INTRODUCTORY STATISTICS

A Student-Centered Approach

Ann Cannon
Cornell College

Daren Starnes
The Lawrenceville School, Emeritus

Josh Tabor
The Potter's School

Austin • Boston • New York • Plymouth

PROGRAM DIRECTOR: **Elizabeth Simmons**
PROGRAM MANAGER: **Lucinda Bingham**
DIRECTOR OF DEVELOPMENT: **Barbara Yien**
DIRECTOR OF CONTENT: **Catriona Kaplan**
DEVELOPMENT EDITOR: **Marita Bley**
CONTENT DEVELOPMENT MANAGER: **Juli Moore**
LEAD CONTENT DEVELOPER: **Aaron Gladish**
SENIOR MEDIA EDITOR: **Doug Newman**
ASSOCIATE MEDIA EDITOR: **Zachary Nothstine**
ASSOCIATE EDITOR: **Nathan Livingston**
SENIOR MARKETING MANAGER: **Leah Christians**
MARKETING ASSISTANT: **Libby Phillips**
MARKET DEVELOPMENT MANAGERS: **Elizabet Cabrera, Kelly Lowery**
SENIOR DIRECTOR, CONTENT MANAGEMENT ENHANCEMENT: **Tracey Kuehn**
EXECUTIVE MANAGING EDITOR: **Michael Granger**
SENIOR MANAGER, PUBLISHING SERVICES: **Andrea Cava**
EXECUTIVE CONTENT PROJECT MANAGER: **Vivien Weiss**
ASSISTANT DIRECTOR, PROCESS WORKFLOW: **Susan Wein**
PRODUCTION SUPERVISOR: **Lawrence Guerra**
DIRECTOR OF DESIGN, CONTENT MANAGEMENT: **Diana Blume**
SENIOR DESIGN SERVICES MANAGER: **Natasha A. S. Wolfe**
INTERIOR DESIGN: **Tamara Newnam, Jerilyn DiCarlo**
SENIOR COVER DESIGN MANAGER: **John Callahan**
ART MANAGER: **Matthew McAdams**
ILLUSTRATIONS: **Lumina Datamatics, Inc.**
EXECUTIVE PERMISSIONS EDITOR: **Robin Fadool**
PHOTO RESEARCHER: **Cheryl Du Bois**
DIRECTOR OF DIGITAL PRODUCTION: **Keri deManigold**
SENIOR MEDIA PROJECT MANAGER: **Elton Carter**
COMPOSITION: **Lumina Datamatics, Inc.**
PRINTING AND BINDING: **Lakeside Book Company**

Copyright © 2024 by Macmillan Learning. All rights reserved. No part of this book may be reproduced, stored in a retrieval system, or transmitted in any form or by any means, electronic, mechanical, photocopying, recording, or otherwise, except as may be permitted by law or expressly permitted in writing by the Publisher.

ISBN: 978-1-319-37174-6 (Loose-leaf Edition)
ISBN: 978-1-319-52342-8 (Paperback Edition)
ISBN: 978-1-319-54531-4 (International Edition)

Library of Congress Control Number: 2023932317
Printed in the United States of America.

1 2 3 4 5 6   28 27 26 25 24 23

Acknowledgments
Acknowledgments and copyrights appear on the same page as the text and art selections they cover; these acknowledgments and copyrights constitute an extension of the copyright page.

Macmillan Learning
120 Broadway
New York, NY 10271
www.macmillanlearning.com

Macmillan Learning
10900 Stonelake Blvd., Suite 300
Austin, Texas 78759
www.macmillan.learning.com

In 1946, William Freeman founded W. H. Freeman and Company and published Linus Pauling's *General Chemistry*, which revolutionized the chemistry curriculum and established the prototype for a Freeman text. W. H. Freeman quickly became a publishing house where leading researchers can make significant contributions to mathematics and science. In 1996, W. H. Freeman joined Macmillan, and we have since proudly continued the legacy of providing revolutionary, quality educational tools for teaching and learning in STEM.

Brief Contents

About the Authors xix
To the Student xx
Preface xxi
Acknowledgments xxvi

Chapter 1 Collecting Data 1
Section 1.1 Introduction to Data Collection 2
Section 1.2 Sampling: Good and Bad 8
Section 1.3 Simple Random Sampling 15
Section 1.4 Other Sampling Methods 21
Section 1.5 Observational Studies and Experiments 28
Section 1.6 Completely Randomized Designs 35
Section 1.7 Blocking 43
Section 1.8 Data Ethics and the Scope of Inference 49

Chapter 2 Displaying Data with Graphs 61
Section 2.1 Displaying Categorical Data 62
Section 2.2 Displaying Relationships Between Two Categorical Variables 72
Section 2.3 Displaying Quantitative Data: Dotplots 82
Section 2.4 Displaying Quantitative Data: Stemplots 92
Section 2.5 Displaying Quantitative Data: Histograms 100
Section 2.6 Displaying Relationships Between Two Quantitative Variables 112

Chapter 3 Numerical Summaries for Quantitative Data 129
Section 3.1 Measuring Center 130
Section 3.2 Measuring Variability 138
Section 3.3 Boxplots and Outliers 148
Section 3.4 Measuring Location in a Distribution 158
Section 3.5 Relationships Between Two Variables: Correlation 167
Section 3.6 More About Correlation 176

Chapter 4 Probability 198
Section 4.1 Randomness, Probability, and Simulation 199
Section 4.2 Basic Probability Rules 207
Section 4.3 Two-Way Tables and Venn Diagrams 213
Section 4.4 Conditional Probability and Independence 220
Section 4.5 The General Multiplication Rule and Bayes' Theorem 229
Section 4.6 The Multiplication Rule for Independent Events 237
Section 4.7 The Multiplication Counting Principle and Permutations 243
Section 4.8 Combinations and Probability 249

Chapter 5 Discrete Random Variables 261
Section 5.1 Introduction to Random Variables 262
Section 5.2 Analyzing Discrete Random Variables 268
Section 5.3 Binomial Random Variables 277
Section 5.4 Analyzing Binomial Random Variables 285
Section 5.5 Poisson Random Variables 292

Chapter 6 Normal Distributions and Sampling Distributions 304
Section 6.1 Continuous Random Variables 305
Section 6.2 Normal Distributions: Finding Areas from Values 314
Section 6.3 Normal Distributions: Finding Values from Areas 322
Section 6.4 Normal Approximation to the Binomial Distribution and Assessing Normality 327
Section 6.5 Sampling Distributions 336

	Section 6.6	Sampling Distributions: Bias and Variability 343
	Section 6.7	Sampling Distribution of the Sample Proportion 352
	Section 6.8	Sampling Distribution of the Sample Mean and the Central Limit Theorem 359

Chapter 7 Estimating a Parameter 375
- Section 7.1 The Idea of a Confidence Interval 376
- Section 7.2 Factors That Affect the Margin of Error 381
- Section 7.3 Estimating a Population Proportion 388
- Section 7.4 Confidence Intervals for a Population Proportion 395
- Section 7.5 Estimating a Population Mean 399
- Section 7.6 Confidence Intervals for a Population Mean 409
- Section 7.7 Estimating a Population Standard Deviation or Variance 414
- Section 7.8 Confidence Intervals for a Population Standard Deviation or Variance 423

Chapter 8 Testing a Claim 433
- Section 8.1 The Idea of a Significance Test 434
- Section 8.2 Significance Tests and Decision Making 441
- Section 8.3 Testing a Claim About a Population Proportion 448
- Section 8.4 Significance Tests for a Population Proportion 457
- Section 8.5 Testing a Claim About a Population Mean 464
- Section 8.6 Significance Tests for a Population Mean 474
- Section 8.7 Power of a Test 481
- Section 8.8 Significance Tests for a Population Standard Deviation or Variance 488

Chapter 9 Comparing Two Populations or Treatments 504
- Section 9.1 Confidence Intervals for a Difference Between Two Population Proportions 505
- Section 9.2 Significance Tests for a Difference Between Two Population Proportions 514
- Section 9.3 Confidence Intervals for a Difference Between Two Population Means 523
- Section 9.4 Significance Tests for a Difference Between Two Population Means 534
- Section 9.5 Analyzing Paired Data: Confidence Intervals for a Population Mean Difference 545
- Section 9.6 Significance Tests for a Population Mean Difference 555
- Section 9.7 Significance Tests for Two Population Standard Deviations or Variances 565

Chapter 10 Chi-Square and Analysis of Variance (ANOVA) 588
- Section 10.1 Testing the Distribution of a Categorical Variable in a Population 589
- Section 10.2 Chi-Square Tests for Goodness of Fit 596
- Section 10.3 Testing the Relationship Between Two Categorical Variables in a Population 606
- Section 10.4 Chi-Square Tests for Association 615
- Section 10.5 Introduction to Analysis of Variance 625
- Section 10.6 One-Way Analysis of Variance 635

Chapter 11 Linear Regression 651
- Section 11.1 Regression Lines 652
- Section 11.2 The Least-Squares Regression Line 658
- Section 11.3 Assessing a Regression Model 671
- Section 11.4 Confidence Intervals for the Slope of a Population Least-Squares Regression Line 681
- Section 11.5 Significance Tests for the Slope of a Population Least-Squares Regression Line 697
- Section 11.6 Confidence Intervals for a Mean Response and Prediction Intervals in Regression 709

Chapter 12 Multiple Regression 725
- Section 12.1 Introduction to Multiple Regression 726
- Section 12.2 Indicator Variables and Interaction 735
- Section 12.3 Inference for Multiple Regression 745

Chapter 13 **Nonparametric Methods** 767
Section 13.1 The Sign Test 768
Section 13.2 The Wilcoxon Signed Rank Test 776
Section 13.3 The Wilcoxon Rank Sum Test 788
Section 13.4 The Kruskal-Wallis Test 799
Section 13.5 Randomization Tests 809
Section 13.6 Bootstrapping 821

Solutions S-1
Glossary G-1
Notes and Data Sources N-1
Index I-1
Tables T-1

Contents

About the Authors xix
To the Student xx
Preface xxi
Acknowledgments xxvi

1 Collecting Data 1

Statistics Matters How can we prevent malaria? 1

Section 1.1 **Introduction to Data Collection** 2
  Class Activity Hiring Discrimination? 2
  Individuals and Variables 3
  Populations and Samples 4
  Observational Studies and Experiments 4
  Section 1.1 Exercises 6

Section 1.2 **Sampling: Good and Bad** 8
  Class Activity Who Wrote the Federalist Papers? 8
  How to Sample Poorly: Convenience and Voluntary Response Samples 9
  How to Sample Well: Random Samples 10
  Other Sources of Bias 11
  Section 1.2 Exercises 13

Section 1.3 **Simple Random Sampling** 15
  Choosing a Simple Random Sample 15
  Sampling Variability 16
  Inference for Sampling 17
  Section 1.3 Exercises 19

Section 1.4 **Other Sampling Methods** 21
  Stratified Random Sampling 22
  Concept Exploration 23
  Cluster Random Sampling 24
  Systematic Random Sampling 25
  Section 1.4 Exercises 26

Section 1.5 **Observational Studies and Experiments** 28
  Confounding 28
  The Language of Experiments 29
  Experiments: Random Assignment 31
  Section 1.5 Exercises 32

Section 1.6 **Completely Randomized Designs** 35
  Experiments: Other Sources of Variability 36
  Completely Randomized Designs 37
  Statistical Significance 38
  Concept Exploration 38
  Section 1.6 Exercises 40

Section 1.7 **Blocking** 43
  Blocking 43
  The Benefit of Blocking 45
  Matched-Pairs Design 46
  Section 1.7 Exercises 47

vii

Section 1.8 Data Ethics and the Scope of Inference 49
Inference About a Population 49
Inference About Cause and Effect 50
Data Ethics 52
Section 1.8 Exercises 53

Statistics Matters *How can we prevent malaria?* 56
Chapter 1 Review 56
Chapter 1 Review Exercises 58

2 Displaying Data with Graphs 61

Statistics Matters *Where is rental housing affordable?* 61

Section 2.1 Displaying Categorical Data 62
Summarizing Data 62
Bar Charts and Pie Charts 63
Avoid Misleading Graphs 66
Section 2.1 Exercises 67

Section 2.2 Displaying Relationships Between Two Categorical Variables 72
Summarizing Data on Two Categorical Variables 72
Displaying the Relationship Between Two Categorical Variables 74
Describing the Relationship Between Two Categorical Variables 76
Section 2.2 Exercises 79

Section 2.3 Displaying Quantitative Data: Dotplots 82
Making and Interpreting Dotplots 83
Describing Shape 84
Describing and Comparing Distributions 86
Section 2.3 Exercises 88

Section 2.4 Displaying Quantitative Data: Stemplots 92
Making Stemplots 93
Describing Stemplots 95
Comparing Distributions with Stemplots 95
Section 2.4 Exercises 97

Section 2.5 Displaying Quantitative Data: Histograms 100
Making Histograms 101
Describing Histograms 103
Class Activity *How Long Is Your Commute?* 104
Comparing Distributions with Histograms 105
Section 2.5 Exercises 107

Section 2.6 Displaying Relationships Between Two Quantitative Variables 112
Making a Scatterplot 113
Describing a Scatterplot 114
Timeplots 115
Section 2.6 Exercises 118

Statistics Matters *Where is rental housing affordable?* 123
Chapter 2 Review 125
Chapter 2 Review Exercises 126
Chapter 2 Project 128

3 Numerical Summaries for Quantitative Data 129

Statistics Matters How can we predict annual water supply? 129

Section 3.1 Measuring Center 130
The Median 130
The Mean 132
Concept Exploration 133
Comparing the Mean and the Median 133

Section 3.1 Exercises 135

Section 3.2 Measuring Variability 138
The Range 138
The Standard Deviation 139
The Interquartile Range 142

Section 3.2 Exercises 144

Section 3.3 Boxplots and Outliers 148
Identifying Outliers 148
Making and Interpreting Boxplots 149
Comparing Distributions with Boxplots 152

Section 3.3 Exercises 154

Section 3.4 Measuring Location in a Distribution 158
Finding and Interpreting Percentiles 159
Finding and Interpreting Standardized Scores (z-Scores) 160
Comparing Location in Different Distributions 162

Section 3.4 Exercises 163

Section 3.5 Relationships Between Two Variables: Correlation 167
Estimating and Interpreting the Correlation 167
Concept Exploration 168
Correlation and Causation 170

Section 3.5 Exercises 172

Section 3.6 More About Correlation 176
Calculating Correlation 176
Properties of the Correlation 178
Outliers and Correlation 179
Concept Exploration 179

Section 3.6 Exercises 182

Statistics Matters *How can we predict annual water supply?* 187

Chapter 3 Review 188

Chapter 3 Review Exercises 189

Chapter 3 Project 191

Cumulative Review Chapters 1–3 193

4 Probability 198

Statistics Matters Should an athlete who fails a drug test be suspended? 198

Section 4.1 Randomness, Probability, and Simulation 199
The Idea of Probability 199
Concept Exploration 199
Myths About Randomness 201
Simulation 202

Section 4.1 Exercises 203

Section 4.2 **Basic Probability Rules** 207
Probability Models 207
Basic Probability Rules 209
Section 4.2 Exercises 210

Section 4.3 **Two-Way Tables and Venn Diagrams** 213
Two-Way Tables and the General Addition Rule 213
Venn Diagrams and Probability 215
Section 4.3 Exercises 217

Section 4.4 **Conditional Probability and Independence** 220
What Is Conditional Probability? 220
Conditional Probability and Independence 223
Section 4.4 Exercises 225

Section 4.5 **The General Multiplication Rule and Bayes' Theorem** 229
The General Multiplication Rule 229
Tree Diagrams 230
Conditional Probability and Bayes' Theorem 232
Section 4.5 Exercises 234

Section 4.6 **The Multiplication Rule for Independent Events** 237
Calculating Probabilities with the Multiplication Rule for Independent Events 237
Use the Multiplication Rule for Independent Events Wisely 239
Section 4.6 Exercises 240

Section 4.7 **The Multiplication Counting Principle and Permutations** 243
The Multiplication Counting Principle 243
Permutations 244
Section 4.7 Exercises 247

Section 4.8 **Combinations and Probability** 249
Combinations 249
Counting and Probability 251
Section 4.8 Exercises 254

Statistics Matters *Should an athlete who fails a drug test be suspended?* 257
Chapter 4 Review 257
Chapter 4 Review Exercises 258

5 Discrete Random Variables 261

Statistics Matters A jury of peers? 261

Section 5.1 **Introduction to Random Variables** 262
Class Activity Smelling Parkinson's Disease 262
Classifying Random Variables 263
Discrete Random Variables 263
Finding Probabilities for Discrete Random Variables 265
Section 5.1 Exercises 266

Section 5.2 **Analyzing Discrete Random Variables** 268
Displaying Discrete Probability Distributions: Histograms and Shape 268
Measuring Center: The Mean (Expected Value) of a Discrete Random Variable 269
Measuring Variability: The Standard Deviation of a Discrete Random Variable 271
Section 5.2 Exercises 274

Section 5.3	**Binomial Random Variables** 277
	Class Activity Pop Quiz! 277
Binomial Settings 277
Calculating Binomial Probabilities 279
Binomial Distributions and Shape 280
	Section 5.3 Exercises 283
Section 5.4	**Analyzing Binomial Random Variables** 285
	Calculating Binomial Probabilities Involving Several Values 285
The Mean and Standard Deviation of a Binomial Random Variable 287
	Section 5.4 Exercises 289
Section 5.5	**Poisson Random Variables** 292
	Poisson Setting 292
Calculating Poisson Probabilities 293
Finding Poisson Probabilities Involving Several Values 294
Finding Poisson Probabilities over Different Interval Lengths 295
	Section 5.5 Exercises 298
	Statistics Matters *A jury of peers?* 300
	Chapter 5 Review 301
	Chapter 5 Review Exercises 302

6 Normal Distributions and Sampling Distributions 304

Statistics Matters How much salt is too much? 304

Section 6.1	**Continuous Random Variables** 305
	Density Curves 305
Normal Distributions 307
The Empirical Rule 309
Concept Exploration 309
	Section 6.1 Exercises 311
Section 6.2	**Normal Distributions: Finding Areas from Values** 314
	Finding Areas to the Left in a Normal Distribution 314
Finding Areas to the Right in a Normal Distribution 317
Finding Areas Between Two Values in a Normal Distribution 318
	Section 6.2 Exercises 320
Section 6.3	**Normal Distributions: Finding Values from Areas** 322
	Finding Boundary Values in a Normal Distribution 322
Finding the Mean or Standard Deviation from Areas in a Normal Distribution 324
	Section 6.3 Exercises 325
Section 6.4	**Normal Approximation to the Binomial Distribution and Assessing Normality** 327
	The Normal Approximation to the Binomial 327
Concept Exploration 329
Using the Normal Approximation to Calculate Probabilities 330
Normal Probability Plots 331
	Section 6.4 Exercises 334
Section 6.5	**Sampling Distributions** 336
	Parameters and Statistics 336
Sampling Distributions 337
Using Sampling Distributions to Evaluate Claims 339
	Section 6.5 Exercises 340

Section 6.6 Sampling Distributions: Bias and Variability 343
Class Activity How many craft sticks are in the bag? 344
Unbiased Estimators 344
Sampling Variability 346
Putting It All Together: Center and Variability 348

Section 6.6 Exercises 349

Section 6.7 Sampling Distribution of the Sample Proportion 352
Center and Variability 352
Shape 354
Concept Exploration 354
Finding Probabilities Involving $\hat{p}$ 355

Section 6.7 Exercises 356

Section 6.8 Sampling Distribution of the Sample Mean and the Central Limit Theorem 359
Center and Variability 359
Shape 360
Concept Exploration 360
Probabilities Involving $\bar{x}$ 362

Section 6.8 Exercises 364

Statistics Matters *How much salt is too much?* 367
Chapter 6 Review 368
Chapter 6 Review Exercises 369
Cumulative Review Chapters 4–6 371

7 Estimating a Parameter 375

Statistics Matters How can I prevent credit card fraud? 375

Section 7.1 The Idea of a Confidence Interval 376
Class Activity What's the "Mystery Mean"? 376
Interpreting Confidence Intervals 377
Building a Confidence Interval 378
Using Confidence Intervals to Make Decisions 378

Section 7.1 Exercises 379

Section 7.2 Factors That Affect the Margin of Error 381
Concept Exploration 381
Interpreting Confidence Level 382
Factors That Affect the Margin of Error 384
Concept Exploration 384
What the Margin of Error Doesn't Account For 385

Section 7.2 Exercises 386

Section 7.3 Estimating a Population Proportion 388
Conditions for Estimating p 388
Critical Values 390
Calculating a Confidence Interval for p 391

Section 7.3 Exercises 393

Section 7.4 Confidence Intervals for a Population Proportion 395
Putting It All Together: The Four-Step Process 395
Determining the Sample Size 396

Section 7.4 Exercises 397

Section 7.5	**Estimating a Population Mean** 399
	Conditions for Estimating μ 400
	The Problem of Unknown σ 400
	Class Activity Confidence interval BINGO! 401
	t^* Critical Values 402
	Calculating a Confidence Interval for μ 404
	Section 7.5 Exercises 407
Section 7.6	**Confidence Intervals for a Population Mean** 409
	The Normal/Large Sample Condition 409
	Putting It All Together: Confidence Interval for μ 410
	Section 7.6 Exercises 412
Section 7.7	**Estimating a Population Standard Deviation or Variance** 414
	The Sampling Distribution of the Sample Variance 415
	Concept Exploration 415
	Conditions for Estimating σ or σ^2 416
	χ^2 Critical Values 417
	Calculating a Confidence Interval for σ or σ^2 418
	Section 7.7 Exercises 421
Section 7.8	**Confidence Intervals for a Population Standard Deviation or Variance** 423
	The Normal Condition 423
	Putting It All Together: Confidence Interval for σ 424
	Section 7.8 Exercises 426
	Statistics Matters How can I prevent credit card fraud? 428
	Chapter 7 Review 429
	Chapter 7 Review Exercises 431
	Chapter 7 Project 432

8 Testing a Claim 433

Statistics Matters What is normal body temperature? 433

Section 8.1	**The Idea of a Significance Test** 434
	Concept Exploration 434
	Stating Hypotheses 435
	Interpreting P-Values 437
	Making Conclusions Based on P-Values 438
	Section 8.1 Exercises 439
Section 8.2	**Significance Tests and Decision Making** 441
	Making Conclusions Using Significance Levels 442
	Type I and Type II Errors 443
	Section 8.2 Exercises 445
Section 8.3	**Testing a Claim About a Population Proportion** 448
	Conditions for Testing a Claim About p 448
	Calculating the Standardized Test Statistic 449
	Finding the P-Value 451
	Section 8.3 Exercises 454
Section 8.4	**Significance Tests for a Population Proportion** 457
	Putting It All Together: The Four-Step Process 457
	Two-Sided Tests 459
	Section 8.4 Exercises 461

Section 8.5 **Testing a Claim About a Population Mean** 464
Conditions for Testing a Claim About μ 464
Calculating the Standardized Test Statistic 465
Finding P-Values 467
Section 8.5 Exercises 471

Section 8.6 **Significance Tests for a Population Mean** 474
Putting It All Together: Testing a Claim About a Population Mean 474
Two-Sided Tests and Confidence Intervals 475
Using Tests Wisely 476
Section 8.6 Exercises 478

Section 8.7 **Power of a Test** 481
The Power of a Test 481
What Affects the Power of a Test? 482
Concept Exploration 483
Section 8.7 Exercises 485

Section 8.8 **Significance Tests for a Population Standard Deviation or Variance** 488
Conditions for Testing a Claim About σ or σ^2 488
Calculating the Test Statistic and Finding P-Values 489
Putting It All Together: Testing a Claim About a Population Standard Deviation or Variance 492
Section 8.8 Exercises 495

Statistics Matters *What is normal body temperature?* 498
Chapter 8 Review 499
Chapter 8 Review Exercises 501
Chapter 8 Project 503

9 Comparing Two Populations or Treatments 504

Statistics Matters How fast-food drive-thrus make money: Speed, accuracy, and customer service 504

Section 9.1 **Confidence Intervals for a Difference Between Two Population Proportions** 505
Conditions for Estimating $p_1 - p_2$ 505
Calculating a Confidence Interval for $p_1 - p_2$ 506
Putting It All Together: Confidence Interval for $p_1 - p_2$ 508
Section 9.1 Exercises 511

Section 9.2 **Significance Tests for a Difference Between Two Population Proportions** 514
Stating Hypotheses and Checking Conditions for a Test About $p_1 - p_2$ 514
Calculations: Standardized Test Statistic and P-Value 515
Putting It All Together: Significance Test About $p_1 - p_2$ 517
Section 9.2 Exercises 520

Section 9.3 **Confidence Intervals for a Difference Between Two Population Means** 523
Conditions for Estimating $\mu_1 - \mu_2$ 524
Calculating a Confidence Interval for $\mu_1 - \mu_2$ 525
Putting It All Together: Confidence Interval for $\mu_1 - \mu_2$ 527
Section 9.3 Exercises 530

Section 9.4 **Significance Tests for a Difference Between Two Population Means** 534
Stating Hypotheses and Checking Conditions for a Test About $\mu_1 - \mu_2$ 534
Calculations: Standardized Test Statistic and P-Value 536
Putting It All Together: Performing a Significance Test About $\mu_1 - \mu_2$ 538
Section 9.4 Exercises 540

Contents xv

Section 9.5 Analyzing Paired Data: Confidence Intervals for a Population Mean Difference 545

Analyzing Paired Data 546
Putting It All Together: Constructing and Interpreting a Confidence Interval for μ_{diff} 549

Section 9.5 Exercises 551

Section 9.6 Significance Tests for a Population Mean Difference 555

Performing a Significance Test About μ_{diff} 555
Paired Data or Two Samples? 558
Class Activity Get Your Heart Beating! 559

Section 9.6 Exercises 560

Section 9.7 Significance Tests for Two Population Standard Deviations or Variances 565

Hypotheses and Conditions for Testing a Claim About Two Standard Deviations or Variances 565
Calculations: Test Statistic and *P*-Value 566
Putting It All Together: Significance Test Comparing Two Standard Deviations or Variances 569

Section 9.7 Exercises 573

Statistics Matters *How fast-food drive-thrus make money: Speed, accuracy, and customer service* 577

Chapter 9 Review 578

Chapter 9 Review Exercises 581

Chapter 9 Project 583

Cumulative Review Chapters 7–9 584

10 Chi-Square and Analysis of Variance (ANOVA) 588

Statistics Matters How racially and ethnically diverse are STEM workers? 588

Section 10.1 Testing the Distribution of a Categorical Variable in a Population 589

Class Activity The Color of Candy 589
Stating Hypotheses 590
Calculating Expected Counts 591
The Chi-Square Test Statistic 592

Section 10.1 Exercises 594

Section 10.2 Chi-Square Tests for Goodness of Fit 596

Conditions for a Chi-Square Test for Goodness of Fit 597
Calculating *P*-values 597
Putting It All Together: The Chi-Square Test for Goodness of Fit 600

Section 10.2 Exercises 603

Section 10.3 Testing the Relationship Between Two Categorical Variables in a Population 606

Stating Hypotheses 606
Calculating Expected Counts 609
The Chi-Square Test Statistic 610

Section 10.3 Exercises 612

Section 10.4 Chi-Square Tests for Association 615

Conditions for a Chi-Square Test for Association 615
Calculating *P*-Values 617
Putting It All Together: The Chi-Square Test for Association 618

Section 10.4 Exercises 622

Section 10.5 Introduction to Analysis of Variance 625
Stating Hypotheses 626
Analyzing Variation, Testing Means 626
ANOVA Table 628

Section 10.5 Exercises 631

Section 10.6 One-Way Analysis of Variance 635
Conditions for a One-Way ANOVA Test 635
P-Values and the One-Way ANOVA Test 636
Fisher's Least Significant Difference (LSD) Intervals 637

Section 10.6 Exercises 640

Statistics Matters *How racially and ethnically diverse are STEM workers?* 646

Chapter 10 Review 647

Chapter 10 Review Exercises 649

Chapter 10 Project 650

11 Linear Regression 651

Statistics Matters How does engine size affect CO_2 emissions? 651

Section 11.1 Regression Lines 652
Making Predictions 652
Residuals 653
Interpreting a Regression Line 654

Section 11.1 Exercises 655

Section 11.2 The Least-Squares Regression Line 658
Calculating the Equation of the Least-Squares Regression Line 659
Outliers and the Least-Squares Regression Line 662
Concept Exploration 662
Regression to the Mean 664

Section 11.2 Exercises 666

Section 11.3 Assessing a Regression Model 671
Residual Plots 671
Standard Deviation of the Residuals 673
The Coefficient of Determination r^2 675

Section 11.3 Exercises 678

Section 11.4 Confidence Intervals for the Slope of a Population Least-Squares Regression Line 681
Concept Exploration 682
Checking Conditions for Inference About the Slope 684
Constructing a Confidence Interval for the Slope of a Least-Squares Regression Line 688
Putting It All Together: Confidence Intervals for the Slope of a Population Least-Squares Regression Line 689

Section 11.4 Exercises 692

Section 11.5 Significance Tests for the Slope of a Population Least-Squares Regression Line 697
Concept Exploration 697
Stating Hypotheses 698
Calculating the Test Statistic and P-Value 699
Putting It All Together: Testing the Slope of a Population Least-Squares Regression Line 701

Section 11.5 Exercises 705

Section 11.6 Confidence Intervals for a Mean Response and Prediction Intervals in Regression 709

Confidence Intervals for the Mean y Value at a Given Value of x 709
Prediction Intervals for an Individual y Value at a Given Value of x 711
Controlling Interval Width 713

Section 11.6 Exercises 716

Statistics Matters *How does engine size affect CO_2 emissions?* 720
Chapter 11 Review 721
Chapter 11 Review Exercises 722
Chapter 11 Project 724

12 Multiple Regression 725

Statistics Matters Which factors affect the growth of whales? 725

Section 12.1 Introduction to Multiple Regression 726

The Idea of Multiple Regression 726
Assessing a Multiple Regression Model 729

Section 12.1 Exercises 732

Section 12.2 Indicator Variables and Interaction 735

Using a Categorical Explanatory Variable 735
Interpreting Multiple Regression Coefficients 738
Interaction 739

Section 12.2 Exercises 741

Section 12.3 Inference for Multiple Regression 745

Testing a Multiple Regression Model 746
Testing an Individual Coefficient 750

Section 12.3 Exercises 754

Statistics Matters *Which factors affect the growth of whales?* 758
Chapter 12 Review 759
Chapter 12 Review Exercises 759
Chapter 12 Project 761

Cumulative Review Chapters 10–12 762

13 Nonparametric Methods 767

Statistics Matters Can acupuncture help chronic headaches? 767

Section 13.1 The Sign Test 768

Median as Measure of Center 768
Hypotheses and Test Statistic for a Sign Test 769
P-Values 770
Putting It All Together: The Sign Test 771

Section 13.1 Exercises 773

Section 13.2 The Wilcoxon Signed Rank Test 776

Ranks and Signed Ranks 777
Test Statistic and *P*-Value 779
Putting It All Together: The Wilcoxon Signed Rank Test 781

Section 13.2 Exercises 784

Section 13.3   The Wilcoxon Rank Sum Test 788

Ranks for Two Samples or Groups 788
Test Statistic and P-Value 790
Putting It All Together: The Wilcoxon Rank Sum Test 793

Section 13.3 Exercises 795

Section 13.4   The Kruskal-Wallis Test 799

Stating Hypotheses and Ranking the Data 799
Test Statistic and P-Value 801
Putting It All Together: The Kruskal-Wallis Test 804

Section 13.4 Exercises 806

Section 13.5   Randomization Tests 809

Randomization Distributions 809
P-Values from Randomization Distributions 811
Putting It All Together: Randomization Tests 813

Section 13.5 Exercises 817

Section 13.6   Bootstrapping 821

Bootstrap Distributions 822
Confidence Intervals from Bootstrap Distributions 823
Putting It All Together: Bootstrap Confidence Intervals 825

Section 13.6 Exercises 828

Statistics Matters *Can acupuncture help chronic headaches?* 832

Chapter 13 Review 833

Chapter 13 Review Exercises 835

Solutions S-1
Glossary G-1
Notes and Data Sources N-1
Index I-1

Tables
Table A: Standard Normal Probabilities T-1
Table B: t Distribution Critical Values T-3
Table C: Chi-Square Distribution Critical Values T-4
Table D: F Distribution Critical Values T-5
Table E: Wilcoxon Signed-Rank Test Critical Values T-13

About the Authors

Ann Cannon is the Watson M. Davis Professor of Mathematics and Statistics at Cornell College in Mount Vernon, Iowa, where she has taught statistics for 30 years. She earned her MA and PhD in statistics from Iowa State University, and her BA in mathematics from Grinnell College. Ann is a Fellow of the American Statistical Association (ASA) and won the Mu Sigma Rho (national statistics honor society) William D. Warde Statistics Education Award. Ann has been very involved with the Statistics and Data Science Education Section of the ASA, serving on the executive committee as member-at-large, secretary/treasurer, and chair. She has also served on the ASA/MAA Joint Committee on Undergraduate Statistics, and as the Secretary/Treasurer and Chair of the Iowa Chapter of the ASA. Ann is currently associate editor for the *Journal of Statistics and Data Science Education*. She has been involved with the AP® Statistics Reading for over 20 years, serving as Reader, Table Leader, Question Leader, and Assistant Chief Reader. Ann is coauthor of *STAT2: Modeling with Regression and ANOVA* (now in its second edition), a textbook designed for the college statistics course following the introductory statistics course. In her spare time, Ann enjoys playing the French horn (particularly in pit orchestras for musical theater), reading, and traveling.

Daren Starnes has taught a variety of statistics courses — including Introductory Statistics, AP® Statistics, and Mathematical Statistics — for 25 years. He earned his MA in mathematics from the University of Michigan and his BS in mathematics from the University of North Carolina at Charlotte. Daren has been a Reader, Table Leader, and Question Leader for the AP® Statistics exam for over 20 years. As a College Board consultant since 1999, Daren has led hundreds of workshops for AP® Statistics teachers throughout the United States and overseas. He frequently presents in-person and online sessions about statistics teaching and learning for high school and college faculty. Daren is an active member of the American Statistical Association (ASA), the National Council of Teachers of Mathematics (NCTM), the American Mathematical Association of Two-Year Colleges (AMATYC), and the International Association for Statistical Education (IASE). He served on the ASA/NCTM Joint Committee on the Curriculum in Statistics and Probability for six years. While on the committee, he edited the *Guidelines for Assessment and Instruction in Statistics Education (GAISE) Report: A Pre-K–12 Curriculum Framework*. Daren is coauthor of *The Practice of Statistics* (now in its seventh edition), the best-selling textbook for AP® Statistics, and of *Statistics and Probability with Applications* (now in its fourth edition), a popular choice for high school introductory statistics. Daren and his wife Judy enjoy traveling, rambling walks, jigsaw puzzles, and spending time with their three sons and seven grandchildren.

Josh Tabor has enjoyed teaching Introductory and AP® Statistics for more than 26 years. He received a BS in mathematics from Biola University, in La Mirada, California. In recognition of his outstanding work as an educator, Josh was named one of five finalists for Arizona Teacher of the Year in 2011. He is a past member of the AP® Statistics Development Committee (2005–2009), as well as an experienced Table Leader, Question Leader, and Exam Leader at the AP® Statistics Reading. In 2013, Josh was named to the SAT® Mathematics Development Committee. Josh is a member of the American Statistical Association (ASA) and was a reviewer for the ASA's *Pre-K–12 Guidelines for Assessment and Instruction in Statistics Education II (GAISE II)*. Each year, Josh leads many workshops and frequently speaks at local, national, and international conferences. In addition to teaching and speaking, he has authored articles in *The American Statistician, The Mathematics Teacher, STATS Magazine,* and *The Journal of Statistics Education*. Josh is coauthor of *The Practice of Statistics* (now in its seventh edition), the best-selling textbook for AP® Statistics, and of *Statistics and Probability with Applications* (now in its fourth edition), a popular choice for high school introductory statistics. Combining his love of statistics and sports, Josh teamed with Christine Franklin to write *Statistical Reasoning in Sports,* an innovative textbook for statistical literacy courses. Outside of work, Josh enjoys gardening, traveling, and playing board games with his family.

To the Student

During the global Covid-19 pandemic, researchers collected mountains of data. These data were used to assess the quality of diagnostic tests, to model the spread and impact of the disease, and to test the effectiveness of vaccines and emerging therapeutics. The researchers relied on statistics to guide the collection, analysis, and interpretation of the data. Careful use of statistics enabled them to get clear answers to challenging questions in the face of uncertainty.

Statistics is also essential for tracking the economy, measuring climate change, gauging people's views about important societal issues, and performing research in many other fields of study. Companies, governments, and other organizations regularly use statistics to make critical decisions that affect us all. In short, statistics matters!

We want to help you develop the statistical thinking skills you will need to make informed, data-based decisions in your major, career, and daily life. That was our primary motivation for writing *Introductory Statistics: A Student-Centered Approach*.

What makes this book student-centered? In addition to the intriguing contexts included in the examples and exercises, we structured the book to make it clear what you need to know and be able to do.

- Each section starts with two or three focused Learning Goals, followed by short narrative passages leading to a Worked Example for each Learning Goal.

- Examples are written in a Problem/Solution format, so they will be a good match for the Exercises at the end of the section. Furthermore, the solution to each example is written in a different font to make it clear what an ideal response looks like. Additional comments, reminders, and suggestions in the solution are placed in adjacent Instructor Talk boxes.

- Other features, such as How To boxes, Summary boxes, and Caution icons, help you to master the skills needed for the course and avoid common mistakes.

- A What Did You Learn? grid wraps up each section by explicitly aligning the examples and selected exercises with the Learning Goals presented at the beginning of the section.

- Each chapter ends with a bulleted summary of key ideas and a set of review exercises that address each of the Learning Goals. In addition, a cumulative review is found at the end of Chapters 3, 6, 9, and 12 that will help you prepare for in-class exams.

We hope that using *Introductory Statistics: A Student-Centered Approach* enhances your statistics course and builds a foundation of statistical literacy that will serve you for the rest of your life.

Sincerely,
Ann Cannon, Daren Starnes, and Josh Tabor

Preface

Introductory Statistics: A Student-Centered Approach is a comprehensive text and digital program that reimagines the process of learning statistics. What makes this text student-centered? It blends concise narrative, engaging examples, real data, and innovative technology with diverse applications chosen to illustrate the importance of statistics in a wide variety of college majors and careers. These features help achieve our primary goal — to help students make informed decisions based on statistical information in their daily lives.

A Student-Centered Approach: Content, Structure, Pedagogy, and Technology

Content

Introductory Statistics: A Student-Centered Approach includes all of the standard topics in a modern introductory statistics course, including a full treatment of descriptive statistics, data collection methods, probability, random variables, sampling distributions, and traditional inference procedures (including nonparametric methods), as well as more contemporary simulation-based inference techniques. The statistical content aligns closely with the recommendations from the American Statistical Association's 2016 *Guidelines for Assessment and Instruction in Statistics Education* (GAISE):

1. Teach statistical thinking.
   - Teach statistics as an investigative process of problem solving and decision making.
   - Give students experience with multivariable thinking.
2. Focus on conceptual understanding.
3. Integrate real data with a context and purpose.
4. Foster active learning.
5. Use technology to explore concepts and analyze data.
6. Use assessments to improve and evaluate student learning.

To maintain a clear focus on statistical thinking, *Introductory Statistics: A Student-Centered Approach* de-emphasizes tedious, formula-based calculations that tend to obscure meaning for students.

One of this book's hallmark features is its **focus on applications.** We selected **real data and studies** to pique students' interest and to showcase the relevance of statistics in the world around them. Browse the chapter content to get a sense of the wide variety of applications from different fields of study that students will encounter.

Structure

Introductory Statistics: A Student-Centered Approach is designed to be **easy to read and easy to use.** Most students don't realize the full value of what their textbook has to offer. For starters, many students do not read the textbook! Many introductory statistics textbooks contain excessive amounts of narrative, making it hard for students to extract the signal from the noise in lengthy exposition or in examples that don't clearly show what is required in a model solution. Students want and deserve *clarity* from their textbook about what's needed to be successful in the course. *Introductory Statistics: A Student-Centered Approach* is written with this goal in mind.

Unlike in other textbooks, every chapter in *Introductory Statistics: A Student-Centered Approach* is organized into **short sections** containing bite-sized chunks of related content. Every section is organized around two or three specific **Learning Goals** stated at the beginning of the section. For each Learning Goal, there is a subsection that includes an illustrative **worked example.** At the end of each section, students will find a group of **exercises** aligned to each Learning Goal. The writing style offers clear, concise explanations that encourage students to read. There is **minimal narrative** between section elements. Our focus on the Learning Goals makes it clear what students need to know and be able to do to achieve success in introductory statistics.

The short-section structure also gives instructors greater flexibility to organize the course based on the desired content coverage, the needs of their students, and the class schedule. As a general rule, at least two sections can be comfortably completed per 50-minute class session.

Pedagogy

Introductory Statistics: A Student-Centered Approach embraces the philosophy that students learn statistics best by *doing* statistics. Our hundreds of worked examples show students how to do statistics well. All examples are in **Problem/Solution format,** with the model student solution shown in a different font and the instructor's voice added in comment boxes adjacent to the solution. Students can use the model solutions as a guide when crafting their responses to the exercises, developing good habits that will serve them well when solving other statistics problems.

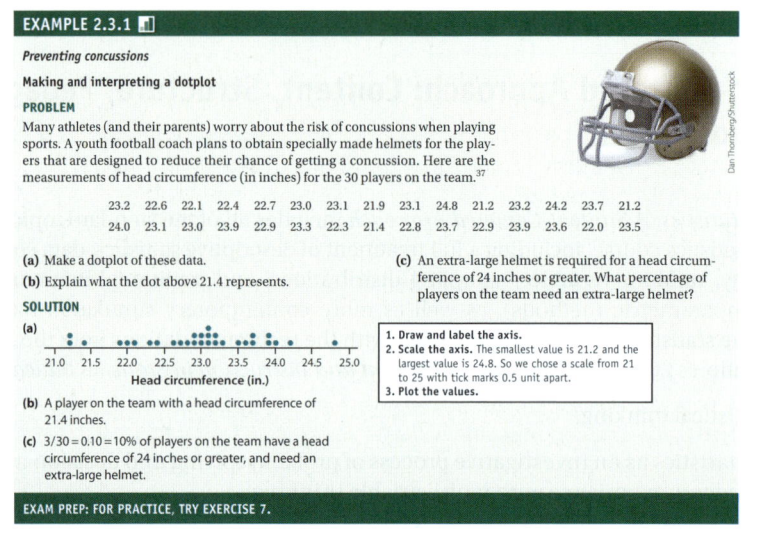

The **Statistics Matters** feature that bookends each chapter gives students an opportunity to use what they have learned to apply statistical thinking in a real-world setting. The text includes several engaging, collaborative **Class Activities** to introduce new content. (Additional Class Activities are suggested in the Instructor's Resources.) The book also features numerous **Concept Explorations,** many involving the use of dynamic **online statistical applets** that help students develop conceptual understanding. **Chapter Projects** at the end of most chapters give students an opportunity to explore large, multivariable data sets using software, and to apply what they have learned in the chapter in an engaging context.

Introductory Statistics: A Student-Centered Approach gives students regular opportunities to practice their skills, deepen their conceptual understanding, and develop their statistical communication using real data and studies. Each section of the book ends with 25 to 30 **exercises** grouped into five distinct categories:

- *Building Concepts and Skills* exercises assess the basic knowledge students should have after reading the section.

- *Mastering Concepts and Skills* exercises reinforce the Learning Goals as shown in the examples. Odd- and even-numbered exercises are paired. Selected exercises are explicitly linked to the corresponding worked example with an icon.

- *Applying the Concepts* exercises ask students to apply multiple Learning Goals in a new context or to apply what they learned in the section in a new way. Odd- and even-numbered exercises are paired.

- *Extending the Concepts* exercises challenge students to explore statistical concepts and methods that go beyond what they learned in the section.

- *Cumulative Review* exercises revisit what students learned in previous sections.

The *Chapter Review* exercises found at the end of each chapter aim to help students review important concepts and skills described by the Learning Goals in each section. For most exercises, the relevant section is noted in parentheses after the exercise title. *Cumulative Reviews* at the end of Chapters 3, 6, 9, and 12 include multiple-choice and free-response questions that are intended to help students review important concepts and skills in the preceding three chapters and prepare for in-class exams.

The design of *Introductory Statistics: A Student-Centered Approach* highlights the book's distinctive structure and important features. Key terms are called out in bold font and explained in **definition boxes**. Step-by-step instructions explaining how to create graphs, perform calculations, or use a statistical process correctly are detailed in **How To boxes**. On-the-spot **Summary boxes** give a quick recap of the key things to remember about important statistical techniques. **Caution icons** alert students to common mistakes. Occasional **Think About It** vignettes encourage students to contemplate subtle statistical issues. A **What Did You Learn?** grid at the end of each section summarizes the connection between the Learning Goals, examples, and exercises. At the end of each chapter, students will find a well-structured **Chapter Review** that includes a bulleted summary of key content.

Technology

Technology is a vital tool for modern statistics. Introductory statistics students benefit from intuitive, easy-to-use technology that allows them to focus on understanding the output. To meet this need, we developed a set of simple applets that are freely accessible to students and instructors for analyzing data, computing probabilities, exploring concepts, and performing inference via simulation or traditional methods — on any mobile device. You can view the applets at www.stapplet.com.

The **Tech Corner** features use these applets, along with the TI-83/84 graphing calculator, which is familiar to many students from their high school mathematics courses. Each Tech Corner also comes with a digital supplement in Achieve that gives instructions for commonly used software: CrunchIt!, Excel, Google Sheets, JMP, Minitab, and R. Screenshots from these software packages are included along with the detailed instructions. Every data set is provided in a format suitable for each technology option, and is marked with a icon at its point of use in the text. Many of the **Chapter Projects,** as well as selected section exercises marked with the Tech icon, ask students to investigate medium and large data sets using technology.

Introductory Statistics: A Student-Centered Approach is available with Macmillan's ground-breaking online learning platform, **Achieve.** The tools in Achieve support learning before, during, and after class for students, and equip instructors with class performance analytics in an easy-to-use interface. Achieve connects the book's innovative structure, pedagogy, and content to rich digital resources that foster understanding and facilitate statistical thinking.

Achieve is the culmination of years of development work put toward creating the most powerful online learning tool for statistics students. It houses all of our assessments, multimedia assets, e-book, and instructor resources in a single user-friendly platform.

Achieve supports educators and students throughout the full range of instruction, including assets suitable for **pre-class** preparation, **in-class** active learning, and **post-class** study and assessment. Achieve is ideal for in-person, flipped, and hybrid classroom designs. The pairing of a powerful platform with outstanding statistics content provides an unrivaled learning experience.

Highlights include:

- A design guided by **learning science** research. Achieve is co-designed through extensive collaboration and testing by both students and faculty.

- A **learning path** of powerful content including **pre-class, in-class,** and **post-class** activities and assessments.

- A detailed **gradebook** with insights for just-in-time teaching and reporting on student achievement by learning objective.

- Easy **integration** and **gradebook sync** with cloud-based **iClicker** classroom engagement solutions.

- Easy integration with your campus **LMS** and availability through **Inclusive Access** programs.

Achieve also includes a **strong focus on practice and assessment.** Robust tutorial-style assessment tools help students develop problem-solving and statistical reasoning skills. Achieve contains thousands of assessment questions designed for both **pre-class** foundational learning and **post-class** homework, which were created by a team of dedicated content developers with subject-matter expertise. For selected questions, our formative assessment environment responds to students' incorrect answers with feedback to guide their study. This Socratic feedback mechanism emulates the office-hours experience, encouraging students to think critically about their identified misconceptions. Students can make the most out of their homework with Achieve's hallmark hints, detailed feedback, and fully worked solutions.

Resources include:

- **LearningCurve Adaptive Quizzes** are formative, adaptive, and conceptually focused quizzes that provide hints and links to the e-book and are personalized to each student's progress.

- **SkillBuilder Tutorial Videos** are brief, worked-out video examples that allow students to visualize how to put their new skills into practice. Many are also paired with Achieve's interactive video assessment features, with conceptual and reflection questions embedded throughout the videos.

- **Reading Quizzes** include all *Building Concepts and Skills* exercises from each section of the book. These quizzes can be used by students as a quick check of their comprehension of the key ideas after they finish reading each section.

- **Desmos Interactive Applets** are statistical simulations powered by the *Desmos Graphing Calculator* and paired with questions to guide students through the usage of and understanding of statistical concepts.

- **Homework Assignments** are carefully selected sets of exercises coded directly from the book with detailed feedback and full solutions, with many including algorithmic variables. Direct links to data files and CrunchIt! web-based statistical software are provided wherever a data set is referenced. Additional editable questions can be added from the Achieve question bank.

- **Corequisite/Prerequisite Support** is provided via **Rapid Review** assignments that pair video and assessment questions to help students review and practice key math and statistics skills.

- **Vocab Coach** allows students to practice and master statistics terminology. Each term defined in the book is included. Students are quizzed on the definition of each term and are also asked to identify the term given its definition. Vocab Coach can be done in class as an iClicker activity, completed outside of class as an adaptive quiz, or assigned for credit.

In addition, the following technology support resources are available:

- All of the **Data Set Files** needed to work through the exercises and examples in the book are available in Excel (.xls and .xlsx), CSV, Minitab, R, JMP, TI, SPSS, PC-Text, and Mac-Text formats.

- **CrunchIt!** is Macmillan's proprietary online statistical software powered by R; it offers computation and graphing functions needed by introductory statistics students. CrunchIt! is preloaded with all of the book's data sets and allows for editing and importing additional data. Direct links to CrunchIt! data sets are available in the e-book and in relevant assessment questions.

- **Online Tech Corners** include step-by-step instructions for CrunchIt!, Excel, Google Sheets, JMP, Minitab, and R. These features are available as separate e-book pages for each topic and software.

- **Software Assistance Videos** for Excel, CrunchIt!, TI calculators, JMP, Minitab, R, and SPSS offer brief, step-by-step instructions for using statistical software to perform the calculations covered in the book.

Achieve also includes the following **Instructor Resources:**

- The **Instructor Solutions Manual** contains full solutions to all exercises in the text.

- **iClicker Question Decks** offer engaging questions related to the key learning objectives in each section of the book. They are designed to work directly with the unique functionality offered in Achieve's powerful iClicker integration, and include Reading Check quizzes, activity tips, a variety of question types, and solution slides.

- **Active Lecture Decks** are demonstration slides and assessment questions for use in class that can be customized and resequenced.

- The **Instructor Guide** provides detailed teaching suggestions for each chapter.

- **Active Learning Activities** are available for each chapter. They provide instructional tips, learning objectives, estimated times, and all materials needed to run activities that can be used in both face-to-face and online class settings.

Acknowledgments

What does it take to build a complete text/media program, from planning the table of contents to drafting the manuscript to polishing the resources in the digital platform — all during a pandemic? *Amazing teamwork!* With so many components to manage, we are indebted to every member of the team for the time, energy, and passion they have invested in making *Introductory Statistics: A Student-Centered Approach* a reality.

To Sarah Seymour, we offer our gratitude for taking a chance on a new title in a challenging time for the publishing industry. Thanks for advocating on our behalf, encouraging us, and coordinating all the different aspects of such a large project. Thanks to Lucinda Bingham for bringing this project across the finish line and championing it with reps and potential adopters. To Andy Dunaway, thanks for supporting our vision and helping us to refine "the story of *SCA*." To Debbie Hardin, thanks for coordinating the development of the program. Thanks to Sara Gordus for her help developing early drafts. Special thanks go to Barbara Yien for her thoughtful commentary and amazing organizational skills, and for shepherding us through the many rounds of reviews, edits, and page proofs.

We also thank Catriona Kaplan and her media team — including Elton Carter, Juli Moore, Aaron Gladish, Doug Newman, and Zachary Nothstine — for their meticulous work on all of the Achieve resources, as well as for proofreading the e-book and user-testing the Online Tech Corners.

We are thankful to Vivien Weiss, Susan Wein, and Paul Rohloff for managing the production aspects of the book. We are also grateful for the many fine people who have worked on various other stages of the process, including Marita Bley, Katrina Mangold, Jennifer Hart, Kaylin Fussell, and Nathan Livingston. Your contributions have been invaluable. Thanks also to Leah Christians and Libby Phillips in marketing and to everyone in the sales department at Macmillan. We look forward to many great years of working together. Thanks to John Samons of Florida State College at Jacksonville for accuracy checking the solutions and page proofs, and to Jill Hobbs for copyediting the manuscript. Thanks also to the people at Lumina Datamatics for their wonderful work implementing a complicated design.

Erica Chauvet and James Bush, both from Waynesburg University, did amazing work writing the solutions manual, creating the short solutions for the back-of-book appendix, and offering suggestions to improve the exercises. Knowing the value of high-quality solutions, we couldn't have asked for a better team to take on this important job. Thanks!

Bob Amar did a great job building the applets at www.stapplet.com, which include many new features for *SCA*. Thanks for helping us make technology freely accessible to students around the globe.

Thanks to all of our students who have, through their encouraging and critical feedback, helped us become better teachers and authors. Thanks also to the many statistics colleagues we have met over the years. We have learned so much from you, whether it be at a conference or talking shop around the lunch table at the AP® Reading. We are also extremely grateful for the people who reviewed the manuscript at various stages. Your comments and suggestions have made a big difference!

Finally, we'd like to express our utmost gratitude and love to our family members: spouses, children, and grandchildren. We know that you have sacrificed time and attention as we worked on this project for several years. Thanks for your encouragement, support, and occasional feedback — it means the world to us.

—*Ann Cannon, Daren Starnes, and Josh Tabor*

Reviewers

We are grateful to colleagues from two-year and four-year colleges and universities who reviewed the manuscript throughout its development:

Gerald Agbegha, *Georgia Gwinnett College*
Carlos Mauricio Amaya, *California State University–Los Angeles*
Wesley Anderson, *Northwest Vista College*
Peter Arvanites, *SUNY Rockland Community College*
Anna Bakman, *Los Angeles Trade Technical Community College*
Wayne Barber, *Chemeketa Community College*
Sarah Bergmann, *Alamance Community College*
Priya Boindala, *Georgia Gwinnett College*
Justine Bonanno-Suquinahua, *Central Piedmont Community College–Central*
Joan Brenneman, *University of Central Oklahoma*
Ryan Carpenter, *Christopher Newport University*
Isabelle Chang, *Temple University*
Hongwei Chen, *Christopher Newport University*
Steve Chung, *California State University–Fresno*
LeAnne Conaway, *Harrisburg Area Community College*
Kristin Cook, *College of Western Idaho*
Tyler Cook, *University of Central Oklahoma*
Jamye Curry Savage, *Georgia Gwinnett College*
Andrew Dellinger, *Elon University*
Ray DeWitt, *Lake Superior State University*
Alok Dhital, *University of New Mexico–Gallup*
Douglas Donohue, *Pearl River Community College*
Jason Droesch, *North Idaho College*
Michelle Duda, *Columbus State Community College*
Daniel Franklin, *Georgia State University*
David French, *Tidewater Community College–Norfolk*
Sydia Gayle-Fenner, *Central Piedmont Community College–Central*
Jeremiah Gilbert, *San Bernardino Valley College*
Ernest Gobert, *Oklahoma City Community College*
Helena Grant, *Cedar Valley College*
Al Groccia, *Valencia College*
Lara Guidroz, *McNeese State University*
David Gurney, *Southeastern Louisiana University*
Rhonda Hatcher, *Texas Christian University*
Andrew Henley, *Guilford Technical Community College–Jamestown*

Linda Hoang, *Cosumnes River College*
Rodney Holke-Farnam, *Hawkeye Community College*
Xiushan Jiang, *College of Charleston*
Victoria Kotlyar, *East Los Angeles College*
Meesook Helena Lee, *McNeese State University*
Sara Lenertz, *Inver Hills Community College*
Sara Lenhart, *Christopher Newport University*
Josh Lewis, *Bakersfield College*
Habib Maagoul, *Northern Essex Community College*
Alexandra Macedo, *El Paso Community College*
lobna Mazzawi, *Everett Community College*
Meghan McIntyre, *Wake Technical Community College*
Cindy McNab, *Northeast State Community College*
Mehdi Mirfattah, *East Los Angeles College*
Paul Peeders, *Madison College-Truax*
Diana Pell, *Riverside Community College*
Jennifer Peters, *Western University*
Chris Petrie, *Eastern Florida State-Titusville*
Matthew Pragel, *Harrisburg Area Community College*
Wendy Royston, *Northeast State Community College*
Mukta Sharma, *Yuba College*
Nigie She, *Bakersfield College*
Jessica Sherwood, *Cleveland State Community College*
Rory Shipowick, *San Diego Mesa College*
Anita Simic Milas, *Bowling Green State University*
Karen Starin, *Columbus State Community College*
Marty Thomas, *Georgia Gwinnett College*
Michael Totoro, *Nassau Community College*
Lise Trivett, *Cascadia Community College*
Sherry Vaughan, *Thomas Nelson Community College-Hampton*
Sukhitha Vidurupola, *Rogers State University*
Angela Ward, *Tri-County Technical College*
Sophia Waymyers, *Francis Marion University*
John Weber, *Georgia State University Perimeter College-Clarkston*
Lloyd Wehrung, *Jacksonville University*
Carol Weideman, *St. Petersburg College-Gibbs Campus*
Paul Weiss, *Emory University*
Trina Wooten, *Northeast State Community College*
Jeffrey Zahnen, *Daytona State College*

1 Collecting Data

Section 1.1 Introduction to Data Collection
Section 1.2 Sampling: Good and Bad
Section 1.3 Simple Random Sampling
Section 1.4 Other Sampling Methods
Section 1.5 Observational Studies and Experiments
Section 1.6 Completely Randomized Designs
Section 1.7 Blocking
Section 1.8 Data Ethics and the Scope of Inference
Chapter 1 Review
Chapter 1 Review Exercises

Statistics Matters How can we prevent malaria?

Throughout this book, we'll open each chapter with a case study illustrating a real-world application of statistics. In this chapter, we'll focus on the role of statistics in the fight to prevent malaria.[1]

Malaria causes hundreds of thousands of deaths each year, and many of the victims are children. In 2021, the World Health Organization approved the first malaria vaccine, but researchers continue to study other methods of prevention as well. Among the questions they have investigated: Would regularly screening children for the malaria parasite and treating those who test positive reduce the proportion of children who develop the disease? Researchers worked with children in 101 schools in southern Kenya, randomly assigning half of the schools to receive regular screenings with rapid diagnostic tests and follow-up treatments for those testing positive. The remaining schools received no regular screening. Children at all 101 schools were tested for malaria at the end of the study.[2]

If the proportion of children who develop the disease is smaller in the schools that received the screening, can we conclude that the screening caused the decrease in malaria? And can we apply the results of this study to children in other countries?

We'll revisit Statistics Matters *at the end of the chapter, so you can use what you have learned to help answer these questions.*

In this chapter, we'll introduce methods of collecting data. You will learn how to select a sample so the results of a survey can be generalized to a larger group of individuals. You will also learn the difference between an observational study and an experiment, and how to design a study so that you can make cause-and-effect conclusions. Welcome to the world of statistics!

Section 1.1 Introduction to Data Collection

LEARNING GOALS

By the end of this section, you will be able to:

- Identify individuals and variables in a data set, then classify variables as categorical or quantitative.
- Identify the population and the sample in a statistical study.
- Distinguish between an observational study and an experiment.

We live in a world of *data*. Every day, the media report poll results, outcomes of medical studies, and analyses of data on everything from gasoline prices to elections to spending habits to the latest technology. The data are trying to tell us a story. To understand what the data are saying, we use **statistics**.

Statistics is the science and art of collecting, analyzing, and drawing conclusions from data.

Statistics is a tool that will help you make informed decisions based on data in your field of study—and in your daily life. The following activity illustrates one of the many uses of statistics in the real world.

Class Activity

Hiring Discrimination?

An airline has just finished training 25 pilots—15 male and 10 female—to become captains. Unfortunately, only 8 captain positions are available. Airline managers announce that they will use a lottery to determine which pilots will fill the available positions. They write the names of all 25 pilots on identical slips of paper. They place the slips in a hat, mix them thoroughly, and draw them out one at a time until all 8 positions are filled. A day later, managers announce the results of the lottery. Of the 8 captains chosen, 5 are female and 3 are male. Some of the male pilots who weren't selected suspect that the lottery was not carried out fairly. They wonder if there may be grounds to file a grievance with the pilots' union.

Is it plausible (believable) that 5 females could be selected by chance alone? To find out, you and your classmates will use an applet to simulate the lottery process that airline managers said they used.

1. Go to www.stapplet.com and launch the *Hiring Discrimination* activity.
2. Click the button to enter an existing class code and enter the code provided by your instructor. If you are doing this activity on your own, click the button to do the activity independently.
3. Once you are in the applet, click the Generate sample button to randomly select a sample of 8 pilots from the population of all 25 pilots. The applet will indicate which pilots were selected, report the number of females in the sample, and add a dot corresponding to that value on the dotplot. Keep generating samples until there are at least 50 dots on the dotplot.
4. Based on the results of the simulation, is it plausible that 5 or more females could be selected by chance alone? Do these results provide convincing evidence that the lottery was unfair?

The previous activity outlined the steps in the **statistical problem-solving process**.[3] You'll learn more about the details of this process in future sections.

The **statistical problem-solving process** consists of the following steps:

- Ask questions: Clarify the research problem and ask one or more valid statistical questions.
- Collect/consider data: Gather appropriate data and identify the individuals and the variables.
- Analyze data: Use appropriate graphical and numerical methods to analyze the data.
- Interpret results: Draw conclusions based on the data analysis. Be sure to answer the statistical question(s)!

Individuals and Variables

This chapter focuses on the first two steps of the statistical problem-solving process. Subsequent chapters focus on analyzing data and interpreting results.

The following table displays data on several roller coasters from around the world.[4]

Roller coaster	Type	Height (ft)	Design	Speed (mph)	Duration (s)
Copperhead Strike	Steel	82.0	Sit down	50.0	144
Eurostar	Steel	98.9	Inverted	50.2	140
Jungle Trailblazer	Wood	108.3	Sit down	54.1	150
Falcon	Steel	197.5	Wing	73.3	156
Olympia Looping	Steel	106.7	Sit down	49.7	105
Time Traveler	Steel	100.0	Sit down	50.3	117

Most data tables follow this format — each row describes an **individual** and each column holds the values of a **variable**.

> An **individual** is a person, animal, or thing described in a set of data.

> A **variable** is any attribute that can take different values for different individuals.

Sometimes individuals are called *cases* or *observational units*.

For the roller coaster data set, the *individuals* are the six roller coasters. The five *variables* recorded for each coaster are type, height (in feet), design, speed (in miles per hour), and duration (in seconds). Type and design are **categorical** variables. Height, speed, and duration are **quantitative** variables.

> A **categorical** variable takes values that are labels, which place each individual into a particular group, called a category.

> A **quantitative** variable takes number values that are quantities — counts or measurements.

Sometimes categorical variables are called *qualitative variables* and quantitative variables are called *numerical variables*.

⚠ **Not every variable that takes number values is quantitative.** Zip code is one example. Although zip codes are numbers, they are neither counts nor measurements of anything. Instead, they are simply labels for a regional location — which makes zip code a categorical variable.

EXAMPLE 1.1.1

How many Twitter followers do members of Congress have?

Individuals and variables

PROBLEM

The Pew Research Center tracks the social media accounts for the 535 members of Congress. The first 5 rows of the following data table show information about the Twitter accounts for several members of Congress.[5]

Name	Party	Chamber	Number of followers	Number of posts	Average post retweets
Bernie Sanders	I	Senate	21,801,423	8234	3693
Elizabeth Warren	D	Senate	10,106,041	9051	1518
Alexandria Ocasio-Cortez	D	House	7,241,799	4527	8559
Ted Cruz	R	Senate	5,272,289	8601	1235
Nancy Pelosi	D	House	5,209,170	5372	3669
⋮	⋮	⋮	⋮	⋮	⋮

Identify the individuals and variables in this data set. Classify each variable as categorical or quantitative.

(continued)

SOLUTION

Individuals: The 535 members of Congress.

Variables:
- Categorical: party, chamber
- Quantitative: number of followers, number of posts, average post retweets

> We don't consider names to be variables, even though they clearly vary from individual to individual. Instead, we consider names to be identifiers for individuals. An ID number would likewise be considered an identifier, not a variable.

EXAM PREP: FOR PRACTICE, TRY EXERCISE 9.

The choice of data analysis method depends on whether a variable is categorical or quantitative. For that reason, it is important to distinguish between these two types of variables. To make life simpler, we sometimes refer to "categorical data" or "quantitative data" instead of identifying the variable as categorical or quantitative.

Populations and Samples

Suppose our statistical question is "What percentage of drivers in the United States text while driving?" To answer this question, we will survey U.S. drivers. Ideally, we would ask them all (take a **census**). But contacting every driver in the United States wouldn't be practical. It would take too much time and cost too much money. Instead, we put the question to a **sample** chosen to represent the entire **population** of drivers.

> The **population** in a statistical study is the entire group of individuals we want information about.

> A **census** collects data from every individual in the population.

> A **sample** is a subset of individuals in the population from which we collect data.

EXAMPLE 1.1.2

Vaccines and soda bottles

Populations and samples

PROBLEM

Identify the population and sample in each of the following settings.

(a) The health office at a university asks 100 students at the school if they have received a COVID-19 vaccine.

(b) The quality-control manager at a bottling company selects 10 bottles from the bottles filled during a particular hour to determine if the volume of soda is within acceptable limits.

SOLUTION

(a) Population: all students at the university
Sample: the 100 students surveyed

(b) Population: all bottles filled that hour
Sample: the 10 bottles inspected

> To identify the population, consider which individuals could have been selected for the sample. In part (b), the sample was only from bottles that hour, so the population is limited to all bottles produced that hour.

EXAM PREP: FOR PRACTICE, TRY EXERCISE 13.

Observational Studies and Experiments

A sample survey usually aims to gather information about a population without disturbing the population in the process. Sample surveys are one kind of **observational study**. Other examples of observational studies include watching the behavior of animals in the wild and tracking the medical history of patients to look for associations between variables such as diet, exercise, and heart disease.

In contrast to observational studies, **experiments** don't involve just observing individuals or asking them questions. Instead, experimenters actively *impose* some treatment to measure the response. Experiments can answer questions like "Does aspirin reduce the chance of a heart attack?" and "Do plants grow better when classical music is playing?"

An **observational study** observes individuals and measures variables of interest, but does not attempt to influence the responses.

An **experiment** deliberately imposes treatments (conditions) on individuals to measure their responses.

The goal of an observational study can be to describe some group or situation, to compare groups, or to examine relationships between variables. The purpose of an experiment is to determine if the treatment causes a change in the response. An observational study, even one based on a well-collected sample, is a poor way to gauge the effect that changes in one variable have on another variable. To see the response to a change, researchers must impose the change. *When the goal is to understand cause and effect, experiments are the only source of fully convincing data.* For this reason, the distinction between an observational study and an experiment is one of the most important ideas in statistics.

EXAMPLE 1.1.3

Family dinners and background music

Observational studies and experiments

PROBLEM

Determine if each of the following settings describes an observational study or an experiment. Explain your answer.

(a) Researchers at Columbia University randomly selected 1000 teenagers in the United States for a survey. According to an ABC News story about the research, "Teenagers who eat with their families at least five times a week are more likely to get better grades in school."[6]

(b) Does the pace of background music affect how fast people eat? Researchers at Aarhus University in Denmark asked volunteers to eat a piece of chocolate and rate it for taste and other features. They randomly assigned each volunteer to listen to either a slow-paced version or a fast-paced version of the same composition while eating. The researchers secretly recorded the amount of time it took the volunteers to finish the chocolate. The average eating time was shorter for the group who listened to the faster-paced music.[7]

SOLUTION

(a) Observational study. There were no treatments imposed on the teenagers. In other words, teenagers weren't told to eat with their families a certain number of times per week.

(b) Experiment. Treatments were imposed on the volunteers. Some volunteers were assigned to eat with fast-paced music playing and others were assigned to eat with slow-paced music playing.

EXAM PREP: FOR PRACTICE, TRY EXERCISE 17.

Because the background music and chocolate-eating study was a randomized experiment, researchers can conclude that the faster-paced music *caused* the volunteers to eat faster. However, in the observational study about teenagers, researchers should *not* conclude that eating more with their families causes teens' grades to be higher. You'll learn why in Section 1.5.

Section 1.1 What Did You Learn?

Review the learning goals from this section. Then practice what you've learned by working through the exercises.

Learning Goal	Example	Exercises
Identify individuals and variables in a data set, then classify variables as categorical or quantitative.	1.1.1	9–12
Identify the population and the sample in a statistical study.	1.1.2	13–16
Distinguish between an observational study and an experiment.	1.1.3	17–22

Section 1.1 Exercises

Building Concepts and Skills These exercises assess the basic knowledge you should have after reading the section.

1. The four steps of the statistical problem-solving process are _____, _____, _____, and _____.

2. True/False: An individual in a data set must be a human being.

3. Give an example of a quantitative variable and an example of a categorical variable.

4. True/False: Any variable that takes number values is a quantitative variable.

5. What is the difference between a population and a sample?

6. What is a census?

7. How does an experiment differ from an observational study?

8. True/False: Both observational studies and experiments can provide convincing evidence of a cause-and-effect relationship between two variables.

Mastering Concepts and Skills These exercises reinforce the learning goals as shown in the examples.

9. **Refer to Example 1.1.1 Box-Office Hits** *Avengers: Endgame* was the top-earning movie of 2019, based on box-office receipts worldwide. The following table displays data on several popular movies.[8] Identify the individuals and variables in this data set. Classify each variable as categorical or quantitative.

Movie	Year	Rating	Time (min)	Genre	Box office ($)
Avengers: Endgame	2019	PG-13	181	Action	2,795,473,000
Avatar	2009	PG-13	162	Action	2,789,705,275
Titanic	1997	PG-13	194	Drama	2,208,208,395
Star Wars: The Force Awakens	2015	PG-13	136	Adventure	2,053,311,220
Avengers: Infinity War	2018	PG-13	156	Action	2,048,134,200
Jurassic World	2015	PG-13	124	Action	1,648,854,864
The Lion King	2019	PG	118	Adventure	1,641,443,366
Furious 7	2015	PG-13	137	Action	1,518,722,794
Marvel's The Avengers	2012	PG-13	142	Action	1,517,935,897
The Avengers: Age of Ultron	2015	PG-13	141	Action	1,403,013,963
Black Panther	2018	PG-13	120	Action	1,348,258,224
Harry Potter and the Deathly Hallows: Part 2	2011	PG-13	130	Fantasy	1,341,693,157

10. **Hotels** A university's lacrosse team is planning to go to Buffalo for a three-day tournament. The tournament's sponsor provides a list of available hotels, along with some information about each hotel. The following table displays data about hotel options. Identify the individuals and variables in this data set. Classify each variable as categorical or quantitative.

Hotel	Pool	Exercise room?	Internet cost ($/day)	Restaurants	Distance to site (mi)	Room service?	Room rate ($/day)
Comfort Inn	Out	Y	0.00	1	8.2	Y	149
Fairfield Inn & Suites	In	Y	0.00	1	8.3	N	119
Baymont Inn & Suites	Out	Y	0.00	1	3.7	Y	60
Chase Suite Hotel	Out	N	15.00	0	1.5	N	139
Courtyard	In	Y	0.00	1	0.2	Dinner only	114
Hilton	In	Y	10.00	2	0.1	Y	156
Marriott	In	Y	9.95	2	0.0	Y	145

11. **Household Data** Every year, the U.S. Census Bureau collects data from more than 3 million households as part of the American Community Survey (ACS). The following table displays some data from the ACS in a recent year. Identify the individuals and variables in this data set. Classify each variable as categorical or quantitative.

Household	Region	Number of people	Time in dwelling (years)	Response mode	Household income	Internet access?
425	Midwest	5	2–4	Internet	$52,000	Yes
936459	West	4	2–4	Mail	$40,500	Yes
50055	Northeast	2	10–19	Internet	$481,000	Yes
592934	West	4	2–4	Phone	$230,800	No
545854	South	9	2–4	Phone	$33,800	Yes
809928	South	2	30+	Internet	$59,500	Yes
110157	Midwest	1	5–9	Internet	$80,000	Yes
999347	South	1	<1	Mail	$8400	No

12. **Car Buyers** A new-car dealer keeps records on car buyers for future marketing purposes. The following table gives information on the previous four buyers. Identify the individuals and variables in this data set. Classify each variable as categorical or quantitative.

Buyer's name	Zip code	Sex	Buyer's distance from dealer (mi)	Car model	City fuel economy (mpg)	Price
P. Smith	27514	M	13	Explorer	21	$32,675
K. Ewing	27510	M	10	Mustang	21	$39,500
L. Suarez	27516	F	2	Fusion	43	$38,400
S. Reice	27243	F	4	F-150	19	$56,000

13. **Refer to Example 1.1.2 Sampling Artifacts** An archaeological dig turns up large numbers of pottery shards, broken stone tools, and other artifacts. Students working on the project classify each artifact and assign it a number. The counts in different categories are important for understanding the site, so the project director chooses 2% of the artifacts at random and checks the students' work. Identify the population and the sample in this setting.

14. **Sampling Hardwood** A furniture maker buys hardwood in large batches. The supplier is supposed to dry the wood before shipping it (wood that isn't dry won't hold its size and shape). The furniture maker chooses five pieces of wood from each batch and tests their moisture content. If any piece exceeds 12% moisture content, the entire batch is sent back. Identify the population and the sample in this setting.

15. **Sampling Orange Juice** How much vitamin C does orange juice contain? A nutrition magazine measures the amount of vitamin C in 50 half-gallon containers of a popular brand of orange juice from 10 different grocery stores and concludes that the containers produced by this company do not have as much vitamin C as advertised. Identify the population and the sample in this setting.

16. **Sampling Envelopes** A large retailer prepares its customers' monthly credit card bills using a machine that folds the bills, stuffs them into envelopes, and seals the envelopes for mailing. To ensure that the envelopes are completely sealed, inspectors choose 40 envelopes at random from the 1000 stuffed each hour for visual inspection. Identify the population and the sample in this setting.

17. **Refer to Example 1.1.3 Indulgent Veggies** Can using indulgent names make college students more likely to eat their veggies? Each day, a cafeteria at Stanford University randomly labeled a vegetable dish in one of four ways: basic ("beets"), healthy restrictive ("lighter-choice beets with no added sugar"), healthy positive ("high-antioxidant beets"), and indulgent ("tangy lime-seasoned beets"). The number of students choosing the veggie was highest when it had the indulgent label, followed by basic, healthy positive, and healthy restrictive.[9] Is this an observational study or an experiment? Explain your answer.

18. **Chocolate and Babies** A University of Helsinki (Finland) study wanted to determine if a mother's consumption of chocolate during pregnancy had an effect on her infant's temperament at age 6 months. Researchers began by asking 305 healthy pregnant women to report their chocolate consumption. When the infants were 6 months old, the researchers asked the mothers to rate their infants' temperament, including smiling, laughter, and fear. The babies born to women who had eaten chocolate daily during pregnancy were found to be more active and "positively reactive" — a measure that the investigators said encompasses traits like smiling and laughter.[10] Is this an observational study or an experiment? Explain your answer.

19. **Social Media and Mood** How does social media use affect the mood and well-being of college students? A total of 143 students at the University of Pennsylvania helped answer this question. At the beginning of the study, they all took a survey that measured their mood and well-being. Then, half were chosen at random to continue their normal social media practices. The other half were limited to 10 minutes per day for each of Facebook, Snapchat, and Instagram. Social media use was verified by screenshots of battery use. The group that had limited social media use had significant decreases in loneliness and depression compared to the other group.[11] Is this an observational study or an experiment? Explain your answer.

20. **Effects of Child Care** A study of child care followed 1364 infants through their sixth year in school. Later, the researchers published an article in which they stated that "the more time children spent in child care from birth to age four-and-a-half, the more adults tended to rate them, both at age four-and-a-half and at kindergarten, as less likely to get along with others, as more assertive, as disobedient, and as aggressive."[12] Is this an observational study or an experiment? Explain your answer.

21. **Cleaning Pacifiers** Do parents who "clean" their babies' pacifiers by sucking on them help protect their children from allergies later in life? A cohort of 184 infants was followed for 36 months, with allergy tests being conducted at 18 and 36 months. Pacifier use and parental cleaning methods were recorded at 6 months in interviews with parents. Pacifier-using children whose parents cleaned their pacifiers by sucking on them were less likely to have eczema at both 18 and 36 months than were pacifier-using children whose parents didn't use this method.[13] Is this an observational study or an experiment? Explain your answer.

22. **Stopping Strokes** Aspirin prevents blood from clotting and so helps prevent strokes. The Second European Stroke Prevention Study asked if adding another anticlotting drug named dipyridamole would help. Patients who had already had a stroke were randomly assigned to receive either aspirin only, dipyridamole only, both aspirin and dipyridamole, or a pill with no active ingredient, and were followed for 2 years.[14] Is this an observational study or an experiment? Explain your answer.

Applying the Concepts These exercises ask you to apply multiple learning goals in a new context or to apply what you learned in this section in a new way.

23. **Coffee and Longevity** A long-term study of more than 400,000 older people in eight states recorded whether participants drank coffee, whether they were alive at the end of the study, and their age when they died

(if applicable). The researchers found that coffee drinkers tend to live longer than non-coffee drinkers.[15]

(a) Identify the variables in this study. Classify each variable as categorical or quantitative.

(b) Identify the population and the sample in this setting.

(c) Is this an observational study or an experiment? Explain your answer.

24. **Quitting Smoking** In an effort to reduce health care costs, General Motors sponsored a study to help employees stop smoking. In the study, 439 volunteer subjects were randomly assigned to receive up to $750 for quitting for a year, while the other 439 volunteer subjects were simply encouraged to stop smoking. After one year, people who received the financial incentive were 3 times more likely to have quit smoking.[16]

(a) Identify the variables in this study. Classify each variable as categorical or quantitative.

(b) Identify the population and the sample in this setting.

(c) Is this an observational study or an experiment? Explain your answer.

25. **Sampling Customers** A department store mails a customer-satisfaction survey to people who make credit card purchases at the store. This month, 45,000 people made credit card purchases. Surveys are mailed to 1000 of these people, chosen at random, and 137 people return the survey.

(a) Identify the population in this setting.

(b) Is the sample the 1000 people who were sent the surveys or the 137 people who responded? Explain your answer.

26. **Sampling Readers** To get feedback from people who read a local weekly newspaper, the newspaper's publisher inserts surveys into 1000 randomly selected copies of the paper. Of the 1000 surveys, 189 are returned to the newspaper's office.

(a) Identify the population in this setting.

(b) Is the sample the 1000 people who were given the surveys or the 189 people who responded? Explain your answer.

Extending the Concepts These exercises challenge you to explore statistical concepts and methods that go beyond what you learned in this section.

27. **Quantigorical?** In most data sets, age is classified as a quantitative variable. Explain how age could be classified as a categorical variable.

28. **Fertilizers** Do organic fertilizers work as well as chemical fertilizers when growing vegetables? Briefly explain how you could address this question with an observational study and with an experiment.

Section 1.2 Sampling: Good and Bad

LEARNING GOALS

By the end of this section, you will be able to:

- Describe how convenience sampling and voluntary response sampling can lead to bias.
- Explain how random sampling can help to avoid bias.
- Describe other sources of bias in a sample survey.

Many statistical studies use information from a sample to make a conclusion about an entire population. To ensure that these conclusions are accurate, we must be mindful of how the sample is selected.

Class Activity

Who Wrote the Federalist Papers?

In this activity, you will learn how statistics can be used to help identify the author of an anonymous text.

The Federalist Papers are a series of 85 essays supporting the ratification of the U.S. Constitution. When newspapers in New York published the essays in 1787 and 1788, the identity of the authors was a secret known to just a few people. Later, the authors were identified as Alexander Hamilton, James Madison, and John Jay. The authorship of 73 of the essays is fairly certain, leaving 12 in dispute. Thanks in some part to statistical analysis, most scholars now believe that these 12 essays were written by Madison alone or in collaboration with Hamilton.[17]

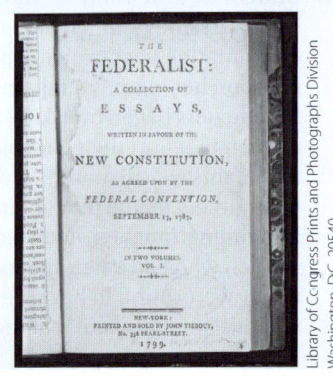

> We can use statistics in several ways to help determine the authorship of a text. One method is to estimate the average word length in a disputed text and compare it to the average word lengths of works where the authorship is not in dispute. The following paragraph is the opening of Federalist Paper No. 51, one of the disputed essays. The theme of this essay is the separation of powers among the three branches of government.
>
> To what expedient, then, shall we finally resort, for maintaining in practice the necessary partition of power among the several departments, as laid down in the Constitution? The only answer that can be given is, that as all these exterior provisions are found to be inadequate, the defect must be supplied, by so contriving the interior structure of the government as that its several constituent parts may, by their mutual relations, be the means of keeping each other in their proper places. Without presuming to undertake a full development of this important idea, I will hazard a few general observations, which may perhaps place it in a clearer light, and enable us to form a more correct judgment of the principles and structure of the government planned by the convention.
>
> 1. Choose 5 words from this passage. Count the number of letters in each of the words you selected and find the average word length.
> 2. Your instructor will draw and label a horizontal axis for a class dotplot. Plot the average word length you obtained in Step 1 on the graph.
> 3. Your instructor will show you how to use a random number generator to select a random sample of 5 words from the 130 words in the opening passage. Count the number of letters in each of the words you selected and find the average word length.
> 4. Your instructor will draw and label another horizontal axis with the same scale for a comparative class dotplot. Plot the average word length you obtained in Step 3 on the graph.
> 5. How do the dotplots compare? Can you think of any reasons why they might be different? Discuss with your classmates.

How to Sample Poorly: Convenience and Voluntary Response Samples

Suppose we want to know how much time students at a university spent doing homework last week. We might go to the university library and ask the first 30 students we see about their homework time. The sample we get is called a **convenience sample**.

> A **convenience sample** consists of individuals from the population who are easy to reach.

Convenience samples tend to produce poor estimates of the values we want to know because the members of the sample often differ from the population in ways that affect their responses. For example, students who are in the library probably spend more time doing homework than a typical student does. If we use this group as our sample, our estimate for the average amount of time spent doing homework will be too high. In fact, if we were to repeat this sampling process again and again, we would almost always overestimate the average time spent doing homework in the population of all students. This predictable overestimation is due to **bias** in the sampling method.

> The design of a statistical study shows **bias** if it is very likely to underestimate or very likely to overestimate the value you want to know.

Many websites have polls that ask visitors to express their opinions. These polls, along with other call-in or write-in polls, are not reliable because they use **voluntary response samples** (sometimes called *self-selected* samples).

> A **voluntary response sample** consists of people who choose to be in the sample by responding to a general invitation.

Voluntary response sampling leads to bias when the members of the sample differ from the population in ways that affect their responses. An advice columnist once asked her followers, "If you had it to do over again, would you have children?" She received nearly 10,000 responses from parents, with almost 70% saying "NO!"[18] Can it be true that 70% of parents regret having children? Not at all. People who feel strongly about an issue, particularly people with strong negative feelings, are more likely to take the trouble to respond. These results are misleading—the percentage of parents who said "No" in her sample is much higher than the proportion who would say "No" in the population of all parents.

EXAMPLE 1.2.1

Do you sleep with a dog?

Convenience and voluntary response samples

PROBLEM

Do most U.S. adult females sleep with a dog? In a study of U.S. adult females, 55% of respondents said they sleep with a dog.[19] This was based on a study from researchers at Canisius College, whose report included the following description of their sampling method[20]:

> Anyone over 18 years of age was eligible to participate in our online survey, but the results in this paper are restricted to female participants residing in the United States who were between 18 and 69 years old. We emailed personalized invitations to 1229 individuals who had previously participated in studies conducted by the Canisius Canine Research Team (CCRT) . . . Initial and reminder emails included a link to the survey that we asked recipients to share with their friends via email and social media sites.

Explain why this sampling method is biased. Is the percentage of women in the sample who sleep with a dog likely greater than or less than the percentage of all adult women in the United States who sleep with a dog?

SOLUTION

The researchers obtained a convenience sample by sending personalized invitations to individuals who had previously completed studies by the Canine Research Team. These people, along with the friends they shared the survey with, are more likely to own (and sleep with) a dog than the rest of the population. As a result, the percentage of women in the sample who sleep with a dog is likely greater than the percentage of all women in the United States who sleep with a dog.

EXAM PREP: FOR PRACTICE, TRY EXERCISE 9.

Bias is not just bad luck in one sample. It's the result of a bad study design that will consistently miss the truth about the population in the same direction.

How to Sample Well: Random Samples

In convenience sampling, the researcher chooses easy-to-reach members of the population. In voluntary response sampling, people decide whether to join the sample. Both sampling methods suffer from bias due to personal choice. The best way to avoid this problem is to let chance choose the sample. That's the idea behind a **random sample**.

> A **random sample** consists of individuals from the population who are selected for the sample using a chance process.

If the advice columnist mentioned earlier had surveyed a random sample of parents instead of using a voluntary response sample, bitter parents would be less likely to be overrepresented in the sample. In other words, the random sample should be much more representative of the population. In fact, when *Newsday* asked the same question to a random sample of parents, 91% said they *would* have kids again.[21]

EXAMPLE 1.2.2

Do you sleep with a dog?

Random samples

PROBLEM

Explain how the researchers from Example 1.2.1 could have avoided bias in their sampling method.

SOLUTION

The researchers could have avoided bias by sending surveys to a random sample of U.S. adult females. If they had used random sampling, dog owners would be much less likely to be overrepresented in the sample and the sample would be more representative of all U.S. females.

EXAM PREP: FOR PRACTICE, TRY EXERCISE 13.

Other Sources of Bias

Even when samples are selected at random, there are other potential sources of bias in a survey. Some of these include *undercoverage, nonresponse,* and *response bias.*

Undercoverage

Some samples are selected by randomly calling landline telephones. Unfortunately, this means people who only use a cell phone (or no phone at all) cannot be included in the sample. These types of samples suffer from **undercoverage.**

> **Undercoverage** occurs when some members of the population are less likely to be chosen or cannot be chosen for the sample.

Undercoverage leads to bias when the "undercovered" individuals differ from the population in ways that affect their responses. For example, randomly dialing landlines is likely to underrepresent young people because many younger people exclusively use a cell phone. If a landline survey is about a tax increase that will help reduce the cost of college, it seems likely that the proportion in the sample who favor the proposal will be less than the proportion in the entire population.

To avoid the bias caused by undercoverage, every member of the population should be given the same chance to be selected. Ideally, samples would be chosen from a *sampling frame* that lists every member of the population. If the sampling frame doesn't contain every member of the population, undercoverage will occur.

Nonresponse

Even if every member of the population is equally likely to be selected for a sample, not all members of the population are equally likely to provide a response. Some people are rarely at home and cannot be reached by pollsters on the phone or in person. Other people see an unfamiliar phone number on their caller ID and never pick up the phone or quickly hang up when they don't recognize the voice of the caller. These are examples of **nonresponse**, another major source of bias in surveys.

> **Nonresponse** occurs when an individual chosen for the sample can't be contacted or refuses to participate.

Nonresponse leads to bias when the individuals who can't be contacted or who refuse to participate differ from the population in ways that affect their responses. Consider a study that uses a mail-in survey to estimate the average number of hours that people work in a typical week. Because people who work very long hours are less likely to respond than people who work fewer hours, the estimated mean from the sample is likely to be less than the mean for the entire population.

How bad is nonresponse? According to the Pew Research Group, the response rate for phone surveys declined from 36% in 1997 to only 9% in 2012. Since 2012, however, the response rate has held steady at about 9%.[22]

The best way to reduce the effect of nonresponse is to follow up with people who don't respond the first time. In the mail-in survey about work hours, the researchers could call people who didn't reply to the mail-in survey or visit the participants at their home. In other types of surveys, polling companies encourage people to respond by offering a small payment as an incentive.

(!) **Don't use the term "voluntary response" instead of "nonresponse" to explain why certain individuals don't respond after being selected for a sample survey.** Think about it this way: Nonresponse can occur only *after* a sample has been selected. In a voluntary response sample, every individual has opted to take part, so there won't be any nonresponse.

Response Bias

Even when a researcher can avoid undercoverage and nonresponse, it is still possible for bias to affect the results. Characteristics of the interviewer, wording of the question, lack of anonymity, and many other factors can lead to **response bias**.

> **Response bias** occurs when there is a consistent pattern of inaccurate responses to a survey question.

A badly worded question can lead to misleading results. The Roper polling organization once asked a random sample of U.S. adults: "Does it seem possible or does it seem impossible to you that the Nazi extermination of the Jews never happened?" Only 65% said it was impossible that the Holocaust never happened, making it seem as if nearly one-third of respondents expressed doubt about the Holocaust's reality. But the wording of the question included a double negative, making it difficult to understand. When the question was revised to remove the double negative, the percentage who were certain that the Holocaust occurred increased dramatically.[23]

Despite involving a high degree of athleticism and competition, cheerleading is not currently recognized as a sport by the National Collegiate Athletic Association (NCAA) or by U.S. federal Title IX guidelines. Do students consider cheerleading a sport? It depends on who's asking. When a cheerleader surveyed students at her school, 80% said "Yes" when she was wearing her cheerleading outfit. When she was dressed in regular clothes, only 28% said "Yes."[24] It appears that people changed their answer when she was wearing her cheerleading outfit so they wouldn't offend her.

EXAMPLE 1.2.3

Did you wash your hands?

Other sources of bias

PROBLEM
What percentage of Americans wash their hands after using the bathroom? The answer to this question depends on how you collect the data. In a telephone survey of 1006 U.S. adults, 96% said they always wash their hands after using a public restroom. An observational study of 6028 adults in public restrooms told a different story: Only 85% of those observed washed their hands after using the restroom. Explain why the results of the two studies are so different.[25]

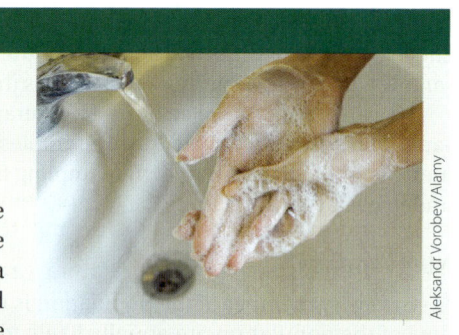

SOLUTION
When asked in person, many people might lie about always washing their hands because they want to appear as if they have good hygiene. When people are only observed and not asked directly, the percentage who wash their hands is smaller — and much closer to the truth.

EXAM PREP: FOR PRACTICE, TRY EXERCISE 17.

In the preceding example, the 85% figure was obtained by stationing an "observer" in the bathroom who was brushing teeth or putting on make-up. Imagine how much lower the percentage of hand washers might be if no one else was present in the bathroom to encourage good behavior!

Completely avoiding bias when conducting a survey is likely impossible. But there are ways that thoughtful researchers can keep bias to a minimum. Selecting a random sample is a good start. And taking steps to avoid undercoverage, nonresponse, and response bias can make surveys even better. Finally, it is a good idea to test out the survey on a small group of people before administering it to the entire sample. This type of pilot study can help uncover any problems with question wording or other unforeseen issues.

Section 1.2 What Did You Learn?

Review the learning goals from this section. Then practice what you've learned by working through the exercises.

Learning Goal	Example	Exercises
Describe how convenience sampling and voluntary response sampling can lead to bias.	1.2.1	9–12
Explain how random sampling can help to avoid bias.	1.2.2	13–16
Describe other sources of bias in a sample survey.	1.2.3	17–22

Section 1.2 Exercises

Building Concepts and Skills These exercises assess the basic knowledge you should have after reading the section.

1. True/False: Convenience samples usually produce reliable estimates.

2. What is bias?

3. Give another name for a voluntary response sample.

4. State a benefit of using a random sample rather than a convenience or voluntary response sample.

5. When does undercoverage lead to bias?

6. How does nonresponse differ from voluntary response?

7. What are two ways that researchers can minimize nonresponse?

8. List three factors that can lead to response bias.

Mastering Concepts and Skills These exercises reinforce the learning goals as shown in the examples.

9. **Refer to Example 1.2.1 Infested Hemlocks** A forester studying the woolly adelgid, an invasive species of insect that infests and weakens hemlock trees, wants to estimate the proportion of hemlocks in a large forest that are infested. The forester starts at a roadside rest area and walks along the road, using the first 50 hemlocks encountered as the sample. Explain why this sampling method is biased. Is the proportion of infected hemlock trees in the sample likely to be greater than or less than the proportion of all hemlock trees in the forest that are infested?

10. **Car Bolts** The quality-control department at an automobile factory checks the tightness of motor-mounting bolts installed by assembly-line workers by sampling the final 25 cars produced by the assembly line each day. Explain why this sampling method is biased. Is the proportion of bolts in the sample that are improperly tightened likely to be greater than or less than the proportion of all bolts in cars that day that are improperly tightened?

11. **Online Reviews** Many websites include customer reviews of products, restaurants, hotels, and so on. The manager of a hotel was upset to see that 26% of reviewers on a travel website gave the hotel a "1 star" rating—the lowest possible rating. Explain why this sampling method is biased. Is the percentage of all the hotel's customers who would give the hotel 1 star in a review likely to be greater than or less than 26%?

12. **Boaty McBoatface** Britain's Natural Environment Research Council invited the public to name its new $300 million polar research ship. To vote on the name, people simply needed to visit a website and record their choice. Ignoring names suggested by the council, more than 124,000 people voted for "Boaty McBoatface," which ended up having more than 3 times as many votes as the second-place name.[26] Explain why this sampling method is biased. Is the proportion who voted for Boaty McBoatface in the sample likely to be greater than or less than the proportion of all British residents who prefer the name Boaty McBoatface?

13. **Refer to Example 1.2.2 Infested Hemlocks** Explain how to avoid the bias you described in Exercise 9.

14. **Car Bolts** Explain how to avoid the bias you described in Exercise 10.

15. **Online Reviews** Explain how to avoid the bias you described in Exercise 11.

16. **Boaty McBoatface** Explain how to avoid the bias you described in Exercise 12.

17. **Refer to Example 1.2.3 Immigration Reform** A news organization wants to know what percentage of U.S. residents support a "pathway to citizenship" for people who came to the United States illegally. The news organization randomly selects registered voters for the survey. Explain how undercoverage might lead to bias in this study. Is the percentage from the sample likely to be greater than or less than the percentage for all U.S. residents?

18. **Eating on Campus** The director of student life at a university wants to know what percentage of students eat regularly on campus. To find out, the director selects a random sample of 300 students who live in the dorms. Explain how undercoverage might lead to bias in this study. Is the percentage from the sample likely to be greater than or less than the percentage for all students?

19. **Driving Distance** Researchers began a survey of drivers by randomly dialing telephone numbers in the United States. Of 45,956 calls to these numbers, 5029 were completed. The goal of the survey was to estimate how far people drive, on average, per day.[27] Explain how nonresponse might lead to bias in this study. Is the average from the sample likely to be greater than or less than the average for all U.S. drivers?

20. **Weight Loss** Three hundred people participated in a free 12-week weight-loss course at a community health clinic. After one year, administrators emailed each of the 300 participants to see how much weight they had lost since the end of the course. Only 56 participants responded to the survey. The average weight loss for this sample was 13.6 pounds. Describe how nonresponse might lead to bias in this study. Is the mean amount of weight loss in the sample likely to be greater than or less than the average for all 300 participants in the course?

21. **Seat Belts** A study in El Paso, Texas, looked at seat belt use by drivers. Drivers were observed at randomly chosen convenience stores. After they left their cars, they were invited to answer questions about seat belt use. In all, 75% said they always used seat belts, yet only 61.5% were wearing seat belts when they pulled into the stores' parking lots.[28] Explain why the two percentages are so different.

22. **Homeless Care** A researcher asked 80 randomly selected people if free health care should be provided to people who are homeless. Half of the people were shown a picture of a homeless woman with a small child. When shown the picture, 67.5% agreed that free health care should be provided to people experiencing homelessness. When the picture was not shown, only 45% agreed with this statement.[29] Explain why the two percentages are so different.

Applying the Concepts These exercises ask you to apply multiple learning goals in a new context or to apply what you learned in this section in a new way.

23. **Literary Digest** One of the most famous flops in survey history occurred in 1936. To predict the outcome of the presidential election between Republican Alf Landon and Democrat Franklin D. Roosevelt, the magazine *Literary Digest* sent more than 10 million "ballots" to its subscribers. It also sent "ballots" to registered owners of an automobile or telephone. Approximately 2.4 million of the ballots were returned, with a large majority (57%) favoring Landon. The election did turn out to be a landslide—but for Roosevelt (61%) instead of Landon.[30]
    (a) Explain how undercoverage might have led to bias in this survey.
    (b) Explain how nonresponse might have led to bias in this survey.
    (c) If the magazine followed up with people who didn't return their ballots and was able to obtain responses, would this eliminate the bias described in part (a) or (b)? Explain your reasoning.

24. **Red Lights** Researchers asked a sample of 880 drivers the following question: "Recalling the last 10 traffic lights you drove through, how many of them were red when you entered the intersections?" Of the 880 respondents, 171 admitted that at least one light had been red. The drivers were selected by randomly dialing residential (landline) telephone numbers.[31]
    (a) Explain how undercoverage might have led to bias in this survey.
    (b) Explain how response bias might affect the results of this survey.
    (c) If the researchers also randomly dialed cell phone numbers, would this eliminate the bias described in part (a) or (b)? Explain your reasoning.

25. **National Health Insurance** Researchers select a random sample of 1200 adult Americans and ask each person the following question:
    In light of the huge national deficit, should the government at this time spend additional money to establish a national system of health insurance?
    (a) Explain how the wording of this question could result in bias. Is the percentage from the sample who say yes likely to be greater than or less than the true percentage of adult Americans who would say they favor establishing a national system of health insurance?
    (b) Create two new questions about establishing a national system of health insurance. Write one that is unbiased and one that is biased in the opposite direction as the original question.

26. **Citizens United** Researchers select a random sample of 1200 adult Americans and ask each person the following question:
    Do you approve or disapprove of the U.S. Supreme Court decision in Citizens United *that let corporations and wealthy individuals spend unlimited amounts on campaigns?*
    (a) Explain how the wording of this question could result in bias. Is the percentage from the sample who say they approve likely to be greater than or less than the true percentage of adult Americans who would say they approve of the *Citizens United* decision?
    (b) Create two new questions about the *Citizens United* ruling. Write one that is unbiased and one that is biased in the opposite direction as the original question.

Extending the Concepts These exercises challenge you to explore statistical concepts and methods that go beyond what you learned in this section.

27. **Randomized Response** When asked sensitive questions, many people give untruthful responses—especially if the survey is not anonymous. To encourage honest answers, researchers developed the "randomized response" method. For example, suppose you want to ask students if they have cheated on an exam this year. Before they answer, have each student privately flip a coin. If the coin lands on heads, the student must answer truthfully. If it lands on tails, the student must answer "Yes."
    (a) Explain why this method might encourage students to answer the question honestly.
    (b) Suppose that 100 students used this method to answer the question about cheating on an exam this year. Estimate the proportion of students who have cheated if 63 students said "Yes" and 37 students said "No."

Cumulative Review These exercises revisit what you learned in previous sections.

28. **Nightlights (1.1)** Does leaving a light on while an infant sleeps affect the child's risk of becoming nearsighted? A study published in the journal *Nature* followed 1220 children for several years. For each child, researchers recorded whether the child slept in darkness, with a nightlight, or in a fully lit room. Researchers also recorded whether the child went on to develop nearsightedness.[32]
    (a) Identify the individuals and variables in this study. Classify each variable as categorical or quantitative.
    (b) Identify the population and sample in this setting.
    (c) Is this an observational study or an experiment? Explain your answer.

Section 1.3 Simple Random Sampling

LEARNING GOALS

By the end of this section, you will be able to:
- Describe how to obtain a simple random sample using a random number generator.
- Explain the concept of sampling variability and the effect of changing sample size.
- Use simulation to test a claim about a population proportion.

In Section 1.2, you learned that choosing a sample at random is a good way to avoid bias. In this section, we start with the details of how to select one type of random sample.

Choosing a Simple Random Sample

At the beginning of each class session, Professor Hebert randomly selects five students to present homework problems on the whiteboard. To make the selection, she uses a deck of index cards with each student's name written on one card. She shuffles the cards well and selects five of them. The students whose names are on the selected cards are the ones chosen to present the homework problems. This type of sample is called a **simple random sample (SRS).**

> A **simple random sample (SRS)** of size n is a sample chosen in such a way that every group of n individuals in the population has an equal chance of being selected as the sample.

In an SRS, each sample of size n has the same chance to be chosen. This implies that each member of the population has the same chance of being chosen for the sample.

Of course, Professor Hebert doesn't have to use index cards to choose an SRS. She can also use technology, which is especially helpful when the population is large.

HOW TO Select an SRS with Technology

1. **Label.** Give each individual in the population a distinct integer label from 1 to N, where N is the number of individuals in the population.
2. **Randomize.** Use a random number generator to obtain n *different* integers from 1 to N.
3. **Select.** Choose the individuals that correspond to the randomly selected integers.

When choosing an SRS, we make the selections *without replacement*. That is, once an individual is selected for the sample, that individual cannot be selected again. Some random number generators sample *with replacement* by default, while others give you the choice of sampling with or without replacement. If you aren't given a choice, it is important that you explain that repeated numbers should be ignored when selecting an SRS. If you are given a choice, choose without replacement.

You can use a graphing calculator to generate random numbers, as shown in the Tech Corner at the end of this section. There are also many random number generators available on the internet, including those at www.random.org. You can even ask Siri!

EXAMPLE 1.3.1

What do we do when the landfill is full?

Selecting an SRS

PROBLEM

When landfills run out of room, they are sometimes converted into parks. Does the presence of a landfill underneath affect the trees that grow in the park? One such park has 283 trees, and researchers want to closely inspect a sample of 20 trees. Describe how to select an SRS of 20 trees using a random number generator.

(continued)

SOLUTION

Give each tree a distinct integer label from 1 to 283. Randomly generate 20 different integers from 1 to 283. Inspect the 20 trees that are labeled with the generated integers.

> Remember to address the possibility of repeated integers by stating "ignore repeated integers" or "generate 20 different integers."

EXAM PREP: FOR PRACTICE, TRY EXERCISE 7.

Sampling Variability

In the preceding example, researchers selected an SRS of $n = 20$ trees from a park that was built on top of a landfill. Based on their sample of 20 trees, they estimated that the mean height of all the trees in the park is 18.6 feet. Is this estimate likely to be correct? Unfortunately, the answer is "no," even though the researchers used an unbiased sampling method. This isn't a contradiction. Instead, it illustrates the concept of **sampling variability.**

> The fact that different random samples of the same size from the same population produce different estimates is called **sampling variability.**

Here is a dotplot showing the mean height for each of 50 different SRSs of size $n = 20$ trees in this park. The original sample, which produced an estimated mean height of 18.6 feet, is shown in red.

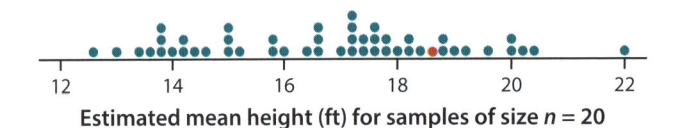

Estimated mean height (ft) for samples of size $n = 20$

The estimated means vary from 12.6 feet to 22.0 feet, even though the samples were all the same size and from the same population. Note that the dots in this dotplot are not tree heights, but rather *mean* tree heights for samples of size $n = 20$.

We can reduce sampling variability by using a larger sample size. Here are the mean heights for each of 50 SRSs of size $n = 80$ trees.

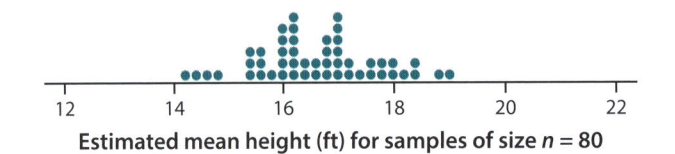

Estimated mean height (ft) for samples of size $n = 80$

Notice that the sample means are now more consistent, varying from just 14.2 feet to 19.0 feet. In other words, our estimates are typically closer to the true mean when using a larger sample size.

EXAMPLE 1.3.2

What's the median income in Virginia?

Sampling variability

PROBLEM

In the American Community Survey, the U.S. Census Bureau interviews households across the country. Based on a random sample of 49,309 Virginia households in 2019, the Census Bureau estimated the median household income in Virginia to be $76,456.[33]

(a) Do you think that the median income of *all* Virginia households was equal to $76,456 in 2019? Explain your answer.

(b) If the sample size was only 1000 households, what effect would this have on the estimated median income?

SOLUTION

(a) No. Because different random samples of 49,309 Virginia households will produce different medians, it is unlikely this sample provides a median that is exactly correct.

(b) Because the sample size is smaller, the estimated median income is less likely to be close to the true median income of all Virginia households.

EXAM PREP: FOR PRACTICE, TRY EXERCISE 11.

⚠ **Increasing the sample size does not reduce bias.** In the example about washing your hands after going to the bathroom from Section 1.2, increasing the number of people in the survey won't change the fact that some respondents will lie about their behavior.

Inference for Sampling

The purpose of obtaining a sample is to give us information about a larger population. The process of drawing conclusions about a population on the basis of sample data is called *inference* because we *infer* information about the population from what we know about the sample.

On March 11, 2021, newly inaugurated President Joe Biden approved a $1.9 trillion relief package, sending $1400 checks to most Americans. Shortly after Biden signed the relief package, Marist University asked a random sample of 1309 U.S. adults if they approved of the job Joe Biden was doing as president. In the sample, 52% said that they approved.[34]

Do these data provide convincing evidence that a majority of *all* U.S. adults approved of Biden's job performance in March 2021? Although more than half of the sample approved of his job performance, it is possible that only 50% (or less) of all U.S. adults approved of his job performance and that the greater-than-50% estimate from the sample was due to sampling variability.

Here are the results of 100 simulated SRSs of size $n = 1309$, assuming that exactly 50% of all U.S. adults approved of Biden's job performance. Each dot on the dotplot represents the percentage who approved of Biden in one simulated sample.

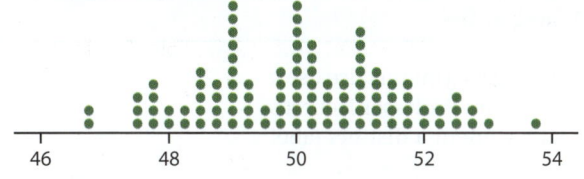

Simulated percent who approved of the president's job performance

In 100 samples of $n = 1309$ from a population where 50% approve of the president's job performance, the estimated job approval varied from about 47% to about 54%, with 11 of the 100 simulated samples producing estimates of 52% or greater. Because it is somewhat likely to get estimates of 52% or higher due to sampling variability alone, the actual estimate of 52% does not provide convincing evidence that more than 50% of all U.S. adults approved of Biden's job performance in March 2021.

EXAMPLE 1.3.3

Is the last Kiss the best Kiss?

Testing a claim

PROBLEM
Do people have a preference for the last thing they taste? Psychologists at the University of Michigan designed a study to find out. The researchers gave 22 students five different flavors of Hershey's Kisses® (milk chocolate, dark chocolate, crème, caramel, and almond) in random order and asked the students to rate each one. Participants were not told how many Kisses they would

(continued)

taste. However, when the fifth and final Kiss was presented, participants were informed that it would be their last one. Of the 22 students in the sample, 14 (64%) gave the final Kiss the highest rating. This is greater than 20% — the expected percentage if people don't have a special preference for the last thing they taste.[35]

To determine if these data provide convincing evidence that more than 20% of students like these prefer the last Kiss they taste, 100 simulated SRSs were selected. Each dot in the graph shows the simulated number of students (out of 22) who preferred the last Kiss, assuming each student had a 20% chance of choosing the last Kiss.

(a) Explain how the graph illustrates the concept of sampling variability.

(b) In the simulation, what percentage of the samples resulted in 14 or more students preferring the last Kiss?

(c) In the actual study, 14 students preferred the last Kiss. Based on your answer to part (b), is there convincing evidence that students like these have a preference for the last Kiss they taste? Explain your reasoning.

SOLUTION

(a) If researchers take many random samples of size 22 from a population of students who show no preference for what they taste last, the number of students in a sample who prefer the last Kiss varies from about 1 to 10.

(b) 0/100 = 0% of the simulated samples resulted in 14 or more students preferring the last Kiss.

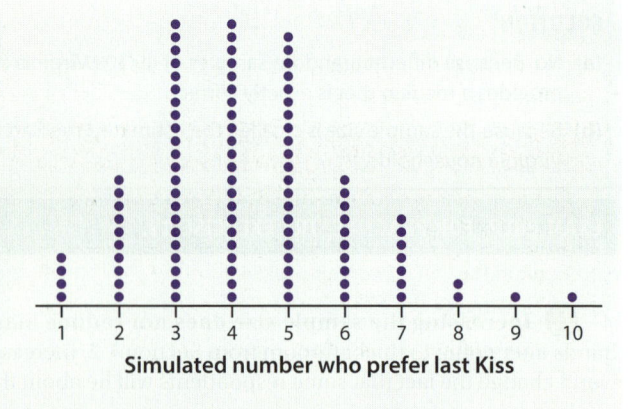

(c) Because it is unlikely to get 14 or more students who prefer the last Kiss due to sampling variability alone, there is convincing evidence that students like these prefer the last thing they taste.

EXAM PREP: FOR PRACTICE, TRY EXERCISE 15.

TECH CORNER Choosing an SRS on a TI-83/84

In addition to using a website to generate random integers, you can also use a graphing calculator. Let's select an SRS of 5 students from Professor Hebert's class using a TI-83/84. She has already given each student a distinct label from 1 to 35.

1. Press MATH, then select PROB (PRB) and randInt(.

2. Complete the command randInt(1,35) and press ENTER until you have chosen 5 *different* numbers. If you are using OS 2.55 or later, leave the entry for *n* blank, as shown in the screen shot.

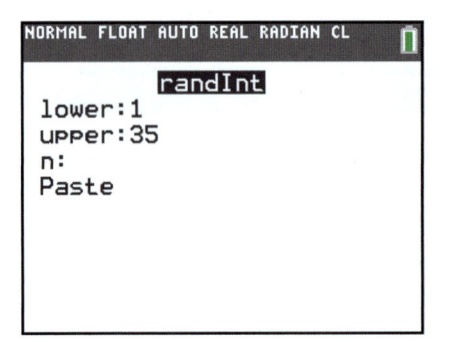

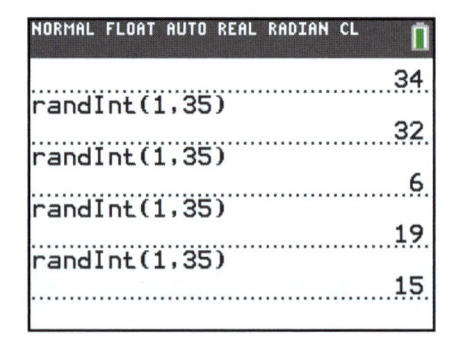

Note: If you have a TI-84 Plus CE or TI-84 Plus C Silver Edition, you can use the command RandIntNoRep(1,35,5) to get a random sample of 5 *different* integers from 1 to 35. If you have a TI-84 with OS 2.55 or later, you can use the command RandIntNoRep(1,35) to sort the integers from 1 to 35 in random order. The first 5 numbers listed give the labels of the chosen students.

Section 1.3 What Did You Learn?

Review the learning goals from this section. Then practice what you've learned by working through the exercises.

Learning Goal	Example	Exercises
Describe how to obtain a simple random sample using a random number generator.	1.3.1	7–10
Explain the concept of sampling variability and the effect of changing sample size.	1.3.2	11–14
Use simulation to test a claim about a population proportion.	1.3.3	15–18

Section 1.3 Exercises

Building Concepts and Skills These exercises assess the basic knowledge you should have after reading the section.

1. What is a simple random sample?
2. True/False: In an SRS, individuals are selected with replacement.
3. What is sampling variability?
4. True/False: Increasing the sample size helps to reduce sampling variability.
5. True/False: Increasing the sample size helps to reduce bias.
6. What is inference?

Mastering Concepts and Skills These exercises reinforce the learning goals as shown in the examples.

7. **Refer to Example 1.3.1 Sampling Pines** To gather data on a 1200-acre pine forest in Louisiana, the U.S. Forest Service laid a grid of 1410 equally spaced circular plots over a map of the forest. The Forest Service wants to visit a sample of 10% of these plots.[36] Describe how to select an SRS of 10% of the plots.

8. **Sampling Gravestones** The local genealogical society in Coles County, Illinois, has compiled records on all 55,914 gravestones in cemeteries in the county for the years 1825 to 1985. Historians plan to use these records to learn about African Americans in Coles County's history. They first choose an SRS of 395 records to check their accuracy by visiting the actual gravestones.[37] Describe how to select an SRS of 395 gravestones.

9. **Football Games** Most National Football League (NFL) games last more than 3 hours, but how much actual playing time is there in a typical NFL game? In 2019, researchers Kirk Goldsberry and Katherine Rowe investigated this question as part of a sports analytics class they taught at the University of Texas using an SRS of 7 games.[38] Describe how to select an SRS of 7 of the 256 regular season games in 2019.

10. **High Finance** A financial research firm wants to study the relationship between the salary of a company's CEO and the 12-month return on the company's stock for the 2800 companies listed on the New York Stock Exchange (NYSE).[39] Describe how to select an SRS of 200 NYSE companies.

11. **Refer to Example 1.3.2 Sampling Pines** Refer to Exercise 7. Based on the sample of plots, the U.S. Forest Service estimated that the mean volume of lumber-quality pine was 2196 ft^3/acre.
    (a) Do you think that the mean volume of lumber-quality pine in the entire forest is equal to 2196 ft^3/acre? Explain your answer.
    (b) If the sample size was decreased to 5% of the plots, what effect would this have on the estimated mean?

12. **Sampling Gravestones** Refer to Exercise 8. Based on the sample, the historians estimated that 0.5% of the gravestones in Coles County during this period are for African Americans.
    (a) Do you think that 0.5% is the true percentage of all gravestones in Coles County during this period that are for African Americans? Explain your answer.
    (b) If the sample size was increased to 1000 gravestones, what effect would this have on the estimated percentage?

13. **Football Games** Refer to Exercise 9. Based on the sample of games, the researchers estimated that NFL games include 17.71 minutes of actual playing time, on average.
    (a) Do you think that the mean amount of playing time in all 2019 NFL games is 17.71 minutes? Explain your answer.
    (b) If the sample size was increased to 20 games, what effect would this have on the estimated mean?

14. **High Finance** Refer to Exercise 10. Based on the sample of companies, the firm estimated that the average 12-month return for companies listed on the NYSE is 9.8%.
    (a) Do you think that the average 12-month return for all NYSE companies is 9.8%? Explain your answer.
    (b) If the sample size was decreased to 100 companies, what effect would this have on the estimated average?

15. **Refer to Example 1.3.3** **Sleep** Sleep Awareness Week begins in the spring with the release of the National Sleep Foundation's annual poll of U.S. sleep habits and ends with the beginning of daylight saving time, when most people lose an hour of sleep.[40] In the foundation's random sample of 1029 U.S. adults, 48% reported that they "often or always" got enough sleep during the previous 7 nights.

To determine if these data provide convincing evidence that fewer than half of all U.S. adults would say they often or always get enough sleep, 100 simulated SRSs of size $n = 1029$ were selected. Each dot in the graph shows the percentage of adults who said they often or always get enough sleep in a simulated SRS, assuming that each adult has a 50% chance of saying they often or always get enough sleep.

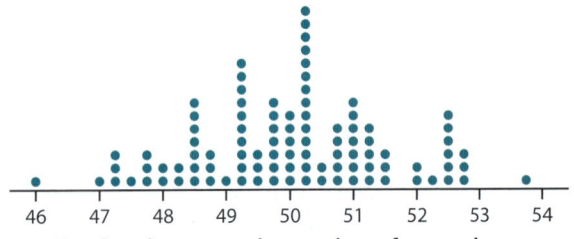

(a) Explain how the graph illustrates the concept of sampling variability.

(b) In the simulation, what percentage of the samples resulted in 48% or fewer of the adults saying they often or always get enough sleep?

(c) In the actual study, 48% of the adults said they often or always got enough sleep. Based on your answer to part (b), is there convincing evidence that fewer than half of all U.S. adults would say they often or always get enough sleep? Explain your reasoning.

16. Kissing the Right Way According to a study, the majority of couples prefer to tilt their heads to the right when kissing. In the study, a researcher observed a random sample of 124 kissing couples and found that 83/124 (66.9%) of the couples tilted to the right.[41]

To determine if these data provide convincing evidence that couples are more likely to tilt their heads to the right, 100 simulated SRSs of size $n = 124$ were selected. Each dot in the graph shows the number of couples who tilted to the right in a simulated SRS, assuming that each couple has a 50% chance of tilting to the right.

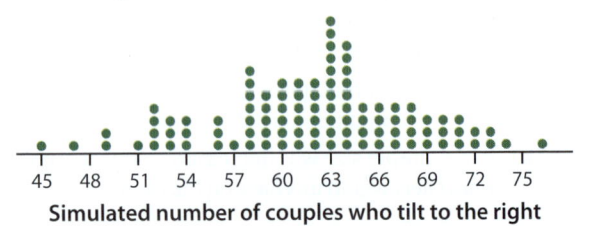

(a) Explain how the graph illustrates the concept of sampling variability.

(b) In the simulation, what percentage of the samples resulted in 83 or more couples tilting to the right?

(c) In the actual study, 83 couples tilted to the right. Based on your answer to part (b), is there convincing evidence that couples prefer to kiss this way? Explain your reasoning.

17. Atmospheric Gases The Pew Research Center and *Smithsonian* magazine recently quizzed a random sample of 1006 U.S. adults on their knowledge of science.[42] One of the questions asked, "Which gas makes up most of the Earth's atmosphere: hydrogen, nitrogen, carbon dioxide, or oxygen?" Only 20% were able to correctly answer "nitrogen."

If people were just guessing, about 25% should get the correct answer. To determine if these data provide convincing evidence that fewer than 25% of all U.S. adults know the answer, 100 simulated SRSs of size $n = 1006$ were selected. Each dot in the graph shows the percentage of adults who got the correct answer in a simulated SRS, assuming that each adult has a 25% chance of answering correctly.

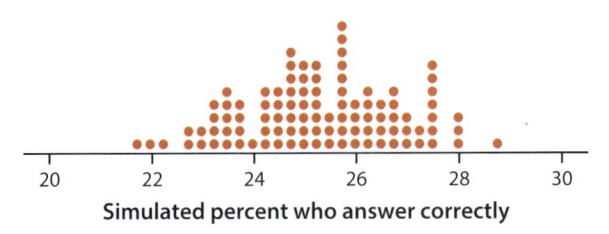

(a) Explain how the graph illustrates the concept of sampling variability.

(b) In the simulation, what percentage of the samples resulted in 20% or fewer of the adults knowing the answer?

(c) In the actual study, 20% of the adults answered correctly. Based on your answer to part (b), is there convincing evidence that fewer than 25% of all U.S. adults would correctly answer "nitrogen"?

18. Weekend Birthdays Over the years, the percentage of births that are planned cesarean sections has been rising. Because doctors can schedule these deliveries, there might be more children born during the week and fewer born on the weekend than if births were uniformly distributed throughout the week. To investigate, researchers selected an SRS of 73 young people. Of these young people, 24 were born on Friday, Saturday, or Sunday.[43]

To determine if these data provide convincing evidence that fewer than 3/7 of all young people were born on Friday, Saturday, or Sunday, 100 simulated SRSs of size $n = 73$ were selected. Each dot in the graph shows the number of people who were born on Friday, Saturday, or Sunday in a simulated SRS, assuming that each person has a 3/7 chance of being born on one of these three days.

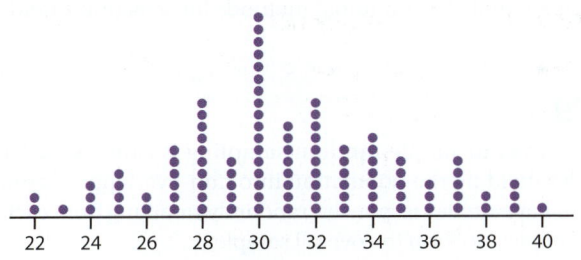

Simulated number who were born on Friday, Saturday, or Sunday

(a) Explain how the graph illustrates the concept of sampling variability.

(b) In the simulation, what percentage of the samples resulted in 24 or fewer people born on Friday, Saturday, or Sunday?

(c) In the actual study, 24 of the young people were born on Friday, Saturday, or Sunday. Based on your answer to part (b), is there convincing evidence that fewer than 3/7 of all young people were born on Friday, Saturday, or Sunday? Explain your reasoning.

Applying the Concepts These exercises ask you to apply multiple learning goals in a new context or to apply what you learned in this section in a new way.

19. **Campus Commute** A college administrator wants to estimate the mean distance that students at a large community college live from campus. To create this estimate, the administrator obtains a list of all students from the registrar's database.

    (a) Describe how you could select an SRS of 100 students.

    (b) Will your sample result be exactly the same as the true population mean? Explain your answer.

    (c) Which would be more likely to yield a sample result close to the true population mean: an SRS of 50 students or an SRS of 100 students? Explain your answer.

20. **Online Banking** The technology director at a bank wants to estimate the proportion of its checking account customers who prefer to deposit checks in their account using a smartphone rather than making deposits in person at the local branch.

    (a) Describe how the director could select an SRS of 100 account holders.

    (b) Will the sample result be exactly the same as the true population proportion? Explain your answer.

    (c) Which would be more likely to yield a sample result close to the true population proportion: an SRS of 50 account holders or an SRS of 100 account holders? Explain your answer.

Extending the Concepts These exercises challenge you to explore statistical concepts and methods that go beyond what you learned in this section.

21. **Flight Arrivals** An airline claims that 95% of its flights arrive on time. To check this claim, an SRS of 25 flights was selected and only 22 were on time. Do these data provide convincing evidence that fewer than 95% of all this airline's flights arrive on time?

    (a) Go to www.stapplet.com and open the *One Categorical Variable, Single Group* applet. Enter the variable name (Flight status), the category names (On-time, Late), and the frequencies (22, 3). Click the Begin analysis button and describe what you see.

    (b) Scroll down to the "Perform Inference" section and choose "Simulate sample count." Keep "On-time" as the category that indicates success and leave the third drop-down menu as "hypothesized value." Enter 95 as the hypothesized percent. *Note:* If the applet asks for a hypothesized *proportion*, enter 0.95 or adjust the preferences. Enter 100 for the number of samples to add and click "Add samples." Describe what you see.

    (c) Based on the results of the simulation, is there convincing evidence that fewer than 95% of all this airline's flights arrive on time? Explain your reasoning.

Cumulative Review These exercises revisit what you learned in previous sections.

22. **Fitness and Confidence (1.1, 1.2)** A fitness center located in a mall offers an exercise program for all mall employees who choose to participate. The program assesses each participant's fitness using a treadmill test and also administers a personality questionnaire. There is a moderately strong relationship between fitness score and self-confidence score.

    (a) Is this an experiment or an observational study? Explain your answer.

    (b) Can we apply these results to all employees in the mall? Explain your answer.

Section 1.4 Other Sampling Methods

LEARNING GOALS

By the end of this section, you will be able to:

- Describe how to obtain a stratified random sample.
- Describe how to obtain a cluster random sample.
- Describe how to obtain a systematic random sample.

In Section 1.2, you learned that selecting a sample at random is a good way to avoid bias in the sampling process. In Section 1.3, you learned how to select a simple random sample (SRS) from a population. In this section, you will learn about three additional methods for selecting a random sample and when they should be used.

Stratified Random Sampling

One of the most commonly used alternatives to simple random sampling is called **stratified random sampling**. This method involves dividing the population into non-overlapping groups (**strata**) of individuals who are expected to have similar responses, randomly sampling from each of these groups, and combining these "subsamples" to form the overall sample.

Strata are groups of individuals in a population that share characteristics thought to be associated with the variables being measured in a study.

Stratified random sampling selects a sample by choosing an SRS from each stratum and combining the SRSs into one overall sample.

Stratified random sampling works best when the individuals within each stratum are similar (homogeneous) with respect to what is being measured and when there are large differences between strata. For example, in a study of sleep habits among university students, the population of students might be divided into three strata: students who live in the dorms, students who live with their parents, and students who live off campus but not with their parents. After all, it is reasonable to think that students who live in the dorms might have different sleeping habits than other students.

HOW TO Select a Stratified Random Sample

1. **Form strata.** Divide the population into groups that share characteristics associated with the variable(s) being measured.
2. **Sample from every stratum.** Select an SRS within each stratum.
3. **Combine.** Combine the selected individuals from each stratum into one overall sample.

How many individuals should be selected from each stratum? It depends on several factors. The most straightforward method is to keep the sample size within each stratum roughly proportional to the size of the strata in the population. For example, if 30% of the students at a university live in the dorms and we want a stratified random sample of 500 students, we would randomly select an SRS of $(0.30)(500) = 150$ students who live in the dorms.

EXAMPLE 1.4.1

How are you planning to vote?

Stratified random sampling

PROBLEM

A news organization in California wants to estimate support for a ballot proposition that would implement a tax increase to pay for additional solar power-generating stations. Using voter registration data, the organization determines that 24% of registered voters in California are Republicans, 46% are Democrats, 24% are Independents, and 6% are registered with other parties.[44]

(a) Explain why stratifying by party makes sense in this context.

(b) Describe how to select a stratified random sample of 1000 registered voters in California.

SOLUTION

(a) Stratifying by party makes sense because opinions about this proposition are likely to be similar within each party but very different between parties.

(b) Select an SRS of 240 Republicans, an SRS of 460 Democrats, an SRS of 240 Independents, and an SRS of 60 others. Then, combine these four groups into one overall sample.

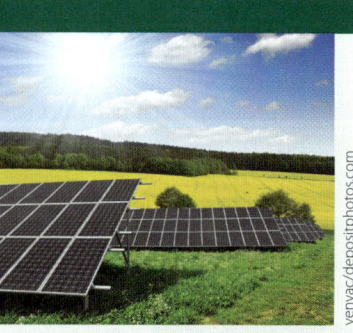

EXAM PREP: FOR PRACTICE, TRY EXERCISE 7.

The following Concept Exploration illustrates the benefit of choosing appropriate strata.

> **CONCEPT EXPLORATION**
>
> **Investigating stratified random sampling using the *Sampling Sunflowers* applet**
>
> In this exploration, you'll learn the advantages of using a properly designed stratified random sample.
>
> A farmer grows sunflowers for making sunflower oil. The field is arranged in a grid pattern, with 10 rows and 10 columns as shown in the figure.
>
> Irrigation ditches run along the top and bottom of the field. The farmer would like to estimate the average number of healthy plants per square in the field. It would take too much time to count the plants in all 100 squares, so the farmer will accept an estimate based on a sample of 10 squares.
>
> 1. Go to www.stapplet.com and launch the *Sampling Sunflowers* activity. Then click the button to do this activity independently.
> 2. Under Method 1, click the button to generate a sample. This will produce a simple random sample (SRS) of 10 plots from the field. The plots chosen are marked with an X.
> 3. Under Method 2, click the button to generate a sample. This will produce a stratified random sample of 10 plots from the field, with one plot chosen at random from each *row*.
> 4. Under Method 3, click the button to generate a sample. This will produce a stratified random sample of 10 plots from the field, with one plot chosen at random from each *column*.
> 5. Which of these three methods do you think will work the best? That is, which method is most likely to give an estimate close to the true mean number of healthy plants in the field? Make sure you think carefully about this before moving on.
> 6. Click the Reveal results button below the "If we did a census . . ." heading. This will reveal the number of healthy plants in each of the squares and the true mean number of healthy plants. It will also calculate the sample means for each of your three samples. Which method produced a sample mean closest to the true mean?
> 7. To see which method is best overall, generate an additional 100 samples for each method using the "Quick add" option at the bottom of the applet. Which method gives estimates that tend to be closest to the true mean? Did this match your answer from Step 5?

The following dotplots show the mean number of healthy plants in 100 samples using each of the three sampling methods in the activity: simple random sampling, stratified random sampling with rows of the field as strata, and stratified random sampling with columns of the field as strata. Notice that all three distributions are centered at about 102.46, the true mean number of healthy plants in all squares of the field. That makes sense because random sampling helps to avoid bias — consistently overestimating or consistently underestimating the true mean.

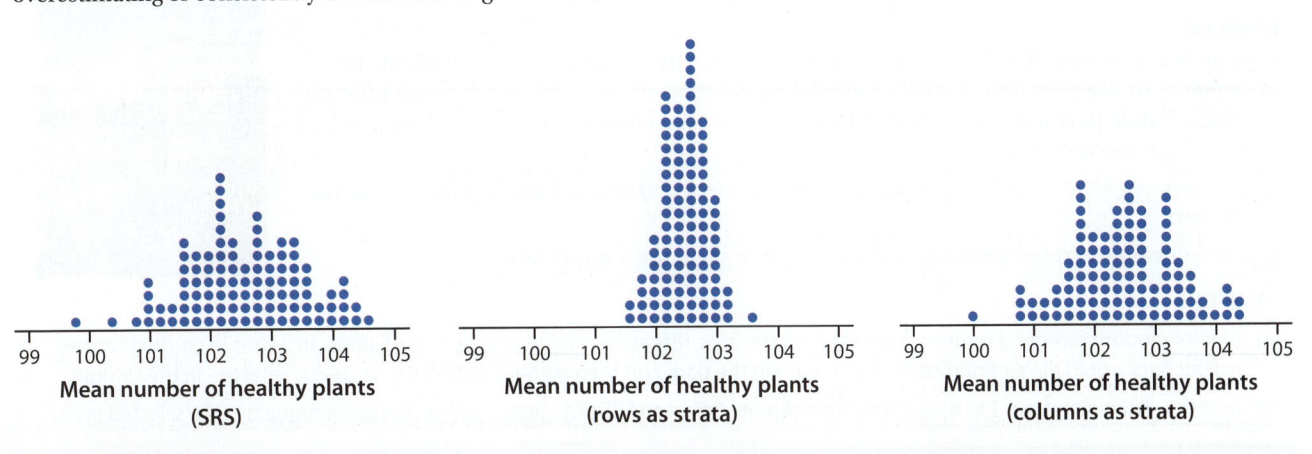

One other detail stands out in the graphs: There is much less variability in the estimates when we use the rows as strata — even though the sample sizes were the same with all three methods.

When strata are chosen wisely, the result is an estimate that is more precise than in a simple random sample. In this case, stratifying by row worked the best because the squares in a particular row are all the same distance from the irrigation ditches—and distance from the water source is likely to be associated with the health of the plants. With the other two methods, it is more likely to get a majority of plots near the ditches (and a large mean) or get a majority of plots far from the ditches (and a small mean).

Cluster Random Sampling

Both simple random sampling and stratified random sampling are hard to use when populations are large and spread out over a wide area. In that situation, we might prefer to use **cluster random sampling.** This method involves dividing the population into non-overlapping groups (**clusters**) of individuals that are "near" one another, then randomly selecting whole clusters to form the overall sample.

> A **cluster** is a group of individuals in the population that are located near each other.

> **Cluster random sampling** selects a sample by randomly choosing clusters and including each member of the selected clusters in the sample.

The main criterion for forming clusters is location. However, cluster random sampling works best when each cluster includes a variety of individuals. This is the opposite of stratified random sampling, where we want to form strata that contain individuals with similar characteristics.

> **HOW TO** Select a Cluster Random Sample
>
> 1. **Form clusters.** Divide the population into groups of individuals located near each other.
> 2. **Select clusters.** Select an SRS of clusters.
> 3. **Combine.** Every individual from each of the selected clusters is included in the sample.

Be sure you understand the difference between strata and clusters. We want each stratum to contain similar individuals and for large differences to exist between strata. For a cluster random sample, we would *like* each cluster to look just like the population, but on a smaller scale. Unfortunately, cluster random samples don't offer the statistical advantage of better information about the population that stratified random samples do. But they are more efficient!

EXAMPLE 1.4.2

How many saguaros are in the park?

Cluster random sampling

PROBLEM
Saguaro National Park near Tucson, Arizona, is famous for its saguaro cactus. To track the health of saguaros in the park and estimate the total number of saguaros, researchers divided the 143-square-mile park into plots of 4 hectares each and used cluster random sampling to select 45 plots for inspection.[45]

(a) Explain why cluster random sampling might be preferred to simple random sampling in this context.

(b) Describe how to select a cluster random sample of saguaros using 45 plots.

SOLUTION

(a) Cluster random sampling is much more efficient because researchers need to visit only 45 plots. In a simple random sample, researchers would have to number each saguaro in the park and then go back to find the saguaros selected in the sample.

(b) Number all the plots from 1 to the total number of plots. Select an SRS of 45 plots and "survey" every saguaro in the selected plots.

EXAM PREP: FOR PRACTICE, TRY EXERCISE 11.

Systematic Random Sampling

Many news organizations conduct exit polls of voters on election day. That is, they survey voters as they leave the polling place, asking questions about how they voted and why they voted that way, along with other demographic questions. To avoid bias, pollsters should select the sample at random. But it is impossible to select a traditional SRS in this context, as this would require knowing which voters will show up, numbering them all, and then being able to identify the randomly selected voters as they leave the polling place. As an alternative, news organizations often use **systematic random sampling**.

> **Systematic random sampling** selects a sample from an ordered arrangement of the population by randomly selecting one of the first k individuals and choosing every kth individual thereafter.

For example, at each polling place the interviewer might be instructed to interview every 20th voter as they leave. To choose a starting point, the interviewer would generate a random number from 1 to 20. If the random number was 6, the 6th voter would be interviewed, followed by the 26th voter, the 46th voter, and so on.

EXAMPLE 1.4.3

How's the market doing?

Systematic random sampling

PROBLEM
A professor of finance wants to do in-depth research on a random sample of 50 companies listed on the New York Stock Exchange. The professor has an alphabetized list of all 2800 stocks on the exchange. Describe how the professor could use systematic random sampling to select the sample.

SOLUTION
Because there are 2800 stocks and the professor wants a random sample of $n = 50$ stocks, the professor should select every $2800/50 = 56$th stock in the alphabetized list, starting with a randomly selected stock from among the first 56 in the list.

> Note that selecting a systematic random sample in this context is easier than selecting an SRS, as the professor doesn't have to number the stocks from 1 to 2800 at the beginning of the selection process.

EXAM PREP: FOR PRACTICE, TRY EXERCISE 15.

Most large-scale sample surveys use *multistage sampling,* which combines two or more sampling methods. For example, the U.S. Census Bureau carries out a monthly Current Population Survey (CPS) of about 60,000 households. Researchers start by choosing a stratified random sample of neighborhoods in 756 of the 2007 geographic areas in the United States. Then they divide each neighborhood into clusters of four nearby households and select a cluster random sample to interview.

Section 1.4 What Did You Learn?

Review the learning goals from this section. Then practice what you've learned by working through the exercises.

Learning Goal	Example	Exercises
Describe how to obtain a stratified random sample.	1.4.1	7–10
Describe how to obtain a cluster random sample.	1.4.2	11–14
Describe how to obtain a systematic random sample.	1.4.3	15–18

Section 1.4 Exercises

Building Concepts and Skills These exercises assess the basic knowledge you should have after reading the section.

1. True/False: In a stratified random sample, it is ideal if the members of a stratum have similar characteristics.

2. What is the main benefit of using a stratified random sample instead of an SRS?

3. True/False: In a cluster random sample, you should randomly select individuals from each cluster.

4. What is the main benefit of using a cluster random sample instead of an SRS?

5. If you want to select a sample of size $n = 100$ from a list of 1000 individuals using systematic random sampling, you should select every _____ individual.

6. If you are selecting every 25th person on a roster using systematic random sampling, how do you determine who is the first person in the sample?

Mastering Concepts and Skills These exercises reinforce the learning goals as shown in the examples.

7. **Refer to Example 1.4.1 Hotel Rooms** A hotel has 30 floors with 40 rooms per floor. Half of the rooms on each floor face the water; the remaining half face the golf course. There is an extra charge for the rooms with a water view. The hotel manager wants to select 120 rooms and survey the registered guest in each of the selected rooms about their overall satisfaction with the property.
   (a) The manager decides to use a stratified random sample and stratify by floor and side of hotel, resulting in $30 \times 2 = 60$ strata. Explain why stratifying by floor and side makes sense in this context.
   (b) Describe how to select a stratified random sample of 120 rooms using floor and side as strata.

8. **Go Blue!** Michigan Stadium, also known as "The Big House," seats more than 100,000 fans for a football game. The University of Michigan Athletic Department wants to survey fans about concessions that are sold during games. Tickets are most expensive for seats on the sidelines. The cheapest seats are in the end zones (where one of the authors sat as a student). Here is a map of the stadium.

   (a) A member of the athletic department proposes that they use a stratified random sample and stratify by region of the stadium (sideline, corner, endzone). Explain why stratifying by region makes sense in this context.
   (b) Describe how to select a stratified random sample of fans using regions of the stadium as strata.

9. **No Tipping** The owner of a large restaurant is considering a new "no tipping" policy and wants to survey a sample of employees about the change. The policy would add 20% to the cost of food and beverages and the additional revenue would be distributed equally among servers and kitchen staff. Describe how to select a stratified random sample of approximately 30 employees. Explain your choice of strata and why stratified random sampling might be preferred in this context.

10. **Campus Parking** The director of student life at a university wants to estimate the proportion of undergraduate students who regularly park a car on campus. Describe how to select a stratified random sample of approximately 100 students. Explain your choice of strata and why stratified random sampling might be preferred in this context.

11. **Refer to Example 1.4.2 Hotel Rooms** Refer to Exercise 7. An assistant manager suggests using a cluster random sample instead of a stratified random sample, using floors as clusters.
    (a) Explain why cluster random sampling might be preferred to stratified random sampling in this context.
    (b) Describe how to select a cluster random sample of 120 rooms.

12. **Go Blue!** Refer to Exercise 8. Another member of the athletic department suggests using a cluster random sample instead of a stratified random sample, using numbered sections (1–44) as clusters.
    (a) Explain why cluster random sampling might be preferred to stratified random sampling in this context.
    (b) Describe how to select a cluster random sample of fans using numbered sections as clusters.

13. **Orientation** Each year, new students at a small college are assigned to a 15-person orientation group. These groups participate in a variety of activities on campus before the rest of the students arrive. On the last day of orientation week, each group meets at a different place on campus for a picnic. To evaluate the orientation program, college administrators select a cluster random sample by randomly selecting 6 orientation groups and surveying each of the students in these 6 groups during the picnic. Explain a benefit of using a cluster random sample instead of an SRS in this context.

14. **Rural Internet** Laying fiber-optic cable is expensive. Internet service providers want to make sure that if they extend their lines to less dense suburban or rural areas, there will be sufficient demand so the work will

be cost-effective. They decide to conduct a survey to determine the proportion of households in a rural subdivision that would buy the service. They select a cluster random sample by randomly selecting five blocks in the subdivision and surveying each family who lives on the selected blocks. Explain a benefit of using a cluster random sample instead of an SRS in this context.

15. **Refer to Example 1.4.3 iPhones** At a factory, 1000 iPhones are produced each day. Management would like to ensure that the phones' display screens meet the company's quality standards before shipping them to retail stores. Because it takes about 10 minutes to inspect an individual phone's display screen, managers decide to inspect a sample of 20 phones from the day's production. Describe how management could use systematic random sampling to select the sample.

16. Airline Boarding Airlines frequently alter their boarding procedures in an effort to reduce boarding time and increase passenger satisfaction. After trying a new boarding system on a flight with 240 passengers, airline executives want to survey 20 randomly selected passengers as they exit the plane about their boarding experience. Describe how the executives could use systematic random sampling to select the sample.

17. Dead Trees In Rocky Mountain National Park, many mature pine trees along Highway 34 are dying due to infestation by pine beetles. Scientists would like to use a sample of size 200 to estimate the proportion of the approximately 5000 pine trees along the highway that have been infested.
   (a) Explain why it wouldn't be practical for the scientists to obtain an SRS in this context.
   (b) Describe how scientists could use systematic random sampling to select the sample.

18. Mall Shoppers The marketing director at a large mall wants to estimate the proportion of weekday shoppers who travel more than 10 miles to visit the mall. On a typical weekday, about 1200 shoppers visit the mall and the director would like to survey a sample of 100 shoppers.
   (a) Explain why it wouldn't be practical for the director to obtain an SRS in this context.
   (b) Describe how the director could use systematic random sampling to select the sample.

Applying the Concepts These exercises ask you to apply multiple learning goals in a new context or to apply what you learned in this section in a new way.

19. Rock On! At the end of a sold-out concert, the concert promoter wants to know how satisfied fans are with the show. The concert venue is rectangular, with 50 rows and 40 seats in each row.
   (a) Describe how to select a stratified random sample of 200 fans. Explain your choice of strata and a benefit of using a stratified random sample in this context.
   (b) Describe how to select a cluster random sample of 200 fans. Explain your choice of clusters and a benefit of using a cluster random sample in this context.
   (c) Describe how to select a systematic random sample of 200 fans and explain a benefit of using a systematic random sample in this context.

20. Wood Screws A factory runs 24 hours a day, producing 30,000 wood screws per day over three 8-hour shifts — day, evening, and overnight. In the last stage of manufacturing, the screws are packaged in boxes containing 10 screws each. The quality control manager inspects a random sample of 300 screws each day.
   (a) Describe how to select a stratified random sample of 300 screws. Explain your choice of strata and a benefit of using a stratified random sample in this context.
   (b) Describe how to select a cluster random sample of 300 screws. Explain your choice of clusters and a benefit of using a cluster random sample in this context.
   (c) Describe how to select a systematic random sample of 300 screws and explain a benefit of using a systematic random sample in this context.

21. Sunflowers Refer to the *Sampling Sunflowers* Concept Exploration in this section. If you were to use a cluster random sample to select 20 squares, would it be better to use rows as clusters or columns as clusters? Explain your reasoning.

22. Campus Parking Refer to Exercise 10. Identify a variable that would *not* be a good choice to stratify by in the survey about parking on campus. Explain your reasoning.

Extending the Concepts These exercises challenge you to explore statistical concepts and methods that go beyond what you learned in this section.

23. Is It an SRS? A corporation employs 2000 assembly-line workers and 500 engineers. A stratified random sample of 200 assembly-line workers and 50 engineers gives every individual in the population the same chance to be chosen for the sample. Is it an SRS? Explain your answer.

24. Wood Screws Refer to Exercise 20. Explain how you could combine two or more sampling methods to increase efficiency but still have a precise estimate of the quality of the screws.

25. Airline Boarding Refer to Exercise 16. Instead of waiting for passengers to exit the airplane, one executive suggests obtaining a systematic random sample by placing a survey in every 12th seat before passengers board. If the plane has 40 rows of 6 seats in each row, explain a possible problem with using systematic random sampling in this context.

Cumulative Review These exercises revisit what you learned in previous sections.

26. **Shiny Pennies (1.2, 1.3)** Professor Lee has a large collection of pennies that she uses for activities with her statistics students. To estimate the average age of the pennies in her collection, she asks Luke to select a random sample of 20 pennies. To avoid getting his hands dirty, Luke selects 20 of the shiniest pennies from the collection.

(a) Explain how Luke's sampling method is biased.

(b) Is the estimated average age from Luke's sample likely to be greater than or less than the average age of the pennies in the entire collection? Explain your reasoning.

(c) If Luke were to increase the sample size to 50 shiny pennies, would this reduce the bias described in part (a)? Explain your reasoning.

Section 1.5 Observational Studies and Experiments

LEARNING GOALS

By the end of this section, you will be able to:

- Explain the concept of confounding and how it limits the ability to make cause-and-effect conclusions.
- Explain the purpose of a control group in an experiment.
- Describe how to randomly assign treatments in an experiment and explain the purpose of random assignment.

In Section 1.1, you learned how to distinguish observational studies from experiments. The goal of an observational study can be to describe some group or situation, to compare groups, or to examine relationships between variables. Unfortunately, with observational studies it is very difficult to identify cause-and-effect relationships.

Confounding

Is taking a vitamin D supplement good for you? Hundreds of observational studies have looked at the relationship between the vitamin D concentration in a person's blood and various health outcomes.[46] In one recent observational study, patients with high vitamin D concentrations were less likely to develop COVID-19 than patients with low vitamin D concentrations.[47] Other observational studies have shown that people with higher vitamin D concentrations have lower rates of cardiovascular disease, better cognitive function, and less risk of diabetes than people with lower concentrations of vitamin D.

In the observational studies involving vitamin D and diabetes, the **explanatory variable** is vitamin D concentration in the blood and the **response variable** is diabetes status — whether or not the person developed diabetes.

A **response variable** measures an outcome of a study.

An **explanatory variable** may help explain or predict changes in a response variable.

Unfortunately, it is very difficult to show that taking vitamin D *causes* a lower risk of diabetes using an observational study. As indicated in the table, there are many possible differences between the group of people with high vitamin D concentration and the group of people with low vitamin D concentration. Any of these differences could be causing the difference in diabetes risk between the two groups.

Variable	Group 1	Group 2
Vitamin D concentration (explanatory)	**High vitamin D concentration**	**Low vitamin D concentration**
Quality of diet	Better diet	Worse diet
Amount of exercise	More exercise	Less exercise
⋮	⋮	⋮
Amount of vitamin supplementation	More likely to take other vitamins	Less likely to take other vitamins
Diabetes status (response)	**Less likely to have diabetes**	**More likely to have diabetes**

For example, it is possible that people who have healthier diets eat lots of foods that are high in vitamin D. Likewise, it is possible that people with healthier diets are less likely to develop diabetes. Vitamin D concentration may not have anything to do with diabetes status, even though there is an association between these two variables. In this case, we say there is **confounding** between vitamin D concentration and diet because we cannot tell which variable is causing the change in diabetes status.

> **Confounding** occurs when two variables are associated in such a way that their effects on a response variable cannot be distinguished from each other.

Likewise, because sun exposure increases vitamin D concentration, it is possible that people who exercise a lot outside have higher concentrations of vitamin D. If people who exercise a lot are also less likely to get diabetes, then amount of exercise and vitamin D concentration are confounded—we can't say which variable is the cause of the smaller diabetes risk. In other words, exercise is a confounding variable because it is related to both vitamin D concentration and diabetes status.

EXAMPLE 1.5.1

Smoking and ADHD

Confounding

PROBLEM
In a study involving more than 4700 children, researchers from Cincinnati Children's Hospital Medical Center found that those children whose mothers smoked during pregnancy were more than twice as likely to develop attention-deficit/hyperactivity disorder (ADHD) as children whose mothers had not smoked.[48] Based on this study, is it reasonable to conclude that a mother's smoking during pregnancy causes an increase in the risk of ADHD in her children? Explain your reasoning.

SOLUTION
No. It is possible that women who smoked during pregnancy had less healthy diets and that eating poorly during pregnancy increases the risk of ADHD in children. If both of these are true, then we would see a relationship between smoking and ADHD even if smoking has no effect on ADHD.

> Notice that the solution describes how diet might be associated with the explanatory variable (smoking status) and how diet might be associated with the response variable (ADHD status). You need to describe both relationships to justify confounding.

EXAM PREP: FOR PRACTICE, TRY EXERCISE 9.

The easiest way to identify confounding in an observational study is to think about other variables that are associated with the explanatory variable that might cause a change in the response variable. In the smoking and ADHD study, there are many potential differences between the group of mothers who smoked during pregnancy and the group of mothers who didn't. Any of these differences could be the cause of the change in the response variable.

The Language of Experiments

To determine if taking vitamin D actually causes a reduction in diabetes risk, researchers in Norway performed an experiment. The researchers randomly assigned 500 **experimental units** (**subjects**) with prediabetes to one of two **treatments**. Half were given a high dose of vitamin D and half were given a **placebo**—a pill that looked exactly like the vitamin D supplement but contained no active ingredient.

> A **treatment** is a specific condition applied to the individuals in an experiment.

> An **experimental unit** is the object to which a treatment is randomly assigned. When the experimental units are human beings, they are often called **subjects**.

> A **placebo** is a treatment that has no active ingredient but is otherwise like other treatments.

The experiment in Norway avoided confounding by letting chance decide who took vitamin D and who didn't. That way, people with healthier diets were split about evenly between the two groups. So were people who exercised a lot and people who took other vitamins. After 5 years, approximately 40% of the people in each group were diagnosed with diabetes.[49] In other words, the association between vitamin D concentration and diabetes status disappeared when comparing two groups that were roughly the same to begin with. *Conducting a well-designed experiment is the best way to avoid confounding.*

To learn what it means for an experiment to be "well designed," let's start with a poorly designed experiment. The students in a physiology class want to investigate whether caffeine affects pulse rates. One of the students proposes the following plan:

1. Have each student measure their initial pulse rate.
2. Have each student drink 12 ounces of cola.
3. Wait for 15 minutes.
4. Have each student measure their final pulse rate.
5. Compare the initial and final pulse rates.

This proposed experiment has a number of problems. If the cola contains sugar as well as caffeine, the class wouldn't be able to tell if any increase in pulse rates was due to the presence of caffeine or the presence of sugar. (Can you say confounding?) Likewise, other events that occurred between the initial and final pulse rate measurements could be the cause of an increase in pulse rates, such as a fire alarm, an increase in room temperature, or an exciting physiology lesson!

The solution to these problems is to include *two groups* in the experiment: one group that receives the caffeine and a **control group** that doesn't receive the caffeine.

> In an experiment, a **control group** is a group used to provide a baseline for comparing the effects of other treatments. Depending on the purpose of the experiment, a control group may be given an inactive treatment, an active treatment, or no treatment at all.

In all other ways, these two groups should be treated exactly the same, so the only difference is the caffeine. This means that one group could get regular cola with caffeine, while the control group gets caffeine-free cola. Both groups would get the same amount of sugar, so sugar consumption would no longer be confounded with caffeine intake. Likewise, both groups would experience the same events during the experiment, so what happens during the experiment won't be confounded with caffeine intake either. *Using a design that compares two or more treatments is the first step in designing a good experiment.*

EXAMPLE 1.5.2

How can we encourage energy conservation?

Control groups

PROBLEM
Many utility companies have introduced programs to encourage energy conservation among their customers. An electric company considers placing small digital displays in households to show current electricity use along with a projected monthly cost. Will the displays reduce electricity use? One cheaper approach is to give customers a chart and information about monitoring their electricity use from their outside meter. Would this method work almost as well? The company decides to conduct an experiment to compare these two approaches (display versus chart) with a control group of customers who receive information about energy consumption but no help in monitoring electricity use. Explain why it is necessary to include the control group.

SOLUTION
A control group would show how much electricity customers tend to use with no help in monitoring their electricity usage. This would serve as a baseline to determine how much electricity is saved by using the display or chart.

> If there wasn't a control group, we wouldn't be able to tell if using the display or chart actually saved electricity. Perhaps the weather this year is less extreme than in the past, allowing households to use less electricity without any monitoring. If this was the case, use of the displays and charts would be confounded with the weather—we wouldn't know if electricity use was lower because of the weather or because of the use of the displays and charts.

EXAM PREP: FOR PRACTICE, TRY EXERCISE 13.

In the caffeine experiment, the physiology class used a control group to help prevent confounding. But even when the study includes a control group, confounding is still possible. If the students in the physiology class know what type of soda they are receiving, the caffeine-consuming group will expect their pulse rates to increase more. And the expectation of an increase in pulse rate may actually increase a subject's pulse rate, apart from the effects of the caffeine itself. This is an example of the **placebo effect.**

> The **placebo effect** describes the fact that some subjects in an experiment will respond favorably to any treatment, even an inactive treatment (placebo).

To avoid the placebo effect, it is important that each treatment look, feel, and taste just like the other treatments, so the subjects will not know which treatment they are receiving. In the caffeine experiment, if neither the students nor the people running the experiment know who is getting which type of cola, the experiment is **double-blind.** Other experiments are **single-blind.**

> In a **double-blind** experiment, neither the subjects nor those who interact with them and measure the response variable know which treatment a subject is receiving.

> In a **single-blind** experiment, either the subjects or the people who interact with them and measure the response variable don't know which treatment a subject is receiving.

The electricity monitoring experiment cannot be double blind because the subjects must know which method to use. It is possible, however, that the experiment could be single blind if the people measuring electricity usage don't know which treatment the household received.

Experiments: Random Assignment

Comparison alone isn't enough to produce results we can trust. Many other variables affect pulse rates besides caffeine. Some of these variables, such as caffeine tolerance, weight, and recent caffeine consumption, describe characteristics of the subjects. Other variables, such as sugar content, amount of soda consumed, and temperature, describe characteristics of the experimental process. To conduct a well-designed experiment, we want to treat every subject the same way and to create groups that are roughly equivalent at the beginning of an experiment. To create these groups, we use **random assignment.**

> In an experiment, **random assignment** means that treatments are assigned to experimental units (or experimental units are assigned to treatments) using a chance process.

Suppose there are 20 subjects in the caffeine experiment. To randomly assign treatments, the professor could number the students from 1 to 20 and randomly generate 10 *different* integers from 1 to 20. The subjects with these labels will receive the cola with caffeine. The remaining 10 subjects will receive the cola without caffeine. The professor could also write the letter "A" on 10 slips of paper, write the letter "B" on 10 slips of paper, shuffle the slips, and hand one to each subject. Subjects who get an "A" slip will receive the cola with caffeine and students who get a "B" slip will receive the cola without caffeine.

Random assignment should distribute the students who already have a caffeine tolerance in roughly equal numbers to each group. It should also roughly balance out the number of students who have consumed caffeine recently and make the average weight of subjects in each group about the same. In general, random assignment should create two groups that are roughly equivalent with respect to *every* variable that might affect the response, whether or not you identified the variable in advance.

EXAMPLE 1.5.3

Is the vaccine effective?

Random assignment

PROBLEM
Before making a vaccine publicly available, pharmaceutical companies must conduct well-designed experiments (called clinical trials) to demonstrate a vaccine's effectiveness and safety. Some of the most important clinical trials in recent years were for the various COVID-19 vaccines. In Pfizer's clinical trial of its COVID-19

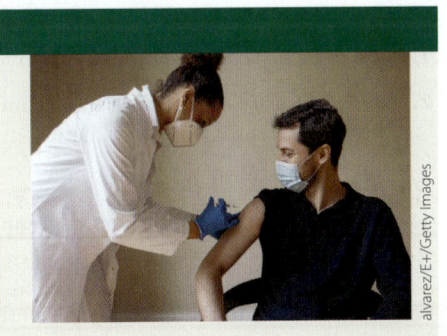

(*continued*)

vaccine, half of the 43,548 volunteer subjects were randomly assigned to receive 2 shots of the vaccine 21 days apart. The other half of the volunteers were randomly assigned to receive 2 shots of saline placebo 21 days apart. After several months, Pfizer determined that its vaccine was 95% effective in preventing COVID-19.[50]

(a) Describe how Pfizer could have randomly assigned the 43,548 subjects to the two treatments, using groups of roughly equal size.

(b) Explain the purpose of random assignment in this experiment.

SOLUTION

(a) Number the subjects from 1 to 43,548. Then, randomly generate 21,774 different integers from 1 to 43,548. Assign the subjects who correspond with these integers to the vaccine. Assign the remaining 21,774 subjects to the placebo.

(b) The purpose of the random assignment was to help avoid confounding by creating two groups of subjects who were roughly equivalent at the beginning of the experiment.

> With random assignment, age, health, occupation, and other variables that affect how likely a subject is to develop COVID-19 should be about the same, on average, in the two groups.

EXAM PREP: FOR PRACTICE, TRY EXERCISE 17.

Random assignment works best when there are many subjects in the experiment, as in the Pfizer clinical trial. If only 20 subjects participated in the clinical trial, the two treatment groups are likely to be somewhat different, even if they were randomly assigned. For example, if 2 of the 20 subjects have underlying health conditions, then there is a 50% chance that these two subjects will be randomly assigned to the same treatment. In the actual trial, there were 43,548 subjects. With this many subjects, it is almost impossible that one of the treatment groups will end up with all of the subjects who have underlying health conditions. The idea that we should use enough subjects to create roughly equivalent groups is called *replication*.

Section 1.5 What Did You Learn?

Review the learning goals from this section. Then practice what you've learned by working through the exercises.

Learning Goal	Example	Exercises
Explain the concept of confounding and how it limits the ability to make cause-and-effect conclusions.	1.5.1	9–12
Explain the purpose of a control group in an experiment.	1.5.2	13–16
Describe how to randomly assign treatments in an experiment and explain the purpose of random assignment.	1.5.3	17–20

Section 1.5 Exercises

Building Concepts and Skills These exercises assess the basic knowledge you should have after reading the section.

1. Two variables are confounded when _____.

2. True/False: Confounding is more common in observational studies than in experiments.

3. "Subject" is another name for _____.

4. What is the purpose of a control group?

5. What is the placebo effect?

6. True/False: If the subjects in an experiment know which treatment they are receiving, the experiment can't be single blind or double blind.

7. True/False: We randomly assign treatments in an experiment so we can generalize the results to a larger population.

8. What does replication mean in the context of an experiment?

Mastering Concepts and Skills These exercises reinforce the learning goals as shown in the examples.

9. **Refer to Example 1.5.1 Social Media and Marriage** In a study conducted by researchers at the University of Texas, people were asked about their social media use and satisfaction with their marriage. Of the heavy social media users, 32% had thought seriously about leaving

their spouse. Only 16% of non-social media users had thought seriously about leaving their spouse.[51] Based on this study, it is reasonable to conclude that using social media makes people more likely to consider leaving their spouse? Explain your reasoning.

10. **Vegetables and Eye Health** In a long-term study, researchers collected information about the diets of participants and whether or not they developed an eye condition called age-related macular degeneration (AMD). People who ate lots of green leafy vegetables, which are high in lutein and zeaxanthin, had a lower risk of AMD.[52] Based on this study, is it reasonable to conclude that eating green leafy vegetables causes a reduction in the risk of AMD? Explain your reasoning.

11. **Breakfast** In a study of 294 British high school students, each student was asked to complete a 7-day food journal. Researchers used these journals to classify the students as rare, occasional, or frequent breakfast eaters. The mean score on an important national exam was significantly lower for those students who rarely ate breakfast compared to frequent eaters.[53] Based on this study, is it reasonable to conclude that eating breakfast more often causes higher test scores? Explain your reasoning.

12. **Screen Time** In a study of early childhood development, researchers studied a sample of 2441 mothers and children. When the children were 24 and 36 months old, their mothers reported how much screen time their children got in a typical day. Then, at ages 36 months and 60 months, the children were assessed on various developmental milestones. The study found that higher amounts of screen time were associated with lower scores on the developmental test.[54] Based on this study, is it reasonable to conclude that more screen time causes impaired development in young children? Explain your reasoning.

13. **Refer to Example 1.5.2 Rainfall and Biomass** The changing climate might affect the amount of rain that falls during different seasons, which in turn affects plant growth. Researchers studied 18 plots of open grassland, each with an area of 70 square meters. One response variable was total plant biomass produced in a plot over the course of a year. Kenwyn Suttle and his coworkers at the University of California at Berkeley wanted to compare the effects of three treatments: adding water equal to 20% of annual rainfall during January to March, adding water equal to 20% of annual rainfall during April to June, and not adding water.[55] Explain why it was necessary to include a control group that didn't get additional water.

14. **Career Education** Surveys of students who dropped out of high school revealed that many of them saw little connection between what they were studying in school and their future plans. To change this perception, researchers developed a program called CareerStart, in which teachers show students how the topics they learn about are used in specific careers. Seven of the 14 schools in Forsyth County, North Carolina, were selected at random to use CareerStart along with the district's standard curriculum. The other seven schools just followed the standard curriculum. Researchers followed both groups of students for several years, collecting data on students' attendance, behavior, standardized test scores, level of engagement in school, and whether or not the students graduated from high school. They found that students at schools that used CareerStart typically did better in each of these areas.[56] Explain why it was necessary to include a control group of schools that didn't use the CareerStart program.

15. **Shark Repellents** Many people like to surf, but there are dangers involved, including the possibility of a shark attack. In response, companies have produced shark repellents in various forms such as magnetic bracelets and scented wax. But do they work? Researchers tested 5 different products in a clever experiment. They built 5 surfboards, each utilizing one of the deterrents. A sixth board had no deterrent but was otherwise identical to the other boards. Then they attached raw tuna about 30 centimeters below each board, where a surfer's foot might be, and recorded the behavior of sharks that came near the board. The shark ate the bait in 96% of the trials with the sixth board, but only 40% of the time with the most effective deterrent.[57] Explain why it was necessary to include the sixth board with no deterrent.

16. **Dead Jellyfish** Research by Andrew Sweetman of Norway's International Research Institute of Stavanger focused on whether deep-sea scavengers consume dead jellyfish. His team lowered platforms piled with dead jellyfish and other platforms piled with dead mackerel more than 4000 feet into Norway's largest fjord; they found that hagfish, crabs, and other scavengers consumed the jellyfish in a few hours—faster than they consumed the mackerel.[58] Explain why it was necessary to include platforms containing mackerel in the experiment, even though the study was focused on the consumption of jellyfish.

17. **Refer to Example 1.5.3 Pizza Prices** The cost of a meal might affect how customers evaluate and appreciate food. To investigate, researchers worked with an Italian all-you-can-eat buffet to perform an experiment. A total of 139 customers were randomly assigned to pay either $4 or $8 for the buffet and then asked to rate the quality of the pizza on a 9-point scale. Customers who paid $8 rated the pizza 11% higher than those who paid only $4.[59]

    (a) Describe how the researchers could have randomly assigned the customers to the two treatments, using groups of roughly equal size.

    (b) Explain the purpose of random assignment in this experiment.

18. **Power and Speech** Recent research suggests that people's speech patterns are influenced by how much power they think they have in a particular negotiation. In one study, researchers randomly assigned 161 college students to two groups. The students in one group were told they had higher status or inside information, giving them

the sense that they had more power than those they were negotiating with. Students in the other group were told something that lowered their perceived power. All students were given the same statement to read. The voices of the students in the "high power" group tended to go up in pitch and become more variable in loudness than the voices of students in the "low-power" group.[60]

(a) Describe how the researchers could have randomly assigned the students to the two treatments, using groups of roughly equal size.

(b) Explain the purpose of random assignment in this experiment.

19. **Effects of Layoffs** Workers who survive a layoff of other employees at their location may suffer from "survivor guilt." A study used as subjects 120 students who were offered an opportunity to earn extra course credit by doing proofreading. Each student worked in the same cubicle as another student, who was an accomplice of the experimenters. At a break midway through the work, one of three things happened:

- *Treatment 1*: The accomplice was told to leave; it was explained that this was because they performed poorly.
- *Treatment 2*: It was explained that unforeseen circumstances meant there was only enough work for one person. By "chance," the accomplice was chosen to be laid off.
- *Treatment 3*: Both students continued to work after the break.

The students' work performance after the break was compared with their performance before the break. Overall, subjects worked harder when told the other student's dismissal was random.[61]

(a) Describe how the researchers could have randomly assigned the students to the three treatments, using groups of equal size.

(b) Explain the purpose of random assignment in this experiment.

20. **Class Notes** University students are increasingly taking class notes on a laptop rather than writing their notes by hand. Students argue that laptops are more efficient, but professors typically disagree. Which method is better for developing conceptual understanding? In an experiment, 109 university students all watched the same lecture. Half were randomly assigned to take notes on a laptop, while the remaining students took notes by hand. Before taking a test to measure conceptual understanding of the lecture, researchers assigned half of each group to review their notes, while the other half weren't given an opportunity for this review. Students who took notes by hand did significantly better on conceptual questions.[62]

(a) Describe how the researchers could have randomly assigned the students to the four treatments, using groups of roughly equal size.

(b) Explain the purpose of random assignment in this experiment.

Applying the Concepts These exercises ask you to apply multiple learning goals in a new context or to apply what you learned in this section in a new way.

21. **Reversing Cognitive Decline** As people age, they are at an increased risk of cognitive decline, including dementia. Can cognitive decline be slowed down or even reversed? In a recent experiment involving 160 older adults, half were randomly assigned to a heart-healthy DASH (Dietary Approaches to Stop Hypertension) diet, with the other half receiving no prescribed diet. Likewise, half of the adults in each diet group were randomly assigned to follow a specific exercise regimen and the other half were not told to exercise. After 6 months, the improvement in cognitive function was recorded for the subjects in each of the four groups. Subjects who were assigned the DASH diet and exercise regimen did best.[63]

(a) Why was it necessary to perform an experiment rather than simply asking older adults about their diet and exercise habits?

(b) Explain the purpose of the control group in this experiment.

(c) Describe how the researchers could have randomly assigned the older adults to the four treatments, using groups of equal size.

(d) Explain the purpose of random assignment in this experiment.

22. **Meal Portions** Over time, the portion sizes of restaurant meals have been increasing. Can offering a to-go box prevent overeating? In a recent experiment involving 53 women, researchers randomly assigned 27 to be informed before the meal that their remaining food would be packaged up in a to-go box. The researchers did not tell the remaining 26 women anything about a to-go box. Furthermore, they randomly assigned each of the 53 women a different portion size: regular, 25% more than regular, 50% more than regular, and 75% more than regular. At the end of each meal, they recorded the total number of calories consumed for the women in the eight groups. Women who were given information about the to-go box tended to eat less, especially at the higher portion sizes.[64]

(a) Why was it necessary to perform an experiment rather than simply asking women if they typically get to-go containers?

(b) Explain the purpose of the control group in this experiment.

(c) Describe how the researchers could have randomly assigned the women to the eight treatments, using groups of roughly equal size.

(d) Explain the purpose of random assignment in this experiment.

23. **Cocoa Effects** A study involved 27 healthy people aged 18 to 72 who consumed a cocoa beverage containing 900 milligrams of flavonols (a class of flavonoids) daily for 5 days. Using a finger cuff,

researchers measured blood flow on the first and fifth days of the study. After 5 days, researchers measured what they called "significant improvement" in blood flow and the function of the cells that line the blood vessels.[65] What flaw in the design of this experiment makes it impossible to say if the cocoa really caused the improved blood flow? Explain your answer.

24. **Comparing Treatments** A large study used records from Canada's national health care system to compare the effectiveness of two ways to treat prostate disease. The two treatments were traditional surgery and a new method that does not require surgery. The records described many cases where doctors had chosen the method of treatment for their patients. The study found that patients treated by the new method were significantly more likely to die within 8 years. What flaw in this study makes it impossible to determine if the new method is more dangerous? Explain your reasoning.[66]

25. **Pain Relief** The extracts of avocado and soybean oils have been shown to slow cell inflammation in test tubes. Will taking avocado and soybean unsaponifiables (ASU; i.e., components of the oily mixture) help relieve pain for subjects with joint stiffness due to arthritis? In an experiment, 345 people were randomly assigned to receive either 300 milligrams of ASU daily or a daily placebo for 3 years.[67] Could this be a single- or double-blind experiment? Explain your answer. Why is blinding an important consideration in this experiment?

26. **Placebo Effect** In a recent study, researchers had volunteers rate the pain of an electric shock before and after taking a new medication. However, half of the subjects were told the medication cost $2.50 per dose, while the other half were told the medication cost $0.10 per dose. In reality, both medications were placebos. Of the "cheap" placebo users, 61% experienced pain relief, while 85% of the "expensive" placebo users experienced pain relief.[68] Explain how the results of this study support the idea of a placebo effect.

Extending the Concepts These exercises challenge you to explore statistical concepts and methods that go beyond what you learned in this section.

27. **Breakfast** Refer to Exercise 11. Design an experiment to investigate if eating breakfast more often causes higher test scores. Explain how your experiment avoids the problem you described in Exercise 11.

28. **Vegetables and Eye Health** Refer to Exercise 10. A well-designed experiment avoids confounding, but a *poorly* designed experiment may not. Design an experiment about eating leafy green vegetables and developing AMD that is subject to confounding, and explain why it is.

29. **"Natural" Experiment** A recent study compared the rate of crashes involving 16- to 18-year-old drivers in two adjacent Virginia counties. For two consecutive years, there were more crashes per person for drivers in this age group in Chesterfield County, where school starts at 7:20 a.m., than in Henrico County, where school starts at 8:45 a.m. The study was described as a "natural experiment," because the two counties are nearly identical in socioeconomic characteristics and in the percentage of roads with traffic congestion. Explain why it is harder to establish a cause-and-effect conclusion with this type of "natural experiment" than with an experiment that uses random assignment.[69]

Cumulative Review These exercises revisit what you learned in previous sections.

30. **Fee Increases (1.2, 1.3)** Do the majority of students at a university support an increase in mandatory fees to pay for upgrades to the student union building? To find out, an administrator surveys 100 students as they enter the student union.

    (a) What type of sample did the administrator obtain?

    (b) Explain why the administrator's sampling method is biased. Is the result from the sample likely to be greater than or less than the percentage of all students who support the increase in fees?

    (c) Would increasing the sample size to 500 students account for the bias described in part (b)? Explain your reasoning.

Section 1.6 Completely Randomized Designs

LEARNING GOALS

By the end of this section, you will be able to:

- Identify other sources of variability in an experiment and explain the benefits of keeping these variables the same for all experimental units.
- Outline an experiment that uses a completely randomized design.
- Use simulation to determine if the difference between two means or two proportions in an experiment is statistically significant.

In Section 1.5, we described a physiology class's experiment to determine if caffeine affects pulse rates. In this section, we finish designing the caffeine experiment and explore what it means to say that the results of an experiment are statistically significant.

Experiments: Other Sources of Variability

Although random assignment should create groups of experimental units that are roughly equivalent at the beginning of an experiment, other variables might also have an impact on the response variable. In the physiology class's caffeine experiment, the sugar content, the amount of soda consumed, and the temperature of the room during the experiment will likely affect pulse rates.

Because sugar almost certainly affects pulse rates, it is important that each subject receive the same amount of sugar. If one group received regular cola with caffeine and the other group received caffeine-free *diet* cola with no sugar, then sugar consumption and caffeine intake would be confounded. *One reason to keep other variables the same for each subject is to prevent confounding, making it easier to determine if one treatment is more effective than another.*

We should also make sure that each student has the same amount of cola to drink. If each student was able to drink as much or as little as they wanted, the changes in pulse rates would be more variable than they would be otherwise. This increase in variability makes it harder to find convincing evidence that caffeine affects pulse rates.

The dotplots in Figure 1.1(a) show the results of an experiment in which the amount of cola consumed was the same for all participating students. From these graphs, it seems clear that caffeine does increase pulse rates. The dotplots in Figure 1.1(b) show the results of an experiment in which the students were able to choose the amount of soda they drank. Notice that the centers of the distributions haven't changed, but the distributions on the right are more variable. The increased overlap in the graphs makes the evidence supporting the effect of caffeine less convincing.

FIGURE 1.1 Dotplots showing the results of the caffeine experiment when (a) the amount of soda was kept the same and (b) when the amount of soda was allowed to vary.

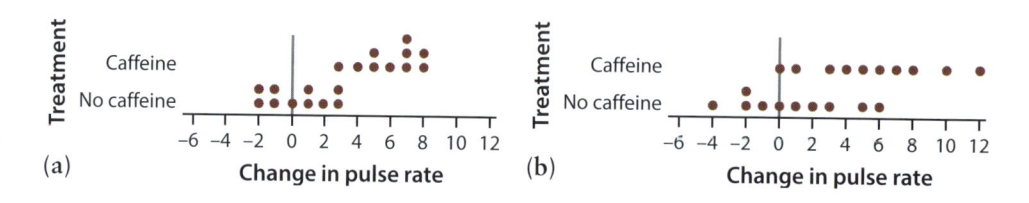

The second reason we keep other variables the same is to reduce the variability in the response variable, making it easier to determine if one treatment is more effective than another.

EXAMPLE 1.6.1

Is it harder to learn while multitasking?

Other sources of variability

PROBLEM

Researchers in Canada performed an experiment with university students to examine the effects of multitasking on student learning. The 40 participants in the study were asked to attend a lecture and take notes with their laptops. Half of the participants were randomly assigned to complete other online tasks not related to the lecture during that time. These tasks were meant to imitate typical student web browsing during classes. The remaining students simply took notes with their laptops. At the end of the lecture, students were given a test about the content of the lecture.[70] Identify one variable that the researchers should keep the same for all subjects. Provide two reasons why it is important for the researchers to keep this variable the same.

SOLUTION

All the students should be given the same test. If groups were given different tests, one test might be easier than the other and we wouldn't know if higher scores were due to the treatment or the ease of the test. Also, if there were different tests for each student, the test scores are likely to have more variability, making it harder to tell if the multitasking was harmful.

> Subjects should also attend the same lecture and be treated the same in every other way, except for the treatment.

EXAM PREP: FOR PRACTICE, TRY EXERCISE 7.

Completely Randomized Designs

The diagram in Figure 1.2 presents the details of the caffeine experiment from Section 1.5: random assignment, the sizes of the groups and which treatment they receive, and the response variable. This type of experimental design is called a **completely randomized design**.

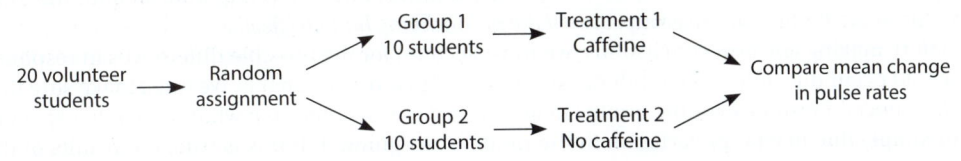

```
20 volunteer → Random → Group 1 → Treatment 1  ↘
students      assignment  10 students  Caffeine     Compare mean change
                        ↘ Group 2 → Treatment 2  ↗   in pulse rates
                          10 students  No caffeine
```

FIGURE 1.2 Outline of a completely randomized design to investigate the effects of caffeine on pulse rates.

> In a **completely randomized design**, the experimental units are assigned to the treatments (or the treatments to the experimental units) completely at random.

Other types of experimental design are also possible. However, in all types of experimental design, random assignment to treatments is essential. Furthermore, there is no requirement that treatment groups be exactly the same size, although there are some statistical advantages if they are roughly equal in size.

EXAMPLE 1.6.2

What distracts drivers?

Completely randomized designs

PROBLEM

Is talking on a cell phone while driving more distracting than talking to a passenger? David Strayer and his colleagues at the University of Utah used 48 undergraduate students as subjects in an experiment. The researchers randomly assigned half of the subjects to drive in a simulator while talking on a cell phone and the other half to drive in the simulator while talking to a passenger. One response variable was whether or not the driver stopped at a rest area that was specified by researchers before the simulation started.[71] Outline a completely randomized design for this experiment.

SOLUTION

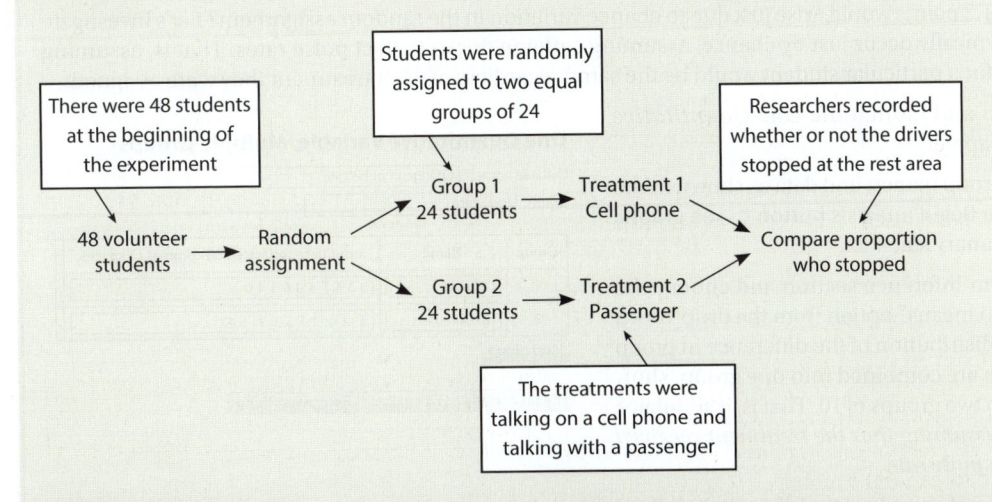

```
There were 48 students       Students were randomly       Researchers recorded
at the beginning of          assigned to two equal        whether or not the drivers
the experiment               groups of 24                 stopped at the rest area

48 volunteer → Random → Group 1 → Treatment 1 ↘
students      assignment  24 students  Cell phone    Compare proportion
                        ↘ Group 2 → Treatment 2 ↗   who stopped
                          24 students  Passenger

                          The treatments were
                          talking on a cell phone and
                          talking with a passenger
```

EXAM PREP: FOR PRACTICE, TRY EXERCISE 11.

Statistical Significance

In the experiment about multitasking, the group that was assigned to multitask did 11 percentage points worse on the test (55% versus 66%) than the control group. Because both groups were relatively small, it is unlikely that the two groups were exactly equivalent to begin with, even though the subjects were randomly assigned. By chance, it is possible that the multitasking group had fewer strong test takers. If so, the mean test score of the multitasking group might be smaller than the mean test score for the control group, *even if the multitasking had no effect.*

Before making any firm conclusions, we must account for the possible differences in response that could occur due *only* to the random assignment of treatments. That is, we must make sure that the difference in response for the groups in an experiment is larger than what we would expect to occur simply due to chance variation in the random assignment. If this is true, the results of the experiment are **statistically significant**.

When an observed difference in responses between the groups in an experiment is so large that it is unlikely to be explained by chance variation in the random assignment, we say that the result is **statistically significant.**

In the multitasking experiment, the researchers determined that the difference in means of 11 percentage points was statistically significant. In other words, a difference in mean test scores of 11 percentage points was too large to occur simply due to chance variation in the random assignment. Note that some *statistically significant* differences may not be considered *practically important*. The real-world importance of experimental results depends greatly on the context of the research. You'll learn more about this in Chapter 8.

CONCEPT EXPLORATION

Drawing conclusions from the caffeine experiment

In this exploration, you will use a simulation to determine if a difference in means is statistically significant.

Here are the results of the physiology class's experiment.

	Change in pulse rate (Final − Initial)										Mean change
Caffeine	8	3	5	1	4	0	6	1	4	0	3.2
No caffeine	3	−2	4	−1	5	5	1	2	−1	4	2.0

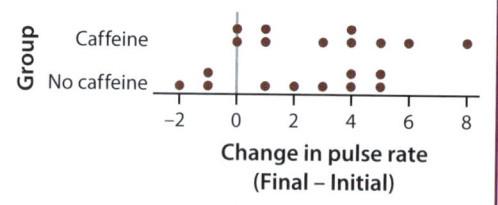

The mean change in pulse rate for the caffeine-consuming group was 1.2 points greater than the mean change for the no-caffeine group. This suggests that caffeine does increase pulse rates. But is the difference statistically significant? Or is it plausible that a difference of 1.2 points would arise just due to chance variation in the random assignment? Let's investigate by seeing what differences typically occur just by chance, assuming caffeine doesn't affect pulse rates. That is, assuming that the change in pulse rate for a particular student would be the same regardless of what treatment they were assigned.

1. Go to www.stapplet.com and launch the *One Quantitative Variable, Multiple Groups* applet.

2. Enter the variable name, group names, and data as shown in the screenshot. Then click the Begin analysis button to see graphs of the data along with summary statistics.

3. Scroll down to the Perform Inference section and choose the "Simulate difference in two means" option from the drop-down menu. This simulates the distribution of the difference in group means when the 20 values are combined into one group, shuffled, and redistributed into two groups of 10. That is, it simulates the random assignment *assuming that the treatment received doesn't affect the change in pulse rate.*

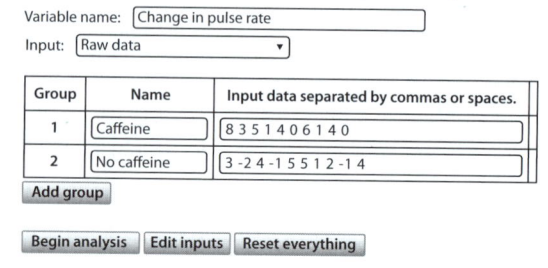

4. Enter 100 for the "Number of samples to add" and click the Add trials button. How many of the simulated random assignments resulted in a difference in means of 1.2 or greater?

5. Based on your answer in Step 4, is the difference of 1.2 statistically significant?

The dotplot in Figure 1.3 shows that getting a difference in means of 1.2 isn't that unusual. In 19 of the 100 trials, we obtained a difference of 1.2 or more simply due to chance variation in the random assignment. Because a difference of 1.2 or greater is somewhat likely to occur by chance alone, the results of the physiology's class experiment aren't statistically significant and do not provide convincing evidence that caffeine affects pulse rate.

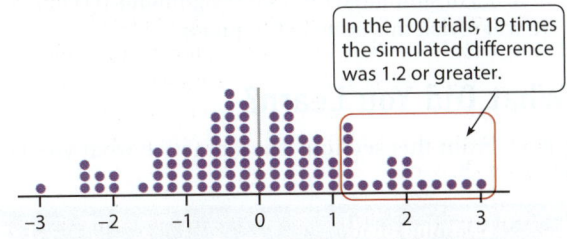

FIGURE 1.3 Dotplot showing the differences in means that occurred in 100 simulated random assignments, assuming that caffeine has no effect on pulse rates.

In the 100 trials, 19 times the simulated difference was 1.2 or greater.

Simulated difference in mean change in pulse rate

EXAMPLE 1.6.3

What distracts drivers?

Determining statistical significance

PROBLEM

Here are the results of the distracted-driver experiment from Example 1.6.2.

		Treatment		
		Cell phone	Passenger	Total
Response	Stopped at rest area	12	21	33
	Didn't stop	12	3	15
	Total	24	24	48

(a) Calculate the proportion of students who stopped at the rest area in each of the two groups, and the difference in proportions (Passenger − Cell phone).

One hundred trials of a simulation were performed using the *One Categorical Variable, Multiple Groups* applet to see what differences in proportions would occur due only to chance variation in the random assignment, assuming that the type of distraction doesn't matter. That is, 33 "stoppers" and 15 "non-stoppers" were randomly assigned to two groups of 24. The dotplot displays the results of the simulation.

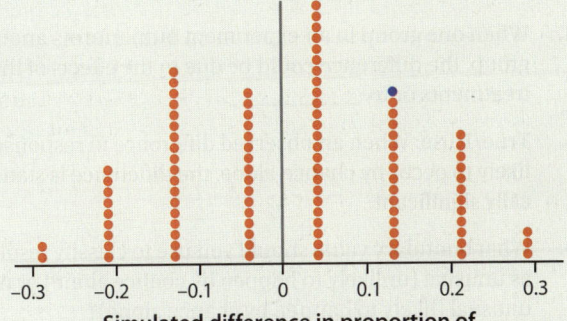

(b) There is a blue dot at a difference of 0.125. Explain what this dot represents.

(c) Based on the results of the simulation, is the difference from part (a) statistically significant? Explain your reasoning.

Simulated difference in proportion of students who stopped (Passenger − Cell phone)

SOLUTION

(a) Passenger: 21/24 = 0.875; Cell phone = 12/24 = 0.500; Difference = 0.875 − 0.500 = 0.375.

(b) One simulated random assignment resulted in a difference in proportions (Passenger − Cell phone) of 0.125.

(c) Because a difference of 0.375 or greater never occurred in the simulation, the difference is statistically significant. It is extremely unlikely to get a difference this big simply due to chance variation in the random assignment.

EXAM PREP: FOR PRACTICE, TRY EXERCISE 15.

In the caffeine example, we said that a difference in means of 1.2 *was not unusual* because a difference this big or bigger occurred in 19% of the simulated random assignments by chance alone. In the distracted-drivers example, we said that a difference in proportions of 0.375 *was unusual* because a difference this big or bigger occurred in 0% of the simulated random assignments by chance alone. So the boundary between "not unusual" and "unusual" must be somewhere between 0% and 19%. For now, we recommend using a boundary of 5%, meaning that differences that would occur in less than 5% of simulated random assignments by chance alone are considered statistically significant. We will revisit this issue in Chapter 8.

Section 1.6 What Did You Learn?

Review the learning goals from this section. Then practice what you've learned by working through the exercises.

Learning Goal	Example	Exercises
Identify other sources of variability in an experiment and explain the benefits of keeping these variables the same for all experimental units.	1.6.1	7–10
Outline an experiment that uses a completely randomized design.	1.6.2	11–14
Use simulation to determine if the difference between two means or two proportions in an experiment is statistically significant.	1.6.3	15–18

Section 1.6 Exercises

Building Concepts and Skills These exercises assess the basic knowledge you should have after reading the section.

1. We should keep other variables the same in an experiment to prevent _____ and reduce _____.

2. Keeping other variables the same in an experiment makes it easier to _____.

3. True/False: In a completely randomized design, researchers consider characteristics of the experimental units when assigning treatments.

4. When one group in an experiment outperforms another group, the difference could be due to the effects of the treatments or to _____.

5. True/False: When an observed difference in response is likely to occur by chance alone, the difference is statistically significant.

6. What boundary value should you use to classify results as unusual (unlikely to happen by chance alone) or not unusual (likely to happen by chance alone)?

Mastering Concepts and Skills These exercises reinforce the learning goals as shown in the examples.

7. **Refer to Example 1.6.1** **Résumé Names** In an experiment about racial discrimination in hiring, researchers sent similar résumés in response to help-wanted ads in two major cities. Half of the résumés were assigned "White-sounding" names (e.g., Emily Walsh or Greg Baker) and half were assigned "Black-sounding" names (e.g., Lakisha Washington or Jamal Jones). *Results*: Applicants with "White-sounding" names had a 1-in-10 response rate but applicants with "Black-sounding" names had a 1-in-15 response rate.[72]

   (a) Identify one variable that researchers kept the same in this experiment.

   (b) Explain how keeping this variable the same helps prevent confounding.

8. **Pay to Play?** Are politicians more available to people who donate money to their campaigns? Researchers had members of a political organization attempt to schedule an appointment with members of Congress to discuss the banning of a particular chemical. Callers were identified as either a "local campaign donor" or a "local constituent," with the identification determined at random for each member of Congress contacted. Otherwise, the protocol for the calls was identical, including the details of the meeting request, which were delivered by email. *Results*: "Donors" were more successful in obtaining meetings and were given better access to higher-level staff, including the member of Congress.[73]

   (a) Identify one variable that researchers kept the same in this experiment.

   (b) Explain how keeping this variable the same helps prevent confounding.

9. **Vitamin D and Asthma** Researchers randomly assigned 408 adults with persistent asthma to take one of two treatments: an inhaled medicine with a vitamin D supplement, or the inhaled medicine with a placebo. *Results*: After 7 months, the vitamin D users were no less likely to experience reduced air flow, hospitalization, or the need for additional medicine.[74]

(a) Identify one variable that researchers kept the same in this experiment.

(b) Other than to help avoid confounding, explain another benefit of keeping this variable the same.

10. **Apple Calories** In an experiment to determine how calories in solid foods affect appetite, researchers randomly assigned volunteers to one of four treatments: 150 calories of apple slices, 150 calories of applesauce, 150 calories of apple juice, or 150 calories of apple juice with added fiber. Fifteen minutes later, researchers provided each participant with a large bowl of pasta and measured how many calories each person consumed. *Results:* The apple slice group ate 190 fewer calories, on average, compared to the other three groups.[75]

    (a) Identify one variable that researchers kept the same in this experiment.

    (b) Other than to help avoid confounding, explain another benefit of keeping this variable the same.

11. **Refer to Example 1.6.2 Fish Oil** To see if fish oil can help reduce blood pressure, researchers recruited 14 male participants with high blood pressure and randomly assigned each to a 4-week diet that included fish oil or a 4-week diet that included a mixture of oils that approximated the types of fat in a typical diet. At the end of the 4 weeks, researchers measured each participant's blood pressure again and recorded the reduction in diastolic blood pressure.[76] Outline a completely randomized design for this experiment.

12. **Echinacea** To see if the popular herbal supplement echinacea can help reduce the length of a cold, researchers randomly assigned 350 volunteers to take echinacea or a placebo. Each volunteer stated that they had a cold or that they thought they were coming down with a cold and had symptoms start at most 36 hours before receiving the treatment. For each volunteer, researchers recorded the length of the illness.[77] Outline a completely randomized design for this experiment.

13. **Cancer** In an experiment to compare treatments for prostate cancer, 731 men with localized prostate cancer were randomly assigned to one of two groups. A group of 364 men was assigned to have surgery and a group of 367 men was assigned to be observed only. After 20 years, researchers noted whether or not each subject was still alive.[78] Outline a completely randomized design for this experiment.

14. **Crime Metaphors** A study examined the impact of metaphors on the way people think about complex issues. In one part of the study, 243 subjects were divided randomly into two groups. A group of 110 people was assigned to read a passage that described crime in a fictional city as a "beast" ravaging the city. A second group of 133 people was assigned to read an identical passage with one change — the word "beast" was replaced with the word "virus." After reading the passage, all the subjects were asked what the town should do in response to crime. Responses were categorized as either suggesting more enforcement or more social programs.[79] Outline a completely randomized design for this experiment.

15. **Refer to Example 1.6.3 Fish Oil** Refer to Exercise 11. The reductions in blood pressure are shown here. Note that a negative value means that the subject's blood pressure *increased*.

Fish oil	8	12	10	14	2	0	0
Mixture	-6	0	1	2	-3	-4	2

    (a) Calculate the mean reduction for the two groups and the difference in means (Fish oil − Mixture).

    One hundred trials of a simulation were performed to see what differences in means would occur due only to chance variation in the random assignment, assuming that the type of oil doesn't matter. The dotplot displays the results of the simulation.

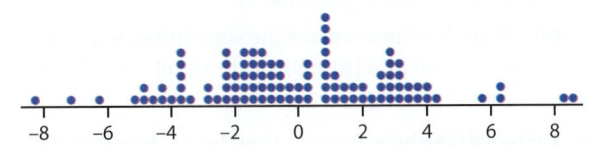

    Simulated difference in mean reduction (Fish oil − Mixture)

    (b) There is one dot at a difference of 5.8. Explain what this dot represents.

    (c) Based on the results of the simulation, is the difference from part (a) statistically significant? Explain your reasoning.

16. **Echinacea** Refer to Exercise 12. The average duration of the cold was 6.34 days for the volunteers who received the echinacea and 6.87 days for the volunteers who received the placebo.

    (a) Calculate the difference in the mean duration for the two groups (Placebo − Echinacea).

    One hundred trials of a simulation were performed to see what differences in means would occur due only to chance variation in the random assignment, assuming that the treatment doesn't matter. The dotplot displays the results of the simulation.

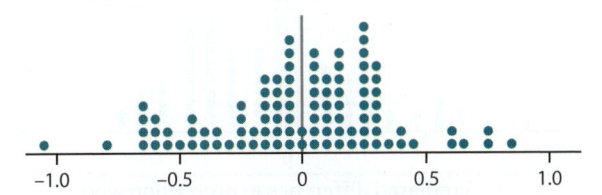

    Simulated difference in mean duration (Placebo − Echinacea)

    (b) There is one dot at a difference of 0.85. Explain what this dot represents.

    (c) Based on the results of the simulation, is the difference from part (a) statistically significant? Explain your reasoning.

17. **Cancer** Refer to Exercise 13. After 20 years, 141 of the 364 men assigned to surgery were still alive and 122 of the 367 men assigned to observation were still alive.

    (a) Calculate the proportion of subjects who were still alive in each of the two treatment groups, and the difference in proportions (Surgery − Observe).

One hundred trials of a simulation were performed to see what differences in proportions would occur due only to chance variation in the random assignment, assuming that the type of treatment doesn't matter. The dotplot displays the results of the simulation.

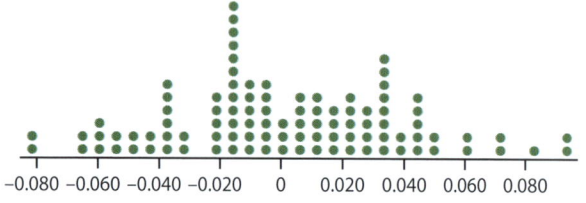

Simulated difference in proportion who are alive (Surgery − Observe)

(b) There is one dot at a difference of 0.082. Explain what this dot represents.

(c) Based on the results of the simulation, is the difference from part (a) statistically significant? Explain your reasoning.

18. **Crime Metaphors** Refer to Exercise 14. Here are the results of the experiment.

		Treatment		
		"Beast"	"Virus"	Total
Response	Enforcement	80	72	152
	Social	30	61	91
	Total	110	133	243

(a) Calculate the proportion of subjects who suggested "enforcement" in each of the two treatment groups, and the difference in proportions (Beast − Virus).

One hundred trials of a simulation were performed to see what differences in proportions would occur due only to chance variation in the random assignment, assuming that the type of metaphor doesn't matter. The dotplot displays the results of the simulation.

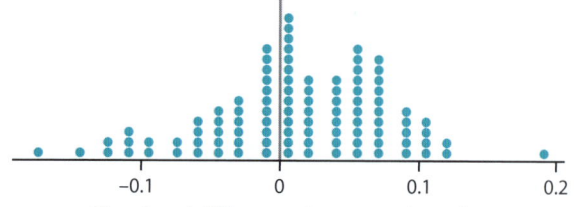

Simulated difference in proportion who suggest enforcement (Beast − Virus)

(b) There is one dot at a difference of 0.19. Explain what this dot represents.

(c) Based on the results of the simulation, is the difference from part (a) statistically significant? Explain your reasoning.

Applying the Concepts These exercises ask you to apply multiple learning goals in a new context or to apply what you learned in this section in a new way.

19. **The Physicians' Health Study** Does regularly taking aspirin help protect people against heart attacks? Does regularly taking beta-carotene help prevent cancer? The Physicians' Health Study I was a medical experiment that helped answer these questions. The subjects in this experiment were 21,996 male physicians. Each of the physicians was assigned to one of the following combinations: aspirin and beta-carotene, aspirin and placebo, placebo and beta-carotene, or placebo and placebo. The physicians were followed for several years and researchers recorded whether each of the physicians had a heart attack or developed cancer.[80]

(a) Identify one variable that researchers kept the same in this experiment and explain a benefit of keeping this variable the same.

(b) Outline a completely randomized design for this experiment.

(c) The difference in the proportion of subjects who developed cancer was not statistically significant. Explain what it means that this difference was not statistically significant. (*Note:* The difference in heart-attack rates *was* statistically significant.)

20. **Sinus Infections** In an experiment, 166 adults from the St. Louis area were recruited and randomly assigned to receive one of two treatments for a sinus infection. Half of the subjects received an antibiotic (amoxicillin) and the other half received a placebo. At different stages during the next month, all subjects took the Sino-Nasal Outcome Test (SNOT).[81]

(a) Identify one variable that researchers kept the same in this experiment and explain a benefit of keeping this variable the same.

(b) Outline a completely randomized design for this experiment.

(c) After 10 days, the difference in average SNOT scores was not statistically significant. Explain what it means that this difference was not statistically significant.

Extending the Concepts These exercises challenge you to explore statistical concepts and methods that go beyond what you learned in this section.

21. **Magnet Treatment** Research has shown that magnetic fields can affect living tissue in humans. To investigate if magnets can be used to help patients with chronic pain, researchers conducted an experiment. A doctor identified a painful site on each patient and asked the patient to rate the pain on a scale from 0 (mild pain) to 10 (severe pain). Then, the doctor selected a sealed envelope containing a magnet at random from a box with a mixture of active and inactive magnets. The chosen magnet was applied to the site of the pain for 45 minutes. After being treated, each patient was again asked to rate the level of pain from 0 to 10 and the improvement in pain level was recorded.[82] Here are the improvements:

Active	0 0 0 0 1 1 1 4 4 5 5 5 5 6 6 6 6 6 6 7 7 7 8 8 8 10 10 10 10
Inactive	0 0 0 0 0 0 0 0 0 1 1 1 1 1 2 3 4 4 5

(a) Open the *One Quantitative Variable, Multiple Groups* applet at www.stapplet.com. Then enter the variable name, group names, and data. Click the Begin analysis button and calculate the difference in the mean improvement (Active − Inactive).

(b) In the Perform Inference section of the applet, simulate the difference in means at least 100 times. What percentage of the simulated random assignments gave a difference at least as large as the one you calculated in part (a)?

(c) Based on your answer to part (b), is the difference from part (a) statistically significant? Explain your reasoning.

22. **Head Lice** A study published in the *New England Journal of Medicine* compared two medicines to treat head lice: an oral medication called ivermectin and a topical lotion containing malathion. Researchers used 376 households in seven areas around the world as experimental units. Of the households randomly assigned to ivermectin, 171 were free from head lice after 2 weeks and 14 were not. In the malathion group, 151 were free from head lice after 2 weeks and 40 were not.[83]

(a) Open the *One Categorical Variable, Multiple Groups* applet at www.stapplet.com. Then enter the group names, variable name, category names, and counts. Click the Begin analysis button and calculate the difference in the proportions who were free from head lice (Ivermectin − Malathion).

(b) In the Perform Inference section of the applet, simulate the difference in proportions at least 100 times. What percentage of the simulated experiments gave a difference at least as large as the one you calculated in part (a)?

(c) Based on your answer to part (b), is the difference from part (a) statistically significant? Explain your reasoning.

Cumulative Review These exercises revisit what you learned in previous sections.

23. **Voice Vote (1.2)** On the popular reality-TV show *The Voice*, viewers can vote for their favorite contestant by voting online, by using a smartphone app, by using the Xfinity X1 set-top box, or by streaming a song on Apple Music. Each method is limited to 10 votes per episode. Explain why the winning contestant may *not* be the one that the majority of the show's viewers prefer.[84]

Section 1.7 Blocking

LEARNING GOALS

By the end of this section, you will be able to:
- Design an experiment that uses blocking.
- Explain the benefits of using blocking in an experiment.
- Design an experiment that uses matched pairs.

In Sections 1.5 and 1.6, you learned how to set up an experiment using a completely randomized design. These simply designed experiments nicely incorporate the principles of comparison and random assignment. Yet, as with sampling, there are times when the simplest method doesn't yield the most precise results. In this section, you will learn how to use a different type of experimental design and why this design can be beneficial.

Blocking

A university's school of agriculture has been developing a new variety of blackberry plant that is supposed to have a greater yield than the current bestselling blackberry plant. To put the new variety to the test, agriculture students perform an experiment using two different plots of land. The smaller plot of land is more fertile and has space for four plants. The larger plot of land is less fertile and has space for six plants.

Because of the difference in fertility, the blackberry yields are likely to be more variable than if all the plants were in equally fertile soil. This additional variability may make it difficult to detect a difference in the average yield of the two varieties. To account for this source of variability, we form **blocks** based on the fertility of the plots.

> A **block** is a group of experimental units that are known before the experiment to be similar in some way that is expected to affect the response to treatments.

Within each block, the agriculture students will randomly assign half of the spaces to receive the new variety and half to receive the current variety, as shown in the figure.

44 CHAPTER 1 Collecting Data

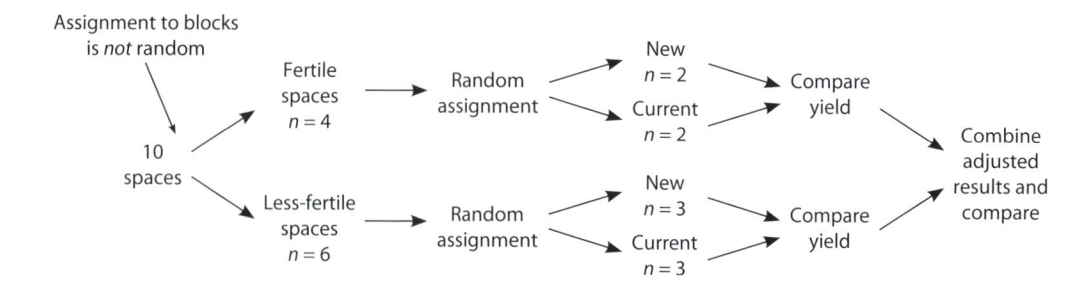

```
                                                      New
Assignment to blocks                                  n = 2      Compare
  is not random        Fertile      Random                       yield
                       spaces    → assignment →       Current                Combine
                       n = 4                          n = 2                  adjusted
     10                                                                      results and
     spaces                                           New                    compare
                       Less-fertile  Random           n = 3      Compare
                       spaces     → assignment →                 yield
                       n = 6                          Current
                                                      n = 3
```

In general, blocking works best when blocks are formed using variables that are the most strongly associated with the response variable(s) in the experiment. In the blackberry experiment, it is reasonable to think that the fertility of the plots is more strongly associated with yield than with other variables in the plots such as proximity to other plants or sun exposure.

HOW TO | Conduct an Experiment Using Blocking

1. **Form blocks.** Divide the experimental units into groups (blocks) that share characteristics thought to be associated with the response variable(s) in the experiment.

2. **Randomly assign treatments within blocks.** Within each block, randomly assign the members of the block to treatments in roughly equal numbers, as in a completely randomized design.

3. **Record the results.** At the end of the experiment, record the value of the response variable for each experimental unit.

We'll talk about how to analyze the results of an experiment with blocking in the next subsection. For now, we will focus on the design of the experiment.

EXAMPLE 1.7.1

Are nurse practitioners as effective as doctors?

Blocking

PROBLEM
Nurse practitioners are nurses with advanced qualifications who often act much like primary care physicians. Are they as effective as doctors in treating patients with chronic conditions? An experiment was conducted with 1316 patients who had been diagnosed with asthma, diabetes, or high blood pressure. One response variable was a measure of the patient's overall health.[85] Design an experiment that uses blocking to investigate this question. Explain your choice of blocks.

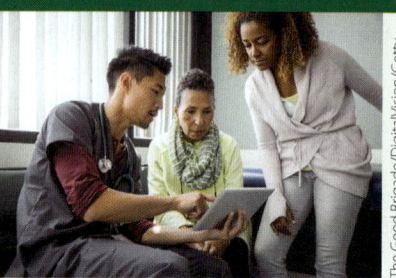

SOLUTION
Form blocks based on chronic condition (asthma, diabetes, or high blood pressure). It is reasonable to think that there is an association between a patient's chronic condition and their overall health. That is, people with certain chronic conditions are likely to be healthier than people with other chronic conditions.

    1. Form blocks.

Randomly assign half of the subjects with asthma to a nurse practitioner and the other half to a doctor. Do the same for the subjects with diabetes and the subjects with high blood pressure.

    2. Randomly assign treatments within blocks.

After the patients have been treated by the nurse practitioner or the doctor for a certain length of time, measure each subject's overall health.

    3. Record the results.

EXAM PREP: FOR PRACTICE, TRY EXERCISE 5.

The Benefit of Blocking

Using an experimental design that incorporates appropriate blocking makes it easier to determine if one treatment is more effective than the other. To find out why, let's explore the results of the blackberry experiment described earlier. The dotplots show the yield (in pounds) for each of the five new variety plants and the five current variety plants.

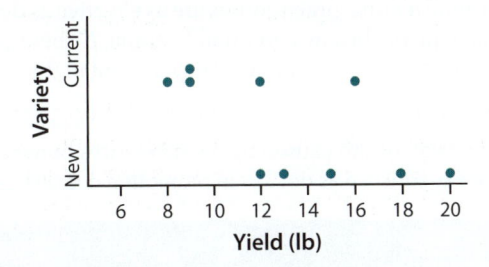

There is some evidence that the new variety results in higher yields, but the evidence isn't especially convincing. Enough overlap occurs in the two distributions that the differences might simply be due to the chance variation in the random assignment.

If we compare the results for the two varieties within each block, however, a different story emerges. Among the four plants in the fertile plot (indicated by the blue dots), the new variety was the clear winner. Likewise, among the six plants in the less fertile plot (indicated by the red squares), the new variety was the clear winner.

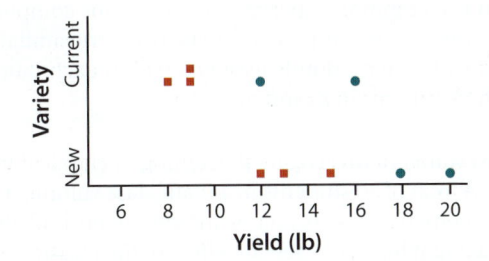

The overlap in the first set of dotplots was due almost entirely to the variation in fertility — the plants in the fertile plot had a higher average yield than the plants in the less fertile plot, regardless of which variety was planted. In fact, the average yield was 5.5 pounds greater in the fertile plot (16.5 versus 11). To account for the variation created by the difference in fertility, let's subtract 5.5 from the yields in the fertile plot to "even the playing field." Here are the adjusted yields (in pounds):

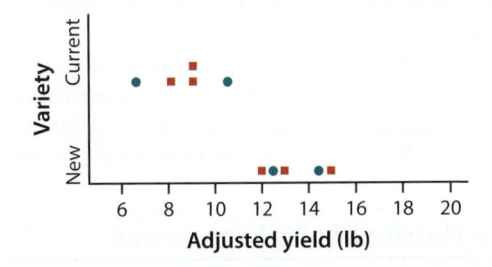

Because we accounted for the variation in yield due to the difference in fertility, the variation in each of the distributions has been reduced. There is no longer any overlap between the two distributions, meaning that the evidence in favor of the new variety is much stronger. *When blocks are formed wisely, it is easier to find convincing evidence that one treatment is more effective than another.*

EXAMPLE 1.7.2

Are nurse practitioners as effective as doctors?

Benefit of blocking

PROBLEM
In Example 1.7.1, we designed an experiment to determine if nurse practitioners are as effective as doctors in treating patients with chronic conditions. Blocks were formed based on type of chronic condition (asthma, diabetes, or high blood pressure). Explain why this experimental design is preferable to a completely randomized experiment in this context.

SOLUTION
This experimental design accounts for the variability in overall health caused by the differences between the chronic conditions. This makes it easier to determine if nurse practitioners are as effective as doctors in promoting overall health of their patients.

EXAM PREP: FOR PRACTICE, TRY EXERCISE 9.

Matched-Pairs Design

A common type of experimental design for comparing two treatments is a **matched-pairs design**. The idea is to create blocks by matching pairs of similar experimental units. The random assignment of subjects to treatments is done within each matched pair. Just as with other forms of blocking, matching helps account for the variation caused by the variable(s) used to form the pairs.

> A **matched-pairs design** is a common experimental design for comparing two treatments that uses blocks of size 2. In some matched-pairs designs, two very similar experimental units are paired and the two treatments are randomly assigned within each pair. In others, each experimental unit receives both treatments in a random order.

Suppose a psychologist wants to investigate if listening to classical music while taking a math exam affects performance. A total of 30 students in a math class volunteer to take part in the experiment. The difference in mathematical ability among the volunteers is likely to create additional variation in the test scores, making it harder to see the effect of the classical music. To account for this variation, we could pair the students by their grade in the class—the two students with the highest grades are paired together, the two students with the next highest grades are paired together, and so on. Within each pair, one student is randomly assigned to take the math exam while listening to classical music, and the other member of the pair is assigned to take the math exam in silence.

Sometimes, each "pair" in a matched-pairs design consists of just one experimental unit that gets both treatments in random order. In the classical music experiment, we could have each student take a math exam in both conditions. To decide the order, we might flip a coin for each student. If the coin lands on heads, the student takes a math exam with classical music playing today and a similar math exam without music playing tomorrow. If it lands on tails, the student does the opposite—no music today and classical music tomorrow.

Randomizing the order of treatments is important to avoid confounding. Suppose everyone did the classical music treatment on the first day and the no-music treatment on the second day, but the testing environment was warmer on the second day. We wouldn't know if any difference in the mean exam score was due to the difference in treatment or the difference in room temperature.

HOW TO | Conduct a Matched-Pairs Experiment

1. **Form pairs.** Create pairs of experimental units that share characteristics thought to be associated with the response variable(s) in the experiment. In some cases, the "pair" is a single experimental unit that is assigned both treatments.

2. **Randomly assign treatments within pairs.** Within each pair, randomly assign one member to receive the first treatment and the other member to the second treatment. When a single experimental unit forms the pair, randomly assign the order in which the treatments are given.

3. **Record the results.** At the end of the experiment, record the value of the response variable for each experimental unit. When a single experimental unit forms the pair, record the value of the response variable after each treatment is given.

You will learn how to analyze the results of a matched-pairs experiment in Sections 9.5 and 9.6.

EXAMPLE 1.7.3

Will an additive improve my mileage?

Matched-pairs designs

PROBLEM
A consumer organization wants to know if using a certain fuel additive increases the fuel efficiency (in miles per gallon) of cars. A total of 20 cars of different types are available for testing. Design an experiment that uses a matched-pairs design to investigate this question. Explain your method of pairing.

SOLUTION
Give each car both treatments. It is reasonable to think that some cars are more fuel efficient than others, so using each car as its own "pair" accounts for the variation in fuel efficiency in the experimental units. For each car, randomly assign the order in which the treatments are given by flipping a coin. Heads indicates using the additive first and no additive second. Tails indicates using no additive first and then the additive second. For each car, record the fuel efficiency (miles per gallon) after using each treatment.

When using each experimental treatment twice in a matched-pairs experiment, it is common to include a "wash-out" period between the treatments to make sure the first treatment doesn't affect the second treatment. In this case, cars that get the additive first might have some remnants of the additive left in the gas tank when it is filled up for the second treatment. The solution is to have every car use no-additive gasoline in between the trials.

EXAM PREP: FOR PRACTICE, TRY EXERCISE 13.

In the preceding example, it is also possible to form pairs of two similar cars. For example, we could pair the two most fuel-efficient cars, the next two most fuel-efficient cars, and so on. This is less ideal, however, because there will still be some differences between the members of each pair that may cause additional variation in the results. Using the same car twice creates perfectly matched "pairs," and it also doubles the number of pairs used in the experiment. Both of these features make it easier to find convincing evidence that the gas additive is effective, if it really is effective.

Section 1.7 What Did You Learn?

Review the learning goals from this section. Then practice what you've learned by working through the exercises.

Learning Goal	Example	Exercises
Design an experiment that uses blocking.	1.7.1	5–8
Explain the benefits of using blocking in an experiment.	1.7.2	9–12
Design an experiment that uses matched pairs.	1.7.3	13–16

Section 1.7 Exercises

Building Concepts and Skills These exercises assess the basic knowledge you should have after reading the section.

1. True/False: The experimental units in each block should be representative of all the experimental units in the experiment.

2. Once experimental units are in blocks, researchers use _____ to determine which units get which treatment.

3. True/False: Blocking can be beneficial because it allows us to account for a source of variability.

4. What are the two ways to form pairs in a matched-pairs experiment?

Mastering Concepts and Skills These exercises reinforce the learning goals as shown in the examples.

5. **Refer to Example 1.7.1 GMAT Prep** The Graduate Management Admission Test (GMAT) is required for many MBA programs. Are synchronous or asynchronous GMAT preparation courses more effective? You recruit 100 business majors to participate in the experiment, some of whom have already taken the

GMAT once, and others who haven't taken it at all. Design an experiment that uses blocking to investigate this question. Explain your choice of blocks.

6. **Testing Mice** Many medical treatments are tried on animals before they are tested on humans. Researchers plan to test the effectiveness of a drug to treat Alzheimer's disease using mice from three different suppliers. At the end of the experiment, the researchers will measure the accumulation of amyloid protein in the brains of the mice. Design an experiment that uses blocking to investigate this question. Explain your choice of blocks.

7. **Corn Varieties** An agriculture researcher wants to compare the yield of five corn varieties: A, B, C, D, and E. The field in which the experiment will be carried out increases in fertility from north to south. The researcher divides the field into 25 plots of equal size, arranged in 5 east–west rows of 5 plots each, as shown in the diagram. Design an experiment that uses blocking to investigate this question. Explain your choice of blocks.

8. **Citrus Fertilizers** A citrus farmer wants to know which of three fertilizers (A, B, and C) is most effective for increasing the amount of fruit on his trees. The farmer is willing to use 30 orange trees, 30 lemon trees, and 30 lime trees from the orchard in an experiment. Design an experiment that uses blocking to investigate this question. Explain your choice of blocks.

9. **Refer to Example 1.7.2** **GMAT Prep** Refer to Exercise 5. Explain why the randomized block design is preferable to a completely randomized experiment in this context.

10. **Testing Mice** Refer to Exercise 6. Explain why the randomized block design is preferable to a completely randomized experiment in this context.

11. **Corn Varieties** Refer to Exercise 7. Explain why the randomized block design is preferable to a completely randomized experiment in this context.

12. **Citrus Fertilizers** Refer to Exercise 8. Explain why the randomized block design is preferable to a completely randomized experiment in this context.

13. **Refer to Example 1.7.3** **Caffeine Dependence** What are the effects of caffeine withdrawal? Researchers recruited 11 volunteers who were diagnosed as being caffeine dependent to serve as subjects. The response variable was a subject's score on a test of depression.[86] Describe how you could use a matched-pairs design to investigate this question. Explain your method of pairing.

14. **Omelet Size** Does our perception of how much we eat affect subsequent food consumption? In a matched-pairs experiment, researchers told volunteers they were eating either a 2-egg omelet or a 4-egg omelet. In reality, all were 3-egg omelets. Volunteers who were told they had a 4-egg omelet ate significantly fewer calories the rest of the day.[87] Describe how you could use a matched-pairs design to investigate this question. Explain your method of pairing.

15. **Valve Surgery** Is the success rate of a new noninvasive method for replacing heart valves using a cardiac catheter more effective than traditional open-heart surgery? Medical researchers have 40 patients, varying in age from 55 to 75, who need valve replacement. One of several response variables will be the percentage of blood that flows backward — in the wrong direction — through the valve on each heartbeat. Describe how you could use a matched-pairs design to investigate this question. Explain your method of pairing.

16. **Nitrogen in Tires** Do nitrogen-filled tires maintain pressure better than air-filled tires? Consumers Union designed an experiment using 31 different types of tires. All tires were inflated to the same pressure and placed outside for a year. At the end of the year, Consumers Union measured the pressure in each tire.[88] Describe how you could use a matched-pairs design to investigate this question. Explain your method of pairing.

Applying the Concepts These exercises ask you to apply multiple learning goals in a new context or to apply what you learned in this section in a new way.

17. **Fresh Bread** Who doesn't love the smell of freshly baked bread? Does the smell of fresh bread also encourage people to eat more? Researchers performed a matched-pairs experiment where 30 volunteers were asked to fill out a survey in a room that either had no smell or was infused with the smell of freshly baked bread. After filling out the survey, volunteers were given as much vegetable soup as they wanted. The amount of soup consumed was significantly greater after subjects were exposed to the bread odor.[89]

    (a) Describe how you could use a completely randomized design to investigate this question.

    (b) Describe how you could use a matched-pairs design to investigate this question. Explain your method of pairing.

    (c) Which design do you prefer? Explain your answer.

18. **Two Apples** An apple a day keeps the doctor away. But will two apples a day lower cholesterol? Researchers performed a matched-pairs experiment in which 40 volunteers ate 2 apples per day for 8 weeks or drank apple juice that contained the same amount of sugar and calories for 8 weeks. At the end of the 8 weeks, researchers measured the change in each subject's cholesterol level, along with changes in other variables. The reduction in cholesterol was significantly greater with the apples.[90]

    (a) Describe how you could use a completely randomized design to investigate this question.

(b) Describe how you could use a matched-pairs design to investigate this question. Explain your method of pairing.

(c) Which design do you prefer? Explain your answer.

Extending the Concepts These exercises challenge you to explore statistical concepts and methods that go beyond what you learned in this section.

19. **Latin Square** Refer to Exercise 7. In addition to blocking by horizontal row, researchers might want to ensure that each treatment occurs only once in each vertical column. Such a design is called a Latin square. Copy the grid onto your paper and assign each plot a treatment (A, B, C, D, E) so that each treatment occurs only once in each row and only once in each column.

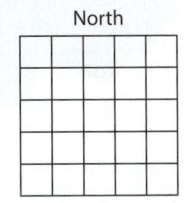

North

20. **Matched Quadruplets** A total of 20 people have agreed to participate in a study of the effectiveness of four weight-loss treatments (A, B, C, and D). The researcher first calculates how overweight each subject is by comparing the subject's current weight with their "ideal" weight. These values are shown in the following table.

Bimbaum	35	Hemandez	25	Moses	25	Smith	29
Brown	34	Jackson	33	Nevesky	39	Stall	33
Brunk	30	Kendall	28	Obrach	30	Tran	35
Cruz	34	Loren	32	Rodriguez	30	Wilansky	42
Deng	24	Mann	28	Santiago	27	Williams	22

(a) Design an experiment that uses blocks of size 4 to investigate which of these treatments works the best. Explain your choice of blocks.

(b) Carry out the random assignment for your design in part (a).

Cumulative Review These exercises revisit what you learned in previous sections.

21. **Rock Songs (1.2)** Are rock-and-roll songs longer now than they were 50 years ago? One way to answer this question would be to record the lengths of the *Billboard* Top 40 rock-and-roll songs from this year and from 50 years ago.

(a) Explain why the method described would not adequately answer this question.

(b) Describe how you would collect data to better answer this question.

Section 1.8 Data Ethics and the Scope of Inference

LEARNING GOALS

By the end of this section, you will be able to:
- Identify the population about which we can make inferences from a sample.
- Determine when it is appropriate to make an inference about cause and effect.
- Evaluate if a statistical study has been carried out in an ethical manner.

Researchers who conduct statistical studies often want to draw conclusions (make inferences) from the data they produce. Here are two examples:

- The U.S. Census Bureau carries out a monthly Current Population Survey of about 60,000 households. One goal is to use data from these randomly selected households to estimate the percentage of unemployed individuals in the population of all U.S. households.[91]
- Scientists performed an experiment involving 21 volunteer subjects between the ages of 18 and 25. Each subject was randomly assigned to one of two treatments: sleep deprivation for one night or unrestricted sleep. The scientists hoped to show that sleep deprivation causes a decrease in performance 2 days later.[92]

The inferences that can be made from a study depend on how the study was designed.

Inference About a Population

In the Census Bureau's sample survey, the households that were selected were *chosen at random* from the population of interest — all U.S. households. Because the sample was randomly selected from all U.S. households, the sample should be representative of this population and the estimated unemployment percentage should be close to the truth. The Census Bureau is safe in making an *inference about the population of all U.S. households* based on the results of the sample.

Note that the Census Bureau should *not* feel safe in making an inference about the percentage of unemployed individuals worldwide, as only U.S. households were eligible to be selected for the sample. Likewise, because the Census Bureau doesn't consider people in prison to be part of a household, it shouldn't make an inference about the population of prisoners based on its sample of U.S. households.

Inference About a Population

Random selection of individuals allows inference about the population from which the sample was selected.

EXAMPLE 1.8.1

A little something sweet

Inference about a population

PROBLEM
The National Health and Nutrition Examination Survey is a long-term research program that uses random sampling methods to examine health and nutrition in adults and children in the United States. One random sample of 2437 young adults (aged 18–25 years) found that 35% reported eating foods with added sugars (e.g., soda, brownies) at least 5 times per week.[93]

(a) Can we generalize this result to all U.S. adults? Explain your answer.

(b) What is the largest population to which we can generalize this result?

SOLUTION

(a) No. Adults older than age 25 weren't part of the population from which the sample was selected.

(b) All young adults (aged 18–25 years) in the United States.

EXAM PREP: FOR PRACTICE, TRY EXERCISE 5.

Inference About Cause and Effect

In the sleep deprivation experiment described at the beginning of this section, subjects were *randomly assigned* to the sleep deprivation and unrestricted sleep treatments. Random assignment helps ensure that the two groups of subjects are as similar as possible before the treatments are imposed. If the unrestricted group did better than the sleep-deprived group, and the difference in performance is statistically significant, the scientists can safely conclude that sleep deprivation caused the decrease in performance. That is, they can make an *inference about cause and effect*.

Inference About Cause and Effect

Random assignment of individuals to groups allows for inference about cause and effect.

Because the sleep researchers didn't randomly select their subjects from any larger population, they shouldn't generalize their results to all people or even to all people between the ages of 18 and 25. *When using volunteers, we should apply the results to only people like the ones in the study*. Determining whether a person is "like" those in an experiment isn't easy, but doctors and other researchers have to make this decision all the time in the real world.

EXAMPLE 1.8.2

Sweets and periodontal disease

Inference about cause and effect

PROBLEM
In Example 1.8.1, you read about a study involving 2437 randomly selected young adults from the United States. In addition to questions about food consumption, researchers asked questions about periodontal disease. The young adults who

reported eating added sugars more than 5 times per week had a 73% higher risk of developing periodontal disease than did those who reported eating no added sugars, and the increase was statistically significant. Based on this study, is it reasonable to say that eating foods with added sugars causes periodontal problems? Explain your answer.

SOLUTION

No, because the researchers didn't randomly assign the young adults to eat a specific number of foods with added sugar per week.

> Although it is certainly possible that eating lots of foods with added sugars causes periodontal disease, such disease could also be due to other variables. If people who eat a lot of foods with added sugars also drink a lot of soda, and the acid in the soda is what causes periodontal disease, then we would see the association between eating sugary food and periodontal disease even if sugar has no effect.

EXAM PREP: FOR PRACTICE, TRY EXERCISE 9.

The following chart summarizes the appropriate scope of inference based on the design of a study.

		Were individuals randomly assigned to groups?	
		Yes	**No**
Were individuals randomly selected?	Yes	Inference about the population: YES Inference about cause and effect: YES	Inference about the population: YES Inference about cause and effect: NO
	No	Inference about the population: NO Inference about cause and effect: YES	Inference about the population: NO Inference about cause and effect: NO

The Challenges of Establishing Causation

A well-designed experiment can tell us that changes in the explanatory variable cause changes in the response variable. However, in some cases it isn't practical or even ethical to do an experiment. Consider these questions:

- Does going to church regularly help people live longer?
- Does smoking cause lung cancer?

To answer these cause-and-effect questions, we need to perform a randomized comparative experiment. Unfortunately, we can't randomly assign people to attend church or to smoke cigarettes. The best data we have about these and many other cause-and-effect questions come from observational studies.

Doctors had long observed that most patients with lung cancer were smokers. Comparison of smokers and similar nonsmokers showed a very strong association between smoking and death from lung cancer. Could the association be due to some other variable? It might be that smokers live unhealthy lives in other ways (e.g., diet, alcohol, lack of exercise) and that one of these variables confounds the relationship between smoking and lung cancer.

Despite the challenges, it is sometimes possible to build a strong case for causation in the absence of experiments. The evidence that smoking causes lung cancer is about as strong as nonexperimental evidence can be. There are several criteria for establishing causation when we can't do an experiment:

- *The association is strong.* The association between smoking and lung cancer is very strong.
- *The association is consistent.* Many studies of different kinds of people in many countries link smoking to lung cancer. That reduces the chance that some other variable specific to one group or one study explains the association.
- *Larger values of the explanatory variable are associated with stronger responses.* People who smoke more cigarettes per day or who smoke over a longer period get lung cancer more often. People who stop smoking reduce their risk.
- *The alleged cause is plausible.* Experiments with animals show that tars from cigarette smoke do cause cancer.
- *The alleged cause precedes the effect in time.* Lung cancer develops after years of smoking. As shown in the graph, the number of people dying of lung cancer rose as smoking became more common, with a lag of about 30 years. And as smoking rates decreased, the number of people dying of lung cancer 30 years later decreased as well.[94]

Cigarette sales and lung cancer mortality for men in the United States

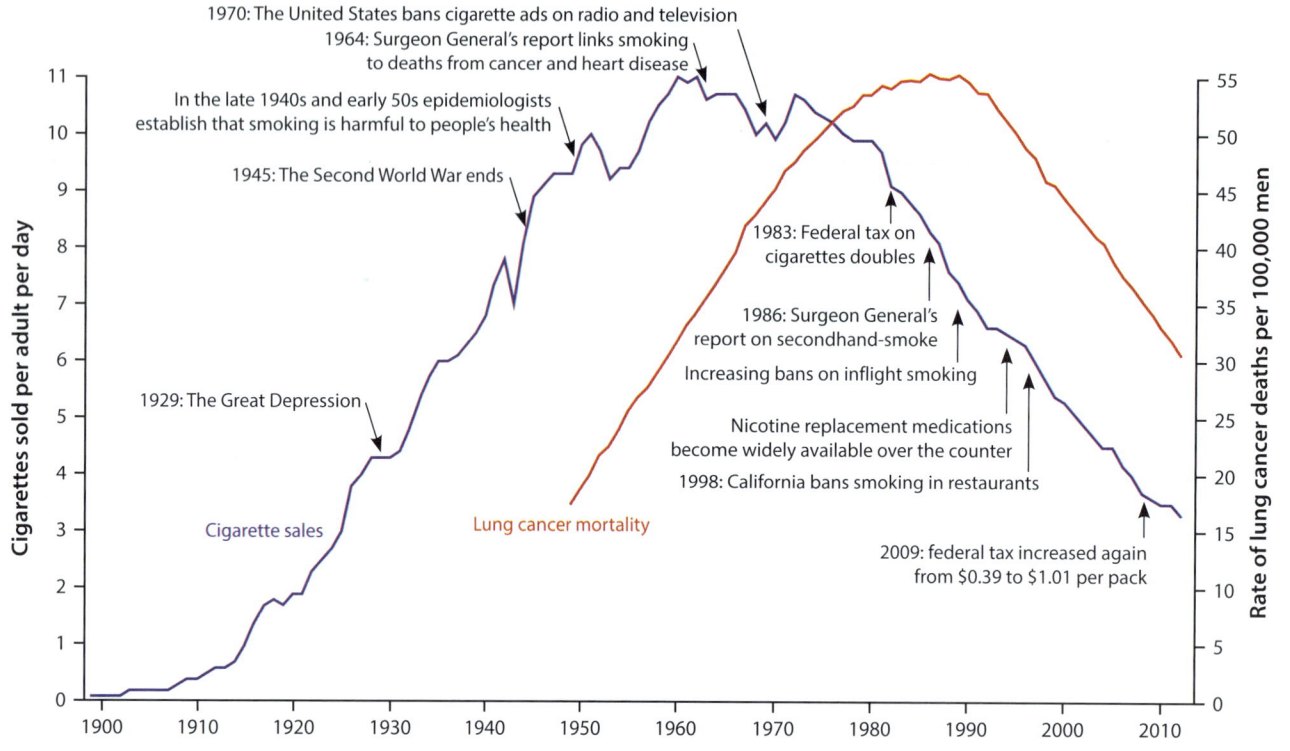

- 1970: The United States bans cigarette ads on radio and television
- 1964: Surgeon General's report links smoking to deaths from cancer and heart disease
- In the late 1940s and early 50s epidemiologists establish that smoking is harmful to people's health
- 1945: The Second World War ends
- 1929: The Great Depression
- 1983: Federal tax on cigarettes doubles
- 1986: Surgeon General's report on secondhand-smoke
- Increasing bans on inflight smoking
- Nicotine replacement medications become widely available over the counter
- 1998: California bans smoking in restaurants
- 2009: federal tax increased again from $0.39 to $1.01 per pack

Data sources: International Smoking Statistics (2017); WHO Cancer Mortality Database (IARC). The death rate from lung cancer is age-standardized.
OurWorldinData.org – Research and data to make progress against the world's largest problems.

Licensed under CC-BY by the author Max Roser.

Data Ethics

Although randomized experiments are the best way to make an inference about cause and effect, in some cases it isn't ethical to do an experiment. For example, does texting while driving increase the risk of having an accident? Although a well-designed experiment would help answer this question, it would be unethical to randomly assign an individual to text while driving!

The most complicated ethical issues arise when we collect data from people. Trials of new medical treatments, for example, can do harm as well as good to their subjects. Likewise, ethical issues must be considered when administering a sample survey, even though no treatments are imposed.

Some basic standards of data ethics must be applied in all studies that gather data from human subjects, whether they are observational studies or experiments. These standards are summarized here.

Basic Principles of Data Ethics

- All planned studies must be reviewed in advance by an *institutional review board* charged with protecting the safety and well-being of the subjects.
- All individuals who are subjects in a study must give their *informed consent* before data are collected.
- All individual data must be kept *confidential.* Only statistical summaries for groups of subjects may be made public.

The law requires that studies carried out or funded by the U.S. federal government obey these principles.[95] The purpose of an institutional review board is not to decide if a proposed study will produce valuable information or if it is statistically sound. Instead, the board's purpose is, in the words of one university's board, "to protect the rights and welfare of human subjects (including patients) recruited to participate in research activities." An institutional review board would certainly reject an experiment that required subjects to smoke cigarettes or text while driving.

Both words in the phrase "informed consent" are important. Subjects must be *informed* in advance about the nature of a study and any risk of harm it may bring. People who are asked to answer survey questions should be told what kinds of questions the survey will ask and approximately how much of their time it will take. Experimenters must tell subjects the nature and purpose of the study and outline possible risks. Subjects must then *consent,* or agree, to participate in writing.

It is important to protect individuals' privacy by keeping all data about them *confidential.* The report of an opinion poll may say what percentage of the 1200 respondents believed that legal immigration should be increased. It may not report what *you* said about this or any other issue. Confidentiality is not the same as *anonymity.* Anonymity means that the names of individuals are not known, even to the director of the study.

EXAMPLE 1.8.3

The Tuskegee Syphilis Study

Data ethics

PROBLEM

In 1932, the U.S. Public Health Service and the Centers for Disease Control and Prevention recruited 600 male Black sharecroppers from Macon County, Alabama, to participate in a medical study. The goal of the study was to observe the natural progression of untreated syphilis, which roughly two-thirds of the men in the study had in latent form. The men in the study were told they were being given free health care and that the study would last 6 months. Instead, the study lasted 40 years, the men were never told about their diagnosis, and they were given placebo treatments, even though effective treatments for syphilis had been developed by 1947.[96] Which principle of data ethics did this study violate? Explain your answer.

SOLUTION

Informed consent. The subjects in this study weren't informed about the true nature of the study, their diagnosis, and the planned use of placebo treatments. If they were informed of these facts, they most certainly would not have consented to participate in the study.

> This terribly unethical study led to the creation of the government's Office for Human Research Protections (OHRP) and laws requiring institutional review boards to prevent such studies from ever occurring again.

EXAM PREP: FOR PRACTICE, TRY EXERCISE 13.

Section 1.8 What Did You Learn?

Review the learning goals from this section. Then practice what you've learned by working through the exercises.

Learning Goal	Example	Exercises
Identify the population about which we can make inferences from a sample.	1.8.1	5–8
Determine when it is appropriate to make an inference about cause and effect.	1.8.2	9–12
Evaluate if a statistical study has been carried out in an ethical manner.	1.8.3	13–16

Section 1.8 Exercises

Building Concepts and Skills These exercises assess the basic knowledge you should have after reading the section.

1. When is it appropriate to use data from a sample to make an inference about a population?

2. When is it appropriate to make an inference about cause and effect?

3. All human subjects in a study should give _____ before the study begins.

4. True/False: Confidentiality means that researchers do not know the identity of the subjects in an experiment.

Mastering Concepts and Skills These exercises reinforce the learning goals as shown in the examples.

5. **Refer to Example 1.8.1** **Naps** A long-term study analyzed data from 3462 randomly selected adults in Lausanne, Switzerland. In the sample, 10.7% reported that they typically take 6 or 7 naps per week.[97]

(a) Can we generalize this result to all adults in Switzerland? Explain your answer.

(b) What is the largest population to which we can generalize this result?

6. **Social Media** A study on teen social media uses analyzed data from 11,872 randomly selected adolescents (aged 13–15 years) in the United Kingdom. In the sample, 20.8% reported using social media 5 or more hours per day, on average.[98]

   (a) Can we generalize this result to teenagers in the United States? Explain your answer.

   (b) What is the largest population to which we can generalize this result?

7. **Resistance and Memory** To study strength training and memory, researchers randomly assigned 46 young adult volunteers to two groups. After both groups were shown 90 pictures, one group had to bend and extend one leg against heavy resistance 60 times. The other group stayed relaxed, while the researchers used the same exercise machine to bend and extend their legs with no resistance. Two days later, each subject was shown 180 pictures — the original 90 pictures plus 90 new pictures and asked to identify which pictures were shown 2 days earlier.[99]

   (a) Can we generalize the results of this study to all adults in the United States? Explain your answer.

   (b) What is the largest population to which we can generalize this result?

8. **Coffee Grounds** Many gardeners add used coffee grounds to their garden, hoping to recycle a waste product and improve their soil. Does it work? Researchers grew 5 different plants (broccoli, leek, radish, viola, and sunflower) in three different types of soil (sandy, sandy clay loam, and loam). Then they randomly assigned each plant to be treated with used (but uncomposted) coffee grounds in 5 different concentrations (0%, 2.5%, 5%, 10%, and 25%).[100]

   (a) Can we generalize the results of this study to all types of plants? Explain your answer.

   (b) What is the largest population to which we can generalize this result?

9. **Refer to Example 1.8.2 Naps and Cardiovascular Disease** In Exercise 5, you read about a study involving 3462 randomly selected adults from Lausanne, Switzerland. In addition to questions about napping, researchers asked questions about cardiovascular disease. A greater proportion of adults who reported taking 6 or 7 naps weekly had cardiovascular events (e.g., heart attacks) than did adults who reported taking fewer naps, and the difference was statistically significant. Based on this study, is it reasonable to say that napping frequently causes cardiovascular problems? Explain your answer.

10. **Social Media and Sleep** In Exercise 6, you read about a study involving 11,872 randomly selected 13- to 15-year-olds in the United Kingdom. In addition to questions about social media use, researchers asked questions about sleep patterns. The teens with the highest reported social media use had more difficulty falling asleep and staying asleep than the teens who reported average social media use, and the difference was statistically significant. Based on this study, is it reasonable to say that increased social media use causes sleep problems? Explain your answer.

11. **Resistance and Memory** In Exercise 7, you read about a study involving 46 young adult volunteers who were randomly assigned to do leg presses with no resistance or with heavy resistance. The members of the heavy-resistance group had significantly better recall of the pictures they were shown at the beginning of the study. Based on this study, is it reasonable to say that increased resistance caused better recall? Explain your answer.

12. **Coffee Grounds** In Exercise 8, you read about a study that applied used coffee grounds in different concentrations to 5 different types of plants in 3 different types of soil. Overall, the plants with higher concentrations of coffee grounds had better soil water retention, but worse growth compared to plants with lower concentrations of coffee grounds. Based on this study, is it reasonable to say that increased concentration of coffee grounds caused the increased soil water retention and worse growth? Explain your answer.

13. **Refer to Example 1.8.3 Facebook Emotions** In cooperation with researchers from Cornell University, Facebook randomly selected almost 700,000 users for an experiment in "emotional contagion." Users' news feeds were manipulated (without their knowledge) to selectively show postings from their friends that were either more positive or more negative in tone, and the emotional tone of their own subsequent postings was measured. The researchers found evidence that people who read emotionally negative postings were more likely to post messages with a negative tone and those reading positive messages were more likely to post messages with a positive tone.[101] Which principle of data ethics did this study violate? Explain your answer.

14. **Stuttering** Beginning in 1939, two researchers at the University of Iowa studying stuttering performed an experiment on 22 orphans. The 10 orphans who already stuttered were divided into two groups, with one group being told their speech was fine and the other being told that their speech was "bad." The remaining 12 orphans had no speech difficulties at the beginning of the study, but half of them were manipulated to believe they were beginning to develop a stutter. The other half were treated as though they had no speech difficulties. All 22 orphans were told they were simply receiving speech therapy and had no knowledge of the purpose of the experiment.[102] Which principle of data ethics did this study violate? Explain your answer.

15. **Malaria in Prison** The University of Chicago and the U.S. Army conducted experiments that tested various malaria treatments in the 1940s. The subjects of the experiment were prisoners at the Stateville Penitentiary near Joliet, Illinois. The prisoners gave informed consent, but were offered reduced prison sentences

to participate.[103] Does the fact that the prisoners gave informed consent make this an ethical experiment? Explain your answer.

16. **Willowbrook Hepatitis Studies** In the 1960s, children entering the Willowbrook State School, an institution for children with intellectual disabilities on Staten Island in New York, were deliberately infected with hepatitis. The researchers justified their actions by arguing that almost all children in the institution quickly became infected anyway. The studies showed for the first time that different strains of hepatitis existed. This finding contributed to the development of effective vaccines.[104] Despite these valuable results, explain why the study was unethical.

Applying the Concepts These exercises ask you to apply multiple learning goals in a new context or to apply what you learned in this section in a new way.

17. **Foster Care** Do abandoned children placed in foster homes do better than similar children placed in an institution? The Bucharest Early Intervention Project involved 136 young children abandoned at birth and living in orphanages in Bucharest, Romania. Half of the children were randomly assigned to be placed in foster homes. The other half remained in the orphanages. The outcomes for children placed in foster homes were better than for the children who remained in the orphanages, and the difference is statistically significant.[105]
    (a) Based on this study, is it reasonable to say that placement in foster care was the cause of the improvement? Explain your answer.
    (b) What is the largest population to which we can generalize these results?
    (c) The children in this study were too young to provide informed consent. Does this make this study unethical? Explain your answer.

18. **Red Lipstick** Does a waitress who wears red lipstick receive better tips? To find out, Lexie wrote "Red" on 30 slips of paper and "Neutral" on 30 slips of paper, and mixed them well. Each time she got ready for work, she randomly selected one piece of paper to determine what lipstick color she would wear that day. The mean daily tip amount was $23 greater when she was wearing red lipstick, and the difference is statistically significant.[106]
    (a) Based on this study, is it reasonable to say that wearing red lipstick was the cause of the increase in tips? Explain your answer.
    (b) What is the largest population to which we can generalize these results?
    (c) The customers whom Lexie served didn't provide consent to participate in her study. Was the study ethical? Explain your reasoning.

19. **Reflected Glory** In a classic study, researchers investigated the tendency of sports fans to "bask in reflected glory" by associating themselves with winning teams. In the study, researchers called randomly selected students at a major university with a highly ranked football team. They randomly assigned half the students to answer questions about a recent game the team lost, and asked the other half about a recent game the team won. If the students were able to correctly identify the winner (showing they were fans of the team), they were asked to describe the game. Students were significantly more likely to identify themselves with the team by their use of the word *we* in the description ("We won the game" versus "They won the game") when describing a win.[107] What conclusions can we draw from this study? Explain your answer.

20. **Berry Good** Eating blueberries and strawberries might improve heart health, according to a long-term study of 93,600 women who volunteered to take part. These berries are high in anthocyanins due to their pigment. Women who reported consuming the most anthocyanins had a significantly smaller risk of heart attack compared to the women who reported consuming the least.[108] What conclusion can we draw from this study? Explain your answer.

Extending the Concepts These exercises challenge you to explore statistical concepts and methods that go beyond what you learned in this section.

21. **Tax Returns** A government agency takes a random sample of income tax returns to obtain information on the average income of people in different occupations. Only the incomes and occupations are recorded from the returns, not the names. Should this study require informed consent? Explain your reasoning.

22. **Religious Behavior** A social psychologist attends public meetings of a religious group to study the behavior patterns of members. Should this study require informed consent? Explain your reasoning. Would your answer change if the social psychologist pretends to be converted and attends private meetings to study the behavior patterns of members?

23. **Review Boards** Government regulations require that institutional review boards consist of at least 5 people, including at least 1 scientist, 1 nonscientist, and 1 person from outside the institution. Most boards are larger, but many contain just 1 outsider.
    (a) Why should review boards contain people who are not scientists?
    (b) Do you think that 1 outside member is enough? How would you choose that member? For example, would you prefer a medical doctor? A religious leader? An activist for patients' rights?

Cumulative Review These exercises revisit what you learned in previous sections.

24. **Assigning Credit (1.5)** An article in the *New York Times Magazine* on Manhattan District Attorney Cyrus Vance, Jr., cites a substantial decrease in crime during Vance's tenure. But the author cautions, "It's hard to know how much one person or policy can affect the crime rate, and consequently how much credit or blame should be assigned when things go well or don't. . . . The sun comes up, the roosters preen." Explain how confounding makes it difficult to give Vance credit for the decrease in crime in Manhattan.[109]

Statistics Matters How can we prevent malaria?

At the beginning of the chapter, we described an experiment that investigated the effect of regular screening for malaria. Researchers worked with children in 101 schools in Kenya, randomly assigning students at 51 of the schools to receive regular screenings and follow-up treatments and students at the remaining schools to receive no regular screening.

1. Why was it necessary to include a group of schools that didn't receive the screening? Does excluding some schools from screening raise any ethical concerns?
2. Describe how you could randomly assign the 101 schools into groups of 51 and 50.
3. What is the purpose of random assignment in this experiment?
4. If the researchers found statistically significant evidence that the proportion with malaria is smaller for children in schools who are regularly screened, would it be reasonable to say that screening caused the reduction in malaria? Would these results apply to all schools in Africa? Explain your answers.
5. Unfortunately, the results of the study were not statistically significant. Explain what this means in the context of this study.

Chapter 1 Review

The Statistical Problem-Solving Process

- **Statistics** is the science and art of collecting, analyzing, and drawing conclusions from data.
- The **statistical problem-solving process** involves four steps: ask questions, collect/consider data, analyze data, and interpret results.
- An **individual** is a person, animal, or thing described in a set of data. A **variable** is any attribute that can take different values for different individuals.
  - A **categorical variable** takes values that are labels, which place each individual into a particular group, called a category.
  - A **quantitative variable** takes number values that are quantities—counts or measurements.
- There are many **data collection methods**, including sample surveys, observational studies, and experiments.
- The **scope of inference** determines how we interpret results.
  - **Random selection** of individuals allows inference about the population from which the sample was selected.
  - **Random assignment** of individuals to groups allows for inference about cause and effect.
  - All data collection methods should follow basic principles of **data ethics**. This includes the use of an institutional review board, informed consent, and confidentiality.

Sampling and Surveys

- The **population** in a statistical study is the entire group of individuals we want information about. A **census** collects data from every individual in the population.
- A **sample** is a subset of individuals in the population from which we collect data.
- A **random sample** consists of individuals from the population who are selected for the sample using a chance process. We can use the data collected from a random sample to make inferences about the population from which the sample was selected.
- A **simple random sample (SRS)** of size n is chosen in such a way that every group of n individuals in the population has an equal chance of being selected as the sample.
- The fact that different random samples of the same size from the same population produce different estimates is called **sampling variability**. Sampling variability can be reduced by increasing the sample size. We can use simulation to determine the values of an estimate that are likely to occur due to sampling variability.
- To use **stratified random sampling**, divide the population into non-overlapping groups of individuals (**strata**) that are similar in some way that might affect their responses. Then choose a separate SRS from each stratum and combine these SRSs to form the sample. When strata are "similar within but different between," stratified random

samples tend to give more precise estimates of unknown population values than do simple random samples.

- To use **cluster random sampling**, divide the population into non-overlapping groups of individuals that are located near each other, called **clusters**. Randomly select some of these clusters. All the individuals in the chosen clusters are included in the sample. Ideally, clusters are "different within but similar between." Cluster random sampling saves time and money by collecting data from entire groups of individuals that are close together.
- To use **systematic random sampling**, select a value of k based on the population size and desired sample size, randomly select a value from 1 to k to identify the first individual in the sample, and choose every kth individual thereafter. Systematic random sampling can be easier to conduct than other sampling methods.
- The design of a statistical study shows **bias** if it is very likely to underestimate or very likely to overestimate the value you want to know.
  - A **convenience sample** consists of individuals from the population who are easy to reach. This method of selecting a sample is biased because the individuals chosen are typically not representative of the population.
  - A **voluntary response sample** consists of people who choose to be in the sample by responding to a general invitation. This method of selecting a sample is biased because the individuals in the sample are typically not representative of the population.
  - **Undercoverage** occurs when some members of the population are less likely to be chosen for the sample. Sampling methods that suffer from undercoverage can show bias if the individuals less likely to be included in the sample differ in relevant ways from the other members of the population.
  - **Nonresponse** occurs when an individual chosen for the sample can't be contacted or refuses to participate. Sampling methods that suffer from nonresponse can show bias if the individuals who don't respond differ in relevant ways from the other members of the population.
  - **Response bias** occurs when there is a consistent pattern of inaccurate responses to a survey question. This kind of bias can be caused by the wording of questions, characteristics of the interviewer, lack of anonymity, and other factors.

Observational Studies

- An **observational study** observes individuals and measures variables of interest but does not attempt to influence the responses.
- A **response variable** measures an outcome of a study. An **explanatory variable** may help explain or predict changes in a response variable.
- **Confounding** occurs when two variables are associated in such a way that their effects on a response variable cannot be distinguished from each other.
- Observational studies cannot definitively show a cause-and-effect relationship between an explanatory variable and a response variable because of potential confounding.

Experiments

- An **experiment** deliberately imposes treatments (conditions) on individuals to measure their responses. We can use the results of a well-designed experiment to make inferences about cause and effect.
- A **treatment** is a specific condition applied to the individuals in an experiment. An **experimental unit** is the object to which a treatment is randomly assigned. When the units are human beings, they are often called **subjects**.
- The **placebo effect** describes the fact that some subjects in an experiment will respond favorably to any treatment, even an inactive treatment. A **placebo** is a treatment that has no active ingredient but is otherwise like other treatments.
- In a **double-blind** experiment, neither the subjects nor those who interact with them and measure the response variable know which treatment a subject received. In a **single-blind** experiment, either the subjects or the people who interact with them and measure the response variable don't know which treatment a subject is receiving.
- In an experiment, a **control group** is used to provide a baseline for comparing the effects of other treatments. Depending on the purpose of the experiment, a control group may be given a placebo, an active treatment, or no treatment at all.
- **Random assignment** means that treatments are assigned to experimental units (or experimental units are assigned to treatments) using a chance process. The purpose of random assignment is to help avoid confounding by creating treatment groups that are roughly equivalent at the beginning of the experiment.
- In a **completely randomized design**, the experimental units are assigned to the treatments (or treatments to the experimental units) completely at random.
- In an experiment, it is important to keep all variables the same, other than the explanatory variable. That way, these additional variables won't be confounded with the explanatory variable or add variability to the response variable, making it easier to determine if one treatment is more effective than another.
- A **randomized block design** forms groups (**blocks**) of experimental units that are similar with respect to a variable that is expected to affect the response. Treatments are assigned at random within each block. Responses are then compared within each block and combined with the responses of other blocks after accounting for the differences between the blocks. When blocks are chosen wisely, it is easier to determine if one treatment is more effective than another.
- A **matched-pairs design** is a form of randomized block design that is commonly used for comparing two treatments. In some matched-pairs designs, each subject receives both treatments in a random order. In other matched-pairs designs, two very similar subjects are paired, and the two treatments are randomly assigned within each pair.
- When an observed difference in responses between the groups in an experiment is so large that it is unlikely to be explained by chance variation in the random assignment, we say that the result is **statistically significant**. We can use simulation to investigate which differences are likely to occur due to the chance variation in random assignment.

Chapter 1 Review Exercises

These exercises will help you review important concepts and skills described by the learning goals in each section. For most exercises, the relevant section is noted in parentheses after the exercise title.

1. **Nurses (1.1, 1.3)** A recent random sample of $n = 805$ adult U.S. residents found that the proportion who rated the honesty and ethical standards of nurses as high or very high was 0.85. This was 0.15 higher than the proportion recorded for doctors, the next highest-ranked profession.[110]
   (a) Identify the sample and the population in this setting.
   (b) Do you think that the proportion of all U.S. residents who would rate the honesty and ethical standards of nurses as high or very high is exactly 0.85? Explain your answer.
   (c) What is the benefit of increasing the sample size in this context?

2. **Campus Parking (1.2, 1.3)** Finding a parking spot can be a major problem on college campuses. The administration at a large community college wants to estimate the proportion of students who are satisfied with the current parking situation.
   (a) Administrators survey the first 50 students to arrive in the school's main parking lot one morning. Explain why this sampling method is biased. Is the proportion of satisfied students in the sample likely to be greater than or less than the proportion of all students who are satisfied with the current parking situation?
   (b) Describe how to select a simple random sample of 50 students from the 8420 students who attend the college.
   (c) Explain why the method in part (b) helps to avoid the problem you identified in part (a).
   (d) Describe another source of bias that could affect the results of the survey, even when selecting an SRS of students.

3. **Campus Parking (1.3)** Refer to Exercise 2. In the SRS of 50 students, $17/50 = 34\%$ said they were satisfied with the current parking situation on campus. To determine if these data provide convincing evidence that fewer than half of all students are satisfied with the current parking situation, 100 simulated SRSs of size $n = 50$ were selected. Each dot in the graph shows the percentage of students who are satisfied with the current parking situation in a simulated SRS, assuming that each student has a 50% chance of being satisfied.

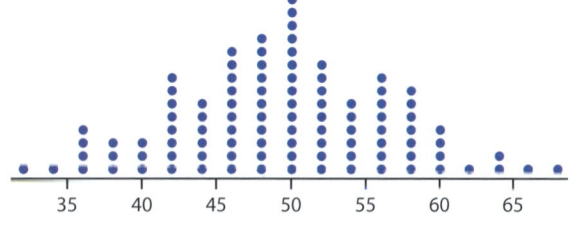

   (a) Explain how the graph illustrates the concept of sampling variability.
   (b) In the simulation, what percentage of the samples resulted in 34% or fewer of the students being satisfied?
   (c) In the actual study, 34% of the students were satisfied. Based on your answer to part (b), is there convincing evidence that fewer than half of all students at the college are satisfied with the current parking situation? Explain your reasoning.

4. **Five-Second Rule (1.1, 1.5)** The "five-second rule" states that a piece of food is safe to eat if it has been on the floor for less than 5 seconds. The rule is based on the belief that bacteria need time to transfer from the floor to the food. But does it work? Researchers from Rutgers University put the five-second rule to the test. They dropped pieces of food onto four different surfaces—stainless steel, ceramic tile, wood, and carpet—and waited for four different lengths of time—less than 1 second, 5 seconds, 30 seconds, and 300 seconds. Finally, they used bacteria prepared two different ways—in a tryptic soy broth or peptone buffer. Once the bacteria were ready, the researchers spread them out on the different surfaces and started dropping food. After waiting the designated amount of time, the food was removed, blended, and plated on tryptic soy agar to allow bacteria to grow.[111]
   (a) Identify the individuals in this experiment.
   (b) What are the explanatory variable(s)? Are they categorical or quantitative?
   (c) What are the response variable(s)? Are they categorical or quantitative?

5. **Organic Food and Cancer (1.1, 1.5)** A study of 68,946 French volunteers found that the respondents who reported eating the most organic food had a smaller risk of developing cancer than the other respondents. Furthermore, the difference was determined to be statistically significant.[112]
   (a) Is this an observational study or an experiment? Explain your answer.
   (b) Explain how confounding prevents a cause-and-effect conclusion in this context.

6. **Mask Up! (1.5, 1.6)** A recent study of 1000 students at the University of Michigan investigated how to prevent catching the common cold. Researchers randomly assigned students to three different cold prevention methods for 6 weeks. Some wore masks, some wore masks and used hand sanitizer, and others took no precautions. The two groups who used masks reported significantly fewer cold symptoms than those who did not wear a mask.[113]

(a) Explain why it was necessary to include a control group that didn't wear a mask.

(b) What is the largest population to which we can generalize this result?

(c) Based on the study, is it reasonable to say that wearing a mask causes fewer cold symptoms? Explain your answer.

(d) What principles should the researchers have followed to make sure this study was done ethically?

7. **Portion Sizes (1.5, 1.6)** To see if changing food portion sizes would influence subjects' food consumption one day later, researchers randomly assigned 75 volunteer subjects to two groups. One group was served small portions of food and the other group was served large portions of food. The next day, all of the subjects were presented with the same food options and allowed to serve themselves, with researchers recording how much food they consumed (in grams).[114]

    (a) Outline a completely randomized design for this experiment. Include details about how to perform the random assignment.

    (b) What is the purpose of random assignment in this experiment?

    (c) Identify one variable the researchers kept the same during the experiment. Explain why it was important to keep that variable the same.

8. **Portion Sizes (1.6)** Refer to Exercise 7. The group that was given the small portions the day before served themselves an average of 144.66 grams of food, while the group that was given the large portions the day before served themselves an average of 189.91 grams of food.

    (a) Calculate the difference in the mean weight of food for the two groups (Large – Small).

    One hundred trials of a simulation were performed to see what differences in means would occur due only to chance variation in the random assignment, assuming that the previous day's portion size doesn't matter. The dotplot displays the results of the simulation.

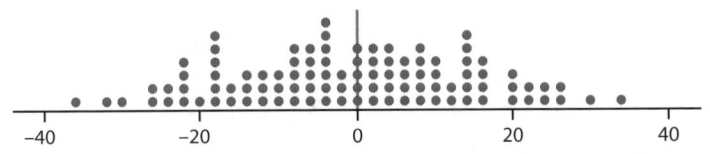

    (b) There is one dot at a difference of 30. Explain what this dot represents.

    (c) Based on the results of the simulation, is the difference from part (a) statistically significant? Explain your reasoning.

9. **Expense Reports (1.4)** An educational publisher has 200 employees who travel around the country trying to sell the company's products. Of these employees, 120 cover densely populated regions, while the remaining 80 cover less densely populated regions that require more travel. Each of these employees takes roughly 25 trips per year. At the end of each sales trip, the employee submits an expense report detailing hotel, meal, and other travel costs. The company accountant wants to do a detailed audit of 100 expense reports to estimate the proportion of expense reports that are accurate.

    (a) Describe how to select a stratified random sample of 100 expense reports. Explain your choice of strata and a benefit of using a stratified random sample in this context.

    (b) Describe how to select a cluster random sample of 100 expense reports. Explain your choice of clusters and a benefit of using a cluster random sample in this context.

    (c) Describe how to select a systematic random sample of 100 expense reports and explain a benefit of using a systematic random sample in this context.

10. **Microwaved Popcorn (1.7)** Some popcorn lovers want to determine if it is better to use the "popcorn button" on their microwave oven or use the amount of time recommended on the bag of popcorn. To measure how well each method works, they will count the number of unpopped kernels remaining after popping. To obtain the experimental units, they go to the store and buy 10 bags each of 4 different varieties of microwave popcorn (butter, cheese, natural, and kettle corn), for a total of 40 bags.

    (a) Describe a randomized block design for this experiment. Justify your choice of blocks.

    (b) Explain why a randomized block design might be preferable to a completely randomized design for this experiment.

11. **E-readers (Chapter 1)** According to researchers at Harvard Medical School, using a light-emitting electronic book (e-reader) before bedtime adversely affects overall health, alertness, and sleep rhythm.[115] The human resources department at a large company wants to determine what proportion of its 5289 employees use e-readers and if the conclusion of the Harvard study applies to its employees.

    (a) Describe how to select a sample of 100 employees to estimate the proportion of all employees who use an e-reader.

    (b) Explain how the human resources department can design a study using these 100 employees to determine how using e-readers before bed affects work performance.

12. **Nut Allergies (Chapter 1)** Can eating nuts during pregnancy help children avoid nut allergies? Researchers studied more than 8000 children who were born in the early 1990s to mothers who volunteered to be part of the Nurses' Health Study II. Children whose mothers ate the most nuts during pregnancy (at least 5 times per week) were significantly less likely to develop nut allergies than children whose mothers ate the least amount of nuts during pregnancy (less than once per month).[116]

    (a) Based on the study, is it reasonable to say that eating nuts during pregnancy caused the reduced risk of nut allergies in children? Explain your answer.

    (b) What is the largest population to which we can generalize these results?

    (c) Would it be ethical to conduct an experiment to answer this question? Explain your reasoning.

2 Displaying Data with Graphs

Section 2.1 Displaying Categorical Data
Section 2.2 Displaying Relationships Between Two Categorical Variables
Section 2.3 Displaying Quantitative Data: Dotplots
Section 2.4 Displaying Quantitative Data: Stemplots
Section 2.5 Displaying Quantitative Data: Histograms
Section 2.6 Displaying Relationships Between Two Quantitative Variables
Chapter 2 Review
Chapter 2 Review Exercises
Chapter 2 Project

Statistics Matters Where is rental housing affordable?

Each year, the National Low Income Housing Coalition (NLIHC) publishes data on the affordability of rental housing in the United States. The usual guideline is that rental expense, to be considered affordable, should not exceed 30% of a household's annual income. In 2020, a household needed an annual income of $49,830 to afford the rent for a typical two-bedroom dwelling. Of course, rent (and the income required to afford it) varies considerably from state to state. In 2020, about 36% of all U.S. households lived in rental housing.[1]

The following table shows part of the 2020 data set from NLIHC. The location of each state is coded by U.S. Census region: 1=Northeast, 2=Midwest, 3=South, and 4=West. The annual household income needed to afford a typical two-bedroom rental in each state is rated as low (less than $35,360), medium (between $35,360 and $49,830), or high (greater than $49,830).

State	Census region	Annual household income to afford two-bedroom rental	Rating of two-bedroom rental cost	Annual median household income	Percent of renter households
Alabama	3	$32,110	Low	$66,123	31
California	4	$76,879	High	$90,909	45
Connecticut	1	$54,956	High	$101,816	34
Florida	3	$50,807	High	$68,669	35
Illinois	2	$44,310	Medium	$85,252	34

How can we display these data in ways that illuminate important aspects of rental households and affordability of rental housing in the 50 states? Relative to annual income, are some regions more affordable than others?

We'll revisit **Statistics Matters** *at the end of the chapter, so you can use what you have learned to help answer these questions.*

In Chapter 1, you learned about the structure of a data set and about methods for collecting data. Now it's time to learn how to *analyze* data. In this chapter, we'll focus on displaying data with tables and graphs to facilitate analysis.

The motto for this chapter is simple: Always plot your data! A well-constructed graph helps us describe patterns in data and identify important departures from those patterns.

Section 2.1 Displaying Categorical Data

LEARNING GOALS
By the end of this section, you will be able to:
- Summarize the distribution of a variable with a frequency table or a relative frequency table.
- Make and interpret bar charts and pie charts of categorical data.
- Recognize ways in which graphs of categorical data can be misleading.

In this section, you'll learn how to create tables that summarize the distribution of a variable, make bar charts and pie charts, and interpret graphs of categorical data. Let's begin by revisiting some key terms from Chapter 1.

Each year, the U.S. Census Bureau collects data from a random sample of more than 3.5 million U.S. households as part of the American Community Survey (ACS). The table displays some data from the ACS in a recent year.[2]

Household	Region	People in household	Time in dwelling	Response mode	Household income	Internet access?
1	South	3	20–29 years	Phone	$272,000	Yes
2	Midwest	3	30+ years	Internet	$54,600	Yes
5	South	1	10–19 years	Mail	$49,900	Yes
6	South	2	2–4 years	Phone	$297,000	Yes
7	Northeast	4	5–9 years	Internet	$130,000	Yes
11	Midwest	3	2–4 years	Internet	$82,000	Yes
14	Midwest	3	10–19 years	Internet	$57,000	Yes
16	West	3	2–4 years	Mail	$36,800	Yes
17	South	2	2–4 years	Mail	$133,000	Yes
25	Midwest	1	20–29 years	Mail	$80,000	No

The *individuals* in this data set are households. Region, time in dwelling, response mode, and internet access status are *categorical variables*. Number of people in the household and household income are *quantitative variables*. Because the sample was randomly selected, the Census Bureau can use these data to make inferences about the population of U.S. households.

Summarizing Data

A variable typically takes values that *vary* from one individual to another. The **distribution** of a variable describes the pattern of variation of these values.

> The **distribution** of a variable tells us what values the variable takes and how often it takes each value.

We can summarize a variable's distribution with a **frequency table** or a **relative frequency table**.

> A **frequency table** shows the number of individuals having each data value.

> A **relative frequency table** shows the proportion or percentage of individuals having each data value.

Some people use the terms *frequency distribution* and *relative frequency distribution* instead. To make either kind of table, start by tallying the number of times that the variable takes each value.

EXAMPLE 2.1.1

Would you rather be rich, happy, famous, or healthy?

Frequency and relative frequency tables

PROBLEM

College admissions departments are continually honing their message to appeal to high school students. Knowing about potential students' long-term goals and priorities helps colleges tailor information about the benefits of attending their institutions. One resource available to recruiters is Census at School, a project sponsored by the American Statistical Association that uses online surveys to collect data about secondary school students. Here is one of the survey questions:

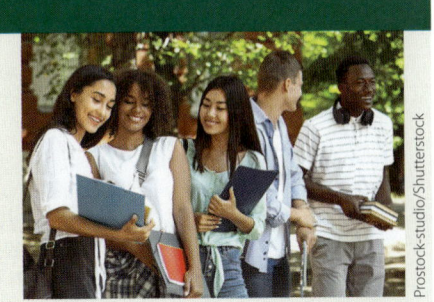

Which would you prefer to be? Select one.

_____ Rich _____ Happy _____ Famous _____ Healthy

We selected a random sample of 75 students who completed the survey in a recent year. Here are the data on preferred status for these students.[3]

Rich	Healthy	Happy	Famous	Rich	Rich	Happy	Rich	Rich	Healthy
Rich	Happy	Rich	Rich	Happy	Happy	Rich	Happy	Rich	Happy
Happy	Happy	Happy	Happy	Happy	Happy	Rich	Healthy	Healthy	Happy
Rich	Happy	Happy	Happy	Rich	Happy	Happy	Happy	Rich	Happy
Rich	Famous	Happy	Rich	Happy	Healthy	Famous	Happy	Rich	Healthy
Happy	Happy	Healthy	Happy	Healthy	Happy	Happy	Happy	Happy	Happy
Happy	Rich	Happy	Healthy	Rich	Happy	Healthy	Happy	Happy	Rich
Rich	Happy	Happy	Rich	Rich					

Summarize the distribution of preferred status with a frequency table and a relative frequency table.

SOLUTION

Frequency table

Preferred status	Frequency
Famous	3
Happy	39
Healthy	10
Rich	23
Total	75

The frequency table shows the *number* of students who chose each status. To create the frequency table, count how many students said "Famous," how many said "Happy," and so on.

Relative frequency table

Preferred status	Relative frequency
Famous	3/75 = 0.040 or 4.0%
Happy	39/75 = 0.520 or 52.0%
Healthy	10/75 = 0.133 or 13.3%
Rich	23/75 = 0.307 or 30.7%
Total	75/75 = 1.000 or 100%

The relative frequency table shows the *proportion* or *percentage* of students who chose each status. Note that in statistics, a proportion is a value between 0 and 1 that is equivalent to a percentage.

EXAM PREP: FOR PRACTICE, TRY EXERCISE 7.

The same process can be used to summarize the distribution of a quantitative variable. However, it would be hard to make a frequency table or a relative frequency table for a quantitative variable that takes many different values, such as household income in the American Community Survey data set. We'll look at a better option for summarizing quantitative variables like this in Section 2.5.

Bar Charts and Pie Charts

A frequency table or relative frequency table summarizes a variable's distribution with numbers. For instance, the Current Population Survey conducted by the U.S. Census Bureau collected data on the highest educational level achieved by U.S. 25- to 34-year-olds in 2020. The relative frequency table summarizes the data.[4] Note: The total is 100.1% due to *roundoff error*.

Level of education	Percent
No high school diploma	6.0
High school graduate	26.1
Some college, no degree	16.1
Associate's degree	10.8
Bachelor's degree	29.1
Advanced degree	12.0

To display the distribution more clearly, you can use a graph. You can make a **bar chart** or a **pie chart** for categorical data. Figure 2.1 shows a bar chart and a pie chart of the data on the education level of U.S. 25- to 34-year-olds in 2020.

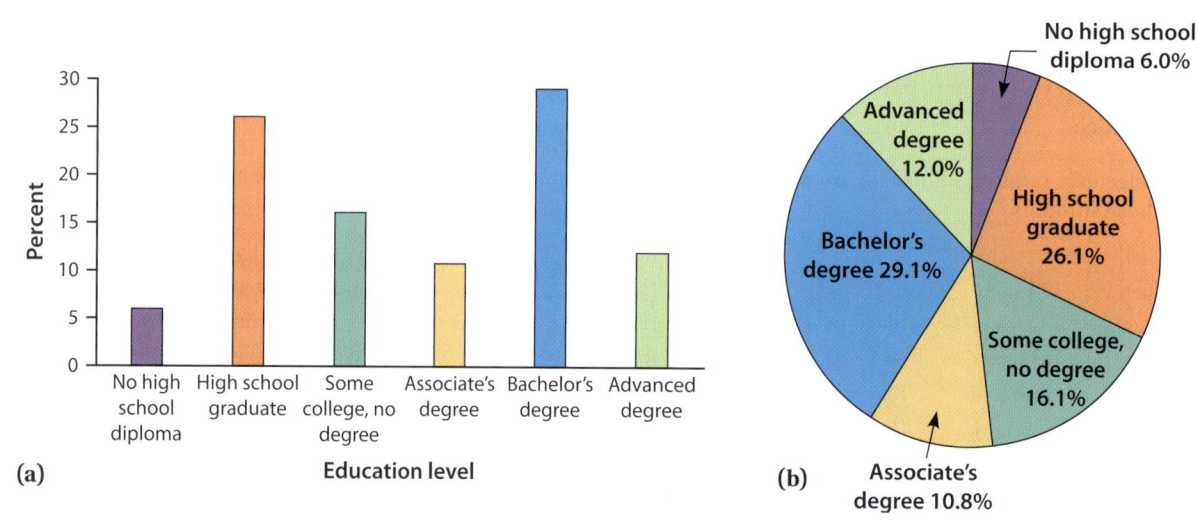

FIGURE 2.1 (a) Bar chart and (b) pie chart of the distribution of education level attained in 2020 by people aged 25 to 34 in the United States.

A **bar chart** shows each category as a bar. The heights of the bars show the category frequencies or relative frequencies.

A **pie chart** shows each category as a sector of a circle. The areas of the sectors are proportional to the category frequencies or relative frequencies.

Bar charts are sometimes called *bar graphs*. Pie charts are sometimes referred to as *circle graphs*. We'll discuss graphs for quantitative data beginning in Section 2.3.

It is fairly straightforward to make a bar chart by hand (see the How To box). For details on making bar charts and pie charts with technology, see the Tech Corner at the end of the section.

HOW TO Make a Bar Chart

1. **Draw and label the axes.** Put the name of the categorical variable under the horizontal axis. To the left of the vertical axis, indicate whether the graph shows the frequency (count) or relative frequency (percentage or proportion) of individuals in each category.

2. **"Scale" the axes.** Write the names of the categories at equally spaced intervals under the horizontal axis. On the vertical axis, start at 0 and place tick marks at equal intervals until you equal or exceed the largest frequency or relative frequency in any category.

3. **Draw bars** above the category names. Make the bars equal in width and leave gaps between them. Be sure that the height of each bar corresponds to the frequency or relative frequency of individuals in that category.

Making a graph is not an end in itself. The real purpose of a graph is to help you interpret the data. When you look at a graph, always ask, "What do I see?" In Figure 2.1, the bar chart and pie chart reveal that the most common level of education for U.S. 25- to 34-year-olds in 2020 was a bachelor's degree (29.1%), followed closely by high school graduate (26.1%). There were much smaller percentages of people with some college but no degree (16.1%), advanced degrees (12.0%), associate's degrees (10.8%), and without high school diplomas (6.0%).

EXAMPLE 2.1.2

Would you rather be rich, happy, famous, or healthy?

Making and interpreting a bar chart

PROBLEM
Here is a frequency table of the preferred status data for the 75 students in Example 2.1.1. Make a bar chart to display the data. Describe what a college admissions department should conclude about potential students' hopes.

Preferred status	Frequency
Famous	3
Happy	39
Healthy	10
Rich	23
Total	75

SOLUTION

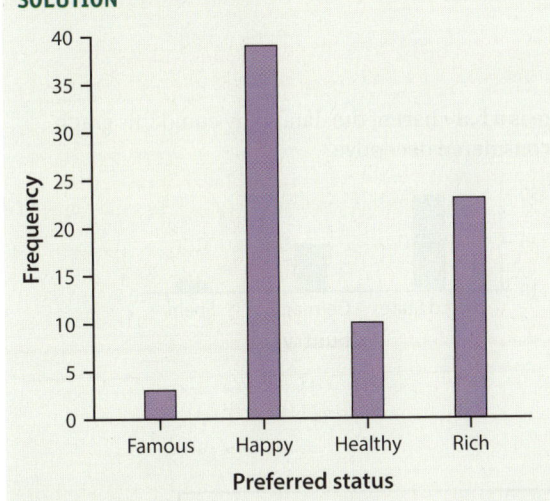

1. Draw and label the axes.
2. "Scale" the axes. The largest frequency is 39. So we chose a vertical scale from 0 to 40, with tick marks 5 units apart. The order of the categories on the *x*-axis is up to you. We arranged the categories alphabetically.
3. Draw bars.

A majority of students (39 out of 75) said they would rather be happy than famous, healthy, or rich. The students' second choice for preferred status was rich (23), followed by healthy (10). The least popular status was famous, chosen by only 3 of the 75 students. The college admissions department may want to make the idea of future happiness a prominent part of its materials for high school students.

EXAM PREP: FOR PRACTICE, TRY EXERCISE 11.

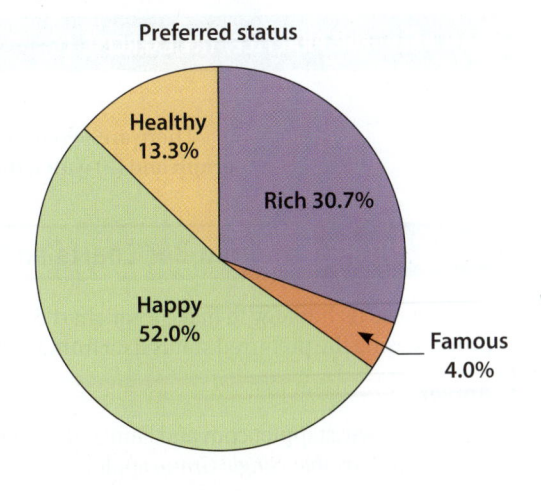

Here is a pie chart of the preferred status data from Example 2.1.2. Notice that the pie shows the *relative frequency* for each response category. For instance, the "Famous" slice makes up 4.0% of the graph because the relative frequency for this category is 3/75 = 0.040 = 4.0%.

Use a pie chart when you want to emphasize each category's relation to the whole. Each slice of the pie shows the count or percentage of individuals in that category.

⚠ **A bar chart or pie chart that displays a distribution of categorical data must include *all* individuals in the data set.** This might require including an "other" category in some cases. See Exercises 13 and 14.

Bar charts are more flexible than pie charts. Both types of graphs can display the distribution of a categorical variable, but a bar chart can also compare any set of quantities that are measured in the same units. See Exercises 23 and 24.

Avoid Misleading Graphs

Bar charts are a bit dull to look at. It is tempting to replace the bars with pictures or to use special 3-D effects to make the graphs seem more interesting. Don't do it! Our eyes react to the area of the bars as well as to their height. When all bars have the same width, the area (width × height) varies in proportion to the height, and our eyes receive the right impression about the quantities being compared.

EXAMPLE 2.1.3

Who doesn't like shopping?

Beware the pictograph!

PROBLEM

Harris Interactive asked adults from several countries if they like or dislike shopping for clothes. The percentage of adults in the United States, Germany, and Spain who said that they dislike shopping for clothes is shown in the table.[5]

Country	Percent
United States	30
Germany	20
Spain	12

(a) Here's a clever graph of the data that uses pictures instead of the more traditional bars. How is this graph misleading?

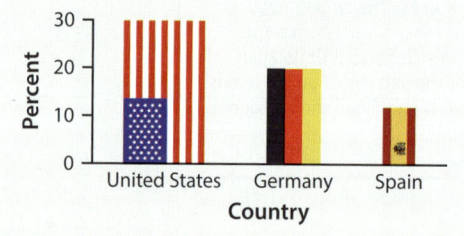

(b) Here is a bar chart of the data. Why could this graph be considered deceptive?

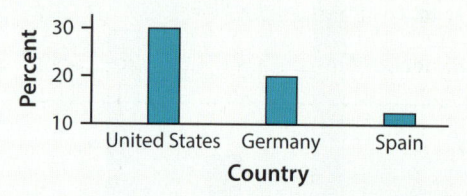

SOLUTION

(a) The areas of the flags make it appear that the percentage of U.S. adults who dislike shopping for clothes is more than 6 times as large as the percentage of Spanish adults who dislike shopping for clothes, when it is really only 2.5 times as large (30% versus 12%).

> Although the heights of the pictures are accurate, our eyes respond to the area of the pictures.

(b) By starting the vertical scale at 10 instead of 0, it looks like the percentage of German adults who dislike shopping for clothes is about 5 times as large as the percentage of Spanish adults who like to shop for new clothes, when it is really less than twice as large (20% versus 12%).

> By starting the vertical scale at a number other than zero, we get a distorted impression of how the relative frequencies for the three countries compare.

EXAM PREP: FOR PRACTICE, TRY EXERCISE 15.

There are two important lessons to be learned from this example: (1) **Beware the pictograph** and (2) **watch those scales.**

TECH CORNER Making Bar Charts and Pie Charts

You can use technology to make a bar chart or pie chart for categorical data. We'll illustrate this process using the preferred status data from Examples 2.1.1 and 2.1.2.

Applet

1. Go to www.stapplet.com and launch the *One Categorical Variable, Single Group* applet.

2. Type Preferred status in the Variable name box.

3. You can choose to input data as Raw data or as Counts in categories.

- Summarized data: Choose to input data as Counts in categories. Type the category names and frequencies. Click on the + button to add rows to the frequency table.

  One Categorical Variable, Single Group

  Variable name: Preferred status
  Input data as: Counts in categories

Category Name	Frequency
1 Famous	3
2 Happy	39
3 Healthy	10
4 Rich	23

  [Begin analysis] [Edit inputs] [Reset everything]

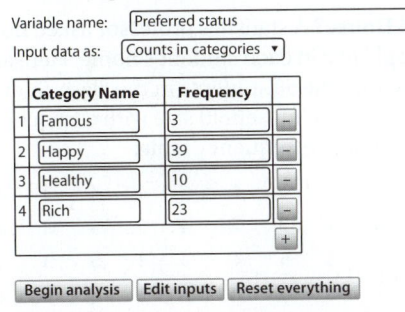

- Raw data: Choose to input data as Raw data. Type the individual data values or copy and paste them from a document or spreadsheet. Be sure to separate the data values with commas or spaces.

  One Categorical Variable, Single Group

  Variable name: Preferred status
  Input data as: Raw data

  Input observations (category names) separated by spaces or commas. (Category names cannot have spaces in them.)
  Rich Healthy Happy Famous Rich Rich Happy Rich Rich Healthy Rich Happy Rich Rich Happy

  [Begin analysis] [Edit inputs] [Reset everything]

4. Click the Begin analysis button. A bar chart of the data should be displayed. To get a pie chart, change the plot type.

Graph Distribution
Plot type: Bar chart   Label bar chart with: Frequency

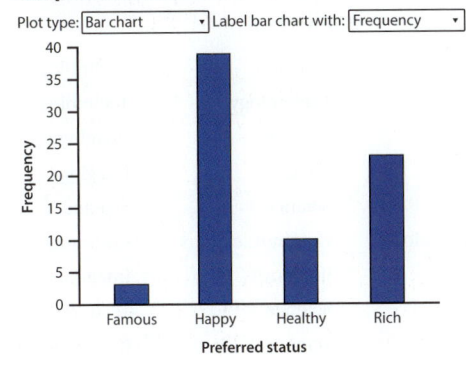

TI-83/84
The TI-83/84 does not produce bar charts or pie charts.

Detailed instructions for using CrunchIt!, Excel, Google Sheets, JMP, Minitab, and R are available in **Achieve**.

Section 2.1 What Did You Learn?

Review the learning goals from this section. Then practice what you've learned by working through the exercises.

Learning Goal	Example	Exercises
Summarize the distribution of a variable with a frequency table or a relative frequency table.	2.1.1	7–10
Make and interpret bar charts and pie charts of categorical data.	2.1.2	11–14
Recognize ways in which graphs of categorical data can be misleading.	2.1.3	15–18

Section 2.1 Exercises

Building Concepts and Skills These exercises assess the basic knowledge you should have after reading the section.

1. The distribution of a variable tells us what _____ the variable takes and _____ it takes each value.

2. Explain the difference between a frequency table and a relative frequency table.

3. Name two types of graphs used to display distributions of categorical data.

4. What is the first step in making a bar chart?

5. True/False: The vertical axis in a bar chart always shows relative frequencies.

6. Why should you "beware the pictograph"?

Mastering Concepts and Skills These exercises reinforce the learning goals as shown in the examples.

7. **Refer to Example 2.1.1 CubeSat Missions** A CubeSat is a miniature satellite used for space research that can be easily deployed from a launch vehicle or from the International Space Station. More than 1000 CubeSats have been launched since specifications for these

satellites were jointly developed by Cal Poly San Luis Obispo and Stanford University in 1999. The table shows the type of mission for each CubeSat launched in November and December 2019. Summarize the distribution of mission type with a frequency table and a relative frequency table.[6]

Science	Science	Technology
Technology	Technology	Technology
Imaging	Imaging	Imaging
Imaging	Imaging	Imaging
Science	Science	Science
Communications	Communications	Communications
Imaging	Imaging	Imaging
Technology	Science	Technology
Technology	Science	Communications
Imaging	Imaging	Imaging
Imaging	Imaging	Technology
Science	Education	Technology
Technology	Education	Imaging
Science	Education	Technology
Technology	Technology	Imaging
Technology	Communications	Imaging
Education		

8. **Raptor Species** The Macbride Raptor Migration Research Station (MRMRS) was established in autumn 1992 as a collaboration between Cornell College and the University of Iowa. Researchers working at this site observed and documented the autumn migration of hawks, eagles, and vultures over a 12-year period. A random sample of 40 of these birds was selected. Their species are noted in the table. Summarize the distribution of species with a frequency table and a relative frequency table.

Red Tailed	Peregrine Falcon	Sharp-Shinned
Red Tailed	Red Tailed	Red Tailed
Bald Eagle	Sharp-Shinned	Red Tailed
Red Tailed	Cooper's Hawk	Red Tailed
Red Tailed	Cooper's Hawk	Red Tailed
Sharp-Shinned	Cooper's Hawk	Red Tailed
Red Tailed	Red Tailed	Sharp-Shinned
Red Tailed	Red Tailed	Red Tailed
Red Tailed	Red Tailed	Red Tailed
Sharp-Shinned	Red Tailed	Red Tailed
Sharp-Shinned	American Kestral	Sharp-Shinned
Sharp-Shinned	Sharp-Shinned	Sharp-Shinned
Red Tailed	Sharp-Shinned	Red Tailed
Sharp-Shinned		

9. **Sleep Statistics** A statistics professor asked how much sleep (in hours) students got on the night before their first exam. Here are the responses from the 50 students in the class. Summarize the distribution of sleep amount with a frequency table and a relative frequency table.

9	8	6	7	7	8	4	7	7	8
8	8	6	7	8	8	7	7	6	8
9	7	6	5	7	8	8	7	9	6
6	6	8	9	5	8	7	7	7	7
2	4	8	3	6	5	5	8	7	3

10. **Crowded House?** A statistics professor asked how many people live in each student's home. Here are the responses from the 50 students in the class. Summarize the distribution of household size with a frequency table and a relative frequency table.

3	5	3	2	4	6	4	4	3	5
4	4	2	2	7	4	3	4	3	3
5	3	5	5	4	4	4	5	3	3
3	4	3	3	4	3	2	6	2	4
8	6	2	4	5	10	3	6	4	3

11. **Refer to Example 2.1.2 Radio Stations** Nielsen Audio, the rating service for radio audiences, places U.S. radio stations into categories that describe the kinds of programs they broadcast. The frequency table summarizes the distribution of station formats in a recent year.[7] Make a bar chart to display the data. Describe what you see.

Format	Count of stations
Adult contemporary	1667
All sports	725
Classic hits	1068
Country	2179
News/talk/information	1996
Oldies	348
Religious	3155
Rock	1032
Spanish language	1183
Variety	1120
Other formats	978

12. **Birthdays** The frequency table summarizes the distribution of day of the week for all babies born in a single week in the United States. Make a bar chart to display the data. Describe what you see.

Day	Births
Sunday	7043
Monday	10,859
Tuesday	11,689
Wednesday	11,652
Thursday	11,776
Friday	11,584
Saturday	8115

13. **Car Colors** The popularity of colors for cars and light trucks changes over time. Silver passed green in 2000 to become the most popular color worldwide, then gave way to shades of white in 2007. Here is a relative

frequency table that summarizes data on the colors of vehicles sold worldwide in 2020.[8]

Color	Percent of vehicles	Color	Percent of vehicles
Black	19	Red	5
Blue	7	Silver	9
Brown/beige	3	White	38
Gray	15	Yellow/gold	2
Green	1	Other	?

(a) What percentage of vehicles falls in the "Other" category?

(b) Make a bar chart to display the data. Describe what you see.

(c) Would it be appropriate to make a pie chart of these data? Explain your answer.

14. **Spam Email** Email spam is a big annoyance. Here is a relative frequency table that summarizes data on the most common types of spam.[9]

Type of spam	Percent	Type of spam	Percent
Adult	19	Leisure	6
Financial	20	Products	25
Health	7	Scams	9
Internet	7	Other	?

(a) What percentage of spam falls in the "Other" category?

(b) Make a bar chart to display the data. Describe what you see.

(c) Would it be appropriate to make a pie chart of these data? Explain your answer.

15. **Refer to Example 2.1.3 Who Buys iMacs?** When Apple, Inc., introduced the iMac, the company wanted to know whether this new computer was expanding its market share. Was the iMac mainly being bought by previous Macintosh owners, or was it being purchased by first-time computer buyers and by previous PC users who were switching over? To find out, Apple hired a firm to conduct a survey of 500 randomly selected iMac customers. The firm categorized each customer as a new computer purchaser, a previous PC owner, or a previous Macintosh owner.[10] The following pictograph summarizes the survey results. Explain how this graph is misleading.

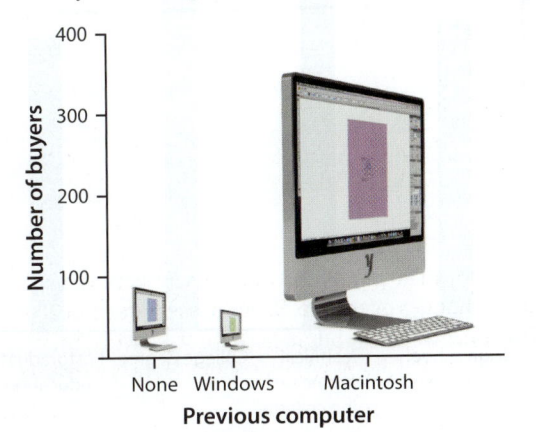

16. **Social Media** The Pew Research Center surveyed a random sample of U.S. teens and adults about their use of social media. The following pictograph displays some results. Explain how this graph is misleading.

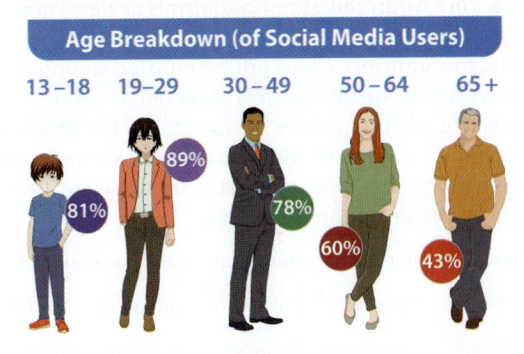

17. **Support the Court?** In 2005, Terry Schiavo, a woman who had been in a persistent vegetative state for 15 years, died following a series of controversial court decisions to remove her feeding tube. CNN reported the results of a survey about the court's decision. The network initially posted on its website a bar chart similar to the one shown here about agreement with the court's decision based on the respondents' political affiliation. Explain how this graph is misleading. (*Note:* When notified about the misleading nature of its graph, the network posted a corrected version.)

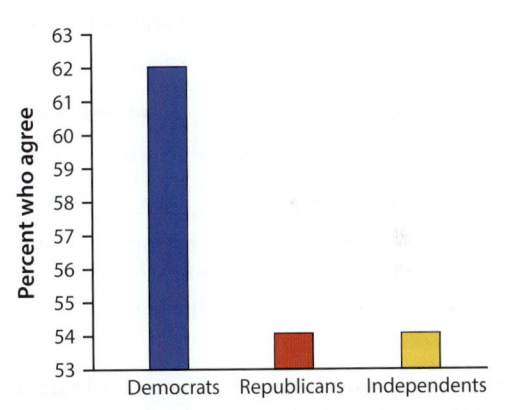

18. **Binge-Watching** Do you "binge-watch" television series by viewing multiple episodes of a series at one sitting? A survey of 800 people who binge-watch were asked how many episodes is too many to watch in one viewing session. Here is a bar chart displaying the results.[11] Explain how this graph is misleading.

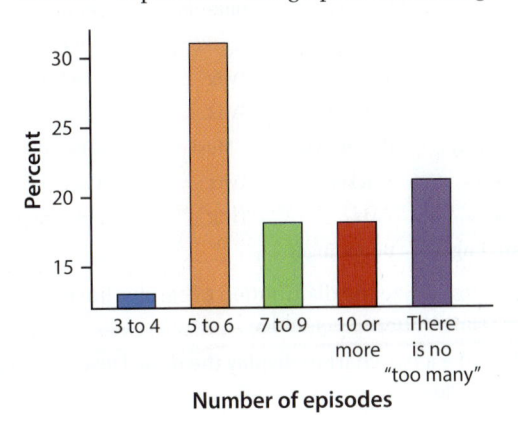

Applying the Concepts These exercises ask you to apply multiple learning goals in a new context or to apply what you learned in this section in a new way.

19. **New York Squirrels** How do squirrels in New York City's Central Park respond to humans? In October 2018, the Squirrel Census project enlisted volunteers to record data on all squirrels observed in Central Park. One of the variables measures how the squirrels responded to humans. We took a random sample of 75 of the more than 3000 squirrels observed. Here are the data on their reactions to humans.[12]

Approach	Run from	Indifferent	Indifferent	Run from
Indifferent	Approach	Indifferent	Indifferent	Indifferent
Approach	Indifferent	Approach	Run from	Run from
Indifferent	Indifferent	Indifferent	Run from	Indifferent
Indifferent	Indifferent	Run from	Indifferent	Indifferent
Indifferent	Run from	Run from	Indifferent	Indifferent
Run from	Run from	Approach	Approach	Run from
Indifferent	Approach	Indifferent	Indifferent	Indifferent
Indifferent	Approach	Indifferent	Indifferent	Indifferent
Indifferent	Run from	Approach	Approach	Indifferent
Indifferent	Indifferent	Run from	Approach	Indifferent
Indifferent	Approach	Indifferent	Run from	Approach
Run from	Approach	Run from	Indifferent	Indifferent
Indifferent	Approach	Indifferent	Run from	Indifferent
Indifferent	Indifferent	Approach	Approach	Indifferent

(a) Summarize the distribution of response to humans with a relative frequency table.

(b) Make a bar chart to display the data. Describe what you see.

20. **Halloween Candy** Professor Cannon bought some candy for her introductory statistics class to eat on Halloween. She offered the students an assortment of Snickers®, Milky Way®, Butterfinger®, Twix®, and 3 Musketeers® candies. Each student was allowed to choose one option. Here are the data on the type of candy selected.

Twix	Snickers	Butterfinger	Butterfinger
3 Musketeers	Snickers	Snickers	Butterfinger
Twix	Twix	Twix	Snickers
Snickers	Milky Way	Twix	Twix
Milky Way	Butterfinger	3 Musketeers	Milky Way
Snickers	Snickers	Twix	Twix
Snickers	Twix	Twix	Butterfinger
Butterfinger	Butterfinger		

(a) Summarize the distribution of candy chosen with a relative frequency table.

(b) Make a bar chart to display the data. Describe what you see.

21. **College Majors** More than 3 million first-year students enroll in U.S. colleges and universities each year. The pie chart displays data on the percentage of first-year students who plan to major in several disciplines.[13] About what percentage of first-year students plan to major in business? In education?

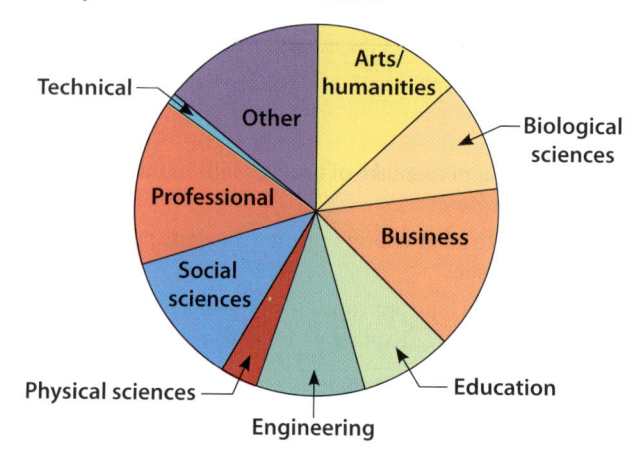

22. **Family Origins** Here is a pie chart of Census Bureau data showing the ethnicity of more than 20 million Asian Americans in the United States.[14] About what percentage of Asian Americans have Chinese ethnicity? Korean ethnicity?

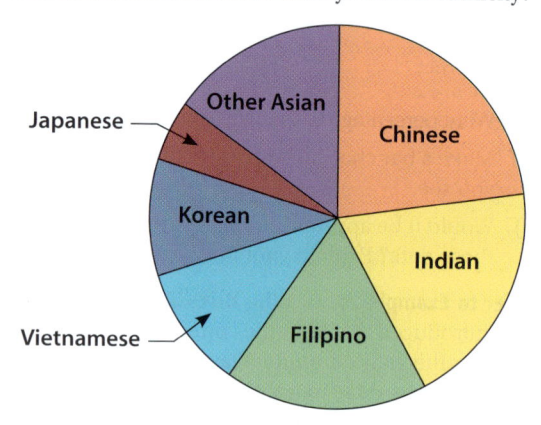

23. **Lotteries and Education** A recent Gallup Poll asked respondents about their highest level of education and whether they had bought a state lottery ticket in the last 12 months.[15] Here is a bar chart of the data.

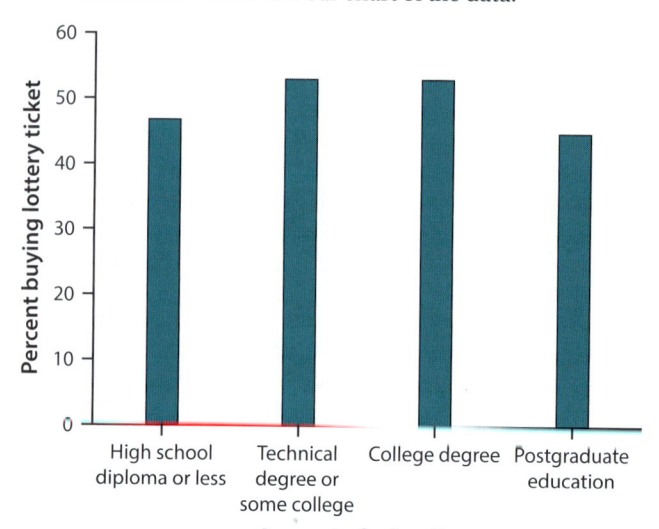

(a) Describe what this graph reveals about lottery ticket buying habits among the different education groups.

(b) Explain why it is not appropriate to make a pie chart for this data set.

24. **Moviegoers** The bar chart displays data on the percentage of people in several age groups who went to a movie during a recent year.[16]

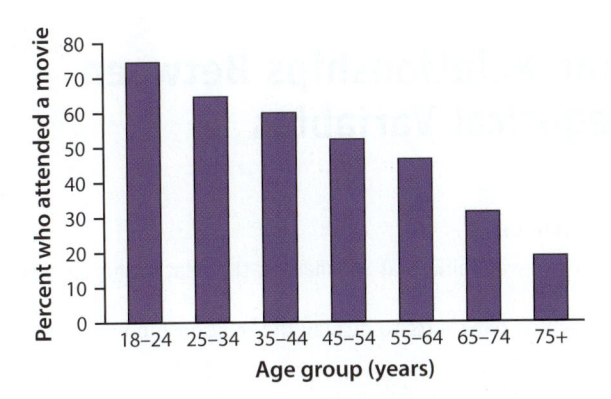

(a) Describe what the graph reveals about movie attendance among different age groups.

(b) Explain why it is not appropriate to make a pie chart for this data set.

25. **Tech Household Languages** What languages do U.S. households speak? The *American Community Survey* data set includes data on several variables for 3000 randomly selected households.[17] Household language (*HHL*) is recorded as follows: 1 = English only, 2 = Spanish, 3 = Other Indo-European languages, 4 = Asian and Pacific Island languages, or 5 = Other language.

(a) Use technology to make a relative frequency table and a bar chart of the household language data.

(b) Describe what you see.

26. **Tech Superpowers** The online survey in Example 2.1.1 also asked which superpower students would choose to have — fly, freeze time, invisibility, super strength, or telepathy. The *Online Survey* data set includes data for 75 randomly selected students who completed the survey.

(a) Use technology to make a relative frequency table and a bar chart of superpower preference (*Superpower*) for the 75 students in the sample.

(b) Describe what you see.

Extending the Concepts These exercises challenge you to explore statistical concepts and methods that go beyond what you learned in this section.

27. **Choropleth Maps** A *choropleth map* is a graphical representation of data by geographic region in which values are depicted by color. For instance, here is a choropleth map that shows the rate of COVID-19 infection for each county in the United States on January 24, 2022.[18] Describe what you see.

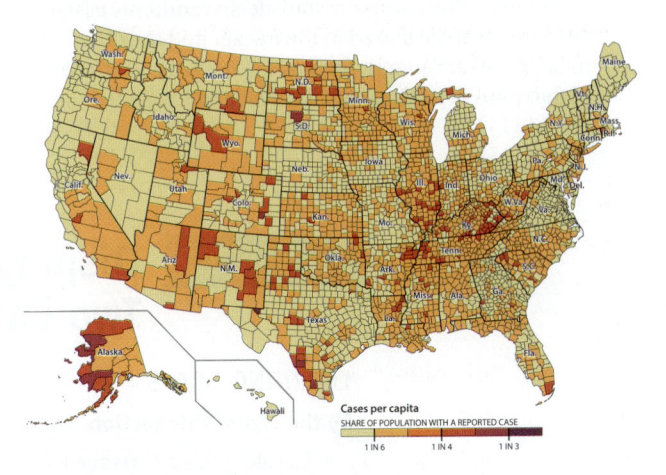

28. **Pareto Charts** It is often more revealing to arrange the bars in a bar chart from tallest to shortest, moving from left to right. Some people refer to this type of bar chart as a *Pareto chart,* named after Italian economist Vilfredo Pareto. Make a Pareto chart for the data in Exercise 13. How is this graph more revealing than one with the bars ordered alphabetically?

Cumulative Review These exercises revisit what you learned in previous sections.

29. **Skyscrapers (1.1)** Here is some information about the tallest buildings in the world (completed by 2020).[19] Identify the individuals and variables in this data set. Classify each variable as categorical or quantitative.

Building	Country	Height (m)	Floors	Use	Year completed
Burj Khalifa	United Arab Emirates	828.0	163	Mixed	2010
Shanghai Tower	China	632.0	121	Mixed	2015
Makkah Royal Clock Tower	Saudi Arabia	601.0	120	Hotel	2012
Ping An Finance Center	China	598.9	115	Office	2017
Lotte World Tower	South Korea	555.7	123	Mixed	2017
One World Trade Center	United States	541.3	94	Office	2014
Guangzhou CTF Finance Center	China	530.0	111	Mixed	2016
Tianjin CTF Finance Centre	China	530.0	97	Mixed	2019
China Zun Tower	China	528.0	108	Mixed	2019
Taipei 101	Taiwan	508.0	101	Office	2004

30. **Batteries on Ice (1.8)** Will storing batteries in a freezer make them last longer? A company that produces batteries takes a random sample of 100 batteries from its warehouse. The company statistician randomly assigns 50 batteries to be stored in the freezer and the other 50 to be stored at room temperature for 3 years. At the end of that time period, each battery's charge is tested. After 3 years, batteries stored in the freezer had a higher average charge, and the difference between the groups was statistically significant.[20]

   (a) Does this study allow for inference about a population? Explain your answer.
   (b) Does this study allow for inference about cause and effect? Explain your answer.

Section 2.2 Displaying Relationships Between Two Categorical Variables

LEARNING GOALS
By the end of this section, you will be able to:
- Calculate relative frequencies from a two-way table that summarizes the relationship between two categorical variables.
- Make side-by-side bar charts or segmented bar charts to display the relationship between two categorical variables.
- Describe the association between two categorical variables.

You can summarize the distribution of a single categorical variable with a frequency table or a relative frequency table. To display the distribution, make a bar chart or a pie chart. In this section, you will learn how to summarize, display, and describe the relationship between two categorical variables.

Summarizing Data on Two Categorical Variables

Yellowstone National Park staff surveyed a random sample of 1526 winter visitors to the park. They asked each person whether they belonged to an environmental club (like the Sierra Club). Respondents were also asked whether they owned, rented, or had never used a snowmobile. The data set looked like this:

Respondent	Environmental club	Snowmobile use
1	No	Own
2	No	Rent
3	Yes	Never
4	Yes	Rent
5	No	Never
⋮	⋮	⋮

It is common to summarize data on two categorical variables with a **two-way table** (sometimes called a *contingency table*). Here is a two-way table that summarizes the Yellowstone survey responses.[21]

		Environmental club	
		No	Yes
Snowmobile use	Never	445	212
	Rent	497	77
	Own	279	16

Section 2.2 Displaying Relationships Between Two Categorical Variables

A **two-way table** is a table of frequencies (or relative frequencies) that summarizes the relationship between two categorical variables for some group of individuals.

In this book, we follow the convention that values of the explanatory variable are summarized in the columns of a two-way table and values of the response variable are summarized in the rows of the table. When there isn't a clear explanatory/response relationship (e.g., between hair color and eye color), it doesn't matter which variable is in the rows and which is in the columns. For the Yellowstone survey, we suspect that environmental club membership may help predict or explain snowmobile use. From what you learned in Chapter 1, that makes environmental club membership the explanatory variable and snowmobile use the response variable.

It's easier to grasp the information in a two-way table if row totals and column totals are included, as in the table shown here.

		Environmental club		
		No	Yes	Total
Snowmobile use	Never used	445	212	657
	Snowmobile renter	497	77	574
	Snowmobile owner	279	16	295
	Total	1221	305	1526

To understand the relationship between environmental club membership and snowmobile use, we need to answer questions like these:

- What percentage of environmental club members in the sample never use snowmobiles? To answer this question, focus on the "Yes" column:

$$\frac{212}{305} = 0.695 = 69.5\%$$

- What percentage of *non*-environmental club members in the sample never use snowmobiles? To answer this question, focus on the "No" column:

$$\frac{445}{1221} = 0.364 = 36.4\%$$

EXAMPLE 2.2.1

Who survived on the *Titanic*?

Calculating relative frequencies from two-way tables

PROBLEM
In 1912, the luxury liner *Titanic*, on its first voyage across the Atlantic, struck an iceberg and sank. Some passengers got off the ship in lifeboats, but many died. The two-way table summarizes data on adult passengers who survived and who died, by class of travel.

		Class of travel			
		First	Second	Third	Total
Survival status	Survived	197	94	151	442
	Died	122	167	476	765
	Total	319	261	627	1207

(a) What percentage of first-class adult passengers survived?
(b) What percentage of second-class adult passengers survived?
(c) What percentage of third-class adult passengers survived?

(continued)

SOLUTION

(a) $\dfrac{197}{319} = 0.618 = 61.8\%$ — To answer this question, focus on the "First" column.

(b) $\dfrac{94}{261} = 0.360 = 36.0\%$ — To answer this question, focus on the "Second" column.

(c) $\dfrac{151}{627} = 0.241 = 24.1\%$ — To answer this question, focus on the "Third" column.

EXAM PREP: FOR PRACTICE, TRY EXERCISE 7.

Displaying the Relationship Between Two Categorical Variables

Let's return to the data from the Yellowstone National Park survey of 1526 randomly selected winter visitors.

		Environmental club		
		No	Yes	Total
Snowmobile use	Never used	445	212	657
	Snowmobile renter	497	77	574
	Snowmobile owner	279	16	295
	Total	1221	305	1526

To explore the relationship between environmental club membership and snowmobile use, we start by calculating the relative frequency of never used, snowmobile renter, and snowmobile owner for each value of the explanatory variable (within each column of the two-way table). The following table summarizes the distribution of snowmobile use separately for environmental club members and non-club members. Note that the percentages in each column should add to 100% (up to the limits of rounding).

Snowmobile use	Not environmental club members	Environmental club members
Never	$\dfrac{445}{1221} = 0.364$ or 36.4%	$\dfrac{212}{305} = 0.695$ or 69.5%
Rent	$\dfrac{497}{1221} = 0.407$ or 40.7%	$\dfrac{77}{305} = 0.252$ or 25.2%
Own	$\dfrac{279}{1221} = 0.229$ or 22.9%	$\dfrac{16}{305} = 0.052$ or 5.2%

⚠ **Be sure to use relative frequencies (percentages or proportions) instead of frequencies (counts) when analyzing data on two categorical variables.** Doing so makes it easier to compare groups of different sizes, as in this setting. There are many more people who never use snowmobiles among the non–environmental club members in the sample (445) than among the environmental club members (212). But these two groups have very different sizes (1221 versus 305). The *percentage* who never use snowmobiles is much higher among environmental club members (69.5% versus 36.4%).

Figure 2.2 compares the distribution of snowmobile use for these two groups with (a) a **side-by-side bar chart** and (b) a **segmented bar chart.** Notice that you can create the segmented bar chart by stacking the bars in the side-by-side bar chart for each of the two environmental club membership categories (no and yes).

A **side-by-side bar chart** displays the distribution of a categorical variable for each value of another categorical variable. The bars are grouped together based on the values of one of the categorical variables and placed next to each other.

A **segmented bar** chart displays the distribution of a categorical variable as portions (segments) of a rectangle, with the area of each segment proportional to the percentage of individuals in the corresponding category.

Section 2.2 Displaying Relationships Between Two Categorical Variables 75

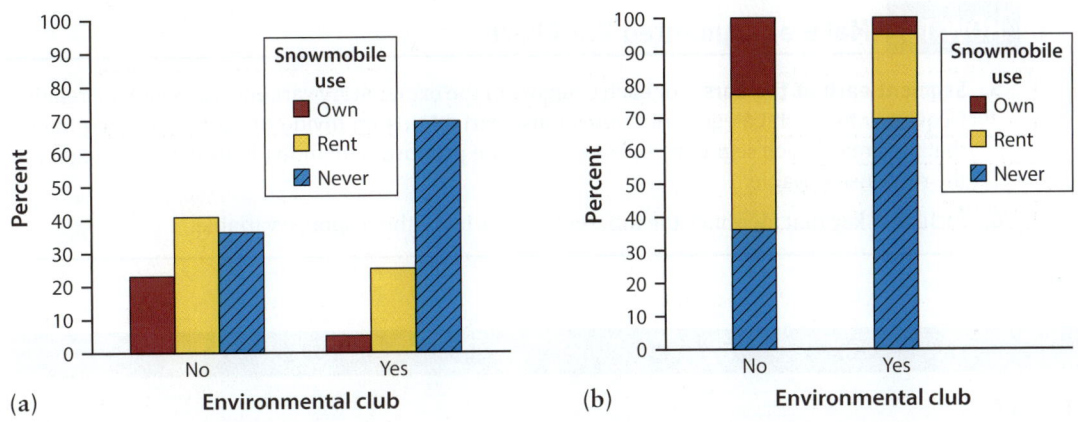

FIGURE 2.2 (a) Side-by-side bar chart and (b) segmented bar chart displaying the distribution of snowmobile use among environmental club members and among non–environmental club members from the 1526 randomly selected winter visitors to Yellowstone National Park.

You can make a side-by-side bar chart using what you learned in Section 2.1. The challenge is deciding how to arrange the bars — within categories of the explanatory variable, as in Figure 2.2(a), or within categories of the response variable, as in the following graph. However you organize the bars, be sure to include a key that describes what each color or type of shading on the graph represents.

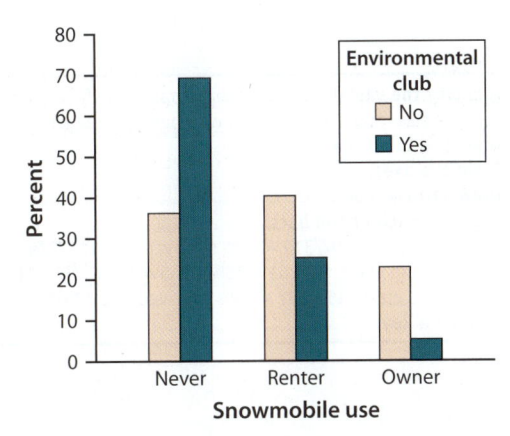

It is a little tedious to make a segmented bar chart by hand (see the How To box). For details on making side-by-side or segmented bar charts with technology, see the Tech Corner at the end of the section.

> **HOW TO** Make a Segmented Bar Chart
>
> 1. **Identify the variables.** Determine which variable is the explanatory variable and which is the response variable. If there is no explanatory/response relationship, choose either variable as "explanatory" and follow the remaining steps.
> 2. **Draw and label the axes.** Put the name of the explanatory variable under the horizontal axis. To the left of the vertical axis, indicate whether the graph shows the percentage (or proportion) of individuals in each category of the response variable.
> 3. **Scale the axes.** Write the names of the categories of the explanatory variable at equally spaced intervals under the horizontal axis. On the vertical axis, start at 0% (or 0) and place tick marks at equal intervals until you reach 100% (or 1).
> 4. **Draw "100%" bars** above each of the category names for the explanatory variable on the horizontal axis so that each bar ends at the top of the graph. Make the bars equal in width and leave gaps between them.

(continued)

HOW TO | Make a Segmented Bar Chart

5. **Segment each of the bars.** For each category of the explanatory variable, calculate the relative frequency for each category of the response variable. Then divide the corresponding bar so that the area of each segment corresponds to the proportion of individuals in each category of the response variable.

6. **Include a key** that identifies the different categories of the response variable.

EXAMPLE 2.2.2

Who survived on the Titanic?

Making a segmented bar chart

PROBLEM

Example 2.2.1 discussed the fate of adult passengers on the *Titanic*. Here, once again, is the two-way table that summarizes data about adult passengers who survived and who died, by class of travel.

Make a segmented bar chart to display the relationship between survival status and class of travel for adult passengers on the *Titanic*.

		Class of travel			
		First	Second	Third	Total
Survival status	Survived	197	94	151	442
	Died	122	167	476	765
	Total	319	261	627	1207

SOLUTION

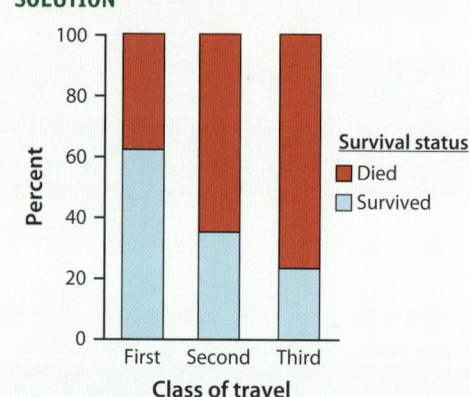

1. **Identify the variables.** Use class of travel for the explanatory variable because class might help predict whether a passenger survived.
2. **Draw and label the axes.**
3. **Scale the axes.**
4. **Draw "100%" bars.**
5. **Segment each of the bars.**
   - In first class, 197/319 = 62% survived and 122/319 = 38% died.
   - In second class, 94/261 = 36% survived and 167/261 = 64% died.
   - In third class, 151/627 = 24% survived and 476/627 = 76% died.
6. **Include a key.**

EXAM PREP: FOR PRACTICE, TRY EXERCISE 11.

As noted in Section 2.1, the next step after making a graph is to describe what you see.

Describing the Relationship Between Two Categorical Variables

Once we make a side-by-side bar chart or a segmented bar chart for data on two categorical variables, the final step is to describe the nature of any **association** between the variables.

> There is an **association** between two variables if knowing the value of one variable helps us predict the value of the other. If knowing the value of one variable does not help us predict the value of the other, then there is no association between the variables.

Both of the following graphs show a clear association between environmental club membership and snowmobile use in the random sample of 1526 winter visitors to Yellowstone National Park. The environmental club members were much less likely to rent (about 25% versus 41%) or own (about 5% versus 23%) snowmobiles than non-club members and were more likely to never use a snowmobile (about 70% versus 36%). Knowing whether a person in the sample is an environmental club member helps us predict that individual's snowmobile use.

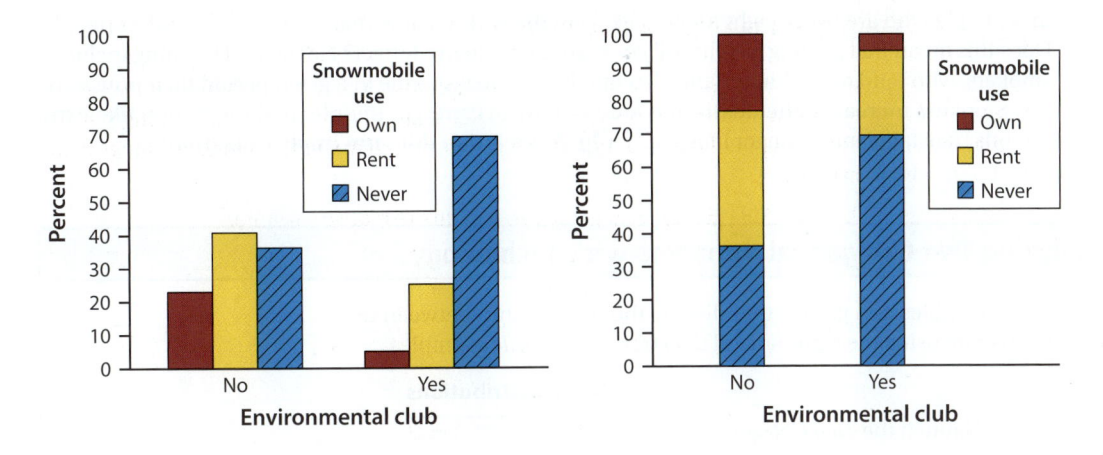

Can we say that there is an association between environmental club membership and snowmobile use in the *population* of winter visitors to Yellowstone National Park? Making this determination requires formal inference, which will have to wait until Chapter 10.

> **THINK ABOUT IT**
>
> **What would the graphs look like if there was *no association* between environmental club membership and snowmobile use in the sample?** If there was no association, then knowing a person's environmental club status wouldn't help predict their snowmobile use. In other words, the distribution of snowmobile use would be the same for both environmental club members and non-members. The segments of the same color—red, yellow, and blue—would be the same height for both the "Yes" and "No" groups, as in this segmented bar chart.

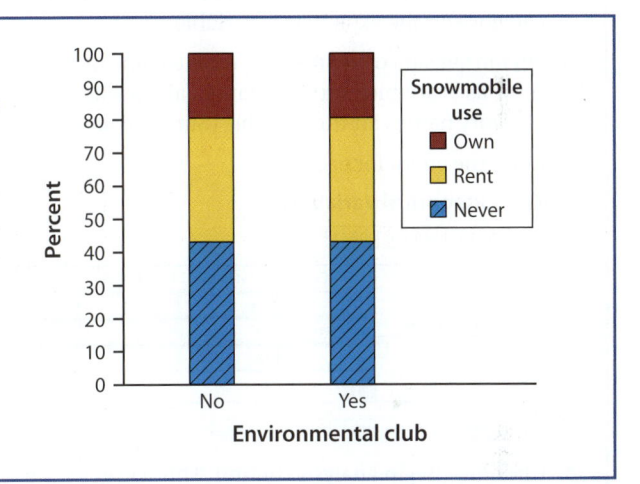

EXAMPLE 2.2.3

Who survived on the Titanic?

Describing association

PROBLEM
Refer to Examples 2.2.1 and 2.2.2. Describe what the segmented bar chart reveals about the association between survival status and class of travel for adult passengers on the *Titanic*.

SOLUTION
There is a clear association between survival status and class of travel on the *Titanic*. Knowing a passenger's class of travel helps us predict whether the passenger survived. Passengers in first class were the most likely to survive (about 62%), followed by second-class passengers (about 36%), and then third-class passengers (about 24%).

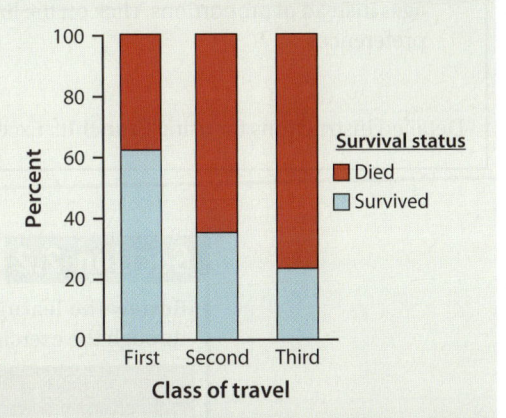

EXAM PREP: FOR PRACTICE, TRY EXERCISE 15.

It may be true that being in a higher class of travel on the *Titanic* increased a passenger's chance of survival. However, as you learned in Chapter 1, there isn't always a cause-and-effect relationship between two variables even if they are clearly associated. For example, a recent study proclaimed that people who are overweight are less likely to die within a few years compared to other people. Does this mean that gaining weight will *cause* someone to live longer? Not at all. The study included smokers, who tend to be thinner and also much more likely to die in a given period than nonsmokers. Smokers increased the death rate among non-overweight people, making it look like extra pounds translated into a longer lifespan.[22] **!** **Association does not imply causation!**

TECH CORNER: Analyzing Two Categorical Variables with Technology

You can use technology to create tables and graphs that display the relationship between two categorical variables. We'll illustrate using the summarized data from the *Titanic* examples.

Applet

1. Go to www.stapplet.com and launch the *Two Categorical Variables* applet.
2. At the top of the table, enter the name of the explanatory variable (Class of travel), along with the category names for this variable (First, Second, and Third). Press the "+" button in the upper right to get a third category for the explanatory variable.
3. On the left side of the table, enter the name of the response variable (Survival status), along with the category names for this variable (Survived and Died).
4. Enter the count for each cell.

Two Categorical Variables

Input data as: Two-way table ▼

	Explanatory variable: Class of travel		
Response variable: Survival status	First	Second	Third
Survived	197	94	151
Died	122	167	476

[Begin analysis] [Edit inputs] [Reset everything]

5. Click the "Begin analysis" button. This will generate a segmented bar chart and a table showing the proportion who survived and died within each class of travel. Use the drop-down menu to get a side-by-side bar graph. *Note:* To have the graph display percentages instead of proportions, click on the link to adjust preferences.

Graph Distributions

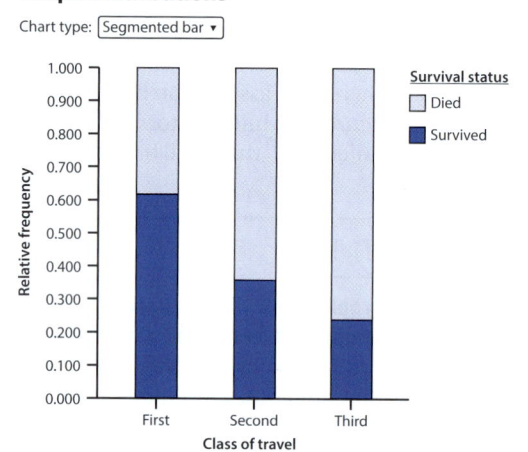

Chart type: Segmented bar ▼

[Segmented bar chart showing Relative frequency (0.000 to 1.000) by Class of travel (First, Second, Third), with Survival status categories Died and Survived.]

Summary Statistics

		Class of travel			
		First	Second	Third	Total
Survival status	Survived	197 (0.618)	94 (0.36)	151 (0.241)	442 (0.366)
	Died	122 (0.382)	167 (0.64)	476 (0.759)	765 (0.634)
	Total	319 (1)	261 (1)	627 (1)	1207 (1)

Proportions in table show the distribution of Survival status for each category of Class of travel.

Note: If you have raw data that isn't already summarized in a two-way table, choose the raw data option in the drop-down menu at the top of the applet.

TI-83/84

The TI-83/84 does not produce side-by-side or segmented bar charts.

Detailed instructions for using CrunchIt!, Excel, Google Sheets, JMP, Minitab, and R are available in *Achieve*.

Section 2.2 What Did You Learn?

Review the learning goals from this section. Then practice what you've learned by working through the exercises.

Learning Goal	Example	Exercises
Calculate relative frequencies from a two-way table that summarizes the relationship between two categorical variables.	2.2.1	7–10
Make side-by-side bar charts or segmented bar charts to display the relationship between two categorical variables.	2.2.2	11–14
Describe the association between two categorical variables.	2.2.3	15–20

Section 2.2 Exercises

Building Concepts and Skills These exercises assess the basic knowledge you should have after reading the section.

1. A two-way table is a table of frequencies (or relative frequencies) that summarizes the relationship between _____ for some group of individuals.

2. Why is it important to use relative frequencies instead of frequencies when analyzing the relationship between two categorical variables?

3. True/False: There is only one way to organize the bars in a side-by-side bar chart.

4. True/False: Each bar in a segmented bar chart represents one category of the response variable.

5. What does it mean if two variables have an association?

6. True/False: If two variables have an association, there is not necessarily a cause-and-effect relationship between them.

Mastering Concepts and Skills These exercises reinforce the learning goals as shown in the examples.

7. **Refer to Example 2.2.1 Car Accident Perceptions** Two researchers asked 150 people to recall the details of a car accident they watched on video. They selected 50 people at random and asked, "About how fast were the cars going when they smashed into each other?" For another 50 randomly selected people, they replaced the words "smashed into" with "hit." They did not ask the remaining 50 people — the control group — to estimate speed at all. A week later, the researchers asked all 150 participants if they saw any broken glass at the accident (there wasn't any). The two-way table summarizes each group's response to the broken glass question.[23]

		Treatment			
		"Smashed into"	"Hit"	Control	Total
Response	Yes	16	7	6	29
	No	34	43	44	121
	Total	50	50	50	150

   (a) What percentage of people who received the "smashed into" wording said that they saw broken glass at the accident?

   (b) What percentage of people who received the "hit" wording said that they saw broken glass at the accident?

8. **Python Nests** How does the temperature of a water python nest influence egg hatching? Researchers assigned newly laid eggs to one of three water temperatures: hot, neutral, or cold. Hot duplicates the extra warmth provided by the mother python, and cold duplicates the absence of the mother. The two-way table summarizes the data on nest temperature and hatching status.[24]

		Nest temperature			
		Cold	Neutral	Hot	Total
Hatching status	Hatched	16	38	75	129
	Didn't hatch	11	18	29	58
	Total	27	56	104	187

   (a) What percentage of python eggs assigned to cold water hatched?

   (b) What percentage of python eggs assigned to hot water hatched?

9. **New York Squirrels** Do adult and juvenile Eastern gray squirrels in New York's Central Park exhibit different behaviors toward humans? That is one of many questions investigated by 323 volunteer squirrel sighters in October 2018.[25] The two-way table summarizes the data for 2898 squirrel sightings in the park.

		Age		
		Juvenile	Adult	Total
Behavior toward humans	Approach	111	756	867
	Indifferent	138	1241	1379
	Run away	81	571	652
	Total	330	2568	2898

   (a) What percentage of juvenile squirrels ran away from humans?

   (b) What percentage of adult squirrels ran away from humans?

10. **Napping and Heart Disease** In a long-term study of 3462 randomly selected adults from Lausanne, Switzerland, researchers investigated the relationship between weekly napping frequency and whether a person experienced a major cardiovascular disease (CVD) event, such as a heart attack or stroke. The two-way table summarizes the data.[26]

		Napping frequency				
		None	1–2 weekly	3–5 weekly	6–7 weekly	Total
CVD event	Yes	93	12	22	28	155
	No	1921	655	389	342	3307
	Total	2014	667	411	370	3462

    (a) What percentage of people who took no naps experienced a major CVD event?

    (b) What percentage of people who took 6–7 naps per week experienced a major CVD event?

11. **Refer to Example 2.2.2 Car Accident Perceptions** Refer to Exercise 7. Make a segmented bar chart to display the relationship between question wording and whether the subject saw broken glass at the accident.

12. **Python Nests** Refer to Exercise 8. Make a segmented bar chart to display the relationship between nest temperature and hatching success.

13. **New York Squirrels** Refer to Exercise 9. Make a side-by-side bar chart to display the relationship between squirrel age and behavior toward humans.

14. **Napping and Heart Disease** Refer to Exercise 10. Make a side-by-side bar chart to display the relationship between napping frequency and CVD status.

15. **Refer to Example 2.2.3 CubeSat Missions** A CubeSat is a miniature satellite used for space research that can be easily deployed from a launch vehicle or from the International Space Station. More than 1000 CubeSats have been launched since specifications for these satellites were jointly developed by Cal Poly San Luis Obispo and Stanford University in 1999. The side-by-side bar chart displays data on the types of missions undertaken by CubeSats launched in 2017 and 2019. Describe the association between mission type and year for these launches.[27]

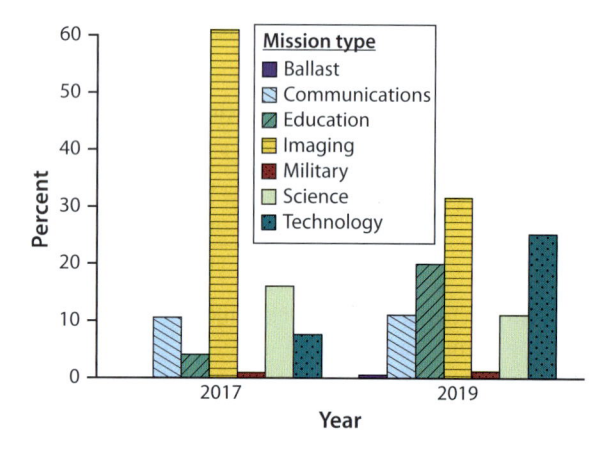

16. **Far from Home** A survey asked first-year college students, "How many miles is this college from your permanent home?" Students selected from the following options: 5 or fewer, 6 to 10, 11 to 50, 51 to 100, 101 to 500, or over 500. The side-by-side bar chart shows the percentage of students at public and private 4-year colleges who chose each option.[28] Describe the association between distance from home and type of college for students who completed the survey.

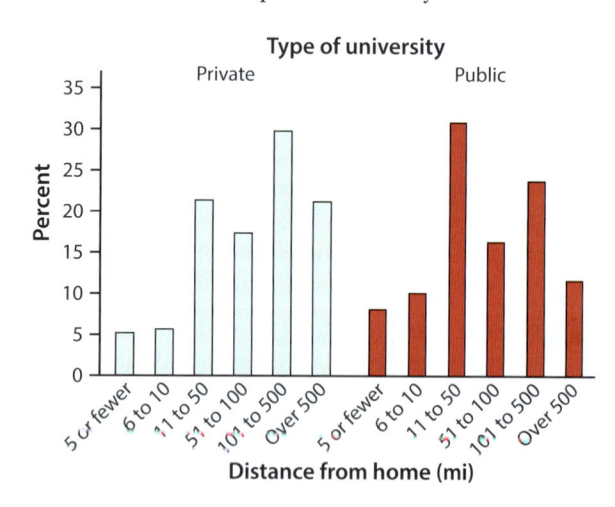

17. **Car Accident Perceptions** Refer to Exercises 7 and 11. Use the graph from Exercise 11 to describe the association between question wording and whether the subject saw broken glass.

18. **Python Nests** Refer to Exercises 8 and 12. Use the graph from Exercise 12 to describe the association between nest temperature and hatching success.

19. **New York Squirrels** Refer to Exercises 9 and 13. Use the graph from Exercise 13 to describe the association between squirrel age and behavior toward humans.

20. **Napping and Heart Disease** Refer to Exercises 10 and 14. Use the graph from Exercise 14 to describe the association between napping frequency and CVD status.

Applying the Concepts These exercises ask you to apply multiple learning goals in a new context or to apply what you learned in this section in a new way.

21. **College Tuition** A Quinnipiac University poll asked a random sample of U.S. adults, "Would you support or oppose major new spending by the federal government that would help undergraduates pay tuition at public colleges without needing loans?" The two-way table shows the responses, grouped by age.[29]

		Age				
		18–34	35–49	50–64	65+	Total
Response	Support	91	161	272	332	856
	Oppose	25	74	211	255	565
	Don't know	4	13	20	51	88
	Total	120	248	503	638	1509

(a) Make a segmented bar chart to display the relationship between age and opinion about government assistance for paying undergraduate tuition for the members of the sample.

(b) Use the graph from part (a) to describe the association between these two variables.

22. **National Languages** The Pew Research Center conducts surveys about a variety of topics in many different countries. One such survey investigated how adult residents of different countries feel about the importance of speaking the national language. Separate random samples of residents of Australia, the United Kingdom, and the United States were asked, "How important do you think it is to be able to speak English?" The two-way table summarizes the responses to this question.[30]

		Country			
		Australia	U.K.	U.S.	Total
Opinion about speaking English	Very important	690	1177	702	2569
	Somewhat important	250	242	221	713
	Not very important	40	28	50	118
	Not at all important	20	13	30	63
	Total	1000	1460	1003	3463

(a) Make a segmented bar chart to display the relationship between country and opinion about speaking English for the members of the sample.

(b) Use the graph from part (a) to describe the association between these two variables.

23. **Tech Household Languages** Is there a relationship between Census Bureau region and household language for U.S. households? The *American Community Survey* data set includes data on several variables for 3000 randomly selected households.[31] Household language (*HHL*) is recorded as follows: 1 = English only, 2 = Spanish, 3 = Other Indo-European languages, 4 = Asian and Pacific Island languages, or 5 = Other language. Census Bureau region (*Region*) is recorded as 1 = Northeast, 2 = Midwest, 3 = South, and 4 = West.

(a) Use technology to make a two-way table and a segmented bar chart that summarizes data on these two variables.

(b) Describe the association between region and household language for the households in the sample.

24. **Tech Internet Access** Is there a relationship between Census Bureau region and internet access for U.S. households? The *American Community Survey* data set includes data on several variables for 3000 randomly selected households.[32] Census Bureau region (*Region*) is recorded as 1 = Northeast, 2 = Midwest, 3 = South, and 4 = West. Internet access (*Access*) is coded as 1 = Yes with subscription to an internet service, 2 = Yes without a subscription to an internet service, and 3 = No.

(a) Use technology to make a two-way table and a segmented bar chart that summarizes data on these two variables.

(b) Describe the association between region and internet access for the households in the sample.

Extending the Concepts These exercises challenge you to explore statistical concepts and methods that go beyond what you learned in this section.

25. **Reading for Pleasure** Common Sense Media surveyed a random sample of more than 1600 U.S. tweens (ages 8–12) and teens (ages 13–18) about how often they read for pleasure. The pie charts summarize the data.[33] Describe similarities and differences between the distributions for tweens and teens.

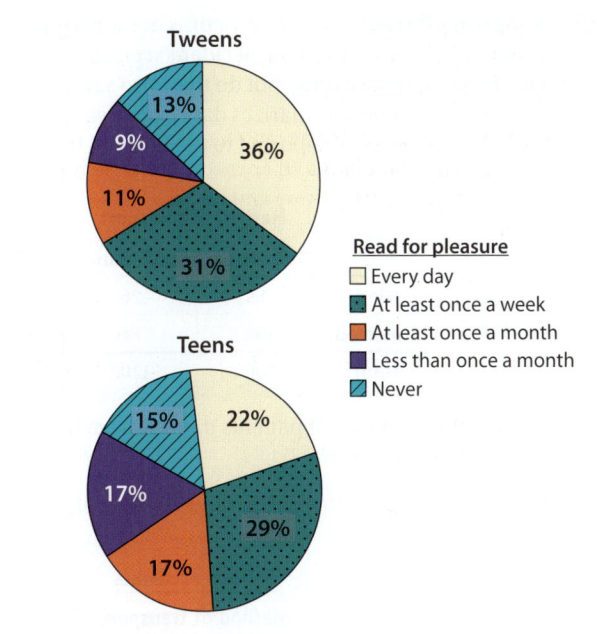

Tweens: 36% Every day, 31% At least once a week, 11% At least once a month, 9% Less than once a month, 13% Never

Teens: 22% Every day, 29% At least once a week, 17% At least once a month, 17% Less than once a month, 15% Never

Read for pleasure
- Every day
- At least once a week
- At least once a month
- Less than once a month
- Never

26. **Mosaic Plots** A *mosaic plot* resembles a segmented bar chart except that the width of the bars corresponds to the proportion of individuals in each category of the explanatory variable. Here is an example of the mosaic plot for the *Titanic* data in Example 2.2.2.

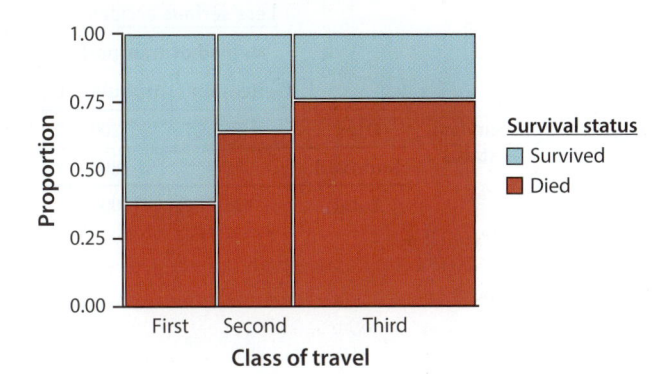

Following this example, make a mosaic plot for the python egg data in Exercise 8.

27. **No Association** Copy and complete the following two-way table so that it shows no association between dominant hand and dominant eye in a sample of 200 people.

		Dominant hand		
		Left	Right	Total
Dominant eye	Left			60
	Right			140
	Total	30	170	200

28. **Simpson's Paradox** Accident victims are sometimes taken by helicopter from the accident scene to a hospital. Helicopters save time. But do they also save lives? The two-way table summarizes data from a sample of patients who were transported to the hospital by helicopter or by ambulance.[34]

		Method of transport		
		Helicopter	Ambulance	Total
Survival status	Died	64	260	324
	Survived	136	840	976
	Total	200	1100	1300

(a) What percentage of patients died with each method of transport?

Here are the same data broken down by severity of accident:

Serious accidents

		Method of transport		
		Helicopter	Ambulance	Total
Survival status	Died	48	60	108
	Survived	52	40	92
	Total	100	100	200

Less serious accidents

		Method of transport		
		Helicopter	Ambulance	Total
Survival status	Died	16	200	216
	Survived	84	800	884
	Total	100	1000	1100

(b) Calculate the percentage of patients who died with each method of transport for the serious accidents. Then calculate the percentage of patients who died with each method of transport for the less serious accidents. What do you notice?

(c) See if you can explain how the result in part (a) is possible given the result in part (b).

Note: This is an example of *Simpson's paradox*, which states that an association between two variables that holds for each value of a third variable can be changed or even reversed when the data for all values of the third variable are combined.

Cumulative Review These exercises revisit what you learned in previous sections.

29. **Confident about computing (1.1, 1.2, 1.3)** Google and Gallup teamed up to survey a random sample of 1673 U.S. students in grades 7–12. One of the questions was "How confident are you that you could learn computer science if you wanted to?" Overall, 54% of students said they were very confident.[35]

    (a) Identify the population and the sample.

    (b) Explain why it was better to randomly select the students rather than putting the survey question on a website and inviting students to answer the question.

    (c) Do you expect that the percentage of *all* U.S. students in grades 7–12 who would say "very confident" is exactly 54%? Explain your answer.

    (d) The report also broke the results down by self-reported gender. For this question, 62% of boys and 48% of girls said they were very confident. Which of the three estimates (54%, 62%, 48%) do you expect is closest to the corresponding population percentage? Explain your answer.

Section 2.3 Displaying Quantitative Data: Dotplots

LEARNING GOALS

By the end of this section, you will be able to:
- Make and interpret dotplots of quantitative data.
- Describe the shape of a distribution of quantitative data.
- Describe or compare distributions of quantitative data with dotplots.

You can use a bar chart or pie chart to display categorical data. A **dotplot** is the simplest graph for displaying quantitative data.

> A **dotplot** shows each data value as a dot above its location on a number line.

In April 2014, managers for the city of Flint, Michigan, decided to save money by using water from the Flint River rather than continuing to buy water sourced from Lake Huron. Soon after, Flint residents noticed that the water coming out of their taps looked, tasted, and smelled bad. Some residents developed rashes, hair loss, and itchy skin. Authorities insisted that drinking water from the Flint River was safe.

As part of its regular water quality testing program, city officials measured lead levels in 71 water samples collected from randomly selected Flint dwellings between January and June 2015. Here are the data (in parts per billion).[36] Figure 2.3 is a dotplot of the data.

0	0	0	0	0	0	0	0	0	0	0	0
0	1	1	1	1	2	2	2	2	2	2	2
2	2	2	2	3	3	3	3	3	3	3	3
3	3	3	4	4	5	5	5	5	5	5	5
5	6	6	6	6	7	7	7	8	8	9	10
10	11	13	18	20	21	22	29	42	42	104	

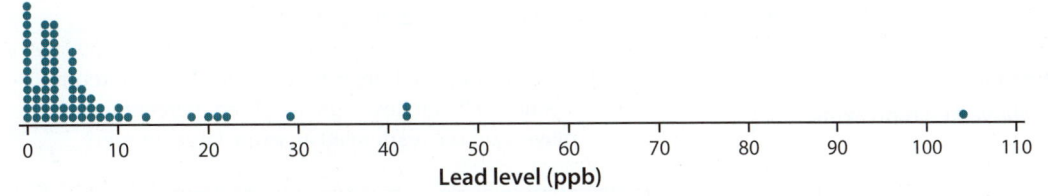

FIGURE 2.3 Dotplot of lead levels in water samples from 71 randomly selected Flint, Michigan, dwellings in January to June, 2015.

You'll learn how to make and interpret dotplots in this section.

Making and Interpreting Dotplots

You can make a dotplot by hand for small sets of quantitative data (see the How To box). For details on making dotplots with technology, see the Tech Corner at the end of the section.

HOW TO Make a Dotplot

1. **Draw and label the axis.** Draw a horizontal axis and put the name of the quantitative variable underneath it.
2. **Scale the axis.** Find the smallest and largest values in the data set. Start the horizontal axis at a number equal to or less than the smallest value and place tick marks at equal intervals until you equal or exceed the largest value.
3. **Plot the values.** Mark a dot above the location on the horizontal axis corresponding to each data value. Try to make all the dots the same size and space them out equally as you stack them.

Remember what we said in Section 2.1: Making a graph is not an end in itself. When you look at a graph, always ask, "What do I see?"

Refer to Figure 2.3. U.S. Environmental Protection Agency (EPA) regulations state that if more than 10% of water samples collected by the city have lead levels greater than 15 parts per billion, action must be taken. The dotplot shows that $8/71 = 0.113 = 11.3\%$ of the water samples in Flint were above this threshold. However, city officials decided to omit two water samples from their analysis: one with 20 parts per billion of lead that came from a business, and one with 104 parts per billion that came from a home using a water filter. With those two values excluded, only $6/69 = 0.087 = 8.7\%$ of the remaining water samples had lead levels greater than 15 parts per billion. So authorities determined that no action was needed.

Note: Flint residents worked with scientists from Virginia Tech University to retest the lead level of water in the city's dwellings. Their result: Far more than 10% of the water samples had lead levels greater than 15 parts per billion. Authorities finally agreed that Flint's water contained dangerously high lead levels and switched back to buying water sourced from Lake Huron. By the end of 2016, the amount of lead in Flint's water was back to a safe level: More than 90% of the water samples had lead levels of at most 12 parts per billion. However, several state and local officials have been charged with crimes for endangering the health of Flint residents.

EXAMPLE 2.3.1

Preventing concussions

Making and interpreting a dotplot

PROBLEM

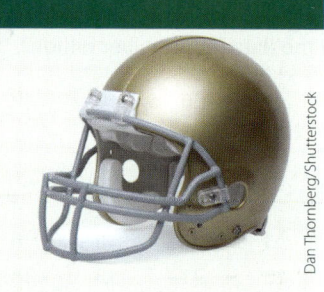

Many athletes (and their parents) worry about the risk of concussions when playing sports. A youth football coach plans to obtain specially made helmets for the players that are designed to reduce their chance of getting a concussion. Here are the measurements of head circumference (in inches) for the 30 players on the team.[37]

23.2	22.6	22.1	22.4	22.7	23.0	23.1	21.9	23.1	24.8	21.2	23.2	24.2	23.7	21.2
24.0	23.1	23.0	23.9	22.9	23.3	22.3	21.4	22.8	23.7	22.9	23.9	23.6	22.0	23.5

(a) Make a dotplot of these data.

(b) Explain what the dot above 21.4 represents.

(c) An extra-large helmet is required for a head circumference of 24 inches or greater. What percentage of players on the team need an extra-large helmet?

SOLUTION

(a)

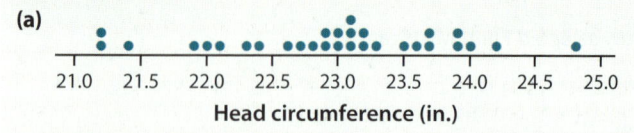

[Dotplot showing Head circumference (in.) on x-axis from 21.0 to 25.0]

1. Draw and label the axis.
2. **Scale the axis.** The smallest value is 21.2 and the largest value is 24.8. So we chose a scale from 21 to 25 with tick marks 0.5 unit apart.
3. Plot the values.

(b) A player on the team with a head circumference of 21.4 inches.

(c) $3/30 = 0.10 = 10\%$ of players on the team have a head circumference of 24 inches or greater, and need an extra-large helmet.

EXAM PREP: FOR PRACTICE, TRY EXERCISE 7.

Describing Shape

When you describe the shape of a dotplot or other graph of quantitative data, focus on the main features. Look for major *peaks*, not for minor ups and downs in the graph. Look for *clusters* of values and obvious *gaps*. Decide if the distribution is **roughly symmetric** or clearly **skewed.**

A distribution is **roughly symmetric** if the right side of the graph (containing the half of the observations with larger values) is approximately a mirror image of the left side.

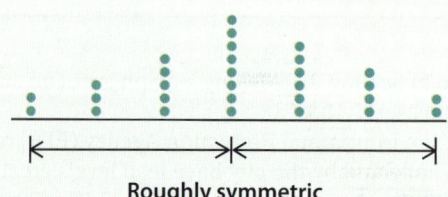

Roughly symmetric

A distribution is **skewed to the right** if the right side of the graph (containing the half of the observations with larger values) is much longer than the left side.

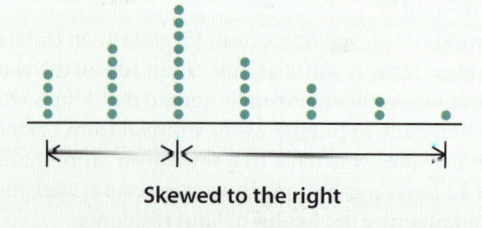

Skewed to the right

A distribution is **skewed to the left** if the left side of the graph (containing the half of the observations with smaller values) is much longer than the right side.

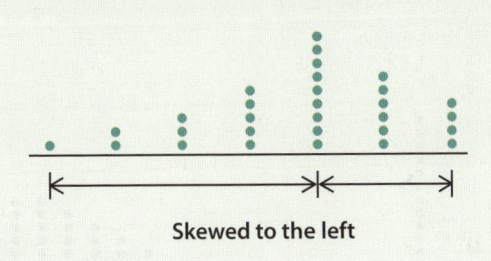

Skewed to the left

For ease, we sometimes say "left-skewed" instead of "skewed to the left" and "right-skewed" instead of "skewed to the right." ⚠ **The direction of skewness is toward the long tail, not the direction where most observations are clustered.** The following drawing is a cute but corny way to help you keep this straight. To avoid danger, Mr. Starnes skis on the gentler slope — in the direction of the skewness.

Skewed to the left!

EXAMPLE 2.3.2

Head circumference and quiz scores

Describing the shape of a distribution

PROBLEM

The following dotplots display two different sets of quantitative data. Graph (a) shows the head circumference measurements (in inches) for the 30 players on a youth football team from Example 2.3.1. Graph (b) shows the scores on a 20-point quiz for each of the 25 students in a statistics class. Describe the shape of each distribution.

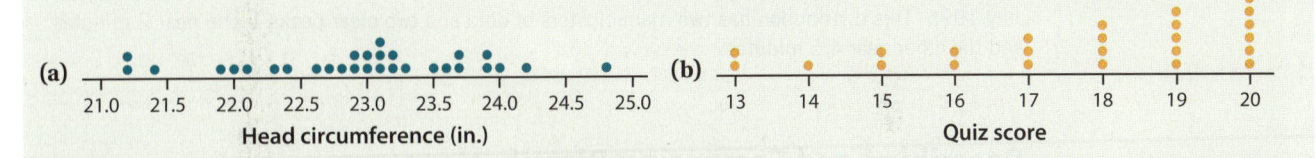

(a) Head circumference (in.) — 21.0 to 25.0

(b) Quiz score — 13 to 20

SOLUTION

(a) The distribution of head circumference is roughly symmetric with a single peak at 23.1 inches, one main cluster of dots between 21.9 and 24.2 inches, a gap from 21.4 to 21.9 inches, and a gap from 24.2 to 24.8 inches.

(b) The distribution of quiz scores is skewed to the left, with a single peak at 20 (a perfect score).

EXAM PREP: FOR PRACTICE, TRY EXERCISE 11.

Some quantitative variables have distributions with predictable shapes. In the following figure, graph (a) is a dotplot of the results of 100 rolls of a 6-sided die. We can describe the shape of the distribution of die rolls as *approximately uniform* because the frequencies are about the same for all possible rolls. Graph (b) is a dotplot of the wing lengths of 100 houseflies.[38] As this graph illustrates, repeated careful measurements of the same attribute for a large group of similar individuals often have distributions that are roughly symmetric and single-peaked. (In Chapter 6, we will describe such distributions as *approximately normal*.)

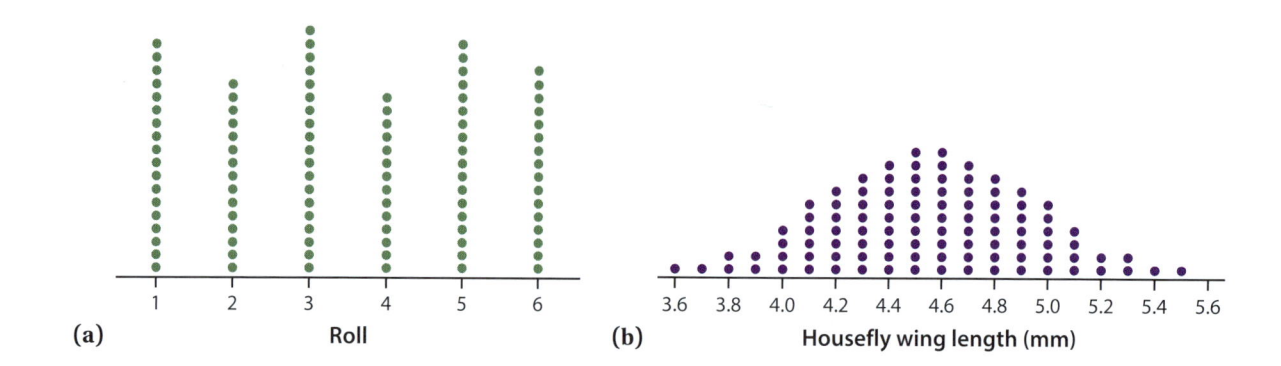

Some people refer to graphs with a single peak as *unimodal*. Figure 2.4 is a dotplot of the duration (in minutes) of 263 eruptions of the Old Faithful geyser in July 1995, when the Starnes family made its first trip to Yellowstone National Park. We describe this graph as *bimodal* because it has two clear peaks. Although we could continue the pattern with "trimodal" for three peaks, and so on, it's more common to refer to distributions with more than two clear peaks as *multimodal*. When you examine a graph of quantitative data, describe any pattern you see as clearly as you can.

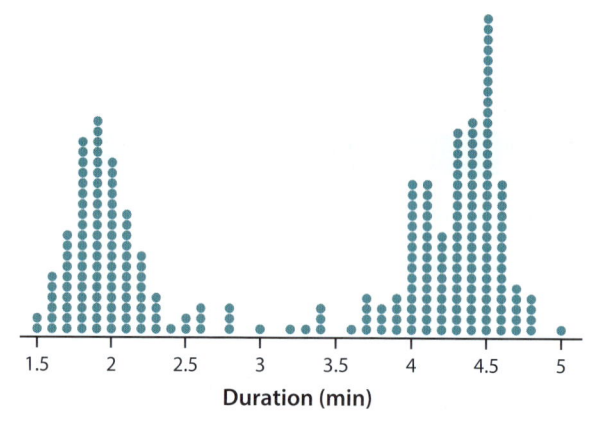

FIGURE 2.4 Dotplot displaying the duration, in minutes, of 263 eruptions of the Old Faithful geyser in July 1995. This distribution has two main clusters of data and two clear peaks—one near 2 minutes and the other near 4.5 minutes.

Describing and Comparing Distributions

Here is a general strategy for describing a distribution of quantitative data.

> **HOW TO** **Describe the Distribution of a Quantitative Variable**
>
> In any graph, look for the *overall pattern* and for clear *departures* from that pattern.
> - You can describe the overall pattern of a distribution by its **shape, center,** and **variability.**
> - An important kind of departure is an **outlier,** a value that falls outside the overall pattern.

We will discuss more formal ways to measure center and variability and to identify outliers in Chapter 3. For now, just use the middle value in the ordered data set (the *median*) to describe the center and the minimum and maximum values to describe the variability. Let's practice with the dotplot of lead levels in water samples from 71 randomly selected Flint, Michigan, dwellings.

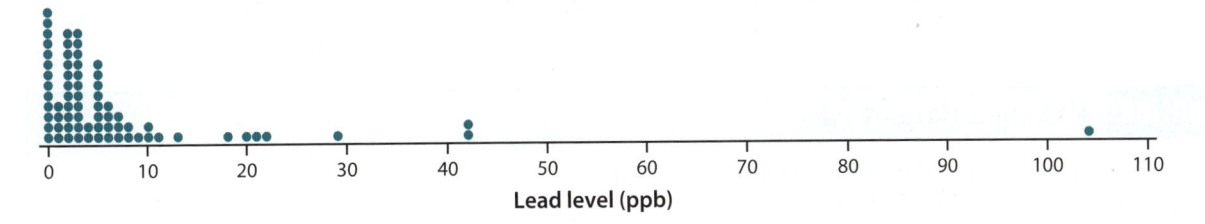

Shape: The distribution of lead level is skewed to the right, with a single peak at 0 ppb.
Outliers: There is one obvious outlier at 104 ppb. The two lead levels of 42 ppb may also be outliers.
Center: The middle value (median) is a lead level of 3 ppb.
Variability: The lead levels vary from 0 to 104 ppb.

Some of the most interesting statistical questions involve comparing two or more groups. Do high school graduates earn more, on average, than students who do not graduate from high school? Which of two popular diets leads to greater long-term weight loss? As the following example suggests, you should always discuss shape, outliers, center, and variability whenever you compare distributions of a quantitative variable.

EXAMPLE 2.3.3

Households in South Africa and the United Kingdom

Comparing distributions with dotplots

PROBLEM
How do the numbers of people living in households in the United Kingdom and South Africa compare? To help answer this question, we used an online tool to select 50 households at random from each country.[39] Here are dotplots of the number of people in each household. Compare the distributions of household size for these two countries.

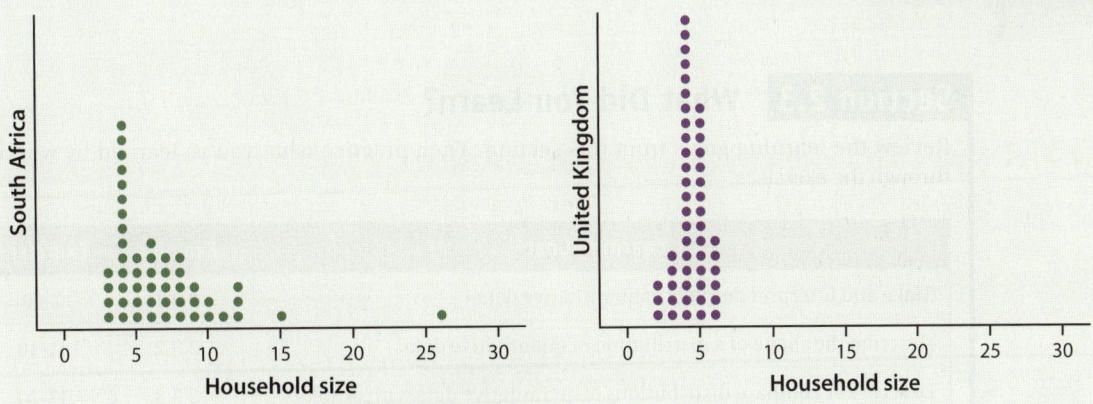

SOLUTION

Shape: The South Africa distribution of household size is skewed to the right and single-peaked, while the U.K. distribution of household size is roughly symmetric and single-peaked.

Outliers: There are no obvious outliers in the U.K. dotplot. The large value in the South Africa dotplot—a household with 26 people—appears to be an outlier.

Center: Household sizes for the South Africa sample tend to be larger (median = 6) than for the U.K. sample (median = 4).

Variability: The household sizes for the South Africa sample vary more (from 3 to 26 people) than for the U.K. sample (from 2 to 6 people).

> Organizing the comparison by characteristic helps ensure that you don't forget to discuss shape, outliers, center, and variability. Also be sure to include context. In this case, the variable of interest is household size.

EXAM PREP: FOR PRACTICE, TRY EXERCISE 17.

When comparing distributions of quantitative data, it's not enough just to list values for the center and variability of each distribution. You have to explicitly *compare* these values, using expressions like "greater than," "less than," or "about the same as."

TECH CORNER: Making a Dotplot

You can use technology to make a dotplot. We'll illustrate using the head circumference data from Example 2.3.1.

Applet

1. Go to www.stapplet.com and launch the *One Quantitative Variable, Single Group* applet.
2. Enter Head circumference (inches) as the Variable name.
3. Select Raw data as the input method.
4. Enter the data manually, or copy and paste them from a document or spreadsheet. Be sure to separate the data values with commas or spaces.

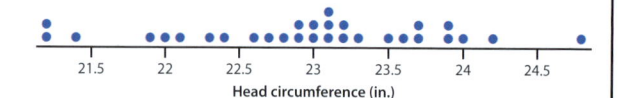

5. Click the Begin analysis button. A dotplot of the data should appear.

Note: You can use the *One Quantitative Variable, Multiple Groups* applet at www.stapplet.com to make parallel dotplots for comparing the distribution of a variable in two or more groups, like the one shown in Example 2.3.3.

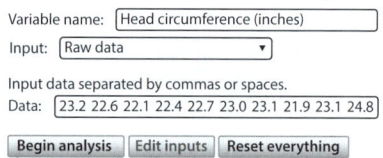

TI-83/84

The TI-83/84 does not produce dotplots.

Detailed instructions for using CrunchIt!, Excel, Google Sheets, JMP, Minitab, and R are available in Achieve.

Section 2.3 What Did You Learn?

Review the learning goals from this section. Then practice what you've learned by working through the exercises.

Learning Goal	Example	Exercises
Make and interpret dotplots of quantitative data.	2.3.1	7–10
Describe the shape of a distribution of quantitative data.	2.3.2	11–16
Describe or compare distributions of quantitative data with dotplots.	2.3.3	17–20

Section 2.3 Exercises

Building Concepts and Skills These exercises assess the basic knowledge you should have after reading the section.

1. A dotplot is the simplest graph for displaying _____ data.

2. What is the first step in making a dotplot?

3. True/False: The following dotplot is skewed to the right because most of the values are tightly clustered on the right side of the graph.

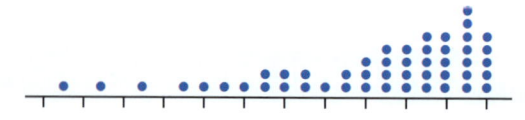

4. Is the distribution of Old Faithful eruption duration in Figure 2.4 roughly symmetric, skewed to the right, or skewed to the left?

5. When describing the distribution of a quantitative variable, look for the _____ and for _____ from that pattern.

6. What four aspects of a distribution should you discuss when describing or comparing graphs of quantitative data?

Mastering Concepts and Skills These exercises reinforce the learning goals as shown in the examples.

7. **Refer to Example 2.3.1 Fuel Efficiency** The EPA is in charge of determining and reporting fuel economy ratings for cars. Here are the EPA estimates of highway gas mileage in miles per gallon (mpg) for a random sample of 21 model-year 2020 midsize cars.[40]

25	30	27	31	38	26	28	40	25	28	30
31	30	30	34	30	31	31	32	48	31	

(a) Make a dotplot of these data.
(b) Explain what the dot at 48 (for the Toyota Prius) represents.
(c) What percentage of the car models in the sample get more than 35 mpg on the highway?

8. **Women's Soccer** How good was the 2019 U.S. women's soccer team? With players like Carli Lloyd, Alex Morgan, and Megan Rapinoe, the team put on an impressive showing en route to winning the 2019 Women's World Cup. Here are data on the number of goals scored by the team in games played in the 2019 season.[41]

1	1	2	2	1	5	6	3	5	3	13	3
2	2	2	2	2	3	4	3	2	1	3	6

(a) Make a dotplot of these data.
(b) Explain what the dot at 13 represents.
(c) In what percentage of games did the team score 4 or more goals?

9. **Sleep Statistics** A statistics professor asked how much sleep students got on the night prior to their first exam. Here are the responses (in hours) from the 50 students in the class.

9	8	6	7	7	8	4	7	7	8
8	8	6	7	8	8	7	7	6	8
9	7	6	5	7	8	8	7	9	6
6	6	8	9	5	8	7	7	7	7
2	4	8	3	6	5	5	8	7	3

(a) Make a dotplot of these data.
(b) What percentage of students in class reported getting less than 7 hours of sleep on the night prior to the exam?

10. **Frozen Pizza** Here are the number of calories per serving for 16 brands of frozen cheese pizza from *Consumer Reports* product reviews.[42]

340	340	310	320	310	360	350	330
260	380	340	320	310	360	350	330

(a) Make a dotplot of these data.
(b) What percentage of the frozen pizzas have fewer than 330 calories per serving?

11. **Refer to Example 2.3.2 Longevity and Dice**

(a) How old is the oldest person you know? Prudential Insurance Company asked 400 people to place a blue sticker on a huge wall next to the age of the oldest person they have ever known. An image of the graph is shown here. Describe the shape of the distribution.

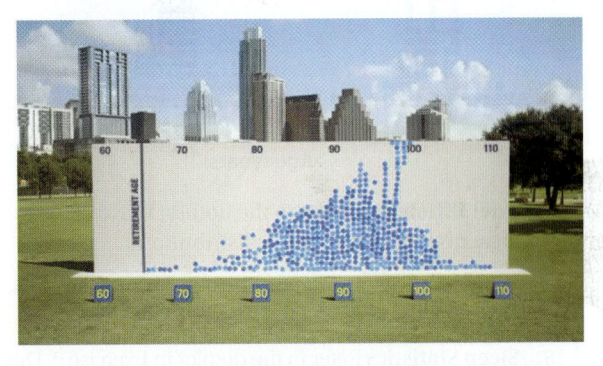

(b) The dotplot shows the results of rolling a pair of 6-sided dice and finding the sum of the up faces 100 times. Describe the shape of the distribution.

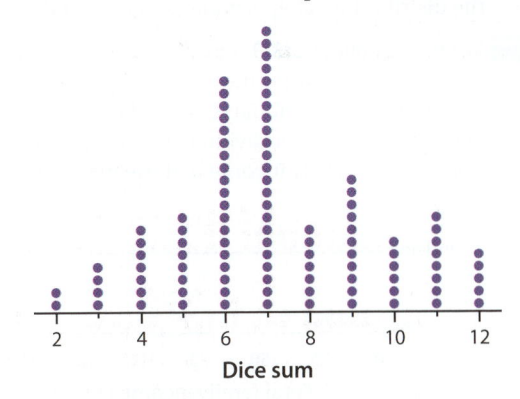

12. **Amusement Park Rides and Family Ages**

(a) Knoebels Amusement Park in Elysburg, Pennsylvania, has earned acclaim for being a lower-cost, family-friendly entertainment venue. Knoebels does not charge for general admission or parking, but it does charge customers for each ride they take. The dotplot shows the cost of each ride in a sample of 22 rides in a recent year. Describe the shape of the distribution.[43]

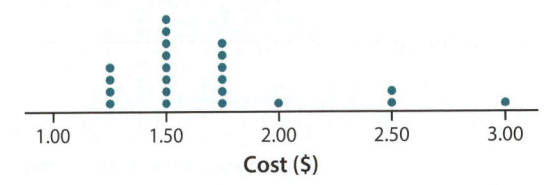

(b) Statistics instructor Paul Myers collected data on the ages (in years) of family members for each student in his class. The dotplot displays the data. Describe the shape of the distribution.

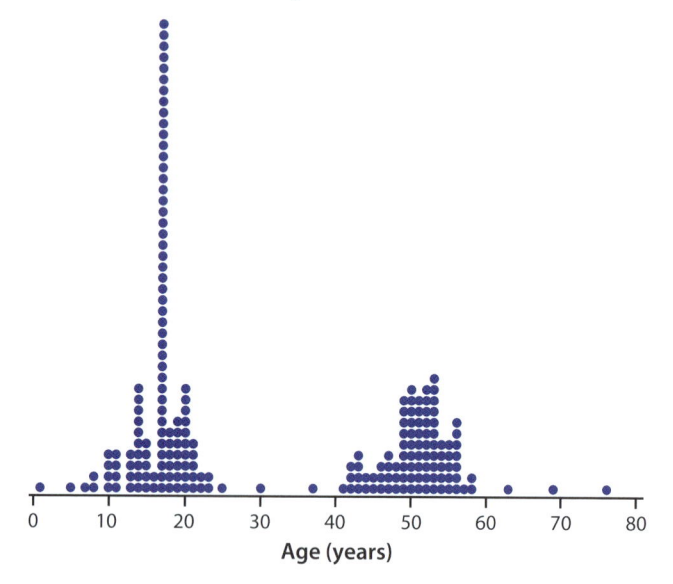

13. Fuel Efficiency Refer to the dotplot in Exercise 7. Describe the shape of the distribution.

14. Women's Soccer Refer to the dotplot in Exercise 8. Describe the shape of the distribution.

15. Sleep Statistics Refer to the dotplot in Exercise 9. Describe the distribution (shape, outliers, center, and variability).

16. Frozen Pizza Refer to the dotplot in Exercise 10. Describe the distribution (shape, outliers, center, and variability).

17. Refer to Example 2.3.3 Family Incomes The following parallel dotplots show the total family income of randomly chosen individuals from Indiana (38 individuals) and New Jersey (44 individuals). Compare the distributions of total family income in these two samples.

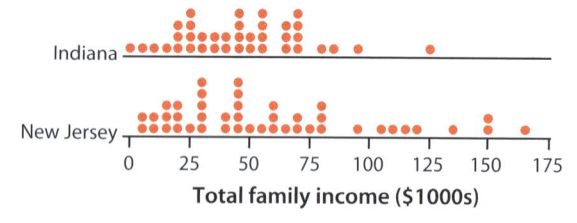

18. Stream Health Nitrates are organic compounds that are a main ingredient in fertilizers. When fertilizers run off into streams, the nitrates can have a toxic effect on fish. An ecologist studying nitrate pollution in two streams measures nitrate concentrations at 42 places in Stony Brook and 42 places in Mill Brook. The parallel dotplots display the data. Compare the distributions of nitrate concentration in the two streams.

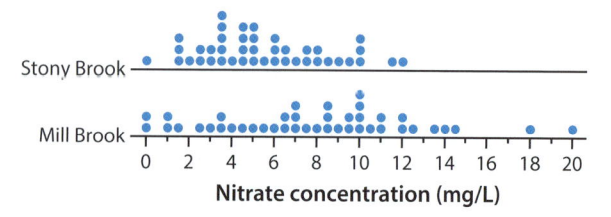

19. Sugary Cereals Researchers collected data on 76 brands of cereal at a local supermarket.[44] For each brand, they recorded the values of several variables, including sugar (grams per serving), calories per serving, and the shelf in the store on which the cereal was located (1 = bottom, 2 = middle, 3 = top). Here are parallel dotplots of the data on sugar content by shelf.

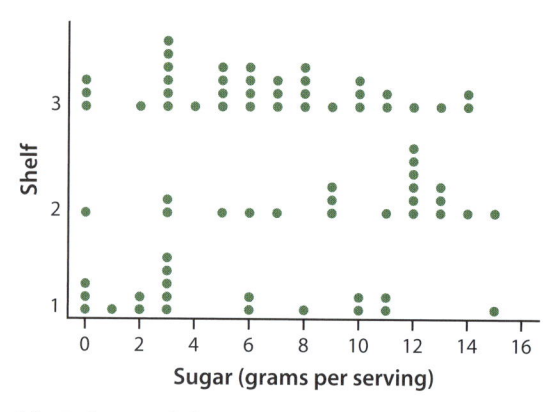

(a) Is the variability in sugar content of the cereals on the three shelves similar or different? Justify your answer.

(b) Critics claim that supermarkets tend to put sugary kids' cereals on lower shelves, where the kids can better see them. Do the data from this study support this claim? Justify your answer.

20. Creativity and Motivation Do external rewards—like money, praise, fame, and grades—promote creativity? Researcher Teresa Amabile recruited 47 experienced creative writers who were college students and divided them at random into two groups. The students in one group were given a list of statements about external reasons (E) for writing, such as public recognition, making money, or pleasing their parents. Students in the other group were given a list of statements about internal reasons (I) for writing, such as expressing yourself and enjoying playing with words. Both groups were then instructed to write a poem about laughter. Each student's poem was rated separately by 12 different poets using a creativity scale.[45] These ratings were averaged to obtain an overall creativity score for each poem. Parallel dotplots of the two groups' creativity scores are shown here.

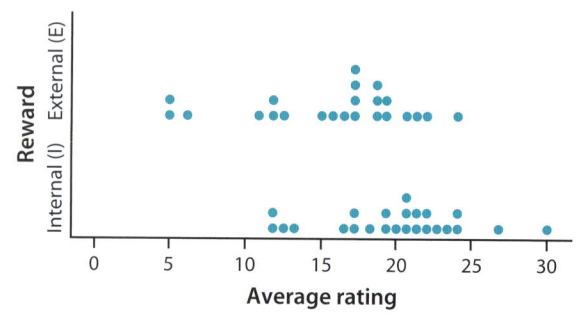

(a) Is the variability in creativity scores similar or different for the two groups? Justify your answer.

(b) What do you conclude about whether external rewards promote creativity? Justify your answer.

Applying the Concepts These exercises ask you to apply multiple learning goals in a new context or to apply what you learned in this section in a new way.

21. **Fuel Efficiency** The dotplot shows the difference in EPA mileage ratings (Highway − City) for each of the 21 model-year 2020 midsize cars from Exercise 7.

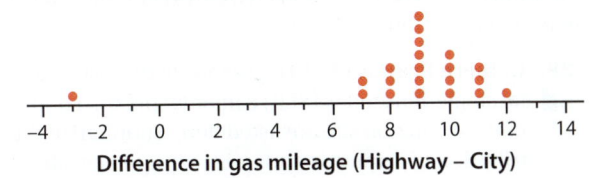

    (a) The dot above −3 is for the Toyota Prius. Explain what this dot represents.
    (b) What percentage of these car models get fuel efficiency of at least 10 mpg more on the highway than in the city?
    (c) Describe the distribution.

22. **Women's Soccer** The dotplot shows the difference in the number of goals scored in each game (U.S. women's team − Opponent) from Exercise 8.

    (a) Explain what the dot above −2 represents.
    (b) In what percentage of its games did the 2019 U.S. women's team score more goals than its opponent?
    (c) Describe the distribution.

23. **Interviewer Effect** For their final project, two statistics students designed an experiment to investigate the effect of the interviewer on people's responses to the question, "How many hours a week do you work out?" Interviewer 1 wore jeans and a school T-shirt, while Interviewer 2 wore athletic attire with the school logo. The two students were each randomly assigned three 15-minute time slots during a 90-minute period to serve as the interviewer. They asked the question of people who passed by their location near campus. Here are parallel dotplots of the data obtained by each interviewer.[46]

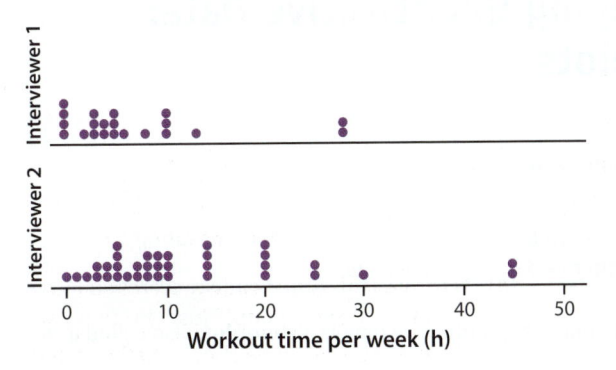

    (a) Explain what the dot above 30 in Interviewer 2's graph represents.
    (b) What percentage of the people surveyed by each interviewer reported working out for 20 or more hours per week?
    (c) Describe the shapes of the two dotplots. Does either distribution have any outliers?
    (d) How does the variability in the reported workout times compare for the two groups? Explain your answer.
    (e) Do these data provide some evidence that the interviewer's clothing affects responses about workout time for people like the participants in this study? Justify your answer.

24. **Calcium in Yogurt** Which is healthier: plain yogurt or flavored yogurt? The *Nutrition Action Healthletter* provided data on calories, saturated fat, protein, calcium, and total sugars for many popular brands of yogurt.[47] Here are parallel dotplots that compare the calcium content as a percentage of the recommended daily value (%DV) of 1000 milligrams for several brands of plain and flavored yogurt.

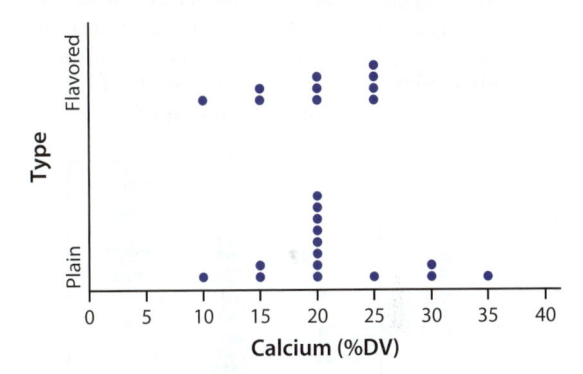

    (a) Explain what the dot above 10 in the flavored yogurt graph represents.
    (b) What percentage of the yogurts of each type have 15% or less of the recommended daily value of calcium?
    (c) Describe the shapes of the two dotplots. Does either distribution have any outliers?
    (d) How does the variability in calcium content compare for the two groups? Explain your answer.
    (e) Does one type of yogurt tend to have more calcium than the other? Justify your answer.

25. **Tech Alcohol and Calories** How many calories are in a 12-ounce serving of beer? How about a 6-ounce serving of wine? Does the number of calories depend on the type of beverage? The **Beer and Wine** data set contains data for 140 different brands of beer and wine.[48]

    (a) Use technology to make a dotplot of the calorie data (*Calories*) for all of the beverages. Describe the distribution (shape, outliers, center, and variability).
    (b) Use technology to make a comparative dotplot of the calories data for beer and wine. What similarities and differences do you see for the two distributions?

26. **Tech Possums** How big are Australian mountain brushtail possums? Does size depend on the sex of the possum? Zoologists in Australia captured a random sample of 104 mountain brushtail possums and recorded the values of 13 different variables.[49] The *Possum* data set gives the data on the *sex* of each possum (m or f) and its total length (in centimeters) in the column labeled *totlngth*.

    (a) Use technology to make a dotplot of the total length data for all of the possums. Describe the distribution (shape, outliers, center, and variability).

    (b) Use technology to make a comparative dotplot of the total length data for male and female possums. What similarities and differences do you see for the two distributions?

Extending the Concepts These exercises challenge you to explore statistical concepts and methods that go beyond what you learned in this section.

27. **Vitamin C and Teeth** Researchers performed an experiment with 60 guinea pigs to investigate the effect of vitamin C on tooth growth. Each animal was randomly assigned to receive one of three dose levels of vitamin C (0.5, 1, or 2 mg/day). The response variable was the length of odontoblast cells (responsible for tooth growth), in micrometers (μm). Here is a special type of dotplot that displays the data.[50]

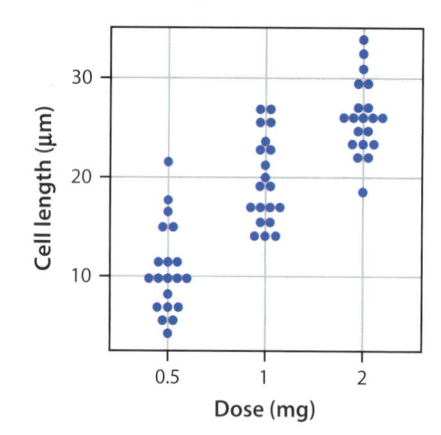

    (a) How does this dotplot differ from the ones shown in this section?

    (b) Do these data provide evidence that vitamin C helps teeth grow in guinea pigs similar to the ones used in this experiment? Justify your answer.

Cumulative Review These exercises revisit what you learned in previous sections.

28. **Camera Brands (2.1)** The brands of the last 45 digital single-lens reflex (SLR) cameras sold on a popular internet auction site are listed here. Summarize the distribution of camera brands with a frequency table and a relative frequency table.

Canon	Sony	Canon	Nikon	Fujifilm
Nikon	Canon	Sony	Canon	Canon
Nikon	Canon	Nikon	Canon	Canon
Canon	Nikon	Fujifilm	Canon	Nikon
Nikon	Canon	Canon	Canon	Canon
Olympus	Canon	Canon	Canon	Nikon
Olympus	Sony	Canon	Canon	Sony
Canon	Nikon	Sony	Canon	Fujifilm
Nikon	Canon	Nikon	Canon	Sony

29. **Marital Status (2.2)** The bar chart compares the marital status of U.S. adult residents (18 years old or older) in 1980 and 2020.[51] Compare the distributions of marital status for these two years.

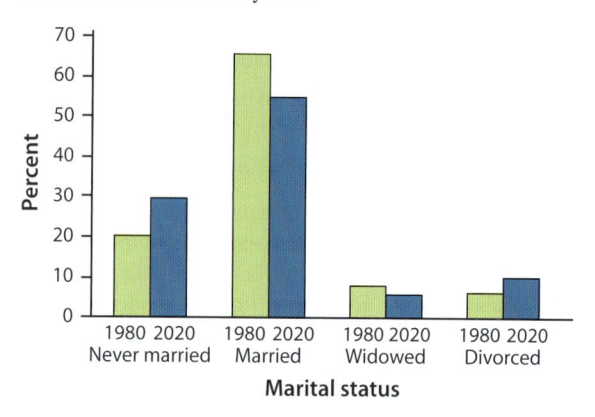

Section 2.4 Displaying Quantitative Data: Stemplots

LEARNING GOALS

By the end of this section, you will be able to:

- Make stemplots of quantitative data.
- Describe distributions of quantitative data with stemplots.
- Compare distributions of quantitative data with stemplots.

Another simple type of graph for displaying quantitative data is a **stemplot**, also called a *stem-and-leaf plot*.

Section 2.4 Displaying Quantitative Data: Stemplots

> A **stemplot** shows each data value separated into two parts: a *stem*, which consists of the leftmost digit(s), and a *leaf*, the final digit. The stems are ordered from least to greatest and arranged in a vertical column. The leaves are arranged in increasing order out from the stems.

Researchers measured the hemoglobin levels (in grams per deciliter) of 41 patients with sickle-cell disease. Here are the data.[52]

7.2	7.7	8.0	8.1	8.1	8.3	8.4	8.4	8.5	8.6	8.7
9.1	9.1	9.1	9.2	9.8	10.0	10.1	10.3	10.4	10.6	
10.7	10.9	11.1	11.3	11.5	11.6	11.7	11.8	11.9	12.0	
12.0	12.1	12.1	12.3	12.6	12.6	13.3	13.3	13.8	13.9	

Figure 2.5 shows a stemplot of the data on hemoglobin levels.

```
 7 | 27
 8 | 011344567
 9 | 11128
10 | 0134679
11 | 1356789
12 | 0011366
13 | 3389
```

KEY: 13 | 9 is a patient with sickle-cell disease and a hemoglobin level of 13.9 g/dL.

FIGURE 2.5 Stemplot showing the hemoglobin level (in grams per deciliter) of 41 patients with sickle-cell disease.

You'll learn how to make and interpret stemplots in this section.

Making Stemplots

Stemplots give us a quick picture of a distribution that includes the actual data values in the graph. You can make a stemplot by hand for small sets of quantitative data (see the How To box). For details on making stemplots with technology, see the Tech Corner at the end of the section.

HOW TO Make a Stemplot

1. **Make stems.** Separate each observation into a stem (all but the final digit of each data value) and a leaf (the final digit of each data value). Write the stems in a vertical column, with the smallest at the top. Draw a vertical line at the right of this column. Do not skip any stems, even if there is no data value for a particular stem.
2. **Add leaves.** Write each leaf in the row to the right of its stem.
3. **Order leaves.** Arrange the leaves in increasing order out from the stem.
4. **Add a key.** Provide a key that explains in context what the stems and leaves represent.

EXAMPLE 2.4.1

Votes in the Electoral College

Making a stemplot

PROBLEM
To become president of the United States, a candidate does not have to receive a majority of the popular vote. The candidate does, however, have to win a majority of the 538 electoral votes that are cast in the Electoral College. The table shows the number of electoral votes for each of the 50 states and the District of Columbia. Make a stemplot of these data.

(continued)

State	EV	State	EV	State	EV	State	EV
Alabama	9	Illinois	19	Montana	4	Rhode Island	4
Alaska	3	Indiana	11	Nebraska	5	South Carolina	9
Arizona	11	Iowa	6	Nevada	6	South Dakota	3
Arkansas	6	Kansas	6	New Hampshire	4	Tennessee	11
California	54	Kentucky	8	New Jersey	14	Texas	40
Colorado	10	Louisiana	8	New Mexico	5	Utah	6
Connecticut	7	Maine	4	New York	28	Vermont	3
Delaware	3	Maryland	10	North Carolina	16	Virginia	13
District of Columbia	3	Massachusetts	11	North Dakota	3	Washington	12
Florida	30	Michigan	15	Ohio	17	West Virginia	4
Georgia	16	Minnesota	10	Oklahoma	7	Wisconsin	10
Hawaii	4	Mississippi	6	Oregon	8	Wyoming	3
Idaho	4	Missouri	10	Pennsylvania	19		

SOLUTION

```
0 | 3333333344444445566666667788899
1 | 000001111234566799
2 | 8
3 | 0
4 | 0
5 | 4
```
KEY: 1|5 is a state with 15 electoral votes.

1. **Make stems.** The number of electoral votes varies from 3 to 54. Think of single-digit numbers like 3 as 03, and use the first digit of each value as the stem. We need stems from 0 to 5.
2. **Add leaves.** Place a 9 on the 0 stem for Alabama's 9 electoral votes, a 3 on the 0 stem for Alaska, a 1 on the 1 stem for Arizona, and so on.
3. **Order leaves.**
4. **Add a key.**

EXAM PREP: FOR PRACTICE, TRY EXERCISE 7.

We can get a better picture of the electoral vote data by *splitting stems*. In Figure 2.6(a), the leaves from 0 to 9 are placed on the "0" stem. Figure 2.6(b) shows another stemplot of the same data. This time, leaves 0 through 4 are placed on one stem, while leaves 5 through 9 are placed on another stem. Now we can see the shape of the distribution more clearly.

(a)
```
0 | 3333333344444445566666667788899
1 | 000001111234566799
2 | 8
3 | 0
4 | 0
5 | 4
```
KEY: 1|5 is a state with 15 electoral votes.

(b)
```
0 | 33333334444444
0 | 556666667788899
1 | 000001111234
1 | 566799
2 |
2 | 8
3 | 0
3 |
4 | 0
4 |
5 | 4
```
KEY: 1|5 is a state with 15 electoral votes.

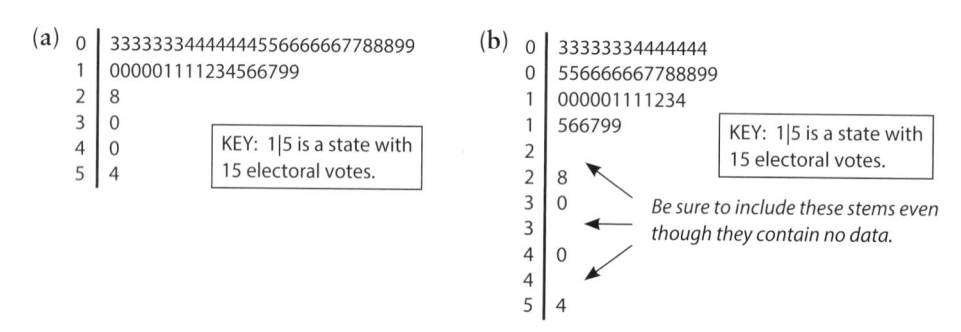

Be sure to include these stems even though they contain no data.

FIGURE 2.6 Two stemplots showing the electoral vote data. The graph in (b) improves on the graph in (a) by splitting the stems.

Here are a few tips to consider before making a stemplot:
- There is no magic number of stems to use. Too few or too many stems will make it difficult to see the distribution's shape. Five stems is a good minimum.
- If you split stems, make sure that each stem is assigned an equal number of possible leaf digits.
- When the data have too many digits, you can get more flexibility by rounding or truncating the data. See Exercise 26.

Describing Stemplots

Making a stemplot is just the beginning. We also need to describe what we see. The stemplot of hemoglobin levels from Figure 2.5 has an approximately uniform shape, with values that vary from 7.2 to 13.9 grams per deciliter. According to the Mayo Clinic, a normal hemoglobin level is between 12.0 and 17.5 g/dL. As you can see from the graph, 30 of the 41 patients with sickle-cell disease have lower than normal hemoglobin levels.

```
 7 | 27
 8 | 011344567
 9 | 11128
10 | 0134679
11 | 1356789
12 | 0011366
13 | 3389
```

KEY: 13 | 9 is a patient with sickle-cell disease and a hemoglobin level of 13.9 g/dL.

EXAMPLE 2.4.2

Votes in the Electoral College

Describing a stemplot

PROBLEM

Use the stemplot in Figure 2.6(b) and the data table from Example 2.4.1 to answer these questions.

(a) What percentage of states (plus the District of Columbia) get 10 or more electoral votes?

(b) Describe the distribution of electoral votes.

SOLUTION

(a) 22/51 = 0.431. About 43.1% of the 50 states plus the District of Columbia get 10 or more electoral votes.

(b) Shape: The distribution of electoral votes is skewed to the right with a single peak on the 05–09 stem and three clear gaps—from 19 to 28, from 30 to 40, and from 40 to 54.

Outliers: 40 (Texas) and 54 (California) appear to be outliers.

Center: The middle value is 8 electoral votes.

Variability: The number of electoral votes in a state (or DC) varies from 3 to 54.

Note that "skewed right" always means skewed toward the larger values, regardless of the orientation of the graph. Imagine rotating the graph 90 degrees counterclockwise to see the shape of the graph more clearly.

EXAM PREP: FOR PRACTICE, TRY EXERCISE 11.

Comparing Distributions with Stemplots

You can use a *back-to-back stemplot* with common stems to compare the distribution of a quantitative variable in two groups. The leaves on each side are placed in order leading out from the common stem.

EXAMPLE 2.4.3

Coastal oaks and acorn size

Comparing distributions with stemplots

PROBLEM

Of the many species of oak trees in the United States, 28 grow on the Atlantic Coast and 11 grow in California. How does the distribution of acorn size compare for oak trees in these two regions? Here are data on the average volumes of acorns (in cubic centimeters) for these 39 oak species:[53]

(continued)

Atlantic Coast	0.3	0.4	0.6	0.8	0.9	0.9	1.1	1.1	1.1	1.1	1.1	1.2	1.4	1.6
	1.8	1.8	1.8	2.0	2.5	3.0	3.4	3.6	3.6	4.8	6.8	8.1	9.1	10.5
California	0.4	1.0	1.6	2.0	2.6	4.1	5.5	5.9	6.0	7.1	17.1			

(a) Make a back-to-back stemplot for these data.

(b) Which of the two regions' oak tree species typically has the larger acorns?

(c) Are the shapes of the acorn size distributions similar or different in the two regions? Justify your answer.

SOLUTION

(a)

```
Atlantic Coast |    | California
        998643 |  0 | 4
   88864211111 |  1 | 06
            50 |  2 | 06
          6640 |  3 |
             8 |  4 | 1
               |  5 | 59
             8 |  6 | 0
               |  7 | 1
             1 |  8 |
             1 |  9 |
             5 | 10 |
               | 11 |
               | 12 |
               | 13 |
               | 14 |
               | 15 |
               | 16 |
               | 17 | 1
```

KEY: 2 | 6 = An oak species whose acorn volume is 2.6 cm³.

1. **Make stems.** The acorn volumes vary from 0.3 cubic centimeter (cm³) to 17.1 cm³. Use the whole-number part of each value as the stem and the decimal part as the leaf. So we need stems from 0 to 17.
2. **Add leaves.**
3. **Order leaves.** For each species, make sure the leaves increase in order going out from the stem.
4. **Add a key.**

(b) California oak tree species tend to have larger acorn volumes (middle value = 4.1 cm³) than Atlantic Coast oak tree species (middle value = 1.7 cm³).

(c) The shapes of the distributions of acorn size in the two regions are quite different. For Atlantic Coast oak tree species, the distribution is skewed to the right and single-peaked. For California oak tree species, the distribution has two distinct clusters (0.4 to 2.6 cm³ and 4.1 to 7.1 cm³), a large gap from 7.1 cm³ to 17.1 cm³, and no clear peak.

Remember: "Skewed right" means that the long tail of the graph is in the direction of the larger values.

EXAM PREP: FOR PRACTICE, TRY EXERCISE 17.

TECH CORNER Making a Stemplot

You can use technology to make a stemplot. We'll illustrate using the electoral vote data from Example 2.4.1.

Applet

1. Go to www.stapplet.com and launch the *One Quantitative Variable, Single Group* applet.
2. Enter Electoral votes as the Variable name.
3. Select Raw data as the input method.
4. Enter the data manually, or copy and paste them from a document or spreadsheet. Be sure to separate the data values with commas or spaces.
5. Click the Begin analysis button.
6. Change the Graph type to Stemplot.
7. Split the stems in two to get a better picture of the distribution.

Note: You can use the *One Quantitative Variable, Multiple Groups* applet at www.stapplet.com to make a back-to-back stemplot like the one shown in Example 2.4.3.

Graph Distribution

Graph type: Stemplot
Split stems: Yes, into two groups
Shift stem: 0 additional decimal places to the left, truncating as needed.
Collapse groups of empty stems? No

```
0 | 33333334444444
0 | 556666667788899
1 | 000001111234
1 | 566799
2 |
2 | 8
3 | 0
3 |
4 | 0
4 |
5 | 4
```

Electoral votes
KEY: 5 | 4 = 54

TI-83/84

The TI-83/84 does not produce stemplots.

Detailed instructions for using CrunchIt!, Excel, Google Sheets, JMP, Minitab, and R are available in Achieve.

Section 2.4 What Did You Learn?

Review the learning goals from this section. Then practice what you've learned by working through the exercises.

Learning Goal	Example	Exercises
Make stemplots of quantitative data.	2.4.1	7–10
Describe distributions of quantitative data with stemplots.	2.4.2	11–16
Compare distributions of quantitative data with stemplots.	2.4.3	17–20

Section 2.4 Exercises

Building Concepts and Skills These exercises assess the basic knowledge you should have after reading the section.

1. A stemplot is also called a _____ plot.

2. True/False: If there is no data value for a particular stem, you should omit that stem from the stemplot.

3. If there are multiple leaves on a given stem in a stemplot, how should the leaves be arranged?

4. What should be included in the key for a stemplot?

5. True/False: When splitting stems, you should make sure that an equal number of possible leaf digits are assigned to each stem.

6. In a back-to-back stemplot, how should the leaves on the left side of the stem be ordered?

Mastering Concepts and Skills These exercises reinforce the learning goals as shown in the examples.

7. **Refer to Example 2.4.1 Kids' Heart Rates** Here are the resting heart rates (in beats per minute) of 19 middle school students. Make a stemplot of these data.

   71 104 76 88 78 71 68 86 70 90
   74 76 69 68 88 96 68 82 120

8. **Beans and Protein** Beans and other legumes are a great source of protein. The following data give the protein content of 31 different varieties of beans, in grams per 100 grams of cooked beans.[54] Make a stemplot of these data.

   7.5   8.2   8.9   9.3   7.1   8.3   8.7   9.5   8.2   9.1   9.0   9.0
   9.7   9.2   8.9   8.1   9.0   7.8   8.0   7.8   7.0   7.5   13.5 8.3
   6.8   16.6 10.6 8.3   7.6   7.7   8.1

9. **Arizona Heat** Here are the high temperature readings in degrees Fahrenheit for Phoenix, Arizona, for each day in July in a recent year. Make a stemplot of these data using split stems.[55]

   111 107 115 108 106 109 111 113 104 103 97
   99   104 110 109 100 105 107 102 101   84   93
   101 105 99   102 104 108 106 106 109

10. **Salmon Health** As part of a study on salmon health, researchers measured the pH of 25 salmon fillets.[56] Here are the data. Make a stemplot of these data using split stems.

    6.34 6.39 6.53 6.36 6.39 6.25 6.45 6.38 6.33 6.26 6.24 6.37 6.32
    6.31 6.48 6.26 6.42 6.43 6.36 6.44 6.22 6.52 6.32 6.32 6.48

11. **Refer to Example 2.4.2 Age Distribution in States** Here is a stemplot of the percentage of residents aged 65 and older in the 50 states and the District of Columbia.[57]

    ```
    11 | 18
    12 | 16
    13 | 9
    14 | 23
    15 | 34444677789999
    16 | 1344455699
    17 | 00112225567
    18 | 12477
    19 | 49
    20 | 56
    ```
    KEY: 13 | 9 = 13.9% of the state's residents are 65 or older.

    (a) In what percentage of states plus the District of Columbia are more than 18% of residents aged 65 and older?
    (b) Describe the distribution.

12. **South Carolina Counties** Here is a stemplot of the areas of the 46 counties in South Carolina. Note that the data have been rounded to the nearest 10 square miles (mi^2).

    ```
     3 | 9999
     4 | 0116689
     5 | 01115566778
     6 | 47899
     7 | 01245579
     8 | 0011
     9 | 13
    10 | 8
    11 | 233
    12 | 2
    ```
    KEY: 6 | 4 represents a county with an area of 640 square miles (rounded to the nearest 10 mi^2).

    (a) What percentage of South Carolina counties have areas of less than 495 mi^2?
    (b) Describe the distribution.

13. **Kids' Heart Rates** Refer to the stemplot in Exercise 7.
    (a) According to the American Heart Association, a resting pulse rate greater than 100 beats per minute is considered high for this age group. What percentage of these middle school students have high resting heart rates by this criterion?
    (b) Describe the shape of the distribution.
    (c) Are there any outliers?

14. **Beans and Protein** Refer to the stemplot in Exercise 8.
    (a) Bean varieties with at least 9 grams of protein per 100 grams of cooked beans are classified as "high protein." What percentage of these bean varieties are "high protein"?
    (b) Describe the shape of the distribution.
    (c) Are there any outliers?

15. **Arizona Heat** Refer to the stemplot in Exercise 9. Describe the distribution.

16. **Salmon Health** Refer to the stemplot in Exercise 10. Describe the distribution.

17. **Refer to Example 2.4.3 Long Jumps** Will children jump farther if they have a target in front of them? To find out, a gym teacher conducted an experiment with 29 twelve-year-old children. The teacher wrote the names of all 29 children on identical slips of paper, put them in a hat, mixed them up, and drew out 15 slips. Each of the 15 selected students did a standing long jump from behind a starting line. The remaining 14 students did a similar standing long jump from behind the starting line, but with a target line placed 200 centimeters from the starting line. The distance from the starting line to the back of each student's closest foot was measured (in centimeters). Here are the data.[58]

No target	146	190	109	181	155	167	154	171
	157	156	128	157	167	162	137	
Target	199	167	147	180	185	170	171	139
	154	126	179	158	181	152		

(a) Make a back-to-back stemplot to compare the jump distances for the two groups.
(b) Are the shapes of the two distributions similar or different? Explain your answer.
(c) Do these data provide some evidence that children like the ones in this experiment will jump farther if they have a target in front of them? Justify your answer.

18. **Basketball Scores** Here are the number of points scored by the winning teams in the first round of the National Collegiate Athletic Association (NCAA) Men's and Women's Basketball tournaments in a recent year.

Men	95	70	79	99	83	72	79	101
	69	82	86	70	79	69	69	70
	95	70	77	61	69	68	69	72
	89	66	84	77	50	83	63	58
Women	80	68	51	80	83	75	77	100
	96	68	89	80	67	84	76	70
	98	81	79	89	98	83	72	100
	101	83	66	76	77	84	71	77

(a) Make a back-to-back stemplot to compare the points scored in men's and women's first-round games.
(b) Are the shapes of the two distributions similar or different? Explain your answer.
(c) Did men's or women's winning teams tend to score more points in their first-round games? Justify your answer.

19. **Finch Evolution** Biologists Peter and Rosemary Grant spent many years collecting data on finches in a remote part of the Galápagos Islands. Their research team caught and measured all of the birds in more than 20 generations of finches. The back-to-back stemplot shows the beak depths (in millimeters) of 89 finches captured the year before a drought occurred and 89 finches captured the year after the drought occurred.[59] Compare the distributions of beak depth in these two years. (*Note:* These data provide evidence that the finches evolved in response to the drought.)

```
   Before drought       After drought
                 2 | 6 |
                 8 | 6 |              KEY: 10 | 0 = 10.0 mm
               411 | 7 | 1             beak depth
                98 | 7 | 9
             44420 | 8 | 044
       9999977765555 | 8 | 778
       444321111100000 | 9 | 0011123344
    99999888887777655 | 9 | 5666666777789999
   4444333221111111000 | 10 | 0002222223333334444
         8766655555 | 10 | 5555566666777778999
                440 | 11 | 0000111134444
                  7 | 11 | 5667
```

20. **Camera Batteries** Researchers decided to investigate which of two popular brands of camera batteries lasts longer. They took separate random samples of 40 camera batteries of each brand, and recorded how long each battery lasted (in hours of use). The back-to-back stemplot displays the data. Compare the two distributions.

```
    Brand X        Brand Y
     86200 | 6 | 9
   99988640 | 7 | 35              KEY: 6 | 9 = Battery
  887531110 | 8 | 248              life of 69 hours
  966422200 | 9 | 014559
    8753111 | 10 | 00258899
         42 | 11 | 001133679
            | 12 | 14666788
            | 13 | 35
            | 14 | 6
```

Applying the Concepts These exercises ask you to apply multiple learning goals in a new context or to apply what you learned in this section in a new way.

21. **Seat Belts** Each year, the National Highway Traffic Safety Administration (NHTSA) conducts an observational study of seat belt use in all 50 states. Trained observers station themselves at randomly selected locations along roadways in each state, and then record data on seat belt use by people in passing vehicles. Here is a stemplot of the percentage of people who were wearing seat belts in each state during a recent year.[60]

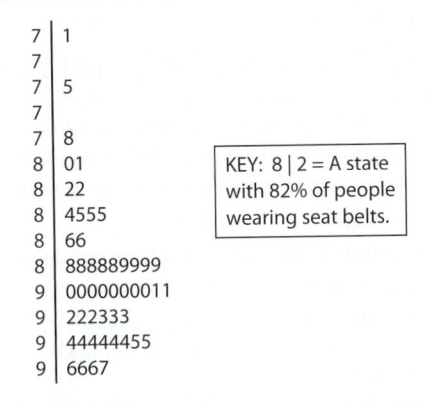

```
7 | 1
7 |
7 | 5
7 |
7 | 8
8 | 01
8 | 22
8 | 4555
8 | 66
8 | 888889999
9 | 0000000011
9 | 222333
9 | 44444455
9 | 6667
```
KEY: 8 | 2 = A state with 82% of people wearing seat belts.

(a) Why did we split stems when making this stemplot?
(b) In how many states was at least 95% seat belt use observed?
(c) Describe the shape of the distribution. Are there any outliers?

Some states enforce a seat belt violation as a primary offense, while other states enforce a seat belt violation only as a secondary offense. The back-to-back stemplot shows the percentage of people who were wearing seat belts in primary enforcement versus secondary enforcement states during the NHTSA study.

```
Primary enforcement       Secondary enforcement
                    7 | 1
                    7 |
                    7 | 5
                    7 |
                    7 | 8
                1 | 8 | 0
                2 | 8 | 2
               55 | 8 | 45
                    8 | 66
             9888 | 8 | 88999
        100000000 | 9 | 1
           333222 | 9 |
          5544444 | 9 | 4
             7666 | 9 |
```
KEY: 8 | 2 = A state with 82% of people wearing seat belts.

(d) Compare the distributions of seat belt use in primary and secondary enforcement states.

22. **Traffic Fatalities** Each year, the National Highway Traffic Safety Administration (NHTSA) collects data on traffic deaths in all 50 states. Here is a stemplot of the traffic fatality rate (deaths per 100,000 population) in each state during a recent year.[61]

```
0 | 55
0 | 6677
0 | 88888999
1 | 0000011111
1 | 222333
1 | 444445555
1 | 667777
1 | 99
2 | 00
2 | 2
```
KEY: 1 | 9 = A state with 1.9 traffic deaths per 100,000 people.

(a) Why did we split stems when making this stemplot?
(b) In how many states was the traffic fatality rate less than 1 death per 100,000 people?
(c) Describe the shape of the distribution. Are there any outliers?

Some states enforce a seat belt violation as a primary offense, while other states enforce a seat belt violation only as a secondary offense. The back-to-back stemplot shows the traffic fatality rate in primary enforcement versus secondary enforcement states during the NHTSA study.

```
Primary enforcement      Secondary enforcement
               55 | 0 |
               76 | 0 | 67
            98888 | 0 | 899
          1111000 | 1 | 001
             3322 | 1 | 23
          5544444 | 1 | 55
             7776 | 1 | 67
                9 | 1 | 9
                0 | 2 | 0
                2 | 2 |
```
KEY: 1 | 9 = A state with 1.9 traffic deaths per 100,000 population.

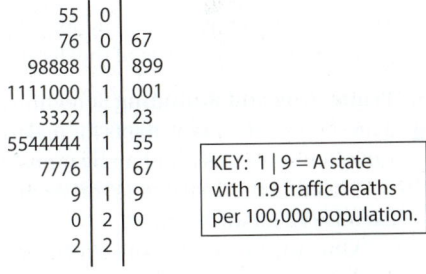

(d) Compare the distributions of traffic fatality rates in primary and secondary enforcement states.

23. **Tech Alcohol Content** What is the alcohol content (measured as percent alcohol) in a 12-ounce serving of beer? How about a 6-ounce serving of wine? Does alcohol content depend on the type of beverage? The *Beer and Wine* data set contains data for 140 different brands of beer and wine. Note that there are some missing data values.[62]

(a) Use technology to make a stemplot of the alcohol content data (*Pct_Alcohol*) for all 110 beverages with a recorded value. Describe the distribution (shape, outliers, center, and variability).
(b) Use technology to make a back-to-back stemplot comparing the alcohol content for beer and wine. What similarities and differences do you see for the two distributions?

24. **Tech Possums** How wide are the skulls of mountain brushtail possums? Does skull width depend on the sex of the possum? Zoologists in Australia captured a random sample of 104 mountain brushtail possums and recorded the values of 13 different variables.[63] The *Possum* data set gives data on the *sex* of each possum (m or f) and its skull width (in millimeters) in the column labeled *skullw*.

(a) Use technology to make a stemplot of the skull width data for all of the possums. Describe the distribution (shape, outliers, center, and variability).

(b) Use technology to make a comparative stemplot of the skull width data for male and female possums. What similarities and differences do you see for the two distributions?

Extending the Concepts These exercises challenge you to explore statistical concepts and methods that go beyond what you learned in this section.

25. **Five-Way Stem Splitting** Sometimes, the variability in a data set is so small that splitting stems in two doesn't produce a stemplot that shows the shape of the distribution well. We can often solve this problem by splitting the stem into five parts, each consisting of two leaf values: 0 and 1, 2 and 3, 4 and 5, and so on. See Exercises 21 and 22 for an example.

    Here are the weights, in ounces, of 36 navel oranges selected from a large shipment to a grocery store.

    5.7 5.4 5.8 5.3 4.6 4.9 5.6 5.3 5.5 5.5 5.4 5.8
    5.3 5.5 5.5 5.4 5.8 5.9 5.4 5.1 5.0 5.5 5.7 4.9
    5.0 5.3 5.1 5.2 5.7 5.6 5.8 4.5 5.2 5.4 5.7 5.6

    Make a stemplot of the data by splitting stems into five parts. Describe the shape of the distribution.

26. **Truncating and Rounding** Sometimes, the values in a data set have too many digits for making a stemplot. We can often solve this problem by *rounding* or by *truncating* [deleting the final digit(s) of] each data value. See Exercise 12 for an example.

    One way to measure the effectiveness of baseball pitchers is to use their *earned run average,* which measures how many earned runs opposing teams score, on average, every nine innings pitched. The overall earned run average for all pitchers in the major leagues in 2019 was 4.49. Here are the earned run averages for all 20 players who pitched at least 15 innings for the World Champion Washington Nationals in 2019.[64]

    3.32 3.25 2.92 3.85 4.50 4.05 4.54 4.86 3.91 6.36
    5.48 3.30 6.23 4.05 6.66 1.44 4.71 5.14 10.13 4.02

    (a) Truncate (delete) the hundredths place of each data value and make a stemplot, using the ones digit as the stem and the tenths digit as the leaf.

    (b) Make another stemplot of the data by rounding each data value to the nearest tenth.

    (c) How do the shapes of the graphs in (a) and (b) compare?

Cumulative Review These exercises revisit what you learned in previous sections.

27. **Healthy Tomatoes (1.5, 1.6)** Agricultural scientists for a chemical company want to determine if a newly developed fertilizer produces heavier tomatoes than the fertilizer the company currently manufactures. They have 24 healthy young tomato plants growing in individual pots.

    (a) Outline a completely randomized design for this experiment.

    (b) Describe how you would carry out the random assignment for your design in part (a).

28. **Comparing Tuition (2.3)** The dotplot shows the 2019 tuition for the 63 largest colleges and universities in North Carolina.[65] Describe the overall pattern of the distribution and identify any clear departures from the pattern.

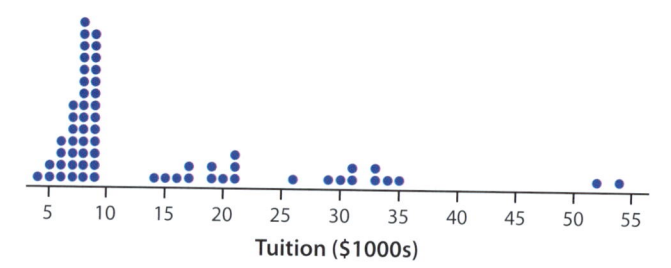

Section 2.5 Displaying Quantitative Data: Histograms

LEARNING GOALS

By the end of this section, you will be able to:
- Make histograms of quantitative data.
- Describe distributions of quantitative data with histograms.
- Compare distributions of quantitative data with histograms.

You can use a dotplot or stemplot to display quantitative data. Both types of graphs show every individual data value. However, for large data sets, this can make it difficult to see the overall pattern in the graph. We often get a cleaner picture of the distribution by grouping nearby values together. Doing so allows us to make a new type of graph: a **histogram.**

A **histogram** shows each interval as a bar. The heights of the bars show the frequencies or relative frequencies of values in each interval.

Figure 2.7 shows a dotplot and a histogram of the duration (in minutes) of 263 eruptions of the Old Faithful geyser in July 1995, when the Starnes family made its first visit to Yellowstone National Park. Notice how the histogram groups nearby values together.

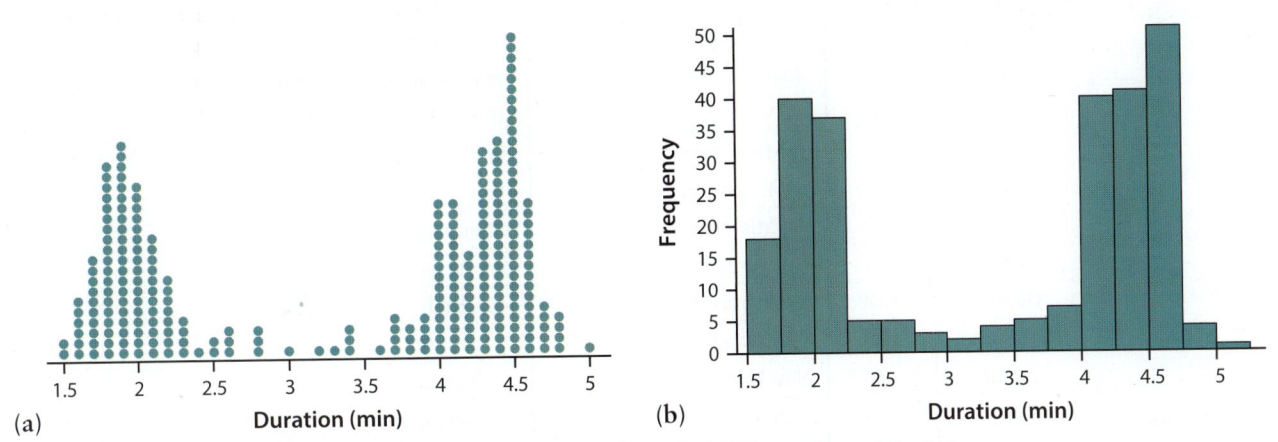

FIGURE 2.7 (a) Dotplot and (b) histogram of the duration (in minutes) of 263 eruptions of the Old Faithful geyser in July 1995.

Making Histograms

You can make a histogram by hand, even for fairly large sets of quantitative data (see the How To box). For details on making histograms with technology, see the Tech Corner at the end of the section.

HOW TO Make a Histogram

1. **Choose equal-width intervals** that span the data. Five intervals is a good minimum.
2. **Make a table** that shows the frequency (count) or relative frequency (percentage or proportion) of data values in each interval.
3. **Draw and label the axes.** Put the name of the quantitative variable under the horizontal axis. To the left of the vertical axis, indicate whether the graph shows the frequency (count) or relative frequency (percentage or proportion) of data values in each interval.
4. **Scale the axes.** Place equally spaced tick marks at the smallest value in each interval along the horizontal axis. On the vertical axis, start at 0 and place equally spaced tick marks until you equal or exceed the largest frequency or relative frequency in any interval.
5. **Draw bars** above the intervals. Make the bars equal in width and leave no gaps between them. Make sure that the height of each bar corresponds to the frequency or relative frequency of data values in that interval. An interval with no data values will appear as a bar of height 0 on the graph.

You can choose intervals of unequal widths when making a histogram, but such graphs are beyond the scope of this book.

EXAMPLE 2.5.1

How many U.S. residents are foreign-born?

Making a histogram

PROBLEM

How does the percentage of foreign-born residents in your state compare to the rest of the country? The table presents the data for all 50 states in a recent year.[66] Make a frequency histogram to display the data.

(continued)

State	Percent	State	Percent	State	Percent
Alabama	3.6	Louisiana	4.2	Ohio	4.8
Alaska	8.0	Maine	3.9	Oklahoma	6.1
Arizona	13.4	Maryland	15.4	Oregon	9.7
Arkansas	5.1	Massachusetts	17.3	Pennsylvania	7.0
California	26.7	Michigan	7.0	Rhode Island	13.7
Colorado	9.5	Minnesota	8.4	South Carolina	5.6
Connecticut	14.8	Mississippi	2.1	South Dakota	4.1
Delaware	10.0	Missouri	4.3	Tennessee	5.5
Florida	21.1	Montana	2.3	Texas	17.1
Georgia	10.3	Nebraska	7.4	Utah	8.6
Hawaii	19.3	Nevada	19.8	Vermont	4.7
Idaho	5.8	New Hampshire	6.4	Virginia	12.7
Illinois	13.9	New Jersey	23.4	Washington	14.9
Indiana	5.3	New Mexico	9.6	West Virginia	1.6
Iowa	5.6	New York	22.4	Wisconsin	5.1
Kansas	7.2	North Carolina	8.4	Wyoming	3.1
Kentucky	4.4	North Dakota	4.1		

SOLUTION

Interval	Frequency
0 to <5	13
5 to <10	20
10 to <15	8
15 to <20	5
20 to <25	3
25 to <30	1

1. **Choose equal-width intervals** that span the data. The data vary from 1.6% to 26.7%. We chose intervals of width 5, beginning at 0: 0 to <5, 5 to <10, and so on. This choice results in more than the minimum of five intervals.
2. **Make a table.** This type of table is sometimes called a *grouped frequency table*.
3. **Draw and label the axes.**
4. **Scale the axes.** The scale on the horizontal axis matches the intervals we chose in Step 1. The highest frequency in an interval is 20, so we scale the vertical axis from 0 to 20, placing tick marks every 5 units.
5. **Draw bars.**

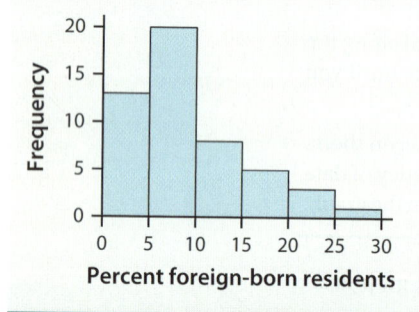

EXAM PREP: FOR PRACTICE, TRY EXERCISE 7.

The convention in this book is to include the left endpoint of an interval and exclude the right endpoint when making histograms. For instance, Delaware's value of 10.0% foreign-born residents was placed in the 10 to <15 bar, not the 5 to <10 bar, in Example 2.5.1.

THINK ABOUT IT

What are we actually doing when we make a histogram? The dotplot in part (a) shows the foreign-born resident data. We grouped the data values into intervals of width 5, beginning with 0 to <5, as indicated by the dashed lines. Then we counted the number of values in each interval. The dotplot in part (b) shows the results of that process. Finally, we drew bars of the appropriate height for each interval to complete the histogram.

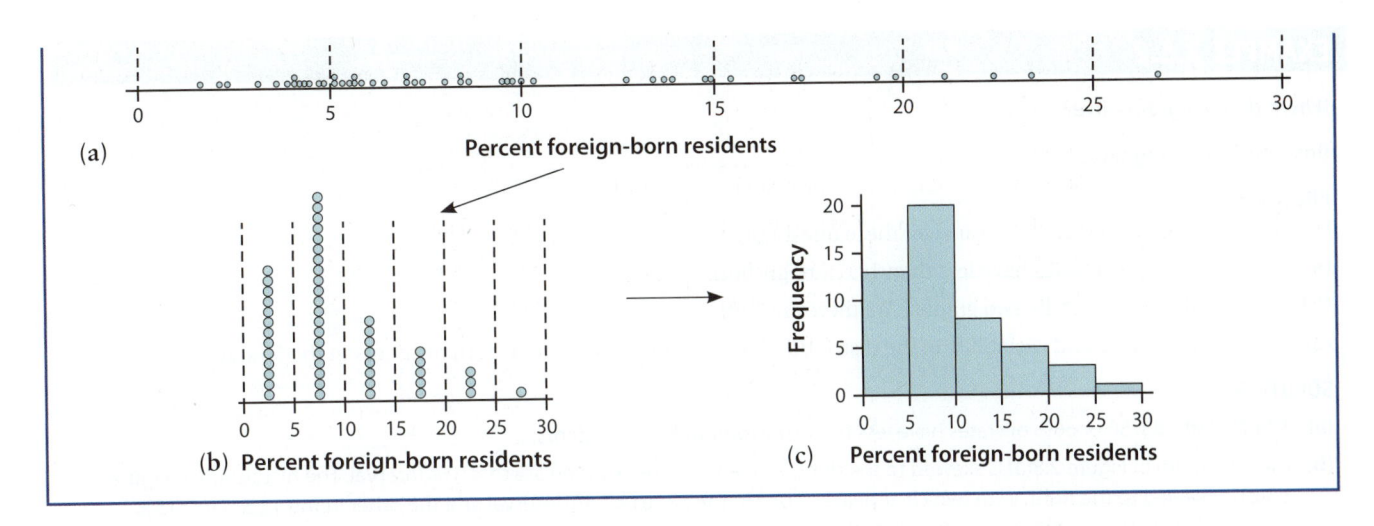

Figure 2.8 shows two different histograms of the foreign-born resident data. Graph (a) uses the intervals of width 5 from the example. Graph (b) uses intervals half as wide: 0 to <2.5, 2.5 to <5, and so on. ❗ **The choice of intervals in a histogram can affect the appearance of a distribution.** Histograms with more intervals show more detail but may have a less clear overall pattern.

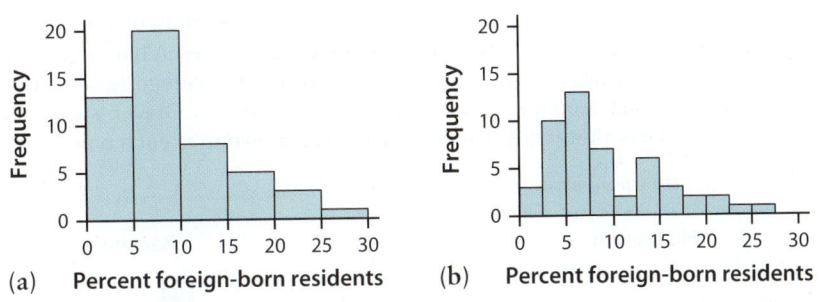

FIGURE 2.8 (a) Frequency histogram of the percentage of foreign-born residents in the 50 states with intervals of width 5, from the previous example. (b) Frequency histogram of the data with intervals of width 2.5.

Describing Histograms

After making a histogram, it's time to describe what we see. The histogram from Figure 2.7 shows that the distribution of Old Faithful eruption duration has two distinct clusters: eruptions lasting about 1.5 to 2.5 minutes and eruptions lasting about 3.5 to 5 minutes. Awareness of these "short" and "long" eruptions is key to predicting when Old Faithful will erupt next.

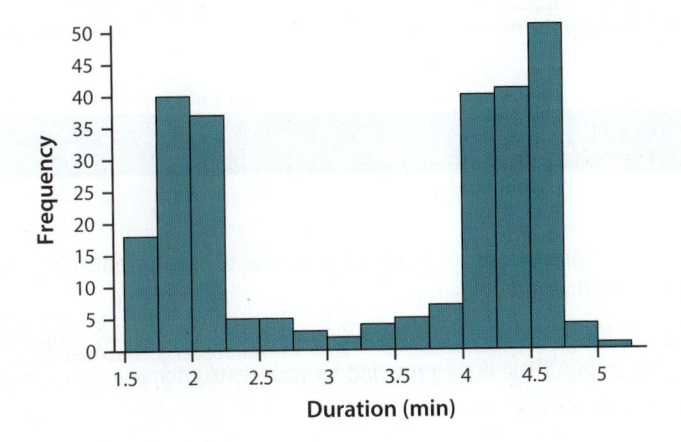

EXAMPLE 2.5.2

Where do immigrants live?

Describing a histogram

PROBLEM

Use the histograms in Figure 2.8 to answer these questions.
(a) What percentage of states have less than 10% foreign-born residents?
(b) Describe the shapes of the two graphs. Are there any outliers?
(c) Estimate the center and variability of the distribution of percentage of foreign-born residents in the 50 states.

SOLUTION

(a) $(13+20)/50 = 33/50$ or 66% of states have less than 10% foreign-born residents.
(b) The histogram in Figure 2.8(a) is skewed to the right, with a single peak in the 5% to <10% interval. The histogram in Figure 2.8(b) is skewed to the right, with two clear peaks—one in the 5% to <7.5% interval and the other in the 12.5% to <15% interval. Neither graph shows any obvious outliers.
(c) From the histogram in Figure 2.8(b), the center of the distribution is in the 5% to <7.5% interval, and the data vary from at least 0% to at most 27.4%.

EXAM PREP: FOR PRACTICE, TRY EXERCISE 11.

⚠ **Don't confuse histograms and bar charts.** Although the two types of graphs look similar, their details and uses are different. A histogram displays the distribution of a *quantitative variable*. Its horizontal axis identifies intervals of values that the variable takes. A bar graph displays the distribution of a *categorical variable*. Its horizontal axis identifies the categories. Be sure to draw bar graphs with blank space between the bars to separate the categories. Draw histograms with no space between bars for adjacent intervals. For comparison, here is one of each type of graph from earlier examples:

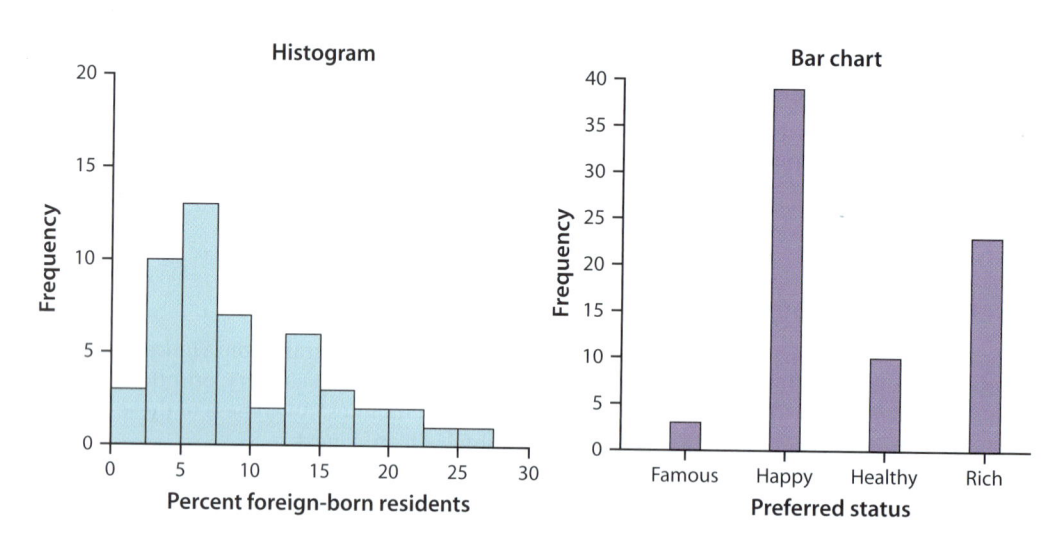

Class Activity

How Long Is Your Commute?

How long did it take you to get to statistics class today? Was your travel time similar to or different from that of the other students in your class? Let's collect some data to find out.

1. Go to www.stapplet.com and launch the *collaborative* version of the *One Quantitative Variable, Single Group* applet.
2. Click on "Enter an Existing Class Code," then input the code provided by your instructor.

3. Consider your answer to this question: How long (in minutes) did it take you to travel to statistics class today (from wherever you were previously)?
4. Type your data value into the box next to "Enter a number into the plot:" and click "Add to Plot."
5. The applet will display a dot for each student's data value on a class dotplot. Your instructor will use the applet to make a stemplot and a histogram of the distribution of travel time. Which type of graph displays the distribution best?
6. Using the graph that was chosen in Step 5, describe the distribution of travel time to class.
7. Suggest a categorical variable that might help explain the variability in travel times for members of the class.
8. Where does your travel time fall in the distribution?

Your instructor will download these data for further analysis in Chapter 3.

Comparing Distributions with Histograms

Histograms can also be used to compare the distribution of a quantitative variable in two or more groups. It's a good idea to use relative frequencies (percentages or proportions) when comparing distributions, especially if the groups have different sizes. Be sure to use the same intervals when making comparative histograms, so the graphs can be drawn using a common horizontal axis scale.

EXAMPLE 2.5.3

Are future earnings related to education?

Comparing distributions with histograms

PROBLEM

Is it true that students who earn an associate's degree or a bachelor's degree make more money than students who attend college but do not earn a degree? To find out, we took a random sample of 500 U.S. residents aged 18 and older who had attended college from a recent Current Population Survey.[67] The educational attainment and annual income of each person were recorded. Here are relative frequency histograms of the income data for the 327 college graduates and the 173 nongraduates. Compare the distributions.

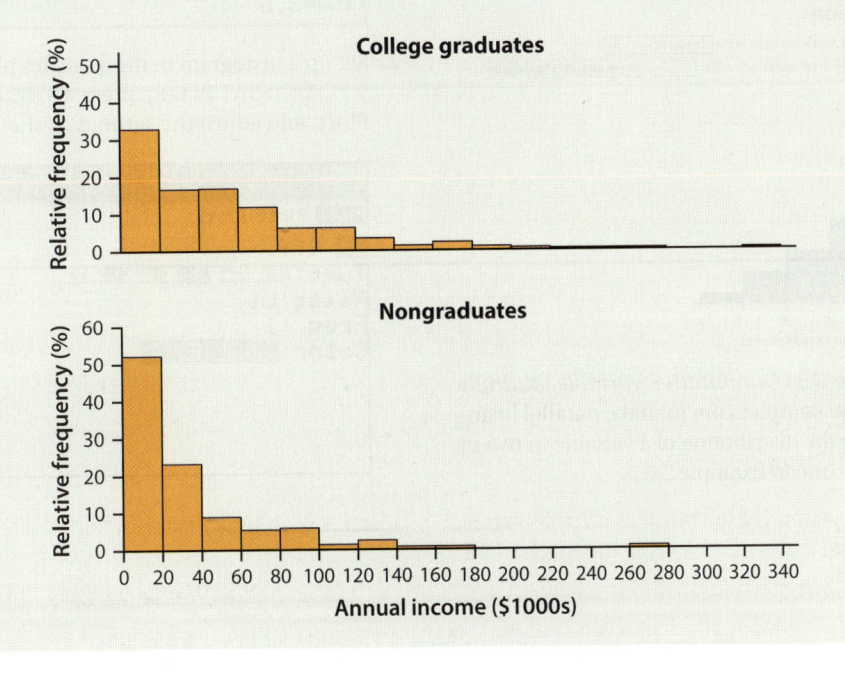

(continued)

SOLUTION

Shape: Both distributions of annual income are skewed to the right, with a single peak between $0 and $20,000.

Outliers: There are possible high outlier incomes in both distributions.

Center: The center of the distribution is larger for people who earned a college degree (median $40K–60K versus $0–20K), indicating that college graduates typically have higher incomes than nongraduates.

Variability: The annual incomes of college graduates vary more (from at least $0 to at most $340K) than the annual incomes of nongraduates (from at least $0 to at most $280K). A noticeably higher percentage of college graduates than nongraduates have annual incomes of at least $100,000.

EXAM PREP: FOR PRACTICE, TRY EXERCISE 17.

TECH CORNER   Making a Histogram

You can also use technology to make a histogram. The technology's default choice of intervals is a good starting point, but you should adjust the intervals to fit with common sense. We'll illustrate using data on the percentage of foreign-born residents in the states from Example 2.5.1.

Applet

1. Go to www.stapplet.com and launch the *One Quantitative Variable, Single Group* applet.
2. Enter Percent foreign-born residents as the Variable name.
3. Select Raw data as the input method.
4. Enter the data manually, or copy and paste them from a document or spreadsheet. Be sure to separate the data values with commas or spaces.
5. Click the Begin analysis button.
6. Change the Graph type to a histogram. You can adjust the interval width or boundary value if desired.

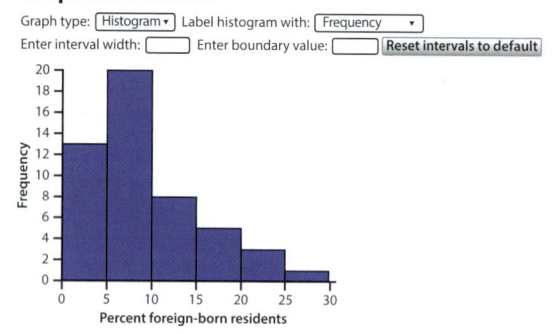

Graph Distribution

TI-83/84

1. Enter the percentage of foreign-born residents in each state in your statistics list editor. Press STAT and choose Edit…; then type the values into list L_1.

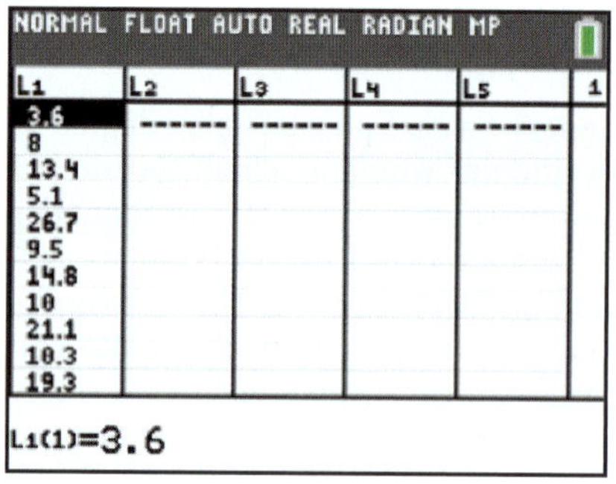

2. Set up a histogram in the statistics plots menu. Press 2nd Y= (STAT PLOT), press ENTER or 1 to go into Plot1, and adjust the settings as shown.

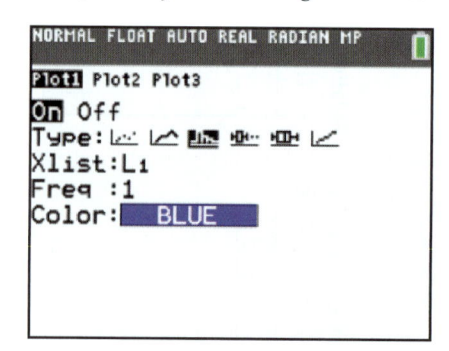

Note: You can use the *One Quantitative Variable, Multiple Groups* applet at www.stapplet.com to make parallel histograms for comparing the distribution of a variable in two or more groups, like the one in Example 2.5.3.

3. Use ZoomStat to let the calculator choose intervals and make a histogram. Press ZOOM and choose 9: ZoomStat. Press TRACE and use the left and right arrow keys to examine the intervals.

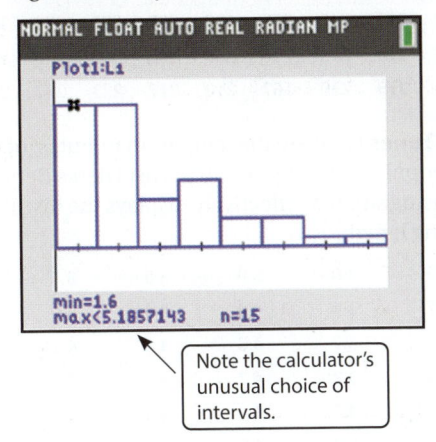

Note the calculator's unusual choice of intervals.

4. Adjust the intervals to match those from the example, then graph the histogram. Press WINDOW and enter the values shown, then press GRAPH. Press TRACE and use the left and right arrow keys to examine the intervals.

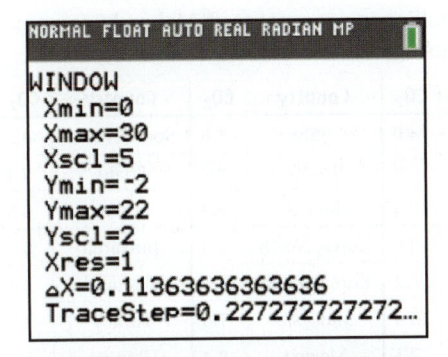

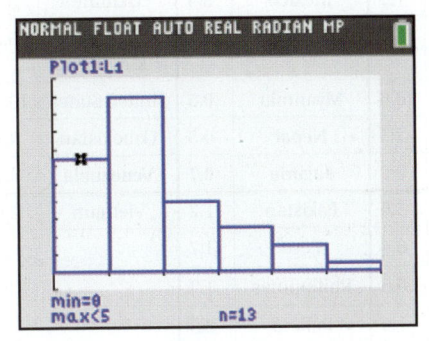

Detailed instructions for using CrunchIt!, Excel, Google Sheets, JMP, Minitab, and R are available in Achieve.

Section 2.5 What Did You Learn?

Review the learning goals from this section. Then practice what you've learned by working through the exercises.

Learning Goal	Example	Exercises
Make histograms of quantitative data.	2.5.1	7–10
Describe distributions of quantitative data with histograms.	2.5.2	11–16
Compare distributions of quantitative data with histograms.	2.5.3	17–20

Section 2.5 Exercises

Building Concepts and Skills These exercises assess the basic knowledge you should have after reading the section.

1. True/False: Dotplots, stemplots, and histograms show every individual value in a set of quantitative data.

2. The first step in making a histogram is to choose _____ that span the data. _____ intervals is a good minimum.

3. True/False: In a histogram, you should leave no gaps between adjacent bars.

4. In this book, if you see a histogram with bars above the intervals from 0–5 and from 5–10 on a number line, in which of the two bars is a data value of 5 included?

5. True/False: The shape of the distribution shown in a histogram will be the same no matter what widths you choose for your intervals.

6. When comparing distributions of quantitative data with histograms, should you scale the vertical axis using frequencies or relative frequencies? Explain your answer.

Mastering Concepts and Skills These exercises reinforce the learning goals as shown in the examples.

7. **Refer to Example 2.5.1 Carbon Dioxide Emissions** Burning fuels in power plants and motor vehicles emits carbon dioxide (CO_2), which contributes to global warming. The table displays CO_2 emissions in metric tons per person from 48 countries with populations of

at least 20 million in a recent year.[68] Make a histogram of the data using intervals of width 2, starting at 0.

Country	CO_2	Country	CO_2	Country	CO_2
Algeria	4.0	Italy	5.6	South Africa	8.2
Argentina	4.0	Japan	8.7	Spain	5.4
Australia	16.3	Kenya	0.3	Sudan	0.5
Bangladesh	0.6	Korea, North	1.5	Tanzania	0.2
Brazil	2.2	Korea, South	11.9	Thailand	4.1
Canada	15.4	Malaysia	7.8	Turkey	4.9
China	7.1	Mexico	3.4	Ukraine	5.1
Colombia	2.0	Morocco	2.0	United Kingdom	5.5
Congo	0.6	Myanmar	0.5	United States	16.1
Egypt	2.5	Nepal	0.5	Uzbekistan	3.3
Ethiopia	0.2	Nigeria	0.7	Venezuela	4.1
France	5.0	Pakistan	1.2	Vietnam	2.6
Germany	8.4	Peru	1.7		
Ghana	0.5	Philippines	1.3		
India	1.9	Poland	8.5		
Indonesia	2.3	Romania	3.9		
Iran	9.4	Russia	11.5		
Iraq	5.6	Saudi Arabia	17.0		

8. **Work Commutes** How long do people travel each day to get to work? The following table gives the average travel times to work (in minutes) in a recent year for workers in each state and the District of Columbia who are at least 16 years old and don't work at home.[69] Make a histogram of the travel times using intervals of width 2 minutes, starting at 16 minutes.

State	Travel time to work (min)	State	Travel time to work (min)	State	Travel time to work (min)
AL	24.9	LA	25.7	OK	21.9
AK	19.1	ME	24.2	OR	23.9
AZ	25.7	MD	33.2	PA	27.2
AR	21.7	MA	30.2	RI	25.2
CA	29.8	MI	24.6	SC	25.0
CO	25.8	MN	23.7	SD	17.2
CT	26.6	MS	24.8	TN	25.2
DE	26.3	MO	23.9	TX	26.6
DC	30.8	MT	18.3	UT	21.9
FL	27.8	NE	18.8	VT	23.3
GA	28.8	NV	24.6	VA	28.7
HI	27.5	NH	27.5	WA	28.0
ID	21.1	NJ	32.2	WV	25.9
IL	29.2	NM	22.3	WI	22.2
IN	23.8	NY	33.6	WY	17.9
IA	19.3	NC	24.8		
KS	19.4	ND	17.3		
KY	23.6	OH	23.7		

9. **Home Runs** Here are the number of home runs hit by each of the 30 Major League Baseball teams during the 2019 season. Make a histogram that effectively displays the distribution of number of home runs hit.

220	249	213	245	256	182	227	223	224	149
288	162	220	279	146	250	307	242	306	257
215	163	219	239	167	210	217	223	247	231

10. **Country Tunes** Here are the lengths, in minutes, of the 50 most popular songs by country artist Dierks Bentley. Make a histogram that effectively displays the distribution of song length.

4.2	4.0	3.9	3.8	3.7
4.7	3.4	4.0	4.4	5.0
4.6	3.7	4.6	4.4	4.1
3.0	3.2	4.7	3.5	3.7
4.3	3.7	4.8	4.4	4.2
4.7	6.2	4.0	7.0	3.9
3.4	3.4	2.9	3.3	4.0
4.2	3.2	3.4	3.7	3.5
3.4	3.7	3.9	3.7	3.8
3.1	3.7	3.6	4.5	3.7

11. **Refer to Example 2.5.2 Stock Returns** The return on a stock is the change in its market price plus any dividend payments made. Return is usually expressed as a percentage of the beginning price. The figure shows a histogram of the distribution of monthly percent return for the U.S. stock market (total return on all common stocks) in 273 consecutive months.[70]

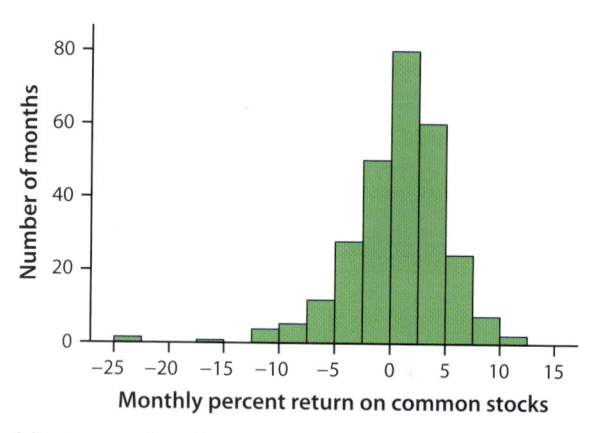

(a) A return less than zero means that stocks lost value in that month. About what percentage of all months had returns less than zero?

(b) Describe the shape of the distribution. Are there any outliers?

(c) Estimate the center and variability of the distribution.

12. **Cereal Calories** Researchers collected data on calories per serving for 77 brands of breakfast cereal. The following histogram displays the data.[71]

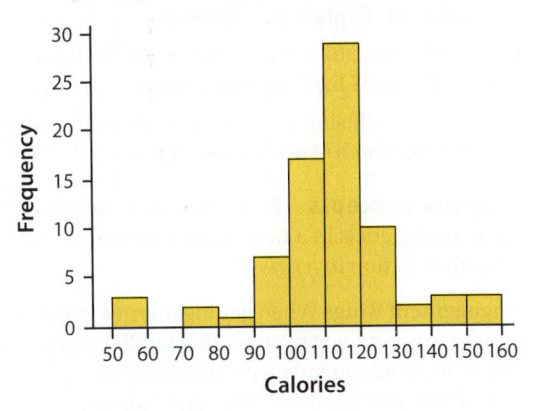

    (a) About what percentage of the cereal brands have 130 or more calories per serving?
    (b) Describe the shape of the distribution. Are there any outliers?
    (c) Estimate the center and variability of the distribution.

13. **Carbon Dioxide Emissions** Refer to the data set and the histogram from Exercise 7.
    (a) What percentage of countries had CO_2 emissions of at least 10 metric tons per person?
    (b) Describe the shape of the distribution. Are there any outliers?
    (c) Estimate the center and variability of the distribution.

14. **Work Commutes** Refer to the data set and the histogram from Exercise 8.
    (a) In what percentage of states is the average travel time at least 20 minutes?
    (b) Describe the shape of the distribution. Are there any outliers?
    (c) Estimate the center and variability of the distribution.

15. **Home Runs** Refer to the histogram from Exercise 9. Describe the distribution.

16. **Country Tunes** Refer to the histogram from Exercise 10. Describe the distribution.

17. **Refer to Example 2.5.3 Households and Income** Here are histograms that display the distributions of household size (number of people) for low-income and high-income households.[72] Low-income households had annual incomes less than $15,000, and high-income households had annual incomes of at least $100,000. Compare the distributions.

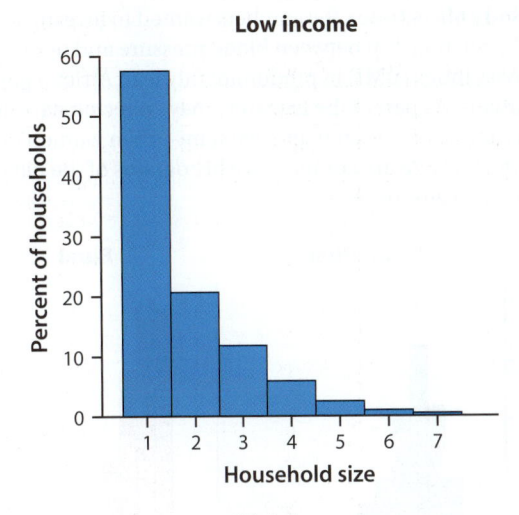

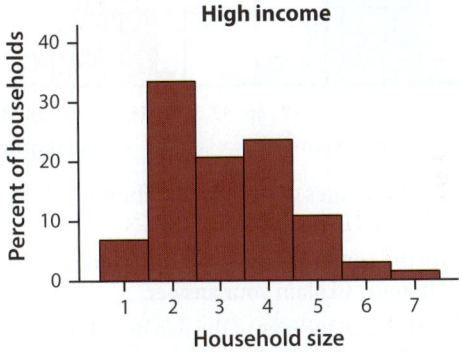

18. **Writing Style** Numerical data can distinguish different types of writing and, sometimes, even individual authors. Here are histograms that display the distribution of word length in Shakespeare's plays and in articles from *Popular Science* magazine.[73] Compare the distributions.

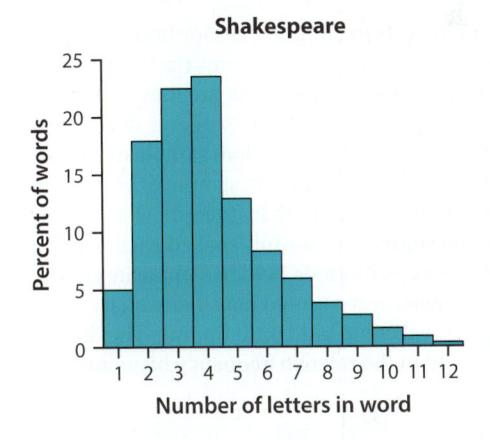

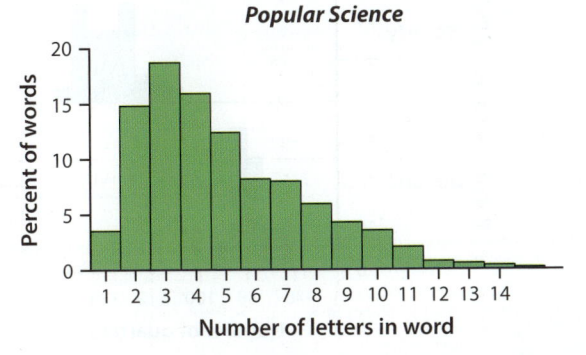

19. **Body Mass Index** Researchers wanted to investigate the relationship between blood pressure and body mass index (BMI) in predominantly lean African populations. As part of the research, they collected data from a sample of 338 rural and 290 semi-urban women in Ghana. Here are comparative histograms of the data for the two groups.[74]

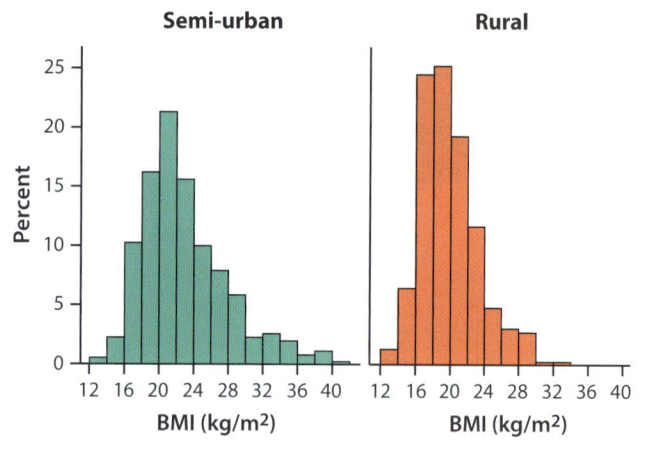

(a) Are the shapes of the two distributions similar or different? Explain your answer.

(b) Are the centers of the two distributions similar or different? Explain your answer.

(c) Are the variabilities of the two distributions similar or different? Explain your answer.

(d) According to the World Health Organization, a person with a BMI less than 18 is considered underweight. Compare the percentages of rural and semi-urban women in the sample who are underweight.

20. **Paper Towels** In commercials for Bounty paper towels, the manufacturer claims that they are the "quicker picker-upper" — but are they also the stronger picker-upper? Two statistics students decided to investigate. They selected a random sample of 30 Bounty paper towels and a random sample of 30 generic paper towels and measured their strength when wet. To do this, the students uniformly soaked each paper towel with 4 ounces of water, held two opposite edges of the paper towel, and counted how many quarters each paper towel could hold until ripping. The data are displayed in the relative frequency histograms.[75]

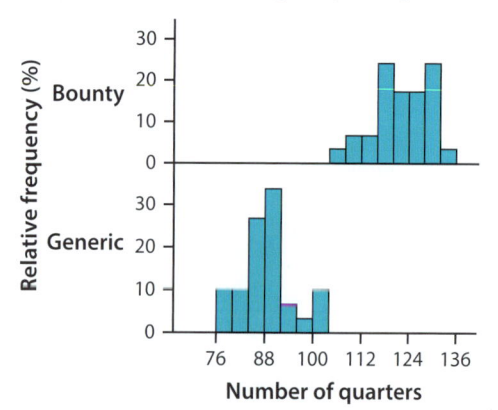

(a) Are the shapes of the two distributions similar or different? Explain your answer.

(b) Are the centers of the two distributions similar or different? Explain your answer.

(c) Are the variabilities of the two distributions similar or different? Explain your answer.

(d) What conclusion should the students make from their research? Explain your answer.

Applying the Concepts These exercises ask you to apply multiple learning goals in a new context or to apply what you learned in this section in a new way.

21. **Engagement Rings** When getting married, some people choose to buy an engagement ring for their partner. How much do people spend on engagement rings, on average? To find out, the *New York Times* and *Morning Consult* surveyed a random sample of 1640 U.S. adults who bought an engagement ring. Here is a histogram of the data.[76]

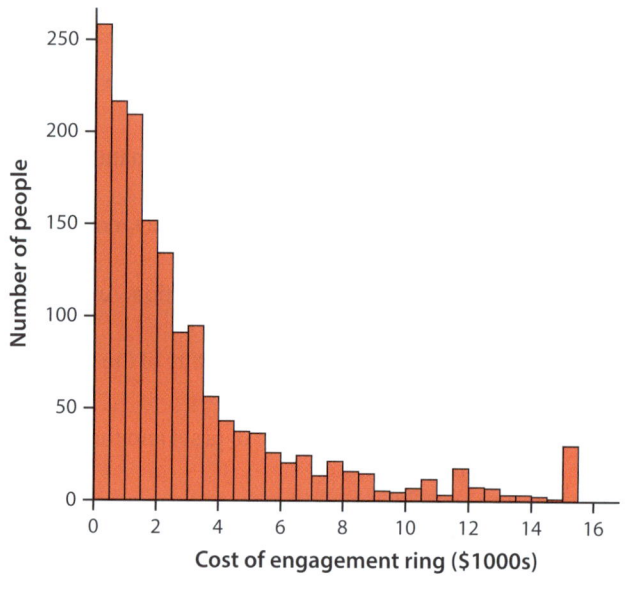

(a) About what percentage of people reported spending at least $5000 on an engagement ring?

(b) Describe the shape of the distribution.

(c) Estimate the center and variability of the distribution.

22. **U.S. President Ages** The histogram displays the ages of the first 46 U.S. presidents when they took office. Teddy Roosevelt was the youngest to take office (42) and Joe Biden was the oldest (78).

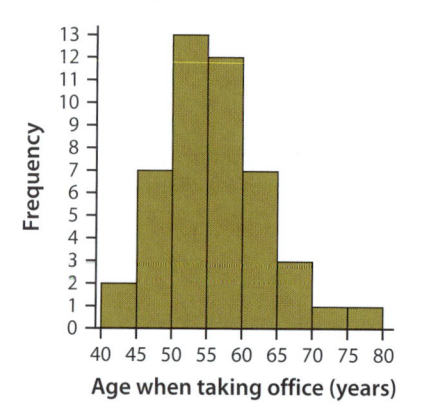

(a) What percentage of U.S. presidents took office before the age of 60?
(b) Describe the shape of the distribution.
(c) Estimate the center and variability of the distribution.

23. **Irises** U.S. botanist Edgar Anderson collected data on iris flowers of three different species: *Setosa*, *Versicolor*, and *Virginia*.[77] British statistician Ronald Fisher used these data to develop a way of classifying an individual iris based on measurements (in centimeters) of sepal length, sepal width, petal length, and petal width. Here are histograms of the data on sepal length. Compare the three distributions.

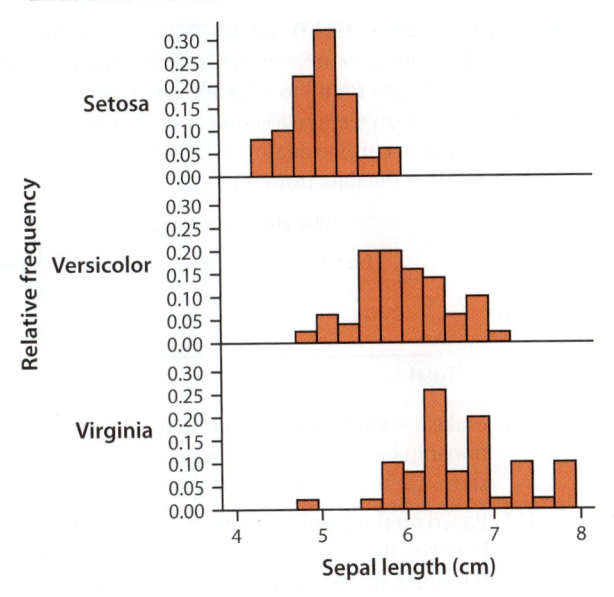

24. **Population Pyramids** The histograms show the distribution of the population's age in Australia and Ethiopia in a recent year from the U.S. Census Bureau's international database.[78] (This type of graph is referred to as a *population pyramid*.)

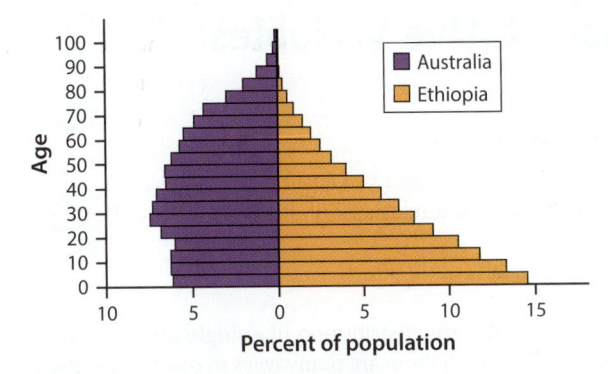

(a) The total population of Australia at this time was about 26 million. Ethiopia's population was about 111 million. Why did we use percentages rather than counts on the horizontal axis?
(b) What important differences do you see between the age distributions?

25. **Tech Carbohydrates in Beer and Wine** How many carbohydrates are in a 12-ounce serving of beer? How about a 6-ounce serving of wine? Does the amount of carbohydrates (in grams) per serving depend on the type of beverage? The **Beer and Wine** data set contains data for 140 different brands of beer and wine. Note that there are some missing data values.[79]

(a) Use technology to make a histogram of the carbohydrate data (*Carbs*) for all of the beverages. Describe the distribution (shape, outliers, center, and variability).
(b) Use technology to make histograms comparing the carbohydrates for beer and wine. What similarities and differences do you see for the two distributions?

26. **Tech Household Income and Location** How much do U.S. households earn in a year? Does it depend on the location of the household? The *American Community Survey* data set includes data on several variables for 3000 randomly selected households.[80] Household income in the past 12 months is recorded in column *HINCP*. Census Bureau region (*Region*) is recorded as 1 = Northeast, 2 = Midwest, 3 = South, and 4 = West.

(a) Use technology to make a histogram of the household income for all of the households. Describe the distribution (shape, outliers, center, and variability).
(b) Use technology to make histograms comparing household incomes across the four regions. What similarities and differences do you see among the four distributions?

Extending the Concepts These exercises challenge you to explore statistical concepts and methods that go beyond what you learned in this section.

27. **Six-Sided Die** Imagine rolling a fair, 6-sided die 60 times. Draw a plausible graph of the distribution of die rolls. Should you use a bar chart or histogram to display the data?

28. **Frequency Polygon** A *frequency polygon* is another graphical way to represent quantitative data. Data are still sorted into equal-width intervals, but instead of drawing a bar for each interval, the frequency in that interval is plotted as a single point. The y value of the point represents the frequency and the x value is the midpoint of the interval. The points are then joined by line segments. Each end of the plot has a line segment going from the last point to the x-axis at what would be the midpoint of the next higher (or lower) interval. Here is the histogram for the presidents' ages when they took office from Exercise 22, along with the frequency polygon for the same data set.

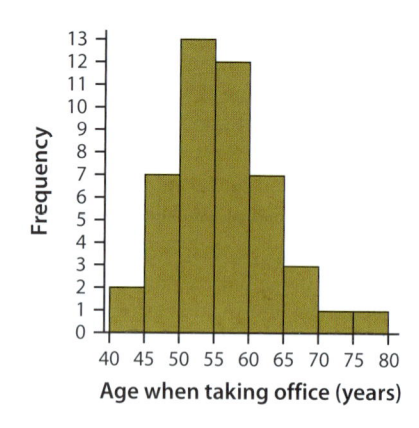

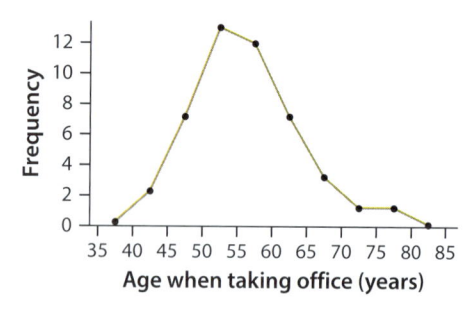

Refer to Exercise 12. Create a frequency polygon of the calories per serving for 77 brands of breakfast cereal.

29. **Exam Grades** The table gives the distribution of grades earned by students taking the AP® Calculus AB and AP® Statistics exams in a recent year.[81]

	1	2	3	4	5	Total number of exams
Calculus AB	55,063	70,088	61,950	56,206	57,352	300,659
Statistics	45,994	42,366	58,321	40,379	32,332	219,392

(a) Make an appropriate graphical display to compare the grade distributions for AP® Calculus AB and AP® Statistics.

(b) Compare the two distributions of exam grades.

Cumulative Review These exercises revisit what you learned in previous sections.

30. **Saunas and Cardiac Death (2.2)** Researchers followed a random sample of 2315 middle-aged men from eastern Finland for up to 30 years. They recorded how often each man went to a sauna and whether he suffered sudden cardiac death (SCD). The two-way table summarizes the data from the study.[82]

		Weekly sauna frequency			
		1 or fewer	2–3	4 or more	Total
SCD	Yes	61	119	10	190
	No	540	1394	191	2125
	Total	601	1513	201	2315

(a) Make a segmented bar chart to display the relationship between weekly sauna frequency and SCD status for the members of the sample.

(b) Based on the graph, is there an association between these variables? Explain your reasoning. If there is an association, briefly describe it.

Section 2.6 Displaying Relationships Between Two Quantitative Variables

LEARNING GOALS

By the end of this section, you will be able to:
- Make a scatterplot to display the relationship between two quantitative variables.
- Describe the direction, form, and strength of a relationship displayed in a scatterplot, and identify outliers.
- Describe a timeplot of a quantitative variable.

In Sections 2.3 to 2.5, you learned how to display the distribution of a single quantitative variable using dotplots, stemplots, and histograms. Although there are many ways to display the distribution of one quantitative variable, a **scatterplot** is the best way to display the relationship between two quantitative variables.

A **scatterplot** shows the relationship between two quantitative variables measured on the same individuals. The values of one variable appear on the horizontal axis, and the values of the other variable appear on the vertical axis. Each individual in the data set appears as a point in the graph.

Section 2.6 Displaying Relationships Between Two Quantitative Variables   113

Figure 2.9 is a scatterplot showing the relationship between income per person (in dollars per year) and the life expectancy (in years) for 175 countries.[83] We expect that the wealth of a country may help us predict the health of its citizens, so income per person is the explanatory variable and life expectancy is the response variable. Note that income per person is plotted on the horizontal axis, life expectancy is plotted on the vertical axis, and each dot represents a country.

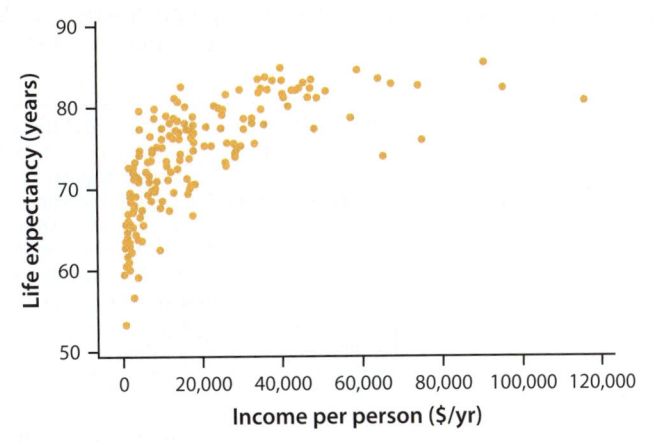

FIGURE 2.9 Scatterplot showing the relationship between income per person and life expectancy in 175 of the world's countries.

Making a Scatterplot

It is straightforward to make a scatterplot by hand (see the How To box). For details on making scatterplots with technology, see the Tech Corner at the end of the section.

HOW TO   Make a Scatterplot

1. **Identify the variables and label the axes.** Put the name of the explanatory variable under the horizontal axis and the name of the response variable to the left of the vertical axis. If there is no explanatory variable, either variable can go on the horizontal axis.

2. **Scale the axes.** Place equally spaced tick marks along the horizontal axis beginning at a number equal to or less than the smallest value of the explanatory variable and continuing until you equal or exceed the largest value. Do the same for the response variable along the vertical axis.

3. **Plot individual data values.** For each individual, plot a point directly above that individual's value for the explanatory variable and directly to the right of that individual's value for the response variable.

EXAMPLE 2.6.1

How much is that truck worth?

Making a scatterplot

PROBLEM
Can we predict the price of a Ford F-150 if we know how many miles it has on the odometer? A random sample of 16 Ford F-150 SuperCrew 4 × 4's was selected from among those listed for sale at autotrader.com. The number of miles driven and price (in dollars) were recorded for each of the trucks.[84] Here are the data.

(continued)

Miles driven	70,583	129,484	29,932	29,953	24,495	75,678	8359	4447
Price ($)	21,994	9500	29,875	41,995	41,995	28,986	31,891	37,991
Miles driven	34,077	58,023	44,447	68,474	144,162	140,776	29,397	131,385
Price ($)	34,995	29,988	22,896	33,961	16,883	20,897	27,495	13,997

Make a scatterplot to display the relationship between miles driven and price.

SOLUTION

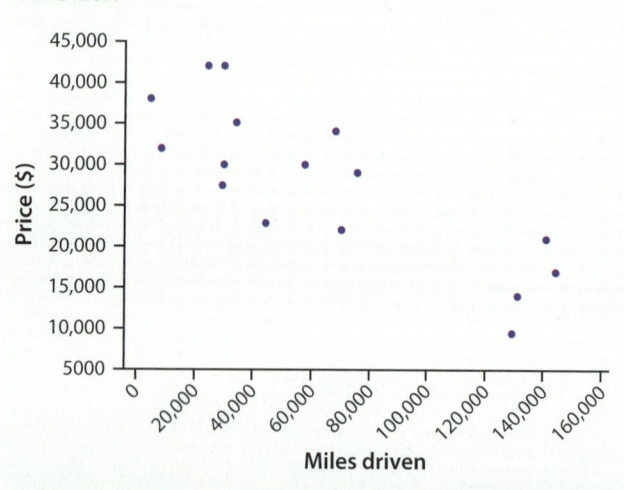

1. **Identify the variables and label the axes.** The explanatory variable is miles driven because a car's mileage should help predict its price.
2. **Scale the axes.** The miles driven values vary from 4447 to 144,162, so we scale the horizontal axis from 0 to 160,000 in increments of 20,000 miles. The prices vary from $9500 to $41,995, so we scale the vertical axis from 5000 to 45,000 in increments of 5000.
3. **Plot individual data values.**

EXAM PREP: FOR PRACTICE, TRY EXERCISE 7.

Describing a Scatterplot

When we analyzed the distribution of one quantitative variable, we looked for the overall pattern (shape, center, and variability) and clear departures from that pattern (outliers). We follow the same general strategy when describing the relationship between two quantitative variables. In this case, however, we use *direction, form,* and *strength* to describe the overall pattern.

HOW TO Describe a Scatterplot

To describe a scatterplot, be sure to use both variable names and address the following four characteristics:

- **Direction:** A scatterplot can show a positive association, a negative association, or no association. In a *positive association,* values of one variable tend to increase as the values of the other variable increase. In a *negative association,* values of one variable tend to decrease as the values of the other variable increase. There is *no association* between two variables if knowing the value of one variable does not help us predict the value of the other variable.
- **Form:** A scatterplot can show a linear form or a nonlinear form. The form is linear if the overall pattern follows a straight line. Otherwise, the form is nonlinear.
- **Strength:** A scatterplot can show a weak, moderate, or strong association. An association is strong if the points don't deviate much from the form identified. An association is weak if the points deviate quite a bit from the form identified.
- **Outliers:** Individual points that fall outside the overall pattern of the relationship.

EXAMPLE 2.6.2

How long a life, how much for that truck?

Describing a scatterplot

PROBLEM
Describe the relationships shown in the scatterplots.

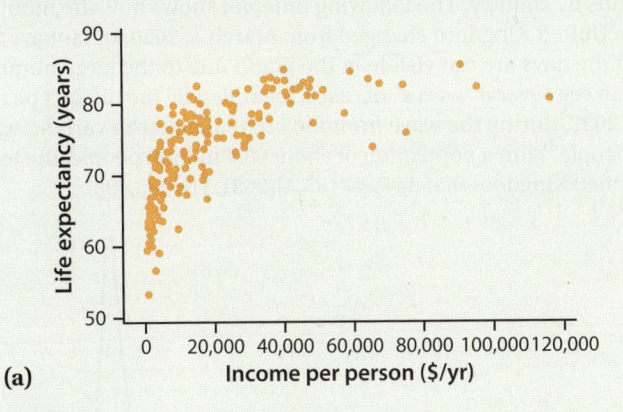

(a) *[Scatterplot of Life expectancy (years) vs. Income per person ($/yr)]*

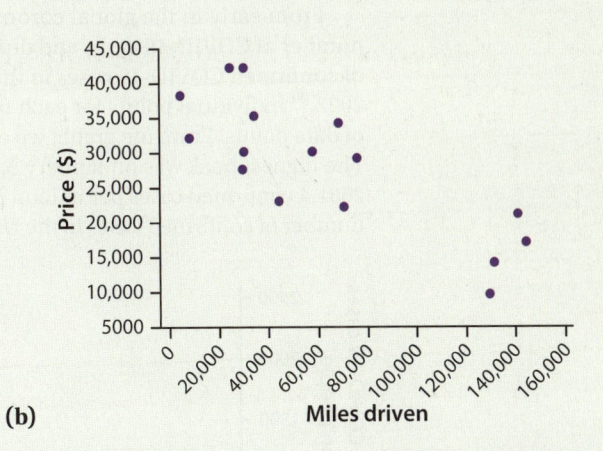

(b) *[Scatterplot of Price ($) vs. Miles driven]*

SOLUTION

(a) Direction: There is a positive association between income per person and life expectancy in these 175 countries. As income per person increases, life expectancy also tends to increase.

Form: There is a curved (nonlinear) pattern in the scatterplot.

Strength: Because the points do not vary too much from the curved pattern, the association is moderately strong.

Outliers: There are two countries with incomes per person between $60,000 and $80,000 per year and life expectancies of only about 75 years that make them stand out below the curved pattern.

> The two outliers are United Arab Emirates, with income per person of $65,300 per year and life expectancy of 73.8 years, and Brunei, with income per person of $75,100 per year and life expectancy of 75.8 years.

(b) Direction: There is a negative association between miles driven and price in this sample of Ford F-150 trucks. As miles driven increases, price tends to decrease.

Form: There is a linear pattern in the scatterplot.

Strength: Because the points do not vary too much from the linear pattern, the association is moderately strong.

Outliers: There don't appear to be any trucks that depart from the linear pattern.

EXAM PREP: FOR PRACTICE, TRY EXERCISE 11.

Even when there is a clear relationship between two variables in a scatterplot, *the direction of the association only describes the overall trend*—not the relationship for each pair of points. For example, in some pairs of Ford F-150 trucks, the truck with more miles driven has a higher price than the truck with fewer miles driven.

Remember the cautionary note we offered in Section 2.2 when describing relationships between two categorical variables: *Association does not imply causation.* Increasing the income per person in a country is not likely to *cause* an increase in life expectancy. Richer countries tend to have better access to health care, nutritious food, and clean water. It is likely these other factors play a larger role in life expectancy. ⚠ **Even a strong association between two variables can sometimes be explained by other variables lurking in the background.** While it may be true that driving a Ford F-150 more miles decreases its value, the truck's price is also affected by its overall condition.

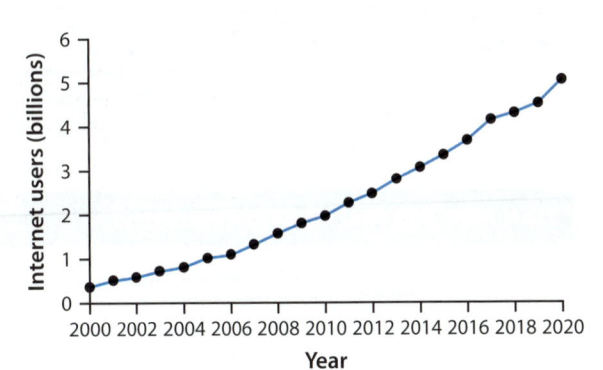

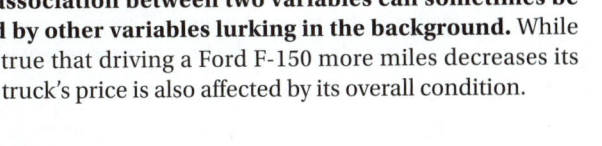

Timeplots

One of the graphs most commonly used by the media is a special type of scatterplot known as a **timeplot.** For instance, here is a timeplot showing how the number of internet users worldwide changed from 2000 to 2020.[85] The graph shows a steady increase in the number of internet users over this time period.

A **timeplot** is a scatterplot with consecutive points connected by a line segment. The graph shows how a single quantitative variable changes over time.

To describe a timeplot, identify major trends and when these trends change. Don't worry about every little wiggle in the graph.

From early in the global coronavirus pandemic, several organizations collected data on the number of COVID-19 cases and deaths by country. The following timeplot shows how the number of confirmed COVID-19 cases in the United Kingdom changed from March 1, 2020, to January 26, 2022.[86] Individual points for each of the days are not visible in the graph due to the large number of data points. From the graph, we can see several "waves" of cases, identified by the distinct peaks. The highest peak was on January 5, 2022, during the wave from the original omicron variant, with 2681.4 confirmed cases per million people. With a population of about 68.4 million people, the total number of confirmed cases in the United Kingdom that day was $(68.4)(2681.4) = 183{,}408$.

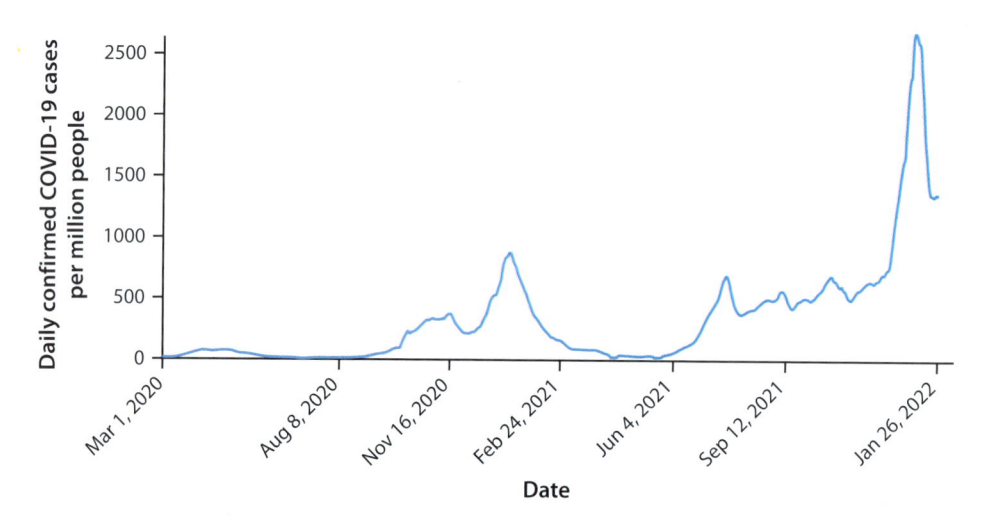

EXAMPLE 2.6.3

Grand Canyon visitors

Describing a timeplot

PROBLEM
The timeplot shows the number of monthly visitors to Grand Canyon National Park from January 2014 to December 2019. Describe what you see.

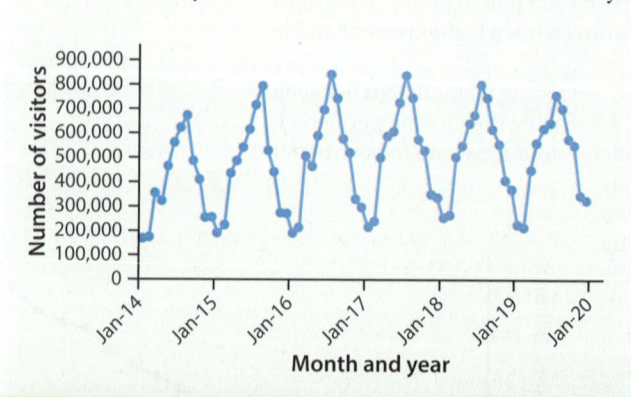

SOLUTION
The number of monthly visitors to Grand Canyon National Park follows a cyclical yearly pattern, with the lowest attendance in the winter months and the highest attendance in the summer months. There seem to be slightly more people visiting the park in more recent years.

EXAM PREP: FOR PRACTICE, TRY EXERCISE 17.

Section 2.6 Displaying Relationships Between Two Quantitative Variables 117

⚠ **Don't assume the trend you see in a timeplot will continue.** For example, the number of monthly visitors to Grand Canyon National Park decreased substantially from prior years starting in March 2020 due to the COVID-19 pandemic.

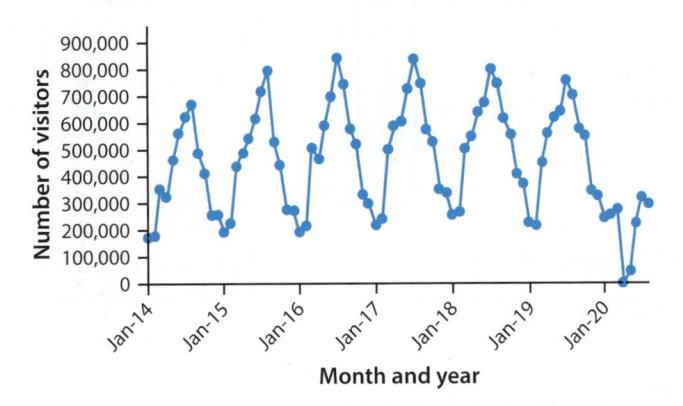

TECH CORNER Making a Scatterplot with Technology

You can use technology to make a scatterplot. We'll illustrate using the Ford F-150 mileage and price data from Example 2.6.1.

Applet

1. Go to www.stapplet.com and launch the *Two Quantitative Variables* applet.
2. Enter the name of the explanatory variable (Miles driven) and the values of the explanatory variable in the first row of boxes. Then, enter the name of the response variable (Price) and the values of the response variable in the second row of boxes. Don't include a comma *within* any data values as you enter them.

Two Quantitative Variables

Variable	Name	Observations (separated by commas or spaces) Keep individuals in the same order.
Explanatory	Miles driven	70583 129484 29932 29953 24495 75678 8359 4447 34077 5
Response	Price	21994 9500 29875 41995 41995 28986 31891 37991 34995 2

[Begin analysis] [Edit inputs] [Reset everything]

3. Click the Begin analysis button to see the scatterplot.

TI-83/84

1. Enter the miles driven values in L_1 and the prices in L_2. Press STAT, choose Edit..., and type the values into L_1 and L_2.

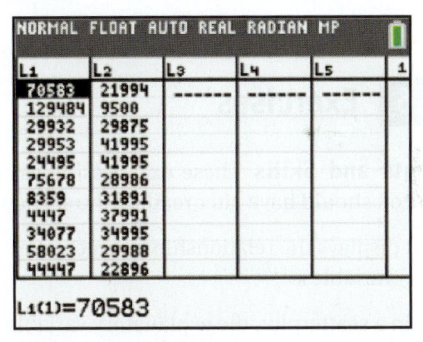

2. Set up a scatterplot in the statistics plots menu. Press 2nd Y= (STAT PLOT), press ENTER or 1 to go into Plot 1, and adjust the settings as shown.

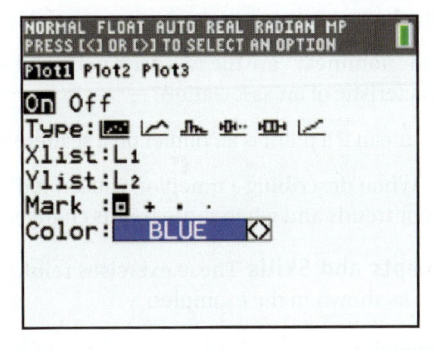

(continued)

3. Use ZoomStat to let the calculator choose an appropriate window. Press ZOOM and choose 9:ZoomStat. Press TRACE and use the left and right arrow keys to examine the intervals.

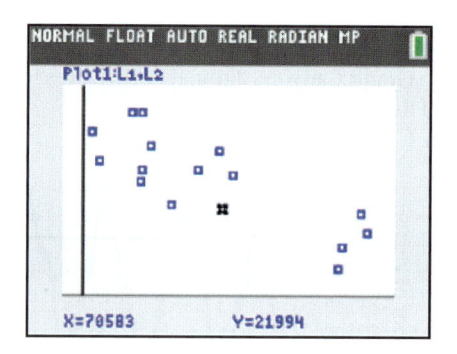

Detailed instructions for using CrunchIt!, Excel, Google Sheets, JMP, Minitab, and R are available in Achieve.

Section 2.6 What Did You Learn?

Review the learning goals from this section. Then practice what you've learned by working through the exercises.

Learning Goal	Example	Exercises
Make a scatterplot to display the relationship between two quantitative variables.	2.6.1	7–10
Describe the direction, form, and strength of a relationship displayed in a scatterplot, and identify outliers.	2.6.2	11–16
Describe a timeplot of a quantitative variable.	2.6.3	17–20

Section 2.6 Exercises

Building Concepts and Skills These exercises assess the basic knowledge you should have after reading the section.

1. A scatterplot displays the relationship between two _____ variables.

2. True/False: In a scatterplot, the explanatory variable can be placed on the horizontal or the vertical axis.

3. To describe the direction of an association, say that there is a _____, _____, or _____ association.

4. "Linear" and "nonlinear" are the two ways to describe which characteristic of an association?

5. What does it mean if a point is an outlier on a scatterplot?

6. True/False: When describing a timeplot, you should identify major trends and when those trends change.

Mastering Concepts and Skills These exercises reinforce the learning goals as shown in the examples.

7. **Refer to Example 2.6.1 Golf Putts** How well do professional golfers putt from various distances to the hole? The data show various distances to the hole (in feet) and the percentage of putts made at each distance for a sample of golfers.[87] Make a scatterplot that shows how the percentage of putts made relates to the distance of the putt.

Distance (ft)	Percent made	Distance (ft)	Percent made
2	93.3	12	25.7
3	83.1	13	24.0
4	74.1	14	31.0
5	58.9	15	16.8
6	54.8	16	13.4
7	53.1	17	15.9
8	46.3	18	17.3
9	31.8	19	13.6
10	33.5	20	15.8

8. **Candy Sugar and Calories** Is there a relationship between the amount of sugar (in grams) and the number of calories in candy? Here are the data from a sample of 12 types of candy.[88] Make a scatterplot to display the relationship between amount of sugar and the number of calories.

Name	Sugar (g)	Calories	Name	Sugar (g)	Calories
Butterfinger Minis	45	450	Reese's Pieces	61	580
Junior Mints	107	570	Skittles	87	450
M&M'S®	62	480	Sour Patch Kids	92	490

Name	Sugar (g)	Calories	Name	Sugar (g)	Calories
Milk Duds	44	370	SweeTarts	136	680
Peanut M&M'S®	79	790	Twizzlers	59	460
Raisinets	60	420	Whoppers	48	350

9. **Roller Coasters** Many people like to ride roller coasters. Amusement parks attract visitors by offering roller coasters that have a variety of speeds and elevations. The table shows data for nine roller coasters.[89] Make a scatterplot to show the relationship between height and maximum speed.

Roller coaster	Height (ft)	Maximum speed (mph)
Apocalypse	100	55
Bullet	196	83
Corkscrew	70	55
Flying Turns	50	24
Goliath	192	66
Hidden Anaconda	152	65
Iron Shark	100	52
Stinger	131	50
Wild Eagle	210	61

10. **Elevation and Temperature** The table presents data on the elevation (in feet) and average January temperature (in degrees Fahrenheit) for 10 cities and towns in Colorado.[90] Make a scatterplot to show the relationship between elevation and average temperature in January.

City	Elevation (ft)	Average January temperature (°F)
Limon	5452	27
Denver	5232	31
Golden	6408	29
Flagler	5002	29
Eagle	6595	21
Vail	8220	18
Glenwood Springs	7183	25
Rifle	5386	26
Grand Junction	4591	29
Dillon	9049	16

11. **Refer to Example 2.6.2 Income and Prayer** The Pew Research Center regularly conducts surveys in the United States and many other countries. The scatterplot shows the relationship between per-capita wealth (per-capita gross domestic product, adjusted for purchasing power parity, in thousands of U.S. dollars) and the percentage of residents who say that they pray daily.[91] For example, in the United States the per-capita wealth is $55,800 and 56% of U.S. residents say that they pray daily. Describe the relationship.

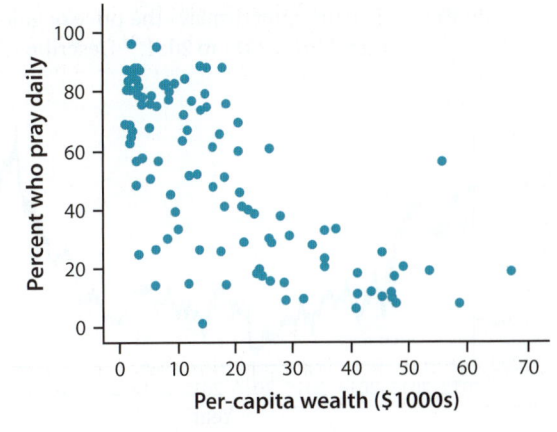

12. **Coin Flips** When a coin is flipped onto a table, it is very unlikely to land on its edge. What if you glued several pennies together and then flipped the stack? The students in Professor Chauvet's college statistics class collected data by repeatedly flipping stacks up to 27 pennies long.[92] The scatterplot shows the relationship between number of pennies in the stack and edge percentage. Describe the relationship.

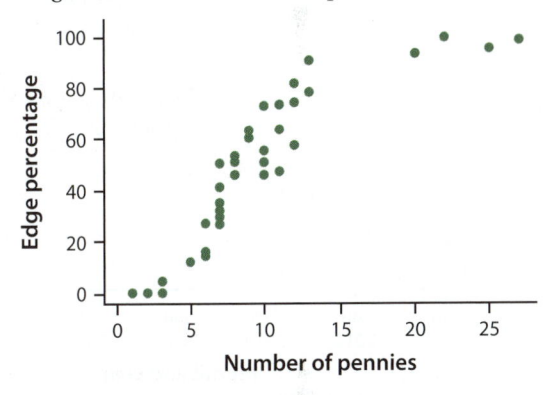

13. **Golf Putts** Refer to Exercise 7. Describe the relationship between distance to the hole and percentage of putts made for the sample of professional golfers.

14. **Candy Sugar and Calories** Refer to Exercise 8. Describe the relationship between amount of sugar and number of calories in the 12 types of candy.

15. **Roller Coasters** Refer to Exercise 9. Describe the relationship between the height and maximum speed of the roller coasters.

16. **Elevation and Temperature** Refer to Exercise 10. Describe the relationship between elevation and average January temperature for these Colorado towns.

17. **Refer to Example 2.6.3 Mortgage Rates** The timeplot displays the interest rate on a 30-year fixed mortgage in the United States from 1985 to 2021.[93] Describe what you see.

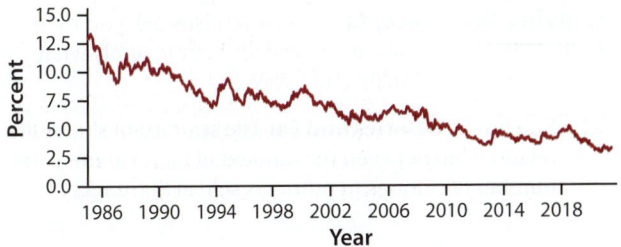

18. Gold Prices The timeplot displays the price of gold (in dollars per ounce) from 2011 to 2021.[94] Describe what you see.

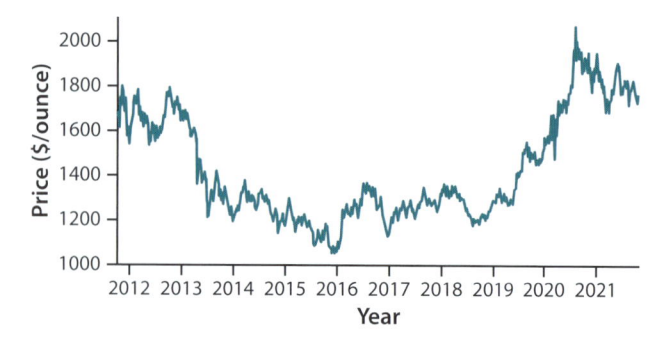

19. Volcanic Carbon Dioxide The Mauna Loa Observatory in Hawaii has been collecting data on CO_2 levels in the atmosphere since the 1950s. Here is a timeplot of the CO_2 level, in parts per million (ppm), recorded at Mauna Loa from January 2010 to May 2021.[95] Describe what you see.

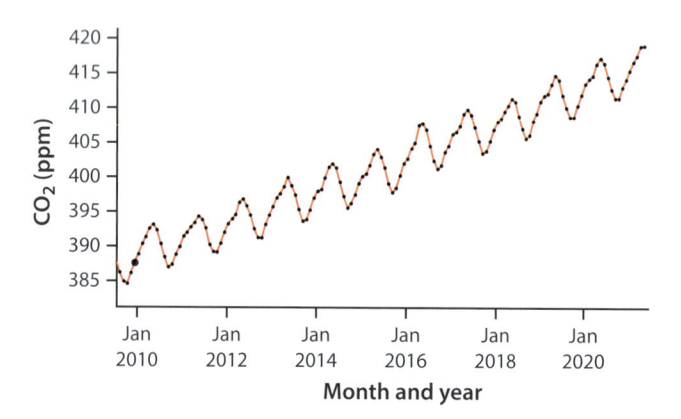

20. Volcanic Carbon Dioxide Refer to Exercise 19. Here is a timeplot of the CO_2 level, in parts per million (ppm), recorded at the Mauna Loa Observatory from the 1970s to 2021. Describe what you see.

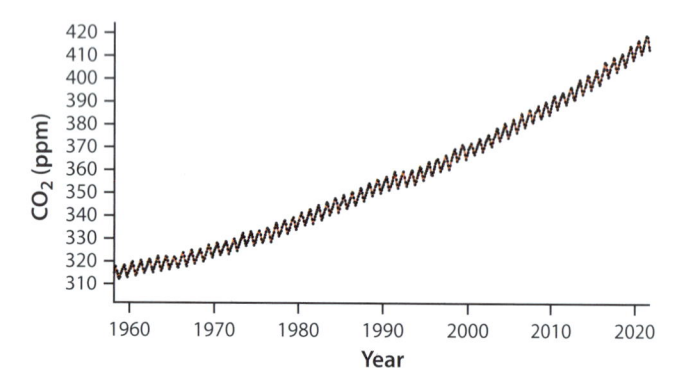

Applying the Concepts These exercises ask you to apply multiple learning goals in a new context or to apply what you learned in this section in a new way.

21. Starbucks Calories and Fat The scatterplot shows the relationship between the amount of fat (in grams) and number of calories in products sold at Starbucks.[96]

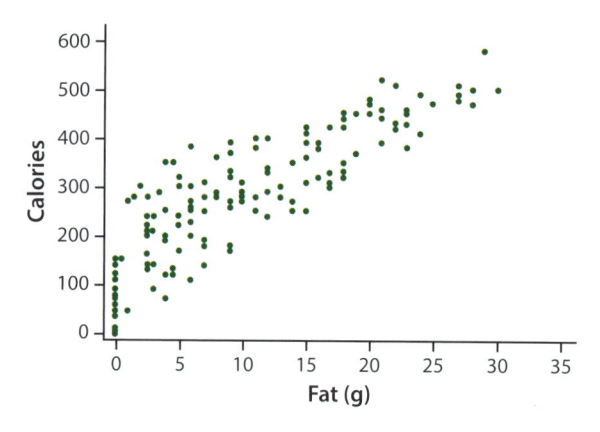

(a) Describe the relationship between fat and calories for these products.

(b) Starbucks Caramelized Apple Pound Cake has 12 grams of fat and 400 calories, while the Pumpkin Cream Cheese Muffin has 14 grams of fat and 350 calories. Explain why this is inconsistent with the association you described in part (a).

How do the nutritional characteristics of food products differ from drink products at Starbucks? The scatterplot shown here enhances the previous scatterplot by plotting the drink products with purple dots and the food products with orange squares.

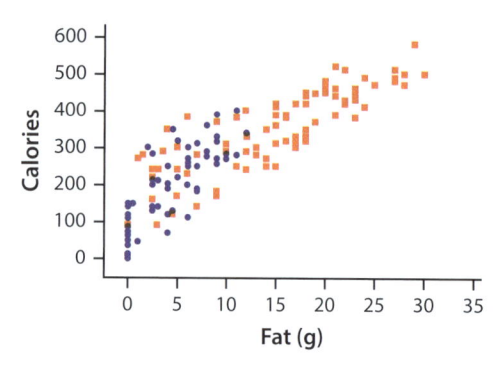

(c) How are the relationships between fat and calories the same for the two types of products? How are the relationships different?

22. Olympic Athletes The scatterplot shows the relationship between height (in inches) and weight (in pounds) for members of the U.S. Olympic Track and Field team.[97]

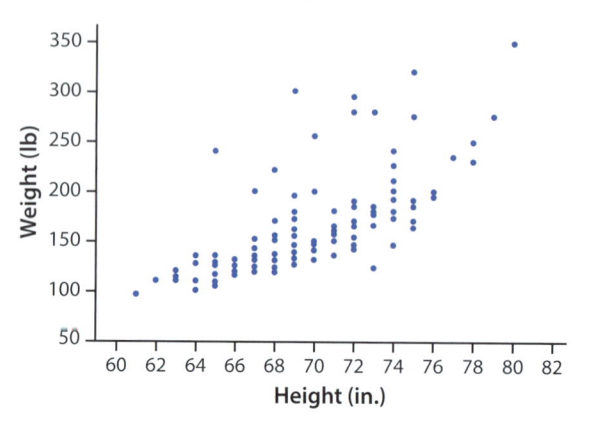

(a) Describe the relationship between height and weight for these athletes.

(b) Discus thrower Kelsey Card is 70 inches tall and weighs 255 pounds, while 400-meter runner Phyllis Francis is 71 inches tall and weighs 158 pounds. Explain why this is inconsistent with the association you described in part (a).

Athletes who participate in the shot put, discus throw, and hammer throw tend to have different physical characteristics than other track and field athletes. The scatterplot shown here enhances the previous scatterplot by plotting these athletes with orange squares and the remaining athletes with purple dots.

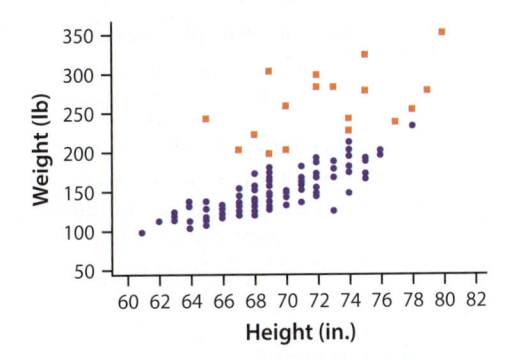

(c) How are the relationships between height and weight the same for the two groups of athletes? How are the relationships different?

23. **Olympic Long Jump** The table shows the Olympic gold medal–winning distance (in meters) in the men's long jump beginning in 1948. Make a timeplot to display how the winning distance has changed since 1948. Describe what you see.

Year	Distance (m)	Year	Distance (m)
1948	7.83	1988	8.72
1952	7.57	1992	8.67
1956	7.83	1996	8.50
1960	8.12	2000	8.55
1964	8.07	2004	8.59
1968	8.90	2008	8.34
1972	8.24	2012	8.31
1976	8.35	2016	8.38
1980	8.54	2020	8.41
1984	8.54		

24. **Olympic 100 Meters** The table shows the Olympic gold medal–winning time (in seconds) in the women's 100-meter dash beginning in 1948. Make a timeplot to display how the winning time has changed since 1948. Describe what you see.

Year	Time (s)	Year	Time (s)
1948	11.90	1988	10.54
1952	11.50	1992	10.82
1956	11.50	1996	10.94
1960	11.00	2000	11.12
1964	11.40	2004	10.93
1968	11.00	2008	10.78
1972	11.07	2012	10.75
1976	11.08	2016	10.71
1980	11.06	2020	10.61
1984	10.97		

25. **Tech Carbohydrates and Calories in Beer and Wine** What is the relationship between the amount of carbohydrates (in grams) and number of calories in a serving of beer? How about in a serving of wine? Does the relationship differ depending on the type of beverage? The **Beer and Wine** data set contains data for 140 different brands of beer and wine. Note that there are some missing data values.[98]

(a) Use technology to make a scatterplot for all of the beverages with amount of carbs (*Carbs*) as the explanatory variable and number of calories (*Calories*) as the response variable. Describe the relationship.

(b) Modify the graph in part (a) to distinguish the points for beer and wine using separate colors or plotting symbols. What do you notice?

26. **Tech Alcohol and Carbohydrates in Beer and Wine** What is the relationship between the alcohol content and amount of carbohydrates (in grams) in a serving of beer? How about in a serving of wine? Does the relationship differ depending on the type of beverage? The **Beer and Wine** data set contains data for 140 different brands of beer and wine. Note that there are some missing data values.[99]

(a) Use technology to make a scatterplot for all of the beverages with alcohol content (*Pct_Alcohol*) as the explanatory variable and amount of carbs (*Carbs*) as the response variable. Describe the relationship.

(b) Modify the graph in part (a) to distinguish the points for beer and wine using separate colors or plotting symbols. What do you notice?

Extending the Concepts These exercises challenge you to explore statistical concepts and methods that go beyond what you learned in this section.

27. **Comparative Timeplots** In many cases, two or more timeplots are shown on the same graph to compare trends. The following graph shows the percentage of U.S. adults who use the internet by level of education.[100] Compare the trends shown in the timeplots.

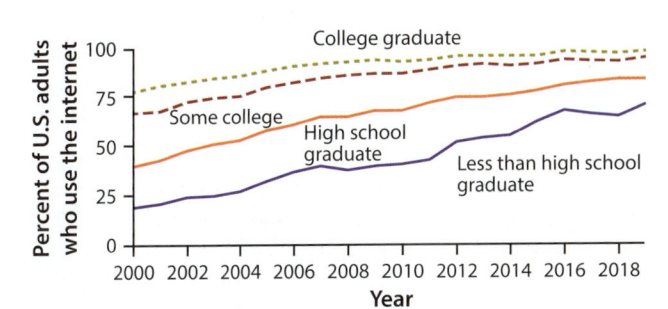

28. **Moving Averages** In the Grand Canyon timeplot (Example 2.6.3), the most obvious feature was the yearly cycles in attendance. But is attendance generally increasing? The ups-and-downs in the graph make it hard to tell. To make it easier to see overall trends, we can plot a *moving average*. The 12-month moving average was calculated using the current month's value along with the values of the previous 11 months. Here is the Grand Canyon timeplot again, with the 12-month moving average added. What does the graph of moving averages show that is hard to see in the original timeplot? What does the graph of moving averages hide?

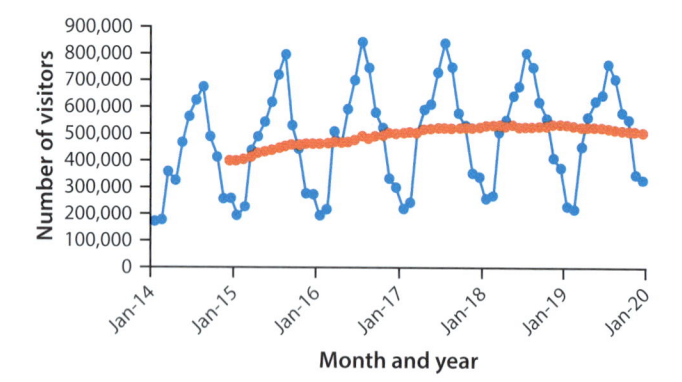

29. **Bubble Charts** The following modified scatterplot is called a *bubble chart* (or *bubble plot*). It displays data on the relationship between life expectancy (in years) and income per person (in dollars) for the world's countries in 2020.[101] The size of each bubble is proportional to the population of that country. The color of each bubble identifies the region of the world in which that country is located. Describe what you see.

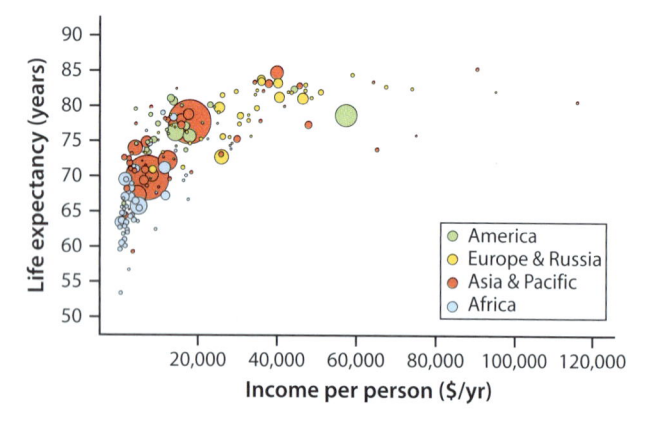

Note: Visit gapminder.org/tools to see an interactive version of this bubble chart with current data.

Cumulative Review These exercises revisit what you learned in previous sections.

30. **Big Diamonds (2.5)** Here are the weights (in milligrams) of 58 diamonds from a nodule carried up to the earth's surface in surrounding rock. These data represent a population of diamonds formed in a single event deep in the earth.[102]

13.8	3.7	33.8	11.8	27.0	18.9	19.3	20.8	25.4	23.1
7.8	10.9	9.0	9.0	14.4	6.5	7.3	5.6	18.5	1.1
11.2	7.0	7.6	9.0	9.5	7.7	7.6	3.2	6.5	5.4
7.2	7.8	3.5	5.4	5.1	5.3	3.8	2.1	2.1	4.7
3.7	3.8	4.9	2.4	1.4	0.1	4.7	1.5	2.0	0.1
0.1	1.6	3.5	3.7	2.6	4.0	2.3	4.5		

    (a) Make a histogram to display these data.

    (b) Describe the shape, center, and variability of the distribution. Are there any obvious outliers?

31. **Murrelet Calls (1.1, 1.8)** Marbled murrelets are seabirds that are commonly found off the northern Pacific Coast of the United States. These birds have experienced notable population declines in recent years due to loss of nesting habitats in old-growth forests as a result of tree cutting. Researchers wondered if they could attract murrelets to potential breeding sites by broadcasting murrelet calls nearby. To find out, the researchers played murrelet calls at 14 randomly assigned sites and played no calls at another 14 randomly assigned sites. A year later, researchers found many more murrelets present during the breeding season, on average, at the sites where they played murrelet calls, and the difference was statistically significant.[103]

    (a) Is this an observational study or an experiment? Explain your answer.

    (b) Based on the study, is it reasonable to say that playing murrelet calls was the cause of increased number of murrelets present during breeding season? Why or why not?

    (c) What is the largest population to which we can generalize these results?

Statistics Matters Where is rental housing affordable?

At the beginning of this chapter, we described a study of affordable rental housing in the United States by the National Low Income Housing Coalition (NLIHC). Here are a few rows of the data table once again. The location of each state is coded by Census region: 1 = Northeast, 2 = Midwest, 3 = South, and 4 = West. The annual household income needed to afford a two-bedroom rental in each state is rated as low (less than $35,360), medium (between $35,360 and $49,830), or high (greater than $49,830).

State	Census region	Annual household income to afford two-bedroom rental	Rating of two-bedroom rental cost	Annual median household income	Percentage of renter households
Alabama	3	$32,110	Low	$66,123	31
California	4	$76,879	High	$90,909	45
Connecticut	1	$54,956	High	$101,816	34
Florida	3	$50,807	High	$68,669	35
Illinois	2	$44,310	Medium	$85,252	34

1. Identify the individuals and variables in this data set. Classify each variable as categorical or quantitative.

2. The two-way table summarizes data on the cost of two-bedroom rental housing in each state and the region in which that state is located.

		Region				
		Midwest	Northeast	South	West	Total
Cost of rental housing	Low	8	0	6	3	17
	Medium	4	5	8	4	21
	High	0	4	2	6	12
	Total	12	9	16	13	50

   (a) Calculate the percentage of states in each region where the cost of two-bedroom rental housing is rated low, medium, and high.
   (b) Make a graph to display the relationship between region and cost of two-bedroom rental housing.
   (c) Describe the association between region and cost of two-bedroom rental housing in the 50 states.

3. How much annual household income ($) is needed to afford two-bedroom rental housing in each state? Here are the data.

AL	32,110	HI	80,613	MA	73,890	NM	34,047	SD	31,701
AK	52,147	ID	34,511	MI	36,227	NY	67,653	TN	35,550
AZ	43,892	IL	44,310	MN	42,705	NC	36,751	TX	43,478
AR	29,514	IN	33,940	MS	30,977	ND	33,647	UT	41,251
CA	76,879	IA	32,151	MO	33,424	OH	33,267	VT	48,597
CO	55,016	KS	34,185	MT	35,112	OK	33,132	VA	49,167
CT	54,956	KY	31,183	NE	33,838	OR	50,687	WA	63,352
DE	45,669	LA	36,356	NV	42,592	PA	39,992	WV	31,135
FL	50,807	ME	41,156	NH	48,726	RI	44,023	WI	35,913
GA	39,758	MD	58,366	NJ	61,762	SC	35,984	WY	35,663

Make a histogram that displays the data effectively. What does the shape of the distribution reveal about the cost of affordable rental housing in the states?

4. Here are parallel dotplots of the percentage of households that live in rental housing in each state by region of the country. Compare the distributions.

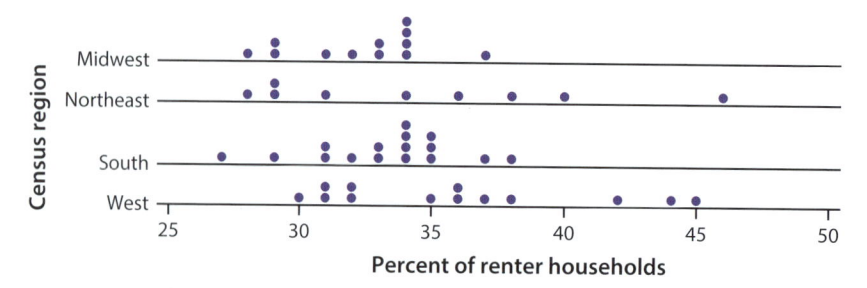

5. The scatterplot displays data on the annual household income ($) needed to afford a two-bedroom rental and the median annual household income ($) in each state, along with the region where each state is located.

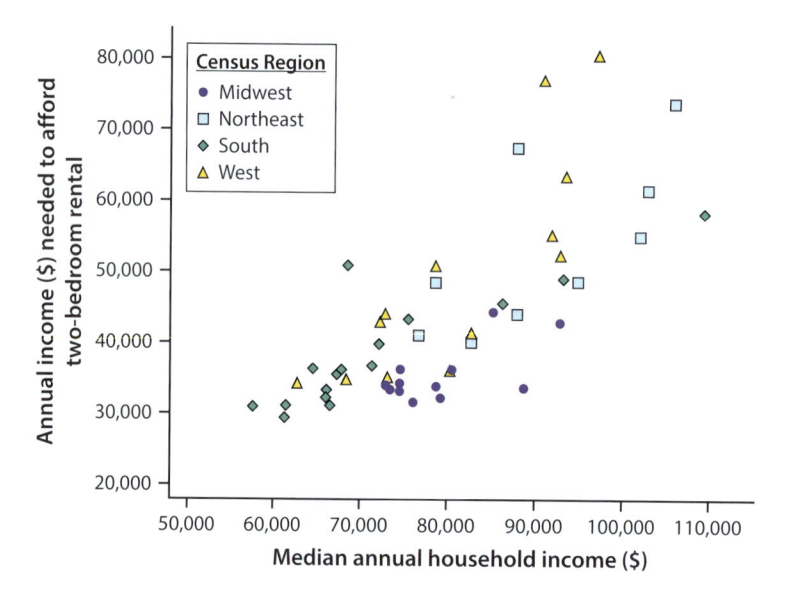

(a) Describe the overall relationship between the annual income needed to afford a two-bedroom rental and the median annual household income in a state.

(b) Relative to annual income, which regions appear to be the most affordable? Least affordable? Explain your answers.

Extension: Use technology to further investigate the data. Write a summary of what you learn.

Postscript: According to the NLIHC, more than 10.8 million of the roughly 44 million U.S. households that lived in rental housing had incomes less than 30% of the annual median household income ($80,320 × 0.30 = $24,096) in 2020. These households can afford monthly rents of only about $600. Many of these low-income households are people aged 65 and older, people with disabilities, single-adult caregivers, or people in school. Racial disparities also exist: 6% of White households, 20% of Black households, 18% of American Indian or Alaska Native households, and 14% of Latino households are low-income renters. Some government assistance is available to low-income households, but there is a lack of affordable rental housing for these households in many states and counties.[104]

Chapter 2 Review

Summarizing One-Variable Data

- The **distribution** of a variable describes what values the variable takes and how often it takes each value.
- You can use a **frequency table** or a **relative frequency table** to summarize the distribution of a variable.

Displaying One-Variable Data

Categorical Data

- **Bar charts** and **pie charts** can be used to display the distribution of a categorical variable.
- Use frequencies or relative frequencies to describe a distribution of categorical data.
- Beware of graphs that mislead the eye. Look at the scales to see if they have been distorted to create a particular impression. Avoid making graphs that replace the bars of a bar chart with pictures whose height and width both change.

Quantitative Data

- You can use a **dotplot, stemplot,** or **histogram** to display the distribution of a quantitative variable. A dotplot displays individual values on a number line. Stemplots separate each observation into a stem and a one-digit leaf. Histograms plot the counts (frequencies) or percentages (relative frequencies) of values in equal-width intervals.
- Histograms are for quantitative data; bar charts are for categorical data. In both types of graphs, be sure to use relative frequencies when comparing data sets of different sizes.
- When examining any graph of quantitative data, look for an *overall pattern* and for clear *departures* from that pattern. **Shape, center,** and **variability** describe the overall pattern of the distribution of a quantitative variable. **Outliers** are observations that lie outside the overall pattern of a distribution. Always look for outliers and try to explain them.
- Some distributions have simple shapes, such as **roughly symmetric, skewed to the left,** or **skewed to the right.** The number of peaks is another aspect of overall shape.
- When comparing distributions of quantitative data, be sure to compare shape, outliers, center, and variability.

Analyzing Relationships Between Two Variables

- Two variables have an **association** if knowing the value of one variable helps us predict the value of the other. If knowing the value of one variable does not help us predict the value of the other, there is no association between the variables.
- Even when two variables have an association, you shouldn't automatically conclude that there is a cause-and-effect relationship between them. The relationship may be better explained by other variables lurking in the background.

Displaying Two-Variable Data

Categorical Data

- A **two-way table** summarizes data on the relationship between two categorical variables for some group of individuals.
- A **segmented bar chart** displays the possible values of a categorical variable as segments of a rectangle, with the area of each segment proportional to the percentage of individuals in the corresponding category.
- You can use a **side-by-side bar chart** or a segmented bar chart to display the relationship between two categorical variables.

Quantitative Data

- A **scatterplot** displays the relationship between two quantitative variables measured on the same individuals. The values of the explanatory variable appear on the horizontal axis, and the values of the response variable appear on the vertical axis. If there is no explanatory variable, either variable can go on the horizontal axis. Each individual in the data set appears as a point on the graph.
- To describe a scatterplot, look for the overall pattern and for striking departures from that pattern. You can describe the overall pattern of a scatterplot by the **direction, form,** and **strength** of the relationship. An important kind of departure is an **outlier,** an individual value that falls outside the overall pattern of the relationship.
  - *Direction:* A relationship has a **positive association** when values of one variable tend to increase as the values of the other variable increase, a **negative association** when values of one variable tend to decrease as the values of the other variable increase, or **no association** when knowing the value of one variable doesn't help predict the value of the other variable.
  - *Form:* A scatterplot can show a linear form or a nonlinear form. The form is linear if the overall pattern follows a straight line. Otherwise, the form is nonlinear.
  - *Strength:* A scatterplot can show a weak, moderate, or strong association. An association is strong if the points don't deviate much from the form identified. An association is weak if the points deviate quite a bit from the form identified.
- A **timeplot** is a special type of scatterplot with consecutive points connected by a line segment. The graph shows how a single quantitative variable changes over time.

Chapter 2 Review Exercises

These exercises will help you review important concepts and skills described by the learning goals in each section. For most exercises, the relevant section is noted in parentheses after the exercise title.

1. **Disc Dogs (2.1)** Here is a list of the breeds of dogs that won the World Canine Disc Championships over a 45-year period.[105]

Whippet	Australian shepherd	Australian shepherd
Whippet	Border collie	Border collie
Whippet	Australian shepherd	Border collie
Mixed breed	Mixed breed	Australian shepherd
Mixed breed	Mixed breed	Border collie
Other purebred	Mixed breed	Border collie
Labrador retriever	Border collie	Other purebred
Mixed breed	Border collie	Border collie
Mixed breed	Australian shepherd	Border collie
Border collie	Border collie	Border collie
Mixed breed	Australian shepherd	Mixed breed
Mixed breed	Border collie	Australian shepherd
Labrador retriever	Mixed breed	Australian shepherd
Labrador retriever	Australian shepherd	Mixed breed
Mixed breed	Australian shepherd	Other purebred

   (a) Summarize the distribution of dog breed with a frequency table.

   (b) Make a bar chart to display the data. Describe what you see.

2. **Cell Phone Features (2.1)** In a survey of more than 2000 U.S. teenagers by Harris Interactive, 47% said that "their social life would end or be worsened without their cell phone."[106] One survey question asked the teens how important it is for their phone to have certain features. The following figure displays data on the percentage who indicated that a particular feature is vital.

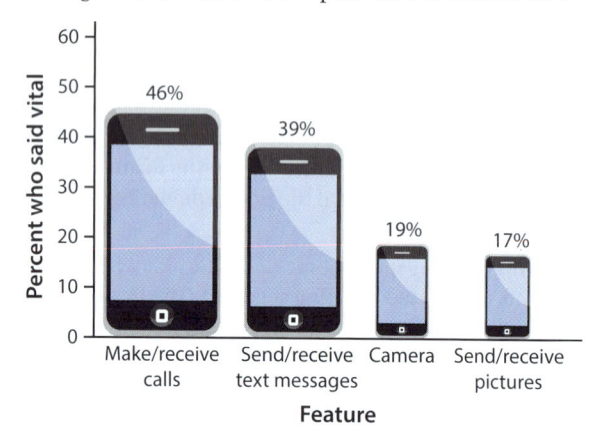

   (a) Explain how the graph gives a misleading impression.

   (b) Would it be appropriate to make a pie chart to display these data? Why or why not?

3. **School Values (2.2)** Researchers carried out a survey of 4th-, 5th-, and 6th-grade students in Michigan. Students were asked if good grades, athletic ability, or being popular was most important to them. The two-way table summarizes the survey data.[107]

		Grade level			
		4th grade	5th grade	6th grade	Total
Most important	Grades	49	50	69	168
	Athletic	24	36	38	98
	Popular	19	22	28	69
	Total	92	108	135	335

   (a) Identify the explanatory and response variables in this context.

   (b) What percentage of fifth-grade students said that grades were most important to them?

   (c) Make a segmented bar chart to show the relationship between grade level and which goal was most important to students.

   (d) Based on the graph, is there an association between these variables? Explain your reasoning. If there is an association, briefly describe it.

4. **Family Income (2.3)** The dotplots show the total family income of 40 randomly chosen individuals each from Connecticut, Indiana, and Maine, based on U.S. Census data. Compare the distributions of family income in these three states.

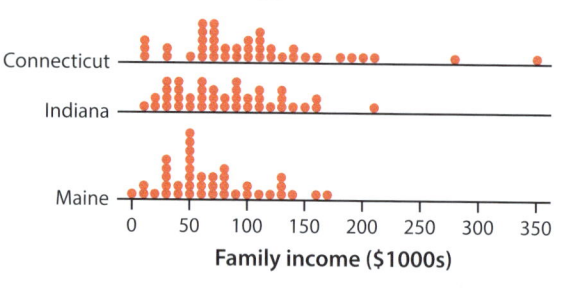

5. **Earth Density (2.4)** In 1798, the English scientist Henry Cavendish measured the density of the earth several times by careful work with a torsion balance. The variable recorded was the density of the earth as a multiple of the density of water. Here are Cavendish's 29 measurements:[108]

   ```
   5.50  5.61  4.88  5.07  5.26  5.55  5.36  5.29  5.58  5.65
   5.57  5.53  5.62  5.29  5.44  5.34  5.79  5.10  5.27  5.39
   5.42  5.47  5.63  5.34  5.46  5.30  5.75  5.68  5.85
   ```

   (a) Make a stemplot of the data.

   (b) Describe the shape of the distribution. Are there any outliers?

   (c) Estimate the center and variability of the distribution.

6. **Music and Memory (2.5)** Two statistics students studied the impact of different types of background music on students' ability to remember words from a list they were allowed to study for 5 minutes. Here are data on the number of words remembered by one group of students who listened to Beethoven's Fifth Symphony:[109]

11	12	23	15	14	15	14	15
10	14	15	9	11	13	25	11
13	13	12	20	17	23	11	12
12	11	20	20	12	12	19	13
15	10	14	11	7	17	13	18

(a) Make a histogram that effectively displays the distribution of number of words remembered.

(b) Describe the distribution.

7. **Crawling Babies (2.6)** At what age do babies learn to crawl? Does it take longer to learn in the winter, when babies are often bundled in clothes that restrict movement? There might even be an association between babies' crawling age and the average temperature during the month when they first try to crawl (around 6 months after birth). Data were collected from parents who reported the birth month and the age at which their child was first able to creep or crawl a distance of 4 feet within 1 minute. The table shows birth month, the average temperature (in degrees Fahrenheit) 6 months after birth, and the average crawling age (in weeks) for a sample of 414 infants.[110]

Birth month	Average temperature 6 months after birth (°F)	Average crawling age (weeks)
January	66	29.84
February	73	30.52
March	72	29.70
April	63	31.84
May	52	28.58
June	39	31.44
July	33	33.64
August	30	32.82
September	33	33.83
October	37	33.35
November	48	33.38
December	57	32.32

(a) Make a scatterplot to display the relationship between average temperature 6 months after birth and average crawling age.

(b) Describe the relationship.

8. **Rating High Schools (Chapter 2)** The nonprofit group Public Agenda conducted telephone interviews with three randomly selected groups of parents of high school children. There were 202 Black parents, 202 Hispanic parents, and 201 White parents. One question asked, "Are the high schools in your state doing an excellent, good, fair, or poor job, or don't you know enough to say?" Here are the survey results:[111]

		Parents' race/ethnicity			
		Black	Hispanic	White	Total
High school rating	Excellent	12	34	22	68
	Good	69	55	81	205
	Fair	75	61	60	196
	Poor	24	24	24	72
	Don't know	22	28	14	64
	Total	202	202	201	605

(a) Make an appropriate graph to display the distribution of high school rating for all respondents to the survey question. Describe what you see.

(b) Here is a segmented bar chart that displays the relationship between parents' race/ethnicity and high school rating. Describe the association between these two variables.

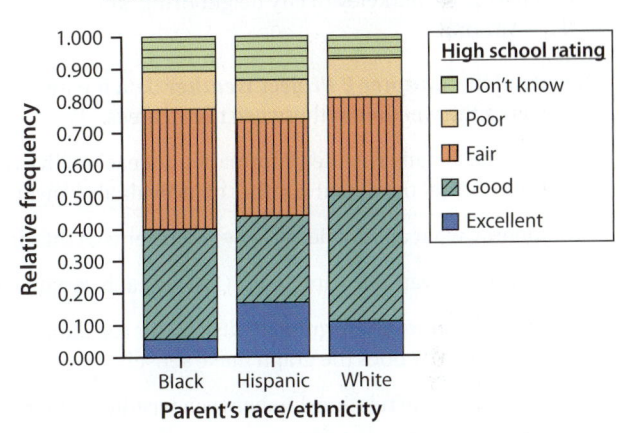

9. **Chess and Reading (Chapter 2)** Many chess advocates believe that chess play develops general intelligence, analytical skill, and the ability to concentrate. Can improving chess-playing skills result in improved reading skills? To investigate, researchers conducted a study involving 53 students who participated in a regional chess tournament. All the students participated in a comprehensive chess program. Researchers measured their reading performances before and after the program. The table shows the difference (Post − Pre) in reading score for each of the students. A negative value indicates that the student's reading performance decreased.[112]

28	3	27	−4	30	42	14	−4	30	28	−19	0	15	−3
−17	16	−12	16	−4	−2	−7	0	1	9	16	12	−3	2
14	10	7	−4	1	−2	−5	0	0	11	11	5	1	−12
16	3	2	22	4	−13	−14	4	−7	9	8			

(a) Make an appropriate graph to display the distribution of difference (Post − Pre) in students' reading scores.

(b) Describe the shape of the distribution.

(c) Did students tend to have higher reading scores after participating in the chess program? Justify your answer.

(d) A scatterplot of the pre-test score and post-test score for each of the 53 students in the program is shown. Describe the relationship between these two variables.

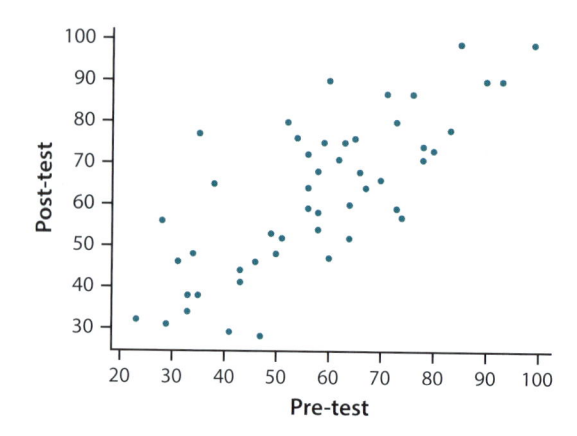

Chapter 2 Project: Weather

Explore a large data set using software and apply what you've learned in this chapter to a real-world scenario.

Researchers collected daily weather data for an entire year from the National Oceanic and Atmospheric Administration (NOAA) National Centers for Environmental Information for two U.S. metro areas: San Francisco and Chicago. The data set contains daily weather information from five different weather stations: San Francisco Downtown, San Francisco International Airport, Berkeley (a city neighboring San Francisco), Chicago Midway Airport (downtown Chicago), and Chicago O'Hare Airport.

Download the **Chapter 2 Project Weather** data file into the software that you are using. The following questions will help you explore the weather in these two metro areas.

1. If we have daily weather information for each of five different weather stations, how many rows of data should we have? Is that the actual number of individuals (rows) in the data set?

2. Identify what an individual (row) represents in this data set.

3. How many variables are there? Classify each variable as categorical or quantitative.

4. Make an appropriate graph to display the fog data. Describe what you see. Think about what you know about the weather. Why does the graph make sense?

5. Investigate the relationship between weather station location and fog status.
    (a) Identify the explanatory variable and the response variable. Explain your reasoning.
    (b) Summarize the relationship with a two-way table.
    (c) Calculate appropriate percentages to describe the relationship.
    (d) Make an appropriate graph to display the relationship. Thinking about what you know about the weather, why does the graph make sense?
    (e) Is there an association between location and fog status? Justify your answer.

6. Make an appropriate graph to compare the distribution of maximum temperature (*Max_Temp*) for the five weather stations. Describe the similarities and differences between the five distributions using your knowledge of the locations.

7. Explore the relationship between average wind speed (*Avg_Wind_Speed*) and average temperature (*Avg_Temp*).
    (a) Identify the explanatory variable and the response variable. Explain your reasoning.
    (b) Make an appropriate graph to display the relationship. Describe what you see. Thinking about what you know about the weather, why does the graph make sense?

8. Investigate whether minimum temperature (*Min_Temp*) is a good predictor of maximum temperature (*Max_Temp*).
    (a) Make an appropriate graph to display the relationship. Describe what you see.
    (b) Modify the graph to include weather station location. How does the relationship differ across weather stations? Thinking about what you know about the weather, why does the graph make sense?

3 Numerical Summaries for Quantitative Data

Section 3.1 Measuring Center
Section 3.2 Measuring Variability
Section 3.3 Boxplots and Outliers
Section 3.4 Measuring Location in a Distribution
Section 3.5 Relationships Between Two Variables: Correlation

Section 3.6 More About Correlation
Chapter 3 Review
Chapter 3 Review Exercises
Chapter 3 Project
Cumulative Review Chapters 1–3

Statistics Matters   How can we predict annual water supply?

Water is an increasingly scarce resource. People need safe water for drinking, bathing, and cooking. But much of the world's water is used for agriculture, business, and electricity. How do resource managers balance these competing demands when making decisions about water allocation? It all starts with predicting the available water supply.

Colorado researchers collected data on the annual water supply (in acre-feet) at several locations in the state over a 25-year period. Here are the data for one of those locations.[1]

17,985	29,430	29,829	35,684	48,460
50,057	41,274	31,559	31,027	34,487
33,289	43,669	35,951	24,772	47,662
49,924	53,650	59,373	40,209	50,722
60,970	66,293	77,738	75,741	71,217

Based on these data, how much available water should the local resource manager budget for the following year? Would knowing the amount of winter snowfall help improve the prediction?

We'll revisit Statistics Matters *at the end of the chapter, so you can use what you have learned to help answer these questions.*

In Chapter 2, you learned how to display quantitative data with graphs. You also explored how to use these graphs to describe and compare distributions, and to analyze the relationship between two quantitative variables. In this chapter, we'll focus on numerical summaries for quantitative data.

Section 3.1 Measuring Center

LEARNING GOALS

By the end of this section, you will be able to:

- Find and interpret the median of a distribution of quantitative data.
- Calculate the mean of a distribution of quantitative data.
- Compare the mean and median of a distribution, and choose the more appropriate measure of center in a given setting.

Let's return to a familiar context from Section 2.3. Recall that city managers in Flint, Michigan, switched the city's water source from Lake Huron to the Flint River to save money. After the switch, Flint residents noticed a change in the look, smell, and taste of the water. Some residents developed rashes, hair loss, and itchy skin. But authorities insisted that drinking water from the Flint River was safe.

Here once again are the lead levels (in parts per billion) in 71 water samples collected from randomly selected Flint dwellings after the switch, along with a dotplot of the data.[2]

0	0	0	0	0	0	0	0	0	0	0	0
0	1	1	1	1	2	2	2	2	2	2	2
2	2	2	2	3	3	3	3	3	3	3	3
3	3	3	4	4	5	5	5	5	5	5	5
5	6	6	6	6	7	7	7	8	8	9	10
10	11	13	18	20	21	22	29	42	42	104	

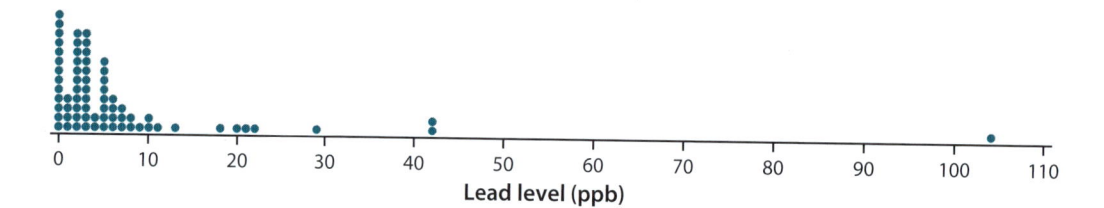

This distribution is right-skewed and single-peaked. The dwelling with a lead level of 104 ppb appears to be an outlier. How much lead does a typical Flint dwelling have in its water?

The *mode* of a distribution is the most frequently occurring data value. For the lead level data, the mode is 0 ppb. But that value isn't representative of how much lead a typical Flint dwelling has in its water. We want to report a value that is in the "center" of the distribution. The mode is often not a good measure of the center because it can fall anywhere in a distribution. Also, a distribution can have multiple modes, or no mode at all.

How should we describe where this (or some other) distribution of quantitative data is centered? The two most common ways to measure the center are the *median* and the *mean*.

The Median

In Chapter 2, we advised you to simply use the "middle value" in an ordered quantitative data set to describe its center. That's the idea of the **median.**

> The **median** is the midpoint of a distribution—the number such that about half the observations are smaller and about half are larger. To find the median, arrange the data values from smallest to largest.
>
> - If the number n of data values is odd, the median is the middle value in the ordered list.
> - If the number n of data values is even, use the average of the two middle values in the ordered list as the median.

You can find the median by hand for small sets of data. For instance, here are data on the population density (in number of people per square kilometer) for all seven countries in Central America.[3]

Country	Belize	Costa Rica	El Salvador	Guatemala	Honduras	Nicaragua	Panama
Population density (people per km^2)	17	100	308	158	82	48	52

To find the median, start by sorting the data values from smallest to largest:

17   48   52   ⓐ2   100   158   308

Because there are $n = 7$ data values (an odd number), the median is the middle value in the ordered list: 82.

Interpretation: About half of Central American countries have a population density less than 82 people per km^2 and about half have a population density greater than 82 people per km^2.

Here is a dotplot of the population density data. You can confirm that the median is 82 by "counting inward" from the minimum and maximum values.

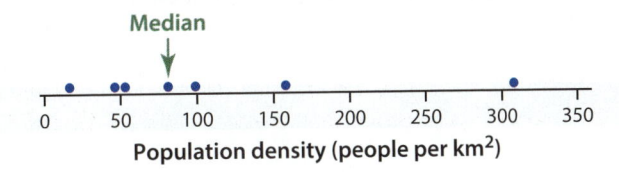

EXAMPLE 3.1.1

More chips please

Finding and interpreting the median

PROBLEM

Have you ever noticed that bags of chips seem to contain lots of air and not enough chips? A group of chip enthusiasts collected data on the percentage of air in each of 14 popular brands of chips. Here are their data, along with a dotplot.[4] Find and interpret the median.

Brand	Percent air	Brand	Percent air
Cape Cod	46	Popchips	45
Cheetos	59	Pringles	28
Doritos	48	Ruffles	50
Fritos	19	Stacy's Pita Chips	50
Kettle Brand	47	Sun Chips	41
Lays	41	Terra	49
Lays Baked	39	Tostitos Scoops	34

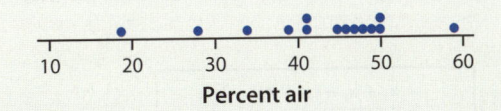

SOLUTION

19 28 34 39 41 41 ㊺ ㊻ 47 48 49 50 50 59

The median is $\dfrac{45 + 46}{2} = 45.5\%$ air. About half of the chip brands have less than 45.5% air, and about half have more.

Sort the data values from smallest to largest (or use the dotplot). Because there are $n = 14$ data values (an even number), use the average of the middle two values in the ordered list as the median.

EXAM PREP: FOR PRACTICE, TRY EXERCISE 9.

The Mean

The most commonly used measure of center is the **mean**.

The **mean** of a distribution of quantitative data is the average of all the individual data values. To find the mean, add all the values and divide by the total number of data values.

If the n data values are $x_1, x_2, \ldots, x_n$, the sample mean $\bar{x}$ (pronounced "x-bar") is given by the formula

$$\bar{x} = \frac{\text{sum of data values}}{\text{number of data values}} = \frac{x_1 + x_2 + \ldots + x_n}{n} = \frac{\sum x_i}{n}$$

The Σ (capital Greek letter sigma) in the formula is short for "add them all up." The subscripts on the observations x_i are just a way of keeping the n data values distinct. They do not necessarily indicate the order or any other special facts about the data.

Actually, the notation $\bar{x}$ refers to the mean of a *sample*. Most of the time, the data we'll encounter can be thought of as a sample from some larger population. When we need to refer to a *population* mean, we'll use the symbol μ (Greek letter mu, pronounced "mew").

EXAMPLE 3.1.2

More chips please

Finding the mean

PROBLEM

Here are the data on percent air in the 14 bags of chips from the preceding example, along with a dotplot.

Brand	Percent air	Brand	Percent air
Cape Cod	46	Popchips	45
Cheetos	59	Pringles	28
Doritos	48	Ruffles	50
Fritos	19	Stacy's Pita Chips	50
Kettle Brand	47	Sun Chips	41
Lays	41	Terra	49
Lays Baked	39	Tostitos Scoops	34

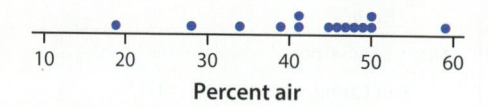

(a) Calculate the mean percent air in the bag for these 14 brands of chips.

(b) The bag of Fritos chips, with only 19% air, is a possible outlier. Calculate the mean percent air in the bag for the other 13 brands of chips. What do you notice?

SOLUTION

(a) $\bar{x} = \dfrac{46 + 59 + 48 + 19 + 47 + \ldots + 34}{14} = \dfrac{596}{14} = 42.57\%$ air

$$\bar{x} = \frac{x_1 + x_2 + \ldots + x_n}{n} = \frac{\sum x_i}{n}$$

(b) $\bar{x} = \dfrac{46 + 59 + 48 + 47 + \ldots + 34}{13} = \dfrac{577}{13} = 44.38\%$ air

Note: It would be incorrect to say the bag of Fritos decreased the mean percent air by 44.38 − 42.57 = 1.81%. This change is a $\dfrac{44.38 - 42.57}{44.38} \times 100 = 4.1\%$ decrease.

The bag of Fritos decreased the mean percent air by 1.81 percentage points.

EXAM PREP: FOR PRACTICE, TRY EXERCISE 13.

The preceding example illustrates an important weakness of the mean as a measure of center: **The mean is not *resistant* to extreme values, such as outliers**.

A statistic is **resistant** if it is not affected much by extreme observations.

The median *is* a resistant measure of center. In the preceding example, the median percent air in all 14 bags of chips is 45.5. If we remove the possible outlier bag of Fritos, the median percent air in the remaining 13 bags is nearly the same (46).

Why is the mean so sensitive to extreme values? The following concept exploration provides some insight.

CONCEPT EXPLORATION

Interpreting the mean

In this exploration, you will investigate a physical interpretation of the mean of a distribution.

1. Stack five pennies at the 6-inch mark on a 12-inch ruler. Place a pencil under the ruler to make a "seesaw" on a desk or table. Move the pencil until the ruler balances. What is the relationship between the location of the pencil and the mean of the five data values: 6, 6, 6, 6, 6?

2. Move one penny off the stack to the 8-inch mark on your ruler. Now move one other penny so that the ruler balances again without moving the pencil. Where did you put the other penny? What is the mean of the five data values represented by the pennies now?

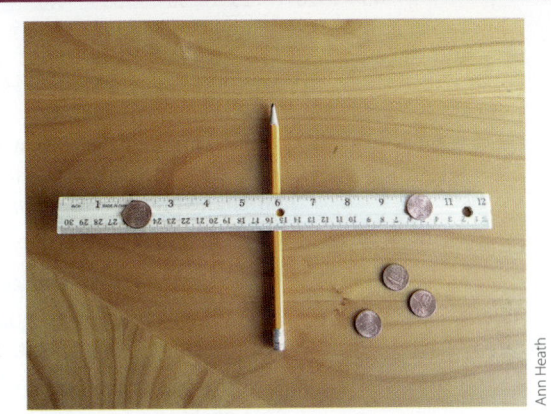

3. Move one more penny off the stack to the 2-inch mark on your ruler. Now move both remaining pennies from the 6-inch mark so that the ruler still balances with the pencil in the same location. Is the mean of the data values still 6?

4. Why is the mean called the "balance point" of a distribution?

The concept exploration gives a physical interpretation of the mean as the balance point of a distribution. For the data on percent air in each of 14 brands of chips, the dotplot balances at $\bar{x} = 42.57\%$.

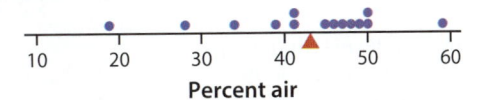

Comparing the Mean and the Median

Which measure — the mean or the median — should we report as the center of a distribution? That depends on both the shape of the distribution and whether there are any outliers.

Shape: Figure 3.1 shows the mean and median for dotplots with three different shapes. Notice how these two measures of center compare in each case. The mean is pulled in the direction of the long tail in a skewed distribution.

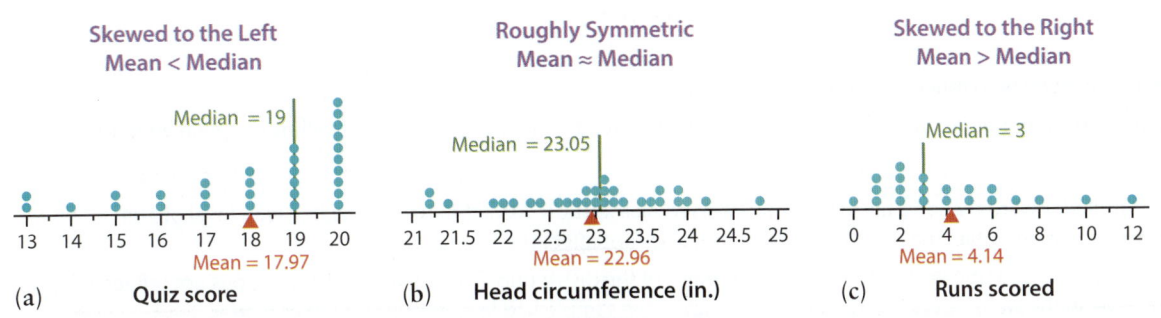

(a) Skewed to the Left — Mean < Median
Median = 19, Mean = 17.97
Quiz score

(b) Roughly Symmetric — Mean ≈ Median
Median = 23.05, Mean = 22.96
Head circumference (in.)

(c) Skewed to the Right — Mean > Median
Median = 3, Mean = 4.14
Runs scored

FIGURE 3.1 Dotplots that show the relationship between the mean and median in distributions with different shapes: (a) scores on a 20-point statistics quiz, (b) head circumference (in inches) for 30 players on a youth football team, and (c) runs scored by a softball team in 21 games played.

134 CHAPTER 3 Numerical Summaries for Quantitative Data

Outliers: We noted earlier that the median is a resistant measure of center, but the mean is not. The dotplot on the left shows the U.S. Environmental Protection Agency (EPA) estimates of city gas mileage in miles per gallon (mpg) for a sample of 21 model year 2021 midsize cars. If we remove the clear outlier — the Toyota Prius, with its 51 mpg in the city — the mean falls from $\bar{x} = 22.52$ to $\bar{x} = 21.10$ mpg, but the median stays at 21 mpg. The dotplot on the right shows the data without the outlier.

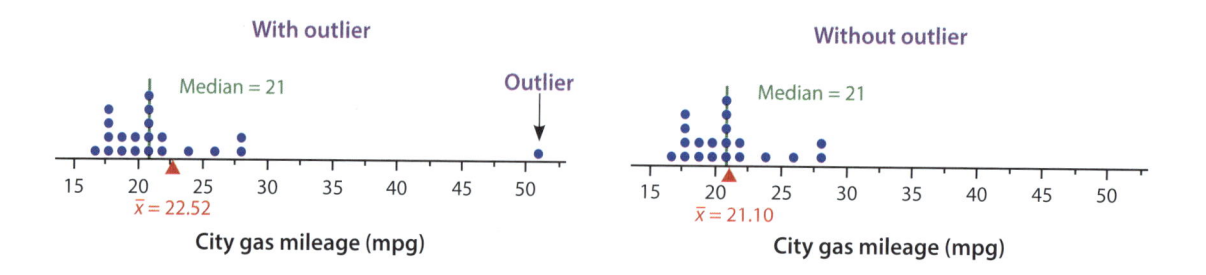

You can investigate how the mean and the median compare for a data set. Go to www.stapplet.com, click the link for Resources for *Introductory Statistics: A Student-Centered Approach*, and launch the *Mean vs. Median* applet.

Choosing a measure of center

- If a distribution of quantitative data is roughly symmetric and has no outliers, use the mean as the measure of center.
- If the distribution is strongly skewed or has outliers, use the median as the measure of center.

EXAMPLE 3.1.3

How much lead is in the water?

Comparing the mean and median

PROBLEM

At the beginning of the section, we presented data on the lead levels (in parts per billion) in water samples taken from 71 dwellings in Flint, Michigan, after the city switched its water source from Lake Huron to the Flint River. Here is a dotplot of the data with the mean and median marked.

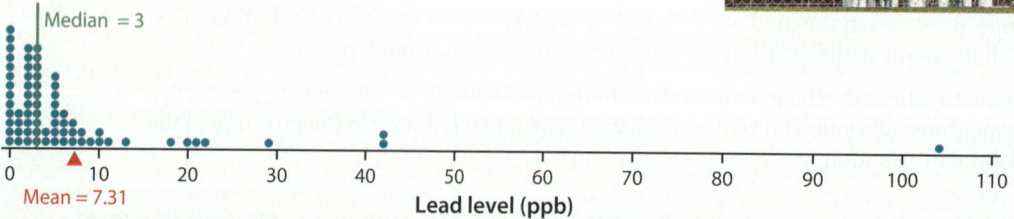

(a) Explain why the mean is so much larger than the median.
(b) Which measure of center better describes the typical amount of lead in the water in a Flint dwelling? Explain your answer.

SOLUTION

(a) The mean is pulled toward the long tail in this right-skewed distribution. Also, the dwelling with a lead level of 104 ppb in the water is an apparent outlier that inflates the mean but does not affect the median as much.
(b) The median lead level of 3 ppb better summarizes the center of the distribution. The mean of 7.31 ppb does not reflect a typical lead level in the water — only 15 of the 71 dwellings in the sample had lead levels this high or higher.

EXAM PREP: FOR PRACTICE, TRY EXERCISE 17.

For medium and large sets of quantitative data, it is preferable to calculate the mean and median using technology. See the Tech Corner in Section 3.2 for details.

Section 3.1 What Did You Learn?

Review the learning goals from this section. Then practice what you've learned by working through the exercises.

Learning Goals	Example	Exercises
Find and interpret the median of a distribution of quantitative data.	3.1.1	9–12
Calculate the mean of a distribution of quantitative data.	3.1.2	13–16
Compare the mean and median of a distribution, and choose the more appropriate measure of center in a given setting.	3.1.3	17–20

Section 3.1 Exercises

Building Concepts and Skills These exercises assess the basic knowledge you should have after reading the section.

1. The two most common ways to measure center are the _____ and the _____.

2. Create a quantitative data set with 10 values for which the mode is not a good measure of the center of the distribution.

3. True/False: The median of a quantitative data set is always one of the individual data values.

4. What is the formula for finding the mean of a distribution of quantitative data?

5. A sample mean is denoted by _____. A population mean is denoted by _____.

6. True/False: The median is a resistant measure of center.

7. What is the physical interpretation of the mean of a distribution?

8. When should you use the median rather than the mean to measure the center of a distribution of quantitative data?

Mastering Concepts and Skills These exercises reinforce the learning goals as shown in the examples.

9. **Refer to Example 3.1.1 Fuel Efficiency** The EPA is in charge of determining and reporting fuel economy ratings for cars in the United States. Here are the EPA estimates of highway gas mileage in miles per gallon (mpg) for a random sample of 21 model year 2020 midsize cars.[5] Find and interpret the median.

25 30 27 31 38 26 28 40 25 28 30 31 30 30 34 30 31 31 32 48 31

10. **Heart Rates** Here are the resting heart rates (in beats per minute) of 19 college students. Find and interpret the median.

71 104 76 88 78 71 68 86 70 90 74 76 69 68 88 96 68 82 120

11. **Copper Mining** In March 2021, a Canadian company found substantial deposits of copper at shallow depths in Arizona's Copper World region, which previously yielded about 440,000 tons of high-quality copper from 1874 to 1969. The company drilled test holes in several randomly selected locations at each of the old mine sites, and measured the amount of copper (as a percentage) in the rock extracted from each hole. Here are the data from the 28 test holes drilled at the Broad Top Butte site, along with a dotplot.[6] Find and interpret the median.

0.00 0.75 1.43 0.43 0.19 0.52 0.00 0.00 0.20 1.38 0.00 0.28 0.71 0.30
0.30 0.44 0.59 0.91 0.33 0.39 0.00 0.70 0.37 0.38 0.67 0.00 0.38 0.30

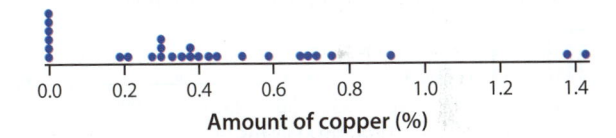

12. **Soccer Stars** How good was the 2019 U.S. Women's National Soccer Team? With players like Carli Lloyd, Alex Morgan, and Megan Rapinoe, the team put on an impressive showing en route to winning the 2019 Women's World Cup. Here are data on the number of goals scored by the team in 24 games played in the 2019 season, along with a dotplot.[7] Find and interpret the median.

1 1 2 2 1 5 6 3 5 3 13 3 2 2 2 2 2 3 4 3 2 1 3 6

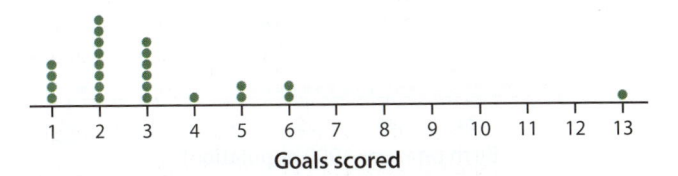

13. **Refer to Example 3.1.2 Fuel Efficiency** Refer to Exercise 9.
   (a) Calculate the mean fuel efficiency in the sample of 21 model year 2020 midsize cars.
   (b) The Toyota Prius, with its 48 mpg fuel efficiency, is a possible outlier in the distribution. Recalculate the median and the mean fuel efficiency for the other 20 car models. What do you notice?

14. **Heart Rates** Refer to Exercise 10.
    (a) Calculate the mean resting heart rate for all 19 college students.
    (b) The student with a resting heart rate of 120 beats per minute is a possible outlier in the distribution. Recalculate the median and the mean resting heart rate for the other 18 students. What do you notice?

15. **Copper Mining** Refer to Exercise 11.
    (a) Calculate the mean percent copper in the rock extracted from all 28 drill holes.
    (b) The test holes that yielded rock with 1.38% and 1.43% copper are apparent outliers. How did these two drill holes affect the mean? Justify your answer with an appropriate calculation.

16. **Soccer Stars** Refer to Exercise 12.
    (a) Calculate the mean number of goals scored by the U.S. Women's National Soccer Team in the 24 games it played during the 2019 season.
    (b) The one game when the team scored 13 goals is a possible outlier in the distribution. How did this game affect the mean? Justify your answer with an appropriate calculation.

17. **Refer to Example 3.1.3 Electoral College** To become president of the United States, a candidate does not have to receive a majority of the popular vote, but that person does have to win a majority of the 538 electoral votes that are cast in the Electoral College. Here is a dotplot of the number of electoral votes in 2021 for each of the 50 states and the District of Columbia.

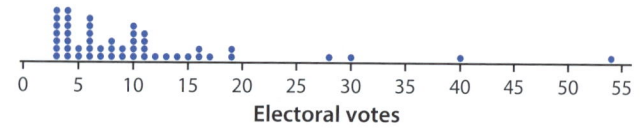

    (a) Explain how the mean and the median compare without doing any calculations.
    (b) Which measure of center better describes a typical number of electoral votes? Explain your answer.

18. **Birth Rates in Africa** One of the important factors in determining population growth rates is the birth rate in a country. The dotplot shows the birth rates per 1000 individuals for 54 African nations in a recent year.[8]

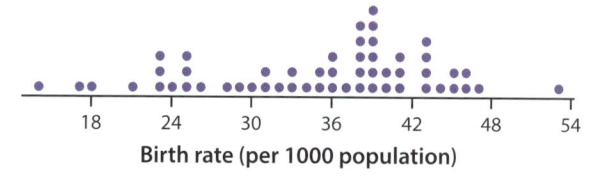

    (a) Explain how the mean and the median compare without doing any calculations.
    (b) Which measure of center better describes the typical birth rate for these 54 African countries? Explain your answer.

19. **Student Sleep** Researchers used data from the American Time Use Survey to investigate how much sleep high school students get per night. The histogram summarizes the data on reported sleep time (in hours per night) from the more than 2700 students who participated in the survey.[9] Explain why the mean is an appropriate measure of center for this distribution.

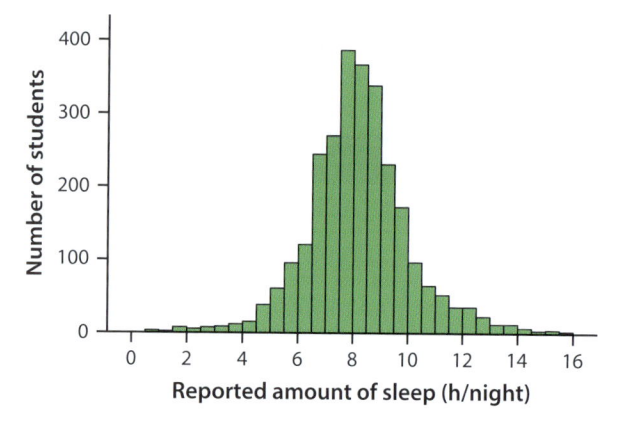

20. **Lightning Strikes** The histogram displays data from a study of lightning storms in Colorado.[10] It shows the distribution of time after midnight (in hours) until the first lightning flash for that day occurred. Explain why the mean is an appropriate measure of center for this distribution.

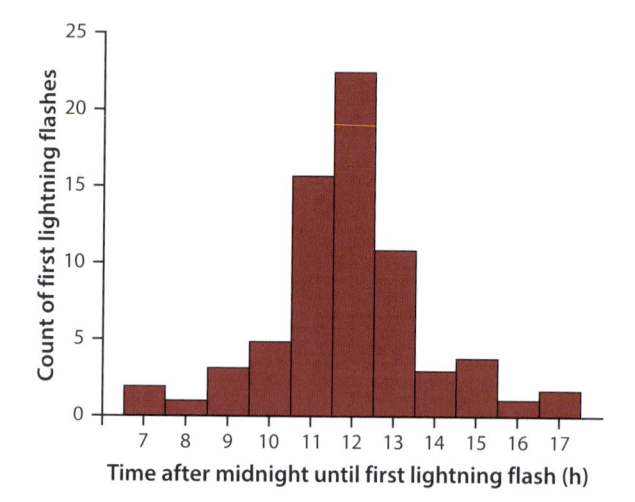

Applying the Concepts These exercises ask you to apply multiple learning goals in a new context or to apply what you learned in this section in a new way.

21. **Work Commutes** How long do people typically spend traveling to work? The answer may depend on where they live. Here are the travel times (in minutes) of 20 randomly chosen workers in New York state.[11]

    10 30 5 25 40 20 10 15 30 20 15 20 85 15 65 15 60 60 40 45

    (a) Find and interpret the median.
    (b) Calculate the mean of the distribution. Give a physical interpretation of the mean.
    (c) Is the mean or the median a more appropriate summary of the center of the distribution? Explain your answer.

22. **Frozen Pizza** Here are the number of calories per serving for 16 brands of frozen cheese pizza.[12]

    340 340 310 320 310 360 350 330 260 380 340 320 310 360 350 330

    (a) Find and interpret the median.

(b) Calculate the mean of the distribution. Give a physical interpretation of the mean.

(c) Is the mean or the median a more appropriate summary of the center of the distribution? Explain your answer.

23. **Teens and Fruit** We all know that fruit is good for us. Here is a histogram of the number of servings of fruit per day that 74 seventeen-year-olds from Pennsylvania claimed they consumed.[13] Find the mean and the median. Show your method clearly.

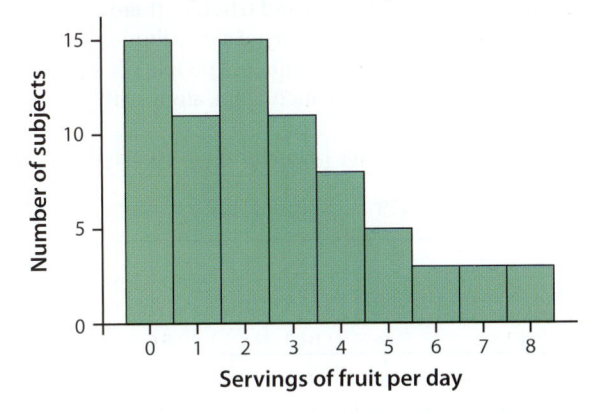

24. **Shakespeare** The following histogram shows the distribution of lengths of words used in Shakespeare's plays.[14] Find the mean and the median. Show your method clearly.

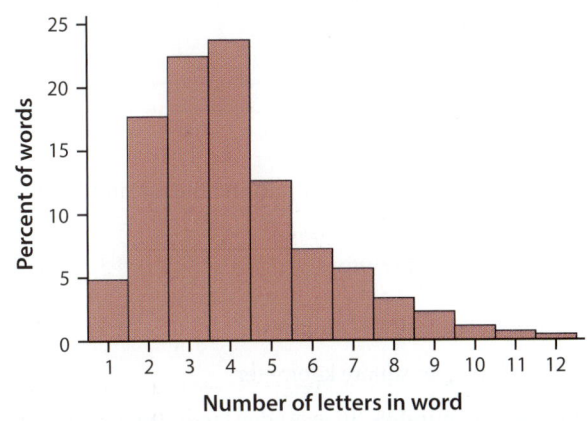

25. **College Loans** In 2021, college student loan debt in the United States reached $1.71 trillion. At the time, there were 45.3 million U.S. borrowers with student loan debt. More than 3 million had loan debts greater than $100,000; about 900,000 had loan debts exceeding $200,000.[15]

(a) Find the mean amount of student loan debt among U.S. borrowers in 2021.

(b) Do you think the median amount of student loan debt was greater than or less than the mean amount? Justify your answer.

26. **Home Prices** The mean and median selling prices of existing single-family homes sold in the United States in February 2021 were $349,900 and $416,000.[16]

(a) Which of these numbers is the mean and which is the median? Explain your reasoning.

(b) Write a sentence to describe how an unethical politician could use these statistics to argue that February 2021 home prices were too high.

Extending the Concepts These exercises challenge you to explore statistical concepts and methods that go beyond what you learned in this section.

Another measure of center for a quantitative data set is the *trimmed mean*. To calculate the trimmed mean, order the data set from lowest to highest, remove the same number of data values from each end, and calculate the mean of the remaining values. For a data set with 10 values, for example, we can calculate the 10% trimmed mean by removing the maximum and minimum value. Why? Because that's one value trimmed from each "end" of the data set out of 10 values, and $1/10 = 0.10$ or 10%.

27. **Student Shoes** How many pairs of shoes does a typical college student have? To find out, an Introductory Statistics instructor surveyed all 20 students in a small section. Here is a dotplot of the number of pairs of shoes each student reported having.[17]

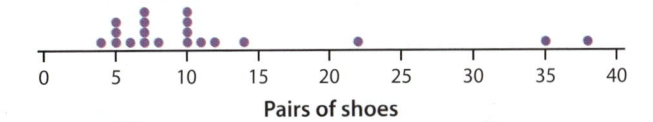

(a) Calculate the mean of the distribution.

(b) Calculate the 10% trimmed mean.

(c) Why is the trimmed mean a better summary of the center of this distribution than the mean?

Cumulative Review These exercises revisit what you learned in previous sections.

28. **Climate Change Insurance (1.1, 1.8, 2.1)** Qualtrics conducted a survey of 1070 randomly selected U.S. homeowners. Respondents were asked, "How much extra would you be willing to spend per year on insurance policies that focus on coverage for climate change–induced risks?" The pie chart summarizes the responses.[18]

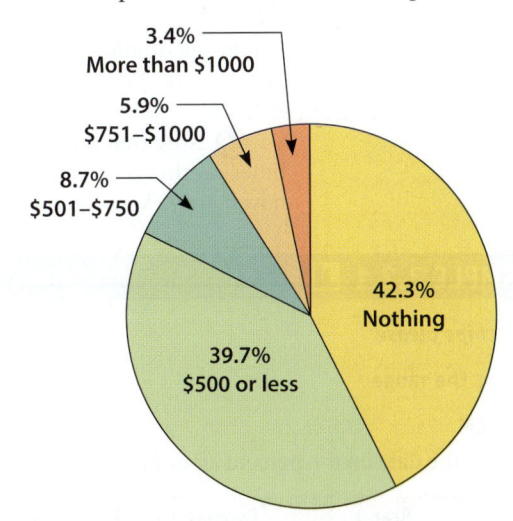

(a) Identify the variable displayed in the pie chart. Is the variable categorical or quantitative? Justify your answer.

(b) Describe what you see in the pie chart.

(c) Would it be appropriate to generalize the findings of this survey to the population of U.S. homeowners? Why or why not?

Section 3.2 Measuring Variability

LEARNING GOALS

By the end of this section, you will be able to:
- Find the range of a distribution of quantitative data.
- Calculate and interpret the standard deviation of a distribution of quantitative data.
- Find and interpret the interquartile range of a distribution of quantitative data.

You can use the mean or the median to describe the center of a distribution of quantitative data. Which statistic you choose depends on the shape of the distribution and whether there are any outliers. A natural follow-up question is: How do we measure the variability of a distribution?

In Section 3.1, we presented data on the population density (in number of people per square kilometer) for the seven countries in Central America. Here are the data once again, along with a dotplot.[19]

Country	Belize	Costa Rica	El Salvador	Guatemala	Honduras	Nicaragua	Panama
Population density (people per km²)	17	100	308	158	82	48	52

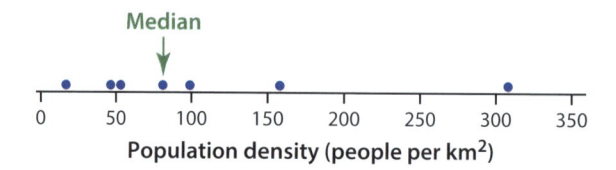

How can we describe the variability of the population densities in these seven countries?

The three most common ways to measure the variability of a distribution of quantitative data are the *range, standard deviation,* and *interquartile range*.

The Range

The simplest measure of variability is the **range**.

> The **range** of a distribution is the distance between the minimum value and the maximum value. That is,
>
> range = maximum − minimum

For the distribution of population density in Central American countries,

range = 308 − 17 = 291 people per square kilometer

! **Note that the range of a data set is a single number.** In everyday language, people sometimes say things like "The data values range from 17 to 308." Be sure to use the term *range* correctly, now that you know its statistical definition.

EXAMPLE 3.2.1

More chips please

Finding the range

PROBLEM

Here are the data on the percent air in 14 bags of chips from Section 3.1, along with a dotplot.[20]

Brand	Percent air	Brand	Percent air	Brand	Percent air
Cape Cod	46	Lays	41	Stacy's Pita Chips	50
Cheetos	59	Lays Baked	39	Sun Chips	41
Doritos	48	Popchips	45	Terra	49
Fritos	19	Pringles	28	Tostitos Scoops	34
Kettle Brand	47	Ruffles	50		

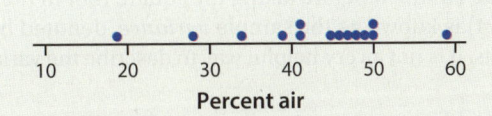

Find the range of the distribution.

SOLUTION

range = 59 − 19 = 40% air

range = maximum − minimum

EXAM PREP: FOR PRACTICE, TRY EXERCISE 9.

The range is *not* a resistant measure of variability. It depends on only the maximum and minimum values, which may be outliers. Look again at the data on the percent air in bags of chips. Without the possible outlier at 19%, the range of the distribution would decrease from 40% air to 59% − 28% = 31% air.

The following graph illustrates another problem with the range as a measure of variability. The parallel dotplots show the lengths (in inches) of a sample of 11 nails produced by each of two machines.[21] Both distributions are symmetric, centered at 70 millimeters (mm), and have a range of 70.2 − 69.8 = 0.4 mm. But the lengths of the nails made by Machine B clearly vary more from the center of 70 mm than the nails made by Machine A.

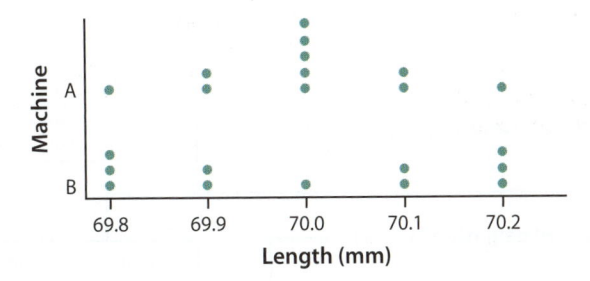

The Standard Deviation

To avoid the problem illustrated by the nail example, we prefer a measure of variability that uses more than just the minimum and maximum values of a quantitative data set. If we summarize the center of a distribution with the mean, then we should use the **standard deviation** to describe the variation of data values around the mean.

> The **standard deviation** measures the typical distance of the values in a distribution from the mean.
>
> To find the sample standard deviation s_x of a quantitative data set with n values:
>
> 1. Find the mean of the distribution.
> 2. Calculate the *deviation* of each value from the mean: deviation = value − mean.
> 3. Square each deviation.
> 4. Add all the squared deviations and divide by $n − 1$.
> 5. Take the square root to return to the original units.

If the values in a data set are given by $x_1, x_2, \ldots, x_n$, we can write the formula for calculating the standard deviation as

$$s_x = \sqrt{\frac{(x_1 - \bar{x})^2 + (x_2 - \bar{x})^2 + \ldots + (x_n - \bar{x})^2}{n-1}} = \sqrt{\frac{\sum (x_i - \bar{x})^2}{n-1}}$$

The notation s_x refers to the standard deviation of a *sample*. Most of the time, the data we'll encounter can be thought of as a sample from some larger population. When we need to refer to a *population* standard deviation, we'll use the symbol σ (lowercase Greek letter sigma). The population standard deviation is calculated by dividing the sum of squared deviations by n instead of $n − 1$ before taking the square root.

The value obtained before taking the square root in the standard deviation calculation (at the end of Step 4) is known as the sample *variance,* denoted by s_x^2. Because variance is measured in squared units, it is not a very helpful way to describe the variability of a distribution.

EXAMPLE 3.2.2

How much football does a football game really include?

Calculating and interpreting standard deviation

PROBLEM

Most National Football League (NFL) games broadcast on TV last more than 3 hours, but how much actual "game time" is there in a typical NFL game? Researchers Kirk Goldsberry and Katherine Rowe investigated this question as part of a sports analytics class they taught at the University of Texas. They carefully watched a random sample of seven games from the 2019 NFL season and recorded the amount of time from when the ball was snapped to when the referee blew the whistle. Here are the game time values (in minutes), excluding any overtime periods, along with a dotplot:[22]

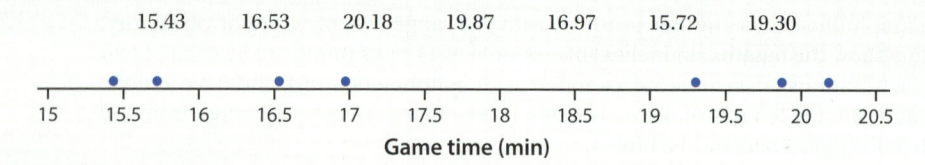

15.43   16.53   20.18   19.87   16.97   15.72   19.30

Calculate and interpret the standard deviation.

SOLUTION

$$\bar{x} = \frac{15.43 + 16.53 + 20.18 + 19.87 + 16.97 + 15.72 + 19.30}{7} = \frac{124.00}{7} = 17.714 \text{ minutes}$$

1. Find the mean of the distribution.

Value x_i	Deviation from mean $x_i - \bar{x}$	Squared deviation $(x_i - \bar{x})^2$
15.43	15.43 − 17.714 = −2.284	$(-2.284)^2 = 5.217$
16.53	16.53 − 17.714 = −1.184	$(-1.184)^2 = 1.402$
20.18	20.18 − 17.714 = 2.466	$(2.466)^2 = 6.081$
19.87	19.87 − 17.714 = 2.156	$(2.156)^2 = 4.648$
16.97	16.97 − 17.714 = −0.744	$(-0.744)^2 = 0.554$
15.72	15.72 − 17.714 = −1.994	$(-1.994)^2 = 3.976$
19.30	19.30 − 17.714 = 1.586	$(1.586)^2 = 2.515$
		Sum = 24.393

2. Calculate the *deviation* of each value from the mean: deviation = value − mean.

3. Square each deviation.

$$s_x^2 = \frac{24.393}{7-1} = 4.066$$

4. Add all the squared deviations and divide by $n - 1$. This gives the sample *variance.*

$$s_x = \sqrt{4.066} = 2.016 \text{ minutes}$$

5. Take the square root to return to the original units.

Interpretation: The actual game time in NFL games typically varies from the mean by about 2.016 minutes.

EXAM PREP: FOR PRACTICE, TRY EXERCISE 13.

More important than the details of calculating s_x are the properties of the standard deviation as a measure of variability:

- **s_x is always greater than or equal to 0.** $s_x = 0$ only when there is no variability—that is, when all values in a distribution are the same.
- **Greater variation from the mean results in larger values of s_x.** For instance, the lengths of nails produced by Machine A have a standard deviation of 0.110 mm, while the lengths of nails

produced by Machine B have a standard deviation of about 0.167 mm. That's about 52% more variability in the lengths of nails produced by Machine B!

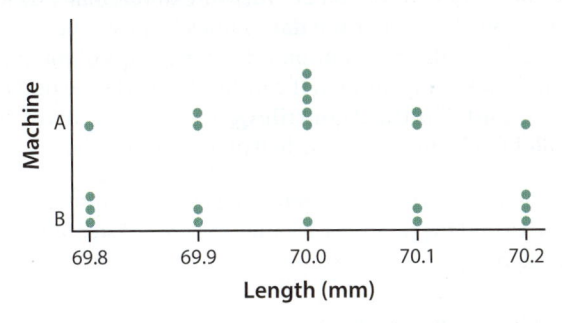

- **s_x is not resistant.** The use of squared deviations makes s_x even more sensitive than $\bar{x}$ to extreme values in a distribution. In Example 3.2.1, the distribution of percent air in the 14 bags of chips has $\bar{x} = 42.571\%$ and $s_x = 10.181\%$. If we omit the minimum value of 19% (Fritos), the mean increases to 44.385% air and the standard deviation decreases to 7.901% air.
- **s_x measures variation about the mean.** It should be used only when the mean is chosen as the measure of center.
- **The standard deviation provides extra information for roughly symmetric, single-peaked, mound-shaped distributions.** The histogram shows data on the percentage of butterfat in milk from 100 three-year-old Ayershire cows.[23] The distribution is roughly symmetric, single-peaked, and mound-shaped. It has a mean of $\bar{x} = 4.173\%$ and a standard deviation of $s_x = 0.291\%$. For distributions like this one, about 68% of data values fall within 1 standard deviation of the mean, about 95% of values fall within 2 standard deviations of the mean, and about 99.7% of values fall within 3 standard deviations of the mean. So we can say that about 95% of the cows in this data set have milk with butterfat content between $4.173 - 2(0.291) = 3.591\%$ and $4.173 + 2(0.291) = 4.755\%$. (If you count frequencies, it's actually 94 of the 100 values.) We will revisit this interesting result, known as the *empirical rule*, in Chapter 6 when we discuss *normal distributions*.

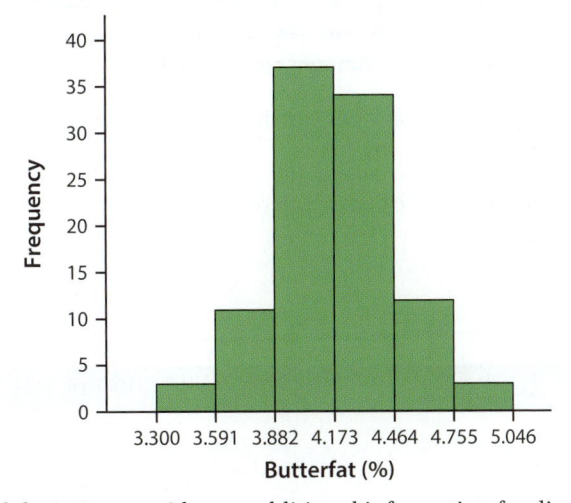

Does the standard deviation provide any additional information for distributions that are *not* roughly symmetric, single-peaked, and mound-shaped? Yes. A result known as *Chebyshev's inequality* (see Exercise 32) tells us that at least 75% of the values in any distribution of quantitative data are within 2 standard deviations of the mean. This inequality also shows that a value at least 5 standard deviations from the mean is unusual in *any* distribution.

> **THINK ABOUT IT**
>
> **Why is the standard deviation calculated in such a complex way?** Add up the deviations from the mean in the preceding example. You should get a sum of 0 (up to rounding error). Why? Because the mean is the balance point of the distribution. We square the deviations to avoid the positive and negative deviations balancing each other out and adding to 0. It might seem strange to "average" the squared deviations by dividing by $n - 1$. The reason for doing this is difficult to explain before Chapter 6. It's easier to understand why we take the square root: to return to the original units (minutes).

The Interquartile Range

We can avoid the impact of extreme values on our measure of variability by focusing on the middle of the distribution. Here's the idea: Order the data values from smallest to largest. Then find the **quartiles**, the values that divide the distribution into four groups of roughly equal size. The **first quartile** Q_1 lies one-fourth of the way through the ordered list. The second quartile is the median, which is halfway through the list. The **third quartile** Q_3 lies three-fourths of the way through the list. The first and third quartiles mark out the middle half of the distribution.

> The **quartiles** of a distribution divide the ordered data set into four groups having roughly the same number of values. To find the quartiles, arrange the data values left to right from smallest to largest and find the median.
>
> The **first quartile** Q_1 is the median of the data values that are to the left of the median in the ordered list.
>
> The **third quartile** Q_3 is the median of the data values that are to the right of the median in the ordered list.

For example, here are the amounts collected each hour by a charity at a local store:

$$\$19, \$26, \$25, \$37, \$31, \$28, \$22, \$22, \$29, \$34, \$39, \$31$$

The dotplot displays the data. Because there are 12 data values, the quartiles divide the distribution into 4 groups of 3 values each.

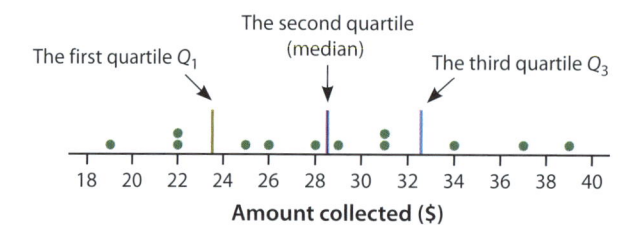

The **interquartile range (IQR)** measures variability using the quartiles. It is simply the range of the middle half of the distribution.

> The **interquartile range (IQR)** is the distance between the first and third quartiles of a distribution. In symbols,
>
> $$IQR = Q_3 - Q_1$$

EXAMPLE 3.2.3

More chips please

Finding and interpreting the *IQR*

PROBLEM
Here again are the data on the percent air in 14 bags of chips.

Brand	Percent air	Brand	Percent air
Cape Cod	46	Popchips	45
Cheetos	59	Pringles	28
Doritos	48	Ruffles	50
Fritos	19	Stacy's Pita Chips	50
Kettle Brand	47	Sun Chips	41
Lays	41	Terra	49
Lays Baked	39	Tostitos Scoops	34

Find the interquartile range. Interpret this value.

SOLUTION

```
19  28  34  39  41  41  ⑤  ㊻  47  48  49  50  50  59
                      Median = 45.5
```
Sort the data values from smallest to largest and find the median.

```
│19  28  34  ㉟  41  41  45 │ 46  47  48  49  50  50  59
           Q₁ = 39      Median = 45.5
```
Find the first quartile Q_1, which is the median of the data values to the left of the median in the ordered list.

```
 19  28  34  ㉟  41  41  45 │ 46  47  48  ㊾  50  50  59 │
           Q₁ = 39      Median = 45.5      Q₃ = 49
```
Find the third quartile Q_3, which is the median of the data values to the right of the median in the ordered list.

$IQR = 49 − 39 = 10\%$ air

$IQR = Q_3 − Q_1$

Interpretation: The middle half of the distribution of percent air in these 14 bags of chips has a range of 10%.

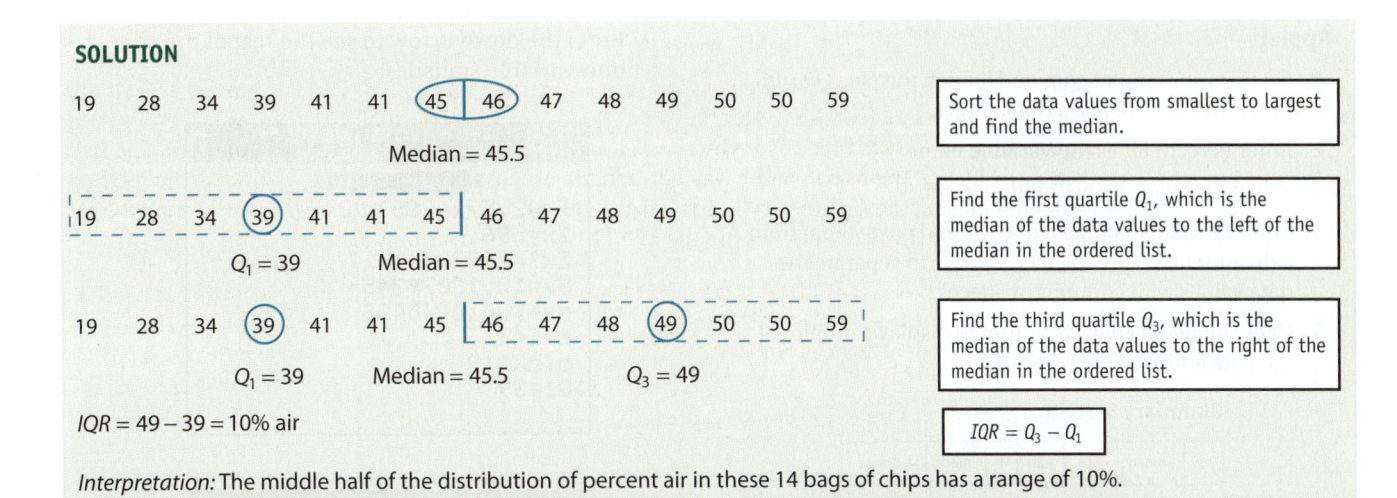

EXAM PREP: FOR PRACTICE, TRY EXERCISE 17.

The quartiles and the interquartile range are *resistant* because they are not affected by a few extreme values. For the air in chips data, Q_3 would still be 49 and the *IQR* would still be 10 if the maximum were 69 rather than 59.

Choosing Measures of Center and Variability

The median and *IQR* are usually better choices than the mean and standard deviation for describing a skewed distribution or a distribution with outliers. Use $\bar{x}$ and s_x for roughly symmetric distributions that don't have outliers.

Here once again is a dotplot of the lead levels (in parts per billion) in 71 water samples taken from randomly selected Flint, Michigan, dwellings after the city switched its water supply from Lake Huron to the Flint River. Numerical summaries of the data set are also provided.[24] In Section 3.1, we decided that the median of 3 ppb provides a better measure of center than the mean of 7.31 ppb because the distribution of lead level is right-skewed and has an apparent outlier at 104 ppb. We should therefore use the interquartile range of $IQR = 7 − 2 = 5$ ppb as our corresponding measure of variability.

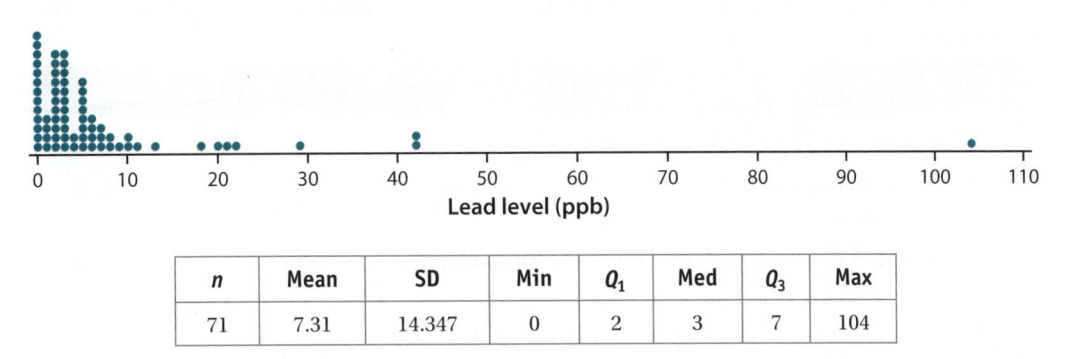

n	Mean	SD	Min	Q_1	Med	Q_3	Max
71	7.31	14.347	0	2	3	7	104

TECH CORNER Calculating Measures of Center and Variability

You can use technology to calculate measures of center and variability for a distribution of quantitative data. That will allow you to concentrate on choosing the right numerical summaries and interpreting results. We'll illustrate using data on the percent air in bags of chips from Example 3.2.1 and Example 3.2.3.

(continued)

Applet

1. Go to www.stapplet.com and launch the *One Quantitative Variable, Single Group* applet.
2. Enter Percent air as the Variable name.
3. Select Raw data as the input method.
4. Enter the data manually, or copy and paste them from a document or spreadsheet. Be sure to separate the data values with commas or spaces.
5. Click the Begin analysis button to display a dotplot of the data and summary statistics.

Summary Statistics

n	mean	SD	min	Q_1	med	Q_3	max
14	42.571	10.181	19	39	45.5	49	59

TI-83/84

1. Type the values into list L_1.
2. Calculate numerical summaries using one-variable statistics.
   - Press STAT, arrow over to the CALC menu, and choose "1-Var Stats."

   OS 2.55 or later: In the dialog box, press 2nd 1 (L_1) and ENTER to specify L_1 as the List. Leave FreqList blank. Arrow down to Calculate and press ENTER.

   Older OS: Press 2nd 1 (L_1) and ENTER.

- Press the down arrow to see the rest of the one-variable statistics.

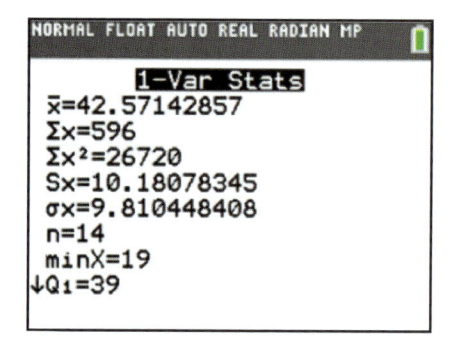

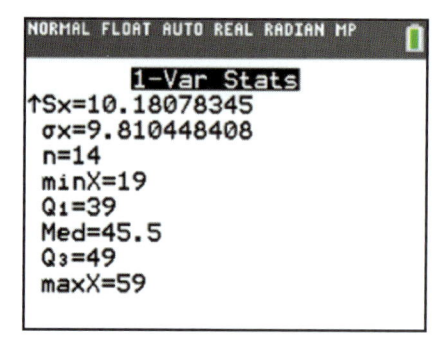

```
NORMAL FLOAT AUTO REAL RADIAN MP
         1-Var Stats
  x̄=42.57142857
  Σx=596
  Σx²=26720
  Sx=10.18078345
  σx=9.810448408
  n=14
  minX=19
 ↓Q₁=39
```

```
NORMAL FLOAT AUTO REAL RADIAN MP
         1-Var Stats
 ↑Sx=10.18078345
  σx=9.810448408
  n=14
  minX=19
  Q₁=39
  Med=45.5
  Q₃=49
  maxX=59
```

Note: Neither the applet nor the TI-83/84 compute the interquartile range (*IQR*) directly. But they do provide Q_1 and Q_3, so you can calculate $IQR = Q_3 - Q_1$.

Detailed instructions for using CrunchIt!, Excel, Google Sheets, JMP, Minitab, and R are available in Achieve.

Section 3.2 What Did You Learn?

Review the learning goals from this section. Then practice what you've learned by working through the exercises.

Learning Goal	Example	Exercises
Find the range of a distribution of quantitative data.	3.2.1	9–12
Calculate and interpret the standard deviation of a distribution of quantitative data.	3.2.2	13–16
Find and interpret the interquartile range of a distribution of quantitative data	3.2.3	17–20

Section 3.2 Exercises

Building Concepts and Skills These exercises assess the basic knowledge you should have after reading the section.

1. The simplest measure of variability is the _____.

2. True/False: In the dotplot near the beginning of this section, the range of the lengths of nails made by Machine A is from 69.8 mm to 70.2 mm.

3. What is the formula for calculating the sample standard deviation of a quantitative data set?

4. True/False: The standard deviation should be interpreted as the typical distance of the values in a distribution from the mean.

5. Explain why the standard deviation of a distribution of quantitative data can never be negative.

6. True/False: For any distribution of quantitative data, about 95% of the data values will be within 2 standard deviations of the mean.

7. How do you find the first quartile of a quantitative data set?

8. For a skewed distribution or a distribution with outliers, use the _____ to describe variability.

Mastering Concepts and Skills These exercises reinforce the learning goals as shown in the examples.

9. **Refer to Example 3.2.1 Fuel Efficiency** The EPA is in charge of determining and reporting fuel economy ratings for cars in the United States. Here are the EPA estimates of highway gas mileage in miles per gallon (mpg) for a random sample of 21 model year 2020 midsize cars.[25] Find the range of the distribution.

25 30 27 31 38 26 28 40 25 28 30 31 30 30 34 30 31 31 32 48 31

10. **Heart Rates** Here are the resting heart rates (in beats per minute) of 19 college students. Find the range of the distribution.

71 104 76 88 78 71 68 86 70 90 74 76 69 68 88 96 68 82 120

11. **Copper Mining** In March 2021, a Canadian company found substantial deposits of copper at shallow depths in Arizona's Copper World region, which previously yielded about 440,000 tons of high-quality copper from 1874 to 1969. The company drilled test holes in several randomly selected locations at each of the old mine sites, and measured the amount of copper (as a percentage) in the rock extracted from each hole. Here are the data from the 28 test holes drilled at the Broad Top Butte site, along with a dotplot.[26] Find the range of the distribution.

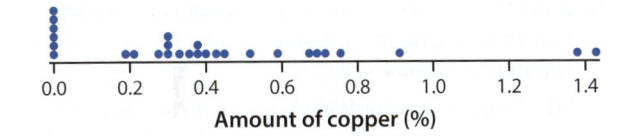

0.00 0.75 1.43 0.43 0.19 0.52 0.00 0.00 0.20 1.38 0.00 0.28 0.71 0.30
0.30 0.44 0.59 0.91 0.33 0.39 0.00 0.70 0.37 0.38 0.67 0.00 0.38 0.30

12. **Soccer Stars** Carli Lloyd, Alex Morgan, and Megan Rapinoe were members of the powerhouse 2019 U.S. Women's National Soccer Team that went on to win the 2019 Women's World Cup. Here are data on the number of goals scored by the team in 24 games played in the 2019 season, along with a dotplot.[27] Find the range of the distribution.

1 1 2 2 1 5 6 3 5 3 13 3 2 2 2 2 2 3 4 3 2 1 3 6

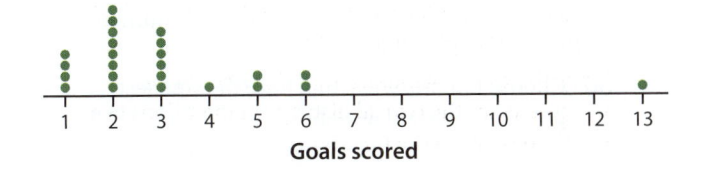

13. **Refer to Example 3.2.2 Golf Simulator** Zufan is a member of the golf team and is thinking about getting a new brand of 7-iron that advertises greater consistency. Before making a decision, Zufan hits 10 shots with the club in a golf simulator. Here are the distances (in yards) for each of the 10 shots.[28] Calculate and interpret the standard deviation.

159 146 153 159 161 142 165 163 154 148

14. **Varying Metabolism** A person's metabolic rate is the rate at which the body consumes energy. Metabolic rate is important in studies of weight gain, dieting, and exercise. Here are the metabolic rates of 7 people who took part in a study on dieting.[29] (The units are calories per 24 hours. These are the same calories used to describe the energy content of foods.) Calculate and interpret the standard deviation.

1792 1666 1362 1614 1460 1867 1439

15. **ASA on Instagram** The American Statistical Association (ASA) has an Instagram account (@amstatnews) to post updates on new statistical publications. Here is a dotplot of the number of Instagram "likes" for 10 posts selected at random.[30] Calculate and interpret the standard deviation.

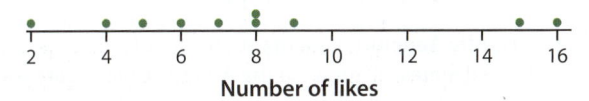

16. **Digital Photos** How much storage space do digital photos use? Here is a dotplot of the file sizes (to the nearest tenth of a megabyte) for 18 randomly selected photos in Noemi's cloud storage.[31] Calculate the standard deviation. Interpret this value.

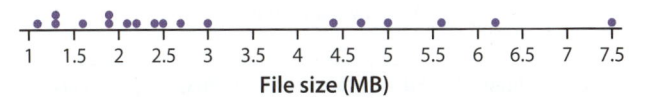

17. **Refer to Example 3.2.3 Fuel Efficiency** Refer to Exercise 9. Find the interquartile range. Interpret this value.

18. **Heart Rates** Refer to Exercise 10. Find the interquartile range. Interpret this value.

19. **Copper Mining** Refer to Exercise 11.
   (a) Find and interpret the *IQR*.
   (b) Which measure best describes the variability of the distribution: the range, the interquartile range, or the standard deviation? Explain your answer.

20. **Soccer Stars** Refer to Exercise 12.
   (a) Find and interpret the *IQR*.
   (b) Which measure best describes the variability of the distribution: the range, the interquartile range, or the standard deviation? Explain your answer.

Applying the Concepts These exercises ask you to apply multiple learning goals in a new context or to apply what you learned in this section in a new way.

21. **Electoral College** To become president of the United States, a candidate has to win a majority of the 538 electoral votes that are cast in the Electoral College. Here is a dotplot of the number of electoral votes in 2021 for each of the 50 states and the District of Columbia, along with numerical summaries.

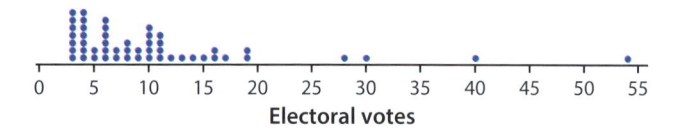

n	Mean	SD	Min	Q_1	Med	Q_3	Max
51	10.549	9.558	3	4	8	12	54

   (a) Find the range of the distribution.
   (b) Interpret the standard deviation.
   (c) Find and interpret the interquartile range.
   (d) Which measure best summarizes the variability of the distribution? Justify your answer.

22. **Atlantic Acorns** Of the many species of oak trees in the United States, 28 grow on the Atlantic Coast. Here is a dotplot of data on the average volumes of acorns (in cubic centimeters) for each of these oak species, along with numerical summaries.[32]

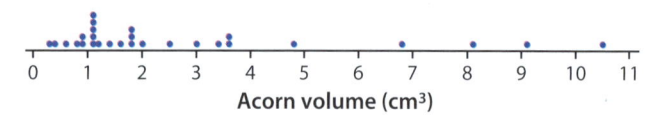

n	Mean	SD	Min	Q_1	Med	Q_3	Max
28	2.729	2.727	0.3	1.1	1.7	3.5	10.5

   (a) Find the range of the distribution.
   (b) Interpret the standard deviation.
   (c) Find and interpret the interquartile range.
   (d) Which measure best summarizes the variability of the distribution? Justify your answer.

23. **Electoral College** Refer to Exercise 21. Find the percentage of data values that fall within 1, 2, and 3 standard deviations of the mean. Explain why these percentages do not match the approximately 68%–95%–99.7% of values specified by the empirical rule.

24. **Atlantic Acorns** Refer to Exercise 22. Find the percentage of data values that fall within 1, 2, and 3 standard deviations of the mean. Explain why these percentages do not match the approximately 68%–95%–99.7% of values specified by the empirical rule.

25. **Estimating SD** The dotplot shows the number of shots per game taken by National Hockey League (NHL) player Sidney Crosby in 81 regular season games in a recent season.[33] Is the standard deviation of this distribution closest to 2, 5, or 10? Explain your answer without doing any calculations.

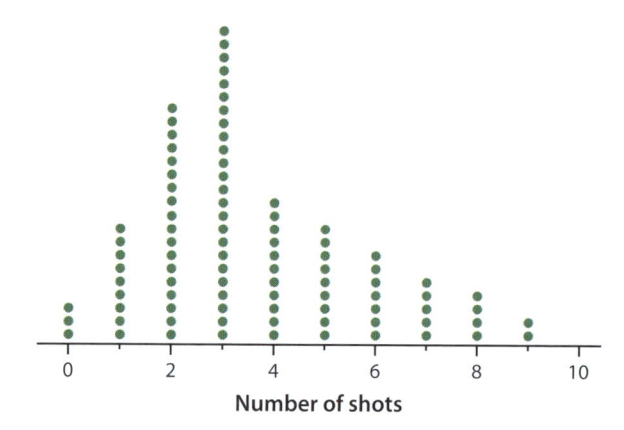

26. **Comparing SD** Which of the following distributions has a larger standard deviation? Justify your answer.

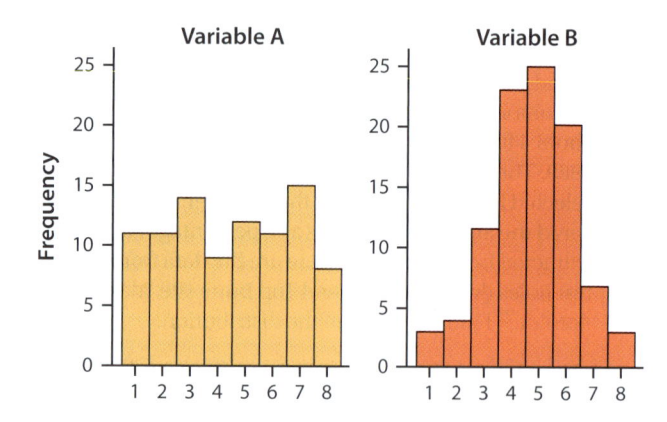

27. **Stats Homework** An Introductory Statistics student's first 14 online homework scores had a mean of 85 and a standard deviation of 8.

   (a) Suppose this student makes an 85 on the next online homework assignment. Would the standard deviation of the student's 15 online homework scores be greater than, equal to, or less than 8? Justify your answer.

   (b) Suppose instead that this student does not complete the next assignment, and earns a score of 0. Would the standard deviation of the student's 15 online homework scores be greater than, equal to, or less than 8? Justify your answer.

28. **SD contest** This is a standard deviation contest. You must choose four numbers from the whole numbers 0 to 10, with repeats allowed.

   (a) Choose four numbers that have the smallest possible standard deviation.

(b) Choose four numbers that have the largest possible standard deviation.

(c) Is more than one choice possible in either part (a) or (b)? Explain your reasoning.

29. **Tech** **Cholesterol Levels** The National Health and Nutrition Examination Survey (NHANES) is an ongoing research program conducted by the National Center for Health Statistics. Researchers selected a random sample of more than 7000 U.S. residents for the most recent survey and recorded lab measurements of their blood cholesterol levels. The **NHANES** data set gives the cholesterol levels in mg/dL (*LBXTC*) and in mmol/L (*LBDTCSI*).[34]

(a) Use technology to make an appropriate graph of the cholesterol levels (in mg/dL). Describe the shape of the distribution.

(b) Calculate the mean and the median cholesterol level. Which measure would you choose to describe the center of the distribution? Explain your answer.

(c) Calculate the range, standard deviation, and interquartile range. Which measure would you choose to describe the variability of the distribution? Explain your answer.

30. **Tech** **Possum Heads** Zoologists in Australia captured a random sample of 104 mountain brushtail possums and recorded the values of 13 different variables, including head length (in millimeters).[35] The **Possum** data set gives the data on head length in the column labeled *hdlngth*.

(a) Use technology to make an appropriate graph of the head length data for all of the possums. Describe the shape of the distribution.

(b) Calculate the mean and the median head length. Which measure would you choose to describe the center of the distribution? Explain your answer.

(c) Calculate the range, standard deviation, and interquartile range. Which measure would you choose to describe the variability of the distribution? Explain your answer.

Extending the Concepts These exercises challenge you to explore statistical concepts and methods that go beyond what you learned in this section.

31. **Student Sleep** Researchers used data from the American Time Use Survey to investigate how much sleep high school students get per night. The histogram summarizes the data on reported sleep time (in hours per night) from the more than 2700 students who participated in the survey.[36]

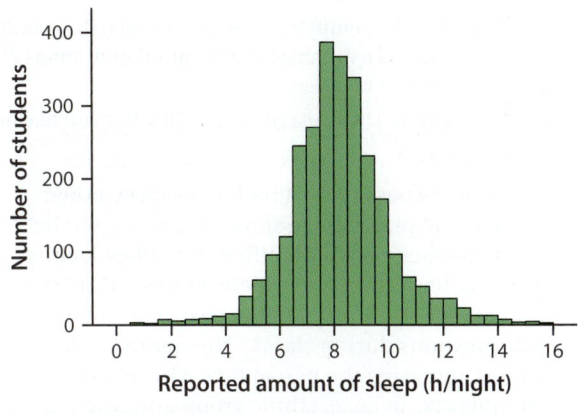

(a) Describe the shape of the distribution.

(b) About what percentage of the values in this distribution should fall within 2 standard deviations of the mean? Justify your answer.

(c) Estimate the mean and the standard deviation of the distribution.

32. **Chebyshev's Inequality** An interesting result known as *Chebyshev's inequality* says that in *any* distribution of quantitative data, at least $100\left(1-\dfrac{1}{k^2}\right)$% of the values are within k standard deviations of the mean. If $k=2$, for example, Chebyshev's inequality tells us that at least

$$100\left(1-\dfrac{1}{2^2}\right)=75\%$$

of the values in any distribution are within 2 standard deviations of the mean. For roughly symmetric, single-peaked, mound-shaped distributions, we know that about 95% of the values are within 2 standard deviations of the mean.

(a) Make a table that shows what percentage of data values must fall within 1, 2, 3, 4, and 5 standard deviations of the mean in any distribution.

(b) Explain why values 5 or more standard deviations from the mean in any distribution should be considered unusual.

33. **Stock Volatility** Many stock traders pay close attention to how much the price of a particular stock varies from day to day, a factor known as its *volatility*. One common way to measure volatility is with the standard deviation of the stock's price over many days or months. Another way is with the *coefficient of variation (CV)*, defined as

$$CV = \dfrac{SD}{mean}$$

The coefficient of variation measures variability relative to the mean of the distribution.

Over a 60-month period, the price of Apple stock at the start of trading on the first day of the month has a mean of $57.86 and a standard deviation of $32.15.[37]

(a) Interpret the standard deviation.

(b) Calculate the coefficient of variation. Explain what this value represents.

34. Variable Words Identify four words for which the number of syllables is more variable than the number of letters.[38]

Cumulative Review These exercises revisit what you learned in previous sections.

35. College Success (2.2) A national survey asked 95,505 first-year college students about specific academic behaviors identified by college faculty as being important for student success. One question asked, "How often in the past year did you ask questions during class?" The figure is a bar chart comparing the percentage who answered "frequently" by race/ethnic group and whether the respondent is a first-generation college student.[39] Describe what you see.

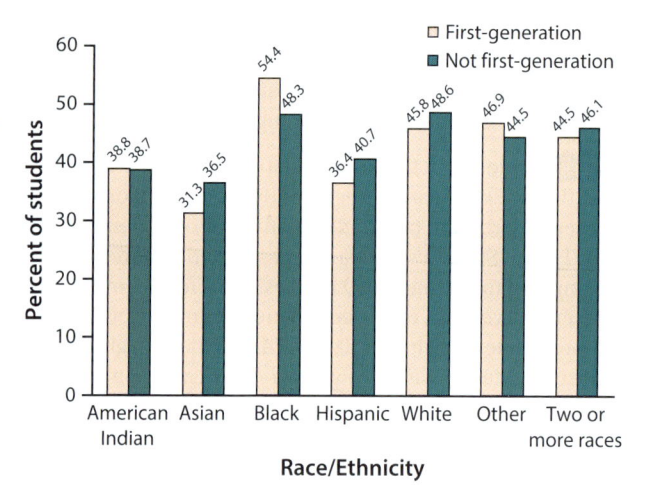

Section 3.3 Boxplots and Outliers

LEARNING GOALS

By the end of this section, you will be able to:

- Use the 1.5 × *IQR* rule to identify outliers.
- Make and interpret boxplots of quantitative data.
- Compare distributions of quantitative data with boxplots.

LeBron James emerged as a superstar in the National Basketball Association (NBA) during his rookie season (2003–2004). He maintained a consistent level of excellence over the first 16 years of his professional career, reaching the NBA Finals 8 consecutive times and winning 3 NBA championships. The dotplot shows the average number of points per game that LeBron scored in each of these 16 seasons.[40]

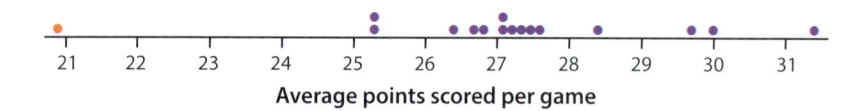

LeBron's 20.9 points per game average in his rookie season stands out (in red) from the rest of the distribution. Should this value be classified as an *outlier*?

Identifying Outliers

Besides serving as a measure of variability, the interquartile range (*IQR*) is often used as a ruler for identifying outliers.

HOW TO Identify Outliers: The 1.5 × *IQR* Rule

Call an observation an outlier if it falls more than 1.5 × *IQR* above the third quartile or more than 1.5 × *IQR* below the first quartile. That is,

Low outliers $< Q_1 - 1.5 \times IQR$ High outliers $> Q_3 + 1.5 \times IQR$

EXAMPLE 3.3.1

An NBA Legend

Identifying outliers

PROBLEM

Here are data on the average number of points per game that LeBron James scored in each of his first 16 NBA seasons. Use the $1.5 \times IQR$ rule to identify any outliers in the distribution.

20.9 27.2 31.4 27.3 30.0 28.4 29.7 26.7 27.1 26.8 27.1 25.3 25.3 26.4 27.5 27.4

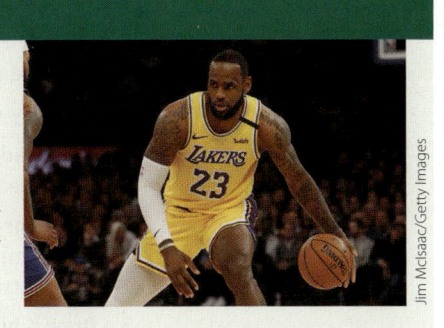

SOLUTION

20.9 25.3 25.3 26.4 | 26.7 26.8 27.1 27.1 || 27.2 27.3 27.4 27.5 | 28.4 29.7 30.0 31.4

$Q_1 = 26.55$ Median = 27.15 $Q_3 = 27.95$

Find the interquartile range (IQR) using the method described in Section 3.2. Then calculate the upper and lower cutoff values for outliers.

$IQR = Q_3 - Q_1 = 27.95 - 26.55 = 1.40$

Outliers $< Q_1 - 1.5 \times IQR = 26.55 - 1.5 \times 1.40 = 24.45$

Outliers $> Q_3 + 1.5 \times IQR = 27.95 + 1.5 \times 1.40 = 30.05$

LeBron James's rookie-season average of 20.9 points per game is a low outlier because it is less than 24.45. The season when he averaged 31.4 points per game is a high outlier because it is greater than 30.05.

EXAM PREP: FOR PRACTICE, TRY EXERCISE 7.

It is important to identify outliers in a distribution for several reasons:

1. **They might be inaccurate data values.** Maybe someone recorded a value as 10.1 instead of 101. Perhaps a measuring device broke down. Or maybe someone gave a silly response, like the student in a class survey who claimed to study 30,000 minutes per night!
2. **They can indicate a remarkable occurrence.** For example, in a graph of career earnings of professional tennis players, Serena Williams is likely to be an outlier.
3. **They can heavily influence the values of some summary statistics,** such as the mean, range, and standard deviation.

Making and Interpreting Boxplots

You can use a dotplot, stemplot, or histogram to display the distribution of a quantitative variable. Another graphical option for quantitative data is a **boxplot** (sometimes called a *box-and-whisker plot*). A boxplot summarizes the data by displaying the locations of five important values within the distribution, known as its **five-number summary.**

> The **five-number summary** of a distribution of quantitative data consists of the minimum, the first quartile Q_1, the median, the third quartile Q_3, and the maximum.

> A **boxplot** is a visual representation of the five-number summary.

It is possible to make a boxplot by hand for relatively small quantitative data sets. For details on making boxplots with technology, see the Tech Corner at the end of the section.

> **HOW TO** Make a Boxplot
>
> 1. **Find the five-number summary** for the distribution.
> 2. **Identify outliers** using the $1.5 \times IQR$ rule.
> 3. **Draw and label the axis.** Draw a horizontal axis and put the name of the quantitative variable underneath it.
> 4. **Scale the axis.** Look at the smallest and largest values in the data set. Start the horizontal axis at a number less than or equal to the smallest value and place tick marks at equal intervals until you equal or exceed the largest value.
> 5. **Draw a box** that spans from the first quartile (Q_1) to the third quartile (Q_3).
> 6. **Mark the median** with a vertical line segment that's the same height as the box.
> 7. **Mark any outliers** with a special symbol such as an asterisk (*).
> 8. **Draw whiskers** — lines that extend from the ends of the box to the smallest and largest data values that are *not* outliers.

To see the steps for making a boxplot in action, let's start with a dotplot. The following figure shows a dotplot of LeBron James's average points per game for each of 16 seasons. We have labeled the first quartile, the median, and the third quartile and marked them with vertical orange line segments. The process of testing for outliers with the $1.5 \times IQR$ rule is illustrated above the dotplot. At the bottom of the figure is the completed boxplot. Because the minimum and maximum values are outliers, we draw the whiskers to the smallest and largest data values that are not outliers.

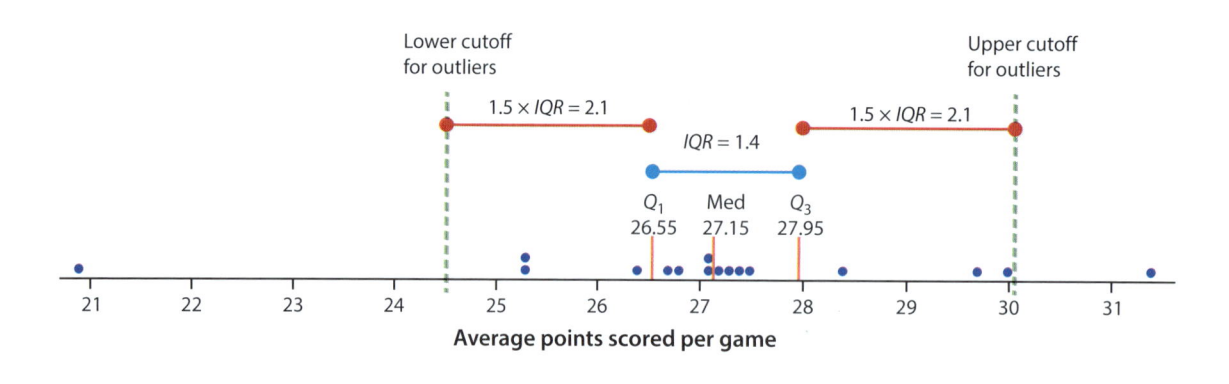

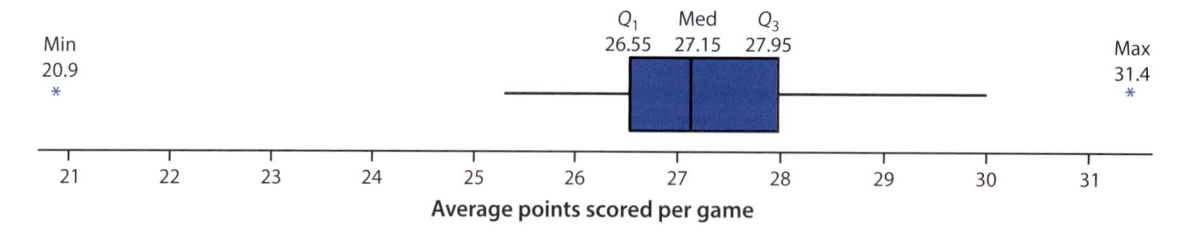

We can see from the boxplot that the distance from the end of the left whisker to the median is smaller than the distance from the median to the end of the right whisker. So we describe LeBron James's average points scored per game distribution as slightly right-skewed with one low outlier and one high outlier.

EXAMPLE 3.3.2

How big are the large fries?

Making and interpreting a boxplot

PROBLEM

According to nutrition information provided by Burger King, the serving size for its large fries is 173 grams.[41] Two young researchers wondered if the company was exaggerating the serving size. To find out, they went to several Burger King restaurants in their area and ordered a total of 15 large fries. The weights of the 15 orders (in grams) are shown here.[42]

178 176 173 172 179 165 179 181 186 184 181 180 183 183 187

(a) Make a boxplot to display the data.

(b) Does the graph in part (a) support the researchers' suspicion that Burger King is exaggerating the serving size of its large fries? Explain your reasoning.

SOLUTION

(a)   165 172 173 (176) 178 179 179 (180) 181 181 183 (183) 184 186 187

    Min Q_1 Med Q_3 Max

$IQR = Q_3 - Q_1 = 183 - 176 = 7$

Low outliers $< Q_1 - 1.5 \times IQR = 176 - 1.5 \times 7 = 165.5$

High outliers $> Q_3 + 1.5 \times IQR = 183 + 1.5 \times 7 = 193.5$

The order of large fries that weighed 165 grams is an outlier.

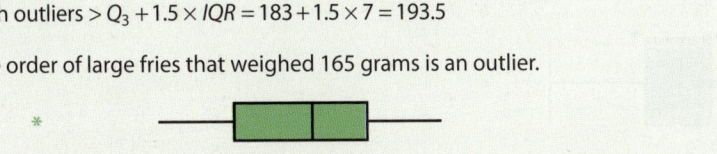

1. Find the five-number summary.
2. Identify outliers.
3. Draw and label the axis.
4. Scale the axis.
5. Draw a box from Q_1 to Q_3.
6. Mark the median.
7. Mark any outliers.
8. Draw whiskers.

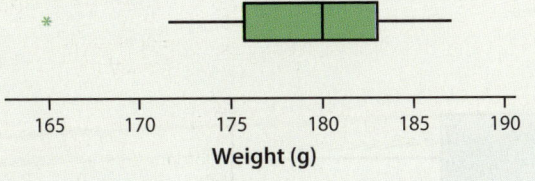

(b) No. From the boxplot, $Q_1 = 176$, so at least 75% of the orders of large fries that the researchers bought from local Burger King restaurants weighed 176 grams or more. Only the outlier (165 grams) and one other order (172 grams) of large fries weighed less than the advertised 173-gram serving size.

EXAM PREP: FOR PRACTICE, TRY EXERCISE 11.

Boxplots provide a quick summary of the center and variability of a distribution. The median is displayed as a vertical line in the central box, the interquartile range is the length of the box, and the range is the length of the entire plot, including outliers.

Boxplots do not display each individual value in a distribution. **Also, boxplots don't show gaps, clusters, or peaks.** For instance, the dotplot below displays the duration (in minutes) of 263 eruptions of the Old Faithful geyser in July 1995, when the Starnes family made its first visit to Yellowstone National Park. The distribution of eruption durations is clearly bimodal (two-peaked). But a boxplot of the data hides this important information about the shape of the distribution.

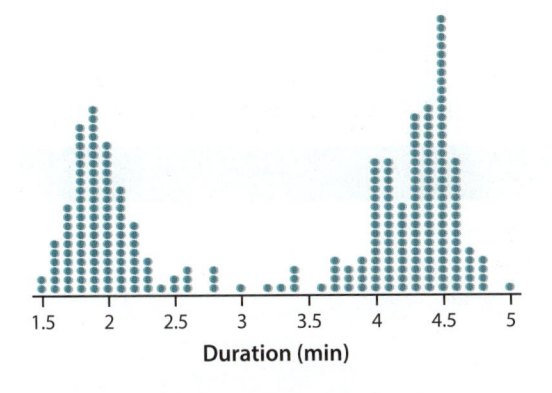

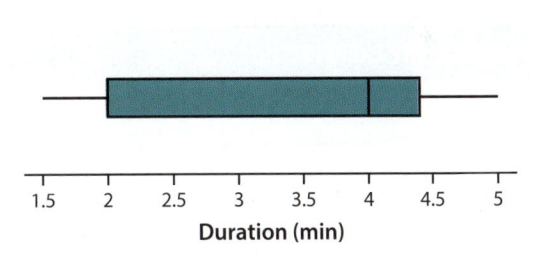

Comparing Distributions with Boxplots

Boxplots are especially effective for comparing the distribution of a quantitative variable in two or more groups.

EXAMPLE 3.3.3

Who is the GOAT?

Comparing distributions with boxplots

PROBLEM

Which NBA player is the Greatest of All Time: LeBron James or Michael Jordan? Here are data on the average points scored per game by each player in the seasons they played through 2019.[43]

| James | 20.9 | 27.2 | 31.4 | 27.3 | 30.0 | 28.4 | 29.7 | 26.7 | 27.1 | 26.8 | 27.1 | 25.3 | 25.3 | 26.4 | 27.5 | 27.4 |
| Jordan | 28.2 | 22.7 | 37.1 | 35.0 | 32.5 | 33.6 | 31.5 | 30.1 | 32.6 | 26.9 | 30.4 | 29.6 | 28.7 | 22.9 | 20.0 | |

Here are numerical summaries and parallel boxplots of the data. Compare the distributions.

Player	$\bar{x}$	s_x	Min	Q_1	Med	Q_3	Max	IQR
James	27.16	2.33	20.90	26.55	27.15	27.95	31.40	1.40
Jordan	29.45	4.76	20.00	26.90	30.10	32.60	37.10	5.70

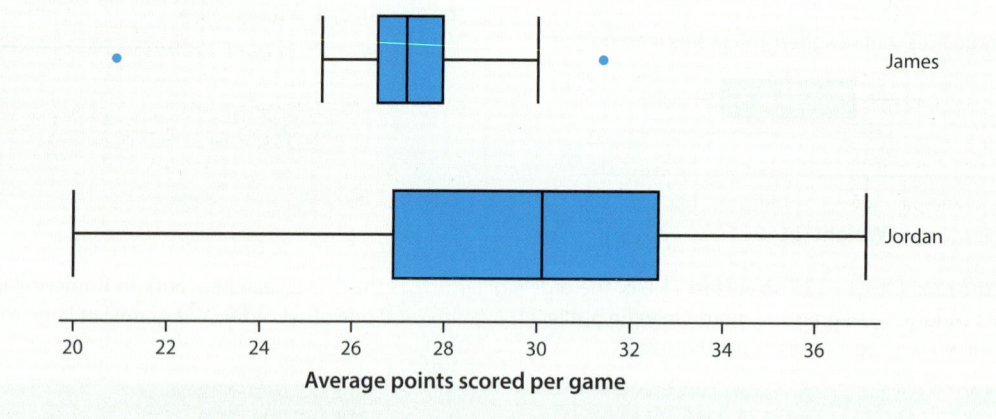

SOLUTION

Shape: LeBron James's distribution of average points per game is slightly right-skewed. Michael Jordan's distribution is slightly left-skewed.

Outliers: The seasons in which James averaged 20.9 and 31.4 points per game (ppg) are outliers. Jordan's distribution has no outliers.

Center: Jordan's average points per game over 15 seasons has a median of 30.1 ppg, almost 3 points higher than James's median of 27.15 ppg. In fact, James averaged more points per game than Jordan's median in only one season!

Variability: There is much more variation in Jordan's average points per game from season to season (IQR = 5.7 ppg) than in James's (IQR = 1.4 ppg).

Michael Jordan is the GOAT!

> Remember to compare shape, outliers, center, and variability!

> When boxplots are provided, we typically use the median and IQR to compare center and variability.

EXAM PREP: FOR PRACTICE, TRY EXERCISE 15.

TECH CORNER Making a Boxplot

You can use technology to make a boxplot. We'll illustrate using data from Example 3.3.3, "Who is the GOAT?"

Applet

1. Go to www.stapplet.com and launch the *One Quantitative Variable, Multiple Groups* applet.
2. Enter Average points scored per game as the Variable name.
3. Select Raw data as the input method.
4. Name Group 1 "James" and Group 2 "Jordan." Enter the data for each group manually, or copy and paste them from a document or spreadsheet. Be sure to separate the data values with commas or spaces.
5. Click the Begin analysis button. Parallel dotplots of the data should appear. Change the graph type to boxplot.

Graph Distributions

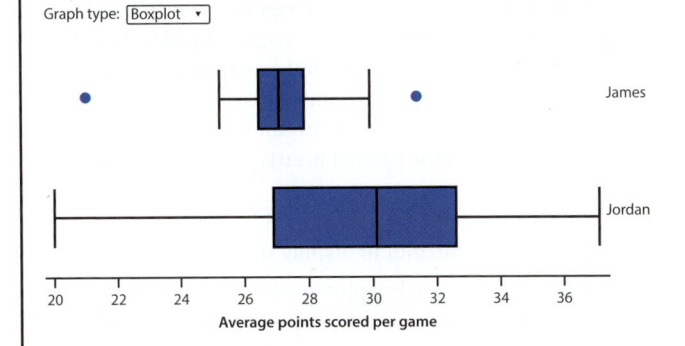

Graph type: Boxplot

TI-83/84

1. Enter the scoring data for James in list L_1 and for Jordan in list L_2.
2. Set up two statistics plots: Plot1 to show a boxplot of James's data and Plot2 to show a boxplot of Jordan's data. The setup for Plot1 is shown. When you define Plot2, be sure to change L_1 to L_2.

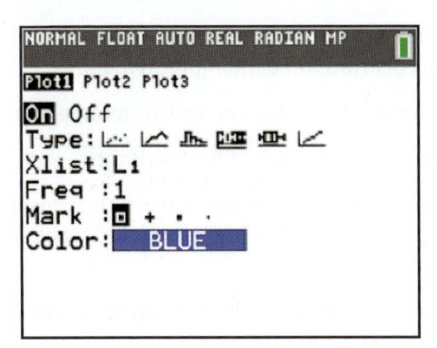

Note: The calculator offers two types of boxplots: one that shows outliers and one that doesn't. We'll always use the type that identifies outliers.

3. Press ZOOM and select ZoomStat to display the parallel boxplots. Then press TRACE and use the arrow keys to view the five-number summary.

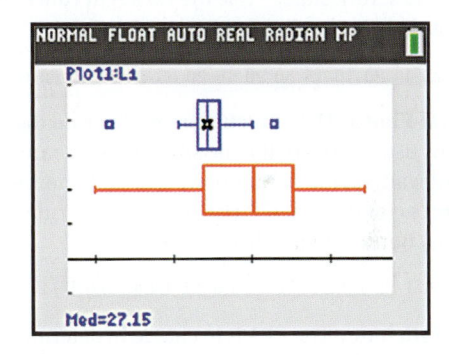

Detailed instructions for using CrunchIt!, Excel, Google Sheets, JMP, Minitab, and R are available in Achieve.

Section 3.3 What Did You Learn?

Review the learning goals from this section. Then practice what you've learned by working through the exercises.

Learning Goals	Example	Exercises
Use the $1.5 \times IQR$ rule to identify outliers.	3.3.1	7–10
Make and interpret boxplots of quantitative data.	3.3.2	11–14
Compare distributions of quantitative data with boxplots.	3.3.3	15–18

Section 3.3 Exercises

Building Concepts and Skills These exercises assess the basic knowledge you should have after reading the section.

1. True/False: The $1.5 \times IQR$ rule says that an outlier is any data value that is more than $1.5 \times IQR$ above or below the median.

2. Give three reasons why it is important to identify outliers in a distribution of quantitative data.

3. What are the five values in the five-number summary?

4. In a boxplot, the width of the box is the _____.

5. True/False: When there are outliers, the whiskers in a boxplot extend to the lower/upper cutoff values for outliers.

6. What important information about the shape of a distribution could a boxplot hide?

Mastering Concepts and Skills These exercises reinforce the learning goals as shown in the examples.

7. **Refer to Example 3.3.1 New York Commutes** How long do people typically spend traveling to work? The answer may depend on where they live. Here are the travel times (in minutes) of 20 randomly chosen workers in New York state.[44] Use the $1.5 \times IQR$ rule to identify any outliers in the distribution.

   10 30 5 25 40 20 10 15 30 20 15 20 85 15 65 15 60 60 40 45

8. **Digital Photos** How much storage space do digital photos use? Here are the file sizes (to the nearest tenth of a megabyte) for 18 randomly selected photos in Noemi's cloud storage.[45] Use the $1.5 \times IQR$ rule to identify any outliers in the distribution.

   2.4 2.7 1.6 1.3 6.2 1.3 5.6 1.1 2.2 1.9 2.1 4.4 4.7 3.0 1.9 2.5 7.5 5.0

9. **Fuel Efficiency** The EPA is in charge of determining and reporting fuel economy ratings for cars in the United States. The dotplot displays the EPA estimates of highway gas mileage in miles per gallon (mpg) for a random sample of 21 model year 2020 midsize cars.[46] Use the $1.5 \times IQR$ rule to identify any outliers in the distribution.

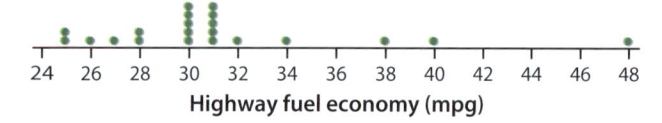

10. **July Heat** The dotplot displays the high temperature readings in degrees Fahrenheit for Phoenix, Arizona, for each day in July in a recent year.[47] Use the $1.5 \times IQR$ rule to identify any outliers in the distribution.

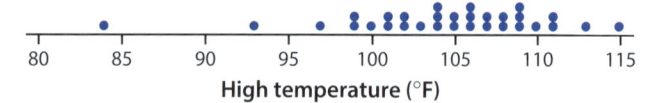

11. **Refer to Example 3.3.2 Copper Mining** In March 2021, a Canadian company found substantial deposits of copper at shallow depths in Arizona's Copper World region, which previously yielded about 440,000 tons of high-quality copper from 1874 to 1969. The company drilled test holes in several randomly selected locations at each of the old mine sites, and measured the amount of copper (as a percentage) in the rock extracted from each hole. Here are the data from the 28 test holes drilled at the Broad Top Butte site.[48]

    0.00 0.75 1.43 0.43 0.19 0.52 0.00 0.00 0.20 1.38 0.00 0.28 0.71 0.30
    0.30 0.44 0.59 0.91 0.33 0.39 0.00 0.70 0.37 0.38 0.67 0.00 0.38 0.30

    (a) Make a boxplot to display the data.

    (b) A copper mine today has about 0.6% copper in its rock. Based on the boxplot in part (a), what percentage of the rock at Broad Top Butte should the company estimate is more than 0.6% copper: less than 25%, between 25% and 50%, between 50% and 75%, or more than 75%? Explain your answer.

12. **Daily Texts** According to blogger Kenneth Burke, adults ages 18 to 24 send an average of 64 texts per day.[49] Dr. Williams suspected that this value was exaggerated, and collected data from a class of Introductory Statistics students on the number of texts they had sent or received in the past 24 hours. Here are the data:

    0 7 1 29 25 8 5 1 25 98 9 0 268 118 72 0 92 52 14 3 3 44 5 42

    (a) Make a boxplot to display the data.

    (b) Based on the boxplot in part (a), what percentage of Dr. Williams's students sent more than 64 texts in the previous day: less than 25%, between 25% and 50%, between 50% and 75%, or more than 75%? Do the data confirm Dr. Williams's suspicion? Explain your answer.

13. **New York Commutes** Refer to Exercise 7.

    (a) Make a boxplot to display the data.

    (b) Which measure of variability—the IQR or standard deviation—would you report for these data? Use the graph from part (a) to help justify your choice.

14. **Digital Photos** Refer to Exercise 8.

    (a) Make a boxplot to display the data.

    (b) Which measure of variability—the IQR or standard deviation—would you report for these data? Use the graph from part (a) to help justify your choice.

15. **Refer to Example 3.3.3 Overthinking It?** Athletes often comment that they try not to "overthink it" when competing in their sport. Is it possible to "overthink it"? To investigate, researchers put some golfers to the test. They recruited 40 experienced golfers and allowed them some time to practice their putting. After practicing, they randomly assigned the golfers in equal numbers to two groups. Golfers in one group were asked to write a detailed description of their putting technique (which could lead to "overthinking it"). Golfers in the other group were asked to do an unrelated verbal task for the same amount of time. After completing their

tasks, each golfer attempted putts from a fixed distance until they made three putts in a row. The boxplots summarize the distribution of the number of putts required for the golfers in each group to make three putts in a row.[50] Compare the distributions.

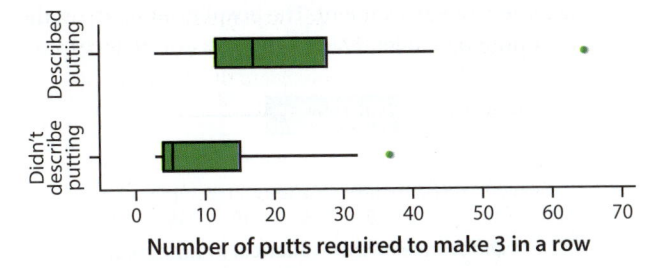

16. **Fast-Food Sandwiches** The following boxplots summarize data on the amount of fat (in grams) in 12 beef sandwiches and 9 chicken or fish sandwiches available on McDonald's menu in 2020.[51] Compare the distributions of fat content for the two types of sandwiches.

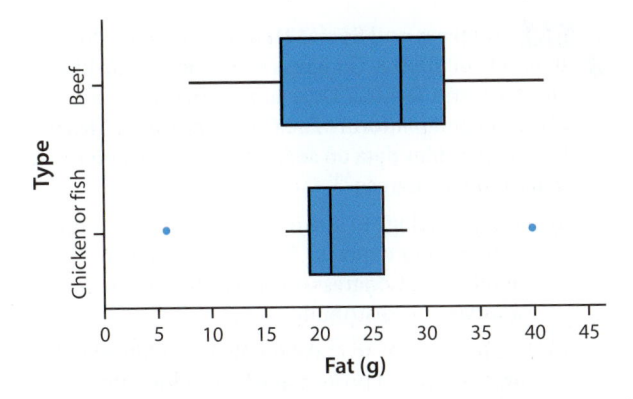

17. **Weight-Loss Strategies** Which dieting strategy works best? Researchers used data from the U.K. television show "How to Lose Weight Well" to investigate. In each episode, 6 participants are assigned in pairs to adopt 3 dieting strategies: crashers, shape shifters, and life changers. The crashers follow extreme diets for 1–2 weeks, such as replacing meals with bone broth or baby food. The shape shifters follow moderately extreme diets for 4–6 weeks, like the blood sugar diet that involves consuming at most 800 calories per day. The life changers follow less extreme diets for 12–16 weeks, like low-carbohydrate or Mediterranean diets. Here are boxplots that summarize the percentage of body weight lost by 137 participants in the TV show over a 5-year period.[52]

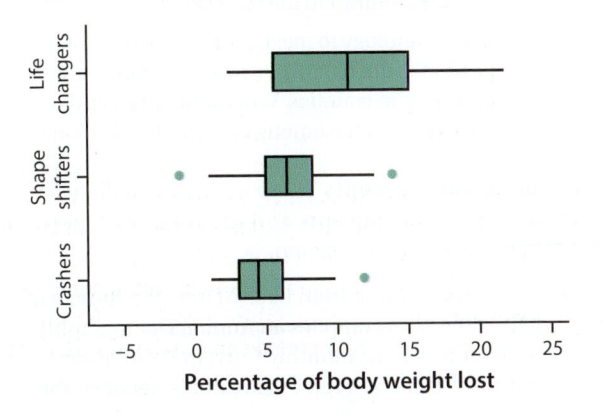

(a) From the boxplots, what can you say about the percentage of people in each of the three dieting groups who lost less than 5% of their body weight?

(b) Compare the distributions of percentage of body weight lost for the three dieting groups.

18. **Fridge Energy Costs** *Consumer Reports* magazine rated different types of refrigerators, including those with bottom freezers, those with top freezers, and those with side freezers. One of the variables measured was annual energy cost (in dollars). The following boxplots show the energy cost distributions for each of these types of refrigerators.[53]

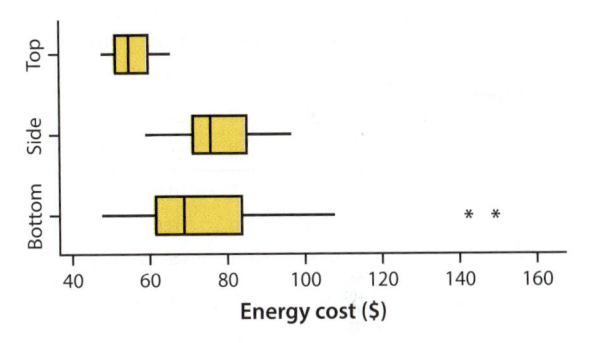

(a) From the boxplots, what can you say about the percentage of each type of freezer that costs more than $60 per year to operate?

(b) Compare the energy cost distributions for the three types of refrigerators.

Applying the Concepts These exercises ask you to apply multiple learning goals in a new context or to apply what you learned in this section in a new way.

19. **Tablet Ratings** In a recent year, *Consumer Reports* rated many tablet computers for performance and quality. Based on several variables, the magazine gave each tablet an overall rating, where higher scores indicate better ratings. The overall ratings of the tablets produced by Apple and Samsung are given here, along with numerical summaries of the data for each group.[54]

Apple   87 87 87 87 86 86 86 86 84 84 84 84 83 83 83 83 81 79 76 73
Samsung 88 87 87 86 86 86 86 86 84 84 83 83 77 76 76 75 75 75 75 74 71 62

Group name	n	Mean	SD	Min	Q_1	Med	Q_3	Max
Apple	20	83.45	3.762	73	83	84	86	87
Samsung	23	79.87	6.737	62	75	83	86	88

(a) Make parallel boxplots to compare the distributions.

(b) Based on your graphs in part (a), which company's tablet would you recommend buying? Give appropriate evidence to support your answer.

20. **Creativity and Motivation** Do external rewards — things like money, praise, fame, and grades — promote creativity? Researcher Teresa Amabile recruited 47 experienced creative writers who were college students and divided them at random into two groups. The students in one

group were given a list of statements about external reasons (E) for writing, such as public recognition, making money, or pleasing their parents. Students in the other group were given a list of statements about internal reasons (I) for writing, such as expressing themselves and enjoying playing with words. Both groups were then instructed to write a poem about laughter. Twelve different poets separately rated each student's poem using a creativity scale.[55] The researchers averaged these ratings to obtain an overall creativity score for each poem. The overall creativity scores for the students in each group are given here, along with numerical summaries of the data.

	12.0	12.0	12.9	13.6	16.6	17.2
Internal	17.5	18.2	19.1	19.3	19.8	20.3
	20.5	20.6	21.3	21.6	22.1	22.2
	22.6	23.1	24.0	24.3	26.7	29.7
	5.0	5.4	6.1	10.9	11.8	12.0
External	12.3	14.8	15.0	16.8	17.2	17.2
	17.4	17.5	18.5	18.7	18.7	19.2
	19.5	20.7	21.2	22.1	24.0	

Group name	n	Mean	SD	Min	Q_1	Med	Q_3	Max
Internal	24	19.883	4.44	12	17.35	20.4	22.4	29.7
External	23	15.739	5.253	5	12	17.2	19.2	24

(a) Make parallel boxplots to compare the distributions.

(b) Based on your graphs in part (a), what do you conclude about whether external rewards promote creativity? Explain your answer.

21. **Nitrogen-Filled Tires** Consumers Union designed an experiment to test whether nitrogen-filled tires would maintain pressure better than air-filled tires. The researchers obtained two tires from each of several brands and then randomly assigned one tire in each pair to be filled with air and the other to be filled with nitrogen. All tires were inflated to the same pressure and placed inside a warehouse for a year. At the end of the year, the researchers measured the pressure (in pounds per square inch) in each tire. The boxplot summarizes data on the difference in pressure loss (Air − Nitrogen) for the tires in each pair.[56]

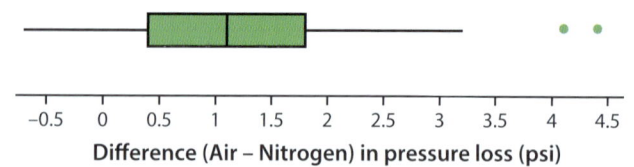

(a) Estimate the median and the *IQR* from the boxplot.

(b) Do these data provide strong evidence that nitrogen-filled tires tend to lose less pressure than air-filled tires? Justify your answer.

22. **Music and Math Tests** Does music hinder academic performance? Researchers investigated this question by having student volunteers complete two equivalent 50-question arithmetic tests with questions like 2+16 and 4+9. One test was given with music playing and the other test was given without music playing, with the order of the two tests determined at random. The amount of time (in seconds) to complete each test was recorded for each student. The boxplot summarizes the distribution of difference in time (Music − Silence).[57]

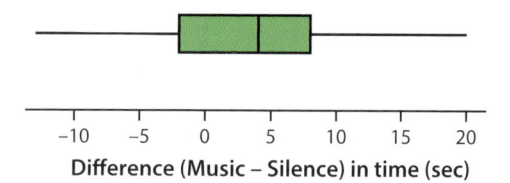

(a) Estimate the median and the *IQR* from the boxplot.

(b) Do these data provide strong evidence that music hinders performance on simple arithmetic tasks for students like the ones in this study? Justify your answer.

23. **Tech Congress and Social Media** How much do members of the U.S. Congress use social media like Facebook and Twitter? Does usage differ by party affiliation and platform? The **Congress Social Media** data set provides data on social media use by members of the 116th Congress.[58]

(a) Use technology to make parallel boxplots of the total number of posts *(Total_Posts)* made by members of Congress on Facebook and on Twitter. Compare the distributions.

(b) Use technology to make parallel boxplots of the total number of posts made by Democratic versus Republican members of Congress on Facebook and on Twitter. What similarities and differences do you see among the four distributions?

24. **Tech Air Quality** How does the air quality compare in different U.S. states? Is air quality similar in bordering states? The **Air Quality** data set provides data.[59]

(a) Use technology to make parallel boxplots of the median Air Quality Index *(Median_AQI)* for the counties in Alabama and Alaska. Compare the distributions.

(b) A "good day" is defined as one with an AQI value of 0 through 50. Calculate the proportion of good days in each county *(Good_Days / Days_w-AQI)*, and add this column to the data set.

(c) Use technology to make parallel boxplots of the proportion of good days in Arizona, California, Nevada, and Oregon counties. What similarities and differences do you see among the four distributions?

Extending the Concepts These exercises challenge you to explore statistical concepts and methods that go beyond what you learned in this section.

25. **Income and Education** Each March, the Bureau of Labor Statistics compiles an Annual Demographic Supplement to its monthly Current Population Survey.[60] Data on 71,067 individuals between the

ages of 25 and 64 who were employed full-time were collected in one of these surveys. The parallel boxplots compare the distributions of income for people with five levels of education. This figure is a variation of the boxplot idea: Because large data sets often contain very extreme observations, we omitted the individuals in each category with the top 5% and bottom 5% of incomes. Also, the whiskers are drawn all the way to the maximum and minimum values of the remaining data for each distribution.

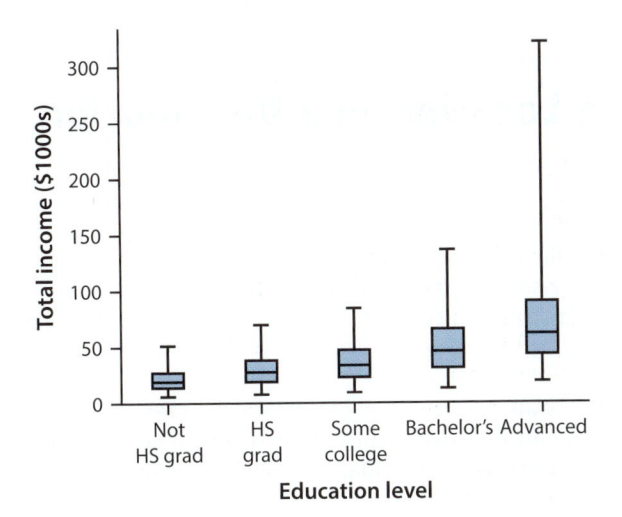

(a) What shapes do the distributions of income have?
(b) Explain how you know that there are outliers in the group that earned an advanced degree.
(c) How does the typical income change as education level increases? Why does this make sense?
(d) Describe how the variability in income changes as education level increases.

26. **Measuring Skewness** Here is a boxplot of the number of electoral votes in 2021 for each of the 50 states and the District of Columbia, along with summary statistics. You can see that the distribution is skewed to the right with four high outliers. How might we compute a numerical measure of skewness?

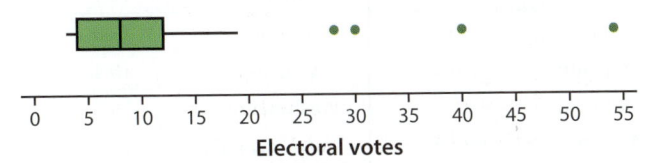

n	Mean	SD	Min	Q_1	Med	Q_3	Max
51	10.549	9.653	3	4	8	12	54

(a) One simple formula for calculating skewness is $\dfrac{\text{maximum} - \text{median}}{\text{median} - \text{minimum}}$. Compute this value for the electoral vote data. Explain why this formula should yield a value greater than 1 for a right-skewed distribution.

(b) Based solely on the summary statistics provided, define a formula for a different statistic that measures skewness. Compute the value of this statistic for the electoral vote data. What values of the statistic might indicate that a distribution is skewed to the right? Explain your reasoning.

27. **Outlier Detection** Some people identify any value more than 2 (or 3) standard deviations from the mean as an outlier in a quantitative data set. This method makes sense for roughly symmetric, single-peaked, mound-shaped distributions. Here are a histogram and numerical summaries of data on the percentage of butterfat in milk from 100 three-year-old Ayrshire cows.[61]

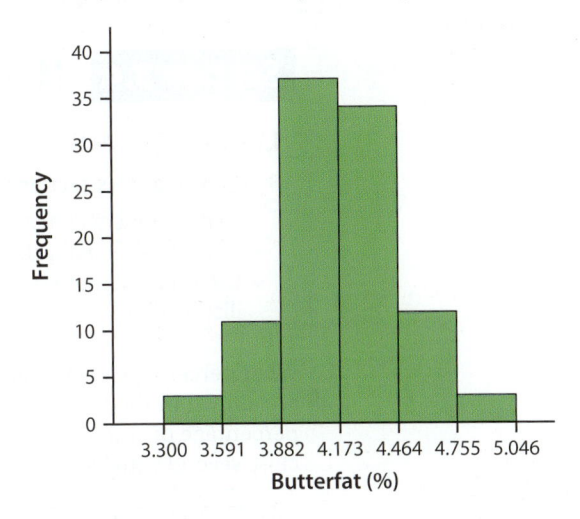

n	Mean	SD	Min	Q_1	Med	Q_3	Max
100	4.173	0.291	3.52	3.97	4.17	4.37	4.91

(a) About what percentage of values does the 2 SD rule identify as outliers?
(b) About what percentage of values does the 3 SD rule identify as outliers?
(c) Are there any outliers in the distribution according to the $1.5 \times IQR$ rule? Explain how you know.

Cumulative Review These exercises revisit what you learned in previous sections.

28. **Cherry Blossoms (2.6)** The timeplot shows the number of days after the start of the year when the cherry trees reached their peak bloom level in Kyoto, Japan, from 812 to 2021.[62] The thick orange line shows the 10-year moving average. Describe what you see.

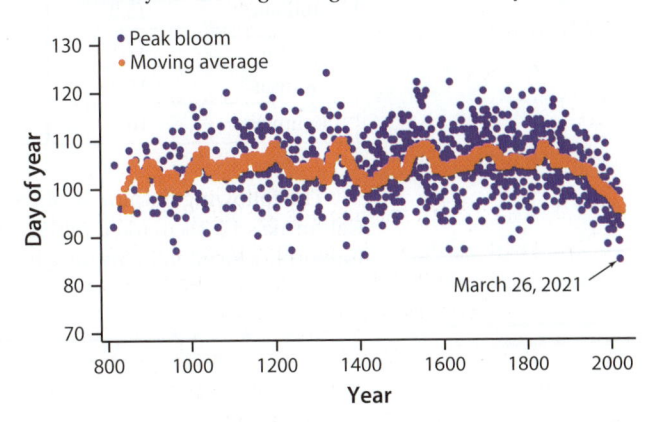

29. **AARP and Medicare (1.2)** To find out what proportion of all U.S. adults support proposed Medicare legislation to help pay medical costs, AARP conducted a survey of its members (people older than age 50 who pay membership dues). One of the questions was "Even if this plan won't affect you personally either way, do you think it should be passed so that seniors with low incomes or people with high drug costs can be helped?" Of the respondents, 75% answered yes.[63]

(a) Describe how undercoverage might lead to bias in this study. Is 75% likely greater than or less than the percentage of all U.S. adults who would answer "yes"?

(b) Describe how the wording of the question might lead to bias in this study. Is 75% likely greater than or less than the percentage of all U.S. adults who support proposed Medicare legislation to help pay medical costs?

Section 3.4 Measuring Location in a Distribution

LEARNING GOALS

By the end of this section, you will be able to:

- Find and interpret a percentile in a distribution of quantitative data.
- Find and interpret a standardized score (z-score) in a distribution of quantitative data.
- Use percentiles or standardized scores (z-scores) to compare the location of values in different distributions.

What percentage of residents in each U.S. state are age 65 or older? The data for a recent year are shown in the following table.[64] You might be surprised to see that Maine has the highest percentage of residents ages 65 and older with 20.6%. Florida, a popular state for retirees, is a close second with 20.5%.

State	Percent age 65 and older	State	Percent age 65 and older	State	Percent age 65 and older
Alabama	16.9	Louisiana	15.4	Ohio	17.1
Alaska	11.8	Maine	20.6	Oklahoma	15.7
Arizona	17.5	Maryland	15.4	Oregon	17.6
Arkansas	17.0	Massachusetts	16.5	Pennsylvania	18.2
California	14.3	Michigan	17.2	Rhode Island	17.2
Colorado	14.2	Minnesota	15.9	South Carolina	17.7
Connecticut	17.2	Mississippi	15.9	South Dakota	16.7
Delaware	18.7	Missouri	16.9	Tennessee	16.4
Florida	20.5	Montana	18.7	Texas	12.6
Georgia	13.9	Nebraska	15.8	Utah	11.1
Hawaii	18.4	Nevada	15.7	Vermont	19.3
Idaho	15.9	New Hampshire	18.1	Virginia	15.4
Illinois	15.6	New Jersey	16.2	Washington	15.5
Indiana	15.8	New Mexico	17.5	West Virginia	19.9
Iowa	17.1	New York	16.5	Wisconsin	16.9
Kansas	15.9	North Carolina	16.3	Wyoming	16.4
Kentucky	16.4	North Dakota	15.4		

The following dotplot displays the data on the percentage of residents ages 65 and older, with California's 14.3% marked in red. The distribution is roughly symmetric. How should we describe California's location in the distribution?

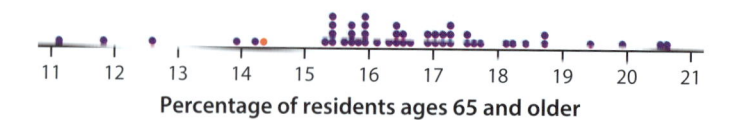

Finding and Interpreting Percentiles

One way to describe California's location in the distribution of percentage of residents ages 65 and older is to calculate its **percentile.** Recall that the three quartiles (Q_1, median, Q_3) divide a distribution of quantitative data into four groups of roughly equal size. The idea of a percentile is similar: The 99 percentiles divide a distribution into 100 groups of roughly equal size. This idea makes sense if a quantitative data set contains a large number of values, but isn't as helpful for smaller data sets.

> An individual's **percentile** is the percentage of values in a distribution that are less than the individual's data value.

From the dotplot, we see that 5 of the 50 data values (10%) are less than California's 14.3% of residents ages 65 and older. So California is at the 10th percentile of the distribution. There are four states with 15.4% of residents ages 65 and older: Louisiana, Maryland, North Dakota, and Virginia. By our definition, all four states are at the 14th percentile of the distribution because 7 out of 50 states have smaller data values.

Which state is at the 78th percentile of the distribution? The one with $(0.78)(50) = 39$ data values less than its own percentage of residents ages 65 and older. Alternatively, we can start with the maximum on the dotplot and count down $50 - 39 = 11$ values to find the 78th percentile. It's Oregon, with 17.6% of residents ages 65 and older.

Here are a few important notes about percentiles:

- ⚠ **Be careful with your language when describing percentiles.** Percentiles are specific locations in a distribution, so an observation isn't "in" the 78th percentile. Rather, it is "at" the 78th percentile.
- Some people define percentile as the percentage of values in a distribution that are *less than or equal to* an individual data value. Using this alternative definition of percentile, it is possible for an individual to fall at the 100th percentile.
- Percentiles are usually reported as whole numbers. Consider a quantitative data set with 43 values. How should we report the percentile for the individual with 30 of the 43 values in the distribution less than their data value? Because $30/43 = 0.698$, we say that this individual is at the 69th percentile of the distribution. We can't say the individual is at the 70th percentile because only 69.8% of the values in the data set are less than this individual's data value.

EXAMPLE 3.4.1

How much lead is in the water?

Finding and interpreting percentiles

PROBLEM

In April 2014, city managers in Flint, Michigan, decided to save money by using water from the Flint River rather than continuing to buy water sourced from Lake Huron. After the switch, Flint residents noticed a change in the look, smell, and taste of the water. Some residents developed rashes, hair loss, and itchy skin. City authorities insisted that drinking water from the Flint River was safe.

As part of its regular water quality testing program, city officials measured lead levels (in parts per billion) in 71 water samples collected from randomly selected Flint dwellings between January and June 2015. Here is a dotplot of the data.[65]

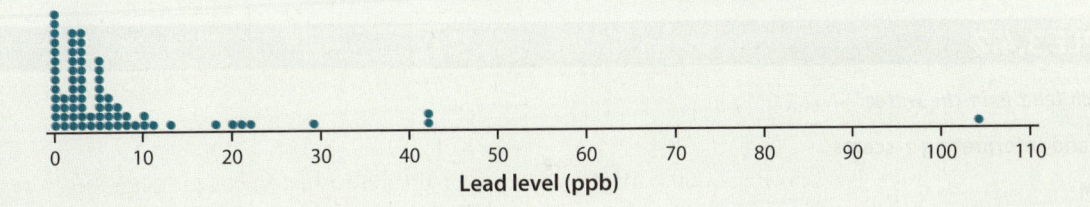

(a) Find the percentile for the water sample with a lead level of 9 ppb. Interpret this value.
(b) EPA regulations require action to be taken if the 90th percentile of lead level for the water samples taken exceeds 15 ppb. Based on these data, should action have been taken?

(continued)

SOLUTION

(a) $58/71 = 0.817$, so this water sample is at the 81st percentile in the distribution of lead level. About 81% of the Flint dwellings tested had water samples with lead levels less than 9 ppb.

> It is easier to start by counting how many values are at or above 9 ppb!

(b) $(0.90)(71) = 63.9$, so the 90th percentile is the data value that has 64 lead levels less than it. Because 64 of the values are less than 20, the 90th percentile is 20 ppb. This exceeds 15 ppb, so action should have been taken in Flint, Michigan, based on these data.

EXAM PREP: FOR PRACTICE, TRY EXERCISE 7.

The median of a distribution is roughly the 50th percentile. The first quartile Q_1 is roughly the 25th percentile of a distribution because it separates the lowest 25% of values from the upper 75%. Likewise, the third quartile Q_3 is roughly the 75th percentile.

Finding and Interpreting Standardized Scores (z-Scores)

A percentile is one way to describe an individual's location in a distribution of quantitative data. Another way is to give the **standardized score (z-score)** for the individual's location.

The **standardized score (z-score)** for an individual value in a distribution tells us how many standard deviations away from the mean the value falls, and in what direction. To find the standardized score (z-score), compute

$$z = \frac{\text{value} - \text{mean}}{\text{standard deviation}}$$

Values larger than the mean have positive z-scores. Values smaller than the mean have negative z-scores. Notice that *standardized scores have no units*. That's because the units of measurement in the numerator and denominator of the z-score formula cancel each other.

Let's return to the data on the percentage of state residents ages 65 and older. Here is the dotplot once again, with California's 14.3% marked in red. The table provides numerical summaries for these data.

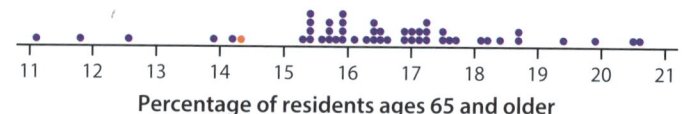

n	Mean	SD	Min	Q_1	Med	Q_3	Max
50	16.50	1.88	11.1	15.7	16.45	17.5	20.6

Where does California's 14.3% fall within the distribution? Its standardized score (z-score) is

$$z = \frac{\text{value} - \text{mean}}{\text{SD}} = \frac{14.3 - 16.50}{1.88} = -1.17$$

That is, California's percentage of residents ages 65 and older is 1.17 standard deviations below the mean for all 50 states.

EXAMPLE 3.4.2

How much lead is in the water?

Finding and interpreting z-scores

PROBLEM

Refer to Example 3.4.1. Numerical summaries of the data are shown in the following table. The water sample from LeeAnne Walters's home had a lead level of 104 ppb. Find and interpret the standardized score (z-score) for this value.

n	Mean	SD	Min	Q_1	Med	Q_3	Max
71	7.31	14.347	0	2	3	7	104

SOLUTION

$$z = \frac{104 - 7.31}{14.347} = 6.74$$

$$z = \frac{\text{value} - \text{mean}}{\text{SD}}$$

The water sample from LeeAnne Walters's home had a lead level 6.74 standard deviations above the mean for all 71 dwellings tested. That's very unusual!

According to Chebyshev's inequality (Section 3.2), a value at least 5 standard deviations from the mean is unusual in any distribution.

EXAM PREP: FOR PRACTICE, TRY EXERCISE 11.

Note: The 104 ppb lead level at LeeAnne Walters's home is even more exceptional since the water was filtered before being tested! After getting no action from local or state officials, LeeAnne contacted environmental researchers at Virginia Tech University, and worked with them to conduct more thorough water-quality testing that ultimately helped resolve the Flint water crisis.

Can we use the standardized score (z-score) of an individual data value to find its corresponding percentile, or vice versa? The answer is "no" for most distributions. For roughly symmetric, single-peaked, mound-shaped distributions, the answer is "yes." We will discuss the connection between z-scores and percentiles for *normal distributions* in Chapter 6.

THINK ABOUT IT

What happens if we standardize *all* the values in a distribution of quantitative data? Here are a dotplot and numerical summaries of the lead levels in the 71 Flint, Michigan, water samples.

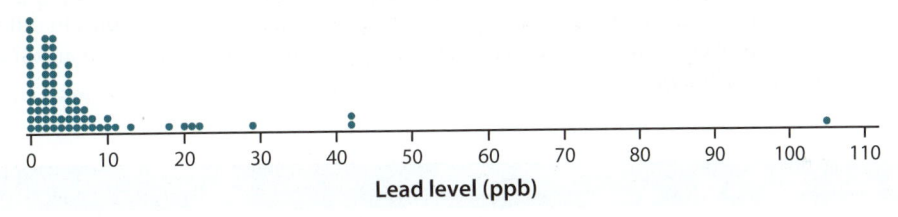

n	Mean	SD	Min	Q_1	Med	Q_3	Max
71	7.31	14.347	0	2	3	7	104

We calculate the standardized score for each water sample using

$$z = \frac{\text{value} - 7.31}{14.347}$$

In other words, we subtract 7.31 from each data value and then divide by 14.347. What effect do these transformations have on the shape, center, and variability of the distribution and on measures of location?

We start by subtracting 7.31 from each observation. Here are a dotplot and numerical summaries of the transformed data, which represent deviations from the mean lead level. Subtracting a constant did not change the shape of the distribution or measures of variability (range, standard deviation, *IQR*). Measures of center and location decreased by 7.31.

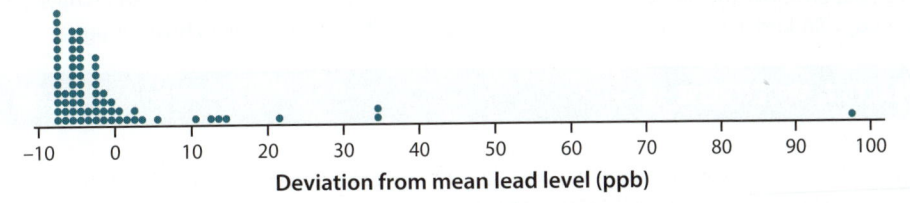

n	Mean	SD	Min	Q_1	Med	Q_3	Max
71	0	14.347	−7.31	−5.31	−4.31	−0.31	96.69

(continued)

Now we divide each deviation by 14.347 to get standardized scores. Here are a dotplot and numerical summaries of the z-scores. Dividing by a constant did not change the shape of the distribution—but the measures of variability, center, and location were all divided by 14.347.

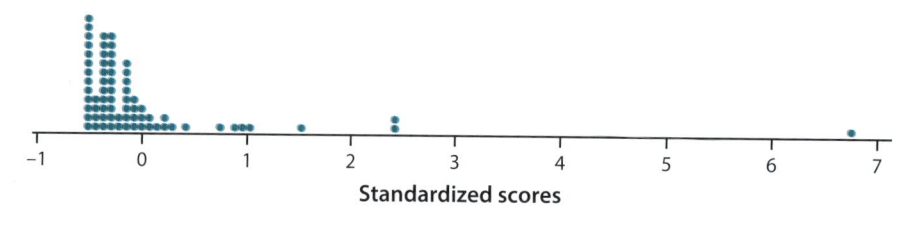

n	Mean	SD	Min	Q_1	Med	Q_3	Max
71	0	1	−0.51	−0.37	−0.3	−0.022	6.739

We would get equivalent results no matter what the original distribution looks like: The distribution of z-scores always has the same shape as the original distribution, a mean of 0, and a standard deviation of 1.

Comparing Location in Different Distributions

How can we compare the locations of values in different distributions of quantitative data? For instance, Shanice and Deiondre are both applying to nursing schools. Shanice takes the National League of Nursing (NLN) Pre-Admission Exam and earns a score of 111. Deiondre takes the Test of Essential Academic Skills (TEAS) Exam and earns a score of 88. Which person did better relative to others taking their respective exams? Shanice's 111 is at the 70th percentile of the NLN score distribution, while Deiondre's 88 is at the 80th percentile of the TEAS score distribution.[66] So Deiondre performed better relative to fellow exam-takers.

Percentiles are one option for comparing the location of individuals in different distributions. Standardized scores (z-scores) are another option if we know the mean and standard deviation of each distribution.

EXAMPLE 3.4.3

Growing like a beanstalk

Comparing location in different distributions

PROBLEM

Jordan (Mr. Tabor's daughter) was 55 inches tall at age 9. The distribution of height for 9-year-old girls has mean 52.5 inches and standard deviation 2.5 inches. Zayne (Mr. Starnes's grandson) was 58 inches tall at age 11. The distribution of height for 11-year-old boys has mean 56.5 inches and standard deviation 3.0 inches.[67] Who is taller relative to other children of their sex and age, Jordan or Zayne? Justify your answer.

SOLUTION

Jordan: $\dfrac{55-52.5}{2.5} = 1.0$ Zayne: $\dfrac{58-56.5}{3.0} = 0.5$

> The standardized heights tell us where each child stands (pun intended!) in the distribution of height for their age group.

Jordan is 1 standard deviation above the mean height of 9-year-old girls, while Zayne is one-half standard deviation above the mean height of 11-year-old boys. So Jordan is taller relative to girls her age than Zayne is relative to boys his age.

EXAM PREP: FOR PRACTICE, TRY EXERCISE 15.

Section 3.4 What Did You Learn?

Review the learning goals from this section. Then practice what you've learned by working through the exercises.

Learning Goal	Example	Exercises
Find and interpret a percentile in a distribution of quantitative data.	3.4.1	7–10
Find and interpret a standardized score (z-score) in a distribution of quantitative data.	3.4.2	11–14
Use percentiles or standardized scores (z-scores) to compare the location of values in different distributions.	3.4.3	15–18

Section 3.4 Exercises

Building Concepts and Skills These exercises assess the basic knowledge you should have after reading the section.

1. An individual's percentile is the percentage of values in a distribution that are _____ the individual's data value.

2. Give an example of when a high percentile would be a bad thing in a health-related setting.

3. The third quartile of a distribution is at roughly the _____ percentile.

4. What is the formula for calculating a standardized score?

5. A z-score of –2 means that the data value is _____ standard deviations _____ the mean of the distribution.

6. Name two ways of comparing location in different distributions of quantitative data.

Mastering Concepts and Skills These exercises reinforce the learning goals as shown in the examples.

7. **Refer to Example 3.4.1 House of Representatives** The U.S. House of Representatives has 435 voting members. The number of representatives that each state has is based on its population when the national census is taken every 10 years. A dotplot of the number of representatives from each of the 50 states in 2021 is shown. The red point on the graph is for the state of Ohio, which has 15 representatives.

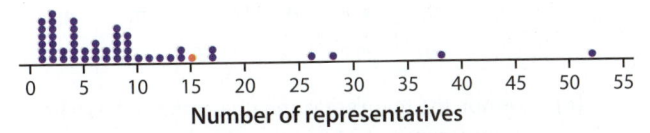

    (a) Find the percentile for Ohio. Interpret this value.
    (b) South Carolina is at the 52nd percentile of the distribution. How many representatives did South Carolina have in 2021?

8. **Play Ball!** The dotplot shows the number of wins for each of the 30 Major League Baseball (MLB) teams in the 2019 season. The red point on the graph is for the Seattle Mariners, which won 68 games.[68]

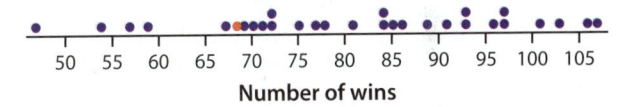

    (a) Find the percentile for the Seattle Mariners. Interpret this value.
    (b) The Washington Nationals' number of wins is at the 70th percentile of the distribution. How many games did Washington win?

9. **Finch Beaks** Biologists Peter and Rosemary Grant spent many years collecting data on finches in a remote part of the Galápagos Islands. Their research team caught and measured all of the birds in more than 20 generations of finches. Here are the beak lengths (in millimeters) of 73 *Fortis* finches measured by the Grants.[69]

    9.05 9.15 9.25 9.25 9.35 9.45 9.55 9.85 9.85 9.85 9.95 9.95
    9.95 10.05 10.05 10.05 10.05 10.05 10.15 10.15 10.15 10.25 10.25
    10.25 10.35 10.35 10.35 10.45 10.45 10.45 10.55 10.65 10.65 10.65 10.65
    10.65 10.65 10.75 10.75 10.85 10.85 10.95 10.95 10.95 10.95 10.95 11.05
    11.05 11.05 11.05 11.05 11.15 11.15 11.15 11.15 11.15 11.25 11.25 11.25
    11.25 11.35 11.35 11.45 11.55 11.55 11.65 11.75 11.75 11.85 11.95 12.05 12.65

    (a) Find the percentile of the finch with a beak length of 11.85 mm.
    (b) What was the beak length of the finch at the 28th percentile of the distribution?

10. **Unlocked Phones** The "sold" listings on a popular auction website included 20 sales of used "unlocked" phones of one popular model. Here are the sales prices (in dollars).

    450 415 495 300 325 430 370 400 325 400
    235 330 304 415 355 405 449 355 425 299

(a) Find the percentile of the phone that sold for $330.

(b) What was the sales price of the phone at the 75th percentile of the distribution?

11. **Refer to Example 3.4.2 House of Representatives** Refer to Exercise 7. In 2021, the number of representatives in a state had mean 8.7 and standard deviation 9.69. Calculate and interpret the standardized score (z-score) for Ohio, which had 15 representatives.

12. Play Ball! Refer to Exercise 8. During the 2019 season, the mean number of wins for MLB teams was 81, with a standard deviation of 15.9 wins. Find and interpret the standardized score (z-score) for the Washington Nationals, which won 93 games (and the World Series!).

13. Stocks The Dow Jones Industrial Average (DJIA) is a commonly used index of the overall strength of the U.S. stock market. In 2019, the mean daily change in the DJIA for the days that the stock markets were open was 20.94 points, with a standard deviation of 206.77 points.

(a) Find the standardized score (z-score) for the change in the DJIA on May 7, 2019, which was −473.39 points. Interpret this value.

(b) The standardized score for December 12, 2019, was $z = 1.03$. Find the change in the DJIA for that date.

14. Household Incomes How do household incomes compare in different states? Here are summary statistics for the median household income in the 50 states and the District of Columbia in a recent year.[70]

n	Mean	SD	Min	Q_1	Med	Q_3	Max
51	69,438.53	11,744.28	44,092	60,010	69,913	78,002	95,310

(a) Find and interpret the z-score for North Carolina, with a median household income of $60,300.

(b) New Jersey had a standardized score of 1.53. Find New Jersey's median household income for that year.

15. **Refer to Example 3.4.3 SAT Versus ACT** Some students who are applying for college take both the SAT and the ACT. One such student, Alejandra, scored 1280 on the SAT and 27 on the ACT. In the year when Alejandra took these tests, the distribution of SAT scores had a mean of 1059 and a standard deviation of 210. The distribution of ACT scores had a mean of 20.7 and a standard deviation of 5.9.[71] Which of Alejandra's two test scores was better, relatively speaking? Justify your answer.

16. Generational GPA Keti and one of her parents, Fatima, graduated from the same high school. Keti's high school GPA (4.2) was higher than Fatima's high school GPA (3.9). In the year when Fatima graduated, the mean GPA at the school was 2.8 with a standard deviation of 0.6. In the year when Keti graduated, the mean GPA was 3.2 with a standard deviation of 0.7. Who had the better GPA, relatively speaking? Explain your reasoning.

17. Gymnastics Scores Simone Biles won the gold medal in women's artistic gymnastics at the 2019 World Championships. Her overall score in the all-around competition was 58.999. More than 40 years earlier, Romanian gymnast Nadia Comaneci took the world by storm with the first perfect 10. With an overall score of 79.275, Comaneci also won the all-around gold medal.[72] Because the scoring systems have changed, these two scores aren't directly comparable. In 2019, the 23 gymnasts who completed the all-around competition had a mean score of 54.719 points and a standard deviation of 1.800 points. In 1976, the top 24 gymnasts in the all-around competition had a mean score of 76.527 points and a standard deviation of 1.327 points. Which gymnast, Biles or Comaneci, had a better performance, relatively speaking? Explain your reasoning.

18. Batting Averages Three landmarks of baseball achievement are Ty Cobb's batting average of 0.420 in 1911, Ted Williams's 0.406 in 1941, and George Brett's 0.390 in 1980. These batting averages cannot be compared directly because the distribution of major league batting averages has changed over the years. The distributions are quite symmetric, except for outliers such as Cobb, Williams, and Brett. While the mean batting average has been held roughly constant by rule changes and the balance between hitting and pitching, the standard deviation has dropped over time. Here are the facts:[73]

Decade	Mean	Standard deviation
1910s	0.266	0.0371
1940s	0.267	0.0326
1980s	0.261	0.0317

Who had the best performance for the decade he played? Explain your reasoning.

Applying the Concepts These exercises ask you to apply multiple learning goals in a new context or to apply what you learned in this section in a new way.

19. Head Circumferences Many athletes (and their parents) worry about the risk of concussions when playing sports. A youth football coach plans to obtain specially made helmets for the players that are designed to reduce the chance of getting a concussion. Here are a dotplot and numerical summaries of the head circumference (in inches) of each player on the team.[74]

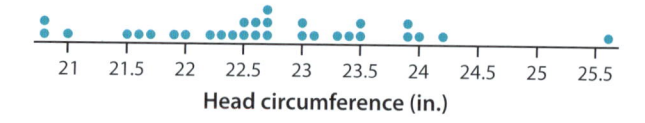

n	Mean	SD	Min	Q_1	Med	Q_3	Max
30	22.697	1.07	20.8	22	22.65	23.4	25.6

(a) Connor, the team's starting quarterback, has a head circumference of 24.0 inches. Find and interpret Connor's percentile.

(b) Find the standardized score (z-score) for Connor's head circumference. Interpret this value.

(c) In the distribution of players' heights, Connor's z-score is 0.87. Is Connor's head circumference relatively large for his height? Explain your reasoning.

20. **Birth Rates** One of the important factors in determining population growth rates is the birth rate in a country. Here is a dotplot that shows the birth rates per 1000 individuals for 54 African nations in a recent year, along with numerical summaries.[75]

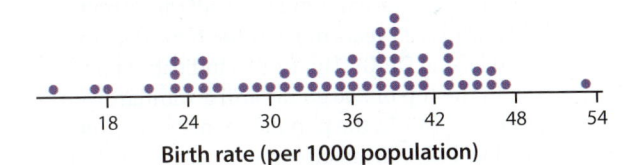

    Birth rate (per 1000 population)

n	Mean	SD	Min	Q_1	Med	Q_3	Max
54	34.907	8.574	14	29	37.5	41	53

    (a) The African country of Chad had a birth rate of 44 per 1000 people. Find and interpret Chad's percentile.
    (b) Find the standardized score (z-score) for Chad. Interpret this value.
    (c) In the distribution of income per person, Chad's z-score was -0.70. Was Chad more unusual with respect to birth rate or income per person compared to other African countries? Explain your reasoning.

21. **California Speed Limits** According to the *Los Angeles Times*, speed limits on California highways are set at the 85th percentile of vehicle speeds on those stretches of road. Explain to someone who knows little statistics what that means.

22. **Cholesterol Level** Dinesh told a friend, "My doctor says my cholesterol level is at the 90th percentile. That means I'm better off than about 90% of people like me." How should Dinesh's friend, who has taken statistics, respond to this statement?

23. **Tech Carbs in Beer and Wine** How do the carbohydrate contents in 12-ounce servings of beer and 6-ounce servings of wine compare? The **Beer and Wine** data set contains data for 140 different brands of beer and wine. Note that there are some missing data values.[76]

    (a) Use technology to make a comparative dotplot of the carbohydrate data *(Carbs)* for beer and wine. Then calculate numerical summaries for each type of beverage.

    Bell's Kalamazoo Stout has 23 grams of carbs in a 12-ounce serving.

    (b) Find its percentile in the beer distribution.
    (c) Calculate its standardized score.

    A 6-ounce serving of Barefoot Pink Moscato wine has 18 grams of carbs.

    (d) Which is more unusual in terms of carbohydrate content relative to other beverages of its type: Bell's Kalamazoo Stout or Barefoot Pink Moscato? Justify your answer.

24. **Tech Congress and Social Media** How much do members of the U.S. Congress use Twitter? Does usage differ by party affiliation? The **Congress Social Media** data set provides data on social media use by members of the 116th Congress.[77]

    (a) Use technology to make a comparative dotplot of the total number of posts *(Total_Posts)* made on Twitter by Democratic versus Republican members of Congress. Then calculate numerical summaries for each party.

    California Representative Ted Liu, a Democrat, made 6698 Twitter posts.

    (b) Find Liu's percentile in the Democratic distribution.
    (c) Calculate Liu's standardized score.

    Tennessee Representative Tim Burchett, a Republican, made 5608 Twitter posts.

    (d) Who is more unusual in terms of Twitter posts relative to other members of their party: Ted Liu or Tim Burchett? Justify your answer.

Extending the Concepts These exercises challenge you to explore statistical concepts and methods that go beyond what you learned in this section.

25. **Presidential Ages** The following table summarizes the ages of the first 46 U.S. presidents at inauguration. Theodore Roosevelt was the youngest to take office (42) and Joe Biden was the oldest (78). Copy and complete the table, filling in the columns for relative frequency, cumulative frequency, and cumulative relative frequency.

Age	Frequency	Relative frequency	Cumulative frequency	Cumulative relative frequency
40–44	2	2/46 = 0.0435 = 4.35%	2	2/46 = 0.0435 = 4.35%
45–49	7	7/46 = 0.1522 = 15.22%	9	9/46 = 0.1957 = 19.57%
50–54	13			
55–59	12			
60–64	7			
65–69	3			
70–74	1			
75–79	1			

    The following figure shows a *cumulative relative frequency graph* (also known as an *ogive*, pronounced "o-jive") for the presidential age at inauguration data. The leftmost point is plotted at a height of 0% at age 40, indicating that none of the first 46 U.S. presidents took office before they were 40 years old; the next point is plotted at a height of 4.35% at age 45, indicating that 4.35% of the first 46 U.S. presidents took office before they were 45 years old; and so on. Consecutive points are connected with line segments to form the completed graph.

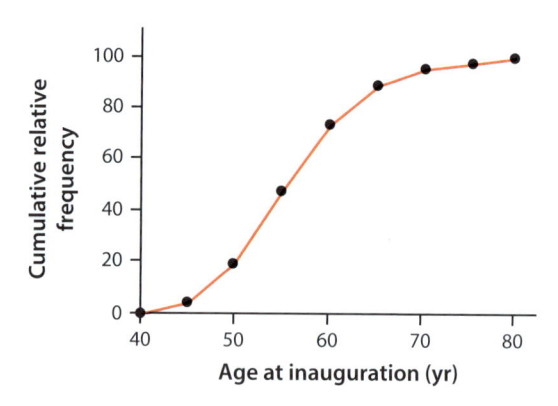

A cumulative relative frequency graph can be used to estimate the location of an individual value in a distribution or to estimate a specified percentile of the distribution.

26. Presidential Ages Use the cumulative relative frequency graph of U.S. presidents' ages when they took office to help you answer each question.

(a) Was Barack Obama, who was first inaugurated at age 47, unusually young?

(b) Estimate and interpret the 65th percentile of the distribution.

27. Comparing Distributions with Ogives Nitrates are organic compounds that are a substantial component of agricultural fertilizers. When those fertilizers run off into streams, the nitrates can have a toxic effect on animals that live in those streams. An ecologist studying nitrate pollution in two streams collects data on nitrate concentrations at 42 places on Stony Brook and 42 places on Mill Brook. Here are cumulative relative frequency graphs of the data for each stream.

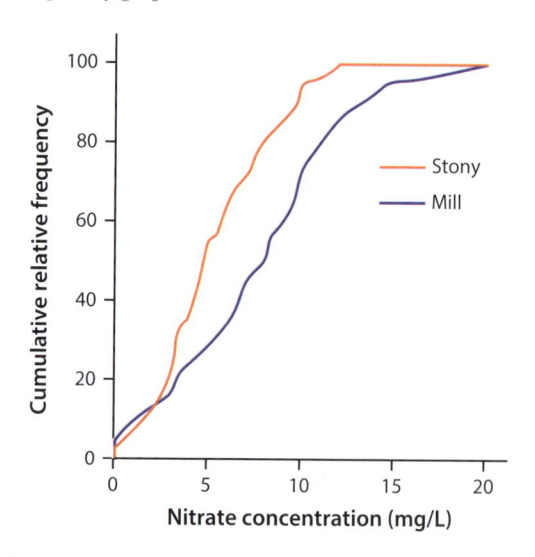

(a) Find and compare the median nitrate concentrations for the two streams.

(b) Which stream has more variability in its nitrate concentrations? Justify your answer with appropriate numerical evidence from the graph.

28. Bone Density People with low bone density have a high risk of broken bones. Currently, the most common method for testing bone density is dual-energy X-ray absorptiometry (DEXA). A patient who undergoes a DEXA test usually gets bone density results in grams per square centimeter (g/cm^2) and in standardized units.

Judy, who is 25 years old, has her bone density measured using DEXA. Her results indicate a bone density in the hip of $948\,g/cm^2$ and a standardized score of $z = -1.45$. In the population of 25-year-old women like Judy, the mean bone density in the hip is $956\,g/cm^2$.[78]

(a) Judy has not taken a statistics class in a few years. Explain in simple language what the standardized score tells her about her bone density.

(b) Use the information provided to calculate the standard deviation of bone density in the population of 25-year-old women.

Cumulative Review These exercises revisit what you learned in previous sections.

29. Separated at Birth (2.6) A researcher studied a group of identical twins who had been separated and adopted at birth. In each case, one twin (Twin A) was adopted by a low-income family and the other (Twin B) by a high-income family. Both twins were given an IQ test as adults.[79] Here are their scores.

| Twin A | 120 | 99 | 99 | 94 | 111 | 97 | 99 | 94 | 104 | 114 | 113 | 100 |
| Twin B | 128 | 104 | 108 | 100 | 116 | 105 | 100 | 100 | 103 | 124 | 114 | 112 |

Make a scatterplot to display the relationship between IQ scores for these sets of twins. Describe what you see.

30. Ketchup or Mustard? (1.6, 2.2) Do the characteristics of the interviewer make a difference? To investigate, a researcher asked 90 mall shoppers if they preferred ketchup or mustard on their hot dogs using three different approaches. The researcher asked 30 shoppers while wearing a full-body ketchup bottle costume, 30 shoppers while wearing a full-body mustard bottle costume, and 30 shoppers while wearing no costume.[80] Here are the results.

		Costume			
		None	Mustard	Ketchup	Total
Preference	Ketchup	9	5	13	27
	Mustard	21	25	17	63
	Total	30	30	30	90

(a) Use the information in the two-way table to make a segmented bar chart to display the association between costume and preference.

(b) Based on the graph in part (a), describe the association between costume and preference.

(c) Is it reasonable to draw a cause-and-effect conclusion based on this study? Explain your answer.

Section 3.5 Relationships Between Two Variables: Correlation

LEARNING GOALS
By the end of this section, you will be able to:
- Estimate the correlation between two quantitative variables from a scatterplot.
- Interpret the correlation.
- Distinguish correlation from causation.

As you learned in Section 2.6, a scatterplot is used to display the relationship between two quantitative variables measured on the same individuals. We describe the association between the variables using direction, form, and strength. Figure 3.2 is a scatterplot of data on the number of miles driven and price (in dollars) for a random sample of 16 Ford F-150 SuperCrew 4 × 4's from autotrader.com.[81] There is a moderately strong, negative, linear relationship between miles driven and price in this sample of Ford F-150 trucks.

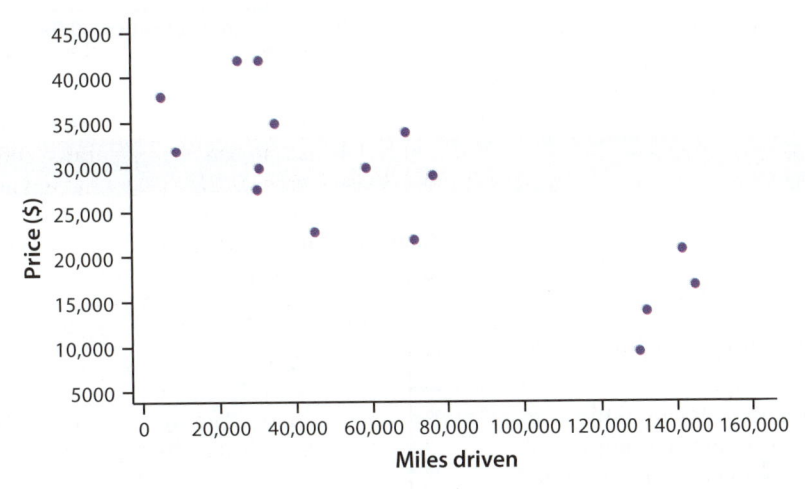

FIGURE 3.2 Scatterplot of miles driven and price for 16 randomly selected Ford F-150 trucks listed on autotrader.com.

To quantify the strength of a linear relationship like the one shown in Figure 3.2, we use the **correlation r**. (Some people refer to r as the "correlation coefficient.") You will learn how to calculate the correlation in the next section. This section focuses on how to understand and interpret the correlation.

> The **correlation r** is a measure of the strength and direction of a linear relationship between two quantitative variables.

Here are some important properties of the correlation r:
- The correlation r is a value between -1 and 1 ($-1 \leq r \leq 1$).
- The correlation r indicates the direction of a linear relationship by its sign: $r > 0$ for a positive association and $r < 0$ for a negative association.
- The extreme values $r = -1$ and $r = 1$ occur *only* in the case of a perfect linear relationship, when the points lie exactly along a straight line.
- If the linear relationship is strong (very little scatter from the linear form), the correlation r will be close to 1 or -1. If the linear relationship is weak, the correlation r will be close to 0.

Estimating and Interpreting the Correlation

Figure 3.3 shows six scatterplots and their corresponding correlations. The value of r describes the direction and strength of the linear relationship in each scatterplot.

FIGURE 3.3 How correlation measures the direction and strength of a linear relationship.

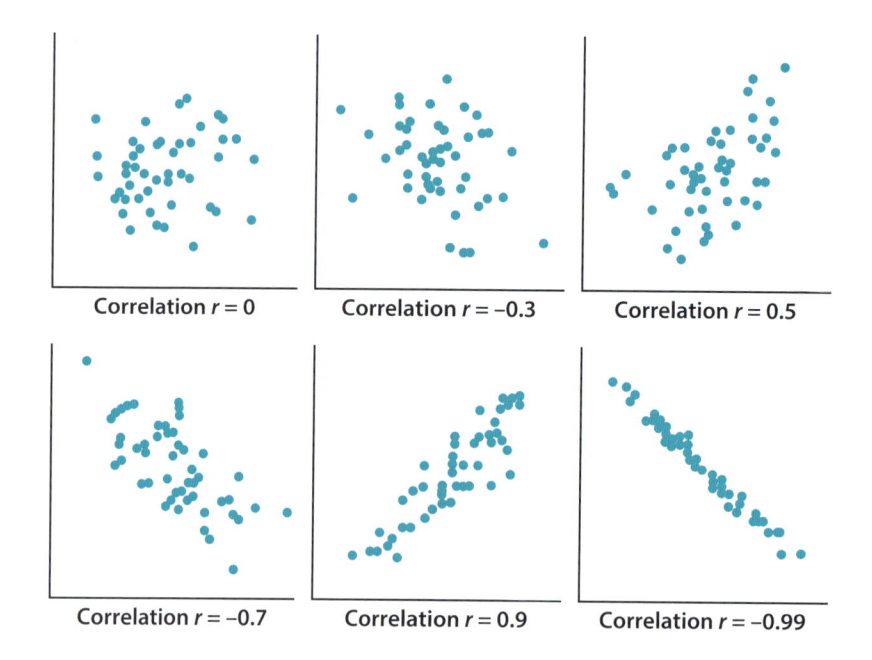

Correlation r = 0

Correlation r = −0.3

Correlation r = 0.5

Correlation r = −0.7

Correlation r = 0.9

Correlation r = −0.99

CONCEPT EXPLORATION

Guess the correlation

In this exploration, you will use an applet to test your skills at estimating correlation from a scatterplot.

1. Go to www.stapplet.com, and launch the *Guess the Correlation* applet (under Activities).[82]

2. Enter your estimate for the correlation shown in the scatterplot and click the Reveal answer button. The applet will calculate the difference between your guess and the actual correlation. How close were you?

3. Click the Do it again! button and repeat Step 2 several times. Are your estimates improving over time?

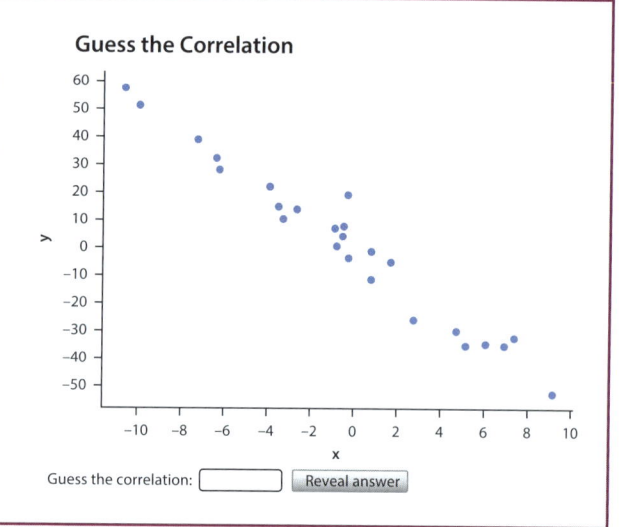

EXAMPLE 3.5.1

How can we help the manatees? Does seat location matter?

Estimating correlation

PROBLEM

For each of the following relationships, is $r > 0$ or is $r < 0$? Closer to $r = 0$ or $r = \pm 1$? Explain your reasoning.

(a) Manatees are large, gentle, slow-moving sea creatures found along the coast of Florida. Many manatees are injured or killed by boats. Here is a scatterplot showing the relationship between the number of boats registered in Florida (in thousands) and the number of manatees killed by boats over a 39-year period.[83]

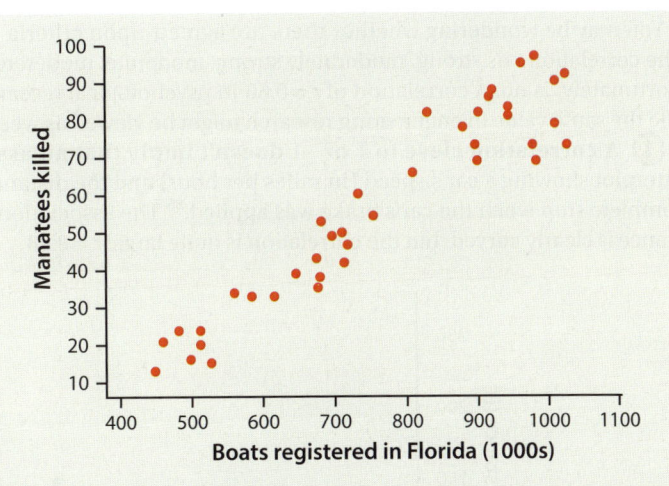

(b) To see if seat location affects student performance, an Introductory Statistics instructor randomly assigned students to seat locations in the classroom for a particular unit and recorded the score for each student on the end-of-unit exam. The explanatory variable in this experiment is which row the student was assigned, where Row 1 is closest to the front and Row 7 is the farthest away. Here is a scatterplot showing the relationship between row and exam score.[84]

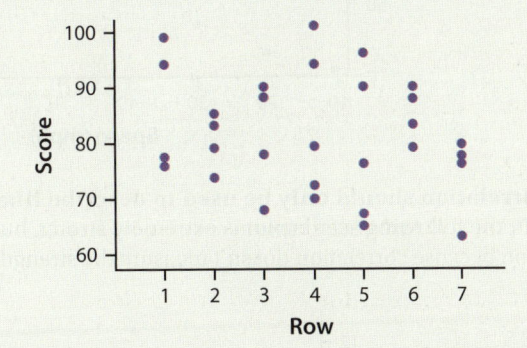

SOLUTION

(a) Because the relationship between number of boats registered and number of manatees killed is positive, $r > 0$. Also, r is closer to 1 than to 0 because the relationship is strong—there isn't much scatter from the linear pattern.

(b) Because the relationship between row and exam score is negative, $r < 0$. Also, r is closer to 0 than to -1 because the relationship is weak—there is a lot of scatter from the linear pattern.

EXAM PREP: FOR PRACTICE, TRY EXERCISE 7.

Now that you know how to estimate the correlation, here is an example of how to interpret the correlation.

EXAMPLE 3.5.2

Can sports teams buy success?

Interpreting correlation

PROBLEM

Does money buy wins in the NBA? Here is a scatterplot showing the relationship between payroll (in millions of dollars) and number of wins for NBA teams in a recent season.[85] The correlation is $r = 0.64$. Interpret this value.

SOLUTION

The correlation of $r = 0.64$ indicates that the linear relationship between team payroll (in millions of dollars) and number of wins for this NBA season is moderately strong and positive.

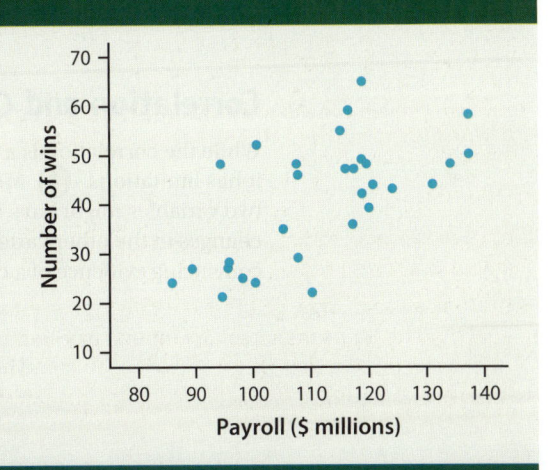

EXAM PREP: FOR PRACTICE, TRY EXERCISE 11.

You may be wondering whether there are agreed-upon criteria for labeling a particular value of the correlation r as strong, moderately strong, moderate, moderately weak, or weak. The answer, unfortunately, is no. A correlation of $r = 0.60$ in psychological research may be considered strong, while the same value in engineering research might be viewed as weak.

⚠️ **A correlation close to 1 or –1 doesn't imply that an association is linear.** Here is a scatterplot showing a car's speed (in miles per hour) and the distance (in feet) needed to come to a complete stop when the car's brake was applied.[86] The association between speed and stopping distance is clearly curved, but the correlation is quite large: $r = 0.98$.

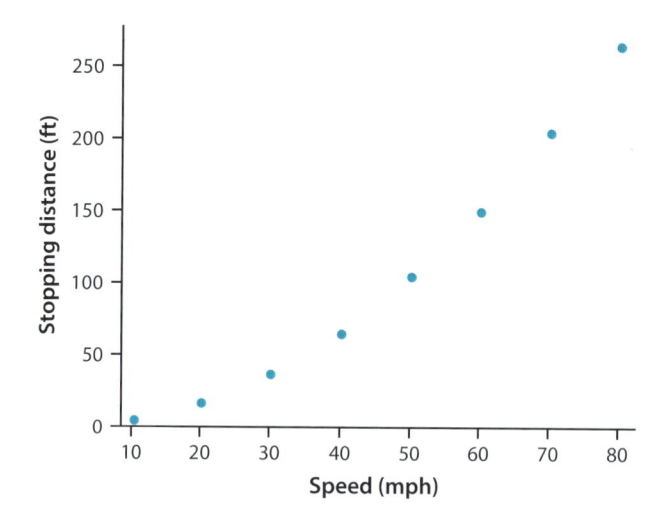

⚠️ **Correlation should only be used to describe linear relationships.** The association displayed in the following scatterplot is extremely strong, but the correlation is $r = 0$. This isn't a contradiction because correlation doesn't measure the strength of nonlinear relationships.

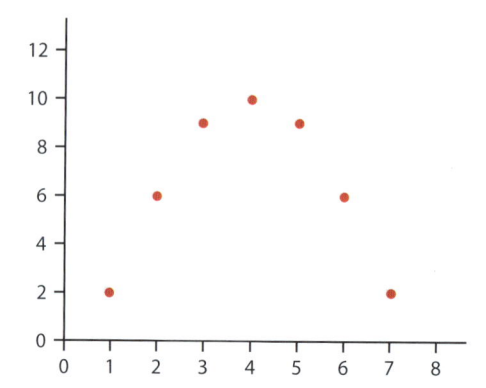

Correlation *alone* doesn't provide any information about form. To determine the form of an association, you must look at a scatterplot.

Correlation and Causation

While the correlation is a good way to measure the strength and direction of a linear relationship, it has limitations. ⚠️ **Most importantly, correlation doesn't imply causation.** In many cases, two variables might have a strong correlation, but changes in one variable won't necessarily cause changes in the other variable. Recall from Chapter 1 that only well-designed experiments can give convincing evidence of a cause-and-effect relationship.

EXAMPLE 3.5.3

If I eat more chocolate, will I win a Nobel Prize?

Distinguishing correlation and causation

PROBLEM
Most people love chocolate for its great taste. But does it also make you smarter? A scatterplot like this one appeared in the *New England Journal of Medicine*.[87] Each point on the scatterplot represents a country. The explanatory variable is the chocolate consumption per resident. The response variable is the number of Nobel Prizes per 10 million residents of that country.

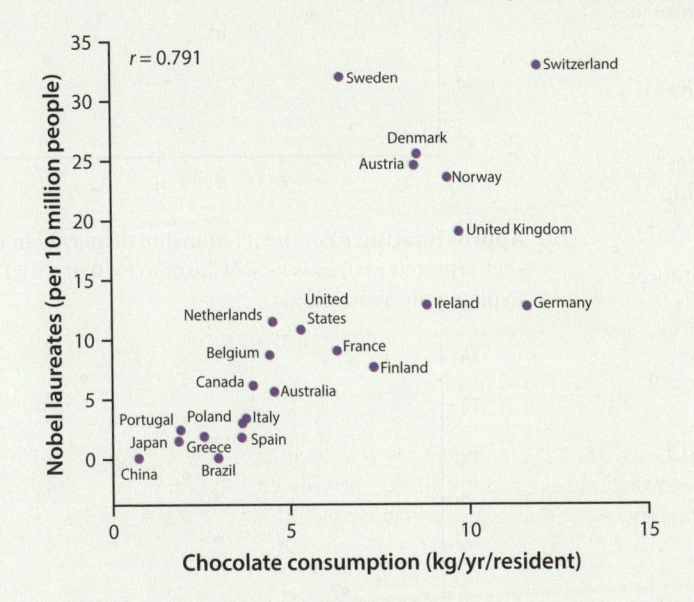

If people in the United States started eating more chocolate, could we expect more Nobel Prizes to be awarded to residents of the United States? Explain your answer.

SOLUTION
No. Even though there is a strong correlation between chocolate consumption and the number of Nobel laureates in a country, you should not infer causation. It is possible that both of these variables are changing due to another variable, such as income per person. Maybe countries with higher income per person tend to have greater chocolate consumption and a higher rate of Nobel laureates.

EXAM PREP: FOR PRACTICE, TRY EXERCISE 15.

Section 3.5 What Did You Learn?

Review the learning goals from this section. Then practice what you've learned by working through the exercises.

Learning Goal	Example	Exercises
Estimate the correlation between two quantitative variables from a scatterplot.	3.5.1	7–10
Interpret the correlation.	3.5.2	11–14
Distinguish correlation from causation.	3.5.3	15–18

Section 3.5 Exercises

Building Concepts and Skills These exercises assess the basic knowledge you should have after reading the section.

1. The correlation r should only be used for _____ relationships.

2. What does the correlation r measure?

3. What is the largest possible value for the correlation? The smallest possible value?

4. Draw a scatterplot where the correlation is very close to $r = -1$.

5. True/False: If the correlation between two variables is close to $r = 1$, the association must be quite linear.

6. True/False: If the correlation between two variables is close to $r = 1$, there is a cause-and-effect relationship between the two variables.

Mastering Concepts and Skills These exercises reinforce the learning goals as shown in the examples.

7. **Refer to Example 3.5.1 Estimating r** For the relationship displayed in the scatterplot, is $r > 0$ or is $r < 0$? Closer to $r = 0$ or $r = \pm 1$? Explain your reasoning.

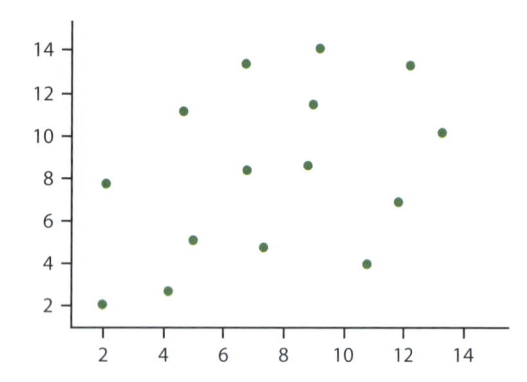

8. **Guessing r** For the relationship displayed in the scatterplot, is $r > 0$ or is $r < 0$? Closer to $r = 0$ or $r = \pm 1$? Explain your reasoning.

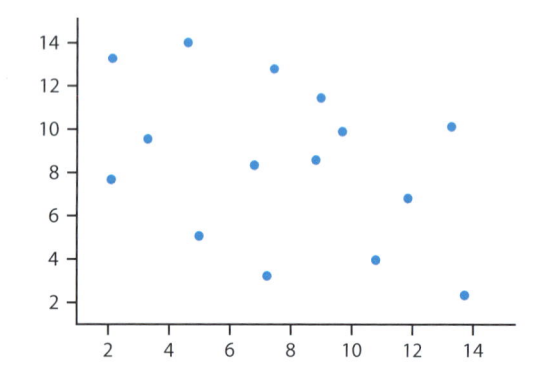

9. **Predicting r** For the relationship displayed in the scatterplot, is $r > 0$ or is $r < 0$? Closer to $r = 0$ or $r = \pm 1$? Explain your reasoning.

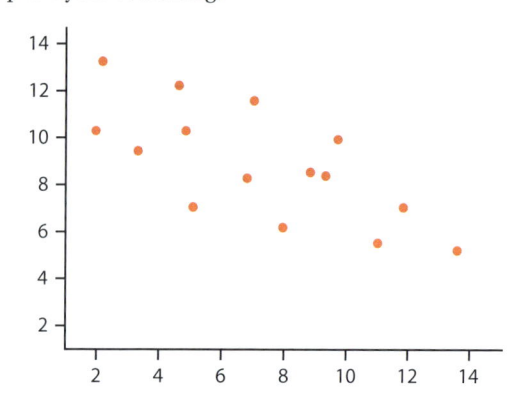

10. **Approximating r** For the relationship displayed in the scatterplot, is $r > 0$ or is $r < 0$? Closer to $r = 0$ or $r = \pm 1$? Explain your reasoning.

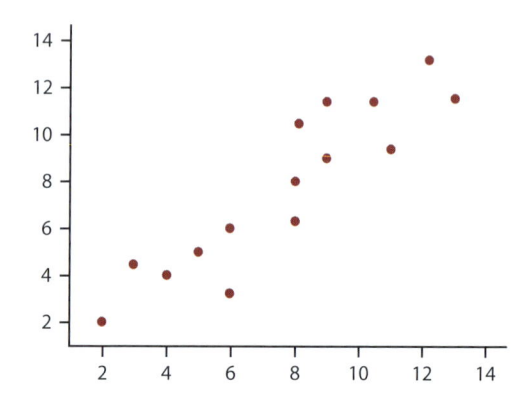

11. **Refer to Example 3.5.2 Crying and IQ** Infants who cry easily may be more easily stimulated than others. This may be a sign of a higher intelligence quotient (IQ). Child development researchers explored the relationship between the crying of infants 4 to 10 days old and their IQ test scores at age 3. A snap of a rubber band on the sole of the foot caused the infants to cry. The researchers recorded the crying and measured its intensity by the number of peaks in the most active 20 seconds. The relationship between these variables is linear, with a correlation of $r = 0.45$.[88] Interpret this value.

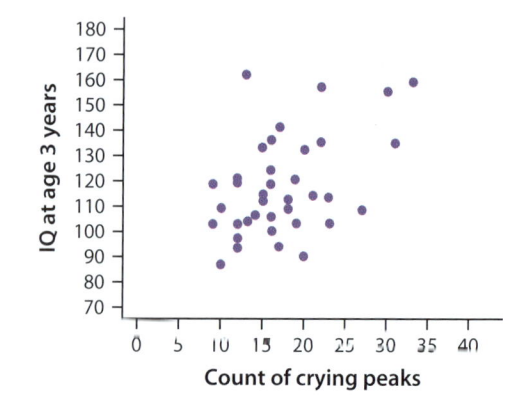

12. **Points and Turnovers** There is a linear relationship between the number of turnovers and the number of points scored for players in a recent NBA season. The correlation for these data is $r = 0.92$. Interpret this value.[89]

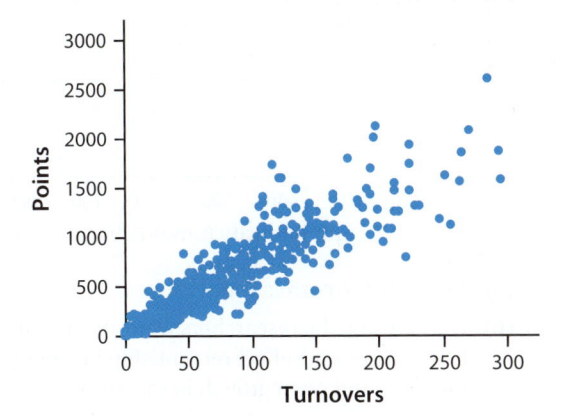

13. **Sparrowhawks** One of nature's patterns connects the percentage of adult birds in a colony that return from the previous year and the number of new adults that join the colony. Research on 13 colonies of sparrowhawks shows a linear relationship between these two variables, with a correlation of $r = -0.75$.[90] Interpret this value.

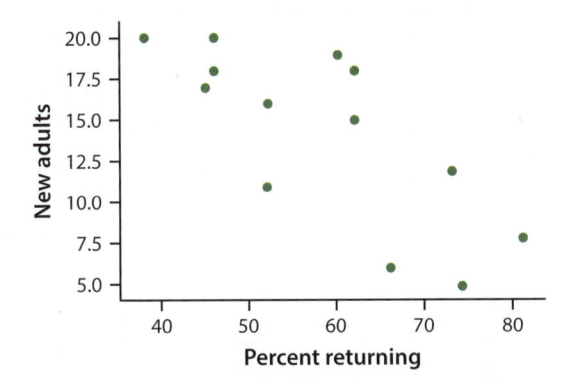

14. **Temperature and Wind Speed** Is there a relationship between temperature and wind speed? Here is a scatterplot showing the average temperature (in degrees Fahrenheit) and average wind speed (in miles per hour) for 365 consecutive days at O'Hare International Airport in Chicago.[91] The correlation for these data is $r = -0.219$. Interpret this value.

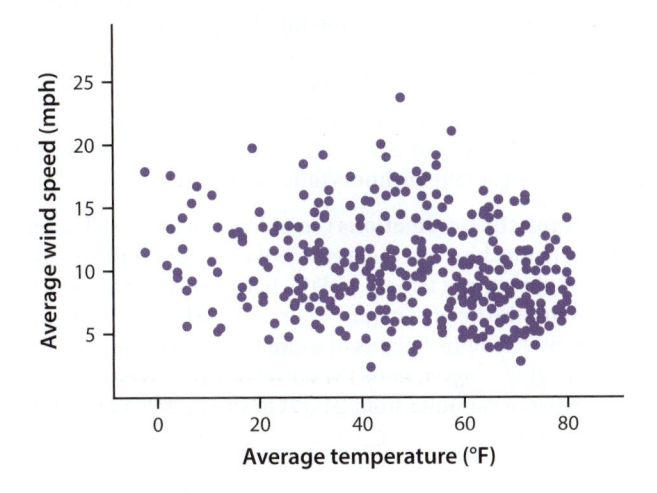

15. **Refer to Example 3.5.3 Pets and Cheese** The scatterplot shows the relationship between consumption of mozzarella cheese per person and money spent on pets each year over a 10-year period. The correlation is $r = 0.93$.[92] Does the strong correlation between these two variables suggest that spending more money on pets causes people to consume more mozzarella cheese? Explain your reasoning.

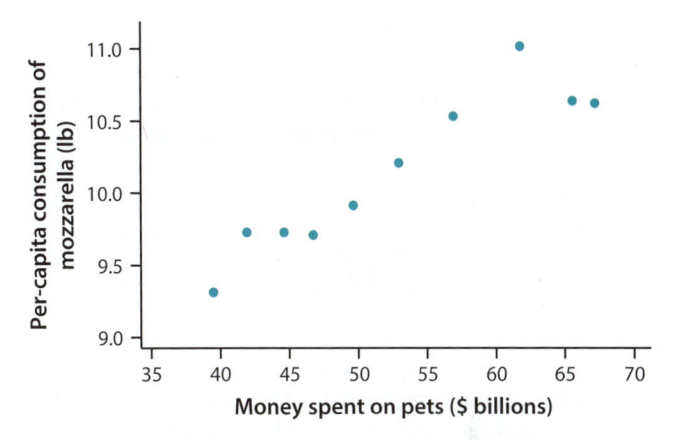

16. **Skiing and Entanglement?** Here is a scatterplot showing the relationship between total revenue generated by skiing facilities in the United States and the number of people who died by becoming tangled in their bedsheets for 10 recent years. The correlation for these data is $r = 0.97$.[93] Does the strong correlation between these two variables suggest that an increase in skiing revenue causes more people to die by becoming tangled in their bedsheets? Explain your reasoning.

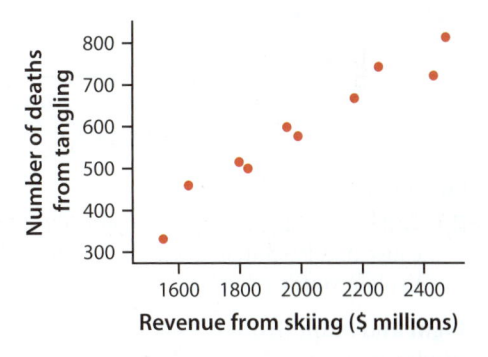

17. **Crying and IQ** Refer to Exercise 11. Does the fact that $r = 0.45$ suggest that making an infant cry will increase their IQ later in life? Explain your reasoning.

18. **Points and Turnovers** Refer to Exercise 12. Does the fact that $r = 0.92$ suggest that an increase in turnovers will cause NBA players to score more points? Explain your reasoning.

Applying the Concepts These exercises ask you to apply multiple learning goals in a new context or to apply what you learned in this section in a new way.

19. **Crowd Support** Do MLB teams that have bigger crowds win more games? Here is a scatterplot showing the average attendance at an MLB team's games and the number of wins for that team.[94]

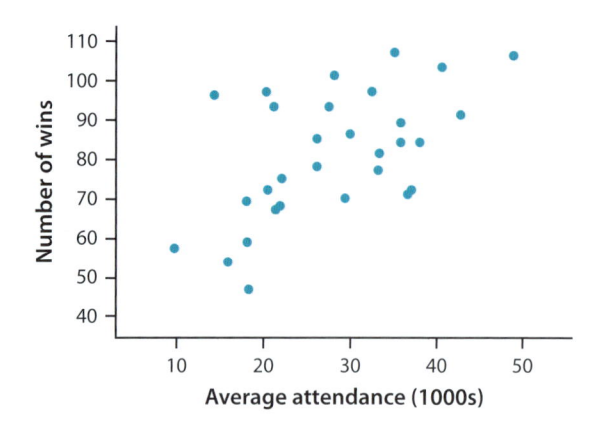

(a) Interpret the value $r = 0.557$.

(b) Would increasing attendance by giving away thousands of tickets to each game cause a team to win more games? Explain your reasoning.

20. Math and Income Is there a relationship between a country's per-capita income and its students' achievement in mathematics? Here is a scatterplot of per-capita gross domestic product versus average score on an international mathematics achievement test for 26 countries in Europe and Asia.[95]

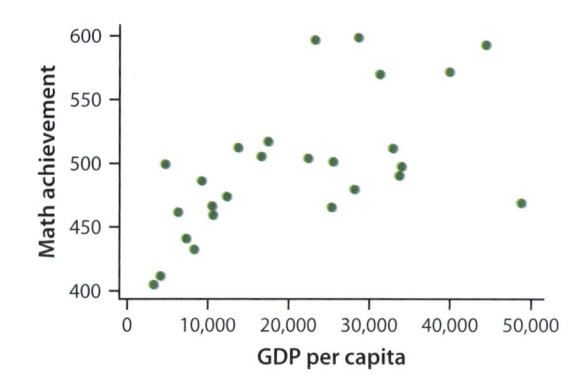

(a) Interpret the value $r = 0.59$.

(b) Will increasing per-capita income in a country cause higher student achievement in mathematics? Explain your reasoning.

21. Light and Plant Growth Meadowfoam seed oil is used in making various skin care products. Researchers interested in maximizing the productivity of meadowfoam plants designed an experiment to investigate the effect of different light intensities on plant growth. The researchers planted 120 meadowfoam seedlings in individual pots, placed 10 pots on each of 12 trays, and put all the trays into a controlled enclosure. Two trays were randomly assigned to each light intensity level (measured in micromoles striking a square meter per second): 150, 300, 450, 600, 750, and 900. The number of flowers produced by each plant was recorded. Here is a scatterplot of data on the average number of flowers per plant on each tray and the corresponding light intensity level.[96]

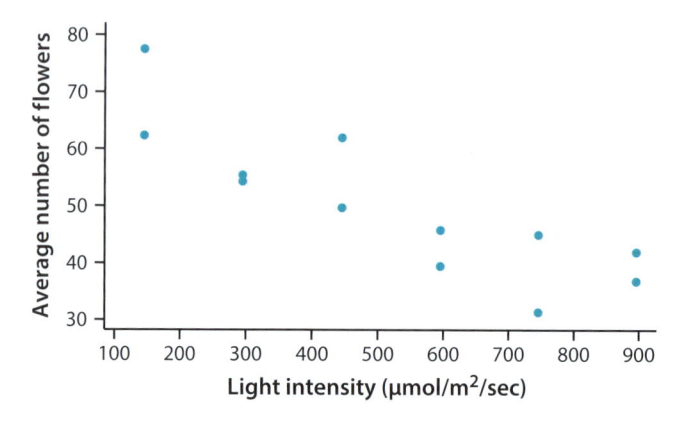

(a) Estimate the correlation.

(b) Explain why the researchers were able to establish a cause-and-effect relationship between light intensity and plant growth in this study.

22. Sweet Flowers Does adding sugar to the water in a vase help flowers stay fresh? To find out, researchers went to a flower shop and randomly selected 12 carnations. They prepared 12 identical vases with the same amount of water in each vase. Then they put 1 tablespoon of sugar in 3 vases, 2 tablespoons of sugar in 3 vases, and 3 tablespoons of sugar in 3 vases. In the remaining 3 vases, they added no sugar. After the vases were prepared, the researchers randomly assigned 1 carnation to each vase and observed how many hours each flower continued to look fresh. Here is a scatterplot of data on the amount of sugar and freshness time.[97]

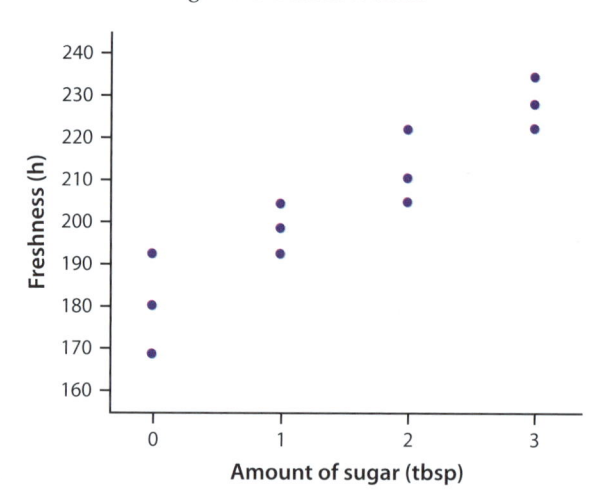

(a) Estimate the correlation.

(b) Explain why the researchers were able to establish a cause-and-effect relationship between amount of sugar and freshness time in this study.

23. Rank the Correlations Consider each of the following relationships in a large group of cisgender individuals: the heights of fathers and the heights of their adult sons, the heights of husbands and the heights of their wives, and the heights of women at age 18 and their heights at age 4. Rank the correlations between these pairs of variables from largest to smallest. Explain your reasoning.

24. **Match the Correlations** Suppose that a middle school physical education teacher collects data about the students in a class. Some of these variables include the number of pull-ups, push-ups, and sit-ups a student can do in 1 minute, and the student's weight. The teacher then calculates the correlation for the following relationships:

    - Number of pull-ups and number of push-ups
    - Number of pull-ups and number of sit-ups
    - Number of pull-ups and weight

    The correlations for these relationships are $r = 0.9$, $r = -0.5$, and $r = 0.3$. Which correlation goes with which relationship? Explain your reasoning.

Extending the Concepts These exercises challenge you to explore statistical concepts and methods that go beyond what you learned in this section.

25. **Teachers and Researchers** A college newspaper interviews a psychologist about student ratings of the teaching of faculty members. The psychologist says, "The evidence indicates that the correlation between the research productivity and teaching rating of faculty members is close to zero." The paper reports this as "Professor McDaniel said that good researchers tend to be poor teachers, and vice versa." Explain why the paper's report is wrong.

26. **1970 Draft Lottery** What should a nation do if there are not enough volunteers for military service? This question arose in the United States during the Vietnam War era. Beginning in 1969, a draft lottery was used to choose draftees by random selection.

    Because taking a simple random sample (SRS) of all eligible individuals was impractical, the draft lottery selected birth dates in a random order. People born on the first date chosen were drafted first, then those born on the second date chosen, and so on. The 366 dates were placed into identical small capsules, which were put in a bowl and publicly drawn one by one.

    The scatterplot displays the results of the 1970 draft lottery. We have plotted the variable "day of year" on the horizontal axis, with Day 1 = January 1 and Day 366 = December 31. The variable "draft number" is plotted on the vertical axis.

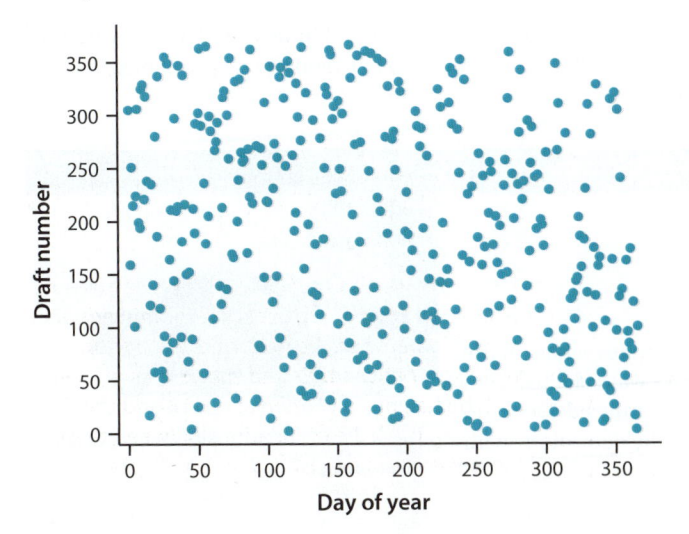

    (a) The correlation is $r = -0.226$. Explain how this value gives some evidence that the draft lottery was not a truly random process.

    Here are side-by-side boxplots of the draft lottery results for each birth month.

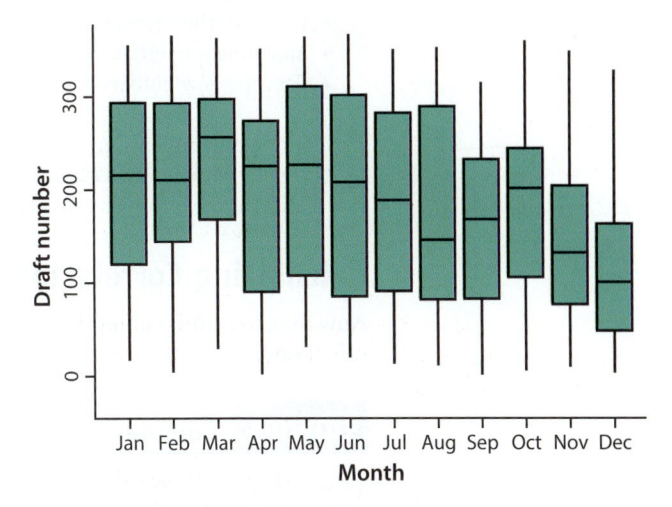

    (b) Describe what the boxplots reveal about the relationship between the day of the year on which people were born and the draft lottery number from the drawing.

Cumulative Review These exercises revisit what you learned in previous sections.

27. **Teenage Heights (3.4)** According to the National Center for Health Statistics, the distribution of heights for 15-year-old males has a mean of 170 centimeters (cm) and a standard deviation of 7.5 cm. Paul is a 15-year-old male and is 179 cm tall.

    (a) Find the z-score corresponding to Paul's height. Explain what this value means.

    (b) Paul's height puts him at the 88th percentile among 15-year-old males. Explain what this means to someone who knows little statistics.

28. **Stock Prices (3.1)** The business magazine *Forbes* reports that 4567 companies sold their first stock to the public between 1990 and 2000. The *mean* change in the stock price of these companies since the first stock was issued was +111%. The *median* change was −31%. Explain how this could happen.[98]

176 CHAPTER 3 Numerical Summaries for Quantitative Data

Section 3.6 More About Correlation

LEARNING GOALS

By the end of this section, you will be able to:

- Calculate the correlation between two quantitative variables.
- Apply the properties of the correlation.
- Describe how outliers influence the correlation.

In Section 3.5, you learned that the correlation r measures the strength and direction of the linear relationship between two quantitative variables. In this section, you will learn how to calculate the correlation and explore its properties.

Calculating Correlation

Now that you understand the meaning and limitations of the correlation r, let's look at how it is calculated.

HOW TO Calculate the Correlation r

Suppose that we have data on variables x and y for n individuals. The values for the first individual are x_1 and y_1, the values for the second individual are x_2 and y_2, and so on. The correlation r between x and y is

$$r = \frac{\sum\left(\frac{x_i - \bar{x}}{s_x}\right)\left(\frac{y_i - \bar{y}}{s_y}\right)}{n-1}$$

To calculate this quantity:

1. Find the mean $\bar{x}$ and the standard deviation s_x of the explanatory variable. Then calculate the corresponding z-score for each individual: $z_{x_i} = \dfrac{x_i - \bar{x}}{s_x}$.

2. Find the mean $\bar{y}$ and the standard deviation s_y of the response variable. Then calculate the corresponding z-score for each individual: $z_{y_i} = \dfrac{y_i - \bar{y}}{s_y}$.

3. Multiply the two z-scores for each individual: $z_{x_i} z_{y_i}$.

4. Add the z-score products for all the individuals and divide by $n-1$.

Notice that the correlation formula can be abbreviated as $r = \dfrac{\sum z_{x_i} z_{y_i}}{n-1}$.

Both variables must be quantitative to calculate the correlation. If one or both of the variables are categorical, we can consider the *association* or *relationship* between the two variables, but not the correlation.

EXAMPLE 3.6.1

Are truck price and mileage closely related?

Calculating correlation

PROBLEM

Can we predict the price of a Ford F-150 if we know how many miles it has on the odometer? A random sample of 16 Ford F-150 SuperCrew 4×4's was selected from among those listed for sale at autotrader.com. The number of miles driven and price (in dollars) were recorded for each of the trucks. Here are the data, along with a scatterplot.[99] Calculate the correlation.

Miles driven	70,583	129,484	29,932	29,953	24,495	75,678	8359	4447
Price ($)	21,994	9500	29,875	41,995	41,995	28,986	31,891	37,991
Miles driven	34,077	58,023	44,447	68,474	144,162	140,776	29,397	131,385
Price ($)	34,995	29,988	22,896	33,961	16,883	20,897	27,495	13,997

SOLUTION

1. Find the mean $\bar{x}$ and the standard deviation s_x. Calculate the z-score for each individual.

2. Find the mean $\bar{y}$ and the standard deviation s_y. Calculate the z-score for each individual.

3. Multiply the two z-scores.

$\bar{x} = 63{,}979.50$, $s_x = 47{,}876.83$

$\bar{y} = 27{,}833.69$, $s_y = 9570.42$

x_i	y_i	$z_{x_i} = \dfrac{x_i - \bar{x}}{s_x}$	$z_{y_i} = \dfrac{y_i - \bar{y}}{s_y}$	$z_{x_i} z_{y_i}$
70,583	21,994	$\dfrac{70{,}583 - 63{,}979.50}{47{,}876.83} = 0.138$	$\dfrac{21{,}994 - 27{,}833.69}{9570.42} = -0.610$	$(0.138)(-0.610) = -0.084$
129,484	9500	$\dfrac{129{,}484 - 63{,}979.50}{47{,}876.83} = 1.368$	$\dfrac{9500 - 27{,}833.69}{9570.42} = -1.916$	$(1.368)(-1.916) = -2.621$
29,932	29,875	$\dfrac{29{,}932 - 63{,}979.50}{47{,}876.83} = -0.711$	$\dfrac{29{,}875 - 27{,}833.69}{9570.42} = 0.213$	$(-0.711)(0.213) = -0.151$
⋮	⋮	⋮	⋮	⋮
29,397	27,495	$\dfrac{29{,}397 - 63{,}979.50}{47{,}876.83} = -0.722$	$\dfrac{27{,}495 - 27{,}833.69}{9570.42} = -0.035$	$(-0.722)(-0.035) = 0.025$
131,385	13,997	$\dfrac{131{,}385 - 63{,}979.50}{47{,}876.83} = 1.408$	$\dfrac{13{,}997 - 27{,}833.69}{9570.42} = -1.446$	$(1.408)(-1.446) = -2.036$

$$r = \dfrac{-0.084 + -2.621 + -0.151 + \ldots + 0.025 + -2.036}{16 - 1} = -0.815$$

4. Add the z-score products for all the individuals and divide by $n - 1$.

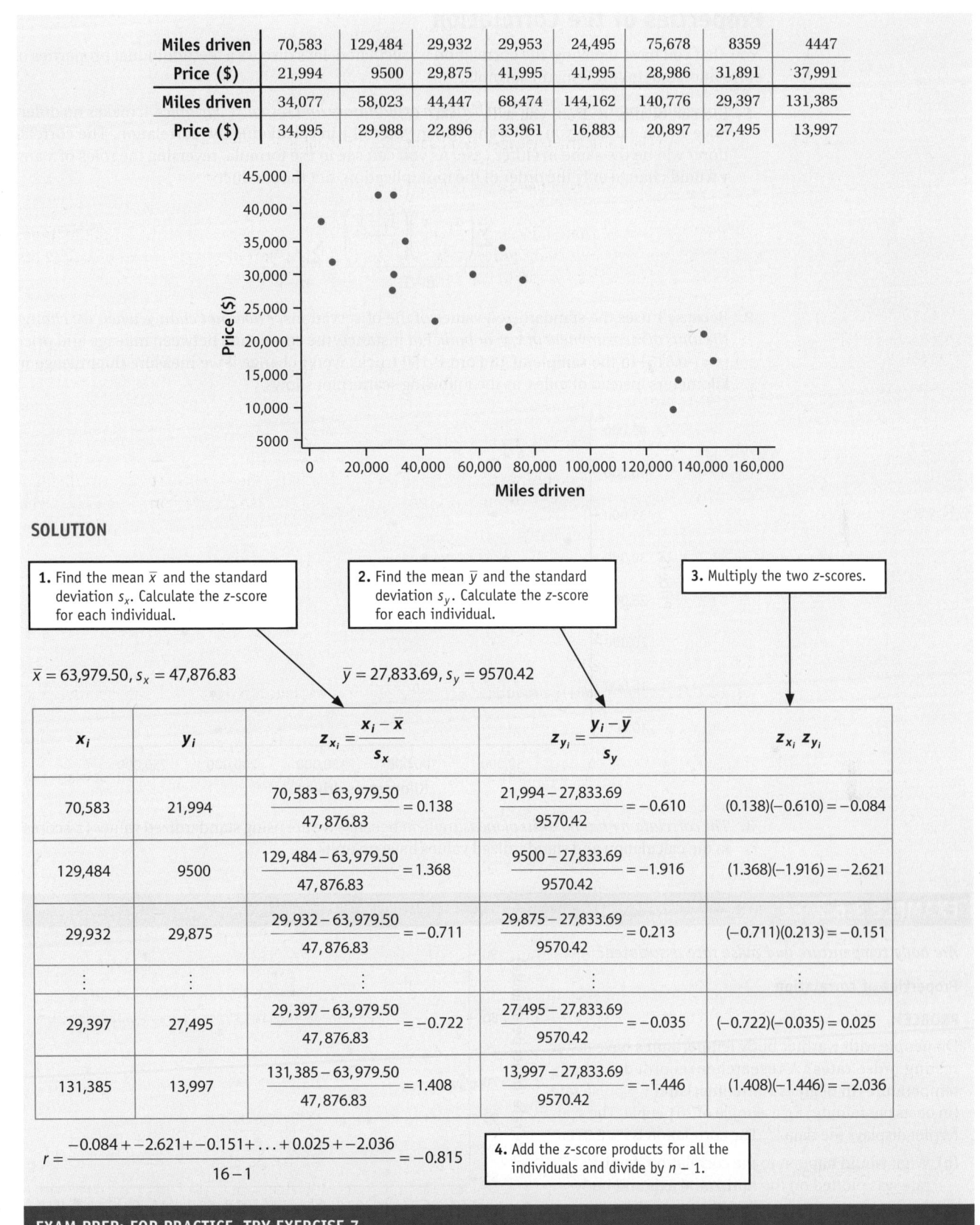

EXAM PREP: FOR PRACTICE, TRY EXERCISE 7.

As Example 3.6.1 illustrates, using the formula to calculate the correlation r is a bit tedious. In practice, you should use technology to find r. See the Tech Corner at the end of the section for details.

Properties of the Correlation

Now that you have seen how the correlation is calculated, let's discuss a few additional properties of correlation that follow from the formula.

1. *Correlation makes no distinction between explanatory and response variables.* It makes no difference which variable you call x and which you call y in calculating the correlation. The correlation r will be the same in either case. As you can see in the formula, reversing the roles of x and y would change only the order of the multiplication, not the product.

$$r = \frac{\sum \left(\frac{x_i - \bar{x}}{s_x}\right)\left(\frac{y_i - \bar{y}}{s_y}\right)}{n-1} = \frac{\sum z_{x_i} z_{y_i}}{n-1}$$

2. Because r uses the standardized values of the observations, *r does not change when we change the units of measurement of x, y, or both*. For instance, the correlation between mileage and price ($r = -0.815$) in the sample of 16 Ford F-150 trucks won't change if we measure the mileage in kilometers instead of miles, as the following scatterplot shows.

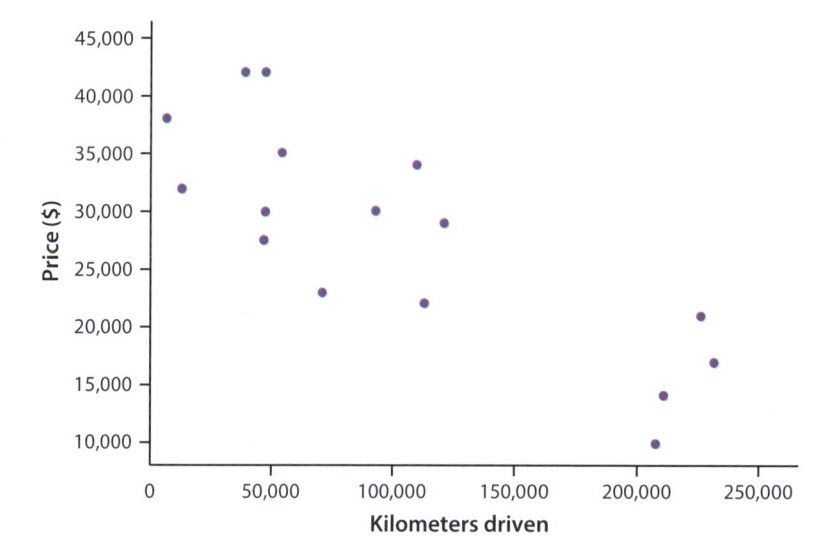

3. *The correlation r has no units of measurement* because we are using standardized values (z-scores) in the calculation and standardized values have no units.

EXAMPLE 3.6.2

Are body temperature and pulse rate associated?

Properties of correlation

PROBLEM

Do people with warmer body temperatures have faster resting pulse rates? A researcher recorded $x =$ body temperature (in degrees Fahrenheit) and $y =$ pulse rate (in beats per minute) for a sample of 20 people. The scatterplot displays the data.[100] The correlation is $r = 0.376$.

(a) What would happen to the correlation if pulse rate was plotted on the horizontal axis and body temperature was plotted on the vertical axis? Explain your answer.

(b) What would happen to the correlation if body temperature was measured in degrees Celsius instead of degrees Fahrenheit? Explain your answer.

(c) Wenona claims that the correlation between body temperature and pulse rate is $r = 0.376$ beats per degree Fahrenheit. Is this correct?

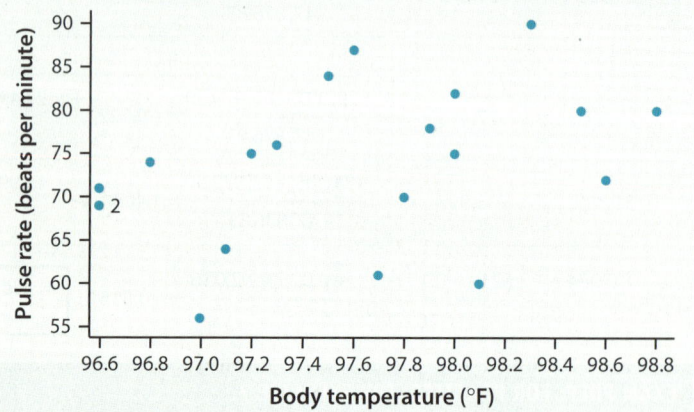

The 2 on the graph indicates that two people in the sample had body temperatures of 96.6°F and pulse rates of 69 beats per minute.

SOLUTION

(a) The correlation would still be $r = 0.376$ because the correlation makes no distinction between explanatory and response variables.

(b) The correlation would still be $r = 0.376$ because the correlation doesn't change when we change the units of either variable.

(c) No. The correlation doesn't have units, so including "beats per degree Fahrenheit" is incorrect.

EXAM PREP: FOR PRACTICE, TRY EXERCISE 11.

Outliers and Correlation

While the correlation is a good way to measure the strength and direction of a linear relationship, it has some limitations. In Section 3.5, you learned that correlation doesn't reveal the form of the relationship, nor does it imply a cause-and-effect relationship. The following exploration illustrates another limitation.

CONCEPT EXPLORATION

How do outliers affect correlation?

In this exploration, you will investigate the influence of outliers on the correlation r.

1. Go to www.stapplet.com, click the link for Resources for *Introductory Statistics: A Student-Centered Approach*, and launch the *Correlation* applet. You should see a scatterplot with 10 points in the lower left corner of the graph that displays a correlation of approximately $r = 0.40$ (labeled as "corr(x_1, y_1)" in the panel to the left of the graph).

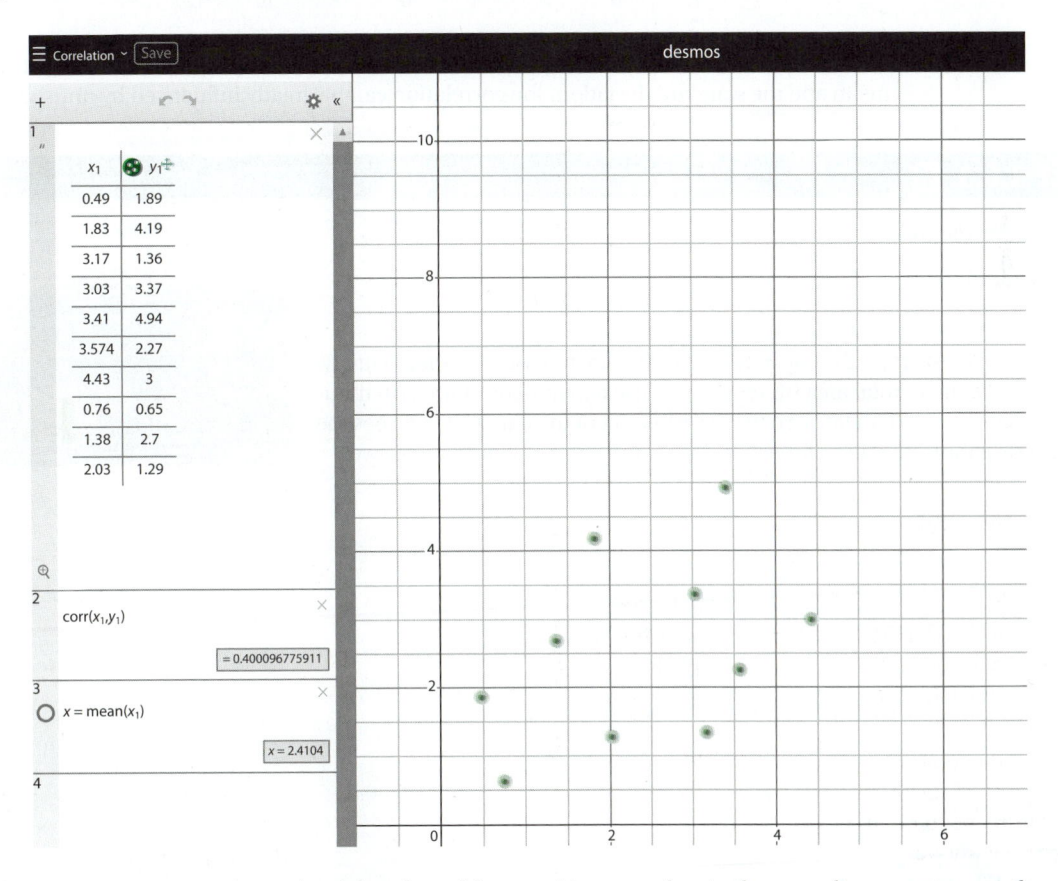

x_1	y_1
0.49	1.89
1.83	4.19
3.17	1.36
3.03	3.37
3.41	4.94
3.574	2.27
4.43	3
0.76	0.65
1.38	2.7
2.03	1.29

2. If you were to move one point to the right edge of the graphing area, but in the same linear pattern as the rest of the points, what do you think will happen to the correlation? Move the point to see if you were correct.

3. Click on the point you just moved, and drag it up and down along the right edge of the graphing area. What happens to the correlation?

(continued)

4. Now, click the circle to the left of $x = \text{mean}(x_1)$ in the left-hand panel to display a vertical line at $\bar{x}$. Move the point from Step 3 onto the vertical $\bar{x}$ line, and then drag the point up and down on this line. What happens to the correlation? Do outliers in the vertical direction have more or less influence on the correlation than outliers in the horizontal direction?

5. Briefly summarize how outliers influence the value of the correlation.

⚠ The correlation is not a resistant measure of strength. In the following scatterplot, the correlation is $r = -0.13$. But when the unusual point in the lower right corner is excluded, the correlation becomes $r = 0.72$.

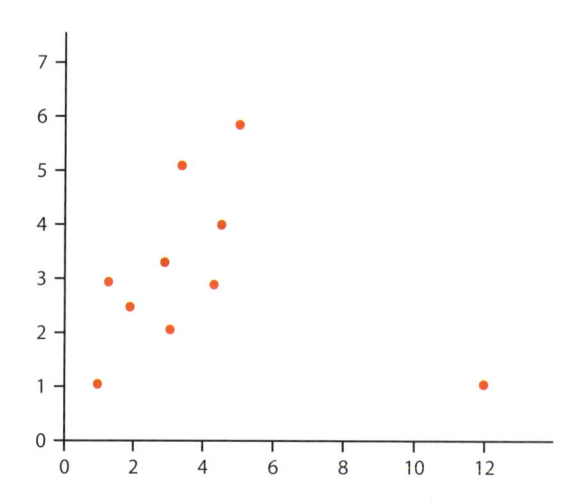

The formula for correlation involves the mean and standard deviation of both variables. Like the mean and the standard deviation, the correlation can be greatly influenced by unusual points.

EXAMPLE 3.6.3

Enough forest for the trees?

Outliers and correlation

PROBLEM

In the article "Do Forests Have the Capacity for 1 Trillion Extra Trees?" author Matthew Russell analyzed data on total area (in thousands of acres) and forest area (in thousands of acres) for the 50 U.S. states. Here is a scatterplot of the data.[101] How does the point for Alaska in the upper right corner affect the correlation? Explain your answer.

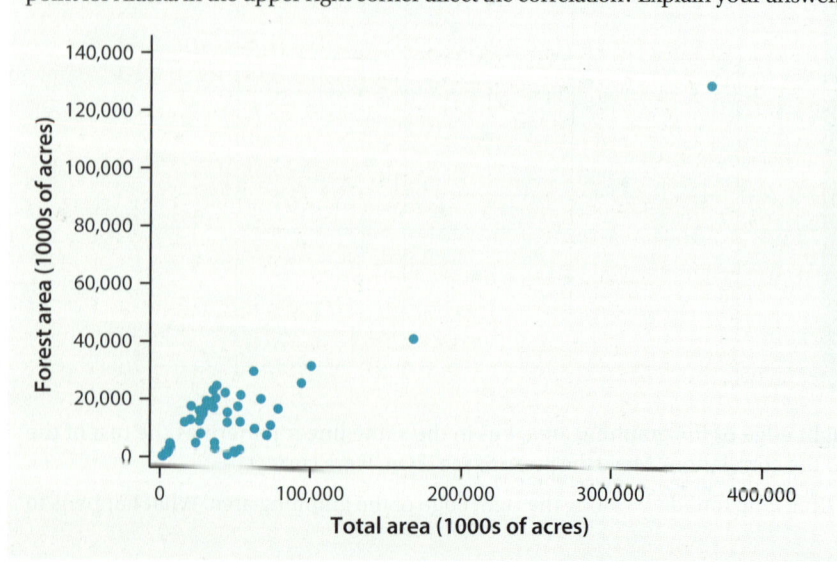

SOLUTION

The scatterplot shows a positive, linear association between total area and forest area in a state. Because the outlier for Alaska is in the same pattern as the rest of the points, it makes the correlation closer to 1.

> With Alaska, the correlation is $r = 0.902$.
> Without Alaska, the correlation is $r = 0.644$.

EXAM PREP: FOR PRACTICE, TRY EXERCISE 15.

TECH CORNER   Calculating Correlation

You can use technology to calculate the correlation between two quantitative variables. We'll illustrate using the Ford F-150 price and mileage data from Example 3.6.1.

Applet

1. Go to www.stapplet.com and launch the *Two Quantitative Variables* applet.
2. Enter the name of the explanatory variable (Miles driven) and the values of the explanatory variable in the first row of boxes. Then, enter the name of the response variable (Price) and the values of the response variable in the second row of boxes. Don't include a comma *within* any data values as you enter them.
3. Click the Begin analysis button to see the scatterplot.
4. Click the Calculate correlation button under the scatterplot to get the correlation.

TI-83/84

The TI-83/84 cannot calculate correlation directly from two lists of data, except as part of calculating the equation of a regression line.

1. Enter the miles driven data into L_1 and the price data into L_2.
2. Make sure the Stat Diagnostics are turned on.
   - **OS 2.55 or later:** Press MODE and set STAT DIAGNOSTICS to ON.
   - **Older OS:** Press 2nd 0 (CATALOG), scroll down to DiagnosticOn, and press ENTER. Press ENTER again to execute the command. The screen should say "Done."
3. To calculate the correlation, press STAT, arrow over to the CALC menu, and choose "LinReg(a+bx)."
   - **OS 2.55 or later:** In the dialog box, enter the following: Xlist:L_1, Ylist:L_2, FreqList (leave blank), Store RegEQ (leave blank), and choose Calculate.
   - **Older OS:** Finish the command to read LinReg(a+bx) L_1,L_2 and press ENTER.

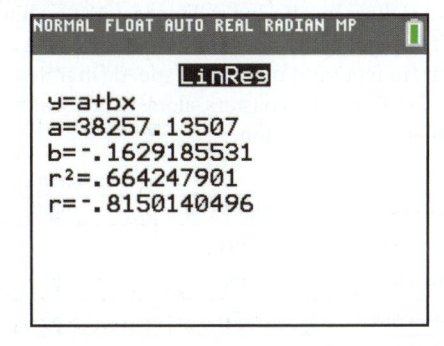

```
NORMAL FLOAT AUTO REAL RADIAN MP
            LinReg
y=a+bx
a=38257.13507
b=-.1629185531
r²=.664247901
r=-.8150140496
```

Note: The correlation is labeled as *r* in the output. You will learn about *a*, *b*, and r^2 in Chapter 11.

Two Quantitative Variables

Variable	Name	Observations (separated by commas or spaces) Keep individuals in the same order.
Explanatory	Miles driven	70583 129484 29932 29953 24495 75678 8359 4447 34077
Response	Price	21994 9500 29875 41995 41995 28986 31891 37991 34995

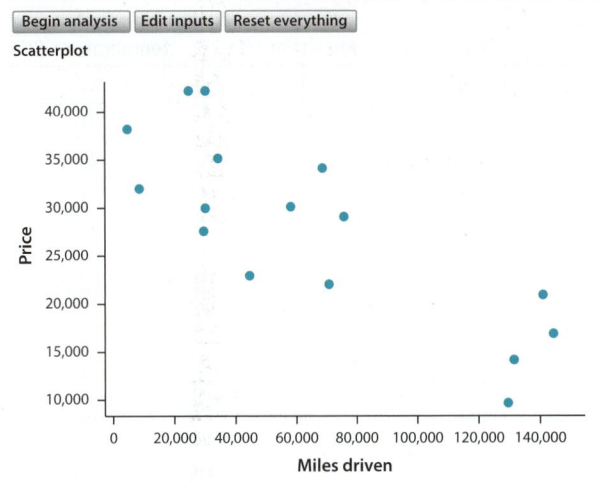

Begin analysis | Edit inputs | Reset everything

Calculate Correlation

Calculate correlation $r = -0.815$

Detailed instructions for using CrunchIt!, Excel, Google Sheets, JMP, Minitab, and R are available in Achieve.

Section 3.6 What Did You Learn?

Review the learning goals from this section. Then practice what you've learned by working through the exercises.

Learning Goal	Example	Exercises
Calculate the correlation between two quantitative variables.	3.6.1	7–10
Apply the properties of the correlation.	3.6.2	11–14
Describe how outliers influence the correlation.	3.6.3	15–18

Section 3.6 Exercises

Building Concepts and Skills These exercises assess the basic knowledge you should have after reading the section.

1. What is the formula for calculating the correlation r?

2. True/False: To calculate the correlation between two variables, both variables must be quantitative.

3. What happens to the correlation if you switch the explanatory and response variables?

4. What happens to the correlation if you change the units of either variable (or both variables)?

5. True/False: The correlation is measured in both the units of the response variable and the units of the explanatory variable.

6. Describe the effect on the correlation of a point that is separated from the rest of the points, but follows the same overall pattern.

Mastering Concepts and Skills These exercises reinforce the learning goals as shown in the examples.

7. **Refer to Example 3.6.1 Roller Coasters** Many people like to ride roller coasters. Amusement parks try to attract visitors by offering roller coasters that have a variety of speeds and elevations. Here are data on the height (in feet) and maximum speed (in miles per hour) for nine roller coasters, along with a scatterplot.[102] Calculate the correlation.

Roller coaster	Height (ft)	Maximum speed (mph)
Apocalypse	100	55
Bullet	196	83
Corkscrew	70	55
Flying Turns	50	24
Goliath	192	66
Hidden Anaconda	152	65
Iron Shark	100	52
Stinger	131	50
Wild Eagle	210	61

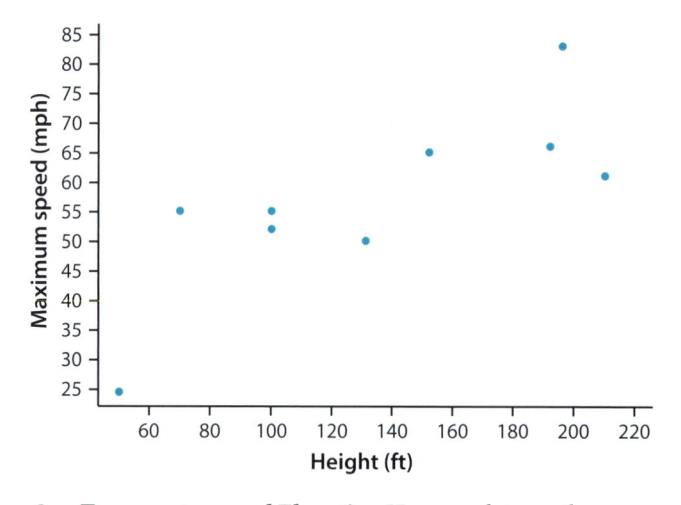

8. **Temperature and Elevation** Here are data on the elevation (in feet) and average January temperature (in degrees Fahrenheit) for 10 cities and towns in Colorado, along with a scatterplot.[103] Calculate the correlation.

City	Elevation (ft)	Average January temperature (°F)
Limon	5452	27
Denver	5232	31
Golden	6408	29
Flagler	5002	29
Eagle	6595	21
Vail	8220	18
Glenwood Springs	7183	25
Rifle	5386	26
Grand Junction	4591	29
Dillon	9049	16

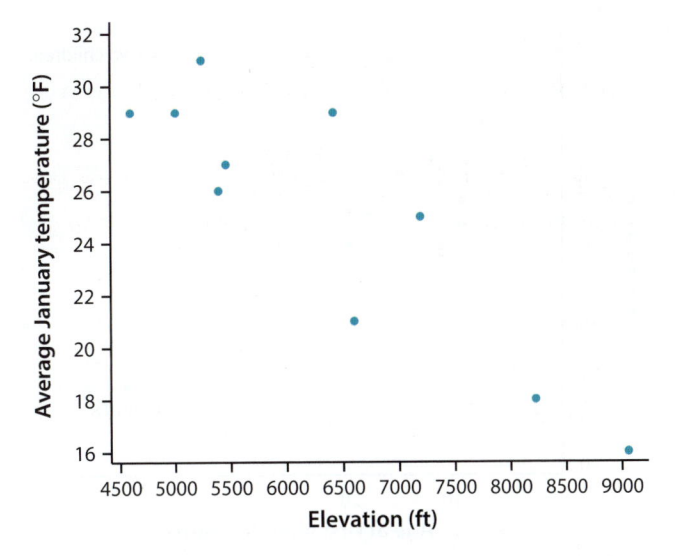

a different species. Here are data on the lengths (in centimeters) of the femur (a leg bone) and the humerus (a bone in the upper arm) for five specimens in which both bones have been preserved.[104]

Length of femur (cm)	38	56	59	64	74
Length of humerus (cm)	41	63	70	72	84

(a) Make a scatterplot using length of femur as the explanatory variable. Do you think that all five specimens come from the same species? Explain your reasoning.

(b) Calculate the correlation. Explain how your value for r matches your graph in part (a).

11. **Refer to Example 3.6.2** **Metabolic Rate** Metabolic rate, the rate at which the body consumes energy, is important in studies of weight gain, dieting, and exercise. We have data on the lean body mass and resting metabolic rate for 12 people who are subjects in a study of dieting. Lean body mass, given in kilograms, is a person's weight leaving out all fat. Metabolic rate is measured in calories burned per 24 hours. The scatterplot shows the relationship between metabolic rate and lean body mass. The correlation is $r = 0.88$.

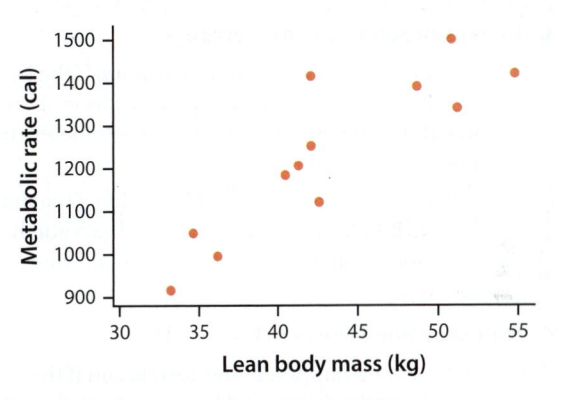

(a) What would happen to the correlation if metabolic rate was plotted on the horizontal axis and lean body mass was plotted on the vertical axis? Explain your answer.

(b) What would happen to the correlation if lean body mass was measured in pounds instead of kilograms? Explain your answer.

(c) Laura claims that the correlation between metabolic rate and lean body mass is $r = 0.88$ cal/kg. Is this correct? Explain your answer.

12. **Crickets** In a famous study published in 1948, communications engineer George Washington Pierce investigated the relationship between air temperature and the rate at which crickets chirp. The scatterplot displays his data on air temperature (in degrees Fahrenheit) and the chirp rate of the striped ground cricket (in chirps per minute). The correlation is $r = 0.835$.[105]

9. **Energy Use** Jasmin is concerned about how much energy she uses to heat her home. She keeps a record of the amount of natural gas her furnace consumes for each month from October to May. Because the months are not equally long, Jasmin divides each month's consumption by the number of days in the month to get the average number of cubic feet of gas used per day. She wants to see if there is a relationship between gas consumption and the average temperature (in degrees Fahrenheit) for each month. The data are given in the table.

Month	Average temperature (°F)	Average fuel consumption (ft³/day)
October	49.4	520
November	38.2	610
December	27.2	870
January	28.6	850
February	29.5	880
March	46.4	490
April	49.7	450
May	57.1	250

(a) Make a scatterplot using average temperature as the explanatory variable.

(b) Calculate the correlation. Explain how your value for r matches your graph in part (a).

10. **Dinosaur Bones** *Archaeopteryx* is an extinct animal that had feathers like a bird, but teeth and a long bony tail like a reptile. Because the known specimens differ greatly in size, some scientists think they are different species rather than individuals from the same species. However, if the specimens belong to the same species and differ in size because some are younger than others, there should be a positive linear relationship between the lengths of a pair of bones from all individuals. An outlier from this relationship would suggest

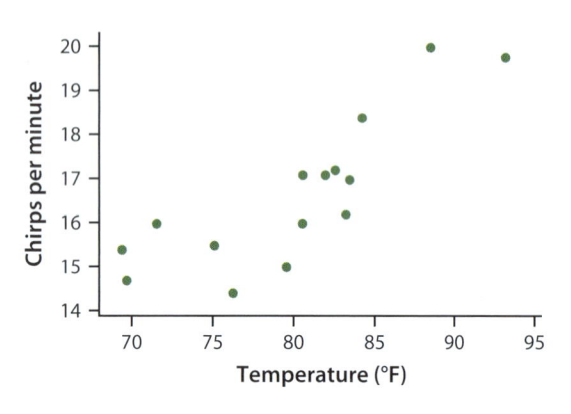

(a) What would happen to the correlation if chirp rate was plotted on the horizontal axis and air temperature was plotted on the vertical axis? Explain your answer.

(b) What would happen to the correlation if air temperature was measured in degrees Celsius instead of degrees Fahrenheit? Explain your answer.

(c) Mariko claims that the correlation between chirp rate and temperature is $r = 0.835$ chirps per minute per degree Fahrenheit. Is this correct? Explain your answer.

13. **Roller Coasters** Refer to Exercise 7.
    (a) What would happen to the correlation if the height of the coasters was measured in meters and the maximum speed was measured in kilometers per hour? Explain your answer.
    (b) What would happen to the correlation if maximum speed was plotted on the horizontal axis and height was plotted on the vertical axis? Explain your answer.

14. **Dinosaur Bones** Refer to Exercise 10.
    (a) What would happen to the correlation if the lengths of the femur and humerus bones in all five specimens were measured in inches instead of centimeters? Explain your answer.
    (b) What would happen to the correlation if length of humerus was plotted on the horizontal axis and length of femur was plotted on the vertical axis? Explain your answer.

15. **Refer to Example 3.6.3 First Speech and Aptitude** Does the age at which a child begins to talk predict a score on a later test of mental ability? A study of the development of young children recorded the age in months at which each of 21 children spoke their first word and that child's Gesell Adaptive Score, the result of an aptitude test taken much later.[106] The scatterplot shows the relationship between Gesell score and age at first word. Two outliers, Child 18 and Child 19, are identified on the scatterplot.

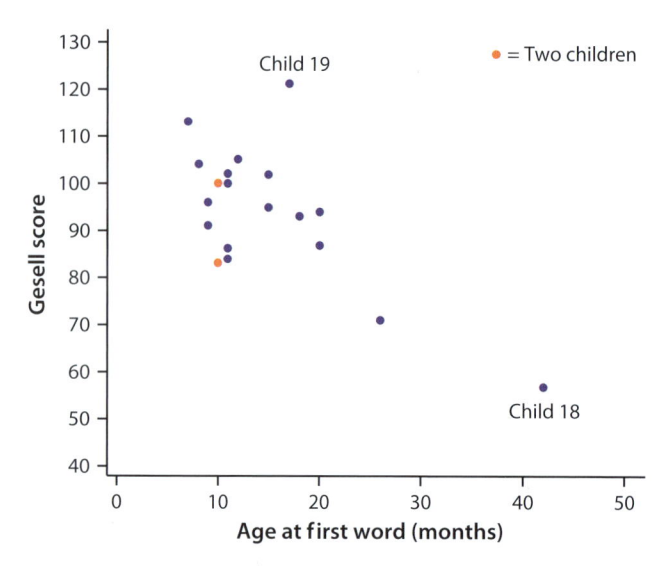

(a) What effect does Child 18 have on the correlation? Explain your answer.

(b) What effect does Child 19 have on the correlation? Explain your answer.

16. **Animal Brains** The scatterplot shows the average brain weight in grams versus average body weight in kilograms for 96 species of mammals.[107] There are many small mammals whose points overlap at the lower left.

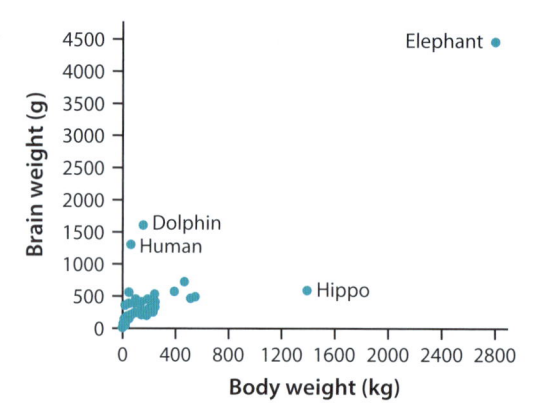

(a) How does the point for the elephant affect the correlation? Explain your answer.

(b) How does the point for the hippo affect the correlation? Explain your answer.

17. **Roller Coasters** Refer to Exercise 7. How does the point for the Flying Turns coaster affect the correlation? Justify your answer.

18. **Dinosaur Bones** Refer to Exercise 10. How does the specimen with femur length 38 cm and humerus length 41 cm affect the correlation? Justify your answer.

Applying the Concepts These exercises ask you to apply multiple learning goals in a new context or to apply what you learned in this section in a new way.

Section 3.6 More About Correlation

19. **Stand Mixers** Is there a relationship between the weight and the price of stand mixers? Here are data for a sample of 11 stand mixers, along with a scatterplot.[108]

Weight (lb)	23	28	19	17	25	26	21	32	16	17	8
Price ($)	180	250	300	150	300	370	400	350	200	150	30

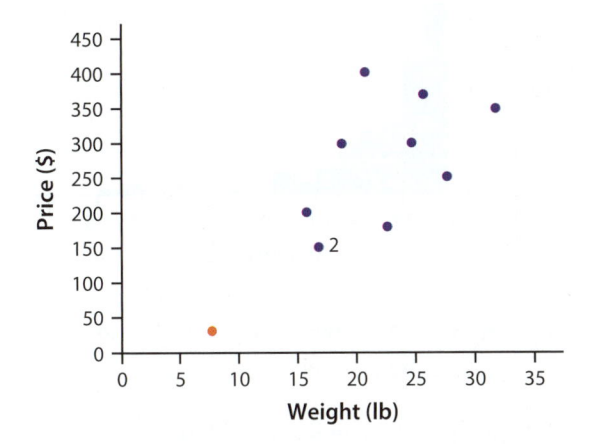

(a) Calculate the correlation. Interpret this value.

(b) The point highlighted in red is a stand mixer from Walmart. How does this point affect the correlation? Explain your reasoning.

20. **Candy Calories** Is there a relationship between the amount of sugar and the number of calories in candy? Here are data from a sample of 12 types of candy, along with a scatterplot.[109]

Name	Sugar (g)	Calories	Name	Sugar (g)	Calories
Butterfinger Minis	45	450	Reese's Pieces	61	580
Junior Mints	107	570	Skittles	87	450
M&M'S	62	480	Sour Patch Kids	92	490
Milk Duds	44	370	SweeTarts	136	680
Peanut M&M'S	79	790	Twizzlers	59	460
Raisinets	60	420	Whoppers	48	350

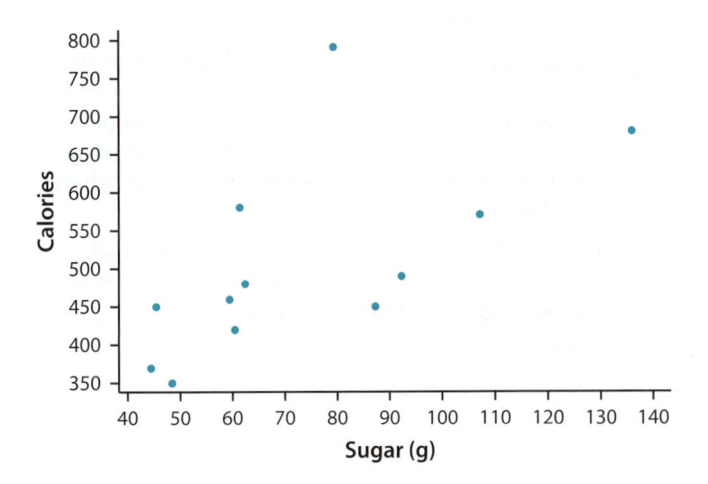

(a) Calculate the correlation. Interpret this value.

(b) Peanut M&M'S, with 79 grams of sugar and 790 calories, are an outlier in the data set. How does this point affect the correlation? Explain your answer.

21. **Spot the Mistake** Each of the following statements contains an error. Explain what is wrong in each case.

(a) There is a high correlation between the birth sex of U.S. workers and their income.

(b) We found a high correlation ($r = 1.09$) between SAT scores and first-year college grade-point average (GPA) for a sample of college students.

22. **Spot the Mistake II** Each of the following statements contains an error. Explain what is wrong in each case.

(a) The correlation between the distance traveled by a hiker and the time spent hiking is $r = 0.9$ m/sec.

(b) There is a correlation of 0.54 between the position a football player plays and their weight.

23. **Tech** **Forests and Trees** How strong is the relationship between the number of trees in a state and its total area? Its forest area? The *Forest and Trees* data set contains data on total area (in thousands of acres), forest area (in thousands of acres), and number of trees for all 50 U.S. states.[110]

(a) Use technology to make a scatterplot with total area *(Total_area)* as the explanatory variable and number of trees *(Num_trees)* as the response variable. Calculate the correlation.

(b) Identify any obvious outliers in the graph from part (a). Determine the effect of each outlier on the correlation.

(c) Make a new graph with the explanatory and response variables reversed. Calculate the correlation. Explain why this result makes sense.

(d) Now make a scatterplot with forest area *(Forest_area)* as the explanatory variable and number of trees as the response variable. Calculate the correlation. Which variable has a stronger linear relationship with the number of trees in a state: total area or forest area?

24. **Tech** **Carbs and Alcohol** How strong is the relationship between the alcohol content and amount of carbohydrates in a serving of beer or wine? Does it depend on the type of beverage? The *Beer and Wine* data set contains data for 140 different brands of beer and wine. Note that there are some missing data values.[111]

(a) Use technology to make a scatterplot for all of the beers with alcohol content *(Pct_Alcohol)* as the explanatory variable and amount of carbs *(Carbs)* as the response variable. Then calculate the correlation.

(b) Identify any obvious outliers in the graph from part (a). Determine the effect of each outlier on the correlation.

(c) Make a new graph with the explanatory and response variables reversed. Calculate the correlation. Explain why this result makes sense.

(d) Now make a scatterplot for all of the wines with alcohol content as the explanatory variable and amount of carbs as the response variable. Calculate the correlation. For which type of beverage is the relationship between alcohol content and carbs stronger: beer or wine?

Extending the Concepts These exercises challenge you to explore statistical concepts and methods that go beyond what you learned in this section.

25. **Wooden Benches** A carpenter sells handmade wooden benches at a craft fair every week. Over the past year, the carpenter has varied the price of the benches from $80 to $120 and recorded the average weekly profit made at each selling price. The prices of the bench and the corresponding average profits are shown in the table.

Price	$80	$90	$100	$110	$120
Average profit	$2400	$2800	$3000	$2800	$2400

(a) Make a scatterplot to show the relationship between price and average profit.
(b) Calculate the correlation for these data.
(c) Explain why the correlation has the value found in part (b) even though there is a strong relationship between price and average profit.

26. **Tough Grader?** Pietro and Elaine are both college English professors. Students think that Elaine is a harder grader, so Elaine and Pietro decide to grade the same 10 essays and see how their scores compare. The correlation was $r = 0.98$, but Elaine's scores were always lower than Pietro's. Draw a scatterplot that illustrates this situation. *Hint:* Include the line $y = x$ on your scatterplot.

27. **Mutual Funds** Investment reports often include correlations. Commenting on one such report that includes a table of correlations among mutual funds, a reporter says, "Two funds can have perfect correlation, yet different returns. For example, Fund A and Fund B may be perfectly correlated, yet Fund A increases in value by $2 whenever Fund B increases in value by $1." Write a brief explanation, for someone who knows little statistics, of how this can happen. Include a sketch to illustrate your explanation.

Cumulative Review These exercises revisit what you learned in previous sections.

28. **Mercury in Tuna (2.5, 3.2, 3.3)** What is the typical mercury concentration in canned tuna sold in stores? A study conducted by Defenders of Wildlife set out to answer this question. Defenders collected a sample of 164 cans of tuna from stores across the United States. They sent the selected cans to a laboratory that is often used by the EPA for mercury testing. A histogram and summary statistics provide information about the mercury concentration in the sampled cans (in parts per million, ppm).[112]

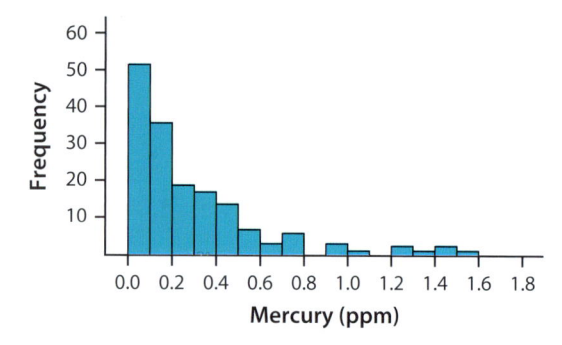

n	Mean	SD	Min	Q_1	Med	Q_3	Max
164	0.285	0.3	0.012	0.072	0.18	0.38	1.5

(a) Interpret the standard deviation.
(b) Determine whether there are any outliers.
(c) Describe the shape, center, and variability of the distribution.

29. **Cartoons (1.4, 1.5, 1.6)** Can watching fast-paced cartoons reduce the mental functioning of children? Sixty 4-year-old children were randomly assigned to three groups of 20 children each. One group was shown a fast-paced cartoon, one group was shown an educational cartoon, and one group was given art supplies and instructed to draw pictures. All of the children spent 9 minutes watching the cartoon or drawing. One of the executive function (decision making) tasks at the end of the study was a delay of gratification. If children could wait long enough, they could have 10 pieces of snack food, but if they couldn't wait long enough, they would get only 2 pieces of snack food. The response variable was how long, in seconds, the children were able to wait.[113]

(a) Describe how researchers could have carried out the random assignment.
(b) Explain the purpose of random assignment in this experiment.
(c) The average waiting time for children in the educational cartoon group was 257.20 seconds compared to an average of 146.15 seconds for the fast-paced cartoon group. The 111.05-second difference was found to be statistically significant. Explain what that means.
(d) Based on the study, is it reasonable to say that watching fast-paced cartoons caused a decrease in waiting times? Explain your answer.

Statistics Matters How can we predict annual water supply?

As we discussed at the beginning of the chapter, water resource managers must balance competing demands when making decisions about water allocation. They start by predicting the available water supply.

Colorado researchers collected data on the annual water supply (in acre-feet) at several locations in the state over a 25-year period. Here are the data for one of those locations, along with a histogram.[114]

17,985	29,430	29,829	35,684	48,460
50,057	41,274	31,559	31,027	34,487
33,289	43,669	35,951	24,772	47,662
49,924	53,650	59,373	40,209	50,722
60,970	66,293	77,738	75,741	71,217

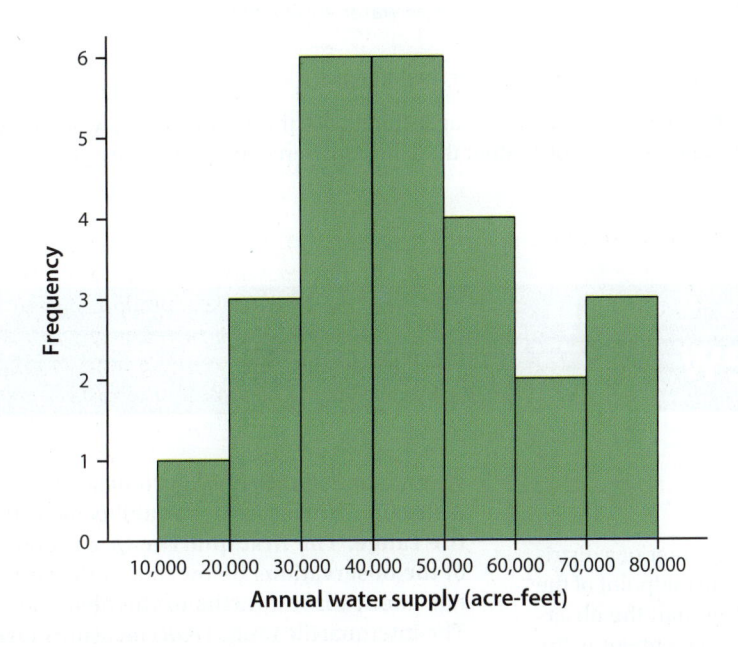

1. Find and interpret the median.
2. (a) Show that there are no outliers in the data set.
   (b) Make a boxplot to display the distribution of annual water supply at this location.
3. Use technology to calculate the mean and the standard deviation. Explain why it would be reasonable to report the mean as a typical annual water supply at this location.
4. The local resource manager decides to use the 20th percentile of the distribution as the predicted amount of water available next year, rather than the median or the mean.
   (a) Find the 20th percentile.
   (b) Give a reason why the resource manager might make this decision.

Researchers also collected data on the winter snowfall at this location each year. They recorded the data as the equivalent amount of water in the snow, in inches. Here is a scatterplot that displays the relationship between snowfall and available water supply the following year at this location over the 25-year period.

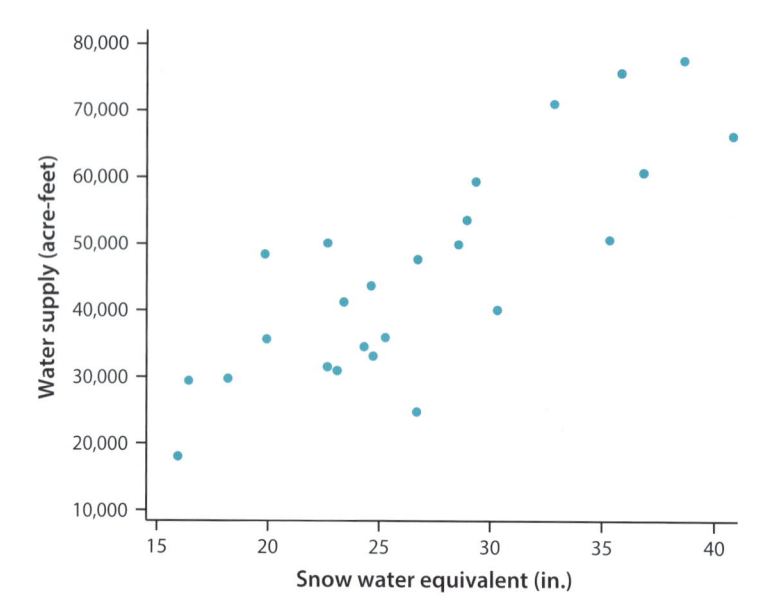

5. The correlation is $r = 0.819$. Interpret this value.

6. Explain why the local resource manager should use the amount of snowfall in a given year to predict annual water supply, rather than the 20th percentile from Question 4.

Chapter 3 Review

Numerical Summaries for One Quantitative Variable

- The **median** and the **mean** measure the center of a distribution in different ways. The median is the midpoint of the distribution, the number such that about half the observations are smaller and half are larger. The mean is the average of the observations:

$$\bar{x} = \frac{\sum x_i}{n}$$

- The simplest measure of variability for a distribution of quantitative data is the **range,** which is the distance from the minimum value to the maximum value.

- When you use the mean to describe the center of a distribution, measure variability using the **standard deviation.** The standard deviation gives the typical distance of the values in a distribution from the mean. In symbols, the sample standard deviation is given by

$$s_x = \sqrt{\frac{\sum (x_i - \bar{x})^2}{n-1}}$$

The value obtained before taking the square root is known as the sample *variance,* denoted by s_x^2. The standard deviation s_x is zero when there is no variability and gets larger as variability from the mean increases.

- When you use the median to describe the center of a distribution, measure its variability using the **interquartile range.** The **first quartile** Q_1 has about one-fourth of the observations below it, and the **third quartile** Q_3 has about three-fourths of the observations below it. The interquartile range (IQR) measures variability in the middle half of the distribution and is found by calculating $IQR = Q_3 - Q_1$.

- The median is a **resistant** measure of center because it tends not to be affected by extreme observations. The mean is not resistant. Among measures of variability, the IQR is resistant, but the standard deviation and range are not.

- The mean and standard deviation are good descriptions for roughly symmetric distributions with no outliers. The median and IQR are a better description for skewed distributions or distributions with outliers.

- According to the $1.5 \times IQR$ rule, an observation is an **outlier** if it is smaller than $Q_1 - 1.5 \times IQR$ or larger than $Q_3 + 1.5 \times IQR$.

- **Boxplots** are based on the **five-number summary** of a distribution, consisting of the minimum, Q_1, the median, Q_3, and the maximum. The box shows the variability in the middle half of the distribution. The median is marked within the box. Lines extend from the box to the smallest and the largest observations that are not outliers. Outliers are plotted with special symbols. Boxplots are helpful for comparing the center (median) and variability (range, IQR)

of multiple distributions. Boxplots aren't as useful for identifying the shape of a distribution because they do not display peaks, clusters, gaps, and other interesting features.

Describing Location in a Distribution

- Two ways of describing an individual data value's location in a distribution of quantitative data are **percentiles** and **standardized scores (z-scores)**.
- An individual's percentile is the percent of values in a distribution that are less than the individual's data value.
- To standardize any data value, subtract the mean of the distribution and then divide the difference by the standard deviation. The resulting standardized score (z-score)

$$z = \frac{\text{value} - \text{mean}}{\text{standard deviation}}$$

measures how many standard deviations the data value lies above or below the mean of the distribution.

- We can also use percentiles and z-scores to compare the location of individuals in different distributions of quantitative data.

Numerical Summaries for Two Quantitative Variables: Correlation

- The **correlation** r is a measure of the strength and direction of a linear relationship between two quantitative variables. The correlation takes values between -1 and 1, where positive values indicate a positive association and negative values indicate a negative association. Values closer to 1 and -1 indicate a stronger linear relationship and values closer to 0 indicate a weaker linear relationship.
- Calculate the correlation using the formula

$$r = \frac{\sum \left(\frac{x_i - \bar{x}}{s_x}\right)\left(\frac{y_i - \bar{y}}{s_y}\right)}{n-1}$$

or with technology.

- The correlation makes no distinction between explanatory and response variables. Correlation has no units. Correlation isn't affected by changes in the units of the explanatory or response variable.
- The correlation isn't resistant — outliers can greatly influence the value of the correlation.
- Correlation does not imply causation!

Chapter 3 Review Exercises

These exercises will help you review important concepts and skills described by the learning goals in each section. For most exercises, the relevant section is noted in parentheses after the exercise title.

1. **Music and Memory (3.1)** For a final project in their Introductory Statistics class, two students studied the impact of different types of background music on students' ability to remember words from a list they were allowed to study for 5 minutes. Here are data on the number of words remembered by one group of students who listened to Beethoven's Fifth Symphony, along with a dotplot.[115]

11	12	23	15	14	15	14	15
10	14	15	9	11	13	25	11
13	13	12	20	17	23	11	12
12	11	20	20	12	12	19	13
15	10	14	11	7	17	13	18

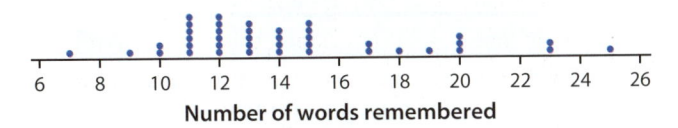

   (a) Find and interpret the median.
   (b) Calculate the mean.
   (c) Explain why the mean is larger than the median for this distribution.

2. **Music and Memory (3.2)** Refer to Exercise 1.
   (a) Find the range of the distribution.
   (b) The standard deviation is 4.05. Interpret this value.
   (c) Find and interpret the interquartile range.

3. **Music and Memory (3.3)** Refer to Exercises 1 and 2.
   (a) Identify any outliers in the distribution.
   (b) Make a boxplot of the data.
   (c) Describe the distribution of number of words remembered.

4. **Reaction Times (3.3)** Researchers suspect that athletes typically have a faster reaction time than non-athletes do. To test this theory, the researchers gave an online reflex test to 33 student athletes and 30 non-athletes at one institution. The following parallel boxplots display the reaction times (in milliseconds) for the two groups of students.

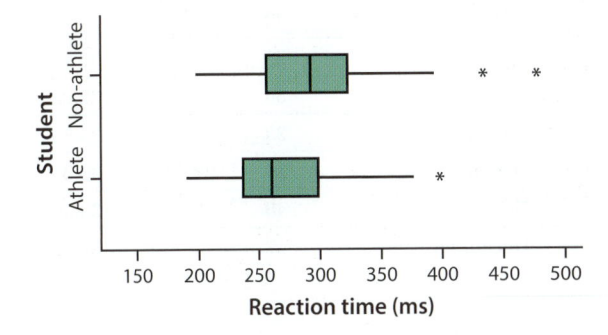

   (a) Compare the distributions.
   (b) What do the data suggest about the researchers' suspicion? Explain your answer.

5. **Iowa Homes (3.4)** The following dotplot gives the sale prices for 40 houses in Ames, Iowa, sold in the same month. The mean sale price was $203,388, with a standard deviation of $87,609.

   (a) Find the percentile of the house represented by the red dot.
   (b) Calculate and interpret the standardized score (z-score) for the house represented by the red dot, which sold for $234,000.
   (c) The same house sold 5 years earlier for $212,500. During the month when this occurred, the mean sales price of homes in Ames, Iowa, was $191,223, with a standard deviation of $76,081. In which of the two years did the house sell for more money, relatively speaking? Explain your answer.

6. **Crawling (3.5, 3.6)** At what age do babies learn to crawl? Does it take longer to learn in the winter, when babies are often bundled in clothes that restrict their movement? There might even be an association between babies' crawling age and the average temperature during the month when they first try to crawl (around 6 months after birth). Data were collected from parents who reported the birth month and the age at which their child was first able to creep or crawl a distance of 4 feet within 1 minute. Here are the data on the average temperature (in degrees Fahrenheit) 6 months after birth, and the average crawling age (in weeks) for a sample of 414 infants, along with a scatterplot.[116]

Birth month	Average temperature 6 months after birth (°F)	Average crawling age (weeks)
January	66	29.84
February	73	30.52
March	72	29.70
April	63	31.84
May	52	28.58
June	39	31.44
July	33	33.64
August	30	32.82
September	33	33.83
October	37	33.35
November	48	33.38
December	57	32.32

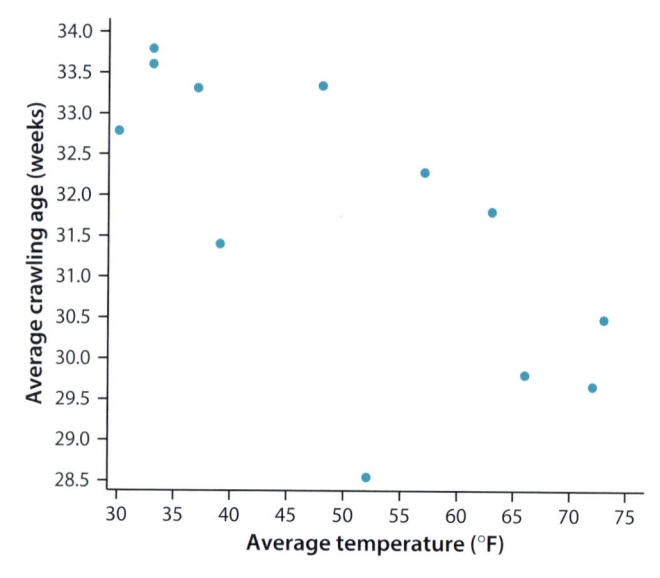

   (a) Calculate and interpret the correlation.
   (b) What effect does the point for babies born in May have on the correlation? Explain your answer.
   (c) How would the correlation be affected if the average crawling age were reported in months instead of weeks? Explain your answer.

7. **Home Prices (Chapter 3)** Ronaldo has been offered a new job that would require moving to a faraway town. Before making a decision, Ronaldo wants to know more about home prices in the town. A local realtor provides detailed data about each of 43 homes that sold in the past several months. Here is a histogram of the data on sales price (in $1000s), along with numerical summaries.[117]

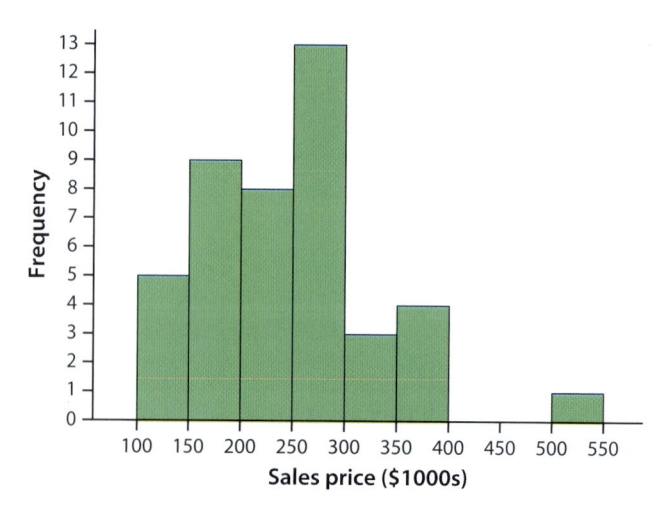

n	Mean	SD	Min	Q_1	Med	Q_3	Max
43	242.302	79.242	107	182	242	285	503

(a) What is a typical home price in this town? Justify your answer.

(b) Explain why the interquartile range is a better measure of the variability of home prices in this town than the standard deviation. Then interpret the *IQR*.

(c) Identify any outliers in the distribution.

(d) Here are parallel boxplots of the sales price for the homes in each region of town. Compare the distributions.

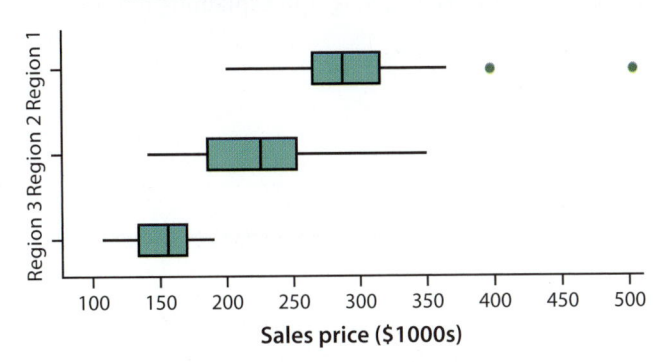

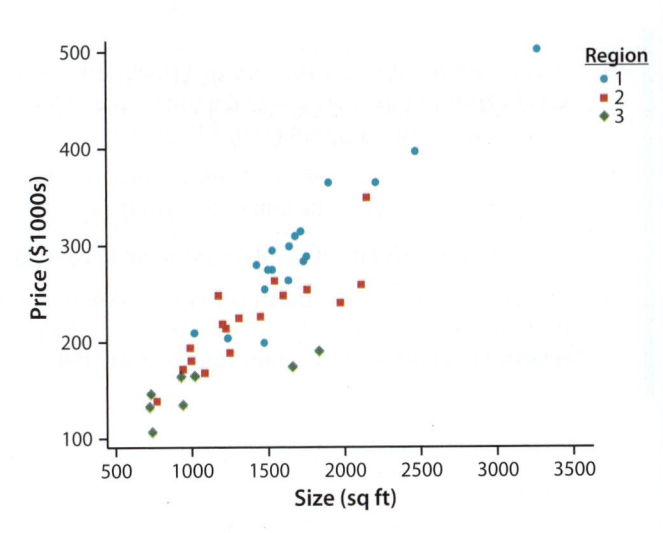

(a) The correlation between price and size for all 43 houses is $r = 0.893$. Interpret this value.

(b) How does the house with 3269 square feet of space that sold for $503,000 affect the correlation? Explain your answer.

(c) If Ronaldo is looking for a house with about 1600 square feet of space, in which region of town would such a house be most affordable? Justify your answer.

8. **Home Prices** Refer to Exercise 7. The scatterplot displays data on the sales price and size (in square feet) of recently sold homes in the town, along with the region in which each home is located.

Chapter 3 Project: Informed Consent

When researchers conduct experiments using human subjects, it is important that the participants understand what they are agreeing to and what risks might be involved. Typically, participants are given an *informed consent form* that explains the experiment and the possible risks. They are asked to read the form and then sign it to confirm that they understand the risks and agree to participate. The challenge for researchers is to create a form that contains all the appropriate information in a way that can be readily understood by a layperson.

Researchers from Switzerland and Tanzania were concerned about how understandable typical informed consent forms are. They designed an experiment to test four different ways of explaining a study and its risks to possible participants. Prospective participants in a randomized clinical trial comparing the efficacy, safety and acceptability of a new chewable drug against particular soil-transmitted infections were randomly assigned to four groups. All four groups were given the informed consent form. Group 1, the control group, was given no additional information about the experiment. Group 2 participated in an oral information session. Group 3 had the oral information session and watched a slideshow. Group 4 had the oral information session and watched a theatrical presentation. After the information was presented, participants were asked to respond to a questionnaire to test how much they had understood about the clinical trial.[118]

The data file **Chapter 3 Project Informed Consent** contains data for all participants in the experiment on two variables: *Group* and *Score* (the score on the questionnaire out of 10).

1. How many participants were there in this experiment? How many were assigned to each treatment?

2. Compute the mean score on the final questionnaire for all participants in the experiment.

3. Compute the mean score on the final questionnaire for each of the four groups. How do these means compare to the overall mean computed in Question 2?

4. Compute the standard deviation for each of the four groups. Which group has the smallest standard deviation? Which group has the largest standard deviation?

(*continued*)

5. Participant number 4 was in group 3 (including a slideshow) and had a final score of 5. Participant number 10 was in group 1 (control) and had a final score of 5. Compute the z-score for each participant. Which one scored better relative to their own group? Explain your answer.

6. Use your answers to Questions 3 and 4 to discuss what you have learned about the scores of the subjects in the control group compared to the other three groups.

7. Create comparative boxplots of the scores in the four groups. Describe what you see.

8. The goal of informed consent is to make sure that participants truly understand the risks they may face by participating in a study. Use your answers to the previous questions to write a paragraph explaining to future researchers how best to communicate these risks.

Cumulative Review Chapters 1–3

This cumulative review is designed to help you review the important concepts and skills from Chapters 1, 2, and 3, and to prepare you for an in-class exam on this material.

Section I: Multiple Choice *Select the best answer for each question.*

1. A sportswriter wants to know how strongly Albuquerque residents support the local minor league baseball team, the Isotopes. The sportswriter stands outside the stadium before a game and interviews the first 20 people who enter the stadium, asking them to rate their enthusiasm for the team on a 1 (lowest) to 5 (highest) scale. Which of the following best describes the results of this survey?
   (a) Because this is a random sample, there will be some sampling variability, but we can expect it to produce quite accurate results.
   (b) This is a random sample, but the size of the sample is too small to produce reliable results.
   (c) This is a voluntary response sample and is likely to underestimate support for the team.
   (d) This is a convenience sample and is likely to overestimate the level of support for the team.

2. The bar chart summarizes responses of dog owners to the question "Where in the car do you let your dog ride?" Which of the following statements is true?

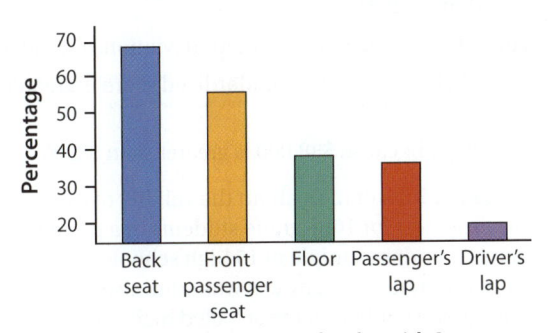

   (a) A majority of owners do not allow their pets to ride in the front passenger seat.
   (b) Roughly twice as many pets are allowed to sit in the front passenger seat as in the passenger's lap.
   (c) The vertical scale of this graph exaggerates the difference between the percentage who let their dogs ride in the driver's lap versus a passenger's lap.
   (d) These data could also be presented in a pie chart.

3. Many professional schools require applicants to take a standardized test. Suppose that 1000 students take such a test. Several weeks after the test, Petra receives a score report: Petra got a 63, which corresponds to the 73rd percentile. This means that
   (a) Petra did worse than about 63% of the test-takers.
   (b) Petra did worse than about 73% of the test-takers.
   (c) Petra did better than about 63% of the test-takers.
   (d) Petra did better than about 73% of the test-takers.

Questions 4 and 5 refer to the following setting. In a statistics class with 136 students, the professor records how much money (in dollars) each student has in their possession during the first class of the semester. The histogram displays the distribution.

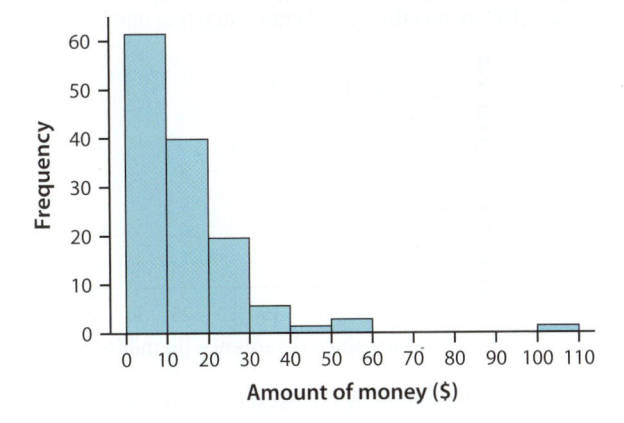

4. The percentage of students with less than $20 in their possession is closest to
   (a) 45%.   (b) 60%.   (c) 75%.   (d) 100%.

5. Which of the following statements about this distribution can be made based on the graph?
   (a) The distribution is left-skewed.
   (b) The range of the distribution is at least $90.
   (c) The standard deviation of the distribution is more than $60.
   (d) The mean is likely less than the median.

6. Some news organizations maintain a database of customers who have volunteered to share their opinions on a variety of issues. Suppose that one of these databases includes 9000 registered voters in California. To measure the amount of support for a controversial ballot issue, 1000 registered voters in California are randomly selected from the database and asked their opinion. Which of the following is the largest population to which the results of this survey should be generalized?
   (a) The 1000 people in the sample
   (b) The 9000 registered voters from California in the database
   (c) All registered voters in California
   (d) All California residents

7. You record the age, employment status, and earned income of a sample of 1463 adults. The number and type of variables you have recorded is
   (a) 3 quantitative and 0 categorical.
   (b) 3 quantitative and 1 categorical.
   (c) 2 quantitative and 1 categorical.
   (d) 2 quantitative and 2 categorical.

8. Biologists assess how closely related similar species are by measuring the number of years since the two species diverged from a common ancestor. Researchers Daniel Bolnick and Thomas Near compared the years since divergence for 12 different pairs of sunfish species to the hatching success of eggs produced by a "cross" between the two species.[119] A scatterplot of their results is shown. A hatching success of 100 means the hybrid eggs hatched as often as single-species eggs hatched. A number less than 100 means the hybrid eggs hatch less often.

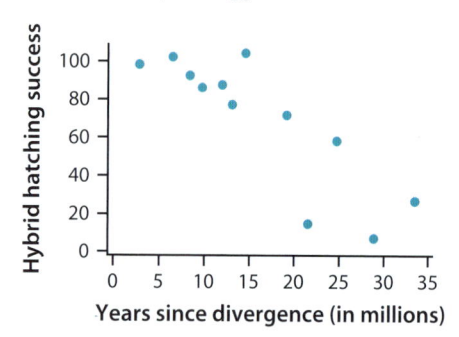

Which of the following is closest to the correlation between these two variables?

(a) −0.7   (b) −0.2   (c) 0.2   (d) 0.7

9. A new headache remedy was given to a group of 25 subjects who had headaches. Four hours after taking the new remedy, 20 of the subjects reported that their headaches had disappeared. From this information, you conclude

(a) that the remedy is effective for the treatment of headaches.
(b) nothing because the sample size is too small.
(c) nothing because there is no control group for comparison.
(d) that the new treatment is better than aspirin.

Questions 10 and 11 refer to the following setting. Forty students took a statistics exam on which the maximum score was 50 points. The exam scores are displayed in the dotplot.

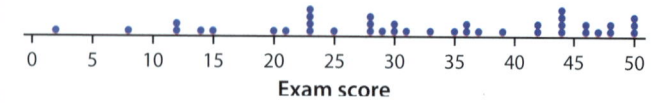

10. Which of the following are the correct median and interquartile range of this distribution?

(a) Median = 31; interquartile range = 21
(b) Median = 32; interquartile range = 21
(c) Median = 32; interquartile range = 23 to 44
(d) Median = 33; interquartile range = 21

11. The standard deviation of this distribution of exam scores is about 13. Which of the following is a correct interpretation of this value?

(a) Every student's exam score is within 13 points of the mean.
(b) The gap between the highest and lowest scores on the exam is 13 points.
(c) The exam scores typically vary from each other by 13 points.
(d) The exam scores typically vary from the mean by 13 points.

12. Consider an experiment to investigate the effectiveness of different insecticides in controlling pests and their impact on the productivity of tomato plants. What is the best reason for randomly assigning treatments (spraying or not spraying) to the farms?

(a) Random assignment allows researchers to generalize conclusions about the effectiveness of the insecticides to all farms.
(b) Random assignment will tend to balance out all other variables, such as soil fertility, so that they are not confounded with the treatment effects.
(c) Random assignment eliminates the effects of other variables, like soil fertility.
(d) Random assignment eliminates chance variation in the responses.

13. In 1965, the mean price of a new car was $2650 and the standard deviation was $1000. In 2021, the mean price was $41,000 and the standard deviation was $9500. If a Ford Mustang cost $2300 in 1965 and $38,000 in 2021, in which year was it more expensive relative to other cars?

(a) 1965, because the standardized score is greater than in 2021
(b) 1965, because the standard deviation is smaller
(c) 2021, because the standardized score is greater than in 1965
(d) 2021, because $38,000 is greater than $2300

14. To gather information about the validity of a new standardized test for 10th-grade students in a particular state, a random sample of 15 high schools was selected from the state. The new test was administered to every 10th-grade student in the selected high schools. What kind of sample is this?

(a) A simple random sample
(b) A stratified random sample
(c) A cluster random sample
(d) A systematic random sample

Questions 15 and 16 refer to the following setting. A survey was designed to study how business operations vary by size. Companies were classified as small, medium, or large. Questionnaires were sent to 200 randomly selected businesses of each size. Because not all questionnaires were returned, researchers decided to investigate the relationship between the response rate and the size of the business. The data are given in the following two-way table.

		Business size		
		Small	Medium	Large
Response?	Yes	125	81	40
	No	75	119	160

15. What percentage of all small companies receiving questionnaires responded?

(a) 12.5% (b) 20.8% (c) 50.8% (d) 62.5%

16. Which of the following conclusions is supported by the data?

(a) Small companies appear to have a higher response rate than medium or large companies.

(b) Exactly the same number of companies responded as didn't respond.

(c) Overall, more than half of companies responded to the survey.

(d) If we combined the medium and large companies, then their response rate would be approximately equal to that of the small companies.

17. A local news agency conducted a survey about unemployment by randomly dialing phone numbers until it had gathered responses from 1000 adults in the state. In the survey, 19% of those who responded said they were not currently employed. In reality, only 6% of the adults in the state were not currently employed at the time of the survey. Which of the following best explains the difference in the two percentages?

(a) The difference is due to sampling variability. We shouldn't expect the results of a random sample to match the truth about the population every time.

(b) The difference is due to response bias. Adults who are employed are likely to lie and say that they are unemployed.

(c) The difference is due to undercoverage. The survey included only adults and did not include teenagers who are eligible to work.

(d) The difference is due to nonresponse. Adults who are employed are less likely to be available to participate in the survey than adults who are unemployed.

18. To investigate if standing up while studying affects performance in an algebra class, a teacher assigns half of the 30 students in the class to stand up while studying and assigns the other half to not stand up while studying. To determine who receives which treatment, the teacher identifies the two students who did best on the last exam and randomly assigns one to stand and one to not stand. The teacher does the same for the next two highest-scoring students and continues in this manner until each student is assigned a treatment. Which of the following best describes this plan?

(a) This is an observational study.

(b) This is a completely randomized experiment.

(c) This is an experiment with blocking, but not a matched-pairs design.

(d) This is a matched-pairs design.

19. Here are the weights (in pounds) of 10 randomly chosen chimpanzees:

96   110   118   118   122   125   126   129   130   145

Which of the following statements about this data set is true?

(a) If the value 96 were removed from the data set, the mean of the remaining 9 weights would be less than the mean of all 10 weights.

(b) If the value 96 were removed from the data set, the standard deviation of the remaining 9 weights would be less than the standard deviation of all 10 weights.

(c) If the value 96 were removed from the data set, the IQR of the remaining 9 weights would be less than the IQR of all 10 weights.

(d) The chimpanzee with a weight of 96 pounds is not considered an outlier by the $1.5 \times IQR$ rule.

20. Which of the following is *not* a benefit of keeping other variables the same in an experiment?

(a) Keeping other variables the same helps prevent confounding.

(b) Keeping other variables the same reduces variability in the response variable.

(c) Keeping other variables the same makes it easier to get statistically significant results if one treatment is more effective than the other.

(d) Keeping other variables the same eliminates the need for random assignment.

21. Researchers conducted an experiment investigating the effect of a new weed killer to prevent weed growth in onion crops. They used two chemicals, the standard weed killer (S) and the new chemical (N), and tested both at high and low concentrations on 50 test plots. They then recorded the number of weeds that grew in each plot. Here are some boxplots of the results. Which of the following is a correct statement about the results of this experiment?

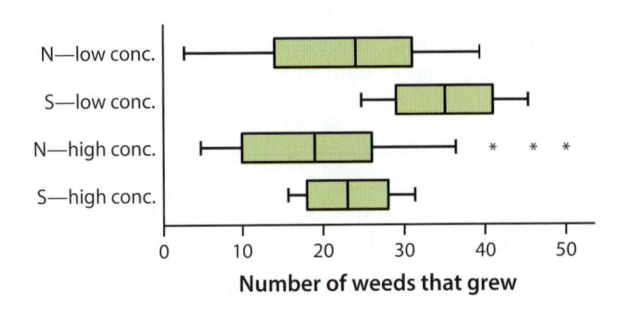

(a) At both high and low concentrations, the new chemical provides better weed control than the standard weed killer.

(b) For both chemicals, a smaller number of weeds typically grew at lower concentrations than at higher concentrations.

(c) The results for the standard weed killer are more variable than those for the new chemical.

(d) The low concentration of the new chemical was much more effective than the high concentration of the standard weed killer.

22. Which of the following is required in a statistical study to ensure that it is conducted in an ethical manner?

    I. Informed consent by participants
    II. Advance approval of the study by an institutional review board
    III. Anonymity of data from all participants

    (a) I only
    (b) II only
    (c) I and II only
    (d) I, II, and III

Section II: Free Response

1. A medical researcher is interested in determining if omega-3 fish oil can help reduce cholesterol in adults. The researcher examines the health records of 200 people in a large medical clinic, classifies them according to whether they take omega-3 fish oil, and records their latest cholesterol readings. An analysis of the data reveals that the mean cholesterol reading for those who are taking omega-3 fish oil is 18 points less than the mean for those who are not taking omega-3 fish oil.

   (a) Is this an observational study or an experiment? Justify your answer.
   (b) Explain the concept of confounding in the context of this study and give one example of a variable that could be confounded with whether people take omega-3 fish oil.
   (c) Researchers find that the 18-point difference in the mean cholesterol readings of the two groups is statistically significant. Can they conclude that omega-3 fish oil is the cause? Why or why not?

2. A researcher is interested in how many contacts older adults have in their smartphones. Here are data on the number of contacts for a random sample of 30 adults older than age 65 with smartphones in a large city.

   7 20 24 25 25 28 28 30 32 35 42 43 44 45 46
   47 48 48 50 51 72 75 77 78 79 83 87 88 135 151

   Display and describe the distribution using appropriate graphical and numerical summaries.

3. On April 17, 2021, The *New York Times* published a scatterplot like the one that follows. The graph displays data on the percentage of residents in each state who had received at least one COVID-19 vaccine shot and the percentage of votes in the state cast for Donald Trump in the 2020 presidential election.[120]

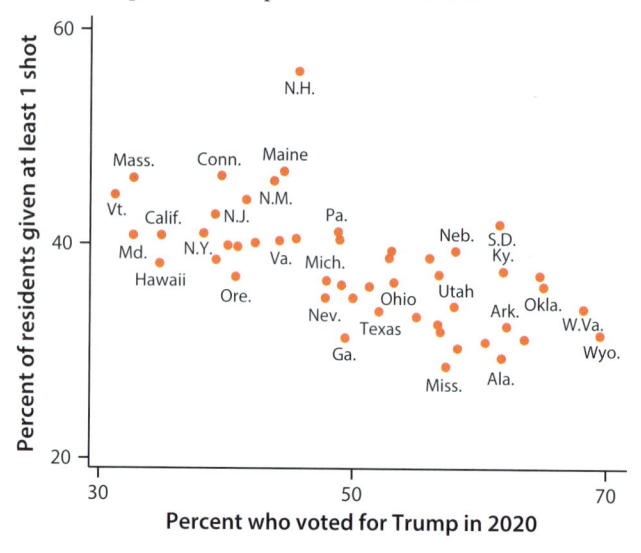

   (a) Identify the individuals and variables in this data set. Classify each variable as categorical or quantitative.
   (b) Describe the relationship displayed in the scatterplot.
   (c) What effect does New Hampshire have on the correlation? Explain your answer.
   (d) How would the correlation change if the explanatory variable was changed to the percentage who voted for Joe Biden in 2020? Explain your answer.

4. A study among the Pima people of Arizona investigated the relationship between a mother's diabetic status and the number of birth defects in her children. The two-way table summarizes the data.

		Diabetic status	
	Nondiabetic	Prediabetic	Diabetic
Number of birth defects — None	754	362	38
Number of birth defects — One or more	31	13	9

   Display and describe the relationship between these two variables using an appropriate graph and numerical summaries.

5. As someone who speaks English as a second language, researcher Sanda is convinced that people find more grammar and spelling mistakes in essays when they think the writer is a non-native English speaker. To test this, Sanda recruits a group of 60 volunteers and randomly assigns them into two groups of 30. Both groups are given the same paragraph to read. One group is told that the author of the paragraph is someone

whose native language is not English. The other group is told nothing about the author. The subjects are asked to count the number of spelling and grammar mistakes in the paragraph. While the two groups find about the same number of *real* mistakes in the passage, the number of things that are *incorrectly* identified as mistakes is more interesting. Here are a dotplot and some numerical summaries of the data on "mistakes" found.

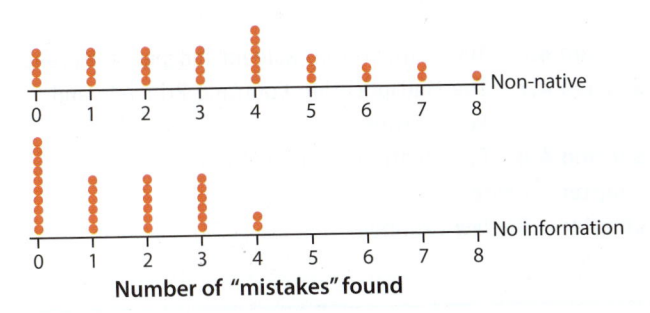

Group	n	Mean	SD
Non-native	30	3.23	2.25
No information	30	1.47	1.33

(a) Compare the distribution of number of "mistakes" found in the two groups.

(b) Describe how Sanda could have carried out the random assignment in this study.

(c) Calculate the difference in mean number of "mistakes" found for the two groups (Non-native − No information).

Five hundred trials of a simulation were performed to see what differences in means would occur due only to chance variation in the random assignment, assuming that providing additional information does not affect the number of incorrectly identified mistakes. The dotplot displays the results of the simulation.

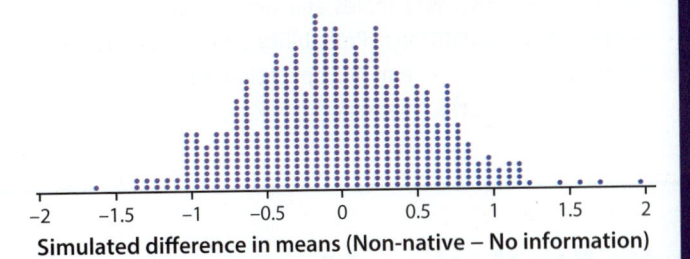

(d) There is a dot at a difference of 1.433. Explain what this dot represents.

(e) Based on the results of the simulation, is the difference from part (a) statistically significant? Explain your reasoning.

4 Probability

Section 4.1 Randomness, Probability, and Simulation
Section 4.2 Basic Probability Rules
Section 4.3 Two-Way Tables and Venn Diagrams
Section 4.4 Conditional Probability and Independence
Section 4.5 The General Multiplication Rule and Bayes' Theorem
Section 4.6 The Multiplication Rule for Independent Events
Section 4.7 The Multiplication Counting Principle and Permutations
Section 4.8 Combinations and Probability
Chapter 4 Review
Chapter 4 Review Exercises

▌▌ Statistics Matters Should an athlete who fails a drug test be suspended?

The National Collegiate Athletic Association (NCAA) has a comprehensive drug-testing program for its student athletes.[1] The main goal of these programs is to reduce the use of banned substances by students who play interscholastic sports. It is not practical to test every athlete for drug use regularly. Instead, an independent drug-testing agency administers tests to randomly selected student athletes at unannounced times during the school year. Athletes who test positive face an immediate loss of one year of NCAA eligibility, and may face other consequences that have long-term effects on their lives and careers.

Drug tests aren't perfect. Sometimes the test results suggest that athletes took a banned substance when they did not. This is known as a *false positive*. At other times, drug test results indicate that athletes are "clean" when they did, in fact, take a banned substance. This is called a *false negative*. Clearly, given the high stakes, athletes and athletic organizations want to ensure that the drug tests are as reliable as possible.

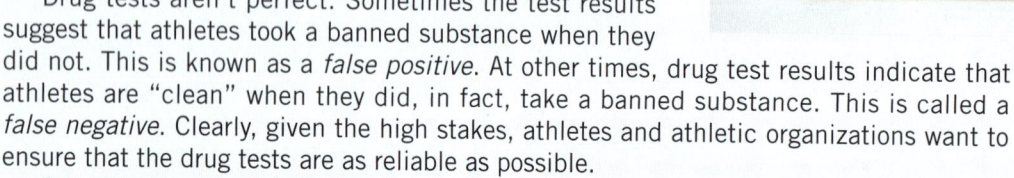

Suppose that 16% of athletes in a certain NCAA sport have taken a banned substance. Further suppose that the drug test used by the independent agency has a false positive rate of 5% and a false negative rate of 10%. If a randomly chosen athlete in this sport tests positive, what's the probability that the student really took a banned substance? Should an athlete who tests positive be suspended from athletic competition for a year?

We'll revisit Statistics Matters *at the end of the chapter, so you can use what you have learned to help answer these questions.*

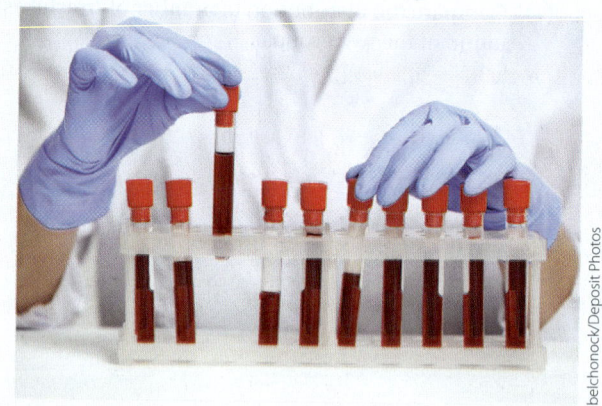

Section 4.1 Randomness, Probability, and Simulation

Accurately interpreting the results of drug and disease testing—including the risks of false positives and false negatives—requires an understanding of probability. The goal of this chapter is to develop your ability to calculate, interpret, and apply probabilities in daily life.

Section 4.1 Randomness, Probability, and Simulation

LEARNING GOALS
By the end of this section, you will be able to:
- Interpret probability as a long-run relative frequency.
- Dispel common myths about randomness.
- Use simulation to model chance behavior.

Chance is all around us. Two children play rock-paper-scissors to determine who gets the last slice of pizza. A coin toss decides which team gets to take the kick-off in a soccer match. People of all ages play games of chance involving cards, dice, or spinners. Chance also plays a large role in the genetic traits that children inherit from their birth parents, such as biological sex, hair and eye color, blood type, handedness, dimples, or whether or not they can roll their tongues.

In this chapter, we will explore the mathematics of chance behavior, otherwise known as *probability*.

The Idea of Probability

Why does soccer require a coin toss to determine which team gets to take the kick-off? Many people would agree that tossing a coin seems a "fair" way to decide. But what exactly does "fair" mean in this context? The following exploration should help shed some light on this question.

CONCEPT EXPLORATION

What is probability?

If you toss a fair coin, what's the probability that it shows heads? It's 1/2, or 0.5, right? But what does a probability of 1/2 really mean? Let's investigate that question by flipping a coin several times.

1. Obtain a coin and flip it once. Record whether you get heads or tails.
2. Flip your coin a second time. Record whether you get heads or tails. What proportion of your first two flips is heads?
3. Flip your coin 8 more times, so that you have 10 flips in all. Record whether you get heads or tails on each flip in a table like the one that follows.

Flip	1	2	3	4	5	6	7	8	9	10
Result (H or T)										
Proportion of heads										

4. Calculate the proportion of heads after each flip and record these values in the bottom row of the table. For instance, suppose you got tails on the first flip and heads on the second flip. Then your proportions of heads would be $0/1 = 0$ after the first flip and $1/2 = 0.50$ after the second flip.
5. Go to www.stapplet.com and launch *The Idea of Probability* applet (found under "Concepts").
6. Keep the probability of heads as 0.5. Set the Number of flips at 10 and click "Flip!" What proportion of the flips were heads?
7. Relaunch the applet and flip the coin 10 more times. What proportion of heads did you get this time? Repeat this process several more times. What do you notice?

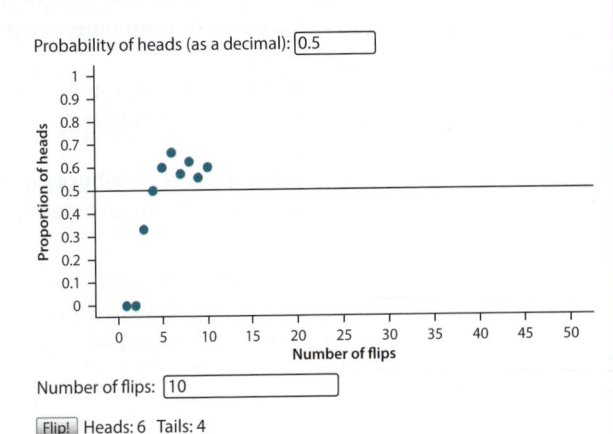

(continued)

8. What if you flip the coin 100 times? Relaunch the applet and have it do 100 flips. Is the proportion of heads exactly equal to 0.5? Close to 0.5?
9. Keep clicking "Flip!" What happens to the proportion of heads?
10. Based on this exploration, write a sentence that explains what the following statement means: "If you flip a fair coin, the probability of heads is 0.5."

Extension: If you flip a coin, it can land heads or tails. If you "toss" a thumbtack, it can land with the point sticking up or with the point down. Does that mean that the probability of a tossed thumbtack landing point up is 0.5? How can you find out? Discuss with your classmates.

Figure 4.1 shows some results from the Concept Exploration. The proportion of flips that land heads varies from 0.30 to 1.00 in the first 10 tosses. As we make more and more flips, however, the proportion of heads gets closer to 0.5 and stays there.

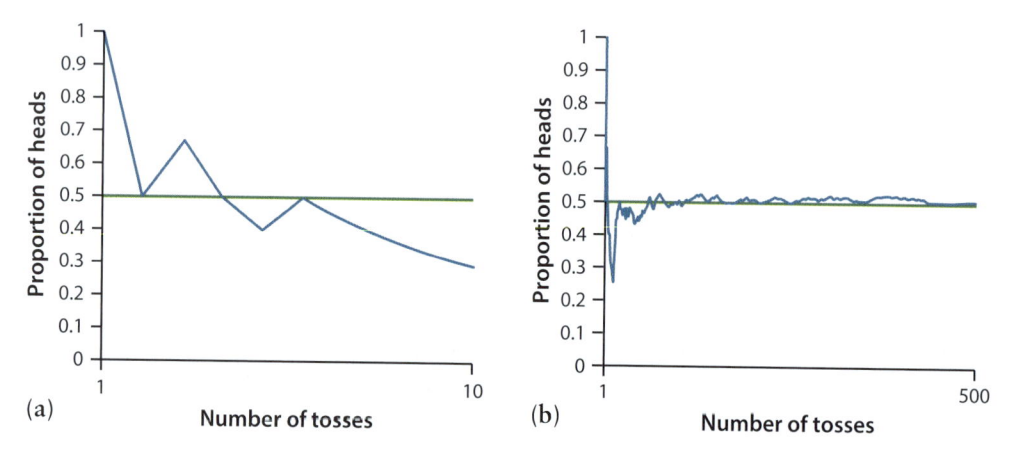

FIGURE 4.1 (a) The proportion of heads in the first 10 tosses of a coin. (b) The proportion of heads in the first 500 tosses of a coin.

When we watch coin tosses or the results of random sampling and random assignment closely, a remarkable fact emerges: *Chance behavior is unpredictable in the short run but has a regular and predictable pattern in the long run.* This is the basis for the idea of **probability**.

The **probability** of any outcome of a chance process is a number between 0 and 1 that describes the proportion of times the outcome would occur in a very large number of repetitions.

Outcomes that never occur have probability 0. An outcome that happens on every repetition has probability 1. An outcome that happens half the time in a very long series of repetitions has probability 0.5.

The fact that the proportion of heads in many tosses eventually closes in on 0.5 is guaranteed by the **law of large numbers**.

The **law of large numbers** says that if we observe more and more repetitions of any chance process, the proportion of times that a specific outcome occurs approaches its probability.

Life-insurance companies, casinos, and other businesses that make decisions based on probability rely on the long-run predictability of chance behavior. The law of large numbers helps ensure that these businesses make sizable profits from a high volume of transactions.

EXAMPLE 4.1.1

Coffee drinkers

Interpreting probability

PROBLEM

According to *The Book of Odds*, the probability that a randomly selected U.S. adult drinks coffee on a given day is 0.56.

(a) Interpret this probability as a long-run relative frequency.
(b) If a researcher randomly selects 100 U.S. adults, will exactly 56 of them drink coffee that day? Explain your answer.

SOLUTION

(a) If you take a very large random sample of U.S. adults, about 56% of them will drink coffee that day.

> The chance process consists of randomly selecting a U.S. adult and recording whether the person drinks coffee that day.

(b) Probably not. With only 100 randomly selected adults, the number who drink coffee that day is likely to differ from 56.

> Probability describes what happens in many, many repetitions (way more than 100) of a chance process.

EXAM PREP: FOR PRACTICE, TRY EXERCISE 5.

Myths About Randomness

The idea of probability is that randomness is predictable in the long run. Unfortunately, our intuitions about randomness lead us to think that chance behavior should also be predictable in the short run.

Suppose you toss a coin 6 times and get tails each time. Some people might think that the next toss is more likely to be heads. While it is true that, in the long run, heads will appear half the time, it is a myth that future outcomes must make up for an imbalance like six straight tails.

Coins and dice have no memories. A coin doesn't know that the first six outcomes were tails, and it can't try to get a head on the next toss to even things out. Of course, things do even out *in the long run*. That's the law of large numbers in action. After 10,000 tosses, the results of the first six tosses don't matter. They are overwhelmed by the results of the next 9994 tosses.

EXAMPLE 4.1.2

Baby number 8

Dispelling myths about randomness

PROBLEM

A heterosexual couple have produced seven children, all of whom were born biologically female. Their doctor mentions that their next baby is much more likely to be born male. Explain why the doctor is wrong.

SOLUTION

The doctor's claim is based on misapplying the law of large numbers to a small number of repetitions. This couple's next baby is just as likely to be born female as male.

> There is about a 1/2 probability that any child born to the couple will be male. (A small percentage of children are intersex at birth.)

EXAM PREP: FOR PRACTICE, TRY EXERCISE 9.

When asked to predict the biological sex — male (M) or female (F) — of the next seven babies born in a local hospital, most people will guess something like M-F-M-F-M-F-F. Few people would say F-F-F-M-M-M-F because this sequence of outcomes doesn't "look random." In reality, these two sequences of births are equally likely. "Runs" consisting of several of the same outcome in a row are surprisingly common in chance behavior.

Simulation

We can model chance behavior and estimate probabilities with a **simulation**.

> A **simulation** imitates a chance process in a way that models real-world outcomes.

You already have some experience with simulations from Chapter 1. In the "Hiring Discrimination?" activity (Section 1.1), you used an applet to imitate a random lottery to choose which pilots would become captains. Section 1.3 showed you how to use a simulation of sampling variability to test a claim about a population proportion. The "Drawing conclusions from the caffeine experiment" Concept Exploration (Section 1.6) asked you to determine whether the results of an experiment are statistically significant by using an applet to mimic the random assignment of subjects to treatments.

These simulations used different methods to imitate a chance process. Even so, the same basic strategy was followed in each simulation.

HOW TO Perform a Simulation

- Describe how to use a chance process to perform one repetition of the simulation. Tell what you will record at the end of each repetition.
- Perform many repetitions.
- Use the results of your simulation to answer the question of interest.

EXAMPLE 4.1.3

How many boxes of cereal to collect them all?

Using simulations to estimate probabilities

PROBLEM

In an attempt to increase sales, a breakfast cereal company decides to offer a promotion. Each box of cereal will contain a collectible comic book featuring a popular superhero: The Flash, Batman, Superman, Wonder Woman, or Green Lantern. The company claims that each of the 5 comic books is equally likely to appear in any box of cereal.

A superfan decides to keep buying boxes of the cereal until they have all 5 superhero comic books. The fan is surprised when it takes 23 boxes to get the full set of comic books. Does this outcome provide convincing evidence that the 5 comic books are not equally likely to be included in each box? To help answer this question, we want to perform a simulation to estimate the probability that it will take 23 or more boxes to get a full set of the superhero collectible comic books.

(a) Describe how to use a random number generator to perform one repetition of the simulation.

We carried out 100 repetitions of the simulation and noted the results. The dotplot shows the number of cereal boxes it took to get all 5 superhero comic books in each repetition.

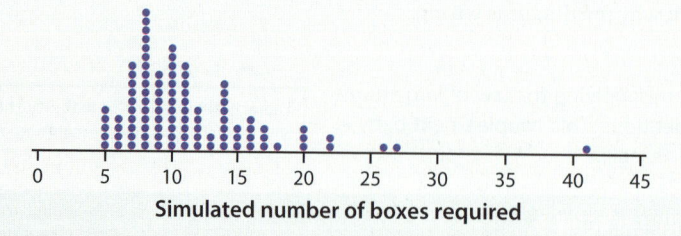

(b) Explain what the dot at 18 represents.

(c) Use the results of the simulation to estimate the probability that it will take 23 or more boxes to get a full set of comic books.

(d) Based on the actual result of 23 boxes and your answer to part (c), is there convincing evidence that the 5 superhero comic books are not equally likely to be found in each box? Explain your reasoning.

SOLUTION

(a) Let 1 = The Flash; 2 = Batman; 3 = Superman; 4 = Wonder Woman; 5 = Green Lantern. Generate a random integer from 1 to 5 to simulate buying one box of cereal and looking at which comic book is inside. Keep generating random integers until all five labels from 1 to 5 appear. Record the number of boxes it takes.

(b) One repetition where it took 18 boxes to get all 5 superhero comic books.

(c) Because 3 of the 100 dots are 23 or greater, the probability ≈ 3/100 = 0.03.

(d) Because it is unlikely that it would take 23 or more boxes to get a full set by chance alone when the superhero comic books are equally likely to be included in each box, this result provides convincing evidence that the superhero comic books are not equally likely to appear in each box of cereal.

> Recall from Section 1.6 that researchers often identify results that occur less than 5% of the time by chance alone as "statistically significant."

EXAM PREP: FOR PRACTICE, TRY EXERCISE 13.

It took our superfan 23 boxes to complete the set of 5 superhero comic books. Does that mean the company lied about how the comic books were distributed? Not necessarily. Our simulation says that it's unlikely (less than a 5% chance) for someone to have to buy at least 23 boxes to get a full set of superhero comic books if each comic book is equally likely to appear in a box of cereal. However, it is still possible that the company was telling the truth and the superfan was just very unlucky.

Section 4.1 What Did You Learn?

Review the learning goals from this section. Then practice what you've learned by working through the exercises.

Learning Goal	Example	Exercises
Interpret probability as a long-run relative frequency.	4.1.1	5–8
Dispel common myths about randomness.	4.1.2	9–12
Use simulation to model chance behavior.	4.1.3	13–16

Section 4.1 Exercises

Building Concepts and Skills These exercises assess the basic knowledge you should have after reading the section.

1. The probability of any outcome of a chance process is a number between _____ and _____.

2. Describe the law of large numbers.

3. True/False: A common myth is that chance behavior is predictable in the short run.

4. List the three steps in performing a simulation.

Mastering Concepts and Skills These exercises reinforce the learning goals as shown in the examples.

5. **Refer to Example 4.1.1 Cystic Fibrosis** It is not uncommon for the husband and wife in a cisgender married couple to carry the gene for cystic fibrosis without having the disease themselves. Suppose we select one of these couples at random. According to the laws of genetics, there is a 0.25 probability that any child they have will develop cystic fibrosis.

    (a) Explain what this probability means.

    (b) If researchers randomly select 4 such couples who each have one child, is one of these couples guaranteed to have a child who develops cystic fibrosis? Explain your answer.

6. **Streaming Commercials** If Lucretia tunes into her favorite streaming radio channel at a randomly selected time, there is a 0.20 probability that a commercial will be playing.

    (a) Explain what this probability means.

    (b) If Lucretia tunes into this channel at 5 randomly selected times, will there be exactly 1 time when a commercial is playing? Explain your answer.

7. **Red Light!** Pedro drives the same route to work on Monday through Friday. The route includes one traffic light. According to the local traffic department, there is a 55% probability that the light will be red. Explain what this probability means.

8. **Chance of Rain** A local weather forecast says that there is a 20% chance of rain tomorrow. Explain what this probability means.

9. **Refer to Example 4.1.2 Softball Stats** A very good professional softball player gets a hit about 35% of the time over an entire season. After the player failed to hit safely in six straight at-bats, a TV commentator said, "She is due for a hit by the law of averages." Explain why the commentator is wrong.

10. **Life Insurance** Jake is an insurance salesperson who is able to complete a sale in about 15% of calls involving life insurance policies. One day, Jake fails to make a sale on the first 10 such calls. A colleague tells Jake not to worry because he is "due for a sale." Explain why the colleague is wrong.

11. **Coin Tosses** Imagine tossing a coin 6 times and recording heads (H) or tails (T) on each toss. Which of the following outcomes is more likely: HTHTTH or TTTHHH? Explain your answer.

12. **Dice Rolls** Imagine rolling a die 12 times and recording the result of each roll. Which of these outcomes is more likely: 1 2 3 4 5 6 6 5 4 3 2 1 or 1 5 4 5 2 4 3 3 6 1 2 6? Explain your answer.

13. **Refer to Example 4.1.3 Train Arrivals** New Jersey Transit claims that its 8:00 a.m. train from Princeton to New York City has probability 0.9 of arriving on time on a randomly selected day. Assume for now that this claim is true. Ariella takes the 8:00 a.m. train to work 20 days in a certain month, and is surprised when the train arrives late in New York on 4 of the 20 days. Should Ariella be surprised? To help answer this question, we want to perform a simulation to estimate the probability that the train would arrive late on 4 or more of 20 days if New Jersey Transit's claim is true.

   (a) Describe how to use a random number generator to perform one repetition of the simulation.

   The dotplot shows the number of days on which the train arrived late in 100 repetitions of the simulation.

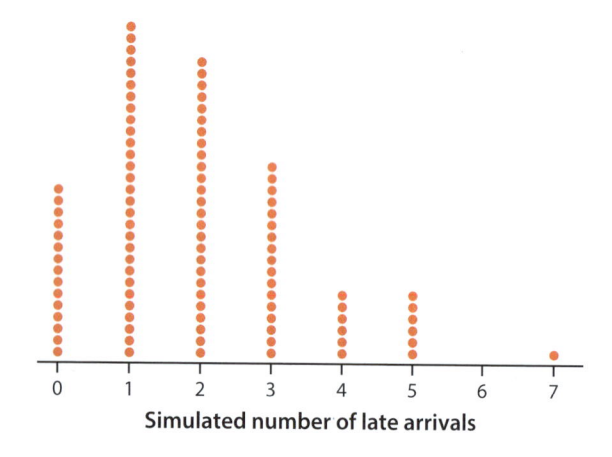

   (b) Explain what the dot at 7 represents.

   (c) Use the results of the simulation to estimate the probability that the train will arrive late on 4 or more of 20 days.

   (d) Based on the actual result of 4 late arrivals in 20 days and your answer to part (c), is there convincing evidence that New Jersey Transit's claim is false? Explain your reasoning.

14. **Double Fault!** A professional tennis player claims to make 90% of their second serves. In a recent match, the player missed 5 of 20 second serves. Is this a surprising result if the player's claim is true? Assume that the player has a 0.10 probability of missing each second serve. We want to carry out a simulation to estimate the probability that this player would miss 5 or more of 20 second serves.

    (a) Describe how to use a random number generator to perform one repetition of the simulation.

    The dotplot displays the number of second serves missed by the player out of 20 second serves in 100 simulated matches.

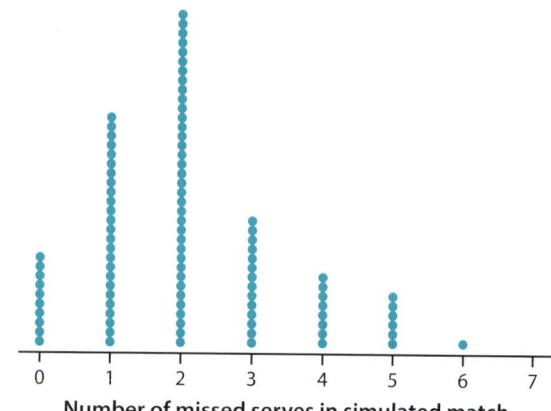

    (b) Explain what the dot at 6 represents.

    (c) Use the results of the simulation to estimate the probability that the player would miss 5 or more of 20 second serves in a match.

    (d) Based on the actual result of 5 missed second serves and your answer to part (c), is there convincing evidence that the player's claim is false? Explain your reasoning.

15. **Bull's-Eye!** In an archery competition, each player continues to shoot until they miss the center of the target twice. Quincy is one of the archers. Based on past experience, Quincy has a 0.60 probability of hitting the center of the target on each shot. Should we be surprised if Quincy stays in the competition for at least 10 shots?

    (a) Describe how you would design a simulation to help answer this question.

    (b) Carry out one repetition of the simulation you described in part (a).

    (c) In 3 out of 100 repetitions of the simulation, Quincy stayed in the competition for at least 10 shots. Based on this result, how would you answer the question of interest? Explain your reasoning.

16. **Color Blindness** About 7% of men in the United States have some form of red–green color blindness. Suppose we randomly select one U.S. adult male at a time until we find one who is red–green color-blind. Should we be surprised if it takes us 20 or more men?

    (a) Describe how you would design a simulation to help answer this question.

(b) Carry out one repetition of the simulation you described in part (a).

(c) In 24 out of 100 repetitions of the simulation, it took 20 or more randomly selected men to find one who is red–green color blind. Based on this result, how would you answer the question of interest?

Applying the Concepts These exercises ask you to apply multiple learning goals in a new context or to apply what you learned in this section in a new way.

17. **Virtual Three Pointers** The figure shows the results of a virtual basketball player shooting many 3-point shots. Explain what this graph tells you about chance behavior in the short run and in the long run.

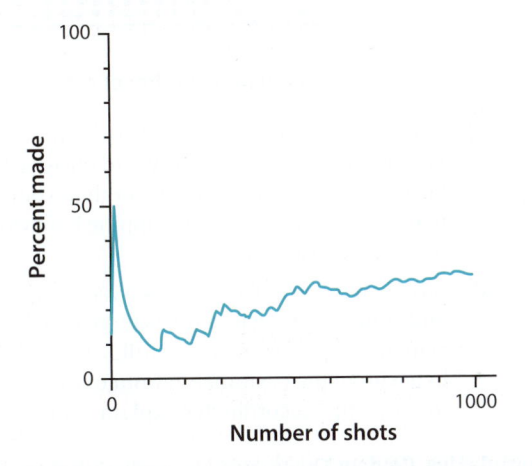

18. **Coin Tosses** The figure shows the results of two different sets of 5000 coin tosses. Explain what this graph tells you about chance behavior in the short run and in the long run. (Note that the horizontal axis is on a *logarithmic scale*.)

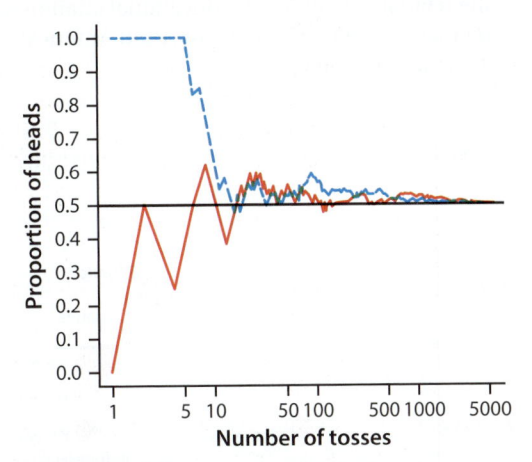

19. **Random Screening?** At a certain airport, security officials claim that they randomly select passengers for extra screening at the gate before boarding some flights. One such flight had 76 passengers — 12 in first class and 64 in economy class. Some passengers were surprised when none of the 10 passengers chosen for screening was seated in first class. Should they be surprised? We want to perform a simulation to estimate the probability that no first-class passengers would be chosen in a truly random selection.

(a) Describe how you would carry out this simulation. *Note:* It is important to avoid selecting the same passenger more than once!

(b) Perform one repetition of your simulation from part (a).

(c) We performed 100 trials of the simulation. In 15 trials, none of the 10 passengers chosen was seated in first class. Does this result provide convincing evidence that the security officials did not carry out a truly random selection? Explain your reasoning.

20. **Scrabble** In the game of Scrabble, the first player draws 7 letter tiles at random from a bag containing 100 tiles. There are 42 vowels, 56 consonants, and 2 blank tiles in the bag. Anise draws first and is surprised to discover that all 7 tiles are vowels. Should Anise be surprised? We want to perform a simulation to estimate the probability that a player will randomly select 7 vowels from the bag.

(a) Describe how you would carry out this simulation. *Note:* It is important to avoid selecting the same tile more than once!

(b) Perform one repetition of your simulation from part (a).

(c) We performed 1000 trials of the simulation. In 2 trials, all 7 tiles were vowels. Does this result give convincing evidence that the bag of tiles was not well mixed when Anise selected the 7 tiles? Explain your reasoning.

21. **Who Recycles?** Researchers want to investigate whether a majority of students at the local community college regularly recycle. To find out, they survey a simple random sample (SRS) of 100 students at the college about their recycling habits. Suppose that 55 students in the sample say that they regularly recycle. Is this convincing evidence that more than half of the students at the college would say they regularly recycle? The dotplot shows the results of taking 200 SRSs of 100 students from a population in which the true proportion who recycle is 0.50.

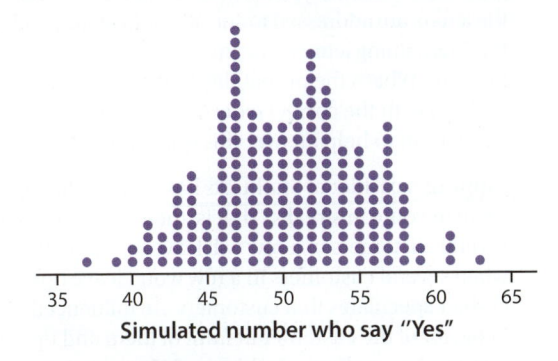

(a) Explain why the sample result (55 out of 100 said "Yes") does not give convincing evidence that more than half of the college's students would say that they regularly recycle.

(b) Suppose instead that 63 students in the researchers' sample had said "Yes." Explain why this result would give convincing evidence that a majority of the college's students would say that they regularly recycle.

22. **Conserving Water** A recent study reported that two-thirds of young adults turn off the water while brushing their teeth.[2] Researchers suspect that the true proportion is lower at their local university. To find out, they ask an SRS of 60 students at the university if they usually brush with the water off. Suppose that 36 students in the sample say "Yes." The dotplot shows the results of taking 200 SRSs of 60 students from a population in which the true proportion who brush with the water off is 2/3.

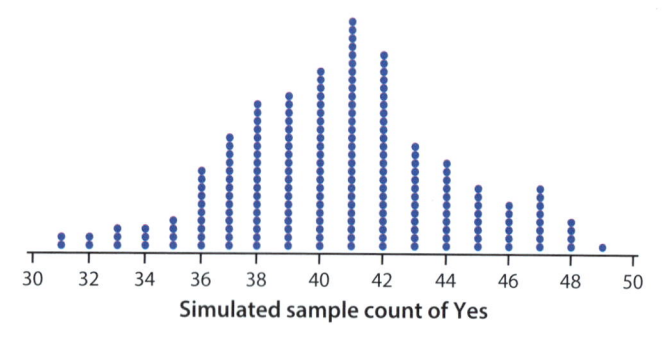

(a) Explain why the sample result (36 out of 60 said "Yes") does not give convincing evidence that fewer than two-thirds of the university's students would say that they brush their teeth with the water off.

(b) Suppose instead that 32 of the 60 students in the researchers' sample had said "Yes." Explain why this result would give convincing evidence that fewer than two-thirds of the university's students would say that they brush their teeth with the water off.

Extending the Concepts These exercises challenge you to explore statistical concepts and methods that go beyond what you learned in this section.

23. **Donor Letters** Charities often send out letters to previous donors asking for additional contributions. At one such charity, a volunteer assists with putting letters into envelopes. Unfortunately, the volunteer does not realize that the letters are addressed to specific individuals, and places the 5 remaining letters into envelopes for these 5 donors at random. What's the probability that none of the 5 donors ends up with the correct letter? Design and carry out a simulation to help answer this question.

24. **Tipping** A cashier at Starbucks has noticed that tips seem to come in streaks. That is, there are times when several customers in a row will leave tips and other times when several customers in a row won't leave tips. The cashier speculates that customers are influenced by the behavior of the customer in front of them and tip (or don't tip) accordingly. Is this true? Here are the results of 50 consecutive customers, where "T" represents one of the 25 customers who left a tip and "N" represents one of the 25 customers who didn't leave a tip.

N TTTT NN T NNNN TTT N T NNNN TTTT NNN TT N T NNN TT NN TTT NN T N TT N T

(a) A run consists of one or more consecutive occurrences of the same outcome. How many runs did the cashier observe in the tipping behavior of these 50 customers?

(b) Given that there will be 25 T's and 25 N's, what is the smallest possible number of runs? Would an outcome with this number of runs suggest that customers are influenced by the previous customer? Explain your answer.

We used technology to simulate the number of runs in 100 different sets of 50 customers—half of whom left a tip and half of whom did not—assuming that one customer's behavior did not influence another's. The dotplot shows the results.

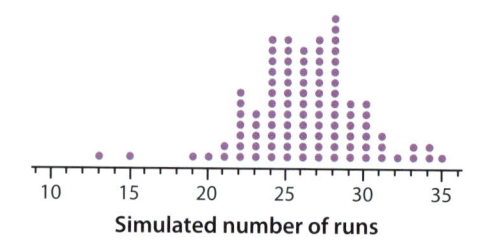

(c) Use the results of the simulation to estimate the probability that the cashier would observe a number of runs as small as or smaller than your answer to part (a) if one customer's tipping behavior does not influence another's.

(d) Based on the actual result observed by the cashier and your answer to part (c), is there convincing evidence that customers are influenced by the behavior of the customer in front of them and tip (or don't tip) accordingly? Explain your reasoning.

Cumulative Review These exercises revisit what you learned in previous sections.

25. **Income and Education (2.6)** The *New York Times* presented a scatterplot like the one that follows, displaying the relationship between educational attainment and median household income for Asian Americans by Asian origin group.[3]

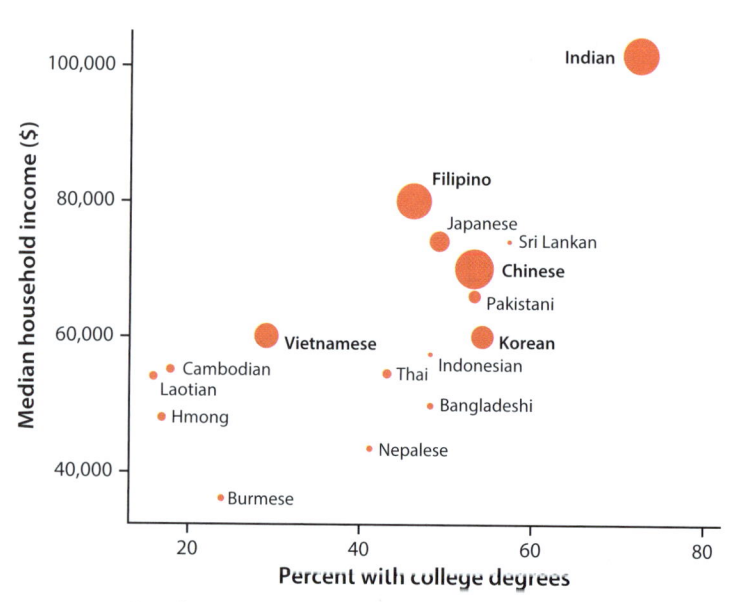

Note: The size of the bubbles is proportional to the number of Asian Americans in each group.

Use the graph to help answer these questions.

(a) Do Asian Americans from all Asian origin groups have similar levels of education? Explain your answer.

(b) What is the relationship between educational attainment and household income for Asian Americans from different Asian origin groups?

26. **Morning Classes (1.8, 2.3)** Professor Swango has two statistics classes, one that meets at 8:00 a.m. and one that meets at 10:00 a.m. The professor wonders if student performance differs in the two classes. Here are the scores earned by the students in both classes on a recent quiz.

10 a.m.	24	28	28	24	11	19	9	28
	24	28	26	23	23	27	20	
8 a.m.	22	21	22	24	23	21	22	18
	21	17	22	23	23	22	21	20

(a) Make parallel dotplots of the data for the two classes.

(b) Use your graph from part (a) to compare the quiz score distributions.

(c) Can Professor Swango conclude that any difference in student performance for the two classes is caused by the times that the classes meet? Explain your answer.

Section 4.2 Basic Probability Rules

LEARNING GOALS

By the end of this section, you will be able to:

- Give a probability model for a chance process with equally likely outcomes and use it to find the probability of an event.
- Use the complement rule to find probabilities.
- Use the addition rule for mutually exclusive events to find probabilities.

In Section 4.1, we used simulation to imitate chance behavior. Do we always have to repeat a chance process — tossing a coin, rolling a die, using a random number generator — many times to determine the probability of a particular outcome? Fortunately, the answer is no.

Probability Models

Many board games involve rolling dice. Imagine rolling two fair, 6-sided dice — one that's red and one that's blue. How do we develop a **probability model** for this chance process? Figure 4.2 displays the set of all possible outcomes in the **sample space**. Because the dice are fair, each of these 36 outcomes will be equally likely and have probability 1/36.

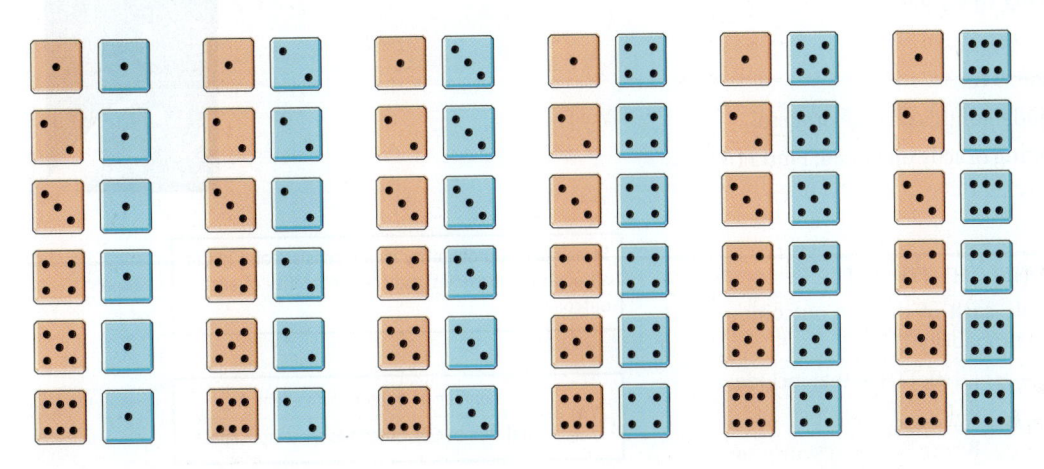

FIGURE 4.2 The 36 possible outcomes of rolling two dice, one red and one blue.

A **probability model** is a description of some chance process that consists of two parts: a list of all possible outcomes and the probability of each outcome.

The list of all possible outcomes is called the **sample space**.

What if the two dice were actually the same color? Figure 4.2 (with the colors of the dice adjusted) would still show the sample space for the chance process of rolling two fair, 6-sided dice. Note that 1 on the first die and 6 on the second die is a different outcome than 6 on the first die and 1 on the second die, even if we can't tell the two dice apart.

A probability model does more than just assign a probability to each outcome: It allows us to find the probability of an **event.**

An **event** is a set of outcomes from some chance process.

Events are usually designated by capital letters, like A, B, C, and so on. For rolling two 6-sided dice, we can define event A as getting a sum of 5. We write the probability of event A as $P(A)$ or $P(\text{sum is 5})$.

It is fairly straightforward to find the probability of an event in the case of equally likely outcomes. There are 4 outcomes in event A:

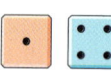

$$P(A) = \frac{\text{number of outcomes with sum of 5}}{\text{total number of outcomes when rolling two dice}} = \frac{4}{36} = 0.111$$

Finding Probabilities: Equally Likely Outcomes

If all outcomes in the sample space are equally likely, the probability that event A occurs can be found using the formula

$$P(A) = \frac{\text{number of outcomes in event A}}{\text{total number of outcomes in sample space}}$$

EXAMPLE 4.2.1

Heads or tails?

Probability models: Equally likely outcomes

PROBLEM

Suppose you flip a fair coin 3 times.
(a) Give a probability model for this chance process.
(b) Define event A as getting exactly one head. Find $P(A)$.

SOLUTION

(a) The sample space is
HHH HHT HTH THH HTT THT TTH TTT
Because the coin is fair, each of these 8 outcomes will be equally likely and have probability 1/8.

(b) There are 3 outcomes — HTT, THT, TTH — that include exactly one head. So $P(A) = \frac{3}{8} = 0.375$.

Remember: A probability model consists of a list of all possible outcomes and the probability of each outcome.

$$P(A) = \frac{\text{number of outcomes in event A}}{\text{total number of outcomes in sample space}}$$

EXAM PREP: FOR PRACTICE, TRY EXERCISE 9.

Basic Probability Rules

Our work so far suggests that a valid probability model must obey two commonsense rules:

- **The probability of any event is a number between 0 and 1, inclusive.** This rule follows from the definition of probability: the proportion of times the event would occur in many repetitions of the chance process.
- **All possible outcomes together must have probabilities that add up to 1.** Any time we observe a chance process, some outcome must occur.

Here's one more rule that follows from the previous two:

- **The probability that an event does *not* occur is 1 minus the probability that the event does occur.** Earlier, we found that the probability of getting a sum of 5 when rolling two fair, 6-sided dice is 4/36. What's the probability that the sum is *not* 5?

$$P(\text{sum is not 5}) = 1 - P(\text{sum is 5}) = 1 - \frac{4}{36} = \frac{32}{36} = 0.889$$

We refer to the event "not A" as the **complement** of A and denote it by A^C. For that reason, this handy result is known as the **complement rule**. Using the complement rule in this setting is much easier than counting all 32 possible ways to get a sum that isn't 5.

> The **complement rule** says that $P(A^C) = 1 - P(A)$, where event A^C is the **complement** of event A — that is, the event that A does not happen.

EXAMPLE 4.2.2

Can you avoid the blue M&M'S?

Complement rule

PROBLEM

Suppose you tear open the corner of a bag of M&M'S® Milk Chocolate Candies, pour one candy into your hand, and observe the color. According to Mars, Incorporated, the maker of M&M'S, the probability model for a bag is shown here.[4]

Color	Blue	Orange	Green	Yellow	Red	Brown
Probability	0.207	0.205	0.198	0.135	0.131	0.124

(a) Explain why this is a valid probability model.
(b) Find the probability that you don't get a blue M&M.

SOLUTION

(a) The probability of each outcome is a number between 0 and 1, and
$0.207 + 0.205 + 0.198 + 0.135 + 0.131 + 0.124 = 1$.
(b) $P(\text{not blue}) = 1 - P(\text{blue}) = 1 - 0.207 = 0.793$

$P(A^C) = 1 - P(A)$

EXAM PREP: FOR PRACTICE, TRY EXERCISE 13.

What's the probability that you get a green M&M or a red M&M? It's

$$P(\text{green or red}) = P(\text{green}) + P(\text{red}) = 0.198 + 0.131 = 0.329$$

Notice that the events "getting a green" and "getting a red" have no outcomes in common — that is, there are no M&M'S that are both green and red. We say that these two events are **mutually exclusive**. As a result, this formula is known as the **addition rule for mutually exclusive events.**

> Two events A and B are **mutually exclusive** if they have no outcomes in common and so can never occur together — that is, if $P(A \text{ and } B) = 0$.

> The **addition rule for mutually exclusive events** A and B says that
> $$P(A \text{ or } B) = P(A) + P(B)$$

EXAMPLE 4.2.3

Who wins the Blazin' Bonus?

Addition rule for mutually exclusive events

PROBLEM
Buffalo Wild Wings ran a promotion called the Blazin' Bonus, in which for every $25 gift card purchased, the customer also received a "Bonus" gift card for $5, $15, $25, or $100. According to the company, here are the probabilities for each Bonus gift card:

Blazin' Bonus	$5	$15	$25	$100
Probability	0.890	0.098	0.010	0.002

Suppose we randomly select a customer who purchases a $25 gift card.

(a) Find the probability that the customer gets a Bonus gift card worth $25 or more.

(b) Find the probability that the customer gets a Bonus gift card worth less than $25.

SOLUTION
(a) $P(\$25 \text{ or more}) = P(\$25 \text{ or } \$100) = P(\$25) + P(\$100) = 0.010 + 0.002 = 0.012$

(b) $P(\text{less than } \$25) = 1 - P(\$25 \text{ or more}) = 1 - 0.012 = 0.988$

> It is OK to add the two probabilities in part (a) because the events "$25 gift card" and "$100 gift card" are mutually exclusive.

EXAM PREP: FOR PRACTICE, TRY EXERCISE 17.

Note that you could also find the probability in part (b) of Example 4.2.3 using the addition rule for mutually exclusive events:

$$P(\text{less than } \$25) = P(\$5 \text{ or } \$15) = P(\$5) + P(\$15) = 0.890 + 0.098 = 0.988$$

Section 4.2 What Did You Learn?

Review the learning goals from this section. Then practice what you've learned by working through the exercises.

Learning Goal	Example	Exercises
Give a probability model for a chance process with equally likely outcomes and use it to find the probability of an event.	4.2.1	9–12
Use the complement rule to find probabilities.	4.2.2	13–16
Use the addition rule for mutually exclusive events to find probabilities.	4.2.3	17–20

Section 4.2 Exercises

Building Concepts and Skills These exercises assess the basic knowledge you should have after reading the section.

1. A probability model is a description of some chance process that consists of two parts: a list of _____ (called the sample space) and the probability of _____.

2. True/False: An event is any collection of outcomes from the sample space.

3. What formula is used to find the probability that event A occurs in a sample space with equally likely outcomes?

4. List two rules that a valid probability model must obey.

5. Define the complement of event A.

6. The complement rule says that $P(A^C) = $ _____.

7. True/False: Two events A and B are mutually exclusive if $P(A \text{ and } B) = 0$.

8. State the addition rule for mutually exclusive events.

Mastering Concepts and Skills These exercises reinforce the learning goals as shown in the examples.

9. **Refer to Example 4.2.1 Four-Sided Dice** A fair, 4-sided die is equally likely to show a 1, 2, 3, or 4 when rolled. Imagine rolling two fair, 4-sided dice and recording the number showing on each die.
    (a) Give a probability model for this chance process.
    (b) Define event A as getting a sum of 5. Find $P(A)$.

10. **Coin Toss** Imagine tossing a fair coin 4 times.
    (a) Give a probability model for this chance process.
    (b) Define event B as getting exactly three tails. Find $P(B)$.

11. **Rock-Paper-Scissors** You're likely familiar with the game "rock-paper-scissors." Two players face each other and, at the count of 3, choose to make a fist (rock), extend a hand with an open palm (paper), or make a "V" with their index and middle fingers (scissors). The winner is determined by these rules: Rock breaks scissors; paper covers rock; and scissors cut paper. If both players choose the same hand shape, then the game is a tie. Suppose that Player 1 and Player 2 are both equally likely to choose rock, paper, or scissors.
    (a) Give a probability model for this chance process.
    (b) Find the probability that Player 1 wins on a single play of the game.

12. **Who's Paying?** Ari, Bettina, Charley, Daniela, and Ezekiel go to the bagel shop for lunch every Thursday. Each time, they randomly pick 2 members of the group to pay for lunch by drawing names from a hat.
    (a) Give a probability model for this chance process.
    (b) Find the probability that Charley or Daniela, or both, end up paying for lunch.

13. **Refer to Example 4.2.2 Blood Types** Human blood can be type O, A, B, or AB, but the distribution of the types varies by race. Here is the distribution of blood types of Black people in the United States.[5]

Blood type	O	A	B	AB
Probability	0.51	0.26	0.19	?

    Suppose we choose one Black person in the United States at random.
    (a) What is the probability that the chosen person has type AB blood?
    (b) Find the probability that the chosen person does not have type AB blood.

14. **Canada's Languages** Canada has two official languages, English and French. Here is the distribution of first language among Canadians, combining various languages from the broad Asia/Pacific region. "Other" includes all other languages, including the many indigenous languages spoken by First Nations peoples.[6]

Language	English	French	Asian/Pacific	Other
Probability	0.58	0.21	0.09	?

    Suppose we choose one Canadian at random.
    (a) What is the probability that this person's first language would be classified as "Other"?
    (b) Find the probability that this Canadian's first language is not English.

15. **Household Size** In U.S. government data, a household consists of all occupants of a dwelling unit. Choose a U.S. household at random and count the number of people it contains. Here is the assignment of probabilities for the outcome.[7]

Number of people	1	2	3	4	5	6	7+
Probability	0.28	0.35	0.15	?	0.06	0.02	0.01

    (a) What probability should replace "?" in the table?
    (b) Find the probability that the chosen household contains more than 1 person.

16. **Road Trips** The U.S. National Household Travel Survey gathers data on the time of day when people begin a trip in their car or other vehicle. Choose a trip at random and record the time at which the trip started. Here is an assignment of probabilities for the outcome.[8]

Time of day	10 p.m.–12:59 a.m.	1 a.m.–5:59 a.m.	6 a.m.–8:59 a.m.	9 a.m.–12:59 p.m.
Probability	0.023	0.019	0.166	0.254
Time of day	1 p.m.–3:59 p.m.	4 p.m.–6:59 p.m.	7 p.m.–9:59 p.m.	
Probability	0.221	?	0.098	

    (a) What probability should replace "?" in the table?
    (b) Find the probability that the chosen trip did not begin between 9 a.m. and 12:59 p.m.

17. **Refer to Example 4.2.3 Blood Types** Refer to Exercise 13. At a particular blood bank, type O and type AB blood are in high demand from donors.
    (a) Find the probability that a randomly chosen Black person in the United States has one of these high-demand blood types.
    (b) Find the probability that a randomly chosen Black person in the United States does not have one of these high-demand blood types.

18. **Canada's Languages** Refer to Exercise 14.
    (a) Find the probability that the chosen Canadian's first language is English or French.
    (b) Find the probability that the chosen Canadian's first language is neither English nor French.

19. **Household Size** Refer to Exercise 15. What's the probability that a randomly selected U.S. household has 5 or fewer people?

20. **Road Trips** Refer to Exercise 16. What's the probability that a randomly chosen trip began between 6 a.m. and 9:59 p.m.?

Applying the Concepts These exercises ask you to apply multiple learning goals in a new context or to apply what you learned in this section in a new way.

21. **Education Levels** Choose a young adult (aged 25 to 29) at random. The probability is 0.065 that the person chosen did not complete high school, 0.29 that the person has a high school diploma but no further education, and 0.387 that the person has at least a bachelor's degree.[9]
    (a) What must be the probability that a randomly chosen young adult has some education beyond high school but does not have a bachelor's degree?
    (b) Find the probability that the young adult completed high school. Which probability rule did you use to find the answer?
    (c) Find the probability that the young adult has further education beyond high school. Which probability rule did you use to find the answer?

22. **GMAT Prep** A company that offers courses to prepare students for the Graduate Management Admission Test (GMAT) has the following information about its customers: 20% are currently undergraduate students in business, 15% are currently undergraduate students in other fields of study, and 60% are college graduates who are currently employed. Choose a customer at random.
    (a) What is the probability that the customer is a college graduate who is not currently employed?
    (b) Find the probability that the customer is currently an undergraduate. Which probability rule did you use to find the answer?
    (c) Find the probability that the customer is not an undergraduate business student. Which probability rule did you use to find the answer?

Extending the Concepts These exercises challenge you to explore statistical concepts and methods that go beyond what you learned in this section.

23. **Rock-Paper-Scissors-Lizard-Spock!** The characters on the television show *Big Bang Theory* played a more complicated version of the traditional rock-paper-scissors game described in Exercise 11. Do some online research to find out how the game works. Suppose that Player 1 and Player 2 are both equally likely to choose rock, paper, scissors, lizard, or Spock. Find the probability that Player 2 wins on a single play of the game.

Cumulative Review These exercises revisit what you learned in previous sections.

24. **Truly Random? (1.3, 4.1)** Chester claims he can act as a random number generator. Shaka has doubts. She tells Chester to write down 50 random digits, suspecting that he won't write down zero as often as he should. Chester includes only 2 zeros. Do these data provide convincing evidence that Chester produces fewer zeros than the expected 10%?

    Shaka simulated 100 samples of 50 random digits, assuming that zero has a 10% chance of occurring each time. The dotplot shows the number of zeros in each simulated sample.

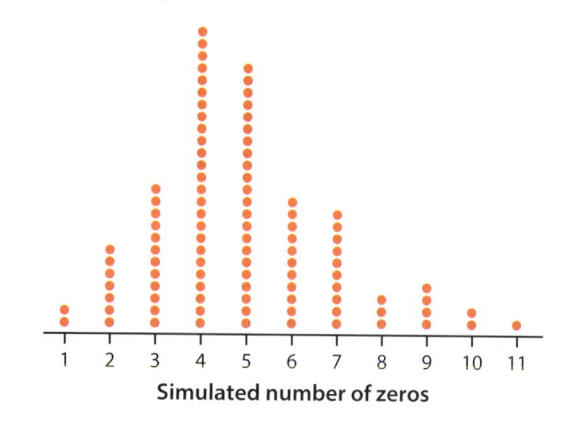

    (a) Explain how the graph illustrates sampling variability.
    (b) In the simulation, what percentage of the samples resulted in 2 or fewer zeros?
    (c) Based on Chester's result of only 2 zeros and your answer to part (b), is there convincing evidence that Chester doesn't choose zero often enough when he's trying to generate random numbers? Explain your reasoning.

25. **Snoring and Age (1.6, 2.2)** Pediatricians were interested in determining if a relationship exists between age and snoring frequency in children. Data from a random sample of 398 Australian children aged 4 to 12 with some history of snoring are summarized in the two-way table.[10]

		Age			
		< 7 years	7–9 years	> 9 years	Total
Snoring frequency	Infrequent	77	70	100	247
	Habitual	58	42	51	151
	Total	135	112	151	398

    (a) Calculate the percentage of children in each age group who had habitual snoring.
    (b) Describe the association between the two variables.
    (c) Explain why we can't conclude, using these data, that getting older *causes* a decrease in snoring frequency.

Section 4.3 Two-Way Tables and Venn Diagrams

LEARNING GOALS

By the end of this section, you will be able to:
- Use a two-way table to find probabilities.
- Calculate probabilities with the general addition rule.
- Use a Venn diagram to find probabilities.

So far, you have learned how to model chance behavior and some basic rules for finding the probability of an event. But what if you're interested in finding probabilities involving two events? For instance, survey results from the Pew Research Center suggest that 69% of U.S. adults use Facebook, 40% use Instagram, and 32% use both.[11] Suppose we select a U.S. adult at random. What's the probability that the person uses Facebook *or* uses Instagram? How about P(does not use Facebook *and* does not use Instagram)? In this section, we will introduce several methods to help you answer questions like these.

Two-Way Tables and the General Addition Rule

There are two different uses of the word "or" in everyday life. In a restaurant, if you are asked if you want "soup or salad," the server expects you to choose one or the other, but not both. However, if you order coffee and are asked if you want "cream or sugar," it's OK to ask for one or the other or both.

Mutually exclusive events A and B cannot both happen at the same time. For such events, "A or B" means that only event A happens or only event B happens. You can find P(A or B) with the addition rule for mutually exclusive events from Section 4.2:

$$P(A \text{ or } B) = P(A) + P(B)$$

How can we find P(A or B) when the two events are *not* mutually exclusive? Now we have to deal with the fact that "A or B" means one or the other or both. For instance, "uses Facebook or uses Instagram" in the scenario just described includes U.S. adults who do both.

When you're trying to find probabilities involving two events, like P(A or B), a two-way table can display the sample space in a way that makes probability calculations easier.

EXAMPLE 4.3.1

Who can roll their tongue or raise one eyebrow?

Two-way tables and probability

PROBLEM
Students in a college statistics class wanted to find out how common it is for young adults to be able to roll their tongues or to raise one eyebrow. They recorded data on whether each of the 200 people in the class could do each of these two things. The two-way table summarizes the data.

		Roll their tongue?		
		Yes	No	Total
Raise one eyebrow?	Yes	42	20	62
	No	107	31	138
	Total	149	51	200

Suppose we choose a student from the class at random. Define event A as student can roll their tongue and event B as student can raise one eyebrow.

(a) Find $P(B)$. Describe this probability in words.
(b) Find P(roll tongue and raise one eyebrow).
(c) Find $P(A \text{ or } B)$.

(continued)

SOLUTION

(a) $P(B) = P(\text{raise one eyebrow}) = \dfrac{62}{200} = 0.31$

There is a 31% probability that a randomly selected student from this class can raise one eyebrow.

> Because each student in the class is equally likely to be selected,
> $$P(B) = \dfrac{\text{number of students who can raise one eyebrow}}{\text{total number of students in the class}}$$

(b) $P(\text{roll tongue and raise one eyebrow}) = \dfrac{42}{200} = 0.21$

> The number of students who can roll their tongue and raise one eyebrow is in the "Yes," "Yes" cell of the two-way table.

(c) $P(A \text{ or } B) = P(\text{roll tongue or raise one eyebrow})$
$= \dfrac{107 + 20 + 42}{200} = \dfrac{169}{200} = 0.845$

> In statistics, "or" means one or the other or both. So "roll tongue or raise one eyebrow" includes (i) roll tongue but not raise one eyebrow; (ii) raise one eyebrow but not roll tongue; and (iii) roll tongue and raise one eyebrow.

EXAM PREP: FOR PRACTICE, TRY EXERCISE 7.

When we found $P(\text{roll tongue and raise one eyebrow})$ in part (b) of Example 4.3.1, we could have described this as either $P(A \text{ and } B)$ or $P(B \text{ and } A)$. Why? Because "A and B" describes the same event as "B and A." Likewise, $P(A \text{ or } B)$ is the same as $P(B \text{ or } A)$.

Part (c) of the example reveals an important fact about finding the probability $P(A \text{ or } B)$: We can't use the addition rule for mutually exclusive events unless events A and B have no outcomes in common. In this case, events A and B share 42 outcomes—there are 42 students who can roll their tongue and raise one eyebrow. If we simply added the probabilities of A and B, we'd get $149/200 + 62/200 = 211/200$. This is clearly wrong, because the probability is greater than 1. As Figure 4.3 illustrates, outcomes common to both events are counted twice when we add the probabilities of these two events.

		Roll their tongue?		
		Yes	No	Total
Raise one eyebrow?	Yes	42	20	62
	No	107	31	138
	Total	149	51	200

$P(A \text{ and } B) = \dfrac{42}{200}$

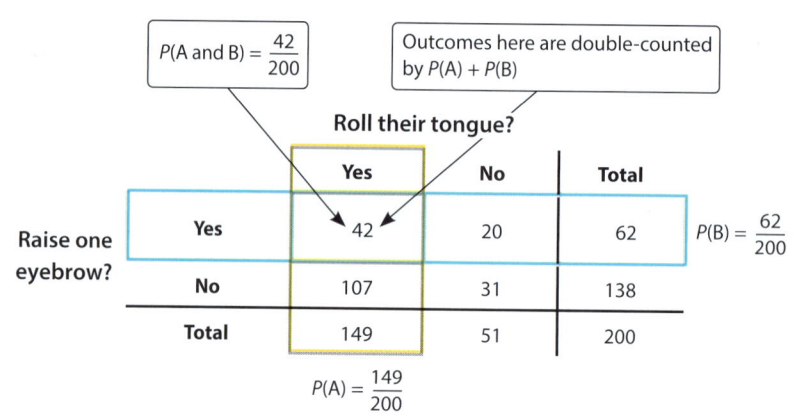

Outcomes here are double-counted by $P(A) + P(B)$

$P(B) = \dfrac{62}{200}$

$P(A) = \dfrac{149}{200}$

FIGURE 4.3 Two-way table showing events A and B from the "roll tongue" and "raise one eyebrow" example. These events are *not* mutually exclusive, so we can't find $P(A \text{ or } B)$ by just adding the probabilities of the two events.

We can fix the double-counting problem illustrated in the two-way table by subtracting the probability $P(\text{roll tongue and raise one eyebrow})$ from the sum. That is,

$P(\text{roll tongue or raise one eyebrow})$
$= P(\text{roll tongue}) + P(\text{raise one eyebrow}) - P(\text{roll tongue and raise one eyebrow})$
$= 149/200 + 62/200 - 42/200$
$= 169/200$

This result is known as the **general addition rule.**

If A and B are any two events resulting from some chance process, the **general addition rule** says that

$$P(A \text{ or } B) = P(A) + P(B) - P(A \text{ and } B)$$

EXAMPLE 4.3.2

Facebook or Instagram?

General addition rule

PROBLEM

A survey about social media use suggests that 69% of U.S. adults use Facebook, 40% use Instagram, and 32% use both. Suppose we select a U.S. adult at random. What's the probability that the person uses Facebook or Instagram?

SOLUTION

Let F = uses Facebook and I = uses Instagram.

$P(F \text{ or } I) = P(F) + P(I) - P(F \text{ and } I)$
$= 0.69 + 0.40 - 0.32$
$= 0.77$

$P(A \text{ or } B) = P(A) + P(B) - P(A \text{ and } B)$

EXAM PREP: FOR PRACTICE, TRY EXERCISE 11.

As Example 4.3.2 suggests, it is sometimes easier to designate events with letters that relate to the context, like F for "uses Facebook" and I for "uses Instagram."

THINK ABOUT IT

What happens if we use the general addition rule for two mutually exclusive events A and B? In that case, $P(A \text{ and } B) = 0$, and the formula reduces to

$$P(A \text{ or } B) = P(A) + P(B) - 0 = P(A) + P(B)$$

In other words, the addition rule for mutually exclusive events is just a special case of the general addition rule.

Venn Diagrams and Probability

We have seen that two-way tables can be used to illustrate the sample space of a chance process involving two events. You can use **Venn diagrams,** like the one shown in Figure 4.4, for the same purpose.

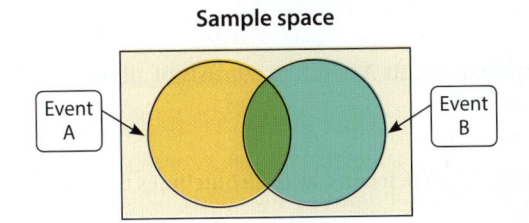

FIGURE 4.4 A typical Venn diagram that shows the sample space and the relationship between two events A and B.

A **Venn diagram** consists of one or more circles surrounded by a rectangle. Each circle represents an event. The region inside the rectangle represents the sample space of the chance process.

In Example 4.3.1, we looked at data from a class activity about who in a large class of college students can roll their tongue or raise one eyebrow. The chance process was selecting a student in the class at random. Our events of interest were A: student can roll their tongue and B: student can raise one eyebrow. Here is the two-way table that summarizes the data.

		Roll their tongue?		Total
		Yes	No	
Raise one eyebrow?	Yes	42	20	62
	No	107	31	138
	Total	149	51	200

The Venn diagram in Figure 4.5 displays the sample space in a slightly different way. There are four distinct regions in the Venn diagram, which correspond to the four cells in the two-way table.

FIGURE 4.5 The completed Venn diagram for the large class of college students. The circles represent the two events A = roll tongue and B = raise one eyebrow.

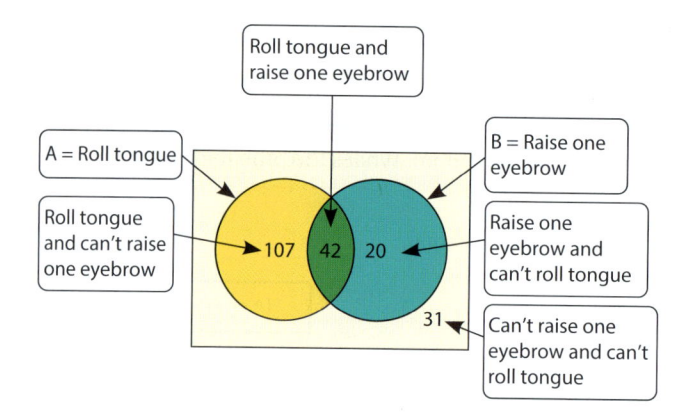

Statisticians have developed some standard vocabulary and notation to make our work with Venn diagrams a bit easier.

- We introduced the *complement* of an event earlier. In Figure 4.6(a), the complement A^C contains the outcomes that are not in A.
- Figure 4.6(b) shows the event "A and B" in green. You can see why this event is also called the **intersection** of A and B. The corresponding notation is $A \cap B$.
- The event "A or B" is shown in green in Figure 4.6(c). This event is also known as the **union** of A and B. The corresponding notation is $A \cup B$.

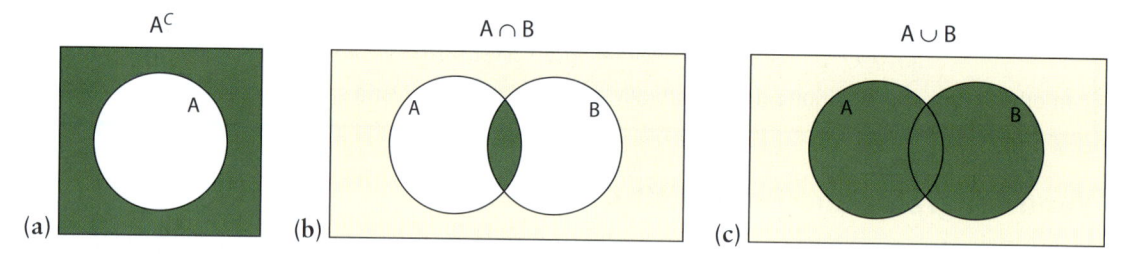

FIGURE 4.6 The green shaded region in each Venn diagram shows (a) the *complement* A^C of event A, (b) the *intersection* of events A and B, and (c) the *union* of events A and B.

The event "A and B" is called the **intersection** of events A and B. It consists of all outcomes that are common to both events, and is denoted $A \cap B$.

The event "A or B" is called the **union** of events A and B. It consists of all outcomes that are in event A or event B, or both, and is denoted $A \cup B$.

Here's a way to keep the symbols straight: $\cup$ for **u**nion; $\cap$ for **in**tersection. With this new notation, we can rewrite the general addition rule in symbols as follows:

$$P(A \cup B) = P(A) + P(B) - P(A \cap B)$$

EXAMPLE 4.3.3

Who reads the paper?

Venn diagrams and probability

PROBLEM

In a large apartment complex, 20% of residents read *USA Today*, 12% read the *New York Times*, and 3% read both papers. Suppose we select a resident of the apartment complex at random and record which of the two papers the person reads.

(a) Make a Venn diagram to display the sample space of this chance process using the events A: reads *USA Today* and B: reads *New York Times*.

(b) Find the probability that the person reads exactly one of the two papers.

SOLUTION

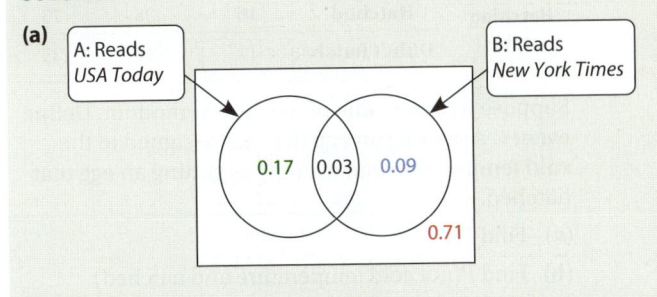

(a)

A: Reads *USA Today* — 0.17 | 0.03 | 0.09 — B: Reads *New York Times*
0.71

Start with the intersection. Because 3% of residents read both papers, $P(A \cap B) = 0.03$.

The 20% of residents who read *USA Today* includes the 3% who read both papers. So 20% − 3% = 17% read only *USA Today*.

By similar reasoning, 12% − 3% = 9% read only the *New York Times*.

A total of 0.17 + 0.03 + 0.09 = 0.29 (or 29%) of residents read at least one of the two papers. By the complement rule, 1 − 0.29 = 0.71 (or 71%) read neither paper.

(b) P(reads exactly one of the two papers) = 0.17 + 0.09 = 0.26

EXAM PREP: FOR PRACTICE, TRY EXERCISE 15.

Section 4.3 What Did You Learn?

Review the learning goals from this section. Then practice what you've learned by working through the exercises.

Learning Goal	Example	Exercises
Use a two-way table to find probabilities.	4.3.1	7–10
Calculate probabilities with the general addition rule.	4.3.2	11–14
Use a Venn diagram to find probabilities.	4.3.3	15–18

Section 4.3 Exercises

Building Concepts and Skills These exercises assess the basic knowledge you should have after reading the section.

1. Explain what "A or B" means in statistics.
2. True/False: "A and B" describes the same event as "B and A."
3. Complete the general addition rule: $P(A \text{ or } B) =$ _____.
4. Sketch a Venn diagram with events A and B that are not mutually exclusive.
5. The event "A and B" is called the _____ of events A and B. It consists of all outcomes that are common to both events.
6. Copy your Venn diagram from Exercise 4 and shade the region that represents $A \cup B$.

Mastering Concepts and Skills These exercises reinforce the learning goals as shown in the examples.

7. **Refer to Example 4.3.1 School Breakfast** Administrators at a new elementary school in a large city are considering a proposal to provide breakfast at school. Before making a decision, the administrators want to know how many children regularly eat breakfast at home. They conducted a survey, asking, "Do you usually eat breakfast at home?" Administrators also recorded whether the student typically walks to school. All 595 students in the school responded to the survey. The resulting data are summarized in the two-way table.[12]

		Walk to school?		
		Yes	No	Total
Usually eat breakfast at home?	Yes	110	190	300
	No	165	130	295
	Total	275	320	595

Suppose we select a student from the school at random. Define event W as getting a student who typically walks to school and event B as getting a student who eats breakfast at home regularly.

(a) Find $P(B^C)$. Describe this probability in words.

(b) Find P(doesn't typically walk to school and doesn't eat breakfast at home regularly).

(c) Find $P(W \text{ or } B^C)$.

8. **Playing Cards** A standard deck of playing cards (with jokers removed) consists of 52 cards in four suits — clubs, diamonds, hearts, and spades. Each suit has 13 cards, with denominations ace, 2, 3, 4, 5, 6, 7, 8, 9, 10, jack, queen, and king. The jack, queen, and king are referred to as "face cards." Imagine that we shuffle the deck thoroughly and deal one card. Let's define events F: getting a face card and H: getting a heart. The two-way table summarizes the sample space for this chance process.

		Face card	Not face card	Total
Suit	Heart	3	10	13
	Not heart	9	30	39
	Total	12	40	52

(a) Find $P(H^C)$. Describe this probability in words.

(b) Find P(face card and not a heart).

(c) Find $P(F \text{ or } H^C)$.

9. *Titanic* **Casualties** In 1912, the *Titanic* struck an iceberg and sank on its first voyage. Some passengers got off the ship safely in lifeboats, but many others did not and died. The following two-way table gives information about adult passengers who survived and who died, by class of travel.

		First	Second	Third
Survived?	Yes	197	94	151
	No	122	167	476

Suppose we randomly select one of the adult passengers who sailed on the *Titanic*. Define event S as getting a person who survived and event F as getting a passenger in first class.

(a) Find $P(F^C)$.

(b) Find P(not in first class and survived).

(c) Find P(not in first class or survived).

10. **Python Nests** How is the hatching of water python eggs influenced by the temperature of the snake's nest? Researchers simulated three environments typical of natural nest conditions by randomly assigning newly laid eggs to incubators set at one of three temperatures: hot, neutral, or cold.

The following two-way table summarizes the data on nest temperature and hatching status.[13]

		Cold	Neutral	Hot
Hatching status	Hatched	16	38	75
	Didn't hatch	11	18	29

Suppose we select one of the eggs at random. Define event C as getting an egg that was assigned to the cold temperature and event H as getting an egg that hatched.

(a) Find $P(H)$.

(b) Find P(not cold temperature and hatched).

(c) Find $P(H^C \text{ or } C)$.

11. **Refer to Example 4.3.2** **Taco Time!** A survey of all students at a large community college revealed that, in the last month, 38% of them had dined at Taco Bell, 16% had dined at Chipotle, and 9% had dined at both. Suppose we select a student from the college at random. What's the probability that the student has dined at Taco Bell or Chipotle in the last month?

12. **Streaming Apps** According to recent survey data, Pandora and Spotify are the music-streaming apps most widely used by young adults. The responses reveal that 26% use Pandora, 48% use Spotify, and 14% use both.[14] Suppose we select a survey respondent at random. What's the probability that the person uses Pandora or Spotify?

13. **Mac or PC?** A recent census on computer use at a major university revealed that 60% of its students mainly used Macs. The rest mainly used PCs. At the time of the census, 67% of the school's students were undergraduates and the rest were graduate students. In the census, 23% of respondents were graduate students who said that they used Macs as their main computers. Suppose we select a student at random from among those who were part of the census. What's the probability that this person is a graduate student or mainly uses a Mac?

14. **Senate Demographics** Suppose that 50% of U.S. senators are Republicans and the rest are Democrats or Independents. Twenty-five percent of the senators are female, and 42% are male Republicans. Let's say we select one of these senators at random. What is the probability that this person is male or a Republican?

15. **Refer to Example 4.3.3** **Taco Time!** Refer to Exercise 11.

(a) Construct a Venn diagram to represent the outcomes of this chance process using the events T: dined at Taco Bell and C: dined at Chipotle.

(b) Find the probability that the chosen student dined at Chipotle but not Taco Bell in the last month.

16. **Streaming Apps** Refer to Exercise 12.
    (a) Construct a Venn diagram to represent the outcomes of this chance process using the events P: uses Pandora and S: uses Spotify.
    (b) Find the probability that the chosen respondent uses Spotify but not Pandora.

17. **Mac or PC?** Refer to Exercise 13.
    (a) Construct a Venn diagram to represent the outcomes of this chance process using the events G: is a graduate student and M: mainly uses a Mac.
    (b) Find $P(G^C \cap M^C)$. Describe this probability in words.

18. **Senate Demographics** Refer to Exercise 14.
    (a) Construct a Venn diagram to represent the outcomes of this chance process using the events M: is male and R: is a Republican.
    (b) Find $P(M^C \cap R^C)$. Describe this probability in words.

Applying the Concepts These exercises ask you to apply multiple learning goals in a new context or to apply what you learned in this section in a new way.

19. **Cell Phones** The Pew Research Center asked a random sample of U.S. adults about their age and cell phone ownership. The two-way table summarizes the data.[15]

		Age				
		18–29	30–49	50–64	65+	Total
Phone ownership	None	2	10	21	33	66
	Cell phone, not smartphone	11	28	57	103	199
	Smartphone	298	451	291	151	1191
	Total	311	489	369	287	1456

    Suppose we select one of the survey respondents at random. What's the probability that:
    (a) The person is not aged 18 to 29 and does not own a smartphone?
    (b) The person is aged 18 to 29 or owns a smartphone?
    (c) What do you notice about the probabilities in parts (a) and (b)? Why does this make sense?

20. **School Values** Researchers carried out a survey of fourth-, fifth-, and sixth-grade students in Michigan. Students were asked whether getting good grades, having athletic ability, or being popular was most important to them. The two-way table summarizes the survey data.[16]

		Grade level			
		4th grade	5th grade	6th grade	Total
Most important	Grades	49	50	69	168
	Athletic	24	36	38	98
	Popular	19	22	28	69
	Total	92	108	135	335

    Suppose we select one of these students at random. What's the probability that:
    (a) The student is a sixth grader or a student who rated good grades as important?
    (b) The student is not a sixth grader and did not rate good grades as important?
    (c) What do you notice about the probabilities in parts (a) and (b)? Why does this make sense?

21. **Color Disks** A jar contains 36 disks, consisting of 9 each of four colors—red, green, blue, and yellow. Each set of disks of the same color is numbered from 1 to 9. Suppose you draw one disk at random from the jar. Define events B: get a blue disk and E: get a disk with the number 8.
    (a) Create a two-way table that describes the sample space in terms of events B and E.
    (b) Find $P(B)$ and $P(E)$.
    (c) Find the probability of getting a blue 8.
    (d) Explain why $P(B \cup E) \neq P(B) + P(E)$. Then use the general addition rule to compute $P(B \cup E)$.

22. **Red Jacks** Suppose you shuffle a standard deck of playing cards and deal one card. (Refer to Exercise 8 for the make-up of a deck of cards.) Define events J: get a jack and R: get a red card.
    (a) Create a two-way table that describes the sample space in terms of events J and R.
    (b) Find $P(J)$ and $P(R)$.
    (c) Find the probability of getting a red jack.
    (d) Explain why $P(J \cup R) \neq P(J) + P(R)$. Then use the general addition rule to compute $P(J \cup R)$.

Extending the Concepts These exercises challenge you to explore statistical concepts and methods that go beyond what you learned in this section.

23. **Mutually Exclusive vs. Complementary** Classify each of the following statements as true or false. Justify your answer.
    (a) If one event is the complement of another event, the two events are mutually exclusive.
    (b) If two events are mutually exclusive, one event is the complement of the other.

24. **University of Venn** At a large university, 71% of the students are undergraduates, 69% live in the dorms, and 42% are in-state residents. Further, 54% of the students are undergraduates who live in the dorms, 28% are undergraduates who are in-state residents, and 21% are in-state residents who live in the dorms. Finally, 13% of the students are undergraduates who live in the dorms and are in-state residents. Pick a student at random from this university. What's the probability that the chosen person is a graduate student who does not live in the dorm and is not an in-state resident? (*Hint:* It might be helpful to make a Venn diagram with three circles.)

Cumulative Review These exercises revisit what you learned in previous sections.

25. **Bone Density (1.6)** Fractures of the spine are common and serious events among people with advanced osteoporosis (low mineral density in the bones). Researchers wondered if taking a medication called strontium ranelate might help prevent fractures. They recruited 1649 women with osteoporosis who had previously had at least one fracture for an experiment. The participants were assigned to take either strontium ranelate or a placebo each day. All of the participants were also taking calcium supplements and receiving standard medical care. One response variable was the number of new fractures over 3 years.[17]

    (a) Describe a completely randomized design for this experiment.

    (b) Explain why it is important to keep the calcium supplements and medical care the same for all people in the experiment.

    (c) The participants who took strontium ranelate had statistically significantly fewer new fractures, on average, than the participants who took a placebo over a 3-year period. Explain what this means to someone who knows little about statistics.

26. **Living Arrangements (2.2)** The U.S. Census Bureau's Current Population Survey collected data on the living arrangements of children in 2020. Here is a side-by-side bar chart that summarizes the data based on children's ages. Note that according to the Census Bureau, two parents can be of the same gender or different genders.[18] Describe the association between the two variables.

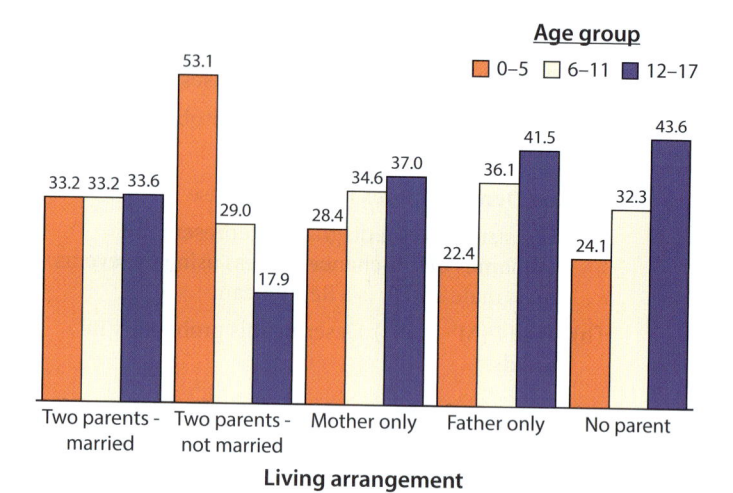

Section 4.4 Conditional Probability and Independence

LEARNING GOALS

By the end of this section, you will be able to:
- Find and interpret conditional probabilities using two-way tables.
- Use the conditional probability formula to calculate probabilities.
- Determine whether two events are independent.

The probability of an event can change if we know that some other event has occurred. For instance, suppose you toss a fair coin twice. The probability of getting two heads is 1/4 because the sample space consists of the four equally likely outcomes

HH   HT   TH   TT

But if you know the first toss was a tail, the probability of getting two heads changes to 0.

This idea is the key to many applications of probability.

What Is Conditional Probability?

Let's return to the college statistics class from Section 4.3. Here is the two-way table we used to find probabilities involving events A: can roll their tongue and B: can raise one eyebrow for a randomly selected student, along with a summary of our previous results.

		Roll their tongue?		
		Yes	No	Total
Raise one eyebrow?	Yes	42	20	62
	No	107	31	138
	Total	149	51	200

$P(A) = P(\text{roll tongue}) = 149/200$

$P(B) = P(\text{raise one eyebrow}) = 62/200$

$P(A \cap B) = P(\text{roll tongue and raise one eyebrow}) = 42/200$

$P(A \cup B) = P(\text{roll tongue or raise one eyebrow}) = 169/200$

Now let's turn our attention to some other interesting probability questions.

1. **If we know that a randomly selected student can raise one eyebrow, what is the probability that the student can roll their tongue?** There are 62 students in the class who can raise one eyebrow. We can restrict our attention to this group, since we are told that the chosen student can raise one eyebrow. Because there are 42 students who can roll their tongue among the 62 students who can raise one eyebrow, the desired probability is

   P(roll tongue *given* raise one eyebrow) = 42/62 = 0.677

		Roll their tongue?		
		Yes	No	Total
Raise one eyebrow?	Yes	42	20	62
	No	107	31	138
	Total	149	51	200

2. **If we know that a randomly selected student can roll their tongue, what's the probability that the student can raise one eyebrow?** This time, our attention is focused on the 149 students who can roll their tongues. Because 42 of the 149 tongue rollers in the class can raise one eyebrow,

   P(raise one eyebrow *given* roll tongue) = 42/149 = 0.282

		Roll their tongue?		
		Yes	No	Total
Raise one eyebrow?	Yes	42	20	62
	No	107	31	138
	Total	149	51	200

These two questions may sound alike, but they're asking about two very different things. Each of these probabilities is an example of a **conditional probability**.

The probability that one event happens given that another event is known to have happened is called a **conditional probability**. The conditional probability that event B happens given that event A has happened is denoted by $P(B|A)$.

With this new notation, we can restate the answers to the two questions just posed as follows:

P(roll tongue | raise one eyebrow) = $P(A|B)$ = 42/62 = 0.677

and

P(raise one eyebrow | roll tongue) = $P(B|A)$ = 42/149 = 0.282

EXAMPLE 4.4.1

Who rides snowmobiles in Yellowstone?

Conditional probabilities and two-way tables

PROBLEM

Yellowstone National Park rangers surveyed a random sample of 1526 winter visitors to the park. Each person was asked whether they owned, rented, or had never used a snowmobile. Respondents were also asked whether or not they belonged to an environmental organization (such as the Sierra Club). The two-way table summarizes the survey responses.[19]

Suppose we randomly select one of the survey respondents. Define events E: environmental club member, S: snowmobile owner, and N: never used a snowmobile.

(a) Find $P(N|E)$. Describe this probability in words.

(continued)

(b) Given that the chosen person is not a snowmobile owner, what's the probability that they are an environmental club member? Write your answer as a probability statement using correct notation for the events.

		Environmental club?		
		No	Yes	Total
Snowmobile use	Never used	445	212	657
	Renter	497	77	574
	Owner	279	16	295
	Total	1221	305	1526

SOLUTION

(a) $P(N|E) = P(\text{never used} | \text{environmental club member})$

$= 212/305 = 0.695$

Given that the randomly chosen person is an environmental club member, there is about a 69.5% probability that they have never used a snowmobile.

To answer part (a), only consider values in the "Yes" column.

(b) $P(\text{environmental club member} | \text{not snowmobile owner})$

$= P(E|S^c) = \dfrac{212+77}{657+574} = \dfrac{289}{1231} = 0.235$

To answer part (b), only consider values in the "Never used" and "Renter" rows.

EXAM PREP: FOR PRACTICE, TRY EXERCISE 5.

Let's look more closely at how conditional probabilities are calculated. From the following two-way table, we see that

$$P(\text{roll tongue} | \text{raise one eyebrow}) = \dfrac{42}{62} = \dfrac{\text{number who can roll their tongue and raise one eyebrow}}{\text{number who can raise one eyebrow}}$$

		Roll their tongue?		
		Yes	No	Total
Raise one eyebrow?	Yes	42	20	62
	No	107	31	138
	Total	149	51	200

What if we focus on probabilities instead of numbers of students? Notice that

$$\dfrac{P(\text{roll tongue and raise one eyebrow})}{P(\text{raise one eyebrow})} = \dfrac{\frac{42}{200}}{\frac{62}{200}} = \dfrac{42}{62} = P(\text{roll tongue} | \text{raise one eyebrow})$$

This observation leads to a general formula for computing a conditional probability.

Calculating Conditional Probabilities

To find the conditional probability $P(A|B)$, use the formula

$$P(A|B) = \dfrac{P(A \text{ and } B)}{P(B)} = \dfrac{P(A \cap B)}{P(B)} = \dfrac{P(\text{both events occur})}{P(\text{given event occurs})}$$

By the same reasoning,

$$P(B|A) = \dfrac{P(B \text{ and } A)}{P(A)} = \dfrac{P(B \cap A)}{P(A)}$$

EXAMPLE 4.4.2

Who reads the paper?

Calculating conditional probability

PROBLEM

In Example 4.3.3, we classified the residents of a large apartment complex based on the events A: reads *USA Today* and B: reads the *New York Times*. The completed Venn diagram is reproduced here.

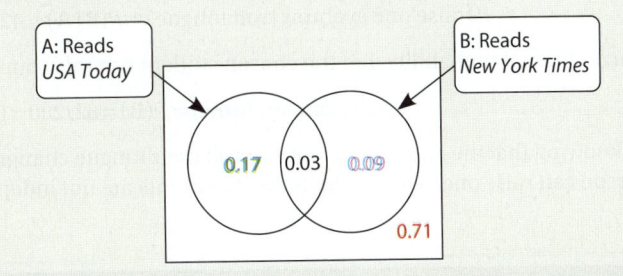

A: Reads USA Today — B: Reads New York Times

0.17 0.03 0.09

0.71

What's the probability that a randomly selected resident who reads *USA Today* also reads the *New York Times*?

SOLUTION

$$P(\text{reads New York Times} \mid \text{reads USA Today}) = \frac{0.03}{0.17 + 0.03} = 0.15$$

$$P(B \mid A) = \frac{P(B \text{ and } A)}{P(A)}$$

EXAM PREP: FOR PRACTICE, TRY EXERCISE 9.

Conditional Probability and Independence

Suppose you toss a fair coin twice. Define events A: first toss is a head and B: second toss is a head. We know that $P(A) = 1/2$ and $P(B) = 1/2$.

- What's $P(B \mid A)$? It's the conditional probability that the second toss is a head given that the first toss was a head. The coin has no memory, so $P(B \mid A) = 1/2$.
- What's $P(B \mid A^C)$? It's the conditional probability that the second toss is a head given that the first toss was not a head. Getting a tail on the first toss does not change the probability of getting a head on the second toss, so $P(B \mid A^C) = 1/2$.

In this case, $P(B \mid A) = P(B \mid A^C) = P(B)$. Knowing whether or not the first toss was a head does not change the probability that the second toss is a head. We say that A and B are **independent events**.

> A and B are **independent events** if knowing whether or not one event has occurred does not change the probability that the other event will happen. In other words, events A and B are independent if
>
> $$P(A \mid B) = P(A \mid B^C) = P(A)$$
>
> Alternatively, events A and B are independent if
>
> $$P(B \mid A) = P(B \mid A^C) = P(B)$$

Let's contrast the coin-toss scenario with our earlier tongue rolling/eyebrow raising example. In that case, the chance process involved randomly selecting a student from a college statistics class. The events of interest were A: can roll their tongue and B: can raise one eyebrow. Are these two events independent?

		Roll their tongue?		Total
		Yes	No	
Raise one eyebrow?	Yes	42	20	62
	No	107	31	138
	Total	149	51	200

- Suppose that the chosen student can roll their tongue. We can see from the two-way table that $P(\text{raise one eyebrow} \mid \text{roll tongue}) = P(B \mid A) = 42/149 = 0.282$.
- Suppose that the chosen student cannot roll their tongue. From the two-way table, we see that $P(\text{raise one eyebrow} \mid \text{not roll tongue}) = P(B \mid A^C) = 20/51 = 0.392$.

Knowing that the chosen student can roll their tongue changes (reduces) the probability that the student can raise one eyebrow. So these two events are not independent.

Another way to determine whether two events A and B are independent is to compare $P(A \mid B)$ to $P(A)$ or $P(B \mid A)$ to $P(B)$. For the tongue rolling/eyebrow raising scenario,

$$P(\text{raise one eyebrow} \mid \text{roll tongue}) = P(B \mid A) = 42/149 = 0.282$$

The unconditional probability that the chosen student can raise one eyebrow is

$$P(\text{raise one eyebrow}) = P(B) = 62/200 = 0.31$$

Again, knowing that the chosen student can roll their tongue changes (reduces) the probability that this person can raise one eyebrow. So these two events are not independent.

EXAMPLE 4.4.3

Are more math professors left-handed?

Checking for independence

PROBLEM

Is there a relationship between subject taught and handedness? To find out, we chose an SRS of 100 college professors at an education conference. The two-way table summarizes the relationship between the subject taught and the dominant hand of each professor. Suppose we choose one of the professors in the sample at random. Are the events "teaches math" and "left-handed" independent? Justify your answer.

		Subject taught		
		Math	Not math	Total
Dominant hand	Right	39	51	90
	Left	7	3	10
	Total	46	54	100

SOLUTION

$P(\text{left-handed} \mid \text{teaches math}) = 7/46 = 0.152$

$P(\text{left-handed}) = 10/100 = 0.10$

Because these probabilities are not equal, the events "teaches math" and "left-handed" are not independent.

Knowing that the professor's subject is math increases the probability that the professor is left-handed.

> Does knowing whether the professor teaches math change the probability of left-handedness?

EXAM PREP: FOR PRACTICE, TRY EXERCISE 13.

In Example 4.4.3, we could have also determined that the two events are not independent by showing that

$$P(\text{left-handed} \mid \text{teaches math}) = 7/46 = 0.152 \neq P(\text{left-handed} \mid \text{doesn't teach math}) = 3/54 = 0.056$$

Or we could have focused on whether knowing that the chosen professor is left-handed changes the probability that the professor teaches math. Because

$$P(\text{teaches math} \mid \text{left-handed}) = 7/10 = 0.70 \neq P(\text{teaches math}) = 46/100 = 0.46$$

the events "teaches math" and "left-handed" are not independent.

THINK ABOUT IT

Is there a connection between independence of events and association between two variables? Yes! In the preceding example, we found that the events "teaches math" and "left-handed" were not independent for the sample of 100 professors. Knowing a professor's subject helped us predict their dominant hand. Applying what you learned in Section 2.2, there is an association between subject taught and handedness for the professors in the sample. The segmented bar chart shows this association in picture form.

Does that mean there is an association between subject taught and handedness in the larger population of professors at the conference? Maybe or maybe not. We'll discuss this issue further in Chapter 10.

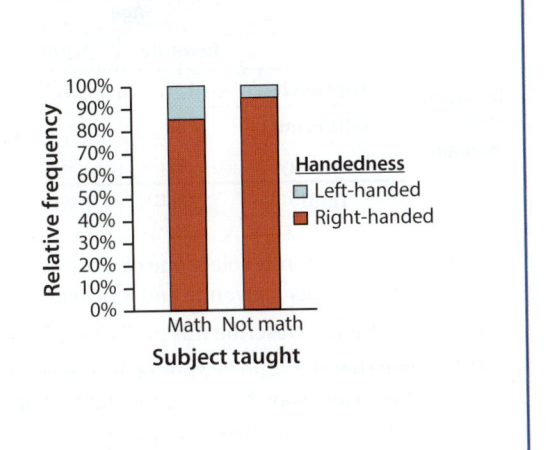

Section 4.4 What Did You Learn?

Review the learning goals from this section. Then practice what you've learned by working through the exercises.

Learning Goal	Example	Exercises
Find and interpret conditional probabilities using two-way tables.	4.4.1	5–8
Use the conditional probability formula to calculate probabilities.	4.4.2	9–12
Determine whether two events are independent.	4.4.3	13–16

Section 4.4 Exercises

Building Concepts and Skills These exercises assess the basic knowledge you should have after reading the section.

1. True/False: $P(A|B)$ is the conditional probability that event B happens given that event A has happened.

2. What is the formula for calculating the conditional probability $P(A|B)$?

3. Explain what it means for two events to be independent.

4. Events A and B are independent if $P(A|B) = $ _____ = _____.

Mastering Concepts and Skills These exercises reinforce the learning goals as shown in the examples.

5. **Refer to Example 4.4.1 Car Accident Recall** Two researchers asked 150 people to recall the details of a car accident they watched on video. They selected 50 people at random and asked, "About how fast were the cars going when they smashed into each other?" For another 50 randomly selected people, they replaced the words "smashed into" with "hit." The remaining 50 people — the control group — were not asked to estimate speed. A week later, all subjects were asked if they saw any broken glass at the accident (there wasn't any). The two-way table summarizes each group's response to the broken glass question.[20]

		Wording			
		"Smashed into"	"Hit"	Control	Total
Broken glass?	Yes	16	7	6	29
	No	34	43	44	121
	Total	50	50	50	150

Suppose we select one of the subjects from this experiment at random. Define events S: "smashed into" wording and Y: "Yes" response.

(a) Find $P(S|Y)$. Describe this probability in words.

(b) Given that the subject did not receive the "smashed into" wording, find the probability that the person said "No" when asked about seeing broken glass at the accident. Write your answer as a probability statement using correct notation for the events.

6. **Gray Squirrels** Do adult and juvenile Eastern gray squirrels in New York's Central Park exhibit different behaviors toward humans? That is one of many questions investigated by 323 volunteer squirrel sighters. Here are the data for 2898 squirrel sightings in the park.[21]

	Age		
Behavior toward humans		Juvenile	Adult	Total
Approach	111	756	867
Indifferent	138	1241	1379
Run away	81	571	652
Total	330	2568	2898

Suppose we randomly select one of these squirrel sightings. Define events J: juvenile and R: run away.

(a) Find $P(R|J)$. Describe this probability in words.

(b) Given that the squirrel spotted by the volunteer did not run away, find the probability that it was an adult. Write your answer as a probability statement using correct notation for the events.

7. **Titanic Casualties** The *Titanic* struck an iceberg and sank on its first voyage across the Atlantic in 1912. Some passengers got off the ship safely in lifeboats, but many others did not and died. The two-way table gives information about adult passengers who survived and who died, by class of travel.

		Class		
		First	Second	Third
Survived?	Yes	197	94	151
	No	122	167	476

Suppose we randomly select one of the adult passengers who sailed on the *Titanic*.

(a) Given that the person selected was in first class, what's the probability that they survived?

(b) If the person selected survived, what's the probability that they were not a third-class passenger?

8. **Python Nests** How is the hatching of water python eggs influenced by the temperature of the snake's nest? Researchers simulated three environments typical of natural nest conditions by randomly assigning newly laid eggs to incubators set at one of three temperatures: hot, neutral, or cold. The following two-way table summarizes the data on nest temperature and hatching status.[22]

		Nest temperature		
		Cold	Neutral	Hot
Hatching status	Hatched	16	38	75
	Didn't hatch	11	18	29

Suppose we select one of the eggs at random.

(a) Given that the chosen egg was assigned to the hot temperature, what is the probability that it hatched?

(b) If the chosen egg hatched, what is the probability that it was not assigned to the hot temperature?

9. **Refer to Example 4.4.2 Taco Time!** A survey of all students at a large community college asked them about eating habits in the past month. Suppose we select one of these students at random. The Venn diagram summarizes the data from the survey using events T: dined at Taco Bell and C: dined at Chipotle. If a student who dined at Chipotle is selected at random, what is the probability that the student also dined at Taco Bell?

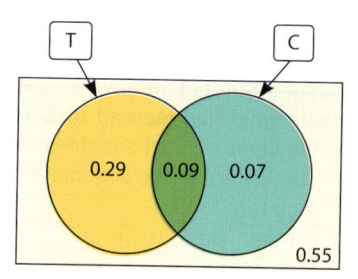

10. **Pet Ownership** Researchers recorded data on pet ownership for randomly selected households in a large city. The Venn diagram summarizes the data using events C: has a cat and D: has a dog. If a household that owns a cat is selected at random, what is the probability that the household owns a dog?

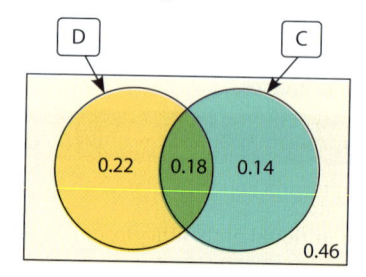

11. **Mac or PC?** A recent census on computer use at a major university revealed that 60% of its students mainly used Macs. The rest mainly used PCs. At the time of the census, 67% of the school's students were undergraduates and the rest were graduate students. In the census, 23% of respondents were graduate students who said that they used Macs as their primary computers. Suppose we randomly select a student from the census who uses a Mac. Find the probability that this person is a graduate student.

12. **Social Media** Survey results from the Pew Research Center suggest that 69% of U.S. adults use Facebook, 40% use Instagram, and 32% use both. Suppose we randomly select a U.S. adult who uses Facebook. Find the probability that the person also uses Instagram.

13. **Refer to Example 4.4.3 Big Papi** Baseball star David Ortiz—nicknamed "Big Papi"—was known for his ability to deliver hits in high-pressure situations. Here is a two-way table of his hits, walks, and outs in all of his regular-season and post-season plate appearances from 1997 through 2014.[23] Suppose we choose a plate appearance at random.

		At-bat			
		Hit	Walk	Out	Total
Season	Regular	2023	1474	5034	8531
	Post	87	57	208	352
	Total	2110	1531	5242	8883

Are the events "hit" and "post season" independent? Justify your answer.

14. **Playing Cards** A standard deck of playing cards (with jokers removed) consists of 52 cards in four suits — clubs, diamonds, hearts, and spades. Each suit has 13 cards, with denominations ace, 2, 3, 4, 5, 6, 7, 8, 9, 10, jack, queen, and king. The jack, queen, and king are referred to as "face cards." Imagine that we shuffle the deck thoroughly and deal one card. Let's define events F: getting a face card and H: getting a heart. The two-way table summarizes the sample space for this chance process.

		Card		
		Face card	Not face card	Total
Suit	Heart	3	10	13
	Not heart	9	30	39
	Total	12	40	52

Are the events "heart" and "face card" independent? Justify your answer.

15. **Titanic Casualties** Refer to Exercise 7. Are the events "survived" and "first class" independent? Justify your answer.

16. **Python Nests** Refer to Exercise 8. Are the events "hot temperature" and "hatched" independent? Justify your answer.

Applying the Concepts These exercises ask you to apply multiple learning goals in a new context or to apply what you learned in this section in a new way.

17. **Cell Phones** The Pew Research Center asked a random sample of U.S. adults about their age and cell phone ownership. The two-way table summarizes the data.[24]

		Age				
		18–29	30–49	50–64	65+	Total
Phone ownership	None	2	10	21	33	66
	Cell phone, not smartphone	11	28	57	103	199
	Smartphone	298	451	291	151	1191
	Total	311	489	369	287	1456

Suppose we select one of the survey respondents at random. Define events S: smartphone owner and O: aged 65 or older.

(a) Find $P(O|S)$. Describe this probability in words.

(b) Given that the chosen person does not have a smartphone, find the probability that the person is aged 65 or older.

(c) Are events S and O independent? Justify your answer.

18. **School Values** Researchers carried out a survey of fourth-, fifth-, and sixth-grade students in Michigan. Students were asked whether getting good grades, having athletic ability, or being popular was most important to them. The two-way table summarizes the survey data.[25]

		Grade level			
		4th grade	5th grade	6th grade	Total
Most important	Grades	49	50	69	168
	Athletic	24	36	38	98
	Popular	19	22	28	69
	Total	92	108	135	335

Suppose we select one of the survey respondents at random. Define events A: athletic ability is most important and F: fifth-grade.

(a) Find $P(F|A)$. Describe this probability in words.

(b) Given that the chosen student did not rank athletic ability as most important, find the probability that the student was a fifth grader.

(c) Are events A and F independent? Justify your answer.

19. **Annual Income** Here is the distribution of the adjusted gross income (in thousands of dollars) reported on individual U.S. federal income tax returns in a prior year. Given that a randomly selected return shows an income of at least $50,000, what is the probability that the income is at least $100,000?

Income	<25	25–49	50–99	100–499	≥500
Probability	0.431	0.248	0.215	0.100	0.006

20. **Language Learners** Researchers selected students in grades 9 to 12 at random and asked if they are studying a language other than English. Here is the distribution of results. What is the probability that a student is studying Spanish given that they are studying some language other than English?

Language	Spanish	French	German	All others	None
Probability	0.26	0.09	0.03	0.03	0.59

21. **Basketball Players** Suppose we select an adult at random. Define events T: person is more than 6 feet tall and B: person is a professional basketball player. Rank the following probabilities from smallest to largest. Explain your reasoning.

$P(T)$ $P(B)$ $P(T|B)$ $P(B|T)$

22. **Choosing to Teach** Suppose we select an adult at random. Define events A: person has earned a college degree and T: person has chosen teaching as their career. Rank the following probabilities from smallest to largest. Explain your reasoning.

$P(A)$ $P(T)$ $P(A|T)$ $P(T|A)$

23. **Independent Dice?** Suppose you roll two fair, 6-sided dice — one red and one blue. Are the events "sum is 7" and "blue die shows a 4" independent? Justify your answer. (See Figure 4.2 for the sample space of this chance process.)

24. **Independent Dice?** Suppose you roll two fair, 6-sided dice — one red and one blue. Are the events "sum is 8" and "blue die shows a 4" independent? Justify your answer. (See Figure 4.2 for the sample space of this chance process.)

Extending the Concepts These exercises challenge you to explore statistical concepts and methods that go beyond what you learned in this section.

25. **What's Your Superpower?** Researchers took a random sample of children from the United Kingdom and the United States. Each student's country of origin was recorded, along with which superpower they would most like to have: the ability to fly, the ability to freeze time, invisibility, superstrength, or telepathy (the ability to read minds). The data are summarized in the following segmented bar chart.[26]

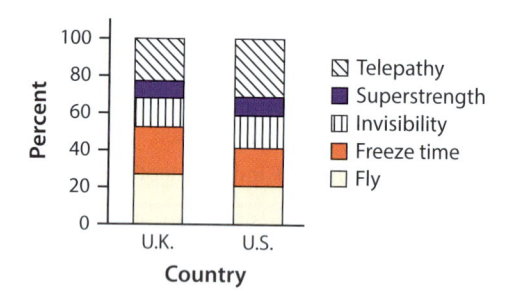

Suppose we choose one of the students from the sample at random.

(a) Explain why the events "U.K." and "fly" are not independent.

(b) Name two events that do appear to be independent.

(c) Use your work from part (a) or (b) to explain whether there is an association between country and superpower preference for the students in the sample.

26. **Independence and Association** The two-way table summarizes data from an experiment comparing the effectiveness of three different diets (A, B, and C) for weight loss. Researchers randomly assigned 300 volunteer subjects to the three diets. The response variable was whether each subject lost weight over a 1-year period.

		Diet			
		A	B	C	Total
Lost weight?	Yes		60		180
	No		40		120
	Total	90	100	110	300

(a) Suppose we randomly select one of the subjects from the experiment. Show that the events "diet B" and "lost weight" are independent.

(b) Copy and complete the table so that there is no association between type of diet and whether a subject lost weight.

(c) Copy and complete the table so that there is an association between type of diet and whether a subject lost weight.

Cumulative Review These exercises revisit what you learned in previous sections.

27. **Observation or Experiment? (1.1)** For each of the following studies, determine whether it is an observational study or an experiment. Explain your answer.

(a) A researcher wants to study the effect of sales on consumers' expectations. The researcher makes up two different histories of the store price of a video game for the past year. Students in an economics course view one or the other price history on a computer. Some students see a steady price, while others see regular sales that temporarily cut the price. Then the students are asked what price they would expect to pay for the video game.

(b) A team of researchers wanted to study the effect that living in public housing had on family stability (defined as the consistency of family structure) in low-income households. They obtained a list of applicants accepted for public housing, together with a list of families who applied but were rejected by housing authorities. Then they took a random sample from each list. Researchers followed up with the families in both groups several times over a period of 20 years.

(c) A doctor at a veterans' hospital examines all of its patient records over a 9-year period and finds that twice as many men as women fell out of their hospital beds during their stay. The doctor asserts that this is evidence men are clumsier than women.

28. **Standard Deviation (3.2)** For the four small data sets given here, indicate which has the smallest standard deviation and which has the largest without performing any calculations. Justify your answer.
    Set 1: 1, 1, 1, 9, 9, 9
    Set 2: 1, 1, 1, 6, 6, 6, 9, 9, 9
    Set 3: 3, 4, 5, 6, 7, 8, 9
    Set 4: 3, 4, 5, 6, 6, 6, 7, 8, 9

Section 4.5 The General Multiplication Rule and Bayes' Theorem

LEARNING GOALS

By the end of this section, you will be able to:
- Use the general multiplication rule to calculate probabilities.
- Use a tree diagram to model a chance process involving a sequence of outcomes.
- Calculate conditional probabilities using Bayes' Theorem.

Suppose that A and B are two events resulting from the same chance process. We can find the probability $P(A \text{ or } B)$ with the general addition rule:

$$P(A \text{ or } B) = P(A) + P(B) - P(A \text{ and } B)$$

How do we find the probability that both events happen, $P(A \text{ and } B)$?

The General Multiplication Rule

About 55% of high school students participate on a school athletic team at some level. Roughly 6% of these athletes go on to play on a team at a college that belongs to the NCAA.[27] What percentage of high school students play a sport in high school *and* go on to play on an NCAA team? About 6% of 55%, or roughly 3.3%.

Let's restate the situation in probability language. Suppose we select a high school student at random. What's the probability that the student plays a sport in high school and goes on to play on an NCAA team? The given information suggests that

$$P(\text{high school sport}) = 0.55 \text{ and } P(\text{NCAA team} \mid \text{high school sport}) = 0.06$$

By the logic just stated,

$$P(\text{high school sport and NCAA team})$$
$$= P(\text{high school sport}) \cdot P(\text{NCAA team} \mid \text{high school sport})$$
$$= (0.55)(0.06) = 0.033$$

This is an example of the **general multiplication rule**.

> For any chance process, the probability that events A and B both occur can be found using the **general multiplication rule**:
>
> $$P(A \text{ and } B) = P(A) \cdot P(B \mid A)$$

The general multiplication rule says that for both of two events to occur, first one must occur. Then, given that the first event has occurred, the second must occur. To confirm that this result is correct, start with the conditional probability formula

$$P(B \mid A) = \frac{P(B \text{ and } A)}{P(A)}$$

The numerator gives the probability we want because $P(B \text{ and } A)$ is the same as $P(A \text{ and } B)$. Multiply both sides of this equation by $P(A)$ to get

$$P(A) \cdot P(B \mid A) = P(A \text{ and } B)$$

EXAMPLE 4.5.1

Internet access

The general multiplication rule

PROBLEM

While internet access has become an important part of many people's lives, not everyone accesses the internet at home in the same way. Some U.S. adults have broadband internet access at home, some have smartphone access via cellular data,

(continued)

and some have both kinds of access. Based on a recent Pew Research Center survey, 96% of U.S. adults aged 18 to 29 own a smartphone and 28% of those who own smartphones don't have broadband at home.[28] Find the probability that a randomly selected U.S. adult aged 18 to 29 owns a smartphone and doesn't have broadband at home.

SOLUTION

P(owns smartphone and no broadband)

$= P$(owns smartphone) $\cdot P$(no broadband | owns smartphone)

$= (0.96)(0.28)$

$= 0.269$

$P(A \text{ and } B) = P(A) \cdot P(B \mid A)$

EXAM PREP: FOR PRACTICE, TRY EXERCISE 5.

Tree Diagrams

Shannon hits the snooze button on her alarm on 60% of school days. If she hits snooze, there is a 0.70 probability that she makes it to her first class on time. If she doesn't hit snooze and gets up right away, there is a 0.90 probability that she makes it to class on time. Suppose we select a school day at random and record whether Shannon hits the snooze button and whether she arrives in class on time. Figure 4.7 shows a **tree diagram** for this chance process.

FIGURE 4.7 A tree diagram displaying the sample space of randomly choosing a school day, noting if Shannon hits the snooze button and whether or not she gets to her first class on time.

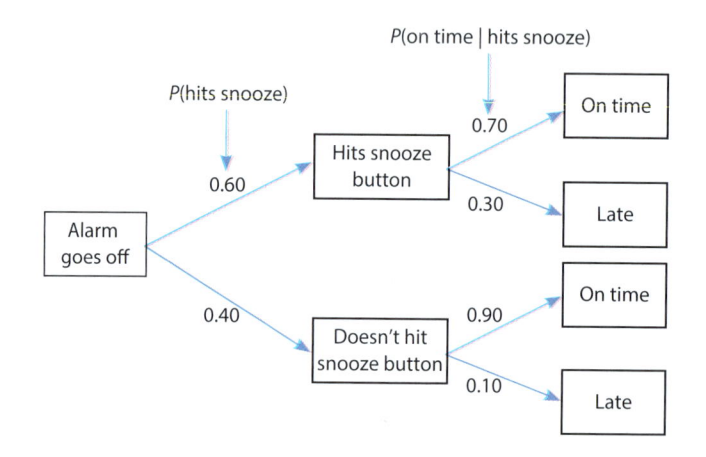

A **tree diagram** shows the sample space of a chance process involving multiple stages. The probability of each outcome is shown on the corresponding branch of the tree. All probabilities after the first stage are conditional probabilities.

There are only two possible outcomes at the first "stage" of this chance process: Shannon hits the snooze button or she doesn't. The first set of branches in the tree diagram displays these outcomes with their probabilities. The second set of branches shows the two possible results at the next "stage" of the process — Shannon gets to her first class either on time or late — and the probability of each result based on whether or not she hit the snooze button. Note that the probabilities on the second set of branches are *conditional* probabilities, like P(on time | hits snooze) = 0.70.

What is the probability that Shannon hits the snooze button and is late for class on a randomly selected school day? The general multiplication rule provides the answer:

P(hits snooze and late) = P(hits snooze) $\cdot P$(late | hits snooze)

$= (0.60)(0.30)$

$= 0.18$

In essence, this calculation amounts to multiplying probabilities along the branches of the tree diagram.

What's the probability that Shannon is late to class on a randomly selected school day? Figure 4.8 illustrates two ways this can happen: Shannon hits the snooze button *and* is late OR she doesn't hit snooze *and* is late.

FIGURE 4.8 Tree diagram with blue arrows showing the two possible ways that Shannon can be late to class on a randomly selected day.

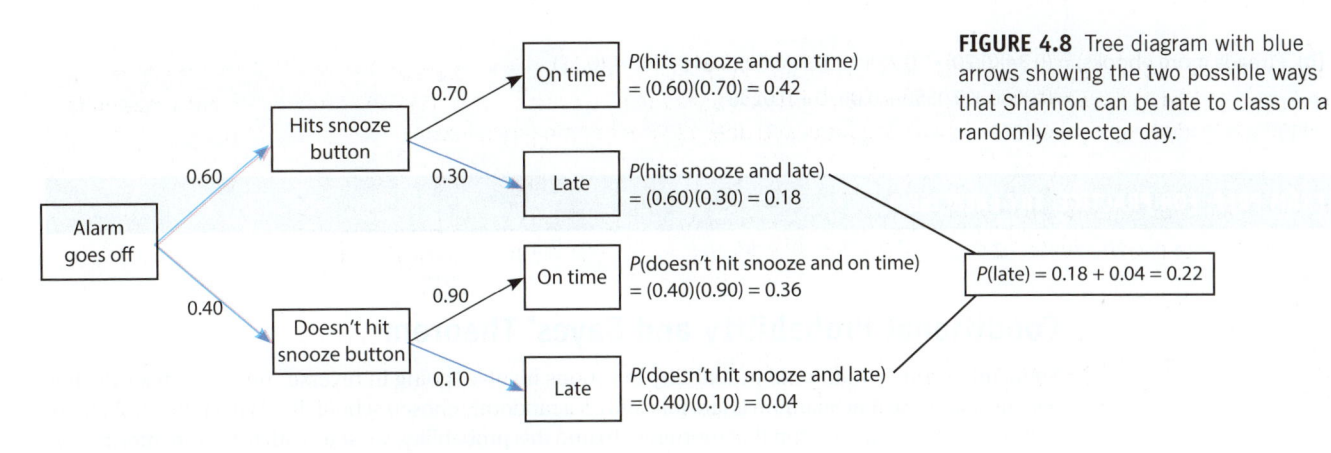

Because these outcomes are mutually exclusive,

$$P(\text{late}) = P(\text{hits snooze and late}) + P(\text{doesn't hit snooze and late})$$

The general multiplication rule tells us that

$$P(\text{doesn't hit snooze and late}) = P(\text{doesn't hit snooze}) \cdot P(\text{late} \mid \text{doesn't hit snooze})$$
$$= (0.40)(0.10)$$
$$= 0.04$$

So

$$P(\text{late}) = 0.18 + 0.04 = 0.22$$

There is about a 22% chance that Shannon will be late to class.

EXAMPLE 4.5.2

Print books and ebooks

Tree diagrams

PROBLEM

Harris Interactive reported that 20% of people aged 18 to 36, 25% of people aged 37 to 48, 21% of people aged 49 to 67, and 17% of people aged 68 and older read more ebooks than print books. According to the U.S. Census Bureau, 34% of adults are aged 18 to 36, 22% are aged 37 to 48, 30% are aged 49 to 67, and 14% are aged 68 and older.[29] Suppose we select one U.S. adult at random and record which age group the person is from and whether they read more ebooks or print books.

(a) Draw a tree diagram to model this chance process.

(b) Find the probability that the person reads more ebooks than print books.

SOLUTION

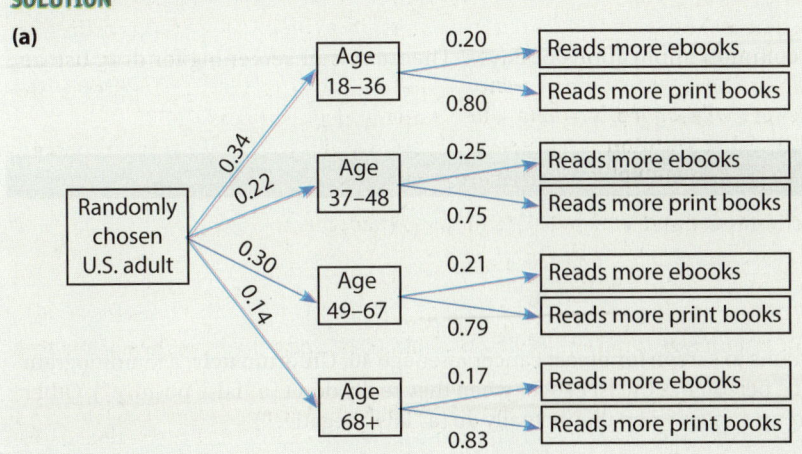

(a)

(*continued*)

(b) P(reads more ebooks) = (0.34)(0.20) + (0.22)(0.25) + (0.30)(0.21) + (0.14)(0.17)

= 0.0680 + 0.0550 + 0.0630 + 0.0238

= 0.2098

EXAM PREP: FOR PRACTICE, TRY EXERCISE 9.

Conditional Probability and Bayes' Theorem

Some interesting conditional probability questions involve "going in reverse" on a tree diagram. For instance, suppose that Shannon is late for class on a randomly chosen school day. What is the probability that she hit the snooze button that morning? To find this probability, we start with the given information that Shannon is late, which is displayed on the second set of branches in Figure 4.8, and ask whether she hit the snooze button, which is shown on the first set of branches. We can use the information from the tree diagram and the conditional probability formula to perform the required calculation:

$$P(\text{hit snooze} \mid \text{late}) = \frac{P(\text{hit snooze and late})}{P(\text{late})}$$

$$= \frac{P(\text{hit snooze}) \cdot P(\text{late} \mid \text{hit snooze})}{P(\text{late})}$$

$$= \frac{(0.60)(0.30)}{(0.60)(0.30) + (0.40)(0.10)}$$

$$= \frac{0.18}{0.22}$$

$$= 0.818$$

When Shannon is late, there is a 0.818 probability that she hit the snooze button that morning.

Let's recap what we just learned:

- On any randomly selected school day, the probability that Shannon hits the snooze button is 0.60. This is known as a *prior* probability.
- Given that Shannon is late to class on a randomly selected day, the probability that she hit snooze increases to 0.818. This is called the *posterior* probability, because it has been updated based on the additional information provided—in this case, that Shannon was late to class.

This alternative way of thinking about probability was developed by the Reverend Thomas Bayes in the mid-1700s. For that reason, the formula for finding a conditional probability when going in reverse on a tree diagram is known as **Bayes' Theorem**.

Bayes' Theorem states that for any two events A and B with nonzero probabilities, the probability of event A given that event B has subsequently occurred is

$$P(A \mid B) = \frac{P(A) \cdot P(B \mid A)}{P(B)}$$

One of the most common applications of Bayes' Theorem is in screening for drug use and disease.

EXAMPLE 4.5.3

How reliable are mammograms?

Bayes' Theorem

PROBLEM

Many women choose to have annual mammograms to screen for breast cancer after age 40. Unfortunately, a mammogram isn't foolproof. Sometimes, the test suggests that a person has breast cancer when they really don't (a "false positive"). Other times, the test says that a person doesn't have breast cancer when they actually do (a "false negative").

Suppose that we know the following information about breast cancer and mammograms in a particular population:

- Of the women aged 40 or older in this population, 1% have breast cancer.
- For women who have breast cancer, the probability of a negative mammogram is 0.03.
- For women who don't have breast cancer, the probability of a positive mammogram is 0.06.

A randomly selected woman aged 40 or older from this population tests positive for breast cancer in a mammogram. Find the probability that she actually has breast cancer.

SOLUTION

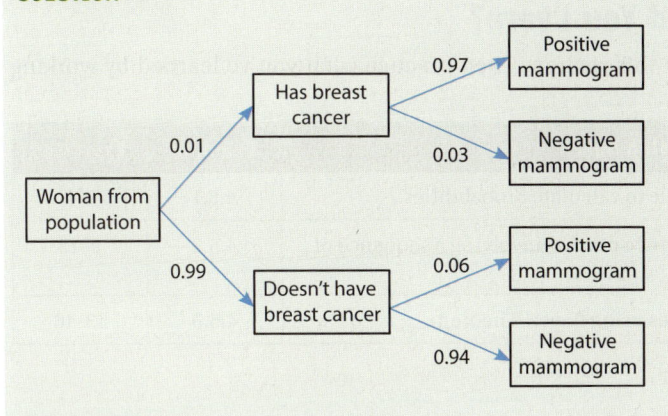

Start by making a tree diagram to summarize the possible outcomes.

- Because 1% of women in this population have breast cancer, 99% don't.
- Of those women who do have breast cancer, 3% would test negative on a mammogram. The remaining 97% would (correctly) test positive.
- Among the women who don't have breast cancer, 6% would test positive on a mammogram. The remaining 94% would (correctly) test negative.

$$P(A|B) = \frac{P(A) \cdot P(B|A)}{P(B)}$$

$P(\text{breast cancer} | \text{positive mammogram})$

$$= \frac{P(\text{breast cancer}) \cdot P(\text{positive mammogram} | \text{breast cancer})}{P(\text{positive mammogram})}$$

$$= \frac{(0.01)(0.97)}{(0.01)(0.97) + (0.99)(0.06)}$$

$$= \frac{0.0097}{0.0691}$$

$$= 0.14$$

EXAM PREP: FOR PRACTICE, TRY EXERCISE 13.

Given that a randomly selected woman from the population has a positive mammogram, there is only about a 14% chance that she has breast cancer. Are you surprised by this result? Most people are. Sometimes, a two-way table that includes counts is more convincing.[30]

To make calculations easier, we'll suppose that there are exactly 10,000 women aged 40 or older in this population, and that exactly 100 have breast cancer (that's 1% of the women in this population).

- How many of those 100 women would have a positive mammogram? That would be 97% of 100, or 97 of them. That leaves 3 who would test negative.
- How many of the 9900 women who don't have breast cancer would have a positive mammogram? That would be 6% of them, or (9900)(0.06) = 594 women. The remaining 9900 − 594 = 9306 would test negative.
- In total, 97 + 594 = 691 women would have positive mammograms and 3 + 9306 = 9309 women would have negative mammograms.

This information is summarized in the two-way table.

		Has breast cancer?		
		Yes	No	Total
Mammogram result	Positive	97	594	691
	Negative	3	9306	9309
	Total	100	9900	10,000

Given that a randomly selected woman has a positive mammogram, the two-way table shows that the conditional probability

$$P(\text{breast cancer} \mid \text{positive mammogram}) = 97/691 = 0.14$$

This example illustrates an important fact when considering proposals for widespread screening for serious diseases or illegal drug use: If the condition being tested is uncommon in the population, many positive screening results will be false positives. The best remedy is to retest any individual who tests positive.

Section 4.5 What Did You Learn?

Review the learning goals from this section. Then practice what you've learned by working through the exercises.

Learning Goal	Example	Exercises
Use the general multiplication rule to calculate probabilities.	4.5.1	5–8
Use a tree diagram to model a chance process involving a sequence of outcomes.	4.5.2	9–12
Calculate conditional probabilities using Bayes' Theorem.	4.5.3	13–16

Section 4.5 Exercises

Building Concepts and Skills These exercises assess the basic knowledge you should have after reading the section.

1. State the general multiplication rule.

2. A tree diagram shows the sample space of a chance process involving multiple _____.

3. True/False: The probability associated with each branch in a tree diagram is a conditional probability.

4. Bayes' Theorem says that the probability of event A given that event B has subsequently occurred is
   $P(A \mid B) = $ _____.

Mastering Concepts and Skills These exercises reinforce the learning goals as shown in the examples.

5. **Refer to Example 4.5.1 Coffee with Cream** Employees at a local coffee shop recorded the drink orders of all the customers on a Saturday. They found that 64% of customers ordered a hot drink, and 80% of these customers added cream to their drink. Find the probability that a randomly selected Saturday customer ordered a hot drink and added cream to the drink.

6. **Health Clubs** Suppose that 10% of adults belong to health clubs, and 40% of these health club members go to the club at least twice a week. Find the probability that a randomly selected adult belongs to a health club and goes there at least twice a week.

7. **Chocolates** Suppose a candy maker offers a special box of 20 chocolate candies that look alike. In reality, 14 of the candies have soft centers and 6 have hard centers. Choose 2 of the candies from a special box at random. Find the probability that both candies have soft centers.

8. **Working Students** An Introductory Statistics class with 30 students includes 20 who work part-time or full-time jobs while attending school and 10 who do not work. Choose 2 of the students in the class at random. Find the probability that both work part-time or full-time jobs while attending school.

9. **Refer to Example 4.5.2 Credit Cards and Gas** In a recent month, 88% of automobile drivers filled their vehicles with regular gasoline, 2% purchased midgrade gas, and 10% bought premium gas.[31] Of those who bought regular gas, 28% paid with a credit card. Of customers who bought midgrade and premium gas, 34% and 42%, respectively, paid with a credit card. Suppose we select a customer at random.
   (a) Draw a tree diagram to model this chance process.
   (b) Find the probability that the customer paid with a credit card.

10. **Lactose Intolerance** Lactose intolerance means that a person has difficulty digesting dairy products that contain lactose (milk sugar). This condition is particularly common among people of African and Asian ancestry. Among people who identify as White alone, Black alone, or Asian alone in the United States, 80% identify as White alone, 14% identify as Black alone, and 6% identify as Asian alone. Moreover, according to the National Institutes of Health (NIH), 15% of White people, 70% of Black people, and 90% of Asian people are lactose intolerant.[32] Suppose we select a person from these populations at random.

(a) Draw a tree diagram to model this chance process.

(b) Find the probability that the person is lactose intolerant.

11. Chocolates Refer to Exercise 7.

(a) Draw a tree diagram to model this chance process.

(b) Find the probability that one of the chocolates has a soft center and the other one doesn't.

12. Working Students Refer to Exercise 8.

(a) Draw a tree diagram to model this chance process.

(b) Find the probability that one of the students works a part-time or full-time job while attending school and the other does not.

13. Refer to Example 4.5.3 First Serves Tennis great Serena Williams made 59% of her first serves in a certain season. When Williams made her first serve, she won 75% of the points. When Williams missed her first serve and had to serve again, she won only 49% of the points.[33] While watching a game from that season in which Williams is serving, you get distracted and miss seeing her serve, but look up in time to see Williams win the point. What's the probability that she missed her first serve?

14. Metal Detector A prospector uses a homemade metal detector to look for valuable metal objects on a beach. The machine isn't perfect—it detects only 98% of the metal objects over which it passes, and it detects 4% of the nonmetallic objects over which it passes. Suppose that 25% of the objects that the machine passes over are metal. If the machine gives a signal when it passes over an object, find the probability that the prospector has found a metal object.

15. Credit Cards and Gas Refer to Exercise 9. Given that the customer paid with a credit card, find the probability that they bought premium gas.

16. Lactose Intolerance Refer to Exercise 10. Given that the chosen person is lactose intolerant, what is the probability that they are of Asian descent?

Applying the Concepts These exercises ask you to apply multiple learning goals in a new context or to apply what you learned in this section in a new way.

17. HIV Testing Enzyme immunoassay (EIA) tests are used to screen blood specimens for the presence of antibodies to human immunodeficiency virus (HIV), the virus that causes acquired immunodeficiency syndrome (AIDS). Antibodies indicate the presence of the virus. The test is quite accurate but is not always correct. A *false positive* occurs when the test gives a positive result but no HIV antibodies are actually present in the blood. A *false negative* occurs when the test gives a negative result but HIV antibodies really are present in the blood. Here are the approximate probabilities of positive and negative EIA outcomes when the blood tested does and does not actually contain antibodies to HIV.[34]

		Test result	
		+	−
Truth	Antibodies present	0.9985	0.0015
	Antibodies absent	0.006	0.994

Suppose that 1% of a large population carries antibodies to HIV in their blood. Imagine choosing a person from this population at random.

(a) Draw a tree diagram to model this chance process.

(b) Find the probability that the EIA test result is positive.

(c) Given that the EIA test is positive, find the probability that the person actually has HIV antibodies.

18. Drug Tests Many employers require prospective employees to take a drug test. A positive result on this test could suggest that the prospective employee uses illegal drugs. However, not all people who test positive use illegal drugs. The test result could be a *false positive*. A negative test result could be a *false negative* if the person really does use illegal drugs. Suppose that 4% of prospective employees use illegal drugs, and that the drug test has a false positive rate of 5% and a false negative rate of 10%.[35] Imagine choosing a prospective employee at random.

(a) Draw a tree diagram to model this chance process.

(b) Find the probability that the drug test result is positive.

(c) Given that the drug test result is positive, find the probability that the prospective employee actually uses illegal drugs.

19. Sensitivity and Specificity The *sensitivity* of a diagnostic test is its probability of correctly diagnosing an individual with the tested-for condition. The *specificity* of a diagnostic test is its probability of correctly diagnosing an individual without the tested-for condition. An at-home screening test for colorectal cancer has a sensitivity of 92.3% and a specificity of 89.8%.[36] Assume that 4% of a certain population has colorectal cancer. If an individual from this population gets a positive result on the at-home screening test, what is the probability that they have colorectal cancer? Show your work.

20. COVID-19 Testing According to a May 13, 2020, article in the *New York Times,*

> … the predictive value of an antibody test with 90 percent accuracy could be as low as 32 percent if the base rate of infection in the population is 5 percent. Put another way, there is an almost 70 percent probability in that case that the test will *falsely* indicate a person has antibodies.
>
> The reason for this is a simple matter of statistics. The lower prevalence there is of a trait in a studied population — here, coronavirus infection — the more likely that a test will return a false positive. While a more accurate test will help, it can't change the statistical reality when the base rate of infection is very low.[37]

Assume that "90 percent accuracy" means the test has a 0.90 probability of giving a positive result for a person who has antibodies and a 0.90 probability of giving a negative result for a person who does not have antibodies. Use the information provided to confirm the 32% probability quoted in the article that a person has antibodies given that they test positive.

Extending the Concepts These exercises challenge you to explore statistical concepts and methods that go beyond what you learned in this section.

21. **HIV Testing** Refer to Exercise 17. Many of the positive results from EIA tests are false positives. It is therefore common practice to perform a second EIA test on another blood sample from a person whose initial specimen tests positive. Assume that the false positive and false negative rates remain the same for a person's second test. Find the probability that a person who gets a positive result on both EIA tests actually has HIV antibodies.

22. **Potential Donors** Tree diagrams can organize problems having more than two stages. The following figure shows probabilities for a charity calling potential donors by telephone.[38] Each person called is either a recent donor, a past donor, or a new prospect. At the next stage, the person called either does or does not pledge to contribute, with conditional probabilities that depend on the donor class to which the person belongs. Finally, those who make a pledge either do or don't make a contribution. Suppose we randomly select a person who is called by the charity.

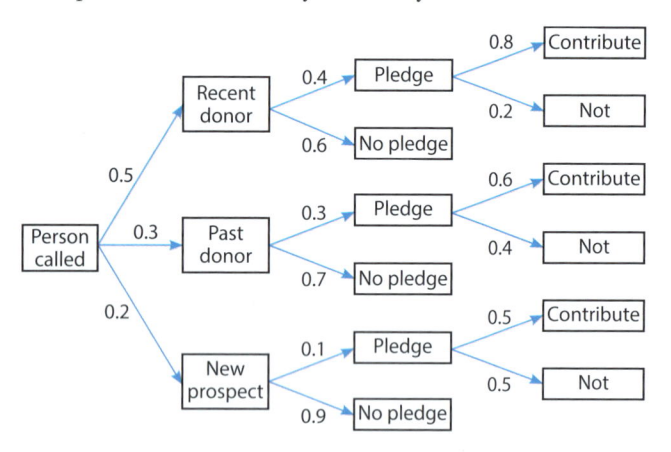

   (a) What is the probability that the person contributed to the charity?
   (b) Given that the person contributed, find the probability that they are a recent donor.

23. **Playing Cards** A standard deck of playing cards consists of 52 cards, with 13 cards in each of four suits: spades, diamonds, clubs, and hearts. Suppose you shuffle the deck thoroughly and deal 5 cards face-up onto a table.
   (a) What is the probability of dealing five spades in a row?
   (b) Find the probability that all five cards on the table have the same suit.

Cumulative Review These exercises revisit what you learned in previous sections.

24. **Water and Wealth (2.6)** Water is an expensive commodity in the U.S. Southwest. An article in the *Arizona Daily Star* reported on the relationship between average household water consumption per year and median household income for 186 different census tracts served by Tucson Water. The scatterplot summarizes this relationship.[39]

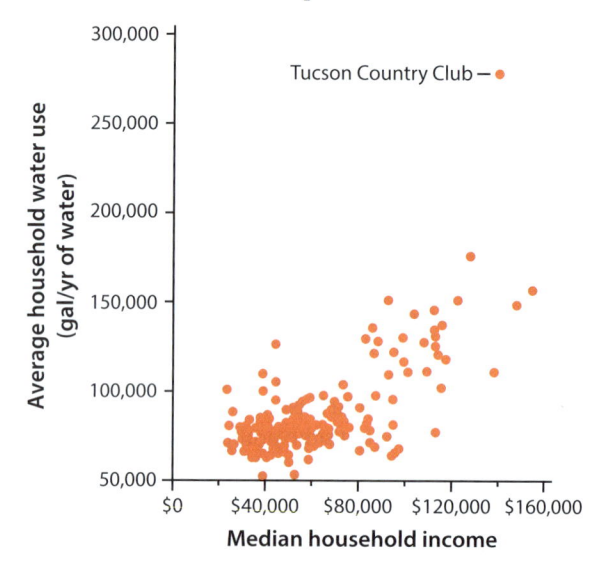

   (a) Use the scatterplot to describe the relationship between median household income and average household water use.
   (b) The single dot in the upper right region of the scatterplot is from the Tucson Country Club Estates development. Describe the characteristics of this neighborhood.

25. **Sports Cars (3.3)** Some people dream of having a sporty car, but worry that it might use too much gas. The U.S. Environmental Protection Agency (EPA) lists most such vehicles in its "two-seater" or "minicompact" categories. The figure shows boxplots for both city and highway gas mileages for these two groups of cars.[40] Compare the distributions.

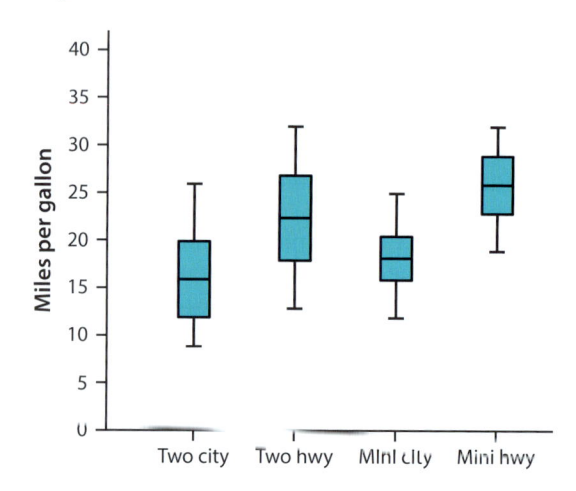

Section 4.6 The Multiplication Rule for Independent Events

LEARNING GOALS

By the end of this section, you will be able to:
- Use the multiplication rule for independent events to calculate probabilities.
- Calculate P(at least one) using the complement rule and the multiplication rule for independent events.
- Determine whether it is appropriate to use the multiplication rule for independent events in a given setting.

What happens to the general multiplication rule in the special case when events A and B are independent? In that case, $P(B|A) = P(B)$ because knowing that event A occurred doesn't change the probability that event B occurs. We can simplify the general multiplication rule as follows:

$$P(A \text{ and } B) = P(A) \cdot P(B|A)$$
$$= P(A) \cdot P(B)$$

This result is known as the **multiplication rule for independent events**.

> The **multiplication rule for independent events** says that if A and B are independent events, the probability that A and B both occur is
>
> $$P(A \text{ and } B) = P(A) \cdot P(B)$$

⚠ Note that this rule applies *only* to independent events.

Calculating Probabilities with the Multiplication Rule for Independent Events

Suppose that Pedro drives the same route to work on Monday through Friday. Pedro's route includes one traffic light. The probability that the light will be green is 0.42, that it will be yellow is 0.03, and that it will be red is 0.55.

1. **What's the probability that the light is green on Monday and red on Tuesday?** Let event A be a green light on Monday and event B be a red light on Tuesday. These two events are independent because knowing whether or not the light was green on Monday doesn't help us predict the color of the light on Tuesday. By the multiplication rule for independent events,

    P(green on Monday and red on Tuesday)
    $= P(A \text{ and } B)$
    $= P(A) \cdot P(B)$
    $= (0.42)(0.55)$
    $= 0.231$

    There is about a 23% probability that the light will be green on Monday and red on Tuesday.

2. **What's the probability that Pedro finds the light red on Monday through Friday?** We can extend the multiplication rule for independent events to more than two events.

    P(red Monday *and* red Tuesday *and* red Wednesday *and* red Thursday *and* red Friday)
    $= P(\text{red Monday}) \cdot P(\text{red Tuesday}) \cdot P(\text{red Wednesday}) \cdot P(\text{red Thursday}) \cdot P(\text{red Friday})$
    $= (0.55)(0.55)(0.55)(0.55)(0.55)$
    $= (0.55)^5$
    $= 0.0503$

    There is about a 5% probability that Pedro will encounter a red light on all five days in a workweek.

EXAMPLE 4.6.1

What factors led to the Challenger disaster?

Multiplication rule for independent events

PROBLEM

On January 28, 1986, Space Shuttle *Challenger* exploded on takeoff, killing all seven crew members aboard. Afterward, scientists and statisticians helped analyze what went wrong. They determined that the failure of O-ring joints in the shuttle's booster rockets caused the explosion. Experts estimated that the probability that an individual O-ring joint would function properly under the cold conditions that day was 0.977. But there were six O-ring joints, and all six had to function properly for the shuttle to launch safely. Assuming that O-ring joints succeed or fail independently, find the probability that the shuttle would launch safely under similar conditions.

SOLUTION

P(O-ring 1 works and O-ring 2 works and . . . and O-ring 6 works)

$= P$(O-ring 1 works) $\cdot P$(O-ring 2 works) $\cdot \ldots \cdot P$(O-ring 6 works)

$= (0.977)(0.977)(0.977)(0.977)(0.977)(0.977)$

$= (0.977)^6$

$= 0.87$

> For the shuttle to launch safely, all six O-ring joints must function properly.

EXAM PREP: FOR PRACTICE, TRY EXERCISE 5.

The multiplication rule for independent events can also be used to help find P(at least one). In Example 4.6.1, the shuttle would *not* launch safely under similar conditions if 1, 2, 3, 4, 5, or all 6 O-ring joints failed — that is, if *at least one* O-ring failed. In other words, the only possible number of O-ring failures excluded is 0. So the events "at least one O-ring joint fails" and "no O-ring joints fail" are complementary events. By the complement rule,

$$P(\text{at least one O-ring fails}) = 1 - P(\text{no O-ring fails})$$
$$= 1 - 0.87$$
$$= 0.13$$

That's a very high chance of failure! As a result of this analysis following the *Challenger* disaster, NASA made important safety changes to the design of the shuttle's booster rockets.

EXAMPLE 4.6.2

Trading speed for accuracy in HIV testing?

Finding the probability of "at least one"

PROBLEM

Many people who come to clinics to be tested for HIV (the virus that causes AIDS) don't come back to learn their test results. Clinics now offer "rapid HIV tests" that give a result while the client waits. In one clinic, use of rapid tests increased the percentage of clients who learned their test results from 69% to 99.7%.

The trade-off for fast results is that rapid tests are often less accurate than slower laboratory tests. Applied to people who have no HIV antibodies, one rapid test has a probability of about 0.008 of producing a false positive (i.e., of falsely indicating that HIV antibodies are present).[41] If a clinic tests 200 randomly selected people who are free of HIV antibodies, what is the chance that at least one false positive will occur? Assume that test results for different individuals are independent.

SOLUTION

P(no false positives) $= P$(all 200 tests negative)

$\qquad = (0.992)(0.992) \ldots (0.992)$

$\qquad = (0.992)^{200}$

$\qquad = 0.201$

P(at least one false positive) $= 1 - 0.201 = 0.799$

> Start by finding P(no false positives).

> The probability that any individual test result is negative is $1 - 0.008 = 0.992$.

> By the complement rule, P(at least one false positive) $= 1 - P$(no false positives).

EXAM PREP: FOR PRACTICE, TRY EXERCISE 9.

Use the Multiplication Rule for Independent Events Wisely

To find the probability of "A or B," we can use the general addition rule:

$$P(A \text{ or } B) = P(A) + P(B) - P(A \text{ and } B)$$

In the special case when A and B are *mutually exclusive* (have no outcomes in common), the addition rule simplifies to

$$P(A \text{ or } B) = P(A) + P(B)$$

To find the probability of "A and B," we can use the general multiplication rule:

$$P(A \text{ and } B) = P(A) \cdot P(B|A)$$

In the special case when A and B are *independent*, the multiplication rule simplifies to

$$P(A \text{ and } B) = P(A) \cdot P(B)$$

! **Resist the temptation to use these simple rules when the conditions that justify them are not met.**

EXAMPLE 4.6.3

Are there many college students aged 55 and older?

Beware lack of independence!

PROBLEM

Government data show that 5.4% of adults are full-time college students and that 37.8% of adults are age 55 or older.[42] If we randomly select an adult, is

$P(\text{full-time college student and age 55 or older}) = (0.054)(0.378) = 0.02$?

Why or why not?

SOLUTION

No, because being a full-time college student and being 55 or older are not independent events. Knowing that the chosen adult is a full-time college student makes it much less likely that they are aged 55 or older.

> For these events to be independent, 37.8% of full-time college students would have to be aged 55 or older.

EXAM PREP: FOR PRACTICE, TRY EXERCISE 13.

The multiplication rule $P(A \text{ and } B) = P(A) \cdot P(B)$ gives us another way to determine whether two events are independent. Let's return to the tongue rolling and eyebrow raising example from earlier in the chapter. The following two-way table summarizes data from a college statistics class.

		Roll their tongue?		
		Yes	No	Total
Raise one eyebrow?	Yes	42	20	62
	No	107	31	138
	Total	149	51	200

Our events of interest were A: can roll their tongue and B: can raise one eyebrow. Are these two events independent? No, because

$$P(A \text{ and } B) = P(\text{roll tongue and raise one eyebrow}) = \frac{42}{200} = 0.21$$

is not equal to

$$P(A) \cdot P(B) = P(\text{roll tongue}) \cdot P(\text{raise one eyebrow}) = \frac{149}{200} \cdot \frac{62}{200} = 0.231$$

If the ability to roll one's tongue and the ability to raise one eyebrow were independent, then about 23.1% of the students would be able to do both. But only 21% of the students could do both—less than expected if these events were independent.

THINK ABOUT IT

Is there a connection between mutually exclusive and independent? Let's start with a new chance process. Choose a U.S. resident at random. Define event A: the person is younger than age 10 and event B: the person has a driver's license. It's fairly clear that these two events are mutually exclusive (can't happen together)! Are they also independent?

If you know that event A has occurred, does this change the probability that event B happens? Of course! If we know the person is younger than age 10, the probability that the person has a driver's license is zero. Because $P(B|A) \neq P(B)$, the two events are not independent.

Two mutually exclusive events (with nonzero probabilities) can *never* be independent, because if one event happens, the other event is guaranteed not to happen.

Section 4.6 What Did You Learn?

Review the learning goals from this section. Then practice what you've learned by working through the exercises.

Learning Goal	Example	Exercises
Use the multiplication rule for independent events to calculate probabilities.	4.6.1	5–8
Calculate P(at least one) using the complement rule and the multiplication rule for independent events.	4.6.2	9–12
Determine whether it is appropriate to use the multiplication rule for independent events in a given setting.	4.6.3	13–16

Section 4.6 Exercises

Building Concepts and Skills These exercises assess the basic knowledge you should have after reading the section.

1. The general multiplication rule says that $P(A \text{ and } B) =$ _____. If A and B are _____ events, the multiplication rule simplifies to $P(A \text{ and } B) = P(A) \cdot P(B)$.

2. $P(\text{at least one}) = 1 -$ _____

3. The general addition rule says that $P(A \text{ or } B) =$ _____. If A and B are _____ events, the addition rule simplifies to $P(A \text{ or } B) = P(A) + P(B)$.

4. True/False: Mutually exclusive events are never independent.

Mastering Concepts and Skills These exercises reinforce the learning goals as shown in the examples.

5. **Refer to Example 4.6.1 Holiday Lights** A string of holiday lights contains 20 lights. The lights are wired in series, so that if any light fails, the whole string will go dark. Each light has probability 0.98 of working for a 3-year period. The lights fail independently of each other. Find the probability that the string of lights will remain bright for 3 years.

6. **A Perfect Game** In baseball, a perfect game is when a pitcher doesn't allow any hitters to reach base in all 9 innings. Historically, Major League Baseball (MLB)

pitchers throw a perfect inning — an inning where no hitters reach base — about 40% of the time.[43] So to throw a perfect game, a pitcher must have 9 perfect innings in a row. Find the probability that an MLB pitcher throws 9 perfect innings in a row, assuming the pitcher's performance in any one inning is independent of his performance in other innings. (You'll check this assumption in Exercise 20.)

7. **Lie Detector** From experience, we know that a certain lie detector will show a positive reading (indicating a lie) 10% of the time when a person is telling the truth. Suppose that a random sample of 5 people take a lie detector test. Find the probability of observing no positive readings if all 5 people are telling the truth.

8. **Airline Reservation** An airline estimates that the probability a randomly selected call to its reservation phone line results in a reservation being made is 0.31. Find the probability that none of 4 randomly selected calls to the phone line results in a reservation.

9. **Refer to Example 4.6.2 O-Negative** People with type O-negative blood are known as "universal donors," because any patient can receive a transfusion of O-negative blood. Only 7% of the U.S. population has O-negative blood.[44] If we choose 10 Americans at random, what is the probability that at least one of them has O-negative blood?

10. **Lottery Odds** If you buy 1 ticket in Canada's Lotto 6/49 lottery game, the probability that you will win a prize is 0.15.[45] Given the nature of lotteries, the probability of winning is independent from one drawing to the next. If you buy one ticket for each of five consecutive drawings, what is the probability that you will win at least one prize?

11. **Dice Rolling** Suppose that you roll a fair, 6-sided die 10 times. What's the probability that you get at least one 6?

12. **On-Time Shipments** A shipping company claims that 90% of its shipments arrive on time. Suppose this claim is true. If we take a random sample of 20 shipments made by the company, what's the probability that at least one of them arrives late?

13. **Refer to Example 4.6.3 Flight Arrivals** An airline reports that 85% of its flights into New York City's LaGuardia Airport arrive on time. To find the probability that its next 4 flights into LaGuardia Airport all arrive on time, can we multiply (0.85)(0.85)(0.85)(0.85)? Why or why not?

14. **Late Shows** Some TV shows begin after their scheduled times when earlier programs run late. According to a network's records, about 3% of its shows start late. To find the probability that 3 consecutive shows on this network start on time, can we multiply (0.97)(0.97)(0.97)? Why or why not?

15. **Newspapers** In a large apartment complex, 20% of residents read *USA Today*, 12% read the *New York Times*, and 3% read both papers. Suppose we select a resident of the apartment complex at random. Are the events A: reads *USA Today* and B: reads *New York Times* independent? Justify your answer.

16. **Pet Ownership** In one large city, 40% of all households own a dog, 32% own a cat, and 18% own both a dog and a cat. Suppose we randomly select a household. Are the events A: owns a cat and B: owns a dog independent? Justify your answer.

Applying the Concepts These exercises ask you to apply multiple learning goals in a new context or to apply what you learned in this section in a new way.

17. **Emergency Calls** Many fire stations handle more emergency calls for medical help than for fires. At one fire station, 81% of incoming calls are for medical help. Suppose we choose 4 incoming calls to the station at random.
    (a) Find the probability that all 4 calls are for medical help.
    (b) What's the probability that at least one of the calls is not for medical help?
    (c) Explain why the calculation in part (a) may not be valid if we choose 4 consecutive calls to the station.

18. **Broken Links** Internet sites often vanish or move, so that references to them can't be followed. In fact, 13% of internet sites referred to in major scientific journals are lost within two years of publication.[46] Suppose we randomly select 7 internet references from scientific journals.
    (a) Find the probability that all 7 references still work two years later.
    (b) What's the probability that at least one of them doesn't work two years later?
    (c) Explain why the calculation in part (a) may not be valid if we choose 7 internet references from one article in the same journal.

Extending the Concepts These exercises challenge you to explore statistical concepts and methods that go beyond what you learned in this section.

19. **Mutually Exclusive vs. Independent** The two-way table summarizes data on the price paid and method of ticket purchase for the passengers on a trolley tour. Imagine choosing a passenger at random. Define event A: paid full price and event B: purchased online.[47]

		Amount paid		
		Full price	Reduced price	Total
Purchase method	Online			10
	In person			40
	Total	20	30	50

(a) Copy and complete the two-way table so that events A and B are mutually exclusive.
(b) Copy and complete the two-way table so that events A and B are independent.

20. **A Perfect Game** Refer to Exercise 6. In the previous 22 seasons, there have been 9 perfect MLB games. In each of these seasons, there were 30 teams, each of which played 162 games. Suppose we randomly select a game from the previous 22 seasons.
    (a) What is the probability of a perfect game?
    (b) How do your answers to part (a) and Exercise 6 compare? What does this imply about the assumption of independence that you made in Exercise 6?

21. **Geometric Probability** You are tossing a pair of fair, 6-sided dice in a board game. Tosses are independent. You land in a danger zone that requires you to roll doubles (both faces showing the same number of spots) before you are allowed to play again.
    (a) What is the probability of rolling doubles on a single toss of the dice?
    (b) What is the probability that you do not roll doubles on the first toss, but you do on the second toss?
    (c) What is the probability that the first two tosses are not doubles and the third toss is doubles? This is the probability that the first doubles occurs on the third toss.
    (d) Do you see the pattern? What is the probability that the first doubles occurs on the kth toss?

    Note: This type of problem, which involves performing repeated, independent trials of the same chance process until a "success" occurs, is known as *geometric probability*.

22. **Dice Rolling** Which is more likely: rolling a fair, 6-sided die 11 times and getting a 1 on each roll *or* rolling the same die 12 times and getting the outcome 154524336126? Justify your answer.

Cumulative Review These exercises revisit what you learned in previous sections.

23. **Desktop or Laptop? (4.5)** A computer company makes desktop and laptop computers at factories in three states — California, Texas, and New York. The California factory produces 40% of the company's computers, the Texas factory makes 25%, and the remaining 35% of computers are manufactured in New York. Of the computers made in California, 75% are laptops. Of those made in Texas and New York, 70% and 50%, respectively, are laptops. All computers are first shipped to a distribution center in Missouri before being sent out to stores. Suppose we select a computer at random from the distribution center.[48]
    (a) Find the probability that the computer is a laptop.
    (b) Given that the computer is a laptop, what is the probability that it was made in Texas?

24. **Clothing Matters (1.5, 1.6, 2.2)** Two young researchers suspect that people are more likely to agree to participate in a survey if the interviewers are dressed up. To test this, the researchers went to the local grocery store on two consecutive Saturday mornings at 10 a.m. On the first Saturday, they wore casual clothing (tank tops and jeans). On the second Saturday, they dressed in button-down shirts and nicer slacks. Each day, they asked every fifth person who walked into the store to participate in a survey. Their response variable was whether or not the person agreed to participate. Here are their results.[49]

		Clothing	
		Casual	Nice
Participation	Yes	14	27
	No	36	23

    (a) Calculate the difference in the proportion of subjects who agreed to participate in the survey in the two groups (Casual − Nice).
    (b) Assume the study design is equivalent to randomly assigning shoppers to the "casual" or "nice" groups. A total of 100 repetitions of a simulation were performed to see what differences in proportions would occur due only to chance variation in this random assignment. Use the results of the simulation in the following dotplot to determine whether the difference in proportions from part (a) is statistically significant. Explain your reasoning.

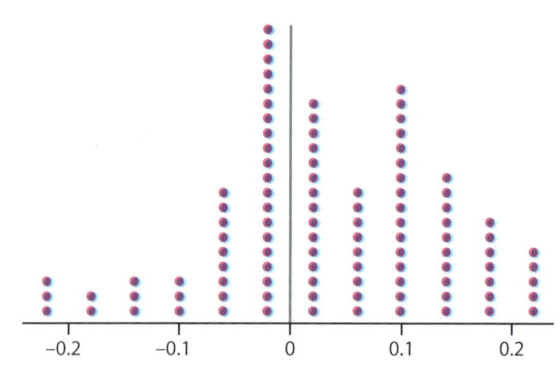

    (c) What flaw in the design of this experiment would prevent the researchers from drawing a cause-and-effect conclusion about the impact of an interviewers' attire on nonresponse in a survey?

Section 4.7 The Multiplication Counting Principle and Permutations

LEARNING GOALS

By the end of this section, you will be able to:
- Use the multiplication counting principle to determine the number of ways to complete a process involving several steps.
- Use factorials to count the number of permutations of a group of individuals.
- Compute the number of permutations of n individuals taken r at a time.

Finding the probability of an event often involves counting the number of possible outcomes of some chance process. In this section, we will show you two techniques for determining the number of ways that a multistep process can happen when the order of the steps matters.

The Multiplication Counting Principle

A college with a large culinary arts program runs a restaurant on campus. Enrolled students and faculty at the college are invited to have one meal per month at the restaurant for free, in exchange for their feedback about the food and service. The restaurant offers a three-course dinner menu. The table shows the options for each course in a recent month. Each diner must choose one appetizer, one main dish, and one dessert.

Appetizer	Main dish	Dessert
Butternut squash soup	Grilled pork chop	Chocolate cake
Green salad	Ribeye steak	Apple and cranberry tart
Caesar salad	Roasted chicken breast	
	Poached salmon	

How many different meals can be ordered? We could try to list all possible orders:

Soup–Pork–Cake, Soup–Pork–Tart, Soup–Steak–Cake, Soup–Steak–Tart, ... However, it might be easier to display all of the options in a diagram, like the one shown here.

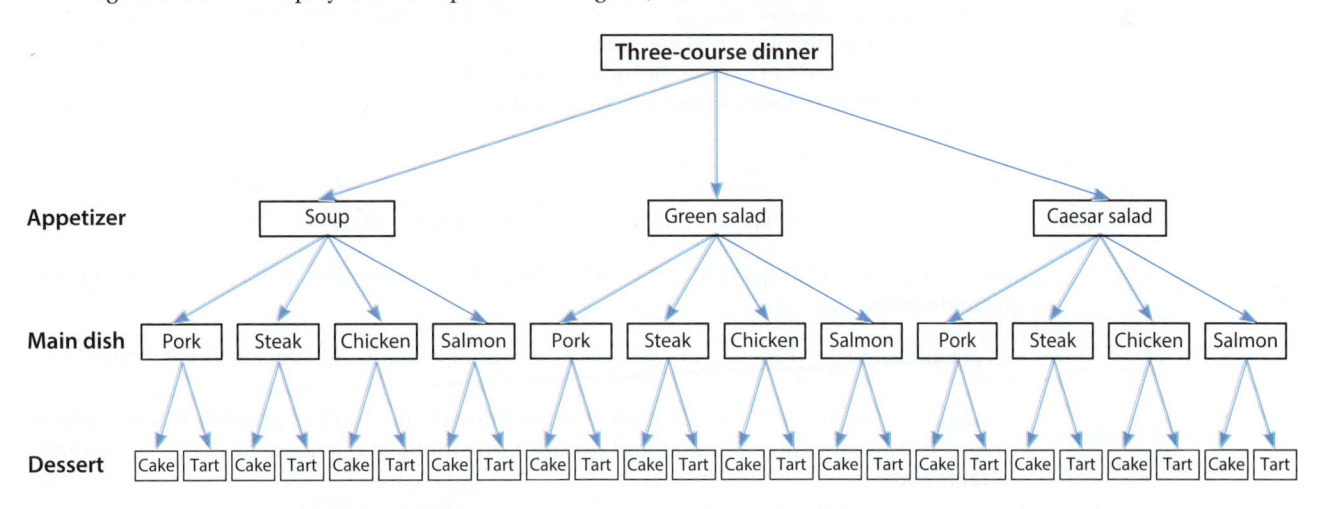

From the diagram, we can see that for each of the three choices of appetizer, there are four choices of main dish, and for each of those main dish choices, there are two dessert choices. So there are

$$\underbrace{3}_{\text{Appetizer}} \cdot \underbrace{4}_{\text{Main dish}} \cdot \underbrace{2}_{\text{Dessert}} = 24$$

different meals that can be ordered from the three-course dinner menu. This is an example of the **multiplication counting principle.**

For a process involving multiple (r) steps, suppose that there are n_1 ways to do Step 1, n_2 ways to do Step 2, ..., and n_r ways to do Step r. The total number of different ways to complete the process is

$$n_1 \cdot n_2 \cdot \ldots \cdot n_r$$

This result is called the **multiplication counting principle**.

EXAMPLE 4.7.1

How many license plates in the Golden State?

The multiplication counting principle

PROBLEM
The standard license plate for California passenger cars has one digit, followed by three letters, and then three more digits. The first digit cannot be a 0. The first and third letters cannot be I, O, or Q. How many possible license plates are there?

SOLUTION
By the multiplication counting principle, there are

9 ·	23 ·	26 ·	23 ·	10 ·	10 ·	10
Digit	Letter	Letter	Letter	Digit	Digit	Digit
Not 0	Not I, O, Q		Not I, O, Q			

There are 26 letters and 10 digits from 0 to 9.

Notice the restrictions: The first digit can't be 0 and the first and third letters can't be I, O, or Q.

= 123,786,000 possible license plates

EXAM PREP: FOR PRACTICE, TRY EXERCISE 7.

Permutations

The multiplication counting principle can also help us determine how many ways there are to arrange a group of people, animals, or things. For example, suppose you have 5 framed photographs of different family members that you want to arrange in a line on top of your dresser. In how many ways can you do this? Let's count the options moving from left to right across the dresser. There are 5 options for the first photo, 4 options for the next photo, and so on. By the multiplication counting principle, there are

$$\underbrace{5}_{\text{Photo \#1}} \cdot \underbrace{4}_{\text{Photo \#2}} \cdot \underbrace{3}_{\text{Photo \#3}} \cdot \underbrace{2}_{\text{Photo \#4}} \cdot \underbrace{1}_{\text{Photo \#5}} = 120$$

different photo arrangements. We call arrangements like this, where the order matters, **permutations.**

A **permutation** is a distinct arrangement of some group of individuals where order matters.

Expressions like $5 \cdot 4 \cdot 3 \cdot 2 \cdot 1$ occur often enough in counting problems that mathematicians invented a special name and notation for them. We write $5 \cdot 4 \cdot 3 \cdot 2 \cdot 1 = 5!$, read as "**5 factorial.**"

For any positive integer n, we define $n!$ (read "n **factorial**") as

$$n! = n(n-1)(n-2) \ldots \cdot 3 \cdot 2 \cdot 1$$

That is, n factorial is the product of the integers starting with n and going down to 1.

EXAMPLE 4.7.2

How many different batting orders?

Permutations and factorials

PROBLEM

The manager of a softball team has picked 9 players to start an upcoming playoff game. How many different ways are there for the manager to arrange these 9 players to make up the team's batting order?

SOLUTION

By the multiplication counting principle, there are

$$\underset{\text{Batter \#1}}{9} \cdot \underset{\text{Batter \#2}}{8} \cdot \underset{\text{Batter \#3}}{7} \cdot \underset{\text{Batter \#4}}{6} \cdot \underset{\text{Batter \#5}}{5} \cdot \underset{\text{Batter \#6}}{4} \cdot \underset{\text{Batter \#7}}{3} \cdot \underset{\text{Batter \#8}}{2} \cdot \underset{\text{Batter \#9}}{1} = 9!$$

= 362,880 possible batting orders

EXAM PREP: FOR PRACTICE, TRY EXERCISE 11.

In the example, note that Louisa batting first (Batter #1) and Winifred batting second (Batter #2) leads to a different set of arrangements than Winifred batting first (Batter #1) and Louisa batting second (Batter #2). This illustrates why we say that "order matters" for permutations.

So far, we have shown how to count the number of distinct arrangements of *all* the individuals in a group of people, animals, or things. Sometimes, we want to determine how many ways there are to select and arrange only *some* of the individuals in a group.

Dr. Cannon likes to get the students in her online statistics class involved in the action. But she doesn't want to play favorites. Each day, Dr. Cannon puts the names of all 28 students in a hat and mixes them up. She then draws out 3 names on camera. The student whose name is picked first gets to monitor the chat for the day. The second student selected is in charge of producing an outline of class notes. The third student picked shares the ebook when requested. In how many different ways can Dr. Cannon fill these three jobs?

By the multiplication counting principle, there are

$$\underset{\text{Job 1}}{28} \cdot \underset{\text{Job 2}}{27} \cdot \underset{\text{Job 3}}{26} = 19{,}656$$

ways for Dr. Cannon to fill the three different jobs. We can describe this result as the number of permutations of 28 people taken 3 at a time. In symbols, we'll write this as $_{28}P_3$.

> **Denoting Permutations:** $_nP_r$
>
> The notation $_nP_r$ represents the number of different permutations of r individuals selected from the entire group of n individuals.

You can use technology to calculate factorials and permutations. See the Tech Corner at the end of the section for details.

EXAMPLE 4.7.3

How many ways can you arrange the batting order?

Finding the number of permutations

PROBLEM

A college softball team has 15 players. How many different ways are there for the team's manager to select and arrange 9 of these players to make up the team's batting order?

(continued)

SOLUTION

There are

$$\underset{\text{Batter \#1}}{15} \cdot \underset{\text{Batter \#2}}{14} \cdot \underset{\text{Batter \#3}}{13} \cdot \underset{\text{Batter \#4}}{12} \cdot \underset{\text{Batter \#5}}{11} \cdot \underset{\text{Batter \#6}}{10} \cdot \underset{\text{Batter \#7}}{9} \cdot \underset{\text{Batter \#8}}{8} \cdot \underset{\text{Batter \#9}}{7}$$

$$= {}_{15}P_9$$

$$= 1{,}816{,}214{,}400 \text{ possible batting orders}$$

EXAM PREP: FOR PRACTICE, TRY EXERCISE 15.

With a little clever math, we can rewrite $_{15}P_9$ as follows:

$$_{15}P_9 = 15 \cdot 14 \cdot 13 \cdot 12 \cdot 11 \cdot 10 \cdot 9 \cdot 8 \cdot 7$$

$$= \frac{15 \cdot 14 \cdot 13 \cdot 12 \cdot 11 \cdot 10 \cdot 9 \cdot 8 \cdot 7 \cdot 6 \cdot 5 \cdot 4 \cdot 3 \cdot 2 \cdot 1}{6 \cdot 5 \cdot 4 \cdot 3 \cdot 2 \cdot 1}$$

$$= \frac{15!}{6!}$$

$$= \frac{15!}{(15-9)!}$$

This method leads to a general formula for $_nP_r$.

HOW TO Calculate Permutations

You can calculate the number of permutations of n individuals taken r at a time (where $r \leq n$) using the multiplication counting principle or with the formula

$$_nP_r = \frac{n!}{(n-r)!}$$

By definition, $0! = 1$.

THINK ABOUT IT

Why do we define 0! = 1? To make the formula for $_nP_r$ work for all possible values of r. Consider $_{28}P_{28}$. This is the number of different arrangements of 28 individuals taken 28 at a time. By the multiplication counting principle, there are $28 \cdot 27 \cdot 26 \cdot \ldots \cdot 3 \cdot 2 \cdot 1 = 28!$ such arrangements. Using the formula for $_nP_r$, we get

$$_{28}P_{28} = \frac{28!}{(28-28)!} = \frac{28!}{0!}$$

If we define $0! = 1$, the values obtained by the two methods will agree.

TECH CORNER Calculating factorials and permutations

You can use an applet or a TI-83/84 to compute the number of distinct permutations of a group of individuals. We'll illustrate this process using the number of different batting orders possible from Examples 4.7.2 and 4.7.3.

Applet

1. Go to www.stapplet.com and launch the *Counting Methods* applet.

2. To evaluate 9! from Example 4.7.2, enter $n = 9$. Then click "Calculate n!."

Counting Methods

n = 9   r =

[Calculate n!] [Calculate nPr] [Calculate nCr]

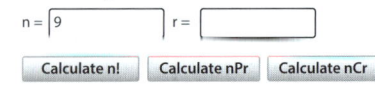

9! = 362880

3. To compute the number of possible batting orders for 9 players on a college softball team with 15 players on the roster from Example 4.7.3, enter $n = 15$ and $r = 9$. Then click "Calculate $_nP_r$."

Counting Methods

$n =$ [15] $r =$ [9]

[Calculate n!] [Calculate nPr] [Calculate nCr]

$_{15}P_9 = 1816214400$

TI-83/84

1. We can use the factorial command to evaluate 9! from Example 4.7.2. Type 9. Then press MATH, arrow to PROB, choose !, and press ENTER.

2. We can use the $_nP_r$ command to compute the number of possible batting orders for 9 players on a college softball team with 15 players on the roster from Example 4.7.3. Type 15. Then press MATH, arrow to PROB, and choose $_nP_r$. Complete the command $_{15}P_9$ (older OS: 15 $_nP_r$ 9) and press ENTER.

```
NORMAL FLOAT AUTO REAL RADIAN MP
9!
                              362880
15P9
                         1816214400
```

Detailed instructions for using CrunchIt!, Excel, Google Sheets, JMP, Minitab, and R are available in Achieve.

Section 4.7 What Did You Learn?

Review the learning goals from this section. Then practice what you've learned by working through the exercises.

Learning Goal	Example	Exercises
Use the multiplication counting principle to determine the number of ways to complete a process involving several steps.	4.7.1	7–10
Use factorials to count the number of permutations of a group of individuals.	4.7.2	11–14
Compute the number of permutations of n individuals taken r at a time.	4.7.3	15–18

Section 4.7 Exercises

Building Concepts and Skills These exercises assess the basic knowledge you should have after reading the section.

1. If you can choose from 2 appetizers and 3 entrees for a dinner, there are $2 \cdot 3 = 6$ possible dinners. This is an example of the _____ principle.

2. Arrangements of individuals where the order matters are called _____.

3. True/False: 3! represents "3 factorial" and is equal to $3 \cdot 2 \cdot 1$.

4. The notation $_nP_r$ stands for the number of permutations of _____ individuals taken _____ at a time.

5. Give the formula for calculating $_nP_r$.

6. True/False: $0! = 0$.

Mastering Concepts and Skills These exercises reinforce the learning goals as shown in the examples.

7. **Refer to Example 4.7.1** **New Jersey Plates** A standard-issue 2020 New Jersey license plate had one letter, then two digits, followed by three letters.

These plates were not allowed to have letters D, T, or X in the fourth position, or the letters I, O, or Q in *any* position. With these restrictions, how many different license plates were possible in 2020?

8. **National Insurance Code** In the United Kingdom, every resident is issued a national insurance "number" for the national health and social security systems. The format of the "number" is two letters, followed by six digits, and then a third letter. The first two letters cannot be D, F, I, Q, U, or V, and the second letter also cannot be O. The last letter is either A, B, C, or D. All

six digits can be 0 through 9, with no restrictions. How many national insurance numbers can be issued with these restrictions?

9. **Three-Letter Call Signs** In 1912, the U.S. government began issuing licenses to radio stations. Each station was given a unique three-letter "call sign." By international agreement, the United States received rights to all call signs beginning with the letters W, N, and K. Radio stations in the western United States were given call signs starting with K. Stations in the East were given call signs starting with W. (N was reserved for use by the U.S. Navy.)

   (a) How many three-letter call signs (like WGO) start with the letter W?

   (b) How many three-letter call signs starting with W or K were available for U.S. radio stations?

10. **Four-Letter Call Signs** Refer to Exercise 9. By 1922, there were more applications for radio station licenses than the number of three-letter call signs available. A radio station in New Orleans applied for and was granted the call sign WAAB.

    (a) How many four-letter call signs starting with W are possible?

    (b) How many four-letter call signs starting with W or K are possible?

11. **Refer to Example 4.7.2 Collecting Donations** Suppose a charity has donation pickups at 6 locations scheduled on a given day. In how many different orders can the charity complete the pickups?

12. **Bookshelf** Shay has 10 science fiction novels on a bedroom bookshelf. How many different ways are there for Shay to arrange the books on the shelf?

13. **Seat Assignments** An adjunct instructor is teaching an evening Introductory Statistics class with 30 students in a room with 30 individual tables. On the first day of class, the instructor places identical slips of paper numbered 1 to 30 in a hat. Each of the 30 students draws a slip from the hat upon entering the classroom to determine their assigned seat. How many possible seating assignments are there?

14. **Line Up** Near the end of an athletic practice, a coach makes the 15 defensive players an offer. They can run sprints, as usual, or they can make every possible arrangement with all 15 players in a single-file line. How many such arrangements are there? Should the players take the offer? Explain your answer.

15. **Refer to Example 4.7.3 Playlist Shuffle** Declan has 100 songs on his playlist. He is going for a run and has time to listen to 8 songs while he runs. Declan decides to use the random shuffle feature to determine which songs will play, and in what order. (This random shuffle feature does not allow any song to be played twice.) How many different lists of 8 songs are possible?

16. **Random Tests** Dr. Vellman has a test bank with 75 multiple-choice questions covering the material in Section 4.7 in this book. The test bank comes with a random question generator that will select and arrange questions to make different versions of a quiz. How many different versions of a 10-question multiple-choice quiz on this lesson could Dr. Vellman make?

17. **Penalty Kicks** A soccer team has 11 players on the field at the end of a scoreless game. According to league rules, the coach must select 5 of the players and designate an order in which they will take penalty kicks. How many different ways are there for the coach to do this?

18. **Padlocks** "Letter lock" padlocks open when a correct sequence of 3 letters is selected on the lock's dial. If the dial has 20 letters on it and letters cannot be repeated, how many different sequences of letters are possible?

Applying the Concepts These exercises ask you to apply multiple learning goals in a new context or to apply what you learned in this section in a new way.

19. **ATM PINs** Many banks require customers who use the automated teller machine (ATM) to enter a four-digit personal identification number (PIN) before they begin a transaction. PINs can include repeated digits.

    (a) How many possible four-digit PINs are there?

    (b) How many four-digit PINs contain no 3s?

20. **Random Music** The pentatonic musical scale contains 5 notes in an octave: C, D, E, G, and A. Noriko decides to look for new musical themes by playing random sequences of 4 notes from the pentatonic scale. The same note can be played more than once in a sequence.

    (a) How many possible four-note sequences can be played?

    (b) How many possible four-note sequences contain no Gs?

21. **We All Scream** The local ice cream shop in Dontrelle's town is called 21 Choices. Why? Because it offers 21 different flavors of ice cream. Dontrelle likes all but three of the flavors that 21 Choices offers: bubble gum, butter pecan, and pistachio.

    (a) A 21 Choices "basic sundae" comes in three sizes—small, medium, or large—and includes one flavor of ice cream and one of 12 toppings. Dontrelle has enough money for a small or medium basic sundae. How many different sundaes could Dontrelle order that include only flavors that he likes?

    (b) Dontrelle could order a cone with three scoops of ice cream instead of a sundae. For variety, Dontrelle prefers to have three different flavors. Dontrelle considers the order of the flavors on the cone to be important. How many three-scoop cones with three different flavors that Dontrelle likes are possible at 21 Choices?

22. **Yahtzee!** Suppose you roll five fair, 6-sided dice.

    (a) How many different possible outcomes are there?

    (b) In how many outcomes do all five dice show the same number (a Yahtzee)?

Extending the Concepts These exercises challenge you to explore statistical concepts and methods that go beyond what you learned in this section.

23. **ATM PINs** Refer to Exercise 19. How many four-digit PINs contain at least two digits that are the same?

24. **Yahtzee!** Refer to Exercise 22.
    (a) Find the probability of getting a Yahtzee on one roll of five fair, 6-sided dice.
    (b) What is the probability that all five dice show a different number?

Cumulative Review These exercises revisit what you learned in previous sections.

25. **Bagel Nutrition (2.6, 3.5, 3.6)** The scatterplot shows the relationship between calories and total carbohydrates (in grams) for 19 varieties of bagels sold by a national chain of bagel stores.[50]

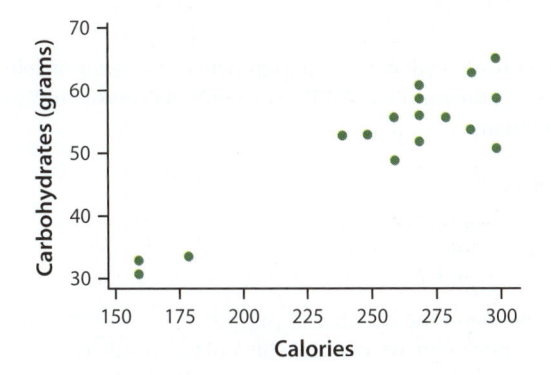

    (a) The correlation for these data is $r = 0.915$. Interpret this value.
    (b) How would the correlation be affected if carbohydrates were measured in ounces instead of grams? (1 gram = 0.035 ounce.)
    (c) The three data points in the lower left represent three special low-calorie bagels. How would the correlation be affected if these three bagels were removed from the data set?

26. **Workplace Injuries (4.2, 4.4)** Pick a nonfatal workplace injury in private industry at random. The table gives the probability model for the day of the week that the injury took place.[51]

Day	Sunday	Monday	Tuesday	Wednesday
Probability	0.06	0.18	0.17	0.17
Day	Thursday	Friday	Saturday	
Probability	x	0.15	0.09	

(a) Find the probability that the randomly chosen injury took place on a Thursday.
(b) Find the probability that the injury took place on a weekday (Monday through Friday).
(c) Given that an injury took place on a weekday, what is the probability that it took place on a Monday?

Section 4.8 Combinations and Probability

LEARNING GOALS
By the end of this section, you will be able to:
- Compute the number of combinations of n individuals taken r at a time.
- Use combinations to calculate probabilities.
- Use the multiplication counting principle and combinations to calculate probabilities.

Recall from Section 4.7 that a permutation is a distinct arrangement of some group of individuals where order matters. Sometimes, we're just interested in finding how many ways there are to choose some number of individuals from a group, but we don't care about the order in which the individuals are selected. For instance, how many ways are there to randomly select 6 winning numbers from 1 to 49 in a lottery drawing? This section focuses on counting the number of possible outcomes when order doesn't matter.

Combinations

Dr. Cannon gives regular, written feedback about homework solutions to students in her online statistics class. Each day, she randomly selects 3 of the 28 students on the class roster. Dr. Cannon then does a digital critique of those students' most recent homework assignments through the online learning management system. In how many different ways can Dr. Cannon choose 3 students from the class of 28? It's tempting to say that there are

$$28 \cdot 27 \cdot 26 = 19{,}656$$

ways for her to do this. That's not correct, however.

Suppose Dr. Cannon randomly selects Lucretia, Tim, and Kiran — in that order. That's really no different from getting Tim, then Lucretia, then Kiran. Or Kiran, then Lucretia, then Tim. How many selections consist of these same 3 students? We can list all the possibilities:

Lucretia-Tim-Kiran   Lucretia-Kiran-Tim   Tim-Lucretia-Kiran
Tim-Kiran-Lucretia   Kiran-Lucretia-Tim   Kiran-Tim-Lucretia

Or we could use the multiplication counting principle: There are 3 possibilities for the first pick, 2 options for the second pick, and only 1 option for the last pick. So there are

$$3 \cdot 2 \cdot 1 = 3! = 6$$

arrangements that consist of these 3 students. This same argument applies for any 3 students whom Dr. Cannon selects.

The order in which Dr. Cannon chooses the 3 students doesn't matter. To avoid counting the same group of 3 students 6 times, we have to divide our original (wrong) answer by $3! = 6$:

$$\text{number of ways to choose 3 homework papers out of 28} = \frac{28 \cdot 27 \cdot 26}{3 \cdot 2 \cdot 1} = 3276$$

This result gives the number of **combinations** of 28 students taken 3 at a time, which we'll write as $_{28}C_3$. Some people prefer to write this as $\binom{28}{3}$ instead.

A **combination** is a set of individuals chosen from some group in which the order of selection doesn't matter. The notation $_nC_r$ represents the number of different combinations of r individuals chosen from the entire group of n individuals.

We can rewrite the preceding answer as

$$_{28}C_3 = \frac{28 \cdot 27 \cdot 26}{3 \cdot 2 \cdot 1} = \frac{_{28}P_3}{3!}$$

which shows an important connection between the number of permutations and the number of combinations in this setting. With a little fancy math, we can also think of this result as

$$_{28}C_3 = \frac{_{28}P_3}{3!} = \frac{\left(\frac{28!}{(28-3)!}\right)}{3!} = \frac{\left(\frac{28!}{25!}\right)}{3!} = \frac{28!}{3!25!} = \frac{28!}{3!(28-3)!}$$

HOW TO Calculate Combinations

You can calculate the number of combinations of n individuals taken r at a time (where $r \leq n$) using the multiplication counting principle, or with either the formula

$$_nC_r = \frac{_nP_r}{r!}$$

or the formula

$$_nC_r = \frac{n!}{r!(n-r)!}$$

EXAMPLE 4.8.1

Which six to pick?

Combinations

PROBLEM

In the New Jersey "Pick Six" lotto game, a player chooses 6 different numbers from 1 to 49. The 6 winning numbers for the lottery are selected at random. If the player matches all 6 numbers, they win the jackpot, which starts at $2 million. How many different possible sets of winning numbers are there for the New Jersey Pick Six lotto game?

SOLUTION

There are

$$_{49}C_6 = \frac{_{49}P_6}{6!} = \frac{49 \cdot 48 \cdot 47 \cdot 46 \cdot 45 \cdot 44}{6 \cdot 5 \cdot 4 \cdot 3 \cdot 2 \cdot 1} = 13{,}983{,}816$$

different sets of winning numbers for this game.

> All that matters is which 6 numbers are picked—the order of selection doesn't matter. That calls for combinations!
>
> $$_nC_r = \frac{_nP_r}{r!}$$

EXAM PREP: FOR PRACTICE, TRY EXERCISE 5.

You can also calculate the solution to Example 4.8.1 using the second formula in the "How To" box:

$$_{49}C_6 = \frac{49!}{6!(49-6)!} = \frac{49!}{6!\,43!} = \frac{49 \cdot 48 \cdot 47 \cdot 46 \cdot 45 \cdot 44 \cdot 43!}{6!\,43!} = \frac{49 \cdot 48 \cdot 47 \cdot 46 \cdot 45 \cdot 44}{6 \cdot 5 \cdot 4 \cdot 3 \cdot 2 \cdot 1}$$
$$= 13{,}983{,}816$$

It is easier to use technology to calculate combinations. See the Tech Corner at the end of the section for details.

Counting and Probability

The focus of this chapter is probability. Recall that when a chance process results in equally likely outcomes, the probability that event A occurs is

$$P(A) = \frac{\text{number of outcomes in event A}}{\text{total number of outcomes in sample space}}$$

You can use the multiplication counting principle and what you have learned about permutations and combinations to help count the number of outcomes.

Consider New Jersey's Pick Six lotto game from Example 4.8.1. What's the probability that a player wins the jackpot by matching all 6 winning numbers? Because the winning numbers are randomly selected, any set of 6 numbers from 1 to 49 is equally likely to be chosen. So we can use the formula just given to calculate

$$P(\text{win the jackpot}) = \frac{\text{number of ways to choose all 6 winning numbers}}{\text{total number of ways to choose 6 numbers from 1 to 49}}$$
$$= \frac{_6C_6}{_{49}C_6}$$
$$= \frac{1}{13{,}983{,}816}$$
$$= 0.0000000715$$

Your calculator may give the probability in scientific notation as 7.15E-08, which is 7.15×10^{-8}. However you write it, the player's chance of winning the jackpot is very small!

EXAMPLE 4.8.2

Randomly selecting students

From counting to probability

PROBLEM

There are 28 students in Dr. Cannon's online statistics class—16 first-year students and 12 second-year students. Suppose Dr. Cannon randomly selects 3 students to provide with written feedback about their most recent homework assignments. Use combinations to find the probability that all 3 students selected are second-years.

(continued)

SOLUTION

P(3 second-years selected)

$$= \frac{\text{\# of ways to choose 3 second-years out of 12}}{\text{Total \# of ways to choose 3 students from the class}}$$

$$= \frac{{}_{12}C_3}{{}_{28}C_3}$$

$$= \frac{\frac{12 \cdot 11 \cdot 10}{3 \cdot 2 \cdot 1}}{\frac{28 \cdot 27 \cdot 26}{3 \cdot 2 \cdot 1}}$$

$$= \frac{220}{3276}$$

$$= 0.067$$

> Because each subgroup of 3 students in the class is equally likely to be chosen, you can use the formula
>
> $$P(A) = \frac{\text{number of outcomes in event A}}{\text{total number of outcomes in sample space}}$$
>
> where event A is that 3 second-year students are chosen.

EXAM PREP: FOR PRACTICE, TRY EXERCISE 9.

We could also have used the general multiplication rule from Section 4.5 to find the desired probability in the example:

P(3 second-years selected)

$= P$(1st student a second-year *and* 2nd student a second-year *and* 3rd student a second-year)

$= P$(1st student a second-year) $\cdot P$(2nd student a second-year | 1st student a second-year) $\cdot$

P(3rd student a second-year | 1st two students are second-years)

$$= \frac{12}{28} \cdot \frac{11}{27} \cdot \frac{10}{26}$$

$$= 0.067$$

Note that these two different methods give the same result.

While it is sometimes possible to use another method to solve a probability question, we recommend using permutations when order matters and combinations when order doesn't matter.

EXAMPLE 4.8.3

Did airline managers conduct an unfair lottery?

Finding probabilities with combinations

PROBLEM

Let's return to the *Hiring Discrimination?* activity in Section 1.1. An airline has just finished training 25 junior pilots — 15 male and 10 female — to become captains. Unfortunately, only 8 captain positions are available. Airline managers announce that they will use a lottery process to determine which pilots will fill the available positions. They write the names of all 25 pilots on identical slips of paper, place them in a hat, mix them well, and draw them out one at a time until all 8 captains are selected.

A day later, managers announce the results of the lottery. Of the 8 captains chosen, 5 are female and 3 are male. Some of the male pilots who were not selected suspect that the lottery was carried out unfairly.[52]

(a) How many possible groups of 8 captains can be selected?

(b) Find the number of ways in which a fair lottery can result in 5 female and 3 male pilots being selected.

(c) Find the probability that a fair lottery would result in the selection of 5 female and 3 male pilots.

(d) The probability that a fair lottery would result in the selection of 5 or more female pilots is 0.1201. Is there convincing evidence that the lottery was carried out unfairly? Explain your reasoning.

SOLUTION

(a) $_{25}C_8 = \dfrac{25!}{8!17!} = 1{,}081{,}575$

> Each of these 1,081,575 possible sets of 8 pilots is equally likely to be selected.

(b) $_{10}C_5 \cdot {_{15}C_3}$

$= \left(\dfrac{10 \cdot 9 \cdot 8 \cdot 7 \cdot 6}{5 \cdot 4 \cdot 3 \cdot 2 \cdot 1} \right) \cdot \left(\dfrac{15 \cdot 14 \cdot 13}{3 \cdot 2 \cdot 1} \right)$

$= (252)(455)$

$= 114{,}660$

> The number of ways to select 5 of the 10 female pilots is $_{10}C_5$. The number of ways to select 3 of the 15 male pilots is $_{15}C_3$. Now use the multiplication counting principle to find the number of ways in which the lottery yields 5 female pilots and 3 male pilots.

(c) $P(\text{5 female and 3 male pilots selected in fair lottery})$

$= \dfrac{\text{number of ways to get 5 female and 3 male pilots}}{\text{total number of ways to select 8 pilots}}$

$= \dfrac{{_{10}C_5} \cdot {_{15}C_3}}{{_{25}C_8}}$

$= \dfrac{114{,}660}{1{,}081{,}575}$

$= 0.106$

(d) Because it is somewhat likely to select 5 or more female pilots by chance alone, there is not convincing evidence that the lottery was carried out unfairly.

> Recall from Section 1.6 that researchers often identify results that occur at least 5% of the time by chance alone as "not statistically significant."

EXAM PREP: FOR PRACTICE, TRY EXERCISE 13.

Part (c) of Example 4.8.3 involves finding the probability of randomly selecting a fixed number of individuals from each of two subgroups in the population — in this case, female and male pilots. The solution involves calculating two combinations of the form $_nC_r$ in the numerator (corresponding to the female and male subgroups) and one combination of the form $_nC_r$ in the denominator (corresponding to the population of 25 pilots who finished training):

$$P(\text{5 female and 3 male pilots selected in fair lottery}) = \dfrac{{_{10}C_5} \cdot {_{15}C_3}}{{_{25}C_8}}$$

Notice two things:

- The sum of the n's in the numerator equals the n in the denominator: $10+15=25$.
- The sum of the r's in the numerator equals the r in the denominator: $5+3=8$.

We could have used the same approach in Example 4.8.2, which focused on Dr. Cannon's class with 16 first-year students and 12 second-year students. The probability that she randomly selects 3 second-year students *and* 0 first-year students to provide feedback about their latest homework assignments is

$$P(\text{3 second-years and 0 first-years selected}) = \dfrac{{_{12}C_3} \cdot {_{16}C_0}}{{_{28}C_3}} = \dfrac{{_{12}C_3} \cdot 1}{{_{28}C_3}} = \dfrac{{_{12}C_3}}{{_{28}C_3}}$$

THINK ABOUT IT

How did we calculate the probability of selecting 5 *or more* female pilots in part (d) of Example 4.8.3? We used combinations and the addition rule for mutually exclusive events.

$P(\text{5 or more female}) = P(\text{5 female and 3 male}) + P(\text{6 female and 2 male})$
$\qquad\qquad\qquad\qquad + P(\text{7 female and 1 male}) + P(\text{8 female and 0 male})$

$= \dfrac{{_{10}C_5} \cdot {_{15}C_3}}{{_{25}C_8}} + \dfrac{{_{10}C_6} \cdot {_{15}C_2}}{{_{25}C_8}} + \dfrac{{_{10}C_7} \cdot {_{15}C_1}}{{_{25}C_8}} + \dfrac{{_{10}C_8} \cdot {_{15}C_0}}{{_{25}C_8}}$

$= 0.10601 + 0.02039 + 0.00166 + 0.00004$

$= 0.1281$

TECH CORNER: Calculating combinations

You can use an applet or a TI-83/84 to compute the number of combinations of n individuals taken r at a time. We'll illustrate this process using the data from Example 4.8.2, describing the number of ways in which Dr. Cannon can randomly select 3 of her 28 students to provide homework assignment feedback.

Applet

1. Go to www.stapplet.com and launch the *Counting Methods* applet.
2. Enter $n = 28$ and $r = 3$. Then click "Calculate nCr."

   Counting Methods

   n = 28   r = 3

   [Calculate n!] [Calculate nPr] [Calculate nCr]

   $_{28}C_3 = 3276$

TI 83/84

1. Type 28. Then press MATH, arrow to PROB, and choose $_nC_r$.
2. Complete the command $_{28}C_3$ (Older OS: 28 $_nC_r$ 3) and press ENTER.

   ```
   NORMAL FLOAT AUTO REAL RADIAN MP
   
   28C3
                            3276
   ```

Detailed instructions for using CrunchIt!, Excel, Google Sheets, JMP, Minitab, and R are available in Achieve.

Section 4.8 What Did You Learn?

Review the learning goals from this section. Then practice what you've learned by working through the exercises.

Learning Goal	Example	Exercises
Compute the number of combinations of n individuals taken r at a time.	4.8.1	5–8
Use combinations to calculate probabilities.	4.8.2	9–12
Use the multiplication counting principle and combinations to calculate probabilities.	4.8.3	13–16

Section 4.8 Exercises

Building Concepts and Skills These exercises assess the basic knowledge you should have after reading the section.

1. True/False: With permutations, order matters. With combinations, order doesn't matter.

2. The notation $_nC_r$ represents the number of combinations of _____ individuals taken _____ at a time.

3. One way to compute $_nC_r$ is to first calculate $_nP_r$ and then divide the result by _____.

4. Another formula for computing combinations is _____.

Mastering Concepts and Skills These exercises reinforce the learning goals as shown in the examples.

5. **Refer to Example 4.8.1 Papa's Pizza** On Monday nights, Papa's Pizza offers a special 50% off deal on its medium two-topping pizzas. All pizzas come with sauce and cheese. The customer can choose any 2 of Papa's 25 toppings to complete the order. How many different possible pizzas are there, assuming that the customer must choose exactly 2 different toppings?

6. **Two Scoops** The local ice cream shop in Dontrelle's town is called 21 Choices. Why? Because it offers 21 different flavors of ice cream. Dontrelle's sister Emogene wants to order a small cup with 2 scoops of different flavors. In how many ways can she do this? (Note that the order in which the flavors are placed in the cup doesn't matter to Emogene.)

7. **Five-Card Hands** A standard deck has 52 unique playing cards. Suppose you shuffle the deck well and deal out 5 cards. How many different possible sets of 5 cards are there?

8. **Random Raffle** At the end of a weeklong seminar, the presenter decides to give away signed copies of their book to 4 randomly selected people in the audience.

How many different ways can this be done if 30 people are present at the seminar?

9. **Refer to Example 4.8.2 Good Batteries** Suppose that you have a drawer containing 8 AAA batteries, but only 6 of them are good. You have to choose 4 for your graphing calculator. If you randomly select 4 batteries, what is the probability that all 4 of them will work? Use combinations to help answer this question.

10. **Scrabble** In the game of Scrabble, each player begins by drawing 7 letter tiles from a bag initially containing 100 tiles. There are 42 vowels, 56 consonants, and 2 blank tiles in the bag. Suppose you are playing Scrabble and get to go first. If you randomly select 7 tiles from the bag, what's the probability that all of them are vowels? Use combinations to help answer this question.

11. **Parking Tickets** At a local high school, 95 students have permission to park on campus. Each month, the student council holds a "golden ticket parking lottery." The 3 lucky winners are given reserved parking spots next to the main entrance. Last month, the winning tickets were drawn by a student council member who is in Dr. Wilder's statistics class. When all 3 golden tickets went to members of that class, some people thought the lottery had been rigged. There are 30 students in the statistics class, all of whom are eligible to park on campus.

    (a) Use combinations to find the probability that a fair lottery would result in all 3 golden tickets going to members of Dr. Wilder's statistics class.

    (b) Based on your answer to part (a), do you think the lottery was carried out fairly? Explain your reasoning.

12. **Airport Security** At a certain airport, security officials claim that they randomly select passengers for extra screening at the gate before boarding some flights. One such flight had 76 passengers — 12 in first class and 64 in economy class. Some passengers were surprised when none of the 10 passengers chosen for screening was seated in first class.

    (a) Use combinations to find the probability that all 10 passengers chosen at random for security screening are in economy class.

    (b) Based on your answer to part (a), should passengers be suspicious about this result? Explain your reasoning.

13. **Refer to Example 4.8.3 You're Fired!** Recently, a company with 30 employees had to fire 12 employees. The company claimed that it selected the employees who were fired at random. Of the 12 employees fired, 6 were aged 60 or older (50%). However, only 10 of the company's 30 employees were aged 60 or older (33%).

    (a) How many possible groups of 12 employees can be selected to be fired?

    (b) Find the number of ways in which a random selection of 12 employees from the company would result in 6 people aged 60 or older and 6 people younger than age 60 being selected.

    (c) Find the probability that a random selection of 12 employees from the company would result in the selection of 6 people aged 60 or older and 6 people younger than age 60.

    (d) The probability that a random selection of 12 employees from the company would result in at least 6 people aged 60 or older being selected is 0.118. Is this convincing evidence that the company did not select the employees who were fired at random? Explain your reasoning.

14. **Soccer Teams** At a university's annual picnic, 22 students in the mathematics/statistics department decide to play a soccer game. Sixteen of the 22 students are math majors and 6 are statistics majors. To divide into two teams of 11, one of the professors put all the players' names into a hat and drew out 11 players to form team Bayes, with the remaining 11 players forming team Newton. The players were surprised when 5 of the statistics majors ended up on team Bayes.

    (a) How many possible groups of 11 students can be selected for team Bayes?

    (b) Find the number of ways in which a random selection of 11 students for team Bayes would result in 5 statistics majors and 6 math majors being selected.

    (c) Find the probability that a random selection of 11 students for team Bayes would result in 5 statistics majors and 6 math majors being selected.

    (d) The probability that a random selection of 11 students for team Bayes and 11 students for team Newton would result in at least 5 statistics majors being on the same team is 0.1486. Is there convincing evidence that the professor didn't mix the names well before drawing them out of the hat? Explain your reasoning.

15. **Random Assignment** Researchers recruited 20 volunteers with elevated cholesterol levels — 8 high and 12 borderline high — to take part in an experiment involving a new drug and a placebo. They randomly assigned the subjects into two groups of 10 people each. To their surprise, 6 of the 8 subjects with high cholesterol levels were randomly assigned to the new drug treatment.

    (a) Find the probability that the random assignment would put 6 subjects with high cholesterol levels and 4 subjects with borderline high cholesterol levels in the new drug treatment group purely by chance.

    (b) The probability that the random assignment would put at least 6 subjects with high cholesterol levels in one treatment group is 0.1698. Should the researchers be surprised by the results of the random assignment? Explain your reasoning.

16. **Three of a Kind** A standard deck has 52 playing cards — 4 cards each of 13 different denominations (ace, two, three, . . . , ten, jack, queen, and king). You and a friend decide to play a game in which you each start with 5 cards. Suppose you shuffle the deck well and deal out 5 cards each to you and your friend. You are surprised when you look at your hand and see 3 aces (along with 2 non-aces).

    (a) Find the probability of dealing yourself 3 aces and 2 non-aces in a 5-card hand from a well-shuffled deck.

(b) The probability of dealing 3 or more aces to either player in a 5-card hand is 0.0035. Should you be surprised by your hand? Explain your reasoning.

Applying the Concepts These exercises ask you to apply multiple learning goals in a new context or to apply what you learned in this section in a new way.

17. **Random Playlist** Janine wants to set up a playlist with 8 songs. She has 50 songs to choose from, including 15 songs by Megan Thee Stallion. Janine's device won't allow any song to appear more than once in a playlist.
    (a) How many different sets of 8 songs are possible for Janine's playlist? Assume that the order of the songs doesn't matter.
    (b) Janine would like to have 2 songs by Megan Thee Stallion and 6 songs by other artists on her playlist. How many different sets of 8 songs for Janine's playlist meet this requirement?
    (c) Suppose Janine lets her device select an 8-song playlist at random. Find the probability that exactly 2 of the songs on the playlist are by Megan Thee Stallion.

18. **Starting Lineup** A certain basketball team has 12 players. Five of these players are classified as guards and 7 of the players are classified as forwards/centers.
    (a) If the coach wanted to choose 5 starters at random by drawing names from a hat, how many possible groups of 5 starters are there?
    (b) To avoid an unbalanced lineup, the coach wants to choose 2 guards and 3 forwards/centers. To do this, the coach places the names of the 5 guards in one hat and the names of the 7 forwards/centers in a second hat. Then, the coach will randomly select 2 guards from the first hat and 3 forwards/centers from the second hat. How many different lineups are possible?
    (c) If the coach chooses 5 starters at random by drawing names from a single hat, find the probability that the selection results in a balanced lineup with 2 guards and 3 forwards/centers.

Exercises 19 and 20 refer to the following setting. The game of Keno is played with 80 balls numbered 1 through 80. During each play of the game, the casino selects 20 of the balls at random. The player picks between 1 and 20 numbers from 1 to 80. For the player to win, all or almost all of the numbers they picked must be among those selected by the casino.

19. **Pick-5 Keno** Suppose that a player picks 5 numbers between 1 and 80.
    (a) Find the probability that all 5 numbers the player picked are among the 20 balls selected at random by the casino. (In this case, the player wins $500 for each $1 bet.)
    (b) The player wins money if they match at least 3 of the numbers selected by the casino. Find the probability that this happens.

20. **Pick-8 Keno** Suppose that a player picks 8 numbers between 1 and 80.
    (a) Find the probability that all 8 numbers the player picked are among the 20 balls selected at random by the casino. (In this case, the player wins $15,000 for each $1 bet.)
    (b) The player wins money if they match at least 4 of the numbers selected by the casino. Find the probability that this happens.

Extending the Concepts These exercises challenge you to explore statistical concepts and methods that go beyond what you learned in this section.

21. **Papa's Pizza** Refer to Exercise 5. On Tuesday nights, Papa's Pizza allows a customer to choose from 0 to 8 toppings for the same price on its medium pizzas. How many different pizzas are possible? (*Note:* A topping cannot be chosen more than once, like double pepperoni.)

22. **Name That Lock** Picture one of those round locks with the numbers from 0 to 39 marked on a circular dial. It is common to refer to these as "combination locks." Explain why a better name might be "permutation locks."

23. **Sampling Beads** A bag contains 20 red beads, 7 white beads, and 5 blue beads. Suppose we choose 6 beads from the bag at random. What's the probability that 3 red beads, 2 white beads, and 1 blue bead are selected?

Cumulative Review These exercises revisit what you learned in previous sections.

24. **Home Runs (3.3)** The dotplot shows the number of home runs hit by each of the 30 MLB teams in a recent year.[53] Construct a boxplot for this distribution. Be sure to identify any outliers.

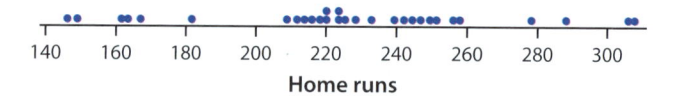

25. **Work Choices (4.4)** The University of Chicago's General Social Survey asked a representative sample of adults this question: "Which of the following statements best describes how your daily work is organized? (1) I am free to decide how my daily work is organized. (2) I can decide how my daily work is organized, within certain limits. (3) I am not free to decide how my daily work is organized." Here is a two-way table of the responses for three levels of education.[54]

		Education level		
		Less than high school	High school	Bachelor's
	1	31	161	81
Response	2	49	269	85
	3	47	112	14

(a) Choose an individual from the sample at random. If the individual's education level was high school, what is the probability that their response was "I am free to decide how my daily work is organized"?
(b) Show that the events "response 3" and "less than high school" are not independent.

Statistics Matters Should an athlete who fails a drug test be suspended?

The chapter-opening Statistics Matters described the NCAA's drug-testing program for student athletes. Suppose that 16% of athletes in a certain NCAA sport have taken a banned substance. Further suppose that the drug test used by the independent testing agency has a false positive rate of 5% and a false negative rate of 10%. Use what you have learned in this chapter to help answer the following questions.

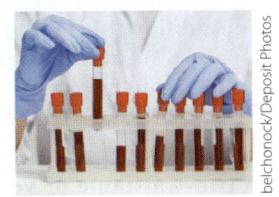

1. What is the probability that a randomly chosen athlete in this sport tests positive for a banned substance?

2. If two athletes in this sport are randomly selected, what is the probability that at least one of them tests positive?

3. If a randomly chosen athlete in this sport tests positive, what is the probability that the student did not take a banned substance? Based on your answer, do you think that an athlete who tests positive should be immediately declared ineligible from athletic competition for a year? Why or why not?

4. If a randomly chosen athlete in this sport tests negative, what is the probability that the student took a banned substance? Explain why it makes sense for the drug-testing process to be designed so that this probability is less than the one you found in Question 3.

5. The NCAA testing protocol involves splitting the initial urine specimen into two vials, one labeled sample A and the other labeled sample B. If an athlete's sample A tests positive, the athlete is notified and given the option to have someone present while sample B is tested. Find the probability that a randomly chosen athlete in this sport who gets a positive test result for both sample A and sample B took a banned substance. Assume that the false positive and false negative rates are the same for both samples. Based on your answer, do you think that an athlete who gets a positive test result for both samples should be suspended from athletic competition for a year? Why or why not?

Chapter 4 Review

Randomness, Probability, and Simulation

- Chance behavior is unpredictable in the short run but has a regular and predictable pattern in the long run.
- The long-run relative frequency of a chance outcome is its **probability**. A probability is a number between 0 (never occurs) and 1 (always occurs).
- The **law of large numbers** says that in many repetitions of the same chance process, the proportion of times that a particular outcome occurs will approach its probability.
- **Simulation** can be used to imitate chance behavior and to estimate probabilities. To perform a simulation:
  - Describe how to use a chance process to perform one repetition of the simulation. Tell what you will record at the end of each repetition.
  - Perform many repetitions.
  - Use the results of your simulation to answer the question of interest.

Probability Models and Probability Rules

- A **probability model** describes a chance process by listing all possible outcomes in the **sample space** and giving the probability of each outcome. A valid probability model requires that all possible outcomes have probabilities that add up to 1.
- An **event** is a collection of possible outcomes from the sample space.
- To find the probability that an event occurs, we can use some basic rules:
  - If all outcomes in the sample space are equally likely,
  $$P(A) = \frac{\text{number of outcomes in event A}}{\text{total number of outcomes in sample space}}$$
  - **Complement rule:** $P(A^C) = 1 - P(A)$, where event A^C is the **complement** of event A — that is, the event that A does not happen.

- **General addition rule:** For any two events A and B,

$$P(A \text{ or } B) = P(A) + P(B) - P(A \text{ and } B)$$

- **Addition rule for mutually exclusive events:** Events A and B are **mutually exclusive** if they have no outcomes in common and so can never occur together—that is, if $P(A \text{ and } B) = 0$. If A and B are mutually exclusive, $P(A \text{ or } B) = P(A) + P(B)$.

- A **two-way table** or **Venn diagram** can be used to display the sample space and to help find probabilities for a chance process involving two events.

- The event "A or B" is known as the **union** of A and B, denoted by $A \cup B$. It consists of all outcomes in event A, event B, or both.

- The event "A and B" is known as the **intersection** of A and B, denoted by $A \cap B$. It consists of all outcomes that are common to both events.

Conditional Probability, Multiplication Rules, and Independence

- A **conditional probability** describes the probability that one event happens given that another event is already known to have happened. To calculate the conditional probability $P(A|B)$ that event A occurs given that event B has occurred, use the formula

$$P(A|B) = \frac{P(A \text{ and } B)}{P(B)} = \frac{P(\text{both events occur})}{P(\text{given event occurs})}$$

- Use the **general multiplication rule** to calculate the probability that events A and B both occur:

$$P(A \text{ and } B) = P(A) \cdot P(B|A)$$

- When a chance process involves multiple stages, a **tree diagram** can be used to display the sample space and to help answer questions involving conditional probability.

- When knowing whether or not one event has occurred does not change the probability that another event happens, we say that the two events are **independent.** For independent events A and B,

$$P(A|B) = P(A|B^C) = P(A) \text{ and}$$
$$P(B|A) = P(B|A^C) = P(B)$$

- In the special case of independent events, the multiplication rule becomes

$$P(A \text{ and } B) = P(A) \cdot P(B)$$

Counting and Probability

- When a chance process involves several steps, the **multiplication counting principle** can be used to determine the total number of possible outcomes. For a process involving r steps, if there are n_1 ways to do Step 1, n_2 ways to do Step 2, . . . , and n_r ways to do Step r, the total number of different ways to complete the process is $n_1 \cdot n_2 \cdot \ldots \cdot n_r$.

- The multiplication counting principle also helps us count the number of distinct arrangements of some group of individuals. We call such arrangements, where the order of selection matters, **permutations.**

- The number of permutations of an entire group of n individuals is n **factorial:**

$$n! = n(n-1)(n-2) \ldots \cdot 3 \cdot 2 \cdot 1$$

- The number of permutations of r individuals selected from a group of n individuals (where $r \leq n$) is denoted by $_nP_r$ and can be computed using the multiplication counting principle, with technology, or with the formula

$$_nP_r = \frac{n!}{(n-r)!}$$

- A **combination** is a set of individuals chosen from some group in which the order of selection doesn't matter. If there are n individuals, $_nC_r$ represents the number of different combinations of r individuals chosen from the entire group of n individuals. You can calculate $_nC_r$ using the multiplication counting principle, with technology, or with the formula

$$_nC_r = \frac{_nP_r}{r!} = \frac{n!}{r!(n-r)!}$$

- The multiplication counting principle, permutations, and combinations are useful tools in helping to calculate probabilities.

Chapter 4 Review Exercises

These exercises will help you review important concepts and skills described by the learning goals in each section. For most exercises, the relevant section is noted in parentheses after the exercise title.

1. **Checked Bags (4.1)** An airline claims that there is a 0.25 probability that a randomly selected solo traveler will check baggage. An agent at the check-in counter records whether each solo traveler checks baggage.

    (a) Explain what probability 0.25 means in this setting.

    The agent randomly selected 12 solo travelers and found that 7 of them checked baggage for their flight. Does this outcome provide convincing evidence that more than 25% of solo travelers check baggage? To help answer this question, we want to perform a simulation to estimate the probability that 7 or more out of 12 randomly selected solo travelers will check baggage if the airline's claim is true.

    (b) Describe how to use a random number generator to perform one repetition of the simulation.

The dotplot shows the number of solo travelers who check baggage in each of 100 simulated random samples of size $n = 12$, assuming that the airline's claim is true.

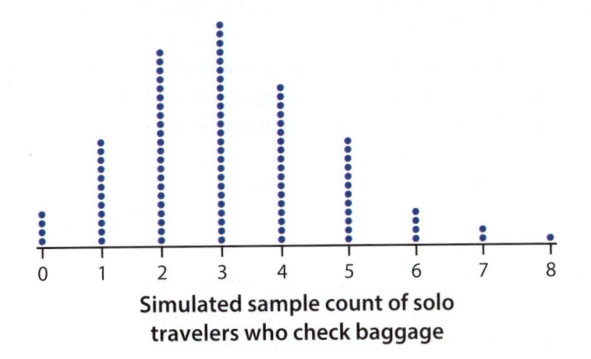

Simulated sample count of solo travelers who check baggage

(c) Use the results of the simulation to estimate the probability that 7 or more of 12 randomly selected solo travelers check baggage.

(d) Based on the actual result of 7 solo travelers who checked baggage and your answer to part (c), is there convincing evidence that more than 25% of solo travelers check baggage? Explain your reasoning.

2. **Vehicle Types (4.2, 4.4)** Suppose we choose a new vehicle sold in the United States in a recent year at random. The probability distribution for the type of vehicle chosen is given here.[55]

Vehicle type	Passenger car	Pickup truck	SUV	Crossover	Minivan
Probability	0.28	0.18	0.08	?	0.05

(a) What is the probability that the vehicle is a crossover?

(b) Given that the vehicle is not a passenger car, what is the probability that it is a pickup truck?

(c) What is the probability that the vehicle is a pickup truck, SUV, or minivan?

3. **Astrology (4.3, 4.4)** The General Social Survey (GSS) asked a random sample of adults their opinion about whether astrology is very scientific, sort of scientific, or not at all scientific. Here is a two-way table of counts for people in the sample who had three levels of higher education.[56]

		Degree held			
		Associate's	Bachelor's	Master's	Total
Opinion about astrology	Not at all scientific	169	256	114	539
	Very or sort of scientific	65	65	18	148
	Total	234	321	132	687

Suppose one person who completed the survey is selected at random.

(a) Find the probability that the person thinks astrology is not at all scientific.

(b) Find the probability that the person has an associate's degree or thinks astrology is not at all scientific.

(c) Find the probability that the person has an associate's degree, given that the person thinks astrology is not at all scientific.

4. **Mushroom Pizza (4.3, 4.4)** The chef at a local pizza shop gives you the following information about the pizzas currently in the oven: 3 of the 7 are thick-crust pizzas, and 2 of the 3 thick-crust pizzas have mushrooms. Of the remaining 4 pizzas, 2 have mushrooms. Suppose that a pizza is selected at random from the oven.

(a) Make a Venn diagram to model this chance process.

(b) Are the events "getting a thick-crust pizza" and "getting a pizza with mushrooms" independent? Justify your answer.

(c) The chef adds an eighth pizza to the oven. This pizza has a thick crust and its only topping is cheese. Now are the events "getting a thick-crust pizza" and "getting a pizza with mushrooms" independent? Justify your answer.

5. **Matching Socks (4.5 or 4.8)** Gabriel is not one for pairing up socks before he throws them into his sock drawer. One dark morning, his drawer contains 8 blue socks, 6 brown socks, and 4 gray socks. He pulls 2 socks at random from his drawer without looking at them.

(a) What is the probability that both socks are blue?

(b) What is the probability that the 2 socks are the same color?

6. **COVID-19 Testing (4.5)** The first at-home test for COVID-19 approved by the U.S. Food and Drug Administration (FDA) gives results in about 15 minutes. This test has a 3% false positive rate and a 5% false negative rate.[57] Suppose that 2% of the U.S. population was infected with COVID-19 when the test was released, and that we give the test to a randomly selected person from the population.

(a) Explain what this test's false positive rate and false negative rate mean.

(b) Make a tree diagram to display the sample space of this chance process.

(c) What's the probability that the randomly selected person tests negative?

(d) If the person tests negative, what is the probability that they don't have COVID-19?

7. **Many-Sided Die (4.6, 4.7)** The outcome of chance events in a fantasy role-playing game is determined by rolling polyhedral dice with anywhere from 4 to 20 sides. A 16-sided die has faces with the numbers from

1 to 16. Suppose you roll a 16-sided die 5 times and observe the number on the top of the die.

(a) How many possible outcomes are there for these 5 rolls?

(b) In how many of the outcomes in part (a) do the 5 rolls produce 5 different numbers?

(c) What is the probability that the 5 rolls produce 5 different numbers?

8. **Lucky Penny? (4.6)** Harris Interactive reported that 33% of U.S. adults believe that finding and picking up a penny is good luck.[58] Assuming that responses from different individuals are independent, what is the probability of randomly selecting 10 U.S. adults and finding at least one person who believes that finding and picking up a penny is good luck?

9. **Talent Show (4.8)** Fourteen musical acts have asked to perform at a college's winter talent show. Eight of them play acoustic music and 6 play electric music. Unfortunately, there is only room on the program for 10 acts — the others will have to wait until the spring show.

(a) How many different groups of 10 acts can be selected to play in the winter show?

(b) If the 10 musical acts for the winter show are randomly selected, what is the probability that 5 acoustic acts and 5 electric acts get to play in the show?

10. **Smoking and Lung Cancer (Chapter 4)** Researchers are interested in the relationship between cigarette smoking and lung cancer. Suppose an adult is randomly selected from a particular large population. The table shows the probabilities of some events related to this chance process.

Event	Probability
Smokes	0.25
Smokes and gets cancer	0.08
Does not smoke and does not get cancer	0.71

(a) Find the probability that the individual gets cancer given that they are a smoker.

(b) Find the probability that the individual smokes or gets cancer.

(c) Two adults are selected at random. Find the probability that at least one of the two smokes.

5 Discrete Random Variables

Section 5.1 Introduction to Random Variables
Section 5.2 Analyzing Discrete Random Variables
Section 5.3 Binomial Random Variables
Section 5.4 Analyzing Binomial Random Variables
Section 5.5 Poisson Random Variables
Chapter 5 Review
Chapter 5 Review Exercises

Statistics Matters A jury of peers?

Are accused criminals in the United States entitled to a "jury of their peers"? Sort of. The Sixth Amendment to the U.S. Constitution begins, "In all criminal prosecutions, the accused shall enjoy the right to a speedy and public trial, by an impartial jury of the State and district wherein the crime shall have been committed. . . ." There is no mention of a "jury of your peers" in the Constitution or any of its amendments. However, an 1879 U.S. Supreme Court decision said that a jury should be chosen from a group "composed of the peers or equals [of the accused]; that is, of his neighbors, fellows, associates, persons having the same legal status in society as he holds."[1]

To meet the Sixth Amendment requirement of impartiality, most courts start by randomly selecting a large jury pool from the citizens who live in the court's jurisdiction. The jurors for a given trial are then chosen from the jury pool in a process known as *voir dire*. Each prospective juror answers a set of questions posed by the judge and the lawyers for both the prosecution and the defense. Depending on their answers, prospective jurors are excluded or seated on the jury.

In one case that made it all the way to the U.S. Supreme Court, a defense lawyer in Michigan challenged the process of selecting the jury pool in the trial of his client, a Black man. Here are the facts:[2]

- About 7.28% of the jury-eligible population in the court's jurisdiction were Black people.
- The jury pool had between 60 and 100 members, only 3 of whom were Black people.

Is it plausible (believable) that a jury pool with so few Black people could be chosen by chance alone?

We'll revisit Statistics Matters *at the end of the chapter, so you can use what you have learned to help answer these questions.*

262 CHAPTER 5 Discrete Random Variables

In Chapter 4, you learned several methods for finding probabilities. In this chapter, you'll learn how to analyze probabilities involving numerical outcomes of a chance process (called *random variables*), like the number of Black citizens in a jury pool. Random variables are an essential tool for statistical inference.

Section 5.1 Introduction to Random Variables

LEARNING GOALS

By the end of this section, you will be able to:
- Classify a random variable as discrete or continuous.
- Verify that the probability distribution of a discrete random variable is valid.
- Calculate probabilities involving a discrete random variable.

Class Activity

Smelling Parkinson's Disease

Joy Milne of Perth, United Kingdom, noticed a "subtle musky odor" on her husband Les that she had never encountered before. At first, Joy thought the smell might be from Les's sweat after long hours of work. But when Les was diagnosed with Parkinson's disease 6 years later, Joy suspected the odor might be a result of the disease.

Scientists were intrigued by Joy's claim and designed an experiment to test her ability to "smell Parkinson's." Joy was presented with 12 different shirts, each worn by a different person, some of whom had Parkinson's disease and some of whom did not. The shirts were given to Joy in a random order, and she had to decide whether each shirt was or was not worn by a patient with Parkinson's disease. Joy identified 11 of the 12 shirts correctly.

Although the researchers wanted to believe that Joy could detect Parkinson's disease by smell, it is possible that she was just a lucky guesser. You and your classmates will perform a simulation to determine which explanation is more believable.[3]

1. Go to www.stapplet.com and launch the "Can You Smell Parkinson's?" activity. Click the button to enter an existing class code and enter the code provided by your instructor. (If you are doing this activity on your own, click the button to do the activity independently.)

2. Once the activity launches, make your guess about whether each of the 12 shirts was worn by a person with Parkinson's disease. Just click on "Yes" or "No" beneath each shirt. After you make all 12 guesses, the applet will indicate whether each of your guesses was correct and will add a dot corresponding to your number of correct guesses on the dotplot.

3. Check whether you and your classmates have performed at least 50 repetitions of the activity. If not, repeat Step 2 as needed. (If you are doing the activity independently, use the Quick Add feature to have the applet perform 50 more repetitions.)

4. How often were 11 or more shirts correctly identified by chance alone? Based on this result, which seems more believable: Joy is just a lucky guesser, or Joy really can smell Parkinson's disease? Explain your reasoning.

In the *Smelling Parkinson's Disease* activity, suppose we define the **random variable** X = the number of correct identifications (out of 12) made by someone who is just guessing. The value of X will vary from one repetition of this chance process to another, but it will always be a whole number between 0 and 12.

A **random variable** takes numerical values that describe the outcomes of a chance process.

We use capital, italic letters (like X or Y) to represent random variables. There are two main types of random variables: *discrete* and *continuous*.

Classifying Random Variables

The variable X in the *Smelling Parkinson's Disease* activity is a **discrete random variable.**

A **discrete random variable** takes a set of individual values with gaps between them on the number line. The set of possible values for a discrete random variable can be finite or infinite.

We can list the possible values of X = the number of correct identifications made by someone who is just guessing as 0, 1, 2, ..., 11, 12. Note that there are gaps between consecutive values on the number line. For example, a gap exists between $X = 1$ and $X = 2$ because X cannot take values such as 1.2 or 1.84.

Now suppose we want to randomly select a number between 0 and 10, allowing *any* number between 0 and 10 as the outcome (like 0.84522 or 9.1111119). Some calculator and computer random number generators will do this. The sample space of this chance process is the entire interval of values between 0 and 10 on the number line. If we define Y = the outcome of such a random number generator, then Y is a **continuous random variable.**

A **continuous random variable** takes any value in a particular interval on the number line. The set of possible values for a continuous random variable is infinite.

Most discrete random variables result from counting something, like the number of siblings for a randomly selected person. Continuous random variables typically result from measuring something, like the height of a randomly selected student or the time it takes that student to run a mile.

EXAMPLE 5.1.1

Feet, gas pumps, and tennis rackets

Discrete versus continuous random variables

PROBLEM

Classify each of the following random variables as discrete or continuous.
(a) X = the foot length of a randomly selected student at your institution
(b) G = the number of pumps (out of 12) in use at a gas station at a randomly selected time of day
(c) T = the string tension of a randomly selected tennis racket

SOLUTION
(a) Continuous — X could take any value in an interval from about 9 inches to 15 inches.
(b) Discrete — G could take any of the values 0, 1, 2, ..., 12, but not values like 2.387465819.
(c) Continuous — T could take any value in an interval from about 50 pounds to 70 pounds.

EXAM PREP: FOR PRACTICE, TRY EXERCISE 9.

This chapter focuses on discrete random variables. We will discuss continuous random variables in Chapter 6.

Discrete Random Variables

Suppose you toss a fair coin 3 times. Let X = the number of heads obtained. We know that X is a discrete random variable because its possible values are 0, 1, 2, and 3. How likely is X to take each of those values?

The sample space for this chance process is

HHH HHT HTH THH HTT THT TTH TTT

Because there are 8 equally likely outcomes, the probability is 1/8 for each possible outcome. Let's group the possible outcomes by the number of heads obtained:

$X = 0$: TTT $X = 1$: HTT THT TTH $X = 2$: HHT HTH THH $X = 3$: HHH

We can summarize the **probability distribution** of X in a table:

Number of heads	0	1	2	3
Probability	1/8	3/8	3/8	1/8

The **probability distribution** of a random variable gives its possible values and their probabilities.

The probability distribution of X = the number of heads obtained in 3 tosses of a fair coin is valid because all the probabilities are between 0 and 1, and their sum is 1:

$$1/8 + 3/8 + 3/8 + 1/8 = 8/8 = 1$$

Probability Distribution of a Discrete Random Variable

The probability distribution of a discrete random variable X lists the values x_i and their probabilities p_i:

Value	x_1	x_2	x_3	...
Probability	p_1	p_2	p_3	...

For the probability distribution to be valid, the probabilities p_i must satisfy two requirements:

1. Every probability p_i is a number between 0 and 1, inclusive.
2. The sum of the probabilities is 1: $p_1 + p_2 + p_3 + \ldots = 1$.

Remember that the number of possible values of a discrete random variable can be either finite or infinite. For instance, let Y = the number of times a person plays the lottery until they win. The smallest possible value of Y is 1 and there is theoretically no largest possible value of Y.

EXAMPLE 5.1.2

Assessing newborn babies

Valid probability distributions

PROBLEM

In 1952, Dr. Virginia Apgar suggested five criteria for measuring a baby's health at birth: skin color, heart rate, muscle tone, breathing, and response when stimulated. She developed a 0–1–2 scale, with 2 being the best score, to rate a newborn on each of the five criteria. A baby's Apgar score is the sum of the ratings on each of the five scales, which gives a whole-number value from 0 to 10. Apgar scores are still used today to evaluate the health of newborns. Although this procedure was later named for Dr. Apgar, the acronym APGAR also represents the five scales: Appearance, Pulse, Grimace, Activity, and Respiration.

Which Apgar scores are typical of newborns? To find out, researchers recorded the Apgar scores of more than 2 million babies at birth in a single year.[4] Imagine selecting one of these newborns at random. (That's our chance process.) Define the random variable X = Apgar score of a randomly selected newborn. The table gives the probability distribution of X.

Apgar score	0	1	2	3	4	5	6	7	8	9	10
Probability	0.001	0.006	0.007	0.008	0.012	0.020	0.038	0.099	0.319	0.437	0.053

Show that this is a valid probability distribution.

SOLUTION

- The probabilities are all between 0 and 1, inclusive.
- The sum of the probabilities is $0.001 + 0.006 + 0.007 + 0.008 + 0.012 + 0.020 + 0.038 + 0.099 + 0.319 + 0.437 + 0.053 = 1$.

EXAM PREP: FOR PRACTICE, TRY EXERCISE 13.

Finding Probabilities for Discrete Random Variables

We can use the probability distribution of a discrete random variable to find the probability of an event. Here again is the probability distribution of X = the number of heads in 3 tosses of a fair coin.

Number of heads	0	1	2	3
Probability	1/8	3/8	3/8	1/8

What's the probability that we get at most 2 heads in 3 tosses of the coin? In symbols, we want to find $P(X \leq 2)$. We know that

$$P(X \leq 2) = P(X = 0 \text{ or } X = 1 \text{ or } X = 2)$$

Because the events $X = 0$, $X = 1$, and $X = 2$ are mutually exclusive, we can add their probabilities to get the answer:

$$P(X \leq 2) = P(X = 0) + P(X = 1) + P(X = 2)$$
$$= 1/8 + 3/8 + 3/8$$
$$= 7/8$$

Alternatively, we could use the complement rule from Section 4.2:

$$P(X \leq 2) = 1 - P(X > 2)$$
$$= 1 - P(X = 3)$$
$$= 1 - 1/8$$
$$= 7/8$$

EXAMPLE 5.1.3

Finding probabilities with discrete random variables

Assessing newborn babies

PROBLEM
Refer to Example 5.1.2. The table gives the probability distribution of X = Apgar score of a randomly selected newborn once again.

Apgar score	0	1	2	3	4	5	6	7	8	9	10
Probability	0.001	0.006	0.007	0.008	0.012	0.020	0.038	0.099	0.319	0.437	0.053

Doctors decided that Apgar scores of 7 or higher indicate a healthy baby. What's the probability that a randomly selected newborn is considered healthy?

SOLUTION

$P(X \geq 7) = 0.099 + 0.319 + 0.437 + 0.053$
$ = 0.908$

> The probability of choosing a healthy baby is
> $P(X \geq 7) = P(X = 7) + P(X = 8) + P(X = 9) + P(X = 10)$

EXAM PREP: FOR PRACTICE, TRY EXERCISE 17.

Note that the probability of randomly selecting a newborn whose Apgar score is *at least* 7 is not the same as the probability that the baby's Apgar score is *greater than* 7. The latter probability is

$$P(X > 7) = P(X = 8) + P(X = 9) + P(X = 10)$$
$$= 0.319 + 0.437 + 0.053$$
$$= 0.809$$

The outcome $X = 7$ is included in "at least 7" but is not included in "greater than 7." **Be sure to consider whether to include the boundary value in your calculations when dealing with discrete random variables.**

Section 5.1 What Did You Learn?

Review the learning goals from this section. Then practice what you've learned by working through the exercises.

Learning Goal	Example	Exercises
Classify a random variable as discrete or continuous.	5.1.1	9–12
Verify that the probability distribution of a discrete random variable is valid.	5.1.2	13–16
Calculate probabilities involving a discrete random variable.	5.1.3	17–20

Section 5.1 Exercises

Building Concepts and Skills These exercises assess the basic knowledge you should have after reading the section.

1. A random variable takes _____ values that describe the outcomes of a chance process.

2. The possible values of a discrete random variable have _____ between consecutive values on a number line.

3. True/False: A discrete random variable can take only a finite number of possible values.

4. A continuous random variable can take an infinite set of possible values in a(n) _____ on the number line.

5. Most _____ random variables result from counting something. Most _____ random variables result from measuring something.

6. What two requirements must be met for the probability distribution of a discrete random variable to be valid?

7. True/False: To find probabilities involving several values of a discrete random variable, you can use the addition rule for mutually exclusive events.

8. True/False: For every discrete random variable X, $P(X \leq 4) = P(X < 4)$.

Mastering Concepts and Skills These exercises reinforce the learning goals as shown in the examples.

9. **Refer to Example 5.1.1 Discrete or Continuous?** Classify each of the following random variables as discrete or continuous.
   (a) X = the pH of a water sample that has been randomly selected from a stream
   (b) Y = the number of correct answers on a recent multiple-choice exam for a randomly selected student in your class
   (c) W = the exact amount of sleep that a randomly selected student from your school got last night

10. **Discrete or Continuous?** Classify each of the following random variables as discrete or continuous.
    (a) H = the number of homework exercises assigned by your statistics instructor on a randomly selected class day
    (b) R = the temperature of the meat in a randomly chosen location in a cooked turkey
    (c) X = the reported score of a randomly selected high school senior who took the SAT Math test

11. **Discrete or Continuous?** Classify each of the following random variables as discrete or continuous.
    (a) R = the number of times you have to roll a fair, 6-sided die to get a 1
    (b) T = winning time in the men's 100-meter dash at a randomly selected international track meet

12. **Discrete or Continuous?** Classify each of the following random variables as discrete or continuous.
    (a) X = number of birthdays had by a randomly selected adult
    (b) Y = head circumference of a randomly selected person

13. **Refer to Example 5.1.2 Benford's Law** Faked numbers on tax returns, invoices, or expense account claims often display patterns that aren't present in legitimate records. Some patterns, like too many round numbers, are obvious and easily avoided by a clever crook. Others are more subtle. It is a striking fact that the first digits of numbers in legitimate records often follow a model known as Benford's law.[5] Call the first digit of a randomly chosen record X for short. Here is the probability distribution of X according to Benford's law (note that a first digit can't be 0). Show that this is a valid probability distribution.

First digit	1	2	3	4	5
Probability	0.301	0.176	0.125	0.097	0.079
First digit	6	7	8	9	
Probability	0.067	0.058	0.051	0.046	

14. **Working Out** Suppose we choose a person aged 19 to 25 years at random and ask, "In the past seven days, how many times did you go to an exercise or fitness center to work out?" Call the response Y for short. Based on a large

sample survey, here is the probability distribution of Y.[6] Show that this is a valid probability distribution.

Days	0	1	2	3	4	5	6	7
Probability	0.68	0.05	0.07	0.08	0.05	0.04	0.01	0.02

15. **College Costs** El Dorado Community College considers students to be attending full-time if they are taking between 12 and 18 units. The tuition charge for a student is $50 per unit. Suppose we choose a full-time El Dorado Community College student at random. The probability distribution of the random variable T = tuition charge (in dollars) for the chosen student is shown here. Describe the event $T = 850$ in words. Then find its probability.

Tuition charge ($)	600	650	700	750	800	850	900
Probability	0.25	0.10	0.05	0.30	0.10	???	0.15

16. **Ferry Fares** A small car ferry runs every half hour from one side of a large river to the other. The ferry can hold a maximum of 5 cars, and the charge is $5 per car. The probability distribution of the random variable X = money collected (in dollars) on a randomly selected ferry trip is shown here. Describe the event $X = 5$ in words. Then find its probability.

Money collected ($)	0	5	10	15	20	25
Probability	0.02	???	0.08	0.16	0.27	0.42

17. **Refer to Example 5.1.3 Benford's Law** Refer to Exercise 13. What's the probability that the first digit of a randomly chosen record is at least 6?

18. **Working Out** Refer to Exercise 14. What's the probability that a randomly selected person says they went to an exercise/fitness center to work out on at most 2 days?

19. **College Costs** Refer to Exercise 15.
    (a) Write the event "tuition charge at most $800" in terms of T. What is the probability of this event?
    (b) Find $P(T < 800)$. Explain why this answer is different from the one you found in part (a).

20. **Ferry Fares** Refer to Exercise 16.
    (a) Write the event "at least $10 is collected" in terms of X. What is the probability of this event?
    (b) Find $P(X > 10)$. Explain why this answer is different from the one you found in part (a).

Applying the Concepts These exercises ask you to apply multiple learning goals in a new context or to apply what you learned in this section in a new way.

21. **Languages** Imagine selecting a U.S. college student at random. Define the random variable X = number of languages spoken by the student. The table gives the probability distribution of X.[7]

Languages	1	2	3	4	5
Probability	0.630	0.295	0.065	0.008	0.002

(a) Is X a discrete or a continuous random variable? Explain your answer.

(b) Explain in words what $P(X \geq 3)$ means. What is this probability?

(c) Find the probability that a randomly selected student speaks fewer than 3 languages.

22. **Skee Ball** Ana is a dedicated Skee Ball player who always rolls for the 50-point slot. The probability distribution of Ana's score X on a roll of the ball is shown here.

Score	10	20	30	40	50
Probability	0.32	0.27	0.19	0.15	0.07

(a) Is X a discrete or a continuous random variable? Explain your answer.

(b) Explain in words what $P(X > 20)$ means. What is this probability?

(c) Find the probability that Ana scores at most 20 on a randomly selected roll.

Extending the Concepts These exercises challenge you to explore statistical concepts and methods that go beyond what you learned in this section.

23. **Coin Tosses** Suppose you toss a fair coin 4 times. Let X = the number of heads you get.
    (a) Find the probability distribution of X.
    (b) Find $P(X \leq 3)$. Describe this probability in words.

24. **Pair-a-Dice** Suppose you roll a pair of fair, 6-sided dice. Let T = the sum of the spots showing on the up-faces.
    (a) Find the probability distribution of T.
    (b) Find $P(T \geq 5)$. Describe this probability in words.

Cumulative Review These exercises revisit what you learned in previous sections.

25. **Phone Alerts (4.6)** According to a recent survey, 67% of cell-phone owners find themselves checking their phone for messages, alerts, or calls—even when they don't notice their phone ringing or vibrating.[8] Suppose we randomly select 5 cell-phone owners at random.
    (a) Find the probability that all 5 cell-phone owners check their phones even when the phone doesn't ring or vibrate.
    (b) What's the probability that at least one of them doesn't check their phone?

26. **Fluoride (1.5)** In an experiment to measure the effect of fluoride "varnish" on the incidence of tooth cavities, thirty-four 10-year-old girls whose parents gave informed consent for them to participate were randomly assigned to two groups. One group was given fluoride varnish annually for 4 years, along with standard dental hygiene; the other group followed only the standard dental hygiene regimen. The mean number of cavities in the two groups was compared at the end of the treatments.[9]
    (a) Is this experiment subject to the placebo effect? Explain your reasoning.
    (b) Describe how you could alter this experiment to make it double-blind.
    (c) Explain the purpose of the random assignment in this experiment.

Section 5.2 Analyzing Discrete Random Variables

LEARNING GOALS

By the end of this section, you will be able to:

- Make a histogram to display the probability distribution of a discrete random variable and describe its shape.
- Calculate and interpret the mean (expected value) of a discrete random variable.
- Calculate and interpret the standard deviation of a discrete random variable.

When we analyzed distributions of quantitative data in Chapters 2 and 3, we made it a point to discuss their shape, center, and variability. We'll do the same with probability distributions of random variables.

Displaying Discrete Probability Distributions: Histograms and Shape

In Section 5.1, we considered the discrete random variable X = Apgar score of a randomly selected newborn. The table gives the probability distribution of X once again.

Apgar score	0	1	2	3	4	5	6	7	8	9	10
Probability	0.001	0.006	0.007	0.008	0.012	0.020	0.038	0.099	0.319	0.437	0.053

We can display this probability distribution graphically using a histogram. Values of the variable go on the horizontal axis and probabilities go on the vertical axis. There is one bar in the histogram for each value of X. The height of each bar gives the probability for the corresponding value of the variable.

Figure 5.1 shows a histogram of the probability distribution of X. This distribution is skewed to the left with a single peak at an Apgar score of 9.

FIGURE 5.1 Histogram of the probability distribution of the random variable X = Apgar score of a randomly selected newborn.

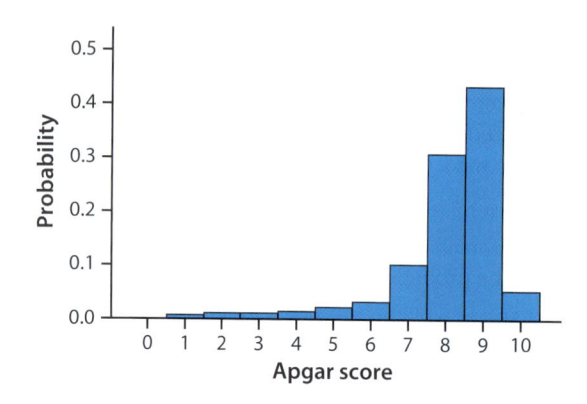

There's another way to think about the graph in Figure 5.1. The histogram models the population distribution of the quantitative variable "Apgar score of a newborn." So we can interpret our earlier result, $P(X > 7) = 0.809$, as saying that about 81% of all newborns have Apgar scores greater than 7.

EXAMPLE 5.2.1

Jeep tours revenue

Displaying a probability distribution

PROBLEM

Pete's Jeep Tours offers a popular half-day trip in a tourist area. There must be at least 2 passengers for the trip to run, and the vehicle will hold up to 6 passengers. Pete charges $150 per passenger. Let C = the total amount of money that Pete collects on a randomly selected trip. The probability distribution of C is given in the table.

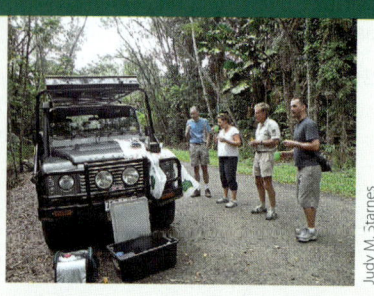

Total collected ($)	300	450	600	750	900
Probability	0.15	0.25	0.35	0.20	0.05

Make a histogram of the probability distribution. Describe its shape.

SOLUTION

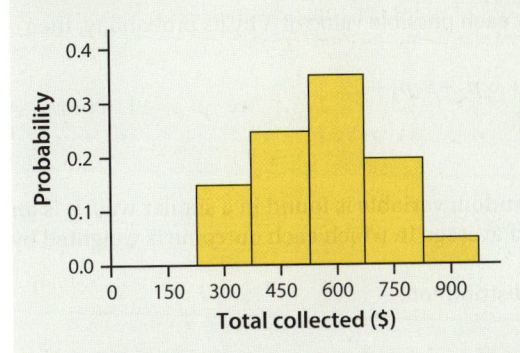

Remember: Values of the variable go on the horizontal axis and probabilities go on the vertical axis. Don't forget to properly label and scale each axis!

The graph is roughly symmetric and has a single peak at $600.

EXAM PREP: FOR PRACTICE, TRY EXERCISE 9.

Measuring Center: The Mean (Expected Value) of a Discrete Random Variable

In Chapter 3, you learned about summarizing the center of a distribution of quantitative data with either the mean or the median. For random variables, the mean is typically used to summarize the center of a probability distribution. Because a probability distribution can model the population distribution of a quantitative variable, we label the mean of a random variable X as μ_X.

To find the mean of a quantitative data set, we compute the sum of the individual observations and divide by the total number of data values. How do we find the *mean of a discrete random variable*?

Consider the random variable C = the total amount of money that Pete collects on a randomly selected Jeep tour from Example 5.2.1. The probability distribution of C is given in the table.

Total collected ($)	300	450	600	750	900
Probability	0.15	0.25	0.35	0.20	0.05

What's the average amount of money that Pete collects on his Jeep tours?

Imagine a hypothetical 100 trips. According to the probability distribution, Pete should collect $300 on 15 of these 100 trips, $450 on 25 trips, $600 on 35 trips, $750 on 20 trips, and $900 on 5 trips. Pete's average amount collected for these trips is

$$\mu_C = \frac{300 \cdot 15 + 450 \cdot 25 + 600 \cdot 35 + 750 \cdot 20 + 900 \cdot 5}{100}$$
$$= \frac{300 \cdot 15}{100} + \frac{450 \cdot 25}{100} + \frac{600 \cdot 35}{100} + \frac{750 \cdot 20}{100} + \frac{900 \cdot 5}{100}$$
$$= 300(0.15) + 450(0.25) + 600(0.35) + 750(0.20) + 900(0.05)$$
$$= \$562.50$$

That is, the **mean of the discrete random variable** C is $\mu_C = \$562.50$. This is also known as the **expected value** of C, denoted by $E(C)$.

The **mean (expected value) of a discrete random variable** is its average value over many, many repetitions of the same chance process.

Suppose that X is a discrete random variable with probability distribution

Value	x_1	x_2	x_3	...
Probability	p_1	p_2	p_3	...

To find the mean (expected value) of X, multiply each possible value of X by its probability, then add all the products:

$$\mu_X = E(X) = x_1 p_1 + x_2 p_2 + x_3 p_3 + \ldots$$
$$= \sum x_i p_i$$

The mean (expected value) of any discrete random variable is found in a similar way. It is an average of the possible outcomes, but a weighted average in which each outcome is weighted by its probability.

Recall that the mean is the balance point of a distribution.

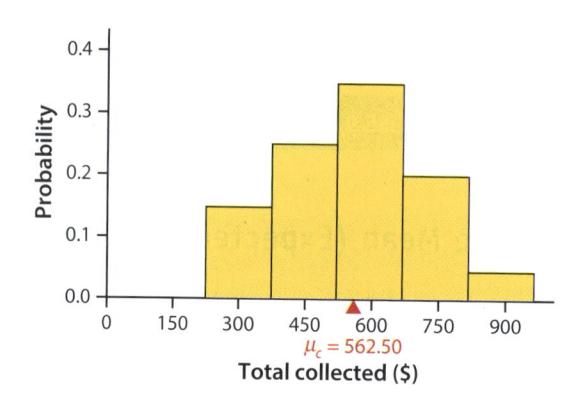

For Pete's distribution of money collected on a randomly selected Jeep tour, the histogram balances at $\mu_C = 562.50$. How do we interpret this value? If we randomly select many, many Jeep tours, Pete will collect about $562.50 per trip, on average.

EXAMPLE 5.2.2

Assessing newborn babies

Finding and interpreting the mean

PROBLEM

Earlier, we defined the random variable X to be the Apgar score of a randomly selected newborn. The table gives the probability distribution of X once again.

Apgar score	0	1	2	3	4	5	6	7	8	9	10
Probability	0.001	0.006	0.007	0.008	0.012	0.020	0.038	0.099	0.319	0.437	0.053

Calculate and interpret the expected value of X.

SOLUTION

$E(X) = \mu_X = 0(0.001) + 1(0.006) + 2(0.007) + \ldots + 10(0.053)$
$= 8.128$

$$E(X) = \mu_X = x_1 p_1 + x_2 p_2 + x_3 p_3 + \ldots$$

If many, many newborns are randomly selected, their average Apgar score will be about 8.128.

EXAM PREP: FOR PRACTICE, TRY EXERCISE 13.

Notice that the mean Apgar score, 8.128, is not a possible value of the random variable X because it is not a whole number between 0 and 10. The decimal value of the mean shouldn't bother you if you think of the expected value as an average over many, many repetitions of the chance process.

⚠ **Don't round the mean (expected value) of a random variable to the nearest integer!**

> **THINK ABOUT IT**
>
> **How can we find the median of a discrete random variable?** In Section 3.1, we defined the median as "the midpoint of a distribution, the number such that about half the observations are smaller and about half are larger." The median of a discrete random variable is the 50th percentile of its probability distribution. We can find the median by adding a cumulative probability row to the probability distribution table, and then locating the smallest value for which the cumulative probability equals or exceeds 0.5. For the distribution of amount of money collected on Pete's Jeep tours, we see that the median is $600.
>
> | Total collected ($) | 300 | 450 | 600 | 750 | 900 |
> |---|---|---|---|---|---|
> | Probability | 0.15 | 0.25 | 0.35 | 0.20 | 0.05 |
> | Cumulative probability | 0.15 | 0.40 | 0.75 | 0.95 | 1.00 |

Measuring Variability: The Standard Deviation of a Discrete Random Variable

Since we're using the mean as our measure of center for a discrete random variable, it shouldn't surprise you that we'll use the standard deviation as our measure of variability. In Section 3.2, we defined the standard deviation s_x of a distribution of quantitative data as the typical distance of the values in the data set from the mean. To get the standard deviation, we started by "averaging" the squared deviations from the mean and then took the square root:

$$s_x = \sqrt{\frac{(x_1 - \bar{x})^2 + (x_2 - \bar{x})^2 + \ldots + (x_n - \bar{x})^2}{n-1}}$$

We can modify this approach to calculate the **standard deviation of a discrete random variable** X. Start by finding a weighted average of the squared deviations $(x_i - \mu_X)^2$ of the values of the variable X from its mean μ_X. The probability distribution gives the appropriate weight for each squared deviation. We call this weighted average the *variance* of X. Then take the square root to get the standard deviation. Because a probability distribution can model the population distribution of a quantitative variable, we label the variance of a random variable X as σ_X^2 and the standard deviation as σ_X.

> The **standard deviation of a discrete random variable** measures how much the values of the variable typically vary from the mean in many, many repetitions of the same chance process.
> Suppose that X is a discrete random variable with probability distribution
>
> | Value | x_1 | x_2 | x_3 | ... |
> |---|---|---|---|---|
> | Probability | p_1 | p_2 | p_3 | ... |
>
> and that μ_X is the mean of X. Then the standard deviation of X is
>
> $$\sigma_X = \sqrt{(x_1 - \mu_X)^2 p_1 + (x_2 - \mu_X)^2 p_2 + (x_3 - \mu_X)^2 p_3 + \ldots}$$
> $$= \sqrt{\sum (x_i - \mu_X)^2 p_i}$$

Let's return to the random variable C = the total amount of money that Pete collects on a randomly selected Jeep tour. The left two columns of the following table give the probability distribution. Recall that the mean of C is $\mu_C = 562.50$. The third column of the table shows the squared deviation of each value from the mean. The fourth column gives the weighted squared deviations.

Total Collected ($)	Probability	Squared Deviation from the Mean	Weighted Squared Deviation
300	0.15	$(300-562.50)^2$	$(300-562.50)^2 (0.15)$
450	0.25	$(450-562.50)^2$	$(450-562.50)^2 (0.25)$
600	0.35	$(600-562.50)^2$	$(600-562.50)^2 (0.35)$
750	0.20	$(750-562.50)^2$	$(750-562.50)^2 (0.20)$
900	0.05	$(900-562.50)^2$	$(900-562.50)^2 (0.05)$
			Sum = 26,718.75

You can see from the fourth column that the sum of the weighted squared deviations is 26,718.75. That's the *variance* of C. The standard deviation of C is the square root of the variance:

$$\sigma_C = \sqrt{(300-562.50)^2(0.15) + (450-562.50)^2(0.25) + \ldots + (900-562.50)^2(0.05)}$$
$$= \sqrt{26{,}718.75}$$
$$= \$163.46$$

How do we interpret this value? If many, many Jeep tours are randomly selected, the amount of money that Pete collects will typically vary from the mean of \$562.50 by about \$163.46.

EXAMPLE 5.2.3

Assessing newborn babies

Finding and interpreting the standard deviation

PROBLEM

Earlier we defined the random variable X to be the Apgar score of a randomly selected newborn. The table gives the probability distribution of X once again.

Apgar score	0	1	2	3	4	5	6	7	8	9	10
Probability	0.001	0.006	0.007	0.008	0.012	0.020	0.038	0.099	0.319	0.437	0.053

In Example 5.2.2, we calculated the mean Apgar score of a randomly chosen newborn to be $\mu_X = 8.128$. Calculate and interpret the standard deviation of X.

SOLUTION

$$\sigma_X = \sqrt{(0-8.128)^2(0.001) + (1-8.128)^2(0.006) + \ldots + (10-8.128)^2(0.053)}$$
$$= \sqrt{2.066}$$
$$= 1.437$$

$$\boxed{\sigma_X = \sqrt{(x_1 - \mu_X)^2 p_1 + (x_2 - \mu_X)^2 p_2 + (x_3 - \mu_X)^2 p_3 + \cdots}}$$

If many, many newborns are randomly selected, the babies' Apgar scores will typically vary from the mean of 8.128 by about 1.437 units.

EXAM PREP: FOR PRACTICE, TRY EXERCISE 17.

Technology can be a big help when you are analyzing discrete random variables.

TECH CORNER | Analyzing Discrete Random Variables

You can use technology to graph the probability distribution of a discrete random variable and to calculate its mean and standard deviation. We'll illustrate this process using the random variable $X =$ Apgar score of a randomly selected newborn, whose probability distribution is shown in Example 5.2.3.

Applet

1. Go to www.stapplet.com and launch the *Discrete Random Variables* applet.

2. Enter "Apgar score" as the variable name. Then input the values of the variable and the corresponding probabilities. Click the "+" sign in the bottom right corner of the table to add rows.

   Discrete Random Variables

   Variable name: Apgar score

	Value	Probability (decimal from 0 to 1)
1	0	.001
2	1	.006
3	2	.007
4	3	.008
5	4	.012
6	5	.020
7	6	.038
8	7	.099
9	8	.319
10	9	.437
11	10	.053

3. Click the Begin analysis button. A histogram of the probability distribution should appear, along with the mean and standard deviation.

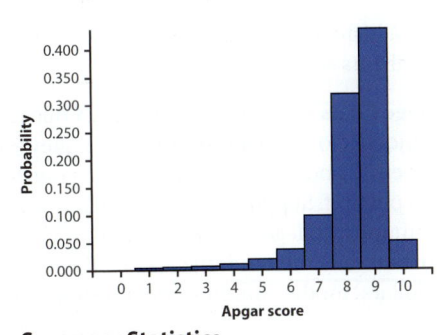

   Summary Statistics

Mean	SD
8.128	1.437

TI-83/84

1. Enter the values of the random variable in list L_1 and the corresponding probabilities in list L_2.

2. To graph a histogram of the probability distribution:
   - Set up a statistics plot to be a histogram with Xlist: L_1 and Freq: L_2.
   - Adjust your window settings as follows:
     Xmin $= -0.5$, Xmax $= 10.5$, Xscl $= 1$, Ymin $= -0.1$, Ymax $= 0.5$, Yscl $= 0.1$
   - Press GRAPH.

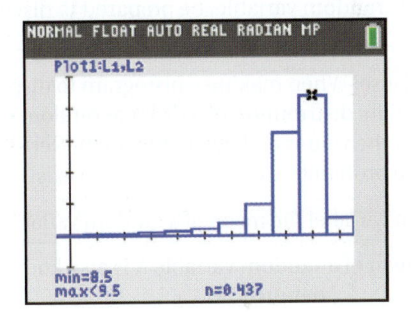

3. To calculate the mean and standard deviation of the random variable, use one-variable statistics with the values in L_1 and the probabilities (relative frequencies) in L_2. Press STAT, arrow over to the CALC menu, and choose 1-Var Stats.

   OS 2.55 or later: In the dialog box, specify List: L_1 and FreqList: L_2. Then choose Calculate.

   Older OS: Execute the command 1-Var Stats L_1, L_2.

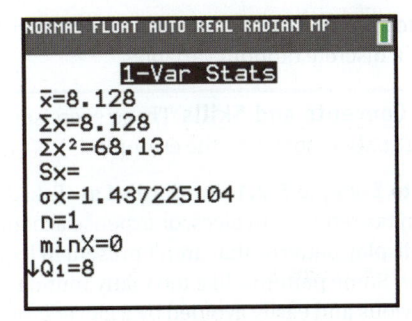

Note: In the 1-Var Stats screen, the calculator labels the mean as $\bar{x}$, but the correct notation for the mean of a random variable is μ_X. Also, be sure to clear the FreqList before trying to calculate summary statistics for one-variable quantitative data!

Detailed instructions for using CrunchIt!, Excel, Google Sheets, JMP, Minitab, and R are available in Achieve.

Section 5.2 What Did You Learn?

Review the learning goals from this section. Then practice what you've learned by working through the exercises.

Learning Goal	Example	Exercises
Make a histogram to display the probability distribution of a discrete random variable and describe its shape.	5.2.1	9–12
Calculate and interpret the mean (expected value) of a discrete random variable.	5.2.2	13–16
Calculate and interpret the standard deviation of a discrete random variable.	5.2.3	17–20

Section 5.2 Exercises

Building Concepts and Skills These exercises assess the basic knowledge you should have after reading the section.

1. When analyzing the probability distribution of a discrete random variable, be prepared to discuss _____, _____, and _____.

2. True/False: When making a histogram to display the probability distribution of a discrete random variable, you put the values of the variable on the vertical axis and the probabilities on the horizontal axis.

3. Why do we label the mean of a random variable X as μ_X?

4. The mean of a random variable X is also known as its _____, denoted by _____.

5. What formula is used to calculate the mean of a discrete random variable?

6. True/False: For a discrete random variable that takes only whole-number values, the expected value may not be a whole number.

7. The standard deviation of a discrete random variable, denoted by _____, measures how much the values of the variable typically vary from the _____.

8. What formula is used to calculate the standard deviation of a discrete random variable?

Mastering Concepts and Skills These exercises reinforce the learning goals as shown in the examples.

9. **Refer to Example 5.2.1 Benford's Law** Faked numbers on tax returns, invoices, or expense account claims often display patterns that aren't present in legitimate records. Some patterns, like too many round numbers, are obvious and easily avoided by a clever crook. Others are more subtle. It is a striking fact that the first digits of numbers in legitimate records often follow a model known as Benford's law.[10] Call the first digit of a randomly chosen record X for short. Here is the probability distribution of X according to Benford's law (note that a first digit can't be 0).

First digit	1	2	3	4	5
Probability	0.301	0.176	0.125	0.097	0.079
First digit	6	7	8	9	
Probability	0.067	0.058	0.051	0.046	

Make a histogram of the probability distribution. Describe its shape.

10. **Working Out** Suppose we choose a person aged 19 to 25 years at random and ask, "In the past seven days, how many times did you go to an exercise or fitness center to work out?" Call the response Y for short. Based on a large sample survey, here is the probability distribution of Y.[11]

Days	0	1	2	3	4	5	6	7
Probability	0.68	0.05	0.07	0.08	0.05	0.04	0.01	0.02

Make a histogram of the probability distribution. Describe its shape.

11. **College Costs** El Dorado Community College considers students to be attending full-time if they are taking between 12 and 18 units. The tuition charge for a student is $50 per unit. Suppose we choose a full-time El Dorado Community College student at random. The probability distribution of the random variable T = tuition charge (in dollars) for the chosen student is shown here.

Tuition charge ($)	600	650	700	750	800	850	900
Probability	0.25	0.10	0.05	0.30	0.10	0.05	0.15

Make a histogram of the probability distribution. Describe its shape.

12. **Ferry Fares** A small car ferry runs every half hour from one side of a large river to the other. The ferry can hold a maximum of 5 cars, and the charge is $5 per car. The probability distribution of the random variable X = money collected (in dollars) on a randomly selected ferry trip is shown here.

Money collected ($)	0	5	10	15	20	25
Probability	0.02	0.05	0.08	0.16	0.27	0.42

Make a histogram of the probability distribution. Describe its shape.

13. **Refer to Example 5.2.2 Benford's Law** Refer to Exercise 9. Calculate and interpret the expected value of X.

14. **Working Out** Refer to Exercise 10. Calculate and interpret the expected value of Y.

15. **College Costs** Refer to Exercise 11. Calculate and interpret the mean of T.

16. **Ferry Fares** Refer to Exercise 12. Compute and interpret the mean of X.

17. **Refer to Example 5.2.3 Benford's Law** Refer to Exercises 9 and 13. Calculate and interpret the standard deviation of X.

18. **Working Out** Refer to Exercises 10 and 14. Calculate and interpret the standard deviation of Y.

19. **College Costs** Refer to Exercises 11 and 15. Calculate and interpret the standard deviation of T.

20. **Ferry Fares** Refer to Exercises 12 and 16. Calculate and interpret the standard deviation of X.

Applying the Concepts These exercises ask you to apply multiple learning goals in a new context or to apply what you learned in this section in a new way.

21. **Languages** Imagine selecting a U.S. college student at random. Define the random variable X = number of languages spoken by the student. The table gives the probability distribution of X.[12]

Languages	1	2	3	4	5
Probability	0.630	0.295	0.065	0.008	0.002

(a) Make a histogram of the probability distribution. Describe its shape.
(b) Calculate and interpret the mean of X.
(c) Calculate and interpret the standard deviation of X.

22. **Skee Ball** Ana is a dedicated Skee Ball player who always rolls for the 50-point slot. The probability distribution of Ana's score X on a roll of the ball is shown here.

Score	10	20	30	40	50
Probability	0.32	0.27	0.19	0.15	0.07

(a) Make a histogram of the probability distribution. Describe its shape.
(b) Calculate and interpret the mean of X.
(c) Calculate and interpret the standard deviation of X.

23. **Life Insurance** A life insurance company sells a term insurance policy to 21-year-old males that pays $100,000 if the insured dies within the next 5 years. The probability that a randomly chosen male will die each year can be found in mortality tables. The company collects a premium of $250 each year as payment for the insurance. The amount Y that the company earns on a randomly selected policy of this type is $250 per year, less the $100,000 that it must pay if the insured dies. Here is the probability distribution of Y.

Age at death	21	22	23
Profit	−$99,750	−$99,500	−$99,250
Probability	0.00183	0.00186	0.00189
Age at death	24	25	26 or older
Profit	−$99,000	−$98,750	$1250
Probability	0.00191	0.00193	0.99058

(a) Explain why the company suffers a loss of $98,750 on such a policy if a customer dies at age 25.
(b) Calculate the expected value of Y. Explain what this result means for the insurance company.
(c) Calculate the standard deviation of Y. Explain what this result means for the insurance company.

24. **Fire Insurance** Suppose a homeowner spends $300 for a home insurance policy that will pay out $200,000 if the home is destroyed by fire. Let P = the profit made by the company on a randomly selected policy. Based on previous data, the probability that a home in this area will be destroyed by fire is 0.0002. Here is the probability distribution of P.

Profit	−$199,700	$300
Probability	0.0002	0.9998

(a) Explain why the company loses $199,700 if the policy holder's home is destroyed by fire.
(b) Calculate the expected value of P. Explain what this result means for the insurance company.
(c) Calculate the standard deviation of P. Explain what this result means for the insurance company.

Exercises 25 and 26 examine how Benford's law (Exercise 9) can be used to detect fraud.

25. **Benford's Law** A not-so-clever employee decided to fake a monthly expense report. This employee believed that the first digits of expense amounts should be equally likely to be any of the numbers from 1 to 9. In that case, the first digit Y of a randomly selected expense amount would have the probability distribution shown in the histogram.

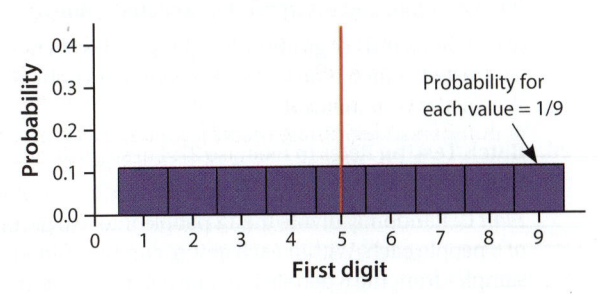

(a) According to the histogram, what is $P(Y > 6)$? According to Benford's law (see Exercise 9), what proportion of first digits in the employee's expense amounts should be greater than 6? How could this information be used to detect a fake expense report?

(b) Explain why the mean of the random variable Y is located at the solid red line in the figure.

(c) According to Benford's law, the expected value of the first digit is $\mu_X = 3.441$. Explain how this information could be used to detect a fake expense report.

26. Benford's Law

(a) Using the histogram from Exercise 25, calculate the standard deviation σ_Y. It gives us an idea of how much variation we would expect in the employee's expense records if the employee assumed that first digits from 1 to 9 were equally likely.

(b) The standard deviation of first digits of randomly selected expense amounts that follow Benford's law is $\sigma_X = 2.46$. Would using standard deviations be a good way to detect fraud? Explain your answer.

Extending the Concepts These exercises challenge you to explore statistical concepts and methods that go beyond what you learned in this section.

Exercises 27–29 explore the potential benefits of using batch testing to detect which people in a population have a disease.[13]

27. Batch Testing Suppose that 12 people need to be given a blood test for a certain disease. Assume that each person has a 5% probability of having the disease, with this chance occurring independently for each person. Consider two different plans for conducting the tests:

Plan A: Give an individual blood test to each person.

Plan B: Combine blood samples from all 12 people into one batch and test that batch.

- If at least one person has the disease, then the batch test result will be positive, and all 12 people must be tested individually.
- If no one has the disease, then the batch test result will be negative, and no additional tests will be needed.

Let X = the total number of tests needed with plan B (batch testing).

(a) Determine the probability distribution of X.

(b) If you implement plan B once, what is the probability that the number of tests needed will be smaller than are needed with plan A?

(c) Calculate and interpret the expected value of X.

(d) If thousands of groups of 12 people need to be tested, which plan — A or B — would be better? Justify your answer.

28. Batch Testing Refer to Exercise 27. Let's consider a third plan.

Plan C: Randomly divide the 12 people into two groups of 6 people each. Within each group, combine blood samples from the 6 people into one batch. Test both batches.

- A batch will test positive only if at least one person in the group has the disease. If a batch tests positive, all 6 people in that group must be tested individually.
- A batch will test negative if no one in the group has the disease. Any batch that tests negative requires no additional testing.

Let Y = the total number of tests needed with plan C (batch testing on two subgroups).

(a) Determine the probability distribution of Y.

(b) Calculate the expected value of Y.

(c) If thousands of groups of 12 people need to be tested, which plan — B or C — would be better? Justify your answer.

29. Batch Testing Refer to Exercises 27 and 28. Let's consider one more plan.

Plan D: Randomly divide the 12 people into three groups of 4 people each. Within each group, combine blood samples from the 4 people into one batch. Test all three batches.

- A batch will test positive only if at least one person in the group has the disease. If a batch tests positive, all 4 people in that group must be tested individually.
- A batch will test negative if nobody in the group has the disease. Any batch that tests negative requires no additional testing.

Let W = the total number of tests needed with Plan D (batch testing on three sub-groups).

(a) Determine the probability distribution of W.

(b) Calculate the expected value of W.

(c) If thousands of groups of 12 people need to be tested, which plan — C or D — would be better? Justify your answer.

Cumulative Review These exercises revisit what you learned in previous sections.

30. Score Change (3.4) A large set of test scores has mean 60 and standard deviation 18. Suppose each score is doubled, and then 5 is subtracted from the result.

(a) Find the mean of the new scores.

(b) Find the standard deviation of the new scores.

31. Late for Work (4.5) Some days, Omari drives to work. The rest of the time Omari rides a bike. Suppose we choose a workday at random. The table gives the probabilities of several events involving Omari.

Event	Probability
Drives to work	0.20
Drives to work and is late	0.05
Late for work, given that he bikes	0.30

(a) Find the probability that Omari is late for work, given that Omari drives.

(b) Draw a tree diagram to summarize the probabilities.

(c) On a randomly selected workday, Omari was late for work. What is the probability that Omari biked?

Section 5.3 Binomial Random Variables

LEARNING GOALS

By the end of this section, you will be able to:
- Determine whether the conditions for a binomial setting are met.
- Calculate probabilities involving a single value of a binomial random variable.
- Make a histogram to display a binomial distribution and describe its shape.

When the same chance process is repeated several times, we are often interested in how many times a particular outcome occurs. Here's an activity that illustrates this idea.

Class Activity

Pop Quiz!

It's time for a pop quiz! We hope you are ready. The quiz consists of 10 multiple-choice questions. Each question has five answer choices, labeled A through E. Now for the bad news: You will not get to see the questions. You have to guess the answer for each one!

1. On a blank piece of paper or document, number from 1 to 10. Then guess the answer to each question: A, B, C, D, or E. Do not look at anyone else's paper! You have 2 minutes.

2. Now it's time to grade the quizzes. Your instructor will display the answer key. The correct answer for each of the 10 questions was determined randomly so that A, B, C, D, or E was equally likely to be chosen.

3. How did you do on your quiz? Make a class dotplot that shows the number of correct answers for each student in your class. As a class, describe what you see.

Binomial Settings

In the *Pop Quiz!* activity, each student is performing repeated *trials* of the same chance process: A trial consists of answering a single multiple-choice question with a randomly generated correct answer. We're interested in the number of times that a specific event occurs: getting a correct answer (which we'll call a "success"). Knowing the outcome of one question (right or wrong answer) tells us nothing about the outcome of any other question. That is, the trials are independent. The number of trials is fixed in advance: $n = 10$. And a student's probability of getting a "success" is the same on each trial: $p = 1/5 = 0.2$. When these conditions are met, we have a **binomial setting.**

> A **binomial setting** arises when we perform n independent trials of the same chance process and count the number of times that a particular outcome (called a "success") occurs.
>
> The four conditions for a binomial setting are:
> - **B**inary? The possible outcomes of each trial can be classified as "success" or "failure."
> - **I**ndependent? Trials are independent. That is, knowing the result of one trial must not tell us anything about the result of any other trial.
> - **N**umber? The number of trials n of the chance process is fixed in advance.
> - **S**ame probability? There is the same probability of success p on each trial.

The boldface letters in the box give you a helpful way to remember the conditions for a binomial setting: Just check the BINS!

When checking the Binary condition, note that there can be more than two possible outcomes per trial. In the *Pop Quiz!* activity, each question (trial) had five possible answer choices: A, B, C, D, or E. If we define "success" as guessing the correct answer to a question, then "failure" occurs when the student guesses any of the four incorrect answer choices.

EXAMPLE 5.3.1

Is blood type like a card game?

Identifying binomial settings

PROBLEM

Determine whether the given scenario describes a binomial setting. Justify your answer.

(a) Genetics says that the genes children receive from their birth parents are independent from one child to another. Each child of a particular set of birth parents has probability 0.25 of having type O blood. Suppose these parents have 5 children. Count the number of children with type O blood.

(b) Shuffle a standard deck of 52 playing cards. Turn over the first 10 cards, one at a time. Record the number of aces you observe.

SOLUTION

(a) • Binary? "Success" = has type O blood. "Failure" = doesn't have type O blood.
 • Independent? Knowing one child's blood type tells you nothing about another child's blood type because each of them inherits genes independently from their parents.
 • Number? $n = 5$
 • Same probability? $p = 0.25$
 This is a binomial setting.

> Check the BINS! A trial consists of observing the blood type for one of these parents' children.

> All the conditions are met, and we are counting the number of successes (children with type O blood).

(b) • Binary? "Success" = get an ace. "Failure" = don't get an ace.
 • Independent? No. If the first card you turn over is an ace, the next card is less likely to be an ace because you're not replacing the top card in the deck. If the first card isn't an ace, the second card is more likely to be an ace.
 This is not a binomial setting because the independent condition is not met.

> Check the BINS! A trial consists of turning over a card from the deck.

> To check for independence, you could also write P(2nd card ace | 1st card ace) = 3/51 and P(2nd card ace | 1st card not ace) = 4/51. Because the two probabilities are not equal, the trials are not independent.

EXAM PREP: FOR PRACTICE, TRY EXERCISE 7.

THINK ABOUT IT

What's the difference between the Independent condition and the Same probability condition? The Independent condition involves *conditional* probabilities. In part (b) of Example 5.3.1,

$$P(\text{2nd card ace | 1st card ace}) = 3/51 \neq P(\text{2nd card ace | 1st card not ace}) = 4/51$$

so the trials are not independent. The Same probability of success condition is about *unconditional* probabilities. Because $P(x\text{th card in a shuffled deck is an ace}) = 4/52$, this condition is met in part (b) of the example. Be sure you understand the difference between these two conditions. *When sampling is done without replacement, the Independent condition is violated.* The Same probability condition would be violated if someone added a few aces to the deck between two of the trials.

The blood type scenario in part (a) of Example 5.3.1 is a binomial setting. If we let X = the number of children with type O blood, then X is a **binomial random variable.**

The count of successes X in a binomial setting is a **binomial random variable**. The possible values of X are 0, 1, 2, ..., n.

In the *Pop Quiz!* activity at the beginning of the section, the binomial random variable is X = the number of correct answers. Note that X is a discrete random variable because it can take only whole-number values from 0 to 10. This is true in general: Binomial random variables are discrete random variables.

Calculating Binomial Probabilities

Let's return to the scenario from part (a) of Example 5.3.1:

> Genetics says that the genes children receive from their birth parents are independent from one child to another. Each child of a particular set of birth parents has probability 0.25 of having type O blood. Suppose these parents have 5 children. Count the number of children with type O blood.

In this binomial setting, a child with type O blood is a "success" (S) and a child with another blood type is a "failure" (F). The count X of children with type O blood is a binomial random variable with $n = 5$ trials and probability of success $p = 0.25$ on each trial.

- What's $P(X = 0)$? That is, what's the probability that *none* of the 5 children has type O blood? The probability that any *one* of this couple's children doesn't have type O blood is $1 - 0.25 = 0.75$ (complement rule). By the multiplication rule for independent events (Section 4.6),

$$P(X = 0) = P(\text{FFFFF}) = (0.75)(0.75)(0.75)(0.75)(0.75) = (0.75)^5 = 0.23730$$

- How about $P(X = 1)$? There are several different ways in which exactly 1 of the 5 children could have type O blood. For instance, the first child born might have type O blood, while the remaining 4 children don't have type O blood. The probability that this happens is

$$P(\text{SFFFF}) = (0.25)(0.75)(0.75)(0.75)(0.75) = (0.25)^1(0.75)^4$$

Alternatively, Child 2 could be the one that has type O blood. The corresponding probability is

$$P(\text{FSFFF}) = (0.75)(0.25)(0.75)(0.75)(0.75) = (0.25)^1(0.75)^4$$

There are three more possibilities to consider — the situations in which Child 3, Child 4, and Child 5 are the only ones who inherit type O blood. Of course, the probability will be the same for each of those cases. In all, there are five different ways in which exactly one child would have type O blood, each with the same probability of occurring. As a result,

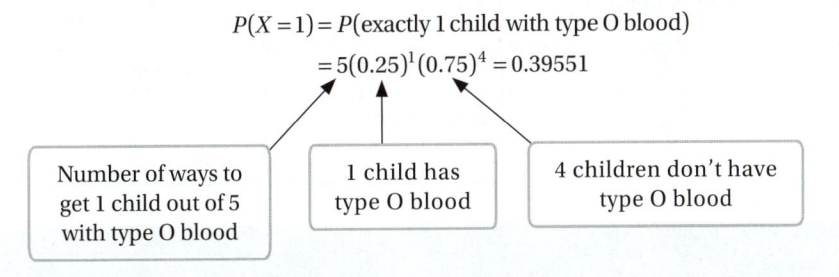

$$P(X = 1) = P(\text{exactly 1 child with type O blood})$$
$$= 5(0.25)^1(0.75)^4 = 0.39551$$

- Number of ways to get 1 child out of 5 with type O blood
- 1 child has type O blood
- 4 children don't have type O blood

Where did the 5 come from in this formula? It's the number of ways to choose which one of the 5 children inherits type O blood. In Section 4.8, we wrote this as $_5C_1$, the number of combinations of 5 individuals taken 1 at a time. So we can rewrite the formula as

$$P(X = 1) = {}_5C_1(0.25)^1(0.75)^4 = 0.39551$$

The pattern of this calculation works for any binomial probability.

Binomial Probability Formula

Suppose that X is a binomial random variable with n trials and probability p of success on each trial. The probability of getting exactly r successes in n trials (for $r = 0, 1, 2, ..., n$) is

$$P(X = r) = {}_nC_r(p)^r(1-p)^{n-r}$$

where

$$_nC_r = \frac{n!}{r!(n-r)!}$$

The binomial probability formula looks complicated, but it is easier to understand if you know what each part of the formula represents:

$$P(X=r) = \binom{\text{number of ways}}{\text{to get } r \text{ successes in } n \text{ trials}} \binom{\text{success}}{\text{probability}}^{\text{number of successes}} \binom{\text{failure}}{\text{probability}}^{\text{number of failures}}$$

EXAMPLE 5.3.2

How many correct on the pop quiz?

Using the binomial probability formula

PROBLEM

To introduce a statistics class to binomial distributions, Professor Desai does the *Pop Quiz!* activity, in which each student in the class guesses an answer from A through E on each of 10 multiple-choice questions. Professor Desai determined the "correct" answer for each of the 10 questions randomly so that A, B, C, D, or E was equally likely to be chosen. Let X = the number of questions that a specific student answers correctly.

Find $P(X = 4)$. Describe this probability in words.

SOLUTION

X is a binomial random variable with $n = 10$ and $p = 1/5 = 0.2$.

$P(X = 4) = {}_{10}C_4 (0.2)^4 (0.8)^6$

$\quad\quad\quad\quad = 210(0.2)^4 (0.8)^6$

$\quad\quad\quad\quad = 0.088$

The probability that this student answers exactly 4 questions correctly on the pop quiz is 0.088.

> $P(X = r) = {}_nC_r (p)^r (1-p)^{n-r}$
>
> Use one of the methods presented in Section 4.8 to find ${}_{10}C_4$:
>
> $${}_{10}C_4 = \frac{{}_{10}P_4}{4!} = \frac{10!}{6!\,4!} = 210$$
>
> or, with technology,
>
> $${}_{10}C_4 = 210$$

EXAM PREP: FOR PRACTICE, TRY EXERCISE 11.

You can also use technology to calculate binomial probabilities. See the Tech Corner prior to Example 5.3.3 for details.

Binomial Distributions and Shape

What does the probability distribution of a binomial random variable look like? The table shows the possible values and corresponding probabilities for X = the number of children with type O blood from earlier in the section. This is a binomial random variable with $n = 5$ and $p = 0.25$. The probability distribution of X is called a **binomial distribution.**

Number with type O blood	0	1	2	3	4	5
Probability	0.23730	0.39551	0.26367	0.08789	0.01465	0.00098

The probability distribution of a binomial random variable is a **binomial distribution**. Any binomial distribution is completely specified by two numbers: the number of trials n of the chance process and the probability p of success on each trial.

A graph of the probability distribution of X = the number of children with type O blood is shown in Figure 5.2. This binomial distribution with $n = 5$ and $p = 0.25$ has a clear right-skewed shape with a single peak at $X = 1$.

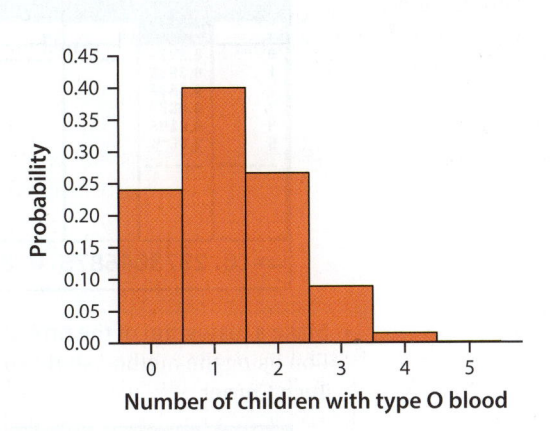

FIGURE 5.2 Histogram showing the probability distribution of the binomial random variable X = number of children with type O blood in a family with 5 children from the same birth parents.

Figure 5.3 shows two more binomial distributions with different shapes. The binomial distribution with $n = 5$ and $p = 0.51$ in Figure 5.3(a) is roughly symmetric. The binomial distribution with $n = 5$ and $p = 0.8$ in Figure 5.3(b) is skewed to the left. In general, when n is small, the probability distribution of a binomial random variable will be roughly symmetric if p is close to 0.5, right-skewed if p is much less than 0.5, and left-skewed if p is much greater than 0.5.

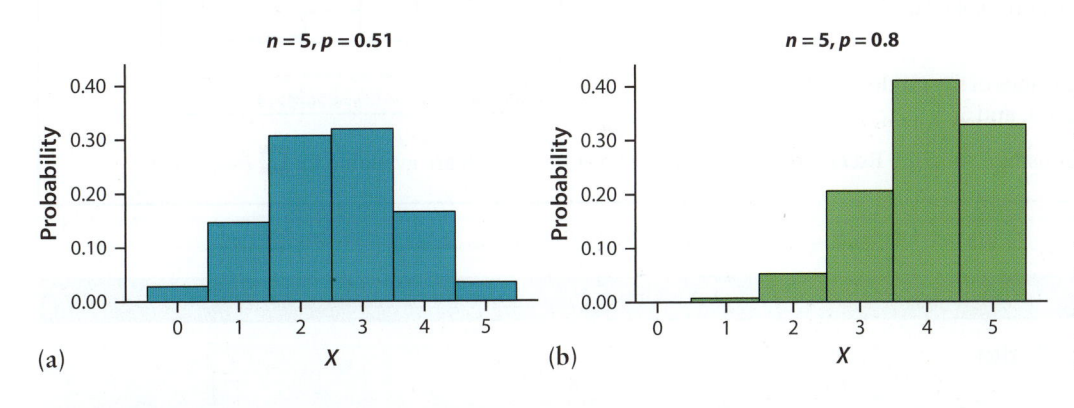

FIGURE 5.3 (a) Probability histogram for the binomial random variable X with $n = 5$ and $p = 0.51$. This binomial distribution is roughly symmetric. (b) Probability histogram for the binomial random variable X with $n = 5$ and $p = 0.8$. This binomial distribution has a left-skewed shape.

TECH CORNER Analyzing Binomial Distributions

You can use technology to graph a binomial distribution and to calculate binomial probabilities. We'll illustrate this process using the familiar binomial random variable X = the number of children with type O blood, with $n = 5$ and $p = 0.25$. Suppose we want to find the probability that exactly 3 of the family's 5 children have type O blood and to make a histogram of the probability distribution.

Applet

1. Go to www.stapplet.com and launch the *Binomial Distributions* applet.
2. Enter $n = 5$ and $p = 0.25$. Then click "Plot distribution" to see a graph of the corresponding binomial distribution.
3. Choose "exactly" from the drop-down menu, type "3," and click "Go!" to find the probability of *exactly* 3 successes. The desired probability is given and illustrated in the graph.

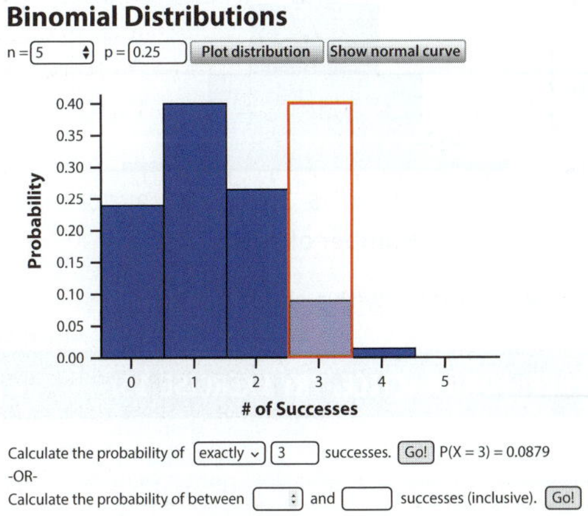

(continued)

TI-83/84

The TI-83/84 command binompdf(n,p,r) computes $P(X = r)$ in a setting with n trials and success probability p on each trial.

1. To calculate $P(X = 3)$, press `2nd` `VARS` (DISTR) and choose binompdf(.

   OS 2.55 or later: In the dialog box, enter these values: trials:5, p:0.25, x value:3. Choose Paste, and then press `ENTER`.

   Older OS: Complete the command binompdf(5,0.25,3) and press `ENTER`.

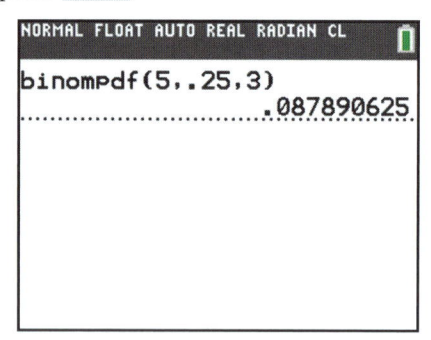

   ```
   NORMAL FLOAT AUTO REAL RADIAN CL
   binompdf(5,.25,3)
                      .087890625
   ```

2. To graph the binomial probability distribution for $n = 5$ and $p = 0.25$:

   - Type the possible values of the random variable X into list L_1: 0, 1, 2, 3, 4, and 5.

   - Highlight L_2 with the cursor. Enter the command binompdf(5,0.25,L_1) and press `ENTER`.

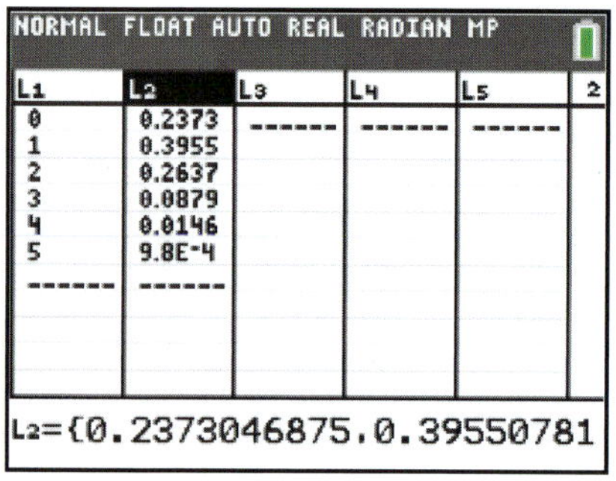

   ```
   NORMAL FLOAT AUTO REAL RADIAN MP
   L1   L2      L3   L4   L5    2
   0    0.2373  ---- ---- ----
   1    0.3955
   2    0.2637
   3    0.0879
   4    0.0146
   5    9.8E-4
   ----------
   L2={0.2373046875,0.39550781
   ```

   - Make a histogram of the probability distribution using the method shown in the Section 5.2 Tech Corner.

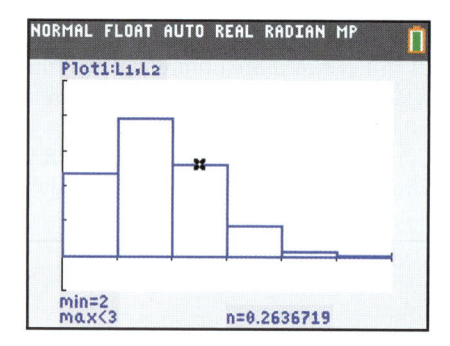

Detailed instructions for using CrunchIt!, Excel, Google Sheets, JMP, Minitab, and R are available in Achieve.

EXAMPLE 5.3.3

Graphing a binomial distribution

Coin tosses

PROBLEM

Imagine that you toss a fair coin 10 times. Let Y = the number of heads you get. Make a histogram of the probability distribution of Y. Describe its shape.

SOLUTION

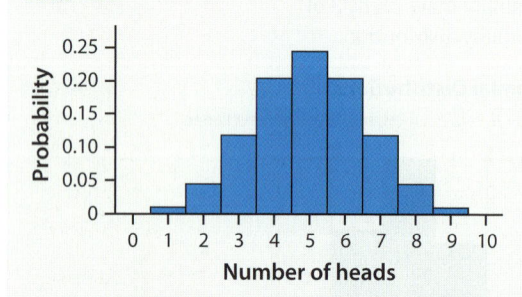

> We used technology to generate the graph of the binomial distribution with $n = 10$ and $p = 0.5$.

The graph is symmetric with a single peak at $Y = 5$.

EXAM PREP: FOR PRACTICE, TRY EXERCISE 15.

Section 5.3 What Did You Learn?

Review the learning goals from this section. Then practice what you've learned by working through the exercises.

Learning Goal	Example	Exercises
Determine whether the conditions for a binomial setting are met.	5.3.1	7–10
Calculate probabilities involving a single value of a binomial random variable.	5.3.2	11–14
Make a histogram to display a binomial distribution and describe its shape.	5.3.3	15–18

Section 5.3 Exercises

Building Concepts and Skills These exercises assess the basic knowledge you should have after reading the section.

1. A binomial setting arises when we perform n _____ trials of the same chance process and count the number of times that a particular outcome, called a _____, occurs.

2. Name the four conditions for a binomial setting. (*Hint:* BINS!)

3. True/False: The possible values of a binomial random variable are $1, 2, \ldots, n$.

4. What formula should you use to calculate the probability of getting exactly r successes in n trials for a binomial random variable X with probability p of success on each trial?

5. A binomial distribution is completely specified by two numbers: _____ and _____.

6. True/False: The shape of a binomial distribution can be left-skewed, symmetric, or right-skewed.

Mastering Concepts and Skills These exercises reinforce the learning goals as shown in the examples.

7. **Refer to Example 5.3.1 Baby Elk** Biologists estimate that a randomly selected baby elk has a 44% probability of surviving to adulthood. Assume this estimate is correct. Suppose researchers choose 7 baby elk at random to monitor. Let X = the number that survive to adulthood. Determine whether this is a binomial setting. Justify your answer.

8. **Bull's-Eye!** Lawrence likes to shoot a bow and arrow. On any shot, Lawrence has about a 10% probability of hitting the bull's-eye. As a challenge one day, Lawrence decides to keep shooting until he gets a bull's-eye. Let X = the number of shots Lawrence takes. Determine whether this is a binomial setting. Justify your answer.

9. **Student Names** Put the first names of all the students in your statistics class in a hat. Mix them up, and draw 4 names without looking. Let Y = the number of students whose first names have more than six letters. Determine whether this is a binomial setting. Justify your answer.

10. **Rolling Doubles** When rolling two fair, 6-sided dice, the probability of rolling doubles is 1/6. Suppose you roll the dice 4 times. Let W = the number of times you roll doubles. Determine whether this is a binomial setting. Justify your answer.

11. **Refer to Example 5.3.2 Baby Elk** Refer to Exercise 7. Find the probability that exactly 4 of the baby elk survive to adulthood.

12. **Rolling Doubles** Refer to Exercise 10. Find the probability that you roll doubles twice.

13. **Train Arrivals** According to New Jersey Transit, the 8:00 a.m. weekday train from Princeton to New York City has a 90% chance of arriving on time on a given day. Suppose this claim is true. Choose 6 weekdays at random. Let Y = the number of days on which the train arrives on time.
    (a) Explain why Y is a binomial random variable.
    (b) Find $P(Y = 4)$. Describe this probability in words.

14. **Red Light!** Pedro drives the same route to work on Monday through Friday. Pedro's route includes one traffic light. According to the local traffic department, there is a 55% probability that the light will be red when Pedro arrives at the intersection on a given workday. Suppose we choose 10 of Pedro's workdays at random and let Y = the number of times that the light is red.
    (a) Explain why Y is a binomial random variable.
    (b) Find $P(Y = 8)$. Describe this probability in words.

15. **Refer to Example 5.3.3 Baby Elk** Refer to Exercise 7. Make a histogram of the probability distribution of X. Describe its shape.

16. **Rolling Doubles** Refer to Exercise 10. Make a histogram of the probability distribution of W. Describe its shape.

17. **Train Arrivals** Refer to Exercise 13. Make a histogram of the probability distribution of Y. Describe its shape.

18. **Red Light!** Refer to Exercise 14. Make a histogram of the probability distribution of Y. Describe its shape.

Applying the Concepts These exercises ask you to apply multiple learning goals in a new context or to apply what you learned in this section in a new way.

19. **Bag Check** Thousands of travelers pass through the airport in Guadalajara, Mexico, each day. Before leaving the airport, each passenger must go through the customs inspection area. Customs agents want to be sure that passengers are not bringing illegal items into the country, but they do not have time to search every traveler's luggage. Instead, they require each person to press a button. Either a red or a green bulb lights up. If a red light flashes, the passenger will be searched by customs agents. A green light means it is OK for the passenger to "go ahead." Customs agents claim that the light has probability 0.30 of showing red on any push of the button. Assume for now that this claim is true.

    Suppose we watch 20 passengers press the button. Let R = the number who get a red light.

    (a) Explain why R is a binomial random variable.
    (b) Find the probability that exactly 6 of the 20 passengers get a red light.
    (c) Make a histogram of the probability distribution of R. Describe its shape.

20. **Cranky Mower** A company has developed an "easy-start" mower that cranks the engine with the push of a button. The company claims that the probability the mower will start on any push of the button is 0.9. Assume for now that this claim is true.

    On the next 10 uses of the mower, let Y = the number of times it starts on the first push of the button.

    (a) Explain why Y is a binomial random variable.
    (b) Find the probability that the mower starts exactly 9 times in 10 uses.
    (c) Make a histogram of the probability distribution of Y. Describe its shape.

Extending the Concepts These exercises challenge you to explore statistical concepts and methods that go beyond what you learned in this section.

21. **Exploring Shape** In this exercise, you will use the *Binomial Distributions* applet at www.stapplet.com to investigate what happens to the shape of a binomial distribution as the sample size increases. For the specified value of p, use the applet to make a graph of the probability distribution and describe its shape for each of these sample sizes: $n = 4$, $n = 20$, $n = 100$.

    (a) $p = 0.2$
    (b) $p = 0.8$

A *geometric setting* arises when we perform independent trials of the same chance process and record the number of trials it takes to get one success. On each trial, the probability p of success must be the same. In a geometric setting, if we define the random variable X to be the number of trials needed to get the first success, then X is called a *geometric random variable*. The probability distribution of X is a *geometric distribution* with probability of success p on any trial. Note that the possible values of X are 1, 2, 3, ... Exercises 22 and 23 introduce you to geometric random variables.

22. **Cranky Mower** To start an old lawn mower, Jordyn has to pull its cord and hope for some luck. On any particular pull, the mower has a 20% chance of starting. Let X be the number of pulls it takes Jordyn to start the mower.

    (a) Explain why X is a geometric random variable.
    (b) Find the probability that it takes Jordyn only 1 pull to start the mower.
    (c) Find the probability that it takes Jordyn exactly 2 pulls to start the mower.
    (d) Find the probability that it takes Jordyn exactly 3 pulls to start the mower.
    (e) Generalize the results in parts (a)–(d) to find the probability it takes Jordyn exactly r pulls to start the mower.

23. **Cranky Mower** Refer to Exercise 22.

    (a) Explain why it would be difficult to make a histogram of the probability distribution of X.
    (b) What shape would the histogram have? Justify your answer.

Cumulative Review These exercises revisit what you learned in previous sections.

24. **Gluten-Free Diets** (1.5, 1.6) Thirty percent of U.S. individuals say they're trying to reduce or eliminate gluten in their diets.[14] Yet only about 1% of the population has an autoimmune response to gluten.

    (a) Many people wonder if a gluten-free diet offers benefits to more than just those individuals with a diagnosed medical condition. This relationship can be difficult to establish because of the placebo effect. Explain why.
    (b) Suppose you have 50 volunteers with gastrointestinal problems that are often associated with "gluten sensitivity." There are dietary supplements available that contain gluten and similar ones that do not. Outline a completely randomized design for an experiment to test the impact of a gluten-free diet on people with these gastrointestinal problems.

25. **Super Bowl Ratings** (2.3, 2.4, 2.5, 3.1, 3.2) The Nielsen Company collects data on the television-viewing habits of U.S. households. One measurement of a show's popularity is the "share," which looks at only the households in which at least one television is turned on, and calculates the percentage of those televisions that are tuned to a particular show. The Nielsen share scores for every NFL Super Bowl from 1967 through 2020 follow.[15] Make an appropriate graph to display the distribution of Nielsen shares. Describe what you see.

    71 68 62 63 62 69 64 63 61 66 69 65 61 67 66 71 64 61
    65 69 62 61 68 61 63 63 68 62 66 70 63 71 69 73 63 67
    74 67 73 78 72 73 72 74 75 69 70 68 79 72 70 68 67 69

Section 5.4 Analyzing Binomial Random Variables

LEARNING GOALS

By the end of this section, you will be able to:

- Use a formula to calculate probabilities involving several values of a binomial random variable.
- Use technology to calculate probabilities involving several values of a binomial random variable.
- Calculate and interpret the mean and standard deviation of a binomial random variable.

In Section 5.3, you learned how to check the conditions for a binomial setting and how to calculate probabilities involving a single value of a binomial random variable. You also used technology to graph the probability distribution of a binomial random variable. In this section, you will learn two ways to find probabilities involving several values of a binomial random variable. Then we will show you how to find the mean and standard deviation of a binomial random variable.

Calculating Binomial Probabilities Involving Several Values

In Section 5.3, we considered this scenario: According to the science of genetics, the genes children receive from their birth parents are independent from one child to another. Each child of a particular set of birth parents has probability 0.25 of having type O blood. Suppose these parents have 5 children. Let X = the number of children with type O blood.

Recall the binomial probability formula from Section 5.3:

$$P(X=r) = {}_nC_r(p)^r(1-p)^{n-r}$$

We can use this formula to find the probability that exactly 3 of the 5 children will have type O blood:

$$P(X=3) = {}_5C_3(0.25)^3(0.75)^2 = 0.08789$$

What if we want to find the probability that *at least* 2 of the couple's 5 children have type O blood? In symbols, that's $P(X \geq 2)$. We could compute this probability by using the binomial probability formula four times:

$$\begin{aligned}P(X \geq 2) &= P(X=2) + P(X=3) + P(X=4) + P(X=5) \\ &= {}_5C_2(0.25)^2(0.75)^3 + {}_5C_3(0.25)^3(0.75)^2 + {}_5C_4(0.25)^4(0.75)^1 + {}_5C_5(0.25)^5(0.75)^0 \\ &= 0.26367 + 0.08789 + 0.01465 + 0.00098 \\ &= 0.36719\end{aligned}$$

A clever alternative is to use the complement rule:

$$\begin{aligned}P(X \geq 2) &= 1 - P(X \leq 1) \\ &= 1 - [P(X=0) + P(X=1)] \\ &= 1 - [{}_5C_0(0.25)^0(0.75)^5 + {}_5C_1(0.25)^1(0.75)^4] \\ &= 1 - [0.23730 + 0.39551] \\ &= 1 - 0.63281 \\ &= 0.36719\end{aligned}$$

Notice that this strategy requires us to add only *two* binomial probabilities!

Here's a helpful tip to avoid making mistakes on probability questions that involve several values of a binomial random variable: Write out the possible values of the variable, circle the ones for which you want to find the probability, and cross out the rest. In this scenario, X can take values from 0 to 5 and we want to find $P(X \geq 2)$:

$$\cancel{0} \quad \cancel{1} \quad \boxed{2 \quad 3 \quad 4 \quad 5}$$

Crossing out the values for 0 and 1 shows why $P(X \geq 2) = 1 - P(X \leq 1)$.

EXAMPLE 5.4.1

Finding binomial probabilities involving several values by formula

Pop quiz

PROBLEM

Let's return to Example 5.3.2 involving Professor Desai's class and the *Pop Quiz!* activity. Recall that each student in the class guessed an answer from A through E on each of the 10 multiple-choice questions. Professor Desai determined the "correct" answer for each of the 10 questions randomly so that A, B, C, D, and E were equally likely to be chosen. Let X = the number of correct answers that a student gets on the quiz.

Octaviah is one of the students in this class. Find the probability that Octaviah gets more than 2 correct answers on the pop quiz.

SOLUTION

$P(X > 2) = 1 - P(X \leq 2)$
$\quad\quad\quad = 1 - [P(X = 0) + P(X = 1) + P(X = 2)]$
$\quad\quad\quad = 1 - [_{10}C_0(0.2)^0(0.8)^{10} + {}_{10}C_1(0.2)^1(0.8)^9 + {}_{10}C_2(0.2)^2(0.8)^8]$
$\quad\quad\quad = 1 - [0.1074 + 0.2684 + 0.3020]$
$\quad\quad\quad = 1 - 0.6778$
$\quad\quad\quad = 0.3222$

You want to find $P(X > 2)$.

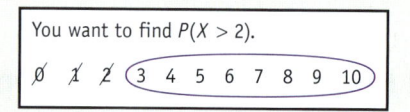

0̸ 1̸ 2̸ ③ 4 5 6 7 8 9 10

EXAM PREP: FOR PRACTICE, TRY EXERCISE 5.

It is tedious to use the binomial probability formula for calculations involving many values of a binomial random variable. Technology is a more practical alternative.

TECH CORNER | Calculating Binomial Probabilities Involving Several Values

You can use technology to calculate probabilities involving several values of a binomial random variable. Let's use technology to confirm our answer to Example 5.4.1. Recall that we were trying to find the probability that Octaviah gets more than 2 correct answers on Professor Desai's pop quiz, and that $n = 10$ and $p = 0.2$.

Applet

1. Go to www.stapplet.com and launch the *Binomial Distributions* applet.

2. Enter $n = 10$ and $p = 0.2$. Then click "Plot distribution." A graph of the appropriate binomial distribution is shown.

3. To find $P(X > 2)$, choose "more than" from the drop-down menu, type 2 in the adjacent box, and click "Go!" The desired probability is illustrated in the graph and displayed to the right of the Go! button.

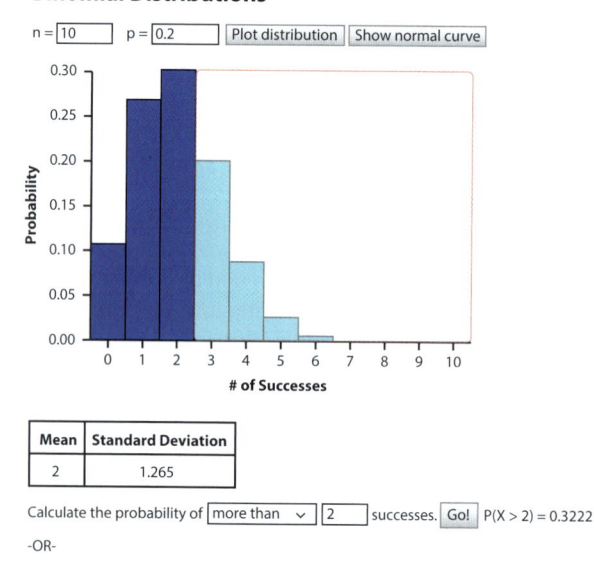

Binomial Distributions

$n = 10$ $p = 0.2$ Plot distribution Show normal curve

Mean	Standard Deviation
2	1.265

Calculate the probability of [more than ▽] [2] successes. Go! $P(X > 2) = 0.3222$

-OR-

Calculate the probability of between [ ] and [ ] successes (inclusive). Go!

TI-83/84

The TI-83/84 command binomcdf(n,p,r) computes $P(X \leq r)$ in a setting with n trials and success probability p on each trial. Note that $P(X > 2) = 1 - P(X \leq 2)$.

- Press 2nd VARS (DISTR) and choose binomcdf(.

  OS 2.55 or later: In the dialog box, enter these values: trials:10, p:0.2, x value:2. Choose Paste, and then press ENTER. Subtract this result from 1 to get the answer.

  Older OS: Complete the command binomcdf(10,0.2,2) and press ENTER. Subtract this result from 1 to get the answer.

Note: The "c" in the TI-83/84 binomcdf command stands for "cumulative." That's because binomcdf(n,p,r) finds $P(X \leq r)$, which is a cumulative probability.

Detailed instructions for using CrunchIt!, Excel, Google Sheets, JMP, Minitab, and R are available in Achieve.

EXAMPLE 5.4.2

Free lunch?

Finding binomial probabilities involving several values with technology

PROBLEM

A local fast-food restaurant is running a "Draw a three, get it free" lunch promotion. After each customer orders, a touchscreen display shows the message "Press here to win a free lunch." A computer program then simulates one card being drawn from a standard deck. If the chosen card is a 3, the customer's order is free. (Note that the probability of drawing a 3 from a standard deck of playing cards is 4/52.) Otherwise, the customer must pay the bill.

(a) On the first day of the promotion, 250 customers place lunch orders. Find the probability that fewer than 10 of them win a free lunch.

(b) In fact, only 9 customers won a free lunch. Does this result give convincing evidence that the computer program is flawed?

SOLUTION

(a) Let Y = the number of customers who win a free lunch. Y has a binomial distribution with $n = 250$ and $p = 4/52$.

$P(Y < 10) = P(Y \leq 9)$

= Applet/binomcdf(trials: 250, p: 4/52, x value: 9)

= 0.00613

(b) Because it is unlikely (probability = 0.00613) that fewer than 10 customers would win a free lunch if the computer program is working properly, there is convincing evidence that the computer program is flawed.

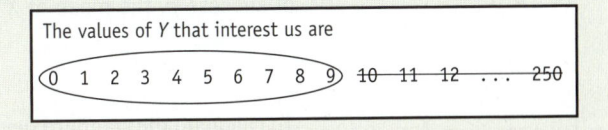

Recall our guideline from Chapter 1: Classify an observed result as unusual if it happens less than 5% of the time by chance alone.

EXAM PREP: FOR PRACTICE, TRY EXERCISE 9.

The Mean and Standard Deviation of a Binomial Random Variable

Let's return to the birth parents with 5 children one more time. Recall that X = the number of children with type O blood. We determined that X is a binomial random variable with $n = 5$ and $p = 0.25$. Its probability distribution is shown in the histogram and the table.

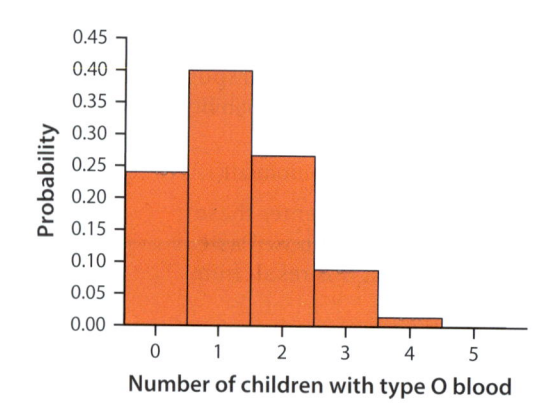

Number with type O blood	0	1	2	3	4	5
Probability	0.23730	0.39551	0.26367	0.08789	0.01465	0.00098

Because X is a discrete random variable, we can calculate its mean (expected value) using the formula

$$\mu_X = E(X) = x_1 p_1 + x_2 p_2 + x_3 p_3 + \ldots$$

from Section 5.2. We get

$$\mu_X = (0)(0.23730) + (1)(0.39551) + \ldots + (5)(0.00098)$$
$$= 1.25$$

Interpretation: If we randomly select many, many families like this one with 5 children, the average number of children per family with type O blood will be about 1.25.

Did you think about why the mean of X is 1.25? Because each child has a 0.25 probability of inheriting type O blood, we would expect 25% of the 5 children to have this blood type. In other words,

$$\mu_X = 5(0.25) = 1.25$$

This method can be used to find the mean of any *binomial* random variable.

Calculating the Mean of a Binomial Random Variable

If a count X of successes has a binomial distribution with number of trials n and probability of success p, the mean of X is

$$\mu_X = np$$

To calculate the standard deviation of X, we use the formula

$$\sigma_X = \sqrt{(x_1 - \mu_X)^2 p_1 + (x_2 - \mu_X)^2 p_2 + (x_3 - \mu_X)^2 p_3 + \ldots}$$

from Section 5.2 with $\mu_X = 1.25$. We get

$$\sigma_X = \sqrt{(0-1.25)^2(0.23730) + (1-1.25)^2(0.39551) + \ldots + (5-1.25)^2(0.00098)}$$
$$= \sqrt{0.9375}$$
$$= 0.968$$

Interpretation: If we randomly select many, many families like this one with 5 children, the number of children per family with type O blood will typically vary from the mean of 1.25 by about 0.968 children.

There is a simple formula for the standard deviation of a binomial random variable, but it isn't easy to explain. For our family with $n = 5$ children and $p = 0.25$ of type O blood, the variance of X is

$$5(0.25)(1 - 0.25) = 0.9375$$

To get the standard deviation, we just take the square root:

$$\sigma_X = \sqrt{5(0.25)(1 - 0.25)} = 0.968$$

This method works for any binomial random variable.

Calculating the Standard Deviation of a Binomial Random Variable

If a count X of successes has a binomial distribution with number of trials n and probability of success p, the standard deviation of X is

$$\sigma_X = \sqrt{np(1-p)}$$

⚠ Remember that these formulas for the mean and standard deviation work *only* for binomial distributions.

EXAMPLE 5.4.3

Pop quiz

Mean and SD of a binomial distribution

PROBLEM

Let's return to Example 5.4.1 involving Professor's Desai's class and the *Pop Quiz!* activity. Recall that X = the number of correct answers a student gets on the quiz.

(a) Calculate and interpret the mean of X.
(b) Calculate and interpret the standard deviation of X.

SOLUTION

The random variable X has a binomial distribution with $n = 10$ and $p = 1/5 = 0.2$.

(a) $\mu_X = 10(0.2) = 2$.

If many, many students took the pop quiz, we'd expect them to get about 2 answers correct, on average.

> The mean of a binomial random variable is $\mu_X = np$.

(b) $\sigma_X = \sqrt{10(0.2)(0.8)} = 1.265$

If many, many students took the quiz, their scores would typically vary from the mean of 2 by about 1.265 correct answers.

> The standard deviation of a binomial random variable is $\sigma_X = \sqrt{np(1-p)}$.

EXAM PREP: FOR PRACTICE, TRY EXERCISE 13.

Section 5.4 What Did You Learn?

Review the learning goals from this section. Then practice what you've learned by working through the exercises.

Learning Goal	Example	Exercises
Use a formula to calculate probabilities involving several values of a binomial random variable.	5.4.1	5–8
Use technology to calculate probabilities involving several values of a binomial random variable.	5.4.2	9–12
Calculate and interpret the mean and standard deviation of a binomial random variable.	5.4.3	13–16

Section 5.4 Exercises

Building Concepts and Skills These exercises assess the basic knowledge you should have after reading the section.

1. What helpful tip is provided in the section to avoid making mistakes on probability questions that involve several values of a binomial random variable?

2. The two ways to find binomial probabilities involving several values are by _____ and using _____.

3. True/False: The formula $\mu_X = np$ can be used to find the mean of any discrete random variable X.

4. What is the formula for the standard deviation of a binomial random variable?

Mastering Concepts and Skills These exercises reinforce the learning goals as shown in the examples.

5. **Refer to Example 5.4.1 Baby Elk** Biologists estimate that a randomly selected baby elk has a 44% probability of surviving to adulthood. Assume this estimate is correct. Suppose researchers choose 7 baby elk at random to monitor. Let X = the number who survive to adulthood. Find the probability that fewer than 3 of the elk survive to adulthood.

6. **Rolling Doubles** When rolling two fair, 6-sided dice, the probability of rolling doubles is 1/6. Suppose you roll the dice 4 times. Let W = the number of times you roll doubles. Find the probability that you roll doubles more than twice.

7. **Smelling Parkinson's** The Section 5.1 Class Activity described an experiment to test whether Joy Milne could smell Parkinson's disease. Joy was presented with 12 shirts, each worn by a different person, some of whom had Parkinson's disease and some of whom did not. The shirts were given to Joy in random order, and she had to decide whether or not each shirt was worn by a patient with Parkinson's disease. Joy identified 11 of the 12 shirts correctly. If we assume that Joy was just guessing, her probability of correctly identifying each shirt was 1/2.

    (a) Find the probability that Joy would identify at least 11 shirts correctly by random guessing.

    (b) Based on Joy's results and your answer to part (a), is there convincing evidence that Joy really can smell Parkinson's disease? Explain your reasoning.

    Note: The researchers later discovered that Joy had correctly identified all 12 shirts. Her one "mistake" was a person who was diagnosed with Parkinson's disease a few months later.

8. **Winning Soda** As a special promotion for its 20-ounce bottles of soda, a soft drink company printed a message on the inside of each cap. Some of the caps said, "Please try again," while others said, "You're a winner!" The company advertised the promotion with the slogan "1 in 6 wins a prize." Suppose the company is telling the truth and that every 20-ounce bottle of soda it fills has a 1-in-6 chance of being a winner. Grayson's statistics class wonders if the company's claim holds true at a nearby convenience store. To find out, all 30 students in the class go to the store and each buys one 20-ounce bottle of the soda.

    (a) Find the probability that two or fewer students would win a prize if the company's claim is true.

    (b) Two of the students in Grayson's class got caps that say, "You're a winner!" Does this result give convincing evidence that the company's 1-in-6 claim is false? Explain your reasoning.

9. **Refer to Example 5.4.2 Last Kiss** Do people have a preference for the last thing they taste? Researchers at the University of Michigan designed a study to find out. The researchers gave 22 students five different Hershey's Kisses (milk chocolate, dark chocolate, crème, caramel, and almond) in random order and asked the student to rate each candy. Participants were not told how many Kisses they would taste. However, when the fifth and final Kiss was presented, participants were told that it would be their last one.[16]

    Assume that the participants in the study don't have a special preference for the last thing they taste—that is, the probability of preferring the last Kiss tasted is $p = 0.20$. Let X = the number of participants who choose the last Kiss. Of the 22 students, 14 gave the final Kiss the highest rating.

    (a) Compute $P(X \geq 14)$ with technology.

    (b) Based on the results of the study and your answer to part (a), do the data give convincing evidence that the participants have a preference for the last thing they taste?

10. **Diet Cola** The makers of a diet cola claim that its taste is indistinguishable from the taste of the full-calorie version of the same cola. To investigate this claim, a researcher prepared small samples of each type of soda in identical cups. Then the researcher had volunteers taste each cola in random order and try to identify which was the diet cola and which was the regular cola.[17]

    If we assume that the volunteers couldn't tell the difference, each one was guessing with a 1/2 chance of being correct. Let X = the number of volunteers who correctly identify the colas. Of the 30 volunteers, 23 made correct identifications.

    (a) Compute $P(X \geq 23)$ with technology.

    (b) Based on the results of the study and your answer to part (a), do the data give convincing evidence that the volunteers can taste the difference between the diet and regular colas?

11. **Train Arrivals** According to New Jersey Transit, the 8:00 a.m. weekday train from Princeton to New York City has a 90% chance of arriving on time on a given day. Suppose this claim is true. Choose 6 weekdays at random. Let Y = the number of days on which the train arrives on time. Find $P(Y \leq 4)$. Describe this probability in words.

12. **Red Light!** Pedro drives the same route to work on Monday through Friday. Pedro's route includes one traffic light. According to the local traffic department, there is a 55% probability that the light will be red when Pedro arrives at the intersection on a given workday. Suppose we choose 10 of Pedro's workdays at random and let X = the number of times that the light is red. Find $P(X \leq 2)$. Describe this probability in words.

13. **Refer to Example 5.4.3 Random Dialing** When a polling company calls a telephone number at random,

there is only a 9% probability that the call reaches a live person and the survey is successfully completed.[18] Suppose the random digit dialing machine makes 15 calls. Let X = the number of calls that result in a completed survey.

(a) Calculate and interpret the mean of X.

(b) Calculate and interpret the standard deviation of X.

14. **Lie Detectors** A federal report finds that lie detector tests given to truthful persons have probability 0.2 of suggesting that the person is deceptive.[19] A company asks 12 job applicants about thefts from previous employers, using a lie detector to assess their truthfulness. Suppose that all 12 answer truthfully. Let X = the number of people whom the lie detector identifies as being deceptive.

(a) Calculate and interpret the mean of X.

(b) Calculate and interpret the standard deviation of X.

15. **Train Arrivals** Refer to Exercise 11.

(a) Calculate and interpret μ_Y.

(b) Calculate and interpret σ_Y.

16. **Red Light!** Refer to Exercise 12.

(a) Calculate and interpret μ_X.

(b) Calculate and interpret σ_X.

Applying the Concepts These exercises ask you to apply multiple learning goals in a new context or to apply what you learned in this section in a new way.

17. **Bag Check** Thousands of travelers pass through the airport in Guadalajara, Mexico, each day. Mexican customs agents want to be sure that passengers are not bringing in illegal items, but do not have time to search every traveler's luggage. Instead, customs requires each person to press a button. Either a red or a green bulb lights up. If the light is red, the passenger will be searched by customs agents. Green means "go ahead." Customs agents claim that the light has probability 0.30 of showing red on any push of the button. Assume for now that this claim is true.

Suppose we watch 20 passengers press the button. Let R = the number who get a red light. Exercise 19 in Section 5.3 asked you to show that R is a binomial random variable.

(a) Find the probability that at most 3 people out of 20 would get a red light if the agents' claim is true.

(b) Suppose that only 3 of the 20 passengers get a red light after pressing the button. Does this give convincing evidence that the customs agents' claimed value of $p = 0.3$ is too high? Explain your reasoning.

(c) Calculate and interpret the expected value of R.

(d) Calculate and interpret the standard deviation of R.

18. **Cranky Mower** A company has developed an "easy-start" mower that cranks the engine with the push of a button. The company claims that the probability the mower will start on any push of the button is 0.9. Assume for now that this claim is true.

On the next 10 uses of the mower, let Y = the number of times it starts on the first push of the button. Exercise 20 in Section 5.3 asked you to show that Y is a binomial random variable.

(a) Find the probability of at most 6 starts in 10 attempts if the company's claim is true.

(b) Suppose that the mower only starts on 6 of the 10 attempts. Does this give convincing evidence that the company's claim is exaggerated? Explain your reasoning.

(c) Calculate and interpret the expected value of Y.

(d) Calculate and interpret the standard deviation of Y.

Extending the Concepts These exercises challenge you to explore statistical concepts and methods that go beyond what you learned in this section.

19. **Random Dialing** Refer to Exercise 13. Note that X = the number of calls that result in a completed survey. Let Y = the number of calls that *don't* result in a completed survey.

(a) Use technology to find $P(Y \geq 13)$ and $P(X \leq 2)$. Explain the relationship between these two values.

(b) Find the mean of Y. How is it related to the mean of X? Explain why this makes sense.

(c) Find the standard deviation of Y. How is it related to the standard deviation of X? Explain why this makes sense.

20. **Diet Cola** Refer to Exercise 10.

(a) Calculate the mean and standard deviation of X.

(b) Of the 30 volunteers, 23 made correct identifications. Find the standardized score (z-score) corresponding to $X = 23$.

(c) Based on your answer to part (b), explain why the observed result of 23 correct identifications is surprising if the 30 volunteers are just guessing.

In the Section 5.3 Extending the Concepts exercises, we introduced geometric random variables. Refer to the introductory paragraph there for important background information, and to Exercises 22 and 23 in Section 5.3 for context.

For a geometric random variable X with probability p of success on each trial, the mean and standard deviation are

$$\mu_X = \frac{1}{p} \quad \text{and} \quad \sigma_X = \frac{\sqrt{1-p}}{p}$$

21. **Start That Mower** To start an old lawn mower, Jordyn has to pull its cord and hope for some luck. On any particular pull, the mower has a 20% chance of starting. Let X be the number of pulls it takes Jordyn to start the mower. In Exercise 22 of Section 5.3, you showed that X is a geometric random variable.

(a) Find the probability that it takes Jordyn more than 6 pulls to start the mower.

(b) Calculate and interpret the mean of X.

(c) Calculate and interpret the standard deviation of X.

Cumulative Review These exercises revisit what you learned in previous sections.

22. **Buying Stock (4.5, 5.2)** You purchase a hot stock for $1000. The stock either gains 30% or loses 25% each day, each with probability 0.5. Its returns on consecutive days are independent of each other. You plan to sell the stock after two days.

    (a) What are the possible values of the stock after two days, and what is the probability for each value?

    (b) What is the probability that the stock is worth more after two days than the $1000 you paid for it?

    (c) What is the mean value of the stock after two days?

Comment: You should see that the two criteria in parts (b) and (c) give different answers to the question "Should I invest?"

23. **Which 'Wich? (4.7)** Sammi's Sandwich Shop lets you design your own sandwich. There are 5 choices for bread, 6 choices for meat, and 4 choices for cheese. You can also choose to include (or not include) each of the following items by request: lettuce, tomato, hot peppers, mayonnaise, or mustard.

    (a) A standard sandwich has 1 type of meat and 1 type of cheese, in addition to bread and any requested items. How many different standard sandwiches can be created?

    (b) A "doubles" sandwich contains 2 different kinds of meat and 2 different kinds of cheese, in addition to bread and any requested items. How many different "doubles" sandwiches can be created?

Section 5.5 Poisson Random Variables

LEARNING GOALS

By the end of this section, you will be able to:
- Calculate probabilities involving a single value of a Poisson random variable.
- Find probabilities involving several values of a Poisson random variable.
- Find probabilities for a Poisson random variable over different interval lengths.

In Sections 5.3 and 5.4, you learned how to use a binomial distribution to calculate probabilities involving the number of successes in a fixed number of trials n of the same chance process with constant probability of success p on each trial. What if we want to find probabilities involving the number of successes X in a *continuous interval* of space or time for a chance process with a constant success rate? We can't use a binomial distribution because there is no fixed number of trials. Instead, we can model the random variable X with a *Poisson distribution*.

Poisson Setting

A factory that manufactures yarn is concerned about the number of imperfections in the yarn it produces. A quality control supervisor selects a 100-foot length of yarn and counts X = the number of imperfections. Because the supervisor is observing a continuous 100-foot segment of yarn and not a fixed number of trials, the probability distribution of X is a *Poisson distribution*.

Other settings in which the Poisson distribution might be used include:

- The number of spam calls you receive on your cell phone in a 24-hour time period.
- The number of prairie dog burrow entrances in a hectare of land.
- The number of customers who arrive at a shop in a 2-hour time interval.

All of these examples have several properties in common. To consider those properties, we need to take the interval of interest (100 feet of yarn, 24 hours, 1 hectare, etc.) and imagine dividing it up into much smaller subintervals. These subintervals need to be the same length, and small enough that the likelihood of two or more successes happening in any one subinterval is (essentially) 0. The probability of a success must be the same for all the subintervals, and a success occurring in one subinterval needs to be independent of a success occurring in any other subinterval. Situations meeting these conditions are examples of a **Poisson setting**.

A **Poisson setting** arises when we count the number of successes that occur over a given continuous interval of space or time. For the Poisson setting to apply, it must be possible to divide the given interval into small, equally sized subintervals for which:
- The probability of two or more successes in any one subinterval is (essentially) 0.
- The probability of success in each subinterval is the same.
- The occurrence of a success in any given subinterval is independent of the occurrence of a success in any other subinterval.

These properties ensure that successes occur at random and at a constant rate over the interval of space or time.

The yarn factory scenario described previously is a Poisson setting. First, the supervisor is observing a fixed length of yarn: 100 feet. Second, it is reasonable to believe that if we divided that 100 feet of yarn into many very small subintervals (say, 0.1 mm in length), each subinterval would be the same length (0.1 mm), would be too short to have 2 or more imperfections, and would have the same probability of having an imperfection. Finally, it is reasonable to assume that imperfections happen randomly along a piece of yarn, so the occurrence of an imperfection in one subinterval would be independent of the occurrence of an imperfection in another subinterval. If we let X = the number of imperfections in a 100-foot length of yarn, then X is a **Poisson random variable.**

The count of successes X in a Poisson setting is a **Poisson random variable**. The possible values of X are 0, 1, 2, ...

Note: Even though we are observing a continuous interval of space or time, a Poisson random variable is a discrete random variable because it is the result of counting successes. The sample space consists of all non-negative integers.

Calculating Poisson Probabilities

As in a binomial setting, we often want to calculate the probability that a Poisson random variable takes a particular value r. Suppose we want to find the probability that the quality control supervisor finds exactly 2 imperfections in a 100-foot-long piece of yarn. Given that the average number of imperfections per 100 feet of yarn produced in this factory is 0.8, the desired probability is

$$P(X=2) = \frac{(0.8)^2 e^{-0.8}}{2!} = 0.144$$

The pattern of this calculation works for any Poisson setting.

Poisson probability formula

Suppose that X is a Poisson random variable and that μ is the population mean number of successes in intervals of a given size. The probability of getting exactly r successes (r = 0, 1, 2, . . .) in an interval of the given size is

$$P(X = r) = \frac{\mu^r e^{-\mu}}{r!}$$

where $e \approx 2.71828$.

Note: Some books and software use the symbol λ (the Greek letter lambda) instead of μ in the Poisson probability formula.

EXAMPLE 5.5.1

Babies born in a 24-hour period

Calculating Poisson probabilities

PROBLEM

The number of babies born in a 24-hour period at St. Luke's Hospital in Cedar Rapids, Iowa, can be modeled by a Poisson distribution with a mean of 6.58 births per 24-hour period.[20] What's the probability that there will be exactly 4 births in a given 24-hour period?

SOLUTION

Let X = the number of births in a given 24-hour period. X has a Poisson distribution with $\mu = 6.58$ births per 24-hour period.

$$P(X=4) = \frac{(6.58)^4 e^{-6.58}}{4!} = 0.108 \qquad \boxed{P(X=r) = \frac{\mu^r e^{-\mu}}{r!}}$$

EXAM PREP: FOR PRACTICE, TRY EXERCISE 7.

The probability distribution of the Poisson random variable X = the number of births in a given 24-hour period is called a **Poisson distribution.**

The probability distribution of a Poisson random variable is a **Poisson distribution.** Any Poisson distribution is completely specified by one number: the mean number of successes μ over the given interval of space or time.

Finding Poisson Probabilities Involving Several Values

In Example 5.5.1, we computed the probability that St. Luke's Hospital would have exactly 4 births in a given 24-hour period. What if we wanted to know the probability of having *fewer than* 4 births in the given 24-hour period? We can simply add the corresponding Poisson probabilities for 0, 1, 2, and 3 births. (Recall from Section 4.7 that $0! = 1$.)

$$P(X<4) = P(X=0) + P(X=1) + P(X=2) + P(X=3)$$
$$= \frac{(6.58)^0 e^{-6.58}}{0!} + \frac{(6.58)^1 e^{-6.58}}{1!} + \frac{(6.58)^2 e^{-6.58}}{2!} + \frac{(6.58)^3 e^{-6.58}}{3!}$$
$$= 0.0014 + 0.0091 + 0.0300 + 0.0659$$
$$= 0.1064$$

There is about an 11% chance of having fewer than 4 births in a given 24-hour period at St. Luke's Hospital.

EXAMPLE 5.5.2

Hurricanes in Hawaii

Finding Poisson probabilities involving several values

PROBLEM

The number of hurricanes that come within 250 nautical miles of Hawaii in a 1-year period can be modeled by a Poisson distribution with mean 0.45.[21] Let X = the number of hurricanes that come within 250 nautical miles of Hawaii in a given 1-year period. Find $P(X \geq 2)$. Describe this probability in words.

SOLUTION

$P(X \geq 2) = 1 - P(X \leq 1)$
$= 1 - [P(X = 0) + P(X = 1)]$
$= 1 - \left(\dfrac{0.45^0 e^{-0.45}}{0!} + \dfrac{0.45^1 e^{-0.45}}{1!} \right)$
$= 1 - (0.6376 + 0.2869)$
$= 0.0755$

> Because
> $P(X \geq 2) = P(X = 2) + P(X = 3) + \ldots$
> has an infinite number of terms, use the complement rule:
> $P(X \geq 2) = 1 - P(X \leq 1)$

There is only about a 7.55% probability that at least 2 hurricanes will come within 250 nautical miles of Hawaii in a given 1-year period.

EXAM PREP: FOR PRACTICE, TRY EXERCISE 11.

It is tedious to use the Poisson probability formula for calculations involving many values of a Poisson random variable. Technology is a more practical alternative. See the Tech Corner at the end of this section for details.

Finding Poisson Probabilities over Different Interval Lengths

In Example 5.5.2, we found that there is a 7.55% probability of having at least 2 hurricanes within 250 nautical miles of Hawaii in a given 1-year period. What if we wanted to find the probability of having at least 2 hurricanes within 250 nautical miles of Hawaii in a specific 2-year period? We can still use a Poisson distribution if we adjust the mean of the distribution appropriately.

The mean number of hurricanes that come within 250 nautical miles of Hawaii in a 1-year period is 0.45. How about in a 2-year period? Because successes occur at a constant rate over time or space in a Poisson setting, the mean number of hurricanes that come within 250 nautical miles of Hawaii in a 2-year period is 2(0.45) = 0.90. Using this strategy allows us to calculate Poisson probabilities for different interval lengths.

HOW TO Adjust a Poisson distribution for different interval lengths

1. Let μ_1 represent the mean of a Poisson distribution for a fixed interval of length s_1.
2. Let s_2 represent the new interval length.
3. Compute the mean of the Poisson distribution for the new interval length as $\mu_2 = \left(\dfrac{s_2}{s_1} \right) \mu_1$.
4. Compute the probabilities for intervals of length s_2 using the Poisson distribution with mean μ_2.

EXAMPLE 5.5.3

Hurricanes in Hawaii

Using different interval sizes to compute Poisson probabilities

PROBLEM
Refer to Example 5.5.2. Find the probability of having at least 2 hurricanes within 250 nautical miles of Hawaii in a given 5-year period.

SOLUTION

$\mu_2 = \left(\dfrac{5}{1} \right)(0.45) = 2.25$

> $\mu_2 = \left(\dfrac{s_2}{s_1} \right) \mu_1$

(continued)

Using Poisson probability formula:

$P(X \geq 2) = 1 - P(X \leq 1)$
$= 1 - [P(X = 0) + P(X = 1)]$
$= 1 - \left(\dfrac{(2.25)^0 e^{-2.25}}{0!} + \dfrac{(2.25)^1 e^{-2.25}}{1!} \right)$
$= 1 - (0.1054 + 0.2371)$
$= 0.6575$

Using technology: Applet / 1 − Poissoncdf(μ:2.25, x value:1) = 0.6575

EXAM PREP: FOR PRACTICE, TRY EXERCISE 15.

While it is fairly unusual for 2 or more hurricanes to occur within 250 nautical miles of Hawaii in a given 1-year period (probability 0.0755), it is quite likely that 2 or more hurricanes will occur within a particular 5-year period (probability 0.6575).

THINK ABOUT IT

Where does the Poisson probability formula come from? Let X be a Poisson random variable with mean number of successes μ for a specified interval size. If we divide that interval into n subintervals, then we would expect $\dfrac{\mu}{n}$ successes in each subinterval. As the number of subintervals gets larger, the size of the subintervals gets smaller. Eventually, the size of the subintervals will be small enough so that at most one success will occur within any subinterval. At that point, each subinterval will have two possible outcomes (success or failure) with probability of success $\dfrac{\mu}{n}$. Now we have a fixed number of subintervals (trials), each with the same probability of success, and the occurrence of a success in one subinterval is independent of the occurrence of a success in any other subinterval. Does this sound familiar? It should: This is a binomial setting. In fact, the Poisson probability formula is exactly the binomial probability formula with $p = \dfrac{\mu}{n}$ when we take the limit as $n \to \infty$. While the mathematics is beyond the scope of this course, it is true that

$$P(X = r) = \lim_{n \to \infty} {}_nC_r \left(\dfrac{\mu}{n} \right)^r \left(1 - \dfrac{\mu}{n} \right)^{n-r} = \dfrac{\mu^r e^{-\mu}}{r!}$$

TECH CORNER — Calculating probabilities for a Poisson random variable

You can use technology to calculate probabilities involving Poisson random variables. We'll illustrate this process using the Poisson random variable X = the number of births at St. Luke's Hospital from Example 5.5.1. Suppose we want to find the probability of having exactly 4 births in a 24-hour period, and the probability of having fewer than 4 births in a 24-hour period. Recall that the mean number of births in a 24-hour period at this hospital is $\mu = 6.58$.

Applet

(a) To calculate $P(X = 4)$:

1. Go to www.stapplet.com and launch the *Poisson Distributions* applet.
2. Enter the mean: 6.58.
3. Choose "exactly" from the drop-down menu, type "4," and click "Go!" to find the probability of *exactly* 4 successes.

Poisson Distributions

Mean = 6.58

Calculate the probability of [exactly ▼] [4] successes. [Go!] P(X = 4) = 0.1084
-OR-
Calculate the probability of between [ ] and [ ] successes (inclusive). [Go!]

(b) To calculate $P(X < 4)$:

1. Go to www.stapplet.com and launch the *Poisson Distributions* applet.

2. Enter the mean: 6.58.

3. Choose "less than" from the drop-down menu, type 4 in the adjacent box, and click "Go!"

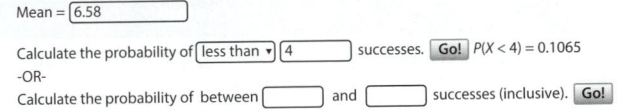

Poisson Distributions

Mean = 6.58

Calculate the probability of [less than ▾] [4] successes. [Go!] $P(X < 4) = 0.1065$
-OR-
Calculate the probability of between [ ] and [ ] successes (inclusive). [Go!]

TI-83/84

(a) To calculate $P(X = 4)$:

The TI-83/84 command poissonpdf(μ,r) computes $P(X = r)$ in a Poisson setting with mean μ successes in a given interval.

- Press 2nd VARS (DISTR) and choose poissonpdf(.

  OS 2.55 or later: In the dialog box, enter these values: μ: 6.58, x value: 4. Choose Paste, and then press ENTER.

  Older OS: Complete the command poissonpdf(6.58,4) and press ENTER.

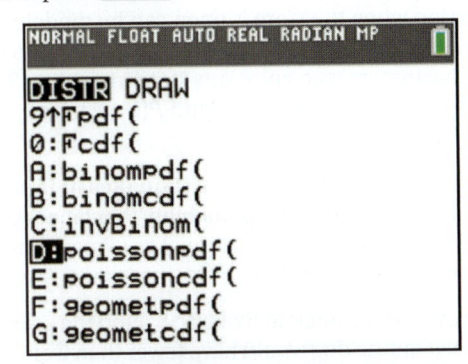

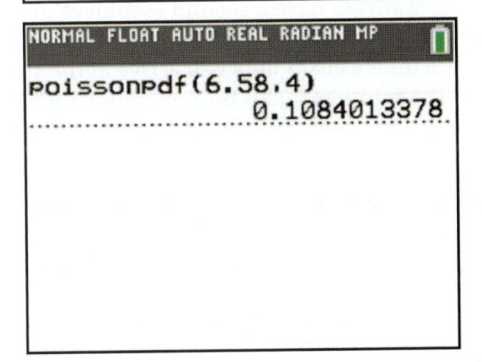

(b) To calculate $P(X < 4)$:

The TI-83/84 command poissoncdf(μ,r) computes $P(X \leq r)$ in a Poisson setting with mean μ successes in a given interval. We need to use the fact that $P(X < 4) = P(X \leq 3)$.

- Press 2nd VARS (DISTR) and choose poissoncdf(.

  OS 2.55 or later: In the dialog box, enter these values: μ: 6.58, x value: 3. Choose Paste, and then press ENTER.

  Older OS: Complete the command poissoncdf(6.58,3) and press ENTER.

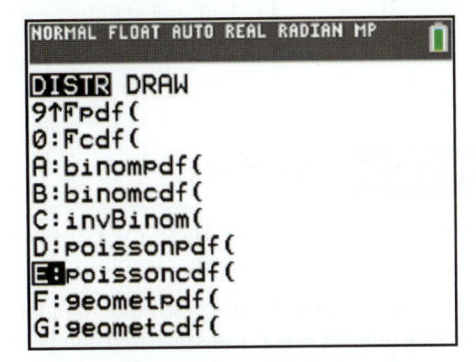

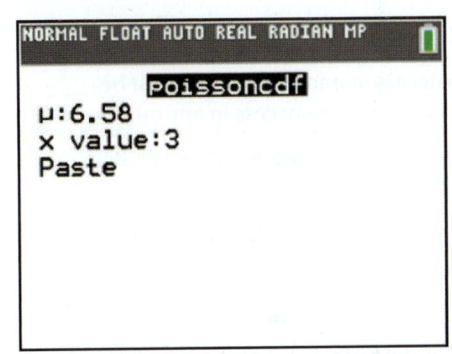

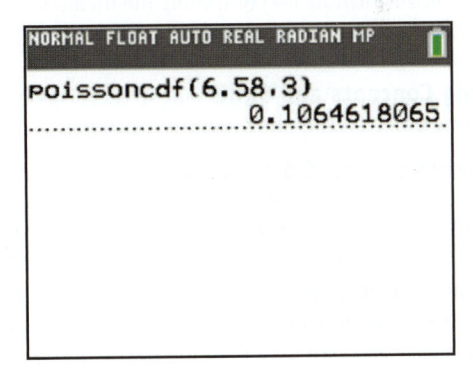

Detailed instructions for using CrunchIt!, Excel, Google Sheets, JMP, Minitab, and R are available in Achieve.

Section 5.5 What Did You Learn?

Review the learning goals from this section. Then practice what you've learned by working through the exercises.

Learning Goal	Example	Exercises
Calculate probabilities involving a single value of a Poisson random variable.	5.5.1	7–10
Find probabilities involving several values of a Poisson random variable.	5.5.2	11–14
Find probabilities for a Poisson random variable over different interval lengths.	5.5.3	15–18

Section 5.5 Exercises

Building Concepts and Skills These exercises assess the basic knowledge you should have after reading the section.

1. True/False: In a Poisson setting, the random variable counts the number of successes in a fixed number of trials.

2. In a Poisson setting, the probability of success must be _____ across all subintervals and the occurrence of a success in one subinterval must be _____ of the occurrence of success in any other subinterval.

3. True/False: The possible values of a Poisson random variable are 0, 1, 2, ...

4. State the Poisson probability formula.

5. A Poisson distribution is completely specified by one number: _____.

6. What is the formula for computing the mean of a Poisson random variable for an interval of a different length?

Mastering Concepts and Skills These exercises reinforce the learning objectives as shown in the examples.

7. **Refer to Example 5.5.1 Earthquakes in Japan** Based on data from 1952–2019, Japan experiences an average of 0.746 earthquake of magnitude 7.0 or greater per year.[22] A Poisson distribution can be used to model the occurrence of earthquakes over time. What is the probability that Japan will experience exactly 1 earthquake of magnitude 7.0 or greater in the next year?

8. **Prairie Dogs** In one study of prairie dog habitats, researchers found that the number of burrow entrances can be modeled by a Poisson distribution with mean 0.737 entrance per hectare.[23] (*Note:* One hectare = 2.47 acres.) What is the probability of finding 2 burrow entrances in one specific hectare of ground in this habitat?

9. **Auto Insurance** Based on prior experience, the number of collision claims filed with an auto insurance company in a 30-day period can be modeled by a Poisson distribution with a mean of 4.9.[24] Let X = the number of collision claims filed with this company in the next 30 days. Find $P(X = 0)$. Describe this probability in words.

10. **Customer Counts** The number of customers who enter a particular store during a specified 30-minute period when the store is open can be modeled by a Poisson distribution with mean 3. Let Y = the number of customers who enter this store in the first 30 minutes it is open tomorrow. Find $P(Y = 5)$. Describe this probability in words.

11. **Refer to Example 5.5.2 Earthquakes in Japan** Refer to Exercise 7. Find the probability that Japan will experience 2 or fewer earthquakes of magnitude 7 or greater in the next year.

12. **Prairie Dogs** Refer to Exercise 8. Find the probability that a prairie dog would have fewer than 2 entrances to choose from in a given hectare of land.

13. **Auto Insurance** Refer to Exercise 9. The insurance company will lose money if at least 12 collision claims are filed in the next 30 days. What is the probability that the company will lose money? Should the insurance company charge more money for its collision coverage? Justify your answer.

14. **Customer Counts** Refer to Exercise 10. If more than 4 customers are likely to arrive in the first 30 minutes that the store is open tomorrow, the owner of the store wants to have a second employee available to help the customers. What is the probability that more than 4 customers will arrive at the store in the first 30 minutes that it is open tomorrow? Should the owner schedule a second employee? Justify your answer.

15. **Refer to Example 5.5.3 Earthquakes in Japan** Refer to Exercise 7. You have been hired for an internship in Japan that will last 6 months. What is the probability that there will be no major earthquakes (magnitude 7 or greater) during your time there?

16. **Prairie Dogs** Refer to Exercise 8. A biologist is interested in a particular plot of land in prairie dog territory that measures 0.5 hectare. What is the probability that there are no burrow entrances on that plot of land?

17. **Auto Insurance** Refer to Exercise 9. Find the probability that the company will have 50 or fewer collision claims filed over the next 365 days.

18. **Customer Counts** Refer to Exercise 10. The store owner considers morning business to be "slow" if fewer than 9 customers arrive between 9:00 a.m. and 12:00 noon. Find the probability that tomorrow morning's business in the store will be slow.

Applying the Concepts These exercises ask you to apply multiple learning objectives in a new context or to apply what you learned in this section in a new way.

19. **Meteors** According to an article published in the journal *Cosmos*, there have been 95 meteors that fell to the earth over a 95-year period.[25] Let X = the number of meteors that fall to the earth over a 1-year period. Then X can be modeled by a Poisson distribution with a mean of 1.
    (a) Calculate the probability that there will be exactly 0 meteors that fall to the earth in the next year.
    (b) Find the probability that at least 2 meteors will fall to the earth in the next year. Is this a likely event? Justify your answer.
    (c) What is the probability that there will be at least 2 meteors that fall to the earth in the next 5 years?

20. **Rare Elephants** A recent study found an average of approximately 0.64 African Forest elephant per square kilometer in Gabon.[26] We can model X = the number of African Forest elephants in a particular square kilometer in Gabon using a Poisson distribution.
    (a) Calculate the probability that there will be exactly 2 elephants in a particular square kilometer section of Gabon.
    (b) You are setting out on a safari to find African Forest elephants and will be looking in a specific square kilometer section of land in Gabon. What is the probability that you will see at least one African Forest elephant? Is this a likely event? Justify your answer.
    (c) One square kilometer is equivalent to 0.386 square mile. What is the probability that there will be at least one African Forest elephant in a particular 1 square mile plot of land in Gabon? Interpret your answer.

Extending the Concepts These exercises challenge you to explore statistical concepts and methods that go beyond what you learned in this section.

21. **Poisson or Binomial?** A store manager wants to learn more about the customers who frequent their store. In each of the following scenarios, which distribution — binomial or Poisson — should the manager use to model the random variable X? Explain your reasoning.
    (a) The manager takes a random sample of 10 customers and observes X = the number of customers who make a purchase. Assume that the probability an individual customer makes a purchase is constant.
    (b) The manager observes X = the number of customers who make a purchase over a 4-hour period.

22. **Wood Screws** A supervisor in a factory that produces 1-inch wood screws thinks that the number of defective screws produced per hour is described by a Poisson distribution. Do the data support this conclusion? The supervisor counts X = the number of defective screws for a random sample of 50 different hours of production. The number of defective screws found is summarized in the table.

Number of defective screws	0	1	2	3	4	5	6+
Frequency	11	17	13	6	2	1	0

    (a) Calculate the average number of defective screws per hour for this sample of 50 hours.
    (b) Compute the relative frequency for each value of X in the sample.
    (c) Using your answer to part (a), compute the probability for each value of X in the table using the Poisson distribution.
    (d) Compare your answers in parts (b) and (c). Does it seem appropriate to use the Poisson distribution to describe the random variable X? Justify your answer.

23. **Analyzing Poisson Distributions** Suppose that the random variable X has a Poisson distribution.
    (a) Use technology to make a graph of the probability distribution for $\mu = 0.5$, 1, 4, and 20. Describe how the shape of the distribution changes as μ increases.

    A Poisson random variable X has mean $\mu_X = \mu$ and standard deviation $\sigma_X = \sqrt{\mu}$.

    (b) Explain why the value of the mean makes sense in a Poisson setting.
    (c) What happens to the standard deviation as μ increases? Explain why this makes sense in a Poisson setting.

Cumulative Review These exercises revisit what you learned in previous sections.

24. **Earthquakes in Japan (2.5, 3.1, 3.2)** Here is a histogram that displays the intensity of all earthquakes of magnitude 5.0 or larger in Japan between 1952 and 2019.[27]

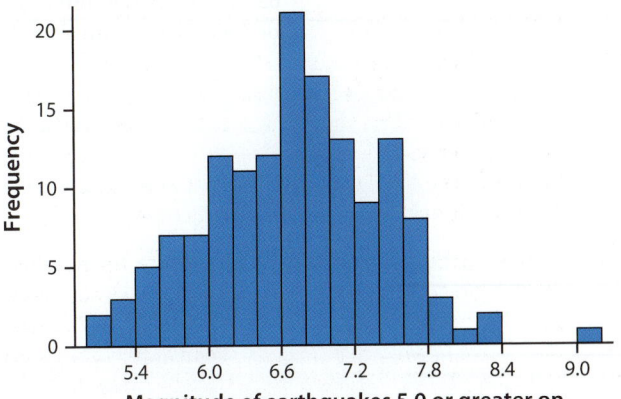

(a) What percentage of these 147 earthquakes measured 7.0 or higher on the Richter scale?
(b) Describe the shape of the distribution. Are there any possible outliers?
(c) Estimate the median of the distribution.
(d) Estimate the interquartile range of the distribution.

25. **Earthquakes in Japan (4.5, 5.3)** Refer to Exercise 24. Suppose we select 5 of these earthquakes at random for further study. Let X = the number of selected earthquakes that measured 7.0 or higher on the Richter scale.

(a) Explain why X is not a binomial random variable.
(b) Use the methods described in Chapter 4 to find $P(X=0)$.
(c) Now use the binomial probability formula to calculate $P(X=0)$. Is this result close to the answer from part (b)?

Statistics Matters A jury of peers?

At the beginning of the chapter, we described what the U.S. Constitution actually says about whether accused criminals in the United States are entitled to a "jury of their peers." To meet the Sixth Amendment requirement of impartiality, most courts start by randomly selecting a large jury pool from the citizens who live in the court's jurisdiction. In one case that made it all the way to the U.S. Supreme Court, a defense attorney in Michigan challenged the process of selecting the jury pool in the trial of his client, a Black man. Here are the facts:[28]

- About 7.28% of the jury-eligible population in the court's jurisdiction were Black people.
- The jury pool had between 60 and 100 members, only 3 of whom were Black people.

For now, assume that the court carried out a proper random selection process to obtain a jury pool with 100 members. Let X = the number of Black citizens in the jury pool.

1. Is X a discrete or a continuous random variable? Explain your answer.
2. What probability distribution does X have? Justify your answer.
3. Use technology to make a graph of the probability distribution. Describe its shape.
4. Calculate and interpret the mean of X.
5. Calculate and interpret the standard deviation of X.
6. Find the probability of getting 3 or fewer Black people in a jury pool with 100 members from a population in which 7.28% of citizens are Black.
7. Based on the actual number of Black citizens in the jury pool and your answer to Question 6, is there convincing evidence that the court did not carry out the random selection process correctly? Explain your reasoning.
8. From the given information, the jury pool may have included fewer than 100 members: 60, 61, 62, ..., or 99. Without doing any calculations, explain why the probability in Question 6 would be even higher with a smaller jury pool.

Chapter 5 Review

Analyzing Random Variables

- A **random variable** takes numerical values determined by the outcomes of a chance process. There are two types of random variables: *discrete* and *continuous*.
- The **probability distribution** of a random variable gives its possible values and their probabilities.
- A **discrete random variable** takes a fixed set of individual values with gaps between them on the number line. The set of possible values for a discrete random variable can be either finite or infinite.
  - A valid probability distribution assigns each of these values a probability between 0 and 1 such that the sum of all the probabilities is exactly 1.
  - We can display the probability distribution as a histogram, with the values of the variable on the horizontal axis and the probabilities on the vertical axis.
  - The probability of any event is the sum of the probabilities of all the values that make up the event.
- A **continuous random variable** takes any value in a particular interval on the number line. The set of possible values for a continuous random variable is infinite.
- We can describe the *shape* of a probability distribution's graph in the same way as we did a distribution of quantitative data—by identifying symmetry or skewness and any major peaks.
- Use the mean to summarize the *center* of a probability distribution. The **mean of a random variable** μ_X is the average value of the variable after many, many repetitions of the chance process. It is also known as the **expected value** of the random variable, $E(X)$.
  - The mean is the balance point of the probability distribution's graph.
  - If X is a discrete random variable, the mean is the average of the values of X, each weighted by its probability:
    $$\mu_X = E(X) = x_1 p_1 + x_2 p_2 + x_3 p_3 + \ldots = \sum x_i p_i$$
- Use the standard deviation to summarize the *variability* of a probability distribution. The **standard deviation of a random variable** σ_X measures how much the values of the variable typically vary from the mean in many, many repetitions of the chance process.
  - If X is a discrete random variable, the standard deviation of X is
    $$\sigma_X = \sqrt{(x_1 - \mu_X)^2 p_1 + (x_2 - \mu_X)^2 p_2 + \ldots} = \sqrt{\sum (x_i - \mu_X)^2 p_i}$$

Binomial Random Variables

- A **binomial setting** arises when we perform n independent trials of the same chance process and count the number of times that a particular outcome (a "success") occurs. The conditions for a binomial setting are as follows:
  - **B**inary? The possible outcomes of each trial can be classified as "success" or "failure."
  - **I**ndependent? Trials must be independent. That is, knowing the result of one trial must not tell us anything about the result of any other trial.
  - **N**umber? The number of trials n of the chance process must be fixed in advance.
  - **S**ame probability? There is the same probability of success p on each trial.

  Remember to check the BINS!

- The count of successes X in a binomial setting is a special type of discrete random variable known as a **binomial random variable**. Its probability distribution is a **binomial distribution**. Any binomial distribution is completely specified by two numbers: the number of trials n of the chance process and the probability of success p on each trial. The possible values of X are the whole numbers $0, 1, 2, \ldots, n$.
- To calculate probabilities involving a binomial random variable, use the binomial probability formula or technology.
  - *Binomial probability formula:* The probability of getting exactly r successes in n trials ($r = 0, 1, 2, \ldots, n$) is
    $$P(X = r) = {}_nC_r (p)^r (1-p)^{n-r}$$
    where
    $${}_nC_r = \frac{n!}{r!(n-r)!}$$
- The mean and the standard deviation of a binomial random variable are
  $$\mu_X = np \quad \text{and} \quad \sigma_X = \sqrt{np(1-p)}$$

Poisson Random Variables

- A **Poisson setting** arises when we count the number of successes that occur over a given continuous interval of space or time. For the Poisson setting to apply, it must be possible to divide the given interval into small, equally sized subintervals for which:
  - The probability of two or more successes in any one subinterval is (essentially) 0.
  - The probability of success in each subinterval is the same.
  - The occurrence of a success in any given subinterval is independent of the occurrence of a success in any other subinterval.

  These properties ensure that successes occur at random and at a constant rate over the interval of space or time.

- The count of successes X in a Poisson setting is a special type of discrete random variable known as a **Poisson random variable**. Its probability distribution is a **Poisson distribution**. Any Poisson distribution is completely specified by one number: the mean number of successes μ in intervals of a given length. The possible values of X are the whole numbers $0, 1, 2, \ldots$.

- To calculate probabilities involving a Poisson random variable, use the Poisson probability formula or technology.
  - *Poisson probability formula:* The probability of getting exactly r successes ($r = 0, 1, 2, \ldots$) in an interval of the given size is
  $$P(X = r) = \frac{\mu^r e^{-\mu}}{r!}$$
  where $e \approx 2.71828$.

- To calculate probabilities involving a Poisson distribution for a different interval length than the one given, use the fact that successes occur at a constant rate over space or time to calculate the mean number of successes for the new interval length. If μ_1 is the mean of a Poisson distribution for an interval of length s_1, and s_2 is the new interval length, use a Poisson distribution with mean $\mu_2 = \left(\dfrac{s_2}{s_1}\right)\mu_1$ to compute probabilities involving the number of successes over the new interval length.

Chapter 5 Review Exercises

These exercises will help you review important concepts and skills described by the learning objectives in each section. For most exercises, the relevant section is noted in parentheses after the exercise title.

1. **Televisions (5.1)** If $X =$ the number of televisions in a randomly selected U.S. household, the probability distribution of X is given here.[29]

Number of TVs	0	1	2	3	4	5
Probability	0.026	0.254	0.331	0.226	0.104	0.059

   (a) Is X a discrete or a continuous random variable? Explain your answer.
   (b) Verify that this probability distribution is legitimate.
   (c) Find $P(X < 2)$. Describe this probability in words.
   (d) What is the probability that a randomly selected household has at least 2 televisions?

2. **Televisions (5.2)** Refer to Exercise 1.
   (a) Make a histogram to display the probability distribution of X. Describe its shape.
   (b) Calculate and interpret the mean of X.
   (c) Calculate and interpret the standard deviation of X.

3. **ESP (5.3, 5.4)** To test whether someone has extrasensory perception (ESP), choose 1 of 4 cards at random — a star, wave, cross, or circle. Ask the person to identify the card without seeing it. Do this a total of 20 times and see how many cards the person identifies correctly. Let $X =$ the number of correct identifications, assuming that the person does not have ESP and is just guessing for each card.

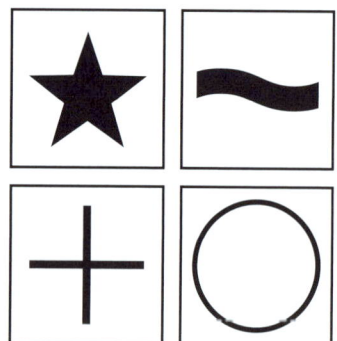

   (a) Explain why X is a binomial random variable.
   (b) Use the binomial probability formula to calculate $P(X = 5)$.
   (c) Find the probability that a person without ESP identifies 8 or more cards correctly.
   (d) Alec makes 8 correct identifications out of 20 cards. Based on this result and your answer to part (c), is there convincing evidence that Alec has ESP? Explain your reasoning.

4. **ESP (5.3, 5.4)** Refer to Exercise 3.
   (a) Make a graph of the probability distribution of X. Describe its shape.
   (b) Calculate the expected value of X.
   (c) Calculate and interpret the standard deviation of X.

5. **Web Server (5.5)** The number of requests received in a 1-second time period by a certain web server can be modeled by a Poisson distribution with mean 23.3.
   (a) Find the probability that this server receives exactly 30 requests in the next second.
   (b) If the server receives 40 or more requests in a second, the website will crash. What is the probability that this will happen?
   (c) Let $Y =$ the number of requests received by this server in a given 5-second period. Describe the probability distribution of Y.
   (d) Find $P(Y < 110)$. Interpret this probability in words.

6. **Keno (Chapter 5)** In a game of 4-Spot Keno, the player picks 4 numbers from 1 to 80. The casino randomly selects 20 winning numbers from 1 to 80. The table shows the possible outcomes of the game and their probabilities, along with the amount of money (payout) that the player wins for a $1 bet. For instance, if the player bets $1 and matches 2 of the 4 numbers

they pick, they will receive their $1 back as a payout. Let X = the payout for a single $1 bet.

Matches	0	1	2	3	4
Payout	$0	$0	$1	$3	$120
Probability	0.308	0.433	0.213	0.043	0.003

(a) Find the probability that a player wins a payout of $1 or more on a $1 bet in 4-Spot Keno.

(b) Calculate and interpret μ_X.

(c) If the player places a $1 bet on 4-Spot Keno for 10 consecutive days, what's the probability that they win a payout of $1 or more on at least 5 days?

7. **Chat Agents (Chapter 5)** A company that provides phone, television, and internet service has 5 people available at all times to chat with customers who are having issues with their service. Based on company records, the number of customer contacts received in a 1-minute period by an individual chat agent can be modeled by a Poisson distribution with a mean of 1.2.

(a) What is the probability that a particular chat agent will receive at least 1 customer contact in the next minute?

(b) Find the probability that at least 4 of the 5 chat agents do not receive any customer contacts in the next minute.

6 Normal Distributions and Sampling Distributions

Section 6.1 Continuous Random Variables
Section 6.2 Normal Distributions: Finding Areas from Values
Section 6.3 Normal Distributions: Finding Values from Areas
Section 6.4 Normal Approximation to the Binomial Distribution and Assessing Normality
Section 6.5 Sampling Distributions
Section 6.6 Sampling Distributions: Bias and Variability
Section 6.7 Sampling Distribution of the Sample Proportion
Section 6.8 Sampling Distribution of the Sample Mean and the Central Limit Theorem
Chapter 6 Review
Chapter 6 Review Exercises
Cumulative Review Chapters 4–6

Statistics Matters How much salt is too much?

Excessive sodium intake can cause high blood pressure, heart disease, and stroke. Restaurant food is one of the biggest sources of sodium in the standard American diet. Researchers analyzed more than 3000 meals ordered from Chipotle restaurants via Grubhub and recorded three variables for each meal: the amount of sodium (mg), the number of calories, and the amount of saturated fat (g).[1]

Here is a graph of the distribution of amount of sodium (mg) in meals from Chipotle, along with a vertical line showing the recommended daily allowance (RDA) for sodium.

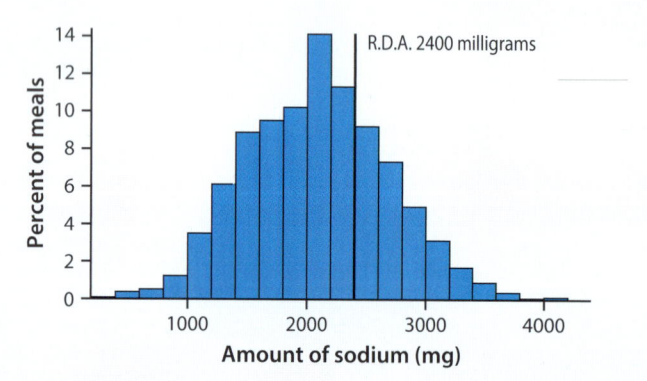

What is the probability that a randomly selected Chipotle meal exceeds the RDA for sodium? What is the probability that the mean amount of sodium in a random sample of 50 Chipotle meals exceeds the RDA?

We'll revisit Statistics Matters at the end of the chapter, so you can use what you have learned to help answer these questions.

In Chapter 1, you learned that statistical inference uses data from a sample to draw conclusions about a population. In Chapters 4 and 5, you developed rules for probability and random variables. In this chapter, we'll combine these ideas to build an understanding of *sampling distributions*—the foundation for the formal inference procedures introduced in Chapters 7–13. Along the way, you'll learn about a very important type of probability distribution—the *normal distribution*.

Section 6.1 Continuous Random Variables

LEARNING GOALS

By the end of this section, you will be able to:
- Use a density curve to model the probability distribution of a continuous random variable.
- Draw a normal distribution with a given mean and standard deviation.
- Use the empirical rule to estimate probabilities in a normal distribution.

In Chapter 5, you learned how to distinguish discrete and continuous random variables and how to analyze discrete random variables. In this chapter, you will learn how to analyze continuous random variables, with an emphasis on random variables that follow a normal distribution.

Density Curves

Recall that discrete random variables often result from *counting* something, like the number of heads in 5 coin tosses or the number of Democrats in a random sample of 100 registered voters. Continuous random variables often result from *measuring* something, like the distance a car can drive on a single tank of gas or the weight of a randomly selected newborn baby.

Selena works at a bookstore in the Denver International Airport. To get to work, she can either take the airport train from the main terminal or use a moving walkway that will allow her to get from the main terminal to the bookstore in 4 minutes. She wonders if it will be faster to walk or take the train to work. Let the random variable X = journey time (in minutes) using the airport train.

In Chapter 5, you learned that we can use a histogram to display the probability distribution of a discrete random variable. When we want to display the probability distribution of a continuous random variable, like X = journey time, we use a **density curve** similar to the one shown here.

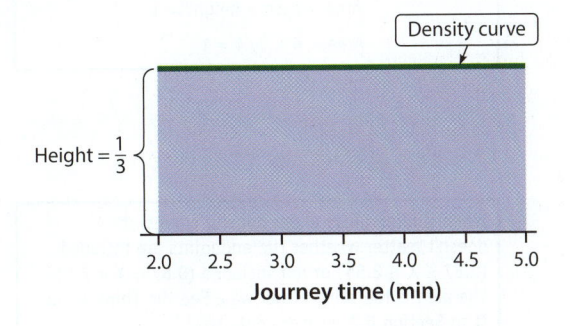

A **density curve** models the probability distribution of a continuous random variable with a curve that:
- is always on or above the horizontal axis.
- has an area of exactly 1 underneath it.

The area under the curve and above any interval of values on the horizontal axis gives the probability that the random variable falls in that interval.

You might wonder why this density curve is drawn at a height of 1/3. Recall that in a probability histogram for a discrete random variable, the heights of the bars add to 1, or 100%. Likewise, we want the rectangular area under the density curve between 2 minutes and 5 minutes to equal 1, or 100%:

$$\text{area} = \text{base} \times \text{height} = 3 \times 1/3 = 1.00 = 100\%$$

Note that this type of density curve is called a *uniform density curve* because it has constant height.

To find the probability that Selena takes less than 4 minutes to get to work when using the airport train, we calculate the area under the density curve between 2 and 4 minutes (shown in red).

$P(X < 4) = \text{base} \times \text{height} = 2 \times 1/3 = 2/3 = 0.667$

The probability that Selena gets to work in less than 4 minutes when using the airport train is 0.667. Because the train will be faster on about two-thirds of the days, she should plan to take the train instead of walking.

EXAMPLE 6.1.1

Uniform distributions

Density curves

PROBLEM

Suppose you use a random number generator to produce a number between 0 and 4 (like 0.84522 or 3.1111119). Let $X =$ the output of the random number generator. The random variable X is continuous and follows a uniform probability distribution because each of the possible values of X is equally likely.

(a) Draw a density curve to model the probability distribution of X. Be sure to include scales on both axes.

(b) What is the probability that X will take a value between 0.87 and 2.55?

(c) Find the 65th percentile of the probability distribution.

SOLUTION

(a)

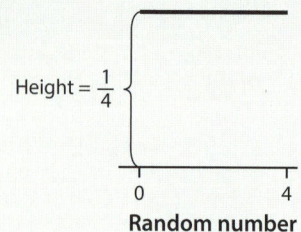

The height of the curve needs to be 1/4 so that:

$$\text{Area} = \text{base} \times \text{height} = 1$$

$$\text{Area} = 4 \times 1/4 = 1$$

(b)

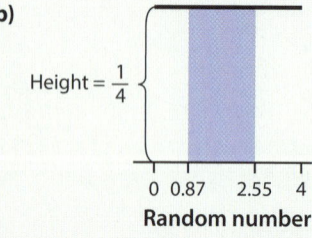

When finding areas under density curves, it doesn't matter whether the endpoints are included ($0.87 \leq X \leq 2.55$) or not included ($0.87 < X < 2.55$). The area is the same either way. See the Think About It in Section 6.2 for more details.

$P(0.87 < X < 2.55) = (2.55 - 0.87) \times 1/4 = 0.42$

(c)

The 65th percentile is the value, x, such that 65% of the area under the density curve is to the left of x. In other words, 65% of the possible random numbers between 0 and 4 are less than x.

$0.65 = (x - 0) \times 1/4 \rightarrow x = 2.60$

EXAM PREP: FOR PRACTICE, TRY EXERCISE 7.

Density curves have many possible shapes. As with the distribution of a quantitative variable, we start our analysis by looking for rough symmetry or clear skewness. We then identify any obvious peaks. Figure 6.1 shows three density curves with distinct shapes.

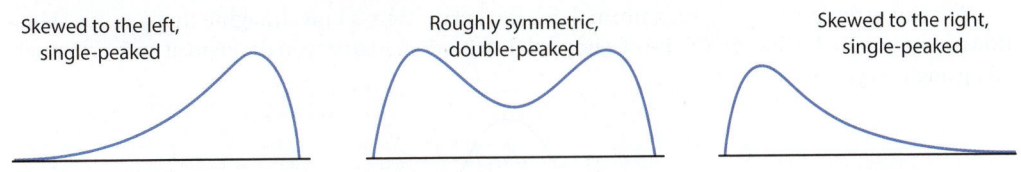

FIGURE 6.1 Density curves with different shapes. Some people refer to graphs with a single peak as *unimodal* and to graphs with two clear peaks as *bimodal*.

The mean μ of a density curve is the point at which the density curve would balance if it was made of solid material. The median of a density curve is the *equal-areas point*, the point that divides the area under the curve in half.

There is no easy way to estimate the standard deviation σ of a density curve in general. But there is one family of density curves for which we can estimate the standard deviation by eye: normal distributions.

Normal Distributions

The histogram in Figure 6.2(a) shows the distribution of total length (in centimeters, cm) for 104 mountain brushtail possums in Australia.[2] The histogram is roughly symmetric, and both tails fall off from a single center peak.

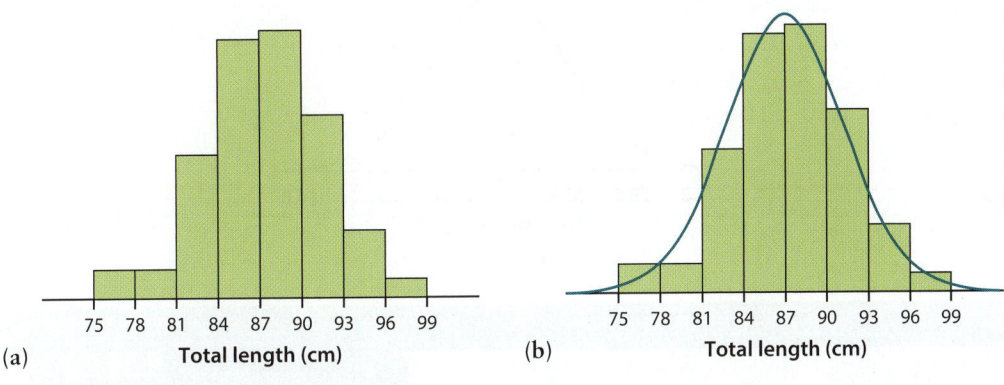

FIGURE 6.2 (a) Histogram showing the distribution of total length (cm) for Australian mountain brushtail possums. (b) A normal density curve models the distribution.

If we let X = the total length of a randomly selected mountain brushtail possum, then X is a continuous random variable. The density curve shown in Figure 6.2(b) is a good model for the probability distribution of X. This density curve is called a **normal curve**. The distributions described by normal curves are called **normal distributions**.

> A **normal distribution** is described by a symmetric, single-peaked, bell-shaped density curve called a **normal curve**. Any normal distribution is completely specified by two numbers: its mean μ and its standard deviation σ.

Look at the two normal distributions in Figure 6.3. They illustrate several important facts.

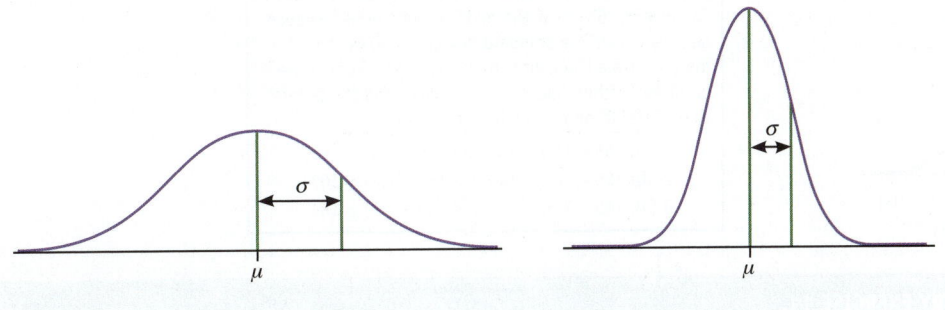

FIGURE 6.3 Two normal distributions showing the mean μ and the standard deviation σ.

- **Shape:** All normal distributions have the same overall shape: symmetric, single-peaked, and bell-shaped.

- **Center:** The mean μ is located at the midpoint of the symmetric density curve and is the same as the median.
- **Variability:** The standard deviation σ measures the variability (width) of a normal distribution.

You can estimate σ by eye on a normal density curve. Here's how: Imagine that you are skiing down a mountain that has the shape of a normal distribution. At first, you descend at an increasingly steep angle as you leave the peak.

Fortunately, instead of skiing straight down the mountain, you find that the slope begins to get flatter rather than steeper as you go out and down.

The points at which these changes of curvature take place are located at a distance σ on either side of the mean μ. (If you have studied calculus, you know these as "inflection points.") You can feel the change in curvature as you run a pencil along a normal curve, which will allow you to estimate the standard deviation.

The total length X for a randomly selected mountain brushtail possum is modeled well by a normal distribution with mean $\mu = 87.1$ cm and standard deviation $\sigma = 4.3$ cm. The density curve shows this distribution, with the points 1, 2, and 3 standard deviations from the mean labeled on the horizontal axis.

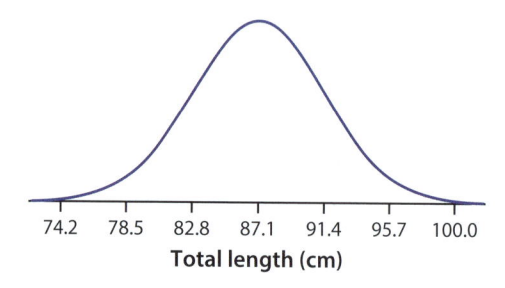

74.2 78.5 82.8 87.1 91.4 95.7 100.0
Total length (cm)

EXAMPLE 6.1.2

Stop the car!

Drawing a normal curve

PROBLEM
Many studies on automobile safety suggest that when automobile drivers make emergency stops, their stopping distances follow an approximately normal distribution. Suppose that for one model of car traveling at 62 mph under typical conditions on dry pavement, the mean stopping distance is $\mu = 155$ feet with a standard deviation of $\sigma = 3$ feet. Sketch the normal curve that approximates the probability distribution of $X =$ stopping distance. Label the mean and the points that are 1, 2, and 3 standard deviations from the mean.

SOLUTION

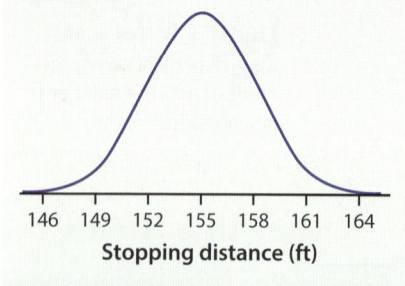

146 149 152 155 158 161 164
Stopping distance (ft)

The mean (155) is at the midpoint of the bell-shaped density curve. The standard deviation (3) is the distance from the center to the change-of-curvature points on either side. Label the mean and the points that are 1, 2, and 3 SDs from the mean:

1 SD: $155 - 1(3) = 152$ and $155 + 1(3) = 158$
2 SD: $155 - 2(3) = 149$ and $155 + 2(3) = 161$
3 SD: $155 - 3(3) = 146$ and $155 + 3(3) = 164$

EXAM PREP: FOR PRACTICE, TRY EXERCISE 11.

⚠ **Remember that μ and σ alone do not specify the appearance of most distributions.** The shape of density curves, in general, does not reveal σ. These are special properties of normal distributions.

Why are normal distributions important in statistics? Here are three reasons:

1. Normal distributions are good descriptions of some distributions of real data. Distributions that are often close to normal include the following:
   - Scores on tests taken by many people (such as the GRE)
   - Repeated careful measurements of the same quantity (like the diameter of a tennis ball)
   - Characteristics of biological populations (such as lengths of crickets and yields of corn)
2. Normal distributions can approximate the results of many kinds of chance outcomes, such as the number of heads in many tosses of a fair coin.
3. Many of the inference methods described in Chapters 7–12 are based on normal distributions.

The Empirical Rule

Because a normal curve isn't a rectangle, calculating the area under a normal curve is different than multiplying base × height. The following exploration will shed some light on how to do it.

Concept Exploration

What's so special about normal distributions?

In this exploration, you will use an applet to discover an interesting property of normal distributions. We'll start with the stopping distance context from Example 6.1.2.

1. Go to www.stapplet.com and launch the *Normal Distributions* applet.
2. Choose "Calculate an area under the normal curve" from the Operation menu at the top. Enter 155 for the mean and 3 for the standard deviation. Then click the Plot distribution button.
3. Use the applet to answer the following questions about the probability distribution of X = stopping distance.
   a. What is the probability that a randomly selected car has a stopping distance between 152 and 158 feet? That is, what proportion of the area under the normal curve lies within 1 standard deviation of the mean?
   b. What is the probability that a randomly selected car has a stopping distance between 149 and 161 feet? That is, what proportion of the area under the normal curve lies within 2 standard deviations of the mean?
   c. What is the probability that a randomly selected car has a stopping distance between 146 and 164 feet? That is, what proportion of the area under the normal curve lies within 3 standard deviations of the mean?
4. The stopping distance Y for a different model of car follows a normal distribution with mean 170 feet and standard deviation 4 feet. What is the probability that Y is within 1 standard deviation of the mean? Within 2 standard deviations of the mean? Within 3 standard deviations of the mean?
5. Change the mean to 0 and the standard deviation to 1. This is called the *standard normal distribution*, based on the fact that standardized scores (z-scores) have a mean of 0 and a standard deviation of 1 (as you learned in Chapter 3). What proportion of the area under the standard normal curve is within 1 standard deviation of the mean? Within 2 standard deviations of the mean? Within 3 standard deviations of the mean?
6. Summarize by completing this sentence: For any normal distribution, the area under the curve within 1, 2, and 3 standard deviations of the mean is about ___%, ___%, and ___%.

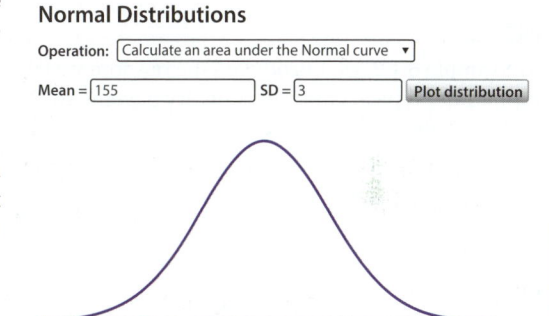

Although there are many different normal distributions, they all have some properties in common. In particular, all normal distributions obey **the empirical rule**. (*Empirical* means "learned from experience or by observation.")

In any normal distribution with mean μ and standard deviation σ, the **empirical rule** states that:
- About 68% of the values fall within 1σ of the mean μ.
- About 95% of the values fall within 2σ of the mean μ.
- About 99.7% of the values fall within 3σ of the mean μ.

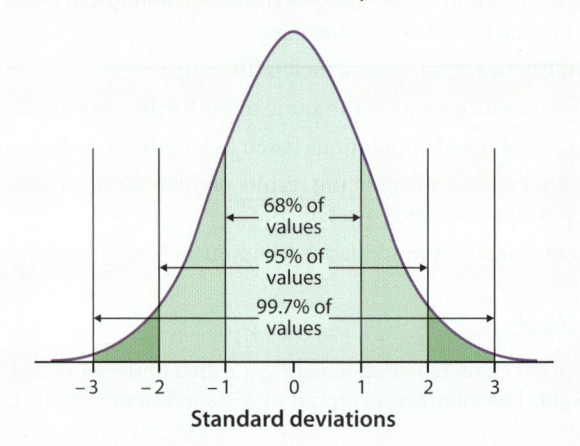

Some people refer to the empirical rule as the *68-95-99.7 rule*. By remembering these three numbers, you can quickly estimate the probability that a random variable falls in a specified interval in a normal distribution.

EXAMPLE 6.1.3

Stop the car!

The empirical rule

PROBLEM

In Example 6.1.2, we introduced the random variable X = stopping distance (in feet) of a certain model of car traveling at 62 mph under typical conditions on dry pavement. The probability distribution of X is approximately normal with $\mu = 155$ feet and $\sigma = 3$ feet.

(a) What is the probability that a car in these conditions will be able to stop in less than 149 feet?

(b) Calculate $P(152 \leq X \leq 161)$.

SOLUTION

(a) Sketch a normal curve with the mean and the values 1, 2, and 3 standard deviations from the mean labeled. Then shade the area of interest.

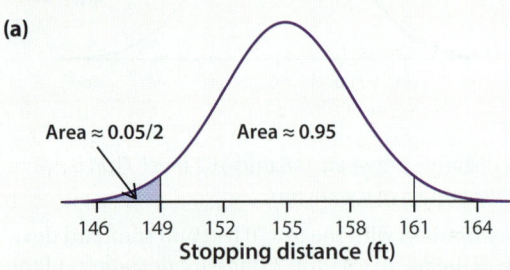

By the empirical rule, about 95% of all stopping distances fall between 149 and 161 feet. The other 5% of distances are outside this interval. Because normal distributions are symmetric, half of these distances are less than 149 feet and half are greater than 161 feet.

$P(X < 149) \approx 0.05/2 = 0.025$

(b) By the empirical rule, about 68% of all stopping distances fall between 152 and 158 feet. About 95% − 68% = 27% of stopping distances are between 1 and 2 standard deviations from the mean. Because normal distributions are symmetric, half of these stopping distances are between 158 and 161 feet.

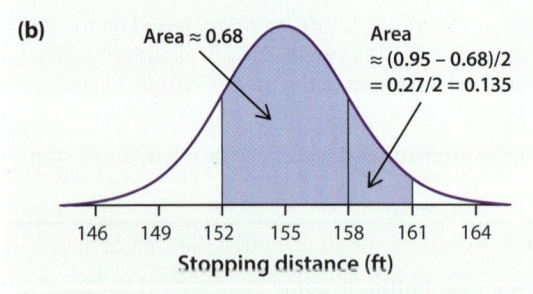

$P(152 < X < 161) \approx 0.68 + 0.27/2 = 0.68 + 0.135 = 0.815$

EXAM PREP: FOR PRACTICE, TRY EXERCISE 15.

If we think of the normal curves in Example 6.1.3 as models for distributions of quantitative data, we can say that these cars will stop in less than 149 feet about 2.5% of the time and will stop in between 152 and 161 feet about 81.5% of the time.

⚠ **The empirical rule applies *only* to normal distributions.** There are other rules for non-normal distributions, however. See Exercise 32 in Section 3.2, for example.

Section 6.1 What Did You Learn?

Review the learning goals from this section. Then practice what you've learned by working through the exercises.

Learning Goal	Example	Exercises
Use a density curve to model the probability distribution of a continuous random variable.	6.1.1	7–10
Draw a normal distribution with a given mean and standard deviation.	6.1.2	11–14
Use the empirical rule to estimate probabilities in a normal distribution.	6.1.3	15–18

Section 6.1 Exercises

Building Concepts and Skills These exercises assess the basic knowledge you should have after reading the section.

1. True/False: The area under a density curve and above an interval of values on the horizontal axis gives the probability that a random variable falls in that interval.

2. The _____ of a density curve is its balance point. The _____ of a density curve is the equal-areas point.

3. Normal curves are completely specified by two values: the _____ and the _____.

4. True/False: The standard deviation of a normal distribution is half the distance between the mean and the maximum.

5. The empirical rule is also called the _____ rule.

6. True/False: The empirical rule applies to the distribution of any continuous random variable.

Mastering Concepts and Skills These exercises reinforce the learning goals as shown in the examples.

7. **Refer to Example 6.1.1 Biking Accidents** The department of parks and recreation collected data on the locations of accidents along a 3-mile bike path in a tourist area. The department determined that X = distance from the start of the bike path to where a randomly selected accident occurred can be modeled by a uniform probability distribution on the interval from 0 miles to 3 miles.

   (a) Draw a density curve to model the probability distribution of X. Be sure to include scales on both axes.

   (b) Aaliyah's property adjoins the bike path between the 0.8-mile mark and the 1.1-mile mark. What is the probability that a randomly selected accident occurred along this section of the path?

   (c) Find the 70th percentile of this distribution.

8. **Waiting for the Bus** Jayden takes the same bus to work every morning. The random variable Y = amount of time (in minutes) that Jayden has to wait for the bus to arrive can be modeled by a uniform probability distribution on the interval from 0 minutes to 10 minutes.

   (a) Draw a density curve to model the probability distribution of Y. Be sure to include scales on both axes.

   (b) What is the probability that Jayden waits more than 7 minutes for the bus?

   (c) Find the 38th percentile of this distribution.

9. **Reaction Time** An internet reaction time test asks subjects to click their mouse as soon as a light flashes on the screen. The light is programmed to go on at a randomly selected time from 2 to 5 seconds after the subject clicks "start." The density curve models the probability distribution of the amount of time the subject has to wait for the light to flash.

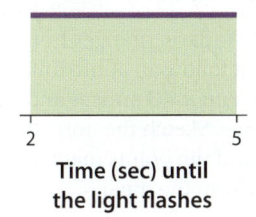

   Time (sec) until the light flashes

   (a) What height must the density curve have? Justify your answer.

   (b) Calculate the probability that the light flashes between 2.5 and 4 seconds after the subject clicks "start."

   (c) Find the first quartile Q_1 of this distribution.

10. **Class Dismissal** Professor Olsen does not always let his statistics class out at the scheduled time. In fact, he seems to end class according to his own "internal clock." The density curve models the probability distribution of the amount of time after the scheduled end of class (in minutes) when Professor Olsen dismisses the class on a randomly selected day. (A negative value indicates he dismissed his class early.)

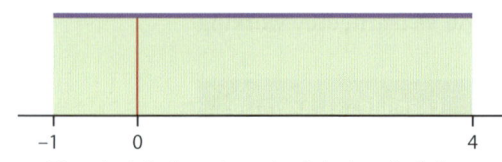

Time (min) after the scheduled end of class when Professor Olsen dismisses class

   (a) What height must the density curve have? Justify your answer.
   (b) Calculate the probability that Professor Olsen ends class within 1 minute (before or after) of the scheduled end of class on a randomly selected day.
   (c) Find the third quartile Q_3 of this distribution.

11. **Refer to Example 6.1.2 Potato Chips** The probability distribution of W = weight of a randomly selected 9-ounce bag of a particular brand of potato chips can be modeled by a normal distribution with mean $\mu = 9.12$ ounces and standard deviation $\sigma = 0.05$ ounce. Sketch the normal density curve. Label the mean and the points that are 1, 2, and 3 standard deviations from the mean.

12. **Rafa Serves!** Rafael Nadal is one of the most accomplished players in tennis history. The probability distribution of S = Nadal's first serve speed (in miles per hour) in a recent year can be modeled by a normal distribution with mean 115 mph and standard deviation 6 mph. Sketch the normal density curve. Label the mean and the points that are 1, 2, and 3 standard deviations from the mean.

13. **Car Batteries** An automaker has found that the life of its batteries varies from car to car according to an approximately normal distribution with mean $\mu = 48$ months and standard deviation $\sigma = 8$ months. Sketch the normal density curve. Label the mean and the points that are 1, 2, and 3 standard deviations from the mean.

14. **Oranges** Mandarin oranges from a certain grove have weights that follow an approximately normal distribution with mean $\mu = 3$ ounces and standard deviation $\sigma = 0.5$ ounce. Sketch the normal density curve. Label the mean and the points that are 1, 2, and 3 standard deviations from the mean.

15. **Refer to Example 6.1.3 Potato Chips** Refer to Exercise 11. Use the empirical rule to answer the following questions.
   (a) What is the probability that a randomly selected bag weighs less than 9.07 ounces?
   (b) What is $P(8.97 \leq W \leq 9.17)$?

16. **Rafa Serves!** Refer to Exercise 12. Use the empirical rule to answer the following questions.
   (a) What is the probability that a first serve is faster than 121 mph?
   (b) What is $P(109 \leq S \leq 133)$?

17. **Car Batteries** Refer to Exercise 13. Use the empirical rule to answer the following questions.
   (a) What is the probability that a battery will last between 40 and 64 months?
   (b) Would you be surprised if a battery lasted at most 32 months? Explain your reasoning.

18. **Oranges** Refer to Exercise 14. Use the empirical rule to answer the following questions.
   (a) What is the probability that a randomly selected orange weighs between 2 and 3.5 ounces?
   (b) Would you be surprised if a randomly selected orange weighed more than 4 ounces? Explain your reasoning.

Applying the Concepts These exercises ask you to apply multiple learning goals in a new context or to apply what you learned in this section in a new way.

19. **Tomato Slices** A particular variety of tomato is prized for its use on hamburgers because of its flavor and size. The probability distribution of X = diameter (cm) of a randomly selected tomato of this variety can be modeled by a normal distribution with $\mu = 10$ cm and $\sigma = 1$ cm.
   (a) Sketch the normal density curve. Label the mean and the points that are 1, 2, and 3 standard deviations from the mean.
   (b) What is the probability that a randomly selected tomato will have a diameter greater than 9 cm?
   (c) About 95% of these tomatoes have diameters between what two values?

20. **Cereal Boxes** A company's cereal boxes advertise that each box contains 9.65 ounces of cereal. In fact, the probability distribution of Y = amount of cereal in a randomly selected box can be modeled by a normal distribution with mean $\mu = 9.71$ ounces and standard deviation $\sigma = 0.03$ ounce.
   (a) Sketch the normal density curve. Label the mean and the points that are 1, 2, and 3 standard deviations from the mean.
   (b) What is the probability that a randomly selected box will be less than the advertised weight?
   (c) About 68% of these boxes have weights between what two values?

21. **Normal Curve** Estimate the mean and standard deviation of the normal density curve in the figure.

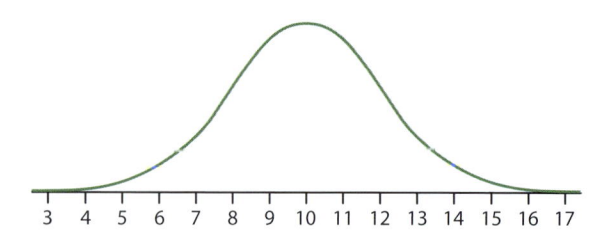

22. Normal Curve Estimate the mean and standard deviation of the normal density curve in the figure.

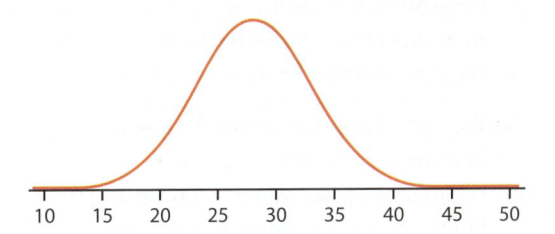

Extending the Concepts These exercises challenge you to explore statistical concepts and methods that go beyond what you learned in this section.

23. Unusual Density Curve The figure shows a density curve that models the probability distribution of a continuous random variable.

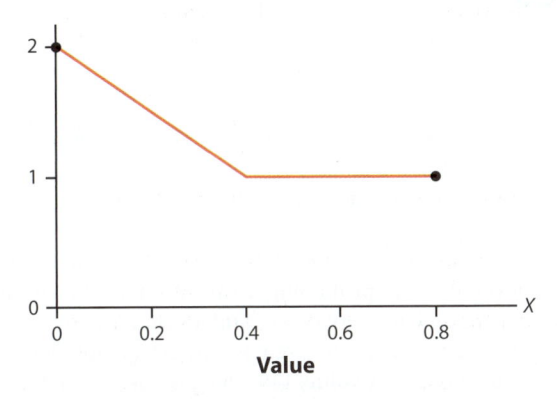

(a) Show that this is a valid density curve.
(b) What is the probability that this random variable takes a value between 0 and 0.2?
(c) The median of the density curve is between 0.2 and 0.4. Explain why.
(d) Is the mean of the density curve less than, equal to, or greater than the median of the density curve? Justify your answer.

24. Find the Center The figure shows a density curve that models the probability distribution of a continuous random variable. Identify the location of the mean and median by letter. Justify your answers.

25. Uniform Distributions What percentage of values in a uniform probability distribution are within 1 standard deviation of the mean? Within 2 standard deviations? For any uniform probability distribution, the standard deviation is $\sigma = \sqrt{\dfrac{(b-a)^2}{12}}$, where a is the minimum value in the distribution and b is the maximum value. Suppose that X is a uniform random variable that takes values from 0 to 10.

(a) What is the mean of X? Explain your reasoning.
(b) Use the formula to calculate the standard deviation of X.
(c) What percentage of the values of X are within 1 standard deviation of the mean? How does this compare with the empirical rule?
(d) What percentage of the values of X are within 2 standard deviations of the mean? How does this compare with the empirical rule?

Cumulative Review These exercises revisit what you learned in previous sections.

26. Double Cream (2.5, 3.1, 3.2) Do Double Stuf Oreo cookies really have twice as much cream as regular Oreo cookies? Two researchers decided to investigate this important question. The researchers obtained separate random samples of 45 regular Oreo cookies and 45 Double Stuf Oreo cookies, and then weighed the amount of cream (in grams) on each cookie. Then they doubled the weight of cream on each regular Oreo cookie.[3] Here are comparative histograms and summary statistics of the data.

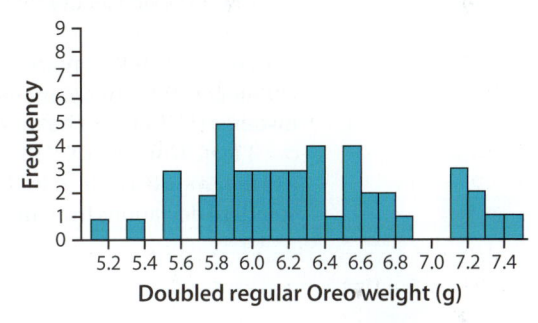

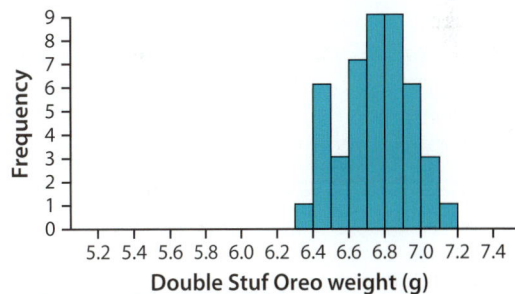

	n	Mean	SD	Min	Q_1	Med	Q_3	Max
Doubled regular Oreos (g)	45	6.301	0.565	5.162	5.881	6.294	6.676	7.46
Double Stuf Oreos (g)	45	6.742	0.184	6.37	6.62	6.73	6.875	7.15

(a) Compare the distributions of cream weight for the two types of cookies.
(b) What conclusion should the researchers make about whether Double Stuf Oreos have twice as much cream as regular Oreos? Justify your answer.

27. **Comparing Quarterbacks (3.4)** In the modern National Football League (NFL), teams emphasize the passing game much more than teams in the 1970s. This makes it very difficult to compare the performances of quarterbacks from these eras—even when they both played for the same team. In 1979, Terry Bradshaw of the Pittsburgh Steelers threw for a career-high 3724 yards. In 2018, the Steelers' Ben Roethlisberger threw for a career-high 5129 yards.[4] Whose performance was better, relatively speaking?

    (a) In 1979, the top quarterback for each of the 28 NFL teams threw for an average of 2789 yards with a standard deviation of 752 yards. Calculate and interpret the z-score for Terry Bradshaw's performance in 1979.

    (b) In 2018, the top quarterback for each of the 32 NFL teams threw for an average of 3552 yards with a standard deviation of 939 yards. Calculate and interpret the z-score for Ben Roethlisberger's performance in 2018.

    (c) Based on your answers to parts (a) and (b), which quarterback had a better performance, relatively speaking? Explain your reasoning.

Section 6.2 Normal Distributions: Finding Areas from Values

LEARNING GOALS

By the end of this section, you will be able to:

- Find the probability that a value falls to the left of a boundary in a normal distribution.
- Find the probability that a value falls to the right of a boundary in a normal distribution.
- Find the probability that a value falls between two boundaries in a normal distribution.

Let's return to the stopping distance context from Section 6.1. Recall that when drivers of a certain car model make an emergency stop on dry pavement, the probability distribution of X = stopping distance (in feet) can be modeled by a normal curve with $\mu = 155$ feet and a standard deviation of $\sigma = 3$ feet. What is $P(X \leq 149.5)$? Here is a normal curve with the area of interest shaded. We can't use the empirical rule to find this area because the boundary value, 149.5 feet, is not exactly 1, 2, or 3 standard deviations from the mean.

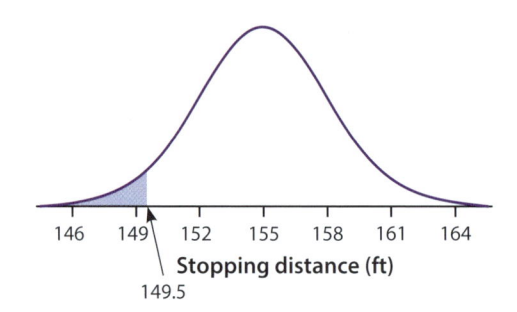

Finding Areas to the Left in a Normal Distribution

As the empirical rule suggests, all normal distributions are the same if we measure them in units of size σ from the mean μ. Changing to these units requires us to standardize, just as we did in Section 3.4:

$$z = \frac{\text{value} - \text{mean}}{\text{standard deviation}} = \frac{x - \mu}{\sigma}$$

Fortunately, subtracting a constant and dividing by a constant don't change the shape of a distribution. If the random variable we standardize has an approximately normal distribution, then so does the new variable Z. This new distribution of standardized values can be modeled with a normal curve having mean $\mu = 0$ and standard deviation $\sigma = 1$. It is called the **standard normal distribution**.

The **standard normal distribution** is the normal distribution with mean 0 and standard deviation 1.

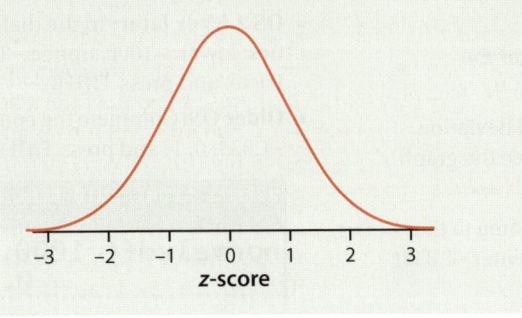

Because all normal distributions are the same when we standardize, we can find areas under any normal curve using the standard normal distribution. For the stopping distance scenario, we want to find $P(X \leq 149.5)$ using a normal distribution with $\mu = 155$ and $\sigma = 3$, as shown in Figure 6.4(a). We start by standardizing the boundary value:

$$z = \frac{\text{value} - \text{mean}}{\text{standard deviation}} = \frac{149.5 - 155}{3} = -1.833$$

A stopping distance of 149.5 feet is 1.833 standard deviations less than the mean. Figure 6.4(b) shows the standard normal distribution with the area to the left of $z = -1.833$ shaded. Notice that the shaded areas in the two graphs are the same.

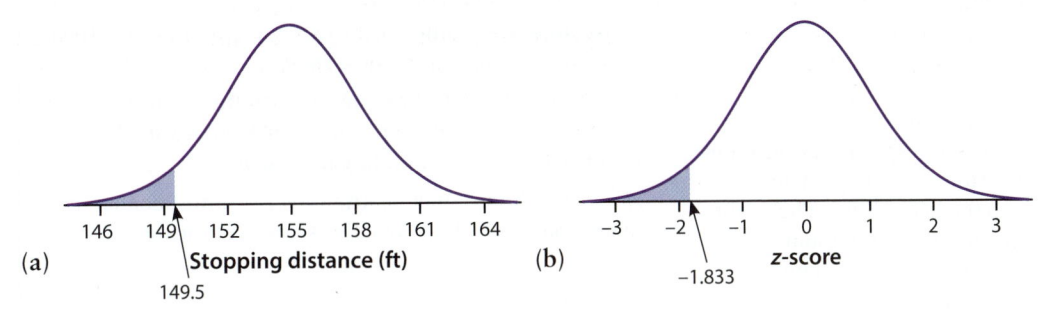

FIGURE 6.4 (a) Normal distribution showing the probability that a car stops in at most 149.5 feet when emergency braking on dry pavement. (b) The corresponding probability in the standard normal distribution.

Once we have standardized the boundary value, we can use Table A in the back of the book or technology to find the relevant area. In Table A, the entry for each z-score is the area under the curve *to the left* of z. To find the area to the left of $z = -1.833$, round to two decimal places (-1.83), locate -1.8 in the left-hand column of the table, then locate the remaining digit 3 as .03 in the top row. The entry to the right of -1.8 and under .03 is .0336. This is the area we seek. There is a 0.0336 probability that a car in this situation will stop in at most 149.5 feet.

z	.02	.03	.04
−1.9	.0274	.0268	.0262
−1.8	.0344	.0336	.0329
−1.7	.0427	.0418	.0409

TECH CORNER Finding areas from values in a normal distribution

You can use technology to find areas from values in *any* normal distribution. Let's use an applet and the TI-83/84 to confirm our earlier result about the probability of stopping in at most 149.5 feet. With technology, we can perform the area calculation using the standard normal distribution or the "unstandardized" normal distribution with mean 155 and standard deviation 3. Recall that for 149.5 feet, $z = \dfrac{149.5 - 155}{3} = -1.833$.

(continued)

Applet

1. Go to www.stapplet.com and launch the *Normal Distributions* applet.
2. Select the option to "Calculate an area under the Normal curve."
3. Enter 0 for the mean and 1 for the standard deviation. Then click the Plot distribution button to see the graph of the standard normal distribution.
4. To find $P(Z \leq -1.833)$, select "Calculate the area to the left of a value" from the drop-down menu, enter -1.833, and click "Calculate area."

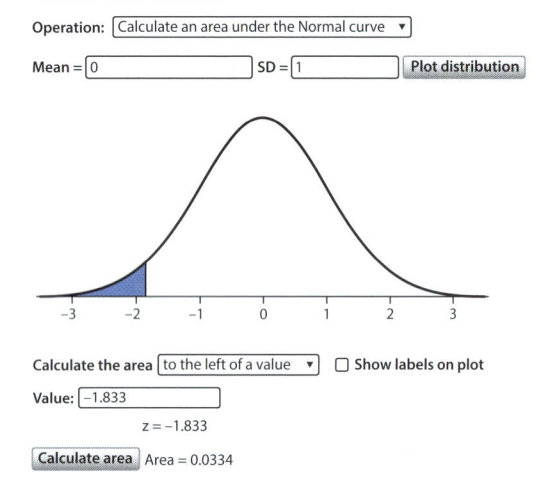

Normal Distributions

Operation: Calculate an area under the Normal curve
Mean = 0   SD = 1   [Plot distribution]

Calculate the area [to the left of a value] ☐ Show labels on plot
Value: -1.833
z = -1.833
[Calculate area]   Area = 0.0334

5. To calculate $P(X \leq 149.5)$ without standardizing, enter 155 for the mean and 3 for the standard deviation. Then calculate the area to the left of 149.5. The result should be the same as the result from the standard normal distribution, other than a small difference because of rounding the z-score.

TI-83/84

1. Press 2nd VARS (Distr) and choose normalcdf(.
   - **OS 2.55 or later:** In the dialog box, enter these values: lower: -1000, upper: -1.833, μ: 0, σ: 1. Choose Paste, and press ENTER.
   - **Older OS:** Complete the command normalcdf(-1000, -1.833, 0, 1) and press ENTER.

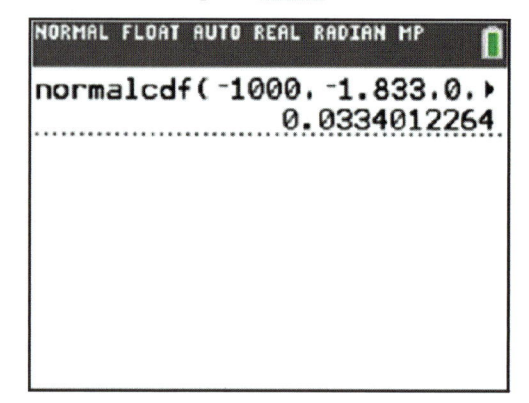

```
NORMAL FLOAT AUTO REAL RADIAN MP

normalcdf(-1000,-1.833,0,▶
                  0.0334012264
```

2. To calculate $P(X \leq 149.5)$ without standardizing, press 2nd VARS (Distr) and choose normalcdf(.
   - **OS 2.55 or later:** In the dialog box, enter these values: lower: -1000, upper: 149.5, μ: 155, σ: 3. Choose Paste, and press ENTER.
   - **Older OS:** Complete the command normalcdf(-1000, 149.5, 155, 3) and press ENTER.

The result should be the same as the result from the standard normal distribution, other than a small difference because of rounding the z-score.

Note: We chose -1000 as the lower bound because it is many, many standard deviations less than the mean.

Detailed instructions for using CrunchIt!, Excel, Google Sheets, JMP, Minitab, and R are available in Achieve.

As the Tech Corner illustrates, it is possible to find $P(X \leq 149.5)$ directly from the original (unstandardized) normal distribution using technology or by standardizing the boundary value and using Table A or technology. *Ask your instructor which method should be used in your class.*

How to Find Areas in Any Normal Distribution

Step 1: Draw a normal distribution with the horizontal axis labeled and scaled using the mean and standard deviation, the boundary value(s) clearly identified, and the area of interest shaded.

Step 2: Perform calculations—show your work! Do one of the following:

(i) Standardize the boundary value(s) and use technology or Table A to find the desired area under the standard normal curve.

(ii) Use technology to find the desired area without standardizing. Label the inputs you used for the applet or calculator.

Section 6.2 Normal Distributions: Finding Areas from Values 317

EXAMPLE 6.2.1

Nelly's driver

Finding area to the left in a normal distribution

PROBLEM

Nelly Korda was one of the top golfers on the Ladies Professional Golf Association (LPGA) tour in 2021. When she hits her driver, the probability distribution of X = distance the ball travels can be modeled by a normal curve with mean 274 yards and standard deviation 8 yards.[5] What is the probability that a shot with her driver goes less than 285 yards?

SOLUTION

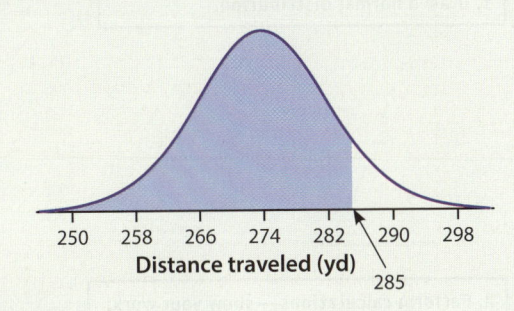

1. **Draw a normal distribution.** Be sure to:
   - Scale the horizontal axis.
   - Label the horizontal axis with the variable name.
   - Clearly identify the boundary value(s).
   - Shade the area of interest.

2. **Perform calculations—show your work!**
   (i) Standardize and use technology or Table A.
   (ii) Use technology without standardizing.

(i) $z = \dfrac{285 - 274}{8} = 1.375$

Using technology: $P(Z < 1.375)$ = Applet/normalcdf(lower: −1000, upper: 1.375, mean: 0, SD: 1) = 0.9154

Using Table A: $P(Z < 1.38) = 0.9162$

(ii) $P(X < 285)$ = Applet/normalcdf(lower: −1000, upper: 285, mean: 274, SD: 8) = 0.9154

Note that the slightly different answer obtained when using Table A occurs because we had to round the z-score to 2 decimal places.

EXAM PREP: FOR PRACTICE, TRY EXERCISE 9.

THINK ABOUT IT

What is the probability that one of Nelly Korda's drives go exactly 285 yards? Because a point on the number line has no width, there is no area directly above the point 285.000000000. . . under the normal density curve in Example 6.2.1. So, the answer to our question based on the normal distribution is 0. One more thing: The areas under the curve with $X < 285$ and $X \leq 285$ are the same. According to the normal model, the probability that a randomly selected drive will travel at most 285 yards is the same as the probability that it will travel less than 285 yards.

Finding Areas to the Right in a Normal Distribution

Finding areas to the right of a boundary line is very similar to finding areas to the left, especially when using technology. When using Table A, the values in the table give areas to the left. Because the total area is 1 under the standard normal curve, to find the area to the right, simply subtract the area to the left from 1. The following example shows you what we mean.

EXAMPLE 6.2.2

Nelly's driver

Finding area to the right in a normal distribution

PROBLEM

When professional golfer Nelly Korda hits her driver, the probability distribution of X = distance the ball travels can be modeled by a normal curve with $\mu = 274$ and $\sigma = 8$. On a specific hole, Korda would need to hit the ball at least 260 yards to have a clear second shot that avoids a large group of trees. Is she likely to have a clear second shot?

SOLUTION

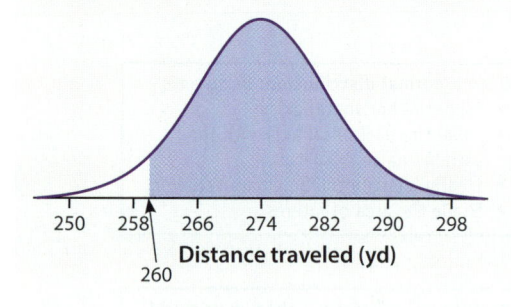

1. Draw a normal distribution.

2. Perform calculations—show your work!

(i) $z = \dfrac{260 - 274}{8} = -1.750$

Using technology: $P(Z \geq -1.750)$ = Applet/normalcdf(lower: −1.750, upper: 1000, mean: 0, SD: 1) = 0.9599
Using Table A: $P(Z \geq -1.75) = 1 - 0.0401 = 0.9599$

(ii) $P(X \geq 260)$ = Applet/normalcdf(lower: 260, upper: 1000, mean: 274, SD: 8) = 0.9599

Because $P(X \geq 260)$ is very large (about 0.96), Korda is likely to have a clear shot.

EXAM PREP: FOR PRACTICE, TRY EXERCISE 13.

Finding Areas Between Two Values in a Normal Distribution

Finding the area between two boundaries is very similar to finding the area to the left or to the right of a boundary, especially when using technology. When using Table A, it's a little more complicated.

Let's use Table A to find the probability that the stopping distance for a car is between 150 and 162 feet. Recall that when drivers of this car model make an emergency stop on dry pavement, the probability distribution of X = stopping distance (in feet) is approximately normal with $\mu = 155$ feet and a standard deviation of $\sigma = 3$ feet. We want to find $P(150 < X < 162)$, as shown by the figure on the left.

Start by standardizing both boundary values:

$$z = \dfrac{150 - 155}{3} = -1.67 \quad \text{and} \quad z = \dfrac{162 - 155}{3} = 2.33$$

The figure on the right shows the corresponding area of interest under the standard normal curve.

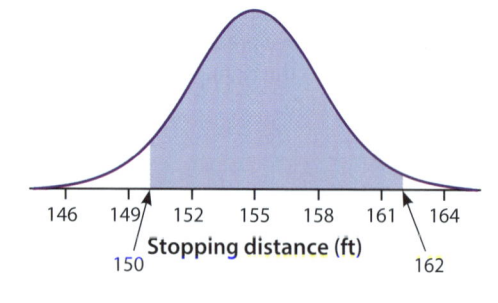

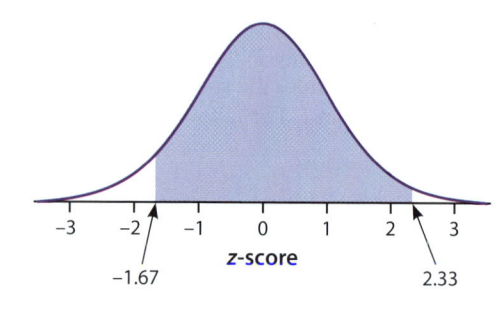

Using Table A makes this process a bit trickier because it shows only the area to the left of a given z-score. The visual offers one way to think about the calculation.

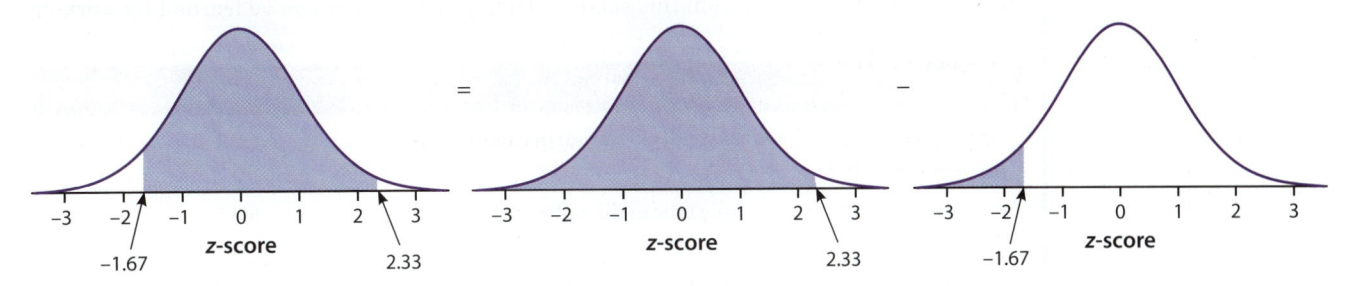

area between $z = -1.67$ and $z = 2.33$
$\qquad$ = (area to the left of $z = 2.33$) − (area to the left of $z = -1.67$)
$\qquad$ = 0.9901 − 0.0475
$\qquad$ = 0.9426

Using technology is easier because you can enter the lower and upper boundaries and the technology does the subtraction of areas for you. Using standardized scores to three decimal places, Applet/normalcdf(lower: −1.667, upper: 2.333, mean: 0, SD: 1) gives 0.9424. Using the unstandardized values, Applet/normalcdf(lower: 150, upper: 162, mean: 155, SD: 3) gives 0.9424.

EXAMPLE 6.2.3

Nelly's driver

Finding area between two values in a normal distribution

PROBLEM
When professional golfer Nelly Korda hits her driver, the probability distribution of X = distance the ball travels can be modeled by a normal curve with $\mu = 274$ and $\sigma = 8$. On another golf hole, Korda can drive the ball onto the green if she hits the ball between 275 and 295 yards. What is the probability that she accomplishes this goal?

SOLUTION

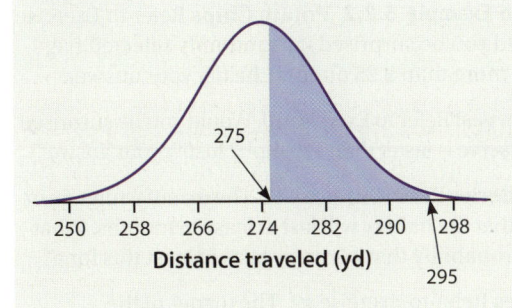

1. Draw a normal distribution.

2. Perform calculations — show your work!

(i) $z = \dfrac{275 - 274}{8} = 0.125$ and $z = \dfrac{295 - 274}{8} = 2.625$

Using technology: $P(0.125 < Z < 2.625)$ = Applet/normalcdf(lower: 0.125, upper: 2.625, mean: 0, SD: 1) = 0.4459

Using Table A: $P(0.13 < Z < 2.63) = 0.9957 - 0.5517 = 0.4440$

(ii) $P(275 < X < 295)$ = Applet/normalcdf(lower: 275, upper: 295, mean: 274, SD: 8) = 0.4459

EXAM PREP: FOR PRACTICE, TRY EXERCISE 17.

Section 6.2 What Did You Learn?

Review the learning goals from this section. Then practice what you've learned by working through the exercises.

Learning Goal	Example	Exercises
Find the probability that a value falls to the left of a boundary in a normal distribution.	6.2.1	9–12
Find the probability that a value falls to the right of a boundary in a normal distribution.	6.2.2	13–16
Find the probability that a value falls between two boundaries in a normal distribution.	6.2.3	17–20

Section 6.2 Exercises

Building Concepts and Skills These exercises assess the basic knowledge you should have after reading the section.

1. What is the formula for calculating a standardized score (z-score)?

2. The standard normal distribution has a mean of _____ and a standard deviation of _____.

3. Before doing any calculations, what is the first step when finding areas in any normal distribution?

4. True/False: When finding area in a normal distribution, it matters whether the boundary value is included or not — for example, $X < 120$ versus $X \leq 120$.

5. True/False: The entry for each z-score in Table A is the area under the standard normal curve to the left of z.

6. In the standard normal curve, the area to the left of $z = 0.53$ is 0.7019. What is the area to the right of $z = 0.53$?

7. If Z = a randomly selected value from the standard normal distribution, find each of the following probabilities.
   (a) $P(Z < 2.46)$
   (b) $P(Z > -1.66)$
   (c) $P(0.50 < Z < 1.79)$

8. If Z = a randomly selected value from the standard normal distribution, find each of the following probabilities.
   (a) $P(Z < -1.39)$
   (b) $P(Z > 2.15)$
   (c) $P(-1.11 < Z < -0.32)$

Mastering Concepts and Skills These exercises reinforce the learning goals as shown in the examples.

9. **Refer to Example 6.2.1 Potato Chips** The probability distribution of W = weight of a randomly selected 9-ounce bag of a particular brand of potato chips can be modeled by a normal distribution with mean $\mu = 9.12$ ounces and standard deviation $\sigma = 0.05$ ounce. What is the probability that a randomly selected bag is less than the advertised weight of 9 ounces?

10. **Rafa Serves!** The probability distribution of S = Rafael Nadal's first serve speed (in miles per hour) in a recent year can be modeled by a normal distribution with mean 115 mph and standard deviation 6 mph. What is the probability that a first serve is less than 100 mph?

11. **Car Batteries** An automaker has found that the life of its batteries varies from car to car according to an approximately normal distribution with mean $\mu = 48$ months and standard deviation $\sigma = 8$ months. What is the probability that a battery lasts less than 50 months?

12. **Oranges** Mandarin oranges from a certain grove have weights that follow an approximately normal distribution with mean $\mu = 3$ ounces and standard deviation $\sigma = 0.5$ ounce. What is the probability that a randomly selected orange weighs less than 3.7 ounces?

13. **Refer to Example 6.2.2 Potato Chips** Refer to Exercise 9. Would you be surprised if a randomly selected bag weighs more than 9.25 ounces? Justify your answer.

14. **Rafa Serves!** Refer to Exercise 10. Would you be surprised if a first serve is faster than 120 mph? Justify your answer.

15. **Car Batteries** Refer to Exercise 11. The manufacturer guarantees that the battery will last at least 36 months. What is the probability that a battery lasts at least this long?

16. **Oranges** Refer to Exercise 12. The owner of the grove will try to sell all oranges that weigh at least 2.75 ounces. What is the probability that a randomly selected orange meets this criterion?

17. **Refer to Example 6.2.3 Potato Chips** Refer to Exercise 9. What is the probability that a randomly selected bag of chips weighs between 9.1 and 9.15 ounces?

18. **Rafa Serves!** Refer to Exercise 10. What is the probability that a first serve is between 110 and 125 mph?

19. **Car Batteries** Refer to Exercise 11. What is the probability that a battery lasts between 55 and 60 months?

20. **Oranges** Refer to Exercise 12. What is the probability that a randomly selected orange weighs between 1.8 and 2.2 ounces?

Applying the Concepts These exercises ask you to apply multiple learning goals in a new context or to apply what you learned in this section in a new way.

21. **GRE Scores** Many graduate schools require applicants to take the GRE. The probability distribution of $G =$ score on the GRE Quantitative Reasoning section for a randomly selected test-taker is approximately normal with mean 153.4 and standard deviation 9.4.[6] A particular graduate program recommends that applicants have a quantitative reasoning score of at least 160.
    (a) What is the probability that a randomly selected test-taker gets a recommended score?
    (b) What is the probability that a randomly selected test-taker scores between 150 and 155?

22. **Egg Weights** In the United States, the Department of Agriculture defines standard egg sizes. A "large" egg, for example, weighs between 57 and 64 grams. The probability distribution of $Y =$ weight of a randomly selected egg produced by the hens of a particular farmer is approximately normal with mean 55.8 grams and standard deviation 7.5 grams.
    (a) What is the probability that a randomly selected egg from this farm weighs less than 50 grams?
    (b) What is the probability that a randomly selected egg from this farm would be classified as "large"?

23. **Cup Lids** At fast-food restaurants, the lids for drink cups are made with a small amount of flexibility, so they can be stretched across the mouth of the cup and then snugly secured. When lids are too small or too large, customers can get frustrated, especially if they end up spilling their drinks. At one restaurant, large drink cups require lids with a diameter of between 3.95 and 4.05 inches. The restaurant's lid supplier claims that the diameter of the large lids follows a normal distribution with mean 3.98 inches and standard deviation 0.02 inch. Assume that the supplier's claim is true.
    (a) What percentage of large lids are too small to fit?
    (b) What percentage of large lids are too big to fit?
    (c) What percentage of large lids have diameters between 3.95 and 4.05 inches?
    (d) Compare your answers to parts (a) and (b). Does it make sense for the lid manufacturer to try to make one of these values larger than the other? Why or why not?

24. **Bottling Soda** A bottling company fills bottles labeled "2 liters" with lemonade. Of course, there is some variation from the target volume. In fact, the distribution of volume is approximately normal with a mean of 2.04 liters and a standard deviation of 0.03 liter.
    (a) About what percentage of bottles are underfilled (i.e., less than 2 liters)?
    (b) About what percentage of bottles have a volume greater than 2.10 liters?
    (c) About what percentage of bottles are within 0.05 liter of the target volume (i.e., between 1.95 and 2.05 liters)?
    (d) Based on your answers to parts (a), (b), and (c), do you think this bottling company needs to adjust its filling machine? Why or why not?

Extending the Concepts These exercises challenge you to explore statistical concepts and methods that go beyond what you learned in this section.

25. **NFL Receivers** In the 2020 NFL season, nearly 500 players caught at least one pass. The mean number of catches was 23.6 and the standard deviation was 24.8.[7] Assume (for now) that the distribution of number of catches is approximately normal.
    (a) What is the probability of randomly selecting a player who has fewer than 0 catches, according to the normal model?
    (b) What does your answer to part (a) indicate about the assumption of normality?

26. **z-Scores** Table A includes z-scores from –3.49 to 3.49. However, there are many instances where a z-score is outside of these boundaries. Find the following probabilities using the standard normal distribution.
    (a) $P(Z < -4.38)$
    (b) $P(Z > -4.38)$
    (c) $P(Z < 8.39)$
    (d) $P(Z > 8.39)$

Cumulative Review These exercises revisit what you learned in previous sections.

27. **Gamers (2.1)** The Pew Research Center asked a random sample of 1996 U.S. adults if they play video games. In the sample, 165 identified themselves as "gamers," 760 said they play video games but don't identify as a gamer, and 1071 said they do not play video games.[8] Here is a graph of these data.

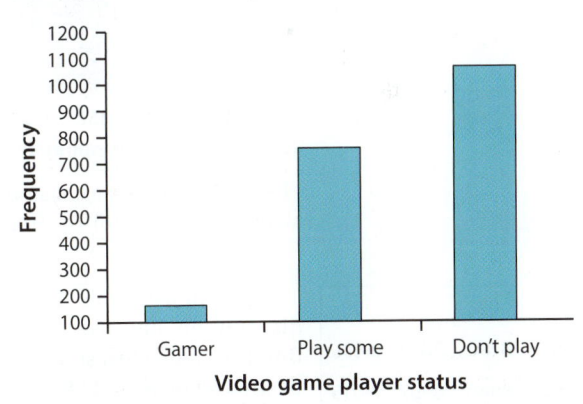

    (a) Explain what is potentially deceptive about this graph.
    (b) Make a bar chart for these data that isn't deceptive.

28. **Standard Deviations (3.2, 6.1)** Continuous random variables A, B, and C all take values between 0 and 10. Density curves that model the probability distributions of each variable, drawn on the same horizontal and vertical scales, are shown here. Rank the standard deviations of the three variables from smallest to largest. Justify your answer.

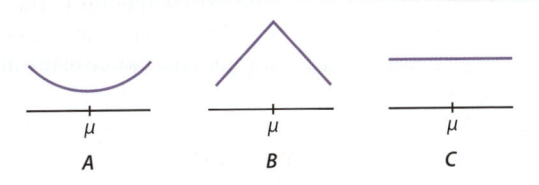

Section 6.3 Normal Distributions: Finding Values from Areas

LEARNING GOALS

By the end of this section, you will be able to:
- Find the value corresponding to a given probability (area) in a normal distribution.
- Find the mean or standard deviation of a normal distribution given a boundary value and corresponding probability (area).

When the probability distribution of a continuous random variable can be modeled with a normal curve, we can use the methods described in Section 6.2 to find the probability that the random variable falls in a particular interval of values. We do this by finding the appropriate area in a normal distribution. What if we want to find the boundary value in a normal distribution corresponding to a given probability? That will require us to reverse the process to go from an area to a value.

Finding Boundary Values in a Normal Distribution

Let's return to the stopping distance scenario from Sections 6.1 and 6.2. Recall that when drivers of a certain car model make an emergency stop on dry pavement, the probability distribution of $X =$ stopping distance (in feet) can be modeled by a normal curve with $\mu = 155$ feet and a standard deviation of $\sigma = 3$ feet. What is the 90th percentile of this distribution? That is, what is the value of X such that the probability of stopping in less than x feet is 0.90?

Figure 6.5(a) shows what we are trying to find: the stopping distance x with 90% of the area to its left under the normal curve. Figure 6.5(b) shows the standard normal distribution with the corresponding area shaded.

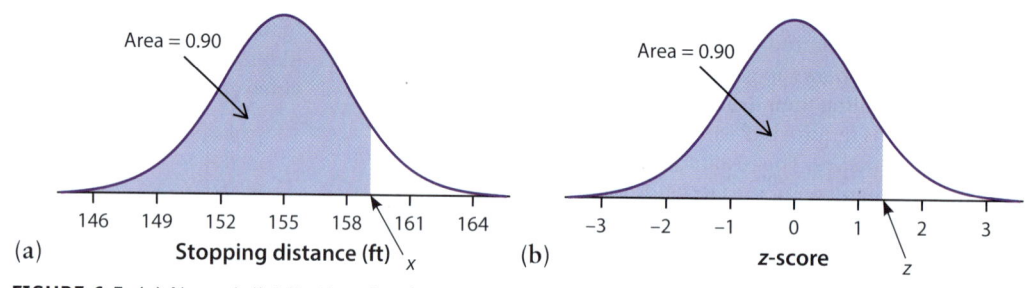

FIGURE 6.5 (a) Normal distribution showing the 90th percentile of the probability distribution of $X =$ stopping distance. (b) The 90th percentile in the standard normal distribution.

z	.07	.08	.09
1.1	.8790	.8810	.8830
1.2	.8980	.8997	.9015
1.3	.9147	.9162	.9177

We can use Table A to find the z-score with an area of 0.90 to its left in a standard normal distribution. Because Table A gives the area to the left of a specified z-score, all we have to do is find the value closest to 0.90 in the middle of the table. From the table excerpt you see that the desired value is $z = 1.28$. To get the corresponding stopping distance x, use the z-score formula to "unstandardize":

$$z = \frac{x - \mu}{\sigma}$$

$$1.28 = \frac{x - 155}{3}$$

$$1.28(3) + 155 = x$$

$$158.84 = x$$

The 90th percentile of the probability distribution of $X =$ stopping distance is 158.84 feet.

We can also use technology to find the 90th percentile of the standard normal distribution and the 90th percentile of the unstandardized distribution of stopping distances.

TECH CORNER Finding values from areas in any normal distribution

You can use technology to find values from areas in *any* normal distribution. Let's use an applet and the TI-83/84 to calculate the 90th percentile of the standard normal distribution and the 90th percentile of the probability distribution of X = stopping distance using the "unstandardized" normal distribution with mean 155 and standard deviation 3.

Applet

1. Go to www.stapplet.com and launch the *Normal Distributions* applet.
2. Select the option to "Calculate a value corresponding to an area."
3. Enter 0 for the mean and 1 for the standard deviation. Then click the Plot distribution button to see the graph of the standard normal distribution.
4. To calculate the 90th percentile in the standard normal distribution, select "Calculate boundary value(s) for a left-tail area" from the drop-down menu, enter 0.90, and click "Calculate value(s)."

Normal Distributions

Operation: Calculate a value corresponding to an area

Mean = 0   SD = 1   Plot distribution

Calculate boundary value(s) for a left-tail ▼ area of 0.90
Calculate value(s) Value = 1.282 (z = 1.282)

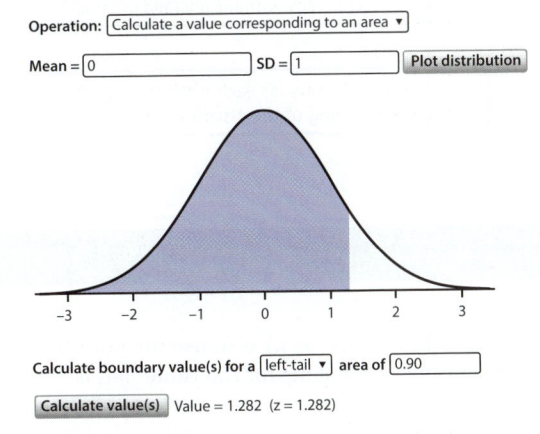

5. To calculate the 90th percentile without standardizing, enter 155 for the mean and 3 for the standard deviation. Then calculate the boundary value for a left-tail area of 0.90.

TI-83/84

1. To calculate the 90th percentile in the standard normal distribution, press 2nd VARS (Distr) and choose invNorm(.
   - **OS 2.55 or later:** In the dialog box, enter these values: area: 0.90, μ: 0, σ: 1. Choose Paste, and press ENTER.
   - **Older OS:** Complete the command invNorm(0.90, 0, 1) and press ENTER.

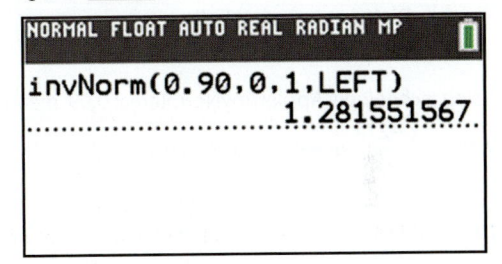

```
NORMAL FLOAT AUTO REAL RADIAN MP
invNorm(0.90,0,1,LEFT)
                    1.281551567
```

Note: The most recent TI-84 Plus CE OS has added an option for specifying area in the LEFT, CENTER, or RIGHT of the distribution. Choose LEFT in this case.

2. To calculate the 90th percentile without standardizing, press 2nd VARS (Distr) and choose invNorm(.
   - **OS 2.55 or later:** In the dialog box, enter these values: area: 0.90, μ: 155, σ: 3. Choose Paste, and press ENTER.
   - **Older OS:** Complete the command invNorm(0.90, 155, 3) and press ENTER.

Detailed instructions for using CrunchIt!, Excel, Google Sheets, JMP, Minitab, and R are available in Achieve.

You can find the 90th percentile using the standard normal distribution with technology, the standard normal distribution with Table A, or the unstandardized normal distribution with technology. *Ask your instructor which method should be used in your class.*

How to Find Values from Areas in Any Normal Distribution

Step 1: Draw a normal distribution with the horizontal axis labeled and scaled using the mean and standard deviation, the area of interest shaded and labeled, and the unknown boundary value clearly marked.

Step 2: Perform calculations—show your work! Do one of the following:

(i) Use technology or Table A to find the value of z with the appropriate area under the standard normal curve, then unstandardize to return to the original distribution.

(ii) Use technology to find the desired value without standardizing. Label the inputs you used for the applet or calculator.

EXAMPLE 6.3.1

Nelly's driver

Finding a value from an area in a normal distribution

PROBLEM

When professional golfer Nelly Korda hits her driver, the probability distribution of $X =$ distance the ball travels can be modeled by a normal curve with $\mu = 274$ and $\sigma = 8$. What is the value of the 99th percentile in this distribution?

SOLUTION

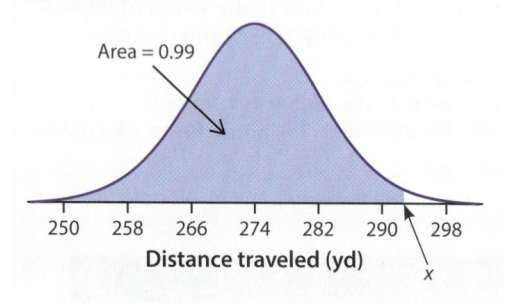

Area = 0.99

250 258 266 274 282 290 298
Distance traveled (yd)
X

1. **Draw a normal distribution.** The 99th percentile of this distribution is the boundary value such that 99% of the area under the curve is less than this value.

(i) *Using technology:* Applet/invNorm(area: 0.99, mean: 0, SD: 1) = 2.326;

$2.326 = \dfrac{x - 274}{8}$ gives $x = 292.61$ yards

Using Table A: 0.99 area to the left gives $z = 2.33$; $2.33 = \dfrac{x - 274}{8}$ gives $x = 292.64$ yards

2. **Perform calculations — show your work!**
   (i) Use technology or Table A and unstandardize.
   (ii) Use technology without unstandardizing.

Note that the answers for each method may vary slightly due to the rounding of the z-scores.

(ii) Applet/invNorm(area: 0.99, mean: 274, SD: 8) gives $x = 292.61$ yards

EXAM PREP: FOR PRACTICE, TRY EXERCISE 5.

Another approach to finding the 99th percentile in Example 6.3.1 is to use the interpretation of the z-score. A standardized score of $z = 2.326$ means we are looking for the value that is 2.326 standard deviations above the mean:

$$\text{mean} + 2.326(\text{SD}) = 274 + 2.326(8) = 292.61$$

So 1% of Korda's drives exceed 292.61 yards.

Finding the Mean or Standard Deviation from Areas in a Normal Distribution

You have seen how to find the value corresponding to a given probability in a normal distribution with known mean and standard deviation. It is also possible to find the mean or standard deviation of a normal distribution using a boundary value and the corresponding probability.

EXAMPLE 6.3.2

Exam time

Finding the standard deviation from an area in a normal distribution

PROBLEM

A statistics professor keeps track of how long each student takes to complete an exam. The distribution of time to finish (in minutes) is approximately normal with a mean of 47 minutes. If 12% of the students are still working on the exam at the end of the 60-minute class, find the standard deviation of the distribution.

SOLUTION

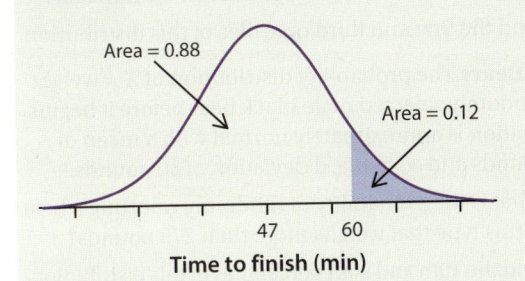

Area = 0.88

Area = 0.12

47   60

Time to finish (min)

1. **Draw a normal distribution.** Because we don't know the standard deviation, we can only label the mean and the boundary value, along with the corresponding areas.

(i) *Using technology:* Applet/invNorm(area: 0.88, mean: 0, SD: 1) gives $z = 1.175$;

$$1.175 = \frac{60 - 47}{\sigma}$$

$1.175\sigma = 13$

$\sigma = 11.06$ minutes

2. **Perform calculations—show your work!** Because we are trying to find the standard deviation, we need to use the z-score formula and fill in the other three values. We know the boundary value is 60 and the mean is 47. The last value we need is the z-score for an area to the left = 1 − 0.12 = 0.88. Substitute these three values into the z-score formula and solve for σ.

Using Table A: 0.88 area to the left gives $z = 1.175$:

$$1.175 = \frac{60 - 47}{\sigma}$$

$1.175\sigma = 13$

$\sigma = 11.06$ minutes

We used 1.175 from Table A because 0.88 is exactly halfway between the area for $z = 1.17$ and $z = 1.18$.

Note that you can't use method (ii) to solve this problem because this method requires a value of σ.

EXAM PREP: FOR PRACTICE, TRY EXERCISE 9.

Example 6.3.2 showed you how to find the standard deviation if you are given the mean of a normal distribution and the probability to the left or right of a given value. If you are given the standard deviation instead of the mean, you can use a similar approach to find the mean of a normal distribution. What if you don't know the mean or standard deviation? With two different probabilities and their corresponding boundary values in a normal distribution, you can solve a system of equations to find μ and σ. See Exercise 18.

Section 6.3 What Did You Learn?

Review the learning goals from this section. Then practice what you've learned by working through the exercises.

Learning Goal	Example	Exercises
Find the value corresponding to a given probability (area) in a normal distribution.	6.3.1	5–8
Find the mean or standard deviation of a normal distribution given a boundary value and corresponding probability (area).	6.3.2	9–12

Section 6.3 Exercises

Building Concepts and Skills These exercises assess the basic knowledge you should have after reading the section.

1. Before doing any calculations, what is the first step when finding values from areas in a normal distribution?

2. What formula do you use to solve for the standard deviation when given a percentile and the mean of a normal distribution?

3. Find the 40th percentile of the standard normal distribution.

4. Find the value in the standard normal distribution with area 0.34 to its right.

Mastering Concepts and Skills These exercises reinforce the learning goals as shown in the examples.

5. **Refer to Example 6.3.1 Potato Chips** The probability distribution of W = weight of a randomly selected 9-ounce bag of a particular brand of potato chips can be modeled by a normal distribution with mean $\mu = 9.12$ ounces and standard deviation $\sigma = 0.05$ ounce. What is the value of the 80th percentile in this distribution?

6. **Rafa Serves!** The probability distribution of S = Rafael Nadal's first serve speed (in miles per hour) can be modeled by a normal distribution with mean 115 mph and standard deviation 6 mph. What is the value of the 15th percentile in this distribution?

7. **Car Batteries** An automaker has found that the life of its batteries varies from car to car according to an approximately normal distribution with mean $\mu = 48$ months and standard deviation $\sigma = 8$ months. What is the value of the 30th percentile in this distribution?

8. **Oranges** Mandarin oranges from a certain grove have weights that follow an approximately normal distribution with mean $\mu = 3$ ounces and standard deviation $\sigma = 0.5$ ounce. What is the value of the 65th percentile in this distribution?

9. **Refer to Example 6.3.2 NBA Heights** The probability distribution of H = height for a randomly selected National Basketball Association (NBA) player is approximately normal with a mean of 78.4 inches.[9] If $P(H > 84) = 0.057$, find the standard deviation of H.

10. **Helmet Sizes** The army reports that the probability distribution of C = head circumference for a randomly selected soldier is approximately normal with a mean of 22.8 inches. If $P(C < 22) = 0.234$, find the standard deviation of C.

11. **Sub Shop** The lengths of foot-long sub sandwiches at a local sub shop follow an approximately normal distribution with unknown mean μ and standard deviation 0.2 inch. If 20% of these sandwiches are shorter than 11.7 inches, find the mean length μ.

12. **Rally Time** A tennis ball machine fires balls a distance that is approximately normally distributed. The mean distance, μ, can be set to different values but the standard deviation is always 1.2 feet. If 5% of balls go farther than 70 feet, find μ.

Applying the Concepts These exercises ask you to apply multiple learning goals in a new context or to apply what you learned in this section in a new way.

13. **Birth Weights** Researchers in Norway analyzed data on the birth weights of 400,000 newborns over a 6-year period. The probability distribution of B = birth weight of a randomly selected newborn is approximately normal with a mean of 3668 grams and a standard deviation of 511 grams.[10] Babies who weigh less than 2500 grams at birth are classified as "low birth weight."

   (a) What is the probability of randomly selecting a newborn who would be identified as low birth weight?

   (b) Find the first and third quartiles of this distribution.

14. **Black Bears** The probability distribution of X = weight of a randomly selected male black bear before it begins hibernation is approximately normal with a mean of 250 pounds and a standard deviation of 50 pounds.

   (a) What is the probability of randomly selecting a bear of this type that weighs more than 275 pounds?

   (b) Find the 10th and 90th percentiles of this distribution.

Exercises 15 and 16 refer to this setting. At fast-food restaurants, the lids for drink cups are made with a small amount of flexibility, so they can be stretched across the mouth of the cup and then snugly secured. When lids are too small or too large, customers can get frustrated, especially if they end up spilling their drinks. At one restaurant, large drink cups require lids with a diameter of between 3.95 and 4.05 inches. The restaurant's lid supplier claims that the diameter of the large lids follows a normal distribution with mean 3.98 inches and standard deviation 0.02 inch. Assume that the supplier's claim is true. The supplier is considering two changes to reduce the percentage of its large-cup lids that are too small to 1%: (1) adjusting the mean diameter of its lids or (2) altering the production process to decrease the standard deviation of the lid diameters.

15. **Cap the Mean**
   (a) If the standard deviation remains $\sigma = 0.02$ inch, at what value should the supplier set the mean diameter of its large-cup lids so that only 1% are too small to fit?
   (b) What effect will the change in part (a) have on the percentage of lids that are too large?

16. **Cap the SD**
   (a) If the mean diameter stays at $\mu = 3.98$ inches, what value of the standard deviation will result in only 1% of lids that are too small to fit?
   (b) What effect will the change in part (a) have on the percentage of lids that are too large?

Extending the Concepts These exercises challenge you to explore statistical concepts and methods that go beyond what you learned in this section.

17. **Birth Weights** Refer to Exercise 13.
   (a) What birth weights would be considered outliers according to the $1.5 \times IQR$ rule from Section 3.3?
   (b) Based on your answer to part (a), what is the probability of randomly selecting a newborn whose weight would be considered an outlier?

18. **Flight Times** An airline flies the same route at the same time each day. The probability distribution of X = flight time (in minutes) of a randomly selected flight on this route can be modeled by a normal distribution having unknown mean and standard deviation. Based on historical data, $P(X > 60) = 0.15$ and $P(X > 75) = 0.03$. Use these probabilities to determine the mean and standard deviation of the probability distribution of X.

Cumulative Review These exercises revisit what you learned in previous sections.

19. **Chamois Leather (3.3)** Chamois leather is smooth and absorbent, making it a popular choice for use in cleaning. Does the temperature of the water affect how much it can absorb? A researcher used ninety 3-inch-by-3-inch squares of chamois leather and randomly assigned 30 to be used with hot water, 30 with room-temperature water, and 30 with cold water.[11] After soaking each piece with the appropriate-temperature water, the researcher carefully measured how much water was absorbed (in milliliters) by wringing out each piece of leather over a graduated cylinder. The boxplots display the distribution of amount of water absorbed for each of these temperatures. Compare these distributions.

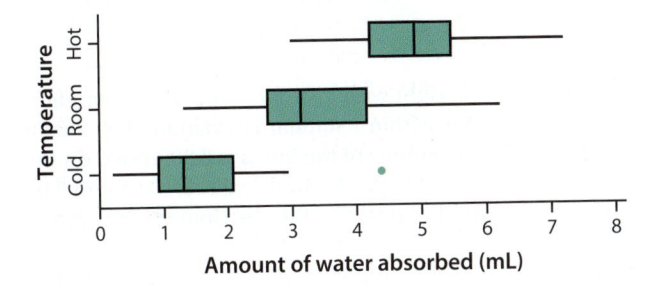

20. **Fuel Efficiency (2.5, 3.1, 3.2)** Here are city fuel efficiency estimates (in miles per gallon, mpg) for 50 randomly selected vehicles produced in 2021.[12]

25	22	12	14	15
15	15	25	22	18
29	15	19	14	17
53	16	16	29	18
17	26	23	24	19
27	25	28	27	13
15	20	18	18	18
33	44	22	13	22
17	17	15	15	15
20	12	31	25	25

(a) Use technology to make a histogram of the data and describe its shape.

(b) Which measures of center and variability would you use to summarize the distribution — the mean and standard deviation or the median and IQR? Explain your answer.

Section 6.4 Normal Approximation to the Binomial Distribution and Assessing Normality

LEARNING GOALS

By the end of this section, you will be able to:
- Determine when it is appropriate to use a normal approximation to a binomial distribution.
- When appropriate, use a normal distribution to approximate a binomial probability.
- Use a normal probability plot to determine if a distribution is approximately normal.

In Sections 5.3 and 5.4, you learned about a discrete probability distribution, called a binomial distribution. The binomial random variable is the number of successes in a fixed number of trials of the same chance process, such as the number of heads in 4 flips of a coin or the number of international students in a random sample of 100 students at a university. In this section, you'll learn when and how to use a normal distribution to approximate a binomial distribution and how to evaluate if a distribution of quantitative data is approximately normal.

The Normal Approximation to the Binomial

Suppose you roll a fair 6-sided die 10 times and count the number of fives. In this case, X = number of fives is a binomial random variable with $n = 10$ and $p = 1/6$. Figure 6.6 shows a graph of the probability distribution of X.

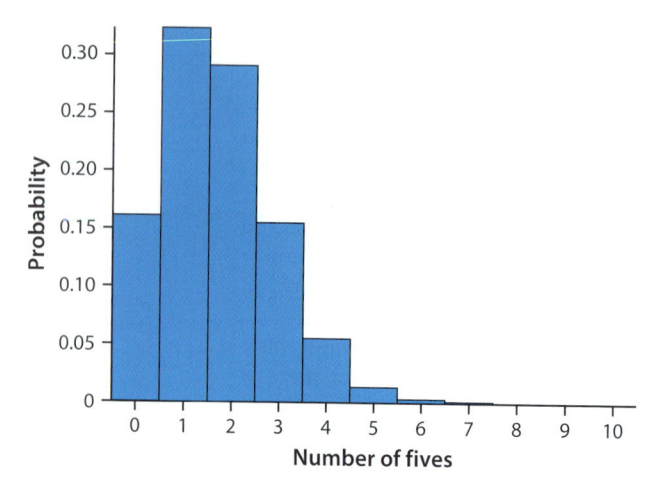

FIGURE 6.6 Histogram showing a binomial distribution with $n = 10$ and $p = 1/6$.

The distribution is single-peaked and skewed to the right with mean $\mu_X = 1.667$ and standard deviation $\sigma_X = 1.179$. In addition, the probability that X is within 1 standard deviation of the mean is 0.61 and the probability that X is within 2 standard deviations of the mean is 0.98. Based on the shape of the distribution (skewed) and the "disagreement" with the empirical rule (0.61 and 0.98 aren't that close to 0.68 and 0.95, respectively), we shouldn't describe this distribution as approximately normal.

Figure 6.7 shows a graph of the probability distribution of $Y =$ number of fives in 100 rolls of the die.

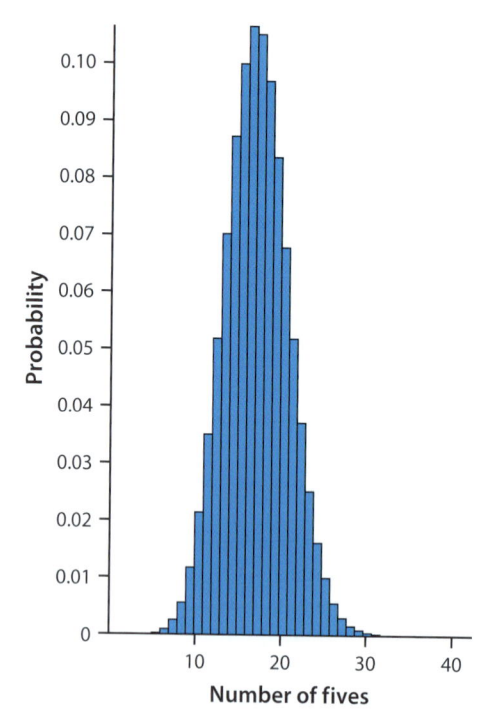

FIGURE 6.7 Histogram showing a binomial distribution with $n = 100$ and $p = 1/6$.

This distribution is single-peaked and roughly symmetric with mean $\mu_Y = 16.667$ and standard deviation $\sigma_Y = 3.727$. The probability that Y is within 1 standard deviation of the mean is 0.72 and the probability that Y is within 2 standard deviations of the mean is 0.96. Because of the shape and the close agreement with the empirical rule, it is safe to call this distribution approximately normal.

Section 6.4 Normal Approximation to the Binomial Distribution and Assessing Normality

Concept Exploration

Normal approximation to the binomial distribution

In this exploration, you'll use an applet to explore the shape of a binomial distribution for different values of n and p.

1. Go to www.stapplet.com and launch the *Binomial Distributions* applet.
2. Enter $n = 10$ and $p = 0.1$. Click the Plot distribution button and describe the shape of the probability distribution of $X =$ the number of successes.
3. Keeping $n = 10$, gradually increase the value of p to $p = 0.90$ by clicking the up arrow in the box displaying the value of p. (Alternatively, you can type in new values of p and click the Plot distribution button.) Describe the change in shape as you move from $p = 0.10$ to $p = 0.90$.
4. Keeping $p = 0.90$, gradually increase the value of n to $n = 100$ by clicking the up arrow in the box displaying the value of n. (Alternatively, you can type in new values of n and click the Plot distribution button.) Describe the change in shape as the sample size increases.

In the Concept Exploration, you learned that the shape of a binomial distribution will be closer to normal when n is larger and when p is closer to 0.5. In general, we can use a normal distribution to approximate a binomial distribution when the **Large Counts condition** is met.

> Let X be the number of successes in a binomial setting with n trials and probability p of success on each trial. The **Large Counts condition** says that the shape of the probability distribution of X is approximately normal if
>
> $$np \geq 10 \quad \text{and} \quad n(1-p) \geq 10$$

This condition is called "large counts" because np is the expected (mean) count of successes and $n(1-p)$ is the expected (mean) count of failures in a binomial setting. The binomial distribution with $n = 10$ and $p = 1/6$ shown in Figure 6.6 is clearly skewed to the right. It is no surprise that the Large Counts condition is not satisfied in this case:

$$10(1/6) = 1.67 < 10 \quad \text{and} \quad 10(1 - 1/6) = 8.33 < 10$$

However, the binomial distribution with $n = 100$ and $p = 1/6$ shown in Figure 6.7 appears approximately normal. This is confirmed by the Large Counts condition:

$$100(1/6) = 16.67 \geq 10 \quad \text{and} \quad 100(1 - 1/6) = 83.33 \geq 10$$

EXAMPLE 6.4.1

Checking for bias

The Large Counts condition

PROBLEM

One way of checking the effect of undercoverage, nonresponse, and other sources of bias in a sample survey is to compare the sample with known facts about the population. According to the Census Bureau, about 16.5% of American adults identify themselves as Hispanic or Latino.[13] Suppose we select a random sample of 1500 American adults and let X be the number of people in the sample who identify themselves as Hispanic or Latino. Is the probability distribution of X approximately normal? Justify your answer.

SOLUTION

Yes, because $1500(0.165) = 247.5$ and $1500(1 - 0.165) = 1252.5$ are both at least 10.

> Check that np and $n(1 - p) \geq 10$.

EXAM PREP: FOR PRACTICE, TRY EXERCISE 5.

Using the Normal Approximation to Calculate Probabilities

When the Large Counts condition is met, we can use a normal distribution to calculate probabilities involving $X =$ the number of successes in a binomial distribution with n trials and probability p of success on each trial. Recall from Section 5.4 that the mean and standard deviation of a binomial random variable X are

$$\mu_X = np \quad \text{and} \quad \sigma_X = \sqrt{np(1-p)}$$

The normal distribution we use to approximate the binomial distribution should have this same mean and standard deviation.

EXAMPLE 6.4.2

Checking for bias

Normal approximation to a binomial distribution

PROBLEM

According to the Census Bureau, about 16.5% of American adults identify themselves as Hispanic or Latino. Suppose we select a random sample of 1500 American adults and let X be the number of people in the sample who identify themselves as Hispanic or Latino. In Example 6.4.1, we showed that the probability distribution of X is approximately normal.

(a) Calculate the mean and standard deviation of X.

(b) Use a normal distribution to calculate the probability that the sample will have at most 200 people who identify as Hispanic or Latino.

SOLUTION

(a) $\mu_X = 1500(0.165) = 247.5$ and $\sigma_X = \sqrt{1500(0.165)(1-0.165)} = 14.38$

> $\mu_X = np$ and $\sigma_X = \sqrt{np(1-p)}$

(b)

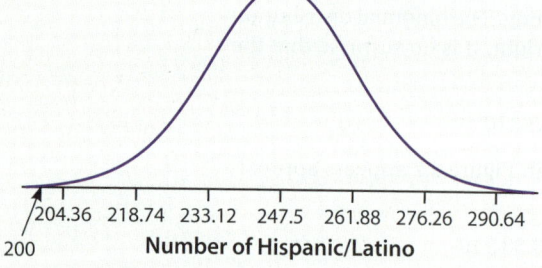

> 1. Draw a normal distribution.

204.36 218.74 233.12 247.5 261.88 276.26 290.64
200
Number of Hispanic/Latino

(i) $z = \dfrac{200 - 247.5}{14.38} = -3.303$

> 2. Perform calculations — show your work!

Using technology: $P(Z \leq -3.303) =$ Applet/normalcdf(lower: −1000, upper: −3.303, mean: 0, SD: 1) = 0.0005

Using Table A: $P(Z \leq -3.30) = 0.0005$

(ii) $P(X \leq 200) =$ Applet/normalcdf(lower: −1000, upper: 200, mean: 247.5, SD: 14.38) = 0.0005

EXAM PREP: FOR PRACTICE, TRY EXERCISE 11.

If we use technology to calculate this probability using a binomial distribution, we get $P(X \leq 200) = 0.0004$. This is very close to the approximated value from the example using a normal distribution. With a probability this small, a random sample of 1500 U.S. adults that has 200 or fewer people who identify as Hispanic or Latino should be investigated for bias.

You may be wondering why we use normal approximations when technology will calculate exact binomial probabilities. As you will discover shortly, inference about the true proportion p of successes in a population is based on the normal approximation to the binomial distribution.

We can also use a normal approximation for some distributions of quantitative data, which is helpful when performing inference about the mean μ of a population distribution. To determine if it is plausible that a sample came from a normally distributed population, we graph the sample data with a dotplot, stemplot, histogram, boxplot, or a *normal probability plot*.

Normal Probability Plots

A graph called a **normal probability plot** (or a *normal quantile plot*) provides a good assessment of whether a distribution of quantitative data is approximately normal.

> A **normal probability plot** is a scatterplot of the ordered pair (data value, expected z-score) for each of the individuals in a quantitative data set. That is, the x-coordinate of each point is the actual data value and the y-coordinate is the expected z-score corresponding to the percentile of that data value in a standard normal distribution.

Figure 6.8 shows histograms and normal probability plots for three different distributions, each with $n = 200$:

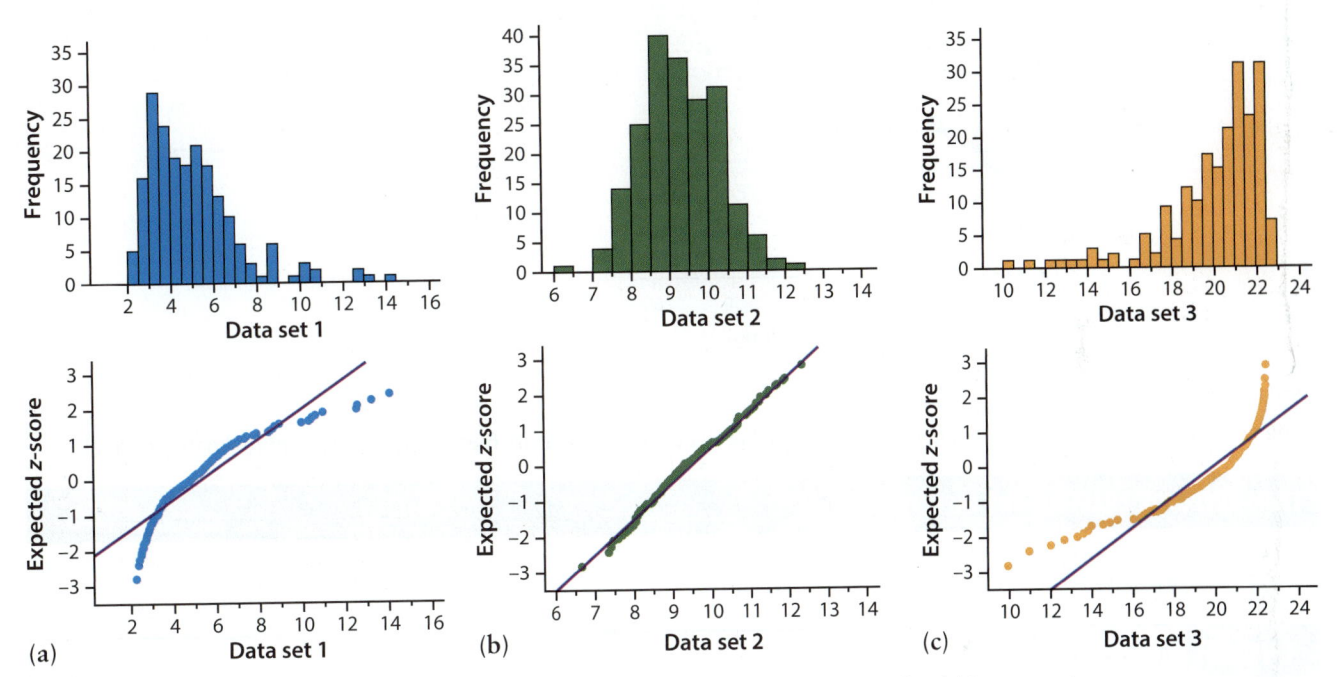

FIGURE 6.8 Histogram and normal probability plot for (a) a distribution that is skewed to the right, (b) a distribution that is approximately normal, and (c) a distribution that is skewed to the left.

HOW TO Assess Normality with a Normal Probability Plot

If the points on a normal probability plot lie close to a straight line, the data are approximately normally distributed. A nonlinear form in a normal probability plot indicates a non-normal distribution.

When examining a normal probability plot, look for shapes that show clear departures from normality. Don't be concerned about minor wiggles in the plot. We used technology to generate three different random samples of size 20 from a normal distribution, shown here.

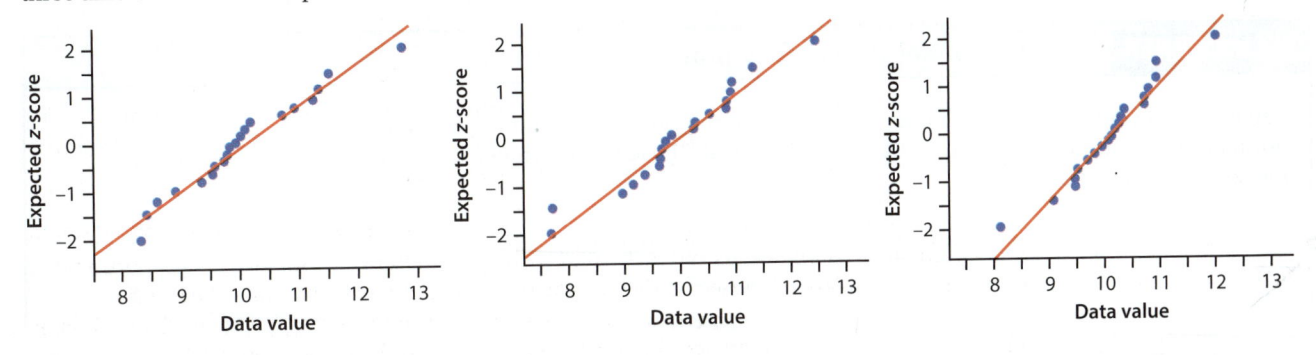

Although none of these plots is perfectly linear, it is reasonable to believe that each sample came from a normally distributed population.

EXAMPLE 6.4.3

Driving in the city

Interpreting normal probability plots

PROBLEM

The U.S. Environmental Protection Agency (EPA) tests vehicles every year and estimates the fuel efficiency for each vehicle model in the city and on the highway. A random sample of 50 model-year 2021 vehicles was selected and the city fuel efficiency estimates (in miles per gallon, mpg) were recorded.[14] A normal probability plot of the data is shown. Use the graph to determine if the distribution of fuel efficiency estimates is approximately normal.

SOLUTION

The normal probability plot is clearly curved, indicating that the distribution of city fuel efficiency is not approximately normal.

EXAM PREP: FOR PRACTICE, TRY EXERCISE 15.

THINK ABOUT IT

Can we determine skewness from a normal probability plot? Look at the normal probability plot of the city fuel efficiency data in Example 6.4.3. Imagine all the points falling down onto the horizontal axis. The resulting dotplot would have many values stacked up between 10 and 20 mpg, fewer values between 20 and 30 mpg, and a few values between 30 and 60 mpg. The distribution would be skewed to the right due to the greater variability in the upper half of the data set. The dotplot of the data confirms our answer.

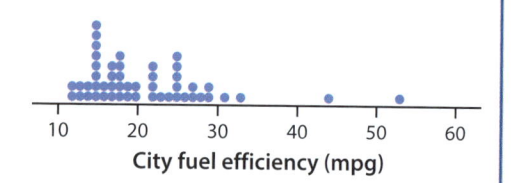

TECH CORNER Making normal probability plots

You can (and should!) use technology to make a normal probability plot to assess the normality of a distribution of quantitative data. Here are the head circumference values (in inches) from Section 2.3:

23.2 22.6 22.1 22.4 22.7 23.0 23.1 21.9 23.1 24.8 21.2 23.2 24.2 23.7 21.2
24.0 23.1 23.0 23.9 22.9 23.3 22.3 21.4 22.8 23.7 22.9 23.9 23.6 22.0 23.5

Section 6.4 Normal Approximation to the Binomial Distribution and Assessing Normality   333

Applet

1. Go to www.stapplet.com and launch the *One Quantitative Variable, Single Group* applet.

2. Enter the variable name "Head circumference (in.)" in the first box. Copy-and-paste or type the data values in the second box, making sure there are spaces or commas between them.

   One Quantitative Variable, Single Group
   Variable name: [Head circumference (in.)]
   Input: [Raw data]
   Input data separated by commas or spaces.
   Data: [23.2 22.6 22.1 22.4 22.7 23.0 23.1 21.9 23.1 24.8]

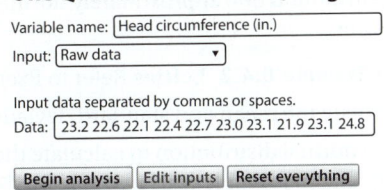

   [Begin analysis] [Edit inputs] [Reset everything]

3. Click the Begin analysis button. This will display a dotplot of the head circumferences and summary statistics. In the drop-down menu, choose Normal probability plot.

Graph Distribution
Graph type: [Normal probability plot ▼]

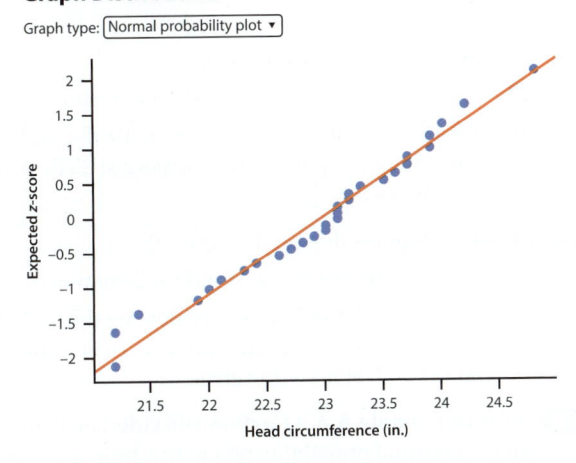

TI-83/84

1. Enter the data values in list L_1.

2. Set up a normal probability plot in the statistics plot menu. The normal probability plot is the sixth option in the list of graph types. Make sure the Data List is L_1 and the Data Axis is X as shown in the screenshot.

3. Use ZoomStat to see the finished graph.

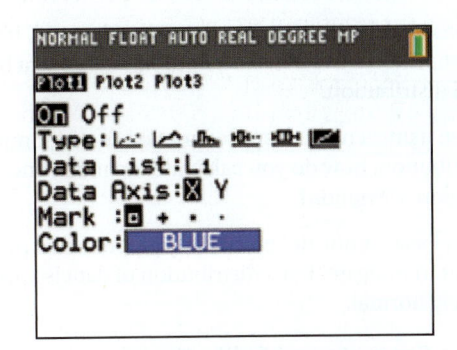

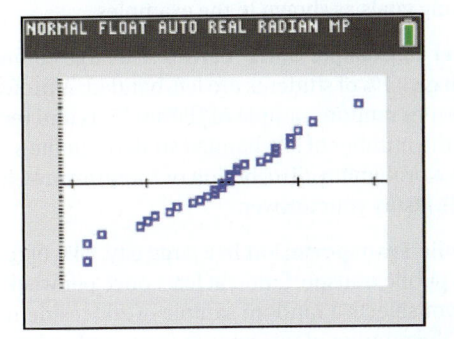

Detailed instructions for using CrunchIt!, Excel, Google Sheets, JMP, Minitab, and R are available in **Achieve**.

Section 6.4   What Did You Learn?

Review the learning goals from this section. Then practice what you've learned by working through the exercises.

Learning Goals	Example	Exercises
Determine when it is appropriate to use a normal approximation to a binomial distribution.	6.4.1	5–10
When appropriate, use a normal distribution to approximate a binomial probability.	6.4.2	11–14
Use a normal probability plot to determine if a distribution is approximately normal.	6.4.3	15–18

Section 6.4 Exercises

Building Concepts and Skills These exercises assess the basic knowledge you should have after reading the section.

1. The distribution of a binomial random variable will be more normal when n is _____ and p is _____.

2. True/False: If either np or $n(1-p)$ is at least 10, it is appropriate to use a normal approximation to a binomial distribution.

3. When using a normal approximation to a binomial distribution, how do you calculate the mean and the standard deviation?

4. True/False: A normal probability plot must be perfectly linear to indicate that a distribution of data is approximately normal.

Mastering Concepts and Skills These exercises reinforce the learning goals as shown in the examples.

5. **Refer to Example 6.4.1** **Lefties** At a large community college, 11% of students are left-handed. A professor selects a random sample of 100 students and records L = the number of left-handed students in the sample. Is the probability distribution of L approximately normal? Justify your answer.

6. **Public Transportation** In a large city, 34% of residents use public transportation at least once per week. The mayor selects a random sample of 200 residents. Let T = the number of residents in the sample who use public transportation at least once per week. Is the probability distribution of T approximately normal? Justify your answer.

7. **Roulette** On U.S. roulette wheels, there are 38 equally likely spaces where the ball can land. Of these spaces, 18 are red, 18 are black, and 2 are green. To test if a particular wheel is fair, regulators spin it 100 times and record G = the number of times the ball lands in a green space. Is the probability distribution of G approximately normal? Justify your answer.

8. **Random Numbers** An online random number generator is supposed to generate the integers 0 to 9 with equal probability. To test if this generator is working correctly, you use it to generate 50 integers from 0 to 9 and record F = the number of fives. Is the probability distribution of F approximately normal? Justify your answer.

9. **Flash Drives** A company that manufactures flash drives claims that 2% of the flash drives it produces are defective. A consumer organization tests a random sample of 1000 flash drives from this company and records X = the number of defective flash drives in the sample. Assume the manufacturer's claim is correct. Is the probability distribution of X approximately normal? Justify your answer.

10. **Expense Reports** The accounting department at a large corporation claims that 84% of all expense reports have at least one error. To test this claim, a supervisor randomly selects 100 expense reports and records E = the number of expense reports with at least one error. Assume the department's claim is correct. Is the probability distribution of E approximately normal? Justify your answer.

11. **Refer to Example 6.4.2** **Lefties** Refer to Exercise 5.
    (a) Calculate the mean and standard deviation of L.
    (b) Use a normal distribution to calculate the probability that the sample will have at least 15 students who are left-handed.

12. **Public Transportation** Refer to Exercise 6.
    (a) Calculate the mean and standard deviation of T.
    (b) Use a normal distribution to calculate the probability that the sample will have at most 60 residents who use public transportation at least once per week.

13. **Flash Drives** Refer to Exercise 9.
    (a) Calculate the mean and standard deviation of X.
    (b) Use a normal distribution to calculate the probability that the sample will have at most 25 defective flash drives.

14. **Expense Reports** Refer to Exercise 10.
    (a) Calculate the mean and standard deviation of E.
    (b) Use a normal distribution to calculate the probability that the sample will have at least 80 expense reports with at least one error.

15. **Refer to Example 6.4.3** **Carbon Dioxide** The figure shows a normal probability plot of the emissions of carbon dioxide per person in 48 countries.[15] Use the graph to determine if this distribution of carbon dioxide emissions is approximately normal.

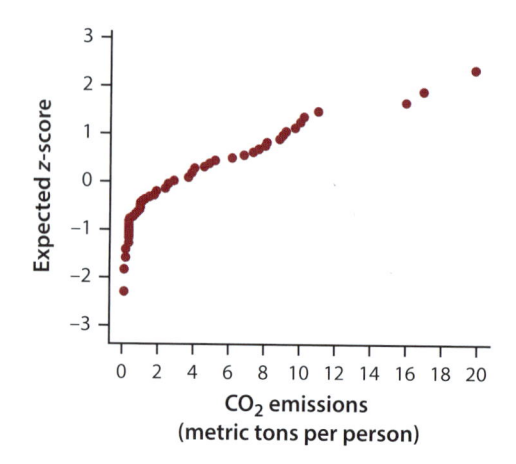

16. Runners' Heart Rates The figure shows a normal probability plot of the heart rates of 200 runners after 6 minutes of exercise on a treadmill.[16] Use the graph to determine if this distribution of heart rates is approximately normal.

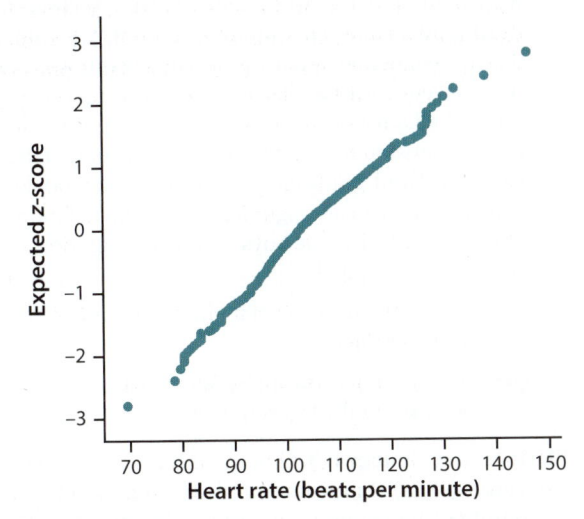

17. Earth's Density In 1798, the English scientist Henry Cavendish used a torsion balance to measure the density of the earth several times. The variable he recorded was the density of the earth as a multiple of the density of water. A normal probability plot of the data is shown.[17] Use the graph to determine if this distribution of density measurements is approximately normal.

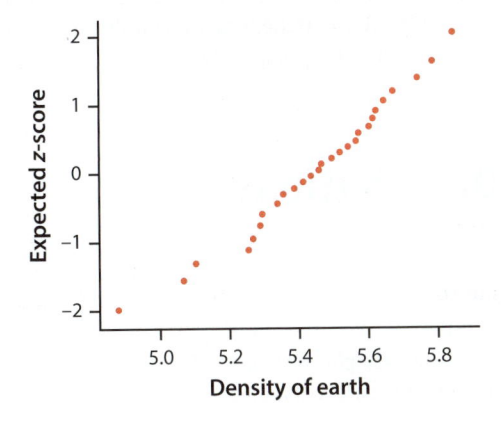

18. Land Areas The normal probability plot displays data on the areas (in thousands of square miles) of each of the 50 states. Use the graph to determine if the distribution of land area is approximately normal.

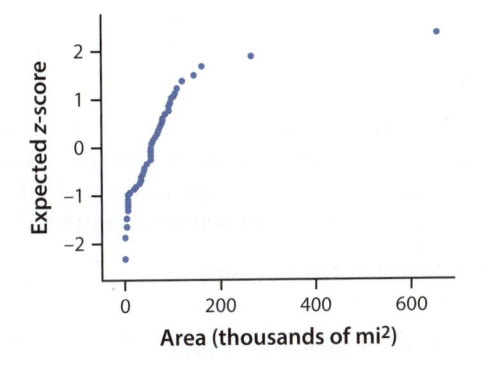

Applying the Concepts These exercises ask you to apply multiple learning goals in a new context or to apply what you learned in this section in a new way.

19. Hip Dysplasia Dysplasia is a malformation of the hip socket that is very common in certain dog breeds and causes arthritis as a dog gets older. According to the Orthopedic Foundation for Animals, 11.6% of all Labrador retrievers have hip dysplasia.[18] A veterinarian will test a random sample of 100 Labrador retrievers and record X = the number of Labs with dysplasia in the sample.

(a) Justify that it is appropriate to use the normal approximation to the binomial distribution in this case.

(b) Calculate the mean and standard deviation of X.

(c) Use a normal distribution to determine the probability that 8 or fewer of the Labs in the sample have dysplasia.

20. Denver Quarters In 2020, 49.5% of quarters were minted by the U.S. Mint in Denver, Colorado (the rest were produced in Philadelphia).[19] Suppose we select a random sample of 50 quarters produced in 2020. Let D = the number of quarters in the sample that were minted in Denver. (You can identify Denver as the mint by the small "D" to the lower right of George Washington.)

(a) Justify that it is appropriate to use the normal approximation to the binomial distribution in this case.

(b) Calculate the mean and standard deviation of D.

(c) Use a normal distribution to determine the probability that 30 or more of the quarters in the sample were minted in Denver.

21. Hip Dysplasia The veterinarian in Exercise 19 selected a random sample of 100 Labs from a particular state and found 8 with hip dysplasia. Based on your answer to part (c) of Exercise 19, does the sample result provide convincing evidence that the percentage of Labs with hip dysplasia in this state is less than 11.6%? Explain your reasoning.

22. Denver Quarters Refer to Exercise 20. In a random sample of 50 quarters from 2021, 30 were minted in Denver. Based on your answer to part (c) of Exercise 20, does the sample result provide convincing evidence that the percentage of quarters minted in Denver in 2021 is greater than 49.5%, the percentage in 2020? Explain your reasoning.

23. Lefties Refer to Exercises 5 and 11. Use a binomial distribution to calculate the probability that the sample will have at least 15 students who are left-handed. How does this compare with your answer to Exercise 11(b)? Which value is more accurate?

24. Public Transportation Refer to Exercises 6 and 12. Use a binomial distribution to calculate the probability that the sample will have at most 60 residents who use public transportation at least once per week. How does this compare with your answer to Exercise 12(b)? Which value is more accurate?

Extending the Concepts These exercises challenge you to explore statistical concepts and methods that go beyond what you learned in this section.

25. **Hip Dysplasia** Refer to Exercise 19. One way to see if the normal approximation is reasonable is to calculate the probability that X is within 1 standard deviation of the mean.

    (a) What values of X are within 1 standard deviation of the mean? Remember that X can only take integer values.

    (b) Use a binomial distribution to calculate the probability that X is within 1 standard deviation of the mean. How does this compare to the empirical rule?

26. **Uninsured Drivers** Suppose that 20% of licensed drivers in a certain state do not have auto insurance. A researcher selects a random sample of 50 licensed drivers from this state. Let X = the number of drivers in the sample who are uninsured.

    (a) Use a binomial distribution to find $P(X \leq 12)$.

    (b) Show that the probability distribution of X is approximately normal. Then use the appropriate normal distribution to approximate the probability that 12 or fewer drivers in the sample are uninsured.

    When using a normal distribution to approximate probabilities involving a binomial random variable, you can get more accurate results using a *continuity correction*. Any value less than 12.5 in the normal distribution corresponds to 12 or fewer uninsured drivers.

    (c) Use the normal distribution from part (b) to find the probability of getting a value less than 12.5. Compare with the answers to part (a) and part (b).

Cumulative Review These exercises revisit what you learned in previous sections.

27. **Summer Reading (1.5, 1.6)** A group of educational researchers studied the impact of summer reading with a randomized experiment involving second- and third-graders in North Carolina. Students were randomly assigned to either a group that was mailed one book a week for 10 weeks or a control group that was not mailed any books. Both groups were given reading comprehension tests at the start and the end of the summer. Third-grade girls who were mailed books showed a statistically significant increase in reading ability, but third-grade boys and second-graders of both sexes did not.[20]

    (a) Explain the purpose of including a control group in this experiment.

    (b) Explain what is meant by "statistically significant increase" in the last sentence.

28. **Internet Charges (3.4)** Some internet service providers (ISPs) charge companies based on how much bandwidth they use in a month. One method that ISPs use to calculate bandwidth is to find the 95th percentile of a company's usage based on random samples of hundreds of 5-minute intervals during a month.

    (a) Explain what "95th percentile" means in this setting.

    (b) Is it possible to determine the z-score for a usage total that is at the 95th percentile? If so, find the z-score. If not, explain why not.

Section 6.5 Sampling Distributions

LEARNING GOALS

By the end of this section, you will be able to:
- Distinguish between a parameter and a statistic.
- Create a sampling distribution using all possible samples from a small population.
- Use the sampling distribution of a statistic to evaluate a claim about a parameter.

To estimate the mean income of U.S. residents with a college degree, the Census Bureau selected a random sample of more than 60,000 people with at least a bachelor's degree. The mean income *in the sample* was $78,862.[21] How close is this estimate to the mean income for *all* members of the population? To find out how an estimate varies from sample to sample, you need to understand *sampling distributions*.

Parameters and Statistics

For the sample of college graduates contacted by the Census Bureau, the mean income was $\bar{x} = \$78{,}862$. The number $78,862 is a **statistic** because it describes this one *sample*. The population that the researchers want to draw conclusions about is all U.S. college graduates. In this case, the **parameter** of interest is the mean income μ of the *population* of all college graduates.

A **statistic** is a number that describes some characteristic of a sample.

A **parameter** is a number that describes some characteristic of the population.

Because we typically can't examine an entire population, the value of a parameter is usually unknown. To estimate the value of a parameter, we use a statistic calculated using data from a random sample of the population.

Remember s and p: statistics come from samples, and parameters come from populations. The notation we use should reflect this distinction. For example, we write μ (the Greek letter mu) for the population mean and $\bar{x}$ for the sample mean. The table lists some additional examples of statistics and their corresponding parameters.

Sample statistic		Population parameter
$\bar{x}$ (the sample mean)	estimates	μ (the population mean)
$\hat{p}$ (the sample proportion)	estimates	p (the population proportion)
s_x (the sample standard deviation)	estimates	σ (the population standard deviation)

EXAMPLE 6.5.1

Instagram your turkey

Parameters and statistics

PROBLEM

Identify the population, the parameter, the sample, and the statistic in each of the following scenarios.

(a) A Pew Research Center poll asked 1502 adults in the United States if they use a variety of social media sites. Of the respondents, 601 said they use Instagram.[22]

(b) Tom is roasting a large turkey breast for a holiday meal. He wants to be sure that the turkey is safe to eat, which requires a minimum internal temperature of 165°F. Tom uses a thermometer to measure the temperature of the turkey breast at four randomly chosen locations. The minimum reading is 170°F.

SOLUTION

(a) Population: all adults in the United States. Parameter: p = the proportion of all adults in the United States who use Instagram. Sample: the 1502 U.S. adults surveyed. Statistic: the sample proportion who use Instagram, $\hat{p} = 601/1502 = 0.40$.

(b) Population: all possible locations in the turkey breast. Parameter: the true minimum temperature in all possible locations. Sample: the four randomly chosen locations. Statistic: the sample minimum, 170°F.

EXAM PREP: FOR PRACTICE TRY EXERCISE 9.

While most parameters and statistics are denoted by special symbols (such as p for the population proportion and $\hat{p}$ for the sample proportion), some parameters and statistics do not have their own symbol. To communicate clearly, include the words *all*, *true*, or *population* in your description of a parameter and the word *sample* in your description of a statistic.

Sampling Distributions

In Section 1.3, you learned about sampling variability—the fact that different random samples of the same size from the same population produce different values of a statistic. The statistics that come from these samples form a **sampling distribution**.

> The **sampling distribution** of a statistic is the distribution of values taken by the statistic in all possible samples of the same size from the same population.

Recall that a distribution describes the possible values of a variable and how often these values occur. The easiest way to picture a distribution is with a graph, such as a dotplot or histogram.

EXAMPLE 6.5.2

Just how tall are their sons?

Sampling distributions

PROBLEM

John and Carol have four grown sons. Their heights (in inches) are 71, 75, 72, and 68. List all 6 possible simple random samples (SRSs) of size $n = 2$ from this population of 4 sons, calculate the mean height for each sample, and display the sampling distribution of the sample mean on a dotplot.

SOLUTION

Sample 1: 71, 75; $\bar{x} = 73$ Sample 4: 75, 72; $\bar{x} = 73.5$
Sample 2: 71, 72; $\bar{x} = 71.5$ Sample 5: 75, 68; $\bar{x} = 71.5$
Sample 3: 71, 68; $\bar{x} = 69.5$ Sample 6: 72, 68; $\bar{x} = 70$

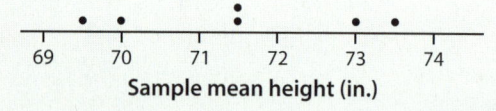

EXAM PREP: FOR PRACTICE, TRY EXERCISE 13.

Every statistic has its own sampling distribution. For example, Figure 6.9 shows the sampling distribution of the sample range of height for SRSs of size $n = 2$ from John and Carol's four sons.

Sample 1: 71, 75; sample range = 4   Sample 4: 75, 72; sample range = 3
Sample 2: 71, 72; sample range = 1   Sample 5: 75, 68; sample range = 7
Sample 3: 71, 68; sample range = 3   Sample 6: 72, 68; sample range = 4

FIGURE 6.9 Dotplot showing the sampling distribution of the sample range of height for SRSs of size $n = 2$.

⚠ **Be specific when you use the word "distribution."** There are three different types of distributions in this setting:

1. The distribution of height in the population (the four heights):

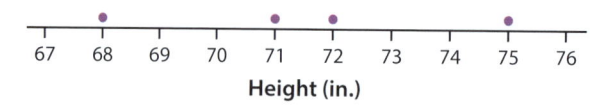

2. The distribution of height in a particular sample (two of the heights):

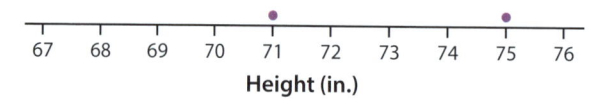

3. The sampling distribution of the sample range for all possible samples (the six sample ranges):

Notice that the first two distributions consist of heights (data values), while the third distribution consists of ranges (statistics). *Lesson:* Never say "the distribution" without providing an additional description that clarifies which distribution you are referring to.

Using Sampling Distributions to Evaluate Claims

Sampling distributions are the foundation for the methods of statistical inference you will learn about in Chapters 7–13. The sampling distribution of a statistic reveals how much the statistic tends to vary from its corresponding parameter and which values of the statistic should be considered unusual.

EXAMPLE 6.5.3

Should the government forgive student loans?

Evaluating a claim

PROBLEM

In a Harris poll of 1015 randomly selected U.S. adults, 55% supported a proposal to have the federal government forgive all student loan debt.[23] Does this provide convincing evidence that more than half of all U.S. adults would support the proposal?

To determine if a sample proportion of $\hat{p} = 0.55$ provides convincing evidence that the population proportion is greater than $p = 0.50$, we simulated 100 SRSs of size $n = 1015$ from a population in which $p = 0.50$. Here are the results of the simulation:

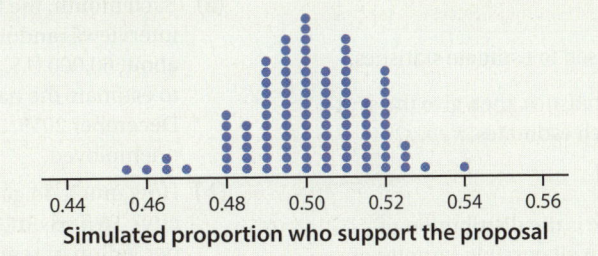

Simulated proportion who support the proposal

(a) There is one dot on the graph at 0.54. Explain what this dot represents.
(b) Would it be surprising to get a sample proportion of 0.55 or greater in an SRS of size 1015 when $p = 0.50$? Explain your reasoning.
(c) Based on the actual sample and your answer to part (b), is there convincing evidence that more than half of all U.S. adults would support the proposal? Explain your reasoning.

SOLUTION

(a) In one simulated SRS of size $n = 1015$, 54% of the respondents supported the proposal.
(b) Yes. In the 100 simulated samples, there were 0 samples with a sample proportion of 0.55 or greater.
(c) Because it is unlikely to get a sample proportion of 0.55 or more by chance alone when $p = 0.50$, there is convincing evidence that more than half of all U.S. adults would support the proposal.

EXAM PREP: FOR PRACTICE, TRY EXERCISE 17.

What if only 52% of the respondents in the actual survey supported the proposal? Because getting a sample proportion of 0.52 or greater by chance alone is not surprising when $p = 0.50$ (it happened 15/100 times in the simulation), there wouldn't be convincing evidence that more than half of U.S. adults would support the proposal.

We used 100 simulated samples to produce the dotplot of sample proportions in this example. Because the simulation didn't include *all* possible samples of size 1015, the dotplot is only an approximation of the actual sampling distribution of $\hat{p}$. Thankfully, the simulated sampling distribution should be a good approximation as long as we use a large number of samples in the simulation.

Section 6.5 What Did You Learn?

Review the learning goals from this section. Then practice what you've learned by working through the exercises.

Learning Goal	Example	Exercises
Distinguish between a parameter and a statistic.	6.5.1	9–12
Create a sampling distribution using all possible samples from a small population.	6.5.2	13–16
Use the sampling distribution of a statistic to evaluate a claim about a parameter.	6.5.3	17–20

Section 6.5 Exercises

Building Concepts and Skills These exercises assess the basic knowledge you should have after reading the section.

1. Parameters describe _____ and statistics describe _____.

2. True/False: Parameters are used to estimate statistics.

3. Identify the following three statistics, then give the symbol for the parameter that each estimates: $\bar{x}, \hat{p}, s_x$.

4. What is sampling variability?

5. The _____ of a statistic is the distribution of values taken by that statistic in all possible samples of the same size from the same population.

6. What are the three different types of distributions described in this section?

7. A sampling distribution reveals how much the _____ tends to vary from _____ and which values of the statistic should be considered _____.

8. When a simulation doesn't include all possible samples of a given size from the population, the resulting distribution of the statistic is a(n) _____ sampling distribution.

Mastering Concepts and Skills These exercises reinforce the learning goals as shown in the examples.

9. **Refer to Example 6.5.1 Smoking and Height** Identify the population, the parameter, the sample, and the statistic in each scenario.

   (a) From a large group of people who each signed a card saying they intended to quit smoking, 1000 people were randomly selected. It turned out that 210 (21%) of the selected individuals had not smoked over the past 6 months.

   (b) A pediatrician wants to know the 75th percentile for the distribution of height of 10-year-old males, so she randomly selects fifty 10-year-old male patients and calculates that the 75th percentile of their heights is 56 inches.

10. **Unemployment and Gas Prices** Identify the population, the parameter, the sample, and the statistic in each scenario.

    (a) Each month, the Current Population Survey interviews randomly selected individuals in about 60,000 U.S. households. One of its goals is to estimate the national unemployment rate. In December 2019, 3.5% of those interviewed were unemployed.

    (b) How much do gasoline prices vary in a large city? To find out, a reporter records the price per gallon of regular unleaded gasoline at 10 randomly selected gas stations in the city on the same day. The range of the prices for these 10 stations is 25 cents.

11. **Tea and Screening** Identify the population, the parameter, the sample, and the statistic in each scenario.

    (a) On Tuesday, the bottles of iced tea filled in a plant were supposed to contain an average of 20 ounces of iced tea. Quality-control inspectors selected 50 bottles at random from the day's production. These bottles contained an average of 19.6 ounces of iced tea.

    (b) On a New York to Denver flight, 8% of the 125 passengers were selected for random security screening before boarding. According to the Transportation Security Administration, 10% of passengers at this airport are supposed to be chosen for random screening.

12. **Bearings and Cows** Identify the population, the parameter, the sample, and the statistic in each scenario.

    (a) A production run of ball bearings is supposed to have a mean diameter of 2.5000 cm. An inspector randomly selects 100 bearings from the container and calculates a mean diameter of 2.5009 cm.

    (b) A farmer has 1000 dairy cows and wants to estimate the minimum daily milk production for dairy cows in the herd. For 20 randomly chosen dairy cows, the minimum production was 6.3 gallons.

Exercises 13–16 refer to the following scenario. An investor has 5 stocks in a portfolio. Here is the net return for each of the five stocks during the previous 12 months:

$$8\% \quad 12\% \quad -5\% \quad -20\% \quad 25\%$$

13. Refer to Example 6.5.2 Sample Means List all 10 possible SRSs of size $n = 2$ from this population of 5 stocks, calculate the mean return for each sample, and display the sampling distribution of the sample mean on a dotplot.

14. Sample Ranges List all 10 possible SRSs of size $n = 3$ from this population of 5 stocks, calculate the range of return for each sample, and display the sampling distribution of the sample range on a dotplot.

15. Sample Proportions List all 10 possible SRSs of size $n = 2$ from this population of 5 stocks, calculate the proportion with a positive return for each sample, and display the sampling distribution of the sample proportion on a dotplot.

16. Sample Medians List all 10 possible SRSs of size $n = 3$ from this population of 5 stocks, calculate the median return for each sample, and display the sampling distribution of the sample median on a dotplot.

17. Refer to Example 6.5.3 Herd Immunity During the Covid-19 pandemic, experts estimated that a community would achieve "herd immunity" when more than 70% of the community members had been vaccinated or recovered from Covid-19.[24] The mayor of a small town surveys a random sample of 50 town residents and finds that 37 have been immunized or have recovered from Covid-19.

To determine if a sample proportion of $\hat{p} = 37/50 = 0.74$ provides convincing evidence that the population proportion is greater than $p = 0.70$, we simulated 100 SRSs of size $n = 50$ from a population in which $p = 0.70$. Here are the results of the simulation:

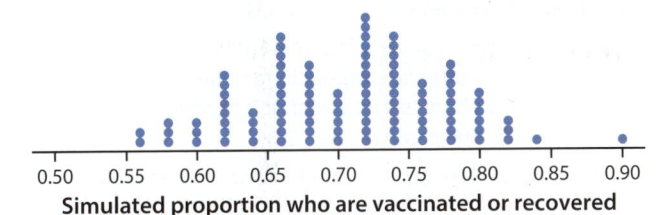

(a) There is one dot on the graph at 0.90. Explain what this dot represents.

(b) Would it be surprising to get a sample proportion of 0.74 or greater in an SRS of size 50 when $p = 0.70$? Explain your reasoning.

(c) Based on the actual sample and your answer to part (b), is there convincing evidence that more than 70% of town residents have been vaccinated or have recovered from Covid-19? Explain your reasoning.

18. The Homework Box At the beginning of her statistics class, Professor Chauvet shows the students a box filled with black and white beads. She claims that the proportion of black beads in the box is $p = 0.50$. To determine the number of homework exercises she will assign that evening, she invites a student to select an SRS of $n = 30$ beads from the box. The number of black beads selected will be the number of homework exercises assigned. When the student selects 19 black beads ($\hat{p} = 19/30 = 0.63$), the class groans and suggests that Professor Chauvet included more than 50% black beads in the box.

To determine if a sample proportion of $\hat{p} = 0.63$ provides convincing evidence that Professor Chauvet cheated, the class simulated 100 SRSs of size $n = 30$, assuming that she was telling the truth. That is, they sampled from a population with 50% black beads. For each sample, they recorded the sample proportion of black beads. The dotplot shows the results of the simulation.

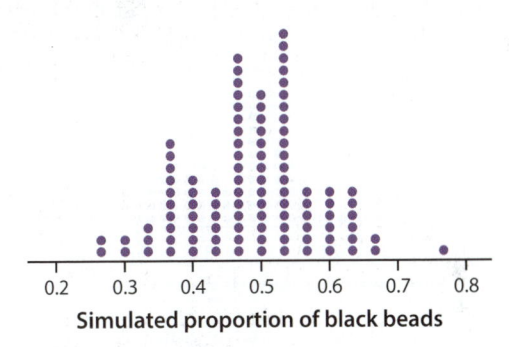

(a) There is one dot on the graph at $\hat{p} = 0.77$. Explain what this dot represents.

(b) Would it be unusual to get a sample proportion of 0.63 or greater in a sample of size 30 when $p = 0.50$? Explain your reasoning.

(c) Based on the actual sample and your answer to part (b), is there convincing evidence that Professor Chauvet lied about the contents of the box? Explain your reasoning.

19. Cold Cabin During the winter months, outside temperatures at the Starneses' cabin in Colorado can stay well below freezing (32°F, or 0°C) for weeks at a time. To prevent the pipes from freezing, Mrs. Starnes sets the thermostat at 50°F. When set at 50°F, the manufacturer claims that the thermostat allows variation in home temperature that follows a normal distribution with $\mu = 50°F$ and $\sigma = 3°F$. Mrs. Starnes programs her digital thermostat to take an SRS of $n = 10$ readings during a 24-hour period and calculates the mean of the results to be $\bar{x} = 49.2°F$.

Suppose that the thermostat is working properly and that the temperatures in the cabin vary according to a normal distribution with mean $\mu = 50°F$ and standard deviation $\sigma = 3°F$. The dotplot shows the distribution of the sample mean in 100 simulated SRSs of size $n = 10$ from this population.

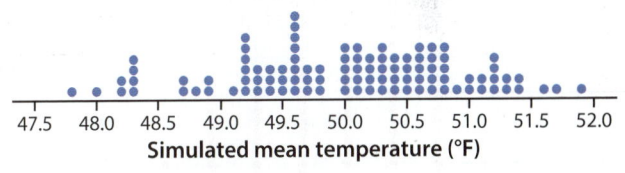

(a) There is one dot on the graph at 47.8. Explain what this dot represents.

(b) Would it be surprising to get a sample mean of 49.2 or smaller in an SRS of size 10 when $\mu = 50$? Explain your reasoning.

(c) Based on the actual sample and your answer to part (b), is there convincing evidence that the thermostat keeps the average temperature colder than the manufacturer claims? Explain your reasoning.

20. **Cold Cabin** Refer to Exercise 19. Mrs. Starnes programs her digital thermostat to take an SRS of $n = 10$ readings during a 24-hour period and calculates the standard deviation of the results to be $s_x = 5°F$.

    Suppose that the thermostat is working properly and that the temperatures in the cabin vary according to a normal distribution with mean $\mu = 50°F$ and standard deviation $\sigma = 3°F$. The dotplot shows the distribution of the sample standard deviation in 100 simulated SRSs of size $n = 10$ from this population.

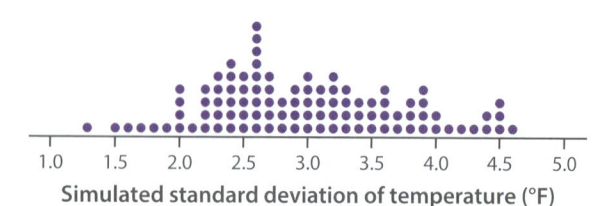

    (a) There is one dot on the graph at 4.6. Explain what this dot represents.

    (b) Would it be surprising to get a sample standard deviation of $s_x = 5$ or greater in an SRS of size 10 when $\sigma = 3$? Explain your reasoning.

    (c) Based on the actual sample and your answer to part (b), is there convincing evidence that the thermostat is more variable than the manufacturer claims? Explain your reasoning.

Applying the Concepts These exercises ask you to apply multiple learning goals in a new context or to apply what you learned in this section in a new way.

21. **Instant Winners** A fast-food restaurant promotes certain food items by giving a game piece with each item. Advertisements proclaim that "25% of the game pieces are Instant Winners!" To test this claim, a frequent diner collects 20 game pieces and gets only 1 instant winner.

    (a) Identify the population, the parameter, the sample, and the statistic in this context.

    Suppose the advertisements are correct and $p = 0.25$. The dotplot shows the distribution of the sample proportion of instant winners in 100 simulated SRSs of size $n = 20$.

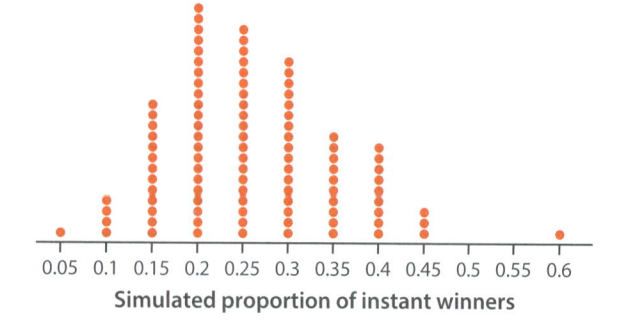

(b) Would it be unusual to get a sample proportion of $\hat{p} = 1/20 = 0.05$ or less in a sample of size $n = 20$ when $p = 0.25$? Explain your reasoning.

(c) Based on the actual sample and your answer to part (b), is there convincing evidence that fewer than 25% of all game pieces are instant winners? Explain your reasoning.

22. **Giraffe Weights** The weights of male giraffes in the wild follow a normal distribution with a mean of 2628 pounds and a standard deviation of 541 pounds.[25] Do male giraffes in zoos weigh more than 2628 pounds, on average? A random sample of $n = 10$ male giraffes in zoos gives a sample mean of 3004 pounds.

    (a) Identify the population, the parameter, the sample, and the statistic in this context.

    Suppose that the weights of male giraffes in zoos follow the same distribution as male giraffes in the wild. The dotplot shows the distribution of the sample mean weight (in pounds) in 100 simulated SRSs of size $n = 10$.

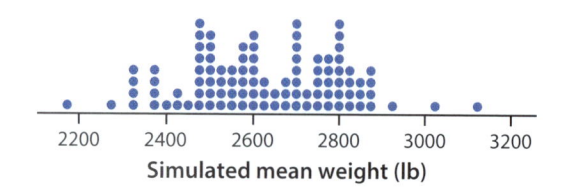

    (b) Would it be unusual to get a sample mean of 3004 or greater in a sample of size $n = 10$ when sampling from a normal population with $\mu = 2628$ and $\sigma = 541$? Explain your reasoning.

    (c) Based on the actual sample and your answer to part (b), is there convincing evidence that male giraffes in zoos weigh more than 2628 pounds, on average? Explain your reasoning.

23. **Instant Winners** Refer to Exercise 21.

    (a) Make a frequency bar chart of the population distribution, assuming that the fast-food restaurant's claim is true and that it produces a total of 1,000,000 game pieces.

    (b) Make a frequency bar chart for the distribution of sample data for one possible SRS of 20 game pieces from this population.

24. **Giraffe Weights** Refer to Exercise 22.

    (a) Make a graph of the population distribution of weight, assuming that weights of male giraffes in zoos follow the same distribution as male giraffes in the wild.

    (b) Sketch a dotplot of the distribution of sample data for one possible SRS of 10 giraffes from this population.

Extending the Concepts These exercises challenge you to explore statistical concepts and methods that go beyond what you learned in this section.

25. **Internet Equity** A school superintendent is concerned about equity within her district. In particular,

she worries that the proportion of South High School students with internet access at home is less than the proportion of North High School students with internet access at home. To investigate, she selects SRSs of size $n = 50$ from each school and finds $\hat{p}_S = 36/50 = 0.72$ and $\hat{p}_N = 46/50 = 0.92$.

To determine if a difference in proportions of $\hat{p}_S - \hat{p}_N = -0.20$ provides convincing evidence that South High School has a smaller proportion of students with internet access at home, we simulated two random samples of size $n = 50$ from populations having the same proportion of students with internet access. Then, we subtracted the simulated sample proportions. Here are the results from repeating this process 100 times.

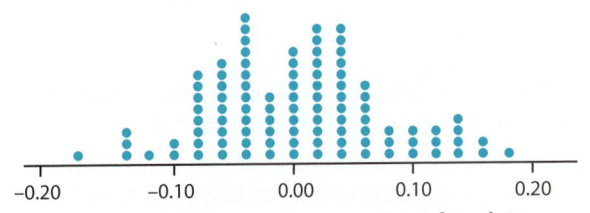

Simulated difference in proportion of students with internet access at home (North – South)

(a) There are 10 dots at 0. Explain what these dots represent.

(b) Would it be unusual to get a difference in sample proportions of –0.20 or smaller when there is no difference in the population proportions? Explain your reasoning.

(c) Based on the actual samples and your answer to part (b), is there convincing evidence that South High School has a smaller proportion of students with internet access at home? Explain your reasoning.

26. **Sample Means** Refer to Exercise 13.
    (a) List all 10 possible SRSs of size $n = 3$ from this population of 5 stocks, calculate the mean return for each sample, and display the sampling distribution of the sample mean on a dotplot.
    (b) How does the variability of the graph in part (a) compare to the variability of the graph in Exercise 13? What does this indicate about the effect of increasing sample size?

Cumulative Review These exercises revisit what you learned in previous sections.

27. **Sampling Tomatoes (4.8, 6.5)** Zach runs a roadside stand during the summer, selling produce from his farm. On a single day in mid-August, he harvests 300 tomatoes. Suppose Zach wants to select a simple random sample of 25 tomatoes from the day's pick to estimate the mean weight of all 300 tomatoes.
    (a) How many possible samples of 25 tomatoes could be selected from the 300 tomatoes in the day's pick?
    (b) What does this say about the practicality of creating the sampling distribution of the sample mean for all samples of size 25 from this population?

28. **Veteran Leadership (2.6, 3.5)** Do basketball teams with veteran players win more games? The scatterplot shows the relationship between the average age of starters and the number of wins for the 30 NBA teams in the 2018 regular season.

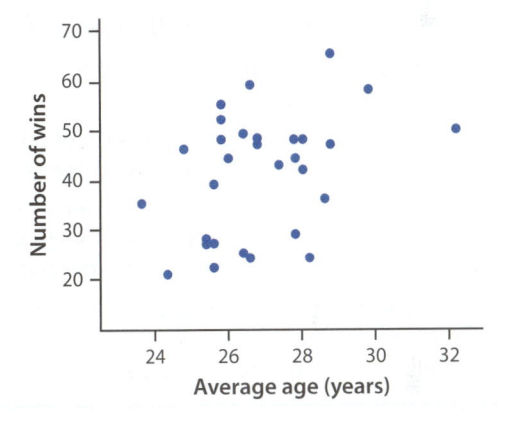

    (a) Describe the association between average age and number of wins.
    (b) The correlation for these data is $r = 0.39$. Interpret this value.
    (c) Based on these data, would starting a 60-year-old player cause a team to win more games?

Section 6.6 Sampling Distributions: Bias and Variability

LEARNING GOALS

By the end of this section, you will be able to:
- Determine if a statistic is an unbiased estimator of a population parameter.
- Describe the relationship between sample size and the variability of a statistic.

Sampling distributions tell us how much the value of a statistic typically varies from the parameter it is trying to estimate and what values of the statistic are likely to happen by chance alone. Two important characteristics of a sampling distribution are its center and variability. The following activity illustrates why these characteristics are important.

Class Activity

How many craft sticks are in the bag?

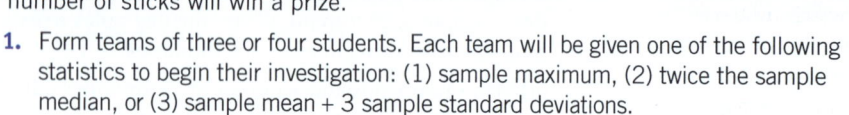

In this activity, you will create a statistic for estimating the total number of craft sticks in a bag (N). The sticks are numbered 1, 2, 3, . . . , N. Near the end of the activity, your instructor will select a random sample of $n = 7$ sticks and read the number on each stick to the class. The team that has the best estimate for the total number of sticks will win a prize.

1. Form teams of three or four students. Each team will be given one of the following statistics to begin their investigation: (1) sample maximum, (2) twice the sample median, or (3) sample mean + 3 sample standard deviations.

2. Before your instructor provides the actual sample of sticks, use simulation to investigate the sampling distribution of the statistic you were provided. For the simulation, assume that there are $N = 100$ sticks in the bag and that you will be selecting random samples of size $n = 7$.
   - Use technology to select an SRS of 7 integers from 1 to 100:
     - With a TI-83/84 calculator, use the command RandIntNoRep(lower: 1, upper: 100, n: 7). [With older OS, use the command RandInt(lower: 1, upper: 100, n: 7) and verify that there are no repeated numbers.]
     - With an internet browser, go to www.random.org/integer-sets. Enter 1 for the number of sets and 7 as the number of unique values in each set. Enter 1 and 100 for the upper and lower boundaries and click "Get Sets."
   - Calculate and record the value of your statistic for this simulated sample.
   - Repeat the process at least 9 more times, recording the value of your statistic each time.
   - Add the 10 values of your statistic to the class dotplot for that statistic. Make sure that the three dotplots are all on the same scale.

3. As a class, discuss the quality of each of the statistics. Do any of them consistently overestimate or consistently underestimate the truth? Are some of the statistics more variable than others?

4. In your group, spend about 10–15 minutes creating a few other statistics (or modifying the original statistics) to estimate the total number of sticks. Create a simulated sampling distribution for each one (as in Step 2) to determine which statistic you will use for the competition.

5. Your instructor will now draw a sample of $n = 7$ sticks from the bag. On a piece of paper, write the names of your group members, your group's estimate for the number of sticks in the bag (a number), and the statistic you used to calculate your estimate (a formula). The closest estimate wins!

Unbiased Estimators

In the craft sticks activity, the goal was to estimate the maximum value in a population, with the assumption that the members of the population are numbered 1, 2, . . . , N. Two possible statistics that might be used to estimate N are the sample maximum (max) and twice the sample median ($2 \times$ median).

Assuming that the population has $N = 100$ members and we use SRSs of size $n = 7$, Figure 6.10 shows simulated sampling distributions of the sample maximum and twice the sample median, along with a red line at $N = 100$.

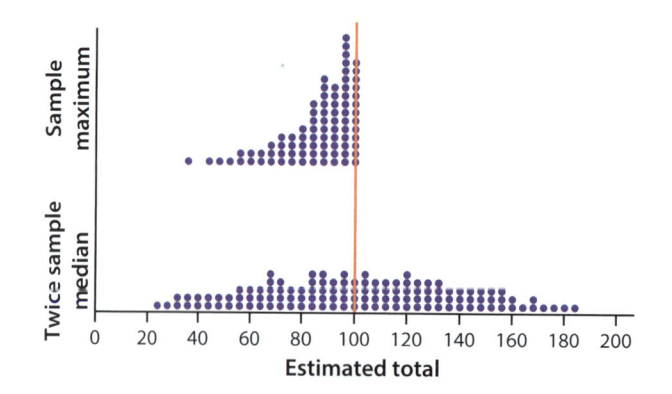

FIGURE 6.10 Simulated sampling distributions of the sample maximum and twice the sample median for samples of size $n = 7$ from a population with $N = 100$.

Section 6.6 Sampling Distributions: Bias and Variability

These simulated sampling distributions look quite different. The sampling distribution of the sample maximum is skewed left, while the sampling distribution of twice the sample median is roughly symmetric.

The values of the sample maximum are consistently less than the population maximum N. However, the values of twice the sample median aren't consistently less than or consistently greater than the population maximum N. It appears that twice the sample median might be an **unbiased estimator** of the population maximum, while the sample maximum is clearly biased.

> A statistic used to estimate a parameter is an **unbiased estimator** if the mean of its sampling distribution is equal to the value of the parameter being estimated.

The use of the word "bias" here is consistent with its use in Chapter 1. The design of a statistical study shows bias if it is very likely to underestimate or very likely to overestimate the value you want to know. Recall the Federalist Papers activity from Section 1.2 in which the estimates from student-selected samples were consistently too large. ⚠ **Don't trust an estimate that comes from a biased sampling method.**

EXAMPLE 6.6.1

Just how tall are their sons?

Biased and unbiased estimators

PROBLEM

In Section 6.5, we created the sampling distribution of the sample range for samples of size $n=2$ from the population of John and Carol's four sons, who have heights of 71, 75, 72, and 68 inches. Is the sample range an unbiased estimator of the population range? Explain your answer.

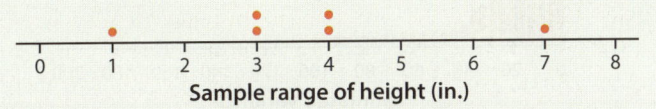

SOLUTION

The mean of the sampling distribution of the sample range is $(1+3+3+4+4+7)/6 = 3.67$ inches. The actual range of the population is $75-68=7$ inches. Because the mean of the sampling distribution of the sample range (3.67) is not equal to the value it is trying to estimate (7), the sample range is a biased estimator of the population range.

> It makes sense that the sample range is a biased estimator of the population range. After all, the sample range can't be larger than the population range, but it can be (and usually is) smaller.

EXAM PREP: FOR PRACTICE, TRY EXERCISE 5.

Here is the sampling distribution of the sample mean height for samples of size $n=2$ from the population of four sons in Example 6.6.1.

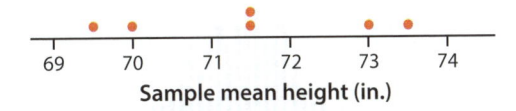

The mean of the sampling distribution of the sample mean is

$$\mu_{\bar{x}} = \frac{69.5+70+71.5+71.5+73+73.5}{6} = 71.5$$

The mean of the population distribution is

$$\mu = \frac{68+71+72+75}{4} = 71.5$$

Because these values are equal, the sample mean $\bar{x}$ is an unbiased estimator of the population mean μ for this population. We will investigate other populations in the exercises and in Section 6.7.

THINK ABOUT IT

Why do we divide by $n - 1$ when calculating the sample standard deviation? In Chapter 3, you learned to calculate the standard deviation of a sample of quantitative data using the formula

$$s_x = \sqrt{\frac{\sum(x_i - \bar{x})^2}{n-1}}$$

What if you divided by n instead of $n - 1$? Let's simulate the sampling distributions of two statistics that can be used to estimate the variance of a population, where the variance is the square of the standard deviation (variance = standard deviation^2):

Statistic 1: $\dfrac{\sum(x_i - \bar{x})^2}{n-1}$ Statistic 2: $\dfrac{\sum(x_i - \bar{x})^2}{n}$

The simulated sampling distributions shown in the dotplots are based on 1000 SRSs of size $n = 3$ from a population with variance = 25. The mean of each distribution is indicated by a blue line segment.

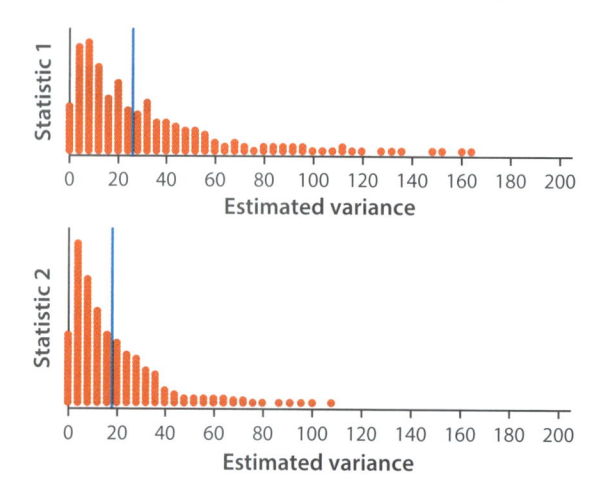

We can see that Statistic 2 is a *biased* estimator of the population variance because the mean of its simulated sampling distribution is clearly less than the value of the population variance (25). Dividing by n instead of $n - 1$ gives estimates that are consistently too small. Statistic 1, however, is unbiased because the mean of its simulated sampling distribution is equal to the population variance (25). That's why we divide by $n - 1$ when calculating the sample variance—and when calculating the sample standard deviation.

Sampling Variability

Another possible statistic that could be used in the craft sticks activity is twice the sample mean. Figure 6.11 shows simulated sampling distributions of twice the sample mean and twice the sample median.

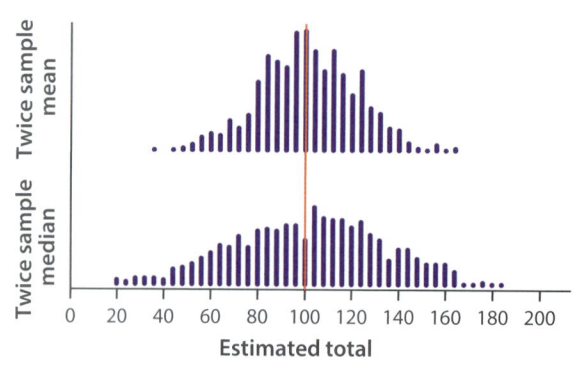

FIGURE 6.11 Simulated sampling distributions of twice the sample mean and twice the sample median for samples of size $n = 7$ from a population with $N = 100$.

Both statistics appear to be unbiased estimators because the mean of each sampling distribution is around 100. However, the sampling distribution of twice the sample mean (standard deviation ≈ 20) is less variable than the sampling distribution of twice the sample median (standard deviation ≈ 32). In general, we prefer statistics that are less variable because they produce estimates that tend to be closer to the true value of the parameter.

For some parameters, there is an obvious choice for a statistic. For example, to estimate the proportion of successes in a population p, we use the proportion of successes in the sample, $\hat{p}$. Fortunately, $\hat{p}$ is an unbiased estimator of p. And as you learned in Section 1.3, we can reduce the variability of an estimate by increasing the sample size.

Figure 6.12 shows simulated sampling distributions for $\hat{p}$ = the proportion of students in the sample who live off-campus when taking SRSs of size $n = 10$ and SRSs of size $n = 50$ from a population in which the proportion of all students who live off-campus is $p = 0.70$.

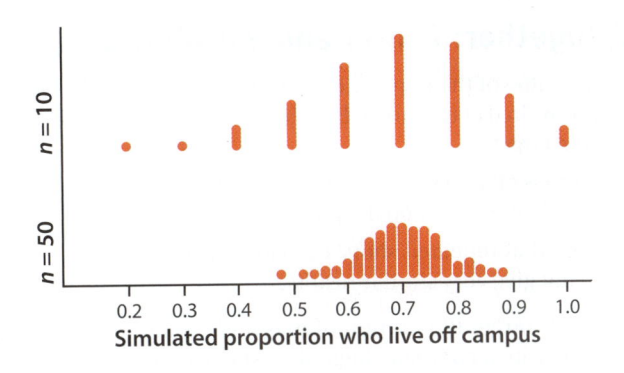

FIGURE 6.12 Simulated sampling distributions of the sample proportion $\hat{p}$ for samples of size $n = 10$ and samples of size $n = 50$ from a population with $p = 0.7$.

As expected, both simulated sampling distributions have means near $p = 0.70$. Also, the sampling distribution of $\hat{p}$ is much more variable when the sample size is $n = 10$, compared with $n = 50$. In a small sample, it is plausible that the sample proportion could be much smaller or much larger than the parameter, just by chance. However, when the sample size gets bigger, we expect the sample proportion to be closer to the value of the parameter.

Decreasing Sampling Variability

The sampling distribution of any statistic will have less variability when the sample size is larger.

EXAMPLE 6.6.2

The lifetime of batteries: Hours or days?

Sampling variability

PROBLEM

For quality control purposes, workers at a battery factory regularly select random samples of batteries to estimate the mean lifetime of the batteries they produce. Here is a simulated sampling distribution of $\bar{x}$, the sample mean lifetime (in hours) for 1000 random samples of size $n = 100$ from a population of AAA batteries.

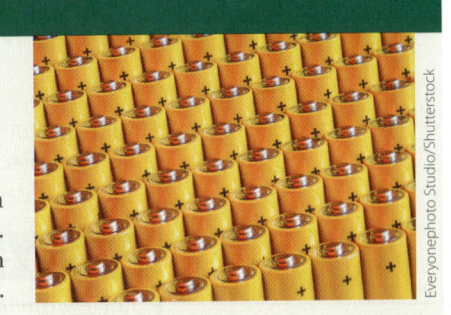

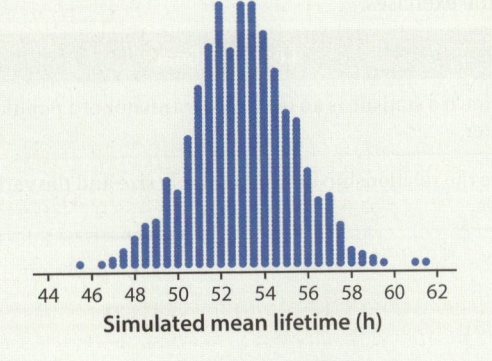

(continued)

(a) What would happen to the sampling distribution of the sample mean $\bar{x}$ if the sample size were $n = 50$ instead? Explain your answer.

(b) How will this change in sample size affect the estimate?

SOLUTION

(a) The sampling distribution of the sample mean $\bar{x}$ will be more variable because the sample size is smaller.

(b) The estimated mean lifetime is less likely to be close to the true mean lifetime. In other words, the estimate will be less precise.

EXAM PREP: FOR PRACTICE, TRY EXERCISE 9.

Putting It All Together: Center and Variability

We can think of the true value of the population parameter as the bull's-eye on a target and of the sample statistic as an arrow fired at the target. Both bias and variability describe what happens when we take many shots at the target.

- *Bias* means that our aim is off and we consistently miss the bull's-eye in the same direction. That is, our sample statistics do not center on the population parameter.
- *High variability* means that repeated shots are widely scattered on the target. In other words, repeated samples do not give very similar results.

Figure 6.13 shows this target illustration of bias and variability. Notice that low variability (shots are close together) can accompany high bias (shots are consistently away from the bull's-eye in one direction). And low or no bias (shots center on the bull's-eye) can accompany high variability (shots are widely scattered). Ideally, we'd like our estimates to be *accurate* (unbiased) and *precise* (have low variability).

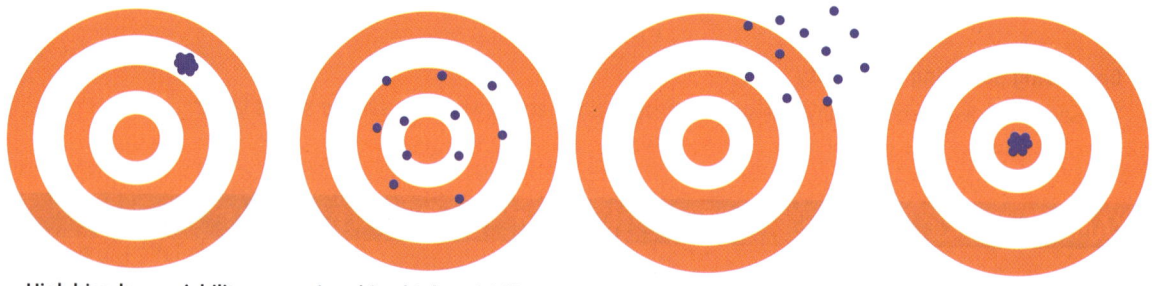

High bias, low variability   Low bias, high variability   High bias, high variability   The ideal: no bias, low variability

FIGURE 6.13 A visual representation of bias and variability. The center of the target represents the value of the parameter and the dots represent possible values of the statistic.

Section 6.6 What Did You Learn?

Review the learning goals from this section. Then practice what you've learned by working through the exercises.

Learning Goal	Example	Exercises
Determine if a statistic is an unbiased estimator of a population parameter.	6.6.1	5–8
Describe the relationship between sample size and the variability of a statistic.	6.6.2	9–12

Section 6.6 Exercises

Building Concepts and Skills These exercises assess the basic knowledge you should have after reading the section.

1. A statistic used to estimate a parameter is an unbiased estimator if _____.

2. True/False: A statistic that is very likely to overestimate the parameter or very likely to underestimate the parameter is called a biased estimator.

3. True/False: The variability of a statistic increases as the sample size increases.

4. Ideally, we'd like our estimators to have _____ bias and _____ variability.

Mastering Concepts and Skills These exercises reinforce the learning goals as shown in the examples.

5. **Refer to Example 6.6.1** Sample Mean Exercise 13 from Section 6.5 featured the following population of 5 stock returns.

    8% 12% −5% −20% 25%

    Here is the sampling distribution of the sample mean return for SRSs of size $n = 2$ from this population of 5 stocks. Is the sample mean an unbiased estimator of the population mean? Explain your answer.

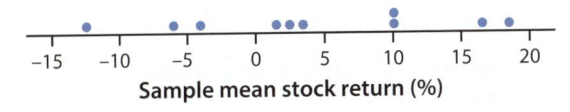

6. **Sample Proportion** Exercise 15 from Section 6.5 featured the following population of 5 stock returns.

    8% 12% −5% −20% 25%

    Here is the sampling distribution of the sample proportion with a positive return for SRSs of size $n = 2$ from this population of 5 stocks. Is the sample proportion an unbiased estimator of the population proportion? Explain your answer.

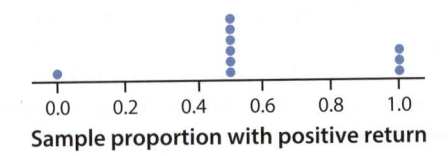

Exercises 7 and 8 refer to the following scenario. The manager of a grocery store records the total amount spent (in dollars) for each customer who makes a purchase at the store during a week. The values in the table summarize the distribution of amount spent for this population.

N	Mean	SD	Min	Q_1	Med	Q_3	Max
749	29.85	24.63	0.20	12.29	22.96	39.93	153.73

7. **Sample Median** To investigate if the sample median is an unbiased estimator of the population median, 1000 SRSs of size $n = 10$ were selected from the population described. The sample median for each of these samples was recorded on the dotplot. The mean of the simulated sampling distribution is indicated by an orange line segment. Does the sample median appear to be an unbiased estimator of the population median? Explain your reasoning.

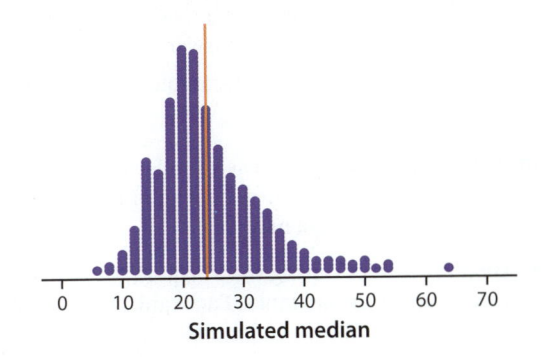

8. **Sample Minimum** To investigate if the sample minimum is an unbiased estimator of the population minimum, 1000 SRSs of size $n = 10$ were selected from the population described. The sample minimum for each of these samples was recorded on the dotplot. The mean of the simulated sampling distribution is indicated by an orange line segment. Does the sample minimum appear to be an unbiased estimator of the population minimum? Explain your reasoning.

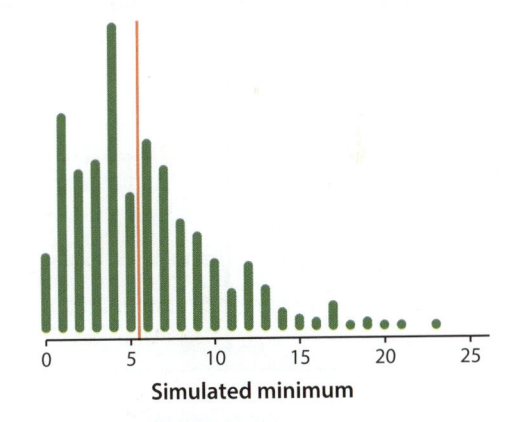

9. **Refer to Example 6.6.2** Sample Mean Refer to Exercise 5.
    (a) What would happen to the sampling distribution of the sample mean if the sample size were $n = 4$ instead? Explain your answer.
    (b) How will this change in sample size affect the estimate?

10. **Sample Proportion** Refer to Exercise 6.
    (a) What would happen to the sampling distribution of the sample proportion if the sample size were $n = 4$ instead? Explain your answer.
    (b) How will this change in sample size affect the estimate?

11. **Sample Median** Refer to Exercise 7.
    (a) What would happen to the sampling distribution of the sample median if the sample size were $n = 5$ instead? Explain your answer.
    (b) How will this change in sample size affect the estimate?

12. **Sample Minimum** Refer to Exercise 8.
    (a) What would happen to the sampling distribution of the sample minimum if the sample size were $n = 5$ instead? Explain your answer.
    (b) How will this change in sample size affect the estimate?

Applying the Concepts These exercises ask you to apply multiple learning goals in a new context or to apply what you learned in this section in a new way.

13. **German Tanks** During World War II, the Allies captured many German tanks. Each tank had a serial number on it. Allied commanders wanted to know how many tanks the Germans had so that they could allocate their forces appropriately. They sent the serial numbers of the captured tanks to a group of mathematicians in Washington, D.C., and asked for an estimate of the total number of German tanks, assuming they were numbered from 1 to N.

    Here are simulated sampling distributions for three statistics that the mathematicians considered, using samples of size $n = 7$. The blue line marks N, the total number of German tanks. The shorter red line segments mark the mean of each simulated sampling distribution.

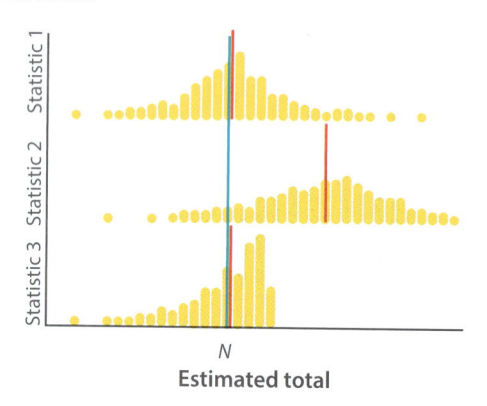

    (a) Do any of these statistics appear to be unbiased? Explain your answer.
    (b) Which of these statistics do you think is best? Explain your reasoning.
    (c) Explain how the Allies could get a more precise estimate of the number of German tanks using the statistic you chose in part (b).

14. **Bias and Variability** The histograms show sampling distributions for four different statistics intended to estimate the same parameter.

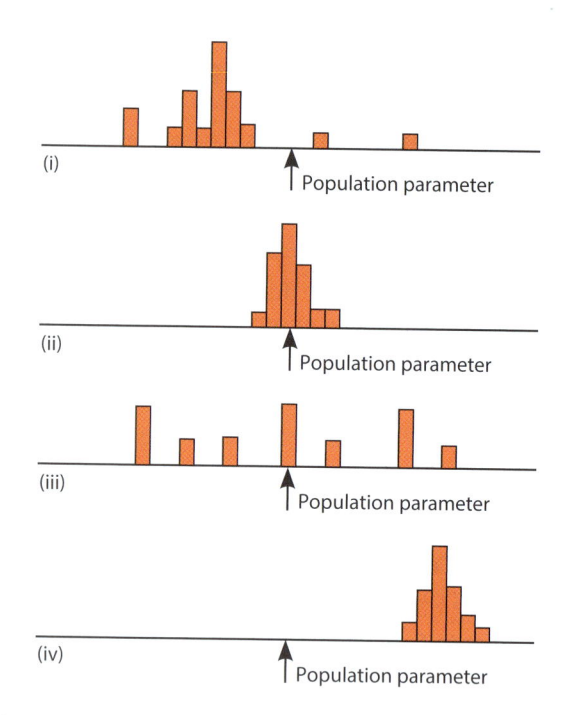

    (a) Do any of these statistics appear to be unbiased? Explain your answer.
    (b) Which of these statistics do you think is best? Explain your reasoning.
    (c) Explain how you could get a more precise estimate of the population parameter using the statistic you chose in part (b).

15. **Cholesterol Levels in Young Adults** A study of the health of young adults plans to measure the blood cholesterol levels of an SRS of 18- to 25-year-olds. The researchers will report the mean $\bar{x}$ from their sample as an estimate of the mean cholesterol level μ in this population. Explain to someone who knows little about statistics what it means to say that $\bar{x}$ is an unbiased estimator of μ.

16. **Election Polling** A polling organization plans to ask a random sample of likely voters who they will vote for in an upcoming election. The researchers will report the sample proportion $\hat{p}$ that favors the incumbent as an estimate of the population proportion p that favors the incumbent. Explain to someone who knows little about statistics what it means to say that $\hat{p}$ is an unbiased estimator of p.

17. **Cholesterol Levels in Young Adults** Refer to Exercise 15. The sample mean $\bar{x}$ is an unbiased estimator of the population mean μ no matter what size SRS the study chooses. Explain to someone who knows nothing about statistics why a large random sample will give more reliable results than a small random sample.

18. **Election Polling** Refer to Exercise 16. The sample proportion $\hat{p}$ is an unbiased estimator of the population proportion p no matter what size random sample the

polling organization chooses. Explain to someone who knows nothing about statistics why a large random sample will give more trustworthy results than a small random sample.

19. **Housing Prices** In a residential neighborhood, the distribution of house values is unimodal and skewed to the right with a median of $200,000 and an *IQR* of $100,000. For which of the following sample sizes, $n = 10$ or $n = 100$, is the sample median more likely to be greater than $250,000? Explain your reasoning.

20. **Basements** In a particular city, 74% of houses have basements. For which of the following sample sizes, $n = 10$ or $n = 100$, is the sample proportion of houses with a basement more likely to be less than 0.70? Explain your reasoning.

Extending the Concepts These exercises challenge you to explore statistical concepts and methods that go beyond what you learned in this section.

21. **Sample Mean** Refer to Exercises 5 and 9. Construct the sampling distribution of the sample mean return for SRSs of size $n = 4$ from this population of 5 stocks. How does the variability in this distribution compare to the variability of the sampling distribution shown in Exercise 5? Does this agree with your answer to Exercise 9?

22. **Sample Proportion** Refer to Exercises 6 and 10. Construct the sampling distribution of the sample proportion with positive returns for SRSs of size $n = 4$ from this population of 5 stocks. How does the variability in this distribution compare to the variability of the sampling distribution shown in Exercise 6? Does this agree with your answer to Exercise 10?

23. **Sample Mean and Midrange** Refer to the population described in Exercises 7 and 8. The manager of the grocery store would like to estimate the true mean amount spent by all customers each week using a random sample of 10 customers. The manager is considering two different statistics to estimate the population mean: (1) the sample mean $\bar{x}$ and (2) the sample midrange = (minimum + maximum)/2.

    (a) Describe one advantage of using the sample midrange instead of the sample mean as the estimate of the population mean.

    (b) To investigate if the sample midrange is an unbiased estimator of the population mean, 1000 SRSs of size 10 were selected from the population of 749 customers, which has a mean amount spent of $29.85. The sample midrange for each of these samples was recorded on the dotplot. Does the sample midrange appear to be an unbiased estimator of the population mean? Explain your reasoning.

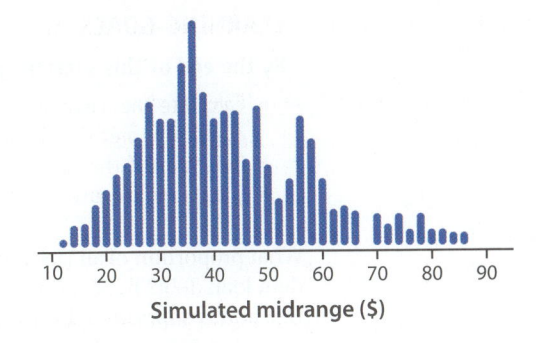

Cumulative Review These exercises revisit what you learned in previous sections.

24. **Student Housing (4.3, 4.4)** There are 104 students in Professor Negroponte's statistics class, 49 math majors and 55 statistics majors. Sixty of the students live in the dorms and the rest live off-campus. Twenty of the math majors live off-campus. Choose a student at random from this class. Let event M = the student is a math major and event D = the student lives in the dorms.

    (a) Construct a Venn diagram to represent the outcomes of this chance process using the events M and D.

    (b) Find each of the following probabilities and describe them in words.
    (i) $P(M \cup D)$ (ii) $P(M^C \cap D)$ (iii) $P(D|M)$

25. **Student Housing (5.3, 5.4)** Refer to Exercise 24. At the beginning of each day that Professor Negroponte's class meets, he randomly selects a member of the class to present the solution to a homework problem. Suppose the class meets 40 times during the semester and the selections are made with replacement. Let X = the number of times a statistics major is selected to present a solution.

    (a) Is X a binomial random variable? Justify your answer.

    (b) Calculate the mean and standard deviation of X.

    (c) For the first 10 meetings of the class, Professor Negroponte selects only 1 statistics major to solve a problem. Is there convincing evidence that his selection process is not really random? Support your answer with an appropriate probability calculation.

Section 6.7 Sampling Distribution of the Sample Proportion

LEARNING GOALS

By the end of this section, you will be able to:

- Calculate the mean and standard deviation of the sampling distribution of a sample proportion $\hat{p}$ and interpret the standard deviation.
- Determine if the sampling distribution of $\hat{p}$ is approximately normal.
- If appropriate, use a normal distribution to calculate probabilities involving $\hat{p}$.

What proportion of all U.S. adults support a proposal to have the federal government forgive all student loan debt? In Section 6.5, you read about a Harris poll in which 558 of 1015 randomly selected U.S. adults supported this proposal.[26] The sample proportion $\hat{p} = 558/1015 = 0.55$ is the statistic that we use to estimate the unknown population proportion p. Because a random sample of 1015 U.S. adults is unlikely to perfectly represent all U.S. adults, we can only say that "about" 55% of all U.S. adults support the student debt proposal.

To determine if the sample proportion of 0.55 provides convincing evidence that more than half of all U.S. adults would support the proposal, we need to know about the **sampling distribution of the sample proportion** $\hat{p}$.

The **sampling distribution of the sample proportion** $\hat{p}$ describes the distribution of values taken by the sample proportion $\hat{p}$ in all possible samples of the same size from the same population.

The dotplot shows a simulated sampling distribution of $\hat{p}$ = the sample proportion of U.S. adults who support the proposal in random samples of size $n = 1015$, assuming that $p = 0.50$. That is, assuming that exactly 50% of all U.S. adults support the proposal.

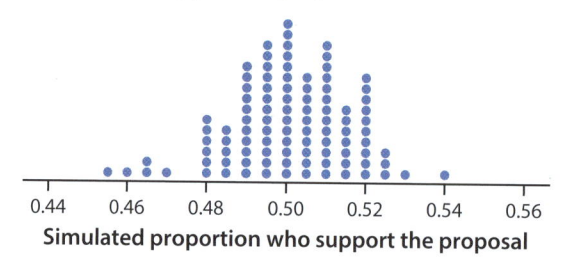

The distribution is approximately normal with a mean of about 0.50 and a standard deviation of about 0.015. By the end of this section, you should be able to anticipate the shape, center, and variability of distributions like this one without having to perform a simulation.

Center and Variability

When we select random samples of size n from a population with proportion of successes p, the value of $\hat{p}$ will vary from sample to sample. As with other types of distributions, there are formulas that describe the center and variability of the sampling distribution of $\hat{p}$.

HOW TO   Calculate $\mu_{\hat{p}}$ and $\sigma_{\hat{p}}$

Let $\hat{p}$ be the proportion of successes in an SRS of size n selected from a large population with proportion of successes p. Then:

- The mean of the sampling distribution of $\hat{p}$ is $\mu_{\hat{p}} = p$.

- The standard deviation of the sampling distribution of $\hat{p}$ is $\sigma_{\hat{p}} = \sqrt{\dfrac{p(1-p)}{n}}$.

Here are some important facts about the mean and standard deviation of the sampling distribution of the sample proportion $\hat{p}$:

- The sample proportion $\hat{p}$ is an unbiased estimator of the population proportion p. In other words, the mean of the sampling distribution $\mu_{\hat{p}}$ is equal to the population proportion p.

Section 6.7 Sampling Distribution of the Sample Proportion

- The standard deviation of the sampling distribution of $\hat{p}$ describes the typical distance between $\hat{p}$ and the population proportion p.
- The sampling distribution of $\hat{p}$ is less variable for larger samples. This is indicated by the n in the denominator of the standard deviation formula. Quadrupling the sample size cuts the standard deviation of the sampling distribution in half.
- The formula for the standard deviation of the sampling distribution of $\hat{p}$ requires that the observations are independent. In practice, we are safe assuming independence when sampling without replacement as long as the sample size is less than 10% of the population size.

EXAMPLE 6.7.1

Sampling employees

Mean and SD of the sampling distribution of $\hat{p}$

PROBLEM

In a large company, 12% of the employees are managers. The human resources department selects an SRS of 50 employees for a survey about work conditions. Let $\hat{p}$ = the proportion of managers in the sample.

(a) Calculate the mean and standard deviation of the sampling distribution of $\hat{p}$.
(b) Interpret the standard deviation from part (a).

SOLUTION

(a) $\mu_{\hat{p}} = 0.12$ and $\sigma_{\hat{p}} = \sqrt{\dfrac{0.12(1-0.12)}{50}} = 0.046$

$\boxed{\mu_{\hat{p}} = p \text{ and } \sigma_{\hat{p}} = \sqrt{\dfrac{p(1-p)}{n}}}$

(b) In SRSs of size $n = 50$, the sample proportion of managers typically varies from the true proportion by about 0.046.

EXAM PREP: FOR PRACTICE, TRY EXERCISE 7.

THINK ABOUT IT

Is the sampling distribution of $\hat{p}$ (the sample proportion of successes) related to the binomial distribution of X (the sample count of successes)? Yes! To calculate the proportion of successes in a sample, divide the number of successes by the sample size.

$$\hat{p} = \frac{\text{number of successes in sample}}{\text{sample size}} = \frac{X}{n}$$

The histogram on the left shows the probability distribution of X = the number of children with type O blood in random samples of size $n = 5$, assuming $p = 0.25$. The histogram on the right shows the sampling distribution of $\hat{p}$ = the sample proportion of children with type O blood in random samples of size $n = 5$, assuming $p = 0.25$. The distributions are exactly the same, other than the scale on the horizontal axis.

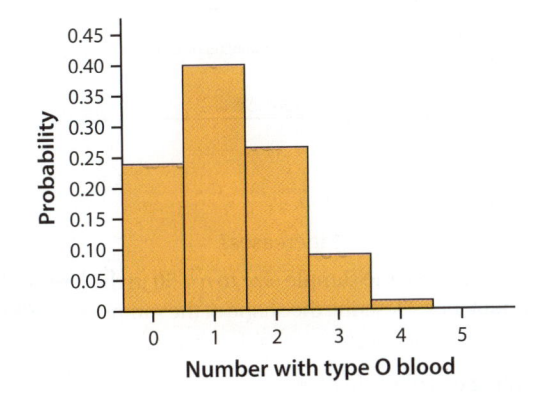

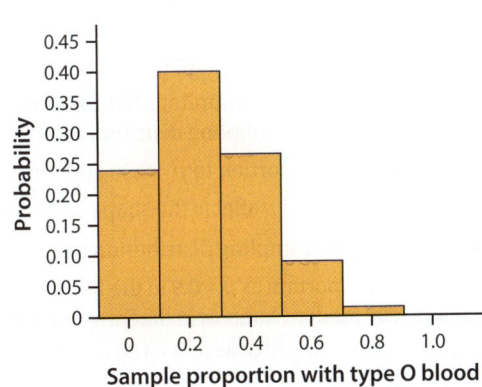

(continued)

Also, the formulas for the mean and standard deviation of the sampling distribution of $\hat{p}$ are derived from the binomial formulas you learned in Section 5.4.

$$\mu_{\hat{p}} = \frac{\mu_X}{n} = \frac{np}{n} = p$$

$$\sigma_{\hat{p}} = \frac{\sigma_X}{n} = \frac{\sqrt{np(1-p)}}{n} = \sqrt{\frac{np(1-p)}{n^2}} = \sqrt{\frac{p(1-p)}{n}}$$

Shape

Both the sample size and the proportion of successes in the population affect the shape of the sampling distribution of the sample proportion $\hat{p}$. The following Concept Exploration helps you explore the effects of these two factors.

Concept Exploration

Shape of the sampling distribution of $\hat{p}$

In this exploration, you will use an applet to investigate the shape of the sampling distribution of the sample proportion $\hat{p}$ for different sample sizes (n) and different proportions of success in the population (p).

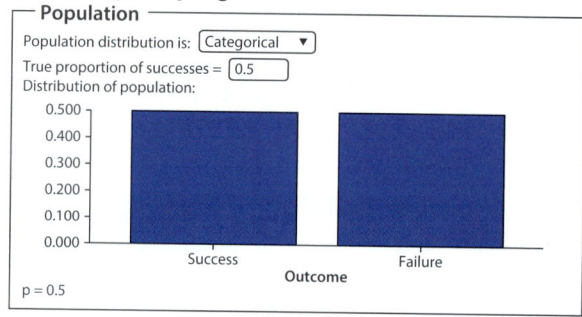

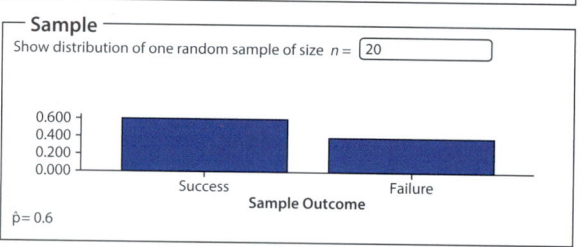

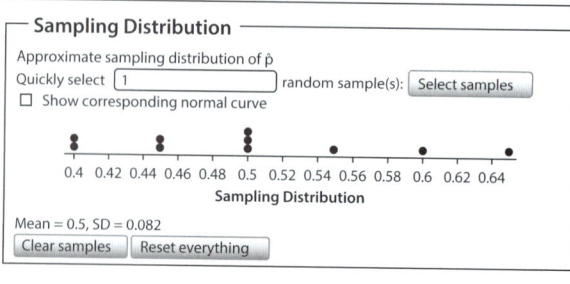

1. Go to www.stapplet.com and launch the *Simulating Sampling Distributions* applet.

2. In the box labeled "Population," choose "Categorical" from the drop-down menu. Keep the true proportion of successes as 0.5.

3. In the box labeled "Sample," change the sample size to $n = 20$ and click the Select sample button. The applet will graph the distribution of the sample and display the value of the sample proportion of successes $\hat{p}$. Was your sample proportion of successes close to the true proportion of successes, $p = 0.50$?

4. In the box labeled "Sampling Distribution," notice that the value of $\hat{p}$ from Step 3 is displayed on a dotplot. Click the Select samples button 9 more times, so that you have a total of 10 sample proportions. Look at the dotplot of your $\hat{p}$ values. Does the distribution have a recognizable shape?

5. To better see the shape of the sampling distribution of $\hat{p}$, enter 990 in the box for quickly selecting random samples and click the Select samples button. You have now selected a total of 1000 random samples of size $n = 20$ from a population with $p = 0.5$. Describe the shape of the simulated sampling distribution of $\hat{p}$ shown in the dotplot. Click the button to show the corresponding normal curve. How well does it fit?

6. How does the shape of the sampling distribution of $\hat{p}$ change if the true proportion of successes changes?

    a. In the Population box, change the true proportion to $p = 0.1$. Keeping $n = 20$ in the Sample box, quickly select 1000 random samples in the Sampling Distribution box. Describe the shape of the sampling distribution of $\hat{p}$.

    b. Now change the true proportion to $p = 0.90$ and repeat Step 6a.

    c. Describe how the value of p affects the shape of the sampling distribution of $\hat{p}$.

7. How does the shape of the sampling distribution of $\hat{p}$ change if the sample size increases?

    a. Keeping the true proportion as $p = 0.9$ in the Population box, increase the sample size to $n = 50$ in the Sample box. Then quickly select 1000 random samples in the Sampling Distribution box. Describe the shape of the sampling distribution of $\hat{p}$.

    b. Repeat Step 7a with sample sizes of $n = 100$ and $n = 500$.

    c. Describe how the value of n affects the shape of the sampling distribution of $\hat{p}$.

As you learned in the Concept Exploration, the shape of the sampling distribution of $\hat{p}$ will be closer to normal when the value of p is closer to 0.5 and when the sample size is larger. This is the same lesson you learned in Section 6.4 when exploring the normal approximation to the binomial distribution. It is no surprise that the **Large Counts condition** is the same as well.

Let $\hat{p}$ be the proportion of successes in a random sample of size n from a population with proportion of successes p. The **Large Counts condition** says that the sampling distribution of $\hat{p}$ will be approximately normal when

$$np \geq 10 \quad \text{and} \quad n(1-p) \geq 10$$

EXAMPLE 6.7.2

Sampling employees

Shape of the sampling distribution of $\hat{p}$

PROBLEM

In a large company, 12% of the employees are managers. The human resources department selects an SRS of 50 employees. Let $\hat{p}$ = the proportion of managers in the sample. Would it be appropriate to use a normal distribution to model the sampling distribution of $\hat{p}$ for samples of size 50? Justify your answer.

SOLUTION

No. Because $np = 50(0.12) = 6 < 10$, the sampling distribution of $\hat{p}$ is not approximately normal.

> If either np or $n(1 - p)$ is less than 10, the sampling distribution of $\hat{p}$ will not be approximately normal.

EXAM PREP: FOR PRACTICE, TRY EXERCISE 11.

Finding Probabilities Involving $\hat{p}$

When the Large Counts condition is met, we can use a normal distribution to calculate probabilities involving $\hat{p}$ = the proportion of successes in a random sample of size n.

EXAMPLE 6.7.3

How far from home do you attend college?

Normal calculations involving $\hat{p}$

PROBLEM

Suppose that 35% of all first-year university students attend school within 50 miles of their home. A polling organization asks an SRS of 1500 first-year university students how far away their home is. Let $\hat{p}$ be the sample proportion who live within 50 miles of their home.

(a) Describe the shape, center, and variability of the sampling distribution of $\hat{p}$.

(b) Find the probability that the SRS of 1500 students will give a result within 2 percentage points of the true value.

SOLUTION

(a) Shape: Approximately normal because $1500(0.35) = 525 \geq 10$ and $1500(1 - 0.35) = 975 \geq 10$

Center: $\mu_{\hat{p}} = 0.35$

Variability: $\sigma_{\hat{p}} = \sqrt{\dfrac{0.35(1 - 0.35)}{1500}} = 0.0123$

> Check that np and $n(1 - p) \geq 10$.

> $\mu_{\hat{p}} = p$ and $\sigma_{\hat{p}} = \sqrt{\dfrac{p(1-p)}{n}}$

(continued)

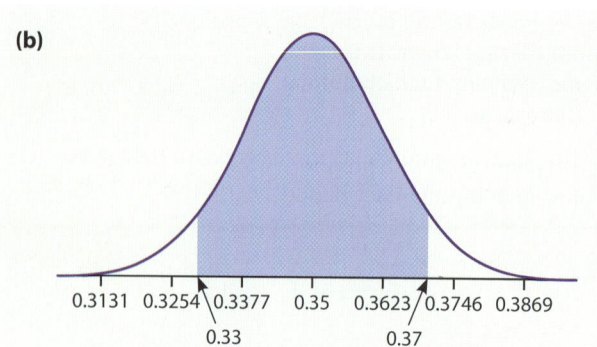

(b)

0.3131 0.3254 0.3377 0.35 0.3623 0.3746 0.3869
              0.33   0.37
Sample proportion who live within 50 miles

1. Draw a normal distribution.

(i) $z = \dfrac{0.33 - 0.35}{0.0123} = -1.626$ and $z = \dfrac{0.37 - 0.35}{0.0123} = 1.626$

2. Perform calculations—show your work!

Using technology: $P(-1.626 < Z < 1.626)$ = Applet/normalcdf(lower: −1.626, upper: 1.626, mean: 0, SD: 1) = 0.8961
Using Table A: $P(-1.63 < Z < 1.63) = 0.9484 - 0.0516 = 0.8968$

(ii) $P(0.33 < \hat{p} < 0.37)$ = Applet/normalcdf(lower: 0.33, upper: 0.37, mean: 0.35, SD: 0.0123) = 0.8961

EXAM PREP: FOR PRACTICE, TRY EXERCISE 15.

Section 6.7 What Did You Learn?

Review the learning goals from this section. Then practice what you've learned by working through the exercises.

Learning Goal	Example	Exercises
Calculate the mean and standard deviation of the sampling distribution of a sample proportion $\hat{p}$ and interpret the standard deviation.	6.7.1	7–10
Determine if the sampling distribution of $\hat{p}$ is approximately normal.	6.7.2	11–14
If appropriate, use a normal distribution to calculate probabilities involving $\hat{p}$.	6.7.3	15–18

Section 6.7 Exercises

Building Concepts and Skills These exercises assess the basic knowledge you should have after reading the section.

1. True/False: The sampling distribution of a sample proportion describes the values in a sample and how often those values occur.

2. What is the formula for the mean of the sampling distribution of the sample proportion? For the standard deviation?

3. The standard deviation of the sampling distribution of $\hat{p}$ measures how far the _____ typically vary from the _____.

4. True/False: Doubling the sample size cuts the standard deviation of the sampling distribution of $\hat{p}$ in half.

5. Why do we check the Large Counts condition?

6. For the Large Counts condition to be met, both _____ and _____ must be at least 10.

Mastering Concepts and Skills These exercises reinforce the learning goals as shown in the examples.

7. **Refer to Example 6.7.1 Orange Skittles®** The makers of Skittles claim that 20% of these candies are orange. You select a random sample of 30 Skittles from a large bag. Let $\hat{p}$ = the proportion of orange Skittles in the sample.
   (a) Calculate the mean and the standard deviation of the sampling distribution of $\hat{p}$.
   (b) Interpret the standard deviation from part (a).

8. **Woodpeckers** In a small forest, 63% of the pine trees have visible woodpecker damage. A researcher selects

a random sample of 20 pine trees and visually inspects each tree for woodpecker damage. Let $\hat{p}$ = the proportion of pine trees with woodpecker damage in the sample.

(a) Calculate the mean and the standard deviation of the sampling distribution of $\hat{p}$.

(b) Interpret the standard deviation from part (a).

9. **On-Time Shipping?** A large mail-order company advertises that it ships 90% of its orders within 3 working days. You select an SRS of 100 of the orders received in the past week for an audit. Let $\hat{p}$ = the proportion of orders in the sample that were shipped within 3 working days.

(a) Calculate the mean and the standard deviation of the sampling distribution of $\hat{p}$.

(b) Interpret the standard deviation from part (a).

10. **Registered Voters** In a congressional district, 55% of registered voters are Democrats. A polling organization selects a random sample of 500 registered voters from this district. Let $\hat{p}$ = the proportion of Democrats in the sample.

(a) Calculate the mean and the standard deviation of the sampling distribution of $\hat{p}$.

(b) Interpret the standard deviation from part (a).

11. **Refer to Example 6.7.2 Orange Skittles** Refer to Exercise 7. Would it be appropriate to use a normal distribution to model the sampling distribution of $\hat{p}$ = the proportion of orange Skittles in the sample? Justify your answer.

12. **Woodpeckers** Refer to Exercise 8. Would it be appropriate to use a normal distribution to model the sampling distribution of $\hat{p}$ = the proportion of pine trees with woodpecker damage in the sample? Justify your answer.

13. **On-Time Shipping?** Refer to Exercise 9. Would it be appropriate to use a normal distribution to model the sampling distribution of $\hat{p}$ = the proportion of orders in the sample that were shipped within 3 working days? Justify your answer.

14. **Registered Voters** Refer to Exercise 10. Would it be appropriate to use a normal distribution to model the sampling distribution of $\hat{p}$ = the proportion of Democrats in the sample? Justify your answer.

15. **Refer to Example 6.7.3 Motorcycle Ownership** In the United States, 8% of households own a motorcycle.[27] You plan to send surveys to an SRS of 500 households. Let $\hat{p}$ be the proportion of households in the sample that own a motorcycle.

(a) Describe the shape, center, and variability of the sampling distribution of $\hat{p}$.

(b) Find the probability that more than 10% of the households in the sample own a motorcycle.

16. **Second Jobs** According to the National Center for Education Statistics, 18% of teachers work at non-school second jobs during the school year.[28] You plan to select an SRS of 200 teachers. Let $\hat{p}$ = the proportion of teachers in the sample who work at non-school second jobs during the school year.

(a) Describe the shape, center, and variability of the sampling distribution of $\hat{p}$.

(b) Find the probability that fewer than 10% of the teachers in the sample work a non-school second job during the school year.

17. **Kickstarter** The fundraising site Kickstarter regularly tracks the success rate of projects that seek funding on its site. Recently, the percentage of projects that were successfully funded was 37.5%.[29] You plan to select a random sample of 50 Kickstarter projects. Let $\hat{p}$ be the proportion of projects in the sample that were successfully funded.

(a) Describe the shape, center, and variability of the sampling distribution of $\hat{p}$.

(b) What is the probability that fewer than 30% of the projects were successfully funded?

18. **French Speakers** Quebec is the only province in Canada where the one official language is French. According to a recent census, 78.1% of Quebec residents identify French as their first language.[30] You plan to select an SRS of 165 Quebec residents. Let $\hat{p}$ be the proportion of residents in the sample who identify French as their first language.

(a) Describe the shape, center, and variability of the sampling distribution of $\hat{p}$.

(b) What is the probability that more than 80% of the residents in the sample identify French as their first language?

Applying the Concepts These exercises ask you to apply multiple learning goals in a new context or to apply what you learned in this section in a new way.

19. **Jury Duty** What proportion of U.S. residents receive a jury summons each year? A polling organization plans to survey a random sample of 500 U.S. residents to find out. Let $\hat{p}$ be the proportion of residents in the sample who received a jury summons in the previous 12 months. According to the National Center for State Courts, 15% of U.S. residents receive a jury summons each year.[31] Suppose that this claim is true.

(a) Calculate the mean and standard deviation of the sampling distribution of $\hat{p}$. Interpret the standard deviation.

(b) Justify that the sampling distribution of $\hat{p}$ is approximately normal.

(c) Calculate the probability that at most 13% of the residents in the sample received a jury summons in the previous 12 months.

(d) Of the poll respondents, 13% said they received a jury summons in the previous 12 months. Based on your answer to part (c), does this poll give convincing evidence that fewer than 15% of all U.S. residents received a jury summons in the previous 12 months? Explain your reasoning.

20. **Cereal Milk** A *USA Today* poll asked a random sample of 1012 U.S. adults what they do with the milk in their cereal bowl after they have eaten the cereal. Let $\hat{p}$ be the proportion of people in the sample who drink the cereal milk. A spokesman for the dairy industry claims that 70% of all U.S. adults drink the cereal milk. Suppose this claim is true.

    (a) Calculate the mean and standard deviation of the sampling distribution of $\hat{p}$. Interpret the standard deviation.

    (b) Justify that the sampling distribution of $\hat{p}$ is approximately normal.

    (c) Calculate the probability that at most 67% of the people in the sample drink the cereal milk.

    (d) Of the poll respondents, 67% said that they drink the cereal milk. Based on your answer to part (c), does this poll give convincing evidence that fewer than 70% of all U.S. adults drink the cereal milk? Explain your reasoning.

21. **Jury Duty** Refer to Exercise 19. What sample size would be required to reduce the standard deviation of the sampling distribution to one-half the value you found in part (a)? Justify your answer.

22. **Cereal Milk** Refer to Exercise 20. What sample size would be required to reduce the standard deviation of the sampling distribution to one-half the value you found in part (a)? Justify your answer.

Extending the Concepts These exercises challenge you to explore statistical concepts and methods that go beyond what you learned in this section.

23. **10% Condition for Proportions** In this section, you learned that the formula for the standard deviation of the sampling distribution of $\hat{p}$ is valid when the sample size is less than 10% of the population size. What if the sample size is larger than 10% of the population size? In this case, we use the *finite population correction factor* in the formula for the standard deviation, where N is the population size:

    $$\sigma_{\hat{p}} = \sqrt{\frac{p(1-p)}{n}} \sqrt{\frac{N-n}{N-1}}$$

    Imagine that you are choosing an SRS from a population of size $N = 1000$ where $p = 0.60$.

    (a) Calculate the standard deviation of the sampling distribution of $\hat{p}$ for samples of size $n = 10$, with and without the finite population correction factor. How do these values compare?

    (b) Calculate the standard deviation of the sampling distribution of $\hat{p}$ for samples of size $n = 500$, with and without the finite population correction factor. How do these values compare?

    (c) Based on your answers to parts (a) and (b), explain why we can use the formula $\sigma_{\hat{p}} = \sqrt{\frac{p(1-p)}{n}}$ when the sample size is less than 10% of the population size.

    (d) Calculate the standard deviation of the sampling distribution of $\hat{p}$ for samples of size $n = 1000$ using the finite population correction factor. Why does the value of the standard deviation make sense in this situation?

24. **Second Jobs** Refer to Exercise 16. Use the appropriate binomial distribution to find the probability that fewer than 10% of the teachers in the sample work a non-school second job during the school year. How does this answer compare to the answer from Exercise 16? Which value is more accurate?

Cumulative Review These exercises revisit what you learned in previous sections.

25. **Malaria Vaccine (1.5, 1.8)** A recent study tested a new malaria vaccine using 450 children in Burkina Faso as subjects. The children were randomly assigned to three groups: a low-dose vaccine group, a high-dose vaccine group, and a placebo vaccine group. At the end of 1 year, the proportion of children who developed malaria was recorded.[32]

    (a) Explain why it was important to include a group that received a placebo vaccine.

    (b) The proportion of children who developed malaria was significantly smaller in the two vaccine groups. Based on the design of this study, can we conclude that the vaccine was the cause? Explain your answer.

    (c) What is the largest population to which we can apply the results of this study?

26. **Working Hard (2.1, 3.1)** The pie chart shows the distribution of hours in the workweek for a sample of 1270 full-time workers in the United States.

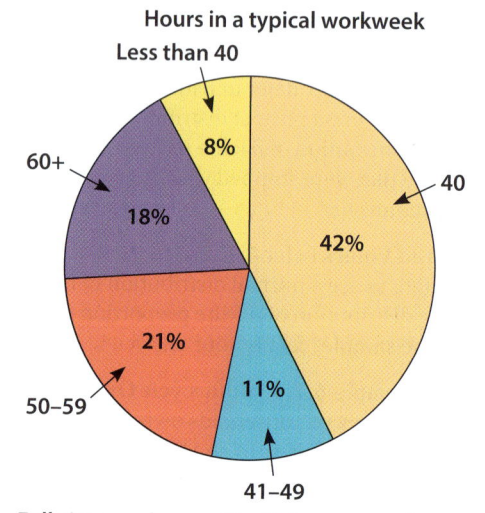

    Hours in a typical workweek

    - Less than 40: 8%
    - 40: 42%
    - 41–49: 11%
    - 50–59: 21%
    - 60+: 18%

    Full-time workers put in 47 hours a week, on average

    (a) Estimate the median number of work hours per week in this sample. Explain your reasoning.

    (b) These data are a "categorical" version of a quantitative variable. Taking into account that the legend says, "Full-time workers put in an average of 47 hours a week," would you say that the distribution of the underlying quantitative variable is symmetric, skewed right, or skewed left? Explain your reasoning.

Section 6.8 Sampling Distribution of the Sample Mean and the Central Limit Theorem

LEARNING GOALS
By the end of this section, you will be able to:
- Calculate the mean and standard deviation of the sampling distribution of a sample mean and interpret the standard deviation.
- Determine if the sampling distribution of $\bar{x}$ is approximately normal.
- If appropriate, use a normal distribution to calculate probabilities involving $\bar{x}$.

When sample data are categorical, we often use the proportion of successes in the sample to make an inference about the population proportion. When sample data are quantitative, we often use the sample mean $\bar{x}$ to estimate the mean μ of a population. But the value of $\bar{x}$ will vary from sample to sample. To understand how much $\bar{x}$ varies from μ and which values of $\bar{x}$ are likely to happen by chance, you need to understand the **sampling distribution of the sample mean $\bar{x}$**.

> The **sampling distribution of the sample mean $\bar{x}$** describes the distribution of values taken by the sample mean $\bar{x}$ in all possible samples of the same size from the same population.

As in Section 6.7, we'll start by investigating the center and variability of the sampling distribution of the sample mean.

Center and Variability

As with the sampling distribution of $\hat{p}$, there are formulas that describe the center and variability of the sampling distribution of $\bar{x}$.

> **HOW TO** Calculate $\mu_{\bar{x}}$ and $\sigma_{\bar{x}}$
>
> Let $\bar{x}$ be the mean of an SRS of size n selected from a large population with mean μ and standard deviation σ. Then:
> - The **mean** of the sampling distribution of $\bar{x}$ is $\mu_{\bar{x}} = \mu$.
> - The **standard deviation** of the sampling distribution of $\bar{x}$ is $\sigma_{\bar{x}} = \dfrac{\sigma}{\sqrt{n}}$.

The behavior of $\bar{x}$ in repeated samples is much like that of the sample proportion $\hat{p}$:

- The sample mean $\bar{x}$ is an unbiased estimator of the population mean μ. In other words, the mean of the sampling distribution $\mu_{\bar{x}}$ is equal to the mean of the population μ.
- The standard deviation of the sampling distribution $\sigma_{\bar{x}}$ describes the typical distance between the sample mean $\bar{x}$ and the population mean μ.
- The distribution of $\bar{x}$ is less variable for larger samples. This is indicated by the n in the denominator of the standard deviation formula. Quadrupling the sample size cuts the standard deviation of the sampling distribution in half.
- The formula for the standard deviation of the sampling distribution of $\bar{x}$ requires that the observations be independent. In practice, we are safe assuming independence when we are sampling without replacement as long as the sample size is less than 10% of the population size.

These facts about the mean and standard deviation of $\bar{x}$ are true *no matter the shape of the population distribution.*

EXAMPLE 6.8.1

Seen any good movies recently?

Mean and standard deviation of the sampling distribution of $\bar{x}$

PROBLEM

The number of movies viewed in the last year by students at a university is skewed to the right with a mean of 25.3 movies and a standard deviation of 15.8 movies. Suppose we select an SRS of 100 students from this school and calculate the mean number of movies viewed by the members of the sample.

(a) Calculate the mean and standard deviation of the sampling distribution of $\bar{x}$.

(b) Interpret the standard deviation from part (a).

SOLUTION

(a) $\mu_{\bar{x}} = 25.3$ and $\sigma_{\bar{x}} = \dfrac{15.8}{\sqrt{100}} = 1.58$

$$\mu_{\bar{x}} = \mu \text{ and } \sigma_{\bar{x}} = \dfrac{\sigma}{\sqrt{n}}$$

(b) In SRSs of size $n = 100$, the sample mean number of movies typically varies from the true mean by about 1.58 movies.

EXAM PREP: FOR PRACTICE, TRY EXERCISE 9.

Shape

The shape of the sampling distribution of the sample mean $\bar{x}$ depends on the shape of the population distribution. In the following Concept Exploration, you will explore what happens when you sample from a normal population and from a non-normal population.

Concept Exploration

Shape of the sampling distribution of $\bar{x}$

In this exploration, you will use an applet to investigate the shape of the sampling distribution of the sample mean $\bar{x}$ for different sample sizes (n) and different population distribution shapes.

1. Go to www.stapplet.com and launch the *Simulating Sampling Distributions* applet.

Sampling from a normal population

2. In the box labeled "Population," there are four choices for the shape of the population distribution when the data are quantitative: normal, uniform, skewed, and bimodal. Select normal.

3. In the box labeled "Sample," change the sample size to $n = 2$ and click the Select sample button. The applet will graph the distribution of the sample and display the value of the sample mean $\bar{x}$. Was your sample mean close to the population mean $\mu = 10$?

4. In the box labeled "Sampling Distribution," notice that the value of $\bar{x}$ from Step 3 is displayed on a dotplot. Click the Select samples button 9 more times, so that you have a total of 10 sample means. Look at the dotplot of your $\bar{x}$ values. Does the distribution have a recognizable shape?

5. To better see the shape of the sampling distribution of $\bar{x}$, enter 990 in the box for quickly selecting random samples and click the Select samples button. You have now selected a total of 1000 random samples of size $n = 2$. Describe the

Simulating Sampling Distributions

Population
Population distribution is: Normal
Distribution of population:

$\mu = 10, \sigma = 2$

Sample
Show distribution of one random sample of size $n = 2$

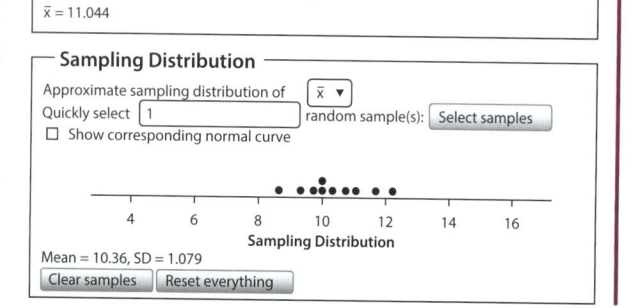

Sample Data
$\bar{x} = 11.044$

Sampling Distribution
Approximate sampling distribution of $\bar{x}$
Quickly select 1 random sample(s): Select samples
☐ Show corresponding normal curve

Sampling Distribution
Mean = 10.36, SD = 1.079
Clear samples | Reset everything

Section 6.8 Sampling Distribution of the Sample Mean and the Central Limit Theorem

shape of the simulated sampling distribution of $\bar{x}$ shown in the dotplot. Click the button to show the corresponding normal curve. How well does it fit?

6. Change the sample size to $n = 10$ in the Sample box and quickly select 1000 random samples in the Sampling Distribution box. Describe the shape of the sampling distribution of $\bar{x}$.

7. Repeat Step 6 with samples of size $n = 30$ and $n = 100$.

8. What have you learned about the shape of the sampling distribution of $\bar{x}$ when the population distribution has a normal shape?

Sampling from a non-normal population

9. In the Population box, change the population shape to uniform and repeat Steps 3–7.

10. In the Population box, change the population shape to skewed and repeat Steps 3–7.

11. In the Population box, change the population shape to bimodal and repeat Steps 3–7.

12. What have you learned about the shape of the sampling distribution of $\bar{x}$ when the population distribution has a non-normal shape?

The screenshots show approximate sampling distributions of $\bar{x}$ for samples of size $n = 2$ and samples of size $n = 30$ from three different populations.

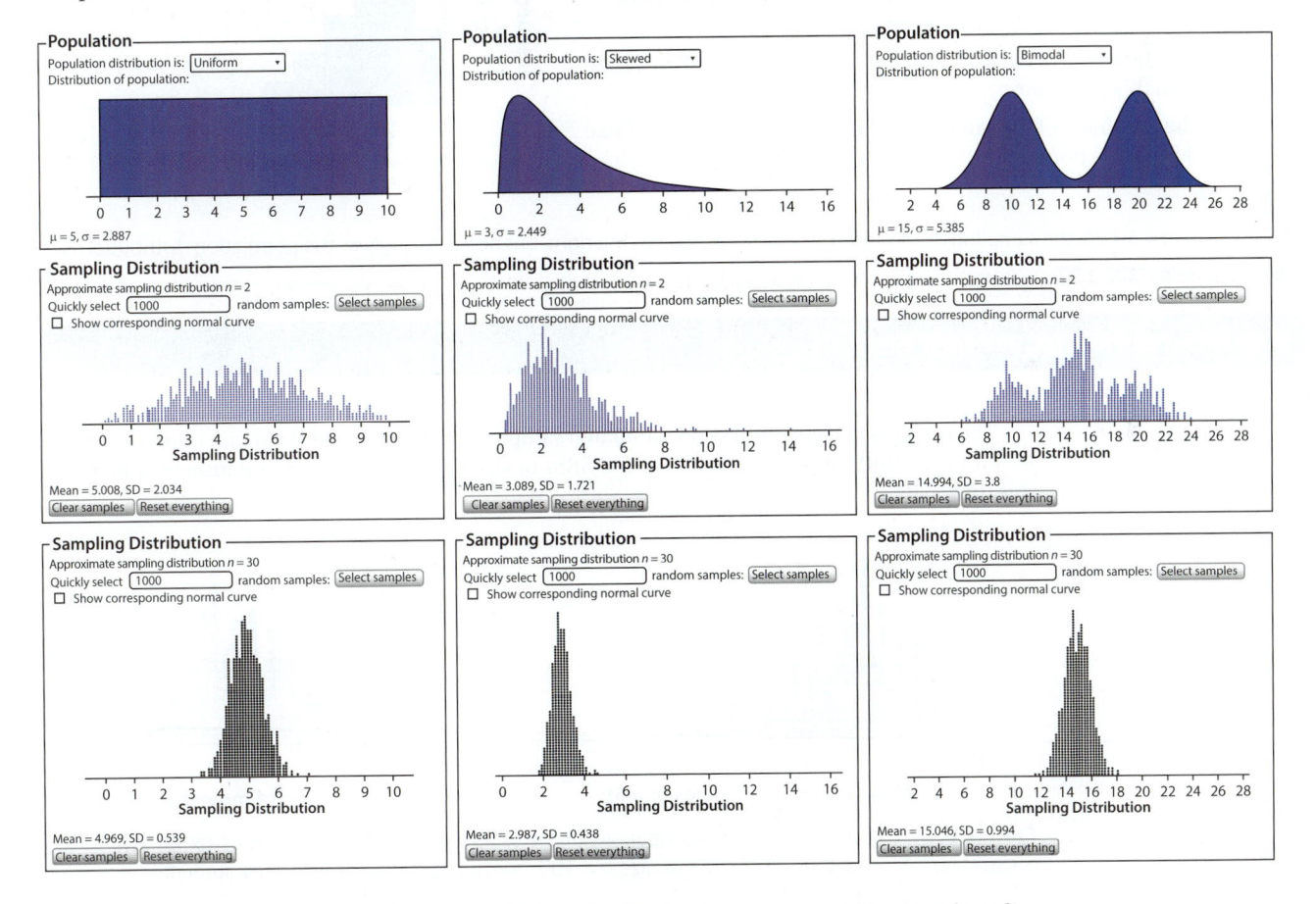

It is a remarkable fact that when the population distribution is non-normal, the sampling distribution of $\bar{x}$ looks more like a normal distribution as the sample size increases. This is true no matter what shape the population distribution has, as long as the population has a finite mean μ and standard deviation σ, and the observations in the sample are independent. This famous fact of probability theory is called the **central limit theorem (CLT)**.

> Select an SRS of size n from any population with mean μ and standard deviation σ. The **central limit theorem (CLT)** says that when n is sufficiently large, the sampling distribution of the sample mean $\bar{x}$ is approximately normal.

362 CHAPTER 6 Normal Distributions and Sampling Distributions

The sample size required for the sampling distribution of $\bar{x}$ to be close to normal depends on the shape of the population distribution. If the population distribution is approximately normal, the distribution of $\bar{x}$ will also be approximately normal, even with a small sample size. However, a larger sample size is required if the shape of the population distribution is far from normal.

> **Describing the Shape of the Sampling Distribution of a Sample Mean**
> - If the population distribution is normal, the sampling distribution of $\bar{x}$ will also be normal, no matter what the sample size n is.
> - If the population distribution is not normal, the sampling distribution of $\bar{x}$ will be approximately normal when the sample size is sufficiently large ($n \geq 30$ in most cases).

EXAMPLE 6.8.2

Seen any good movies recently?

Shape of the sampling distribution of $\bar{x}$

PROBLEM

The number of movies viewed in the last year by students at a university is skewed to the right with a mean of 25.3 movies and a standard deviation of 15.8 movies, as shown in the histogram.

(a) Describe the shape of the sampling distribution of $\bar{x}$ for SRSs of size $n = 5$ from this population.

(b) Describe the shape of the sampling distribution of $\bar{x}$ for SRSs of size $n = 100$ from this population.

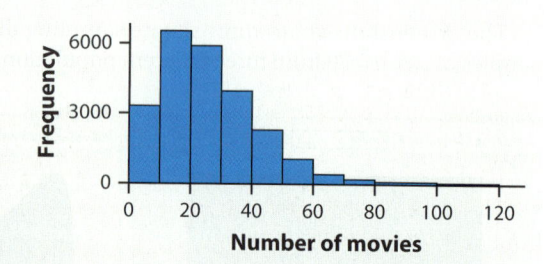

SOLUTION

(a) Because $n = 5 < 30$, the sampling distribution of $\bar{x}$ will be skewed to the right, but not as strongly as the population distribution is.

(b) Because $n = 100 \geq 30$, the sampling distribution of $\bar{x}$ will be approximately normal.

EXAM PREP: FOR PRACTICE, TRY EXERCISE 13.

The dotplots in Figure 6.14 show simulated sampling distributions of $\bar{x}$ = mean number of movies for (a) 200 SRSs of size $n = 5$ and (b) 200 SRSs of size $n = 100$ from the population of students at this university.

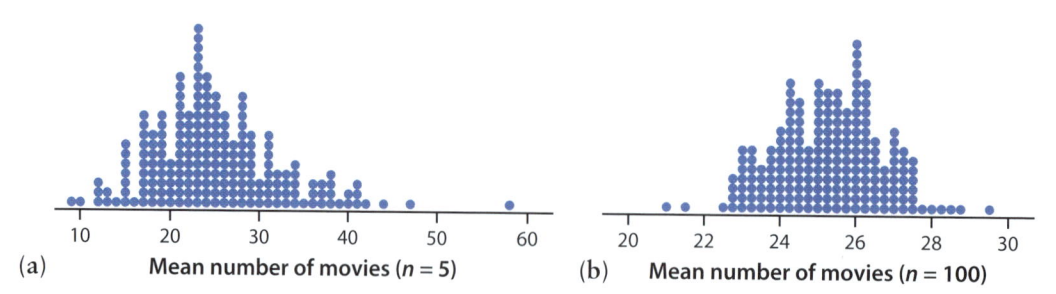

FIGURE 6.14 Simulated sampling distributions of the sample mean number of movies for (a) 200 SRSs of size $n = 5$ and (b) 200 SRSs of size $n = 100$ from a population of university students.

As expected, the simulated sampling distribution of $\bar{x}$ for SRSs of size $n = 5$ is skewed to the right, but not as strongly as the population distribution is. The simulated sampling distribution of $\bar{x}$ for SRSs of size $n = 100$ is approximately normal — thanks to the central limit theorem.

Probabilities Involving $\bar{x}$

We can do probability calculations involving $\bar{x}$ if the shape of the population distribution is approximately normal or if the sample size is large ($n \geq 30$).

Section 6.8 Sampling Distribution of the Sample Mean and the Central Limit Theorem

EXAMPLE 6.8.3

Are these peanuts underweight?

Probabilities involving $\bar{x}$

PROBLEM

A snack-foods company uses a machine to place dry-roasted, shelled peanuts in jars labeled "16 ounces." The distribution of weight in the jars is approximately normal with a mean of 16.1 ounces and a standard deviation of 0.15 ounce. Let $\bar{x}$ be the mean weight of the contents in 10 randomly selected jars.

(a) Describe the shape, center, and variability of the sampling distribution of $\bar{x}$ for samples of size 10. Include a justification for your description of the shape.

(b) Find the probability that the mean weight of the contents in 10 randomly selected jars is less than the advertised weight of 16 ounces.

SOLUTION

(a) Shape: Approximately normal because the population distribution is approximately normal

Center: $\mu_{\bar{x}} = 16.1$ ounces

Variability: $\sigma_{\bar{x}} = \dfrac{0.15}{\sqrt{10}} = 0.047$ ounce

> Don't forget to divide the standard deviation of the population by $\sqrt{n}$!

(b)

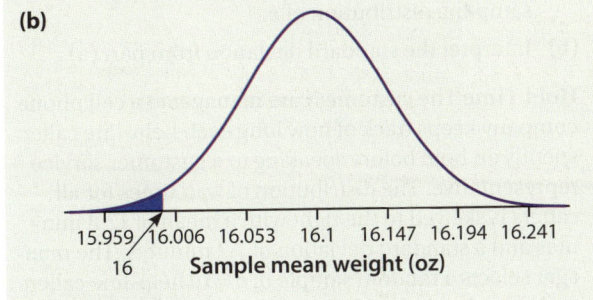

> 1. Draw a normal distribution.

> If we weren't told that the population distribution of peanut weight was approximately normal, we wouldn't be able to find this probability using a normal distribution because $n = 10 < 30$.

15.959 16.006 16.053 16.1 16.147 16.194 16.241
  16
Sample mean weight (oz)

> 2. Perform calculations — show your work!

(i) $z = \dfrac{16 - 16.1}{0.047} = -2.128$

Using technology: $P(Z < -2.128)$ = Applet/normalcdf(lower: −1000, upper: −2.128, mean: 0, SD: 1) = 0.0167

Using Table A: $P(Z < -2.13) = 0.0166$

(ii) $P(\bar{x} < 16)$ = Applet/normalcdf(lower: −1000, upper: 16, mean: 16.1, SD: 0.047) = 0.0167

EXAM PREP: FOR PRACTICE, TRY EXERCISE 17.

Averages are less variable than individuals, so randomly selecting a single jar that is under the advertised weight of 16 ounces is more likely than randomly selecting a sample of 10 jars with an average weight less than 16 ounces. As you can see in the figure, the area under the purple curve to the left of 16 is much greater than the area under the blue curve to the left of 16.

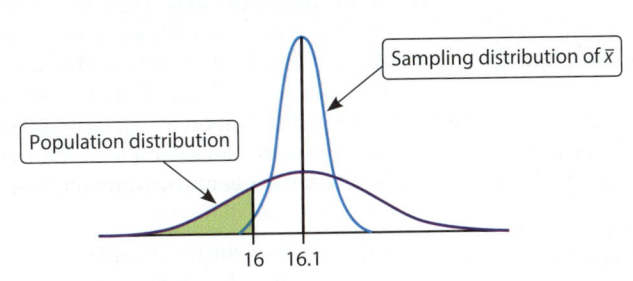

Sampling distribution of $\bar{x}$

Population distribution

16 16.1

The fact that averages of several observations are less variable than individual observations is important in many settings. For example, it is common practice in science and medicine to repeat a measurement several times and report the average of the results.

Section 6.8 What Did You Learn?

Review the learning goals from this section. Then practice what you've learned by working through the exercises.

Learning Goal	Example	Exercises
Calculate the mean and standard deviation of the sampling distribution of a sample mean and interpret the standard deviation.	6.8.1	9–12
Determine if the sampling distribution of $\bar{x}$ is approximately normal.	6.8.2	13–16
If appropriate, use a normal distribution to calculate probabilities involving $\bar{x}$.	6.8.3	17–20

Section 6.8 Exercises

Building Concepts and Skills These exercises assess the basic knowledge you should have after reading the section.

1. True/False: When data are quantitative, we often use μ to estimate $\bar{x}$.

2. What is the formula for the mean of the sampling distribution of $\bar{x}$? For the standard deviation?

3. The _____ of the sampling distribution of $\bar{x}$ measures how far the values of the sample mean typically vary from the population mean.

4. True/False: Increasing the sample size decreases the variability of the sampling distribution of $\bar{x}$.

5. True/False: When sampling from a non-normal population, the central limit theorem states that the population distribution becomes more normal as the sample size increases.

6. The sampling distribution of the sample mean will have an approximately normal distribution when the sample size is large or when _____.

7. For most non-normal populations, we can apply the central limit theorem when the sample size is at least _____.

8. Averages of several observations are _____ variable than individual observations.

Mastering Concepts and Skills These exercises reinforce the learning goals as shown in the examples.

9. **Refer to Example 6.8.1 Song Lengths** David has about 10,000 songs in his collection. The distribution of play time for these songs is heavily skewed to the right with a mean of 225 seconds and a standard deviation of 60 seconds. David chooses a random sample of $n = 5$ songs from his collection and records $\bar{x}$ = sample mean play time.

   (a) Calculate the mean and standard deviation of the sampling distribution of $\bar{x}$.

   (b) Interpret the standard deviation from part (a).

10. **Hold Time** The customer care manager at a cell phone company keeps track of how long each help-line caller spends on hold before speaking to a customer service representative. The distribution of wait times for all callers is skewed to the right with a mean of 12.8 minutes and a standard deviation of 7.2 minutes. The manager selects a random sample of $n = 10$ help-line callers and records $\bar{x}$ = sample mean wait time.

    (a) Calculate the mean and standard deviation of the sampling distribution of $\bar{x}$.

    (b) Interpret the standard deviation from part (a).

11. **Birth Weights** The birth weights of babies born full term vary according to an approximately normal distribution with a mean of 3.4 kilograms (kg) and standard deviation of 0.5 kg.[33] A large city hospital selects a random sample of 15 full-term babies born in the previous six months and records $\bar{x}$ = sample mean birth weight.

    (a) Calculate the mean and standard deviation of the sampling distribution of $\bar{x}$.

    (b) Interpret the standard deviation from part (a).

12. **Grinding Auto Parts** A grinding machine in an auto-parts factory prepares axles with a mean diameter of 40.125 millimeters (mm). The machine has some variability, so the axle diameter varies according to a normal distribution with a standard deviation of 0.002 mm. A supervisor selects a random sample of 4 axles each hour for quality control purposes and records $\bar{x}$ = sample mean diameter.

    (a) Calculate the mean and standard deviation of the sampling distribution of $\bar{x}$.

    (b) Interpret the standard deviation from part (a).

13. **Refer to Example 6.8.2 Song Lengths** Refer to Exercise 9, where $\bar{x}$ is the sample mean play time (in seconds).
(a) Describe the shape of the sampling distribution of $\bar{x}$ for SRSs of size $n = 5$ from the population of songs in David's collection. Justify your answer.
(b) Describe the shape of the sampling distribution of $\bar{x}$ for SRSs of size $n = 100$ from the population of songs in David's collection. Justify your answer.

14. Hold Times Refer to Exercise 10, where $\bar{x}$ is the sample mean wait time (in minutes).
(a) Describe the shape of the sampling distribution of $\bar{x}$ for SRSs of size $n = 10$ from the population of help-line callers. Justify your answer.
(b) Describe the shape of the sampling distribution of $\bar{x}$ for SRSs of size $n = 50$ from the population of help-line callers. Justify your answer.

15. Birth Weights Refer to Exercise 11, where $\bar{x}$ is the sample mean birth weight (in kilograms).
(a) Describe the shape of the sampling distribution of $\bar{x}$ for SRSs of size $n = 15$ from the population of all full-term babies at this hospital. Justify your answer.
(b) Describe the shape of the sampling distribution of $\bar{x}$ for SRSs of size $n = 50$ from the population of all full-term babies at this hospital. Justify your answer.

16. Grinding Auto Parts Refer to Exercise 12, where $\bar{x}$ is the sample mean diameter (in millimeters).
(a) Describe the shape of the sampling distribution of $\bar{x}$ for SRSs of size $n = 4$ from the population of all axles produced in an hour. Justify your answer.
(b) Describe the shape of the sampling distribution of $\bar{x}$ for SRSs of size $n = 50$ from the population of all axles produced in an hour. Justify your answer.

17. **Refer to Example 6.8.3 Finch Beaks** One dimension of bird beaks is "depth"—the height of the beak where it arises from the bird's head. During a study on one island in the Galápagos archipelago, researchers found that the beak depth of medium ground finches follows an approximately normal distribution with mean $\mu = 9.5$ mm and standard deviation $\sigma = 1.0$ mm.[34] Let $\bar{x}$ be the sample mean beak depth in 15 randomly selected medium ground finches.
(a) Describe the shape, center, and variability of the sampling distribution of $\bar{x}$ for samples of size 15. Include a justification for your description of the shape.
(b) What is the probability that the mean beak depth $\bar{x}$ in 15 randomly selected medium ground finches is within 0.5 mm of the population mean?

18. Cola Bottling A company uses a machine to fill plastic bottles with cola. The volume in a bottle follows an approximately normal distribution with mean $\mu = 298$ milliliters (mL) and standard deviation $\sigma = 3$ mL. Let $\bar{x}$ be the sample mean volume in an SRS of 16 bottles.
(a) Describe the shape, center, and variability of the sampling distribution of $\bar{x}$ for samples of size 16. Include a justification for your description of the shape.
(b) What is the probability that the sample mean volume $\bar{x}$ in 16 randomly selected bottles is within 1 mL of the population mean volume?

19. Lightning Strikes The number of lightning strikes on a square kilometer of open ground in a year has a mean of 6 and standard deviation of 2.4. The National Lightning Detection Network (NLDN) uses automatic sensors to watch for lightning in a random sample of fifty 1-square-kilometer plots of land for one year. Let $\bar{x}$ be the average number of lightning strikes in the sample.
(a) Describe the shape, center, and variability of the sampling distribution of $\bar{x}$ for samples of size $n = 50$ from this population. Include a justification for your description of the shape.
(b) Calculate the probability that the average number of lightning strikes per square kilometer $\bar{x}$ in 50 randomly selected square kilometers is less than 5.

20. Insurance Claims An insurance company claims that in the population of homeowners, the mean annual loss from fire is $250 with a standard deviation of $5000. The distribution of loss is strongly right-skewed: Many homeowners have $0 loss, but a few have large losses. Let $\bar{x}$ be the average loss in a random sample of 10,000 homeowners.
(a) Describe the shape, center, and variability of the sampling distribution of $\bar{x}$ for samples of size $n = 10,000$ from this population. Include a justification for your description of the shape.
(b) Suppose that the insurance company charges $300 for each policy. Calculate the probability that the mean average loss $\bar{x}$ for 10,000 randomly selected homeowners is less than $300.

Applying the Concepts These exercises ask you to apply multiple learning goals in a new context or to apply what you learned in this section in a new way.

21. Air Conditioning Your company has a contract to perform preventive maintenance on thousands of air-conditioning units in a large city. Based on service records from the past year, the time (in hours) that a technician requires to complete the work follows a strongly right-skewed distribution with $\mu = 1$ hour and $\sigma = 1.5$ hours. As a promotion, your company will provide service to a random sample of 70 air-conditioning units free of charge. You plan to budget an average of 1.1 hours per unit for a technician to complete the work. Will this be enough time?
(a) Describe the shape, center, and variability of the sampling distribution of $\bar{x}$ for samples of size $n = 70$ from this population. Include a justification for your description of the shape.
(b) Calculate the probability that the average maintenance time $\bar{x}$ for 70 randomly selected units exceeds 1.1 hours.
(c) Based on your answer to part (b), did the company budget enough time? Explain your answer.

22. **Fuel Efficiency** Driving styles differ, so there is variability in the fuel efficiency of the same model automobile when driven by different people. The manufacturer of a certain car model claims that its fuel efficiency varies according to a normal distribution with $\mu = 23.6$ miles per gallon (mpg) and $\sigma = 2.5$ mpg.[35] To test this claim, a consumer organization selects an SRS of 25 owners of this model and calculates the sample mean fuel efficiency to be 23.2 mpg. Does this provide convincing evidence that the manufacturer is overstating the mean fuel efficiency?
    (a) Describe the shape, center, and variability of the sampling distribution of $\bar{x}$ for samples of size $n = 25$ from this population, assuming the manufacturer's claim is true. Include a justification for your description of the shape.
    (b) Calculate the probability that the average fuel efficiency $\bar{x}$ for 25 randomly selected owners is at most 23.2 mpg.
    (c) Based on your answer to part (b), is there convincing evidence that the manufacturer is overstating the mean fuel efficiency? Explain your answer.

23. **Finch Beaks** In Exercise 17, you calculated the probability that $\bar{x}$ would estimate the true mean beak depth within ±0.5 mm of μ in samples of size 15.
    (a) If you randomly selected one finch instead of 15, would it be more likely, less likely, or equally likely to have a beak depth within 0.5 mm of μ? Explain your reasoning without doing any calculations.
    (b) Calculate the probability of the event described in part (a) to confirm your answer.

24. **Cola Bottling** In Exercise 18, you calculated the probability that $\bar{x}$ would estimate the true mean volume of cola within ±1 mL of μ in samples of size 16.
    (a) If you randomly selected one bottle instead of 16, would it be more likely, less likely, or equally likely to contain a volume of cola within ±1 mL of μ? Explain your reasoning without doing any calculations.
    (b) Calculate the probability of the event described in part (a) to confirm your answer.

25. **Song Lengths** Refer to Exercise 9. How many songs would you need to randomly select to get a standard deviation of the sampling distribution of $\bar{x}$ that is half the value you found in Exercise 9?

26. **Hold Times** Refer to Exercise 10. How many help-line callers would you need to randomly select to get a standard deviation of the sampling distribution of $\bar{x}$ that is half the value you found in Exercise 10?

27. **Central Limit Theorem** Asked what the central limit theorem says, a student replies, "As you select larger and larger samples from a population, the graph of the sample values looks more and more normal." Is the student right? Explain your answer.

28. **Central Limit Theorem** Asked what the central limit theorem says, a student replies, "As you select larger and larger samples from a population, the variability of the sampling distribution of the sample mean decreases." Is the student right? Explain your answer.

Extending the Concepts These exercises challenge you to explore statistical concepts and methods that go beyond what you learned in this section.

29. **10% Condition for Means** In this section, you learned that the formula for the standard deviation of the sampling distribution of $\bar{x}$ is valid when the sample size is less than 10% of the population size. What if the sample size is larger than 10% of the population size? In this case, we use the *finite population correction factor* in the formula for the standard deviation, where N is the population size:
$$\sigma_{\bar{x}} = \frac{\sigma}{\sqrt{n}} \sqrt{\frac{N-n}{N-1}}$$
Imagine that you are choosing an SRS from a population of size $N = 1000$ where $\mu = 60$ and $\sigma = 5$.
    (a) Calculate the standard deviation of the sampling distribution of $\bar{x}$ for samples of size $n = 10$, with and without the finite population correction factor. How do these values compare?
    (b) Calculate the standard deviation of the sampling distribution of $\bar{x}$ for samples of size $n = 500$, with and without the finite population correction factor. How do these values compare?
    (c) Based on your answers to parts (a) and (b), explain why we can use the formula $\sigma_{\bar{x}} = \sigma/\sqrt{n}$ when the sample size is less than 10% of the population size.
    (d) Calculate the standard deviation of the sampling distribution of $\bar{x}$ for samples of size $n = 1000$ using the finite population correction factor. Why does the value of the standard deviation make sense in this situation?

30. **Insurance Claims** Refer to Exercise 20. If the company wants to be 90% certain that the mean loss from fire in an SRS of 10,000 homeowners is less than the amount it charges for the policy, how much should the company charge?

Cumulative Review These exercises revisit what you learned in previous sections.

31. **Devices and Sleep (5.1, 5.2)** A National Sleep Foundation survey of 1103 parents asked, among other questions, how many electronic devices (TVs, video games, smartphones, computers, tablets, and so on) children had in their bedrooms.[36] Let $X =$ the number of devices in a randomly chosen child's bedroom. Here is the probability distribution of X.

Number of devices	0	1	2	3	4	5
Probability	0.28	0.27	0.18	0.16	0.07	0.04

(a) Show that this is a legitimate probability distribution.

(b) What is the probability that a randomly chosen child has at least one electronic device in their bedroom?

(c) Calculate the expected value and standard deviation of X.

32. **Carpet Flaws (5.5, 6.8)** The number of flaws in a 1-square-yard piece carpet is a Poisson random variable with $\mu = 1.6$ flaws and $\sigma = 1.26$ flaws.

(a) Calculate the probability that a randomly selected 1-square-yard piece of carpet has more than 2 flaws.

(b) Suppose you randomly select ten 1-square-yard pieces of carpet and calculate $\bar{x}$ = mean number of flaws. Explain why you shouldn't use a normal distribution to do calculations involving $\bar{x}$.

(c) Calculate $P(\bar{x} > 2)$ for a random sample of fifty 1-square-yard pieces of carpet. How does this compare to your answer from part (a)?

Statistics Matters How much salt is too much?

As discussed at the beginning of this chapter, researchers analyzed more than 3000 meals ordered from Chipotle restaurants using the online site Grubhub. The researchers recorded three variables for each meal: the amount of sodium (mg), the number of calories, and the amount of saturated fat (g).[37]

The probability distribution of X = amount of sodium (mg) in a randomly selected meal from Chipotle can be modeled by a normal distribution with mean $\mu = 1993$ mg and standard deviation $\sigma = 593$ mg.

1. The recommended daily allowance (RDA) for sodium is 2400 mg. What is the probability that a randomly selected Chipotle meal exceeds the RDA for sodium?

2. What is the 90th percentile of the probability distribution of X?

Suppose you randomly select 50 meals from Chipotle and calculate $\bar{x}$ = mean amount of sodium in the sample and $\hat{p}$ = proportion of meals in the sample that exceed the RDA for sodium.

3. Round your answer from Question 1 to 2 decimal places and use it as the value of p. Then find the probability that more than 40% of meals in the sample exceed the RDA for sodium.

4. What is the probability that the mean amount of sodium in the sample exceeds the RDA for sodium?

The probability distribution of Y = amount of saturated fat (g) in a randomly selected meal from Chipotle can be modeled by a density curve that is skewed to the left with mean $\mu = 13.9$ g and standard deviation $\sigma = 5.5$ g.

5. The RDA for saturated fat is 20 g. Explain why you can't calculate the probability that a randomly selected meal from Chipotle exceeds the RDA for saturated fat with the given information.

6. Explain why you can calculate the probability that the mean amount of saturated fat in 50 randomly selected Chipotle meals exceeds the RDA for saturated fat. Then calculate this probability.

Chapter 6 Review

Continuous Random Variables and Density Curves

- We model the distribution of a continuous random variable with a **density curve**. A density curve always remains on or above the horizontal axis and has total area 1 underneath it. An area under a density curve and above any interval of values on the horizontal axis estimates the probability that the random variable falls in that interval.

- The mean and the median of a density curve can be located by eye. The mean μ is the balance point of the curve. The median divides the area under the curve in half. The standard deviation σ cannot be located by eye on most density curves.

Normal Distributions

- A **normal distribution** is described by a symmetric, single-peaked, bell-shaped density curve called a **normal curve**. Any normal distribution is completely specified by two numbers: its mean and its standard deviation. The mean μ is the center of the curve and the standard deviation σ is the distance from μ to the change-of-curvature points on either side.

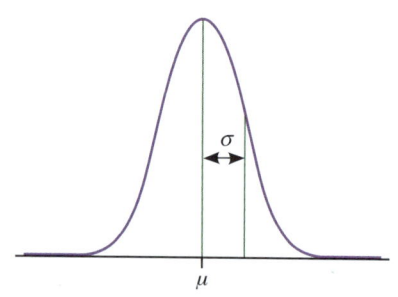

- The **empirical rule** (also called the 68–95–99.7 rule) describes the approximate probability that the value of a random variable falls within 1, 2, and 3 standard deviations of the mean in a normal distribution.

- All normal distributions are the same when values are standardized. If X follows a normal distribution with mean μ and standard deviation σ, we standardize using the following formula:

$$z = \frac{\text{value} - \text{mean}}{\text{standard deviation}} = \frac{x - \mu}{\sigma}$$

The resulting distribution is called the **standard normal distribution** and has mean $\mu = 0$ and standard deviation $\sigma = 1$.

- To find the area in a normal distribution corresponding to given boundary values:
  - **Step 1: Draw a normal distribution** with the horizontal axis labeled and scaled using the mean and standard deviation, the boundary value(s) clearly identified, and the area of interest shaded.
  - **Step 2: Perform calculations—show your work!** Do one of the following:
    (i) Standardize the boundary value(s) and use technology or Table A to find the desired area under the standard normal curve.
    (ii) Use technology to find the desired area without standardizing. Label the inputs you used for the applet or calculator.

- To find the value in a normal distribution corresponding to a given percentile (area):
  - **Step 1: Draw a normal distribution** with the horizontal axis labeled and scaled using the mean and standard deviation, the area of interest shaded and labeled, and the unknown boundary value clearly marked.
  - **Step 2: Perform calculations—show your work!** Do one of the following:
    (i) Use technology or Table A to find the value of z with the appropriate area under the standard normal curve, then unstandardize to return to the original distribution.
    (ii) Use technology to find the desired value without standardizing. Label the inputs you used for the applet or calculator.

- You can find the mean or standard deviation of a normal distribution using one or more percentiles by solving for the missing value in the z-score formula.

- You can use the **normal approximation to the binomial distribution** when the Large Counts condition is met. If X is the number of successes in a binomial setting with n trials and p probability of success on each trial, the **Large Counts condition** says that the shape of the probability distribution of X is approximately normal if $np \geq 10$ and $n(1-p) \geq 10$. The probability distribution of X has mean $\mu_X = np$ and standard deviation $\sigma_X = \sqrt{np(1-p)}$.

- A **normal probability plot** is a scatterplot of the ordered pair (data value, expected z-score) for each of the individuals in a quantitative data set. That is, the x-coordinate of each point is the actual data value and the y-coordinate is the expected z-score corresponding to the percentile of that data value in a standard normal distribution. If the points on a normal probability plot lie close to a straight line, the data are approximately normally distributed. A nonlinear form in a normal probability plot indicates a non-normal distribution.

Sampling Distributions

- A **parameter** is a number that describes some characteristic of the population. A **statistic** is a number that describes some characteristic of a sample. We use statistics to estimate parameters.

- The **sampling distribution** of a statistic is the distribution of values taken by the statistic in all possible samples of the same size from the same population.

- To determine a sampling distribution, list all possible samples of a particular size, calculate the value of the statistic for each sample, and graph the distribution of the statistic. If there are many possible samples, use simulation to approximate the sampling distribution: Repeatedly select random samples of a particular size, calculate the value of the statistic for each sample, and graph the distribution of the statistic.

- We can use sampling distributions to determine which values of a statistic are likely to happen by chance alone and how much a statistic typically varies from the parameter it is trying to estimate.
- A statistic used to estimate a parameter is an **unbiased estimator** if the mean of its sampling distribution is equal to the value of the parameter being estimated. That is, the statistic doesn't consistently overestimate or consistently underestimate the value of the parameter when many random samples are selected.
- The sampling distribution of any statistic will have less variability when the sample size is larger. That is, the statistic will be a more precise estimator of the parameter with larger sample sizes.
- Let $\hat{p}$ = the proportion of successes in a random sample of size n from a large population with proportion of successes p. The **sampling distribution of a sample proportion $\hat{p}$** describes the distribution of values taken by the sample proportion $\hat{p}$ in all possible samples of the same size from the same population.
  - The **mean** of the sampling distribution of $\hat{p}$ is $\mu_{\hat{p}} = p$. The mean describes the average value of $\hat{p}$ in repeated random samples.
  - The **standard deviation** of the sampling distribution of $\hat{p}$ is $\sigma_{\hat{p}} = \sqrt{\dfrac{p(1-p)}{n}}$. The standard deviation describes how far the values of $\hat{p}$ typically vary from p in repeated random samples.
  - The shape of the sampling distribution of $\hat{p}$ will be approximately normal when the **Large Counts condition** is met: $np \geq 10$ and $n(1-p) \geq 10$.
- Let $\bar{x}$ = the mean of a random sample of size n from a large population with mean μ and standard deviation σ. The **sampling distribution of a sample mean $\bar{x}$** describes the distribution of values taken by the sample mean $\bar{x}$ in all possible samples of the same size from the same population.
  - The **mean** of the sampling distribution of $\bar{x}$ is $\mu_{\bar{x}} = \mu$. The mean describes the average value of $\bar{x}$ in repeated random samples.
  - The **standard deviation** of the sampling distribution of $\bar{x}$ is $\sigma_{\bar{x}} = \dfrac{\sigma}{\sqrt{n}}$. The standard deviation describes how far the values of $\bar{x}$ typically vary from μ in repeated random samples.
  - The shape of the sampling distribution of $\bar{x}$ will be approximately normal when the population distribution is approximately normal or when the sample size is large ($n \geq 30$). The fact that the sampling distribution of $\bar{x}$ becomes approximately normal—even when the population is non-normal—as the sample size increases is called the **central limit theorem (CLT)**.
- When the sampling distribution of $\hat{p}$ or $\bar{x}$ is approximately normal, you can use z-scores and Table A or technology to perform probability calculations involving the statistic.
- To determine which sampling distribution to use, consider whether the variable of interest is categorical or quantitative. If it is categorical, use the sampling distribution of a sample proportion $\hat{p}$. If it is quantitative, use the sampling distribution of a sample mean $\bar{x}$.

Chapter 6 Review Exercises

These exercises will help you review important concepts and skills described by the learning goals in each section. For most exercises, the relevant section is noted in parentheses after the exercise title.

1. **Horse Pregnancies (6.1)** Bigger animals tend to carry their young longer before birth. The distribution of H = length of horse pregnancy from conception to birth is approximately normal with mean 336 days and standard deviation 6 days.
   (a) Sketch the normal curve that models the distribution of horse pregnancy length. Label the mean and the points that are 1, 2, and 3 standard deviations from the mean.
   (b) Use the empirical rule to estimate the probability that a randomly selected horse pregnancy is longer than 342 days.

2. **Copper Rockfish (6.2)** The length X of randomly selected commercially raised copper rockfish can be modeled by a normal distribution with mean $\mu = 16.4$ inches and standard deviation $\sigma = 2.1$ inches.[38] Calculate the probability that a randomly selected commercially raised copper rockfish is:
   (a) less than 10 inches long.
   (b) greater than 17 inches long.
   (c) between 15 and 20 inches long.

3. **Copper Rockfish (6.3)** Refer to Exercise 2.
   (a) Find the 25th percentile of the distribution of X.
   (b) Suppose that the length Y of wild copper rockfish can be modeled by a normal distribution with mean μ and standard deviation $\sigma = 3$ inches. If 10% of wild copper rockfish are greater than 19 inches long, what is the mean length of wild copper rockfish?

4. **U.S. Kids (6.4)** According to the U.S. Census Bureau, 24% of U.S. residents are younger than age 18 years. Suppose we select a random sample of 500 U.S. residents. Let X = the number of people in the sample who are younger than age 18.

(a) Show that the probability distribution of X is approximately normal.

(b) Calculate the mean and standard deviation of the appropriate normal distribution.

(c) Use this normal distribution to calculate the probability that the number of people younger than age 18 in a random sample of size 500 is between 100 and 110.

5. **Whale Length (6.4)** The following normal probability plot shows the distribution of length (in meters, m) for a sample of 202 North Atlantic right whales.[39] Use the graph to determine if this distribution of total length is approximately normal.

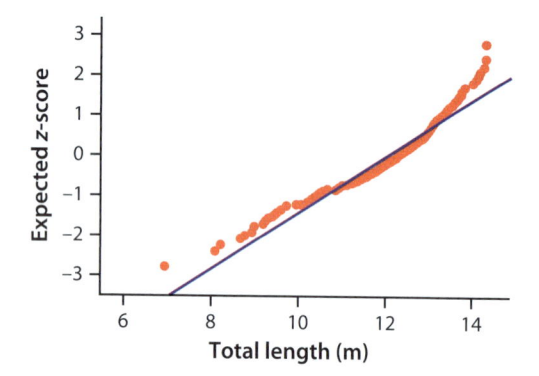

6. **Bad Eggs (6.5)** People who eat eggs that are contaminated with the *Salmonella* bacterium can get food poisoning. A large egg producer randomly selects 200 eggs from all the eggs shipped in one day. The laboratory reports that $9/200 = 0.045$ of these eggs had *Salmonella* contamination. Identify the population, the parameter, the sample, and the statistic.

7. **Children's Books (6.5, 6.6)** An author has written 5 children's books. The numbers of pages in these books are 64, 66, 71, 73, and 76.
   (a) List all 10 possible SRSs of size $n = 3$, calculate the median number of pages for each sample, and display the sampling distribution of the sample median on a dotplot.
   (b) Show that the sample median is a biased estimator of the population median for this population.
   (c) Describe how the variability of the sampling distribution of the sample median would change if the sample size was increased to $n = 4$.

8. **Five-Second Rule (6.5, 6.7)** A report claimed that 20% of people subscribe to the "five-second rule." That is, they would eat a piece of food that fell onto the kitchen floor if it was picked up within 5 seconds. Assume this figure is true for the population of U.S. adults. Let $\hat{p}$ = the proportion of people who subscribe to the five-second rule in an SRS of size 80 from this population.
   (a) Calculate the mean and the standard deviation of the sampling distribution of $\hat{p}$.
   (b) Interpret the standard deviation from part (a).
   (c) Justify that it is appropriate to use a normal distribution to model the sampling distribution of $\hat{p}$.

(d) Calculate the probability that at most 10% of the people in the sample subscribe to the five-second rule.

(e) In an actual SRS of size 80, only 10% subscribed to the five-second rule. Based on your answer to part (d), does this result provide convincing evidence that the proportion of all U.S. adults who subscribe to the five-second rule is less than 0.20? Explain your reasoning.

9. **Copper Rockfish (6.8)** Refer to Exercise 2. Let $\bar{x}$ = the mean length in a random sample of 10 commercially raised copper rockfish.
   (a) Calculate the mean and standard deviation of the sampling distribution of $\bar{x}$.
   (b) Interpret the standard deviation from part (a).
   (c) What is the probability that the mean length is greater than 17 inches in a random sample of 10 commercially raised copper rockfish?
   (d) Would your answers to any of parts (a), (b), or (c) be affected if the distribution of length in the population of commercially raised copper rockfish was distinctly non-normal? Explain your answer.

10. **Gypsy Moths (Chapter 6)** The gypsy moth is a serious threat to oak and aspen trees. A state agriculture department places traps throughout the state to detect the moths. Each month, an SRS of 50 traps is inspected, the number of moths in each trap is recorded, and the mean number of moths is calculated. Based on years of data, the distribution of moth counts is strongly skewed with mean 0.5 and standard deviation 0.7.
    (a) If possible, calculate the probability that a randomly selected trap has at least one moth. If not, explain why this probability cannot be calculated.
    (b) Calculate the probability that the mean number of moths in a sample of size 50 is at least 0.6 moth.
    (c) In a recent month, the mean number of moths in an SRS of size 50 was 0.6. Based on this result, should officials in the state agricultural department be worried that the moth population is getting larger in their state? Explain your reasoning.

11. **Ketchup (Chapter 6)** A fast-food restaurant has just installed a new automatic ketchup dispenser for use in preparing its burgers. The amount of ketchup dispensed by the machine can be modeled by a normal distribution with mean 1.05 fluid ounces and standard deviation 0.08 fluid ounce.
    (a) If the restaurant's goal is to put between 1 and 1.2 ounces of ketchup on each burger, what is the probability of achieving this goal on a randomly selected burger?
    (b) Suppose that the manager adjusts the machine's settings so that the mean amount of ketchup dispensed is 1.1 ounces. How much does the machine's standard deviation have to be reduced to ensure that at least 99% of the restaurant's burgers have between 1 and 1.2 ounces of ketchup on them?

Cumulative Review Chapters 4–6

This cumulative review is designed to help you review the important concepts and skills from Chapters 4, 5, and 6, and to prepare you for an in-class exam on this material.

Section I: Multiple Choice *Select the best answer for each question.*

1. An ecologist studying starfish populations records values for each of the following variables from randomly selected plots on a rocky coastline.

    X = The number of starfish in the plot
    Y = The total weight of starfish in the plot
    W = The percentage of area in the plot that is covered by barnacles (a popular food for starfish)

    How many of these are continuous random variables and how many are discrete random variables?

    (a) Three continuous
    (b) Two continuous, one discrete
    (c) One continuous, two discrete
    (d) Three discrete

2. A study of voting chose 663 registered voters at random shortly after an election. Of these, 72% said they had voted in the election. Election records show that only 56% of registered voters voted in the election. Which of the following statements is true about these percentages?

    (a) 72% and 56% are both statistics.
    (b) 72% is a statistic and 56% is a parameter.
    (c) 72% is a parameter and 56% is a statistic.
    (d) 72% and 56% are both parameters.

3. In the town of Tower Hill, the number of cell phones in a randomly selected household is a random variable C with the following probability distribution:

Number of cell phones	0	1	2	3	4	5
Probability	0.1	0.1	0.25	0.3	0.2	???

    What is the probability that a randomly selected household has either 0 cell phones or 5 cell phones?

    (a) 0.005 (b) 0.05 (c) 0.10 (d) 0.15

4. Based on the normal probability plot of heights shown here, what is the best description of the shape of the distribution of height?

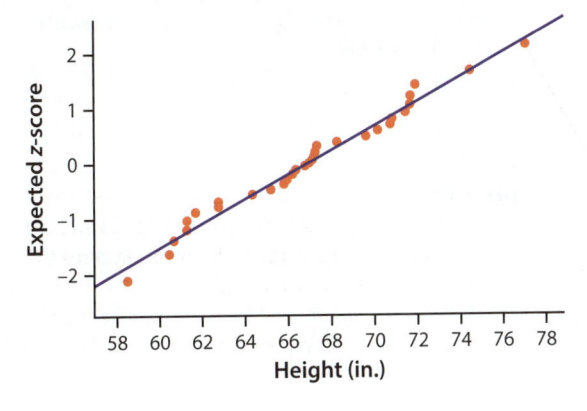

    (a) Skewed left
    (b) Symmetric but not approximately normal
    (c) Approximately normal
    (d) Exactly normal

5. A spinner has three equally sized regions: blue, red, and green. Jonny spins the spinner 3 times and gets 3 blues in a row. If he spins the spinner 297 more times, how many more blues is he most likely to get?

    (a) 97 (b) 99 (c) 100 (d) 101

6. The alumni department at a small college has 5 employees. The ages of these employees are 23, 34, 37, 42, and 58. Suppose you select a random sample of 4 employees and calculate the sample minimum age. Which of the following shows the sampling distribution of the sample minimum age?

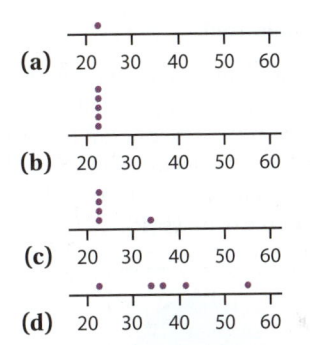

7. Suppose that a student is randomly selected from a large university. The probability that the student is a graduate student is 0.22. The probability that the student is older than age 25 is 0.30. If the probability that the student is older than age 25 or a graduate student is 0.36, what is the probability that the student is older than age 25 and a graduate student?

    (a) 0.060 (b) 0.066 (c) 0.080 (d) 0.160

8. The density curve shown models the probability distribution of a continuous random variable that is equally likely to take any value in the interval from 0 to 2. What is the probability of randomly selecting a value between 0.5 and 1.2?

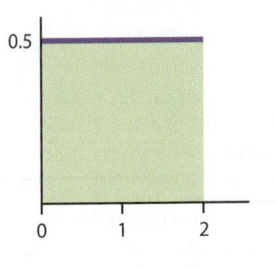

    (a) 0.25 (b) 0.35 (c) 0.50 (d) 0.70

9. When three-digit telephone area codes were first put in place in the United States and Canada, the first number could be any digit from 2 through 9, the second number could be only 0 or 1, and the third number could be any digit from 0 through 9. How many possible area codes can be created with this format?

    (a) 24
    (b) 126
    (c) 140
    (d) 160

10. A school has 25 students in each of grades 1 through 12. The principal randomly selects 1 student from each grade and records whether or not the student is more than 5 feet tall. Let X = the number of students in the sample who are more than 5 feet tall. Which of the following requirements for a binomial setting is *violated* in this case?

    (a) There are two possible outcomes for each trial.
    (b) The trials are independent.
    (c) The probability of success is the same for each trial.
    (d) The number of trials is fixed.

11. The Venn diagram displays the sample space for the chance process of randomly selecting a faculty member from a college and recording whether or not the faculty member regularly uses the internet as a source of news (I) and whether or not the faculty member regularly uses print media as a source of news (P).

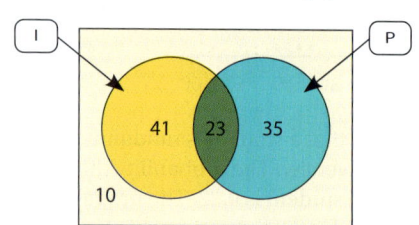

    Which of the following two-way tables conveys the same information?

    (a)
	Print news		
	P	P^C	Total
Internet news — I	23	17	40
Internet news — I^C	35	10	45
Total	58	27	85

    (b)
	Print news		
	P	P^C	Total
Internet news — I	23	41	64
Internet news — I^C	35	10	45
Total	58	51	109

    (c)
	Print news		
	P	P^C	Total
Internet news — I	10	41	51
Internet news — I^C	35	23	58
Total	45	64	109

    (d)
	Print news		
	P	P^C	Total
Internet news — I	23	18	41
Internet news — I^C	41	10	51
Total	64	28	92

12. Suppose you roll a fair 6-sided die 8 times. What is the probability of getting at least one six?

    (a) 0.167
    (b) 0.233
    (c) 0.767
    (d) 0.833

13. For the normal distribution shown here,

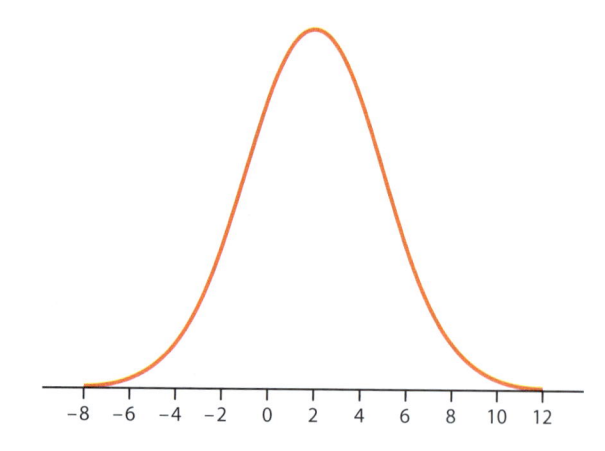

    the standard deviation is closest to:

    (a) 1.
    (b) 2.
    (c) 3.
    (d) 6.

14. On weekdays, the average number of emails received by a marketing manager is 143 per day. Use a Poisson distribution to calculate the probability that the marketing manager gets at most 125 emails on a randomly selected weekday.

    (a) 0.01
    (b) 0.07
    (c) 0.87
    (d) 0.95

15. When packing for a business trip, Chien chooses 3 of his 10 dress shirts and 4 of his 12 ties. How many different shirt and tie combinations can Chien pack?

   (a) 12
   (b) 120
   (c) 59,400
   (d) 8,553,600

16. The figure shows the probability distribution of a discrete random variable X. Which of the following best describes this random variable?

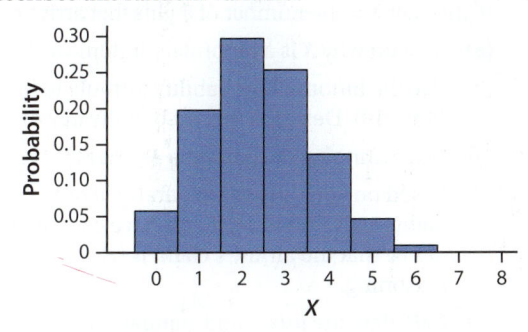

   (a) Binomial with $n = 8$, $p = 0.1$
   (b) Binomial with $n = 8$, $p = 0.3$
   (c) Binomial with $n = 8$, $p = 0.8$
   (d) Normal with mean 2, standard deviation 1

17. A species of cockroach has weights that are approximately normally distributed with a mean of 50 grams. After measuring the weights of many of these cockroaches, a lab assistant reports that 14% of the cockroaches weigh more than 55 grams. Based on this report, what is the approximate standard deviation of weight for this species of cockroaches?

   (a) 4.6
   (b) 5.0
   (c) 6.2
   (d) 14.0

18. How do people express laughter on Reddit? Researchers studied millions of posts on Reddit and isolated comments that used a textual form of laughter such as "lol" or "haha" (emojis weren't included). Of the comments with a textual form of laughter, 55.8% used some form of "lol."[40] If you randomly select 300 comments on Reddit that use a textual form of laughter, what is the probability that fewer than half use a variation of "lol"?

   (a) 0.02
   (b) 0.04
   (c) 0.06
   (d) 0.08

Questions 19 and 20 refer to the following scenario.
Debabu is a fine basketball player, but his free-throw shooting could use some work. For the past three seasons, he has made only 56% of his free throws. His coach sends him to a summer clinic to work on his shot, and when he returns, his coach has him step to the free-throw line and take 50 shots. He makes 34 shots. Is this result convincing evidence that Debabu's shooting has improved? We want to perform a simulation to estimate the probability that a 56% free-throw shooter would make 34 or more in a sample of 50 shots.

19. Which of the following is a correct way to perform the simulation?

   (a) Let integers from 1 to 34 represent making a free throw and integers from 35 to 50 represent missing a free throw. Generate 50 random integers from 1 to 50. Count the number of made free throws. Repeat this process many times.

   (b) Let integers from 1 to 34 represent making a free throw and integers from 35 to 50 represent missing a free throw. Generate 50 random integers from 1 to 50 with no repeats allowed. Count the number of made free throws. Repeat this process many times.

   (c) Let integers from 1 to 56 represent making a free throw and integers from 57 to 100 represent missing a free throw. Generate 50 random integers from 1 to 100. Count the number of made free throws. Repeat this process many times.

   (d) Let integers from 1 to 56 represent making a free throw and integers from 57 to 100 represent missing a free throw. Generate 50 random integers from 1 to 100 with no repeats allowed. Count the number of made free throws. Repeat this process many times.

20. The dotplot displays the number of made shots in 100 simulated sets of 50 free throws by someone with 0.56 probability of making a free throw.

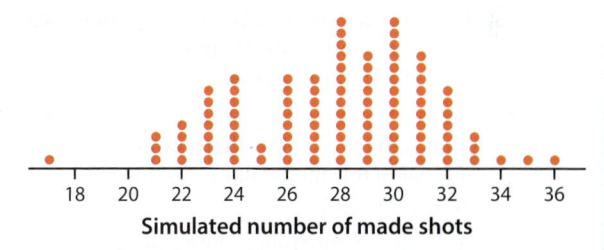

   Which of the following is an appropriate statement about Debabu's free-throw shooting, based on this dotplot?

   (a) If Debabu were still only a 56% shooter, the probability that he would make at least 34 of his shots is about 0.03.

   (b) If Debabu were still only a 56% shooter, the probability that he would make at least 34 of his shots is about 0.97.

   (c) If Debabu is now shooting better than 56%, the probability that he would make at least 34 of his shots is about 0.03.

   (d) If Debabu is now shooting better than 56%, the probability that he would make at least 34 of his shots is about 0.97.

Section II: Free Response

1. Prior to the lockdowns and social distancing associated with the Covid-19 pandemic, about 44% of all reported injuries in U.S. households occurred at home. With many more people working from home during the pandemic, researchers surveyed 2009 U.S. households in June 2020, roughly 3 months after lockdowns began.[41] Researchers asked whether anyone in the household had an injury that occurred at home or ingested something that could make them sick while at home. They also classified the households as urban, suburban, or rural. Here are the results:

		Type of incident			
		Both	Injury or ingestion, but not both	Neither	Total
Type of household	Urban	171	83	404	658
	Suburban	80	137	750	967
	Rural	18	81	285	384
	Total	269	301	1439	2009

Randomly select one of the 2009 households surveyed.

   (a) What is the probability that the selected household had at least one incident?
   (b) Given that the selected household is classified as urban, what is the probability that the household had both types of incidents?
   (c) Are the events "selecting an urban household" and "selecting a household with both types of incidents" independent? Justify your answer.

2. According to the truecar.com website, the probability distribution of X = asking price of a randomly selected new car of a certain model is approximately normal with mean \$35,987 and standard deviation \$607.50.[42]

   (a) The factory invoice price for this model is \$35,625. What is the probability that a randomly selected new car of this model has an asking price below factory invoice?
   (b) According to the truecar.com site, any asking price between \$34,772 and \$36,225 for this car model is considered "good" or "great." What is the probability that a randomly selected new car of this model has a "good" or "great" asking price?
   (c) The manufacturer's suggested retail price (MSRP) for this car model is at the 98th percentile of the probability distribution of asking price. Find the MSRP.

3. Let Y denote the number of broken eggs in a randomly selected carton of one dozen "store-brand" eggs at a local supermarket. The probability distribution of Y is as follows:

Number of broken eggs	0	1	2	3	4
Probability	0.77	0.11	0.08	0.03	0.01

   (a) What is the probability that at least 2 eggs in a randomly selected carton are broken?
   (b) Make a histogram of the probability distribution. Describe its shape.
   (c) Calculate and interpret the mean of Y.
   (d) Calculate and interpret the standard deviation of Y.

4. An airline claims that its 7:00 a.m. New York to Los Angeles flight has an 85% chance of arriving on time on a randomly selected day. Assume for now that this claim is true. Suppose we select a random sample of 20 of these flights. Let X = the number of flights that arrive on time.

   (a) Explain why X is a binomial random variable.
   (b) Use the binomial probability formula to find $P(X = 19)$. Describe this probability in words.
   (c) Use technology to compute $P(X \leq 14)$.
   (d) Based on your answer to part (c), if only 14 of the 20 flights arrive on time, do we have convincing evidence that the airline's claim is false? Explain your reasoning.
   (e) Calculate the mean and standard deviation of X.
   (f) Would it be reasonable to use the normal approximation to the binomial distribution to calculate the probability in part (c)? Justify your answer.

5. Three machines—A, B, and C—are used to produce a large quantity of identical parts at a factory. Machine A produces 60% of the parts, while Machines B and C produce 30% and 10% of the parts, respectively. Historical records indicate that 1% of the parts produced by Machine A are defective, compared with 3% for Machine B and 4% for Machine C. Suppose we randomly select a part produced by one of these three machines.

   (a) Draw a tree diagram to represent this chance process.
   (b) What is the probability that the part is defective?
   (c) If the part is inspected and found to be defective, what is the probability that it was produced by Machine C?

6. The amount that households pay service providers for access to the internet varies quite a bit, but the mean monthly fee is \$50 and the standard deviation is \$20.[43] The distribution is not normal: Many households pay a lower rate for slower access, but some pay much more for faster connections. A sample survey asks an SRS of 60 households with internet access how much they pay per month. Let $\bar{x}$ be the mean amount paid by the members of the sample.

   (a) Explain what it means that $\bar{x}$ is an unbiased estimator of μ.
   (b) Calculate the mean and standard deviation of the sampling distribution of $\bar{x}$. Interpret the standard deviation.
   (c) What is the shape of the sampling distribution of $\bar{x}$? Explain your answer.
   (d) Find the probability that the average amount paid by the sample of households is less than \$45.

7 Estimating a Parameter

Section 7.1	The Idea of a Confidence Interval	Section 7.7	Estimating a Population Standard Deviation or Variance
Section 7.2	Factors That Affect the Margin of Error		
Section 7.3	Estimating a Population Proportion	Section 7.8	Confidence Intervals for a Population Standard Deviation or Variance
Section 7.4	Confidence Intervals for a Population Proportion		
		Chapter 7 Review	
Section 7.5	Estimating a Population Mean	Chapter 7 Review Exercises	
Section 7.6	Confidence Intervals for a Population Mean	Chapter 7 Project	

▌▘ Statistics Matters How can I prevent credit card fraud?

Fraud warning!
Your phone buzzes with a fraud alert from your credit card company. Someone has charged your card $512 for building supplies in a different state. When you check your wallet, you discover that you still have the card. The thief must have stolen your credit card number and used it to make the purchase. Fortunately, you had asked your card company to notify you whenever a charge occurs without the card being physically present.

For credit card holders and the banks that issue them, time is of the essence when fraud happens. The banks need to cancel stolen card numbers as soon as possible to prevent additional fraudulent charges—and to reduce customer stress. Knowing this, many banks try hard to minimize the time required for customers to connect to a service representative.

A large bank decided to study the call response times in its credit card fraud department. The bank's goal was to have a representative answer an incoming call in less than 30 seconds. The histogram shows the distribution of response time in a random sample of 241 calls to the bank's credit card fraud department.[1] We can use the data in the sample to estimate the mean response time and the proportion of calls that are answered in less than 30 seconds. But how confident should we be in these estimates?

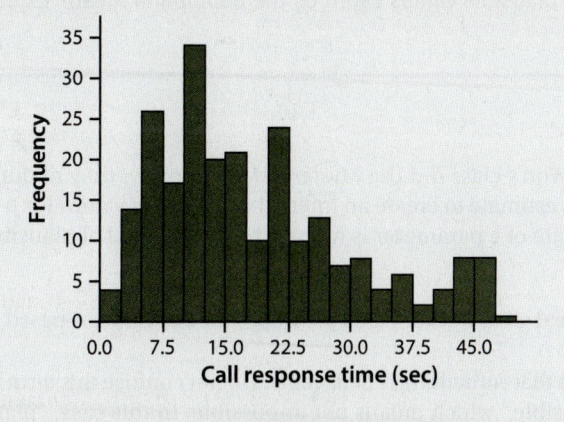

We'll revisit Statistics Matters *at the end of the chapter, so you can use what you have learned to help answer this question.*

CHAPTER 7   Estimating a Parameter

Chapter 7 begins the formal study of *statistical inference*—using information from a sample to draw conclusions about a population parameter, such as the population proportion p or the population mean μ. This is an important transition from Chapter 6, where you were given information about a population and asked questions about the distribution of a sample statistic, such as the sample proportion $\hat{p}$ or the sample mean $\bar{x}$.

Section 7.1 The Idea of a Confidence Interval

LEARNING GOALS

By the end of this section, you will be able to:
- Interpret a confidence interval.
- Determine the point estimate and margin of error from a confidence interval.
- Use confidence intervals to make decisions.

How long does a battery last on the newest iPhone, on average? What proportion of college students have been vaccinated for Covid-19? How much does income vary among residents in a large city?

It wouldn't be practical to determine the lifetime of *every* iPhone battery or to ask *all* college students about their vaccine status or to survey *every* resident in the city. Instead, we choose a random sample of individuals (iPhone batteries, undergraduates, residents) to represent the population and use the sample data to estimate the value of a population parameter. The following activity illustrates this process.

Class Activity

What's the "Mystery Mean"?

In this activity, each team of three or four students will use data from a random sample to estimate the mean μ of a normally distributed population with a standard deviation of $\sigma = 20$.[2]

1. Before class, your instructor chose a value of μ and used technology to select a random sample of $n = 16$ values from a normal population with mean μ and standard deviation $\sigma = 20$. Your instructor will now give you the mean $\bar{x}$ of these 16 values.
2. How confident are you that the sample mean $\bar{x}$ provided by your instructor is equal to the population mean μ? Explain your answer.
3. How much will the sample mean typically vary from the population mean in SRSs of size $n = 16$ from a population with $\sigma = 20$? *Hint:* You'll need to use what you learned in Chapter 6.
4. Use your answer from Question 3 to create an interval of *plausible* values for the population mean μ.
5. How confident are you that your interval of plausible values captures the population mean? Explain your reasoning.
6. Share your team's results with the class.

When Professor Woo's class did the *Mystery Mean* activity, they obtained a sample mean of $\bar{x} = 23.8$ and used this estimate to create an interval of plausible values for μ that went from 13.8 to 33.8. When the estimate of a parameter is reported as an interval of plausible values, it is called a **confidence interval.**

A **confidence interval** gives a set of plausible values for a parameter based on sample data.

"Plausible" means that something is believable. Do not confuse this term with "probable," which means likely, or "possible," which means not impossible. In this case, "plausible" means that we shouldn't be surprised if any one of the values in the interval is equal to the value of the parameter. We use an interval of plausible values rather than a single-value estimate to increase our confidence that we have a correct value for the parameter.

Interpreting Confidence Intervals

We construct confidence intervals so that we know *how much* confidence we have that the interval successfully captures the parameter. The most commonly used **confidence level** is 95%. You will learn more about how to interpret confidence levels in Section 7.2.

> The **confidence level** C gives the long-run capture rate of confidence intervals calculated with C% confidence.

The Gallup polling organization recently asked a random sample of U.S. adults about the most important financial problem facing their family today. The highest percentage of respondents (14%) selected "health care costs."[3] Based on this sample, the 95% confidence interval for the population proportion is 0.10 to 0.18. That is, we are 95% confident that the interval from 0.10 to 0.18 captures the proportion of *all* U.S. adults who would say that health care costs are the biggest financial problem facing their family today.

HOW TO Interpret a Confidence Interval

To interpret a C% confidence interval for an unknown parameter, say, "We are C% confident that the interval from _____ to _____ captures the [parameter in context]."

⚠ **When interpreting a confidence interval, make sure that you are describing the parameter and not the statistic.** It's wrong to say that we are 95% confident that the interval from 0.10 to 0.18 captures the proportion of U.S. adults who *said* health care costs were the biggest financial problem facing their family today. The "proportion who *said* health care costs" is the sample proportion, which is known to be 0.14. The interval gives plausible values for the proportion of *all* U.S. adults who *would say* health care costs if asked. Recall from Chapter 6 that parameter descriptions should include the word *all*, *true*, or *population*.

EXAMPLE 7.1.1

What's your blood pressure?

Interpreting a confidence interval

PROBLEM

The National Health and Nutrition Examination Survey (NHANES) is an ongoing research program conducted by the National Center for Health Statistics. Researchers selected a random sample of more than 8000 U.S. residents for study in the most recent survey. One of the many variables they recorded was blood pressure. Based on the sample, a 95% confidence interval for the mean systolic blood pressure of U.S. residents age 8 and older is 121.1 to 122.1 millimeters of mercury (mmHg).[4] Interpret this confidence interval.

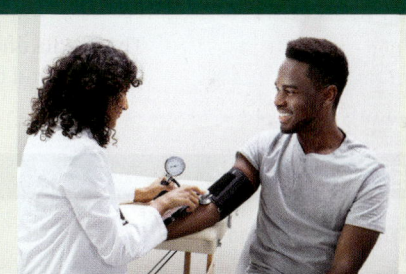

SOLUTION

We are 95% confident that the interval from 121.1 to 122.1 mmHg captures the mean systolic blood pressure of all U.S. residents age 8 and older.

EXAM PREP: FOR PRACTICE, TRY EXERCISE 9.

Sometimes, confidence intervals are reported in the form

(lower boundary, upper boundary)

For example, the confidence interval in the preceding example could be reported as (121.1, 122.1) instead of "121.1 to 122.1."

Another way to interpret a confidence interval is by describing the parameter before giving the endpoints of the interval: "We are 95% confident that the mean systolic blood pressure of all U.S. residents age 8 and older is between 121.1 mmHg and 122.1 mmHg." Either interpretation is correct, as long as it includes a statement about the confidence level, the endpoints of the interval, and a description of the parameter in context.

Building a Confidence Interval

To create an interval of plausible values for a parameter, we typically need two components: a **point estimate** to use as the midpoint of the interval, and a **margin of error** to account for sampling variability.

> A **point estimate** is a single-value estimate of a population parameter.

> The **margin of error** of an estimate describes how far, at most, we expect the point estimate to vary from the population parameter.

The point estimate is our best guess for the value of the parameter based on data from a sample. Unfortunately, the point estimate is unlikely to be correct because of the variability introduced by random sampling. We include the margin of error to account for this sampling variability. In a $C\%$ confidence interval, the distance between the point estimate and the true parameter value will be less than the margin of error in about $C\%$ of all samples.

In general, the structure of a confidence interval is

$$\text{point estimate} \pm \text{margin of error}$$

We can visualize a $C\%$ confidence interval like this:

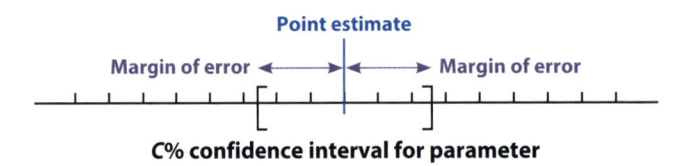

```
                        Point estimate
    Margin of error ←——————→←——————→ Margin of error
   └─┴─┴─┴─┴─┴─┴─[─┴─┴─┴─┴─┴─┴─]─┴─┴─┴─┴─┴─┘
              C% confidence interval for parameter
```

Earlier, you learned that the 95% confidence interval for the proportion of all U.S. adults who would say that health care costs are the biggest financial problem facing their family today is 0.10 to 0.18. This interval could also be expressed as

$$0.14 \pm 0.04$$

The media often presents confidence intervals in a similar way, but with the point estimate in the headline and the margin of error in the fine print.

EXAMPLE 7.1.2

What's your blood pressure?

Determining the point estimate and margin of error

PROBLEM

In Example 7.1.1, you learned that the 95% confidence interval for the mean systolic blood pressure of all U.S. residents age 8 and older is 121.1 to 122.1 mmHg. Calculate the point estimate and margin of error used to create this confidence interval.

SOLUTION

$$\text{point estimate} = \frac{121.1 + 122.1}{2} = 121.6 \text{ mmHg}$$

The point estimate is the midpoint of the interval.

$$\text{margin of error} = 122.1 - 121.6 = 0.5 \text{ mmHg}$$

The margin of error is the distance from the point estimate to the endpoints of the interval. You can calculate the margin of error using the lower endpoint as well: $121.6 - 121.1 = 0.5$ mmHg.

EXAM PREP: FOR PRACTICE, TRY EXERCISE 13.

Using Confidence Intervals to Make Decisions

In addition to estimating the value of a population parameter, we can use confidence intervals to make decisions about a parameter.

EXAMPLE 7.1.3

What's your blood pressure?

Using confidence intervals to make decisions

PROBLEM

In the preceding examples, you learned that the 95% confidence interval for the mean systolic blood pressure of all U.S. residents age 8 and older is 121.1 to 122.1 mmHg. According to the Centers for Disease Control and Prevention, "normal" systolic blood pressure is less than 120 mmHg.[5] Based on this interval, is there convincing evidence that the mean systolic blood pressure of all U.S. residents age 8 and older is greater than 120 mmHg? Explain your reasoning.

SOLUTION

Because all of the plausible values for the population mean (121.1 to 122.1) are greater than 120, there is convincing evidence that the mean systolic blood pressure of all U.S. residents age 8 and older is greater than 120 mmHg.

> Even though all the values in the interval are greater than 120, it is certainly possible that some individuals have a systolic blood pressure less than 120. The confidence interval gives plausible values for only the *mean* systolic blood pressure.

EXAM PREP: FOR PRACTICE, TRY EXERCISE 17.

What if the confidence interval was 119.1 mmHg to 124.1 mmHg instead? Because this interval includes some values less than 120, there would *not* be convincing evidence that the mean systolic blood pressure of all U.S. residents age 8 and older is greater than 120 mmHg. It is plausible that the mean systolic blood pressure could be as low as 119.1 mmHg.

Section 7.1 What Did You Learn?

Review the learning goals from this section. Then practice what you've learned by working through the exercises.

Learning Goal	Example	Exercises
Interpret a confidence interval.	7.1.1	9–12
Determine the point estimate and margin of error from a confidence interval.	7.1.2	13–16
Use confidence intervals to make decisions.	7.1.3	17–20

Section 7.1 Exercises

Building Concepts and Skills These exercises assess the basic knowledge you should have after reading the section.

1. In this chapter, we use information from a(n) _____ to draw conclusions about a(n) _____ parameter. In Chapter 6, we used information about a(n) _____ to answer questions about a(n) _____ statistic.

2. True/False: A confidence interval gives all the possible values for a parameter.

3. True/False: The confidence level gives the long-run capture rate of intervals calculated using that confidence level.

4. How do you interpret a confidence interval?

5. The structure of a confidence interval is _____ ± _____.

6. What is a point estimate?

7. Why do we include a margin of error when estimating a population parameter?

8. True/False: If a proposed value for a parameter is not captured in a confidence interval for that parameter, there is convincing evidence that the proposed value is incorrect.

Mastering Concepts and Skills These exercises reinforce the learning goals as shown in the examples.

9. **Refer to Example 7.1.1 Atmospheric Gases** The Pew Research Center and *Smithsonian* magazine recently quizzed a random sample of 1006 U.S. adults on their knowledge of science.[6] One of the questions asked, "Which gas makes up most of the Earth's atmosphere: hydrogen, nitrogen, carbon dioxide, or oxygen?" A 95% confidence interval for the proportion who would correctly answer nitrogen is 0.175 to 0.225. Interpret this confidence interval.

10. **Control Groups** The Pew Research Center and *Smithsonian* magazine recently quizzed a random sample of 1006 U.S. adults on their knowledge of science.[7] One of the questions asked, "Which is the better way to determine whether a new drug is effective in treating a disease? If a scientist has a group of 1000 volunteers with the disease to study, should she (a) give the drug to all of them and see how many get better or (b) give the drug to half of them and not to the other half, and compare how many in each group get better?" A 95% confidence interval for the proportion who would correctly answer (b) is 0.723 to 0.777. Interpret this confidence interval.

11. **Flint Water** In Chapters 2 and 3, you read about the dangerous lead levels in the drinking water in Flint, Michigan. Using the data collected by the Michigan Department of Environmental Quality in 2015, the 99% confidence interval for the mean lead level in Flint tap water in that year is 2.8 to 11.8 parts per billion (ppb).[8] Interpret this confidence interval.

12. **Median Income** According to the U.S. Census Bureau, the 90% confidence interval for Alabama's median household income in 2019 is $51,134 to $52,334.[9] Interpret this confidence interval.

13. **Refer to Example 7.1.2 Atmospheric Gases** Calculate the point estimate and margin of error used to create the confidence interval in Exercise 9.

14. **Control Groups** Calculate the point estimate and margin of error used to create the confidence interval in Exercise 10.

15. **Flint Water** Calculate the point estimate and margin of error used to create the confidence interval in Exercise 11.

16. **Median Income** Calculate the point estimate and margin of error used to create the confidence interval in Exercise 12.

17. **Refer to Example 7.1.3 Atmospheric Gases** Refer to Exercise 9. If people guess one of the four choices at random, about 25% should get the answer correct. Does the interval in Exercise 9 provide convincing evidence that fewer than 25% of all U.S. adults would answer this question correctly? Explain your reasoning.

18. **Control Groups** Refer to Exercise 10. If people guess one of the two choices at random, about 50% should get the answer correct. Does the interval in Exercise 10 provide convincing evidence that more than 50% of all U.S. adults would answer this question correctly? Explain your reasoning.

19. **Flint Water** Does the interval in Exercise 11 provide convincing evidence that the mean lead level in Flint tap water exceeds 5 ppb, the threshold for safe drinking water established by the Food and Drug Administration?[10] Explain your reasoning.

20. **Median Income** Does the interval in Exercise 12 provide convincing evidence that Alabama's median household income in 2019 was greater than $50,000? Explain your reasoning.

Applying the Concepts These exercises ask you to apply multiple learning goals in a new context or to apply what you learned in this section in a new way.

21. **International Travel** A recent survey from Morning Consult asked a random sample of U.S. adults if they had ever traveled internationally. Based on the sample, the 95% confidence interval for the true proportion of U.S. adults who have traveled internationally is 0.436 to 0.478.[11]
    (a) Interpret the confidence interval.
    (b) Calculate the point estimate and margin of error used to create this confidence interval.
    (c) Based on the confidence interval, is it plausible that a majority of U.S. adults have traveled internationally? Explain your reasoning.

22. **New Year's Resolutions** At the end of 2019, Morning Consult surveyed a random sample of U.S. adults about their New Year's resolutions. Among the 655 people in the sample who made a resolution, 402 said they had kept their resolution. The 95% confidence interval for the proportion of all resolution-making U.S. adults who kept their resolution is 0.576 to 0.651.[12]
    (a) Interpret the confidence interval.
    (b) Calculate the point estimate and margin of error used to create this confidence interval.
    (c) Based on the confidence interval, is there convincing evidence that a majority of U.S. resolution makers kept their resolutions? Explain your reasoning.

23. **Bottling Cola** A particular type of cola advertises that individual cans contain 12 ounces of the beverage. Each hour, a supervisor selects 10 cans at random, measures their contents, and computes a 95% confidence interval for the true mean volume. For one particular hour, the 95% confidence interval is 11.97 ounces to 12.05 ounces.
    (a) Does the confidence interval provide convincing evidence that the true mean volume is different than 12 ounces? Explain your reasoning.
    (b) Does the confidence interval provide convincing evidence that the true mean volume is 12 ounces? Explain your reasoning.

24. **Fun Size Candy** A candy bar manufacturer sells a "fun size" version that is advertised to weigh 17 grams. A hungry researcher selected a random sample of 44 fun-size bars and found a 95% confidence interval for the true mean weight to be 16.945 grams to 17.395 grams.
    (a) Does the confidence interval provide convincing evidence that the true mean weight is different than 17 grams? Explain your reasoning.
    (b) Does the confidence interval provide convincing evidence that the true mean weight is 17 grams? Explain your reasoning.

Extending the Concepts These exercises challenge you to explore statistical concepts and methods that go beyond what you learned in this section.

25. **International Travel** Refer to Exercise 21. The proportion of respondents in the sample who had traveled internationally is 0.457. How confident are you that the proportion of all U.S. adults who have traveled internationally is 0.457? Explain your answer.

26. **Crabby Crabs** Biologists worried about the effect of noise pollution on crabs selected a sample of 34 shore crabs and randomly divided them into two groups. They exposed one group to 7.5 minutes of ship noise and exposed the other group to 7.5 minutes of ambient harbor noise. Because animals under stress typically increase their oxygen consumption, the researchers measured the amount of oxygen consumption for these two groups of crabs. A 95% confidence interval for the *difference* in mean oxygen consumption (Ship noise − Ambient noise) is 33.3 to 101.8 μmoles per hour.[13]

    (a) Interpret the confidence interval.

    (b) Does the confidence interval give convincing evidence that crabs have a greater mean oxygen consumption in the presence of ship noise? Explain your reasoning.

Cumulative Review These exercises revisit what you learned in previous sections.

27. **NBA Player Heights (3.3)** The Golden State Warriors of the National Basketball Association (NBA) have won several NBA Championships in recent years, including 2015, 2017, 2018, and 2021. They also won the NBA Championship in 1975. Here are the heights (in inches) of all the players on the Warriors and their opponents in the NBA finals in 1975 and 2015.[14]

1975	72	72	73	74	75	75	75	75
	76	76	76	76	76	77	78	78
	78	79	79	79	79	79	80	81
	81	83						
2015	75	75	75	76	77	78	78	78
	78	78	79	79	79	79	80	80
	80	80	80	81	81	82	82	82
	82	83	84	84	85	85		

    Draw parallel boxplots using the data from each year. Use the plots to compare the distribution of height for the two years.

28. **Flossing (1.2, 1.3)** A dentist wants to estimate the proportion of U.S. adults who floss their teeth regularly, so she surveys 100 randomly selected patients. When asked directly by the dentist, 83% of the patients in the sample claim to floss their teeth regularly.

    (a) Explain how undercoverage might affect the estimate.

    (b) Explain how response bias might affect the estimate.

    (c) Would increasing the sample size to 200 patients address the issues in (a) or (b)? Explain your reasoning.

Section 7.2 Factors That Affect the Margin of Error

LEARNING GOALS

By the end of this section, you will be able to:

- Interpret a confidence level.
- Describe how the confidence level and sample size affect the margin of error.
- Explain how practical issues like nonresponse, undercoverage, and response bias can affect the interpretation of a confidence interval.

In Section 7.1, you learned that the confidence level gives the long-run capture rate of the method used to calculate a confidence interval. The following Concept Exploration will help you investigate the idea of confidence level in more depth.

CONCEPT EXPLORATION

Investigating confidence level with an applet

In this exploration, you will use an applet to learn what it means to say that we are "95% confident" that our confidence interval captures the parameter value.

1. Go to www.stapplet.com and launch the *Simulating Confidence Intervals* applet.
2. In the box labeled "Population," keep the default setting for the normal population distribution.

(continued)

3. In the box labeled "Sample," keep the default settings for sample size ($n = 30$), confidence level (95%), and calculation method (t distribution, which you'll learn about in Section 7.5). Click the Go! button. The applet will display the distribution of the sample and calculate the sample mean, sample standard deviation, and endpoints of the 95% confidence interval. Does the interval include the population mean, $\mu = 10$?

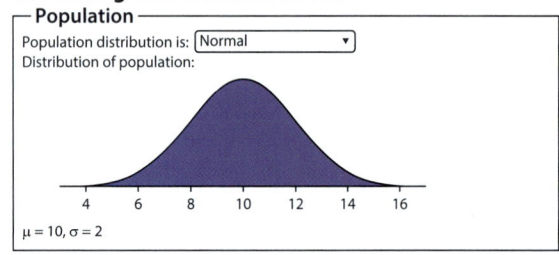

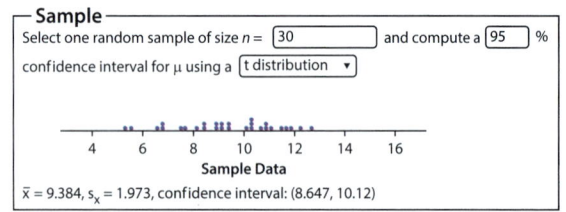

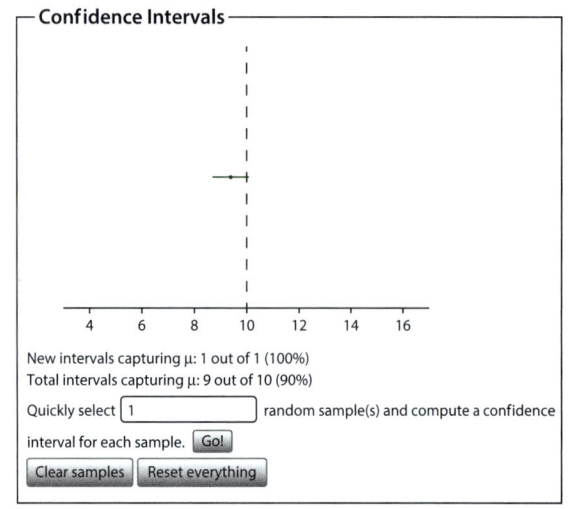

Simulating Confidence Intervals

Population
Population distribution is: Normal
Distribution of population:

[Normal distribution plot centered at 10, range 4 to 16]
$\mu = 10$, $\sigma = 2$

Sample
Select one random sample of size $n = 30$ and compute a 95 % confidence interval for μ using a t distribution

[Sample Data dotplot, range 4 to 16]
Sample Data
$\bar{x} = 9.384$, $s_x = 1.973$, confidence interval: (8.647, 10.12)

Confidence Intervals

[Single confidence interval displayed]

New intervals capturing μ: 1 out of 1 (100%)
Total intervals capturing μ: 9 out of 10 (90%)
Quickly select 1 random sample(s) and compute a confidence interval for each sample. Go!
Clear samples   Reset everything

4. In the box labeled "Confidence Intervals," notice that the confidence interval is displayed as a horizontal line segment, along with a vertical line segment at $\mu = 10$. If the interval includes $\mu = 10$, the interval is green. Otherwise, it is red. (Note that you can hover over the interval with your mouse to see the endpoints.) Click the Go! button 9 times to generate 9 additional random samples and their corresponding 95% confidence intervals. Look in the lower-left corner to determine how many of the intervals captured $\mu = 10$.

5. Change the number of intervals to 50 and click the Go! button many times to generate at least 1000 confidence intervals. What do you notice about the total percentage of the intervals that capture μ?

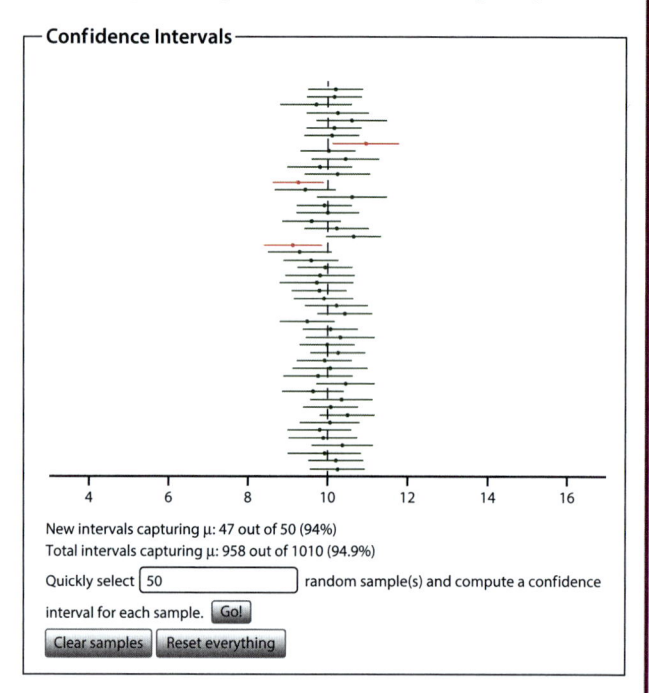

Confidence Intervals

[Plot of ~50 confidence intervals, range 4 to 16]

New intervals capturing μ: 47 out of 50 (94%)
Total intervals capturing μ: 958 out of 1010 (94.9%)
Quickly select 50 random sample(s) and compute a confidence interval for each sample. Go!
Clear samples   Reset everything

6. Click the button to Clear samples. Then repeat Step 5 using a 99% confidence level.

7. Click the button to Clear samples. Then repeat Step 5 using an 80% confidence level.

8. Summarize what you have learned about the relationship between confidence level and capture rate after taking many samples.

We will investigate the effect of changing the sample size later.

Interpreting Confidence Level

In the preceding section, we claimed to be "95% confident that the interval from 0.10 to 0.18 captures the proportion of all U.S. adults who would say that health care costs are the biggest financial problem facing their family today."[15] The claim of "95% confidence" means that we are using a method that successfully captures the parameter in 95% of all possible samples. That is, if we were to select many random samples and construct a 95% confidence interval using each sample, about 95% of those intervals would capture p, the proportion of all U.S. adults who would answer "health care costs."

HOW TO Interpret a Confidence Level

To interpret the confidence level C, say, "If we were to select many random samples from a population and construct a C% confidence interval using each sample, about C% of the intervals would capture the [parameter in context]."

⚠️ **Don't interpret the confidence *interval* when you are asked to interpret a confidence *level* — and vice versa.** Confidence levels describe the process used to make an interval, not any one particular interval.

EXAMPLE 7.2.1

Do you eat fast food?

Interpreting confidence level

PROBLEM

The National Center for Health Statistics (NCHS) asked a random sample of 3626 U.S. adults about a variety of topics. According to the study, the 95% confidence interval for the true proportion of adults who eat fast food on a given day is 0.350 to 0.382.[16] Interpret the confidence level.

SOLUTION

If the NCHS selected many random samples of U.S. adults and constructed a 95% confidence interval using each sample, about 95% of these intervals would capture the true proportion of U.S. adults who eat fast food on a given day.

EXAM PREP: FOR PRACTICE, TRY EXERCISE 7.

⚠️ **The confidence level does *not* tell us the probability that a particular confidence interval captures the population parameter.** Once a particular confidence interval is calculated, its endpoints are fixed. And because the value of a parameter is also a constant, a particular confidence interval either includes the parameter (probability = 1) or doesn't include the parameter (probability = 0). To illustrate, we simulated 50 random samples of size $n = 30$ from a population where $\mu = 10$ and calculated a 95% confidence interval for μ with each sample. As seen in Figure 7.1, no individual 95% confidence interval has a 95% probability of capturing the true parameter value.

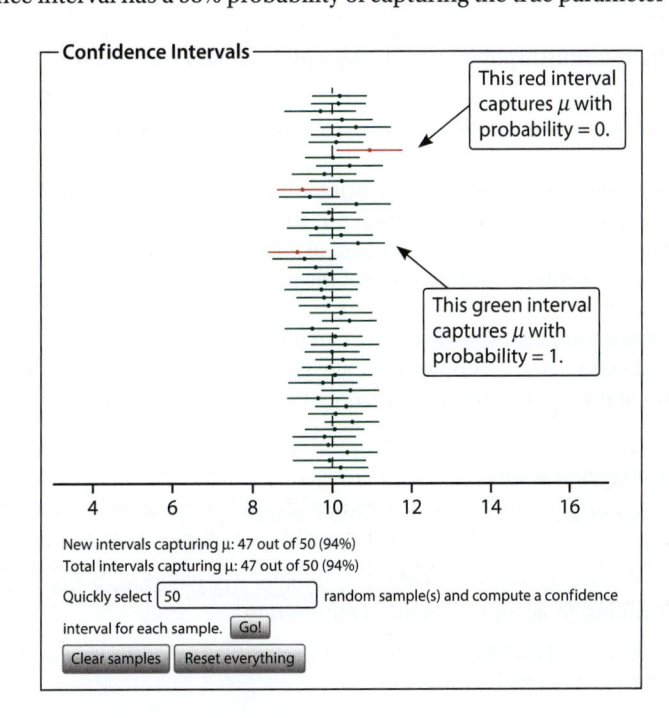

FIGURE 7.1 Confidence intervals from 50 simulated random samples showing that the probability a particular 95% confidence interval captures the true parameter value is either 0 (red intervals) or 1 (green intervals) and not 0.95.

Factors That Affect the Margin of Error

In general, we prefer narrow confidence intervals — that is, confidence intervals with a small margin of error. After all, knowing that tomorrow's high temperature will be between −40°F and 200°F won't help you pick out your clothes in the morning! But knowing that tomorrow's high temperature will be between 80°F and 82°F would be quite useful. To reduce the margin of error in a confidence interval, we can change two factors: the sample size and the confidence level. In the following Concept Exploration, you will investigate how each of these factors affects the margin of error.

CONCEPT EXPLORATION

Exploring margin of error with an applet

In this exploration, you will use the applet from the preceding Concept Exploration to investigate the relationships among the confidence level, the sample size, and the margin of error.

1. Go to www.stapplet.com and launch the *Simulating Confidence Intervals* applet.

Part 1: Adjusting the Confidence Level

2. In the box labeled "Population," keep the default setting for the normal population distribution.
3. In the box labeled "Sample," keep the default settings for sample size ($n = 30$), confidence level (95%), and calculation method (t distribution). Click the Go! button. The applet will display the distribution of the sample and calculate the sample mean, sample standard deviation, and endpoints of the 95% confidence interval.
4. In the box labeled "Confidence Intervals," change the number of intervals to 50, click the Go! button, and note the width of the intervals.
5. In the box labeled "Sample," increase the confidence level gradually by clicking the up arrow until the confidence level is 99%. What happens to the width of the intervals?
6. Now click the down arrow to reduce the confidence level to 80%. What happens to the width of the intervals?
7. Summarize what you learned about the relationship between the confidence level and the margin of error for a fixed sample size.

Part 2: Adjusting the Sample Size

8. Click the Reset everything button, change the confidence level back to 95% in the box labeled "Sample," and click the Go! button. Then quickly generate 50 confidence intervals in the box labeled "Confidence Intervals."
9. In the box labeled "Sample," decrease the sample size to $n = 10$. Then quickly generate 50 confidence intervals in the box labeled "Confidence Intervals." What happens to the width of the confidence intervals?
10. Repeat Step 9 with samples of size $n = 100$.
11. Summarize what you learned about the relationship between the sample size and the margin of error for a fixed confidence level.

As the Concept Exploration illustrates, the price we pay for greater confidence is a wider interval. If we're satisfied with 80% confidence, then our interval of plausible values for the parameter will be much narrower than if we insist on 95% or 99% confidence. But intervals constructed at an 80% confidence level will capture the true value of the parameter much less often than intervals that use a 99% confidence level.

The Concept Exploration also shows that we can get a more precise estimate of a parameter by increasing the sample size. Larger samples typically yield narrower confidence intervals. However, larger samples carry higher costs: They require more time and more money to obtain.

Decreasing the margin of error

In general, we prefer an estimate with a small margin of error. The margin of error gets smaller when:

- *The confidence level decreases.* To obtain a smaller margin of error from the same data, you must be willing to accept a smaller capture rate.
- *The sample size n increases.* In general, increasing the sample size n reduces the margin of error for any fixed confidence level.

EXAMPLE 7.2.2

Do you eat fast food?

Changing sample size and confidence level

PROBLEM

In Example 7.2.1, you read about a random sample of 3626 U.S. adults who were surveyed by the NCHS. According to the study, the 95% confidence interval for the true proportion of adults who eat fast food on a given day is 0.350 to 0.382.

(a) Explain what would happen to the width of the interval if the confidence level were increased to 99%.

(b) Explain what would happen to the width of the interval if the sample size were increased to 5000.

SOLUTION

(a) The confidence interval would be wider because increasing the confidence level increases the margin of error.

> To be more confident that our interval captures the true proportion, we need a wider interval.

(b) The confidence interval would be narrower because increasing the sample size typically decreases the margin of error.

> Recall from Chapter 6 that increasing the sample size decreases the standard deviation of the sampling distribution of $\hat{p}$ because we have more information about the population.

EXAM PREP: FOR PRACTICE, TRY EXERCISE 11.

What the Margin of Error Doesn't Account For

When we calculate a confidence interval, we include the margin of error because we expect the value of the point estimate to vary somewhat from the parameter value. However, the margin of error accounts for *only* the variability due to random sampling. It does not account for practical difficulties, such as undercoverage and nonresponse in a sample survey. These problems can result in estimates that are much farther from the parameter value than the margin of error would suggest. Remember this unpleasant fact when reading the results of an opinion poll or other sample survey. ⚠ **The margin of error does *not* account for any sources of bias in the data collection process.**

EXAMPLE 7.2.3

Do you eat fast food?

Bias and the margin of error

PROBLEM

In the preceding examples, you read about a random sample of 3626 U.S. adults who were surveyed by the NCHS. According to the study, the 95% confidence interval for the true proportion of adults who eat fast food on a given day is 0.350 to 0.382.

(a) Why did the NCHS include a margin of error with its estimate?

(b) Describe one potential source of bias in the NCHS survey that is not accounted for by the margin of error.

SOLUTION

(a) To account for sampling variability—the fact that different samples of size 3626 from the population of U.S. adults will give different estimates for the proportion of U.S. adults who eat fast food on a given day.

(b) The margin of error doesn't account for the fact that some respondents might lie and say they didn't eat fast food on the day in question, knowing that fast food is generally considered unhealthy.

EXAM PREP: FOR PRACTICE, TRY EXERCISE 15.

Section 7.2 What Did You Learn?

Review the learning goals from this section. Then practice what you've learned by working through the exercises.

Learning Goal	Example	Exercises
Interpret a confidence level.	7.2.1	7–10
Describe how the confidence level and sample size affect the margin of error.	7.2.2	11–14
Explain how practical issues like nonresponse, undercoverage, and response bias can affect the interpretation of a confidence interval.	7.2.3	15–18

Section 7.2 Exercises

Building Concepts and Skills These exercises assess the basic knowledge you should have after reading the section.

1. How do you interpret a confidence level?

2. True/False: When you calculate a 95% confidence interval, there is a 0.95 probability that this interval captures the value of the parameter.

3. What is the benefit of decreasing the confidence level? What is the drawback?

4. What is the benefit of increasing the sample size? What is the drawback?

5. True/False: The margin of error in a confidence interval accounts for sampling variability in the data collection process.

6. True/False: The margin of error in a confidence interval accounts for bias in the data collection process.

Mastering Concepts and Skills These exercises reinforce the learning goals as shown in the examples.

7. **Refer to Example 7.2.1 Exercise Estimates** How accurately do people estimate the amount of exercise they get? Researchers randomly selected 3806 adults in New York City and asked them to estimate the number of minutes per week they spent doing moderate exercise. Based on the sample, a 90% confidence interval for the mean amount of time doing moderate exercise per week is 517 minutes to 542 minutes.[17] Interpret the confidence level.

8. **Casinos** The Gallup polling organization asked a random sample of 1025 U.S. adults how often they participated in various activities during the previous year. The 99% confidence interval for the mean number of times people report visiting a gambling casino during the previous year is 1.65 to 3.35.[18] Interpret the confidence level.

9. **Online Dating** A recent study conducted by the Pew Research Center asked a random sample of U.S. adults about their experience with online dating websites and apps. Based on this survey, the 95% confidence interval for the proportion of all U.S. adults who have used an online dating website or app is 0.279 to 0.321.[19] Interpret the confidence level.

10. **Weight Loss** A Gallup poll asked a random sample of U.S. adults, "Would you like to lose weight?" Based on this poll, the 95% confidence interval for the proportion of all U.S. adults who want to lose weight is 0.56 to 0.62. Interpret the confidence level.[20]

11. **Refer to Example 7.2.2 Exercise Estimates** Refer to Exercise 7.
    (a) Explain what would happen to the width of the interval if the confidence level were increased to 99%.
    (b) Explain what would happen to the width of the interval if the sample size were increased to 10,000 adults.

12. **Casinos** Refer to Exercise 8.
    (a) Explain what would happen to the width of the interval if the confidence level were decreased to 90%.
    (b) Explain what would happen to the width of the interval if the sample size were increased to 2000 adults.

13. **Online Dating** Refer to Exercise 9.
    (a) Explain what would happen to the width of the interval if the confidence level were changed to 90%.
    (b) Explain what would happen to the width of the interval if the sample size was half as big as in the actual study.

14. **Weight Loss** Refer to Exercise 10.
    (a) Explain what would happen to the width of the interval if the confidence level were changed to 99%.
    (b) Explain what would happen to the width of the interval if the sample size was half as big as in the actual study.

15. **Refer to Example 7.2.3 Exercise Estimates** Refer to Exercise 7.
    (a) Why did the researchers include a margin of error with their estimate?

(b) Describe one potential source of bias in the study that is not accounted for by the margin of error.

16. **Casinos** Refer to Exercise 8.
    (a) Why did Gallup include a margin of error with its estimate?
    (b) Describe one potential source of bias in the study that is not accounted for by the margin of error.

17. **Online Dating** Refer to Exercise 9. In the description of its methodology, the Pew Research Center says that in addition to sampling variability (what it calls "sampling error"), "one should bear in mind that question wording and practical difficulties in conducting surveys can introduce error or bias into the findings of opinion polls." Give an example of a practical difficulty that could affect the accuracy of the 95% confidence interval.

18. **Weight Loss** Refer to Exercise 10. In the description of its methodology, Gallup states that the 3 percentage point margin of error for this poll includes only sampling variability (what it calls "sampling error"). What other potential sources of error (Gallup calls these "nonsampling errors") could affect the accuracy of the 95% confidence interval?

Applying the Concepts These exercises ask you to apply multiple learning goals in a new context or to apply what you learned in this section in a new way.

19. **California Traffic** People love living in California for many reasons, but traffic isn't one of them. Based on a random sample of 572 employed California adults, a 90% confidence interval for the average travel time to work for all employed California adults is 23 minutes to 26 minutes.[21]
    (a) Interpret the confidence level.
    (b) Name two actions you could take to reduce the margin of error. What drawbacks do these actions have?
    (c) Describe one source of bias in this survey. Does the stated margin of error account for this possible bias? Explain your answer.

20. **California Employment** Each month the government releases unemployment statistics. The published unemployment rate doesn't include people who choose not to be employed, such as retirees. Based on a random sample of 1000 California adults, a 99% confidence interval for the proportion of all California adults who say they are employed is 0.532 to 0.612.[22]
    (a) Interpret the confidence level.
    (b) Name two actions you could take to reduce the margin of error. What drawbacks do these actions have?
    (c) Describe one source of bias in this survey. Does the stated margin of error account for this possible bias? Explain your answer.

21. **Which Confidence Level?** The figure shows the result of taking 25 SRSs from a normal population and constructing a confidence interval for μ using each sample. Which confidence level — 80%, 90%, 95%, or 99% — do you think was used? Explain your reasoning.

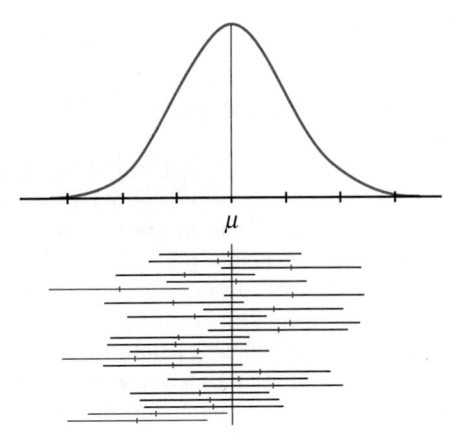

22. **Which Confidence Level?** The figure shows the result of taking 25 SRSs from a normal population and constructing a confidence interval for μ using each sample. Which confidence level — 80%, 90%, 95%, or 99% — do you think was used? Explain your reasoning.

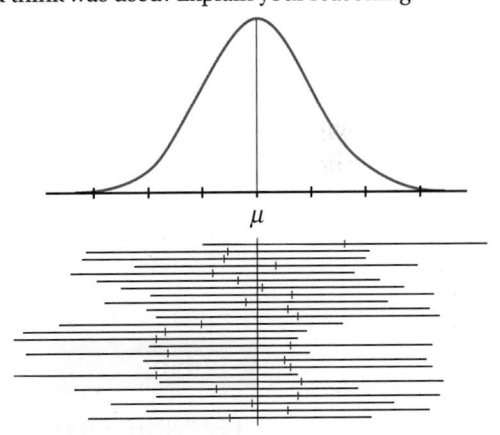

Extending the Concepts These exercises challenge you to explore statistical concepts and methods that go beyond what you learned in this section.

23. **Sample Size and Confidence** Does increasing the sample size make you more confident? You can investigate this question by going to www.stapplet.com and launching the *Simulating Confidence Intervals* applet.[23]
    (a) In the box labeled "Population," choose the Categorical population. Leave the true proportion of successes = 0.5. In the box labeled "Sample," change the sample size to $n = 125$ and leave the confidence level as 95%. Click the Go! button to generate one confidence interval. In the box labeled "Confidence Intervals," change the number of samples to 100 and click the Go! button 10 times. What percentage of the intervals captured the population proportion $p = 0.5$?
    (b) Keeping the 95% confidence level, increase the sample size to $n = 200$ and repeat the process from part (a). What percentage of the intervals captured the population proportion $p = 0.5$?
    (c) Keeping the 95% confidence level, increase the sample size to $n = 500$ and repeat the process from part (a). What percentage of the intervals captured the population proportion $p = 0.5$?
    (d) Based on your answers, does increasing the sample size increase your confidence that a confidence interval will capture the parameter? Explain your reasoning.

Cumulative Review These exercises revisit what you learned in previous sections.

24. **Taxes and Championships (2.6, 3.6)** Do states with lower income tax rates attract better professional athletes? Using data from 2000–2019, Ty Schalter at www.fivethirtyeight.com investigated the relationship between x = average state income tax rate and y = "championship points" per team-year.[24] Three championship points were awarded for winning the title and one point for being runner-up. For example, California had the most championship points (55) but also the most team-years (301), giving it a y-value of 55/301 = 0.183. Here is a scatterplot summarizing this relationship.

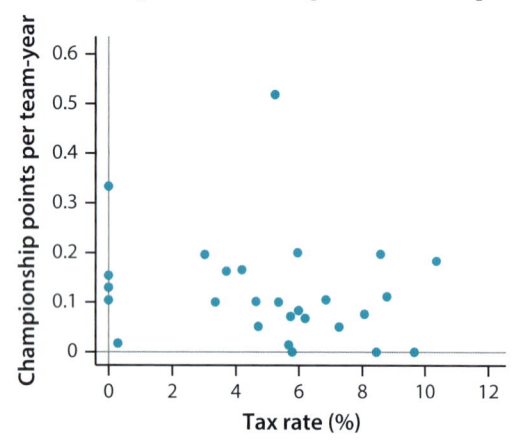

(a) Describe the association shown in the scatterplot.

(b) The correlation for these data is $r = -0.20$. The state of Massachusetts has an average income tax rate of 5.28% and 0.519 championship points. If the point for Massachusetts was removed, what would happen to the value of the correlation? Explain your reasoning.

25. **Educational Attainment (5.4, 6.6)** According to a recent Current Population Survey from the U.S. Census Bureau, 89.8% of adult U.S. residents have earned a high school diploma. Suppose we select a random sample of 120 American adults and record the proportion $\hat{p}$ of individuals in our sample who have a high school diploma.[25]

(a) Describe the shape, center, and variability of the sampling distribution of $\hat{p}$.

(b) Find the probability that the sample proportion of residents who have earned a high school diploma in a random sample of 120 residents is at least 0.95.

(c) If the sample size was 20, it would not be appropriate to use a normal distribution to perform probability calculations. Instead, use a binomial distribution to estimate the probability that the sample proportion is at least 0.95 in a sample of size 20.

Section 7.3 Estimating a Population Proportion

LEARNING GOALS

By the end of this section, you will be able to:

- Check the Random and Large Counts conditions for constructing a confidence interval for a population proportion.
- Determine the critical value for calculating a C% confidence interval for a population proportion.
- Calculate a C% confidence interval for a population proportion.

In Sections 7.1 and 7.2, you learned that a confidence interval can be used to estimate an unknown population parameter. We are often interested in estimating the proportion p of some outcome in the population. Here are some examples:

- What proportion of U.S. adults are unemployed right now?
- What proportion of college students have cheated on a test?
- What proportion of pine trees in a national park are infested with beetles?

In this section, you will learn how to calculate a confidence interval for a proportion.

Conditions for Estimating p

Before calculating a confidence interval for a population proportion p, you should check two important conditions:

> **HOW TO** Check the Conditions for Constructing a Confidence Interval for p
>
> To construct a confidence interval for the population proportion p, check that the following conditions are met:
>
> - **Random:** The data come from a random sample from the population of interest.
> - **Large Counts:** Both $n\hat{p}$ and $n(1-\hat{p})$ are at least 10.

In Chapter 1, you learned that the scope of inference is limited to the members of the sample if the sample wasn't selected at random from the population of interest. This means that you can't generalize to a larger population if the Random condition isn't met. Remember that the margin of error in a confidence interval doesn't account for bias in the data collection process.

The method we use to calculate a confidence interval for p requires that the sampling distribution of $\hat{p}$ be approximately normal. In Section 6.6, you learned that this will be true whenever np and $n(1-p)$ are both at least 10. Because we don't know the value of p, we use $\hat{p}$ when checking the Large Counts condition.

The method we use also requires that the observations within the sample be independent. If we are sampling at random without replacement, we can view the observations as independent as long as the sample size is less than 10% of the population size. In this and subsequent chapters, we will present only contexts where this condition is met so you don't have to specifically check for it.

EXAMPLE 7.3.1

Can you name the branches of government?

Checking conditions to construct a confidence interval for p

PROBLEM

In a random sample of 1416 U.S. adults conducted by the Annenberg Foundation, only 510 could name all three branches of government.[26] Check if the conditions are met for calculating a confidence interval for p = the proportion of all U.S. adults who could name all three branches of government.

SOLUTION

Random? The Annenberg Foundation selected a random sample of U.S. adults. ✓

Large Counts? $1416\left(\dfrac{510}{1416}\right) = 510 \geq 10$ and $1416\left(1 - \dfrac{510}{1416}\right) = 906 \geq 10$ ✓

> To check the Large Counts condition, make sure both $n\hat{p}$ and $n(1-\hat{p})$ are at least 10.

EXAM PREP: FOR PRACTICE, TRY EXERCISE 7.

Did you notice something about the values of $n\hat{p}$ and $n(1-\hat{p})$ in Example 7.3.1? The value $n\hat{p}$ will always equal the observed count of successes in the sample and the value $n(1-\hat{p})$ will always equal the observed count of failures in the sample.

THINK ABOUT IT

What happens if one of the conditions is violated? If the data come from a convenience sample or if other sources of bias are present in the data collection process, there's no reason to calculate a confidence interval for p. You should have *no* confidence in an interval when the Random condition is violated.

To explore violations of the Large Counts condition, we performed a simulation with $p = 0.3$ and $n = 10$. This violates the Large Counts condition, as $np = 10(0.3) = 3$, which is clearly less than 10. Here are 100 "95%" confidence intervals based on samples of size 10 from this population. Only 81 of them captured the true proportion—much less than the advertised capture rate of 95%!

When the Large Counts condition is violated, the actual capture rate will almost always be smaller than the stated confidence level. When we increased the sample size to $n = 50$ so that the Large Counts condition was met, 94% of the "95%" confidence intervals captured the true proportion $p = 0.3$. That's much better!

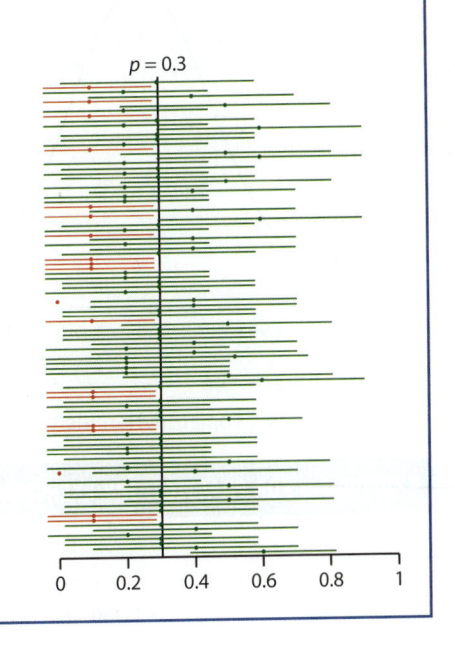

Critical Values

You already know that confidence intervals are made up of two parts, the point estimate and the margin of error:

$$\text{point estimate} \pm \text{margin of error}$$

When we are calculating a confidence interval for a population proportion p, the point estimate we use is the sample proportion $\hat{p}$.

The margin of error is more complicated because it is based on two factors: the confidence level and how much the sample proportion typically varies from the population proportion. Fortunately, we can use the facts about the sampling distribution of $\hat{p}$ from Section 6.6 as a foundation:

- The standard deviation of the sampling distribution of $\hat{p}$ is $\sigma_{\hat{p}} = \sqrt{\dfrac{p(1-p)}{n}}$.
- When the Large Counts condition is met, the sampling distribution of $\hat{p}$ is approximately normal.
- According to the empirical rule, $\hat{p}$ will be within 2 standard deviations of p in about 95% of all samples.
- Therefore, in about 95% of all samples, p will be within 2 standard deviations of $\hat{p}$.

Of course, we don't always go out 2 standard deviations in each direction. The number of standard deviations, called a **critical value**, depends on the confidence level.

> The **critical value** is a multiplier that makes the interval wide enough to have the stated capture rate.

When the Large Counts condition is met, we can use the standard normal distribution to find the critical value z^* for a specific confidence level. In the following example, we calculate a more precise critical value for a 95% confidence interval.

EXAMPLE 7.3.2

What's the z?

Finding a critical value

PROBLEM

Find the critical value z^* for a 95% confidence interval. Assume the Large Counts condition is met.

SOLUTION

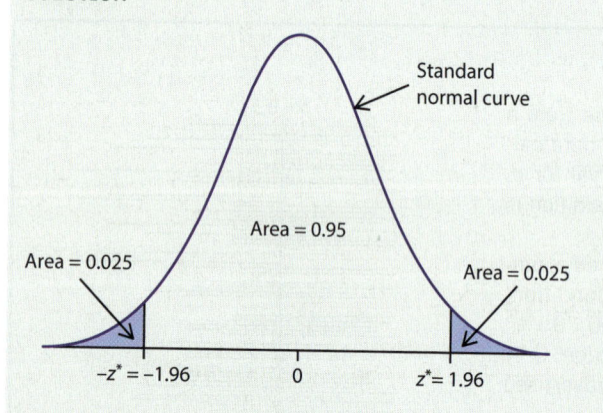

Standard normal curve

Area = 0.95
Area = 0.025 (left tail)
Area = 0.025 (right tail)
$-z^* = -1.96$ 0 $z^* = 1.96$

To find the critical value for 95% confidence, we find the boundaries that capture the middle 95% under the standard normal curve.

With 95% of the area between the two boundaries, there will be 5%/2 = 2.5% of the area in each tail.

Using Table A, search the body of the table for an area closest to 0.025. A z-score of −1.96 corresponds to an area of 0.025.

Using Table A: $z^* = 1.96$
Using technology: Applet/invNorm(left-tail area: 0.025, mean: 0, SD: 1) gives $z = -1.960$, so $z^* = 1.960$.

EXAM PREP: FOR PRACTICE, TRY EXERCISE 11.

Critical values are always reported as positive numbers. If you get a negative z-score, such as −1.960, just drop the negative sign and report $z^* = 1.960$.

Calculating a Confidence Interval for p

In the previous subsection, we reminded you that the standard deviation of the sampling distribution of $\hat{p}$ is $\sigma_{\hat{p}} = \sqrt{\dfrac{p(1-p)}{n}}$. Unfortunately, we don't know the value of p if we are trying to estimate it using a confidence interval. So we replace p with $\hat{p}$ to calculate the **standard error of $\hat{p}$** or $\text{SE}_{\hat{p}}$.

> The **standard error of $\hat{p}$** is an estimate of the standard deviation of the sampling distribution of $\hat{p}$.
>
> $$\text{SE}_{\hat{p}} = \sqrt{\dfrac{\hat{p}(1-\hat{p})}{n}}$$
>
> The standard error of $\hat{p}$ estimates how much $\hat{p}$ typically varies from p.

The *standard error* of an estimate and the *margin of error* of an estimate are related, but they aren't the same thing. The standard error estimates how much we expect $\hat{p}$ to vary from p, on average. The margin of error estimates how much we expect $\hat{p}$ to vary from p, at most. We use the standard error and the critical value to calculate the margin of error.

HOW TO Calculate a Confidence Interval for p

When the Random and Large Counts conditions are met, a C% confidence interval for the population proportion p is

point estimate ± margin of error

$$\hat{p} \pm z^* \sqrt{\dfrac{\hat{p}(1-\hat{p})}{n}}$$

where z^* is the critical value for the standard normal curve with C% of its area between $-z^*$ and z^*.

EXAMPLE 7.3.3

Can you name the branches of government?

Calculating a confidence interval for p

PROBLEM

In Example 7.3.1, you read about a random sample of 1416 U.S. adults surveyed by the Annenberg Foundation where only 510 could name all three branches of government. Calculate a 90% confidence interval for p = the proportion of all U.S. adults who could name all three branches of government. Note that we verified the conditions in Example 7.3.1.

SOLUTION

$\hat{p} = 510/1416 = 0.36$ | Identify the value of $\hat{p}$ for this sample.

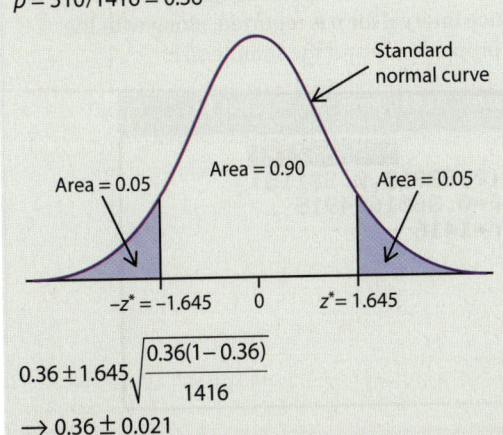

Standard normal curve: Area = 0.05 (left), Area = 0.90 (middle), Area = 0.05 (right); $-z^* = -1.645$, $z^* = 1.645$

Calculate the critical value z^* for 90% confidence using the standard normal curve.

$0.36 \pm 1.645 \sqrt{\dfrac{0.36(1-0.36)}{1416}}$

→ 0.36 ± 0.021

→ 0.339 to 0.381

$\hat{p} \pm z^* \sqrt{\dfrac{\hat{p}(1-\hat{p})}{n}}$

EXAM PREP: FOR PRACTICE, TRY EXERCISE 15.

We can be 90% confident that the interval from 0.339 to 0.381 captures the proportion of all U.S. adults who could name all three branches of government. Because all of the plausible values are less than 0.50, there is convincing evidence that a majority of U.S. adults cannot name all three branches.

TECH CORNER: Confidence Intervals for a Population Proportion

You can use technology to calculate a confidence interval for a population proportion. We'll illustrate using the *"Can you name the branches of government?"* examples from earlier in this section. Recall that there were 1416 U.S. adults in the sample and only 510 (36%) of them could name all three branches of government.

Applet

1. Go to www.stapplet.com and launch the *One Categorical Variable, Single Group* applet.

2. Enter the variable name "Response" and the Category names "Correct" and "Incorrect." Then, enter 510 for the frequency of "Correct" and $1416 - 510 = 906$ for "Incorrect."

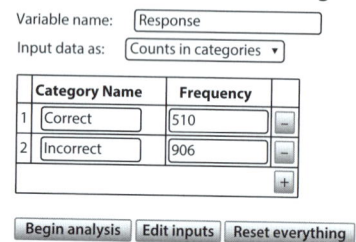

3. Click the Begin analysis button. Then, scroll down to the "Perform Inference" section.

4. Choose 1-sample z-interval for the procedure, "Correct" as the category to indicate as success, and 90% for the confidence level. Then click the Perform inference button.

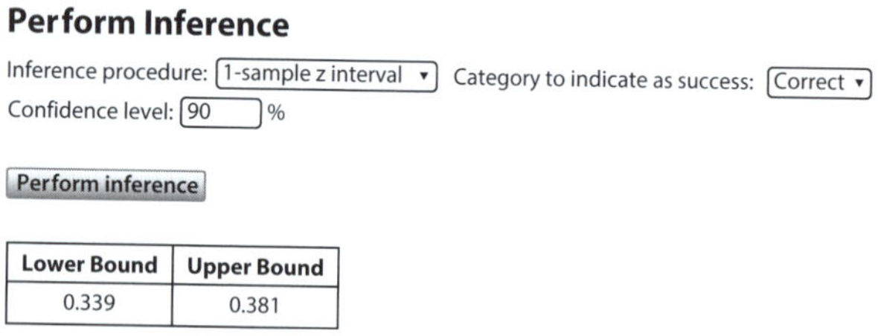

Lower Bound	Upper Bound
0.339	0.381

Note: If you have raw data instead of summary statistics, choose "Raw data" from the Input menu and enter the data. Then follow the directions in Steps 3 and 4.

TI-83/84

1. Press STAT, then choose TESTS and 1−PropZInt.

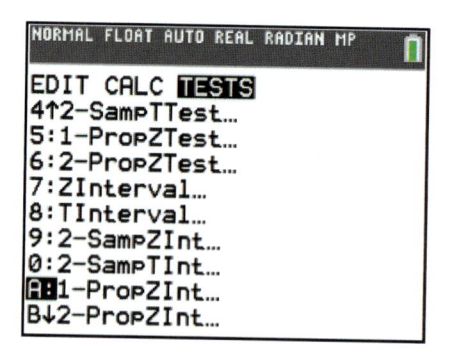

2. When the 1−PropZInt screen appears, enter $x = 510$, $n = 1416$, and confidence level = 0.9. Note that x is the *number* of successes, not the proportion of successes ($\hat{p}$). The value of x must be a whole number or the calculator will return an error message.

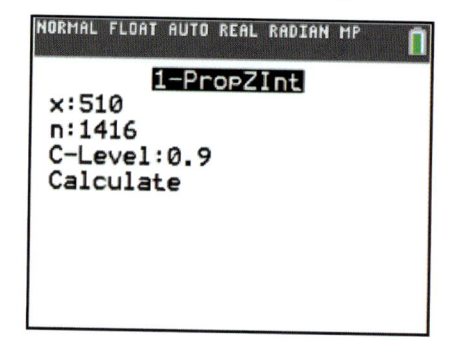

3. Highlight "Calculate" and press ENTER. The 90% confidence interval for p is reported, along with the sample proportion $\hat{p}$ and the sample size.

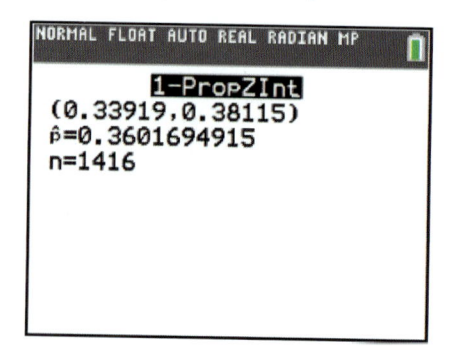

Detailed instructions for using CrunchIt!, Excel, Google Sheets, JMP, Minitab, and R are available in Achieve.

Section 7.3 What Did You Learn?

Review the learning goals from this section. Then practice what you've learned by working through the exercises.

Learning Goal	Example	Exercises
Check the Random and Large Counts conditions for constructing a confidence interval for a population proportion.	7.3.1	7–10
Determine the critical value for calculating a C% confidence interval for a population proportion.	7.3.2	11–14
Calculate a C% confidence interval for a population proportion.	7.3.3	15–18

Section 7.3 Exercises

Building Concepts and Skills These exercises assess the basic knowledge you should have after reading the section.

1. What is the Random condition?

2. Why do we use $\hat{p}$ in the Large Counts condition rather than p?

3. True/False: When the Large Counts condition is violated, the capture rate is less than the confidence level.

4. What is a critical value?

5. The _____ of $\hat{p}$ is an estimate of the standard deviation of the sampling distribution of $\hat{p}$.

6. What is the formula for a confidence interval for a population proportion?

Mastering Concepts and Skills These exercises reinforce the learning goals as shown in the examples.

7. **Refer to Example 7.3.1 Languages** What proportion of U.S. adults speak a language other than English at home? The American Community Survey asked a random sample of 2746 U.S. adults this question and found that 560 of the respondents spoke a language other than English at home.[27] Check if the conditions for calculating a confidence interval for p are met, where $p =$ the proportion of all U.S. adults who speak a language other than English at home.

8. **Crowdfunding** What proportion of U.S. adults contribute to crowdfunding campaigns such as GoFundMe? In a random sample of 1535 U.S. adults, 487 reported that they give to a crowdfunding campaign in a typical year.[28] Check if the conditions for calculating a confidence interval for p are met, where $p =$ the proportion of all U.S. adults who contribute to a crowdfunding campaign in a typical year.

9. **Air Quality** Los Angeles County is known for having bad air quality. In 2020, air quality was recorded on 183 days. Of these days, the air quality was very unhealthy on 2 days.[29] Check if the conditions for calculating a confidence interval for p are met, where $p =$ the proportion of all days in Los Angeles County in 2020 when the air quality was very unhealthy.

10. **Sponges** NSF International, a public health organization, wanted to learn more about the presence of germs in the homes of college students. It randomly selected 20 off-campus college households in Michigan and tested a set of items from each household. All 20 of the kitchen sponges tested contained yeast and mold. Check if the conditions for calculating a confidence interval for p are met, where $p =$ the true proportion of kitchen sponges from off-campus college households in Michigan that contain yeast and mold.[30]

11. **Refer to Example 7.3.2 Finding z^*** Find the critical value z^* for a 98% confidence interval. Assume the Large Counts condition is met.

12. **Finding z^*** Find the critical value z^* for a 96% confidence interval. Assume the Large Counts condition is met.

13. **Finding z^*** Find the critical value z^* for a 93% confidence interval. Assume the Large Counts condition is met.

14. **Finding z^*** Find the critical value z^* for a 99.9% confidence interval. Assume the Large Counts condition is met.

15. **Refer to Example 7.3.3 Languages** Refer to Exercise 7. Calculate a 99% confidence interval for the proportion of all U.S. adults who speak a language other than English at home.

16. **Crowdfunding** Refer to Exercise 8. Calculate a 90% confidence interval for the proportion of all U.S. adults who contribute to a crowdfunding campaign in a typical year.

17. **Smelling Covid** Dogs are known to have a very good sense of smell and are sometimes used to detect drugs, bombs, and diseases like cancer. Can they also smell Covid-19? Researchers in Germany trained dogs to identify Covid-19 from samples of human saliva. After training, they presented the dogs with 1012 randomized samples, some that were positive for Covid-19 and others that were negative. The dogs correctly identified 93.8% of the samples.[31] Calculate a 95% confidence interval for the true proportion of samples that these dogs would correctly identify. Assume the conditions are met.

18. **Spoilers** What proportion of American adults think it is OK to post spoilers about a new movie on social media within 48 hours of its release? According to a Morning Consult poll of 2200 randomly selected U.S. adults, 36% think it is OK.[32] Calculate a 95% confidence interval for the proportion of all U.S. adults who think it is OK to post spoilers about a new movie on social media within 48 hours of its release. Assume the conditions are met.

Applying the Concepts These exercises ask you to apply multiple learning goals in a new context or to apply what you learned in this section in a new way.

19. **Cell Phones** A recent poll of 738 randomly selected customers of a major U.S. cell-phone carrier found that 170 of them had walked into something or someone while talking on a cell phone.[33]
    (a) Show that the conditions for calculating a confidence interval for a proportion are met.
    (b) Calculate a 90% confidence interval for the proportion of all customers who have walked into something or someone while talking on a cell phone.
    (c) Interpret the interval from part (b).

20. **Cheating** What proportion of students are willing to report cheating by other students? A student project put this question to an SRS of 172 undergraduates at a large university: "You witness two students cheating on a quiz. Do you go to the professor?" Only 19 answered "Yes."[34]
    (a) Show that the conditions for calculating a confidence interval for a proportion are met.
    (b) Calculate a 99% confidence interval for the proportion of all undergraduate students at this university who would answer "Yes."
    (c) Interpret the interval from part (b).

21. **Cell Phones** Refer to Exercise 19. Calculate and interpret the standard error of $\hat{p}$, where $\hat{p}$ = the sample proportion of customers who have walked into something or someone while talking on a cell phone.

22. **Cheating** Refer to Exercise 20. Calculate and interpret the standard error of $\hat{p}$ where $\hat{p}$ = the sample proportion of students who answer "Yes."

Extending the Concepts These exercises challenge you to explore statistical concepts and methods that go beyond what you learned in this section.

23. **Equality** Have efforts to promote equality for women gone far enough in the United States? A poll on this issue by the cable network MSNBC contacted 1019 adults. A newspaper article about the poll said, "Results have a margin of sampling error of plus or minus 3 percentage points."[35]
    (a) The news article said that 65% of men, but only 43% of women, believe that efforts to promote equality have gone far enough. Explain why we do not have enough information to give confidence intervals for men and women separately.
    (b) Would the margin of error for women alone be less than 0.03, about equal to 0.03, or greater than 0.03? Why? (You can see that the news article's statement about the margin of error for poll results is a bit misleading.)

24. **Sponges** In Exercise 10, you read about a study to estimate p = the true proportion of kitchen sponges from college households in Michigan that contain yeast and mold. In the study, $\hat{p} = 20/20$. Because the Large Counts condition isn't met, we shouldn't use the methods of this section to construct a confidence interval for p. In situations where the Large Counts condition isn't met, we can use the "+4 interval." To construct a +4 interval, simply add 2 successes and 2 failures to the sample and compute the interval using the formula from this section.
    (a) After adding 2 sponges with yeast and mold and 2 sponges without yeast and mold to the sample, what is the value of $\hat{p}$? What is the sample size n?
    (b) Use your answers from part (a) to construct a 95% confidence interval for p.

Cumulative Review These exercises revisit what you learned in previous sections.

25. **Rolling Doubles (4.1, 4.6)** You are playing a board game with some friends. Each turn begins with rolling two dice. In this game, rolling "doubles" — the same number on both dice — is especially beneficial. The probability of rolling doubles is 1/6.
    (a) What is the probability of rolling doubles 3 times in a row?
    (b) After you roll doubles on your last three turns, one of your friends says, "No way you'll roll doubles this time. It would be nearly impossible." Explain what is wrong with your friend's reasoning.
    (c) Given that you have rolled doubles 3 times in a row, what is the probability that you roll doubles on your fourth roll?

26. **People Younger Than 18 (2.5)** Here are the percentages of people younger than age 18 in each of the 50 states, according to a recent U.S. Census.[36]

20.7	22.3	23.6	24.2	25.1	20.7	22.6	23.6
24.4	25.2	20.9	22.6	23.6	24.4	25.5	21.3
22.9	23.7	24.4	25.5	21.3	22.9	23.7	24.6
25.5	21.7	23.2	23.7	24.7	25.7	21.8	23.4
23.8	24.8	26.4	22.0	23.4	23.9	24.8	27.3
22.3	23.5	23.9	24.9	27.4	22.3	23.5	24.0
25.0	31.5						

    (a) Make a histogram of these data.
    (b) Describe the distribution.

Section 7.4 Confidence Intervals for a Population Proportion

LEARNING GOALS

By the end of this section, you will be able to:
- Use the four-step process to construct and interpret a confidence interval for a population proportion.
- Determine the sample size required to obtain a *C*% confidence interval for a population proportion with a specified margin of error.

In Section 7.3, you learned how to check the conditions and calculate a confidence interval for a population proportion. In this section, you will use a four-step process to construct and interpret a confidence interval for a population proportion.

Putting It All Together: The Four-Step Process

When you are constructing and interpreting a confidence interval, use the following four-step process.

HOW TO   Use the Four-Step Process: Confidence Intervals

State: State the parameter you want to estimate and the confidence level.
Plan: Identify the appropriate inference method and check the conditions.
Do: If the conditions are met, perform the calculations.
Conclude: Interpret your interval in the context of the problem.

A confidence interval for a population proportion is often called a **one-sample *z* interval for a proportion**.

A **one-sample *z* interval for a proportion** is a confidence interval used to estimate a population proportion *p*.

EXAMPLE 7.4.1

Do you get enough sleep?

Confidence interval for *p*

PROBLEM

Sleep Awareness Week begins in the spring with the release of the National Sleep Foundation's annual poll of U.S. sleep habits and ends with the beginning of daylight-saving time, when most people lose an hour of sleep.[37] In the foundation's random sample of 1029 U.S. adults, 48% reported that they "often or always" got enough sleep during the previous 7 nights. Construct and interpret a 95% confidence interval for the proportion of all U.S. adults who would report they often or always got enough sleep during the previous 7 nights.

SOLUTION

State: 95% CI for *p* = the proportion of all U.S. adults who would report they often or always got enough sleep during the previous 7 nights.

> **State:** State the parameter you want to estimate and the confidence level.

Plan: One-sample *z* interval for *p*.

> **Plan:** Identify the appropriate inference method and check the conditions.

(continued)

Random? Random sample of 1029 U.S. adults ✓
Large Counts? 1029(0.48) ≈ 494 ≥ 10 and 1029(1− 0.48) ≈ 535 ≥ 10 ✓

> Because these are the observed counts of successes and failures, they should be rounded to the nearest integer.

Do:

$$0.48 \pm 1.96 \sqrt{\frac{0.48(1-0.48)}{1029}}$$

→ 0.48 ± 0.031
→ 0.449 to 0.511

Using technology: 0.450 to 0.511

> **Do:** If the conditions are met, perform the calculations.

> When the calculations from technology differ from the by-hand calculations, we'll use the values from technology in the conclusion.

Conclude: We are 95% confident that the interval from 0.450 to 0.511 captures the proportion of all U.S. adults who would report they often or always got enough sleep during the previous 7 nights.

> **Conclude:** Interpret your interval in the context of the problem.

EXAM PREP: FOR PRACTICE, TRY EXERCISE 5.

Notice that we didn't make a conclusion about the proportion of U.S. adults who *reported* that they often or always got enough sleep. That would be the sample proportion, $\hat{p} = 0.48$. Instead, the conclusion is about the true proportion of all U.S. adults who *would report* that they often or always got enough sleep if we did a census.

Does this interval provide convincing evidence that a majority (more than 50%) of U.S. adults would report that they often or always got enough sleep during the previous 7 nights? Although there are plausible values in the interval greater than 0.5, the interval also includes many values less than 0.5. Thus, the interval does not provide convincing evidence that the majority of U.S. adults would report that they often or always got enough sleep during the previous 7 nights.

Determining the Sample Size

When planning a study, we may want to choose a sample size that allows us to estimate a population proportion within a given margin of error. The formula for the margin of error (*ME*) in a confidence interval for *p* is

$$ME = z^* \sqrt{\frac{\hat{p}(1-\hat{p})}{n}}$$

The margin of error calculation includes the sample proportion $\hat{p}$. Unfortunately, we won't know the value of $\hat{p}$ until *after* the study has been conducted. This means we have to guess the value of $\hat{p}$ when choosing *n*. Here are two ways to do this:

1. Use a guess for $\hat{p}$ based on a preliminary study or past experience with similar studies.
2. Use $\hat{p} = 0.5$ as the guess. The margin of error *ME* is largest when $\hat{p} = 0.5$, so this guess yields an upper bound for the sample size that will result in a given margin of error. If we get a $\hat{p}$ other than 0.5 when we do our study, the margin of error will be smaller than planned.

Once you have a guess for $\hat{p}$, the formula for the margin of error can be solved to give the required sample size *n*.

HOW TO Calculate Sample Size for a Desired Margin of Error

To determine the sample size *n* that will yield a *C*% confidence interval for a population proportion *p* with a maximum margin of error *ME*, solve the following inequality for *n*:

$$z^* \sqrt{\frac{\hat{p}(1-\hat{p})}{n}} \leq ME$$

where $\hat{p}$ is a guessed value for the sample proportion. The margin of error will always be less than or equal to *ME* if you use $\hat{p} = 0.5$.

EXAMPLE 7.4.2

Who has a tattoo?

Determining sample size

PROBLEM

Suppose that you want to estimate $p=$ the true proportion of students at your school who have a tattoo with 90% confidence and a margin of error of no more than 0.07. Determine how many students should be surveyed.

SOLUTION

$1.645\sqrt{\dfrac{0.5(1-0.5)}{n}} \leq 0.07$ — Because we don't have a value of $\hat{p}$ to use from a preliminary or previous study, use $\hat{p} = 0.5$ as the guessed value of $\hat{p}$. Then solve the inequality for n.

$\sqrt{\dfrac{0.5(1-0.5)}{n}} \leq \dfrac{0.07}{1.645}$ — Divide each side by 1.645.

$\dfrac{0.5(1-0.5)}{n} \leq \left(\dfrac{0.07}{1.645}\right)^2$ — Square both sides.

$0.5(1-0.5) \leq n\left(\dfrac{0.07}{1.645}\right)^2$ — Multiply both sides by n.

$\dfrac{0.5(1-0.5)}{\left(\dfrac{0.07}{1.645}\right)^2} \leq n$ — Divide both sides by $\left(\dfrac{0.07}{1.645}\right)^2$.

$138.0625 \leq n$ — Answer the question, making sure to follow the inequality sign when rounding.

We need to survey at least 139 students.

EXAM PREP: FOR PRACTICE, TRY EXERCISE 9.

Even though 138.0625 is closer to 138 than 139, we have to use at least 139 students to *guarantee* that the margin of error will be less than or equal to 0.07. If $\hat{p} = 0.5$ and $n = 138$, the margin of error is 0.07002, which is greater than the desired margin of error of 0.07. In general, we round to the next highest integer when solving for sample size to make sure the margin of error is less than or equal to the desired value.

Section 7.4 What Did You Learn?

Review the learning goals from this section. Then practice what you've learned by working through the exercises.

Learning Goal	Example	Exercises
Use the four-step process to construct and interpret a confidence interval for a population proportion.	7.4.1	5–8
Determine the sample size required to obtain a $C\%$ confidence interval for a population proportion with a specified margin of error.	7.4.2	9–12

Section 7.4 Exercises

Building Concepts and Skills These exercises assess the basic knowledge you should have after reading the section.

1. Identify the names of each step in the four-step process for constructing and interpreting a confidence interval.

2. What is the formula for the margin of error when calculating a confidence interval for a population proportion?

3. What are the two different options for picking a value of $\hat{p}$ to use in a sample size calculation?

4. **True/False:** You should always round your answer up to the next integer when calculating the required sample size to guarantee a certain margin of error in a confidence interval for a proportion.

Mastering Concepts and Skills These exercises reinforce the learning goals as shown in the examples.

5. **Refer to Example 7.4.1 Unknown Callers** What do people do when an unknown number calls their cell phone? In a Pew Research Center survey, only 19% of the 10,211 randomly selected U.S. adults would answer the phone to see who it is.[38] Construct and interpret a 95% confidence interval for the proportion of all U.S. adults who would answer their cell phone when an unknown number calls.

6. **Paint Flaws** A large automobile manufacturing plant produces 1200 new cars every day. A quality-control inspector checks a random sample of 90 cars from one day's production and finds that 12 of them have minor paint flaws. Construct and interpret a 99% confidence interval for the proportion of all cars produced that day with minor paint flaws.

7. **Pesticides** The U.S. Department of Agriculture (USDA) regularly tests produce labeled as "organic" for chemical residue, usually from pesticides. In a recent random sample of 409 items labeled as organic, 21.3% contained some chemical residue. Construct and interpret a 95% confidence interval for the true proportion of organic produce that contains chemical residue.[39]

8. **Soft Drinks** A Gallup poll of 1015 randomly selected adults found that 69% would oppose a law that limited the size of soft drinks and other sugary beverages served in restaurants.[40] Construct and interpret a 95% confidence interval for the proportion of all adults who would oppose a law limiting the size of soft drinks.

9. **Refer to Example 7.4.2 PTC** The substance PTC is found in foods like cabbage and broccoli. Some people find that PTC has a strong bitter taste, but others find it to be tasteless. The ability to taste PTC is inherited. About 75% of Italians can taste PTC, for example. You want to estimate the proportion of Americans of Italian descent who can taste PTC. How large a sample must you test to estimate the proportion of PTC tasters in this population within 0.04 with 90% confidence? Answer this question using the 75% estimate as the guessed value for $\hat{p}$.

10. **Education Vouchers** A national opinion poll found that 44% of all U.S. adults agree that parents should be given vouchers that can be applied toward their children's education at any public or private school of their choice. The result was based on a small sample. How large an SRS is required to obtain a margin of error of at most 0.03 in a 99% confidence interval? Answer this question using the previous poll's result as the guessed value for $\hat{p}$.

11. **Mayoral Polling** Edgar Martinez and Ingrid Gustafson are the candidates for mayor in a large city. We want to estimate the proportion p of all registered voters in the city who plan to vote for Gustafson with 95% confidence and a margin of error no greater than 0.03. How large a random sample do we need?

12. **Nightclub** A college student organization wants to start a nightclub for students who are younger than age 21. To assess support for the idea, the organization will select an SRS of students and ask each if they would patronize this type of establishment. What sample size is required to obtain a 90% confidence interval with a margin of error of at most 0.04?

Applying the Concepts These exercises ask you to apply multiple learning goals in a new context or to apply what you learned in this section in a new way.

13. **Yankees in Connecticut** Connecticut is located between New York and Boston, so baseball fans in the state typically root for either the Yankees or the Red Sox. In a recent poll of 803 randomly selected baseball fans in Connecticut, 44% said their favorite team was the Yankees.[41]

    (a) Construct and interpret a 95% confidence interval for the proportion of all baseball fans in Connecticut who would say their favorite team is the Yankees.

    (b) Based on the interval, is it plausible that the Yankees are the favorite team for a majority of Connecticut baseball fans? Explain your reasoning.

    (c) If the researchers wanted to reduce the margin of error to at most 2%, about how many *additional* baseball fans do they have to randomly select? Use the value of $\hat{p}$ from the initial study.

14. **Life on Mars** A recent CNN poll found that in a random sample of 508 U.S. residents, 31% answered "Yes" to the question "Do you think life has ever existed on Mars?"[42]

    (a) Construct and interpret a 99% confidence interval for the proportion of all U.S. residents who believe life ever existed on Mars.

    (b) Based on the interval, is it plausible that more than 35% of U.S. residents think life ever existed on Mars? Explain your reasoning.

    (c) If the polltakers wanted to reduce the margin of error to at most 4%, about how many *additional* U.S. residents do they have to randomly select? Use the value of $\hat{p}$ from the initial study.

15. **Pesticides** Refer to Exercise 7. How large a sample would be required to cut the margin of error in half?

16. **Soft Drinks** Refer to Exercise 8. How large a sample would be required to cut the margin of error in half?

17. **Tech City Trees** The data set *NYCTrees* contains detailed information about a random sample of 8253 trees on New York City streets.[43] The variable *sidewalk* indicates whether there is damage to the sidewalk next

to the tree, due to the growth of the tree and its root system. Calculate and interpret a 90% confidence interval for the proportion of all living trees on New York City streets that are next to damaged sidewalks. *Note:* Do not include dead trees in your calculations.

18. **Tech** **Teens and Sleep** Medical experts suggest that teenagers should be getting between 8 and 10 hours of sleep a night, on average. A random sample of 446 teens, aged 14 to 18, were asked the question, "On an average school night, how many hours of sleep do you get?" For the variable *outcome*, the researchers recorded a 1 if the teen got at least 7 hours of sleep and recorded a 0 if the teen got less than 7 hours of sleep. The data are stored in the *LosingSleep* data set.[44] Calculate and interpret a 95% confidence interval for the proportion of all teenagers between 14 and 18 who get at least 7 hours of sleep.

Extending the Concepts These exercises challenge you to explore statistical concepts and methods that go beyond what you learned in this section.

19. **Gambling** In college athletics, gambling is an issue of great concern. Because of this, the National Collegiate Athletic Association (NCAA) surveyed randomly selected student athletes about gambling-related behaviors. Of the 5594 Division I male athletes in the survey, 3547 reported participation in some form of gambling. This includes playing cards, betting on games of skill, buying lottery tickets, and betting on sports. A report of this study cited a 1% margin of error.[45]
    (a) The confidence level was not stated in the report. Use what you have learned to find the confidence level, assuming that the NCAA selected an SRS.
    (b) Describe one potential source of bias in the NCAA study that is not accounted for by the margin of error.

20. **Paint Flaws** Use your confidence interval from Exercise 6 to calculate a 99% confidence interval for the *total number* of cars produced that day with minor paint flaws.

21. **Sample Size** According to a Common Sense Media survey of randomly selected U.S. teens, a 95% confidence interval for the proportion of all teens who would report watching TV every day is 0.5382 to 0.6025. "Watching TV" includes watching programming from a broadcast or cable network on a TV set, watching shows that were recorded on a DVR, and streaming shows to an internet-connected device with a subscription service such as Netflix.[46] What sample size was used to calculate this interval?

Cumulative Review These exercises revisit what you learned in previous sections.

22. **Bone Density (6.2)** Osteoporosis is a condition in which the bones become brittle due to loss of minerals. To diagnose osteoporosis, an elaborate apparatus measures bone mineral density (BMD) in standardized form. The standardization is based on a population of healthy young adults. The World Health Organization (WHO) criterion for osteoporosis is a BMD score that is 2.5 standard deviations below the mean for young adults. BMD measurements in a population of people similar in age roughly follow a normal distribution.
    (a) What percentage of healthy young adults have osteoporosis by the WHO criterion?
    (b) People aged 70 to 79 are, of course, not young adults. The mean BMD in this age group is about -2 on the standard scale for young adults. Suppose that the standard deviation is the same as for young adults. What percentage of this older population has osteoporosis?

23. **Home Sales (6.8)** The distribution of prices for home sales in a certain New Jersey county is skewed right with a mean of $290,000 and a standard deviation of $145,000.
    (a) Suppose you select an SRS of 100 home sales from this large population. What is the probability that the mean of the sample is greater than $325,000?
    (b) Suppose you select an SRS of 5 home sales from this population. Explain why you cannot calculate the probability that the mean of this sample is greater than $325,000.

Section 7.5 Estimating a Population Mean

LEARNING GOALS

By the end of this section, you will be able to:
- State and check the Random and Normal/Large Sample conditions for constructing a confidence interval for a population mean.
- Determine critical values for calculating a C% confidence interval for a population mean.
- Calculate a C% confidence interval for a population mean.

In Section 7.3, you learned how to check conditions and calculate a confidence interval for a population proportion p. In this section, you will learn how to check conditions and calculate a confidence interval for a population mean μ.

Conditions for Estimating μ

As with confidence intervals for a population proportion, certain conditions need to be met to construct a confidence interval for a population mean.

> **HOW TO** **Check the Conditions for Constructing a Confidence Interval for μ**
>
> To construct a confidence interval for the population mean μ, check that the following conditions are met:
> - **Random:** The data come from a random sample from the population of interest.
> - **Normal/Large Sample:** The data come from a normally distributed population or the sample size is large ($n \geq 30$).

The methods for calculating a confidence interval for a mean work best when the population distribution is normal. When the shape of the population distribution isn't normal (or is unknown), having a large sample size helps to ensure that the methods we use will be approximately correct. What constitutes "large" depends on how far from normal the population distribution is. In most cases, however, sample sizes of at least 30 are sufficient.

EXAMPLE 7.5.1

Put a ring on it?

Checking conditions for constructing a confidence interval for μ

PROBLEM

When getting married, some people choose to buy an engagement ring for their partner. How much do people spend on engagement rings, on average? To find out, the *New York Times* and Morning Consult surveyed a random sample of 1640 U.S. adults who bought an engagement ring.[47] Check if the conditions for calculating a confidence interval for μ are met.

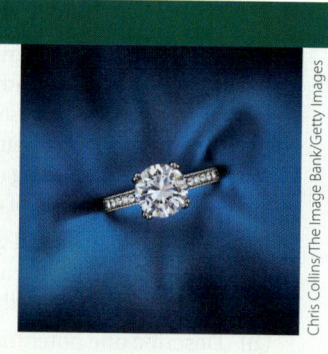

SOLUTION

Random? Random sample of U.S. adults who have bought an engagement ring. ✓

Normal/Large Sample? We don't know the shape of the population distribution, but the sample size is large: $n = 1640 \geq 30$. ✓

EXAM PREP: FOR PRACTICE, TRY EXERCISE 7.

The Problem of Unknown σ

In Section 7.3, we presented the formula for a confidence interval for a population proportion:

$$\hat{p} \pm z^* \sqrt{\frac{\hat{p}(1-\hat{p})}{n}}$$

In more general terms, this is

$$\text{statistic} \pm \text{critical value} \times \text{standard error of statistic}$$

A confidence interval for a population mean has a formula with the same structure. Using $\bar{x}$ as the point estimate for μ and $\sigma/\sqrt{n}$ as the standard deviation of the sampling distribution of $\bar{x}$ gives

$$\bar{x} \pm z^* \frac{\sigma}{\sqrt{n}}$$

Unfortunately, if we don't know the true value of μ, we rarely know the true value of σ, either. We can use s_x as an estimate for σ, but things don't work out as nicely as we might like.

Class Activity

Confidence interval BINGO!

In this activity, your class will investigate the problem caused by replacing σ with s_x when calculating a confidence interval for μ, and learn how to fix it.

A homesteading family wants to estimate the mean weight (in grams) of all tomatoes grown on their farm. To do so, they select a random sample of 4 tomatoes, calculate the mean weight (in grams), and use the sample mean $\bar{x}$ to create a 99% confidence interval for the population mean μ. Suppose that the weights of all tomatoes on the farm are normally distributed, with a mean of 10 ounces and a standard deviation of 2 ounces.

Let's use an applet to simulate taking a random sample of $n = 4$ tomatoes and calculating a 99% confidence interval for μ using three different methods.

Method 1 (assuming σ is known)

$$\bar{x} \pm z^* \frac{\sigma}{\sqrt{n}} \rightarrow \bar{x} \pm 2.576 \frac{2}{\sqrt{4}}$$

1. Go to www.stapplet.com and launch the *Simulating Confidence Intervals* applet.

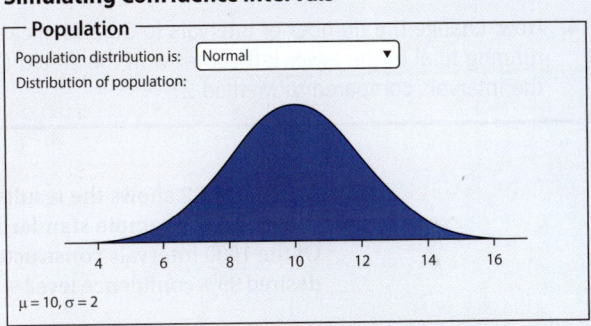

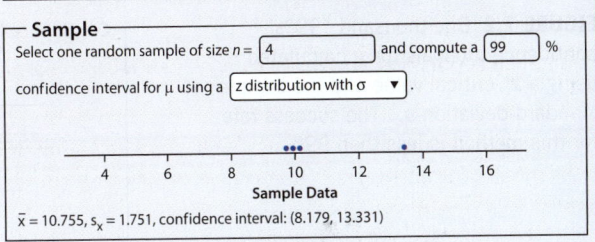

2. In the box labeled "Population," keep the default setting for the normal population distribution.

3. In the box labeled "Sample," change the sample size to $n = 4$, change the confidence level to 99%, and change the method to "z distribution with σ." Click the Go! button to select one random sample and calculate the corresponding 99% confidence interval.

4. In the box labeled "Confidence Intervals," the confidence interval is displayed as a horizontal line segment, along with a vertical line segment at $\mu = 10$. If the interval includes $\mu = 10$, the interval is green. Otherwise, it is red.

5. In the box labeled "Confidence Intervals," quickly generate 1 confidence interval at a time by pressing the Go! button, shouting out "BINGO!" whenever you get an interval that misses $\mu = 10$ (i.e., a red interval). Stop when your instructor calls time.

6. How well did Method 1 work? Compare the running total in the lower-left corner with the stated confidence level of 99%.

Method 2 (using s_x as an estimate for σ)

$$\bar{x} \pm z^* \frac{s_x}{\sqrt{n}} \rightarrow \bar{x} \pm 2.576 \frac{s_x}{\sqrt{4}}$$

1. Click the Clear samples button at the bottom of the applet. Then, in the box labeled "Sample," change the method to "z distribution with sx." Keep everything else the same.

2. In the box labeled "Confidence Intervals," quickly generate 1 confidence interval at a time by pressing the Go! button, shouting out "BINGO!" whenever you get an interval that misses $\mu = 10$ (i.e., a red interval). Stop when your instructor calls time.

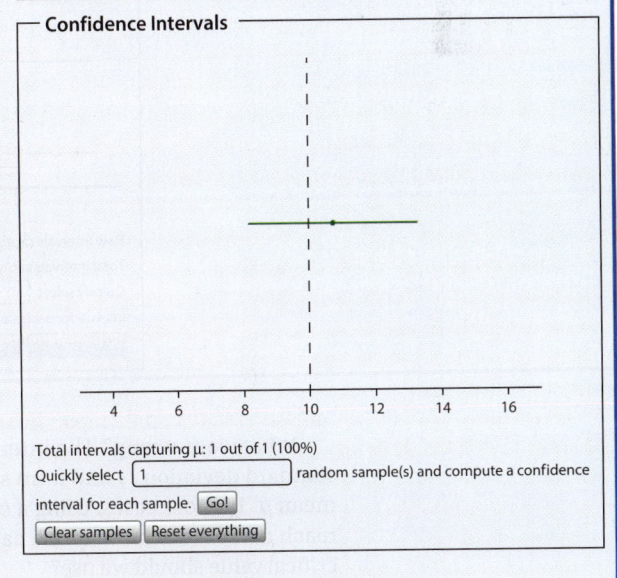

3. Did Method 2 work as well as Method 1? Discuss with your classmates.

4. Now, change the number of intervals to 50 and click "Go!" until you have more than 1000 intervals. Compare the running total in the lower-left corner with the stated confidence level of 99%. What do you notice about the width of the intervals that missed?

(continued)

To increase the capture rate of the intervals to 99%, we need to make the intervals wider. We can do this by using a different critical value, called a t^* critical value. You'll learn how to find this number soon.

Method 3 (using s_x as an estimate for σ and a t^* critical value instead of a z^* critical value)

$$\bar{x} \pm t^* \frac{s_x}{\sqrt{n}} \rightarrow \bar{x} \pm ??? \frac{s_x}{\sqrt{4}}$$

1. Click the Clear samples button at the bottom of the applet. Then, in the box labeled "Sample," change the method to "t distribution." Keep everything else the same.
2. In the box labeled "Confidence Intervals," reset the number of samples to 1 and quickly generate 1 confidence interval at a time, shouting out "BINGO!" whenever you get an interval that misses $\mu = 10$ (i.e., a red interval). Stop when your instructor calls time.
3. Did Method 3 work better than Method 2? How does it compare to Method 1? Discuss with your classmates.
4. Now, change the number of intervals to 50 and click "Sample" until you have more than 1000 intervals. Compare the running total in the lower-left corner with the stated confidence level of 99%. What do you notice about the width of the intervals compared to Method 2?

Figure 7.2 shows the results of repeatedly constructing confidence intervals using a z^* critical value and the sample standard deviation s_x, as described in Method 2 of the preceding activity. Of the 1000 intervals constructed, only 91.7% captured the population mean. That's far below our desired 99% confidence level!

FIGURE 7.2 One thousand "99%" confidence intervals for μ calculated using a z^* critical value and the sample standard deviation s_x. The success rate for this method is less than 99%.

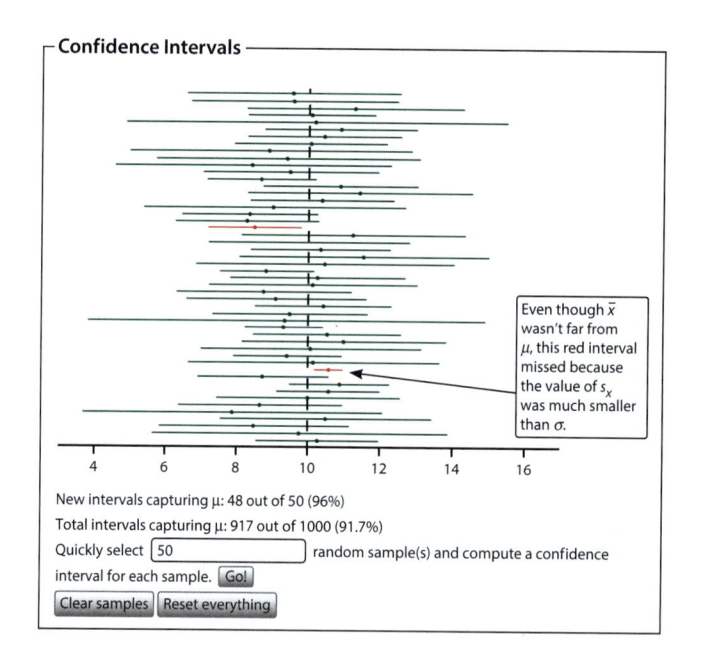

What went wrong? The intervals that missed (those in red) came from samples with a small standard deviation s_x and from samples in which the sample mean $\bar{x}$ was far from the population mean μ. In those cases, using a critical value of $z^* = 2.576$ didn't produce long enough intervals to reach $\mu = 10$. To achieve a 99% capture rate, we need to multiply by a larger critical value. But what critical value should we use?

t^* Critical Values

When calculating a confidence interval for a population mean, we use a t^* critical value rather than a z^* critical value whenever we use s_x to estimate σ. This means that we almost always use a t^* critical value when constructing a confidence interval for μ because we seldom know the value of σ.

The critical value is denoted t^* because it comes from a **t distribution**, not the standard normal distribution. The critical value t^* has the same interpretation as z^*: It measures how many standard errors we need to extend from the point estimate to get the desired level of confidence.

A **t distribution** is described by a symmetric, single-peaked, bell-shaped density curve. Any t distribution is completely specified by its *degrees of freedom* (df). When performing inference about a population mean based on a random sample of size n using the sample standard deviation s_x to estimate the population standard deviation σ, use a t distribution with df $= n-1$.

Figure 7.3 shows two different t distributions, along with the standard normal distribution. Because the t distributions have more area in the tails than the standard normal distribution, t^* critical values will always be larger than z^* critical values for a specified level of confidence. As the degrees of freedom increase, the t distributions have less area in the tails and approach the standard normal distribution. This makes sense because the value of s_x will typically be closer to σ as the sample size increases.

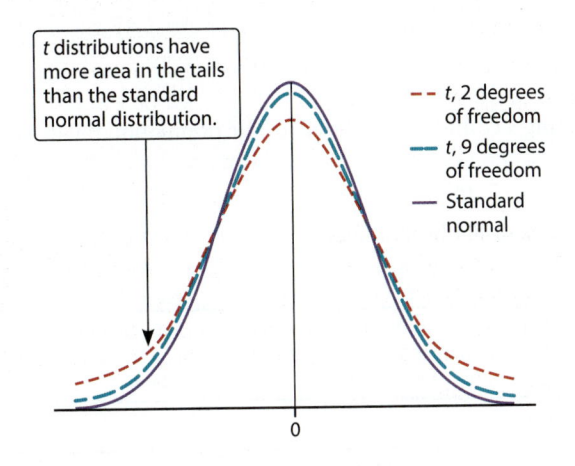

FIGURE 7.3 Density curves for the t distributions with 2 and 9 degrees of freedom and the standard normal distribution. All are symmetric with center 0. The t distributions have more variability than the standard normal distribution does, but approach the standard normal distribution as the df increases.

You will learn more about t distributions in Chapter 8. For now, we will focus on how to calculate the critical value t^* for various sample sizes and confidence levels. As with calculating z^*, you can use technology or a table to find t^*. See the Tech Corner at the end of the section for instructions about using technology.

HOW TO Find t^* Using Table B

1. Using Table B in the back of the book, find the correct confidence level at the bottom of the table.
2. On the left side of the table, find the correct number of *degrees of freedom* (df). For this type of confidence interval, df $= n-1$.
3. In the body of the table, find the value of t^* that corresponds to the confidence level and df.
4. If the correct df isn't listed, use the greatest df available that is less than the correct df.

In the BINGO! activity, we calculated 99% confidence intervals with $n = 4$, so df $= 4-1 = 3$. Here is an excerpt from Table B that shows how to find t^* in this case.

df	.02	.01	.005	.0025
1	15.89	31.82	63.66	127.3
2	4.849	6.965	9.925	14.09
3	3.482	4.541	5.841	7.453
⋮	⋮	⋮	⋮	⋮
z^*	2.054	2.326	2.576	2.807
	96%	98%	99%	99.5%

Tail probability p (top header); **Confidence level C** (bottom)

For 99% confidence and 3 degrees of freedom, $t^* = 5.841$. That is, the interval should extend 5.841 standard errors on both sides of the point estimate to have a capture rate of 99%. This t^* critical value is more than twice as large as the z^* critical value for 99% confidence ($z^* = 2.576$).

The bottom row of Table B gives z^* critical values. That's because the t distributions approach the standard normal distribution as the degrees of freedom approaches infinity.

If the correct df isn't listed, use the greatest df available that is less than the correct df. "Rounding up" to a larger df will result in confidence intervals that are too narrow. The intervals won't be wide enough to capture the true population value as often as suggested by the confidence level.

EXAMPLE 7.5.2

Would you like some t?*

Finding t^* critical values

PROBLEM

What critical value t^* should be used in constructing a confidence interval for the population mean in each of the following scenarios? Assume the conditions are met.

(a) A 95% confidence interval based on an SRS of size $n = 12$
(b) A 90% confidence interval from a random sample of 48 observations

SOLUTION

(a) df = 12 − 1 = 11;
Using Table B: $t^* = 2.201$
Using technology: Applet/invT(area:0.025, df:11) = −2.201, so $t^* = 2.201$

> In Table B, use the column for 95% confidence and the row corresponding to df = 12 − 1 = 11. Directions for using technology are in the Tech Corner at the end of the section.

(b) df = 48 − 1 = 47;
Using Table B: with df = 40, $t^* = 1.684$
Using technology: Applet/invT(area:0.05, df:47) = −1.678, so $t^* = 1.678$

> There is no df = 47 row in Table B, so we use the more conservative df = 40. Technology can handle the full df = 47.

EXAM PREP: FOR PRACTICE, TRY EXERCISE 11

You are now ready to calculate a confidence interval for a population mean.

Calculating a Confidence Interval for μ

Because we almost never know the population standard deviation σ, we must use the sample standard deviation s_x as an estimate for σ. This means that we must estimate the standard deviation of the sampling distribution of $\bar{x}$ with the **standard error of $\bar{x}$**.

The **standard error of $\bar{x}$** is an estimate of the standard deviation of the sampling distribution of $\bar{x}$.

$$SE_{\bar{x}} = \frac{s_x}{\sqrt{n}}$$

The standard error of $\bar{x}$ estimates how much $\bar{x}$ typically varies from μ.

Now that we know how to calculate the t^* critical value and the standard error of $\bar{x}$, we can calculate a confidence interval for μ.

HOW TO Calculate a Confidence Interval for μ

When the Random and Normal/Large Sample conditions are met, a $C\%$ confidence interval for the unknown population mean μ is

$$\bar{x} \pm t^* \frac{s_x}{\sqrt{n}}$$

where t^* is the critical value for a t distribution with df $= n - 1$ and $C\%$ of its area between $-t^*$ and t^*.

EXAMPLE 7.5.3

Put a ring on it?

Calculating a confidence interval for μ

PROBLEM

In Example 7.5.1, we verified that the conditions were met for constructing a confidence interval for μ = the true mean cost of an engagement ring for all U.S. adults who have bought an engagement ring. After adjusting for inflation, the mean cost in the sample of 1640 was $3329 with a standard deviation of $4739.[48] Calculate a 99% confidence interval for μ.

SOLUTION

df = 1640 − 1 = 1639

Using technology: $t^* = 2.579$

$3329 \pm 2.579 \dfrac{4739}{\sqrt{1640}}$

→ 3329 ± 302

→ $3027 to $3631

> In Table B, there is no row for df = 1639, so we would use the more conservative df = 1000 to get $t^* = 2.581$.

$$\bar{x} \pm t^* \dfrac{s_x}{\sqrt{n}}$$

EXAM PREP: FOR PRACTICE, TRY EXERCISE 15.

For decades, diamond jewelry marketers popularized the idea that you should spend three months' salary for an engagement ring. For U.S. adults, the typical three-month salary is a little more than $15,000.[49] Because all the plausible values for the mean cost of an engagement ring are much less than $15,000, the three-month "rule" doesn't seem to hold true anymore (if it ever did!).

TECH CORNER   Critical Values and Confidence Intervals for a Population Mean

Calculating t^* critical values

You can use technology to calculate t^* critical values. We'll illustrate using the example with $n = 12$ and 95% confidence.

Applet

1. Go to www.stapplet.com and launch the *t Distributions* applet.
2. Choose "Calculate a value corresponding to an area" from the drop-down menu, enter 11 for the degrees of freedom, and click "Plot distribution."
3. Using the drop-down menu below the curve, choose "central" and enter an area of 0.95. Then click "Calculate value(s)" and check the box to show labels on plot.

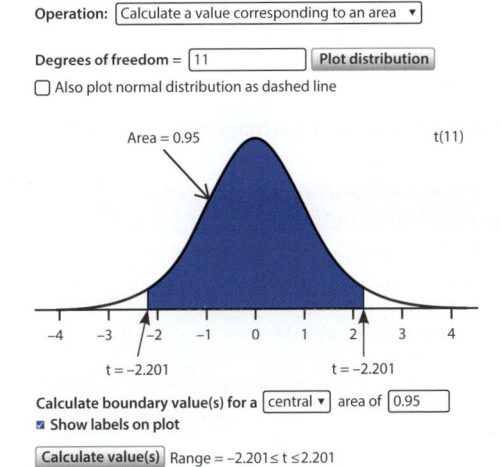

TI-83/84

1. Press 2nd VARS (DISTR) and choose invT(.
2. Because we want a central area of 0.95, each tail has an area of 0.025.
   - **OS 2.55 or later:** In the dialog box, enter area: 0.025 and df: 11. Choose Paste, and then press ENTER.
   - **Older OS:** Complete the command invT(0.025,11).

```
NORMAL FLOAT AUTO REAL RADIAN MP
invT(.025,11)
              -2.200985143
```

Note: The t^* critical value is positive 2.201.

Calculating a confidence interval for μ

You can also use technology to calculate a confidence interval for a population mean. We'll illustrate using the engagement ring example. Recall that the mean was $3329, the standard deviation was $4739, and the sample size was 1640. *Note:* Technology uses the exact df, so the results obtained with technology may vary slightly from results obtained with Table B.

(continued)

Applet

1. Go to www.stapplet.com and launch the *One Quantitative Variable, Single Group* applet.

2. Enter the variable name "Cost" and choose "Mean and standard deviation" from the Input menu. Then enter the mean, standard deviation, and sample size.

    One Quantitative Variable, Single Group

    Variable name: Cost
    Input: Mean and standard deviation

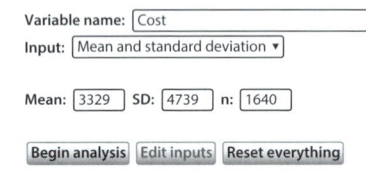

    Mean: 3329   SD: 4739   n: 1640

    [Begin analysis] [Edit inputs] [Reset everything]

3. Click the Begin analysis button. In the "Perform Inference" section, select the "1-sample t interval for μ" for the inference procedure and 99% for the confidence level. Then click the Perform inference button.

    Perform Inference

    Inference procedure: 1-sample t interval for μ Confidence level: 99 %

    [Perform inference]

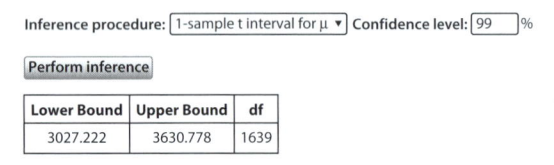

Lower Bound	Upper Bound	df
3027.222	3630.778	1639

Note: If you have raw data instead of summary statistics, choose "Raw data" from the Input menu and enter the data. Then follow the directions in Step 3.

TI-83/84

1. Press STAT, then choose TESTS and TInterval.

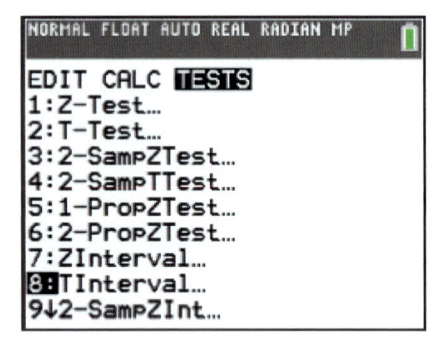

2. When the TInterval screen appears, choose Stats as the input method. Then enter the mean, standard deviation, and sample size, along with 0.99 as the confidence level.

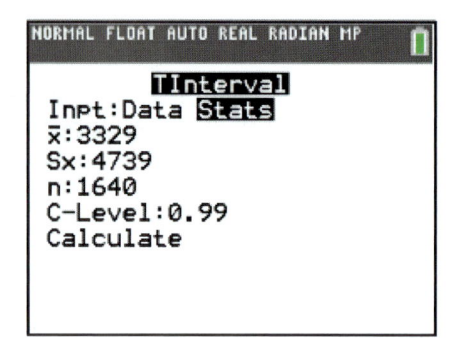

3. Highlight "Calculate" and press ENTER. The 99% confidence interval for μ is reported, along with the sample mean $\bar{x}$, sample standard deviation s_x, and sample size n.

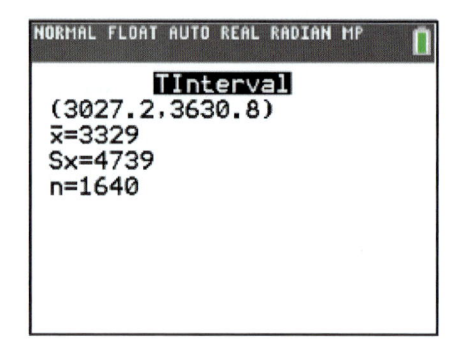

Note: If you have the raw data entered in a list, change the input method to Data and enter the name of the list where the data are stored (e.g., L_1).

Detailed instructions for using CrunchIt!, Excel, Google Sheets, JMP, Minitab, and R are available in Achieve.

Section 7.5 What Did You Learn?

Review the learning goals from this section. Then practice what you've learned by working through the exercises.

Learning Goal	Example	Exercises
State and check the Random and Normal/Large Sample conditions for constructing a confidence interval for a population mean.	7.5.1	7–10
Determine critical values for calculating a C% confidence interval for a population mean.	7.5.2	11–14
Calculate a C% confidence interval for a population mean.	7.5.3	15–18

Section 7.5 Exercises

Building Concepts and Skills These exercises assess the basic knowledge you should have after reading the section.

1. In the Normal/Large Sample condition, the data should come from a normally distributed _____ or the sample size should be at least _____.

2. True/False: When calculating a confidence interval for a population mean, we usually know the population standard deviation.

3. Whenever we use s_x to estimate _____, we need to use a(n) _____ critical value rather than a z^* critical value.

4. How do you calculate the degrees of freedom when estimating a population mean?

5. True/False: When you are using Table B, if the correct degrees of freedom are not included in the table, you should always round up to the next larger degrees of freedom listed.

6. What is the formula for a confidence interval for a population mean?

Mastering Concepts and Skills These exercises reinforce the learning goals as shown in the examples.

7. **Refer to Example 7.5.1 Live Shows** How often do U.S. adults attend live music or theater events in a given year, on average? Gallup asked a random sample of 1025 U.S. adults about this and other leisure activities. Check if the conditions for calculating a confidence interval for μ are met.[50]

8. **Possums** How long are Australian possums, on average? Zoologists in Australia captured a random sample of 104 possums and recorded the values of 9 different variables, including total length (in centimeters).[51] Check if the conditions for calculating a confidence interval for μ are met.

9. **Presidents and Longevity** To estimate the average age at which U.S. presidents have died, a historian obtains a list of all U.S. presidents who have died and their ages at death. Check if the conditions for calculating a confidence interval for μ are met.

10. **Stock Prices** To estimate the average price of all 30 stocks in the Dow Jones Industrial Average, a finance major obtains the prices of all 30 stocks at the close of trading. Check if the conditions for calculating a confidence interval for μ are met.

11. **Refer to Example 7.5.2 Finding t^*** Which critical value t^* should be used when constructing a confidence interval for the population mean in each of the following scenarios? Assume the conditions are met.
    (a) A 95% confidence interval based on $n = 10$ randomly selected observations
    (b) A 90% confidence interval based on a random sample of 77 individuals

12. **Finding t^*** Which critical value t^* should be used when constructing a confidence interval for the population mean in each of the following scenarios? Assume the conditions are met.
    (a) A 90% confidence interval based on $n = 12$ randomly selected observations
    (b) A 99% confidence interval based on a random sample of size 58

13. **Finding t^*** Which critical value t^* should be used when constructing a confidence interval for the population mean in each of the following scenarios? Assume the conditions are met.
    (a) A 99% confidence interval from an SRS of 20 observations
    (b) A 95% confidence interval based on $n = 85$ randomly selected individuals

14. **Finding t^*** Which critical value t^* should be used when constructing a confidence interval for the population mean in each of the following scenarios? Assume the conditions are met.
    (a) A 95% confidence interval from an SRS of 30 observations
    (b) A 90% confidence interval based on $n = 162$ randomly selected individuals

15. **Refer to Example 7.5.3 Tootsie Pops** Many people have asked the question, but few have been patient enough to collect the data: How many licks does it take to get to the center of a Tootsie Pop? Researcher Corey Heid decided to find out.[52] He instructed a sample of 92 students to lick a Tootsie Pop along the non-banded side until they could taste the chocolate center. The mean number of licks was 356.1 with a standard deviation of 185.7 licks. Calculate a 99% confidence interval for the true mean number of licks to get to the center of a Tootsie Pop. Assume the conditions are met.

16. **Bunions** A bunion on the big toe is fairly uncommon in youth and often requires surgery as treatment. Doctors used X rays to measure the angle (in degrees) of deformity on the big toe in a random sample of 37 patients younger than age 21 who came to a medical center for surgery to correct a bunion. The angle is a measure of the seriousness of the deformity. For these 37 patients, the mean angle of deformity was 24.76 degrees and the standard deviation was 6.34 degrees. Calculate a 90% confidence interval for the mean angle of deformity in the population of patients like these. Assume the conditions are met.[53]

17. **Live Shows** Refer to Exercise 7. In the sample of 1025 adults, the mean number of live music and theater events attended was 3.8 with a standard deviation of 6.95. Note that the conditions are met.
    (a) Calculate a 95% confidence interval for the mean number of live music and theater events U.S. adults attend in a given year.
    (b) Interpret the confidence level.

18. **Possums** Refer to Exercise 8. In the sample of 104 possums, the mean total length was 87.1 cm with a standard deviation of 4.3 cm. Note that the conditions are met.
    (a) Calculate a 95% confidence interval for the mean total length of Australian possums.
    (b) Interpret the confidence level.

Applying the Concepts These exercises ask you to apply multiple learning goals in a new context or to apply what you learned in this section in a new way.

19. **Fishing** Nick is an avid angler, but gets frustrated when his fishing line breaks when he is reeling in a big catch. To test whether his favorite brand of fishing line lives up to its "6 pound" claim, Nick randomly selected 30 pieces of fishing line, attached each to a bucket and filled the bucket with water until the line broke. Then, he measured the weight of the bucket of water. The mean weight was 6.44 pounds with a standard deviation of 0.75 pound.[54]
    (a) Verify that the conditions are met for constructing a confidence interval for μ = true mean breaking strength of this type of fishing line.
    (b) Construct a 99% confidence interval for μ.
    (c) Based on your interval from part (b), is there convincing evidence that the mean breaking strength is different than 6 pounds? Explain your reasoning.

20. **Bone Mineral Loss** Breastfeeding mothers secrete calcium into their milk. Some of this calcium may come from their bones, so mothers may lose bone mineral. Researchers measured the percent change in bone mineral content (BMC) of the spines of 47 randomly selected mothers during 3 months of breastfeeding. The mean change in BMC was −3.587% and the standard deviation was 2.506%.[55]
    (a) Verify that the conditions are met for constructing a confidence interval for μ = true mean change in BMC for breastfeeding mothers.
    (b) Construct a 99% confidence interval for μ.
    (c) Based on your interval from part (b), do these data give convincing evidence that nursing mothers lose bone mineral, on average? Explain your reasoning.

21. **Blood Pressure** A medical study randomly selected 27 adults and measured the seated systolic blood pressure of each adult. In the sample, $\bar{x} = 114.9$ and $s_x = 9.3$. What is the standard error of the mean? Interpret this value.

22. **Commute Times** A study of commuting times reports the travel times to work of a random sample of 20 employed adults in New York State. The mean is $\bar{x} = 31.25$ minutes and the standard deviation is $s_x = 21.88$ minutes. What is the standard error of the mean? Interpret this value.

Extending the Concepts These exercises challenge you to explore statistical concepts and methods that go beyond what you learned in this section.

23. **House Appraisals** The mayor of a small town wants to estimate the average property value for the houses built there in the last year. She randomly selects 15 houses and pays an appraiser to determine the value of each house. The mean value of these houses is $183,100 with a standard deviation of $29,200. Assume the conditions are met.
    (a) Calculate a 90% confidence interval for the mean value of new houses in this town.
    (b) There are a total of 300 new houses in the mayor's town and the town levies a 1% property tax on each house. Use your confidence interval from part (a) to calculate a 90% confidence interval for the total amount of tax revenue generated by the new houses.

24. **Bunions** Refer to Exercise 16. Researchers omitted 1 patient with a deformity angle of 50 degrees from the analysis due to a measurement issue. What effect would including this outlier have on the confidence interval? Explain your answer without doing any calculations.

25. **Board Lengths** A lumber mill takes long pieces of wood and cuts them into shorter lengths for easier use. The machine the workers use measures and cuts the wood and can be set for a variety of board lengths. The owner knows from long experience that the standard deviation of board lengths for a particular setting on this machine is $\sigma = 0.5$ inch. On a particular day, the machine is set to cut boards to a length of 24 inches. The owner selects a random sample of 30 boards cut that day and finds the sample mean to be $\bar{x} = 24.1$ inches. Find a 95% confidence interval for μ = the mean length of all boards cut that day. *Hint:* You should use a z^* critical value rather than a t^* critical value because you don't need to estimate σ with s_x.

Cumulative Review These exercises revisit what you learned in previous sections.

26. **Mail Call! (1.3)** Even though electronic communication through text messaging and social media has become increasingly important, a recent study found that 41% of U.S. residents still look forward to opening their U.S. Post Office mailbox each day.[56] A social scientist wonders if the proportion of people in their 20s who look forward to opening their "real" mailbox is less than 41%. He selects a random sample of 50 people in their 20s and finds that only 14 of them say that they look forward to opening their mailbox each day. To determine if these data provide convincing evidence that fewer than 41% of people in their 20s look forward to opening their mailbox each day, we simulated 100 random samples. Each dot in the graph shows the number of people in their 20s who look forward to opening their mailbox each day in a random sample of 50 people, assuming that the true percentage is 41%.

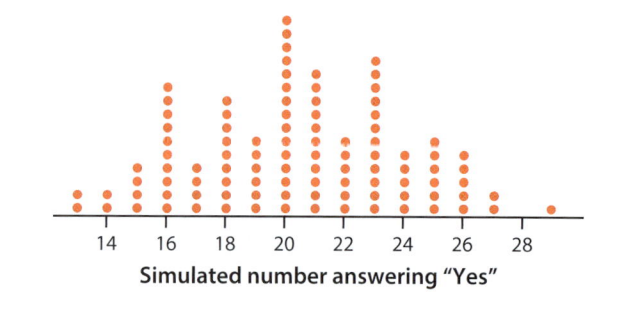

(a) Explain how the graph illustrates the concept of sampling variability.

(b) In the simulation, what percentage of the samples resulted in 14 or fewer people who said "Yes"?

(c) In the actual study, 14 people said "Yes." Based on your answer to part (b), is there convincing evidence that fewer than 41% of people in their 20s look forward to opening their mailbox each day? Explain your reasoning.

27. **Mail Call (5.3, 5.4)** Refer to Exercise 26. Let $X =$ the number of people in a random sample of size 50 who say they look forward to opening their mailbox each day. Assume the true proportion of people who look forward to opening their mailbox is 0.41.

(a) Explain why we can use a binomial distribution to model the distribution of X.

(b) Calculate the mean and standard deviation of the distribution of X. Interpret the standard deviation.

(c) The mean and the standard deviation of the simulated number answering "Yes" in Exercise 26 are 20.46 and 3.54, respectively. How do these compare to your results in part (b)?

Section 7.6 Confidence Intervals for a Population Mean

LEARNING GOALS

By the end of this section, you will be able to:

- Use sample data to check the Normal/Large Sample condition.
- Use the four-step process to construct and interpret a confidence interval for a population mean.

In Section 7.5, you learned how to check the conditions and calculate a confidence interval for a mean. In this section, we will revisit the Normal/Large Sample condition and practice using the four-step process for constructing and interpreting a confidence interval.

The Normal/Large Sample Condition

In Section 7.5, we introduced the Normal/Large Sample condition for constructing a confidence interval for a population mean:

- **Normal/Large Sample:** The data come from a normally distributed population or the sample size is large ($n \geq 30$).

But what if the sample size is small ($n < 30$) and the shape of the population distribution is unknown? In this case, we graph the sample data and ask this question: "Is it plausible that these data came from a normally distributed population?" Outliers or strong skewness in the sample data might indicate that the population distribution is non-normal. However, if the distribution of sample data is roughly symmetric or has only moderate skewness with no outliers, it is believable that the data came from a normally distributed population.

HOW TO Check the Normal/Large Sample Condition

There are three ways that the Normal/Large Sample condition can be met:

1. The data come from a normally distributed population.
2. The sample size is large ($n \geq 30$).
3. When the sample size is small and the shape of the population distribution is unknown, a graph of the sample data shows no strong skewness or outliers.

It can be difficult to determine if a graph shows "strong" skewness or only "moderate" skewness. The following example should help you make this distinction.

EXAMPLE 7.6.1

Are the conditions met?

The Normal/Large Sample condition

PROBLEM

Determine if the Normal/Large Sample condition is met in each of the following scenarios:

(a) How much force does it take to pull wood apart? The stemplot shows the force (in pounds) required to pull apart a random sample of 20 pieces of Douglas fir.

```
23 | 0
24 | 0
25 |
26 | 5
27 |          Key: 31|3 = 313
28 | 7        pounds of force
29 |
30 | 259
31 | 399
32 | 033677
33 | 0236
```

(b) Suppose you want to estimate the mean SAT Math score for the incoming class at a small college. The boxplot summarizes the distribution of SAT Math scores for a random sample of 12 incoming students.

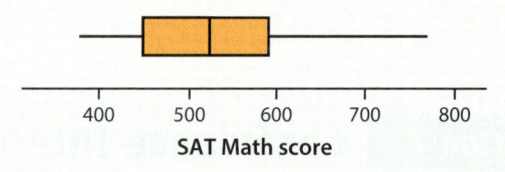

SOLUTION

(a) No. The sample size is small (20 < 30) and the stemplot is strongly skewed to the left with possible outliers.

(b) Yes. Although the sample size is small, the boxplot is not strongly skewed and there are no outliers.

> It is *not* plausible that this sample came from a normal population because of the strong skewness and outliers.

> Even though there is slight skewness in the boxplot, it is plausible that this sample came from a normal population.

EXAM PREP: FOR PRACTICE, TRY EXERCISE 5.

Putting It All Together: Confidence Interval for μ

When you construct and interpret a **one-sample t interval for a mean**, follow the four-step process from Section 7.4: State, Plan, Do, and Conclude.

A **one-sample t interval for a mean** is a confidence interval used to estimate a population mean μ.

EXAMPLE 7.6.2

How much football is really in a football game?

Constructing a confidence interval for μ

PROBLEM

Most National Football League (NFL) games on TV last more than 3 hours, but how much actual "game time" is there in a typical NFL game? Researchers Kirk Goldsberry and Katherine Rowe investigated this question as part of a sports analytics class they taught at the University of Texas. They carefully watched a random sample of 7 games from the 2019 NFL season and recorded the amount of time from when the ball was snapped to when the referee blew the whistle. Here are the game time values (in minutes), excluding any overtime periods:[57]

15.43 16.53 20.18 19.07 16.97 15.72 19.30

Construct and interpret a 99% confidence interval for μ = the mean amount of game time in all 2019 NFL games.

SOLUTION

State: 99% CI for μ = the mean amount of game time in all 2019 NFL games.

> **State:** State the parameter you want to estimate and the confidence level.

Plan: One-sample t interval for μ
Random? Random sample of seven 2019 NFL games ✓
Normal/Large Sample? The sample size is small, but the dotplot doesn't show any outliers or strong skewness. ✓

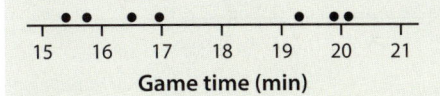

```
      • •   • •       •  • •
   ┬───┬───┬───┬───┬───┬───┬
  15  16  17  18  19  20  21
           Game time (min)
```

> **Plan:** Identify the appropriate inference method and check the conditions.

Do: For these data, $\bar{x} = 17.71$, $s_x = 2.02$, and $n = 7$.
With 99% confidence and df $= 7 - 1 = 6$, $t^* = 3.707$.

$$17.71 \pm 3.707 \frac{2.02}{\sqrt{7}}$$

→ 17.71 ± 2.83
→ 14.88 to 20.54

Using technology: 14.89 to 20.54

> **Do:** If the conditions are met, perform the calculations. *Use the sample data to calculate the mean and standard deviation.*

Conclude: We are 99% confident that the interval from 14.89 minutes to 20.54 minutes captures the mean amount of game time in all 2019 NFL games.

> **Conclude:** Interpret your interval in the context of the problem. *Make sure that your conclusion is about a population mean and includes units when appropriate.*

EXAM PREP: FOR PRACTICE, TRY EXERCISE 9.

⚠ **Make sure to include the graph of sample data when the sample size is small and you are checking the Normal/Large Sample condition.** In the preceding examples, we used a dotplot, a stemplot, and a boxplot to address the Normal/Large Sample condition. You can also use histograms or normal probability plots to assess normality. Each of these graphs has strengths and weaknesses, so we recommend following the advice of your instructor when choosing a graph. In particular, be careful when describing the shape of a distribution based on a boxplot. Boxplots hide modes and gaps, making it impossible to determine whether a distribution is approximately normal based on the boxplot alone. However, because boxplots clearly show skewness and outliers, they can be helpful for identifying important departures from normality.

THINK ABOUT IT

Is it possible to determine the sample size needed to achieve a specified margin of error when estimating a population mean? Yes, but it's complicated. The margin of error (*ME*) in the confidence interval for μ is

$$ME = t^* \frac{s_x}{\sqrt{n}}$$

Unfortunately, there are two problems with using this formula to determine the sample size.

1. We don't know the sample standard deviation s_x because we haven't produced the data yet.
2. The critical value t^* depends on the sample size n that we choose.

The second problem is more serious. To get the correct value of t^*, we have to know the sample size. But that's precisely what we're trying to find!

One alternative is to come up with a reasonable estimate for the *population* standard deviation σ from a similar study that was done in the past or from a small-scale preliminary study. By pretending that σ is known, we can use a z^* critical value rather than a t^* critical value.

Section 7.6 What Did You Learn?

Review the learning goals from this section. Then practice what you've learned by working through the exercises.

Learning Goal	Example	Exercises
Use sample data to check the Normal/Large Sample condition.	7.6.1	5–8
Use the four-step process to construct and interpret a confidence interval for a population mean.	7.6.2	9–12

Section 7.6 Exercises

Building Concepts and Skills These exercises assess the basic knowledge you should have after reading the section.

1. When the sample size is small and the population distribution's shape is unknown, how do you check the Normal/Large Sample condition?

2. True/False: A graph of sample data must be very close to normal to satisfy the Normal/Large Sample condition.

3. List the types of graphs you can use to check the Normal/Large Sample condition when the sample size is small and the population distribution's shape is unknown.

4. Identify one challenge with determining the sample size required when calculating a confidence interval for a population mean.

Mastering Concepts and Skills These exercises reinforce the learning goals as shown in the examples.

5. **Refer to Example 7.6.1 Medical Jargon** Judy is interested in estimating the reading level of a medical journal using average word length. She records the word lengths in a random sample of 100 words. The histogram displays the data. Determine if the Normal/Large Sample condition is met.

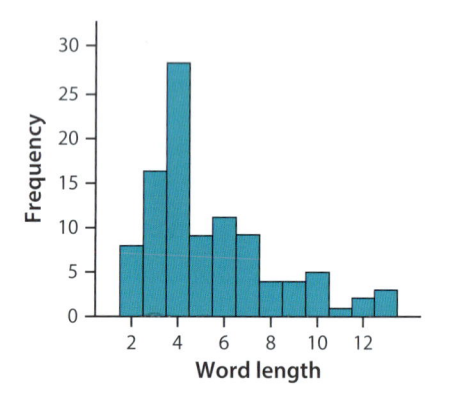

6. **Velvetleaf** The invasive weed known as velvetleaf is often found in U.S. cornfields, where it produces lots of seeds. How many seeds do velvetleaf plants produce? The histogram shows the counts from a random sample of 28 plants that came up in a cornfield when no herbicide was used.[58] Determine if the Normal/Large Sample condition is met.

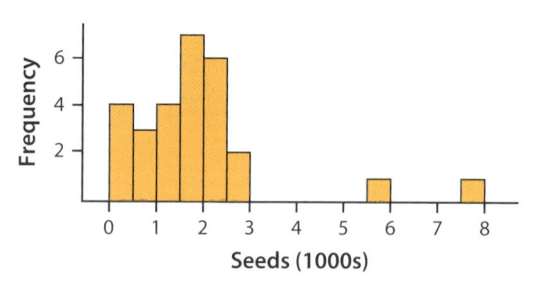

7. **Base Salaries** What is the mean base salary for full-time employees of community colleges in New York City? Researchers selected a random sample of 20 full-time NYC community college employees and recorded the base salary for each employee.[59] The results are summarized in the boxplot. Determine if the Normal/Large Sample condition is met.

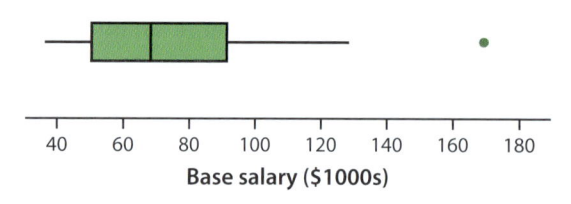

8. **Dissolved Oxygen** The level of dissolved oxygen (DO) in a river shows the water's ability to support aquatic life. A researcher collects water samples from 15 randomly chosen locations along a stream and measures the DO. The results are summarized in the boxplot. Determine if the Normal/Large Sample condition is met.

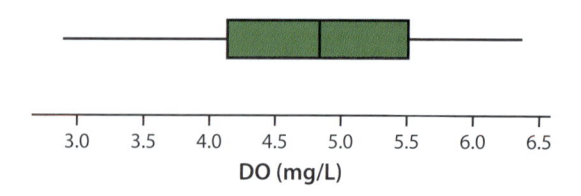

9. **Refer to Example 7.6.2 Iridium in Asteroids** Did an asteroid strike create a dust cloud that led to the extinction of dinosaurs? Researchers took rock samples in Gubbio, Italy, from various depths and measured the concentration of iridium (a rare metal that is more common in asteroids) in them. The deeper the sample, the older the rocks are. A sudden increase in iridium concentration at some point in time would lend

support to the asteroid hypothesis. Here we analyze just those samples from a depth of 347 meters to estimate the mean amount of iridium in a particular time period. Here are the iridium measurements in parts per billion (ppb) for the 5 samples from that depth:[60]

   290   450   620   710   875

Construct and interpret a 90% confidence interval for μ = the mean amount of iridium in all rock samples at a depth of 347 meters in Gubbio, Italy.

10. **Vitamin Testing** Several years ago, the U.S. Agency for International Development provided 238,300 metric tons of corn–soy blend (CSB) for emergency relief in countries throughout the world. CSB is a low-cost fortified food. As part of a study to evaluate appropriate vitamin C levels in this food, measurements were taken on packages of CSB produced in a factory. The following data are the amounts of vitamin C, measured in milligrams per 100 grams (mg/100 g) of blend, for a random sample of 8 packages from one production run:[61]

   26   31   23   22   11   22   14   31

Construct and interpret a 95% confidence interval for μ = the mean amount of vitamin C in CSB packages from this production run.

11. **Healing Newts** Biologists studying the healing of skin wounds measured the rate at which new cells closed a cut made in the skin of an anesthetized newt. Here are data from a random sample of 18 newts, measured in micrometers (millionths of a meter) per hour:[62]

29 27 34 40 22 28 14 35 26 35 12 30 23 18 11 22 23 33

Construct and interpret a 99% confidence interval for μ = the true mean healing rate.

12. **Crabby Crabs** Researchers designed an experiment to determine how noise affects stress level in crabs. They selected a random sample of crabs and wanted to estimate, before running the experiment, the mean weight of crabs in the population. Here are the weights (g) of the crabs in the sample:[63]

22.7	34.6	36.0	40.1	47.5	49.6	50.7	54.4	57.4	59.7
60.5	61.3	67.9	68.5	84.9	84.9	56.0	25.3	29.2	32.0
39.4	41.9	43.8	50.8	53.6	53.0	55.3	59.8	63.6	62.3
57.3	58.8	72.7	74.5						

Construct and interpret a 99% confidence interval for μ = the mean weight of all crabs in the population.

Applying the Concepts These exercises ask you to apply multiple learning goals in a new context or to apply what you learned in this section in a new way.

13. **Video Screens** A manufacturer of high-resolution video terminals must control the tension on the mesh of fine wires that lies behind the surface of the viewing screen. Too much tension will tear the mesh, and too little will allow wrinkles. The tension is measured by an electrical device with output readings in millivolts (mV). Some variation is inherent in the production process. Here are the tension readings from a random sample of 20 screens from a single day's production:

269.5 297.0 269.6 283.3 304.8 280.4 233.5 257.4 317.5 327.4
264.7 307.7 310.0 343.3 328.1 342.6 338.8 340.1 374.6 336.1

Construct and interpret a 90% confidence interval for μ = the mean tension of all the screens produced on this day.

14. **Elephant Size** How tall are adult male African elephants that lived through droughts during their first two years of life? Researchers measured the shoulder height (in centimeters) of a random sample of 14 mature (age 12 or older) male African elephants that lived through droughts during their first two years of life.[64] Here are the data:

200.00	272.91	217.57	294.15	296.84
212.00	257.00	251.39	266.75	237.19
265.85	212.00	220.00	225.00	

Construct and interpret a 90% confidence interval for μ = the mean shoulder height for all mature male African elephants that lived through droughts during their first two years of life.

15. **Video Screens** Refer to Exercise 13. The manufacturer's goal is to produce screens with an average tension of 300 mV. Based on the interval from Exercise 13, is there convincing evidence that the screens produced this day don't meet the manufacturer's goal? Explain your reasoning.

16. **Elephant Size** Refer to Exercise 14. The mean shoulder height for all mature male African elephants is 360 cm.[65] Based on the interval from Exercise 14, is there convincing evidence that the mean shoulder height is smaller for mature male African elephants that went through a drought during their first two years of life? Explain your reasoning.

17. Tech **Boston Commutes** How far do workers commute in Boston? The data set **BostonCommutes** includes the variable *distance*, which records the length of commute (in miles) for a random sample of 500 workers in Boston.[66] Construct and interpret a 90% confidence interval for the mean distance that all workers in Boston commute.

18. Tech **City Trees** The data set **NYCTrees** contains detailed information about a random sample of 8253 trees on New York City streets.[67] One of the measurements taken for each standing tree (i.e., not just a stump) is the diameter at breast height (*DBH*), in inches. Breast height was defined as 54 inches above ground. Construct and interpret a 99% confidence interval for the mean diameter at breast height of all trees on New York City streets.

Extending the Concepts These exercises challenge you to explore statistical concepts and methods that go beyond what you learned in this section.

19. **Estimating BMI** The body mass index (BMI) of all individuals in a certain population is normally distributed with a standard deviation of about 7.5. How large a sample would be needed to estimate the mean BMI in this population to within ±1 with 99% confidence?

20. **Tax Time** How much time do people spend filling out their tax forms, on average? A tax preparation software company wants to estimate the mean amount of time that customers use its software at a 90% confidence level with a margin of error of at most 30 minutes. A pilot study that recorded the amount of "active" time customers spent using the software gave a standard deviation of 154 minutes. How many customers does the software company need to study to meet the goal?

21. **Vitamin Testing** Refer to Exercise 10. What percentage of the data values are within the 95% confidence interval that you calculated? Why does this percentage differ from 95%?

22. **Diving Judges** A spectator wanted to know whether judges at a diving competition were scoring dives with similar strictness. She kept track of all scores given to each diver during one competition. Here are the scores given by two of the judges for one randomly selected competitor's 11 dives:

Dive	Judge A	Judge B
1	5.5	6.5
2	6.0	6.0
3	4.0	4.5
4	4.0	3.0
5	5.5	5.5
6	6.0	6.5
7	5.0	5.0
8	6.0	5.5
9	6.0	6.0
10	5.0	5.5
11	4.5	4.5

(a) Calculate the difference in score (Judge A − Judge B) for each dive.

(b) Graph the differences using a dotplot. Based on the graph, is the Normal/Large Sample condition satisfied? Explain your answer.

(c) Calculate the mean and standard deviation of the differences. Then, calculate a 90% confidence interval for the true mean difference in score.

(d) Interpret the interval from part (c). Based on the interval, is there convincing evidence that one judge is scoring more strictly than the other, on average? Explain your reasoning.

Cumulative Review These exercises revisit what you learned in previous sections.

23. **Arizona Lottery (4.7, 4.8)** In an Arizona lottery game called "The Pick," players pick 6 numbers from 1 to 44. Then lottery officials draw 6 winning numbers from 1 to 44. If a player matches all 6 numbers, they win a share of the jackpot. If they match any 5 of the winning numbers, they win $2000.

(a) How many different ways are there to choose 6 numbers from 1 to 44?

(b) What is the probability of winning the jackpot if you play 1 time?

(c) How many different ways are there to choose 5 of the 6 winning numbers and 1 of 38 non-winning numbers?

(d) What is the probability of winning $2000 if you play one time?

24. **Smoking (4.5)** British government statistics classify adults in the workforce as "managerial and professional" (37% of the population), "intermediate" (35%), or "routine and manual" (28%). A survey finds that 10.2% of adults in managerial and professional occupations smoke, 15.7% of the intermediate group smoke, and 25.5% in routine and manual occupations smoke.[68]

(a) Use a tree diagram to find the percentage of all British adults in the workforce who smoke.

(b) Find the percentage of adult smokers in the workforce who have routine and manual occupations.

Section 7.7 Estimating a Population Standard Deviation or Variance

LEARNING GOALS

By the end of this section, you will be able to:

- State and check the Random and Normal conditions for constructing a confidence interval for a population standard deviation or variance.
- Determine the critical values for calculating a C% confidence interval for a population standard deviation or variance.
- Calculate a C% confidence interval for a population standard deviation or variance.

In Sections 7.3–7.6, you learned how to construct and interpret a confidence interval for a population proportion or a population mean. In this section, you will learn how to check conditions and calculate a confidence interval for a population standard deviation σ or a population variance σ^2. We begin by exploring the sampling distribution of the sample variance s_x^2.

The Sampling Distribution of the Sample Variance

In Section 7.1, you learned that confidence intervals are typically presented in the following form:

$$\text{point estimate} \pm \text{margin of error}$$

Confidence intervals built in this way have the point estimate as the midpoint of the interval. This works well when the sampling distribution of the statistic is symmetric, as is the case for proportions and means when the conditions are met. Is the sampling distribution of the sample variance s_x^2 symmetric? The following Concept Exploration will help you investigate this question.

CONCEPT EXPLORATION

Investigating the Sampling Distribution of the Sample Variance

In this exploration, you will investigate the sampling distribution of the sample variance s_x^2 when sampling from a normally distributed population.

1. Go to www.stapplet.com and launch the *Simulating Sampling Distributions* applet.
2. In the box labeled "Population," choose the normal population distribution. Note that $\sigma = 2$, so $\sigma^2 = 2^2 = 4$.
3. In the box labeled "Sample," change the sample size to $n = 10$ and press the Select sample button. The applet will graph the distribution of the sample and display the value of the sample mean $\bar{x}$.
4. In the box labeled "Sampling Distribution," choose s^2 in the first drop-down menu. Notice that the value of s^2 is now shown in the box labeled "Sample." Now have the applet quickly select 1000 random samples and display the simulated sampling distribution of s^2. Describe the shape of the sampling distribution.
5. Underneath the sampling distribution, the applet displays the mean and standard deviation of the simulated sampling distribution. How does the mean of the sampling distribution compare with the value of $\sigma^2 = 4$?
6. Click the Clear samples button at the bottom of the applet. Then, in the box labeled "Sample," change the sample size to $n = 20$ and repeat Step 4. How does the sample size affect the shape of the sampling distribution of s^2?

Here is an approximate sampling distribution of s_x^2 using 1000 random samples of size $n = 20$ from a normally distributed population with $\sigma_x^2 = 4$.

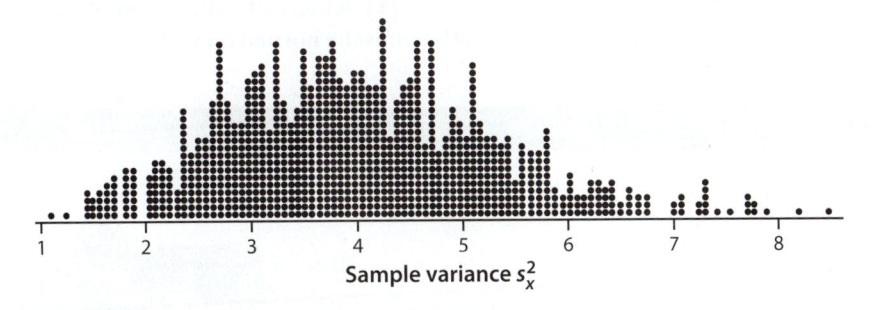

Unlike the sampling distribution of $\hat{p}$ or $\bar{x}$, the sampling distribution of s_x^2 is right skewed, even when the population distribution is normal. This means we cannot use z^* or t^* critical values when calculating a confidence interval for σ^2. Fortunately, when the conditions for inference are met, the statistic

$$\chi^2 = \frac{(n-1)s_x^2}{\sigma^2}$$

follows a **chi-square distribution** (*chi* is pronounced "kye," like "rye").

> A **chi-square distribution** is defined by a density curve that takes only non-negative values and is skewed to the right. A particular chi-square distribution is specified by its *degrees of freedom*. When performing inference about a population standard deviation or variance based on a random sample of size n, use a chi-square distribution with df $= n - 1$.

Figure 7.4 shows the density curves for three members of the chi-square family of distributions. As you can see, the chi-square distributions are skewed to the right, but become less skewed as the sample size (and degrees of freedom) increases.

FIGURE 7.4 The density curves for three members of the chi-square family of distributions.

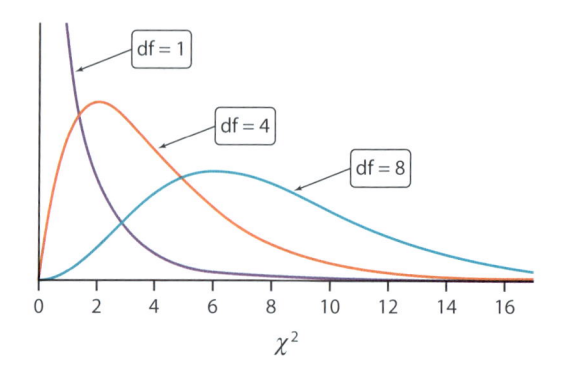

Conditions for Estimating σ or σ^2

Just as for the confidence intervals discussed earlier in this chapter, certain conditions need to be met to construct a confidence interval for a population standard deviation or population variance.

> **HOW TO** **Check the Conditions for Constructing a Confidence Interval for σ or σ^2**
>
> To construct a confidence interval for a population standard deviation σ or population variance σ^2, check that the following conditions are met:
> - **Random:** The data come from a random sample from the population of interest
> - **Normal:** The data come from a normally distributed population.

When the Random condition is satisfied, we can generalize about the population from which the sample was selected. We check the Normal condition to make sure that the sampling distribution of $\dfrac{(n-1)s_x^2}{\sigma^2}$ follows a χ^2 distribution with degrees of freedom df $= n-1$. Note that there is no Large Sample override to the Normal condition in this case. ⚠ **When estimating a population standard deviation or variance, the population distribution must be normal no matter the sample size.**

EXAMPLE 7.7.1

How variable is butterfat in milk from Canadian cows?

Checking conditions

PROBLEM

Researchers wanted to estimate the variability of butterfat content in the milk produced by Canadian cows. To do so, they measured the percentage of butterfat in milk from a random sample of 10 pure-bred Canadian cows.[69] Here are the data, along with a boxplot of the data:

3.92 4.95 4.47 4.28 4.07 4.10 4.38 3.98 4.46 5.05

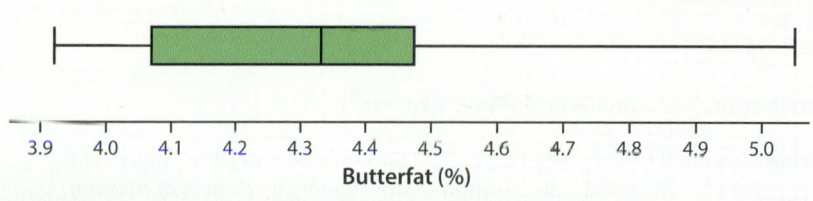

Check whether the conditions for calculating a confidence interval for σ or σ^2 are met.

SOLUTION

Random? Random sample of Canadian cows. ✓

Normal? The boxplot is not strongly skewed and there are no outliers. ✓

> Even though there is slight skewness in the boxplot, it is plausible that this sample came from a normally distributed population.

EXAM PREP: FOR PRACTICE, TRY EXERCISE 5.

Recall that a graph of the sample data does not have to show exact symmetry to satisfy the Normal condition. Samples that come from a normal population distribution are likely to show some skewness due to sampling variability.

χ^2 Critical Values

Because chi-square distributions are not symmetric, finding the critical values for a confidence interval for a population standard deviation or variance is different than finding the value of z^* or t^*. Instead of having one critical value, as we did for means and proportions, we will now have two. We label these χ_L^2 and χ_U^2 for the lower and upper critical values.

Suppose we want to calculate a 95% confidence interval for the population standard deviation or variance based on a random sample of size $n = 15$ from a normally distributed population. To find the upper and lower critical values for the confidence interval, we use a chi-square distribution with df $= 15 - 1 = 14$. With 95% of the area under the chi-square density curve between χ_L^2 and χ_U^2, there is an area of $(100\% - 95\%)/2 = 2.5\%$ or 0.025 to the left of χ_L^2 and an area of 0.025 to the right of χ_U^2, as shown in Figure 7.5.

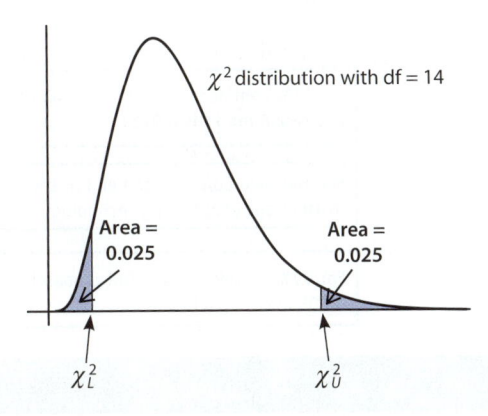

FIGURE 7.5 Upper and lower chi-square critical values for a 95% confidence interval.

Just as when we calculated z^* and t^*, we can use technology or a table to calculate the upper and lower critical values from a chi-square distribution. See the Tech Corner at the end of the section for instructions about using technology.

Here is an excerpt from Table C (in the back of the book) that we can use to find the upper critical value χ_U^2 in this setting. Note that Table C shows the area to the *right* of a given critical value. In this case, the area to the right of χ_U^2 is 0.025. For 95% confidence and df $= 14$, $\chi_U^2 = 26.119$.

	Tail probability p		
df	.05	.025	.01
13	22.362	24.736	27.688
14	23.685	26.119	29.141
15	24.996	27.488	30.578

To find the lower critical value χ_L^2, look for the area to the right of $1 - 0.025 = 0.975$. For 95% confidence and df $= 14$, $\chi_L^2 = 5.629$.

	Tail probability p		
df	.99	.975	.95
13	4.107	5.009	5.892
14	4.660	5.629	6.571
15	5.229	6.262	7.261

Finding χ_U^2 and χ_L^2 Using Table C

1. Using Table C, find the row corresponding to df = $n-1$ on the left side of the table. If the correct df isn't listed, use the greatest df available that is less than the correct df.
2. For a C% confidence interval, the percentage of the area to the right of χ_U^2 is $(100-C)/2$. Convert this value to a decimal that represents the tail probability p. Find the upper critical value χ_U^2 in the corresponding column of the table.
3. For a C% confidence interval, the area to the right of χ_L^2 is $1-p$. Find the lower critical value χ_L^2 in the corresponding column of the table.

EXAMPLE 7.7.2

Is this mission critical?

Finding χ^2 critical values

PROBLEM

Which critical values χ_L^2 and χ_U^2 should be used in constructing a confidence interval for a population standard deviation or variance in the following scenarios? Assume the conditions are met.

(a) A 95% confidence interval based on a random sample of size $n=10$
(b) A 99% confidence interval based on a random sample of size $n=30$

SOLUTION

(a) df = $10-1=9$,

Using Table C: $\chi_L^2 = 2.700$ and $\chi_U^2 = 19.023$

Using technology: $\chi_L^2 = 2.700$ and $\chi_U^2 = 19.023$

> For 95% confidence, the tail probability p for χ_U^2 is 0.025 and for χ_L^2 is 0.975.
>
> See the Tech Corner at the end of the section for instructions about using technology.

(b) df = $30-1=29$,

Using Table C: $\chi_L^2 = 13.121$ and $\chi_U^2 = 52.336$

Using technology: $\chi_L^2 = 13.121$ and $\chi_U^2 = 52.336$

> For 99% confidence, the tail probability p for χ_U^2 is 0.005 and for χ_L^2 is 0.995.

EXAM PREP: FOR PRACTICE, TRY EXERCISE 9.

In both parts of Example 7.7.2, there was a row in Table C for the correct df. If you are using Table C and the correct df isn't listed, use the greatest df available that is less than the correct df. Alternatively, see the Tech Corner at the end of the section.

Now we are ready to calculate a confidence interval for the population standard deviation or variance.

Calculating a Confidence Interval for σ or σ^2

In most real-world settings, we are interested in calculating a confidence interval for the population standard deviation σ rather than the population variance σ^2. Fortunately, once we have a confidence interval for σ^2, it is easy to calculate a confidence interval for σ.

Recall that when the Random and Normal conditions are met, $\dfrac{(n-1)s_x^2}{\sigma^2}$ follows an approximate χ^2 distribution with df = $n-1$. This means that C% of the $\dfrac{(n-1)s_x^2}{\sigma^2}$ values will be between χ_L^2 and χ_U^2. In other words,

$$\chi_L^2 \leq \frac{(n-1)s_x^2}{\sigma^2} \leq \chi_U^2$$

for C% of samples. Solving each side of this inequality for σ^2 yields a confidence interval for the population variance:

$$\frac{(n-1)s_x^2}{\chi_U^2} \leq \sigma^2 \leq \frac{(n-1)s_x^2}{\chi_L^2}$$

While the computation of this interval is different from those you have encountered before, its meaning is the same—an interval of plausible values for the population variance based on sample data.

How do we find a confidence interval for the population standard deviation? Just take the square root of all three expressions:

$$\sqrt{\frac{(n-1)s_x^2}{\chi_U^2}} \leq \sigma \leq \sqrt{\frac{(n-1)s_x^2}{\chi_L^2}}$$

HOW TO Calculate a Confidence Interval for σ or σ^2

When the Random and Normal conditions are met, a C% confidence interval for the population variance σ^2 is

$$\frac{(n-1)s_x^2}{\chi_U^2} \leq \sigma^2 \leq \frac{(n-1)s_x^2}{\chi_L^2}$$

where the area between χ_L^2 and χ_U^2 is $\dfrac{C}{100}$ in a chi-square distribution with df $= n-1$.

Therefore, a C% confidence interval for the population standard deviation σ is

$$\sqrt{\frac{(n-1)s_x^2}{\chi_U^2}} \leq \sigma \leq \sqrt{\frac{(n-1)s_x^2}{\chi_L^2}}$$

EXAMPLE 7.7.3

How variable is butterfat in milk from Canadian cows?

Calculating a confidence interval for σ

PROBLEM
Because milk from many cows is combined before it reaches consumers, it would help farmers to know how much variability there is in the percentage of butterfat in milk from different cows. The standard deviation of the butterfat percentage in samples of milk from 10 randomly selected Canadian cows is 0.3862%. We checked the conditions for these data in Example 7.7.1. Calculate a 95% confidence interval for the population standard deviation.

SOLUTION
For these data, $s_x = 0.3862$, $n = 10$, and df $= 10 - 1 = 9$.

Using Table C: $\chi_L^2 = 2.700$ and $\chi_U^2 = 19.023$

Using technology: $\chi_L^2 = 2.700$ and $\chi_U^2 = 19.023$

$$\sqrt{\frac{(10-1)(0.3862)^2}{19.023}} \leq \sigma \leq \sqrt{\frac{(10-1)(0.3862)^2}{2.700}}$$

→ $0.2656 \leq \sigma \leq 0.7051$

→ 0.2656% to 0.7051%

Start by calculating the sample standard deviation and the appropriate degrees of freedom. Then find the critical values.

$$\sqrt{\frac{(n-1)s_x^2}{\chi_U^2}} \leq \sigma \leq \sqrt{\frac{(n-1)s_x^2}{\chi_L^2}}$$

EXAM PREP: FOR PRACTICE, TRY EXERCISE 13.

We are 95% confident that the interval from 0.2656% to 0.7051% captures the standard deviation of the butterfat percentage in milk from all Canadian cows.

TECH CORNER: Critical Values and Confidence Intervals for a Population Standard Deviation

Calculating χ^2 critical values

You can use technology to calculate χ_L^2 and χ_U^2. We'll illustrate using the example with $n = 10$ and 95% confidence.

Applet

1. Go to www.stapplet.com and launch the χ^2 *Distributions* applet.
2. Choose "Calculate a value corresponding to an area" from the drop-down menu, enter 9 for the degrees of freedom, and click "Plot distribution."
3. Using the drop-down menu below the curve, choose "central" and enter an area of 0.95. Then click "Calculate value(s)" and check the box to show labels on plot.

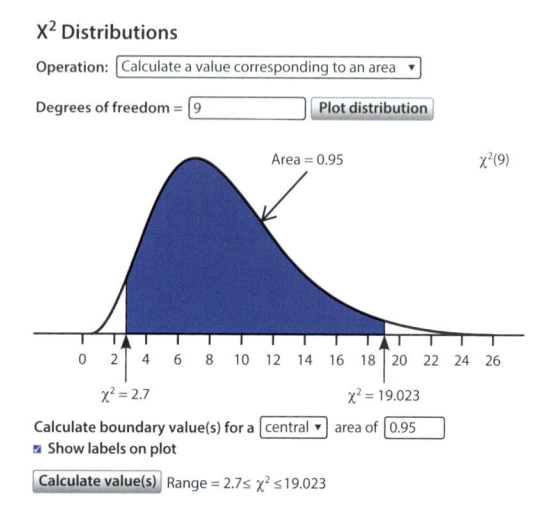

TI-83/84

The inverse χ^2 command is not currently available on the TI-83/84.

Calculating a confidence interval for a population standard deviation

You can also use technology to calculate a confidence interval for a population standard deviation or variance.

We'll illustrate with the butterfat data from the previous examples. Here again are the data:

3.92 4.95 4.47 4.28 4.07 4.10 4.38 3.98 4.46 5.05

Applet

1. Go to www.stapplet.com and launch the *One Quantitative Variable, Single Group* applet.
2. Enter the variable name "Butterfat (%)" and enter the data values, separated by commas or spaces.
3. Click the Begin analysis button. In the "Perform Inference" section, select the "1-sample χ^2 interval for σ" for the inference procedure and 95% for the confidence level. Then click the Perform inference button.

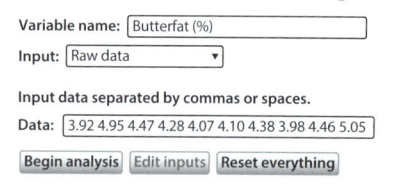

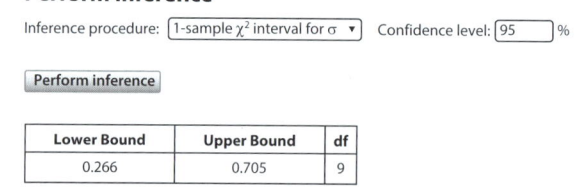

Note: If you have summary statistics instead of raw data, choose "Mean and standard deviation" from the Input menu and enter the values. Then follow the directions in Step 3.

TI-83/84

This type of confidence interval is not currently available on the TI-83/84.

Detailed instructions for using CrunchIt!, Excel, Google Sheets, JMP, Minitab, and R are available in Achieve.

Section 7.7 What Did You Learn?

Review the learning goals from this section. Then practice what you've learned by working through the exercises.

Learning Goal	Example	Exercises
State and check the Random and Normal conditions for constructing a confidence interval for a population standard deviation or variance.	7.7.1	5–8
Determine the critical values for calculating a C% confidence interval for a population standard deviation or variance.	7.7.2	9–12
Calculate a C% confidence interval for a population standard deviation or variance.	7.7.3	13–16

Section 7.7 Exercises

Building Concepts and Skills These exercises assess the basic knowledge you should have after reading the section.

1. True/False: The Normal condition for calculating a confidence interval for a population standard deviation or variance is satisfied whenever $n \geq 30$.

2. How do you calculate the degrees of freedom when estimating a population standard deviation or variance?

3. When computing a 90% confidence interval for a population variance, what right-tail probability p should you use to find χ_L^2? To find χ_U^2?

4. What is the formula for a confidence interval for a population standard deviation?

Mastering Concepts and Skills These exercises reinforce the learning goals as shown in the examples.

5. **Refer to Example 7.7.1 Corn Plants** Just how much variability is there in the height of mature corn plants? You plan to select a random sample of 20 corn plants from a large field and measure the height of each plant. From past experience, the heights of this variety of corn are normally distributed. Check if the conditions for calculating a confidence interval for σ are met.

6. **Screws** A manufacturer of screws is concerned about the variability in the length of the screws being produced in the factory. The quality-control supervisor plans to select a random sample of 30 screws produced in one day and measure their length. From past experience, the lengths of screws produced in the factory are normally distributed. Check if the conditions for calculating a confidence interval for σ are met.

7. **Vitamin Testing** In Exercise 10 from Section 7.6, you found a confidence interval for the mean amount of vitamin C in packages of corn–soy blend (CSB) used for emergency relief in countries throughout the world.[70] In that exercise, we were most interested in the mean amount of vitamin C in all packages produced. It is also important to know just how variable the vitamin C content is. Here is a boxplot of the vitamin C measurements in milligrams per 100 grams (mg/100 g) from a random sample of 8 packages from one production run. Check if the conditions for calculating a confidence interval for σ are met.

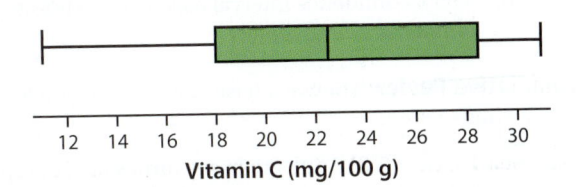

8. **Dissolved Oxygen** The level of dissolved oxygen (DO) in a river shows the water's ability to support aquatic life. If the level of DO varies too much, aquatic life could be endangered even if the mean DO level is acceptable. A researcher collects water samples from 15 randomly chosen locations along a stream and measures the DO. Here is a boxplot of the distribution of DO in milligrams per liter (mg/L). Check if the conditions for calculating a confidence interval for σ are met.

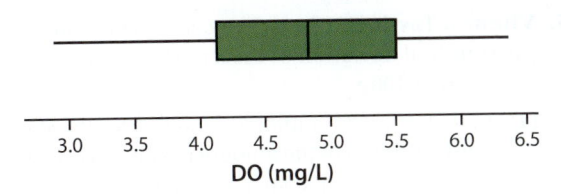

9. **Refer to Example 7.7.2 Finding χ_L^2 and χ_U^2** Which critical values χ_L^2 and χ_U^2 should be used when constructing a confidence interval for the population standard deviation in each of the following scenarios? Assume the conditions are met.
   (a) A 95% confidence interval based on $n = 5$ randomly selected observations
   (b) A 90% confidence interval based on a random sample of 22 individuals

10. **Finding χ_L^2 and χ_U^2** Which critical values χ_L^2 and χ_U^2 should be used when constructing a confidence interval for the population standard deviation in each of the following scenarios? Assume the conditions are met.
    (a) A 90% confidence interval based on $n = 18$ randomly selected observations
    (b) A 99% confidence interval based on a random sample of 25 individuals

11. **Finding χ_L^2 and χ_U^2** Which critical values χ_L^2 and χ_U^2 should be used when constructing a confidence interval for the population standard deviation in each of the following scenarios? Assume the conditions are met.
    (a) A 99% confidence interval based on an SRS of 20 observations
    (b) A 95% confidence interval based on $n = 103$ randomly selected individuals

12. **Finding χ_L^2 and χ_U^2** Which critical values χ_L^2 and χ_U^2 should be used when constructing a confidence interval for the population standard deviation in each of the following scenarios? Assume the conditions are met.
    (a) A 95% confidence interval based on an SRS of 16 observations
    (b) A 90% confidence interval based on $n = 45$ randomly selected individuals

13. **Refer to Example 7.7.3 Corn Plants** Refer to Exercise 5. After selecting the random sample of 20 corn plants, the sample standard deviation was determined to be 8 inches. Calculate a 95% confidence interval for the standard deviation of the height of all corn plants in this field.

14. Screws Refer to Exercise 6. After selecting the random sample of 30 screws, the sample standard deviation was determined to be 0.1 inch. Calculate a 99% confidence interval for the standard deviation of the length of screws produced that day.

15. Vitamin Testing Refer to Exercise 7. The sample standard deviation of the 8 packages is $s_x = 7.19$ mg/100 g.

(a) Calculate a 90% confidence interval for the standard deviation of the amount of vitamin C in all packages from that day's production run.

(b) Interpret the confidence level.

16. Dissolved Oxygen Refer to Exercise 8. The sample standard deviation of the 15 DO values is $s_x = 0.94$ mg/L.

(a) Calculate a 95% confidence interval for the standard deviation of DO in the entire stream.

(b) Interpret the confidence level.

Applying the Concepts These exercises ask you to apply multiple learning goals in a new context or to apply what you learned in this section in a new way.

17. Fishing In Exercise 19 from Section 7.5, you found a confidence interval for the true mean breaking strength of fishing line advertised as "6 pound." In addition to making sure the mean breaking strength meets or exceeds the advertised weight, it is important that the breaking strength is consistent. It would certainly be a problem if some pieces of line could hold 10 pounds, but other pieces could hold only 2 pounds. Here is a dotplot of the 30 breaking strength values from the study.[71]

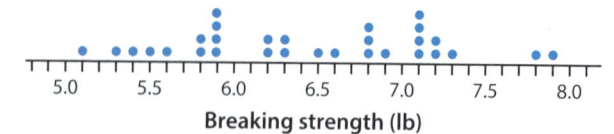

(a) Verify that the conditions are met for calculating a confidence interval for σ = the standard deviation of the breaking strength for this type of fishing line.

(b) For this sample, $\bar{x} = 6.44$ pounds and $s_x = 0.75$ pound. Calculate a 99% confidence interval for σ.

(c) Is there convincing evidence that the true standard deviation is less than 1 pound? Explain your reasoning.

18. Video Screens In Exercise 13 from Section 7.6, you estimated the mean amount of tension on the mesh of fine wires that lie behind the surface of video screens. Some variation is inherent in the production process, but the manufacturer would like to keep this variation to a minimum. Here is a boxplot of the tension measurements, in millivolts (mV), from 20 screens randomly selected from those produced on a particular day.

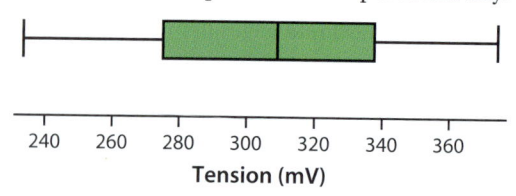

(a) Verify that the conditions are met for calculating a confidence interval for σ = the standard deviation of the amount of tension in all screens produced that day,

(b) For this sample, $\bar{x} = 306.32$ mV and $s_x = 36.21$ mV. Calculate a 90% confidence interval for σ.

(c) The goal for the manufacturer is to produce screens with a standard deviation of at most 25 mV. Based on the interval, is there convincing evidence that the screens produced this day don't meet the manufacturer's goal? Explain your reasoning.

Extending the Concepts These exercises challenge you to explore statistical concepts and methods that go beyond what you learned in this section.

19. Corn Plants Refer to Exercise 13.

(a) Compute a 95% confidence interval for σ^2 = the variance of the height of all corn plants in this field.

(b) What are the units for the confidence interval computed in part (a)?

(c) Compare the length of the confidence interval computed in part (a) to the one you computed in Exercise 13. You should have found that the interval in part (a) is longer. Explain why.

20. Engagement Rings In Section 7.5, you worked through examples dealing with the cost of engagement rings. In the random sample of 1640 engagement rings, the mean cost was $3329 with a standard deviation of $4739.[72] Explain why the conditions for constructing a confidence interval for σ are not met.

21. Exact Critical Values Use technology to find the exact values of χ_L^2 and χ_U^2 to be used when constructing a confidence interval for the population standard deviation in each of the following scenarios. Assume the conditions are met.

(a) A 90% confidence interval based on $n = 72$ randomly selected observations

(b) A 99% confidence interval based on a random sample of 35 individuals

Cumulative Review These exercises revisit what you learned in previous sections.

22. Sea Turtles (6.3) Adult green sea turtles have carapace (shell) lengths that follow a normal distribution with mean 98.5 cm and standard deviation 5.17 cm. What shell length would put a green sea turtle in the largest 8% of its species?

23. **Full House (4.8)** A standard deck has 52 playing cards — 4 cards each of 13 different denominations (ace, two, three, ..., ten, jack, queen, and king). You and a friend decide to play a game in which you each start with 5 cards. Suppose you shuffle the deck well and deal out 5 cards to you and your friend. You are surprised when you look at your hand and see three aces along with two kings. Should you be surprised? Find the probability of dealing yourself three aces and two kings in a 5-card hand from a well-shuffled deck.

Section 7.8 Confidence Intervals for a Population Standard Deviation or Variance

LEARNING GOALS

By the end of this section, you will be able to:
- Make a graph of sample data and check the Normal condition.
- Use the four-step process to construct and interpret a confidence interval for a population standard deviation.

In Section 7.7, you learned how to check the conditions and calculate a confidence interval for a population standard deviation or variance using summarized data. In this section, we will revisit the Normal condition and practice using the four-step process for constructing and interpreting a confidence interval for σ.

The Normal Condition

In Section 7.7, we introduced the Normal condition for constructing a confidence interval for a population standard deviation or variance:

- **Normal:** The data come from a normally distributed population.

There are two ways to assess this condition. It is possible, but rare, that we have prior knowledge that the population has a normal distribution. In most cases, we assess normality using a graph of the sample data. For dotplots, stemplots, histograms, and boxplots, check that there is no strong skewness or outliers. For normal probability plots, check that the graph has a reasonably linear form. Each of these graphs has strengths and weaknesses when assessing normality, so ask your instructor which graph you should use.

EXAMPLE 7.8.1

How variable are brains?

The Normal condition

PROBLEM

Amyloid-β (Abeta) is a protein fragment that has been linked to Alzheimer's disease. Researchers collected brain tissue from a random sample of Catholic priests and measured the concentration of Abeta (pmol/g) in the tissue. How much variability is there in the amount of Abeta in brain tissue for the population of Catholic priests? Here are the data from this study.[73] Determine if the Normal condition is met.

114	41	276	0	16	228	927	0	211	829	1561	0	276	959	16
24	325	49	537	73	33	16	8	276	537	0	569	772	0	260
423	780	1610	0	309	512	797	24	57	106	407	390	1154	138	634
919	1415	390	1024	1154	195	715	1496	407	1171	439	894			

(continued)

SOLUTION

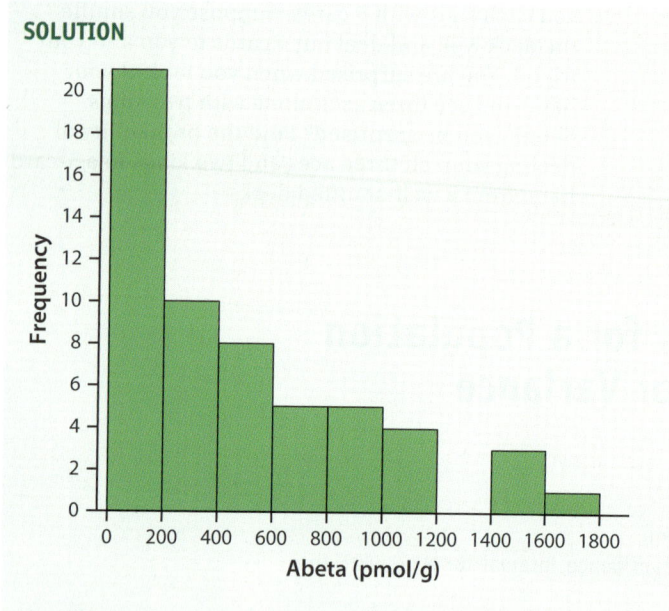

No. A histogram of the data shows a clear right skew. It is not plausible that these data came from a normal population.

> Even though $n \geq 30$, we still have to check for normality. There is no large sample size override when constructing a confidence interval for σ or σ^2.

EXAM PREP: FOR PRACTICE, TRY EXERCISE 5.

Here is a normal probability plot of the brain data from Example 7.8.1. The graph shows a clear curved pattern, so the Normal condition is not met. It would not be appropriate to calculate a confidence interval for the population standard deviation or variance in this case.

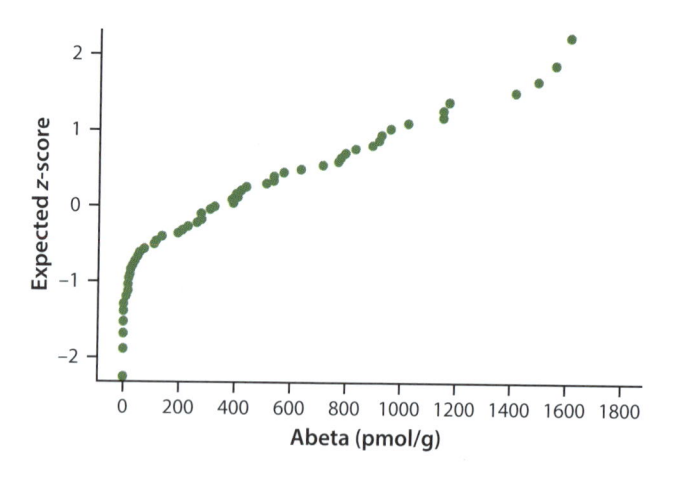

Putting It All Together: Confidence Interval for σ

Because standard deviations are easier to interpret and have the same units as the data, it is much more common to construct confidence intervals for a population standard deviation than for a population variance. When you construct and interpret a **one-sample confidence interval for a standard deviation (variance)**, follow the four-step process from Sections 7.4 and 7.6: State, Plan, Do, and Conclude.

> A **one-sample confidence interval for a standard deviation (variance)** is a confidence interval used to estimate a population standard deviation σ (or a population variance σ^2).

EXAMPLE 7.8.2

How well does it fit?

Constructing a confidence interval for σ

PROBLEM

Clothing manufacturers need information about body size to create clothing that fits appropriately. For a new jacket it plans to introduce, a company measured the shoulder-to-hip ratio for a random sample of 38 undergraduate male students.[74] Here are the data:

1.352	1.349	1.431	1.405	1.307	1.363	1.218	1.136	1.333	1.218	1.457
1.436	1.409	1.247	1.545	1.249	1.643	1.269	1.310	1.127	1.342	1.368
1.277	1.186	1.464	1.315	1.428	1.099	1.284	1.249	1.363	1.377	1.475
1.259	1.402	1.182	1.087	1.322						

Construct and interpret a 95% confidence interval for σ = the standard deviation of shoulder-to-hip ratios in all undergraduate male students.

SOLUTION

State: 95% CI for σ = the standard deviation of shoulder-to-hip ratio in all undergraduate male students.

Plan: One-sample confidence interval for σ

Random? Random sample of undergraduate male students. ✓

Normal? The histogram does not show strong skewness or outliers. ✓

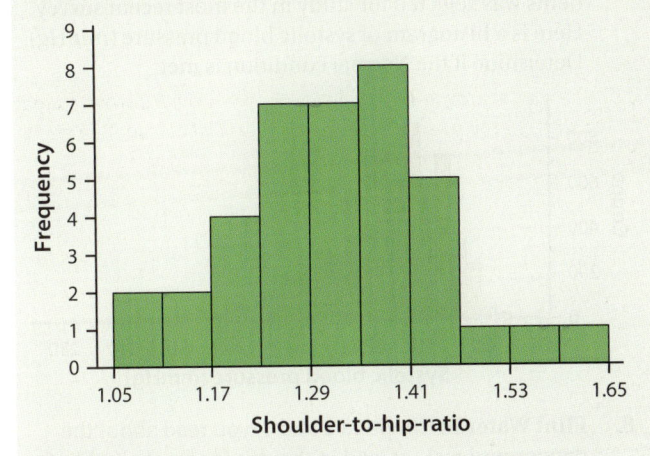

Do: For these data, $s_x = 0.1215$, $n = 38$, and df = 38 − 1 = 37.

Using technology: With df = 37 and 95% confidence, $\chi_L^2 = 22.106$ and $\chi_U^2 = 55.668$.

$$\sqrt{\frac{(37)(0.1215)^2}{55.668}} \leq \sigma \leq \sqrt{\frac{(37)(0.1215)^2}{22.106}}$$

→ $0.099 \leq \sigma \leq 0.157$

Conclude: We are 95% confident that the interval from 0.099 to 0.157 captures the standard deviation of shoulder-to-hip ratios in all undergraduate male students.

State: State the parameter you want to estimate and the confidence level.

Plan: Identify the appropriate inference method and check the conditions.

Do: If the conditions are met, perform the calculations. *Use technology to calculate the standard deviation of the sample data and the critical values.*

When using Table C in this context, we would have to use df = 30 to calculate the critical values and n − 1 = 30 in the formula for the confidence interval. Read the discussion after the example for more details.

Conclude: Interpret your interval in the context of the problem. *Make sure that your conclusion is in context, about a population standard deviation, and includes units, when appropriate.*

EXAM PREP: FOR PRACTICE, TRY EXERCISE 9.

If we wanted to use Table C in Example 7.8.2, we would need to use df = 30 instead of df = 37 because there is no row for df = 37. And because the critical values for df = 30 (46.979 and 16.791) are smaller than the critical values for df = 37 (55.668 and 22.106), we would need to use a smaller value for n − 1 in the formula for the confidence interval to compensate (30 rather than 37). If we used df = 30 for the critical values but $n-1=37$ in the formula, the endpoints of the interval would be larger than they should be, and our method would be biased. Using technology is not only more straightforward, but the resulting interval (0.099 to 0.157) is also narrower than when using Table C (0.097 to 0.162).

Section 7.8 What Did You Learn?

Review the learning goals from this section. Then practice what you've learned by working through the exercises.

Learning Goal	Example	Exercises
Make a graph of sample data and check the Normal condition.	7.8.1	5–8
Use the four-step process to construct and interpret a confidence interval for a population standard deviation.	7.8.2	9–12

Section 7.8 Exercises

Building Concepts and Skills These exercises assess the basic knowledge you should have after reading the section.

1. True/False: A graph of sample data must be very close to normal to satisfy the Normal condition for constructing a confidence interval for a population standard deviation or variance.

2. List the types of graphs you can use to check the Normal condition.

3. Why do we typically calculate confidence intervals for the population standard deviation σ, rather than the population variance σ^2?

4. In the *Conclude* step, the interpretation should always be in _____ and include _____ when appropriate.

Mastering Concepts and Skills These exercises reinforce the learning goals as shown in the examples.

5. **Refer to Example 7.8.1 Cracker Fiber** Would eating crackers made from different types of fiber before meals affect the number of calories digested? Researchers designed an experiment to answer this question, using 12 volunteer subjects. They compared four types of fiber, with each subject receiving each type of fiber in a random order. The sample values of digested calories when the subjects ate bran (one type of fiber) crackers are given here in calories per day.[75] Determine if the Normal condition is met.

2047 2658 1752 1669 2207 1707 2767 2280 2293 2357 2003 2288

6. **CEO Compensation** How much variability is there in CEO compensation in the United States? Here are the salaries, in millions of dollars, for a random sample of 20 U.S. CEOs in a recent year.[76] Determine if the Normal condition is met.

21.80 11.20 6.69 12.12 14.48 1.72 18.95 19.39 15.16 10.21
8.41 11.84 9.67 1.50 8.30 29.14 17.95 22.51 16.43 31.33

7. **Blood Pressure** The National Health and Nutrition Examination Survey (NHANES) is an ongoing research program conducted by the National Center for Health Statistics. A random sample of more than 8000 U.S. residents was selected for study in the most recent survey.[77] Here is a histogram of systolic blood pressure (mmHg). Determine if the Normal condition is met.

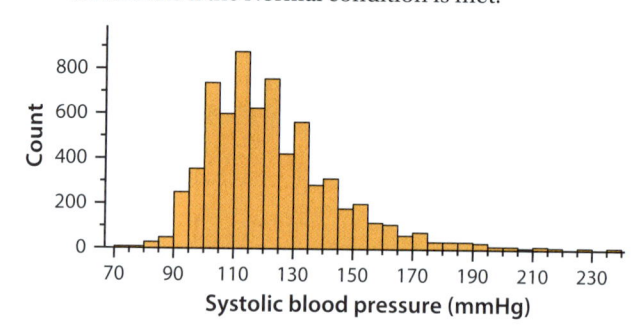

8. **Flint Water** In Chapters 2 and 3, you read about the dangerous levels of lead in the drinking water in Flint, Michigan. Here are the 71 observations collected by the Michigan Department of Environmental Quality in 2015.[78] Determine if the Normal condition is met.

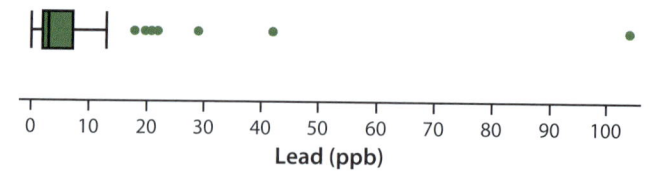

9. **Refer to Example 7.8.2 Red-Tail Hawks** A biology professor spent many years studying birds of prey. He and his students captured, measured, and released such birds as they were migrating through their state. One variable of interest was the tail length. Here are the tail lengths, in millimeters, for a random sample of 10 red-tail hawks they observed over the years.[79]

238 215 220 212 220 233 249 198 226 228

Construct and interpret a 95% confidence interval for σ = the standard deviation of tail length in all red-tail hawks migrating through the area.

10. **Football Games** Most NFL games on TV last more than 3 hours, but the actual "game time" is much shorter. How much does game time vary? Researchers Kirk Goldsberry and Katherine Rowe investigated this question as part of a sports analytics class they taught at the University of Texas. They carefully watched a random sample of 7 games from the 2019 NFL season and recorded the amount of time from when the ball was snapped to when the referee blew the whistle. Here are the game time values (in minutes), excluding any overtime periods:[80]

    15.43   16.53   20.18   19.87   16.97   15.72   19.30

    Construct and interpret a 90% confidence interval for σ = the standard deviation of game time for all 2019 NFL games.

11. **Crabby Crabs** In Exercise 12 from Section 7.6, you found a confidence interval for the mean weight of crabs using a random sample of crabs that were to be used in an experiment about noise and stress level. Before running the experiment, the researchers wanted to estimate how variable crab weights are in the population. Here is a histogram of weight in grams (g) for the 34 crabs in the sample. The sample standard deviation for these data is $s_x = 15.36$ g.[81]

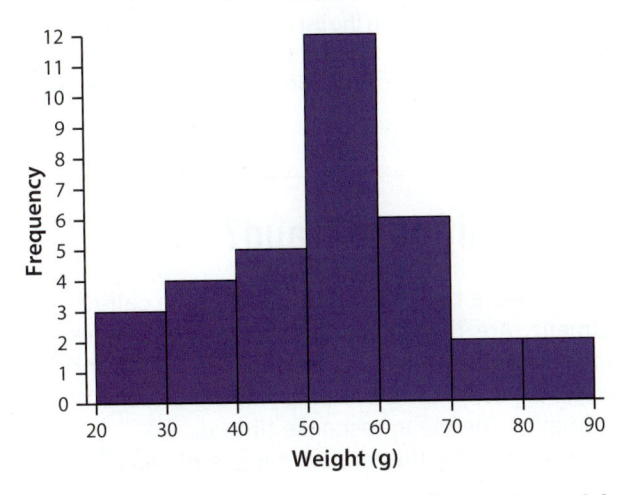

    Construct and interpret a 99% confidence interval for σ = the standard deviation of weight for all crabs in the population.

12. **Possums** In Exercises 8 and 18 from Section 7.5, you found a confidence interval for the mean length of an Australian possum. How much do the lengths of Australian possums typically vary? Here is a histogram of the distribution of total length (in centimeters) for a random sample of 104 possums. The sample standard deviation for these data is $s_x = 4.31$ cm.[82]

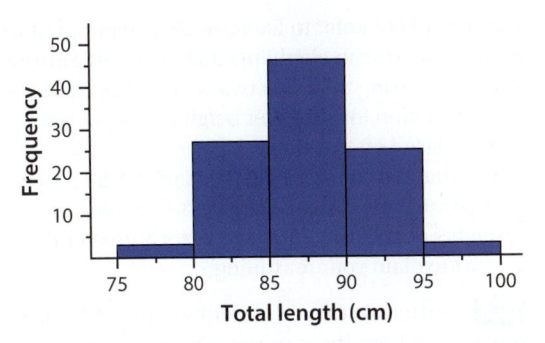

    Construct and interpret a 95% confidence interval for σ = the standard deviation of total length for all Australian possums.

Applying the Concepts These exercises ask you to apply multiple learning goals in a new context or to apply what you learned in this section in a new way.

13. **Iridium in Asteroids** Did an asteroid strike create a dust cloud that led to the extinction of dinosaurs? Researchers took rock samples in Gubbio, Italy, from various depths and measured the concentration of iridium (a rare metal that is more common in asteroids) in them. The deeper the sample, the older the rocks are. A sudden increase in iridium concentration at some point in time would lend support for the asteroid hypothesis. Here we analyze just those samples from a depth of 347 meters to see how variable the iridium measurements are within a particular time period. Here are the iridium measurements in parts per billion (ppb) for the 5 samples from that depth:[83]

    290   450   620   710   875

    Construct and interpret a 95% confidence interval for σ = the standard deviation of amount of iridium for all rock samples at a depth of 347 meters.

14. **Elephant Size** How variable are the heights of adult male African elephants that lived through droughts during their first two years of life? Researchers measured the shoulder height (in centimeters) of a random sample of 14 mature (age 12 or older) male African elephants that lived through droughts during their first two years of life.[84] Here are the data:

200.00	272.91	217.57	294.15	296.84
212.00	257.00	251.39	266.75	237.19
265.85	212.00	220.00	225.00	

    Construct and interpret a 90% confidence interval for σ = the standard deviation of shoulder height for all mature male African elephants that lived through droughts during their first two years of life.

15. **Iridium in Asteroids** Refer to Exercise 13. Researchers expect that the true standard deviation for iridium samples at a constant depth is 200 ppb. Based on the interval, is there convincing evidence that the true standard deviation is not 200 ppb? Explain your reasoning.

16. **Elephant Size** Refer to Exercise 14. The population of adult male African elephants that do not live through droughts during their first two years of life has a standard deviation for shoulder heights of approximately 17 cm. Based on the interval, is there convincing evidence that the standard deviation of shoulder height for mature male African elephants that live through droughts during their first two years is greater than 17 cm? Explain your reasoning.

17. **Tech** **South Pole CO_2** A random sample of 100 carbon dioxide readings (in parts per million, or ppm) taken between 1988 and 2016 from the South Pole (CO_2) are stored in the data set **SouthPole**.[85] If it is appropriate, calculate and interpret a 99% confidence interval for the standard deviation of all possible carbon dioxide readings from the South Pole during this time period. If it is not appropriate, explain why not.

18. **Tech** **Hot Coffee** How much variability is there in the price (in dollars) of hot coffee at different restaurants and coffee shops? A university statistics class recorded hot coffee prices (*HotPrice*) at randomly selected restaurants, cafes, and coffee shops near the university. The data are in the data set **Coffee**.[86] If it is appropriate, calculate and interpret a 95% confidence interval for the standard deviation of hot coffee prices in all restaurants, cafes, and coffee shops near the university. If it is not appropriate, explain why not.

Extending the Concepts These exercises challenge you to explore statistical concepts and methods that go beyond what you learned in this section.

19. **Red-Tail Hawks** Refer to Exercise 9. Construct and interpret a 90% confidence interval for σ^2 = the variance of tail length in all red-tail hawks migrating through the area.

Cumulative Review These exercises revisit what you learned in previous sections.

20. **College Class Years (4.2)** A liberal arts college classifies each student as being either a first-year, sophomore, junior, senior, or other. Here is the distribution of class year for a recent academic year. Suppose we choose a student at random.

Year	First-year	Sophomore	Junior	Senior	Other
Probability	0.29	0.24	?	0.25	0.02

   (a) What is the probability that this student is a junior?
   (b) What is the probability that this student is either a first-year or belongs in the "other" category?

21. **Dogs and Cats (6.5)** For each of the following scenarios, determine whether the value measured is a statistic or a parameter.
   (a) The manager at a large animal shelter records how long each of the cats currently at the shelter has been there. He then computes the average length of stay.
   (b) A town council is considering creating a dog park in their town. One council member gets a list of local dog owners from a veterinarian. She calls a random sample of the people on the list and asks whether they would take their dog to the dog park if the town built it. She computes the proportion of owners who say "Yes."

Statistics Matters How can I prevent credit card fraud?

At the beginning of this chapter, we described a bank that was studying the call-response times in its credit card fraud department. Are the bank's customer service representatives generally meeting the goal of answering incoming calls in less than 30 seconds? We can approach this question in two ways: by estimating the proportion p of calls that are answered within 30 seconds or by estimating the mean response time μ.

Here are a histogram and summary statistics for the random sample of calls.[87]

n	Mean	SE Mean	StDev	Min	Q_1	Med	Q_3	Max
241	18.353	0.758	11.761	1	9	16	25	49

1. Based on the sample data, a 95% confidence interval for the proportion of all calls to the bank's credit card fraud department that are answered in less than 30 seconds is 0.783 to 0.877. Interpret this interval.

2. Calculate the point estimate and the margin of error for the interval in Question 1.

3. Explain two ways that the bank could reduce the margin of error for the confidence interval in Question 1, along with any drawbacks to these actions.

4. Construct and interpret a 95% confidence interval for the mean response time for all calls to the bank's credit card fraud department.

5. In this context, what does it mean to be 95% confident?

6. Is the credit card fraud department meeting its goal of answering calls in less than 30 seconds? Explain your reasoning.

7. Explain why it wouldn't be a good idea to calculate a confidence interval for the population standard deviation using these data.

Chapter 7 Review

The Idea of a Confidence Interval

- A **confidence interval** gives a set of plausible values for a parameter based on sample data. Confidence intervals can be used to make decisions about the value of a parameter.
  - To interpret a C% confidence interval for an unknown parameter, say, "We are C% confident that the interval from _____ to _____ captures the [parameter in context]."
- The **confidence level** C gives the long-run capture rate of the method for calculating the confidence interval.
  - To interpret the confidence level C, say, "If we were to select many random samples from a population and construct a C% confidence interval using each sample, about C% of the intervals would capture the [parameter in context]."
- The typical structure of a confidence interval is

    point estimate ± margin of error

- A **point estimate** is a single-value estimate of a population parameter calculated from sample data.
- The **margin of error** of an estimate describes how far, at most, we expect the point estimate to vary from the population parameter. In a C% confidence interval, the distance between the point estimate and the true parameter value will be less than the margin of error in about C% of all samples.
  - The margin of error is affected by the confidence level and the sample size. Assuming everything else remains the same, increasing the confidence level C increases the margin of error and increasing the sample size decreases the margin of error.
  - The margin of error does not account for problems in the data collection process, such as undercoverage, nonresponse, or poorly worded questions. The margin of error only accounts for variability due to random sampling.

- When constructing and interpreting a confidence interval, follow the **four-step process:**
  - **State:** State the parameter you want to estimate and the confidence level.
  - **Plan:** Identify the appropriate inference method and check the conditions.
  - **Do:** If the conditions are met, perform the calculations.
  - **Conclude:** Interpret your interval in the context of the problem.

Confidence Intervals for a Proportion

- There are two **conditions** for constructing a confidence interval for a population proportion p:
  - **Random:** The data come from a random sample from the population of interest.
  - **Large Counts:** Both $n\hat{p}$ and $n(1-\hat{p})$ are at least 10. That is, the observed number of successes and the observed number of failures are both at least 10.
- When the conditions are met, a **one-sample z interval for p** is

$$\hat{p} \pm z^* \sqrt{\frac{\hat{p}(1-\hat{p})}{n}}$$

where $\hat{p}$ is the proportion of successes in the sample, z^* is the critical value, and n is the sample size.
  - The **critical value** z^* is a multiplier that makes the interval wide enough to have the stated capture rate. On a standard normal curve, C% of the area will be between $-z^*$ and z^*. Use Table A or technology to find the value of z^*.

- The **standard error of $\hat{p}$** is an estimate of the standard deviation of the sampling distribution of $\hat{p}$ and estimates how much $\hat{p}$ typically varies from p.

$$SE_{\hat{p}} = \sqrt{\frac{\hat{p}(1-\hat{p})}{n}}$$

- To determine the sample size n that will yield a $C\%$ confidence interval for a population proportion p with a maximum margin of error ME, solve the following inequality for n:

$$z^*\sqrt{\frac{\hat{p}(1-\hat{p})}{n}} \leq ME$$

where $\hat{p}$ is a guessed value for the sample proportion. The margin of error will always be less than or equal to ME if you use $\hat{p} = 0.5$.

Confidence Intervals for a Mean

- There are two **conditions** for constructing a confidence interval for μ:
  - **Random:** The data come from a random sample from the population of interest.
  - **Normal/Large Sample:** The data come from a normally distributed population or the sample size is large ($n \geq 30$). When the sample size is small and the shape of the population distribution is unknown, a graph of the sample data shows no strong skewness or outliers.

- When the conditions are met, a **one-sample t interval for μ** is

$$\bar{x} \pm t^* \frac{s_x}{\sqrt{n}}$$

where $\bar{x}$ is the sample mean, t^* is the critical value, s_x is the sample standard deviation, and n is the sample size.

  - The **critical value t^*** is a multiplier that makes the interval wide enough to have the stated capture rate.

On the graph of a t distribution with $n-1$ degrees of freedom, $C\%$ of the area will be between $-t^*$ and t^*. Use technology or Table B to find the value of t^*.

- When calculating a confidence interval for a population mean, we use a t^* critical value rather than a z^* critical value whenever we use s_x to estimate σ.

- The **standard error of $\bar{x}$** is an estimate of the standard deviation of the sampling distribution of $\bar{x}$ and estimates how much $\bar{x}$ typically varies from μ.

$$SE_{\bar{x}} = \frac{s_x}{\sqrt{n}}$$

Confidence Intervals for a Standard Deviation or Variance

- There are two **conditions** for constructing a confidence interval for σ or σ^2:
  - **Random:** The data come from a random sample from the population of interest.
  - **Normal:** The data come from a normally distributed population. When the shape of the population distribution is unknown, a graph of the sample data shows no strong skewness or outliers.

- When the conditions are met, a **one-sample confidence interval for σ** is

$$\sqrt{\frac{(n-1)s_x^2}{\chi_U^2}} \leq \sigma \leq \sqrt{\frac{(n-1)s_x^2}{\chi_L^2}}$$

where χ_L^2 and χ_U^2 are the critical values, s_x is the sample standard deviation, and n is the sample size.

- The **critical values χ_L^2 and χ_U^2** come from a chi-square distribution with $n-1$ degrees of freedom. $C\%$ of the area will be between χ_L^2 and χ_U^2. Use technology or Table C to find these values.

	Comparing confidence intervals		
	Confidence interval for p	Confidence interval for μ	Confidence interval for σ
Name (TI-83/84)	One-sample z interval for p (1–PropZInt)	One-sample t interval for μ (TInterval)	One-sample confidence interval for σ (not available)
Conditions	• **Random:** The data come from a random sample from the population of interest. • **Large Counts:** Both $n\hat{p}$ and $n(1-\hat{p})$ are at least 10. That is, the number of successes and the number of failures in the sample are both at least 10.	• **Random:** The data come from a random sample from the population of interest. • **Normal/Large Sample:** The data come from a normally distributed population or the sample size is large ($n \geq 30$). When the sample size is small and the shape of the population distribution is unknown, a graph of the sample data shows no strong skewness or outliers.	• **Random:** The data come from a random sample from the population of interest. • **Normal:** The data come from a normally distributed population. When the sample size is small and the shape of the population distribution is unknown, a graph of the sample data shows no strong skewness or outliers.
Formula (degrees of freedom)	$\hat{p} \pm z^*\sqrt{\dfrac{\hat{p}(1-\hat{p})}{n}}$	$\bar{x} \pm t^*\dfrac{s_x}{\sqrt{n}}$ (df = $n-1$)	$\sqrt{\dfrac{(n-1)s_x^2}{\chi_U^2}} \leq \sigma \leq \sqrt{\dfrac{(n-1)s_x^2}{\chi_L^2}}$ (df = $n-1$)

Chapter 7 Review Exercises

These exercises will help you review important concepts and skills described by the learning goals in each section. For most exercises, the relevant section is noted in parentheses after the exercise title.

1. **Paying College Athletes (7.1, 7.2)** A recent ABC News/*Washington Post* poll asked a random sample of U.S. residents, "Beyond any scholarships they receive, do you support or oppose paying salaries to college athletes?" The 95% confidence interval for the proportion of people who support salaries is 0.295 to 0.365.[88]
   (a) Interpret the confidence interval.
   (b) Interpret the confidence level.
   (c) Calculate the point estimate and the margin of error used to create this confidence interval.
   (d) Name two ways that the pollsters could reduce the margin of error. What are the drawbacks of these actions?

2. **Critical Values (7.3, 7.5)** Calculate the relevant critical value for each of the following scenarios, assuming the conditions are met:
   (a) You want to calculate a 94% confidence interval for a population proportion based on a sample of size 1200.
   (b) You want to calculate a 99% confidence interval for a population mean based on a sample of size 8.
   (c) You want to calculate a 90% confidence interval for a population mean using a sample of size 95.

3. **Extreme Poverty (7.1, 7.3, 7.4)** In the past 20 years, has the proportion of the world's population living in extreme poverty almost halved, remained more or less the same, or almost doubled? When a random sample of 1000 U.S. adults were asked this question in the Gapminder Misconception Study, only 5% got it right — the proportion living in extreme poverty has been nearly cut in half.[89]
   (a) Verify that the conditions have been met for calculating a confidence interval for p, the true proportion of U.S. adults who would correctly answer this question.
   (b) Calculate a 95% confidence interval for p.
   (c) Gapminder points out that one-third of chimpanzees would get this question correct when asked to choose from three bananas with the answer choices written on them. Does the interval in part (b) provide convincing evidence that fewer than one-third of U.S. adults would correctly answer this question? Explain your reasoning.

4. **E-cigarettes (7.2)** The dean of students of a large university wants to estimate the proportion of her students who have smoked an e-cigarette within the previous 30 days. She conducts a survey by randomly selecting students entering one of the dining halls for lunch and asking them to fill out a questionnaire and return it to her. Describe one potential source of bias in the study that is not accounted for by the margin of error.

5. **Gamers (7.4)** You want to estimate, with 99% confidence, the proportion of students at a college who play video games at least once a day. How large a sample should you select to ensure that the margin of error is at most 0.05?

6. **Christmas Trees (7.5, 7.6)** The National Christmas Tree Association periodically conducts surveys to learn more about consumer preferences for real versus fake Christmas trees. In a random sample of 2020 U.S. adults in 2018, 531 bought a real tree. The mean cost of a real tree was $78 with a standard deviation of $81.[90]
   (a) Verify that the conditions for constructing a confidence interval for μ are met.
   (b) Construct a 90% confidence interval for the mean cost of all real Christmas trees bought in 2018.
   (c) Interpret the interval from part (b).

7. **Fuel Efficiency (7.5, 7.6)** A consumer advocacy group randomly selected 20 owners of a particular model of pickup truck. Each owner was asked to report the number of miles per gallon (mpg) for their most recent tank of gas. Here are the mpg values for these 20 owners:

15.8	13.6	15.6	19.1	22.4	15.6	22.5	17.2	19.4	22.6
19.4	18.0	14.6	18.7	21.0	14.8	22.6	21.5	14.3	20.9

   Construct and interpret a 95% confidence interval for $\mu=$ the true mean fuel efficiency for this model of pickup truck.

8. **Fuel Efficiency (7.7, 7.8)** Refer to Exercise 7. Construct and interpret a 95% confidence interval for $\sigma=$ the standard deviation of fuel efficiency for all pickup trucks of this model.

9. **Number 51? (Chapter 7)** Should Washington, D.C., become a separate state? That's the question Gallup asked of a random sample of 1018 U.S. adults. In the sample, only 29% favored making Washington, D.C., a separate state.[91]
   (a) Construct and interpret a 95% confidence interval for the proportion of all U.S. adults who favor making Washington, D.C., a separate state.
   (b) Explain what is meant by "95% confident" in this context.

10. **Bottled Water (Chapter 7)** A certain brand of bottled water claims to match the body's natural pH of 7.4. To test this claim, Benni selected a random sample of these bottles from a local grocery store and measured the pH of each bottle using an electronic pH meter.[92] Here are the results:

7.4	7.4	7.3	7.3	7.3	7.3	7.2	7.3	7.3
7.2	7.2	7.3	7.4	7.4	7.5	7.3	7.4	7.4

    (a) Construct and interpret a 90% confidence interval for the true mean pH for this brand of bottled water from this grocery store.
    (b) Does the confidence interval from part (a) provide convincing evidence that the true mean pH is different than 7.4? Explain your reasoning.
    (c) Construct and interpret a 90% confidence interval for the standard deviation of pH level for this brand of bottled water from this grocery store.

Chapter 7 Project: Towed Vehicles

Explore a large data set using software and apply what you've learned in this chapter to a real-world scenario.

Like many cities, Baltimore, Maryland, shares public data on its Open Data website data.baltimorecity.gov. The data set **BaltimoreTow** contains a random sample of 190 vehicles listed as "Reclaimed by Owner" from the more than 250,000 vehicles towed by the city between January 1, 2010, and January 15, 2021. For each vehicle towed, more than 40 variables were recorded. In this project, we will concentrate on four of them: *TagState* (the state in which the vehicle is licensed), *TotalPaid* (how much money the owner had to pay to retrieve the vehicle), *VehicleYear* (the manufacturing year of the vehicle), and *TowYear* (the year in which the vehicle was towed).

Download this data set into the technology that you are using. The following questions will help you begin to explore this information.

1. Baltimore is a major metropolitan area located not far from Washington, D.C., the capital of the United States. Due to its proximity to Washington, we might expect the vehicles towed by the city of Baltimore to include a relatively high proportion of vehicles with license plates from states other than Maryland. Construct and interpret a 95% confidence interval for the proportion of reclaimed vehicles that have Maryland license plates. *Note: Ignore vehicles with a missing value for TagState.*

2. How much should vehicle owners expect to pay to retrieve their vehicle once it has been towed by the city? Construct and interpret a 95% confidence interval for the mean total paid to retrieve the vehicle.

3. A total paid of $100 in prior years is different than a total paid of $100 today, due to inflation. How would the interval from Question 2 change if you accounted for inflation before doing calculations?

4. Is it appropriate to construct a 95% confidence interval for the population standard deviation of the total paid to retrieve a vehicle? If so, construct and interpret the interval. If not, explain why not.

5. What can we say about the ages of vehicles when they were towed? Create a new variable that measures the age of the vehicle at the time it was towed. In other words, create a new variable *TowAge = TowYear − VehicleYear*. What is the sample mean age at towing?

6. Construct and interpret a 95% confidence interval for the population mean age at towing.

7. According to *Car and Driver* magazine, the average age of vehicles on the road is 11.9 years.[93] Based on your interval in Question 6, is there convincing evidence that the mean age of reclaimed vehicles is different from the mean age of all vehicles? Why might this be?

8 Testing a Claim

Section 8.1 The Idea of a Significance Test
Section 8.2 Significance Tests and Decision Making
Section 8.3 Testing a Claim About a Population Proportion
Section 8.4 Significance Tests for a Population Proportion
Section 8.5 Testing a Claim About a Population Mean
Section 8.6 Significance Tests for a Population Mean
Section 8.7 Power of a Test
Section 8.8 Significance Tests for a Population Standard Deviation or Variance

Chapter 8 Review
Chapter 8 Review Exercises
Chapter 8 Project

▐▐ Statistics Matters What is normal body temperature?

A fever often indicates that a person is ill. But what temperature should the thermometer show for a healthy person?

Researchers conducted a study to determine whether the "accepted" value for normal body temperature, 98.6°F, is accurate. They used an oral thermometer to measure the temperatures of a random sample of 130 healthy people aged 18 to 40. The dotplot shows the temperature readings.[1] We have added a vertical line at 98.6°F for reference.

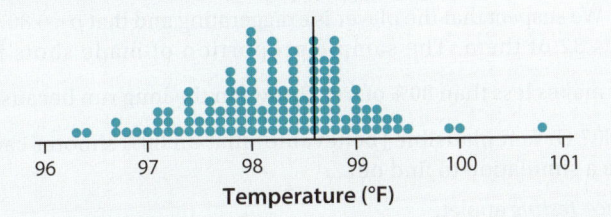

Exploratory data analysis revealed several interesting facts about this data set:

- The mean temperature is $\bar{x} = 98.25°F$.
- The standard deviation of the temperature readings is $s_x = 0.73°F$.
- 62.3% of the temperature readings are less than 98.6°F.

Do these data provide convincing evidence that "normal" body temperature in the population of healthy 18- to 40-year-olds is not 98.6°F?

We'll revisit Statistics Matters *at the end of the chapter, so you can use what you have learned to help answer these questions.*

In Chapter 7, you learned how to use a confidence interval to estimate a population parameter based on sample data. This chapter shows you how to use sample data to test a claim about a population parameter.

Section 8.1 The Idea of a Significance Test

LEARNING GOALS

By the end of this section, you will be able to:

- State appropriate hypotheses for a significance test about a population parameter.
- Interpret a *P*-value in context.
- Make an appropriate conclusion for a significance test based on a *P*-value.

Confidence intervals are one of the two most common types of statistical inference. You use a confidence interval when your goal is to estimate a population parameter, such as the population proportion p, the population mean μ, or the population standard deviation σ. You use the second common type of inference, called a **significance test**, when you have a different goal: to test a claim about a population parameter.

A **significance test** is a formal procedure for using observed data to decide between two competing claims (called *hypotheses*). The claims are usually statements about population parameters.

A significance test is sometimes referred to as a *test of significance, a hypothesis test*, or a *test of hypotheses*.

The following Concept Exploration illustrates the reasoning of a significance test.

CONCEPT EXPLORATION

I'm a great free-throw shooter!

In this exploration, you will perform a simulation to test a claim about a population proportion.

A basketball player claims to be an 80% free-throw shooter. That is, the player claims that $p = 0.80$, where p is the true proportion of free throws the player will make in the long run. We suspect that the player is exaggerating and that $p < 0.80$.

Suppose the player shoots 50 free throws and makes 32 of them. The sample proportion of made shots is $\hat{p} = \dfrac{32}{50} = 0.64$. This result gives *some* evidence that the player makes less than 80% of free throws in the long run because $0.64 < 0.80$. But does it give *convincing* evidence that $p < 0.80$? Or is it plausible (believable) that an 80% shooter can have a performance this poor by chance alone? You can use a simulation to find out.

1. Go to www.stapplet.com and launch the *Logic of Significance Testing* applet.
2. Click the Shoot button 50 times to simulate a sample of 50 shots by an 80% free-throw shooter. The sample proportion $\hat{p}$ of made shots will be displayed on a dotplot in the Simulation Results section.
3. Use the Quick Add section of the applet to perform 399 more repetitions of the simulation, so that you have a total of 400 simulated samples of 50 free throws. The 400 values of $\hat{p}$ = the sample proportion of made free throws will be displayed on the dotplot. Note that this is a simulated sampling distribution of $\hat{p}$.
4. Based on the simulated sampling distribution, how likely is it for an 80% free-throw shooter to make 64% or less of their shots when shooting 50 free throws? Use the applet to count the number and percent of dots less than or equal to 0.64.
5. Based on your answer to Question 4, does the observed $\hat{p} = 0.64$ result give convincing evidence that the player is exaggerating? Or is it plausible (believable) that an 80% shooter can have a performance this poor by chance alone?

In the Concept Exploration, the shooter made only 32 of 50 free-throw attempts ($\hat{p} = 32/50 = 0.64$). There are two possible explanations for why the shooter made only 64% of shots:

1. The player's claim is true ($p = 0.80$). That is, the player really is an 80% free-throw shooter and the poor performance happened by chance alone.
2. The player's claim is false ($p < 0.80$). That is, the player really makes less than 80% of free throws in the long run.

If Explanation 1 is plausible, then we don't have convincing evidence that the shooter is exaggerating—the player's poor performance could have occurred purely by chance. However, if it is unlikely for an 80% free-throw shooter to get a proportion of 0.64 or less in 50 attempts, then we can rule out Explanation 1.

We used the applet to simulate 400 sets of 50 shots, assuming that the player is really an 80% free-throw shooter. Figure 8.1 shows a dotplot of the results. Each dot on the graph represents the simulated proportion $\hat{p}$ of made shots in one sample of 50 free throws.

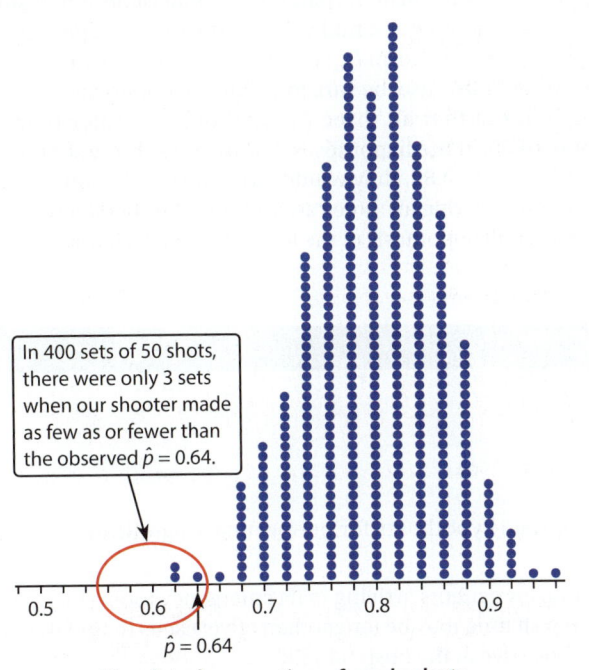

FIGURE 8.1 Dotplot of a simulated sampling distribution of $\hat{p}$, the proportion of free throws made by an 80% shooter in a sample of 50 shots.

In 400 sets of 50 shots, there were only 3 sets when our shooter made as few as or fewer than the observed $\hat{p} = 0.64$.

$\hat{p} = 0.64$

Simulated proportion of made shots

The simulation shows that it would be very unlikely for an 80% free-throw shooter to make 32 or fewer free throws in 50 shots ($\hat{p} \leq 0.64$) just by chance. This small probability ($\approx 3/400 = 0.0075$) leads us to rule out Explanation 1. The observed result gives us convincing evidence that Explanation 2 is correct: The player makes less than 80% of free throws in the long run.

Stating Hypotheses

A significance test starts with a careful statement of the claims we want to compare. In the free-throw shooter context, the player claims a long-run proportion of made free throws of $p = 0.80$. This is the claim we seek evidence *against*. We call it the **null hypothesis**, abbreviated H_0. Usually, the null hypothesis is a statement of "no difference." For the free-throw shooter, no difference from the player's claim gives H_0: $p = 0.80$.

The claim we hope or suspect to be true instead of the null hypothesis is called the **alternative hypothesis**. We abbreviate the alternative hypothesis as H_a. In this case, we believe the player might be exaggerating, so our alternative hypothesis is H_a: $p < 0.80$.

> The claim that we weigh evidence *against* in a significance test is called the **null hypothesis (H_0)**. The claim that we are trying to find evidence *for* is the **alternative hypothesis (H_a)**.

In the free-throw shooter context, our hypotheses are

$$H_0: p = 0.80$$
$$H_a: p < 0.80$$

where p is the true proportion of free throws the player will make in the long run. The alternative hypothesis is **one-sided** ($p < 0.80$) because we suspect the player makes less than 80% of all free throws. If you suspect that the true value of a parameter is different from the null value—either greater than *or* less than—use a **two-sided** alternative hypothesis. For simplicity, we sometimes refer to a significance test with a one-sided alternative hypothesis as a "one-sided test" and a significance test with a two-sided alternative hypothesis as a "two-sided test."

The alternative hypothesis is **one-sided** if it states that a parameter is *greater than* the null value or if it states that the parameter is *less than* the null value.

The alternative hypothesis is **two-sided** if it states that the parameter is *different from* the null value (it could be either greater than or less than the null value).

The null hypothesis usually has the form H_0: parameter = null value. A one-sided alternative hypothesis has either the form H_a: parameter < null value or the form H_a: parameter > null value. A two-sided alternative hypothesis has the form H_a: parameter ≠ null value. To determine the correct form of H_a, read the problem carefully. Also, be sure to define the population parameter in context.

Note: Some people insist that all three possibilities — greater than, less than, and equal to — should be accounted for in the hypotheses. For the free-throw shooter example, because the alternative hypothesis is $H_a: p < 0.80$, they would write the null hypothesis as $H_0: p \geq 0.80$. Despite the mathematical appeal of covering all three cases, we use the claimed value $p = 0.80$ when carrying out the test. So we'll use a null hypothesis of the form $H_0: p = 0.80$ in this book.

EXAMPLE 8.1.1

ER wait times and food insufficiency

Stating hypotheses

PROBLEM
For each of the following scenarios, state appropriate hypotheses for performing a significance test. Be sure to define the parameter of interest.

(a) A hospital advertises that the average wait time for patients arriving in its emergency room (ER) is 15 minutes. Local health advocates suspect that the average ER wait time may be longer than advertised. To find out, they record the wait times for a random sample of 500 patients who arrive at the hospital's ER.

(b) Results of a government survey suggest that 14% of U.S. households with children experienced food insufficiency (defined as sometimes or often not having enough to eat) in a given month.[2] Researchers wonder whether this result applies to their large city. They ask a random sample of 1000 households with children in the city whether they experienced food insufficiency during that month.

SOLUTION

(a) $H_0: \mu = 15$
$H_a: \mu > 15$

where μ = the mean wait time (in minutes) for all patients arriving at this hospital's ER.

> H_a is one-sided because health advocates suspect the average ER wait time is *longer* than advertised.

(b) $H_0: p = 0.14$
$H_a: p \neq 0.14$

where p = the proportion of all households with children in this large city that would report food insufficiency that month.

> The researchers wonder if the proportion of all households with children that experienced food insufficiency in their large city *differs from* (is either greater than or less than) the claimed proportion of $p = 0.14$ based on the government survey, so H_a is two-sided.

EXAM PREP: FOR PRACTICE, TRY EXERCISE 7.

⚠ **The hypotheses should express the belief or suspicion we have *before* we see the data.** It is cheating to look at the data first and then frame the alternative hypothesis to fit what the data show. For example, the data from the study of food insufficiency showed that $\hat{p} = 0.12$ for a random sample of 1000 households with children in the large city. Researchers should not change the alternative hypothesis to $H_a: p < 0.14$ after looking at the data.

⚠ **Hypotheses refer to a population, not to a sample.** After all, the goal of inference is to make some conclusion about the population based on sample data! Be sure to state H_0 and H_a in terms of population parameters. It is *never* correct to write a hypothesis about a sample statistic, such as $H_0: \hat{p} = 0.14$ or $H_a: \bar{x} > 15$. Also, do not refer to the sample when defining the parameter. For example, "the proportion of households *who said* they had food insufficiency" describes the sample, not the population.

Interpreting *P*-Values

It might seem strange to state a null hypothesis and then try to find evidence against it. Maybe it would help to think about how a criminal trial works in the United States. The defendant is "innocent until proven guilty." That is, the null hypothesis is innocence and the prosecution must offer convincing evidence against this hypothesis and in favor of the alternative hypothesis: guilt. That's exactly how significance tests work, although in statistics we deal with evidence provided by data and use a probability to say how strong the evidence is.

In the free-throw shooter scenario presented at the beginning of the section, a player who claimed to be an 80% free-throw shooter made only $\hat{p} = 32/50 = 0.64$ of shots in a random sample of 50 free throws. This is evidence *against* the null hypothesis that $p = 0.80$ and *in favor of* the alternative hypothesis $p < 0.80$. But is the evidence convincing? To answer this question, we have to know how likely it is for an 80% shooter to make 64% or less of their free throws by chance alone in a random sample of 50 attempts. This probability is called a **P-value**.

> The **P-value** of a test is the probability, assuming the null hypothesis H_0 is true, of getting evidence for the alternative hypothesis H_a as strong as or stronger than the observed evidence by chance alone.

We used simulation to estimate the *P*-value for our free-throw shooter: $3/400 = 0.0075$. How do we interpret this *P*-value? Assuming that the player makes 80% of free throws in the long run, there is about a 0.0075 probability of getting a sample proportion of 0.64 or less by chance alone in a random sample of 50 shots.

EXAMPLE 8.1.2

Do teens get enough calcium?

Interpreting a P-value

PROBLEM
Calcium is a vital nutrient for healthy bones and teeth. The National Institutes of Health (NIH) recommends a calcium intake of 1300 milligrams (mg) per day for teenagers. The NIH is concerned that teenagers aren't getting enough calcium, on average. Is this true? Researchers decide to perform a test of

$$H_0: \mu = 1300$$
$$H_a: \mu < 1300$$

where μ is the mean daily calcium intake in the population of teenagers. They ask a random sample of 20 teens to record their food and drink consumption for 1 day. The researchers then compute the calcium intake for each teen. Data analysis reveals that $\bar{x} = 1198$ mg and $s_x = 411$ mg. Researchers performed a significance test and obtained a *P*-value of 0.1405.

(a) Explain why the data give some evidence for H_a.
(b) Interpret the *P*-value.

SOLUTION

(a) The sample mean of 1198 mg is less than 1300 mg.

(b) Assuming that the mean daily calcium intake in the teen population is 1300 mg, there is a 0.1405 probability of getting a sample mean of 1198 mg or less by chance alone in a random sample of 20 teens.

> Be sure to include the condition—assuming that the null hypothesis is true—when interpreting a *P*-value.

EXAM PREP: FOR PRACTICE, TRY EXERCISE 13.

The *P*-value is the probability, given that the null hypothesis H_0 is true, of getting evidence for H_a as strong as or stronger than the observed result by chance alone. In other words, the *P*-value is a conditional probability. For the teens and calcium example, the *P*-value = $P(\bar{x} \leq 1198 \mid \mu = 1300) = 0.1405$.

When H_a is two-sided (parameter ≠ null value), values of the sample statistic less than or greater than the null value both count as evidence for H_a. Suppose we want to perform a test of $H_0: p = 0.5$ versus $H_a: p \neq 0.5$ based on a simple random sample (SRS) with a sample proportion of $\hat{p} = 0.65$. This result gives some evidence for $H_a: p \neq 0.5$ because $0.65 \neq 0.5$. In this case, evidence for H_a as strong as or stronger than the observed result includes any value of $\hat{p}$ greater than or equal to 0.65 as well as any value of $\hat{p}$ less than or equal to 0.35. Why? Because $\hat{p}=0.35$ is just as different from the null value of $p=0.5$ as is $\hat{p}=0.65$. For this scenario, the P-value is equal to the conditional probability $P(\hat{p} \leq 0.35 \text{ or } \hat{p} \geq 0.65 \mid p=0.5)$.

Making Conclusions Based on P-Values

The final step in performing a significance test is to make a conclusion about the competing claims being tested. We make a decision based on the strength of evidence in favor of the alternative hypothesis (and against the null hypothesis) as measured by the P-value. Small P-values give convincing evidence for H_a because they say that the observed result is unlikely to occur when H_0 is true. Larger P-values fail to give convincing evidence for H_a because they say that the observed result is likely to occur by chance alone when H_0 is true.

HOW TO   Make a Conclusion in a Significance Test

- If the P-value is small, reject H_0 and conclude there is convincing evidence for H_a (in context).
- If the P-value is not small, fail to reject H_0 and conclude there is not convincing evidence for H_a (in context).

The wording of a significance test conclusion may seem unusual at first, but it's consistent with what happens in a U.S. criminal trial. Once the jury has weighed the evidence against the null hypothesis of innocence, they return one of two verdicts: "guilty" (reject H_0) or "not guilty" (fail to reject H_0). A not-guilty verdict doesn't guarantee that the defendant is innocent; it just says that there's not convincing evidence of guilt. Likewise, a fail-to-reject H_0 decision in a significance test doesn't guarantee that H_0 is true.

How small does a P-value have to be for us to reject H_0? In Chapter 1, we recommended using a boundary of 5% when determining if a result is *statistically significant*. Keep following this recommendation for now—that is, view a P-value less than 0.05 as "small." We'll consider this boundary value more carefully in Section 8.2.

In the free-throw shooter activity, the estimated P-value was 0.0075. Because the P-value is small (less than 0.05), we reject $H_0: p=0.80$. We have convincing evidence that the player makes fewer than 80% of their free throws in the long run.

EXAMPLE 8.1.3

Can we tell if teens get enough calcium?

Making a conclusion based on a P-value

PROBLEM

In Example 8.1.2, researchers collected data on the daily calcium intake for a random sample of 20 teens. Data analysis revealed that $\bar{x}=1198$ mg and $s_x=411$ mg. The researchers used these data to perform a test of

$$H_0: \mu = 1300$$
$$H_a: \mu < 1300$$

where μ is the mean daily calcium intake in the population of teenagers. The resulting P-value is 0.1405. What conclusion would you make?

SOLUTION

The P-value of 0.1405 is not small (not less than 0.05), so we fail to reject H_0. We don't have convincing evidence that teens are getting fewer than 1300 mg of calcium per day, on average.

EXAM PREP: FOR PRACTICE, TRY EXERCISE 19.

Be careful how you write conclusions when the P-value is not small. Don't conclude that the null hypothesis is true just because we didn't find convincing evidence for the alternative hypothesis. For example, it would be incorrect to conclude that teens *are* getting 1300 mg of calcium per day, on average. We found *some* evidence that the teens weren't getting enough calcium, but the evidence wasn't convincing enough to reject H_0. ⚠ **Never "accept H_0" or conclude that H_0 is true!** In fact, the 90% confidence interval for μ = the mean daily calcium intake in the population of teenagers is 1039.1 to 1356.9 milligrams. You can see that 1300 is just one of the many plausible values for μ based on the sample data.

Section 8.1 What Did You Learn?

Review the learning goals from this section. Then practice what you've learned by working through the exercises.

Learning Goal	Example	Exercises
State appropriate hypotheses for a significance test about a population parameter.	8.1.1	7–12
Interpret a *P*-value in context.	8.1.2	13–18
Make an appropriate conclusion for a significance test based on a *P*-value.	8.1.3	19–24

Section 8.1 Exercises

Building Concepts and Skills These exercises assess the basic knowledge you should have after reading the section.

1. The goal of a significance test is to test a _____ about a _____.

2. What are the two possible explanations when a basketball player who claims to be an 80% free-throw shooter makes only 64% of free throws taken in 50 attempts?

3. The claim we weigh evidence against in a significance test is called the _____ hypothesis. The claim we are trying to find evidence for is the _____ hypothesis.

4. True/False: Hypotheses should never be stated in terms of sample statistics, like $\hat{p}$ or $\bar{x}$.

5. Define *P*-value.

6. True/False: If the *P*-value is not small, you would accept H_0 and conclude that there is not convincing evidence for H_a.

Mastering Concepts and Skills These exercises reinforce the learning goals as shown in the examples.

7. **Refer to Example 8.1.1 Pineapple Weights** At the Hawaii Pineapple Company, managers are interested in the size of the pineapples grown in the company's fields. Last year, the mean weight of the pineapples harvested from one large field was 31 ounces. A new irrigation system was installed in this field after the growing season. Managers wonder if the mean weight of pineapples grown in the field this year will be greater than it was last year. They select a random sample of 50 pineapples from this year's crop. State appropriate hypotheses for performing a significance test. Be sure to define the parameter of interest.

8. **Sturdy Boards** Lumber companies dry freshly cut wood in kilns before selling it. A certain percentage of the boards become "checked," which means that cracks develop at the ends of the boards during drying. The current drying procedure is known to produce cracks in 16% of the boards. The drying supervisor at a lumber company wants to test a new method to determine if fewer boards crack. The supervisor uses the new method on a random sample of 200 boards. State appropriate hypotheses for performing a significance test. Be sure to define the parameter of interest.

9. **Southpaws** A media report claims that 10% of all adults in the United States are left-handed. A professor wonders if this figure applies to the large community college where the professor works. To investigate, the professor chooses an SRS of 100 students from the college and records whether or not each student is left-handed. State appropriate hypotheses for performing a significance test. Be sure to define the parameter of interest.

10. **Students' Attitudes** The Survey of Study Habits and Attitudes (SSHA) is a psychological test that measures students' attitudes toward school and study habits. Scores range from 0 to 200, with higher scores indicating more positive attitudes and better study habits. The mean score for U.S. college students is about 115. A researcher wonders if older students' attitudes toward school differ from the college population as a whole. The researcher gives the SSHA to an SRS of 45 of the more than 1000 students at a large college who are at least 30 years of age. State appropriate hypotheses for performing a significance test. Be sure to define the parameter of interest.

11. **Thermostat Check** During the winter months, outside temperatures at the Starnes family cabin in Colorado can stay well below freezing (32°F, or 0°C) for weeks at a time. To prevent the pipes from freezing, Mrs. Starnes sets the thermostat at 50°F. The manufacturer claims that the thermostat allows variation in home temperature that follows an approximately normal distribution

with standard deviation $\sigma = 3°F$. Mrs. Starnes suspects that the manufacturer is overstating the consistency of the thermostat. To investigate, she programs a digital thermometer to take an SRS of $n = 10$ readings during a 24-hour period. State appropriate hypotheses for performing a significance test. Be sure to define the parameter of interest.

12. **New Golf Club** An avid golfer would like to improve their game. Based on years of experience, the golfer has established that the distance balls travel when hit with their current 3-iron follows an approximately normal distribution with standard deviation $\sigma = 15$ yards. The golfer is hoping that a new 3-iron will make their shots more consistent (less variable). To find out, the golfer hits 50 shots on a driving range with the new 3-iron. State appropriate hypotheses for performing a significance test. Be sure to define the parameter of interest.

13. **Refer to Example 8.1.2 Pineapple Weights** In the study of pineapple weights from Exercise 7, the sample mean weight from this year's crop is 31.4 ounces and the sample standard deviation is 2.5 ounces. A significance test yields a P-value of 0.1317.
    (a) Explain why the data give some evidence for H_a.
    (b) Interpret the P-value.

14. **Sturdy Boards** In the study of a new lumber drying method from Exercise 8, the sample proportion of cracked boards is $22/200 = 0.11$. A significance test yields a P-value of 0.027.
    (a) Explain why the data give some evidence for H_a.
    (b) Interpret the P-value.

15. **Southpaws** In the professor's study of left-handedness at a community college from Exercise 9, the sample proportion of left-handed students is $16/100 = 0.16$. A significance test yields a P-value of 0.0455.
    (a) Explain why the data give some evidence for H_a.
    (b) Interpret the P-value.

16. **Students' Attitudes** In the study of older students' attitudes toward school from Exercise 10, the sample mean SSHA score is 122.7 and the sample standard deviation is 29.8. A significance test yields a P-value of 0.09.
    (a) Explain why the data give some evidence for H_a.
    (b) Interpret the P-value.

17. **Thermostat Check** In the study of cabin temperatures from Exercise 11, the standard deviation of the 10 temperature readings is 5°F. A significance test yields a P-value of 0.003.
    (a) Explain why the data give some evidence for H_a.
    (b) Interpret the P-value.

18. **New Golf Club** In the study of 3-iron consistency from Exercise 12, the standard deviation of distance traveled for the sample of 50 shots is 13.9 yards. A significance test yields a P-value of 0.25.
    (a) Explain why the data give some evidence for H_a.
    (b) Interpret the P-value.

19. **Refer to Example 8.1.3 Pineapple Weights** Refer to Exercises 7 and 13. What conclusion would you make?

20. **Sturdy Boards** Refer to Exercises 8 and 14. What conclusion would you make?

21. **Southpaws** Refer to Exercises 9 and 15. What conclusion would you make?

22. **Students' Attitudes** Refer to Exercises 10 and 16. What conclusion would you make?

23. **Thermostat Check** Refer to Exercises 11 and 17. What conclusion would you make?

24. **New Golf Club** Refer to Exercises 12 and 18. What conclusion would you make?

Applying the Concepts These exercises ask you to apply multiple learning goals in a new context or to apply what you learned in this section in a new way.

25. **Kissing Tilt** According to an article in the *San Gabriel Valley Tribune*, the majority of couples prefer to tilt their heads to the right when kissing. In the study, a researcher observed a random sample of 124 kissing couples and found that $83/124$ ($\hat{p} = 0.669$) of the couples tilted to the right.[3] Do these data provide convincing evidence that more than 50% of kissing couples prefer to tilt their heads to the right?
    (a) State appropriate hypotheses for performing a significance test. Be sure to define the parameter of interest.
    (b) Explain why the data give some evidence for H_a.
    (c) The P-value for the test in part (a) is 0.0001. Interpret the P-value.
    (d) What conclusion would you make?

26. **Honest Loaf** The mean weight of loaves of bread produced at a bakery is supposed to be 1 pound. The quality control supervisor at the bakery is concerned that employees are making loaves that are too light. When the supervisor weighs a random sample of 50 bread loaves, they find that the mean weight is $\bar{x} = 0.975$ pound.
    (a) State appropriate hypotheses for performing a significance test. Be sure to define the parameter of interest.
    (b) Explain why the data give some evidence for H_a.
    (c) The P-value for the test in part (a) is 0.0806. Interpret the P-value.
    (d) What conclusion would you make?

27. **Flawed Hypotheses** A researcher suspects that the mean birth weight of babies whose mothers did not see a doctor before delivery is less than 3000 grams. The researcher states the hypotheses as

$$H_0: \bar{x} = 3000$$
$$H_a: \bar{x} < 3000$$

Explain what's wrong with the stated hypotheses. Then give the correct hypotheses.

28. Flawed Hypotheses A change is made that should improve student satisfaction with the parking situation at a university. Before the change, 37% of students approve of the parking that's provided. A researcher states the hypotheses as

$$H_0: p > 0.37$$
$$H_a: p = 0.37$$

Explain what's wrong with the stated hypotheses. Then give the correct hypotheses.

Extending the Concepts These exercises challenge you to explore statistical concepts and methods that go beyond what you learned in this section.

29. Shakespeare's Work? Statistics can help decide the authorship of literary works. Sonnets by William Shakespeare contain an average of $\mu = 6.9$ new words (words not used in the poet's other works) and a standard deviation of $\sigma = 2.7$ new words. The number of new words is approximately normally distributed. Scholars expect sonnets by other authors to contain more new words. A scholarly sleuth discovers a manuscript with many new sonnets, and a debate erupts over whether the manuscript is really Shakespeare's work. Some scholars select a random sample of 5 sonnets from the manuscript and count the number of new words in each. The mean number of new words in these 5 sonnets is $\bar{x} = 9.2$.

The dotplot shows the results of simulating 200 random samples of size 5 from a normal distribution with a mean of 6.9 and a standard deviation of 2.7, and calculating the mean number of new words for each sample.

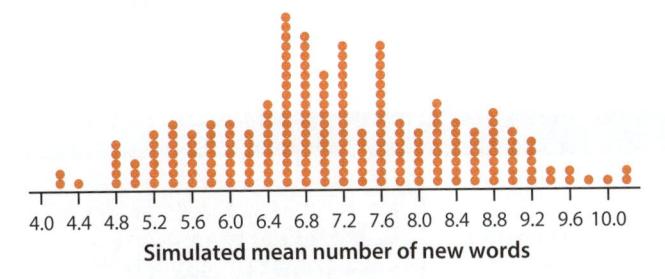

(a) State appropriate hypotheses for performing a significance test. Be sure to define the parameter of interest.

(b) Explain why the data give some evidence for H_a.

(c) Use the simulation results to estimate the P-value of the test in part (a). Interpret the P-value.

(d) What conclusion would you make?

Cumulative Review These exercises revisit what you learned in previous sections.

30. Tornadoes (2.5, 3.1, 3.2) In any year, every U.S. state has at least a few tornadoes. The histogram shows the distribution of the total number of tornadoes reported in each state over a 60-year period.[4]

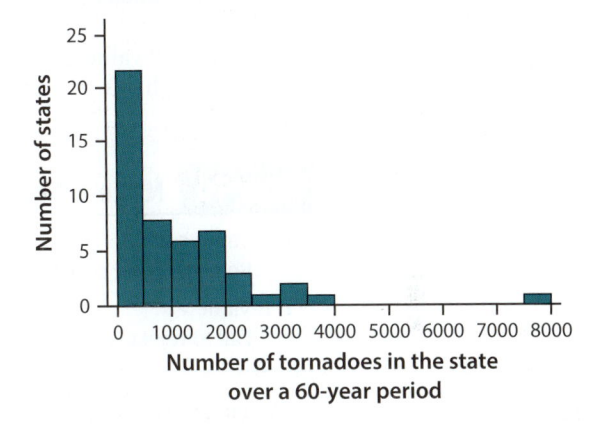

(a) Describe the shape of this distribution. Identify any obvious outliers.

(b) Which measures of center and variability should be used to summarize this distribution? Explain your answer.

31. Explaining Confidence (7.2) Here is an explanation from the Associated Press concerning one of its opinion polls. Explain what is wrong with the statement.

> For a poll of 1600 adults, the variation due to sampling error is no more than 3 percentage points either way. The error margin is said to be valid at the 95% confidence level. This means that, if the same questions were repeated in 20 polls, the results of at least 19 surveys would be within 3 percentage points of the results of this survey.

Section 8.2 Significance Tests and Decision Making

LEARNING GOALS

By the end of this section, you will be able to

- Make an appropriate conclusion in a significance test using a significance level.
- Interpret a Type I error and a Type II error in context.
- Give a consequence of a Type I error and a Type II error in a given setting.

There are two types of conclusions you can make in a significance test:

P-value small → Reject H_0 → Convincing evidence for H_a (in context)
P-value not small → Fail to reject H_0 → Not convincing evidence for H_a (in context)

In Section 8.1, we encouraged you to view a P-value less than 0.05 as small. Choosing this boundary value means we are requiring evidence for H_a so strong that sample results as extreme or more extreme would happen less than 5% of the time by chance alone when H_0 is true.

Sometimes, it may be preferable to use a different boundary value—like 0.01 or 0.10—when making a conclusion in a significance test. By the end of this section, you will understand why.

Making Conclusions Using Significance Levels

To determine if a P-value should be considered small, we compare it to a boundary value called the **significance level**. We denote it by α, the Greek letter alpha.

> The **significance level** α is the value that we use as a boundary to decide if an observed result is unlikely to happen by chance alone when the null hypothesis is true.

When our P-value is less than the chosen significance level α in a significance test, we reject the null hypothesis H_0 and conclude there is convincing evidence in favor of the alternative hypothesis H_a.

HOW TO   Make a Conclusion in a Significance Test

If P-value $< \alpha$:
  Reject H_0 and conclude there is convincing evidence for H_a (in context).

If P-value $> \alpha$:
  Fail to reject H_0 and conclude there is not convincing evidence for H_a (in context).

What if the P-value is *exactly equal* to the significance level α? This almost never happens, but if it does, ask a statistician for advice!

Significance at the $\alpha = 0.05$ level is often expressed by the statement "The results were significant at the 5% level ($P < 0.05$)." Here, P stands for the P-value. *The P-value is more informative than a statement about significance* because it describes the strength of evidence for the alternative hypothesis. For example, both an observed result with P-value = 0.03 and an observed result with P-value = 0.0003 are significant at the 5% level. But the P-value of 0.0003 gives much stronger evidence against H_0 and in favor of H_a than the P-value of 0.03. This is why we always include the P-value in our conclusions, and not just a statement about significance.

EXAMPLE 8.2.1

Do the company's new batteries last longer?

Making a conclusion using a significance level

PROBLEM

A company has developed a new deluxe AAA battery that is supposed to last longer than its regular AAA battery. However, these new batteries are more expensive to produce, so the company would like to be convinced that they really do last longer. Based on years of experience, the company knows that its regular AAA batteries last for 30 hours of continuous use, on average. The company selects an SRS of 50 new deluxe batteries and uses them continuously until they are completely drained. The sample mean lifetime is $\bar{x} = 33.9$ hours. A significance test is performed using the hypotheses

$$H_0: \mu = 30$$
$$H_a: \mu > 30$$

where μ is the true mean lifetime (in hours) of the new deluxe AAA batteries. The resulting P-value is 0.0729. What conclusion would you make at the $\alpha = 0.01$ level?

SOLUTION

Because the P-value of $0.0729 > \alpha = 0.01$, we fail to reject H_0. We do not have convincing evidence that the company's deluxe AAA batteries last longer than 30 hours, on average.

> P-value $> \alpha \rightarrow$ Fail to reject $H_0 \rightarrow$ Not convincing evidence for H_a (in context)

EXAM PREP: FOR PRACTICE, TRY EXERCISE 7.

When a researcher plans to make a conclusion based on a significance level, ⚠ **α should be stated *before* the data are collected.** Otherwise, a deceptive user of statistics might choose α *after* the data have been analyzed in an attempt to manipulate the conclusion. This is just as inappropriate as choosing an alternative hypothesis after looking at the data. *If no significance level is provided, use* $\alpha = 0.05$ *as you did in Section 8.1.*

Note: In this book, we use *P*-values and significance levels to make conclusions in significance tests. An alternative method based on significance levels, known as the *rejection region approach,* was commonly used before technology became readily available. We illustrate this alternative method in the Think About It in Section 8.3.

Type I and Type II Errors

When we make a conclusion from a significance test based on sample data, we hope our conclusion will be correct. But sometimes the data may lead us to an incorrect conclusion. There are two types of errors we can make: a **Type I error** or a **Type II error**.

> A **Type I error** occurs if we reject H_0 when H_0 is true. That is, the data give convincing evidence that H_a is true when it really isn't.

> A **Type II error** occurs if we fail to reject H_0 when H_a is true. That is, the data do not give convincing evidence that H_a is true when it really is.

The possible outcomes of a significance test are summarized in Figure 8.2.

		Truth about the population	
		H_0 true	H_a true
Conclusion based on sample	Reject H_0	Type I error	Correct conclusion
	Fail to reject H_0	Correct conclusion	Type II error

FIGURE 8.2 The two types of errors in significance tests.

The truth about the population is either that H_0 is true or H_a is true. It must be one or the other! If H_0 is true:
- Our conclusion is correct if we don't find convincing evidence that H_a is true.
- We make a Type I error if we find convincing evidence that H_a is true.

If H_a is true:
- Our conclusion is correct if we find convincing evidence that H_a is true.
- We make a Type II error if we do not find convincing evidence that H_a is true.

Here's a helpful reminder to keep the two types of errors straight: "Fail to" goes with Type II.

It is important to be able to describe Type I and Type II errors in the context of a problem. In the batteries example, the company performed a test of

$$H_0: \mu = 30$$
$$H_a: \mu > 30$$

where μ is the true mean lifetime (in hours) of its new deluxe AAA batteries.
- A Type I error occurs if the company finds convincing evidence that its new AAA batteries last longer than 30 hours, on average, when their true mean lifetime really is 30 hours.
- A Type II error occurs if the company doesn't find convincing evidence that the new batteries last longer than 30 hours, on average, when their true mean lifetime really is greater than 30 hours.

Be sure to discuss the evidence and the truth when you interpret Type I and Type II errors.

Of course, only one error is possible at a time, depending on the conclusion we make. In Example 8.2.1, it is possible the battery company made a Type II error because it didn't find convincing evidence that the batteries last longer than 30 hours, on average.

EXAMPLE 8.2.2

Can we make fast food faster?

Type I and Type II errors

PROBLEM

The manager of a fast-food restaurant wants to reduce the proportion of drive-thru customers who have to wait longer than 2 minutes to receive their food after placing an order. Based on store records, the proportion of customers who had to wait longer than 2 minutes was 0.63. To reduce this proportion, the manager assigns an additional employee to assist with drive-thru orders. During the next month, the manager will collect a random sample of 250 drive-thru times and test the following hypotheses at the $\alpha = 0.10$ significance level:

$$H_0: p = 0.63$$
$$H_a: p < 0.63$$

where $p =$ the proportion of all drive-thru customers at this restaurant who have to wait longer than 2 minutes to receive their food. Describe a Type I error and a Type II error in this setting.

SOLUTION

Type I error: The manager finds convincing evidence that the true proportion of drive-thru customers who have to wait longer than 2 minutes has decreased from 0.63, when it is really still 0.63.

> A Type I error occurs if a test finds convincing evidence that H_a is true when it really isn't.

Type II error: The manager doesn't find convincing evidence that the true proportion of drive-thru customers who have to wait longer than 2 minutes has decreased from 0.63, when it really is less than 0.63.

> A Type II error occurs if a test does not find convincing evidence that H_a is true when it really is.

EXAM PREP: FOR PRACTICE, TRY EXERCISE 11.

Which is more serious: a Type I error or a Type II error? That depends on the situation. For the batteries example, a Type I error would result in the company spending lots of money to produce the new AAA batteries when they don't last any longer than the older, cheaper type. A Type II error would lead the company not to produce the new AAA batteries even though they last longer, on average. This is a potential missed opportunity for the company to increase customer satisfaction, which might also lead to increased sales. It is important to consider the possible consequences of the two types of errors prior to carrying out a significance test (and when choosing the significance level).

EXAMPLE 8.2.3

Can we make fast food faster?

Consequences of Type I and Type II errors

PROBLEM

Refer to Example 8.2.2. Give a consequence of a Type I error and a Type II error in this setting.

SOLUTION

Type I error: The restaurant assigns an extra employee to the drive-thru even though service time has not improved. This increases the restaurant's costs but brings no benefit.

Type II error: The restaurant decides not to keep the extra employee even though drive-thru service time has improved. This decision could lead to customers becoming upset by slow service, which might decrease sales.

EXAM PREP: FOR PRACTICE, TRY EXERCISE 17.

The most common significance levels are $\alpha = 0.05$, $\alpha = 0.01$, and $\alpha = 0.10$. Which is the best choice for a given significance test? That depends on whether a Type I error or a Type II error is more serious.

In the fast-food example, a Type I error occurs if the true proportion of customers who have to wait at least 2 minutes remains $p = 0.63$, but we get a sample proportion $\hat{p}$ small enough to yield a

P-value less than $\alpha=0.10$. When H_0 is true, this will happen 10% of the time just by chance. In other words, $P(\text{Type I error})=\alpha$.

> **Determining Type I Error Probability**
>
> The probability of making a Type I error in a significance test is equal to the significance level α.

We can decrease the probability of making a Type I error in a significance test by using a smaller significance level. For instance, the restaurant manager could use $\alpha=0.05$ instead of $\alpha=0.10$. But there is a trade-off between $P(\text{Type I error})$ and $P(\text{Type II error})$: As one decreases, the other increases, assuming everything else remains the same. If we make it more difficult to reject H_0 by decreasing α, we increase the probability that we don't find convincing evidence for H_a when it is true. That's why it is important to consider the possible consequences of each type of error before choosing a significance level.

If the fast-food restaurant manager really wants to avoid a Type I error, using $\alpha=0.01$ is the best choice. This small significance level will make it harder to find convincing evidence for H_a. Alternatively, if the manager really wants to avoid a Type II error, the manager should use a larger value of α, such as $\alpha=0.10$.

Section 8.2 What Did You Learn?

Review the learning goals from this section. Then practice what you've learned by working through the exercises.

Learning Goal	Example	Exercises
Make an appropriate conclusion in a significance test using a significance level.	8.2.1	7–10
Interpret a Type I error and a Type II error in context.	8.2.2	11–16
Give a consequence of a Type I error and a Type II error in a given setting.	8.2.3	17–22

Section 8.2 Exercises

Building Concepts and Skills These exercises assess the basic knowledge you should have after reading the section.

1. When the P-value is less than the significance level α in a significance test, we say that the result is _____ at the α level.

2. True/False: When a researcher plans to make a conclusion in a significance test based on a significance level, α should be stated before the data are collected.

3. Define Type II error.

4. True/False: A Type I error is always worse than a Type II error.

5. The probability of making a Type I error in a significance test is equal to _____.

6. As $P(\text{Type I error})$ decreases, $P(\text{Type II error})$ _____, assuming everything else remains the same.

Mastering Concepts and Skills These exercises reinforce the learning goals as shown in the examples.

7. **Refer to Example 8.2.1 Clean Water** The Environmental Protection Agency has determined that safe drinking water should contain no more than 1.3 milligrams (mg) of copper per liter (L) of water, on average. To test water from a new source, you collect water in small bottles at each of 30 randomly selected locations. The mean copper content of your bottles is 1.35 mg/L and the standard deviation is 0.21 mg/L. You perform a test of

$$H_0: \mu = 1.3$$
$$H_a: \mu > 1.3$$

where μ is the true mean copper content of the water from the new source. The test yields a P-value of 0.1012. What conclusion would you make at the $\alpha=0.01$ level?

8. **Grapefruit Juice** The labels on the bottles of a company's grapefruit juice say that they contain 180 milliliters (mL) of liquid. A quality-control supervisor suspects that the true mean is less than that. The supervisor selects a random sample of 40 bottles and measures the volume of liquid in each bottle. The mean volume of liquid in the bottles is 179.5 mL and the standard deviation is 1.3 mL. The supervisor performs a test of

$$H_0: \mu = 180$$
$$H_a: \mu < 180$$

where μ is the mean amount of liquid in all of the company's bottles of grapefruit juice. The test yields a P-value of 0.0098. What conclusion should the quality-control supervisor make at the $\alpha = 0.05$ level?

9. **Unknown Numbers** A telemarketing company claims that 1 in 5 U.S. adults will answer their cell phone when a call is received from an unknown number. What do the data say? In a Pew Research Center survey of 10,211 randomly selected U.S. adult cell phone owners, 19% said that they would answer the phone when an unknown number calls.[5] We use these data to perform a test of

$$H_0: p = 0.20$$
$$H_a: p \neq 0.20$$

where $p =$ the proportion of all U.S. adults with cell phones who would answer if an unknown number calls. The test yields a P-value of 0.0115.

(a) What conclusion should we make at the $\alpha = 0.05$ level?

(b) Would the conclusion from part (a) change if we used a 1% significance level instead? Explain your reasoning.

10. **Blood Type and Covid** According to a CNN Health report, "People with blood type O may be less vulnerable to COVID-19...." The article was based on a Danish study of 7422 people who tested positive for Covid. Only 38.4% of those people had type O blood, while 41.7% of the Danish population have type O blood.[6] To check the conclusion cited in the article, we perform a test of

$$H_0: p = 0.417$$
$$H_a: p < 0.417$$

where $p =$ the proportion of all Danish people who test positive for Covid who have type O blood. The test yields a P-value of 0.000000004.

(a) What conclusion should we make at the $\alpha = 0.01$ level?

(b) Would the conclusion from part (a) change if we used a 5% significance level instead? Explain your reasoning.

11. **Refer to Example 8.2.2 Mean Income** You are thinking about opening a restaurant and are searching for a good location. From your research, you know that the mean income of the households near the restaurant must be more than $85,000 to support the type of upscale restaurant you wish to open. You decide to select a simple random sample of 50 households near one potential site. Based on the mean income of this sample, you will perform a test of

$$H_0: \mu = 85{,}000$$
$$H_a: \mu > 85{,}000$$

where μ is the mean income in the population of households near the restaurant.[7] Describe a Type I error and a Type II error in this scenario.

12. **Highway Speeds** A city manager is trying to determine if it is worth the cost to install a speed sensor and traffic camera on a highway near the city. The manager will install these devices if there is convincing evidence that more than 20% of cars are speeding. The city's police department selects a random sample of 100 cars on the highway and measures their speed. The resulting data are used to perform a test of

$$H_0: p = 0.20$$
$$H_a: p > 0.20$$

where p is the proportion of all cars on this highway that are speeding. Describe a Type I error and a Type II error in this scenario.

13. **First Responders** Several cities have begun to monitor emergency response times because accident victims with life-threatening injuries generally need medical attention within 8 minutes. In one city, emergency personnel took more than 8 minutes to arrive at 22% of all calls involving life-threatening injuries last year. The city manager then issued guidelines for improving response time to local first responders. After 6 months, the city manager selects an SRS of 400 calls involving life-threatening injuries and examines the response times. Based on the sample data, the city manager performs a test of

$$H_0: p = 0.22$$
$$H_a: p < 0.22$$

where p is the true proportion of calls involving life-threatening injuries during this 6-month period for which emergency personnel took more than 8 minutes to arrive. Describe a Type I error and a Type II error in this scenario.

14. **High Blood Pressure?** A company markets a computerized device for detecting high blood pressure. The device measures an individual's blood pressure once per hour at a randomly selected time throughout a 12-hour period. Then it calculates the mean systolic (top number) pressure for the sample of measurements. Based on the sample results, the device performs a test of

$$H_0: \mu = 130$$
$$H_a: \mu > 130$$

where μ is the person's true mean systolic pressure. Describe a Type I error and a Type II error in this scenario.

15. **Thermostat Check** During the winter months, outside temperatures at the Starnes family cabin in Colorado can stay well below freezing (32°F, or 0°C) for weeks at a time. To prevent the pipes from freezing, Mrs. Starnes sets the thermostat at 50°F. The manufacturer claims that the thermostat allows variation in home temperature that follows an approximately normal distribution with standard deviation $\sigma = 3$°F. To test this claim, Mrs. Starnes programs her digital thermometer to take an SRS of $n = 10$ readings during a 24-hour period. Based on the data, Mrs. Starnes performs a test of

$$H_0: \sigma = 3$$
$$H_a: \sigma > 3$$

where σ is the true standard deviation of the temperatures in the cabin with this thermostat setting. Describe a Type I error and a Type II error in this scenario.

16. **New Golf Club** An avid golfer would like to improve their game. Based on years of experience, the golfer has established that the distance balls travel when hit with the current 3-iron follows an approximately normal distribution with standard deviation $\sigma = 15$ yards. The golfer is hoping that a new 3-iron will make their shots more consistent (less variable). To find out, the golfer hits 50 shots on a driving range with the new 3-iron. Based on the data, the golfer performs a test of
$$H_0: \sigma = 15$$
$$H_a: \sigma < 15$$
where σ is the true standard deviation of the distance balls travel when the golfer hits the new 3-iron. Describe a Type I error and a Type II error in this scenario.

17. **Refer to Example 8.2.3 Mean Income** Refer to Exercise 11. Give a consequence of each type of error in this scenario.

18. **Highway Speeds** Refer to Exercise 12. Give a consequence of each type of error in this scenario.

19. **First Responders** Refer to Exercise 13.
    (a) Give a consequence of each type of error in this scenario.
    (b) Which type of error do you think is more serious? Justify your answer.

20. **High Blood Pressure?** Refer to Exercise 14.
    (a) Give a consequence of each type of error in this scenario.
    (b) Which type of error do you think is more serious? Justify your answer.

21. **Thermostat Check** Refer to Exercise 15.
    (a) Give a consequence of each type of error in this scenario.
    (b) Which type of error do you think is more serious? Justify your answer.

22. **New Golf Club** Refer to Exercise 16.
    (a) Give a consequence of each type of error in this scenario.
    (b) Which type of error do you think is more serious? Justify your answer.

Applying the Concepts These exercises ask you to apply multiple learning goals in a new context or to apply what you learned in this section in a new way.

23. **Road Repairs** Members of the city council want to know if a majority of city residents support a 1% increase in the sales tax to fund road repairs. To investigate, they survey a random sample of 300 city residents and use the results to test the hypotheses
$$H_0: p = 0.50$$
$$H_a: p > 0.50$$
where p is the proportion of all city residents who support a 1% increase in the sales tax to fund road repairs. In the sample, $\hat{p} = 158/300 = 0.527$. The resulting P-value is 0.18.
    (a) What conclusion should the council make at $\alpha = 0.05$?
    (b) Which type of error — a Type I error or a Type II error — could the council have made in part (a)? Explain your answer.
    (c) Give a consequence of the type of error you chose in part (b).

24. **Light Bulbs** A contract between a manufacturer and a consumer of light bulbs specifies that the mean lifetime of the bulbs must be at least 1000 hours. An ordinary testing procedure is difficult because 1000 hours is more than 41 days! Because the lifetime of a bulb decreases as the voltage applied increases, a common procedure is to perform an accelerated lifetime test in which the bulbs are lit using 400 volts (compared to the usual 110 volts). At 400 volts, a 1000-hour bulb is expected to last only 3 hours. This is a well-known procedure, and both sides have agreed that the results from the accelerated test will be a valid indicator of the bulb's lifetime. The manufacturer will test the hypotheses
$$H_0: \mu = 3$$
$$H_a: \mu < 3$$
where μ = the true mean lifetime of the bulbs using $\alpha = 0.05$. A random sample of 100 bulbs yields an average lifetime of $\bar{x} = 2.90$ hours. The resulting P-value is 0.04.
    (a) What conclusion should the manufacturer make at $\alpha = 0.05$?
    (b) Which type of error — a Type I error or a Type II error — could the manufacturer have made in part (a)? Explain your answer.
    (c) Give a consequence of the type of error you chose in part (b).

25. **First Responders** Refer to Exercises 13 and 19. Which significance level — 0.10, 0.05, or 0.01 — would you choose for the test? Justify your answer.

26. **High Blood Pressure?** Refer to Exercises 14 and 20. Which significance level — 0.10, 0.05, or 0.01 — would you choose for the test? Justify your answer.

27. **Statistical Significance**
    (a) Explain why an observed result that is statistically significant at the 1% level must always be significant at the 5% level.
    (b) If a result is significant at the 5% level, what can you say about its significance at the 1% level?

28. **Statistical Significance** Asked to explain the meaning of "statistically significant at the 5% level," a student says, "This means the probability that the null hypothesis is true is less than 0.05." Is this explanation correct? Why or why not?

Extending the Concepts These exercises challenge you to explore statistical concepts and methods that go beyond what you learned in this section.

The *power* of a test is the probability that the test will find convincing evidence for H_a at significance level α when a specific alternative value of the parameter is true. We will discuss power in more detail in Section 8.7.

29. **Power** A drug manufacturer claims that fewer than 10% of patients who take its new drug for treating Alzheimer's disease will experience nausea. A researcher plans to give the drug to a random sample of $n=125$ patients with Alzheimer's disease, and to use the resulting data to perform a test of

$$H_0: p = 0.10$$
$$H_a: p < 0.10$$

where $p =$ the true proportion of patients with Alzheimer's disease who will experience nausea when taking this new drug. Suppose that the manufacturer's claim about its new drug is correct, and only 7% of all patients will experience nausea. The power of this test at the $\alpha = 0.05$ significance level is 0.27.

(a) Explain what "power = 0.27" means in this scenario.
(b) Describe a Type II error in this setting.
(c) Find the probability of a Type II error.
(d) Name one way to increase the power of the test.

Cumulative Review These exercises revisit what you learned in previous sections.

30. **Paul Bunyan (1.2)** Bangor, Maine, is one of several towns that claim to be the birthplace of the legendary lumberjack Paul Bunyan. Bangor is proud of its 31-foot fiberglass statue of Paul, and some townspeople think a second statue should be made of Paul's equally legendary sidekick, Babe the Blue Ox. The *Bangor Daily News* conducted an online poll, asking visitors to its website, "Should Bangor add Babe the Blue Ox to the Paul Bunyan statue site?" Of the 1123 people who responded, 864 said "Yes." Explain why this sampling method is biased. Is the sample proportion who said "Yes" likely greater than or less than the proportion of all Bangor residents who think that a statue should be added for Babe the Blue Ox?

31. **Short Novels (6.7)** In a city library, 30% of the novels have fewer than 400 pages. Suppose you randomly select 50 novels from the library. Let $\hat{p}$ be the proportion of novels in the sample that have fewer than 400 pages.

(a) Calculate the mean and standard deviation of the sampling distribution of $\hat{p}$. Interpret the standard deviation.
(b) Justify that the sampling distribution of $\hat{p}$ is approximately normal.
(c) Find the probability that more than 40% of the novels in the sample have fewer than 400 pages.

Section 8.3 Testing a Claim About a Population Proportion

LEARNING GOALS

By the end of this section, you will be able to:

- Check the Random and Large Counts conditions for performing a significance test about a population proportion.
- Calculate the standardized test statistic for a significance test about a population proportion.
- Find the P-value for a one-sided significance test about a population proportion.

In Sections 8.1 and 8.2, you saw that a significance test can be used to test a claim about an unknown population parameter. We are often interested in testing a claim about the proportion p of some outcome in a population. For example, a large car dealership claims that the proportion of all car buyers who purchase an extended warranty is $p = 0.60$, but an investigative reporter suspects that the true proportion is somewhat lower. This lesson shows you how to check conditions and perform calculations for a significance test about a population proportion.

Conditions for Testing a Claim About p

In Section 7.3, we introduced two conditions that should be met before we construct a confidence interval for a population proportion. We called them the Random and Large Counts conditions. These same conditions must be verified before carrying out a significance test about a population proportion.

The Large Counts condition for proportions requires that both np and $n(1-p)$ be at least 10. Because we assume H_0 is true when performing a significance test, we use the parameter value specified by the null hypothesis (denoted p_0) when checking the Large Counts condition. In this case, the Large Counts condition says that the *expected* count of successes np_0 and the *expected* count of failures $n(1-p_0)$ are both at least 10.

HOW TO Check Conditions for Performing a Significance Test About p

To perform a test about a population proportion p, check that the following conditions are met:
- **Random:** The data come from a random sample from the population of interest.
- **Large Counts:** Both np_0 and $n(1-p_0)$ are at least 10, where p_0 is the value of p specified by H_0.

Random sampling allows us to make an inference about the population based on sample data. If the data come from a convenience sample or a voluntary response sample, there's no point carrying out a significance test for p. The same is true if other sources of bias are present in the data-collection process. When the Large Counts condition is met, the sampling distribution of $\hat{p}$ is approximately normal. If this condition is violated, a P-value calculated from a normal distribution will not be accurate.

EXAMPLE 8.3.1

Can we make fast food faster?

Checking conditions

PROBLEM

In Section 8.2, you read that the proportion of all drive-thru customers at a fast-food restaurant who had to wait longer than 2 minutes after placing their order was $p = 0.63$. The restaurant's manager wants to reduce this proportion, and assigns an additional employee to assist with drive-thru orders. During the next month, the manager collects data on wait times from a random sample of 250 drive-thru orders, and finds that only $141/250 = 0.564$ of the customers have to wait longer than 2 minutes. The manager would like to carry out a test at the $\alpha = 0.10$ significance level of

$$H_0: p = 0.63$$
$$H_a: p < 0.63$$

where p = the proportion of all drive-thru customers at this restaurant who have to wait longer than 2 minutes to receive their food. Check if the conditions for performing the significance test are met.

SOLUTION

Random? Random sample of 250 drive-thru orders ✓

Large Counts? $250(0.63) = 157.5 \geq 10$ and $250(1-0.63) = 92.5 \geq 10$ ✓

> Large Counts condition: $np_0 \geq 10$ and $n(1-p_0) \geq 10$. Be sure to use the null value p_0, not the sample proportion $\hat{p}$, when checking this condition!

EXAM PREP: FOR PRACTICE, TRY EXERCISE 5.

Calculating the Standardized Test Statistic

In the fast-food example, the sample proportion of drive-thru customers who had to wait more than 2 minutes for their order is $\hat{p} = \dfrac{141}{250} = 0.564$. Because this result is less than 0.63, there is *some* evidence against $H_0: p = 0.63$ and in favor of $H_a: p < 0.63$.

But do we have *convincing* evidence that the proportion of all customers who have to wait longer than 2 minutes has decreased? To answer this question, we want to know if it's unlikely to get a sample proportion of 0.564 or less by chance alone when the null hypothesis is true. In other words, we are looking for a P-value.

Suppose for now that the null hypothesis $H_0: p = 0.63$ is true. Consider the sample proportion $\hat{p}$ of customers who have to wait more than 2 minutes for their drive-thru orders in a random sample of size $n = 250$. You learned in Section 6.7 that the sampling distribution of $\hat{p}$ will have mean

$$\mu_{\hat{p}} = p = 0.63$$

and standard deviation

$$\sigma_{\hat{p}} = \sqrt{\dfrac{p_0(1-p_0)}{n}} = \sqrt{\dfrac{0.63(0.37)}{250}} = 0.03054$$

Because the Large Counts condition is met, the sampling distribution of $\hat{p}$ will be approximately normal. Figure 8.3 displays this distribution. We have added the manager's sample result, $\hat{p} = \dfrac{141}{250} = 0.564$.

FIGURE 8.3 Normal approximation to the sampling distribution of the sample proportion $\hat{p}$ of drive-thru customers who have to wait at least 2 minutes in random samples of 250 orders when $p = 0.63$.

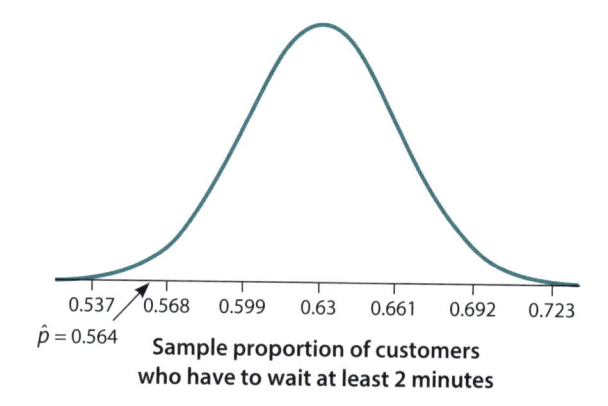

$\hat{p} = 0.564$

0.537   0.568   0.599   0.63   0.661   0.692   0.723

Sample proportion of customers who have to wait at least 2 minutes

To assess how far the statistic ($\hat{p} = 0.564$) is from the null value ($p_0 = 0.63$), standardize the statistic:

$$z = \dfrac{\hat{p} - p_0}{\sqrt{\dfrac{p_0(1-p_0)}{n}}} = \dfrac{0.564 - 0.63}{0.03054} = -2.161$$

This value is called the **standardized test statistic**.

> A **standardized test statistic** measures how far a sample statistic is from what we would expect if the null hypothesis H_0 were true, in standard deviation units. That is,
>
> $$\text{standardized test statistic} = \dfrac{\text{statistic} - \text{null value}}{\text{standard deviation (error) of statistic}}$$

The standardized test statistic says how far the sample result is from the null value, and in which direction, on a standardized scale. In this case, the fast-food manager's sample proportion of $\hat{p} = 0.564$ is 2.161 standard deviations less than the null value of $p = 0.63$.

EXAMPLE 8.3.2

How popular is online dating?

Calculating the standardized test statistic

PROBLEM

A Morning Consult poll revealed that 176 respondents in a random sample of 487 U.S. adults aged 18 to 29 had ever used an online dating app or service.[8] Do these data provide convincing evidence that more than 1 in 3 U.S. adults aged 18 to 29 have used an online dating app or service? To find out, we want to perform a test of

$$H_0: p = 1/3$$
$$H_a: p > 1/3$$

where p = the proportion of all U.S. adults aged 18 to 29 who have ever used an online dating app or service. Calculate the standardized test statistic.

SOLUTION

Statistic: $\hat{p} = \dfrac{176}{487} = 0.3614$

$z = \dfrac{0.3614 - 1/3}{\sqrt{\dfrac{(1/3)(2/3)}{487}}} = 1.314$

First calculate the sample statistic.

$$\text{standardized test statistic} = \dfrac{\text{statistic} - \text{null value}}{\text{standard deviation (error) of statistic}}$$

$$z = \dfrac{\hat{p} - p_0}{\sqrt{\dfrac{p_0(1 - p_0)}{n}}}$$

EXAM PREP: FOR PRACTICE, TRY EXERCISE 9.

Finding the *P*-Value

When the Random and Large Counts conditions are met, you can use the standardized test statistic to find the *P*-value for a significance test. Let's return to the fast-food drive-thru example. Our test of

$$H_0: p = 0.63$$
$$H_a: p < 0.63$$

based on the observed sample result of $\hat{p} = 0.564$ gave a standardized test statistic of $z = -2.161$. The *P*-value is the probability of getting a sample proportion less than or equal to $\hat{p} = 0.564$ by chance alone when $H_0: p = 0.63$ is true. The shaded area in Figure 8.4(a) shows this probability. Figure 8.4(b) shows the corresponding area to the left of $z = -2.161$ in the standard normal distribution.

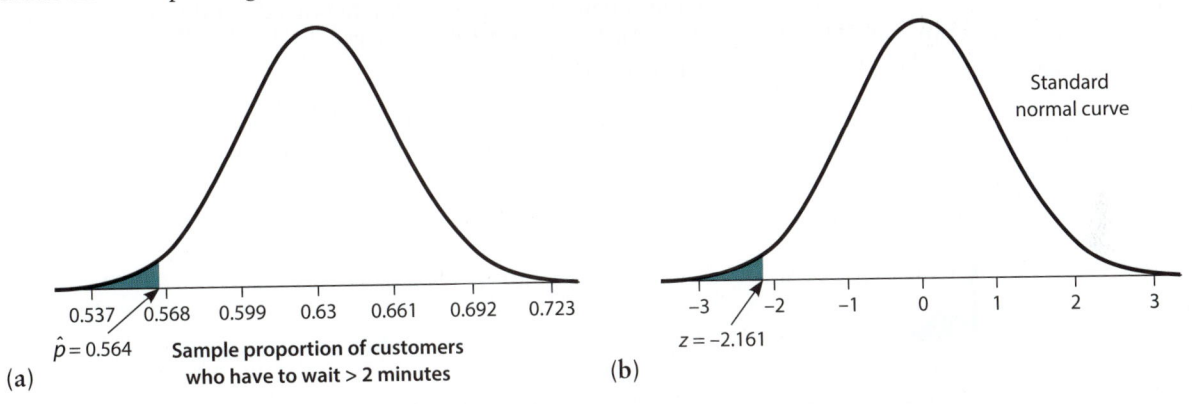

FIGURE 8.4 The shaded area shows the *P*-value for the fast-food example about the proportion of drive-thru customers who had to wait more than 2 minutes to receive their orders (a) on the approximately normal sampling distribution of $\hat{p}$ and (b) on the standard normal curve.

We can find the *P*-value from the standardized test statistic using Table A or technology. Table A gives $P(z \leq -2.16) = 0.0154$ using the *z*-score rounded to 2 decimal places. The *Normal Distributions* applet and the TI-83/84 command normalcdf(lower: −1000, upper: −2.161, μ: 0, σ: 1) give a *P*-value of 0.0153.

You learned how to make a conclusion for a statistical test based on a *P*-value and a significance level in Section 8.2. In the fast-food scenario, the manager chose a significance level of $\alpha = 0.10$. Because the *P*-value of 0.0153 is less than $\alpha = 0.10$, we reject H_0. We have convincing evidence that the proportion of all drive-thru customers who have to wait more than 2 minutes is now less than 0.63. ⚠ **We cannot conclude that assigning an additional employee to the drive-thru *caused* the decrease in wait times because this study was not a randomized experiment.** It could be that customer orders were simpler during the data collection period, for example.

Calculating the Standardized Test Statistic and P-Value in a Test About a Population Proportion p

Suppose the Random and Large Counts conditions are met. To perform a test of $H_0: p = p_0$, compute the standardized test statistic

$$z = \frac{\hat{p} - p_0}{\sqrt{\dfrac{p_0(1 - p_0)}{n}}}$$

Find the P-value by calculating the probability of getting a z statistic this large or larger in the direction specified by the alternative hypothesis H_a in a standard normal distribution.

EXAMPLE 8.3.3

How popular is online dating?

Finding the P-value

PROBLEM

In Example 8.3.2, we started to perform a test of

$$H_0: p = 1/3$$
$$H_a: p > 1/3$$

where $p =$ the proportion of all U.S. adults aged 18 to 29 who have ever used an online dating app or service. Morning Consult took a random sample of 487 U.S. adults aged 18 to 29 and found that 176 had ever used an online dating app or service. Note that the Random condition is met because the data came from a random sample of U.S. adults aged 18 to 29, and the Large Counts condition is met because $np_0 = 487(1/3) = 162.33 \geq 10$ and $n(1 - p_0) = 487(2/3) = 324.67 \geq 10$.

The resulting standardized test statistic is $z = 1.314$. Find the P-value.

SOLUTION

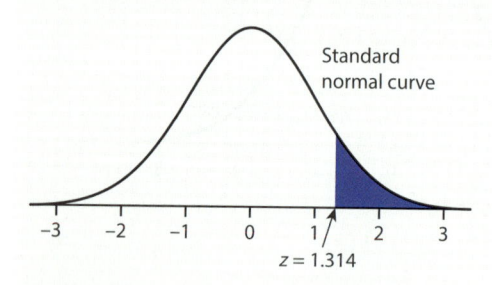

Using Table A: P-value = 1 − 0.9049 = 0.0951
Using technology: Applet/normalcdf(lower: 1.314, upper: 1000, mean: 0, SD: 1) = 0.0944

> The P-value from Table A is obtained using $z = 1.31$.

EXAM PREP: FOR PRACTICE, TRY EXERCISE 13.

For the online dating survey, because the P-value of 0.0944 is greater than our default significance level of $\alpha = 0.05$, we fail to reject H_0. We don't have convincing evidence that the proportion of all U.S. adults aged 18 to 29 who have ever used an online dating app or service is greater than 1/3.

THINK ABOUT IT

What is the *rejection region* approach? Instead of making a conclusion in a significance test based on a P-value, some people prefer to establish criteria for rejecting H_0 or failing to reject H_0 based on *critical values* of the test statistic. In the online dating examples, we wanted to perform a test of $H_0: p = 1/3$ versus $H_a: p > 1/3$ at the $\alpha = 0.05$ significance level. Which values of the z test statistic would give us convincing evidence for H_a in this upper-tailed test? Because the

Large Counts condition is met, we can use the standard normal distribution to answer this question. Using technology, the critical value z^* corresponding to a right-tail area of 0.05 is $z^* = 1.645$, as shown in the following figure.

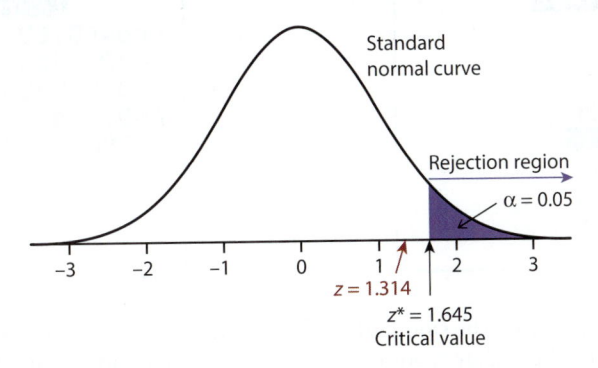

If H_0 is true, a value of z greater than 1.645 will occur less than 5% of the time by chance alone. That is, a value of $z > 1.645$ will lead us to reject H_0. For that reason, we call $z > 1.645$ the *rejection region*. A value of $z < 1.645$ will lead us to fail to reject H_0. (If the test statistic is exactly equal to the critical value, ask a statistician for advice.)

With the decision-making criteria established, all we need to do is see where our test statistic falls. The calculated value of $z = 1.314$ does not fall in the rejection region, so we fail to reject H_0. This is consistent with our decision using the P-value approach in Example 8.3.3.

As with confidence intervals, you can use technology to do all of the necessary calculations in a significance test for a population proportion. The following Tech Corner provides the details.

TECH CORNER Significance Tests About a Population Proportion

We'll illustrate the process of performing calculations in a significance test about a proportion using the fast-food scenario in Example 8.3.1, where we tested $H_0: p = 0.63$ versus $H_a: p < 0.63$. Recall that in a random sample of 250 drive-thru orders, 141 customers had to wait longer than 2 minutes to get their order.

Applet

1. Go to www.stapplet.com and launch the *One Categorical Variable, Single Group* applet.

2. Enter the variable name "Wait more than 2 minutes?" and choose to input data as counts in categories. Enter the category name "Yes" with a frequency of 141 and the category name "No" with a frequency of $250 - 141 = 109$.

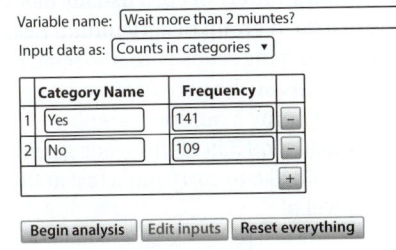

3. Click the Begin analysis button. Then, scroll down to the "Perform Inference" section.

4. Choose 1-sample z-test for the procedure, "Yes" as the category to indicate as success, $p <$ for the alternative hypothesis, and 0.63 for the hypothesized proportion. Then, click the Perform inference button.

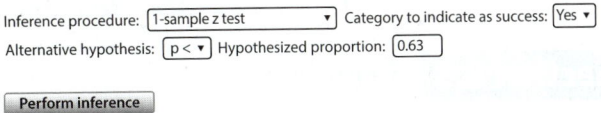

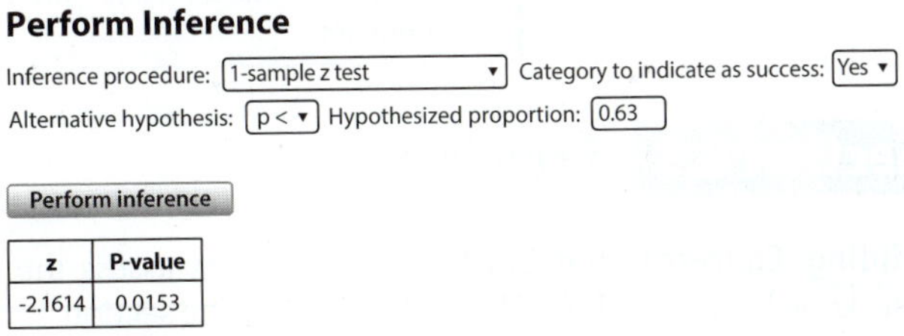

z	P-value
-2.1614	0.0153

TI-83/84

1. Press STAT, then choose TESTS and 1-PropZTest.

2. When the 1-PropZTest screen appears, enter $p_0 = 0.63$, $x = 141$, and $n = 250$. Specify the alternative hypothesis as prop $< p_0$. Note that x is the *number* of successes, not the *proportion* of successes ($\hat{p}$). The values of x and n must be whole numbers or the calculator will return an error message.

(continued)

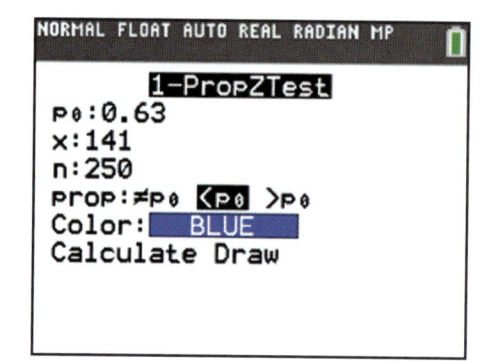

```
NORMAL FLOAT AUTO REAL RADIAN MP
        1-PropZTest
  p₀:0.63
  x:141
  n:250
  prop:≠p₀  <p₀  >p₀
  Color:  BLUE
  Calculate Draw
```

```
NORMAL FLOAT AUTO REAL RADIAN MP
        1-PropZTest
  prop<0.63
  z=-2.161438103
  p=0.0153307017
  p̂=0.564
  n=250
```

3. Highlight "Calculate" and press ENTER. The standardized test statistic and P-value for the test are reported, along with the sample proportion $\hat{p}$ and the sample size n.

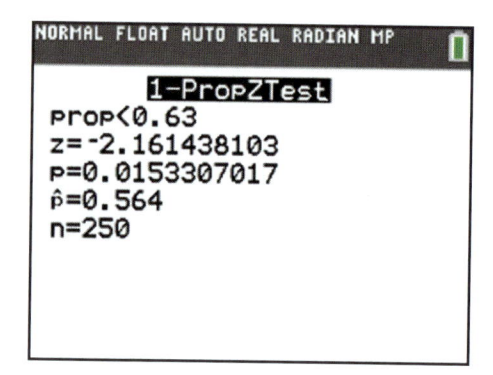

Note: If you select the Draw option, you will get a picture of the standard normal distribution with the area of interest shaded and the standardized test statistic and P-value labeled.

Detailed instructions for using CrunchIt!, Excel, Google Sheets, JMP, Minitab, and R are available in Achieve.

The standardized test statistic and P-value calculated using the "full technology" approach described in the Tech Corner are generally more accurate than the values we obtained earlier in the section when we rounded while doing our calculations. For that reason, we will use the values from the full technology approach when reporting results "Using technology" in Section 8.4.

Section 8.3 What Did You Learn?

Review the learning goals from this section. Then practice what you've learned by working through the exercises.

Learning Goal	Example	Exercises
Check the Random and Large Counts conditions for performing a significance test about a population proportion.	8.3.1	5–8
Calculate the standardized test statistic for a significance test about a population proportion.	8.3.2	9–12
Find the P-value for a one-sided significance test about a population proportion.	8.3.3	13–16

Section 8.3 Exercises

Building Concepts and Skills These exercises assess the basic knowledge you should have after reading the section.

1. State the Large Counts condition for a significance test about a population proportion.

2. Why does the Large Counts condition need to be met?

3. The general formula for a standardized test statistic is _____.

4. Give the specific formula for a standardized test statistic when performing a test of $H_0: p = p_0$.

Mastering Concepts and Skills These exercises reinforce the learning goals as shown in the examples.

5. **Refer to Example 8.3.1 Corn Growth** The germination rate of seeds is defined as the proportion of seeds that sprout and grow when properly planted and watered. A certain variety of corn usually has a germination rate of 0.80. A group of agricultural researchers wants to see if spraying seeds with a special, nontoxic chemical will increase the germination rate of this variety of corn. They spray a random sample of 400 seeds with the chemical, and 339 of the seeds germinate. The researchers would like to carry out a test at the $\alpha = 0.01$ significance level of

$$H_0: p = 0.80$$
$$H_a: p > 0.80$$

where p = the true proportion of seeds of this variety of corn that will germinate when sprayed with the chemical. Check if the conditions for performing the significance test are met.

6. **Rural Internet** A June 2020 report says that 78% of U.S. students in grades K–12 have reliable internet

access at home.[9] Researchers believe the proportion is smaller in their large, rural school district. To investigate, they choose a random sample of 120 K–12 students in the school district and find that 85 have reliable internet access at home. The researchers would like to carry out a test at the $\alpha = 0.05$ significance level of

$$H_0: p = 0.78$$
$$H_a: p < 0.78$$

where p = the proportion of all K–12 students in this large, rural school district who have reliable internet access at home. Check if the conditions for performing the significance test are met.

7. **Fire the Coach?** A college president says, "More than two-thirds of the alumni believe I should fire the football coach." The president's statement is based on 200 emails received from alumni in the past three months. The college's athletic director wants to perform a test of $H_0: p = 2/3$ versus $H_a: p > 2/3$, where p = the proportion of all the college's alumni who favor firing the coach. Check if the conditions for performing the significance test are met.

8. **Coin Toss** You want to determine if a coin is fair. You toss it 10 times and record the proportion of tosses that land heads. You would like to perform a test of $H_0: p = 0.5$ versus $H_a: p \neq 0.5$, where p = the proportion of all tosses of the coin that would land heads. Check if the conditions for performing the significance test are met.

9. **Refer to Example 8.3.2 Corn Growth** Refer to Exercise 5. Calculate the standardized test statistic.

10. **Rural Internet** Refer to Exercise 6. Calculate the standardized test statistic.

11. **Working Students** According to the National Center for Education Statistics, 49% of full-time students in 2-year colleges are employed.[10] An administrator at a large, rural 2-year college suspects that the proportion of full-time students who are employed is less than the national figure. The administrator would like to carry out a test at the 5% significance level of

$$H_0: p = 0.49$$
$$H_a: p < 0.49$$

where p = the proportion of all full-time students at this 2-year college who are employed. The administrator selects a random sample of 200 full-time students from the college and finds that 91 of them are employed.

(a) Verify that the conditions for performing the test are met.

(b) Calculate the standardized test statistic.

12. **Green Tea** Two young researchers decided to investigate whether green labeling makes consumers believe that a product is natural. They took a random sample of 40 people, and served each person two cups of tea — one in a green cup and one in a clear cup. (Although the participants did not know it, both cups were filled with the same type of tea.) The researchers asked each person, "Which cup of tea do you believe has a more natural flavor?" Twenty-nine of the 40 people said the one in the green cup.[11] Based on these data, the researchers want to carry out a test at the 1% significance level of

$$H_0: p = 0.50$$
$$H_a: p > 0.50$$

where p = the true proportion of people who would identify the green cup of tea as having the more natural flavor.

(a) Verify that the conditions for performing the test are met.

(b) Calculate the standardized test statistic.

13. **Refer to Example 8.3.3 Corn Growth** Refer to Exercises 5 and 9. Find the *P*-value.

14. **Rural Internet** Refer to Exercises 6 and 10. Find the *P*-value.

15. **Working Students** Refer to Exercise 11.

(a) Find the *P*-value.

(b) What conclusion should the administrator make?

16. **Green Tea** Refer to Exercise 12.

(a) Find the *P*-value.

(b) What conclusion should the researchers make?

Applying the Concepts These exercises ask you to apply multiple learning goals in a new context or to apply what you learned in this section in a new way.

17. **Singing Last** On TV shows that feature singing competitions, contestants often wonder if there is an advantage in performing last. To investigate this question, researchers selected a random sample of 600 college students and showed each student the audition videos of 12 different singers. Each student viewed the videos in random order. We would expect approximately 1/12 of the students to prefer the last singer seen, assuming order doesn't matter. In this study, 59 of the 600 students preferred the last singer they viewed. Do the data provide convincing evidence that performing last offers an advantage?

(a) State appropriate hypotheses for performing a significance test. Be sure to define the parameter of interest.

(b) Check that the conditions for performing the test are met.

(c) Calculate the standardized test statistic.

(d) Find the *P*-value. What conclusion would you make?

18. **Potato Chips** A company that makes potato chips requires each shipment of potatoes to meet certain quality standards. If the company finds convincing evidence that more than 8% of the potatoes in the shipment have "blemishes," the truck will be sent back to the supplier to get another load of potatoes.

Otherwise, the entire truckload will be used to make potato chips. A supervisor selects a random sample of 500 potatoes from a truck and finds that 52 of the potatoes have blemishes. Do the data provide convincing evidence that the truck should be sent back to the supplier?

(a) State appropriate hypotheses for performing a significance test. Be sure to define the parameter of interest.

(b) Check that the conditions for performing the test are met.

(c) Calculate the standardized test statistic.

(d) Find the P-value. What conclusion would you make?

19. **Singing Last** Refer to Exercise 17. Which type of error — a Type I error or a Type II error — could you have made in part (d)? Explain your answer.

20. **Potato Chips** Refer to Exercise 18. Which type of error — a Type I error or a Type II error — could you have made in part (d)? Explain your answer.

21. **Corn Growth** Refer to Exercise 13.
    (a) Interpret the P-value.
    (b) What conclusion would you make?

22. **Rural Internet** Refer to Exercise 14.
    (a) Interpret the P-value.
    (b) What conclusion would you make?

Extending the Concepts These exercises challenge you to explore statistical concepts and methods that go beyond what you learned in this section.

23. **Green Tea** Refer to Exercise 12 and the Think About It feature in this section.
    (a) Find the critical value and the rejection region.
    (b) What conclusion would you make? Explain your reasoning.

24. **Cell-Phone Passwords** A consumer organization suspects that fewer than half of parents know their child's cell-phone password. The Pew Research Center asked a random sample of 1060 parents if they knew their child's cell-phone password; 551 reported that they knew the password.
    (a) State appropriate hypotheses for testing the consumer organization's belief. Be sure to define the parameter of interest.
    (b) Check that the conditions for performing the test are met.
    (c) Explain why it isn't necessary to carry out the significance test based on the sample result.
    Let's go ahead and carry out the test anyway to see what will happen.

(d) Calculate the standardized test statistic and P-value.

(e) How is the conclusion of the test consistent with your answer to part (c)?

Cumulative Review These exercises revisit what you learned in previous sections.

25. **Under 18 (5.3, 5.4, 8.1)** According to the U.S. Census Bureau, 22.4% of U.S. residents are younger than age 18 years. Suppose we select a random sample of 10 U.S. residents. Let $X =$ the number of people in the sample who are younger than age 18.

    (a) Explain why it is reasonable to use the binomial distribution for probability calculations involving X.

    (b) Explain why the probability distribution of X is not approximately normal.

    (c) Find the probability that at most 1 person in the sample is younger than age 18.

    (d) Researchers want to perform a test of $H_0: p = 0.224$ versus $H_a: p < 0.224$, where p is the proportion of all North Carolina residents who are younger than age 18. They select a random sample of 10 North Carolina residents; only 1 person is younger than 18. Based on your answer to part (c), what conclusion should the researchers make?

26. **Cheese Nutrition (3.5, 3.6)** The scatterplot shows the relationship between $x =$ protein content in grams and $y =$ total fat content in grams for 1 ounce of 25 different types of cheese.[12]

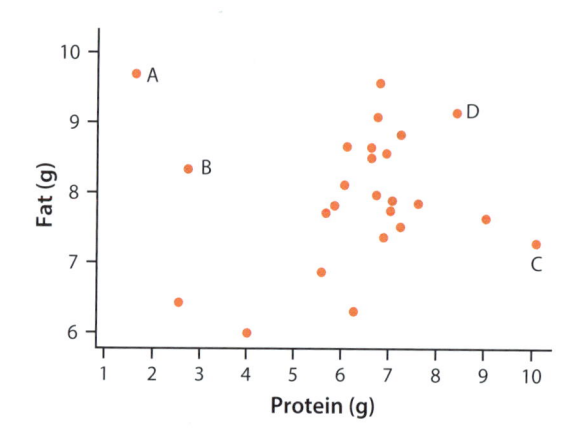

    (a) Is the correlation between these two variables close to −1, close to +1, or close to 0? Explain your answer.

    (b) What effect do points A and B have on the correlation?

    (c) If you want to buy a cheese that has a high protein content relative to its fat content, would you choose the cheese whose point is indicated by A, B, C, or D? Explain your reasoning.

Section 8.4 Significance Tests for a Population Proportion

LEARNING GOALS

By the end of this section, you will be able to:

- Use the four-step process to perform a one-sided significance test about a population proportion.
- Calculate the *P*-value for a two-sided significance test about a population proportion.
- Use the four-step process to perform a two-sided significance test about a population proportion.

In Section 8.3, you learned how to check conditions and perform calculations for a **one-sided test** about a population proportion. We begin this lesson by showing you how to use the four-step process to carry out a one-sided significance test for p. Then we discuss **two-sided tests**.

A significance test involving a one-sided alternative hypothesis is called a **one-sided test.** A significance test involving a two-sided alternative hypothesis is called a **two-sided test.**

Putting It All Together: The Four-Step Process

To perform a significance test, we state hypotheses, check conditions, calculate a test statistic and *P*-value, and make a conclusion in the context of the problem. The four-step process is ideal for organizing our work.

> **HOW TO** Use the Four-Step Process: Significance Tests
>
> **State:** State the hypotheses, parameter(s), significance level, and evidence for H_a.
> **Plan:** Identify the appropriate inference method and check the conditions.
> **Do:** If the conditions are met, perform calculations.
> - Calculate the test statistic.
> - Find the *P*-value.
>
> **Conclude:** Make a conclusion about the hypotheses in the context of the problem.

We are now ready to use the four-step process to carry out a **one-sample *z* test for a proportion**.

A **one-sample *z* test for a proportion** is a significance test of the null hypothesis that a population proportion p is equal to a specified value.

In all of this section's examples, "Using technology" indicates that we are getting results from the full technology approach described in the Section 8.3 Tech Corner.

EXAMPLE 8.4.1

Who watches Survivor?

One-sided test for a proportion

PROBLEM

Advertisers want to know about the viewership of various reality shows before investing their marketing dollars. According to Nielsen ratings, *Survivor* was one of the most-watched shows in the United States during every week that it aired. An avid *Survivor* fan (your textbook author, Mr. Starnes) claims that 35% of all U.S. adults have watched *Survivor*. A skeptical editor believes this figure is too high. The editor asks a random sample of 200 U.S. adults if they have watched *Survivor*; 60 say, "Yes." Is there convincing evidence at the $\alpha = 0.05$ significance level to confirm the editor's belief?

(continued)

SOLUTION

State:

$H_0: p = 0.35$
$H_a: p < 0.35$

where p = the proportion of all U.S. adults who have watched *Survivor*.
Use $\alpha = 0.05$.

The evidence for H_a is: $\hat{p} = \dfrac{60}{200} = 0.30 < 0.35$

> **State:** State the hypotheses, parameter(s), significance level, and evidence for H_a.

Plan: One-sample z test for p
Random? Random sample of 200 U.S. adults ✓
Large Counts? $200(0.35) = 70 \geq 10$ and $200(1 - 0.35) = 130 \geq 10$ ✓

> **Plan:** Identify the appropriate inference method and check the conditions. *Be sure to use the null value p_0 when checking the Large Counts condition.*

Do: $\hat{p} = \dfrac{60}{200} = 0.30$

- $z = \dfrac{0.30 - 0.35}{\sqrt{\dfrac{0.35(1-0.35)}{200}}} = -1.482$

- *P*-value:

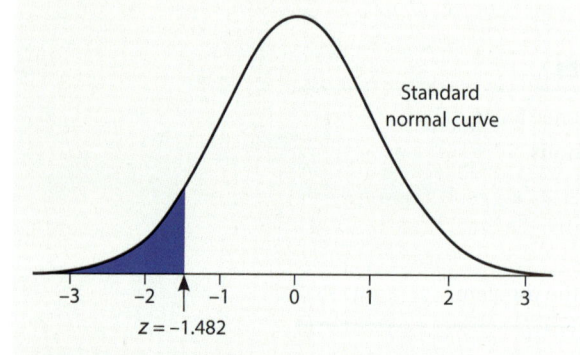

> **Do:** If the conditions are met, perform calculations:
> - Calculate the test statistic.
> - Find the *P*-value.

[Standard normal curve with shaded left tail at $z = -1.482$]

Using Table A: *P*-value = 0.0694
Using technology: $z = -1.482$, *P*-value = 0.0691

> When the calculations from technology differ from the by-hand calculations, we'll use the values from technology in the conclusion.

Conclude: Because the *P*-value of 0.0691 is greater than $\alpha = 0.05$, we fail to reject H_0. There is not convincing evidence that the proportion of all U.S. adults who have watched *Survivor* is less than 0.35.

> **Conclude:** Make a conclusion about the hypotheses in the context of the problem.

EXAM PREP: FOR PRACTICE, TRY EXERCISE 5.

Example 8.4.1 reminds us why significance tests are important. The sample proportion of adults who have watched *Survivor* is $\hat{p} = 60/200 = 0.30$. This result gives some evidence against $H_0: p = 0.35$ and in favor of $H_a: p < 0.35$. To see whether such an outcome is unlikely to occur by chance alone when H_0 is true, we had to carry out a significance test. The *P*-value told us that a sample proportion of 0.30 or smaller would occur in about 7% of all random samples of 200 U.S. adults from a population in which 35% have watched *Survivor*. So we can't rule out sampling variability as a plausible explanation for getting a sample proportion of $\hat{p} = 0.30$. Of course, we could have made a Type II error in this case by failing to reject H_0 when $H_a: p < 0.35$ is true.

> **THINK ABOUT IT**
>
> **What happens when the data don't support H_a?** Suppose the skeptical editor in Example 8.4.1 had obtained a sample proportion of $\hat{p} = 76/200 = 0.38$. This sample result doesn't even give *some* evidence to support the alternative hypothesis $H_a: p < 0.35$ because $\hat{p} = 0.38$ is *greater than* 0.35! There's no need to continue with a significance test.

Two-Sided Tests

When the appropriate conditions are met, the *P*-value in a one-sided test about a population proportion is the area in one tail of a standard normal distribution—the tail specified by H_a. (For that reason, a one-sided test is sometimes called a *one-tailed test*.) In a two-sided test, the alternative hypothesis has the form $H_a: p \neq p_0$. The *P*-value in such a test is the probability of getting a sample proportion as far or farther from p_0 than the observed value of $\hat{p}$, in either direction, assuming the null hypothesis is true. As a result, you have to find the area in both tails of a standard normal distribution to get the *P*-value. (For that reason, a two-sided test is sometimes called a *two-tailed test*.)

EXAMPLE 8.4.2

What about a two-sided test?

Finding the *P*-value

PROBLEM
Suppose that you want to perform a test of

$$H_0: p = 0.70$$
$$H_a: p \neq 0.70$$

An SRS of size 50 from the population of interest yields 39 successes. Note that the Random and Large Counts conditions are met. Calculate the standardized test statistic and find the *P*-value.

SOLUTION

$\hat{p} = \dfrac{39}{50} = 0.78$

$z = \dfrac{0.78 - 0.70}{\sqrt{\dfrac{0.70(1-0.70)}{50}}} = 1.234$

$$z = \dfrac{\hat{p} - p_0}{\sqrt{\dfrac{p_0(1-p_0)}{n}}}$$

Standard normal curve

$z = -1.234$ $z = 1.234$

> As the graph suggests, the *P*-value for a two-sided test is twice the area in one tail of the standard normal distribution.

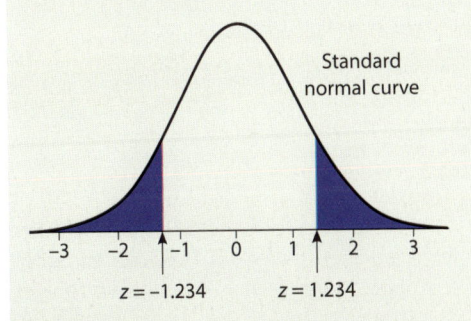

Using Table A: *P*-value = 0.1093 + 0.1093 = 2(0.1093) = 0.2186

Using technology: $z = 1.234$, *P*-value = 0.2170

> Recall that "Using technology" means we are getting the *P*-value from the full technology approach described at the end of Section 8.3.

EXAM PREP: FOR PRACTICE, TRY EXERCISE 9.

Now you are ready to perform a two-sided test about a population proportion.

EXAMPLE 8.4.3

Do most college students vape?

Two-sided test about a proportion

PROBLEM
According to the University of Michigan's *Monitoring the Future* study, 39.6% of college students used e-cigarettes (vaped) in a recent year.[13] Public health researchers wonder if this national result applies to their own local two- and four-year colleges and universities. To investigate, the researchers collect data from a random sample of 750 students attending these local institutions; 331 of them vaped that year. Is there convincing evidence that the proportion of all college students enrolled in these local institutions who vaped that year differs from the national result?

SOLUTION

State:

$H_0: p = 0.396$

$H_a: p \neq 0.396$

where p = the proportion of all students in local two- and four-year colleges and universities who vaped in a recent year.
Use $\alpha = 0.05$ because no significance level was stated.

The evidence for H_a is: $\hat{p} = \dfrac{331}{750} = 0.4413 \neq 0.396$

> **State:** State the hypotheses, parameter(s), significance level, and evidence for H_a.

Plan: One-sample z test for p

Random? Random sample of 750 college students ✓

Large Counts? $750(0.396) = 297 \geq 10$ and $750(1 - 0.396) = 453 \geq 10$ ✓

> **Plan:** Identify the appropriate inference method and check the conditions.

Do: $\hat{p} = \dfrac{331}{750} = 0.4413$

- $z = \dfrac{0.4413 - 0.396}{\sqrt{\dfrac{0.396(1 - 0.396)}{750}}} = 2.537$

- P-value:

> **Do:** If the conditions are met, perform calculations:
> - Calculate the test statistic.
> - Find the P-value.

[Standard normal curve with shaded tails at $z = -2.537$ and $z = 2.537$]

Using Table A: P-value $= 2(0.0055) = 0.0110$

Using technology: $z = 2.539$, P-value $= 0.0111$

Conclude: Because the P-value of 0.0111 is less than $\alpha = 0.05$, we reject H_0. There is convincing evidence that the proportion of all college students enrolled in these local institutions who vaped that year differs from the national result of 0.396.

> **Conclude:** Make a conclusion about the hypotheses in the context of the problem.

EXAM PREP: FOR PRACTICE, TRY EXERCISE 13.

How do we interpret the P-value for the two-sided test in Example 8.4.3? If the proportion of all students in local two- and four-year colleges and universities who vaped in a recent year is the same as the national result ($p_0 = 0.396$), there is a 0.0111 probability of getting a sample proportion at least as far from p_0 in either direction as 0.4413 by chance alone. In symbols, we can summarize the P-value as $P(\hat{p} \leq 0.3507 \text{ or } \hat{p} \geq 0.4413 \mid p = 0.396) = 0.0111$.

Confidence Intervals Give More Information

The result of a significance test begins with the decision to reject H_0 or fail to reject H_0. In the public health researchers' vaping study, for instance, the data led us to reject $H_0: p = 0.396$ because we found convincing evidence that the proportion of college students enrolled in local two- and four-year institutions who vaped in a recent year differs from the national value. We're left wondering what the actual proportion p might be. A confidence interval can shed some light on this issue.

You learned how to calculate a confidence interval for a population proportion in Section 7.3. A 95% confidence interval for p is

$$0.4413 \pm 1.96 \sqrt{\frac{(0.4413)(1-0.4413)}{750}} \rightarrow 0.4413 \pm 0.0355 \rightarrow 0.4058 \text{ to } 0.4768$$

This interval gives the values for p that are plausible based on the sample data. We would not be surprised if the proportion of all college students in these local institutions who vaped in a recent year was any value between 0.4058 and 0.4768. However, we would be surprised if the true proportion was 0.396 because this value is not contained in the confidence interval.

There is a link between confidence intervals and *two-sided* tests. The 95% confidence interval (0.4058, 0.4768) gives an approximate set of p_0's that should not be rejected by a two-sided test at the $\alpha = 0.05$ significance level. Any p_0 value outside the interval should be rejected as implausible.

Section 8.4 What Did You Learn?

Review the learning goals from this section. Then practice what you've learned by working through the exercises.

Learning Goal	Example	Exercises
Use the four-step process to perform a one-sided significance test about a population proportion.	8.4.1	5–8
Calculate the P-value for a two-sided significance test about a population proportion.	8.4.2	9–12
Use the four-step process to perform a two-sided significance test about a population proportion.	8.4.3	13–16

Section 8.4 Exercises

Building Concepts and Skills These exercises assess the basic knowledge you should have after reading the section.

1. In the four-step process for a significance test, what four things should you include in the State step?

2. A significance test for a population proportion is often referred to as a _____.

3. The P-value in a two-sided test about a population proportion requires you to find the area in _____ of a standard normal distribution.

4. Explain why a confidence interval for a population proportion gives more information than a two-sided test of $H_0: p = p_0$.

Mastering Concepts and Skills These exercises reinforce the learning goals as shown in the examples.

5. **Refer to Example 8.4.1 Drug Side Effects** A drug manufacturer claims that fewer than 10% of patients who take its new drug for treating Alzheimer's disease will experience nausea. To test this claim, researchers conduct an experiment. They give the new drug to a random sample of 300 patients with Alzheimer's disease whose families have given informed consent for the patients to participate in the study. In all, 25 of the subjects experience nausea. Do these data give convincing evidence at the $\alpha = 0.01$ significance level to support the drug manufacturer's claim?

6. **Campus Parking** A university makes a change to improve student satisfaction with parking on campus. Before the change, 37% of students approved of the parking provided by the university. After the change, the university's administration surveys an SRS of 200 students at the school. In all, 83 students say that they approve of the new parking arrangement. Do these data give convincing evidence at the $\alpha = 0.05$ significance level that the change was effective?

7. **Sexting** A research article claims that 60% of U.S. college students have engaged in electronic sharing of sexual content, known as sexting.[14] The counseling staff at a large college in the United States worries that the actual figure might be higher at their institution. To find out, they administer an anonymous survey to a random sample of 250 students at their college. All 250 respond, and 167 admit to sexting. Do these data give convincing evidence at the 5% significance level that the counselor's belief is correct?

8. **Coffee Challenge** Do coffee drinkers prefer fresh-brewed coffee or instant coffee? A researcher claims that only half of all coffee drinkers prefer fresh-brewed coffee. To test this claim, we ask a random sample of 50 coffee drinkers in a small city to take part in a study. Each person tastes two unmarked cups — one containing instant coffee and one containing fresh-brewed coffee — and says which they prefer. Of the 50 participants, 36 choose the fresh-brewed coffee. Do these results give convincing evidence at the 1% significance level that coffee drinkers favor fresh-brewed coffee over instant coffee?

9. **Refer to Example 8.4.2 Two-Sided Test** Suppose that you want to perform a test of $H_0: p = 0.7$ versus $H_a: p \neq 0.7$. An SRS of size 80 from the population of interest yields 59 successes. Calculate the standardized test statistic and P-value. Note that the Random and Large Counts conditions are met.

10. **Two-Tailed Test** Suppose that you want to perform a test of $H_0: p = 0.65$ versus $H_a: p \neq 0.65$. An SRS of size 80 from the population of interest yields 41 successes. Calculate the standardized test statistic and P-value. Note that the Random and Large Counts conditions are met.

11. **Another Two-Sided Test** Suppose that you want to perform a test of $H_0: p = 0.25$ versus $H_a: p \neq 0.25$. A random sample of size 60 from the population of interest yields 8 successes.
    (a) Explain why the conditions for inference are met.
    (b) Calculate the standardized test statistic and P-value.

12. **Another Two-Tailed Test** Suppose that you want to perform a test of $H_0: p = 0.45$ versus $H_a: p \neq 0.45$. A random sample of size 60 from the population of interest yields 33 successes.
    (a) Explain why the conditions for inference are met.
    (b) Calculate the standardized test statistic and P-value.

13. **Refer to Example 8.4.3 Wealth Goals** In a recent year, 84.3% of first-year college students responding to a national survey identified "being very well-off financially" as an important personal goal.[15] A state university finds that 157 of an SRS of 200 of its first-year students say that this goal is important. Is there convincing evidence at the $\alpha = 0.05$ significance level that the proportion of all first-year students at this university who think being very well-off financially is important differs from the national value of 84.3%?

14. **Driving Tests** A state's Department of Motor Vehicles (DMV) claims that 60% of people pass their driving test on the first attempt. An investigative reporter examines an SRS of the DMV records for 125 people who took their driving test for the first time; 86 of them passed the test. Is there convincing evidence at the $\alpha = 0.05$ significance level that the DMV's claim is incorrect?

15. **Mendel's Peas** Gregor Mendel (1822–1884), an Austrian monk, is considered the father of genetics. Mendel studied the inheritance of various traits in pea plants. One such trait is whether the pea is smooth or wrinkled. Mendel predicted a ratio of 3 smooth peas for every 1 wrinkled pea, or 75% smooth peas. In one experiment, he observed 423 smooth and 133 wrinkled peas.
    (a) Do these data provide convincing evidence that Mendel's prediction is incorrect? Assume that the Random condition is met.
    (b) Interpret the P-value.

16. **Coin Spins** When a fair coin is flipped, we all know that the probability the coin lands on heads is 0.50. However, what if a coin is spun? According to the article "Euro Coin Accused of Unfair Flipping" in the *New Scientist* journal, two Polish math professors and their students spun a Belgian euro coin 250 times. It landed heads 140 times. One of the professors concluded that the coin was minted asymmetrically. A representative from the Belgian mint indicated the result was just chance.[16] Assume that the conditions for inference are met.
    (a) Do these data provide convincing evidence that this euro coin is not equally likely to land on heads or tails when spun? Assume that the Random condition is met.
    (b) Interpret the P-value.

Applying the Concepts These exercises ask you to apply multiple learning goals in a new context or to apply what you learned in this section in a new way.

17. **Selling Upgrades** A software company is trying to decide whether to produce an upgrade of one of its programs. Customers would have to pay $100 for the upgrade. For the upgrade to be profitable, the company has to sell it to more than 20% of its customers. You contact a random sample of 60 customers and find that 16 would be willing to pay $100 for the upgrade.
    (a) Which would be a more serious mistake in this setting — a Type I error or a Type II error? Justify your answer.

(b) Do the sample data give convincing evidence at the $\alpha = 0.01$ significance level that more than 20% of the company's customers are willing to purchase the upgrade?

18. **Flu Vaccine** A drug company has developed a new vaccine for preventing the flu. The company claims that fewer than 5% of adults who use its vaccine will get the flu. To test this claim, researchers give the vaccine to a random sample of 1000 adults. Of these subjects, 43 get the flu.

    (a) Which would be a more serious mistake in this setting—a Type I error or a Type II error? Justify your answer.

    (b) Do these data give convincing evidence at the $\alpha = 0.01$ significance level to support the company's claim?

19. **Wealth Goals** Refer to Exercise 13 and Section 7.4.

    (a) Construct and interpret a 95% confidence interval for the proportion of all first-year students at the university who would identify being very well-off financially as an important personal goal. Assume the conditions for inference are met.

    (b) Explain why the interval in part (a) provides more information than the test in Exercise 13.

20. **Driving Tests** Refer to Exercise 14 and Section 7.4.

    (a) Construct and interpret a 95% confidence interval for the proportion of all people who pass their driving test on the first attempt. Assume the conditions for inference are met.

    (b) Explain why the interval in part (a) provides more information than the test in Exercise 14.

21. **Weight Loss** A Gallup poll found that 55% of the people in its sample said "Yes" when asked, "Would you like to lose weight?" Gallup announced: "For results based on the total sample of national adults, one can say with 95% confidence that the margin of (sampling) error is ±3 percentage points."[17] Does this interval provide convincing evidence that the actual proportion of U.S. adults who would say they want to lose weight differs from 0.60? Justify your answer.

22. **Who Tweets?** The Pew Research Center asked a random sample of U.S. adults, "Do you ever use Twitter?" In this sample, 22% said "Yes." According to Pew, the margin of error for the 95% confidence interval is ±2.9 percentage points.[18] Does this interval provide convincing evidence that the actual proportion of U.S. adults who would say they use Twitter differs from 0.20? Justify your answer.

23. **Tech** **Household Languages** What percentage of U.S. households speak a language other than English? The *American Community Survey* data set contains detailed information about a random sample of 3000 U.S. households.[19] The Census Bureau records household language (HHL) as 1 = English only, 2 = Spanish, 3 = other Indo-European languages, 4 = Asian and Pacific Island languages, or 5 = other language. A researcher claims that 21.5% of U.S. households speak a language other than English. Do these data give convincing evidence to refute the researcher's claim?

24. **Tech** **High Cholesterol** What percentage of U.S. residents have high cholesterol levels? The National Health and Nutrition Examination Survey (NHANES) is an ongoing research program conducted by the National Center for Health Statistics. Researchers selected a random sample of more than 7000 U.S. residents for study in the most recent survey. The *NHANES* data set gives the cholesterol levels in mg/dL ($LBXTC$).[20] A media report claims that more than 1 in 4 U.S. residents have high cholesterol levels—that is, have total cholesterol levels of 200 mg/dL or higher. Do these data provide convincing evidence to support the report's claim? (*Note:* You should create a new variable that indicates whether or not each person has a total cholesterol level of 200 mg/dL or higher.)

Extending the Concepts These exercises challenge you to explore statistical concepts and methods that go beyond what you learned in this section.

When the Large Counts condition is violated, you shouldn't perform a one-sample z test for a proportion. But you can carry out a significance test using the binomial distribution.

25. **Cranky Mower** A company has developed an "easy-start" mower that cranks the engine with the push of a button. The company claims that the probability this mower model will start on any push of the button is 0.9. A consumer testing agency suspects that this claim is exaggerated. To test the claim, an agency researcher selects a random sample of 20 of these mowers and attempts to start each one by pushing the button. Only 15 of the mowers start.

    (a) State appropriate hypotheses for performing a significance test. Be sure to define the parameter of interest.

    (b) Show that the Large Counts condition is not met.

    (c) Assuming that the null hypothesis from part (a) is true, calculate the probability that 15 or fewer of the 20 randomly selected mowers would start.

    (d) Based on your result in part (c), what conclusion would you make?

26. **Flu Vaccine** Refer to Exercise 18 and the Think About It in Section 8.3.

    (a) Find the critical value and the rejection region.

    (b) What conclusion would you make? Explain your reasoning.

Cumulative Review These exercises revisit what you learned in previous sections.

27. **Miles Driven (6.8)** The service department of a large automobile dealership records the odometer readings of cars that it repairs and determines that the distribution of miles driven per year by all of its customers has a mean of 14,000 miles and a standard deviation

of 4000 miles. The distribution is skewed to the right. Suppose a random sample of 12 cars is selected from the dealership's service records. Let $\bar{x}$ = the mean number of miles driven per year for the cars in the sample.

(a) What is the mean of the sampling distribution of $\bar{x}$?

(b) Calculate and interpret the standard deviation of the sampling distribution of $\bar{x}$.

(c) Describe the shape of the sampling distribution of $\bar{x}$.

28. **Gas Mileage (6.2, 6.3)** In its 2021 *Fuel Economy Guide*, the Environmental Protection Agency gives data on car models that have gasoline engines and are not hybrids. The combined city and highway gas mileage for these vehicles is approximately normal with mean 23.2 miles per gallon (mpg) and standard deviation 5.0 mpg.[21]

(a) The Chevrolet Malibu with a standard four-cylinder engine has a combined gas mileage of 29 mpg. What percentage of all vehicles have worse gas mileage than the Malibu?

(b) How high must a vehicle's gas mileage be to fall in the top 10% of all these car models?

Section 8.5 Testing a Claim About a Population Mean

LEARNING GOALS

By the end of this section, you will be able to:

- Check the Random and Normal/Large Sample conditions for performing a significance test about a population mean.
- Calculate the standardized test statistic for a significance test about a population mean.
- Find the *P*-value for a significance test about a population mean.

You learned how to perform a significance test about a population proportion in Sections 8.3 and 8.4. Now we'll examine the details of testing a claim about a population mean μ.

Conditions for Testing a Claim About μ

In Section 7.5, we introduced conditions that should be met before we construct a confidence interval for a population mean. We called them the Random and Normal/Large Sample conditions. These same conditions must be verified before carrying out a significance test.

> **HOW TO** Check the Conditions for Performing a Significance Test About μ
>
> To perform a test about a population mean μ, check that the following conditions are met:
> - **Random:** The data come from a random sample from the population of interest.
> - **Normal/Large Sample:** The data come from a population that is approximately normally distributed or the sample size is large ($n \geq 30$). When the sample size is small and the shape of the population distribution is unknown, a graph of the sample data shows no strong skewness or outliers.

EXAMPLE 8.5.1

Does this radio station play enough music?

Checking conditions

PROBLEM

A "classic rock" radio station claims to play an average of 50 minutes of music every hour. However, it seems that every time you tune into this station, a commercial is playing. To investigate the station's claim, you randomly select 12 different hours during the next week and record how many minutes of music the station plays in each of those hours. Here are the data:

44 49 45 51 49 53 49 44 47 50 46 48

You would like to perform a test at the $\alpha = 0.05$ significance level of

$$H_0: \mu = 50$$
$$H_a: \mu < 50$$

where μ = true mean amount of music played (in minutes) during each hour by this station. Check that the conditions for performing the test are met.

SOLUTION
Random? Random sample of 12 different hours

Normal/Large Sample? The sample size is small, but the dotplot doesn't show any outliers or strong skewness. ✓

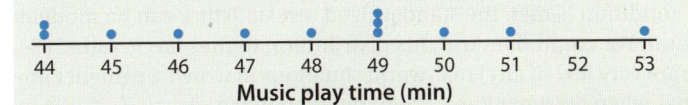

> Because the sample size is less than 30, graph the sample data to see if it is plausible that they came from an approximately normally distributed population.

EXAM PREP: FOR PRACTICE, TRY EXERCISE 7.

We used a dotplot to check the Normal/Large Sample condition in Example 8.5.1 because it is an easy graph to make by hand. You can also make a stemplot, histogram, boxplot, or normal probability plot to check this condition. Figure 8.5 shows a boxplot and a normal probability plot of the music play time data. The boxplot is roughly symmetric and has no outliers. The normal probability plot is fairly linear, as we would expect if the data came from an approximately normally distributed population of music play times during each hour on the radio station.

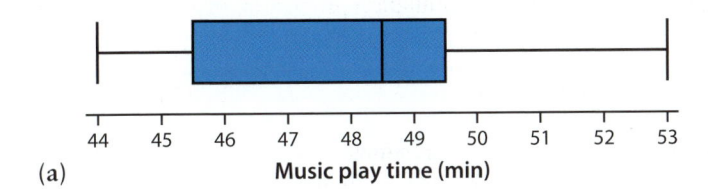

FIGURE 8.5 (a) A boxplot and (b) a normal probability plot of the music play times (in minutes) in a random sample of 12 hours of broadcasting by a classic rock radio station. The boxplot does not show any strong skewness or outliers, and the normal probability plot is fairly linear, so it is plausible that the population distribution of music play times during each hour on the radio station is approximately normal.

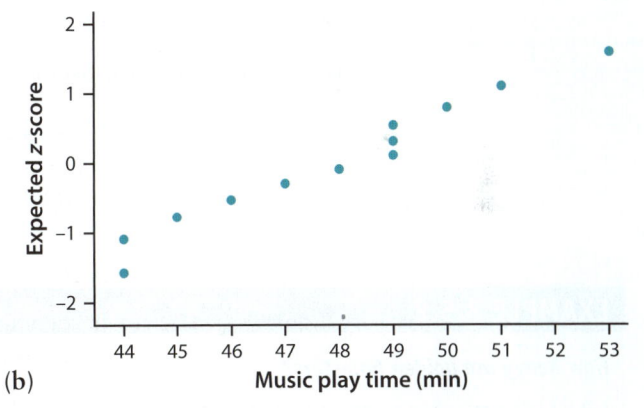

Calculating the Standardized Test Statistic

In the radio station example, the sample mean amount of music played during 12 randomly selected hours was $\bar{x} = 47.917$ minutes. Because this result is less than 50 minutes, there is *some* evidence against $H_0: \mu = 50$ and in favor of $H_a: \mu < 50$. But do we have *convincing* evidence that the true mean amount of music played during each hour by this station is less than 50 minutes? To answer this question, we want to know if it is unlikely to get a sample mean of 47.9 minutes or less by chance alone when the null hypothesis is true. As with proportions, we start by calculating a standardized test statistic.

Suppose for now that the null hypothesis $H_0: \mu = 50$ is true. Consider the sample mean amount of music played $\bar{x}$ (in minutes) for a random sample of $n = 12$ hours. You learned in Section 6.8 that the sampling distribution of $\bar{x}$ will have mean

$$\mu_{\bar{x}} = \mu = 50 \text{ minutes}$$

and standard deviation

$$\sigma_{\bar{x}} = \frac{\sigma}{\sqrt{n}}$$

To assess how far the sample mean $\bar{x}$ is from the null hypothesis value μ_0, we standardize the statistic:

$$\text{standardized test statistic} = \frac{\text{statistic} - \text{null value}}{\text{standard deviation (error) of statistic}}$$

In an ideal world where we know the population standard deviation σ, our standardized test statistic would be

$$z = \frac{\bar{x} - \mu_0}{\frac{\sigma}{\sqrt{n}}}$$

When the Normal/Large Sample condition is met, the standardized test statistic z can be modeled by the standard normal distribution. We could then use this distribution to find the P-value. (See Exercise 27.) Unfortunately, there are very few (if any) real-world situations in which we might know the population standard deviation σ when we don't know the population mean μ!

Because the population standard deviation σ is usually unknown, we use the sample standard deviation s_x in its place. For the radio station data, $s_x = 2.811$. Our resulting estimate of $\sigma_{\bar{x}}$ is the standard error (SE) of the mean:

$$SE_{\bar{x}} = \frac{s_x}{\sqrt{n}} = \frac{2.811}{\sqrt{12}} = 0.8115$$

When we use the sample standard deviation s_x to estimate the unknown population standard deviation σ, the standardized test statistic is denoted by t rather than z (you will learn why shortly). So the formula becomes

$$t = \frac{\bar{x} - \mu_0}{\frac{s_x}{\sqrt{n}}}$$

The standardized test statistic for the radio station data is

$$t = \frac{47.917 - 50}{\frac{2.811}{\sqrt{12}}} = \frac{47.917 - 50}{0.8115} = -2.567$$

EXAMPLE 8.5.2

How heavy are golden hamsters?

Calculating the standardized test statistic

PROBLEM

According to the Animal Diversity Web, adult golden hamsters have an average weight of 4 ounces.[22] To test this claim, an animal researcher selects and weighs a random sample of 30 golden hamsters. The mean weight is $\bar{x} = 4.033$ ounces with standard deviation $s_x = 0.199$ ounce. The researcher wants to perform a test at the 10% significance level of

$$H_0: \mu = 4$$
$$H_a: \mu \neq 4$$

where μ = the mean weight (in ounces) in the population of adult golden hamsters. Calculate the standardized test statistic.

SOLUTION

$$t = \frac{4.033 - 4}{\frac{0.199}{\sqrt{30}}} = 0.908 \qquad \boxed{t = \frac{\bar{x} - \mu_0}{\frac{s_x}{\sqrt{n}}}}$$

EXAM PREP: FOR PRACTICE, TRY EXERCISE 13.

Remember that the standardized test statistic tells us how far the sample result is from the null value, and in which direction, on a standardized scale. In Example 8.5.2, the mean weight $\bar{x} = 4.033$ ounces of the 30 golden hamsters in the sample is 0.908 standard errors greater than the null value of $\mu = 4$ ounces.

Finding P-Values

When the Normal/Large Sample condition is met and the null hypothesis H_0 is true, the standardized test statistic

$$t = \frac{\bar{x} - \mu_0}{\frac{s_x}{\sqrt{n}}}$$

can be modeled by a t distribution. As you learned in Section 7.5, we specify a particular t distribution by giving its degrees of freedom (df). When we perform inference about a population mean μ using a t distribution, the appropriate degrees of freedom are found by subtracting 1 from the sample size n, making df $= n - 1$.

The t distributions and the t inference procedures were invented by William S. Gosset (1876–1937). Gosset worked for the Guinness brewery, and his goal was to help the company make better beer. He used his new t procedures to find the best varieties of barley and hops. Because Gosset published his work under the pen name "Student," you will often see the t distribution called "Student's t" in his honor.

Figure 8.6 compares the density curves of the standard normal distribution and the t distributions with 2 and 9 degrees of freedom once again. The figure illustrates these facts about the t distributions:

- The t distributions are similar in shape to the standard normal distribution. They are symmetric about 0, single-peaked, and bell-shaped.
- The t distributions have more variability than the standard normal distribution. It is more likely to get an extremely large value of t (say, greater than 3) than an extremely large value of z because the t distributions have more area in the tails of the distribution.
- As the degrees of freedom increase, the t distributions approach the standard normal distribution.

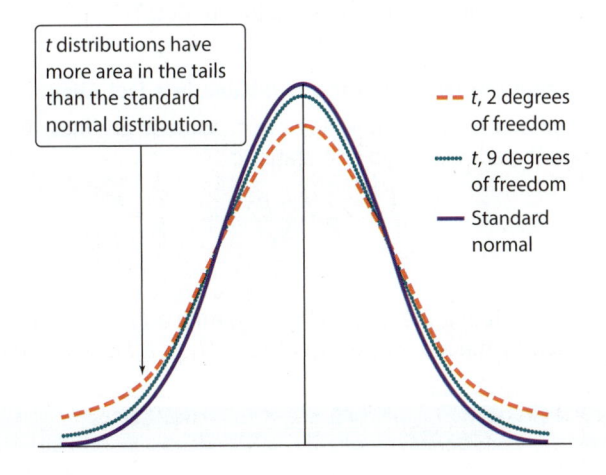

FIGURE 8.6 Density curves for the t distributions with 2 and 9 degrees of freedom and the standard normal distribution. All are symmetric with center 0. The t distributions have more variability compared to the standard normal distribution.

When the appropriate conditions are met, we can use Table B or technology to find a P-value from the appropriate t distribution when performing a test about a population mean. In the radio station example, we planned to carry out a test of

$$H_0: \mu = 50$$
$$H_a: \mu < 50$$

where μ = the true mean amount of music played (in minutes) during each hour by this station. In $n = 12$ randomly selected hours, the radio station played an average of $\bar{x} = 47.917$ minutes of music. The P-value is the probability of getting a result this small or smaller by chance alone when $H_0: \mu = 50$ is true. Earlier, we calculated the standardized test statistic to be $t = -2.567$. So we estimate the P-value by finding $P(t \leq -2.567)$ in a t distribution with df $= 12 - 1 = 11$. The shaded area in Figure 8.7 shows this probability.

FIGURE 8.7 The shaded area shows the P-value for the radio station example as the area to the left of $t = -2.567$ in a t distribution with 11 degrees of freedom.

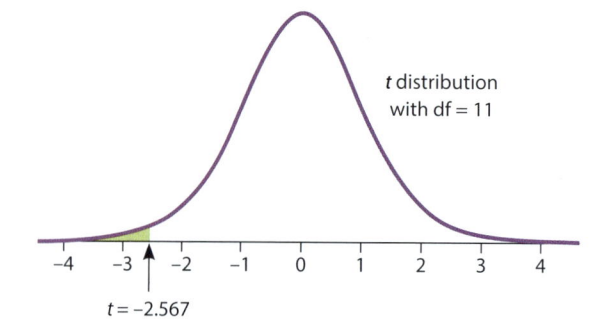

Table B shows only *positive* t-values and areas in the *right* tail of the t distributions. But the t distributions are symmetric around their center of 0, so $P(t \leq -2.567) = P(t \geq 2.567)$. Go to the df = 11 row. The value $t = 2.567$ falls between the values 2.328 and 2.718. Now look at the top of the corresponding columns in Table B. You see that the "Upper-tail probability p" is between 0.02 and 0.01. Therefore, the P-value for this test is between 0.01 and 0.02.

	Upper-tail probability p		
df	.02	.01	.005
10	2.359	2.764	3.169
11	2.328	2.718	3.106
12	2.303	2.681	3.055

As you can see, Table B gives only an interval of possible P-values for a significance test. We can still make a conclusion from the test in the same way as if we had a single probability. Because the P-value of between 0.01 and 0.02 is less than the stated $\alpha = 0.05$ significance level, we reject $H_0: \mu = 50$. We have convincing evidence that the classic rock radio station is playing fewer than 50 minutes of music per hour, on average.

You can also calculate the P-value from the t statistic using technology. The *t Distributions* applet and the TI-83/84 command tcdf(lower: –1000, upper: –2.567, df: 11) give $P(t \leq -2.567) = 0.0131$. See the first part of the Tech Corner at the end of this section for details.

Calculating the Standardized Test Statistic and P-Value in a Test About a Population Mean μ

Suppose the Random and Normal/Large Sample conditions are met. To perform a test of $H_0: \mu = \mu_0$, compute the standardized test statistic

$$t = \frac{\bar{x} - \mu_0}{\frac{s_x}{\sqrt{n}}}$$

Find the P-value by calculating the probability of getting a t statistic this large or larger in the direction specified by the alternative hypothesis H_a in a t distribution with df $= n - 1$.

EXAMPLE 8.5.3

How heavy are golden hamsters?

Finding the P-value

PROBLEM

In Example 8.5.2, an animal researcher decided to perform a test at the 10% significance level of

$$H_0: \mu = 4$$
$$H_a: \mu \neq 4$$

where μ = the mean weight (in ounces) in the population of adult golden hamsters. The researcher's random sample of 30 adult golden hamsters had mean weight $\bar{x} = 4.033$ ounces and standard deviation $s_x = 0.199$ ounce. Note that the Random condition is met with the random sample of golden hamsters, and the Normal/Large Sample condition is met because $n = 30 \geq 30$.

The resulting standardized test statistic is $t = 0.908$. Find the P-value.

SOLUTION

t distribution with df = 29

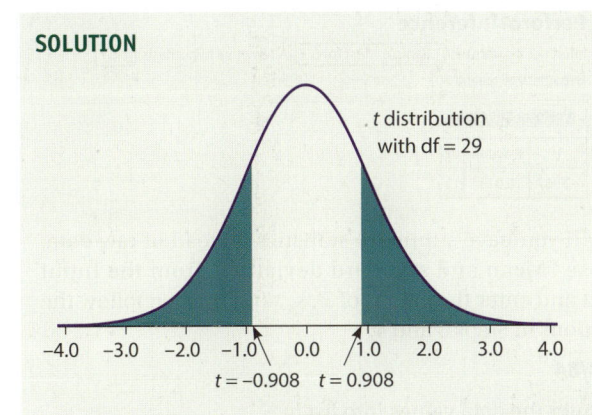

$t = -0.908$ $t = 0.908$

Using Table B: The *P*-value is between 2(0.15) = 0.30 and 2(0.20) = 0.40.
Using technology: Applet/tcdf(lower: 0.908, upper: 1000, df: 29) × 2 = 0.3714

EXAM PREP: FOR PRACTICE, TRY EXERCISE 17.

If the mean weight in the population of golden hamsters is 4 ounces, there is a 0.3714 probability of getting a sample mean weight as unusual (in either direction) as $\bar{x} = 4.033$ ounces by chance alone. Because 0.3714 is greater than $\alpha = 0.10$, we fail to reject H_0. We do not have convincing evidence that the mean weight in the population of adult golden hamsters differs from 4 ounces.

In addition to giving only an interval of possible *P*-values for a significance test, Table B has another limitation. It includes probabilities only for *t* distributions with degrees of freedom from 1 to 30 and then skips to df = 40, 50, 60, 80, 100, and 1000. The bottom row—labeled z^*—gives probabilities for the standard normal distribution. **If the df you need isn't provided in Table B, use the next smaller df that is available.** It's not fair "rounding up" to a larger df, which is like pretending that your sample size is larger than it really is. Doing so would give you a smaller *P*-value than is true and would make you more likely to incorrectly reject H_0 when it's true (i.e., make a Type I error). Of course, "rounding down" to a smaller df will give you a larger *P*-value than is correct, which makes you more likely to commit a Type II error!

Given the limitations of Table B, our advice is to use technology to find *P*-values when carrying out a significance test about a population mean. The standardized test statistic and *P*-value calculated using the "full technology" approach described in the second part of the Tech Corner are generally more accurate than the values we obtained earlier in the section when we rounded while doing our calculations. For that reason, we will use the values from the full technology approach when reporting results "Using technology" in Section 8.6.

TECH CORNER   *P*-Values and Significance Tests for a Population Mean

Calculating *P*-values

You can use technology to find the *P*-value from the standardized test statistic in a significance test for a population mean. We'll illustrate this process using the radio station data from Example 8.5.1, where $t = -2.567$ and df = 11.

Applet

1. Go to www.stapplet.com and launch the *t Distributions* applet.

2. Choose "Calculate an area under the t curve" from the drop-down menu, enter 11 for the degrees of freedom, and click "Plot distribution."

3. Using the drop-down menu below the curve, choose "to the left of a value," and enter −2.567. Then click "Calculate area."

t Distributions

Operation: Calculate an area under the t curve
Degrees of freedom = 11   [Plot distribution]
☐ Also plot normal distribution as dashed line

Calculate the area: to the left of a value
Value: −2.567
[Calculate area]   Area = 0.0131

(continued)

TI-83/84

1. Press 2nd VARS (DISTR) and choose tcdf(.
2. We want to find the area to the left of $t = -2.567$ under the t distribution curve with df $=11$.
   - **OS 2.55 or later:** In the dialog box, enter lower: -1000, upper: -2.567, and df: 11. Choose Paste, and then press ENTER.
   - **Older OS:** Complete the command tcdf(-1000, -2.567, 11) and press ENTER.

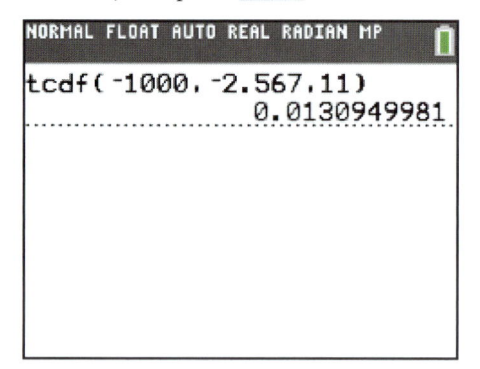

```
NORMAL FLOAT AUTO REAL RADIAN MP
tcdf(-1000,-2.567,11)
              0.0130949981
```

Performing a significance test for μ

You can also use technology to perform all the calculations for a significance test about a population mean. We'll illustrate this process using the radio station data from Example 8.5.1. Here are the data again:

44 49 45 51 49 53 49 44 47 50 46 48

Applet

1. Got to www.stapplet.com and launch the *One Quantitative Variable, Single Group* applet.
2. Enter "Music play time (min)" as the variable name. Then input the data. Be sure to separate the data values with commas or spaces as you type them.

   One Quantitative Variable, Single Group
   Variable name: Music play time (min)
   Input: Raw data

   Input data separated by commas or spaces.
   Data: 44 49 45 51 49 53 49 44 47 50

   [Begin analysis] [Edit inputs] [Reset everything]

3. Click the Begin analysis button.
4. Scroll down to the "Perform Inference" section. Choose the 1-sample t test for μ for the procedure, $\mu <$ for the alternative hypothesis, and enter 50 for the hypothesized mean. Then click the Perform inference button.

Perform Inference

Inference procedure: 1-sample t test for μ Alternative hypothesis: $\mu <$
Hypothesized mean: 50

[Perform inference]

t	P-value	df
-2.567	0.013	11

Note: If you have summary statistics instead of raw data, choose "Mean and standard deviation" from the Input menu and enter the values of $\bar{x}$, s_x, and n. Then follow the directions in Steps 3 and 4.

TI-83/84

1. Enter the data values into list L_1.
2. Press STAT, then choose TESTS and T-Test.
3. When the T-Test screen appears, choose Data as the input method. Then enter $\mu_0 = 50$, List: L_1, Freq: 1, and choose $\mu < \mu_0$.
4. Highlight "Calculate" and press ENTER.

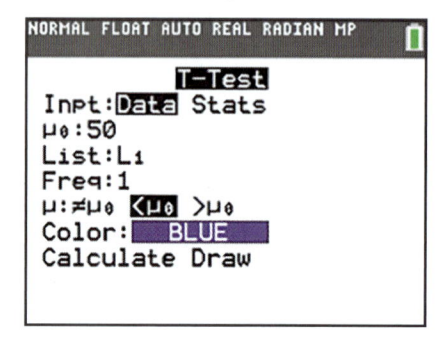

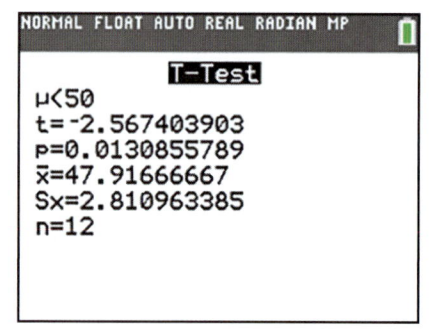

```
NORMAL FLOAT AUTO REAL RADIAN MP
         T-Test
Inpt:Data Stats
μ0:50
List:L1
Freq:1
μ:≠μ0 <μ0 >μ0
Color: BLUE
Calculate Draw
```

```
NORMAL FLOAT AUTO REAL RADIAN MP
         T-Test
μ<50
t=-2.567403903
p=0.0130855789
x̄=47.91666667
Sx=2.810963385
n=12
```

Notes:
- If you select the Draw option, you will get a picture of the appropriate t distribution with the area of interest shaded and the standardized test statistic and P-value labeled.
- If you have summary statistics instead of raw data, change the input method to Stats and enter the appropriate values.

Detailed instructions for using CrunchIt!, Excel, Google Sheets, JMP, Minitab, and R are available in Achieve.

Section 8.5 What Did You Learn?

Review the learning goals from this section. Then practice what you've learned by working through the exercises.

Learning Goal	Example	Exercises
Check the Random and Normal/Large Sample conditions for performing a significance test about a population mean.	8.5.1	7–12
Calculate the standardized test statistic for a significance test about a population mean.	8.5.2	13–16
Find the P-value for a significance test about a population mean.	8.5.3	17–20

Section 8.5 Exercises

Building Concepts and Skills These exercises assess the basic knowledge you should have after reading the section.

1. What are the three ways that the Normal/Large Sample condition for performing a significance test about a population mean can be met?

2. True/False: If the sample size is small and the shape of the population distribution is unknown, a graph of the sample data must look approximately normal to satisfy the Normal/Large Sample condition.

3. Give the specific formula for the standardized test statistic when performing a test of $H_0: \mu = \mu_0$.

4. When the Normal/Large Sample condition is met in a test of $H_0: \mu = \mu_0$, the standardized test statistic can be modeled by the t distribution with how many degrees of freedom?

5. The t distributions are _____ about 0, _____-peaked, and _____-shaped.

6. True/False: It is more likely to get an extremely large value of t than an extremely large value of z because the t distributions have more area in the tails of the distribution.

Mastering Concepts and Skills These exercises reinforce the learning goals as shown in the examples.

7. **Refer to Example 8.5.1 Salmon Fillets** As part of a study on salmon health, researchers measured the pH of a random sample of 25 salmon fillets at a fish processing plant.[23] Here are the data:

    6.34 6.39 6.53 6.36 6.39 6.25 6.45 6.38 6.33
    6.26 6.24 6.37 6.32 6.31 6.48 6.26 6.42
    6.43 6.36 6.44 6.22 6.52 6.32 6.32 6.48

    One concern is that the process of filleting salmon may result in fillets that are too acidic and unpleasant to eat. To avoid this problem, plant managers set a goal of producing salmon fillets with an average pH of at least 6.40. Do these data provide convincing evidence that the plant managers are failing to meet their goal? The researchers want to perform a test at the $\alpha = 0.05$ significance level of $H_0: \mu = 6.40$ versus $H_a: \mu < 6.40$, where μ is the mean pH level in the population of all salmon fillets at the plant. Check if the conditions for performing the test are met.

8. **Fancy Fonts** Does the use of fancy type fonts slow down the reading of text on a computer screen? Adults can read four paragraphs of text in the commonly used Times New Roman font in an average time of 22 seconds. Researchers asked a random sample of 24 adults to read these same four paragraphs in the ornate font named 𝒢𝒾𝑔𝒾 (Gigi). Here are their times, in seconds:[24]

    23.2 21.2 28.9 27.7 29.1 27.3 16.1 22.6 25.6 34.2 23.9 26.8
    20.5 34.3 21.4 32.6 26.2 34.1 31.5 24.6 23.0 28.6 24.4 28.1

    The researchers want to perform a test at the $\alpha = 0.05$ significance level of $H_0: \mu = 22$ versus $H_a: \mu > 22$, where μ is the mean time to read the four paragraphs of text in Gigi font in the population of adults. Check if the conditions for performing the test are met.

9. **Pineapple Weights** At the Hawaii Pineapple Company, managers are interested in the size of the pineapples grown in the company's fields. Last year, the mean weight of the pineapples harvested from one large field was 31 ounces. A new irrigation system was installed in this field after the growing season. Managers wonder if the mean weight of pineapples grown in the field this year will be greater than it was last year. They select an SRS of 50 pineapples from this year's crop. The managers want to perform a test of $H_0: \mu = 31$ versus $H_a: \mu > 31$, where μ is the mean weight of all pineapples grown in the field this year. Check if the conditions for performing the test are met.

10. **Students' Attitudes** The Survey of Study Habits and Attitudes (SSHA) is a psychological test that measures students' attitudes toward school and study habits. Scores range from 0 to 200, with higher scores indicating more positive attitudes and better study habits. The mean score for U.S. college students is about 115. A researcher wonders if older students' attitudes toward school differ from the attitudes of the college population as a whole.

The researcher gives the SSHA to an SRS of 45 of the more than 1000 students at a large college who are at least 30 years of age. The researcher wants to perform a test of $H_0: \mu = 115$ versus $H_a: \mu \neq 115$, where μ is the mean SSHA score in the population of students at this college who are at least 30 years old. Check if the conditions for performing the test are met.

11. **Competitive Prices** A retailer entered into an exclusive agreement with a supplier that guaranteed to provide all products at competitive prices. To be sure the supplier honored the terms of the agreement, the retailer had an audit performed on a random sample of 25 invoices. The percentage of purchases on each invoice for which an alternative supplier offered a lower price than the original supplier was recorded.[25] For example, a data value of 38 means that the price would be lower with a different supplier for 38% of the items on the invoice. The retailer would like to determine if there is convincing evidence that the mean percentage of purchases for which an alternative supplier offered lower prices is greater than 50% in the population of this company's invoices. A histogram and some numerical summaries of the data are shown here.

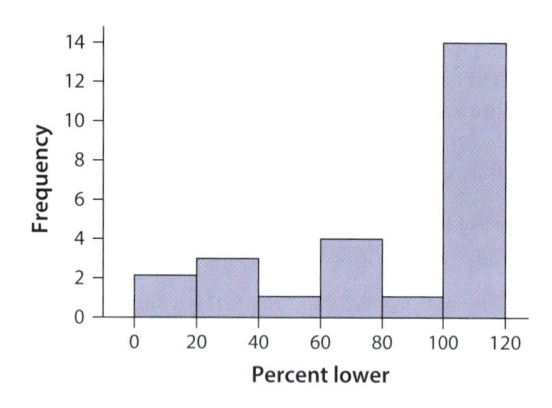

n	Mean	SD	Min	Q_1	Med	Q_3	Max
25	77.76	32.6768	0	68	100	100	100

(a) State appropriate hypotheses for the retailer to test. Be sure to define your parameter.

(b) Check if the conditions for performing the test in part (a) are met.

12. **Battery Life** A tablet computer manufacturer claims that its batteries last an average of 11.5 hours when playing videos. The quality-control department randomly selects 20 tablets from each day's production and tests the fully charged batteries by playing a video repeatedly until the battery dies. The quality-control department will discard the batteries from that day's production run if they find convincing evidence that the mean battery life is less than 11.5 hours. Here are a dotplot and summary statistics of the data from one day:

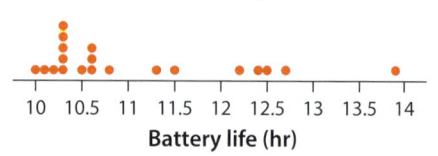

n	Mean	SD	Min	Q_1	Med	Q_3	Max
20	11.07	1.097	10	10.3	10.6	11.85	13.9

(a) State appropriate hypotheses for the quality-control department to test. Be sure to define the parameter of interest.

(b) Check if the conditions for performing the test in part (a) are met.

13. **Refer to Example 8.5.2 Salmon Fillets** Refer to Exercise 7. The sample mean pH level is 6.367 and the sample standard deviation is 0.087. Calculate the standardized test statistic.

14. **Fancy Fonts** Refer to Exercise 8. The sample mean time is 26.5 seconds and the sample standard deviation is 4.73 seconds. Calculate the standardized test statistic.

15. **Pineapple Weights** Refer to Exercise 9. The weights of the 50 randomly selected pineapples have mean 31.935 ounces and standard deviation 2.394 ounces. Calculate the standardized test statistic.

16. **Students' Attitudes** Refer to Exercise 10. The SSHA scores of the 45 randomly selected older students have mean 125.7 and standard deviation 29.8. Calculate the standardized test statistic.

17. **Refer to Example 8.5.3 Salmon Fillets** Refer to Exercise 13. Find the P-value.

18. **Fancy Fonts** Refer to Exercise 14. Find the P-value.

19. **Pineapple Weights** Refer to Exercise 15.
    (a) Find the P-value.
    (b) What conclusion would you make?

20. **Students' Attitudes** Refer to Exercise 16.
    (a) Find the P-value.
    (b) What conclusion would you make?

Applying the Concepts These exercises ask you to apply multiple learning goals in a new context or to apply what you learned in this section in a new way.

21. **Gas Prices** Donatella reads that the average price of regular gas in her state is $3.06 per gallon. To see if the average price of gas is different in her city, she selects 10 gas stations at random and records the price per gallon for regular gas at each station. Here are the data:

    3.13 3.01 3.09 3.05 2.97 2.99 3.05 2.98 3.09 3.02

    Do these data provide convincing evidence at the 5% significance level that the average price per gallon of regular gas in Donatella's city is different from $3.06?

    (a) State appropriate hypotheses for performing a significance test. Be sure to define the parameter of interest.
    (b) Check that the conditions for performing the test are met.
    (c) Calculate the standardized test statistic.
    (d) Find the P-value. What conclusion would you make?

22. **Construction Zones** Every road has one at some point — a construction zone that has much lower speed limits. To see if drivers obey these lower speed limits, a police officer uses a radar gun to measure the speed (in miles per hour [mph]) of a random sample of 10 drivers in a 25-mph construction zone. Here are the data:

    27 33 32 21 30 30 29 25 27 34

    Do these data provide convincing evidence at the 5% significance level that the average speed of drivers in this construction zone is greater than the posted speed limit?
    (a) State appropriate hypotheses for performing a significance test. Be sure to define the parameter of interest.
    (b) Check that the conditions for performing the test are met.
    (c) Calculate the standardized test statistic.
    (d) Find the P-value. What conclusion would you make?

23. **Salmon Fillets** Refer to Exercise 17.
    (a) Interpret the P-value.
    (b) What conclusion would you make?

24. **Fancy Fonts** Refer to Exercise 18.
    (a) Interpret the P-value.
    (b) What conclusion would you make?

25. **Supermarket Shoppers** A marketing consultant observes 50 consecutive shoppers at a supermarket, recording how much each shopper spends in the store. Explain why it would not be wise to use these data to carry out a significance test about the mean amount spent by all shoppers at this supermarket.

26. **Presidential Ages** Joe is writing a report on the backgrounds of American presidents. He looks up the ages of all the presidents when they entered office. Because Joe took a statistics course, he uses these numbers to perform a significance test about the mean age of all U.S. presidents when they took office. Explain why this makes no sense.

Extending the Concepts These exercises challenge you to explore statistical concepts and methods that go beyond what you learned in this section.

27. **What If σ Is Known?** Several restaurants have started using automated soft drink machines. One such machine is programmed to dispense liquid according to an approximately normal distribution with mean μ ounces and standard deviation $\sigma = 0.2$ ounce. A quality-control inspector decides to test the machine by setting μ to 18 and recording the amount of liquid dispensed for a random sample of 10 cups. The sample mean is 17.85 ounces. The inspector would like to perform a test at the $\alpha = 0.05$ significance level to determine if the machine is dispensing something other than 18 ounces of liquid per cup, on average.
    (a) State appropriate hypotheses for this test. Be sure to define the parameter of interest.
    (b) Check that the conditions for inference are met.
        Because σ is known, the standardized test statistic is a z statistic and the P-value comes from the standard normal distribution.
    (c) Calculate the standardized test statistic and P-value.
    (d) What conclusion would you make?

28. **Construction Zones** Refer to Exercise 22 and the Think About It in Section 8.3.
    (a) Find the critical value and the rejection region.
    (b) What conclusion would you make? Explain your reasoning.

29. **Gas Prices** Refer to Exercise 21 and the Think About It in Section 8.3.
    (a) Find the critical value and the rejection region.
    (b) What conclusion would you make? Explain your reasoning.

Cumulative Review These exercises revisit what you learned in previous sections.

30. **Spoofing (1.1, 1.5)** To collect information such as passwords, online criminals use "spoofing" to direct internet users to fraudulent websites. In one study of internet fraud, students were warned about spoofing and then asked to log into their university account starting from the university's home page. In some cases, the log-in link led to the genuine dialog box. In others, the box looked genuine but, in fact, was linked to a different site that recorded the ID and password the student entered. The box that appeared for each student was determined at random. An alert student could detect the fraud by looking at the true internet address displayed in the browser status bar, but most just entered their ID and password.
    (a) Is this an observational study or an experiment? Explain your answer.
    (b) Identify the explanatory and response variables.

31. **Digital Photos (7.6)** Rafiq notices that the file sizes of photographs taken with a digital camera vary, depending on the image. Rafiq selects a random sample of 49 photo files and finds that the mean size is 8.05 megabytes and the standard deviation is 1.96 megabytes. Calculate and interpret a 99% confidence interval for the mean file size of all photographs taken with Rafiq's camera.

Section 8.6 Significance Tests for a Population Mean

LEARNING GOALS

By the end of this section, you will be able to:
- Use the four-step process to perform a significance test about a population mean.
- Use a confidence interval to make a conclusion about a two-sided test for a population mean.
- Understand how to wisely use and interpret the results of significance tests.

In Section 8.5, you learned how to check conditions and perform calculations for a test about a population mean. This lesson shows you how to use the four-step process to carry out a significance test about μ, and how to use a confidence interval to make a conclusion about a two-sided test for μ. Then we offer some advice about using significance tests wisely.

Putting It All Together: Testing a Claim About a Population Mean

We are now ready to use the four-step process to carry out a **one-sample t test for a mean**.

A **one-sample t test for a mean** is a significance test of the null hypothesis that a population mean μ is equal to a specified value.

EXAMPLE 8.6.1

Are these subs too short?

Performing a significance test about μ

PROBLEM
Two young researchers noticed that the lengths of the "6-inch" sub sandwiches they get at their favorite restaurant seemed shorter than advertised. To investigate, they randomly selected 24 different times during the next month and ordered a "6-inch" sub. Here are the actual lengths (in inches) of each of the 24 sandwiches:

| 4.50 | 4.75 | 4.75 | 5.00 | 5.00 | 5.00 | 5.50 | 5.50 | 5.50 | 5.50 | 5.50 | 5.50 |
| 5.75 | 5.75 | 5.75 | 6.00 | 6.00 | 6.00 | 6.00 | 6.00 | 6.50 | 6.75 | 6.75 | 7.00 |

Do these data give convincing evidence at the $\alpha = 0.10$ level that the "6-inch" sandwiches at this restaurant are shorter than advertised, on average?

SOLUTION

State:

$H_0: \mu = 6$
$H_a: \mu < 6$

where μ = the mean length of all "6-inch" subs from this restaurant.
Use $\alpha = 0.10$.
The evidence for H_a is: $\bar{x} = 5.6771 < 6$

Plan: One-sample t test for μ
Random? They randomly selected 24 times to buy a sub. ✓
Normal/Large Sample? The sample size is small, but the dotplot doesn't show any outliers or strong skewness. ✓

> Follow the four-step process!
>
> **State:** State the hypotheses, parameter(s), significance level, and evidence for H_a.
>
> **Plan:** Identify the appropriate inference method and check the conditions.

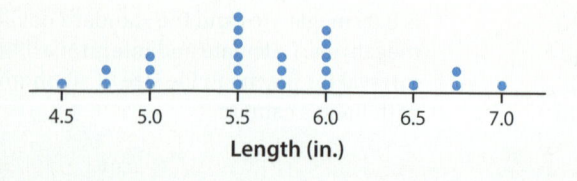

Do: $\bar{x} = 5.6771$ inches, $s_x = 0.6572$ inch

- $t = \dfrac{5.6771 - 6}{\dfrac{0.6572}{\sqrt{24}}} = -2.407$

- P-value: df = 24 − 1 = 23

> **Do:** If the conditions are met, perform calculations: *Use technology to compute the sample mean and standard deviation.*
> - Calculate the test statistic.
> - Find the P-value.

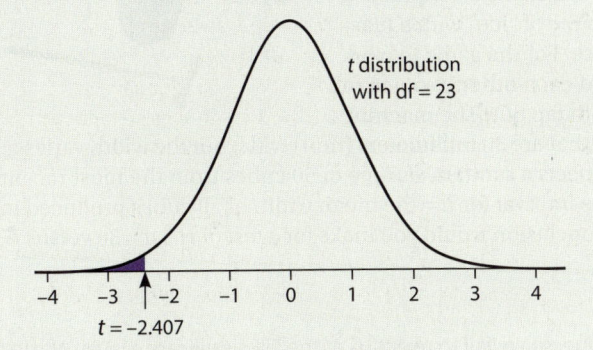

t distribution with df = 23

$t = -2.407$

Using Table B: The P-value is between 0.01 and 0.02
Using technology: $t = -2.407$, P-value = 0.0123 using df = 23

Conclude: Because the P-value of 0.0123 is less than $\alpha = 0.10$, we reject H_0. There is convincing evidence that the mean length of all "6-inch" subs at this restaurant is less than 6 inches.

> **Conclude:** Make a conclusion about the hypotheses in the context of the problem.

EXAM PREP: FOR PRACTICE, TRY EXERCISE 5.

The P-value from Example 8.6.1 tells us there is about a 0.012 probability of getting a random sample of 24 subs with a mean length of 5.6771 inches or less if $H_0: \mu = 6$ is true. This unlikely result allows us to rule out the explanation that the observed result happened by chance alone.

Two-Sided Tests and Confidence Intervals

You learned in Section 8.4 that a confidence interval gives more information than a significance test does — it provides the entire set of plausible values for the parameter based on sample data. The connection between two-sided tests and confidence intervals is even stronger for means than it was for proportions. That's because both inference methods for means use the same standard error of $\bar{x}$ in the calculations:

$$\text{standardized test statistic: } t = \frac{\bar{x} - \mu_0}{\dfrac{s_x}{\sqrt{n}}} \qquad \text{confidence interval: } \bar{x} \pm t^* \dfrac{s_x}{\sqrt{n}}$$

The link between two-sided tests and confidence intervals for a population mean allows us to make a conclusion directly from a confidence interval.

- If a 99% confidence interval for μ does not capture the null value μ_0, we can reject $H_0: \mu = \mu_0$ in a two-sided test at the 1% significance level ($\alpha = 0.01$).
- If a 99% confidence interval for μ captures the null value μ_0, then we should fail to reject $H_0: \mu = \mu_0$ in a two-sided test at the 1% significance level.

The same logic applies for other confidence levels, but *only* for a two-sided test.

EXAMPLE 8.6.2

Who broke the ice?

Confidence intervals and two-sided tests

PROBLEM

In the children's game Don't Break the Ice, small plastic ice cubes are squeezed into a square frame. Each child takes turns tapping out a cube of "ice" with a plastic hammer, hoping that the remaining cubes don't collapse. For the game to work correctly, the cubes must be big enough so that they hold each other in place in the plastic frame, but not so big that they are too difficult to tap out. The machine that produces the plastic cubes is designed to make cubes that are 30 millimeters (mm) wide, but the width varies a little. To ensure that the machine is working well, a supervisor inspects a random sample of 50 cubes from the most recent hour of production and measures their width. The 95% confidence interval for $\mu =$ the mean width of all cubes produced in the last hour is 29.997 to 30.043 mm. Based on this interval, what conclusion would you make for a test of $H_0: \mu = 30$ versus $H_a: \mu \neq 30$ at the $\alpha = 0.05$ significance level? Explain your reasoning.

SOLUTION

The 95% confidence interval contains $\mu = 30$ as a plausible value, so we fail to reject H_0 at the 5% significance level. We do not have convincing evidence that the mean width of all plastic ice cubes produced in this hour differs from 30 mm.

EXAM PREP: FOR PRACTICE, TRY EXERCISE 9.

For the random sample of 50 cubes in Example 8.6.2, the mean is $\bar{x} = 30.02$ mm and the standard deviation is $s_x = 0.08$ mm. The standardized test statistic is

$$t = \frac{\bar{x} - \mu_0}{\frac{s_x}{\sqrt{n}}} = \frac{30.02 - 30}{\frac{0.08}{\sqrt{50}}} = 1.768$$

The corresponding P-value from a t distribution with $50 - 1 = 49$ degrees of freedom is 0.0833. Because the P-value of $0.0833 > \alpha = 0.05$, we fail to reject H_0. This is consistent with our conclusion based on the 95% confidence interval in the example.

Would a 90% confidence interval for μ include 30 mm as a plausible value for the parameter? Only if a two-sided test would fail to reject $H_0: \mu = 30$ at the 10% significance level. Because the P-value of $0.0833 < \alpha = 0.10$, we would reject H_0. At the 10% significance level, we *do* have convincing evidence that the mean width of all plastic cubes produced this hour differs from 30 mm. So the 90% confidence interval would *not* contain 30 mm. You can check that the interval is 30.001 to 30.039 mm.

Using Tests Wisely

Significance tests are widely used in reporting the results of research in many fields. New drugs require convincing evidence of their effectiveness and safety before their manufacturers are allowed to market them. Courts ask about statistical significance when hearing discrimination cases. Marketers want to know whether a new ad campaign significantly outperforms the old one, and medical researchers want to know whether a new therapy performs significantly better than the existing treatments. In all these uses, statistical significance is valued because it points to an effect that is unlikely to occur by chance alone.

Carrying out a significance test is often quite simple, especially if you use technology. Using tests wisely, however, is not so simple. Here are some points to keep in mind when using or interpreting significance tests.

⚠ **The foolish user of statistics who feeds the data to a calculator or computer without performing exploratory analysis will often be embarrassed.** Plot your data and examine them carefully. Is the difference you are seeking visible in your graphs? If not, ask yourself whether the difference is large enough to be practically important. Are there outliers or other departures from a consistent pattern? A few outliers can produce highly significant results if you blindly apply common significance tests. Outliers can also destroy the significance of otherwise convincing data.

Statistical Significance and Practical Importance

When a null hypothesis of no effect or no difference can be rejected at the usual significance levels ($\alpha = 0.10$, $\alpha = 0.05$, or $\alpha = 0.01$), there is convincing evidence of a difference. But that difference may be very small. When large samples are used, even tiny deviations from the null hypothesis may be statistically significant.

EXAMPLE 8.6.3

Can we cut healing time?

Statistical significance versus practical importance

PROBLEM

Researchers want to test a new antibacterial cream, "Formulation NS," to see how it affects the healing rate of small cuts. Previous research shows that with no medication, the mean healing time (defined as the time for the scab to fall off) is 7.6 days. The researchers want to determine if Formulation NS speeds healing. They make minor cuts on a random sample of 250 college students who have given informed consent to participate in the study, and apply Formulation NS to the wounds. The researchers want to perform a test at the $\alpha = 0.05$ significance level of

$$H_0: \mu = 7.6$$
$$H_a: \mu < 7.6$$

where μ = the mean healing time (in days) in the population of college students whose cuts are treated with Formulation NS.

The mean healing time for these participants is $\bar{x} = 7.5$ days and the standard deviation is $s_x = 0.9$ day. After confirming that the conditions for inference are met, the researchers carry out a one-sample t test for μ, and find that $t = -1.757$ and P-value $= 0.0401$. Explain why this result is statistically significant, but not practically important.

SOLUTION

Because the P-value of 0.0401 is less than $\alpha = 0.05$, the observed result is statistically significant. The researchers have convincing evidence that Formulation NS reduces the average healing time to less than 7.6 days in the population of college students whose cuts are treated with Formulation NS. The result is not practically important because the observed sample mean was $\bar{x} = 7.5$ days. Having scabs fall off one-tenth of a day sooner, on average, is no big deal!

EXAM PREP: FOR PRACTICE, TRY EXERCISE 13.

Remember the wise saying: ⚠ **Statistical significance is not the same thing as practical importance.** The remedy for attaching too much importance to statistical significance is to give a confidence interval for the parameter in which you are interested. A confidence interval provides a set of plausible values for the parameter rather than simply asking if the observed result is unlikely to occur by chance alone when H_0 is true. Confidence intervals are not used as often as they should be, whereas significance tests are perhaps overused.

Beware of Multiple Analyses

Statistical significance should mean that you have found a difference that you were looking for. The reasoning behind statistical significance works well if you decide what difference you are seeking, design a study to search for it, and use a significance test to weigh the evidence you get. In other settings, statistical significance may have little meaning. Here's one such example.

Might the radiation from cell phones be harmful to users? Many studies have found little or no connection between using cell phones and various illnesses. Here is part of a news account of one study:

> A hospital study that compared brain cancer patients and a similar group without brain cancer found no statistically significant difference in cell phone use for the two groups. But when 20 distinct types of brain cancer were considered separately, a significant difference in cell phone use was found for one rare type. Puzzlingly, however, this risk appeared to decrease rather than increase with greater mobile phone use.[26]

Think for a moment. Suppose that the 20 null hypotheses for these 20 significance tests are all true. Then each test has a 5% chance of making a Type I error — that is, of showing statistical significance at the $\alpha = 0.05$ level. That's what $\alpha = 0.05$ means: Results this extreme occur only 5% of the time by chance alone when the null hypothesis is true. Therefore, we expect about 1 of 20 tests to give a significant result just by chance. Running one test and reaching the $\alpha = 0.05$ level is reasonably good evidence that you have found something; running 20 tests and reaching that level only once is not.

Searching data for patterns is certainly a legitimate endeavor. Performing every conceivable significance test on a data set with many variables until you obtain a statistically significant result is not. This unfortunate practice is known by many names, including data dredging and P-hacking. To learn more about the pitfalls of multiple analyses, do an internet search for the XKCD comic about jelly beans causing acne.

Section 8.6 What Did You Learn?

Review the learning goals from this section. Then practice what you've learned by working through the exercises.

Learning Goal	Example	Exercises
Use the four-step process to perform a significance test about a population mean.	8.6.1	5–8
Use a confidence interval to make a conclusion about a two-sided test for a population mean.	8.6.2	9–12
Understand how to wisely use and interpret the results of significance tests.	8.6.3	13–16

Section 8.6 Exercises

Building Concepts and Skills These exercises assess the basic knowledge you should have after reading the section.

1. A significance test for a population mean (when σ is unknown) is often referred to as a _____.

2. True/False: If a 95% confidence interval for μ does not capture the null value μ_0, we should fail to reject $H_0: \mu = \mu_0$ in a two-sided test at the $\alpha = 0.05$ significance level.

3. Statistical significance is not the same thing as _____ importance.

4. What is P-hacking?

Mastering Concepts and Skills These exercises reinforce the learning goals as shown in the examples.

5. **Refer to Example 8.6.1 Dissolved Oxygen** The level of dissolved oxygen (DO) in a stream or river is an important indicator of the water's ability to support aquatic life. A researcher measures the DO level at 15 randomly chosen locations along a stream. Here are the results in milligrams per liter (mg/L):

   | 4.53 | 5.04 | 3.29 | 5.23 | 4.13 | 5.50 | 4.83 | 4.40 |
   | 5.42 | 6.38 | 4.01 | 4.66 | 2.87 | 5.73 | 5.55 | |

   A mean DO level below 5 mg/L puts aquatic life at risk. Do we have convincing evidence at the $\alpha = 0.10$ significance level that aquatic life in this stream is at risk?

6. **Women and Calcium** The recommended daily allowance (RDA) of calcium for women between the ages of 18 and 24 years is 1200 milligrams (mg). Researchers who were involved in a large-scale study of women's bone health suspected that their participants had significantly lower calcium intakes than the RDA. To test this suspicion, the researchers measured the daily calcium intake of a random sample of 36 women from the study who fell within the desired age range. The sample mean was 856.2 mg and the standard deviation was 306.7 mg. Do these data give convincing evidence at the $\alpha = 0.01$ significance level that the researchers' suspicion is correct?

7. **Reading level** A school librarian purchases a novel for the library. The publisher claims that the book is written at a fifth-grade reading level, but the librarian suspects that the reading level is higher than that. The librarian selects a random sample of 40 pages and uses a standard readability test to assess the reading level of each page. The mean reading level of these pages is 5.4 with a standard deviation of 0.8. Do these data give convincing evidence at the $\alpha = 0.05$ significance level that the average reading level of this novel is greater than 5?

8. **Cola Bottles** Bottles of a popular cola are supposed to contain 300 milliliters (mL) of cola. Some variation occurs from bottle to bottle because the filling machinery is not perfectly precise. An inspector measures the contents (in mL) of six randomly selected bottles from a single day's production. Here are the data:

   | 299.4 | 297.7 | 301.0 | 298.9 | 300.2 | 297.0 |

   Do these data give convincing evidence at the 10% significance level that the mean amount of cola in all the bottles filled that day differs from the target value of 300 mL?

9. Refer to Example 8.6.2 Radon Detectors Radon is a colorless, odorless gas that is naturally released by rocks and soils and may concentrate in tightly closed houses. It is slightly radioactive, so there is some concern that it may pose a health hazard. Radon detectors are sold to homeowners, but the detectors may be inaccurate. Researchers placed a random sample of 11 detectors in a chamber and exposed them to 105 picocuries per liter of radon over 3 days. A graph of the radon readings from the 11 detectors shows no strong skewness or outliers. The 90% confidence interval for μ = the true mean reading in the population of radon detectors (in picocuries per liter) is 99.61 to 110.03. Based on this interval, what conclusion would you make for a test of $H_0: \mu = 105$ versus $H_a: \mu \neq 105$ at the $\alpha = 0.10$ significance level? Explain your reasoning.

10. Pill Quality A drug manufacturer forms tablets by compressing grains of material that contain the active ingredient and fillers. To assure quality, the hardness of a sample from each batch of tablets produced is measured. The target value for the hardness is $\mu = 11.5$ newtons. Researchers select a random sample of 20 tablets and measure their hardness. A graph of the data is roughly symmetric with no outliers. The 95% confidence interval for μ = the true mean hardness in the population of tablets is 11.472 to 11.561 newtons. What conclusion would you make for a test of $H_0: \mu = 11.5$ versus $H_a: \mu \neq 11.5$ at the $\alpha = 0.05$ significance level? Explain your reasoning.

11. Internet Speed How long does it take for a chunk of information to travel from one server to another and back? According to internettrafficreport.com, the average response time is 200 milliseconds (about one-fifth of a second). Researchers wonder if this claim is true, so they collect data on response times (in milliseconds) for a random sample of 14 servers in Europe. A graph of the data reveals no strong skewness or outliers.

(a) State appropriate hypotheses for a significance test in this setting. Be sure to define the parameter of interest.

(b) Explain why the conditions for inference are met.

(c) The 95% confidence interval for the true mean response time is 158.22 to 189.64 milliseconds. Based on this interval, what conclusion would you make for a test of the hypotheses in part (a) at the 5% significance level? Explain your answer.

12. Water Intake A blogger claims that U.S. adults drink, on average, 40 ounces of water per day. Researchers wonder if this claim is true, so they ask a random sample of 24 U.S. adults about their daily water intake. A graph of the data shows a roughly symmetric shape with no outliers.

(a) State appropriate hypotheses for a significance test in this setting. Be sure to define the parameter of interest.

(b) Explain why the conditions for inference are met.

(c) The 90% confidence interval for the mean daily water intake is 30.35 to 36.92 ounces. Based on this interval, what conclusion would you make for a test of the hypotheses in part (a) at the 10% significance level? Explain your answer.

13. Refer to Example 8.6.3 MCAT Scores A national company offers a review course to help students prepare for the Medical College Admission Test (MCAT). Note that MCAT scores vary from 472 to 528. The company wants to know if using a smartphone app in addition to its prep course will help increase participants' scores more than just taking the prep course. On average, people taking the prep course increase their scores by 10 points. To investigate use of the smartphone app, the company has a random sample of 500 students use the app along with the regular course and measures their improvement. Then the company performs a test of $H_0: \mu = 10$ versus $H_a: \mu > 10$, where μ is the true mean improvement in the MCAT score for students who take this company's prep course and use the smartphone app. For the sample of 500 students, the average improvement was $\bar{x} = 10.4$ with a standard deviation of $s_x = 5$. The standardized test statistic is $t = 1.79$ with a P-value of 0.0371. Explain why this result is statistically significant, but not practically important.

14. Music and Mazes A researcher wishes to determine if people are able to complete a certain pencil-and-paper maze more quickly while listening to classical music. Suppose previous research has established that the mean time needed for people to complete a certain maze without music is 40 seconds. The researcher, therefore, decides to test the hypotheses $H_0: \mu = 40$ versus $H_a: \mu < 40$, where μ = the true average time (in seconds) needed to complete the maze while listening to classical music. To do so, the researcher has a random sample of 10,000 people complete the maze with classical music playing. The mean time for these people is $\bar{x} = 39.92$ seconds, and the P-value of the significance test is 0.0002. Explain why this result is statistically significant, but not practically important.

15. ESP Tests A researcher looking for evidence of extrasensory perception (ESP) tests 500 people. Four of these people do significantly better (P-value < 0.01) than random guessing.

(a) Is it proper to conclude that these four people have ESP? Explain your answer.

(b) What should the researcher now do to test whether any of these four subjects have ESP?

16. Echinacea A medical experiment investigated whether taking the herb Echinacea could help reduce cold symptoms. The study measured 50 different response variables usually associated with colds, such as low-grade fever, congestion, frequency of coughing, and so on. At the end of the study, the participants taking Echinacea displayed significantly better responses at the $\alpha = 0.05$ level than those taking a placebo for three of the 50 response variables studied. Should we be convinced that Echinacea helps reduce cold symptoms? Why or why not?

Applying the Concepts These exercises ask you to apply multiple learning goals in a new context or to apply what you learned in this section in a new way.

17. **Cooling Reactions** A chemical production process uses water-cooling to carefully control the temperature of the system so that the correct products are obtained from the chemical reaction. The system must be maintained at a temperature of 120°F. A technician suspects that the average temperature is different from 120°F and checks the process by taking temperature readings at 12 randomly selected times over a 4-hour period. Here are the data:

120.1	124.2	122.4	124.4	120.8	121.4
121.8	119.6	120.2	121.5	118.7	122.0

    (a) Is there convincing evidence at the $\alpha=0.05$ significance level that the true mean temperature μ of the system is different from 120°F?

    (b) The 95% confidence interval for μ is 120.33°F to 122.52°F. Explain how this interval is consistent with, but gives more information than, the test in part (a).

18. **Candy Bags** A machine is supposed to fill bags with an average of 19 ounces of candy. The manager of the candy factory wants to be sure that the machine does not consistently underfill or overfill the bags. The manager selects a random sample of 75 bags of candy produced that day and weighs each bag. The mean weight of these bags is 19.28 ounces and the standard deviation is 0.81 ounce.

    (a) Is there convincing evidence at the $\alpha=0.01$ significance level that the mean amount of candy μ in all bags filled that day differs from 19 ounces?

    (b) The 99% confidence interval for μ is 19.033 to 19.527 ounces. Explain how this interval is consistent with, but gives more information than, the test in part (a).

19. **Tests and Confidence Intervals** The P-value for a two-sided test of the null hypothesis $H_0: \mu=10$ is 0.06.

    (a) Does the 95% confidence interval for μ include 10? Why or why not?

    (b) Does the 90% confidence interval for μ include 10? Why or why not?

20. **Tests and Confidence Intervals** The P-value for a two-sided test of the null hypothesis $H_0: \mu=15$ is 0.03.

    (a) Does the 99% confidence interval for μ include 15? Why or why not?

    (b) Does the 95% confidence interval for μ include 15? Why or why not?

21. **Tech Household Size** How many people are in a typical U.S. household? The *American Community Survey* data set contains detailed information about a random sample of 3000 U.S. households.[27] The U.S. Census Bureau records the number of people (NP) in each surveyed household. According to the Census Bureau's website (www.census.gov), the average number of people per household was 2.52 in the year when this sample was taken. Do these data provide convincing evidence at the $\alpha=0.01$ significance level that the Census Bureau's claim was incorrect?

22. **Tech High Cholesterol** What is the average cholesterol level of U.S. adults age 20 or older? The National Health and Nutrition Examination Survey (NHANES) is an ongoing research program conducted by the National Center for Health Statistics. Researchers selected a random sample of more than 7000 U.S. residents for study in the most recent survey. The **NHANES** data set gives the cholesterol levels in mg/dL (*LBXTC*).[28] According to the U.S. Centers for Disease Control and Prevention (CDC), the mean cholesterol level for adults age 20 or older was 191 mg/dL in the year when this sample was taken. Do these data provide convincing evidence at the $\alpha=0.01$ significance level that the CDC's claim was incorrect?

Extending the Concepts These exercises challenge you to explore statistical concepts and methods that go beyond what you learned in this section.

23. **Confidence Intervals and One-Sided Tests** There is a connection between a one-sided test and a confidence interval for a population mean—but it's a little complicated. Consider a one-sided test at the $\alpha=0.05$ significance level of $H_0: \mu=10$ versus $H_a: \mu>10$ based on an SRS of $n=20$ observations.

    (a) Complete this statement: We should reject H_0 if the standardized test statistic $t > $ _____. In other words, we should reject H_0 if the sample mean is more than _____ standard errors greater than $\mu=10$.

    (b) Find the t^* critical value for a 90% confidence level in this setting. The resulting interval is $\bar{x} \pm t^* (SE_{\bar{x}})$.

    (c) Suppose that the sample mean $\bar{x}$ leads us to reject H_0. Use parts (a) and (b) to explain why the 90% confidence interval *cannot* contain $\mu=10$.

24. **One-Sided Confidence Interval** A pharmaceutical company has developed a new drug to reduce cholesterol levels in patients. The company's goal is to estimate with 95% confidence a *lower bound* for $\mu=$ the true mean reduction in cholesterol (in milligrams per deciliter, mg/dL) for patients who use this drug. To accomplish this goal, the company gives the drug to a random sample of 30 patients with high cholesterol levels.

    (a) Use the appropriate t distribution to find the value of t^* with area $= 0.95$ to its right.

    (b) For the 30 patients in the sample, the mean reduction in cholesterol is 25 mg/dL and the standard deviation is 10 mg/dL. Use the t^* value from part (a) to calculate an estimate of the lower bound for μ with 95% confidence.

    (c) Interpret your interval from part (b).

Cumulative Review These exercises revisit what you learned in previous sections.

25. **Homework Habits (1.8)** Researchers in Spain interviewed 7725 13-year-olds about their homework habits—how much time they spent per night on homework and whether they got help from their parents or not—and then had them take a test with 24 math questions and 24 science questions. They found that students who spent between 90 and 100 minutes on their homework did only a little better on the test than those who spent 60 to 70 minutes on their homework. Beyond 100 minutes, students who spent more time did worse than those who spent less time. The researchers concluded that 60 to 70 minutes per night is the optimal amount of time for students to spend on homework.[29] Is it appropriate to conclude that students who reduce their homework time from 120 minutes to 70 minutes will likely improve their performance on tests such as those used in this study? Why or why not?

26. **Getting Ahead (7.1, 7.2)** A recent CBS/*New York Times* poll asked a random sample of Americans, "Which comes closer to your view: In today's economy, everyone has a fair chance to get ahead in the long run, or in today's economy, it's mainly just a few people at the top who have a chance to get ahead?" The 95% confidence interval for the percentage of people who would choose "it's mainly just a few people at the top" is 58% to 64%.[30]

    (a) Interpret the confidence interval.

    (b) Interpret the confidence level.

Section 8.7 Power of a Test

LEARNING GOALS

By the end of this section, you will be able to:

- Interpret the power of a significance test.
- Describe how changes in sample size, significance level, and alternative parameter value affect the power of a test.

Researchers often perform a significance test in hopes of finding convincing evidence *for* the alternative hypothesis. Why? Because H_a states the claim about the population parameter that they believe is true. For instance, a drug manufacturer claims that fewer than 10% of patients who take its new drug for treating Alzheimer's disease will experience nausea. To test this claim, researchers want to carry out a test of

$$H_0: p = 0.10$$
$$H_a: p < 0.10$$

where p is the proportion of all patients who would experience nausea when taking the new Alzheimer's drug. They plan to give the new drug to a random sample of patients with Alzheimer's disease whose families have given informed consent for the patients to participate in the study.

Suppose that H_a is true—the true proportion who will experience nausea is less than 10%. As you learned in Section 8.2, there are two possible outcomes for the significance test:

- A correct conclusion if the test rejects H_0 and finds convincing evidence for H_a.
- A Type II error if the test fails to reject H_0 and doesn't find convincing evidence for H_a.

This section investigates the factors that give researchers the best chance of finding convincing evidence for the alternative hypothesis (and avoiding a Type II error) when H_a is true.

The Power of a Test

Suppose that the true proportion of subjects like the ones in the Alzheimer's study who would experience nausea after taking the drug is $p = 0.08$. Researchers would make a Type II error if they failed to find convincing evidence for $H_a: p < 0.10$ based on the sample data. How likely is the significance test to *avoid* a Type II error in this case? We refer to this probability as the **power** of the test.

> The **power** of a test is the probability that the test will find convincing evidence for H_a when a specific alternative value of the parameter is true.

Like a *P*-value, the power of a test is a *conditional* probability: Power = P(reject H_0 | parameter = some specific alternative value). In other words, power is the probability that we find convincing evidence the alternative hypothesis is true, given that the alternative hypothesis really is true. To interpret the power of a test in a given setting, just interpret the relevant conditional probability. Start by assuming that a specific alternative parameter value is true. That's quite different from interpreting a P-value, when we start by assuming that the null hypothesis is true.

Let's return to the Alzheimer's study. Suppose the researchers decide to perform a test of $H_0: p = 0.10$ versus $H_a: p < 0.10$ at the $\alpha = 0.05$ significance level based on data from a random sample of 300 patients with Alzheimer's disease. Advanced calculations reveal that the power of this test to detect $p = 0.08$ is 0.29. *Interpretation:* If the proportion of all patients with Alzheimer's disease who would experience side effects when taking the new drug is $p = 0.08$, there is a 0.29 probability that the researchers will find convincing evidence for $H_a: p < 0.10$.

EXAMPLE 8.7.1

Can we tell if the new batteries last longer?

Interpreting the power of a test

PROBLEM
A company has developed a new deluxe AAA battery that is supposed to last longer than its regular AAA battery. The new batteries are more expensive to produce, however, so the company would like to be convinced that they really do last longer. Based on years of experience, the company knows that its regular AAA batteries last for 30 hours of continuous use, on average. The company plans to select an SRS of 50 new batteries and use them continuously until they are completely drained. Then the company will perform a test at the $\alpha = 0.05$ significance level of

$$H_0: \mu = 30$$
$$H_a: \mu > 30$$

where μ is the true mean lifetime (in hours) of the new deluxe AAA batteries. The power of the test to detect that $\mu = 31$ hours is 0.762. Interpret this value.

SOLUTION
If the true mean lifetime of the company's new AAA batteries is $\mu = 31$ hours, there is a 0.762 probability that the company will find convincing evidence for $H_a: \mu > 30$.

EXAM PREP: FOR PRACTICE, TRY EXERCISE 5.

The company in Example 8.7.1 has a good chance (power = 0.762) of rejecting H_0 if the true mean lifetime of its new AAA batteries is 31 hours. That is, $P(\text{reject } H_0 \mid \mu = 31) = 0.762$. What's the probability that the company makes a Type II error in this case? It's $P(\text{fail to reject } H_0 \mid \mu = 31) = 1 - 0.762 = 0.238$. We can generalize this relationship between the power of a significance test and the probability of a Type II error.

> **Relating Power and Type II Error**
>
> The power of a test to detect a specific alternative parameter value is the probability of avoiding a Type II error for that alternative:
>
> Power = 1 − P(Type II error) and P(Type II error) = 1 − Power

The power of the test in the Alzheimer's study to detect $p = 0.08$ is only 0.29. In other words, researchers have a 1 − 0.29 = 0.71 probability of making a Type II error by failing to find convincing evidence for $H_a: p < 0.10$ when $p = 0.08$. So what can researchers do to decrease the probability of making a Type II error and increase the power of the test?

What Affects the Power of a Test?

Here is a Concept Exploration that will help you answer this question.

CONCEPT EXPLORATION

A great free-throw shooter?

In this Concept Exploration, you will use an applet to investigate the factors that affect the power of a test.

A basketball player claims to be an 80% free-throw shooter. That is, the player claims that $p = 0.80$, where p is the true proportion of free throws the player will make in the long run. Suppose the player is exaggerating and really makes fewer than 80% of free throws ($p < 0.80$). We have the player shoot $n = 50$ free throws and record the sample proportion $\hat{p}$ of made shots. Then we use the sample result to perform a test at the $\alpha = 0.05$ significance level of

$$H_0: p = 0.80$$
$$H_a: p < 0.80$$

where p = the true proportion of free throws the shooter makes in the long run.

1. Go to www.stapplet.com and launch the *Power* applet.

2. Select proportions for the type of test. Enter 0.80 for the null hypothesis value p_0; enter 0.66 for the true proportion p (indicating that the player is actually a 66% free-throw shooter); choose less than for the type of alternative hypothesis; enter 50 for the sample size n; and leave the significance level as $\alpha = 0.05$. Choose to calculate and plot α, the Type I error probability, then click "Calculate." Once the graphs appear, click the box to "Show rejection region."

   The null sampling distribution is shown with a dashed line. The true sampling distribution is shown with a solid line.
   $\alpha = 0.05$

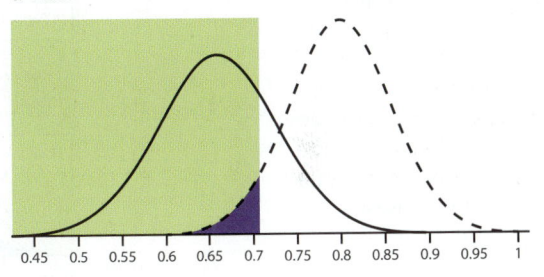

   - The curve on the right shows the sampling distribution of the sample proportion $\hat{p}$ for random samples of size $n = 50$ when $H_0: p = 0.80$ is true. We refer to this as the *null distribution*. A value of $\hat{p}$ that falls along the horizontal axis within the green shaded region (the *rejection region*) is far enough below 0.80 that we should reject $H_0: p = 0.80$. (You can hover over the green region to see the values of $\hat{p}$ that would lead to a reject H_0 conclusion.) The corresponding area under the null distribution curve (the blue shaded region) is equal to the significance level, $\alpha = 0.05$. Note that this is also the probability of making a Type I error—rejecting H_0 when H_0 is true.
   - The curve on the left shows the true sampling distribution of the sample proportion $\hat{p}$ for random samples of size $n = 50$ when $p = 0.66$. We refer to this as the *alternative distribution*. A value of $\hat{p}$ that falls along the horizontal axis within the green region is far enough below 0.80 that it would lead to a correct rejection of $H_0: p = 0.80$. In other words, the corresponding area under the alternative distribution curve represents the *power* of the test. A value of $\hat{p}$ that falls along the horizontal axis outside the green region in the alternative distribution would lead to a Type II error because it is not far enough below 0.80 to reject $H_0: p = 0.80$.

3. Use the applet to calculate and plot β, the probability of making a Type II error, using the same settings as in Step 2.

4. Use the applet to calculate and plot the power of the test using the same settings as in Step 2. Complete this interpretation of the power: If the shooter makes 66% of their free throws in the long run, there is a _____ probability of finding convincing evidence for $H_a: p < 0.80$ in a sample of 50 free throws.

5. *Sample size:* Gradually increase n from 50 to 100 by clicking the up arrow in the Sample size box. (Alternatively, you can type in new values of n and click the Calculate button.) Does the power increase or decrease from the value in Step 4? Explain why this makes sense.

6. *Significance level:* Reset the sample size to $n = 50$.
   (a) Change the significance level to $\alpha = 0.01$. How does this affect the probability of a Type I error? How about the probability of a Type II error? Does the power of the test increase or decrease from the value in Step 4?
   (b) Make a guess about what will happen to the power of the test if you change the significance level to $\alpha = 0.10$. Use the applet to test your conjecture.
   (c) Explain what the results in parts (a) and (b) tell you about the relationship between Type I error probability, Type II error probability, and power.

7. *Difference between null and alternative parameter value:* Reset the sample size to $n = 50$ and the significance level to $\alpha = 0.05$. Will we be more likely to detect that the player is exaggerating if they are really a 60% free-throw shooter or if they are really a 70% free-throw shooter? Use the applet to test your conjecture.

As Step 5 of the Concept Exploration confirms, we get better information about the true proportion of free throws that the player makes from a random sample of 100 shots than from a random sample of 50 shots. The power of the test to detect that $p = 0.66$ increases from 0.758 to 0.941 when the sample size increases from $n = 50$ to $n = 100$.

Will it be easier to reject H_0 if $\alpha = 0.05$ or if $\alpha = 0.10$? When α is larger, it is easier to reject H_0 because the P-value doesn't need to be as small. Step 6 of the Concept Exploration shows that the power of the test to detect that $p = 0.66$ increases from 0.758 to 0.843 when the significance level increases from $\alpha = 0.05$ to $\alpha = 0.10$. Increasing α means that we are less likely to make a Type II error — failing to reject H_0 when H_a is true. Unfortunately, increasing α also makes it more likely that we will make a Type I error — rejecting H_0 when H_0 is true.

Figure 8.8 illustrates the connection between Type I and Type II error probabilities and the power of a test. Note that $P(\text{Type I error}) = \alpha = 0.10$ and that $P(\text{Type II error}) = 1 - \text{Power} = 1 - 0.843 = 0.157$. You can see that increasing the Type I error probability α would decrease the Type II error probability and increase the power of the test. By the same logic, decreasing the chance of a Type I error results in a higher chance of a Type II error and less power.

FIGURE 8.8 The connection between Type I error, Type II error, and power. Because $P(\text{Type I error}) = \alpha$, as the significance level α increases: $P(\text{Type I error})$ increases, $P(\text{Type II error})$ decreases, and power increases.

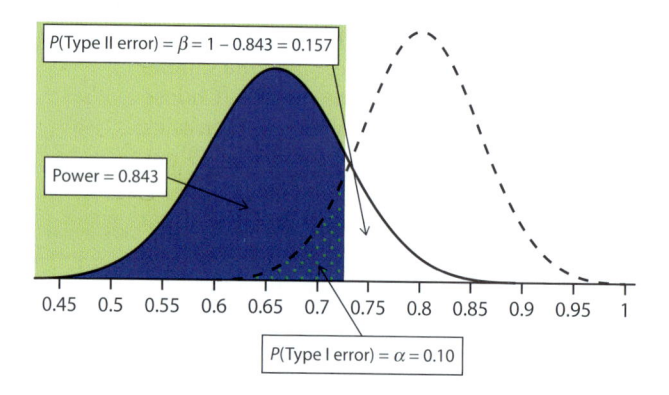

Step 7 of the Concept Exploration shows that it is easier to detect large differences between the null and alternative parameter values than to detect small differences. When $n = 50$ and $\alpha = 0.05$, the power of the test to detect $p = 0.60$ is 0.939, whereas the power of the test to detect $p = 0.70$ is only 0.543.

The difference between the null parameter value and the specific alternative parameter value of interest is often referred to as the *effect size*. For the basketball player setting with a null value of $p = 0.80$, it is much easier to detect if the true proportion of free throws the player makes is $p = 0.60$ (an effect size of 0.20, or 20 percentage points) than if the true proportion is $p = 0.70$ (an effect size of 0.10, or 10 percentage points).

Increasing the Power of a Significance Test

The power of a significance test to detect an alternative value of the parameter when H_0 is false and H_a is true will be greater when:

- The sample size n is larger.
- The significance level α is larger.
- The null and alternative parameter values are farther apart (larger effect size).

There are other ways to increase the power of a test, such as using a more sophisticated data collection method. Section 1.4 (stratified random sampling) and Section 1.7 (blocking) describe two of these strategies. In an experiment, power will increase if you have fewer sources of variability by keeping other variables constant. Our best advice for maximizing the power of a test is to collect data wisely, choose as high an α level (Type I error probability) as you are willing to risk, and use as large a sample size as you can afford.

In practice, researchers should consult with statisticians to determine the required sample size to detect an alternative parameter value when planning a study. This is analogous to calculating the sample size needed to achieve a desired margin of error when researchers plan to construct a confidence interval.

EXAMPLE 8.7.2

Can we tell if the new drug reduces nausea?

What affects the power of a test

PROBLEM
The researchers in the Alzheimer's experiment want to test the drug manufacturer's claim that fewer than 10% of patients who take its new drug for treating Alzheimer's disease will experience nausea. That is, they want to carry out a test of

$$H_0: p = 0.10$$
$$H_a: p < 0.10$$

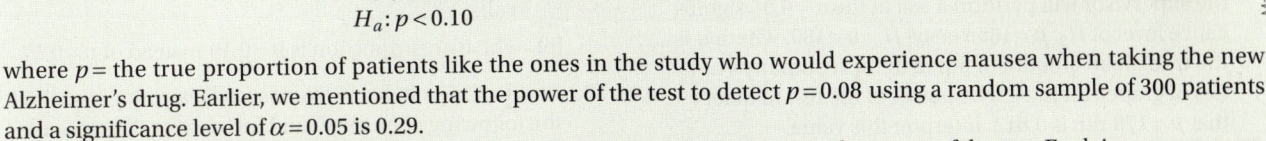

where $p =$ the true proportion of patients like the ones in the study who would experience nausea when taking the new Alzheimer's drug. Earlier, we mentioned that the power of the test to detect $p = 0.08$ using a random sample of 300 patients and a significance level of $\alpha = 0.05$ is 0.29.

Determine whether each of the following changes would increase or decrease the power of the test. Explain your answers.

(a) Use $\alpha = 0.01$ instead of $\alpha = 0.05$.
(b) If the true proportion is $p = 0.06$ instead of $p = 0.08$.
(c) Use $n = 200$ instead of $n = 300$.

SOLUTION
(a) Decrease; using a smaller significance level makes it harder to reject H_0 when H_a is true.
(b) Increase; it is easier to detect a bigger difference between the null and alternative parameter values.
(c) Decrease; a smaller sample size gives less information about the true proportion p.

EXAM PREP: FOR PRACTICE, TRY EXERCISE 9.

Section 8.7 What Did You Learn?

Review the learning goals from this section. Then practice what you've learned by working through the exercises.

Learning Goal	Example	Exercises
Interpret the power of a significance test.	8.7.1	5–8
Describe how changes in sample size, significance level, and alternative parameter value affect the power of a test.	8.7.2	9–12

Section 8.7 Exercises

Building Concepts and Skills These exercises assess the basic knowledge you should have after reading the section.

1. The power of a test is the probability that the test will find _____ for H_a when a specific alternative value of the parameter is true.

2. Power $= P($ _____ $H_0 \mid$ parameter $=$ some specific alternative value$)$

3. What is the relationship between the power of a test and the probability of a Type II error?

4. Name three ways to increase the power of a test.

Mastering Concepts and Skills These exercises reinforce the learning goals as shown in the examples.

5. **Refer to Example 8.7.1 Mean Income** You are thinking about opening a restaurant and are searching for a suitable location. From research you have done, you know that the mean household income of the people living near the restaurant must be more than $85,000 to support the type of restaurant you wish to open. You decide to take a simple random sample of 50 households near one potential site. Using the sample data, you will perform a test at the $\alpha = 0.05$ significance level

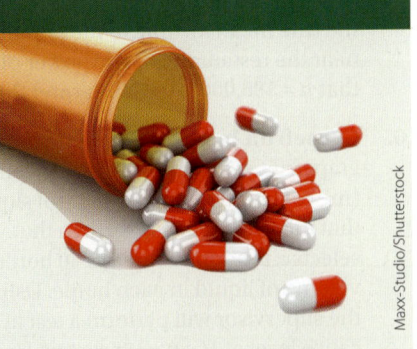

of $H_0: \mu = \$85{,}000$ versus $H_a: \mu > \$85{,}000$, where μ is the true mean income in the population of households near the restaurant.[31] The power of the test to detect that $\mu = \$86{,}000$ is 0.64. Interpret this value.

6. **Grapefruit Juice** The labels on bottles of a company's grapefruit juice say that they contain 180 milliliters (mL) of liquid. A quality-control supervisor suspects that the true mean is less than that. The supervisor selects a random sample of 40 bottles and measures the volume of liquid in each bottle. Using the sample data, the supervisor will perform a test at the $\alpha = 0.05$ significance level of $H_0: \mu = 180$ versus $H_a: \mu < 180$, where μ is the true mean amount of liquid in this company's bottles of grapefruit juice. The power of the test to detect that $\mu = 178$ mL is 0.812. Interpret this value.

7. **First Responders** Several cities have begun to monitor emergency response times because accident victims with life-threatening injuries generally need medical attention within 8 minutes. In one city, emergency personnel took more than 8 minutes to arrive at 22% of all calls involving life-threatening injuries last year. The city manager then issued guidelines for improving response time to local first responders. After 6 months, the city manager selects an SRS of 400 calls involving life-threatening injuries and examines the response times. Using the sample data, the city manager will perform a test at the 5% significance level of $H_0: p = 0.22$ versus $H_a: p < 0.22$, where p is the proportion of all calls involving life-threatening injuries during this 6-month period for which emergency personnel took more than 8 minutes to arrive. The power of the test to detect that $p = 0.17$ is 0.80. Interpret this value.

8. **Potato Chips** A company that makes potato chips requires each shipment of potatoes to meet certain quality standards. If the company finds convincing evidence that more than 8% of the potatoes in the shipment have "blemishes," the truck will be sent back to the supplier to get another load of potatoes. Otherwise, the entire truckload will be used to make potato chips. To make the decision, a supervisor will inspect a random sample of 500 potatoes from the shipment. Using the sample data, the supervisor will perform a test at the 5% significance level of $H_0: p = 0.08$ versus $H_a: p > 0.08$, where p is the proportion of all potatoes with blemishes in a given truckload. The power of the test to detect that $p = 0.11$ is 0.764. Interpret this value.

9. **Refer to Example 8.7.2 Mean Income** Refer to Exercise 5. Determine if each of the following changes would increase or decrease the power of the test. Explain your answers.
   (a) Use a random sample of 30 households instead of 50 households.
   (b) Try to detect that $\mu = \$85{,}500$ instead of $\mu = \$86{,}000$.
   (c) Change the significance level to $\alpha = 0.10$.

10. **Grapefruit Juice** Refer to Exercise 6. Determine if each of the following changes would increase or decrease the power of the test. Explain your answers.
    (a) Use a random sample of 100 bottles instead of 40 bottles.
    (b) Try to detect that $\mu = 177$ mL instead of $\mu = 178$ mL.
    (c) Change the significance level to $\alpha = 0.01$.

11. **First Responders** Refer to Exercise 7. Determine if each of the following changes would increase or decrease the power of the test. Explain your answers.
    (a) Change the significance level to $\alpha = 0.01$.
    (b) Select a random sample of 500 calls instead of 400 calls.
    (c) The true proportion is $p = 0.19$ instead of $p = 0.17$.

12. **Potato Chips** Refer to Exercise 8. Determine if each of the following changes would increase or decrease the power of the test. Explain your answers.
    (a) Change the significance level to $\alpha = 0.10$.
    (b) Select a random sample of 250 potatoes instead of 500 potatoes.
    (c) The true proportion is $p = 0.10$ instead of $p = 0.11$.

Applying the Concepts These exercises ask you to apply multiple learning goals in a new context or to apply what you learned in this section in a new way.

13. **Plywood Floors** A company that manufactures plywood claims that the mean breaking strength of its 1-inch plywood flooring is 500 pounds. A construction engineer suspects that the manufacturer is exaggerating the breaking strength of its plywood. The engineer measures the breaking strength of a random sample of 40 pieces of the company's 1-inch plywood flooring, and uses the data to perform a test at the $\alpha = 0.05$ significance level of

    $$H_0: \mu = 500$$
    $$H_a: \mu < 500$$

    where μ is the true mean breaking strength (in pounds) of this company's 1-inch plywood flooring.
    (a) Find the probability of a Type I error.
    (b) The power of the test to detect that $\mu = 495$ is 0.738. Interpret this value.
    (c) Find the probability of a Type II error for the test in part (b).
    (d) Describe two ways to increase the power of the test in part (b).

14. **Campus Parking** A university makes a change to improve student satisfaction with parking on campus. Before the change, 37% of students approved of the parking provided by the university. After the change, the university's administration surveys an SRS of 200 students at the school. Using the sample data, they would like to perform a test at the $\alpha = 0.05$ significance level of

    $$H_0: p = 0.37$$
    $$H_a: p > 0.37$$

    where p is the proportion of all students at the university who are satisfied with the parking situation after the change.

(a) Find the probability of a Type I error.

(b) The power of the test to detect that $p=0.45$ is 0.75. Interpret this value.

(c) Find the probability of a Type II error for the test in part (b).

(d) Describe two ways to increase the power of the test in part (b).

15. **Error Probabilities and Power** You read that a statistical test at the 1% significance level has probability 0.14 of making a Type II error when a specific alternative parameter value is true. What is the power of the test against this alternative?

16. **Power and Error** A scientist calculates that a test at the 5% significance level has probability 0.23 of making a Type II error when a specific alternative parameter value is true. What is the power of the test against this alternative?

17. **Mean Income** Refer to Exercises 5 and 9.

    (a) Explain one disadvantage of using $\alpha=0.10$ instead of $\alpha=0.05$ when performing the test.

    (b) Explain one disadvantage of taking a random sample of 50 households instead of 30 households.

18. **Potato Chips** Refer to Exercises 8 and 12.

    (a) Explain one disadvantage of using $\alpha=0.10$ instead of $\alpha=0.05$ when performing the test.

    (b) Explain one disadvantage of taking a random sample of 500 potatoes instead of 250 potatoes from the shipment.

Extending the Concepts These exercises challenge you to explore statistical concepts and methods that go beyond what you learned in this section.

19. **Power Curves** As mentioned, researchers should consider the power of a test to detect an alternative parameter value when planning a study. Most statistical software can produce *power curves* to help researchers determine the required sample size for various alternative parameter values or effect sizes. Suppose a researcher wants to perform a test at the $\alpha=0.05$ significance level of $H_0: p=0.60$ versus $H_a: p>0.60$. The graph shows power curves for different alternative parameter values and sample sizes.

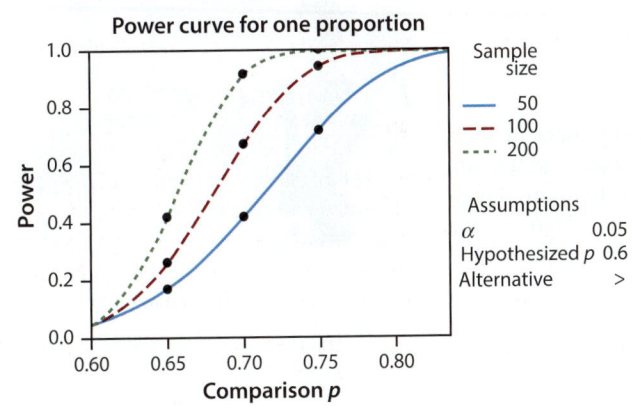

    (a) Estimate the power of the test to detect that $p=0.65$ using a sample size of $n=100$.

    (b) If the researcher wants to have a good chance to detect if $p=0.70$, what sample size would you recommend? Explain your answer.

    (c) Estimate the power of the test to detect that $p=0.75$ for $n=50$, 100, and 200. Which sample size would you recommend that the researcher use for the study, taking into account both power and cost?

It is possible (but not pleasant) to calculate the power of a test by hand. The following exercise gives you the basic idea. Our advice is to use technology when calculating power.

20. **Potato Chips** Refer to Exercise 8.

    (a) Suppose that $H_0: p=0.08$ is true. Describe the shape, center, and variability of the sampling distribution of $\hat{p}$ in random samples of size 500.

    (b) Use the sampling distribution from part (a) to find the value of $\hat{p}$ with an area of 0.05 to the right of it. If the proportion of defective potatoes in the sample is greater than this value of $\hat{p}$, the supervisor will reject $H_0: p=0.08$ at the 5% significance level.

    (c) Now suppose that $p=0.11$. Describe the shape, center, and variability of the sampling distribution of $\hat{p}$ in random samples of size 500.

    (d) Use the sampling distribution from part (c) to find the probability of getting a sample proportion greater than the value you found in part (b). This result is the power of the test to detect $p=0.11$.

Cumulative Review These exercises revisit what you learned in previous sections.

21. **Homework Check (4.3)** There are 35 students in Dr. Feng's Introductory Statistics class. One day, 24 students completed the Concept Exploration prior to class, 14 students finished the online homework assignment, and 8 students did both. Suppose we randomly select a student from the class.

    (a) Make a two-way table that describes the sample space of this chance process.

    (b) What is the probability that a randomly chosen student did not do either task?

22. **Food Safety (7.2)** "Do you feel confident or not confident that the food available at most grocery stores is safe to eat?" When a Gallup poll asked this question, 87% of the sample said they were confident. Gallup announced the poll's margin of error for 95% confidence as ±3 percentage points.[32] Which of the following sources of error are included in this margin of error? Explain.

    (i) Gallup dialed landline telephone numbers at random, so it missed all people without landline phones, including people whose only phone is a cell phone.

    (ii) Some people whose numbers were chosen never answered the phone in several calls, or answered but refused to participate in the poll.

    (iii) Other samples will provide different estimates due to chance variation in the random selection of telephone numbers.

Section 8.8 Significance Tests for a Population Standard Deviation or Variance

LEARNING GOALS

By the end of this section, you will be able to:

- State and check the Random and Normal conditions for performing a significance test about a population standard deviation or variance.
- Calculate the standardized test statistic and find the *P*-value for a significance test about a population standard deviation or variance.
- Use the four-step process to perform a significance test about a population standard deviation or variance.

So far in this chapter, you have learned how to perform a significance test about a population proportion or a population mean. In this section, we examine the details of testing a claim about a population standard deviation or variance.

Conditions for Testing a Claim About σ or σ^2

In Section 7.7, we introduced the conditions that should be met before we construct a confidence interval for a population standard deviation or variance. We called them the Random and Normal conditions. These same conditions must be verified before carrying out a significance test for a population standard deviation σ or a population variance σ^2.

HOW TO Check the Conditions for a Significance Test About σ or σ^2

To perform a test about a population standard deviation σ or a population variance σ^2, check that the following conditions are met:

- **Random:** The data come from a random sample from the population of interest.
- **Normal:** The data come from a population that is approximately normally distributed.

⚠ There is no Large Sample override to the Normal condition for tests about σ or σ^2. No matter how large the sample size *n* is, the data must appear to have come from an approximately normal population distribution. There are two ways this condition can be met:

1. The population is known to be approximately normally distributed.
2. The data from the sample do not show any strong skewness or outliers.

EXAMPLE 8.8.1

How variable are resting heart rates?

Checking conditions

PROBLEM

One way that doctors assess whether their patients are healthy is by checking their resting pulse rates (measured in beats per minute, beats/min). How variable are resting pulse rates among humans? A common belief is that the standard deviation of the resting pulse rate among humans is $\sigma = 6.67$ beats/min.[33] Here are the data on the resting heart rates for a random sample of 20 people:[34]

```
69  69  76  81  68  78  68  70  79  85
70  79  72  70  78  83  73  61  75  74
```

We would like to perform a test at the $\alpha = 0.05$ significance level of

$$H_0: \sigma = 6.67$$
$$H_a: \sigma \neq 6.67$$

where σ = the true standard deviation of resting heart rate (in beats/min) in humans. Check if the conditions for performing the test are met.

SOLUTION

Random? Random sample of 20 people ✓

Normal?

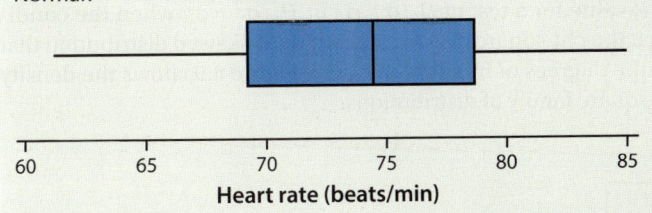

> Appropriate graphs for assessing the Normal condition include dotplots, stemplots, histograms, boxplots, and normal probability plots.

The boxplot is not strongly skewed and there are no outliers. ✓

EXAM PREP: FOR PRACTICE, TRY EXERCISE 5.

Calculating the Test Statistic and Finding *P*-Values

The Animal Diversity Web suggests that the standard deviation of the weight of golden hamsters is 0.15 ounce.[35] An animal researcher would like to perform a two-sided test of this claim. We can state hypotheses for the significance test in terms of either the population standard deviation or the population variance:

$H_0: \sigma = 0.15$
$H_a: \sigma \neq 0.15$

where σ = the population standard deviation of weight (in ounces) for golden hamsters

$H_0: \sigma^2 = (0.15)^2 = 0.0225$
$H_a: \sigma^2 \neq (0.15)^2 = 0.0225$

where σ^2 = the population variance of weight (in ounces) for golden hamsters

The original claim was about the population standard deviation, so we will use the hypotheses involving σ for this test.

The animal researcher took a random sample of 30 golden hamsters and found the weight (in ounces) of each hamster. A graph of the data reveals no strong skewness or outliers, so the Random and Normal conditions are met. The standard deviation of weight in the sample is $s_x = 0.199$ ounce. Because $s_x = 0.199 \neq 0.15$, the sample result gives *some* evidence against H_0 and in favor of H_a. But do we have *convincing* evidence that the population standard deviation σ of golden hamster weight is different from 0.15 ounce? To answer this question, we want to know if it is unlikely to get a sample standard deviation of 0.199 ounce or more extreme (in either direction) when the null hypothesis is true.

When testing a claim about a population proportion or mean, we compute a standardized test statistic to measure the size of the difference between the sample statistic and the null value:

$$\text{standardized test statistic} = \frac{\text{statistic} - \text{null value}}{\text{standard deviation (error) of statistic}}$$

When testing a claim about a population standard deviation or variance, the test statistic has a different form. Instead of measuring the difference between the sample statistic and the null value, the test statistic is based on their *ratio*. The formula for the test statistic is

$$\chi^2 = \frac{(n-1)s_x^2}{\sigma_0^2}$$

where σ_0^2 is the population variance according to the null hypothesis. Note that *the test statistic is the same whether you are performing a test about a population variance or standard deviation.* The only difference is how you state the hypotheses and write your conclusion.

The test statistic for the golden hamsters is

$$\chi^2 = \frac{(n-1)s_x^2}{\sigma_0^2} = \frac{(30-1)(0.199)^2}{(0.15)^2} = \frac{(29)(0.0396)}{0.0225} = 51.04$$

How unusual is it to get a test statistic of 51.04 or more extreme when the null hypothesis is true? To answer that question, we need to find the P-value.

In Section 7.7, you learned that the statistic $\chi^2 = \dfrac{(n-1)s_x^2}{\sigma^2}$ has approximately a chi-square distribution with df $= n-1$ when the population distribution is approximately normal. We can use this chi-square distribution to find the P-value for a test of $H_0: \sigma = \sigma_0$ or $H_0: \sigma^2 = \sigma_0^2$ when the conditions for inference are met. Recall that the chi-square distribution is a right-skewed distribution that becomes somewhat less skewed as the degrees of freedom increase. Figure 8.9 shows the density curves for three members of the chi-square family of distributions.

FIGURE 8.9 The density curves for three members of the chi-square family of distributions.

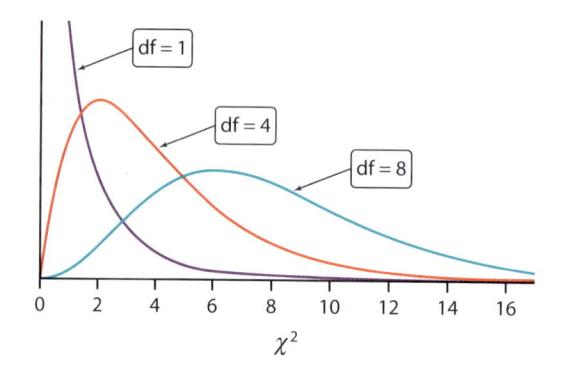

For the hamster example, we calculated the test statistic to be $\chi^2 = 51.04$ with a sample of size 30. Because conditions are met, the P-value can be obtained from a chi-square distribution with df $= n - 1 = 30 - 1 = 29$. Start by plotting the test statistic on a sketch of the chi-square distribution with 29 degrees of freedom, as shown in Figure 8.10. Recall that we are performing a test of $H_0: \sigma = 0.15$ versus $H_a: \sigma \neq 0.15$. Because this is a two-sided test, and the chi-square distribution is not symmetric, we first find the area in the tail with the test statistic, and then multiply that area by 2. (If the test were one-sided, we would find the area in the tail specified by the alternative hypothesis.)

FIGURE 8.10 The right-tail area for the significance test about the standard deviation of weight in the population of golden hamsters. Because this test is two-sided, we multiply the right-tail area by 2 to get the P-value.

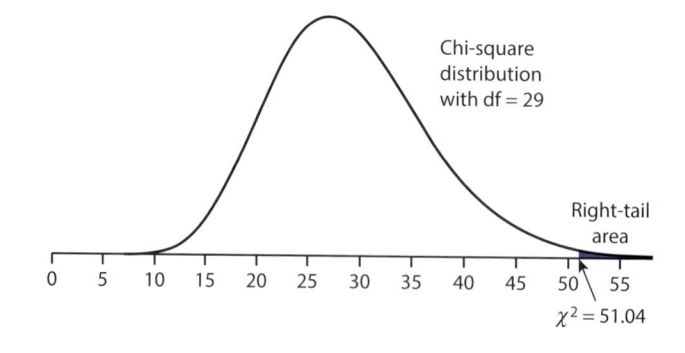

We can use Table C or technology to find the P-value from the chi-square test statistic using the appropriate chi-square distribution. See the first part of the Tech Corner at the end of the section for instructions about using technology. In this case, the test statistic is in the upper tail, so we start by finding the area under the curve to the right of $\chi^2 = 51.04$. Here is an excerpt from Table C that shows upper-tail areas for several chi-square distributions. In the row for 29 df, we see that our test statistic of 51.04 falls between 49.588 and 52.336. Reading to the top of those two columns, we see that the corresponding right-tail area is between 0.005 and 0.01. Because this is a two-sided test, the P-value is twice the area in the upper tail, so it is between $2(0.005) = 0.01$ and $2(0.01) = 0.02$. This interval of values is consistent with the result from technology: Applet/χ^2cdf(lower: 51.04, upper: 1000, df: 29) × 2 = 0.0139.

Section 8.8 Significance Tests for a Population Standard Deviation or Variance

df	Tail probability p			
	.05	.025	.01	.005
27	40.113	43.194	46.963	49.645
28	41.337	44.461	48.278	50.993
29	42.557	45.722	49.588	52.336
30	43.773	46.979	50.892	53.672

Because the P-value of 0.0139 is less than our default significance level of $\alpha = 0.05$, we reject H_0. These data give convincing evidence that the standard deviation of weight in the population of golden hamsters is different from the 0.15-ounce value claimed by the Animal Diversity Web.

Calculating the Standardized Test Statistic and P-Value in a Test About a Population Standard Deviation σ or a Population Variance σ^2

Suppose the Random and Normal conditions are met. To perform a test of $H_0: \sigma = \sigma_0$ or $H_0: \sigma^2 = \sigma_0^2$, compute the standardized test statistic

$$\chi^2 = \frac{(n-1)s_x^2}{\sigma_0^2}$$

Find the P-value by calculating the probability of getting a χ^2 statistic this large or larger in the direction specified by the alternative hypothesis H_a in a chi-square distribution with df $= n - 1$.

EXAMPLE 8.8.2

How variable are resting heart rates?

Test statistic and P-value

PROBLEM

In Example 8.8.1, we started to perform a test at the $\alpha = 0.05$ significance level of

$$H_0: \sigma = 6.67$$
$$H_a: \sigma \neq 6.67$$

where σ = the true standard deviation of resting heart rate (in beats per minute) in humans. Here are the data once again on the resting heart rates for a random sample of 20 people.[36]

69	69	76	81	68	78	68	70	79	85
70	79	72	78	78	83	73	61	75	74

We already showed that the conditions for inference are met.

(a) Calculate the test statistic.
(b) Find the P-value.

SOLUTION

(a) $n = 20$, $s_x = 5.983$

$$\chi^2 = \frac{(20-1)(5.983)^2}{(6.67)^2} = 15.29$$

Use technology to calculate the sample standard deviation.

$$\chi^2 = \frac{(n-1)s_x^2}{\sigma_0^2}$$

(continued)

(b) df = 20 − 1 = 19

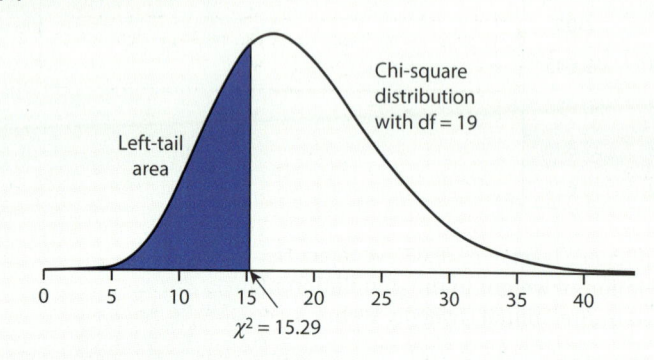

> This test is two-sided. Start by finding the area in the tail with the test statistic, $\chi^2 = 15.29$. Then multiply the left-tail area by 2 to get the *P*-value.

Chi-square distribution with df = 19

Left-tail area

$\chi^2 = 15.29$

Using Table C: *P*-value > 2(0.10) = 0.20
Using technology: Applet/χ^2cdf(lower: 0, upper: 15.29, df: 19) × 2 = 0.5920

EXAM PREP: FOR PRACTICE, TRY EXERCISE 9.

If the standard deviation of resting heart rates in the population of humans is $\sigma = 6.67$ beats/min, there is a 0.592 probability of getting a sample standard deviation as unusual (in either direction) as $s_x = 5.983$ beats/min by chance alone. Because 0.592 is greater than $\alpha = 0.05$, we fail to reject H_0. We do not have convincing evidence that the population standard deviation of resting heart rates in the human population differs from the claimed 6.67 beats/min.

A 95% confidence interval for the population standard deviation σ in Example 8.8.2 is 4.55 to 8.74 beats/min. Note that the null value of $\sigma = 6.67$ is a plausible value of the parameter, which would lead us to fail to reject H_0. So the 95% confidence interval and the two-sided test at the $\alpha = 0.05$ significance level yield consistent decisions. But the confidence interval gives us more information about the parameter σ.

Putting It All Together: Testing a Claim About a Population Standard Deviation or Variance

We are now ready to use the four-step process to carry out a **one-sample test for a standard deviation (variance)**.

> A **one-sample test for a standard deviation (variance)** is a significance test of the null hypothesis that a population standard deviation σ (or a population variance σ^2) is equal to a specified value.

EXAMPLE 8.8.3

Is the water temperature consistent enough?

Four-step process

PROBLEM

A chemical production process uses water-cooling to carefully control the temperature of the system so that the correct products are obtained from the chemical reaction. The mean temperature of the system must be maintained at 120°F (see Section 8.6, Exercise 17). But just maintaining the mean temperature is not enough: There also can't be too much variability in the temperature of the system. For the process to work properly, the variance of the distribution of temperatures cannot be larger than 1 (°F)^2. A technician took temperature readings at 12 randomly selected times over a 4-hour period. Here are the data:

| 120.1 | 124.2 | 122.4 | 124.4 | 120.8 | 121.4 |
| 121.8 | 119.6 | 120.2 | 121.5 | 118.7 | 122.0 |

Is there convincing evidence at the 1% significance level that the variance in temperature during this time period is too large?

SOLUTION

State:

$H_0: \sigma^2 = 1$

$H_a: \sigma^2 > 1$

where σ^2 = the true variance of the chemical production process temperature in (°F)^2 over this 4-hour period.

Use $\alpha = 0.01$.

The evidence for H_a is: $s_x^2 = (1.716)^2 = 2.945 > 1$.

Plan: One-sample test for σ^2

Random? Random sample of 12 temperatures

Normal? The dotplot does not show any outliers or strong skewness.

> **State:** State the hypotheses, parameter(s), significance level, and evidence for H_a.

> **Plan:** Identify the appropriate inference method and check the conditions.

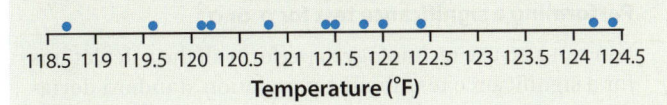

Do: $n = 12, s_x = 1.716$

- $\chi^2 = \dfrac{(12-1)(1.716)^2}{1} = 32.391$

- P-value: df = 12 − 1 = 11

> **Do:** If the conditions are met, perform calculations:
> - Calculate the test statistic.
> - Find the P-value.

> Note that this is a one-sided test because the alternative hypothesis is $H_a: \sigma^2 > 1$. The P-value is the area to the right of $\chi^2 = 32.391$ under the chi-square curve.

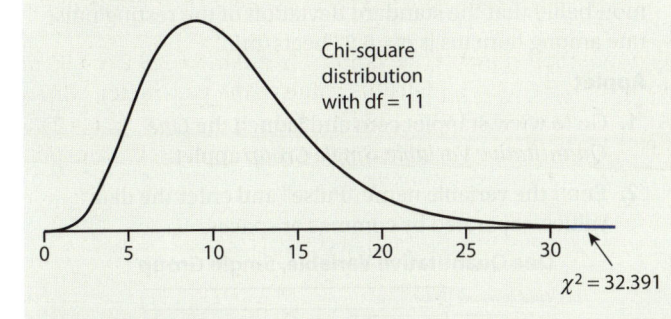

Chi-square distribution with df = 11

$\chi^2 = 32.391$

Using Table C: P-value is less than 0.005

Using technology: $\chi^2 = 32.383$, P-value < 0.001

Conclude: Because the P-value of <0.001 is less than $\alpha = 0.01$, we reject H_0. There is convincing evidence that the variance of the process temperature is larger than 1 (°F)^2.

> **Conclude:** Make a conclusion about the hypotheses in the context of the problem.

EXAM PREP: FOR PRACTICE, TRY EXERCISE 13.

What if the df you need isn't listed in Table C? In Section 7.7, we told you to use the greatest df available in the table that is less than the correct df when finding critical values using a chi-square distribution. There is an added complication when performing a significance test for a population standard deviation or variance. In the formula for the test statistic

$$\chi^2 = \dfrac{(n-1)s_x^2}{\sigma_0^2}$$

you should substitute the df you are using from Table C as the value of $(n-1)$, *not* the actual sample size minus 1.

Given the limitations of Table C, our advice is to use technology to find P-values when carrying out a significance test about a population standard deviation or variance. The test statistic and P-value calculated using the "full technology" approach described in the second part of the Tech Corner are generally more accurate than the values obtained when we round while doing our calculations.

TECH CORNER: P-Values and Significance Tests for a Population Standard Deviation or Variance

Calculating P-values

You can use technology to find the P-value from the chi-square test statistic in a significance test for a population standard deviation or variance. We'll illustrate this process using the pulse data from Example 8.8.2, where $\chi^2 = 15.29$ and df = 19.

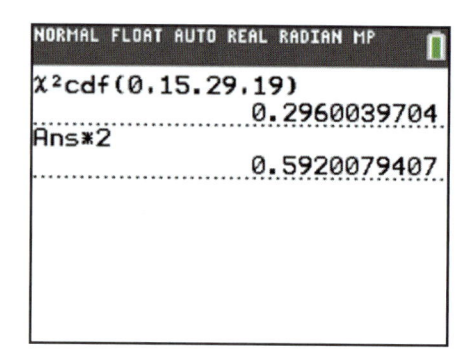

Applet

1. Go to www.stapplet.com and launch the χ^2 *Distributions* applet.

2. Choose "Calculate an area under the χ^2 curve" from the drop-down menu, enter 19 for the degrees of freedom, and click "Plot distribution."

3. Using the drop-down menu below the curve, choose "to the left of a value," and enter 15.29. Then click "Calculate area."

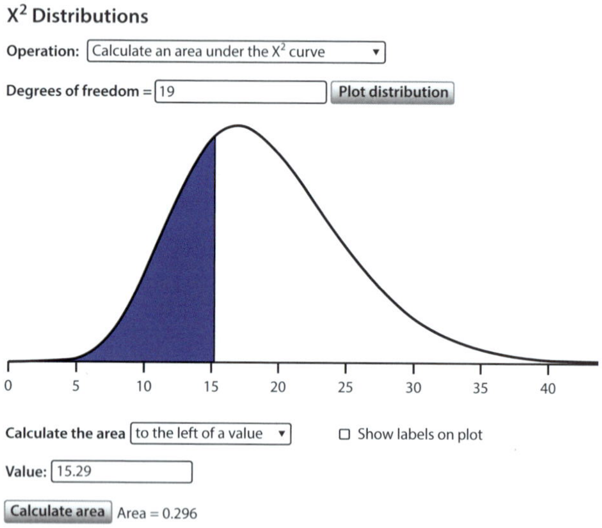

```
NORMAL FLOAT AUTO REAL RADIAN MP
χ²cdf(0,15.29,19)
                     0.2960039704
Ans*2
                     0.5920079407
```

To get the P-value for this two-sided test, multiply the left-tail area by 2: $0.296 \times 2 = 0.592$.

TI-83/84

1. Press `2nd` `VARS` (DISTR) and choose χ^2cdf(.

2. We want to find the area to the left of 15.29 under the χ^2 distribution curve with df = 19.
   - **OS 2.55 or later:** In the dialog box, enter lower: 0, upper: 15.29, and df: 19. Choose Paste, and then press `ENTER`.
   - **Older OS:** Complete the command χ^2cdf(0,15.29,19) and press `ENTER`.

3. To find the P-value for this two-sided test, multiply the left-tail area by 2.

Performing a significance test for σ or σ^2

You can also use technology to perform all the calculations for a significance test about a population standard deviation or variance. We'll illustrate this process using the pulse data from Example 8.8.2. Here are the data once again:

69	69	76	81	68	78	68	70	79	85
70	79	72	78	78	83	73	61	75	74

Recall that we want to perform a two-sided test of the common belief that the standard deviation of the resting pulse rate among humans is $\sigma = 6.67$ beats/min.

Applet

1. Go to www.stapplet.com and launch the *One Quantitative Variable, Single Group* applet.

2. Enter the variable name "Pulse" and enter the data values, separated by commas or spaces.

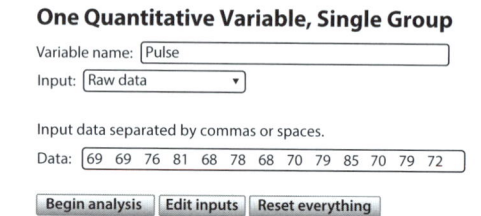

3. Click the Begin analysis button.

4. Scroll down to the "Perform Inference" section. Choose the 1-sample χ^2 test for σ for the procedure, $\sigma \neq$ for the alternative hypothesis, and enter 6.67 for the hypothesized standard deviation. Then click the Perform inference button.

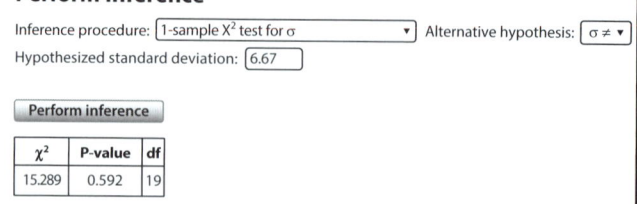

χ^2	P-value	df
15.289	0.592	19

TI 83/84

This type of significance test is not currently available on the TI-83/84.

Detailed instructions for using CrunchIt!, Excel, Google Sheets, JMP, Minitab, and R are available in Achieve.

Section 8.8 What Did You Learn?

Review the learning goals from this section. Then practice what you've learned by working through the exercises.

Learning Goal	Example	Exercises
State and check the Random and Normal conditions for performing a significance test about a population standard deviation or variance.	8.8.1	5–8
Calculate the standardized test statistic and find the P-value for a significance test about a population standard deviation or variance.	8.8.2	9–12
Use the four-step process to perform a significance test about a population standard deviation or variance.	8.8.3	13–16

Section 8.8 Exercises

Building Concepts and Skills These exercises assess the basic knowledge you should have after reading the section.

1. What are the two conditions that must be met to perform a significance test for a population standard deviation or variance?

2. State the formula for the test statistic in a significance test about a population standard deviation or variance.

3. When conditions are met, find the P-value in a test about a population standard deviation or variance using a _____ distribution with _____ df.

4. True/False: For a two-sided test about a population variance or standard deviation, find the area beyond the test statistic in the tail where the test statistic is located, and multiply this area by 2.

Mastering Concepts and Skills These exercises reinforce the learning goals as shown in the examples.

5. **Refer to Example 8.8.1 Salmon Fillets** As part of a study on salmon health, researchers measured the pH of a random sample of 25 salmon fillets at a fish processing plant.[37] Here are the data:

   6.34 6.39 6.53 6.36 6.39 6.25 6.45 6.38 6.33 6.26 6.24 6.37 6.32
   6.31 6.48 6.26 6.42 6.43 6.36 6.44 6.22 6.52 6.32 6.32 6.48

   The researchers want to perform a test at the $\alpha = 0.05$ significance level of $H_0: \sigma = 0.10$ versus $H_a: \sigma \neq 0.10$, where σ is the standard deviation of the pH level in the population of all salmon fillets. Check if the conditions for performing the test are met.

6. **Fancy Fonts** Does the use of fancy type fonts change the variability in reading time of text on a computer screen? Adults can read four paragraphs of text in the common Times New Roman font in an average time of 22 seconds with a standard deviation of 5.69 seconds.[38] Researchers asked a random sample of 24 adults to read these same four paragraphs in the ornate font named *Gigi* (Gigi). Here are their times in seconds:

   23.2 21.2 28.9 27.7 29.1 27.3 16.1 22.6 25.6 34.2 23.9 26.8
   20.5 34.3 21.4 32.6 26.2 34.1 31.5 24.6 23.0 28.6 24.4 28.1

   The researchers want to perform a test at the $\alpha = 0.05$ significance level of $H_0: \sigma = 5.69$ versus $H_a: \sigma \neq 5.69$, where σ is the standard deviation of the time (in seconds) to read the four paragraphs of text in Gigi font in the population of adults. Check if the conditions for performing the test are met.

7. **GRE Scores** The Educational Testing Service reports that the standard deviation for the quantitative reasoning portion of all GRE exams is 9.44.[39] A national chain of GRE preparation schools wants to know if using its regular program will help decrease the variation in student scores. To investigate this question, the company takes a random sample of 150 students and finds a sample standard deviation of 8.46. GRE scores are known to be approximately normally distributed.

   (a) State appropriate hypotheses for the company to test.

   (b) Check that the conditions for performing the test are met.

8. **Laundry Detergent** A machine is supposed to fill bottles of laundry detergent with an average of 64 ounces and a variance of at most 3 ounces². If there is too much variability, too many bottles will either overfill and overflow, or underfill and make consumers angry. A worker takes a random sample of 22 bottles in a particular hour. The sample variance is 3.9 square ounces and it is known that the amount of detergent in the bottles is approximately normally distributed.

   (a) State appropriate hypotheses for a test about the variability of the bottle-filling machine.

   (b) Check that the conditions for performing the test are met.

9. **Refer to Example 8.8.2 Salmon Fillets** Refer to Exercise 5.
   (a) Calculate the test statistic.
   (b) Find the P-value.

10. **Fancy Fonts** Refer to Exercise 6.
    (a) Calculate the test statistic.
    (b) Find the P-value.

11. **GRE Scores** Refer to Exercise 7.
    (a) Calculate the test statistic.
    (b) Find the P-value using technology.
    (c) What conclusion would you make?

12. **Laundry Detergent** Refer to Exercise 8.
    (a) Calculate the test statistic.
    (b) Find the P-value using technology.
    (c) What conclusion would you make?

13. **Refer to Example 8.8.3 Fishing Line** A certain brand of fishing line is advertised as having a breaking strength of 6 pounds. In addition to making sure that the mean breaking strength meets or exceeds the advertised weight, it is important that the breaking strength is consistent. It would certainly be a problem if some pieces of line could hold 10 pounds, but other pieces could hold only 2 pounds. Consistency of breaking strength implies a small standard deviation. An angler randomly selects 30 pieces of fishing line, attaches each to a bucket, and fills the bucket with water until the line breaks. Here is a dotplot of the breaking strength (in pounds) for the 30 trials.[40]

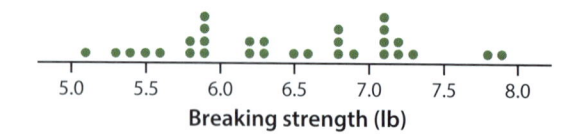

    For this sample, $\bar{x} = 6.44$ pounds and $s_x = 0.75$ pound. Is there convincing evidence at the $\alpha = 0.05$ significance level that the true standard deviation of the breaking strength is greater than 0.5 pound?

14. **Video Screens** A manufacturer of high-resolution video terminals must control the tension on the mesh of fine wires that lies behind the surface of the viewing screen. Too much tension will tear the mesh, and too little will allow wrinkles. The tension is measured by an electrical device with output readings in millivolts (mV). Some variation is inherent in the production process, but the manufacturer would like to keep this variation to a minimum. Here is a boxplot of the tension measurements, in millivolts (mV), from 20 randomly selected screens produced on a particular day.

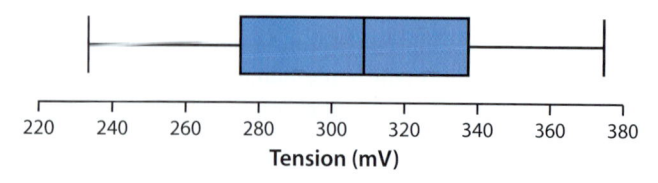

    For this sample, $\bar{x} = 306.32$ mV and $s_x = 36.21$ mV. The goal for the manufacturer is to produce screens with a standard deviation of at most 25 mV. Is there convincing evidence at the $\alpha = 0.01$ significance level that the screens produced this day don't meet the manufacturer's goal?

15. **Red-Tail Hawks** A biology professor spent many years studying birds of prey. The professor and their students captured, measured, and released such birds as they were migrating through their state. One variable of interest was the tail length. Here are the tail lengths, in millimeters (mm), for a random sample of 10 red-tail hawks they observed over the years:[41]

    238   215   220   212   220   233   249   198   226   228

    According to one source, σ = the standard deviation of tail length in all red-tail hawks is 11.8 mm.[42] Is there convincing evidence at the $\alpha = 0.05$ significance level that the population of red-tailed hawks migrating through this particular area has a different standard deviation of tail length?

16. **Asteroid Strike?** Did an asteroid strike create a dust cloud that led to the extinction of most dinosaurs? Researchers took rock samples in Gubbio, Italy, from various depths and measured them for the concentration of iridium (a rare metal that is more common in asteroids). The deeper the sample, the older the rocks were. A sudden increase in iridium at some point in time would lend support for the asteroid hypothesis. In this exercise, we analyze just those samples from a depth of 347 meters to see how variable the iridium measurements are within a particular time period. Here are the iridium measurements in parts per billion (ppb) for the 5 randomly selected samples from that depth:[43]

    290   450   620   710   875

    Researchers expect that the true variance for iridium samples at a constant depth is 40,000 ppb^2. Is there convincing evidence at the $\alpha = 0.10$ significance level that the true variance is not 40,000 ppb^2?

Applying the Concepts These exercises ask you to apply multiple learning goals in a new context or to apply what you learned in this section in a new way.

17. **Elephant Height** How variable are the heights of adult male African elephants that lived through droughts during their first two years of life? Researchers measured the shoulder height (in centimeters, cm) of a random sample of 14 such elephants.[44] Here are the data:

    | 200.00 | 272.91 | 217.57 | 294.15 | 296.84 | 212.00 |
    | 257.00 | 251.39 | 266.75 | 237.19 | 265.85 | 212.00 |
    | 220.00 | 225.00 | | | | |

    According to the Animal Diversity Web, the standard deviation of shoulder height for all mature male African elephants is $\sigma = 13.3$ cm. Is there convincing evidence at the 5% significance level that, in the population of adult male African elephants that lived through droughts during their first two years of life, shoulder height has a larger standard deviation than 13.3 cm?

18. **SAT Math** The National Center for Education Statistics reports that the standard deviation of scores on the math portion of the SAT exam is 107.[45] A random sample of 18 incoming students to a liberal arts college was taken, and their math SAT scores were recorded. Here are the data:

480	550	760	540	600	480	610	620	550
780	570	800	720	610	590	470	670	550

Is there convincing evidence at the 5% significance level that the standard deviation of SAT math scores in the population of incoming students at this college is different from the national value of 107?

19. **Elephant Height** Refer to Exercise 17. The power of the test to detect that $\sigma = 20$ is 0.738.
    (a) Interpret the power.
    (b) Would the power of the test to detect that $\sigma = 25$ be larger or smaller than 0.738? Justify your answer.
    (c) Name two things the researchers could do to increase the power of the test.

20. **SAT Math** Refer to Exercise 18. The power of the test to detect that $\sigma = 150$ is 0.662.
    (a) Interpret the power.
    (b) Would the power of the test to detect that $\sigma = 125$ be larger or smaller than 0.662? Justify your answer.
    (c) Name two things the researchers could do to increase the power of the test.

21. **Salmon Fillets** Refer to Exercises 5 and 9.
    (a) Interpret the P-value.
    (b) What conclusion would you make?
    (c) The 95% confidence interval for σ is 0.0680 to 0.1212. Explain how this interval is consistent with, but gives more information than, your conclusion in part (b).

22. **Fancy Fonts** Refer to Exercises 6 and 10.
    (a) Interpret the P-value.
    (b) What conclusion would you make?
    (c) The 95% confidence interval for σ is 3.67 to 6.63. Explain how this interval is consistent with, but gives more information than, your conclusion in part (b).

23. **Tech Coffee Prices** How much variability is there in the price of hot coffee at different restaurants and coffee shops? A university statistics class recorded hot coffee prices at randomly selected restaurants, cafes, and coffee shops in their city. Their data are found in the data set *Coffee*.[46] One professor at the college claims that the variance of the price of hot coffee (*HotPrice*) at all restaurants and coffee shops in the city is 0.25 dollar². Is there convincing evidence that the variance of hot coffee prices in this city is larger than the professor's claim?

24. **Tech College Tuition** How much variability is there in the tuition charged at public colleges and universities in the United States? Researchers took a random sample of such schools and recorded the tuition amounts. Their data are found in the data set *Tuition*.[47] Is there convincing evidence that the standard deviation of tuition at all public colleges and universities is different from $5000?

Extending the Concepts These exercises challenge you to explore statistical concepts and methods that go beyond what you learned in this section.

25. **Elephant Height** Refer to Exercise 17 and the Think About It in Section 8.3.
    (a) Find the critical value and the rejection region.
    (b) What conclusion would you make? Explain your reasoning.

26. **SAT Math** Refer to Exercise 18 and the Think About It in Section 8.3.
    (a) Find the critical value and the rejection region.
    (b) What conclusion would you make? Explain your reasoning.

27. **Normal Approximation** In situations where the sample size is larger than 50, a chi-square distribution with df $= n - 1$ can be approximated by a normal distribution with mean $n - 1$ and standard deviation $\sqrt{2(n-1)}$. For a significance test, this means that the χ^2 statistic calculated in this section can be standardized by subtracting the mean and dividing by the standard deviation:

$$z = \frac{\chi^2 - (n-1)}{\sqrt{2(n-1)}}$$

The standard normal distribution can then be used to find the P-value for a significance test.

A machine is supposed to fill bags with a standard deviation of 0.7 ounce of candy. The manager wants to be sure that there isn't too much variability among the bags. The manager selects a random sample of 75 bags of candy produced that day and weighs each bag. The standard deviation is 0.81 ounce. Assume that the weights of the bags are approximately normally distributed.

   (a) What are the hypotheses the manager wishes to test?
   (b) Calculate the χ^2 statistic.
   (c) Compute the standardized test statistic z.
   (d) Find the P-value based on the value of z you computed in part (c).
   (e) The P-value based on the chi-square distribution is 0.027. How does this compare to the P-value you computed in part (d)?

Cumulative Review These exercises revisit what you learned in previous sections.

28. **Eyestrain (1.7)** A researcher wants to find the best computer monitor setup to reduce eyestrain. There are four different setups they wish to test. The researcher recruits 16 volunteers to help with the test: 4 children (younger than age 18), 4 young adults (age 18–25), 4 adults (age 26–60), and 4 older adults (age 61+).

    (a) Design an experiment that uses blocking to investigate this question. Explain your choice of blocks.

    (b) Explain why the randomized block design is preferable to a completely randomized experiment in this context.

29. **Computer Chips (5.5)** A manufacturer of computer chips knows from experience that the number of defective chips produced per day follows a Poisson distribution with mean 3.2 defectives per day.

    (a) What is the probability of producing no defective chips on a particular day?

    (b) What is the probability that more than 5 defective chips are produced on a particular day?

Statistics Matters What is normal body temperature?

At the beginning of the chapter, we described a study investigating whether "normal" human body temperature is really 98.6°F. Knowing the answer to that question is crucial in determining whether someone has a fever—a common symptom of illness. Researchers used an oral thermometer to measure the temperatures of a random sample of 130 healthy men and women ages 18 to 40. The dotplot shows the temperature readings.[48] We have added a vertical line at 98.6°F for reference.

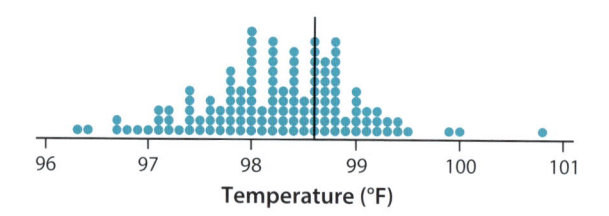

Exploratory data analysis revealed several interesting facts about this data set:

- The mean temperature is $\bar{x} = 98.25°F$.
- The standard deviation of the temperature readings is $s_x = 0.73°F$.
- 62.3% of the temperature readings are less than 98.6°F.

1. If "normal" body temperature really is 98.6°F, we would expect that half of all healthy 18- to 40-year-olds will have a body temperature less than 98.6°F. Do the data from this study give convincing evidence at the $\alpha = 0.05$ significance level that the proportion of all healthy 18- to 40-year-olds with a body temperature less than 98.6°F differs from 0.5?

2. Based on the conclusion in Question 1, which type of error could have been made: a Type I error or a Type II error? Explain your answer.

3. Do the data give convincing evidence at the 5% significance level that the mean body temperature in all healthy 18- to 40-year-olds is not 98.6°F?

4. A 95% confidence interval for the population mean is 98.123°F to 98.377°F. Explain how the confidence interval is consistent with, but gives more information than, the significance test in Question 3.

5. To determine who has a fever, it would be helpful to know how much the body temperatures of healthy people vary. A report claims that the standard deviation of body temperature in healthy 18- to 40-year-olds is 0.8°F. The researchers in this study suspect that the actual value is somewhat less.

    (a) State appropriate hypotheses for a significance test to determine whether the researchers' suspicion is correct.

(b) Check that the conditions for performing the test in part (a) are met.

(c) Calculate the appropriate test statistic. What distribution should be used to find the P-value?

(d) The P-value of the test is 0.083. Interpret this value in context. What conclusion would you make?

Postscript: An interesting study published in 2020 confirmed that the mean oral temperature of adults is lower than the 98.6°F value that was established in the 1800s. By examining large data sets from three time periods (1860–1940, 1971–1975, and 2007–2017), the researchers found convincing evidence that average adult body temperature has decreased over time. The mean temperature of the more than 150,000 adults in the 2007–2017 data set was 98.0°F.[49]

Chapter 8 Review

The Idea of a Significance Test

- A **significance test** is a procedure for using observed data to decide between two competing claims, called hypotheses. The hypotheses are usually statements about a parameter, such as the population proportion p, the population mean μ, or the population standard deviation σ.
- The claim we weigh evidence *against* in a statistical test is called the **null hypothesis (H_0)**. The null hypothesis often has the form H_0: parameter = null value.
- The claim about the population that we are trying to find evidence *for* is the **alternative hypothesis (H_a)**.
  - A **one-sided** alternative hypothesis has the form H_a: parameter < null value or H_a: parameter > null value.
  - A **two-sided** alternative hypothesis has the form H_a: parameter ≠ null value.
- A **standardized test statistic** typically measures how far a sample statistic is from what we would expect if the null hypothesis H_0 were true, in standardized units. That is,

$$\text{standardized test statistic} = \frac{\text{statistic} - \text{null value}}{\text{standard deviation (error) of statistic}}$$

- The ***P*-value** of a test is the probability of getting evidence for the alternative hypothesis H_a as strong as or stronger than the observed evidence by chance alone when the null hypothesis H_0 is true.
- Small *P*-values are evidence against the null hypothesis and for the alternative hypothesis because they say that the observed result is unlikely to occur when H_0 is true. To determine if a *P*-value should be considered small, we compare it to the **significance level α**.

- We make a conclusion in a significance test based on the *P*-value.
  - If *P*-value < α: Reject H_0 and conclude there is convincing evidence for H_a (in context).
  - If *P*-value > α: Fail to reject H_0 and conclude there is not convincing evidence for H_a (in context).
- When we make a conclusion in a significance test, we can make two kinds of mistakes.
  - A **Type I error** occurs if we find convincing evidence that H_a is true when it really isn't.
  - A **Type II error** occurs if we do not find convincing evidence that H_a is true when it really is.
- The probability of making a Type I error is equal to the significance level α. There is a trade-off between $P(\text{Type I error})$ and $P(\text{Type II error})$: As one decreases, the other increases, assuming everything else remains the same. So it is important to consider the possible consequences of each type of error before choosing a significance level.

How Significance Tests Work

- When you perform a significance test, follow the four-step process:

  State: State the hypotheses, parameter(s), significance level, and evidence for H_a.

  Plan: Identify the appropriate inference method and check the conditions.

  Do: If the conditions are met, perform calculations:
  - Calculate the test statistic.
  - Find the *P*-value.

  Conclude: Make a conclusion about the hypotheses in the context of the problem.

Significance Tests for a Population Proportion

- The conditions for performing a significance test of $H_0: p = p_0$ are:
  - **Random:** The data come from a random sample from the population of interest.
  - **Large Counts:** Both np_0 and $n(1-p_0)$ are at least 10.
- The standardized test statistic for a **one-sample z test for p** is

$$z = \frac{\hat{p} - p_0}{\sqrt{\frac{p_0(1-p_0)}{n}}}$$

- When the conditions are met, the standardized test statistic has approximately a standard normal distribution. You can use Table A or technology to find the P-value.
- Confidence intervals provide additional information that significance tests do not—namely, a set of plausible values for the population proportion p based on sample data. A 95% confidence interval for p gives information about the parameter that is generally consistent with a two-sided test of $H_0: p = p_0$ at the $\alpha = 0.05$ significance level.

Significance Tests for a Population Mean

- The conditions for performing a significance test of $H_0: \mu = \mu_0$ are:
  - **Random:** The data come from a random sample from the population of interest.
  - **Normal/Large Sample:** The data come from an approximately normally distributed population or the sample size is large ($n \geq 30$). When the sample size is small and the shape of the population distribution is unknown, a graph of the sample data shows no strong skewness or outliers.
- The standardized test statistic for a **one-sample t test for μ** is

$$t = \frac{\bar{x} - \mu_0}{\frac{s_x}{\sqrt{n}}}$$

- When the conditions are met, the standardized test statistic can be modeled by a t distribution with $n-1$ degrees of freedom (df). You can use Table B or technology to find the P-value.
- Confidence intervals provide additional information that significance tests do not—namely, a set of plausible values for the population mean μ based on sample data. A 95% confidence interval for μ gives information about the parameter that is consistent with a two-sided test of $H_0: \mu = \mu_0$ at the $\alpha = 0.05$ significance level.

Significance Tests for a Population Standard Deviation or Variance

- The conditions for performing a significance test of $H_0: \sigma = \sigma_0$ or $H_0: \sigma^2 = \sigma_0^2$ are:
  - **Random:** The data come from a random sample from the population of interest.
  - **Normal:** The data come from a population that is approximately normally distributed. When the shape of the population distribution is unknown, a graph of the sample data shows no strong skewness or outliers.

  There is *no* large sample size override to the Normal condition for this test.
- The test statistic for a **one-sample test for σ (or σ^2)** is

$$\chi^2 = \frac{(n-1)s_x^2}{\sigma_0^2}$$

- When the conditions are met, the test statistic can be modeled by a chi-square distribution with $n-1$ degrees of freedom (df). You can use Table C or technology to find the P-value.

Using Tests Wisely

- Very small deviations from the null hypothesis can be highly significant (small P-value) when a test is based on a large sample. A statistically significant result may not be practically important.
- Many tests that are run at once will likely produce some significant results by chance alone, even if all the null hypotheses are true. Beware of P-hacking.
- The **power** of a test is the probability that the test will find convincing evidence for H_a when a specific alternative value of the parameter is true. In other words, the power of a test is the probability of avoiding a Type II error. For a specific alternative, Power $= 1 - P(\text{Type II error})$.
- We can increase the power of a significance test (decrease the probability of a Type II error) by increasing the sample size, increasing the significance level, or increasing the difference that is important to detect between the null and alternative parameter values (known as the *effect size*). Wise choices when collecting data, such as controlling for other variables and using blocking in experiments or stratified random sampling, can also increase power.

	Comparing significance tests		
	Significance test for p	**Significance test for μ**	**Significance test for σ (or σ^2)**
Name (TI-83/84)	One-sample z test for p (1-PropZTest)	One-sample t test for μ (TTest)	One-sample test for σ (or σ^2) (not available)
Conditions	• **Random:** The data come from a random sample from the population of interest. • **Large Counts:** Both np_0 and $n(1-p_0)$ are at least 10. That is, the expected number of successes and the expected number of failures in the sample are both at least 10.	• **Random:** The data come from a random sample from the population of interest. • **Normal/Large Sample:** The data come from an approximately normally distributed population or the sample size is large ($n \geq 30$). When the sample size is small and the shape of the population distribution is unknown, a graph of the sample data shows no strong skewness or outliers.	• **Random:** The data come from a random sample from the population of interest. • **Normal:** The data come from an approximately normally distributed population. When the shape of the population distribution is unknown, a graph of the sample data shows no strong skewness or outliers.
Formula	$z = \dfrac{\hat{p} - p_0}{\sqrt{\dfrac{p_0(1-p_0)}{n}}}$ P-value from standard normal distribution	$t = \dfrac{\bar{x} - \mu_0}{\dfrac{s_x}{\sqrt{n}}}$ P-value from t distribution with df $= n-1$	$\chi^2 = \dfrac{(n-1)s_x^2}{\sigma_0^2}$ P-value from chi-square distribution with df $= n-1$

Chapter 8 Review Exercises

These exercises will help you review important concepts and skills described by the learning goals in each section. For most exercises, the relevant section is noted in parentheses after the exercise title.

1. **Stating Hypotheses (8.1)** For each of the following scenarios, state appropriate hypotheses for a significance test. Be sure to define the parameter of interest.
   (a) The website popcorn.org claims that U.S. adults consume an average of 47 quarts of popcorn per year. Researchers wonder if this claim is valid in their state.
   (b) According to the Humane Society, 47% of U.S. households own at least one dog. You suspect that the proportion of households that own at least one dog in your county is greater than the national proportion.
   (c) "Ping" and "jitter" are both important to the quality of an internet connection. Ping measures the response time (in milliseconds) when a request is made. Jitter measures the standard deviation in response times. For Zoom meetings, jitter should be at most 40 milliseconds.[50] A frequent Zoom user is concerned that their internet connection does not meet the jitter standard.

2. **Mean Gravel (8.1, 8.2)** The company that is providing gravel for a road project claims to put an average of 20 cubic meters (m^3) of gravel in a truckload. To test this claim, the project manager will measure the volume of gravel in a random sample of 50 truckloads.
   (a) State hypotheses for a significance test to determine whether the company is delivering less gravel than it claims. Be sure to define the parameter of interest.
   (b) Describe a Type I error and a Type II error in this scenario.
   (c) Give a consequence of each type of error. Which type of error is more serious?

3. **Finding P-Values (8.3, 8.5, 8.7)** Find the P-value in each of the following scenarios. Assume that the conditions for inference are met.
   (a) A test of $H_0: p = 0.7$ versus $H_a: p > 0.7$ based on a random sample of size $n = 40$ yields a standardized test statistic of 1.83.
   (b) A test of $H_0: \mu = 30$ versus $H_a: \mu \neq 30$ based on a random sample of size $n = 20$ yields a standardized test statistic of -2.35.
   (c) A test of $H_0: \sigma = 1.5$ versus $H_a: \sigma < 1.5$ based on a random sample of size $n = 25$ yields a test statistic of 15.62.

4. **Roulette (8.3)** A U.S. roulette wheel has 18 red slots among its 38 slots. To test if a particular roulette wheel is fair, you spin the wheel 50 times and the ball lands in a red slot 31 times.
   (a) State appropriate hypotheses for performing a significance test. Be sure to define the parameter of interest.
   (b) Check that the conditions for performing the significance test are met.
   (c) The P-value for this test is 0.0384. Interpret the P-value.

(d) What conclusion would you make at the $\alpha = 0.05$ significance level?

(e) The casino manager uses your data to produce a 99% confidence interval for p and gets (0.443, 0.797). The manager says that this interval provides convincing evidence that the wheel is fair. How do you respond?

5. **Underemployment (8.4)** Economists often track employment trends by measuring the proportion of people who are "underemployed," meaning they are either unemployed or would like to work full-time but are only working part-time. At the beginning of 2020, 7.0% of U.S. adults were underemployed. The mayor of Carrboro suspects that the situation in this small city is worse than in the rest of the country. A staff member selects an SRS of 150 adult Carrboro residents and finds that 14 of them are underemployed. Do these data give convincing evidence at the 10% significance level that the proportion of underemployed workers in Carrboro is greater than elsewhere in the country?

6. **Aspirin (8.5)** The makers of Aspro brand aspirin want to be sure that their tablets contain the right amount of active ingredient (acetylsalicylic acid). When the production process is working properly, Aspro tablets contain an average of 320 milligrams (mg) of active ingredient. A quality-control inspector selects a random sample of 30 tablets from a batch in production. The amount of active ingredient in the 30 selected tablets has mean 319 mg and standard deviation 3 mg. The inspector will perform a test at the $\alpha = 0.05$ significance level of $H_0: \mu = 320$ versus $H_a: \mu \neq 320$, where $\mu =$ the mean amount of active ingredient in all tablets from this production batch.

(a) Explain why the sample data give some evidence for H_a.

(b) Check that the conditions are met for carrying out the test.

(c) Calculate the standardized test statistic and P-value.

(d) What conclusion would you make?

7. **Sweet Corn (8.6)** A certain variety of sweet corn is known to produce individual ears with a mean weight of 8 ounces. A farmer is testing a new fertilizer designed to produce larger ears of corn, as measured by their weight. The farmer selects a random sample of 10 ears of corn and determines their weight (in ounces). The data are as follows:

8.30 7.35 9.05 8.05 8.25 8.75 9.00 7.60 8.45 8.55

Do these data provide convincing evidence that the mean weight for all ears of corn grown using the new fertilizer is greater than 8 ounces?

8. **Unemployment (8.7)** Refer to Exercise 5.

(a) The power of the test to detect the fact that 11% of adult Carrboro residents are underemployed is 0.70. Interpret this value.

(b) Explain two ways that you can increase the power of the test.

9. **Sweet Corn (8.8)** Refer to Exercise 7. The farmer is concerned about a possible increase in the variability of the weight for individual ears of corn produced using the new fertilizer. The grocery store that purchases this farmer's corn will terminate the contract if there is convincing evidence at the $\alpha = 0.01$ significance level that $\sigma > 0.5$ ounce.

(a) Explain why the sample data give the farmer a reason to worry.

(b) Perform the significance test. What conclusion would you make?

10. **Signature Verification (Chapter 8)** When a petition is submitted to government officials to put a political candidate's name on a ballot, a certain number of valid voters' signatures are required. Rather than check the validity of all the signatures, officials often randomly select a sample of signatures for verification and perform a significance test to see if the true proportion of signatures is greater than the required value. Suppose a petition has 30,000 signatures and 18,000 valid signatures are required for a candidate to be on the ballot, so at least 60% of the signatures on this petition must be valid. The officials select a random sample of 300 signatures and find that 159 are valid.

(a) Do the data provide convincing evidence at the $\alpha = 0.01$ significance level that the true proportion of valid signatures on the petition is less than 0.60?

(b) Which type of error could you have made in part (a): a Type I error or a Type II error? Explain your answer.

11. **Vertical Leaps (Chapter 8)** Student researchers saw an article on the internet claiming that the average vertical jump for college students was 15 inches. They wondered if this claim is valid for the students at their community college, so they selected a random sample of 20 students. After contacting these students several times, they finally convinced them to allow their vertical jumps to be measured for research purposes. Here are the data (in inches):

11.0 11.5 12.5 26.5 15.0 12.5 22.0 15.0 13.5 12.0
23.0 19.0 15.5 21.0 12.5 23.0 20.0 8.5 25.5 20.5

(a) Carry out an appropriate significance test at the 5% significance level.

(b) A 95% confidence interval for the mean vertical jump of all students at this community college is 14.488 to 19.512 inches. Explain how the confidence interval gives more information than the test in part (a).

12. **Vertical Leaps (Chapter 8)** Refer to Exercise 11. Do the data provide convincing evidence that the vertical jump height of all students at this community college has variance less than 36 square inches?

Chapter 8 Project: Air Quality

Explore a large data set using software and apply what you've learned in this chapter to a real-world scenario.

The quality of the air around us has an enormous effect on our lives. Pollution in the air can cause health problems for individuals, as well as affect the climate by causing more acid rain, depleting the ozone layer around the earth, and damaging plant life. The data set **AirQuality2020** contains a random sample of 100 counties from throughout the United States and Puerto Rico. The variables in the data set represent several different measures of air quality. One way the Environmental Protection Agency (EPA) reports the air quality of a particular place is by calculating the Air Quality Index (AQI). The AQI varies from 0 to 500 and measures the amount of pollution in the air. Larger numbers indicate more pollution.

Download this data set into the technology that you are using. The following questions will help you begin to analyze the data. We will return to this data set for the Chapter 9 Project.

1. Instead of reporting an actual value for the AQI, EPA often reports categories that the general public can better understand. For instance, if the AQI is between 101 and 150, the EPA will report the air quality as being unhealthy for medically sensitive people. The variable *Sensitive* measures whether that particular county had any measured days in 2020 with an AQI between 101 and 150. Note that 0 = No "unhealthy for sensitive people" days and 1 = At least one "unhealthy for sensitive people" day. Do we have convincing evidence at the $\alpha = 0.10$ level that more than one-fourth of the counties in the United States and Puerto Rico experienced "unhealthy for sensitive individuals" air days in 2020?

2. How could the power of the test in Question 1 be increased?

3. Construct a 95% confidence interval for the true proportion of counties that experienced "unhealthy for sensitive individuals" air days in 2020. What additional information does this interval provide compared to the significance test in Question 1?

4. Days with an AQI less than 100 are considered to be either good or moderate (not unhealthy) days. One variable measured was the maximum AQI (*Max AQI*) over the course of the year in each of these counties. Do we have convincing evidence at the 10% significance level that the mean maximum AQI is less than 100 for all counties in the United States and Puerto Rico in 2020?

5. Interpret the *P*-value from the significance test in Question 4.

6. Which type of error could you have made in the significance test in Question 4?

7. Construct a 95% confidence interval for the population mean value of the maximum AQI for all counties in the United States and Puerto Rico. What additional information does this interval provide compared to the significance test in Question 4?

8. Graph the data for the *Max AQI* variable. Identify the counties that are outliers for this variable. Redo Questions 4 and 7 without these data values. What effect do these observations have on the confidence interval and significance test? Explain your answer.

9 Comparing Two Populations or Treatments

Section 9.1	Confidence Intervals for a Difference Between Two Population Proportions	**Section 9.6**	Significance Tests for a Population Mean Difference
Section 9.2	Significance Tests for a Difference Between Two Population Proportions	**Section 9.7**	Significance Tests for Two Population Standard Deviations or Variances
Section 9.3	Confidence Intervals for a Difference Between Two Population Means	**Chapter 9 Review**	
Section 9.4	Significance Tests for a Difference Between Two Population Means	**Chapter 9 Review Exercises**	
		Chapter 9 Project	
Section 9.5	Analyzing Paired Data: Confidence Intervals for a Population Mean Difference	**Cumulative Review Chapters 7–9**	

Statistics Matters: How fast-food drive-thrus make money: Speed, accuracy, and customer service

Consumers spend billions of dollars each year in the drive-thru lanes of U.S. fast-food restaurants. Having quick, accurate, and friendly service at a drive-thru window translates directly into restaurant revenue. As a result, industry executives, stockholders, and analysts closely follow the ratings of fast-food drive-thru lanes that appear each year in *QSR*, a publication that reports on the quick-service restaurant industry.

The 2021 *QSR* drive-thru study (conducted during the Covid pandemic) involved visits to a random sample of restaurants in 10 major fast-food chains in all 50 U.S. states. During each visit, the researcher ordered a modified main item (for example, a hamburger with no pickles), a side item, and a drink. If any item was not received as ordered, or if the restaurant failed to give the correct change or supply a straw and a napkin, the order was considered "inaccurate." Service time, which is the time from when the car stopped at the speaker to when the entire order was received, was measured for each visit. In addition, *QSR* conducted a survey about customer service experiences at these fast-food drive-thrus.[1]

Here are some results from the study:

- In the 159 visits to McDonald's, 91.2% of the orders were accurate. In the 163 visits to Wendy's, 85.3% of the orders were accurate.
- McDonald's average service time for 159 drive-thru visits was 311 seconds with a standard deviation of 182 seconds. Burger King's service time for 145 drive-thru visits had a mean of 359 seconds and a standard deviation of 194 seconds.
- The two top-rated restaurants for customer service in the survey were Chick-Fil-A and KFC. In a random sample of 1002 people, 525 rated Chick-Fil-A's service as "very friendly." In a separate random sample of 1000 people, 444 rated KFC's service as "very friendly."

Section 9.1 Confidence Intervals for a Difference Between Two Population Proportions

In 2021, was there a significant difference in order accuracy at McDonald's versus Wendy's drive-thrus? How much better was the average service time at McDonald's than at Burger King restaurants? Do McDonald's drive-thru service times vary significantly more than Burger King's? Is Chick-Fil-A the undisputed winner for customer service?

We'll revisit Statistics Matters at the end of the chapter, so you can use what you have learned to help answer these questions.

In Chapter 7, you learned how to estimate a single population parameter (proportion, mean, standard deviation) using a confidence interval. In Chapter 8, you learned how to test a claim about a single population parameter with a significance test. This chapter shows you how to perform inference for comparing two population parameters.

Section 9.1 Confidence Intervals for a Difference Between Two Population Proportions

LEARNING GOALS

By the end of this section, you will be able to:
- Check the Random and Large Counts conditions for constructing a confidence interval for a difference between two proportions.
- Calculate a $C\%$ confidence interval for a difference between two proportions.
- Use the four-step process to construct and interpret a confidence interval for the difference between two proportions.

In Sections 7.3 and 7.4, you learned how to calculate and interpret a confidence interval for a population proportion p. What if we want to estimate the difference $p_1 - p_2$ between the proportions of successes in Population 1 and Population 2? For instance, perhaps we want to estimate the difference in the proportions of all young adult (ages 18 to 29) Democrats and Republicans in the United States who feel that politics has become too polarized. The ideal strategy is to take a separate random sample from each population and to use the difference $\hat{p}_1 - \hat{p}_2$ between the sample proportions as our point estimate. To make inferences about $p_1 - p_2$, we have to first know about the sampling distribution of $\hat{p}_1 - \hat{p}_2$. Here are the details.

> **HOW TO** Describe the Sampling Distribution of $\hat{p}_1 - \hat{p}_2$
>
> Let $\hat{p}_1$ be the proportion of successes in a simple random sample (SRS) of size n_1 selected from a large population with proportion of successes p_1. Let $\hat{p}_2$ be the proportion of successes in an SRS of size n_2 selected from a large population with proportion of successes p_2. Then:
> - The **shape** of the sampling distribution of $\hat{p}_1 - \hat{p}_2$ is approximately normal if $n_1 p_1$, $n_1(1-p_1)$, $n_2 p_2$, and $n_2(1-p_2)$ are all at least 10.
> - The **mean** of the sampling distribution of $\hat{p}_1 - \hat{p}_2$ is $\mu_{\hat{p}_1 - \hat{p}_2} = p_1 - p_2$.
> - The **standard deviation** of the sampling distribution of $\hat{p}_1 - \hat{p}_2$ is
>
> $$\sigma_{\hat{p}_1 - \hat{p}_2} = \sqrt{\frac{p_1(1-p_1)}{n_1} + \frac{p_2(1-p_2)}{n_2}}$$

This section focuses on constructing confidence intervals for a difference between two proportions. You will learn how to perform significance tests about $p_1 - p_2$ in Section 9.2.

Conditions for Estimating $p_1 - p_2$

In Section 7.3, we introduced two conditions that should be met before we construct a confidence interval for a population proportion: Random and Large Counts. Now that we are comparing *two* proportions, we have to modify these conditions slightly.

> **HOW TO** Check Conditions for Constructing a Confidence Interval for $p_1 - p_2$
>
> To construct a confidence interval for a difference in population proportions $p_1 - p_2$, check that the following conditions are met:
>
> - **Random:** The data come from independent random samples from the two populations of interest or from two groups in a randomized experiment.
> - **Large Counts:** The counts of "successes" and "failures" in each sample or group — $n_1\hat{p}_1$, $n_1(1-\hat{p}_1)$, $n_2\hat{p}_2$, and $n_2(1-\hat{p}_2)$ — are all at least 10.

Recall from Chapter 1 that the Random condition is important for determining the scope of inference. Random sampling allows us to generalize our results to the populations of interest, and random assignment in an experiment permits us to make cause-and-effect conclusions.

The method we use to calculate a confidence interval for $p_1 - p_2$ requires that the sampling distribution of $\hat{p}_1 - \hat{p}_2$ be approximately normal. This will be true whenever n_1p_1, $n_1(1-p_1)$, n_2p_2, and $n_2(1-p_2)$ are all at least 10. Because we don't know the value of p_1 or p_2 when we are estimating their difference, we use $\hat{p}_1$ and $\hat{p}_2$ when checking the Large Counts condition.

EXAMPLE 9.1.1

Who prefers brand-name clothing?

Checking conditions

PROBLEM

A Harris Interactive survey asked independent random samples of adults from the United States and Germany about the importance of brand names when buying clothes. Of the 2309 U.S. adults surveyed, 26% said brand names were important, compared with 22% of the 1058 German adults surveyed.[2] Let p_1 = the proportion of all U.S. adults who think brand names are important when buying clothes and p_2 = the proportion of all German adults who think brand names are important when buying clothes. Check if the conditions for calculating a confidence interval for $p_1 - p_2$ are met.

SOLUTION

Random? Independent random samples of 2309 U.S. adults and 1058 German adults. ✓

Large Counts? $2309(0.26) = 600.34 \rightarrow 600$

$2309(1 - 0.26) = 2309(0.74) = 1708.66 \rightarrow 1709$

$1058(0.22) = 232.76 \rightarrow 233$

$1058(1 - 0.22) = 1058(0.78) = 825.24 \rightarrow 825$

All counts are at least 10. ✓

> Be sure to check that the counts of "successes" and "failures" in each sample or group — $n_1\hat{p}_1$, $n_1(1-\hat{p}_1)$, $n_2\hat{p}_2$, and $n_2(1-\hat{p}_2)$ — are *all* at least 10. Because these are the observed counts of successes and failures, they should be rounded to the nearest integer.

EXAM PREP: FOR PRACTICE, TRY EXERCISE 5.

Calculating a Confidence Interval for $p_1 - p_2$

When the appropriate conditions are met, we can calculate a confidence interval for $p_1 - p_2$ using the familiar formula

$$\text{point estimate} \pm \text{margin of error}$$

or, in slightly expanded form,

$$\text{statistic} \pm \text{critical value} \cdot \text{standard error of statistic}$$

The observed difference in sample proportions $\hat{p}_1 - \hat{p}_2$ is our point estimate for $p_1 - p_2$. We can find the critical value z^* for the given confidence level using either technology or Table A.

Section 9.1 Confidence Intervals for a Difference Between Two Population Proportions

The standard deviation of the sampling distribution of $\hat{p}_1 - \hat{p}_2$ is

$$\sigma_{\hat{p}_1-\hat{p}_2} = \sqrt{\frac{p_1(1-p_1)}{n_1} + \frac{p_2(1-p_2)}{n_2}}$$

Because we don't know the values of the parameters p_1 and p_2, we replace them with the sample proportions $\hat{p}_1$ and $\hat{p}_2$. The result is the *standard error* of $\hat{p}_1 - \hat{p}_2$:

$$SE_{\hat{p}_1-\hat{p}_2} = \sqrt{\frac{\hat{p}_1(1-\hat{p}_1)}{n_1} + \frac{\hat{p}_2(1-\hat{p}_2)}{n_2}}$$

This is the last piece we need to calculate the confidence interval.

HOW TO Calculate a Confidence Interval for $p_1 - p_2$

When the Random and Large Counts conditions are met, a C% confidence interval for the difference $p_1 - p_2$ between two population proportions is

$$(\hat{p}_1 - \hat{p}_2) \pm z^* \sqrt{\frac{\hat{p}_1(1-\hat{p}_1)}{n_1} + \frac{\hat{p}_2(1-\hat{p}_2)}{n_2}}$$

where z^* is the critical value for the standard normal distribution with C% of its area between $-z^*$ and z^*.

EXAMPLE 9.1.2

Who prefers brand-name clothing?

Calculating a confidence interval for $p_1 - p_2$

PROBLEM

Recall from Example 9.1.1 that in a random sample of 2309 U.S. adults, 26% said brand names are important when buying clothes, and in an independent random sample of 1058 German adults, 22% said brand names are important when buying clothes.

(a) Calculate a 95% confidence interval for the difference (U.S. − German) in the proportions of all U.S. adults and all German adults who think brand names are important when buying clothes.

(b) Based on the interval from part (a), is there convincing evidence of a difference in the population proportions? Explain your reasoning.

SOLUTION

(a) $(0.26 - 0.22) \pm 1.960 \sqrt{\dfrac{0.26(0.74)}{2309} + \dfrac{0.22(0.78)}{1058}}$

$\rightarrow 0.04 \pm 0.031 \rightarrow 0.009$ to 0.071

$(\hat{p}_1 - \hat{p}_2) \pm z^* \sqrt{\dfrac{\hat{p}_1(1-\hat{p}_1)}{n_1} + \dfrac{\hat{p}_2(1-\hat{p}_2)}{n_2}}$

(b) Yes, because the confidence interval (0.009, 0.071) does not include 0 as a plausible value for the difference in the population proportions of U.S. and German adults who think brand names are important when buying clothes.

EXAM PREP: FOR PRACTICE, TRY EXERCISE 9.

How should we interpret the confidence interval from Example 9.1.2? We are 95% confident that the interval from 0.009 to 0.071 captures the difference (U.S. − German) in the proportions of all U.S. adults and all German adults who think brand names are important when buying clothes. The interval suggests that the percentage of all adults who think brand names are important when buying clothes is between 0.9 and 7.1 percentage points higher in the United States than in Germany.

Notice that we have added a second sentence to our usual interpretation of the confidence interval. It describes what the interval tells us about the difference between the two population proportions — which is larger and by how much — in plain language. Be sure to include a similar sentence anytime you interpret a confidence interval for the difference between two parameters.

What would happen to the confidence interval in Example 9.1.2 if we had switched the order of the subtraction to (German − U.S.)? The resulting 95% confidence interval would be −0.071 to −0.009. Both intervals are correct, as long as the order of subtraction is clearly defined.

The researchers in Example 9.1.2 selected independent random samples from the two populations they wanted to compare. In practice, it's common to take one random sample that includes individuals from both populations of interest and then to separate the chosen individuals into two groups. The inference procedures for comparing two proportions are still valid in such situations, provided that the two groups can be viewed as independent random samples from their respective populations of interest. (See Exercises 13 and 14.)

THINK ABOUT IT

Why percentage points higher, not percent higher? It would *not* be correct to say that the importance of brand names when buying clothes for U.S. adults is between 0.9 and 7.1 *percent* higher than for German adults. To see why, suppose that $p_1 = 0.26$ and $p_2 = 0.22$. The difference $p_1 - p_2 = 0.26 - 0.22 = 0.04$, or 4 percentage points. But the proportion of U.S. adults who think brand names are important when buying clothes is $0.04/0.22 = 0.182$, or 18.2% higher than the corresponding proportion of German adults.

Putting It All Together: Confidence Interval for $p_1 - p_2$

We are now ready to use the four-step process to construct and interpret a **two-sample z interval for a difference between two proportions.** You can use technology to perform the calculations in the Do step. See the Tech Corner at the end of the section for details.

A **two-sample z interval for a difference between two proportions** is a confidence interval used to estimate a difference in the proportions of successes for two populations or treatments.

Examples 9.1.1 and 9.1.2 involved inference about $p_1 - p_2$ using data that were produced by random sampling. In such cases, the parameters p_1 and p_2 are the proportions of successes in the corresponding populations. However, many important statistical results come from randomized experiments. For example, how much does a new treatment for pancreatic cancer increase the 5-year survival rate compared to the current standard-of-care treatment? In this case, we want to estimate the value of $p_1 - p_2$, where p_1 and p_2 are the true proportions of success for individuals like the ones in the experiment who receive Treatment 1 or Treatment 2. Fortunately, we can still use the sampling distribution of $\hat{p}_1 - \hat{p}_2$ to perform inference about $p_1 - p_2$. See the Think About It feature in Section 9.2 for more details.

Most experiments on people use recruited volunteers as subjects. When participants are not randomly selected, researchers should not generalize the results of their experiment to some larger populations of interest. But the researchers can still make cause-and-effect conclusions that apply to individuals like the ones who took part in the experiment, as long as the treatments were randomly assigned. This same logic applies to experiments on animals or things.

EXAMPLE 9.1.3

How can lower back pain be treated?

Confidence interval for $p_1 - p_2$

PROBLEM

Patients with lower back pain are often given nonsteroidal anti-inflammatory drugs (NSAIDs) like naproxen to help ease their pain. Researchers wondered if taking Valium along with the naproxen would affect pain relief. To find out, they recruited 112 patients with severe lower back pain and randomly assigned them

Section 9.1 Confidence Intervals for a Difference Between Two Population Proportions

to one of two treatments: (1) naproxen and Valium or (2) naproxen and placebo. After 1 week, 39 of the 57 subjects who took naproxen and Valium reported reduced lower back pain, compared with 43 of the 55 subjects in the naproxen and placebo group.[3] Construct and interpret a 99% confidence interval for the difference in the proportion of patients like the ones in this study who would report reduced lower back pain after taking naproxen and Valium versus after taking naproxen and placebo for a week.

SOLUTION

State: 99% CI for $p_{NV} - p_{NP}$, where p_{NV} = true proportion of patients like these who would report reduced lower back pain after taking naproxen and Valium for a week and p_{NP} = true proportion of patients like these who would report reduced lower back pain after taking naproxen and placebo for a week.

| **State:** State the parameters you want to estimate and the confidence level. *Be sure to indicate the order of subtraction.* |

Plan: Two-sample z interval for $p_{NV} - p_{NP}$

Random? Randomly assigned patients to take naproxen and Valium or naproxen and placebo. ✓

Large Counts? 39, 57 − 39 = 18, 43, and 55 − 43 = 12 are all at least 10. ✓

| **Plan:** Identify the appropriate inference method and check conditions. |

| These are just the observed counts of successes and failures. |

Do: $\hat{p}_{NV} = \dfrac{39}{57} = 0.684$, $\hat{p}_{NP} = \dfrac{43}{55} = 0.782$

$$(0.684 - 0.782) \pm 2.576\sqrt{\dfrac{0.684(0.316)}{57} + \dfrac{0.782(0.218)}{55}}$$

$\rightarrow -0.098 \pm 0.214 \rightarrow -0.312$ to 0.116

Using technology: -0.3114 to 0.1162

| **Do:** If the conditions are met, perform calculations. |

| As in Chapters 7 and 8, when the calculations from technology differ from the by-hand calculations, we'll use the values from technology in the conclusion. |

Conclude: We are 99% confident that the interval from -0.3114 to 0.1162 captures the difference ($NV - NP$) in the true proportions of patients like the ones in this study who would report reduced pain after taking naproxen and Valium versus after taking naproxen and a placebo for a week. The interval suggests that the true percentage of patients like these who would report reduced pain after taking naproxen and Valium is between 31.14 percentage points lower and 11.62 percentage points higher than after taking naproxen and placebo.

| **Conclude:** Interpret your interval in the context of the problem. *Don't forget the second sentence when you interpret the interval!* |

EXAM PREP: FOR PRACTICE, TRY EXERCISE 13.

Because the confidence interval in Example 9.1.3 includes 0 as a plausible value for $p_{NV} - p_{NP}$, we don't have convincing evidence that taking Valium along with naproxen affects pain relief for patients like the ones in the study. Keep in mind that 0 is just one of many plausible values for $p_{NV} - p_{NP}$ based on the sample data. ⚠ **Never suggest that you believe the difference between the true proportions is 0 just because 0 is in the interval!**

TECH CORNER Confidence Intervals for a Difference Between Two Proportions

You can use technology to calculate a confidence interval for $p_1 - p_2$. We'll illustrate this process using Examples 9.1.1 and 9.1.2. Recall that of the 2309 U.S. adults surveyed, $2309(0.26) = 600.34 \rightarrow 600$ think brand names are important when buying clothes. Of the 1058 German adults surveyed, $1058(0.22) = 232.76 \rightarrow 233$ think brand names are important when buying clothes.

Applet

1. Go to www.stapplet.com and launch the *One Categorical Variable, Multiple Groups* applet.

2. Choose to input data as "Two-way table." Enter "Brand names important?" as the variable name. Type the group names in the boxes at the top of the columns and the category names in the boxes at the left of the rows, as shown.

(continued)

3. Input the number of successes and failures for each group, as shown.

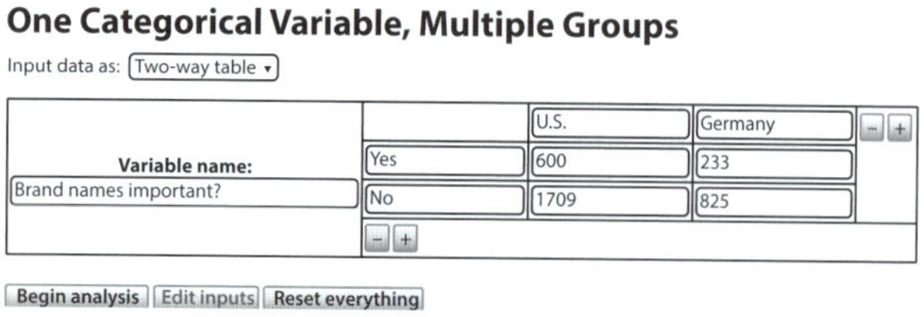

One Categorical Variable, Multiple Groups

Input data as: Two-way table

		U.S.	Germany
Variable name: Brand names important?	Yes	600	233
	No	1709	825

Begin analysis | Edit inputs | Reset everything

4. Click the Begin analysis button. Then scroll down to the "Perform Inference" section. Choose the 2-sample z interval $(p_1 - p_2)$ for the procedure, 95% for the confidence level, and Yes as the category to indicate as success. Then click the Perform inference button. *Note:* If the output is in percentages, change it to proportions by clicking on the preferences link at the bottom of the applet.

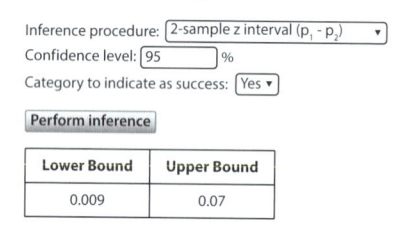

Perform Inference

Inference procedure: 2-sample z interval $(p_1 - p_2)$
Confidence level: 95 %
Category to indicate as success: Yes

Perform inference

Lower Bound	Upper Bound
0.009	0.07

Note: If you have raw data instead of summary statistics, choose "Raw data" from the Input menu and enter the data for each group. Then follow the directions in Step 4.

TI-83/84

1. Press STAT, then choose TESTS and 2-PropZInt.
2. When the 2-PropZInt screen appears, enter the values shown. Note that the values of x1, n1, x2, and n2 must all be integers!

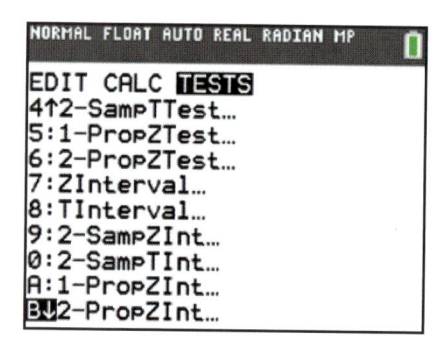

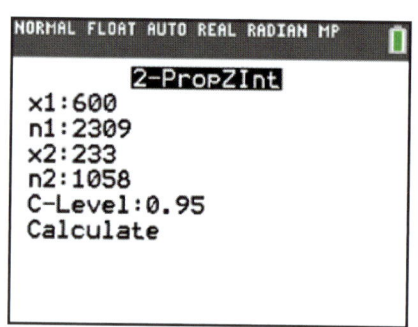

3. Highlight "Calculate" and press ENTER. The 95% confidence interval for $p_1 - p_2$ is reported, along with the sample proportions and the sample sizes.

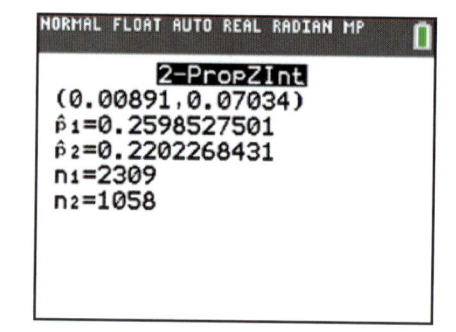

Detailed instructions for using CrunchIt!, Excel, Google Sheets, JMP, Minitab, and R are available in Achieve.

Section 9.1 What Did You Learn?

Review the learning goals from this section. Then practice what you've learned by working through the exercises.

Learning Goal	Example	Exercises
Check the Random and Large Counts conditions for constructing a confidence interval for a difference between two proportions.	9.1.1	5–8
Calculate a $C\%$ confidence interval for a difference between two proportions.	9.1.2	9–12
Use the four-step process to construct and interpret a confidence interval for the difference between two proportions.	9.1.3	13–16

Section 9.1 Exercises

Building Concepts and Skills These exercises assess the basic knowledge you should have after reading the section.

1. The shape of the sampling distribution of $\hat{p}_1 - \hat{p}_2$ is approximately normal if _____, _____, _____, and _____ are all at least 10.

2. What are the two conditions that need to be satisfied to construct a confidence interval for $p_1 - p_2$?

3. What is the formula for calculating a confidence interval for $p_1 - p_2$?

4. True/False: When a confidence interval for $p_1 - p_2$ contains 0, the interval provides convincing evidence that there is no difference in the true proportions.

Mastering Concepts and Skills These exercises reinforce the learning goals as shown in the examples.

5. **Refer to Example 9.1.1 Instagram Rising** Do people use Instagram more now than in years past? A 2021 Pew Research Center survey took a random sample of 1502 U.S. adults and found that 40% of the sample used Instagram. A similar survey in 2016 took a random sample of 1520 U.S. adults and found that 32% of U.S. adults used Instagram.[4] Let p_1 = the proportion of all U.S. adults who used Instagram in 2021 and p_2 = the proportion of all U.S. adults who used Instagram in 2016. Check if the conditions for calculating a confidence interval for $p_1 - p_2$ are met.

6. **Student Employment** A sample survey is given to independent random samples of 500 high school students and 550 college students. Researchers want to estimate the difference in the proportions of high school students and college students who are employed. In all, 20.4% of the high school students and 45.1% of the college students surveyed are employed.[5] Let p_1 = the proportion of all high school students who are employed and p_2 = the proportion of all college students who are employed. Check if the conditions for calculating a confidence interval for $p_1 - p_2$ are met.

7. **Birth Defects** The book and movie *A Civil Action* tells the story of a major legal battle that took place in the small town of Woburn, Massachusetts. A town well that supplied water to East Woburn residents was contaminated by industrial chemicals. During the period that residents drank water from this well, 16 of the 414 babies born had birth defects. On the west side of Woburn, 3 of the 228 babies born during the same time period had birth defects. Let p_1 = the proportion of all babies born with birth defects in West Woburn and p_2 = the proportion of all babies born with birth defects in East Woburn. Check if the conditions for calculating a confidence interval for $p_1 - p_2$ are met.

8. **Mice and Acorns** Mice populations in the wild rise and fall depending on the abundance of acorns, their favorite food. Experimenters studied two similar forest areas in a year when the acorn crop failed. They added hundreds of thousands of acorns to one area to imitate an abundant acorn crop, while leaving the other area untouched. The next spring, 54 of the 72 mice trapped in the first area were in breeding condition, versus 10 of the 17 mice trapped in the second area.[6] Let p_1 = the true proportion of mice like these in the acorn-supplemented area that are in breeding condition and p_2 = the true proportion of mice like these in the unsupplemented area that are in breeding condition. Check if the conditions for calculating a confidence interval for $p_1 - p_2$ are met.

9. **Refer to Example 9.1.2 Instagram Rising** Refer to Exercise 5.
   (a) Calculate a 90% confidence interval for the difference (2021 – 2016) in the proportions of all U.S. adults who use Instagram.
   (b) Based on the interval from part (a), is there convincing evidence of a change in the proportions of U.S. adults who use Instagram from 2016 to 2021? Explain your reasoning.

10. **Student Employment** Refer to Exercise 6.
    (a) Calculate a 99% confidence interval for the difference (High school – College) in the proportions of all students who are employed.
    (b) Based on the interval from part (a), is there convincing evidence of a difference in the proportions of high school students and college students who are employed? Explain your reasoning.

11. **Treating Prostate Cancer** In an experiment to compare treatments for prostate cancer, 731 men with localized prostate cancer were randomly assigned either to have surgery or to be observed only. After 20 years, 141 of the 364 men assigned to surgery were still alive, compared with 122 of the 367 men assigned to observation only.[7] Let p_S = the true proportion of men with localized prostate cancer like the ones in this study who survive for 20 years after surgery and p_O = the true proportion of men with localized prostate cancer like the ones in this study who survive for 20 years with observation only. Researchers want to construct a 95% confidence interval for $p_S - p_O$.
    (a) Show that the conditions for calculating the confidence interval are met.
    (b) Calculate the confidence interval.
    (c) Based on the interval from part (b), is there convincing evidence that surgery helps increase 20-year survival compared to observation only for men with localized prostate cancer like the ones in this study? Explain your reasoning.

12. **Aspirin and Dementia** In an experiment that investigated the effectiveness of low-dose aspirin in preventing dementia, researchers recruited 19,114 participants aged 70 or older who were free from cardiovascular disease, physical disability, and dementia. Participants were randomly assigned to take either low-dose aspirin

or a placebo daily. Of the 9525 participants who took low-dose aspirin, 283 developed dementia within 5 years. Of the 9589 participants who took placebo, 292 developed dementia within 5 years.[8] Let p_A = the true proportion of people like the ones in this study taking low-dose aspirin who develop dementia within 5 years and p_{PL} = the true proportion of people like the ones in this study taking placebo daily who develop dementia within 5 years. Researchers want to construct a 95% confidence interval for $p_A - p_{PL}$.

(a) Show that the conditions for calculating the confidence interval are met.

(b) Calculate the confidence interval.

(c) Based on the interval from part (b), is there convincing evidence that low-dose aspirin helps reduce the risk of developing dementia within 5 years compared to placebo for people like the ones in this study? Explain your reasoning.

13. **Refer to Example 9.1.3 Christmas Trees** An association of Christmas tree growers in Indiana wants to know if there is a difference in preference for natural trees between urban and rural households. To investigate, the association sponsored a survey of a random sample of Indiana households that had a Christmas tree last year. Of the 160 rural households surveyed, 64 had a natural tree. Of the 261 urban households surveyed, 89 had a natural tree.[9] Due to the sampling method used in this survey, it is reasonable to consider these as independent random samples of rural and urban households in Indiana. Construct and interpret a 95% confidence interval for the difference (Rural − Urban) in the proportions of all rural and urban Indiana households that had a natural tree last year.

14. **Partisan Politics** The Harvard Youth Poll asked a random sample of young adults (ages 18 to 29) in the United States, "Do you agree or disagree with this statement: Politics has become too partisan?" Of the 1083 respondents who identified as Democrats, 57% said "Agree." Of the 564 respondents who identified as Republicans, 59% said "Agree."[10] Due to the sampling method used in this survey, it is reasonable to consider these as independent random samples from the populations of 18- to 29-year-old Democrats and Republicans in the United States. Construct and interpret a 95% confidence interval for the difference (Democrat − Republican) in the proportions of all U.S. young adult Democrats and Republicans who would say that politics has become too partisan.

15. **I Want Candy!** In an experiment carried out at Cambridge University, researchers wanted to determine if moving candy closer to people in a waiting room would increase the proportion of subjects who ate the candy. They randomly assigned subjects to a waiting room with a bowl of candy placed near the seating location (20 centimeters, cm) or far from the seating location (70 cm). Of the 61 subjects assigned to sit near the bowl, 39 of them ate the candy, while only 24 of the 61 subjects assigned to sit far from the bowl ate the candy.[11] Construct and interpret a 90% confidence interval for the difference in the proportion of subjects like the ones in this study who would eat the candy when it is placed nearby and the proportion who would eat the candy when it is placed far away.

16. **Cockroaches** The pesticide diazinon is commonly used to treat infestations of the German cockroach, *Blattella germanica*. A study investigated the persistence of this pesticide on various types of surfaces. Researchers applied a 0.5% emulsion of diazinon to glass and plasterboard. After 14 days, they randomly assigned 72 cockroaches to two groups of 36, placed one group on each surface, and recorded the number that died within 48 hours. On the glass, 18 cockroaches died; on the plasterboard, 25 died.[12] Construct and interpret a 90% confidence interval for the difference in the proportions of cockroaches like the ones in this study that would die within 48 hours on glass treated with diazinon and on plasterboard treated with diazinon, respectively.

Applying the Concepts These exercises ask you to apply multiple learning goals in a new context or to apply what you learned in this section in a new way.

17. **New Heart Valve** For patients needing an aortic-valve replacement, there are two commonly used approaches—traditional surgery and a new method called transcatheter aortic-valve replacement (TAVR). Researchers wanted to know if the new TAVR method produces better outcomes, so they randomly assigned patients to the two approaches and then recorded the number of patients who had negative outcomes (death, stroke, or rehospitalization). Of the 454 patients assigned to undergo traditional surgery, 68 had negative outcomes. Of the 496 patients assigned to the TAVR method, only 42 had negative outcomes.[13]

(a) Construct and interpret a 99% confidence interval for the difference (Surgery − TAVR) in the true proportion of patients like the ones in this study who would have negative outcomes after undergoing traditional surgery and the true proportion who would have negative outcomes after undergoing TAVR.

(b) Based on your interval, is there convincing evidence that a smaller proportion of subjects like these will have negative outcomes with the new TAVR method? Explain your reasoning.

18. **Quitting Smoking** Nicotine patches are often used to help smokers quit. Does giving medicine to fight depression also help? A randomized double-blind experiment assigned 244 smokers to receive nicotine patches and another 245 smokers to receive both a patch and an antidepressant drug. A year later, 40 subjects in the nicotine patch group still abstained from smoking, as did 87 subjects in the patch-plus-drug group.[14]

(a) Construct and interpret a 99% confidence interval for the difference in the true proportion of smokers like the ones in this study who would abstain when using an antidepressant and a nicotine patch and the true proportion who would abstain when using only a patch.

(b) Based on your interval, is there convincing evidence that an antidepressant plus a nicotine patch is more effective than the patch alone in helping subjects like these abstain from smoking? Explain your reasoning.

19. **Christmas Trees** Refer to Exercise 13. A 95% confidence interval for the difference (Rural − Urban) in the population proportions is −0.036 to 0.154.
   (a) Interpret the confidence level.
   (b) Does the confidence interval provide convincing evidence that the two population proportions are different? Explain your reasoning.
   (c) Does the confidence interval provide convincing evidence that the two population proportions are equal? Explain your reasoning.

20. **Partisan Politics** Refer to Exercise 14. A 95% confidence interval for the difference (Democrat − Republican) in the population proportions is −0.071 to 0.029.
   (a) Interpret the confidence level.
   (b) Does the confidence interval provide convincing evidence that the two population proportions are different? Explain your reasoning.
   (c) Does the confidence interval provide convincing evidence that the two population proportions are equal? Explain your reasoning.

21. **Tech City Trees** Does a tree's location—on the curb or offset from the curb—help predict whether there is damaged sidewalk next to the tree (perhaps due to the growth of the tree and its root system)? The data set **NYCTrees** contains detailed information about a random sample of 8253 trees on New York City streets. The location of each tree (*curb_loc*) and the corresponding sidewalk status (*sidewalk*) were recorded.[15] Calculate and interpret a 90% confidence interval for the difference (On curb − Offset) in the proportions of all live trees on the curb and offset from the curb that are next to damaged sidewalk in New York City. *Note:* The Random condition is met when we take a single random sample and split it into two groups based on a categorical variable.

22. **Tech Cautious Squirrels** Do adult and juvenile Eastern gray squirrels in New York's Central Park exhibit different behavior toward humans? The **Squirrels** data set includes data on 3019 squirrel sightings in the park. For 2868 of the sightings, each squirrel's *age*—adult or juvenile—and whether the squirrel approached humans (*approach*) were recorded.[16] Calculate and interpret a 90% confidence interval for the difference in the proportions of all adult squirrels and all juvenile squirrels in Central Park that would approach humans. Assume that the Random condition is met.

Extending the Concepts These exercises challenge you to explore statistical concepts and methods that go beyond what you learned in this section.

23. **Jelly Beans** A candy maker offers child- and adult-size bags of jelly beans containing different color mixes. The company claims that the child mix has 30% red jelly beans, while the adult mix contains 10% red jelly beans. Assume that the candy maker's claim is true. Suppose we take a random sample of 50 jelly beans from the child mix and a separate random sample of 100 jelly beans from the adult mix. Let $\hat{p}_C$ and $\hat{p}_A$ be the sample proportions of red jelly beans from the child and adult mixes, respectively.
   (a) What is the shape of the sampling distribution of $\hat{p}_C - \hat{p}_A$? Why?
   (b) Find the mean of the sampling distribution.
   (c) Calculate the standard deviation of the sampling distribution.
   (d) Find the probability that the proportion of red jelly beans is greater in the sample from the child mix than in the sample from the adult mix.

24. **SE Versus ME for Proportions** In a random sample of 125 workers from Company 1, 35 admitted to using sick leave when they weren't really ill. In a separate random sample of 68 workers from Company 2, 17 admitted that they had used sick leave when they weren't ill. The difference in the sample proportions of workers from the two companies who admitted to using sick leave when they weren't ill is $\hat{p}_1 - \hat{p}_2 = 0.28 - 0.25 = 0.03$.
   (a) The standard error of $\hat{p}_1 - \hat{p}_2$ is 0.066. Interpret this value.
   (b) The margin of error for a 95% confidence interval for $p_1 - p_2$ is 0.13. Interpret this value.

Cumulative Review These exercises revisit what you learned in previous sections.

25. **Acupuncture and Pregnancy (1.5)** A study reported in the medical journal *Fertility and Sterility* sought to determine whether the ancient Chinese art of acupuncture could help infertile women become pregnant.[17] A total of 160 healthy women who planned to have in vitro fertilization (IVF) were recruited for the study. Half of the subjects (80) were randomly assigned to receive acupuncture 25 minutes before implanting the embryo and again 25 minutes after the implant. The remaining 80 women were assigned to a control group and instructed to lie still for 25 minutes after the embryo transfer. Here are the results: 34 of the 80 women in the acupuncture group got pregnant, compared to 21 out of 80 in the control group.
   (a) Explain the purpose of the control group in this experiment.
   (b) Describe how researchers could have randomly assigned the women to the two groups.
   (c) What is the purpose of random assignment in this experiment?

26. **Acupuncture and Pregnancy (1.6, 1.8, 8.2)** Refer to Exercise 25.
   (a) The results of the study were statistically significant. Explain what "statistically significant" means in the context of this study.
   (b) Which type of error might have occurred in this study, a Type I error or a Type II error? Justify your answer.
   (c) Based on the results of this study, would it be reasonable to say that acupuncture caused an increase in the pregnancy rate for women like the ones in this study? Explain your answer.

Section 9.2 Significance Tests for a Difference Between Two Population Proportions

LEARNING GOALS

By the end of this section, you will be able to:

- State hypotheses and check conditions for performing a significance test about a difference between two proportions.
- Calculate the standardized test statistic and *P*-value for a significance test about a difference between two proportions.
- Use the four-step process to perform a significance test about a difference between two proportions.

An observed difference between two sample proportions can reflect an actual difference in the population proportions, or it may just be due to chance variation in random sampling or random assignment. In Section 1.7, we used simulation to determine which explanation makes more sense in a randomized experiment. This section shows you how to perform a significance test about a difference between two proportions.

Stating Hypotheses and Checking Conditions for a Test About $p_1 - p_2$

In a test for comparing two proportions, the null hypothesis has the general form

$$H_0: p_1 - p_2 = \text{hypothesized value}$$

We'll focus on situations in which the hypothesized difference is 0. Then the null hypothesis says that there is no difference between the two population proportions:

$$H_0: p_1 - p_2 = 0$$

The null hypothesis can also be written in the equivalent form $H_0: p_1 = p_2$. The alternative hypothesis says what kind of difference we expect.

The conditions for performing a significance test about $p_1 - p_2$ are the same as those for constructing a confidence interval for a difference in proportions that you learned in Section 9.1.

> **HOW TO** Check Conditions for Performing a Significance Test About $p_1 - p_2$
>
> To perform a test about a difference in population proportions $p_1 - p_2$, check that the following conditions are met:
>
> - **Random:** The data come from independent random samples from the two populations of interest or from two groups in a randomized experiment.
> - **Large Counts:** The counts of "successes" and "failures" in each sample or group — $n_1 \hat{p}_1$, $n_1(1-\hat{p}_1)$, $n_2 \hat{p}_2$, and $n_2(1-\hat{p}_2)$ — are all at least 10.

EXAMPLE 9.2.1

Misperceptions about the world

Stating hypotheses and checking conditions

PROBLEM

Where does the majority of the world's population live: in low-income countries, middle-income countries, or high-income countries? When a random sample of 500 South Korean adults was asked this question in the Gapminder Misconception Study, only 39% got it right — the majority of the world's population lives in middle-income countries. When a random sample of 1000 U.S. adults was asked this question, only 36% answered correctly.[18] Do these data give convincing evidence of a difference in the population proportions who would answer correctly in these two countries?

(a) State appropriate hypotheses for performing a significance test. Be sure to define the parameters of interest.
(b) Do the data provide *some* evidence of a difference in the population proportions? Justify your answer.
(c) Check that the conditions for performing the test are met.

SOLUTION

(a) $H_0: p_1 - p_2 = 0$

$H_a: p_1 - p_2 \neq 0$

where p_1 = the proportion of all South Korean adults who would answer the question correctly and p_2 = the proportion of all U.S. adults who would answer the question correctly.

> H_a is two-sided because we're asked if there is convincing evidence of a *difference* in the population proportions (meaning that $p_1 - p_2$ could be either less than or greater than 0).

(b) Yes, because $\hat{p}_1 - \hat{p}_2 = 0.39 - 0.36 = 0.03 \neq 0$.

(c) Random? The data came from independent random samples of adults from the two countries. ✓

Large Counts? $500(0.39) = 195$, $500(0.61) = 305$, $1000(0.36) = 360$, $1000(0.64) = 640$ are all at least 10. ✓

EXAM PREP: FOR PRACTICE, TRY EXERCISE 5.

Calculations: Standardized Test Statistic and *P*-Value

If the conditions are met, we can proceed with calculations. To do a test of $H_0: p_1 - p_2 = 0$, standardize $\hat{p}_1 - \hat{p}_2$ to get a z statistic:

$$\text{standardized test statistic} = \frac{\text{statistic} - \text{null value}}{\text{standard deviation (error) of statistic}}$$

$$z = \frac{(\hat{p}_1 - \hat{p}_2) - 0}{\sqrt{\dfrac{p_1(1-p_1)}{n_1} + \dfrac{p_2(1-p_2)}{n_2}}}$$

Unfortunately, we don't know the value of p_1 or p_2. You might be tempted to replace p_1 and p_2 in the denominator with the corresponding sample proportions. Don't do it! Here's why.

A significance test begins by assuming that $H_0: p_1 - p_2 = 0$ is true. In that case, $p_1 = p_2$. This means we should not substitute different values for p_1 and p_2. Instead, we should use a common value, which we call p_C. To estimate p_C, we combine (or "pool") the data from the two samples as if they came from one larger sample. This *combined sample proportion* is

$$\hat{p}_C = \frac{\text{number of successes in both samples combined}}{\text{number of individuals in both samples combined}} = \frac{X_1 + X_2}{n_1 + n_2}$$

In other words, $\hat{p}_C$ gives the overall proportion of successes in the combined samples. Use $\hat{p}_C$ in place of both p_1 and p_2 in the denominator of the standardized test statistic:

$$z = \frac{(\hat{p}_1 - \hat{p}_2) - 0}{\sqrt{\dfrac{\hat{p}_C(1-\hat{p}_C)}{n_1} + \dfrac{\hat{p}_C(1-\hat{p}_C)}{n_2}}}$$

When the Large Counts condition is met, this z statistic will have approximately the standard normal distribution. We can find the appropriate *P*-value using either technology or Table A.

EXAMPLE 9.2.2

Misperceptions about the world

Calculating the standardized test statistic and P-value

PROBLEM

Refer to Example 9.2.1. The two-way table summarizes the data from the Gapminder Misconception Study, in which independent random samples of South Korean and U.S. adults were asked, "Where does the majority of the world's population live: in low-income countries, middle-income countries, or high-income countries?

		Country		
		South Korea	United States	Total
Answered correctly?	Yes	195	360	555
	No	305	640	945
	Total	500	1000	1500

(a) Calculate the standardized test statistic.
(b) Find the *P*-value.

SOLUTION

(a) $\hat{p}_1 = \dfrac{195}{500} = 0.39$, $\hat{p}_2 = \dfrac{360}{1000} = 0.36$

$\hat{p}_C = \dfrac{195 + 360}{500 + 1000} = \dfrac{555}{1500} = 0.37$

$z = \dfrac{(0.39 - 0.36) - 0}{\sqrt{\dfrac{0.37(0.63)}{500} + \dfrac{0.37(0.63)}{1000}}} = 1.134$

> First calculate the sample proportions and the combined sample proportion. The "Total" column in the two-way table makes it easy to see that the overall (combined) proportion of successes in the two samples is $\hat{p}_C = 555/1500$.
>
> $$z = \dfrac{(\hat{p}_1 - \hat{p}_2) - 0}{\sqrt{\dfrac{\hat{p}_C(1 - \hat{p}_C)}{n_1} + \dfrac{\hat{p}_C(1 - \hat{p}_C)}{n_2}}}$$

(b) *P*-value:

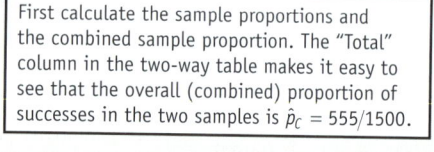

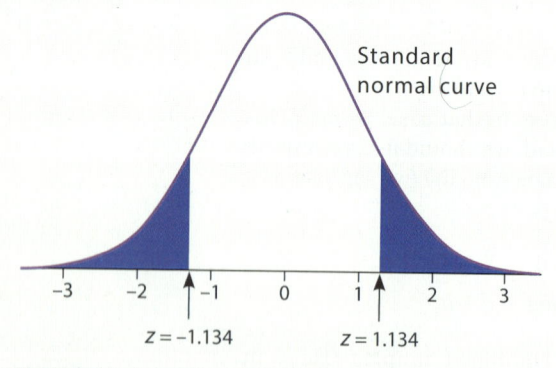

[Standard normal curve with shaded areas below $z = -1.134$ and above $z = 1.134$]

Using Table A: *P*-value = 2(0.1292) = 0.2584
Using technology: Applet/normalcdf(lower: 1.134, upper: 1000, mean: 0, SD: 1) × 2 = 0.2568

EXAM PREP: FOR PRACTICE, TRY EXERCISE 9.

What does the *P*-value in Example 9.2.2 tell us? If there is no difference in the population proportions of South Korean and U.S. adults who would answer the question correctly, and we repeated the random sampling process many times, we would get a difference in sample proportions as large as or larger than 0.03 in either direction about 26% of the time. Because the *P*-value of 0.26 is greater than our default significance level of $\alpha = 0.05$, we fail to reject H_0. We don't have convincing evidence of a difference in the proportion of all South Korean adults and the proportion of all U.S. adults who would correctly answer the question about where the majority of the world's population lives.

Section 9.2 Significance Tests for a Difference Between Two Population Proportions

We can get additional information about the difference between the population proportions by constructing a confidence interval. Technology gives the 95% confidence interval for $p_1 - p_2$ as -0.022 to 0.082. That is, we are 95% confident that the interval from -0.022 to 0.082 captures the difference (South Korean – U.S.) in the proportions of all adults in the two countries who would correctly answer the question about where the majority of the world's population lives. The interval suggests that the percentage of all South Korean adults who would answer the question correctly is between 2.2 percentage points lower and 8.2 percentage points higher than the percentage of all U.S. adults who would answer the question correctly. This is consistent with our "fail to reject H_0" conclusion, because 0 is included in the interval of plausible values for $p_1 - p_2$.

Putting It All Together: Significance Test About $p_1 - p_2$

We are now ready to use the four-step process to carry out a **two-sample z test for a difference between two proportions.** You can use technology to perform the calculations in the Do step. See the Tech Corner at the end of the section for details.

> A **two-sample z test for a difference between two proportions** is a significance test of the null hypothesis that the difference in the proportions of successes for two populations or treatments is equal to 0.

EXAMPLE 9.2.3

Does a cholesterol-reducing drug lower the risk of heart attacks?

Significance test for $p_1 - p_2$

PROBLEM

The Helsinki Heart Study recruited middle-aged men with high cholesterol levels but no history of other serious medical problems to investigate whether a cholesterol-reducing drug could lower the risk of heart attacks. The volunteer subjects were assigned at random to one of two treatments: 2051 men took the drug gemfibrozil to reduce their cholesterol levels, and a control group of 2030 men took a placebo. During the next 5 years, 56 men in the gemfibrozil group and 84 men in the placebo group had heart attacks.[19] Do the results of this study give convincing evidence at the 1% significance level that gemfibrozil is effective in preventing heart attacks?

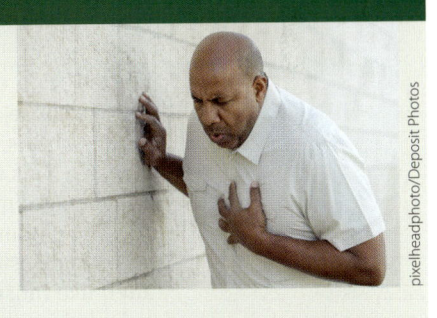

SOLUTION

State:

$H_0: p_G - p_{PL} = 0$

$H_a: p_G - p_{PL} < 0$

where p_G = the true heart attack rate for middle-aged men like the ones in this study who take gemfibrozil and p_{PL} = the true heart attack rate for middle-aged men like the ones in this study who take a placebo.

Use $\alpha = 0.01$.

The evidence for H_a is: $\hat{p}_G - \hat{p}_{PL} = \dfrac{56}{2051} - \dfrac{84}{2030} = 0.0273 - 0.0414 = -0.0141 < 0$

> **State:** State the hypotheses, parameter(s), significance level, and evidence for H_a.

Plan: Two-sample z test for $p_G - p_{PL}$

Random? Volunteer subjects were randomly assigned to receive gemfibrozil or a placebo. ✓

Large Counts? 56, 2051 − 56 = 1995, 84, 2030 − 84 = 1946 are all at least 10. ✓

> **Plan:** Identify the appropriate inference method and check the conditions.

Do: $\hat{p}_G = 0.0273$, $\hat{p}_{PL} = 0.0414$, $\hat{p}_C = \dfrac{56 + 84}{2051 + 2030} = \dfrac{140}{4081} = 0.0343$

- $z = \dfrac{(0.0273 - 0.0414) - 0}{\sqrt{\dfrac{0.0343(0.9657)}{2051} + \dfrac{0.0343(0.9657)}{2030}}} = -2.475$

> **Do:** If the conditions are met, perform calculations.
> - Calculate the test statistic.
> - Find the P-value.

(continued)

- P-value:

Standard normal curve

z = −2.475

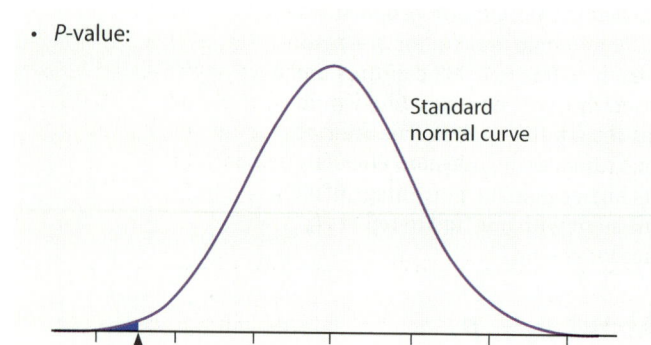

Using Table A: P-value = 0.0066

Using technology: z = −2.470, P-value = 0.0068

Conclude: Because the P-value of 0.0068 is less than $\alpha = 0.01$, we reject H_0. There is convincing evidence of a lower heart attack rate for middle-aged men like the ones in this study who take gemfibrozil than for those who take only a placebo.

Conclude: Make a conclusion about the hypotheses in the context of the problem.

EXAM PREP: FOR PRACTICE, TRY EXERCISE 13.

We chose a 1% significance level in Example 9.2.3 to reduce the chance of making a Type I error—that is, finding convincing evidence that gemfibrozil reduces heart attack risk when it really doesn't. This error could have serious consequences if an ineffective drug was given to lots of middle-aged men with high cholesterol levels!

The random assignment in the Helsinki Heart Study allowed researchers to draw a cause-and-effect conclusion. They could say that gemfibrozil reduces the rate of heart attacks for middle-aged men like those who took part in the experiment. But because the subjects were not randomly selected from a larger population, researchers could not generalize the findings of this study any further. No conclusions could be drawn about the effectiveness of gemfibrozil in preventing heart attacks in all middle-aged men, in men of other ages, or in women.

THINK ABOUT IT

Why do the inference methods for random sampling work for randomized experiments? Confidence intervals and tests for $p_1 - p_2$ are based on the sampling distribution of $\hat{p}_1 - \hat{p}_2$. But in most experiments, researchers don't select subjects at random from any larger populations. They do randomly assign subjects to treatments. As we did in Section 1.7, we can use simulation to see what would happen if the random assignment were repeated many times, assuming no treatment effect—that is, assuming that $H_0: p_1 - p_2 = 0$ is true.

We used the *One Categorical Variable, Multiple Groups* applet to randomly reassign the 4081 subjects in the Helsinki Heart Study to the two groups 500 times, assuming the drug received (gemfibrozil or placebo) doesn't affect whether or not each individual has a heart attack. Figure 9.1 shows the value of $\hat{p}_G - \hat{p}_{PL}$ in the 500 simulated trials. This distribution (sometimes referred to as a *randomization distribution* of $\hat{p}_G - \hat{p}_{PL}$) has an approximately normal shape with mean 0 and standard deviation 0.0058. This matches well with the distribution we used to perform calculations in Example 9.2.3.

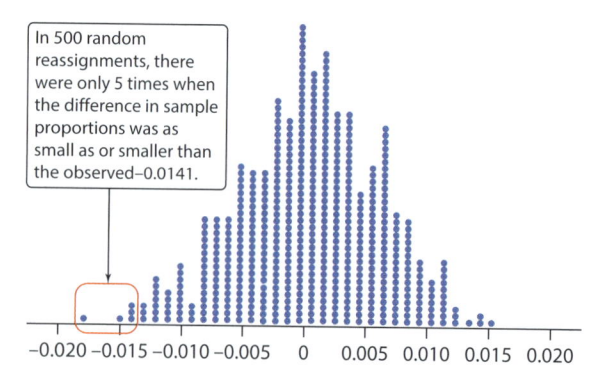

In 500 random reassignments, there were only 5 times when the difference in sample proportions was as small as or smaller than the observed −0.0141.

Simulated difference in sample proportions of subjects who have heart attacks ($\hat{p}_G - \hat{p}_{PL}$)

FIGURE 9.1 Dotplot of the values of $\hat{p}_G - \hat{p}_{PL}$ from each of 500 simulated random reassignments of subjects to treatment groups in the Helsinki Heart Study, assuming the null hypothesis of no treatment effect is true.

Section 9.2 Significance Tests for a Difference Between Two Population Proportions

In the Helsinki Heart Study, the difference in the proportions of subjects who had a heart attack in the gemfibrozil and placebo groups was $0.0273 - 0.0414 = -0.0141$. How likely is it that a difference this large or larger would happen purely by chance when H_0 is true? Figure 9.1 provides a rough answer: 5 of the 500 random reassignments yielded a difference in proportions less than or equal to -0.0141. That is, our estimate of the P-value is 0.01. This is quite close to the 0.0068 P-value that we calculated in Example 9.2.3, suggesting that it's OK to use inference methods for random sampling to analyze randomized experiments. We will discuss *randomization tests* like this in more detail in Section 13.5.

TECH CORNER Significance Tests for a Difference Between Two Proportions

You can use technology to perform the calculations for a significance test of $H_0: p_1 - p_2 = 0$. We'll illustrate this process using the Gapminder Misconception Study data from Examples 9.2.1 and 9.2.2. Recall that of the 500 South Korean adults surveyed, $500(0.39) = 195$ correctly answered the question about where the majority of the world's population lives. Of the 1000 U.S. adults surveyed, $1000(0.36) = 360$ correctly answered the question.

Applet

1. Go to www.stapplet.com and launch the *One Categorical Variable, Multiple Groups* applet.

2. Choose to input data as "Two-way table." Enter "Answered correctly?" as the variable name. Type the group names in the boxes at the top of the columns and the category names in the boxes at the left of the rows, as shown.

3. Input the number of successes and failures for each group, as shown.

One Categorical Variable, Multiple Groups

Input data as: Two-way table

	South Korea	U.S.
Yes	195	360
No	305	640

Variable name: Answered correctly?

4. Click the Begin analysis button. Then scroll down to the "Perform Inference" section. Choose the 2-sample z test $(p_1 - p_2)$ for the procedure, $p_1 - p_2 \neq 0$ for the alternative hypothesis, and "Yes" as the category to indicate as success. Then, click the Perform inference button. *Note:* If the output is shown as percentages, change to proportions by clicking on the preferences link at the bottom of the applet.

Perform Inference

Inference procedure: 2-sample z test $(p_1 - p_2)$
Alternative hypothesis: $p_1 - p_2 \neq 0$
Category to indicate as success: Yes

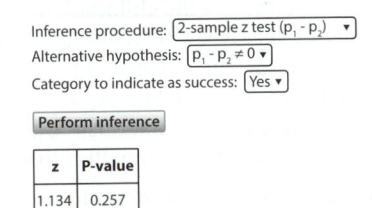

z	P-value
1.134	0.257

Note: If you have raw data instead of summary statistics, choose "Raw data" from the Input menu and enter the data for each group. Then follow the directions in Step 4.

TI-83/84

1. Press STAT, then choose TESTS and 2-PropZTest.

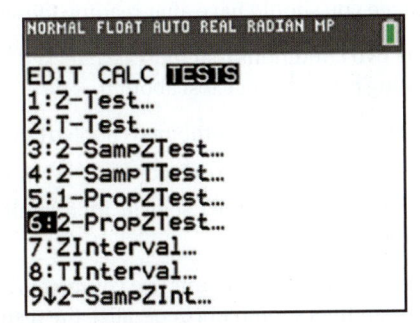

```
NORMAL FLOAT AUTO REAL RADIAN MP

EDIT CALC TESTS
1:Z-Test…
2:T-Test…
3:2-SampZTest…
4:2-SampTTest…
5:1-PropZTest…
6:2-PropZTest…
7:ZInterval…
8:TInterval…
9↓2-SampZInt…
```

2. When the 2-PropZTest screen appears, enter the values shown. Note that the values of x1, n1, x2, and n2 must all be integers!

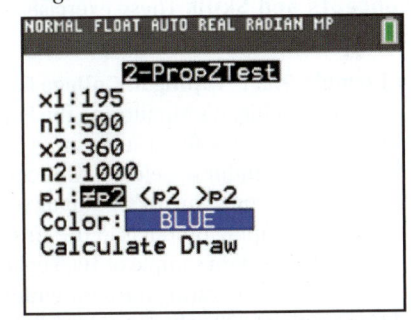

```
NORMAL FLOAT AUTO REAL RADIAN MP

2-PropZTest
x1:195
n1:500
x2:360
n2:1000
p1:≠p2  <p2  >p2
Color: BLUE
Calculate  Draw
```

3. Highlight "Calculate" and press ENTER.

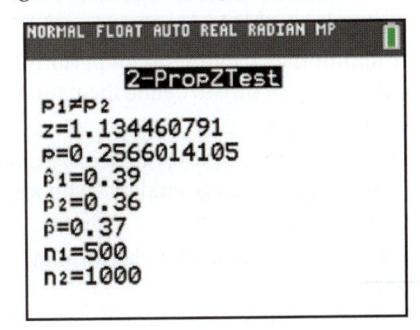

```
NORMAL FLOAT AUTO REAL RADIAN MP

2-PropZTest
p1≠p2
z=1.134460791
p=0.2566014105
p̂1=0.39
p̂2=0.36
p̂=0.37
n1=500
n2=1000
```

Detailed instructions for using CrunchIt!, Excel, Google Sheets, JMP, Minitab, and R are available in Achieve.

Section 9.2 What Did You Learn?

Review the learning goals from this section. Then practice what you've learned by working through the exercises.

Learning Goal	Example	Exercises
State hypotheses and check conditions for performing a significance test about a difference between two proportions.	9.2.1	5–8
Calculate the standardized test statistic and P-value for a significance test about a difference between two proportions.	9.2.2	9–12
Use the four-step process to perform a significance test about a difference between two proportions.	9.2.3	13–16

Section 9.2 Exercises

Building Concepts and Skills These exercises assess the basic knowledge you should have after reading the section.

1. State the two conditions that must be checked before performing a significance test about $p_1 - p_2$.

2. Give the formula for the combined (pooled) sample proportion.

3. What is the formula for the z statistic for a significance test about $p_1 - p_2$?

4. True/False: When experiments use volunteers as subjects, researchers should not generalize the results of an experiment beyond individuals like the ones in the experiment.

Mastering Concepts and Skills These exercises reinforce the learning goals as shown in the examples.

5. **Refer to Example 9.2.1 Vaping in College** Each year, the University of Michigan's Monitoring the Future study administers a survey about substance use and related behaviors to randomly selected U.S. college students. One survey question asks students whether they have used any vaping products in the current year. In 2019, 43.7% of a random sample of 1005 college students said "Yes." In 2020 (during the pandemic), 39.6% of a random sample of 800 college students said "Yes."[20] Researchers want to know if college students' vaping habits changed over this 1-year period.
    (a) State appropriate hypotheses for performing a significance test. Be sure to define the parameters of interest.
    (b) Do the data provide *some* evidence of a change in the proportion of college students who were vaping from 2019 to 2020? Justify your answer.
    (c) Check that the conditions for performing the test are met.

6. **Milk or Juice?** Companies that market products targeted at children are naturally interested in children's decision-making behaviors. As part of one study, researchers randomly selected children in different age groups and compared their ability to sort new products into correct product categories (in this case, milk or juice). The table summarizes the data.[21]

Age group	n	Number who sorted correctly
4- to 5-year-olds	50	10
6- to 7-year-olds	53	28

Researchers want to know if a greater proportion of 6- to 7-year-olds than 4- to 5-year-olds can sort correctly.
    (a) State appropriate hypotheses for performing a significance test. Be sure to define the parameters of interest.
    (b) Do the data provide *some* evidence that a greater proportion of 6- to 7-year-olds than 4- to 5-year-olds can sort correctly? Justify your answer.
    (c) Check that the conditions for performing the test are met.

7. **Preschool Effects** Researchers recruited 123 children from low-income families in Michigan who had never attended preschool. They randomly assigned 62 of the children to attend preschool (paid for by the study budget) and the other 61 to serve as a control group who would not attend preschool. One response variable of interest was the children's need for social services as adults. Over a 10-year period, 38 children in the preschool group and 49 children in the control group needed social services.[22] Researchers want to know if attending preschool reduces the need for social services as an adult for children like these.
    (a) State appropriate hypotheses for performing a significance test. Be sure to define the parameters of interest.
    (b) Do the data provide *some* evidence that preschool reduces the need for social services as an adult for children like these? Justify your answer.
    (c) Check that the conditions for performing the test are met.

8. **Rocket Fish!** French researchers set an ambitious goal to farm fish on the moon. They performed an experiment to determine whether certain varieties of fish can be safely transported on a rocket into space. They randomly assigned 400 European seabass eggs into two groups of 200 eggs each. All of the eggs in both groups were placed in a dish filled with seawater. The first group of eggs was then placed in a vibration chamber designed to simulate a typical takeoff, while the second group of eggs was kept in similar environmental conditions with no vibrations. In the vibrations group, 76% of the eggs went on to hatch, compared to 82% of the eggs in the control group.[23] Researchers want to know if the vibrations from a rocket launch will affect the proportion of European seabass eggs that hatch.
   (a) State appropriate hypotheses for performing a significance test. Be sure to define the parameters of interest.
   (b) Do the data provide *some* evidence that vibrations affect the proportion of European seabass eggs like the ones in this study that hatch? Justify your answer.
   (c) Check that the conditions for performing the test are met.

9. **Refer to Example 9.2.2 Vaping in College** Refer to Exercise 5.
   (a) Calculate the standardized test statistic.
   (b) Find the *P*-value.

10. **Milk or Juice?** Refer to Exercise 6.
    (a) Calculate the standardized test statistic.
    (b) Find the *P*-value.

11. **Preschool Effects** Refer to Exercise 7.
    (a) Calculate the standardized test statistic and the *P*-value.
    (b) What conclusion would you make?

12. **Rocket Fish!** Refer to Exercise 8.
    (a) Calculate the standardized test statistic and the *P*-value.
    (b) What conclusion would you make?

13. **Refer to Example 9.2.3 Low-Birth-Weight Babies** Babies born weighing less than 1500 grams (about 3.3 pounds) are classified as very low birth weight (VLBW). A long-term study followed 242 randomly selected VLBW babies to age 20 years, along with a group of 233 randomly selected babies from the same population who had normal birth weight. At age 20 years, 179 of the VLBW group and 193 of the normal-birth-weight group had graduated from high school.[24] Do these data provide convincing evidence at the 1% significance level that the graduation rate among VLBW babies is less than that for normal-birth-weight babies?

14. **Financial Incentives and Smoking** In an effort to reduce health care costs, General Motors sponsored a study to help its employees stop smoking. In the study, half of the subjects (439) were randomly assigned to receive $750 if they agreed to quit smoking for a year, while the other half (439) were simply encouraged to use traditional methods to stop smoking. None of the 878 volunteers knew that there was a financial incentive when they first signed up. At the end of one year, 15% of those in the financial rewards group had quit smoking, while only 5% in the traditional group had quit smoking.[25] Do the results of this study give convincing evidence at the $\alpha = 0.01$ significance level that a financial incentive helps employees like the ones in this study stop smoking?

15. **Homeless Care** A researcher asked 80 randomly selected people if free health care should be provided to people experiencing homelessness. Half of the respondents, determined at random, were shown a picture of a homeless woman with a small child. When shown the picture, 67.5% agreed that free health care should be provided to the homeless. When the picture was not shown, only 45% agreed with this statement.[26] Do these results give convincing evidence that showing the picture affects people's opinions about free health care for those experiencing homelessness?

16. **Risky Behavior** Each year, the Youth Risk Behavior Surveillance Survey (YRBSS) administers a survey to randomly selected U.S. high school students in each state. One survey question asks if, within the past 30 days, the student has ridden with a driver who had been drinking alcohol. Among students who completed the YRBSS in a recent year, 351 of 1909 randomly selected Arizona students said "Yes," compared to 275 of 1329 randomly selected California students. Do these data provide convincing evidence of a difference in the population proportions for these two states?

Applying the Concepts These exercises ask you to apply multiple learning goals in a new context or to apply what you learned in this section in a new way.

17. **Peanut Allergies** A recent study of peanut allergies explored the relationship between early exposure to peanuts and the subsequent development of an allergy to peanuts. Infants (4 to 11 months old) who had shown evidence of other kinds of allergies were randomly assigned to one of two groups: Group 1 consumed a baby-food form of peanut butter, and Group 2 avoided peanut butter. At 5 years old, 10 of 307 children in the peanut-consumption group were allergic to peanuts, and 55 of 321 children in the peanut-avoidance group were allergic to peanuts.[27]
    (a) Does this study provide convincing evidence at the $\alpha = 0.05$ significance level of a difference in the development of peanut allergies in infants like the ones in this study who consume or avoid peanut butter?
    (b) Based on your conclusion in part (a), which mistake—a Type I error or a Type II error—could you have made? Explain your answer.
    (c) Should you generalize the result in part (a) to all infants? Why or why not?

18. **Controlling Cholesterol** Which of two popular drugs — Lipitor or Pravachol — is more effective at helping lower "bad cholesterol"? Researchers designed an experiment to find out. The experiment included about 4000 people with heart disease as subjects. These volunteers were randomly assigned to one of two treatment groups: Lipitor or Pravachol. At the end of the study, researchers compared the proportion of subjects in each group who died, had a heart attack, or suffered other serious consequences within two years. For the 2063 subjects using Pravachol, the proportion was 0.263. For the 2099 subjects using Lipitor, the proportion was 0.224.[28]

   (a) Does this study provide convincing evidence at the $\alpha = 0.05$ significance level of a difference in the effectiveness of Lipitor and Pravachol for subjects like the ones in this study?

   (b) Based on your conclusion in part (a), which mistake — a Type I error or a Type II error — could you have made? Explain your answer.

   (c) Should you generalize the result in part (a) to all people with heart disease? Why or why not?

19. **Peanut Allergies** Refer to Exercise 17.

   (a) Construct and interpret a 95% confidence interval for the difference between the true proportions. If you already defined parameters and checked conditions in Exercise 17, you don't need to do those tasks again here.

   (b) Explain how the confidence interval provides more information than did the test in Exercise 17.

20. **Controlling Cholesterol** Refer to Exercise 18.

   (a) Construct and interpret a 95% confidence interval for the difference between the true proportions. If you already defined parameters and checked conditions in Exercise 18, you don't need to do those tasks again here.

   (b) Explain how the confidence interval provides more information than did the test in Exercise 18.

21. **Vaping in College** Refer to Exercise 9.

   (a) Interpret the *P*-value.

   (b) What conclusion would you make at the 1% significance level?

22. **Milk or Juice?** Refer to Exercise 10.

   (a) Interpret the *P*-value.

   (b) What conclusion would you make at the 1% significance level?

Extending the Concepts These exercises challenge you to explore statistical concepts and methods that go beyond what you learned in this section.

23. **Texting and Driving** Does providing additional information affect responses to a survey question? Researchers decided to investigate this issue by asking different versions of a question about texting and driving. They randomly divided 50 mall shoppers into two groups of 25. The first group was asked Version A and the second group was asked Version B. Here are the actual questions:

   - **Version A:** A lot of people text and drive. Are you one of them?

   - **Version B:** About 6000 deaths occur per year due to texting and driving. Knowing the potential consequences, do you text and drive?

   Of the 25 shoppers assigned to Version A, 16 admitted to texting and driving. Of the 25 shoppers assigned to Version B, only 12 admitted to texting and driving.

   (a) State appropriate hypotheses for performing a significance test. Be sure to define the parameters of interest.

   (b) Explain why you should not use the methods introduced in Section 9.2 to calculate the *P*-value.

   (c) We performed 100 repetitions of a simulation to see what differences in proportions (Version A – Version B) would occur due only to chance variation in the random assignment, assuming that the question asked doesn't matter. A dotplot of the results is shown here. What is the estimated *P*-value?

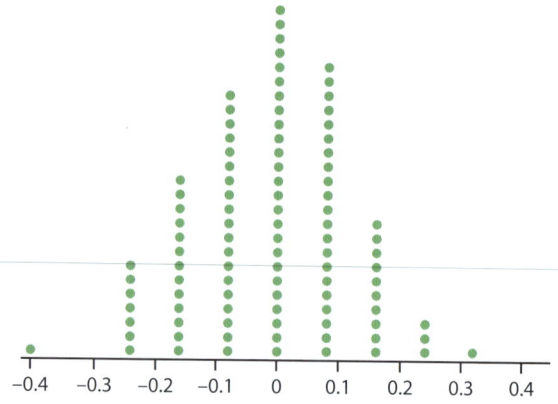

   (d) What conclusion would you make?

24. **Rejection Region** Refer to the Think About It in Section 8.3 and to Exercise 22.

   (a) Find the critical value and the rejection region.

   (b) What conclusion would you make? Explain your reasoning.

Cumulative Review These exercises revisit what you learned in previous sections.

25. **Car of the Year (2.1)** Every year, *Motor Trend* magazine gives a car manufacturer the coveted "Car of the Year" award.

   (a) The pie chart provides the distribution of winning manufacturer over a 70 year period.[29] In about what percentage of years did General Motors win this award? Justify your answer.

Car of the year

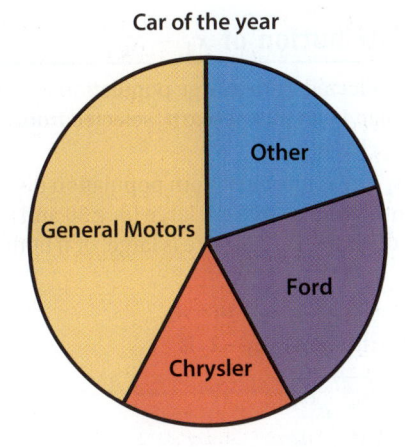

(b) The pictograph shows the same data about *Motor Trend's* Car of the Year as in part (a). Explain how this graph is misleading.

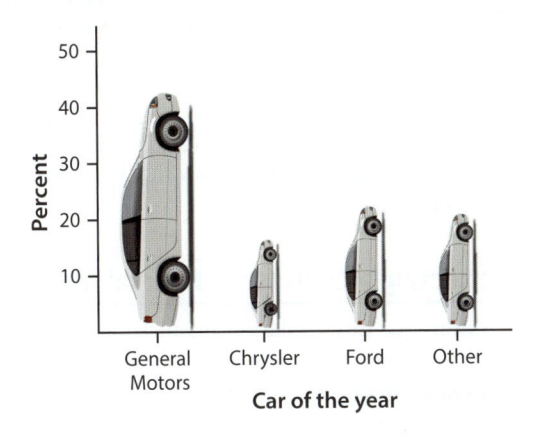

26. Watching TV (5.1, 5.2, 6.8) Choose a young person (ages 19 to 25) at random and ask, "In the past seven days, how many days did you watch television?" Call the response X for short. Here is the probability distribution of X.

Days	0	1	2	3	4	5	6	7
Probability	0.04	0.03	0.06	0.08	0.09	0.08	0.05	???

(a) What is the probability that $X = 7$? Justify your answer.

(b) Calculate the mean and standard deviation of the random variable X.

(c) Suppose that you asked 100 randomly selected young people (ages 19 to 25) to respond to the question and found the mean $\bar{x}$ of their responses. Describe the sampling distribution of $\bar{x}$.

(d) Would a sample mean $\bar{x} = 4.96$ be surprising? Explain your reasoning.

Section 9.3 Confidence Intervals for a Difference Between Two Population Means

LEARNING GOALS
By the end of this section, you will be able to:
- Check the Random and Normal/Large Sample conditions for constructing a confidence interval for a difference between two means.
- Calculate a $C\%$ confidence interval for a difference between two means.
- Use the four-step process to construct and interpret a confidence interval for the difference between two means.

In Section 9.1, you learned how to construct and interpret a confidence interval for a difference $p_1 - p_2$ between two population proportions. What if we want to estimate the difference $\mu_1 - \mu_2$ between two population means? For instance, maybe we want to estimate the difference in the average annual income of U.S. adults with college degrees and U.S. adults who attended college but did not earn a degree. The ideal strategy is to take a separate random sample from each population and to use the difference $\bar{x}_1 - \bar{x}_2$ between the sample means as our point estimate. To make inferences about $\mu_1 - \mu_2$, we have to first know about the sampling distribution of $\bar{x}_1 - \bar{x}_2$. Here are the details.

HOW TO: Describe the Sampling Distribution of $\bar{x}_1 - \bar{x}_2$

Let $\bar{x}_1$ be the sample mean in an SRS of size n_1 selected from a large population with mean μ_1 and standard deviation σ_1. Let $\bar{x}_2$ be the sample mean in an SRS of size n_2 selected from a large population with mean μ_2 and standard deviation σ_2. Then:

- The **shape** of the sampling distribution of $\bar{x}_1 - \bar{x}_2$ is normal if both population distributions are normal. It is approximately normal if both sample sizes are large ($n_1 \geq 30$ and $n_2 \geq 30$), or if one population distribution is normal and the other sample size is large, or if both population distributions are approximately normal.
- The **mean** of the sampling distribution of $\bar{x}_1 - \bar{x}_2$ is $\mu_{\bar{x}_1 - \bar{x}_2} = \mu_1 - \mu_2$.
- The **standard deviation** of the sampling distribution of $\bar{x}_1 - \bar{x}_2$ is

$$\sigma_{\bar{x}_1 - \bar{x}_2} = \sqrt{\frac{\sigma_1^2}{n_1} + \frac{\sigma_2^2}{n_2}}$$

This section focuses on constructing confidence intervals for a difference between two population means. You will learn how to perform significance tests about $\mu_1 - \mu_2$ in Section 9.4.

Conditions for Estimating $\mu_1 - \mu_2$

In Section 7.5, we introduced two conditions that should be met before we construct a confidence interval for a population mean: Random and Normal/Large Sample. Now that we are comparing *two* means, we have to modify these conditions slightly.

HOW TO: Check Conditions for Constructing a Confidence Interval for $\mu_1 - \mu_2$

To construct a confidence interval for a difference in population means $\mu_1 - \mu_2$, check that the following conditions are met:

- **Random:** The data come from independent random samples from the two populations of interest or from two groups in a randomized experiment.
- **Normal/Large Sample:** For each sample, the data come from a normally distributed population or the sample size is large ($n \geq 30$). For each sample, if the population distribution has unknown shape and $n < 30$, a graph of the sample data shows no strong skewness or outliers.

As we mentioned in Section 7.6, it is sometimes difficult to determine if a graph of quantitative data shows "strong" skewness or only "moderate" skewness when the sample size is small ($n < 30$). Example 7.6.1 and the following example should help you make this distinction.

EXAMPLE 9.3.1

Windy City or Big Apple?

Checking conditions

PROBLEM

A recent college graduate is considering a job offer that would allow them to live in either Chicago or New York City. To help make the decision, they want to estimate the typical cost of living in each city. Here are data on the monthly rents (in dollars) for independent random samples of 10 one-bedroom apartments in each city:[30]

| Chicago | 1495 | 1100 | 2000 | 1295 | 2000 | 1400 | 1250 | 1180 | 1000 | 1090 |
| New York | 2800 | 1450 | 2250 | 2075 | 2400 | 2400 | 1700 | 900 | 1470 | 1800 |

Let μ_1 = the mean monthly rent (\$) of all one-bedroom apartments in Chicago and μ_2 = the mean monthly rent (\$) of all one-bedroom apartments in New York City. Check if the conditions for calculating a 99% confidence interval for $\mu_1 - \mu_2$ are met.

SOLUTION

Random? Independent random samples of 10 one-bedroom apartments in Chicago and New York City. ✓

Normal/Large Sample? Both sample sizes are small, and the Chicago boxplot is moderately skewed to the right. But neither boxplot shows any strong skewness or outliers. ✓

> Because the sample sizes are less than 30, graph both samples of data to see if it is reasonable to assume that they came from normally distributed populations.

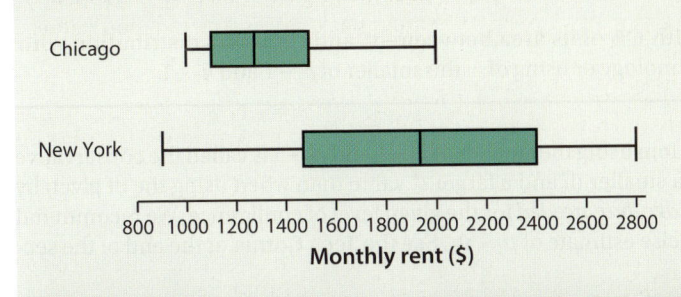

[Boxplots of Chicago and New York monthly rent ($), with x-axis from 800 to 2800]

EXAM PREP: FOR PRACTICE, TRY EXERCISE 5.

We used boxplots to check the Normal/Large Sample condition in Example 9.3.1 because they clearly display any outliers or strong skewness. You can also use dotplots, stemplots, histograms, or normal probability plots to assess normality. Just be sure to include graphs of sample data when the sample size is small ($n < 30$) and you are checking the Normal/Large Sample condition.

Calculating a Confidence Interval for $\mu_1 - \mu_2$

When conditions are met, we can calculate a confidence interval for $\mu_1 - \mu_2$ using the familiar formula

$$\text{point estimate} \pm \text{margin of error}$$

or, in slightly expanded form,

$$\text{statistic} \pm \text{critical value} \times \text{standard error of statistic}$$

The observed difference in sample means $\bar{x}_1 - \bar{x}_2$ is our point estimate for $\mu_1 - \mu_2$.

The standard deviation of the sampling distribution of $\bar{x}_1 - \bar{x}_2$ is

$$\sigma_{\bar{x}_1 - \bar{x}_2} = \sqrt{\frac{\sigma_1^2}{n_1} + \frac{\sigma_2^2}{n_2}}$$

Because we usually don't know the values of the population standard deviations σ_1 and σ_2, we replace them with the sample standard deviations s_1 and s_2. The result is the *standard error* of $\bar{x}_1 - \bar{x}_2$:

$$SE_{\bar{x}_1 - \bar{x}_2} = \sqrt{\frac{s_1^2}{n_1} + \frac{s_2^2}{n_2}}$$

We can find the critical value t^* for the given confidence level by using either technology or Table B. But what df should we use? There are two practical options.

Option 1 (Technology): Use the t distribution with degrees of freedom calculated by technology using the following formula. Note that the df given by this formula is usually *not* a whole number. Statisticians B. L. Welch and F. E. Satterthwaite discovered this fairly remarkable formula in the 1940s:

$$df = \frac{\left(\frac{s_1^2}{n_1} + \frac{s_2^2}{n_2}\right)^2}{\frac{1}{n_1 - 1}\left(\frac{s_1^2}{n_1}\right)^2 + \frac{1}{n_2 - 1}\left(\frac{s_2^2}{n_2}\right)^2}$$

Option 2 (Conservative approach): Use the t distribution with degrees of freedom equal to the *smaller* of $n_1 - 1$ and $n_2 - 1$. This option was more popular in the days when Table B and a four-function calculator were the main calculation tools.

HOW TO Calculate a Confidence Interval for $\mu_1 - \mu_2$

When the Random and Normal/Large Sample conditions are met, a C% confidence interval for the difference $\mu_1 - \mu_2$ between two population means is

$$(\bar{x}_1 - \bar{x}_2) \pm t^* \sqrt{\frac{s_1^2}{n_1} + \frac{s_2^2}{n_2}}$$

where t^* is the critical value with C% of its area between $-t^*$ and t^* in the t distribution with degrees of freedom given by technology or using df = the smaller of $n_1 - 1$ and $n_2 - 1$.

Calculating the degrees of freedom using the smaller of $n_1 - 1$ and $n_2 - 1$ is called the conservative approach because it will result in a smaller df and a larger t^* value than when using the df given by technology, *making the interval wider than needed* for the given level of confidence. We recommend using technology to get a more precise estimate of $\mu_1 - \mu_2$. See the Tech Corner at the end of the section for details.

EXAMPLE 9.3.2

Windy City or Big Apple?

Calculating a confidence interval for $\mu_1 - \mu_2$

PROBLEM

Recall from Example 9.3.1 that independent random samples of 10 one-bedroom apartments in Chicago and New York City were selected, and their monthly rents (in dollars) were recorded. The table shows summary statistics for the two samples of apartments.

City	Sample size	Mean	SD
Chicago	$n_1 = 10$	$\bar{x}_1 = 1381.00$	$s_1 = 357.99$
New York	$n_2 = 10$	$\bar{x}_2 = 1924.50$	$s_2 = 567.11$

(a) Calculate a 99% confidence interval for the difference (Chicago − New York) in the mean monthly rent of all one-bedroom apartments in the two cities.

(b) Based on the interval from part (a), can the prospective employee conclude that the mean monthly rent for one-bedroom apartments is higher in one city than in the other? Explain your reasoning.

SOLUTION

(a) *Using technology:*

df = 15.19; −1167.33 to 80.33

Conservative approach:

df = smaller of (10 − 1, 10 − 1) = 9; $t^* = 3.250$

> See the Tech Corner at the end of the section for details on how to obtain this interval.

> Find the value of t^* using Table B or Applet/invT (area: 0.005, df = 9).

$$(1381.00 - 1924.50) \pm 3.250 \sqrt{\frac{357.99^2}{10} + \frac{567.11^2}{10}}$$

$\rightarrow -543.50 \pm 689.25 \rightarrow -1232.75 \text{ to } 145.75$

> $(\bar{x}_1 - \bar{x}_2) \pm t^* \sqrt{\frac{s_1^2}{n_1} + \frac{s_2^2}{n_2}}$

(b) No, the employee cannot conclude that the mean monthly rent for one-bedroom apartments is higher in one city than in the other because the confidence interval (−1167.33, 80.33) includes 0 as a plausible value for the difference in the mean monthly rents of all one-bedroom apartments in Chicago and in New York City.

EXAM PREP: FOR PRACTICE, TRY EXERCISE 11.

We are 99% confident that the interval from −$1167.33 to $80.33 captures the difference in the mean monthly rents of all one-bedroom apartments in Chicago and in New York City. The interval suggests that the mean monthly rent of all one-bedroom apartments in Chicago is between $1167.33 less and $80.33 more than the mean monthly rent of all one-bedroom apartments in New York City.

In Example 9.3.2, the 99% confidence interval for $\mu_1 - \mu_2$ using the conservative approach with df = 9 and $t^* = 3.250$ is −$1232.75 to $145.75. This interval has margin of error $689.25. Technology gives the 99% confidence interval as −$1167.33 to $80.33. This interval has margin of error $623.83, which is $65.42 smaller than the result found when using the conservative approach. The confidence interval obtained using technology is narrower because it uses df = 15.19 instead of df = 9, which results in a smaller critical value: $t^* = 2.941$ instead of $t^* = 3.250$.

Putting It All Together: Confidence Interval for $\mu_1 - \mu_2$

We are now ready to use the four-step process to construct and interpret a **two-sample t interval for a difference between two means.** Remember that you can use technology to perform the calculations in the Do step.

> A **two-sample t interval for a difference between two means** is a confidence interval used to estimate a difference in the means of two populations or treatments.

Examples 9.3.1 and 9.3.2 involved inference about $\mu_1 - \mu_2$ using data that were produced by random sampling. In such cases, the parameters μ_1 and μ_2 are the means of the corresponding populations. However, many important statistical results involve comparing two means based on data from randomized experiments. Then the parameters μ_1 and μ_2 are the true mean responses for individuals like the ones in the experiment who receive Treatment 1 or Treatment 2. Fortunately, we can still use the sampling distribution of $\bar{x}_1 - \bar{x}_2$ to perform inference about $\mu_1 - \mu_2$. See the Think About It feature in Section 9.4 for more details.

EXAMPLE 9.3.3

Do portion sizes affect food consumption?

Confidence interval for $\mu_1 - \mu_2$

PROBLEM

In a study published in the *American Journal of Clinical Nutrition*, researchers wanted to know if changing food portion sizes would influence subjects' food consumption one day later. The volunteer subjects were randomly assigned to two groups, with one group being served small portions of food and the other group being served large portions of food. The next day all of the subjects were allowed to serve themselves, and the researchers recorded how much food they consumed (in grams). Here is a summary of the results:[31]

Treatment	n	Mean	SD
Small portion	38	144.66	72.36
Large portion	37	189.91	55.62

Construct and interpret a 95% confidence interval for the difference in the true mean amounts of food consumed by people like the ones in this study one day after being given large versus small portions.

SOLUTION

State: 95% CI for $\mu_L - \mu_S$, where μ_L = true mean amount of food consumed (grams) by people like the ones in this study who are given large portions the previous day and μ_S = true mean amount of food consumed (grams) by people like the ones in this study who are given small portions the previous day.

> **State:** State the parameters you want to estimate and the confidence level. *Be sure to indicate the order of subtraction.*

Plan: Two-sample t interval for $\mu_L - \mu_S$

Random? Randomly assigned volunteers to receive small portions or large portions of food. ✓

Normal/Large Sample? Both sample sizes are large: $n_L = 37 \geq 30$ and $n_S = 38 \geq 30$. ✓

> **Plan:** Identify the appropriate inference method and check the conditions.

(continued)

Do: *Using technology:*
df = 69.30; 15.57 to 74.93

Conservative approach:
df = smaller of (37 − 1, 38 − 1) = 36;
Use df = 30 on Table B → $t^* = 2.042$

$$(189.91 - 144.66) \pm 2.042 \sqrt{\frac{55.62^2}{37} + \frac{72.36^2}{38}}$$

→ 45.25 ± 30.384 → 14.866 to 75.634

Conclude: We are 95% confident that the interval from 15.57 to 74.93 captures the difference $(\mu_L - \mu_S)$ in the true mean amounts of food consumed (grams) by people like the ones in this study who are given large portions the previous day and those who are given small portions the previous day. The interval suggests that the true mean amount of food consumed by people like these is between 15.57 grams and 74.93 grams higher after being given large portions the previous day than after being given small portions the previous day.

Do: If the conditions are met, perform calculations.

You could use technology to find the critical value t^* for df = 36: Applet/invT(area:0.025, df:36) = 2.028

$$(189.91 - 144.66) \pm 2.028 \sqrt{\frac{55.62^2}{37} + \frac{72.36^2}{38}}$$

→ 45.25 ± 30.176 → 15.074 to 75.426

Conclude: Interpret your interval in the context of the problem. *Don't forget the second sentence when you interpret the interval!*

EXAM PREP: FOR PRACTICE, TRY EXERCISE 15.

Example 9.3.3 gives an important reminder if you are using the conservative approach with Table B: If the degrees of freedom you need is not listed in the table (df = 36), be sure to use the next lowest df that does appear there (df = 30). Of course, doing so leads to a wider interval than if we had used df = 36 and the corresponding $t^* = 2.028$, which reduces the margin of error from 30.384 to 30.176. Technology yields an even more precise estimate — with df = 69.30, the margin of error is only 29.68.

We chose the order of subtraction in Example 9.3.3 to estimate $\mu_L - \mu_S$. What if we had reversed the order and estimated $\mu_S - \mu_L$ instead? The resulting 95% confidence interval from technology is −74.93 to −15.57. This interval suggests that the true mean amount of food consumed by people like the ones in this study is between 15.57 grams and 74.93 grams lower after being given small portions the previous day than after being given large portions the previous day. Note that this is equivalent to the conclusion we reached in the example.

The Pooled Two-Sample t Procedures

Most technology offers a choice of two-sample t procedures. One is often labeled "unequal" variances or "unpooled"; the other is called "equal" variances or "pooled." (Recall that the variance is the square of the standard deviation.) The unequal variance procedure uses our formula for the two-sample t interval, with df calculated using the complicated formula developed by Satterthwaite and Welch that we presented earlier. *This approach is valid whether or not the population variances are equal.*

The other choice is a special version of the two-sample t procedures that assumes the two population distributions have equal variances. This procedure combines (*pools*) the two sample variances to estimate the common population variance σ^2. The formula for the pooled estimate of σ^2 is

$$s_p^2 = \frac{(n_1 - 1)s_1^2 + (n_2 - 1)s_2^2}{(n_1 - 1) + (n_2 - 1)}$$

If the two population variances are equal and both population distributions are normal, we can use a t distribution with df $= (n_1 - 1) + (n_2 - 1) = n_1 + n_2 - 2$ and s_p^2 to perform calculations. This method offers more degrees of freedom than does Option 1 (technology), which leads to narrower confidence intervals and smaller P-values.

The pooled t procedures were widely used before technology made it easy to carry out the two-sample t procedures we present in this book. In practice, population variances are rarely equal, so our two-sample t procedures are almost always more accurate than the pooled procedures. Our advice: *Don't use the pooled two-sample t procedures unless a statistician says it is OK to do so.*

Section 9.3 Confidence Intervals for a Difference Between Two Population Means 529

TECH CORNER Confidence Intervals for a Difference Between Two Means

You can use technology to calculate a confidence interval for $\mu_1 - \mu_2$. We'll illustrate using the data from Examples 9.3.1 and 9.3.2, which dealt with monthly rents of one-bedroom apartments in Chicago and New York City. Here are the summary statistics once again:

City	Sample size	Mean	SD
Chicago	$n_1 = 10$	$\bar{x}_1 = 1381.00$	$s_1 = 357.99$
New York	$n_2 = 10$	$\bar{x}_2 = 1924.50$	$s_2 = 567.11$

Applet

1. Go to www.stapplet.com and launch the *One Quantitative Variable, Multiple Groups* applet.
2. Enter "Monthly rent ($)" as the variable name.
3. Choose to input Mean and standard deviation. Enter the name, mean, standard deviation, and sample size for both groups, as shown.

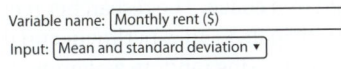

4. Click the Begin analysis button. Scroll down to the "Perform Inference" section. Choose the 2-sample t interval ($\mu_1 - \mu_2$) for the procedure, "No" for the Conservative degrees of freedom, and 99% for the confidence level. Then click the Perform inference button.

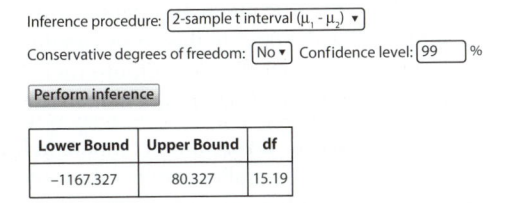

Note: If you have raw data instead of summary statistics, choose "Raw data" from the Input menu in Step 3. Then enter the values from each of the two samples and proceed as in Step 4.

TI-83/84

1. Press STAT, then choose TESTS and 2-SampTInt.
2. When the 2-SampTInt screen appears, choose Stats as the input method. Then enter the summary statistics, as shown. Be sure to use 0.99 for the confidence level and choose "No" for Pooled.

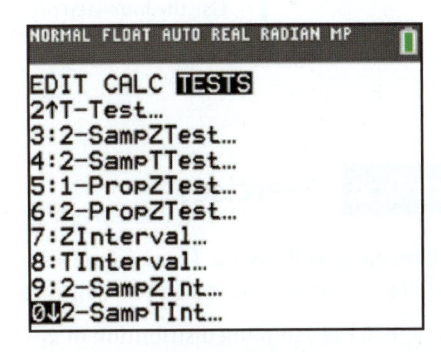

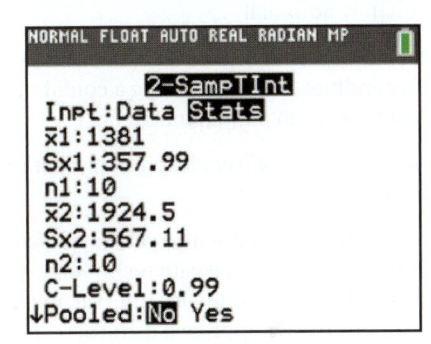

3. Highlight "Calculate" and press ENTER.

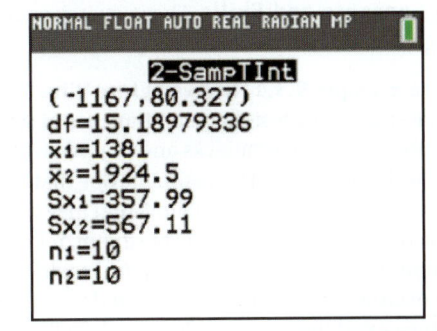

Note: To calculate a confidence interval for a difference in means from raw data, start by entering the data into lists L_1 and L_2. Then choose Data as the input method in Step 2.

Detailed instructions for using CrunchIt!, Excel, Google Sheets, JMP, Minitab, and R are available in Achieve.

Section 9.3 What Did You Learn?

Review the learning goals from this section. Then practice what you've learned by working through the exercises.

Learning Goal	Example	Exercises
Check the Random and Normal/Large Sample conditions for constructing a confidence interval for a difference between two means.	9.3.1	5–10
Calculate a C% confidence interval for a difference between two means.	9.3.2	11–14
Use the four-step process to construct and interpret a confidence interval for the difference between two means.	9.3.3	15–18

Section 9.3 Exercises

Building Concepts and Skills These exercises assess the basic knowledge you should have after reading the section.

1. The shape of the sampling distribution of $\bar{x}_1 - \bar{x}_2$ is approximately normal if _____.

2. In what two ways can the data be collected to satisfy the Random condition for constructing a confidence interval for a difference in population means?

3. What is the formula for calculating a confidence interval for $\mu_1 - \mu_2$?

4. True/False: The confidence interval for a difference in population means obtained with technology will always be narrower than the confidence interval obtained with the more conservative approach using df = smaller of $n_1 - 1$ and $n_2 - 1$.

Mastering Concepts and Skills These exercises reinforce the learning goals as shown in the examples.

5. **Refer to Example 9.3.1 Red vs. White Wine** Observational studies suggest that moderate use of alcohol by adults reduces heart attacks and that red wine may have special benefits. One reason may be that red wine contains polyphenols, antioxidants that may reduce the risk of heart attacks. In an experiment, researchers randomly assigned healthy participants to drink half a bottle of either red or white wine each day for 2 weeks. They measured the level of polyphenols in the participants' blood before and after the 2-week period. Here are the percent changes in polyphenol levels for the subjects in each group:[32]

Red wine	3.5	8.1	7.4	4.0	0.7	4.9	8.4	7.0	5.5
White wine	3.1	0.5	-3.8	4.1	-0.6	2.7	1.9	-5.9	0.1

Let μ_1 = the true mean percent change in polyphenol levels for subjects like the ones in this study after drinking red wine daily for 2 weeks, and μ_2 = the true mean percent change in polyphenol levels for subjects like the ones in this study after drinking white wine daily for 2 weeks. Check if the conditions for calculating a confidence interval for $\mu_1 - \mu_2$ are met.

6. **Encouragement Effect** A researcher wondered if people would perform better in the barbell curl if they were encouraged by a coach. The researcher recruited 31 subjects to participate in an experiment and then split them into two groups using random assignment. Group 1 received positive encouragement during the exercise, while Group 2 served as a control group that received no encouragement. The researcher recorded the number of barbell curl repetitions each participant was able to complete before setting the bar down. Here are the data.[33]

With encouragement	50	11	60	5	40	45	11	30
	8	50	11	15	9	40	75	
No encouragement	43	19	33	1	0	2	6	61
	21	11	40	15	15	18	37	22

Let μ_1 = the true mean number of barbell curls that subjects like the ones in this study can do with encouragement and μ_2 = the true mean number of barbell curls that subjects like the ones in this study can do with no encouragement. Check if the conditions for calculating a confidence interval for $\mu_1 - \mu_2$ are met.

7. **Family Income** How do incomes compare in Indiana and New Jersey? The parallel dotplots show the total family income of independent random samples of 38 Indiana residents and 44 New Jersey residents. Let μ_I = the mean total family income for all Indiana residents and μ_{NJ} = the mean total family income for all New Jersey residents. Check if the conditions for calculating a confidence interval for $\mu_I - \mu_{NJ}$ are met.

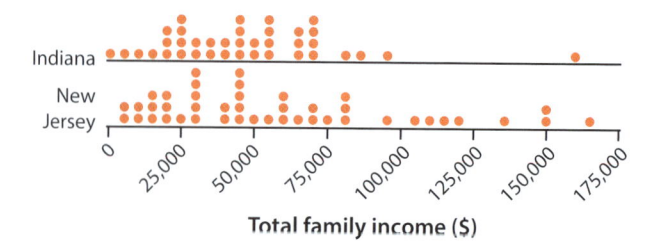

8. **Household Sizes** How do the numbers of people living in households in the United Kingdom and South Africa

compare? To help answer this question, we used an online random data selector to choose separate random samples of 36 students from South Africa and 31 students from the United Kingdom. Here are dotplots of the household sizes reported by the students in the survey. Let μ_{SA} = the mean household size for all students in South Africa and μ_{UK} = the mean household size for all students in the United Kingdom. Check if the conditions for calculating a confidence interval for $\mu_{SA} - \mu_{UK}$ are met.

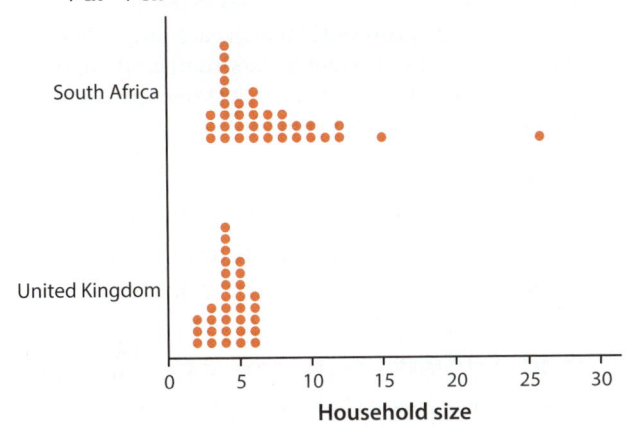

9. **Overthinking** Athletes often comment that they try not to "overthink it" when competing in their sport. Is it possible to "overthink"? To investigate, researchers put some golfers to the test. They recruited 40 experienced golfers and allowed them some time to practice their putting. After practicing, they randomly assigned the golfers in equal numbers to two groups. Golfers in one group had to write a detailed description of their putting technique (which could lead to "overthinking it"). Golfers in the other group had to do an unrelated verbal task for the same amount of time. After completing their tasks, each golfer was asked to attempt putts from a fixed distance until they made 3 putts in a row. The boxplots summarize the distribution of the number of putts required for the golfers in each group to make 3 putts in a row.[34] Let μ_1 = the true mean number of putts required by subjects like the ones in this study after describing their putting and μ_2 = the true mean number of putts required by subjects like the ones in this study after not describing their putting. Check if the conditions for calculating a confidence interval for $\mu_1 - \mu_2$ are met.

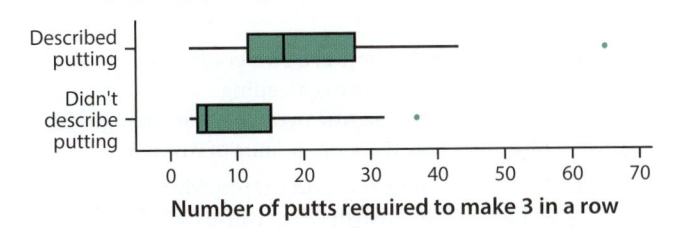

10. **Word Length** Sanderson is interested in comparing the mean word length in articles from a medical journal and from an airline's in-flight magazine. Sanderson counts the number of letters in the first 400 words of an article in the medical journal and in the first 100 words of an article in the airline magazine, and then uses statistical software to produce the histograms shown. Note that J stands for journal and M for magazine. Check if the conditions for calculating a confidence interval for $\mu_J - \mu_M$ are met.

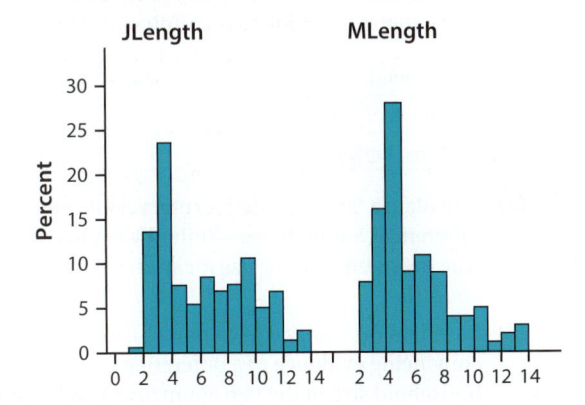

11. **Refer to Example 9.3.2 Red vs. White Wine** Refer to Exercise 5. Here are summary statistics for the two groups:

Group	Sample size	Mean	SD
Red wine	$n_1 = 9$	$\bar{x}_1 = 5.50$	$s_1 = 2.517$
White wine	$n_2 = 9$	$\bar{x}_2 = 0.233$	$s_2 = 3.292$

(a) Calculate a 90% confidence interval for the difference (Red wine – White wine) in the true mean percent change in polyphenol levels for subjects like the ones in this study under the two conditions.

(b) Based on the interval from part (a), is there convincing evidence of a difference in the true means? Justify your answer.

12. **Encouragement Effect** Refer to Exercise 6. Here are summary statistics for the two groups:

Group	Sample size	Mean	SD
With encouragement	$n_1 = 15$	$\bar{x}_1 = 30.667$	$s_1 = 22.363$
No encouragement	$n_2 = 16$	$\bar{x}_2 = 21.5$	$s_2 = 17.232$

(a) Calculate a 90% confidence interval for the difference (Encouragement – No encouragement) in the true mean number of barbell curl repetitions that subjects like the ones in this study can do under the two conditions.

(b) Based on the interval from part (a), is there convincing evidence of a difference in the true means? Justify your answer.

13. **Family Income** Refer to Exercise 7. Here are summary statistics for the two samples:

State	n	Mean	SD
Indiana	38	$47,400	$29,400
New Jersey	44	$58,100	$41,900

(a) Calculate a 99% confidence interval for the difference (Indiana – New Jersey) in the mean total family income for all residents in the two states.

(b) Based on the interval from part (a), is there convincing evidence of a difference in the mean total family income in the two states? Justify your answer.

14. **Household Sizes** Refer to Exercise 8. Here are summary statistics for the two samples:

Country	n	Mean	SD
South Africa	36	7.06	4.38
United Kingdom	31	4.26	1.18

(a) Calculate a 99% confidence interval for the difference (South Africa − United Kingdom) in the true mean household size for students in the two countries.

(b) Based on the interval from part (a), is there convincing evidence of a difference in the average household size in the two countries? Justify your answer.

15. Refer to Example 9.3.3 **Squirrel Weights** In many parts of the northern United States, two color variants of the Eastern gray squirrel — gray and black — are found in the same habitats. A scientist wonders if there is a difference in the sizes of the two color variants. The scientist collects separate random samples of 40 squirrels of each color from a large forest and weighs them. The 40 black squirrels have a mean weight of $\bar{x}_B = 20.3$ ounces and a standard deviation of $s_B = 2.1$ ounces. The 40 gray squirrels have a mean weight of $\bar{x}_G = 19.2$ ounces and a standard deviation of $s_G = 1.9$ ounces. Construct and interpret a 95% confidence interval for the difference in mean weight of all black and gray squirrels in this forest.

16. **Beta Blockers** In a study of heart surgery, one concern was the effect of drugs called beta blockers on the pulse rate of patients during surgery. The available subjects were randomly split into two groups of 30 patients each. One group received a beta blocker; the other group received a placebo. The pulse rate of each patient at a critical point during the operation was recorded. The treatment group had a mean pulse rate of 65.2 beats per minute (bpm) and a standard deviation of 7.8 bpm. For the control group, the mean pulse rate was 70.3 bpm and the standard deviation was 8.3 bpm. Construct and interpret a 99% confidence interval for the true difference in mean pulse rates for patients like the ones in this study who receive a beta blocker and those who receive a placebo.

17. **Note Taking** In a recent study at the University of California, researchers wanted to compare the amount of information recorded by students who write notes with pencil and paper and those who type notes on a laptop. They conducted an experiment in which 109 volunteer subjects were randomly assigned to one of these two strategies for taking notes. One of the variables they measured was the number of words written. The 55 students assigned to writing notes with pencil and paper had a mean of 390.7 words and a standard deviation of 143.9 words. The 54 students assigned to typing notes on a laptop had a mean of 548.7 words and a standard deviation of 252.7 words. Construct and interpret a 95% confidence interval for the true difference in the mean number of words students like the ones in this study write versus type in their notes.[35]

18. **Tropical Flowers** Different varieties of the tropical flower *Heliconia* are fertilized by different species of hummingbirds. Researchers believe that over time, the lengths of the flowers and the forms of the hummingbirds' beaks have evolved to match each other. Here are data on the lengths (in millimeters, mm) for random samples of two color varieties of the same species of flower on the island of Dominica:[36]

H. caribaea red

41.90	42.01	41.93	43.09	41.17	41.69	39.78	40.57
39.63	42.18	40.66	37.87	39.16	37.40	38.20	38.07
38.10	37.97	38.79	38.23	38.87	37.78	38.01	

H. caribaea yellow

| 36.78 | 37.02 | 36.52 | 36.11 | 36.03 | 35.45 | 38.13 | 37.10 |
| 35.17 | 36.82 | 36.66 | 35.68 | 36.03 | 34.57 | 34.63 | |

Construct and interpret a 95% confidence interval for the difference in the true mean lengths of these two varieties of flowers on the island of Dominica.

Applying the Concepts These exercises ask you to apply multiple learning goals in a new context or to apply what you learned in this section in a new way.

19. **Breastfeeding and Calcium** Breastfeeding mothers secrete calcium into their milk. Some of this calcium may come from their bones, so mothers may lose bone mineral. Researchers compared a random sample of 47 breastfeeding women with a random sample of 22 women of similar age who were neither pregnant nor lactating. They measured the percent change in the bone mineral content (BMC) of the women's spines over 3 months. The table summarizes the data.[37]

Group	n	Mean	SD
Breastfeeding	47	−3.59	2.51
Not pregnant or lactating	22	0.31	1.30

Graphs of the data for both groups revealed no strong skewness or outliers.

(a) Calculate and interpret a 95% confidence interval for the difference (Breastfeeding − Not pregnant or lactating) in the true mean percent changes in BMC for these two populations of women.

(b) Does your interval in part (a) provide convincing evidence that breastfeeding causes a decrease in bone mineral content, on average? Justify your answer.

20. **Temperature and Surgery** Researchers wondered whether maintaining a patient's body temperature close to normal by heating the patient during surgery

would affect wound infection rates. Patients were assigned at random to two groups: the normothermic group (patients' core temperatures were maintained at near normal levels, 36.5°C, with heating blankets) and the hypothermic group (patients' core temperatures were allowed to decrease to about 34.5°C). If keeping patients warm during surgery alters the chance of infection, patients in the two groups should have hospital stays of different lengths. Here are summary statistics on their hospital stays (in number of days) for the two groups:[38]

Group	n	Mean	SD
Normothermic	104	12.1	4.4
Hypothermic	96	14.7	6.5

(a) Construct and interpret a 95% confidence interval for the difference in the true mean lengths of hospital stays for normothermic and hypothermic patients like the ones in this study.

(b) Does your interval in part (a) provide convincing evidence that keeping patients like the ones in this study warm during surgery affects the average length of patients' hospital stays? Justify your answer.

21. **Bird Eggs** A researcher wants to see if birds that build larger nests lay larger eggs. She selects independent random samples of small nests and large nests. Then she weighs one egg (chosen at random if there is more than one egg) from each nest. A 95% confidence interval for the difference (Large − Small) between the mean masses (in grams) of eggs in small and large nests is 1.6 ± 2.0.

(a) Does the interval provide convincing evidence of a difference in the mean egg mass of birds with small nests and birds with large nests? Explain your reasoning.

(b) Does the interval provide convincing evidence that the mean egg mass of birds with small nests and birds with large nests is the same? Explain your reasoning.

22. **Reaction Times** Researchers wanted to know if student athletes (students on at least one varsity team) have faster reaction times than non-athletes. They took separate random samples of 33 athletes and 30 non-athletes from their school and tested their reaction times using an online reaction test, which measured the time (in seconds) between when a green light went on and the subject pressed a key on the computer keyboard. A 95% confidence interval for the difference (Non-athlete − Athlete) in the mean reaction times was 0.018 ± 0.034 second.

(a) Does the interval provide convincing evidence of a difference in the mean reaction times of athletes and non-athletes at this school? Explain your reasoning.

(b) Does the interval provide convincing evidence that the mean reaction time of athletes and the mean reaction time of non-athletes at this school are the same? Explain your reasoning.

23. **Note Taking** Refer to Exercise 17.
(a) Interpret the confidence level.
(b) Would a 99% confidence interval be narrower or wider than the 95% confidence interval you calculated in Exercise 17? Explain why without doing any calculations.

24. **Tropical Flowers** Refer to Exercise 18.
(a) Interpret the confidence level.
(b) Would a 90% confidence interval be narrower or wider than the 95% confidence interval you calculated in Exercise 18? Explain why without doing any calculations.

25. Tech **City Trees** How do the average diameters of trees on the curb and trees offset from the curb in New York City compare? The data set *NYCTrees* contains detailed information about a random sample of 8253 trees on New York City streets.[39] Each tree's location (*curb_loc*) is described as either on the curb or offset from the curb. One of the measurements taken for each standing tree (i.e., not just a stump) was the diameter at breast height (*DBH*), in inches. Breast height was defined as 54 inches above ground. Due to the sampling method used in this survey, it is reasonable to consider these as independent random samples of trees on the curb and trees offset from the curb in New York City. Calculate and interpret a 95% confidence interval for the difference in the mean DBH in these two populations of trees.

26. Tech **Computers and Household Income** How do the average incomes of households with computers and households without computers compare? The *American Community Survey* data set contains detailed information from a random sample of 3000 U.S. households.[40] Each household is classified based on whether it has a desktop or laptop computer (*LAPTOP* = 1) or does not (*LAPTOP* = 2). The variable *HINCP* records household income in the past 12 months. Due to the sampling method used in this survey, it is reasonable to consider these as independent random samples of U.S households with and without a computer when these data were collected. Calculate and interpret a 95% confidence interval for the difference in the mean incomes in these two populations of U.S. households.

Extending the Concepts These exercises challenge you to explore statistical concepts and methods that go beyond what you learned in this section.

27. **Dogs and Cats** Dogs vary in size a great deal more than cats do. The weights of all dogs in a certain city have a mean of 28 pounds and a standard deviation of 14 pounds. The weights of all cats in the same city have a mean of 9.5 pounds and a standard deviation of 2 pounds. Suppose you take independent random samples of 40 dogs and 30 cats from this city. Let $\bar{x}_1 - \bar{x}_2$ be the difference (Dogs − Cats) in the sample mean weights.

(a) Is the shape of the sampling distribution of $\bar{x}_1 - \bar{x}_2$ approximately normal? Explain your answer.

(b) Find the mean of the sampling distribution.

(c) Calculate the standard deviation of the sampling distribution.

(d) Find the probability that the mean weight for the sample of 40 dogs is at least 20 pounds greater than the mean weight for the sample of 30 cats.

28. **SE vs. ME for Means** In a random sample of 125 workers from Company 1, the mean number of sick days taken in the previous year was 4.8 days with a standard deviation of 2.1 days. In a separate random sample of 68 employees from Company 2, the mean number of sick days taken in the previous year was 2.5 days with a standard deviation of 1.1 days. The difference in the sample mean number of sick days from the two companies is $\bar{x}_1 - \bar{x}_2 = 4.8 - 2.5 = 2.3$.

(a) The standard error of $\bar{x}_1 - \bar{x}_2$ is 0.23. Interpret this value.

(b) Using technology, the margin of error for a 95% confidence interval for $\mu_1 - \mu_2$ is 0.45. Interpret this value.

Cumulative Review These exercises revisit what you learned in previous sections.

29. **Fever and Flu (4.5, 5.3)** A large community college is in the midst of a bad flu season. The probability that a randomly selected student has the flu is 0.35. Given that a student has the flu, there is a 0.90 probability they have a high fever. Given that a student does not have the flu, there is a 0.12 probability they have a high fever (as a result of some other ailment).

(a) Find the probability that a randomly selected student has a high fever.

(b) Suppose that a randomly selected student has a high fever. Find the probability that the student has the flu.

(c) Suppose we randomly select 5 students. Find the probability that exactly 3 of them have the flu and a high fever.

30. **Mobile Phone Quality (7.4)** A quality-control inspector is testing mobile phones made during a single day at a factory to determine the proportion of phones with minor defects. The inspector selects an SRS of 200 phones and finds that 12 of them have minor defects. Calculate and interpret a 99% confidence interval for the proportion of all phones produced that day with minor defects.

Section 9.4 Significance Tests for a Difference Between Two Population Means

LEARNING GOALS

By the end of this section, you will be able to:

- State hypotheses and check conditions for performing a significance test about a difference between two means.
- Calculate the standardized test statistic and P-value for a significance test about a difference between two means.
- Use the four-step process to perform a significance test about a difference between two means.

An observed difference between two sample means can reflect an actual difference in the population means, or it may just be due to chance variation in random sampling or random assignment. In Section 1.7, we used simulation to determine which explanation makes more sense in a randomized experiment. This section shows you how to perform a significance test about a difference between two means.

Stating Hypotheses and Checking Conditions for a Test About $\mu_1 - \mu_2$

In a test comparing two means, the null hypothesis has the general form

$$H_0: \mu_1 - \mu_2 = \text{hypothesized value}$$

We'll focus on situations in which the hypothesized difference is 0. Then the null hypothesis says that there is no difference between the two population means:

$$H_0: \mu_1 - \mu_2 = 0$$

The null hypothesis can also be written in the equivalent form $H_0: \mu_1 = \mu_2$. The alternative hypothesis says what kind of difference we expect.

The conditions for performing a significance test about $\mu_1 - \mu_2$ are the same as those for constructing a confidence interval for a difference in means that you learned in Section 9.3.

Section 9.4 Significance Tests for a Difference Between Two Population Means

HOW TO Check Conditions for Performing a Significance Test About $\mu_1 - \mu_2$

To perform a significance test about a difference in population means $\mu_1 - \mu_2$, check that the following conditions are met:

- **Random:** The data come from independent random samples from the two populations of interest or from two groups in a randomized experiment.
- **Normal/Large Sample:** For each sample, the data come from a normally distributed population or the sample size is large ($n \geq 30$). For each sample, if the population distribution has unknown shape and $n < 30$, a graph of the sample data shows no strong skewness or outliers.

EXAMPLE 9.4.1

Do calcium supplements reduce blood pressure?

Stating hypotheses and checking conditions

PROBLEM

Does increasing the amount of calcium in our diet reduce blood pressure? Researchers designed a randomized comparative experiment to find out. The subjects were 21 healthy men who volunteered to participate. They were randomly assigned to two groups: 10 of the men received a calcium supplement for 12 weeks, while the control group of 11 men received a placebo pill that looked identical. The experiment was double-blind. The response variable was the decrease in systolic (top number) blood pressure for a subject after 12 weeks, in millimeters of mercury (mmHg). An increase appears as a negative number. Here are the data:[41]

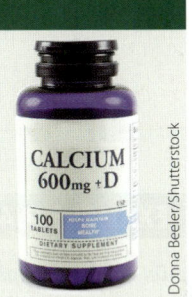

Group 1 (calcium)	7	−4	18	17	−3	−5	1	10	11	−2	
Group 2 (placebo)	−1	12	−1	−3	3	−5	5	2	−11	−1	−3

The researchers want to know if a calcium supplement reduces blood pressure more than a placebo, on average, for healthy men like the ones in this study.

(a) State appropriate hypotheses for performing a significance test. Be sure to define the parameters of interest.

(b) Do the data provide *some* evidence that a calcium supplement reduces blood pressure more than a placebo, on average, for healthy men like the ones in this study? Justify your answer.

(c) Check that the conditions for performing the test are met.

SOLUTION

(a) $H_0: \mu_1 - \mu_2 = 0$

$H_a: \mu_1 - \mu_2 > 0$

where μ_1 = the true mean decrease in systolic blood pressure for healthy men like the ones in this study who take a calcium supplement and μ_2 = the true mean decrease in systolic blood pressure for healthy men like the ones in this study who take a placebo.

> H_a is one-sided because researchers want to know if a calcium supplement reduces blood pressure *more than* a placebo, on average.

(b) Yes, because $\bar{x}_1 - \bar{x}_2 = 5.0 - (-0.273) = 5.273 > 0$.

> Use technology to calculate the sample means.

(c) Random? The 21 subjects were randomly assigned to the calcium or placebo treatments. ✓

Normal/Large Sample? The sample sizes are small, but the dotplots show no strong skewness and no outliers. ✓

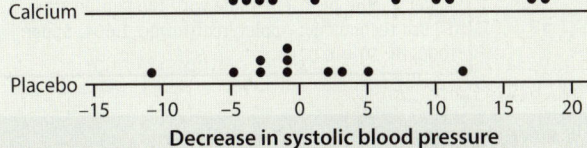

EXAM PREP: FOR PRACTICE, TRY EXERCISE 5.

Calculations: Standardized Test Statistic and *P*-Value

If the conditions are met, we can proceed with calculations. To do a test of $H_0: \mu_1 - \mu_2 = 0$, start by standardizing $\bar{x}_1 - \bar{x}_2$:

$$\text{standardized test statistic} = \frac{\text{statistic} - \text{null value}}{\text{standard deviation (error) of statistic}}$$

Because we usually don't know the population standard deviations, we use the *standard error* of $\bar{x}_1 - \bar{x}_2$ in the denominator of the standardized test statistic:

$$t = \frac{(\bar{x}_1 - \bar{x}_2) - 0}{\sqrt{\dfrac{s_1^2}{n_1} + \dfrac{s_2^2}{n_2}}}$$

To find the *P*-value, use the *t* distribution with degrees of freedom given by technology or equal to the smaller of $n_1 - 1$ and $n_2 - 1$ (conservative approach). See the Tech Corner at the end of the section for details about using technology to calculate the test statistic and *P*-value.

EXAMPLE 9.4.2

Do calcium supplements reduce blood pressure?

Calculating the standardized test statistic and *P*-value

PROBLEM

Refer to Example 9.4.1, which describes a randomized experiment to investigate whether calcium supplements reduce blood pressure more than a placebo, on average.

(a) Calculate the standardized test statistic.
(b) Find the *P*-value.

SOLUTION

(a) $\bar{x}_1 = 5.0$, $s_1 = 8.743$, $n_1 = 10$,
$\bar{x}_2 = -0.273$, $s_2 = 5.901$, $n_2 = 11$

$$t = \frac{[5.0 - (-0.273)] - 0}{\sqrt{\dfrac{8.743^2}{10} + \dfrac{5.901^2}{11}}} = \frac{5.273}{3.288} = 1.604$$

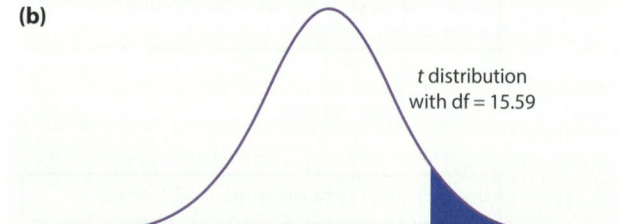

> First, calculate the sample statistics from the raw data using technology.

$$t = \frac{(\bar{x}_1 - \bar{x}_2) - 0}{\sqrt{\dfrac{s_1^2}{n_1} + \dfrac{s_2^2}{n_2}}}$$

(b)

t distribution with df = 15.59

t = 1.604

Using technology: *P*-value = 0.0644 using df = 15.59
Conservative approach: df = smaller of (10 − 1, 11 − 1) = 9
Table B gives a *P*-value between 0.05 and 0.10.

> To get a more precise *P*-value with the conservative df, use technology: Applet/tcdf(lower: 1.604, upper: 1000, df: 9) = 0.0716.

EXAM PREP: FOR PRACTICE, TRY EXERCISE 9.

Using the conservative df = 9 from Example 9.4.2, the P-value of the test is 0.0716. The actual P-value is 0.0644 when using df = 15.59 obtained from technology. Notice that technology gives smaller, more accurate P-values for a significance test about a difference between two population means. That's because calculators and software use the more complicated formula from Section 9.3 to obtain a larger number of degrees of freedom.

What does the P-value tell us? If $H_0: \mu_1 - \mu_2 = 0$ is true, there is a 0.0644 probability of getting a difference (Calcium − Placebo) in mean blood pressure reduction for the two groups of 5.273 or greater just by the chance involved in the random assignment. Because the P-value of 0.0644 is greater than our default significance level of $\alpha = 0.05$, we fail to reject H_0. We do not have convincing evidence that a calcium supplement reduces blood pressure more than a placebo, on average, for healthy men like the ones in this study.

Why didn't researchers find a significant difference in the calcium and blood pressure experiment? The difference in mean systolic blood pressures for the two groups was 5.273 mmHg. This seems like a fairly large difference. With the small group sizes of 10 and 11, however, this difference wasn't large enough to find convincing evidence for the one-sided alternative $H_a: \mu_1 - \mu_2 > 0$. We suspect that larger groups might show a similar difference in mean blood pressure reduction, which would indicate that calcium has a significant effect. If so, then the researchers in this experiment made a Type II error—not finding convincing evidence for H_a when it is actually true.

In fact, later studies involving more subjects showed that an increase in calcium intake slightly reduces both systolic and diastolic blood pressures in healthy people.[42] *Sample size strongly affects the power of a test.* It is easier to detect a difference in the effectiveness of two treatments if both are applied to large numbers of subjects.

THINK ABOUT IT

Why do the inference methods for random sampling work for randomized experiments? Confidence intervals and tests for $\mu_1 - \mu_2$ are based on the sampling distribution of $\bar{x}_1 - \bar{x}_2$. But in most experiments, researchers don't select subjects at random from any larger populations. They do randomly assign subjects to treatments. As in Section 1.7, we can use simulation to see what would happen if the random assignment were repeated many times, assuming no treatment effect—that is, assuming that $H_0: \mu_1 - \mu_2 = 0$ is true.

We used the *One Quantitative Variable, Multiple Groups* applet at www.stapplet.com to randomly reassign the 21 subjects in the calcium and blood pressure experiment to the two groups 1000 times, assuming the drug received (calcium or placebo) *doesn't affect* each individual's change in systolic blood pressure. Figure 9.2 shows the value of $\bar{x}_1 - \bar{x}_2$ in the 1000 simulated trials. This distribution (sometimes referred to as a *randomization distribution* of $\bar{x}_1 - \bar{x}_2$) has an approximately normal shape with mean 0 and standard deviation 3.288. This matches well with the standard error we used when calculating the t statistic in Example 9.4.2.

64 of the 1000 random reassignments resulted in a difference in sample means of 5.273 or greater.

Simulated difference in sample mean decrease in systolic blood pressure ($\bar{x}_1 - \bar{x}_2$)

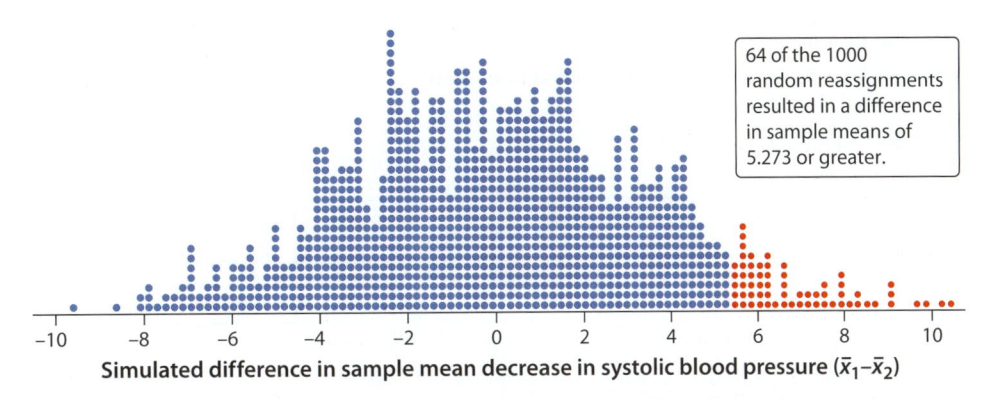

FIGURE 9.2 Dotplot of the values of $\bar{x}_1 - \bar{x}_2$ from each of 1000 simulated random reassignments of subjects to treatment groups in the calcium and blood pressure experiment, assuming the null hypothesis of no treatment effect is true.

In the actual experiment, the difference between the mean decreases in systolic blood pressure in the calcium and placebo groups was 5.000 − (−0.273) = 5.273. How likely is it that a difference this large or larger would happen purely by chance when H_0 is true? Figure 9.2 provides a rough answer: 64 of the 1000 random reassignments yielded a difference in means greater than or equal to 5.273. That is, our estimate of the P-value is 0.064. This is quite close to the 0.0644 P-value that we calculated in Example 9.4.2, suggesting that it's OK to use inference methods for random sampling to analyze randomized experiments. We will discuss *randomization tests* like this in more detail in Section 13.5.

Putting It All Together: Performing a Significance Test About $\mu_1 - \mu_2$

We are now ready to use the four-step process to carry out a **two-sample t test for a difference between two means**. Remember that you can use technology to perform the calculations in the Do step.

A **two-sample t test for a difference between two means** is a significance test of the null hypothesis that the difference in the means of two populations or treatments is equal to a specified value (usually 0).

EXAMPLE 9.4.3

Where do the big trees grow?

Significance test for $\mu_1 - \mu_2$

PROBLEM

The Wade Tract Preserve in Georgia is an old-growth forest of longleaf pines that has survived in a relatively undisturbed state for hundreds of years. One question of interest to foresters who study the area is "How do the sizes of longleaf pine trees in the northern and southern halves of the forest compare?" To find out, researchers took random samples of 30 trees from each half and measured the diameter at breast height (DBH) in centimeters. The table shows summary statistics for the two samples of trees.[43]

Forest location	n	Mean	SD
Northern half	30	23.70	17.50
Southern half	30	34.53	14.26

Do these data give convincing evidence of a difference in the population means at the 5% significance level?

SOLUTION

State:

$H_0: \mu_N - \mu_S = 0$

$H_a: \mu_N - \mu_S \neq 0$

where μ_N = the mean DBH (cm) of all trees in the northern half of the forest and μ_S = the mean DBH (cm) of all trees in the southern half of the forest

Use $\alpha = 0.05$.

The evidence for H_a is: $\bar{x}_N - \bar{x}_S = 23.70 - 34.53 = -10.83 \neq 0$

Plan: Two-sample t test for $\mu_N - \mu_S$

Random? The data came from independent random samples of trees from the northern and southern halves of the forest. ✓

Normal/Large Sample? Both sample sizes are large:

$n_N = 30 \geq 30$ and $n_S = 30 \geq 30$. ✓

Do:

- $t = \dfrac{(23.70 - 34.53) - 0}{\sqrt{\dfrac{17.50^2}{30} + \dfrac{14.26^2}{30}}} = -2.628$

- P-value:

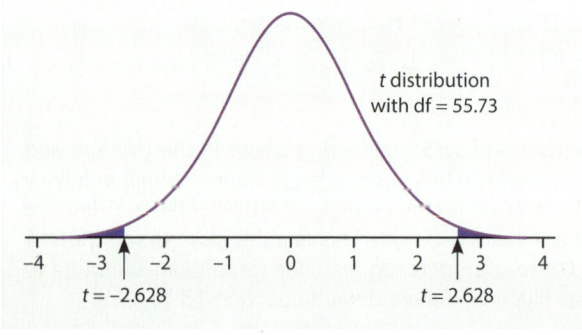

t distribution with df = 55.73

$t = -2.628$ $t = 2.628$

> **State:** State the hypotheses, parameter(s), significance level, and evidence for H_a.

> **Plan:** Identify the appropriate inference method and check the conditions.

> **Do:** If the conditions are met, perform calculations.
> • Calculate the test statistic.
> • Find the P-value.

Using technology: $t = -2.628$; *P*-value $= 0.011$ using df $= 55.73$

Conservative approach: df $=$ smaller of $(30 - 1, 30 - 1) = 29$;

Table B gives a *P*-value between $2(0.005) = 0.01$ and $2(0.01) = 0.02$

Conclude: Because the *P*-value of 0.011 is less than $\alpha = 0.05$, we reject H_0. These data give convincing evidence of a difference in the mean DBH of all trees in the northern and southern halves of the forest.

> Use technology to get a more precise *P*-value: Applet/ tcdf(lower: -1000, upper: -2.628, df: 29) $\times 2 = 0.0136$.
>
> **Conclude:** Make a conclusion about the hypotheses in the context of the problem.

EXAM PREP: FOR PRACTICE, TRY EXERCISE 13.

We can get additional information about the difference between the population means in Example 9.4.3 by constructing a confidence interval. Technology gives the 95% confidence interval for $\mu_N - \mu_S$ as -19.087 to -2.573. That is, we are 95% confident that the interval from -19.087 to -2.573 captures the difference (North − South) in the mean DBH (in cm) of all the northern trees and all the southern trees in the Wade Tract Forest Preserve. The interval suggests that the mean DBH of the trees in the northern half of the forest is between 2.573 cm and 19.087 cm smaller than the mean DBH of the trees in the southern half of the forest. The 95% confidence interval is consistent with our "reject H_0" conclusion at the 5% significance level, because 0 is not included in the interval of plausible values for $\mu_N - \mu_S$.

TECH CORNER: Significance Tests for a Difference Between Two Means

You can use technology to perform the calculations for a significance test of $H_0: \mu_1 - \mu_2 = 0$. We'll illustrate using the calcium and blood pressure data from Examples 9.4.1 and 9.4.2. Here are the data once again:

Group 1 (calcium)	7	−4	18	17	−3	−5	1	10	11	−2
Group 2 (placebo)	−1	12	−1	−3	3	−5	5	2	−11	−1	−3

Applet

1. Go to www.stapplet.com and launch the *One Quantitative Variable, Multiple Groups* applet.
2. Enter "Decrease in systolic blood pressure" as the variable name.
3. Choose to input raw data. Type the group names "Calcium" and "Placebo" in the boxes at the left. Enter the data for each group manually, or copy and paste them from a document or spreadsheet. Be sure to separate the data values with commas or spaces.

One Quantitative Variable, Multiple Groups

Variable name: Decrease in systolic blood pressure
Input: Raw data

Group	Name	Input data separated by commas or spaces.
1	Calcium	7 −4 18 17 −3 −5 1 10 11 −2
2	Placebo	−1 12 −1 −3 3 −5 5 2 −11 −1 −3

Add group

Begin analysis | Edit inputs | Reset everything

4. Click the Begin analysis button. Scroll down to the "Perform Inference" section. Choose the 2-sample *t* test $(\mu_1 - \mu_2)$ for the procedure, "No" for the Conservative degrees of freedom, and $\mu_1 - \mu_2 > 0$ for the alternative hypothesis. Then click the Perform inference button.

Perform Inference

Inference procedure: 2-sample t test $(\mu_1 - \mu_2)$ | Conservative degrees of freedom: No
Alternative hypothesis: $\mu_1 - \mu_2 > 0$

Perform inference

t	P-value	df
1.604	0.064	15.591

Note: If you have summary statistics instead of raw data, choose "Mean and standard deviation" from the Input menu in Step 3 and enter the name, mean, standard deviation, and sample size for each group. Then follow the directions in Step 4.

(continued)

TI-83/84

1. Enter the decrease in systolic blood pressure data for the calcium group in list L_1 and for the placebo group in list L_2.
2. Press STAT, then choose TESTS and 2-SampTTest.

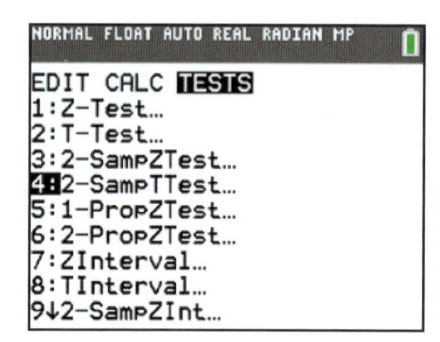

```
NORMAL FLOAT AUTO REAL RADIAN MP
EDIT CALC TESTS
1:Z-Test…
2:T-Test…
3:2-SampZTest…
4:2-SampTTest…
5:1-PropZTest…
6:2-PropZTest…
7:ZInterval…
8:TInterval…
9↓2-SampZInt…
```

3. When the 2-SampTTest screen appears, choose Data as the input method and enter the values shown. Be sure to choose $\mu_1 > \mu_2$ as the alternative hypothesis and "No" for Pooled.

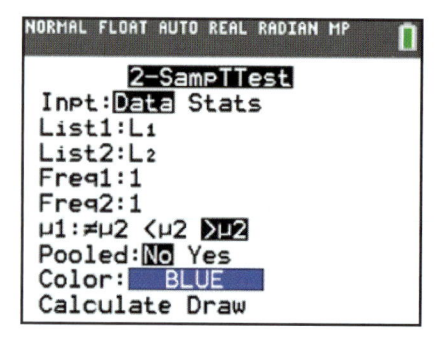

```
NORMAL FLOAT AUTO REAL RADIAN MP
2-SampTTest
Inpt: Data Stats
List1: L1
List2: L2
Freq1: 1
Freq2: 1
μ1: ≠μ2  <μ2  >μ2
Pooled: No  Yes
Color:    BLUE
Calculate  Draw
```

4. Highlight "Calculate" and press ENTER.

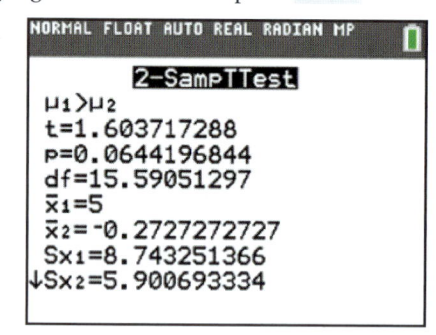

```
NORMAL FLOAT AUTO REAL RADIAN MP
2-SampTTest
μ1>μ2
t=1.603717288
p=0.0644196844
df=15.59051297
x̄1=5
x̄2=-0.2727272727
Sx1=8.743251366
↓Sx2=5.900693334
```

Note: To perform a significance test for a difference in means from summary statistics, start with Step 2 and choose Stats as the input method in Step 3.

Detailed instructions for using CrunchIt!, Excel, Google Sheets, JMP, Minitab, and R are available in **Achieve**.

Section 9.4 What Did You Learn?

Review the learning goals from this section. Then practice what you've learned by working through the exercises.

Learning Goal	Example	Exercises
State hypotheses and check conditions for performing a significance test about a difference between two means.	9.4.1	5–8
Calculate the standardized test statistic and P-value for a significance test about a difference between two means.	9.4.2	9–12
Use the four-step process to perform a significance test about a difference between two means.	9.4.3	13–16

Section 9.4 Exercises

Building Concepts and Skills These exercises assess the basic knowledge you should have after reading the section.

1. Use appropriate notation to write out the null hypothesis that there is no difference between two population means.

2. To satisfy the Normal/Large Sample condition for performing a significance test for a difference in means: For each sample, the data come from a _____ population or the sample size is _____ (_____). For each sample, if the population distribution has unknown shape and $n < 30$, a graph of the sample data shows no _____ or _____.

3. What is the formula for the t statistic for a test of no difference between two population means?

4. True/False: Using technology for a two-sample t test for a difference between two means results in a smaller df and a larger P-value than the more conservative approach using df = smaller of $n_1 - 1$ and $n_2 - 1$.

Mastering Concepts and Skills These exercises reinforce the learning goals as shown in the examples.

5. **Refer to Example 9.4.1 Work Hours** Has the mean number of hours worked in a week changed in the

United States over time? One of the questions on the General Social Survey (GSS) survey asks respondents how many hours they work each week. Responses from random samples of U.S. employees in 1978 and 2018 are summarized in the table.[44]

Year	n	Mean	SD
2018	1381	41.28	6.59
1978	855	40.81	8.40

Researchers want to know if there is a difference in the mean number of hours worked per week for all U.S. employees in 1978 and 2018.

(a) State appropriate hypotheses for performing a significance test. Be sure to define the parameters of interest.

(b) Do the data provide *some* evidence of a difference in the mean number of hours worked per week for all U.S. employees in 1978 and 2018? Justify your answer.

(c) Check that the conditions for performing the test are met.

6. **Sleep Deprivation** Does sleep deprivation linger for more than a day? Researchers designed a study using 21 volunteer subjects between the ages of 18 and 25. All 21 participants took a computer-based visual discrimination test at the start of the study. Then the subjects were randomly assigned into two groups. The 11 subjects in one group were deprived of sleep for an entire night in a laboratory setting. The 10 subjects in the other group were allowed unrestricted sleep for the night. Both groups were allowed as much sleep as they wanted for the next two nights. On Day 4, all the subjects took the same visual discrimination test on the computer. Researchers recorded the improvement in time (measured in milliseconds) from Day 1 to Day 4 on each subject's tests. Here are the data:[45]

Sleep deprivation	−14.7	−10.7	−10.7	2.2	2.4	4.5
	7.2	9.6	10.0	21.3	21.8	
Unrestricted sleep	−7.0	11.6	12.1	12.6	14.5	18.6
	25.2	30.5	34.5	45.6		

The researchers want to know if sleep deprivation decreases the true mean improvement time on the visual discrimination task compared to unrestricted sleep for subjects like the ones in this study.

(a) State appropriate hypotheses for performing a significance test. Be sure to define the parameters of interest.

(b) Do the data provide *some* evidence that sleep deprivation decreases the true mean improvement time on the visual discrimination task compared to unrestricted sleep for subjects like the ones in this study? Justify your answer.

(c) Check that the conditions for performing the test are met.

7. **Creativity and Motivation** Do external rewards—such as money, praise, fame, and grades—promote or inhibit creativity? Researcher Teresa Amabile recruited 47 experienced creative writers who were college students and divided them at random into two groups. The students in one group were given a list of statements about external reasons (E) for writing, such as public recognition, making money, or pleasing their parents. Students in the other group were given a list of statements about internal reasons (I) for writing, such as expressing yourself and enjoying playing with words. Both groups were then instructed to write a poem about laughter. Each student's poem was rated separately by 12 different poets using a creativity scale. These ratings were averaged to obtain an overall creativity score for each poem. Here are the data:[46]

Internal	12.0	12.0	12.9	13.6	16.6	17.2
	17.5	18.2	19.1	19.3	19.8	20.3
	20.5	20.6	21.3	21.6	22.1	22.2
	22.6	23.1	24.0	24.3	26.7	29.7
External	5.0	5.4	6.1	10.9	11.8	12.0
	12.3	14.8	15.0	16.8	17.2	17.2
	17.4	17.5	18.5	18.7	18.7	19.2
	19.5	20.7	21.2	22.1	24.0	

Dr. Amabile wants to know if there is a difference in the true mean creativity scores for students like the ones in this study when given internal versus external reasons for writing.

(a) State appropriate hypotheses for performing a significance test. Be sure to define the parameters of interest.

(b) Do the data provide *some* evidence of a difference in the true mean creativity scores for students like the ones in this study when given internal versus external reasons for writing? Justify your answer.

(c) Check that the conditions for performing the test are met.

8. **Happy Customers** As the Spanish-speaking population in the United States has grown, businesses have increased their focus on providing an English or Spanish option for customer service. One study interviewed a random sample of customers leaving a bank. Customers were asked if they spoke English or Spanish when talking to bank representatives. Due to the sampling method used in this survey, it is reasonable to consider these as independent random samples of customers who speak each of the two languages. Each customer rated the importance of several aspects of bank service on a 10-point scale. Here are summary results for the importance of "reliability" (the accuracy of account records, etc.):[47]

Group	n	Mean	SD
English	92	6.37	0.60
Spanish	86	5.91	0.93

Researchers want to know if there is a difference in the mean reliability ratings of customers who speak English and customers who speak Spanish when talking to bank representatives.

(a) State appropriate hypotheses for performing a significance test. Be sure to define the parameters of interest.

(b) Do the data provide *some* evidence of a difference in the mean reliability ratings of all customers who speak English and all customers who speak Spanish when talking to bank representatives? Justify your answer.

(c) Check that the conditions for performing the test are met.

9. Refer to Example 9.4.2 Work Hours Refer to Exercise 5.

(a) Calculate the standardized test statistic.

(b) Find the *P*-value.

10. Sleep Deprivation Refer to Exercise 6.

(a) Calculate the standardized test statistic.

(b) Find the *P*-value.

11. Creativity and Motivation Refer to Exercise 7.

(a) Calculate the standardized test statistic and *P*-value.

(b) What conclusion would you make?

12. Happy Customers Refer to Exercise 8.

(a) Calculate the standardized test statistic and *P*-value.

(b) What conclusion would you make?

13. Refer to Example 9.4.3 Education and Income Is it true that students who earn an associate's degree or bachelor's degree make more money than students who attend college but do not earn a degree? To find out, researchers took a random sample of 500 U.S. residents aged 18 and older who had attended college from a recent Current Population Survey.[48] They recorded the educational attainment and annual income of each person. Due to the sampling method used in this survey, it is reasonable to consider these as independent random samples of students who earned degrees and those who did not. Here are summary statistics for the two groups:

Group	n	Mean	SD
Earned degree	327	$49,454.80	$51,257.10
No degree	173	$29,299.20	$38,298.00

Do these data provide convincing evidence that the mean annual income is higher in the population of college graduates than in the population of college attendees who did not graduate?

14. Music and Plant Growth Researchers designed an experiment to determine whether plants grow better if they are exposed to classical music or to metal music. They selected 10 bean seeds and planted each in a Styrofoam cup. They randomly assigned half of these cups to be exposed to metal music each night, while the other half were exposed to classical music each night. They recorded the amount of growth, in millimeters, for each plant after 2 weeks. Here are the data:[49]

Metal	22	36	73	57	3
Classical	87	78	124	121	19

Do these data provide convincing evidence of a difference in the true mean growth of plants like the ones in this study that are exposed to classical music versus metal music?

15. Fish Oil To see if fish oil can help reduce blood pressure, researchers recruited 14 men with high blood pressure and performed an experiment. They randomly assigned 7 of the men to a 4-week diet that included fish oil. They assigned 7 other men to a 4-week diet that included a mixture of oils that approximated the types of fat in a typical diet. At the end of the 4 weeks, they measured each volunteer's blood pressure (in mmHg) again and recorded the reduction in diastolic blood pressure. These differences are shown in the table. Note that a negative value means that the subject's blood pressure increased.[50]

Fish oil	8	12	10	14	2	0	0
Regular oil	−6	0	1	2	−3	−4	2

Do these data provide convincing evidence at the 5% significance level that fish oil helps reduce blood pressure more than regular oil, on average, for men like the ones in this study?

16. Shoe Prices Researchers who planned to go shopping for shoes wondered how the prices at Payless compared to the prices at Famous Footwear. From each company, they randomly selected 30 pairs of shoes and recorded the price of each pair. The table shows summary statistics for the two samples of shoes.[51]

Store	Mean	SD
Famous Footwear	$42.67	$16.33
Payless	$20.00	$7.31

Do these data provide convincing evidence at the 5% significance level that shoes cost less, on average, at Payless than at Famous Footwear?

Applying the Concepts These exercises ask you to apply multiple learning goals in a new context or to apply what you learned in this section in a new way.

17. Reading Activities An educator believes that new reading activities in the classroom will help elementary school students improve their reading ability. She recruits 44 third-grade students and randomly assigns them to two groups. One group of 21 students does these new activities for an 8-week period. A control group of 23 third-graders follows the same curriculum without the activities. At the end of the 8 weeks, all students are given the Degree of Reading Power (DRP) test, which measures the aspects of reading ability that the treatment is designed to improve. Parallel boxplots and summary statistics for the groups are shown here.[52]

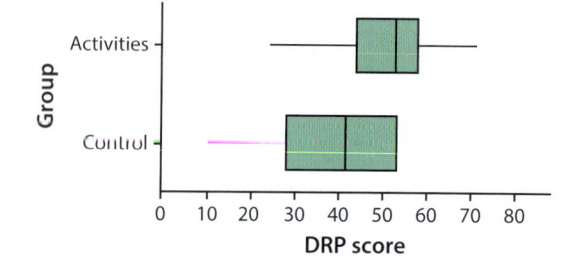

Group	n	Mean	SD
Activities	21	51.48	11.01
Control	23	41.52	17.15

(a) Is there convincing evidence at the $\alpha = 0.05$ significance level that the true mean DRP score is greater for students like the ones in this study who do the reading activities?

(b) Can we conclude that the new reading activities caused an increase in the mean DRP score? Justify your answer.

(c) Based on your conclusion in part (a), which type of error — a Type I error or a Type II error — could you have made? Explain your answer.

18. **Possible Bias?** As someone who speaks English as a second language, researcher Sanda is convinced that people find more grammar and spelling mistakes in essays when they think the writer is a non-native English speaker. To test this, Sanda recruits a group of 60 volunteers and randomly assigns them into two groups of 30. Both groups are given the same paragraph to read. One group is told that the author of the paragraph is someone whose native language is not English. The other group is told nothing about the author. The subjects are asked to count the number of spelling and grammar mistakes in the paragraph. While the two groups found about the same number of *real* mistakes in the passage, the number of things that were *incorrectly* identified as mistakes was more interesting. Here are a dotplot and some numerical summaries of the data on "mistakes" found.

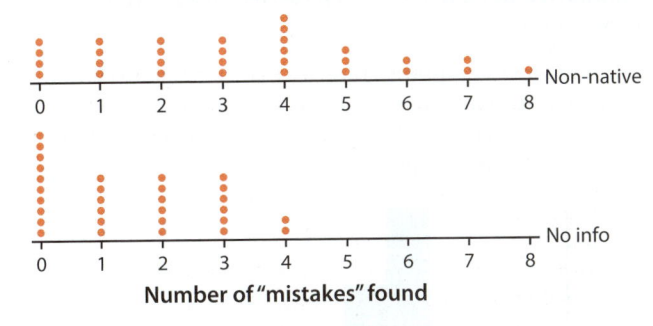

Number of "mistakes" found

Group	Mean	SD
Non-native	3.23	2.25
No info	1.47	1.33

(a) Is there convincing evidence at the $\alpha = 0.05$ significance level that the mean number of "mistakes" found will be greater when people like these volunteers are told that the author is someone whose native language is not English?

(b) Can we conclude that telling volunteers the author is someone whose native language is not English caused an increase in the mean number of "mistakes" found? Justify your answer.

(c) Based on your conclusion in part (a), which type of error — a Type I error or a Type II error — could you have made? Explain your answer.

19. **Reading Activities** Refer to Exercise 17.
    (a) Construct and interpret a 95% confidence interval for the difference between the true means. If you already defined parameters and checked conditions in Exercise 17, you don't need to do those tasks again here.
    (b) Explain how the confidence interval provides more information than the test in Exercise 17.

20. **Possible Bias?** Refer to Exercise 18.
    (a) Construct and interpret a 95% confidence interval for the difference between the true means. If you already defined parameters and checked conditions in Exercise 18, you don't need to do those tasks again here.
    (b) Explain how the confidence interval provides more information than the test in Exercise 18.

21. **Work Hours** Refer to Exercise 9.
    (a) Interpret the *P*-value.
    (b) What conclusion would you make?

22. **Sleep Deprivation** Refer to Exercise 10.
    (a) Interpret the *P*-value.
    (b) What conclusion would you make?

23. Tech **Glass Forensics** Glass can be used as evidence in a criminal investigation if it can be correctly identified. Researchers were interested in whether glass types can be determined by certain characteristics. They designed a study that used a total of 214 randomly selected specimens of glass of several different types, including headlamps and vehicle windows that were float processed (made by pouring molten glass onto molten tin). The **Glass** data set contains measurements of various characteristics of these glass specimens, including the percentage of aluminum by weight (*Aluminum*).[53] Extract the data for the specimens labeled "Headlamp" and "Vehicle window float processed" in the *Type* column for analysis. Due to the sampling method used in this survey, it is reasonable to view these specimens as independent random samples from the corresponding populations of these two types of glass. Do the data provide convincing evidence at the 5% significance level of a difference in the mean amount of aluminum contained in all headlamps and all vehicle windows that are float processed?

24. Tech **Possum Length** Does the size of Australian possums differ by sex? Zoologists in Australia captured a random sample of 104 possums and recorded the values of 13 different variables, including total length (*totlngth*, in centimeters) and *sex*, in the **Possum** data set.[54] Due to the sampling method used in this survey, it is reasonable to view the male and female possums in this data set as independent random samples from the corresponding populations. Do the data provide convincing evidence at the 5% significance level of a difference in the mean total length of male and female Australian possums?

Extending the Concepts These exercises challenge you to explore statistical concepts and methods that go beyond what you learned in this section.

25. **Overthinking** Athletes often comment that they try not to "overthink it" when competing in their sport. Is it possible to "overthink"? To investigate, researchers put some golfers to the test. They recruited 40 experienced golfers and allowed them some time to practice their putting. After practicing, they randomly assigned the golfers in equal numbers to two groups. Golfers in one group had to write a detailed description of their putting technique (which could lead to "overthinking it"). Golfers in the other group had to do an unrelated verbal task for the same amount of time. After completing their tasks, each golfer was asked to attempt putts from a fixed distance until they made three putts in a row. Here are the data, along with boxplots and summary statistics of the distribution of the number of putts required for the golfers in each group to make three putts in a row:[55]

Described putting	26	17	12	8	28	24	37	12	3	65
	15	43	3	16	29	5	17	28	25	11
Didn't describe putting	19	16	6	37	3	5	20	5	3	4
	4	3	9	5	4	11	32	6	5	15

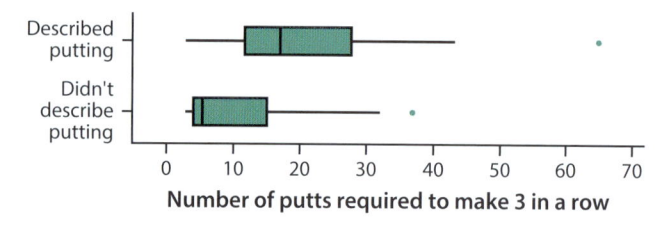

Group Name	n	mean	SD
1: Described putting	20	21.2	15.07
2: Didn't describe putting	20	10.6	9.832

Let μ_1 = the true mean number of putts required by subjects like the ones in this study after describing their putting and μ_2 = the true mean number of putts required by subjects like the ones in this study after not describing their putting. The Normal/Large Sample condition for inference about $\mu_1 - \mu_2$ is not met because $n_1 = n_2 = 20 < 30$ and both distributions are clearly skewed right with a high outlier.

We used the *One Quantitative Variable, Multiple Groups* applet to randomly reassign the 40 subjects to the two groups 500 times, assuming the treatment received doesn't affect each golfer's putting. A dotplot of the simulated difference in the mean number of putts required to make three in a row (Described putting − Didn't describe putting) is shown here.

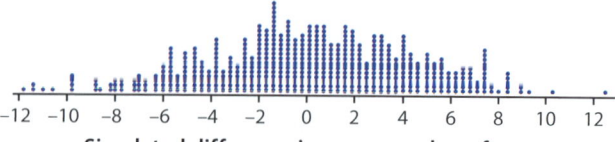

(a) State appropriate hypotheses for testing whether there is convincing evidence of overthinking it when putting by golfers like the ones in this study.

(b) What is the estimated *P*-value?

(c) What conclusion would you make?

26. **Hotel Toilets** A company that makes hotel toilets claims that its new pressure-assisted toilet reduces the average amount of water used by more than 0.5 gallon per flush when compared to its current model. To test this claim, the company randomly selects 30 toilets of each type and measures the amount of water that is used when each toilet is flushed once. For the current-model toilets, the mean amount of water used is 1.64 gallons with a standard deviation of 0.29 gallon. For the new toilets, the mean amount of water used is 1.09 gallons with a standard deviation of 0.18 gallon.

(a) Carry out an appropriate significance test. What conclusion would you make? (Note that the null hypothesis is *not* $H_0: \mu_1 - \mu_2 = 0$.)

(b) Based on your conclusion in part (a), could you have made a Type I error or a Type II error? Explain your answer.

27. **Rejection Region** Refer to Exercise 18 and the Think About It in Section 8.3.

(a) Find the critical value and the rejection region.

(b) What conclusion would you make? Explain your reasoning.

Cumulative Review These exercises revisit what you learned in previous sections.

28. **Europe's Economies (2.5, 3.1)** The histogram shows the distribution of per-capita gross domestic product (GDP), expressed in thousands of U.S. dollars, for 39 countries in Europe.[56]

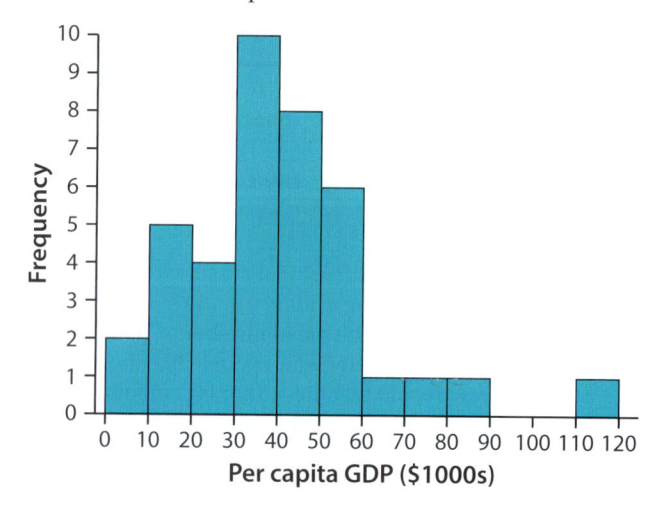

(a) In what percentage of countries is per-capita GDP less than $20,000?

(b) Describe the shape of the distribution and identify any possible outliers.

(c) Is it possible to find the exact value of the median GDP from this histogram? Explain your reasoning.

29. **Price Cuts (1.5)** Stores advertise price reductions to attract customers. What type of price cut is most attractive? Experiments with more than one factor allow insight into interactions between the factors. A study of the attractiveness of advertised price discounts focused on two factors: percentage of all goods on sale (25%, 50%, 75%, or 100%) and whether the discount was stated precisely (e.g., "60% off") or as a range (e.g., "40% to 70% off"). Subjects rated the attractiveness of the sale on a scale of 1 to 7.

   (a) List the treatments for this experiment, assuming researchers will use all combinations of the two factors.
   (b) Describe how you would randomly assign 200 volunteer subjects to treatments.
   (c) Explain the purpose of the random assignment in part (b).
   (d) The figure shows the mean ratings for the eight treatments formed from the two factors.[57] Based on these results, write a careful description of how percentage of goods on sale and precise discount versus range of discounts influence the attractiveness of a sale.

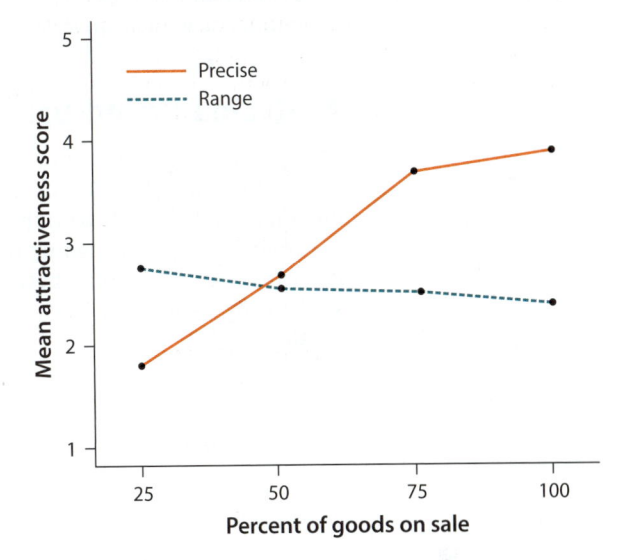

Section 9.5 Analyzing Paired Data: Confidence Intervals for a Population Mean Difference

LEARNING GOALS

By the end of this section, you will be able to:

- Use a graph to analyze the distribution of differences in a set of paired data.
- Calculate the mean and standard deviation of the differences in a set of paired data, and interpret the mean difference.
- Use the four-step process to construct and interpret a confidence interval for a mean difference.

Sections 9.3 and 9.4 showed you how to perform inference about the difference between two means when data come from two independent random samples or two groups in a randomized experiment. What if we want to compare means in a setting that involves measuring a quantitative variable twice for the same individual or for two very similar individuals?

For instance, trace metals found in wells affect the taste of drinking water, and high concentrations can pose a health risk. Researchers measured the concentration of zinc (in milligrams per liter, mg/L) near the top and the bottom of 10 randomly selected wells in a large region. The data are provided in the table.[58]

Well	1	2	3	4	5	6	7	8	9	10
Top	0.415	0.238	0.390	0.410	0.605	0.609	0.632	0.523	0.411	0.612
Bottom	0.430	0.266	0.567	0.531	0.707	0.716	0.651	0.589	0.469	0.723

Notice that these two groups of zinc concentrations did *not* come from independent random samples of measurements made at the top of some wells and at the bottom of other wells. The data were obtained from *pairs* of measurements — one measurement at the top of a well and the other measurement at the bottom of the *same* well. This set of zinc concentrations is an example of **paired data**.

Paired data result from recording two values of the same quantitative variable for each individual or for each pair of similar individuals.

This section focuses on how to analyze paired data and how to construct and interpret a confidence interval for a *mean difference*.

Analyzing Paired Data

The graph in Figure 9.3 shows parallel dotplots of the zinc concentrations from the observational study of wells. We can see that the zinc concentration tends to be lower, on average, at the top of wells ($\bar{x}_T = 0.4845$) than at the bottom of wells ($\bar{x}_B = 0.5649$). There is a similar amount of variability in zinc concentration at the top and bottom of these wells: $s_T = 0.1312$ and $s_B = 0.1468$. But with so much overlap between the groups, the difference in means does not appear to be statistically significant. A test of $H_0: \mu_B - \mu_T = 0$ versus a two-sided alternative yields $t = 1.29$ and P-value $= 0.213$.

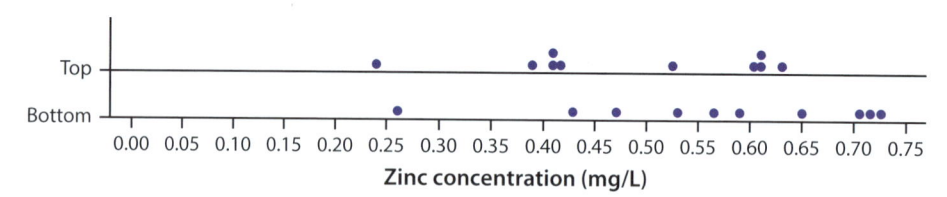

FIGURE 9.3 Parallel dotplots of the zinc concentrations at the top and bottom of 10 randomly selected wells in a large region.

The previous analysis ignores the fact that these are *paired* data. Let's look at the difference in zinc concentrations at the top and bottom of each well.

Well	1	2	3	4	5	6	7	8	9	10
Top	0.415	0.238	0.390	0.410	0.605	0.609	0.632	0.523	0.411	0.612
Bottom	0.430	0.266	0.567	0.531	0.707	0.716	0.651	0.589	0.469	0.723
Difference (Bottom − Top)	0.015	0.028	0.177	0.121	0.102	0.107	0.019	0.066	0.058	0.111

The dotplot in Figure 9.4 displays these differences. Notice that all the differences are positive!

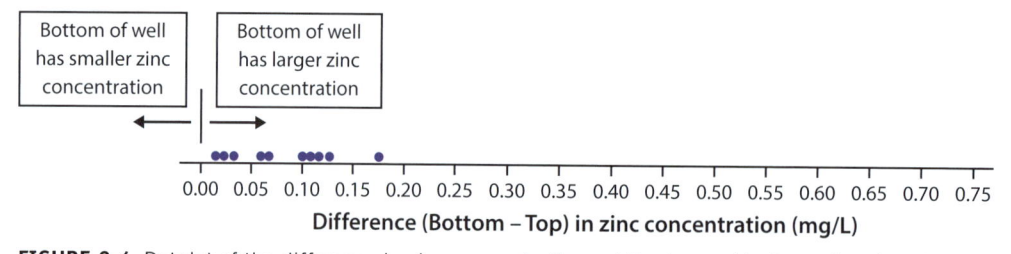

FIGURE 9.4 Dotplot of the difference in zinc concentrations at the top and bottom of each well.

For all 10 wells in the sample, the zinc concentration is greater at the bottom of the well than at the top of the well. This graph provides stronger evidence that zinc concentration tends to be larger at the bottom than at the top of wells in this region. Because we used the same scale in Figures 9.3 and 9.4, you can also see that there is much less variability in the paired differences than in the unpaired zinc concentrations at the top and bottom of the wells.

Analyzing Paired Data Graphically

To analyze paired data, start by computing the difference for each pair. Then use a graph to see if the differences are consistently greater than or consistently less than 0.

Section 9.5 Analyzing Paired Data: Confidence Intervals for a Population Mean Difference

EXAMPLE 9.5.1

Do cross-fertilized or self-fertilized plants grow taller?

Analyzing paired data with graphs

PROBLEM

Charles Darwin, author of *On the Origin of Species* (1859), designed an experiment to compare the effects of cross-fertilization and self-fertilization on the size of plants. He planted pairs of very similar seedling *Zea mays* (maize) plants, one cross-fertilized and one self-fertilized, in each of 15 pots at the same time. He randomly assigned plants to pots. After a period of time, Darwin measured the heights (in inches) of all the plants. Here are the data:[59]

Pair	1	2	3	4	5	6	7	8	9	10	11	12	13	14	15
Cross	23.5	12.0	21.0	22.0	19.1	21.5	22.1	20.4	18.3	21.6	23.3	21.0	22.1	23.0	12.0
Self	17.4	20.4	20.0	20.0	18.4	18.6	18.6	15.3	16.5	18.0	16.3	18.0	12.8	15.5	18.0

Make a dotplot of the difference in height (Cross – Self) for each pair of maize plants. What does the graph reveal?

SOLUTION

Pair	1	2	3	4	5	6	7	8	9	10	11	12	13	14	15
Difference (Cross – Self)	6.1	–8.4	1.0	2.0	0.7	2.9	3.5	5.1	1.8	3.6	7.0	3.0	9.3	7.5	–6.0

Start by computing the difference (Cross – Self) for each pair of plants.

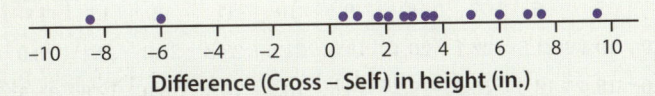

There are 13 positive differences and 2 negative differences, meaning that in 13 of the 15 pairs of maize plants, the cross-fertilized plant grew taller. The difference values vary from –8.4 (self-fertilized plant was 8.4 inches taller than the cross-fertilized plant in that pair) to 9.3 (cross-fertilized plant was 9.3 inches taller than the self-fertilized plant in that pair).

EXAM PREP: FOR PRACTICE, TRY EXERCISE 7.

THE MEAN AND STANDARD DEVIATION OF THE DIFFERENCES

Now that we have looked at a graph of the differences for paired data, it's time to calculate numerical summaries. Here again are the data on zinc concentration (in mg/L) at the top and bottom of 10 randomly selected wells in a large region, along with a rescaled dotplot of the differences (Bottom – Top).

Well	1	2	3	4	5	6	7	8	9	10
Top	0.415	0.238	0.390	0.410	0.605	0.609	0.632	0.523	0.411	0.612
Bottom	0.430	0.266	0.567	0.531	0.707	0.716	0.651	0.589	0.469	0.723
Difference (Bottom – Top)	0.015	0.028	0.177	0.121	0.102	0.107	0.019	0.066	0.058	0.111

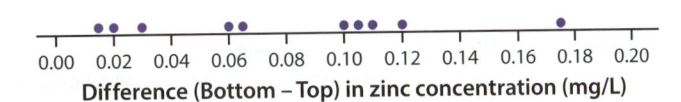

The *mean difference* for these data is

$$\bar{x}_{\text{diff}} = \bar{x}_{B-T} = \frac{0.015 + 0.028 + \ldots + 0.111}{10} = \frac{0.804}{10} = 0.0804 \text{ mg/L}$$

This value tells us that the zinc concentration at the bottom of a well is 0.0804 mg/L greater, on average, than the zinc concentration at the top of the well for this random sample of 10 wells.

The standard deviation of the differences is

$$s_{\text{diff}} = s_{B-T} = \sqrt{\frac{(0.015 - 0.0804)^2 + (0.028 - 0.0804)^2 + \ldots + (0.111 - 0.0804)^2}{10 - 1}} = 0.0523$$

This value tells us that the difference in zinc concentrations at the top and bottom of wells in this region typically varies from the mean difference by about 0.0523 mg/L. Notice that s_{diff} is much smaller than the standard deviations we computed earlier when we (incorrectly) viewed the two groups of zinc concentration measurements at the top and bottom of these 10 wells as unrelated: $s_T = 0.1312$ and $s_B = 0.1468$. ❗ **Remember: The proper method of analysis depends on how the data are produced.**

Analyzing Paired Data Numerically

When analyzing paired data, use the mean difference $\bar{x}_{diff}$ and the standard deviation of the differences s_{diff} as summary statistics.

For practical purposes, you can use technology to find the mean and standard deviation of the differences.

EXAMPLE 9.5.2

Do cross-fertilized or self-fertilized plants grow taller?

Analyzing paired data numerically

PROBLEM

Refer to Example 9.5.1. Here again are the data on the heights (in inches) of the cross-fertilized and self-fertilized maize plants in each pair from Darwin's study, along with the differences (Cross − Self).

Pair	1	2	3	4	5	6	7	8	9	10	11	12	13	14	15
Cross	23.5	12.0	21.0	22.0	19.1	21.5	22.1	20.4	18.3	21.6	23.3	21.0	22.1	23.0	12.0
Self	17.4	20.4	20.0	20.0	18.4	18.6	18.6	15.3	16.5	18.0	16.3	18.0	12.8	15.5	18.0
Difference (Cross − Self)	6.1	−8.4	1.0	2.0	0.7	2.9	3.5	5.1	1.8	3.6	7.0	3.0	9.3	7.5	−6.0

Calculate the mean and standard deviation of the difference (Cross − Self) in height for each pair of maize plants. Interpret the mean difference.

SOLUTION

$$\bar{x}_{diff} = \bar{x}_{C-S} = \frac{6.1 + (-8.4) + \ldots + (-6.0)}{15} = \frac{39.1}{15} = 2.60667$$

$$s_{diff} = s_{C-S} = \sqrt{\frac{(6.1 - 2.60667)^2 + (-8.4 - 2.60667)^2 + \ldots + (-6.0 - 2.60667)^2}{15 - 1}} = 4.713$$

The cross-fertilized maize plant in each pair grew 2.60667 inches taller, on average, than the self-fertilized maize plant.

EXAM PREP: FOR PRACTICE, TRY EXERCISE 11.

THINK ABOUT IT

What's the relationship between the mean difference and the difference in the means? For the observational study of 10 randomly selected wells in a large region, the mean zinc concentrations are $\bar{x}_B = 0.5649$ at the bottom of the wells and $\bar{x}_T = 0.4845$ mg/L at the top of the wells. The difference in the mean zinc concentrations is

$$\bar{x}_B - \bar{x}_T = 0.5649 - 0.4845 = 0.0804 \text{ mg/L}$$

But that is exactly the same value as the mean difference: $\bar{x}_{diff} = \bar{x}_{B-T} = 0.0804$ mg/L. This result holds in general: *The mean difference is equal to the difference in the means.*

So why do we need a different method of analysis? Because the standard error of the mean difference is *not* equal to the standard error of the difference in means. The standard error of the mean difference is

$$SE_{\bar{x}_{diff}} = \frac{s_{diff}}{\sqrt{n_{diff}}} = \frac{0.0523}{\sqrt{10}} = 0.0165 \text{ mg/L}$$

which is less than the standard error of the difference in means

$$SE_{\bar{x}_B - \bar{x}_T} = \sqrt{\frac{s_B^2}{n_B} + \frac{s_T^2}{n_T}} = \sqrt{\frac{0.1468^2}{10} + \frac{0.1312^2}{10}} = 0.0623 \text{ mg/L}$$

Using the appropriate standard error when data are paired typically results in narrower confidence intervals and smaller P-values. We'll discuss how to choose the proper method of inference in Section 9.6.

Putting It All Together: Constructing and Interpreting a Confidence Interval for μ_{diff}

When paired data come from a random sample or a randomized experiment, the statistic $\bar{x}_{diff}$ is a point estimate for the population mean difference μ_{diff}. Before constructing a confidence interval for μ_{diff}, we also have to check that the Normal/Large Sample condition is met.

HOW TO Check Conditions for Constructing a Confidence Interval for μ_{diff}

To construct a confidence interval for a population mean difference μ_{diff}, check that the following conditions are met:
- **Random:** Paired data come from a random sample from the population of interest or from a randomized experiment.
- **Normal/Large Sample:** Paired data come from a normally distributed population of differences or the sample size is large ($n_{diff} \geq 30$). When the sample size is small and the shape of the population distribution of differences is unknown, a graph of the sample differences shows no strong skewness or outliers.

If these conditions are met, we can safely calculate a confidence interval for a population mean difference. The method is the same as in Sections 7.5 and 7.6, except that the data are just the sample of differences within each pair. Note that the Normal/Large Sample condition is *not* met for Darwin's study of maize plants because the sample size is small ($n_{diff} = 15 < 30$) and the difference of −8.4 inches in the growth of the cross-fertilized plant and the self-fertilized plant in Pair 2 is an outlier.

HOW TO Calculate a Confidence Interval for μ_{diff}

When the Random and Normal/Large Sample conditions are met, a C% confidence interval for the population mean difference μ_{diff} is

$$\bar{x}_{diff} \pm t^* \frac{s_{diff}}{\sqrt{n_{diff}}}$$

where t^* is the critical value for a t distribution with df $= n_{diff} - 1$ and C% of its area between $-t^*$ and t^*.

As with any inference procedure, follow the four-step process when you construct and interpret a **paired t interval for a mean difference.** You can use technology to perform the calculations in the Do step with either raw data or summary statistics of the differences. See the Tech Corner in Section 7.5 for details.

> A **paired t interval for a mean difference** is a confidence interval used to estimate a population (true) mean difference for paired data.

This procedure is sometimes called a *one-sample t interval for a mean difference* to remind you that it is basically the same method you learned in Section 7.5.

EXAMPLE 9.5.3

Does zinc sink?

Confidence intervals for μ_{diff}

PROBLEM

The data on zinc concentration (in mg/L) at the top and bottom of 10 randomly selected wells in a large region are shown again in the table, along with the differences we calculated earlier.

Well	1	2	3	4	5	6	7	8	9	10
Top	0.415	0.238	0.390	0.410	0.605	0.609	0.632	0.523	0.411	0.612
Bottom	0.430	0.266	0.567	0.531	0.707	0.716	0.651	0.589	0.469	0.723
Difference (Bottom − Top)	0.015	0.028	0.177	0.121	0.102	0.107	0.019	0.066	0.058	0.111

Construct and interpret a 95% confidence interval for the mean difference in zinc concentrations at the top and bottom of all wells in this region.

SOLUTION

State: 95% CI for μ_{diff} = mean difference (Bottom − Top) in zinc concentrations at the top and bottom of all wells in this large region.

> **State:** State the parameter you want to estimate and the confidence level. *Be sure to indicate the order of subtraction.*

Plan: Paired *t* interval for μ_{diff}

Random? Paired data come from zinc concentration measurements at the top and bottom of a random sample of 10 wells in the region. ✓

Normal/Large Sample? The sample size is small ($n_{\text{diff}} = 10 < 30$), but the dotplot of differences doesn't show any strong skewness or outliers. ✓

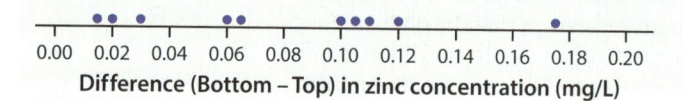

> **Plan:** Identify the appropriate inference method and check the conditions.

```
         • • •       • •         • • •  •
                                              •
├──┼──┼──┼──┼──┼──┼──┼──┼──┼──┼──┤
0.00  0.02  0.04  0.06  0.08  0.10  0.12  0.14  0.16  0.18  0.20
      Difference (Bottom − Top) in zinc concentration (mg/L)
```

Do: $\bar{x}_{\text{diff}} = 0.0804$, $s_{\text{diff}} = 0.0523$, $n_{\text{diff}} = 10$

With 95% confidence and df = 10 − 1 = 9, $t^* = 2.262$

$$0.0804 \pm 2.262 \frac{0.0523}{\sqrt{10}}$$

→ 0.0804 ± 0.0374

→ 0.0430 to 0.1178

Using technology: 0.043 to 0.118

> **Do:** If the conditions are met, perform calculations.

> See the Section 7.5 Tech Corner for details on calculating the confidence interval with technology.

Conclude: We are 95% confident that the interval from 0.043 to 0.118 mg/L captures the mean difference (Bottom − Top) in the zinc concentrations at the top and bottom of all wells in this large region. The interval suggests that the zinc concentration is between 0.043 and 0.118 mg/L greater, on average, at the bottom than at the top of wells in this region.

> **Conclude:** Interpret your interval in the context of the problem. *Don't forget the second sentence when you interpret the interval!*

EXAM PREP: FOR PRACTICE, TRY EXERCISE 15.

The 95% confidence interval in Example 9.5.3 tells us that all the plausible values for μ_{diff} are positive. This gives convincing evidence that zinc concentrations are greater at the bottom than at the top of wells in this large region, on average.

Section 9.5 What Did You Learn?

Review the learning goals from this section. Then practice what you've learned by working through the exercises.

Learning Goal	Example	Exercises
Use a graph to analyze the distribution of differences in a set of paired data.	9.5.1	7–10
Calculate the mean and standard deviation of the differences in a set of paired data, and interpret the mean difference.	9.5.2	11–14
Use the four-step process to construct and interpret a confidence interval for a mean difference.	9.5.3	15–18

Section 9.5 Exercises

Building Concepts and Skills These exercises assess the basic knowledge you should have after reading the section.

1. What is the definition of paired data?

2. To analyze paired data graphically, use a graph to see if the _____ are consistently greater than or consistently less than 0.

3. When analyzing paired data numerically, use the _____ and the _____ as summary statistics.

4. State the Normal/Large Sample condition for constructing a confidence interval for a population mean difference.

5. True/False: The method for calculating a paired t interval for a mean difference is the same as the method for calculating a one-sample t interval for a mean from Section 7.5.

6. What is the formula for calculating a confidence interval for μ_{diff}?

Mastering Concepts and Skills These exercises reinforce the learning goals as shown in the examples.

7. **Refer to Example 9.5.1 Fuel Efficiency** An interested buyer is considering 21 different model-year 2020 midsize cars for purchase. Here are the U.S. Environmental Protection Agency (EPA) estimates of city gas mileage and highway gas mileage in miles per gallon (mpg) for each car:[60]

Model	City	Highway
Alfa Romeo Giulia	24	33
BMW 540i	22	30
Buick Regal AWD	21	29
Ford Fusion FWD	21	31
Honda Civic	32	42
Infiniti Q50 AWD	19	27
Jaguar XF AWD	25	34
Kia Forte	31	41
Kia Stinger RWD	22	29
Lexus GS350	20	28
Lexus LS500	25	33
Lincoln MKZ FWD	20	31
Mercedes-Benz E350	23	32
Mini JCW Countryman	26	33
Nissan Altima SR/Platinum	27	37
Nissan Maxima	20	30
Subaru Impreza	28	36
Toyota Avalon	22	32
Toyota Camry XSE	22	32
Toyota Prius	43	48
Volvo S90 AWD	21	31

Make a dotplot of the difference (Highway − City) in gas mileage for each car model. What does the graph reveal?

8. **Women's Soccer** How good was the 2019 U.S. women's soccer team? With players like Carli Lloyd, Alex Morgan, and Megan Rapinoe, the team put on an impressive showing en route to winning the 2019 Women's World Cup. Here are data on the number of goals scored by the U.S. team and its opponent in games played during the 2019 season:[61]

U.S.	Opponent	U.S.	Opponent
1	3	3	2
1	0	6	0
5	3	2	2
6	0	2	2
3	0	1	0
5	0	13	0
3	0	3	0
3	0	2	0
4	0	2	1
3	0	2	1
2	0	2	1
1	1	2	0

Make a dotplot of the difference (U.S. – Opponent) in goals scored for each game. What does the graph reveal?

9. **Express Checkout** Researchers investigated which line was faster in the supermarket: the express lane or the regular lane. They randomly selected 15 times during a week, went to the same store, and bought the same item. One of the researchers used the express lane and the other used the closest regular lane. To decide which lane each of them would use, they flipped a coin. They entered their lanes at the same time, paid in the same way, and recorded the time in seconds it took them to complete the transaction. The table displays the data.[62]

Time in express lane (sec)	Time in regular lane (sec)
337	342
226	472
502	456
408	529
151	181
284	339
150	229
357	263
349	332
257	352
321	341
383	397
565	694
363	324
85	127

(a) Calculate the difference (Regular lane – Express lane) in transaction time for each visit to the supermarket.

(b) A dotplot of the differences from part (a) is shown here. What does the graph reveal about the difference in transaction times?

10. **Internet Speed** Ramon has found that his computer's internet connection is slower when he is farther from his wireless modem, which is located in his living room. To examine this difference, he randomly selects 14 times during the day and uses an online "speed test" to determine the download speeds to his computer in his bedroom and living room at each time (choosing which location to test first by flipping a coin). Here are the data, with download speeds in megabits per second (Mbps):[63]

Time	1	2	3	4	5	6	7
Bedroom	13.5	15.5	18.4	14.8	14.9	12.1	9.8
Living room	16.6	24.1	25.0	20.4	29.7	12.5	22.2
Time	8	9	10	11	12	13	14
Bedroom	16.0	11.1	14.3	15.6	10.5	15.6	11.3
Living room	17.6	26.7	18.5	28.7	15.7	22.8	27.0

(a) Calculate the difference (Living room – Bedroom) in download speeds for each time.

(b) A dotplot of the differences from part (a) is shown here. What does the graph reveal about the difference in download speeds?

11. **Refer to Example 9.5.2 Fuel Efficiency** Refer to Exercise 7. Calculate the mean and standard deviation of the difference (Highway – City) in gas mileages. Interpret the mean difference.

12. **Women's Soccer** Refer to Exercise 8. Calculate the mean and standard deviation of the difference (U.S. – Opponent) in goals scored. Interpret the mean difference.

13. **Express Checkout** Refer to Exercise 9.

(a) Calculate the mean and standard deviation of the difference (Regular lane – Express lane) in transaction times.

(b) Interpret the mean difference.

(c) Interpret the standard deviation of the differences.

14. **Internet Speed** Refer to Exercise 10.

(a) Calculate the mean and standard deviation of the difference (Living room – Bedroom) in download speeds.

(b) Interpret the mean difference.

(c) Interpret the standard deviation of the differences.

15. **Refer to Example 9.5.3 Express Checkout** Refer to Exercises 9 and 13. Construct and interpret a 95% confidence interval for the true mean difference in transaction times.

16. **Internet Speed** Refer to Exercises 10 and 14. Construct and interpret a 95% confidence interval for the true mean difference in download speeds.

17. **Piano Lessons** Do piano lessons improve the spatial-temporal reasoning of preschool children? A study measured the spatial-temporal reasoning of a random sample of 34 preschool children before and after 6 months of piano lessons. The differences (After – Before) in the reasoning scores have mean 3.618 and standard deviation 3.055.[64]

(a) Construct and interpret a 90% confidence interval for the mean difference in reasoning scores for all preschool students who take 6 months of piano lessons.

(b) Based on your interval from part (a), can you conclude that taking 6 months of piano lessons would cause an increase in preschool students' average reasoning scores? Why or why not?

18. **Annual Fee?** A bank wonders whether eliminating the annual credit card fee for customers who charge at least $2400 in a year will increase the amount they

decide to charge on its credit cards. The bank makes this change for an SRS of 200 of its credit card customers. It then compares these customers' charges for this year with their charges for last year. The mean increase in the sample is $332 and the standard deviation is $108.

(a) Construct and interpret a 99% confidence interval for the true mean increase in the amount spent by this bank's credit card customers when the annual fee is dropped.

(b) Based on the interval from part (a), can you conclude that dropping the annual fee would cause an increase in the average amount spent by this bank's credit card customers? Why or why not?

Applying the Concepts These exercises ask you to apply multiple learning goals in a new context or to apply what you learned in this section in a new way.

19. **Caffeine Dependence** Researchers designed an experiment to study the effects of caffeine withdrawal. They recruited 11 volunteers who were diagnosed as being caffeine dependent to serve as subjects. Each subject was barred from coffee, colas, and other substances with caffeine for the duration of the experiment. During one 2-day period, subjects took capsules containing their normal caffeine intake. During another 2-day period, they took placebo capsules. The order in which subjects took the caffeine capsules and the placebo was randomized. At the end of each 2-day period, a test for depression was given to all 11 subjects. Researchers wanted to know whether being deprived of caffeine would lead to an increase in depression.[65] The table contains data on the subjects' scores on the depression test. Higher scores correspond to more symptoms of depression.

Subject	1	2	3	4	5	6	7	8	9	10	11
Depression (caffeine)	5	5	4	3	8	5	0	0	2	11	1
Depression (placebo)	16	23	5	7	14	24	6	3	15	12	0

(a) Make a dotplot of the difference (Placebo − Caffeine) in depression test scores for each subject. What does the graph reveal?

(b) Construct and interpret a 90% confidence interval for the true mean difference in depression test scores for subjects like the ones in this study.

(c) Does the interval in part (b) give convincing evidence that caffeine deprivation causes an increase in the average depression test score for subjects like these? Justify your answer.

20. **Groovy Tires** Researchers were interested in comparing two methods for estimating tire mileage. The first method used the amount of weight lost by a tire. The second method used the amount of wear in the grooves of the tire. Researchers obtained a random sample of 16 tires. They used both methods to estimate the total distance traveled by each tire. The table provides the two estimates (in thousands of miles) for each tire.[66]

Tire	Weight	Groove
1	45.9	35.7
2	41.9	39.2
3	37.5	31.1
4	33.4	28.1
5	31.0	24.0
6	30.5	28.7
7	30.9	25.9
8	31.9	23.3
9	30.4	23.1
10	27.3	23.7
11	20.4	20.9
12	24.5	16.1
13	20.9	19.9
14	18.9	15.2
15	13.7	11.5
16	11.4	11.2

(a) Make a dotplot of the differences in the estimates of tire mileage for the two methods (Weight − Groove). What does the graph reveal?

(b) Construct and interpret a 95% confidence interval for the true mean difference (Weight − Groove) in the estimates from these two methods in the population of tires.

(c) Does your interval in part (b) give convincing evidence of a difference in the average estimated tire mileage between the two methods? Justify your answer.

21. **Fuel Efficiency** Refer to Exercises 7 and 11. Explain why it would not be appropriate to construct a confidence interval for the mean difference (Highway − City) in gas mileage for all model-year 2020 midsize car models.

22. **Women's Soccer** Refer to Exercises 8 and 12. Explain why it would not be appropriate to construct a confidence interval for the mean difference (U.S. − Opponent) in goals scored for all games played by the team in 2019.

23. **Chewing Gum** Researchers designed an experiment to investigate whether students can improve their short-term memory by chewing the same flavor of gum while studying for and taking an exam.[67] After recruiting 30 volunteers, they randomly assigned 15 to chew gum while studying a list of 40 words for 90 seconds. Immediately after the 90-second study period — and while chewing the same gum — the participants wrote down as many words as they could remember. The remaining 15 volunteers followed the same procedure without chewing gum. Two weeks later, each of the 30 participants did the opposite treatment. Researchers recorded the number of words each volunteer correctly remembered for each test.

(a) Explain why it was important for researchers to randomly assign the order in which each participant did the task with and without chewing gum.

(b) Verify that the conditions for constructing a paired t interval for a mean difference are satisfied.

(c) The 95% confidence interval for the true mean difference (Gum − No gum) in number of words remembered is −0.67 to 1.54. Interpret the confidence interval and the confidence level.

(d) Based on the interval, is there convincing evidence that chewing gum helps subjects like the ones in this study with short-term memory? Explain your answer.

24. **Stressful Puzzles** Do people get stressed out when other people watch them work? To find out, researchers recruited 30 volunteers to take part in an experiment.[68] They randomly assigned 15 of the participants to complete a word search puzzle while researchers stood close by and visibly took notes. The remaining 15 were assigned to complete a word search puzzle while researchers stood at a distance. After each participant completed the word search, they completed a second word search under the opposite treatment. Researchers recorded the amount of time each volunteer required to complete each puzzle.

(a) Explain why it was important for researchers to randomly assign the order in which each participant did the task with researchers close by and standing at a distance.

(b) Verify that the conditions for constructing a one-sample t interval for a mean difference are satisfied.

(c) The 95% confidence interval for the true mean difference (Close by − At a distance) in amount of time needed to complete the puzzle is −12.7 seconds to 119.4 seconds. Interpret the confidence interval and the confidence level.

(d) Based on the interval, is there convincing evidence that standing close by causes subjects like the ones in this study to take longer to complete a word search? Explain your answer.

25. **Tech Hot vs. Iced Coffee** How much difference is there in the prices of hot coffee and iced coffee at different establishments? A university statistics class recorded coffee prices at randomly selected local restaurants, cafes, and coffee shops. The data can be found in the data set *Coffee*.[69] Construct and interpret a 95% confidence interval for the mean difference in the prices of hot coffee (*HotPrice*) and iced coffee (*IcedPrice*) at all establishments near this university.

26. **Tech Student Sleep** How much more sleep, on average, do students get on non-school nights than on school nights? The *Census At School* data set includes data from a random sample of 75 U.S. high school students who took part in an online survey. Respondents were asked how many hours of sleep per night they usually get when they have school the next day (*Sleep_Hours_Schoolnight*) versus when they don't have school the next day (*Sleep_Hours_Non_Schoolnight*). Construct and interpret a 95% confidence interval for the mean difference in the amount of sleep per night U.S. high school students get on school nights versus non-school nights.

Extending the Concepts These exercises challenge you to explore statistical concepts and methods that go beyond what you learned in this section.

27. **Darwin's Data** Refer to the data from Darwin's study of cross-fertilized versus self-fertilized maize plants in Example 9.5.1.

(a) Make a scatterplot of the paired data with cross-fertilized plant height on the horizontal axis and self-fertilized plant height on the vertical axis. Use the same scale on both axes.

(b) Because paired data result from recording two values of the same quantitative variable for each individual or for each pair of similar individuals, we expect a scatterplot of paired data to have a positive correlation. Is that true for Darwin's data?

(c) Calculate the mean and standard deviation of height for the 15 cross-fertilized plants and for the 15 self-fertilized plants (as if the data were unpaired).

(d) Confirm that the difference in mean height for the two types of plants is equal to the mean difference in height $\bar{x}_{diff} = \bar{x}_{C-S} = 2.60667$ inches.

(e) Compare the standard deviation of height for the two types of plants [from part (c)] to $s_{diff} = 4.713$ inches. Did pairing help reduce the variability in height in this study?

28. **Groovy Tires** Refer to Exercise 20.

(a) A student doesn't notice that the data are paired, and uses a two-sample t interval to estimate the difference (Weight − Groove) in the mean tire mileage for the two methods with 95% confidence. Calculate this interval.

(b) Compare the interval in part (a) to the correct paired t interval from Exercise 20. Identify a consequence of using the incorrect method.

Cumulative Review These exercises revisit what you learned in previous sections.

29. **Growing Tomatoes (8.1, 8.2)** A chemical company is developing a new fertilizer for tomatoes. The company wants to know if tomatoes grown with its new fertilizer are larger, on average, than tomatoes grown with its current fertilizer. From prior experience, the company knows the true mean weight of tomatoes grown with the current fertilizer is $\mu = 1.2$ lb. Researchers apply the new fertilizer to a random sample of young tomato plants and measure the mean weight of the 35 tomatoes produced by the plants. The result is $\bar{x} = 1.33$ lb.

(a) State the null and alternative hypotheses for this test.

(b) Describe a Type II error in the context of this study and give a possible consequence.

(c) At the end of the study, the researchers report a P-value of 0.062. Interpret this value.

(d) What conclusion should researchers make about the effectiveness of the new fertilizer?

30. **Advanced Degree (5.3, 6.4)** In 2020, about 13% of U.S. adults aged 25 and older had a master's or doctoral degree.[70] Suppose we select a random sample of 500 U.S. adults aged 25 and older and let X be the number of people who have a master's or doctoral degree.

(a) Explain why this is a binomial setting.

(b) Show that the probability distribution of X is approximately normal.

(c) Calculate the mean and standard deviation of the appropriate normal distribution.

(d) Use this normal distribution to calculate the probability that a random sample of 500 U.S. adults aged 25 and older will contain 70 or more people who have a master's or doctoral degree.

Section 9.6 Significance Tests for a Population Mean Difference

LEARNING GOALS

By the end of this section, you will be able to:

- Use the four-step process to perform a significance test about a mean difference.
- Determine whether you should use two-sample t procedures for inference about $\mu_1 - \mu_2$ or paired t procedures for inference about μ_{diff} in a given setting.

In Section 9.5, you learned to analyze paired data by looking at the difference within each pair. You also saw how to construct and interpret a confidence interval for a population mean difference μ_{diff}. This section shows you how to perform a significance test about μ_{diff}. The second half of the section explores inference about a difference between two means when the data come from two independent random samples or treatment groups versus inference about a mean difference when the data are paired.

Performing a Significance Test About μ_{diff}

When paired data come from a random sample or a randomized experiment, we may want to perform a significance test about the true mean difference μ_{diff}. The null hypothesis has the general form

$$H_0: \mu_{\text{diff}} = \text{hypothesized value}$$

We'll focus on situations where the hypothesized value is 0. Then the null hypothesis says that the true mean difference is 0:

$$H_0: \mu_{\text{diff}} = 0$$

The alternative hypothesis says what kind of difference we expect.

The conditions for performing a significance test about a population mean difference are the same as those for constructing a confidence interval for μ_{diff} in Section 9.5.

HOW TO Check Conditions for Performing a Significance Test About μ_{diff}

To perform a significance test about a population mean difference μ_{diff}, check that the following conditions are met:

- **Random:** Paired data come from a random sample from the population of interest or from a randomized experiment.
- **Normal/Large Sample:** Paired data come from a normally distributed population of differences or the sample size is large ($n_{\text{diff}} \geq 30$). When the sample size is small and the shape of the population distribution of differences is unknown, a graph of the sample differences shows no strong skewness or outliers.

When these conditions are met, we can carry out a test of $H_0: \mu_{\text{diff}} = 0$. The standardized test statistic is

$$t = \frac{\bar{x}_{\text{diff}} - 0}{\frac{s_{\text{diff}}}{\sqrt{n_{\text{diff}}}}}$$

We can use technology or Table B to find the P-value from the t distribution with df $= n_{\text{diff}} - 1$. The method is the same as in Sections 8.5 and 8.6, except that the data are just the sample of differences within each pair.

Consumers Union designed an experiment to test whether nitrogen-filled tires would maintain their pressure better than air-filled tires. Its researchers obtained two tires from each of several brands and then randomly assigned one tire in each pair to be filled with air and the other to be filled with nitrogen. All tires were inflated to the same pressure and placed outside for a year. At the end of the year, the researchers measured the pressure in each tire. The pressure loss (in pounds per square inch, psi) during the year for the tires of each brand is shown in the table.[71]

Brand	Air	Nitrogen	Brand	Air	Nitrogen
BF Goodrich Traction T/A HR	7.6	7.2	Pirelli P6 Four Seasons	4.4	4.2
Bridgestone HP50 (Sears)	3.8	2.5	Sumitomo HTR H4	1.4	2.1
Bridgestone Potenza G009	3.7	1.6	Yokohama Avid H4S	4.3	3.0
Bridgestone Potenza RE950	4.7	1.5	BF Goodrich Traction T/A V	5.5	3.4
Bridgestone Turanza EL400	2.1	1.0	Bridgestone Potenza RE950_P195	4.1	2.8
Continental Premier Contact H	4.9	3.1	Continental ContiExtreme Contact	5.0	3.4
Cooper Lifeliner Touring SLE	5.2	3.5	Continental ContiPro Contact	4.8	3.3
Dayton Daytona HR	3.4	3.2	Cooper Lifeliner Touring SLE_T	3.2	2.5
Falken Ziex ZE-512	4.1	3.3	General Exclaim UHP	6.8	2.7
Fuzion Hrl	2.7	2.2	Hankook Ventus V4 H105	3.1	1.4
General Exclaim	3.1	3.4	Michelin Energy MXV4 Plus_S8	2.5	1.5
Goodyear Assurance Tripletred	3.8	3.2	Michelin Pilot Exalto A/S	6.6	2.2
Hankook Optimo H418	3.0	0.9	Michelin Pilot HX MXM4	2.2	2.0
Kumho Solus KH16	6.2	3.4	Pirelli P6 Four Seasons Plus	2.5	2.7
Michelin Energy MXV4 Plus	2.0	1.8	Sumitomo HTR+	4.4	3.7
Michelin Pilot XGT H4	1.1	0.7			

A dotplot of the difference (Air-filled − Nitrogen-filled) in pressure loss for each pair of tires is shown below. We can see that 28 of the 31 differences are positive. That is, the air-filled tire lost more pressure than the nitrogen-filled tire for all but three brands. The mean difference in pressure loss is $\bar{x}_{\text{diff}} = 1.252$ psi, meaning that the air-filled tires lost 1.252 psi more than the nitrogen-filled tires, on average. The standard deviation of the differences is $s_{\text{diff}} = 1.202$ psi.

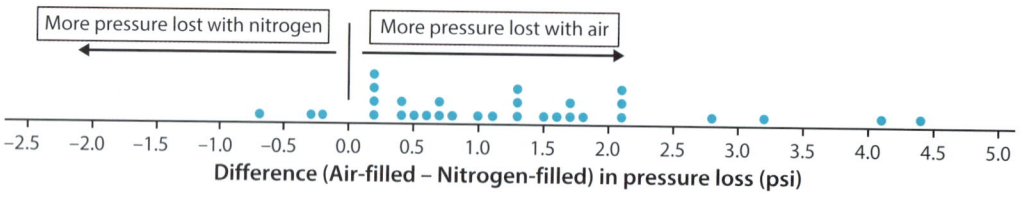

These data provide *some* evidence that filling tires with nitrogen instead of air reduces pressure loss. But do they give *convincing* evidence that the true mean difference μ_{diff} in pressure loss (Air-filled − Nitrogen-filled) is positive for tire brands like the ones in this study? To answer that question, we need to perform a significance test.

As with any inference procedure, be sure to follow the four-step process when performing a **paired t test for a mean difference.** You can use technology to perform the calculations in the Do step using raw data or summary statistics of the differences. See the Tech Corner in Section 8.5 for details.

A **paired t test for a mean difference** is a significance test of the null hypothesis that a population mean difference is equal to a specified value, usually 0.

Section 9.6 Significance Tests for a Population Mean Difference

This procedure is sometimes called a *one-sample t test for a mean difference* to remind you that it is basically the same method you learned in Section 8.5.

EXAMPLE 9.6.1

Does filling tires with nitrogen reduce pressure loss?

Performing a significance test about μ_{diff}

PROBLEM

Refer to the tire experiment carried out by Consumers Union. Do the data give convincing evidence at the $\alpha = 0.05$ significance level that air-filled tires lose more pressure, on average, than nitrogen-filled tires for brands like the ones in this study?

SOLUTION

State:

$H_0: \mu_{\text{diff}} = 0$

$H_a: \mu_{\text{diff}} > 0$

where μ_{diff} = the true mean difference (Air-filled − Nitrogen-filled) in pressure loss for tire brands like the ones in this study.

Use $\alpha = 0.05$.

The evidence for H_a is: $\bar{x}_{\text{diff}} = 1.252 > 0$.

State: State the hypotheses, parameter(s), significance level, and evidence for H_a.

Plan: Paired t test for μ_{diff}

Random? Random assignment of one tire in each pair to be filled with nitrogen and one to be filled with air. ✓

Normal/Large Sample? The number of differences in the sample is large: ($n_{\text{diff}} = 31 \geq 30$). ✓

Plan: Identify the appropriate inference method and check the conditions.

Do: $\bar{x}_{\text{diff}} = 1.252$ psi, $s_{\text{diff}} = 1.202$ psi, $n_{\text{diff}} = 31$

- $t = \dfrac{1.252 - 0}{\dfrac{1.202}{\sqrt{31}}} = 5.799$

- P-value: df $= 31 - 1 = 30$

Do: If the conditions are met, perform calculations:
- Calculate the test statistic.
- Find the P-value.

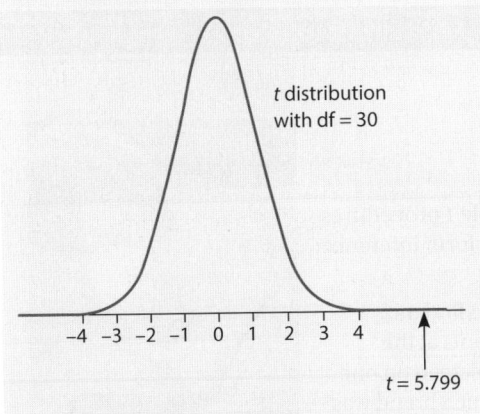

t distribution with df = 30

$t = 5.799$

Using Table B: P-value < 0.0005
Using technology: $t = 5.797$, P-value ≈ 0

Conclude: Because the P-value of approximately 0 is less than $\alpha = 0.05$, we reject H_0. There is convincing evidence that air-filled tires lose more pressure, on average, than nitrogen-filled tires for brands like the ones in this study.

Conclude: Make a conclusion about the hypotheses in the context of the problem. *We can make a cause-and-effect conclusion because this was a randomized experiment.*

EXAM PREP: FOR PRACTICE, TRY EXERCISE 5.

The significance test in Example 9.6.1 led to a simple decision: reject H_0. But we know that a confidence interval gives more information than a test — it provides the entire set of plausible values for the parameter based on the data. A *two-sided* significance test about a population mean using an $\alpha = 0.05$ significance level gives equivalent results to a 95% confidence interval for μ. For the *one-sided* test in the example with $\alpha = 0.05$, a 90% confidence interval gives equivalent information.

The 90% confidence interval for μ_{diff} is

$$\bar{x}_{\text{diff}} \pm t^* \frac{s_{\text{diff}}}{\sqrt{n_{\text{diff}}}} \to 1.252 \pm 1.697 \frac{1.202}{\sqrt{31}} \to 1.252 \pm 0.366 \to 0.886 \text{ to } 1.618$$

We are 90% confident that the interval from 0.886 to 1.618 psi captures the true mean difference (Air-filled – Nitrogen-filled) in pressure loss for tire brands like the ones in this study. The interval suggests that air-filled tires lose, on average, 0.886 to 1.618 psi more pressure than nitrogen-filled tires in a one-year period. Because the null value of 0 is not contained in the interval, we would reject H_0, which is consistent with the decision made when performing the significance test.

In general, there are two ways that an experiment with two treatments can yield paired data:

1. Each experimental unit can be given both treatments in a random order.
2. The researcher can form pairs of similar experimental units and randomly assign each treatment to exactly one member of every pair.

As you learned in Section 1.7, these are both known as *matched-pairs* designs. The nitrogen in tires experiment involved randomly assigning the air and nitrogen treatments to one tire in each pair. We will show an example of the other type of matched-pairs design at the end of the section.

Paired Data or Two Samples?

In Sections 9.3 and 9.4, we used two-sample t procedures to perform inference about the difference $\mu_1 - \mu_2$ between two population means. These methods require data that come from independent random samples from the populations of interest or from two groups in a randomized experiment. In Sections 9.5 and 9.6, we use paired t procedures to perform inference about the population mean difference μ_{diff}. These methods require *paired data* that come from a random sample from the population of interest or from a randomized experiment. *The proper inference method depends on how the data were produced.*

EXAMPLE 9.6.2

Sunscreen, video games, and adolescent romance

Two samples or paired data?

PROBLEM

In each of the following scenarios, decide whether you should use two-sample t procedures to perform inference about a difference in means or paired t procedures to perform inference about a mean difference. Explain your choice.

(a) Which of two brands of sunscreen is more effective at preventing sunburn? To find out, researchers conducted an experiment with a group of volunteer college students at the New Jersey shore. The researchers applied the same amounts of Brand A sunscreen on one arm and Brand B sunscreen on the other arm of each teen. Which arm got which brand was randomly determined for each volunteer. After the college students sunbathed for an hour, researchers compared the amount of redness on each person's left and right arms.

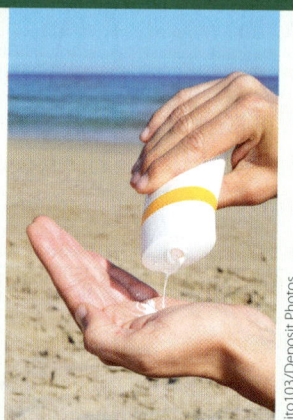

(b) Do experienced video game players earn higher scores when they play with someone present to cheer them on or when they play alone? Fifty young adults with experience playing a particular video game volunteer for a study. Researchers randomly assign 25 of them to play the game alone and the other 25 to play the game with a supporter present. They record each player's score.

(c) How do young adults look back on adolescent romance? Investigators interviewed a random sample of 40 couples in their mid-20s. They interviewed each partner separately. They asked each participant about a romantic relationship that lasted at least 2 months when they were aged 15 or 16. One response variable was a measure on a numerical scale of how much the attractiveness of the adolescent partner mattered. Researchers want to find out how much the older and younger partner in a couple differ on this measure.

SOLUTION

(a) Paired t procedures. In this experiment, the same quantitative variable—skin redness—is recorded twice for each volunteer: on one arm with Brand A sunscreen applied and on the other arm with Brand B sunscreen applied.

(b) Two-sample t procedures. The data come from a randomized experiment in which young adults are randomly assigned to one of two conditions: playing a video game alone or playing with a supporter present.

(c) Paired t procedures. The same quantitative variable—a rating of the importance of attractiveness—is recorded for the two individuals in each couple.

EXAM PREP: FOR PRACTICE, TRY EXERCISE 9.

When designing an experiment to compare two treatments, a completely randomized design may not be the best option. A matched-pairs design might be a better choice, as the following class activity shows.

Class Activity

Get Your Heart Beating!

Are standing pulse rates higher, on average, than sitting pulse rates? In this activity, you will perform two experiments to try to answer this question.

Experiment 1: Completely randomized design

1. Your instructor will randomly assign half of the students in your class to stand and the other half to sit. Once the two groups have been formed, students should stand or sit as required. Then each student should measure their pulse and share their data anonymously using technology.

2. Analyze the data for this completely randomized design. Make parallel dotplots and calculate the mean pulse rate for each group. Is there *some* evidence that standing pulse rates are higher, on average? Explain your answer.

Experiment 2: Matched-pairs design

1. To produce paired data in this setting, each student should receive both treatments in random order. Because each participant already sat or stood in Step 1, they just need to do the opposite now. Everyone should measure their pulse again in the new position. Then each participant should calculate the difference (Standing − Sitting) in pulse rate and share this value anonymously using technology.

2. Analyze the data for the matched-pairs design. Make a dotplot of these differences and calculate their mean. Is there *some* evidence that standing pulse rates are higher, on average? Explain your answer.

3. Which design provides more convincing evidence that standing pulse rates are higher, on average, than sitting pulse rates? Justify your answer.

From the *Get Your Heart Beating!* activity, we see that the type of experimental design helps us identify the proper inference procedure for comparing means. A completely randomized design requires two-sample t procedures, while a matched-pairs design calls for paired t procedures. Let's see what this looks like for an introductory statistics class with 24 students who did the activity.

Figure 9.5 shows a dotplot of the pulse rates for the completely randomized design. The mean pulse rate for the standing group is $\bar{x}_1 = 74.83$ bpm; the mean for the sitting group is $\bar{x}_2 = 68.33$ bpm. So the average pulse rate is 6.5 bpm higher in the standing group. However, the variability in pulse rates for the two groups creates a lot of overlap in the dotplots. A two-sample t test of $H_0: \mu_1 - \mu_2 = 0$ versus $H_a: \mu_1 - \mu_2 > 0$ yields $t = 1.42$ and a P-value of 0.09. Because the P-value is greater than our default $\alpha = 0.05$ significance level, we fail to reject H_0. These data do not provide convincing evidence that standing pulse rates are higher, on average, than sitting pulse rates for people like the students in this class.

FIGURE 9.5 Parallel dotplots of the pulse rates for the standing and sitting groups in a statistics class's completely randomized experiment.

What about the class's matched-pairs design? Figure 9.6 shows a dotplot of the difference (Standing − Sitting) in pulse rate for each of the 24 students. We can see that 21 of the 24 students recorded a positive difference, indicating that their standing pulse rate was higher. The mean difference is $\bar{x}_{diff} = 6.83$ bpm. A paired t test of $H_0: \mu_{diff} = 0$ versus $H_a: \mu_{diff} > 0$ gives $t = 6.483$ and a P-value of approximately 0. Because the P-value is less than our default $\alpha = 0.05$ significance level, we reject H_0. These data provide *very* convincing evidence that standing pulse rates are higher, on average, than sitting pulse rates for people like the students in this class.

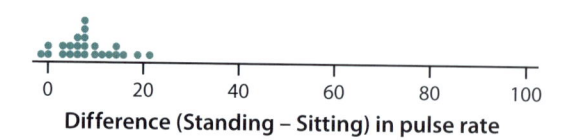

FIGURE 9.6 Dotplot of the difference in pulse rate (Standing − Sitting) for each student in a statistics class's matched-pairs design.

Let's take one more look at Figures 9.5 and 9.6. Notice that we used the same scale for both graphs. The matched-pairs design reduced the variability in the response variable by accounting for a big source of variability—the differences between individual students. In other words, the matched-pairs design has more power to detect the fact that standing causes an increase in the average pulse rate. With the larger amount of variability in the completely randomized design, we could not make such a conclusion.

Section 9.6 What Did You Learn?

Review the learning goals from this section. Then practice what you've learned by working through the exercises.

Learning Goal	Example	Exercises
Use the four-step process to perform a significance test about a mean difference.	9.6.1	5–8
Determine whether you should use two-sample t procedures for inference about $\mu_1 - \mu_2$ or paired t procedures for inference about μ_{diff} in a given setting.	9.6.2	9–12

Section 9.6 Exercises

Building Concepts and Skills These exercises assess the basic knowledge you should have after reading the section.

1. State the null hypothesis for performing a significance test about a population mean difference, assuming the null hypothesis is that there is no difference.

2. True/False: The conditions for performing a significance test about a population mean difference are the same as the conditions for constructing a confidence interval for μ_{diff}.

3. What is the formula for the t statistic for a test of the null hypothesis that the true mean difference is 0?

4. To analyze data from a completely randomized experimental design, use _____ t procedures; for a matched-pairs experimental design, use _____ t procedures.

Mastering Concepts and Skills These exercises reinforce the learning goals as shown in the examples.

5. Refer to Example 9.6.1 Better Barley Does drying barley seeds in a kiln increase the yield of barley? A famous experiment by William S. Gosset (who discovered the t distributions) investigated this question. Gosset marked 11 pairs of adjacent plots in a large field. For each pair, he planted regular barley seeds in one plot and kiln-dried seeds in the other. He used a coin flip to determine which plot in each pair got the regular barley seed and which got the kiln-dried seed. The table displays the data on barley yield (in pounds per acre) for each plot.[72]

Plot	Regular	Kiln
1	1903	2009
2	1935	1915
3	1910	2011
4	2496	2463
5	2108	2180
6	1961	1925
7	2060	2122
8	1444	1482
9	1612	1542
10	1316	1443
11	1511	1535

Do these data provide convincing evidence that drying barley seeds in a kiln increases the yield of barley, on average?

6. Faster Food Many people think it's faster to order at the drive-thru than to order inside at fast-food restaurants. Two researchers decided to investigate by visiting a local fast-food restaurant on 10 randomly selected days during lunch. Each time they went, they flipped a coin to determine which of them would go inside and which would use the drive-thru. Both researchers ordered the same item, paid with the same amount of cash, and recorded how long it took (in seconds) to wait in line, pay, and receive their item. Here are their data:[73]

Visit	Drive-thru	Inside
1	325	170
2	608	110
3	90	52
4	519	158
5	216	66
6	263	128
7	559	81
8	154	163
9	449	64
10	512	120

Do these data provide convincing evidence of a difference in the average service time inside and at the drive-thru for this local fast-food restaurant?

7. Baseball Pitches During the 2021 season, Major League Baseball (MLB) announced that it would begin stricter enforcement of a rule banning pitchers from applying sticky substances to baseballs. It was an open secret that some pitchers were using various substances to try to increase the spin rate of their pitches, which in turn made the ball harder to hit. Researchers wondered if the enhanced enforcement would decrease the average spin rate (in revolutions per minute, rpm) of balls thrown by MLB pitchers. They collected data on the spin rates of a random sample of 44 pitchers before and after the stricter enforcement was announced, and calculated the change (After − Before) in spin rate for each pitcher.[74] A dotplot of the differences is shown here. The mean change is −74.114 rpm and the standard deviation is 79.948 rpm.

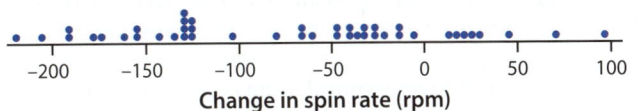

(a) Do the data provide convincing evidence at the $\alpha = 0.01$ significance level of a decrease in the average spin rate of all MLB pitchers from before the stricter enforcement of sticky substances was announced and after this announcement?

(b) Can we conclude that the result in part (a) is due to the announced change in enforcement by MLB officials? Why or why not?

8. Friday the 13th Does Friday the 13th have an effect on people's shopping behavior? Researchers collected data on the number of shoppers at a random sample of 45 grocery stores on Friday the 6th and Friday the 13th in the same month. A dotplot of the difference (Friday the 6th − Friday the 13th) in the number of shoppers at each store on these 2 days is shown here. The mean difference is −46.5 and the standard deviation of the differences is 178.[75]

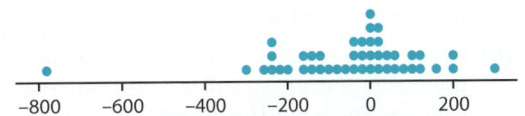

(a) Do these data provide convincing evidence at the $\alpha = 0.05$ significance level of a difference in the average number of shoppers at grocery stores on these 2 days?

(b) Can we conclude that the result in part (a) is because Friday the 13th does not affect people's behavior? Why or why not?

9. Refer to Example 9.6.2 Two Samples or Paired Data? For each scenario, decide whether you should use two-sample t procedures to perform inference about a difference in means or paired t procedures to perform inference about a mean difference. Explain your choice in each case.

(a) Before exiting the water, scuba divers remove their fins. A maker of scuba equipment advertises a new style of fins that is supposed to be faster to remove.

A consumer advocacy group suspects that the time to remove the new fins may be no different than the time required to remove the old fins, on average. They recruit 20 experienced scuba divers to test the new fins. Each diver flips a coin to determine if they wear the new fin on the left foot and the old fin on the right foot, or vice versa. The researchers record the time it takes each diver to remove each type of fin.

(b) To study the health of aquatic life, scientists gathered a random sample of 60 White Piranha fish from a tributary of the Amazon River during one year. They compared the average length of these fish to a random sample of 82 White Piranha from the same tributary a decade ago.

(c) Can a wetsuit deter shark attacks? A researcher designs a new wetsuit with color variations that are suspected to deter shark attacks. To test this idea, she fills two identical drums with bait and covers one in the standard black neoprene wetsuit and the other in the new suit. Over a period of one week, she selects 16 two-hour time periods and randomly assigns 8 of them to the drum in the black wetsuit. She assigns the other 8 time periods to the drum with the new suit. During each time period, she submerges the appropriate drum in waters that sharks frequent, and records the number of times a shark bites the drum.

10. **Two Samples or Paired Data?** For each scenario, decide whether you should use two-sample t procedures to perform inference about a difference in means or paired t procedures to perform inference about a mean difference. Explain your choice in each case.[76]

(a) To compare the average weight gain of pigs fed two different diets, researchers used nine pairs of pigs. The pigs in each pair were littermates. A coin toss was used to decide which pig in each pair got Diet A and which got Diet B.

(b) Researchers select separate random samples of professors at two-year colleges and four-year colleges. They compare the average salaries of professors at the two types of institutions.

(c) To test the effects of a new fertilizer, researchers treat 100 plots with the new fertilizer, and treat another 100 plots with another fertilizer. They use a computer's random number generator to determine which plots get which fertilizer.

11. **Paired Data or Two Samples?** For each scenario, decide whether you should use two-sample t procedures to perform inference about a difference in means or paired t procedures to perform inference about a mean difference. Explain your choice in each case.[77]

(a) To test the wear characteristics of two tire brands, A and B, researchers randomly assign Brand A or Brand B tires to each of 50 cars of the same make and model.

(b) To test the effect of background music on productivity, researchers observe factory workers. For one month, each subject works without music. For another month, each subject works while listening to music on a portable music player. The month in which each subject listens to music is determined by a coin toss.

12. **Paired Data or Two Samples?** For each scenario, decide whether you should use two-sample t procedures to perform inference about a difference in means or paired t procedures to perform inference about a mean difference. Explain your choice in each case.

(a) Which of two brands of ski wax works better? To find out, researchers randomly assign 15 skiers to race downhill with one brand of wax and 15 different skiers to race downhill with the other brand of wax. The researchers compare the average times of the skiers in the two groups.

(b) A local taco shop is considering a switch to a new tortilla. To investigate whether customers like the new tortilla better than the one the taco shop is currently using, the shop owners select a random sample of 20 regular customers. They ask each customer to try both tortillas and record a taste score for each. They randomly determine the order in which the customers try the two tortillas.

Applying the Concepts These exercises ask you to apply multiple learning goals in a new context or to apply what you learned in this section in a new way.

13. **Insomnia Drug** Researchers carried out an experiment with 10 randomly selected patients who frequently experienced insomnia to investigate the effectiveness of a drug that was designed to increase sleep time. The data show the number of additional hours of sleep gained by each subject after taking the drug.[78] (A negative value indicates that the subject got less sleep after taking the drug.)

1.9 0.8 1.1 0.1 −0.1 4.4 5.5 1.6 4.6 3.4

(a) Why should you use paired t procedures to perform inference about a mean difference in this scenario?

(b) Is there convincing evidence at the $\alpha = 0.05$ significance level that patients with insomnia would get more sleep, on average, when taking the drug?

(c) Can we conclude that the drug is effective at increasing the average amount of sleep time? Justify your answer.

(d) Based on your conclusion in part (b), which type of error — a Type I error or a Type II error — could you have made? Explain your answer.

14. **Warming Up** Researchers conducted an experiment to determine if athletes run faster after warming up. They randomly selected 30 athletes at a large school and had them run a 100-meter sprint with and without warming up (one week apart), randomly determining which they ran first. For each runner, researchers calculated the difference (No warm-up − Warm-up) in 100-meter sprint times, in seconds. The mean difference is 1.293 seconds and the standard deviation is 0.438 second.[79]

(a) Why should you use paired t procedures to perform inference about a mean difference in this scenario?

(b) Is there convincing evidence at the $\alpha = 0.05$ significance level that athletes from this school run faster (have shorter times), on average, when they warm up?

(c) Can we conclude that athletes at this school will run faster by warming up? Justify your answer.

(d) Based on your conclusion in part (b), which type of error — a Type I error or a Type II error — could you have made? Explain your answer.

15. **Insomnia Drug** Refer to Exercise 13.

    (a) Construct and interpret a 90% confidence interval for the true mean amount of additional sleep that patients with insomnia like the ones in this study get when taking the drug. If you already defined parameters and checked conditions in Exercise 13, you don't need to do those tasks again here.

    (b) Explain how the confidence interval provides more information than the test in Exercise 13.

16. **Warming Up** Refer to Exercise 14.

    (a) Construct and interpret a 90% confidence interval for the true mean difference (No warm-up − Warm-up) in 100-meter sprint times for athletes at this school. If you already defined parameters and checked conditions in Exercise 14, you don't need to do those tasks again here.

    (b) Explain how the confidence interval provides more information than the test in Exercise 14.

17. **Tech Coastal Home Sales** How do the list price and the sales price of homes along the coast compare? The *Coastal Sales* data set contains detailed information about a random sample of dwellings sold along the South Carolina coast during the third quarter of 2020.[80] Two of the variables recorded for each dwelling were the list price (*List_Price*) and the sales price (*Sales_Price*), both in dollars.

    (a) Do these data provide convincing evidence at the 1% significance level of a nonzero mean difference in the list price and the sales price in the population of South Carolina coastal dwellings sold during the third quarter of 2020?

    (b) Calculate and interpret a 99% confidence interval for the population mean difference.

18. **Tech Texting Teens** How do the number of text messages sent and received each day by high school students compare? The *Census At School* data set includes data from a random sample of 75 U.S. high school students who took part in an online survey. Two of the variables recorded for each student were the number of text messages they sent (*Text_Messages_Sent_Yesterday*) and received (*Text_Messages_Received_Yesterday*) on the previous day.

    (a) Do the data provide convincing evidence at the 1% significance level of a nonzero mean difference in the number of text messages sent and received the previous day in the population of U.S. high school students who took part in the online survey?

    (b) Calculate and interpret a 99% confidence interval for the population mean difference.

Exercises 19 and 20 refer to the following scenario. Coaching companies claim that their courses can raise the SAT scores of high school students. Of course, the scores of students who retake the SAT without paying for coaching generally improve. A random sample of students who took the SAT twice included 427 who were coached and 2733 who were uncoached.[81] Here are summary statistics for their SAT Verbal scores on the first and second tries, as well as for their gain in score from the first try to the second try.

	Try 1		Try 2		Gain	
	n	$\bar{x}$	s_x	$\bar{x}$	s_x	$\bar{x}$	s_x
Coached	427	500	92	529	97	29	59
Uncoached	2733	506	101	527	101	21	52

19. **Coaching and SAT Scores** Do the scores of students who are coached increase significantly?

    (a) You could use the information in the Coached row to carry out either a two-sample t test comparing Try 1 with Try 2 for coached students or a paired t test using Gain. Which is the correct test? Why?

    (b) Carry out the proper test. What do you conclude?

20. **Coaching and SAT Scores** What we really want to know is whether coached students improve more than uncoached students do, and whether any advantage is large enough to be worth paying for. Use the information given previously to answer the following questions.

    (a) How much more do the scores of coached students increase, on average, compared to uncoached students? Construct and interpret a 99% confidence interval.

    (b) Does the interval in part (a) give convincing evidence that the scores of coached students increase more, on average, than those of uncoached students? Explain your reasoning.

    (c) Based on your work, do you think coaching courses are worth paying for? Why or why not?

Extending the Concepts These exercises challenge you to explore statistical concepts and methods that go beyond what you learned in this section.

21. **Treating Arthritis** A pharmaceutical company has developed a new drug to reduce arthritis pain. To test the drug's effectiveness, researchers perform an experiment. They recruit 10 volunteer subjects with a history of arthritis pain who are not currently taking any medication. Each person takes the company's new drug for a 3-month period and the company's current arthritis drug for a different 3-month period, with a 3-month "washout" period in between when the subject takes no medication. Researchers randomly assign the order in which each person takes the two drugs. They ask participants to record the number of hours that they are free from arthritis pain each day during the experiment. The table shows the number of pain-free hours per day reported by each person on days

when they took the two drugs, along with the difference (New drug − Current drug) for each person.

Subject	1	2	3	4	5	6	7	8	9	10
New drug	3.5	5.7	2.9	2.4	9.9	16.7	6.0	4.0	20.9	3.3
Current drug	2.0	3.6	2.6	2.6	7.3	14.9	6.6	2.0	8.5	3.4
Difference	1.5	2.1	0.3	−0.2	2.6	1.8	−0.6	2.0	12.4	−0.1

(a) State appropriate hypotheses for performing a test about the true mean difference.

(b) Calculate the mean and standard deviation of the differences.

(c) Explain why it is not appropriate to carry out a paired t test in this case.

(d) We used the *One Quantitative Variable, Single Group* applet at www.stapplet.com to redo the random assignment of the new drug and current drug order for each participant 500 times, assuming no treatment effect. The simulation uses the same two values for the number of pain-free hours per day as in the original experiment for each person, but randomly assigns the new drug and current drug labels to those two values. Then it calculates the sample mean difference (New drug − Current drug) for this random reassignment of treatment order to individuals. Note that the difference value for each person will either be the same as in the original experiment or the *opposite* of that observed difference if the random assignment reverses the two drug labels. The dotplot shows the mean difference for each of the 500 simulated random assignments. Use the results of the simulation to estimate the P-value. What conclusion would you make?

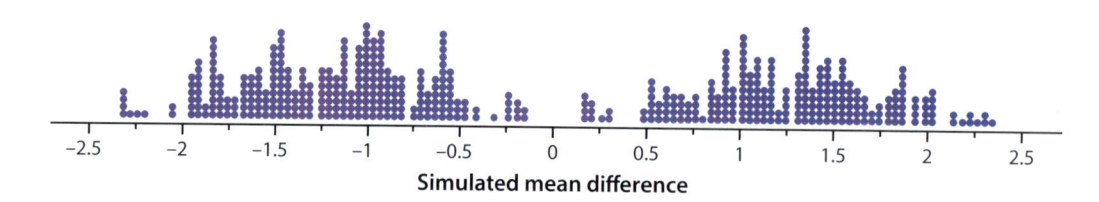

Cumulative Review
These exercises revisit what you learned in previous sections.

22. **Catching Food (1.5, 2.2, 3.1)** Tempe thinks she's pretty skilled at "mouth-catching" (no hands) bits of snack food that have been thrown to her from across a room. She wonders if she's better at catching some foods than others. To find out, a friend tosses 150 pieces of each of three different kinds of snack food (in random order) and records how often Tempe catches the food in her mouth. Here are the results:

Snack food

		Goldfish	Grapes	Gummies	Total
Caught?	Yes	90	100	115	305
	No	60	50	35	145
	Total	150	150	150	450

(a) Identify the explanatory and response variables.

(b) Use the information in the two-way table to create a segmented bar chart that shows the relationship between snack food type and Tempe's catching success.

(c) Is there an association between the two variables? Explain your reasoning. If an association does exist, briefly describe it.

23. **Which Inference Method? (Chapters 7, 8, 9)** For each scenario, state which inference procedure from Chapter 7, 8, or 9 you would use. Be specific. For example, you might say, "Two sample z test for the difference between two proportions." You do not have to carry out any procedures.

(a) Drowning in bathtubs is a major cause of death in children younger than age 5 years. Researchers asked a random sample of parents many questions related to bathtub safety. Overall, 85% of the sample said they used baby bathtubs for infants. Estimate the percentage of all parents of young children who use baby bathtubs.

(b) How seriously do people view speeding in comparison with other annoying behaviors? A large random sample of adults was asked to rate a number of behaviors on a scale of 1 (no problem at all) to 5 (very severe problem). Do speeding drivers get a higher average rating than noisy neighbors?

(c) You have data from interviews with a random sample of students who failed to graduate from a particular college in 7 years and also from a random sample of students who entered at the same time and did graduate within 7 years. You will use these data to estimate the difference in the percentages of students from rural backgrounds among dropouts and graduates.

(d) A bank wants to know if either of two incentive plans will lead to a larger increase in the average amount spent on its credit cards. The bank offers incentive A to one group of current credit card customers and incentive B to another group of current credit card customers, with the offers determined at random. The amount charged by customers in the two groups during the following 6 months is compared.

Section 9.7 Significance Tests for Two Population Standard Deviations or Variances

LEARNING GOALS

By the end of this section, you will be able to:

- State hypotheses and check conditions for a significance test comparing two standard deviations or variances.
- Calculate the test statistic and *P*-value for a significance test comparing two standard deviations or variances.
- Use the four-step process to perform a significance test comparing two standard deviations or variances.

In Sections 9.1–9.4, you learned how to compare proportions and means from two different populations or treatments using confidence intervals and significance tests about a *difference* in parameters: $p_1 - p_2$ or $\mu_1 - \mu_2$. When we compare standard deviations or variances for two populations or treatments, we instead use a *ratio* for performing inference.

Hypotheses and Conditions for Testing a Claim About Two Standard Deviations or Variances

In a test comparing two population standard deviations or variances, the null hypothesis has the general form

$$H_0: \frac{\sigma_1}{\sigma_2} = \text{hypothesized value} \quad \text{or} \quad H_0: \frac{\sigma_1^2}{\sigma_2^2} = \text{hypothesized value}$$

We'll restrict ourselves to situations in which the hypothesized ratio is 1. Then the null hypothesis is

$$H_0: \frac{\sigma_1}{\sigma_2} = 1 \quad \text{or} \quad H_0: \frac{\sigma_1^2}{\sigma_2^2} = 1$$

which is equivalent to saying the two population standard deviations or variances are the same:

$$H_0: \sigma_1 = \sigma_2 \quad \text{or} \quad H_0: \sigma_1^2 = \sigma_2^2$$

The alternative hypothesis says how we expect the standard deviations or variances to compare, and can be either one-sided or two-sided. We will always write the hypotheses in the format without ratios, for reasons that will become clear shortly.

In Section 8.8, we introduced two conditions that should be met before we perform a significance test about a population standard deviation or variance: Random and Normal. Now that we are comparing *two* standard deviations or variances, we have to modify these conditions slightly.

HOW TO Check the Conditions for a Significance Test Comparing Two Standard Deviations or Variances

To perform a test comparing two population standard deviations or variances, check that the following conditions are met:

- **Random:** The data come from independent random samples from the two populations of interest or from two groups in a randomized experiment.
- **Normal:** For each sample, the data come from a normally distributed population.

⚠ **As with the tests for σ or σ^2 described in Section 8.8, there is no Large Sample override to the Normal condition for tests comparing two standard deviations or variances.** No matter how large the sample sizes are, the data from both samples must appear to have come from normal population distributions.

EXAMPLE 9.7.1

Does a fungus affect tadpole gut length?

Stating hypotheses and checking conditions

PROBLEM

When tadpoles are exposed to the fungus *Batrachrochytrium dendrobatidis* (BD), they suffer from damage to their mouthparts, which then reduces their food consumption. Biologists wondered whether this reduction in food consumption would result in changes to the gut length of the tadpole. In particular, they were curious whether the variability in gut length would be different for tadpoles that were exposed to the fungus and for tadpoles that were not exposed to the fungus.

For this study, biologists randomly assigned 14 tadpoles to be exposed to the fungus and 13 tadpoles to be controls, with no exposure to the fungus. They measured gut length (in mm) for each tadpole 46 days after the treatment.[82] Here are the data:

BD	191.40	142.92	169.81	152.18	171.33	153.92	181.08	124.90	173.06	207.01	143.77	220.35	130.00	195.13
Control	177.92	181.58	154.29	217.67	185.88	249.29	196.90	202.16	210.86	215.60	174.16	222.83	228.74	

(a) State appropriate hypotheses for performing a significance test. Be sure to define the parameters of interest.

(b) Do the data provide *some* evidence that the standard deviation of gut length is different for tadpoles like the ones in this study that are exposed to BD and those that are not exposed to BD? Justify your answer.

(c) Check if the conditions for performing a test comparing the standard deviations are met.

SOLUTION

(a) $H_0: \sigma_{BD} = \sigma_{Control}$

$H_a: \sigma_{BD} \neq \sigma_{Control}$

where σ_{BD} = the standard deviation of the gut length for all tadpoles like the ones in the experiment that are exposed to BD and $\sigma_{Control}$ = the standard deviation of the gut length for all tadpoles like the ones in the experiment that are not exposed to BD.

This is a two-sided test because the researchers wondered if the standard deviation of gut length differs for tadpoles exposed to BD and tadpoles not exposed to BD.

(b) $s_{BD} = 28.72 \neq 26.18 = s_{Control}$

Use technology to calculate the standard deviation for each group.

(c) Random? The tadpoles were randomly assigned to the fungus exposure and control groups. ✓

Normal? Neither dotplot shows any strong skewness or outliers. ✓

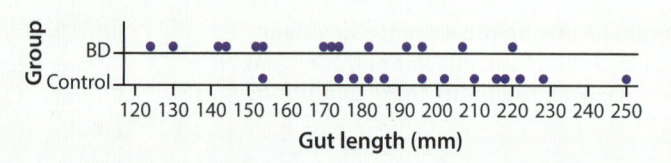

Always check a graph (dotplot, histogram, boxplot, normal probability plot) to see if it is plausible that both populations are normally distributed.

EXAM PREP: FOR PRACTICE, TRY EXERCISE 5.

Calculations: Test Statistic and *P*-Value

If the Random and Normal conditions are met, we can proceed with calculations. To test $H_0: \sigma_1 = \sigma_2$ or $H_0: \sigma_1^2 = \sigma_2^2$, we compute the ratio of sample variances as our test statistic:

$$F = \frac{s_L^2}{s_S^2}$$

where s_L^2 is the larger sample variance and s_S^2 is the smaller sample variance. When the conditions are met, the *F* statistic follows a new type of distribution called an ***F* distribution**.

Section 9.7 Significance Tests for Two Population Standard Deviations or Variances

An **F distribution** is defined by a density curve that takes only non-negative values and is skewed to the right. A particular F distribution is specified by its *numerator degrees of freedom* and its *denominator degrees of freedom*. When performing inference comparing two population standard deviations or variances, the numerator degrees of freedom is $df_{Num} = n_L - 1$, where n_L is the size of the sample with the larger variance, and the denominator degrees of freedom is $df_{Denom} = n_S - 1$, where n_S is the size of the sample with the smaller variance.

Figure 9.7 shows the density curves for five members of the F family of distributions (denoted $F_{df_{Num}, df_{Denom}}$). As you can see, the F distributions, like the chi-square distributions, are skewed to the right, but will be less skewed as the sample sizes (and degrees of freedom) increase.

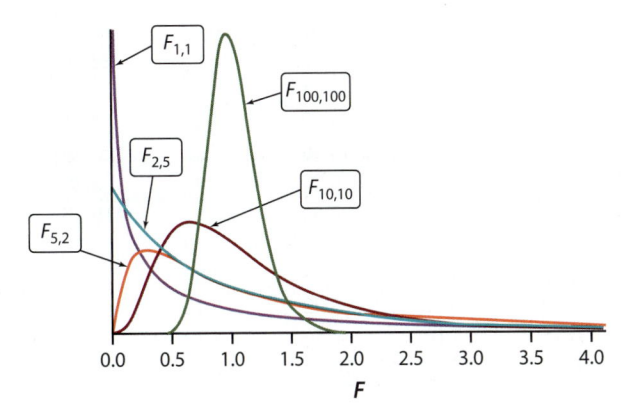

FIGURE 9.7 Density curves for five members of the F family of distributions.

To find the P-value from the F statistic in a test of $H_0: \sigma_1 = \sigma_2$, we start by remembering that the formula for the test statistic always puts the larger sample variance in the numerator. This means that the value of the test statistic will always be greater than the null value of 1. Consequently, we always calculate the area under the F density curve *to the right* of the test statistic. If the test is two-sided, we double this area to get the P-value. If the test is one-sided, the area to the right of the test statistic *is* the P-value. ⚠ If $H_a: \sigma_1 < \sigma_2$ but $s_1 > s_2$, **the sample data provide no evidence for the alternative hypothesis and no test should be performed.**

Use either technology or Table D to find the area to the right of the F test statistic in the appropriate F distribution. As with the t distributions, when using the table we can only put bounds on the P-value, so using technology is preferred. To illustrate the use of Table D in a two-sided test, imagine that we have calculated $F = 5.06$ with $n_L = 8$ and $n_S = 7$. Choose the column in the table with $df_{Num} = 8 - 1 = 7$ and the subset of the rows with $df_{Denom} = 7 - 1 = 6$.

TABLE D F distribution critical values

		\multicolumn{9}{c	}{Degrees of freedom in the numerator}
	p	1	2	3	4	5	6	7	8	9
1	.100	39.86	49.50	53.59	55.83	57.24	58.20	58.91	59.44	59.86
	.050	161.45	199.50	215.71	224.58	230.16	233.99	236.77	238.88	240.54
	.025	647.79	799.50	864.16	899.58	921.85	937.11	948.22	956.66	963.28
	.010	4052.18	4999.50	5403.35	5624.58	5763.65	5858.99	5928.36	5981.07	6022.47
	.001	405284	500000	540379	562500	576405	585937	592873	598144	602284
2	.100	8.53	9.00	9.16	9.24	9.29	9.33	9.35	9.37	9.38
	.050	18.51	19.00	19.16	19.25	19.30	19.33	19.35	19.37	19.38
	.025	38.51	39.00	39.17	39.25	39.30	39.33	39.36	39.37	39.39
	.010	98.50	99.00	99.17	99.25	99.30	99.33	99.36	99.37	99.39
	.001	998.50	999.00	999.17	999.25	999.30	999.33	999.36	999.37	999.39
	.100	5.54	5.46	5.39	5.34	5.31	5.28	5.27	5.25	5.24
	.050	10.13	9.55	9.28	9.12	9.01	8.94	8.89	8.85	8.81

(continued)

	p	1	2	3	4	5	6	7	8	9
3	.025	17.44	16.04	15.44	15.10	14.88	14.73	14.62	14.54	14.47
	.010	34.12	30.82	29.46	28.71	28.24	27.91	27.67	27.49	27.35
	.001	167.03	148.50	141.11	137.10	134.58	132.85	131.58	130.62	129.86
4	.100	4.54	4.32	4.19	4.11	4.05	4.01	3.98	3.95	3.94
	.050	7.71	6.94	6.59	6.39	6.26	6.16	6.09	6.04	6.00
	.025	12.22	10.65	9.98	9.60	9.36	9.20	9.07	8.98	8.90
	.010	21.20	18.00	16.69	15.98	15.52	15.21	14.98	14.80	14.66
	.001	74.14	61.25	56.18	53.44	51.71	50.53	49.66	49.00	48.47
5	.100	4.06	3.78	3.62	3.52	3.45	3.40	3.37	3.34	3.32
	.050	6.61	5.79	5.41	5.19	5.05	4.95	4.88	4.82	4.77
	.025	10.01	8.43	7.76	7.39	7.15	6.98	6.85	6.76	6.68
	.010	16.26	13.27	12.06	11.39	10.97	10.67	10.46	10.29	10.16
	.001	47.18	37.12	33.20	31.09	29.75	28.83	28.16	27.65	27.24
6	.100	3.78	3.46	3.29	3.18	3.11	3.05	3.01	2.98	2.96
	.050	5.99	5.14	4.76	4.53	4.39	4.28	4.21	4.15	4.10
	.025	8.81	7.26	6.60	6.23	5.99	5.82	5.70	5.60	5.52
	.010	13.75	10.92	9.78	9.15	8.75	8.47	8.26	8.10	7.98
	.001	35.51	27.00	23.70	21.92	20.80	20.03	19.46	19.03	18.69
7	.100	3.59	3.26	3.07	2.96	2.88	2.83	2.78	2.75	2.72
	.050	5.59	4.74	4.35	4.12	3.97	3.87	3.79	3.73	3.68
	.025	8.07	6.54	5.89	5.52	5.29	5.12	4.99	4.90	4.82
	.010	12.25	9.55	8.45	7.85	7.46	7.19	6.99	6.84	6.72
	.001	29.25	21.69	18.77	17.20	16.21	15.52	15.02	14.63	14.33

Degrees of freedom in the denominator

The value 5.06 is not one of the values listed in the appropriate part of the table. But we do notice that $4.21 < 5.06 < 5.70$. This means that the area under the curve to the right of $F = 5.06$ is between 0.025 and 0.05.

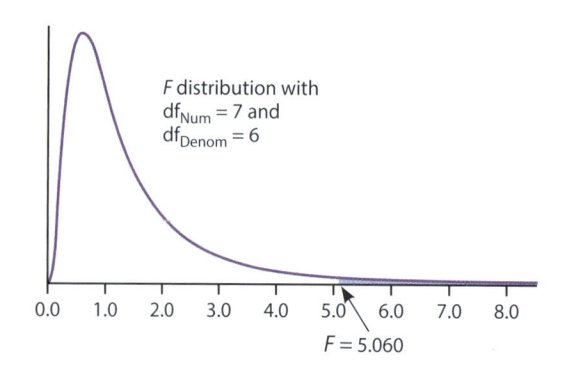

F distribution with $df_{Num} = 7$ and $df_{Denom} = 6$

$F = 5.060$

p	F critical value
.100	3.01
.050	4.21
.025	5.70
.010	8.26
.001	19.46

Because this is a two-sided test, we double these values to get $0.05 < P\text{-value} < 0.10$. For details on finding the P-value from the F statistic with technology, see the Tech Corner at the end of the section.

EXAMPLE 9.7.2

Does a fungus affect tadpole gut length?

Calculating the test statistic and P-value

PROBLEM

Refer to Example 9.7.1. The following table gives the summary statistics for the gut lengths of tadpoles exposed to BD and tadpoles not exposed to BD in the biologists' experiment. Do these data provide convincing evidence that the true standard deviations of gut length differ for tadpoles like the ones in this study that are exposed to the fungus and those that are not exposed?

Group	n	$\bar{x}$	s_x
BD	14	168.35	28.72
Control	13	201.38	26.18

(a) Calculate the test statistic.
(b) Find the P-value.

SOLUTION

(a) $F = \dfrac{s_L^2}{s_S^2} = \dfrac{28.72^2}{26.18^2} = 1.20$

Remember: The larger sample variance goes in the numerator.

(b) P-value:

$df_{Num} = 14 - 1 = 13$ and $df_{Denom} = 13 - 1 = 12$

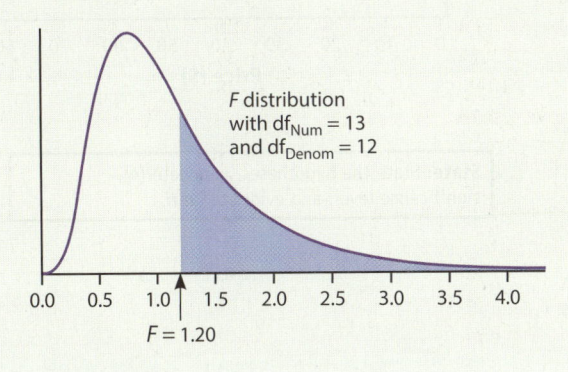

$df_{Num} = n_L - 1$; $df_{Denom} = n_S - 1$

F distribution with $df_{Num} = 13$ and $df_{Denom} = 12$

$F = 1.20$

See the Tech Corner at the end of the section for details about using technology.

Using technology: P-value $= 2(0.379) = 0.758$
Using Table D: P-value $> 2(0.10) = 0.20$

Using the table: The table does not include $df_{Num} = 13$. As with the t distributions, use the next lower value: $df_{Num} = 12$.

EXAM PREP: FOR PRACTICE, TRY EXERCISE 11.

With a P-value of $0.758 > \alpha = 0.05$ in the tadpole experiment, we fail to reject the null hypothesis. We do not have convincing evidence of a difference in the standard deviations of the gut lengths of tadpoles like the ones in this study that are exposed to BD and the gut lengths of tadpoles like the ones in this study that are not exposed to BD.

Putting It All Together: Significance Test Comparing Two Standard Deviations or Variances

We are now ready to use the four-step process to perform a **two-sample F test comparing two standard deviations (variances).** You can use technology to perform the calculations in the Do step. See the Tech Corner at the end of the section for details.

A **two-sample F test comparing two standard deviations (variances)** is a significance test of the null hypothesis that the standard deviations (variances) for two populations or treatments are equal.

EXAMPLE 9.7.3

Do prices vary less at Payless?

Significance test comparing two standard deviations

PROBLEM

Researchers who planned to go shopping for shoes wondered how the prices at Payless compared to the prices at Famous Footwear. To investigate this question, they randomly selected 30 pairs of shoes from each company and recorded the price of each pair. In Section 9.4, Exercise 16 asked you to compare the mean prices of shoes at these two stores. The researchers also wondered whether the prices at Payless are less variable than those at Famous Footwear.

(*continued*)

Do the data provide convincing evidence at the $\alpha = 0.05$ significance level that the standard deviation of shoe prices is smaller at Payless than at Famous Footwear? Here are the summary statistics and boxplots of data from the two samples:

Store	Mean	SD
Famous Footwear	$42.67	$16.33
Payless	$20.00	$7.31

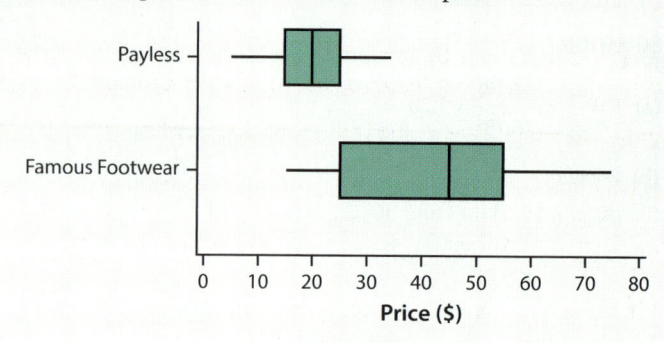

SOLUTION

State:

$H_0: \sigma_{FF} = \sigma_P$

$H_a: \sigma_{FF} > \sigma_P$

where σ_{FF} = the true standard deviation of shoe prices ($) at Famous Footwear and σ_P = the true standard deviation of shoe prices ($) at Payless. Use $\alpha = 0.05$.

The evidence for H_a is: $s_{FF} = 16.33 > 7.31 = s_P$.

Plan: Two-sample F test comparing two standard deviations

Random? The data came from independent random samples of shoes from Payless stores and Famous Footwear stores. ✓

Normal? Neither boxplot is strongly skewed or has outliers. ✓

Do:

- $F = \dfrac{s_{FF}^2}{s_P^2} = \dfrac{16.33^2}{7.31^2} = 4.99$

- P-value: $df_{Num} = 30 - 1 = 29$ and $df_{Denom} = 30 - 1 = 29$

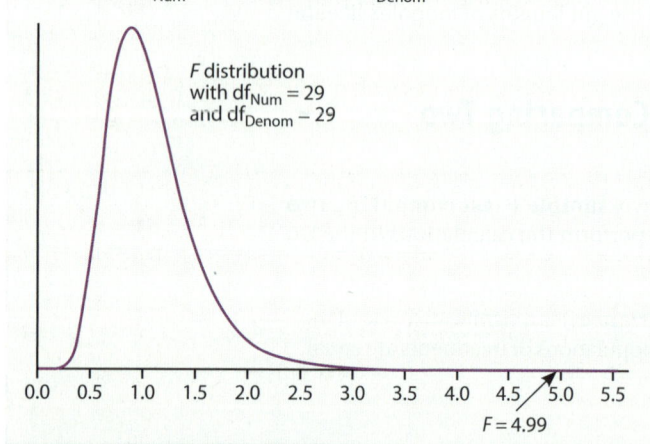

Using technology: $F = 4.99$, P-value ≈ 0
Using Table D: P-value < 0.001

> **State:** State the hypotheses, parameter(s), significance level, and evidence for H_a.

> **Plan:** Identify the appropriate inference method and check the conditions.

> **Do:** If the conditions are met, perform calculations.
> - Calculate the test statistic.
> - Find the P-value.

> Using the table: While the table includes $df_{Denom} = 29$, it does not include $df_{Num} = 29$. As with the t distributions, use the next lower value: $df_{Num} = 25$.

Conclude: Because the P-value of approximately 0 is less than $\alpha = 0.05$, we reject H_0. The data provide convincing evidence that the standard deviation of shoe prices is smaller at Payless than at Famous Footwear.

> **Conclude:** Make a conclusion about the hypotheses in the context of the problem.

EXAM PREP: FOR PRACTICE, TRY EXERCISE 15.

Section 9.7 Significance Tests for Two Population Standard Deviations or Variances

THINK ABOUT IT

How do we construct a 95% confidence interval for σ_1/σ_2? We can get additional information about the ratio of two population standard deviations by constructing a confidence interval. When we compute a confidence interval for the ratio of standard deviations, it does not matter which population standard deviation we put in the numerator. In Examples 9.7.1 and 9.7.2, we used the ratio $\dfrac{s_{BD}}{s_{Control}}$ in the test statistic because $s_{BD} > s_{Control}$, where s_{BD} is the sample standard deviation of the gut lengths of the tadpoles exposed to BD and $s_{Control}$ is the sample standard deviation of the gut lengths of the tadpoles not exposed to BD. Here we consider a confidence interval for the corresponding ratio of true standard deviations: $\dfrac{\sigma_{BD}}{\sigma_{Control}}$.

Technology gives the 95% confidence interval for $\dfrac{\sigma_{BD}}{\sigma_{Control}}$ as 0.610 to 1.948. That is, we are 95% confident that the interval from 0.610 to 1.948 captures the ratio (BD/Control) of the standard deviations of the gut lengths for tadpoles exposed and not exposed to BD. This interval suggests that the standard deviation of the gut lengths for tadpoles like those in the experiment that are exposed to BD is between 0.61 times and 1.948 times the standard deviation of gut lengths for tadpoles like those in the experiment that are not exposed to BD. This is consistent with our earlier "fail to reject H_0" conclusion, because 1 is included in the interval of plausible values for $\dfrac{\sigma_{BD}}{\sigma_{Control}}$.

Computing the interval by hand requires a bit more work. We first compute the confidence interval for the ratio of the variances. A 95% confidence interval for $\dfrac{\sigma_1^2}{\sigma_2^2}$ is given by

$$\frac{1}{F^*_{n_1-1,n_2-1}}\left(\frac{s_1^2}{s_2^2}\right) \text{ to } F^*_{n_2-1,n_1-1}\left(\frac{s_1^2}{s_2^2}\right)$$

where $F^*_{a,b}$ is the 97.5th percentile (with area 0.025 to the right) in an F distribution with $df_{Num} = a$ and $df_{Denom} = b$.

The confidence interval for the ratio of the true variances $\dfrac{\sigma^2_{BD}}{\sigma^2_{Control}}$ for tadpoles exposed to BD and tadpoles not exposed to BD is

$$\frac{1}{F^*_{13,12}}\left(\frac{s_1^2}{s_2^2}\right) \text{ to } F^*_{12,13}\left(\frac{s_1^2}{s_2^2}\right) \rightarrow \frac{1}{3.239}\left(\frac{28.72^2}{26.18^2}\right) \text{ to } 3.153\left(\frac{28.72^2}{26.18^2}\right) \rightarrow 0.372 \text{ to } 3.794$$

Taking the square root of each endpoint gives the interval 0.610 to 1.948 for the ratio of the standard deviations.

TECH CORNER **P-Values and Significance Tests for Comparing Two Standard Deviations or Variances**

Calculating P-Values

You can use technology to find the P-value from the F test statistic in a significance test comparing two standard deviations or variances. We'll illustrate this process using the shoe store data from Example 9.7.3, where $F = 4.99$, $df_{Num} = 29$, and $df_{Denom} = 29$.

Applet

1. Go to www.stapplet.com and launch the *F Distributions* applet.
2. Choose "Calculate an area under the F curve" from the drop-down menu, enter 29 for the numerator degrees of freedom, enter 29 for the denominator degrees of freedom, and click "Plot distribution."
3. Choose the option to calculate the area to the right of a value.
4. Enter the value of F, 4.99, and click "Calculate area."

(continued)

TI-83/84

1. Press 2nd VARS (DISTR) and choose Fcdf(.
2. We want to find the area to the right of $F = 4.99$ under the F distribution curve with numerator df = 29 and denominator df = 29.

   - **OS 2.55 or later:** In the dialog box, enter lower: 4.99, upper: 1000, dfNumer: 29, dfDenom: 29. Choose Paste, and then press ENTER.
   - **Older OS:** Complete the command Fcdf(4.99,1000,29,29) and press ENTER.

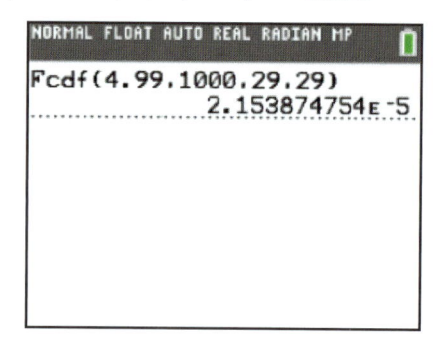

```
NORMAL FLOAT AUTO REAL RADIAN MP
Fcdf(4.99,1000,29,29)
                 2.153874754E-5
```

Performing a Significance Test for Comparing Two Standard Deviations

You can use technology to perform the calculations for a significance test of $H_0: \sigma_1 = \sigma_2$. We'll illustrate using the shoe store data from Example 9.7.3. The data for the two samples are given here:

Famous Footwear	15	25	25	25	25	25	25	25	35	35
	35	35	35	35	45	45	45	45	45	45
	45	45	55	55	65	65	65	65	75	75

Payless	5	5	15	15	15	15	15	15	15	15
	15	15	15	15	15	25	25	25	25	25
	25	25	25	25	25	25	25	25	35	35

Applet

1. Go to www.stapplet.com and launch the *One Quantitative Variable, Multiple Groups* applet.
2. Enter "Price" as the variable name.
3. Choose to input raw data. Type the group names "Famous Footwear" and "Payless" in the boxes at the left. Enter the data for each group manually, or copy and paste them from a document or spreadsheet. Be sure to separate the data values with commas or spaces.

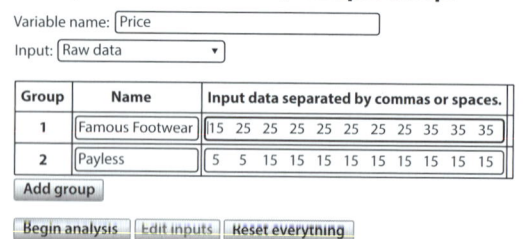

One Quantitative Variable, Multiple Groups
Variable name: Price
Input: Raw data

Group	Name	Input data separated by commas or spaces.
1	Famous Footwear	15 25 25 25 25 25 25 25 35 35 35
2	Payless	5 5 15 15 15 15 15 15 15 15

Add group

Begin analysis | Edit inputs | Reset everything

4. Click the Begin analysis button. Scroll down to the "Perform Inference" section. Choose the F test for σ_1^2/σ_2^2 for the procedure, and $\sigma_1^2 > \sigma_2^2$ for the alternative hypothesis. Then click the Perform inference button.

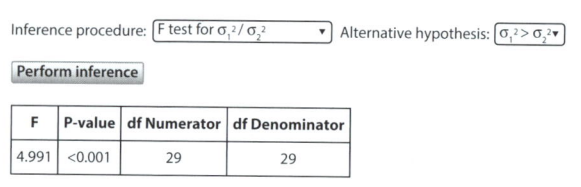

Perform Inference

Inference procedure: F test for σ_1^2/σ_2^2 Alternative hypothesis: $\sigma_1^2 > \sigma_2^2$

Perform inference

F	P-value	df Numerator	df Denominator
4.991	<0.001	29	29

Note: If you have summary statistics instead of raw data, choose "Mean and standard deviation" from the Input menu in Step 3 and enter the name, mean, standard deviation, and sample size for each group. Then follow the directions in Step 4.

TI-83/84

1. Enter the shoe prices for Famous Footwear in list L_1 and for Payless in list L_2.
2. Press STAT, then choose TESTS and 2-SampFTest.

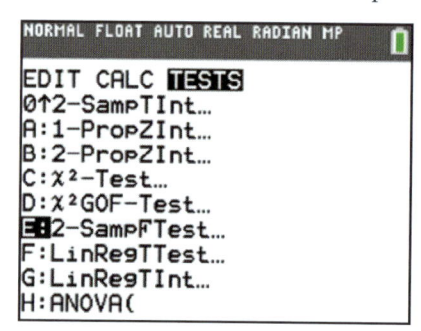

```
NORMAL FLOAT AUTO REAL RADIAN MP
EDIT CALC TESTS
0↑2-SampTInt...
A:1-PropZInt...
B:2-PropZInt...
C:χ²-Test...
D:χ²GOF-Test...
E:2-SampFTest...
F:LinRegTTest...
G:LinRegTInt...
H:ANOVA(
```

3. When the 2-SampFTest screen appears, choose Data as the input method and enter the values shown. Be sure to choose $\sigma_1 > \sigma_2$ as the alternative hypothesis.

```
NORMAL FLOAT AUTO REAL RADIAN MP
         2-SampFTest
Inpt:Data Stats
List1:L1
List2:L2
Freq1:1
Freq2:1
σ1:≠σ2  <σ2  >σ2
Color: BLUE
Calculate  Draw
```

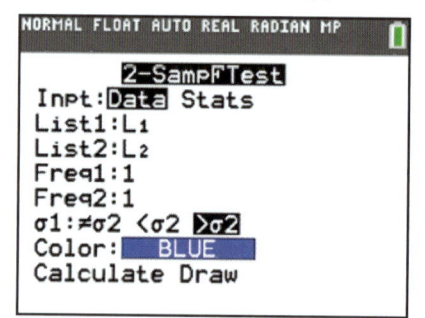

4. Highlight "Calculate" and press ENTER.

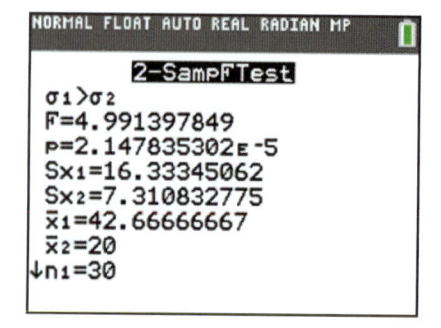

```
NORMAL FLOAT AUTO REAL RADIAN MP
         2-SampFTest
σ1>σ2
F=4.991397849
p=2.147835302E-5
Sx1=16.33345062
Sx2=7.310832775
x̄1=42.66666667
x̄2=20
↓n1=30
```

Detailed instructions for using CrunchIt!, Excel, Google Sheets, JMP, Minitab, and R are available in Achieve.

Section 9.7 What Did You Learn?

Review the learning goals from this section. Then practice what you've learned by working through the exercises.

Learning Goals	Example	Exercises
State hypotheses and check conditions for a significance test comparing two standard deviations or variances.	9.7.1	5–10
Calculate the test statistic and P-value for a significance test comparing two standard deviations or variances.	9.7.2	11–14
Use the four-step process to perform a significance test comparing two standard deviations or variances.	9.7.3	15–18

Section 9.7 Exercises

Building Concepts and Skills These exercises assess the basic knowledge you should have after reading the section.

1. Use appropriate notation to write out the null hypothesis that two population standard deviations are the same.

2. State the two conditions that must be checked before performing a significance test comparing two standard deviations or variances.

3. The F test statistic has $df_{Num} =$ _____ and $df_{Denom} =$ _____.

4. True/False: It doesn't matter which variance appears in the numerator of the formula for the F test statistic.

Mastering Concepts and Skills These exercises reinforce the learning goals as shown in the examples.

5. **Refer to Example 9.7.1 Crabs and Noise** Researchers collected 34 crabs and randomly allocated them to one of two treatments: 7.5 minutes of ship noise or 7.5 minutes of ambient harbor noise. The researchers measured the total amount of oxygen consumed by the crabs during treatment. Do the data provide convincing evidence at the 10% significance level that the standard deviation of oxygen consumption is different for crabs like the ones in the study when exposed to ship noise as compared to ambient harbor noise?[83] Here are the oxygen consumption levels for the crabs in each group:

Ambient	22.7	34.6	36.0	40.1	47.5	49.6	50.7	54.4	56.0
	57.4	59.7	60.5	61.3	67.9	68.5	84.9	84.9	
Ship	25.3	29.2	32.0	39.4	41.9	43.8	50.8	53.0	53.6
	55.3	57.3	58.8	59.8	63.6	62.3	72.7	74.5	

(a) State appropriate hypotheses for performing a significance test. Be sure to define the parameters of interest.

(b) Do the data provide *some* evidence of a difference in standard deviations between the oxygen consumption of crabs like the ones in the study exposed to ship noise and crabs like the ones in the study exposed to ambient harbor noise? Justify your answer.

(c) Check that the conditions for performing the test are met.

6. **Sleep Deprivation** Does sleep deprivation affect consistency in cognitive tasks? Researchers designed a study using 21 volunteer subjects between the ages of 18 and 25. All 21 participants took a computer-based visual discrimination test at the beginning of the study. Then the researchers randomly assigned the subjects into two groups. The 11 subjects in one group were deprived of sleep for an entire night in a laboratory setting. The 10 subjects in the other group were allowed unrestricted sleep for the night. Both groups were allowed as much sleep as they wanted for the next two nights. On Day 4, all the subjects took the same visual discrimination test on the computer. Researchers recorded the improvement in time (measured in milliseconds) from Day 1 to Day 4 on each subject's tests. Do the data provide convincing evidence at the 5% significance level that the standard deviation of improvement is different between people like the ones in the study who had sleep deprivation and people like the ones in the study who had unrestricted sleep? Here are the data:[84]

Sleep deprivation	−14.7	−10.7	−10.7	2.2	2.4	4.5
	7.2	9.6	10.0	21.3	21.8	
Unrestricted sleep	−7.0	11.6	12.1	12.6	14.5	18.6
	25.2	30.5	34.5	45.6		

(a) State appropriate hypotheses for performing a significance test. Be sure to define the parameters of interest.

(b) Do the data provide *some* evidence that the standard deviation of improvement in time from Day 1 to Day 4 is different between people like the ones in the study who had sleep deprivation and people like the ones in the study who had unrestricted sleep? Justify your answer.

(c) Check that the conditions for performing the test are met.

7. **Red vs. White Wine** Observational studies suggest that moderate use of alcohol by adults reduces heart attacks and that red wine may have special benefits. One reason may be that red wine contains polyphenols, antioxidants that may reduce the risk of heart attacks. In an experiment, researchers randomly assigned healthy participants to drink half a bottle of either red or white wine each day for 2 weeks. They measured the level of polyphenols in the participants' blood before and after the 2-week period. Do the data provide convincing evidence that the standard deviation of percent change in polyphenol levels is smaller for red wine than for white wine in people like those in the study? Parallel boxplots and summary statistics for the groups are shown.[85]

Group	n	Mean	SD
Red	9	5.50	2.517
White	9	0.23	3.290

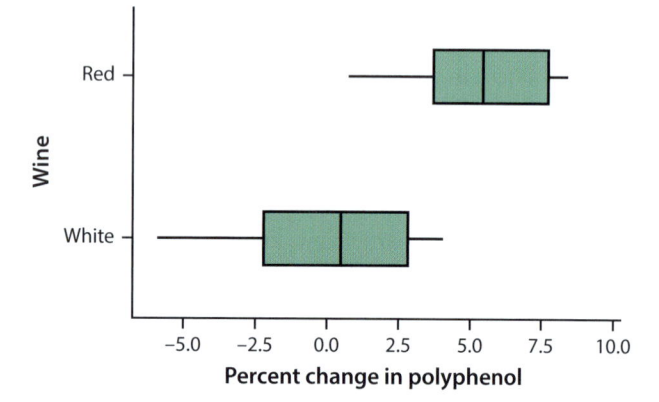

(a) State appropriate hypotheses for performing a significance test. Be sure to define the parameters of interest.
(b) Do the data provide *some* evidence that the standard deviation of percent change in polyphenol levels is smaller for red wine than for white wine in people like those in the study? Justify your answer.
(c) Check that the conditions for performing the test are met.

8. **Friend Effect** Does whether you are buying from a friend or a stranger affect the consistency with which you offer a price for the item? A researcher randomly divided 15 subjects into two groups. Both groups received a description of an item they might wish to purchase. One group was told to imagine buying that item from a friend whom they expected to see again. The other group was told to imagine buying that item from a stranger. Do the data provide convincing evidence that the standard deviation of prices offered is smaller when buying from a friend than when buying from a stranger for people like those in the study? Normal probability plots and summary statistics for the two groups are shown.[86]

Group	n	Mean	SD
Friend	8	281.88	18.31
Stranger	7	211.40	46.40

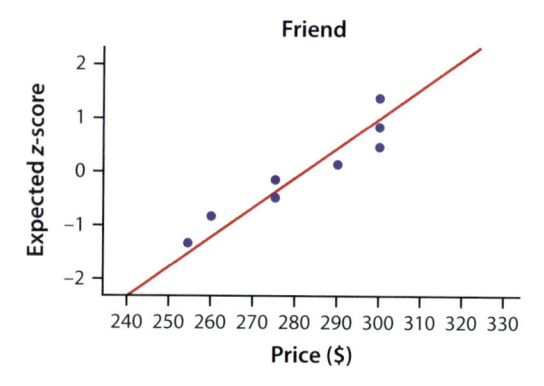

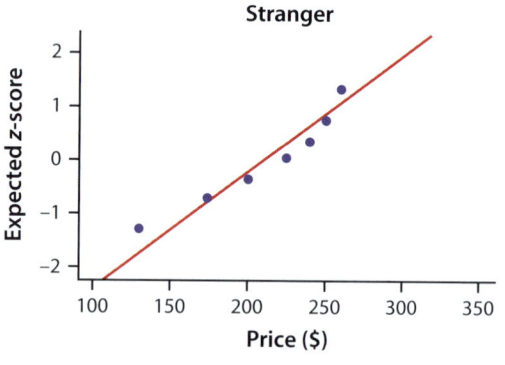

(a) State appropriate hypotheses for performing a significance test. Be sure to define the parameters of interest.
(b) Do the data provide *some* evidence that the standard deviation of prices offered is smaller when buying from a friend than when buying from a stranger for people like those in the study? Justify your answer.
(c) Check that the conditions for performing the test are met.

9. **Family Income** How do incomes compare in Indiana and New Jersey? The parallel dotplots show the total family income of independent random samples of 38 Indiana residents and 44 New Jersey residents. Is there convincing evidence that the standard deviations of family incomes in Indiana and New Jersey are different? Check if the conditions for conducting the appropriate significance test are met.

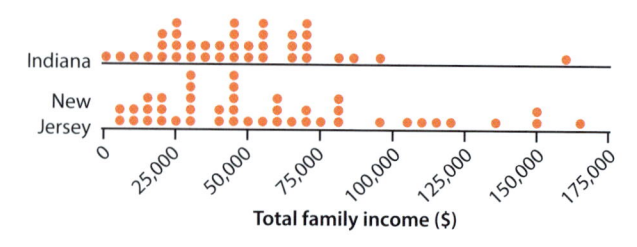

10. **Word Consistency** Sanderson is interested in comparing the word length in articles from a medical journal and from an airline's in-flight magazine. Sanderson counts the number of letters in the first 400 words of an article in the medical journal and in the first 100 words of an article in the airline magazine, and then uses statistical software to produce the histograms shown. Note that *J* stands for journal and

M for magazine. Is there convincing evidence that the standard deviations of word lengths in the two types of publications are different? Check if the conditions for conducting the appropriate significance test are met.

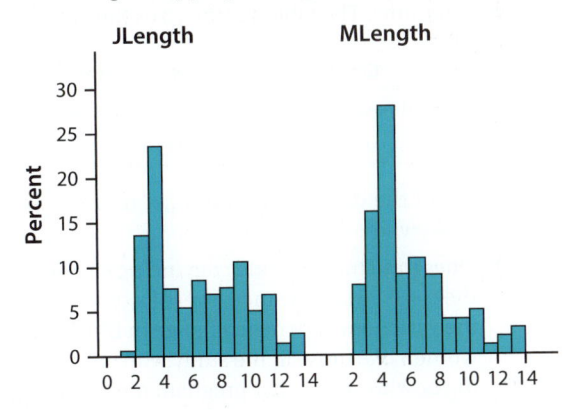

11. **Refer to Example 9.7.2 Crabs and Noise** Refer to Exercise 5.
   (a) Calculate the test statistic.
   (b) Find the *P*-value.

12. **Sleep Deprivation** Refer to Exercise 6.
   (a) Calculate the test statistic.
   (b) Find the *P*-value.

13. **Red vs. White Wine** Refer to Exercise 7.
   (a) Calculate the test statistic and the *P*-value.
   (b) What conclusion would you make?

14. **Friend Effect** Refer to Exercise 8.
   (a) Calculate the test statistic and the *P*-value.
   (b) What conclusion would you make?

15. **Refer to Example 9.7.3 Encouragement Effect** A researcher wondered if people would perform differently in the barbell curl if they were encouraged by a coach. The researcher recruited 31 subjects to participate in an experiment and then split them into two groups using random assignment. Group 1 received positive encouragement during the exercise, while Group 2 served as a control group that received no encouragement. The researcher recorded the number of barbell curl repetitions each participant was able to complete before setting the bar down. Here are the data:[87]

With encouragement	50	11	60	5	40	45	11	30
	8	50	11	15	9	40	75	
No encouragement	43	19	33	1	0	2	6	61
	21	11	40	15	15	18	37	22

Do these data provide convincing evidence at the $\alpha = 0.01$ significance level that the standard deviation of the number of barbell curl repetitions that people like the ones in this study can do differs depending on whether they receive encouragement or not?

16. **Maintaining Balance** As we get older, can we continue to maintain our balance just as well as when we were younger? Researchers selected a random sample of nine elderly people and a separate random sample of eight young people. All subjects were told to stand barefoot on a "force platform" that measures their movement and given a hand-held button. They were asked to maintain a stable upright position while also pressing the button every time they heard a randomly timed specific noise. One variable measured was the amount of side-to-side sway (in mm) each subject demonstrated. Here are the data:[88]

Elderly	14	41	18	11	16	24	18	21	37
Young	17	10	16	22	12	14	12	18	

Do these data provide convincing evidence at the $\alpha = 0.05$ significance level that the standard deviation of the amount of sway differs in the populations of elderly and young people? *Note:* Advanced methods indicate that the Normal condition is met for these data.

17. **Reading Activities** An educator believes that new reading activities in the classroom will help elementary school students improve the consistency in their reading ability. She recruits 44 third-grade students and randomly assigns them to two groups. One group of 21 students does the new activities for an 8-week period. A control group of 23 third-graders follows the same curriculum without the activities. At the end of the 8 weeks, all students are given the Degree of Reading Power (DRP) test, which measures the aspects of reading ability that the treatment is designed to improve. Parallel boxplots and summary statistics for the groups are shown here.[89]

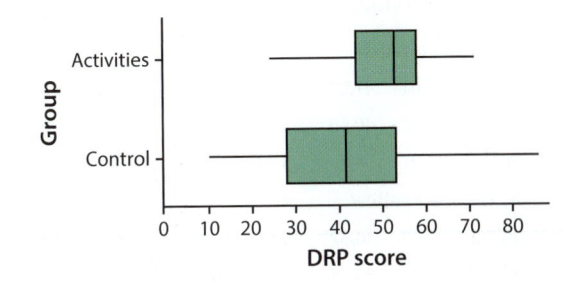

Group	*n*	Mean	SD
Activities	21	51.48	11.01
Control	23	41.52	17.15

Do these data provide convincing evidence at the 5% significance level that the standard deviation in DRP score is less for students like the ones in this study who do the activities than for students who do not do the activities?

18. **Squirrel Weights** In many parts of the northern United States, two color variants of the Eastern gray squirrel — gray and black — are found in the same habitats. A scientist wonders if the weights of all gray Eastern gray squirrels are less variable than the weights of all black Eastern gray squirrels. The scientist collects separate random samples of 40 squirrels of each color from a large forest and weighs them. Parallel dotplots and summary statistics for the groups are shown here.

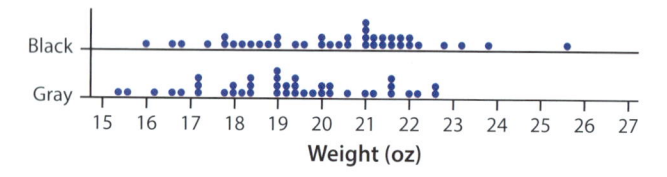

Group	n	Mean	SD
Black	40	20.3	2.1
Gray	40	19.2	1.9

Do these data provide convincing evidence at the 5% significance level that the standard deviation of the weights of all gray Eastern gray squirrels is less than the standard deviation of the weights of all black Eastern gray squirrels in this large forest?

Applying the Concepts These exercises ask you to apply multiple learning goals in a new context or to apply what you learned in this section in a new way.

19. **Tropical Flowers** Different varieties of the tropical flower *Heliconia* are fertilized by different species of hummingbirds. Researchers believe that over time, the lengths of the flowers and the forms of the hummingbirds' beaks have evolved to match each other. Here are data on the lengths (in mm) for random samples of two color varieties of the same species of flower on the island of Dominica:[90]

H. caribaea red

41.90	42.01	41.93	43.09	41.17	41.69	39.78	40.57
39.63	42.18	40.66	37.87	39.16	37.40	38.20	38.07
38.10	37.97	38.79	38.23	38.87	37.78	38.01	

H. caribaea yellow

| 36.78 | 37.02 | 36.52 | 36.11 | 36.03 | 35.45 | 38.13 | 37.10 |
| 35.17 | 36.82 | 36.66 | 35.68 | 36.03 | 34.57 | 34.63 | |

(a) Does this study provide convincing evidence at the $\alpha = 0.05$ significance level of a difference in the variances of the lengths of these two species of flower on the island of Dominica?

(b) Based on your conclusion in part (a), which mistake — a Type I error or a Type II error — could you have made? Explain your answer.

20. **Breastfeeding and Calcium** Breastfeeding mothers secrete calcium into their milk. Some of this calcium may come from their bones, so mothers may lose bone mineral. Researchers compared a random sample of 47 breastfeeding women with a random sample of 22 women of similar age who were neither pregnant nor lactating. They measured the percent change in the bone mineral content (BMC) of the women's spines over 3 months. The table summarizes the data.[91]

Group	n	Mean	SD
Breastfeeding	47	−3.59	2.51
Not pregnant nor lactating	22	0.31	1.30

Graphs of the data for both groups revealed no strong skewness or outliers.

(a) Does this study provide convincing evidence at the $\alpha = 0.05$ significance level of a difference in the variances of percent change in BMC between women like those in the study who are breastfeeding or who are neither pregnant nor lactating?

(b) Based on your conclusion in part (a), which mistake — a Type I error or a Type II error — could you have made? Explain your answer.

21. **Crabs and Noise** Refer to Exercises 5 and 11.

(a) Interpret the *P*-value.

(b) What conclusion would you make?

22. **Sleep Deprivation** Refer to Exercises 6 and 12.

(a) Interpret the *P*-value.

(b) What conclusion would you make?

23. **Tech Possum Length** Does the size of Australian possums differ by sex? Zoologists in Australia captured a random sample of 104 possums and recorded the values of 13 different variables, including foot length (*footlgth*, in mm) and *sex*, in the **Possum** data set.[92] Due to the sampling method used in this survey, it is reasonable to view the male and female possums in this data set as independent random samples from the corresponding populations. Does this study provide convincing evidence of a difference in the standard deviations of the foot length of all male and all female possums?

24. **Tech Glass Forensics** Glass can be used as evidence in a criminal investigation if it can be correctly identified. Researchers were interested in whether glass types can be determined by certain characteristics. They designed a study that used a total of 214 randomly selected specimens of glass of several different types, including headlamps and vehicle windows that were float processed (made by pouring molten glass onto molten tin). The **Glass** data set contains measurements of various characteristics of these glass specimens, including the percentage of aluminum by weight (*Aluminum*).[93] Extract the data for the specimens labeled "Headlamp" and "Vehicle window float processed" in the *Type* column for analysis. Due to the sampling method used in this survey, it is reasonable to view these specimens as independent random samples from the corresponding populations of these two types of glass. Does this study provide convincing evidence of a difference in the standard deviations of the amount of aluminum in headlamps and float-processed glass from vehicle windows?

Extending the Concepts These exercises challenge you to explore statistical concepts and methods that go beyond what you learned in this section.

25. **Tropical Flowers** In Exercise 19, you performed a significance test to determine if there is convincing evidence that the standard deviation of the length of *H. caribaea* red flowers is different from the standard deviation of the length of *H. caribaea* yellow flowers on the island of Dominica. Summary statistics of the data from the two samples are given here.

Group	n	SD
Red	23	1.786
Yellow	15	0.975

   (a) Use the method described in the Think About It feature in this section to compute a 95% confidence interval for $\dfrac{\sigma_R}{\sigma_Y}$.

   (b) Explain how the interval you calculated in part (a) is consistent with, but gives more information than, the significance test in Exercise 19.

Cumulative Review These exercises revisit what you learned in previous sections.

26. **Tadpoles (9.4)** Refer to Example 9.7.1. Recall that biologists wondered whether a reduction in food consumption would result in changes to the gut length of the tadpoles. In particular, one question was whether the mean gut length would increase for those tadpoles exposed to BD, allowing them to increase their nutrient absorption from the limited amount of food. The data are given here once again (in mm):

BD	191.40 142.92 169.81 152.18 171.33 153.92 181.08
	124.90 173.06 207.01 143.77 220.35 130.00 195.13
Control	177.92 181.58 154.29 217.67 185.88 249.29 196.90
	202.16 210.86 215.60 174.16 222.83 228.74

   (a) State appropriate hypotheses for performing a significance test. Be sure to define the parameters of interest.

   (b) Do the data provide *some* evidence that the mean gut length of tadpoles like the ones in this study that are exposed to BD is larger than the mean gut length for tadpoles not exposed to BD? Justify your answer.

   (c) What does your answer to part (b) tell you about possible values for the *P*-value in this significance test?

27. **Liquid Detergent (6.2, 6.8)** The amount of liquid detergent poured into "138-ounce" bottles at a certain factory is normally distributed with mean 139.4 ounces and standard deviation 0.58 ounce.

   (a) What is the probability that a randomly selected bottle contains less than 138 ounces of detergent?

   (b) What is the probability that the mean amount of detergent in a sample of 5 randomly selected bottles will be less than 138 ounces?

Statistics Matters How fast-food drive-thrus make money: Speed, accuracy, and customer service

At the beginning of the chapter, we described a study about drive-thru service at fast-food restaurants. Here is a summary of the details from the 2021 *QSR* study:

- In the 159 visits to McDonald's, 91.2% of the orders were accurate. In the 163 visits to Wendy's, 85.3% of the orders were accurate.
- McDonald's average service time for 159 drive-thru visits was 311 seconds with a standard deviation of 182 seconds. Burger King's average service time for 145 drive-thru visits was 359 seconds with a standard deviation of 194 seconds.
- The two top-rated restaurants for customer service in the survey were Chick-Fil-A and KFC. In a random sample of 1002 people, 525 rated Chick-Fil-A's service as "very friendly." In a separate random sample of 1000 people, 444 rated KFC's service as "very friendly."

1. Construct and interpret a 95% confidence interval for the difference in the true proportions of accurate drive-thru orders at McDonald's and Wendy's restaurants in 2021.

2. Do the data provide convincing evidence that a greater proportion of all Chick-Fil-A customers than KFC customers in 2021 would rate their restaurant's service as "very friendly"?

3. Should you use two-sample *t* procedures or paired *t* procedures to compare the mean drive-thru service times for all McDonald's and Burger King restaurants in 2021? Explain your answer.

4. Do the data provide convincing evidence at the $\alpha = 0.05$ significance level of a difference in the mean drive-thru service times at all McDonald's and Burger King restaurants in 2021?

5. A 95% confidence interval for the difference in the population mean drive-thru service times (McDonald's – Burger King) is –90.57 seconds to –5.43 seconds. Explain how the confidence interval is consistent with, but gives more information than, the significance test in Question 4.

6. *Challenge:* Explain why it would not be appropriate to use these data to carry out a test of $H_0: \sigma_M = \sigma_{BK}$ versus $H_a: \sigma_M \neq \sigma_{BK}$, where σ_M is the standard deviation of service time at all McDonald's restaurant drive-thrus in 2021 and σ_{BK} is the standard deviation of service time at all Burger King restaurant drive-thrus in 2021.

Postscript: Chick-Fil-A drive-thrus had, by far, the longest average service time among the 10 fast-food restaurants in the 2021 *QSR* study: 541 seconds (about 9 minutes).

Chapter 9 Review

Comparing Two Populations or Treatments

- Inference methods for comparing proportions, means, or standard deviations/variances of two populations or treatments build on the procedures for inference about one parameter introduced in Chapters 7 and 8.
- Be sure to follow the four-step process whenever you construct a confidence interval or perform a significance test for a difference in proportions, a difference in means, a mean difference, or a ratio of two standard deviations/variances.

Inference About Two Proportions

- Let $\hat{p}_1$ be the proportion of successes in an SRS of size n_1 selected from a large population with proportion of successes p_1. Let $\hat{p}_2$ be the proportion of successes in an SRS of size n_2 selected from a large population with proportion of successes p_2. The sampling distribution of $\hat{p}_1 - \hat{p}_2$ has the following properties:
  - **Shape:** Approximately normal if $n_1 p_1$, $n_1(1-p_1)$, $n_2 p_2$, and $n_2(1-p_2)$ are all at least 10.
  - **Center:** The mean is $\mu_{\hat{p}_1 - \hat{p}_2} = p_1 - p_2$.
  - **Variability:** The standard deviation is

  $$\sigma_{\hat{p}_1 - \hat{p}_2} = \sqrt{\frac{p_1(1-p_1)}{n_1} + \frac{p_2(1-p_2)}{n_2}}$$

- Before estimating or testing a claim about $p_1 - p_2$, check that these conditions are met:
  - **Random:** The data come from independent random samples from the two populations of interest or from two groups in a randomized experiment.
  - **Large Counts:** The counts of "successes" and "failures" in each sample or group—$n_1 \hat{p}_1$, $n_1(1-\hat{p}_1)$, $n_2 \hat{p}_2$, and $n_2(1-\hat{p}_2)$—are all at least 10.

- When the Random and Large Counts conditions are met, a $C\%$ confidence interval for $p_1 - p_2$ is

$$(\hat{p}_1 - \hat{p}_2) \pm z^* \sqrt{\frac{\hat{p}_1(1-\hat{p}_1)}{n_1} + \frac{\hat{p}_2(1-\hat{p}_2)}{n_2}}$$

where z^* is the critical value for the standard normal distribution with $C\%$ of its area between $-z^*$ and z^*. This is called a **two-sample z interval for a difference between two proportions.**

- To estimate the standard deviation of the sampling distribution of $\hat{p}_1 - \hat{p}_2$ in a significance test of $H_0: p_1 - p_2 = 0$, use the *combined (pooled) sample proportion*

$$\hat{p}_C = \frac{\text{number of successes in both samples combined}}{\text{number of individuals in both samples combined}} = \frac{X_1 + X_2}{n_1 + n_2}$$

- The standardized test statistic for a **two-sample z test for a difference between two proportions** is

$$z = \frac{(\hat{p}_1 - \hat{p}_2) - 0}{\sqrt{\frac{\hat{p}_C(1-\hat{p}_C)}{n_1} + \frac{\hat{p}_C(1-\hat{p}_C)}{n_2}}}$$

When the Random and Large Counts conditions are met, the standardized test statistic has approximately a standard normal distribution. You can use Table A or technology to find the P-value.

Inference About Two Means

- Let $\bar{x}_1$ be the sample mean in an SRS of size n_1 selected from a large population with mean μ_1 and standard deviation σ_1. Let $\bar{x}_2$ be the sample mean in an SRS of size n_2

selected from a large population with mean μ_2 and standard deviation σ_2. The sampling distribution of $\bar{x}_1 - \bar{x}_2$ has the following properties:

- **Shape:** Normal if both population distributions are normal; approximately normal if both sample sizes are large ($n_1 \geq 30$ and $n_2 \geq 30$), or if one population distribution is normal and the other sample size is large, or if both population distributions are approximately normal.
- **Center:** The mean is $\mu_{\bar{x}_1 - \bar{x}_2} = \mu_1 - \mu_2$.
- **Variability:** The standard deviation is

$$\sigma_{\bar{x}_1 - \bar{x}_2} = \sqrt{\frac{\sigma_1^2}{n_1} + \frac{\sigma_2^2}{n_2}}$$

- Before estimating or testing a claim about $\mu_1 - \mu_2$, check that these conditions are met:
  - **Random:** The data come from independent random samples from the two populations of interest or from two groups in a randomized experiment.
  - **Normal/Large Sample:** For each sample, the data come from a normally distributed population or the sample size is large ($n \geq 30$). For each sample, if the population distribution has unknown shape and $n < 30$, a graph of the sample data shows no strong skewness or outliers.
- When the Random and Normal/Large Sample conditions are met, a $C\%$ confidence interval for $\mu_1 - \mu_2$ is

$$(\bar{x}_1 - \bar{x}_2) \pm t^* \sqrt{\frac{s_1^2}{n_1} + \frac{s_2^2}{n_2}}$$

where t^* is the critical value with $C\%$ of its area between $-t^*$ and t^* in the t distribution with degrees of freedom given by technology or using df = the smaller of $n_1 - 1$ and $n_2 - 1$. This is called a **two-sample t interval for a difference between two means**.

- A significance test of $H_0: \mu_1 - \mu_2 = 0$ is called a **two-sample t test for a difference between two means**. The standardized test statistic is

$$t = \frac{(\bar{x}_1 - \bar{x}_2) - 0}{\sqrt{\frac{s_1^2}{n_1} + \frac{s_2^2}{n_2}}}$$

When the Random and Normal/Large Sample conditions are met, find the P-value using the t distribution with degrees of freedom given by technology or using the smaller of $n_1 - 1$ and $n_2 - 1$ (conservative approach).

Inference About a Mean Difference: Paired Data

- **Paired data** result from recording two values of the same quantitative variable for each individual or for each pair of similar individuals.
- To analyze paired data, start by calculating the difference for each pair. Then use a graph to see if the differences are consistently greater than or consistently less than 0.
- When analyzing paired data, use the mean difference $\bar{x}_{\text{diff}}$ and the standard deviation of the differences s_{diff} as summary statistics.

- Before estimating or testing a claim about μ_{diff}, check that these conditions are met:
  - **Random:** Paired data come from a random sample from the population of interest or from a randomized experiment.
  - **Normal/Large Sample:** Paired data come from a normally distributed population of differences or the sample size is large ($n_{\text{diff}} \geq 30$). When the sample size is small and the shape of the population distribution of differences is unknown, a graph of the sample differences shows no strong skewness or outliers.
- When the Random and Normal/Large Sample conditions are met, a $C\%$ confidence interval for the true mean difference μ_{diff} is

$$\bar{x}_{\text{diff}} \pm t^* \frac{s_{\text{diff}}}{\sqrt{n_{\text{diff}}}}$$

where t^* is the critical value for a t distribution with df $= n_{\text{diff}} - 1$ and $C\%$ of its area between $-t^*$ and t^*. This is called a **paired t interval for a mean difference**.

- A significance test of $H_0: \mu_{\text{diff}} = 0$ is called a **paired t test for a mean difference**. The standardized test statistic is

$$t = \frac{\bar{x}_{\text{diff}} - 0}{\frac{s_{\text{diff}}}{\sqrt{n_{\text{diff}}}}}$$

When the Random and Normal/Large Sample conditions are met, use technology or Table B to find the P-value from a t distribution with df $= n_{\text{diff}} - 1$.

- The proper inference method depends on how the data were produced. For paired data, use paired t procedures for μ_{diff}. For quantitative data that come from independent random samples from two populations or from two groups in a randomized experiment, use two-sample t procedures for $\mu_1 - \mu_2$.

Significance Tests for Two Population Standard Deviations or Variances

- In a test comparing two population standard deviations or variances:
  - The null hypothesis typically states that the ratio of the parameters is 1, which is equivalent to saying that the two parameters are equal:

$$H_0: \sigma_1 = \sigma_2 \quad \text{or} \quad H_0: \sigma_1^2 = \sigma_2^2$$

  - The alternative hypothesis says how we expect the standard deviations or variances to compare, and can be either one-sided or two-sided.
- The conditions for performing a significance test comparing two population standard deviations or variances are:
  - **Random:** The data come from independent random samples from the two populations of interest or from two groups in a randomized experiment.
  - **Normal:** For each sample, the data come from a normally distributed population.

    There is *no* large sample size override to the Normal condition for this test.

- The test statistic for a **two-sample F test comparing two standard deviations (or variances)** is

$$F = \frac{s_L^2}{s_S^2}$$

where s_L^2 is the larger sample variance and s_S^2 is the smaller sample variance.

- When the Random and Normal conditions are met, the test statistic can be modeled by an **F distribution** with numerator degrees of freedom $df_{Num} = n_L - 1$, where n_L is the size of the sample with the larger variance, and denominator degrees of freedom $df_{Denom} = n_S - 1$, where n_S is the size of the sample with the smaller variance. You can use technology or Table D to find the P-value.

Inference methods for comparing proportions, means, and variances

	Difference in proportions $p_1 - p_2$	Difference in means $\mu_1 - \mu_2$	Mean difference μ_{diff}	Ratio of SDs or variances $\frac{\sigma_1}{\sigma_2}$ or $\frac{\sigma_1^2}{\sigma_2^2}$
Conditions	• **Random:** The data come from independent random samples from the two populations of interest or from two groups in a randomized experiment. • **Large Counts:** The counts of "successes" and "failures" in each sample or group — $n_1\hat{p}_1, n_1(1-\hat{p}_1), n_2\hat{p}_2, n_2(1-\hat{p}_2)$ — are all at least 10.	• **Random:** The data come from independent random samples from the two populations of interest or from two groups in a randomized experiment. • **Normal/Large Sample:** For each sample, the data come from a normally distributed population or the sample size is large ($n \geq 30$). For each sample, if the population distribution has unknown shape and $n < 30$, a graph of the sample data shows no strong skewness or outliers.	• **Random:** Paired data come from a random sample from the population of interest or from a randomized experiment. • **Normal/Large Sample:** Paired data come from a normally distributed population of differences or the sample size is large ($n_{diff} \geq 30$). When the sample size is small and the shape of the population distribution of differences is unknown, a graph of the sample differences shows no strong skewness or outliers.	• **Random:** The data come from independent random samples from the two populations of interest or from two groups in a randomized experiment. • **Normal:** For each sample, the data come from a normally distributed population.

Significance tests

	Two-sample z test for $p_1 - p_2$ (2-PropZTest)	Two-sample t test for $\mu_1 - \mu_2$ (2-SampTTest)	Paired t test for μ_{diff} (T-Test)	Two-sample F test comparing SDs or variances (2-SampFTest)
Name (TI-83/84)				
Null hypothesis	$H_0: p_1 - p_2 = 0$	$H_0: \mu_1 - \mu_2 = 0$	$H_0: \mu_{diff} = 0$	$H_0: \sigma_1 = \sigma_2$ or $H_0: \sigma_1^2 = \sigma_2^2$
Formula	$z = \dfrac{(\hat{p}_1 - \hat{p}_2) - 0}{\sqrt{\dfrac{\hat{p}_C(1-\hat{p}_C)}{n_1} + \dfrac{\hat{p}_C(1-\hat{p}_C)}{n_2}}}$ where $\hat{p}_C = \dfrac{X_1 + X_2}{n_1 + n_2}$ P-value from standard normal distribution	$t = \dfrac{(\bar{x}_1 - \bar{x}_2) - 0}{\sqrt{\dfrac{s_1^2}{n_1} + \dfrac{s_2^2}{n_2}}}$ P-value from t distribution with df from technology or smaller of $n_1 - 1$ and $n_2 - 1$	$t = \dfrac{\bar{x}_{diff} - 0}{\dfrac{s_{diff}}{\sqrt{n_{diff}}}}$ P-value from t distribution with $df = n_{diff} - 1$	$F = \dfrac{s_L^2}{s_S^2}$ where L = larger variance, S = smaller variance P-value from F distribution with $df_{Num} = n_L - 1$ and $df_{Denom} = n_S - 1$

	Confidence intervals			
Name (TI-83/84)	Two-sample z interval for $p_1 - p_2$ (2-PropZInt)	Two-sample t interval for $\mu_1 - \mu_2$ (2-Samp TInt)	Paired t interval for μ_{diff} (T-Interval)	See the Think About It in Section 9.7 for information about a confidence interval for $\dfrac{\sigma_1}{\sigma_2}$ or $\dfrac{\sigma_1^2}{\sigma_2^2}$
Formula (degrees of freedom)	$(\hat{p}_1 - \hat{p}_2) \pm z^* \sqrt{\dfrac{\hat{p}_1(1-\hat{p}_1)}{n_1} + \dfrac{\hat{p}_2(1-\hat{p}_2)}{n_2}}$	$(\bar{x}_1 - \bar{x}_2) \pm t^* \sqrt{\dfrac{s_1^2}{n_1} + \dfrac{s_2^2}{n_2}}$ df from technology or smaller of $n_1 - 1, n_2 - 1$	$\bar{x}_{\text{diff}} \pm t^* \dfrac{s_{\text{diff}}}{\sqrt{n_{\text{diff}}}}$ df $= n_{\text{diff}} - 1$	

Chapter 9 Review Exercises

These exercises will help you review important concepts and skills described by the learning goals in each section. For most exercises, the relevant section is noted in parentheses after the exercise title.

1. **Restless Legs (9.2)** Restless legs syndrome (RLS) causes a powerful urge to move your legs—so much so that it becomes uncomfortable to sit or lie down. Sleep is difficult. Researchers undertook a randomized trial of the drug pramipexole to determine its effectiveness in treating RLS. They randomly assigned patients to one of two groups: One group was treated with pramipexole, the other with a placebo. Of the 193 subjects in the pramipexole group, 81.9% reported "much improved" symptoms. In comparison, 54.3% of the 92 subjects in the placebo group reported "much improved" symptoms.[94] Does this experiment provide convincing evidence at the $\alpha = 0.05$ significance level of a difference (Pramipexole – Placebo) in the true proportion of individuals with RLS like the ones in this study who will experience much improved symptoms if they take pramipexole versus a placebo?

2. **Restless Legs (9.1)** Refer to Exercise 1.
    (a) Construct and interpret a 95% confidence interval for the difference (Pramipexole – Placebo) in the true proportions of individuals with RLS like the ones in this study who will experience much improved symptoms. If you already defined parameters and checked conditions in Exercise 1, you don't need to do those tasks again here.
    (b) Explain how the confidence interval from part (b) provides more information than the significance test in Exercise 1.

3. **Water Fleas (9.3)** *Daphnia pulicaria* is a water flea—a small crustacean that lives in lakes and is a major food supply for many species of fish. When fish are present in the lake water, they release chemicals called kairomones that induce water fleas to grow long tail spines that make them more difficult for the fish to eat. One study of this phenomenon compared the relative length of tail spines in two populations of *D. pulicaria*: one grown when kairomones were present and one grown when they were not. Here are summary statistics on the relative tail spine lengths, measured as a percentage of the entire length of the water flea, for separate random samples from the two populations:[95]

    Relative Tail Spine Length

	n	Mean	SD
Fish kairomone present	214	37.26	4.68
Fish kairomone absent	152	30.67	4.19

    Construct and interpret a 99% confidence interval for the difference in the mean relative tail spine lengths of all *D. pulicaria* with fish kairomone present and absent.

4. **Caffeine and Pulse Rate (9.4)** A physiology class performed an experiment to investigate whether drinking a caffeinated beverage would increase pulse rates. Twenty students in the class volunteered to take part in the experiment. All of the students measured their initial pulse rates. Then the professor randomly assigned the students into two groups of 10. Each student in the first group drank 12 ounces of cola with caffeine. Each student in the second group drank 12 ounces of caffeine-free cola. All students then measured their pulse rates again. The table displays the change in pulse rate for the students in both groups.[96]

    Change in Pulse Rate (Final pulse rate – Initial pulse rate)

Caffeine	8	3	5	1	4	0	6	1	4	0
No caffeine	3	–2	4	–1	5	5	1	2	–1	4

    Do these data give convincing evidence at the 1% significance level that drinking caffeine increases pulse rates, on average, for people like the ones in this study?

5. **Reaction Time (9.5)** Does exercise improve reaction time? Researchers used an online tool to measure the reaction time of 15 randomly selected college students before and after each student had run two laps on a track. They measured reaction time in milliseconds (ms), with smaller numbers indicating faster reactions. Here are the data, along with the difference (Before − After) in reaction time for each student and a dotplot of the differences:[97]

Student	1	2	3	4	5	6	7	8
Before	275	329	297	476	314	366	317	313
After	263	358	264	426	288	334	328	286
Difference (Before − After)	12	−29	33	50	26	32	−11	27
Student	9	10	11	12	13	14	15	
Before	254	317	292	314	299	295	259	
After	275	324	278	276	287	276	256	
Difference (Before − After)	−21	−7	14	38	12	19	3	

(a) What does the graph reveal?

(b) Calculate the mean and standard deviation of the differences in reaction time, and interpret the mean difference.

(c) Do the data provide some evidence that exercise improves the mean reaction time of all college students? Explain your answer.

6. **Reaction Time (9.5, 9.6)** Refer to Exercise 5.

(a) Calculate and interpret a 95% confidence interval for the population mean difference (Before − After) in reaction time.

(b) Based on the confidence interval in part (a), do the data provide convincing evidence that exercise is associated with a decrease in the mean reaction time of all college students? Explain your answer.

7. **Stock Volatility (9.7)** An investor is comparing the volatility of two stocks, A and B. She wants to know if there is a difference in the variability in return on investment for these two stocks as measured by the percent increase or decrease in the price of the stock from its date of purchase. The investor takes a random sample of 50 annualized daily returns over the past 5 years for each stock. Graphs of the data suggest that it is reasonable to believe that the corresponding population distributions are normal. Here are summary statistics for the daily returns of each stock:[98]

Stock	Mean	SD
A	11.8%	12.9%
B	7.1%	9.6%

Do the data provide convincing evidence of a difference in the standard deviations of the daily returns of these two stocks at the $\alpha = 0.01$ significance level?

8. **Subliminal Messages (Chapter 9)** A "subliminal" message is below our threshold of awareness but may nonetheless influence us. Can subliminal messages help students learn math? A group of 18 students who had failed the mathematics part of the City University of New York Skills Assessment Test agreed to participate in a study to find out. All received a daily subliminal message, flashed on a screen too rapidly to be consciously read. The treatment group of 10 students (assigned at random) was exposed to "Each day I am getting better in math." The control group of 8 students was exposed to a neutral message, "People are walking on the street." All 18 students participated in a summer program designed to improve their math skills, and all took the assessment test again at the end of the program. The following table gives data on the students' scores before and after the program.[99]

Treatment group			Control group		
Pre-test	Post-test	Difference	Pre-test	Post-test	Difference
18	24	6	18	29	11
18	25	7	24	29	5
21	33	12	20	24	4
18	29	11	18	26	8
18	33	15	24	38	14
20	36	16	22	27	5
23	34	11	15	22	7
23	36	13	19	31	12
21	34	13			
17	27	10			

(a) Do the data provide convincing evidence that participating in the summer program improves the assessment test performance of students like the ones in this study who are only exposed to a neutral message?

(b) How much more, on average, do students like the ones in this study improve their assessment test scores when exposed to the subliminal message about getting better in math versus when exposed to a neutral message? Calculate a 95% confidence interval to help answer this question.

(c) Can we generalize the results of this study to the population of all students who failed the mathematics part of the City University of New York Skills Assessment Test? Why or why not?

Chapter 9 Project: Gas Prices

Explore a large data set using software and apply what you've learned in this chapter to a real-world scenario.

When buying or leasing a new car, one of the factors that customers consider is the type of fuel it uses. Some people prefer vehicles that use diesel, while others favor vehicles that use regular unleaded gasoline. Which of these two types of fuel costs more at the pump, on average? Does the answer to this question depend on the state where you live? Is the variation in price per gallon greater for one type of fuel than the other?

Researchers collected data on the price per gallon (in dollars) of diesel and regular unleaded gasoline from a random sample of gas stations in 6 states: Colorado, Illinois, Indiana, Kansas, Missouri, and Ohio. The file **GasPrices** contains data from a total of 82 gas stations.

Download this data set into the technology that you are using. The following questions will help you begin to analyze the data.

1. Start by calculating the difference (*Diesel – Unleaded*) in gas prices for all 82 stations, and store these values in a new column titled *Difference*. Then create a graph that displays the distribution of difference in gas prices. What does the graph reveal?

2. Construct and interpret a 95% confidence interval for the true mean difference. Does the interval provide convincing evidence of a difference in the mean prices per gallon of diesel and regular unleaded gasoline at gas stations in these 6 states?

3. Create a graph that compares the difference (*Diesel – Unleaded*) in gas prices for each of the 6 states. Write a few sentences comparing the 6 distributions.

4. Construct and interpret a 95% confidence interval for the difference in the proportion of all gas stations in Missouri and the proportion of all gas stations in the other 5 states at which the price of diesel is higher than the price of unleaded.

5. We might expect the mean difference (*Diesel – Unleaded*) in gas prices to be the same for adjacent states. Use your graph from Question 3 to choose 2 adjacent states for which the conditions for inference are met. Then carry out an appropriate test to see if the data provide convincing evidence to contradict this expectation.

6. We might also expect the variability of the difference (*Diesel – Unleaded*) in gas prices to be the same for adjacent states. Use your graph from Question 3 to choose 2 adjacent states for which the conditions for inference are met. Then carry out an appropriate test to see if the data provide convincing evidence to contradict this expectation.

Cumulative Review Chapters 7–9

This cumulative review is designed to help you review important concepts and skills from Chapters 7, 8, and 9, and to prepare you for an in-class exam on this material.

Section I: Multiple Choice *Select the best answer for each question.*

1. In a Gallup poll of randomly selected U.S. adults, 75% said they would vote for a law that imposed term limits on members of the U.S. Congress.[100] The poll's margin of error was 4 percentage points at the 95% confidence level. Why did Gallup include the margin of error with its estimate?
   (a) To account for sampling variability
   (b) To account for undercoverage
   (c) To account for nonresponse
   (d) To account for possible recording errors when the responses were collected

2. Experiments on learning in animals sometimes measure how long it takes mice to find their way through a maze. The mean time is 18 seconds for one particular maze. A researcher thinks that a loud noise will cause the mice to complete the maze in less time. The researcher measures how long each of 10 mice takes with a noise stimulus. What are the appropriate hypotheses for a significance test?
   (a) $H_0: \mu = 18; H_a: \mu \neq 18$
   (b) $H_0: \mu = 18; H_a: \mu > 18$
   (c) $H_0: \mu < 18; H_a: \mu = 18$
   (d) $H_0: \mu = 18; H_a: \mu < 18$

3. Each dotplot shows the distribution of a sample selected at random from a different population. For which of the three samples would it be safe to construct a one-sample t interval for the population mean?

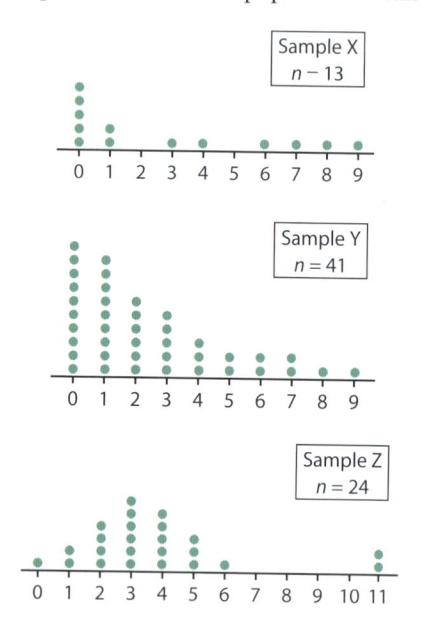

   Sample X, $n = 13$
   Sample Y, $n = 41$
   Sample Z, $n = 24$

   (a) Sample X only
   (b) Sample Y only
   (c) Samples Y and Z
   (d) None of the samples

4. A random sample of 100 likely voters in a small city produced 59 voters in favor of Candidate A. The value of the standardized test statistic for performing a test of $H_0: p = 0.5$ versus $H_a: p > 0.5$ is
   (a) $z = \dfrac{0.59 - 0.5}{\sqrt{\dfrac{0.59(0.41)}{100}}}$
   (b) $z = \dfrac{0.59 - 0.5}{\sqrt{\dfrac{0.5(0.5)}{100}}}$
   (c) $z = \dfrac{0.5 - 0.59}{\sqrt{\dfrac{0.59(0.41)}{100}}}$
   (d) $z = \dfrac{0.5 - 0.59}{\sqrt{\dfrac{0.5(0.5)}{100}}}$

5. To estimate the difference in the proportions of students at Community College A and Community College B who have a full-time job, researchers selected a random sample of 100 students from each institution. A 90% confidence interval for $p_A - p_B$ is -0.06 ± 0.10. Based on this interval, which conclusion is best?
   (a) Because –0.06 is in the interval, there is convincing evidence of a difference in the population proportions.
   (b) Because 0 is in the interval, there is convincing evidence that the population proportions are equal.
   (c) Because 0 is in the interval, there is not convincing evidence of a difference in the population proportions.
   (d) Because most of the interval is negative, there is convincing evidence that a greater proportion of students at Community College B have full-time jobs.

6. One reason for using a t distribution instead of the standard normal distribution to find critical values when calculating a confidence interval for a population mean is that
   (a) z^* can be used only for large samples.
   (b) z^* requires that you know the population standard deviation σ.
   (c) z^* requires that you can regard your data as an SRS from the population.
   (d) z^* will lead to a wider interval than t^*.

7. Many television viewers express doubts about the validity of certain ads. In an attempt to answer their critics, Timex Corporation wishes to estimate the proportion p of all consumers who believe what is shown about Timex watches in the company's ads. Which of the following is the smallest number of consumers that Timex can survey to guarantee a margin of error of 0.05 or less at the 99% confidence level?

(a) 600
(b) 650
(c) 700
(d) 750

8. A significance test is performed to test the null hypothesis $H_0: \mu = 2$ versus the alternative hypothesis $H_a: \mu \neq 2$. A sample of size 28 produced a test statistic of $t = 2.051$. Assuming all conditions for inference are met, which of the following intervals contains the P-value for this test?

   (a) $0.01 \leq P\text{-value} < 0.02$
   (b) $0.02 \leq P\text{-value} < 0.025$
   (c) $0.025 \leq P\text{-value} < 0.05$
   (d) $0.05 \leq P\text{-value} < 0.10$

9. A Gallup poll found that only 28% of U.S. adults expect to inherit money or valuable possessions from a relative. The poll's margin of error was ±3 percentage points at a 95% confidence level. This means that

   (a) Gallup can be 95% confident that between 25% and 31% of the sample expect an inheritance.
   (b) there is a 95% probability that the percentage of all adults who expect an inheritance is between 25% and 31%.
   (c) the poll used a method that gets an answer within 3% of the true population percentage in about 95% of all possible samples.
   (d) if Gallup takes another poll on this issue, there is a 95% probability that the result of the second poll will be between 25% and 31%.

10. A fresh-fruit distributor claims that only 4% of its Macintosh apples are bruised. A buyer for a grocery store chain suspects that the true proportion p is greater than 4%. The buyer selects a random sample of 30 apples to test the null hypothesis $H_0: p = 0.04$ against the alternative hypothesis $H_a: p > 0.04$. Which of the following statements about conditions for performing a one-sample z test for the population proportion is correct?

    (a) The test should not be performed because the Random condition has not been met.
    (b) The test should not be performed because the Large Counts condition has not been met.
    (c) We cannot determine if the conditions have been met until we have the sample proportion $\hat{p}$.
    (d) All conditions for performing the test have been met.

11. An ecologist is studying differences in the populations of a certain species of lizard on two different islands. The ecologist collects independent random samples of lizards from the two islands, weighs the lizards (in grams), and then releases them. The table provides summary statistics for the data.

Island	n	Mean (g)	SD (g)
A	24	46.5	5.97
B	30	44.2	4.24

    Which of the following is the correct formula for the standardized test statistic to test the null hypothesis that the mean weights of lizards on the two islands are equal?

    (a) $t = \dfrac{(46.5 - 44.2) - 0}{\left(\dfrac{5.97}{\sqrt{24}} + \dfrac{4.24}{\sqrt{30}}\right)^2}$

    (b) $t = \dfrac{(46.5 - 44.2) - 0}{\sqrt{\dfrac{5.97}{24} + \dfrac{4.24}{30}}}$

    (c) $t = \dfrac{(46.5 - 44.2) - 0}{\sqrt{\dfrac{5.97^2}{24} + \dfrac{4.24^2}{30}}}$

    (d) $t = \dfrac{(46.5 - 44.2) - 0}{\sqrt{\dfrac{5.97^2}{24}} + \sqrt{\dfrac{4.24^2}{30}}}$

12. Vigorous exercise helps people live several years longer (on average). Whether mild-intensity activities like slow walking extend life is not clear. Suppose that the added life expectancy from regular slow walking is just 2 months. A significance test is more likely to find convincing evidence of an increase in mean life expectancy with regular slow walking if it is based on

    (a) a very large random sample and a 5% significance level is used.
    (b) a very large random sample and a 1% significance level is used.
    (c) a very small random sample and a 5% significance level is used.
    (d) a very small random sample and a 1% significance level is used.

13. Some people say that more babies are born in September than in any other month. To test this claim, researchers take a random sample of 150 college students and find that 21 of them were born in September. Researchers are interested in whether the proportion born in September is greater than 1/12 — what you would expect if September was no different from any other month. Thus, the null hypothesis is $H_0: p = 1/12$. The P-value from an appropriate significance test is 0.0056. Which of the following statements best describes what the P-value measures?

    (a) The probability that September birthdays are no more common than birthdays in any other month is 0.0056.
    (b) The probability that the proportion of September birthdays in the population is equal to 1/12 is 0.0056.
    (c) 0.0056 is the probability of getting a sample with a proportion of September birthdays that is 21/150 or higher, given that the true proportion is 1/12.
    (d) 0.0056 is the probability of getting a sample with a proportion of September birthdays that is 21/150 or higher, given that the true proportion is greater than 1/12.

14. After a name-brand drug has been sold for several years, the Food and Drug Administration (FDA) will allow other companies to produce a generic equivalent. The FDA will permit the generic drug to be sold as long as there isn't convincing evidence that it is less effective than the name-brand drug. For a proposed generic drug intended to lower blood pressure, the following hypotheses will be used:

$$H_0: \mu_G = \mu_N$$
$$H_a: \mu_G < \mu_N$$

where μ_G = true mean reduction in blood pressure using the generic drug and μ_N = true mean reduction in blood pressure using the brand-name drug. In the context of this situation, which of the following describes a Type I error?

(a) The FDA finds convincing evidence that the generic drug is less effective, when in reality it is less effective.

(b) The FDA finds convincing evidence that the generic drug is less effective, when in reality it is equally effective.

(c) The FDA fails to find convincing evidence that the generic drug is less effective, when in reality it is less effective.

(d) The FDA fails to find convincing evidence that the generic drug is less effective, when in reality it is equally effective.

15. Which of the following is a valid way for the Normal condition to be met when constructing a confidence interval for a population standard deviation or variance?

  I. The population distribution is normal.
  II. The sample size is at least 30.
  III. The sample size is less than 30, but a graph of the sample data suggests that it is reasonable to believe that the population distribution is normal.

(a) I only
(b) I and II only
(c) I and III only
(d) I, II, and III

16. A quality-control inspector will measure the salt content (in milligrams, mg) in a random sample of bags of potato chips from an hour of production. Which of the following would result in the smallest margin of error in estimating the true mean salt content μ?

(a) 90% confidence; $n = 25$
(b) 90% confidence; $n = 50$
(c) 95% confidence; $n = 25$
(d) 95% confidence; $n = 50$

17. A researcher plans to conduct a significance test at the $\alpha = 0.01$ significance level. The researcher designs the study to have a power of 0.90 to detect a particular alternative value of the parameter. The probability that the researcher will commit a Type II error for the particular alternative value of the parameter is

(a) 0.01.
(b) 0.10.
(c) 0.90.
(d) 0.99.

18. A total of 35 people from a random sample of 125 workers from Company A admitted to using sick leave when they weren't really ill. A total of 17 employees from a random sample of 68 workers from Company B admitted that they had used sick leave when they weren't ill. What would be a 95% confidence interval for the difference in the proportions of workers at the two companies who would admit to using sick leave when they weren't ill?

(a) $0.03 \pm 1.96 \sqrt{\dfrac{(0.28)(0.72)}{125} + \dfrac{(0.25)(0.75)}{68}}$

(b) $0.03 \pm 1.645 \sqrt{\dfrac{(0.28)(0.72)}{125} + \dfrac{(0.25)(0.75)}{68}}$

(c) $0.03 \pm 1.96 \sqrt{\dfrac{(0.269)(0.731)}{125} + \dfrac{(0.269)(0.731)}{68}}$

(d) $0.03 \pm 1.645 \sqrt{\dfrac{(0.269)(0.731)}{125} + \dfrac{(0.269)(0.731)}{68}}$

19. After checking that conditions are met, you perform a significance test of $H_0: \mu = 1$ versus $H_a: \mu \neq 1$. You obtain a P-value of 0.022. Which of the following must be true?

(a) A 95% confidence interval for μ will include the value 1.
(b) A 95% confidence interval for μ will include the value 0.
(c) A 99% confidence interval for μ will include the value 1.
(d) A 99% confidence interval for μ will include the value 0.

20. Are TV commercials louder than the programs they precede or follow? To find out, researchers collected data on 50 randomly selected commercials in a given week. With the television's volume at a fixed setting, they determined whether or not the maximum loudness of each commercial exceeded the maximum loudness in the first 30 seconds of regular programming that followed. Assuming conditions for inference are met, the most appropriate method for answering the question of interest is

(a) a one-sample z test for a proportion.
(b) a two-sample z test for a difference in proportions.
(c) a paired t test for a mean difference.
(d) a two-sample t test for a difference in means.

Section II: Free Response

1. Members at a popular fitness club currently pay a $40 per month membership fee. The owner of the club wants to raise the fee to $50 but is concerned that some members will leave the gym if the fee increases. To investigate, the owner surveys a random sample of the club members. The resulting 95% confidence interval for the proportion of all members who would quit if the fee was raised to $50 is 0.18 ± 0.075.

   (a) One of the conditions for calculating the confidence interval is that $n\hat{p} \geq 10$ and $n(1-\hat{p}) \geq 10$. Explain why it is necessary to check this condition.

   (b) Interpret the confidence interval.

   (c) According to the club's accountant, the fee increase will be worthwhile if fewer than 20% of the members quit. Based on the confidence interval, can the owner be confident that the fee increase will be worthwhile? Explain your answer.

2. A researcher wants to determine if a 5-week crash diet is effective over a long period of time. The researcher selects a random sample of 15 five-week crash dieters, and records each person's weight (in pounds) before starting the diet and 1 year after the diet is concluded. Here are the data:[101]

Dieter	1	2	3	4	5	6	7	8
Before	158	185	176	172	164	234	258	200
After	163	182	188	150	161	220	235	191
Dieter	9	10	11	12	13	14	15	
Before	228	246	198	221	236	255	231	
After	228	237	209	220	222	268	234	

   Do the data provide convincing evidence that 5-week crash dieters weigh less, on average, 1 year after finishing the diet?

3. A Pew Research Center study asked a random sample of 1520 U.S. adults how many books they had read in the previous 12 months. The average number of books read by the members of the sample (including those who reported reading zero books) was 12 books, with a standard deviation of 18 books.[102]

   (a) Construct and interpret a 90% confidence interval for the mean number of books read by all U.S. adults in the previous 12 months.

   (b) Interpret the confidence level.

   (c) Explain why it would not be appropriate to calculate a confidence interval for the population standard deviation in this setting.

4. AZT was the first drug that seemed effective in delaying the onset of acquired immunodeficiency syndrome (AIDS). Evidence of AZT's effectiveness came from a large randomized comparative experiment. The subjects were 870 volunteers who were infected with human immunodeficiency virus (HIV, the virus that causes AIDS), but who did not yet have AIDS. The study assigned 435 of the subjects at random to take 500 mg of AZT each day and the other 435 subjects to take a placebo. At the end of the study, 38 of the subjects who took the placebo and 17 of the subjects who took AZT had developed AIDS.[103] Researchers want to determine whether the data provide convincing evidence that taking AZT lowers the proportion of infected people like the ones in this study who will develop AIDS in a given period of time.

   (a) If the results of the study are statistically significant, is it reasonable to conclude that AZT is the cause of the decrease in the proportion of people like these who will develop AIDS? Explain your answer.

   (b) State appropriate hypotheses for performing a significance test. Be sure to define any parameters you use.

   (c) Describe a Type I error and a Type II error in this context, and give a possible consequence of each.

   (d) Based on your answer to part (c), should researchers use a significance level of $\alpha = 0.01$, 0.05, or 0.10 when performing the test? Explain your answer.

   (e) Carry out the test using the significance level you chose in part (d).

5. The body's natural electrical field helps wounds heal. If diabetes changes this field, it might explain why people with diabetes heal more slowly. A study of this idea compared randomly selected normal mice and randomly selected mice bred to spontaneously develop diabetes. The investigators attached sensors to the right hip and front feet of the mice and measured the difference in electrical potential (in millivolts) between these locations. Graphs of the data for each group were consistent with normal population distributions. The table provides summary statistics for the two groups of mice.[104]

Group	n	Mean	SD
Mice with diabetes	24	13.090	4.839
Normal mice	18	10.022	2.915

   (a) Construct and interpret a 95% confidence interval for the difference in the population means for normal mice and mice with diabetes.

   (b) The P-value of a two-sided test for a difference between the two population means is 0.015. Interpret this value.

   (c) Explain why the confidence interval in part (a) is consistent with, but gives more information than, the significance test in part (b).

6. Refer to Exercise 5. Do the data provide convincing evidence at the $\alpha = 0.05$ significance level that the difference in electrical potential varies more (has a larger standard deviation) in mice with diabetes than in normal mice?

10 Chi-Square and Analysis of Variance (ANOVA)

Section 10.1 Testing the Distribution of a Categorical Variable in a Population
Section 10.2 Chi-Square Tests for Goodness of Fit
Section 10.3 Testing the Relationship Between Two Categorical Variables in a Population
Section 10.4 Chi-Square Tests for Association
Section 10.5 Introduction to Analysis of Variance
Section 10.6 One-Way Analysis of Variance
Chapter 10 Review
Chapter 10 Review Exercises
Chapter 10 Project

▌▪ Statistics Matters How racially and ethnically diverse are STEM workers?

Careers in science, technology, engineering, and mathematics (STEM) are prized for their high salaries, good job security, and pleasant working conditions. But how racially and ethnically diverse are workers in STEM fields? To investigate, the Pew Research Center used data from the American Community Survey to compare the distribution of race and ethnicity in STEM jobs to the distribution of race and ethnicity in all jobs. Here is a side-by-side bar chart of their results.[1]

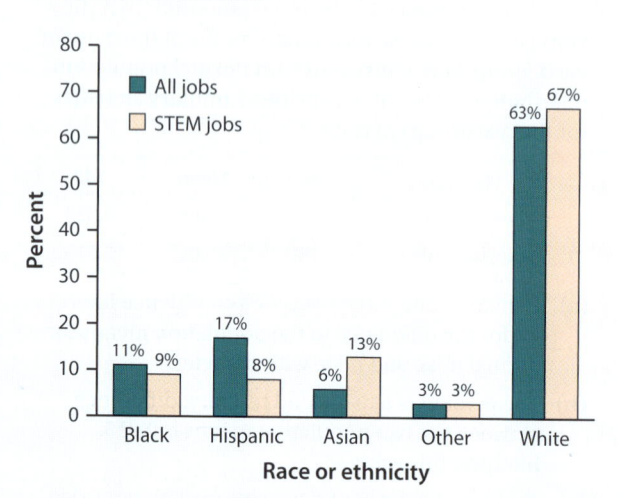

Do these data provide convincing evidence that the distribution of race and ethnicity in STEM jobs is different than the distribution in all jobs? Has the distribution of race and ethnicity in STEM jobs changed over time?

We'll revisit Statistics Matters *at the end of the chapter, so you can use what you have learned to help answer these questions.*

In Chapter 9, you learned how to perform inference about two parameters, such as two population means or two population proportions. This chapter introduces more general inference methods for comparing several means or comparing several proportions, in addition to inference about the distribution of a categorical variable.

Section 10.1 Testing the Distribution of a Categorical Variable in a Population

LEARNING GOALS

By the end of this section, you will be able to:

- State hypotheses for a test about the distribution of a categorical variable.
- Calculate expected counts for a test about the distribution of a categorical variable.
- Calculate the test statistic for a test about the distribution of a categorical variable.

In Sections 8.3 and 8.4, you learned how to perform a one-sample z test for a population proportion. This test is appropriate when the data come from a single sample and can be divided into two categories: success and failure. In this section, you will again encounter categorical data from a single sample, but the variable can have *two or more* categories. You'll learn how to state hypotheses, calculate expected counts, and calculate the test statistic for a test about the distribution of a categorical variable in this section. You'll learn how to do the full test in Section 10.2.

CLASS ACTIVITY

The Color of Candy

Mars, Inc., is famous for its milk chocolate candies. Here's what the company's Consumer Affairs Department claims about the distribution of color for M&M'S® Milk Chocolate Candies produced at its Hackettstown, New Jersey, factory:

Brown: 12.5%   Red: 12.5%   Yellow: 12.5%
Green: 12.5%   Orange: 25%   Blue: 25%

The purpose of this activity is to investigate whether the distribution of color in a large bag of M&M'S Milk Chocolate Candies differs from the distribution of color claimed by the Hackettstown factory.

1. Your class will select a random sample of 60 M&M'S Milk Chocolate Candies from a large bag and count the number of candies of each color. Make a table that summarizes these *observed counts*. Does the distribution of color in the sample match the claimed distribution? Which colors are closest to expected? Farthest from expected?

2. How can you tell if the sample data give convincing evidence against the company's claim? Each team of three or four students should discuss this question and devise a formula for a test statistic that measures the difference between the observed and expected color distributions. The test statistic should yield a single number when the observed and expected values are plugged in. Also, larger differences between the observed and expected distributions should result in a larger value for the statistic.

3. Each team will share its proposed test statistic with the class. Your instructor will then reveal how the *chi-square test statistic* χ^2 is calculated. We'll discuss it in more detail shortly.

4. Discuss as a class: If your sample is consistent with the company's claim, will the value of χ^2 be large or small? If your sample is not consistent with the company's claim, will the value of χ^2 be large or small?

5. Compute the value of the χ^2 test statistic for the class's data.

   We can use simulation to determine if your class's chi-square test statistic is large enough to provide convincing evidence that the distribution of color in the large bag differs from the company's claim.

6. Go to www.stapplet.com and launch the *M&M's/Skittles/Froot Loops* applet. Keep the color distribution you are comparing to as M&M's Milk Chocolate. Enter the class's observed counts in the table and click the Begin analysis button. The applet will display the distribution of color with a bar chart and calculate the value of the chi-square statistic for these data.

(continued)

7. To see which values of the chi-square statistic are likely to happen by chance alone when taking samples of the same size from the claimed population, enter 1 for the number of samples to simulate and click the Simulate button. A graph of the simulated sample will appear, along with the chi-square test statistic for that sample. Was the simulated chi-square statistic smaller or larger than the chi-square statistic from the actual sample?

8. Keep clicking the Simulate button to generate more simulated chi-square statistics. To speed up the process, change the number of samples to 100 and click the Simulate button again and again.

9. Describe the shape, center, and variability of the sampling distribution of the simulated chi-square test statistic. Click the box to plot the approximate χ^2 density curve. Does it fit well?

10. To estimate the P-value, use the counter at the bottom of the applet to count the percentage of dots greater than or equal to the observed chi-square statistic from Step 5. Based on the P-value, what conclusion would you make?

Stating Hypotheses

As with any test, we begin by stating hypotheses. The null hypothesis in a test about the distribution of a categorical variable should state a claim about the distribution of the variable in the population of interest. In the case of *The Color of Candy* activity, the categorical variable we're measuring is color and the population of interest is the large bag of M&M'S® Milk Chocolate Candies. The appropriate null hypothesis is

H_0: The distribution of color in the large bag of M&M'S Milk Chocolate Candies is the same as the claimed distribution.

The alternative hypothesis for this test is that the categorical variable does *not* have the specified distribution. For the M&M'S, our alternative hypothesis is

H_a: The distribution of color in the large bag of M&M'S Milk Chocolate Candies *is different than* the claimed distribution.

Although we usually write the hypotheses in words, we can also write them in symbols. For example, here are the hypotheses for *The Color of Candy* activity:

H_0: $p_{brown} = 0.125$, $p_{red} = 0.125$, $p_{yellow} = 0.125$, $p_{green} = 0.125$, $p_{orange} = 0.25$, $p_{blue} = 0.25$

H_a: At least two of these proportions differ from the values stated by the null hypothesis.

where p_{color} = the true proportion of M&M'S Milk Chocolate Candies of that color in the large bag.

Why don't we write the alternative hypothesis as "H_a: At least one of these proportions differs from the values stated by the null hypothesis" instead? If the stated proportion in one category is wrong, then the stated proportion in at least one other category must be wrong because the sum of the proportions must be 1.

EXAMPLE 10.1.1

Are more NHL players born earlier in the year?

Stating hypotheses

PROBLEM

In his book *Outliers,* Malcolm Gladwell introduces the idea of the relative-age effect with an example from hockey. Many National Hockey League (NHL) players come from Canada. Because January 1 is the cut-off birth date for youth leagues in Canada, children born earlier in the year compete against players who may be as much as 12 months younger. Gladwell argues that these "relatively old" players tend to be bigger, stronger, and more coordinated and hence get more playing time and more coaching, and have a better chance of being successful.

To see if the birth dates of NHL players are uniformly distributed across the four quarters of the year, a random sample of 80 NHL players was selected and their birthdays were recorded.[2] The table summarizes the distribution of birthday for these 80 players.

Birthday	Jan–Mar	Apr–Jun	Jul–Sep	Oct–Dec
Number of players	32	20	16	12

Do these data provide convincing evidence that the true proportions of NHL players born in each quarter are not the same?

(a) State appropriate hypotheses for a test that addresses this question.
(b) Calculate the proportion of players in the sample who were born in each quarter.
(c) Explain how the proportions in part (b) give some evidence for the alternative hypothesis.

SOLUTION

(a) H_0: The true proportions of NHL players born in each quarter are the same.
H_a: The true proportions of NHL players born in each quarter are not all the same.

The hypotheses could also be stated symbolically:
H_0: $p_{Jan-Mar} = p_{Apr-Jun} = p_{Jul-Sep} = p_{Oct-Dec} = 0.25$
H_a: At least 2 of the proportions $\neq 0.25$, where $p_{quarter}$ = the proportion of all NHL players born in the specified quarter.

(b) Jan–Mar: 32/80 = 0.40
Apr–Jun: 20/80 = 0.25
Jul–Sep: 16/80 = 0.20
Oct–Dec: 12/80 = 0.15

(c) There is evidence for H_a because the sample proportions are not all the same.

EXAM PREP: FOR PRACTICE, TRY EXERCISE 5.

Calculating Expected Counts

Andre's class did "The Color of Candy" activity. The table summarizes the data from the class's sample of M&M'S® Milk Chocolate Candies.

Color	Brown	Red	Yellow	Green	Orange	Blue	Total
Observed count	12	3	7	9	9	20	60

To begin their analysis, Andre's class compared the observed counts from their sample with the counts that would be expected if the manufacturer's claim is true. To calculate the expected counts, they started with the manufacturer's claim:

Brown: 12.5%   Red: 12.5%   Yellow: 12.5%
Green: 12.5%   Orange: 25%   Blue: 25%

Assuming that the claimed distribution is true, 12.5% of all M&M'S Milk Chocolate Candies in the large bag should be brown. For random samples of 60 candies, the expected count of brown M&M'S is $(60)(0.125) = 7.5$.

HOW TO Calculate Expected Counts

The expected count for category i in the distribution of a categorical variable is

$$np_i$$

where n is the overall sample size and p_i is the proportion for category i specified by the null hypothesis.

Using the same method, Andre's class came up with these expected counts for the remaining five colors:

Red: $(60)(0.125) = 7.5$
Yellow: $(60)(0.125) = 7.5$
Green: $(60)(0.125) = 7.5$
Orange: $(60)(0.25) = 15.0$
Blue: $(60)(0.25) = 15.0$

Note that most of these expected counts are not integers. Like the expected values from Section 5.2, the expected counts here represent the *average* number of M&M'S® Milk Chocolate Candies of each color in many, many random samples of size 60.

EXAMPLE 10.1.2

Are more NHL players born earlier in the year?

Calculating expected counts

PROBLEM

A random sample of 80 NHL players was selected to determine if their birthdays are uniformly distributed across the four quarters of the year. The table summarizes birthday data for these 80 players.

Birthday	Jan–Mar	Apr–Jun	Jul–Sep	Oct–Dec
Number of players	32	20	16	12

Calculate the expected counts for a test of the null hypothesis that the true proportions of NHL players born in each quarter are the same.

SOLUTION

Jan–Mar: $(80)(0.25) = 20$
Apr–Jun: $(80)(0.25) = 20$
Jul–Sep: $(80)(0.25) = 20$
Oct–Dec: $(80)(0.25) = 20$

> If the proportion of players born in each of the four quarters is the same, then $100\%/4 = 25\% = 0.25$ of NHL players should be born in each quarter.

EXAM PREP: FOR PRACTICE, TRY EXERCISE 9.

The Chi-Square Test Statistic

Here is a graph comparing the observed and expected counts of M&M'S® Milk Chocolate Candies for Andre's class.

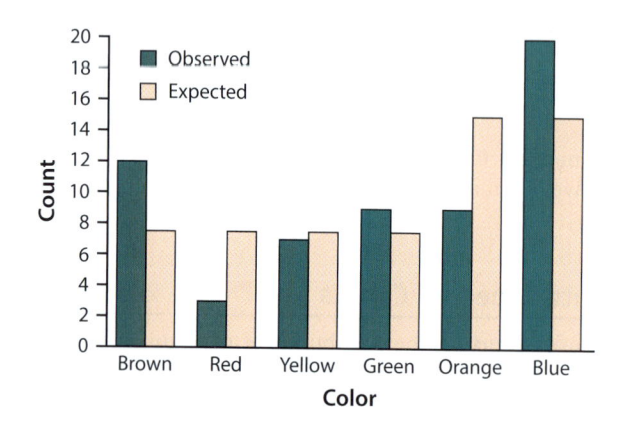

Because the observed counts are different from the expected counts, there is some evidence that the claimed distribution is not correct. However, it may be that the claimed distribution is correct and the differences found by Andre's class were due to sampling variability.

To assess if these differences are larger than what is likely to happen by chance alone when H_0 is true, we calculate a test statistic and P-value. For a test about the distribution of a categorical variable, we use the **chi-square test statistic**.

The **chi-square test statistic** is a measure of how different the observed counts are from the expected counts, relative to the expected counts. The formula for the statistic is

$$\chi^2 = \sum \frac{(\text{observed count} - \text{expected count})^2}{\text{expected count}}$$

where the sum is over all categories of the categorical variable.

The table shows the observed and expected counts for Andre's class data.

Color	Brown	Red	Yellow	Green	Orange	Blue	Total
Observed count	12	3	7	9	9	20	60
Expected count	7.5	7.5	7.5	7.5	15.0	15.0	60

For these data, the chi-square test statistic is

$$\chi^2 = \sum \frac{(\text{observed count} - \text{expected count})^2}{\text{expected count}}$$

$$= \frac{(12-7.5)^2}{7.5} + \frac{(3-7.5)^2}{7.5} + \frac{(7-7.5)^2}{7.5} + \frac{(9-7.5)^2}{7.5} + \frac{(9-15.0)^2}{15.0} + \frac{(20-15.0)^2}{15.0}$$

$$= 2.7 + 2.7 + 0.03 + 0.30 + 2.4 + 1.67$$

$$= 9.8$$

⚠️ **Make sure to use the observed and expected *counts*, not the observed and expected *proportions*, when calculating the chi-square test statistic.** In addition to giving an incorrect value for the test statistic, using proportions instead of counts removes valuable information about sample size from the calculation.

EXAMPLE 10.1.3

Are more NHL players born earlier in the year?

Calculating the chi-square test statistic

PROBLEM

To see if NHL player birthdays are uniformly distributed across the four quarters of the year, we recorded the birthdays of a random sample of 80 NHL players. The table shows the observed and expected counts for a test of the null hypothesis that the true proportions of NHL players born in each quarter are the same. Calculate the value of the chi-square test statistic.

Birthday	Jan–Mar	Apr–Jun	Jul–Sep	Oct–Dec
Observed count	32	20	16	12
Expected count	20	20	20	20

SOLUTION

$$\chi^2 = \frac{(32-20)^2}{20} + \frac{(20-20)^2}{20} + \frac{(16-20)^2}{20} + \frac{(12-20)^2}{20}$$

$$\chi^2 = 7.2 + 0 + 0.8 + 3.2$$

$$\chi^2 = 11.2$$

$$\chi^2 = \sum \frac{(\text{observed count} - \text{expected count})^2}{\text{expected count}}$$

EXAM PREP: FOR PRACTICE, TRY EXERCISE 13.

Larger values of χ^2 arise when the differences between the observed counts and the expected counts are larger. Thus, the larger the value of χ^2, the stronger the evidence for the alternative hypothesis (and against the null hypothesis). But how large does the value of χ^2 need to be for the evidence to be convincing? We can use simulation to get an idea.

The dotplot in Figure 10.1 shows the results of a simulation in which 100 random samples of size 80 were selected from a population of NHL players where the proportion of players born in each quarter was the same. For each simulated sample of 80 players, the chi-square test statistic was calculated.

FIGURE 10.1 Dotplot showing values of the chi-square test statistic in 100 simulated samples of size $n = 80$ from a population of NHL players where players are equally likely to be born in each of the four quarters.

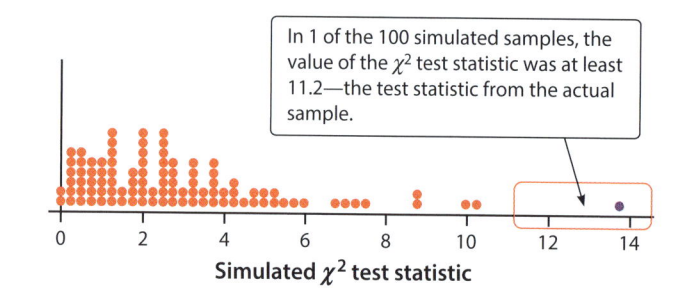

In 1 of the 100 simulated samples, the value of the χ^2 test statistic was at least 11.2—the test statistic from the actual sample.

Because there is only one value greater than or equal to the observed chi-square test statistic of 11.2, the P-value is approximately $1/100 = 0.01$. Because it is unlikely to get differences between the observed and expected counts this large or larger by chance alone, these data provide convincing evidence that the proportion of NHL players born in each quarter is not the same for all quarters. You will learn more about calculating P-values in the next section.

Section 10.1 What Did You Learn?

Review the learning goals from this section. Then practice what you've learned by working through the exercises.

Learning Goal	Example	Exercises
State hypotheses for a test about the distribution of a categorical variable.	10.1.1	5–8
Calculate expected counts for a test about the distribution of a categorical variable.	10.1.2	9–12
Calculate the test statistic for a test about the distribution of a categorical variable.	10.1.3	13–16

Section 10.1 Exercises

Building Concepts and Skills These exercises assess the basic knowledge you should have after reading the section.

1. The null hypothesis in a test about the distribution of a categorical variable should state a claim about the distribution of a _____ in the _____ of interest.

2. To find the expected count for a particular category, multiply the _____ by the _____ specified by the _____.

3. What is the formula for the chi-square test statistic?

4. True/False: Larger values of χ^2 arise when the differences between the observed counts and the expected counts are larger.

Mastering Concepts and Skills These exercises reinforce the learning goals as shown in the examples.

5. **Refer to Example 10.1.1 Fruit Fly Offspring**
Biologists wish to mate pairs of fruit flies having genetic makeup RrNn, indicating that each has one dominant gene (R) and one recessive gene (r) for eye color, along with one dominant (N) and one recessive (n) gene for wing type. The biologists predict a ratio of 9:3:3:1 for red-eyed/straight-winged, red-eyed/curly-winged, white-eyed/straight-winged, and white-eyed/curly-winged offspring, respectively. To test their hypothesis about the distribution of offspring, the biologists randomly select pairs of fruit flies and mate them. Of the 200 offspring, 99 have red eyes/straight wings, 42 have red eyes/curly wings, 49 have white eyes/straight wings, and 10 have white eyes/curly wings.

(a) State appropriate hypotheses for a test that addresses the biologists' claim.

(b) Calculate the proportion of each type of fly in the sample.

(c) Explain how the proportions in part (b) give some evidence for the alternative hypothesis.

6. **Roulette Slots** Roulette is a casino game in which players place bets that certain colors or numbers will come up when a roulette wheel is spun. Casinos are required to verify that their games operate as advertised. In the United States, roulette wheels have 38 equally likely slots — 18 red, 18 black, and 2 green. The managers of a casino spin one of these roulette wheels 200 times. The ball lands on red 85 times, on black 99 times, and on green 16 times.

(a) State appropriate hypotheses for a test of this roulette wheel.

(b) Calculate the proportion of each color in the sample.

(c) Explain how the proportions in part (b) give some evidence for the alternative hypothesis.

7. **Chicken Nuggets** According to McDonalds, Chicken McNuggets® come in four different shapes: Bone, Bell, Boot, and Ball. Are these shapes equally likely? To find out, researchers randomly selected 200 nuggets and identified the shape of each one.[3]

Nugget shape	Bone	Bell	Boot	Ball	Total
Frequency	50	40	59	51	200

(a) State appropriate hypotheses for a test that addresses this question.

(b) Calculate the proportion of each shape in the sample.

(c) Explain how the proportions in part (b) give some evidence for the alternative hypothesis.

8. **Cereal Colors** Are the colors of Kellogg's Froot Loops® cereal equally likely? To find out, researchers randomly selected 120 loops and recorded the color of each loop.

Color	Orange	Yellow	Purple	Red	Blue	Green	Total
Frequency	28	21	16	25	14	16	120

(a) State appropriate hypotheses for a test that addresses this question.

(b) Calculate the proportion of each color in the sample.

(c) Explain how the proportions in part (b) give some evidence for the alternative hypothesis.

9. **Refer to Example 10.1.2 Fruit Fly Offspring** Calculate the expected counts for a test of the null hypothesis stated in Exercise 5.

10. **Roulette Slots** Calculate the expected counts for a test of the null hypothesis stated in Exercise 6.

11. **Chicken Nuggets** Calculate the expected counts for a test of the null hypothesis stated in Exercise 7.

12. **Cereal Colors** Calculate the expected counts for a test of the null hypothesis stated in Exercise 8.

13. **Refer to Example 10.1.3 Fruit Fly Offspring** Refer to Exercises 5 and 9. Calculate the value of the χ^2 test statistic.

14. **Roulette Slots** Refer to Exercises 6 and 10. Calculate the value of the χ^2 test statistic.

15. **Chicken Nuggets** Refer to Exercises 7 and 11. Calculate the value of the χ^2 test statistic.

16. **Cereal Colors** Refer to Exercises 8 and 12. Calculate the value of the χ^2 test statistic.

Applying the Concepts These exercises ask you to apply multiple learning goals in a new context or to apply what you learned in this section in a new way.

17. **Birds and Trees** Researchers want to know if birds prefer particular types of trees when they're searching for seeds and insects. In an Oregon forest, 54% of the trees are Douglas firs, 40% are ponderosa pines, and 6% are other types of trees. At a randomly selected time during the day, the researchers observed 156 red-breasted nuthatches: 70 were seen in Douglas firs, 79 in ponderosa pines, and 7 in other types of trees.[4]

(a) State the hypotheses that the researchers are interested in testing.

(b) Calculate the expected counts for a test of the null hypothesis in part (a).

(c) Calculate the value of the chi-square test statistic for these data.

18. **Seagull Landings** Do seagulls show a preference for where they land? To answer this question, biologists conducted a study in an enclosed outdoor space with a piece of shore made up of 56% sand, 29% mud, and 15% rocks by area. The biologists chose 200 seagulls at random. Each seagull was released into the outdoor space on its own and observed until it landed somewhere on the piece of shore. In all, 128 seagulls landed on the sand, 61 landed in the mud, and 11 landed on the rocks.

(a) State the hypotheses that the biologists are interested in testing.

(b) Calculate the expected counts for a test of the null hypothesis in part (a).

(c) Calculate the value of the chi-square test statistic for these data.

19. **Birds and Trees** Refer to Exercise 17. The dotplot shows the results of a simulation in which 100 random samples of size 156 were selected from a population where 54% of red-breasted nuthatches are in Douglas firs, 40% are in ponderosa pines, and 6% are in other types of trees. For each simulated sample of 156 nuthatches, the chi-square test statistic was

calculated. Based on the χ^2 test statistic from Exercise 17 and the results of the simulation, what conclusion would you make?

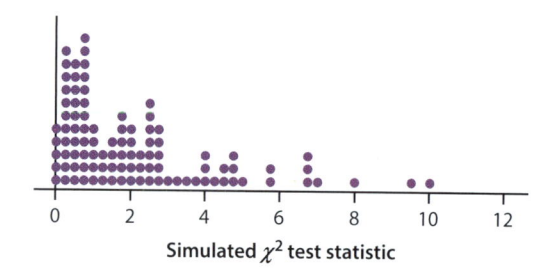

20. **Seagull Landings** Refer to Exercise 18. The dotplot shows the results of a simulation in which 100 random samples of size 200 were selected from a population where 56% of seagulls land on sand, 29% land on mud, and 15% land on rocks. For each simulated sample of 200 seagulls, the chi-square test statistic was calculated. Based on the χ^2 test statistic from Exercise 18 and the results of the simulation, what conclusion would you make?

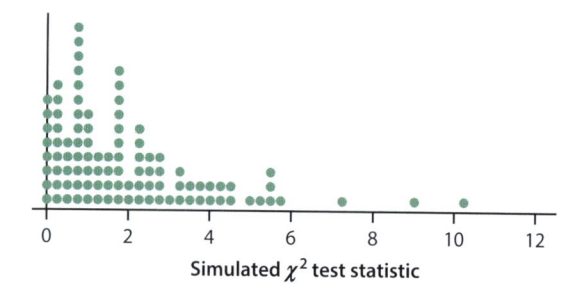

Extending the Concepts These exercises challenge you to explore statistical concepts and methods that go beyond what you learned in this section.

21. **Carrot Weights** A gardener claims that the weights of carrots from their garden are approximately normally distributed with a mean of 60 grams and a standard deviation of 10 grams. To test this claim, you select a random sample of 45 carrots from their garden and weigh them. The table summarizes the distribution of weight for the carrots in the sample.

Weight	Less than 50 g	50–<60 g	60–<70 g	At least 70 g	Total
Frequency	10	16	13	6	45

(a) If the gardener's claim is true, what proportion of the carrots should be in each of the weight categories?

(b) Calculate the expected number of carrots in each of the weight categories if the gardener's claim is true.

(c) Calculate the value of the chi-square test statistic for these data.

Cumulative Review These exercises revisit what you learned in previous sections.

22. **Flight Arrivals (5.3, 5.4)** The probability that a particular flight from Chicago to Amsterdam arrives on time on any given day is 0.70. Suppose we select a random sample of 10 days and let X = the number of on-time arrivals for that flight.

(a) Explain why this is a binomial setting.

(b) Calculate and interpret μ_X and σ_X.

(c) What is the probability that the flight has an on-time arrival on exactly 9 of those days?

23. **Polling Bias? (1.2)** The *New York Times Magazine* conducted a poll of its subscribers and reported for several weeks on the responses of "3244 subscribers who chose to participate." One question posed was, "Are you afraid that there will be another terrorist attack in the United States in your lifetime on the order of the September 11 attacks?" A total of 49% of the respondents said "Yes."[5] Give one reason why this poll might produce biased results and describe the direction of bias.

Section 10.2 Chi-Square Tests for Goodness of Fit

LEARNING GOALS

By the end of this section, you will be able to:
- Check conditions for a test about the distribution of a categorical variable.
- Calculate the P-value for a test about the distribution of a categorical variable.
- Use the four-step process to perform a chi-square test for goodness of fit.

In Section 10.1, you learned how to state hypotheses, calculate expected counts, and calculate the chi-square test statistic for a test about the distribution of a categorical variable. In this section, you will learn how to check conditions, calculate a P-value, and complete the four-step process for this test, called a **chi-square test for goodness of fit**.

A **chi-square test for goodness of fit** is a significance test of the null hypothesis that a categorical variable has a specified distribution in the population of interest.

Conditions for a Chi-Square Test for Goodness of Fit

Like the other significance tests you have learned about, there are conditions that need to be satisfied when performing a chi-square test for goodness of fit.

> **HOW TO** Check the Conditions for a Chi-Square Test for Goodness of Fit
>
> To perform a chi-square test for goodness of fit, check that the following conditions are met:
> - **Random:** The data come from a random sample from the population of interest.
> - **Large Counts:** All expected counts are at least 5.

For Andre's class data given in Section 10.1, the conditions for a chi-square test for goodness of fit are met:
- Random? Andre's class selected a random sample of M&M'S® Milk Chocolate Candies from the large bag.
- Large Counts? All the expected counts are at least 5:

Color	Brown	Red	Yellow	Green	Orange	Blue	Total
Expected count	7.5	7.5	7.5	7.5	15.0	15.0	60

EXAMPLE 10.2.1

Are more NHL players born earlier in the year?

Checking conditions

PROBLEM

To see if NHL player birthdays are uniformly distributed across the four quarters of the year, a random sample of 80 NHL players was selected and their birthdays recorded. The table shows the observed and expected counts for a test of the null hypothesis that the true proportions of NHL players born in each quarter are the same.

Birthday	Jan–Mar	Apr–Jun	Jul–Sep	Oct–Dec
Observed count	32	20	16	12
Expected count	20	20	20	20

Check whether the conditions for performing a chi-square test for goodness of fit are met.

SOLUTION

Random? The data come from a random sample of NHL players. ✓
Large Counts? All expected counts (20, 20, 20, 20) are at least 5. ✓

> Make sure to give the values of the *expected* counts when checking the Large Counts condition.

EXAM PREP: FOR PRACTICE, TRY EXERCISE 7.

Calculating *P*-values

Andre's class did *The Color of Candy* activity in Section 10.1 and got a chi-square test statistic of $\chi^2 = 9.8$. Is this an unusually large value? Or is this a value that is likely to happen by chance alone when the distribution of color in the bag is what the company claims? To investigate, let's use computer software to simulate taking 1000 random samples of size 60 from the population distribution of M&M'S® Milk Chocolate Candies given by Mars, Inc. The dotplot in Figure 10.2 shows the values of the chi-square test statistic for these 1000 samples.

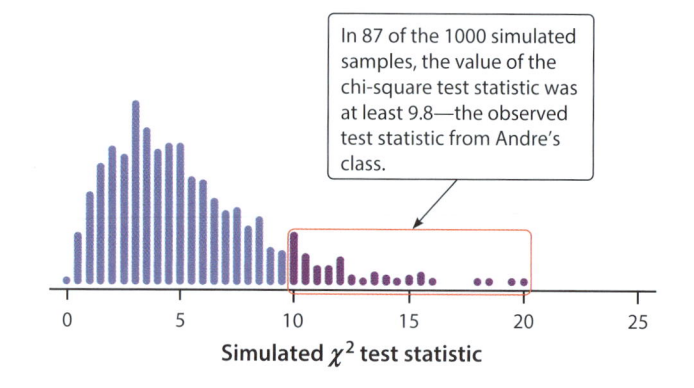

FIGURE 10.2 Dotplot showing values of the chi-square test statistic in 1000 simulated samples of size $n = 60$ from the population distribution of M&M'S® Milk Chocolate Candies stated by the company.

> In 87 of the 1000 simulated samples, the value of the chi-square test statistic was at least 9.8—the observed test statistic from Andre's class.

In the simulation, 87 of the 1000 samples produced a χ^2 statistic of 9.8 or larger. Thus, the *P*-value is about 0.087. There is about a 0.087 probability of getting an observed distribution of color this different or more different from the expected distribution of color by chance alone, assuming that the distribution of color in the bag is the same as the company claims. Because this approximate *P*-value isn't less than 0.05, we don't have convincing evidence that the distribution of color in the large bag of M&M'S® Milk Chocolate Candies is different than the claimed distribution.

As Figure 10.2 suggests, the sampling distribution of the chi-square test statistic is *not* a normal distribution. Instead, it is a right-skewed distribution that allows only values greater than or equal to 0 because χ^2 can never be negative. When the Large Counts condition is met, the sampling distribution of the χ^2 test statistic is modeled well by a **chi-square distribution** with degrees of freedom (df) equal to the number of categories minus 1. As with the *t* distributions, there is a different chi-square distribution for each possible df value. Unlike for *t* tests and intervals, however, the degrees of freedom for a chi-square test for goodness of fit are based on the number of categories, not the sample size.

> A **chi-square distribution** is described by a density curve that takes only non-negative values and is skewed to the right. A particular chi-square distribution is specified by its degrees of freedom.

Figure 10.3 shows the density curves for three members of the chi-square family of distributions. As df increases, the density curves become less skewed, and larger values of χ^2 become more likely.

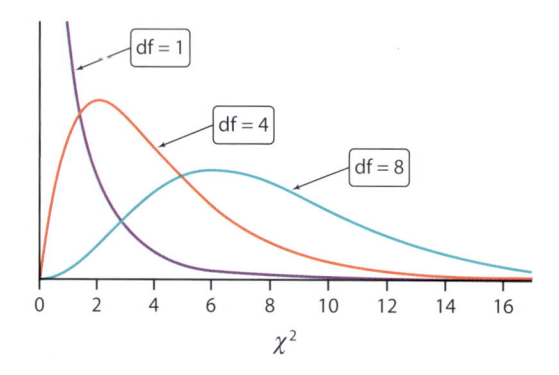

FIGURE 10.3 The density curves for three members of the chi-square family of distributions.

To get *P*-values from a chi-square distribution, we can use technology or Table C. For Andre's class data, the conditions were met and $\chi^2 = 9.8$. Because there are 6 different categories for color, df $= 6 - 1 = 5$. The *P*-value is the probability of getting a value of χ^2 as large as or larger than 9.8 when H_0 is true. Figure 10.4 shows this probability as an area under the chi-square density curve with 5 degrees of freedom.

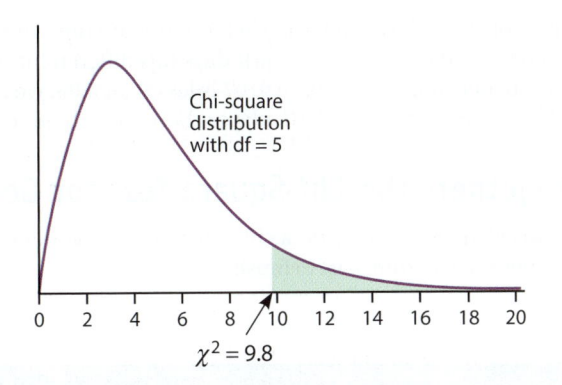

FIGURE 10.4 The *P*-value for a chi-square test for goodness of fit using Andre's M&M'S® Milk Chocolate Candies class data.

Tail probability *p*		
df	.10	.05
4	7.779	9.488
5	9.236	11.071
6	10.645	12.592

For tests of goodness of fit, we always find the area under the chi-square distribution to the right of the χ^2 test statistic. This is because larger values of χ^2 indicate bigger differences between the observed and expected counts.

To find the *P*-value using Table C, look in the df = 5 row. The value $\chi^2 = 9.8$ falls between the critical values 9.236 and 11.071. Looking at the top of the corresponding columns, we find that the right tail area of the chi-square distribution with 5 degrees of freedom is between 0.10 and 0.05. So, the *P*-value for a test based on Andre's data is between 0.05 and 0.10. This is consistent with the value of 0.087 we got from the simulation earlier.

You can also calculate the *P*-value using technology, which gives *P*-value = 0.081. See the Tech Corner at the end of this section for details.

> **Calculating the *P*-value in a chi-square test for goodness of fit**
>
> Suppose the Random and Large Counts conditions are met. Find the *P*-value by calculating the probability of getting a χ^2 statistic as large or larger than the observed value of χ^2 using a χ^2 distribution with df = number of categories – 1.

EXAMPLE 10.2.2

Are more NHL players born earlier in the year?

Calculating *P*-values

PROBLEM

A random sample of 80 NHL players was selected to see if their birthdays are uniformly distributed across the four quarters of the year. The table in Example 10.2.1 shows the observed and expected counts for a test of the null hypothesis that the true proportions of NHL players born in each quarter are the same. The chi-square test statistic for these data is $\chi^2 = 11.2$. Calculate the *P*-value.

SOLUTION

df = 4 – 1 = 3

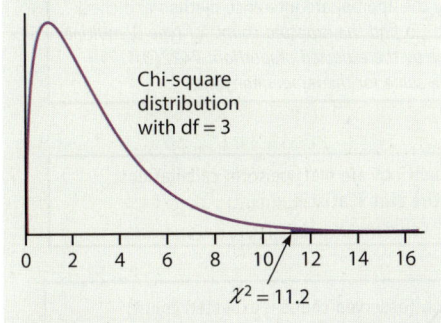

Using Table C: In the df = 3 row, 11.2 is between 9.348 and 11.345, so 0.01 < *P*-value < 0.025.
Using technology: Applet/χ^2cdf(lower: 11.2, upper: 1000, df: 3) gives *P*-value = 0.0107.

EXAM PREP: FOR PRACTICE, TRY EXERCISE 11.

If the true proportions of NHL players born in each quarter are the same, there is only a 0.0107 probability of getting an observed distribution of birthdays this different or more different from the expected distribution by chance alone. Because 0.0107 is less than 0.05, we reject H_0. There is convincing evidence that the true proportions of NHL players born in each quarter are not the same.

Putting It All Together: The Chi-Square Test for Goodness of Fit

Now that we know how to calculate P-values for a chi-square test for goodness of fit, we can use the four-step process to complete a test from start to finish.

EXAMPLE 10.2.3

What's blood got to do with it?

Performing a chi-square test for goodness of fit

PROBLEM

Are people with certain blood types more susceptible to developing Covid-19? A study in Denmark used medical records to identify the blood types of 7422 individuals who tested positive for Covid-19 between March 2020 and July 2020.[6] The table shows the distribution of blood type for these individuals and the distribution of blood type in the population of Denmark.

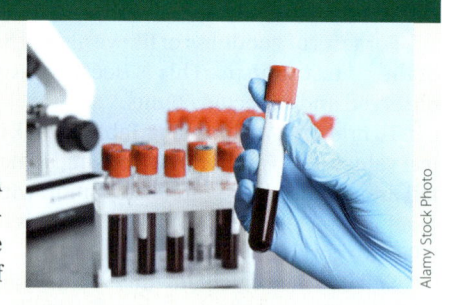

Blood type	O	A	B	AB	Total
Covid-19 positive	2851	3296	897	378	7422
Population	41.7%	42.4%	11.4%	4.5%	100%

Assuming that the 7422 individuals whose data were used in the study are equivalent to a random sample of individuals who tested positive in Denmark during this time period, do these data provide convincing evidence that the distribution of blood type for Danish individuals who test positive for Covid-19 differs from the distribution of blood type in the population of Denmark? Use $\alpha = 0.05$.

SOLUTION

State: H_0: The distribution of blood type for Danish individuals who test positive for Covid-19 is the same as the distribution of blood type in the population of Denmark.

H_a: The distribution of blood type for Danish individuals who test positive for Covid-19 is different than the distribution of blood type in the population of Denmark.

$\alpha = 0.05$

The evidence for H_a is: The percentages in the sample (38.4%, 44.4%, 12.1%, 5.1%) don't all match the population percentages.

Plan: Chi-square test for goodness of fit

Random? Assume the sample of 7422 is equivalent to a random sample. ✓

Large Counts? All expected counts (3095.0, 3146.9, 846.1, 334.0) are ≥ 5.

Do:

$$\chi^2 = \frac{(2851 - 3095.0)^2}{3095.0} + \frac{(3296 - 3146.9)^2}{3146.9}$$
$$+ \frac{(897 - 846.1)^2}{846.1} + \frac{(378 - 334.0)^2}{334.0} = 35.2$$

df = 4 − 1 = 3

> **State:** State the hypotheses, parameter(s), significance level, and evidence for H_a. *If you state the hypotheses in words, there are no parameters to define.*

> The percent of type O in the sample is 2851/7422 = 38.4%. The other percentages are calculated similarly.

> **Plan:** Identify the appropriate inference method and check the conditions. *To find the expected count for type O, multiply the sample size by the expected proportion: 7422(0.417) = 3095.0. Do the same for the remaining types.*

> **Do:** If the conditions are met, perform calculations:
> - Calculate the test statistic.
> - Find the P-value.

$$\chi^2 = \sum \frac{(\text{observed count} - \text{expected count})^2}{\text{expected count}}$$

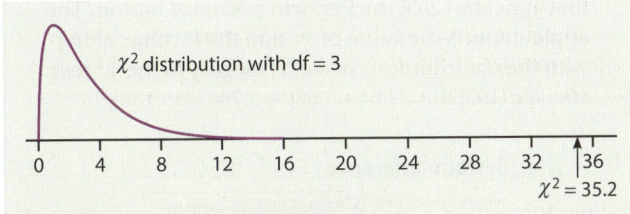

χ^2 distribution with df = 3

0 4 8 12 16 20 24 28 32 36
χ^2 = 35.2

Using Table C: P-value < 0.005
Using technology: $\chi^2 = 35.2$, df $= 3$, P-value ≈ 0
Conclude: Because the P-value of ≈ 0 is $< \alpha = 0.05$, we reject H_0. There is convincing evidence that the distribution of blood type for Danish individuals who test positive for Covid-19 is different than the distribution of blood type in the population of Denmark.

Conclude: Make a conclusion about the hypotheses in the context of the problem.

EXAM PREP: FOR PRACTICE, TRY EXERCISE 15.

When the null hypothesis is rejected in a chi-square test for goodness of fit, it is common to do an informal follow-up analysis to see which categories had the biggest differences between observed and expected counts. In the blood type example, there were 244 fewer individuals with type O blood that tested positive than expected. Conversely, there were 149 more individuals with type A blood who tested positive than expected.

THINK ABOUT IT

Can't we just do a bunch of one-sample z tests? In Example 10.2.3 about blood type, we could perform four separate one-sample z tests, one for each blood type. For example, we could test $H_0: p_O = 0.417$ versus $H_a: p_O \neq 0.417$, $H_0: p_A = 0.424$ versus $H_a: p_A \neq 0.424$, and so on. In addition to being much more tedious, this approach has the problem of requiring multiple tests. If we use $\alpha = 0.05$ for each of the tests, each individual test has a 0.05 probability of making a Type I error. However, the probability of making *at least one* Type I error in the collection of four tests is greater than 0.05. To keep the probability of a Type I error at 0.05, we conduct one overall test—the chi-square test for goodness of fit.

TECH CORNER P-Values and Chi-Square Tests for Goodness of Fit

Calculating P-values

You can use technology to calculate P-values for a chi-square test for goodness of fit. We'll illustrate this process by calculating the P-value from the χ^2 test statistic using the hockey data from Example 10.2.2. Recall that $\chi^2 = 11.2$ and df $= 3$.

Applet

1. Go to www.stapplet.com and launch the χ^2 *Distributions* applet.

2. Choose "Calculate an area under the χ^2 curve" from the drop-down menu, enter 3 for the degrees of freedom, and click "Plot distribution."

3. Choose to calculate the area to the right of a value and enter 11.2 as the value. Click the Calculate area button.

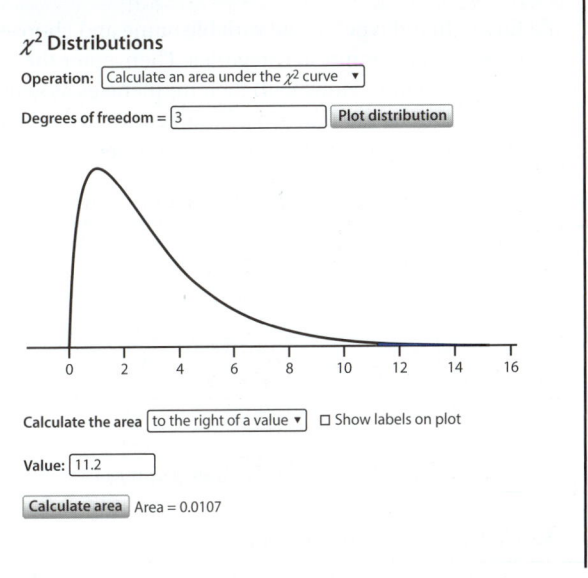

χ^2 **Distributions**
Operation: [Calculate an area under the χ^2 curve ▼]
Degrees of freedom = [3]   [Plot distribution]

0 2 4 6 8 10 12 14 16

Calculate the area [to the right of a value ▼] ☐ Show labels on plot
Value: [11.2]
[Calculate area] Area = 0.0107

(continued)

TI-83/84

1. Press 2nd VARS to get to the distribution menu and arrow down to χ^2cdf(.

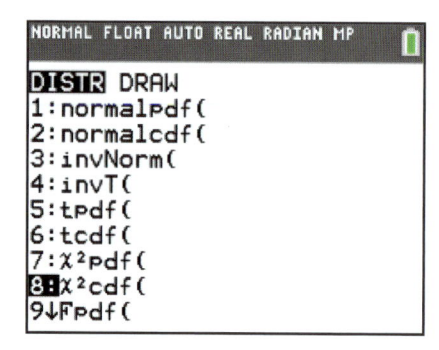

```
NORMAL FLOAT AUTO REAL RADIAN MP

DISTR  DRAW
1:normalpdf(
2:normalcdf(
3:invNorm(
4:invT(
5:tpdf(
6:tcdf(
7:χ²pdf(
8:χ²cdf(
9↓Fpdf(
```

2. Enter lower: 11.2, upper: 1000, df: 3. Click Paste and then press ENTER. **Older OS:** Complete the command χ^2cdf(11.2, 1000, 3) and press ENTER.

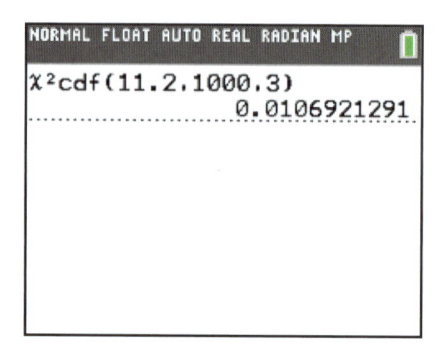

```
NORMAL FLOAT AUTO REAL RADIAN MP

χ²cdf(11.2,1000,3)
             0.0106921291
```

Performing a Chi-Square Goodness of Fit Test

You can also use technology to do all the calculations for a χ^2 test for goodness of fit. We'll illustrate this process with the blood type data from Example 10.2.3.

Applet

1. Go to www.stapplet.com and launch the *One Categorical Variable, Single Group* applet.
2. Enter "Blood type" as the variable name and choose to enter data as counts in categories. Then, enter the four category names along with their frequencies as shown in the screenshot. *Note:* To add categories to the table, click the + button in the lower-right corner.

One Categorical Variable, Single Group

Variable name: Blood type
Input data as: Counts in categories

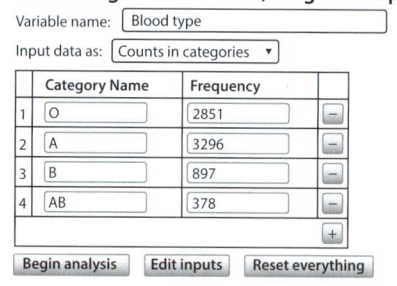

	Category Name	Frequency
1	O	2851
2	A	3296
3	B	897
4	AB	378

Begin analysis   Edit inputs   Reset everything

3. Click the Begin analysis button.
4. Scroll down to the Perform Inference section and choose "Chi-square goodness-of-fit test" from the menu. Then enter the expected counts in the table that appears. Click the Perform inference button. The applet reports the value of χ^2 and the *P*-value, along with the contributions of each category to the χ^2 test statistic (19.236 + 7.064 + 3.062 + 5.796 = 35.159).

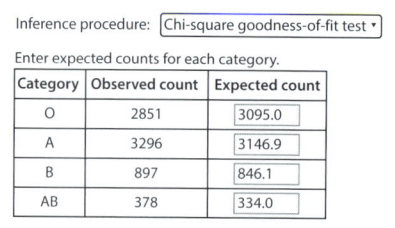

Perform Inference

Inference procedure: Chi-square goodness-of-fit test

Enter expected counts for each category.

Category	Observed count	Expected count
O	2851	3095.0
A	3296	3146.9
B	897	846.1
AB	378	334.0

Perform inference

χ^2	*P*-value	df
35.159	<0.001	3

Category	Contribution
O	19.236
A	7.064
B	3.062
AB	5.796

Note: If you have raw data (e.g., O, A, A, O, AB, B, A, B, . . .) choose to "Input data as: Raw data" at the top of the applet.

TI-84

The chi-square test for goodness of fit isn't available on the TI-83 and some older models of the TI-84.

1. Enter the observed counts into list L_1 and the expected counts into list L_2.

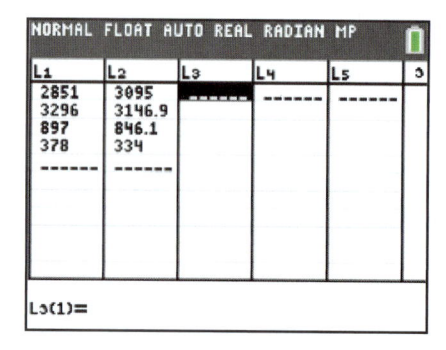

```
NORMAL FLOAT AUTO REAL RADIAN MP

L1      L2      L3    L4    L5    3
2851    3095    ----  ----  ----
3296    3146.9
897     846.1
378     334
----    ----

L5(1)=
```

2. Press STAT, then choose TESTS and χ^2GOF-Test. Enter L_1 for observed, L_2 for expected, and 3 for df. Highlight "Calculate" and press ENTER. The calculator reports the value of χ^2 and the *P*-value, along with the contributions of each category to the χ^2 test statistic (19.236 + 7.064 + 3.062 + 5.796 = 35.159). Note that the calculator reports the *P*-value in scientific notation.

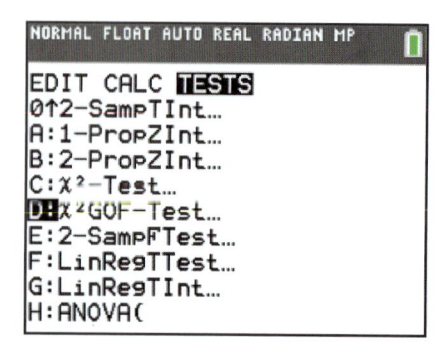

```
NORMAL FLOAT AUTO REAL RADIAN MP

EDIT CALC TESTS
0↑2-SampTInt…
A:1-PropZInt…
B:2-PropZInt…
C:χ²-Test…
D:χ²GOF-Test…
E:2-SampFTest…
F:LinRegTTest…
G:LinRegTInt…
H:ANOVA(
```

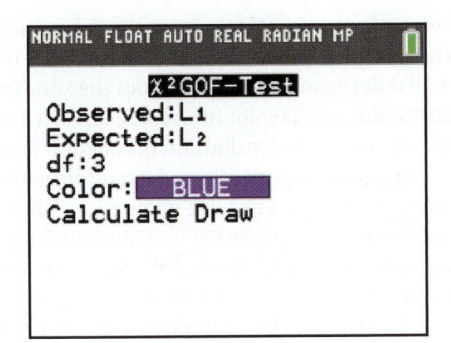

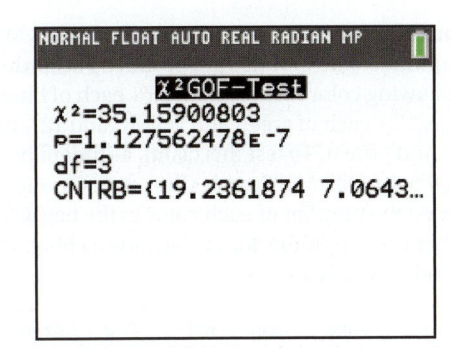

Note: If you select the "Draw" option, you will get a picture of the appropriate χ^2 distribution with the area of interest shaded and the test statistic and *P*-value labeled.

Detailed instructions for using CrunchIt!, Excel, Google Sheets, JMP, Minitab, and R are available in Achieve.

Section 10.2 What Did You Learn?

Review the learning goals from this section. Then practice what you've learned by working through the exercises.

Learning Goal	Example	Exercises
Check conditions for a test about the distribution of a categorical variable.	10.2.1	7–10
Calculate the *P*-value for a test about the distribution of a categorical variable.	10.2.2	11–14
Use the four-step process to perform a chi-square test for goodness of fit.	10.2.3	15–18

Section 10.2 Exercises

Building Concepts and Skills These exercises assess the basic knowledge you should have after reading the section.

1. True/False: To meet the Large Counts condition, the observed count in each category must be at least 5.

2. What shape do all chi-square distributions have?

3. True/False: In a test for goodness of fit, df = $n - 1$.

4. What happens to the shape of the chi-square distributions as the degrees of freedom increase?

5. We always use the area in the _____ tail of a chi-square distribution to calculate *P*-values for tests for goodness of fit.

6. How do you show evidence for the alternative hypothesis in the "State" step of a test for goodness of fit?

Mastering Concepts and Skills These exercises reinforce the learning goals as shown in the examples.

7. **Refer to Example 10.2.1 Landline Sampling** Will randomly selecting households with landline phones produce a representative sample? According to the Census Bureau, of all U.S. residents aged 18 and older, 13% are 18–24 years old, 35% are 25–44 years old, 35% are 45–64 years old, and 17% are 65 years and older.[7] The table gives the age distribution for a sample of U.S. residents aged 18 and older who were chosen by randomly dialing landline telephone numbers. Check whether the conditions for performing a chi-square test for goodness of fit are met.

Age	18–24	25–44	45–64	65+	Total
Count	17	118	161	91	387

8. **Peanut M&M'S** Mars, Inc., reports that its M&M'S Peanut Chocolate Candies are produced according to the following color distribution: 23% each of blue and orange, 15% each of green and yellow, and 12% each of red and brown. To test this claim, a student bought a small bag of M&M'S Peanut Chocolate Candies and recorded the number of each color in the bag. Check whether the conditions for performing a chi-square test for goodness of fit are met.

Color	Blue	Orange	Green	Yellow	Red	Brown	Total
Count	10	6	9	4	7	2	38

9. **Animal Crackers** Are the different animals in a box of animal crackers equally likely? To investigate, a student opened a box and recorded the number of each type of animal in the box. Check whether the conditions for performing a chi-square test for goodness of fit are met.

Animal	Cow	Horse	Buffalo	Moose	Elephant
Frequency	1	5	4	3	5
Animal	Camel	Goat	Polar bear	Donkey	Cat (tail)
Frequency	6	4	3	6	2
Animal	Lion	Cat (no tail)	Wombat	Total	
Frequency	3	3	5	50	

10. **Die Rolls** Josephina made a 6-sided die in her ceramics class and rolled it 60 times to test whether each side was equally likely to show up. The table summarizes the outcomes of her 60 rolls. Check whether the conditions for performing a chi-square test for goodness of fit are met.

Outcome	1	2	3	4	5	6	Total
Frequency	13	11	6	12	10	8	60

11. **Refer to Example 10.2.2 Finding P-values** Calculate the P-value in each of the following settings.
    (a) $\chi^2 = 19.03$, df $= 11$
    (b) $\chi^2 = 19.03$, df $= 3$

12. **More P-values** Calculate the P-value in each of the following settings.
    (a) $\chi^2 = 4.49$, df $= 5$
    (b) $\chi^2 = 4.49$, df $= 1$

13. **Landline Sampling** Refer to Exercise 7.
    (a) Calculate the χ^2 test statistic and P-value.
    (b) Interpret the P-value.
    (c) What conclusion would you make?

14. **Die Rolls** Refer to Exercise 10.
    (a) Calculate the χ^2 test statistic and P-value.
    (b) Interpret the P-value.
    (c) What conclusion would you make?

15. **Refer to Example 10.2.3 Climate and Car Color** Does the warm, sunny weather in Arizona affect a driver's choice of car color? Conventional wisdom suggests that drivers might opt for a lighter color with the hope that it will reflect some of the heat from the sun. To see if the distribution of car color in Oro Valley, near Tucson, is different from the distribution of car color across North America, a researcher selected a random sample of 300 cars in Oro Valley. The table shows the distribution of car color for the sample in Oro Valley and the distribution of car color in North America.[8] Do these data provide convincing evidence that the distribution of car color in Oro Valley differs from the North American distribution? Use $\alpha = 0.01$.

Color	White	Black	Gray	Silver	Red
Oro Valley sample	84	38	31	46	27
North America	23%	18%	16%	15%	10%
Color	Blue	Green	Other	Total	
Oro Valley sample	29	6	39	300	
North America	9%	2%	7%	100%	

16. **Crossing Tomatoes** Genetics researchers in Canada performed an experiment in which they crossed tall, cut-leaf tomatoes with dwarf, potato-leaf tomatoes.[9] Genetic laws state that the characteristics of height and leaf shape should occur in the ratio 9:3:3:1 for tall/cut, tall/potato, dwarf/cut, and dwarf/potato. Do these data provide convincing evidence that the genetic theory is incorrect? Assume that the Random condition is met and use $\alpha = 0.10$.

Characteristics	Tall, cut	Tall, potato	Dwarf, cut	Dwarf, potato	Total
Frequency	926	288	293	104	1611

17. **Pea Plants** Gregor Mendel (1822–1884), an Austrian monk, is considered the father of genetics. Mendel studied the inheritance of various traits in pea plants. One such trait is whether the pea is smooth or wrinkled. Mendel predicted a ratio of 3 smooth peas for every 1 wrinkled pea. In one experiment, he observed 423 smooth and 133 wrinkled peas. Do these data provide convincing evidence at the 5% significance level that Mendel's prediction is incorrect? Assume that the Random condition is met.

18. **Coin Spin** When you flip a fair coin, the probability the coin lands on heads is 0.50. However, what if you spin the coin? According to the article "Euro Coin Accused of Unfair Flipping" in the *New Scientist*, two Polish math professors and their students spun a Belgian euro coin 250 times. It landed on heads 140 times. One of the professors concluded that the coin was minted asymmetrically. A representative from the Belgian mint indicated the result happened by chance alone. Do these data provide convincing evidence at the 5% significance level that this euro coin is not equally likely to land on heads or tails when spun? Assume the Random condition is met.

Applying the Concepts These exercises ask you to apply multiple learning goals in a new context or to apply what you learned in this section in a new way.

19. **Benford's Law** Faked numbers in tax returns, invoices, or expense account claims often display patterns that aren't present in legitimate records. Some patterns are obvious and easily avoided by a clever crook. Others are more subtle. It is a striking fact that the first digits of numbers in legitimate records often follow a model known as Benford's law.[10] Here is the distribution of first digit for variables that follow Benford's law.

First digit	1	2	3	4	5
Proportion	0.301	0.176	0.125	0.097	0.079
First digit	6	7	8	9	Total
Proportion	0.067	0.058	0.051	0.046	1.000

A forensic accountant who is familiar with Benford's law inspects a random sample of 250 invoices from a company that is accused of committing fraud. The table displays the sample data. Is there convincing evidence that the invoices from this company don't follow Benford's law?

First digit	1	2	3	4	5
Frequency	61	50	43	34	25
First digit	6	7	8	9	Total
Frequency	16	7	8	6	250

20. **Race and Ethnicity of Residents** According to the U.S. Census Bureau, the distribution of race and ethnicity of the New York City population is as follows: Hispanic: 26%; Black: 26%; White: 33%; Asian: 13%; Other: 2%. The manager of a large housing complex in the city wonders if the distribution of race and ethnicity of the complex's residents is consistent with the population distribution. To find out, the manager records data from a random sample of 800 residents.[11] Based on the sample data, is there convincing evidence that the distribution of race and ethnicity in the housing complex is different from the distribution in New York City?

Race and ethnicity	Hispanic	Black	White	Asian	Other	Total
Frequency	212	202	270	94	22	800

21. **Benford's Law** Refer to Exercise 19.
    (a) Describe a Type I error and a Type II error in this setting, and give a possible consequence of each. Which do you think is more serious?
    (b) For which first digits were the observed counts much greater than expected? For which first digits were the observed counts much less than expected? Justify your answer.

22. **Race and Ethnicity of Residents** Refer to Exercise 20.
    (a) Describe a Type I error and a Type II error in this setting, and give a possible consequence of each. Which do you think is more serious?
    (b) For which race and ethnicity categories were the observed counts much greater than expected? For which categories were the observed counts much less than expected? Justify your answer.

Extending the Concepts These exercises challenge you to explore statistical concepts and methods that go beyond what you learned in this section.

23. **Climate and Car Color** Refer to Exercise 15. Why do you think the researchers combined several colors (e.g., pink, purple) into a single category called "Other"?

24. **Coin Spin** Refer to Exercise 18. In Section 8.4, Exercise 16, you analyzed these data with a one-sample z test for a proportion. The hypotheses were $H_0: p = 0.5$ and $H_a: p \ne 0.5$, where $p =$ the true proportion of heads.
    (a) Calculate the z statistic and P-value for this test.
    (b) How does the P-value from part (a) compare to the P-value from Exercise 18 in this section? Calculate the value of z^2. How does it compare to the value of χ^2?

25. **Truly Random?** Use a random number generator on a calculator or the internet to generate 200 digits from 0 to 9.
    (a) Create a table showing the observed count for each digit. *Hint:* To obtain the observed counts, make a histogram of the 200 random digits. You may have to adjust the scale to go from −0.5 to 9.5 with bar widths of 1.
    (b) Based on the sample in part (a), is there convincing evidence that the digits from the random number generator are not equally likely? Use $\alpha = 0.05$.
    (c) Assuming that a student's random number generator is working properly, what is the probability that the student will make a Type I error in part (b)?
    (d) Suppose that 25 students in a statistics class independently do this exercise for homework and that all of their random number generators are working properly. Find the probability that at least one of them makes a Type I error.

26. **Mice Cancer Treatments** Researchers conducted an experiment on 36 mice to compare four possible treatments for cancerous tumors: control, drug only, radiation only, and drug plus radiation.[12] The mice had cells from a human cancer cell line implanted. Once a tumor grew to be large enough, the mice were randomly assigned to one of the four groups. The table shows the mean growth (End size − Start size) for each treatment in cubic millimeters (mm^3). Explain why it would not be appropriate to perform a chi-square test for goodness of fit using these data.

Treatment	Control	Drug	Radiation	Both
Mean growth	1997	1736	1808	1126

Cumulative Review These exercises revisit what you learned in previous sections.

27. **Soccer Risks (4.3, 4.4)** A study in Sweden looked at former elite soccer players, people who had played soccer but not at the elite level, and people of the same age who did not play soccer. The two-way table classifies these individuals by whether or not they had arthritis of the hip or knee by their mid-50s.[13]

		Soccer experience			
		Elite	Non-elite	Did not play	Total
Arthritis	Yes	10	9	24	43
	No	61	206	548	815
	Total	71	215	572	858

Suppose we choose one of these players at random.
(a) What is the probability that the player has arthritis?
(b) What is the probability that the player has arthritis, given that they were classified as an elite soccer player?

28. **Soccer Risks (2.1)** Refer to Exercise 27. We suspect that the more serious soccer players have more arthritis later in life.
(a) Make a graph to show the relationship between soccer experience and arthritis for these players.
(b) Based on your graph, is there an association between soccer experience and arthritis for these players? Explain your answer.

Section 10.3 Testing the Relationship Between Two Categorical Variables in a Population

LEARNING GOALS

By the end of this section, you will be able to:
- State hypotheses for a test about the relationship between two categorical variables.
- Calculate expected counts for a test about the relationship between two categorical variables.
- Calculate the test statistic for a test about the relationship between two categorical variables.

In Section 2.2, we considered the relationship between two categorical variables, such as snowmobile use and environmental club membership for a random sample of winter visitors to Yellowstone National Park. Because you hadn't learned about inference yet, we could only talk about the association *in that sample.* In this section, you will learn how to state hypotheses, calculate expected counts, and calculate the test statistic for a test about the relationship between two categorical variables *in a population.*

Stating Hypotheses

In Chapter 2, we explored relationships between two variables to determine if the variables have an **association.**

> There is an **association** between two variables if knowing the value of one variable helps us predict the value of the other. If knowing the value of one variable does not help us predict the value of the other, then there is no association between the variables.

Is there an association between subject taught and handedness for college professors at an education conference? To investigate, we selected a random sample of 100 professors from the conference. The two-way table summarizes the relationship between subject taught and handedness for these professors.

		Subject taught		
		Math	Not math	Total
Dominant hand	Right	39	51	90
	Left	7	3	10
	Total	46	54	100

Section 10.3 Testing the Relationship Between Two Categorical Variables in a Population

In Section 4.4, we analyzed these data by calculating and displaying conditional probabilities. When selecting a professor at random from this sample, $P(\text{left-handed} \mid \text{math}) = 7/46 = 0.152$ and $P(\text{left-handed} \mid \text{not math}) = 3/54 = 0.056$. These probabilities are displayed in Figure 10.5 as blue segments in the segmented bar chart.

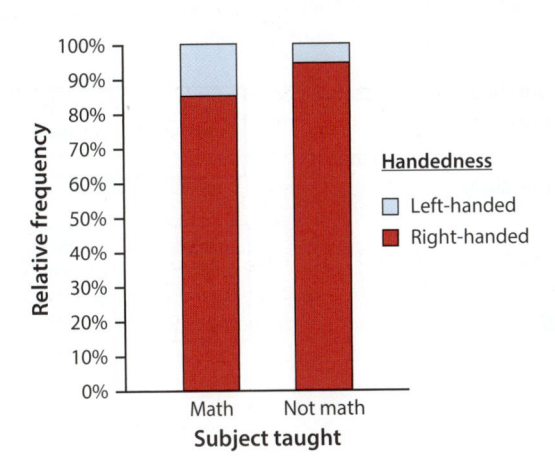

FIGURE 10.5 Segmented bar chart showing the relationship between subject taught and handedness for a sample of 100 college professors.

Because these probabilities differ, we concluded that the events "math" and "left-handed" are not independent. That is, we concluded that there is an association between subject taught and handedness *for the members of the sample*.

Does the association in the sample provide convincing evidence that there is an association between subject taught and handedness for *all* professors at the conference? Or is it plausible that there is no association between these variables in the population of professors at the conference and that the differences observed in the sample were due to sampling variability alone? We can use a significance test to find out.

As with other significance tests, we begin the test for the relationship between two categorical variables by stating hypotheses. The null hypothesis for this test states that there is no association between the variables in the population. The alternative hypothesis states that there is an association. Here are the hypotheses for the subject taught and handedness example:

H_0: There is no association between subject taught and handedness in the population of professors at the conference.

H_a: There is an association between subject taught and handedness in the population of professors at the conference.

EXAMPLE 10.3.1

Do angry people have a greater risk of heart disease?

Stating hypotheses

PROBLEM

A study followed a random sample of 8474 people with normal blood pressure for about 4 years. All the individuals were free of heart disease at the beginning of the study. Each person took the Spielberger Trait Anger Scale test, which measures how prone a person is to sudden anger. Researchers also recorded if each individual developed coronary heart disease. This classification includes people who had heart attacks and those who needed medical treatment for heart disease.[14] The two-way table summarizes the data.

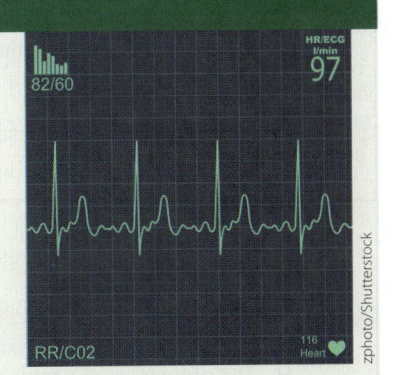

		Anger level			
		Low	Moderate	High	Total
Heart disease status	Yes	53	110	27	190
	No	3057	4621	606	8284
	Total	3110	4731	633	8474

(continued)

(a) State the hypotheses for a test about the relationship between anger level and heart disease status in the population of people with normal blood pressure.
(b) Make a segmented bar chart to show the relationship between anger level and heart disease status in the sample.
(c) Explain how the graph in part (b) gives some evidence for the alternative hypothesis.

SOLUTION

(a) H_0: There is no association between anger level and heart disease status in the population of people with normal blood pressure.

H_a: There is an association between anger level and heart disease status in the population of people with normal blood pressure.

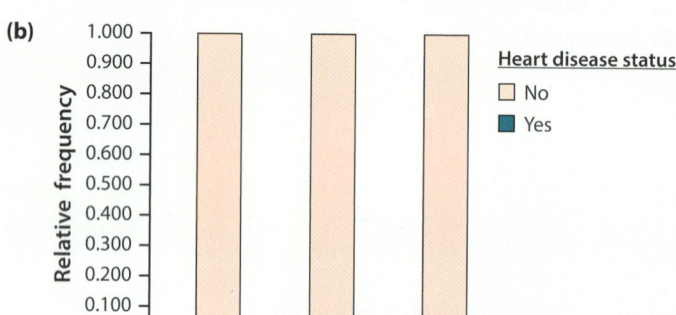

> The null hypothesis for this test states that there is *no* association between the variables in the *population of interest*.

(b)

> To make the segmented bar chart, calculate the proportion of each anger level group that has heart disease. For example, among the low anger group, $53/3110 = 0.017$ have heart disease.

(c) Because the proportion with heart disease is different for the three anger level groups, there is an association between anger level and heart disease status in the sample.

> Knowing a person's anger level helps to predict if the person developed heart disease for the members of the sample.

EXAM PREP: FOR PRACTICE, TRY EXERCISE 5.

Notice that the alternative hypothesis says that the null hypothesis is *not* true. For a test about the relationship between two categorical variables, we don't have one-sided alternative hypotheses like $H_a: p < 0.50$ or $H_a: \mu > 100$.

THINK ABOUT IT

Can we state the hypotheses another way? Yes. If two variables have no association, we can also say the two variables are independent. Using the anger and heart disease example, we can state the hypotheses as follows:

H_0: Anger level and heart disease status are independent in the population of people with normal blood pressure.

H_a: Anger level and heart disease status are not independent in the population of people with normal blood pressure.

Also, when there are only two response categories, we can state the hypotheses in terms of proportions:

H_0: The proportion of people with normal blood pressure who develop heart disease is the same for all three anger levels (e.g., $p_{low} = p_{moderate} = p_{high}$).

H_a: The proportion of people with normal blood pressure who develop heart disease is not the same for all three anger levels (e.g., at least one of the proportions is different).

Note that the alternative hypothesis says that the null hypothesis is not true, just as when we stated the hypotheses using the idea of association.

Calculating Expected Counts

How do we determine if an association between two variables in a sample provides convincing evidence of an association in the population? We measure how different the observed counts are from what we would expect if there was no association between the variables in the population.

For instance, if there is no association between subject taught and handedness in the population of college professors at the conference, then the proportion of math professors who are left-handed should be the same as the proportion of non-math professors who are left-handed. Because 10% (10/100) of the professors in the sample were left-handed, 10% of the math professors should be left-handed and 10% of the non-math professors should be left-handed when H_0 is true.

Observed Counts

		Subject taught		
		Math	Not math	Total
Dominant hand	Right	39	51	90
	Left	7	3	10
	Total	46	54	100

Expected Counts

		Subject taught		
		Math	Not math	Total
Dominant hand	Right	? (90%)	? (90%)	90 (90%)
	Left	? (10%)	? (10%)	10 (10%)
	Total	46	54	100

That is, under the null hypothesis of no association, the expected number of left-handed math professors is 10% of the 46 math professors in the sample:

$$\frac{10}{100} \times 46 = 4.6$$

Don't worry that the expected count isn't an integer—we aren't saying that a sample can have 4.6 left-handed math professors! The expected count tells us that if many samples are selected from a population like this one—but with no association between subject taught and handedness—we would expect an average of 4.6 left-handed math professors in samples of size 100.

Notice that the calculation can be rewritten as

$$\frac{10}{100} \times 46 = \frac{10 \times 46}{100}$$

This is the total number of lefties (row total) multiplied by the total number of math professors (column total), divided by the total number in the sample (table total). This leads to the general rule for computing expected counts in a two-way table.

HOW TO Calculate Expected Counts for a Test About the Relationship Between Two Categorical Variables

For a test about the relationship between two categorical variables, the expected count for a cell in the two-way table is

$$\text{expected count} = \frac{\text{row total} \times \text{column total}}{\text{table total}}$$

You can complete the table of expected counts by using the formula for each cell or by using the formula for some cells and subtracting to find the remaining cells. For example, if the expected count of left-handed math professors is 4.6, we know the expected count of right-handed math professors is $46 - 4.6 = 41.4$. Here is the completed table of expected counts:

Expected Counts

		Subject taught		
		Math	Not math	Total
Dominant hand	Right	41.4 (90%)	48.6 (90%)	90 (90%)
	Left	4.6 (10%)	5.4 (10%)	10 (10%)
	Total	46	54	100

EXAMPLE 10.3.2

Do angry people have a greater risk of heart disease?

Calculating expected counts

PROBLEM

Here are the data from the study of anger and heart disease. Calculate the expected counts for a test of the null hypothesis that there is no association between anger level and heart disease status in the population of people with normal blood pressure.

		Anger level			
		Low	Moderate	High	Total
Heart disease status	Yes	53	110	27	190
	No	3057	4621	606	8284
	Total	3110	4731	633	8474

SOLUTION

Expected Counts

		Anger level			
		Low	Moderate	High	Total
Heart disease status	Yes	$\frac{190 \times 3110}{8474} = 69.73$	$\frac{190 \times 4731}{8474} = 106.08$	$190 - 69.73 - 106.08 = 14.19$	190
	No	$3110 - 69.73 = 3040.27$	$4731 - 106.08 = 4624.92$	$633 - 14.19 = 618.81$	8284
	Total	3110	4731	633	8474

$$\text{expected count} = \frac{\text{row total} \times \text{column total}}{\text{table total}}$$

EXAM PREP: FOR PRACTICE, TRY EXERCISE 9.

Notice that we used subtraction to calculate the expected counts for the cells in the last row and the last column. In other words, we needed to use the formula for expected counts in only two cells. This observation will be useful when we discuss degrees of freedom and calculating *P*-values in the next section.

The Chi-Square Test Statistic

For a test about the relationship between two categorical variables, we use the familiar chi-square test statistic:

$$\chi^2 = \sum \frac{(\text{observed count} - \text{expected count})^2}{\text{expected count}}$$

This time, the sum is over all cells in the two-way table, not including the totals. Here are the observed and expected counts for the example about professors at the education conference:

Observed Counts

		Subject taught		
		Math	Not math	Total
Dominant hand	Right	39	51	90
	Left	7	3	10
	Total	46	54	100

Expected Counts

		Subject taught		
		Math	Not math	Total
Dominant hand	Right	41.4	48.6	90
	Left	4.6	5.4	10
	Total	46	54	100

The chi-square test statistic is

$$\chi^2 = \frac{(39-41.4)^2}{41.4} + \frac{(51-48.6)^2}{48.6} + \frac{(7-4.6)^2}{4.6} + \frac{(3-5.4)^2}{5.4} = 2.58$$

You can use technology to calculate the chi-square test statistic and expected counts. See the Tech Corner in Section 10.4 for details.

EXAMPLE 10.3.3

Do angry people have a greater risk of heart disease?

Calculating the χ^2 test statistic

PROBLEM
Here are the observed and expected counts from the study of anger and heart disease. Calculate the value of the chi-square test statistic.

Observed Counts

		Anger level			
		Low	Moderate	High	Total
Heart disease status	Yes	53	110	27	190
	No	3057	4621	606	8284
	Total	3110	4731	633	8474

Expected Counts

		Anger level			
		Low	Moderate	High	Total
Heart disease status	Yes	69.73	106.08	14.19	190
	No	3040.27	4624.92	618.81	8284
	Total	3110	4731	633	8474

SOLUTION

$$\chi^2 = \frac{(53-69.73)^2}{69.73} + \frac{(110-106.08)^2}{106.08} + \cdots + \frac{(606-618.81)^2}{618.81} = 16.08$$

$$\chi^2 = \sum \frac{(\text{observed count} - \text{expected count})^2}{\text{expected count}}$$

EXAM PREP: FOR PRACTICE, TRY EXERCISE 13.

How unusual is it to get a χ^2 test statistic as large as or larger than 16.08 just by chance? You will learn how to use a chi-square distribution to calculate the *P*-value in the next section. For now, we can estimate the *P*-value with a simulation. The dotplot in Figure 10.6 shows the χ^2 test statistic for 100 simulated random samples of 8474 people with normal blood pressure, assuming there is no association between anger level and heart disease status.

In 0 of the 100 simulated samples, the value of the χ^2 test statistic was at least 16.08—the test statistic from the actual sample.

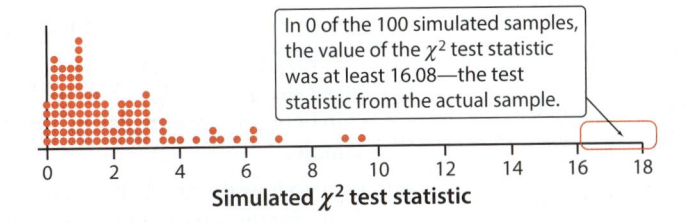

FIGURE 10.6 The χ^2 test statistic from each of 100 simulated random samples of size 8474 selected from a specific population where there is no association between anger and heart disease status.

Because the observed χ^2 test statistic of 16.08 is greater than any of the simulated χ^2 test statistics, the *P*-value is approximately 0. There is almost no chance of getting an association between anger level and heart disease status as strong as or stronger than the one observed in the sample by chance alone when there is no association between these variables in the population. This sample provides convincing evidence of an association between anger level and heart disease status in the population of people with normal blood pressure. Of course, this was not an experiment, so it would be incorrect to conclude that anger *causes* heart disease based on this study. (!) **Remember that association doesn't imply causation!**

Section 10.3 What Did You Learn?

Review the learning goals from this section. Then practice what you've learned by working through the exercises.

Learning Goal	Example	Exercises
State hypotheses for a test about the relationship between two categorical variables.	10.3.1	5–8
Calculate expected counts for a test about the relationship between two categorical variables.	10.3.2	9–12
Calculate the test statistic for a test about the relationship between two categorical variables.	10.3.3	13–16

Section 10.3 Exercises

Building Concepts and Skills These exercises assess the basic knowledge you should have after reading the section.

1. What does it mean if two variables have an association?

2. True/False: The null hypothesis for a test about the relationship between two categorical variables is:
H_0: There is an association between the two variables in the population of interest.

3. What is the formula for calculating expected counts in a test about the relationship between two categorical variables?

4. Which test statistic do we use in a test about the relationship between two categorical variables?

Mastering Concepts and Skills These exercises reinforce the learning goals as shown in the examples.

5. **Refer to Example 10.3.1 Nightlights and Myopia**
Researchers at The Ohio State University wanted to know if there was an association between using a nightlight and myopia (nearsightedness) in children. They surveyed the parents of 1220 randomly selected children and recorded the lighting conditions the children slept in during their first two years of life and whether they had myopia at age 10.[15] The two-way table summarizes the data.

		Lighting condition			
		No light	Nightlight	Fully lit	Total
Myopia status	Yes	83	129	10	222
	No	334	629	35	998
	Total	417	758	45	1220

(a) State the hypotheses for a test about the relationship between lighting condition and myopia status in the population of children.

(b) Make a segmented bar chart to show the relationship between lighting condition and myopia status in the sample.

(c) Explain how the graph in part (b) gives some evidence for the alternative hypothesis.

6. **Gamer Ages** To determine if there is an association between age and playing video games, the Pew Research Center asked randomly selected U.S. adults for their age and if they "ever play video games on a computer, TV, game console, or portable device like a cell phone."[16] The two-way table summarizes the data.

		Age group				
		18–29	30–49	50–64	65+	Total
Play video games?	Yes	887	1217	650	279	3033
	No	429	872	985	840	3126
	Total	1316	2089	1635	1119	6159

(a) State the hypotheses for a test about the relationship between age and playing video games in the population of U.S. adults.

(b) Make a segmented bar chart to show the relationship between age and playing video games in the sample.

(c) Explain how the graph in part (b) gives some evidence for the alternative hypothesis.

7. **Democrats on Twitter** Are self-identified Democrats who use Twitter more likely to describe their political views as liberal than self-identified Democrats who do not use Twitter? The Pew Research Center studied a random sample of 6077 U.S. adults who self-identified as Democrats or Democrat-leaning. Each member of the sample was asked if they use Twitter and how they would describe their political views. The two-way table summarizes the data.[17]

		Use Twitter?		
		No	Yes	Total
Political views	Very conservative	127	39	166
	Conservative	257	56	313
	Moderate	2131	673	2804
	Liberal	1261	599	1860
	Very liberal	539	395	934
	Total	4315	1762	6077

(a) State the hypotheses for a test about the relationship between Twitter use and political views in the population of U.S. adults who self-identify as Democrats.

(b) Make a segmented bar chart to show the relationship between Twitter use and political views in the sample.

(c) Explain how the graph in part (b) gives some evidence for the alternative hypothesis.

8. **Snowmobiles and Environmental Clubs** Yellowstone National Park staff surveyed a random sample of 1526 winter visitors to the park. They asked each person whether they belonged to an environmental club (like the Sierra Club). Respondents were also asked whether they owned, rented, or had never used a snowmobile. Is there an association between club membership and snowmobile use? The two-way table summarizes the data.

		Environmental club		
		No	Yes	Total
Snowmobile use	Never used	445	212	657
	Snowmobile renter	497	77	574
	Snowmobile owner	279	16	295
	Total	1221	305	1526

(a) State the hypotheses for a test about the relationship between club membership and snowmobile use in the population of winter visitors to Yellowstone.

(b) Make a segmented bar chart to show the relationship between club membership and snowmobile use in the sample.

(c) Explain how the graph in part (b) gives some evidence for the alternative hypothesis.

9. **Refer to Example 10.3.2 Nightlights and Myopia** Calculate the expected counts for a test of the hypotheses stated in Exercise 5.

10. **Gamer Ages** Calculate the expected counts for a test of the hypotheses stated in Exercise 6.

11. **Democrats on Twitter** Calculate the expected counts for a test of the hypotheses stated in Exercise 7.

12. **Snowmobiles and Environmental Clubs** Calculate the expected counts for a test of the hypotheses stated in Exercise 8.

13. **Refer to Example 10.3.3 Nightlights and Myopia** Refer to Exercises 5 and 9. Calculate the value of the chi-square test statistic.

14. **Gamer Ages** Refer to Exercises 6 and 10. Calculate the value of the chi-square test statistic.

15. **Democrats on Twitter** Refer to Exercises 7 and 11. Calculate the value of the chi-square test statistic.

16. **Snowmobiles and Environmental Clubs** Refer to Exercises 8 and 12. Calculate the value of the chi-square test statistic.

Applying the Concepts These exercises ask you to apply multiple learning goals in a new context or to apply what you learned in this section in a new way.

17. **Saunas and Cardiac Death** Researchers followed a random sample of 2315 middle-aged men from eastern Finland for up to 30 years. They recorded how often each man went to a sauna and whether or not he suffered sudden cardiac death (SCD).[18] The two-way table summarizes the data from the study.

		Weekly sauna frequency			
		1 or fewer	2-3	4 or more	Total
SCD	Yes	61	119	10	190
	No	540	1394	191	2125
	Total	601	1513	201	2315

(a) State the hypotheses for a test about the relationship between weekly sauna frequency and SCD for middle-aged men from eastern Finland.

(b) Make a graph that displays the relationship between weekly sauna frequency and SCD for the members of the sample. Explain how the graph provides some evidence for the alternative hypothesis.

(c) Calculate the expected counts for a test of the hypotheses stated in part (a).

(d) Calculate the value of the chi-square test statistic.

18. **Home Injuries** Recognizing that many more people were working from home during the Covid pandemic, researchers surveyed a random sample of 2009 U.S. households in June 2020, roughly 3 months after lockdowns began.[19] The researchers asked whether anyone in the household had an injury that had occurred at home or ingested something that could make them sick while at home during the previous 3 months. They also

classified the households as urban, suburban, or rural. Here are the results:

		Type of household			
		Urban	Suburban	Rural	Total
Type of incident	Both	171	80	18	269
	Injury or ingestion, but not both	83	137	81	301
	Neither	404	750	285	1439
	Total	658	967	384	2009

(a) State the hypotheses for a test about the relationship between type of incident and type of household for U.S. households in June 2020.

(b) Make a graph that displays the relationship between type of household and type of incident for the households in the sample. Explain how the graph provides some evidence for the alternative hypothesis.

(c) Calculate the expected counts for a test of the hypotheses stated in part (a).

(d) Calculate the value of the chi-square test statistic.

Extending the Concepts These exercises challenge you to explore statistical concepts and methods that go beyond what you learned in this section.

19. **Saunas and Cardiac Death** In Exercise 17(b), you constructed a graph to display the relationship between weekly sauna frequency and sudden cardiac death for the members of the sample. Assuming the overall proportion who experience SCD remains the same, construct a new graph that shows no association between weekly sauna frequency and SCD.

20. **Subject Taught and Handedness** In this section, we analyzed the relationship between subject taught and handedness for a sample of 100 college professors at an education conference. The chi-square test statistic for these data is $\chi^2 = 2.58$. The dotplot shows the results of a simulation in which 200 random samples of size 100 were selected from a population of professors where there is no association between subject taught and handedness. For each simulated sample of 100 professors, the chi-square test statistic was calculated. Based on the χ^2 test statistic and the results of the simulation, what conclusion would you make?

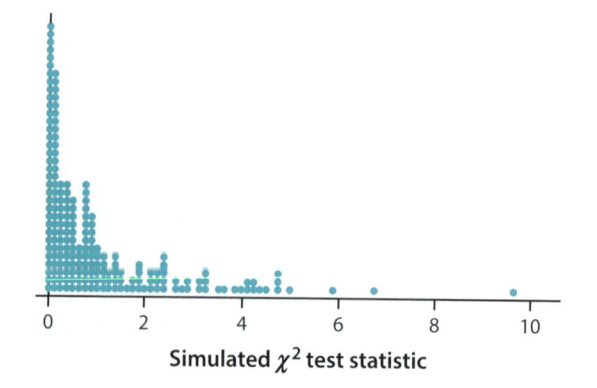

Cumulative Review These exercises revisit what you learned in previous sections.

21. **Butterfly Wings (3.3)** Are male and female butterflies different sizes? Random samples of 16 male and 16 female *Boloria chariclea* butterflies were selected in Greenland and their wing lengths measured.[20] Construct a boxplot for each distribution. Then compare the distributions, making sure to answer the initial question.

Male	Female
18.1	19.1
18.2	18.8
18.4	19.5
18.1	19.0
17.9	18.9
17.8	18.6
17.8	18.9
17.9	18.9
17.7	18.8
18.3	19.1
18.0	18.9
17.7	18.7
17.6	18.6
17.4	18.6
17.8	18.4
17.4	18.2

22. **Encouraging Texts (1.5, 1.6)** Can sending encouraging text messages help patients stick to exercise programs? An experiment in Australia randomly assigned patients with knee osteoarthritis to one of two groups. The first group got basic information about the importance of physical activity, a specific knee-strengthening regimen, and text messages encouraging adherence. The second group got only the basic information, with no specific knee-strengthening regimen or text messages. After 24 weeks, each subject was asked for a pain rating on a scale from 0 (no pain) to 10 (extreme pain). The mean pain rating was significantly smaller for the patients who received the exercise regimen supported by text messages.[21]

(a) Explain the purpose of the random assignment in this experiment.

(b) Explain what it means to say that the mean pain rating was "significantly" smaller.

(c) Based on this study, explain why it is impossible to determine whether the text messages alone are the cause of the decrease in mean pain rating.

Section 10.4 Chi-Square Tests for Association

LEARNING GOALS

By the end of this section, you will be able to:
- Check conditions for a test about the relationship between two categorical variables.
- Calculate the *P*-value for a test about the relationship between two categorical variables.
- Use the four-step process to perform a chi-square test for association.

In Section 10.3, you learned how to state hypotheses, calculate expected counts, and calculate the chi-square test statistic for a test about the relationship between two categorical variables. In this section, you will learn how to check the conditions, calculate the *P*-value, and perform the four-step process for this test, called a **chi-square test for association.**

A **chi-square test for association** is a significance test of the null hypothesis that there is no association between two categorical variables in the population of interest.

Conditions for a Chi-Square Test for Association

As with the chi-square test for goodness of fit, there are two conditions to be satisfied in a chi-square test for association: Random and Large Counts. However, there are now three ways to verify that the Random condition is met:

1. The data come from a single random sample from the population of interest. In this case, the values of two categorical variables are recorded for each individual.
2. The data come from independent random samples from the populations of interest. In this case, the value of one categorical variable is recorded for each individual in each sample.
3. The data come from groups in a randomized experiment. In this case, the value of one categorical variable is recorded for each individual in each group.

In Section 10.3, all the examples and exercises were from a single random sample with two variables recorded for each individual (#1). In this section, we will introduce data collected from independent random samples (#2) and from randomized experiments (#3). All three of these data collection plans result in a two-way table that can be analyzed with a chi-square test for association. *Note*: Some people refer to a test based on a single random sample (#1) as a *test for independence* and a test based on independent random samples (#2) or a randomized experiment (#3) as a *test for homogeneity*. We'll continue to refer to all of these cases as a *test for association*.

> **HOW TO** Check the Conditions for a Chi-Square Test for Association
>
> To perform a chi-square test for association, check that the following conditions are met:
> - **Random:** The data come from a random sample from the population of interest, from independent random samples from the populations of interest, or from groups in a randomized experiment.
> - **Large Counts:** All expected counts are at least 5.

In the anger and heart disease example from Section 10.3, the conditions are satisfied:
- Random? The study used a random sample of size 8474 from the population of people with normal blood pressure. ✓
- Large Counts? The expected counts shown in the table are all at least 5. ✓

Expected Counts

		Anger level			
		Low	Moderate	High	Total
Heart disease status	Yes	69.73	106.08	14.19	190
	No	3040.27	4624.92	618.81	8284
	Total	3110	4731	633	8474

However, the conditions are *not* satisfied in the example about handedness and subject taught among college professors at a conference. Even though these data came from a random sample, one of the expected counts shown in the table (4.6) is less than 5.

Expected Counts

		Subject taught		
		Math	Not math	Total
Dominant hand	Right	41.4	48.6	90
	Left	4.6	5.4	10
	Total	46	54	100

EXAMPLE 10.4.1

How important is speaking English?

Checking conditions

PROBLEM

The Pew Research Center conducts surveys about a variety of topics in many different countries. One such survey investigated how adult residents of different countries feel about the importance of speaking the national language. Independent random samples of adult residents of Australia, the United Kingdom, and the United States were asked, "How important do you think it is to be able to speak English?" The two-way table summarizes the responses to this question.[22] Check whether the conditions for performing a chi-square test for association are met.

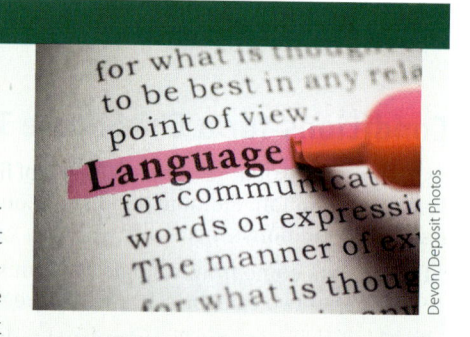

		Country			
		Australia	U.K.	U.S.	Total
Opinion about speaking English	Very important	690	1177	702	2569
	Somewhat important	250	242	221	713
	Not very important	40	28	50	118
	Not at all important	20	13	30	63
	Total	1000	1460	1003	3463

SOLUTION

- **Random?** The Pew Research Center selected independent random samples from Australia, the United Kingdom, and the United States. ✓
- **Large Counts?** All expected counts are ≥ 5 (see the table). ✓

$$\text{expected count} = \frac{\text{row total} \times \text{column total}}{\text{table total}}$$

Expected counts

		Country			
		Australia	U.K.	U.S.	Total
Opinion about speaking English	Very important	741.8	1083.1	744.1	2569
	Somewhat important	205.9	300.6	206.5	713
	Not very important	34.1	49.7	34.2	118
	Not at all important	18.2	26.6	18.2	63
	Total	1000	1460	1003	3463

EXAM PREP: FOR PRACTICE, TRY EXERCISE 5.

Calculating *P*-Values

When the conditions are met, we use a chi-square distribution to calculate the *P*-value for a chi-square test for association. For the chi-square test for goodness of fit, we used df = number of categories − 1. This method for calculating degrees of freedom is correct when there is only one sample and one variable. Because there are two variables in a test for association, we use a different rule for calculating the degrees of freedom.

> **HOW TO** **Calculate the Degrees of Freedom in a Test for Association**
>
> To calculate the *P*-value in a chi-square test for association when the conditions are met, use the chi-square distribution with
>
> $$\text{df} = (\text{number of rows} - 1)(\text{number of columns} - 1)$$
>
> where the number of rows and number of columns do not include the totals.

Here is the two-way table for the anger and heart disease example:

		\multicolumn{3}{c	}{Anger level}	
		Low	Moderate	High	Total
Heart disease status	Yes	53	110	27	190
	No	3057	4621	606	8284
	Total	3110	4731	633	8474

There are 2 rows and 3 columns in this two-way table (not counting the totals). Thus,

$$\text{df} = (2-1)(3-1) = 2$$

To calculate the *P*-value, use Table C or technology as in Section 10.2. Recall that $\chi^2 = 16.08$.

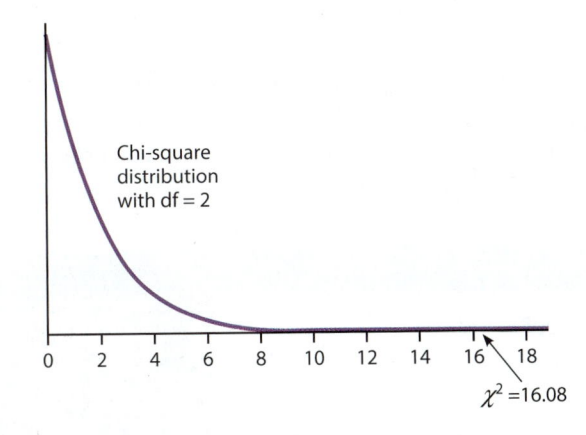

Chi-square distribution with df = 2

$\chi^2 = 16.08$

Using Table C, the *P*-value is < 0.005, as shown in the table excerpt. Technology gives *P*-value = 0.0003.

	Tail probability *p*	
df	.01	.005
1	6.635	7.879
2	9.210	**10.597**
3	11.345	12.838

Because the *P*-value is very small, we reject H_0. There is convincing evidence of an association between anger and heart disease status in the population of people with normal blood pressure. This matches our conclusion from the end of Section 10.3, where we estimated the *P*-value using simulation.

Calculating the P-Value in a Chi-Square Test for Association

Suppose the Random and Large Counts conditions are met. Find the P-value by calculating the probability of getting a χ^2 statistic as large or larger than the observed value of χ^2 using a χ^2 distribution with df = (number of rows − 1)(number of columns − 1).

EXAMPLE 10.4.2

How important is speaking English?

Calculating P-values

PROBLEM

In Example 10.4.1, you checked the conditions for a test of the association between country and opinion about speaking English. The chi-square test statistic for this test is $\chi^2 = 68.57$. Calculate the P-value.

SOLUTION

df = (4 − 1)(3 − 1) = 6

Using Table C: P-value < 0.005

Using technology: Applet/χ^2cdf(lower: 68.57, upper: 1000, df: 6) gives P-value ≈ 0.

> df = (number of rows − 1)(number of columns − 1)

EXAM PREP: FOR PRACTICE, TRY EXERCISE 9.

If there really was no association between country and opinion about speaking English among the residents in Australia, the United Kingdom, and the United States, the probability of getting an association as strong or stronger than the association observed in the sample would be essentially 0. Because the P-value is small, we reject H_0. There is convincing evidence of an association between country and opinion about speaking English.

Putting It All Together: The Chi-Square Test for Association

As with the other significance tests you have learned, be sure to follow the four-step process when performing a chi-square test for association. Like always, you can do the calculations using technology. See the Tech Corner at the end of this section.

EXAMPLE 10.4.3

Does background music influence what customers buy?

Performing a chi-square test for association

PROBLEM

Market researchers suspect that background music may affect the mood and buying behavior of customers. A study compared three randomly assigned treatments in a grocery store: no music, French accordion music, and Italian string music. Under each condition, the researchers recorded whether a customer purchased a bottle of wine and if so, what country the wine came from.[23] Here are the results. Do these data provide convincing evidence at the 1% significance level of an association between type of background music and type of wine purchased?

		Type of background music			
		None	French	Italian	Total
Type of wine purchased	French	30	39	30	99
	Italian	11	1	19	31
	Other	43	35	35	113
	Total	84	75	84	243

SOLUTION

State: H_0: There is no association between type of wine purchased and type of background music at this store.

H_a: There is an association between type of wine purchased and type of background music at this store.

$\alpha = 0.01$

The evidence for H_a is: The proportions who bought French wine (0.357, 0.52, 0.357) are not the same for all three treatments. The same is true for Italian wine (0.131, 0.013, 0.226) and other wine (0.512, 0.467, 0.417).

Plan: Chi-square test for association

Random? Type of background music was randomly assigned. ✓

Large Counts? All expected counts are ≥ 5 (see table). ✓

		Type of background music			Total
		None	French	Italian	
Type of wine purchased	French	34.22	30.56	34.22	99
	Italian	10.72	9.57	10.72	31
	Other	39.06	34.88	39.06	113
	Total	84	75	84	243

Do:

$$\chi^2 = \frac{(30-34.22)^2}{34.22} + \frac{(39-30.56)^2}{30.56} + \cdots + \frac{(35-39.06)^2}{39.06} = 18.28$$

$df = (3-1)(3-1) = 4$

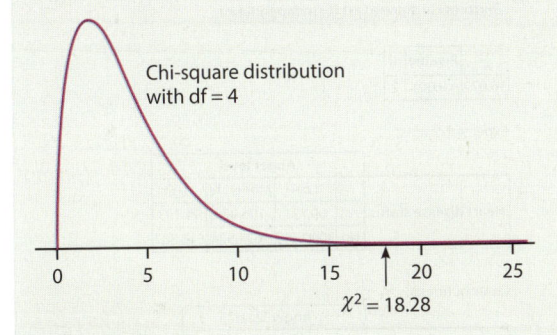

Chi-square distribution with df = 4

$\chi^2 = 18.28$

Using Table C: P-value < 0.005

Using technology: $\chi^2 = 18.28$, df = 4, P-value = 0.0011

Conclude: Because the P-value of 0.0011 is < α = 0.01, we reject H_0. There is convincing evidence of an association between type of wine purchased and type of background music at this store.

State: State the hypotheses, significance level, and evidence for H_a.

Show evidence for H_a with calculations or a graph that demonstrates the distribution of the response variable isn't the same for each category of the explanatory variable.

Plan: Identify the appropriate inference method and check conditions.

Do: If the conditions are met, perform calculations:
- Calculate the test statistic.
- Find the P-value.

Conclude: Make a conclusion about the hypotheses in the context of the problem.

EXAM PREP: FOR PRACTICE, TRY EXERCISE 13.

Because the type of background music was randomly assigned each day and the results of the study were statistically significant, the researchers can conclude that background music does influence what wine shoppers at this store purchase. Furthermore, the association seen in the study makes sense: There were more French bottles purchased than expected when French music was playing (39 > 30.56) and more Italian bottles purchased than expected when Italian music was playing (19 > 10.72).

THINK ABOUT IT

What do you do with raw data? All of the examples so far have presented data in summarized form. That is, all the data sets have been summarized in a two-way table. In practice, we usually start with a data table that has variables in columns and individuals in rows. For example:

Individual	Type of background music	Type of wine purchased
1	French	French
2	French	Other
3	None	French
4	Italian	Other
⋮	⋮	⋮

Although you could try to create a two-way table by hand from raw data, technology is better suited for such a task. The *Two Categorical Variables* applet at www.stapplet.com will create a two-way table from raw data, as will statistical software.

TECH CORNER: Chi-Square Test for Association

You can use technology to perform the calculations for a chi-square test for association. We'll illustrate this process using the anger and heart disease data in Examples 10.3.1–10.3.3.

Applet

1. Go to www.stapplet.com and launch the *Two Categorical Variables* applet.

2. Enter "Anger level" as the explanatory variable. Then, input each of the different levels. Click the + button in the upper right to get an additional column in the table. Then, enter "Heart disease status" as the response variable, along with the different categories. Finally, enter the observed counts in each cell, as shown in the screenshot. Click the Begin analysis button.

Two Categorical Variables

Input data as: Two-way table

Response variable: Heart disease status	Explanatory variable: Anger level		
	Low	Moderate	High
Yes	53	110	27
No	3057	4621	606

Begin analysis | Edit inputs | Reset everything

3. Scroll down to the "Perform Inference" section and click the button to perform the chi-square test for association. (*Note:* The applet calls this a chi-square test for independence.) The test statistic, P-value, degrees of freedom, expected counts, and contributions to the χ^2 statistic will be displayed.

Perform Inference

Perform chi-square test for independence

χ^2	P-value	df
16.077	<0.001	2

Expected counts:

		Anger level		
		Low	Moderate	High
Heart disease status	Yes	69.731	106.076	14.193
	No	3040.269	4624.924	618.807

Contributions:

		Anger level		
		Low	Moderate	High
Heart disease status	Yes	4.014	0.145	11.557
	No	0.092	0.003	0.265

Note: If you have raw data, choose to "Input data as: Raw data" at the top of the applet. Then enter the values of the explanatory variable and the response variable.

TI-83/84

1. Enter the observed counts into matrix [A].
   - Press 2nd X^{-1} (MATRIX), arrow to EDIT, and choose [A].
   - Enter the dimensions of the matrix: 2 × 3.
   - Enter the observed counts in the same order as the two-way table.

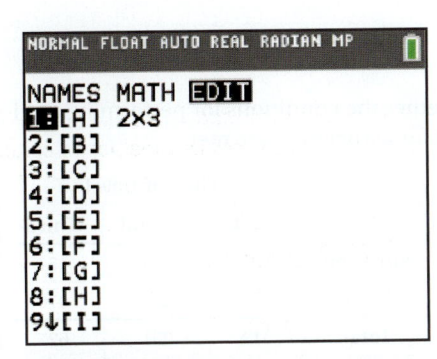

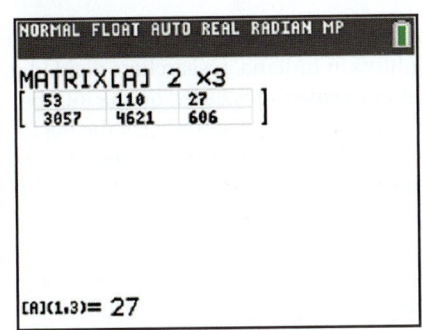

2. Press STAT, arrow to TESTS, and choose χ^2-Test. Then choose [A] for the observed counts and [B] for the expected counts. *Note:* You don't need to enter the expected counts in matrix [B]. This is telling the calculator where to store the expected counts. Highlight "Calculate" and press ENTER.

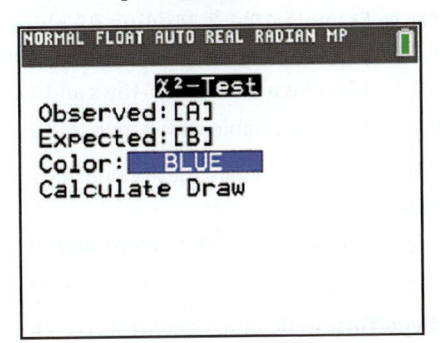

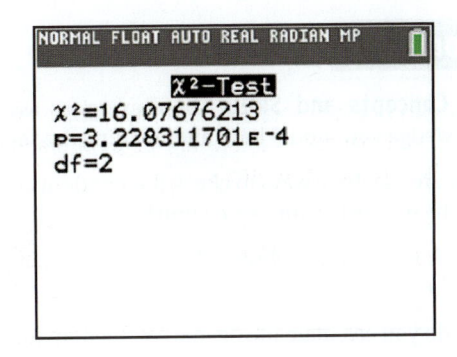

Notes:
- Notice that the *P*-value is expressed in scientific notation. It isn't 3.22; it's 0.000323. *P*-values must be between 0 and 1!
- If you select the "Draw" option, you will get a picture of the appropriate χ^2 distribution with the area of interest shaded and the test statistic and *P*-value labeled.
- To see the expected counts, view matrix [B]. Press 2nd X^{-1} (MATRIX), arrow to [B], and press ENTER. The calculator doesn't have an option to see the contributions to the χ^2 statistic for this test.

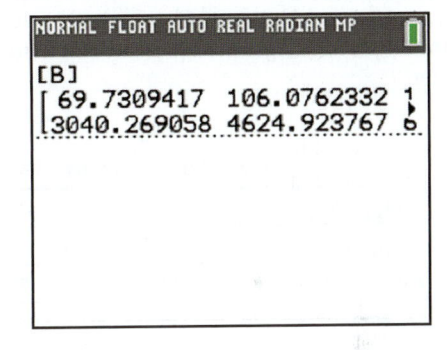

Detailed instructions for using CrunchIt!, Excel, Google Sheets, JMP, Minitab, and R are available in Achieve.

Section 10.4 What Did You Learn?

Review the learning goals from this section. Then practice what you've learned by working through the exercises.

Learning Goal	Example	Exercises
Check conditions for a test about the relationship between two categorical variables.	10.4.1	5–8
Calculate the *P*-value for a test about the relationship between two categorical variables.	10.4.2	9–12
Use the four-step process to perform a chi-square test for association.	10.4.3	13–16

Section 10.4 Exercises

Building Concepts and Skills These exercises assess the basic knowledge you should have after reading the section.

1. What are the three ways to check the Random condition in a chi-square test for association?

2. True/False: Both the observed and expected counts must be at least 5 to meet the Large Counts condition.

3. How do you calculate the degrees of freedom for a chi-square test for association?

4. How do you show evidence for the alternative hypothesis in a chi-square test for association?

Mastering Concepts and Skills These exercises reinforce the learning goals as shown in the examples.

5. **Refer to Example 10.4.1 College Tuition** A random sample of U.S. residents was recently asked the following question: "Would you support or oppose major new spending by the federal government that would help undergraduates pay tuition at public colleges without needing loans?" The two-way table shows the responses, grouped by age.[24] Check whether the conditions for performing a chi-square test for association are met.

		Age				
		18–34	35–49	50–64	65+	Total
Response	Support	91	161	272	332	856
	Oppose	25	74	211	255	565
	Don't know	4	13	20	51	88
	Total	120	248	503	638	1509

6. **Smartphone Ownership** The Pew Research Center regularly conducts surveys about technology use. In early 2021, it surveyed 1502 randomly selected U.S. adults. In addition to demographic questions, each participant was asked if they own a smartphone. The two-way table shows the responses, grouped by race and ethnicity.[25] *Note:* The Pew Research Center describes Black and White adults as those who report being only one race and are not Hispanic. Hispanics can be of any race. Check whether the conditions for performing a chi-square test for association are met.

		Race or ethnicity				
		Black	Hispanic	White	Other	Total
Own a smartphone?	Yes	100	138	896	143	1277
	No	20	24	158	23	225
	Total	120	162	1054	166	1502

7. **Surviving the *Titanic*** In 1912, the luxury liner *Titanic* struck an iceberg and sank. Some passengers got off the ship in lifeboats, but many died. The two-way table gives information about all adult passengers who survived and who died, by class of travel. Check whether the conditions for performing a chi-square test for association are met.

		Class of travel			
		First	Second	Third	Total
Survival status	Survived	197	94	151	442
	Died	122	167	476	765
	Total	319	261	627	1207

8. **Firefighters and CVD** In a long-term study of male firefighters in Indiana, researchers divided the firefighters into five categories based on the number of pushups they could do without stopping at the beginning of the study. At the end of 10 years, researchers recorded whether or not the firefighters had experienced a cardiovascular disease (CVD)–related event since the beginning of the study.[26] The two-way table summarizes the data. Check whether the conditions for performing a chi-square test for association are met.

		Number of pushups					
		0–10	11–20	21–30	31–40	40+	Total
CVD-related event	Yes	8	9	9	10	1	37
	No	67	191	380	275	154	1067
	Total	75	200	389	285	155	1104

9. **Refer to Example 10.4.2 Finding *P*-Values** Calculate the *P*-value in each of the following scenarios.
   (a) $\chi^2 = 17.34$ for a table with 5 rows and 4 columns
   (b) $\chi^2 = 17.34$ for a table with 3 rows and 2 columns

10. **More *P*-Values** Calculate the *P*-value in each of the following scenarios.
    (a) $\chi^2 = 8.03$ for a table with 2 rows and 3 columns
    (b) $\chi^2 = 8.03$ for a table with 5 rows and 3 columns

11. **College Tuition** Refer to Exercise 5. The chi-square test statistic for these data is $\chi^2 = 39.75$. Calculate the *P*-value.

12. **Smartphone Ownership** Refer to Exercise 6. The chi-square test statistic for these data is $\chi^2 = 0.44$. Calculate the *P*-value.

13. **Refer to Example 10.4.3 Shark Repellents** Many people like to surf, but there are dangers involved, including the possibility of a shark attack. In response, companies have produced shark repellents in various forms such as magnetic bands, scented wax, and electrical fields. But do they work? Researchers tested five different products in a clever experiment. They built five surfboards, each utilizing one of the repellents. A sixth board had no repellent but was otherwise identical to the other boards. For each trial, they attached raw tuna about 30 cm below each board, where a surfer's foot might be, and recorded the behavior of sharks that came near the board. The researchers repeated this process nearly 300 times, with the assignment of

repellent determined at random for each trial.[27] The two-way table summarizes the results of the study. Is there convincing evidence at the 5% significance level of an association between repellent and shark response for surfboards like the ones in the study?

		Repellent						
		Control	Wax	Magnet band	Magnet leash	Electric (Rpela)	Electric (Surf+)	Total
Shark response	Bite	48	43	44	47	44	19	245
	No bite	2	7	6	3	5	29	52
	Total	50	50	50	50	49	48	297

14. **Quitting Smoking** It's hard for smokers to quit, but there are a variety of treatments aimed at helping them, including the nicotine patch and bupropion, a drug that is commonly used to fight depression. The two-way table summarizes data from a randomized, double-blind trial that compared four treatments: nicotine patch alone, depression drug alone, nicotine patch with depression drug, and placebo (no treatment). A "success" means that the participant did not smoke for a year following the beginning of the study.[28] Is there convincing evidence at the 5% significance level of an association between treatment and response for subjects like the ones in the study?

		Treatment				
		Nicotine patch	Drug	Patch plus drug	Placebo	Total
Response	Success	40	74	87	25	226
	Failure	204	170	158	135	667
	Total	244	244	245	160	893

15. **Going to the Movies** During the Covid pandemic, attendance at movie theaters was down dramatically. But was attendance at movie theaters going down anyway, considering how many in-home viewing options are available? Gallup polled 1002 randomly selected U.S. adults in 2001 and 1025 randomly selected U.S. adults in 2019 and asked how often these adults went to the movies.[29] The table and graph summarize the results of the two surveys. Is there convincing evidence of an association between year and number of trips to the movie theater? Use $\alpha = 0.01$.

		Year		
		2001	2019	Total
Number of trips to the movie theater	None	261	277	538
	1	70	102	172
	2	110	123	233
	3–5	210	246	456
	6–9	90	82	172
	10+	261	195	456
	Total	1002	1025	2027

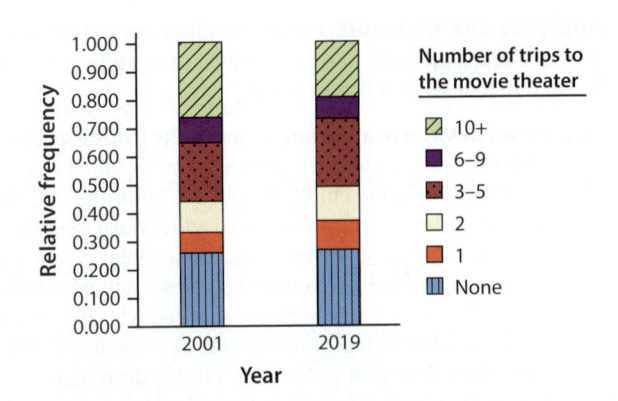

16. **Medieval Bone Fractures** To study the challenges of living in medieval times, archaeologists in Cambridge, England, examined samples of skeletons from three different medieval cemeteries for bone fractures. The cemetery of the All Saints by the Castle parish was a place of burial for the majority of people in the area. The cemetery at the Hospital of St. John the Evangelist was a place of burial for those who were impoverished, chronically ill, or both. The cemetery at the Augustinian friary was used by friars and wealthier townspeople.[30] The table and graph summarize the relationship between cemetery and number of bone fractures. Is there convincing evidence of an association between cemetery and number of fractures? Assume the conditions for inference are met and use $\alpha = 0.10$.

		Cemetery			
		All Saints parish	Hospital of St. John	Augustinian friary	Total
Number of fractures	None	47	113	19	179
	1	21	22	5	48
	2 or more	16	20	4	40
	Total	84	155	28	267

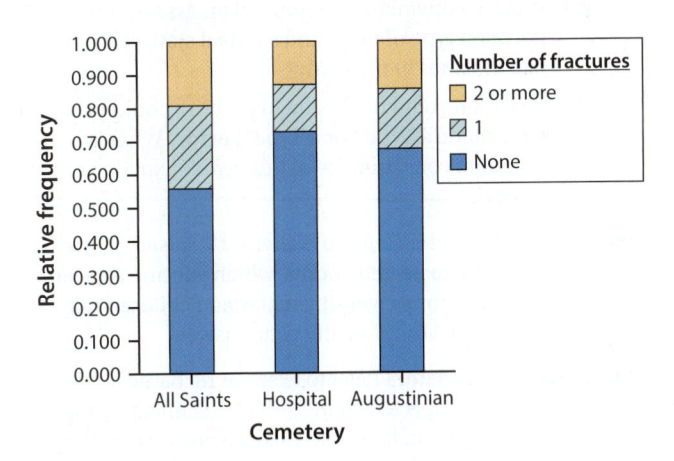

Applying the Concepts These exercises ask you to apply multiple learning goals in a new context or to apply what you learned in this section in a new way.

17. **Malaria Vaccines** Malaria is one of the leading causes of death worldwide. An experiment in Burkina Faso randomly assigned children aged 5–17 months into one of three groups. Group 1 received a low-dose malaria vaccine, Group 2 received a high-dose malaria vaccine, and Group 3 received a control (rabies) vaccine. At the end of 12 months, researchers recorded whether or not each child had at least one episode of clinical malaria.[31] In the low-dose group, 50 of the 146 children had at least one episode of malaria. In the high-dose group, 39 of the 146 children had at least one episode of malaria. In the control group, 106 of the 147 children had at least one episode of malaria.

    (a) Is there convincing evidence of an association between type of vaccine and malaria status for children like the ones in the study?

    (b) Based on your conclusion in part (a), could you have made a Type I or a Type II error? What is a potential consequence of this error? Explain your answer.

18. **Cows and Predators** Cows in Botswana are regularly attacked by predators, including lions. In response, farmers often attempt to kill the predators. However, if predators can be tricked into thinking that the cows have spotted them, attacks might decrease, saving the lives of both cows and predators. Would painting eyes on the rear ends of cows be enough to scare off predators? Researchers randomly assigned cows to one of three treatments: eyes painted on the rear end, two Xs painted on the rear end, or no marks on the rear end. After a month, the researchers counted the number of cows that had been killed by predators in each group.[32] None of the 683 cows with painted eyes was killed by a predator, 4 of the 543 cows with Xs were killed by a predator, and 15 of the 835 cows with no marks were killed by a predator.

    (a) Is there convincing evidence of an association between type of mark and survival status for cows like the ones in the study?

    (b) Based on your conclusion in part (a), could you have made a Type I or a Type II error? What is a potential consequence of this error? Explain your answer.

19. **Malaria Vaccines** Refer to Exercise 17. Based on the observed and expected counts, which vaccine appears most effective for preventing malaria in children like the ones in the study? Justify your answer.

20. **Cows and Predators** Refer to Exercise 18. Based on the observed and expected counts, which treatment(s) appear most effective for helping cows like the ones in the study avoid being killed by predators? Justify your answer.

21. **Tech Python Nests** How is the hatching of water python eggs influenced by the temperature of the snake's nest? Researchers randomly assigned newly laid eggs to one of three water temperatures: hot, neutral, or cold. Hot duplicates the extra warmth provided by the mother python, and cold duplicates the absence of the mother.[33] The variables *Temperature* and *Hatched* are stored in the **Pythons** data set. Is there convincing evidence of an association between temperature and hatching for python eggs like the ones in the study?

22. **Tech Streaming Subscribers** Many households use streaming TV services such as Netflix. Morning Consult asked a random sample of 2200 U.S. adults if they currently subscribe to a streaming service, have subscribed in the past but not now, or have never subscribed. Respondents were also asked for their age.[34] The variables *Age* and *Streaming status* are stored in the **Streaming** data set. Is there convincing evidence of an association between age and streaming status for U.S. adults?

Extending the Concepts These exercises challenge you to explore statistical concepts and methods that go beyond what you learned in this section.

23. **Reflected Glory** In a classic study, researchers investigated the tendency of sports fans to "bask in reflected glory" by associating themselves with winning teams. The researchers called randomly selected students at a major university with a highly ranked football team. Half the students were randomly assigned to answer questions about a recent game the team lost, and the other half were asked about a recent game the team won. If the students were able to correctly identify the winner (showing they were fans of the team), they were asked to describe the game. Researchers recorded if the students identified themselves with the team by their use of the word *we* in the description ("We won the game" versus "They won the game").[35] The table shows the results of the experiment.

		Outcome of game		
		Win	Loss	Total
Association with team	"We"	27	15	42
	"They"	58	68	126
	Total	85	83	168

    (a) Calculate the χ^2 test statistic and P-value for a test of H_0: There is no association between outcome of game and association with team. Note that the conditions are met.

    Because there are only two treatments, each with two outcomes, you can also analyze the data with a two-sample z test for a difference between two proportions. Let p_W = the proportion of students who would associate themselves with the team after a win and p_L = the proportion of students who would associate themselves with the team after a loss.

    (b) Calculate the z test statistic and P-value for a test of $H_0: p_W - p_L = 0$ versus $H_a: p_W - p_L \neq 0$. Note that the conditions are met.

    (c) How do the P-values from part (a) and part (b) compare? How do the values of χ^2 and z^2 compare?

24. **Firefighters and CVD** Refer to Exercise 8. In the study about pushups and cardiovascular disease among firefighters, the Large Counts condition wasn't met when using the original two-way table (shown again here). In cases like this, we can combine two or more columns (or two or more rows) so that the Large Counts condition will be satisfied.

		\multicolumn{6}{c}{Number of pushups}
		0–10	11–20	21–30	31–40	40+	Total
CVD-related event	Yes	8	9	9	10	1	37
	No	67	191	380	275	154	1067
	Total	75	200	389	285	155	1104

(a) Based on your answer to Exercise 8, which two columns should be combined? Explain your reasoning.

(b) After combining the two columns you chose in part (a), show that the Large Counts condition is met.

25. **Mixed Nuts** A company sells deluxe and premium nut mixes, both of which contain only cashews, Brazil nuts, almonds, and peanuts. The premium nuts are much more expensive than the deluxe nuts. A consumer group suspects that the two nut mixes are really the same. To find out, the group selected independent random samples of 20 pounds of each nut mix and recorded the weights of each type of nut in the sample. Here are the data.[36] Explain why we can't use a chi-square test to determine if there is an association between type of nut and type of mix.

		\multicolumn{2}{c}{Type of mix}
		Deluxe	Premium
Type of nut	Cashew	6 lb	5 lb
	Brazil nut	3 lb	4 lb
	Almond	5 lb	6 lb
	Peanut	6 lb	5 lb

Cumulative Review These exercises revisit what you learned in previous sections.

Exercises 26 and 27 refer to the following scenario.
Many chess masters and chess advocates believe that chess play develops general intelligence, analytical skill, and the ability to concentrate. To investigate if chess training increases reading ability, researchers conducted a study. All of the subjects in the study participated in a comprehensive chess program, and their reading performances were measured before and after the program. Summary statistics and boxplots of pre-test scores, post-test scores, and differences in scores are shown here.

Descriptive Statistics: Pretest, Posttest, Post − Pre

Variable	N	Mean	Median	StDev	Min	Max	Q_1	Q_3
Pretest	53	57.70	58.00	17.84	23.00	99.00	44.50	70.50
Posttest	53	63.08	64.00	18.70	28.00	99.00	48.00	76.00
Post − Pre	53	5.38	3.00	13.02	−19.00	42.00	−3.50	14.00

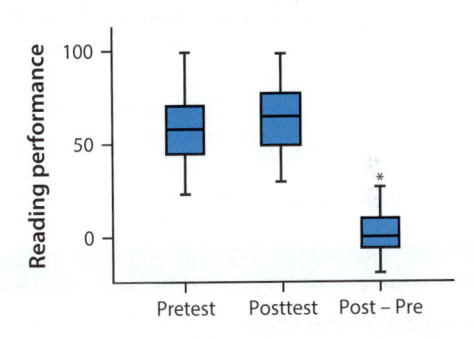

26. **Chess and Reading (3.3)** Compare the distribution of pre-test scores with the distribution of post-test scores. Is there some evidence that chess training helps?

27. **Chess and Reading (1.6, 9.6)** Assume the conditions for inference are met.
(a) Which test, a two-sample t test or a paired t test, would be the appropriate test to use in this context? Explain your reasoning.
(b) If the test in part (a) shows a statistically significant increase in reading scores, could you conclude that participating in the chess program is what caused the increase? Explain your reasoning.

Section 10.5 Introduction to Analysis of Variance

LEARNING GOALS

By the end of this section, you will be able to:
- State the hypotheses for a one-way ANOVA test.
- Compare the variation between groups to the variation within groups.
- Identify the components of an ANOVA table and their relationship to each other.

In Section 9.4, you learned how to perform a two-sample t test for a difference between two means. The next two sections will introduce you to a significance test that is appropriate for comparing the means of *two or more* populations or treatments.

Stating Hypotheses

For a two-sample t test for a difference between two means, the usual null hypothesis is that there is no difference between the two population means. When we are comparing means for three or more populations, the null hypothesis remains that there are no differences among the population means. For example, if we are comparing four population means, the null hypothesis is

$$H_0: \mu_1 = \mu_2 = \mu_3 = \mu_4$$

The alternative hypothesis for a two-sample t test for a difference between two means is either that the two means are not equal (a two-sided alternative) or that one mean is larger than the other (a one-sided alternative). The **one-way analysis of variance (ANOVA) test** only identifies whether there are differences between the population means, not the direction of those differences. Therefore, the alternative hypothesis for this test is

$$H_a: \text{At least one mean is different}$$

Note that the alternative hypothesis does not say that each mean is different from all the others ($\mu_1 \neq \mu_2 \neq \mu_3 \neq \mu_4$). The alternative is true even if just one mean differs from the rest.

A **one-way analysis of variance (ANOVA) test** is a significance test of the null hypothesis that two or more population (or treatment) means are the same.

The test is called "one-way" because we are considering only one explanatory variable. If there were two explanatory variables, it would be called "two-way" ANOVA.

EXAMPLE 10.5.1

Comparing cancer treatments

Stating hypotheses

PROBLEM

Researchers conducted an experiment on 36 mice to compare four possible treatments for cancerous tumors: control, drug only, radiation only, and drug plus radiation.[37] The mice had cells from a human cancer cell line implanted. Once a tumor grew to be large enough, the mice were randomly assigned to one of the four groups. Researchers focused on the amount the tumor grew (in mm³) over the course of the experiment, measured as growth = end size − start size. Do the data provide convincing evidence that the mean tumor growth in mice like the ones in the study is not the same for all four treatments? State appropriate hypotheses for a test that addresses this question.

SOLUTION

$H_0: \mu_{\text{control}} = \mu_{\text{drug}} = \mu_{\text{radiation}} = \mu_{\text{both}}$

H_a: at least one mean is different

where $\mu_{\text{treatment}}$ represents the mean tumor growth in mice like the ones in the study when exposed to the specified treatment.

EXAM PREP: FOR PRACTICE, TRY EXERCISE 5.

Analyzing Variation, Testing Means

The term *analysis of variance* puts the focus on variation. But the hypotheses are about means. How can we make conclusions about population means by analyzing variances?

To illustrate, we selected independent random samples of size 50 from each of four populations and recorded the values of the same quantitative variable for each sample. In all four populations, the variable follows a normal distribution with a standard deviation of $\sigma = 1$. The means of the four population distributions are $\mu_1 = 5$, $\mu_2 = 5$, $\mu_3 = 4.7$, and $\mu_4 = 5.1$. The data from these samples are summarized by the boxplots.

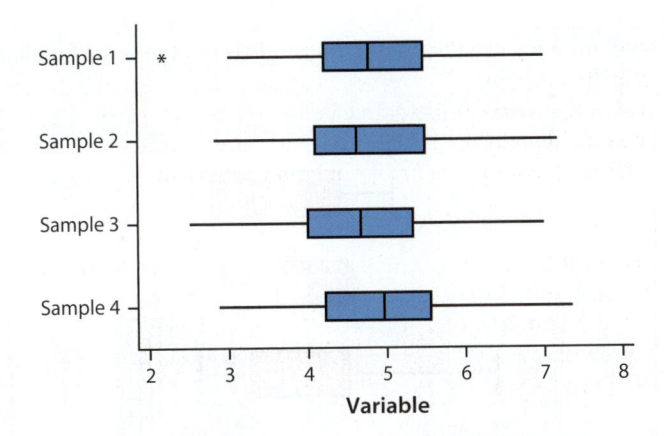

Because the boxplots overlap so much, there is little reason to believe that any of the population means are different. That is, it seems plausible that these samples came from the same population. In general, when there is a lot of overlap in the distributions of the sample data, there isn't strong evidence that at least one of the population means is different.

Now here are four different independent random samples of size 50 from each of four normal populations. The means of the four population distributions are unchanged: $\mu_1 = 5$, $\mu_2 = 5$, $\mu_3 = 4.7$, and $\mu_4 = 5.1$. However, each of these populations has a much smaller standard deviation, $\sigma = 0.10$. The data from these new samples are summarized by the following boxplots.

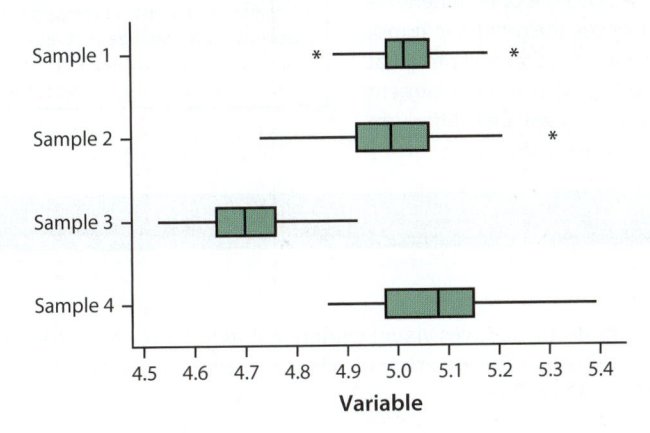

Because the boxplot from Sample 3 overlaps very little with the other boxplots, it seems obvious that Sample 3 comes from a population with a different mean than the other three samples. However, because there is a lot of overlap between the boxplots from Samples 1, 2, and 4, it is difficult to infer that the mean of Population 4 is different than the means of Populations 1 and 2.

When comparing the boxplots in each set, we actually consider two different kinds of variability: the variability *between* the samples (the differences in the sample means) and the variability *within* the samples (the widths of the boxplots). If there is very little overlap, then the variability between the sample means is bigger relative to the amount of variability within the samples. When this happens, we have strong evidence that there are differences among the population means.

If the samples show lots of overlap, then the variability between the sample means is smaller relative to the amount of variability within the samples. In that case, we do not have strong evidence that there are differences among the population means.

EXAMPLE 10.5.2

Comparing cancer treatments

Using variation to compare means

PROBLEM
Again, we consider the experiment on 36 mice to compare four possible treatments for cancerous tumors: control, drug only, radiation only, and drug plus radiation. The boxplots show the distribution of tumor growth (mm^3) for each treatment group.

(continued)

Do the data suggest that the mean tumor growth for mice like the ones in the study is not the same for all four treatments? Explain your answer.

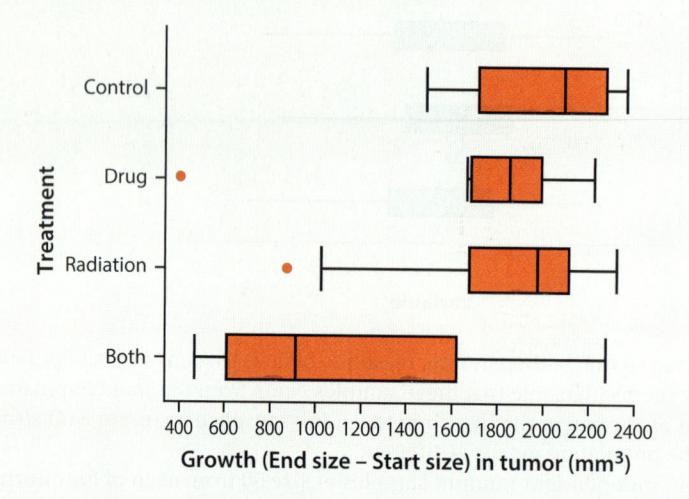

SOLUTION

There is considerable overlap between the control, drug, and radiation groups in the boxplots, suggesting that there may be little or no difference in mean tumor growth for those treatments. However, the growth in tumor size for the group that received both the drug and radiation is centered at a much smaller value and has little overlap with the other three treatment groups. The boxplots suggest that there may be at least one difference among the four treatment means for mice like the ones in the study.

> When we compare more than two groups, the alternative hypothesis encompasses many different possibilities. It could be that only one mean is different. It could be that each mean is different from all other means. Or it could be something in between.

EXAM PREP: FOR PRACTICE, TRY EXERCISE 9.

The graph in Example 10.5.2 gives visual evidence that at least one of the treatment means is different for mice like those in the experiment. The next step is to gather the numerical evidence needed to complete the ANOVA test.

ANOVA Table

Recall that the sample standard deviation is defined to be

$$s_x = \sqrt{\frac{\sum(x_i - \bar{x})^2}{n-1}}$$

If we square the standard deviation, the resulting statistic is called the sample variance:

$$s_x^2 = \frac{\sum(x_i - \bar{x})^2}{n-1}$$

The numerator of the variance $\sum(x_i - \bar{x})^2$ is referred to as the sum of squares total (SST). It measures the variation of the individual values from the mean.

For a one-way ANOVA test, we start by combining all the samples into one larger sample. The SST is computed by using that larger sample with $\bar{x}$ representing the mean of all observations, regardless of sample or treatment. The SST can then be divided into two parts: sum of squares groups (SSG) and sum of squares error (SSE).

$$\text{SST} = \text{SSG} + \text{SSE}$$

The SSG measures the amount of variability between the groups and the SSE measures the amount of variability within the groups.

In a similar fashion, the denominator of the variance, $n-1$, is the total degrees of freedom, df_T. It is the sum of two other degrees of freedom terms, one for groups and one for error:

$$df_T = df_G + df_E$$

If there are I groups of data, then $df_G = I - 1$. Because $df_T = n - 1$, we can determine that $df_E = n - I$.

To compute the final measure of variability between groups and within groups, we find the ratio of the relevant sum of squares (SS) and degrees of freedom (df), and call it the mean square (MS):

$$MSG = \frac{SSG}{df_G} = \frac{SSG}{I-1}$$

$$MSE = \frac{SSE}{df_E} = \frac{SSE}{n-I}$$

To determine whether there is more variability between groups than within groups, we calculate the ratio of the two mean squares:

$$F = \frac{MSG}{MSE}$$

If F is approximately 1, then the variation between the groups is similar to the variation within the groups, suggesting that any observed differences in the sample means may be due to random chance alone. If $F > 1$, the variation between the groups is larger than the variation within the groups and the differences between the sample means may be significant. The larger F is, the stronger the evidence that at least one of the means is different from the other means.

We combine all of this information into a handy table, called the **ANOVA table**.

> The **ANOVA table** summarizes the numerical information needed to conduct a significance test about differences among two or more population or treatment means. The typical format for the table is

Source	df	SS	MS	F
Groups	$df_G = I - 1$	SSG	MSG	F
Error	$df_E = n - I$	SSE	MSE	
Total	$df_T = n - 1$	SST		

> where SST = SSG + SSE, each MS = SS/df, and F = MSG/MSE.

Note that ANOVA tables typically include a P-value to the right of the F statistic. You'll see examples of this in Section 10.6.

EXAMPLE 10.5.3

Comparing cancer treatments

The ANOVA table

PROBLEM

Here is a partial ANOVA table for the experiment on 36 mice comparing four treatments for cancer: control, drug alone, radiation alone, and both drug and radiation.

Source	df	SS	MS	F
Groups			1,150,630	
Error		8,241,477		
Total		11,693,368		

(a) Fill in the empty cells in the ANOVA table.
(b) Is there some evidence that at least one of the treatment means is different? Justify your answer using your calculated value for F.

(continued)

SOLUTION

(a) $df_G = 4 - 1 = 3$
$df_E = 36 - 4 = 32$
$df_T = 36 - 1 = 35$
$11{,}693{,}368 = SSG + 8{,}241{,}477$
$SSG = 3{,}451{,}891$
$MSE = 8{,}241{,}477/32 = 257{,}546$
$F = 1{,}150{,}630/257{,}546 = 4.468$

> $df_G = I - 1$ where I = number of groups
> $df_E = n - I$
> $df_T = n - 1$

> $SST = SSG + SSE$

> $MSE = SSE/df_E$

> $F = MSG/MSE$

The final ANOVA table is:

Source	df	SS	MS	F
Treatments	3	3,451,891	1,150,630	4.468
Error	32	8,241,477	257,546	
Total	35	11,693,368		

(b) Because the F statistic of 4.468 is larger than 1, there is some evidence that the mean tumor growth for mice like the ones in the study is not the same for all four treatments.

EXAM PREP: FOR PRACTICE, TRY EXERCISE 13.

How unusual is it to get an F statistic of 4.468 or larger by chance alone? You will learn more about P-values for one-way ANOVA in Section 10.6. For now, we can estimate the P-value with a simulation. The dotplot in Figure 10.7 shows the F statistic for 100 simulated random assignments, assuming there is no difference in the mean tumor growth for the four treatments.

FIGURE 10.7 The F statistic from each of 100 simulated random assignments assuming there is no difference in the mean tumor growth for the four treatments.

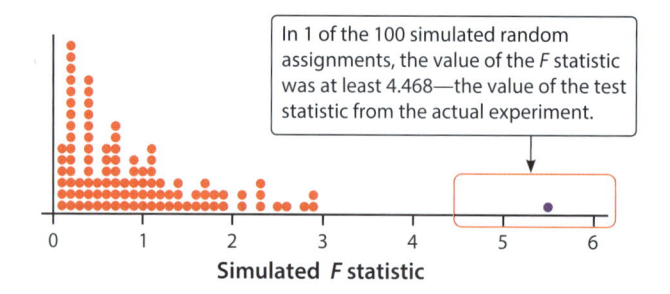

> In 1 of the 100 simulated random assignments, the value of the F statistic was at least 4.468—the value of the test statistic from the actual experiment.

Only 1 of the 100 simulated F statistics was greater than or equal to the observed test statistic of $F = 4.468$, so the P-value is approximately 0.01. Because the P-value of 0.01 is less than $\alpha = 0.05$, we reject H_0. There is convincing evidence that at least one of the mean tumor growth values is different for mice like the ones in the experiment. Which mean (or means) is different than the others? You'll learn how to do a follow-up analysis in Section 10.6.

Section 10.5 What Did You Learn?

Review the learning goals from this section. Then practice what you've learned by working through the exercises.

Learning Goal	Example	Exercises
State the hypotheses for a one-way ANOVA test.	10.5.1	5–8
Compare the variation between groups to the variation within groups.	10.5.2	9–12
Identify the components of an ANOVA table and their relationship to each other.	10.5.3	13–16

Section 10.5 Exercises

Building Concepts and Skills These exercises assess the basic knowledge you should have after reading the section.

1. What is the alternative hypothesis for a one-way ANOVA test?

2. True/False: There is stronger evidence for the alternative hypothesis in a one-way ANOVA test when the amount of variability within groups is larger relative to the amount of variability between groups.

3. What is the relationship between the mean square, the sum of squares, and the degrees of freedom?

4. What is the formula for the F statistic?

Mastering Concepts and Skills These exercises reinforce the learning goals as shown in the examples.

5. **Refer to Example 10.5.1 Effects of Logging** How does logging in a tropical rainforest affect the forest in later years? Researchers compared randomly selected forest plots in Borneo that had never been logged with similar randomly selected plots nearby that had been logged 1 year earlier and 8 years earlier. They counted the number of trees on each plot.[38] Do the data provide convincing evidence that the mean number of trees per plot is not the same for all three levels of logging? State appropriate hypotheses for a test that addresses this question.

6. **Sugary Cereals** Many favorite kids' cereals contain lots of sugar. Do supermarkets deliberately place more sugary cereals on shelves where kids can better see them? To find out, researchers went to the cereal aisle in a large supermarket. They selected independent random samples of cereal from the top shelf, middle shelf, and bottom shelf. For each cereal, they recorded the sugar content (grams per serving).[39] Do the data provide convincing evidence that the mean amount of sugar in cereals is not the same for all three shelves at the grocery store? State appropriate hypotheses for a test that addresses this question.

7. **Frozen Desserts** Researchers were interested in the nutritional attributes of several categories of frozen desserts. They selected a sample of 65 different desserts, recorded the amount of saturated fat per serving (g), and classified each dessert as one of the following: ice cream, frozen yogurt, gelato, sorbet, or dairy free.[40] Do the data provide convincing evidence that the mean amount of saturated fat in frozen desserts is not the same for all five types of frozen desserts? State appropriate hypotheses for a test that addresses this question.

8. **Glass Forensics** Researchers were interested in whether the type of glass can be determined by the various characteristics of the glass. This research was motivated by criminal investigations — glass can be used as evidence if it can be correctly identified. The study used 214 randomly selected specimens of glass of six different types: building windows that were float processed, building windows that were not float processed, vehicle windows that were float processed, containers, tableware, and headlamps. One characteristic measured was the refractive index.[41] Do the data provide convincing evidence that the mean refractive index is not the same for all six types of glass? State appropriate hypotheses for a test that addresses this question.

9. **Refer to Example 10.5.2 Effects of Logging** Refer to Exercise 5. Here are dotplots of the number of trees for the three samples. Do the data suggest that the mean number of trees is not the same for all three levels of logging in Borneo? Explain your answer.

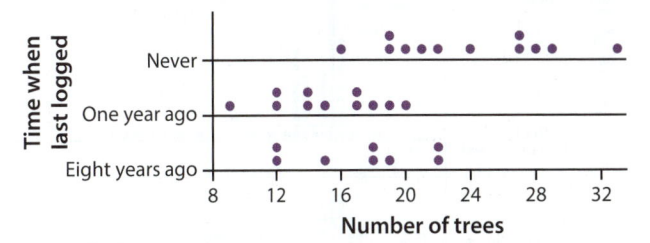

10. **Sugary Cereals** Refer to Exercise 6. Here are boxplots of the amount of sugar (in grams) for the three samples. Do the data suggest that the mean amount of sugar in cereals is not the same for all three shelves at this grocery store? Explain your answer.

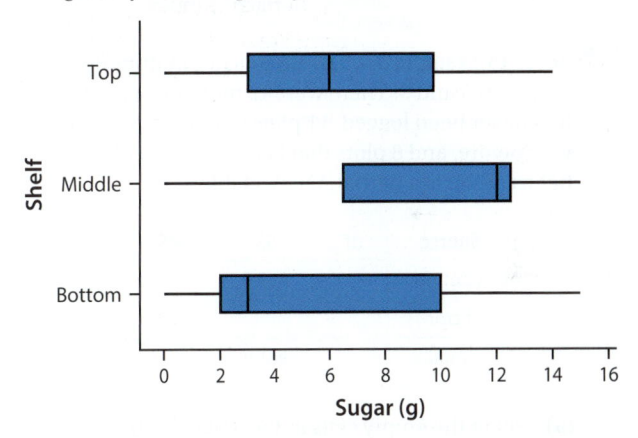

11. **Frozen Desserts** Refer to Exercise 7. Here are boxplots of the amount of saturated fat (g) for the five types of frozen desserts. Do the data suggest that the mean amount of saturated fat is not the same for all five types of desserts? Explain your answer.

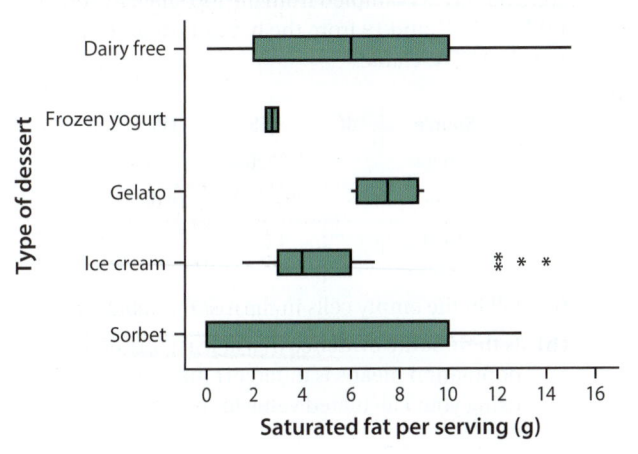

12. **Glass Forensics** Refer to Exercise 8. Dotplots of the refractive index values for the six types of glass are given. Do the data suggest that the mean refractive index is not the same for all six types of glass? Explain your answer.

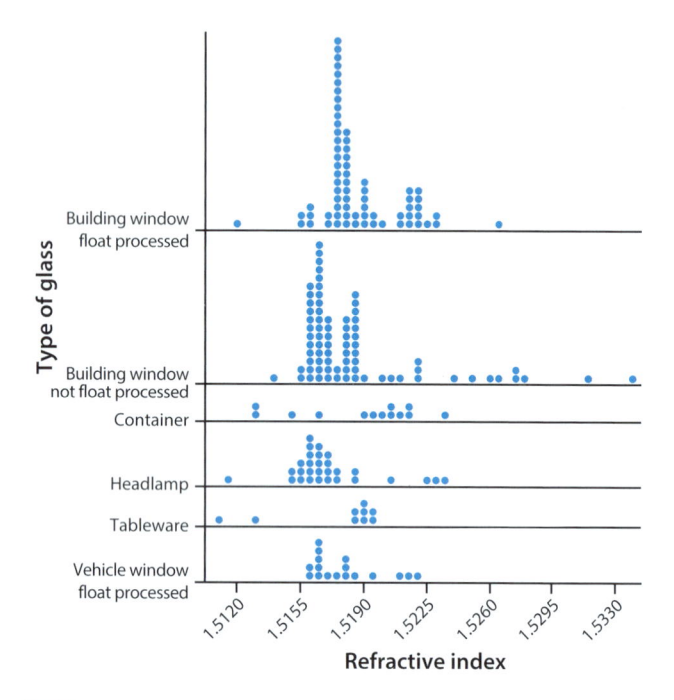

13. **Refer to Example 10.5.3** **Effects of Logging** Refer to Exercises 5 and 9. There were 12 plots surveyed that had never been logged, 11 plots that were logged one year before, and 8 plots that had been logged 8 years before. Here is a partial ANOVA table:

Source	df	SS	MS	F
Groups		455.6		
Error			18.05	
Total		961.0		

(a) Fill in the empty cells in the ANOVA table.
(b) Is there some evidence that at least one of the population means is different? Justify your answer using your calculated value for F.

14. **Sugary Cereals** Refer to Exercises 6 and 10. There were 36 cereals sampled from the top shelf, 21 from the middle shelf, and 19 from the bottom shelf. Here is a partial ANOVA table:

Source	df	SS	MS	F
Groups		220.2		
Error			16.68	
Total		1437.9		

(a) Fill in the empty cells in the ANOVA table.
(b) Is there some evidence that at least one of the population means is different? Justify your answer using your calculated value for F.

15. **Frozen Desserts** Refer to Exercises 7 and 11. The data contained 19 dairy-free desserts, 2 frozen yogurts, 4 gelatos, 31 types of ice cream, and 9 types of sorbet. Here is a partial ANOVA table:

Source	df	SS	MS	F
Groups			21.17	
Error		970.00		
Total				

(a) Fill in the empty cells in the ANOVA table.
(b) Is there some evidence that at least one of the population means is different? Justify your answer using your calculated value for F.

16. **Glass Forensics** Refer to Exercises 8 and 12. The data contained 70 specimens of float-processed building window glass, 76 specimens of non-float-processed building window glass, 13 specimens of container glass, 29 specimens of headlamp glass, 9 specimens of tableware glass, and 17 specimens of float-processed vehicle window glass. Here is a partial ANOVA table:

Source	df	SS	MS	F
Groups			0.000015	
Error		0.001891		
Total				

(a) Fill in the empty cells in the ANOVA table.
(b) Is there some evidence that at least one of the population means is different? Justify your answer using your calculated value for F.

Applying the Concepts These exercises ask you to apply multiple learning goals in a new context or to apply what you learned in this section in a new way.

17. **Applying Fabric Dye** To compare three different methods of dying cloth, researchers took 24 pieces of fabric made of ramie and randomly divided them into three groups. For each group, they used either Method A, Method B, or Method C to apply the same amount of blue dye. Next, they used a colorimeter to measure the lightness of the color on a scale from 0 to 100 where 0 is black and 100 is white. Do the data provide convincing evidence that the mean amount of lightness is not the same for all three methods?

(a) State appropriate hypotheses for a test that addresses this question.
(b) Here are dotplots of the lightness value for the three methods. Do the data suggest that the mean amount of lightness is not the same for all three methods? Explain your answer.

(c) A partial ANOVA table is given here. Fill in the empty cells.

Source	df	SS	MS	F
Groups		1.788		
Error			0.09383	
Total				

(d) Is there some evidence that at least one of the population means is different? Justify your answer using your calculated value for F.

18. **Chamois Absorption** Chamois leather is very absorbent and is often used in cleaning. A researcher wondered whether water temperature affects how much water the chamois leather absorbs. The researcher cut out ninety 3-inch-by-3-inch squares of chamois leather and randomly assigned 30 to hot water, 30 to room-temperature water, and 30 to cold water. Each square was soaked in the assigned water and then wrung out over a graduated cylinder to measure how much water it absorbed.[42] Do the data provide convincing evidence that the mean amount of water absorbed is not the same for all three temperatures?

    (a) State appropriate hypotheses for a test that addresses this question.

    (b) Here are boxplots of the amount of water absorbed for the three temperatures. Do the data suggest that the mean amount of water absorbed is not the same for all three temperatures? Explain your answer.

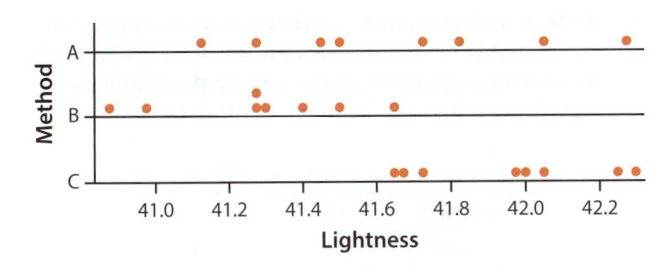

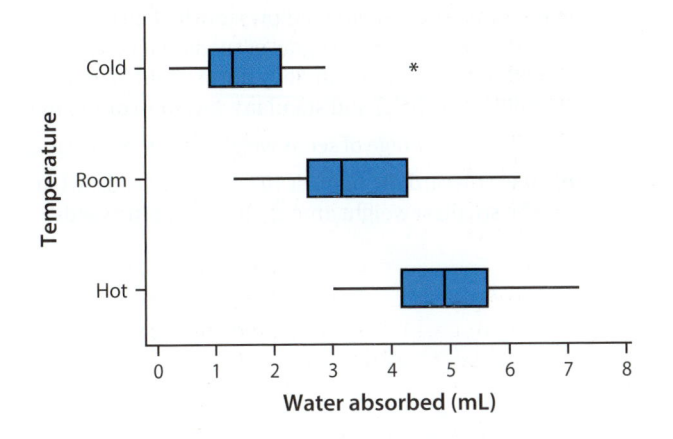

(c) A partial ANOVA table is given here. Fill in the empty cells.

Source	df	SS	MS	F
Groups		182.8		
Error			1.159	
Total				

(d) Is there some evidence that at least one of the population means is different? Justify your answer using your calculated value for F.

19. **More Significant** The boxplots summarize the results of an experiment comparing three treatments.

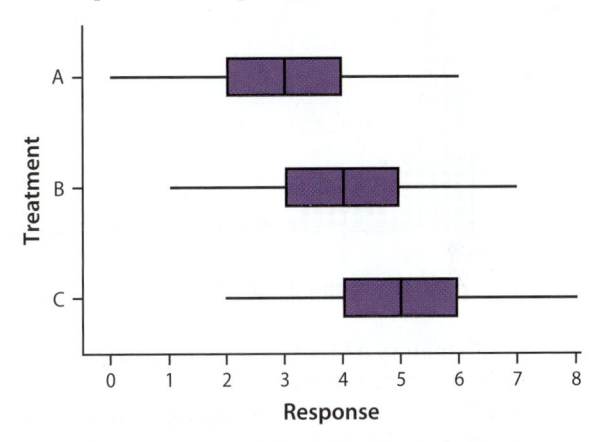

    (a) Keeping the means the same, draw three new boxplots that would result in a smaller P-value than the original data.

    (b) Keeping the variability of the original boxplots the same, draw three new boxplots that would result in a smaller P-value than the original data.

20. **Less Significant** The boxplots summarize the results of an experiment comparing three treatments.

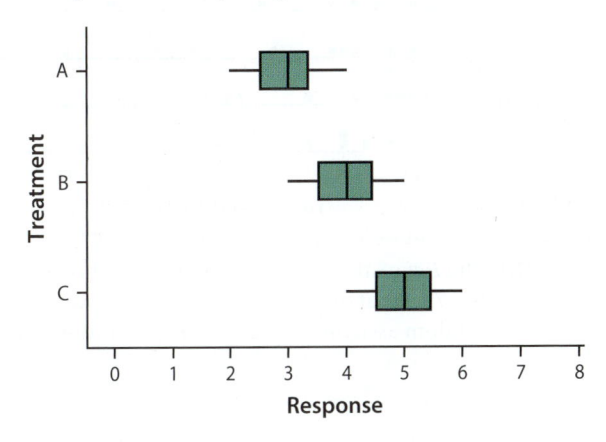

    (a) Keeping the means the same, draw three new boxplots that would result in a larger P-value than the original data.

    (b) Keeping the variability of the original boxplots the same, draw three new boxplots that would result in a larger P-value than the original data.

Extending the Concepts These exercises challenge you to explore statistical concepts and methods that go beyond what you learned in this section.

21. **Effects of Logging** In Exercises 5, 9, and 13, you compared the mean number of trees on 12 plots that hadn't previously been logged, 11 plots that had been logged a year previously, and 8 plots that had been logged 8 years previously. In Exercise 13, you calculated $F = 12.62$. The dotplot shows the results of 200 random samples of 12, 11, and 8 plots, assuming the null hypothesis is true. For each random sample, the F statistic was calculated. Based on the actual F statistic of 12.62 and the results of the simulation, what conclusion would you make?

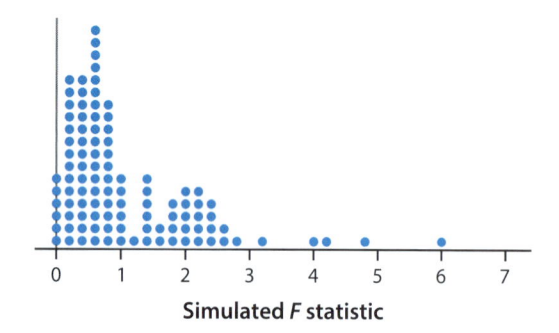

22. **Confirmatory ANOVA** Sometimes ANOVA is used to confirm that an assumption of equal means is reasonable. The experiment described in Example 10.5.1 started with the researchers randomly assigning the mice to the four possible treatments. The purpose of the random allocation was to create groups that were as similar as possible at the beginning of the experiment.

    (a) Here are dotplots of the tumor start sizes. Based on the dotplots, is it plausible that these four samples came from populations with the same mean start size? Explain your answer.

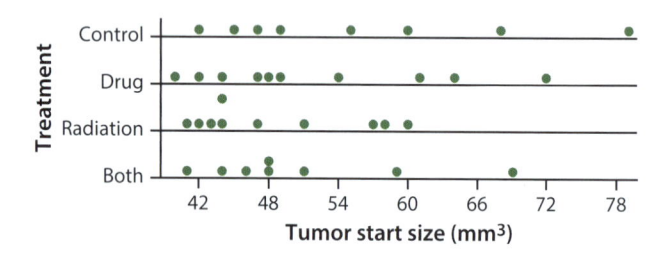

    (b) The F statistic in the ANOVA table for this analysis is 0.73. What does this suggest about how well the random assignment worked? Explain your answer.

23. **ANOVA with Summary Statistics** It is possible to construct an ANOVA table using only the means, standard deviations, and sample sizes. Here are the summary statistics for three independent random samples:

	Mean	SD	n
Sample 1	118	27	16
Sample 2	121	29	14
Sample 3	125	32	18

    (a) Calculate the overall mean $\bar{x}$ using a weighted average of the sample means:

    $$\bar{x} = \frac{n_1 \bar{x}_1 + n_2 \bar{x}_2 + n_3 \bar{x}_3}{n_1 + n_2 + n_3}$$

    (b) Calculate the SSG by finding the sum of the weighted squared deviations of the sample means from the overall mean:

    $$SSG = n_1(\bar{x}_1 - \bar{x})^2 + n_2(\bar{x}_2 - \bar{x})^2 + n_3(\bar{x}_3 - \bar{x})^2$$

    (c) Calculate the SSE by finding the sum of the weighted sample variances:

    $$SSE = (n_1 - 1)s_1^2 + (n_2 - 1)s_2^2 + (n_3 - 1)s_3^2$$

    (d) Construct the complete ANOVA table.

Cumulative Review These exercises revisit what you learned in previous sections.

24. **Ravenous Caterpillars (7.2, 7.4)** In its caterpillar stage, the gypsy moth does considerable damage to hardwoods in the eastern United States. A forester selects a random sample of 200 trees in a certain forest and finds that 124 of them show damage from gypsy moth caterpillars.

    (a) Construct and interpret a 99% confidence interval for the proportion of all trees in this forest that show damage from gypsy moth caterpillars.

    (b) Explain what 99% confidence means in the context of this problem.

25. **Seed Weights (6.2, 6.3)** Biological measurements on the same species often follow a normal distribution quite closely. The weights of seeds of a variety of winged bean are approximately normal with a mean of 525 milligrams (mg) and standard deviation of 110 mg.

    (a) What percentage of seeds weigh more than 500 mg?

    (b) If we discard the lightest 10% of these seeds, what is the smallest weight among the remaining seeds?

Section 10.6 One-Way Analysis of Variance

LEARNING GOALS

By the end of this section, you will be able to:
- State and check the conditions for a one-way ANOVA test.
- Find the *P*-value and use the four-step process to perform a one-way ANOVA test.
- Compare each pair of groups using Fisher's LSD intervals when a one-way ANOVA test is significant.

In Section 10.5, you learned how to state hypotheses for a test comparing means from two or more populations or treatments. You also learned how Analysis of Variance (ANOVA) measures the amount of evidence for the alternative hypothesis that at least one population (treatment) mean is different than the others. In this section, you will learn the details of carrying out a one-way ANOVA test and how to do a follow-up analysis.

Conditions for a One-Way ANOVA Test

Just like for the other statistical tests you have learned about, certain conditions must be met for ANOVA test conclusions to be valid. The first two conditions, Random and Normal/Large Sample, should be familiar from the two-sample *t* test for a difference between two means covered in Section 9.4. The third condition, Equal SD, states that the population distributions have the same standard deviation. Due to sampling variability, the standard deviations in the samples are unlikely to be equal. However, the one-way ANOVA procedure works well as long as the largest sample standard deviation is less than twice the smallest sample standard deviation.

HOW TO Check the Conditions for a One-Way ANOVA Test

To perform a one-way ANOVA test, check that the following conditions are met:
- **Random:** The data come from independent random samples from the populations of interest or from groups in a randomized experiment.
- **Normal/Large Sample:** For each sample, the data come from a normally distributed population or the sample size is large ($n \geq 30$). When the sample size is small and the shape of the population distribution is unknown, a graph of the sample data shows no strong skewness or outliers.
- **Equal SD:** The largest sample standard deviation is less than twice the smallest sample standard deviation.

If the Normal/Large Sample condition is not met, consider using the nonparametric Kruskal-Wallis test described in Section 13.4.

EXAMPLE 10.6.1

Comparing cancer treatments

Checking conditions

PROBLEM

In Section 10.5, we considered data from an experiment comparing four possible treatments for cancerous tumors in mice: control, drug only, radiation only, and drug plus radiation. Recall that the mice had cells from a human cancer cell line implanted. Once a tumor grew to be large enough, the mice were randomly assigned to one of the four groups. The response variable is the amount of growth in the tumor (mm^3) during the experiment. Use the data to check the conditions for a one-way ANOVA test.

(continued)

Control	1522.2	1484.3	2360.9	1896.5	2108.1	2065.1	2295.8	2246.5		
Drug	1949.2	1986.1	1897.0	1678.3	1661.9	1800.0	401.7	1720.5	2221.8	2042.9
Radiation	2014.6	2320.3	1671.5	1902.0	2110.7	2108.4	1926.2	870.1	1024.3	2130.5
Drug + radiation	1862.8	497.5	762.5	1062.2	2273.8	716.5	464.0	1372.5		

SOLUTION

- Random? The mice were randomly assigned to treatments. ✓
- Normal/Large Sample? The normal probability plots for the control, radiation, and drug plus radiation groups are roughly linear. However, the normal probability plot for the drug group has an obvious outlier. The Normal/Large Sample condition is not met.

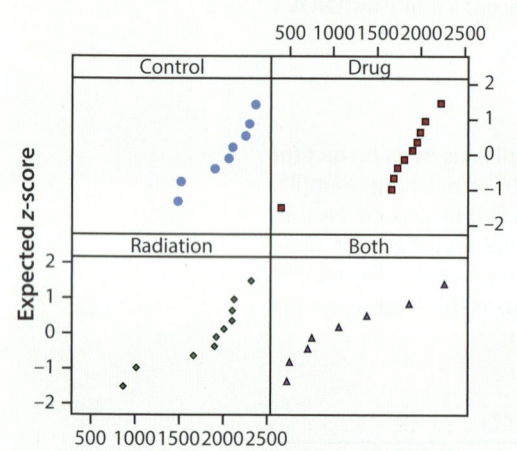

> When using normal probability plots to assess normality, look for roughly linear patterns. Strong curvature or outliers suggest lack of normality. You can also use dotplots, histograms, or boxplots to look for strong skewness or outliers.

[Normal probability plots for Control, Drug, Radiation, and Both; x-axis: Growth (End size − Start size) in tumor (mm^3); y-axis: Expected z-score]

- Equal SD? The sample standard deviations are $s_{Control} = 338$, $s_{Drug} = 501$, $s_{Radiation} = 486$, and $s_{Both} = 661$, and $661 < 2(338) = 676$. ✓

> To check the Equal SD condition, show that the largest SD is less than 2 times the smallest SD.

EXAM PREP: FOR PRACTICE, TRY EXERCISE 5.

P-Values and the One-Way ANOVA Test

In Section 10.5, we introduced $F = MSG/MSE$ as the relevant statistic to consider in the one-way ANOVA setting. When the conditions are met, the F statistic can be modeled by an F distribution with numerator degrees of freedom $df_G = I - 1$ and denominator degrees of freedom $df_E = n - I$. While you can look up the P-value in an F table, the P-value is usually given in the ANOVA table provided by technology. See the Tech Corner at the end of this section.

EXAMPLE 10.6.2

Aphid honeydew

Performing a one-way ANOVA test

PROBLEM

Researchers studying a small type of insect called an aphid were interested in the relationship between aphids and different types of host plants.[43] As part of the study, the researchers measured the amount of "honeydew" (aphid waste — not related to honeydew melons) produced by aphids on independent random samples of three different types of host plants (*Trifolium*, *Pisum*, and *Medicago*). Is there convincing evidence at the $\alpha = 0.05$ significance level of at least one difference in the mean amount of honeydew produced by aphids on these three types of plants?

Trifolium	1.08	2.21	2.63	1.63	3.51	2.53	2.92	0.98	2.39	2.05	0.36	0.74	1.00	0.79	0.55	1.05	1.46	1.09
Pisum	1.03	2.48	1.31	4.33	2.33	2.68	3.34	2.46	2.74	5.83	0.89	1.69	1.52	2.87	3.54	3.76	1.96	
Medicago	1.06	0.88	1.87	1.42	0.39	1.20	0.39	1.41	0.88	0.59	1.08	2.71	1.63	2.98	2.94	2.43	2.34	

SOLUTION

State: $H_0: \mu_{Trifolium} = \mu_{Pisum} = \mu_{Medicago}$

H_a: At least one mean is different

where μ_{Type} is the mean amount of honeydew produced by aphids on the specified type of host plant.

$\alpha = 0.05$

The evidence for H_a is: The dotplots show that the *Pisum* group is somewhat shifted to the right, indicating that a larger mean amount of honeydew may be produced by aphids on this plant.

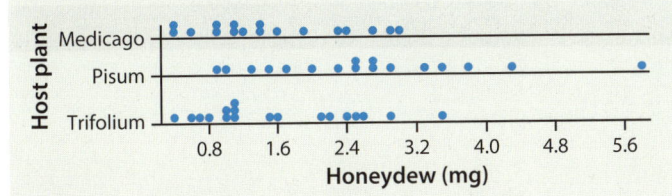

> **State:** State the hypotheses, parameters, significance level, and evidence for H_a.

Plan: One-way ANOVA test for differences in means.
- Random? Three independent random samples of host plants. ✓
- Normal/Large Sample? The dotplots do not show strong skewness or outliers. ✓
- Equal SD? The sample standard deviations are $s_{Trifolium} = 0.912$, $s_{Pisum} = 1.271$, and $s_{Medicago} = 0.865$, and $1.271 < 2(0.865) = 1.73$. ✓

Do: Using technology:

> **Plan:** Identify the appropriate inference method and check conditions.

Source	df	SS	MS	F	P-value
Groups	2	12.82	6.408	6.04	0.005
Error	49	51.95	1.060		
Total	51	64.77			

> **Do:** If the conditions are met, perform calculations:
> - Calculate the test statistic.
> - Find the *P*-value.

The ANOVA table gives $F = 6.04$ and *P*-value = 0.005.

Conclude: Because the *P*-value of $0.005 < \alpha = 0.05$, we reject H_0. There is convincing evidence of at least one difference in the mean amount of honeydew produced by aphids on the three different types of host plants.

> **Conclude:** Make a conclusion about the hypotheses in the context of the problem.

EXAM PREP: FOR PRACTICE, TRY EXERCISE 9.

In this example, the data provided convincing evidence of at least one difference in the mean amount of honeydew produced by aphids on these three types of plants. But this conclusion seems incomplete. Which population means are different? Is one type of host plant different in terms of average honeydew production by aphids, but the other two are the same? If so, which one is different? Or are all three types of host plants different from each other with respect to average honeydew production by aphids? Answering these questions requires a follow-up analysis.

Fisher's Least Significant Difference (LSD) Intervals

There are several methods for performing a follow-up analysis when the results of an ANOVA test are significant. Here we present one of those methods: *Fisher's least significant difference (LSD) intervals*.

HOW TO — Compute Fisher's Least Significant Difference (LSD) Intervals

When we find the one-way ANOVA test result to be significant, we compute Fisher's LSD intervals as follows:

1. Find $s_p^2 = \text{MSE}$, where the MSE comes from the ANOVA table.
2. Calculate t^* using df_E as the degrees of freedom and the stated confidence level (e.g., 95%).
3. Use the formula $(\bar{x}_1 - \bar{x}_2) \pm t^* \sqrt{\dfrac{s_p^2}{n_1} + \dfrac{s_p^2}{n_2}}$ to compute the LSD interval for each pair of means.

Any interval that does not contain 0 provides convincing evidence that the corresponding population means are different.

Fisher's LSD intervals are appropriate only when the results of the ANOVA test are significant. Otherwise, we run into the problem of multiple comparisons described in Section 8.6. That is, each confidence interval on its own has the stated confidence level (e.g., 95%), but taken together, the set of intervals has an overall confidence level that is less than the stated confidence level.

EXAMPLE 10.6.3

Aphid honeydew

Computing Fisher's LSD intervals

PROBLEM

In Example 10.6.2, we found convincing evidence of at least one difference in the mean honeydew output for aphids on three different types of host plants. Compute the appropriate 95% Fisher's LSD intervals and report which differences are significant.

Source	df	SS	MS	F	P-value
Groups	2	12.82	6.408	6.04	0.005
Error	49	51.95	1.060		
Total	51	64.77			

SOLUTION

$s_p^2 = 1.06$

$t^* = 2.010$

Find t^* using a t distribution with $df_E = 49$ degrees of freedom and 95% confidence.

$\bar{x}_{Trifolium} = 1.609$, $\bar{x}_{Pisum} = 2.633$, $\bar{x}_{Medicago} = 1.541$

$n_{Trifolium} = 18$, $n_{Pisum} = 17$, $n_{Medicago} = 17$

Trifolium vs. Pisum:

$(1.609 - 2.633) \pm 2.010 \sqrt{\dfrac{1.06}{18} + \dfrac{1.06}{17}} = -1.024 \pm 0.700 = (-1.724, -0.324)$

$(\bar{x}_1 - \bar{x}_2) \pm t^* \sqrt{\dfrac{s_p^2}{n_1} + \dfrac{s_p^2}{n_2}}$, where $s_p^2 = \text{MSE}$

Trifolium vs. Medicago:

$(1.609 - 1.541) \pm 2.010 \sqrt{\dfrac{1.06}{18} + \dfrac{1.06}{17}} = 0.068 \pm 0.700 = (-0.632, 0.768)$

Pisum vs. Medicago:

$(2.633 - 1.541) \pm 2.010 \sqrt{\dfrac{1.06}{17} + \dfrac{1.06}{17}} = 1.092 \pm 0.710 = (0.382, 1.802)$

Because 0 is included in the second interval, there isn't convincing evidence that the mean honeydew production is different for *Trifolium* and *Medicago*. However, because 0 is not included in the first or third interval, there is convincing evidence that the mean honeydew production is different for *Pisum* when compared to either *Trifolium* or *Medicago*. In fact, the intervals suggest that the mean for *Pisum* is greater than the mean for *Trifolium* and the mean for *Medicago*.

EXAM PREP: FOR PRACTICE, TRY EXERCISE 15.

The phrase "least significant difference" actually refers to the margin of error in the LSD interval. When a difference in means is greater than the corresponding margin of error, the difference is significant. When Fisher first proposed this idea for a follow-up analysis to ANOVA in 1935, the concept of a confidence interval had not yet been introduced. Even after confidence intervals were introduced in 1937, the name "Fisher's LSD" stuck around.

TECH CORNER One-Way ANOVA

You can use technology to display the data graphically, calculate the summary statistics for each group, produce the ANOVA table, and calculate Fisher's LSD intervals. We'll illustrate this process with the *Aphid* data set from Examples 10.6.2 and 10.6.3.

Applet

1. Go to www.stapplet.com and launch the *One Quantitative Variable, Multiple Groups* applet.

2. Enter the variable name and choose "Raw data" from the Input menu. Then enter the name and data for each group. Click the "Add group" button to add the third group.

One Quantitative Variable, Multiple Groups
Variable name: Honeydew
Input: Raw data

Group	Name	Input data separated by commas or spaces.
1	Trifolium	1.08 2.21 2.63 1.63 3.51 2.53 2.92 0.98 2.39 2.0
2	Pisum	1.03 2.48 1.31 4.33 2.33 2.68 3.34 2.46 2.74 5.8
3	Medicago	1.06 0.88 1.87 1.42 0.39 1.20 0.39 1.41 0.88 0.5

Add group

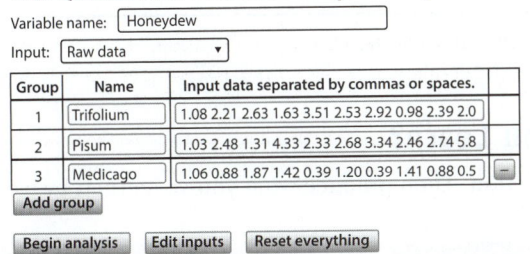

Begin analysis Edit inputs Reset everything

3. Click the Begin analysis button. The result is a set of plots (dotplots by default, but this can be changed to histograms, boxplots, or normal probability plots), and the summary statistics for each group.

Graph Distributions
Graph type: Dotplot

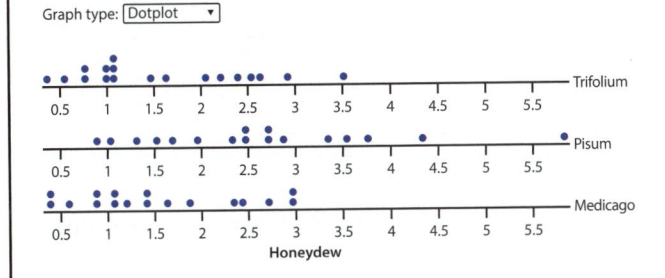

Summary Statistics

Group Name	n	mean	SD	min	Q_1	med	Q_3	max
1: Trifolium	18	1.609	0.912	0.36	0.98	1.275	2.39	3.51
2: Pisum	17	2.633	1.271	0.89	1.605	2.48	3.44	5.83
3: Medicago	17	1.541	0.865	0.39	0.88	1.41	2.385	2.98

Export summary statistics

4. Choose One-way ANOVA as the inference procedure and check the box to include follow-up analysis (Fisher's LSD intervals). Use 95% for the confidence level. Click the Perform inference button.

Perform Inference
Inference procedure: One-way ANOVA ☑ Include follow-up analysis using a confidence level of 95 %
Perform inference

Source	df	Sum of Squares	Mean Square	F-value	P-value
Group	2	12.816	6.408	6.044	0.005
Error	49	51.953	1.06		
Total	51	64.77			

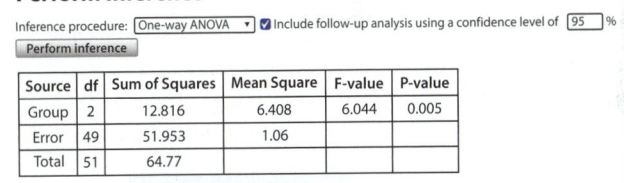

Groups	Fisher's LSD Interval
Group 1 - Group 2	(−1.723, −0.324)
Group 1 - Group 3	(−0.632, 0.768)
Group 2 - Group 3	(0.382, 1.802)

Note: You can also enter raw data "by variable" if you have one column of honeydew values and one column of plant values instead of three columns of honeydew values.

TI-83/84

1. Enter the data for *Trifolium* into list L_1, the data for *Pisum* into list L_2, and the data for *Medicago* into list L_3.

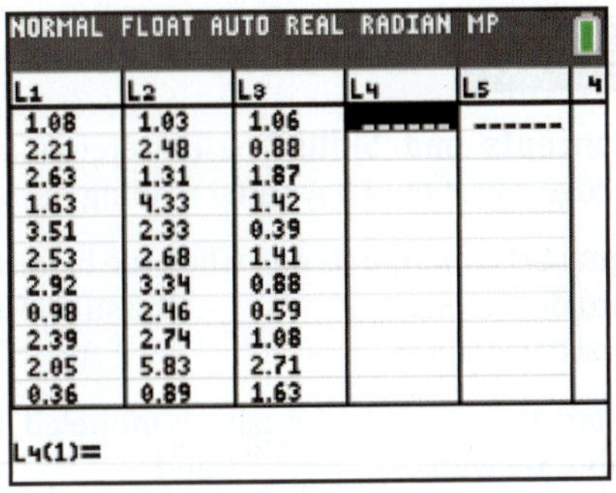

(*continued*)

2. Press STAT, arrow to Tests, and choose ANOVA(. Then enter the lists that contain the data and press ENTER. The output includes all the components of the ANOVA table, along with the (pooled) standard deviation to help calculate Fisher's LSD intervals. Note that the calculator doesn't have an option to calculate these intervals.

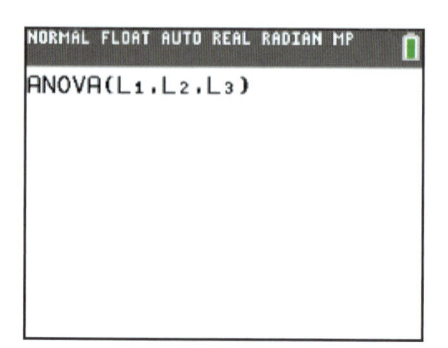

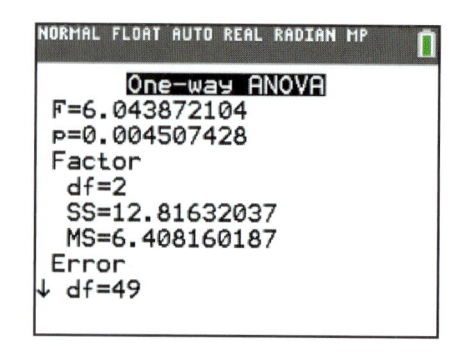

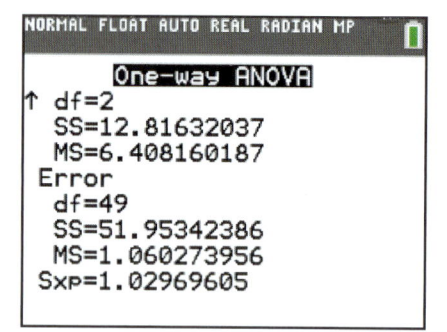

```
NORMAL FLOAT AUTO REAL RADIAN MP
ANOVA(L1,L2,L3)
```

```
NORMAL FLOAT AUTO REAL RADIAN MP
        One-way ANOVA
F=6.043872104
p=0.004507428
Factor
  df=2
  SS=12.81632037
  MS=6.408160187
Error
↓ df=49
```

```
NORMAL FLOAT AUTO REAL RADIAN MP
        One-way ANOVA
↑ df=2
  SS=12.81632037
  MS=6.408160187
Error
  df=49
  SS=51.95342386
  MS=1.060273956
Sxp=1.02969605
```

Detailed instructions for using CrunchIt!, Excel, Google Sheets, JMP, Minitab, and R are available in Achieve.

Section 10.6 What Did You Learn?

Review the learning goals from this section. Then practice what you've learned by working through the exercises.

Learning Goal	Examples	Exercises
State and check the conditions for a one-way ANOVA test.	10.6.1	5–8
Find the P-value and use the four-step process to perform a one-way ANOVA test.	10.6.2	9–14
Compare each pair of groups using Fisher's LSD intervals when a one-way ANOVA test is significant.	10.6.3	15–18

Section 10.6 Exercises

Building Concepts and Skills These exercises assess the basic knowledge you should have after reading the section.

1. The equal SD condition is met when the largest sample standard deviation is _____ the smallest sample standard deviation.

2. The F statistic in the ANOVA table is modeled by an F distribution with _____ and _____ degrees of freedom.

3. True/False: The Normal/Large Sample condition for one-way ANOVA is checked in the same way as the Normal/Large Sample condition for the two-sample t test for a difference in means.

4. True/False: Fisher's LSD intervals should be computed for every one-way ANOVA test.

Mastering Concepts and Skills These exercises reinforce the learning goals as shown in the examples.

5. **Refer to Example 10.6.1 Glass Forensics** Researchers were interested in whether the type of glass can be determined by the various characteristics of the glass. This research was motivated

by criminal investigations — glass can be used as evidence if it can be correctly identified. The study used 214 randomly selected specimens of glass of six different types: building windows that were float processed, building windows that were not float processed, vehicle windows that were float processed, containers, tableware, and headlamps. One characteristic measured was the refractive index.[44] Use the given information to check the conditions for a one-way ANOVA test.

Type	n	Mean	SD
Building window float processed	70	1.5187	0.00227
Building window not float processed	76	1.5186	0.00380
Container	13	1.5189	0.00335
Headlamp	29	1.5171	0.00255
Tableware	9	1.5175	0.00312
Vehicle window float processed	17	1.5180	0.00192

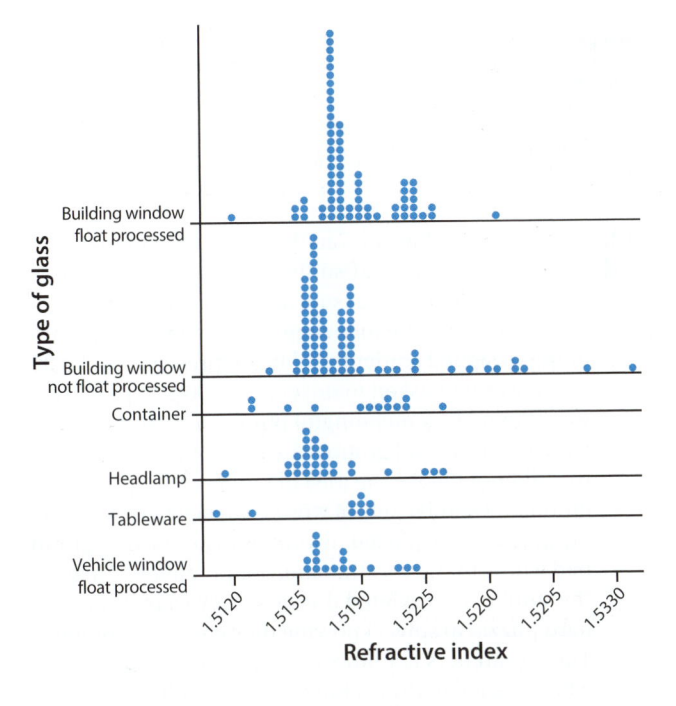

6. **Sugary Cereals** Many favorite kids' cereals contain lots of sugar. Do supermarkets deliberately place more sugary cereals on shelves where kids can better see them? To find out, researchers went to the cereal aisle in a large supermarket. They selected independent random samples of cereal from the top shelf, middle shelf, and bottom shelf. For each cereal, they recorded the sugar content (grams per serving).[45] Use the given information to check the conditions for a one-way ANOVA test.

Shelf	n	Mean	SD
Top	36	6.528	3.836
Middle	21	9.619	4.129
Bottom	19	5.110	4.480

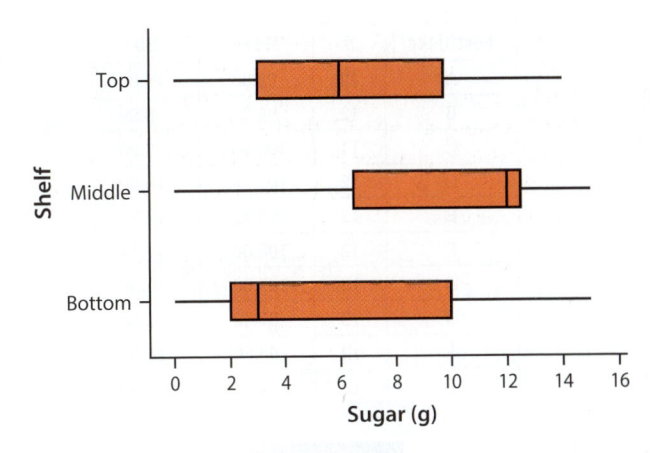

7. **Effects of Logging** How does logging in a tropical rainforest affect the forest in later years? Researchers compared randomly selected forest plots in Borneo that had never been logged with similar randomly selected plots nearby that had been logged 1 year earlier and 8 years earlier. They counted the number of trees on each plot.[46] Use the data to check the conditions for a one-way ANOVA test.

Never	27	22	29	21	19	33	16	20	24	27	28	19
One year earlier	12	12	15	9	20	18	17	14	14	17	19	
Eight years earlier	18	22	15	18	19	22	12	12				

8. **Frozen Desserts** Researchers were interested in the nutritional attributes of several categories of frozen desserts. They selected a sample of 65 different desserts, recorded the number of calories per serving, and classified the dessert as one of the following: ice cream, frozen yogurt, gelato, sorbet, or dairy free.[47] Use the data to check the conditions for a one-way ANOVA test.

| Dairy free | 50 | 160 | 120 | 100 | 230 | 110 | 110 | 350 | 190 | 230 |
	230	240	230	240	350	130	240	220	330	
Frozen yogurt	230	190								
Gelato	210	240	290	300						
Ice cream	110	100	110	150	150	150	120	140	140	150
	160	140	160	190	200	120	160	170	200	150
	190	170	220	170	190	170	160	200	380	250
	350									
Sorbet	160	150	150	170	180	180	280	300	300	

9. **Refer to Example 10.6.2 Fertilized Bean Sprouts** A scientist tested the effects of 10 different fertilizers on mung bean sprouts. Each fertilizer was randomly assigned to 11 or 12 bean sprouts. One week later, the heights of the bean sprouts were measured in millimeters.[48] Use the information given to determine if there is convincing evidence at the $\alpha = 0.10$ significance level that the mean growth for bean sprouts like the ones in the study is not the same for all 10 fertilizers.

Fertilizer	n	Mean	SD
A	12	99.83	15.83
B	12	95.25	8.99
C	11	100.09	8.96
D	12	103.25	16.80
E	12	93.42	11.90
F	12	108.00	14.15
G	12	103.08	16.38
H	12	97.42	12.58
I	12	98.50	15.97
J	12	101.75	8.99

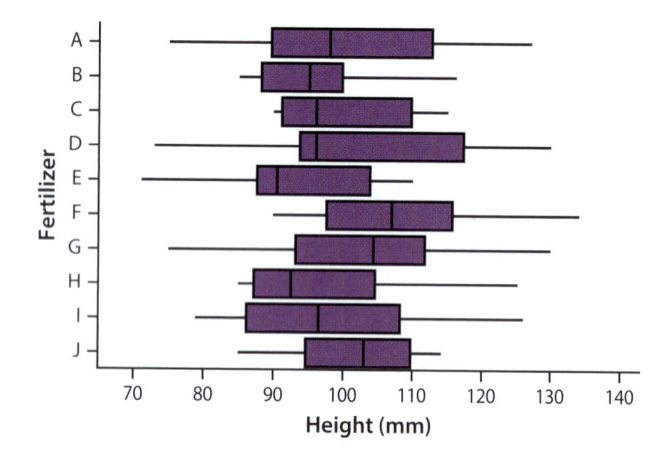

Source	df	SS	MS	F	P-value
Groups	9	1943.51	215.946	1.194	0.3059
Error	109	19,707.1	180.799		
Total	118	21,650.6			

10. Chemicals in Water Random samples of water were selected at three different depths (bottom, mid-depth, and surface) from the Wolf River in Tennessee at a site downstream from an abandoned dump that had been used by the pesticide industry. Typical water testing happens at a single depth, but the typical method could be problematic if the concentration of the chemical varies by depth.[49] Here we analyze hexachlorobenzene (HCB) concentrations (in nanograms per liter, ng/L) in the water samples. Use the information given to determine if there is convincing evidence at the $\alpha = 0.01$ significance level that at least one depth has a different mean concentration of HCB than other depths.

Depth	n	Mean	SD
Surface	10	4.804	0.631
Mid-depth	10	5.330	1.106
Bottom	10	5.839	1.014

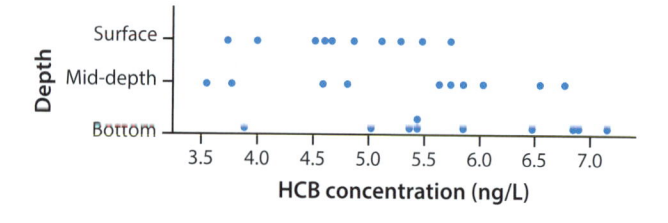

	Analysis of variance				
Source	df	SS	MS	F	P-value
Depth	2	5.357	2.6783	3.03	0.065
Error	27	23.848	0.8833		
Total	29	29.205			

11. Effects of Logging In Exercise 7, you considered a study in which researchers compared randomly selected forest plots in Borneo that had never been logged with similar randomly selected plots that had either been logged 1 year earlier or 8 years earlier. In that exercise, you analyzed the number of trees in each plot. The researchers also measured the number of species in each plot. These species counts are shown in the table. Do the data provide convincing evidence at the 5% significance level that the mean number of species per plot is not the same for all three levels of logging in Borneo?

Never	22	18	22	20	15	21	13	13	19	13	19	15
One year previously	11	11	14	7	18	15	15	12	13	15	8	
Eight years previously	17	18	14	18	15	15	10	12				

12. Money and Behavior Can thinking about money make people more self-sufficient? Kathleen Vohs of the University of Minnesota and colleagues tested this theory with the following experiment.[50] They randomly assigned students to one of three groups. Each group was first asked to unscramble 30 groups of five words to make a meaningful phrase with four of the words. The control group were given groups of words that did not invoke the idea of money. The "play" group had similar sets of words, but stacks of Monopoly money were placed near them. The "prime" group had words that led to thinking about money. Once the students completed this task, they were given a hard puzzle to solve, knowing they could ask for help. The researchers hypothesized that the two groups who were led to think about money would feel more self-sufficient and therefore take longer to ask for help. The values in the table show the amount of time (in seconds) until the subject asked for help. Do the data provide convincing evidence at the 5% significance level that the mean amount of time before students like the ones in the study ask for help is not the same for these three treatments?

Prime	609	444	242	199	174	55	251	466	443
	531	135	241	476	482	362	69	160	
Play	455	100	238	243	500	570	231	380	222
	71	232	219	320	261	290	495	600	67
Control	118	272	413	291	140	104	55	189	126
	400	92	64	88	142	141	373	156	

13. **Counselor Bias** Do high school counselors exhibit bias when recommending students for advanced courses? Researchers at the University of Massachusetts Amherst developed fictional student transcripts that were identical except for the name on them. One transcript represented a student who was strong both academically and behaviorally. Counselors were randomly assigned to one of four groups, with each group seeing a different name on that transcript. The names used were chosen to suggest that the student was either a Black female, Black male, White female, or White male. The counselors were asked to rate the preparedness of the student for advanced coursework.[51] Use the information given to determine if there is convincing evidence that the mean rating given by counselors like the ones in the study is not the same for all four student names.

Treatment	n	Mean	SD
Black female	34	6.912	1.712
Black male	28	8.679	1.090
White female	32	8.375	1.385
White male	30	8.433	1.305

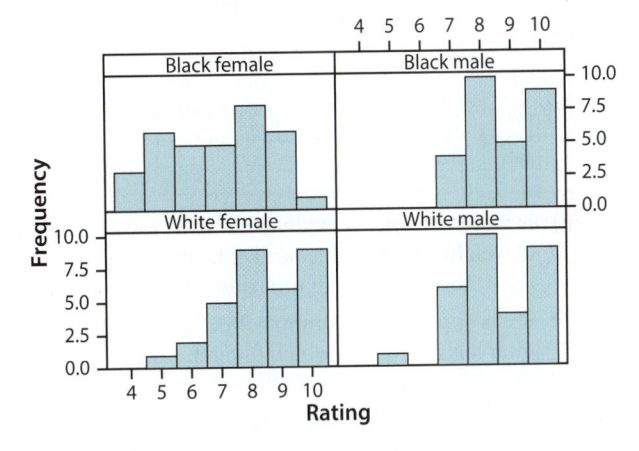

Analysis of variance

Source	df	SS	MS	F	P-value
Treatment	3	62.90	20.965	10.58	0.000
Error	120	237.71	1.981		
Total	123	300.60			

14. **Road Rage** "The phenomenon of road rage has been frequently discussed but infrequently examined." So begins a report based on interviews with 1382 randomly selected drivers.[52] The respondents' answers to interview questions produced scores on an "angry/threatening driving scale" with values between 0 and 19. Do people of different ages have different amounts of road rage? The summary statistics show the means, standard deviations, and sample sizes for three age groups. Although the researchers selected a single random sample, it is reasonable to consider these age groups as independent random samples. Use the information given to determine if there is convincing evidence that the mean road rage score is not the same for the three age groups.

Age group	n	$\bar{x}$	s_x
Younger than 30 years	244	2.22	3.11
30 to 55 years	734	1.33	2.21
Older than 55 years	364	0.66	1.60

Source	df	SS	MS	F	P-value
Groups	2	356.14	178.07	34.8	0.000
Error	1339	6859.65	5.123		
Total	1341	7215.79			

15. **Refer to Example 10.6.3 Effects of Logging** Refer to Exercise 11. Compute the appropriate 95% Fisher's LSD intervals and report which differences are significant.

16. **Money and Behavior** Refer to Exercise 12. Compute the appropriate 95% Fisher's LSD intervals and report which differences are significant.

17. **Counselor Bias** Refer to Exercise 13. Here is computer output including information about the 95% Fisher's LSD intervals. Report the conclusions that can be made from these intervals.

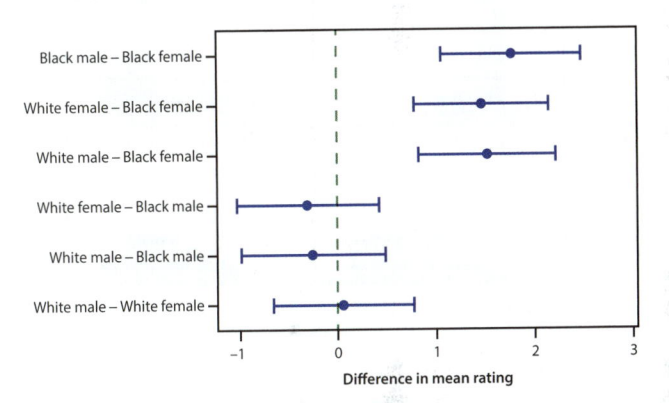

18. **Road Rage** Refer to Exercise 14. The 95% Fisher's LSD intervals are given here. Report the conclusions that can be made from these intervals.

Groups	Fisher's LSD interval
(Younger than 30) − (30 to 55)	(0.56, 1.22)
(Younger than 30) − (Older than 55)	(1.19, 1.927)
(30 to 55) − (Older than 55)	(0.39, 0.96)

Applying the Concepts These exercises ask you to apply multiple learning goals in a new context or to apply what you learned in this section in a new way.

19. **Informed Consent** When experiments are conducted using humans as subjects, it is important that the participants understand what they are agreeing to do, and what risks they might be taking. Typically, participants are given a form that explains the experiment and the possible risks. They are then asked to sign this form, confirming that they understand the risks and agree to participate. Researchers from Switzerland and Tanzania were concerned about how well people understand the typical informed consent form. They designed a randomized experiment to test four different ways of

explaining an experiment and its risks to possible participants in Tanzania. All four groups were given the same informed consent form. For the control group, this was all the information they were given. The second group was also given an oral information session. The third group had the oral information session as well as a slideshow. The fourth group had the oral information session as well as a theatrical presentation. Participants were then asked to respond to a questionnaire to test how much they had understood about the experiment.[53] Use the given information to answer the following questions.

Group	n	Mean	SD
ICF alone	150	4.413	1.466
Oral and slideshow	174	6.690	1.996
Oral and theater	145	7.034	1.701
Oral information	135	6.837	1.626

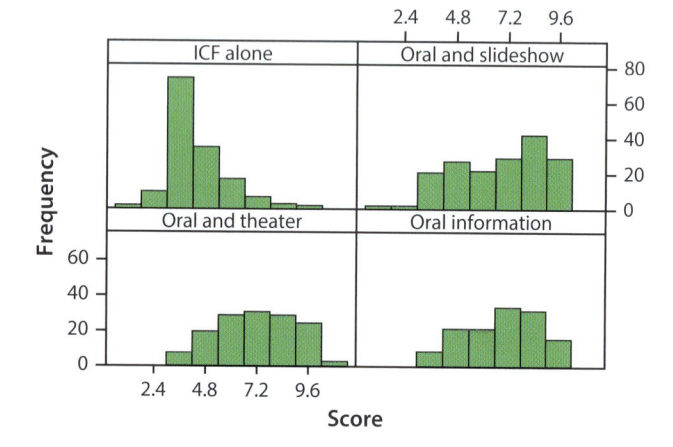

Analysis of variance

Source	df	SS	MS	F	P-value
Group	3	675.3	225.111	75.84	0.000
Error	600	1780.9	2.968		
Total	603	2456.2			

(a) Is there convincing evidence that the mean level of understanding for subjects like the ones in the study is not the same for all four treatments? Use $\alpha = 0.05$.

(b) If it is appropriate, compute 95% Fisher's LSD intervals and report which differences are significant. If it is not appropriate, explain why not.

(c) What is the take-away message from this study about informing potential participants in an experiment? Explain your answer.

20. **Diets** Researchers compared four popular diets in an experiment to determine whether there were differences in mean weight loss after 12 months. They randomly assigned 93 subjects to one of the four diets (Atkins, Ornish, Weight Watchers, Zone). They measured the amount of weight lost (in kilograms, kg) after 12 months.[54] Use the given information to answer the following questions.

Diet	n	Mean	SD
Atkins	21	3.92	6.05
Ornish	20	6.56	9.29
WW	26	4.59	5.39
Zone	26	4.88	6.92

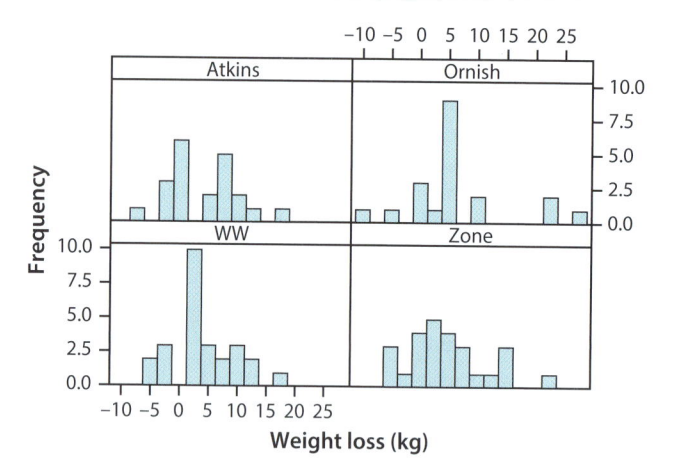

Analysis of variance

Source	df	SS	MS	F	P-value
Diet	3	77.60	25.87	0.54	0.659
Error	89	4293.71	48.24		
Total	92	4371.31			

(a) Is there convincing evidence that the mean amount of weight loss for subjects like the ones in the study is not the same for all four diets? Use $\alpha = 0.05$.

(b) If it is appropriate, compute 95% Fisher's LSD intervals and report which differences are significant. If it is not appropriate, explain why not.

(c) What is the take-away message from this study about these different diet plans? Explain your answer.

21. **Tech Glass Forensics** The researchers from Exercise 5 also measured the amount of *Magnesium* in the glass. The measurement was made in percent of magnesium oxide by weight. Again, the six different kinds of glass were building windows that were float processed, building windows that were not float processed, vehicle windows that were float processed, containers, tableware, and headlamps.[55] The data are stored in the *Glass* data set. Use technology to check the Normal/Large Sample and Equal SD conditions for a one-way ANOVA test.

22. **Tech Hummingbirds** Researchers wanted to study the relationship between varieties of the tropical flower *Heliconia* on the island of Dominica and the different species of hummingbirds that pollinate the flowers. They hypothesized that over time the flowers and birds that pollinate them would have evolved so the length of the hummingbirds' beaks were a good fit for the flowers. To test this hypothesis, they selected independent random samples of three

varieties of the flower, each pollinated by a different species of hummingbird, and measured the *Length* of each flower in millimeters.[56] The data are stored in the **Hummingbirds** data set. Use technology to check the Normal/Large Sample and Equal SD conditions for a one-way ANOVA test.

23. Tech **Cola Caffeine** Researchers selected random samples of fountain cola at three popular restaurant chains: Applebee's, Kentucky Fried Chicken, and McDonald's. They measured the amount of *Caffeine* in each sample (in milligrams per 12 ounces, or mg/12 oz).[57] The data are stored in the **Caffeine** data set.

    (a) Is there convincing evidence that the mean amount of caffeine per serving is not the same at all three restaurant chains?

    (b) If it is appropriate, compute 95% Fisher's LSD intervals and report which differences are significant. If it is not appropriate, explain why not.

    (c) What is the take-away message from this study about the caffeine in fountain colas from these restaurants? Explain your answer.

24. Tech **Fruit Fly Longevity** An experiment was conducted to investigate whether increased sexual activity shortened the lifespan of male fruit flies. Researchers randomly allocated 125 fruit flies to live in a test tube in one of five conditions: alone, with 1 pregnant female, with 8 pregnant females, with 1 virgin female, with 8 virgin females. The *Longevity* of the male fruit fly was measured in days.[58] The data are stored in the **Fruitflies** data set.

    (a) Is there convincing evidence that the mean longevity for male fruit flies like the ones in the study is not the same for all five living conditions?

    (b) If it is appropriate, compute 95% Fisher's LSD intervals and report which differences are significant. If it is not appropriate, explain why not.

    (c) What is the take-away message from this study about the longevity of male fruit flies like the ones in the study? Explain your answer.

Extending the Concepts These exercises challenge you to explore statistical concepts and methods that go beyond what you learned in this section.

25. **ANOVA vs. Two-Sample t Test** All the examples and exercises in Sections 10.5 and 10.6 have three or more populations or treatments. When used for two populations or treatments, a one-way ANOVA test will give the same results as a slightly modified version of the two-sample t test that you learned about in Chapter 9. Recall that the two-sample t test did not include a condition that the standard deviations of the two populations be the same. There is a version of the two-sample t test that does have this requirement, called the *pooled two-sample t test for a difference in means*. The P-value of a two-sided pooled two-sample t test will be identical to the P-value from the ANOVA test.

    In this exercise, we will compare the two tests using the **Caffeine** data set discussed in Exercise 23. Random samples of servings of fountain cola were selected at three popular restaurant chains: Applebee's, Kentucky Fried Chicken, and McDonald's. At Applebee's and McDonald's, the fountain cola was Coca-Cola. At Kentucky Fried Chicken, it was Pepsi. Is there a significant difference in the mean amount of *Caffeine* per serving between Coke and Pepsi?

    (a) The test statistic for the pooled two-sample t test is $t = \dfrac{(\bar{x}_1 - \bar{x}_2) - 0}{\sqrt{\dfrac{s_p^2}{n_1} + \dfrac{s_p^2}{n_2}}}$, where s_p is the pooled standard deviation and $s_p = \sqrt{\dfrac{(n_1-1)s_1^2 + (n_2-1)s_2^2}{(n_1-1)+(n_2-1)}}$. The t statistic then has $n_1 + n_2 - 2$ degrees of freedom. Assume that the conditions are met for this test. Use the information given to compute the test statistic and P-value for a pooled two-sample t test for $H_a: \mu_C \neq \mu_P$.

Brand	n	Mean	SD
Coke	40	41.699	4.461
Pepsi	20	36.101	3.155

    (b) Here is the ANOVA table for the same data. What is the conclusion from this test?

    Analysis of variance

Source	df	SS	MS	F	P-value
Brand	1	417.9	417.95	25.11	0.000
Error	58	965.3	16.64		
Total	59	1383.3			

    (c) How does the t statistic you computed in part (a) compare to the F statistic from the ANOVA table in part (b)? What is the relationship? Note that this relationship will always be true when using the pooled t test.

26. **Transforming Alzheimer's Protein** Amyloid-β is a protein fragment that has been associated with Alzheimer's disease. Data were collected from the autopsies of a sample of priests who donated their bodies to medical research. The priests were divided into three categories: those who had not exhibited any cognitive impairment before death, those who had exhibited mild cognitive impairment, and those who had mild to moderate Alzheimer's disease. The amount of amyloid-β (in picomoles per gram, pmol/g, of tissue from the posterior cingulate cortex) was measured for each subject and recorded in the following table.[59]

No impairment	114	41	276	0	16	228	927	0	211	829	1561
	0	276	959	16	24	325	49	537			
Mild impairment	73	33	16	8	276	537	0	569	772	0	260
	423	780	1610	0	309	512	797	24	57	106	
Alzheimer's disease	407	390	1154	138	634	919	1415	390	1024	1154	195
	715	1496	407	1171	439	894					

(a) Use the given data to show that neither the Normal/Large Sample nor the Equal SD condition for a one-way ANOVA test is met.

(b) Sometimes transforming the data can help with meeting the Normal/Large Sample and/or Equal SD conditions. Take the square root of all the data values. Show that both conditions are now met.

Cumulative Review These exercises revisit what you learned in previous sections.

27. **Blood Alcohol Level (2.6)** An experiment was performed at The Ohio State University to explore the relationship between the number of beers someone drinks and their blood alcohol content (BAC). The researchers assigned 16 student volunteers to drink a specific number of beers. They then measured the volunteers' blood alcohol content 30 minutes after consuming the final beer. Here are the BAC data:

Beers	5	2	9	8	3	7	3	5
BAC	0.10	0.03	0.19	0.12	0.04	0.095	0.07	0.06
Beers	3	5	4	6	5	7	1	4
BAC	0.02	0.05	0.07	0.1	0.085	0.09	0.01	0.05

(a) Make a scatterplot that shows how the number of beers relates to the BAC.

(b) Describe the relationship between number of beers drunk and BAC for these students.

28. **Data Ethics (1.8)** Describe the role of an Institutional Review Board and explain why it is important when an experiment includes human subjects.

Statistics Matters How racially and ethnically diverse are STEM workers?

At the beginning of this chapter, we described a study about the distribution of race and ethnicity in STEM jobs in the United States. According to the Pew Research Center, here is the distribution of race and ethnicity for all jobs in 2019:

Black: 11%   Hispanic: 17%   Asian: 6%   Other: 3%   White: 63%

Pew classifies Black, Asian, and White adults as those who report being only one race and are not Hispanic. Hispanics can be of any race. "Other" includes non-Hispanic American Indian or Alaskan Native, non-Hispanic Native Hawaiian or Pacific Islander, and non-Hispanic two or more major racial groups.

Here is the distribution of race and ethnicity from a random sample of 1022 U.S. adults in STEM jobs in 2019:

Black: 95   Hispanic: 81   Asian: 136   Other: 28   White: 682

1. Do these data provide convincing evidence at the $\alpha = 0.05$ significance level that the distribution of race and ethnicity in STEM jobs in 2019 was different than the distribution of race and ethnicity in all jobs?

2. Interpret the *P*-value from the test in Question 1.

Here is the distribution of race and ethnicity from a random sample of 1086 U.S. adults in STEM jobs in 2016:[60]

Black: 98   Hispanic: 78   Asian: 139   Other: 26   White: 745

3. Make a graph to compare the distribution of race and ethnicity in STEM jobs in 2016 and 2019. Describe what you see.

4. Is there convincing evidence at the 5% significance level that the distribution of race and ethnicity in STEM jobs has changed from 2016 to 2019?

5. Which type of error, Type I or Type II, is possible based on your conclusion to Question 4? Explain your answer.

Suppose we want to perform a test to determine if there are any differences in mean salary for STEM jobs among the same racial and ethnic groups in the United States in 2019.

6. Identify the appropriate test and state the null and alternative hypotheses.

7. What conditions would need to be met to carry out the test?

Chapter 10 Review

Tests About the Distribution of One Categorical Variable

- A test about how well the distribution of one categorical variable "fits" a claimed distribution is called a **chi-square test for goodness of fit.**

- The null hypothesis in a test for goodness of fit is that the distribution of a categorical variable in a population is the same as a claimed distribution. The alternative hypothesis is that the population distribution is different than the claimed distribution.

- There are two conditions for a chi-square test for goodness of fit:
  - **Random:** The data come from a random sample from the population of interest.
  - **Large Counts:** All *expected* counts are at least 5.

- Calculate the **expected count** for each category by multiplying the sample size by the proportion specified by the null hypothesis.

- The **chi-square test statistic** is a measure of how different the observed counts are from the expected counts, relative to the expected counts. The formula for the statistic is

$$\chi^2 = \sum \frac{(\text{observed count} - \text{expected count})^2}{\text{expected count}}$$

  In a chi-square test for goodness of fit, the sum is over all categories of the categorical variable.

- A **chi-square distribution** is described by a density curve that takes only non-negative values and is skewed to the right. A particular chi-square distribution is specified by its degrees of freedom.

- If the conditions are met, find the *P*-value for a chi-square test for goodness of fit using Table C or technology and the chi-square distribution with degrees of freedom = number of categories − 1.

- Be sure to follow the four-step process whenever you perform a chi-square test for goodness of fit.

Tests About the Relationship Between Two Categorical Variables

- Two categorical variables have an **association** if knowing the value of one variable helps us predict the value of the other. If knowing the value of one variable does not help us predict the value of the other, then there is no association between the variables.

- A test about the relationship between two categorical variables is called a **chi-square test for association.**

- The null hypothesis for this test is that there is *no* association between the two variables in the population of interest. The alternative hypothesis is that there *is* an association between the two variables in the population of interest.

- There are two conditions for a chi-square test for association:
  - **Random:** The data come from a random sample from the population of interest, from independent random samples from the populations of interest, or from groups in a randomized experiment.
  - **Large Counts:** All *expected* counts are at least 5.

- Calculate **expected counts** using the formula

$$\text{expected count} = \frac{\text{row total} \times \text{column total}}{\text{table total}}$$

- Use the **chi-square test statistic,** where the sum is over all the cells in the table (not including the totals):

$$\chi^2 = \sum \frac{(\text{observed count} - \text{expected count})^2}{\text{expected count}}$$

- If conditions are met, find the *P*-value using Table C or technology and the chi-square distribution with degrees of freedom = (number of rows − 1)(number of columns − 1).

- Be sure to follow the four-step process whenever you perform a chi-square test for association.

Tests About Two or More Means

- A test that compares two or more means is called a **one-way Analysis of Variance** (**ANOVA**) test.

- The null hypothesis in an ANOVA test is that all population (treatment) means are equal. The alternative hypothesis is that at least one of the population (treatment) means is different than the rest.

- The evidence for the alternative hypothesis is stronger when there is less overlap in the distributions of the response variable for each of the groups. That is, the evidence is stronger when the centers of the distributions are far from each other and the variation within each distribution is small.

- There are three conditions for a one-way ANOVA test:
  - **Random:** The data come from independent random samples from the populations of interest or from groups in a randomized experiment.
  - **Normal/Large Sample:** For each sample, the data come from a normally distributed population or the sample size is large ($n \geq 30$). When the sample size is small and the shape of the population distribution is unknown, a graph of the sample data shows no strong skewness or outliers.
  - **Equal SD:** The largest sample SD is less than twice the smallest sample SD.

- The ***F* statistic** is a ratio of the variability between group means and the variability within groups. A value of *F* greater than 1 suggests that the differences in the means are bigger than expected based on chance alone. The formula for the statistic is

$$F = \frac{\text{MSG}}{\text{MSE}} = \frac{\text{SSG}/\text{df}_G}{\text{SSE}/\text{df}_E}$$

where MSG = mean square for the groups, MSE = mean square for the errors, SSG = sum of squares for the groups, SSE = sum of squares for the errors, df_G = degrees of

- freedom for the groups = number of groups − 1 and df_E = degrees of freedom for the error = n − number of groups.
- If the conditions are met, find the P-value for a one-way ANOVA test using technology and the F distribution with numerator degrees of freedom = number of groups − 1 = $I - 1$ and denominator degrees of freedom = sample size − number of groups = $n - I$.
- An **ANOVA table** summarizes the test by providing the sums of squares and degrees of freedom for the groups and for the errors. It also provides the mean squares, F statistic, and P-value.
- Be sure to follow the four-step process whenever you perform a one-way ANOVA test.

- When the results of an ANOVA test are significant, do a follow-up analysis to identify which means are different than the others. One method of follow-up analysis is to use **Fisher's least significant difference (LSD) intervals**. The formula is

$$(\bar{x}_1 - \bar{x}_2) \pm t^* \sqrt{\frac{s_p^2}{n_1} + \frac{s_p^2}{n_2}}$$

where s_p^2 = MSE and t^* is the t critical value for $C\%$ confidence and df_E = sample size − number of groups = $n - I$.
- If 0 is not included in a Fisher's LSD interval, then there is convincing evidence that the corresponding population (treatment) means are not the same.

	Comparing Tests		
Name	Chi-square test for goodness of fit	Chi-square test for association	One-way ANOVA test for differences in means
Null hypothesis	The stated distribution of a categorical variable in the population is correct.	There is no association between two categorical variables in the population.	$\mu_1 = \mu_2 = \mu_3 = \cdots$
Alternative hypothesis	The stated distribution of a categorical variable in the population is not correct.	There is an association between two categorical variables in the population.	At least one of the means is different.
Conditions	• **Random:** The data come from a random sample from the population of interest. • **Large Counts:** All expected counts are at least 5.	• **Random:** The data come from a random sample from the population of interest, from independent random samples from the populations of interest, or from groups in a randomized experiment. • **Large Counts:** All expected counts are at least 5.	• **Random:** The data come from independent random samples from the populations of interest or from groups in a randomized experiment. • **Normal/Large Sample:** For each sample, the corresponding population distribution is approximately normal or the sample size is large ($n \geq 30$). For each sample, if the population distribution has unknown shape and $n < 30$, a graph of the sample data shows no strong skewness or outliers. • **Equal SD:** The largest sample standard deviation is less than twice the smallest sample standard deviation.
Expected counts	(sample size) × (proportion specified by null)	$\dfrac{\text{row total} \times \text{column total}}{\text{table total}}$	Not applicable
Test statistic	$\chi^2 = \sum \dfrac{(\text{observed count} - \text{expected count})^2}{\text{expected count}}$		$F = \dfrac{\text{MSG}}{\text{MSE}} = \dfrac{\text{SSG}/df_G}{\text{SSE}/df_E}$
Degrees of freedom	Number of categories − 1	(number of rows − 1) × (number of columns − 1)	df_G = number of groups − 1 = $I - 1$ df_E = n − number of groups = $n - I$
TI-83/84 Name	χ^2 GOF-test (not available on TI-83)	χ^2-test	ANOVA

Chapter 10 Review Exercises

These exercises will help you review important concepts and skills described by the learning goals in each section. For most exercises, the relevant section is noted in parentheses after the exercise title.

1. **Plant Colors (10.1, 10.2)** Biologists wish to cross pairs of tobacco plants having genetic makeup Gg, indicating that each plant has one dominant gene (G) and one recessive gene (g) for color. Each offspring plant will receive one gene for color from each parent. The biologists predict that the expected ratio of green (GG) to yellow-green (Gg) to albino (gg) tobacco plants should be 1:2:1. In other words, the biologists predict that 25% of the offspring will be green, 50% will be yellow-green, and 25% will be albino. To test their hypothesis, the biologists mate 84 randomly selected pairs of yellow-green parent plants. Of 84 offspring, 23 plants were green, 50 were yellow-green, and 11 were albino. Do these data provide convincing evidence at the $\alpha = 0.01$ level that the true distribution of offspring is different from what the biologists predict?

2. **Car Accident Perceptions (10.3, 10.4)** Two researchers asked 150 people to recall the details of a car accident they watched on video. They selected 50 people at random and asked, "About how fast were the cars going when they smashed into each other?" For another 50 randomly selected people, they replaced the words "smashed into" with "hit." They did not ask the remaining 50 people — the control group — to estimate speed at all. A week later, the researchers asked all 150 participants if they saw any broken glass at the accident (there wasn't any). The table shows each group's response to the broken glass question.[61]

		Wording			
		"Smashed into"	"Hit"	Control	Total
Response	Yes	16	7	6	29
	No	34	43	44	121
	Total	50	50	50	150

   (a) State the null and alternative hypotheses for a test to determine if there is an association between wording and response for subjects like the ones in the study.

   (b) Calculate the expected counts for a test of the null hypothesis stated in part (a).

   (c) Check that the conditions for this test have been met.

   (d) Calculate the test statistic and P-value for this test. What conclusion would you make at the 5% significance level?

3. **Coffee and Cholesterol (10.5, 10.6)** Does drinking boiled coffee, such as Scandinavian coffee, have an effect on blood cholesterol levels? Does it matter if the coffee has been filtered? To find out, researchers in the Netherlands randomly assigned 64 volunteers to one of three treatments: boiled coffee, boiled and filtered coffee, or no coffee. They measured total cholesterol level (mmol/L) for each volunteer before the study and after 81 days.[62] The increase in total cholesterol for each volunteer is shown in the table (a negative value indicates a decrease in cholesterol).

Boiled	−0.4	−0.3	−0.1	0	0.1	0.2	0.2	0.3	0.4	0.4	0.5
	0.6	0.7	0.7	0.8	0.8	0.9	1	1.1	1.3	1.3	1.5
Filtered	−0.6	−0.5	−0.5	−0.4	−0.3	−0.2	−0.1	0	0	0.1	0.1
	0.2	0.3	0.4	0.4	0.5	0.5	0.6	0.7	0.7	0.8	
None	−0.8	−0.6	−0.6	−0.5	−0.4	−0.3	−0.3	−0.3	−0.2	−0.1	−0.1
	−0.1	0	0	0	0.3	0.3	0.5	0.5	0.8	1	

Source	df	SS	MS	F	P-value
Groups	2	3.9555	1.9778	8.686	0.0005
Error	61	13.8888	0.2277		
Total	63	17.8444			

   (a) Is there convincing evidence at the $\alpha = 0.05$ significance level that the mean change in cholesterol for subjects like the ones in the study is not the same for all three treatments?

   (b) If it is appropriate, compute 95% Fisher's LSD intervals and report which differences are significant. If it is not appropriate, explain why not.

   (c) What is the take-away message from this study about the effect of boiled coffee and coffee filters? Explain your answer.

4. **Napping and Heart Disease (Chapter 10)** In a long-term study of 3462 randomly selected adults from Lausanne, Switzerland, researchers investigated the relationship between weekly napping frequency and whether or not a person experienced an event related to cardiovascular disease (CVD), such as a heart attack. The table summarizes the results.[63]

		Napping frequency				
		None	1–2 weekly	3–5 weekly	6–7 weekly	Total
CVD event status	Yes	93	12	22	28	155
	No	1921	655	389	342	3307
	Total	2014	667	411	370	3462

   (a) Do these data provide convincing evidence of an association between napping frequency and CVD event status for all adults in Lausanne?

   (b) Based on your answer to part (a), is it reasonable to conclude that taking more naps caused a change in the CVD status of these adults?

Chapter 10 Project: Trees of New York

Explore a large data set using software and apply what you've learned in this chapter to a real-world scenario.

The city of New York shares many data sets on its website data.cityofnewyork.us. The data set **NYCTrees** contains a random sample of 8253 trees on New York City streets. The data set includes the values of many variables for each tree, including details about location, variety, and health. The variable *Borough* lists which of the five boroughs (Bronx, Brooklyn, Manhattan, Queens, Staten Island) each tree is located in. The variable *Health* classifies the health of each tree as good, fair, or poor. The variable *DBH* gives the diameter of a tree at breast height (roughly 4.5 feet off the ground) in inches.

Are trees on New York City streets proportionally distributed across the five boroughs? That is, if a borough has 30% of the land area, does it also have 30% of the street trees? The table shows the land area of each borough.

Borough	Area (square miles)
Bronx	42.10
Brooklyn	70.82
Manhattan	22.83
Queens	108.53
Staten Island	58.37
Total	302.65

1. Graph the distribution of *Borough* for the trees in the sample and report the number of trees in the sample for each borough.
2. Calculate the percentage of total land area occupied by each borough.
3. Conduct an appropriate test to determine if trees on New York City streets are proportionally distributed across the five boroughs.
4. If there is convincing evidence that trees on New York City streets are not proportionally distributed across the five boroughs, which boroughs have more street trees than expected? Fewer?

Are trees on New York City streets equally healthy in the five boroughs?

5. Create a two-way table and a graph to summarize the relationship between *Borough* and *Health* in the sample.
6. Conduct an appropriate test to determine if there is an association between *Borough* and *Health* for all trees on New York City streets.
7. If there is convincing evidence of an association, which boroughs typically had healthier street trees? Less healthy?

Are trees on New York City streets equally large in the five boroughs?

8. Create a graph to compare the distribution of *DBH* in each of the five boroughs. Describe what you see.
9. Conduct an appropriate test to determine if the mean *DBH* for trees on New York City streets is different in any of the five boroughs.
10. If there is convincing evidence of a difference in mean *DBH*, use Fisher's LSD intervals to identify which means are significantly different.

Further exploration: Do the five boroughs have different distributions of street-tree varieties (variable *Species*)? Is the proportion of street-trees that are alive (variable *Status*) the same in the five boroughs? Are there other interesting relationships between variables?

11 Linear Regression

Section 11.1 Regression Lines
Section 11.2 The Least-Squares Regression Line
Section 11.3 Assessing a Regression Model
Section 11.4 Confidence Intervals for the Slope of a Population Least-Squares Regression Line
Section 11.5 Significance Tests for the Slope of a Population Least-Squares Regression Line
Section 11.6 Confidence Intervals for a Mean Response and Prediction Intervals in Regression
Chapter 11 Review
Chapter 11 Review Exercises
Chapter 11 Project

▌▀ Statistics Matters How does engine size affect CO_2 emissions?

According to the U.S. Environmental Protection Agency (EPA), 29% of all greenhouse gas emissions in the United States come from transportation.[1] Although cars and trucks emit several different greenhouse gases, the volume of carbon dioxide (CO_2) emitted is much greater than the other gases. The scatterplot shows the relationship between x = engine size (in liters) and y = CO_2 emissions (in grams per mile) for a random sample of 176 gas-powered cars and trucks produced in 2021.

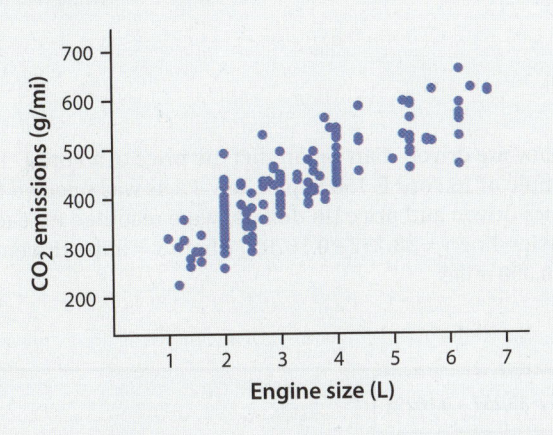

How can we use engine size to predict CO_2 emissions, and how accurate will these predictions be? Do the sample data provide convincing evidence of an association between engine size and CO_2 emissions in the population of all 2021 gas-powered cars and trucks?

We'll revisit Statistics Matters *at the end of the chapter, so you can use what you have learned to help answer these questions.*

In Chapters 2 and 3, you learned how to use scatterplots to display the relationship between two quantitative variables and how to calculate and interpret the correlation between the two variables. In this chapter, you'll learn how to use a regression line to model a linear relationship between a quantitative explanatory variable and a quantitative response variable. You'll also learn how to assess the model and how to use the model to make inferences about a population regression line.

Section 11.1 Regression Lines

LEARNING GOALS

By the end of this section, you will be able to:

- Make predictions using a regression line, keeping in mind the dangers of extrapolation.
- Calculate and interpret a residual.
- Interpret the slope and y-intercept of a regression line.

When the relationship between two quantitative variables is linear, we can use a **regression line** to model the relationship and make predictions.

A **regression line** is a line that models how a response variable y changes as an explanatory variable x changes. Regression lines are expressed in the form $\hat{y} = b_0 + b_1 x$, where $\hat{y}$ (pronounced "y hat") is the predicted value of y for a given value of x.

We could also express the regression line in the form $y = mx + b$, as we do in algebra. However, statisticians prefer the reordered format $\hat{y} = b_0 + b_1 x$ because it works better when they use several explanatory variables, as you will learn in Chapter 12. Just remember, the slope is always the coefficient of the x variable.

Making Predictions

One common use of a regression line is to make predictions.

EXAMPLE 11.1.1

How much is that truck worth?

Making predictions

PROBLEM

Everyone knows that cars and trucks lose value the more miles they are driven. Can we predict the price of a Ford F-150 if we know how many miles it has on the odometer? A random sample of 16 Ford F-150 SuperCrew 4 × 4s was selected from among those listed for sale at autotrader.com. The number of miles driven and price (in dollars) were recorded for each of the trucks.[2] Here is a scatterplot of the data, along with the regression line $\hat{y} = 38{,}257 - 0.1629x$, where x = miles driven and y = price. Predict the price of a Ford F-150 that has been driven 100,000 miles.

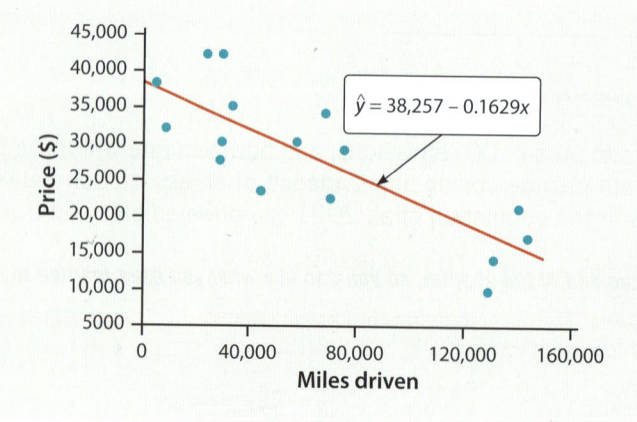

SOLUTION

$\hat{y} = 38{,}257 - 0.1629(100{,}000)$

$\hat{y} = \$21{,}967$

> To predict the price of a Ford F-150 that has been driven 100,000 miles, substitute $x = 100{,}000$ into the equation and simplify.

EXAM PREP: FOR PRACTICE, TRY EXERCISE 9.

Can we predict the price of a Ford F-150 with 300,000 miles on the odometer? We can certainly substitute 300,000 into the equation of the line. The prediction is

$$\hat{y} = 38{,}257 - 0.1629(300{,}000) = -\$10{,}613$$

We predict that we would need to pay someone $10,613 just to take the truck off our hands! This prediction is an **extrapolation**.

> **Extrapolation** is the use of a regression line for prediction outside the interval of x values used to obtain the line. The further we extrapolate, the less reliable the predictions become.

Residuals

Even when we are not extrapolating, our predictions are seldom perfect. For a specific point, the difference between the actual value of y and the predicted value of y is called a **residual**.

> A **residual** is the difference between an actual value of y and the value of y predicted by the regression line. That is,
>
> residual = actual y − predicted y
>
> $= y - \hat{y}$

The scatterplot in Figure 11.1 shows the residual for the F-150 with 70,583 miles and an actual price of $21,994. The predicted price is $\hat{y} = 38{,}257 - 0.1629(70{,}583) = \$26{,}759$, so the residual is $21{,}994 - 26{,}759 = -\$4765$. The negative value means that the price of this truck is $4765 less than predicted, based on the number of miles it has been driven. It also means that the point is below the regression line on the scatterplot.

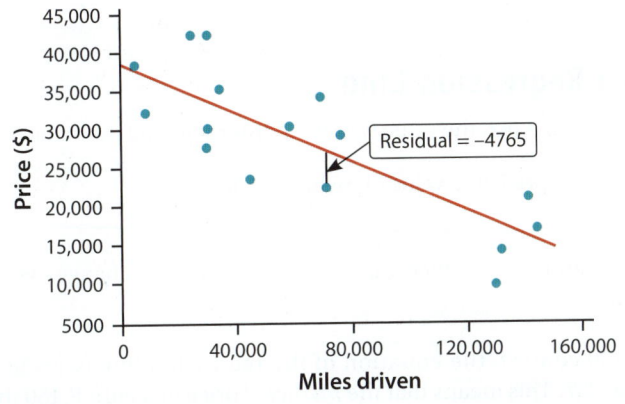

FIGURE 11.1 Scatterplot showing the relationship between miles driven and price for a sample of Ford F-150s, along with the regression line. The residual for one truck is illustrated with a vertical line segment.

EXAMPLE 11.1.2

Studying ponderosa pines

Calculating and interpreting a residual

PROBLEM

The U.S. Forest Service randomly selected ponderosa pine trees in western Montana to investigate the relationships between diameter at breast height (DBH), height, and volume of usable lumber.[3] The scatterplot shows the relationship between $x =$ DBH (in inches) and $y =$ height (in feet) for a random sample of 40 ponderosa pines, along with the regression line $\widehat{height} = 43.5 + 2.62\, DBH$.

(a) Calculate the residual for the ponderosa pine that had a DBH of 25.4 inches and a height of 133 feet.

(b) Interpret the residual from part (a).

SOLUTION

(a) $\widehat{height} = 43.5 + 2.62(25.4) = 110.05$ feet

residual $= 133 - 110.05 = 22.95$ feet

> Find the predicted height using the regression line.
> residual = actual y − predicted y = height − $\widehat{height}$

(b) This ponderosa pine was 22.95 feet taller than predicted by the regression line with $x =$ DBH.

EXAM PREP: FOR PRACTICE, TRY EXERCISE 13.

Interpreting a Regression Line

In the regression line $\hat{y} = b_0 + b_1 x$, the value b_0 is the **y-intercept** and the value b_1 is the **slope**.

The **y-intercept** b_0 is the predicted value of y when $x = 0$.

The **slope** b_1 is the amount by which the predicted value of y changes when x increases by one unit.

In the Ford F-150 context, the equation of the regression line is $\hat{y} = 38{,}257 - 0.1629x$. The y-intercept is $b_0 = 38{,}257$. This means that the *predicted* price of a Ford F-150 that has been driven 0 miles is $38,257. The slope is the coefficient of x, $b_1 = -0.1629$. This means that the *predicted* price of a Ford F-150 decreases by 0.1629 dollar for each increase of 1 mile driven. ⚠ **It is very important to include the word "predicted" (or its equivalent) in the interpretation of the slope and y-intercept.** Otherwise, it may seem that our predictions will be exactly correct.

EXAMPLE 11.1.3

Studying ponderosa pines

Interpreting a regression line

PROBLEM

In Example 11.1.2, the equation of the regression line is $\widehat{height} = 43.5 + 2.62 DBH$.

(a) Interpret the slope of the regression line.

(b) Does the value of the *y*-intercept have meaning in this context? If so, interpret the *y*-intercept. If not, explain why not.

SOLUTION

(a) The predicted height of a ponderosa pine increases by 2.62 feet for each increase of 1 inch in diameter at breast height.

(b) It is not possible for a ponderosa pine to have a diameter of 0 inches, so the *y*-intercept does not have meaning in this context.

EXAM PREP: FOR PRACTICE, TRY EXERCISE 17.

If we interpreted the *y*-intercept before considering whether it had meaning, we'd say: "If a ponderosa pine had a diameter of 0 inches, the predicted height is 43.5 feet." This clearly doesn't make sense!

Section 11.1 What Did You Learn?

Review the learning goals from this section. Then practice what you've learned by working through the exercises.

Learning Goal	Example	Exercises
Make predictions using a regression line, keeping in mind the dangers of extrapolation.	11.1.1	9–12
Calculate and interpret a residual.	11.1.2	13–16
Interpret the slope and *y*-intercept of a regression line.	11.1.3	17–20

Section 11.1 Exercises

Building Concepts and Skills These exercises assess the basic knowledge you should have after reading the section.

1. In the regression line $\hat{y} = b_0 + b_1 x$, what does $\hat{y}$ represent?

2. What is extrapolation? Is it a good idea to extrapolate?

3. What is a residual?

4. What does it mean if a residual is negative?

5. How do you interpret the *y*-intercept of a regression line?

6. Give an example of a situation where it doesn't make sense to interpret the *y*-intercept.

7. How do you interpret the slope of a regression line?

8. Which word is missing from this interpretation of slope? The price of a Ford F-150 decreases by 0.1629 dollar for each increase of 1 mile that the truck is driven.

Mastering Concepts and Skills These exercises reinforce the learning goals as shown in the examples.

9. **Refer to Example 11.1.1 Old Faithful** One of the major attractions in Yellowstone National Park is the Old Faithful geyser. The scatterplot shows the relationship between *x* = the duration of the previous eruption (in minutes) and *y* = wait time until the next eruption (in minutes). The equation of the regression line relating these variables is $\widehat{wait} = 33.35 + 13.29 duration$. Predict the wait time if the previous eruption lasted 4 minutes.

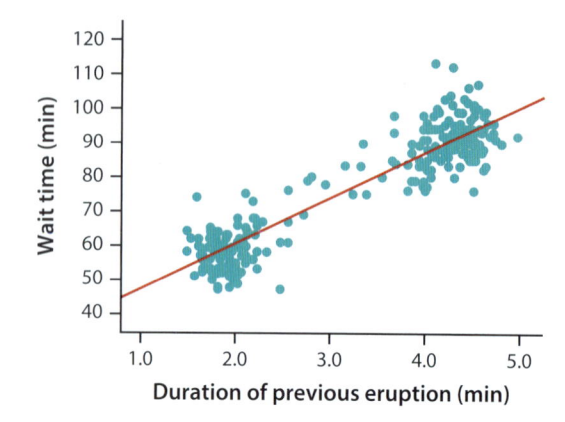

10. Crickets The scatterplot shows the relationship between x = temperature in degrees Fahrenheit and y = chirps per minute for the striped ground cricket. The equation of the regression line relating these variables is $\widehat{chirp} = -0.31 + 0.212\, temp$. Predict the cricket chirp rate when the temperature is 82°F.[4]

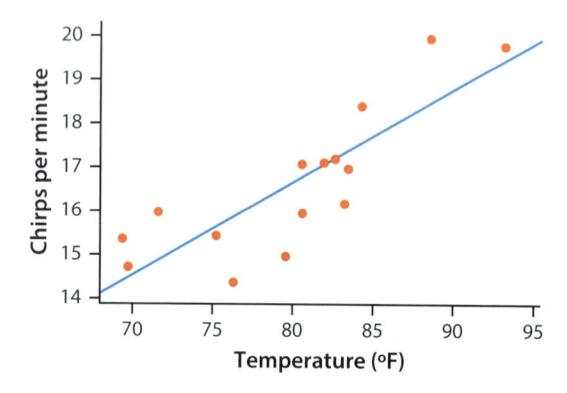

11. Cherry Blossoms Many people look forward to the blossoming of cherry trees for their beautiful flowers. Because the flowers don't last long, it's helpful to be able to predict when the trees will blossom. Here is a scatterplot showing the relationship between x = the average temperature in March (in °C) and y = number of days in April until the first blossom for a 24-year period in Japan.[5] The equation of the regression line is $\hat{y} = 33.1 - 4.69x$. Predict the date of first blossom if the average March temperature was 6°C.

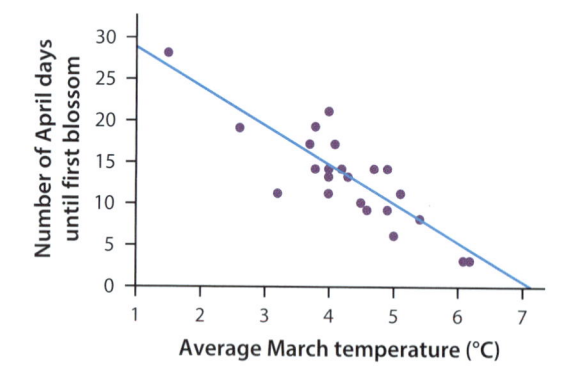

12. Coffee Drinks Single-serve coffee drinks are quite popular. Some people drink them for their caffeine and others for the sugar — and some people for both! Here is a scatterplot showing the relationship between x = calories and y = caffeine content (in milligrams, mg) for 23 different single-serve coffee drinks.[6] The equation of the regression line is $\hat{y} = 201.1 - 0.2264x$. Predict the amount of caffeine in a single-serve coffee drink with 200 calories.

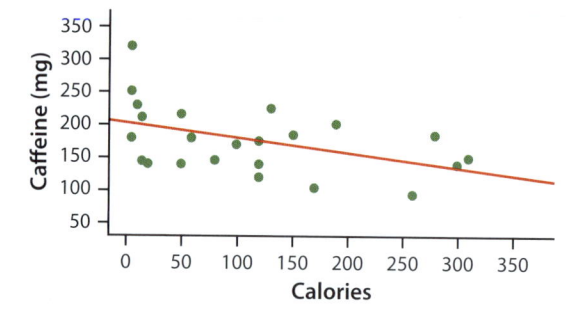

13. Refer to Example 11.1.2 Old Faithful Refer to Exercise 9. The equation of the regression line for predicting y = wait time from x = duration of previous eruption is $\widehat{wait} = 33.35 + 13.29\, duration$.

   (a) In one cycle, it took 62 minutes between eruptions, and the duration of the previous eruption was 2 minutes. Calculate the residual for this cycle.

   (b) Interpret the residual for this cycle.

14. Crickets Refer to Exercise 10. The equation of the regression line for predicting y = chirps per minute from x = temperature in degrees Fahrenheit is $\widehat{chirp} = -0.31 + 0.212\, temp$.

   (a) One observation in these data measured 16.2 chirps per minute at 83.3°F. Calculate the residual for this observation.

   (b) Interpret the residual for this observation.

15. Cherry Blossoms Refer to Exercise 11. The equation of the regression line relating y = number of days in April until the first blossom to x = the average temperature in March (in °C) is $\hat{y} = 33.1 - 4.69x$.

   (a) One year the first blossom occurred on April 11 and the average March temperature was 3.2°C. Calculate the residual for this year.

   (b) Interpret the residual for this year.

16. Coffee Drinks Refer to Exercise 12. The equation of the regression line relating y = caffeine content (in mg) to x = calories is $\hat{y} = 201.1 - 0.2264x$.

   (a) The Dunkin Donuts Mocha Iced Coffee has 280 calories and 185 mg of caffeine. Calculate the residual for this drink.

   (b) Interpret the residual for this drink.

17. Refer to Example 11.1.3 Old Faithful Refer to Exercise 9. The equation of the regression line for predicting y = wait time from x = duration of previous eruption is $\widehat{wait} = 33.35 + 13.29\, duration$.

   (a) Interpret the slope of the regression line.

   (b) Does the value of the y-intercept have meaning in this context? If so, interpret the y-intercept. If not, explain why not.

18. **Crickets** Refer to Exercise 10. The equation of the regression line for predicting y = chirps per minute from x = temperature in degrees Fahrenheit is $\widehat{chirp} = -0.31 + 0.212\,temp$.
    (a) Interpret the slope of the regression line.
    (b) Does the value of the y-intercept have meaning in this context? If so, interpret the y-intercept. If not, explain why not.

19. **Cherry Blossoms** Refer to Exercise 11. The equation of the regression line relating y = number of days in April until the first blossom to x = the average temperature in March (in °C) is $\hat{y} = 33.1 - 4.69x$.
    (a) Interpret the slope of the regression line.
    (b) Does the value of the y-intercept have meaning in this context? If so, interpret the y-intercept. If not, explain why not.

20. **Coffee Drinks** Refer to Exercise 12. The equation of the regression line relating y = caffeine content (in mg) to x = calories is $\hat{y} = 201.1 - 0.2264x$.
    (a) Interpret the slope of the regression line.
    (b) Does the value of the y-intercept have meaning in this context? If so, interpret the y-intercept. If not, explain why not.

Applying the Concepts These exercises ask you to apply multiple learning goals in a new context or to apply what you learned in this section in a new way.

21. **Energy Use** Jasmin is concerned about how much energy she uses to heat her home, so she keeps a record of the natural gas her furnace consumes for each month from October to May. Because the months are not equally long, she divides each month's consumption by the number of days in the month. She wants to see if there is a relationship between x = average temperature (in degrees Fahrenheit) and y = average gas consumption (in cubic feet per day) for each month. Here is a scatterplot along with the regression line $\hat{y} = 1425 - 19.87x$.

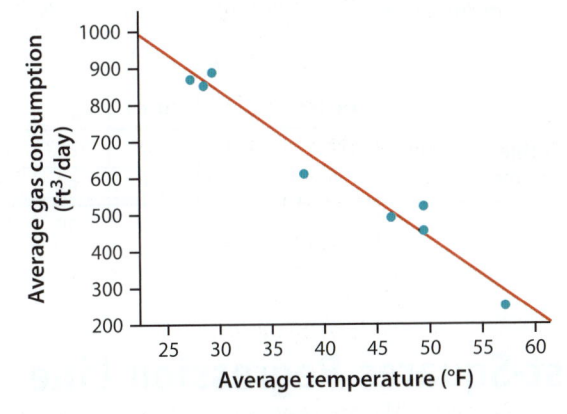

    (a) Calculate and interpret the residual for March, when the average temperature was 49.4°F and Jasmin used an average of 520 cubic feet per day.
    (b) Interpret the slope of the regression line.
    (c) Does the value of the y-intercept have meaning in this context? If so, interpret the y-intercept. If not, explain why not.

22. **Roller Coasters** Roller coasters with larger heights usually go faster than shorter ones. Here is a scatterplot of x = height (in feet) versus y = maximum speed (in miles per hour) for a random sample of 9 roller coasters. The equation of the regression line for this relationship is $\widehat{speed} = 28.17 + 0.2143\,height$.

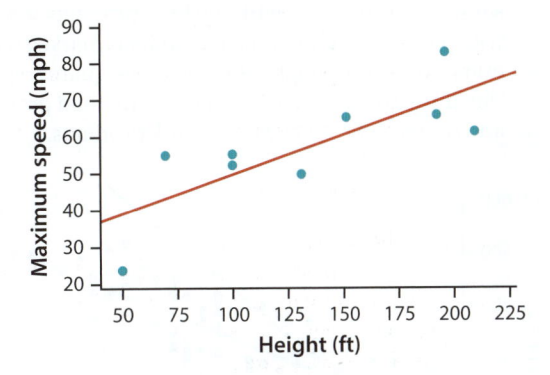

    (a) Calculate and interpret the residual for the Iron Shark, which has a height of 100 feet and a top speed of 52 miles per hour.
    (b) Interpret the slope of the regression line.
    (c) Does the value of the y-intercept have meaning in this context? If so, interpret the y-intercept. If not, explain why not.

23. **Energy Use** Refer to Exercise 21. Suppose that the average temperature in the current month is 10 degrees warmer than the previous month. Approximately how much less gas per day should Jasmin expect to use this month than in the previous month? Explain your reasoning.

24. **Roller Coasters** Refer to Exercise 22. Suppose a park owner plans to increase the height of a roller coaster by 20 feet during the off-season. Approximately how much faster is the top speed expected to be? Explain your reasoning.

Extending the Concepts These exercises challenge you to explore statistical concepts and methods that go beyond what you learned in this section.

25. **Starbucks** The scatterplot shows the relationship between the amount of fat (in grams) and number of calories in products sold at Starbucks.[7] The equation of the least-squares regression line is $\widehat{calories} = 118 + 15\,fat$.

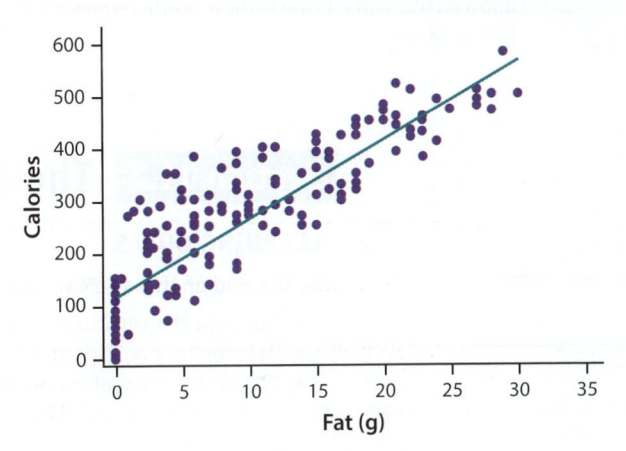

(a) Interpret the slope of the regression line.
(b) Does the value of the y-intercept have meaning in this context? If so, interpret the y-intercept. If not, explain why not.

Is the relationship between fat and calories the same for drink products and for food products? Here is the same scatterplot, but with the food products marked with purple dots and the drink products marked with orange dots. The equations of the least-squares regression lines are $\widehat{calories} = 170 + 11.8\,fat$ for food products and $\widehat{calories} = 88 + 24.5\,fat$ for drink products.

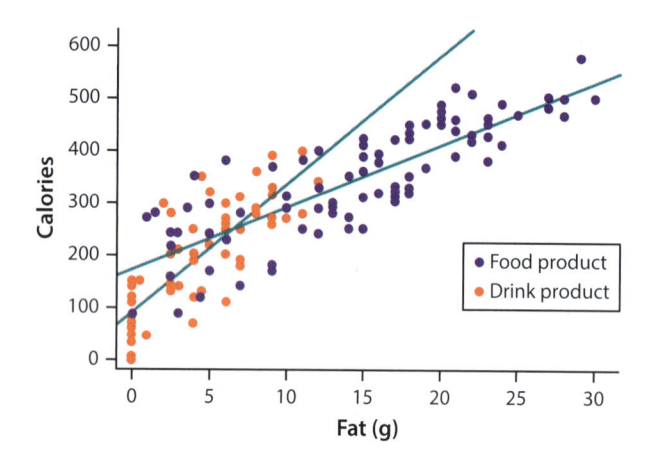

(c) Explain how the relationship between fat and calories differs for food products and drink products.
(d) According to the scatterplot, when is the predicted number of calories per gram of fat the same for food products and drink products? Explain your reasoning.

26. **Pythagorean Winning Percentage** Prediction models come in many different forms, not just linear. Bill James, one of the founders of baseball analytics (called *Sabermetrics*), proposed the following formula for predicting a baseball team's winning percentage based on RS = runs scored and RA = runs allowed. It is called the "Pythagorean" formula because the denominator looks like the Pythagorean theorem from geometry.

$$\text{predicted winning percentage} = \frac{RS^2}{RS^2 + RA^2}$$

(a) In 2021, the New York Yankees scored 711 runs and allowed 669 runs. What is their predicted winning percentage?

(b) Use your answer from part (a) to calculate the predicted *number* of wins for the Yankees in their 162-game season.
(c) The Yankees actually won 92 games. Calculate and interpret their residual.

Cumulative Review These exercises revisit what you learned in previous sections.

27. **Hot Dogs (3.5, 3.6)** Are hot dogs that are high in calories also high in salt? The scatterplot shows the relationship between calories and salt content (measured as milligrams of sodium) in 17 brands of meat hot dogs.[8]

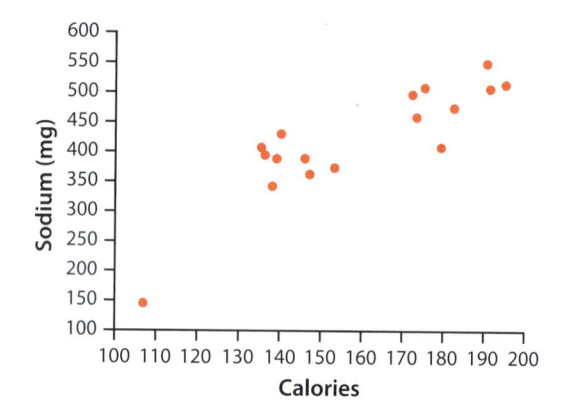

(a) The correlation for these data is $r = 0.87$. Explain what this value means.
(b) What effect does the hot-dog brand with the lowest calorie content have on the correlation? Explain your answer.

28. **Online Banking (10.4)** A recent poll conducted by the Pew Research Center asked a random sample of 1846 internet users if they do any of their banking online. The table summarizes their responses by age.[9] Do these data provide convincing evidence of an association between age and use of online banking among internet users?

		Age				
		18–29	30–49	50–64	65+	Total
Online banking	Yes	265	352	304	167	1088
	No	130	190	249	189	758
	Total	395	542	553	356	1846

Section 11.2 The Least-Squares Regression Line

LEARNING GOALS
By the end of this section, you will be able to:
- Calculate the equation of the least-squares regression line using technology.
- Describe how outliers affect the least-squares regression line.
- Explain the concept of regression to the mean.

A good regression line makes the residuals as small as possible so that the predicted values are close to the actual values. For this reason, statisticians prefer using the **least-squares regression line**.

> The **least-squares regression line** is the line that makes the sum of the squared residuals as small as possible.

The scatterplots in Figure 11.2 show the relationship between the price of a Ford F-150 and the number of miles it has been driven, along with the least-squares regression line. Figure 11.2(a) shows the residuals as vertical line segments. To determine how well the line works, we could add the residuals and hope for a small total.

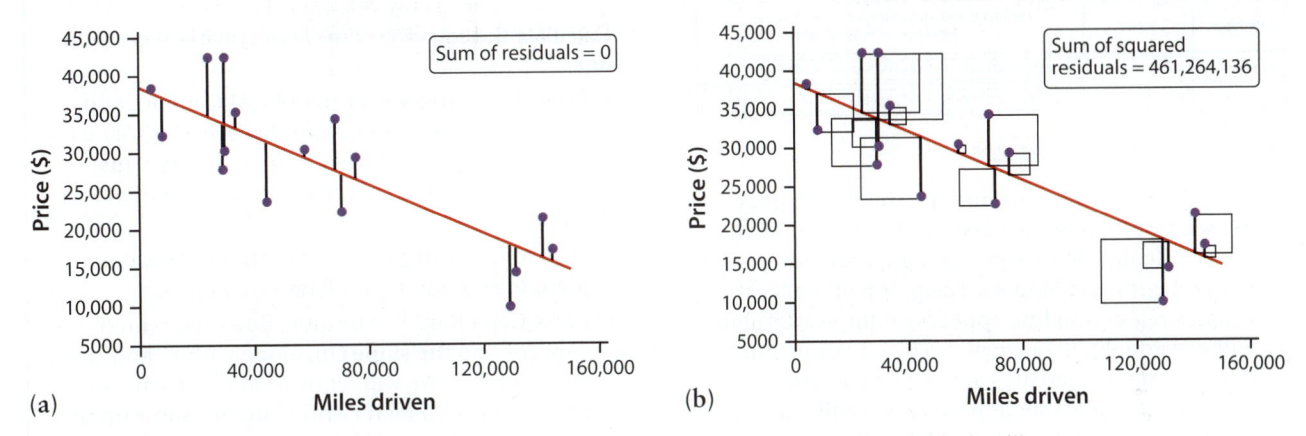

FIGURE 11.2 Scatterplots of the Ford F-150 data with the regression line added. (a) The residuals will add to approximately 0 when using a good regression line. (b) A good regression line should make the sum of squared residuals as small as possible.

If the line is a good fit, some of the residuals will be positive and others negative. If we add these residuals, the positive and negative residuals will cancel each other out and add to approximately 0. Unfortunately, the residuals of some worse-fitting lines (e.g., a horizontal line at $\bar{y} = \$27{,}834$) will also add to 0, making the sum of residuals a flawed measure of quality. To avoid this problem, we square the residuals before adding them, as shown in Figure 11.2(b). Of all the possible lines we could use, the least-squares regression line is the one that makes the sum of the *squares* of the residuals the *least*.

Calculating the Equation of the Least-Squares Regression Line

Technology makes it easy to calculate the equation of the least-squares regression line.

TECH CORNER Calculating the Equation of the Least-Squares Regression Line

You can use technology to calculate the equation of the least-squares regression line. We'll illustrate this process using the Ford F-150 data from Example 11.1.1.

Miles driven	Price ($)	Miles driven	Price ($)
70,583	21,994	34,077	34,995
129,484	9500	58,023	29,988
29,932	29,875	44,447	22,896
29,953	41,995	68,474	33,961
24,495	41,995	144,162	16,883
75,678	28,986	140,776	20,897
8359	31,891	29,397	27,495
4447	37,991	131,385	13,997

(continued)

Applet

1. Go to www.stapplet.com and launch the *Two Quantitative Variables* applet.

2. Enter the name of the explanatory variable (Miles driven) and the values of the explanatory variable in the first row of boxes. Then, enter the name of the response variable [Price ($)] and the values of the response variable in the second row of boxes.

Two Quantitative Variables

Variable	Name	Observations (separated by commas or spaces) Keep individuals in the same order.
Explanatory	Miles driven	70583 129484 29932 29953 24495 75678 8359 4447 34077 580
Response	Price ($)	21994 9500 29875 41995 41995 28986 31891 37991 34995 299

[Begin analysis] [Edit inputs] [Reset everything]

3. Click the Begin analysis button. This will display the scatterplot of price versus miles driven. Then, click "Calculate least-squares regression line" under Regression Models. The graph of the least-squares regression line appears on the scatterplot and a residual plot is created. The equation and other summary statistics are displayed in the Regression Models section. *Note:* We will learn about residual plots and the additional summary statistics in Section 11.3.

Scatterplot

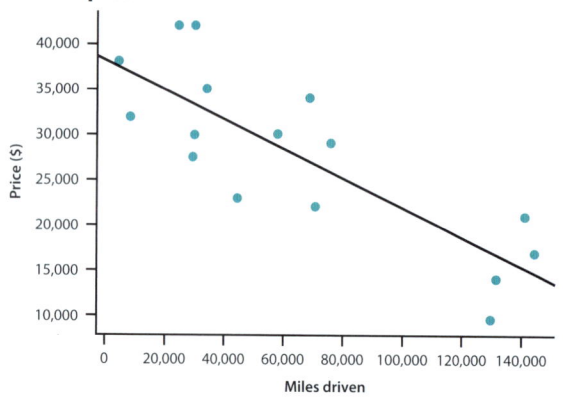

Regression Models

[Calculate least-squares regression line]

Equation	n	s	r^2
$\hat{y} = 38257.13507 - 0.16292x$	16	5740.131	0.664

TI-83/84

1. Enter the miles driven in L_1 and the prices in L_2.
   - Press STAT and choose Edit . . . .
   - Type the x-values into L_1 and the y-values into L_2.

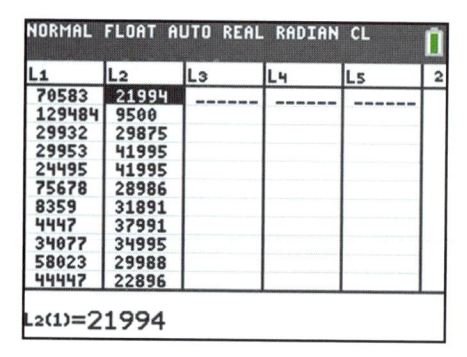

2. Calculate the equation of the least-squares regression line.
   - Press STAT, arrow over to the CALC menu, and choose LinReg(a+bx). Note that the TI-83/84 uses the form $\hat{y} = a + bx$ rather than $\hat{y} = b_0 + b_1x$. Just remember that the slope is the coefficient of x in either form.
   - Adjust the settings as shown. **Older OS:** Complete the command LinReg(a+bx) L_1, L_2.
   - Press Calculate. The output shows the y-intercept (a) and the slope (b), along with r^2 and the correlation r. We will learn what r^2 measures in Section 11.3. *Note:* If r and r^2 do not show up in the output, you need to turn on the Stat Diagnostics. Find this option by pressing MODE or choosing DiagnosticOn in the Catalog (2nd 0).

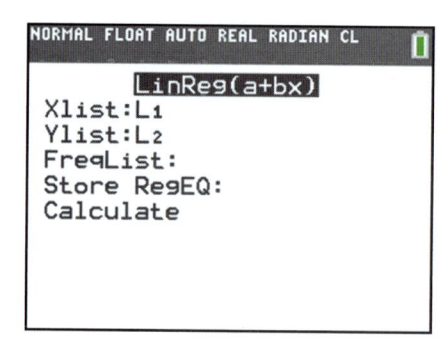

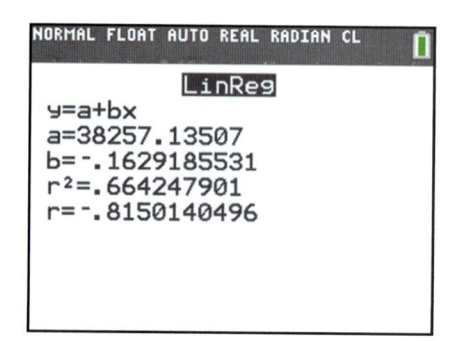

3. *Optional:* Graph the least-squares regression line on a scatterplot.

   - Set up a scatterplot. See the Tech Corner in Section 2.6.
   - Press Y= and enter the equation of the least-squares regression line for Y1.
   - Press ZOOM and choose ZoomStat.

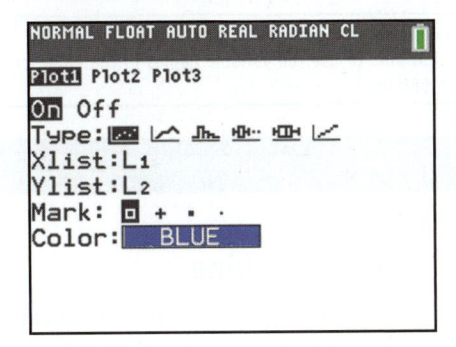

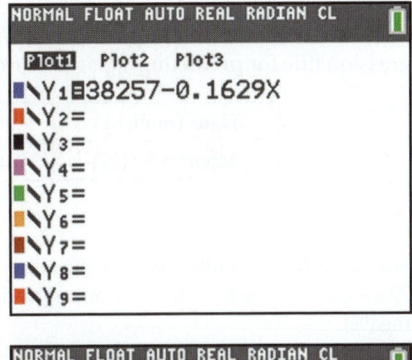

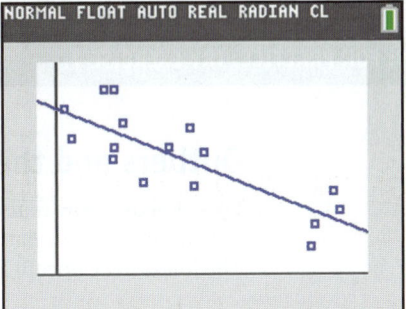

Detailed instructions for using CrunchIt!, Excel, Google Sheets, JMP, Minitab, and R are available in Achieve.

It is very common to use statistical software to calculate least-squares regression lines. The output from statistical software can be mysterious looking at first glance. Fortunately, all software follows a similar format. Here is output for the Ford F-150 data from two statistical software packages: Minitab and JMP. Each set of output records the slope and y-intercept of the least-squares regression line, along with many other statistics that we'll use in subsequent sections. In particular, in Section 11.3 you'll learn about the standard deviation of the residuals and r^2.

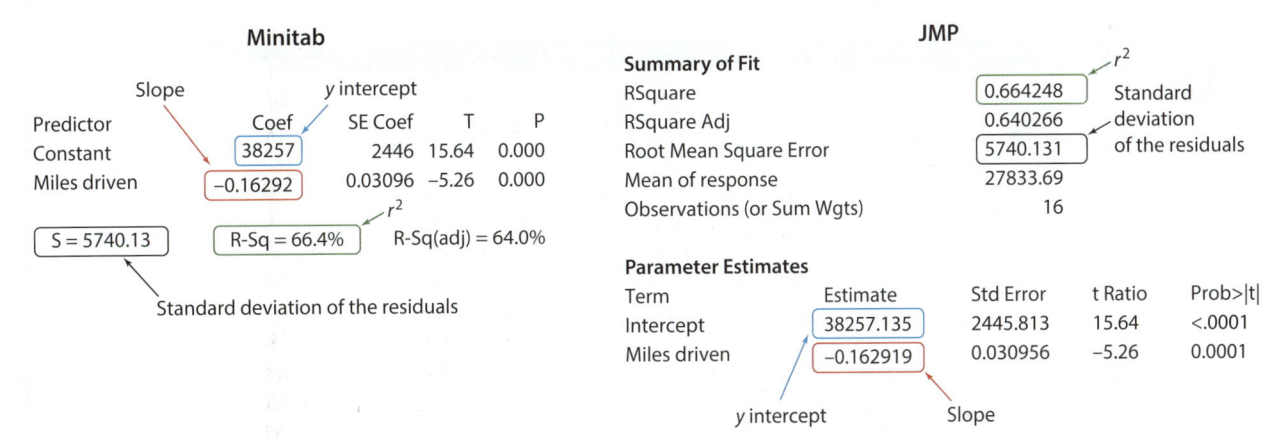

Minitab

Predictor	Coef	SE Coef	T	P
Constant	38257	2446	15.64	0.000
Miles driven	−0.16292	0.03096	−5.26	0.000

S = 5740.13   R-Sq = 66.4%   R-Sq(adj) = 64.0%

- 38257 = y intercept
- −0.16292 = Slope
- S = 5740.13 = Standard deviation of the residuals
- R-Sq = 66.4% = r^2

JMP

Summary of Fit

RSquare	0.664248
RSquare Adj	0.640266
Root Mean Square Error	5740.131
Mean of response	27833.69
Observations (or Sum Wgts)	16

- RSquare = r^2
- Root Mean Square Error = Standard deviation of the residuals

Parameter Estimates

Term	Estimate	Std Error	t Ratio	Prob>|t|
Intercept	38257.135	2445.813	15.64	<.0001
Miles driven	−0.162919	0.030956	−5.26	0.0001

- 38257.135 = y intercept
- −0.162919 = Slope

EXAMPLE 11.2.1

Does diet soda get less sweet over time?

Calculating the least-squares regression line using technology

PROBLEM

People who drink diet sodas may notice a decrease in the sweetness of the soda if it has been sitting around for a few months — particularly in a warm environment. To determine how fast the sweetener aspartame degrades over time, researchers used diet soda of three different acidity levels (pH). They kept some at 20°C, some at 30°C, and the rest at 40°C. The following table shows the time (in months) and the

(continued)

remaining aspartame (%) for a soda with pH = 2.75 stored at 20°C.[10] Use technology to calculate the equation of the least-squares regression line for predicting aspartame from time.

Time (months)	0	1	2	3	4	5
Aspartame (%)	100	90.3	85.3	80.9	74.2	64.5

SOLUTION

Using the *Two Quantitative Variables* applet, the equation is $\hat{y} = 99.0 - 6.58x$, where $x =$ time (months) and $y =$ aspartame (%).

> Make sure to clearly define x and y when stating the equation of a least-squares regression line. Or use the variable names in the equation: *aspartame* $= 99.0 - 6.58$*time*.

EXAM PREP: FOR PRACTICE, TRY EXERCISE 7.

Outliers and the Least-Squares Regression Line

You already know that outliers can greatly affect summary statistics such as the mean, standard deviation, and correlation. Do outliers also affect the equation of a least-squares regression line? The following Concept Exploration will help you find out.

CONCEPT EXPLORATION

Outliers and the least-squares regression line

In this exploration, you will investigate how outliers affect the least-squares regression line.

1. Go to www.stapplet.com, click the link for Resources for *Introductory Statistics: A Student-Centered Approach*, and launch the *Regression* applet.

2. A scatterplot should appear in the Desmos graphing calculator, along with a least-squares regression line (the orange line), a vertical line at $\bar{x}$ (the purple line), and a horizontal line at $\bar{y}$ (the green line). Notice that the least-squares regression line goes through the point $(\bar{x}, \bar{y})$.

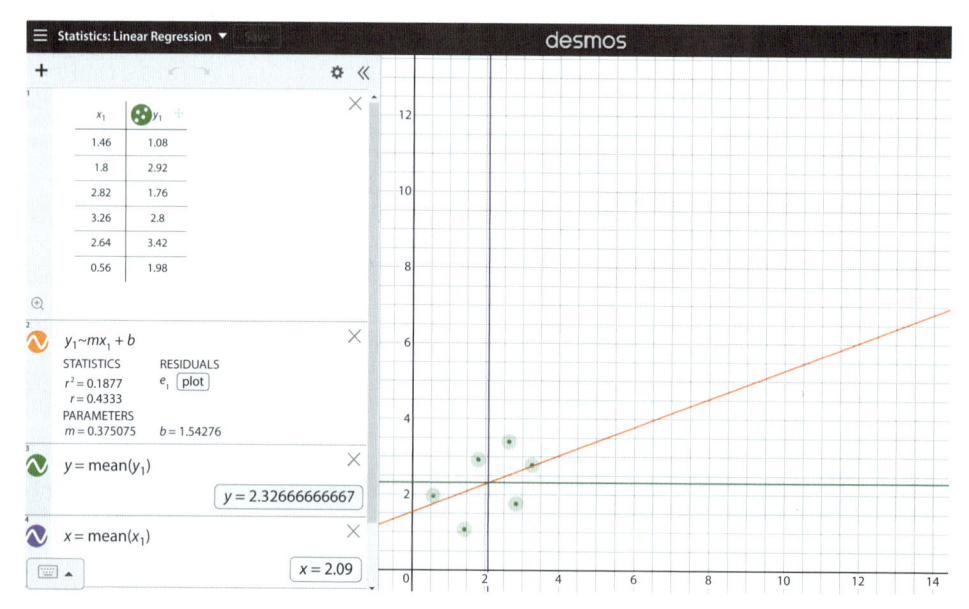

3. There is currently one point on the least-squares regression line. If you were to move this point along the regression line to the right edge of the graphing area, what do you think will happen to the equation of the least-squares regression line? Move this point to see if you were correct.

4. Click on the point you just moved and drag it up and down along the right edge of the graphing area. What happens to the equation of the least-squares regression line? Does the least-squares regression line still go through the point $(\bar{x}, \bar{y})$?

5. Now, move this point so that it is on the vertical $\bar{x}$ line. Drag the point up and down on the $\bar{x}$ line. What happens to the equation of the least-squares regression line? Do outliers in the vertical direction have more or less influence on the least-squares regression line than outliers in the horizontal direction?
6. Briefly summarize how outliers influence the equation of the least-squares regression line.

As you learned in the Concept Exploration, the equation of the least-squares regression line can be greatly influenced by outliers, so the slope and *y*-intercept are not resistant. However, some outliers are more influential than others!

EXAMPLE 11.2.2

First speech and mental aptitude

Describing the effect of outliers

PROBLEM

Does the age at which a child begins to talk predict a later score on a test of mental ability? A study of the development of young children recorded the age in months at which each of 23 children spoke their first word and that child's Gesell Adaptive Score, the result of an aptitude test taken much later.[11] The scatterplot shows the relationship between Gesell score and age at first word, along with the least-squares regression line. Two outliers, Child 18 and Child 19, are identified on the scatterplot.

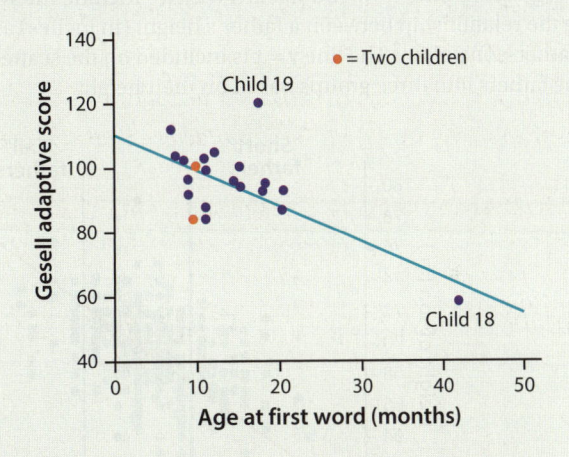

(a) Describe the effect Child 18 has on the equation of the least-squares regression line.
(b) Describe the effect Child 19 has on the equation of the least-squares regression line.

SOLUTION

(a) Because the point for this child is below the line on the right, it is making the slope of the regression line steeper (more negative) and increasing the *y*-intercept.
(b) Because this point is near $\bar{x}$ but above the rest of the points, it pulls the line up a little, which increases the *y*-intercept but doesn't change the slope very much.

EXAM PREP: FOR PRACTICE, TRY EXERCISE 11.

Figure 11.3 shows the effects of removing Child 18 and Child 19 on the correlation and the regression line. The graph adds two more regression lines, one calculated after leaving out Child 18 and the other after leaving out Child 19. You can see that removing the point for Child 18 changes the correlation and moves the line quite a bit. Because of Child 18's extreme position on the age scale, this point has a strong influence on both the correlation and the line. However, removing Child 19 has only a small effect on the correlation and not much effect on the regression line.

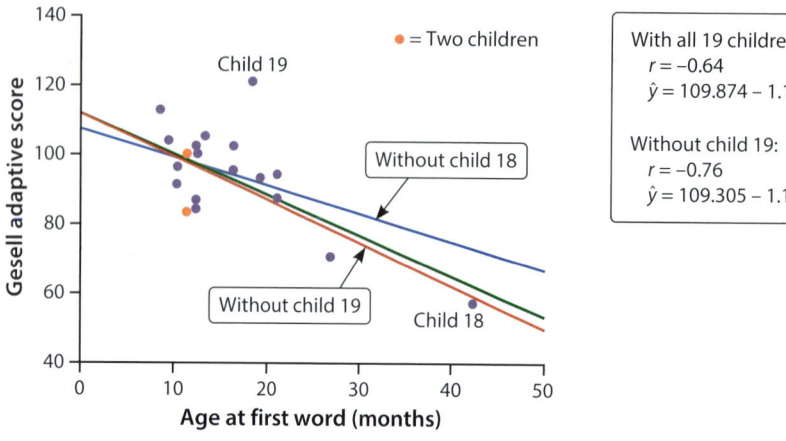

FIGURE 11.3 Three least-squares regression lines for predicting Gesell score using age at first word. The green line is calculated from all the data. The blue line is calculated leaving out Child 18. Child 18 is an influential observation because leaving out this point moves the regression line quite a bit. The red line is calculated leaving out only Child 19.

Regression to the Mean

Earlier in this section, you learned why we include the phrase *least-squares* in the name of the least-squares regression line. Now you'll learn why we include the word *regression*. Here is a scatterplot showing the relationship between a father's height (in inches) and his adult son's height (in inches) for 465 father–son pairs.[12] The line $y = x$ is included on the scatterplot, along with vertical lines separating the fathers into three groups based on their height.

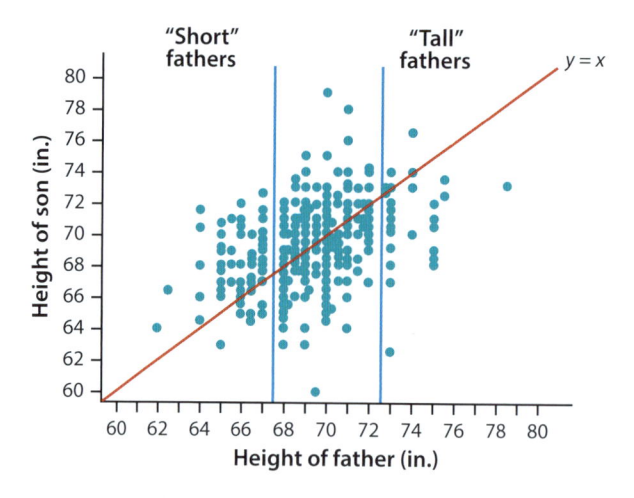

If a father and son are exactly the same height, the point for this pair will be on the line $y = x$. If the son is taller than his father, their point will be above the $y = x$ line. And if the son is shorter than his father, their point will be below the $y = x$ line.

Before looking at the data, it seems reasonable to believe that any particular son would have a 50% chance of being taller than his father. However, consider the sons of the "tall" fathers on the right side of the scatterplot — these sons still tended to be tall, but most were shorter than their fathers. The story is reversed for the sons of "short" fathers on the left side of the graph — these sons tended to be short, but almost all of them were taller than their fathers. In both cases, the heights of the sons tended to *regress* to the overall mean. This phenomenon is called **regression to the mean**.

> The tendency for extreme values of the explanatory variable to be paired with less extreme values of the response variable is called **regression to the mean**.

Here is a scatterplot of the same data showing both the $y = x$ line and the least-squares regression line. Notice how the least-squares regression line predicts that sons of tall fathers will be tall, but not as tall as their fathers, on average. Likewise, the least-squares regression line predicts that sons of short fathers will be short, but not as short as their fathers, on average.

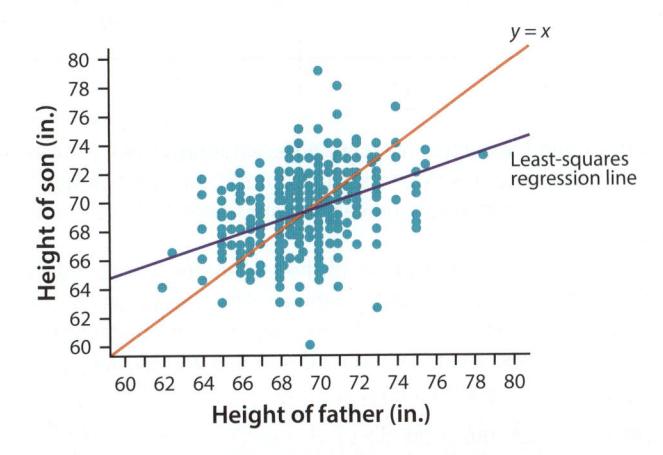

EXAMPLE 11.2.3

Do LPGA golfers regress to the mean?

Regression to the mean

PROBLEM

The scatterplot shows the results of the third and fourth rounds of the 2021 Ladies Professional Golf Association (LPGA) U.S. Open golf tournament.[13] Each dot represents a single golfer, with dots slightly staggered to avoid overlapping points. The line $y = x$ is included on the scatterplot, along with vertical lines separating the golfers into three groups based on the Round 3 score. Remember that lower scores are better in golf!

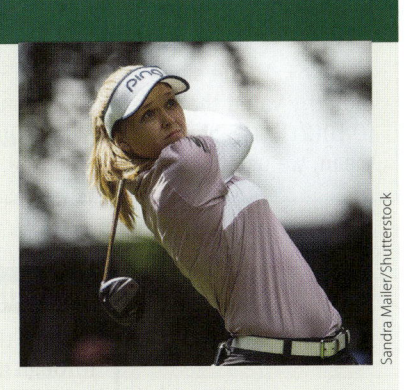

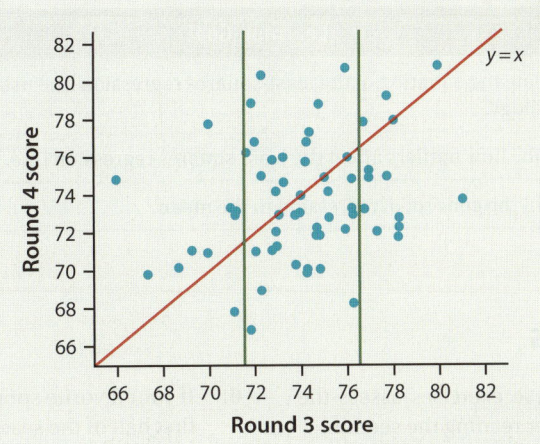

(a) Of the 10 golfers who shot 71 or less in Round 3, what percentage did worse (scored higher) in Round 4?
(b) Of the 13 golfers who shot 77 or more in Round 3, what percentage did better (scored lower) in Round 4?
(c) Explain how your answers to parts (a) and (b) illustrate regression to the mean.

SOLUTION

(a) 9/10 = 90%
(b) 9/13 = 69.2%
(c) Among the golfers who did the best in Round 3, more than half (90%) had a worse (less extreme) performance in Round 4. Likewise, among the golfers who did the worst in Round 3, more than half (69.2%) had a better (less extreme) performance in Round 4.

EXAM PREP: FOR PRACTICE, TRY EXERCISE 15.

If your favorite team or player is off to a great start, don't get too excited! Regression to the mean is likely to occur sooner or later.

> **THINK ABOUT IT**
>
> **Does regression to the mean still occur when the *x* and *y* variables measure different things?** When investigating the relationship between two variables that have roughly the same mean and standard deviation, we can use the line $y = x$ as a reference to illustrate regression to the mean. However, in many instances, we are interested in the relationship between two variables with different means and standard deviations.
>
> Recall that regression to the mean is the tendency for extreme values of the explanatory variable to be paired with less extreme values of the response variable. We can measure "extreme-ness" with standard deviations. For example, an *x* value that is 2 standard deviations greater than $\bar{x}$ could be considered extreme. Regression to the mean says that the predicted value of *y* for this *x* value will be less extreme—that is, *less than* 2 standard deviations greater than $\bar{y}$. How much less?
>
> Here are two additional facts about the least-squares regression line:
>
> 1. The least-squares regression line always passes through the point $(\bar{x}, \bar{y})$.
> 2. The slope of the least-squares regression line is $b_1 = r \dfrac{s_y}{s_x}$.
>
> If $r = 1$, the slope is $b_1 = s_y/s_x$. Thus, an *x* value that is 2 standard deviations greater than $\bar{x}$ would result in a predicted *y* value that is exactly 2 standard deviations greater than $\bar{y}$. But if $r = 0.8$, the predicted *y* value would be only $(0.8)(2) = 1.6$ standard deviations greater than $\bar{y}$. Because the correlation determines the extent of the *regression* to the mean, we use the symbol "*r*" for the correlation.

Section 11.2 What Did You Learn?

Review the learning goals from this section. Then practice what you've learned by working through the exercises.

Learning Goal	Example	Exercises
Calculate the equation of the least-squares regression line using technology.	11.2.1	7–10
Describe how outliers affect the least-squares regression line.	11.2.2	11–14
Explain the concept of regression to the mean.	11.2.3	15–18

Section 11.2 Exercises

Building Concepts and Skills These exercises assess the basic knowledge you should have after reading the section.

1. What does "least-squares" mean in the phrase *least-squares regression line*?

2. Which applet or calculator function can you use to calculate the equation of a least-squares regression line?

3. True/False: Outliers always affect the equation of the least-squares regression line.

4. True/False: It is possible for an outlier to affect the *y*-intercept of a least-squares regression line but not the slope.

5. What is regression to the mean?

6. If your favorite sports team is off to a great start in the first half of the season, what is most likely to happen in the second half of the season: the team will do better than in the first half, the team will do about the same as in the first half, or the team will do worse than in the first half?

Mastering Concepts and Skills These exercises reinforce the learning goals as shown in the examples.

7. **Refer to Example 11.2.1 Candy Calories** Here are data from a sample of 12 types of candy.[14] Use technology to calculate the equation of the least-squares regression line relating y = calories to x = amount of sugar (in grams).

Name	Sugar	Calories
Butterfinger Minis	45	450
Junior Mints	107	570
M&M'S	62	480
Milk Duds	44	370
Peanut M&M'S	79	790
Raisinettes	60	420
Reese's Pieces	61	580
Skittles	87	450
Sour Patch Kids	92	490
SweeTarts	136	680
Twizzlers	59	460
Whoppers	48	350

8. **Buried Change** Drilling down beneath a lake in Alaska yields chemical evidence of past changes in climate. Biological silicon, left by the skeletons of single-celled creatures called diatoms, is a measure of the abundance of life in the lake. Another variable, based on the ratio of certain isotopes, gives an indirect measure of moisture—mostly from snow. As we drill down, we look further into the past.[15] Use technology to calculate the equation of the least-squares regression line relating y = silicon to x = isotope.

Isotope (%)	Silicon (mg/g)
−19.90	97
−19.84	106
−19.46	118
−20.20	141
−20.71	154
−20.80	265
−20.86	267
−21.28	296
−21.63	224
−21.63	237
−21.19	188
−19.37	337

9. **Anscombe's First Two** Statistician Francis Anscombe created four data sets to illustrate certain properties of least-squares regression lines. Here are the first two data sets. The final two data sets are in Exercise 10.

Data Set A

x	10	8	13	9	11	14	6	4	12	7	5
y	8.04	6.95	7.58	8.81	8.33	9.96	7.24	4.26	10.84	4.82	5.68

Data Set B

x	10	8	13	9	11	14	6	4	12	7	5
y	9.14	8.14	8.74	8.77	9.26	8.10	6.13	3.10	9.13	7.26	4.74

(a) Use technology to calculate the least-squares regression line for each data set. What do you notice?

(b) Use technology to make a scatterplot for each data set. What do you notice?

10. **Anscombe's Final Two** Statistician Francis Anscombe created four data sets to illustrate certain properties of least-squares regression lines. The first two data sets are in Exercise 9. Here are the final two data sets:

Data Set C

x	10	8	13	9	11	14	6	4	12	7	5
y	7.46	6.77	12.74	7.11	7.81	8.84	6.08	5.39	8.15	6.42	5.73

Data Set D

x	8	8	8	8	8	8	8	8	8	8	19
y	6.58	5.76	7.71	8.84	8.47	7.04	5.25	5.56	7.91	6.89	12.50

(a) Use technology to calculate the least-squares regression line for each data set. What do you notice?

(b) Use technology to make a scatterplot for each data set. What do you notice?

11. **Refer to Example 11.2.2 Tree Density** In the article "Do Forests Have the Capacity for 1 Trillion Extra Trees?" author Matthew Russell analyzed data on the total area, forest area, and number of trees for each U.S. state. Here is a scatterplot showing y = the number of trees (in billions) and x = forest area (in millions of acres), along with the least-squares regression line, excluding Alaska and Hawaii.[16]

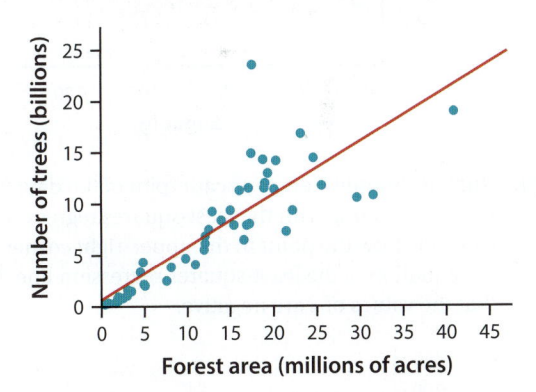

(a) Describe how the point representing Maine (17.6, 23.6) affects the equation of the least-squares regression line.

(b) Describe how the point representing Texas (41.0, 19.0) affects the equation of the least-squares regression line.

12. **Gestation and Life Span** Is there a relationship between the gestational period (time from conception to birth) of an animal and its average lifespan? Here is a scatterplot of the x = gestational period (days) and y = average lifespan (years), along with the least-squares regression line, for 43 species of animals.[17]

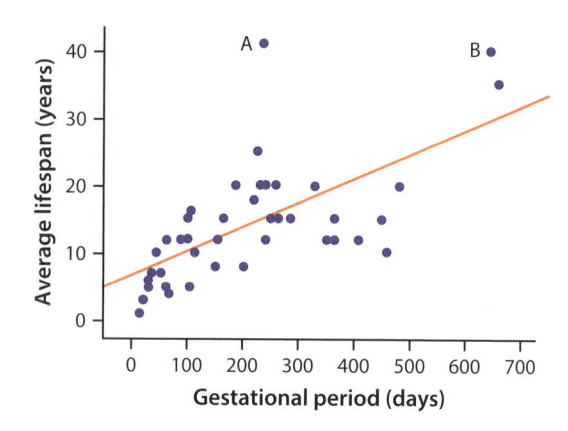

(a) Point A is the hippopotamus. Describe how this point affects the equation of the least-squares regression line.

(b) Point B is the Asian elephant. Describe how this point affects the equation of the least-squares regression line.

13. **Candy Calories** Here is a scatterplot of the data from Exercise 7, along with the least-squares regression line. Describe how the point representing Peanut M&M'S® at (79, 790) affects the least-squares regression line.

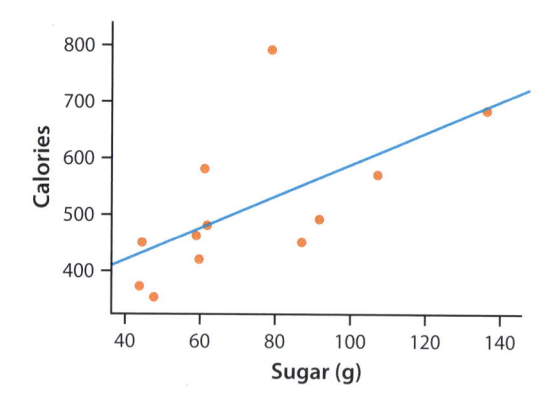

14. **Buried Change** Here is a scatterplot of the data from Exercise 8, along with the least-squares regression line. Describe how the point in the upper-right corner affects the equation of the least-squares regression line. Note that the values of x are negative.

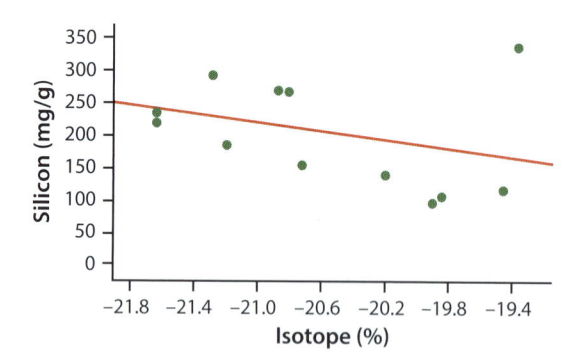

15. **Refer to Example 11.2.3 NBA Teams** The scatterplot shows the relationship between winning percentage before the All-Star break (roughly mid-season) and winning percentage after the All-Star break for the 30 National Basketball Association (NBA) teams in a recent season.[18] Also included on the scatterplot are vertical lines at 40% and 60% and the line $y = x$.

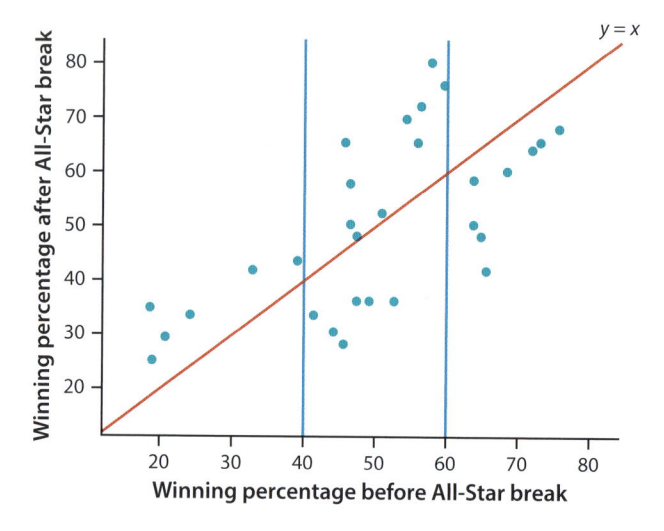

(a) Of the 6 teams that won fewer than 40% of their games before the All-Star break, what percentage did better after the All-Star break?

(b) Of the 8 teams that won more than 60% of their games before the All-Star break, what percentage did worse after the All-Star break?

(c) Explain how your answers to parts (a) and (b) illustrate regression to the mean.

16. **SAT Practice** To get extra practice, students in an SAT preparation class took two versions of the SAT math section on the same day. The scatterplot shows the relationship between scores on the first practice test and scores on the second practice test. The line $y = x$ is included on the scatterplot, along with vertical lines separating the students into three groups based on their score on the first practice exam.

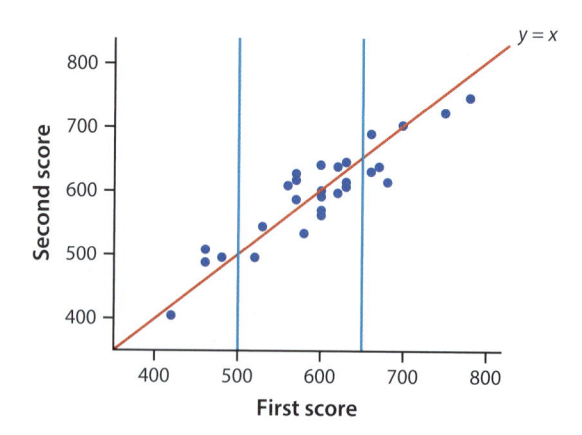

(a) Of the 4 students who scored less than 500 on the first exam, what percentage scored higher on the second exam?

(b) Of the 7 students who scored greater than 650 on the first exam, what percentage scored lower on the second exam?

(c) Explain how your answers to parts (a) and (b) illustrate regression to the mean.

17. **Midterms and Finals** We expect that students who do well on the midterm exam in a course will also do well on the final exam. Gary Smith of Pomona College looked at the relationship between x = midterm score and y = final exam score for all 346 students who took his statistics class over a 10-year period.[19] Assume that both the midterm and the final exam were scored out of 100 points.

    (a) State the equation of the least-squares regression line if each student's score on the final was equal to that student's score on the midterm.

    (b) The actual least-squares regression line for predicting final exam score y from midterm exam score x is $\widehat{final} = 46.6 + 0.41 midterm$. Predict the final exam score of a student who scored 50 on the midterm and a student who scored 100 on the midterm.

    (c) Explain how your answers to part (b) illustrate regression to the mean.

18. **Batting Averages** We expect that a baseball player who has a high batting average in the first month of the season will also have a high batting average in the rest of the season. Using performance data for 66 Major League Baseball (MLB) players from a recent season,[20] a least-squares regression line was calculated to predict y = rest-of-season batting average from x = first-month batting average. *Note:* A player's batting average is the proportion of plate appearances in which he gets a hit. A batting average that exceeds 0.300 is considered very good in MLB.

    (a) State the equation of the least-squares regression line if each player had the same batting average in the rest of the season as he did in the first month of the season.

    (b) The actual equation of the least-squares regression line is $\hat{y} = 0.245 + 0.109x$. Predict the rest-of-season batting average for a player who had a 0.200 batting average in the first month of the season and for a player who had a 0.400 batting average in the first month of the season.

    (c) Explain how your answers to part (b) illustrate regression to the mean.

Applying the Concepts These exercises ask you to apply multiple learning goals in a new context or to apply what you learned in this section in a new way.

19. **Mighty Beans** Beans and other legumes are an excellent source of protein. The table gives data on the total protein and carbohydrate content (each in grams) of a one-half-cup portion of cooked beans for 12 different varieties.[21]

Type	Carbohydrate (g)	Protein (g)
Soybeans	8.5	14
Garbanzos	22.5	7
Black beans	20	8
Adzuki beans	28.5	8.5
Cranberry beans	21.5	8
Great Northern beans	18.5	7
Kidney beans	20	7.5
Navy beans	23.5	7.5
Pinto beans	22.5	7.5
White beans	22.5	8.5
Lima beans	19.5	7
Mung beans	19	7

    (a) Make a scatterplot to show the relationship between protein and carbohydrate content for these bean varieties, using carbohydrate content as the explanatory variable.

    (b) Using technology, calculate the equation of the least-squares regression line relating y = protein content to x = carbohydrate content.

    (c) What effect do you think the observation for soybeans has on the equation of the least-squares regression line? Calculate the equation of the least-squares regression line without this variety to confirm your answer.

20. **Measuring Glucose** People with diabetes measure their fasting plasma glucose (FPG; measured in units of milligrams per milliliter, mg/mL) after fasting for at least 8 hours. Another measurement, made at regular medical checkups, is called HbA1c. It is roughly the percentage of red blood cells that have a glucose molecule attached. HbA1c measures average exposure to glucose over a period of several months. The table gives data on both HbA1c and FPG for 18 people with diabetes five months after they completed a diabetes education class.[22]

HbA1c (%)	FPG (mg/mL)
6.1	141
6.3	158
6.4	112
6.8	153
7.0	134
7.1	95
7.5	96
7.7	78
7.9	148
8.7	172
9.4	200
10.4	271
10.6	103
10.7	172
10.7	359
11.2	145
13.7	147
19.3	255

(a) Make a scatterplot to show the relationship between FPG and HbA1c, using HbA1c as the explanatory variable.

(b) Using technology, calculate the equation of the least-squares regression line relating $y =$ FPG to $x =$ HbA1c.

(c) What effect do you think the subject with HbA1c of 19.3% and FPG of 255 mg/ml has on the equation of the least-squares regression line? Calculate the equation of the least-squares regression line without this subject to confirm your answer.

21. **Tech** **Home Prices** Do larger houses cost more? Use the *Home Sales* data set to investigate this question.[23]

(a) Make a scatterplot showing the relationship between $y =$ Price and $x =$ Size.

(b) Use technology to calculate the equation of the least-squares regression line.

(c) Calculate and interpret the residual for the 1633-square-foot house that sold for $265,000.

(d) Interpret the slope of the least-squares regression line.

22. **Tech** **Frozen Desserts** What is the relationship between saturated fat and calories in frozen desserts? Use the *Frozen Desserts* data set to investigate this question.[24]

(a) Make a scatterplot showing the relationship between $y =$ Calories and $x =$ SatFat.

(b) Use technology to calculate the equation of the least-squares regression line.

(c) Calculate and interpret the residual for Breyer's Natural Vanilla, which has 6 grams of saturated fat and 170 calories.

(d) Interpret the slope of the least-squares regression line.

23. **Car Price and Age** Do Honda CRVs decrease in value as they age? Using a random sample of 191 used Honda CRVs, we calculated a least-squares regression line to predict price (dollars) from age (in years). Use the computer output to determine the equation of the least-squares regression line. Make sure to define the variables in the equation.

```
Predictor    Coefficient    StdError       T       P
Constant         31687.1      304.03   104.224   0.000
Age             -1754.03       52.98   -33.108   0.000
S=2248.08                  R-sq=0.853
```

24. **Car Price and Mileage** Do Honda CRVs decrease in value as they are driven more? Using a random sample of 191 used Honda CRVs, we calculated a least-squares regression line to predict price (dollars) from miles driven. Use the computer output to determine the equation of the least-squares regression line. Make sure to define the variables in the equation.

```
Predictor      Coefficient    StdError       T       P
Constant          30773.4       417.41   73.724  0.000
Miles driven      -0.1497       0.0069  -21.675  0.000
S=3139.8                    R-sq=0.713
```

Extending the Concepts These exercises challenge you to explore statistical concepts and methods that go beyond what you learned in this section.

25. **Reversing the Variables** In Section 3.6, you learned that reversing the explanatory and response variables doesn't affect the correlation. Does reversing the variables affect the equation of the least-squares regression line?

(a) Using the candy data from Exercise 7, calculate the equation of the least-squares regression line relating $y =$ sugar to $x =$ calories.

(b) Are the slope and y-intercept the same as in Exercise 7?

26. **Predicting Test Scores** Each year, students in an elementary school take a standardized math test at the end of the school year. For a class of fourth-graders, the average score was 55.1, with a standard deviation of 12.3. In the third grade, these same students had an average score of 61.7, with a standard deviation of 14.0. The correlation between the two sets of scores is $r = 0.95$. Use the two facts from the Think About It feature in this section to calculate the equation of the least-squares regression line for predicting a fourth-grade score from a third-grade score.

27. **Lean Body Mass and Metabolism** The data and scatterplot show the lean body mass and metabolic rate for a sample of 5 adults. For each person, the lean body mass is the adult's total weight in kilograms less any weight due to fat. The metabolic rate is the number of calories burned in a 24-hour period.

Mass (kg)	33	43	40	55	49
Rate (cal/day)	1050	1120	1400	1500	1700

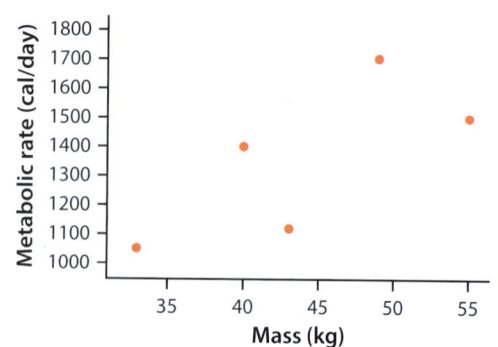

In this context, it makes sense to model the relationship between metabolic rate and body mass with a direct variation function in the form $y = kx$. After all, a person with no lean body mass should burn no calories, and functions in the form $y = kx$ always go through the point (0, 0). But what value of k would be best?

Several different values of k, such as $k = 25$, $k = 26$, and so on, were used to predict the metabolic rate, and the sum of squared residuals (SSR) was calculated for each value of k. Here is a scatterplot showing the relationship between SSR and k. According to the scatterplot, what is the ideal value of k to use for predicting metabolic rate? Explain your answer.

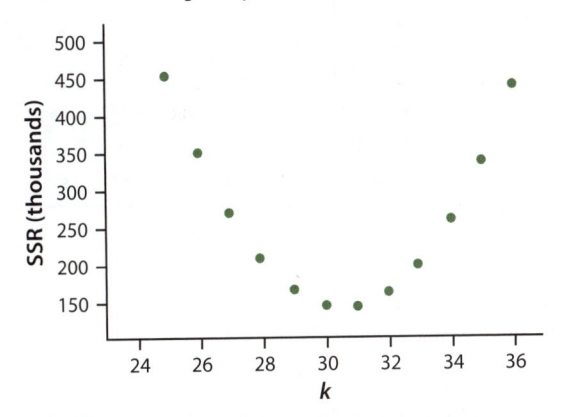

Cumulative Review These exercises revisit what you learned in previous sections.

28. **Accidents and Marijuana (1.8, 2.2)** Researchers in New Zealand interviewed 907 drivers at age 21. The researchers had data on previous traffic accidents and asked the drivers about marijuana use. Here are data on whether these drivers had caused accidents at age 19, broken down by marijuana use at that age:[25]

		\multicolumn{4}{c}{Frequency of marijuana use per year}	
		Never	1–10 times	11–50 times	51+ times	Total
Caused accident	No	393	193	55	106	747
	Yes	59	36	15	50	160
	Total	452	229	70	156	907

(a) Make a segmented bar chart to determine if there is an association between these variables. If there is an association, describe it. If there is no association, explain how you know.

(b) Explain why we can't conclude, using these data, that marijuana use *causes* accidents.

29. **Heavy Diamonds (2.5, 3.1, 3.2)** Here are the weights (in milligrams) of 58 diamonds from a nodule carried up to the earth's surface in surrounding rock. These data represent a population of diamonds formed in a single event deep in the earth.[26] Make a histogram that shows the distribution of weights of these diamonds. Describe what you see. Give appropriate numerical measures of center and variability.

13.8	3.7	33.8	11.8	27.0	18.9	19.3	20.8	25.4	23.1	7.8
10.9	9.0	9.0	14.4	6.5	7.3	5.6	18.5	1.1	11.2	7.0
7.6	9.0	9.5	7.7	7.6	3.2	6.5	5.4	7.2	7.8	3.5
5.4	5.1	5.3	3.8	2.1	2.1	4.7	3.7	3.8	4.9	2.4
1.4	0.1	4.7	1.5	2.0	0.1	0.1	1.6	3.5	3.7	2.6
4.0	2.3	4.5

Section 11.3 Assessing a Regression Model

LEARNING GOALS
By the end of this section, you will be able to:
- Use a residual plot to determine whether a regression model is appropriate.
- Interpret the standard deviation of the residuals.
- Interpret r^2.

Now that you have learned how to calculate a least-squares regression line, it is important to assess how well the line fits the data. We do this by asking two questions:

- Is a line the right model to use, or would a curve be better?
- If a line is the right model to use, how well does it make predictions?

Residual Plots

In Section 11.1, you learned how to calculate and interpret a residual. You can also use residuals to assess whether a regression model is appropriate by making a **residual plot**.

A **residual plot** is a scatterplot that plots the residuals on the vertical axis and the explanatory variable on the horizontal axis.

Figure 11.4(a) is a scatterplot showing the relationship between diameter at breast height (DBH) in inches and volume of usable lumber in board feet for a random sample of 40 ponderosa pines. The resulting residual plot is shown in Figure 11.4(b).

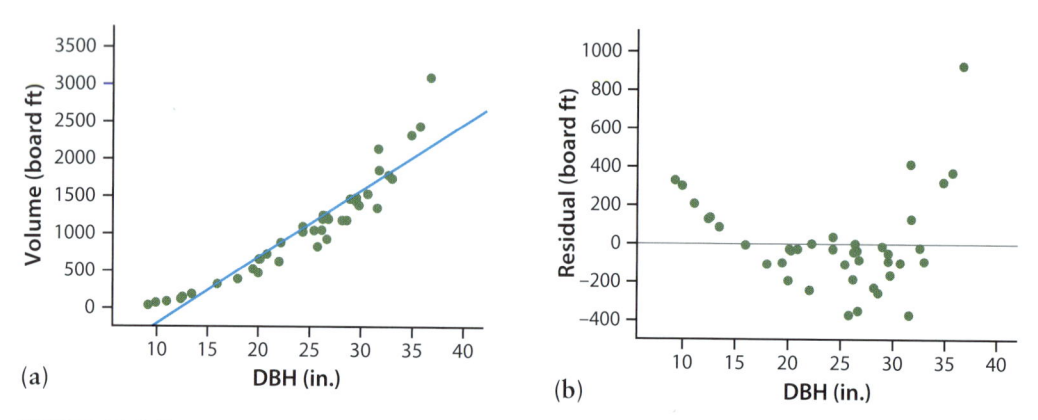

FIGURE 11.4 For a random sample of 40 ponderosa pines, (a) the scatterplot showing the relationship between DBH and volume of usable lumber and (b) the residual plot for the linear model.

The least-squares regression line clearly doesn't fit this association very well. This is especially obvious when looking at the leftover curved pattern in the residual plot! For skinny trees, the actual volume is always greater than the line predicts, resulting in positive residuals. For trees with DBH from roughly 16 to 32 inches, the actual volumes are typically less than the line predicts, resulting in negative residuals. For trees with DBH greater than 32 inches, the actual volumes are typically greater than the line predicts, again resulting in positive residuals. This positive–negative–positive pattern in the residual plot indicates that the linear form of our model doesn't match the form of the association. A curved model might be better in this case.

Figure 11.5 gives a scatterplot (a) showing the Ford F-150 data from Section 11.1, along with the corresponding residual plot (b). Looking at the scatterplot, the line seems to be a good fit for the association. You can "see" that the line is appropriate by the lack of a leftover curved pattern in the residual plot. In fact, the residuals look randomly scattered around the residual = 0 line.

FIGURE 11.5 For a random sample of 16 Ford F-150s, (a) the scatterplot showing the relationship between miles driven and price and (b) the residual plot for the linear model.

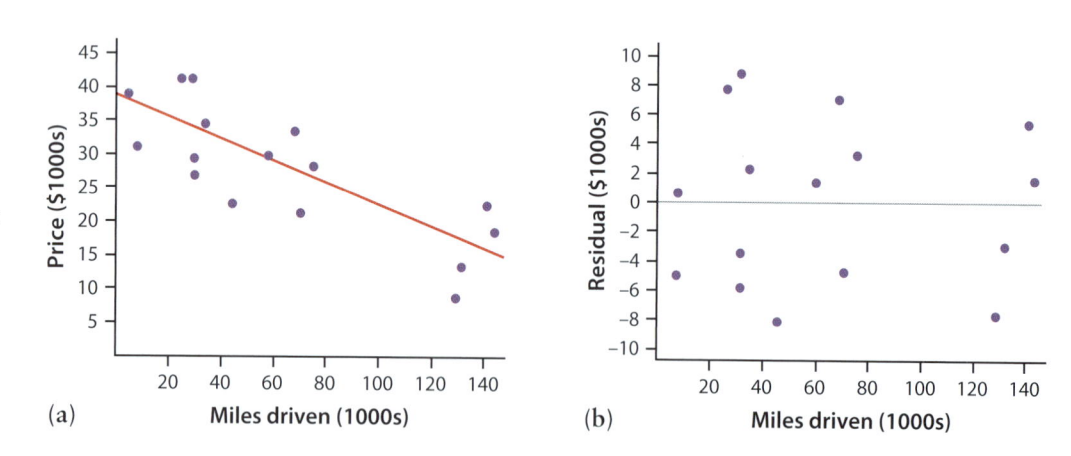

Interpreting a Residual Plot

To determine whether the regression model is appropriate, look at the residual plot.
- If there is no leftover curved pattern in the residual plot, the regression model is appropriate.
- If there is a leftover curved pattern in the residual plot, the regression model is not appropriate.

> **THINK ABOUT IT**
>
> **Why do we look for patterns in residual plots?**
>
> The word "residual" comes from the Latin word *residuum*, meaning "left over." When we calculate a residual, we are calculating what is left over after subtracting the predicted value from the actual value:
>
> $$\text{residual} = \text{actual } y - \text{predicted } y$$
>
> Likewise, when we look at the form of a residual plot, we are looking at the form that is left over after subtracting the form of the model from the form of the association:
>
> $$\text{form of residual plot} = \text{form of association} - \text{form of model}$$
>
> When there is a leftover curved form in the residual plot, the form of the association and the form of the model are not the same. However, if the form of the association and the form of the model are the same, the residual plot should have no obvious form, other than random scatter.

EXAMPLE 11.3.1

Studying ponderosa pines

Interpreting a residual plot

PROBLEM

In Section 11.1, we used a least-squares regression line to model the relationship between the height of a ponderosa pine tree (in feet) and its DBH (in inches). Here is the residual plot for that model. Use the residual plot to determine whether the regression model is appropriate.

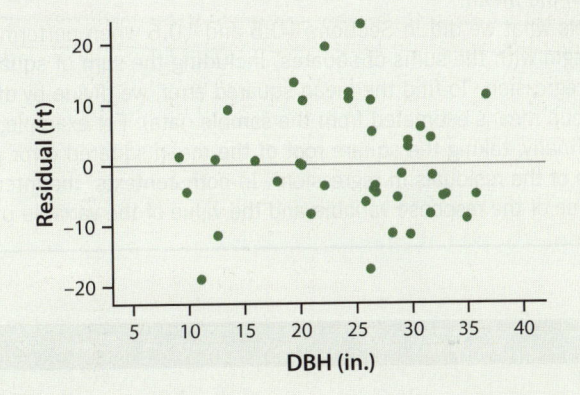

SOLUTION

Because there is no leftover curved pattern in the residual plot, the least-squares regression line is an appropriate model for relating the height of a ponderosa pine tree to its diameter at breast height.

> Random scatter in the residual plot means the form of the model (a line) matches the form of the association (linear).

EXAM PREP: FOR PRACTICE, TRY EXERCISE 7.

Standard Deviation of the Residuals

Once we have all the residuals, we can measure how well the line makes predictions with the **standard deviation of the residuals** s.

> The **standard deviation of the residuals** s measures the size of a typical residual. That is, s measures the typical distance between the actual y values and the predicted y values.

To calculate the standard deviation of the residuals s, we square each of the residuals, add them, divide the sum by $n-2$, and take the square root. We can write the formula for calculating the standard deviation of the residuals as follows:

$$s = \sqrt{\frac{\text{sum of squared residuals}}{\text{number of points} - 2}} = \sqrt{\frac{\sum(y_i - \hat{y}_i)^2}{n-2}}$$

In Section 11.1, we calculated a residual of –4765 dollars for the Ford F-150 with 70,583 miles and an actual price of $21,994. The price of this truck was $4765 less than predicted, based on the number of miles it had been driven. Here are the residuals for all 16 trucks in our sample using the least-squares regression line $\hat{y} = 38{,}257 - 0.1629x$:

| –4765 | –7664 | –3506 | 8617 | 7728 | 3057 | –5004 | 458 |
| 2289 | 1183 | –8121 | 6858 | 2110 | 5572 | –5973 | –2857 |

The standard deviation of the residuals is

$$s = \sqrt{\frac{(-4765)^2 + (-7664)^2 + \cdots + (-2857)^2}{16-2}} = \sqrt{\frac{461{,}264{,}136}{14}} = \$5740$$

The actual price of a Ford F-150 is typically about $5740 different than the price predicted by the least-squares regression line with x = miles driven.

> **THINK ABOUT IT**
>
> **Why do we divide by $n-2$ instead of $n-1$ when calculating the standard deviation of the residuals?** Because we use the sample data to estimate *two* values—the sample slope and the sample y-intercept—when calculating the standard deviation of the residuals. When calculating the standard deviation of a single quantitative variable, we use the sample data to estimate only *one* value—the sample mean.
>
> Dividing by $n-2$ also parallels what we did in Sections 10.5 and 10.6 when performing a one-way ANOVA test. When building the ANOVA table, we begin with the sums of squares, including the sum of squared errors, which is equivalent to the sum of squared residuals in regression. To find the mean squared error, we divide by df = n – (number of groups), which is the same as n – (number of group means estimated from the sample data). For example, if we were using one-way ANOVA with two groups, $df_{error} = n - 2$. Finally, taking the square root of the mean squared error gives s, the standard deviation of the errors (the standard deviation of the residuals in regression). In both contexts, the interpretation is the same: the typical difference between the actual value of the response variable and the value of the variable predicted by the model.

EXAMPLE 11.3.2

Studying ponderosa pines

Interpreting s

PROBLEM

In previous examples, we used a least-squares regression line to model the relationship between the height of a ponderosa pine tree (in feet) and the diameter at breast height (in inches). The standard deviation of the residuals for this model is $s = 9.4$. Interpret this value.

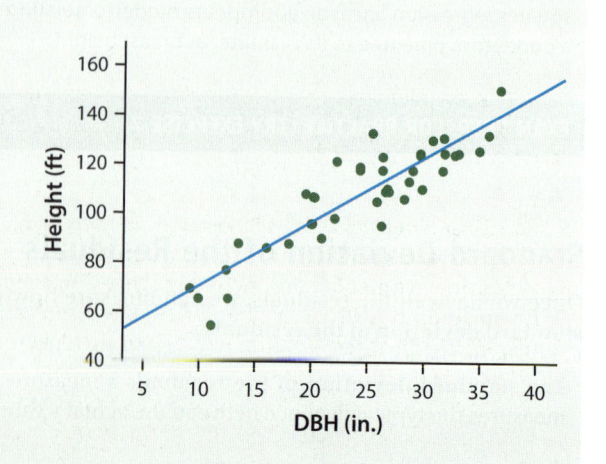

SOLUTION

The actual height of a ponderosa pine tree is typically about 9.4 feet away from the height predicted by the least-squares regression line with x = diameter at breast height.

EXAM PREP: FOR PRACTICE, TRY EXERCISE 11.

The Coefficient of Determination r^2

Besides the standard deviation of the residuals s, we can also use the **coefficient of determination** r^2 to measure how well the regression line makes predictions.

> The **coefficient of determination** r^2 measures the percent reduction in the sum of squared residuals when using the least-squares regression line to make predictions, rather than the mean value of y. In other words, r^2 measures the percentage of the variability in the response variable that is accounted for by the least-squares regression line.

Suppose that we wanted to predict the price of a particular Ford F-150, but we didn't know how many miles it had been driven. Our best guess would be the average price of a Ford F-150 from our sample of 16 trucks, $\bar{y}$ = $27,834. Of course, this prediction is unlikely to be very good, as the prices vary quite a bit from the mean (s_y = $9570). If we knew how many miles the truck had been driven, we could use the least-squares regression line to make a better prediction. How much better are predictions using the least-squares regression line with x = miles driven, rather than using only the average price? The answer is r^2.

The scatterplot in Figure 11.6(a) shows the squared residuals along with the sum of squared residuals (1,373,893,151) when using the average price as the predicted value. The scatterplot in Figure 11.6(b) shows the squared residuals along with the sum of squared residuals (461,264,136) when using the least-squares regression line with x = miles driven to predict the price.

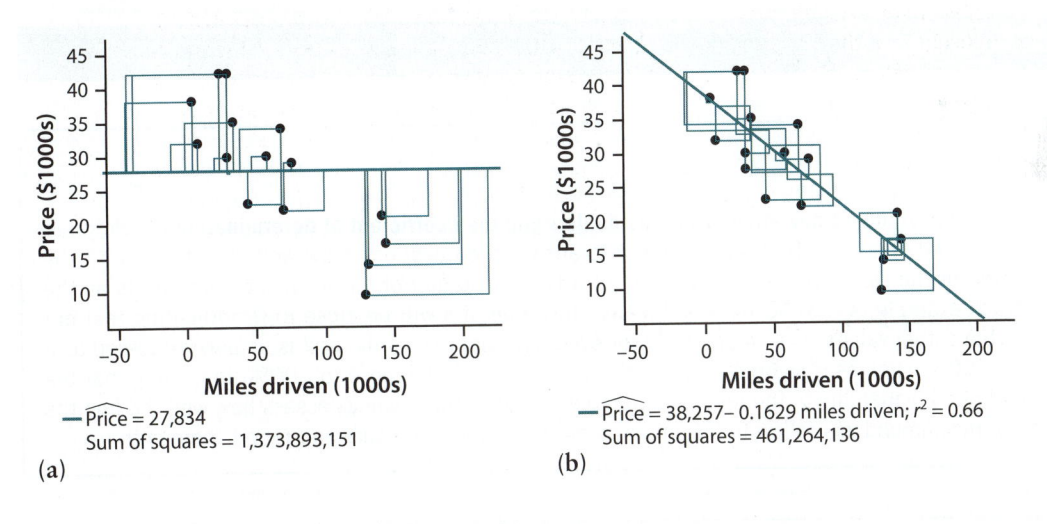

FIGURE 11.6 (a) The sum of squared residuals is 1,373,893,151 if we use the mean price as our prediction for the price of a Ford F-150. (b) The sum of squared residuals from the least-squares regression line is 461,264,136.

(a) —— $\widehat{\text{Price}}$ = 27,834
Sum of squares = 1,373,893,151

(b) —— $\widehat{\text{Price}}$ = 38,257 − 0.1629 miles driven; r^2 = 0.66
Sum of squares = 461,264,136

To find r^2, calculate the percent reduction in the sum of squared residuals:

$$r^2 = \frac{1{,}373{,}893{,}151 - 461{,}264{,}136}{1{,}373{,}893{,}151} = \frac{912{,}629{,}015}{1{,}373{,}893{,}151} = 0.66$$

The sum of squared residuals has been reduced by 66%. That is, 66% of the variability in the price of a Ford F-150 is accounted for by the least-squares regression line with x = miles driven. The remaining 34% is due to other factors, including age, color, condition, and other features of the truck.

If you studied ANOVA in Chapter 10, the three values in the fractions in the calculation are equivalent to the three sums of squares in the ANOVA table. The denominator of the fraction (1,373,893,151) is the sum of squares total (SST). The sum of the squared residuals (461,264,136) is the sum of squares error (SSE). The difference between SST and SSE (912,629,015) is called the sum of squares model (what we called SSG in ANOVA).

The easiest way to calculate the value of r^2 is to square the value of the correlation r. Recall that you learned how to calculate the correlation in Section 3.6. The value of r^2 is also typically provided every time you calculate the equation of a least-squares regression line using technology.

EXAMPLE 11.3.3

Studying ponderosa pines

Interpreting r^2

PROBLEM

In previous examples, we used a least-squares regression line to model the relationship between the height of a ponderosa pine tree (in feet) and its diameter at breast height (in inches). Interpret the value $r^2 = 0.812$ for this model.

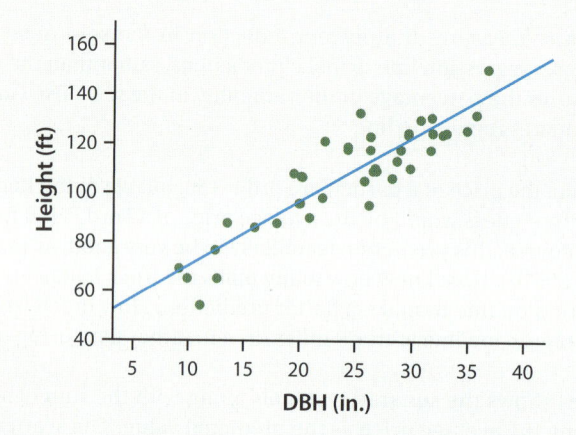

SOLUTION

81.2% of the variability in height is accounted for by the least-squares regression line with $x =$ diameter at breast height.

EXAM PREP: FOR PRACTICE, TRY EXERCISE 15.

THINK ABOUT IT

What's the relationship between the standard deviation of the residuals s and the coefficient of determination r^2? Both are calculated from the sum of squared residuals. Both also attempt to answer the question "How well does the line fit the data?" The standard deviation of the residuals reports the size of a typical prediction error, in the same units as the response variable. In the truck example, $s = 5740$ *dollars*. Ideally, the value of s will be close to 0, indicating that our predictions are very close to the actual values. The value of r^2, however, does not have units and is usually expressed as a percentage between 0% and 100%, such as $r^2 = 66\%$. Ideally, the value of r^2 will be close to 100%, indicating that the model accounts for almost all of the variability in the response variable. Because these values assess how well the line fits the data in different ways, we recommend you follow the example of most statistical software and report them both.

TECH CORNER — Making residual plots, Calculating s and r^2

You can use technology to make a residual plot, calculate r^2, and calculate s. We'll illustrate this process using the Ford F-150 data.

Miles driven	Price ($)	Miles driven	Price ($)	Miles driven	Price ($)
70,583	21,994	8359	31,891	68,474	33,961
129,484	9500	4447	37,991	144,162	16,883
29,932	29,875	34,077	34,995	140,776	20,897
29,953	41,995	58,023	29,988	29,397	27,495
24,495	41,995	44,447	22,896	131,385	13,997
75,678	28,986				

Applet

1. Go to www.stapplet.com and launch the *Two Quantitative Variables* applet.
2. Enter the name of the explanatory variable (Miles driven) and the values of the explanatory variable in the first row of boxes. Then, enter the name of the response variable [Price ($)] and the values of the response variable in the second row of boxes.

Two Quantitative Variables

Variable	Name	Observations (separated by commas or spaces) Keep individuals in the same order.
Explanatory	Miles driven	70583 129484 29932 29953 24495 75678 8359 4447 34077 580
Response	Price ($)	21994 9500 29875 41995 41995 28986 31891 37991 34995 299

[Begin analysis] [Edit inputs] [Reset everything]

3. Click the Begin analysis button. This will display the scatterplot of price versus miles driven. Then, click "Calculate least-squares regression line" under Regression Models. The residual plot appears directly below the scatterplot. The equation of the least-squares regression line, n (number of points), s, and r^2 are displayed in the Regression Models section.

Scatterplot

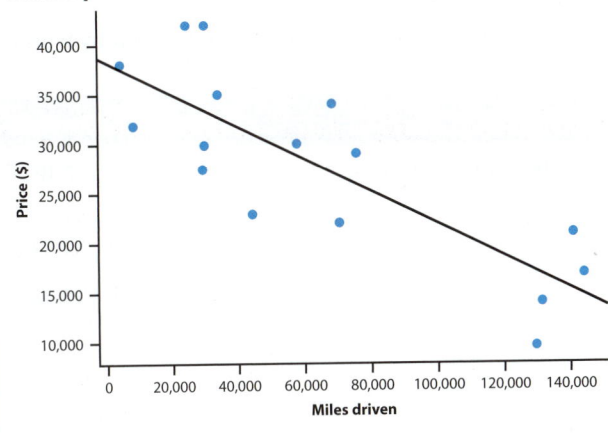

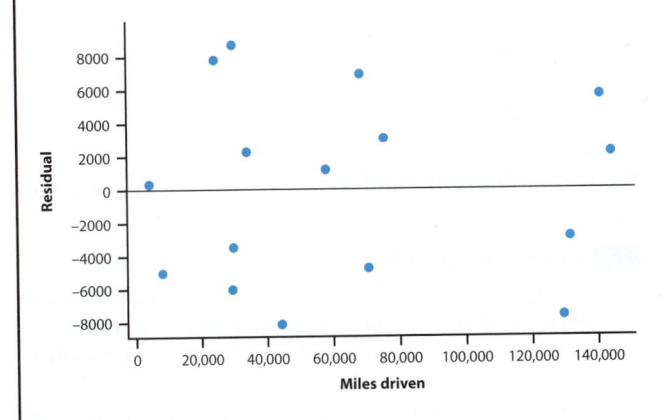

Regression Models

[Calculate least-squares regression line]

Equation	n	s	r^2
$\hat{y} = 38257.13507 - 0.16292x$	16	5740.131	0.664

TI-83/84

1. Enter the miles driven in L_1 and the prices in L_2.
   - Press STAT and choose Edit . . . .
   - Type the x-values into L_1 and the y-values into L_2.

2. Calculate the equation of the least-squares regression line and r^2.
   - Press STAT, arrow over to the CALC menu, and choose LinReg(a+bx).
   - Adjust the settings as shown and press Calculate. In addition to the equation of the least-squares regression line, the calculator displays the value of r^2. **Older OS:** Complete the command LinReg(a+bx) L_1, L_2. *Note:* If r and r^2 do not show up in the output, you need to turn on the Stat Diagnostics. Find this option by pressing MODE or choosing DiagnosticOn in the catalog (2nd 0).

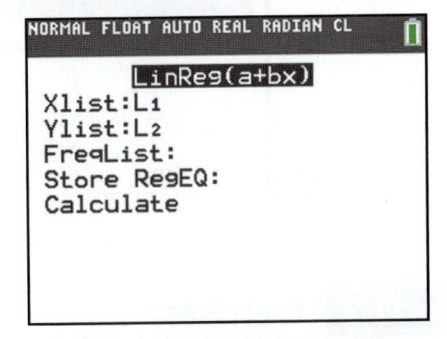

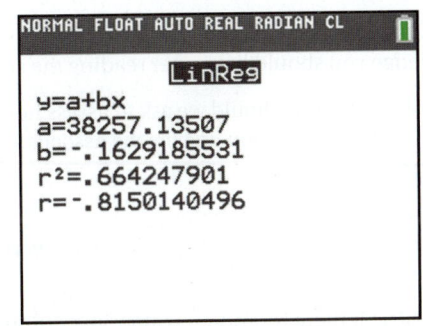

3. To make a residual plot, set up a scatterplot in the statistics plots menu.
   - Press 2nd Y= (STAT PLOT).
   - Press ENTER or 1 to go into Plot1.
   - Adjust the settings as shown. The RESID list is found in the List menu by pressing 2nd STAT. *Note:* You have to calculate the equation of the

(continued)

least-squares regression line using the calculator *before* making a residual plot. Otherwise, the RESID list will include the residuals from a previously calculated least-squares regression line.

- Press ZOOM and choose 9: ZoomStat.

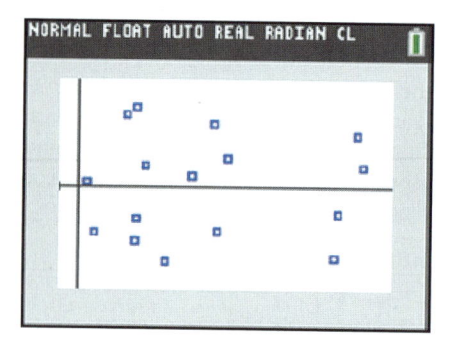

Note: It is possible to calculate the standard deviation of the residuals s on the TI-83/84 using the LinRegTTest option. See the Tech Corner in Section 11.5.

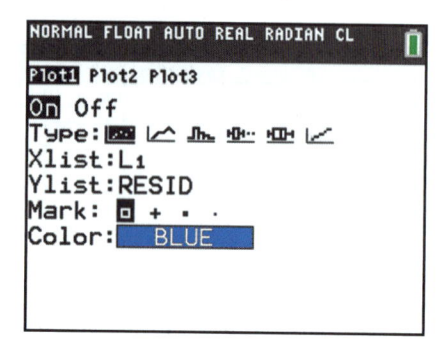

4. Use ZoomStat to let the calculator choose an appropriate window.

Detailed instructions for using CrunchIt!, Excel, Google Sheets, JMP, Minitab, and R are available in Achieve.

Section 11.3 What Did You Learn?

Review the learning goals from this section. Then practice what you've learned by working through the exercises.

Learning Goal	Example	Exercises
Use a residual plot to determine whether a regression model is appropriate.	11.3.1	7–10
Interpret the standard deviation of the residuals.	11.3.2	11–14
Interpret r^2.	11.3.3	15–18

Section 11.3 Exercises

Building Concepts and Skills These exercises assess the basic knowledge you should have after reading the section.

1. What two questions should we ask to assess how well a least-squares regression line fits the association in a scatterplot?

2. In a residual plot, we graph the _____ on the horizontal axis and the _____ on the vertical axis.

3. True/False: If there is random scatter in a residual plot, then the model used to create the residual plot is appropriate.

4. What does the standard deviation of the residuals s measure?

5. What does the coefficient of determination r^2 measure?

6. True/False: A model will make better predictions when the standard deviation of the residuals s is larger and the coefficient of determination r^2 is smaller.

Mastering Concepts and Skills These exercises reinforce the learning goals as shown in the examples.

7. **Refer to Example 11.3.1 Click-Through Rates**
   Companies work hard to have their website listed at the top of an internet search. A linear model was used to predict y = the percentage of people who click on a link for the website from x = the website's position in the results of an internet search on a mobile device (1 = top position, 2 = second position, etc.).[27] Here is a residual plot for this model. Use it to determine if the regression model is appropriate.

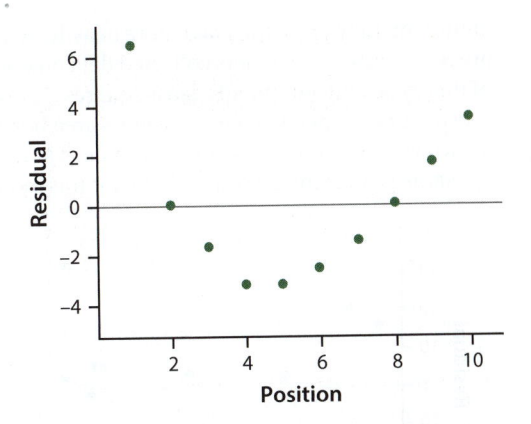

8. **Height Disadvantage?** Very tall basketball players have a reputation of being bad free-throw shooters (even though their shots start closer to the rim!). Is this true for players in the Women's National Basketball Association (WNBA)? Using data from a recent season, a linear model was used to predict y = free-throw percentage from x = height (in inches). Here is the residual plot for that model. Use it to determine whether the regression model is appropriate. *Note: The slope of the least-squares regression line is −0.77, so it appears that taller WNBA players are worse free-throw shooters.*

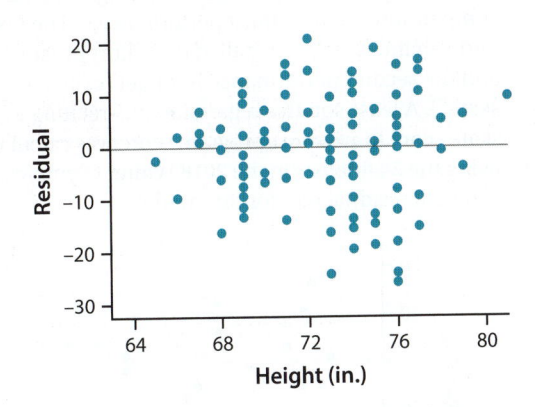

9. **Tablet Batteries** Can you predict the battery life of a tablet from the tablet's price? Using data from a sample of 15 tablets, a linear model was created to predict y = battery life (in hours) from x = price (in dollars).[28] Here is a residual plot for this model. Use it to determine if the regression model is appropriate.

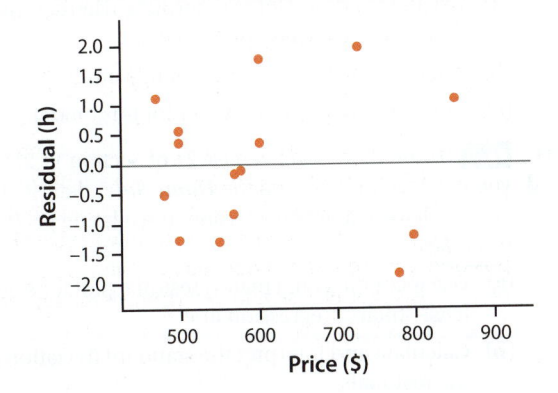

10. **Grill Cooling** An avid barbecuer wondered how fast a gas grill cools once it has been turned off. The barbecuer recorded the grill's thermometer reading at several points in time and used a linear model to predict y = grill temperature (in °F) from time elapsed since the grill was turned off (in minutes). Here is the residual plot for that model. Use it to determine whether the regression model is appropriate.

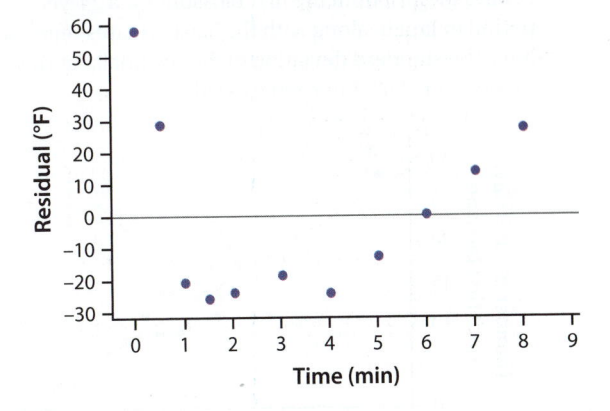

11. **Refer to Example 11.3.2 Old Faithful** One of the major attractions in Yellowstone National Park is the Old Faithful geyser. The scatterplot shows the relationship between x = the duration of the previous eruption (in minutes) and y = wait time until the next eruption (in minutes), along with the least-squares regression line. The standard deviation of the residuals for this model is $s = 6.49$. Interpret this value.

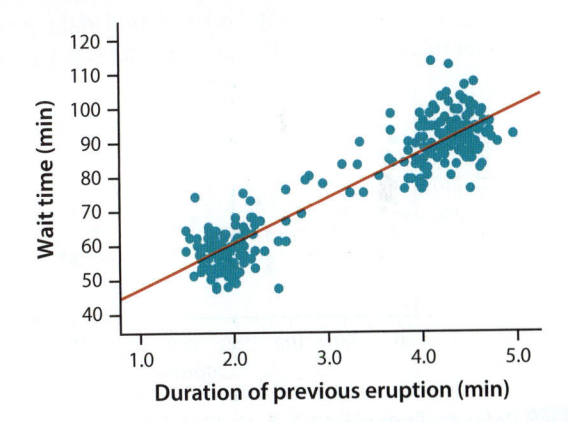

12. **Crickets** The scatterplot shows the relationship between x = temperature (in °F) and y = chirps per minute for the striped ground cricket, along with the least-squares regression line. The standard deviation of the residuals for this model is $s = 0.97$. Interpret this value.[29]

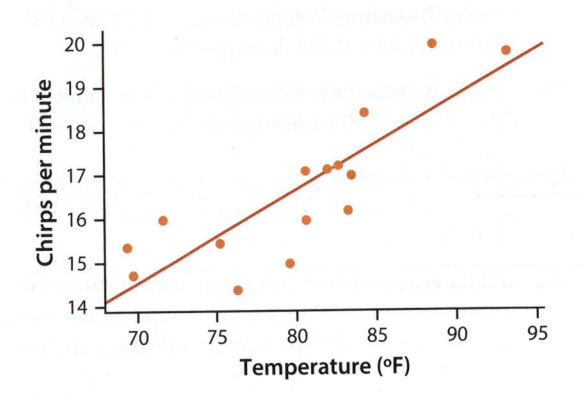

13. **Cherry Blossoms** Many people look forward to the blossoming of cherry trees for their beautiful flowers. Because the flowers don't last long, it's helpful to be able to predict when the trees will blossom. Here is a scatterplot showing the relationship between $x =$ the average temperature in March (in °C) and $y =$ number of days in April until the first blossom for a 24-year period in Japan, along with the least-squares regression line. The standard deviation of the residuals for this model is $s = 3.02$. Interpret this value.

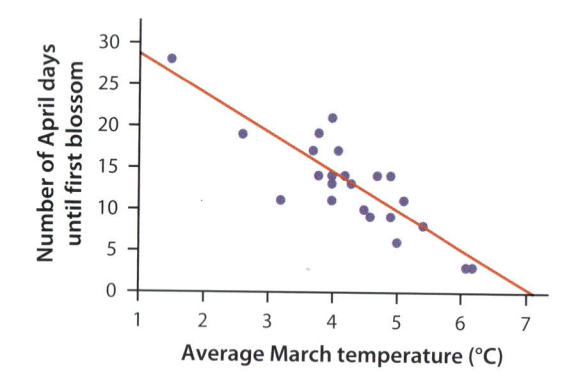

14. **Coffee Drinks** Single-serve coffee drinks are quite popular. Some people drink them for their caffeine and others for the sugar — and some people for both! Here is a scatterplot showing the relationship between $x =$ calories and $y =$ caffeine content (in mg) for 23 different single-serve coffee drinks, along with the least-squares regression line.[30] The standard deviation of the residuals for this model is $s = 46.9$. Interpret this value.

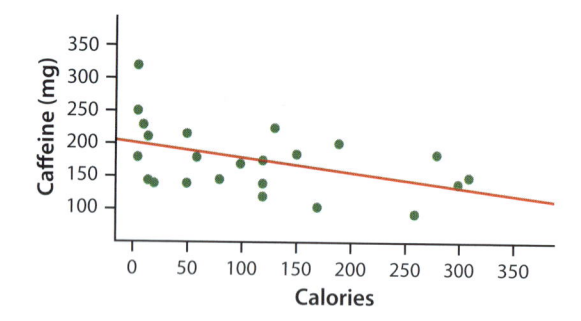

15. **Refer to Example 11.3.3 Old Faithful** Refer to Exercise 11. The value of r^2 for this model is 0.854. Interpret this value.

16. **Crickets** Refer to Exercise 12. The value of r^2 for this model is 0.697. Interpret this value.

17. **Cherry Blossoms** Refer to Exercise 13. The value of r^2 for this model is 0.724. Interpret this value.

18. **Coffee Drinks** Refer to Exercise 14. The value of r^2 for this model is 0.196. Interpret this value.

Applying the Concepts
These exercises ask you to apply multiple learning goals in a new context or to apply what you learned in this section in a new way.

19. **Wildebeest and Fire** Long-term records from the Serengeti National Park in Tanzania show interesting ecological relationships. When wildebeest are more abundant, they graze the grass more heavily, so there are fewer fires. Researchers collected data on one part of this cycle and computed a least-squares regression line relating $y =$ percentage of the grass area burned to $x =$ wildebeest abundance (in thousands of animals) in the same year. Here is a residual plot for this model.[31]

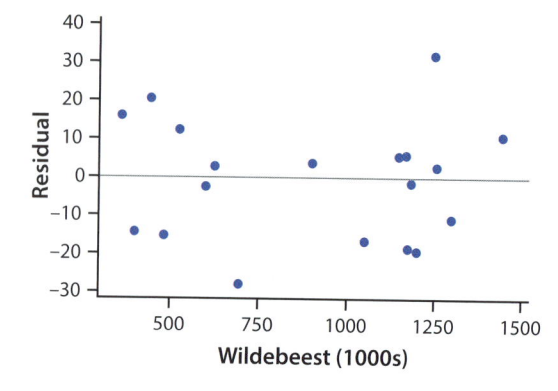

    (a) Use the residual plot to determine whether the regression model is appropriate.
    (b) Interpret the value $s = 15.99$ for this model.
    (c) Interpret the value $r^2 = 0.646$ for this model.

20. **Figure Skating** In Olympic women's figure skating competitions, skaters must perform twice. The first performance is shorter (called the "short program") and the second performance is longer (called the "free skate"). A least-squares regression line relating $y =$ free-skate score to $x =$ short-program score was calculated using the 24 finalists in the 2018 Winter Olympics.[32] Here is a residual plot for this model.

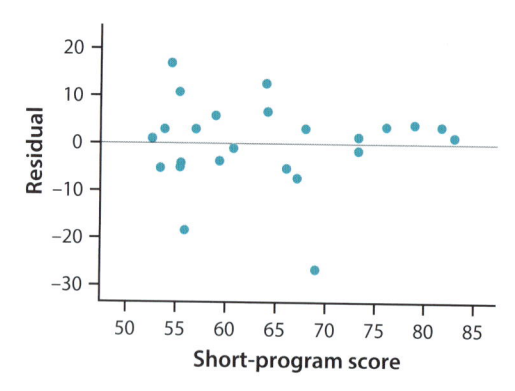

    (a) Use the residual plot to determine whether the regression model is appropriate.
    (b) Interpret the value $s = 9.42$ for this model.
    (c) Interpret the value $r^2 = 0.774$ for this model.

21. **Tech Home Prices** In Exercise 21 of Section 11.2, you used technology and the **Home Sales** data set to create a least-squares regression line relating $y =$ Price to $x =$ Size.

    (a) Use technology to create a residual plot for this least-squares regression line.
    (b) Calculate and interpret the standard deviation of the residuals.
    (c) Calculate and interpret the value of r^2.

22. **Tech Frozen Desserts** In Exercise 22 of Section 11.2, you used technology and the *Frozen Desserts* data set to create a least-squares regression line relating $y =$ Calories to $x =$ SatFat.
    (a) Use technology to create a residual plot for this least-squares regression line.
    (b) Calculate and interpret the standard deviation of the residuals.
    (c) Calculate and interpret the value of r^2.

Extending the Concepts These exercises challenge you to explore statistical concepts and methods that go beyond what you learned in this section.

23. **Old Faithful** Refer to Exercise 11. Suppose that the wait times were recorded in hours instead of minutes. How would this change affect the values of s and r^2? Explain your reasoning.

24. **Grill Cooling** Refer to Exercise 10. There is a negative association between $x =$ time since the grill was turned off and $y =$ temperature. Use this fact along with the residual plot from Exercise 10 to sketch a scatterplot of $x =$ time versus $y =$ temperature.

25. **Form and s** The value of s alone doesn't reveal information about the form of an association. Sketch a scatterplot showing a nonlinear association with a small value of s and a second scatterplot showing a linear association with a large value of s.

Cumulative Review These exercises revisit what you learned in previous sections.

26. **Teenager Height (3.4)** According to the National Center for Health Statistics, the distribution of height for 15-year-old males has a mean of 170 centimeters (cm) and a standard deviation of 7.5 cm. Paul is 15 years old and 179 cm tall.
    (a) Find the z-score corresponding to Paul's height. Explain what this value means.
    (b) Paul's height puts him at the 88th percentile among 15-year-old males. Explain what this means to someone who knows no statistics.

27. **Stair Climber (11.1, 11.2)** Alana's favorite exercise machine is a stair climber. On the "random" setting, the machine changes speeds at regular intervals, so the total number of simulated "floors" she climbs varies from session to session. Alana also exercises for different lengths of time each session. She decides to explore the relationship between the number of minutes she works out on the stair climber and the number of floors it tells her that she's climbed. Alana records climbing time (in minutes) and number of floors climbed for 6 exercise sessions.

Time (min)	Floors
15	73
16	82
18	88
20	103
22	109
25	127

(a) Calculate the equation of the least-squares regression line relating $y =$ number of floors climbed to $x =$ climbing time.
(b) Using the regression equation you calculated in part (a), predict the number of simulated "floors" Alana would climb in 21 minutes.
(c) Would you be willing to use the regression line to predict the number of floors Alana would climb in 35 minutes? Explain your answer.

Section 11.4 Confidence Intervals for the Slope of a Population Least-Squares Regression Line

LEARNING GOALS

By the end of this section, you will be able to:
- Check the conditions for inference about the slope of a least-squares regression line.
- Calculate a $C\%$ confidence interval for the slope of a least-squares regression line.
- Use the four-step process to construct and interpret a confidence interval for the slope of a least-squares regression line.

In Sections 11.1–11.3, you learned how to use a regression line to summarize the relationship between two quantitative variables. If the data are a random sample from a larger population, we can use the slope of the **sample regression line** to estimate the slope of the **population regression line**.

A regression line calculated from every value in the population is called a **population regression line** (or true regression line). The equation of a population regression line is $y = \beta_0 + \beta_1 x + \varepsilon$, where

- y is the value of the response variable for a given value of x.
- β_0 is the population y-intercept.
- β_1 is the population slope.
- ε (the Greek letter epsilon) is the error term. We add the error term ε to the model as a reminder that individual values of the response variable will vary from values predicted by the model.

A regression line calculated from a sample is called a **sample regression line.** The equation of a sample regression line is $\hat{y} = b_0 + b_1 x$, where

- $\hat{y}$ is the predicted y value (or estimated mean y value) for a given value of x.
- b_0 is the sample y-intercept.
- b_1 is the sample slope.

Figure 11.7 shows the relationship between $x =$ duration of an eruption (in minutes) and $y =$ wait time until the next eruption (in minutes) for all 263 eruptions of the Old Faithful geyser during a particular month. Because the scatterplot includes all the eruptions in a particular month, the least-squares regression line shown is the population regression line $y = 33.3 + 13.3x + \varepsilon$.

FIGURE 11.7 Scatterplot of the duration and wait time between eruptions of Old Faithful for all 263 eruptions in a single month. The population least-squares regression line is shown in blue.

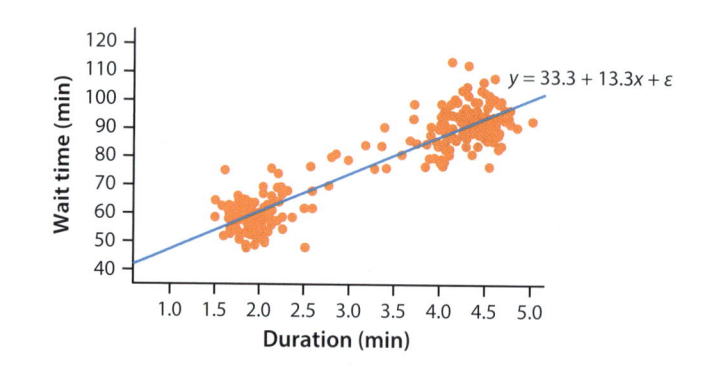

To construct confidence intervals and perform significance tests about the population slope β_1, we need to know about the sampling distribution of the sample slope b_1. The following Concept Exploration will get you started.

CONCEPT EXPLORATION

Sampling from Old Faithful

In this exploration, you will investigate the sampling distribution of the sample slope using the population of eruptions displayed in Figure 11.7.

1. Go to www.stapplet.com and launch the *Old Faithful* applet from the Activities section.

2. In the box initially labeled "Population," the applet will display the population of 263 eruptions. Near the bottom of the box, keep the default sample size of $n = 15$ and click the Go! button. The applet will highlight the 15 randomly selected eruptions and display the slope b_1 of the sample least-squares regression line. Was your sample slope close to the population slope $\beta_1 = 13.3$?

3. In the box labeled "Sampling Distribution," notice that the value of the sample slope from Step 2 is displayed on a dotplot. Click the Select samples button 9 times so that you have a total of 10 sample slopes. Look at the dotplot of sample slopes. Does the distribution have a recognizable shape?

Section 11.4 Confidence Intervals for the Slope of a Population Least-Squares Regression Line

Sampling from Old Faithful

Sample

The scatterplot below displays data for all 263 eruptions of Old Faithful in one month. The population regression line, *Predicted wait time* = 33.3 + 13.3 *Duration*, is shown as a solid line. To investigate the sampling distribution of the sample slope, the applet will select a random sample of *n* eruptions, display the sample regression line as a dashed line, and record the value of the sample slope.

Select one random sample of size $n = 15$ and compute the sample slope.

Sample slope = 11.209

Sampling Distribution

Quickly select 1 random samples: [Select samples]
☐ Show corresponding normal curve

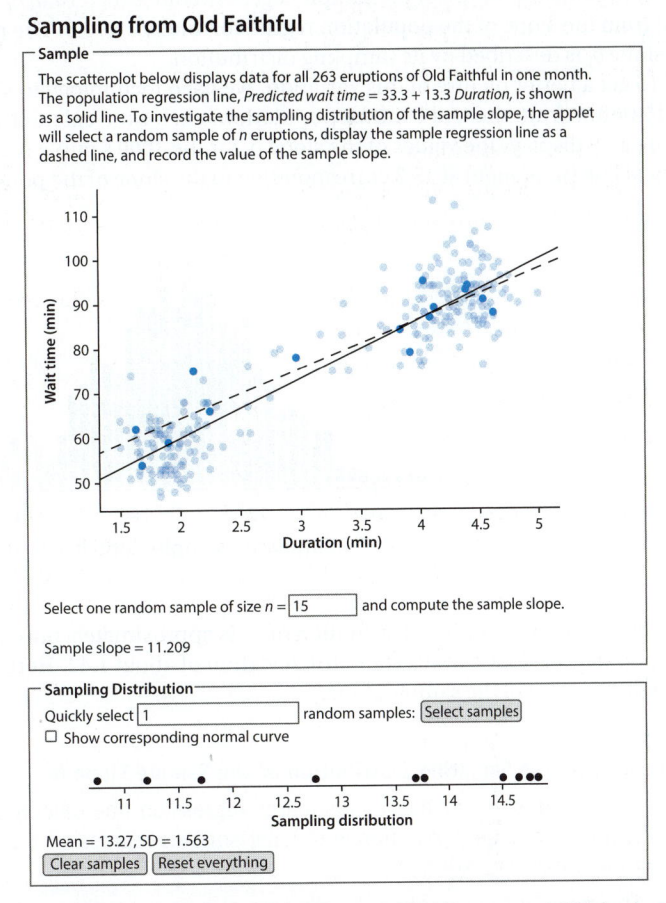

Mean = 13.27, SD = 1.563
[Clear samples] [Reset everything]

4. To better see the shape of the sampling distribution of the sample slope, enter 990 in the box for quickly selecting random samples and click the Select samples button. You have now selected a total of 1000 random samples of size $n = 15$.

5. Describe the shape of the simulated sampling distribution of b_1 shown in the dotplot. Click the button to show the corresponding normal curve. How well does it fit?

6. What is the mean of your simulated sampling distribution? How does it compare to the population slope $\beta_1 = 13.3$?

7. What is the standard deviation of the simulated sampling distribution? Now change the sample size to $n = 50$ in the Sample box and quickly select 1000 random samples in the Sampling Distribution box. What happened to the value of the standard deviation? Explain why this makes sense.

Figure 11.8 shows the results of taking three different SRSs of 15 Old Faithful eruptions from the population described earlier. Each graph displays the selected points and the least-squares regression line for that sample (in purple). The population regression line ($y = 33.3 + 13.3x + \varepsilon$) is also shown (in orange).

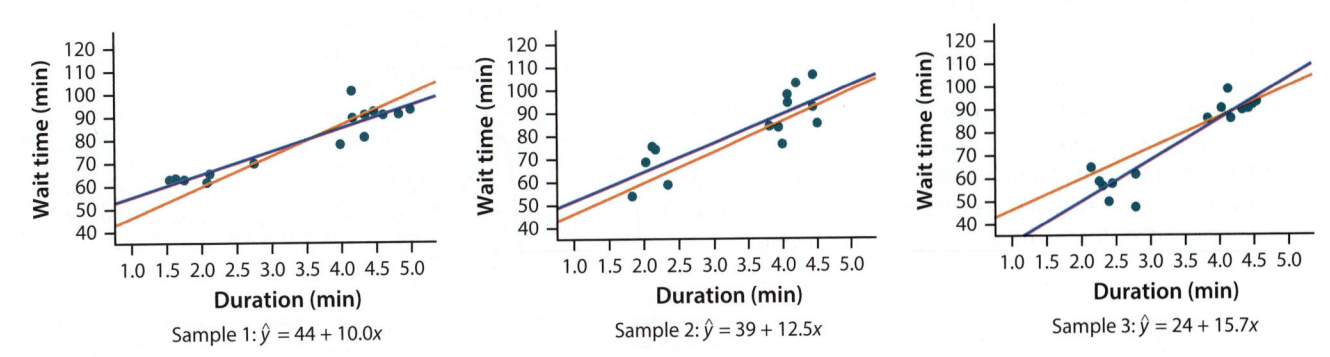

Sample 1: $\hat{y} = 44 + 10.0x$

Sample 2: $\hat{y} = 39 + 12.5x$

Sample 3: $\hat{y} = 24 + 15.7x$

FIGURE 11.8 Scatterplots and least-squares regression lines (in purple) for three different SRSs of 15 Old Faithful eruptions, along with the population regression line (in orange).

Notice that the slopes of the sample regression lines ($b_1 = 10.0$, $b_1 = 12.5$, and $b_1 = 15.7$) vary quite a bit from the slope of the population regression line, $\beta_1 = 13.3$. The pattern of variation in the sample slope b_1 is described by its sampling distribution.

To get a better picture of this variation, we used technology to simulate choosing 1000 SRSs of $n = 15$ points from the Old Faithful population, each time calculating the sample regression line. Figure 11.9 displays the values of the slope b_1 for the 1000 sample regression lines. We have added a vertical line (in orange) at 13.3 corresponding to the slope of the population regression line β_1.

FIGURE 11.9 Dotplot of the sample slope b_1 of the least-squares regression line in 1000 simulated SRSs of $n = 15$ eruptions. The population slope (13.3) is marked with an orange vertical line.

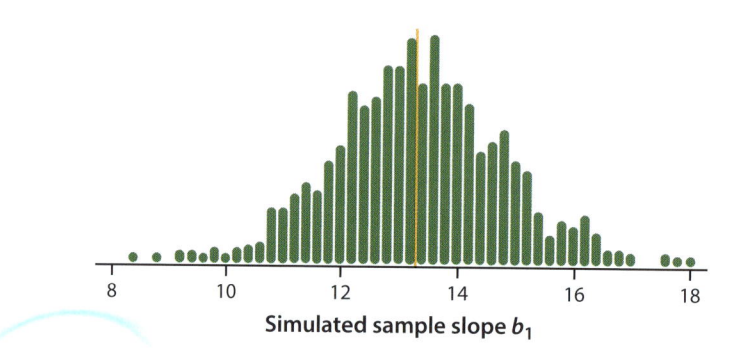

The simulated sampling distribution of b_1 is approximately normal, with a mean of roughly 13.3 (the population slope), and a standard deviation of about 1.42. Here are the details about the sampling distribution of the sample slope.

> **Describing the Sampling Distribution of the Sample Slope b_1**
>
> Let b_1 be the slope of the least-squares regression line calculated from an SRS of size n drawn from a large population with least-squares regression line $y = \beta_0 + \beta_1 x + \varepsilon$. Assuming the conditions are met:
>
> - The **mean** of the sampling distribution of b_1 is $\mu_{b_1} = \beta_1$.
>
> - The **standard deviation** of the sampling distribution of b_1 is $\sigma_{b_1} = \dfrac{\sigma}{\sigma_x \sqrt{n}}$, where σ is the population standard deviation of the residuals and σ_x is the population standard deviation of the explanatory variable.
>
> - The **shape** of the sampling distribution of b_1 is approximately normal.

Checking Conditions for Inference About the Slope

Like inference for proportions and means, certain conditions need to be satisfied to construct confidence intervals and perform significance tests about the slope of a least-squares regression line. Figure 11.10 shows the regression model in picture form *when the conditions are met*.

FIGURE 11.10 The regression model when the conditions for inference are met. The line is the population (true) regression line, which shows how the mean response μ_y changes as the explanatory variable x changes. For any fixed value of x, the observed response y varies according to a normal distribution having mean μ_y and standard deviation σ.

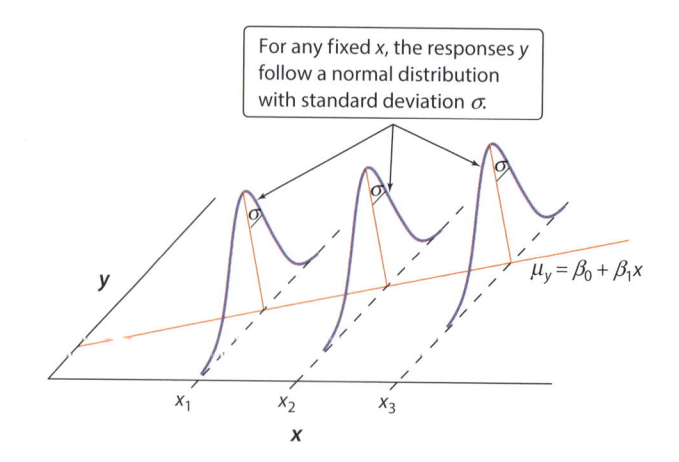

Here are some key observations:
- **Shape:** For each value of x, the distribution of y is normal.
- **Center:** For each value of x, the mean μ_y of the distribution of y falls on the population regression line.
- **Variability:** For each value of x, the distribution of y has the same standard deviation.

Before performing inference about the slope of a least-squares regression line, be sure to check the Random condition and verify that a regression model like the one in Figure 11.10 is reasonably accurate. That is, verify that for each value of x, the distribution of ε is approximately normal with mean 0 and standard deviation σ.

HOW TO Check the Conditions for Inference About the Slope

To perform inference about the slope of population least-squares regression line, check that the following conditions are met:
- **Random:** The data come from a random sample from the population of interest or from a randomized experiment.
- **Linear:** The form of the association between the two variables is linear. Check this condition with a scatterplot or residual plot. The residual plot should have no leftover curved patterns.
- **Equal SD:** For all values of x, the standard deviation of y is the same. Check this condition with a residual plot. Common violations of this condition include a < pattern (residuals tend to grow in size as x increases) or a > pattern (residuals tend to shrink in size as x increases).
- **Normal:** For all values of x, the distribution of y is normal. Check this condition with a graph (e.g., dotplot) of the residuals. There should be no strong skewness or outliers.

Because it is rare to have more than one or two y values for each x value in a regression context, we can't check the Normal or Equal SD conditions for each value of x as we did with ANOVA. Instead, we calculate the residuals and combine them all together. To check the Normal condition, make a graph (e.g., dotplot, boxplot, normal probability plot) of all the residuals. If there is no strong skewness or outliers, it is plausible that the distribution of y is approximately normal for each x.

To check the Equal SD condition, make a residual plot and verify that the amount of variation from the residual = 0 line is roughly the same for all values of x in the data set. For example, here is a scatterplot showing the smoking death rates (per 100,000 people) for 231 countries in 1990 and 2017, along with the least-squares regression line and corresponding residual plot. Thankfully, all but 12 of the countries had a smaller rate in 2017.[33]

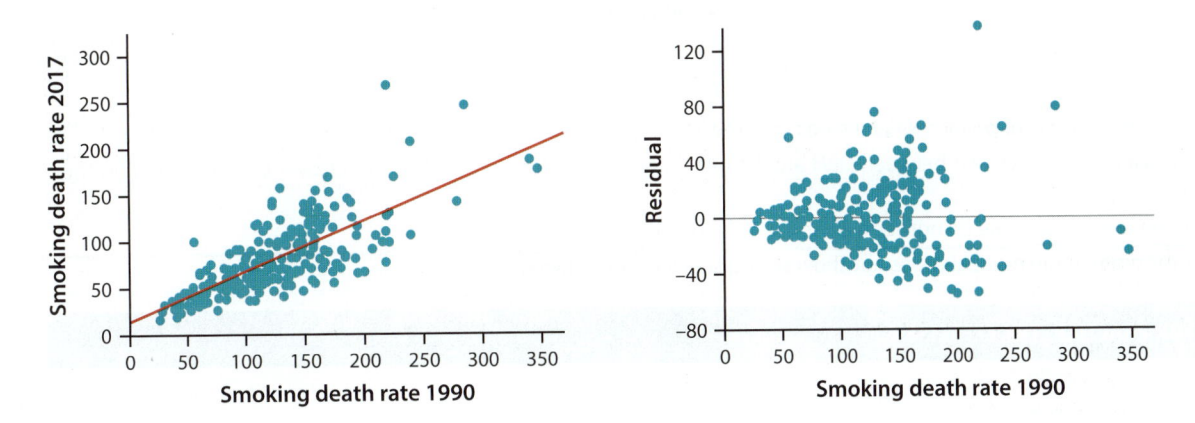

Because there is a < shaped pattern in both the scatterplot and the residual plot, the Equal SD condition is not satisfied for these data. The 2017 death rates are much more variable for countries that had larger death rates in 1990.

You will always see some irregularity when you look for normality and equal standard deviations in the residuals, especially when you have few observations. Don't overreact to minor issues in the graphs when checking the Normal and Equal SD conditions.

EXAMPLE 11.4.1

Studying ponderosa pines

Checking conditions

PROBLEM

The U.S. Forest Service randomly selected ponderosa pine trees in western Montana to investigate the relationships between diameter at breast height (DBH), height, and volume of usable lumber.[34] The scatterplot shows the relationship between $x =$ DBH (in inches) and $y =$ height (in feet) for a random sample of 40 ponderosa pines, along with the least-squares regression line and the corresponding residual plot and histogram of residuals. Check if the conditions for performing inference about the slope of the least-squares regression line are met.

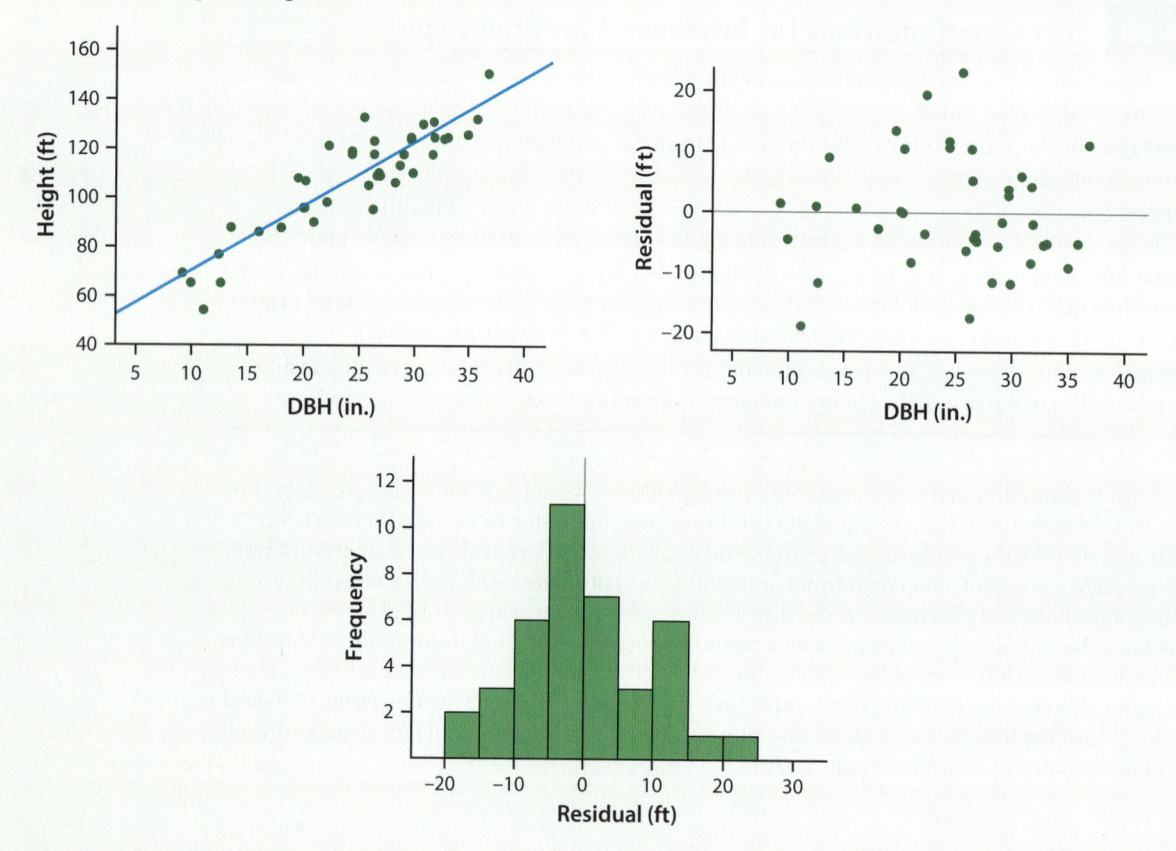

SOLUTION

- Random? A random sample of ponderosa pines was selected. ✓
- Linear? There is a linear association between DBH and height in the scatterplot, and there is no leftover curved pattern in the residual plot. ✓
- Equal SD? In the residual plot, we do not see a clear < pattern or > pattern. ✓
- Normal? A histogram of the residuals does not show strong skewness or outliers. ✓

EXAM PREP: FOR PRACTICE, TRY EXERCISE 5.

THINK ABOUT IT

What if the one of the conditions isn't met? In many cases, performing a transformation of one or both variables can help. For example, here is a scatterplot showing $x =$ diameter at breast height (in inches) and volume of usable lumber (in board feet) from a random sample of 40 ponderosa pines. The linear condition is clearly not met.

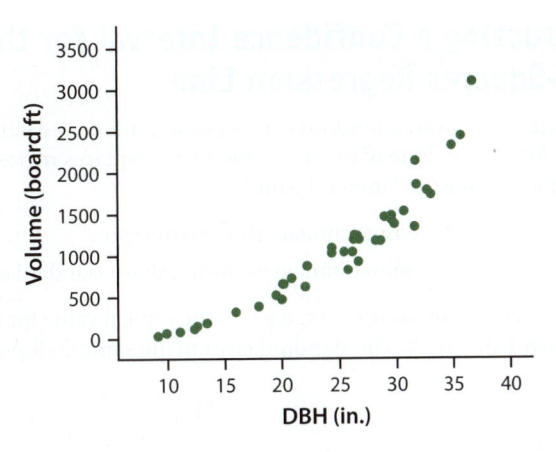

Section 11.4 Confidence Intervals for the Slope of a Population Least-Squares Regression Line

Knowing that volume is a three-dimensional measurement and DBH is a one-dimensional measurement, we can cube-root each of the volumes. Here is the resulting scatterplot. The form of the association is much more linear!

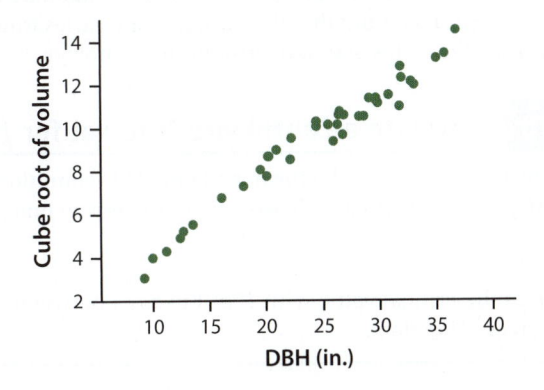

Likewise, here is the scatterplot showing the death rates from smoking in 1990 and 2017 for 231 countries.

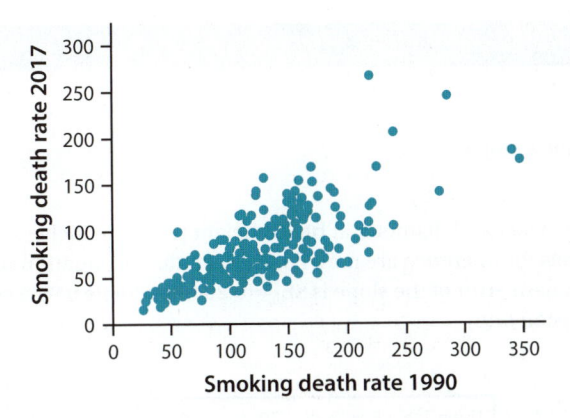

And here is a scatterplot showing $x = \ln$(death rate in 1990) and $y = \ln$(death rate in 2017), where ln is the natural logarithm. Notice how the < shape disappears and the Equal SD condition is now met.

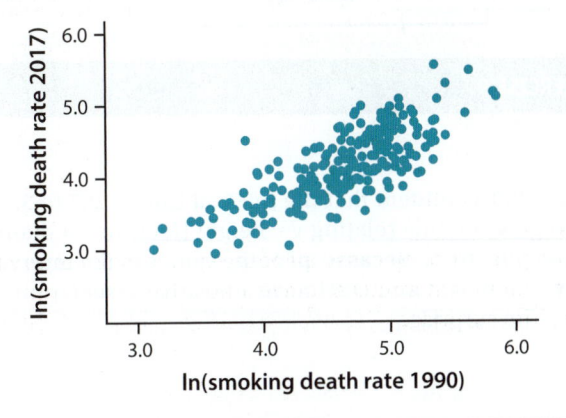

Constructing a Confidence Interval for the Slope of a Least-Squares Regression Line

When data come from a random sample or a randomized experiment, the statistic b_1 is our point estimate for β_1, the slope of the population least-squares regression line. We calculate a confidence interval for β_1 using the familiar formula

$$\text{point estimate} \pm \text{margin of error}$$

$$\text{point estimate} \pm \text{critical value} \times \text{standard error of statistic}$$

When the conditions are met, we find the critical value for the confidence interval using a t distribution with df = $n-2$. The standard error of the sample slope is

$$SE_{b_1} = \frac{s}{s_x \sqrt{n-1}}$$

where s is the standard deviation of the residuals, s_x is the sample standard deviation of the explanatory variable, and n is the sample size. As with other standard errors, the standard error of the slope estimates the typical amount that the sample slope varies from the true slope. Fortunately, we can usually get the value of the standard error from technology.

HOW TO Calculate a Confidence Interval for β_1

When the Random, Normal, Linear, and Equal SD conditions are met, a $C\%$ confidence interval for the slope of the population least-squares regression line is

$$b_1 \pm t^* SE_{b_1}$$

where t^* is the critical value with $C\%$ of its area between $-t^*$ and t^* in the t distribution with $n-2$ degrees of freedom.

EXAMPLE 11.4.2

Studying ponderosa pines

Calculating a confidence interval for a slope

PROBLEM

In Example 11.4.1 about the height in feet and diameter at breast height (DBH) in inches of 40 randomly selected ponderosa pines, we verified that the conditions for inference are met. For these data, the equation of the least-squares regression line is $\widehat{height} = 43.5 + 2.62\,DBH$. The standard error of the slope is $SE_{b_1} = 0.204$. Calculate a 95% confidence interval for the slope of the population least-squares regression line.

SOLUTION

df = $40 - 2 = 38, t^* = 2.024$

$2.62 \pm 2.024(0.204)$

$\rightarrow 2.62 \pm 0.413$

$\rightarrow 2.207$ to 3.033

Using Table B with df = 30, $t^* = 2.042$.

$b_1 \pm t^* SE_{b_1}$

EXAM PREP: FOR PRACTICE, TRY EXERCISE 9.

We are 95% confident that the interval from 2.207 to 3.033 captures the slope of the population regression line relating y = height (feet) to x = diameter at breast height (inches) for ponderosa pine trees. Because all of the values in the interval are positive, there is convincing evidence that height and DBH have a positive association. Thicker ponderosa pines tend to be taller — no surprise.

Section 11.4 Confidence Intervals for the Slope of a Population Least-Squares Regression Line

Putting It All Together: Confidence Intervals for the Slope of a Population Least-Squares Regression Line

We are now ready to use the four-step process to construct and interpret a *t* **interval for the slope of a least-squares regression line**.

> The *t* **interval for the slope of a least-squares regression line** is a confidence interval used to estimate the slope of a population (true) regression line.

When provided with raw data, it is best to use technology to calculate the interval. See the Tech Corner at the end of this section.

EXAMPLE 11.4.3

How much is that truck worth?

Confidence interval for β_1

PROBLEM

Earlier in this chapter, you analyzed the relationship between $y =$ asking price (in dollars) and $x =$ miles driven for a random sample of 16 Ford F-150 SuperCrew 4 × 4s. Here are the data.

Miles driven	70,583	129,484	29,932	29,953	24,495	75,678	8359	4447
Price ($)	21,994	9500	29,875	41,995	41,995	28,986	31,891	37,991
Miles driven	34,077	58,023	44,447	68,474	144,162	140,776	29,397	131,385
Price ($)	34,995	29,988	22,896	33,961	16,883	20,897	27,495	13,997

Calculate and interpret a 95% confidence interval for the slope of the population least-squares regression line relating $y =$ asking price to $x =$ miles driven.

SOLUTION

State: 95% CI for $\beta_1 =$ the slope of the population least-squares regression line relating $y =$ asking price ($) to $x =$ miles driven.

> **State:** State the parameter you want to estimate and the confidence level.

Plan: *t* interval for the slope of a least-squares regression line
- Random? Random sample of 16 Ford F-150 SuperCrew 4 × 4s. ✓
- Linear? There is a linear association between number of miles driven and asking price in the scatterplot, and there is no leftover curved pattern in the residual plot. ✓
- Equal SD? In the residual plot, we do not see a clear < pattern or > pattern. ✓
- Normal? The dotplot of the residuals doesn't show strong skewness or outliers. ✓

> **Plan:** Identify the appropriate inference method and check the conditions. *If the graphs aren't provided, make sure to include them in your response.*

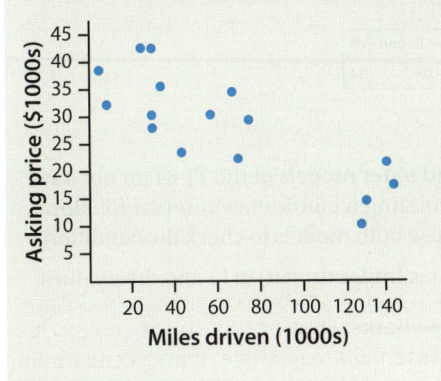

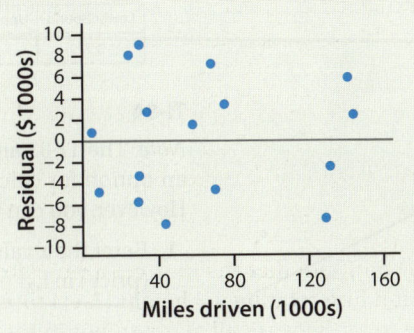

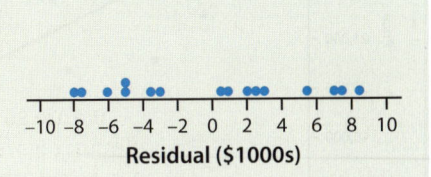

(continued)

Do: −0.229 to −0.097 with df = 14

Conclude: We are 95% confident that the interval from −0.229 to −0.097 captures the slope of the population least-squares regression line relating y = asking price to x = miles driven for Ford F-150 SuperCrew 4 × 4s.

Do: If the conditions are met, perform the calculations. *Use technology when provided with raw data.*

Conclude: Interpret your interval in the context of the problem.

EXAM PREP: FOR PRACTICE, TRY EXERCISE 13.

Because all of the plausible values in the interval are less than 0, these data provide convincing evidence of a negative association between asking price and miles driven for Ford F-150 SuperCrew 4 × 4s. For each additional mile driven, we expect the asking price to go down between about 10 and 23 cents.

TECH CORNER *t* Interval for the Slope of a Least-Squares Regression Line

You can use technology to check the conditions and calculate a *t* interval for the slope of a least-squares regression line. We'll illustrate this process using the Ford F-150 data from Example 11.4.3.

Applet

1. Go to www.stapplet.com and launch the *Two Quantitative Variables* applet.
2. Enter "Miles driven" as the explanatory variable. Then, input the 16 values for this variable. Enter "Price ($)" as the response variable, along with the corresponding 16 values for this variable.

Two Quantitative Variables

Variable	Name	Observations (separated by commas or spaces) Keep individuals in the same order.
Explanatory	Miles driven	70583 129484 29932 29953 24495 75678 8359 4447 34077 580
Response	Price ($)	21994 9500 29875 41995 41995 28986 31891 37991 34995 299

3. Click the Begin analysis button. To check the conditions, click "Calculate least-squares regression line." This will display the scatterplot, residual plot, and dotplot of the residuals, allowing you to check the Normal, Linear, and Equal SD conditions.

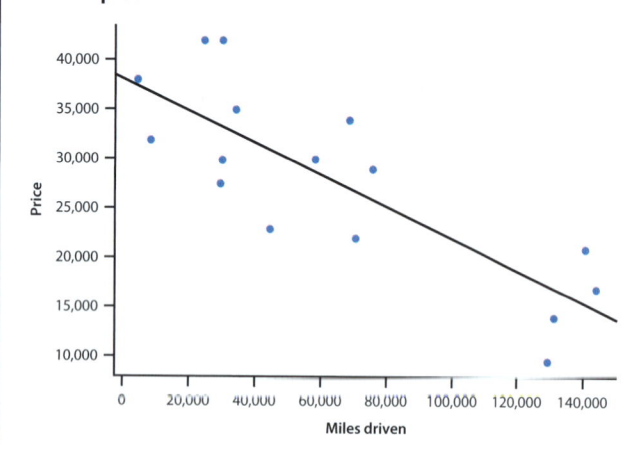

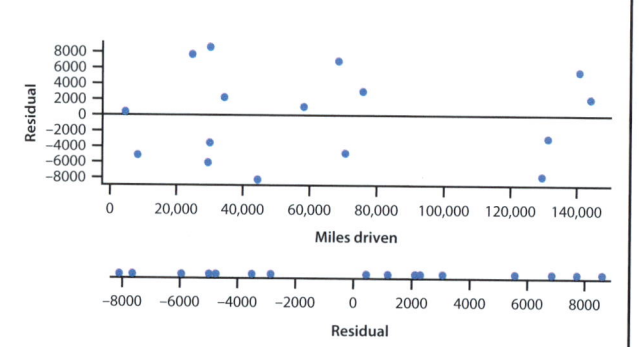

4. To calculate the confidence interval, scroll down to the "Perform Inference" section and choose "*t* interval for slope" from the drop-down menu. For the confidence level, enter 95%. Click the Perform inference button. The confidence interval and degrees of freedom will be displayed.

Perform Inference

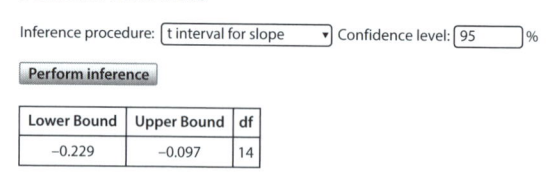

Inference procedure: [t interval for slope]   Confidence level: [95] %

[Perform inference]

Lower Bound	Upper Bound	df
−0.229	−0.097	14

TI-84

Note: The TI-83 and older models of the TI-84 do not have an option for calculating a confidence interval for slope. However, you can use both models to check the conditions.

1. Enter the *x* values (miles driven) in L_1 and the *y* values (price) in L_2.

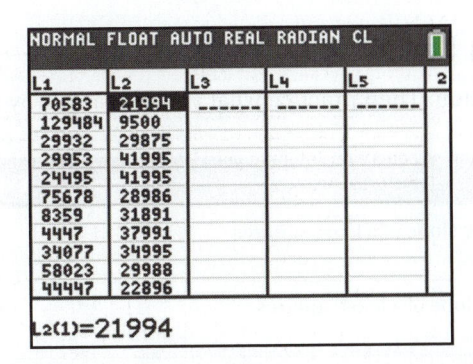

Section 11.4 Confidence Intervals for the Slope of a Population Least-Squares Regression Line

2. Press STAT, arrow to TESTS, and choose LinRegTInt. Adjust the settings as shown. Highlight "Calculate" and press ENTER. The confidence interval, estimated slope (b), and degrees of freedom (df) are reported, along with the standard deviation of the residuals (s), y-intercept (a), r^2, and correlation (r).

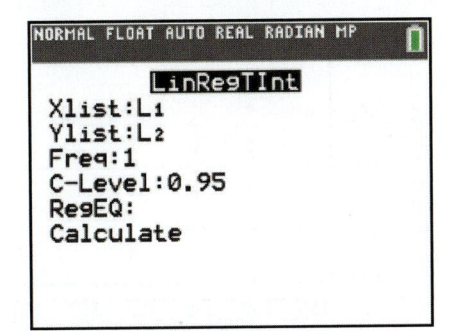

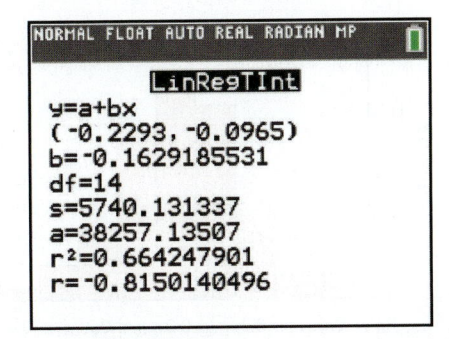

- Press ZOOM and choose 9:ZoomStat.

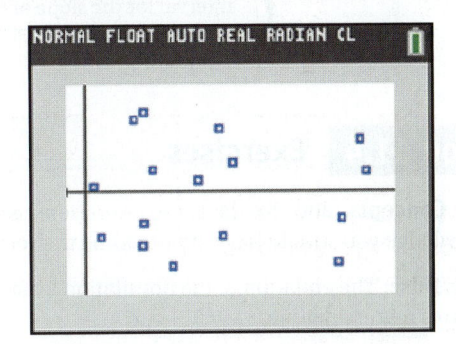

- To make a histogram (or boxplot) of the residuals, adjust the STAT PLOT settings as shown, press ZOOM, and choose 9:ZoomStat.

3. To check the conditions, begin by making a residual plot. The TI-83/84 automatically calculates the residuals when the LinRegTInt operation is performed (or the least-squares regression line is calculated), so you must calculate the interval or calculate the line *before* plotting the residuals.

    - Press 2nd Y= (STAT PLOT).
    - Press ENTER or 1 to go into Plot1.
    - Adjust the settings as shown. The RESID list can be found in the List menu (2nd STAT).

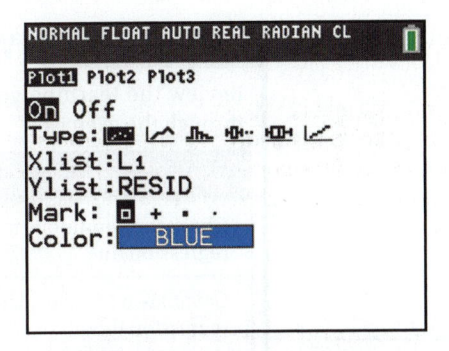

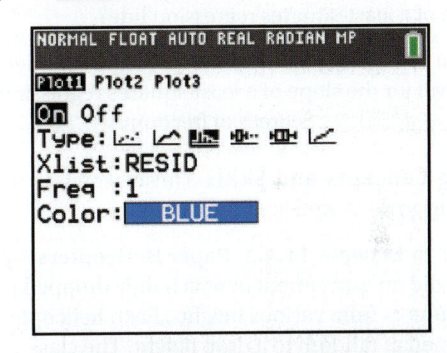

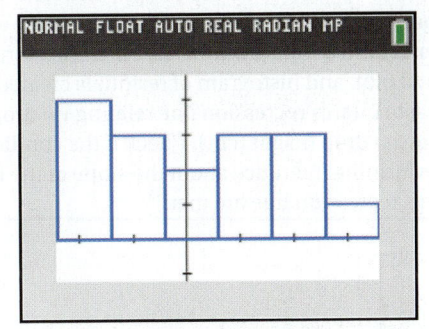

Detailed instructions for using CrunchIt!, Excel, Google Sheets, JMP, Minitab, and R are available in Achieve.

Section 11.4 What Did You Learn?

Review the learning goals from this section. Then practice what you've learned by working through the exercises.

Learning Goal	Example	Exercises
Check the conditions for inference about the slope of a least-squares regression line.	11.4.1	5–8
Calculate a C% confidence interval for the slope of a least-squares regression line.	11.4.2	9–12
Use the four-step process to construct and interpret a confidence interval for the slope of a least-squares regression line.	11.4.3	13–16

Section 11.4 Exercises

Building Concepts and Skills These exercises assess the basic knowledge you should have after reading the section.

1. True/False: The equation of the population least-squares regression line is $\hat{y} = b_0 + b_1 x$.

2. Which graph(s) can you use to check the Normal condition? The Linear condition? The Equal SD condition?

3. What is the formula for a confidence interval for the slope of a least-squares regression line?

4. When calculating the critical value for a confidence interval for the slope of a least-squares regression line, use _____ degrees of freedom.

Mastering Concepts and Skills These exercises reinforce the learning goals as shown in the examples.

5. **Refer to Example 11.4.1 Paper Helicopters** A physics class did an experiment in which they dropped paper helicopters from various heights. Each helicopter was assigned at random to a drop height. The class suspects that helicopters dropped from a greater height will take longer to land on the ground. Here are a scatterplot, residual plot, and histogram of residuals created from the least-squares regression line relating y = drop time (sec) to x = drop height (cm). Check if the conditions for performing inference about the slope of the least-squares regression line are met.[35]

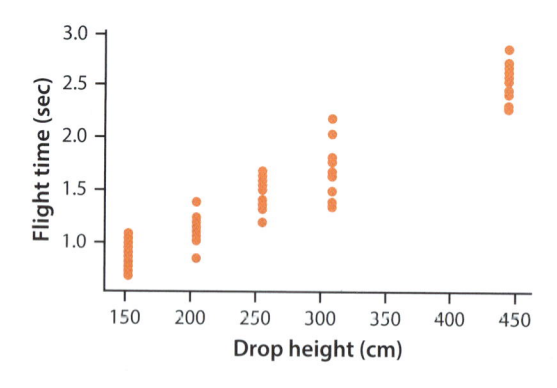

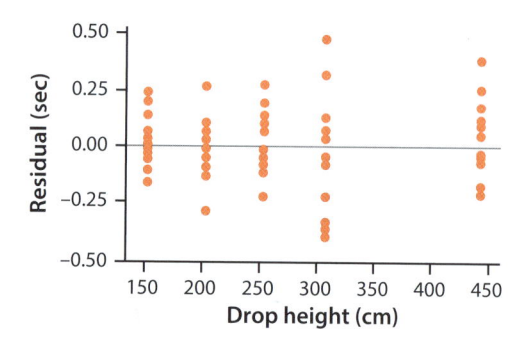

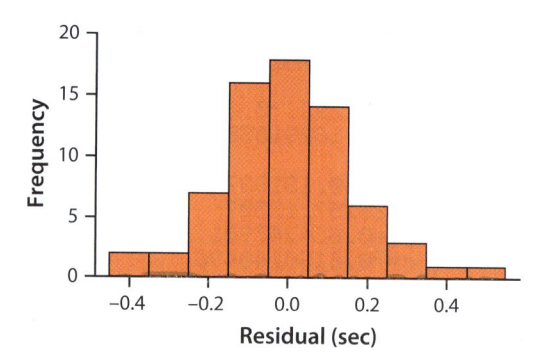

6. **Weeds and Corn Growth** Lamb's-quarter is a common weed that interferes with the growth of corn. An agriculture researcher planted corn at the same rate in 16 small plots of ground and then weeded the plots by hand to allow a fixed number of lamb's-quarter plants to grow in each meter of the corn rows. The decision of how many of these plants to leave in each plot was made at random. No other weeds were allowed to grow. Here are a scatterplot, residual plot, and dotplot of residuals created from the least-squares regression line relating y = corn yield (bushels per acre) to x = number of weeds per meter. Check if the conditions for performing inference about the slope of the least-squares regression line are met.[36]

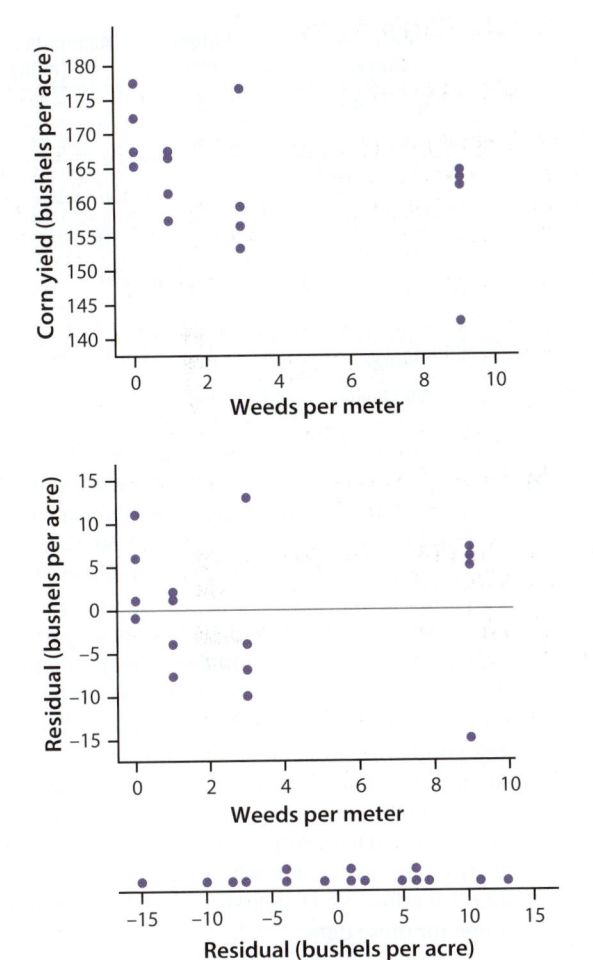

7. **Income and Mortality** What does a country's income per person (measured in adjusted gross domestic product per person in dollars) tell us about the mortality rate for children younger than 5 years of age (per 1000 live births)? A random sample of 14 countries was selected to investigate. Here are a scatterplot, residual plot, and dotplot of residuals created from the least-squares regression line relating y = mortality rate to x = income. Check if the conditions for performing inference about the slope of the least-squares regression line are met.[37]

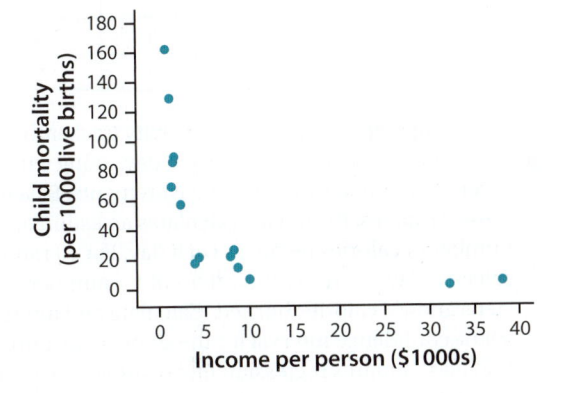

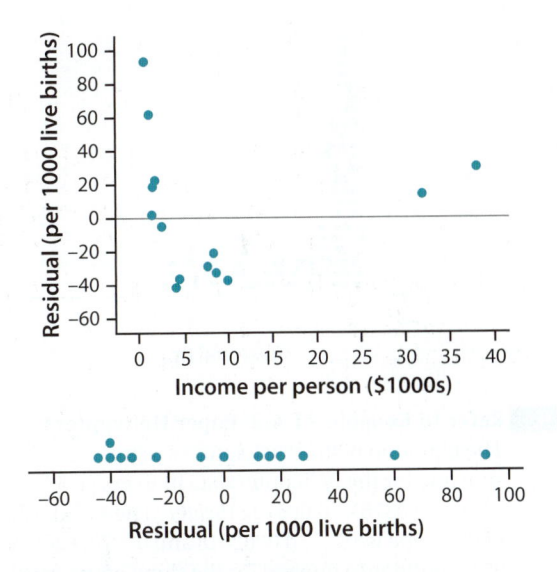

8. **Tall Saguaros** Saguaro National Park near Tucson, Arizona is famous for its saguaro cactus. To track the health of saguaros in the park and estimate the total number of saguaros, researchers randomly select saguaros for inspection every 10 years.[38] Is there a relationship between the height of a saguaro and the elevation where it grows? Here are a scatterplot, residual plot, and dotplot of residuals created from the least-squares regression line relating y = height (meters) to x = elevation (meters). Check if the conditions for performing inference about the slope of the least-squares regression line are met.

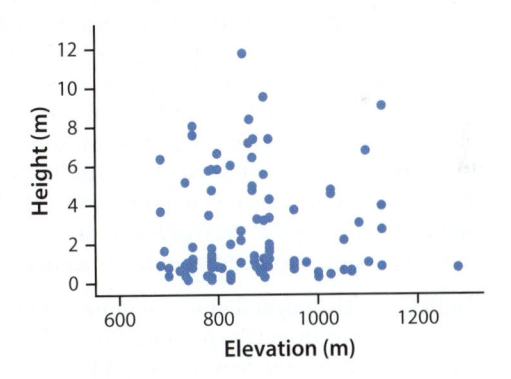

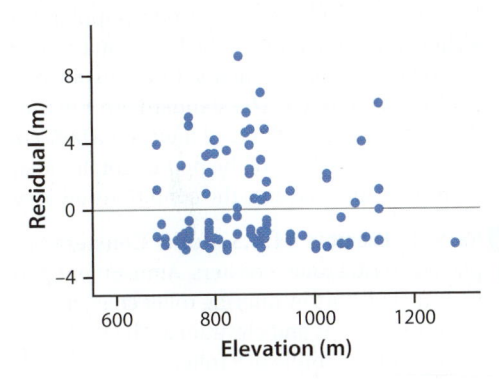

Roller coaster	Height (ft)	Maximum speed (mph)
Apocalypse	100	55
Bullet	196	83
Corkscrew	70	55
Flying Turns	50	24
Goliath	192	66
Hidden Anaconda	152	65
Iron Shark	100	52
Stinger	131	50
Wild Eagle	210	61

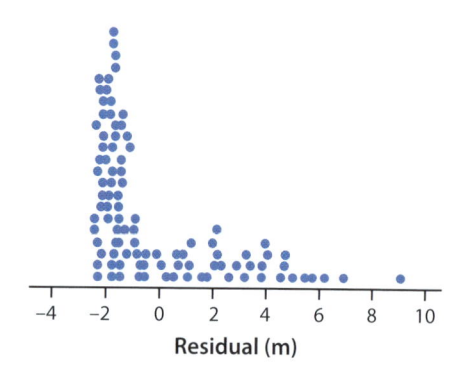

9. **Refer to Example 11.4.2 Paper Helicopters** The equation of the least-squares regression line for the helicopter data in Exercise 5 is $\widehat{time} = -0.03761 + 0.0057244\,height$. The standard error of the slope is $SE_{b_1} = 0.0002018$ and $n = 70$. Calculate a 95% confidence interval for the slope of the true least-squares regression line.

10. **Weeds and Corn Growth** The equation of the least-squares regression line for the weeds and corn data in Exercise 6 is $\widehat{yield} = 166.4500 - 1.0808\,weeds$. The standard error of the slope is $SE_{b_1} = 0.5777$ and $n = 16$. Calculate a 95% confidence interval for the slope of the true least-squares regression line.

11. **Loud Music and Tests** Two psychology students wanted to know if listening to music at a louder volume hurts test performance. To investigate, they recruited 30 volunteers and randomly assigned 10 to listen to music at 30 decibels, 10 to listen to music at 60 decibels, and 10 to listen to music at 90 decibels. While listening to the music, each volunteer took a 10-question math test.[39] The equation of the least-squares regression line relating $y =$ number correct to $x =$ volume (decibels) is $\hat{y} = 9.900 - 0.0483x$ and the standard error of the slope is $SE_{b_1} = 0.0116$. Calculate and interpret a 90% confidence interval for the slope of the true least-squares regression line. Assume the conditions for inference are met.

12. **Flight Costs** Do longer flights cost more money? A frequent flier recorded the distance from Philadelphia to a random sample of 6 cities and the cost of the cheapest flight to that city on a popular discount airline.[40] The equation of the least-squares regression line relating $y =$ cost (dollars) to $x =$ distance (miles) is $\hat{y} = 107.08 + 0.0416x$. The standard error of the slope is $SE_{b_1} = 0.0106$. Calculate and interpret a 90% confidence interval for the slope of the population least-squares regression line. Assume the conditions are met.

13. **Refer to Example 11.4.3 Roller Coasters** Many people like to ride roller coasters. Amusement parks try to attract visitors by offering roller coasters that have a variety of speeds and elevations. The table shows data for a random sample of 9 roller coasters.[41] Calculate and interpret a 95% confidence interval for the slope of the population least-squares regression line relating $y =$ maximum speed (mph) to $x =$ height (ft).

14. **Soda Cans** Researchers wanted to investigate if tapping on a can of soda would reduce the amount of soda expelled after the can has been shaken. For their experiment, they vigorously shook 40 cans of soda and randomly assigned each can to be tapped for 0 seconds, 4 seconds, 8 seconds, or 12 seconds. After opening the cans and waiting for the fizzing to stop, they measured the amount expelled (in milliliters) by subtracting the amount remaining from the original amount in the can.[42] Use the data to calculate and interpret a 95% confidence interval for the slope of the true least-squares regression line relating $y =$ amount expelled (mL) to $x =$ tapping time (sec). *Note:* Due to the large sample size, it isn't surprising to find outliers in the distribution of residuals when the Normal condition is met. Therefore, you can assume the Normal condition is met for these data.

Amount expelled (mL)			
0 sec	4 sec	8 sec	12 sec
110	95	88	80
100	105	84	75
105	105	87	80
105	105	85	75
105	95	79	70
110	90	100	65
107	88	85	71
105	95	85	77
104	94	80	76
106	96	80	75

15. **Tracking Steps** Josh wears an activity tracker to record the number of steps he takes each day, along with several other health-related measurements. Based on these variables, the device calculates an estimate of the number of calories he burns each day. For 10 randomly selected days, here are the values of $x =$ number of steps and $y =$ calories burned. Calculate and interpret a 90% confidence interval for the slope of the population least-squares regression line relating $y =$ calories burned to $x =$ number of steps for Josh.

Number of steps	Calories burned
7997	2569
6318	2328
6620	2549
7708	2519
12,627	2840
10,961	2864
10,819	2771
7715	2596
10,958	2684
9694	2566

16. **Tennis Serves** Are taller tennis players able to serve faster? Physics would suggest this is the case, as taller players have longer arms and can generate more racquet speed. The table shows x = height (in inches) and y = average first serve speed (in miles per hour) for 16 randomly selected male professional tennis players.[43] Calculate and interpret a 99% confidence interval for the slope of the population least-squares regression line relating y = serve speed (mph) to x = height (in.) for professional male tennis players.

Height (in.)	Speed (mph)
75	108
78	122
76	113
78	123
77	127
73	123
77	113
74	120
77	118
80	128
73	116
72	106
78	124
75	119
73	125
71	110

Applying the Concepts These exercises ask you to apply multiple learning goals in a new context or to apply what you learned in this section in a new way.

17. **Light and Plant Growth** Meadowfoam seed oil is used in making various skin care products. Researchers interested in maximizing the productivity of meadowfoam plants designed an experiment to investigate the effect of different light intensities on plant growth. The researchers planted 120 meadowfoam seedlings in individual pots, placed 10 pots on each of 12 trays, and put all the trays into a controlled enclosure. Two trays were randomly assigned to each light intensity level (micromoles per square meter per second): 150, 300, 450, 600, 750, and 900. The number of flowers produced by each plant was recorded and the average number of flowers was recorded for each tray.[44] Here are the data:

Light intensity	150	150	300	300	450	450
Average number of flowers	62.3	77.4	55.3	54.2	49.6	61.9
Light intensity	600	600	750	750	900	900
Average number of flowers	39.4	45.7	31.3	44.9	36.8	41.9

(a) Calculate and interpret a 95% confidence interval for the slope of the true least-squares regression line relating y = average number of flowers to x = light intensity.

(b) Based on the confidence interval from part (a), is there convincing evidence that greater light intensity produces fewer flowers, on average? Explain your reasoning.

18. **Sugary Flowers** Does adding sugar to the water in a vase help flowers stay fresh? To find out, two researchers prepared 12 identical vases with exactly the same amount of water in each vase. They put 1 tablespoon of sugar in 3 vases, 2 tablespoons of sugar in 3 vases, and 3 tablespoons of sugar in 3 vases. In the remaining 3 vases, they put no sugar. After the vases were prepared, the researchers randomly assigned 1 carnation to each vase and observed how many hours each flower continued to look fresh. Here are the data:

Amount of sugar (tbs)	0	0	0	1	1	1
Freshness time (hr)	168	180	192	192	204	204
Amount of sugar (tbs)	2	2	2	3	3	3
Freshness time (hr)	204	210	210	222	228	234

(a) Calculate and interpret a 95% confidence interval for the slope of the true least-squares regression line relating y = freshness time (in hours) to x = amount of sugar (in tablespoons).

(b) Based on the confidence interval from part (a), is there convincing evidence that adding more sugar helps to keep carnations looking fresh longer? Explain your reasoning.

19. Tech **Car Price and Age** Refer to Exercise 23 in Section 11.2. The data set **Honda CRV** contains information about a random sample of 191 used Honda CRVs. Use technology to construct and interpret a 95% confidence interval for the slope of the population least-squares regression line relating y = Price to x = Age (in years).

20. Tech **Car Price and Mileage** Refer to Exercise 24 in Section 11.2. The data set **Honda CRV** contains information about a random sample of 191 used Honda CRVs. Use technology to construct and interpret a 95% confidence interval for the slope of the population least-squares regression line relating y = Price to x = Miles.

Extending the Concepts These exercises challenge you to explore statistical concepts and methods that go beyond what you learned in this section.

21. **Income and Mortality** Refer to Exercise 7. To help meet the conditions for inference about the slope, we can often transform one or both variables using mathematical operations such as square roots and logarithms. Here are the data from Exercise 7:

Income	Mortality
38,003.9	4.4
2475.68	56.4
1202.53	127.5
1382.95	68.5
7858.97	21.3
1830.97	87.5
8199.03	26.3
4523.44	21.6
32,021	2.8
643.39	160.3
10,005.2	7.1
1493.53	84
4016.2	17.6
8826.9	14.5

(a) Using technology, calculate the base-10 logarithm of each of the mortality values and the base-10 logarithm of each of the income values.

(b) Construct a scatterplot showing the relationship between $y = \log(\text{mortality})$ and $x = \log(\text{income})$. What do you notice?

22. **Weeds and Corn Growth** Refer to Exercises 6 and 10. The equation of the least-squares regression line for the weeds and corn data in Exercise 6 is $\hat{y} = 166.4500 - 1.0808x$, where x = weeds per meter and y = corn yield (bushels per acre). The standard error of the slope is $SE_{b_1} = 0.5777$ and $n = 16$. Interpret the value 0.5777.

23. **Teenager Height** Using the health records of every student at a high school, the school nurse created a scatterplot relating y = height (in centimeters) to x = age (in years). After verifying that the conditions for the regression model were met, the nurse calculated the equation of the population regression line to be $y = 105 + 4.2x + \varepsilon$ with a standard deviation of the residuals of $\sigma = 7$ cm.

(a) According to the population regression line, what is the average height of 15-year-old students at this high school?

(b) About what percentage of 15-year-old students at this school are taller than 180 cm?

24. **Teenager Height** Refer to Exercise 23. In addition to calculating the equation of the least-squares regression line, the nurse calculated the standard deviation of age to be $\sigma_x = 1.28$ years. Suppose the nurse selects a random sample of 10 students from the school and calculates the equation of the least-squares regression line for this sample.

(a) Calculate the mean and standard deviation of the sampling distribution of the sample slope for samples of size 10. What shape does the sampling distribution have?

(b) What is the probability that the slope of the least-squares regression line for this sample is less than 4?

Cumulative Review These exercises revisit what you learned in previous sections.

Exercises 25–27 refer to the following scenario. Does the color in which words are printed affect your ability to read them? Do the words themselves affect your ability to name the color in which they are printed? Professor Starnes had his 16 students investigate these questions. Each student performed two tasks in a random order while a partner timed the activity: (1) Read 32 words aloud as quickly as possible, and (2) say the color in which each of 32 words is printed as quickly as possible. Try both tasks for yourself using the following word list.

YELLOW	RED	BLUE	GREEN
RED	GREEN	YELLOW	YELLOW
GREEN	RED	BLUE	BLUE
YELLOW	BLUE	GREEN	RED
BLUE	YELLOW	RED	RED
RED	BLUE	YELLOW	GREEN
BLUE	GREEN	GREEN	BLUE
GREEN	YELLOW	RED	YELLOW

25. **Colorful Words (1.1, 1.6)** Let's review the design of the study.

(a) Explain why this was an experiment and not an observational study.

(b) Explain the purpose of the random assignment in the context of the study.

Here are the data from the experiment. For each student, the time to perform the two tasks is given to the nearest second.

Subject	Words	Colors
1	13	20
2	10	21
3	15	22
4	12	25
5	13	17
6	11	13
7	14	32
8	16	21
9	10	16
10	9	13
11	11	11
12	17	26
13	15	20
14	15	15
15	12	18
16	10	18

26. **Colorful Words (9.5, 9.6)**
    (a) Calculate the difference in time for each student, make a boxplot of the differences, and describe what you see.
    (b) Calculate and interpret the mean difference.
    (c) Explain why it is not safe to use paired t procedures to do inference about the mean difference in the time to complete the two tasks.

27. **Colorful Words (2.6, 11.1, 11.2)** Can we use students' word task time to predict their color task time?
    (a) Make an appropriate scatterplot to help answer this question. Describe what you see.
    (b) Use technology to find the equation of the least-squares regression line.
    (c) Find and interpret the residual for the student who completed the word task in 9 seconds.

    Note: John Ridley Stroop is often credited with the discovery in 1935 of the fact that the color in which "color words" are printed interferes with people's ability to identify the color. The so-called Stroop effect, though, was originally discussed by German researchers in a 1929 paper.

Section 11.5 Significance Tests for the Slope of a Population Least-Squares Regression Line

LEARNING GOALS

By the end of this section, you will be able to:
- State hypotheses for performing a significance test about the slope of a least-squares regression line.
- Calculate the standardized test statistic and P-value for a significance test about the slope of a least-squares regression line.
- Use the four-step process to perform a significance test about the slope of a least-squares regression line.

In Sections 10.3 and 10.4, you learned how to perform a test about the relationship between two *categorical* variables, such as anger level and heart disease status. In this section, you will learn how to state hypotheses, calculate the test statistic and P-value, and carry out a test about the relationship between two *quantitative* variables. The following Concept Exploration gives you a preview of what's to come.

CONCEPT EXPLORATION

Should you sit in front?

Many people believe that students learn better if they sit closer to the front of the classroom. Does sitting closer *cause* higher achievement, or do better students simply choose to sit in the front? To investigate, a statistics instructor randomly assigned students to seat locations in the classroom for a particular unit and recorded the test score for each student at the end of the unit. In this exploration, you will use simulation to determine if these data provide convincing evidence that sitting closer to the front improves test scores.

1. The scatterplot shows the relationship between $x =$ row number (the rows are equally spaced, with row 1 closest to the front and row 7 farthest away) and $y =$ test score, along with the least-squares regression line $\hat{y} = 85.71 - 1.12x$. Do these data provide *some* evidence that sitting closer improves test scores? How do you know?

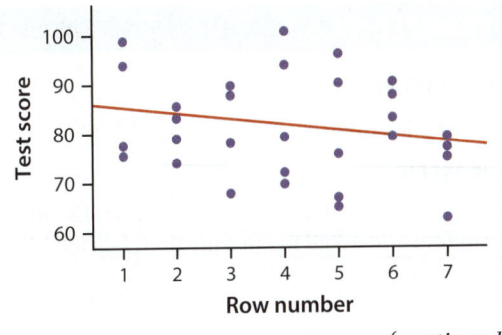

(continued)

Does the slope of $b_1 = -1.12$ provide *convincing* evidence that sitting closer to the front improves test scores for students like these, or is it plausible that the association is due to the chance variation in the random assignment? Here are the data:

Row	1	1	1	1	2	2	2	2	3	3	3	3	4	4	4
Score	76	77	94	99	83	85	74	79	90	88	68	78	94	72	101
Row	4	4	5	5	5	5	5	6	6	6	6	7	7	7	7
Score	70	79	76	65	90	67	96	88	79	90	83	79	76	77	63

2. Go to www.stapplet.com and launch the *Two Quantitative Variables* applet. Enter the Row values as the explanatory variable and the Score values as the response variable. Then click the Begin analysis button to generate a scatterplot of the data.

3. In the Perform Inference section, choose "Simulate sample slope" as the inference procedure. Enter 1 for the number of samples to add and click "Add samples." This randomly reassigns the score to the row values and calculates the resulting slope. The simulated slope is displayed on the dotplot. How does it compare to the actual slope of −1.12?

4. Keep clicking "Add samples" to perform more random reassignments (or increase the number of samples to add in the entry box). Is −1.12 unusual or something that is likely to occur by chance alone? What conclusion should you make based on the instructor's experiment?

Stating Hypotheses

In Section 10.3, you learned how to state hypotheses for a test about the relationship between two *categorical* variables. Stating hypotheses for a test about the relationship between two *quantitative* variables is very similar, except that we can now specify a direction for the relationship in the alternative hypothesis.

In the seating chart exploration, we wanted to test if sitting closer to the front improves test scores for students like the ones in the study. That is, we wanted to know if smaller row numbers are associated with larger test scores (a negative association). Here is one way to state the hypotheses for this test:

H_0: There is no association between row number and test score for students like these.

H_a: There is a negative association between row number and test score for students like these.

We can also state the hypotheses symbolically using β_1, the slope of the true least-squares regression line. If there is no association between these variables, the slope of the true least-squares regression line will be 0. On the other hand, if there is a negative association between these variables, the slope of the true least-squares regression line will be negative. In symbols, the hypotheses are

$$H_0: \beta_1 = 0$$
$$H_a: \beta_1 < 0$$

where β_1 is the slope of the true least-squares regression line relating y = test score to x = row number for students like the ones in this experiment.

As in tests for means and proportions, a two-sided alternative hypothesis is possible. For example, if we had no initial belief about the relationship between seat location and test scores, the alternative hypothesis would be $H_a: \beta_1 \neq 0$ (there *is* an association between row number and test score for students like these). As always, the alternative hypothesis is formulated based on the statistical question being investigated—before the data are collected.

EXAMPLE 11.5.1

Can foot length predict height?

Stating hypotheses

PROBLEM

Are students with longer feet typically taller than their classmates with shorter feet? Fifteen college students were selected at random and asked to measure their foot length and height, both in centimeters. The equation of the least-squares regression line is $\widehat{height} = 102.7 + 2.77 footlength$.

(a) State hypotheses for a test about the relationship between foot length and height in the population of college students.
(b) What is the evidence for H_a?

SOLUTION

(a) $H_0: \beta_1 = 0$ and $H_a: \beta_1 > 0$, where β_1 is the slope of the population least-squares regression line relating $y =$ height (cm) to $x =$ foot length (cm).

> The alternative hypothesis specifies a positive association because we suspect that students with above-average foot length also have above-average height.

(b) The evidence for H_a is: $b_1 = 2.77 > 0$.

EXAM PREP: FOR PRACTICE, TRY EXERCISE 5.

You could also state the hypotheses for this example in words:

H_0: There is no association between foot length and height for college students.

H_a: There is a positive association between foot length and height for college students.

Calculating the Test Statistic and *P*-Value

The standardized test statistic for a test about the slope of a least-squares regression line has the same form as the standardized test statistic for a test about a mean or a test about a proportion:

$$\text{standardized test statistic} = \frac{\text{statistic} - \text{null value}}{\text{standard error of statistic}}$$

The statistic we use is b_1, the slope of the sample least-squares regression line. When the null hypothesis is no association between two variables, the hypothesized slope is 0. We use the standard error of the slope SE_{b_1} in the denominator, which is almost always calculated with technology. Here is the formula:

$$t = \frac{b_1 - 0}{SE_{b_1}}$$

In most cases, we want to test the null hypothesis that the true slope is 0. However, there are occasions when we use a null value different than 0. (See Exercise 28 in this section.)

When the conditions are met for a test about the slope of a least-squares regression line, we use a t distribution with $n - 2$ degrees of freedom to calculate the *P*-value. In the seating chart context, the conditions are met:

- Random? The students were randomly assigned to seats for the experiment. ✓
- Linear? There is a linear association between row number and test score in the scatterplot shown earlier, and there is no leftover curved pattern in the residual plot. ✓

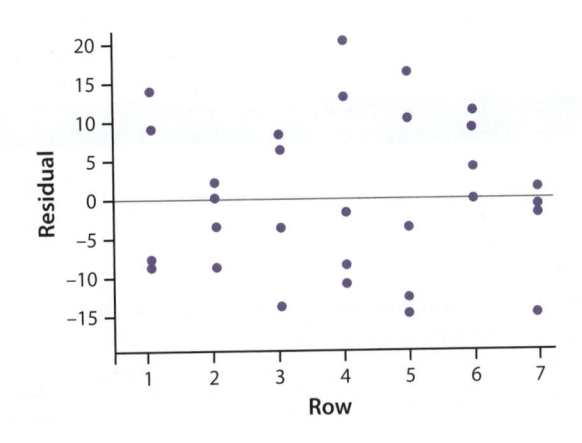

- Equal SD? In the residual plot, we do not see a clear < pattern or > pattern. ✓
- Normal? A dotplot of the residuals does not show strong skewness or outliers. ✓

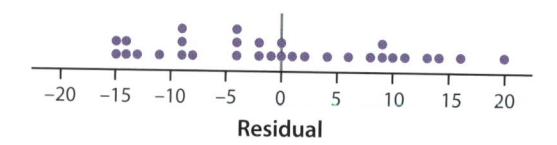

The equation of the least-squares regression line is $\hat{y} = 86 - 1.12x$ and the standard error of the slope is $SE_{b_1} = 0.95$. The value of the standardized test statistic is

$$t = \frac{-1.12 - 0}{0.95} = -1.18$$

Because there were 30 students in the experiment, we should use the t distribution with $30 - 2 = 28$ degrees of freedom to calculate the P-value. We want to find $P(t \leq -1.18)$. Using Table B, the P-value is between 0.10 and 0.15. Using technology, we get a more precise P-value of 0.124.

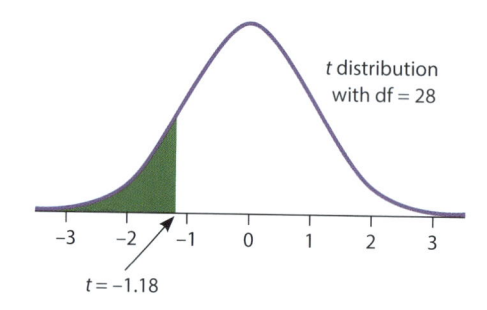

Because this P-value is larger than $\alpha = 0.05$, we fail to reject H_0. These data do not provide convincing evidence that sitting closer to the front of the classroom improves test scores for students like the ones in this experiment.

Calculating the Standardized Test Statistic and P-Value in a Significance Test About the Slope of a Population Regression Line

Suppose the conditions for inference about the slope are met. To perform a test of $H_0: \beta_1 = 0$, compute the standardized test statistic

$$t = \frac{b_1 - 0}{SE_{b_1}}$$

Find the P-value by calculating the probability of getting a t statistic this large or larger in the direction specified by the alternative hypothesis H_a using a t distribution with $n - 2$ degrees of freedom.

EXAMPLE 11.5.2

Can foot length predict height?

Calculating test statistics and P-values

PROBLEM

In Example 11.5.1, the equation of the least-squares regression line is $\widehat{height} = 102.7 + 2.77\, footlength$ and the standard error of the slope is $SE_{b_1} = 0.806$. Calculate the standardized test statistic and P-value for a test of the hypotheses $H_0: \beta_1 = 0$ versus $H_a: \beta_1 > 0$, where β_1 is the slope of the population least-squares regression line relating $y =$ height to $x =$ foot length. Assume the conditions are met.

SOLUTION

- Standardized test statistic: $t = \dfrac{2.77 - 0}{0.806} = 3.437$ $\qquad$ $t = \dfrac{b_1 - 0}{SE_{b_1}}$

- P-value: df = 15 − 2 = 13

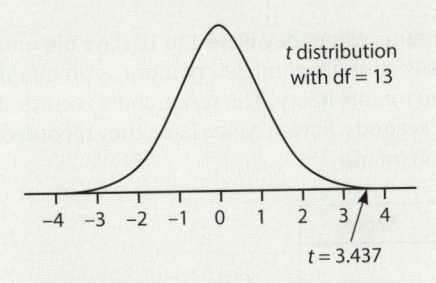

t distribution with df = 13

t = 3.437

Using Table B: 0.001 < P-value < 0.0025
Using technology: P-value = 0.0022

EXAM PREP: FOR PRACTICE, TRY EXERCISE 9.

As we mentioned in Section 11.2, regression calculations are often performed using statistical software. Here is computer output from JMP using the data from Example 11.5.2. The standard error of the slope and the t statistic match the values in the example. Note that JMP reports the two-sided P-value by default. If a test is one-sided and there is evidence for H_a, simply divide the P-value provided in half. Dividing 0.0044 in half gives 0.0022, the value from the example.

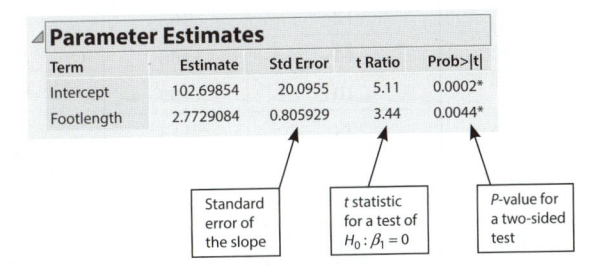

Parameter Estimates

Term	Estimate	Std Error	t Ratio	Prob>|t|
Intercept	102.69854	20.0955	5.11	0.0002*
Footlength	2.7729084	0.805929	3.44	0.0044*

Standard error of the slope — t statistic for a test of $H_0: \beta_1 = 0$ — P-value for a two-sided test

Assuming there is no association between foot length and height in the population of college students, there is a 0.0022 probability of getting a sample slope of 2.77 or greater by chance alone. Because the P-value in this example is less than $\alpha = 0.05$, we reject H_0. These data provide convincing evidence of a positive association between foot length and height in the population of college students.

As we've seen in previous chapters, confidence intervals can provide more information than a significance test does. The 95% confidence interval for the population slope in this context is 1.032 to 4.514. Because all of the values in the interval are positive, the confidence interval gives the same conclusion as the test—plus an interval of plausible values for the slope.

Putting It All Together: Testing the Slope of a Population Least-Squares Regression Line

We are now ready to use the four-step process to perform a t *test for the slope of a least-squares regression line*.

A **t test for the slope of a least-squares regression line** is a significance test of the null hypothesis that there is no linear association between two quantitative variables.

When raw data are provided, as in the seating chart Concept Exploration or the following example, you should use technology to calculate the equation of the least-squares regression line, produce graphs to check conditions, and find the test statistic and P-value. See the Tech Corner at the end of the section for details.

EXAMPLE 11.5.3

Can infant crying predict aptitude later in life?

Performing a test for slope

PROBLEM

Child development researchers explored the relationship between the crying of infants when they were 4 to 10 days old and their scores on an aptitude test later in life, thinking that infants who are more easily stimulated might cry more as an infant and also develop faster. A snap of a rubber band on the sole of the foot caused the infants to cry. The researchers recorded the crying and measured its intensity by the number of peaks in the most active 20 seconds. Several years later, they recorded the aptitude score for the child. The table contains data from a random sample of 38 infants.[45]

Cry count	Score	Cry count	Score
10	87	13	162
12	97	17	94
9	103	19	103
16	106	13	104
18	109	18	109
15	114	18	112
12	119	16	118
20	132	19	120
16	136	22	135
33	159	30	155
20	90	12	94
16	100	12	103
23	103	14	106
27	108	10	109
15	112	23	113
21	114	9	119
12	120	16	124
15	133	31	135
17	141	22	157

Do these data provide convincing evidence at the $\alpha = 0.05$ level of a positive linear relationship between cry count and later aptitude score in the population of infants?

SOLUTION

State:

$H_0: \beta_1 = 0$
$H_a: \beta_1 > 0$

where β_1 = the slope of the population regression line relating y = aptitude score to x = count of crying peaks in the population of infants.
Use $\alpha = 0.05$.
The evidence for H_a is: $b_1 = 1.4929 > 0$.

Plan: *t* test for the slope of a least-squares regression line.
- Random? Random sample of 38 infants. ✓
- Linear? There is a linear pattern in the scatterplot, and there are no leftover curved patterns in the residual plot. ✓

> **State:** State the hypotheses, parameter(s), significance level, and evidence for H_a. Calculate the sample slope using technology.

> **Plan:** Identify the appropriate inference method and check the conditions. *Use technology to create the residual plot and graph of the residuals.*

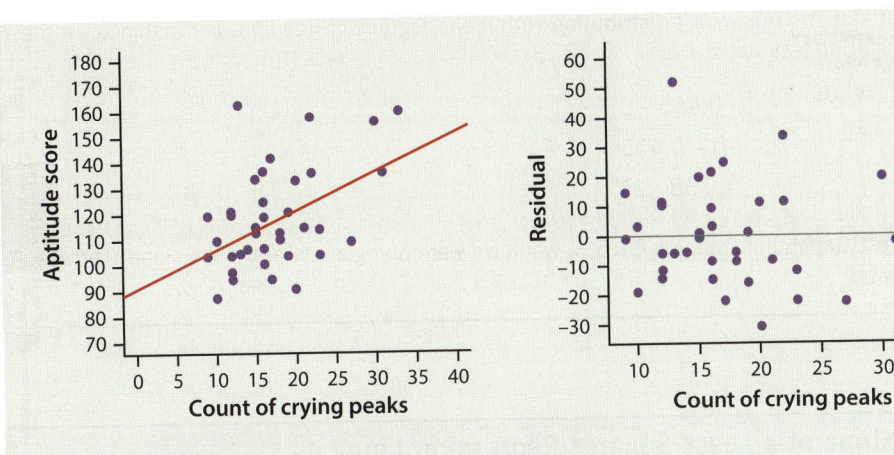

Section 11.5 Significance Tests for the Slope of a Population Least-Squares Regression Line

- **Equal SD?** The residual plot does not show a > or < pattern. ✓
- **Normal?** The histogram of residuals does not show strong skewness or obvious outliers. ✓

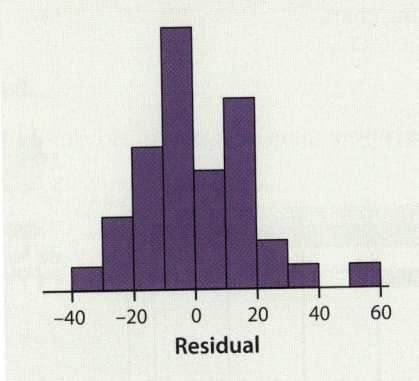

Do: Using technology,
- $t = 3.07$
- P-value $= 0.002$ with df $= 38 - 2 = 36$

Conclude: Because the P-value of $0.002 < \alpha = 0.05$, we reject H_0. There is convincing evidence of a positive linear relationship between the count of crying peaks and later aptitude score in the population of infants.

Do: If the conditions are met, perform the calculations:
- Calculate the test statistic.
- Find the P-value.

Conclude: Make a conclusion about the hypotheses in the context of the problem.

EXAM PREP: FOR PRACTICE, TRY EXERCISE 13.

Based on the results of the crying study, should we ask doctors and parents to make infants cry more so that they'll have greater aptitude scores later in life? Hardly. This observational study gives statistically significant evidence of a positive linear relationship between the two variables. However, we can't conclude that more intense crying as an infant *causes* an increase in aptitude. Maybe infants who cry more are more alert to begin with and tend to score higher on aptitude tests.

THINK ABOUT IT

Could we also do a test for the correlation? Yes. Testing the null hypothesis $H_0: \beta_1 = 0$ is the same as testing the null hypothesis $H_0: \rho = 0$, where ρ (Greek lowercase rho) is the population correlation between the two variables. In both cases, the null hypothesis specifies that there is no association between the variables. Here is the formula for the test statistic with this null hypothesis, where r is the sample correlation:

$$t = \frac{r - 0}{\sqrt{\frac{1 - r^2}{n - 2}}}$$

(continued)

When the conditions are met, this test statistic follows a t distribution with $n - 2$ degrees of freedom. For instance, in the preceding example, $r = 0.455$. The resulting test statistic is

$$t = \frac{0.455 - 0}{\sqrt{\frac{1 - 0.455^2}{38 - 2}}} = 3.07$$

This is exactly the same value as the test statistic in Example 11.5.3, when we were doing a test about the slope. The one-sided P-value (0.002) is the same as well!

TECH CORNER t Test for the Slope of a Least-Squares Regression Line

You can use technology to check conditions and perform the calculations for a t test for the slope of a least-squares regression line. We'll illustrate this process using the seating chart data from the Concept Exploration at the beginning of this section.

Applet

1. Go to www.stapplet.com and launch the *Two Quantitative Variables* applet.

2. Enter "Row number" as the explanatory variable. Next, input the 30 values for this variable. Then, enter "Test score" as the response variable, along with the 30 corresponding values for this variable.

Two Quantitative Variables

Variable	Name	Observations (separated by commas or spaces) Keep individuals in the same order.
Explanatory	Row number	1 1 1 1 2 2 2 2 3 3 3 3 4
Response	Test score	76 77 94 99 83 85 74 79 90 88 68 78 9

Begin analysis | Edit inputs | Reset everything

3. Click the Begin analysis button. To check the conditions, refer to the Tech Corner in Section 11.4. To calculate the standardized test statistic and P-value, scroll down to the "Perform Inference" section and choose "t test for slope" from the drop-down menu. For the alternative hypothesis, choose $\beta < 0$. (Note that the applet uses β to represent the slope of the population regression line, not β_1.) Click the Perform inference button. The test statistic, P-value, and degrees of freedom will be displayed.

Perform Inference

Inference procedure: [t test for slope ▼] Alternative hypothesis: [$\beta < 0$ ▼]

[Perform inference]

t	P-value	df
−1.179	0.124	28

TI-83/84

1. Enter the x values (row number) in L_1 and the y values (test score) in L_2.

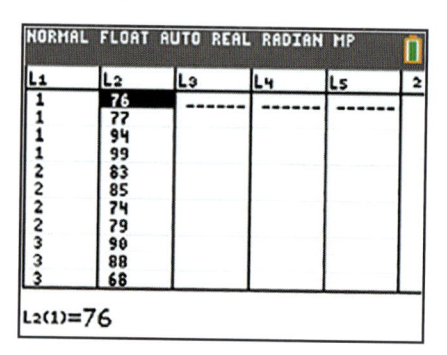

NORMAL FLOAT AUTO REAL RADIAN MP

L1	L2	L3	L4	L5	2
1	76	------	------	------	
1	77				
1	94				
1	99				
2	83				
2	85				
2	74				
2	79				
3	90				
3	88				
3	68				

L2(1)=76

2. To do the calculations, press STAT, arrow to TESTS, choose LinRegTTest, and adjust the settings as shown. Highlight "Calculate" and press ENTER. The test statistic (t), P-value (p), and degrees of freedom (df) are reported, along with the y-intercept (a), slope (b), standard deviation of the residuals (s), r^2, and correlation (r).

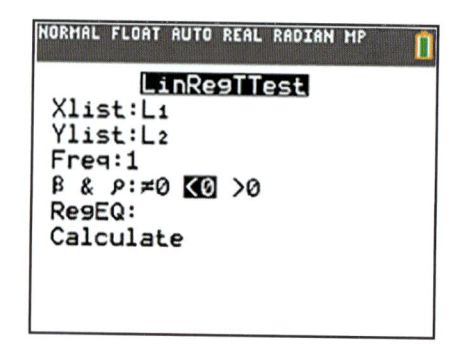

NORMAL FLOAT AUTO REAL RADIAN MP

```
         LinRegTTest
Xlist:L1
Ylist:L2
Freq:1
β & ρ:≠0  <0  >0
RegEQ:
Calculate
```

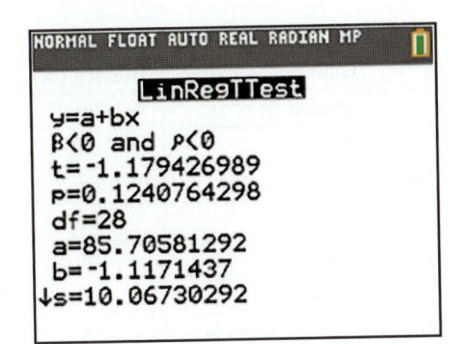

Section 11.5 Significance Tests for the Slope of a Population Least-Squares Regression Line

```
NORMAL FLOAT AUTO REAL RADIAN MP
        LinRegTTest
y=a+bx
β<0 and ρ<0
t=-1.179426989
p=0.1240764298
df=28
a=85.70581292
b=-1.1171437
↓s=10.06730292
```

Notes:
- The symbol ρ is the Greek letter rho, which represents the value of the true correlation. The test for slope and the test for correlation give the same standardized test statistic and *P*-value.
- To check the conditions, refer to the Tech Corner in Section 11.4.

Detailed instructions for using CrunchIt!, Excel, Google Sheets, JMP, Minitab, and R are available in **Achieve**.

Section 11.5 What Did You Learn?

Review the learning goals from this section. Then practice what you've learned by working through the exercises.

Learning Goal	Example	Exercises
State hypotheses for performing a significance test about the slope of a least-squares regression line.	11.5.1	5–8
Calculate the standardized test statistic and *P*-value for a significance test about the slope of a least-squares regression line.	11.5.2	9–12
Use the four-step process to perform a significance test about the slope of a least-squares regression line.	11.5.3	13–16

Section 11.5 Exercises

Building Concepts and Skills These exercises assess the basic knowledge you should have after reading the section.

1. True/False: Hypotheses for a test about the relationship between two quantitative variables must be expressed in words.

2. How do you show the evidence for H_a in a test about the slope of a least-squares regression line?

3. What is the formula for the standardized test statistic in a test about the slope of a least-squares regression line?

4. True/False: The proper degrees of freedom for a test about the slope of a least-squares regression line is $n-1$.

Mastering Concepts and Skills These exercises reinforce the learning goals as shown in the examples.

5. **Refer to Example 11.5.1 Hungry Fish** Does a larger concentration of fish attract more predators? One study looked at kelp perch and their common predator, the kelp bass. The researcher set up 16 large circular pens on sandy ocean bottoms off the coast of southern California. Young perch were selected at random from a large group and placed in the pens. Four pens had 10 perch each, four pens had 20 perch each, 4 pens had 40 perch each, and the final 4 pens had 60 perch each. The researcher then dropped the nets protecting the pens, allowing bass to swarm in, and counted the perch left after 2 hours.[46] The equation of the least-squares regression line is $\hat{y} = 0.120 + 0.00857x$, where y = proportion of perch killed and x = number of perch in the pen.

   (a) State hypotheses for a test about the relationship between number of perch and proportion of perch killed.

   (b) What is the evidence for H_a?

6. **Car Statistics** Do older cars have more miles driven? Researchers selected a random sample of 21 cars from a university parking lot and asked the cars' drivers to report the age (in years) of the vehicles and the number of miles driven. The equation of the least-squares regression line is $\hat{y} = 7288.54 + 11630.6x$, where x = age (years) and y = miles driven.

   (a) State hypotheses for a test about the relationship between age and miles driven for cars in this university parking lot.

   (b) What is the evidence for H_a?

7. **Wine and Heart Health** Is higher per capita wine consumption associated with a lower death rate from heart disease? A researcher from the University of California, San Diego, collected data on average per capita wine consumption and heart disease death rate in a random sample of 19 countries for which data were available.[47] The equation of the least-squares regression line is $\widehat{rate} = 260.6 - 22.97 wine$, where wine consumption is measured as alcohol from wine per person (in liters) and heart disease death rate is the number of deaths from heart disease per 100,000 people.

   (a) State hypotheses for a test about the relationship between wine consumption and heart disease death rate in these countries.

   (b) What is the evidence for H_a?

8. **Swimming and Pulse Rate** Professor Moore recorded the time he took to swim 2000 yards and his pulse rate after swimming on a random sample of 23 days. Does his pulse rate tend to be higher on days that he swims faster (has smaller times)? The equation of the least-squares regression line is $\widehat{pulse} = 484 - 9.81 time$, where pulse rate is in beats per minute and time is in minutes.

   (a) State hypotheses for a test about the relationship between pulse rate and swimming time for days when Professor Moore swims 2000 yards.

   (b) What is the evidence for H_a?

9. **Refer to Example 11.5.2 Hungry Fish** Refer to Exercise 5. The standard error of the slope is $SE_{b_1} = 0.002456$ and the sample size is $n = 16$. Calculate the standardized test statistic and P-value for a test of the hypotheses in Exercise 5. Assume the conditions are met.

10. **Car Statistics** Refer to Exercise 6. The standard error of the slope is $SE_{b_1} = 1249$ and the sample size is $n = 21$. Calculate the standardized test statistic and P-value for a test of the hypotheses in Exercise 6. Assume the conditions are met.

11. **Wine and Heart Health** Refer to Exercise 7. The standard error of the slope is $SE_{b_1} = 3.56$ and the sample size is $n = 19$. Calculate the standardized test statistic and P-value for a test of the hypotheses in Exercise 7. Assume the conditions are met.

12. **Swimming and Pulse Rate** Refer to Exercise 8. The standard error of the slope is $SE_{b_1} = 1.84$ and the sample size is $n = 23$. Calculate the standardized test statistic and P-value for a test of the hypotheses in Exercise 8. Assume the conditions are met.

13. **Refer to Example 11.5.3 Literacy and Birth Rates** The table shows $x =$ female literacy rate (for women 15 years of age or older) and $y =$ birth rate (births per 1000 population) for 20 randomly selected countries.[48] Do these data provide convincing evidence at the $\alpha = 0.05$ significance level that countries with higher female literacy rates have lower birth rates?

Country	Female literacy (%)	Birth rate
Bhutan	39	23
Poland	99	9.8
Sao Tome and Principe	68	38
Venezuela	94	22
Belize	70	35
Malta	94	9.2
Congo Democratic Republic	49	46
Vietnam	89	17
Zimbabwe	80	35
Madagascar	63	38
Central African Republic	39	37
Ghana	58	34
Nicaragua	78	24
Egypt	59	25
Ecuador	84	23
Suriname	89	21
Mongolia	98	20
Macedonia	95	12
Gabon	80	32
Mauritania	47	37

14. **Beer and BAC** Does drinking more cans of beer result in a higher blood alcohol content (BAC)? Sixteen volunteers aged 21 or older with an initial BAC of 0 took part in a study to find out. Each volunteer drank a randomly assigned number of cans of beer and had their BAC measured 30 minutes later.[49] The table shows $x =$ number of beers and $y =$ BAC. Do these data provide convincing evidence at the $\alpha = 0.05$ significance level that volunteers like these will have a higher BAC when consuming more cans of beer?

Beers	BAC	Beers	BAC
5	0.1	3	0.02
2	0.03	5	0.05
9	0.19	4	0.07
8	0.12	6	0.1
3	0.04	5	0.085
7	0.095	7	0.09
3	0.07	1	0.01
5	0.06	4	0.05

15. **Heartbeats** Is there a relationship between a person's body temperature and their resting pulse rate? The table shows $x =$ body temperature (in °F) and $y =$ pulse rate (in beats per minute) for each of 20 randomly selected students at a university.[50] Do these data provide convincing evidence at the 1% significance level of a linear relationship between body temperature and resting pulse rate for students at this university?

Temperature	Pulse rate
97.9	78
97.1	64
98.6	72
98.5	80
98.1	60
97.0	56
97.2	75
97.8	70
98.3	90
97.3	76
98.8	80
98.0	82
97.5	84
98.0	75
97.6	87
97.7	61
96.6	69
96.6	69
96.6	71
96.8	74

16. Toddler Mealtimes Is there an association between the length of time a toddler sits at the lunch table and the number of calories consumed? Researchers collected data on a random sample of 20 toddlers observed over several months. The table shows $x=$ the mean time a child spent at the lunch table (minutes) and $y=$ the mean number of calories consumed.[51] Do these data provide convincing evidence at the 10% significance level of a linear association between time and calories for toddlers?

Time	Calories	Time	Calories
21.4	472	42.4	450
30.8	498	43.1	410
37.7	465	29.2	504
33.5	456	31.3	437
32.8	423	28.6	489
39.5	437	32.9	436
22.8	508	30.6	480
34.1	431	35.1	439
33.9	479	33.0	444
43.8	454	43.7	408

Applying the Concepts These exercises ask you to apply multiple learning goals in a new context or to apply what you learned in this section in a new way.

17. Beavers and Beetles Researchers laid out 23 circular plots, each 4 meters in diameter, at random in an area where beavers were cutting down cottonwood trees. In each plot, they counted the number of stumps from trees cut by beavers and the number of clusters of beetle larvae. Ecologists think that the new sprouts from stumps are more tender than other cottonwood growth, such that beetles prefer them. If so, we would expect more beetle larvae when there are more stumps.[52] Do these data provide convincing evidence of a positive association between number of stumps and number of beetle larvae in this area?

Stumps	Larvae	Stumps	Larvae
2	10	2	25
2	30	1	8
1	12	2	21
3	24	2	14
3	36	1	16
4	40	1	6
3	43	4	54
1	11	1	9
2	27	2	13
5	56	1	14
1	18	4	50
3	40		

18. Lean Body Mass and Metabolism We have data on the lean body mass and resting metabolic rate for a random sample of 12 people who volunteered for a study of dieting. Lean body mass, given in kilograms (kg), is a person's weight leaving out all fat. Metabolic rate is measured in calories burned per 24 hours (cal/24 hr). The researchers believe that lean body mass is positively associated with metabolic rate. Do these data provide convincing evidence of a positive association between lean body mass and metabolic rate for people like these?

Lean body mass (kg)	Metabolic rate (cal/24 hr)
36.1	995
54.6	1425
48.5	1396
42.0	1418
50.6	1502
42.0	1256
40.3	1189
33.1	913
42.4	1124
34.5	1052
51.1	1347
41.2	1204

19. Beavers and Beetles Interpret the P-value from Exercise 17.

20. Lean Body Mass and Metabolism Interpret the P-value from Exercise 18.

21. **Heartbeats** Refer to Exercise 15.
    (a) Calculate and interpret a 99% confidence interval for the slope of the population least-squares regression line relating $y =$ resting pulse rate to $x =$ body temperature (in °F). Assume the conditions are met.
    (b) Explain how the interval from part (a) is consistent with your conclusion in Exercise 15.

22. **Toddler Mealtimes** Refer to Exercise 16.
    (a) Calculate and interpret a 90% confidence interval for the slope of the population least-squares regression line relating $y =$ mean number of calories consumed to $x =$ mean time a child spent at the lunch table (in minutes). Assume the conditions are met.
    (b) Explain how the interval from part (a) is consistent with your conclusion in Exercise 16.

23. Tech **Car Price and Age** Refer to Exercise 23 in Section 11.2. The data set **Honda CRV** contains information about a random sample of 191 used Honda CRVs. Use technology to determine if these data provide convincing evidence of a negative linear association between $y =$ Price and $x =$ Age (years). *Note*: Due to the large sample size, it isn't surprising to find outliers in the distribution of residuals when the Normal condition is met. Therefore, you can assume the Normal condition is met for these data.

24. Tech **Car Price and Mileage** Refer to Exercise 24 in Section 11.2. The data set **Honda CRV** contains information about a random sample of 191 used Honda CRVs. Use technology to determine if these data provide convincing evidence of a negative linear association between $y =$ Price and $x =$ Miles.

Extending the Concepts These exercises challenge you to explore statistical concepts and methods that go beyond what you learned in this section.

25. **Literacy and Birth Rates** Refer to Exercise 13, which provides data about $x =$ female literacy rate (for women 15 years of age or older) and $y =$ birth rate (births per 1000 population) for 20 randomly selected countries. Do these data provide convincing evidence that countries with higher female literacy rates have lower birth rates?
    (a) State the hypotheses for this question in terms of the population correlation coefficient ρ.
    (b) Calculate the sample correlation r for these data.
    (c) Use the formula from the Think About It feature in this section to calculate the value of the test statistic. How does this compare to the value of the test statistic from Exercise 13?
    (d) Calculate the P-value using the test statistic in part (c). How does it compare to the P-value from Exercise 13?

26. **Car Statistics** Refer to the car data in Exercises 6 and 10. In addition to performing tests about the slope of a least-squares regression line, it is possible to perform a test about the y-intercept.

    (a) If $x =$ age (years) and $y =$ miles driven, what should the value of the y-intercept equal? Explain your answer.
    (b) Based on your answer to part (a), state the hypotheses for a test about the y-intercept of the population least-squares regression line. Use β_0 to represent the y-intercept of the population least-squares regression line.
    (c) The equation of the least-squares regression line is $\hat{y} = 7288.54 + 11{,}630.6x$, where $x =$ age (years) and $y =$ miles driven. The standard error of the y-intercept is $SE_{b_0} = 6591$ and the sample size is $n = 21$. Calculate the t statistic and P-value for a test of the hypotheses in part (b), assuming the conditions for inference are met.
    (d) Based on the P-value in part (c), what would you conclude?

27. **Infant Crying and Aptitude** In Example 11.5.3, you used the slope to perform a test about the relationship between $x =$ cry count and $y =$ later aptitude score in the population of infants. It is also possible to classify the values for each variable into categories and use a chi-square test for association. The two-way table summarizes these data:

		Cry count		
		Less than 16	At least 16	Total
Aptitude score	Less than 115	10	12	22
	At least 115	5	11	16
	Total	15	23	38

   (a) Do these data provide convincing evidence of an association between cry count and later aptitude score for infants?
   (b) Describe a disadvantage of using a chi-square test instead of a t test for slope in this context.

28. **Car Statistics** Refer to the car data in Exercises 6 and 10. Although it is most common to test $H_0: \beta_1 = 0$, it is possible to use values other than 0 in the null hypothesis. A national automotive group claims that a typical car is driven 15,000 miles per year.
    (a) State the null and alternative hypotheses for a test of the automotive group's claim.
    (b) Use the slope and standard error of the slope from Exercises 6 and 10 to calculate the test statistic for a test of the hypotheses in part (a).
    (c) Calculate the P-value. What conclusion would you make?

Cumulative Review These exercises revisit what you learned in previous sections.

29. **Shower Time (3.3, 6.2, 6.3, 6.8)** Marcella takes a shower every morning when she gets up. Her time in the shower varies according to a normal distribution with mean 4.5 minutes and standard deviation 0.9 minute.

(a) Find the probability that Marcella's shower lasts between 3 and 6 minutes on a randomly selected day.

(b) If Marcella took a 7-minute shower, would it be classified as an outlier by the $1.5 \times IQR$ rule? Justify your answer.

(c) Find the probability that the *mean* length of her shower times on 10 randomly selected days exceeds 5 minutes.

30. **Inference Procedures (Chapters 7–10)** In each of the following settings, state which inference procedure from Chapters 7–10 you would use. For example, you might say "two-sample z test for the difference between two proportions." You do not have to carry out any procedures.[53]

    (a) What is the average voter turnout during an election? A random sample of 38 cities was asked to report the percentage of registered voters who actually voted in the most recent election.

    (b) Are people who live in urban areas more likely to visit libraries than people who live in rural areas? Separate random samples of 300 urban residents and 300 rural residents were asked if they had been to a library in the previous year.

    (c) Is there a relationship between attendance at religious services and alcohol consumption? A random sample of 1000 adults was asked if they regularly attend religious services and if they drink alcohol daily.

    (d) Separate random samples of 75 undergraduate students and 75 graduate students were asked how much time, on average, they spend watching television each week. We want to estimate the difference in the average amount of TV watched by undergraduate and graduate students.

Section 11.6 Confidence Intervals for a Mean Response and Prediction Intervals in Regression

> **LEARNING GOALS**
>
> By the end of this section, you will be able to:
> - Interpret a confidence interval for the mean y value at a given x value.
> - Interpret a prediction interval for an individual y value at a given x value.
> - Describe the factors that affect the width of a confidence interval for the mean y value and the width of a prediction interval for an individual y value at a given x value.

In Section 11.4, you learned how to find a confidence interval for β_1, the slope of the population (true) least-squares regression line. Knowing the possible values for the slope β_1 helps us understand the relationship between the explanatory and response variables in the population.

Another use of the regression line is to make predictions about the response variable y at specific values of the explanatory variable x. In Section 11.1, you learned how to calculate a predicted value of y for a specific value of x using the regression line computed from data. Now we use the fact that the sample regression line is only an estimate of the population regression line, which means that the predicted value of y we compute from a sample regression line is also only an estimate. In this section, we will explore the idea of a confidence interval for the predicted value of y at a given value of x. In fact, we will see two different intervals, depending on how we interpret the predicted value.

Confidence Intervals for the Mean y Value at a Given Value of x

Throughout this chapter, we have been analyzing data from the U.S. Forest Service about the relationship between $x =$ the diameter at breast height (DBH) and $y =$ the height of ponderosa pines in western Montana. The least-squares regression line calculated from the random sample of 40 trees is $\widehat{height} = 43.5 + 2.62\,DBH$. The scatterplot shows this relationship.

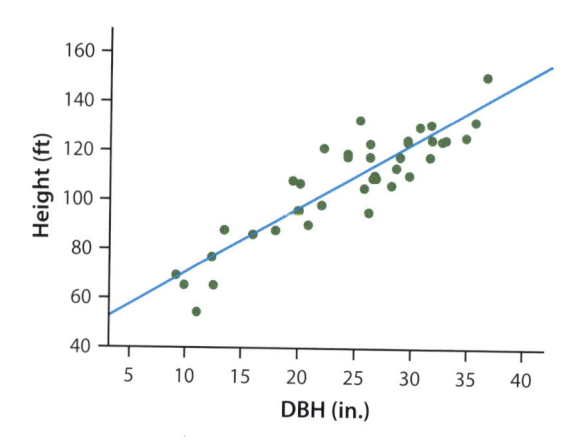

Consider trees that have a DBH of 24 inches. Using the least-squares regression line, we find $\hat{y} = 43.5 + 2.62(24) = 106.4$ feet. Thinking back to the population regression model in Figure 11.10 from Section 11.4, we can interpret the value 106.4 as the estimated mean height of all ponderosa pines in western Montana with a DBH of 24 inches.

When data come from a random sample or a randomized experiment, the statistic $\hat{y}$ is a point estimate for μ_y, the population mean value of y at a given value of x. When our goal is to estimate μ_y, we refer to $\hat{y}$ as $\widehat{\mu_y}$. The confidence interval for μ_y has a familiar form:

$$\text{point estimate} \pm \text{margin of error}$$

$$\text{point estimate} \pm \text{critical value} \times \text{standard error of statistic}$$

The same conditions are required here as for inference about the slope of the regression line. When those conditions are met, the critical value for the confidence interval comes from the t distribution with df $= n - 2$.

The standard error of $\hat{\mu}_y$ is

$$SE_{\hat{\mu}_y} = s\sqrt{\frac{1}{n} + \frac{(x^* - \bar{x})^2}{(n-1)s_x^2}}$$

where s is the standard deviation of the residuals, x^* is the given value of x, n is the sample size, and s_x is the standard deviation of the explanatory variable x. As with other standard errors, $SE_{\hat{\mu}_y}$ estimates the typical amount that $\widehat{\mu_y}$ varies from μ_y at the given value of x.

To compute the standard error for the ponderosa pine trees, we need several pieces of information. We already know that $n = 40$ and we have specified $x^* = 24$. Technology gives $s = 9.428$, $\bar{x} = 24.50$, and $s_x = 7.39$. Putting this all together, we get

$$SE_{\hat{\mu}_y} = 9.428\sqrt{\frac{1}{40} + \frac{(24 - 24.50)^2}{39(7.39^2)}} = 1.494$$

With a sample size of 40, df $= 40 - 2 = 38$. Using technology, we find that $t^* = 2.024$ for a 95% confidence interval. The resulting confidence interval is

$$106.4 \pm 2.024(1.494) \rightarrow 106.4 \pm 3.0 \rightarrow 103.4 \text{ to } 109.4$$

We are 95% confident that the interval from 103.4 to 109.4 feet captures the mean height of all ponderosa pine trees in western Montana that have a DBH of 24 inches.

In practice, we do not typically compute the confidence interval by hand, but rather rely on technology to compute it for us.

EXAMPLE 11.6.1

Professional tennis serves

Interpreting a confidence interval for μ_y

PROBLEM

How fast can 79-inch-tall (6-foot 7-inch) male professional tennis players serve the ball, on average? Physics would suggest that taller players have longer arms and can, therefore, generate more racquet speed. Researchers selected a random sample of 16 male professional tennis players and recorded their height in inches and their average first serve speed in miles per hour.[54] Assume that the conditions for regression inference are met. The confidence interval for the mean y value at $x = 79$ inches is 117.9 to 129.0. Interpret this interval.

SOLUTION

We are 95% confident that the interval from 117.9 to 129.0 mph captures the average first serve speed of all 79-inch-tall professional male tennis players.

> Remember to specify the value of x in your interpretation. This interval is for the mean of y at only that particular value of x.

EXAM PREP: FOR PRACTICE, TRY EXERCISE 5.

We now turn to the second interpretation of $\hat{y}$.

Prediction Intervals for an Individual y Value at a Given Value of x

In the previous subsection, we interpreted $\hat{y}$ as the estimated mean value of y when x takes a specified value. We can also interpret $\hat{y}$ as the estimated value of y for a single observation that has a specified value of x. So, $\hat{y} = 106.4$ feet is the estimated mean height of all ponderosa pines with DBH = 24 inches *and* $\hat{y} = 106.4$ feet is the estimated height of an individual ponderosa pine tree with DBH = 24 inches.

When our goal is to estimate y = the value of the response variable for an individual observation, we will refer to the estimate as $\hat{y}$ instead of $\widehat{\mu_y}$. We refer to the confidence interval for y as a *prediction interval* to emphasize that we are predicting an individual value, not estimating a mean of many values. The prediction interval for y at a given value of x has the same familiar form as the confidence interval for μ_y:

$$\text{point estimate} \pm \text{margin of error}$$

$$\text{point estimate} \pm \text{critical value} \times \text{standard error of statistic}$$

As with a confidence interval for a mean response, the conditions for a prediction interval are the same as those for inference about the slope of the regression line. When those conditions are met, the critical value for the confidence interval comes from the t distribution with df $= n - 2$. The standard error of $\hat{y}$ is

$$\text{SE}_{\hat{y}} = s\sqrt{1 + \frac{1}{n} + \frac{(x^* - \bar{x})^2}{(n-1)s_x^2}}$$

where s is the standard deviation of the residuals, x^* is the given value of x, n is the sample size, and s_x is the standard deviation of the explanatory variable x. As with other standard errors, $\text{SE}_{\hat{y}}$ estimates the typical amount that $\hat{y}$ varies from y at the given value of x.

Recall that the standard error of $\widehat{\mu_y}$ is

$$\text{SE}_{\hat{\mu}_y} = s\sqrt{\frac{1}{n} + \frac{(x^* - \bar{x})^2}{(n-1)s_x^2}}$$

which looks very similar to the standard error of $\hat{y}$. The only difference is the "1 +" under the square root in $\text{SE}_{\hat{y}}$. The result of this difference is that $\text{SE}_{\hat{y}} > \text{SE}_{\hat{\mu}_y}$ and that the $C\%$ prediction interval for y is always wider than the corresponding $C\%$ confidence interval for μ_y at a given value of x.

What is a 95% prediction interval for the height of a single ponderosa pine with DBH = 24 inches? We saw earlier that

$$SE_{\hat{\mu}_y} = 9.428\sqrt{\frac{1}{40} + \frac{(24-24.5)^2}{39(7.39^2)}} = 1.494$$

Thus,

$$SE_{\hat{y}} = 9.428\sqrt{1 + \frac{1}{40} + \frac{(24-24.5)^2}{39(7.39^2)}} = 9.546$$

and the prediction interval is

$$106.4 \pm 2.024(9.546) \rightarrow 106.4 \pm 19.3 \rightarrow 87.1 \text{ to } 125.7$$

We are 95% confident that the interval from 87.1 to 125.7 feet captures the height of a single ponderosa pine tree in western Montana with a DBH of 24 inches. This interval is quite a bit wider than the confidence interval for the *mean* height of all ponderosa pines in western Montana with a DBH of 24 inches, which was 103.4 to 109.4 feet.

Intuitively, it makes sense that a prediction interval for an individual value of y will be wider than the corresponding confidence interval for the mean value of y. The mean value of y for all observations with a fixed x value is a constant. So a confidence interval for μ_y considers only the variability of $\widehat{\mu_y}$ from μ_y. But individual observations with the same value of x will likely have values of y that differ from μ_y. Therefore, a prediction interval for an individual value of y needs to take into consideration both the variability of $\widehat{\mu_y}$ from μ_y *and* the variability of the individual values of y from μ_y. Because of this additional source of variability, prediction intervals are wider than the corresponding confidence intervals. As you learned in Section 6.8, individuals are more variable than averages.

EXAMPLE 11.6.2

Professional tennis serves

Interpreting a prediction interval for y

PROBLEM

How fast can an individual 79-inch-tall male professional tennis player serve the ball? The data from Example 11.6.1 yield a prediction interval for an individual y value with $x = 79$ inches of (109.5, 137.4). Interpret this interval.

SOLUTION

We are 95% confident that the interval from 109.5 to 137.4 mph captures the first serve speed of a single 79-inch-tall professional male tennis player.

> A prediction interval predicts the value for a *specific individual*, not the mean of all individuals.

EXAM PREP: FOR PRACTICE, TRY EXERCISE 9.

As with confidence intervals for μ_y, we will rely on technology to compute prediction intervals for y. Given here are two examples of what computer output might look like for the ponderosa pine data with $x^* = 24$. The interval marked "CI" is the confidence interval for the mean value of y at the given value of x, and the interval marked "PI" is the prediction interval for an individual value of y at the given value of x.

Section 11.6 Confidence Intervals for a Mean Response and Prediction Intervals in Regression

Minitab

Prediction for Height

WORKSHEET 1
Prediction for Height

Regression Equation

Height = 43.47 + 2.618 DBH

Settings

Variable	Setting
DBH	24

Prediction

Fit	SE Fit	95% CI	95% PI
106.309	1.49421	(103.285, 109.334)	(86.9854, 125.633)

95% CI → Confidence interval
95% PI → Prediction interval

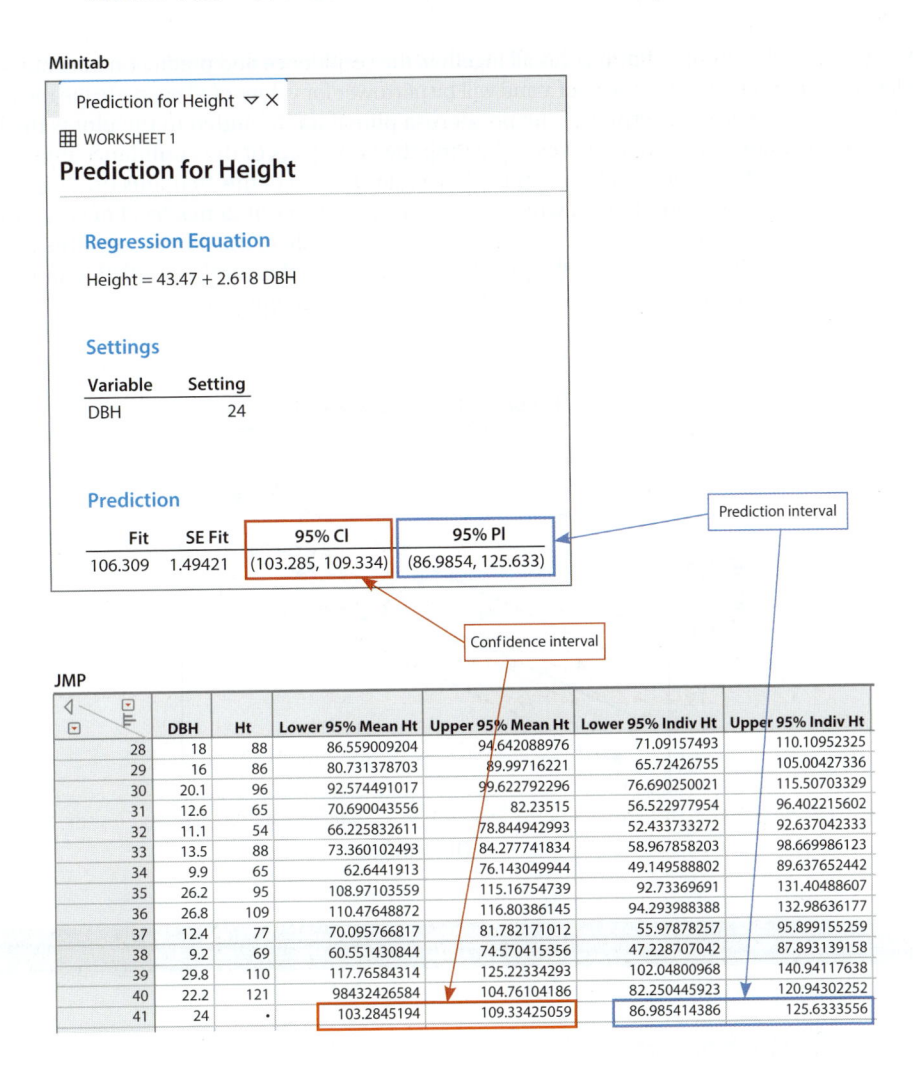

JMP

	DBH	Ht	Lower 95% Mean Ht	Upper 95% Mean Ht	Lower 95% Indiv Ht	Upper 95% Indiv Ht
28	18	88	86.559009204	94.642088976	71.09157493	110.10952325
29	16	86	80.731378703	89.99716221	65.72426755	105.00427336
30	20.1	96	92.574491017	99.622792296	76.690250021	115.50703329
31	12.6	65	70.690043556	82.23515	56.522977954	96.402215602
32	11.1	54	66.225832611	78.844942993	52.433733272	92.637042333
33	13.5	88	73.360102493	84.277741834	58.967858203	98.669986123
34	9.9	65	62.6441913	76.143049944	49.149588802	89.637652442
35	26.2	95	108.97103559	115.16754739	92.73369691	131.40488607
36	26.8	109	110.47648872	116.80386145	94.293988388	132.98636177
37	12.4	77	70.095766817	81.782171012	55.97878257	95.899155259
38	9.2	69	60.551430844	74.570415356	47.228707042	87.893139158
39	29.8	110	117.76584314	125.22334293	102.04800968	140.94117638
40	22.2	121	98.432426584	104.76104186	82.250445923	120.94302252
41	24	.	103.2845194	109.33425059	86.985414386	125.6333556

Controlling Interval Width

We can control the widths of both the confidence interval for a mean response μ_y and the prediction interval for an individual value of y at a particular value of x in three ways: by changing the confidence level, by changing the sample size, or by changing the location of x^* = the value of x chosen. As with the other confidence intervals you've learned about, the price we pay for gaining more confidence is a wider interval for both the confidence interval and the prediction interval. As the confidence level increases, so does the width of the intervals.

Sample size also affects the width of these intervals in the way that we expect. As the sample size gets larger, the width of the intervals gets smaller. Intuitively, this makes sense: If we have more data, we have better information about the population value. Mathematically, the fact that a larger sample size leads to a narrower interval also makes sense. Look again at the formulas for the standard error for both the confidence interval and the prediction interval:

$$SE_{\hat{\mu}_y} = s\sqrt{\frac{1}{n} + \frac{(x^*-\bar{x})^2}{(n-1)s_x^2}} \qquad SE_{\hat{y}} = s\sqrt{1 + \frac{1}{n} + \frac{(x^*-\bar{x})^2}{(n-1)s_x^2}}$$

In both cases, n appears in the denominator of the fractions. If we divide by a larger number, the overall result is a smaller SE. And if the SE is smaller, the width of the interval is also smaller.

How does the specific value of x affect the width of the confidence and prediction intervals? On the face of it, this seems less intuitive than the effect of the confidence level or sample size. In both SE formulas, x^* represents the chosen value of x and the term $(x^*-\bar{x})^2$ measures the squared distance between the x-coordinate of the chosen point and the sample mean of the x variable, $\bar{x}$. The closer the value of x is to $\bar{x}$, the smaller $(x^*-\bar{x})^2$ will be; conversely, the further the value of x is from $\bar{x}$, the larger this term will be. And because $(x^*-\bar{x})^2$ is in the numerator, the standard errors will get larger as x^* gets farther from $\bar{x}$. Putting this all together, the confidence and prediction intervals will be wider for values of x^* far from the mean of x and will be narrower for values of x^* nearer to the mean of x.

Figure 11.11 shows a scatterplot of the ponderosa pine data. Included in the plot is the least-squares regression line, as well as curves indicating the endpoints of the confidence intervals for μ_y (red) and the prediction intervals for y (purple). Notice that 39 of the 40 points (97.5%) are captured by the purple prediction bands, which is very close to the confidence level of 95%. You can also see that the confidence intervals are wider toward the outside of the plot because the red lines are farther apart than in the middle of the plot. It is harder to see this for the prediction intervals in this scatterplot, but they are also wider closer to the edges of the scatterplot.

FIGURE 11.11 Scatterplot showing the relationship between DBH and height of ponderosa pine trees in western Montana, along with the regression line. Also shown are the bands representing the endpoints of the confidence intervals for μ_y (red) and the bands representing the endpoints of the prediction intervals for y (purple).

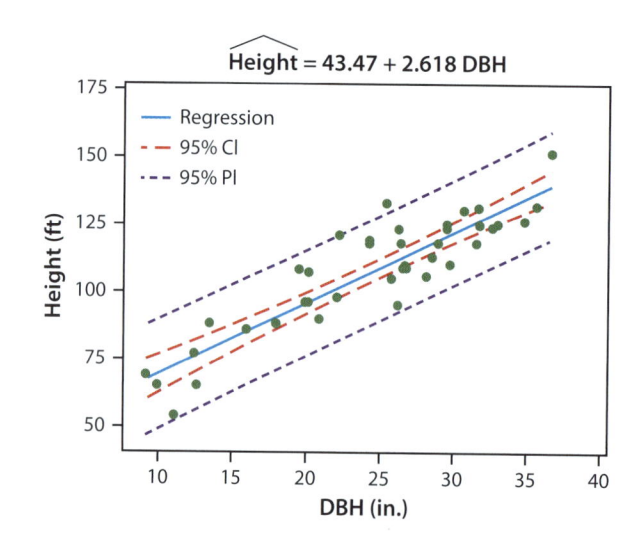

EXAMPLE 11.6.3

Professional tennis serves

Changing confidence level, sample size, and location of x^*

PROBLEM

We saw in Examples 11.6.1 and 11.6.2 that when $x^* = 79$ inches, the 95% confidence interval for μ_y is (117.9, 129.0) and the 95% prediction interval for y is (109.5, 137.4).

(a) Explain what would happen to the width of these intervals if we change the confidence level to 90%.

(b) Explain what would happen to the width of these intervals if we increased the sample size to 35.

(c) Here is a scatterplot comparing the height of the players to the speed of their serves. Explain what would happen to the width of these intervals if we changed x^* to 74 inches.

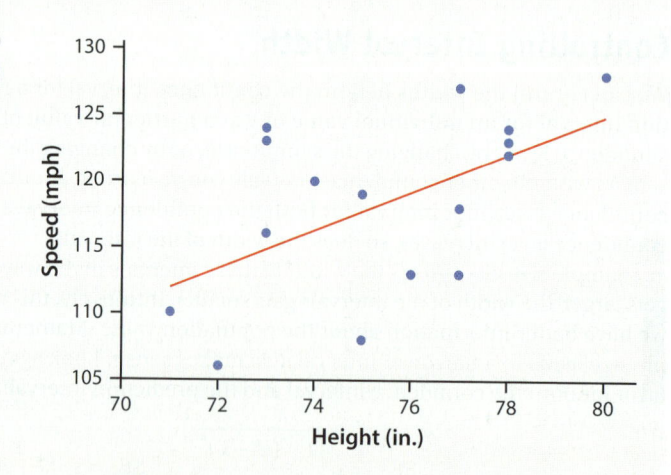

SOLUTION

(a) The width of both intervals will decrease because decreasing the confidence level decreases the margin of error.

(b) Increasing the sample size will decrease the width of both intervals because increasing the sample size decreases the SE, which decreases the margin of error.

(c) Moving x^* from 79 to 74 will decrease the width of both intervals because 74 is closer to the mean value of x than $x^* = 79$. This decreases the SE, which decreases the margin of error.

EXAM PREP: FOR PRACTICE, TRY EXERCISE 13.

Section 11.6 Confidence Intervals for a Mean Response and Prediction Intervals in Regression

TECH CORNER Confidence Intervals for a Mean Response and Prediction Intervals in Regression

Although we do not ask you to compute them in this section, you can use technology to find confidence intervals for a mean response and prediction intervals for an individual response in regression. We'll illustrate this process with the data for the professional men's tennis players used in Examples 11.6.1–11.6.3 by computing the confidence and prediction intervals for $x^* = 79$ inches.

Height (in.)	Speed (mph)	Height (in.)	Speed (mph)
75	108	77	118
78	122	80	128
76	113	73	116
78	123	72	106
77	127	78	124
73	123	75	119
77	113	73	125
74	120	71	110

Applet

1. Go to www.stapplet.com and launch the *Two Quantitative Variables* applet.

2. Enter the name of the explanatory variable (Height) and the values of the explanatory variable in the first row of boxes. Then, enter the name of the response variable (Speed) and the values of the response variable in the second row of boxes.

Two Quantitative Variables

Variable	Name	Observations (separated by commas or spaces) Keep individuals in the same order.
Explanatory	Height (in.)	75 78 76 78 77 73 77 74 77 80 73 72 78 75 73 71
Response	Speed (mph)	108 122 113 123 127 123 113 120 118 128 116 106 124 119 125

[Begin analysis] [Edit inputs] [Reset everything]

3. Click the Begin analysis button. To check the conditions, click "Calculate least-squares regression line." This will display the scatterplot, residual plot, and dotplot of the residuals, allowing you to check the Normal, Linear, and Equal SD conditions.

4. To calculate a confidence interval for the mean of the response variable for a particular value of the explanatory variable, scroll down to the "Perform Inference" section and choose "Confidence interval for mean of response variable" from the drop-down menu. Enter the value of x^* you are using and choose 95% for the confidence interval. Click the Perform inference button. The confidence interval and degrees of freedom will be displayed. You can also click the checkbox to show the interval boundaries on the scatterplot for all values of the explanatory variable.

Perform Inference

Inference procedure: [Confidence interval for mean of response variable ▼]
Explanatory value: [79]   Confidence level: [95]%
☑ Show interval bounds on the scatterplot with regression line

[Perform inference]

Lower Bound	Upper Bound	df
117.898	129.025	14

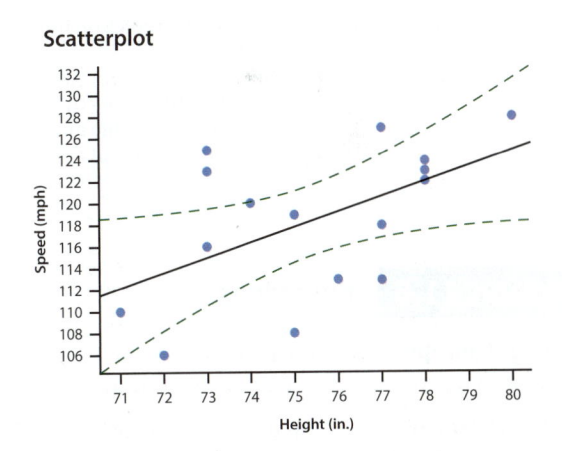

Scatterplot

5. To calculate the prediction interval for an individual response, scroll down to the "Perform Inference" section and choose "Prediction interval for value of response variable" from the drop-down menu. Enter the value of x^* you are using and choose 95% for the confidence interval. Click the Perform inference button. The confidence interval

(continued)

and degrees of freedom will be displayed. You can also click the checkbox to show the interval boundaries on the scatterplot for all values of the explanatory variable.

Perform Inference

Inference procedure: Prediction interval for value of response variable
Explanatory value: 79   Confidence level: 95 %
☑ Show interval bounds on the scatterplot with regression line

[Perform inference]

Lower Bound	Upper Bound	df
109.522	137.401	14

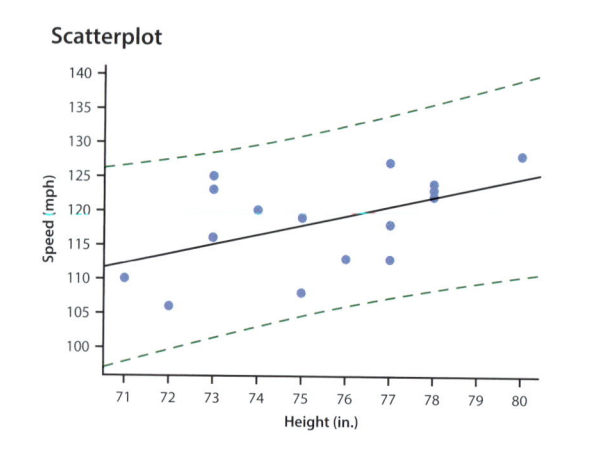

TI-83/84

These computations are not available on the TI-83/84.

Detailed instructions for using CrunchIt!, Excel, Google Sheets, JMP, Minitab, and R are available in Achieve.

Section 11.6 What Did You Learn?

Review the learning goals from this section. Then practice what you've learned by working through the exercises.

Learning Goal	Example	Exercises
Interpret a confidence interval for the mean y value at a given x value.	11.6.1	5–8
Interpret a prediction interval for an individual y value at a given x value.	11.6.2	9–12
Describe the factors that affect the width of a confidence interval for the mean y value and the width of a prediction interval for an individual y value at a given x value.	11.6.3	13–16

Section 11.6 Exercises

Building Concepts and Skills These exercises assess the basic knowledge you should have after reading the section.

1. Which two values can you use $\hat{y}$ to estimate?

2. True/False: A prediction interval for y gives the plausible values of y for all values of x.

3. Which interval, the confidence interval or the prediction interval, will be wider at a given value of x?

4. True/False: Computing a prediction interval for a value of x near $\bar{x}$ will result in a narrower interval than computing a prediction interval for a value of x further away from $\bar{x}$.

Mastering Concepts and Skills These exercises reinforce the learning goals as shown in the examples.

5. **Refer to Example 11.6.1 Loud Music and Tests** Two psychology students wanted to know if listening to music at a louder volume hurts test performance. To investigate, they recruited 30 volunteers and randomly assigned 10 to listen to music at 30 decibels, 10 to listen to music at 60 decibels, and 10 to listen to music at 90 decibels. While listening to the music, each volunteer took a 10-question math test.[55] Let y = score on the math test and x = volume of music (in decibels). Assume the conditions for inference are met. The 95% confidence interval for the mean y value at x = 45 decibels is (7.04, 8.41). Interpret this interval.

6. **Hungry Fish** Does a larger concentration of fish attract more predators? One study looked at kelp perch and their common predator, the kelp bass. The researcher set up 16 large circular pens on sandy ocean bottoms off the coast of southern California. Young perch were selected at random from a large group and were placed in the pens. Four pens had 10 perch each, four pens had 20 perch each, 4 pens had 40 perch each, and the final

4 pens had 60 perch each. The researcher then dropped the nets protecting the pens, allowing bass to swarm in, and counted the perch left after 2 hours.[56] Let $y =$ proportion of perch left and $x =$ number of perch in the pen initially. Assume the conditions for inference are met. The 95% confidence interval for the mean y value at $x = 30$ perch is (0.28, 0.48). Interpret this interval.

7. **Wine and Heart Health** Is higher per capita wine consumption associated with lower death rates from heart disease? A researcher from the University of California, San Diego, collected data on average per capita wine consumption and heart disease death rate in a random sample of 19 countries for which data were available.[57] Let $y =$ heart disease death rate (deaths per 100,000 people) and $x =$ wine consumption (measured as alcohol from wine per person, in liters). Assume the conditions for inference are met. The 90% confidence interval for the mean y value at $x = 2$ liters is (198.23, 231.02). Interpret this interval.

8. **Beer and BAC** Does drinking more cans of beer result in a higher blood alcohol content (BAC)? Sixteen volunteers aged 21 or older with an initial BAC of 0 took part in a study to find out. Each volunteer drank a randomly assigned number of cans of beer and had their BAC measured 30 minutes later.[58] Let $y =$ BAC and $x =$ number of beers. Assume the conditions for inference are met. The 99% confidence interval for the mean y value at $x = 4$ beers is (0.04, 0.08). Interpret this interval.

9. **Refer to Example 11.6.2 Loud Music and Tests** Refer to Exercise 5. The 95% prediction interval for an individual y value with $x = 45$ decibels is (4.46, 10.99). Interpret this interval.

10. **Hungry Fish** Refer to Exercise 6. The 95% prediction interval for an individual y value with $x = 30$ perch in the pen is (−0.04, 0.79). Interpret this interval.

11. **Wine and Heart Health** Refer to Exercise 7. The 90% prediction interval for an individual y value with $x = 2$ liters is (146.72, 285.53). Interpret this interval.

12. **Beer and BAC** Refer to Exercise 8. The 99% prediction interval for an individual y value with $x = 4$ beers is (0.00, 0.12). Interpret this interval.

13. **Refer to Example 11.6.3 Loud Music and Tests** Refer to Exercises 5 and 9. When $x^* = 45$ decibels, the 95% confidence interval for μ_y is (7.04, 8.41) and the 95% prediction interval for y is (4.46, 10.99). Note that $\bar{x} = 60$ decibels.

    (a) Explain what would happen to the width of these intervals if we changed the confidence level to 90%.

    (b) Explain what would happen to the width of these intervals if we increased the sample size to 40.

    (c) Explain what would happen to the width of these intervals if we changed x^* to 55 decibels.

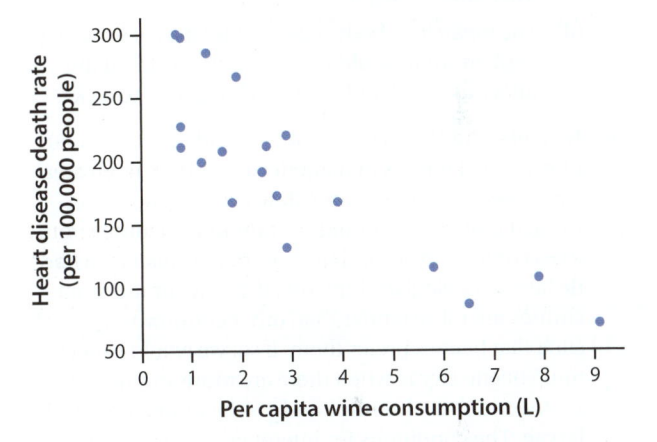

14. **Hungry Fish** Refer to Exercises 6 and 10. When $x^* = 30$ perch, the 95% confidence interval for μ_y is (0.28, 0.48) and the 95% prediction interval for y is (−0.04, 0.79). Note that $\bar{x} = 32.5$ fish.

    (a) Explain what would happen to the width of these intervals if we changed the confidence level to 98%.

    (b) Explain what would happen to the width of these intervals if we increased the sample size to 28.

    (c) Explain what would happen to the width of these intervals if we changed x^* to 50 perch.

15. **Wine and Heart Health** Refer to Exercises 7 and 11. When $x^* = 2$ liters, the 90% confidence interval for μ_y is (198.23, 231.02) and the 90% prediction interval for y is (146.72, 285.53).

    (a) Explain what would happen to the width of these intervals if we changed the confidence level to 95%.

    (b) Explain what would happen to the width of these intervals if we decreased the sample size to 15.

    (c) Here is a scatterplot showing the association between per capita wine consumption and the heart disease death rate. Explain what would happen to the width of these intervals if we changed x^* to 7 liters.

16. **Beer and BAC** Refer to Exercises 8 and 12. When $x^* = 4$ beers, the 99% confidence interval for μ_y is (0.04, 0.08) and the 99% prediction interval for y is (0.00, 0.12).

    (a) Explain what would happen to the width of these intervals if we changed the confidence level to 95%.

    (b) Explain what would happen to the width of these intervals if we decreased the sample size to 10.

    (c) Here is a scatterplot showing the association between the number of beers and BAC. Explain what would happen to the width of these intervals if we changed x^* to 5 beers.

Applying the Concepts These exercises ask you to apply multiple learning goals in a new context or to apply what you learned in this section in a new way.

17. **Lean Body Mass and Metabolism** We have data on the lean body mass and resting metabolic rate for a random sample of 12 people who volunteered for a study of dieting. Let x = lean body mass, given in kilograms (kg), which is a person's weight leaving out all fat, and y = metabolic rate, measured in calories burned per 24 hours (cal/24 hr). The conditions for inference are met and the least-squares regression line computed from these data is $\hat{y} = 201 + 24.03x$.

    (a) Compute the predicted metabolic rate for a person with a lean body mass of 45 kg.

    (b) The 98% confidence interval for the mean y value at $x = 45$ kg is (1203, 1362). Interpret this interval.

    (c) The 98% prediction interval for an individual y value with $x = 45$ kg is (1008, 1557). Interpret this interval.

    (d) The mean lean body mass in this study is $\bar{x} = 43$ kg. Explain what would happen to the width of the intervals in (b) and (c) if we changed x^* to 38 kg.

18. **Beavers and Beetles** Researchers laid out 23 circular plots, each 4 meters in diameter, at random in an area where beavers were cutting down cottonwood trees. In each plot, they counted the number of stumps from trees cut by beavers and the number of clusters of beetle larvae. Ecologists think that the new sprouts from stumps are more tender than other cottonwood growth, such that beetles prefer them. If so, we would expect more beetle larvae when there are more stumps.[59] Let x = the number of stumps and y = the number of beetle larvae. The conditions for inference are met and the least-squares regression line computed from these data is $\hat{y} = -1.29 + 11.89x$.

    (a) Compute the predicted number of beetle larvae for a plot with 4 stumps.

    (b) The 95% confidence interval for the mean y value at $x = 4$ stumps is (41.2, 51.3). Interpret this interval.

    (c) The 95% prediction interval for an individual y value with $x = 4$ stumps is (32.0, 60.6). Interpret this interval.

    (d) The mean number of stumps in this study is $\bar{x} = 2.2$. Explain what would happen to the width of the intervals in (b) and (c) if we changed x^* to 3 stumps.

19. **Roller Coasters** Amusement parks try to attract visitors by offering roller coasters that have a variety of speeds and elevations. A random sample of 9 roller coasters was selected.[60] Let y = maximum speed (mph) and x = height (ft) of these roller coasters. The conditions for inference are met and the least-squares regression line computed from these data is $\widehat{speed} = 28.17 + 0.2143 height$.

    (a) Compute the predicted maximum speed for a roller coaster with a height of 175 feet.

    (b) A student used software to compute the 95% confidence interval and the 95% prediction interval for $x^* = 175$ feet. The two intervals computed are (38.67, 92.70) and (55.24, 76.13). Which interval is the confidence interval and which is the prediction interval? Explain your answer.

20. **Toddler Mealtimes** Researchers collected data on a random sample of 20 toddlers observed over several months. Let x = the mean time a child spent at the lunch table (minutes) and y = the mean number of calories consumed.[61] The conditions for inference are met and the least-squares regression line computed from these data is $\widehat{calories} = 560.7 - 3.077 time$.

    (a) Compute the predicted number of calories consumed for a toddler who sat for 33 minutes, on average.

    (b) A student used software to compute the 90% confidence interval and the 90% prediction interval for $x^* = 33$ minutes. The two intervals computed are (449.9, 468.3) and (417.5, 500.7). Which interval is the confidence interval and which is the prediction interval? Explain your answer

21. **Tech Car Price and Age** Refer to Exercise 23 in Section 11.2. The data set **Honda CRV** contains information about a random sample of 191 used Honda CRVs, including the variables *Price* and *Age*. Note that the conditions for inference are met.

    (a) Use technology to calculate and interpret a 95% confidence interval for the mean price of all used Honda CRVs that are 5 years old.

    (b) Use technology to calculate and interpret a 95% prediction interval for the price of a single used Honda CRV that is 5 years old.

22. **Tech Car Price and Mileage** Refer to Exercise 24 in Section 11.2. The data set **Honda CRV** contains information about a random sample of 191 used Honda CRVs, including the variables *Price* and *Miles*. Note that the conditions for inference are met.

    (a) Use technology to calculate and interpret a 95% confidence interval for the mean price of all used Honda CRVs that have been driven 50,000 miles.

    (b) Use technology to calculate and interpret a 95% prediction interval for the price of a single used Honda CRV that has been driven 50,000 miles.

Extending the Concepts These exercises challenge you to explore statistical concepts and methods that go beyond what you learned in this section.

23. **Explanatory Variable Variability** Consider again the standard error for the confidence and prediction intervals introduced in this section:

$$SE_{\hat{\mu}_y} = s\sqrt{\frac{1}{n} + \frac{(x^* - \bar{x})^2}{(n-1)s_x^2}} \qquad SE_{\hat{y}} = s\sqrt{1 + \frac{1}{n} + \frac{(x^* - \bar{x})^2}{(n-1)s_x^2}}$$

Notice that s_x^2 occurs in the denominator of the last fraction in both standard errors.

(a) What will happen to the SE as the value of s_x^2 increases? How will that affect the width of the confidence and prediction intervals?

(b) All other things being equal, would you prefer a larger or smaller value of s_x^2 for computing confidence and prediction intervals in a regression setting? Justify you answer.

(c) How can you assure that you get the type of value for s_x^2 you picked in (b) when designing an experiment? How is this different than in an observational study? Explain your answer.

Cumulative Review These exercises revisit what you learned in previous sections.

24. **Butterflies (11.1)** Scientists were interested in the effect of climate change on butterflies. They took a random sample of butterflies of the species *Boloria chariclea* over a number of years and measured $y =$ the wing length for the butterfly and $x =$ the average temperature during the larval growing season in the previous summer.[62] The bigger butterflies are, the hardier they are considered to be. If they are getting smaller as the temperature increases with climate change, this could be a problem for the species. Here is a scatterplot of the data along with the regression line $\overline{winglength} = 18.87 - 0.24 AveTemp$.

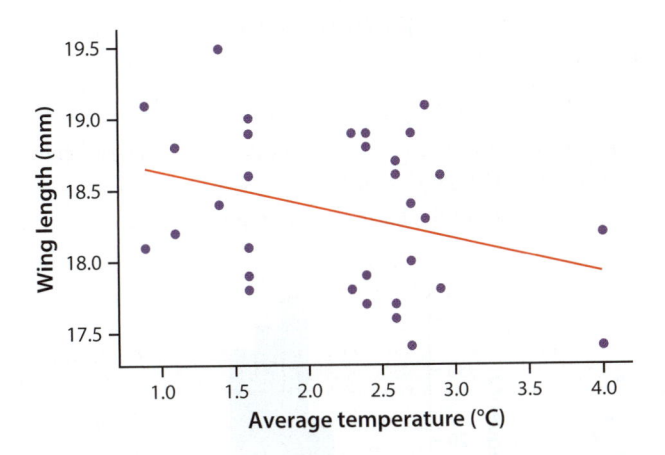

(a) Calculate and interpret the residual for the butterfly with a larval growing season average temperature of 1.6°C and a wing length of 18.1 mm.

(b) Interpret the slope of the regression line.

(c) Does the value of the y-intercept have meaning in this context? If so, interpret the y-intercept. If not, explain why not.

25. **Sea Slugs (10.5)** Sea slugs eat vaucherian seaweed. Researchers wondered how the slug larvae find this seaweed.[63] Do chemicals that leach from the seaweed attract the slug larvae? Researchers collected seawater over a patch of vaucherian seaweed every 5 minutes over a 25-minute period as the tide was coming in (so there were 6 collection times: 0, 5, 10, 15, 20, and 25 minutes). They thought that the concentration of the chemicals would decrease when the seaweed was deeper underwater. The researchers divided each water sample into 6 equal parts and randomly assigned larvae to the final water samples. They recorded the percentage of larvae that metamorphosed for each of the 36 experimental units.

(a) Dotplots of the percentage that metamorphosed for the 6 samples are given. Do the data suggest that the mean percentage that metamorphosed is not the same for all 6 levels of time after the tide starts coming in? Explain your answer.

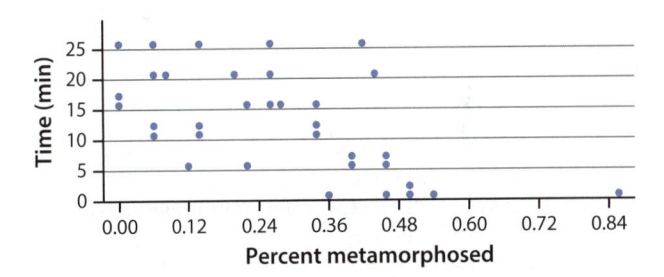

(b) Do the data provide convincing evidence that the mean percentage of larvae that metamorphose is not the same in water collected at different tide levels? State appropriate hypotheses for a test that addresses this question.

(c) There were 36 different experimental units. A partial ANOVA table is given below. Fill in the empty cells. Is there some evidence for at least one difference among the population means? Justify your answer using your calculated value for F.

Source	df	SS	MS	F
Groups		0.6309		
Error			0.02115	
Total		1.2655		

Statistics Matters How does engine size affect CO_2 emissions?

As discussed at the beginning of this chapter, 29% of all greenhouse gas emissions in the United States come from transportation.[64] Although cars and trucks emit several different greenhouse gases, the volume of carbon dioxide (CO_2) emitted is much greater than the other gases. The scatterplot shows the relationship between x = engine size (in liters, L) and y = CO_2 emissions (in grams per mile, g/mi) for a random sample of 176 gas-powered cars and trucks produced in 2021, along with the least-squares regression line $\hat{y} = 244.7 + 54.9x$.

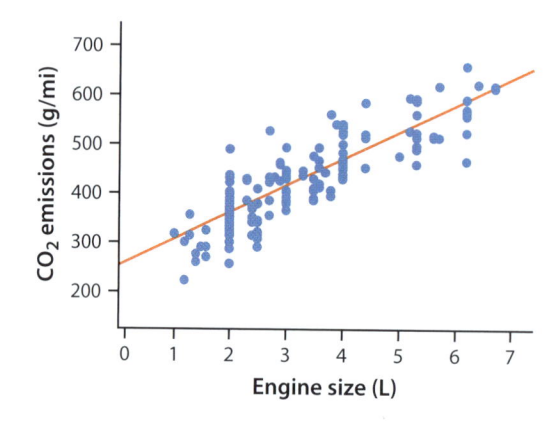

1. Interpret the slope of the least-squares regression line.

2. The 2021 Toyota Sequoia 4WD has an engine size of 5.7 L and CO_2 emissions of 618 g/mi. Calculate and interpret the residual for this vehicle.

3. The standard deviation of the residuals for this model is $s = 45.2$. Interpret this value.

4. The value of r^2 for this model is 0.738. Interpret this value. Can you think of other factors that might account for the unexplained variability in CO_2 emissions?

Here are a residual plot and histogram of the residuals for the least-squares regression line. The standard error of the slope is 2.48.

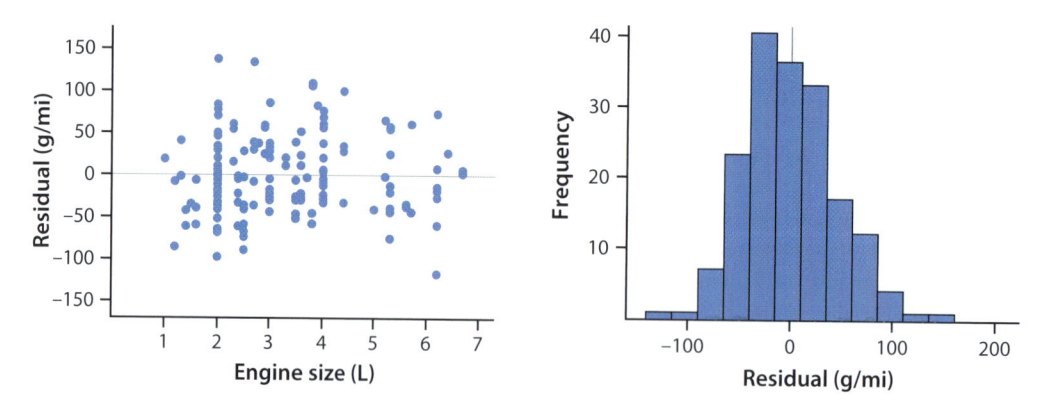

5. Do these data provide convincing evidence of a positive association between engine size and CO_2 emissions in the population of all gas-powered 2021 vehicles?

6. Calculate and interpret a 95% confidence interval for the slope of the population least-squares regression line.

7. For vehicles with an engine size of 4 L, the 95% prediction interval for the CO_2 emission of a single vehicle is (374.60, 553.82). Interpret this interval. Will the 95% confidence interval for the mean CO_2 for vehicles with an engine size of 4 L be wider or narrower than the corresponding prediction interval? Explain your answer.

Chapter 11 Review

Modeling Linear Associations

- A **regression line** is a line that models how a response variable y changes as an explanatory variable x changes. Regression lines are expressed in the form $\hat{y} = b_0 + b_1 x$, where $\hat{y}$ is the predicted value of y for a particular value of x.
- **Extrapolation** is the use of a regression line for prediction outside the interval of x values used to obtain the line. The further we extrapolate, the less reliable the predictions become.
- A **residual** is the difference between an actual value of y and the value of y predicted by the regression line. That is, residual = actual y − predicted $y = y - \hat{y}$.
- The **slope** b_1 of a regression line is the amount by which the *predicted* value of y changes when x increases by one unit. The **y-intercept** b_0 is the *predicted* value of y when $x = 0$.
- The **least-squares regression line** is the line that makes the sum of the squared residuals as small as possible.
- The least-squares regression line is not resistant: Outliers can greatly influence the equation of the line.
- **Regression to the mean** is the tendency for extreme values of the explanatory variable to be paired with less extreme values of the response variable.
- A **residual plot** is a scatterplot that plots the residuals on the vertical axis and the explanatory variable on the horizontal axis. If a residual plot shows no leftover curved patterns, the regression model is appropriate. If a residual plot shows a leftover curved pattern, the regression model is not appropriate.
- The **standard deviation of the residuals s** measures the size of a typical residual. That is, s measures the typical distance between the actual y values and the predicted y values.
- The **coefficient of determination r^2** measures the percent reduction in the sum of squared residuals when using the least-squares regression line to make predictions rather than the mean value of y. In other words, r^2 measures the percentage of the variability in the response variable that is accounted for by the least-squares regression line.

Inference for the Relationship Between Two Quantitative Variables

- A regression line calculated from every value in the population is called a **population regression line** (or true regression line). The equation of a population regression line is $y = \beta_0 + \beta_1 x + \varepsilon$, where y is the value of the response variable for a given value of x, β_0 is the population y-intercept, β_1 is the population slope, and ε is the error term.
- A regression line calculated from a sample is called a **sample regression line**. The equation of a sample regression line is $\hat{y} = b_0 + b_1 x$, where $\hat{y}$ is the predicted y value or the estimated mean y value for a given value of x, b_0 is the sample y-intercept, and b_1 is the sample slope.
- There are four **conditions** for performing inference about the slope of a least-squares regression line:
  - **Random:** The data come from a random sample from the population of interest or from a randomized experiment.
  - **Linear:** The form of the association between the two variables is linear.
  - **Equal SD:** For all values of x, the standard deviation of y is the same.
  - **Normal:** For all values of x, the distribution of y is normal.
- When the conditions are met, the ***t* interval for the slope of a least-squares regression line** is
$$b_1 \pm t^* \operatorname{SE}_{b_1}$$
where SE_{b_1} is the standard error of the slope and t^* is the critical value with $C\%$ of its area between $-t^*$ and t^* in the t distribution with $\mathrm{df} = n - 2$.
- The test for a relationship between two quantitative variables is called the ***t* test for the slope of a least-squares regression line.**
  - The usual null hypothesis for this test is that there is no association between the two variables in the population of interest. That is, $\beta_1 = 0$, where β_1 is the slope of the population least-squares regression line. The usual alternative hypothesis is that there is some kind of association between the two variables in the population of interest. That is, $\beta_1 > 0$ (positive association), $\beta_1 < 0$ (negative association), or $\beta_1 \neq 0$ (an association).
  - The **standardized test statistic** is
$$t = \frac{b_1 - 0}{\operatorname{SE}_{b_1}}$$
where 0 is the hypothesized value of the slope and SE_{b_1} is the standard error of the slope.
  - When the conditions are met, you can find the P-value using a t distribution with $\mathrm{df} = n - 2$.
- Be sure to follow the four-step process whenever you calculate a confidence interval or perform a significance test for β_1, the slope of the population least-squares regression line.
- The **confidence interval for a mean response** gives an interval of plausible values for the mean response μ_y at a given value x^* of the explanatory variable. Use technology to calculate this interval or use the formula
$$\hat{y} \pm t^* s \sqrt{\frac{1}{n} + \frac{(x^* - \bar{x})^2}{(n-1)s_x^2}}$$
where t^* is the critical value with $C\%$ of its area between $-t^*$ and t^* in the t distribution with $\mathrm{df} = n - 2$.
- The **prediction interval for an individual response** gives an interval of plausible values for a single value of the response variable y at a given value x^* of the

explanatory variable. Use technology to calculate this interval or the formula

$$\hat{y} \pm t^* s \sqrt{1 + \frac{1}{n} + \frac{(x^* - \bar{x})^2}{(n-1)s_x^2}}$$

where t^* is the critical value with $C\%$ of its area between $-t^*$ and t^* in the t distribution with df $= n - 2$.

- For a given value of x, the prediction interval for an individual y value will always be wider than the corresponding confidence interval for a mean response.
- All other things being equal, the width of a confidence interval for a mean y value and the width of a prediction interval for an individual y value will be narrower if the confidence level decreases, the sample size increases, or the x value of interest gets closer to $\bar{x}$.

Chapter 11 Review Exercises

1. **Ice Break (11.1)** The Nenana ice classic is a contest held each spring in Nenana, Alaska.[65] For the price of $2.50 per ticket, contestants guess the date and time when the ice will break up on the Tanana River. The winning guess can be worth more than $300,000! The scatterplot shows the relationship between $y =$ ice break day (number of days into the year when the ice breaks) and $x =$ number of years since 1900, along with the least-squares regression line $\hat{y} = 130.49 - 0.0861x$.

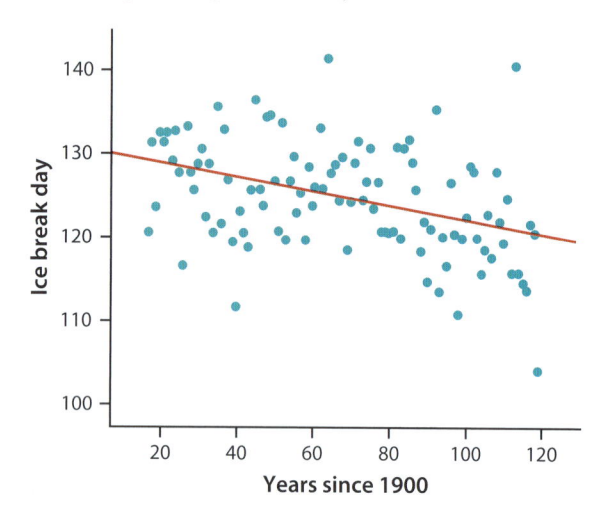

   (a) Predict the date when the ice will break on the Tanana River in the year 2050. How confident are you in this prediction? Explain your answer.
   (b) Calculate and interpret the residual for 2019 when the ice broke on day 104.01 (just after midnight on April 14).
   (c) Interpret the slope of the least-squares regression line.
   (d) Does the y-intercept have meaning in this case? If so, interpret the y-intercept. If not, explain why not.

2. **Crawling (11.2)** At what age do babies learn to crawl? Does it take longer to learn in the winter, when babies are often bundled in clothes that restrict their movement? If so, there would be an association between babies' crawling age and the average temperature during the month when they first try to crawl (approximately 6 months after birth). Data were collected from parents who reported the birth month and the age at which their child was first able to creep or crawl a distance of 4 feet within 1 minute. The table shows birth month, the average temperature (in degrees Fahrenheit) 6 months after birth, and the average crawling age (in weeks) for a sample of 414 infants.[66]

Birth month	Average temperature 6 months after birth (°F)	Average crawling age (weeks)
January	66	29.84
February	73	30.52
March	72	29.70
April	63	31.84
May	52	28.58
June	39	31.44
July	33	33.64
August	30	32.82
September	33	33.83
October	37	33.35
November	48	33.38
December	57	32.32

   (a) Use technology to construct a scatterplot and calculate the equation of the least-squares regression line for predicting average crawling age from average temperature 6 months after birth.
   (b) How does the point representing May affect the equation of the least-squares regression line? Explain your answer.

3. **Quarterbacks (11.2)** Do quarterbacks regress to the mean? The scatterplot shows the completion percentage for the 30 NFL quarterbacks who had at least 160 passing attempts in two recent consecutive seasons.[67] The line $y = x$ is included on the scatterplot, along with vertical lines separating the quarterbacks into three groups based on their completion percentages in Season 1.

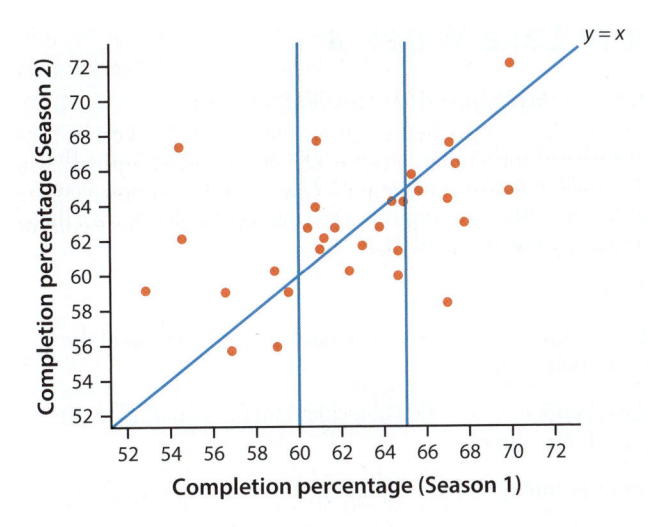

(a) Of the 8 quarterbacks who completed less than 60% of their passes in Season 1, what percentage did better in Season 2?

(b) Of the 9 quarterbacks who completed more than 65% of their passes in Season 1, what percentage did worse in Season 2?

(c) Explain how your answers to parts (a) and (b) illustrate regression to the mean.

4. **Ice Break (11.3)** In Exercise 1, we used a least-squares regression line to model the relationship between $y =$ ice break day (number of days into the year when the ice breaks) and $x =$ number of years since 1900. Here is the residual plot for this model.

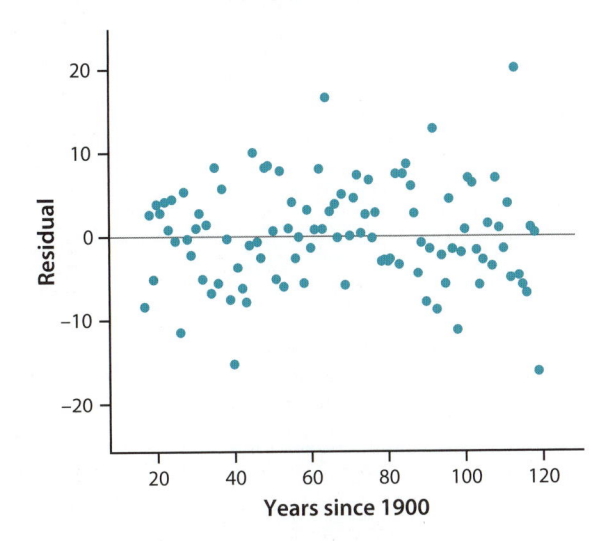

(a) Explain what the residual plot suggests about the appropriateness of the linear model.

(b) The standard deviation of the residuals for this model is $s = 6.1$ days. Interpret this value.

(c) The value of r^2 for this model is $r^2 = 0.15$. Interpret this value.

5. **Big Chickens (11.4)** Growth hormones are often used to increase the weight gain of chickens. In an experiment using 15 chickens, researchers randomly assigned 3 chickens to each of 5 different doses of growth hormone (0, 0.2, 0.4, 0.8, and 1.0 milligrams). They recorded the subsequent weight gain (in ounces) for each chicken. The equation of the least-squares regression line is $\overline{weightgain} = 4.5459 + 4.8323\,hormonedose$. The standard error of the slope is $SE_{b_1} = 1.0164$. Calculate and interpret a 95% confidence interval for the slope of the true least-squares regression line. Assume the conditions are met.

6. **Protein Bars (11.5)** When companies create food products, they try to make them as tasty as possible. This is often done by adding some combination of sugar and fat to the products. Because the makers of protein bars want to keep calories low, they will typically add either mostly sugar or mostly fat, but not both. The table shows $x =$ amount of sugar (tsp) and $y =$ amount of saturated fat (g) in 19 randomly selected types of protein bars.[68] Do these data provide convincing evidence of a negative association between sugar and fat in protein bars? Use $\alpha = 0.05$.

Sugar (tsp)	Saturated fat (g)
3.5	1
4.5	1
3.5	1.5
0	2
0.5	2
1.5	2
2	2.5
2.5	2.5
3	2.5
4.5	2.5
1	3
4	3
4.5	3
1.5	3.5
1.5	3.5
3.5	3.5
0	4
2	4
0	6

7. **Protein Bars (11.6)** Refer to Exercise 6. How much fat should we expect from a protein bar with 1 teaspoon of sugar? We calculated a 95% confidence interval for the mean fat content and a 95% prediction interval for the fat content, both for $x = 1$ teaspoon of sugar. One of these intervals is (0.808, 5.619) and the other is (2.524, 3.902).

(a) Which interval is the prediction interval? Explain how you know.

(b) Interpret both intervals.

Chapter 11 Project: Dissolved Oxygen in Lake Water

Explore a large data set using software and apply what you've learned in this chapter to a real-world scenario.

The concentration of dissolved oxygen (DO) in a lake is very important for the health of the lake and the organisms that depend on it. The data set **Champlain** contains a random sample of 500 measurements of DO from water samples taken at various depths over the past 30 years in Lake Champlain. In addition to measurements of DO (mg/L) and depth (m), the data set includes the temperature of the water when the measurements were obtained (°C).[69]

1. Download the **Champlain** data set into your preferred technology.

2. Create a scatterplot to investigate the relationship between $x = $ *Temp* and $y = $ *DO*. Then calculate the equation of the least-squares regression line, the standard deviation of the residuals, and r^2.

3. Create a scatterplot to investigate the relationship between $x = $ *Depth* and $y = $ *DO*. Then calculate the equation of the least-squares regression line, the standard deviation of the residuals, and r^2.

4. Compare the two associations. Is temperature or depth a better predictor of DO? Explain your reasoning using the values of s and r^2.

5. Do these data provide convincing evidence of a linear relationship between DO and the variable you chose in Question 4? Perform a significance test to answer this question using $\alpha = 0.05$.

6. Calculate and interpret a 95% confidence interval for the slope of the population least-squares regression line relating DO to the variable you chose in Question 4.

Extension: Is there a significant trend in DO over time? Use the data to investigate.

12 Multiple Regression

Section 12.1 Introduction to Multiple Regression
Section 12.2 Indicator Variables and Interaction
Section 12.3 Inference for Multiple Regression
Chapter 12 Review

Chapter 12 Review Exercises
Chapter 12 Project
Cumulative Review Chapters 10–12

Statistics Matters Which factors affect the growth of whales?

Thanks to restrictions on whaling, whale populations are making a comeback around the world. However, whales still face other challenges to their health. One way to measure the health of a whale is by its length. After all, we'd expect healthy whales to grow longer than whales that are not as healthy.

Here is a scatterplot showing the relationship between x = age (years) and y = length (m) for a sample of 145 North Atlantic right whales ages 2 to 20 years.[1]

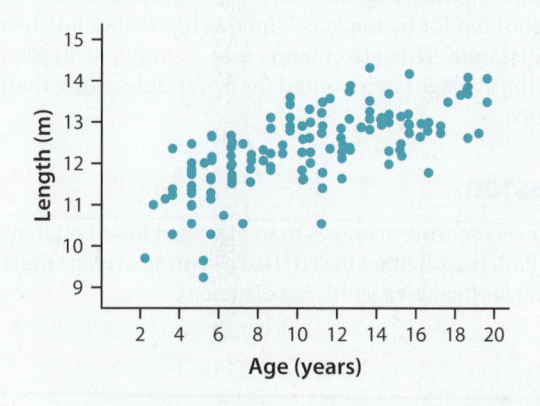

There is a moderately strong, positive linear relationship between age and length for these whales.

After accounting for age, do whales born in more recent years have different lengths than those born in the past? Do whales that have been entangled in fishing gear tend to have smaller offspring?

We'll revisit Statistics Matters at the end of the chapter, so you can use what you have learned to help answer these questions.

In Chapter 11, you learned how to use a least-squares regression line to make predictions and summarize the relationship between a quantitative explanatory variable and a quantitative response variable. In this chapter, you will learn how to use more than one explanatory variable, including both quantitative and categorical explanatory variables, to model a quantitative response variable.

Section 12.1 Introduction to Multiple Regression

LEARNING GOALS

By the end of this section, you will be able to:

- Use a multiple regression model to make predictions.
- Use a multiple regression model to calculate and interpret residuals.
- Interpret the standard deviation of the residuals and R^2 for a multiple regression model.

When there is a linear relationship between two quantitative variables, we use a least-squares regression line to summarize the relationship. To evaluate how well the regression line models the association, we use the standard deviation of the residuals s and the coefficient of determination r^2. Ideally, s is close to 0, indicating that the actual values of y are close to the values predicted by the line. Values of r^2 close to 1 indicate that the linear model accounts for a large percentage of the variability in the response variable.

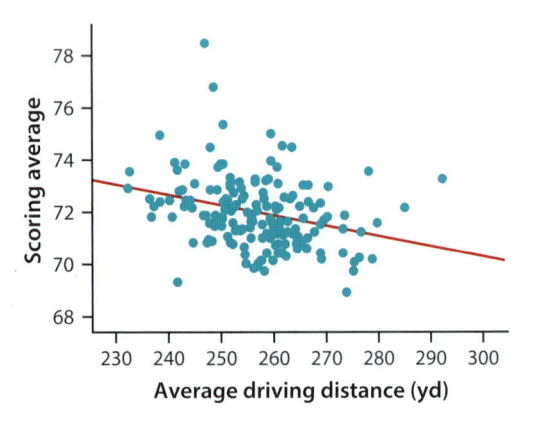

Here is a scatterplot showing the relationship between x = average driving distance (yards) and y = scoring average for 164 Ladies Professional Golf Association (LPGA) golfers in 2021.[2] There is a weak, negative linear association between average driving distance and scoring average. Remembering that lower scores are better in golf, the scatterplot shows that players who hit the ball farther typically have better (lower) scores.

The least-squares regression line for these data is $\hat{y} = 81.9886 - 0.0392x$. The standard deviation of the residuals is $s = 1.30$, indicating that the actual scoring averages are typically 1.30 shots away from the values predicted by the least-squares regression line using x = average driving distance. The value of r^2 is 0.087, indicating that only 8.7% of the variability in scoring average is accounted for by the least-squares regression line using x = average driving distance. This also means that as much as 91.3% of the variability in scoring average is accounted for by variables other than average driving distance.

The Idea of Multiple Regression

Roughly half of the golfers have better (lower) scoring averages than expected based on their average driving distance. For example, Inbee Park (highlighted in red) had a scoring average that is 3.22 shots better than expected after accounting for her average driving distance.

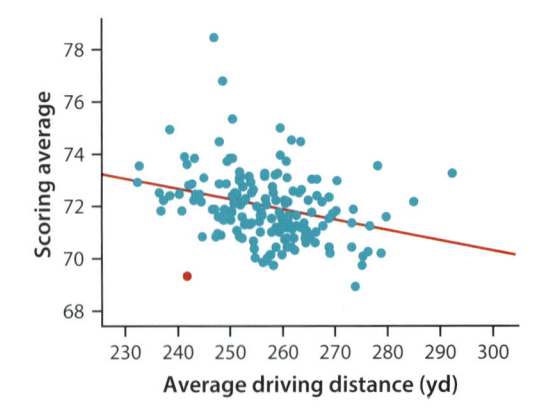

What do the golfers with negative residuals have in common? Perhaps they are good putters. Perhaps they are accurate drivers or play well when close to the green. Maybe all of these. To determine if a different explanatory variable can help account for the previously unaccounted-for variation in scoring average, we can see if any of these variables has an association with the residuals from the original least-squares regression line relating $y =$ scoring average to $x =$ average driving distance.

For example, Figure 12.1 shows a scatterplot of $x =$ driving accuracy (percentage of tee shots that land in the fairway, with bigger values being better) and $y =$ residual from the original model relating scoring average to average driving distance. The point representing Inbee Park is shown in red.

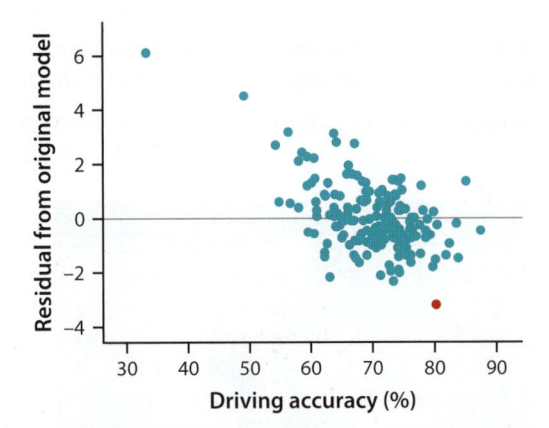

FIGURE 12.1 Scatterplot showing the relationship between $x =$ driving accuracy and $y =$ residual from the original model relating scoring average to average driving distance. The association in the scatterplot indicates that driving accuracy helps account for some of the variability in scoring average that was unaccounted for by average driving distance.

The negative association in the scatterplot shows that players with high values for driving accuracy tended to have negative residuals in the original model. That is, *players with high values for driving accuracy tended to have scoring averages that were better than expected, after accounting for average driving distance.* This makes sense: After accounting for average driving distance, players who are more accurate should have better scores than players who are less accurate.

Figure 12.2 shows scatterplots investigating two additional explanatory variables: putting average (number of putts per hole, on average) and sand saves (percentage of times the player successfully gets out of a sand trap). Players with small values for putting average and players with high values for sand saves tended to have scoring averages that were better (lower) than expected after accounting for average driving distance.

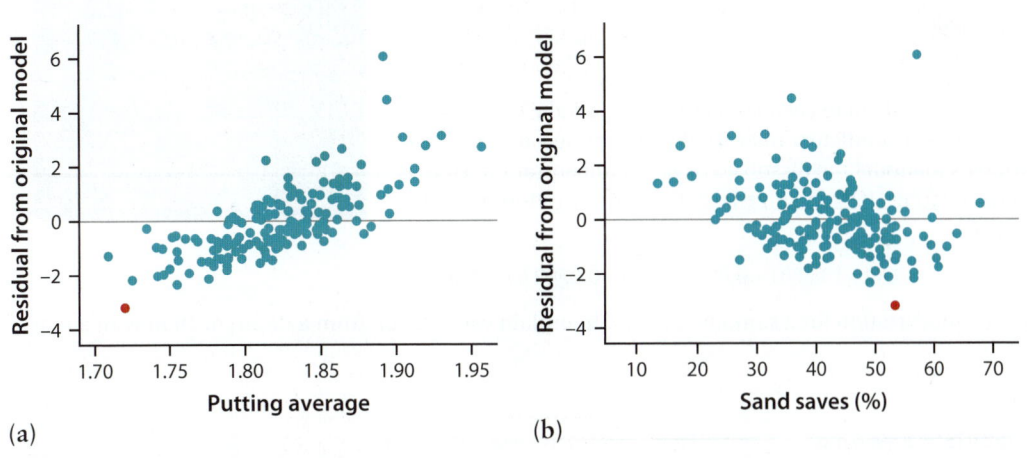

FIGURE 12.2 (a) Scatterplot showing the relationship between $x =$ putting average and $y =$ residual from the original model relating scoring average to average driving distance. (b) Scatterplot showing the relationship between $x =$ sand saves and $y =$ residual from the original model relating scoring average to average driving distance. The associations in the scatterplots indicate that putting average and sand saves help account for some of the variability in scoring average that was unaccounted for by average driving distance.

Now that we know driving accuracy, putting average, and sand saves can account for some of the previously unaccounted-for variability in scoring average, we can create a **multiple regression model** to improve our predictions. Note that a linear regression model with only one explanatory variable is often called a *simple linear regression model* to differentiate it from a *multiple linear regression model*. Because the structure of all the multiple regression models in this book are linear, for simplicity we won't include the word "linear" when referring to a multiple regression model.

A **multiple regression model** is a least-squares regression equation that relates a quantitative response variable to two or more explanatory variables. Multiple regression models are expressed in the form $\hat{y} = b_0 + b_1 x_1 + b_2 x_2 + \ldots$.

Here is the multiple regression model relating y = scoring average to x_1 = average driving distance, x_2 = driving accuracy, x_3 = putting average, and x_4 = sand saves for LPGA golfers in 2021:

$$\hat{y} = 60.6344 - 0.0556 x_1 - 0.1024 x_2 + 18.4071 x_3 - 0.0197 x_4$$

In 2021, Inbee Park's average driving distance was 241.6 yards, her driving accuracy was 80.3%, her putting average was 1.72, and her sand save percentage was 53.33. This gives a predicted scoring average of

$$\hat{y} = 60.6344 - 0.0556(241.6) - 0.1024(80.3) + 18.4071(1.72) - 0.0197(53.33) = 69.59$$

Based on Park's average driving distance, driving accuracy, putting average, and sand saves, the model predicts a scoring average of 69.59 shots.

EXAMPLE 12.1.1

What factors predict the level of dissolved oxygen in water?

Using a multiple regression model

PROBLEM

Is depth a good predictor of the dissolved oxygen concentration in Lake Champlain? Would water temperature be a better predictor? Instead of picking one variable or the other, we can use both variables to model the dissolved oxygen concentration. Here is the multiple regression model relating DO = dissolved oxygen concentration (mg/L) to depth (m) and temperature (°C):

$$\widehat{DO} = 13.5881 - 0.0217\ depth - 0.2583\ temperature$$

Predict the dissolved oxygen concentration for a sample of Lake Champlain water taken from a depth of 10 m with a temperature of 15°C.

SOLUTION

$\hat{y} = 13.5881 - 0.0217(10) - 0.2583(15) = 9.4966$ mg/L

EXAM PREP: FOR PRACTICE, TRY EXERCISE 5.

How good was our prediction for Inbee Park's scoring average? To find out, we calculate the residual using her actual scoring average of 69.3:

$$\text{residual} = y - \hat{y} = 69.3 - 69.59 = -0.29 \text{ shot}$$

Park's actual scoring average was 0.29 shot better (lower) than the value predicted by the multiple regression model with x_1 = average driving distance, x_2 = driving accuracy, x_3 = putting average, and x_4 = sand saves. This is a big improvement from her residual of 3.22 shots when using only average driving distance to make the prediction.

EXAMPLE 12.1.2

What factors predict the level of dissolved oxygen in water?

Calculating and interpreting residuals

PROBLEM

Here is the multiple regression model relating DO = dissolved oxygen concentration (mg/L) to depth (m) and temperature (°C) from Example 12.1.1.

$$\widehat{DO} = 13.5881 - 0.0217\, depth - 0.2583\, temperature$$

One of the water samples used to create the multiple regression model has a depth of 26 m, a temperature of 6.83°C, and a dissolved oxygen concentration of 12.5 mg/L.

(a) Calculate the residual for this water sample.

(b) Interpret the residual from part (a).

SOLUTION

(a) $\hat{y} = 13.5881 - 0.0217(26) - 0.2583(6.83) = 11.2597$ mg/L

Residual = $12.5 - 11.2597 = 1.2403$ mg/L

Find $\hat{y}$ using the regression model.
residual = $y - \hat{y}$

(b) The actual dissolved oxygen concentration for this water sample is 1.2403 mg/L greater than the value predicted by the multiple regression model with x_1 = depth (m) and x_2 = temperature (°C).

When interpreting a residual, remember to include the direction (greater than/less than) in addition to the size of the residual.

EXAM PREP: FOR PRACTICE, TRY EXERCISE 9.

Assessing a Multiple Regression Model

The purpose of including additional explanatory variables in a regression model is to make better predictions. In other words, we include additional explanatory variables to reduce the size of the residuals and to account for other sources of variability in the response variable.

To see how adding driving accuracy, putting average, and sand saves to the model helps improve our predictions of scoring average, we can compare the size of the residuals for the simple linear regression (one-predictor) model using average driving distance to the size of the residuals for the multiple regression (four-predictor) model. Not surprisingly, the boxplots show that the residuals tend to be much smaller in the four-predictor model.

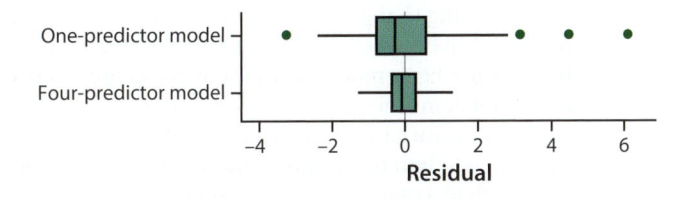

When using all four explanatory variables, the standard deviation of the residuals is $s = 0.560$ shot. The actual scoring averages typically vary by about 0.560 shot from the values predicted by the multiple regression model using x_1 = average driving distance, x_2 = driving accuracy, x_3 = putting average, and x_4 = sand saves. This is an improvement from the original model using only average driving distance, which produced a standard deviation of $s = 1.30$.

The value of R^2 for our four-predictor model is 0.833. That is, 83.3% of the variability in scoring average is accounted for by the multiple regression model using x_1 = average driving distance, x_2 = driving accuracy, x_3 = putting average, and x_4 = sand saves. Again, this is a big improvement from the original one-predictor model, which has an r^2 of 0.087. Note that it is common practice to use uppercase R^2 when describing a multiple regression model.

EXAMPLE 12.1.3

What factors predict the level of dissolved oxygen in water?

Interpreting s and R^2

PROBLEM

In the preceding examples, we used a multiple regression model relating dissolved oxygen concentration (mg/L) to depth (m) and temperature (°C). For this model, $s = 1.06$ and $R^2 = 0.505$. Interpret these values.

SOLUTION

- The actual dissolved oxygen concentrations are typically about 1.06 mg/L away from the concentrations predicted by the multiple regression model using depth (m) and temperature (°C).
- 50.5% of the variability in dissolved oxygen concentration is accounted for by the multiple regression model using depth (m) and temperature (°C).

EXAM PREP: FOR PRACTICE, TRY EXERCISE 13.

In the dissolved oxygen context, if we used a single-predictor model with x = depth, then $s = 1.46$ mg/L and $r^2 = 0.058$. If we used a single-predictor model with x = temperature, then $s = 1.13$ mg/L and $r^2 = 0.434$. Using both variables in the model from the examples, we reduced s to 1.06 mg/L and increased R^2 to 0.505.

In practice, we can use technology to get the values of both s and R^2 (see the Tech Corner at the end of this section). However, it can be helpful to know how these values are calculated. The good news is that the formulas are very similar to the formulas in Chapter 11. In fact, the way we calculate R^2 is exactly the same for simple linear regression and multiple regression.

The standard deviation of the residuals is

$$s = \sqrt{\frac{\sum(y_i - \hat{y}_i)^2}{n - (k+1)}}$$

where k is the number of explanatory variables used in the model and $(k+1)$ is the number of estimated coefficients (the coefficients of the explanatory variables plus the constant term). In our LPGA model, we used four explanatory variables, so $n - (k+1) = 164 - (4+1) = 159$.

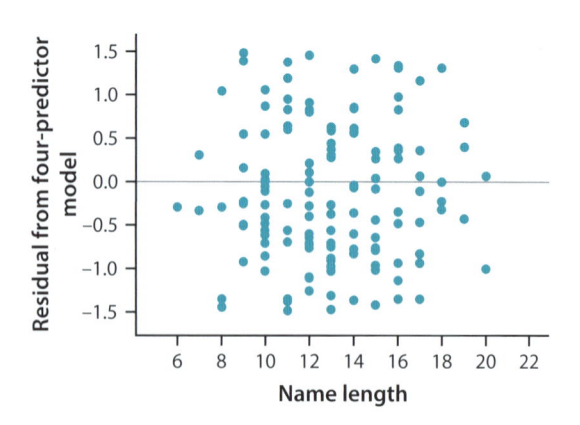

Because of the way s is calculated, we can use changes in s to evaluate whether or not to include a variable in our model. If a new variable helps reduce the residuals by a lot, s will decrease even though the denominator gets reduced by 1. However, if a new variable doesn't help reduce the residuals by much, the value of s may actually increase because of the decrease of 1 in the denominator.

Returning to the LPGA example, here is a scatterplot showing the relationship between x_5 = golfer's name length and y = residual from the four-predictor model. The lack of association in the scatterplot indicates that name length doesn't account for any of the previously unaccounted-for variability. The change in s also confirms this fact: adding x_5 = name length to our model increases s from 0.560 to 0.561. Clearly, the length of a player's name doesn't have anything to do with how well she plays golf!

THINK ABOUT IT

Can you use the same explanatory variable more than once in a multiple regression model? Yes! For example, when the association between x and y looks like a parabola, we can build a multiple regression model in the form $\hat{y} = b_0 + b_1x + b_2x^2$. A model in this form is often called a quadratic model because it has an x^2 term, along with an x term and a constant term.

Here is a scatterplot showing x = carapace (shell) length (cm) and y = clutch size (number of eggs) for a random sample of female gopher tortoises in Okeeheelee County Park, Florida.[3] Also included is a graph of the quadratic model $\hat{y} = -1018.95 + 66.181x - 1.0636x^2$.

In addition to adding a quadratic term, we could add other transformations of an explanatory variable, including a cubic term (x^3), a square-root term ($\sqrt{x}$), or a log term (ln x). See Exercise 22.

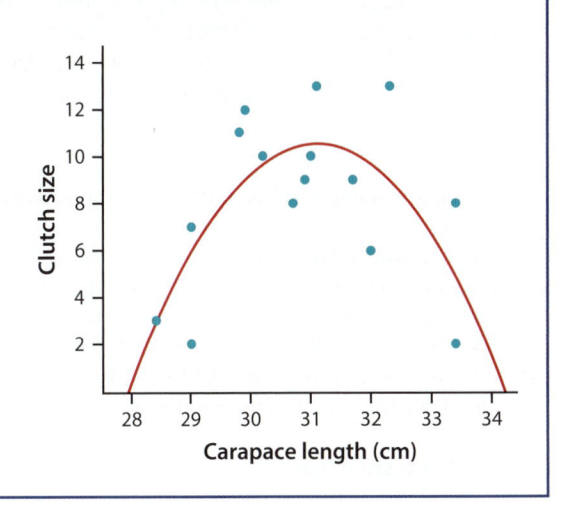

TECH CORNER Creating a Multiple Regression Model

Although we do not ask you to compute a multiple regression model in this section, you can use technology to find the equation of the model and to compute s and R^2. We'll illustrate this process with the carapace length and clutch size data from the preceding Think About It feature. You can also find the data in the **Tortoise** data set.

Clutch size	Carapace length (cm)	Squared carapace length (cm²)
3	28.4	806.56
2	29.0	841.00
7	29.0	841.00
11	29.8	888.04
12	29.9	894.01
10	30.2	912.04
8	30.7	942.49
9	30.9	954.81
10	31.0	961.00
13	31.1	967.21
9	31.7	1004.89
6	32.0	1024.00
13	32.3	1043.29
2	33.4	1115.56
8	33.4	1115.56

Applet

1. Go to www.stapplet.com and launch the *Multiple Regression* applet.
2. Enter the name of explanatory variable 1 (Carapace length) and the values of explanatory variable 1 in the first row of boxes. Then, enter the name of explanatory variable 2 (Squared carapace length) and the values of explanatory variable 2 in the second row of boxes. Make sure both boxes are checked to include these variables in the model. Finally, enter the name of the response variable (Clutch size) and the values of the response variable in the final row of boxes. *Note:* If you have more than two explanatory variables, click the + button in the lower right to add more variables.

(continued)

Multiple Regression

Variable	Name	Observations (separated by commas or spaces) Keep individuals in the same order.	Included in model
Explanatory 1	Carapace length	28.4 29.0 29.0 29.8 29.9 30.2 30.7 30.9 31.0 31.1 31.7 32.0 3	✓
Explanatory 2	Squared carapa	806.56 841.00 841.00 888.04 894.01 912.04 942.49 954.81 9	✓
Response	Clutch size	3 2 7 11 12 10 8 9 10 13 9 6 13 2 8	

[Begin analysis] [Edit inputs] [Reset everything]

3. Click the Begin analysis button. The applet will display the equation of the multiple regression model, along with R^2 and s. The applet will also display other output, including a residual plot and a dotplot of residuals. These will be helpful in Section 12.3 when we're checking the conditions for inference. Finally, the applet allows you to calculate a predicted value by entering the values of the explanatory variables.

Linear Regression Model

Clutch size = −1018.95531 + 66.180967 (Carapace length) + −1.063569 (Squared carapace length)
$S = 2.788$ $R^2 = 0.499$ $R^2(adj) = 0.416$

Predictor	Coef	SE Coef	T	P
Constant	−1018.955	298.811	−3.41	0.005
Carapace length	66.181	19.315	3.426	0.005
Squared carapace length	−1.064	0.312	−3.413	0.005

Analysis of Variance

Source	df	Sum of Squares	Mean Square	F-value	P-value
Regression	2	93.094	46.547	5.986	0.016
Error	12	93.306	7.776		
Total	14	186.4			

Residual Plot

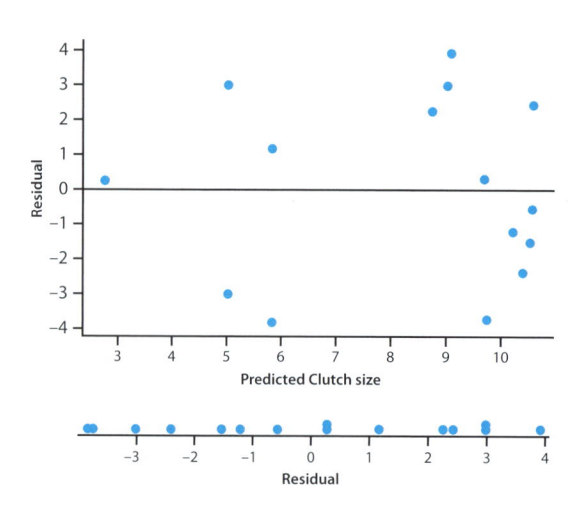

Make a Prediction

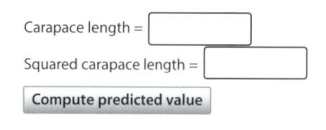

Carapace length = [ ]
Squared carapace length = [ ]
[Compute predicted value]

Note: To investigate what happens when you add or remove a variable, click the Edit inputs button and then adjust the checkboxes to the right of the explanatory variables.

TI-83/84

The TI-83/84 does not perform multiple regression.

Detailed instructions for using CrunchIt!, Excel, Google Sheets, JMP, Minitab, and R are available in Achieve.

Section 12.1 What Did You Learn?

Review the learning goals from this section. Then practice what you've learned by working through the exercises.

Learning Goal	Example	Exercises
Use a multiple regression model to make predictions.	12.1.1	5–8
Use a multiple regression model to calculate and interpret residuals.	12.1.2	9–12
Interpret the standard deviation of the residuals and R^2 for a multiple regression model.	12.1.3	13–16

Section 12.1 Exercises

Building Concepts and Skills These exercises assess the basic knowledge you should have after reading the section.

1. True/False: The goal of a multiple regression model is to account for some of the variability in the response variable that is unaccounted for in a simple linear regression model.

2. A multiple regression model uses _____ explanatory variables to predict the value of a quantitative response variable.

3. Explain how to calculate a residual when using a multiple regression model.

4. **True/False:** When you add a variable to a multiple regression model that helps to account for variability in the response variable, the value of s will increase.

Mastering Concepts and Skills These exercises reinforce the learning goals as shown in the examples.

5. **Refer to Example 12.1.1 Predicting College GPA**
   When the admissions department at a university decides which applicants to accept, it considers many factors. Two of the factors most commonly included in this assessment are high school grade-point average (GPA) and SAT scores. How well do these variables predict a student's GPA in college? Researchers at a large university studied a sample of 224 computer science majors to find out.[4] Here is the multiple regression model relating cGPA = college GPA (after three semesters) to hsGPA = high school GPA and SAT score (out of 1600):

   $$\widehat{cGPA} = -0.1025 + 0.6974\, hsGPA + 0.0003\, SAT$$

   Predict the college GPA for a computer science major who had a 3.05 GPA in high school and scored 1140 on the SAT.

6. **Engineering Glass** The refractive index of glass measures how fast light passes through the glass, with larger values indicating more refraction. For example, the refractive index of water is 1.33, meaning that light travels 1.33 times *slower* in water than in a vacuum.[5] Which factors affect the refractive index of glass? Using a sample of 214 different pieces of glass, researchers measured the amounts of different minerals in the glass (in percent by weight).[6] Here is the multiple regression model relating RI = refractive index to calcium and silicon:

   $$\widehat{RI} = 1.6155 + 0.0016\, calcium - 0.0015\, silicon$$

   Predict the refractive index for a piece of glass that is 8.75% calcium and 71.78% silicon by weight.

7. **NFL Wins** The goal of every team in the National Football League (NFL) is to win games. To win a game, a team must score more points than it allows the other team to score. Using data from the 2020 regular season,[7] we build a multiple regression model relating y = number of wins to x_1 = points scored and x_2 = points allowed:

   $$\hat{y} = 3.4022 + 0.0366 x_1 - 0.0251 x_2$$

   Predict the number of wins for a team that scores 400 points and allows 300 points.

8. **Treadmill Calories** Many exercise bikes, elliptical trainers, and treadmills display basic information like distance, speed, calories burned, and duration of the workout. Here is a multiple regression model relating y = calories burned per hour on a treadmill using x_1 = speed (in miles per hour) and x_2 = incline (in percent).

   $$\hat{y} = -120.22 + 149.96 x_1 + 36.26 x_2$$

   Predict the number of calories burned when a person walks on this treadmill at 3.0 miles per hour with a 4% incline.

9. **Refer to Example 12.1.2 Predicting College GPA**
   Refer to Exercise 5. One student in the data set had a college GPA of 3.73, a high school GPA of 3.89, and an SAT score of 1350.
   (a) Calculate the residual for this student.
   (b) Interpret the residual from part (a).

10. **Engineering Glass** Refer to Exercise 6. One piece of glass in the data set had a refractive index of 1.5167, a calcium content of 10.17%, and a silicon content of 73.88%.
    (a) Calculate the residual for this piece of glass.
    (b) Interpret the residual from part (a).

11. **NFL Wins** Refer to Exercise 7. The Super Bowl champion Tampa Bay Buccaneers had 11 wins, scored 492 points, and allowed 355 points.
    (a) Calculate the residual for the Buccaneers.
    (b) Interpret the residual from part (a).

12. **Treadmill Calories** Refer to Exercise 8. After walking on the treadmill at 3.5 miles per hour at a 2% incline, Anne burned 436 calories per hour.
    (a) Calculate the residual for this workout.
    (b) Interpret the residual from part (a).

13. **Refer to Example 12.1.3 Predicting College GPA**
    Refer to Exercise 5. For this model, $s = 0.71$ and $R^2 = 0.18$. Interpret these values.

14. **Engineering Glass** Refer to Exercise 6. For this model, $s = 0.0014$ and $R^2 = 0.802$. Interpret these values.

15. **NFL Wins** Refer to Exercise 7. For this model, $s = 1.375$ and $R^2 = 0.855$. Interpret these values.

16. **Treadmill Calories** Refer to Exercise 8. For this model, $s = 38$ and $R^2 = 0.99$. Interpret these values.

Applying the Concepts These exercises ask you to apply multiple learning goals in a new context or to apply what you learned in this section in a new way.

17. **Used Car Prices** Which factors affect the price of a used car? We selected a random sample of 191 used Honda CRVs and recorded several variables for each car.[8] Here is the multiple regression model relating y = price to x_1 = age (years) and x_2 = annual miles (miles per year):

    $$\hat{y} = 33{,}957.20 - 1813.42 x_1 - 0.17875 x_2$$

    (a) Predict the price for a 7-year-old Honda CRV that has been driven 7479 miles per year.
    (b) The actual price for the car in part (a) is $18,990. Calculate and interpret the residual for this CRV.
    (c) For this model, $s = 2129$ and $R^2 = 0.868$. Interpret these values.

18. **House Prices** Which factors affect the price of a house? We selected a random sample of 43 recent home sales and recorded several variables for each house.[9] Here is the multiple regression model relating y = price (in $1000s) to x_1 = house size (square feet) and x_2 = yard size (square feet):

$$\hat{y} = 28.048 + 0.1237x_1 + 0.0052x_2$$

  (a) Predict the price for a 1633-square-foot house with a 7117-square-foot yard.
  (b) The actual price for the house in part (a) is $265,000. Calculate and interpret the residual for this house.
  (c) For this model, $s = 33.872$ and $R^2 = 0.826$. Interpret these values.

19. **Tech Yogurt** How do the amounts of saturated fat (g), added sugar (tsp), and protein (g) affect the number of calories in yogurt? Import the **Yogurt** data set into your preferred technology.
  (a) Construct a multiple regression model relating y = Calories to x_1 = SatFat, x_2 = Sugar, and x_3 = Protein.
  (b) Calculate and interpret the residual for the yogurt with 90 calories, 1.5 grams of saturated fat, 0 tsp of added sugar, and 17 grams of protein.
  (c) Identify and interpret the values of s and R^2.

20. **Tech Frozen Desserts** How do the amounts of saturated fat (g), added sugar (tsp), and protein (g) affect the number of calories in frozen desserts like ice cream and gelato? Import the **Frozen Desserts** data set into your preferred technology.
  (a) Construct a multiple regression model relating y = Calories to x_1 = SatFat, x_2 = AddSug, and x_3 = Protein.
  (b) Calculate and interpret the residual for the Haagen Dazs ice cream with 350 calories, 14 grams of saturated fat, 5.5 tsp of added sugar, and 5 grams of protein.
  (c) Identify and interpret the values of s and R^2.

Extending the Concepts These exercises challenge you to explore statistical concepts and methods that go beyond what you learned in this section.

21. **Residual Plots** Besides the standard deviation of the residuals and R^2, we can use a residual plot to assess a multiple regression model. When there is only one explanatory variable, we plot the explanatory variable on the horizontal axis and the residuals on the vertical axis. When there are multiple explanatory variables, however, we use the predicted y values on the horizontal axis. This is because the predicted y values are a combination of all the explanatory variables. Here is the residual plot for the four-predictor model of y = scoring average for LPGA golfers.

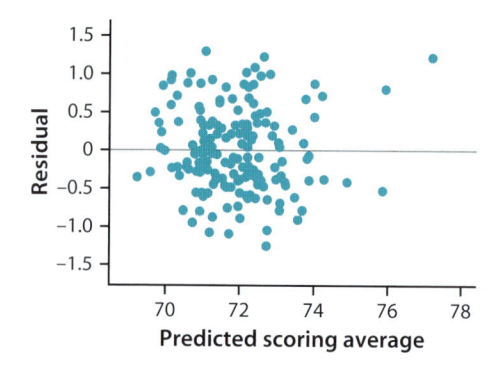

Based on the residual plot, is this multiple regression model appropriate for predicting scoring average? Explain how you know.

22. **Cubic Models** In Chapter 11, we analyzed the relationship between DBH = diameter at breast height (inches), height (feet), and volume of usable lumber (board-feet) for a random sample of ponderosa pines. Here is a scatterplot showing x = DBH and y = volume.

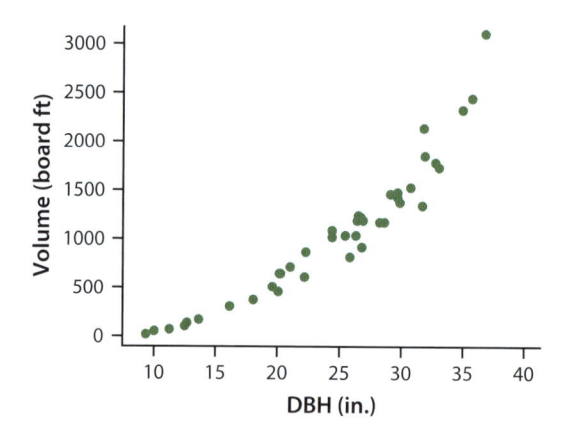

In Chapter 11, we used a cube-root transformation to linearize the association. We can also use multiple regression to fit a cubic model relating y = volume to x = DBH:

$$\hat{y} = -1333.7 + 206.54x - 9.0280x^2 + 0.1760x^3$$

  (a) Calculate and interpret the residual for the ponderosa pine tree with a DBH of 31.7 inches and a volume of 2127 board-feet.
  (b) The model gives $s = 153.3$ and $R^2 = 0.956$. Interpret these values.

23. **LPGA** In this section we introduced a four-predictor model to predict y = scoring average for LPGA golfers. Which of these four explanatory variables is the most important? Because these variables are measured on different scales, we cannot simply compare the coefficients. As in Chapter 3, we can use standardized scores (z-scores) to make these comparisons possible. After standardizing the values of each of the explanatory

variables, we created the following multiple regression model, where z_1 = z-score for average driving distance, z_2 = z-score for driving accuracy, z_3 = z-score for putting average, and z_4 = z-score for sand saves:

$$\hat{y} = 71.92 - 0.5667z_1 - 0.7340z_2 + 0.7980z_3 - 0.1942z_4$$

(a) Predict the scoring average for a golfer who is 1 standard deviation better than average on each of the four variables. Remember that smaller scores are better in golf, so being above average is better for average driving distance, driving accuracy, and sand saves — but not for putting average.

(b) A change of 1 standard deviation in which explanatory variable leads to the biggest change in predicted scoring average? The smallest change? Explain how this could have been predicted using the scatterplots in Figures 12.1 and 12.2.

Cumulative Review These exercises revisit what you learned in previous sections.

24. **Marketing Cereal (10.1, 10.2)** A grocery store sells four different sizes of a popular brand of corn flakes. The table shows the distribution of box size for all sales of this brand at the store during the past several years.

Size	Small	Medium	Large	Jumbo
Frequency	10%	15%	60%	15%

The store manager decides to adjust the pricing of the four sizes and wants to know if the distribution of box size has changed. A month after the price adjustment, the manager selects a random sample of 120 transactions involving this brand of corn flakes and counts how many boxes of each size were sold. Here are the results:

Size	Small	Medium	Large	Jumbo
Frequency	8	24	61	27

(a) State the null and alternative hypotheses for a test to determine if the distribution of box size has changed.

(b) Calculate the expected counts for a test of the null hypothesis stated in part (a).

(c) Check if the conditions for this test have been met.

(d) Calculate the test statistic and P-value for this test. What conclusion would you make?

25. **Oranges (6.8)** Mrs. Tabor keeps track of the number of seeds in each mandarin orange that she eats from the large tree in her backyard. The distribution is skewed to the right with a mean of 8.5 seeds and a standard deviation of 5.9 seeds.

(a) Based on the information provided, is it possible to find the probability that a randomly selected orange has at least 10 seeds? If so, find the probability. If not, explain why not.

(b) Suppose that Mrs. Tabor selects a random sample of 30 oranges from her tree. Is it possible to find the probability that the sample mean number of seeds is at least 10? If so, find the probability. If not, explain why not.

Section 12.2 Indicator Variables and Interaction

LEARNING GOALS

By the end of this section, you will be able to:

- Use a multiple regression model with an indicator variable.
- Interpret the coefficients in a multiple regression model.
- Use a multiple regression model with an interaction term.

In Section 12.1, you learned how to use a multiple regression model to make predictions and calculate residuals. You also learned how to assess a multiple regression model using the standard deviation of the residuals and R^2. In this section, you will learn how to use multiple regression models that include a binary categorical explanatory variable. You will also learn how to interpret coefficients in a multiple regression model and use an interaction term.

Using a Categorical Explanatory Variable

A long-term study of red squirrels in Canada investigated the relationship between food availability and population size.[10] Red squirrels in the Yukon have a diet made up mostly of seeds from white spruce trees. They gather these seeds in the fall, store them underground, and survive on them for the following year. The following scatterplot shows the relationship between cone index in autumn (a measure of food availability) and squirrel density the following spring (number of squirrels per hectare) for plots observed over a number of years.

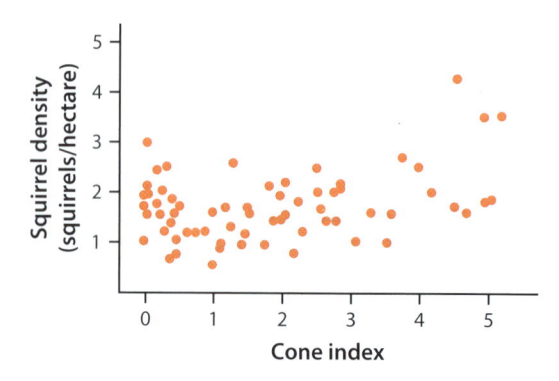

The positive association shown in the scatterplot suggests that squirrels are more likely to breed when they know they have adequate food stored. To confirm this belief, researchers supplemented the food supply in randomly assigned plots during autumn and measured the squirrel density the following spring. Here is a scatterplot that shows the entire data set, with the blue squares indicating plots with food supplementation. The positive association between cone index and squirrel density is still present, but the squirrel density tends to be much higher in the supplemented plots.

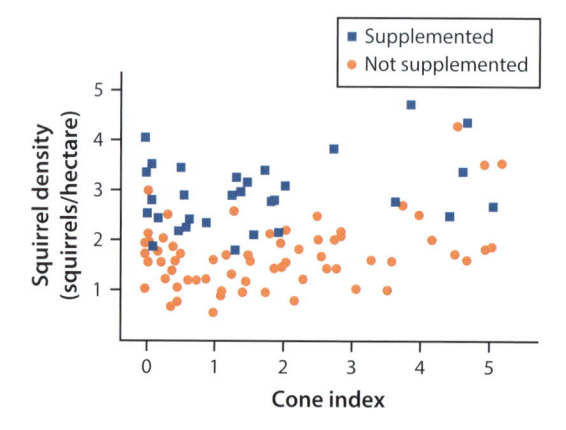

We can use a multiple regression model to describe the relationship between cone index and squirrel density for these two types of plots if we use an **indicator variable** to indicate whether or not a plot was supplemented with food.

An **indicator variable** is a categorical variable with two possible outcomes. These outcomes are coded numerically, so they can be included in regression calculations. A success is reported as a "1" and a failure is recorded as a "0."

Making a prediction when the model includes indicator variables is essentially the same as making a prediction when all the explanatory variables are quantitative. Simply substitute a 1 or a 0, depending on the value of the categorical variable. Here is the multiple regression model relating squirrel density to cone index and food supplementation (1 if yes, 0 if no):

$$\widehat{density} = 1.3810 + 0.1674 \, coneindex + 1.2534 \, supplementation$$

If a plot has a cone index of 2.5 and was supplemented with food, the predicted squirrel density is

$$\widehat{density} = 1.3810 + 0.1674(2.5) + 1.2534(1) = 3.0529 \text{ squirrels/hectare}$$

In contrast, if a plot has a cone index of 2.5 and was *not* supplemented with food, the predicted squirrel density is

$$\widehat{density} = 1.3810 + 0.1674(2.5) + 1.2534(0) = 1.7995 \text{ squirrels/hectare}$$

Why not just fit separate regression lines for each group? Building a multiple regression model with an indicator variable allows us to test if there are differences in the slopes or intercepts for the two groups. You'll learn more about this in Section 12.3.

Section 12.2 Indicator Variables and Interaction 737

EXAMPLE 12.2.1

Pricing Legos: Marvel vs. DC

Indicator variables

PROBLEM

One of the many uses of multiple regression is to model the price of a product. To illustrate, Iowa State University professors Anna Peterson and Laura Ziegler collected data about Lego sets available for sale online at amazon.com.[11] Are Lego sets with a Marvel Comics theme priced differently than Lego sets with a DC Comics theme? Here is a scatterplot showing the relationship between y = price ($) and x = number of pieces, along with separate regression lines for sets of Marvel (purple dots) and DC (orange squares).

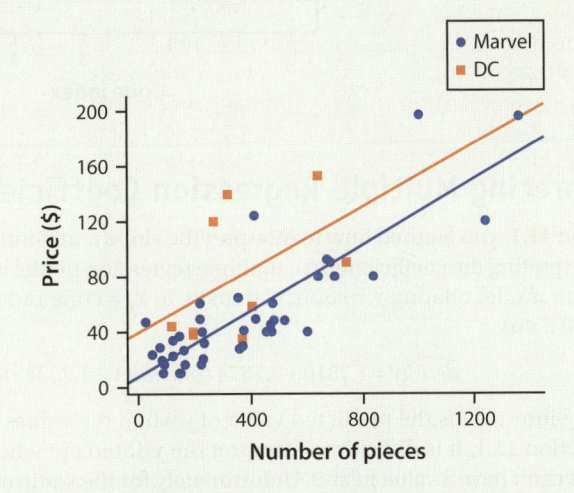

(a) Based on the scatterplot, explain why a multiple regression model for the price of a Lego set should include an indicator variable.

Here is a multiple regression model relating y = price of a Lego set to x_1 = number of pieces and x_2 = Marvel (1 if Marvel, 0 if DC):

$$\hat{y} = 36.81 + 0.1179 x_1 - 30.92 x_2$$

(b) Calculate and interpret the residual of a 1244-piece Marvel set that costs $119.95.

SOLUTION

(a) The scatterplot shows that for any given number of pieces, the Marvel sets generally have lower prices than the DC sets.

(b) $\hat{y} = 36.81 + 0.1179(1244) - 30.92(1) = \$152.56$

residual $= 119.95 - 152.56 = -\$32.61$

The actual cost of this Lego set is $32.61 less than the value predicted by the multiple regression model with x_1 = number of pieces and x_2 = Marvel.

> Find $\hat{y}$ using the regression model.
> residual $= y - \hat{y}$

> When interpreting a residual, remember to include the direction (greater than/less than) in addition to the size of the residual.

EXAM PREP: FOR PRACTICE, TRY EXERCISE 5.

When a multiple regression model includes one quantitative explanatory variable and one indicator variable, the model creates parallel regression lines, one for each value of the indicator variable. Using the squirrel data, if there is no food supplementation (*supplementation* = 0):

$$\widehat{density} = 1.3810 + 0.1674 \, coneindex + 1.2534(0) = 1.3810 + 0.1674 \, coneindex$$

If there is food supplementation (*supplementation* = 1):

$$\widehat{density} = 1.3810 + 0.1674 \, coneindex + 1.2534(1) = 2.6344 + 0.1674 \, coneindex$$

Note that the slopes of these two lines are the same, but the y-intercept differs by 1.2534 — the coefficient of the indicator variable. Here is a scatterplot showing the multiple regression model:

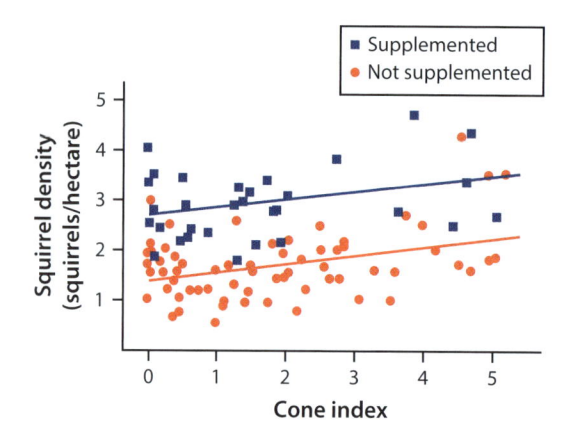

Interpreting Multiple Regression Coefficients

In Section 11.1, you learned how to interpret the slope and y-intercept of a least-squares regression line. Interpreting the coefficients in a multiple regression model is similar. Here again is the multiple regression model relating y = squirrel density to x_1 = cone index and x_2 = food supplementation (1 if yes, 0 if no):

$$\widehat{density} = 1.3810 + 0.1674\, coneindex + 1.2534\, supplementation$$

The y-intercept is the predicted value of y when the values of all explanatory variables are 0. As in Section 11.1, it is difficult to interpret the y-intercept when one or more of the explanatory variables can't have a value near 0. Unfortunately for the squirrels, there were years when the cone index was near 0, so the y-intercept does have meaning in this case: When the cone index is 0 and there is no food supplementation, the predicted squirrel density is 1.3810 squirrels/hectare.

The coefficient of cone index is 0.1674. After accounting for food supplementation, the predicted squirrel density increases by 0.1674 squirrel/hectare for each increase of 1 in cone index. The coefficient of supplemented is 1.2534. After accounting for cone index, the predicted squirrel density increases by 1.2534 squirrels/hectare when the food supply is supplemented.

EXAMPLE 12.2.2

Pricing Legos: Marvel vs. DC

Interpreting coefficients

PROBLEM

Here is the multiple regression model from Example 12.2.1 relating y = price of a Lego set to x_1 = number of pieces and x_2 = Marvel (1 if Marvel, 0 if DC):

$$\hat{y} = 36.81 + 0.1179 x_1 - 30.92 x_2$$

(a) Interpret the coefficients of x_1 and x_2.
(b) If it makes sense, interpret the y-intercept. If not, explain why not.

SOLUTION

(a) x_1: After accounting for whether the set is Marvel or DC, the predicted price of a Lego set increases by $0.1179 for each additional piece in the set.

x_2: After accounting for the number of pieces, the predicted price of a Marvel Lego set is $30.92 less than the predicted price of a DC Lego set.

> When interpreting a coefficient, remember to account for other variables and to include the word "predicted."

(b) Although it is possible for Marvel to equal 0, it doesn't make sense to price a Lego set with 0 pieces.

EXAM PREP: FOR PRACTICE, TRY EXERCISE 9.

Interpreting coefficients can be tricky, especially when the explanatory variables are strongly related to each other. This issue, called *multicollinearity*, is beyond the scope of this book.

Interaction

In the previous examples, we used a multiple regression model with an indicator variable to fit parallel regression lines for two groups. If we want to use a multiple regression model to create regression lines with different slopes, we use an **interaction term** in the model. In general, interaction terms should be included whenever the effect of one explanatory variable on the response variable depends on the value of another explanatory variable.

> An **interaction term** in a multiple regression model is a term created by multiplying two explanatory variables and including the product as an additional explanatory variable in the model. A multiple regression model with two explanatory variables and an interaction term has the following form: $\hat{y} = b_0 + b_1 x_1 + b_2 x_2 + b_3 x_1 x_2$.

Here is a scatterplot showing the relationship between y = highway fuel efficiency (mpg) and x = engine size (liters) for 200 randomly selected model-year 2021 vehicles. Also included are separate regression lines for gas-powered vehicles (teal squares) and hybrid vehicles (green circles).

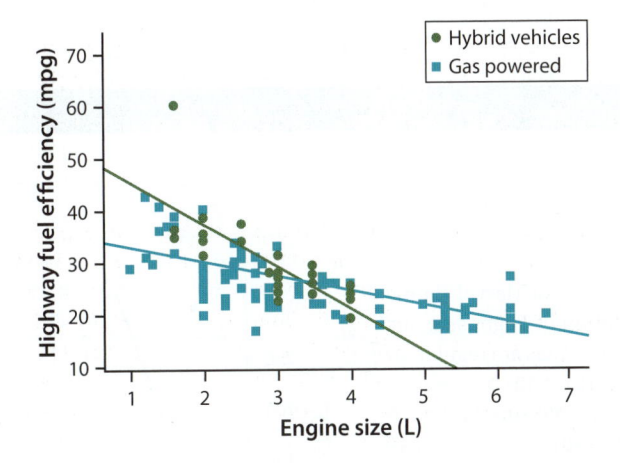

There is a negative association between engine size and highway fuel efficiency for both types of vehicles, but the slope of the regression line for the hybrids is much steeper than that for gas-powered vehicles. To account for the different slopes, we use an interaction term that is the product of the variable engine size and the variable hybrid. Here is the model, where y = highway fuel efficiency, x_1 = engine size, and x_2 = hybrid (1 = hybrid, 0 = not):

$$\hat{y} = 35.1583 - 2.7393 x_1 + 17.6539 x_2 - 5.2562 x_1 x_2$$

The Kia Sorrento hybrid has an engine size of 1.6 liters. Its predicted fuel efficiency is

$$\hat{y} = 35.1583 - 2.7393(1.6) + 17.6539(1) - 5.2562(1.6)(1) = 40.02 \text{ mpg}$$

The Sorrento hybrid has an actual fuel efficiency of 37 mpg, so it gets about 3.02 mpg less than expected based on the multiple regression model with x_1 = engine size and x_2 = hybrid.

To see how using an interaction term creates lines with different slopes, we can separately substitute 0 and 1 into the equation and simplify. If hybrid = 0,

$$\hat{y} = 35.1583 - 2.7393 x_1 + 17.6539(0) - 5.2562 x_1 (0)$$
$$\hat{y} = 35.1583 - 2.7393 x_1$$

If hybrid = 1,

$$\hat{y} = 35.1583 - 2.7393 x_1 + 17.6539(1) - 5.2562 x_1 (1)$$
$$\hat{y} = (35.1583 + 17.6539) + (-2.7393 x_1 - 5.2562 x_1)$$
$$\hat{y} = 52.8122 - 7.9955 x_1$$

When the value of the indicator variable is 1, the coefficient of the indicator variable (17.6539) describes the change in the y-intercept from when the value of the indicator variable is 0. In other words, the y-intercept is 17.6539 greater for hybrids than for gas-powered cars.

Likewise, the coefficient of the interaction term (−5.2562) measures the change in slope from when the value of the indicator variable is 0. The slope is 5.2562 more negative for hybrids than for gas-powered cars.

Here is the scatterplot, along with a graph of the multiple regression model.

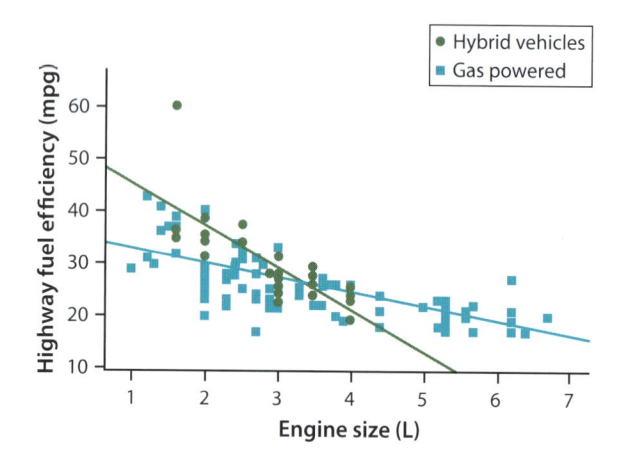

EXAMPLE 12.2.3

Pricing Legos: Classic vs. Duplo

Interaction

PROBLEM

In Examples 12.2.1 and 12.2.2, we compared the price of Marvel-themed and DC-themed Lego sets. What about the "unthemed" Legos sets, such as Duplo (large-size bricks for smaller children) and Classic (regular-size bricks)? Here is a scatterplot showing the relationship between $y =$ price and $x =$ number of pieces for Classic and Duplo sets, along with separate regression lines for sets of Duplo (blue squares) and Classic (orange circles).

(a) Based on the scatterplot, explain why a multiple regression model for the price of a Lego set should include an interaction term.

Here is a multiple regression model relating the price of a Lego set to the number of pieces and Duplo (1 if Duplo, 0 if Classic):

$$\widehat{Price} = 5.37 + 0.0659\, Pieces - 3.53\, Duplo + 0.8943\, Pieces \times Duplo$$

(b) Predict the price of a Classic set with 200 pieces.
(c) State the equation of the regression line for Duplo sets.
(d) Explain how the values −3.53 and 0.8943 affect the regression line for Duplo sets.

SOLUTION

(a) The model should include an interaction term because the slopes of the separate regression lines are different for Classic and Duplo sets.

(b) $\hat{y} = 5.37 + 0.0659(200) - 3.53(0) + 0.8943(200)(0) = \$18.55$

(c) $\hat{y} = 5.37 + 0.0659\, Pieces - 3.53(1) + 0.8943\, Pieces(1)$

$\hat{y} = 1.84 + 0.9602\, Pieces$

(d) −3.53: The y-intercept of the regression line for Duplo bricks is \$3.53 smaller than the y-intercept of the regression line for Classic bricks.

0.8943: The slope of the regression line for Duplo bricks is 0.8943 greater than the slope of the regression line for Classic bricks.

EXAM PREP: FOR PRACTICE, TRY EXERCISE 13.

In this model, the effect of the number of pieces on the price of a set depends on whether the bricks were Duplo or Classic. Because the slope for the Duplo bricks (0.9602) is much bigger than the slope of the Classic sets (0.0659), it appears that an individual Duplo brick is much more expensive than a Classic brick. This makes sense, given how much bigger the Duplo bricks are.

Section 12.2 What Did You Learn?

Review the learning goals from this section. Then practice what you've learned by working through the exercises.

Learning Goal	Example	Exercises
Use a multiple regression model with an indicator variable.	12.2.1	5–8
Interpret the coefficients in a multiple regression model.	12.2.2	9–12
Use a multiple regression model with an interaction term.	12.2.3	13–16

Section 12.2 Exercises

Building Concepts and Skills These exercises assess the basic knowledge you should have after reading the section.

1. When using an indicator variable, we indicate a "success" with a(n) _____ and a "failure" with a(n) _____.

2. True/False: A multiple regression model with one quantitative explanatory variable and one indicator variable creates parallel regression lines.

3. True/False: The coefficient of an indicator variable describes the change in the actual values of the response variable when the value of the indicator variable is equal to 1.

4. Interaction terms should be included whenever the effect of one _____ variable on the _____ variable depends on the value of another _____ variable.

Mastering Concepts and Skills These exercises reinforce the learning goals as shown in the examples.

5. **Refer to Example 12.2.1 Meadowfoam** Meadowfoam seed oil is used in making various skin care products. Researchers interested in maximizing the productivity of meadowfoam plants designed an experiment to investigate the effects of different light intensities on plant growth. The researchers planted 240 meadowfoam seedlings in individual pots, placed 10 pots in each of 24 trays, and put all the trays into a controlled enclosure. They randomly assigned four trays to each light intensity level (micromoles per square meter per second, $\mu mol/m^2/sec$): 150, 300, 450, 600, 750, and 900. At each light intensity level, they randomly assigned two of the trays to receive light treatment 24 days earlier than the remaining trays. They recorded the number of flowers produced by each plant and the average number of flowers for each tray.[12] Here is a scatterplot that shows the relationship between y = average number of flowers and x = light intensity, with green squares indicating trays that were given the treatment early.

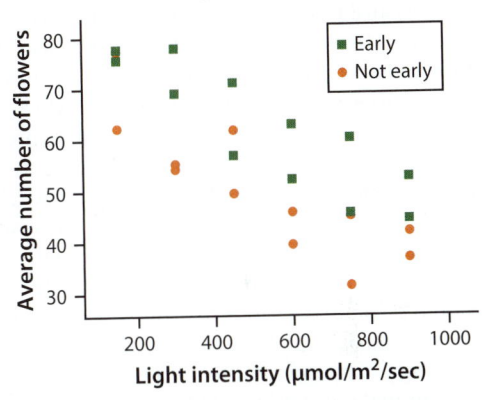

   (a) Based on the scatterplot, explain why a multiple regression model for the average number of flowers should include an indicator variable.

   Here is a multiple regression model relating these variables, where "early" is an indicator variable (1 = early light treatment, 0 = no early treatment):

   $$\widehat{flowers} = 71.31 - 0.0405\ light + 12.16\ early$$

   (b) Calculate and interpret the residual for the tray with an average number of flowers of 62.9 that was exposed to a light intensity of 600 $\mu mol/m^2/sec$ 24 days early.

6. **Butterflies** Greenland is home to the *Boloria chariclea* butterfly. Scientists studying this species of butterfly investigated some of the factors that might affect wing size, including sex and average temperature during the previous summer, when larvae were growing. Here is a scatterplot that shows the relationship between y = wing length (mm) and x = average temperature in the previous summer (°C), with red squares indicating female butterflies.

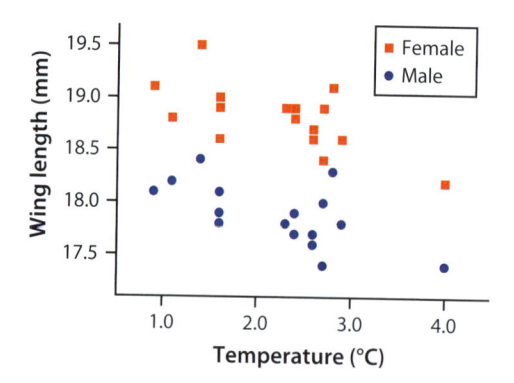

(a) Based on the scatterplot, explain why a multiple regression model for the wing length should include an indicator variable.

Here is a multiple regression model relating these variables, where "female" is an indicator variable (1 = female, 0 = male):

$$\widehat{wing} = 18.40 - 0.2350\ temp + 0.9312\ female$$

(b) Calculate and interpret the residual for the female butterfly with a wing length of 18.6 mm whose larvae grew in an average temperature of 1.6°C.

7. **Wolves and Deer** In areas where deer are common, collisions between vehicles and deer are also common. Researchers noticed that when wolves are introduced into an area with a deer population, vehicle–deer collisions decrease. This is partly due to wolves preying on deer. Wolves also tend to travel near roads, encouraging deer to avoid these areas. Using data from rural counties in Wisconsin over several decades,[13] researchers recorded whether there were wolves present and how many miles residents drove per year (figuring that there would be more collisions in areas where more vehicles were being driven). Here is a multiple regression model relating y = annual number of vehicle-deer collisions, x_1 = annual miles driven (in billions), and x_2 = 5-year wolf presence (1 if wolves have been present in the previous 5 years, 0 otherwise):

$$\hat{y} = 109.8 + 372.82 x_1 - 118.25 x_2$$

Calculate and interpret the residual for the county that had 143 vehicle–deer collisions in a year where wolves were present and residents drove 1.023 billion miles.

8. **MLB Attendance** Attendance at Major League Baseball (MLB) games is affected by many factors. In general, teams that win more games draw more fans. Does a recent appearance in the World Series also boost attendance? Here is a multiple regression model relating y = average attendance to x_1 = number of wins and x_2 = world series (1 if the team has been in the World Series in past 5 years, 0 otherwise) for the 30 MLB teams in a recent (nonpandemic) season:

$$\hat{y} = 3919.2 + 313.7 x_1 + 4546.0 x_2$$

Calculate and interpret the residual for the Boston Red Sox, who won 78 games, have been in a recent World Series, and averaged 35,564 fans.

9. **Refer to Example 12.2.2 Meadowfoam** Refer to Exercise 5.
   (a) Interpret the coefficients of *light* and *early*.
   (b) If it makes sense, interpret the y-intercept. If not, explain why not.

10. **Butterflies** Refer to Exercise 6.
    (a) Interpret the coefficients of *temp* and *female*.
    (b) If it makes sense, interpret the y-intercept. If not, explain why not.

11. **Wolves and Deer** Refer to Exercise 7.
    (a) Interpret the coefficients of x_1 and x_2.
    (b) If it makes sense, interpret the y-intercept. If not, explain why not.

12. **MLB Attendance** Refer to Exercise 8.
    (a) Interpret the coefficients of x_1 and x_2.
    (b) If it makes sense, interpret the y-intercept. If not, explain why not.

13. **Refer to Example 12.2.3 Alcohol** Here is a scatterplot showing the relationship between y = calories and x = percent alcohol for various types of beer and wine, along with separate regression lines for beers (purple squares) and wines (orange circles).

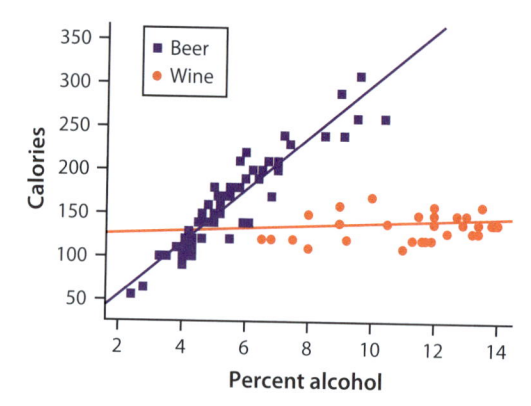

(a) Based on the scatterplot, explain why a multiple regression model for the number of calories should include an interaction term.

Here is a multiple regression model relating calories to percent alcohol and beer (1 = beer, 0 = wine):

$$\widehat{Calories} = 119.4 + 1.6314\ Percent - 126.63\ Beer + 28.75\ Percent \times Beer$$

(b) Predict the number of calories in a wine with 10% alcohol.
(c) State the equation of the regression line for beers.
(d) Explain how the values −126.63 and 28.75 affect the regression line for beers.

14. **Starbucks** Is the relationship between fat and calories the same for drink products and for food products at Starbucks? Here is a scatterplot showing the relationship between y = calories and x = fat (g) for a random sample of Starbucks products, with the food products

marked with green squares and the drink products marked with teal dots.[14]

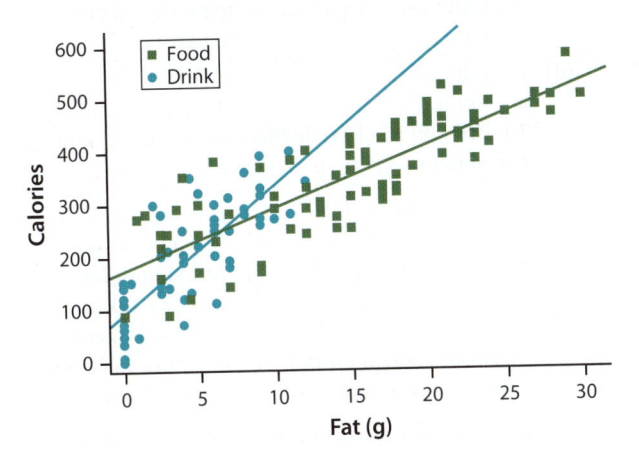

(a) Based on the scatterplot, explain why a multiple regression model for the number of calories should include an interaction term.

Here is a multiple regression model relating calories to fat and food (1 = food, 0 = drink):

$$\widehat{Calories} = 88.26 + 24.45\, Fat + 82.07\, Food - 12.62\, Fat \times Food$$

(b) Predict the number of calories for drinks with 10 grams of fat.

(c) State the equation of the regression line for food products.

(d) Explain how the values 82.07 and −12.62 affect the regression line for food products.

15. Wolves and Deer Refer to Exercise 7. Here is a revised model that includes an interaction term for annual miles driven and wolf presence:

$$\hat{y} = 64.16 + 465.44 x_1 - 52.01 x_2 - 139.12 x_1 x_2$$

(a) Calculate and interpret the residual for the county that had 143 vehicle–deer collisions in a year where wolves were present and residents drove 1.023 billion miles.

(b) State the equation of the regression line for counties with no wolf presence in the previous 5 years.

(c) The standard deviation of the residuals went from 148.1 to 146.9 when including the interaction term. What does this indicate about the model?

16. MLB Attendance Refer to Exercise 8. Here is a revised model that includes an interaction term for number of wins and a recent World Series appearance:

$$\hat{y} = 1154.7 + 348.6 x_1 + 21445.7 x_2 - 197.41 x_1 x_2$$

(a) Calculate and interpret the residual for the Boston Red Sox, who won 78 games, have been in a recent World Series, and averaged 35,564 fans.

(b) State the equation of the regression line for teams that have not made a recent World Series appearance.

(c) The standard deviation of the residuals went from 6160 to 6228 when including the interaction term. What does this indicate about the model?

Applying the Concepts These exercises ask you to apply multiple learning goals in a new context or to apply what you learned in this section in a new way.

17. Used Car Prices Which factors affect the price of a used car? We selected a random sample of 191 used Honda CRVs and recorded several variables for each car.[15] Here is the multiple regression model relating y = price to x_1 = age (years), x_2 = annual miles (miles per year), and x_3 = leather (1 if car has leather seats, 0 if not):

$$\hat{y} = 32{,}855.70 - 1757.65 x_1 - 0.2082 x_2 + 2288.18 x_3$$

(a) Calculate and interpret the residual for the 6-year-old CRV that has been driven 10,171.5 miles per year, has leather seats, and has a price of $20,990.

(b) Interpret the coefficients of x_1, x_2, and x_3.

(c) For this model, $s = 1810$ and $R^2 = 0.905$. Interpret these values.

18. House Prices Which factors affect the price of a house? We selected a random sample of 43 recent home sales and recorded several variables for each house.[16] Here is the multiple regression model relating y = price (in $1000s) to x_1 = house size (square feet), x_2 = yard size (square feet), and x_3 = pool (1 = has pool, 0 = no pool):

$$\hat{y} = 28.097 + 0.1249 x_1 + 0.0051 x_2 - 9.3745 x_3$$

(a) Calculate and interpret the residual for the 1421-square-foot house that has a 6679-square-foot yard, a pool, and a price of $280,000.

(b) Interpret the coefficients of x_1, x_2, and x_3.

(c) For this model, $s = 34.095$ and $R^2 = 0.828$. Interpret these values.

19. Used Car Prices Refer to Exercise 17. Here is a revised model with an interaction term for age and leather seats:

$$\hat{y} = 32131.80 - 1615.47 x_1 - 0.2126 x_2 + 3703.99 x_3 - 289.59 x_1 x_3$$

(a) For this model, $s = 1760$. Based on this value, did including the interaction term help? Explain your answer.

(b) Calculate and interpret the residual for the 6-year-old CRV that has been driven 10,171.5 miles per year, has leather seats, and has a price of $20,990.

20. House Prices Refer to Exercise 18. Here is a revised model with an interaction term for house size and pool:

$$\hat{y} = 18.642 + 0.1333 x_1 + 0.0047 x_2 + 280.34 x_3 - 0.1771 x_1 x_3$$

(a) For this model, $s = 29.064$. Based on this value, did including the interaction term help? Explain your answer.

(b) Calculate and interpret the residual for the 1421-square-foot house that has a 6679-square-foot yard, a pool, and a price of $280,000.

21. Tech Yogurt How do the amounts of saturated fat (g), added sugar (tsp), and protein (g) affect the number of calories in yogurt? Does it matter if it is plant-based? Import the *Yogurt* data set into your preferred technology. Then create the variable *Plant* to indicate if the yogurt is plant-based (1 = yes, 0 = no) and fill in the values for this variable.

(a) Construct a multiple regression model relating $y =$ *Calories* to $x_1 =$ *SatFat*, $x_2 =$ *Sugar*, $x_3 =$ *Protein*, and $x_4 =$ *Plant*.

(b) Interpret the coefficient of x_4.

(c) Did adding the indicator variable for plant-based help the model? Explain your answer.

22. **Tech** **Frozen Desserts** How do the amounts of saturated fat (g), added sugar (tsp), and protein (g) affect the number of calories in frozen desserts like ice cream and gelato? Does it matter if it is a sorbet? Import the **Frozen Desserts** data set into your preferred technology. Then create the variable *Sorbet* to indicate if the dessert is a sorbet (1 = yes, 0 = no) and fill in the values for this variable.

(a) Construct a multiple regression model relating $y =$ *Calories* to $x_1 =$ *SatFat*, $x_2 =$ *AddSug*, $x_3 =$ *Protein*, and $x_4 =$ *Sorbet*.

(b) Interpret the coefficient of x_4.

(c) Did adding the indicator variable for sorbet help the model? Explain your answer.

23. **Tech** **Yogurt** Refer to Exercise 21. Create an interaction term for $x_1 =$ *SatFat* and $x_4 =$ *Plant* and fill in the values for this variable. Did adding the interaction term help the model? Explain your answer.

24. **Tech** **Frozen Desserts** Refer to Exercise 22. Create an interaction term for $x_1 =$ *SatFat* and $x_4 =$ *Sorbet* and fill in the values for this variable. Did adding the interaction term help the model? Explain your answer.

Extending the Concepts These exercises challenge you to explore statistical concepts and methods that go beyond what you learned in this section.

25. **Used Car Prices** Refer to Exercises 17 and 19. In this section, we incorporated categorical variables into a multiple regression model, but only categorical variables with exactly two categories (so they could be coded as 0 and 1). In reality, many categorical variables have more than two categories. For example, CRVs come in four models: LX, EX, EX-L, and Touring. To incorporate a categorical variable with four categories, we create three indicator variables. Here is the multiple regression model relating $y =$ price to $x_1 =$ age (years), $x_2 =$ annual miles (miles per year), $x_3 =$ EX (1 if EX, 0 if not), $x_4 =$ EX-L (1 if EX-L, 0 if not), $x_5 =$ Touring (1 if Touring, 0 if not). Note that if x_3, x_4, and x_5 are all 0, the CRV is an LX.

$$\hat{y} = 31{,}739.2 - 1706.85x_1 - 0.2399x_2 + 1899.89x_3 + 2943.07x_4 + 5741.08x_5$$

(a) Predict the price of a 5-year-old CRV EX that has been driven 12,500 miles per year.

(b) Interpret the coefficient of x_5.

(c) The primary difference between an EX and an EX-L is that the EX-L has leather seats. Based on the model, how much does the predicted price increase for CRVs that have leather seats?

(d) How does your answer to part (c) differ from your answer in Exercise 17? How does knowing that Touring models have leather seats help explain the difference?

26. **Lake Interaction** In Section 12.1, we used a multiple regression model relating DO = dissolved oxygen concentration (mg/L) to depth (m) and temperature (°C) in Lake Champlain. For this model, $s = 1.06$ and $R^2 = 0.505$. It is also possible to create an interaction term between two quantitative explanatory variables. Here is the revised model with the interaction term:

$$\widehat{DO} = 13.18 - 0.0183\,depth - 0.1934\,temp - 0.0063\,depth \times temp$$

$$s = 0.97 \qquad R^2 = 0.582$$

(a) Did the interaction term help? Explain your answer.

(b) Calculate and interpret the residual for the sample with a depth of 26 m, a temperature of 6.83°C, and a dissolved oxygen concentration of 12.5 mg/L.

27. **Aspartame** People who drink diet sodas may notice a decrease in the sweetness of the soda if it has been sitting around for a long time — particularly if it has been in a warm environment. To determine how fast the sweetener aspartame degrades over time, researchers used diet soda of three different acidity levels (pH). They kept some at 20°C, some at 30°C, and the rest at 40°C. The scatterplot shows the time (in months) and the remaining aspartame (%) for a soda with pH = 2.75 stored at 20°C (red squares) or 30°C (blue circles).[17]

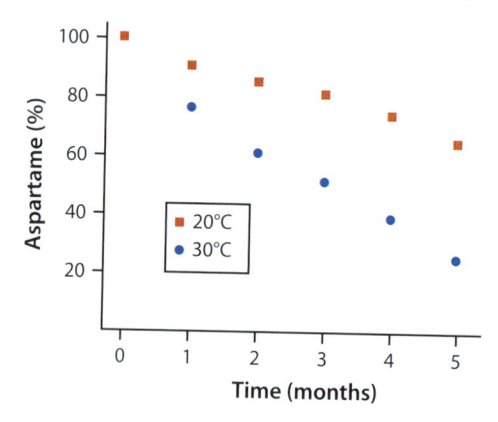

Here is a multiple regression model relating $y =$ aspartame remaining to $x_1 =$ time and $x_2 =$ temp (1 if temperature = 20°C, 0 if temperature = 30°C):

$$\hat{y} = 96.5357 - 14.837x_1 + 8.926x_1x_2$$

(a) Explain why the model needs an interaction term, but not a term with just the indicator variable.

(b) Predict the amount of aspartame remaining after 2.5 months if a diet soda is stored at 20°C.

28. **Engineering Glass** One way to look for multicollinearity (association between explanatory variables) is with a scatterplot matrix, which shows the relationship between each pair of explanatory variables in a multiple regression setting. Shown here is a scatterplot matrix with four variables: refractive index (the response variable) and three possible explanatory variables—sodium content, calcium content, and silicon content.

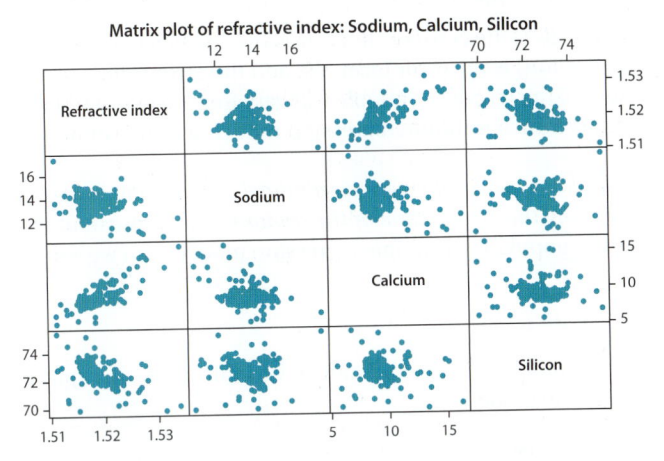

Matrix plot of refractive index: Sodium, Calcium, Silicon

(a) If you were to build a one-predictor model of refractive index, which explanatory variable would you use? Explain your answer.

(b) If you were to a build a two-predictor model of refractive index, which pair of explanatory variables show the most multicollinearity?

(c) In Exercise 6 from Section 12.1, we built a two-predictor model of refractive index using calcium content and silicon content. Explain why we don't need to worry about multicollinearity with this model.

29. **Icicles** We have data on the growth of icicles starting at length 10 centimeters (cm) and at length 20 cm. Suppose icicles that start at 10 cm grow at a rate of 0.15 cm per minute and icicles that start at 20 cm grow at a rate of 0.12 cm per minute. Give a multiple regression model that describes how y = length changes with x_1 = time (min) and x_2 = starting length (1 if 20 cm, 0 if 10 cm).

Cumulative Review These exercises revisit what you learned in previous sections.

30. **Preschool (1.5, 1.6)** Does preschool help low-income children stay in school and find good jobs later in life? The Carolina Abecedarian Project (the name suggests the ABCs) has followed a group of 111 children since 1972. Back then, these individuals were all healthy infants from low-income families in Chapel Hill, North Carolina. All the infants received nutritional supplements and help from social workers. Half were also assigned at random to an intensive preschool program.[18]

   (a) Describe how the researchers could have carried out the random assignment.

   (b) Explain the purpose of random assignment in this experiment.

   (c) Identify one variable that the researchers kept the same for all subjects. Why was it important that the researchers kept this variable the same?

31. **Gummy Bears (10.3, 10.4)** Two researchers wondered if the distribution of color was the same for name-brand gummy bears (Haribo Gold) and store-brand gummy bears (Great Value). To investigate, they randomly selected gummy bears of each brand and counted how many bears were red, green, yellow, orange, or white.[19] Based on the data summarized in the table, is there convincing evidence at the 5% significance level of an association between brand of gummy bear and color?

		Brand		
		Haribo Gold	Great Value	Total
	Red	137	212	349
	Green	53	104	157
Color	Yellow	50	85	135
	Orange	81	127	208
	White	52	94	146
	Total	373	622	995

Section 12.3 Inference for Multiple Regression

LEARNING GOALS
By the end of this section, you will be able to:
- State the hypotheses for a test about the usefulness of a multiple regression model.
- Make a conclusion for a test about the usefulness of a multiple regression model based on a P-value.
- Determine whether an explanatory variable provides additional useful information about the response variable, after accounting for other variables in the model.

In Sections 12.1 and 12.2, you learned the basics of summarizing data from a sample using a multiple regression model. In this section, we'll introduce you to evaluating a multiple regression model with an eye toward describing a population.

Testing a Multiple Regression Model

In Sections 12.1 and 12.2, we used data from a sample to create models of the form $\hat{y} = b_0 + b_1 x_1 + b_2 x_2 + \ldots$. These models describe the relationship between several explanatory variables and the response variable. In this section, we begin thinking of this model as an estimate for a "true" regression model that describes an entire population. The notation we use for the population (true) model is

$$y = \beta_0 + \beta_1 x_1 + \beta_2 x_2 + \ldots + \varepsilon$$

where y is the value of the response variable, β_0 is the constant term, β_i is the coefficient of explanatory variable x_i, and ε (the Greek letter epsilon) is the error term. We add the error term ε to the model as a reminder that individual values of the response variable will vary from values predicted by the model. When we calculate residuals from a multiple regression model, we are estimating these unknown errors.

One question we ask is: *Does the population multiple regression model describe the values of the response variable better than simply using the mean value of the response?* If the mean of the response variable μ_y does just as well as the population multiple regression model, then we would prefer the simpler mean-only model:

$$y = \mu_y + \varepsilon$$

Because this is equivalent to a regression model with no explanatory variables, we can replace μ_y with β_0 in the population multiple regression model:

$$y = \beta_0 + \varepsilon$$

Our null hypothesis is that there is no difference in our ability to predict the value of the response variable between this mean-only model and a more complicated multiple regression model. If the null hypothesis is true, then the coefficients of each of the explanatory variables in the population multiple regression model would be 0:

$$y = \beta_0 + 0 x_1 + 0 x_2 + \ldots + \varepsilon = \beta_0 + \varepsilon$$

Our alternative hypothesis is that the population multiple regression model adds useful information about the values of the response variable. If the alternative hypothesis is true, the value of at least one of the coefficients of the explanatory variables in the model would be nonzero. Here are the hypotheses in symbols, assuming there are k explanatory variables:

$$H_0: \beta_1 = \beta_2 = \ldots = \beta_k = 0$$
$$H_a: \text{At least one of these } \beta_i\text{'s} \neq 0$$

Example 12.1.1 presented a multiple regression model for predicting the amount of dissolved oxygen (DO) in water samples from Lake Champlain, based on the depth of the sample and the water temperature at that depth. The model given was

$$\widehat{DO} = 13.5881 - 0.0217 \, depth - 0.2583 \, temperature$$

Is there convincing evidence that a population multiple regression model using depth and temperature to predict DO is more effective than the mean-only model? We want to test the following hypotheses for the model $y = \beta_0 + \beta_1 \, depth + \beta_2 \, temperature + \varepsilon$:

$$H_0: \beta_1 = \beta_2 = 0$$
$$H_a: \text{At least one of } \beta_1 \text{ or } \beta_2 \neq 0$$

If we reject H_0, we have convincing evidence that including temperature or depth (or both) in the model provides additional useful information about dissolved oxygen levels compared to the model that uses only the mean DO level to make predictions.

EXAMPLE 12.3.1

How old is that carpet?

Stating hypotheses for testing a population multiple regression model

PROBLEM

A group of chemists was interested in whether they could use chemical composition to date pieces of wool carpet. They selected a random sample of 23 wool carpet pieces with known ages ranging from 120 to 1750 years and measured the levels (in grams per 100 grams of protein) of four different amino acids: cysteic acid, cystine, methionine, and tyrosine. Do these data provide convincing evidence that the following population multiple regression model provides additional useful information about age? State the appropriate hypotheses for this test.

$$age = \beta_0 + \beta_1\,cysteic + \beta_2\,cystine + \beta_3\,methionine + \beta_4\,tyrosine + \varepsilon$$

SOLUTION

$H_0: \beta_1 = \beta_2 = \beta_3 = \beta_4 = 0$
$H_a:$ At least one of these β_i's $\neq 0$

EXAM PREP: FOR PRACTICE, TRY EXERCISE 7.

Checking the Conditions

The conditions required for this test are essentially the same as the conditions for performing inference about the slope of a simple linear regression model as described in Section 11.4.

HOW TO — Check the Conditions for Inference About a Population Multiple Regression Model

To perform inference about a population multiple regression model, check that the following conditions are met:

- **Random:** The data come from a random sample from the population of interest or from a randomized experiment.
- **Linear:** The form of the association between the set of explanatory variables used in the model and the response variable is linear. That is, the model is of the form $y = \beta_0 + \beta_1 x_1 + \beta_2 x_2 + \ldots + \beta_k x_k + \varepsilon$, where $k =$ the number of explanatory variables. Check this condition with a plot of the residuals versus the predicted values, which should have no leftover patterns.
- **Equal SD:** For all combinations of values of the explanatory variables, the standard deviation of y is the same. Check this condition with a residual plot. Common violations of this condition include a $<$ pattern (residuals tend to grow in size as the predicted values increase) or a $>$ pattern (residuals tend to shrink in size as predicted values increase).
- **Normal:** For all combinations of values of the explanatory variables, the distribution of y is normal. Check this condition with a graph (e.g., dotplot, normal probability plot) of the residuals. There should be no strong skewness or outliers.

One big difference between checking conditions for simple linear regression and multiple regression is the way we create residual plots. Because there are multiple explanatory variables in multiple regression, we use the predicted y values on the horizontal axis rather than one of the explanatory variables. This is because the predicted y values are a combination of *all* the explanatory variables. Often these predicted y values are called *fitted* values.

Another difference is that we can't use a scatterplot to check the Linear condition as we could when there was only one explanatory variable. With two explanatory variables, we would need a 3-D plot to look for a plane (rather than a line). And with each additional explanatory variable, we would need a plot with an extra dimension. Instead, we rely on a residual plot to identify departures from linearity.

Because the water specimens from Lake Champlain were a random sample, the Random condition has been met. The residual plot created from this model does not show any leftover curved patterns, so the Linear condition is met. There is also a relatively even band of points across the residual plot, indicating that the Equal SD condition is met.

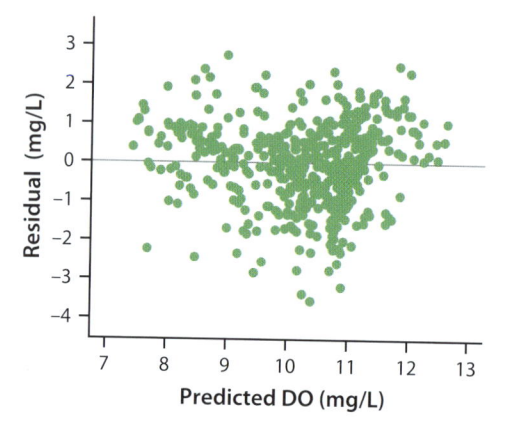

A histogram of the residuals is single-peaked and only slightly left-skewed, so the Normal condition is also met.

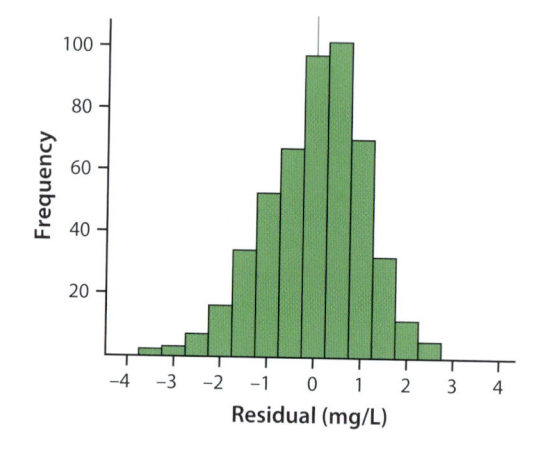

Completing the Test

Now that conditions have been met, we proceed with our test of the multiple regression model. Here again are the hypotheses for a test of the usefulness (sometimes referred to as *utility*) of the population multiple regression model in the form $y = \beta_0 + \beta_1 x_1 + \beta_2 x_2 + \ldots + \beta_k x_k + \varepsilon$:

$H_0: \beta_1 = \beta_2 = \ldots = \beta_k = 0$

$H_a:$ At least one of these β_i's $\neq 0$

We use the sample multiple regression model $\hat{y} = b_0 + b_1 x_1 + b_2 x_2 + \ldots + b_k x_k$ to determine whether we have convincing evidence for the alternative hypothesis. If we find convincing evidence for H_a, we conclude that the population multiple regression model with at least one of the explanatory variables is better than the model that uses only the mean y value. In other words, we conclude that at least one of the explanatory variables in the population multiple regression model provides additional useful information about the response variable.

Note that the test of a population multiple regression model does *not* tell us which explanatory variables are useful—only that at least one of them is useful. This is similar to the conclusion from the one-way ANOVA test discussed in Section 10.5 when the null hypothesis is rejected. In that case, we conclude that at least one mean is different, but we don't know which one is different.

The results of the test of a population multiple regression model are organized in an ANOVA table, similar to the tables discussed in Sections 10.5 and 10.6.

Section 12.3 Inference for Multiple Regression

Source	df	SS	MS	F
Regression	$df_R = k$	SSR	MSR	F
Error	$df_E = n - k - 1$	SSE	MSE	
Total	$df_T = n - 1$	SST		

In this ANOVA table,

- k = the number of explanatory variables
- $df_R + df_E = df_T$
- $SSR + SSE = SST$
- $MSR = SSR/df_R$
- $MSE = SSE/df_E$
- $F = MSR/MSE$

We use the F statistic as the test statistic for an **F test for a population multiple regression model**. When the conditions are met, this test statistic has an F distribution with df_R and df_E degrees of freedom. Fortunately, technology provides a P-value in addition to the F statistic.

> An **F test for a population multiple regression model** is a significance test of the null hypothesis that a population multiple regression model does not provide additional useful information about a response variable compared to a model that uses only the mean of the response variable.

The ANOVA table for the dissolved oxygen model is shown here:

Source	df	SS	MS	F	P-Value
Regression	2	565.2	282.6	253.52	0.000
Error	497	554.0	1.115		
Total	499	1119.3			

Because the P-value of ≈ 0.000 is less than $\alpha = 0.05$, we reject the null hypothesis. We have convincing evidence that either temperature or depth (or both) provides additional useful information about the level of dissolved oxygen. In other words, we have convincing evidence that the population multiple regression model is better than the model that uses only the mean level of dissolved oxygen to make predictions.

EXAMPLE 12.3.2

How old is that carpet?

F test for multiple regression

PROBLEM
Recall from Example 12.3.1 that chemists were interested in using amino acid values to predict the age of samples of wool carpet. The conditions for inference about the population multiple regression model are met. Based on the ANOVA table for this model, what conclusion would you make for a test of the hypotheses in Example 12.3.1 using $\alpha = 0.05$?

Source	df	SS	MS	F	P-Value
Regression	4	8,694,828	2,173,707	1010.54	0.000
Error	18	38,719	2151		
Total	22	8,733,546			

SOLUTION
Because the P-value of approximately $0 < \alpha = 0.05$, we reject H_0. There is convincing evidence that at least one of the explanatory variables (cysteic acid, cystine, methionine, and tyrosine levels) provides additional useful information about the age of wool carpet samples.

EXAM PREP: FOR PRACTICE, TRY EXERCISE 11.

Testing an Individual Coefficient

Finding convincing evidence for the alternative hypothesis is a good start. Unfortunately, the F test only reveals that at least one of the explanatory variables provides additional useful information about the response variable. It doesn't tell us which explanatory variable (or variables) provides this information. For that, we turn to significance tests for individual coefficients.

There are two competing aims when building a multiple regression model: (1) do a good job of modeling the response variable and (2) be as simple as possible. If a multiple regression model has too many explanatory variables, it can be hard to interpret. This is because the explanatory variables are often interrelated. To help balance these competing aims, we can test whether individual explanatory variables provide useful information to the model, after accounting for the other explanatory variables in the model.

In Section 11.5, you learned how to test the slope of a least-squares regression line. In that test, the null hypothesis is $H_0: \beta_1 = 0$, where β_1 is the slope of the population regression line. A **t test for an individual coefficient** is similar. This test can be used for the coefficient of any individual explanatory variable in a population multiple regression model, assuming that all of the other explanatory variables are already in the model. The conditions required for the test are the same as those required for testing the population multiple regression model as a whole.

> A **t test for an individual coefficient** is a significance test of the null hypothesis that an explanatory variable in a population multiple regression model does not provide additional useful information about the response variable, after accounting for other variables in the model.

Computer output for multiple regression typically contains a table with the results of the test for every coefficient in the model, including the y-intercept. Here is the output for the multiple regression model relating the amount of dissolved oxygen in water samples from Lake Champlain to depth and temperature:

```
Coefficients
Term            Coef        SE Coef      T-Value      P-Value
Constant        13.588      0.197        68.95        0.000
Depth           -0.02168    0.00257      -8.44        0.000
Temp            -0.2583     0.0122       -21.19       0.000
```

Each line in the table gives the t statistic and P-value for a significance test for one of the coefficients. We typically ignore the first line, which tests whether the y-intercept is 0 or not. Because the y-intercept is often an extrapolation, we usually don't worry too much about its specific value. For the remaining rows, the null hypothesis is that the coefficient of the specified explanatory variable is 0, and the alternative hypothesis is two-sided. Because the P-values for depth and temperature are both approximately 0, we conclude that each explanatory variable adds useful information about dissolved oxygen levels, even after accounting for the other variable in the model.

EXAMPLE 12.3.3

How old is that carpet?

Testing coefficients

PROBLEM

Recall from Examples 12.3.1 and 12.3.2 that chemists were interested in predicting the age of samples of wool carpet based on the amounts of four different amino acids present in the samples. Computer output for the multiple regression model includes the following table:

```
Term         Coef      SE Coef     T-Value     P-Value
Constant     411       311         1.32        0.203
cysteic      291.6     66.5        4.38        0.000
cystine      12.9      44.1        0.29        0.773
methionine   -1203     445         -2.70       0.015
tyrosine     -74.3     70.5        -1.05       0.306
```

For each of the explanatory variables, is there convincing evidence at the 5% significance level that the variable provides additional useful information, after accounting for the other variables in the model? Explain your answer.

SOLUTION

Both cystine and tyrosine have P-values that are greater than $\alpha = 0.05$. We do not have convincing evidence that either of these variables provides additional useful information about carpet age in a model with the other three variables included. However, the P-values for cysteic and methionine are both less than $\alpha = 0.05$. We have convincing evidence that each of these variables provides additional useful information about carpet age in a model with the other three variables included.

EXAM PREP: FOR PRACTICE, TRY EXERCISE 15.

The results in Example 12.3.3 suggest that either cystine, tyrosine, or both could be eliminated from the model. How do we decide which of those three options is best? Is there a fourth option that is better still?

Choosing Variables for a Final Model

We now have the tools to revise our model and balance the two goals mentioned earlier: usefulness and simplicity. In Example 12.3.3, we discovered that both cysteic acid and methionine are likely to provide additional useful information about carpet age, but perhaps we don't need one or both of cystine and tyrosine. How do we find which model is best? Deciding which variables to include in a model is as much an art as it is a science. It also involves a lot of trial and error!

Let's start by fitting the model using just cysteic acid and methionine (the two amino acids with P-values < 0.05 in the original model). Computer output gives the following results:

```
Regression Equation
age = 216 + 336.5 cysteic - 1195 methionine


Model Summary
    S       R-sq    R-sq(adj)
 45.5571   99.52%    99.48%


Coefficients
Term         Coef    SE Coef   T-Value   P-Value
Constant     216      167       1.30     0.210
cysteic      336.5    39.9      8.43     0.000
methionine  -1195     358      -3.33     0.003
```

How well does this model fit? Based on the individual P-values, there is still convincing evidence that both variables provide additional useful information about carpet age. We also consider the value of the standard deviation of the residuals s and the value of R^2 for the model. Here we see that the actual ages of the pieces of carpet typically vary by about $s = 45.6$ years from the values predicted by the model. This is much better than the standard deviation from the mean-only model ($s_{age} = 630$ years). Also, the value of R^2 indicates that 99.5% of the variability in age is accounted for by this model. Clearly, this model does a very good job of predicting ages. But will adding another variable improve it even more?

The variable tyrosine had the next smallest P-value in the original model: 0.306. What happens if we add that variable back into the model? Here is the output:

```
Regression Equation
age = 454 + 284.7 cysteic - 1130 methionine - 77.0 tyrosine


Model Summary
    S       R-sq    R-sq(adj)
 45.2495   99.55%    99.48%
```

```
Coefficients
Term           Coef     SE Coef    T-Value    P-Value
Constant        454       268        1.69      0.107
cysteic       284.7      60.6        4.69      0.000
methionine    -1130       361       -3.13      0.005
tyrosine      -77.0      68.2       -1.13      0.273
```

Because the P-value for tyrosine $> \alpha = 0.05$, we do not have convincing evidence that tyrosine adds useful information about carpet ages beyond what we gained from including cysteic and methionine.

Before we move on, let's compare the values of s and R^2 for this model and the previous one. Notice that s has decreased from 45.56 to 45.25. This suggests that the typical differences between the actual values and the predicted values are a tiny bit smaller (better) with this model. We also notice that R^2 has increased from 0.9952 to 0.9955. In fact, R^2 will almost always increase when we include additional variables in the model. Even if a new variable has nothing to do with the response, R^2 will never decrease.

For this reason, statisticians often use R^2-adjusted to evaluate whether a variable should be included in a model. R^2-adjusted is similar to R^2, but it takes into consideration the complexity of the model. Its value will increase only if the new variable provides enough extra information to justify the increased complexity of the model. In fact, if the new variable does not help the model at all, R^2-adjusted will get smaller, suggesting that the added complexity is not worth it.

For both of these models, R^2-adjusted is 0.9948. In other words, any additional information that tyrosine adds is balanced out by the greater complexity of the model. The two variable model is less complex and provides about as much useful information.

Would a model with just one explanatory variable be nearly as good? What about the model with cysteic acid, methionine, and cystine in it? Would an interaction term help? See the Chapter 12 Project to keep exploring these data.

THINK ABOUT IT

How does R^2-adjusted account for the complexity of the model? It might help to know how R^2-adjusted is calculated:

$$R^2\text{-adjusted} = 1 - \frac{s^2}{s_y^2}$$

where s_y is the standard deviation of the response variable and s is the standard deviation of the residuals:

$$s = \sqrt{\frac{\sum (y_i - \hat{y}_i)^2}{n - (k + 1)}}$$

As variables are added to the model, predictions will get better and the numerator of s (the sum of squared residuals) will decrease. But so will the denominator. Recall that k is the number of explanatory variables, so increasing k will decrease the value of $n - (k + 1)$. If a variable doesn't provide much additional useful information, the denominator will have a larger decrease relative to the numerator, making s larger. And if s gets larger, R^2-adjusted gets smaller because s_y is a constant.

TECH CORNER   Assessing a Multiple Regression Model

Although we do not ask you to compute test statistics or P-values for multiple regression models in this section, you can use technology to do these calculations. You can also use technology to generate graphs to check the conditions. We'll illustrate this process with the carapace length and clutch size data from the Tech Corner in Section 12.1. You can also find the data in the **Tortoise** data set.

Clutch size	Carapace length (cm)	Squared carapace length (cm²)
3	28.4	806.56
2	29.0	841.00
7	29.0	841.00
11	29.8	888.04
12	29.9	894.01
10	30.2	912.04
8	30.7	942.49
9	30.9	954.81
10	31.0	961.00
13	31.1	967.21
9	31.7	1004.89
6	32.0	1024.00
13	32.3	1043.29
2	33.4	1115.56
8	33.4	1115.56

Applet

1. Go to www.stapplet.com and launch the *Multiple Regression* applet.

2. Enter the name of explanatory variable 1 (Carapace length) and the values of explanatory variable 1 in the first row of boxes. Then, enter the name of explanatory variable 2 (Squared carapace length) and the values of explanatory variable 2 in the second row of boxes. Make sure both boxes are checked to include these variables in the model. Finally, enter the name of the response variable (Clutch size) and the values of the response variable in the final row of boxes. *Note:* If you have more than two explanatory variables, click the + button in the lower right to add more variables.

3. Click the Begin analysis button. The applet will display the equation of the multiple regression model, along with R^2 and s. The applet will also display a table that includes details about each coefficient, an ANOVA table, a residual plot, and a dotplot of residuals.

Linear Regression Model

Clutch size = −1018.95531 + 66.180967 (Carapace length) + −1.063569 (Squared carapace length)
$S = 2.788$ $R^2 = 0.499$ $R^2(adj) = 0.416$

Predictor	Coef	SE Coef	T	P
Constant	−1018.955	298.811	−3.41	0.005
Carapace length	66.181	19.315	3.426	0.005
Squared carapace length	−1.064	0.312	−3.413	0.005

Analysis of Variance

Source	df	Sum of Squares	Mean Square	F-value	P-value
Regression	2	93.094	46.547	5.986	0.016
Error	12	93.306	7.776		
Total	14	186.4			

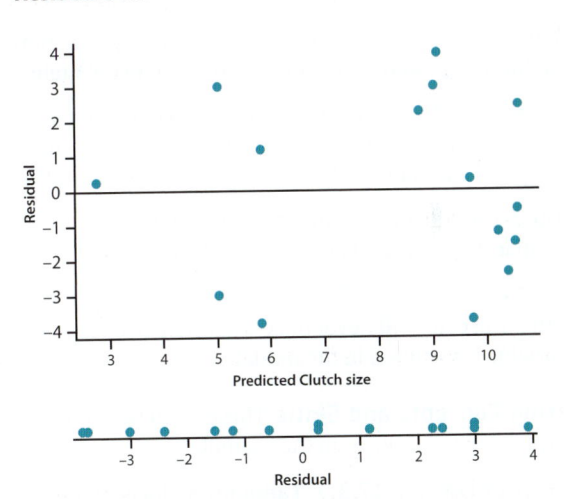

Make a Prediction

Carapace length = [ ]
Squared carapace length = [ ]
Compute predicted value

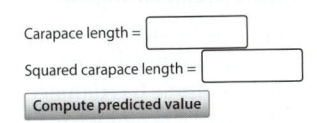

Note: To investigate what happens when you add or remove a variable, click the Edit inputs button and then adjust the checkboxes to the right of the explanatory variables.

TI-83/84

The TI-83/84 does not perform multiple regression.

Detailed instructions for using CrunchIt!, Excel, Google Sheets, JMP, Minitab, and R are available in Achieve.

Section 12.3 What Did You Learn?

Review the learning goals from this section. Then practice what you've learned by working through the exercises.

Learning Goal	Example	Exercises
State the hypotheses for a test about the usefulness of a multiple regression model.	12.3.1	7–10
Make a conclusion for a test about the usefulness of a multiple regression model based on a P-value.	12.3.2	11–14
Determine whether an explanatory variable provides additional useful information about the response variable, after accounting for other variables in the model.	12.3.3	15–18

Section 12.3 Exercises

Building Concepts and Skills These exercises assess the basic knowledge you should have after reading the section.

1. What is the alternative hypothesis for a test about a population multiple regression model with three explanatory variables?

2. You should check the Linear and Equal SD conditions using a graph of _____ versus _____ .

3. True/False: The P-value for the test about a population multiple regression model is found in an ANOVA table.

4. True/False: The t test for an individual coefficient tests the linear relationship between that variable and the response variable, independent of any other variables.

5. The P-value given in computer output for a test of an individual coefficient is typically for a(n) _____-sided alternative hypothesis.

6. True/False: R^2 will never decrease when an additional variable is included in the model.

Mastering Concepts and Skills These exercises reinforce the learning goals as shown in the examples.

7. **Refer to Example 12.3.1 Tadpoles** Biologists wondered whether tadpoles exposed to the *Batrachochytrium dendrobatidis* (Bd) fungus would be able to adjust the relative length of their intestinal tract to help with food absorption.[20] The biologists already had evidence that this happens when the tadpole has damage to its mouth that reduces food intake. The biologists randomly assigned 27 tadpoles to two groups. The first group with 14 tadpoles was exposed to Bd. The second group with 13 tadpoles was the control group. The biologists measured four variables: GutLength (the length of the intestinal tract in mm), MouthDamage (a quantitative variable measuring the level of damage to the mouth), Body (the length of the tadpole in mm), and Bd (an indicator variable denoting treatment with 1 = Bd, 0 = control). Do these data provide convincing evidence that the following population multiple regression model provides additional useful information about gut length? State the appropriate hypotheses for this test.

$GutLength = \beta_0 + \beta_1 \, MouthDamage + \beta_2 \, Body + \beta_3 \, Bd + \varepsilon$

8. **Squirrels** A long-term study of red squirrels in Canada investigated the relationship between food availability and population size.[21] Red squirrels in the Yukon have a diet made up mostly of seeds from white spruce trees. They gather these seeds in the fall, store them underground, and survive on them for the following year. Researchers believe that squirrels are more likely to breed when they know they have adequate food stored. To confirm this, they supplemented the food supply in randomly assigned plots during autumn and measured the squirrel density the following spring. The variables recorded were Cone (cone index, a measure of food availability in autumn), Density (number of squirrels per hectare the following spring), and Supp (1 = supplemented plot, 0 = not a supplemented plot) for plots observed over a number of years. Do these data provide convincing evidence that the following population multiple regression model provides additional useful information about squirrel density? State the appropriate hypotheses for this test.

$Density = \beta_0 + \beta_1 \, Cone + \beta_2 \, Supp + \varepsilon$

9. **Household Income** The American Community Survey is conducted every month by the U.S. Census Bureau. For a random sample of households that participated in the American Community Survey, we recorded three variables: household income (Income), property value (Property), and monthly electricity cost (Electric). Do these data provide convincing evidence that the following population multiple regression model provides additional useful information about household income? State the appropriate hypotheses for this test.

$Income = \beta_0 + \beta_1 \, Property + \beta_2 \, Electric + \varepsilon$

10. **Meadowfoam** Meadowfoam seed oil is used in making various skin care products. Researchers interested in maximizing the productivity of meadowfoam plants designed an experiment to investigate the effect of different light intensities on plant growth. The researchers planted 240 meadowfoam seedlings in individual pots, placed 10 pots in each of 24 trays, and put all the trays into a controlled enclosure. They randomly assigned four trays to each light intensity level (μmol/m^2/sec): 150, 300, 450, 600, 750, and 900. At each light intensity level, they randomly assigned two of the trays to receive the light treatment 24 days earlier than the remaining trays. They recorded the number of flowers produced by each plant and the average number of flowers for each tray, along with the light intensity and whether the plants got early light (1 = early, 0 = not early).[22] Do these data provide convincing evidence that the following population multiple regression model provides additional useful information about average number of flowers? State the appropriate hypotheses for this test.

$$Flowers = \beta_0 + \beta_1\, Light + \beta_2\, Early + \beta_3\, Light \times Early + \varepsilon$$

11. **Refer to Example 12.3.2 Tadpoles** Refer to Exercise 7. The conditions for inference about the multiple regression model are met. Based on the ANOVA table for this model, what conclusion would you make for a test of the hypotheses in Exercise 7 using $\alpha = 0.05$?

Source	df	SS	MS	F	P-Value
Regression	3	14,469	4823	9.38	0.000
Error	23	11,829	514.3		
Total	26	26,298			

12. **Squirrels** Refer to Exercise 8. The conditions for inference about the multiple regression model are met. Based on the ANOVA table for this model, what conclusion would you make for a test of the hypotheses in Exercise 8 using $\alpha = 0.05$?

Source	df	SS	MS	F	P-Value
Regression	2	37.30	18.65	43.74	0.000
Error	93	39.66	0.43		
Total	95	76.96			

13. **Household Income** Refer to Exercise 9. The conditions for inference about the multiple regression model are met. Based on the ANOVA table for this model, what conclusion would you make for a test of the hypotheses in Exercise 9 using $\alpha = 0.05$?

Source	df	SS	MS	F	P-Value
Regression	2	2.53099E+10	1.26549E+10	6.37	0.003
Error	58	1.15148E+11	1.98532E+9		
Total	60	1.40458E+11			

14. **Meadowfoam** Refer to Exercise 10. The conditions for inference about the multiple regression model are met. Based on the ANOVA table for this model, what conclusion would you make for a test of the hypotheses in Exercise 10 using $\alpha = 0.05$?

Source	df	SS	MS	F	P-Value
Regression	3	3467.28	1155.76	26.55	0.000
Error	20	870.66	43.53		
Total	23	4337.94			

15. **Refer to Example 12.3.3 Tadpoles** Refer to Exercises 7 and 11. Computer output includes the following table:

Term	Coef	SE Coef	T-Value	P-Value
Constant	5.2	57.2	0.09	0.929
MouthpartDamage	96.8	45.8	2.11	0.046
Body	6.44	2.75	2.35	0.028
Bd	−25.4	11.2	−2.27	0.033

For each of the explanatory variables, is there convincing evidence at the 5% significance level that the variable provides additional useful information about gut length, after accounting for the other variables in the model? Explain your answer.

16. **Squirrels** Refer to Exercises 8 and 12. Computer output includes the following table:

Term	Coef	SE Coef	T-Value	P-Value
Constant	1.381	0.115	11.98	0.000
Cone	0.1674	0.0436	3.84	0.000
Supp	1.253	0.143	8.77	0.000

For each of the explanatory variables, is there convincing evidence at the 5% significance level that the variable provides additional useful information about squirrel density, after accounting for the other variables in the model? Explain your answer.

17. **Household Income** Refer to Exercises 9 and 13. Computer output includes the following table:

Term	Coef	SE Coef	T-Value	P-Value
Constant	32741	12651	2.59	0.012
Property	0.0775	0.0241	3.21	0.002
Electric	96.6	59.2	1.63	0.108

For each of the explanatory variables, is there convincing evidence at the 5% significance level that the variable provides additional useful information about household income, after accounting for the other variables in the model? Explain your answer.

18. **Meadowfoam** Refer to Exercises 10 and 14. Computer output includes the following table:

Term	Coef	SE Coef	T-Value	P-Value
Constant	71.62	4.34	16.49	0.000
Light	−0.04108	0.00744	−5.52	0.000
Early	11.52	6.14	1.88	0.075
Light*Early	0.0012	0.0105	0.12	0.910

For each of the explanatory variables, is there convincing evidence at the 5% significance level that the variable provides additional useful information about the average number of flowers, after accounting for the other variables in the model? Explain your answer.

Applying the Concepts These exercises ask you to apply multiple learning goals in a new context or to apply what you learned in this section in a new way.

19. **Used Car Prices** Which factors affect the price of a used car? We selected a random sample of 191 used Honda CRVs and recorded several variables for each car.[23] Here is the multiple regression model relating y = price to x_1 = age (years), x_2 = annual miles (miles per year), and x_3 = leather (1 if car has leather seats, 0 if not):

$$\hat{y} = 32{,}855.70 - 1757.65x_1 - 0.2082x_2 + 2288.18x_3$$

(a) Do these data provide convincing evidence that the population multiple regression model $y = \beta_0 + \beta_1 x_1 + \beta_2 x_2 + \beta_3 x_3 + \varepsilon$ provides additional useful information about price? State the appropriate hypotheses for this test.

(b) The conditions for inference about the population multiple regression model are met. Based on the ANOVA table for this model, what conclusion would you make for a test of the hypotheses in part (a)?

Source	df	SS	MS	F	P-Value
Regression	3	6.0054E+9	2.0018E+9	764.62	0.000
Error	187	4.8957E+8	2,618,014		
Total	190	6.4949E+9			

(c) Computer output includes the following table:

Term	Coef	SE Coef	T-Value	P-Value
Constant	32856	484	67.94	0.000
Age	−1757.65	44.6	−39.38	0.000
Miles	−0.2082	0.0313	−6.646	0.000
IndLeather	2288.18	268	8.53	0.000

For each of the explanatory variables, is there convincing evidence that the variable provides additional useful information about price, after accounting for the other variables in the model? Explain your answer.

20. **House Prices** Which factors affect the price of a house? We selected a random sample of 43 recent home sales and recorded several variables for each house.[24] Here is the multiple regression model relating y = price (in $1000s) to x_1 = house size (square feet), x_2 = yard size (square feet), and x_3 = pool (1 = has pool, 0 = no pool):

$$\hat{y} = 28.097 + 0.1249x_1 + 0.0051x_2 - 9.3745x_3$$

(a) Do these data provide convincing evidence that the population multiple regression model $y = \beta_0 + \beta_1 x_1 + \beta_2 x_2 + \beta_3 x_3 + \varepsilon$ provides additional useful information about price? State the appropriate hypotheses for this test.

(b) The conditions for inference about the population multiple regression model are met. Based on the ANOVA table for this model, what conclusion would you make for a test of the hypotheses in part (a)?

Source	df	SS	MS	F	P-Value
Regression	3	218,392	72,797	62.62	0.000
Error	39	45,335	1162		
Total	42	263,727			

(c) Computer output includes the following table:

Term	Coef	SE Coef	T-Value	P-Value
Constant	28.1	16.5	1.70	0.098
Size	0.1249	0.0116	10.74	0.000
Yard	0.00514	0.00192	2.68	0.011
Pool	−9.4	13.5	−0.69	0.493

For each of the explanatory variables, is there convincing evidence that the variable provides additional useful information about price, after accounting for the other variables in the model? Explain your answer.

21. **Tech Starbucks** Is the relationship between fat (g) and calories the same for drink products and for food products at Starbucks? The variables *Calories, Fat*, and *Type* were recorded for a random sample of Starbucks products.[25] Import the **Starbucks** data set into your preferred technology and create an indicator variable *Food* (1 = food, 0 = drink).

(a) Do these data provide convincing evidence that the population multiple regression model $Calories = \beta_0 + \beta_1\ Fat + \beta_2\ Food + \beta_3\ Fat \times Food + \varepsilon$ provides additional useful information about calories? State the appropriate hypotheses for this test.

(b) Verify that the conditions for inference about the population multiple regression model are met.

(c) Based on the ANOVA table for this model, what conclusion would you make for a test of the hypotheses in part (a)?

(d) Based on your answer to part (c), does it make sense to look at the table of t tests for the individual variables in this setting? Explain your reasoning.

22. **Tech Pine Trees** The Department of Biology at Kenyon College conducted an experiment to study the growth of pine trees. Volunteers planted 1000 white pine (*Pinus strobus*) seedlings at the Brown Family Environmental Center. They randomly assigned seedlings to be planted in two grids, distinguished by spacing of either 10 or 15 feet between the seedlings. Students recorded data on a number of variables for each of the trees, including *Hgt0* (height, in cm, at the beginning of the study), *Hgt7* (height, in cm, after 7 years), and *Spacing* (in feet).[26] Import the **Pine Seedlings** data set into your preferred technology. Note that we excluded 222 of the trees from the data set because some of the values were missing.

(a) Do these data provide convincing evidence that the population multiple regression model $Hgt7 = \beta_0 + \beta_1\, Hgt0 + \beta_2\, Spacing + \varepsilon$ provides additional useful information about Hgt7? State the appropriate hypotheses for this test.

(b) Verify that the conditions for inference about the population multiple regression model are met.

(c) Based on the ANOVA table for this model, what conclusion would you make for a test of the hypotheses in part (a)?

(d) Based on your answer to part (c), does it make sense to look at the table of t tests for the individual variables in this setting? Explain your reasoning.

Extending the Concepts These exercises challenge you to explore statistical concepts and methods that go beyond what you learned in this section.

23. **Used Car Prices** In Section 11.4, you learned how to calculate a confidence interval for the slope of a population least-squares regression line. In a multiple regression model, you can calculate a confidence interval for the coefficient of an explanatory variable using the same formula:

$$b_i \pm t^* \, SE_{b_i}$$

where t^* is the critical value for $C\%$ confidence in a t distribution with $df = n - (k+1)$ and k is the number of explanatory variables.

(a) Use the information from Exercise 19 to construct and interpret 95% confidence intervals for the coefficients of age and leather.

(b) Based on the intervals from part (a), is there convincing evidence at the 5% significance level that age and leather each provide additional useful information about price? Explain your reasoning.

24. **Meadowfoam** Refer to Exercises 10, 14, and 18. Does providing early light exposure affect the average number of flowers in meadowfoam plants? We can use a two-sample t test to help answer this question using the following hypotheses:

$$H_0: \mu_E - \mu_C = 0$$
$$H_a: \mu_E - \mu_C \neq 0$$

where μ_E = the average number of flowers for meadowfoam plants like the ones in the study that receive early light and μ_C = the average number of flowers for meadowfoam plants like the ones in the study that do not receive early light.

(a) For this test, $t = 2.38$ and P-value $= 0.027$. What conclusion would you make using $\alpha = 0.01$?

(b) Here is computer output for a multiple regression model using both light and early as explanatory variables. At the 1% significance level, is there convincing evidence that the variable early provides additional useful information about average number of flowers?

Term	Coef	SE Coef	T-Value	P-Value
Constant	71.31	3.27	21.78	0.000
Light	-0.0405	0.0051	-7.89	0.000
Early	12.16	2.63	4.42	0.000

(c) Explain why the P-value for the variable early is smaller in the multiple regression model than in part (a).

Cumulative Review These exercises revisit what you learned in previous sections.

25. **Brownies (1.5, 1.6)** Researchers invited 186 university students to take part in an experiment. While the students were watching a video, they were given brownies to snack on. Unknown to the students, the size and number of brownies differed from student to student. There were three different sizes (8 g, 16 g, or 32 g) and four different quantities (1, 2, 4, or 8 brownies). Researchers measured the total weight of brownies consumed and whether each student finished the brownies.[27]

(a) If researchers wanted to use every possible combination of size and quantity, how many treatments would there be?

(b) Explain how to randomly assign the 186 university students to treatment groups of roughly equal size.

(c) Which variables would be important to control in this experiment? Explain your answer.

26. **Brownies (12.2)** Refer to Exercise 25. To analyze the data, researchers constructed a multiple regression model to predict total weight consumed based on size and quantity. Explain why it would be important to include an interaction term in the model.

Statistics Matters Which factors affect the growth of whales?

Thanks to restrictions on whaling, whale populations are making a comeback around the world. However, whales still face other challenges to their health. One way to measure the health of a whale is by its length. After all, we'd expect healthy whales to grow longer than whales that are not as healthy.

Here is a scatterplot showing the relationship between x = age (years) and y = length (m) for a sample of 145 North Atlantic right whales ages 2–20 years that were observed from 2000 to 2019.[28] Whales whose mothers had been entangled in fishing gear are shown in red. Whales that have been born after the year 2000 are indicated by squares (blue or red).

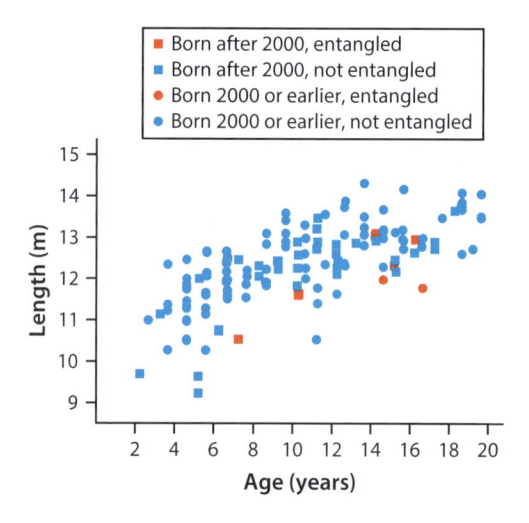

1. Based on the graph, how do the lengths of whales whose mothers had been entangled in fishing gear (indicated by red circles or squares) compare to the lengths of the other whales?

2. Based on the graph, how do the lengths of whales born since 2000 (indicated by blue or red squares) compare to the lengths of the other whales?

Here is the multiple regression model relating y = length (m) to x_1 = age (years), x_2 = entangled (1 = yes, 0 = no), and x_3 = recent (1 = born after 2000, 0 = not):

$$\hat{y} = 10.86 + 0.1567 x_1 - 0.6579 x_2 - 0.4064 x_3$$

$$s = 0.734 \qquad R^2 = 0.724$$

3. Calculate and interpret the residual for the whale with a length of 11.96 m that was 14.6 years old, was born in 1986, and whose mother had been entangled in fishing gear.

4. Interpret the values of s and R^2.

5. Interpret the coefficients of x_1 and x_2.

6. According to the model, how do the lengths of whales born since 2000 compare to the lengths of other whales? Explain your answer.

Chapter 12 Review

Multiple Regression Models

- A **multiple regression model** is a least-squares regression equation that relates a quantitative response variable to two or more explanatory variables. Multiple regression models are expressed in the form $\hat{y} = b_0 + b_1 x_1 + b_2 x_2 + \ldots$ and should be calculated with technology.
- To make predictions using a multiple regression model, substitute the value of each explanatory variable into the equation and compute the result.
- To calculate a residual, subtract the value predicted by the multiple regression model from the actual value:

  $$\text{residual} = \text{actual } y - \text{predicted } y = y - \hat{y}$$

- The standard deviation of the residuals s measures how far the actual values of y typically vary from the values predicted by the multiple regression model. If s decreases when adding an explanatory variable to the model, this variable provides additional useful information about the response variable.
- The value of R^2 measures the percentage of the variability in y that is accounted for by the multiple regression model. The value of R^2 will never decrease when adding an explanatory variable, so changes in R^2 are not the best way to determine whether the newly included variable provides additional useful information about the response variable.
- The coefficient of a quantitative explanatory variable estimates the change in the predicted value for each increase of one unit in the explanatory variable, after accounting for the other variables in the model.
- An **indicator variable** is a categorical variable with two possible outcomes: 1 (success) or 0 (failure).
  - Using an indicator variable in a multiple regression model creates parallel regression lines, one for each value of the indicator variable.
  - The coefficient of an indicator variable estimates the difference in the y-intercepts of the parallel regression lines.
- An **interaction term** in a multiple regression model is a term created by multiplying two explanatory variables and including the product as an additional explanatory variable in the model.
  - A multiple regression model with one quantitative explanatory variable, one indicator variable, and an interaction term creates two regression lines with different slopes and y-intercepts. The model has the form $\hat{y} = b_0 + b_1 x_1 + b_2 x_2 + b_3 x_1 x_2$.
  - The coefficient of the interaction term estimates the difference in the slopes of the two regression lines.

Inference for Multiple Regression

- The form of a **population multiple regression model** is

  $$y = \beta_0 + \beta_1 x_1 + \beta_2 x_2 + \ldots + \varepsilon$$

  where y is the value of the response variable, β_0 is the constant term, β_i is the coefficient of explanatory variable x_i, and ε is the error term.
- An **F test for a population multiple regression model** is a significance test of the null hypothesis that a population multiple regression model does not provide additional useful information about a response variable compared to a model that uses only the mean of the response variable. The alternative hypothesis says that the population model does provide additional useful information about the response variable. The hypotheses, in symbols, for a model with k explanatory variables are

  $$H_0: \beta_1 = \beta_2 = \ldots = \beta_k = 0$$

  $$H_a: \text{At least one of these } \beta_i\text{'s} \neq 0$$

- The F statistic and its corresponding P-value are typically found in the ANOVA table.
- Four conditions must be met to do inference about a population multiple regression model, including tests for individual coefficients: Random, Linear, Equal SD, and Normal.
- A **t test for an individual coefficient** is a significance test of the null hypothesis that an explanatory variable in a population multiple regression model does not provide additional useful information about the response variable, after accounting for the other variables in the model. That is, the null hypothesis is $H_0: \beta_i = 0$ for explanatory variable x_i. The values of the t statistics and their corresponding P-values are typically found in output from technology.
- When building a multiple regression model, we need to balance the desire to make good predictions with the desire to have a simple model. When evaluating whether a variable should be included in the model, consider the results of a test for the coefficient of the variable, as well as the standard deviation of the residuals s and the value of R^2-adjusted.

Chapter 12 Review Exercises

1. **Cheese (12.1)** Which factors affect the number of calories in cheese? We selected a sample of 18 types of sliced cheese and recorded several variables for each.[29] Here is the multiple regression model relating calories to saturated fat (g) and protein (g) in a single serving:

   $$\widehat{\text{calories}} = 7.2629 + 12.9850\, \text{satfat} + 3.8457\, \text{protein}$$

   (a) Predict the price for a sliced cheese with 1.5 g of saturated fat and 7 g of protein.

   (b) The actual number of calories for the cheese in part (a) is 50. Calculate and interpret the residual for this cheese.

   (c) For this model, $s = 4.46$ and $R^2 = 0.760$. Interpret these values.

2. **CO_2 Emissions (12.2)** Here is a scatterplot showing the relationship between $y = CO_2$ emissions (grams/mile) and $x =$ engine size (liters) for a random sample of 200 model-year 2021 vehicles. Also shown are separate regression lines for vehicles that are exclusively front-wheel drive (purple squares) and other types of vehicles, such as all-wheel drive (orange circles).

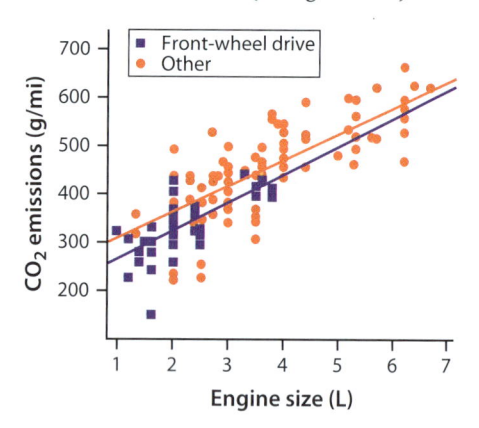

Here is the multiple regression model relating $y = CO_2$ emissions (grams/mile) to $x_1 =$ engine size (liters) and $x_2 =$ front (1 = front-wheel drive, 0 = not):

$$\hat{y} = 250.696 + 53.439 x_1 - 37.776 x_2$$

(a) Calculate and interpret the residual for the front-wheel drive Volkswagen Jetta, with an engine size of 1.4 liters and CO_2 emissions of 260 grams/mile.

(b) Interpret the coefficients of x_1 and x_2.

(c) Based on the graph, do you think adding an interaction term would be worthwhile? Explain your reasoning.

3. **CO_2 Emissions (12.3)** Here is computer output for the multiple regression model in Exercise 2. Note that the conditions for inference are met.

Source	df	SS	MS	F	P-Value
Regression	2	1204254	602127	248.5	0.000
Error	197	477382	2423.26		
Total	199	1681636			

Term	Coef	SE Coef	T-Value	P-Value
Constant	250.6959	10.5090	23.855	0.000
EngSize	53.4388	2.8371	18.835	0.000
Front	-37.7758	9.2880	-4.067	0.000

(a) Do these data provide convincing evidence that the population multiple regression model $y = \beta_0 + \beta_1 x_1 + \beta_2 x_2 + \varepsilon$ provides additional useful information about CO_2 emissions? State the appropriate hypotheses for this test. What conclusion would you make?

(b) For each of the explanatory variables, is there convincing evidence that the variable provides additional useful information about CO_2 emissions, after accounting for the other variables in the model? Explain your answer.

(c) Is adding an interaction term (EngSize*Front) helpful? Use the following computer output to explain your answer.

Source	df	SS	MS	F	P-Value
Regression	3	1204653	401551	165.003	0.000
Error	196	476984	2433.59		
Total	199	1681636			

Term	Coef	SE Coef	T-Value	P-Value
Constant	251.8487	10.9108	23.083	0.000
EngSize	53.1036	2.9617	17.930	0.000
Front	-47.6811	26.2161	-1.819	0.071
EngSize*Front	4.2743	10.5757	0.404	0.687

4. **Nesting Birds (Chapter 12)** Researchers investigated the relationship between nestling mass, measured in grams, and nest humidity index, measured as the ratio of total mass of water in the nest divided by the nest dry mass, for two different groups of great titmice parents.[30] One group was exposed to fleas during egg laying and the other was not. Exposed parents were coded as 1, and unexposed parents were coded as 0. Here is the output from a multiple regression model using humidity and exposed to predict mass:

Source	df	SS	MS	F	P-Value
Regression	2	32.008	16.004	15.51	0.000
Error	34	35.085	1.032		
Total	36	67.092			

Term	Coef	SE Coef	T-Value	P-Value
Constant	18.0848	0.6592	27.43	0.000
Humidity	-5.411	1.377	-3.93	0.000
Exposed	0.8484	0.3587	2.37	0.024

S = 1.01583   R-Sq = 47.7%   R-Sq(adj) = 44.6%

(a) Predict the nestling mass for a nest with a humidity index of 1.2 that was exposed to fleas.

(b) After accounting for humidity, how does the predicted mass change for nests that were exposed to fleas?

(c) What percentage of the variability in mass is accounted for by the multiple regression model?

(d) State the hypotheses for an F test about the population multiple regression model. What conclusion would you make?

(e) Does each of the explanatory variables provide additional useful information about the response variable, after accounting for the other variables? Explain your answer.

Chapter 12 Project: Building a Multiple Regression Model: How Old Is That Carpet?

A group of chemists was interested in determining whether they could use chemical composition to date pieces of wool carpet. They selected a random sample of 23 wool carpet pieces ranging in age from 120 to 1750 years and measured the levels (in grams per 100 grams of protein) of four different amino acids: cysteic acid, cystine, methionine, and tyrosine.

In Section 12.3, you read about three different models to predict age, each with advantages and disadvantages.
- Model 1: four explanatory variables (*cysteic, cystine, methionine,* and *tyrosine*)
- Model 2: two explanatory variables (*cysteic* and *methionine*)
- Model 3: three explanatory variables (*cysteic, methionine,* and *tyrosine*)

In this project, you'll explore several additional models and decide which model you think is best. Start by loading the **Carpet** data set into your preferred technology.

1. Build Model 4 using the following explanatory variables: *cysteic, cystine,* and *methionine*.
   (a) State the equation of the model, along with the values of s, R^2, and R^2-adjusted.
   (b) Create a residual plot and a graph of the residuals and use these graphs to verify that the conditions for inference are met.
   (c) What does the value of the F statistic and its P-value tell you about this model?
   (d) Based on the t statistics and their P-values, are there any variables that don't seem to provide additional useful information about age? Explain your answer.

2. Build Model 5 using the following explanatory variables: *cysteic, methionine,* and an interaction variable created by multiplying the values of *cysteic* and *methionine*.
   (a) State the equation of the model, along with the values of s, R^2, and R^2-adjusted.
   (b) Does Model 5 seem better than Model 4? Explain your reasoning.

3. Create at least two more models, including at least one model with a single explanatory variable. For each model, state the equation of the model, along with the values of s, R^2, and R^2-adjusted.

4. Of the seven (or more) models, which do you think is best? Explain your reasoning.

Cumulative Review Chapters 10–12

This cumulative review is designed to help you review the important concepts and skills from Chapters 10, 11, and 12, and to prepare you for an in-class exam on this material.

Section I: Multiple Choice *Select the best answer for each question.*

Exercises 1–3 refer to the following scenario. Researchers investigating two different drugs to treat migraines in children conducted an experiment. They randomly assigned 328 children ages 9–17 who suffered from migraines to either amitriptyline, topiramate, or placebo. The primary outcome was a reduction of at least 50% in the number of headache days. The table summarizes the results.[31] Is there an association between drug and outcome for children like these?

		Drug			
		Amitriptyline	Topiramate	Placebo	Total
Outcome	At least 50% reduction	69	72	40	181
	Less than 50% reduction	63	58	26	147
	Total	132	130	66	328

1. The alternative hypothesis for the test is:
   (a) H_a: The observed count in each cell is different from the expected count for that cell.
   (b) H_a: The proportion who have at least a 50% reduction is different for all three treatments.
   (c) H_a: There is no association between drug and outcome for children like these.
   (d) H_a: There is an association between drug and outcome for children like these.

2. The expected count for the "Placebo/At least 50% reduction" cell is given by which of the following expressions?
   (a) $\dfrac{(181)(40)}{66}$
   (b) $\left(\dfrac{40}{66}\right)(328)$
   (c) $\dfrac{(181)(66)}{328}$
   (d) $\dfrac{(181)(40)}{328}$

3. The value of the test statistic is $\chi^2 = 1.24$. Which of the following intervals contains the P-value?
   (a) P-value < 0.01
   (b) 0.01 < P-value < 0.05
   (c) 0.05 < P-value < 0.10
   (d) P-value > 0.10

4. The scatterplot shows the relationship between life expectancy (in years) and internet users (per 100 individuals in the population) for 35 countries in North and South America, along with the least-squares regression line. Haiti had a life expectancy of 45 years and 8.37 internet users per 100 people.[32] What effect would removing this point have on the regression line?

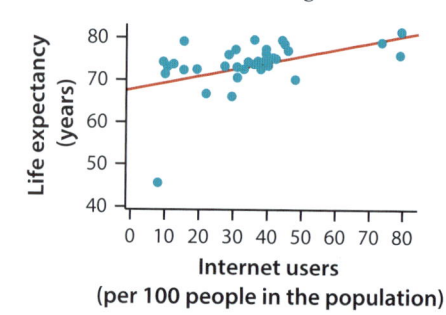

   (a) Slope would increase; y-intercept would increase.
   (b) Slope would increase; y-intercept would decrease.
   (c) Slope would decrease; y-intercept would increase.
   (d) Slope would decrease; y-intercept would decrease.

5. In a certain population there is an approximately linear relationship between y = height of a person (in centimeters) and x = their age (in years, from 5 to 18) described by the equation $\hat{y} = 50.3 + 6.1x$. Which one of the following statements must be true?
   (a) The estimated slope is 6.1, which implies that people in this population between the ages of 5 and 18 are predicted to grow about 6.1 cm in height for each year they grow older.
   (b) The estimated intercept is 50.3 cm. We can conclude from the intercept that the typical height of people in this population at birth is 50.3 cm.
   (c) A 10-year-old person in this population is at least 110.3 cm tall.
   (d) An 8-year-old person in this population who is 115 cm tall is shorter than expected.

Exercises 6–8 refer to the following scenario. A chewing-gum manufacturer wants to find out if any of its four flavors of gum is more popular than the others among regular gum chewers. A random sample of 80 regular gum chewers is asked to identify their favorite flavor of gum. Here are the results:

Flavor	Peppermint	Cinnamon	Wintergreen	Spearmint
Frequency	25	19	22	14

6. Which of the following would be an appropriate null hypothesis for the company to test?
   (a) $H_0: \mu_1 = \mu_2 = \mu_3 = \mu_4$
   (b) H_0: Flavor preference for the population is uniformly distributed across the four flavors.
   (c) H_0: At least one of the four flavor preferences in the population is different from the other three.
   (d) H_0: The observed counts are equal to the expected counts.

7. Which of the following conditions must be met to use a chi-square distribution to test the null hypothesis from Exercise 6?

   I. The sample size is at least 30.

   II. All expected cell counts are at least 5.

   III. Respondents come from a random sample of all people who regularly chew gum.

   (a) I and II only
   (b) II and III only
   (c) III only
   (d) I, II, and III

8. Which of the following is the correct chi-square test statistic for this test?

   (a) $\chi^2 = \dfrac{(25-20)^2}{20} + \dfrac{(19-20)^2}{20} + \dfrac{(22-20)^2}{20} + \dfrac{(14-20)^2}{20}$

   (b) $\chi^2 = \dfrac{(25-20)^2}{25} + \dfrac{(19-20)^2}{19} + \dfrac{(22-20)^2}{22} + \dfrac{(14-20)^2}{14}$

   (c) $\chi^2 = \dfrac{(25-20)}{20} + \dfrac{(19-20)}{20} + \dfrac{(22-20)}{20} + \dfrac{(14-20)}{20}$

   (d) $\chi^2 = \dfrac{(25-20)}{25} + \dfrac{(19-20)}{19} + \dfrac{(22-20)}{22} + \dfrac{(14-20)}{14}$

9. Which of the following is *not* one of the conditions that must be satisfied to perform inference about the slope of a least-squares regression line?

   (a) For each value of x, the distribution of y values has the same standard deviation.
   (b) The data come from a random sample or a randomized experiment.
   (c) The distribution of x and the distribution of y must both be normal.
   (d) The association between x and y is linear.

Exercises 10–13 refer to the following scenario. Scientists measured the activity level of 7 randomly selected fish at different temperatures. Activity level was measured from 0 to 100, where 0 = no activity and 100 = maximal activity. The least-squares regression line summarizing the relationship between y = activity level and x = temperature (°C) is $\hat{y} = 148.62 - 3.2167x$. The standard deviation of the residuals is 4.785 and the standard error of the slope is 0.4533. Assume the conditions for inference are met.

10. Let β_1 = the slope of the population regression line relating y = activity level to x = temperature. Which of the following are the appropriate hypotheses for a test to determine if fish like these tend to be more active in colder water?

    (a) $H_0: \beta_1 = 0,\ H_a: \beta_1 > 0$
    (b) $H_0: \beta_1 = 0,\ H_a: \beta_1 < 0$
    (c) $H_0: \beta_1 \neq 0,\ H_a: \beta_1 = 0$
    (d) $H_0: \beta_1 < 0,\ H_a: \beta_1 > 0$

11. Which of the following is the value of the standardized test statistic for this test?

    (a) $t = -7.10$
    (b) $t = -0.67$
    (c) $t = 327.86$
    (d) $t = 31.06$

12. Which of the following gives a correct interpretation of the value 4.785 in this setting?

    (a) For every 1°C increase in temperature, fish activity is predicted to increase by 4.785 units.
    (b) The typical distance of the temperature readings from their mean is about 4.785°C.
    (c) The typical distance of the activity level ratings from the least-squares line is about 4.785 units.
    (d) The typical distance of the activity-level ratings from their mean is about 4.785 units.

13. Which of the following is a 95% confidence interval for the slope of the population regression line?

    (a) -3.2167 ± 0.4533
    (b) -3.2167 ± 4.785
    (c) $-3.2167 \pm 1.96(0.4533)$
    (d) $-3.2167 \pm 2.571(0.4533)$

Exercises 14–15 refer to the following scenario. Does high-dose zinc or ascorbic acid supplementation reduce the duration of symptoms for patients with Covid-19? To investigate, researchers randomly assigned 214 patients to receive either no supplementation, a high-dose zinc supplement, an ascorbic acid supplement, or both supplements. The primary outcome was the number of days required to reach a 50% reduction in symptoms.[33]

14. Here is a partially complete ANOVA table. What is the value of the F statistic?

Source	df	SS	MS	F
Groups	3	48.93		
Error	210	3619.56		
Total	213	3668.49		

    (a) $F = 0.014$
    (b) $F = 0.946$
    (c) $F = 1.057$
    (d) $F = 73.97$

15. The P-value for a test of $H_0: \mu_1 = \mu_2 = \mu_3 = \mu_4$ is 0.42. Based on this P-value, what conclusion would you make at the $\alpha = 0.05$ level?

    (a) Because the P-value of $0.42 > \alpha = 0.05$, we fail to reject H_0. There is convincing evidence that the mean number of days is the same for all four treatments when given to patients like these.
    (b) Because the P-value of $0.42 > \alpha = 0.05$, we fail to reject H_0. There is not convincing evidence that the mean number of days is different for all four treatments when given to patients like these.

(c) Because the P-value of $0.42 > \alpha = 0.05$, we fail to reject H_0. There is not convincing evidence that the mean number of days is different for at least one of these treatments when given to patients like these.

(d) Because the P-value of $0.42 > \alpha = 0.05$, we reject H_0. There is convincing evidence that the mean number of days is different for at least one of these treatments when given to patients like these.

16. Derek Jeter of the New York Yankees was one of the best baseball players of all time. A linear regression model was calculated to predict y = the number of home runs Jeter hit in a given year from x = his age that year.[34] Based on the residual plot, which of the following is the best conclusion?

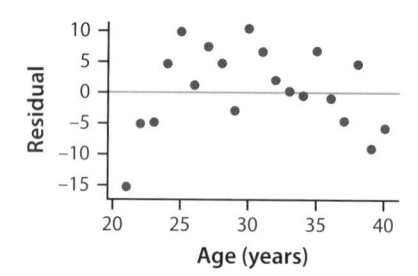

(a) Because there are roughly the same number of positive and negative residuals, the linear model is appropriate.

(b) Because some of the residuals are greater than 10 or less than −10, the linear model is not appropriate.

(c) Because there is a leftover curved pattern in the residual plot, the linear model is not appropriate.

(d) Because there is a leftover curved pattern in the residual plot, the linear model is appropriate.

17. The association between two quantitative variables, x and y, is positive, linear, and strong with $r = 0.9$. One of the x values is 3 standard deviations greater than the mean value of x. What does the least-squares regression line predict about the value of y for this value of x?

(a) The predicted value of y will be more than 3 standard deviations greater than the mean of y.

(b) The predicted value of y will be exactly 3 standard deviations greater than the mean of y.

(c) The predicted value of y will be less than 3 standard deviations greater than the mean of y.

(d) There is not enough information to determine anything about the predicted value of y.

Exercises 18–20 refer to the following scenario. Many consumers are concerned about the amount of energy used by household appliances. Researchers tested 62 different refrigerators and recorded several variables for each, including annual energy cost, cost, capacity, and type (top freezer, bottom freezer, or side-by-side). Here is a multiple regression model relating annual energy cost ($) to price ($) and top (1 if a top freezer, 0 otherwise):

$\text{energy cost} = 56.81 + 0.0074\, price + 0.6831\, top - 0.0094\, price \times top$

18. Predict the energy use for a $1000 refrigerator with a bottom freezer.

(a) $54.81
(b) $55.49
(c) $56.81
(d) $64.21

19. For this model, $R^2 = 0.355$. Interpret this value.

(a) 35.5% of the variability in energy cost is accounted for by the multiple regression model using *price*, *top*, and their interaction.

(b) 35.5% of the predictions from the multiple regression model using *price, top*, and their interaction will be correct.

(c) After accounting for *top*, the predicted energy cost increases by $0.355 for each increase of $1 in *price*.

(d) After accounting for price, the predicted energy cost is $0.355 greater for refrigerators with top freezers.

20. The given multiple regression model creates separate regression lines for refrigerators with top freezers and for refrigerators that have other types of freezers. What is the slope of the regression line for refrigerators with top freezers?

(a) −0.0094
(b) −0.0020
(c) 0.0074
(d) 0.6811

Section II: Free Response

Exercises 1–3 refer to the following scenario. Can exercise improve sleep quality? Researchers randomly assigned 437 participants to one of four treatments: no exercise ($n = 92$), light exercise ($n = 151$), moderate exercise ($n = 99$), or heavy exercise ($n = 95$). At the beginning of the experiment, each participant was given a score on the Medical Outcomes Study Sleep Scale (MOSSS). At the end of the 6-month period, researchers measured improvement in MOSSS scores and whether participants had a significant sleep disturbance in the final 4 weeks of the study.[35]

1. The two-way table summarizes the data on sleep disturbances for these participants.

		Amount of exercise				
		None	Light	Moderate	Heavy	Total
Sleep disturbance?	Yes	41	46	33	31	151
	No	51	105	66	64	286
	Total	92	151	99	95	437

Do these data provide convincing evidence of an association between amount of exercise and whether or not a person has significant sleep disturbance for people like the ones in this study?

2. The table gives the mean improvement in MOSSS score for each group, along with the standard deviation and sample size.[36]

Group	Mean	SD	n
None	2.1	9.9	92
Light	4.0	10.2	151
Moderate	4.1	10.0	99
Heavy	6.3	9.6	95

(a) State the hypotheses for a test to determine if there is convincing evidence that the mean improvement in MOSSS scores is not the same for all four amounts of exercise for people like the ones in this study. Make sure to define your parameter(s).

(b) Verify that the conditions for this test are met.

(c) For this test, the *P*-value is 0.04. Interpret the *P*-value.

(d) What conclusion would you make at the $\alpha = 0.05$ level?

3. The amount of exercise assigned to each group was based on the number of kilocalories of energy expenditure per kilogram of body weight (KKW). The light exercise group was assigned 4 KKW, the moderate exercise group was assigned 8 KKW, and the heavy exercise group was assigned 12 KKW, with the no exercise group assigned 0 KKW. Do improvements in MOSSS scores tend to get larger as the amount of exercise (KKW) increases for people like the ones in this study?

(a) State the hypotheses for a test to answer this question. Make sure to define your parameter(s).

(b) How is this test different from the test in Exercise 2?

(c) For this test, the *P*-value is 0.02. What conclusion would you make at the $\alpha = 0.05$ level?

Exercises 4–5 refer to the following scenario. In Major League Baseball, the goal of a pitcher is to prevent the other team from scoring runs. One of the primary ways to measure the effectiveness of a pitcher is with a statistic called earned run average (ERA). ERA measures how many earned runs the pitcher allows every 9 innings, on average. Lower values of ERA indicate better pitching. To investigate which factors affect ERA, we selected a random sample of 75 pitchers with at least 50 innings pitched in the 2021 season and recorded three other variables in addition to ERA: strikeout rate (SO9), the average number of strikeouts the pitchers get in 9 innings; home run rate (HR9), the average number of home runs the pitchers allow in 9 innings; and walk rate (BB9), the average number of walks the pitchers allow in 9 innings. Larger values are better for strikeout rate, but smaller values are better for home run rate and walk rate.[37]

4. How does strikeout rate (SO9) affect ERA? Here is a scatterplot, along with the regression line $\widehat{ERA} = 6.12 - 0.22(SO9)$. Also shown are a residual plot and a histogram of the residuals.

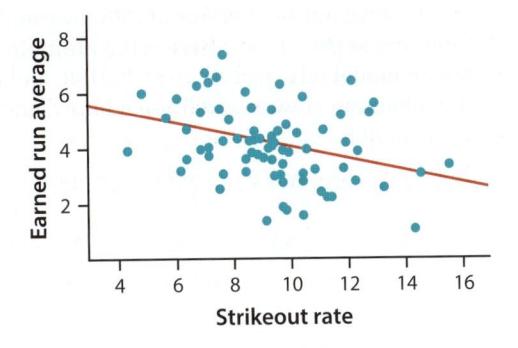

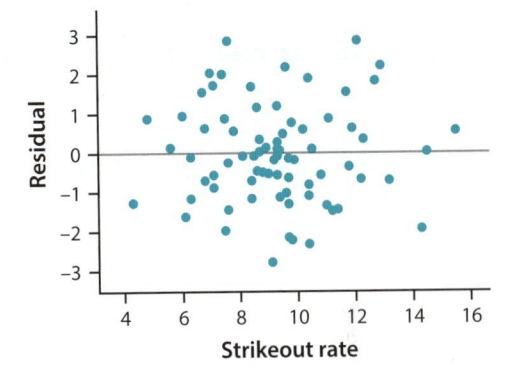

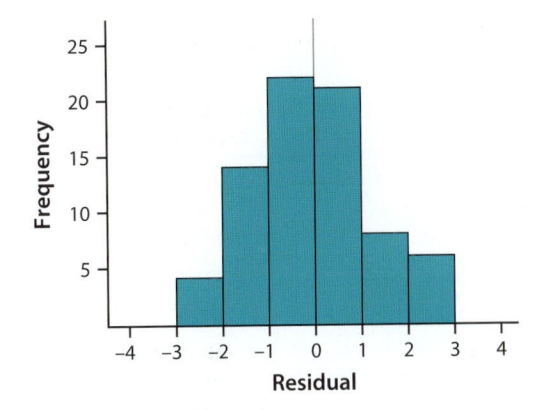

(a) Calculate and interpret the residual for Jacob deGrom, who had a strikeout rate of 14.3 and an ERA of 1.08.

(b) Interpret the slope of the regression line. If the *y*-intercept has meaning in this context, interpret it. If not, explain why not.

(c) Verify that the conditions for performing inference about the slope of the least-squares regression line are met.

(d) A 95% prediction interval for a pitcher with an SO9 of 14.3 is 0.355 to 5.684. Interpret this interval.

(e) Would a 95% confidence interval for the mean ERA among pitchers with an SO9 of 14.3 be wider or narrower than the prediction interval? Explain your answer.

5. Can we improve our predictions of ERA if we add HR9 and BB9 as predictors? Here is the multiple regression model relating ERA to SO9, HR9, and BB9, along with output showing additional information about the model.

$$\widehat{ERA} = 2.73 - 0.19\, SO9 + 1.81\, HR9 + 0.25\, BB9$$

Source	df	SS	MS	F	P-Value
Regression	3	99.867	33.2891	62.475	0.000
Error	71	37.832	0.5328		
Total	74	137.699			

Term	Coef	SE Coef	T-Value	P-Value
Constant	2.7296	0.4564	5.981	0.000
SO9	-0.1939	0.0418	-4.634	0.000
HR9	1.8110	0.1669	10.854	0.000
BB9	0.2547	0.0760	3.353	0.0013

(a) Calculate and interpret the residual for Jacob deGrom, who had a strikeout rate of 14.3, a home run rate of 0.6, a walk rate of 1.1, and an ERA of 1.08.

(b) Interpret the coefficient of SO9.

(c) Do these data provide convincing evidence that the population multiple regression model $y = \beta_0 + \beta_1\, SO9 + \beta_2\, HR9 + \beta_3\, BB9 + \varepsilon$ provides additional useful information about ERA? State the appropriate hypotheses for this test. What conclusion would you make?

(d) For each of the explanatory variables, is there convincing evidence that the variable provides additional useful information about ERA, after accounting for the other variables in the model? Explain your answer.

13 Nonparametric Methods

Section 13.1 The Sign Test
Section 13.2 The Wilcoxon Signed Rank Test
Section 13.3 The Wilcoxon Rank Sum Test
Section 13.4 The Kruskal-Wallis Test
Section 13.5 Randomization Tests
Section 13.6 Bootstrapping
Chapter 13 Review
Chapter 13 Review Exercises

Statistics Matters Can acupuncture help chronic headaches?

You are just getting ready to go out when you suddenly feel another headache coming on. How bad is it going to be? How will it impact your plans? One study of U.S. adults found that 18% of respondents suffered multiple headaches a month.[1] Researchers wondered if acupuncture might help reduce the frequency and severity of headaches, and designed a clinical trial to investigate.[2]

The researchers recruited volunteers who suffered from chronic headaches and randomly assigned them to two groups. The treatment group received up to 12 acupuncture treatments over 3 months from a licensed acupuncturist. The control group did not have acupuncture treatments. Both groups continued to be treated medically for their headaches, as they had been before the study. The researchers followed all of the participants for 1 year. The participants filled out a daily headache diary for three 4-week periods: one before the study officially began, another at 3 months, and another at 1 year.

What is the average number of years that people like those in the study have suffered from chronic headaches? Was there a bigger decrease in headache frequency for those who received acupuncture? If there was an effect, did it last beyond when the acupuncture ended at 3 months?

We'll revisit **Statistics Matters** *at the end of the chapter, so you can use what you have learned to help answer these questions.*

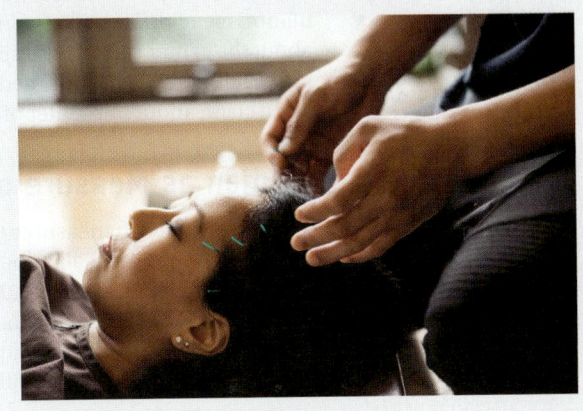

CHAPTER 13 Nonparametric Methods

In Chapters 7–12, you learned a number of statistical inference methods. While the methods were for a variety of parameters, nearly all of them have one thing in common: They require normal distributions. For inference about means with small sample sizes, it is the population from which the observations are drawn that needs to be normally distributed. For regression and ANOVA, the population distribution of the response variable needs to be normal at each value of the explanatory variable. Methods that require some underlying distribution structure, such as normality, are broadly called *parametric methods*. In this chapter, we present the basics of several *nonparametric methods* appropriate for situations in which the conditions for parametric methods aren't met. In-depth discussions of these nonparametric methods are beyond the scope of this text.

Section 13.1 The Sign Test

LEARNING GOALS

By the end of this section, you will be able to:

- State the hypotheses and calculate the test statistic for a sign test.
- Calculate the *P*-value for a sign test.
- Use the four-step process to perform a sign test.

In Sections 8.5, 8.6, and 9.6, we introduced significance tests for a population mean μ and for a population mean difference μ_{diff} in a paired data setting. These methods use inference to assess claims about the center (mean) of a single population distribution. They also require that the population of interest is normally distributed. In this section, we discuss an inference method that can be used to test a claim about the center of a non-normal population distribution.

Median as Measure of Center

The city of Chicago measures the amount of *Escherichia coli* bacteria in random samples of water from Lake Michigan at all its beaches on a daily basis every summer. Two beaches located near each other are Oak Street Beach and Ohio Street Beach.[3] Residents suspect that the bacteria levels might be higher at Ohio Street Beach. Here are the *E. coli* measurements (in colony-forming units per 100 milliliters, CFU/100 mL) in random samples of water from both locations for each day of the week starting July 28, 2019, along with the differences (Oak − Ohio).

Date	7/28/19	7/29/19	7/30/19	7/31/19	8/1/19	8/2/19	8/3/19
Oak Street	85.9	54.5	176.9	87.1	182.4	41.1	42.3
Ohio Street	125.8	61.4	343.5	103.7	184.8	72.0	82.1
Difference	−39.9	−6.9	−166.6	−16.6	−2.4	−30.9	−39.8

The data are paired because we have bacteria levels for both beaches on the same 7 days. To compare the levels, we compute the difference between the two beaches on each day.

If the levels of bacteria are generally the same, we would expect a distribution of differences centered at 0. Our first thought for inference, from Chapter 9, would be a paired *t* test for a mean difference. But these data do not meet the required conditions. There are only seven differences and there is one obvious outlier at −166.6, as the following boxplot shows.

Difference (Oak Street − Ohio Street) in *E. coli*, in cfu/100 mL

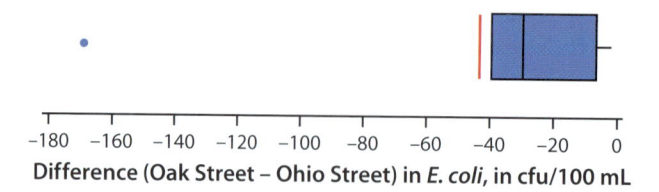

Note the red line segment in the boxplot of the differences showing the mean difference (Oak Street − Ohio Street) in *E. coli* measurements. Because the outlier is on the left side, it's no surprise that the mean is quite a bit smaller than the median. As discussed in Section 3.1, the median is a better measure of center when a distribution is skewed or includes outliers. Because nonparametric tests for the center of a distribution are used primarily when the population distributions are skewed or have outliers, they rely on the median as the measure of center.

Hypotheses and Test Statistic for a Sign Test

The most common use of the **sign test** is for a situation like the bacteria comparison on the two Chicago beaches: We have paired data and we want to evaluate whether the population distribution of differences is centered at 0. This translates to a null hypothesis that the population *median* difference is 0.

> A **sign test** is a nonparametric significance test of the null hypothesis that a population median difference is equal to a specified value, usually 0.

For the beach data, residents are concerned that there might be a higher level of *E. coli* at Ohio Street Beach. We subtracted the Ohio Street Beach value from the Oak Street Beach value, so the residents are asking for a test of the following hypotheses:

$$H_0: \text{Population median difference (Oak - Ohio)} = 0$$
$$H_a: \text{Population median difference (Oak - Ohio)} < 0$$

If the population median difference is 0, then we would expect about half of the nonzero sample differences to be positive and about half to be negative. This leads us to the **test statistic for the sign test.** Simply count the number of differences in the sample that are positive (or negative, depending on the alternative hypothesis).

> Compute the difference for each pair of observations. Classify each difference by its sign (ignore differences of 0). The **test statistic for the sign test** is:
> - X = the number of negative differences when H_a: Population median difference < 0.
> - X = the number of positive differences when H_a: Population median difference > 0.
> - X = the larger of the number of positive or negative differences when H_a: Population median difference ≠ 0.

The alternative hypothesis in the beach context is that the population median difference is less than 0, so we count the number of negative differences. Because there are zero positive differences and seven negative differences, the test statistic is $X = 7$.

EXAMPLE 13.1.1

Shrew heart rates

Stating hypotheses and calculating the test statistic

PROBLEM

Researchers wondered whether the heart rate of shrews differs during different phases of sleep. They obtained a random sample of six shrews and attached heart and brain wave monitors to each one. The researchers measured the average heart rate for each shrew during two different phases of sleep: REM (rapid eye movement) sleep, which is associated with dreaming, and DSW (deep slow wave), which is one type of non-REM sleep.[4]

(a) State the hypotheses the researchers wish to test.

(b) Calculate the test statistic.

Shrew	A	B	C	D	E	F
REM	15.7	21.5	18.3	17.1	22.5	18.9
DSW	11.7	21.1	19.7	18.2	23.2	20.7

SOLUTION

(a) The hypotheses are:

H_0: Population median difference (REM − DSW) = 0

H_a: Population median difference (REM − DSW) ≠ 0

> This is a two-sided test because the researchers wondered if there was a difference in heart rates.

(continued)

(b) Calculating REM − DSW, the differences are 4.0, 0.4, −1.4, −1.1, −0.7, and −1.8. Because two of the differences are positive and four are negative, the test statistic is $X = 4$.

> Because this is a two-sided test, the test statistic is the larger of the number of positive differences and the number of negative differences.

EXAM PREP: FOR PRACTICE, TRY EXERCISE 7.

Why not use a paired t test for the data in Example 13.1.1? The sample is small and the dotplot shows that the data are clearly skewed to the right with a high outlier.

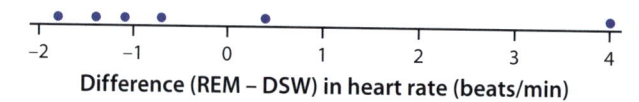

Difference (REM − DSW) in heart rate (beats/min)

P-Values

When the null hypothesis is true, the median difference in the population is 0. Therefore, we expect half of all nonzero differences in the population to be negative and half to be positive. The test statistic X is the count of the number of positive (or negative) differences in the sample. According to the null hypothesis, each nonzero difference has probability 0.5 of being positive and probability 0.5 of being negative. Because there is a fixed number of nonzero differences in the sample, if those differences are independent, X has a binomial distribution with n = number of nonzero differences and $p = 0.5$.

In the beach scenario from earlier, we are testing the following hypotheses:

H_0: Population median difference (Oak − Ohio) = 0

H_a: Population median difference (Oak − Ohio) < 0

We calculated the test statistic for the beach data to be $X = 7$. Because this is a one-sided test, we compute $P(X \geq 7)$ to get the P-value. We observed random samples of water from each of two locations on seven days and all observed differences are nonzero, so X is a binomial random variable with $n = 7$ and $p = 0.5$.

$$P\text{-value} = P(X \geq 7) = \binom{7}{7}(0.5)^7(0.5)^0 = 0.0078$$

With a P-value so small, we reject the null hypothesis. There is convincing evidence that the median difference (Oak − Ohio) in *E. coli* level is less than 0. It appears that the *E. coli* level is higher at Ohio Street Beach, as residents suspected.

Condition and P-value for a sign test

There is only one condition for a sign test:

- **Random:** Paired data come from a random sample from the population of interest or from a randomized experiment.

The P-value is calculated using a binomial distribution with n = the number of nonzero differences and $p = 0.5$.

- For a one-sided test, the P-value is $P(X \geq x)$.
- For a two-sided test, the P-value is $2P(X \geq x)$.

Here, x is the observed value of the test statistic for the sign test.

Why do we need the Random condition? For two reasons: (1) The randomness assures that the differences are independent, so the binomial distribution is appropriate; and (2) the randomness allows us to infer the results to a larger population or to make an inference about cause and effect.

EXAMPLE 13.1.2

Shrew heart rates

Calculating a P-value

PROBLEM

In Example 13.1.1, we started to perform a sign test using the following hypotheses:

H_0: Population median difference (REM − DSW) = 0

H_a: Population median difference (REM − DSW) ≠ 0

We calculated a test statistic of $X = 4$ based on the differences in average heart rate during REM sleep and DSW sleep for a random sample of six shrews.

(a) Calculate the P-value.

(b) Using a significance level of 0.05, what conclusion would you make?

SOLUTION

(a) P-value $= 2P(X \geq 4)$

$$= 2\left[\binom{6}{4}(0.5)^4(0.5)^2 + \binom{6}{5}(0.5)^5(0.5)^1 + \binom{6}{6}(0.5)^6(0.5)^0\right]$$

$$= 2(0.344) = 0.688$$

- We have 6 shrews in our sample and all differences are nonzero, so $n = 6$.
- The test is two-tailed, so the P-value is $2P(X \geq 4)$.

(b) Because the P-value of $0.688 > \alpha = 0.05$, we fail to reject H_0. There is not convincing evidence that the median difference in heart rate for shrews in REM versus DSW sleep (REM − DSW) is different from 0.

You can also use technology to calculate this probability: 2 × (1 − binomcdf(trials:6, p:0.5, x value:3)). Alternatively, you can use the *Binomial Distributions* applet.

EXAM PREP: FOR PRACTICE, TRY EXERCISE 11.

The sign test can also be used to conduct a test about the median of a single population. Simply subtract the null hypothesis value of the population median from each data value, and then proceed with the sign test as if these differences came from paired data.

> **Sign test about the median of a single population**
>
> The null hypothesis is:
>
> H_0: Population median = hypothesized value
>
> The condition is:
>
> - **Random:** The data come from a random sample from the population of interest.
>
> The differences are computed as difference = value − hypothesized median.
>
> To calculate the test statistic and P-value, use a sign test with these computed differences.

In the case of testing the median of a single population, the value in the null hypothesis is not necessarily (in fact, often not) 0. Also, the random condition is slightly different than when using the sign test with paired data, because a randomized experiment is not possible with only one treatment group.

Putting It All Together: The Sign Test

As in previous chapters, we can use the four-step process to complete a sign test.

EXAMPLE 13.1.3

Rent in Chicago

The sign test

PROBLEM

Housing costs are typically one of the largest items in a family's budget. In May 2021, the reported average price for all apartments in Chicago was $1862 per month.[5] Here are the rents ($) for a random sample of 9 apartments that did not have an application fee and were available for rent in May 2021.[6] Because the distribution of the data is strongly left skewed, we focus on the median rather than the mean. Is there convincing evidence that the distribution of rents for apartments with no application fee has a median price less than $1862?

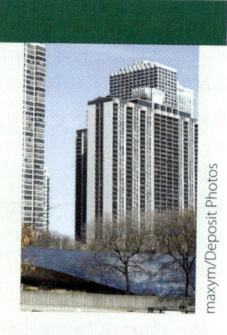

| 975 | 1895 | 1500 | 1800 | 925 | 1600 | 1900 | 1600 | 870 |

SOLUTION

State:

H_0: Population median = 1862

H_a: Population median < 1862

where Population median is the median apartment rent of all apartments in Chicago for which there was no application fee in May 2021. Use $\alpha = 0.05$.

The evidence for H_a is: More than half the values (7 of 9) are less than 1862.

Plan: Sign test

Random? Random sample of apartments. ✓

Do:
- The 9 differences are: −887, 33, −362, −62, −937, −262, 38, −262, and −992
  $X = 7$
- P-value: $n = 9$, $p = 0.5$; Using technology: $P(X \geq 7) = 0.090$

Conclude: Because the P-value of $0.090 > \alpha = 0.05$, we fail to reject H_0. We do not have convincing evidence that the median rent of apartments in Chicago that do not have an application fee was less than $1862 in May 2021.

Follow the four-step process.

State: State the hypotheses, parameter(s), significance level, and evidence for H_a.

Plan: Identify the appropriate inference method and check the conditions. For the sign test, the only condition is randomness.

Do: If the conditions are met, perform calculations:
- *Calculate the test statistic. Compute the differences by subtracting the hypothesized median from all observed values. This is a lower-tailed test, so count the number of negative differences.*
- *Find the P-value.*

Conclude: Make a conclusion about the hypotheses in the context of the problem.

EXAM PREP: FOR PRACTICE, TRY EXERCISE 15.

Section 13.1 What Did You Learn?

Review the learning goals from this section. Then practice what you've learned by working through the exercises.

Learning Goal	Example	Exercises
State the hypotheses and calculate the test statistic for a sign test.	13.1.1	7–10
Calculate the P-value for a sign test.	13.1.2	11–14
Use the four-step process to perform a sign test.	13.1.3	15–18

Section 13.1 Exercises

Building Concepts and Skills These exercises assess the basic knowledge you should have after reading the section.

1. True/False: If the alternative hypothesis in a sign test is that the population median difference is less than 0, the test statistic is the number of positive differences.

2. True/False: To perform a sign test for a population median difference, it must be reasonable to assume that the differences come from a normal population.

3. In performing the sign test, differences that are equal to 0 are _____.

4. What constitutes evidence for H_a: Population median difference > 0 when conducting a sign test?

5. When conducting a sign test for the median of a single population, you would calculate each difference as value − _____.

6. When calculating the P-value for a sign test, you would use the _____ distribution with $n =$ _____ and $p =$ _____.

Mastering Concepts and Skills These exercises reinforce the learning goals as shown in the examples.

7. **Refer to Example 13.1.1 Flight Times** Emirates Airline offers one outbound flight from Dubai, United Arab Emirates, to Doha, Qatar, and one return flight from Doha to Dubai each day. An experienced Emirates pilot suspects that the Dubai-to-Doha outbound flight typically takes longer. To find out, the pilot collects data about these flights on a random sample of 12 days.[7]

Day	1	2	3	4	5	6
Outbound (Dubai → Doha)	75	42	62	63	54	46
Return (Doha → Dubai)	42	37	37	44	42	40
Day	7	8	9	10	11	12
Outbound (Dubai → Doha)	52	50	42	46	43	52
Return (Doha → Dubai)	41	44	41	42	48	48

(a) State the hypotheses the pilot wishes to test.
(b) Calculate the test statistic.

8. **Darwin's Plants** Charles Darwin, author of *On the Origin of Species* (1859), designed an experiment to compare the effects of cross-fertilization and self-fertilization on the size of plants. He planted pairs of very similar seedling plants, one self-fertilized and one cross-fertilized, in each of 15 pots. He randomly assigned the plant pairs to pots. After a period of time, Darwin measured the heights (in inches) of all the plants. We wish to determine if there is convincing evidence that the median difference in heights (Cross-fertilized − Self-fertilized) for plants like the ones in Darwin's study is not 0.

Pair	Cross	Self	Pair	Cross	Self
1	23.5	17.4	9	18.3	16.5
2	12.0	20.4	10	21.6	18.0
3	21.0	20.0	11	23.3	16.3
4	22.0	20.0	12	21.0	18.0
5	19.1	18.4	13	22.1	12.8
6	21.5	18.6	14	23.0	15.5
7	22.1	18.6	15	12.0	18.0
8	20.4	15.3			

(a) State the hypotheses we wish to test.
(b) Calculate the test statistic.

9. **Strong Arms** Is the ability to build biceps strength different using more repetitions (reps) with lighter weights versus with fewer reps and heavier weights? A group of 10 right-handed athletes volunteered to take part in an experiment. A researcher randomly assigned each athlete to use one program with their left arm and the other program with their right arm. The first program consisted of 10 sets of 10 curls with the assigned arm, using weights that were 50% of the maximum weight each athlete could curl. The second program consisted of 10 sets of 5 curls with the other arm, using weights that were 75% of the maximum weight each athlete could curl. At the end of 2 weeks, each athlete did as many curls as possible with each arm (simultaneously) using weights that were 90% of the maximum weight they could curl at the beginning of the experiment. The researcher hypothesized that more reps with less weight tends to yield a different median number of curls for athletes like the ones in the study versus fewer reps with more weight.

Athlete	Number of curls (more reps/less weight)	Number of curls (fewer reps/more weight)
1	18	16
2	14	15
3	23	22
4	22	24
5	7	6
6	9	9
7	12	10
8	9	11
9	18	15
10	22	20

(a) State the hypotheses the researcher wishes to test.
(b) Calculate the test statistic.

10. **Pain Relief** A pharmaceutical company has developed a new drug to reduce arthritis pain. To test the drug's effectiveness, researchers perform an experiment. They recruit 12 volunteer subjects with a history of arthritis pain who are not currently taking any medication. Each subject takes the company's new drug for a 3-month period and the company's current arthritis drug for a different 3-month period, with a 3-month "washout" period in between when the subject takes no medication. The order in which each subject takes the two drugs is randomly assigned. The table shows the average number of hours that subjects are free from arthritis pain each day during the experiment. The company wishes to determine whether the new drug tends to give more hours of pain relief than its current drug for patients with arthritis like the ones in the study.

Subject	1	2	3	4	5	6	7	8	9	10
New drug	3.5	5.7	2.9	2.4	9.9	16.7	6.0	4.0	8.5	3.3
Current drug	2.0	3.6	2.6	2.6	7.3	14.9	6.6	2.0	20.9	3.4

(a) State the hypotheses the company wishes to test.
(b) Calculate the test statistic.

11. **Refer to Example 13.1.2 Flight Times** Refer to Exercise 7.
(a) Calculate the P-value.
(b) Using a significance level of 0.05, what conclusion would you make?

12. **Darwin's Plants** Refer to Exercise 8.
(a) Calculate the P-value.
(b) Using a significance level of 0.05, what conclusion would you make?

13. **Strong Arms** Refer to Exercise 9.
(a) Calculate the P-value.
(b) Using a significance level of 0.05, what conclusion would you make?

14. **Pain Relief** Refer to Exercise 10.
(a) Calculate the P-value.
(b) Using a significance level of 0.05, what conclusion would you make?

15. **Refer to Example 13.1.3 House Prices** According to the U.S. Census Bureau, the median sale price for a single-family home in March 2021 was $330,800.[8] Here are the selling prices (in thousands of dollars) for 15 randomly selected Huntingdon County, Pennsylvania, homes sold in March 2021.[9] Do these data provide convincing evidence at the $\alpha = 0.10$ significance level that the median single-family home price in Huntingdon County is less than the national median?

45	93	98	167	6
1650	140	298	285	15
85	65	185	90	104

16. **Tech Stocks** Many U.S. investors put their money into the stock market. Some investors think that tech stocks are a particularly good place to earn money. Given here is the daily change in price (in $) for the stock labeled with the symbol GOOG (Alphabet, Inc., the owner of Google) for a random sample of 15 days during the first four months of 2021.[10] Is there convincing evidence at the $\alpha = 0.05$ significance level that the median daily change in price is not $0?

8.04	−20.32	15.95	−3.62	−13.66
−40.50	−2.09	1.15	−20.31	13.86
6.81	−2.93	−25.31	2.25	−76.96

17. **Hearing Tests** When testing people's hearing ability, audiologists use lists of 50 words that are designed so the lists are all equally hard to hear. In one study, researchers compared two such lists when played at low volume with a noisy background. The subjects were people of normal hearing and the order in which they heard the lists (denoted L1 and L2) was randomized.[11] The data recorded are the percentage of words correctly identified. Is there convincing evidence that these two lists are not equally hard to hear?

Subject	L1	L2	Subject	L1	L2
1	28	24	13	32	36
2	24	32	14	40	32
3	32	20	15	28	38
4	30	14	16	48	14
5	34	32	17	34	26
6	30	22	18	28	14
7	36	20	19	40	38
8	32	26	20	18	20
9	48	26	21	20	14
10	32	38	22	26	18
11	32	30	23	36	22
12	38	16	24	40	34

18. **Chocolate Boost?** Chocolate contains a chemical called theobromine that is a member of the same alkaloid family as caffeine and nicotine. Researchers wondered if theobromine would have stimulating effects similar to caffeine and nicotine. They designed an experiment with four volunteers in which each volunteer would be subject to both a dose of theobromine and a dose of a placebo (with a washout period in between).[12] Two hours later, they measured how many finger taps the people could do in a specified amount of time. They randomized the order in which the volunteers received the treatment. Is there convincing evidence that people like the volunteers are able to tap more quickly after they receive theobromine than after they receive the placebo?

Subject	Placebo	Theobromine
A	11	20
B	56	71
C	15	41
D	6	32

Applying the Concepts These exercises ask you to apply multiple learning goals in a new context or to apply what you learned in this section in a new way.

19. **Dog Yawns** Many people believe that seeing other people yawn makes them more likely to yawn themselves. What about dogs? To find out, three researchers in Japan performed an experiment using 25 dogs.[13] During different time periods, each dog was with a human who yawned or a human who made mouth movements like a yawn but with no large intake of breath. The order in which each dog experienced these two conditions was determined at random. Researchers recorded the number of yawns for each dog. Do these data give convincing evidence that dogs like the ones in this study tend to yawn more when in the presence of a yawning human than a non-yawning human? A dotplot of the difference (Human yawned − Human didn't yawn) in number of yawns made by the dog is shown.

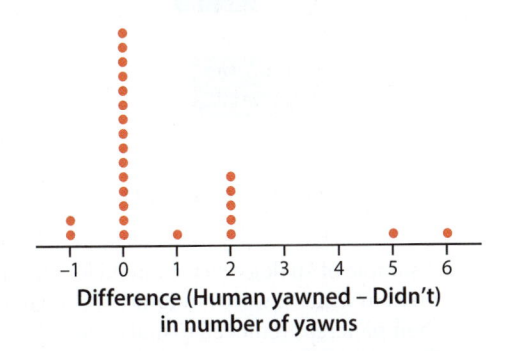

   (a) Explain why it is not safe to perform a one-sample t test based on these data.
   (b) State the hypotheses and check the condition for the sign test.
   (c) Calculate the test statistic and the P-value.
   (d) Using a significance level of 0.05, what conclusion would you make?

20. **Wound Healing** Differences in electric potential occur naturally from point to point on the body's skin. Is the natural electric field best for helping wounds to heal? If so, changing the field will slow healing. To investigate, researchers anesthetized 14 newts and made razor cuts in their hind limbs. For each newt, they let one limb heal naturally and used an electrode to change the electric field in the other limb to half its normal value. They determined which leg received each condition at random. After 2 hours, researchers measured the healing rate in micrometers per hour. Do these data give convincing evidence that cuts tend to heal more slowly (take longer) when the electric field is reduced for newts like the ones in this study? The dotplot shows the difference (Electric field − Natural) in healing rates for each of the newts.[14]

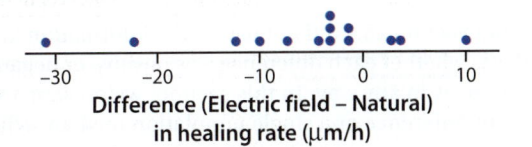

   (a) Explain why it is not safe to perform a one-sample t test based on these data.
   (b) State the hypotheses and check the condition for the sign test.
   (c) Calculate the test statistic and P-value.
   (d) Using a significance level of 0.05, what conclusion would you make?

Extending the Concepts These exercises challenge you to explore statistical concepts and methods that go beyond what you learned in this section.

21. **Spinning Euros** The sign test can also be used to conduct a significance test of $H_0: p = 0.5$ in a setting in which there are only two possible outcomes for each observation and $p =$ the population proportion for one specific category. Simply define one outcome as being a positive outcome and the other as a negative outcome. The test statistic is calculated in the same way as presented earlier in this section. Use this idea in the following scenario.

    When a fair coin is flipped, the probability of it landing heads is 0.5. But what if you spin the coin? When two Polish math professors and their students spun a Belgian euro coin 250 times, it landed heads 140 times. One of the professors concluded that the coin was minted asymmetrically. A representative from the Belgian mint indicated the result was just chance.[15] Conduct the appropriate sign test with $\alpha = 0.05$.

22. **Sign Test Limitations** The sign test is not particularly useful when the sample size is quite small. Here you will explore why that is the case.

    (a) Based on the P-value in Exercise 18, explain why a sign test with $n = 4$ can never provide convincing evidence for H_a when using a 0.05 significance level.
    (b) If $n = 5$, could you reject a one-sided alternative hypothesis? What about a two-sided alternative hypothesis?

23. **Sign Test Versus t Test** If the data meet the conditions for a t test, does it matter whether you use a t test or a sign test? This exercise will help you explore that question.

    Do polyurethane swimsuits help swimmers swim faster, on average? To investigate, Lauren, a statistics student and competitive swimmer, had 8 volunteers from her swim team participate in an experiment. On two consecutive Saturdays, she had each swimmer perform the same warm-up routine and swim a 50-meter freestyle. On one Saturday, the swimmer wore a polyurethane suit; on the other Saturday, the swimmer wore a traditional suit. The order was determined at random for each swimmer. The results for each swimmer are shown in the table.

Swimmer	Traditional suit	Polyurethane suit
Brianna	26.32	25.89
Caitlyn	28.13	28.01
Griff	27.06	27.33
Leighanne	27.97	27.22
Alex	25.24	24.44
Courtney	28.32	28.29
Matt	26.77	26.91
Kirsten	27.66	27.01

(a) State the hypotheses for the paired t test, making sure to define any parameter(s) used.

(b) State and check the conditions for the paired t test.

(c) The P-value for the paired t test is 0.042. Based on this P-value, state an appropriate conclusion using a significance level of 0.05.

(d) State the hypotheses for the sign test, making sure to define any parameter(s) used.

(e) State and check the condition for the sign test.

(f) Calculate the P-value. Using a significance level of 0.05, what conclusion would you make?

(g) Both the t test and the sign test evaluate a claim about the center of the population distribution of differences. The t test evaluates a claim about the mean, whereas the sign test evaluates a claim about the median. But when the population distribution of differences is symmetric (which is required for the t test), the mean and the median are very close to each other and we can compare the two tests. Based on the P-values in part (c) and part (f), which test is less likely to result in a Type II error? Explain your choice and why you think the chosen test has this characteristic.

Cumulative Review These exercises revisit what you learned in previous sections.

24. **House Prices (7.4)** In a poll of 961 U.S. adults taken by the Gallup organization in April 2021, 682 reported thinking that house prices would increase over the next year in their area.[16] Construct a 90% confidence interval for p = the proportion of all U.S. adults who, in April 2021, thought that house prices would increase in their area over the next year.

25. **Comparing Two Groups (9.4, 9.6)** For each of the following situations, determine whether a t test is appropriate.

    (a) A random sample of 65 healthy males and 65 healthy females had their temperature taken for a study of normal human body temperature.[17] Boxplots of the data for each group are given. We are interested in determining if healthy males and females have the same mean normal body temperature.

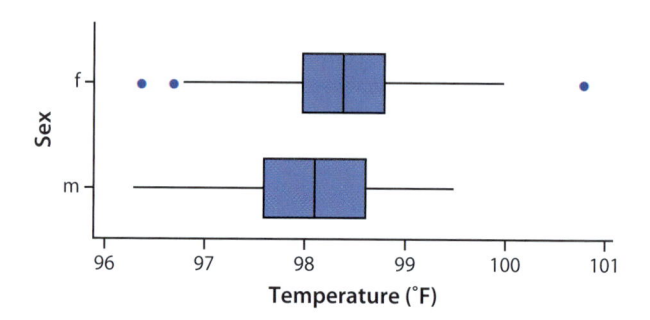

    (b) A sample of 10 dogs was obtained for an experiment in which 5 were randomly allocated to have their pancreas removed to make them diabetic. The other 5 did not receive the surgery. After the treatment (surgery or not), the rate of turnover of lactic acid was measured for each dog.[18] The research question was whether having diabetes changed the rate or not. Dotplots of both groups are given.

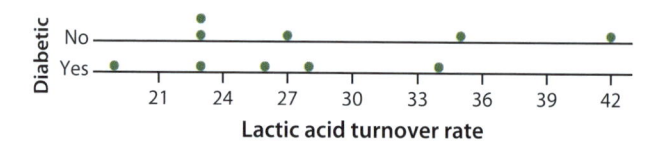

Section 13.2 The Wilcoxon Signed Rank Test

LEARNING GOALS

By the end of this section, you will be able to:
- Calculate the ranks and signed ranks in a sample of differences.
- Check the conditions, calculate the test statistic, and find the P-value for a Wilcoxon signed rank test.
- Use the four-step process to perform a Wilcoxon signed rank test.

You may have noticed that the sign test from Section 13.1 did not use all the information available to us from the observations. All we noted was whether each difference was positive or negative. We did not use the magnitudes of those differences in any way. In this section, we present a second nonparametric test for a population median difference or a single population median. While this second test also does not use the actual values of the differences, it does take their magnitude into consideration by ranking the differences.

Ranks and Signed Ranks

William S. Gosset, a brewer for the Guinness Brewery in Ireland who also discovered the t distributions, ran a famous experiment to determine if barley seeds that were dried in a kiln would result in a greater yield than non-kiln-dried barley seeds. He had 11 pairs of adjacent plots in a large field. For each pair of plots, he used a coin flip to determine which plot would get regular barley seed and which would get kiln-dried seed. The table displays the data on barley yield (in pounds per acre) for each plot and the difference in yield (Kiln − Regular) for each plot.[19]

Plot	1	2	3	4	5	6	7	8	9	10	11
Regular	1903	1935	1910	2496	2108	1961	2060	1444	1612	1316	1511
Kiln	2009	1915	2011	2463	2180	1925	2122	1482	1542	1443	1535
Difference	106	−20	101	−33	72	−36	62	38	−70	127	24

Gossett was interested in testing the following hypotheses:

H_0: True median difference (Kiln − Regular) = 0

H_a: True median difference (Kiln − Regular) > 0

Because the Random condition is met, we can apply the sign test from Section 13.1. Our test statistic is $X = 7$ because this is an upper-tail test, which means we count the number of positive differences. The resulting P-value is 0.274. Because the P-value of $0.274 > \alpha = 0.05$, we fail to reject H_0. The sign test does not provide convincing evidence that the true median difference (Kiln − Regular) in barley yield is greater than 0. In other words, this test does not provide enough evidence to convince the brewer that he should take the extra step to dry the seeds in the kiln before planting them.

But if we look at a dotplot of the differences, all four negative differences are fairly close to 0, while several of the positive differences are quite far from 0. How can we take this information into consideration?

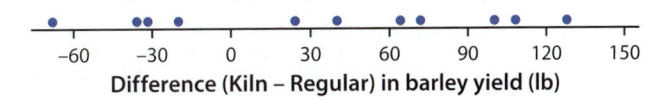

Start by considering the *absolute value* of the nonzero differences and assigning them a rank from the smallest (1) to the largest (11). Ignore any differences that are zero.

Plot	1	2	3	4	5	6	7	8	9	10	11
Difference	106	−20	101	−33	72	−36	62	38	−70	127	24
|Difference|	106	20	101	33	72	36	62	38	70	127	24
Rank	10	1	9	3	8	4	6	5	7	11	2

Next, find each plot for which the original difference was negative. For those plots, put a negative sign in front of the rank, signifying that the rank belongs to a negative difference. For example, the smallest absolute difference was 20 from plot 2. Its rank is 1. But because the original difference was negative, we change the rank to −1 and refer to it as the *signed rank*. Any ranks that come from positive differences remain positive. Here are the signed ranks for all 11 plots:

Plot	1	2	3	4	5	6	7	8	9	10	11
Difference	106	−20	101	−33	72	−36	62	38	−70	127	24
|Difference|	106	20	101	33	72	36	62	38	70	127	24
Rank	10	1	9	3	8	4	6	5	7	11	2
Signed rank	10	−1	9	−3	8	−4	6	5	−7	11	2

CHAPTER 13 Nonparametric Methods

Finding signed ranks

To find the signed ranks:

1. Compute the absolute values of all nonzero differences.
2. Rank these absolute differences. Start with 1 for the smallest absolute difference. If two or more absolute differences are equal, give each of them the same rank: the average of the ranks they would have been given individually.
3. Once all absolute differences have been ranked, put negative signs in front of the ranks that came from negative differences. Leave the ranks for those differences that were positive as positive numbers. These are the signed ranks.

EXAMPLE 13.2.1

Getting the attention of cats

Finding ranks and signed ranks

PROBLEM

Do cats recognize their names, even if it seems like they are ignoring us? Researchers in Japan investigated this question by presenting a random sample of 16 cats with five spoken words separated by 15 seconds each.[20] The first four words were nouns of the same length as the cat's name. The fifth word was the cat's name. Researchers expected that as the cat heard the first four words, it would pay increasingly less attention. However, they thought that when the cat's name was spoken, the cat's attention would increase. Attention was measured by 10 evaluators who could not hear the words spoken to the cat. The evaluators rated the magnitude of response (on a 0–3 scale) for each word spoken to each cat. Here are the mean magnitudes for the 16 cats after hearing noun 4 and after hearing their name.

(a) Find the ranks of the differences.
(b) Compute the signed ranks.

Cat	Cr	Ekcn	Icg	Knt	Kro	Krr	Kucn	Mm	Nk	Okr	Rnk	Sco	Sm	Sr	Te	Um
Noun 4	2.2	1.0	0.6	0.3	1.5	0.7	0.4	1.1	0.6	0.6	0.2	0.1	1.1	2.1	1.1	0.1
Name	1.0	0.5	2.7	0.8	1.6	1.7	2.9	1.7	2.2	0.9	2.2	2.1	2.0	0.5	0.9	0.3

SOLUTION

(a)

Cat	Noun 4	Name	Difference = Name − Noun 4	|Difference|	Rank
Cr	2.2	1.0	−1.2	1.2	10
Ekcn	1.0	0.5	−0.5	0.5	5.5
Icg	0.6	2.7	2.1	2.1	15
Knt	0.3	0.8	0.5	0.5	5.5
Kro	1.5	1.6	0.1	0.1	1
Krr	0.7	1.7	1.0	1.0	9
Kucn	0.4	2.9	2.5	2.5	16
Mm	1.1	1.7	0.6	0.6	7
Nk	0.6	2.2	1.6	1.6	11.5
Okr	0.6	0.9	0.3	0.3	4
Rnk	0.2	2.2	2.0	2.0	13.5
Sco	0.1	2.1	2.0	2.0	13.5
Sm	1.1	2.0	0.9	0.9	8
Sr	2.1	0.5	−1.6	1.6	11.5
Te	1.1	0.9	−0.2	0.2	2.5
Um	0.1	0.3	0.2	0.2	2.5

- Compute the difference for each cat.
- Compute the absolute value of each difference.
- Rank the absolute differences.
- The smallest absolute difference is 0.1, so this difference gets rank 1.
- The next two smallest absolute differences are both 0.2. Because they are the same, these differences both get the average of ranks 2 and 3. That is, both differences are assigned rank 2.5.
- Continue until all absolute differences are ranked.

(b)

Cat	Difference = Name − Noun	|Difference|	Rank	Signed Rank
Cr	−1.2	1.2	10	−10
Ekcn	−0.5	0.5	5.5	−5.5
Icg	2.1	2.1	15	15
Knt	0.5	0.5	5.5	5.5
Kro	0.1	0.1	1	1
Krr	1.0	1.0	9	9
Kucn	2.5	2.5	16	16
Mm	0.6	0.6	7	7
Nk	1.6	1.6	11.5	11.5
Okr	0.3	0.3	4	4
Rnk	2.0	2.0	13.5	13.5
Sco	2.0	2.0	13.5	13.5
Sm	0.9	0.9	8	8
Sr	−1.6	1.6	11.5	−11.5
Te	−0.2	0.2	2.5	−2.5
Um	0.2	0.2	2.5	2.5

Give each rank the sign that matches the associated difference.

EXAM PREP: FOR PRACTICE, TRY EXERCISE 5.

In Example 13.2.1, we had paired data and computed the signed ranks based on the differences within each pair. We can use the same idea when conducting a significance test about the median of a single population. In this case, the null hypothesis becomes H_0: Population median = hypothesized value, and the differences are computed by subtracting the hypothesized median value from each of the observed values. Signed ranks are computed as before based on the absolute values of the differences.

We use the signed ranks to calculate the test statistic for the **Wilcoxon signed rank test**.

A **Wilcoxon signed rank test** is a nonparametric significance test of the null hypothesis that a population median difference is 0 or that a population median is equal to a specified value.

Test Statistic and P-Value

Here again are the signed ranks for the barley experiment:

Plot	1	2	3	4	5	6	7	8	9	10	11
Signed rank	10	−1	9	−3	8	−4	6	5	−7	11	2

Notice that most of the negative ranks are small in magnitude and most of the positive ranks are large in magnitude. Start by computing the sum of the positive ranks W^+, and the sum of the *absolute values* of the negative ranks W^-. For the barley differences:

$$W^+ = 10+9+8+6+5+11+2 = 51$$
$$W^- = 1+3+4+7 = 15$$

Recall that the alternative hypothesis for the barley data was H_a: True median difference (Kiln − Regular) > 0. Large positive differences (and small negative ones) would provide some evidence for H_a. In this case, the sum of the positive ranks (51) is larger than the sum of the negative ranks (15), which gives some evidence for H_a.

The **test statistic W for the Wilcoxon signed rank test** is the *smaller* of W^+ and W^-. For the barley experiment, $W = 15$.

Let W^+ be the sum of the positive signed ranks, and let W^- be the sum of the absolute values of the negative signed ranks. The **test statistic W for the Wilcoxon signed rank test** is the smaller of W^+ and W^-.

The distribution of the test statistic W is known when the conditions are met. For paired data, the differences must come from a random sample or randomized experiment, and the population distribution of differences must be symmetric. Likewise, when testing a median of a single population, the data must come from a random sample and the population distribution must be symmetric. In Gosset's study of barley seeds, the data came from a randomized experiment and the dotplot of differences (shown again here) is reasonably symmetric.

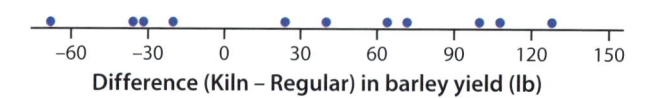

The P-value for this test cannot be calculated with a formula. Although it is best to rely on technology to calculate the P-value (see the Tech Corner at the end of this section), you can use Table E in the back of the book to get bounds for the P-value. When using this table, choose the correct portion of the table based on whether you are conducting a one-tailed test or a two-tailed test and the number of signed ranks n. Find the row with that number under the column labeled "n." Compare the test statistic to the two numbers in the table. If W is larger than both numbers, then P-value > 0.05. If W is between the two numbers, then $0.01 < P$-value < 0.05. If W is smaller than both numbers, then P-value < 0.01.

Conditions and P-value for the Wilcoxon signed rank test

There are two conditions for the Wilcoxon signed rank test:

- **Random:** When testing paired data, the differences come from a random sample from the population of interest or from a randomized experiment. When testing a population median, the data come from a random sample from the population of interest.
- **Symmetry:** When testing paired data, the population distribution of differences is symmetric. When testing a population median, the distribution of the population of interest is symmetric.

To find the P-value for the Wilcoxon signed rank test, use technology or Table E.

Statistical software gives a P-value of 0.06 for the barley experiment. Using the typical significance level of 0.05, we would not reject the null hypothesis, which is the same conclusion that we reached with the sign test. But notice that the P-value of 0.06 with the Wilcoxon signed rank test is much smaller than the P-value of 0.274 with the sign test. In general, the Wilcoxon signed rank test is less likely to result in a Type II error than the sign test because the Wilcoxon signed rank test considers the magnitudes of the differences. This extra power comes at the cost of having an extra condition—namely, that the population distribution of differences is symmetric.

EXAMPLE 13.2.2

Getting the attention of cats

Completing the Wilcoxon signed rank test

PROBLEM

Return to the data from Example 13.2.1. The hypotheses are

H_0: Median difference (name − noun) in cat response $= 0$
H_a: Median difference (name − noun) in cat response > 0

Here are the signed ranks found in that example:

Cat	Signed rank	Cat	Signed rank	Cat	Signed rank
Cr	−10	Kucn	16	Sm	8
Ekcn	−5.5	Mm	7	Sr	−11.5
Icg	15	Nk	11.5	Te	−2.5
Knt	5.5	Okr	4	Um	2.5
Kro	1	Rnk	13.5		
Krr	9	Sco	13.5		

(a) Check the conditions for the Wilcoxon signed rank test.
(b) Calculate the test statistic W.
(c) Find the P-value.
(d) What conclusion would you make?

SOLUTION

(a) Random? Researchers used a random sample of 16 cats. ✓

Symmetric? The boxplot shows that the distribution of differences is reasonably symmetric. ✓

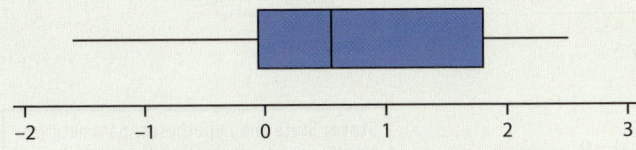

Difference (Name − Noun 4) in attention score

(b) $W^+ = 15+5.5+1+9+16+7+11.5+4+13.5+13.5+8+2.5 = 106.5$

$W^- = 10+5.5+11.5+2.5 = 29.5$

$W = 29.5$

(c) Using technology: P-value $= 0.0246$

Using Table E: $0.01 < P$-value < 0.05

(d) Because the P-value of 0.0246 is less than $\alpha = 0.05$, we reject H_0. We have convincing evidence that cats typically are more attentive to their name than to other nouns.

> W is the smaller of W^+ and W^-.

> Table: There are 16 non-zero differences, so look in the line with $n = 16$. Along that row $23 < 29.5 < 35$. The labels for those two columns are 0.05 and 0.01, so $0.01 < P$-value < 0.05.

EXAM PREP: FOR PRACTICE, TRY EXERCISE 9.

In Example 13.2.2, we report an exact P-value of 0.0246. All exact P-values in this section were found using Minitab statistical software.

Putting It All Together: The Wilcoxon Signed Rank Test

We follow the four-step process when performing a Wilcoxon signed rank test.

EXAMPLE 13.2.3

World life expectancy

The Wilcoxon signed rank test

PROBLEM

Life expectancy is defined as the average age a newborn is expected to reach if current age-based mortality rates stay the same throughout the child's lifetime. Did life expectancy in the world's countries improve between 2011 and 2021? Here is a random sample of 15 countries with their life expectancies in each of those two years. Is there convincing evidence at the $\alpha = 0.01$ significance level that the median change in life expectancy was greater than 0? To help answer this question, perform the appropriate Wilcoxon signed rank test.

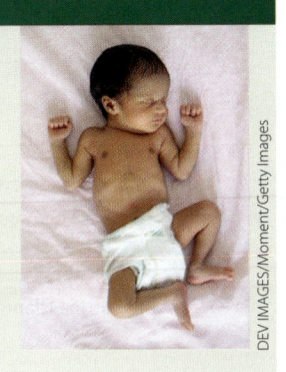

(continued)

CHAPTER 13 Nonparametric Methods

Country	Life expectancy	
	2011	2021
Bahrain	78.15	79.67
Barbados	74.34	78.31
Bosnia and Herzegovina	78.81	77.74
Botswana	58.05	65.24
Burkina Faso	53.70	63.06
Liberia	57.00	65.10
Micronesia	71.52	74.17
Paraguay	76.19	78.13
Philippines	71.66	70.32
Russia	66.29	72.16
Saint Lucia	76.84	78.71
Seychelles	73.52	75.84
Tajikistan	66.03	69.06
Togo	62.71	70.99
Wallis and Futuna Islands	78.98	80.45

SOLUTION

State:

H_0: Population median difference in life expectancy (2021 − 2011) = 0
H_a: Population median difference in life expectancy (2021 − 2011) > 0
Use $\alpha = 0.01$.

The evidence for H_a is: More than half of the countries (13 of 15) showed an increase in life expectancy from 2011 to 2021.

Plan: Wilcoxon signed rank test

Random? The sample of 15 countries was randomly selected. ✓
Symmetric? A dotplot of the differences is not strongly skewed. ✓

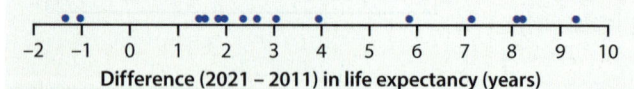

> **State:** State the hypotheses, parameter(s), significance level, and evidence for H_a.

> **Plan:** Identify the appropriate inference method and check the conditions.

Difference (2021 − 2011) in life expectancy (years)

Do:

Country	Difference	Rank	Signed Rank
Bahrain	1.52	4	4
Barbados	3.97	10	10
Bosnia and Herzegovina	−1.07	1	−1
Botswana	7.19	12	12
Burkina Faso	9.36	15	15
Liberia	8.10	13	13
Micronesia	2.65	8	8
Paraguay	1.94	6	6
Philippines	−1.34	2	−2
Russia	5.87	11	11
Saint Lucia	1.87	5	5
Seychelles	2.32	7	7
Tajikistan	3.03	9	9
Togo	8.28	14	14
Wallis and Futuna Islands	1.47	3	3

- $W^+ = 4+10+12+15+13+8+6+11+5+7+9+14+3 = 117$
  $W^- = 1+2 = 3$
  $W = 3$
- *P*-value:

Using technology: *P*-value = 0.001

Using Table E: *P*-value < 0.01

Conclude: Because the *P*-value of $0.001 < \alpha = 0.01$, we reject H_0. We have convincing evidence that the median change in life expectancy (2021 − 2011) for the world's countries was greater than 0.

Do: If the conditions are met, perform the calculations:
- Calculate the test statistic.
- Find the *P*-value.

Conclude: Make a conclusion about the hypotheses in the context of the problem.

EXAM PREP: FOR PRACTICE, TRY EXERCISE 13.

TECH CORNER   Wilcoxon signed rank test

You can use technology to perform a Wilcoxon signed rank test for a population median or a population median difference. We'll illustrate this process using the barley seed experiment data from this section.

Plot	1	2	3	4	5	6	7	8	9	10	11
Regular	1903	1935	1910	2496	2108	1961	2060	1444	1612	1316	1511
Kiln	2009	1915	2011	2463	2180	1925	2122	1482	1542	1443	1535
Difference	106	−20	101	−33	72	−36	62	38	−70	127	24

Applet

1. Go to www.stapplet.com and launch the *One Quantitative Variable, Single Group* applet.

2. Enter the variable name "Difference in yield (Kiln − Regular)" and choose "Raw data" as the "Input." Then type the differences into the "Data" box.

   One Quantitative Variable, Single Group
   Variable name: Difference in yield (Kiln - Regular)
   Input: Raw data
   Input data separated by commas or spaces.
   Data: 106 −20 101 −33 72 −36 62 38 −70 127
   Begin analysis | Edit inputs | Reset everything

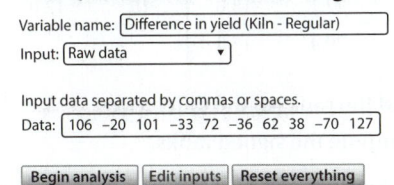

3. Click the Begin analysis button. Then scroll down to the "Perform Inference" section.

4. Choose Wilcoxon signed rank test for paired data for the procedure and choose the appropriate alternative hypothesis (">0" for this example). Click the Perform inference button.

Perform Inference

Inference procedure: Wilcoxon signed rank test for paired data
Alternative hypothesis is that the population median difference (Group 1- Group 2) is >0

Perform inference

W	P-value
15	>0.05

Notes:

1. If you want to test the median of a single population, enter the data values in Step 2, choose Wilcoxon signed rank test for population median as the procedure in Step 4, and enter a value for the hypothesized median.

2. If the sample size is greater than 30, the applet uses a normal approximation for the *P*-value.

TI-83/84

This type of significance test is not currently available on the TI-83/84.

Detailed instructions for using CrunchIt!, Excel, Google Sheets, JMP, Minitab, and R are available in Achieve.

Section 13.2 What Did You Learn?

Review the learning goals from this section. Then practice what you've learned by working through the exercises.

Learning Goal	Example	Exercises
Calculate the ranks and signed ranks in a sample of differences.	13.2.1	5–8
Check the conditions, calculate the test statistic, and find the P-value for a Wilcoxon signed rank test.	13.2.2	9–12
Use the four-step process to perform a Wilcoxon signed rank test.	13.2.3	13–16

Section 13.2 Exercises

Building Concepts and Skills These exercises assess the basic knowledge you should have after reading the section.

1. Before ranking the differences, first take the _____ _____ of them.

2. True/False: When using the Wilcoxon signed rank test for paired data, the conditions are that the differences come from a random sample or randomized experiment, and that the distribution of differences is symmetric.

3. In performing the Wilcoxon signed rank test, differences that equal 0 are _____.

4. The test statistic W is the _____ of W^+ and W^-.

Mastering Concepts and Skills These exercises reinforce the learning goals as shown in the examples.

5. **Refer to Example 13.2.1 Flight Times** Emirates Airline offers one outbound flight from Dubai, United Arab Emirates, to Doha, Qatar, and one return flight from Doha to Dubai each day. An experienced Emirates pilot suspects that the Dubai-to-Doha outbound flight typically takes longer. To find out, the pilot collects data about these flights on a random sample of 12 days.[21]

Day	1	2	3	4	5	6
Outbound (Dubai → Doha)	75	42	62	63	54	46
Return (Doha → Dubai)	42	37	37	44	42	40
Day	7	8	9	10	11	12
Outbound (Dubai → Doha)	52	50	42	46	43	52
Return (Doha → Dubai)	41	44	41	42	48	48

(a) Find the ranks of the differences.
(b) Compute the signed ranks.

6. **Darwin's Plants** Charles Darwin, author of *On the Origin of Species* (1859), designed an experiment to compare the effects of cross-fertilization and self-fertilization on the size of plants. He planted pairs of very similar seedling plants, one self-fertilized and one cross-fertilized, in each of 15 pots. He randomly assigned the pairs of plants to pots. After a period of time, Darwin measured the heights (in inches) of all the plants. We wish to determine if there is convincing evidence that the median difference in heights (Cross-fertilized − Self-fertilized) in plants like the ones in Darwin's study is not 0.

Pair	Cross	Self	Pair	Cross	Self
1	23.5	17.4	9	18.3	16.5
2	12.0	20.4	10	21.6	18.0
3	21.0	20.0	11	23.3	16.3
4	22.0	20.0	12	21.0	18.0
5	19.1	18.4	13	22.1	12.8
6	21.5	18.6	14	23.0	15.5
7	22.1	18.6	15	12.0	18.0
8	20.4	15.3			

(a) Find the ranks of the differences.
(b) Compute the signed ranks.

7. **Strong Arms** Is the ability to build biceps strength different using more repetitions (reps) with lighter weights versus with fewer reps and heavier weights? A group of 10 right-handed athletes volunteered to take part in an experiment. A researcher randomly assigned each athlete to use one program with their left arm and the other program with their right arm. The first program consisted of 10 sets of 10 curls with the assigned arm, using weights that were 50% of the maximum weight each athlete could curl. The second program consisted of 10 sets of 5 curls with the other arm, using weights that were 75% of the maximum weight each athlete could curl. At the end of 2 weeks, each athlete did as many curls as possible with each arm (simultaneously) using weights that were 90% of the maximum weight they could curl at the beginning of the experiment. The researcher hypothesized that more reps with less weight tends to yield a different median number of

curls for athletes like the ones in this study than fewer reps with more weight.

Athlete	Number of curls (more reps/less weight)	Number of curls (fewer reps/more weight)
1	18	16
2	14	15
3	23	22
4	22	24
5	7	6
6	9	9
7	12	10
8	9	11
9	18	15
10	22	20

(a) Find the ranks of the differences.
(b) Compute the signed ranks.

8. **Pain Relief** A pharmaceutical company has developed a new drug to reduce arthritis pain. To test the drug's effectiveness, researchers perform an experiment. They recruit 12 volunteer subjects with a history of arthritis pain who are not currently taking any medication. Each subject takes the company's new drug for a 3-month period and the company's current arthritis drug for a different 3-month period, with a 3-month "washout" period in between, when the subject takes no medication. The order in which each subject takes the two drugs is randomly assigned. The table shows the average number of hours that subjects are free from arthritis pain each day during the experiment. The company wishes to determine whether the new drug tends to give more hours of pain relief than its current drug for patients with arthritis like the ones in this study.

Subject	1	2	3	4	5	6	7	8	9	10
New drug	3.5	5.7	2.9	25.4	9.9	16.7	6.0	4.0	8.5	3.3
Current drug	2.0	3.6	2.6	2.6	7.3	14.9	6.6	2.0	20.9	3.4

(a) Find the ranks of the differences.
(b) Compute the signed ranks.

9. **Refer to Example 13.2.2 Flight Times** Refer to Exercise 5. The hypotheses are

H_0: Population median difference in flight time (Outbound − Return) = 0

H_a: Population median difference in flight time (Outbound − Return) > 0

(a) Check the conditions for the Wilcoxon signed rank test.
(b) Calculate the test statistic W.
(c) Find the P-value.
(d) Using a significance level of 0.05, what conclusion would you make?

10. **Darwin's Plants** Refer to Exercise 6. The hypotheses are

H_0: Population median difference in size (Cross-fertilized − Self-fertilized) = 0

H_a: Population median difference in size (Cross-fertilized − Self-fertilized) ≠ 0

(a) Check the conditions for the Wilcoxon signed rank test.
(b) Calculate the test statistic W.
(c) Find the P-value.
(d) Using a significance level of 0.05, what conclusion would you make?

11. **Strong Arms** Refer to Exercise 7. The hypotheses are

H_0: Population median difference in number of curls (More reps − Fewer reps) = 0

H_a: Population median difference in number of curls (More reps − Fewer reps) ≠ 0

(a) Check the conditions for the Wilcoxon signed rank test.
(b) Calculate the test statistic W.
(c) Find the P-value.
(d) Using a significance level of 0.05, what conclusion would you make?

12. **Pain Relief** Refer to Exercise 8. The hypotheses are

H_0: Population median difference in hours of pain relief (New drug − Old drug) = 0

H_a: Population median difference in hours of pain relief (New drug − Old drug) > 0

(a) Check the conditions for the Wilcoxon signed rank test.
(b) Calculate the test statistic W.
(c) Find the P-value.
(d) Using a significance level of 0.05, what conclusion would you make?

13. **Refer to Example 13.2.3 Diving Judges** A spectator wanted to know whether judges at a diving competition were scoring dives with similar strictness. She kept track of all scores given to each diver during one competition. Here are the scores given by two of the judges for 14 randomly selected dives during the competition. Is there convincing evidence at the $\alpha = 0.05$ level that these two judges are judging the divers differently? To help answer this question, perform the appropriate Wilcoxon signed rank test.

Dive	Judge A	Judge B	Dive	Judge A	Judge B
1	3.5	3.0	8	6.0	6.0
2	4.0	5.0	9	4.5	5.5
3	5.0	5.0	10	5.0	5.5
4	5.5	5.5	11	4.5	5.0
5	3.5	3.5	12	6.0	6.0
6	6.0	6.0	13	5.5	5.0
7	6.0	5.0	14	4.0	4.5

14. **New York City Salaries** In 2019, the median annual salary in the state of New York was $72,108.[22] A random sample of salaries for 22 people who work for the City of New York in 2020 was obtained and the data are given here.[23] Is there convincing evidence that the median salary for New York City workers in 2020 was less than the median salary of all workers in New York state in 2019? To help answer this question, perform the appropriate Wilcoxon signed rank test.

$78,989 $132,217 $61,893 $37,565 $39,868 $66,299 $33,554 $108,106
$62,074 $60,293 $99,763 $85,292 $45,142 $45,000 $44,083 $108,811
$27,230 $73,460 $86,023 $88,080 $50,518 $113,762

15. **Construction Zones** Every road has one at some point — construction zones that have much lower speed limits. To see if drivers obey these lower speed limits, a police officer uses a radar gun to measure the speed (in miles per hour, mph) of a random sample of 10 drivers in a 25 mph construction zone. Here are the data:

    27   33   32   21   30   30   29   25   27   34

Do these data provide convincing evidence at the $\alpha = 0.05$ significance level that the median speed of drivers in this construction zone is greater than the posted speed limit? To help answer this question, perform the appropriate Wilcoxon signed rank test.

16. **Hearing Tests** When testing people's hearing ability, audiologists use lists of 50 words that are designed so the lists are all equally hard to hear. In one study, researchers compared two such lists when played at low volume with a noisy background. The subjects were people of normal hearing and the order in which they heard the lists (denoted L1 and L2) was randomized.[24] The data recorded are the percentage of words correctly identified. Is there convincing evidence at the $\alpha = 0.05$ level that these two lists are not equally hard to hear? To help answer this question, perform the appropriate Wilcoxon signed rank test.

Subject	L1	L2	Subject	L1	L2
1	28	24	13	32	36
2	24	32	14	40	32
3	32	20	15	28	38
4	30	14	16	48	14
5	34	32	17	34	26
6	30	22	18	28	14
7	36	20	19	40	38
8	32	26	20	18	20
9	48	26	21	20	14
10	32	38	22	26	18
11	32	30	23	36	22
12	38	16	24	40	34

Applying the Concepts These exercises ask you to apply multiple learning goals in a new context or to apply what you learned in this section in a new way.

17. **Chicago Rent** In Example 13.1.3, we considered rent prices in Chicago. Recall that in May 2021, the reported average price for all apartments in Chicago was $1862 per month.[25] A random sample of 9 apartments that did not have an application fee and were available for rent in May 2021 was selected, and the rent amounts ($) are given here.[26] In Example 13.1.3, we conducted a sign test to evaluate whether there is convincing evidence that the distribution of rents for apartments with no application fee has a median price less than $1862. Now you will apply the Wilcoxon signed rank test to the same data.

    975   1895   1500   1800   925   1600   1900   1600   870

    (a) Compute the differences, ranks, and signed ranks.
    (b) State the hypotheses and check the conditions.
    (c) Calculate the test statistic W.
    (d) Find the P-value.
    (e) Make a conclusion for the test. How do the P-value and conclusion compare to those in Example 13.1.3?

18. **Shrews** Recall that in Examples 13.1.1 and 13.1.2, researchers wondered whether the heart rate of shrews differs for different phases of sleep. They used a random sample of six shrews and attached heart and brain wave monitors to each one. In particular, they measured the average heart rate for each shrew during two different phases of sleep: REM (rapid eye movement) sleep, which is associated with dreaming, and DSW (deep slow wave), which is one type of non-REM sleep.[27] Here are the data once again:

Shrew	A	B	C	D	E	F
REM	15.7	21.5	18.3	17.1	22.5	18.9
DSW	11.7	21.1	19.7	18.2	23.2	20.7

    (a) Compute the differences, ranks, and signed ranks.
    (b) State the hypotheses and check the conditions.
    (c) Calculate the test statistic W.
    (d) Minitab gives an exact P-value of 0.529 for this test. Make a conclusion for the test. How do the P-value and conclusion compare to those in Example 13.1.2?

19. **Tech College Weight Gain** There is a common belief that students in their first year of college typically gain weight. A professor of nutrition recruited 68 first-year students from a large university for a study on weight gain.[28] Assume that the conditions for inference are met. Use the Wilcoxon signed rank test with the data in the **WeightGain** data file to determine if there is convincing evidence at the 10% significance level that the median weight gain (*Difference*) of first-year students at this university is larger than 0.

20. **Tech Normal Body Temperature?** Most people are taught that "normal" body temperature is 98.6°F or 37°C. This value comes from data collected by Dr. Carl Wunderlich in 1851 from approximately 25,000 patients in Leipzig, Germany.[29] Researchers wondered if this is, in fact, the median body temperature of adults in modern times. They selected a random sample of 130 adults and measured their temperature in degrees Fahrenheit.[30] Use the Wilcoxon signed rank test with the data in the *NormTemp* data file to determine if there is convincing evidence at the 1% significance level that the median body temperature (*Temperature*) of adults is not 98.6°F.

Extending the Concepts These exercises challenge you to explore statistical concepts and methods that go beyond what you learned in this section.

21. **Sign Test, Wilcoxon Signed Rank Test, or *t* Test?** How do these three tests compare on the same data set? We return to the data set from Section 13.1, Exercise 22.

Swimmer	Traditional suit	Polyurethane suit
Brianna	26.32	25.89
Caitlyn	28.13	28.01
Griff	27.06	27.33
Leighanne	27.97	27.22
Alex	25.24	24.44
Courtney	28.32	28.29
Matt	26.77	26.91
Kirsten	27.66	27.01

    Do polyurethane swimsuits help swimmers swim faster, on average? To investigate, Lauren, a statistics student and competitive swimmer, had 8 volunteers from her swim team participate in an experiment. On two consecutive Saturdays, she had each swimmer perform the same warm-up routine and swim a 50-meter freestyle. On one Saturday, the swimmer wore a polyurethane suit; on the other Saturday, the swimmer wore a traditional suit. The order was determined at random for each swimmer. The results for each swimmer are shown in the table. Is there convincing evidence that swimmers swim faster with polyurethane swimsuits?

    (a) What are the conditions required for the sign test, the Wilcoxon signed rank test, and the *t* test? Put them in order from least restrictive to most restrictive.

    (b) In Section 13.1, Exercise 23, you found the *P*-value for the sign test to be 0.145 and the *P*-value for the *t* test to be 0.042. Minitab gives an exact *P*-value of 0.071 for the Wilcoxon signed rank test. Put the *P*-values in order from largest to smallest. How does this order compare to the order from part (a)? While this relationship is not always true, it is generally true.

Cumulative Review These exercises revisit what you learned in previous sections.

22. **Fruit Fly Longevity (9.4)** In Section 10.6, Exercise 24, we considered an experiment that was conducted to investigate whether increased sexual activity shortens the lifespan of male fruit flies. The researchers randomly allocated 125 fruit flies to live in a test tube in one of five conditions: alone, with 1 pregnant female, with 8 pregnant females, with 1 virgin female, or with 8 virgin females. The longevity of the male fruit fly was measured in days.[31] In this exercise we will consider only the 100 fruit flies assigned to live with either pregnant females ($n = 50$) or virgin females ($n = 50$). Is there convincing evidence at the 1% significance level that the mean longevity for male fruit flies living with virgin females is shorter than that for males living with pregnant females?

Treatment	n	$\bar{x}$	s_x
Virgin	50	47.7	16.2
Pregnant	50	64.1	15.0

23. **Increasing Tips (10.4)** Will telling a joke to customers increase the likelihood of getting a tip? Researchers in France collected data to answer this question. A waiter at a famous resort was asked to randomly assign their customers to one of three treatments when leaving the bill at the table for the guest: include a card with a joke on it, include a card with an advertisement for a local restaurant on it, or include no card at all.[32] The response variable was whether the customer left a tip or not. The two-way table shows the results. Is there convincing evidence of an association between type of card and whether or not customers like the ones in this study left a tip?

		Treatment		
		Joke card	Advertisement card	No card
Tip	Yes	30	14	16
	No	42	60	49

Section 13.3 The Wilcoxon Rank Sum Test

LEARNING GOALS

By the end of this section, you will be able to:

- Find the appropriate ranks for two samples or groups.
- Calculate the test statistic and *P*-value for a Wilcoxon rank sum test.
- Use the four-step process to perform a Wilcoxon rank sum test.

In the first two sections of this chapter, we considered nonparametric tests that can be used nearly always with paired data or when the data come from one population. In this section, we introduce a nonparametric version of the two-sample *t* test. Like most nonparametric tests, this test eliminates the need for data to come from normally distributed populations.

Ranks for Two Samples or Groups

When finding ranks for the Wilcoxon signed rank test in Section 13.2, we computed differences, ignored the signs, and ranked the unsigned differences. We then regrouped the ranks based on which type of difference, positive or negative, they came from. When we are comparing data from two independent samples or from two groups in a randomized experiment, we will follow a similar procedure.

Researchers wondered whether procyanidin B-2 (a compound found in apples) would help with hair growth in men who had male pattern baldness. They randomly assigned 19 men to receive procyanidin B-2 for 6 months and 10 men to the control group, which did not receive any treatment. At the end of the 6-month period, they measured the increase in total number of hairs per 0.25 cm^2.[33] The researchers hoped that the typical increase in total number of hairs in treated men like those in the experiment would be larger than the typical increase in total number of hairs in untreated men like those in the experiment. This seems like a clear case to consider a two-sample *t* test. However, both sample sizes are small and the dotplots show that there is a low outlier in the control group, and possibly an outlier or two in the treatment group.

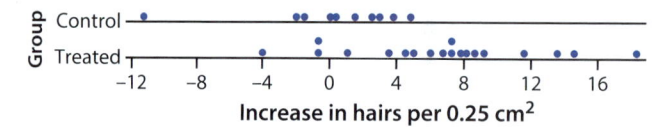

When normality is in question, as it is here, we use a nonparametric method to analyze the data. As discussed in Section 3.1, the median is a better measure of center when a distribution is skewed or has outliers. Like the sign test from Section 13.1 and the Wilcoxon signed rank test from Section 13.2, the **Wilcoxon rank sum test** also uses the median as the measure of center. In this case, the hypotheses are

H_0: True median increase in hairs when untreated = True median increase in hairs when treated

H_a: True median increase in hairs when untreated < True treated median increase in hairs when treated

A **Wilcoxon rank sum test** is a nonparametric significance test of the null hypothesis that two population (or treatment) medians are the same.

While the test statistic is different, the Wilcoxon rank sum test is equivalent to the Mann-Whitney U test. Thus, they are sometimes referred to collectively as the Wilcoxon-Mann-Whitney test.

Under the null hypothesis, the median increase in total number of hairs per 0.25 cm^2 is the same for treated and untreated men like the ones in this experiment. Because the two medians are the same when the null hypothesis is true, if we combine the two samples and rank all of the values, the ranks should be randomly distributed between the two groups. We will exploit this fact when creating the test statistic.

For the experiment comparing procyanidin B-2 to no treatment, the increase in total number of hairs per 0.25 cm^2 for each person in the two groups is as follows:

Control	0.3	1.4	3.0	3.7	−1.5	−2.0	0.0	4.8	2.4	−11.3
Treatment	3.5	5.0	7.3	18.3	14.5	6.7	9.0	−0.7	7.8	−4.0
	6.0	4.5	8.0	11.4	1.0	7.3	8.5	−0.7	13.5	

Let's combine the data from the two groups, put them in numerical order, and then rank them.

Group	Value	Rank	Group	Value	Rank
Control	−11.3	1	Control	4.8	16
Treatment	−4.0	2	Treatment	5.0	17
Control	−2.0	3	Treatment	6.0	18
Control	−1.5	4	Treatment	6.7	19
Treatment	−0.7	5.5	Treatment	7.3	20.5
Treatment	−0.7	5.5	Treatment	7.3	20.5
Control	0.0	7	Treatment	7.8	22
Control	0.3	8	Treatment	8.0	23
Treatment	1.0	9	Treatment	8.5	24
Control	1.4	10	Treatment	9.0	25
Control	2.4	11	Treatment	11.4	26
Control	3.0	12	Treatment	13.5	27
Treatment	3.5	13	Treatment	14.5	28
Control	3.7	14	Treatment	18.3	29
Treatment	4.5	15			

> **Finding ranks for the Wilcoxon rank sum test**
>
> To find the ranks for the Wilcoxon rank sum test:
>
> 1. Combine both samples (groups) and order the values from smallest to largest, keeping track of which sample (group) each value came from.
> 2. Assign ranks for all values, starting with 1 for the smallest value. If two or more values are equal, give each of them the same rank: the average of the ranks they would have been given individually.

EXAMPLE 13.3.1

Pain threshold

Finding ranks

PROBLEM

Do pain thresholds differ for people with blond hair versus people with brown hair? Researchers at the University of Melbourne chose random samples of people with naturally blond hair and people with naturally brown hair in Australia. Each person completed a pain sensitivity test and was given a pain threshold score (the higher the score, the higher the tolerance for pain). Here are the data.[34] Is there convincing evidence at the 1% significance level that the median pain threshold is different in people with blond hair and people with brown hair in Australia?

Blond	62	60	71	55	48	63	57	43	41	52
Brown	42	50	41	37	32	39	51	30	35	

(a) State the hypotheses the researchers wish to test.
(b) Find the ranks for all observed values.

(continued)

SOLUTION

(a) H_0: Population median pain threshold in people with blond hair = Population median pain threshold in people with brown hair
H_a: Population median pain threshold in people with blond hair ≠ Population median pain threshold in people with brown hair

(b)

Group	Pain threshold	Rank	Group	Pain threshold	Rank
Brown	30	1	Brown	50	11
Brown	32	2	Brown	51	12
Brown	35	3	Blond	52	13
Brown	37	4	Blond	55	14
Brown	39	5	Blond	57	15
Blond	41	6.5	Blond	60	16
Brown	41	6.5	Blond	62	17
Brown	42	8	Blond	63	18
Blond	43	9	Blond	71	19
Blond	48	10			

Combine the two groups before ranking, but keep the information about which group each value came from.

EXAM PREP: FOR PRACTICE, TRY EXERCISE 5.

Once we have the ranks for all values in a study, we can build a test statistic and find the relevant P-value.

Test Statistic and P-Value

When the null hypothesis of equal population (or treatment) medians is true, and both samples are combined and then ranked, the ranks will act as if they were randomly allocated to the two samples. Let R_1 be the sum of the ranks in one group and R_2 be the sum of the ranks in the other group. Designate n_1 as the size of the first group and n_2 as the size of the second group. Because there are $n = n_1 + n_2$ total values, we know that $R_1 + R_2$ must be equal to the sum of the first n integers: $\frac{n(n+1)}{2}$.

That means if we know R_1, we know what R_2 must be. Consequently, it does not matter which group is designated as Group 1.

The distribution of R_1 has been well studied and is the basis for the test statistic for the Wilcoxon rank sum test. When the null hypothesis is true, the data come from two independent random samples or two groups in a randomized experiment, both n_1 and n_2 are at least 5, and the two population standard deviations are equal, R_1 has an approximately normal distribution with mean $\mu_1 = \frac{n_1(n_1 + n_2 + 1)}{2}$ and standard deviation $\sigma_1 = \sqrt{\frac{n_1 n_2 (n_1 + n_2 + 1)}{12}}$. We then define the standardized test statistic to be

$$z = \frac{R_1 - \mu_1}{\sigma_1}$$

and compute the P-value using the standard normal distribution.

For the hair growth data, we pick the control group as Group 1. This means that $n_1 = 10$ and $n_2 = 19$. We compute R_1 to be

$$R_1 = 1 + 3 + 4 + 7 + 8 + 10 + 11 + 12 + 14 + 16 = 86$$

and

$$\mu_1 = \frac{10(10+19+1)}{2} = \frac{300}{2} = 150$$

$$\sigma_1 = \sqrt{\frac{10(19)(10+19+1)}{12}} = \sqrt{\frac{5700}{12}} = \sqrt{475} = 21.79$$

Putting all of that information together, we get a test statistic of

$$z = \frac{86 - 150}{21.79} = -2.937$$

Before using a normal distribution to calculate the P-value, we must check the conditions. This was a randomized experiment; the sample sizes of 10 and 19 are both at least 5; and the standard deviations of the two groups are $s_C = 4.56$ and $s_T = 5.53$. We will use the same rule of thumb that we did for ANOVA in Sections 10.5 and 10.6. The two standard deviations will be considered to be close enough if the larger standard deviation is less than twice the smaller standard deviation. Here $5.53 < 2(4.56) = 9.12$, so the standard deviations are close enough. The conditions have been met.

In this case, the alternative hypothesis is that the true median hair growth when untreated is less than the true median hair growth using the treatment for people like the ones in this study. We declared the untreated group to be Group 1. If the untreated group has less hair growth, the ranks for this group will be smaller. The smaller the sum of the ranks, the stronger the evidence against the null hypothesis, so the area to the left of $z = -2.937$ under the standard normal curve gives the P-value. Using technology, we find that the P-value is 0.0017. Because $0.0017 < 0.05$, we reject H_0. The data provide convincing evidence that procyanidin B-2 causes hair growth in men like those in the study.

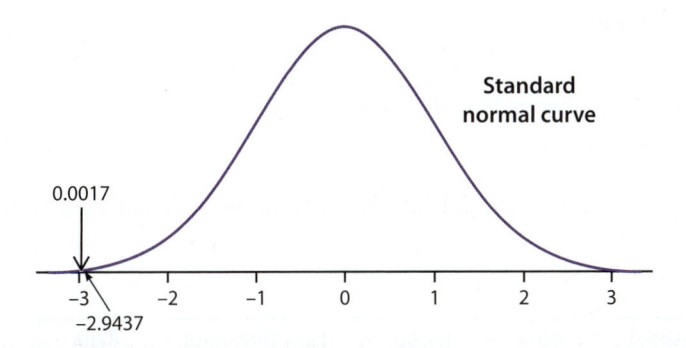

Why did we choose the control group as Group 1? Just for ease of computation. The sample size is smaller, so there are fewer ranks to add to get R_1. Some people might choose the treatment group to be Group 1 because they are thinking about the alternative hypothesis in terms of how the true treatment median compares to the true control median.

THINK ABOUT IT

Does it matter which group is Group 1? What would have happened if we had chosen the treatment group to be Group 1 instead? We would have come to the same conclusion. In this case, $n_1 = 19$, $n_2 = 10$, and $R_1 = 349$ (just add the rest of the ranks that we didn't use in the earlier calculations). This leads to $\mu_1 = 285$ and $\sigma_1 = 21.79$. Putting all that together, we get $z = 2.937$. But this time we have a test statistic based on the treatment group, so the area to the right of $z = 2.937$ under the standard normal curve gives the P-value. The P-value is once again 0.0017, which leads to the same conclusion.

Conditions, test statistic, and P-value for the Wilcoxon rank sum test

Conditions:
- **Random:** The data come from independent random samples from the two populations of interest or from two groups in a randomized experiment.
- **Large Sample:** Each sample size is at least 5.
- **Equal SD:** The larger sample standard deviation is less than twice the smaller sample standard deviation.

(continued)

Test statistic:
- Designate one of the two groups as Group 1 and the other as Group 2.
- Compute R_1 = the sum of the ranks for the values in Group 1.
- Compute $\mu_1 = \dfrac{n_1(n_1 + n_2 + 1)}{2}$ and $\sigma_1 = \sqrt{\dfrac{n_1 n_2 (n_1 + n_2 + 1)}{12}}$.
- Calculate the standardized test statistic $z = \dfrac{R_1 - \mu_1}{\sigma_1}$.

P-value: The standardized test statistic z has a distribution that is approximately standard normal, so use either technology or Table A to compute the *P*-value.

EXAMPLE 13.3.2

Pain threshold

Check conditions, calculate test statistic and *P*-value

PROBLEM

Let's return to the scenario in Example 13.3.1. Recall that we have the following hypotheses:

H_0: Population median pain threshold in people with blond hair = Population median pain threshold in people with brown hair

H_a: Population median pain threshold in people with blond hair ≠ Population median pain threshold in people with brown hair

We previously found the ranks for this data set:

Group	Pain threshold	Rank	Group	Pain threshold	Rank
Brown	30	1	Brown	50	11
Brown	32	2	Brown	51	12
Brown	35	3	Blond	52	13
Brown	37	4	Blond	55	14
Brown	39	5	Blond	57	15
Blond	41	6.5	Blond	60	16
Brown	41	6.5	Blond	62	17
Brown	42	8	Blond	63	18
Blond	43	9	Blond	71	19
Blond	48	10			

(a) Check that the conditions for performing the Wilcoxon rank sum test are met.
(b) Calculate the standardized test statistic.
(c) Find the *P*-value.
(d) What conclusion would you make at the 1% significance level?

SOLUTION

(a) Random? Independent random samples of people with blond hair and people with brown hair. ✓
Large Sample? $n_{Bl} = 10 \geq 5$, $n_{Br} = 9 \geq 5$ ✓
Equal SD? $s_{Bl} = 9.40$, $s_{Br} = 7.28$, $9.40 < 2(7.28) = 14.56$ ✓

(b) Let the people with brown hair be Group 1. Then $n_1 = 9, n_2 = 10$,
$R_1 = 1 + 2 + 3 + 4 + 5 + 6.5 + 8 + 11 + 12 = 52.5$,

$$\mu_1 = \frac{9(9+10+1)}{2} = 90, \text{ and } \sigma_1 = \sqrt{\frac{9(10)(9+10+1)}{12}} = 12.247. \text{ Therefore}$$

$$z = \frac{52.5 - 90}{12.247} = -3.062.$$

- R_1 = sum of all ranks of values that originally came from Group 1.
- $\mu_1 = \dfrac{n_1(n_1 + n_2 + 1)}{2}$
- $\sigma_1 = \sqrt{\dfrac{n_1 n_2(n_1 + n_2 + 1)}{12}}$

(c)

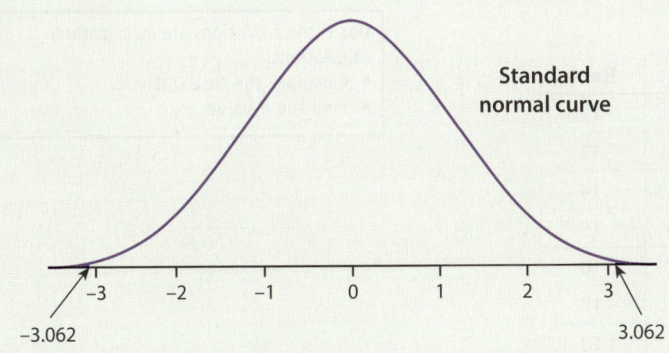

Standard normal curve, with -3.062 and 3.062 marked in the tails.

Using Table A: P-value = 2(0.0011) = 0.0022

Using technology: Applet/normalcdf(lower: –1000, upper: –3.062, mean: 0, SD: 1) × 2 = 0.0022

(d) Because the P-value of 0.0022 is less than $\alpha = 0.01$, we reject H_0. We have convincing evidence of a difference in the median pain threshold of people with blond hair and people with brown hair in Australia.

EXAM PREP: FOR PRACTICE, TRY EXERCISE 9.

Putting It All Together: The Wilcoxon Rank Sum Test

We follow the four-step process when performing a Wilcoxon rank sum test.

EXAMPLE 13.3.3

Does calcium reduce blood pressure?

The Wilcoxon rank sum test

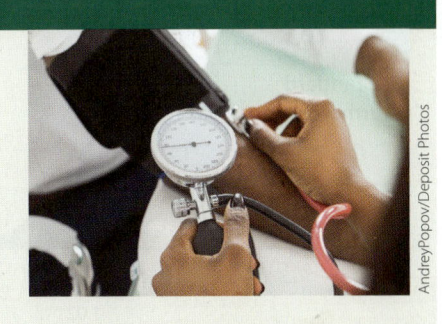

PROBLEM

Does calcium reduce blood pressure? Researchers designed a randomized comparative experiment to find out. The subjects were 21 healthy men who volunteered to take part in the experiment. They were randomly assigned to two groups: 10 of the men received a calcium supplement, while the control group of 11 men received a placebo pill that looked identical to the calcium supplement. The experiment was double-blind. The response variable is the decrease in systolic (top number) blood pressure for a subject after 12 weeks, in millimeters of mercury. An increase appears as a negative number. Here are the data:[35]

Group 1 (calcium)	7	–4	18	17	–3	–5	1	10	11	–2	
Group 2 (placebo)	–1	12	–1	–3	3	–5	5	2	–11	–1	–3

Use the Wilcoxon rank sum test to determine if there is convincing evidence that calcium supplements tend to reduce blood pressure more than a placebo, for healthy men like the ones in this study.

SOLUTION

State:

H_0: True median decrease in systolic blood pressure with calcium = True median decrease in systolic blood pressure without calcium

H_a: True median decrease in systolic blood pressure with calcium > True median decrease in systolic blood pressure without calcium

State: State the hypotheses, parameter(s), significance level, and evidence for H_a.

(continued)

Use $\alpha = 0.05$ because no significance level was given.

The evidence for H_a is: The median decrease for the calcium group = 4 > −1 = median decrease for the placebo group

Plan: Wilcoxon rank sum test

Random? The data come from two groups of men in a randomized experiment, in which one group received a placebo and the other received a calcium supplement. ✓

Large Sample? The sample sizes of 10 and 11 are both ≥ 5. ✓

Equal SD? The SDs are $s_C = 8.74$ and $s_P = 5.90$, so $8.74 < 2(5.90) = 11.80$. ✓

Do:

> **Plan:** Identify the appropriate inference method and check the conditions. *For the Wilcoxon signed rank test, check the Random, Large Sample, and Equal SD conditions.*

> **Do:** If the conditions are met, perform calculations:
> - Calculate the test statistic.
> - Find the P-value.

Group	Value	Rank	Group	Value	Rank
Placebo	−11	1	Calcium	1	12
Placebo	−5	2.5	Placebo	2	13
Calcium	−5	2.5	Placebo	3	14
Calcium	−4	4	Placebo	5	15
Calcium	−3	6	Calcium	7	16
Placebo	−3	6	Calcium	10	17
Placebo	−3	6	Calcium	11	18
Calcium	−2	8	Placebo	12	19
Placebo	−1	10	Calcium	17	20
Placebo	−1	10	Calcium	18	21
Placebo	−1	10			

Let the Calcium group be Group 1. Then $n_1 = 10$ and $n_2 = 11$.

$R_1 = 2.5 + 4 + 6 + 8 + 12 + 16 + 17 + 18 + 20 + 21 = 124.5$

$$\mu_1 = \frac{10(10+11+1)}{2} = 110 \qquad \sigma_1 = \sqrt{\frac{10(11)(10+11+1)}{12}} = 14.20$$

$$z = \frac{124.5 - 110}{14.20} = 1.021$$

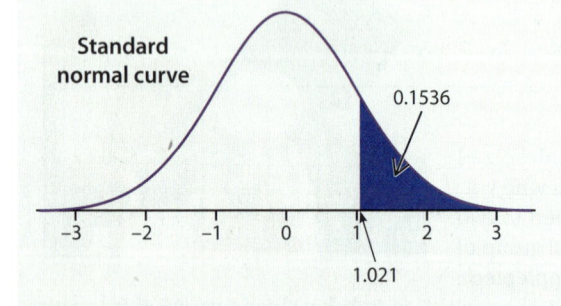

Standard normal curve — shaded area 0.1536 to the right of $z = 1.021$.

Using Table A: P-value = $1 − 0.8461 = 0.1539$

Using technology: P-value = 0.1536

Conclude: Because the P-value of $0.1536 > \alpha = 0.05$, we fail to reject H_0. We do not have convincing evidence that the median decrease in blood pressure is greater with a calcium supplement than without it, for men like the ones in this study.

> **Conclude:** Make a conclusion about the hypotheses in the context of the problem.

EXAM PREP: FOR PRACTICE, TRY EXERCISE 13.

In Section 9.4, we determined that the blood pressure data in Example 13.3.3 met the Normal/Large Sample condition and analyzed the data from this experiment using a two-sample *t* test. The P-value calculated in Example 9.4.2 was 0.0644. In other words, we failed to reject the null hypothesis in both tests. But the P-value was smaller with the two-sample *t* test. This will be true in general because we have added the condition that the underlying populations must be normally distributed. When we relax our conditions by not requiring normality, the *P*-value tends to increase. The result is that when you can use parametric tests, they are preferable.

TECH CORNER — Wilcoxon rank sum test

You can use technology to perform a Wilcoxon rank sum test to compare two population medians. We'll illustrate this process using the hair growth experiment data from this section.

Control	0.3	1.4	3.0	3.7	−1.5	−2.0	0.0	4.8	2.4	−11.3
Treatment	3.5	5.0	7.3	18.3	14.5	6.7	9.0	−0.7	7.8	−4.0
	6.0	4.5	8.0	11.4	1.0	7.3	8.5	−0.7	13.5	

Applet

1. Go to www.stapplet.com and launch the *One Quantitative Variable, Multiple Groups* applet.
2. Enter the variable name "Hair growth" and choose "Raw data" as the "Input."
3. Enter each group name and the values for that group as shown.

One Quantitative Variable, Multiple Groups
Variable name: Hair growth
Input: Raw data

Group	Name	Input data separated by commas or spaces.
1	Control	0.3 1.4 3.0 3.7 −1.5 −2.0 0.0 4.8 2.4 −11.3
2	Treatment	3.5 5.0 7.3 18.3 14.5 6.7 9.0 −0.7 7.8 −4.0 6.0 4.5 8.0

Add group

Begin analysis | Edit inputs | Reset everything

4. Click the Begin analysis button. Then scroll down to the "Perform Inference" section.

5. Choose Wilcoxon rank sum test (Normal approx) for the procedure and choose the appropriate alternative hypothesis ("<" for this example). Then, click the Perform inference button.

Perform Inference
Inference procedure: Wilcoxon rank sum test (Normal approx)
Alternative hypothesis is that population median 1 is [<] population median 2

Perform inference

z	P-value
−2.937	0.002

Note: If your data set is organized by variable (i.e., one column for treatment and one column for growth), you can use the "Raw data by variable" input option.

TI-83/84

This type of significance test is not currently available on the TI-83/84.

Detailed instructions for using CrunchIt!, Excel, Google Sheets, JMP, Minitab, and R are available in Achieve.

Section 13.3 What Did You Learn?

Review the learning goals from this section. Then practice what you've learned by working through the exercises.

Learning Goal	Example	Exercises
Find the appropriate ranks for two samples or groups.	13.3.1	5–8
Calculate the test statistic and *P*-value for a Wilcoxon rank sum test.	13.3.2	9–12
Use the four-step process to perform a Wilcoxon rank sum test.	13.3.3	13–16

Section 13.3 Exercises

Building Concepts and Skills These exercises assess the basic knowledge you should have after reading the section.

1. The null hypothesis for the Wilcoxon rank sum test is _____.

2. The Large Sample condition for the Wilcoxon rank sum test states that both sample sizes must be at least _____.

3. True/False: Once the data have been ordered from smallest to largest, if the fourth and fifth values are the same, they both get the rank of 4.

4. When computing the standardized test statistic, the formula for the mean is $\mu_1 =$ _____ and the formula for the standard deviation is $\sigma_1 =$ _____.

Mastering Concepts and Skills These exercises reinforce the learning goals as shown in the examples.

5. **Refer to Example 13.3.1 Lifeboat Training** The task of launching a lifeboat from a larger vessel is complex. Moreover, when it becomes necessary, this procedure must be completed quickly. Typically, crew learn how to launch a lifeboat using a lecture video and reading material. Researchers wondered if employing a more interactive training using a computer would lead to better knowledge transfer. They designed an experiment and randomly assigned 16 people who had already received some maritime safety training to each treatment. After all the people received the training, they took an exam to test their knowledge of how to launch a lifeboat correctly.[36] Is there convincing evidence at the 5% significance level that the median test score for people like those in this study who use a computer for training is larger than the median test score for people like those in this study who use a lecture video and readings for training?

The data for both groups are shown here:

Computer	5.45	8.92	8.35	8.32	9.78	6.27	7.13	7.09
	5.07	8.39	9.42	8.91	8.52	6.82	8.64	6.25
Lecture	4.56	6.66	5.64	6.44	4.86	0.33	5.70	6.35
	6.75	6.60	3.24	6.17	2.41	1.99	5.22	5.98

(a) State the hypotheses the researchers wish to test.
(b) Find the ranks for all observed values.

6. **Aluminum in Glass** Researchers were interested in whether types of glass can be determined by various characteristics of the glass. This research was motivated by criminal investigations — glass can be used as evidence if it can be correctly identified. The study used a total of 214 randomly selected specimens of glass of several different types, including vehicle windows that were float-processed and glass containers. Here we limit the analysis to the amount of aluminum in just these two types of glass. Is there convincing evidence at the 1% level that there is a different median amount of aluminum in all float-processed vehicle window glass than in all container glass?

The amounts of aluminum in each type of glass measured are shown here:

Container	3.50	1.86	1.56	1.56	1.65	1.83	1.76
	1.58	3.04	3.02	1.40	2.17	1.51	
Window	1.11	1.34	1.38	1.35	1.76	0.83	0.65
	1.22	1.31	1.26	1.28	0.58	1.52	1.63
	1.54	0.75	0.91				

(a) State the hypotheses the researchers wish to test.
(b) Find the ranks for all observed values.

7. **Allergies and Height** Is there a relationship between whether a child has allergies and how tall the child is? Researchers asked a random sample of U.S. school-age children whether they had allergies or not. Due to the sampling method used in this survey, it is reasonable to consider these as independent random samples of children with allergies and without allergies. Researchers also measured each child's height (cm).[37] Is there convincing evidence at the 5% level that all children with allergies have a different median height than all children without allergies?

The heights are shown here:

Allergies	165.0	183.0	170.1	183.0	170.6	162.3	166.0
	171.0	169.0	164.0	174.0	165.0	163.0	
No allergies	181.0	172.7	165.1	173.0	180.0	152.0	175.0
	182.0	182.0	160.0	176.0	175.5	173.0	160.0
	170.0	164.0	160.0				

(a) State the hypotheses the researchers wish to test.
(b) Find the ranks for all observed values.

8. **Swim Faster** Which factors can lead to faster times in the swimming pool? One group of researchers wondered if swimming with goggles would lead to faster times than swimming without goggles. They designed an experiment with one swimmer to investigate. They asked the swimmer to repeatedly swim one lap of 25 meters in a pool, 12 times with goggles and 12 times without. The order of the laps with or without goggles was randomized. Is there convincing evidence at the 5% level that the median time to swim one lap of 25 meters with goggles is less than the median time to swim one lap of 25 meters without goggles for this person?

Goggles	16.55	17.22	17.70	21.53	22.49	22.50
	16.14	16.39	16.40	19.97	19.95	20.32
No goggles	17.77	17.43	18.70	23.78	24.29	24.89
	16.85	17.80	16.81	22.63	22.81	22.31

(a) State the hypotheses the researchers wish to test.
(b) Find the ranks for all observed values.

9. **Refer to Example 13.3.2 Lifeboat Training** Refer to Exercise 5.
   (a) Check that the conditions for performing the Wilcoxon rank sum test are met.
   (b) Calculate the standardized test statistic.
   (c) Find the *P*-value.
   (d) What conclusion would you make?

10. **Aluminum in Glass** Refer to Exercise 6.
    (a) Check that the conditions for performing the Wilcoxon rank sum test are met.
    (b) Calculate the standardized test statistic.
    (c) Find the *P*-value.
    (d) What conclusion would you make?

11. **Allergies and Height** Refer to Exercise 7.
    (a) Check that the conditions for performing the Wilcoxon rank sum test are met.
    (b) Calculate the standardized test statistic.
    (c) Find the P-value.
    (d) What conclusion would you make?

12. **Swim Faster** Refer to Exercise 8.
    (a) Check that the conditions for performing the Wilcoxon rank sum test are met.
    (b) Calculate the standardized test statistic.
    (c) Find the P-value.
    (d) What conclusion would you make?

13. **Refer to Example 13.3.3 Walk the Dogs** A statistician keeps track of the number of steps they walk each day. They wondered whether the number of steps they walk is different on days when they walk the dog than on days when they don't. The statistician kept track of the number of steps they walked on a random sample of days when they walked the dog and a random sample of days when they didn't.[38] Use the Wilcoxon rank sum test to determine if there is convincing evidence at the 5% level that the median number of steps for all days on which the statistician walks the dog is different from the median number of steps for days when they don't walk the dog.

Walk the dog	5528	6937	9374	7576	3642	7090
	7218	5772	6789	7533	11,042	
Don't walk the dog	9456	4988	1889	10,204	4497	14,327
	10,286	3336	6981	6151	2721	3449

14. **Thalidomide** Patients with human immunodeficiency virus (HIV) infection can sometimes experience weight loss. Researchers wondered whether treating these patients with thalidomide might help them gain weight back again. In their study, the researchers randomly assigned 16 HIV-positive patients to receive thalidomide and 16 other HIV-positive patients to receive a placebo. The response variable for each individual was the weight gain (kg) over a 21-day period.[39] Use the Wilcoxon rank sum test to determine if there is convincing evidence at the 5% level that the median weight gain for patients like those in the study who are given thalidomide is greater than the median weight gain for patients like those in the study who receive the placebo.

Thalidomide	9.0	6.0	4.5	2.0	2.5	3.0	1.0	1.5
	1.0	−1.0	−2.0	0.0	−3.0	−3.0	0.5	−2.5
Placebo	2.5	3.5	4.0	1.0	0.5	4.0	1.5	2.0
	−0.5	0.0	2.5	0.5	−1.5	0.0	1.0	3.5

15. **Teaching Methods** Researchers in Rwanda were concerned that students in a Physics course were struggling because of the English-language technical vocabulary. The course was traditionally taught by a lecturer using a blackboard. Researchers wondered whether using a multimedia presentation on a screen in the classroom would lead to a better outcome for the students. The researchers randomly assigned 32 students to one of the two teaching methods for the 4-week course. They collected post-test scores for all students in both groups.[40] Use the Wilcoxon rank sum test to determine if there is convincing evidence at the 5% level that the true median post-test score for students like those in the study who are taught using the multimedia presentation is greater than the true median post-test score for students like those in the study who are taught using the traditional lecture method.

Multimedia	8.20	14.03	17.20	3.73	14.62	14.92	10.68
	10.79	10.72	9.44	11.06	9.02	9.89	
Lecture	8.83	11.21	8.36	7.30	6.23	6.43	15.09
	9.01	7.35	7.48	9.88	6.81	4.80	10.19
	5.93	8.62	6.16	11.59	7.39		

16. **Sodium in Glass** Refer to Exercises 6 and 10. The researchers were interested in not only the amount of aluminum in the glass, but also the amount of sodium. Once again, we limit the analysis to specimens of glass of two types: vehicle windows that were float-processed and glass containers. Use the Wilcoxon rank sum test to determine if there is convincing evidence at the 5% level that there is a different median amount of sodium in all container glass than in all window glass.

Container	14.01	12.73	11.56	11.03	12.64	12.86	13.27
	13.44	13.02	13.00	13.38	12.85	12.97	
Window	13.65	13.33	13.24	12.16	13.14	14.32	13.64
	13.42	12.86	13.04	13.41	14.03	13.53	13.50
	13.33	13.64	14.19				

Applying the Concepts These exercises ask you to apply multiple learning goals in a new context or to apply what you learned in this section in a new way.

17. **Jump for the Target** Will children jump farther if they have a target in front of them? To find out, a gym teacher conducted an experiment with 29 twelve-year-old children. The teacher wrote the names of all 29 children on identical slips of paper, put them in a hat, mixed them up, and drew out 15 slips. Each of the 15 selected students did a standing long jump from behind a starting line. The remaining 14 students did a similar standing long jump from behind the starting line, but with a target line placed 200 centimeters from the starting line. The distance from the starting line to the back of each student's closest foot was measured (in centimeters). Here are the data:[41]

No target	146	190	109	181	155	167	154	171
	157	156	128	157	167	162	137	
Target	199	167	147	180	185	170	171	139
	154	126	179	158	181	152		

(a) Use the Wilcoxon rank sum test to determine if there is convincing evidence at the $\alpha = 0.05$ level that the median distance jumped will be longer when there is a target than when there is no target for children like the ones in the study.

(b) Explain the purpose of the random assignment.

18. **Sensory Deprivation** Researchers designed an experiment to assess whether sensory deprivation for a longer period of time would affect the alpha-wave pattern frequency in the brain. The subjects in their experiment were 20 prisoners, randomly divided into two groups: one group was placed in solitary confinement for 7 days and the other group stayed in their usual cells.[42] Here are the data:

Solitary	9.6	10.4	9.7	10.3	9.2
	9.3	9.9	9.5	9.0	10.9
Usual cell	10.7	10.7	10.4	10.9	10.5
	9.6	10.3	11.1	11.2	10.4

(a) Use the Wilcoxon rank sum test to determine if there is convincing evidence at the $\alpha = 0.05$ level of a difference in the typical alpha-wave pattern frequency between those prisoners in solitary confinement and those prisoners in their usual cells for people like those in the study.

(b) In what way was this experiment unethical?

19. **Tech Meal Prep Effect** Does children's participation in meal preparation affect how many calories children eat? Swiss researchers randomly allocated 47 child–parent pairs to one of two groups. In the first group, the child participated in the meal prep. In the second group, the child was in the kitchen while the parent prepared the meal, but was otherwise occupied (e.g., with a coloring book). The data are in the file **KidsCalories**.[43] Use the Wilcoxon rank sum test to determine if there is convincing evidence at the 5% level that there is a different true median amount of calories consumed by children like those in the study who help their parents prepare the meal compared to children like those in the study who do not help their parents prepare the meal.

20. **Tech Teaching Reading** A teacher wanted to compare two methods for teaching reading: the traditional method and a method that involved extra activities. The teacher randomly assigned 40 third graders to one of the two methods. At the end of the 8-week program, the teacher gave the students a reading test. The scores for the students in both groups are in the file **Reading**.[44] Use the Wilcoxon rank sum test to determine if there is convincing evidence at the 5% level that the true median test score for children like those in the study who engage in the extra activities is greater than the true median test score for children like those in the study who use only the traditional materials.

Extending the Concepts These exercises challenge you to explore statistical concepts and methods that go beyond what you learned in this section.

21. **Outliers** Is the Wilcoxon rank sum test resistant to outliers? This exercise explores that question. In Example 13.3.3, we found that the P-value for testing whether the median decrease was larger when taking a calcium supplement as compared to a placebo was 0.1536. What if the largest blood pressure decrease with the placebo was not actually 12 mm mercury, but rather 22 mm mercury? Here is the data set again, but with 22 substituted for the 12 in the placebo group:

Group 1 (calcium)	7	−4	18	17	−3	−5	1	10	11	−2
Group 2 (placebo)	−1	22	−1	−3	3	−5	5	2	−11	−1	−3

(a) Find the P-value for the Wilcoxon rank sum test with this updated data set. How much did the P-value change? Does the conclusion change?

(b) Assume the conditions are met for using a two-sample t test on the original data (using the original value of 12 in the placebo group). What is the P-value?

(c) Ignoring the violation of the Normal/Large Sample condition, calculate the P-value for the two-sample t test with the value of 22 instead of 12 in the placebo group. Was there much change?

(d) Which test is more resistant to outliers, the two-sample t test or the Wilcoxon rank sum test? Explain your answer.

Cumulative Review These exercises revisit what you learned in previous sections.

22. **Two Samples or Paired? (9.6, 13.1, 13.2, 13.3)** For each scenario, decide whether you should use a two-sample procedure to perform inference about a difference in means or medians or a paired procedure to perform inference about a mean difference or median difference. Explain your choice in each case.

(a) Ford has announced that it will produce an all-electric Ford F-150 pickup truck, and Chevy has announced that it will produce an all-electric Chevy Silverado pickup truck. Which truck travels farther on a full battery charge? Researchers drive a random sample of 10 F-150 prototypes until they stop from lack of power, and the distance driven is measured. A random sample of 10 Silverado prototypes is also driven until they stop from lack of power and the distance driven is measured.

(b) Is the wind stronger at the north rim of the Grand Canyon or at the south rim? Researchers select a set of 15 random times within a specific week and measure the wind speed at each of the 15 times at both the north rim and the south rim.

23. **Himalayan Trekkers (1.5, 10.4)** Acute mountain sickness can be a problem for people in the mountains once they get above 2000 m (6500 feet). Himalayan trekkers often go far above that height, so they are especially susceptible to this illness. Researchers wondered

whether gingko biloba, acetazolamide, or both would help trekkers to avoid acute mountain sickness. They devised a randomized, double-blind, placebo-controlled experiment to evaluate whether any of these treatments would help. The subjects were 487 Western trekkers in the Himalayas.[45]

(a) Explain what *randomized, double-blind,* and *placebo-controlled* mean.

(b) The two-way table summarizes the results of the study. Is there is convincing evidence of an association between medication given and the occurrence of acute mountain sickness in people like those in the study?

		Treatment			
		Placebo	Acetazolamide	Gingko biloba	Both
Acute mountain sickness	Yes	40	14	43	18
	No	79	104	81	108

Section 13.4 The Kruskal-Wallis Test

LEARNING GOALS
By the end of this section, you will be able to:
- State the hypotheses and find the appropriate ranks for a Kruskal-Wallis test.
- Calculate the test statistic and *P*-value for a Kruskal-Wallis test.
- Use the four-step process to perform a Kruskal-Wallis test.

In Section 13.3, we considered a nonparametric test for comparing the medians from two populations or treatments. Here we expand on that idea to include situations in which we want to compare medians from two *or more* populations or treatments. You should recognize ideas in this section as similar to ones from both Section 13.3 (nonparametric Wilcoxon rank sum test comparing two medians) and Sections 10.5 and 10.6 (parametric ANOVA test comparing two or more means).

Stating Hypotheses and Ranking the Data

In Section 10.5, we introduced the ANOVA test for comparing means of two or more populations or treatments. One of the conditions for the ANOVA test is that the populations are normally distributed. But what if this condition is not met or is questionable? Once again we turn to a nonparametric test.

When comparing the medians of two or more populations (treatments), the null hypothesis is that all of the population (treatment) medians are the same:

H_0: All population medians are equal.

As in Section 10.5, where we considered the null hypothesis that all of the population means are the same, the alternative hypothesis is nondirectional. That is, the alternative is simply that at least one of the medians is different:

H_a: At least one median is different.

Like in Section 10.5, if we reject H_0, we cannot say which median (or medians) is different, only that at least one of them is different. A test of these hypotheses is called a **Kruskal-Wallis test**.

> A **Kruskal-Wallis test** is a nonparametric significance test of the null hypothesis that two or more population medians are the same.

Researchers wondered whether the color of the iris in people's eyes affects the speed of flicker that the eye can detect in flickering light. They took a random sample of 19 people, categorized them by the color of their eyes (brown, green, blue), and then tested them to determine the highest frequency (cycles/sec) of flicker they could detect in a flickering light.[46] The hypotheses they were testing are

H_0: $\text{Median}_{\text{Blue}} = \text{Median}_{\text{Brown}} = \text{Median}_{\text{Green}}$

H_a: At least one median is different.

where $\text{Median}_{\text{Color}}$ is the median of the highest frequency of flicker that all people with that color iris can detect.

Here are the data on highest flicker frequency (cycles/sec) from the iris study:

Brown	26.8	27.9	23.7	25.0	26.3	24.5
	24.8	25.7				
Green	26.4	24.2	28.0	26.9	29.1	
Blue	25.7	27.2	29.9	28.5	29.4	28.3

The dotplots show that all three samples are reasonably symmetric and do not have any large outliers. However, because the sample sizes are so small, it is hard to determine whether they all could be from normal distributions. For instance, the values for the green-eyed subjects could come from a uniform distribution. To be safe, we suggest a Kruskal-Wallis test with this data set.

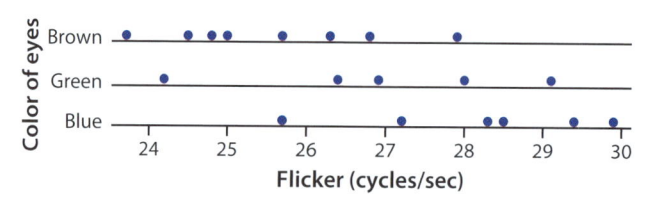

The Kruskal-Wallis test, like the nonparametric tests in Sections 13.2 and 13.3, uses the ranks of the data values rather than the individual values themselves. We find ranks in the same way that we did for the Wilcoxon rank sum test. That is, we combine the measurements from all the samples, and rank them. As before, if there are ties, we give each of the tied data values the same rank: the average of the ranks they would have been given individually.

We rank the values as follows:

Group	Value	Rank	Group	Value	Rank
Brown	23.7	1	Green	26.9	11
Green	24.2	2	Blue	27.2	12
Brown	24.5	3	Brown	27.9	13
Brown	24.8	4	Green	28.0	14
Brown	25.0	5	Blue	28.3	15
Brown	25.7	6.5	Blue	28.5	16
Blue	25.7	6.5	Green	29.1	17
Brown	26.3	8	Blue	29.4	18
Green	26.4	9	Blue	29.9	19
Brown	26.8	10			

Next, we add the ranks for the values in each group separately:

$$R_{\text{Brown}} = 1 + 3 + 4 + 5 + 6.5 + 8 + 10 + 13 = 50.5$$

$$R_{\text{Green}} = 2 + 9 + 11 + 14 + 17 = 53$$

$$R_{\text{Blue}} = 6.5 + 12 + 15 + 16 + 18 + 19 = 86.5$$

These sums will be used to create the test statistic used to evaluate the evidence for the alternative hypothesis.

Finding sums of ranks for the Kruskal-Wallis test

To find the sum of the ranks for each group in the Kruskal-Wallis test:

1. Combine all groups into one large sample and order the values from smallest to largest, keeping track of which group each value came from.
2. Assign ranks for all values, starting with 1 for the smallest value. If two or more values are equal, give each of them the same rank: the average of the ranks they would have been given individually.
3. For each group, sum the ranks of all values.

EXAMPLE 13.4.1

Density of rocks

Stating hypotheses and finding ranks

PROBLEM

Geologists were interested in comparing the density (kg/m^3) of the rock below the earth's surface at three locations in Iran. They selected eight cylindrical sections of subsurface rock (core samples) at random spots within each location and measured their densities. Here are the data.[47] Is there convincing evidence that at least one of the locations has a different population median rock density than the other locations?

Location 1	2811.0	2857.4	2906.7	2847.7	2825.6	2803.6	2916.0	2800.0
Location 2	2854.2	2918.4	2873.4	2861.1	2862.1	2879.8	2932.0	2851.0
Location 3	2817.0	2832.0	2854.5	2804.4	2853.6	2796.4	2860.0	2790.0

(a) State the hypotheses the researchers wish to test.
(b) Find the sum of the ranks for each group.

SOLUTION

(a) H_0: Median$_1$ = Median$_2$ = Median$_3$
H_a: At least one median is different.
where Median$_i$ = the median density of all subsurface rock at location i.

(b)

Location	Value	Rank	Location	Value	Rank
3	2790.0	1	2	2854.2	13
3	2796.4	2	3	2854.5	14
1	2800.0	3	1	2857.4	15
1	2803.6	4	3	2860.0	16
3	2804.4	5	2	2861.1	17
1	2811.0	6	2	2862.1	18
3	2817.0	7	2	2873.4	19
1	2825.6	8	2	2879.8	20
3	2832.0	9	1	2906.7	21
1	2847.7	10	1	2916.0	22
2	2851.0	11	2	2918.4	23
3	2853.6	12	2	2932.0	24

> Combine the three samples before ranking, but keep the information about which value came from which sample.

$R_1 = 3 + 4 + 6 + 8 + 10 + 15 + 21 + 22 = 89$
$R_2 = 11 + 13 + 17 + 18 + 19 + 20 + 23 + 24 = 145$
$R_3 = 1 + 2 + 5 + 7 + 9 + 12 + 14 + 16 = 66$

> Add the ranks for each group separately.

EXAM PREP: FOR PRACTICE, TRY EXERCISE 5.

Test Statistic and *P*-Value

When the null hypothesis of equal population (or treatment) medians is true and the values from all groups are combined and then ranked, the ranks will act as if they were randomly allocated to the different populations (or treatments). Let R_1 be the sum of the ranks in one group, R_2 be the sum of the ranks in the next group, and so on. Designate n_1 as the sample size of the first group, n_2 as the sample size in the second group, and so on. Finally, let $N = n_1 + n_2 + \cdots + n_k$ be the total number of values, and $k =$ the number of groups.

The test statistic for the Kruskal-Wallis test takes into consideration the sums of the ranks of all the groups R_i, the sample sizes of all the groups n_i, and the overall sample size, N. It is defined as

$$H = \left(\frac{12}{N(N+1)} \sum_{i=1}^{k} \frac{R_i^2}{n_i}\right) - 3(N+1)$$

To calculate H for the flickering light data, recall that $R_{\text{Brown}} = 50.5$, $R_{\text{Green}} = 53$, and $R_{\text{Blue}} = 86.5$. The sample sizes are $n_{\text{Brown}} = 8$, $n_{\text{Green}} = 5$, and $n_{\text{Blue}} = 6$. This leads to $N = 8 + 5 + 6 = 19$. The test statistic, then, is

$$H = \left(\frac{12}{19(20)} \left(\frac{50.5^2}{8} + \frac{53^2}{5} + \frac{86.5^2}{6}\right)\right) - 3(20) = 67.188 - 60 = 7.188$$

For very small sample sizes, the distribution of H requires the use of a special table. However, if all sample sizes are 5 or more, and the variability of the samples is similar, then H has approximately a chi-square distribution with $k - 1$ degrees of freedom, where k = the number of samples in the data set. As was the case in an ANOVA test, large values of the test statistic give evidence against the null hypothesis and in favor of the alternative hypothesis. This means we always find the area in the upper tail of the chi-square distribution when calculating the P-value for the Kruskal-Wallis test.

In the flickering light study, we have three groups (Brown, Green, and Blue), so $k = 3$ and all three sample sizes are at least 5. To check whether the variability of the samples is similar enough, use the rule of thumb that the largest standard deviation is less than twice the smallest. The sample standard deviations for the three groups are $s_{\text{Brown}} = 1.365$, $s_{\text{Green}} = 1.843$, and $s_{\text{Blue}} = 1.528$. In this case, the largest SD = $1.843 < 2.73 = 2(1.365)$ = twice the smallest SD, so we can use the chi-square distribution. This means that the P-value is $P(H \geq 7.188)$, where H has a chi-square distribution with df = $k - 1 = 3 - 1 = 2$. Using technology, the P-value = 0.027.

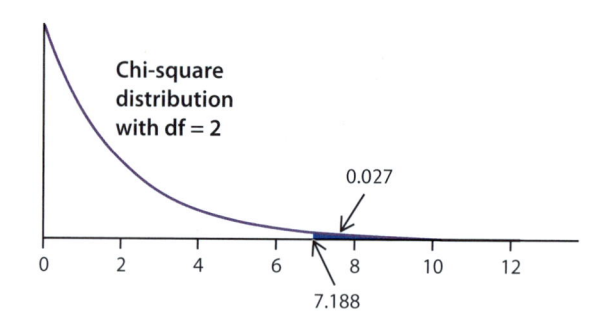

Conditions, test statistic, and P-value for the Kruskal-Wallis test

Conditions:

- **Random:** The data come from independent random samples from two or more populations of interest or from two or more groups in a randomized experiment.
- **Large Sample:** Each sample size is at least 5.
- **Equal SD:** The largest sample standard deviation is less than twice the smallest sample standard deviation.

Test statistic:

- Compute R_i = the sum of the ranks for the values in group i for each group.
- Let n_i = the sample size of group i, and $N = n_1 + n_2 + \ldots + n_k$, where k = number of groups.
- Calculate the test statistic $H = \left(\frac{12}{N(N+1)} \sum_{i=1}^{k} \frac{R_i^2}{n_i}\right) - 3(N+1)$.

P-value: The test statistic H has a distribution that is approximately chi-square with $k - 1$ df, so use either Table C at the back of the book or technology to compute the P-value.

EXAMPLE 13.4.2

Density of rocks

Check conditions, calculate test statistic and P-value

PROBLEM

Recall the hypotheses we stated in Example 13.4.1:

H_0: Median$_1$ = Median$_2$ = Median$_3$

H_a: At least one median is different.

where Median$_i$ = the median density of all subsurface rock at location i.

Here are the data once again, along with the sum of ranks for each location:

Location 1	2811.0	2857.4	2906.7	2847.7	2825.6	2803.6	2916.0	2800.0
Location 2	2854.2	2918.4	2873.4	2861.1	2862.1	2879.8	2932.0	2851.0
Location 3	2817.0	2832.0	2854.5	2804.4	2853.6	2796.4	2860.0	2790.0

$$R_1 = 89, R_2 = 145, \text{ and } R_3 = 66.$$

(a) Check that the conditions for performing the Kruskal-Wallis test are met.

(b) Calculate the test statistic.

(c) Find the *P*-value.

(d) What conclusion would you make?

SOLUTION

(a) Random? The cylindrical sections of subsurface rock were collected at random spots within locations. ✓

Large Sample? $n_1 = 8 \geq 5, n_2 = 8 \geq 5, n_3 = 8 \geq 5$ ✓

Equal SD? The three sample standard deviations are $s_1 = 45.2$, $s_2 = 30.2$, and $s_3 = 28.0$. The largest sample standard deviation is $s_1 = 45.2 < 2(28.0) = 56 = 2$ times the smallest sample standard deviation. ✓

(b) $N = 8 + 8 + 8 = 24$

$$H = \left(\frac{12}{24(25)}\right)\left(\frac{89^2}{8} + \frac{145^2}{8} + \frac{66^2}{8}\right) - 3(25) = 83.255 - 75 = 8.255$$

(c) df = 2

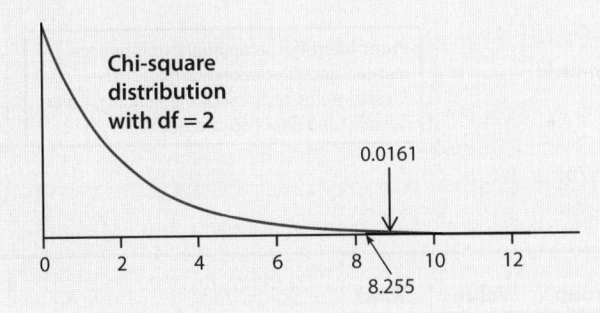

Using Table C: $0.01 < P\text{-value} < 0.025$

Using technology: Applet/χ^2 cdf(lower: 8.255, upper: 1000, df: 2) = 0.0161

(d) Because the *P*-value of 0.0161 is smaller than the typically chosen $\alpha = 0.05$, we reject H_0. We have convincing evidence that at least one of the three sites has a different population median subsurface rock density.

EXAM PREP: FOR PRACTICE, TRY EXERCISE 9.

Several methods can be used to perform a follow-up analysis when the results of the Kruskal-Wallis test are significant. These are beyond the scope of this book.

Putting It All Together: The Kruskal-Wallis Test

As with our other inference procedures, we follow the four-step process when performing a Kruskal-Wallis test.

EXAMPLE 13.4.3

Growing bacteria

The Kruskal-Wallis test

PROBLEM

What is the best way to culture (grow) bacteria so that a patient with an infection can be diagnosed and treated correctly? Researchers designed an experiment to try five different concentrations of tryptone, a nutrient for the bacteria. They randomly assigned six dishes with a form of *Staphylococcus aureus* to each of the concentrations. At the end of the experiment, the researchers counted the number of colony-forming units (CFUs) in each dish.[48] Is there convincing evidence at the 5% level that at least one of the concentrations of tryptone produces a different median number of CFUs of this type of bacteria? To help answer this question, perform the appropriate Kruskal-Wallis test.

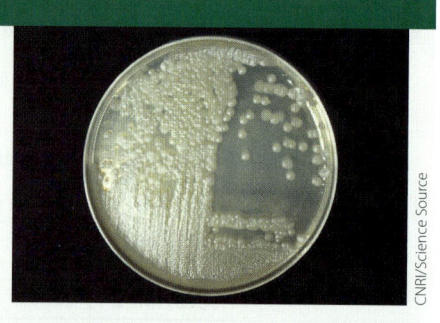

0.6%	10	129	93	146	42	118
0.8%	26	145	98	217	108	99
1.0%	50	156	89	269	96	141
1.2%	52	243	149	284	131	234
1.4%	47	178	113	186	121	172

SOLUTION

State:

H_0: The true median number of CFUs is the same in 0.6%, 0.8%, 1.0%, 1.2%, and 1.4% concentrations of tryptone.

H_a: At least one true median is different.

Use $\alpha = 0.05$.

The evidence for H_a is: The sample median numbers of CFUs are not all the same: Median$_{0.6\%}$ = 105.5, Median$_{0.8\%}$ = 103.5, Median$_{1.0\%}$ = 118.5, Median$_{1.2\%}$ = 191.5, Median$_{1.4\%}$ = 146.5.

Plan: Kruskal-Wallis test

Random? The measurements of CFUs come from 5 groups in a randomized experiment. ✓

Large Sample? The sample sizes are $6 \geq 5$, $6 \geq 5$, $6 \geq 5$, $6 \geq 5$, and $6 \geq 5$. ✓

Equal SD? The standard deviations are $s_{0.6\%} = 53.2$, $s_{0.8\%} = 63.0$, $s_{1.0\%} = 76.5$, $s_{1.2\%} = 86.5$, $s_{1.4\%} = 53.4$, and $86.5 < 2(53.2) = 106.4$. ✓

Do:

> **State:** State the hypotheses, parameter(s), significance level, and evidence for H_a.

> **Plan:** Identify the appropriate inference method and check conditions. *For the Kruskal-Wallis test, check the Random, Large Sample, and Equal SD conditions.*

Group	Value	Rank	Group	Value	Rank	Group	Value	Rank
0.6%	10	1	0.8%	99	11	1.2%	149	21
0.8%	26	2	0.8%	108	12	1.0%	156	22
0.6%	42	3	1.4%	113	13	1.4%	172	23
1.4%	47	4	0.6%	118	14	1.4%	178	24
1.0%	50	5	1.4%	121	15	1.4%	186	25
1.2%	52	6	0.6%	129	16	0.8%	217	26
1.0%	89	7	1.2%	131	17	1.2%	234	27
0.6%	93	8	1.0%	141	18	1.2%	243	28
1.0%	96	9	0.8%	145	19	1.0%	269	29
0.8%	98	10	0.6%	146	20	1.2%	284	30

$R_{0.6\%} = 1+3+8+14+16+20 = 62$
$R_{0.8\%} = 2+10+11+12+19+26 = 80$
$R_{1.0\%} = 5+7+9+18+22+29 = 90$
$R_{1.2\%} = 6+17+21+27+28+30 = 129$
$R_{1.4\%} = 4+13+15+23+24+25 = 104$
$n_{0.6\%} = 6, n_{0.8\%} = 6, n_{1.0\%} = 6, n_{1.2\%} = 6, n_{1.4\%} = 6$
$N = 6+6+6+6+6 = 30$

$$H = \left(\frac{12}{30(31)}\right)\left(\frac{62^2}{6} + \frac{80^2}{6} + \frac{90^2}{6} + \frac{129^2}{6} + \frac{104^2}{6}\right) - 3(31) = 98.497 - 93 = 5.497$$

> **Do:** If the conditions are met, perform calculations:
> - Calculate the test statistic.
> - Find the P-value.

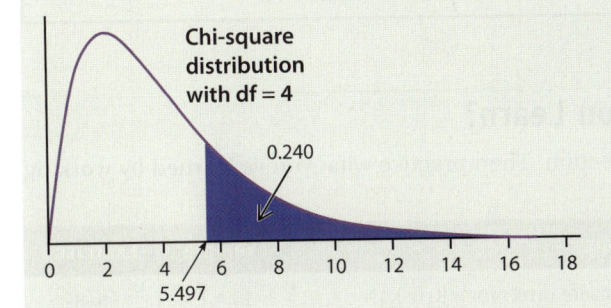

- *P-value:* H has a chi-square distribution with $df = 5 - 1 = 4$.

Chi-square distribution with df = 4

0.240

5.497

Using Table C: P-value > 0.10
Using technology: P-value = 0.240

Conclude: Because the P-value of $0.240 > \alpha = 0.05$, we fail to reject H_0. We do not have convincing evidence that the true median number of bacteria grown is different for at least one concentration level of tryptone.

> **Conclude:** Make a conclusion about the hypotheses in the context of the problem.

EXAM PREP: FOR PRACTICE, TRY EXERCISE 13.

TECH CORNER Kruskal-Wallis test

You can use technology to conduct a Kruskal-Wallis test to compare two or more population medians. We'll illustrate this process using the rock density data from this section.

Location 1	2811.0	2857.4	2906.7	2847.7	2825.6	2803.6	2916.0	2800.0
Location 2	2854.2	2918.4	2873.4	2861.1	2862.1	2879.8	2932.0	2851.0
Location 3	2817.0	2832.0	2854.5	2804.4	2853.6	2796.4	2860.0	2790.0

Applet

1. Go to www.stapplet.com and launch the *One Quantitative Variable, Multiple Groups* applet.
2. Enter the variable name "Rock density" and choose "Raw data" as the "Input."
3. Enter each group name and the values for that group as shown. To add a third group, click the Add group button.

One Quantitative Variable, Multiple Groups

Variable name: Rock density
Input: Raw data

Group	Name	Input data separated by commas or spaces
1	Location 1	2811.0 2857.4 2906.7 2847.7 2825.6 28
2	Location 2	2854.2 2918.4 2873.4 2861.1 2862.1 28
3	Location 3	2817.0 2832.0 2854.5 2804.4 2853.6 27

Add group

Begin analysis Edit inputs Reset everything

(continued)

Perform Inference

Inference procedure: [Kruskal-Wallis test ▾]
[Perform inference]

H	P-value	df
8.255	0.016	2

Detailed instructions for using CrunchIt!, Excel, Google Sheets, JMP, Minitab, and R are available in Achieve.

Note: If your data set is organized by variable (i.e., one column for location and one column for density), you can use the "Raw data by variable" input option.

TI-83/84

This type of significance test is not currently available on the TI-83/84.

Section 13.4 What Did You Learn?

Review the learning goals from this section. Then practice what you've learned by working through the exercises.

Learning Goals	Example	Exercises
State the hypotheses and find the appropriate ranks for a Kruskal-Wallis test.	13.4.1	5–8
Calculate the test statistic and P-value for a Kruskal-Wallis test.	13.4.2	9–12
Use the four-step process to perform a Kruskal-Wallis test.	13.4.3	13–16

Section 13.4 Exercises

Building Concepts and Skills These exercises assess the basic knowledge you should have after reading the section.

1. The null and alternative hypotheses for the Kruskal-Wallis test are _____ and _____.

2. The names of the three conditions for the Kruskal-Wallis test are _____, _____, and _____.

3. True/False: When conditions are met, the P-value for a Kruskal-Wallis test is always calculated as a one-tail area.

4. The P-value is found by using a(n) _____ distribution with _____ df.

Mastering Concepts and Skills These exercises reinforce the learning goals as shown in the examples.

5. **Refer to Example 13.4.1 Wool Shrinkage** How can manufacturers minimize the amount of shrinkage in wool items they produce? Researchers tested three different treatments they could apply during the production process to reduce the amount of shrinkage. Is the median shrinkage the same or is at least one different from the others?[49] The researchers randomly assigned 27 pieces of wool to one of the three treatments and measured the amount of shrinkage (%). Here are the data:

Treatment 1	42.8	46.2	45.6	43.0	47.0	43.8	43.5	45.2	45.7
Treatment 2	33.6	40.4	40.1	35.0	39.4	39.5	36.2	37.9	39.2
Treatment 3	35.6	35.2	36.6	34.8	33.0	32.3	34.8	32.3	35.6

(a) State the hypotheses for a Kruskal-Wallis test in this context.

(b) Find the sum of the ranks for each group.

6. **Iron in Pottery** Can the chemical composition of a piece of ancient pottery help determine where it came from? Researchers took a random sample of ancient pottery pieces from three different sites in Great Britain and measured the amount of iron (measured as percentage of oxide) in each pottery piece's clay.[50] Is the median amount of iron the same in pieces from all three sites, or is at least one site different? Here are the data:

	7.00	7.08	7.09	6.37	7.06
Llanedryn	4.26	5.78	5.49	6.92	6.13
	6.64	6.69	6.44	6.26	
Isle Thorns	1.28	2.39	1.50	1.88	1.51
Ashley Rails	1.12	1.14	0.92	2.74	1.64

(a) State the hypotheses for a Kruskal-Wallis test in this context.

(b) Find the sum of the ranks for each group.

7. **Water Pollution** Is pollution mixed uniformly in the water in a river or does it vary with the depth in the river? Researchers filled 10 containers with randomly sampled water at each of three different depths in the Wolf River in Tennessee, downstream from an abandoned dump site.[51] They measured the amount of hexachlorobenzene (HCB) in each container (nanograms per liter). Is the median amount of pollution the same at all three depths, or is it different in at least one depth? Here are the data:

Surface	3.74	4.61	4.00	4.67	4.87
	5.12	4.52	5.29	5.74	5.48
Mid-depth	6.03	6.55	3.55	4.59	3.77
	4.81	5.85	5.74	6.77	5.64
Bottom	5.44	6.88	5.37	5.44	5.03
	6.48	3.89	5.85	6.85	7.16

(a) State the hypotheses for a Kruskal-Wallis test in this context.

(b) Find the sum of the ranks for each group.

8. **Back Pain** Does how you twist when you lift heavy objects affect the amount of back pain you may experience? Researchers designed an experiment to find out. They randomly assigned 27 people to lift at one of three different twisting angles (0, 30, and 45 degrees). They then measured the amount of pain on an index where higher numbers indicate more pain.[52] Are all three twisting angles the same with respect to the median amount of pain or is at least one of the angles different? Here are the data:

0 degrees	0.97	1.06	1.12	1.43	1.55	1.71	1.95	2.07	2.24
30 degrees	1.08	1.15	1.26	1.62	1.72	1.89	2.16	2.30	2.52
45 degrees	1.13	1.20	1.32	1.70	1.80	1.98	2.26	2.40	2.64

(a) State the hypotheses for a Kruskal-Wallis test in this context.

(b) Find the sum of the ranks for each group.

9. **Refer to Example 13.4.2 Wool Shrinkage** Refer to Exercise 5.

(a) Check that the conditions for performing the Kruskal-Wallis test are met.

(b) Calculate the test statistic.

(c) Find the P-value.

(d) What conclusion would you make at the 5% significance level?

10. **Iron in Pottery** Refer to Exercise 6.

(a) Check that the conditions for performing the Kruskal-Wallis test are met.

(b) Calculate the test statistic.

(c) Find the P-value.

(d) What conclusion would you make at the 1% significance level?

11. **Water Pollution** Refer to Exercise 7.

(a) Check that the conditions for performing the Kruskal-Wallis test are met.

(b) Calculate the test statistic.

(c) Find the P-value.

(d) What conclusion would you make at the $\alpha = 0.01$ significance level?

12. **Back Pain** Refer to Exercise 8.

(a) Check that the conditions for performing the Kruskal-Wallis test are met.

(b) Calculate the test statistic.

(c) Find the P-value.

(d) What conclusion would you make at the $\alpha = 0.05$ significance level?

13. **Refer to Example 13.4.3 Marketing Salaries** Does the salary that a marketing manager receives depend on their geographic location? To investigate, a researcher took random samples of marketing manager salaries (in $) from three regions of the United States:[53] Northeast (NE), Southeast (SE), and West (W). Is there convincing evidence at the 5% level that at least one of these regions has a different population median salary for marketing managers? To help answer this question, perform the appropriate Kruskal-Wallis test.

NE	103,850	69,730	113,050	81,730	116,890
	129,420	91,760	72,210	79,330	
SE	76,480	69,680	92,380	89,360	61,550
	56,630	69,030	62,020		
W	67,100	115,620	85,850	73,050	75,500
	76,940	112,310	84,100		

14. **Wool and pH** In Exercise 5, we considered three treatments to help reduce the shrinkage of wool. The same researchers also investigated how the pH of the water in which the wool was washed affected shrinkage. They randomly assigned the pieces of wool to water with three different pH values. Is there convincing evidence at the 5% level that at least one of these pHs produces a different median amount of shrinkage (%)? To help answer this question, perform the appropriate Kruskal-Wallis test.

pH = 2	42.8	43.0	43.5	33.6	35.0	36.2	35.6	34.8	34.8
pH = 6	46.2	47.0	45.2	40.4	39.4	37.9	35.2	33.0	32.3
pH = 10	45.6	43.8	45.7	40.1	39.5	39.2	36.6	32.3	35.6

15. **Salinity** Is the salinity level the same in different parts of the Bimini Lagoon in the Bahamas? Researchers took random samples of water in several spots at three different locations in the lagoon and measured salinity in parts per thousand.[54] Is there convincing evidence that the median salinity level in the Bimini Lagoon is different in at least one of the locations? To help answer this question, perform the appropriate Kruskal-Wallis test.

Site 1	37.54	37.01	36.71	37.03	37.32	37.01
	37.03	37.70	37.36	36.75	37.45	38.85
Site 2	40.17	40.80	39.76	39.70	40.79	40.44
	39.79	39.38				
Site 3	39.04	39.21	39.05	38.24	38.53	38.71
	38.89	38.66	38.51	40.58		

16. **Espresso** Does the amount of foam in a cup of espresso depend on how it was produced? Researchers measured the amount of foam in nine randomly selected cups of espresso produced with each of three different methods: Bar Machine (BM), Hyper-Espresso Method (HEM), and Espresso System (ES). The amount of foam is measured as the percentage of the amount of liquid in the cup.[55] Is there convincing evidence that the median amount of foam is different for at least one of these methods? To help answer this question, perform the appropriate Kruskal-Wallis test.

BM	36.64	39.65	37.74	35.96	38.52	21.02	24.81	34.18	23.08
HEM	70.84	46.68	73.19	57.78	48.61	72.77	65.04	62.53	54.26
ES	56.19	36.67	35.35	40.11	33.52	37.12	37.33	32.68	48.33

Applying the Concepts These exercises ask you to apply multiple learning goals in a new context or to apply what you learned in this section in a new way.

17. **Bacteria and Temperature** Refer to Example 13.4.3. The researchers took a different strain of bacteria and randomly assigned 10 dishes of bacteria to each of three temperatures (°C). At the end of the experiment, they again measured the number of CFUs of bacteria in each dish. Is there convincing evidence that the median amount of bacteria grown is different for at least one temperature?

27°	14	20	17	29	18	102	109	129	161	158
35°	102	110	106	178	124	136	139	156	233	213
43°	31	62	119	132	142	111	148	97	177	212

(a) State the hypotheses and check the conditions for a Kruskal-Wallis test.
(b) Calculate the test statistic H.
(c) Find the P-value.
(d) What conclusion would you make?

18. **Aluminum in Pottery** Refer to Exercise 6. The same researchers also measured the amount of aluminum in the pieces of pottery. Is there convincing evidence that the median amount of aluminum in the clay of the pottery is different for at least one location?

Llanedryn	14.4	13.8	14.6	11.5	13.8
	10.9	10.7	11.6	11.1	13.4
	12.4	13.1	12.7	12.5	
Isle Thorns	18.3	15.8	18.0	18.0	20.8
Ashley Rails	17.7	18.3	16.7	14.8	19.1

(a) State the hypotheses and check the conditions for a Kruskal-Wallis test.
(b) Calculate the test statistic H.
(c) Find the P-value.
(d) What conclusion would you make?

19. **Tech Electronic Paper Displays** Several types of electronic readers use an electronic display that mimics the look and experience of ink on paper. Does the type of electronic paper display affect reading speed? Researchers randomly assigned 20 people to each of three different electronic paper displays and measured the number of seconds that it took each person to read a specific passage.[56] Use the Kruskal-Wallis test with the data in the **EReader** data file to determine if there is convincing evidence at the 5% significance level that the true median reading time is different for at least one of the electronic paper displays.

20. **Tech Serotonin in Mice** Does serotonin affect the mood of mice in the same way that it does in humans? Researchers genetically altered mice by "knocking out" the expression of a gene that regulates serotonin production. They created three types of mice: minus (mice with two copies of the serotonin negative gene), plus (mice with two copies of the serotonin positive gene), and mixed (one copy of each type of gene).[57] The researchers randomly selected mice from each group and measured the number of contacts the mice made with other mice during an observation time. Use the Kruskal-Wallis test with the data in the **MouseBrain** data file to determine whether there is convincing evidence at the 5% significance level that the true median number of contacts for at least one of the groups of mice is different.

Extending the Concepts These exercises challenge you to explore statistical concepts and methods that go beyond what you learned in this section.

21. **Comparing Kruskal-Wallis and Wilcoxon Rank Sum Tests** Refer to Examples 13.3.1 and 13.3.2. Those examples illustrated the use of a Wilcoxon rank sum test to determine whether there is convincing evidence that the median pain threshold in people with blond hair is different from the median pain threshold in people with brown hair.

Here again are the data:

Blond	62	60	71	55	48	63	57
	43	41	52				
Brown	42	50	41	37	32	39	51
	30	35					

In Example 13.3.2, we calculated $z = -3.062$ and the P-value $= 0.0022$.

(a) Report the test statistic H and the P-value from the appropriate Kruskal-Wallis test.

(b) Compare the values of z^2 and H, and compare the two P-values.

In fact, when you are comparing two independent samples or two groups in a randomized experiment, the Kruskal-Wallis test is identical to the two-sided Wilcoxon rank sum test.

Cumulative Review These exercises revisit what you learned in previous sections.

22. **Bird Watching (5.5)** An experienced bird watcher knows that the number of different bird species they will see at a particular location over a 15-minute period can be modeled by a Poisson distribution with a mean of 5 bird species.

    (a) What is the probability that in one 15-minute period of bird watching, the bird watcher will see exactly 4 different species?

    (b) What is the probability that in one 15-minute period of bird watching, the bird watcher will see between 4 and 7 (inclusive) different bird species?

23. **Late Students (4.6)** Suppose that there are 15 students in a particular class. In this scenario, whether one student is late to class is independent of whether another student is late to class. Also, the probability that any one individual student is late is 0.10. Find the probability that all students will be in class on time for one particular class session.

Section 13.5 Randomization Tests

LEARNING GOALS

By the end of this section, you will be able to:

- Describe how to use note cards to create a randomization distribution for a test about a difference between two proportions or two means.
- Estimate the P-value from a randomization distribution.
- Use the four-step process to perform a randomization test about a difference between two proportions or two means.

In this section, we consider a new method for performing a significance test to compare two population proportions or two population means. There are two crucial differences between the methods from Chapter 9 and the method we present here: how we calculate the test statistic and which conditions need to be met to estimate the P-value.

Randomization Distributions

In Section 1.6, we introduced informal inference using simulation. Now that we know the general structure of significance tests, we can revisit those ideas in more detail.

Let's start with an example about a difference between two proportions. Is the occurrence of side effects from a flu vaccine the same whether you get a full dose or a half dose? Or does a half dose lead to fewer side effects? Researchers set out to study this question by randomly allocating 1259 patients to receive either the full dose (628 patients) or a half dose (631 patients) of the vaccine. Afterward, they measured how many patients in each group experienced headache. Of those receiving a full dose, 37 reported having had a headache, while of those receiving a half dose, 28 reported having a headache.[58] The hypotheses in question are

$$H_0: p_{\text{Full}} - p_{\text{Half}} = 0$$
$$H_a: p_{\text{Full}} - p_{\text{Half}} > 0$$

where p_{Full} = the proportion of people like those in the study who receive the full dose of the vaccine and experience headache, and p_{Half} = the proportion of people like those in the study who receive the half dose of the vaccine and experience headache.

In this new test about a difference in proportions, we use the difference in sample proportions as the test statistic. For this study,

$$\hat{p}_{\text{Full}} - \hat{p}_{\text{Half}} = \frac{37}{628} - \frac{28}{631} = 0.0589 - 0.0444 = 0.0145$$

Because this difference is greater than 0, there is *some* evidence for the alternative hypothesis. But how unlikely is a difference in sample proportions of 0.0145 or greater, if the null hypothesis is true? To answer that question, we need the sampling distribution of $\hat{p}_{\text{Full}} - \hat{p}_{\text{Half}}$ when the null hypothesis is true. We know from Chapter 9 that if the Random and Large Counts conditions are met, this distribution is approximately normal. In this section, we approximate the sampling distribution of $\hat{p}_{Full} - \hat{p}_{Half}$ in a different way—one that only requires the Random condition.

The null hypothesis says that the proportion of all people receiving the flu vaccine who will experience a headache is the same whether they receive a full dose or a half dose. One way to interpret this null hypothesis is that no matter which dose a person is given, they will have the same reaction. That is, a person who develops a headache would get it under either condition. In this experiment, that means 65 of the 1259 participants were destined to get a headache no matter which dose of flu vaccine they received. It is just the luck of random assignment that allocated 37 of them to the full dose and 28 to the half dose.

To approximate the distribution of $\hat{p}_{Full} - \hat{p}_{Half}$, we mimic this random assignment process. Imagine taking 1259 note cards. Write "headache" on 65 of them and write "no headache" on the rest. Shuffle the cards and then deal 628 of them into one pile representing the people who got the full dose, and 631 into another pile representing the people who got the half dose. This simulates randomly assigning the 1259 people to the two treatments. Now count how many people with headaches are in each pile, compute a new sample proportion for each group, and compute the difference in sample proportions. This difference is one possible value in the distribution of $\hat{p}_{Full} - \hat{p}_{Half}$ when $p_{Full} - p_{Half} = 0$.

To build the approximate sampling distribution, repeat this process at least 1000 times. The result is called a **randomization distribution**. Of course, no one wants to do this by hand with a deck of note cards, so we will use technology to create a randomization distribution. Once we have the randomization distribution, we can perform a **randomization test**.

A **randomization distribution** is the approximation of the sampling distribution of a statistic created by repeatedly simulating the possible values of the sample statistic while assuming the null hypothesis is true.

A **randomization test** is a nonparametric significance test that uses a randomization distribution to estimate the *P*-value.

The same general procedure can be used for a test about a difference between two means, when the null hypothesis is $H_0: \mu_1 - \mu_2 = 0$. Begin by writing the sample values from both groups on note cards and shuffling the cards. Then deal two piles the same size as the original group sizes, compute the two sample means, and find their difference. This difference is one possible value in a randomization distribution of the test statistic $\bar{x}_1 - \bar{x}_2$. As with proportions, to better approximate the distribution, repeat this process at least 1000 times using technology.

EXAMPLE 13.5.1

Hair growth

Creating randomization distributions

PROBLEM

In Section 13.3, we considered data from an experiment designed to determine if procyanidin B-2 would result in more hair growth than no treatment for men with male pattern baldness. The researchers randomly assigned 19 men to the treatment group and 10 men to the control group, and then measured the increase in total number of hairs per 0.25 cm² for each person in the two groups. Here are the data once again:

Control	0.3	1.4	3.0	3.7	−1.5	−2.0	0.0	4.8	2.4	−11.3
Treatment	3.5	5.0	7.3	18.3	14.5	6.7	9.0	−0.7	7.8	−4.0
	6.0	4.5	8.0	11.4	1.0	7.3	8.5	−0.7	13.5	

For this experiment, let $\mu_{\text{Treatment}}$ = the true mean increase in total number of hairs per 0.25 cm² in people like those in the study who use procyanidin B-2, and let μ_{Control} = the true mean increase in total number of hairs per 0.25 cm² in people like those in the study who use nothing. The hypotheses are

$$H_0: \mu_{\text{Treatment}} - \mu_{\text{Control}} = 0$$

$$H_a: \mu_{\text{Treatment}} - \mu_{\text{Control}} > 0$$

Describe how to use note cards to simulate one possible value in a randomization distribution of $\bar{x}_{\text{Treatment}} - \bar{x}_{\text{Control}}$ for this experiment.

SOLUTION

Write each of the 29 observed values of the increase in hair growth on a separate note card. Shuffle the cards and deal two piles: one pile of 10 representing the control group, and one pile of 19 representing the treatment group. Compute the mean for each pile and calculate $\bar{x}_{\text{Treatment}} - \bar{x}_{\text{Control}}$.

EXAM PREP: FOR PRACTICE, TRY EXERCISE 5.

P-Values from Randomization Distributions

We used technology to create a randomization distribution of $\hat{p}_{\text{Full}} - \hat{p}_{\text{Half}}$ for the flu vaccine experiment. This graph shows 1000 simulated values of $\hat{p}_{\text{Full}} - \hat{p}_{\text{Half}}$.

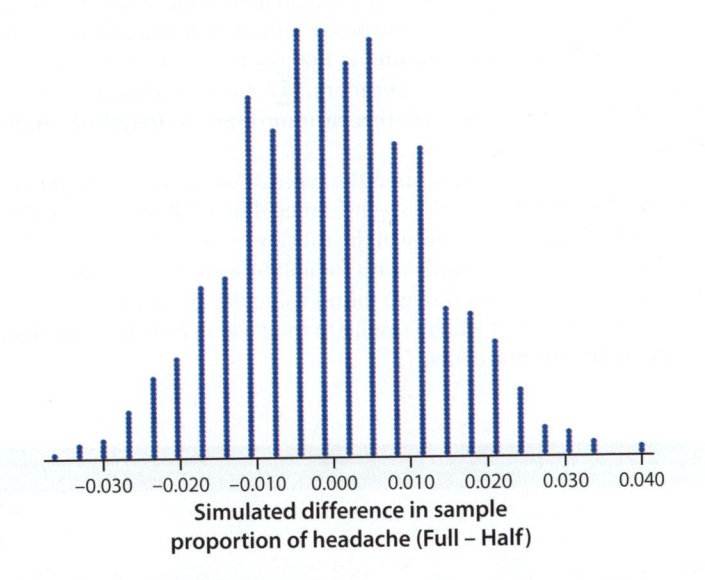

Notice that the center of the distribution is 0, exactly the value specified by the null hypothesis $H_0: p_{\text{Full}} - p_{\text{Half}} = 0$. The distribution also looks approximately normal, which is not surprising. In the full vaccine group, there were 37 people who experienced headache and 591 who did not. In the half dose group, there were 28 people who experienced headache and 603 who did not. All of these numbers are at least 10, so the Large Counts condition is met. But because we only used the fact that the data were randomly assigned to groups in constructing the randomization distribution, the Random condition is the only condition required for a randomization test.

> **HOW TO** Check the Condition for Performing a Randomization Test About $p_1 - p_2$ or $\mu_1 - \mu_2$

To perform a randomization test about a difference in population proportions $p_1 - p_2$ or a difference in population means $\mu_1 - \mu_2$, check that the following condition is met:

- **Random:** The data come from independent random samples from the two populations of interest or from two groups in a randomized experiment.

When the Random condition is met, we estimate the P-value by computing the proportion of dots in a randomization distribution that are in the appropriate tail(s) based on the alternative hypothesis. Because the flu shot example has an upper-tail alternative hypothesis, we need to count the dots greater than or equal to the observed difference of 0.0145.

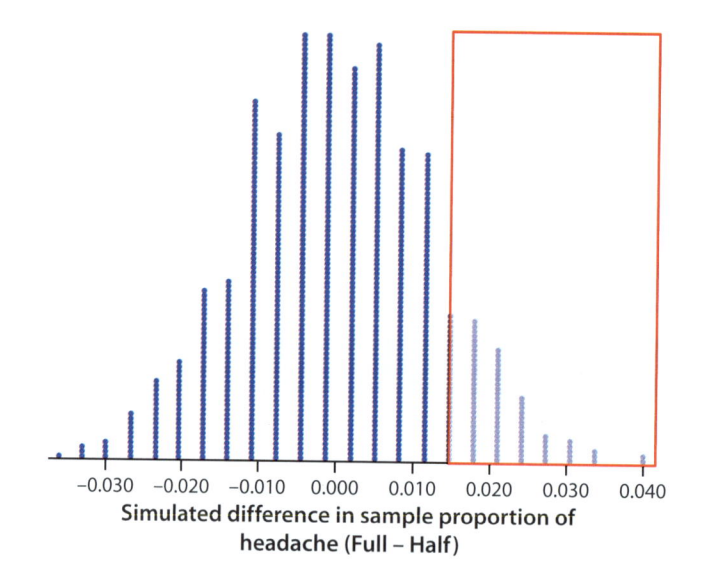

In this case, there are 130 dots greater than or equal to 0.0145, so the P-value is approximately $130/1000 = 0.130$. That is, there is about a 0.130 probability of getting a difference in sample proportions of at least 0.0145 by chance alone, assuming that the two doses are equally likely to produce headaches in people like the ones in the experiment. ⚠ **Note that the P-value from a randomization test will differ each time you create a randomization distribution, but the P-values should be similar.**

Because the P-value of 0.130 is larger than the typical 0.05 significance level, we fail to reject $H_0: p_{Full} - p_{Half} = 0$. We do not have convincing evidence that a full dose of the flu vaccine is more likely to result in a headache than a half dose of the vaccine.

What is the P-value for the two-sample z test for a difference between two proportions for this study? It is 0.122, very close to the P-value we found using the randomization test. In general, the P-value from the randomization test will be similar to the P-value from the traditional inference test when the conditions for both tests are met.

EXAMPLE 13.5.2

Hair growth

P-value from a randomization distribution

PROBLEM

Return to Example 13.5.1. Recall the hypotheses we identified:

$$H_0: \mu_{Treatment} - \mu_{Control} = 0$$
$$H_a: \mu_{Treatment} - \mu_{Control} > 0$$

Here are the data once again:

Control	0.3	1.4	3.0	3.7	−1.5	−2.0	0.0	4.8	2.4	−11.3
Treatment	3.5	5.0	7.3	18.3	14.5	6.7	9.0	−0.7	7.8	−4.0
	6.0	4.5	8.0	11.4	1.0	7.3	8.5	−0.7	13.5	

We used technology to create the following randomization distribution by simulating 100 values of $\bar{x}_{Treatment} - \bar{x}_{Control}$.

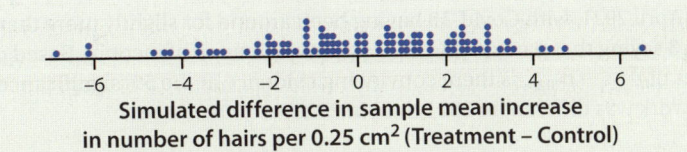

[Dotplot: Simulated difference in sample mean increase in number of hairs per 0.25 cm² (Treatment − Control), x-axis from −6 to 6]

(a) Calculate the value of $\bar{x}_{\text{Treatment}} - \bar{x}_{\text{Control}}$ for the original experiment.
(b) Estimate the P-value using the randomization distribution.
(c) What conclusion would you make at the $\alpha = 0.05$ significance level?

SOLUTION

(a) $\bar{x}_{\text{Treatment}} - \bar{x}_{\text{Control}} = 6.679 - 0.080 = 6.599$
(b) There are no dots in the randomization distribution as large as or larger than 6.599, so the P-value is approximately 0.
(c) Because the P-value of approximately 0 is less than $\alpha = 0.05$, we reject H_0. There is convincing evidence that the mean increase in hair growth is larger for people like those in the study who use procyanidin B-2 than for people like those in the study who do not use the treatment.

EXAM PREP: FOR PRACTICE, TRY EXERCISE 9.

In Example 13.5.2, the randomization distribution was based on only 100 trials so that you could estimate the P-value by counting dots. Typically, we would use at least 1000 trials to get a better estimate of the P-value, as shown here for the hair treatment data from Example 13.5.2. Notice this graph shows more detail about the shape of the randomization distribution.

[Dotplot: Simulated difference in sample mean increase in number of hairs per 0.25 cm² (Treatment − Control), x-axis from −6 to 8, with a red box highlighting values near 7]

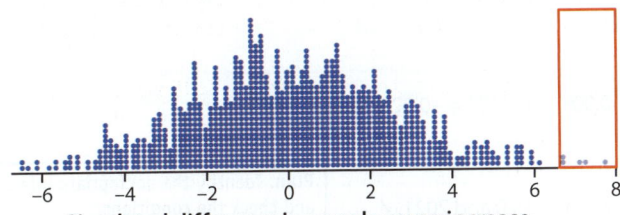

In this case, 5 of the 1000 simulated differences in means were 6.599 or larger, so the P-value is approximately 0.005. This is very similar to the near-zero result from the example.

Putting It All Together: Randomization Tests

As with other inference procedures, we follow the four-step process when performing a randomization test.

EXAMPLE 13.5.3

How did COVID-19 affect retiree income?

Four-step process for randomization test

PROBLEM

The Gallup Company does a yearly survey of retirees asking about their sources of income and how comfortable they are with the amount of income they have. One question asks if part-time work is a major source of income for them. Did that change due to the Covid-19 pandemic? In April 2020, Gallup received 312 responses in its survey of retirees. Of those, 12 indicated that part-time work was

(continued)

a major source of income. In April 2021, with Covid-19 having been around for slightly more than a year, Gallup received 331 responses from retirees, with 3 saying that part-time work was a major source of income. Based on the following randomization distribution of 100 values of $\hat{p}_{2021} - \hat{p}_{2020}$, is there convincing evidence at the 5% significance level that the proportion of retirees relying on part-time work was different in 2021 than in 2020?

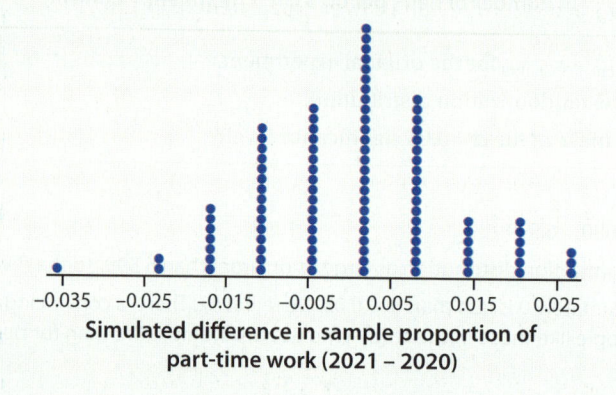

SOLUTION

State:

$H_0: p_{2021} - p_{2020} = 0$

$H_a: p_{2021} - p_{2020} \neq 0$

p_{2021} = the population proportion of retirees in 2021 for whom part-time work was a major source of income

p_{2020} = the population proportion of retirees in 2020 for whom part-time work was a major source of income

$\alpha = 0.05$

The evidence for H_a is: $\hat{p}_{2021} - \hat{p}_{2020} = \dfrac{3}{331} - \dfrac{12}{312} = 0.0091 - 0.0385 = -0.0294 \neq 0$

> **STATE:** State the hypotheses, parameter(s), significance level, and evidence for H_a.

Plan: Randomization test for the difference between two proportions.

Random? Independent random samples of retirees in both 2020 and 2021. ✓

> **Plan:** Identify the appropriate inference method and check the conditions.

Do:
- $\hat{p}_{2021} - \hat{p}_{2020} = 0.0091 - 0.0385 = -0.0294$
- P-value:

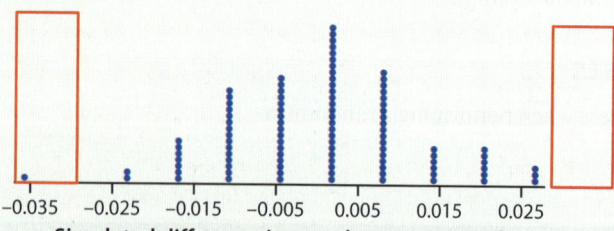

> **Do:** If the conditions are met, perform calculations.
> - Calculate the test statistic.
> - Find the P-value.
>
> *Remember:* The test statistic is the difference in sample proportions that you calculated in the State step as evidence for the alternative hypothesis.

P-value ≈ 0.01

> Because randomization distributions are typically not exactly symmetric, count the number of dots in both tails of the distribution.

Conclude: Because the P-value of approximately 0.01 is less than $\alpha = 0.05$, we reject H_0. There is convincing evidence that the population proportion of retirees relying on part-time work was different in 2021 than in 2020.

> **Conclude:** Make a conclusion about the hypotheses in the context of the problem.

EXAM PREP: FOR PRACTICE, TRY EXERCISE 13.

The major benefit of the randomization test is that it is appropriate for use in a wider variety of studies than the two-sample t test for a difference in means or the two-sample z test for a difference in proportions. The results from the randomization test will be quite similar to the results of the t test or the z test when the conditions are met for these tests (as in the flu example in this section). But randomization tests are also appropriate when the distribution-based conditions aren't met. For example, we can use a randomization test for the hair growth experiment in Example 13.5.1, even though each group has an outlier.

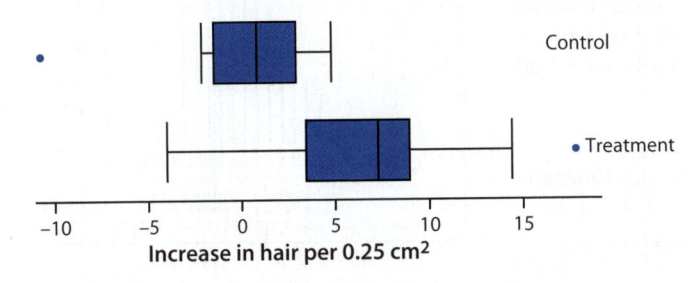

Likewise, a randomization test is also useful in the case of Example 13.5.3. There the data do not meet the Large Counts condition, as there are only 3 successes in the 2021 group.

TECH CORNER   Randomization test

Difference between two means

You can use technology to conduct a randomization test to compare two population means. We'll illustrate this process using the hair growth data from this section.

Control	0.3	1.4	3.0	3.7	−1.5	−2.0	0.0	4.8	2.4	−11.3
Treatment	3.5	5.0	7.3	18.3	14.5	6.7	9.0	−0.7	7.8	−4.0
	6.0	4.5	8.0	11.4	1.0	7.3	8.5	−0.7	13.5	

Applet

1. Go to www.stapplet.com and launch the *One Quantitative Variable, Multiple Groups* applet.

2. Enter the variable name "Hair growth" and choose "Raw data" as the "Input."

3. Enter each group name and the values for that group as shown.

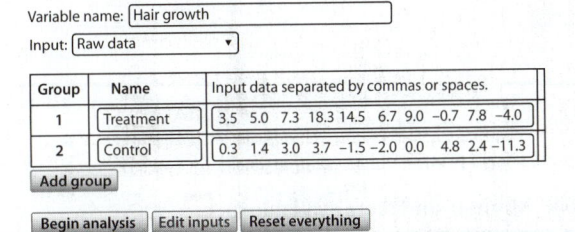

4. Click the Begin analysis button. Then, scroll down to the "Perform Inference" section.

5. Choose Simulate difference in two means for the procedure, enter 1000 for Number of samples to add, and click the Add trials button.

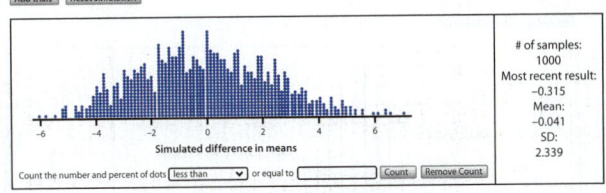

6. To estimate the *P*-value, choose "greater than" in the counter below the dotplot and type the test statistic 6.599. Click the Count button. The estimated *P*-value is the percentage of dots in the selected region.

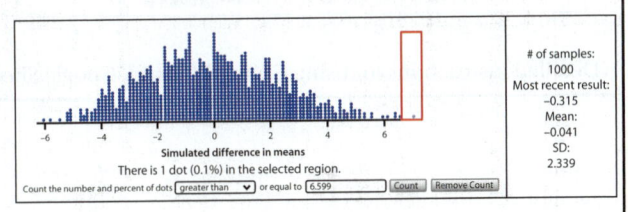

Note: If you are doing a two-sided test, count the number of dots in both tails (e.g., greater than 6.599 and less than −6.599).

(*continued*)

TI-83/84

This type of significance test is not currently available on the TI-83/84.

Difference between two proportions

You can use technology to conduct a randomization test to compare two population proportions. We'll illustrate this process using the flu vaccine data from this section. Recall that 37 out of 628 patients who got the full dose experienced headache, while 28 out of 631 who received the half dose experienced headache.

Applet

1. Go to www.stapplet.com and launch the *Two Categorical Variables* applet.

2. Choose "Two-way table" as the Input and fill in the table as shown.

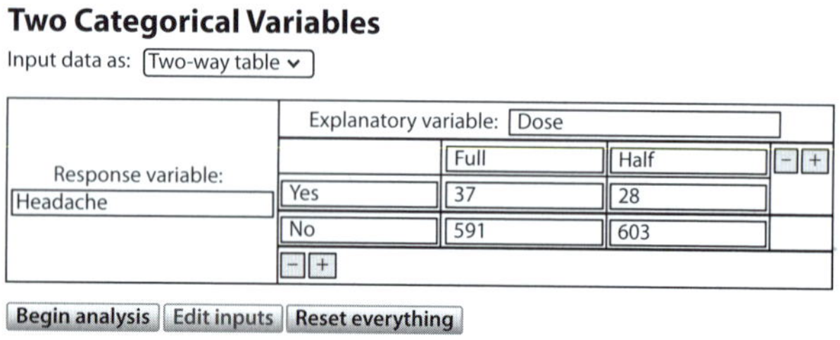

Two Categorical Variables
Input data as: Two-way table

	Explanatory variable: Dose	
Response variable: Headache	Full	Half
Yes	37	28
No	591	603

3. Click the Begin analysis button. Then, scroll down to the "Perform Inference" section.

4. Choose Simulate difference in proportions for the procedure, enter 1000 for Number of samples to add, and click the Add trials button.

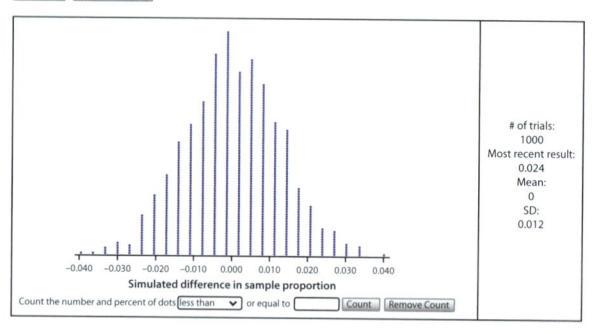

5. To estimate the *P*-value, choose "greater than" in the counter below the dotplot and type the test statistic 0.0145. Click the Count button. The *P*-value is the percentage of dots in the selected region.

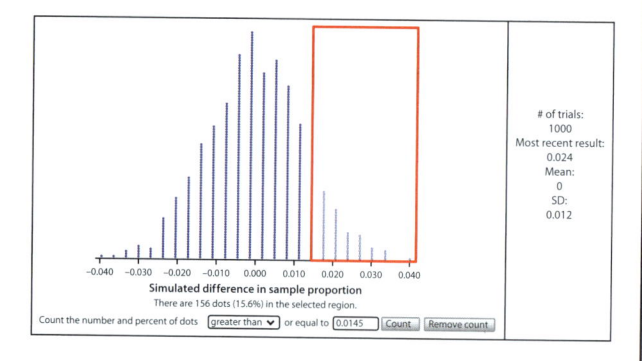

Notes:

1. If you are doing a two-sided test, count the number of dots in both tails (e.g., greater than 0.0145 and less than −0.0145).

2. If the applet is set up to use percentages rather than proportions, click "Adjust color, rounding, and percent/proportion preferences" at the bottom of the page and choose decimals in the first box.

TI-83/84

This type of significance test is not currently available on the TI-83/84.

Detailed instructions for using CrunchIt!, Excel, Google Sheets, JMP, Minitab, and R are available in Achieve.

Section 13.5 What Did You Learn?

Review the learning goals from this section. Then practice what you've learned by working through the exercises.

Learning Goal	Example	Exercises
Describe how to use note cards to create a randomization distribution for a test about a difference between two proportions or two means.	13.5.1	5–8
Estimate the *P*-value from a randomization distribution.	13.5.2	9–12
Use the four-step process to perform a randomization test about a difference between two proportions or two means.	13.5.3	13–16

Section 13.5 Exercises

Building Concepts and Skills These exercises assess the basic knowledge you should have after reading the section.

1. What is the test statistic in a randomization test about a difference between two means?

2. What is the only condition required for a randomization test about a difference between two proportions or two means?

3. True/False: If two people complete a randomization test on the same set of data, they will always get the same *P*-values.

4. True/False: If the Large Counts condition is met, the *P*-value from a randomization test will be close to the *P*-value from a two-sample *z* test for a difference in proportions.

Mastering Concepts and Skills These exercises reinforce the learning goals as shown in the examples.

5. **Refer to Example 13.5.1 Computer Industry** Have perceptions of the computer industry changed over 20 years? In 2001, Gallup randomly selected a sample of 326 U.S. adults and asked them for their view of the industry. Three of those adults answered that they had a very negative view of the industry. In 2021, Gallup asked the question again, this time to a random sample of 1006 U.S. adults.[59] In this survey, 70 responded that they had a very negative view of the industry. The hypotheses the researchers wanted to test are:

$$H_0: p_{2021} - p_{2001} = 0$$
$$H_a: p_{2021} - p_{2001} \neq 0$$

Describe how to use note cards to simulate one possible value of $\hat{p}_{2021} - \hat{p}_{2001}$ in a randomization distribution for this study.

6. **Vaping in College** Each year, the University of Michigan's Monitoring the Future study administers a survey about substance use and related behaviors to randomly selected U.S. college students. One survey question asks students whether they have used any vaping products in the current year. In 2019, 439 of a random sample of 1005 college students said "Yes." In 2020 (during the Covid-19 pandemic), 317 of a random sample of 800 college students said "Yes."[60] Researchers want to know if the proportion of college students who vape decreased over this 1-year period. The hypotheses the researchers wanted to test are:

$$H_0: p_{2020} - p_{2019} = 0$$
$$H_a: p_{2020} - p_{2019} < 0$$

Describe how to use note cards to simulate one possible value of $\hat{p}_{2020} - \hat{p}_{2019}$ in a randomization distribution for this study.

7. **Teaching Methods** Researchers in Rwanda were concerned that students in a Physics course were struggling because of the English-language technical vocabulary. The course was traditionally taught by a lecturer using a blackboard. Researchers wondered whether using a multimedia presentation on a screen in the classroom would create a better outcome for the students. They randomly assigned 32 students to one of the two teaching methods for the 4-week course. They collected post-test scores for all students in both groups.[61]

Lecture	8.83	11.21	8.36	7.30	6.23	6.43	15.09
	9.01	7.35	7.48	9.88	6.81	4.80	10.19
	5.93	8.62	6.16	11.59	7.39		
Multimedia	8.20	14.03	17.20	3.73	14.62	14.92	10.68
	10.79	10.72	9.44	11.06	9.02	9.89	

Researchers would like to know if the true mean post-test score for students like those in the study who are taught using the multimedia presentation is greater than the true mean post-test score for students like those in the study who are taught using the traditional lecture method. The hypotheses the researchers wanted to test are:

$$H_0: \mu_{\text{Lecture}} - \mu_{\text{Multimedia}} = 0$$
$$H_a: \mu_{\text{Lecture}} - \mu_{\text{Multimedia}} < 0$$

Describe how to use note cards to simulate one possible value of $\bar{x}_{\text{Lecture}} - \bar{x}_{\text{Multimedia}}$ in a randomization distribution for this experiment.

8. **Music and Plant Growth** Researchers designed an experiment to determine whether plants grow better if they are exposed to classical music or to metal music. They selected 10 bean seeds and planted each seed in a Styrofoam cup. They assigned half of these cups to be exposed to metal music each night, while the other half were exposed to classical music each night. They recorded the amount of growth, in millimeters, for each plant after 2 weeks. Here are the data:[62]

Metal	22	36	73	57	3
Classical	87	78	124	121	19

   Researchers want to know if there is a difference in true mean growth of plants like the ones in this study that are exposed to classical music versus metal music. The hypotheses the researchers wanted to test are:

   $$H_0: \mu_{Metal} - \mu_{Classical} = 0$$
   $$H_a: \mu_{Metal} - \mu_{Classical} \neq 0$$

   Describe how to use note cards to simulate one possible value of $\bar{x}_{Metal} - \bar{x}_{Classical}$ in a randomization distribution for this experiment.

9. **Refer to Example 13.5.2 Computer Industry** Refer to Exercise 5. Let $\hat{p}_{2021}$ = the proportion of U.S. adults in the 2021 study who had a very negative view of the computer industry and $\hat{p}_{2001}$ = the proportion of U.S. adults in the 2001 study who had a very negative view of the computer industry.

   (a) Calculate the value of $\hat{p}_{2021} - \hat{p}_{2001}$ for the original study.

   (b) Estimate the P-value using the randomization distribution with 100 values given here.

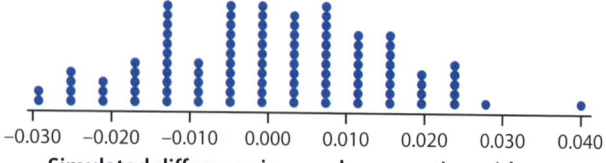

   (c) What conclusion would you make at the $\alpha = 0.05$ significance level?

10. **Vaping in College** Refer to Exercise 6. Let $\hat{p}_{2020}$ = the proportion of U.S. college students in the 2020 study who used vaping products and $\hat{p}_{2019}$ = the proportion of U.S. college students in the 2019 study who used vaping products.

    (a) Calculate the value of $\hat{p}_{2020} - \hat{p}_{2019}$ for the original study.

    (b) Estimate the P-value using the randomization distribution with 100 values given here.

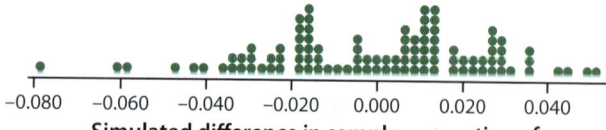

    (c) What conclusion would you make at the $\alpha = 0.05$ significance level?

11. **Teaching Methods** Refer to Exercise 7. Let $\bar{x}_{Lecture}$ = the sample mean post-test score for the students in the lecture group and $\bar{x}_{Multimedia}$ = the sample mean post-test score for the students in the multimedia group.

    (a) Calculate the value of $\bar{x}_{Lecture} - \bar{x}_{Multimedia}$ for the original experiment.

    (b) Estimate the P-value using the randomization distribution with 100 values given here.

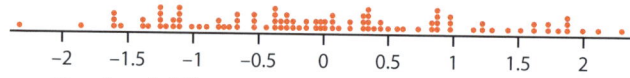

    (c) Interpret the P-value you found in part (b).

12. **Music and Plant Growth** Refer to Exercise 8. Let $\bar{x}_{Metal}$ = the sample mean growth for the plants exposed to metal music and $\bar{x}_{Classical}$ = the sample mean growth for the plants exposed to classical music.

    (a) Calculate the value of $\bar{x}_{Metal} - \bar{x}_{Classical}$ for the original experiment.

    (b) Estimate the P-value using the randomization distribution with 100 values given here.

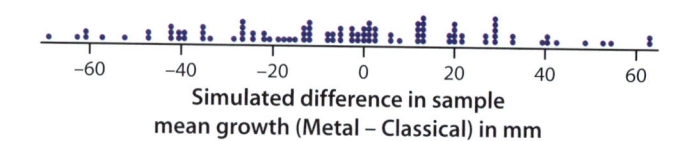

    (c) Interpret the P-value you found in part (b).

13. **Refer to Example 13.5.3 Sleep Deprivation** Do the effects of sleep deprivation linger for more than a day? Researchers designed a study using 21 volunteer subjects between the ages of 18 and 25. All 21 participants took a computer-based visual discrimination test at the start of the study. Then researchers randomly assigned the subjects into two groups. The 11 subjects in one group were deprived of sleep for an entire night in a laboratory setting. The 10 subjects in the other group were allowed unrestricted sleep for the night. Both groups were allowed as much sleep as they wanted for the next two nights. On Day 4, all the subjects took the same visual discrimination test on the computer. Researchers recorded the improvement in time (measured in milliseconds) from Day 1 to Day 4 on each subject's tests. Here are the data:[63]

Sleep deprivation	−14.7	−10.7	−10.7	2.2	2.4	4.5
	7.2	9.6	10.0	21.3	21.8	
Unrestricted sleep	−7.0	11.6	12.1	12.6	14.5	18.6
	25.2	30.5	34.5	45.6		

    Let $\bar{x}_{Deprived}$ = the sample mean improvement in test time for subjects deprived of sleep and $\bar{x}_{Unrestricted}$ = the sample mean improvement in test time for subjects allowed unrestricted sleep. Is there convincing evidence that sleep deprivation decreases the true mean improvement time on the visual discrimination task compared to unrestricted sleep for subjects like the ones in this study? Use a randomization test with 100 values of $\bar{x}_{\text{Deprived}} - \bar{x}_{\text{Unrestricted}}$ in the given randomization distribution.

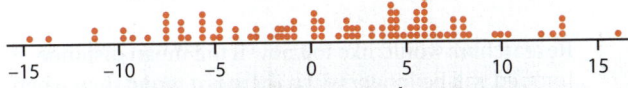

14. **Marketing to Children** Many new products introduced into the market are targeted toward children. The behavior of children with regard to choosing new products is of particular interest to companies that design marketing strategies for these products. In one study, researchers randomly selected children in different age groups and compared their ability to sort new products into correct product categories (in this case, milk or juice). Here is a summary of the data:[64]

Age group	n	Number who sorted correctly
4- to 5-year-olds	50	10
6- to 7-year-olds	53	28

Let $\hat{p}_{\text{younger}}$ = the proportion of the 4- to 5-year-olds who sorted correctly and $\hat{p}_{\text{older}}$ = the proportion of 6- to 7-year-olds who sorted correctly. Is there convincing evidence at the 5% significance level that a greater proportion of 6- to 7-year-olds than 4- to 5-year-olds can sort correctly? Use a randomization test with 100 values of $\hat{p}_{\text{younger}} - \hat{p}_{\text{older}}$ in the given randomization distribution.

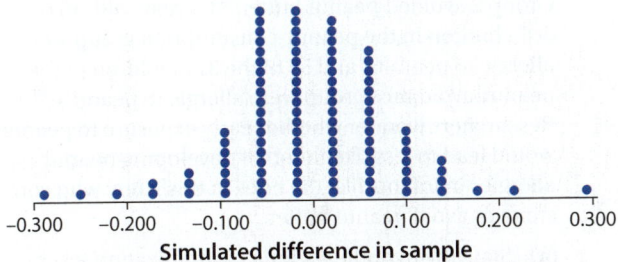

15. **Fish on the Moon** French researchers set an ambitious goal to farm fish on the moon. They performed an experiment to determine whether certain varieties of fish can be safely transported on a rocket into space. The researchers randomly assigned 400 European seabass eggs into two groups of 200 eggs each. They placed all of the eggs in both groups in a dish filled with seawater. They then placed the first group of eggs in a vibration chamber designed to simulate a typical takeoff. They kept the second group of eggs in similar environmental conditions, but with no vibrations. In the vibration group, 76% of the eggs went on to hatch, compared to 82% of the eggs in the control group.[65] Let $\hat{p}_{\text{vibration}}$ = the proportion of eggs receiving the vibration treatment that hatched and $\hat{p}_{\text{control}}$ = the proportion of eggs with no vibrations that hatched. The researchers want to know if the vibrations from a rocket launch will affect the proportion of European seabass eggs that hatch. Use a randomization test with 100 values of $\hat{p}_{\text{vibration}} - \hat{p}_{\text{control}}$ in the given randomization distribution.

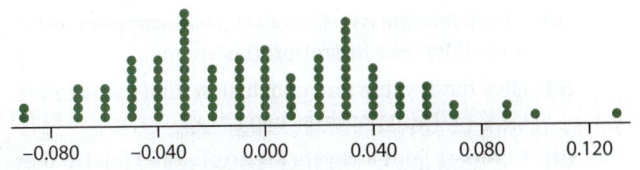

16. **Cloud Seeding** Does seeding clouds affect the amount of rainfall they produce? Researchers in Tasmania randomly assigned some clouds to be seeded and some to be left unseeded, and then measured the amount of rainfall produced by each cloud (in inches). The observed values of rainfall are given here.[66] Let $\bar{x}_{\text{seeded}}$ = the sample mean amount of rainfall for the seeded clouds and $\bar{x}_{\text{unseeded}}$ = the sample mean amount of rainfall for the unseeded clouds. For clouds like the ones in this study, is there convincing evidence at the 1% significance level that the mean amount of rainfall is different in seeded and unseeded clouds? Use a randomization test with 200 values of $\bar{x}_{\text{seeded}} - \bar{x}_{\text{unseeded}}$ in the given randomization distribution.

Seeded	0.81	2.48	0.37	0.42	0.88	1.25	1.11
	1.09	0.79	0.76	0.24	2.35	1.63	1.08
Unseeded	1.44	0.84	0.37	0.64	0.30	0.76	1.08
	0.74	4.06	0.40	2.36	2.23	1.16	6.00

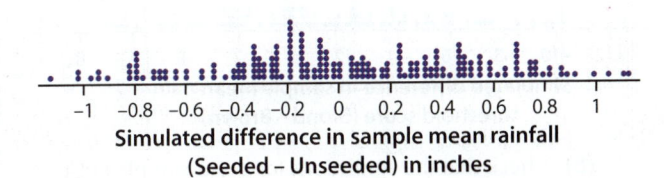

Applying the Concepts These exercises ask you to apply multiple learning goals in a new context or to apply what you learned in this section in a new way.

17. **Trust in Science** How do the people of Nigeria and Ethiopia compare when it comes to trust in science? A global survey asked random samples of people in many nations of the world about their trust in science. Of the 1002 respondents in Nigeria, 439 reported that they trust scientists a little or a lot. Of the 1003 respondents in Ethiopia, 753 reported that they trust scientists a little or a lot.[67]

(a) Use the given randomization distribution with 100 values and a randomization test to determine whether there is convincing evidence at the $\alpha = 0.05$ significance level that the proportions of people who trust scientists are different between Ethiopia and Nigeria.

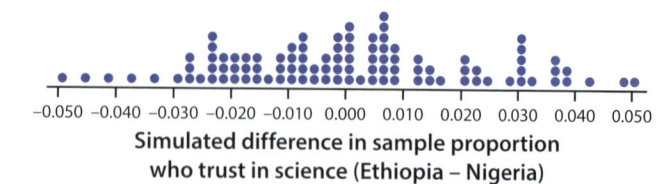

(b) Check that the conditions for a two-sample z test for a difference in proportions are met.

(c) Give the P-value and conclusion for a two-sample z test of the hypotheses in part (a).

(d) Compare your answers to parts (a) and (c). Do they give similar P-values and conclusions?

18. **Hair Color and Pain** Are pain thresholds different for people with blond hair versus people with brown hair? This is the question that researchers at the University of Melbourne asked. They chose random samples of people with naturally blond hair and people with naturally brown hair in Australia. Each person completed a pain sensitivity test and was given a pain threshold score (the higher the score, the higher the tolerance for pain). Here are the data:[68]

Blond	62	60	71	55	48	63	57
	43	41	52				
Brown	42	50	41	37	32	39	51
	30	35					

(a) Use the given randomization distribution with 100 values and a randomization test to determine whether there is convincing evidence at the 1% significance level that the mean pain threshold is different in people with blond hair and people with brown hair in Australia.

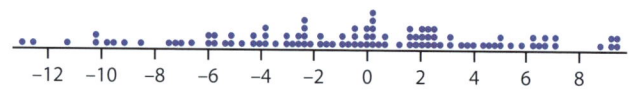

(b) Check that the conditions for a two-sample t test for a difference in means are met.

(c) Give the P-value and conclusion for a two-sample t test of the hypotheses in part (a).

(d) Compare your answers to parts (a) and (c). Do they give similar P-values and conclusions?

19. **Tech** **Jump for the Target** Will children jump farther if they have a target in front of them? To find out, a gym teacher conducted an experiment with 29 twelve-year-old children. The teacher wrote the names of all 29 children on identical slips of paper, put them in a hat, mixed them up, and drew out 15 slips. Each of the 15 selected students did a standing long jump from behind a starting line. The remaining 14 students did a similar standing long jump from behind the starting line, but with a target line placed 200 centimeters from the starting line. The distance from the starting line to the back of each student's closest foot was measured (in centimeters). The data are shown here and in the file **LongJump**:[69]

No target	146	190	109	181	155	167	154	171
	157	156	128	157	167	162	137	
Target	199	167	147	180	185	170	171	139
	154	126	179	158	181	152		

Researchers would like to know if the mean distance jumped will be longer when there is a target than when there is no target for children like the ones in the study.

(a) State the hypotheses for a randomization test in this context.

(b) Describe how to use note cards to simulate one possible value in a randomization distribution for this experiment.

(c) Let $\bar{x}_{\text{Target}}$ = the sample mean distance jumped for children who had a target and $\bar{x}_{\text{NoTarget}}$ = the sample mean distance jumped for the children who did not have a target. Calculate the value of $\bar{x}_{\text{Target}} - \bar{x}_{\text{NoTarget}}$ for the original experiment.

(d) Use technology to find a randomization distribution with at least 1000 values, and estimate the P-value.

(e) What conclusion would you make at the $\alpha = 0.05$ significance level?

20. **Tech** **Peanut Allergies** A recent study of peanut allergies explored the relationship between early exposure to peanuts and the subsequent development of an allergy to peanuts. Infants (4 to 11 months old) who had shown evidence of other kinds of allergies were randomly assigned to one of two groups: Group 1 consumed a baby-food form of peanut butter, and Group 2 avoided peanut butter. At 5 years old, 10 of the 307 children in the peanut-consumption group were allergic to peanuts, and 55 of the 321 children in the peanut-avoidance group were allergic to peanuts.[70] Researchers wonder whether early exposure to peanuts would lead to less likelihood of developing peanut allergies in infants like the ones in this study who consume or avoid peanut butter.

(a) State the hypotheses for a randomization test in this context.

(b) Describe how to use note cards to simulate one possible value in a randomization distribution for this experiment.

(c) Let $\hat{p}_{\text{Peanuts}}$ = the proportion of 5-year-olds in the peanut butter group who developed peanut allergies and $\hat{p}_{\text{NoPeanuts}}$ = the proportion of 5-year-olds in the non-peanut-butter group who developed peanut allergies. Calculate the value of $\hat{p}_{\text{Peanuts}} - \hat{p}_{\text{NoPeanuts}}$ for the original experiment.

(d) Use technology to find a randomization distribution with at least 1000 values, and estimate the P-value.

(e) What conclusion would you make at the $\alpha = 0.05$ significance level?

Extending the Concepts These exercises challenge you to explore statistical concepts and methods that go beyond what you learned in this section.

21. **Randomization Test for Slope** Is there a relationship between a person's body temperature and their resting pulse rate? The table shows x = body temperature (°F) and y = pulse rate (in beats per minute) for each of 20 randomly selected students at a university.[71] Do these data provide convincing evidence of a linear relationship between body temperature and resting pulse rate for students at this university?

Temperature	Pulse rate	Temperature	Pulse rate
97.9	78	98.8	80
97.1	64	98.0	82
98.6	72	97.5	84
98.5	80	98.0	75
98.1	60	97.6	87
97.0	56	97.7	61
97.2	75	96.6	69
97.8	70	96.6	69
98.3	90	96.6	71
97.3	76	96.8	74

The hypotheses for this test are

$$H_0: \beta_1 = 0$$
$$H_a: \beta_1 \neq 0$$

where β_1 = the slope of the population regression line relating heart rate to temperature.

To complete a test to determine if we have convincing evidence against the null hypothesis that the slope is 0, we need to know what the distribution of possible sample slopes is when the population slope is 0—that is, when there is no relationship between pulse rate and temperature in the population. We can simulate this by separating the specific temperatures and heart rates in our sample, randomly reordering the heart rates, and reassigning them to the individual temperatures.

(a) Describe how you could use note cards to simulate one possible value of b_1 in a randomization distribution of the slope in this study.

(b) The sample slope for these data is $b_1 = 4.944$. Use the given randomization distribution with 100 values to estimate the P-value for the randomization test to determine whether there is convincing evidence of a linear relationship between body temperature and resting pulse rate for students at this university.

Simulated slope of least-squares regression line relating heart rate to temperature

(c) What conclusion would you make at the $\alpha = 0.05$ significance level?

Cumulative Review These exercises revisit what you learned in previous sections.

22. **Comfortable Retirement (7.5)** Gallup asked a random sample of 630 non-retirees whether they think they will have enough income to live comfortably in retirement. The responses included 334 people who said yes.[72] Construct and interpret a 99% confidence interval for the proportion of all non-retirees who would say they will have enough income to live comfortably in retirement.

23. **Marker for Alzheimer's Disease (3.3)** Amyloid-β (Abeta) is a protein fragment that has been linked to Alzheimer's disease. Autopsies from a sample of Catholic priests included measurements of Abeta (pmol/g tissue from the posterior cingulate cortex) from three groups: subjects who had exhibited no cognitive impairment before death, subjects who had exhibited mild cognitive impairment, and subjects who had mild to moderate Alzheimer's disease.[73] Construct boxplots for all three groups and compare the distributions.

No impairment	114	41	276	0	16	228	927
	0	211	829	1561	0	276	959
	16	24	325	49	537		
Mild impairment	73	33	16	8	276	537	0
	569	772	0	260	423	780	1610
	0	309	512	797	24	57	106
Alzheimer's disease	407	390	1154	138	634	919	1415
	390	1024	1154	195	715	1496	407
	1171	439	894				

Section 13.6 Bootstrapping

LEARNING GOALS

By the end of this section, you will be able to:
- Describe how to use note cards to create a bootstrap distribution for a population proportion or population mean.
- Use a bootstrap distribution to estimate a confidence interval for a population proportion or population mean.
- Use the four-step process to estimate and interpret a bootstrap confidence interval for a population proportion or population mean.

Sections 13.1–13.5 introduced ways of performing significance tests when some of the conditions required for the traditional inference methods presented in Chapters 8–10 are not met. What about confidence intervals? In this section, we offer an alternative approach to constructing confidence intervals for a population proportion or a population mean that eliminates the Large Counts or Normal/Large Sample condition.

Bootstrap Distributions

When constructing a confidence interval for a population mean or population proportion, the structure of the confidence intervals has been defined as

$$\text{point estimate} \pm \text{margin of error}$$
$$= \text{point estimate} \pm (\text{critical value})(\text{standard error of statistic})$$

where the critical value is z^* or t^*. However, when the Large Counts condition isn't met, we shouldn't use a normal distribution to calculate z^*. Likewise, when the Normal/Large Sample condition isn't met, we shouldn't use a t distribution to calculate t^*. To determine how much the sample statistic is expected to vary, at most, from the population parameter, we can use simulation to approximate the sampling distribution of the statistic.

In Example 13.1.3, we recorded the rent for a random sample of nine apartments in Chicago that did not have an application fee. Here are the data once again:

$975 $1895 $1500 $1800 $925 $1600 $1900 $1600 $870

For this sample, the mean rent is $\bar{x} = \$1451.67$. To construct a confidence interval for the population mean rent μ, we need to know how much $\bar{x}$ varies from μ, at most. Ideally, we would take many different samples of nine apartments from the population of all apartments for rent in Chicago without an application fee and compute the mean for each sample. But if we had the resources to take many, many random samples of nine such apartments, we should have the resources to just measure the whole population and calculate the population mean μ. The next best thing is to *simulate* taking many samples from the population.

We start with the random sample that we do have. Because the apartments were randomly selected, we hope that this sample does a good job of representing the population of all apartments in Chicago with no application fee. To create a "population" to sample from, we replicate the original sample many times to create an approximation of the true population. For the rent data, here is a dotplot of the original sample.

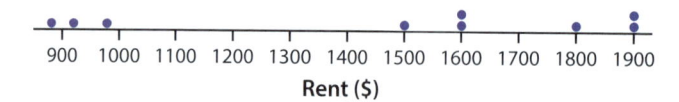

And here is the approximate population after replicating the original sample many times.

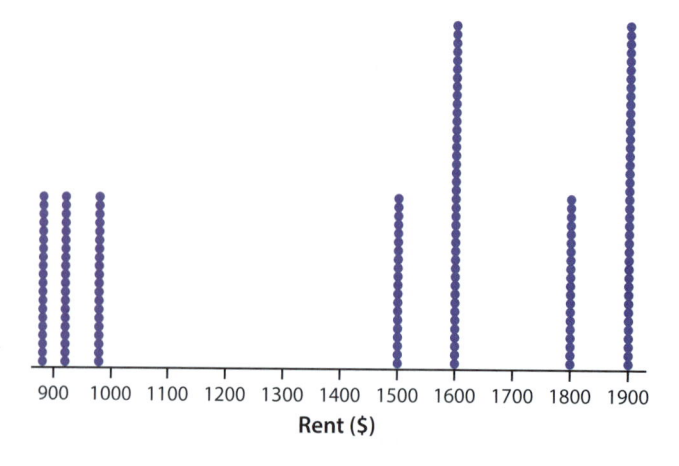

To estimate the sampling distribution of the sample mean rent $\bar{x}$, we select many random samples of size 9 from this approximate population.

Note that selecting a random sample of size 9 from this approximate population is equivalent to selecting a random sample of size 9 *with replacement* from the original sample of nine apartments.

This key observation makes it much easier to simulate the sampling distribution of $\bar{x}$. To perform the simulation, write each of the rent values from the sample on a separate note card. Shuffle the cards, deal one, and record the rent value. Put the card back into the deck and shuffle again. Choose a new card and record the rent value. Continue this process until you have recorded nine rent values. This sample of nine values is called a **bootstrap sample.** Then compute the sample mean from the bootstrap sample. A **bootstrap distribution** consists of the sample means computed from many, many bootstrap samples.

> A **bootstrap sample** is a random sample of the same size as the original sample that is selected with replacement from the original sample.

> A **bootstrap distribution** consists of the collection of sample statistics computed from many bootstrap samples.

Why is it called a "bootstrap" distribution? The phrase "pulling yourself up by your bootstraps" refers to an individual using only their own strength to achieve a goal. Likewise, a bootstrap distribution uses only the original sample to generate the estimated sampling distribution.

EXAMPLE 13.6.1

Can you name the branches of government?

Creating bootstrap samples

PROBLEM

In a random sample of 1416 U.S. adults surveyed by the Annenberg Foundation, only 510 could name all three branches of government.[74] Describe how to use note cards to simulate one possible value in a bootstrap distribution of $\hat{p}$ for this study.

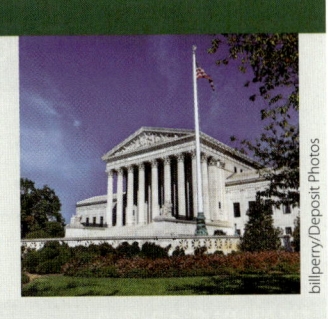

SOLUTION

Take 1416 note cards. Write yes on 510 and no on the remaining 906 cards. Shuffle the cards and deal one. Record the word written on the card. Replace the card in the deck, shuffle the deck, and deal one card. Again record the word written on the card. Continue this process until 1416 words have been written down. Compute $\hat{p}$ = the proportion of cards drawn that say "yes."

EXAM PREP: FOR PRACTICE, TRY EXERCISE 5.

Confidence Intervals from Bootstrap Distributions

We used technology to select 1000 bootstrap samples from the Chicago rent data, calculate the mean of each bootstrap sample, and graph the resulting bootstrap distribution.

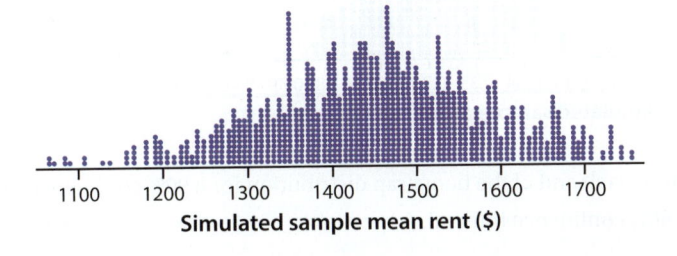

Note that the center of this distribution is approximately $1452, the mean of our original sample and the point estimate for our confidence interval. To finish the computation of the confidence interval, we need to account for the variability in the sampling distribution.

With a bootstrap distribution, this is actually quite simple. If we want a 95% confidence interval, simply eliminate the smallest 2.5% of the dots and the largest 2.5% of the dots. The endpoints of the confidence interval are the minimum and maximum of the remaining dots. This interval will be roughly centered at the sample statistic (also the center of the bootstrap distribution) and covers the middle 95% of the bootstrap distribution. For the bootstrap distribution of 1000 sample mean rents shown previously, we eliminate the smallest 25 points and the largest 25 points. The minimum and maximum of the remaining points give a 95% confidence interval of $1191.20 to $1696.10, as shown in the following figure.

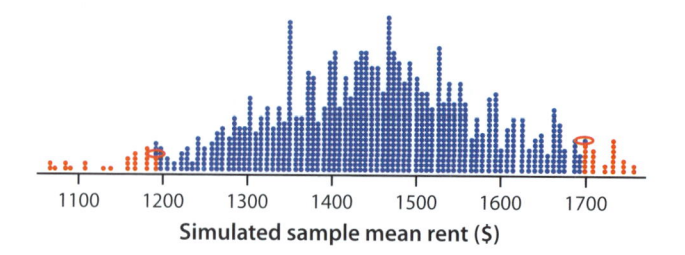

Simulated sample mean rent ($)

We are 95% confident that the mean rent for all apartments without application fees in Chicago is between $1191.20 and $1696.10.

This same idea can be used for any confidence level. If you want a 90% confidence interval, divide the remaining 10% in two, and cut off the bottom and top 5% of the bootstrap distribution. The only condition required for the bootstrap confidence interval is the Random condition. When the Large Counts or Normal/Large Sample condition is also met, the bootstrap confidence interval will be similar to the confidence interval calculated using the methods described in Chapter 7.

HOW TO | **Check the Condition for Constructing a Bootstrap Confidence Interval for p or μ**

To construct a bootstrap confidence interval for a population proportion p or a population mean μ, check that the following condition is met:

- **Random:** The data come from a random sample from the population of interest.

EXAMPLE 13.6.2

Can you name the branches of government?

Estimate a bootstrap confidence interval

PROBLEM

Let's return to Example 13.6.1. In a random sample of 1416 U.S. adults surveyed by the Annenberg Foundation, only 510 could name all three branches of government. The dotplot shows the value of $\hat{p}$ = the sample proportion of people who said yes in 200 bootstrap samples of size 1416.

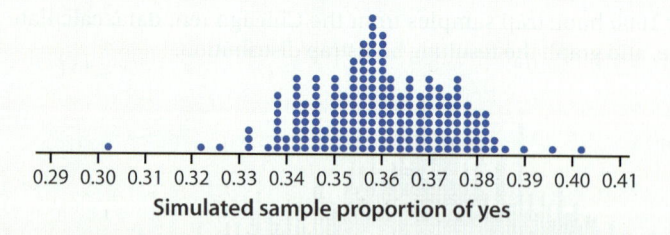

Simulated sample proportion of yes

(a) How many dots should be eliminated from each end of the bootstrap distribution for a 90% confidence interval?

(b) Use the bootstrap distribution to find a 90% confidence interval.

SOLUTION:

(a) For 90% confidence, we need to eliminate 5% of dots from each end of the bootstrap distribution → eliminate 200(0.05) = 10 dots from each end.

(b) The 90% confidence interval for the proportion of people who can name all three branches of government is (0.338, 0.382).

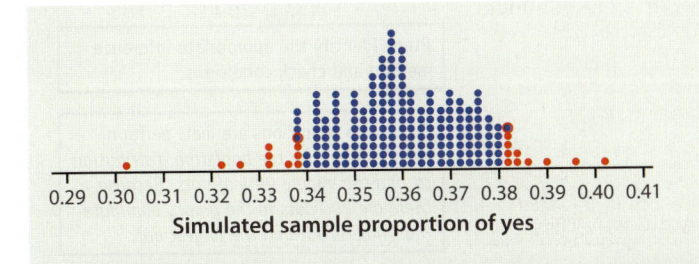

> The 11th dot from the left is located at 0.338 and the 11th dot from the right is located at 0.382.

Simulated sample proportion of yes

EXAM PREP: FOR PRACTICE, TRY EXERCISE 9.

In Example 7.3.3, we were 90% confident that the interval from 0.339 to 0.381 captured the proportion of U.S. adults who can name all three branches of government. That interval is quite close to the bootstrap interval that we just found in Example 13.6.2.

Putting It All Together: Bootstrap Confidence Intervals

One last time, we follow the four-step process when constructing a **bootstrap confidence interval**.

> A **bootstrap confidence interval** is a nonparametric confidence interval for a population parameter estimated with a bootstrap distribution.

EXAMPLE 13.6.3

How much football is really in a football game?

Four-step process for a bootstrap confidence interval

PROBLEM

Most National Football League (NFL) games on TV last more than 3 hours, but how much actual playing time is there in a typical NFL game? Researchers Kirk Goldsberry and Katherine Rowe investigated this question as part of a sports analytics class they taught at the University of Texas. They carefully watched a random sample of 7 games from the 2019 NFL season and recorded the amount of time from when the ball was snapped to when the referee blew the whistle. Here are the game time values (in minutes), excluding any overtime periods.[75]

15.43 16.53 20.18 19.87 16.97 15.72 19.30

Construct and interpret a 99% bootstrap confidence interval for μ = the mean amount of game time (in minutes) in all 2019 NFL games using the given bootstrap distribution with 200 values.

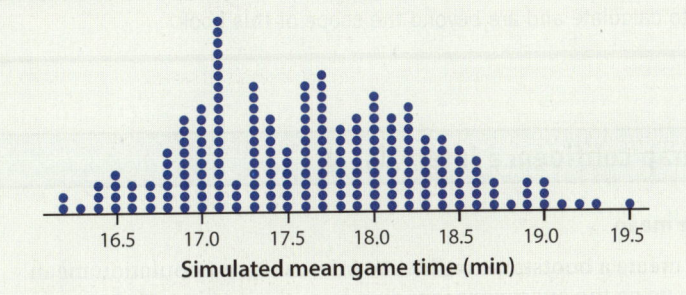

Simulated mean game time (min)

SOLUTION

State: 99% CI for μ = the mean amount of game time (in minutes) in all 2019 NFL games.

> **STATE:** State the parameter you want to estimate and the confidence level.

(continued)

Plan: Bootstrap confidence interval for μ.

Random? Random sample of seven 2019 NFL games. ✓

Do:

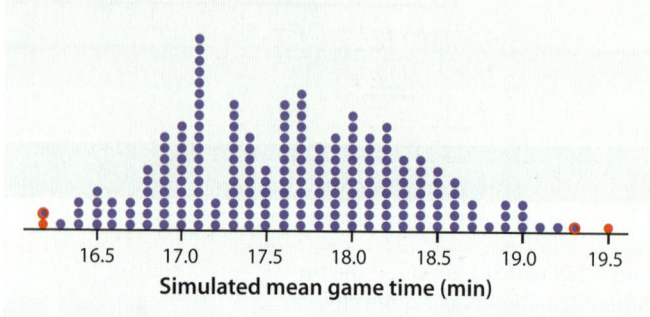

[Dotplot of simulated mean game time from 16.5 to 19.5 minutes]

Simulated mean game time (min)

CI: 16.2 to 19.3

Conclude: We are 99% confident that the interval from 16.2 minutes to 19.3 minutes captures the mean game time for all 2019 NFL games.

Plan: Identify the appropriate inference method and check conditions.

Do: If the conditions are met, perform calculations. *Use the bootstrap distribution to find the middle 99% of dots. Because there are 200 dots in the graph, eliminate the smallest dot and the largest dot.*

Conclude: Make a conclusion about the hypotheses in the context of the problem.

EXAM PREP: FOR PRACTICE, TRY EXERCISE 13.

THINK ABOUT IT

Why do we need bootstrap confidence intervals when we already have other methods? For two reasons: (1) Formula-based confidence intervals don't exist for some parameters, such as a population median. But constructing a bootstrap confidence interval for a median is no harder than constructing a bootstrap confidence interval for a mean. Just record the sample median from each bootstrap sample instead of the sample mean. (2) We can use bootstrap confidence intervals when the distribution-based conditions for inference aren't met. For example, if we have a small sample with an outlier, we shouldn't use a one-sample t interval for μ because the Normal/Large Sample condition isn't met. Because the bootstrap confidence interval has only one condition, it can be used in more situations than our parametric inference methods from Chapters 7–12.

So why don't we always use bootstrap confidence intervals? Because bootstrap intervals do not always achieve the long-run capture rate specified by the confidence level. In Example 7.6.2, we used the NFL game time data to construct a 99% confidence interval of 14.88 minutes to 20.54 minutes using a t distribution. Because the Normal/Large Sample condition was met, we can be 99% confident that this interval actually captures the true mean amount of game time. Note that this interval is wider than the interval we constructed in Example 13.6.3. Because the bootstrap interval of 16.2 minutes to 19.3 minutes is narrower, we are less than 99% confident that this interval captures the true mean amount of game time.

In general, bootstrap intervals tend to have smaller capture rates than the advertised confidence level when sample sizes are small. But when the data meet the additional distribution-based conditions, bootstrap intervals tend to have capture rates close to the confidence level. There are also other types of bootstrap intervals that have better capture rates, but they are more complicated to calculate and are beyond the scope of this book.

TECH CORNER Bootstrap confidence intervals

Confidence interval for one mean

You can use technology to create a bootstrap confidence interval for a population mean. We'll illustrate this process using the apartment rent data from this section. Here are the data in dollars:

975   1895   1500   1800   925   1600   1900   1600   870

(continued)

Applet

1. Go to www.stapplet.com and launch the *One Quantitative Variable, Single Group* applet.
2. Enter the variable name "Rent" and choose "Raw data" as the Input.
3. Enter the values for that group as shown.

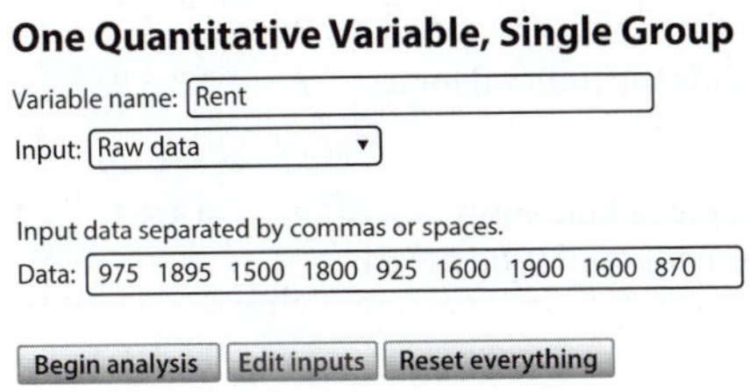

    One Quantitative Variable, Single Group
    Variable name: Rent
    Input: Raw data
    Input data separated by commas or spaces.
    Data: 975 1895 1500 1800 925 1600 1900 1600 870

    [Begin analysis] [Edit inputs] [Reset everything]

4. Click the Begin analysis button. Then, scroll down to the "Perform Inference" section.
5. Choose Simulate sample mean for the procedure, enter 1000 for Number of samples to add, and click the Add samples button.
6. To find the 95% confidence interval, enter 95 in the box to find endpoints for the middle _____ % of simulated means and click the Find endpoints button.

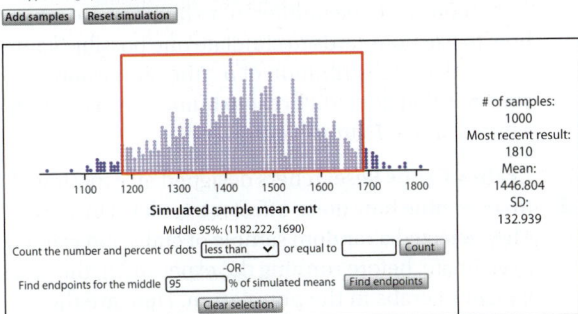

Perform Inference
Inference procedure: Simulate sample mean
Simulates the distribution of the sample mean when selecting samples of the original size *with replacement* from the original sample.

Number of samples to add: 1000
The applet will graph all of your results until you hit "Reset simulation."
[Add samples] [Reset simulation]

Simulated sample mean rent
Middle 95%: (1182.222, 1690)

# of samples: 1000
Most recent result: 1810
Mean: 1446.804
SD: 132.939

TI-83/84

This type of confidence interval is not currently available on the TI-83/84.

Confidence interval for one proportion

You can use technology to create a bootstrap confidence interval for one proportion. We'll illustrate using the survey results about the three branches of government from this section. Recall that 510 U.S. adults could name all three branches of the government, out of a sample of 1416 U.S. adults.

Applet

1. Go to www.stapplet.com and launch the *One Categorical Variable, Single Group* applet.
2. Enter the variable name "Can name branches of government" and choose "Counts in categories" as the Input.
3. Fill in the data table as shown.

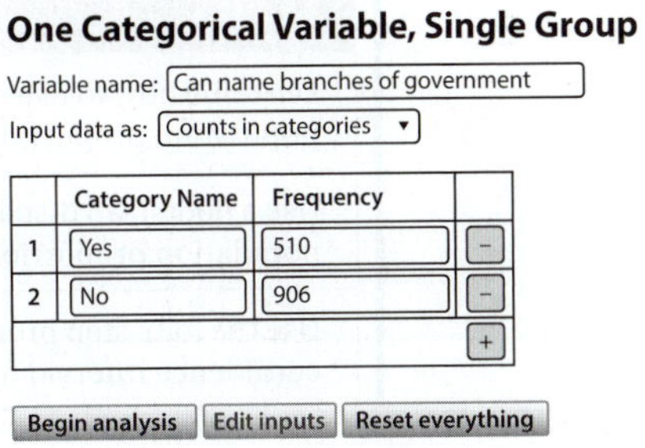

    One Categorical Variable, Single Group
    Variable name: Can name branches of government
    Input data as: Counts in categories

	Category Name	Frequency
1	Yes	510
2	No	906

    [Begin analysis] [Edit inputs] [Reset everything]

4. Click the Begin analysis button. Then, scroll down to the "Perform Inference" section.
5. Choose Simulate sample proportion for the procedure, choose "Yes" as the category to indicate as a success, choose "the observed value" for what the proportion is equal to, and enter 1000 for Number of samples to add. Click the Add samples button.
6. To find the 95% confidence interval, enter 95 in the box to find endpoints for the middle _____ % of simulated results and click the Find endpoints button.

Perform Inference
Inference procedure: Simulate sample proportion   Category to indicate as success: Yes
Simulate the distribution of the sample proportion for samples of the original size assuming that the true proportion is equal to the observed value.
Number of samples to add: 1000
The applet will graph all of your results until you hit "Reset simulation."
[Add samples] [Reset simulation]

Simulated sample proportion of Yes
Middle 95%: (0.336, 0.386)

# of samples: 1000
Most recent result: 0.396
Mean: 0.361
SD: 0.013

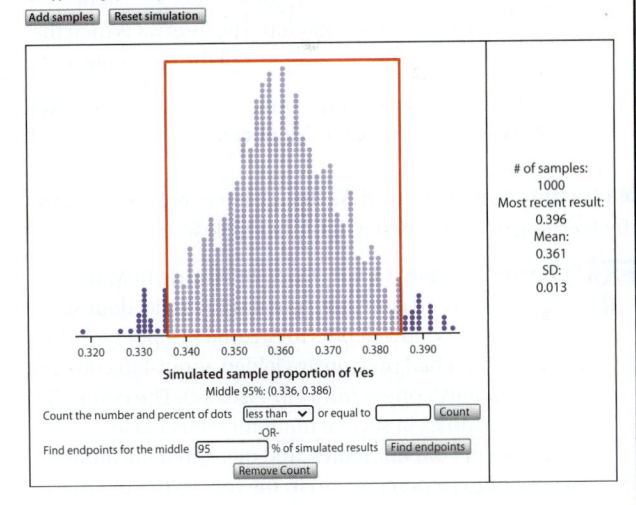

Note: If the applet is set up to use percentages rather than proportions, click "Adjust color, rounding, and percent/proportion preferences" at the bottom of the page and choose decimals in the first box.

TI-83/84

This type of confidence interval is not currently available on the TI-83/84.

Detailed instructions for using CrunchIt!, Excel, Google Sheets, JMP, Minitab, and R are available in Achieve.

Section 13.6 What Did You Learn?

Review the learning goals from this section. Then practice what you've learned by working through the exercises.

Learning Goal	Example	Exercises
Describe how to use note cards to create a bootstrap distribution for a population proportion or population mean.	13.6.1	5–8
Use a bootstrap distribution to estimate a confidence interval for a population proportion or population mean.	13.6.2	9–12
Use the four-step process to estimate and interpret a bootstrap confidence interval for a population proportion or population mean.	13.6.3	13–16

Section 13.6 Exercises

Building Concepts and Skills These exercises assess the basic knowledge you should have after reading the section.

1. What is the only condition required for a bootstrap confidence interval for a population proportion or a population mean?

2. True/False: When selecting a bootstrap sample by hand, write each of the observed values on a separate note card. Shuffle the cards and then select a sample, with replacement, the same size as the original sample.

3. For a 90% confidence interval, eliminate the bottom _____ % of dots and the top _____ % of dots in the bootstrap distribution. The interval is then the _____ and the _____ of the remaining dots.

4. True/False: Two bootstrap confidence intervals created from the same data will be identical.

Mastering Concepts and Skills These exercises reinforce the learning goals as shown in the examples.

5. **Refer to Example 13.6.1 Copper Mining** In March 2021, a Canadian company found substantial deposits of copper at shallow depths in Arizona's Copper World region, which had previously yielded about 440,000 tons of high-quality copper from 1874 to 1969. The company drilled test holes in several randomly selected locations at each of the old mine sites, and measured the amount of copper (as a percentage) in the rock extracted from each hole. Here are the data from the 28 test holes drilled at the Broad Top Butte site.[76] Describe how to use note cards to simulate one possible value in a bootstrap distribution of $\bar{x}$ = the mean percent copper in a random sample of 28 rock samples at the Broad Top Butte site.

0.00 0.75 1.43 0.43 0.19 0.52 0.00 0.00 0.20 1.38 0.00 0.28 0.71 0.30
0.30 0.44 0.59 0.91 0.33 0.39 0.00 0.70 0.37 0.38 0.67 0.00 0.38 0.30

6. **Crowdfunding** What proportion of U.S. adults contribute to crowdfunding campaigns such as GoFundMe? In a random sample of 1535 U.S. adults, 487 reported that they give to a crowdfunding campaign in a typical year.[77] Describe how to use note cards to simulate one possible value in a bootstrap distribution of $\hat{p}$ = the proportion of U.S. adults who report they give to a crowdfunding campaign in a typical year in a random sample of size 1535.

7. **New York Squirrels** How do squirrels in New York City's Central Park respond to humans? In October 2018, the Squirrel Census project enlisted volunteers to record data on all squirrels observed in Central Park. One of the variables measured how the squirrels responded to humans. We selected a random sample of 75 squirrel sightings. In that sample, 3 of the squirrels were reported to be indifferent to humans. Describe how to use note cards to simulate one possible value in a bootstrap distribution of $\hat{p}$ = the proportion of squirrels that are indifferent to humans in a random sample of size 75 from Central Park.

8. **Crabby Crabs** Researchers designed an experiment to determine how noise affects stress level in crabs. They selected a random sample of crabs and wanted to estimate, before running the experiment, the mean weight of crabs in the population. Here are the weights (g) of the crabs in the sample.[78] Describe how to use note cards to simulate one possible value in a bootstrap distribution of $\bar{x}$ = the mean weight of crabs in a random sample of size 34.

22.7 34.6 36.0 40.1 47.5 49.6 50.7 54.4 57.4 59.7
60.5 61.3 67.9 68.5 84.9 84.9 56.0 25.3 29.2 32.0
39.4 41.9 43.8 50.8 53.6 53.0 55.3 59.8 63.6 62.3
57.3 58.8 72.7 74.5

9. **Refer to Example 13.6.2 Copper Mining** Refer to Exercise 5. The dotplot shows the value of the sample mean percent of copper in rock samples for 100 bootstrap samples.

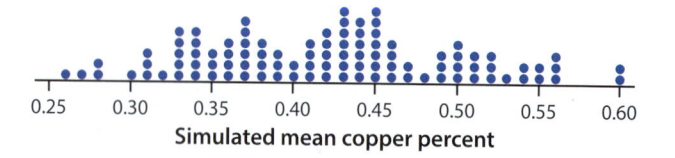

(a) How many dots should be eliminated from each end of the bootstrap distribution for a 90% confidence interval?

(b) Use the bootstrap distribution to find a 90% confidence interval.

10. **Crowdfunding** Refer to Exercise 6. The dotplot shows the value of the sample proportion of people who report that they give to a crowdfunding campaign in a typical year for 200 bootstrap samples.

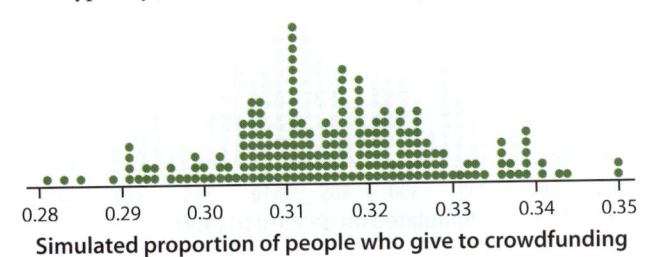

(a) How many dots should be eliminated from each end of the bootstrap distribution for a 95% confidence interval?

(b) Use the bootstrap distribution to find a 95% confidence interval.

11. **New York Squirrels** Refer to Exercise 7. The dotplot shows the value of the sample proportion of squirrels that appear indifferent to humans in Central Park for 200 bootstrap samples.

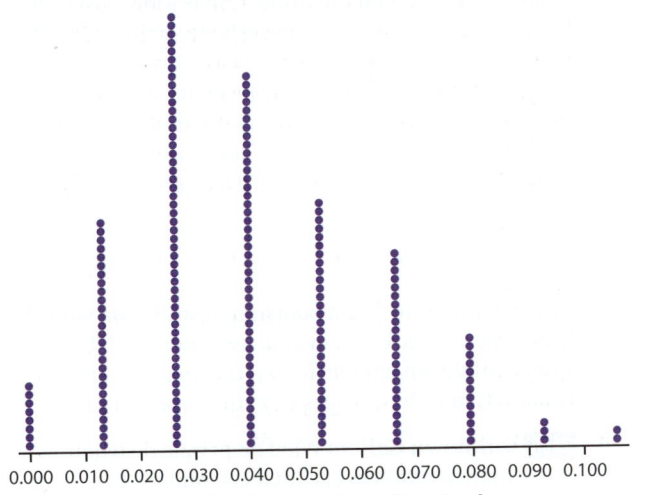

(a) How many dots should be eliminated from each end of the bootstrap distribution for a 99% confidence interval?

(b) Use the bootstrap distribution to find a 99% confidence interval.

12. **Crabby Crabs** Refer to Exercise 8. The dotplot shows the value of the sample mean weight of crabs (g) for 100 bootstrap samples.

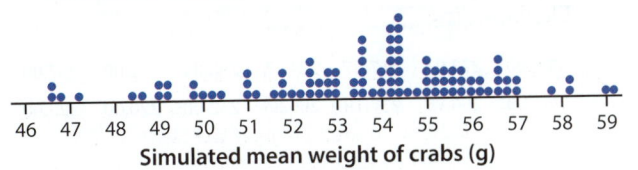

(a) How many dots should be eliminated from each end of the bootstrap distribution for a 90% confidence interval?

(b) Use the bootstrap distribution to find a 90% confidence interval.

13. **Refer to Example 13.6.3 Smelling Covid-19** Dogs are known to have a very good sense of smell and are sometimes used to detect drugs, bombs, and diseases such as cancer. Can they also smell Covid-19? Researchers in Germany trained dogs to identify the Covid-19 virus in samples of human saliva. After training, they presented the dogs with 1012 randomized samples, some that were positive for Covid-19 and others that were negative. The dogs correctly identified 949 of the samples.[79] Estimate a 95% confidence interval for the true proportion of samples that these dogs would correctly identify using the given bootstrap distribution with 200 values.

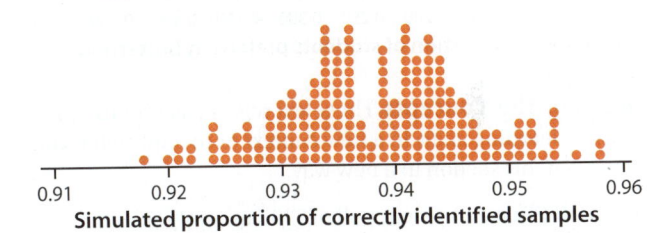

14. **Commute Time** How long do people typically spend traveling to work? The answer may depend on where they live. A random sample of 20 workers in New York state was chosen and their travel times (in minutes) recorded.[80] Estimate a 90% confidence interval for the true mean commute time for workers in New York state using the bootstrap distribution with 100 values.

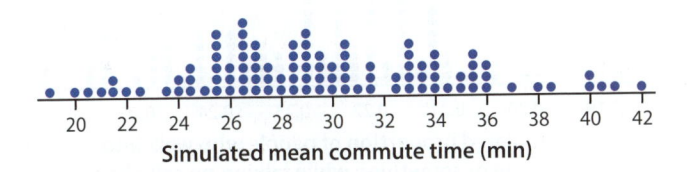

15. **Elephant Size** How tall are adult male African elephants that lived through droughts during their first two years of life? Researchers measured the shoulder height (in cm) of a random sample of 14 mature (age 12 years or older) male African elephants that lived through droughts during their first two years of life.[81] Estimate a 90% confidence interval for the true mean shoulder height of male African elephants who lived through droughts during their first years of life using the bootstrap distribution with 100 values.

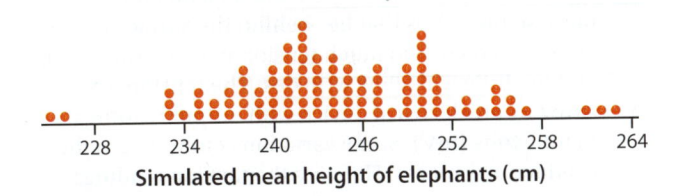

16. **Candy** A random sample of 30 college students was asked what their favorite candy bar is. Butterfinger was the selection for 7 students in the sample. Estimate a 99% confidence interval for the true proportion of college students who would choose Butterfinger as their favorite candy bar using the given bootstrap distribution with 200 values.

0.050 0.100 0.150 0.200 0.250 0.300 0.350 0.400 0.450 0.500
Simulated proportion of students preferring Butterfinger

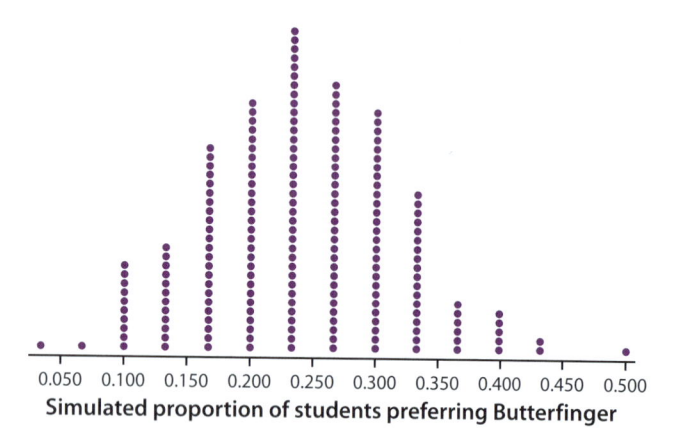

Applying the Concepts These exercises ask you to apply multiple learning goals in a new context or to apply what you learned in this section in a new way.

17. **Cell Phones** A recent poll of 738 randomly selected customers of a major U.S. cell-phone carrier found that 170 of them had walked into something or someone while talking on a cell phone.[82]

    (a) Use the given bootstrap distribution with 100 values to calculate a 90% bootstrap confidence interval for the proportion of all customers of this U.S. cell-phone carrier who would say they have walked into someone or something while talking on a cell phone.

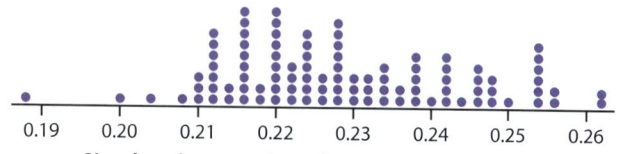

    0.19 0.20 0.21 0.22 0.23 0.24 0.25 0.26
    Simulated proportion of people who walk into someone or something while talking on cell phone

    (b) Check that the conditions are met to calculate a 90% one-sample z interval for a proportion.

    (c) Calculate a 90% one-sample z interval for the proportion of all customers of this U.S. cell-phone carrier who would say they walked into someone or something while talking on a cell phone.

    (d) Compare your answers to parts (a) and (c). Do they give similar intervals?

18. **Video Screens** A manufacturer of high-resolution video terminals must control the tension on the mesh of fine wires that lies behind the surface of the viewing screen. Too much tension will tear the mesh, and too little will allow wrinkles. The tension is measured by an electrical device with output readings in millivolts (mV). Some variation is inherent in the production process. Here are the tension readings from a random sample of 20 screens from a single day's production:

    269.5 297.0 269.6 283.3 304.8 280.4 233.5 257.4 317.5 327.4
    264.7 307.7 310.0 343.3 328.1 342.6 338.8 340.1 374.6 336.1

    (a) Use the given bootstrap distribution with 200 values to calculate a 95% bootstrap confidence interval for the mean tension in the mesh of all high-resolution video terminals produced that day.

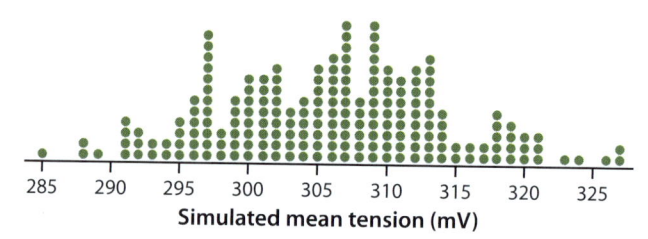

    285 290 295 300 305 310 315 320 325
    Simulated mean tension (mV)

    (b) Check that the conditions are met to calculate a 95% one-sample t interval for a mean.

    (c) Calculate a 95% one-sample t interval for the mean tension in the mesh of high-resolution video terminals.

    (d) Compare your answers to parts (a) and (c). Do they give similar intervals?

19. **Tech Vitamin Testing** Several years ago, the U.S. Agency for International Development provided 238,300 metric tons of corn–soy blend (CSB) for emergency relief in countries throughout the world. CSB is a low-cost fortified food. As part of a study to evaluate vitamin C levels in this food, measurements were taken on packages of CSB produced in a factory. The following data (also in the data file **VitaminC**) are the amounts of vitamin C, measured in milligrams per 100 grams (mg/100 g) of blend, for a random sample of 8 packages from one production run:[83]

    26   31   23   22   11   22   14   31

    Use technology to find a bootstrap distribution with at least 1000 values. Estimate and interpret a 90% bootstrap confidence interval for μ = the mean amount of vitamin C in CSB packages from this production run.

20. **Tech Unknown Callers** What do people do when an unknown number calls their cell phone? In a Pew Research Center survey, only 19% of the 10,211 randomly selected U.S. adults would answer the phone to see who it is.[84] Use technology to find a bootstrap distribution with at least 1000 values. Estimate and interpret a 95% bootstrap confidence interval for the proportion of all U.S. adults who would answer their cell phone when an unknown number calls.

21. **Tech Elephant Size** Refer to Exercise 15. The bootstrap distribution in Exercise 15 was based on 100 bootstrap samples. What is the effect of using more bootstrap samples? The data are given here and in the data file **ElephantSize**:

    200.00 272.91 217.57 294.15 296.84 212.00 257.00
    251.39 266.75 237.19 265.85 212.00 220.00 225.00

(a) Create a bootstrap distribution with at least 1000 bootstrap samples and estimate a 90% confidence interval for the true shoulder height of male African elephants that lived through droughts during their first two years of life.

(b) Are the bootstrap intervals from part (a) and from Exercise 15 the same?

(c) Which interval is more trustworthy? Explain your answer.

22. **Tech Candy** Refer to Exercise 16. The bootstrap distribution in Exercise 16 was based on 200 bootstrap samples. What is the effect of using more bootstrap samples? Recall that 7 respondents in a random sample of 30 college students chose Butterfinger as their favorite candy bar.

(a) Create a bootstrap distribution with at least 1000 bootstrap samples and estimate a 99% confidence interval for the true proportion of college students who would choose Butterfinger as their favorite candy bar.

(b) Are the bootstrap intervals from part (a) and from Exercise 16 the same?

(c) Which interval is more trustworthy? Explain your answer.

Extending the Concepts These exercises challenge you to explore statistical concepts and methods that go beyond what you learned in this section

23. **Bootstrap Confidence Interval for Difference in Two Means** The idea of a bootstrap confidence interval can be extended to other statistics, such as the difference in two population means. In this setting, we have two samples (call them sample A and sample B), one from each of two populations. The statistic of interest is $\bar{x}_A - \bar{x}_B$. To create one bootstrap value of $\bar{x}_A - \bar{x}_B$, take a bootstrap sample from sample A and a bootstrap sample from sample B. Compute the sample means from each bootstrap sample, and then calculate the difference. This difference is one value in the bootstrap distribution. Repeat the process many times to create a bootstrap distribution that can be used to find the confidence interval.

Let's return to the pain threshold data from Section 13.5, Exercise 18. Do pain thresholds differ for people with naturally blond hair and people with naturally brown hair? This is the question that researchers at the University of Melbourne asked. They chose random samples of people with naturally blond hair and people with naturally brown hair in Australia. Each person completed a pain sensitivity test and was given a pain threshold score (the higher the score, the higher the tolerance for pain). Here are the data:[85]

Blond	62	60	71	55	48	63	57
	43	41	52				
Brown	42	50	41	37	32	39	51
	30	35					

(a) Describe how to use note cards to simulate one possible value in a bootstrap distribution of $\bar{x}_{Blond} - \bar{x}_{Brown}$ where $\bar{x}_{Blond}$ = the mean pain threshold score for people with naturally blond hair in Australia and $\bar{x}_{Brown}$ = the pain threshold score for people with naturally brown hair in Australia.

(b) Use the given bootstrap distribution with 200 values to calculate a 95% bootstrap confidence interval for the difference in mean pain threshold (Blond − Brown) for people like those in the experiment.

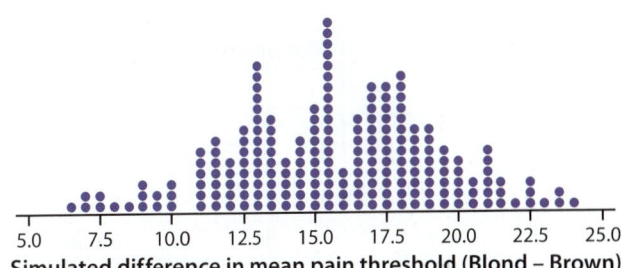

Simulated difference in mean pain threshold (Blond − Brown)

Cumulative Review These exercises revisit what you learned in previous sections.

24. **Marker for Alzheimer's Disease (1.1)** Amyloid-β (Abeta) is a protein fragment that has been linked to Alzheimer's disease. Autopsies from a sample of 57 Catholic priests included measurements of Abeta (pmol/g tissue from the posterior cingulate cortex) from three groups: subjects who had exhibited no cognitive impairment before death, subjects who had exhibited mild cognitive impairment, and subjects who had mild to moderate Alzheimer's disease.[86]

(a) Identify the variables in this study. Classify each as categorical or quantitative.

(b) Identify the population and sample in this setting.

(c) Is this an observational study or an experiment? Explain your answer.

25. **SAT Scores (6.2)** Scores on the math portion of the SAT test are designed to have a normal distribution with $\mu = 500$ and $\sigma = 100$.

(a) What is the probability that a randomly selected test-taker will score at least 640 on the math portion of the SAT?

(b) What is the probability that a randomly selected test-taker will score between 450 and 700 on the math portion of the SAT?

Statistics Matters Can acupuncture help chronic headaches?

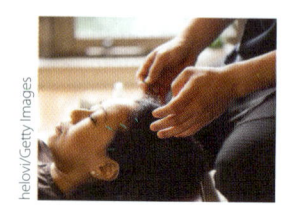

At the beginning of this chapter, we described a study to determine whether acupuncture can help those who suffer from chronic headaches. Here we evaluate how many headaches people like those in the study suffer, on average, before acupuncture treatment, and what conclusions we can draw about acupuncture and chronic headaches.

We will use the volunteers from one particular clinic. The variables measured include chronicity = the number of years the subjects have had chronic headaches, and three frequency variables that measure the number of days with headaches out of 4 weeks at the beginning of the study, at 3 months, and at 1 year. Here are the data:

Chronicity	Group	Begin frequency	3-month frequency	1-year frequency
5	Control	27	0	0
20	Control	23	13	16
31	Control	11	5	13
31	Control	28	28	28
30	Control	26	26	28
25	Control	12	6	11
10	Control	13	11	12
10	Control	9	3	6
9	Control	17	19	23
30	Control	10	9	7
14	Control	16	16	9
9	Control	16	10	17
44	Control	20	0	0
17	Control	8	8	12
39	Control	8	0	10
5	Control	18	8	10
8	Treatment	9	5	8
29	Treatment	18	0	0
25	Treatment	12	5	11
6	Treatment	16	8	9
30	Treatment	11	8	10
20	Treatment	28	26	27
15	Treatment	11	8	3
10	Treatment	22	12	14
45	Treatment	9	20	0
5	Treatment	10	10	7
15	Treatment	8	0	0
6	Treatment	7	9	6
27	Treatment	23	20	17
31	Treatment	28	0	0
42	Treatment	28	28	27
40	Treatment	28	28	28

We start by learning about the population from which these volunteers come.

1. Chronicity measures the number of years that the subjects have suffered from chronic headaches. Use a 95% bootstrap confidence interval to estimate the mean number of years that people like the subjects in this study have suffered from chronic headaches.

   Next, we evaluate the conclusions that we can reach from the study.

2. Compute the change in number of days with headaches from the beginning of the study to the 3-month point, when the treatment ended for each subject.

3. Does the acupuncture seem to help? Using the differences you calculated in Exercise 2, perform the appropriate Wilcoxon rank sum test to determine whether there is convincing evidence of an improvement in the median number of days with headache for people like those in the study when treated with acupuncture.

4. Compute the change in number of days with headaches from the beginning of the study to the 1-year point.

5. Does acupuncture seem to have a lasting effect? Use the differences calculated in Exercise 4 and a randomization test to determine whether there is convincing evidence of an improvement in the mean number of days with headache for people like those in the study when treated with acupuncture.

Chapter 13 Review

Nonparametric Inference

- Nonparametric procedures are typically used when one or more of the conditions for the parametric inference methods of Chapters 7–10 are not met.
- Nonparametric tests often have lower power (produce larger P-values) than the corresponding parametric inference methods.
- Many nonparametric procedures are based on the median rather than the mean because the underlying population distribution is often not symmetric.
- As with parametric inference procedures, be sure to follow the four-step process when using nonparametric procedures.

Sign Test

- The **sign test** is a significance test for the median difference of paired data.
  - The null hypothesis is that the median difference is 0.
  - Make sure the Random condition is met: Paired data come from a random sample from the population of interest or from a randomized experiment.
  - To compute the test statistic:
    - Calculate the difference for each pair.
    - Count the number of positive and negative differences that result from the previous step. Ignore any differences of 0.
  - The test statistic is:
    - For a lower-tail test: X = the number of negative differences.
    - For an upper-tail test: X = the number of positive differences.
    - For a two-tailed test: X = the larger of the number of positive or negative differences.
- The **sign test** is also a significance test for the median of a single population.
  - The null hypothesis is that the population median is equal to a specified value.
  - Make sure the Random condition is met: The data come from a random sample from the population of interest.
  - To compute the test statistic:
    - Subtract the null hypothesis value from all data values.
    - Count the number of positive and negative differences that result from the previous step. Ignore any differences of 0.
  - The test statistic is:
    - For a lower-tail test: X = the number of negative differences.
    - For an upper-tail test: X = the number of positive differences.
    - For a two-tailed test: X = the larger of the number of positive or negative differences.

- When the condition is met, the P-value for both versions of the test is calculated using a binomial distribution with n = the number of nonzero differences and $p = 0.5$. Let x be the observed value of the test statistic.
  - For a one-sided test, the P-value is $P(X \geq x)$.
  - For a two-sided test, the P-value is $2P(X \geq x)$.

Wilcoxon Signed Rank Test

- The **Wilcoxon signed rank test** is a significance test for the median difference for paired data. It typically has more power than the sign test.
  - The null hypothesis is that the median difference is 0.
  - Make sure the conditions are met:
    - Random: The differences come from a random sample from the population of interest or from a randomized experiment.
    - Symmetry: The population distribution of differences is symmetric.
  - Compute the signed ranks:
    - Compute the absolute values of all nonzero differences.
    - Rank these absolute differences. Start with 1 for the smallest absolute difference. If two or more absolute differences are equal, give each of them the same rank: the average of the ranks they would have been given individually.
    - Once all absolute differences have been ranked, put negative signs in front of the ranks that came from negative differences. Leave the ranks for those differences that were positive as positive numbers. These are the signed ranks.
  - Calculate the test statistic:
    - Let W^+ = the sum of the positive ranks and W^- = the sum of the *absolute values* of the negative ranks.
    - The test statistic W = the *smaller* of W^+ and W^-.
- The **Wilcoxon signed rank test** is also a significance test for the median of one population.
  - The null hypothesis is that the population median is equal to a specified value.
  - Make sure the conditions are met:
    - Random: The data come from a random sample from the population of interest.
    - Symmetry: The distribution of the population of interest is symmetric.
  - Compute the signed ranks:
    - Subtract the null hypothesis value from all data values.
    - Compute the absolute values of all nonzero differences.
    - Rank these absolute differences. Start with 1 for the smallest absolute difference. If two or more absolute differences are equal, give each of them the same rank: the average of the ranks they would have been given individually.
    - Once all absolute differences have been ranked, put negative signs in front of the ranks that came from negative differences. Leave the ranks for those differences that were positive as positive numbers. These are the signed ranks.
  - Calculate the test statistic:
    - Let W^+ = the sum of the positive ranks and W^- = the sum of the *absolute values* of the negative ranks.
    - The test statistic W = the *smaller* of W^+ and W^-.
  - When the conditions are met for either version of the test, you can use technology or Table E to find the P-value.

Wilcoxon Rank Sum Test

- The **Wilcoxon rank sum test** is a significance test of the null hypothesis that two population medians are the same.
- Make sure the conditions are met:
  - Random: The data come from independent random samples from the two populations of interest or from two groups in a randomized experiment.
  - Large Sample: Each sample size is at least 5.
  - Equal SD: The larger sample standard deviation is less than twice the smaller sample standard deviation.
- Find the ranks:
  - Combine both samples (groups) and order the values from smallest to largest, keeping track of which sample (group) each value came from.
  - Assign ranks for all values, starting with 1 for the smallest value. If two or more values are equal, give each of them the same rank: the average of the ranks they would have been given individually.
- Calculate the test statistic:
  - Designate one of the two groups as Group 1 and the other as Group 2.
  - Compute R_1 = the sum of the ranks for the values in Group 1.
  - Compute $\mu_1 = \dfrac{n_1(n_1 + n_2 + 1)}{2}$ and $\sigma_1 = \sqrt{\dfrac{n_1 n_2(n_1 + n_2 + 1)}{12}}$.
  - Calculate the standardized test statistic $z = \dfrac{R_1 - \mu_1}{\sigma_1}$.
- When the conditions are met, the standardized test statistic has approximately a standard normal distribution. You can use Table A or technology to find the P-value.

Kruskal-Wallis Test

- The **Kruskal-Wallis test** is a significance test of the null hypothesis that two or more population medians are the same.
- Make sure the conditions are met:
  - Random: The data come from independent random samples from the populations of interest or from groups in a randomized experiment.

- Large Sample: Each sample size is at least 5.
- Equal SD: The largest sample standard deviation is less than twice the smallest sample standard deviation.
- Find the ranks:
  - Combine all groups into one large sample and order the values from smallest to largest, keeping track of which group each value came from.
  - Assign ranks for all values, starting with 1 for the smallest value. If two or more values are equal, give each of them the same rank: the average of the ranks they would have been given individually.
- Calculate the test statistic:
  - Compute R_i = the sum of the ranks for the values in group i for each group.
  - Let n_i = the sample size of group i, and $N = n_1 + n_2 + \cdots + n_k$, where k = number of groups.
  - Calculate the test statistic
  $$H = \left(\frac{12}{N(N+1)} \sum_{i=1}^{k} \frac{R_i^2}{n_i}\right) - 3(N+1).$$
- When the conditions are met, the test statistic has an approximately chi-square distribution with $k-1$ df. You can use Table C or technology to find the P-value.

Randomization Tests

- A **randomization test** is a significance test that uses a randomization distribution to estimate the P-value.
- When performing a randomization test for a difference in means or a difference in proportions, the null hypothesis is that the two means (proportions) are equal.
- Make sure the Random condition is met: The data come from independent random samples from the two populations of interest or from two groups in a randomized experiment.
- Create a **randomization distribution** with technology or by hand:
  - For both samples, write each value on a separate note card.
  - Combine the two samples, shuffle the cards, and deal them into two piles of the same sizes as the original sample sizes.
  - Compute the sample statistic for each pile. Then compute the difference between the two sample statistics.
  - Repeat this process many times to create a randomization distribution of differences between sample statistics.
- When the condition is met, find the P-value by:
  - For a one-sided test, count the dots in the randomization distribution as extreme as or more extreme in the direction of H_a than the value found from the original two samples. The P-value is the number of dots counted divided by the total number of dots in the randomization distribution.
  - For a two-sided test, count the number of extreme dots in both tails of the randomization distribution.

Bootstrap Confidence Intervals

- A **bootstrap confidence interval** is a confidence interval for a population parameter estimated with a bootstrap distribution.
- Make sure the Random condition is met: The data come from a random sample from the population of interest.
- Create a **bootstrap distribution** using technology or by hand:
  - Write each value from the sample on a separate note card.
  - Create a **bootstrap sample** by selecting a sample with replacement from the cards (shuffling the cards between selections) the same size as the original sample.
  - Compute the sample statistic from the bootstrap sample.
  - Repeat this process many times to create a bootstrap distribution of sample statistics.
- When the conditions are met, find a 95% confidence interval by eliminating the smallest 2.5% and the largest 2.5% of the dots in the bootstrap distribution. The endpoints of the 95% confidence interval are the minimum and maximum of the remaining dots in the bootstrap distribution. This same idea can be used for any confidence level. For example, if you want a 90% interval, divide the remaining 10% in two, and cut off the bottom and top 5% of the bootstrap distribution.

Chapter 13 Review Exercises

These exercises will help you review important concepts and skills described by the learning goals in each section. For most exercises, the relevant section is noted in parentheses after the exercise title.

1. **Digging Manioc (13.1, 13.2)** Anthropologists wondered whether a metal tool called a machete would increase the speed with which Machiguenga women in Peru could dig manioc, a root vegetable that is a staple of their diet, when compared with their traditional wooden tool. They asked five women to dig manioc with each tool in a random order.[87] Is there convincing evidence at the 5% significance level that the women work better with the machete than with the traditional tool? The researchers measured the amount of manioc

each woman was able to dig in kilograms per hour with each tool. Here are the data:

Person	Machete	Wooden tool
1	119	39
2	216	114
3	240	150
4	129	51
5	137	60

(a) Explain why neither a paired t test for a mean difference nor a Wilcoxon signed rank test is appropriate to analyze this experiment.

(b) Conduct a sign test to answer the researcher's question.

2. **Sweet Corn (13.2)** Sweet corn of a certain variety is known to produce individual ears of corn with a mean weight of 8 ounces. A farmer is testing a new fertilizer designed to produce larger ears of corn, as measured by their weight. The farmer selects a random sample of 10 ears of corn and determines their weight (in ounces). The data are as follows:

8.30 7.35 9.05 8.05 8.25 8.75 9.00 7.60 8.45 8.55

Do the data provide convincing evidence that the fertilizer produces a median weight greater than 8 ounces? Use a Wilcoxon signed rank test to answer the farmer's question.

3. **Slippery Shoes (13.3)** Researchers wanted to know whether oil-resistant outer soles on safety shoes would be more likely to slip in icy conditions than non-oil-resistant outer soles on safety shoes. They selected a random sample of safety shoes with each kind of outer soles and measured the coefficient of friction as a measure of slipperiness.[88] The larger the value of the coefficient of friction, the less slippery the shoe is. Do the data provide convincing evidence at the 5% significance level that safety shoes with oil-resistant outer soles are more slippery than safety shoes without oil-resistant outer soles? Use a Wilcoxon rank sum test to answer the question. Here are the data:

Non-oil-resistant	0.176	0.131	0.173	0.105	0.169
	0.103	0.068	0.209	0.187	0.239
	0.073	0.152	0.153	0.148	
Oil resistant	0.131	0.112	0.101	0.119	0.178
	0.156	0.120	0.133	0.177	

4. **Rope Failure (13.4)** What might cause a climbing rope to fail sooner than expected? Researchers designed an experiment in which they randomly assigned ropes to be dirty, soaked in water, or both. They then measured the pressure (in pounds) at which the rope failed.[89] Is there convincing evidence at the 5% level that at least one of these treatments produces a different median failure pressure? To help answer this question, perform the appropriate Kruskal-Wallis test.

Both	515	524	417	486	520	520	415	486
Dirt	655	624	699	722	625	620	702	721
Soaked	621	637	664	609	615	640	660	600

5. **Staying Warm (13.5)** When people need to warm their cars up before driving in winter, do they stay in the vehicle while it is warming up, or do they wait in a building for the vehicle to warm up? A random sample of 250 Iowa residents resulted in 65 saying that they wait in their vehicle, and a random sample of 270 Vermont residents resulted in 84 saying that they wait in their vehicle. Is there convincing evidence of a difference between the population proportion of Iowa and Vermont residents who stay in their vehicle?

(a) Let $\hat{p}_{\text{Iowa}}$ = the sample proportion of people in Iowa who stay in their vehicle while it warms up, and $\hat{p}_{\text{Vermont}}$ = the sample proportion of people in Vermont who stay in their vehicle while it warms up. Describe how to use note cards to simulate one possible value of $\hat{p}_{\text{Iowa}} - \hat{p}_{\text{Vermont}}$ in a randomization distribution for this study.

(b) Use a randomization test at the 1% significance level with the 200 values of $\hat{p}_{\text{Iowa}} - \hat{p}_{\text{Vermont}}$ in the given randomization distribution to answer the question of interest.

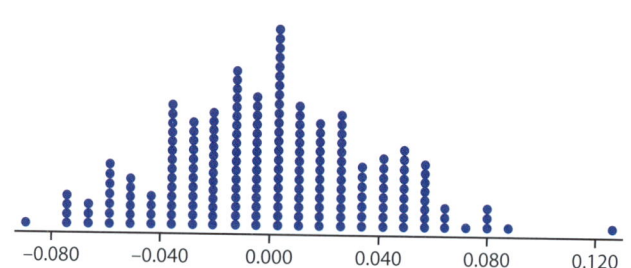

6. **Gas Mileage (13.6)** A consumer advocacy group randomly selected 20 owners of a particular model of pickup truck. Each owner was asked to report the number of miles per gallon (mpg) for their most recent tank of gas. Estimate and interpret a 95% bootstrap confidence interval for μ = the true mean fuel efficiency for this model of pickup truck using the given bootstrap distribution with 200 values.

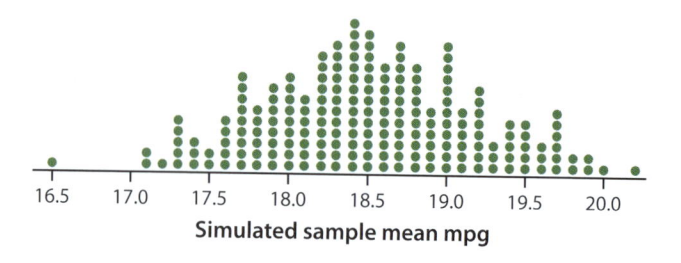

7. **Caffeine and Pulse (Chapter 13)** A physiology class performed an experiment to investigate whether drinking a caffeinated beverage would increase pulse rates. Twenty students in the class volunteered to take part in

the experiment. All of the students measured their initial pulse rates. Then the professor randomly assigned the students into two groups of 10. Each student in the first group drank 12 ounces of cola with caffeine. Each student in the second group drank 12 ounces of caffeine-free cola. All students then measured their pulse rates again. The table displays the change in pulse rate (Final − Initial) for the students in both groups.[90]

Caffeine	8	3	5	1	4	0	6	1	4	0
No caffeine	3	−2	4	−1	5	5	1	2	−1	4

Do these data give convincing evidence at the $\alpha = 0.01$ significance level that drinking caffeine tends to increase pulse rates for people like the ones in this study?

(a) There are two nonparametric methods covered in this chapter that could be used for this significance test. Which are they, and why would they be appropriate?

(b) Choose one of the methods from your answer to part (a) and perform the test.

Solutions

CHAPTER 1
Answers to Section 1.1 Exercises

1. Ask questions, collect/consider data, analyze data, and interpret results
3. *Quantitative:* Age (in years), height, weight, GPA. *Categorical:* Age bracket (0–9, 10–19, 21–29, etc.), eye color, state you live in
5. The population is the entire group of individuals we want information about. A sample is a subset of individuals in the population from which we collect data.
7. In an experiment, a treatment is imposed upon individuals. In an observational study, no treatment is imposed.
9. *Individuals:* The 12 movies. *Categorical:* Year, rating, genre. *Quantitative:* Time (minutes), box office sales (dollars)
11. *Individuals:* The eight households selected by the U.S. Census Bureau. *Categorical:* Region, time in dwelling, response mode, whether or not they have internet access. *Quantitative:* Number of people, household income (dollars)
13. *Population:* All of the artifacts collected on the archaeological dig. *Sample:* The 2% of artifacts that are checked.
15. *Population:* All of the half-gallon containers of a popular brand of orange juice at these 10 stores. *Sample:* The 50 half-gallon containers of orange juice that were measured.
17. Experiment. Treatments were imposed. The cafeteria randomly assigned a label for a vegetable dish: basic, healthy restrictive, healthy positive, or indulgent.
19. Experiment. Treatments were imposed on the students. Students were assigned at random to either continue their normal social media practices or to be limited to 10 minutes per day for each of Facebook, Snapchat, and Instagram.
21. Observational study. There were no treatments imposed on the parents.
23. (a) *Variables:* Whether participants drank coffee and whether they were alive at the end of the study are both categorical; Age at death is quantitative. **(b)** *Population:* All of the older people in eight states. *Sample:* The 400,000 older people who participated in the survey. **(c)** Observational study. There were no treatments imposed on the older people.
25. (a) *Population:* All 45,000 people who made credit card purchases. **(b)** *Sample:* the 137 people who responded. The sample is the subset of the population from which we have obtained data.
27. Age (in years) is quantitative, but age could be classified as categorical if people are asked to report what age bracket they are in (0–9, 10–19, 20–29, etc.).

Answers to Section 1.2 Exercises

1. False. Convenience samples tend to produce poor estimates of the values we want to know because the members of the sample often differ from the population in ways that affect their responses.
3. Self-selected sample
5. Undercoverage leads to bias when the individuals that are less likely to be selected differ from the population in ways that affect their responses.
7. Researchers can minimize nonresponse by following up with people who do not respond the first time. They can also encourage people to respond by offering a small incentive.
9. The sampling method (convenience sampling) is biased because the trees along the road are more likely to be damaged by cars and people, and may be more susceptible to infestation. The proportion of infected hemlock trees in the sample is likely to be greater than the proportion of all hemlock trees in the forest that are infested.
11. This sampling method (voluntary response sampling) is biased because it is likely that the customers who chose to leave reviews feel strongly about the hotel, often due to a bad experience. Customers who had an average or good experience are less likely to leave a review. The percentage of 1s in the sample is likely greater than the percentage of all customers who would give the hotel 1 star.
13. Inspect a random sample of hemlock trees. Then infested trees would be much less likely to be overrepresented in the sample and the sample would be more representative of all hemlock trees.
15. Survey a random sample of hotel customers. Upset customers would be much less likely to be overrepresented in the sample and the sample would be more representative of all hotel customers.
17. This study displays undercoverage because the U.S. residents who are not registered voters can't be part of the sample, but they are more likely to support a "pathway to citizenship" because this group likely includes some people who came to the United States illegally. The percentage from the sample is likely less than the percentage of all U.S. residents who would support a pathway to citizenship.
19. This study shows nonresponse because people who drive more miles per day are less likely to respond to the survey and may even miss the survey call because they are driving. The average from the sample is likely less than the average for all U.S. drivers.
21. When asked in person, people may lie about always wearing their seat belts. When people are only observed and not asked directly, the percent who wear a seat belt will be smaller, and much closer to the truth.
23. (a) People who did not have a subscription to *Literary Digest* and those who did not own an automobile or telephone were not sent a ballot. These people may have been more likely to support the Democratic candidate (Roosevelt), but were underrepresented in the sample. **(b)** 2,400,000 of the ballots were returned, so at least 7,600,000 ballots were not returned. Those people who did not return their ballots may have been more likely to support the Democratic candidate (Roosevelt), but were underrepresented in the sample. **(c)** Following up with people who didn't return their ballots would not eliminate the bias due to undercoverage, as those people without an automobile or phone are still left out of the survey results. Following up will help to eliminate the bias due to nonresponse, as a greater percentage of those selected for the sample will be contacted and surveyed.
25. (a) If they make note of the huge national deficit, Americans might become more concerned about the U.S. government spending additional money on social programs. The percentage from the sample who say "Yes" is likely less than the true percentage of adult Americans who favor spending money to establish a national system of health insurance. **(b)** Answers will vary. *Unbiased:* Should the government establish a national system of health insurance? *Biased:* A national system of health insurance would help the poor and would help many children get the life-saving health care they need. Do you support the establishment of a national system of health insurance?
27. (a) A student does not have to feel guilty about answering "Yes" to the question, as it is possible that their coin landed on tails and they are simply following the protocol. **(b)** If 100 students each flipped a coin, we would expect 50 of them to get tails and answer "Yes" according to the protocol. The other 50 students are (we hope) answering the question honestly. This means that $63 - 50 = 13$ of the 50 admit to cheating. We estimate that the proportion of students who have cheated is $13/50 = 0.26$.

Answers to Section 1.3 Exercises

1. A simple random sample (SRS) of size n is a sample chosen in such a way that every group of n individuals in the population has an equal chance to be selected for the sample.

3. Sampling variability is the fact that different random samples of the same size from the same population produce different estimates.
5. False. Increasing the sample size does not reduce bias.
7. Give each circular plot a distinct integer label from 1 to 1410. Randomly generate 141 different integers from 1 to 1410. Visit the 141 circular plots that are labeled with the generated integers.
9. Give each regular season game a distinct integer label from 1 to 256. Randomly generate 7 different integers from 1 to 256. Record the length of each of the 7 games that are labeled with the generated integers.
11. (a) No. Because different random samples will produce different means, it is unlikely this sample provides a mean volume of lumber-quality pine that is exactly correct. **(b)** Because the sample size is smaller, the estimated mean of lumber-quality pine volume is less likely to be close to the true mean of lumber-quality pine volume.
13. (a) No. Because different random samples will produce different means, it is unlikely this sample provides a mean game time that is exactly correct. **(b)** Because the sample size is larger, the estimated mean game time is more likely to be close to the true mean game time.
15. (a) If researchers take many random samples of size 1029 from a population of U.S. adults where 50% say they often or always get enough sleep, the percentage of U.S. adults in a sample who say they often or always get enough sleep varies from about 46% to 53.8%. **(b)** $11/100 = 11\%$ **(c)** Because researchers are likely to get 48% or fewer U.S. adults who often or always get enough sleep due to sampling variability alone, there is not convincing evidence that less than half of all U.S. adults would say they often or always get enough sleep.
17. (a) If researchers take many random samples of size 1006 from a population of U.S. adults in which 25% would get the answer correct, the percentage of U.S. adults in a sample who answer correctly varies from about 21.75% to about 28.75%. **(b)** $0/100 = 0\%$ **(c)** Because researchers are unlikely to get 20% or fewer adults who answer correctly due to sampling variability alone, there is convincing evidence that fewer than 25% of all U.S. adults would correctly answer "nitrogen."
19. (a) Give each student a distinct integer label from 1 to the total number of students, N. Using a random number generator, generate 100 different integers from 1 to N. Survey the 100 students who are labeled with the generated integers. **(b)** No. Because different random samples will produce different means, it is unlikely that this sample will provide a mean that is *exactly* correct. **(c)** We would expect the sample mean to be closer to the true population mean when using an SRS of 100 students rather than 50 students. This is because increasing the sample size reduces sampling variability.
21. (a) After clicking the Begin analysis button, a bar chart appears. The bar chart shows a frequency of 22 for on-time flights and a frequency of 3 for late flights. Summary statistics also appear, which give the frequency and relative frequency for on-time and late flights. **(b)** Answers will vary.

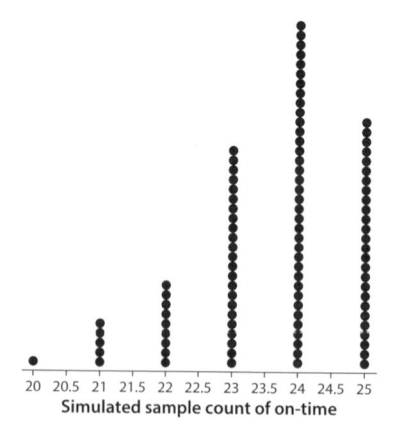

(c) Answers will vary. No. $15/100 = 15\%$ of the simulated samples resulted in 22 or fewer on-time flights. Because we are likely to get 22 or fewer on-time flights due to sampling variability alone, there is not convincing evidence that fewer than 95% of all this airline's flights arrive on time.

Answers to Section 1.4 Exercises

1. True
3. False. In a cluster random sample, you should randomly select the clusters, and every individual from the selected clusters are included in the sample.
5. 10th
7. (a) We might expect the satisfaction to be similar among those on the same floor/side, but different across floors/sides. **(b)** Select an SRS of 2 rooms from each of the 60 floor/side strata. Then, combine these 120 rooms into one overall sample.
9. The opinion of the employees might be the same within each type of employee (servers, kitchen staff), but differ across the different employee types. Use the type of employee as strata. Select an SRS of 15 employees who are servers and an SRS of 15 employees who work in the kitchen. Finally, combine these two SRSs into one overall sample. *Benefit:* A more precise estimate of the proportion who approve.
11. (a) A cluster random sample is more convenient. The manager would need to visit only 3 floors. In a stratified random sample, the manager would need to use rooms on every floor. **(b)** Number all the floors from 1 to 30. Select an SRS of 3 floors and survey the people staying in every room on the selected floors.
13. The administrator would need to visit only six groups. In an SRS, the administrator would have to number all the students and go all over the campus to find and survey the selected students.
15. Because there are 1000 iPhones and the manager wants a random sample of $n = 20$ phones, the manager should select every $1000/20 = 50$th iPhone that comes off the production line, starting with a randomly selected iPhone from among the first 50 that come off the production line.
17. (a) To obtain an SRS, every tree would need to be identified and numbered. This is not practical. **(b)** Because there are 5000 pine trees that are along Highway 34 and we want a random sample of $n = 200$ pine trees, we should select every $5000/200 = 25$th pine tree, starting with a randomly selected pine tree from among the first 25 along Highway 34.
19. (a) *Strata:* Rows. *Method:* Select an SRS of 4 fans from each of the 50 rows. Combine these 200 fans into one overall sample. *Benefit:* This will help provide a more precise estimate of the mean satisfaction rating with the show. **(b)** *Clusters:* Each row. *Method:* Number the rows from 1 to 50. Select an SRS of 4 rows and survey every fan in the randomly selected rows. *Benefit:* This is efficient because the researcher has to visit only 4 rows to obtain the entire sample. *Note:* Alternatively, randomly select 5 columns. **(c)** *Systematic sampling method:* Because there are $50 \times 40 = 2000$ fans in attendance and we want a random sample of $n = 200$ fans, we should select every $2000/200 = 10$th fan who leaves the venue, starting with a randomly selected fan from among the first 10 who leave the venue. *Benefit:* This is easier to implement than an SRS.
21. The columns. Each column serves as a small-scale replica of the population, whereas the rows are systematically different due to their proximity to the water.
23. No. An SRS would give every possible group of $n = 250$ employees the same chance of being selected from the population of 2500 employees. Using the method described, it would be impossible to end up with a group of 250 assembly-line workers.
25. A problem with this method of sampling is that every survey could go to someone sitting in a window seat. Passengers who sit in window seats may have a different boarding experience than passengers who have to sit in the middle seat.

Answers to Section 1.5 Exercises

1. Two variables are confounded when they are associated in such a way that their effects on a response variable cannot be distinguished from each other.

3. a human experimental unit

5. The placebo effect describes the fact that some subjects in an experiment will respond favorably to any treatment, even an inactive treatment.

7. False. We randomly assign treatments in an experiment so we can draw conclusions about cause and effect.

9. No, it is possible that people who are heavy social media users are more dissatisfied with their own lives and that people who are dissatisfied with their own lives are more likely to consider leaving their spouse. If both of these are true, then we would see a relationship between social media use and consideration of leaving a spouse, even if social media use has no effect on whether a person considers leaving their spouse.

11. No. It is possible that the students who ate breakfast in the morning were more conscientious and that conscientious students tend to score better on the national exam. If both of these are true, then we would see a relationship between eating breakfast and the score on the national exam, even if eating breakfast has no effect on the national exam score.

13. A control group would show how much plant biomass is produced without any additional water. This would serve as a baseline to determine how much additional biomass is produced by adding water during different seasons.

15. A control group would show what percentage of the time the shark ate the bait from a surfboard with no deterrent. This would serve as a baseline to determine how much less frequently the shark ate the bait for each of the deterrents.

17. (a) Label each customer with a different integer from 1 to 139. Then, randomly generate 70 different integers from 1 to 139. The customers with these labels will pay $4 for the buffet. The remaining 69 customers will pay $8 for the buffet. **(b)** To avoid confounding by creating two groups of customers who are roughly equivalent at the beginning of the experiment.

19. (a) Label each student with a different integer from 1 to 120. Then, randomly generate 40 different integers from 1 to 120. The students with these labels will be given Treatment 1. Next, randomly generate an additional 40 different integers from 1 to 120 (being careful not to choose the students who have already been selected for Treatment 1). The students with these labels will be given Treatment 2. The remaining 40 students will be given Treatment 3. **(b)** To avoid confounding by creating three groups of students who are roughly equivalent at the beginning of the experiment.

21. (a) It was necessary to perform an experiment rather than simply asking older adults about their diet and exercise habits to avoid confounding and to enable the researchers to draw cause-and-effect conclusions. **(b)** A control group would show the change in cognitive function among older adults who do not follow the DASH diet or a specific exercise regimen. This would serve as a baseline for comparing the effects of the treatments (diet and exercise) on the change in cognitive function. **(c)** Label each adult with a different integer from 1 to 160. Then, randomly generate 40 different integers from 1 to 160. The adults with these labels will be given Treatment 1 (DASH diet, no exercise regimen). Next, randomly generate an additional 40 different integers from 1 to 160, being careful not to choose the adults who have already been selected for Treatment 1. The adults with these labels will be given Treatment 2 (DASH diet, exercise regimen). Next, randomly generate an additional 40 different integers from 1 to 160, being careful not to choose the adults who have already been selected for Treatment 1 or 2. The adults with these labels will be given Treatment 3 (no prescribed diet, no exercise regimen). The remaining 40 adults will be given Treatment 4 (no prescribed diet, exercise regimen). **(d)** To avoid confounding by creating four groups of adults who are roughly equivalent at the beginning of the experiment.

23. There was no control group for comparison purposes. To make a cause-and-effect conclusion possible, we need to randomly assign some subjects to get flavonols and others to get a placebo.

25. This experiment could be double-blind if the treatment (ASU or placebo) assigned to a subject was unknown to both the subject and those responsible for assessing the effectiveness of that treatment. It is important for the subjects to be blinded so the people who receive ASU don't show more improvement or benefits because they have more positive expectations. It is important for the experimenters to be blinded so that they will be unbiased in the way that they interact and assess the subjects.

27. Label the 294 British high school students from 1 to 294. Randomly assign 147 students to eat breakfast regularly. The remaining 147 students are assigned to not eat breakfast. At the end of a predetermined time period, compare the average test scores for each group.

29. There are many other variables that might have an effect on the crash rate of 16- to 18-year-old drivers. These variables could include taking a driver preparation course, driving and texting, number of hours of practice driving, and many others. Only with random assignment could we be comfortable that all of these variables have been equally distributed across the two groups, so that we could be convinced that the difference in crash rates is due to the school starting time.

Answers to Section 1.6 Exercises

1. confounding; variability in the response variable

3. False. In a completely randomized design, the experimental units are assigned to the treatments (or treatments to the experimental units) completely at random.

5. False. When an observed difference in response is *unlikely* to occur by chance alone, the difference is statistically significant.

7. (a) The résumés were similar. **(b)** If one set of résumés was distinctively more impressive than the other, we wouldn't know if the applicant names or the résumé quality was the cause of the difference in the response rates.

9. (a) The amount of time for treatment (7 months). **(b)** If the amount of time for treatment was allowed to vary, the values of the response variables would be more variable, making it harder to determine whether the vitamin D supplement helps.

11.

14 males → Random assignment → Group 1: 7 men → Treatment 1: Fish oil ↘
                                Group 2: 7 men → Treatment 2: Mixture ↗ Compare mean reduction in blood pressure

13.

731 men → Random assignment → Group 1: 364 men → Treatment 1: Surgery ↘
                               Group 2: 367 men → Treatment 2: Observe ↗ Compare proportion of men still alive

15. (a) Fish oil group: 6.57 mmHg; mixture group: −1.14 mmHg. The difference in mean reduction is 6.57 − (−1.14) = 7.71 mmHg. **(b)** One simulated random assignment resulted in a difference in means (Fish oil − Mixture) of 5.8 mmHg. **(c)** Because a difference of means of 7.71 or greater rarely occurred in the simulation (2/100 = 2%), the difference is statistically significant. Researchers are unlikely to get a difference this big simply due to chance variation in the random assignment.

17. (a) Surgery group: 141/364 = 0.387; observation group: 122/367 = 0.332. The difference in proportions (Surgery−Observe) is 0.387 − 0.332 = 0.055. **(b)** One simulated random assignment resulted in a difference of proportions (Surgery−Observe) of 0.082. **(c)** Because a difference of proportions of 0.055 or greater was not unusual in the simulation (7/100 = 7%), the difference is not statistically significant. Researchers are likely to get a difference this big simply due to chance variation in the random assignment.

19. (a) The researchers used only male physicians. *Benefit:* This design reduces variation in the response variable(s). **(b)** Here is a diagram of the experiment.

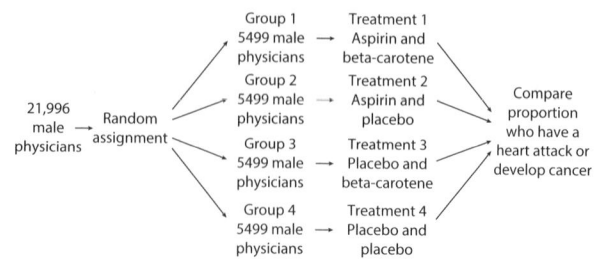

```
                              Group 1        Treatment 1
                              5499 male  →   Aspirin and
                              physicians     beta-carotene
                           ↗                                ↘
                              Group 2        Treatment 2
                              5499 male  →   Aspirin and
                              physicians     placebo
   21,996        Random    ↗                                   Compare
   male      →  assignment                                  →  proportion
   physicians              ↘                                    who have a
                              Group 3        Treatment 3        heart attack or
                              5499 male  →   Placebo and        develop cancer
                              physicians     beta-carotene
                           ↘                                ↗
                              Group 4        Treatment 4
                              5499 male  →   Placebo and
                              physicians     placebo
```

(c) The difference (Beta-carotene − Placebo) in the proportion of doctors who developed cancer was small enough that it could be explained by chance variation in the random assignment to treatments.

21. (a) Active magnets: 5.2414; inactive magnets: 1.0952. The difference in the mean improvement (Active − Inactive) is 5.2414 − 1.0952 = 4.1462. **(b)** Answers will vary. 0/100 = 0% **(c)** Because a difference in means of 4.1462 or greater never occurred in the simulation (0/100 = 0%), the difference is statistically significant. Researchers are unlikely to get a difference this big simply due to chance variation in the random assignment.

23. This is an example of voluntary response sampling, because people chose to be in the sample by responding to a general invitation. It is likely that only those people who feel strongly about their favorite contestant will respond. As a result, the winning contestant may not be the one whom the majority of the show's viewers prefer.

Answers to Section 1.7 Exercises

1. False. A block is a group of experimental units that are known before the experiment *to be similar* in some way, which is expected to affect the response to treatments.
3. True
5. Form blocks based on experience (never taken the GMAT, taken the GMAT before). It is reasonable to think that there is an association between experience and effectiveness of the preparation course. Randomly assign half of the subjects who have taken the GMAT before to the synchronous prep course and the other half to the asynchronous prep course. Do the same for the subjects who have never taken the GMAT. After they have taken the prep courses, rate the effectiveness for each student by recording their GMAT score.
7. Form blocks based on row. It is reasonable to think that there is an association between row and yield because the field increases in fertility from north to south. Randomly assign one plot in each row to variety A, B, C, D, and E. At the end of the growing season, measure the yield of each plot.
9. It accounts for the variability in effectiveness caused by the fact that some students already took the test and some did not. This design makes it easier to determine if the asynchronous course is as effective as the synchronous course in effectively preparing students for the GMAT.
11. It accounts for the variability in yield caused by the fact that the field increases in fertility from north to south. This design makes it easier to determine if any particular varieties of corn can be expected to provide a greater yield than the others.
13. Give each subject both treatments. For each subject, randomly assign the order in which the treatments are given by flipping a coin. Heads indicates getting a caffeine pill that contains the subject's normal daily caffeine consumption on the first day and a placebo on the second day. Tails indicates receiving the placebo first and then the caffeine pill second. For each subject, record the depression score after using each treatment.
15. Match the 40 patients according to age. In other words, the oldest two patients would form a pair, and so on, down to the two youngest patients. For each pair, have one person flip a coin. If the person flips heads, that person gets the new method and the other person gets the surgery. If it is tails, do the opposite. After the surgery, record the percentage of blood that flows backward for each patient.
17. (a) Label each volunteer with a different integer from 1 to 30. Then, randomly generate 15 different integers from 1 to 30. The 15 volunteers will be assigned to the room that smells of freshly baked bread. The remaining 15 volunteers will be assigned to the room with no smell. Record the amount of soup consumed for each volunteer. **(b)** Match the 30 volunteers according to when they have last eaten. In other words, the volunteers who have eaten the most recently will form a pair, down to the two who have not eaten for the longest amount of time. For each pair, have one person flip a coin. If the person flips heads, that person gets the room that smells like bread and the other person gets the room with no smell. If it is tails, do the opposite. Record the amount of soup consumed for each volunteer. **(c)** I prefer the matched-pairs design because it accounts for variation in hunger among the subjects.

19.

B	E	C	A	D
C	A	D	B	E
D	B	E	C	A
E	C	A	D	B
A	D	B	E	C

21. (a) There would be undercoverage because only songs in the Top 40 could be selected. Perhaps these songs are shorter than other songs from those years. **(b)** Select an SRS of rock songs from a numbered list of all rock songs from this current year and a separate SRS of rock songs from a numbered list of all rock songs from 50 years ago. Calculate the average length for each random sample, then compare the averages.

Answers to Section 1.8 Exercises

1. It is appropriate to use data from a sample to make an inference about a population when the sample is selected at random from the population.
3. informed consent
5. (a) No. The sample was not selected randomly from the population of all adults in Switzerland. **(b)** All adults in Lausanne, Switzerland.
7. (a) No. The participants were not randomly selected from the population of all adults in the United States. **(b)** All young adults like those who volunteered.
9. No. The researchers didn't randomly assign adults from Lausanne, Switzerland, to take a specific number of naps per week, so it is not reasonable to say that napping frequently causes cardiovascular problems.
11. Yes. The researchers randomly assigned the volunteers to do leg presses with no resistance or with heavy resistance, so it is reasonable to say that increased resistance caused better recall.
13. Those Facebook users involved in the study did not know that they were going to be subjected to treatments, and they did not provide informed consent before the study was conducted.
15. The ethics of this experiment are questionable. Because the prisoners were offered reduced prison sentences to participate, they may have been coerced into providing informed consent.
17. (a) Yes, because this study involved random assignment of the infants to the treatments (foster care or institutional care). **(b)** Children like the ones in the study. **(c)** Yes, it was unethical. The subjects were not able to give informed consent. They did not know what was happening to them and they were not old enough to understand the ramifications.
19. The sample was selected from the population of all students at a major university with a highly ranked football team, so we cannot generalize this result to all sports fans, but rather only to all students at this university. The students were randomly assigned to the treatment groups (win, loss), so it is reasonable to say that the fact that the team won or lost caused students to identify themselves with the team (or not).
21. Many would consider this to be an appropriate use of collecting data without participants' knowledge because the data are, in effect, anonymous and confidential.

23. Possible answers include: **(a)** A nonscientist might be more likely to consider the subjects as people and not be tempted by the scientific results that might be discovered. **(b)** One might consider at least two outside members. A religious leader might be chosen because we would expect them to help lead the committee in ethical and moral discussions. You might also choose a patient advocate to speak for the subjects involved.

Answers to Chapter 1 Review Exercises

1. (a) The population is all adult U.S. residents and the sample is the 805 adult U.S. residents interviewed. **(b)** Even though the sample size is very large, it is unlikely that the percentage in the entire population would be exactly the same as the percentage in the sample because of sampling variability. However, the number 0.85 should be fairly close to the value for the whole population. **(c)** A larger random sample is more likely to produce a sample result close to the true population value.

2. (a) It is biased because it is a convenience sample. Those who are first to arrive in the main parking lot will certainly find good spots, so the proportion of these students who are satisfied would likely be greater than the proportion of all students who are satisfied with the current parking situation. **(b)** Label the 8420 students who attend the college with the numbers 1 to 8420. Randomly select 50 different numbers from 1 to 8420. The students with those labels will be selected for the sample. **(c)** An SRS is more likely to accurately represent the whole population because each individual is equally likely to be included in the sample, whereas a convenience sample often results in a sample that does not have the same characteristics as the whole population. **(d)** Another source of bias that could affect the results of the survey, even when selecting an SRS of students, is nonresponse. If the students who choose to not respond to the survey differ from the rest of the population with regard to parking satisfaction, the results of the survey will be biased.

3. (a) If researchers take many random samples of size 50 from a population of students for which each student has a 50% chance of being satisfied, the percentage of students in a sample who are satisfied varies from about 32% to 68%. **(b)** $2/100 = 2\%$ **(c)** Because researchers are unlikely to get 34% or fewer who are satisfied with the current parking situation due to sampling variability alone, there is convincing evidence that fewer than half of all students at the college are satisfied with the current parking situation.

4. (a) The individuals are the pieces of food. **(b)** *Explanatory:* Surface (categorical), length of time (quantitative), and bacteria preparation (categorical) **(c)** *Response:* Number of bacteria on the food (quantitative)

5. (a) This is an observational study because no treatment is imposed on the volunteers. **(b)** It is possible that volunteers who eat the most organic food have healthier diets in general and that eating a healthy diet decreases the risk of developing cancer. If both of these are true, then we would see a relationship between eating organic food and smaller risk of developing cancer, even if eating organic food has no effect on the risk of developing cancer.

6. (a) A control group would show how many symptoms students experience with no mask. This would serve as a baseline to determine how many fewer symptoms students may experience while wearing a mask or while using hand sanitizer and wearing a mask. **(b)** All students at the University of Michigan like the ones in the study. **(c)** Yes, because the researchers randomly assigned the students to the treatment groups. **(d)** The researchers should have consulted an IRB, informed the students about the experiment, and received consent from the students who wanted to participate. When publishing results, the outcomes from students must be confidential.

7. (a)

```
                         Group 1         Treatment 1
                       37 volunteers    Small portions     Compare how
     75      Random                                        much food they
  volunteers assignment                                    consumed
                         Group 2         Treatment 2
                       38 volunteers    Large portions
```

Number the volunteers from 1 to 75. Generate 37 different integers from 1 to 75 and assign the corresponding volunteers to the small portions. Assign the remaining 38 volunteers to the large portions. **(b)** To avoid confounding by creating two groups of volunteers who are roughly equivalent at the beginning of the experiment. **(c)** The researchers presented each volunteer with the same food options and allowed the volunteers to serve themselves. This reduces variability in the amount of food the volunteers consume.

8. (a) The difference = $189.91 - 144.66 = 45.25$ grams. **(b)** One simulated random assignment resulted in a difference in means (Large − Small) of 30 grams. **(c)** Because a difference of means of 45.25 or greater never occurred in the simulation ($0/100 = 0\%$), the difference is statistically significant. Researchers are unlikely to get a difference this big simply due to chance variation in the random assignment.

9. (a) *Strata:* Type of region. *Method:* Select an SRS of 60 expense reports among those submitted by employees who cover densely populated regions and an SRS of 40 expense reports among those submitted by employees who cover less densely populated regions. Then, combine these 100 expense reports into one overall sample. *Benefit:* Using the type of region as strata will help provide a more precise estimate of the proportion of accurate expense reports. **(b)** *Clusters:* The employees. *Method:* Number the employees from 1 to 200. Select an SRS of 4 employees and review all 25 expense reports for each of the selected employees. *Benefit:* This sampling method is efficient because the auditor has to pull the accounts of only 4 employees. **(c)** *Systematic method:* Because there are $200 \times 25 = 5000$ expense reports and we want a random sample of $n = 100$ expense reports, we should select every $5000/100 = 50$th expense report that is submitted, starting with a randomly selected expense report from among the first 50 that are submitted. *Benefit:* This method is easier to implement than an SRS. This method also ensures that the sample is selected uniformly as the reports are turned in.

10. (a) Form blocks based on variety, because the number of unpopped kernels is likely to differ by variety. Randomly assign 5 bags of each variety to the popcorn button treatment and 5 to the timed treatment. After popping each of the 40 bags in random order, count the number of unpopped kernels in each bag and compare the results within each variety. **(b)** A randomized block design accounts for the variability in the number of unpopped kernels created by the different varieties of popcorn (butter, cheese, natural, kettle). This makes it easier to determine if using the microwave button is more effective for reducing the number of unpopped kernels.

11. (a) Give each employee a distinct integer label from 1 to 5289. Randomly generate 100 different integers from 1 to 5289. The employees with these labels are selected to be part of the sample. Ask each of the selected employees if they use an e-reader before bedtime. **(b)** The human resources department can carry out an experiment. Label each of the 100 employees with a different integer from 1 to 100. Then, randomly generate 50 different integers from 1 to 100. These 50 employees will be assigned to use an e-reader before bed each night. The remaining 50 employees will not use an e-reader before bed. After a predetermined time period ends, rate the work performance of all 100 employees.

12. (a) No, because the mothers in the study were not randomly assigned to eat nuts at least 5 times per week or less than once per month. **(b)** All mothers like the ones in the study. **(c)** Although the unborn cannot give consent, some would consider this to be an ethical experiment if an institutional review board reviewed it, the mothers gave their informed consent, and individual data were kept confidential.

CHAPTER 2

Answers to Section 2.1 Exercises

1. values; how often

3. bar chart and pie chart

5. False. The vertical axis in a bar chart can display frequencies or relative frequencies.

7.

Frequency table

Mission type	Frequency
Communications	5
Education	4
Imaging	17
Science	9
Technology	14
Total	49

Relative frequency table

Mission type	Relative frequency
Communications	5/49 = 0.102 or 10.2%
Education	4/49 = 0.082 or 8.2%
Imaging	17/49 = 0.347 or 34.7%
Science	9/49 = 0.184 or 18.4%
Technology	14/49 = 0.286 or 28.6%
Total	49/49 = 1.00 or 100%

9.

Frequency table

Hours of sleep	Frequency
2	1
3	2
4	2
5	4
6	8
7	15
8	14
9	4
Total	50

Relative frequency table

Hours of sleep	Relative frequency
2	1/50 = 0.02 or 2%
3	2/50 = 0.04 or 4%
4	2/50 = 0.04 or 4%
5	4/50 = 0.08 or 8%
6	8/50 = 0.16 or 16%
7	15/50 = 0.30 or 30%
8	14/50 = 0.28 or 28%
9	4/50 = 0.08 or 8%
Total	50/50 = 1.00 or 100%

11.

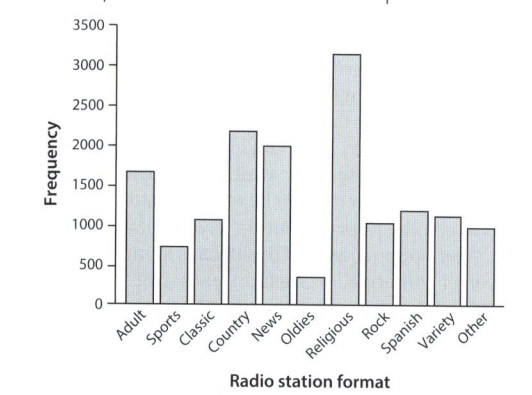

According to Nielsen Audio, the most common radio station formats are religious (3155), country (2179), and news/talk (1996), while the least common radio station formats are sports (725), other formats (978), and oldies (348).

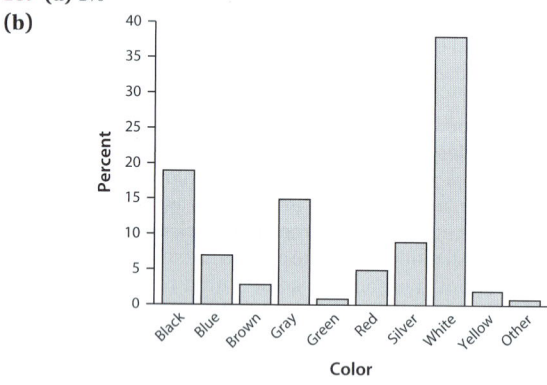

13. **(a)** 1%
(b)

For vehicles sold worldwide in 2020, the most popular colors were white (38%), black (19%), and gray (15%). The least popular colors were yellow/gold (2%), green (1%), and other (1%). **(c)** Yes, we have the percentages from all the categories that make up the whole 100%.

15. The areas of the computers make it appear that the number of buyers who purchase a Mac is more than 9 times as large as the number of buyers who previously had no computer or a Windows computer when the number of buyers is really less than 4 times as many.

17. By starting the vertical scale at 53 instead of 0, it looks like the percentage of Democrats who agree with the decision is almost 10 times higher than for Republicans and Independents. In truth, the percentage of Democrats who agree with the decision (62) is only slightly higher than the percentages of Republicans (54) and Independents (54) who agree.

19. (a)

Relative frequency table

Response	Relative frequency
Approach	17/75 = 0.227 or 22.7%
Indifferent	41/75 = 0.547 or 54.7%
Run from	17/75 = 0.227 or 22.7%
Total	75/75 = 1.00 or 100%

(b)

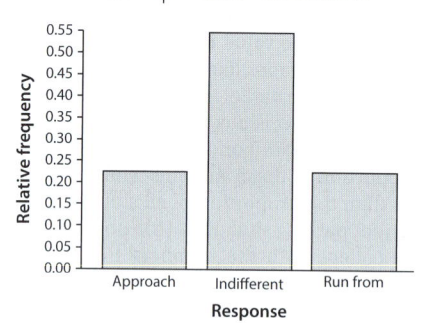

The squirrels in the random sample were most likely to be indifferent (54.7%) and were equally likely to approach (22.7%) or run from the humans (22.7%).

21. For business, slightly more than 1/8 of the circle or about 14%. For education, less than 1/10 of the circle, or about 8%.

23. (a) Those who have a technical degree or some college and those who have a college degree are more likely to buy a lottery ticket (a little more than 50%) than those who have a high school diploma or less (about 47%) or those who have a postgraduate education (about 45%). **(b)** The data do not represent parts of a whole. We have the percentage of each education group who has bought a lottery ticket, rather than the percentage of all lottery ticket buyers who are in each education group.

25. (a)

HHL	Frequency	Relative frequency
1	2398	0.799
2	298	0.099
3	138	0.046
4	138	0.046
5	28	0.009
	3000	1

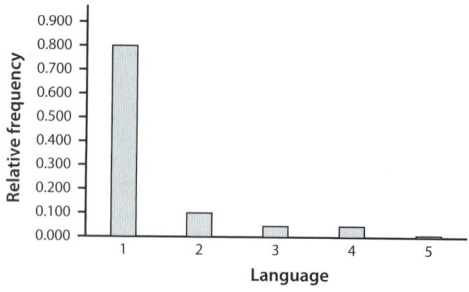

(b) Approximately 80% of the randomly selected households speak English, about 10% speak Spanish, and about 4.6% speak other Indo-European languages and Asian and Pacific Islander languages. Fewer than 1% of the households speak languages other than these.

27. On January 24, 2022, the average number of cases per capita were the highest in northwestern and southwestern Alaska, the west side of the southern tip of Texas, eastern Kentucky, near the Arkansas/Tennessee border, and in scattered parts of Illinois, North Dakota, and New Mexico. The average number of cases was the least in the Northwest, Nevada, and most of the Northeast.

29. *Individuals:* 10 tallest buildings in the world. *Categorical:* Country, use, year completed. *Quantitative:* Height (m), number of floors.

Answers to Section 2.2 Exercises

1. two categorical variables

3. False. You can arrange the bars within categories of the explanatory variable or within categories of the response variable.

5. Two variables have an association if knowing the value of one variable helps us predict the value of the other.

7. (a) $16/50 = 32\%$ (b) $7/50 = 14\%$

9. (a) $81/330 = 24.5\%$ (b) $571/2568 = 22.2\%$

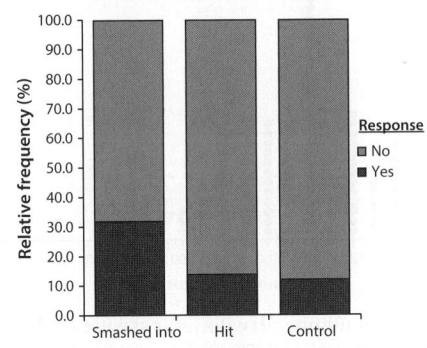

11.

13.

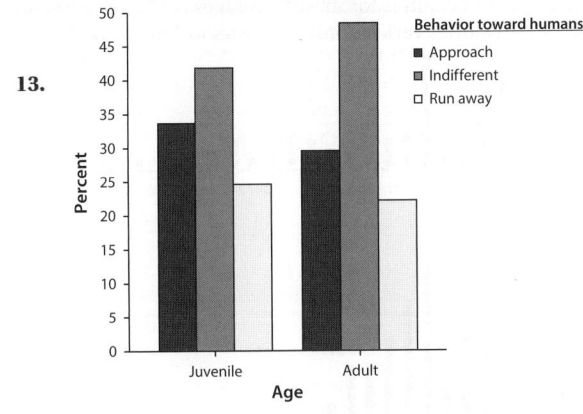

15. There is an association between year and mission type. Over 60% of missions were for imaging in 2017, compared to only about 30% in 2019. The percentage of missions for education and technology increased noticeably from 2017 to 2019, while the percentage of missions for science decreased somewhat.

17. There is an association between wording and whether or not the person recalled seeing broken glass. The subjects were most likely to recall seeing broken glass when they were told the cars "smashed into" each other (32%), followed by 14% who recalled seeing broken glass when they were told the cars "hit" each other, and 12% of the control group who recalled seeing broken glass.

19. There is an association between age and behavior for squirrels. Juvenile squirrels were more likely to run away (34% versus 29%), less likely to be indifferent (42% versus 48%), and more likely to approach (25% versus 22%).

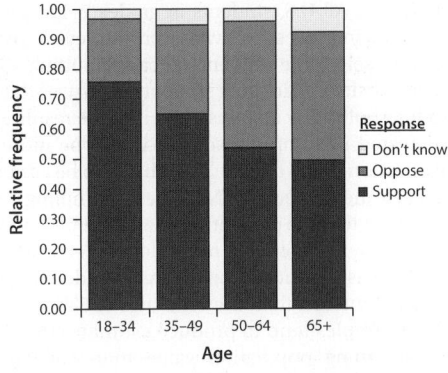

21. (a)

(b) There is an association between age and response. As age increases, the proportion of adults who support the new spending by the federal government that would help undergraduates pay tuition at public colleges without needing loans decreases.

23. (a)

		Region				
		1 = Northeast	2 = Midwest	3 = South	4 = West	Total
Language	1 = English only	181 (0.77)	730 (0.90)	914 (0.85)	573 (0.65)	2398
	2 = Spanish	17 (0.07)	30 (0.04)	96 (0.09)	155 (0.18)	298
	3 = Indo-European	28 (0.12)	30 (0.04)	32 (0.03)	48 (0.05)	138
	4 = Asian and Pacific	9 (0.04)	17 (0.02)	23 (0.02)	89 (0.10)	138
	5 = Other	1 (0.00)	6 (0.01)	6 (0.01)	15 (0.02)	28
	Total	236	813	1071	880	3000

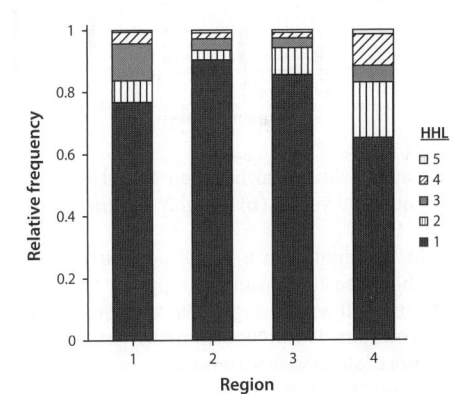

(b) There is an association between region and household language. The region that has the greatest proportion of homes in which the residents speak English only is the Midwest (90%), followed by the South (85%), Northeast (77%), and West (65%).

25. Tweens were more likely than teens to read everyday (36% versus 22%). Teens were more likely than tweens to read at least once a month (17% versus 11%), and less than once a month (17% versus 9%). Both tweens and teens are about as likely to read at least once per week (31% versus 29%) and to never read (13% versus 15%).

27.

		Dominant hand		
		Left	Right	Total
Dominant eye	Left	9	51	60
	Right	21	119	140
	Total	30	170	200

29. (a) *Population:* All U.S. students in grades 7–12. *Sample:* The 1673 U.S. students in grades 7–12 who were surveyed. **(b)** It is better to randomly select the students because putting the survey question on a website and inviting students to answer the question would introduce voluntary response bias into the results. Students who are interested in computer science would be more likely to answer the question and would be more likely to be confident that they could learn the subject. **(c)** No. Due to sampling variability, we can only expect that the percentage of all U.S. students in grades 7–12 who would say "very confident" is close to 54%. **(d)** 54%. The estimate of 54% was produced from a larger random sample from the population of interest than the estimates of 62% and 48%, and larger random samples tend to produce estimates that are closer to the corresponding population value than smaller random samples.

Answers to Section 2.3 Exercises

1. quantitative
3. False. The dotplot is skewed to the left because the left side of the graph (containing the half of the observations with smaller values) is much longer than the right side.
5. overall pattern; departures
7. (a)

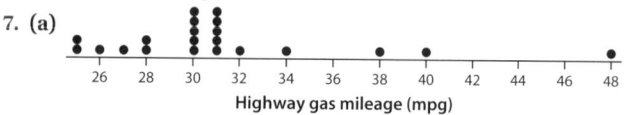

[dotplot: Highway gas mileage (mpg), 26 to 48]

(b) The Toyota Prius gets 48 mpg on the highway.
(c) $3/21 = 0.143$ or 14.3%
9. (a)

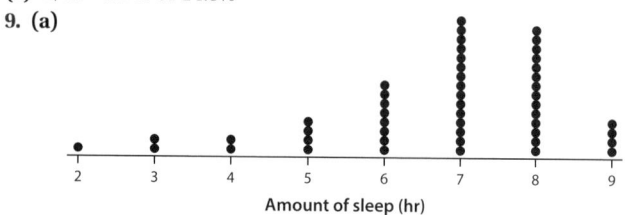

[dotplot: Amount of sleep (hr), 2 to 9]

(b) $17/50 = 0.34 = 34\%$
11. (a) Left-skewed with a peak between 90 and 100 years. There is a small gap around 70 years. **(b)** Roughly symmetric with a single peak at 7.
13. Skewed to the right with a single peak around 30–31 mpg. There is a gap between 40 mpg and 48 mpg.
15. Skewed to the left with a single peak around 7–8 hours. There are no obvious outliers. The median amount of sleep is 7 hours. The number of hours of sleep varies from 2 to 9.
17. *Shape:* Indiana: Roughly symmetric; New Jersey: Right-skewed. *Outliers:* $125,000 in the Indiana dotplot. No obvious outliers for the New Jersey dotplot. *Center:* Indiana has a similar center to New Jersey (around $45,000). *Variability:* The variability for Indiana ($0 to $125,000) is less than the variability for New Jersey ($5000 to $165,000).
19. (a) The variability appears to be similar for all three shelves (from 0 grams to about 14–15 grams). **(b)** The claim tends to hold true for the middle shelf, but not the lowest shelf. The middle shelf has the cereals with the most sugar, as the center of the distribution of sugar per serving on the middle shelf has the greatest median at around 12 grams per serving. It is possible that supermarkets want these cereals at "eye level" for kids. However, the median amount of sugar is the least for cereals on the low shelf (median = 3 grams per serving).
21. (a) The Toyota Prius's city gas mileage is 3 mpg higher than its highway gas mileage. **(b)** $8/21 = 0.38$ or about 38% **(c)** The distribution of difference in gas mileage (Highway − City) is fairly symmetric with a single peak at 9 mpg and a low outlier at −3 mpg. The median difference is 9 mpg and the differences vary from −3 mpg to 12 mpg.

23. (a) Among all the people whom Interviewer 2 surveyed, there was one person who said they worked out 30 hours per week. **(b)** *Interviewer 1:* $2/21 = 0.095$ or about 9.5%. *Interviewer 2:* $9/36 = 0.25$ or 25%. **(c)** *Interviewer 1:* Skewed to the right with 2 possible outliers at about 28 hours per week. *Interviewer 2:* Skewed to the right, with 2 possible outliers at about 45 hours per week. **(d)** The variability of workout times for Interviewer 1's sample is less than for Interviewer 2's sample. Interviewer 1's data vary from 0 to about 28 hours. Interviewer 2's data vary from 0 to about 45 hours. **(e)** Yes. If the interviewer did not affect responses about workout time for people like the ones in the study, we would expect the distributions to have similar shapes, centers, and variability. The people in Interviewer 2's sample reported greater workout times (median = 9 hours) than the people in Interviewer 1's sample (median = 5 hours).
25. (a)

[dotplot: Number of calories, 60 to 300]

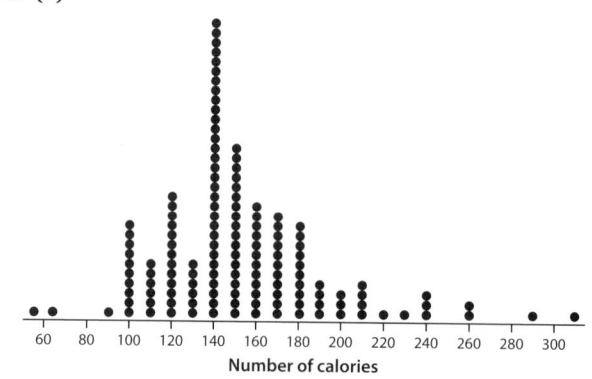

The distribution of number of calories is skewed to the right with a single peak at 140 calories. There are no apparent outliers. The median number of calories for these beverages is 150 calories and the number of calories varies from 55 calories to 310 calories.
(b)

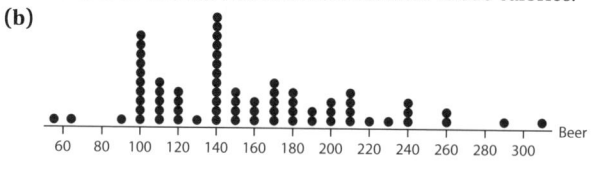

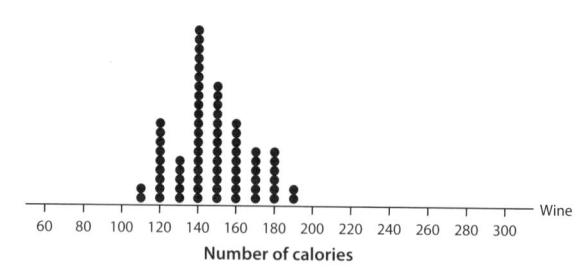

[dotplot: Beer, Number of calories, 60 to 300]

[dotplot: Wine, Number of calories, 60 to 300]

The distribution of the number of calories in the beers is slightly skewed to the right with two clear peaks (100 and 140), while the distribution of the number of calories in the wines is roughly symmetric with a single peak at 140. Both types of beverages have a median of 150 calories. Neither distribution contains any clear outliers. The number of calories in these beers varies from 55 calories to 310 calories, which is much greater than the variability of the number of calories in the wines, which varies from 110 calories to 190 calories.
27. (a) This dotplot is oriented vertically rather than horizontally, and the dots go out from the center line instead of stacking up on a number line. **(b)** Yes, the median cell length increases as the dose of vitamin C increases.
29. U.S. adult residents in 2020 were less likely to be married or widowed and more likely to be divorced or never married than U.S.

adult residents from 1980. The percentage of those who had never married was about 20% in 1980 and about 30% in 2020. The percentage of those who were married was about 65% in 1980 and about 55% in 2020. The percentage of those who were widowed was about 8% in 1980 and about 6% in 2020. The percentage of those who were divorced was about 6% in 1980 and about 10% in 2020.

Answers to Section 2.4 Exercises

1. stem-and-leaf
3. The leaves should be arranged in ascending order coming out from the stem.
5. True
7.
```
 6 | 8889
 7 | 0114668
 8 | 2688
 9 | 06
10 | 4
11 |
12 | 0
```
KEY: 12|0 represents a student with a resting heart rate of 120 bpm.

9.
```
 8 | 4
 8 |
 9 | 3
 9 | 799
10 | 011223444
10 | 556667788999
11 | 0113
11 | 5
```
KEY: 11|5 represents a high temperature in "Phoenix, Arizona," of 115°F.

11. (a) 9/51 = 0.176 or 17.6% **(b)** The shape is roughly symmetric with a single peak around 15–15.9%. There do not appear to be any states that are outliers. The median of the distribution is 16.4%. The percentage of residents aged 65 and older varies from 11.1% to 20.6%.
13. (a) 2/19 = 0.105 or 10.5% **(b)** The shape is skewed to the right with a single peak in the 70–79 bpm stem and a gap from 104 to 120. **(c)** There is one possible outlier—the resting heart rate of 120 bpm.
15. The distribution is slightly left-skewed with a single peak on the 105–109 stem and a gap from 84 to 93. The high temperature of 84°F appears to be an outlier. The median high temperature is 105°F and the high temperatures vary from 84°F to 115°F.
17. (a)
```
      No target    Target
              9 | 10 |
                | 11 |
            8 | 12 | 6
            7 | 13 | 9
            6 | 14 | 7
        77654 | 15 | 248
          772 | 16 | 7
            1 | 17 | 019
            1 | 18 | 015
            0 | 19 | 9
```
KEY: 19|9 represents a student who jumped 199 cm.

(b) Different. *No target:* The shape is fairly symmetric with a single peak in the 150s. *Target:* The shape is skewed to the left with peaks in the 150s and in the 170–180s. **(c)** Yes, the median jump distance of the children who had a target (168.5 cm) is greater than those who did not have a target (157 cm).
19. The shapes of both distributions of beak depth are skewed to the left; however, after the drought, the shape is more clearly skewed left. There do not appear to be any outliers in either distribution. The depths of the finch beaks tend to be greater after the drought (median = 10.3 mm) than before (median = 9.7 mm). Beak depths vary more before the drought (6.2 mm to 11.7 mm) than after (7.1 mm to 11.7 mm).
21. (a) To get a better picture of the distribution of the percentage of observed people who were wearing seat belts. **(b)** Six states **(c)** Skewed to the left with a single peak in the 90–91 stem and gaps from 71% to 75% and from 75% to 78%. The values 71% and 75% are possible outliers. **(d)** The shapes of both distributions of seat belt use are single-peaked and slightly skewed to the left. *Primary enforcement states:* No clear outliers. *Secondary enforcement states:* Two possible outliers (71% and 75%). The median of the percentage of observed people who were wearing seat belts in primary enforcement states was greater (median = 91.5%) than that in secondary enforcement states (median = 86%). In primary enforcement states, seat belt usage varied less (81% to 97%) than in secondary enforcement states (71% to 94%).

23. (a)
```
 2 | 48
 3 | 358
 4 | 0000111222233335566677888889
 5 | 00000122245556889
 6 | 00012455788
 7 | 0002455
 8 | 0005
 9 | 00001256
10 | 045
11 | 035679
12 | 000479
13 | 000244458999999
14 | 0
```
KEY: 14|0 represents a beverage that is 14.0% alcohol.

The distribution of percent alcohol is fairly symmetric and bimodal with a peak in the 4.0–4.9% stem and a peak in the 13.0–13.9% stem. There are no obvious outliers. The median alcohol content is 6.45% alcohol and the alcohol contents vary from 2.4% to 14%.

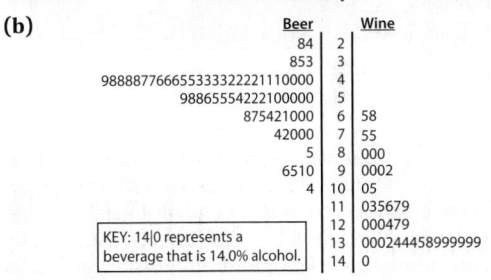

(b)
```
                        Beer      Wine
                         84 |  2 |
                        853 |  3 |
     98888776665533332222111 0000 |  4 |
          98865554222100000 |  5 |
               875421000 |  6 | 58
                  42000 |  7 | 55
                      5 |  8 | 000
                   6510 |  9 | 0002
                      4 | 10 | 05
                        | 11 | 035679
                        | 12 | 000479
                        | 13 | 000244458999999
                        | 14 | 0
```
KEY: 14|0 represents a beverage that is 14.0% alcohol.

The distribution of percent alcohol for beer is single-peaked and skewed to the right and the distribution of percent alcohol for wine is single-peaked and skewed to the left. The median alcohol content for the beers is 5%, which is less than the median for the wines (12%). The alcohol content for beer varies from 2.4% to 10.4%, which is similar to the variability for wine (6.5% to 14%).

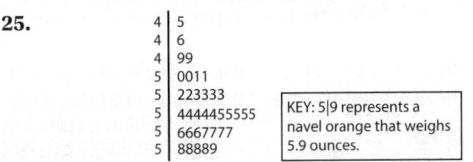

25.
```
4 | 5
4 | 6
4 | 99
5 | 0011
5 | 223333
5 | 4444455555
5 | 6667777
5 | 88889
```
KEY: 5|9 represents a navel orange that weighs 5.9 ounces.

The distribution of navel orange weight is skewed to the left with a single peak in the 5.4–5.5 stem.

27. (a) 24 healthy young tomato plants → Random assignment → Group 1: 12 plants → Treatment 1: New fertilizer; Group 2: 12 plants → Treatment 2: Current fertilizer → Compare weight

(b) Label each tomato plant with a different integer from 1 to 24. Then, randomly generate 12 different integers from 1 to 24. The tomato plants with these labels will be given Treatment 1. The remaining 12 tomato plants will be given Treatment 2.

Answers to Section 2.5 Exercises

1. False. Dotplots and stemplots show every individual value in a set of quantitative data, but histograms do not.
3. True
5. False. The choice of the interval width in a histogram can affect the appearance of the distribution.

7. [histogram: CO₂ (metric tons per person)]

9. [histogram: Number of home runs hit]

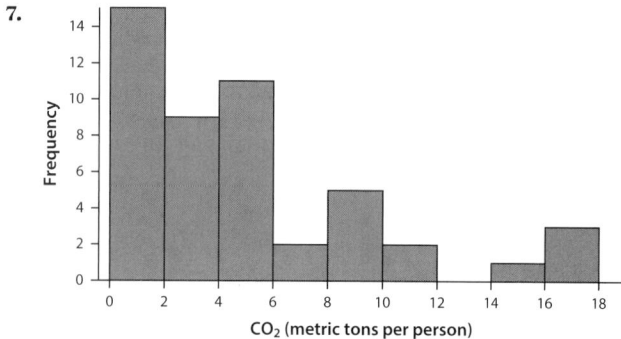

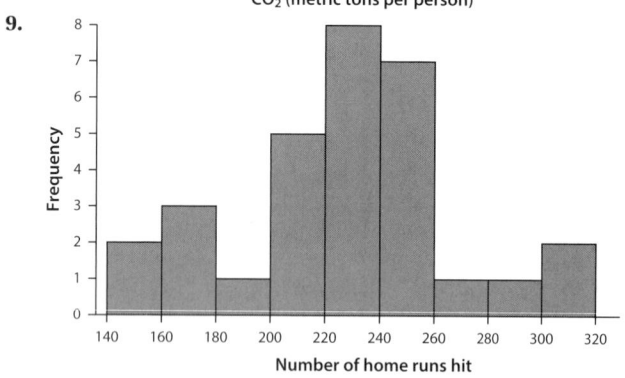

11. (a) $100/273 = 37\%$ (b) *Shape:* roughly symmetric with a single peak in the 0% to <2.5% interval. There is a gap from −22.5% to <−17.5% and there is a gap from −15% to <−12.5%. There are possible outliers in the interval −25% to <−22.5% and the interval −17.5% to <−15%. (c) The median is in the 0 to <2.5% interval. The monthly percent return on common stocks varies from possibly as low as −25% to possibly as high as about 12.5%.

13. (a) $6/48 = 0.125$ or 12.5% (b) *Shape:* Skewed to the right with a single peak in the 0 to <2 interval. There is a gap in the 12 to <14 interval. Canada (15.4 metric tons), the United States (16.1 metric tons), Australia (16.3 metric tons), and Saudi Arabia (17.0 metric tons) are possible outliers. (c) The center (median) is 3.95 metric tons of CO_2 emissions per person and the CO_2 emissions per person vary from 0.2 metric ton per person to 17 metric tons per person.

15. The distribution is fairly symmetric with a single large peak in the 220 to <240 interval, and two minor peaks on both ends of the distribution. There are no obvious outliers. The median number of home runs hit is 223.5 home runs and the number of home runs hit varies from 146 to 307 home runs.

17. *Shape:* Low-income: Strongly skewed to the right. High-income: Slightly skewed to the right. *Outliers:* There are no apparent outliers in either distribution. *Center:* The center is greater for high-income (median = 3) than low-income (median = 1). High-income households tend to have larger household sizes than low-income households. *Variability:* Household size among both low- and high-income households varies from 1 to 7.

19. (a) Similar. Both are single-peaked and skewed to the right. (b) Different. The median BMI for the semi-urban women (20 to <22) is greater than that of the rural women (18 to <20). (c) Different. The BMI measurements vary more (about 12 to <42) among semi-urban women than among rural women (about 12 to <34). (d) A greater percentage of rural women are underweight (33%) compared to semi-urban women (13%).

21. (a) About $337/1640 = 0.205$ or 20.5% (b) Single-peaked and skewed to the right (c) The median cost is a little less than $2000 and the costs vary from about $0 to about $15,500.

23. All three distributions of sepal length have a shape that is fairly symmetric with a single peak. The *Virginia* species is the only distribution that appears to have any outliers. The biggest difference is the centers. The median sepal length is approximately 5 cm for the *Setosa* species, which is less than that of the *Vericolor* species (about 6 cm), which is less than that of the *Virginia* species (about 6.5 cm). Variability is smallest for the *Setosa* species and is largest for the *Virginia* species.

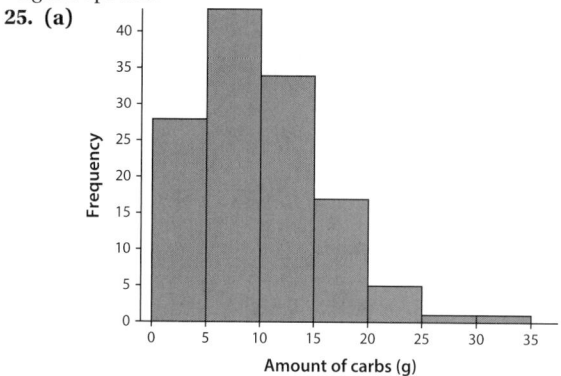

25. (a) [histogram: Amount of carbs (g)]

The distribution of the amount of carbs is single-peaked and skewed to the right. There are no obvious outliers. The median is 8 grams and the amount of carbs varies from 1 to 30 grams.

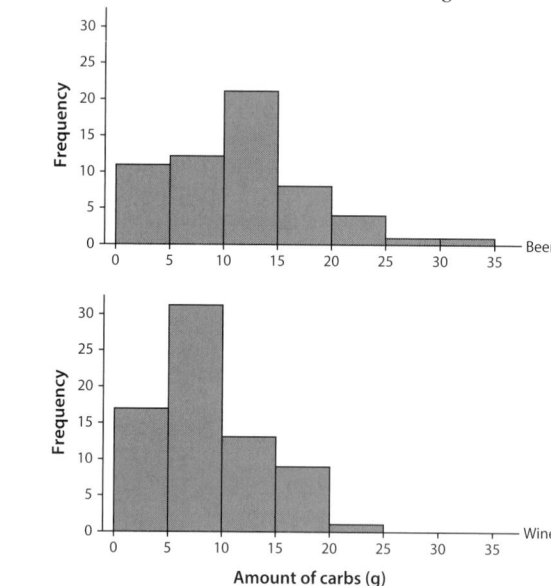

(b) [histograms: Amount of carbs (g) for Beer and Wine]

Both beverages have distributions that are single-peaked and skewed to the right. There are no obvious outliers in either distribution. The median amount of carbs for beers (12 grams) is greater than that for the wines (6 grams). The amount of carbs varies more for beer (2 to 30 grams) than for wine (1 to 22 grams).

27. A histogram should be used because die roll is a quantitative variable. A possible histogram is given here.

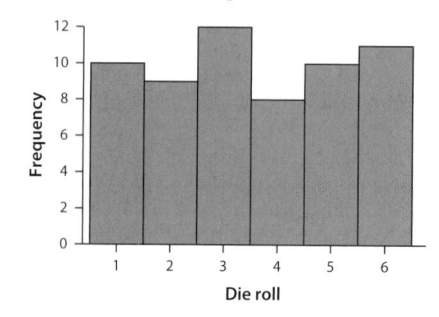

[histogram: Die roll]

29. (a) The histograms are using relative frequencies because many more students take the AP® Calculus AB exam than take the AP® Statistics exam.

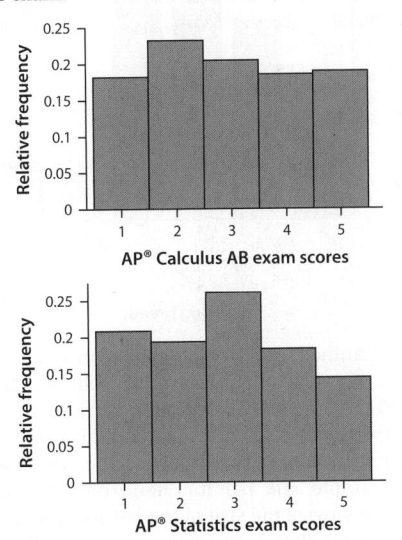

(b) The distribution of AP® Calculus AB exam scores is fairly uniform, while the distribution of AP® Statistics exam scores is single-peaked and fairly symmetric. The most common score on the AP® Calculus exam is 2 and the most common score on the AP® Statistics exam is 3. Neither distribution has any outliers. The median of both distributions is a score of 3. Although scores on both exams vary from 1 to 5, there are more scores close to the center on the AP® Statistics exam, and on the AP® Calculus exam, the scores tend to be more evenly distributed across the scores 1 to 5.

Answers to Section 2.6 Exercises

1. quantitative
3. positive; negative; no
5. In a scatterplot, an outlier is an individual point that falls outside the overall pattern of the relationship.
7.

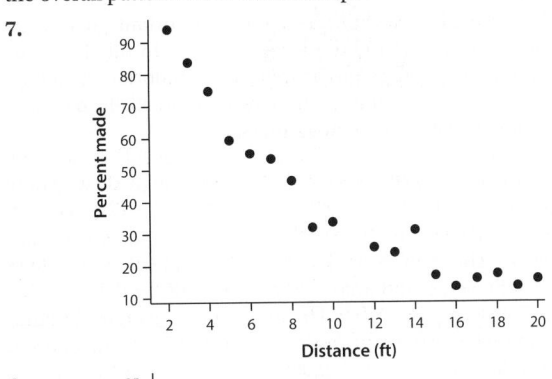

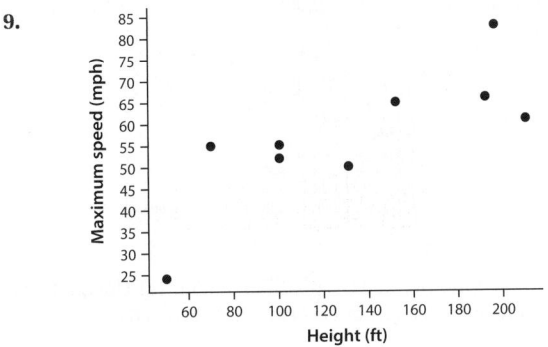

9.

11. *Direction:* Negative association. Countries with greater per-capita wealth tended to have a lower percentage of residents who say that they pray daily, and vice versa. *Form:* There is a curved (nonlinear) pattern in the scatterplot. *Strength:* Fairly weak. *Outliers:* The United States deviates from the overall pattern with a per-capita wealth of $55,800 and 56% of residents who say that they pray daily.
13. *Direction:* Negative association. As distance increases, the percentage of putts made tends to decrease. *Form:* There is a curved (nonlinear) pattern in the scatterplot. *Strength:* Strong. *Outliers:* There is an unusual point with a distance of 14 feet and 31% of putts made.
15. *Direction:* Positive association. Taller roller coasters tend to have a higher maximum speed and shorter roller coasters tend to have a lower maximum speed. *Form:* There is a linear pattern in the scatterplot. *Strength:* Moderate strength. *Outliers:* The Flying Turns coaster, with a height of 50 feet and a maximum speed of 24 mph, falls well below the linear pattern of the remaining points.
17. The mortgage rates have tended to decline steadily from a high of about 13% in 1985 to a low of about 2.6% in 2021.
19. The CO_2 level measurements show seasonal variation as well as a positive long-term trend. Each year, the CO_2 levels appear to rise throughout the winter and spring, but then decline throughout the summer and fall. The CO_2 levels were about 388 ppm in January 2010 and have continued to rise each January thereafter to a peak of about 420 ppm in April 2021.
21. (a) There is a moderately strong, positive, linear relationship between the amount of fat and the number of calories in Starbucks products. No outliers are visible in the scatterplot. **(b)** This is inconsistent because the Pumpkin Cream Cheese Muffin has more grams of fat than the Caramelized Apple Pound Cake, but has *fewer* calories than the Caramelized Apple Pound Cake. Generally, items with more fat have more calories, but these two items are an exception to the overall relationship. **(c)** For both food products and beverage products, there is a moderately strong, positive, linear association between the amount of fat and the number of calories. However, the majority of high-fat items are food products and the majority of low-fat items are drink products.
23.

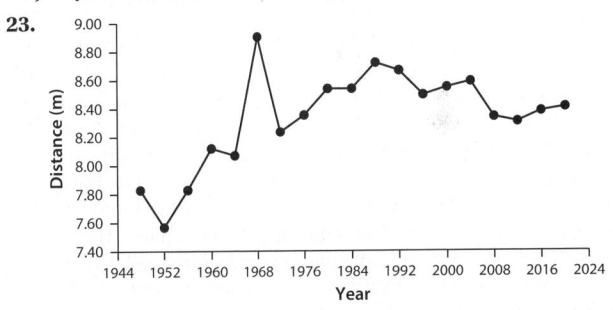

The distance of the winning men's Olympic long jump tended to increase from 1948 to 1988. Bob Beamon's 8.9-meter jump in 1968 is an outlier during this time period and still stands as the record today. After 1988, the winning long jumps decreased slightly from 8.72 meters in 1988 to 8.41 meters in 2020.
25. (a)

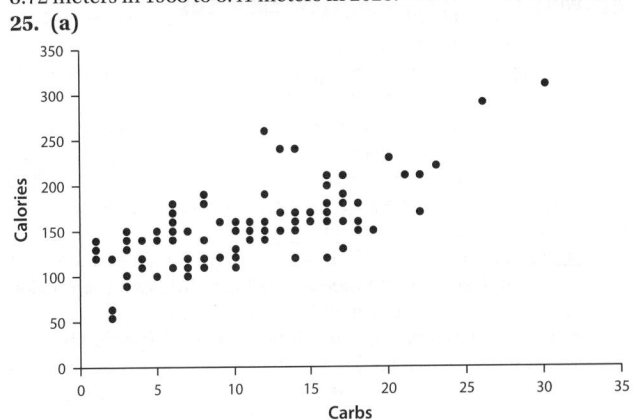

There is a moderately strong, positive, linear relationship between amount of carbs and number of calories for these beverages. There are no obvious outliers in the scatterplot.

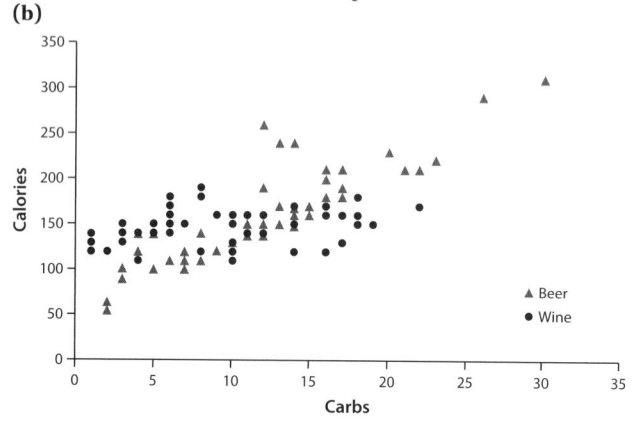

(b)

There is a fairly strong, positive, linear association between the amount of carbs and the number of calories for the beers. There is a weak, positive, linear association between the amount of carbs and the number of calories for the wines. There are three outliers in the beers scatterplot with 12 to 14 grams of carbs and 240 to 260 calories, and no obvious outliers in the wine scatterplot. The amount of carbs in beer varies more than the amount of carbs in wine, and the number of calories tends to increase more rapidly for beer than for wine as the amount of carbs increases.

27. Those who have the most education account for the largest percentage of adults who use the internet. As education level decreases, the percentage of adults who use the internet decreases as well. However, for all four categories of education level, the percentage of U.S. adults who use the internet has been increasing over time. The rate of increase has been fastest for those at the lower education levels, closing the gap in internet usage with people who have attended college.

29. The bubble chart shows the relationship between income per person and life expectancy for all countries. There is a strong, positive, nonlinear relationship between income per person and life expectancy in a country. As income per person increases, life expectancy also increases, but at a decreasing rate. African countries tend to have both low income per person and low life expectancy. A couple of large Asian countries (India and China) have low income per person but moderate life expectancies. The United States has a fairly high income per person (almost $60,000) relative to the other countries, but its life expectancy is only moderately high (about 78 years).

31. (a) This study is an experiment because a treatment is imposed. **(b)** Yes, because the researchers randomly assigned the treatments to the sites. **(c)** All sites like the ones used in the study.

Answers to Chapter 2 Review Exercises

1. (a)

Breed group	Frequency
Whippet	3
Australian shepherd	10
Border collie	13
Mixed breed	13
Other purebred	3
Labrador retriever	3
Total	45

(b)

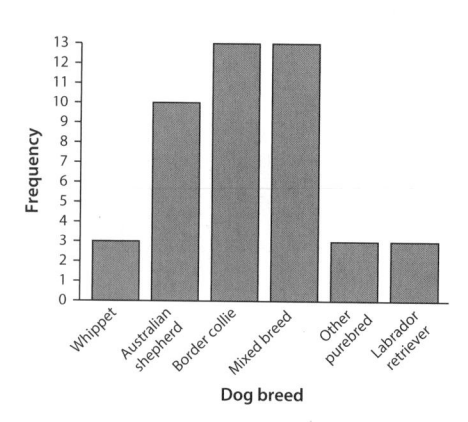

For the World Canine Disc Championships, mixed breed dogs and border collies won the most often (13), followed by Australian shepherds (10). Whippets (3), Labrador retrievers (3), and other purebreds (3) were the least represented in the group of winners.

2. (a) The graph is misleading because the "bars" are different widths. For example, the bar for "send/receive text messages" should be roughly twice the size of the bar for "camera" when it is actually much more than twice as large in area. **(b)** No, because they do not describe parts of the same whole. Teenagers were free to answer in more than one category.

3. (a) *Explanatory:* Grade level of the student. *Response:* Most important goal. Grade level might help predict which goal was most important to them. **(b)** $50/108 = 0.463$ or 46.3%

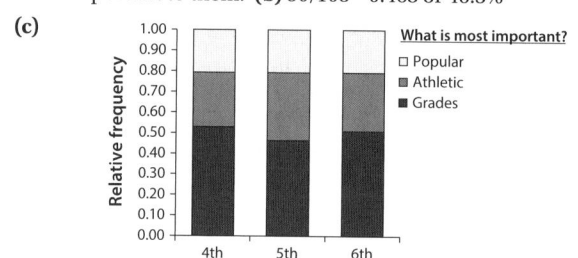

(c)

(d) Yes, because the distribution of most important goal is not exactly the same for each grade level. Students in the 4th grade more often choose good grades as their most important goal than do 5th and 6th graders, while 5th graders are more likely to choose athletic ability than are 4th or 6th graders.

4. *Shape:* All three income distributions are skewed to the right, with the Connecticut distribution being more strongly skewed than the Indiana and Maine distributions. *Outliers:* Connecticut: $280,000 and $350,000; Indiana: $210,000; Maine: No obvious outliers. *Center:* The center is largest for Connecticut (median ≈ $95,000), followed by Indiana (median ≈ $75,000) and then Maine (median ≈ $60,000). *Variability:* The total family incomes for families from Connecticut vary the most ($10,000 to $350,000), followed by Indiana (about $10,000 to $210,000) and then Maine (about $0 to $170,000).

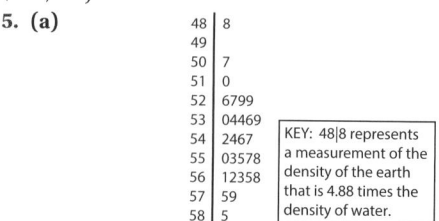

5. (a)

```
48 | 8
49 |
50 | 7
51 | 0
52 | 6799
53 | 04469
54 | 2467
55 | 03578
56 | 12358
57 | 59
58 | 5
```

KEY: 48|8 represents a measurement of the density of the earth that is 4.88 times the density of water.

(b) Roughly symmetric; 4.88 is a possible outlier. **(c)** The median is 5.46. The density measurements vary from 4.88 to 5.85.

6. (a)

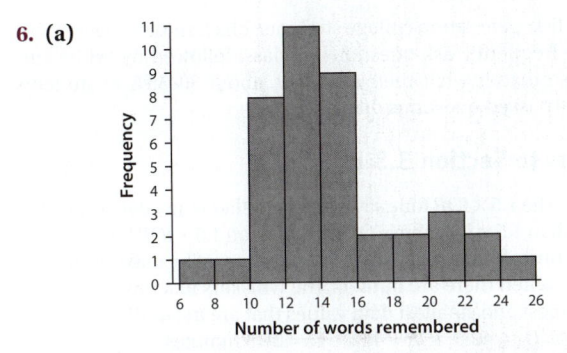

(b) The distribution of number of words remembered is skewed to the right with a single peak in the 12 to <14 interval. There are no apparent outliers. The median number of words remembered is 13 and the number of words remembered varies from 7 to 25 words.

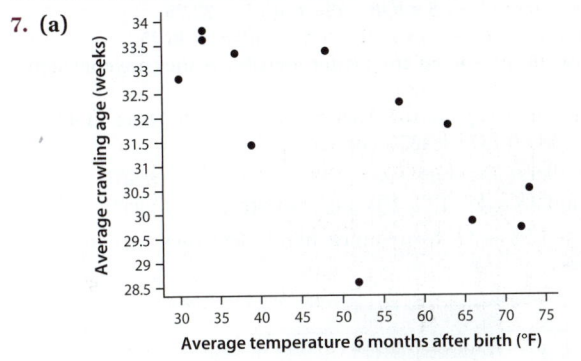

7. (a)

(b) There is a moderately strong, negative, linear relationship between average temperature 6 months after birth and average crawling age for these infants. The birth month of May (or crawling month of November), at point (52, 28.58) departs from the overall linear pattern.

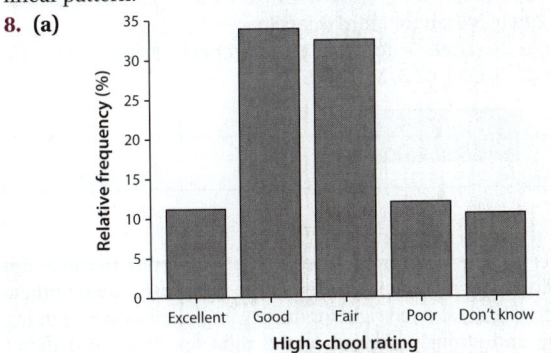

8. (a)

About 34% of parents surveyed rated the high schools in their state as doing a good job. A similar percentage (32%) gave high schools only a fair rating. Roughly equal percentages of parents rated their states' high schools as excellent (11%) or poor (12%). Overall, about 45% of parents surveyed rated high schools as doing an excellent or good job, whereas about 43% rated high schools as fair or poor. It seems like parents are fairly evenly split about how well their states' high schools are doing. **(b)** There is an association between parents' race/ethnicity and their rating of how well their states' high schools are doing because the segmented bar graphs are not identical. Only 40% of Black parents surveyed said that high schools were doing an excellent (6%) or good (34%) job. A slightly higher 44% of Hispanic parents rated their states' high schools favorably: excellent (17%) or good (27%). About 51% of White parents rated their states' high schools favorably: excellent (11%) or good (40%). A higher percentage of Black parents rated their states' high schools as fair (37%) than did White or Hispanic parents (30%).

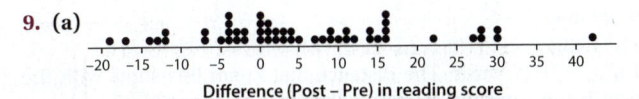

9. (a)

(b) The shape is fairly symmetric with three peaks: −4, 0, and 16. **(c)** Yes. There are many more positive differences than negative differences. **(d)** There is a moderately strong, positive, linear relationship between the pre-test and post-test scores for these 53 students. There are no clear outliers.

CHAPTER 3

Answers to Section 3.1 Exercises

1. mean; median
3. False. If there are an even number of values in the data set, the median will be the average of the two middle values in the ordered list and may have a different value than the individual data values.
5. $\bar{x}; \mu$
7. It is the balance point of the distribution.
9. 30 mpg. About half of the cars get less than 30 mpg and about half get more.
11. 0.375%. About half of the test holes have less than 0.375% copper and about half have more.
13. (a) 31.2381 mpg **(b)** New mean = 30.4 mpg. New median = 30 mpg. The mean decreased when the outlier was removed, but the median remained the same.
15. (a) 0.4268% **(b)** New mean = 0.3515%. The two outliers had increased the mean. Without the outliers, the percent copper in the remaining 26 drill holes was just 0.3515%.
17. (a) The distribution of electoral votes is skewed to the right with several possible high outliers, so the mean is greater than the median. **(b)** The median, because the distribution of electoral votes is skewed to the right and has possible outliers.
19. The distribution of the reported amount of sleep is fairly symmetric, and there are no obvious outliers.
21. (a) 22.5 minutes. About half of the randomly chosen workers in New York State have a travel time of less than 22.5 minutes and about half have a travel time that is more than 22.5 minutes. **(b)** 31.25 minutes. This is the balance point of the distribution. **(c)** The distribution of travel times is skewed to the right and the travel time of 85 minutes is a possible outlier, so the median is a more appropriate summary of the center.
23. The mean is 194/74 = 2.62 servings of fruit per day. The median would be the average of the 37th and 38th values (when written in order). The median is 2 servings.
25. (a) $37,748.34 **(b)** The distribution of student loan debt is likely skewed to the right since there are many borrowers with loan debts of $100,000 or more, so the median < the mean of $37,748.34.
27. (a) 11.65 pairs of shoes per person **(b)** 9.4375 pairs of shoes per person **(c)** The trimmed mean, because it removes the four most extreme values (including the two possible outliers of 35 and 38) so as to give a better measure of the center of the distribution.

Answers to Section 3.2 Exercises

1. range
3. $s_x = \sqrt{\dfrac{\sum (x_i - \bar{x})^2}{n-1}}$
5. Standard deviation measures the typical distance that the observations are away from the mean, and distance is always nonnegative. Also, the formula for calculating the standard deviation involves squaring the deviations of individual data values from the mean, which ensures a nonnegative sum.
7. First, arrange all the values left to right from smallest to largest. Next, find the median. The first quartile is the median of the data values that are to the left of the median.

9. 23 mpg
11. 1.43%
13. $s_x = 7.717$ yards. The distance that Zufan hits shots with the new 7-iron typically varies from the mean by about 7.717 yards.
15. $s_x = 4.472$ likes. The number of likes for ASA Instagram posts typically varies from the mean by about 4.472 likes.
17. $IQR = 31.5 - 28 = 3.5$ mpg. The range of the middle half of the fuel economy ratings for these 21 cars is 3.5 mpg.
19. (a) $IQR = 0.63 - 0.195 = 0.435\%$. The range of the middle half of the percent copper for these 28 test holes is 0.435% copper. **(b)** The interquartile range, because the distribution of percent copper is skewed to the right with two possible outliers.
21. (a) 51 votes **(b)** The number of votes for these 50 states and DC typically varies from the mean by 9.558 votes. **(c)** $IQR = 12 - 4 = 8$ votes. The range of the middle half of the number of electoral votes for these 50 states and DC is 8 electoral votes. **(d)** The interquartile range, because the distribution of number of electoral votes is skewed to the right with at least one possible outlier.
23. Mean ± 1SD: 0.991 to 20.107   47 out of 51 = 92.2%
Mean ± 2SD: −8.567 to 29.665   48 out of 51 = 94.1%
Mean ± 3SD: −18.125 to 39.223   49 out of 51 = 96.1%
These percentages do not match the values specified by the empirical rule because the distribution of electoral college votes is not single-peaked, mound-shaped, and fairly symmetric.
25. SD ≈ 2. The mean is about 3 or 4. The typical distance that the values are away from the mean cannot be 5 or 10 because the data values only vary from 0 to 9.
27. (a) SD < 8. Including an additional score that is equal to the mean decreases the typical distance of the homework scores from the mean. **(b)** SD > 8. Including an additional score (0) that is very far from the mean increases the typical distance of 8 from the mean.
29. (a)

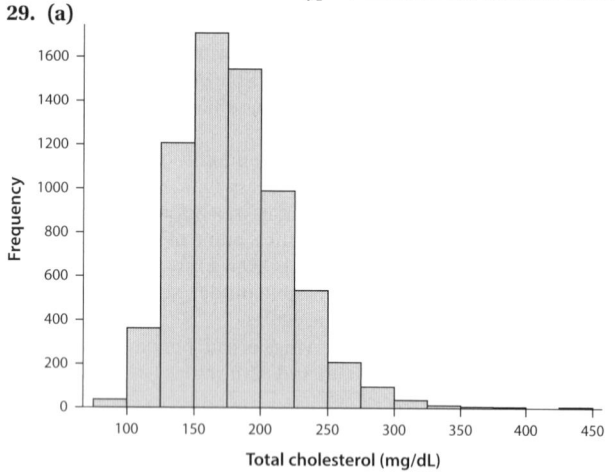

The shape of the distribution of total cholesterol is skewed to the right with a single peak around 165 mg/dL of cholesterol. **(b)** Mean = 179.8946 mg/dL. Median = 176 mg/dL. The median, because the distribution is skewed to the right with some possible outliers. **(c)** Range = 370 mg/dL. Standard deviation = 40.6022 mg/dL. $IQR = 204 - 151 = 53$ mg/dL. The interquartile range, because the distribution is skewed to the right with some possible outliers.
31. (a) The shape of the distribution of reported amount of sleep is fairly symmetric and mound-shaped with a single peak around 8 hours/night. **(b)** About 95%, because the distribution is single-peaked, mound-shaped, and fairly symmetric. **(c)** Mean ≈ 8. SD ≈ 2.33. Note that 99.7% of students got between approximately 1–15 hours of sleep. These values are 7 hours from the mean and represent 3 SD from the mean. So, SD ≈ 7/3 ≈ 2.33 hours.
33. (a) The price of the Apple stock typically varies from the mean by about $32.15. **(b)** CV = 0.556. For the Apple stock, the standard deviation is 55.6% the size of the mean.
35. For all races, the percentage of students who frequently asked questions in class were similar for those who were first-generation and not first-generation college students. Black students were most likely to frequently ask questions in class, followed by White students. No matter what their ethnicity, about 30–55% of students frequently asked questions during class.

Answers to Section 3.3 Exercises

1. False. The $1.5 \times IQR$ Rule says that an outlier is any data value that is more than $1.5 \times IQR$ above Q_3 or more than $1.5 \times IQR$ below Q_1.
3. minimum, first quartile, median, third quartile, maximum
5. False. When there are outliers, the whiskers in a boxplot extend to the largest and smallest data values that are *not* outliers.
7. $Q_1 = 15$, $Q_3 = 42.5$, $IQR = 42.5 - 15 = 27.5$ minutes
Low outliers < $Q_1 - 1.5 \times IQR = 15 - 1.5(27.5) = -26.25$
High outliers > $Q_3 + 1.5 \times IQR = 42.5 + 1.5(27.5) = 83.75$
Because 85 is greater than 83.75, it is considered an outlier.
9. $Q_1 = 28$, $Q_3 = 31.5$, $IQR = 31.5 - 28 = 3.5$ mpg
Low outliers < $Q_1 - 1.5 \times IQR = 28 - 1.5(3.5) = 22.75$
High outliers > $Q_3 + 1.5 \times IQR = 31.5 + 1.5(3.5) = 36.75$
Because 38, 40, and 48 are greater than 36.75, they are considered outliers.
11. (a) Min = 0, $Q_1 = 0.195$, Med = 0.375, $Q_3 = 0.63$, Max = 1.43
$IQR = 0.63 - 0.195 = 0.435\%$ copper
Low outliers < $Q_1 - 1.5 \times IQR = 0.195 - 1.5(0.435) = -0.4575$
High outliers > $Q_3 + 1.5 \times IQR = 0.63 + 1.5(0.435) = 1.2825$
Because 1.38 and 1.43 are more than 1.2825, they are considered outliers.

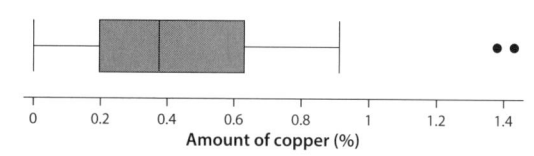

(b) Between 25% and 50% of the rock at Broad Top Butte has a copper content greater than 0.6% because 0.6% is greater than the median, but less than the third quartile.
13. (a) See Exercise 7 for the outlier check. Min = 5, $Q_1 = 15$, Median = 22.5, $Q_3 = 42.5$, Max = 85

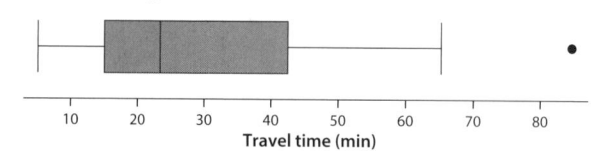

(b) IQR is a more appropriate measure of variability because the distribution of travel time is skewed to the right with a high outlier.
15. *Shape:* Slightly skewed right for those who described their putting technique and strongly skewed to the right for those who didn't describe their putting technique. *Outliers:* There is one outlier of 65 putts among those who described their putting technique, and there is one outlier of 37 putts among those who did not describe their putting technique. *Center:* Those who described their putting technique tended to require more putts to make 3 in a row (median = 17 putts) than those who did not describe their putting technique (median = 5 putts). *Variability:* There is more variation in the number of putts required to make 3 in a row for those who described their putting technique ($IQR = 16$) than for those who did not describe their putting technique ($IQR = 11$).
17. (a) More than 50% of the crashers lost less than 5% of their body weight, about 25% of the shape shifters lost less than 5% of their body weight, and less than 25% of the life changers lost less than 5% of their body weight. **(b)** *Shape:* Fairly symmetric for all three dieting strategies. *Outliers:* There are no outliers among those who followed the life changers diet, there are two outliers among those who followed the shape shifters diet, and there is one outlier among those who followed the crashers diet. *Center:* The crashers tend to have the least percentage of body weight lost (median ≈ 4%), followed by the shape shifters (median ≈ 6.5%).

The life changers tend to have the greatest percentage of body weight lost (median ≈ 11%). *Variability:* There is more variation in the percentage of body weight lost among the life changers ($IQR ≈ 9.5\%$) than for the shape shifters ($IQR ≈ 3\%$) and the crashers ($IQR ≈ 3.5\%$).

19. (a) *Apple:* $IQR = 86 - 83 = 3$
Low outliers $< Q_1 - 1.5 \times IQR = 83 - 1.5(3) = 78.5$
High outliers $> Q_3 + 1.5 \times IQR = 86 + 1.5(3) = 90.5$
There are two data values (73 and 76) that are less than 78.5, and there are no data values greater than 90.5, so this distribution has 2 outliers.
Samsung: $IQR = 86 - 75 = 11$
Low outliers $< Q_1 - 1.5 \times IQR = 75 - 1.5(11) = 58.5$
High outliers $> Q_3 + 1.5 \times IQR = 86 + 1.5(11) = 102.5$
This distribution has no outliers.

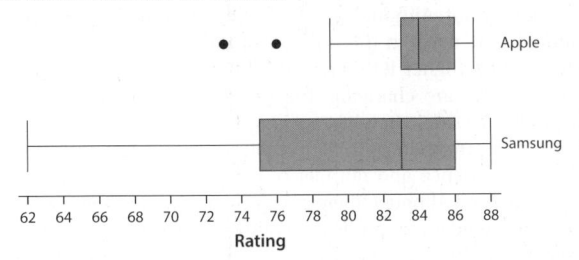

(b) Apple. The median rating for Apple (median = 84) is slightly greater than the median rating for Samsung (median = 83); however, 75% of Apple's ratings are greater than or equal to Samsung's median rating. Apple's two lowest ratings are both outliers and Samsung's lowest two ratings are less than Apple's lowest (outlier) rating.

21. (a) Median ≈ 1.1 psi. $IQR ≈ 1.75 - 0.4 = 1.35$ psi. **(b)** Yes. More than 75% of the differences (Air − Nitrogen) are positive, meaning the amount of pressure lost for air-filled tires tended to be greater than the amount of pressure lost for nitrogen-filled tires.

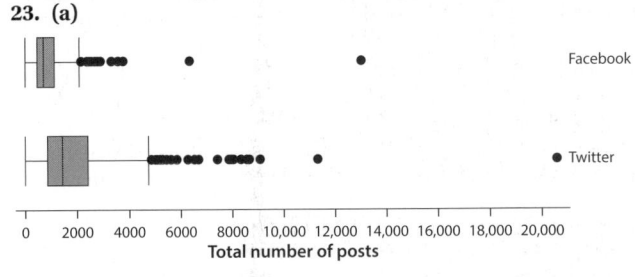

23. (a)

Shape: Skewed to the right for both Facebook and Twitter. *Outliers:* There are many high outliers in the total number of posts made on both Facebook and Twitter. *Center:* The total number of posts tends to be greater for Twitter (median = 1429.5 posts) than for Facebook (median = 697.5 posts). *Variability:* The variability in the total number of posts is greater for Twitter ($IQR = 1554$ posts) than for Facebook ($IQR = 670$ posts).

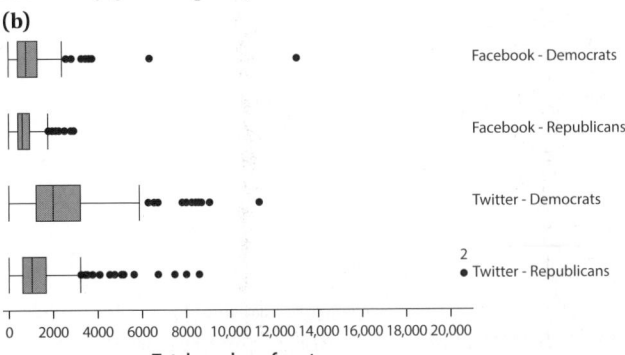

(b)

Similarities: All four distributions are skewed to the right with outliers. *Differences:* The distributions have different centers and variability. Regardless of platform, Republicans tend to post less often (Facebook median = 607.5, Twitter median = 1017) than Democrats (Facebook median = 791, Twitter median = 1982). Republicans also have less variability in the total number of posts regardless of platform (Facebook $IQR = 539$, Twitter $IQR = 1028$) than Democrats (Facebook $IQR = 805$, Twitter $IQR = 1998$).

25. (a) All five income distributions are skewed to the right. Within each level of education, there is much more variability in income for the upper 50% of the distribution than for the lower 50% of the distribution. **(b)** The upper whisker for the advanced degree boxplot is much longer than 1.5 times its IQR, so there is at least one upper outlier in the group who earned an advanced degree. **(c)** As education level rises, the median, quartiles, and extremes rise. Individuals with more education tend to attain jobs that yield more income. **(d)** The variability in income increases as the education level increases.

27. (a) Mean ± 2SD: 3.591 to 4.755. 6 out of 100 or 6% of the values are classified as outliers. **(b)** Mean ± 3SD: 3.3 to 5.046. 0% of the values are classified as outliers. **(c)** $IQR = 4.37 - 3.97 = 0.4$
Low outliers $< Q_1 - 1.5 \times IQR = 3.97 - 1.5(0.4) = 3.37$
High outliers $> Q_3 + 1.5 \times IQR = 4.37 + 1.5(0.4) = 4.97$
Because the minimum is >3.37 and the maximum is <4.97, there are no outliers.

29. (a) People who are 50 years old or less did not have a chance to participate in the survey. Younger people generally don't have as many prescription drugs and might be less willing to support a program like this. 75% is likely greater than the percentage of all U.S. adults who would answer "Yes." **(b)** Including the additional information of "can be helped" might have encouraged respondents to say yes to the program because they liked the idea of helping people. 75% is likely greater than the percentage of all U.S. adults who support this legislation.

Answers to Section 3.4 Exercises

1. less than
3. 75th
5. 2; below
7. (a) $43/50 = 0.86$. 86th percentile. 86% of the states have fewer representatives than Ohio. **(b)** $(0.52)(50) = 26$. There are 26 states that have fewer representatives. South Carolina has 7 representatives.
9. (a) $69/73 = 0.9452$. This beak length is at the 94th percentile. **(b)** $(0.28)(73) = 20.44$. The 28th percentile is the beak length that has 21 beak lengths that are less than it. The beak length of 10.25 mm is at the 28th percentile.
11. $z = 0.65$. The number of representatives for the state of Ohio is 0.65 standard deviation above the mean number of representatives for all 50 states.
13. (a) $z = -2.39$. The change in the DJIA on May 7 was 2.39 standard deviations below the mean change of 20.94 points. **(b)** Let $x = $ the unknown value. $1.03 = \dfrac{x - 20.94}{206.77}$ so $x = 233.913$. On December 12, 2019, the DJIA went up 233.913 points.
15. *SAT:* $z = 1.052$; *ACT:* $z = 1.068$. Alejandra scored better on the ACT because her z-score on that test was greater than her z-score on the SAT.
17. *Biles:* $z = 2.38$; *Comaneci:* $z = 2.07$. Biles had a better performance because her z-score was greater than Comaneci's z-score.
19. (a) $27/30 = 0.9$. Connor's head circumference is at the 90th percentile. 90% of players on the team have a smaller head circumference than Connor. **(b)** $z = 1.22$. Connor's head circumference is 1.22 standard deviations above the mean head circumference. **(c)** Yes. Because Connor's z-score for head circumference ($z = 1.22$) is greater than his z-score for height ($z = 0.87$), Connor's head circumference is relatively large for his height.
21. The speed limit is set at the value that will have 85% of the vehicles on that road traveling at a speed less than the speed limit.

23. (a)

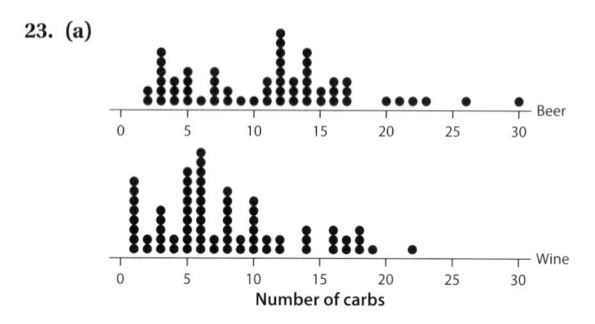

Group name	n	Mean	SD	Min	Q_1	Med	Q_3	Max
1: Beer	58	11.1379	6.3091	2	5	12	14	30
2: Wine	71	7.8732	5.2124	1	5	6	10	22

(b) 55/58 = 0.948 of the beers have fewer carbs than Bell's Kalamazoo Stout. Bell's Kalamazoo Stout is at the 94th percentile with regard to number of carbs. (c) Bell's Kalamazoo Stout's z-score: $z = 1.88$ (d) Barefoot Pink Moscato's z-score: $z = 1.94$. Barefoot Pink Moscato is more unusual because its z-score ($z = 1.94$) is greater than that of Bell's Kalamazoo Stout ($z = 1.88$).

25.

Age	Frequency	Relative frequency	Cumulative frequency	Cumulative relative frequency
40–44	2	2/46 = 0.0435 = 4.35%	2	2/46 = 0.0435 = 4.35%
45–49	7	7/46 = 0.1522 = 15.22%	9	9/46 = 0.1957 = 19.57%
50–54	13	13/46 = 0.2826 = 28.26%	22	22/46 = 0.4783 = 47.83%
55–59	12	12/46 = 0.2609 = 26.09%	34	34/46 = 0.7391 = 73.91%
60–64	7	7/46 = 0.1522 = 15.22%	41	41/46 = 0.8913 = 89.13%
65–69	3	3/46 = 0.0652 = 6.52%	44	44/46 = 0.9565 = 95.65%
70–74	1	1/46 = 0.0217 = 2.17%	45	45/46 = 0.9783 = 97.83%
75–79	1	1/46 = 0.0217 = 2.17%	46	46/46 = 1.00 = 100%

27. (a) The median for Stony Brook is about 4.5% and the median for Mill Brook is about 7.5%. (b) The Mill Brook stream has more variability. The nitrate concentration for the Mill Brook stream varies from 0 to about 20 mg/L, which is greater than that for the Stony Brook stream, which varies from 0 to about 12 mg/L.
29. Here is the scatterplot.

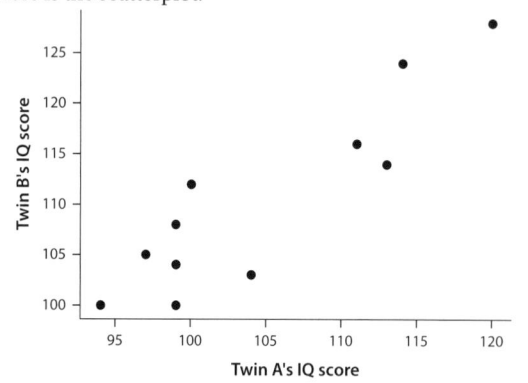

There is a moderately strong, positive, linear relationship between the IQs of twin A and twin B. There are no clear outliers.

Answers to Section 3.5 Exercises

1. linear
3. The largest possible value for the correlation is $r = 1$ and the smallest possible value is $r = -1$.
5. False. It is possible for the correlation to be close to 1 for a nonlinear association if the association is positive and very strong.
7. Because the relationship is positive, $r > 0$. Also, r is closer to 0 than to 1 because the relationship is weak—there is a lot of scatter from the linear pattern.
9. Because the relationship is negative, $r < 0$. Also, r is closer to −1 than to 0 because the relationship is strong—there isn't much scatter from the linear pattern.
11. The correlation of 0.45 indicates that the linear relationship between the count of crying peaks and IQ at age 3 years is somewhat weak and positive.
13. The correlation of −0.75 indicates that the linear relationship between the percentage of adult birds that return from the previous year and the number of new adults that join for 13 colonies of sparrowhawks is moderately strong and negative.
15. Probably not. Although there is a strong, positive correlation, an increase in spending money on pets is not likely to cause more mozzarella cheese to be consumed. It is likely that both of these variables are increasing due to other variables, such as an increase in wealth/prosperity.
17. Probably not. Although there is a somewhat weak, positive correlation, an increase in the number of times an infant cries is not likely to cause higher IQ scores later in life. It is likely that both of these variables are changing due to other variables, such as the babies' alertness or irritability.
19. (a) The correlation of 0.557 indicates that the linear relationship between attendance and number of wins is moderate and positive. (b) Probably not. Although there is a moderate, positive correlation, an increase in attendance is not likely to cause a team to win more games. It is likely that both of these variables are changing due to several other variables, such as how well the players are playing this season.
21. (a) $r \approx -0.85$ (b) They randomly assigned the trays to the light intensity levels.
23. Heights of women at age 4 and at age 18 have the largest correlation since it is reasonable to expect taller children to become taller adults and shorter children to become shorter adults. The next largest would be the correlation between the heights of male parents and their adult children because they share genes. Tall fathers tend to have relatively tall sons and short fathers tend to have relatively short sons. The smallest correlation would be between husbands and their wives. Some tall men may prefer to marry tall women, but this isn't always the case.
25. The paper's report is wrong because the correlation of $r \approx 0$ means that there is no linear association between research productivity and teaching rating.
27. (a) $z = 1.2$. Paul's height is 1.2 standard deviations above the mean height for his age. (b) 88% of 15-year-old males are shorter than Paul.

Answers to Section 3.6 Exercises

1. $r = \dfrac{\sum z_{x_i} z_{y_i}}{n-1}$

3. The correlation stays the same.
5. False. The correlation is unitless.
7. $r = 0.78$
9. (a)

(b) $r = -0.98$. The linear relationship is strong and negative.
11. (a) Still $r = 0.88$. Correlation makes no distinction between explanatory and response variables. (b) Still $r = 0.88$. Correlation doesn't change when we change the units of either variable. (c) No. Correlation doesn't have units.

13. (a) Still $r = 0.78$. Correlation doesn't change when we change the units of either variable. **(b)** Still $r = 0.78$. Correlation makes no distinction between explanatory and response variables.
15. (a) Child 18's values follow the same pattern as the rest of the points. This child makes the correlation closer to -1. **(b)** Child 19's values do not follow the same pattern as the rest of the points, so this child makes the correlation closer to 0.
17. The Flying Turns point makes the overall pattern more linear, thereby increasing the correlation.
19. (a) $r = 0.741$. The linear relationship between weight and price is moderately strong and positive. **(b)** Because the stand mixer from Walmart follows the same pattern as the rest of the points, this mixer makes the correlation closer to 1.
21. (a) Birth sex is a categorical variable. The term "correlation" is not appropriate here. **(b)** The stated correlation is greater than 1, but the largest possible value of the correlation is $r = 1$.
23. (a)

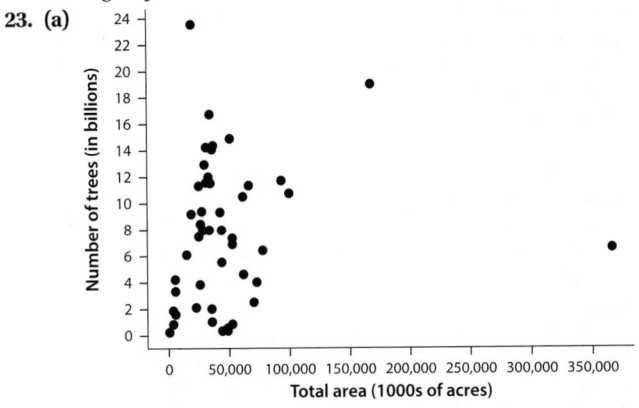

$r = 0.178$ **(b)** The state with a total area of 365,616 thousands of acres and only 6,583,055,894 trees is the most striking outlier. This outlier makes the correlation closer to 0 because it does not follow the linear pattern formed by the rest of the points. The other two potential outliers—Maine (17,579, 23,582,556,256) and Texas (40,970, 19,030,943,451)—do not follow the overall linear pattern either, and weaken the correlation.
(c)

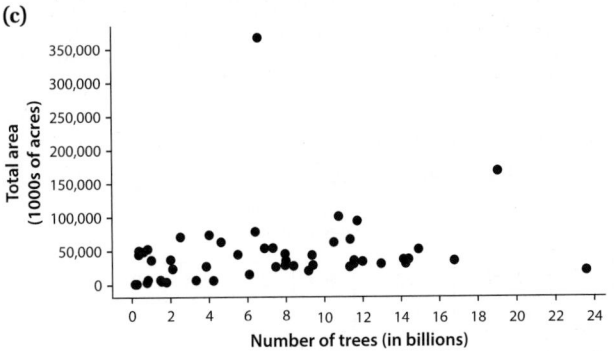

The correlation is the same: $r = 0.178$. Correlation makes no distinction between explanatory and response variables.
(d)

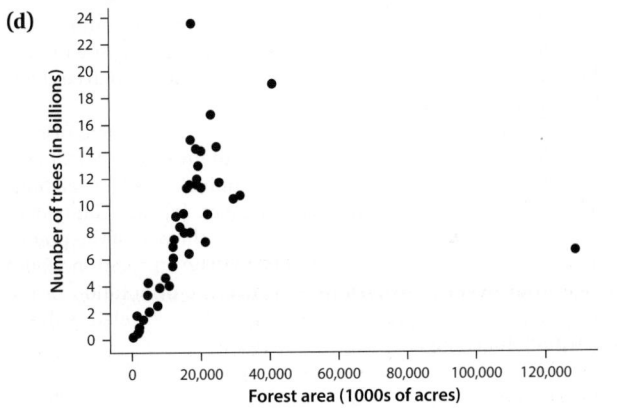

$r = 0.3998$. There is a stronger linear relationship between forest area and number of trees than between total area and number of trees because $0.3998 > 0.178$.
25. (a)

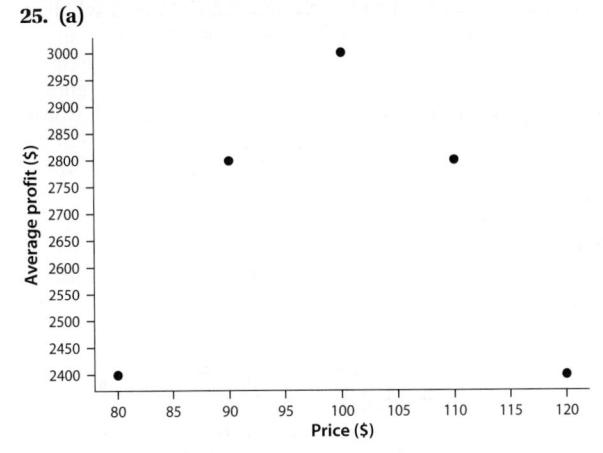

(b) $r = 0$ **(c)** Correlation measures the strength of a *linear* association between two quantitative variables. This plot shows a nonlinear relationship between price and average profit.
27. Here is one possibility.

Fund A ($)	Fund B ($)
10	10
12	11
14	12
16	13
18	14

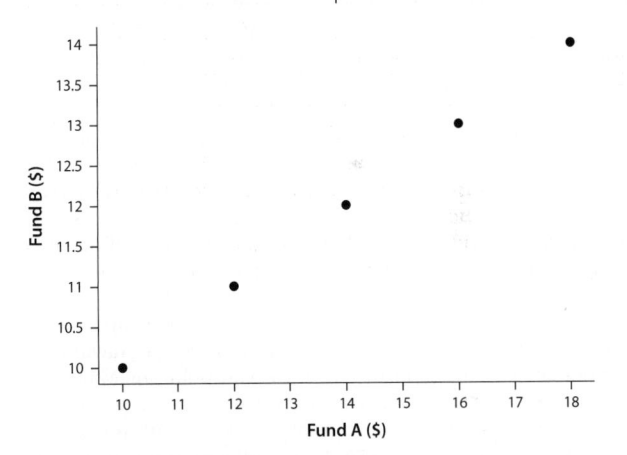

As you can see from the table, for each $2 increase in Fund A, Fund B increases by $1. The graph shows that the correlation is $r = 1$.
29. (a) Label each child with a different integer from 1 to 60. Then, randomly generate 20 different integers from 1 to 60. The children with these labels will be shown a fast-paced cartoon. Next, randomly generate an additional 20 different integers from 1 to 60 (being careful not to choose the children who have already been selected for the first treatment). The children with these labels will be shown an educational cartoon. The remaining 20 children will be given art supplies and instructed to draw pictures. **(b)** Avoid confounding by creating 3 groups of children that are roughly equivalent at the beginning of the experiment. **(c)** The difference in the average waiting time in these 2 groups was large enough that it could not be explained by chance variation in the random assignment to treatments. **(d)** Yes. The researchers randomly assigned the children to the treatments.

Answers to Chapter 3 Review Exercises

1. (a) 13 words. About half of the students remembered fewer than 13 words and about half remembered more. **(b)** 14.3 words **(c)** Mean > median because the distribution of number of words remembered is skewed to the right and there are possible high outliers.
2. (a) Range $= 25 - 7 = 18$ words **(b)** The number of words remembered typically varies from the mean by about 4.05 words. **(c)** $IQR = 16 - 11.5 = 4.5$ words. The range of the middle half of the number of words remembered for these 40 students is 4.5 words.
3. (a) $Q_1 = 11.5, Q_3 = 16, IQR = 16 - 11.5 = 4.5$ words
Low outliers $< Q_1 - 1.5 \times IQR = 11.5 - 1.5(4.5) = 4.75$
High outliers $> Q_3 + 1.5 \times IQR = 16 + 1.5(4.5) = 22.75$
Because 23, 23, and 25 are greater than 22.75, they are considered outliers.
(b)

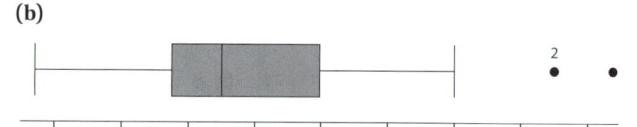

[Boxplot showing Number of words remembered from 8 to 24, with three outliers near 22-25]

(c) The distribution of number of words remembered is skewed to the right with 3 outliers. The median number of words remembered is 13 words and the IQR of the number of words remembered is 4.5.
4. (a) *Shape:* Skewed to the right for both athletes and non-athletes. *Outliers:* Both distributions have high outliers. *Center:* The athletes tend to have a faster reaction time (median ≈ 260 ms) than the non-athletes (median ≈ 295 ms). *Variability:* The athletes' reaction times vary less ($IQR \approx 70$ ms) than non-athletes' reaction times ($IQR \approx 80$ ms). **(b)** The data provide some evidence for the researcher's suspicion. The median of the distribution of reaction time for athletes (around 260 milliseconds) is less than the median for non-athletes (around 295 milliseconds). This suggests that athletes typically have a faster reaction time than non-athletes.
5. (a) $26/40 = 0.65$, so the house represented by the red dot is at the 65th percentile. **(b)** $z = 0.35$. This home has a sale price that is 0.35 standard deviation above the mean sale price. **(c)** 5 years earlier: $z = 0.28$. The house sold for more money, relatively speaking, recently because the z-score for the recent selling price ($z = 0.35$) is greater than the z-score for the selling price 5 years earlier ($z = 0.28$).
6. (a) *Using technology:* $r = -0.70$. The linear relationship between average temperature 6 months after birth and average crawling age is moderately strong and negative. **(b)** The point for May is far below the linear pattern shown by the rest of the points, so the point for May weakens the correlation (makes it closer to 0). **(c)** The correlation would still be $r = -0.70$. The correlation is not affected by changes to the units of measurement.
7. (a) $242,000. This is the median price. **(b)** Because the distribution of home prices is skewed to the right and has an apparent high outlier. The middle half of sale prices for these 43 homes has a range of $103,000. **(c)** $Q_1 = 182, Q_3 = 285, IQR = 285 - 182 = 103$
Low outliers $< Q_1 - 1.5 \times IQR = 182 - 1.5(103) = 27.5$
High outliers $> Q_3 + 1.5 \times IQR = 285 + 1.5(103) = 439.5$
There are no values less than 27.5, but there is one value greater than 439.5, the maximum home price of $503,000. This value is an outlier. **(d)** *Shape:* The distribution of home prices for Region 1 and Region 2 are skewed to the right. The distribution of home prices in Region 3 is somewhat skewed left. *Outliers:* Region 1 has 2 high outliers. Region 2 and Region 3 have no outliers. *Center:* The home prices in Region 1 tend to be the most expensive (median ≈ $275,000), followed by Region 2 (median ≈ $230,000); Region 3 is the least expensive (median ≈ $155,000). *Variability:* The home prices vary the most in Region 1 (about $200,000 to about $500,000), followed by Region 2 (about $140,000 to about $350,000). The home prices in Region 3 varied the least (about $100,000 to about $190,000).
8. (a) The linear relationship between price and size is strong and positive. **(b)** Because the house with 3269 square feet that sold for $503,000 follows the same pattern as the rest of the points, this house makes the correlation closer to 1. **(c)** Region 3. In the past several months, 2 homes that were at least 1600 square feet sold in Region 3 for less than $200,000. In Regions 1 and 2, homes of this size tend to sell for at least $250,000.

Answers to Chapters 1–3 Cumulative Review

Section I: Multiple Choice

1. d
2. c
3. d
4. c
5. b
6. b
7. c
8. a
9. c
10. b
11. d
12. b
13. c
14. c
15. d
16. a
17. d
18. d
19. b
20. d
21. a
22. c

Section II: Free Response

1. (a) This is an observational study. No treatment is imposed. **(b)** Overall lifestyle is one variable that could be confounded with whether or not people take the omega-3 fish oil and also whether they have low cholesterol levels. People who live a healthier lifestyle may be more likely to take omega-3 fish oil and also have lower cholesterol levels than those who live a less healthy lifestyle. **(c)** No. There was no random assignment of the subjects to taking omega-3 fish oil or not, so it is not reasonable to say that taking omega-3 fish oil will decrease cholesterol readings.
2.

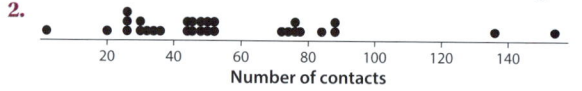

[Dotplot of Number of contacts from 20 to 140]

n	Mean	SD	Min	Q_1	Med	Q_3	Max
30	54.7667	32.6472	7	30	46.5	77	151

$IQR = 77 - 30 = 47$ contacts
Low outliers $< Q_1 - 1.5 \times IQR = 30 - 1.5(47) = -40.5$
High outliers $> Q_3 + 1.5 \times IQR = 77 + 1.5(47) = 147.5$
Because 151 is greater than 147.5, it is considered an outlier. The distribution of number of contacts is skewed to the right with one outlier, the adult who had 151 contacts. The median is 46.5 contacts and the IQR is 47 contacts.
3. (a) Individuals = the 50 states. Variables = Percentage of residents who voted for Trump in 2020 (quantitative) and Percentage of residents given at least 1 shot (quantitative). **(b)** The relationship is moderately strong, negative, and linear. **(c)** New Hampshire does not follow the linear pattern formed by the rest of the points, so it makes the correlation closer to 0. **(d)** The correlation would change from negative to positive because the orientation of the horizontal axis would reverse. The strength of the correlation, however, would not change.

4.

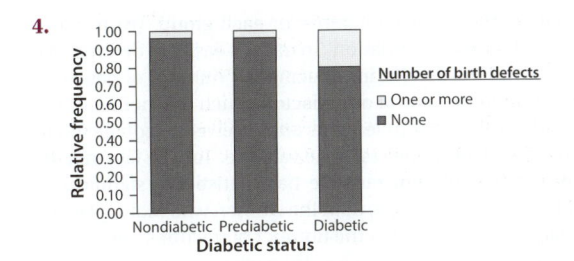

		Diabetic status			
		Nondiabetic	Prediabetic	Diabetic	Total
Number of birth defects	None	754 (0.96)	362 (0.97)	38 (0.81)	1154 (0.96)
	One or more	31 (0.04)	13 (0.03)	9 (0.19)	53 (0.04)
	Total	785 (1)	375 (1)	47 (1)	1207 (1)

The graph and numerical summaries show that the proportion of children with birth defects is greatest among mothers with diabetes (0.19), and is similar among women who do not have diabetes (0.04) or have prediabetes (0.03).

5. (a) The shape of the distribution of number of mistakes is slightly skewed to the right for both the non-native group and the no-information group. There are no obvious outliers in either distribution. The center of the distribution for the non-native group (mean = 3.23) is greater than for the no-information group (mean = 1.47). The variability of the distribution for the non-native group (SD = 2.25) is greater than for the no-information group (SD = 1.33). **(b)** Label each volunteer with a different integer from 1 to 60. Then, randomly generate 30 different integers from 1 to 60. The volunteers with these labels will be told the author is someone whose native language is not English. The remaining 30 volunteers will be told nothing about the author. **(c)** The difference is 3.23 − 1.47 = 1.76 "mistakes." **(d)** One simulated random assignment resulted in a difference of means (Non-native − No information) of 1.433 "mistakes." **(e)** Because a difference of means of 1.76 or greater rarely occurred in the simulation (1/500 = 0.2%), the difference is statistically significant. Researchers are unlikely to get a difference this big simply due to chance variation in the random assignment.

CHAPTER 4
Answers to Section 4.1 Exercises

1. 0; 1
3. True
5. (a) If you select a very large random sample of couples like these, about 25% of their children will develop cystic fibrosis. **(b)** No. This probability describes what happens in many, many repetitions (far more than 4) of a chance process.
7. If you select a very large random sample of times during the day and record the color of the traffic light, about 55% of the times the light will be red.
9. Even after the player failed to hit safely in six straight at-bats, she will continue to have the same 35% probability of getting a hit in the next at-bat.
11. These two sequences of coin tosses are equally likely. Any sequence of 6 coin tosses has a 1/64 probability of occurring.
13. (a) Let 1–9 = on time and 10 = late. Generate a random integer from 1 to 10 to simulate taking a train ride. Repeat this a total of 20 times. Count the number of 10s that occurred. The number of 10s indicates the number of train rides (out of 20) that are late. **(b)** The dot at 7 represents one repetition where there were 7 late trains out of 20. **(c)** Probability ≈ 13/100 = 0.13 **(d)** Because it is likely to have 4 or more late trains by chance alone when 90% of the days the train is on time, this result does not provide convincing evidence that the claim is false.

15. (a) Let 1–6 = hit center and 7–10 = miss center. Generate a random integer from 1 to 10 to simulate the outcome of a single shot. Continue to generate integers until a number between 7 and 10 is obtained twice. Record the number of shots Quincy took. **(b)** Randomly generated integers: **10**, 3, 3, 2, 1, **8**. Quincy stayed in the competition for 6 shots. **(c)** It is unlikely (probability ≈ 0.03) for a 60% shooter to remain in the competition for at least 10 shots, so we should be surprised if Quincy remains in the competition for at least 10 shots.
17. In the short run, the player's percentage of made three-point shots is extremely variable. In the long run, the percentage of made shots approaches a specific value of around 30%.
19. (a) Let 1–12 = first class, 13–76 = economy class. Generate a random integer from 1 to 76 to simulate selecting a single passenger at random for screening. Continue to generate integers until you get 10 unique integers (no repeats). Record the number of first-class passengers in the sample. **(b)** Randomly generated integers: 67, 31, 62, **9**, **1**, 57, 44, **12**, 52, 58. There were 3 first-class passengers. **(c)** It is somewhat likely (probability ≈ 0.15) that none of the 10 randomly selected passengers will be seated in first class. This result does not provide convincing evidence that the TSA officers did not carry out a truly random selection.
21. (a) Assuming that the true proportion of students who would say they regularly recycle is 0.50, the estimated probability that a random sample of 100 students will yield 55 or more who say they regularly recycle is 43/200 = 0.215. Because this result is likely to happen by chance alone, there is not convincing evidence that more than half of the college's students would say they regularly recycle. **(b)** Assuming that the true proportion of students who would say they regularly recycle is 0.50, the estimated probability that a random sample of 100 students will yield 63 or more who say they regularly recycle is 1/200 = 0.005. Because this result is very unlikely to happen by chance alone, we have convincing evidence that a majority of the school's students would say they regularly recycle.
23. Let 1 = Letter 1, 2 = Letter 2, 3 = Letter 3, 4 = Letter 4, and 5 = Letter 5. Generate 5 random integers from 1 to 5, excluding repeats. If the numbers come out in the order 1, 2, 3, 4, 5, the donors will all end up with the correct letter. For no one to get the correct letter, none of the numbers can be in the correct position. The more trials that are conducted, the closer the result should be to the theoretical probability, which is 44/120 = 0.367.
25. (a) No. About 70% of Indian Asian Americans have college degrees, while less than 20% of Laotian, Cambodian, and Hmong Asian Americans have college degrees. **(b)** The relationship between educational attainment and household income is moderately strong, positive, and linear.

Answers to Section 4.2 Exercises

1. all possible outcomes; each outcome

3. $P(A) = \dfrac{\text{number of outcomes in event A}}{\text{total number of outcomes in sample space}}$

5. The complement of A is the event that A does not happen.
7. True
9. (a) *Sample space:* (1, 1), (1, 2), (1, 3), (1, 4), (2, 1), (2, 2), (2, 3), (2, 4), (3, 1), (3, 2), (3, 3), (3, 4), (4, 1), (4, 2), (4, 3), (4, 4). Each of these 16 outcomes has probability 1/16. **(b)** Outcomes with a sum of 5: (1, 4), (2, 3), (3, 2), (4, 1). $P(A) = 4/16 = 0.25$.
11. (a) Using R = rock, P = paper, S = scissors, *Sample space:* RR, RP, RS, PR, PP, PS, SR, SP, SS. Each has probability 1/9. **(b)** 3 outcomes where player 1 wins: RS, PR, SP. $P(\text{Player 1 wins}) = 3/9 = 0.33$.
13. (a) $1 - (0.51 + 0.26 + 0.19) = 1 - 0.96 = 0.04$ **(b)** $1 - 0.04 = 0.96$
15. (a) $1 - (0.28 + 0.35 + 0.15 + 0.06 + 0.02 + 0.01) = 1 - 0.87 = 0.13$ **(b)** $1 - 0.28 = 0.72$
17. (a) $0.51 + 0.04 = 0.55$ **(b)** $1 - 0.55 = 0.45$
19. $1 - (0.02 + 0.01) = 0.97$

21. (a) $1-(0.065+0.29+0.387)=0.258$ (b) $1-0.065=0.935$. Complement rule. (c) $0.258+0.387=0.645$. Addition rule for mutually exclusive events.

23. Using R = rock, P = paper, S = scissors, L = lizard, Sp = Spock, *Sample space:* RR, RP, RS, RL, RSp, PR, PP, PS, PL, PSp, SR, SP, SS, SL, SSp, LR, LP, LS, LL, LSp, SpR, SpP, SpS, SpL, SpSp. Each has probability 1/25. Player 2 wins with PS, RP, LR, SpL, SSp, LS, PL, SpP, RSp, SR. $P(\text{win})=10/25=0.4$.

25. (a) <7 age group, $58/135=0.430$; 7–9 age group, $42/112=0.375$; >9 age group, $51/151=0.338$ (b) As age increases, the proportion of children who habitually snore decreases. (c) We cannot conclude that getting older causes a decrease in snoring frequency because this was an observational study, not an experiment.

Answers to Section 4.3 Exercises

1. Event A happens, or event B happens, or both events happen.
3. $P(A \text{ or } B) = P(A) + P(B) - P(A \text{ and } B)$
5. intersection
7. (a) $P(B^C)=295/595=0.496$. There is about a 49.6% probability that a randomly selected student from this school does not eat breakfast at home regularly. (b) $130/595=0.218$ (c) $405/595=0.681$
9. (a) $888/1207=0.736$ (b) $245/1207=0.203$ (c) $1085/1207=0.899$
11. $0.38+0.16-0.09=0.45$
13. $100-67=33\%$ are graduate students. $0.33+0.60-0.23=0.70$
15. (a) Here is a Venn diagram.

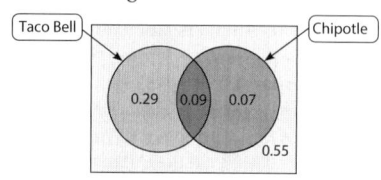

(b) 0.07
17. (a) Here is a Venn diagram.

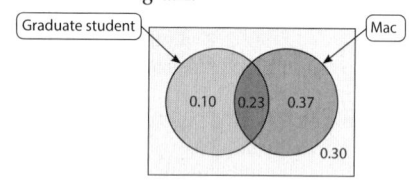

(b) 0.30. There is a 30% probability that a randomly selected student from this major university is an undergraduate student who uses a PC.
19. (a) $252/1456=0.173$ (b) $1-0.173=0.827$ (c) They add to 1. This makes sense because the event "$P(\text{age 18 to 29 or owns a smartphone})$" is the complement of the event "$P(\text{not age 18 to 29 and does not own a smartphone})$."
21. (a)

		Disk number		
		Number 8	Not 8	Total
Disk color	Blue	1	8	9
	Not blue	3	24	27
	Total	4	32	36

(b) $P(B)=9/36=0.25$; $P(E)=4/36=0.111$ (c) $P(B \text{ and } E)=1/36=0.028$ (d) The events "disk is blue" and "disk is the number 8" are not mutually exclusive.

$$P(B \cup E) = 9/36 + 4/36 - 1/36 = 12/36 = 0.333$$

23. (a) True. Complementary events have no outcomes in common. (b) False. Two mutually exclusive events do not necessarily make up the entire sample space
25. (a) Label the women from 1 to 1649. Using a random number generator, generate 825 different random integers from 1 to 1649. Assign the women with those numbers to take strontium ranelate. The other 824 women will take a placebo. At the end of 3 years, compare the number of new fractures for each group. (b) To minimize additional sources of variation. To make it easier to determine if strontium ranelate is effective in reducing the number of new fractures, it is important to control other factors (such as whether or not the women took calcium supplements and received adequate medical care) that might affect the response variable. (c) The fact that the women who took strontium ranelate had statistically significantly fewer new fractures, on average, than the women who took a placebo over a 3-year period means that the number of fractures observed in the strontium ranelate group was less than what we would expect to happen by chance alone if the treatment were not effective.

Answers to Section 4.4 Exercises

1. False. $P(A|B)$ is the conditional probability that event A happens given that event B has happened.
3. When two events are independent, knowing whether or not one event occurred does not change the probability that the other event will happen.
5. (a) $P(S|Y)=P(\text{smashed into}|\text{yes})=16/29=0.552$. Given that the selected subject said "yes," there is a 55.2% probability that they received the "smashed into" wording. (b) $P(Y^C|S^C)=87/100=0.87$
7. (a) $197/319=0.618$ (b) $291/442=0.658$
9. $0.09/0.16=0.5625$
11. $0.23/0.60=0.383$
13. $P(\text{hit}|\text{post-season})=87/352=0.247$

$P(\text{hit})=2110/8883=0.238$

Because the probabilities are not equal, the events "hit" and "post-season" are not independent.
15. $P(\text{survived}|\text{first class})=197/319=0.618$

$P(\text{survived})=442/1207=0.366$

Because the probabilities are not equal, the events "survived" and "first class" are not independent.
17. (a) $P(O|S)=151/1191=0.127$. Given that the respondent is a smartphone user, there is a 12.7% probability that the respondent is age 65 or older. (b) $P(O|S^C)=136/265=0.513$ (c) Because the probabilities in part (a) and part (b) are not equal, the events "smartphone owner" and "65 or older" are not independent.
19. $0.106/0.321=0.330$
21. $P(B) < P(B|T) < P(T) < P(T|B)$. There are very few professional basketball players, so $P(B)$ should be the smallest probability. If you are a professional basketball player, it is quite likely that you are tall, so $P(T|B)$ should be the largest probability. Finally, it's much more likely to be more than 6 feet tall than it is to be a professional basketball player if you're more than 6 feet tall.
23. $P(\text{sum is 7}|\text{blue die is a 4})=1/6$

$P(\text{sum is 7}|\text{blue die is not a 4})=5/30=1/6$

The probabilities are equal, so the events "blue die shows a 4" and "sum is 7" are independent.
25. (a) The bottom segment in each graph shows the percentage of students who said "fly" in each country. $P(\text{fly}|\text{U.K.})$ is on the left and $P(\text{fly}|\text{U.S.})$ is on the right. Because the probabilities are not equal, the events "U.K." and "fly" are not independent. (b) The segments for superstrength look roughly the same size, so the events "superstrength" and "U.K." appear to be independent. (c) There is an association between country and superpower preference because the distribution of superpower is not the same for each country [as shown in part (a)].
27. (a) Experiment. Treatments were imposed on the economics students. (b) Observational study. There were no treatments imposed on the families. (c) Observational study. There were no treatments imposed on the patients.

Answers to Section 4.5 Exercises

1. $P(A \text{ and } B) = P(A) \cdot P(B|A)$
3. False. The probabilities on the branches in the first stage of a tree diagram are not conditional probabilities. The probabilities on the branches *after* the first stage are conditional probabilities.

5. $(0.64)(0.80) = 0.512$
7. $(14/20)(13/19) = 0.479$
9. (a)

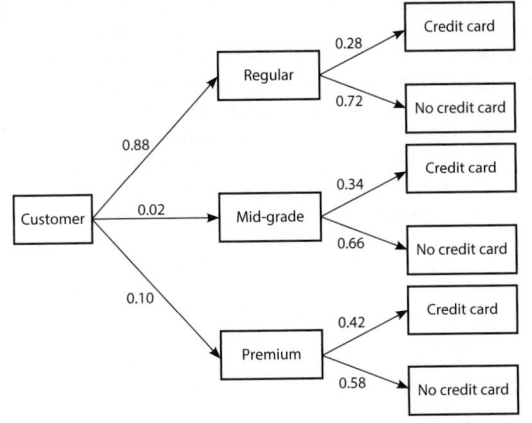

[Tree diagram: Customer → Regular (0.88) → Credit card (0.28), No credit card (0.72); Mid-grade (0.02) → Credit card (0.34), No credit card (0.66); Premium (0.10) → Credit card (0.42), No credit card (0.58)]

(b) $(0.88)(0.28) + (0.02)(0.34) + (0.10)(0.42) = 0.2464 + 0.0068 + 0.0420 = 0.2952$

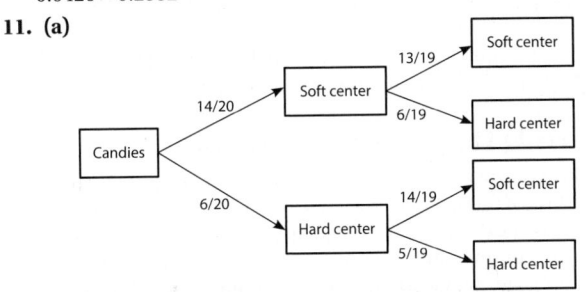

11. (a)

[Tree diagram: Candies → Soft center (14/20) → Soft center (13/19), Hard center (6/19); Hard center (6/20) → Soft center (14/19), Hard center (5/19)]

(b) $(14/20)(13/19) + (6/20)(14/19) = 0.221 + 0.221 = 0.442$

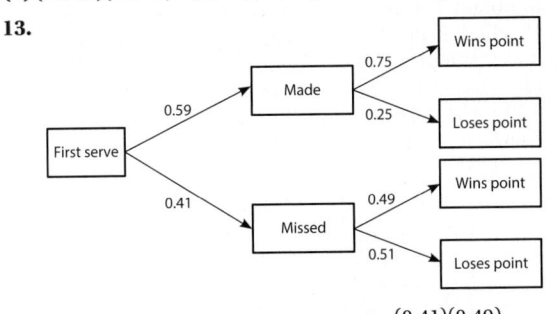

13.

[Tree diagram: First serve → Made (0.59) → Wins point (0.75), Loses point (0.25); Missed (0.41) → Wins point (0.49), Loses point (0.51)]

$P(\text{missed first serve} \mid \text{won point}) = \dfrac{(0.41)(0.49)}{(0.59)(0.75) + (0.41)(0.49)} = 0.312$

15. $P(\text{premium gasoline} \mid \text{credit card}) =$

$\dfrac{(0.10)(0.42)}{(0.88)(0.28) + (0.02)(0.34) + (0.10)(0.42)} = 0.142$

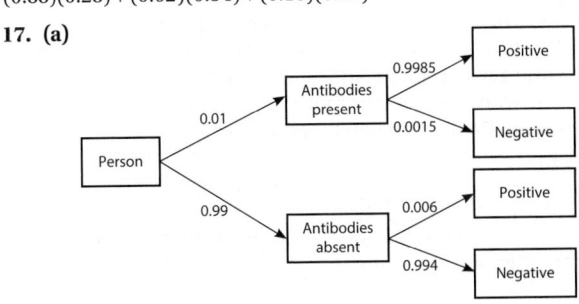

17. (a)

[Tree diagram: Person → Antibodies present (0.01) → Positive (0.9985), Negative (0.0015); Antibodies absent (0.99) → Positive (0.006), Negative (0.994)]

(b) $P(\text{positive}) = (0.01)(0.9985) + (0.99)(0.006) = 0.015925$
(c) $P(\text{antibody} \mid \text{positive}) = (0.01)(0.9985)/0.015925 = 0.627$

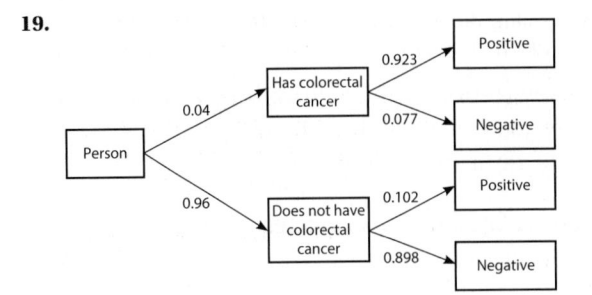

19.

[Tree diagram: Person → Has colorectal cancer (0.04) → Positive (0.923), Negative (0.077); Does not have colorectal cancer (0.96) → Positive (0.102), Negative (0.898)]

$P(\text{has colorectal cancer} \mid \text{positive}) =$

$\dfrac{(0.04)(0.923)}{(0.04)(0.923) + (0.96)(0.102)} = 0.274$

21. $P(\text{has HIV antibody} \mid 2 + \text{tests}) =$

$\dfrac{(0.01)(0.9985)(0.9985)}{(0.01)(0.9985)(0.9985) + (0.99)(0.006)(0.006)} = 0.996$

23.
(a) $\left(\dfrac{13}{52}\right)\left(\dfrac{12}{51}\right)\left(\dfrac{11}{50}\right)\left(\dfrac{10}{49}\right)\left(\dfrac{9}{48}\right) = 0.000495$ **(b)** Because there are four possible suits in which to have all five cards of the same suit, the probability is $4(0.000495) = 0.00198$.

25. *Shape:* The distribution of highway gas mileage appears to be fairly symmetric for all four age groups. *Outliers:* There are no outliers for any of the distributions. *Center:* The typical highway gas mileage for "Mini hwy" (median ≈ 25 mpg) is greater than that for "Two hwy" (median ≈ 22 mpg), which is greater than that for "Mini city" (median ≈ 18 mpg), with "Two city" having the lowest highway gas mileage (median ≈ 16 mpg). *Variability:* The variation in highway gas mileage for "Two city" and "Two hwy" (*IQR* ≈ 8 or 9 mpg) is greater than that for "Mini city" and "Mini hwy" (*IQR* ≈ 5 or 6 mpg).

Answers to Section 4.6 Exercises

1. $P(A) \cdot P(B \mid A)$; independent
3. $P(A) + P(B) - P(A \text{ and } B)$; mutually exclusive
5. $(0.98)^{20} = 0.668$
7. $(0.90)^5 = 0.5905$
9. $(0.93)^{10} = 0.484$
$P(\text{at least one is O}^-) = 1 - 0.484 = 0.516$
11. $(5/6)^{10} = 0.1615$
$P(\text{at least one 6}) = 1 - 0.1615 = 0.8385$
13. No, because the 4 consecutive flights being on time are not independent events. Knowing that one flight is late, perhaps due to bad weather, makes it more likely that other flights will be late.
15. Does $0.03 = (0.20)(0.12)$? $0.03 \neq 0.024$; the events are not independent.
17. (a) $(0.81)^4 = 0.4305$ **(b)** $1 - 0.4305 = 0.5695$ **(c)** The calculation in part (a) might not be valid because the 4 consecutive calls being medical are not independent events.

19. (a)

Amount paid

		Full price	Reduced price	Total
Purchase method	Online	0	10	10
	In person	20	20	40
	Total	20	30	50

(b)

Amount paid

		Full price	Reduced price	Total
Purchase method	Online	4	6	10
	In person	16	24	40
	Total	20	30	50

21. (a) $P(\text{rolling doubles}) = 1/6 = 0.167$ (b) $(5/6)(1/6) = 0.139$
(c) $(5/6)(5/6)(1/6) = 0.116$
(d) $P(\text{first doubles occurs on the } k\text{th toss}) = \left(\dfrac{5}{6}\right)^{k-1}\left(\dfrac{1}{6}\right)$

23. (a) $(0.40)(0.75) + (0.25)(0.70) + (0.35)(0.50) = 0.30 + 0.175 + 0.175 = 0.65$ (b) $(0.25)(0.70)/0.65 = 0.269$

Answers to Section 4.7 Exercises

1. multiplication counting
3. True
5. $_nP_r = \dfrac{n!}{(n-r)!}$
7. $23 \times 10 \times 10 \times 20 \times 23 \times 23 = 24,334,000$ possible license plates
9. (a) $1 \times 26 \times 26 = 676$ possible three-letter radio call signs (b) 676 signs that start with W and 676 signs that start with K: $676 + 676 = 1352$
11. $6 \times 5 \times 4 \times 3 \times 2 \times 1 = 6! = 720$ possible orders to complete the pickups
13. $30 \cdot 29 \cdot 28 \cdots 3 \cdot 2 \cdot 1 = 30! = 2.65 \times 10^{32}$ possible seating arrangements
15. $_{100}P_8 = 7.5 \times 10^{15}$ possible lists
17. $_{11}P_5 = 55,440$ possible ways
19. (a) $10 \times 10 \times 10 \times 10 = 10,000$ possible four-digit passwords
(b) $9 \times 9 \times 9 \times 9 = 6561$ possible four-digit passwords (with no 3s)
21. (a) $2 \times 18 \times 12 = 432$ possible sundaes (b) $_{18}P_3 = 18 \times 17 \times 16 = 4896$ possible three-scoop cones
23. $10 \times 9 \times 8 \times 7 = 5040$ possible four-digit passwords with all four digits different. There are $10,000 - 5040 = 4960$ four-digit passwords that contain at least two of the same digit.
25. (a) The linear relationship between calories and carbohydrates for these 19 varieties of bagels is strong and positive.
(b) $r = 0.915$. The correlation doesn't change when we change the units of either variable. (c) With these points removed, the correlation is less strongly positive and will get closer to 0.

Answers to Section 4.8 Exercises

1. True
3. $r!$
5. $_{25}C_2 = \dfrac{_{25}P_2}{2!} = \dfrac{25 \cdot 24}{2 \cdot 1} = 300$ different possible two-topping pizzas
7. $_{52}C_5 = \dfrac{_{52}P_5}{5!} = \dfrac{52 \cdot 51 \cdot 50 \cdot 49 \cdot 48}{5 \cdot 4 \cdot 3 \cdot 2 \cdot 1} = 2,598,960$ different possible five-card hands
9. $\dfrac{_6C_4}{_8C_4} = \dfrac{15}{70} = 0.214$
11. (a) $\dfrac{_{30}C_3}{_{95}C_3} = \dfrac{4060}{138,415} = 0.029$ (b) No. There is less than a 5% probability that 3 randomly selected students would all be from Mr. Wilder's class in a fair lottery. Because this result is unlikely to happen by chance alone, we have convincing evidence that the lottery was not carried out fairly.
13. (a) $_{30}C_{12} = \dfrac{30!}{12!18!} = 86,493,225$
(b) $_{10}C_6 \cdot _{20}C_6 = (210)(38,760) = 8,139,600$ different ways
(c) $\dfrac{_{10}C_6 \cdot _{20}C_6}{_{30}C_{12}} = \dfrac{8,139,600}{86,493,225} = 0.0941$
(d) No. Because it is somewhat likely (probability = 0.118) that a random selection of 12 employees would result in at least 6 employees age 60 and older being selected, there is not convincing evidence that the company did not select the employees who were fired at random.
15. (a) $\dfrac{_8C_6 \cdot _{12}C_4}{_{20}C_{10}} = \dfrac{13,860}{184,756} = 0.0750$ (b) No. Because it is somewhat likely (probability = 0.1698) that the random assignment would put at least 6 volunteers with high cholesterol levels in one treatment group, the subjects should not be surprised by the results of the random assignment.

17. (a) $_{50}C_8 = 536,878,650$ different sets of 8 songs
(b) $_{15}C_2 \cdot _{35}C_6 = 170,431,800$ different sets of 8 songs
(c) $\dfrac{_{15}C_2 \cdot _{35}C_6}{_{50}C_8} = \dfrac{170,431,800}{536,878,650} = 0.317$
19. (a) $\dfrac{_{20}C_5}{_{80}C_5} = \dfrac{15504}{24,040,016} = 0.000645$
(b) $\dfrac{_{20}C_3 \cdot _{60}C_2}{_{80}C_5} + \dfrac{_{20}C_4 \cdot _{60}C_1}{_{80}C_5} + \dfrac{_{20}C_5}{_{80}C_5} = 0.096672$
21. $_{25}C_8 + _{25}C_7 + _{25}C_6 + _{25}C_5 + _{25}C_4 + _{25}C_3 + _{25}C_2 + _{25}C_1 + _{25}C_0 = 1,807,781$ different pizzas
23. $\dfrac{_{20}C_3 \cdot _7C_2 \cdot _5C_1}{_{32}C_6} = 0.132$
25. (a) $161/542 = 0.297$
(b) $P(\text{Response 3} \mid \text{less than high school}) = 47/127 = 0.370$
$P(\text{Response 3} \mid \text{not less than high school}) = 126/722 = 0.175$
Because the probabilities are not equal, the events "Response 3" and "less than high school" are not independent.

Answers to Chapter 4 Review Exercises

1. (a) If you take a very large random sample of solo travelers, about 25% of them will check baggage. (b) Let 1 = check baggage and 2–4 = does not check baggage. Generate a random integer from 1 to 4 to simulate selecting one solo traveler and seeing if they check baggage. Repeat this a total of 12 times. Record the number of travelers who check baggage. (c) $\approx 3/100 = 3\%$ (d) Because you are unlikely to obtain 7 or more solo travelers out of 12 who check baggage by chance alone when 25% of all solo travelers check baggage, this result provides convincing evidence that more than 25% of all solo travelers check baggage.
2. (a) $1 - 0.59 = 0.41$ (b) $0.18/0.72 = 0.25$ (c) $0.18 + 0.08 + 0.05 = 0.31$
3. (a) $539/687 = 0.785$ (b) $234/687 + 539/687 - 169/687 = 0.879$
(c) $169/539 = 0.314$
4. (a)

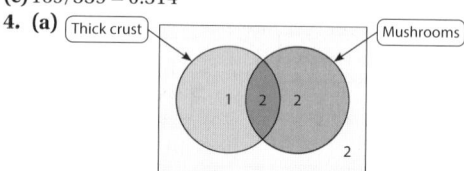

(b) $P(\text{thick crust}) = 3/7 = 0.43$. $P(\text{thick crust} \mid \text{mushrooms}) = 2/4 = 0.50$. Because the probabilities are not equal, the events "thick crust" and "mushrooms" are not independent. (c) $P(\text{thick crust}) = 4/8 = 0.50$. $P(\text{thick crust} \mid \text{mushroom}) = 2/4 = 0.50$. Because the probabilities are equal, the events "thick crust" and "mushrooms" are independent.
5. (a) $(8/18)(7/17) = 0.183$
(b) $P(\text{same color}) = (8/18)(7/17) + (6/18)(5/17) + (4/18)(3/17) = 0.32$
6. (a) The 3% false positive rate means that 3% of people who take this test will get a positive result when they do not have Covid. The 5% false negative rate means that 5% of people who take this test will get a negative result when they do have Covid.
(b)

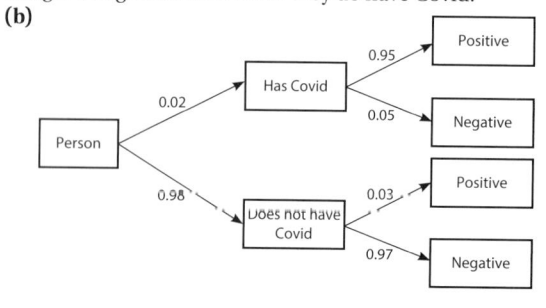

(c) $(0.02)(0.05)+(0.98)(0.97)=0.9516$ (d) P(does not have Covid | test negative) $=\dfrac{(0.98)(0.97)}{(0.02)(0.05)+(0.98)(0.97)}=0.9989$

7. (a) $16\times16\times16\times16\times16=1{,}048{,}576$ possible outcomes for 5 rolls (b) $16\times15\times14\times13\times12=524{,}160$ possible outcomes in which all 5 rolls are different (c) $524{,}160/1{,}048{,}576=0.4999$

8. $P(\text{none})=(1-0.33)^{10}=0.018228$ so $P(\text{at least }1)=1-0.018228=0.981772$

9. (a) $_{14}C_{10}=\dfrac{_{14}P_{10}}{10!}=1001$ different groups of 10 acts
(b) $_6C_5\cdot{}_8C_5=(6)(56)=336$ different groups of 10 acts that have 5 electric and 5 acoustic

$P(\text{5 electric, 5 acoustic})=\dfrac{_6C_5\cdot{}_8C_5}{_{14}C_{10}}=\dfrac{336}{1001}=0.336$

10. (a) $P(\text{cancer}\mid\text{smoker})=0.08/0.25=0.32$
(b) $P(\text{smokes or cancer})=0.25+0.12-0.08=0.29$
(c) $P(\text{at least 1 smokes})=1-(0.75)(0.75)=0.4375$

CHAPTER 5

Answers to Section 5.1 Exercises

1. numerical
3. False. The set of possible values for a discrete random variable can be finite or infinite.
5. discrete; continuous
7. True
9. (a) Continuous (b) Discrete (c) Continuous
11. (a) Discrete (b) Continuous
13. Each of the probabilities is between 0 and 1 (inclusive); the sum of the probabilities $=1$.
15. $T=850$ is the event that a randomly chosen student has a tuition charge of $850. $P(T=850)=0.05$.
17. $P(X\geq 6)=0.067+0.058+0.051+0.046=0.222$
19. (a) $P(T\leq 800)=0.25+0.10+0.05+0.30+0.10=0.80$
(b) $P(T<800)=0.25+0.10+0.05+0.30=0.70$. This answer does not include $T=800$.
21. (a) X is a discrete random variable because it takes a fixed set of numerical values with gaps between them. (b) $P(X\geq 3)$ is the probability that a randomly selected U.S. college student speaks at least three languages. $P(X\geq 3)=0.065+0.008+0.002=0.075$.
(c) $P(X<3)=1-0.075=0.925$
23. (a)

Number of heads	0	1	2	3	4
Probability	1/16	4/16	6/16	4/16	1/16

(b) $P(X\leq 3)=1-1/16=15/16=0.9375$. There is about a 0.9375 probability of getting less than or equal to 3 heads when tossing a fair coin 4 times.
25. (a) $P(\text{all 5 check phones})=(0.67)^5=0.135$ (b) $P(\text{at least 1 doesn't check phone})=1-0.135=0.865$

Answers to Section 5.2 Exercises

1. shape; center; variability
3. Because a probability distribution can model the population distribution of a quantitative variable.
5. $\mu_X=E(X)=x_1p_1+x_2p_2+x_3p_3+\cdots=\sum x_ip_i$
7. σ_X; mean

9.

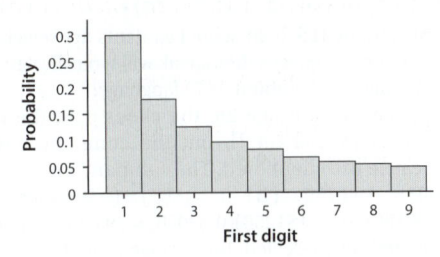

The graph is skewed to the right with a single peak at 1.

11.

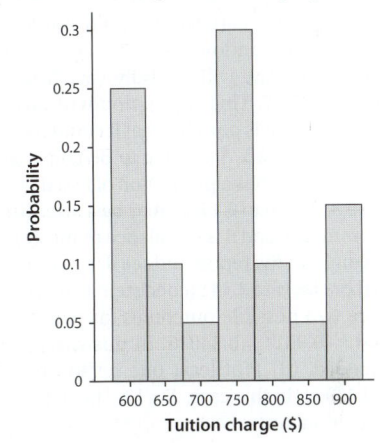

The graph is fairly symmetric with peaks at $600, $750, and $900.
13. $E(X)=\mu_X=1(0.301)+2(0.176)+3(0.125)+\cdots+9(0.046)=3.441$.
If many, many legitimate reports are randomly selected, the average of the first digits in the reports will be about 3.441.
15. $\mu_T=E(T)=600(0.25)+650(0.10)+\cdots+900(0.15)=732.50$.
If many, many students from El Dorado Community College are randomly selected, their average tuition charge will be about $732.50.
17. $\sigma_X=\sqrt{(1-3.441)^2(0.301)+(2-3.441)^2(0.176)+\cdots+(9-3.441)^2(0.046)}$
$=2.462$. If many, many legitimate records are randomly selected, the first digit will typically vary from the mean of 3.441 by about 2.462.
19.
$\sigma_T=\sqrt{(600-732.50)^2(0.25)+(650-732.50)^2(0.10)+\cdots+(900-732.50)^2(0.15)}$
$=102.80$. If many, many students from El Dorado Community College are randomly selected, their tuition typically varies from the mean of $732.50 by about $102.80.
21. (a)

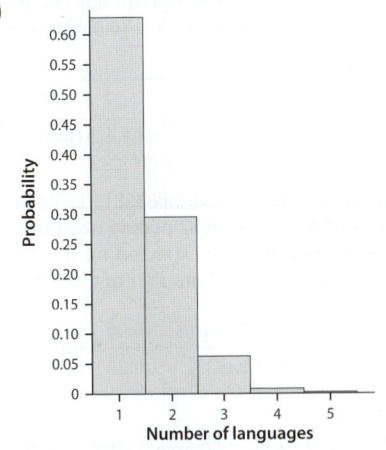

The graph is skewed to the right with a single peak at 1.
(b) $E(X)=\mu_X=1(0.630)+2(0.295)+3(0.065)+4(0.008)+5(0.002)=1.457$. If many, many U.S. high school students are selected at random, the average number of languages they speak will be about 1.457.

(c) $\sigma_X = \sqrt{(1-1.457)^2(0.630)+(2-1.457)^2(0.295)+\cdots+(5-1.457)^2(0.002)}$ = 0.671. If many, many U.S. high school students are selected at random, the number of languages they speak will typically vary from the mean of 1.457 languages by about 0.671 language.

23. (a) If a client dies at age 25, the client paid the company $250 for 5 years ($1250 total), but the company had to pay $100,000 because the client died. The company's overall profit is $1250-100,000=-\$98,750$. **(b)** $\mu_Y = E(Y) = (-\$99,750)(0.00183) + -\$99,500(0.00186)+\cdots+(\$1250)(0.99058) = 303.35$. If many, many policies were randomly selected, the average profit for the company will be about $303.35 per policy. **(c)** $\sigma_Y = \sqrt{(-99750-303.35)^2(0.00183)+(-99500-303.35)^2(0.00186)+\cdots+(1250-303.35)^2(0.99058)} = 9707.57$. If many, many policies are randomly selected, the amount of profit per policy for the company will typically vary from the mean of $303.35 by about $9707.57. This large amount of variability is dangerous for a company, as it is possible that it could have huge losses.

25. (a) $P(Y > 6) = 3/9 = 0.333$. According to Benford's law: $P(Y > 6) = 0.155$. If an expense report has a proportion of first digits greater than 6 that is closer to 0.333 than to 0.155, it may be a fake. **(b)** Because this distribution is symmetric and 5 is its balance point. **(c)** Compute the mean of the first digits in the report and see if it is closer to 5 (suggesting a fake report) or closer to 3.441 (consistent with a truthful report).

27. (a) There are two possible outcomes for plan B: (1) P(no one has the disease) $= (0.95)^{12} = 0.54036$. In this case, $X=1$. Only the batch test is needed. (2) P(at least one person has the disease) $= 1-(0.95)^{12} = 0.45964$. In this case, $X=13$. The batch test is needed as well as 12 individual tests.

X = Total number of tests with plan B	1	13
Probability	0.54036	0.45964

(b) Probability = 0.54036 because plan A will always need 12 tests. **(c)** $E(X) = 1(0.54036) + 13(0.45964) = 6.51568$. If many, many batches are tested, the average number of tests needed will be about 6.516 tests. **(d)** Plan B would be better because the expected number of tests is about 6.5 tests, while plan A will require 12 tests.

29. (a) There are four possible outcomes for Plan D: (1) P(no one has the disease in all three groups) $= (0.95)^4(0.95)^4(0.95)^4 = 0.54036$. In this case, $W = 3$. (2) P(no one has the disease in two groups and at least one person has the disease in one group) $= 3(0.95)^4(0.95)^4[1-(0.95)^4] = 0.36918$. In this case, $W = 7$. (3) P(no one has the disease in one group and at least one person has the disease in two groups) $= 3(0.95)^4[1-(0.95)^4][1-(0.95)^4] = 0.08408$. In this case, $W = 11$. (4) P(at least one person has the disease in all three groups) $= [1-(0.95)^4][1-(0.95)^4][1-(0.95)^4] = 0.00638$. In this case, $W = 15$.

W = total # of tests with Plan D	3	7	11	15
Probability	0.54036	0.36918	0.08408	0.00638

(b) $E(W) = 3(0.54036)+7(0.36918)+11(0.08408)+15(0.00638) = 5.22592$ **(c)** If thousands of groups of 12 people need to be tested, Plan C would be better because the expected number of tests is about 5.179 tests, while the expected number of tests for Plan D is 5.226 tests.

31. (a) $P(\text{late} | \text{drives}) = \dfrac{P(\text{late and drives})}{P(\text{drives})} = \dfrac{0.05}{0.20} = 0.25$

(b)

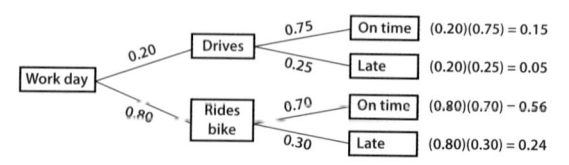

Work day → 0.20 Drives → 0.75 On time (0.20)(0.75) = 0.15; 0.25 Late (0.20)(0.25) = 0.05
Work day → 0.80 Rides bike → 0.70 On time (0.80)(0.70) = 0.56; 0.30 Late (0.80)(0.30) = 0.24

(c) $P(\text{biked} | \text{late}) = \dfrac{(0.80)(0.30)}{(0.20)(0.25)+(0.80)(0.30)} = 0.828$

Answers to Section 5.3 Exercises

1. independent; success
3. False. The possible values of a binomial random variable are $0, 1, 2, \ldots, n$.
5. n; p
7. This is a binomial setting. *Binary?* "Success" = baby elk survives to adulthood. "Failure" = baby elk does not survive. *Independent?* Knowing whether or not one randomly selected elk survives to adulthood tells you nothing about whether or not another randomly selected elk survives to adulthood. *Number?* $n = 7$. *Same probability?* $p = 0.44$
9. No, because the *Independent* condition is not met. If the first person selected has a last name with more than six letters, the next person chosen is less likely to have a last name with more than six letters (and vice versa) because you're not replacing the first person's name back into the hat.
11. $P(X=4) = {}_7C_4(0.44)^4(1-0.44)^3 = 0.2304$
13. (a) This is a binomial setting and Y = the number of successes. *Binary?* "Success" = train arrives on time. "Failure" = train does not arrive on time. *Independent?* Knowing whether or not the train arrives on time one day tells you nothing about whether or not the train will arrive on time another day. *Number?* $n=6$. *Same probability?* $p = 0.90$. **(b)** $P(Y=4) = {}_6C_4(0.90)^4(1-0.90)^2 = 0.0984$. If we randomly select 6 weekdays, there is about a 0.0984 probability that the train will be on time exactly 4 times.
15.

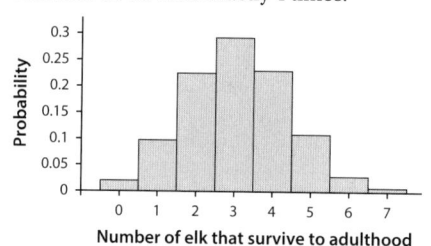

The graph is fairly symmetric with a single peak at $X = 3$.

17.

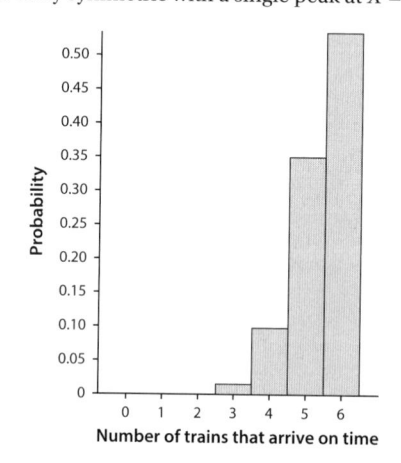

The graph is left-skewed with a single peak at $Y = 6$.
19. (a) R is a binomial random variable because the conditions are met and R = the number of successes. *Binary?* "Success" = the light is red. "Failure" = the light is not red. *Independent?* Knowing whether or not one passenger has a red light tells you nothing about whether or not another passenger gets a red light. *Number?* $n = 20$. *Same probability?* $p = 0.30$. **(b)** $P(R=6) = {}_{20}C_6(0.30)^6(1-0.30)^{14} = 0.1916$

(c)

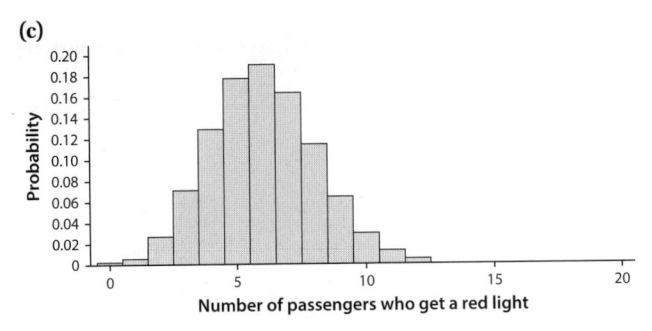

The graph is fairly symmetric with a single peak at $R = 6$.

21. (a) Here are the probability distribution histograms for $p = 0.2$.

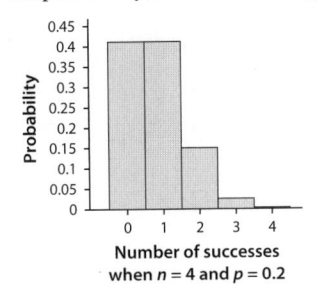

The graph is strongly right-skewed with a single peak that includes $X = 0$ and $X = 1$.

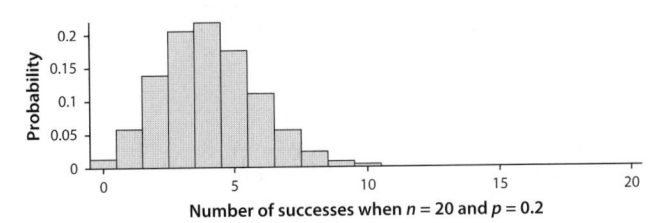

The graph is slightly right-skewed with a single peak at $X = 4$.

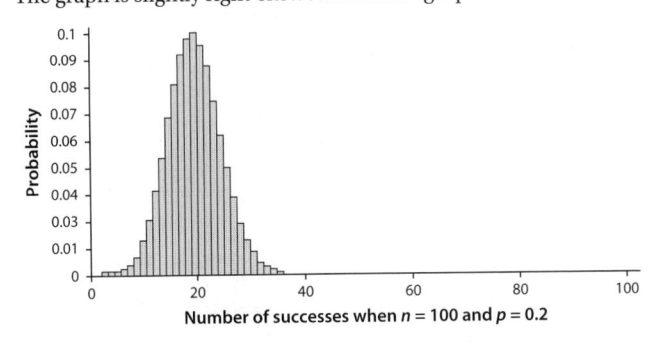

The graph is fairly symmetric and bell-shaped with a single peak at $X = 20$.

(b) Here are the probability distribution histograms for $p = 0.8$.

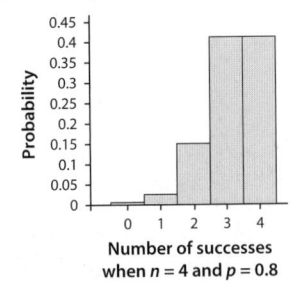

The graph is strongly left-skewed with a single peak that includes $X = 3$ and $X = 4$.

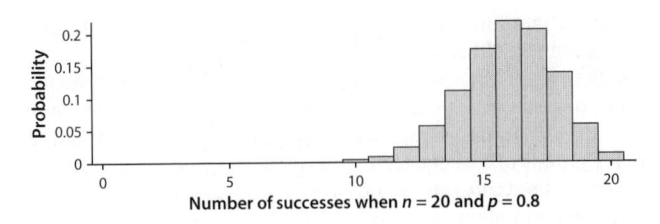

The graph is slightly left-skewed with a single peak at $X = 16$.

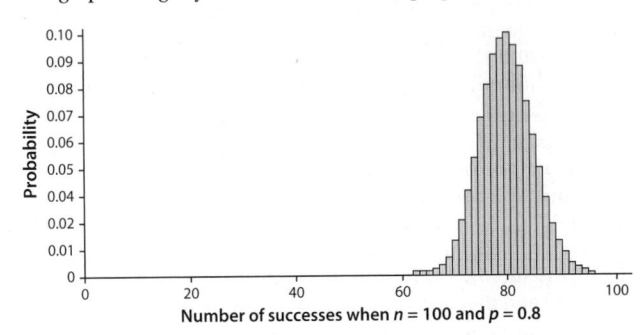

The graph is fairly symmetric and bell-shaped with a single peak at $X = 80$. In both cases, as n increases, the graph becomes more symmetric and bell-shaped.

23. (a) It (theoretically) could take Jordyn an infinite number of tries before the mower starts. **(b)** Skewed to the right. The probability of $X = 1$ will always be the tallest in a geometric distribution because $P(X = 1) = p$ and every outcome thereafter is repeatedly multiplied by a number less than 1—the probability of failure.

25. Answers will vary.

Shape: Slightly right-skewed and double-peaked. *Outliers:* There are two possible upper outliers. (The $1.5 \times IQR$ rule shows that there are no outliers.) *Center:* The median Nielsen share score is 68. *Variability:* The IQR of Nielsen share scores is $71 - 63 = 8$.

Answers to Section 5.4 Exercises

1. Write out the possible values of the variable, circle the ones for which you want to find the probability, and cross out the rest.

3. False. The formula $\mu_X = np$ can only be used to find the mean of a binomial random variable.

5. $P(X < 3) = {}_7C_0(0.44)^0(1-0.44)^7 + {}_7C_1(0.44)^1(1-0.44)^6 + {}_7C_2(0.44)^2(1-0.44)^5 = 0.0173 + 0.0950 + 0.2239 = 0.3362$

7. Let X = the number of shirts that Joy identifies correctly.
(a) $P(X \geq 11) = {}_{12}C_{11}(0.5)^{11}(1-0.5)^1 + {}_{12}C_{12}(0.5)^{12}(1-0.5)^0 = 0.002930 + 0.000244 = 0.003174$ **(b)** Yes. Because it is very unlikely for Joy to correctly identify at least 11 shirts correctly by chance alone, there is convincing evidence that Joy really can smell Parkinson's disease.

9. (a) Tech: $P(X \geq 14) = 0.00001$ **(b)** Yes. It is very unlikely that at least 14 people will choose the last Kiss by chance alone; therefore we have convincing evidence that participants have a preference for the last thing they taste.

11. Tech: $P(Y \leq 4) = 0.1143$. There is a 0.1143 probability that the train will be on time at most 4 of the 6 randomly selected weekdays.

13. (a) $\mu_X = 15(0.09) = 1.35$. If many, many sets of 15 phone calls are made by the random dialing machine, we expect about 1.35 surveys will be completed, on average. **(b)** $\sigma_X = \sqrt{15(0.09)(0.91)} = 1.11$. If many, many sets of 15 phone calls are made by the random dialing

machine, the number of completed surveys would typically vary from the mean of 1.35 by about 1.11 completed surveys.
15. (a) $\mu_Y = 6(0.9) = 5.4$. If many, many groups of 6 weekdays are randomly selected, we expect that the train will be on time about 5.4 of the days, on average. **(b)** $\sigma_Y = \sqrt{6(0.9)(0.1)} = 0.73$. If many, many groups of 6 weekdays are randomly selected, the number of days that the train will be on time would typically vary from the mean of 5.4 by about 0.73 day.
17. (a) *Tech*: $P(R \le 3) = 0.1071$ **(b)** No. It is somewhat likely that 3 or fewer people will get a red light by chance alone, so we do not have convincing evidence against the custom agent's claim. **(c)** $E(R) = 20(0.3) = 6$. If many, many groups of 20 passengers press the button, we expect about 6 of them will get a red light, on average. **(d)** $\sigma_R = \sqrt{20(0.3)(0.7)} = 2.05$. If many, many groups of 20 passengers press the button, the number of passengers who will get the red light would typically vary from the mean of 6 by about 2.05 passengers.
19. (a) $P(Y \ge 13) = 0.8531$, $P(X \le 2) = 0.8531$. In a set of 15 calls, 13 or more not completing the survey is the same as 2 or fewer completing the survey. **(b)** $\mu_Y = 15(0.91) = 13.65$. The sum of the mean of X (1.35) and the mean of Y (13.65) equals 15. In sets of 15 phone calls, we expect about 1.35 surveys to be completed, on average, and an average of 13.65 calls to end with incomplete surveys. **(c)** $\sigma_Y = \sqrt{15(0.91)(0.09)} = 1.11$. In many, many sets of 15 phone calls, the number of incomplete surveys would typically vary from the mean of 13.65 by about 1.11 surveys. This is the same as the standard deviation of X. This makes sense because the success rate for completed surveys is the same as the failure rate for incomplete surveys.
21. (a) $P(X > 6) = 1 - P(X \le 6) = 1 - [0.2 + (0.8)(0.2) + (0.8)^2(0.2) + (0.8)^3(0.2) + (0.8)^4(0.2) + (0.8)^5(0.2)] = 0.262$
(b) $\mu_X = \dfrac{1}{0.2} = 5$; it will take Jordyn, on average, 5 pulls to start the mower.
(c) $\sigma_X = \dfrac{\sqrt{1-0.2}}{0.2} = 4.47$; the number of pulls that it will take to start the mower will typically vary from the mean of 5 by about 4.47 pulls.
23. (a) $5 \times 6 \times 4 \times 2 \times 2 \times 2 \times 2 \times 2 = 3840$ different standard sandwiches **(b)** $5 \times {}_6C_2 \times {}_4C_2 \times 2 \times 2 \times 2 \times 2 \times 2 = 14{,}400$ different double sandwiches

Answers to Section 5.5 Exercises

1. False. The Poisson random variable counts the number of successes that occur over a given continuous interval of space or time.
3. True
5. the mean number of successes μ over the given interval of space or time
7. $P(X=1) = \dfrac{0.746^1 e^{-0.746}}{1!} = 0.354$
9. $P(X=0) = \dfrac{4.9^0 e^{-4.9}}{0!} = 0.007$; there is a 0.007 probability that 0 collision claims are filed with this company in the next 30 days.
11. $P(X \le 2) = \dfrac{0.746^0 e^{-0.746}}{0!} + \dfrac{0.746^1 e^{-0.746}}{1!} + \dfrac{0.746^2 e^{-0.746}}{2!} = 0.960$
13. $P(X \ge 12) = $ Applet / $1 - \text{Poissoncdf}(\mu: 4.9, x \text{ value}: 11) = 0.005$. No, because it is unlikely that the insurance company will lose money in a given 30-day time period.
15. $\mu_2 = \left(\dfrac{6}{12}\right)(0.746) = 0.373$ earthquake in 6 months;
$P(X=0) = \dfrac{0.373^0 e^{-0.373}}{0!} = 0.689$

17. $\mu_2 = \left(\dfrac{365}{30}\right)(4.9) = 59.6167$ collision claims filed in 365 days;
Tech: $P(X \le 50) = $ Applet / Poissoncdf(μ: 59.6167, x value: 50) = 0.117
19. (a) $P(X=0) = \dfrac{1^0 e^{-1}}{0!} = 0.368$
(b) $P(X \ge 2) = 1 - \left[\dfrac{1^0 e^{-1}}{0!} + \dfrac{1^1 e^{-1}}{1!}\right] = 0.264$. This is a fairly likely event with over a 25% chance of occurring.
(c) $\mu_2 = \left(\dfrac{5}{1}\right)(1) = 5$ meteors in 5 years;
$P(X \ge 2) = 1 - \left[\dfrac{5^0 e^{-5}}{0!} + \dfrac{5^1 e^{-5}}{1!}\right] = 0.960$
21. (a) Binomial. The manager is counting the number of successes among a fixed number of trials ($n=10$). **(b)** Poisson. The manager is counting the number of successes that occur over a given continuous interval of time.
23. (a) Here are the probability distributions for $\mu = 0.5$, 1, 4, and 20.

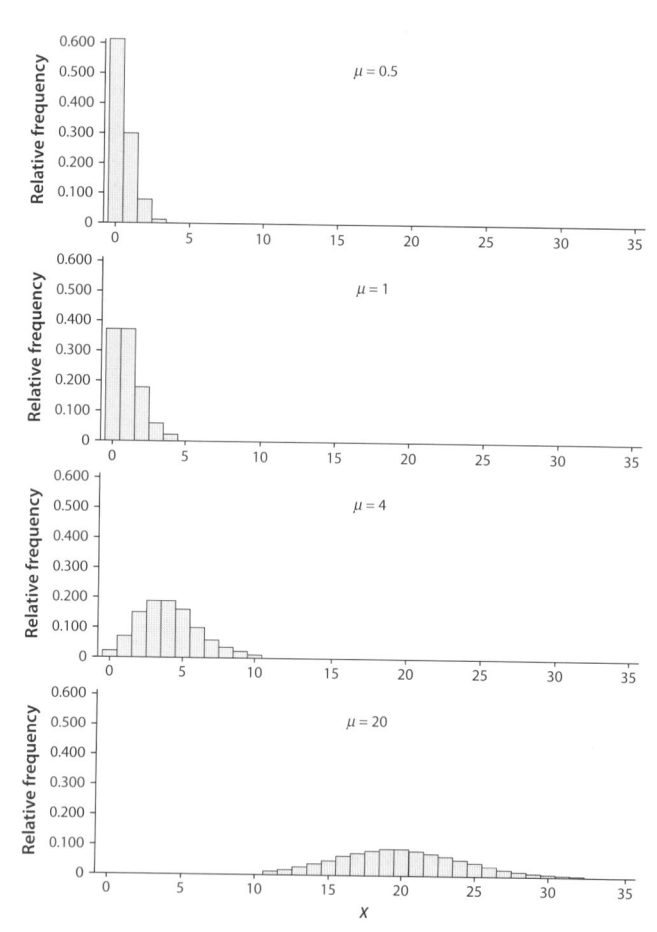

As the mean increases, the shape changes from right-skewed to more and more like that of a normal distribution. **(b)** The mean of a Poisson distribution is the average number of successes that occur over a given interval of space or time, which is μ. **(c)** As μ increases, the standard deviation increases. As μ increases, the number of successes can take on a greater number of possible values.

25. (a) X is not a binomial random variable because the outcomes are not independent. If the first earthquake selected has a magnitude over 7.0, the next earthquake is less likely to have a magnitude over 7.0 (and vice versa) because we are not replacing the first earthquake that was selected.

(b) $P(X=0) = \left(\dfrac{97}{147}\right)\left(\dfrac{96}{146}\right)\left(\dfrac{95}{145}\right)\left(\dfrac{94}{144}\right)\left(\dfrac{93}{143}\right) = 0.121$

(c) $P(X=0) = {}_5C_0\left(\dfrac{50}{147}\right)^0\left(\dfrac{97}{147}\right)^5 = 0.125$

Even though we are sampling without replacement from a finite population, we are only sampling a small fraction of the population, so the binomial probability is close to that which was obtained by using the general multiplication rule.

Answers to Chapter 5 Review Exercises

1. (a) X is a discrete random variable because it takes a fixed set of numerical values with gaps between them. **(b)** The probability distribution is legitimate because each probability is a value between 0 and 1 and the sum of the probabilities equals 1. **(c)** $P(X<2) = 0.026 + 0.254 = 0.28$. There is a 0.28 probability that a randomly selected U.S. household will have fewer than 2 televisions. **(d)** $P(X \geq 2) = 1 - 0.28 = 0.72$
2. (a) The shape is roughly symmetric with a single peak at 2.

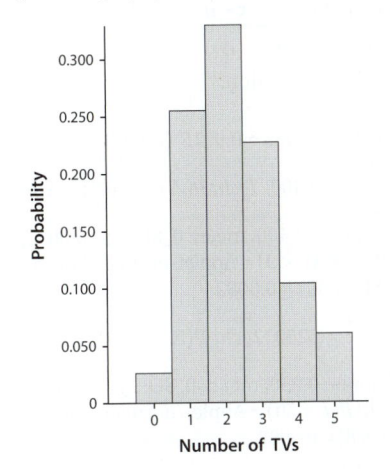

(b) $\mu_X = E(X) = 0(0.026) + 1(0.254) + 2(0.331) + 3(0.226) + 4(0.104) + 5(0.059) = 2.305$. If many, many U.S. households are randomly selected, their average number of televisions will be about 2.305.
(c) $\sigma_X = \sqrt{(0-2.305)^2(0.026) + (1-2.305)^2(0.254) + \cdots + (5-2.305)^2(0.059)} = 1.199$. If many, many U.S. households are randomly selected, the number of televisions in a household typically varies from the mean of 2.305 by about 1.199 televisions.
3. (a) X is a binomial random variable because the following conditions are met and X counts the number of successes. *Binary?* "Success" = correctly identifies the card. "Failure" = incorrectly identifies the card. *Independent?* Knowing whether or not one card is correctly identified tells you nothing about whether or not another card is correctly identified. *Number?* $n=20$. *Same probability?* $p=0.25$ **(b)** $P(X=5) = {}_{20}C_5(0.25)^5(1-0.25)^{15} = 0.202$
(c) *Tech:* $P(X \geq 8)$ = Applet / 1− binomcdf(trials: 20, p: 0.25, x value: 7) = 0.1018 **(d)** Because it is somewhat likely to correctly identify 8 or more out of 20 cards by chance alone, there is not convincing evidence that Alec has ESP.

4. (a)

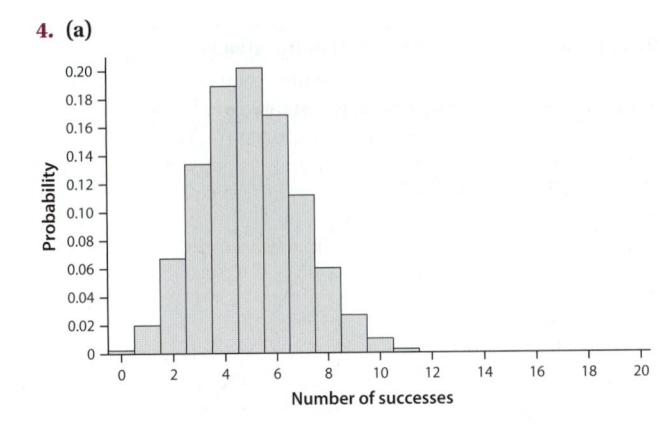

Shape: Single-peaked, fairly symmetric **(b)** $\mu_X = 20(0.25) = 5$
(c) $\sigma_X = \sqrt{20(0.25)(1-0.25)} = 1.94$. If many, many different sets of 20 cards are selected, the number of correctly identified cards would typically vary from the mean of 5 by about 1.94 cards.

5. (a) $P(X=30) = \dfrac{23.3^{30} e^{-23.3}}{30!} = 0.03$

(b) *Tech:* $P(X \geq 40)$ = Applet / 1− Poissoncdf(μ: 23.3, x value: 39) = 0.001 **(c)** Y has a Poisson distribution with a mean of $(5)(23.3) = 116.5$. **(d)** *Tech:* $P(Y<110)$ = Applet / Poissoncdf(μ: 116.5, x value: 109) = 0.261. There is a 0.261 probability that the server will receive fewer than 110 requests in a given 5-second period.
6. (a) $P(\text{wins \$1 or more}) = 0.213 + 0.043 + 0.003 = 0.259$
(b) $\mu_X = (0)(0.308) + (0)(0.433) + (1)(0.213) + (3)(0.043) + (120)(0.003) = \$0.70$. If you were to play many, many games of 4-Spot Keno, you would get, on average, a payout of about \$0.70 per game. **(c)** Let Y = number of \$1 payouts that occur. *Binary?* "Success" = win a payout of \$1. "Failure" = does not win a payout of \$1. *Independent?* Knowing whether or not a \$1 payout is won on one day tells you nothing about whether or not a \$1 payout is won on any other day. *Number?* $n=10$. *Same probability?* $p=0.259$. *Tech:* $P(Y \geq 5)$ = Applet / 1− binomcdf(n: 10, p: 0.259, x value: 4) = 0.0891

7. (a) $P(X \geq 1) = 1 - \dfrac{1.2^0 e^{-1.2}}{0!} = 0.699$ **(b)** Let Y = the number of chat agents who do not receive any customer contacts in the next minute. *Binary?* "Success" = no customer contacts. "Failure" = at least 1 customer contact. *Independent?* Knowing whether one agent receives a customer contact tells you nothing about whether or not another agent receives a customer contact. *Number?* $n=5$. *Same probability?* $p=0.301$. $P(Y \geq 4) = {}_5C_4(0.301)^4(0.699)^1 + {}_5C_5(0.301)^5(0.699)^0 = 0.0312$

CHAPTER 6

Answers to Section 6.1 Exercises

1. True
3. mean (μ); standard deviation (σ)
5. 68-95-99.7
7. (a)

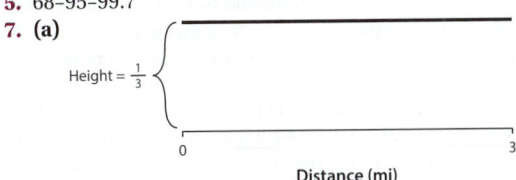

(b) $P(0.8 < X < 1.1) = (1.1 - 0.8)\left(\dfrac{1}{3}\right) = 0.1$ **(c)** $(0.7)(3) = 2.1$ mi

9. (a) The height must be $\frac{1}{3}$ so that the area is 1. Area $= (5-2)\left(\frac{1}{3}\right) = 1$ **(b)** $P(2.5 < Y < 4) = (4-2.5)\left(\frac{1}{3}\right) = 0.5$ **(c)** $0.25 = (Q_1 - 2)\left(\frac{1}{3}\right); Q_1 = 2.75$

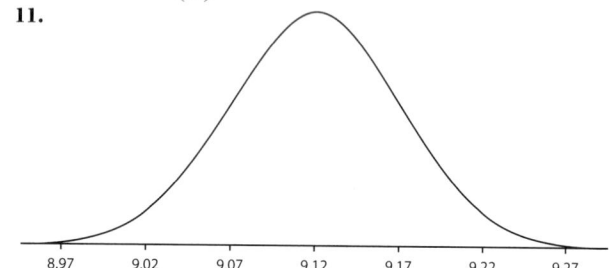

11.

[Normal distribution curve centered at 9.12, with x-axis values 8.97, 9.02, 9.07, 9.12, 9.17, 9.22, 9.27, labeled "Bag of chips weight (oz)"]

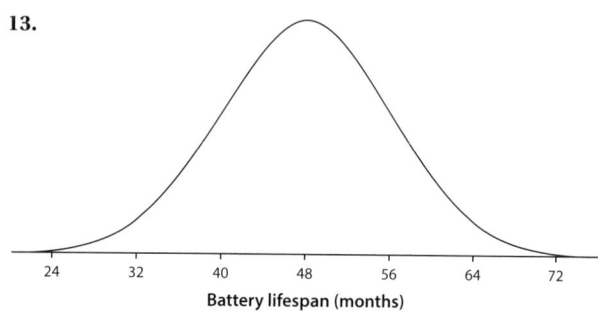

13.

[Normal distribution curve centered at 48, with x-axis values 24, 32, 40, 48, 56, 64, 72, labeled "Battery lifespan (months)"]

15. (a) $P(W < 9.07) = (1 - 0.68)/2 = 0.16$ **(b)** $P(8.97 \le W \le 9.17) = (0.997 - 0.68)/2 + 0.68 = 0.1585 + 0.68 = 0.8385$

17. (a) $P(40 < L < 64) = 0.68 + (0.95 - 0.68)/2 = 0.68 + 0.135 = 0.815$ **(b)** $P(L \le 32) = (1 - 0.95)/2 = 0.025$. Yes, only 2.5% of batteries die this quickly.

19. (a)

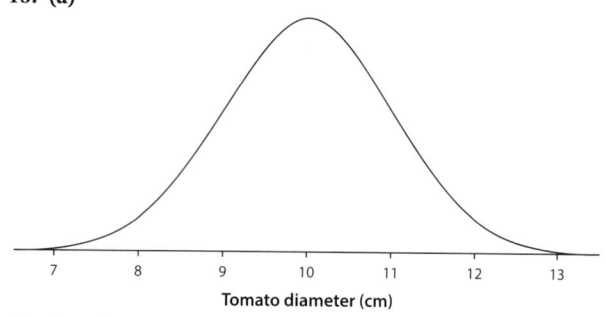

[Normal distribution curve centered at 10, with x-axis values 7, 8, 9, 10, 11, 12, 13, labeled "Tomato diameter (cm)"]

(b) $P(X > 9) = 0.68 + (1 - 0.68)/2 = 0.68 + 0.16 = 0.84$ **(c)** 8 cm and 12 cm

21. The center appears to be at 10, so 10 is the estimate for the mean. The curvature changes at 8 and 12, so 2 is the estimate for the standard deviation.

23. (a) The density curve is entirely above the horizontal axis and the area under the density curve = 1. Split the area under the density curve into a trapezoid (from $X = 0$ to $X = 0.4$) and a rectangle (from $X = 0.4$ to $X = 0.8$). The area of the trapezoid is $(1/2)(2+1)(0.4) = 0.6$. The area of the rectangle is $(0.4)(1) = 0.4$. The total area is $0.6 + 0.4 = 1$. **(b)** Area $= (1/2)(2 + 1.5)(0.2) = 0.35$ **(c)** Because only 35% of the area is to the left of 0.2 [from part (b)], we know the median is > 0.2. And because 60% of the area is to the left of 0.4 [from part (a)], the point with 50% of the area to the left must be between 0.2 and 0.4. **(d)** The mean will be greater than the median due to the right-skewed shape of this distribution.

25. (a) The mean of X is $\frac{a+b}{2} = \frac{0+10}{2} = 5$.

(b) $\sigma = \sqrt{\frac{(10-0)^2}{12}} = 2.887$ **(c)** mean ± 1SD $= 5 \pm 2.887 = 2.113$ to 7.887. Percentage of values within 1 SD of the mean $= \frac{7.887 - 2.113}{10} = 0.5774$ or 57.74%. This is not the same as the empirical rule, which states that about 68% of observations are within 1 SD of the mean in a normal distribution. **(d)** mean ± 2SD $= 5 \pm 2(2.887) = -0.774$ to 10.774, which is beyond the boundaries of 0 to 10. 100% of the values are within 2 SD of the mean. This is not the same as the empirical rule, which states that about 95% of observations are within 2 SD of the mean in a normal distribution.

27. (a) $z = \frac{3724 - 2789}{752} = 1.24$. The number of yards that Terry Bradshaw threw for is 1.24 standard deviations above the mean in 1979. **(b)** $z = \frac{5129 - 3552}{939} = 1.68$. The number of yards that Ben Roethlisberger threw for is 1.68 standard deviations above the mean in 2018. **(c)** Relatively speaking, Ben Roethlisberger had a better performance because his z-score ($z = 1.68$) is greater than Terry Bradshaw's z-score ($z = 1.24$).

Answers to Section 6.2 Exercises

1. $z = \frac{\text{value} - \text{mean}}{\text{standard deviation}} = \frac{x - \mu}{\sigma}$

3. Draw a normal distribution.

5. True

7. (a) 0.9931 **(b)** $1 - 0.0485 = 0.9515$ **(c)** $0.9633 - 0.6915 = 0.2718$

9. (i) $z = \frac{9 - 9.12}{0.05} = -2.400$; Tech: $P(Z < -2.400)$ = Applet/normalcdf(lower: −1000, upper: −2.400, mean: 0, SD: 1) = 0.0082; Table A: $P(Z < -2.40) = 0.0082$ **(ii)** $P(W < 9)$ = Applet/normalcdf(lower: −1000, upper: 9, mean: 9.12, SD: 0.05) = 0.0082

11. (i) $z = \frac{50 - 48}{8} = 0.250$; Tech: $P(Z < 0.250)$ = Applet/normalcdf(lower: −1000, upper: 0.250, mean: 0, SD: 1) = 0.5987; Table A: $P(Z < 0.25) = 0.5987$ **(ii)** $P(L < 50)$ = Applet/normalcdf(lower: −1000, upper: 50, mean: 48, SD: 8) = 0.5987

13. (i) $z = \frac{9.25 - 9.12}{0.05} = 2.600$; Tech: $P(Z > 2.600)$ = Applet/normalcdf(lower: 2.600, upper: 1000, mean: 0, SD: 1) = 0.0047; Table A: $P(Z > 2.60) = 1 - 0.9953 = 0.0047$ **(ii)** $P(W > 9.25)$ = Applet/normalcdf(lower: 9.25, upper: 1000, mean: 9.12, SD: 0.05) = 0.0047. This is surprising because it is very unlikely to randomly select a bag that weighs more than 9.25 ounces.

15. (i) $z = \frac{36 - 48}{8} = -1.500$; Tech: $P(Z \ge -1.500)$ = Applet/normalcdf(lower: −1.500, upper: 1000, mean: 0, SD: 1) = 0.9332; Table A: $P(Z \ge -1.5) = 1 - 0.0668 = 0.9332$ **(ii)** $P(L \ge 36)$ = Applet/normalcdf(lower: 36, upper: 1000, mean: 48, SD: 8) = 0.9332

17. (i) $z = \frac{9.1 - 9.12}{0.05} = -0.400$ and $z = \frac{9.15 - 9.12}{0.05} = 0.600$ Tech: $P(-0.400 < Z < 0.600)$ = Applet/normalcdf(lower: −0.400, upper: 0.600, mean: 0, SD: 1) = 0.3812; Table A: $P(-0.40 < Z < 0.60) = 0.7257 - 0.3446 = 0.3811$ **(ii)** $P(9.1 < W < 9.15)$ = Applet/normalcdf(lower: 9.1, upper: 9.15, mean: 9.12, SD: 0.05) = 0.3812

19. (i) $z = \frac{55 - 48}{8} = 0.875$ and $z = \frac{60 - 48}{8} = 1.500$ Tech: $P(0.875 < Z < 1.500)$ = Applet/normalcdf(lower: 0.875, upper: 1.500, mean: 0, SD: 1) = 0.1240; Table A: $P(0.88 < Z < 1.50) = 0.9332 - 0.8106 = 0.1226$ **(ii)** $P(55 < L < 60)$ = Applet/normalcdf(lower: 55, upper: 60, mean: 48, SD: 8) = 0.1240

21. (a) (i) $z = \dfrac{160 - 153.4}{9.4} = 0.702$; *Tech:* $P(Z \geq 0.702)$ = Applet/normalcdf(lower: 0.702, upper: 1000, mean: 0, SD: 1) = 0.2413; *Table A:* $P(Z \geq 0.70) = 1 - 0.7580 = 0.2420$ **(ii)** $P(G \geq 160)$ = Applet/normalcdf(lower: 160, upper: 1000, mean: 153.4, SD: 9.4) = 0.2413 **(b) (i)** $z = \dfrac{150 - 153.4}{9.4} = -0.362$ and $z = \dfrac{155 - 153.4}{9.4} = 0.170$; *Tech:* $P(-0.362 < Z < 0.170)$ = Applet/normalcdf(lower: −0.362, upper: 0.170, mean: 0, SD: 1) = 0.2088; *Table A:* $P(-0.36 < Z < 0.17) = 0.5675 - 0.3594 = 0.2081$ **(ii)** $P(150 < G < 155)$ = Applet/normalcdf(lower: 150, upper: 155, mean: 153.4, SD: 9.4) = 0.2088

23. (a) (i) $z = \dfrac{3.95 - 3.98}{0.02} = -1.500$; *Tech:* $P(Z < -1.500)$ = Applet/normalcdf(lower: −1000, upper: −1.500, mean: 0, SD: 1) = 6.68%; *Table A:* $P(Z < -1.50) = 6.68\%$ **(ii)** $P(X < 3.95)$ = Applet/normalcdf(lower: −1000, upper: 3.95, mean: 3.98, SD: 0.02) = 6.68% **(b) (i)** $z = \dfrac{4.05 - 3.98}{0.02} = 3.500$; *Tech:* $P(Z > 3.500)$ = Applet/normalcdf(lower: 3.500, upper: 1000, mean: 0, SD: 1) = 0.02%; *Table A:* $P(Z > 3.50) = 100\% - 99.98\% = 0.02\%$ **(ii)** $P(X > 4.05)$ = Applet/normalcdf(lower: 4.05, upper: 1000, mean: 3.98, SD: 0.02) = 0.02% **(c)** 100% − 6.68% − 0.02% = 93.3% **(d)** It makes sense for the lid manufacturer to make the lids so that it is more likely that a lid will be too small than too large. If a lid is too small, the consumer will realize it and select a new lid. If the lid is too large, the consumer may not realize it and may spill their drink.

25. (a) (i) $z = \dfrac{0 - 23.6}{24.8} = -0.952$; *Tech:* $P(Z < -0.952)$ = Applet/normalcdf(lower: −1000, upper: −0.952, mean: 0, SD: 1) = 0.1705; *Table A:* $P(Z < -0.95) = 0.1711$ **(ii)** $P(X < 0)$ = Applet/normalcdf(lower: −1000, upper: 0, mean: 23.6, SD: 24.8) = 0.1706 **(b)** The distribution of number of catches cannot be approximately normal because there is a 0% probability that a randomly selected player would have fewer than 0 catches. No players have fewer than 0 catches.

27. (a) The graph is potentially deceptive because the scale on the vertical axis does not begin at 0.

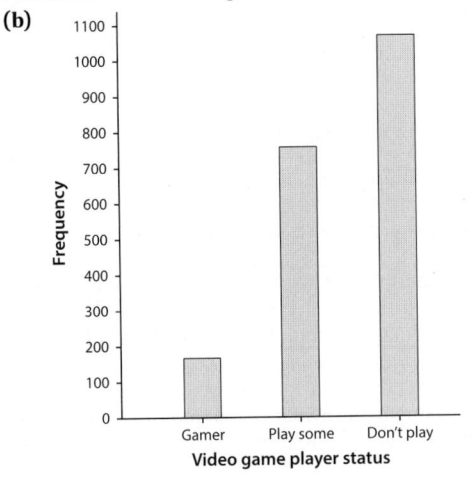

(b) [Bar chart: Frequency vs. Video game player status. Gamer ≈ 170, Play some ≈ 760, Don't play ≈ 1080. Y-axis 0 to 1100.]

Answers to Section 6.3 Exercises

1. The first step is to draw a normal distribution.

3. $z = -0.253$ (−0.25 with Table A)

5. (i) *Tech:* Applet/invNorm(area: 0.84, mean: 0, SD: 1) gives $z = 0.842$; *Table A:* 0.80 area to the left gives $z = 0.84$; $0.842 = \dfrac{x - 9.12}{0.05}$, $x = 9.16$ oz **(ii)** *Tech:* Applet/invNorm(area: 0.80, mean: 9.12, SD: 0.05) = 9.16 oz

7. (i) *Tech:* Applet/invNorm(area: 0.30, mean: 0, SD: 1) gives $z = -0.524$; *Table A:* 0.30 area to the left gives $z = -0.52$; $-0.524 = \dfrac{x - 48}{8}$, $x = 43.81$ months **(ii)** *Tech:* Applet/invNorm(area: 0.30, mean: 48, SD: 8) = 43.80 months

9. *Tech:* Applet/invNorm(area: 0.943, mean: 0, SD: 1) = 1.580; $1.580 = \dfrac{84 - 78.4}{SD}$, SD = 3.54 inches; *Table A:* 0.94 area to the left gives $z = 1.555$, which is the average of 1.55 and 1.56; $1.555 = \dfrac{84 - 78.4}{SD}$, SD = 3.60 inches

11. *Tech:* Applet/invNorm(area: 0.20, mean: 0, SD: 1) = −0.842; $-0.842 = \dfrac{11.7 - \mu}{0.2}$; $\mu = 11.87$ inches; *Table A:* 0.20 area to the left gives $z = -0.84$; $-0.84 = \dfrac{11.7 - \mu}{0.2}$; $\mu = 11.87$ inches

13. (a) (i) $z = \dfrac{2500 - 3668}{511} = -2.286$; *Tech:* $P(Z < -2.286)$ = Applet/normalcdf(lower: −1000, upper: −2.286, mean: 0, SD: 1) = 0.0111; *Table A:* $P(Z < -2.29) = 0.0110$ **(ii)** $P(B < 2500)$ = Applet/normalcdf(lower: −1000, upper: 2500, mean: 3668, SD: 511) = 0.0111 **(b)** Q_1: **(i)** *Tech:* Applet/invNorm(area: 0.25, mean: 0, SD: 1) gives $z = -0.674$; *Table A:* 0.25 area to the left gives $z = -0.67$; $-0.674 = \dfrac{Q_1 - 3668}{511}$, $Q_1 = 3323.6$ g **(ii)** *Tech:* Applet/invNorm(area: 0.25, mean: 3668, SD: 511) = 3323.3 g Q_3: **(i)** *Tech:* Applet/invNorm(area: 0.75, mean: 0, SD: 1) gives $z = 0.674$; *Table A:* 0.75 area to the left gives $z = 0.67$; $0.674 = \dfrac{Q_3 - 3668}{511}$, $Q_3 = 4012.4$ g **(ii)** *Tech:* Applet/invNorm(area: 0.75, mean: 3668, SD: 511) = 4012.7 g

15. (a) *Tech:* Applet/invNorm(area: 0.01, mean: 0, SD: 1) = −2.326; $-2.326 = \dfrac{3.95 - \mu}{0.02} \rightarrow \mu = 3.9965$ inches; *Table A:* 0.01 area to the left gives $z = -2.33$; $-2.33 = \dfrac{3.95 - \mu}{0.02} \rightarrow \mu = 3.9966$ inches **(b)** The percentage of lids that are too large will increase.

17. (a) From Exercise 13(b): $Q_1 = 3323$ g and $Q_3 = 4013$ g; IQR = 4013 − 3323 = 690 g. Low outlier < $Q_1 - 1.5 \times IQR = 3323 - 1.5(690) = 2288$. High outlier > $Q_3 + 1.5 \times IQR = 4013 + 1.5(690) = 5048$. Any birth weight less than 2288 g or greater than 5048 g will be an outlier. **(b) (i)** $z = \dfrac{2288 - 3668}{511} = -2.701$; *Tech:* $P(Z < -2.701)$ = Applet/normalcdf(lower: −1000, upper: −2.701, mean: 0, SD: 1) = 0.0035; *Table A:* $P(Z < -2.70) = 0.0035$ **(ii)** $P(X < 2288)$ = Applet/normalcdf(lower: −1000, upper: 2288, mean: 3668, SD: 511) = 0.0035; **(i)** $z = \dfrac{5048 - 3668}{511} = 2.701$; *Tech:* $P(Z > 2.701)$ = Applet/normalcdf(lower: 2.701, upper: 1000, mean: 0, SD: 1) = 0.0035; *Table A:* $P(Z > 2.70) = 0.0035$ **(ii)** $P(X > 5048)$ = Applet/normalcdf(lower: 5048, upper: 1000, mean: 3668, SD: 511) = 0.0035. There is a 0.0035 + 0.0035 = 0.0070 probability of randomly selecting a newborn whose weight would be considered an outlier.

19. *Shape:* Slightly skewed to the right for all three temperatures. *Outliers:* The cold temperature distribution contains a high outlier. The other distributions do not contain any outliers. *Center:* The median amount of water absorbed is greatest for the hot-temperature distribution (about 4.9 mL); it is less for the room-temperature distribution (about 3.1 mL), and is even less for the cold-temperature distribution (about 1.2 mL). *Variability:* The variability in amount of water absorbed is greatest for the room-temperature distribution ($IQR \approx 1.6$ mL), is slightly less for the hot-temperature distribution ($IQR \approx 1.3$ mL), and is slightly less for the cold-temperature distribution ($IQR \approx 1.2$ mL).

Answers to Section 6.4 Exercises

1. larger; closer to 0.5
3. Mean: $\mu_X = np$, SD: $\sigma_X = \sqrt{np(1-p)}$
5. Yes, because $np = 100(0.11) = 11 \geq 10$ and $n(1-p) = 100(1-0.11) = 89 \geq 10$
7. No, because $np = 100(2/38) = 5.26 < 10$
9. Yes, because $np = 1000(0.02) = 20 \geq 10$ and $n(1-p) = 1000(1-0.02) = 980 \geq 10$
11. (a) $\mu_L = 100(0.11) = 11$, $\sigma_L = \sqrt{100(0.11)(1-0.11)} = 3.13$
(b) (i) $z = \dfrac{15-11}{3.13} = 1.278$; Tech: $P(Z \geq 1.278) =$ Applet/normalcdf(lower: 1.278, upper: 1000, mean: 0, SD: 1) = 0.1006; Table A: $P(Z \geq 1.28) = 1 - 0.8997 = 0.1003$ (ii) $P(L \geq 15) =$ Applet/normalcdf(lower: 15, upper: 1000, mean: 11, SD: 3.13) = 0.1006
13. (a) $\mu_X = 1000(0.02) = 20$, $\sigma_X = \sqrt{1000(0.02)(1-0.02)} = 4.427$
(b) (i) $z = \dfrac{25-20}{4.427} = 1.129$; Tech: $P(Z \leq 1.129) =$ Applet/normalcdf(lower: -1000, upper: 1.129, mean: 0, SD: 1) = 0.8706; Table A: $P(Z \leq 1.13) = 0.8708$ (ii) $P(X \leq 25) =$ Applet/normalcdf(lower: -1000, upper: 25, mean: 20, SD: 4.427) = 0.8706
15. The normal probability plot is clearly curved; the distribution of CO_2 emissions is not approximately normal.
17. The normal probability plot is fairly linear; the distribution of earth density estimates is approximately normal.
19. (a) $np = 100(0.116) = 11.6 \geq 10$ and $n(1-p) = 100(1-0.116) = 88.4 \geq 10$ (b) $\mu_X = 100(0.116) = 11.6$, $\sigma_X = \sqrt{100(0.116)(1-0.116)} = 3.202$ (c) (i) $z = \dfrac{8-11.6}{3.202} = -1.124$; Tech: $P(Z \leq -1.124) =$ Applet/normalcdf(lower: -1000, upper: -1.124, mean: 0, SD: 1) = 0.1305; Table A: $P(Z \leq -1.12) = 0.1314$ (ii) $P(X \leq 8) =$ Applet/normalcdf(lower: -1000, upper: 8, mean: 11.6, SD: 3.202) = 0.1304
21. No, it is not unlikely (probability = 0.1304) that a random sample of 100 Labs would include 8 or fewer dogs with hip dysplasia when selecting from a population of Labs in which 11.6% have hip dysplasia. We do not have convincing evidence that the percentage of Labs with hip dysplasia in this state is less than 11.6%.
23. $P(X \geq 15) = 1 -$ Applet/binomcdf(trials: 100, $p = 0.11$, x value: 14) = 0.1330. The normal approximation calculation gave a probability of 0.1006, which is less accurate than 0.1330.
25. (a) $\mu_X = 100(0.116) = 11.6$, $\sigma_X = \sqrt{100(0.116)(1-0.116)} = 3.202$. Mean ± 1SD $= 11.6 - 3.202 = 8.398$ to $11.6 + 3.202 = 14.802$. The values of X within 1 standard deviation of the mean are $9 \leq X \leq 14$. (b) $P(9 \leq X \leq 14) =$ Applet/binomcdf(trials: 100, $p = 0.11$, x value: 14) $-$ Applet/binomcdf(trials: 100, $p = 0.11$, x value: 8) = 0.6504. The empirical rule says that 68% of observations fall within 1 standard deviation of the mean; 65% is fairly close to 68%.
27. (a) The purpose of the control group is to provide a baseline for comparing the effects of the treatment. Otherwise, we wouldn't be able to tell if the books or something else (e.g., students maturing) caused an increase in reading comprehension. (b) The difference in the reading scores in the third-grade girls' groups was so large that it is unlikely to be explained by chance variation alone in the random assignment to treatments.

Answers to Section 6.5 Exercises

1. populations; samples
3. $\bar{x} =$ sample mean, $\hat{p} =$ sample proportion, $s_x =$ sample standard deviation. $\bar{x}$ estimates μ; $\hat{p}$ estimates p; s_x estimates σ.
5. sampling distribution
7. statistic; its corresponding parameter; unusual
9. (a) *Population:* All people who signed a card saying that they intend to quit smoking. *Parameter:* $p =$ the true proportion of the population who actually quit smoking. *Sample:* The 1000 people who were selected. *Statistic:* The proportion of the sample who actually quit smoking, $\hat{p} = 0.21$. (b) *Population:* All 10-year-old boys. *Parameter:* The true 75th percentile height of the population of all 10-year-old boys. *Sample:* The fifty 10-year-old boys who were selected. *Statistic:* The 75th percentile height of the sample of fifty 10-year-old boys, 56 in.
11. (a) *Population:* All bottles of iced tea filled in a plant on Tuesday. *Parameter:* $\mu =$ the true mean amount of tea in the population of all bottles filled on Tuesday. *Sample:* The 50 bottles that were selected. *Statistic:* The mean amount of tea in the sample of 50 bottles, $\bar{x} = 19.6$ oz. (b) *Population:* All passengers in the airport. *Parameter:* $p =$ the true proportion of the population of all passengers who are chosen for random screening. *Sample:* The 125 passengers on a New York-to-Denver flight who were selected. *Statistic:* The proportion of the sample of 125 passengers who were selected for security screening, $\hat{p} = 0.08$.
13. Sample #1: 8% and 12%; $\bar{x} = 10\%$
Sample #2: 8% and -5%; $\bar{x} = 1.5\%$
Sample #3: 8% and -20%; $\bar{x} = -6\%$
Sample #4: 8% and 25%; $\bar{x} = 16.5\%$
Sample #5: 12% and -5%; $\bar{x} = 3.5\%$
Sample #6: 12% and -20%; $\bar{x} = -4\%$
Sample #7: 12% and 25%; $\bar{x} = 18.5\%$
Sample #8: -5% and -20%; $\bar{x} = -12.5\%$
Sample #9: -5% and 25%; $\bar{x} = 10\%$
Sample #10: -20% and 25%; $\bar{x} = 2.5\%$
Here is the sampling distribution of the sample mean for samples of size $n = 2$.

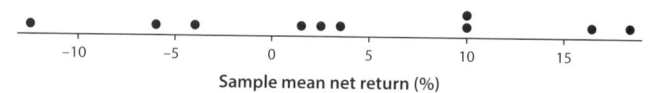

15. Sample #1: 8% and 12%; $\hat{p} = 1$
Sample #2: 8% and -5%%; $\hat{p} = 0.5$
Sample #3: 8% and -20%; $\hat{p} = 0.5$
Sample #4: 8% and 25%; $\hat{p} = 1$
Sample #5: 12% and -5%; $\hat{p} = 0.5$
Sample #6: 12% and -20%; $\hat{p} = 0.5$
Sample #7: 12% and 25%; $\hat{p} = 1$
Sample #8: -5% and -20%; $\hat{p} = 0$
Sample #9: -5% and 25%; $\hat{p} = 0.5$
Sample #10: -20% and 25%; $\hat{p} = 0.5$
Here is the sampling distribution of the sample proportion for samples of size $n = 2$.

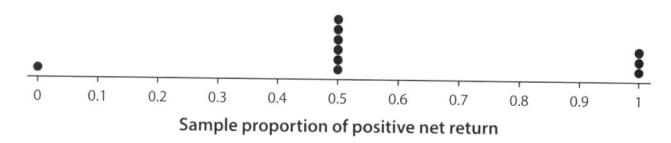

17. (a) In one simulated SRS of size $n = 50$, 90% of the town residents have been immunized or have recovered from Covid-19. (b) No. In the 100 simulated samples, 39 of the 100 SRSs have a sample proportion of 0.74 or greater. (c) No. Because it is likely to get a sample proportion of 0.74 or greater by chance alone when $p = 0.70$, there is not convincing evidence that more than 70% of the town residents have been vaccinated or have recovered from Covid-19.
19. (a) In one simulated SRS of size $n = 10$, the mean temperature was 47.8°F. (b) No. In the 100 simulated samples, 20 of the SRSs had a sample mean of 49.2 or less. (c) No. Because it is somewhat likely to get a sample mean of 49.2 or less by chance alone when $\mu = 50$, there is not convincing evidence that the thermostat keeps the average temperature colder than the manufacturer claims.
21. (a) *Population:* All game pieces. *Parameter:* $p =$ the true proportion of all game pieces that are instant winners. *Sample:* The 20 game pieces collected by the frequent diner. *Statistic:* The proportion of the 20 game pieces that are instant winners, $\hat{p} = 1/20 = 0.05$. (b) Yes. In the 100 simulated samples, only 1 of the SRSs had a sample proportion of 0.05 or less. (c) Yes. Because it is unlikely to get a

sample proportion of 0.05 or less by chance alone when $p = 0.25$, there is convincing evidence that fewer than 25% of all game pieces are instant winners.

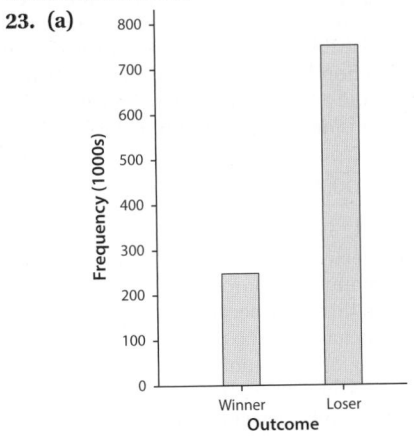

23. (a)

[Bar chart: Frequency (1000s) vs Outcome. Winner ≈ 250, Loser ≈ 750]

(b) Answers may vary. The number of winners should be close to 5, and the total number of winners and losers should be 20.

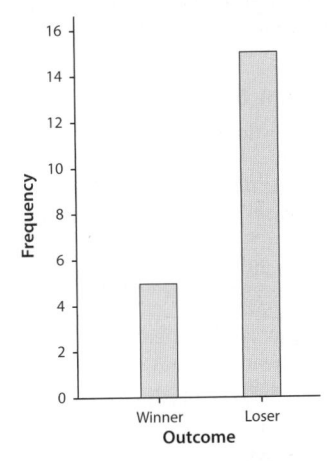

[Bar chart: Frequency vs Outcome. Winner ≈ 5, Loser ≈ 15]

25. (a) In 10 cases of taking a random sample of size $n = 50$ from each high school, the difference in the proportions of students with internet access at home is 0%. This means the proportions of students with internet access were the same for each high school in 10 simulated trials of this sampling process. **(b)** Yes. In the 100 simulated trials of this process, 0 of the trials had a difference in proportions of −0.20 or less. **(c)** Yes. Because it is very unlikely to get a difference in the sample proportions of −0.20 or less by chance alone when there is no difference in the population proportions, there is convincing evidence that South High School has a smaller proportion of students with internet access at home than does North High School.

27. (a) $_{300}C_{25} = 1.95 \times 10^{36}$ different possible samples of 25 tomatoes **(b)** Due to the extremely large number of possible samples, it is not practical to examine the complete sampling distribution of sample means for samples of size 25.

Answers to Section 6.6 Exercises

1. The mean of its sampling distribution is equal to the value of the parameter being estimated.

3. False. The variability of a statistic decreases as the sample size increases.

5. $\mu_{\bar{x}} = \dfrac{10 + 1.5 - 6 + 16.5 + 3.5 - 4 + 18.5 - 12.5 + 10 + 2.5}{10} = 4$ and

$\mu = \dfrac{8 + 12 - 5 - 20 + 25}{5} = 4$. Because $\mu_{\bar{x}}$ is equal to the value it is trying to estimate (μ), the sample mean is an unbiased estimator of the population mean.

7. Yes. The mean of the simulated sampling distribution is very close to 22.96, the value of the population median.

9. (a) The sampling distribution of the sample mean will be less variable because the sample size is larger. **(b)** The estimated mean net return $\bar{x}$ is more likely to be close to the true mean net return μ. The estimate will be more precise.

11. (a) The sampling distribution of the sample median will be more variable because the sample size is smaller. **(b)** The estimated median amount spent is less likely to be close to the true median amount spent. The estimate will be less precise.

13. (a) Statistics 1 and 3 both appear to be unbiased, because the mean of each simulated sampling distribution is very close to N, the estimated total number of German tanks. Statistic 2 appears to be biased, because the mean of its sampling distribution is clearly greater than N. **(b)** Statistic 3. While both statistics 1 and 3 are unbiased, statistic 3 appears to have less variability. **(c)** The Allies could get a more precise estimate by capturing more tanks (increasing the sample size). This way the estimated number of tanks is more likely to be close to the true number of tanks (more precise).

15. If we chose many SRSs and calculated the sample mean $\bar{x}$ for each sample, the distribution of $\bar{x}$ would be centered at the value of μ.

17. A larger random sample will provide more information about the population and, therefore, a more precise estimate. The variability of the distribution of $\bar{x}$ decreases as the sample size increases.

19. $n = 10$. Sample medians obtained from smaller samples are more variable than those obtained from larger samples, so the sample size of $n = 10$ is more likely produce a sample median greater than \$250,000 (further from \$200,000). The larger sample size ($n = 100$) is more likely to produce a sample median that is close to \$200,000, the population median.

21. Sample #1: 8%, 12%, −5%, −20%; $\bar{x} = -1.25\%$
Sample #2: 8%, 12%, −5%, 25%; $\bar{x} = 10\%$
Sample #3: 8%, 12%, −20%, 25%; $\bar{x} = 6.25\%$
Sample #4: 8%, −5%, −20%, 25%; $\bar{x} = 2\%$
Sample #5: 12%, −5%, −20%, 25%; $\bar{x} = 3\%$
Here is the sampling distribution of the sample mean for samples of size $n = 4$.

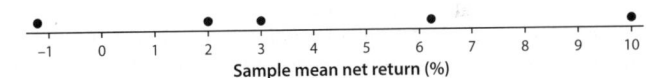

[Dotplot: Sample mean net return (%), values from −1 to 10]

The variability of the sampling distribution of the sample mean for samples of size $n = 4$ is less than that for samples of size $n = 2$. This is consistent with the answer to Exercise 9, where we concluded that the sampling distribution of the sample mean will be less variable for larger sample sizes.

23. (a) The midrange is easier to calculate because it is just the average of the minimum and the maximum rather than the average of all 10 amounts spent. **(b)** No, the sample midrange does not appear to be an unbiased estimator of the population mean. The mean of the sampling distribution is clearly greater than \$29.85, the value of the population mean.

25. (a) This is a binomial setting. *Binary?* "Success" = statistics major is selected. "Failure" = math major is selected. *Independent?* Because they are sampling with replacement, knowing whether or not one randomly selected student is a statistics major tells you nothing about whether or not another randomly selected student is a statistics major. *Number?* $n = 40$. *Same probability?* $p = \dfrac{55}{104} = 0.529$

(b) $\mu_X = np = 40(0.529) = 21.16$, $\sigma_X = \sqrt{np(1-p)} = \sqrt{40(0.529)(0.471)} = 3.16$ **(c)** $P(X \leq 1) = P(X = 0) + P(X = 1) =$ Applet/binomcdf(n: 40, p: 0.529, x: 1) = 0.0066. If the professor were to randomly choose students for the first 10 meetings, there is less than a 1% probability that he would select one statistics major or fewer by chance alone. Because this is unlikely, we have convincing evidence that his selection process is not truly random.

Answers to Section 6.7 Exercises

1. False. The sampling distribution describes the possible values of $\hat{p}$ and how often these values occur.
3. sample proportions; population proportion
5. To determine if the shape of the sampling distribution of the sample proportion is approximately normal.
7. (a) $\mu_{\hat{p}} = 0.20$, $\sigma_{\hat{p}} = \sqrt{\frac{0.20(1-0.20)}{30}} = 0.073$ **(b)** In SRSs of size $n = 30$, the sample proportion of orange Skittles typically varies from the true proportion by about 0.073.
9. (a) $\mu_{\hat{p}} = 0.90$, $\sigma_{\hat{p}} = \sqrt{\frac{0.90(1-0.90)}{100}} = 0.03$ **(b)** In SRSs of size $n = 100$, the sample proportion of orders that were shipped within three working days typically varies from the true proportion by about 0.03.
11. No, because $np = (30)(0.2) = 6 < 10$
13. Yes, because $np = (100)(0.90) = 90 \geq 10$ and $n(1-p) = (100)(1-0.90) = 10 \geq 10$
15. (a) Approximately normal because $np = (500)(0.08) = 40 \geq 10$ and $n(1-p) = (500)(1-0.08) = 460 \geq 10$, $\mu_{\hat{p}} = 0.08$, $\sigma_{\hat{p}} = \sqrt{\frac{0.08(1-0.08)}{500}} = 0.012$ **(b)** (i) $z = \frac{0.10-0.08}{0.012} = 1.667$; Tech: $P(Z > 1.667) = $ Applet/normalcdf(lower: 1.667, upper: 1000, mean: 0, SD: 1) = 0.0478; Table A: $P(Z > 1.67) = 1 - 0.9525 = 0.0475$ (ii) $P(\hat{p} > 0.10) = $ Applet/normalcdf(lower: 0.10, upper: 1000, mean: 0.08, SD: 0.012) = 0.0478
17. (a) Approximately normal because $np = (50)(0.375) = 18.75 \geq 10$ and $n(1-p) = (50)(1-0.375) = 31.25 \geq 10$, $\mu_{\hat{p}} = 0.375$, $\sigma_{\hat{p}} = \sqrt{\frac{0.375(1-0.375)}{50}} = 0.068$ **(b)** (i) $z = \frac{0.30-0.375}{0.068} = -1.103$; Tech: $P(Z < -1.103) = $ Applet/normalcdf(lower: −1000, upper: −1.103, mean: 0, SD: 1) = 0.1350; Table A: $P(Z < -1.10) = 0.1357$ (ii) $P(\hat{p} < 0.30) = $ Applet/normalcdf(lower: −1000, upper: 0.30, mean: 0.375, SD: 0.068) = 0.1350
19. (a) $\mu_{\hat{p}} = 0.15$, $\sigma_{\hat{p}} = \sqrt{\frac{0.15(1-0.15)}{500}} = 0.016$. In SRSs of size $n = 500$, the sample proportion of people who received a jury summons in the previous 12 months typically varies from the true proportion by about 0.016. **(b)** Because $np = (500)(0.15) = 75 \geq 10$ and $n(1-p) = (500)(1-0.15) = 425 \geq 10$, the sampling distribution of $\hat{p}$ is approximately normal. **(c)** (i) $z = \frac{0.13-0.15}{0.016} = -1.250$; Tech: $P(Z \leq -1.250) = $ Applet/normalcdf(lower: −1000, upper: −1.250, mean: 0, SD: 1) = 0.1056; Table A: $P(Z < -1.25) = 0.1056$ (ii) $P(\hat{p} < 0.13) = $ Applet/normalcdf(lower: −1000, upper: 0.13, mean: 0.15, SD: 0.016) = 0.1056 **(d)** Because it is somewhat likely to get a sample proportion of 0.13 or less by chance alone when $p = 0.15$, there is not convincing evidence that fewer than 15% of all U.S. residents received a jury summons in the previous 12 months.
21. To cut the standard deviation in half, the sample size would need to be 4 times as large. Select a random sample of $500(4) = 2000$ U.S. residents.
23. (a) $\sigma_{\hat{p}} = \sqrt{\frac{0.6(1-0.6)}{10}} = 0.1549$, $\sigma_{\hat{p}} = \sqrt{\frac{0.6(1-0.6)}{10}}\sqrt{\frac{1000-10}{1000-1}} = 0.1542$. In this case, the standard deviation of the sampling distribution of $\hat{p}$ is about the same whether the finite population correction factor is used or not. **(b)** $\sigma_{\hat{p}} = \sqrt{\frac{0.6(1-0.6)}{500}} = 0.0219$, $\sigma_{\hat{p}} = \sqrt{\frac{0.6(1-0.6)}{500}}\sqrt{\frac{1000-500}{1000-1}} = 0.0155$. In this case, the standard deviation of the sampling distribution of $\hat{p}$ is much smaller when the finite population correction factor is used. **(c)** When the sample size is less than 10% of the population size [part (a)], the finite population correction factor is not needed because the standard deviation of the sampling distribution is about the same whether the finite population correction factor is used or not. When the sample size is at least 10% of the population size [part (b)], the finite population correction factor is needed to determine the standard deviation of the sampling distribution of $\hat{p}$.
(d) $\sigma_{\hat{p}} = \sqrt{\frac{0.6(1-0.6)}{1000}}\sqrt{\frac{1000-1000}{1000-1}} = 0$. In this case, all 1000 members of the population are selected for the sample. Because this is a census, the true value of the population proportion p will be known. There will be no variability in this value because it is a parameter.
25. (a) It was important to include a group that received a placebo vaccine to serve as a baseline for determining the effectiveness of the low-dose and high-dose vaccines. The control group shows the proportion of children who would develop malaria without the vaccine. **(b)** Because this study involved random assignment of the children to the treatment groups (low-dose vaccine, high-dose vaccine, or placebo), we can conclude that the vaccine was the cause of the decrease in the proportion of children who developed malaria. **(c)** All children like those in the study.

Answers to Section 6.8 Exercises

1. False. We use $\bar{x}$ to estimate μ.
3. standard deviation
5. False. The *sampling distribution* of $\bar{x}$ becomes more normal as the sample size increases, not the population distribution.
7. $n = 30$
9. (a) $\mu_{\bar{x}} = 225$, $\sigma_{\bar{x}} = \frac{60}{\sqrt{5}} = 26.833$ **(b)** In SRSs of size $n = 5$ from David's collection, the sample mean play time of songs typically varies from the true mean by about 26.833 seconds.
11. (a) $\mu_{\bar{x}} = 3.4$, $\sigma_{\bar{x}} = \frac{0.5}{\sqrt{15}} = 0.129$ **(b)** In SRSs of size $n = 15$, the sample mean birth weight typically varies from the true mean by about 0.129 kg.
13. (a) Because $n = 5 < 30$, the sampling distribution will be skewed to the right, but not as strongly as the population distribution. **(b)** Because $n = 100 \geq 30$, the sampling distribution will be approximately normal.
15. (a) The sampling distribution will be approximately normal because the population distribution is approximately normal. **(b)** The sampling distribution will be approximately normal because the population distribution is approximately normal.
17. (a) Approximately normal because the population distribution is approximately normal. $\mu_{\bar{x}} = 9.5$, $\sigma_{\bar{x}} = \frac{1}{\sqrt{15}} = 0.258$
(b) (i) $z = \frac{9-9.5}{0.258} = -1.938$ and $z = \frac{10-9.5}{0.258} = 1.938$; Tech: $P(-1.938 < Z < 1.938) = $ Applet/normalcdf(lower: −1.938, upper: 1.938, mean: 0, SD: 1) = 0.9474; Table A: $P(-1.94 < Z < 1.94) = 0.9738 - 0.0262 = 0.9476$ (ii) $P(9 < \bar{x} < 10) = $ Applet/normalcdf(lower: 9, upper: 10, mean: 9.5, SD: 0.258) = 0.9474
19. (a) Approximately normal because $n = 50 \geq 30$, $\mu_{\bar{x}} = 6$, $\sigma_{\bar{x}} = \frac{2.4}{\sqrt{50}} = 0.339$ **(b)** (i) $z = \frac{5-6}{0.339} = -2.950$; Tech: $P(Z < -2.950) = $ Applet/normalcdf(lower: −1000, upper: −2.950, mean: 0, SD: 1) = 0.0016, Table A: $P(Z < -2.95) = 0.0016$ (ii) $P(\bar{x} < 5) = $ Applet/normalcdf(lower: −1000, upper: 5, mean: 6, SD: 0.339) = 0.0016

21. (a) Approximately normal because $n = 70 \geq 30$, $\mu_{\bar{x}} = 1$, $\sigma_{\bar{x}} = \frac{1.5}{\sqrt{70}} = 0.179$ **(b)** (i) $z = \frac{1.1-1}{0.179} = 0.559$; Tech: $P(Z > 0.559)$ = Applet/normalcdf(lower: 0.559, upper: 1000, mean: 0, SD: 1) = 0.2881; Table A: $P(Z > 0.56) = 1 - 0.7123 = 0.2877$ (ii) $P(\bar{x} > 1.1)$ = Applet/normalcdf(lower: 1.1, upper: 1000, mean: 1, SD: 0.179) = 0.2882 **(c)** It is somewhat likely that the mean maintenance time will exceed 1.1 hours, so no, the company did not budget enough time.

23. (a) Less likely. Individual values vary more than averages do. **(b)** Approximately normal because the population distribution is approximately normal. $\mu_{\bar{x}} = 9.5$, $\sigma_{\bar{x}} = \frac{1}{\sqrt{1}} = 1$ (i) $z = \frac{9-9.5}{1} = -0.5$ and $z = \frac{10-9.5}{1} = 0.5$; Tech: $P(-0.500 < Z < 0.500)$ = Applet/normalcdf(lower: −0.500, upper: 0.500, mean: 0, SD: 1) = 0.3829; Table A: $P(-0.5 < Z < 0.5) = 0.6915 - 0.3085 = 0.3830$ (ii) $P(9 < \bar{x} < 10)$ = Applet/normalcdf(lower: 9, upper: 10, mean: 9.5, SD: 1) = 0.3829. This probability of 0.3829 is much less than the probability calculated in Exercise 17 (0.9474).

25. To cut $\sigma_{\bar{x}}$ in half, select a sample that is 4 times as large. The sample size should be increased to $n = (5)(4) = 20$ songs.

27. No. The graph of the sample will look similar to the shape of the population distribution, regardless of the sample size. The student should say that the graph of the *sampling distribution of the sample mean* ($\bar{x}$) looks increasingly more normal as you take ever larger samples from a population.

29. (a) $\sigma_{\bar{x}} = \frac{5}{\sqrt{10}} = 1.581$, $\sigma_{\bar{x}} = \frac{5}{\sqrt{10}}\sqrt{\frac{1000-10}{1000-1}} = 1.574$

In this case, the standard deviation of the sampling distribution of $\bar{x}$ is about the same whether the finite population correction factor is used or not.

(b) $\sigma_{\bar{x}} = \frac{5}{\sqrt{500}} = 0.224$, $\sigma_{\bar{x}} = \frac{5}{\sqrt{500}}\sqrt{\frac{1000-500}{1000-1}} = 0.158$

In this case, the standard deviation of the sampling distribution of $\bar{x}$ is much smaller when the finite population correction factor is used. **(c)** When the sample size is less than 10% of the population size [part (a)], the finite population correction factor is not needed because the standard deviation of the sampling distribution is about the same whether the finite population correction factor is used or not. However, when the sample size is not less than 10% of the population size [part (b)], the finite population correction factor is needed to determine the standard deviation of the sampling distribution of $\bar{x}$.

(d) $\sigma_{\bar{x}} = \frac{5}{\sqrt{1000}}\sqrt{\frac{1000-1000}{1000-1}} = 0$. In this case, all 1000 members of the population are selected for the sample. Because this is a census, the true value of the population mean μ will be known. There will be no variability in this value because it is a parameter.

31. (a) The probabilities are all between 0 and 1, inclusive, and the sum of the probabilities is 1. **(b)** Using the complement rule, $P(X \geq 1) = 1 - P(X = 0) = 1 - 0.28 = 0.72$. **(c)** $\mu_X = 0(0.28) + 1(0.27) + 2(0.18) + 3(0.16) + 4(0.07) + 5(0.04) = 1.59$ devices

$\sigma_X = \sqrt{(0-1.59)^2(0.28) + (1-1.59)^2(0.27) + \cdots + (5-1.59)^2(0.04)} = 1.422$ devices

Answers to Chapter 6 Review Exercises

1. (a)

(b) $(1 - 0.68)/2 = 0.16$

2. (a) (i) $z = \frac{10-16.4}{2.1} = -3.048$; Tech: $P(Z < -3.048)$ = Applet/normalcdf(lower: −1000, upper: −3.048, mean: 0, SD: 1) = 0.0012; Table A: $P(Z < -3.05) = 0.0011$ (ii) $P(X < 10)$ = Applet/normalcdf(lower: −1000, upper: 10, mean: 16.4, SD: 2.1) = 0.0012 **(b)** (i) $z = \frac{17-16.4}{2.1} = 0.286$; Tech: $P(Z > 0.286)$ = Applet/normalcdf(lower: 0.286, upper: 1000, mean: 0, SD: 1) = 0.3874; Table A: $P(Z > 0.29) = 1 - 0.6141 = 0.3859$ (ii) $P(X > 17)$ = Applet/normalcdf(lower: 17, upper: 1000, mean: 16.4, SD: 2.1) = 0.3875

(c) (i) $z = \frac{15-16.4}{2.1} = -0.667$ and $z = \frac{20-16.4}{2.1} = 1.714$; Tech: $P(-0.667 < Z < 1.714)$ = Applet/normalcdf(lower: −0.667, upper: 1.714, mean: 0, SD: 1) = 0.7043; Table A: $P(-0.67 < Z < 1.71) = 0.9564 - 0.2514 = 0.7050$ (ii) $P(15 < X < 20)$ = Applet/normalcdf(lower: 15, upper: 20, mean: 16.4, SD: 2.1) = 0.7043

3. (a) (i) Tech: Applet/invNorm(area: 0.25, mean: 0, SD: 1) gives $z = -0.674$; Table A: 0.25 area to the left gives $z = -0.67$; $-0.674 = \frac{x-16.4}{2.1}$, $x = 14.985$ inches (ii) Tech: Applet/invNorm(area: 0.25, mean: 16.4, SD: 2.1) = 14.984 inches **(b)** Tech: Applet/invNorm(area: 0.90, mean: 0, SD: 1) = 1.282; $1.282 = \frac{19-\mu}{3} \rightarrow \mu = 15.154$ inches; Table A: 0.9 area to the left gives $z = 1.28$; $1.28 = \frac{19-\mu}{3} \rightarrow \mu = 15.16$ inches

4. (a) Approximately normal because $np = 500(0.24) = 120 \geq 10$ and $n(1-p) = 500(1-0.24) = 380 \geq 10$ **(b)** $\mu_X = 500(0.24) = 120$, $\sigma_X = \sqrt{500(0.24)(1-0.24)} = 9.550$ **(c)** (i) $z = \frac{100-120}{9.550} = -2.094$ and $z = \frac{110-120}{9.550} = -1.047$; Tech: $P(-2.094 < Z < -1.047)$ = Applet/normalcdf(lower: −2.094, upper: −1.047, mean: 0, SD: 1) = 0.1294; Table A: $P(-2.09 \leq Z \leq -1.05) = 0.1469 - 0.0183 = 0.1286$ (ii) $P(100 \leq X \leq 110)$ = Applet/normalcdf(lower: 100, upper: 110, mean: 120, SD: 9.550) = 0.1294

5. The normal probability plot has a curved pattern, indicating that the distribution of total length for the 202 North Atlantic right whales is not approximately normal.

6. *Population:* All eggs shipped in one day. *Parameter:* The proportion p of eggs shipped that day that had *Salmonella*. *Sample:* The 200 eggs examined. *Statistic:* The proportion of eggs in the sample that had *Salmonella*, $\hat{p} = \frac{9}{200} = 0.045$.

7. (a) Sample #1: 64, 66, 71; Median = 66
Sample #2: 64, 66, 73; Median = 66
Sample #3: 64, 66, 76; Median = 66
Sample #4: 64, 71, 73; Median = 71
Sample #5: 64, 71, 76; Median = 71
Sample #6: 64, 73, 76; Median = 73
Sample #7: 66, 71, 73; Median = 71
Sample #8: 66, 71, 76; Median = 71
Sample #9: 66, 73, 76; Median = 73
Sample #10: 71, 73, 76; Median = 73

Here is the sampling distribution of the sample median.

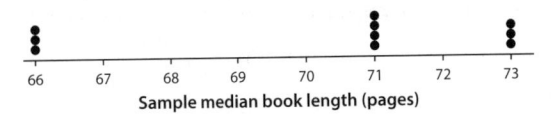

(b) $\mu_{\text{median}} = \frac{66+66+66+71+71+71+71+73+73+73}{10} = 70.1$

The sample median is a biased estimator of the population median. The mean of the sampling distribution is 70.1, which is not the same as the population median (71). **(c)** If n is increased to 4, the sampling distribution of the sample median will be less variable because the sample size is larger.

8. (a) $\mu_{\hat{p}} = 0.20$ and $\sigma_{\hat{p}} = \sqrt{\dfrac{0.20(1-0.20)}{80}} = 0.045$ **(b)** In SRSs of size $n = 80$, the sample proportion of people who subscribe to the five-second rule typically varies from the true proportion by about 0.045. **(c)** Approximately normal because $np = (80)(0.2) = 16 \geq 10$ and $n(1-p) = (80)(1-0.2) = 64 \geq 10$. **(d) (i)** $z = \dfrac{0.1-0.2}{0.045} = -2.222$; Tech: $P(Z < -2.222) =$ Applet/normalcdf(lower: −1000, upper: −2.222, mean: 0, SD: 1) = 0.0131; Table A: $P(Z < -2.22) = 0.0132$ **(ii)** $P(\hat{p} < 0.10) =$ Applet/normalcdf(lower: −1000, upper: 0.1, mean: 0.2, SD: 0.045) = 0.0131 **(e)** Because this result is unlikely to happen by chance alone when $p = 0.20$, there is convincing evidence that the proportion of all U.S. adults who subscribe to the five-second rule is less than 0.20.

9. (a) $\mu_{\bar{x}} = 16.4$ and $\sigma_{\bar{x}} = \dfrac{2.1}{\sqrt{10}} = 0.664$ **(b)** In SRSs of size $n = 10$, the sample mean length typically varies from the true mean by about 0.664 inches. **(c)** Approximately normal, because the population distribution is approximately normal. **(i)** $z = \dfrac{17 - 16.4}{0.664} = 0.904$; Tech: $P(Z > 0.904) =$ Applet/normalcdf(lower: 0.904, upper: 1000, mean: 0, SD: 1) = 0.1830; Table A: $P(Z > 0.90) = 1 - 0.8159 = 0.1841$ **(ii)** $P(\bar{x} > 17) =$ Applet/normalcdf(lower: 17, upper: 1000, mean: 16.4, SD: 0.664) = 0.1831 **(d)** If the distribution of length in the population of commercially raised copper rockfish were distinctly non-normal, the answers to parts (a) and (b) would be the same because the mean and standard deviation do not depend on the shape of the population distribution. We would not be able to answer part (c), because we would not be sure that the sampling distribution is approximately normal.

10. (a) Not possible because the distribution of the population is strongly skewed and $n = 1 < 30$ **(b)** Approximately normal because $n = 50 \geq 30$, $\mu_{\bar{x}} = 0.5$, $\sigma_{\bar{x}} = \dfrac{0.7}{\sqrt{50}} = 0.099$ **(i)** $z = \dfrac{0.6 - 0.5}{0.099} = 1.010$; Tech: $P(Z \geq 1.010) =$ Applet/normalcdf(lower: 1.010, upper: 1000, mean: 0, SD: 1) = 0.1562; Table A: $P(Z \geq 1.01) = 1 - 0.8438 = 0.1562$ **(ii)** $P(\bar{x} \geq 0.6) =$ Applet/normalcdf(lower: 0.6, upper: 1000, mean: 0.5, SD: 0.099) = 0.1562 **(c)** No. Because this result is somewhat likely to happen by chance alone when $\mu = 0.5$, the state agricultural department should not be worried that the moth population is getting larger in their state.

11. (a) (i) $z = \dfrac{1 - 1.05}{0.08} = -0.625$ and $z = \dfrac{1.2 - 1.05}{0.08} = 1.875$; Tech: $P(-0.625 < Z < 1.875) =$ Applet/normalcdf(lower: −0.625, upper: 1.875, mean: 0, SD: 1) = 0.7036; Table A: $P(-0.63 < Z < 1.88) = 0.9699 - 0.2643 = 0.7056$ **(ii)** $P(1 < X < 1.2) =$ Applet/normalcdf(lower: 1, upper: 1.2, mean: 1.05, SD: 0.08) = 0.7036 **(b)** Tech: Applet/invNorm(area: 0.005, mean: 0, SD: 1) gives $= -2.576$; Table A: 0.005 area to the left gives $z = -2.58$. Solving $-2.576 = \dfrac{1 - 1.1}{\sigma}$ gives $\sigma = 0.039$.

Answers to Cumulative Review II: Chapters 4–6

Section I: Multiple Choice

1. b
2. b
3. d
4. c
5. b
6. c
7. d
8. b
9. d
10. c
11. b
12. c
13. c
14. b
15. c
16. b
17. a
18. a
19. c
20. a

Section II: Free Response

1. (a) P(at least one incident) $= 1 - 1439/2009 = 0.284$ **(b)** P(both types | urban) $= 171/658 = 0.260$ **(c)** P(both types) $= 269/2009 = 0.134$. Because P(both types) $\neq P$(both types | urban), these events are not independent.

2. (a) (i) $z = \dfrac{35{,}625 - 35{,}987}{607.5} = -0.596$; Tech: $P(Z < -0.596) =$ Applet/normalcdf(lower: −1000, upper: −0.596, mean: 0, SD: 1) = 0.2756; Table A: $P(Z < -0.60) = 0.2743$ **(ii)** $P(X < 35625) =$ Applet/normalcdf(lower: −100000, upper: 35625, mean: 35987, SD: 607.5) = 0.2756 **(b) (i)** $z = \dfrac{34{,}772 - 35{,}987}{607.5} = -2.000$ and $z = \dfrac{36{,}225 - 35{,}987}{607.5} = 0.392$; Tech: $P(-2.000 < Z < 0.392) =$ Applet/normalcdf(lower: −2.000, upper: 0.392, mean: 0, SD: 1) = 0.6297; Table A: $P(-2.00 < Z < 0.39) = 0.6517 - 0.0228 = 0.6289$ **(ii)** $P(34772 < X < 36225) =$ Applet/normalcdf(lower: 34772, upper: 36225, mean: 35987, SD: 607.5) = 0.6296 **(c) (i)** Tech: Applet/invNorm(area: 0.98, mean: 0, SD: 1) gives $z = 2.054$; Table A: Area of 0.98 to the left gives $z = 2.05$. $2.054 = \dfrac{x - 35{,}987}{607.5}$, $x = \$37{,}234.81$ **(ii)** Tech: Applet/invNorm(area: 0.98, mean: 35987, SD: 607.5) = \$37,234.65

3. (a) "At least 2 eggs" is equivalent to $Y \geq 2$. $P(Y \geq 2) = 0.08 + 0.03 + 0.01 = 0.12$.

(b)

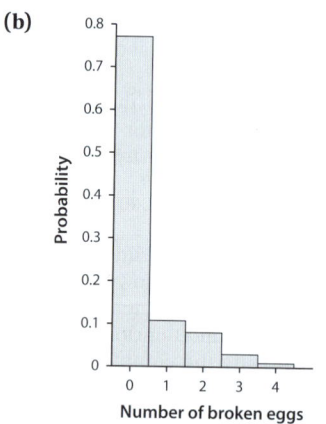

The distribution of number of broken eggs is skewed to the right with a single peak at 0. **(c)** $\mu_Y = 0(0.77) + 1(0.11) + \cdots + 4(0.01) = 0.4$. If many, many cartons of "store brand" eggs were randomly selected, the average number of broken eggs will be about 0.4.

(d) $\sigma_Y = \sqrt{(0-0.4)^2(0.77) + (1-0.4)^2(0.11) + \cdots + (4-0.4)^2(0.01)} = \sqrt{0.7} = 0.837$. If many, many cartons of "store brand" eggs were randomly selected, the number of broken eggs in a carton typically varies from the mean of 0.4 by about 0.837 egg.

4. (a) X is a binomial random variable. *Binary?* "Success" = flight arrives on time. "Failure" = flight arrives late. *Independent?* Knowing whether or not the flight is on time for one randomly selected day tells you nothing about whether or not the flight is on time for another randomly selected day. *Number?* $n = 20$. *Same probability?* $p = 0.85$ **(b)** $P(X = 19) = {}_{20}C_{19}(0.85)^{19}(1 - 0.85)^1 = 0.1368$. There is about a 14% chance that exactly 19 flights will arrive on time.

(c) *Tech:* $P(X \leq 14) =$ Applet/binomcdf(trials: 20, p: 0.85, x value: 14) = 0.0673 (d) Because it is somewhat likely (probability = 0.0673) that 14 or fewer flights will arrive on time by chance alone, we don't have convincing evidence against the airline's claim that 85% of these flights arrive on time. (e) $\mu_X = 20(0.85) = 17$ and $\sigma_X = \sqrt{20(0.85)(1-0.85)} = 1.597$ (f) Because $n(1-p) = 20(1-0.85) = 3 < 10$, it is not reasonable to use the normal approximation.

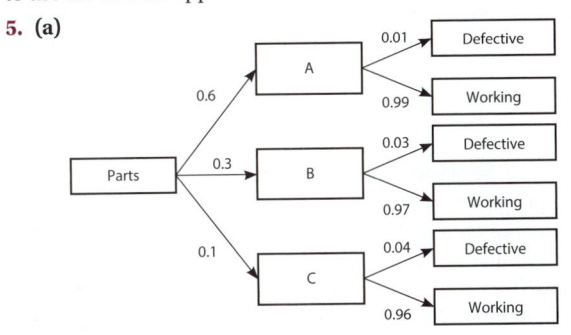

5. (a)

(b) $P(\text{defective}) = (0.60)(0.01) + (0.30)(0.03) + (0.10)(0.04) = 0.006 + 0.009 + 0.004 = 0.019$ **(c)** $P(C \mid \text{defective}) = \dfrac{0.004}{0.019} = 0.211$

6. (a) The mean of the sampling distribution of $\bar{x}$ is equal to μ, the mean amount paid by all members. **(b)** $\mu_{\bar{x}} = \$50$ and $\sigma_{\bar{x}} = \dfrac{20}{\sqrt{60}} = \$2.582$. In SRSs of size $n = 60$, the sample mean amount paid for internet service typically varies from the true mean by about $2.58. **(c)** Approximately normal because $n = 60 \geq 30$ **(d)** (i) $z = \dfrac{45-50}{2.582} = -1.936$; *Tech:* $P(Z < -1.936) =$ Applet/normalcdf(lower: -1000, upper: -1.936, mean: 0, SD: 1) = 0.0264; *Table A:* $P(Z < -1.94) = 0.0262$ (ii) $P(\bar{x} < 45) =$ Applet/normalcdf(lower: -1000, upper: 45, mean: 50, SD: 2.582) = 0.0264

CHAPTER 7
Answers to Section 7.1 Exercises

1. sample; population; population; sample
3. True
5. point estimate ± margin of error
7. We include a margin of error when estimating a population parameter to account for sampling variability.
9. We are 95% confident that the interval from 0.175 to 0.225 captures the proportion of all U.S. adults who would correctly identify nitrogen as the answer.
11. We are 99% confident that the interval from 2.8 ppb to 11.8 ppb captures the mean lead level for all tap water in Flint, Michigan, in 2015.
13. point estimate = $\dfrac{0.175 + 0.225}{2} = 0.20$; margin of error = $0.225 - 0.20 = 0.025$
15. point estimate = $\dfrac{2.8 + 11.8}{2} = 7.3$; margin of error = $11.8 - 7.3 = 4.5$
17. Yes. Because all the plausible values for the population proportion (0.175, 0.225) are less than 0.25, there is convincing evidence that less than 25% of all U.S. adults would answer this question correctly.
19. No. Because some of the plausible values for the population mean (2.8, 11.8) are less than 5 ppb, the interval does not provide convincing evidence that the mean lead level in Flint tap water exceeds 5 ppb.
21. (a) We are 95% confident that the interval from 0.436 to 0.478 captures the true proportion of U.S. adults who have traveled internationally. **(b)** point estimate = $\dfrac{0.436 + 0.478}{2} = 0.457$; margin of error = $0.478 - 0.457 = 0.021$ **(c)** No. None of the plausible values for the population proportion (0.436, 0.478) is greater than 0.50, so it is not plausible that a majority of U.S. adults have traveled internationally.

23. (a) No. Because 12 is one of the plausible values for the true mean volume (11.97, 12.05), there is not convincing evidence that the true mean volume is different than 12 ounces. **(b)** No. Although 12 is one plausible value for the true mean volume (11.97, 12.05), there are many other plausible values besides 12 in the confidence interval. Because any of these values could be the true mean, there is not convincing evidence that the true mean volume is 12 ounces.
25. Not confident. While we can say that we are 95% confident that the *interval* from 0.436 to 0.478 captures the true proportion of U.S. adults who have traveled internationally, it is very unlikely that the true proportion of U.S. adults who have traveled internationally is *exactly* 0.457 due to sampling variability.
27. Here are parallel boxplots.

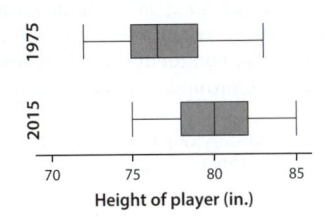

Shape: The distribution of height for the 1975 players appears to be slightly skewed to the right, while the distribution of height for 2015 is fairly symmetric. *Outliers:* Neither distribution has outliers. *Center:* The median height in 1975 (76.5 inches) is less than the median height in 2015 (80 inches). *Variability:* The variability in height is similar for both distributions. (In 1975, range = 11 and $IQR = 4$. In 2015, range = 10 and $IQR = 4$.)

Answers to Section 7.2 Exercises

1. To interpret the confidence level C, say, "If we were to select many random samples from a population and construct a $C\%$ confidence interval using each sample, about $C\%$ of the intervals would capture the [parameter in context]."
3. The benefit of decreasing the confidence level is that the margin of error will decrease. The drawback is that we would have less confidence that we have captured the population parameter.
5. True
7. If the researcher selects many random samples of adults in New York City and constructs a 90% confidence interval for each sample, about 90% of these intervals would capture the true mean amount of time spent doing moderate exercise per week for all adults in New York City.
9. If the Pew Research Center selects many random samples of U.S. adults and constructs a 95% confidence interval for each sample, about 95% of these intervals would capture the true proportion of all U.S. adults who have used an online dating website or app.
11. (a) Wider, because increasing the confidence level increases the margin of error. **(b)** Narrower, because increasing the sample size typically decreases the margin of error.
13. (a) Narrower, because decreasing the confidence level decreases the margin of error. **(b)** Wider, because decreasing the sample size typically increases the margin of error.
15. (a) The researchers included a margin of error with their estimate to account for sampling variability, which is the fact that different samples of 3806 adults in New York City will give different estimates. **(b)** The margin of error does not account for response bias. Some people may exaggerate the number of minutes per week they spend doing moderate exercise.
17. Answers may vary. The wording of the question may affect the response people give about whether or not they have used an online dating website or app. Certain phrasings may embarrass the survey respondent into saying "no," even if the answer is really "yes."
19. (a) If we select many random samples of employed California adults and construct a 90% confidence interval for each sample, about 90% of these intervals would capture the average travel time to work for all employed California adults. **(b)** Decrease the confidence

level. Drawback: We can't be as confident that our interval will capture the true mean. Increase the sample size. Drawback: Larger samples cost more time and money to obtain. **(c)** People who have longer travel times to work might have less time to respond to a survey. This would cause our estimate from the sample to be less than the true mean travel time to work. The bias due to nonresponse is not accounted for by the margin of error, because the margin of error accounts for only variability we expect from random sampling.
21. 4 of the 25 confidence intervals (16%) did not capture the true parameter. So, 84% of the intervals actually did capture the true parameter, which suggests that these were 80% or 90% confidence intervals.
23. (a) Answers may vary, but should be close to 95%. **(b)** Answers may vary, but should be close to 95%. **(c)** Answers may vary, but should be close to 95%. **(d)** No. A 95% confidence level will have a capture rate of about 95%, regardless of the sample size. As the sample size increases, the confidence intervals become narrower, but the capture rate (determined by the confidence level) stays about the same.
25. (a) *Shape:* Approximately normal because $np = 120(0.898) = 107.76 \geq 10$ and $n(1-p) = 120(1-0.898) = 12.24 \geq 10$

Center: $\mu_{\hat{p}} = 0.898$ *Variability:* $\sigma_{\hat{p}} = \sqrt{\dfrac{0.898(1-0.898)}{120}} = 0.028$

(b) (i) $z = \dfrac{0.95 - 0.898}{0.028} = 1.857$; *Tech:* $P(Z > 1.857) =$ Applet/ normalcdf(lower: 1.857, upper: 1000, mean: 0, SD: 1) $= 0.0317$; *Table A:* $P(Z \geq 1.86) = 1 - 0.9686 = 0.0314$ (ii) *Tech:* $P(\hat{p} \geq 0.95) =$ Applet/normalcdf(lower: 0.95, upper: 1000, mean: 0.898, SD: 0.028) $= 0.0316$ **(c)** Let $X =$ the number of residents who have earned a high school diploma. The random variable X has a binomial distribution with $n = 20$ and $p = 0.898$. $P(X \geq 19) = 1 - $ binomcdf$(n = 20, p = 0.898, x$ value $= 18) = 0.3805$

Answers to Section 7.3 Exercises

1. The Random condition states that the data must come from a random sample from the population of interest.
3. True. When the Large Counts condition is violated, the actual capture rate will typically be smaller than the stated capture rate.
5. standard error
7. *Random?* Random sample of U.S. adults. ✓ *Large Counts?* $n\hat{p} = 560 \geq 10$ and $n(1-\hat{p}) = 2186 \geq 10$. ✓ The conditions for inference are met.
9. *Random?* We do not know if the 183 days are a random sample from the population of all days in 2020. *Large Counts?* $n\hat{p} = 2 < 10$, so the sampling distribution of $\hat{p}$ is not approximately normal. The conditions for inference are not met.
11. $z^* = 2.326$
13. $z^* = 1.812$
15. $0.204 \pm 2.576\sqrt{\dfrac{0.204(1-0.204)}{2746}}$; 0.184 to 0.224

17. $0.938 \pm 1.960\sqrt{\dfrac{0.938(1-0.938)}{1012}}$; 0.923 to 0.953

19. (a) *Random?* Random sample of customers of a major U.S. cell-phone carrier. ✓ *Large Counts?* $n\hat{p} = 170 \geq 10$ and $n(1-\hat{p}) = 568 \geq 10$. ✓
(b) $0.230 \pm 1.645\sqrt{\dfrac{0.230(1-0.230)}{738}}$; 0.205 to 0.255 **(c)** We are 90% confident that the interval from 0.205 to 0.255 captures the proportion of all customers of this major U.S. cell-phone carrier who have walked into something or someone while talking on a cell phone.
21. $SE_{\hat{p}} = \sqrt{\dfrac{0.23(1-0.23)}{738}} = 0.0155$. In SRSs of size $n = 738$, the sample proportion of customers who have walked into something or someone while talking on a cell phone will typically vary from the true proportion by about 0.0155.

23. (a) We do not know the sample sizes for the men and for the women. **(b)** The margin of error for women alone would be greater than 0.03 because the sample size for women alone is less than 1019.
25. (a) $P(\text{roll doubles 3 times in a row}) = \left(\dfrac{1}{6}\right)\left(\dfrac{1}{6}\right)\left(\dfrac{1}{6}\right) = \dfrac{1}{216}$ **(b)** The friend misapplied the Law of Large Numbers to a small number of repetitions. Even after rolling doubles three turns in a row, there is still the same 1/6 probability of getting doubles on the next roll. **(c)** Because each roll is independent, the probability of rolling doubles on the fourth roll is 1/6.

Answers to Section 7.4 Exercises

1. State, Plan, Do, Conclude
3. Use a guess for $\hat{p}$ based on a preliminary study or past experience with similar studies or use $\hat{p} = 0.5$ as the guess.
5. S: 95% CI for $p =$ the proportion of all U.S. adults who would answer their cell phone when an unknown number calls. **P:** One-sample z interval for p. *Random?* Random sample of 10,211 U.S. adults. ✓ *Large Counts?* $1940 \geq 10$ and $8271 \geq 10$. ✓ **D:** $0.19 \pm 1.96\sqrt{\dfrac{0.19(1-0.19)}{10,211}}$; 0.182 to 0.198; *Tech:* 0.182 to 0.198. **C:** We are 95% confident that the interval from 0.182 to 0.198 captures the proportion of all U.S. adults who would answer their cell phone when an unknown number calls.
7. S: 95% CI for $p =$ the true proportion of organic produce items that have chemical residue. **P:** One-sample z interval for p. *Random?* Random sample of 409 items. ✓ *Large Counts?* $87 \geq 10$ and $322 \geq 10$. ✓ **D:** $0.213 \pm 1.960\sqrt{\dfrac{0.213(1-0.213)}{409}}$; 0.173 to 0.253; *Tech:* 0.173 to 0.252.
C: We are 95% confident that the interval from 0.173 to 0.252 captures the true proportion of organic produce items that have chemical residue.

9. $1.645\sqrt{\dfrac{0.75(1-0.75)}{n}} \leq 0.04$, $317.11 \leq n$. We need to survey at least 318 Americans of Italian descent.

11. Use $\hat{p} = 0.5$. $1.960\sqrt{\dfrac{0.5(1-0.5)}{n}} \leq 0.03$, $1067.11 \leq n$. We need to survey at least 1068 registered voters.
13. (a) S: 95% CI for $p =$ the proportion of all baseball fans in Connecticut who would say their favorite team is the Yankees. **P:** One-sample z interval for p. *Random?* Random sample of 803 baseball fans in Connecticut. ✓ *Large Counts?* $353 \geq 10$ and $450 \geq 10$. ✓

D: $0.44 \pm 1.960\sqrt{\dfrac{0.44(1-0.44)}{803}}$; 0.406 to 0.474; *Tech:* 0.405 to 0.474.

C: We are 95% confident that the interval from 0.405 to 0.474 captures the proportion of all baseball fans in Connecticut who would say their favorite team is the Yankees. **(b)** No. Because all the plausible values for the population proportion (0.406, 0.474) are less than 0.50, it is not plausible that more than 50% (a majority) of all baseball fans in Connecticut would say their favorite team is the Yankees.

(c) $1.960\sqrt{\dfrac{0.44(1-0.44)}{n}} \leq 0.02$, $2366.43 \leq n$. We need to survey at least 2367 baseball fans in Connecticut. This would require an *additional* $2367 - 803 = 1564$ baseball fans in Connecticut to be randomly selected.
15. The sample size would need to be multiplied by 4 to cut the margin of error in half. A sample of $4(409) = 1636$ is needed.
17. S: 90% CI for $p =$ the proportion of all living trees on New York City streets that are next to damaged sidewalks. **P:** One-sample z interval for p. *Random?* Random sample of 7858 trees on New York

City streets. ✓ *Large Counts?* $2221 \geq 10$ and $5637 \geq 10$. ✓ **D:** *Tech:* 0.274 to 0.291. **C:** We are 90% confident that the interval from 0.274 to 0.291 captures the proportion of all living trees on New York City streets that are next to damaged sidewalks.

19. (a) $0.01 = z^* \sqrt{\dfrac{0.6341(0.3659)}{5594}}$; thus, $z^* = 1.55$. The area between -1.55 and 1.55 under the standard normal curve is 0.8789. The confidence level is about 88%. **(b)** Athletes may not have been truthful in their responses (response bias).

21. $\hat{p} = \dfrac{0.5382 + 0.6025}{2} = 0.57035$; margin of error $= 0.6025 - 0.57035 = 0.03215$; $z^* = 1.960$; $0.03215 = 1.960 \sqrt{\dfrac{0.57035(1-0.57035)}{n}}$, $n = 910.765 \to 911$

A sample size of 911 was used to calculate this interval.

23. (a) *Mean:* $\mu_{\bar{x}} = \$290{,}000$; *SD:* $\sigma_{\bar{x}} = \dfrac{145{,}000}{\sqrt{100}} = \$14{,}500$;

Shape: Because $n = 100 \geq 30$, the sampling distribution of $\bar{x}$ is approximately normal. **(i)** $z = \dfrac{325{,}000 - 290{,}000}{14{,}500} = 2.414$;

Tech: $P(Z > 2.414) =$ Applet/normalcdf(lower: 2.414, upper: 1000, mean: 0, SD: 1) $= 0.0079$; *Table A:* $P(Z > 2.41) = 1 - 0.9920 = 0.0080$ **(ii)** *Tech:* $P(\bar{x} > 325{,}000) =$ Applet/normalcdf(lower: 325,000, upper: 1,000,000, mean: 290,000, SD: 14,500) $= 0.0079$. **(b)** Because $n = 5 < 30$ and the population distribution is skewed to the right, the sampling distribution of $\bar{x}$ is skewed to the right, although not as strongly as the population distribution.

Answers to Section 7.5 Exercises

1. population; 30
3. σ; t^*
5. False. Round down to the largest degrees of freedom listed in Table B that is less than the actual degrees of freedom.
7. *Random?* Random sample of 1025 U.S. adults. ✓ *Normal/Large Sample?* $n = 1025 \geq 30$. ✓
9. *Random?* This is not a random sample, so this condition is not met. Inference is not needed here because we have the age at death for the entire population (all U.S. presidents who have died). *Normal/Large Sample?* $n = 40 \geq 30$. ✓ (There are 40 presidents who have died as of the writing of this solution.)
11. (a) Using df $= 10 - 1 = 9$ gives $t^* = 2.262$. **(b)** *Using tech:* With df $= 77 - 1 = 76$, $t^* = 1.665$; *Using Table B:* With df $= 60$, $t^* = 1.671$.
13. (a) Using df $= 20 - 1 = 19$ gives $t^* = 2.861$. **(b)** *Using tech:* With df $= 85 - 1 = 84$, $t^* = 1.989$; *Using Table B:* With df $= 80$, $t^* = 1.990$.
15. *Using Table B:* With df $= 80$, $356.1 \pm 2.639 \dfrac{185.7}{\sqrt{92}}$; 305.007 to 407.193 licks; *Tech:* 305.164 to 407.036 using df $= 91$
17. (a) *Using Table B:* With df $= 1000$, $3.8 \pm 1.962 \dfrac{6.95}{\sqrt{1025}}$; 3.374 to 4.226 events attended; *Tech:* 3.374 to 4.226 using df $= 1024$. **(b)** If Gallup selects many random samples of adults and constructs a 95% confidence interval for each sample, about 95% of these intervals would capture the mean number of live music and theater events attended in a given year for all adults.
19. (a) *Random?* Random sample of 30 pieces of fishing line. ✓ *Normal/Large Sample?* $n = 30 \geq 30$. ✓ **(b)** $6.44 \pm 2.756 \dfrac{0.75}{\sqrt{30}}$; 6.063 to 6.817 pounds; *Tech:* 6.063 to 6.817 using df $= 29$ **(c)** Yes. Because all of the plausible values for the true mean (6.063, 6.817) are different than 6, there is convincing evidence that the true mean breaking strength is different than 6 pounds.

21. $\text{SE}_{\bar{x}} = \dfrac{9.3}{\sqrt{27}} = 1.79$; if we select many random samples of size 27, the sample mean systolic blood pressure will typically vary from the true mean by about 1.79.

23. (a) $183{,}100 \pm 1.761 \dfrac{29{,}200}{\sqrt{15}}$ \$169,823 to \$196,377; *Tech:* \$169,820.77 to \$196,379.24 using df $= 14$ **(b)** ($169{,}820.77)(0.01)(300) = \$509{,}462.31$ and ($196{,}379.24)(0.01)(300) = \$589{,}137.72$. A 90% confidence interval for the total amount of tax revenue generated by 300 new homes at a rate of 1% per home is \$509,462.31 to \$589,137.72.

25. $24.1 \pm 1.96 \dfrac{0.5}{\sqrt{30}}$; 23.921 to 24.279

27. (a) *B?* "Success" = person looks forward to opening their mailbox each day. "Failure" = person does not look forward to opening their mailbox each day. *I?* Knowing whether or not one person looks forward to opening their mailbox each day tells you nothing about whether or not another randomly selected person looks forward to opening their mailbox each day. *N? n* $= 50$ *S? p* $= 0.41$ **(b)** $\mu_X = 50(0.41) = 20.5$ and $\sigma_X = \sqrt{50(0.41)(1-0.41)} = 3.48$. If many random samples of size 50 were taken, the number of people who say they look forward to opening their mailbox each day would typically vary from the mean by about 3.48 people. **(c)** The mean of the simulated samples (20.46) is very close to the calculated mean (20.5). Also, the standard deviation of the simulated samples (3.54) is very close to the calculated standard deviation (3.48).

Answers to Section 7.6 Exercises

1. Examine a graph of the sample data. Confirm that the graph shows no strong skewness or outliers.
3. dotplot, stemplot, boxplot, histogram, normal probability plot
5. Yes. $n = 100 \geq 30$. ✓
7. No. The sample size is small ($n = 20 < 30$) and the boxplot shows an outlier.
9. S: 90% CI for μ = the mean amount of iridium in all rock samples at a depth of 347 m in Gubbio, Italy. **P:** One-sample t interval for μ. *Random?* Random sample of 5 rock samples. ✓ *Normal/Large Sample?* The sample size is small, but the dotplot doesn't show any outliers or strong skewness. ✓

D: $589 \pm 2.132 \dfrac{227.002}{\sqrt{5}}$; 372.563 to 805.437; *Tech:* 372.578 to 805.422 with df $= 4$. **C:** We are 90% confident that the interval from 372.578 ppb to 805.442 ppb captures the mean amount of iridium in all rock samples at a depth of 347 m in Gubbio, Italy.

11. S: 99% CI for μ = the true mean healing rate (micrometers per hour) of newts. **P:** One-sample t interval for μ. *Random?* Random sample of 18 newts. ✓ *Normal/Large Sample?* The sample size is small, but the dotplot doesn't show any outliers or strong skewness. ✓

D: $25.667 \pm 2.898 \dfrac{8.324}{\sqrt{18}}$; 19.981 to 31.353; *Tech:* 19.980 to 31.353 with df $= 17$. **C:** We are 99% confident that the interval from 19.980 micrometers/hour to 31.353 micrometers/hour captures the true mean healing rate of newts.

13. S: 90% CI for μ = the mean tension of all the screens produced on this day. **P:** One-sample t interval for μ. *Random?* Random sample of 20 screens. ✓ *Normal/Large Sample?* The sample size is small, but the dotplot doesn't show any outliers or strong skewness. ✓

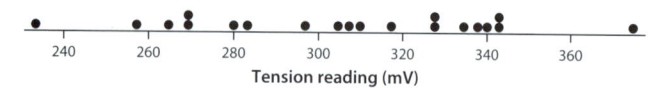

D: $306.32 \pm 1.729 \dfrac{36.209}{\sqrt{20}}$; 292.321 to 320.319; Tech: 292.32 to 320.32 with df = 19. C: We are 90% confident that the interval from 292.32 mV to 320.32 mV captures the mean tension of all the screens produced on this day.

15. No. Because 300 is one of the plausible values for the population mean (292.32, 320.32), it is a plausible value for the mean tension for all screens produced this day. We do not have convincing evidence that the screens do not meet the manufacturer's goal.

17. S: 90% CI for μ = the mean distance that all workers in Boston commute. **P:** One-sample t interval for μ. *Random?* Random sample of 500 workers in Boston. ✓ *Normal/Large Sample?* $n = 500 \geq 30$ ✓ **D:** *Tech:* 11.180 to 12.896 with df = 499. **C:** We are 90% confident that the interval from 11.180 miles to 12.896 miles captures the mean distance that all workers in Boston commute.

19. Because we are told the population standard deviation, use z^* instead of t^*. $2.576 \dfrac{7.5}{\sqrt{n}} \leq 1$; $373.26 \leq n$, select an SRS of 374 individuals.

21. 4 out of the 8 values (only 50%) are within the 95% confidence interval 16.49 to 28.51 mg/100 g. This is because the confidence interval estimates the *mean* vitamin C level and does not tell us anything about *individual* values. Because means are less variable than individual values, we would expect that fewer than 95% of individual values would lie inside the 95% confidence interval for the mean.

23. (a) $_{44}C_6 = 7{,}059{,}052$ ways

(b) $P(\text{winning the jackpot}) = \dfrac{1}{7{,}059{,}052} = 0.000000142$

(c) $(_6C_5)(_{38}C_1) = 228$ ways

(d) $P(\text{winning \$2000}) = \dfrac{228}{7{,}059{,}052} = 0.000032$

Answers to Section 7.7 Exercises

1. False. There is no Large Sample override to the Normal condition for the sampling distribution of the sample variance.

3. For a 90% confidence interval, use $p = 0.95$ to find χ_L^2 and $p = 0.05$ to find χ_U^2.

5. *Random?* Random sample of 20 corn plants. ✓ *Normal?* The heights of the population of this variety of corn are normally distributed. ✓

7. *Random?* Random sample of 8 packages from one production run. ✓ *Normal?* The boxplot is not strongly skewed and there are no outliers. ✓

9. (a) $\chi_L^2 = 0.484$ and $\chi_U^2 = 11.143$ **(b)** $\chi_L^2 = 11.591$ and $\chi_U^2 = 32.671$

11. (a) $\chi_L^2 = 6.844$ and $\chi_U^2 = 38.582$ **(b)** *Table C with df = 100:* $\chi_L^2 = 74.222$ and $\chi_U^2 = 129.561$; *Tech with df = 102:* $\chi_L^2 = 75.946$ and $\chi_U^2 = 131.838$

13. $\sqrt{\dfrac{(20-1)(8)^2}{32.852}} \leq \sigma \leq \sqrt{\dfrac{(20-1)(8)^2}{8.907}}$; 6.084 inches to 11.684 inches

15. (a) $\sqrt{\dfrac{(8-1)(7.19)^2}{14.067}} \leq \sigma \leq \sqrt{\dfrac{(8-1)(7.19)^2}{2.167}}$; 5.072 mg/100 g to 12.923 mg/100 g. **(b)** If many random samples of packages from the production run are selected and a 90% confidence interval is constructed for each sample, about 90% of these intervals would capture the true standard deviation of the amount of vitamin C in all packages from that day's production run.

17. (a) *Random?* Random sample of 30 pieces of fishing line. ✓ *Normal?* The dotplot is not strongly skewed and there are no outliers. ✓

(b) $\sqrt{\dfrac{(30-1)(0.75)^2}{52.336}} \leq \sigma \leq \sqrt{\dfrac{(30-1)(0.75)^2}{13.121}}$; 0.558 pound to 1.115 pounds **(c)** No. Because some of the plausible values for the true standard deviation (0.558, 1.115) are greater than 1 pound, the interval does not provide convincing evidence that the true standard deviation is less than 1 pound. It is plausible that the true standard deviation is as small as 0.558 pound and as great as 1.115 pounds.

19. (a) $\dfrac{(20-1)(8)^2}{32.852} \leq \sigma^2 \leq \dfrac{(20-1)(8)^2}{8.907}$; 37.014 in.^{2} to 136.522 in.^{2} **(b)** The units for this confidence interval are inches^2. **(c)** The interval for the variance is longer than the interval for the standard deviation because the variance is the square of the standard deviation; so, the bounds of the confidence interval for the population variance are the square of the bounds of the confidence interval for the population standard deviation.

21. (a) $\chi_L^2 = 52.6$ and $\chi_U^2 = 91.67$ **(b)** $\chi_L^2 = 16.501$ and $\chi_U^2 = 58.964$

23. $P(\text{3 aces and 2 kings}) \dfrac{(_4C_3)(_4C_2)}{_{52}C_5} = \dfrac{24}{2{,}598{,}960} = 0.00000923$

You should be very surprised to receive 3 aces and 2 kings from a well-shuffled deck because it is extremely unlikely that this hand would be dealt just by chance.

Answers to Section 7.8 Exercises

1. False. To satisfy the Normal condition, a graph of the sample data must simply show no strong skewness or outliers to plausibly come from a population that is normally distributed.

3. Standard deviations are easier to interpret and have the same units as the data.

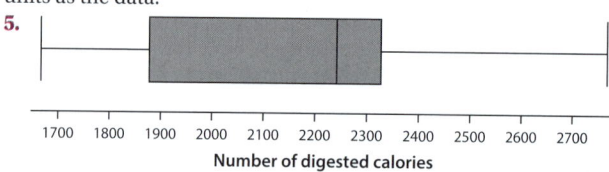

5.

The boxplot shows no strong skewness or outliers. It is plausible that these data came from a normal population.

7. The histogram of the sample data shows strong right-skewness and possibly a few high outliers, so the Normal condition is not met.

9. S: 95% CI for σ = the standard deviation of tail length in all red-tail hawks migrating through the area. **P:** One-sample confidence interval for σ. *Random?* Random sample of 10 red-tail hawks. ✓ *Normal?* A boxplot of the sample data shows no strong skewness or outliers. ✓

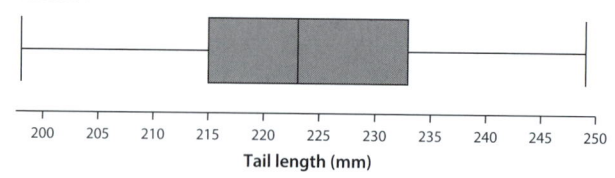

D: *Tech:* $\sqrt{\dfrac{(9)(14.356)^2}{19.023}} \leq \sigma \leq \sqrt{\dfrac{(9)(14.356)^2}{2.7}}$; $9.875 \leq \sigma \leq 26.210$. **C:** We are 95% confident that the interval from 9.875 mm to 26.210 mm captures the standard deviation of tail length in all red-tail hawks migrating through the area.

11. S: 99% CI for σ = the standard deviation of weight for all crabs in the population. **P:** One-sample confidence interval for σ. *Random?* Random sample of 34 crabs. ✓ *Normal?* The histogram of the sample data shows no strong skewness or outliers. ✓

D: *Tech:* $\sqrt{\dfrac{(33)(15.36)^2}{57.648}} \leq \sigma \leq \sqrt{\dfrac{(33)(15.36)^2}{15.815}}$; $11.621 \leq \sigma \leq 22.188$. **C:** We are 99% confident that the interval from 11.621 g to 22.188 g captures the standard deviation of weight for all crabs in the population.

13. S: 95% CI for σ = the standard deviation of amount of iridium for all rock samples at a depth of 347 m. **P:** One-sample confidence interval for σ. *Random?* Random sample of 5 rock samples. ✓ *Normal?* The dotplot of the sample data shows no strong skewness or outliers. ✓

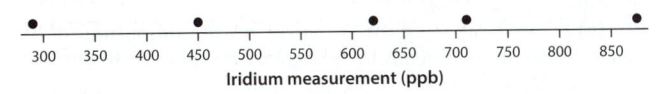

D: $\sqrt{\dfrac{(4)(227.002)^2}{11.143}} \leq \sigma \leq \sqrt{\dfrac{(4)(227.002)^2}{0.484}}$; $136.006 \leq \sigma \leq 652.585$. **C:** We are 95% confident that the interval from 136.006 ppb to 652.585 ppb captures the standard deviation of amount of iridium for all rock samples at a depth of 347 m.

15. No, there is not convincing evidence that the true standard deviation is not 200 ppb because 200 ppb is within the 95% confidence interval.

17. S: 99% CI for σ = the standard deviation of carbon dioxide readings. **P:** One-sample confidence interval for σ. *Random?* Random sample of 100 carbon dioxide readings. ✓ *Normal?* The dotplot of the sample data shows no strong skewness or outliers. ✓

D: $\sqrt{\dfrac{(99)(14.347)^2}{138.987}} \leq \sigma \leq \sqrt{\dfrac{(99)(14.347)^2}{66.51}}$; $12.109 \leq \sigma \leq 17.504$. **C:** We are 99% confident that the interval from 12.109 ppm to 17.504 ppm captures the standard deviation of carbon dioxide readings.

19. S: 95% CI for σ^2 = the variance of tail length in all red-tail hawks migrating through the area. **P:** One-sample confidence interval for σ^2. *Random?* Random sample of 10 red-tail hawks. ✓ *Normal?* A boxplot of the sample data shows no strong skewness or outliers. ✓

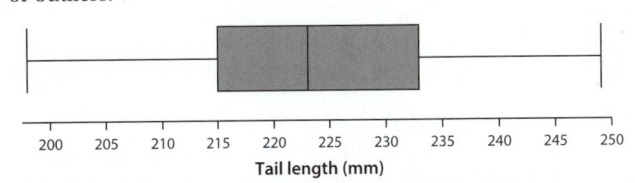

D: $\dfrac{(9)(14.356)^2}{19.023} \leq \sigma^2 \leq \dfrac{(9)(14.356)^2}{2.7}$; $97.506 \leq \sigma^2 \leq 686.982$. **C:** We are 95% confident that the interval from 97.506 mm² to 686.982 mm² captures the variance of tail length in all red-tail hawks migrating through the area.

21. (a) Parameter. The average length of stay describes the population of all cats at the shelter. **(b)** Statistic. The proportion of owners who say "yes" is a number that describes the sample of people on the list.

Answers to Chapter 7 Review Exercises

1. (a) We are 95% confident that the interval from 0.295 to 0.365 captures the true proportion of all U.S. residents who support salaries for college athletes. **(b)** If the pollsters select many random samples of U.S. residents and construct a 95% confidence interval for each sample, about 95% of these intervals would capture the true proportion of all U.S. residents who support salaries for college athletes. **(c)** point estimate = $\dfrac{0.295 + 0.365}{2} = 0.33$ and margin of error = $0.365 - 0.33 = 0.035$ **(d)** Decreasing the confidence level. Drawback: They can't be as confident that their interval will capture the true proportion. Increasing the sample size. Drawback: Larger samples cost more time and money to obtain.

2. (a) $z^* = 1.881$ **(b)** Using $df = 8 - 1 = 7$ gives $t^* = 3.499$. **(c)** *Using tech:* With $df = 95 - 1 = 94$, $t^* = 1.661$; *Using Table B:* With $df = 80$, $t^* = 1.664$.

3. (a) *Random?* Random sample of 1000 U.S. adults. ✓ *Large Counts?* $n\hat{p} = 50 \geq 10$ and $n(1 - \hat{p}) = 950 \geq 10$. ✓

(b) $0.05 \pm 1.960\sqrt{\dfrac{0.05(1 - 0.05)}{1000}}$; 0.036 to 0.064; *Tech:* 0.036 to 0.064

(c) Yes. Because all of the plausible values for the true proportion (0.036, 0.064) are less than 1/3, there is convincing evidence that fewer than one-third of U.S. adults would answer this question correctly.

4. Response bias. Students might lie and answer "no" because they are being surveyed by the dean, and they don't want to get in trouble.

5. Use $\hat{p} = 0.5$, $2.576\sqrt{\dfrac{0.5(1 - 0.5)}{n}} \leq 0.05$, $663.58 \leq n$. We need to sample at least 664 students.

6. (a) *Random?* Random sample of 2020 adults. ✓ *Normal/Large Sample?* $n = 2020 \geq 30$. ✓ **(b)** $78 \pm 1.646 \dfrac{81}{\sqrt{2020}}$; $75.03 to $80.97; *Tech:* $75.03 to $80.97 using $df = 2019$ **(c)** We are 90% confident that the interval from $75.03 to $80.97 captures the mean cost of all real Christmas trees bought in 2018.

7. S: 95% CI for μ = the true mean fuel efficiency (mpg) for this model of pickup truck. **P:** One-sample t interval for μ. *Random?* Randomly selected 20 owners. ✓ *Normal/Large Sample?* The sample size is small, but the dotplot doesn't show any outliers or strong skewness. ✓

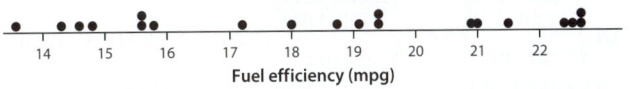

D: $18.48 \pm 2.093 \dfrac{3.116}{\sqrt{20}}$; 17.022 to 19.938; *Tech:* 17.022 to 19.938 using $df = 19$. **C:** We are 95% confident that the interval from 17.022 mpg to 19.938 mpg captures the true mean fuel efficiency for this model of pickup truck.

8. S: 95% CI for σ = the standard deviation of fuel efficiency for all pickup trucks of this model. **P:** One-sample confidence interval for σ. *Random?* Randomly selected 20 owners. ✓ *Normal?* The dotplot of the sample data shows no strong skewness or outliers. ✓

D: $\sqrt{\dfrac{(19)(3.116)^2}{32.852}} \leq \sigma \leq \sqrt{\dfrac{(19)(3.116)^2}{8.907}}$; $2.370 \leq \sigma \leq 4.551$. **C:** We are 95% confident that the interval from 2.370 mpg to 4.551 mpg captures the standard deviation of fuel efficiency values for all pickup trucks of this model.

9. (a) S: 95% CI for p = the proportion of all U.S. adults who favor making Washington, D.C., a separate state. **P:** One-sample z interval for p. *Random?* Random sample of 1018 U.S. adults. ✓ *Large Counts?* $295 \geq 10$ and $723 \geq 10$. ✓ **D:** $0.29 \pm 1.960\sqrt{\dfrac{0.29(1 - 0.29)}{1018}}$; 0.262 to 0.318; *Tech:* 0.262 to 0.318. **C:** We are 95% confident that the interval from 0.262 to 0.318 captures the proportion of all U.S. adults who favor making Washington, D.C., a separate state. **(b)** If Gallup selects many random samples of U.S. adults and constructs a 95% confidence interval for each sample, about 95% of these intervals would capture the true proportion of all U.S. adults who favor making Washington, D.C., a separate state.

10. (a) S: 90% CI for μ = the true mean pH for this brand of bottled water from this grocery store. **P:** One-sample t interval for μ.

Random? Random sample of 18 bottles of water. ✓ *Normal/Large Sample?* The sample size is small, but the dotplot doesn't show any outliers or strong skewness. ✓

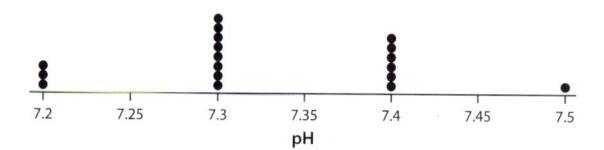

D: $7.328 \pm 1.740 \frac{0.083}{\sqrt{18}}$; 7.294 to 7.362; *Tech:* 7.294 to 7.362 using df = 17. **C:** We are 90% confident that the interval from 7.294 to 7.362 captures the true mean pH for this brand of bottled water from this grocery store. **(b)** Yes. Because all of the plausible values for the true mean (7.294, 7.362) are different than 7.4, there is convincing evidence that the true mean pH is different from 7.4. **(c) S:** 90% CI for σ = the standard deviation of pH level for this brand of bottled water from this grocery store. **P:** *Random?* Random sample of 18 bottles of water. ✓ *Normal/Large Sample?* The sample size is small, but the dotplot doesn't show any outliers or strong skewness. ✓

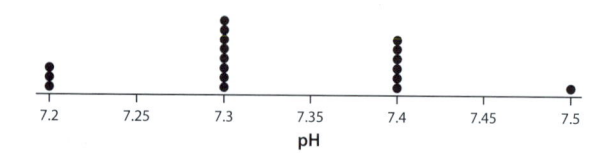

D: $\sqrt{\frac{(17)(0.083)^2}{27.587}} \leq \sigma \leq \sqrt{\frac{(17)(0.083)^2}{8.672}}$; $0.065 \leq \sigma \leq 0.116$. **C:** We are 90% confident that the interval from 0.065 to 0.116 captures the standard deviation of pH level for this brand of bottled water from this grocery store.

CHAPTER 8
Answers to Section 8.1 Exercises

1. claim; parameter
3. null; alternative
5. The *P*-value of a test is the probability, assuming the null hypothesis H_0 is true, of getting evidence for the alternative hypothesis H_a as strong or stronger than the observed evidence by chance alone.
7. $H_0: \mu = 31, H_a: \mu > 31$, where μ = the true mean weight (in ounces) of pineapples grown in the field this year.
9. $H_0: p = 0.10, H_a: p \neq 0.10$, where p = the proportion of all students at the community college who are left-handed.
11. $H_0: \sigma = 3, H_a: \sigma > 3$, where σ = the true standard deviation of the temperature (in °F) in the Starnes family cabin.
13. (a) The evidence for H_a is: $\bar{x} = 31.4 > 31$. **(b)** Assuming that the mean weight of all pineapples grown in the field this year is 31 ounces, there is a 0.1317 probability of getting a sample mean of 31.4 ounces or greater by chance alone in a random sample of 50 pineapples.
15. (a) The evidence for H_a is: $\hat{p} = 0.16 \neq 0.10$. **(b)** Assuming that the proportion of all students at the college who are left-handed is 0.10, there is a 0.0455 probability of getting a sample proportion greater than or equal to 0.16 or less than or equal to 0.04 by chance alone in a random sample of 100 students.
17. (a) The evidence for H_a is: $s_x = 5 > 3$. **(b)** Assuming that the standard deviation of the temperature in the cabin is 3°F, there is a 0.003 probability of getting a sample standard deviation of 5°F or greater by chance alone in a random sample of 10 temperature readings.
19. The *P*-value of 0.1317 is not small (less than 0.05), so we fail to reject H_0. We do not have convincing evidence that the mean weight of the pineapples grown this year is greater than 31 ounces.

21. The *P*-value of 0.0455 is small, so we reject H_0. We have convincing evidence that the proportion of all students at this community college who are left-handed is different from 0.10.
23. The *P*-value of 0.003 is small, so we reject H_0. We have convincing evidence that the true standard deviation of the temperature in the cabin is greater than 3°F.
25. (a) $H_0: p = 0.50, H_a: p > 0.50$, where p = the proportion of all couples who tilt their heads to the right when kissing. **(b)** The evidence for H_a is: $\hat{p} = 0.669 > 0.50$. **(c)** Assuming that the proportion of all couples who tilt their heads to the right when kissing is 0.50, there is a 0.0001 probability of getting a sample proportion of 0.669 or greater by chance alone in a random sample of 124 couples. **(d)** The *P*-value of 0.0001 is small, so we reject H_0. We have convincing evidence that more than 50% of all kissing couples prefer to tilt their heads to the right.
27. The hypotheses should be stated in terms of the population parameter, μ, not the sample statistic, $\bar{x}$. $H_0: \mu = 3000, H_a: \mu < 3000$, where μ = the mean birth weight (in grams) of all babies whose mothers did not see a doctor before delivery.
29. (a) $H_0: \mu = 6.9, H_a: \mu > 6.9$, where μ = the mean number of new words for all sonnets in the new manuscript. **(b)** The evidence for H_a is: $\bar{x} = 9.2 > 6.9$. **(c)** Estimated *P*-value = 13/200 = 0.065. Assuming that the mean number of new words for the population of all sonnets in the new manuscript is 6.9, there is an estimated 0.065 probability of getting a sample mean of 9.2 or greater by chance alone in a random sample of 5 sonnets. **(d)** The *P*-value of 0.065 is not small, so we fail to reject H_0. We do not have convincing evidence that the mean number of new words for all sonnets in the new manuscript is greater than 6.9.
31. One mistake is in saying that 95% of other polls would have results within three percentage points *of the results of this survey*. It should say that other polls would have results within three percentage points *of the true proportion*. Another mistake is saying "at least 19" of the 20 surveys will be within three percentage points when it should say "about 19" of the 20 surveys.

Answers to Section 8.2 Exercises

1. statistically significant
3. A Type II error occurs if we fail to reject H_0 when H_a is true. That is, the data do not give convincing evidence that H_a is true when it really is.
5. the significance level, α
7. Because the *P*-value of $0.1012 > \alpha = 0.01$, we fail to reject H_0. We do not have convincing evidence that the true mean copper content of the water from the new source is greater than 1.3 mg/L.
9. (a) Because the *P*-value of $0.0115 < \alpha = 0.05$, we reject H_0. We have convincing evidence that the proportion of all U.S. adults who will answer their cell phone when a call is received from an unknown number is different than 0.20. **(b)** Yes, because $0.0115 > \alpha = 0.01$. At the 1% significance level, we would fail to reject H_0.
11. *Type I:* You find convincing evidence that the true mean income level in this area is greater than $85,000, when it is really equal to $85,000. *Type II:* You do not find convincing evidence that the true mean income level in this area is greater than $85,000, when it really is greater than $85,000.
13. *Type I:* The city manager finds convincing evidence that the true proportion of calls involving life-threatening injuries for which emergency personnel took more than 8 minutes to arrive during this 6-month period is less than 0.22, when it is really equal to 0.22. *Type II:* The city manager doesn't find convincing evidence that the true proportion of calls involving life-threatening injuries for which emergency personnel took more than 8 minutes to arrive during this 6-month period is less than 0.22, when it really is less than 0.22.
15. *Type I:* Mrs. Starnes finds convincing evidence that the true standard deviation of temperatures in the cabin with this thermostat setting is greater than 3°F, when it is really equal to 3°F. *Type II:* Mrs. Starnes doesn't find convincing evidence that the true standard deviation of temperatures in the cabin with this thermostat setting is greater than 3°F, when it really is greater than 3°F.

17. *Consequence of Type I:* You decide to open a restaurant in this area, but there is not enough money in the area and the restaurant fails. *Consequence of Type II:* You decide not to open a restaurant in this area when it would have been successful.
19. (a) *Consequence of Type I:* The city manager concludes that response times have improved when they really haven't, so no further action will be taken and more patients with life-threatening conditions will be at risk. *Consequence of Type II:* The city manager may conclude that the new guidelines were not helpful in getting first responders to emergencies faster, when they really were. The city manager may take further action that costs the city money unnecessarily. **(b)** The Type I error is more serious. The city manager concludes that the response times have improved and does not spend more time and resources working to further improve response times. In reality, the risk for patients with life-threatening conditions has not improved, so more lives are at risk.
21. (a) *Consequence of Type I:* Mrs. Starnes concludes that the temperatures are more variable than $\sigma = 3°F$ when they really are not that variable. She may buy a new thermostat when the current one is working properly. *Consequence of Type II:* Mrs. Starnes concludes that the temperatures have a standard deviation of $3°F$ when they actually vary more than that. She will think the thermostat is working properly when it is not and will keep the defective thermostat. **(b)** The Type II error is more serious. If the temperatures have a standard deviation of more than $3°F$, the cabin temperature may drop too low and the pipes are more susceptible to freezing and bursting.
23. (a) Because the P-value of $0.18 > \alpha = 0.05$, we fail to reject H_0. We do not have convincing evidence that the proportion of all city residents who support a 1% increase in the sales tax to fund road repairs is greater than 0.50. **(b)** Because we failed to reject the null hypothesis, it is possible that we made a Type II error. **(c)** The city council does not implement the increase in the sales tax and the roads do not get repaired, when in reality more than 50% of the city residents support the increase in sales tax.
25. $\alpha = 0.01$. It would decrease the probability of the more serious Type I error.
27. (a) For an observed result to be statistically significant at the 1% level, the P-value must be less than 1%. If the P-value is less than 1%, it must also be less than 5%, and therefore also significant at the 5% level. **(b)** For a result to be statistically significant at the 5% level, the P-value must be less than 5%. This P-value may or may not be less than 1%, so the result may or may not be significant at the 1% level.
29. (a) If the true proportion of patients with Alzheimer's disease who will experience nausea when taking this new drug is 0.07, there is a 0.27 probability that this test will find convincing evidence that the true proportion is less than 0.10. **(b)** We fail to find convincing evidence that the true proportion of patients with Alzheimer's disease who will experience nausea when taking this new drug is less than 0.10, when in fact the true proportion of patients with Alzheimer's disease who will experience nausea when taking this new drug is 0.07. **(c)** $P(\text{Type II error}) = 1 - \text{Power} = 1 - 0.27 = 0.73$ **(d)** We can increase the power either by increasing the sample size (n) or by increasing the significance level (α).
31. (a) $\mu_{\hat{p}} = 0.30$, $\sigma_{\hat{p}} = \sqrt{\frac{0.30(1-0.30)}{50}} = 0.065$. In SRSs of size $n = 50$, the sample proportion of books that have fewer than 400 pages typically varies from the true proportion by about 0.065. **(b)** Because $np = (50)(0.30) = 15 \geq 10$ and $n(1-p) = (50)(1-0.30) = 35 \geq 10$, the sampling distribution of $\hat{p}$ is approximately normal.
(c) (i) $z = \frac{0.40 - 0.30}{0.065} = 1.538$, *Tech:* Applet/normalcdf(lower: 1.538, upper: 1000, mean: 0, SD: 1) = 0.0620; *Table A:* $P(z > 1.54) = 0.0618$
(ii) *Tech:* $P(\hat{p} > 0.40) = $ Applet/normalcdf(lower: 0.40, upper: 1000, mean: 0.3, SD: 0.065) = 0.0620

Answers to Section 8.3 Exercises

1. The Large Counts condition says that both np_0 and $n(1-p_0)$ are at least 10.
3. standardized test statistic $= \dfrac{\text{statistic} - \text{null value}}{\text{standard deviation (error) of statistic}}$
5. *Random?* Random sample of 400 seeds. ✓ *Large Counts?* $np_0 = 400(0.80) = 320 \geq 10$ and $n(1-p_0) = 400(1-0.80) = 80 \geq 10$. ✓
7. *Random?* The data came from a voluntary response sample. This condition is not satisfied. *Large Counts?* $np_0 = 200\left(\dfrac{2}{3}\right) = 133.33 \geq 10$ and $n(1-p_0) = 200\left(1-\dfrac{2}{3}\right) = 66.67 \geq 10$. ✓
9. $z = \dfrac{0.8475 - 0.80}{\sqrt{\dfrac{0.80(1-0.80)}{400}}} = 2.375$
11. (a) *Random?* The administrator selected a random sample of 200 full-time students from the college. ✓ *Large Counts?* $np_0 = 200(0.49) = 98 \geq 10$ and $n(1-p_0) = 200(1-0.49) = 102 \geq 10$. ✓
(b) $z = \dfrac{0.455 - 0.49}{\sqrt{\dfrac{0.49(1-0.49)}{200}}} = -0.990$
13. *Table A:* $1 - 0.9913 = 0.0087$; *Tech:* Applet/normalcdf(lower: 2.375, upper: 1000, mean: 0, SD: 1) = 0.0088
15. (a) *Table A:* 0.1611; *Tech:* Applet/normalcdf(lower: −1000, upper: −0.990, mean: 0, SD: 1) = 0.1611 **(b)** Because the P-value of $0.1611 > \alpha = 0.05$, we fail to reject H_0. We do not have convincing evidence that the proportion of all full-time students from this college who are employed is less than 0.49.
17. (a) $H_0: p = 1/12$, $H_a: p > 1/12$, where $p = $ the true proportion of college students who prefer the last singer they see. Use $\alpha = 0.05$. **(b)** *Random?* Researchers selected a random sample of 600 college students and each student viewed the videos in a random order. ✓ *Large Counts?* $np_0 = 600\left(\dfrac{1}{12}\right) = 50 \geq 10$ and $n(1-p_0) = 600\left(1-\dfrac{1}{12}\right) = 550 \geq 10$. ✓ **(c)** $z = \dfrac{0.0983 - \left(\dfrac{1}{12}\right)}{\sqrt{\dfrac{\dfrac{1}{12}\left(1-\dfrac{1}{12}\right)}{600}}} = 1.326$
(d) *Table A:* $1 - 0.9082 = 0.0918$; *Tech:* Applet/normalcdf(lower: 1.326, upper: 1000, mean: 0, SD: 1) = 0.0924. Because the P-value of $0.0924 > \alpha = 0.05$, we fail to reject H_0. We do not have convincing evidence that the true proportion of college students who prefer the last singer they see is greater than $1/12$ (i.e., that there is an advantage to performing last).
19. Because we failed to reject the null hypothesis, it is possible that we could have made a Type II error.
21. (a) Assuming that the proportion of all seeds of this variety of corn that will germinate is 0.80, there is a 0.0088 probability of getting a sample proportion of 0.8475 or greater by chance alone in a random sample of 400 seeds. **(b)** Because the P-value of $0.0088 < \alpha = 0.01$, we reject H_0. We have convincing evidence that the proportion of all seeds of this variety of corn that will germinate is greater than 0.80.
23. (a) invNorm(area: 0.99, mean: 0, SD: 1) = 2.326. A value of $z > 2.326$ will lead us to reject H_0. **(b)** Because $z = 2.846 > 2.326$, we reject H_0. We have convincing evidence that the proportion of all people who would identify the green cup of tea as having the more natural flavor is greater than 0.50.

25. (a) It is reasonable to use the binomial distribution for probability calculations involving X because the following conditions are met: *Binary?* "Success" = person is younger than 18 years. "Failure" = person is not younger than 18 years. *Independent?* Knowing whether or not one person is younger than age 18 tells you nothing about whether or not another person is younger than age 18. *Number?* $n = 10$. *Same probability?* $p = 0.224$. **(b)** Because $np = 10(0.224) = 2.24 < 10$ and $n(1-p) = 10(1-0.224) = 7.76 < 10$, the probability distribution of X is not approximately normal. **(c)** $P(X \leq 1) = $ binomcdf(trials: 10, p: 0.224, x-value: 1) $= 0.308$ **(d)** It is likely (probability = 0.308) that a random sample of 10 people would contain at most 1 person younger than age 18 by chance alone. Therefore, because $0.308 > \alpha = 0.05$, the researchers should fail to reject H_0. There is not convincing evidence that the proportion of all North Carolina residents who are younger than age 18 is less than 0.224.

Answers to Section 8.4 Exercises

1. State hypotheses, parameter(s), significance level, and evidence for H_a.

3. both tails

5. S: $H_0: p = 0.10, H_a: p < 0.10$, where $p = $ the proportion of all patients with Alzheimer's disease who would experience nausea from the new drug. Use $\alpha = 0.01$. The evidence for H_a is: $\hat{p} = 25/300 = 0.0833 < 0.10$. **P:** One-sample z test for p. *Random?* Random sample of 300 patients. ✓ *Large Counts?* $np_0 = 30 \geq 10$ and $n(1-p_0) = 270 \geq 10$. ✓ **D:** $z = \dfrac{0.0833 - 0.10}{\sqrt{\dfrac{0.10(1-0.10)}{300}}} = -0.964$; *Table A:* 0.1685; *Tech:* $z = -0.962$, *P*-value $= 0.1680$ **C:** Because the *P*-value of $0.1680 > \alpha = 0.01$, we fail to reject H_0. We do not have convincing evidence to support the drug manufacturer's claim that the proportion of all patients with Alzheimer's disease who would experience nausea from the new drug is less than 0.10.

7. S: $H_0: p = 0.60, H_a: p > 0.60$, where $p = $ the proportion of all students at this college who have engaged in sexting. Use $\alpha = 0.05$. The evidence for H_a is: $\hat{p} = 0.668 > 0.60$. **P:** One-sample z test for p. *Random?* SRS of 250 students. ✓ *Large Counts?* $np_0 = 150 \geq 10$ and $n(1-p_0) = 100 \geq 10$. ✓ **D:** $z = \dfrac{0.668 - 0.6}{\sqrt{\dfrac{0.6(1-0.6)}{250}}} = 2.1947$; *Table A:* $1 - 0.9857 = 0.0143$; *Tech:* $z = 2.1947$, *P*-value $= 0.0141$ **C:** Because the *P*-value of $0.0141 < \alpha = 0.05$, we reject H_0. We have convincing evidence that the proportion of all students at this college who have engaged in sexting is greater than 0.6.

9. $z = \dfrac{0.7375 - 0.7}{\sqrt{\dfrac{0.7(1-0.7)}{80}}} = 0.732$, *Table A:* $2(0.2327) = 0.4654$; *Tech:* $z = 0.732$, *P*-value $= 0.4642$

11. (a) *Random?* A random sample was selected. ✓ *Large Counts?* $np_0 = 60(0.25) = 15 \geq 10$ and $n(1-p_0) = 60(1-0.25) = 45 \geq 10$. ✓ **(b)** $z = \dfrac{0.1333 - 0.25}{\sqrt{\dfrac{0.25(1-0.25)}{60}}} = -2.088$; *Table A:* $2(0.0183) = 0.0366$; *Tech:* $z = -2.087$, *P*-value $= 0.0369$

13. S: $H_0: p = 0.843, H_a: p \neq 0.843$, where $p = $ the proportion of all first-year students at this university who think being very well-off financially is an important personal goal. Use $\alpha = 0.05$. The evidence for H_a is: $\hat{p} = 0.785 \neq 0.843$. **P:** One-sample z test for p. *Random?* We have an SRS of 200 first-year students. ✓ *Large Counts?* $np_0 = 168.6 \geq 10$ and $n(1-p_0) = 31.4 \geq 10$. ✓ **D:** $z = \dfrac{0.785 - 0.843}{\sqrt{\dfrac{0.843(1-0.843)}{200}}} = -2.255$; *Table A:* $2(0.0119) = 0.0238$; *Tech:* $z = -2.255$, *P*-value $= 0.0242$. **C:** Because the *P*-value of $0.0242 < \alpha = 0.05$, we reject H_0. We have convincing evidence that the proportion of first-year students at this university who think being very well-off financially is an important goal is different than 0.843.

15. (a) S: $H_0: p = 0.75, H_a: p \neq 0.75$, where $p = $ the true proportion of peas that are smooth. Use $\alpha = 0.05$. The evidence for H_a is: $\hat{p} = 0.7608 \neq 0.75$. **P:** One-sample z test for p. *Random?* Random sample of peas (Assume). ✓ *Large Counts?* $np_0 = 417 \geq 10$ and $n(1-p_0) = 139 \geq 10$. ✓ **D:** $z = \dfrac{0.7608 - 0.75}{\sqrt{\dfrac{0.75(1-0.75)}{556}}} = 0.588$; *Table A:* $2(0.2776) = 0.5552$; *Tech:* $z = 0.588$, *P*-value $= 0.5568$. **C:** Because the *P*-value of $0.5568 > \alpha = 0.05$, we fail to reject H_0. We do not have convincing evidence that the true proportion of smooth peas differs from 0.75. **(b)** Assuming that the true proportion of peas that are smooth is 0.75, there is a 0.5568 probability of getting a sample proportion at least as far from 0.75 as 0.7608 (in either direction) by chance alone in a random sample of 556 peas.

17. (a) *Type I:* We find convincing evidence that more than 20% of the company's customers are willing to purchase the upgrade, when the true proportion is only 0.20. The company produces the upgrade and it is not profitable. *Type II:* We don't find convincing evidence that more than 20% of the company's customers are willing to purchase the upgrade, when the true proportion really is greater than 0.20. The company doesn't produce the upgrade and misses out on the potential profit. A Type I error is more serious because the company would produce the upgrade and lose money. A Type II error would mean only losing a potential profit. **(b) S:** $H_0: p = 0.20, H_a: p > 0.20$, where $p = $ the proportion of all the company's customers that would be willing to purchase the upgrade. Use $\alpha = 0.01$. The evidence for H_a is: $\hat{p} = 16/60 = 0.2667 > 0.20$. **P:** One-sample z test for p. *Random?* Random sample of 60 customers. ✓ *Large Counts?* $np_0 = 12 \geq 10$ and $n(1-p_0) = 48 \geq 10$. ✓ **D:** $z = \dfrac{0.2667 - 0.20}{\sqrt{\dfrac{0.20(1-0.20)}{60}}} = 1.292$; *Table A:* $1 - 0.9015 = 0.0985$; *Tech:* $z = 1.291$, *P*-value $= 0.0984$ **C:** Because the *P*-value of $0.0984 > \alpha = 0.01$, we fail to reject H_0. We do not have convincing evidence that more than 20% of the company's customers are willing to purchase the upgrade.

19. (a) S: 95% CI for $p = $ the proportion of all first-year students at the university who would identify being well-off financially as an important personal goal. **P:** One-sample z interval for p. Assume the conditions are met. ✓ **D:** $0.785 \pm 1.960 \sqrt{\dfrac{0.785(1-0.785)}{200}} \rightarrow 0.728$ to 0.842; *Tech:* 0.728 to 0.842. **C:** We are 95% confident that the interval from 0.728 to 0.842 captures the proportion of all first-year students at the university who would identify being well-off financially as an important personal goal. **(b)** The significance test in Exercise 13 only allowed us to reject $p = 0.843$ as a plausible value for the parameter. The confidence interval gives the values of p that are plausible based on the sample data (notice that 0.843 is not one of them).

21. Yes. An estimate of 0.55 with a margin of error of 0.03 gives a confidence interval of 0.52 to 0.58. Because 0.60 is not contained in the interval, it is not a plausible value for the population proportion. We have convincing evidence that the actual proportion of U.S. adults who would say they want to lose weight differs from 0.60.

23. S: $H_0: p = 0.215, H_a: p \neq 0.215$, where $p = $ the proportion of U.S. households that speak a language other than English. Use $\alpha = 0.05$. The evidence for H_a is: $\hat{p} = 602/3000 = 0.200667 \neq 0.215$. **P:** One-sample z test for p. *Random?* Random sample of 3000 households. ✓ *Large Counts?* $np_0 = 645 \geq 10$ and $n(1-p_0) = 2355 \geq 10$. ✓

D: *Tech:* $z = -1.911$, *P*-value $= 0.0560$. **C:** Because the *P*-value of $0.0560 > \alpha = 0.05$, we fail to reject H_0. We do not have convincing evidence that the proportion of U.S. households that speak a language other than English is different than 0.215. These data do not refute the researcher's claim.
25. **(a)** $H_0: p = 0.9, H_a: p < 0.9$, where $p =$ the proportion of all mowers of this model that would start on a push of the button. **(b)** *Large Counts?* $np_0 = 18 \geq 10$, but $n(1-p_0) = 2 < 10$. This condition is not met. **(c)** Let $X =$ the number of mowers that start out of 20. X is a binomial random variable with $n = 20$ and $p = 0.9$. *Tech:* $P(X \leq 15) =$ binomcdf(trials: 20, *p*: 0.9, *x*-value: 15) $= 0.043$. **(d)** Because the *P*-value of $0.043 < \alpha = 0.05$, we reject H_0. We have convincing evidence that the proportion of all mowers of this model that would start on a push of the button is less than the company's claim of 0.9.
27. **(a)** $\mu_{\bar{x}} = 14{,}000$ miles **(b)** $\sigma_{\bar{x}} = \dfrac{4000}{\sqrt{12}} = 1154.7$ miles. In SRSs of size $n = 12$ from the population of all cars repaired at this dealership, the sample mean number of miles driven per year typically varies from the true mean by about 1154.7 miles. **(c)** Because $n = 12 < 30$ and the population distribution is skewed to the right, the sampling distribution of $\bar{x}$ will also be skewed to the right, but not quite as strongly as the population.

Answers to Section 8.5 Exercises

1. (1) The data come from a population that is approximately normally distributed, or (2) $n \geq 30$, or (3) a graph of the sample data shows no strong skewness or outliers.
3. $t = \dfrac{\bar{x} - \mu_0}{\dfrac{s_x}{\sqrt{n}}}$
5. symmetric; single; bell
7. *Random?* Random sample of 25 salmon fillets. ✓ *Normal/Large Sample?* The sample size is small ($n < 30$), but the dotplot doesn't show any outliers or strong skewness. ✓

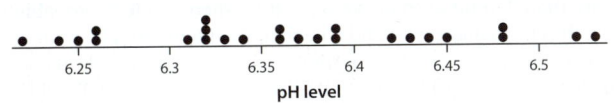

9. *Random?* SRS of 50 pineapples. ✓ *Normal/Large Sample?* $n = 50 \geq 30$. ✓
11. **(a)** $H_0: \mu = 50, H_a: \mu > 50$, where $\mu =$ the true mean percentage of purchases for which an alternative supplier offered lower prices. **(b)** *Random?* Random sample of 25 invoices. ✓ *Normal/Large Sample?* The sample size is small and the histogram shows strong skewness. This condition is not met.
13. $t = \dfrac{6.367 - 6.4}{\dfrac{0.087}{\sqrt{25}}} = -1.897$
15. $t = \dfrac{31.935 - 31}{\dfrac{2.394}{\sqrt{50}}} = 2.762$
17. *Table B:* The *P*-value is between 0.025 and 0.05. *Tech:* Applet/tcdf(lower: -1000, upper: -1.897, df: 24) $= 0.0350$.
19. **(a)** *Table B:* The *P*-value is between 0.0025 and 0.005 using df $= 40$. *Tech:* Applet/tcdf(lower: 2.762, upper: 1000, df: 49) $= 0.004$. **(b)** Because the *P*-value of $0.004 < \alpha = 0.05$, we reject H_0. We have convincing evidence that the mean weight of all pineapples grown in the field this year is greater than 31 ounces.
21. **(a)** $H_0: \mu = \$3.06, H_a: \mu \neq \$3.06$, where $\mu =$ the mean price of regular gasoline at all stations in this city. **(b)** *Random?* Random sample of 10 gas stations in the city. ✓ *Normal/Large Sample?* The sample size is small, but the dotplot doesn't show any outliers or strong skewness. ✓

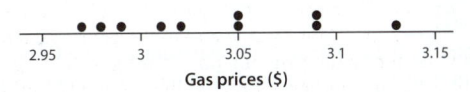

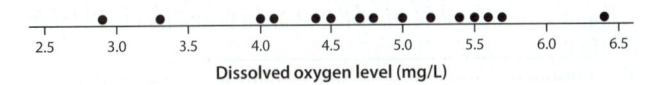

(c) $t = \dfrac{3.038 - 3.06}{\dfrac{0.053}{\sqrt{10}}} = -1.313$ **(d)** *Table B:* The *P*-value is between $2(0.10) = 0.20$ and $2(0.15) = 0.30$. *Tech:* Applet/tcdf(lower: -1000, upper: -1.313, df: 9) $\times 2 = 0.2217$. Because the *P*-value of $0.2217 > \alpha = 0.05$, we fail to reject H_0. We do not have convincing evidence that the true mean price of gas in this city differs from $3.06.
23. **(a)** Assuming that the mean pH of all salmon fillets at this processing plant is 6.40, there is a 0.0350 probability of getting a sample mean of 6.367 or less by chance alone in a random sample of 25 fillets. **(b)** Because the *P*-value of $0.0350 < \alpha = 0.05$, we reject H_0. We have convincing evidence that the mean pH of all salmon fillets at this processing plant is less than 6.40.
25. The Random condition is not satisfied because the marketing consultant has used a convenience sample, which may not be representative of all shoppers. It would not be wise to carry out a significance test about the mean amount spent by all shoppers at the supermarket.
27. **(a)** $H_0: \mu = 18, H_a: \mu \neq 18$, where $\mu =$ the true mean amount of liquid dispensed (ounces) per cup. **(b)** *Random?* Random sample of 10 cups. ✓ *Normal/Large Sample?* The machine is programmed to dispense liquid according to an approximately normal distribution. ✓
(c) $z = \dfrac{17.85 - 18}{\dfrac{0.2}{\sqrt{10}}} = -2.372$; *Table A:* $2(0.0089) = 0.0178$; *Tech:* $z = -2.372$, *P*-value $= 0.0177$ **(d)** Because the *P*-value of $0.0177 < \alpha = 0.05$, we reject H_0. We have convincing evidence that the true mean amount of liquid dispensed per cup is different than 18 ounces.
29. **(a)** The critical value is $t^* = 2.262$ because the total area to the left of $t = -2.262$ and to the right of $t = 2.262$ in a t distribution with df $= 9$ is 0.05. A value of $t < -2.262$ or $t > 2.262$ will lead us to reject H_0. **(b)** Because $t = -1.313$ is not < -2.262 or > 2.262, we fail to reject H_0. We do not have convincing evidence that the true mean price of gas in this city differs from $3.06.
31. S: 99% CI for $\mu =$ the mean file size (in megabytes) for all photographs taken with Rafiq's camera. **P:** One-sample t interval for μ. *Random?* Random sample of 49 photo files. ✓ *Normal/Large Sample?* $n = 49 \geq 30$. ✓ **D:** $8.05 \pm 2.682 \dfrac{1.96}{\sqrt{49}}$; 7.299 to 8.801; *Tech:* 7.299 to 8.801 using df $= 48$. **C:** We are 99% confident that the interval from 7.299 megabytes to 8.801 megabytes captures the mean file size for all photographs taken with Rafiq's camera.

Answers to Section 8.6 Exercises

1. one-sample t test for μ
3. practical
5. S: $H_0: \mu = 5, H_a: \mu < 5$, where $\mu =$ the true mean dissolved oxygen level (mg/L) in this stream. Use $\alpha = 0.10$. The evidence for H_a is: $\bar{x} = 4.771 < 5$. **P:** One-sample t test for μ. *Random?* Water is sampled at 15 randomly chosen locations. ✓ *Normal/Large Sample?* The sample size is small, but the dotplot doesn't show any outliers or strong skewness. ✓

D: $t = \dfrac{4.771-5}{\dfrac{0.940}{\sqrt{15}}} = -0.944$; *Table B:* The *P*-value is between 0.15 and 0.20; *Tech: t* = −0.943, *P*-value = 0.1809 using df = 14. **C:** Because the *P*-value of 0.1809 > α = 0.10, we fail to reject H_0. There is not convincing evidence that the true mean dissolved oxygen level in this stream is less than 5 mg/L (i.e., that aquatic life in this stream is at risk).

7. S: $H_0: \mu = 5$, $H_a: \mu > 5$, where μ = the true mean reading level for all pages of this novel. Use $\alpha = 0.05$. The evidence for H_a is: $\bar{x} = 5.4 > 5$. **P:** One-sample *t* test for μ. *Random?* Random sample of 40 pages from this novel. ✓ *Normal/Large Sample?* $n = 40 \geq 30$. ✓

D: $t = \dfrac{5.4-5}{\dfrac{0.8}{\sqrt{40}}} = 3.162$; *Table B:* Use df = 30. The *P*-value is between 0.001 and 0.0025. *Tech: t* = 3.162, *P*-value = 0.0015 using df = 39. **C:** Because the *P*-value of 0.0015 < α = 0.05, we reject H_0. There is convincing evidence that the mean reading level for all pages of this novel is greater than a fifth-grade level.

9. The 90% confidence interval contains $\mu = 105$ as a plausible value, so we fail to reject H_0 at the $\alpha = 0.10$ significance level. We do not have convincing evidence that the true mean reading in the population of radon detectors is different from the actual value of 105 picocuries per liter.

11. (a) $H_0: \mu = 200$, $H_a: \mu \neq 200$, where μ = the true mean response time (milliseconds) for servers in Europe. **(b)** The 14 servers were randomly selected. Although the sample size is small, a graph of the data reveals no strong skewness or outliers. **(c)** The 95% CI does not contain $\mu = 200$ as a plausible value, so we reject H_0 at the $\alpha = 0.05$ significance level. We have convincing evidence that the true mean response time for servers in Europe is different from the site's claim of 200 milliseconds.

13. With a *P*-value of 0.0371 < 0.05, the company has convincing evidence that the average score increase for students who use the smartphone app in addition to the regular program is greater than 10 points. The result is not practically important because an increase in average MCAT score of 0.4 point is not impressive.

15. (a) No. The researcher tested 500 subjects. At the $\alpha = 0.01$ level, we would expect 500(0.01) = 5 subjects to do significantly better than random guessing by chance alone. **(b)** The researcher should test these four subjects again to see if any of the four subjects have ESP.

17. (a) S: $H_0: \mu = 120$, $H_a: \mu \neq 120$, where μ = the mean temperature of the system (°F) for all times over a 4-hour period. Use $\alpha = 0.05$. The evidence for H_a is: $\bar{x} = 121.425 \neq 120$. **P:** One-sample *t* test for μ. *Random?* Random sample of 12 times over a 4-hour period. ✓ *Normal/Large Sample?* The sample size is small, but the boxplot doesn't show any outliers or strong skewness. ✓

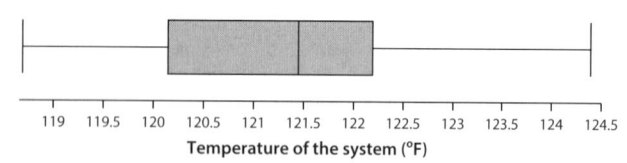

D: $t = \dfrac{121.425 - 120}{\dfrac{1.716}{\sqrt{12}}} = 2.877$; *Table B:* The *P*-value is between 2(0.005) = 0.01 and 2(0.01) = 0.02. *Tech: t* = 2.877, *P*-value = 0.0151 using df = 11. **C:** Because the *P*-value of 0.0151 < α = 0.05, we reject H_0. There is convincing evidence that the mean temperature of the system for all times over the 4-hour period is different from 120°F. **(b)** In the test, the *P*-value of 0.0151 < α = 0.05, so we reject H_0. The 95% confidence interval does not contain $\mu = 120$ as a plausible value, so we reject $H_0: \mu = 120$ at the $\alpha = 0.05$ significance level. The decision about the two-sided test is consistent with the decision made based on the corresponding confidence interval. But the confidence interval gives more information: the set of all plausible values of μ based on the sample data.

19. (a) Yes. Because the *P*-value of 0.06 > α = 0.05, we fail to reject H_0. This means that $\mu = 10$ is plausible and therefore must be contained in the 95% confidence interval. **(b)** No. Because the *P*-value of 0.06 < α = 0.10, we reject H_0. This means that $\mu = 10$ is not plausible and therefore must not be contained in the 90% confidence interval.

21. S: $H_0: \mu = 2.52$, $H_a: \mu \neq 2.52$, where μ = the mean number of people per household in the year the sample was taken. Use $\alpha = 0.01$. The evidence for H_a is: $\bar{x} = 2.4097 \neq 2.52$. **P:** One-sample *t* test for μ. *Random?* Random sample of 3000 households. ✓ *Normal/Large Sample?* $n = 3000 \geq 30$. ✓ **D:** *Tech: t* = −4.341, *P*-value ≈ 0 using df = 2999. **C:** Because the *P*-value of approximately 0 < α = 0.01, we reject H_0. There is convincing evidence that the mean number of people per household in the year the sample was taken is different from 2.52 people per household.

23. (a) 1.729; 1.729 **(b)** Using df = 20 − 1 = 19, the t^* critical value for a 90% confidence level is 1.729. **(c)** If we reject H_0, we know from part (a) that the sample mean $\bar{x}$ is more than 1.729 standard errors greater than $\mu = 10$. If $\bar{x}$ is more than 1.729 standard errors away from the mean, we know from part (b) that the 90% confidence interval does not contain $\mu = 10$.

25. No, because this was an observational study. An experiment would be needed to show a cause-and-effect relationship between study time and test scores.

Answers to Section 8.7 Exercises

1. convincing evidence
3. Power = 1 − *P*(Type II error)
5. If the true mean income in the population of people who live near the restaurant is μ = $86,000, there is a 0.64 probability that I will find convincing evidence for $H_a: \mu > $85,000.
7. If the true proportion of all calls involving life-threatening injuries during this 6-month period for which emergency personnel took more than 8 minutes to arrive is $p = 0.17$, there is a 0.80 probability that the city manager will find convincing evidence for $H_a: p < 0.22$.
9. (a) Decrease. A smaller sample size gives less information about the true mean μ. **(b)** Decrease. It is harder to detect a smaller difference between the null and alternative parameter values. **(c)** Increase. Using a larger significance level makes it easier to reject H_0 when H_a is true.
11. (a) Decrease. Using a smaller significance level makes it harder to reject H_0 when H_a is true. **(b)** Increase. A larger sample size gives more information about the true proportion *p*. **(c)** Decrease. It is harder to detect a smaller difference between the null and alternative parameter values.
13. (a) *P*(Type I error) = 0.05 **(b)** If the mean breaking strength (in pounds) of the company's 1-inch plywood flooring is $\mu = 495$, there is a 0.738 probability that the company will find convincing evidence for $H_a: \mu < 500$. **(c)** *P*(Type II error) = 1 − 0.738 = 0.262. **(d)** The power of the test in part (b) can be increased by either increasing the sample size or using a larger significance level.
15. Power = 1 − 0.14 = 0.86
17. (a) The larger significance level will increase the probability of a Type I error. **(b)** The larger sample size would require more time and money.
19. (a) Power ≈ 0.25 **(b)** $n = 200$. Of the choices, this sample size has the greatest power to detect that $p = 0.70$. **(c)** The power to detect $p = 0.75$ with a sample size of $n = 50$ is ≈ 0.72. The power to detect $p = 0.75$ with a sample size of $n = 100$ is ≈ 0.94. The power to detect $p = 0.75$ with a sample size of $n = 200$ is ≈ 0.99. I recommend the researcher use a sample size of $n = 100$. By doubling the sample size from $n = 100$ to $n = 200$, the power would only increase from about 0.94 to about 0.99. It does not seem that the small increase in power is worth doubling the cost of time and money.

21. (a)

		Finished the online homework?		
		Yes	No	Total
Completed the Concept Exploration?	Yes	8	16	24
	No	6	5	11
	Total	14	21	35

(b) $5/35 = 0.143$

Answers to Section 8.8 Exercises

1. *Random:* The data come from a random sample from the population of interest. *Normal:* The data come from a population that is approximately normally distributed.
3. chi-square; $n-1$
5. *Random?* Random sample of 25 salmon fillets. ✓ *Normal?* The dotplot shows no strong skewness or outliers. It is plausible that these data came from a normal population. ✓

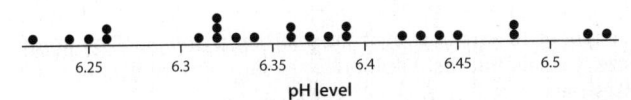

7. (a) $H_0: \sigma = 9.44$, $H_a: \sigma < 9.44$, σ = the standard deviation of GRE scores for all students who use the program. **(b)** *Random?* Random sample of 150 students. ✓ *Normal?* GRE scores are known to be approximately normally distributed. ✓

9. (a) $\chi^2 = \dfrac{(25-1)(0.0871)^2}{(0.10)^2} = 18.207$; **(b)** *Table C:* P-value $> 2(0.10) = 0.20$; *Tech:* $\chi^2 = 18.214$, P-value $= 0.415$

11. (a) $\chi^2 = \dfrac{(150-1)(8.46)^2}{(9.44)^2} = 119.669$; **(b)** *Tech:* $\chi^2 = 119.669$, P-value $= 0.0370$ **(c)** Because the P-value of $0.0370 < \alpha = 0.05$, we reject H_0. There is convincing evidence that the standard deviation of GRE scores for all students who use the program is less than 9.44.

13. S: $H_0: \sigma = 0.5$, $H_a: \sigma > 0.5$, where σ = the standard deviation of breaking strength (in pounds) for all pieces of fishing line. Use $\alpha = 0.05$. The evidence for H_a is: $s_x = 0.75 > 0.5$. **P:** One-sample test for σ. *Random?* Random sample of 30 pieces of fishing line. ✓ *Normal?* The dotplot shows no strong skewness or outliers. It is plausible that these data came from a normal population. ✓
D: $\chi^2 = \dfrac{(30-1)(0.75)^2}{0.5^2} = 65.25$; *Table C:* The P-value is less than 0.005. *Tech:* $\chi^2 = 65.25$, P-value < 0.001 using df $= 29$. **C:** Because the P-value of less than $0.001 < \alpha = 0.05$, we reject H_0. There is convincing evidence that the standard deviation of breaking strength for all pieces of fishing line is greater than 0.5 pound.

15. S: $H_0: \sigma = 11.8$, $H_a: \sigma \neq 11.8$, where σ = the standard deviation of tail length (in mm) for all red-tail hawks migrating through this particular area. Use $\alpha = 0.05$. The evidence for H_a is: $s_x = 14.3562 \neq 11.8$. **P:** One-sample test for σ. *Random?* Random sample of 10 red-tail hawks. ✓ *Normal?* The dotplot shows no strong skewness or outliers. It is plausible that these data came from a normal population. ✓

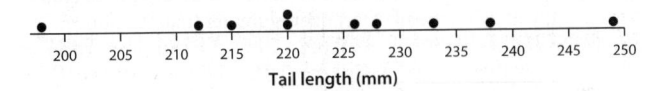

D: $\chi^2 = \dfrac{(10-1)(14.3562)^2}{11.8^2} = 13.322$; *Table C:* The P-value is greater than $2(0.10)$. *Tech:* $\chi^2 = 13.322$, P-value $= 0.2972$ using df $= 9$. **C:** Because the P-value of $0.2972 > \alpha = 0.05$, we fail to reject H_0. There is not convincing evidence that the standard deviation of tail length for all red-tail hawks migrating through this particular area is different than 11.8 mm.

17. S: $H_0: \sigma = 13.3$, $H_a: \sigma > 13.3$, where σ = the standard deviation of shoulder height (in cm) for all mature male elephants that lived through droughts during their first 2 years of life. Use $\alpha = 0.05$. The evidence for H_a is: $s_x = 31.6422 > 13.3$. **P:** One-sample test for σ. *Random?* Random sample of 14 elephants. ✓ *Normal?* The dotplot shows no strong skewness or outliers. It is plausible that these data came from a normal population. ✓

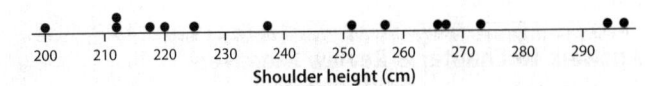

D: $\chi^2 = \dfrac{(14-1)(31.6422)^2}{13.3^2} = 73.582$; *Table C:* The P-value is less than 0.005. *Tech:* $\chi^2 = 73.582$, P-value < 0.001 using df $= 13$. **C:** Because the P-value of less than $0.001 < \alpha = 0.05$, we reject H_0. There is convincing evidence that the standard deviation of shoulder heights for all male African elephants that lived through droughts during their first 2 years of life is greater than 13.3 cm.

19. (a) If the true standard deviation in shoulder height is $\sigma = 20$ cm, there is a 0.738 probability that the researchers will find convincing evidence for $H_a: \sigma > 13.3$ cm. **(b)** Power will be larger than 0.738. It is easier to detect a larger difference between the null and alternative parameter value. **(c)** The power of the test could be increased by either increasing the sample size or using a larger significance level.

21. (a) Assuming that the standard deviation of the pH of all salmon fillets is 0.10, there is a 0.415 probability of getting a sample standard deviation of 0.0871 or more extreme (in either direction) by chance alone in a random sample of 25 salmon fillets. **(b)** Because the P-value of $0.0871 > \alpha = 0.05$, we fail to reject H_0. There is not convincing evidence that the standard deviation of pH for all salmon fillets is different than 0.10. **(c)** The 95% confidence interval contains $\sigma = 0.10$ as a plausible value, so we cannot reject $H_0: \sigma = 0.10$ at the $\alpha = 0.05$ significance level. The confidence interval gives more information than the conclusion in part (b) because it gives an interval of plausible values for the population standard deviation.

23. S: $H_0: \sigma^2 = 0.25$, $H_a: \sigma^2 > 0.25$, where σ^2 = the variance of price (in dollars^2) for all hot coffees in this city. Use $\alpha = 0.05$. The evidence for H_a is: $s_x^2 = (0.5176)^2 = 0.26791 > 0.25$. **P:** One-sample test for σ^2. *Random?* Random sample of 41 hot coffees. ✓ *Normal?* The dotplot shows no strong skewness or outliers. It is plausible that these data came from a normal population. ✓

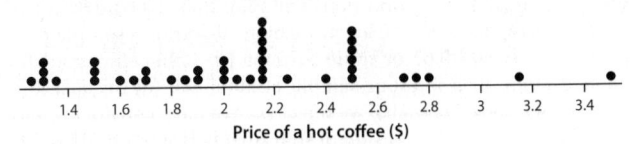

D: *Tech:* $\chi^2 = 42.873$, P-value $= 0.349$ using df $= 40$. **C:** Because the P-value of $0.349 > \alpha = 0.05$, we fail to reject H_0. There is not convincing evidence that the variance of price for all hot coffees in this city is greater than 0.25 dollar^2.

25. (a) A value of $\chi^2 > 22.362$ will lead us to reject H_0. **(b)** Because $\chi^2 = 73.582$ is greater than 22.362, we reject H_0. There is convincing evidence that the standard deviation of shoulder heights for all male African elephants that lived through droughts during their first 2 years of life is greater than 13.3 cm.

27. (a) $H_0: \sigma = 0.7$, $H_a: \sigma > 0.7$, where σ = the standard deviation of weight (in ounces) for all bags of candy produced that day.

(b) $\chi^2 = \dfrac{(75-1)(0.81)^2}{0.7^2} = 99.084$

(c) $z = \dfrac{\chi^2 - (n-1)}{\sqrt{2(n-1)}} = \dfrac{99.084 - (75-1)}{\sqrt{2(75-1)}} = 2.062$

(d) *Table A:* $1 - 0.9803 = 0.0197$; *Tech:* 0.0196 (e) The *P*-value based on the standard normal curve (0.0196) is fairly close to the *P*-value based on the chi-square distribution (0.027).

29. (a) $P(X = 0) = \dfrac{3.2^0 e^{-3.2}}{0!} = 0.0408$

(b) $P(X > 5) = 1 - P(X \leq 5) = 1 - \text{poissoncdf}(\mu: 3.2, x \text{ value}: 5) = 0.1054$

Answers to Chapter 8 Review Exercises

1. (a) $H_0: \mu = 47, H_a: \mu \neq 47$, where $\mu =$ the mean quantity of popcorn (in quarts per year) consumed by all adults in the researcher's state. (b) $H_0: p = 0.47, H_a: p > 0.47$, where $p =$ the proportion of all households in your county that own at least one dog. (c) $H_0: \sigma = 40, H_a: \sigma > 40$, where $\sigma =$ the standard deviation of response times (in milliseconds) for this user's internet connections.
2. (a) $H_0: \mu = 20, H_a: \mu < 20$, where $\mu =$ the mean volume of gravel (m³) in all truckloads sent by the company for this road project. (b) *Type I:* You find convincing evidence that the true mean amount of gravel per truckload is less than 20 m³, when it is really equal to 20 m³. *Type II:* You don't find convincing evidence that the true mean amount of gravel per truckload is less than 20 m³, when it is really less than 20 m³. (c) *Consequence of Type I:* You tell the company that it lied about its claim, even though it delivered the appropriate amount of gravel. This decision would likely upset the gravel company managers and possibly delay completing the road. *Consequence of Type II:* You don't say anything to the company, even though it did not deliver the appropriate amount of gravel. You won't have enough gravel to complete the road. This will lead to a delay and an additional cost, as more gravel would need to be purchased. The more serious error is a Type II error because the road will not be completed on time and will cost more than it should have.
3. (a) *Table A:* $1 - 0.9664 = 0.0336$; *Tech:* Applet/normalcdf(lower: 1.83, upper: 1000, mean: 0, SD: 1) = 0.0336 (b) *Table B:* The *P*-value is between $2(0.01) = 0.02$ and $2(0.02) = 0.04$; *Tech:* Applet/tcdf(lower: −1000, upper: −2.35, df: 19) × 2 = 0.0297 (c) *Table C:* The *P*-value is between 0.05 and 0.10. *Tech:* χ^2cdf(lower: 0, upper: 15.62, df: 24) = 0.0987 using df = 24
4. (a) $H_0: p = \dfrac{18}{38} = 0.474, H_a: p \neq \dfrac{18}{38} = 0.474$, where $p =$ the proportion of all spins that land on red. (b) *Random?* Random sample of 50 spins. ✓ *Large Counts?* $np_0 = 23.7 \geq 10$ and $n(1-p_0) = 26.3 \geq 10$. ✓ (c) Assuming that the proportion of all spins that will land on red is $18/38 = 0.474$, there is a 0.0384 probability of getting a sample proportion of $31/50 = 0.62$ or more extreme (in either direction) by chance alone in a random sample of 50 spins. (d) Because the *P*-value of $0.0384 < \alpha = 0.05$, we reject H_0. We have convincing evidence that the true proportion of spins that will land on red is different than 18/38. It seems that the wheel is unfair. (e) The manager's conclusion is inconsistent with the conclusion in part (d) because the manager used a 99% confidence interval, which is equivalent to a test using $\alpha = 0.01$. If the manager had used a 95% confidence interval, 18/38 would not be considered a plausible value. Also, the manager incorrectly claims that the interval provides convincing evidence that H_0 is true.
5. S: $H_0: p = 0.07, H_a: p > 0.07$, where $p =$ the true proportion of Carrboro residents who are underemployed. Use $\alpha = 0.10$. The evidence for H_a is: $\hat{p} = 14/150 = 0.0933 > 0.07$. **P:** One-sample *z* test for *p*. *Random?* Simple random sample of 150 Carrboro residents. ✓ *Large Counts?* $np_0 = 10.5 \geq 10$ and $n(1-p_0) = 139.5 \geq 10$. ✓

D: $z = \dfrac{0.0933 - 0.07}{\sqrt{\dfrac{0.07(1-0.07)}{150}}} = 1.118$; *Table A: P*-value = 0.1314; *Tech:* $z = 1.120$, *P*-value = 0.1313. **C:** Because the *P*-value of $0.1313 > \alpha = 0.10$, we fail to reject H_0. We do not have convincing evidence that the proportion of Carrboro residents who are underemployed is greater than the national proportion of 0.07.
6. (a) The evidence for H_a is: $\bar{x} = 319 \neq 320$. (b) *Random?* Random sample of 30 tablets from a batch in production. ✓ *Normal/Large Sample?* $n = 30 \geq 30$. ✓ (c) $t = \dfrac{319 - 320}{\dfrac{3}{\sqrt{30}}} = -1.826$; *Table B:* The *P*-value is between $2(0.025) = 0.05$ and $2(0.05) = 0.10$. *Tech:* $t = -1.826$, *P*-value = 0.0782 using df = 29. (d) Because the *P*-value of $0.0782 > \alpha = 0.05$, we fail to reject H_0. There is not convincing evidence that the true mean amount of active ingredient for all aspirin tablets in this production batch is different than 320 mg.
7. S: $H_0: \mu = 8, H_a: \mu > 8$, where $\mu =$ the mean weight for all ears of corn grown using the new fertilizer. Use $\alpha = 0.05$. The evidence for H_a is: $\bar{x} = 8.335 > 8$. **P:** One-sample *t* test for μ. *Random?* Random sample of 10 ears of corn. ✓ *Normal/Large Sample?* The sample size is small, but the dotplot doesn't show any outliers or strong skewness. ✓

D: $t = \dfrac{8.335 - 8}{\dfrac{0.5563}{\sqrt{10}}} = 1.904$; *Table B:* The *P*-value is between 0.025 and 0.05; *Tech:* $t = 1.904$, *P*-value = 0.0446 using df = 9. **C:** Because the *P*-value of $0.0446 < \alpha = 0.05$, we reject H_0. There is convincing evidence that the mean weight of all ears of corn grown using the new fertilizer is greater than 8 ounces.
8. (a) If the true proportion of adult Carrboro residents who are underemployed is $p = 0.11$, there is a 0.70 probability that the researchers will find convincing evidence for $H_a: p > 0.07$. (b) Increase the sample size or use a larger significance level.
9. (a) The sample data give the farmer a reason to worry because $s_x = 0.5563 > 0.50$. (b) **S:** $H_0: \sigma = 0.5, H_a: \sigma > 0.5$, where $\sigma =$ the standard deviation of weight for all ears of corn produced using the new fertilizer. Use $\alpha = 0.01$. The evidence for H_a is: $s_x = 0.5563 > 0.50$. **P:** One-sample test for σ. *Random?* Random sample of 10 ears of corn. ✓ *Normal?* The dotplot shows no strong skewness or outliers. It is plausible that these data came from an approximately normally distributed population. ✓

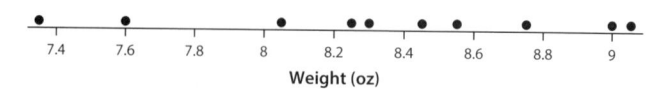

D: $\chi^2 = \dfrac{(10-1)(0.5563)^2}{0.5^2} = 11.141$; *Table C:* The *P*-value is greater than 0.10. *Tech:* $\chi^2 = 11.141$, *P*-value = 0.2662 using df = 9. **C:** Because the *P*-value of $0.2662 > \alpha = 0.01$, we fail to reject H_0. There is not convincing evidence that the standard deviation of weight for all ears of corn produced using the new fertilizer is greater than 0.5 ounce.
10. (a) **S:** $H_0: p = 0.60, H_a: p < 0.60$, where $p =$ the proportion of all signatures that are valid. Use $\alpha = 0.01$. The evidence for H_a is: $\hat{p} = \dfrac{159}{300} = 0.53 < 0.60$. **P:** One-sample *z* test for *p*. *Random?* Random sample of 300 signatures. ✓ *Large Counts?* $np_0 = 180 \geq 10$ and $n(1-p_0) = 120 \geq 10$. ✓

D: $z = \dfrac{0.53 - 0.60}{\sqrt{\dfrac{0.60(1-0.60)}{300}}} = -2.475$; *Table A:* The *P*-value is 0.0066. *Tech:* $z = -2.475$, *P*-value = 0.0067. **C:** Because the *P*-value of $0.0067 < \alpha = 0.01$, we reject H_0. We have convincing evidence that the proportion of all signatures on the petition that are valid is less than 0.60. **(b)** Because we rejected the null hypothesis, it is possible that we made a Type I error.

11. (a) S: $H_0: \mu = 15$, $H_a: \mu \ne 15$, where μ = the true mean vertical jump (in inches) for all students at this community college. Use $\alpha = 0.05$. The evidence for H_a is: $\bar{x} = 17 \ne 15$. **P:** One-sample *t* test for μ. *Random?* Random sample of 20 students. ✓ *Normal/Large Sample?* The sample size is small, but the boxplot doesn't show any outliers or strong skewness. ✓

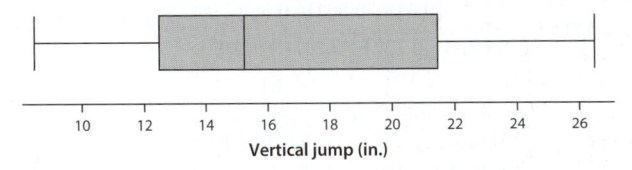

D: $t = \dfrac{\bar{x} - \mu_0}{\dfrac{s_x}{\sqrt{n}}} = \dfrac{17 - 15}{\dfrac{5.368}{\sqrt{20}}} = 1.666$; *Table B:* The *P*-value is between $2(0.05) = 0.10$ and $2(0.10) = 0.20$. *Tech:* $t = 1.666$, *P*-value = 0.1121 with df = 19. **C:** Because the *P*-value of $0.1121 > \alpha = 0.05$, we fail to reject H_0. We do not have convincing evidence that the mean vertical jump in the population of students at this school is different than 15 inches. **(b)** The confidence interval gives more information than the test, as it provides the entire set of plausible values for the true mean vertical jump based on the data (including 15 inches).

12. S: $H_0: \sigma^2 = 36$, $H_a: \sigma^2 < 36$, where σ^2 = the variance of vertical jump height (inches^2) for all students at this community college. Use $\alpha = 0.05$. The evidence for H_a is: $s_x^2 = (5.368)^2 = 28.815 < 36$. **P:** One-sample test for σ^2. *Random?* Random sample of 20 students. ✓ *Normal?* The dotplot shows no strong skewness or outliers. It is plausible that these data came from a normal population. ✓

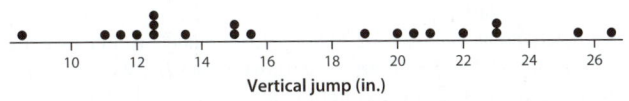

D: $\chi^2 = \dfrac{(20-1)(28.815)}{36} = 15.208$; *Table C:* The *P*-value is greater than 0.10. *Tech:* $\chi^2 = 15.208$, *P*-value = 0.2907 using df = 19. **C:** Because the *P*-value of $0.2907 > \alpha = 0.05$, we fail to reject H_0. There is not convincing evidence that the variance of vertical jump distance for all students at this community college is less than 36 in.^{2}

CHAPTER 9

Answers to Section 9.1 Exercises

1. $n_1\hat{p}_1$; $n_1(1-\hat{p}_1)$; $n_2\hat{p}_2$; $n_2(1-\hat{p}_2)$

3. $(\hat{p}_1 - \hat{p}_2) \pm z^* \sqrt{\dfrac{\hat{p}_1(1-\hat{p}_1)}{n_1} + \dfrac{\hat{p}_2(1-\hat{p}_2)}{n_2}}$

5. *Random?* Independent random samples of 1502 U.S. adults in 2021 and 1520 U.S. adults in 2016. ✓ *Large Counts?* $n_1\hat{p}_1 = 601$, $n_1(1-\hat{p}_1) = 901$, $n_2\hat{p}_2 = 486$, $n_2(1-\hat{p}_2) = 1034$. All are at least 10. ✓

7. *Random?* The data include the whole population of babies born on the east and west sides and are not a random sample. Condition not met. *Large Counts?* $n_1\hat{p}_1 = 16$, $n_1(1-\hat{p}_1) = 398$, $n_2\hat{p}_2 = 3$, $n_2(1-\hat{p}_2) = 225$. These counts are not all at least 10. Condition not met.

9. (a) $(0.40 - 0.32) \pm 1.645 \sqrt{\dfrac{0.40(1-0.40)}{1502} + \dfrac{0.32(1-0.32)}{1520}} \to 0.051$ to 0.109 **(b)** Yes. Because the plausible values for the difference in proportions (0.051, 0.109) are all positive, there is convincing evidence of a change in the proportions of U.S. adults who used Instagram from 2016 to 2021.

11. (a) *Random?* Randomly assigned men to have surgery or to be observed only. ✓ *Large Counts?* $n_S\hat{p}_S = 141$, $n_S(1-\hat{p}_S) = 223$, $n_O\hat{p}_O = 122$, $n_O(1-\hat{p}_O) = 245$. All are at least 10. ✓

(b) $(0.387 - 0.332) \pm 1.960 \sqrt{\dfrac{0.387(1-0.387)}{364} + \dfrac{0.332(1-0.332)}{367}} \to$ -0.014 to 0.124 **(c)** No. Because the confidence interval (−0.014, 0.124) includes 0 as a plausible value for the difference in the population proportions, there is not convincing evidence that surgery helps increase 20-year survival compared to observation only for men with localized prostate cancer like the ones in this study.

13. S: 95% CI for $p_{\text{rural}} - p_{\text{urban}}$, where p_{rural} = the proportion of all rural households in Indiana that had a natural Christmas tree last year and p_{urban} = the proportion of all urban households in Indiana that had a natural Christmas tree last year. **P:** Two-sample *z* interval for $p_{\text{rural}} - p_{\text{urban}}$. *Random?* Independent random samples of rural and urban households in Indiana. ✓ *Large Counts?* $64, 160 - 64 = 96$, $89, 261 - 89 = 172$ are all at least 10. ✓

D: $(0.400 - 0.341) \pm 1.960 \sqrt{\dfrac{0.400(1-0.400)}{160} + \dfrac{0.341(1-0.341)}{261}} \to$ -0.036 to 0.154 *Tech:* -0.036 to 0.154. **C:** We are 95% confident that the interval from −0.036 to 0.154 captures the true difference (Rural − Urban) in the proportion of Indiana households that had a natural Christmas tree last year. The interval suggests that the percentage of rural Indiana households that had a Christmas tree last year is between 3.6 percentage points less than and 15.4 percentage points greater than the proportion of urban Indiana households that had a Christmas tree last year.

15. S: 90% CI for $p_N - p_F$, where p_N = the proportion of all subjects like these near the bowl of candy who eat the candy and $p_F = \ldots$ far from the bowl of candy who eat the candy. **P:** Two-sample *z* interval for $p_N - p_F$. *Random?* People were randomly assigned to a room with candy near the seating location or far from the seating location. ✓ *Large Counts?* $39, 61 - 39 = 22$, $24, 61 - 24 = 37$ are all at least 10. ✓

D: $(0.639 - 0.393) \pm 1.645 \sqrt{\dfrac{0.639(1-0.639)}{61} + \dfrac{0.393(1-0.393)}{61}} \to$ 0.102 to 0.390 *Tech:* 0.102 to 0.390. **C:** We are 90% confident that the interval from 0.102 to 0.390 captures the difference between the true proportions of people like these who would eat the candy when placed in a seating location far from the bowl versus in a seating location near the bowl. The interval suggests that the percentage of people who would eat the candy when placed in a seating location near the bowl is between 10.2 and 39 percentage points greater than the percentage of people like these who would eat the candy when placed in a location far from the bowl.

17. (a) S: 99% CI for $p_S - p_T$, where p_S = the true proportion of patients like these who have traditional surgery and will have negative outcomes and $p_T = \ldots$ TAVR and will have negative outcomes. **P:** Two-sample *z* interval for $p_S - p_T$. *Random?* Patients were randomly assigned to traditional surgery or TAVR. ✓ *Large Counts?* $68, 454 - 68 = 386$, $42, 496 - 42 = 454$ are all at least 10. ✓

D: $(0.150 - 0.085) \pm 2.576 \sqrt{\dfrac{0.150(1-0.150)}{454} + \dfrac{0.085(1-0.085)}{496}} \to$ 0.011 to 0.119 *Tech:* 0.011 to 0.119. **C:** We are 99% confident that the interval from 0.011 to 0.119 captures the difference (Surgery − TAVR) in the true proportion of patients like these who will have negative outcomes with traditional surgery and the proportion who will have negative outcomes with TAVR. The interval suggests that the percentage of patients like these who have traditional surgery and have negative outcomes is between 1.1 percentage points and 11.9 percentage points greater than the percentage who have TAVR and have negative outcomes. **(b)** Yes. Because the plausible values for the true difference (S − T) in proportions (0.011, 0.119) are all positive, there is convincing evidence that a smaller proportion of subjects like these will have negative outcomes with the new TAVR method.

19. (a) If the association selects many random samples of 160 rural households in Indiana and 261 urban households in Indiana and constructs a 95% confidence interval for the difference in the proportions each time, about 95% of these intervals would capture the true difference in the proportions of all rural and urban households that had a natural tree last year. **(b)** No. Because our interval (−0.036, 0.154) includes a difference of 0, it is plausible that there is no difference (Rural − Urban) in the population proportions of all rural and urban households with natural trees. **(c)** No. While it is plausible that the population proportions have a difference of 0, it is also plausible that the population proportions differ by any value in the interval −0.036 to 0.154.

21. S: 90% CI for $p_C - p_O$, where p_C = the proportion of all live trees on the curb that are next to damaged sidewalk and $p_O = \ldots$ offset from the curb that are next to damaged sidewalk. **P:** Two-sample z interval for $p_C - p_O$. *Random?* The problem states that the Random condition is met. ✓ *Large Counts?* 2181, 7529 − 2181 = 5348, 40, 329 − 40 = 289 are all at least 10. ✓ **D:** *Tech:* 0.137 to 0.199. **C:** We are 90% confident that the interval from 0.137 to 0.199 captures the difference in the proportion of all live trees on the curb that are next to damaged sidewalk and the proportion of all live trees offset from the curb that are next to damaged sidewalk. The interval suggests that the percentage of live trees on the curb that are next to damaged sidewalk is between 13.7 and 19.9 percentage points greater than the percentage of all offset trees that are next to damaged sidewalk.

23. (a) Approximately normal, because $n_C p_C = 15$, $n_C(1 - p_C) = 35$, $n_A p_A = 10$, and $n_A(1 - p_A) = 90$ are all ≥ 10. **(b)** $\mu_{\hat{p}_C - \hat{p}_A} = 0.30 - 0.10 = 0.20$ **(c)** $\sigma_{\hat{p}_C - \hat{p}_A} = \sqrt{\dfrac{0.30(1-0.30)}{50} + \dfrac{0.10(1-0.10)}{100}} = 0.0714$ **(d)** We want to calculate $P(\hat{p}_C - \hat{p}_A) > 0$.

(i) $z = \dfrac{0 - 0.20}{0.0714} = -2.801$; *Tech:* Applet/normalcdf(lower: −2.801, upper: 1000, mean: 0, SD: 1) = 0.9975; *Table A:* 1 − 0.0026 = 0.9974 (ii) *Tech:* Applet/normalcdf(lower: 0, upper: 1000, mean: 0.20, SD: 0.0714) = 0.9975.

25. (a) A control group would show the success rate of IVF with no acupuncture. This would serve as a baseline to determine how much acupuncture helps. Otherwise, we wouldn't be able to tell if the acupuncture caused an increase in the pregnancy rate. **(b)** Label each subject with a different integer from 1 to 160. Then, randomly generate 80 different integers from 1 to 160. The subjects with these labels will be in the acupuncture group. The remaining 80 subjects will be in the control group. **(c)** The purpose of random assignment is to create roughly equivalent groups of infertile women at the beginning of the experiment.

Answers to Section 9.2 Exercises

1. *Random:* The data come from independent random samples from the two populations of interest or from two groups in a randomized experiment; *Large Counts:* The counts of "successes" and "failures" in each sample or group — $n_1\hat{p}_1$, $n_1(1-\hat{p}_1)$, $n_2\hat{p}_2$, and $n_2(1-\hat{p}_2)$ — are all at least 10.

3. $z = \dfrac{(\hat{p}_1 - \hat{p}_2) - 0}{\sqrt{\dfrac{\hat{p}_C(1-\hat{p}_C)}{n_1} + \dfrac{\hat{p}_C(1-\hat{p}_C)}{n_2}}}$

5. (a) $H_0: p_1 - p_2 = 0, H_a: p_1 - p_2 \neq 0$, where p_1 = proportion of all U.S. college students who used vaping products in 2019 and $p_2 = \ldots$ in 2020. **(b)** Yes. $\hat{p}_1 - \hat{p}_2 = 0.437 - 0.396 = 0.041 \neq 0$ **(c)** *Random?* Independent random samples from the populations of U.S. college students in 2019 and 2020. ✓ *Large Counts?* 439, 566, 317, and 483 are all at least 10. ✓

7. (a) $H_0: p_P - p_{NP} = 0, H_a: p_P - p_{NP} < 0$, where p_P = true proportion of children like those in the study who would need social services as an adult if they go to preschool and $p_{NP} = \ldots$ don't go to preschool. **(b)** Yes. $\hat{p}_P - \hat{p}_{NP} = 0.613 - 0.803 = -0.19 < 0$ **(c)** *Random?* Children were randomly assigned to attend preschool or not. ✓ *Large Counts?* 38, 24, 49, and 12 are all at least 10. ✓

9. (a) $z = \dfrac{(0.437 - 0.396) - 0}{\sqrt{\dfrac{0.419(1-0.419)}{1005} + \dfrac{0.419(1-0.419)}{800}}} = 1.754$ **(b)** *Table A:* P-value $= 2(0.0401) = 0.0802$; *Tech:* Applet/normalcdf(lower: 1.754, upper: 1000, mean: 0, SD: 1) × 2 = 0.0794

11. (a) $z = \dfrac{(0.613 - 0.803) - 0}{\sqrt{\dfrac{0.707(1-0.707)}{62} + \dfrac{0.707(1-0.707)}{61}}} = -2.315$

Table A: P-value $= 0.0102$; *Tech:* Applet/normalcdf(lower: −1000, upper: −2.315, mean: 0, SD: 1) = 0.0103 **(b)** Because the P-value of 0.0103 is less than $\alpha = 0.05$, we reject H_0. We have convincing evidence that the true proportion of children like those in the study who need social services later is less for those who go to preschool than for those who do not.

13. Because the P-value of 0.0103 is less than $\alpha = 0.05$, we reject H_0. We have convincing evidence that the true proportion of children like those in the study who need social services later is less for those who go to preschool than for those who do not. **S:** $H_0: p_1 - p_2 = 0, H_a: p_1 - p_2 < 0$, where p_1 = proportion of all babies with VLBW who graduate from high school and $p_2 = \ldots$ babies with normal birth weight who graduate from high school. Use $\alpha = 0.01$. The evidence for H_a is: $\hat{p}_1 - \hat{p}_2 = 0.740 - 0.828 = -0.088 < 0$. **P:** Two-sample z test for $p_1 - p_2$. *Random?* Independent random samples of 242 babies with VLBW and 233 babies with normal birth weight. ✓ *Large Counts?* 179, 63, 193, and 40 are all at least 10. ✓

D: $z = \dfrac{(0.740 - 0.828) - 0}{\sqrt{\dfrac{0.783(1-0.783)}{242} + \dfrac{0.783(1-0.783)}{233}}} = -2.326$; *Table A:*

P-value $= 0.0099$; *Tech:* $z = -2.344$, P-value $= 0.0095$. **C:** Because the P-value of 0.0095 is less than $\alpha = 0.01$, we reject H_0. We have convincing evidence that the graduation rate among babies with VLBW is less than the graduation rate for babies with normal birth weight.

15. S: $H_0: p_1 - p_2 = 0, H_a: p_1 - p_2 \neq 0$, where p_1 = proportion of all people who would agree that free health care should be provided to homeless individuals after being shown the picture of a homeless woman with a small child and $p_2 = \ldots$ when no picture is shown. Use $\alpha = 0.05$. The evidence for H_a is: $\hat{p}_1 - \hat{p}_2 = 0.675 - 0.45 = 0.225 \neq 0$. **P:** Two-sample z test for $p_1 - p_2$. *Random?* Random sample of 80 people and respondents were randomly assigned to see a picture of a homeless woman with a child or not. ✓ *Large Counts?* 27, 13, 18, and 22 are all at least 10. ✓

D: $z = \dfrac{(0.675 - 0.45) - 0}{\sqrt{\dfrac{0.563(1-0.563)}{40} + \dfrac{0.563(1-0.563)}{40}}} = 2.029$; *Table A:*

P-value $= 2(0.0212) = 0.0424$; *Tech:* $z = 2.028$, P-value $= 0.0425$. **C:** Because the P-value of 0.0425 is less than $\alpha = 0.05$, we reject H_0. We have convincing evidence that showing the picture affects people's opinions about free health care for individuals experiencing homelessness.

17. (a) S: $H_0: p_1 - p_2 = 0, H_a: p_1 - p_2 \neq 0$, where p_1 = true proportion of infants like these who become allergic to peanut butter if they consume a baby-food form of peanut butter and $p_2 = \ldots$ avoid peanut butter. Use $\alpha = 0.05$. The evidence for H_a is: $\hat{p}_1 - \hat{p}_2 = 0.033 - 0.171 = -0.138 \neq 0$. **P:** Two-sample z test for $p_1 - p_2$. *Random?* Infants were randomly assigned to either consume a baby-food form of peanut butter or to avoid peanut butter. ✓ *Large Counts?* 10, 297, 55, and 266 are all at least 10. ✓

D: $z = \dfrac{(0.033-0.171)-0}{\sqrt{\dfrac{0.104(1-0.104)}{307}+\dfrac{0.104(1-0.104)}{321}}} = -5.663$; *Table A:*

P-value ≈ 2(0) = 0; *Tech:* $z = -5.707$, *P*-value ≈ 0. **C:** Because the *P*-value of approximately 0 is less than $\alpha = 0.05$, we reject H_0. We have convincing evidence that there is a difference in the proportion of infants like these who develop peanut allergies among those who consume peanut butter versus those who avoid peanut butter. **(b)** Because we rejected the null hypothesis, it is possible that we may have made a Type I error. That is, we could have found convincing evidence of a difference in the proportion of infants like these who develop peanut allergies among those who consume peanut butter and those who avoid peanut butter when there is actually no difference in the population proportions. **(c)** No. The infants in this study were not randomly selected from the population of all infants, but rather were infants who have shown evidence of other kinds of allergies.

19. (a) See Exercise 17 for definition of parameters and condition check. **S:** 95% CI for $p_1 - p_2$. **P:** Two-sample z interval for $p_1 - p_2$.

D: $(0.033-0.171) \pm 1.960 \sqrt{\dfrac{0.033(1-0.033)}{307}+\dfrac{0.171(1-0.171)}{321}} \rightarrow$

-0.184 to -0.092 *Tech:* -0.185 to -0.093. **C:** We are 95% confident that the interval from -0.185 to -0.093 captures the true difference between the proportions of infants like these who become allergic to peanut butter if they consume a baby-food form of peanut butter and those who avoid peanut butter. The interval suggests that the percentage of infants who consume peanut butter and develop peanut allergies is between 9.3 and 18.5 percentage points less than the percentage for infants who avoid peanut butter and develop peanut allergies. **(b)** The test in Exercise 17 only allowed us to reject $H_0: p_1 - p_2 = 0$ in favor of a two-sided alternative. The confidence interval gives us the entire set of plausible values for the difference in the true proportion of infants who develop peanut allergies among those who consume peanut butter and those who do not: -0.185 to -0.093 (which does not include a difference of 0 as plausible).

21. (a) Assuming that the difference in the proportions of all U.S. college students who used any vaping products in 2019 and 2020 is 0, there is a 0.0794 probability of getting a difference in sample proportions of 0.041 or more extreme (in either direction) by chance alone in random samples of 1005 and 800 students. **(b)** Because the *P*-value of 0.0794 is greater than $\alpha = 0.01$, we fail to reject H_0. We do not have convincing evidence that there is a difference between the proportions of all U.S. college students in 2019 and 2020 who used vaping products.

23. (a) $H_0: p_A - p_B = 0$, $H_a: p_A - p_B \neq 0$, where p_A = true proportion of subjects like these who would admit to texting and driving when asked Version A of the question and $p_B = \ldots$ Version B of the question. **(b)** The Large Counts condition is not satisfied. The number of subjects who did not admit to texting and driving when asked Version A ($25 - 16 = 9$) is less than 10. **(c)** $\hat{p}_A = \dfrac{16}{25} = 0.64$, $\hat{p}_B = \dfrac{12}{25} = 0.48$, $\hat{p}_A - \hat{p}_B = 0.64 - 0.48 = 0.16$. There are 21 repetitions of the simulation with a difference of proportions less than or equal to -0.16 and 14 repetitions of the simulation with a difference of proportions greater than or equal to 0.16. Estimated *P*-value = $35/100 = 0.35$. **(d)** Because the estimated *P*-value of 0.35 is greater than $\alpha = 0.05$, we fail to reject H_0. We do not have convincing evidence that there is a difference between the true proportions of subjects like these who would admit to texting and driving when asked Version A versus when asked Version B of the question.

25. (a) General Motors has won this award in about 40% of the years. In the pie chart, the section for General Motors is more than one-fourth and less than one-half of the pie, and looks as though it is closer to one-half than it is to one-fourth. **(b)** The areas in the pictograph make it seem as if the number of times that General Motors has won the award is more than 4 times as often as the other companies, which isn't the case.

Answers to Section 9.3 Exercises

1. For each sample, the corresponding population distribution is approximately normal or the sample size is large ($n \geq 30$)

3. $(\bar{x}_1 - \bar{x}_2) \pm t^* \sqrt{\dfrac{s_1^2}{n_1} + \dfrac{s_2^2}{n_2}}$

5. *Random?* Participants were randomly assigned to drink half a bottle of either red or white wine each day for 2 weeks. ✓ *Normal/Large Sample?* Both sample sizes are small, but neither dotplot shows any strong skewness or outliers. ✓

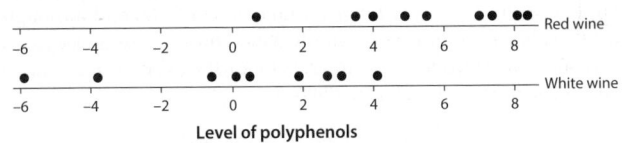

7. *Random?* Independent random samples of residents from Indiana and New Jersey. ✓ *Normal/Large Sample?* Both sample sizes are large. $n_1 = 38 \geq 30$ and $n_2 = 44 \geq 30$. ✓

9. *Random?* Golfers were randomly assigned to describe their putting technique or to do an unrelated verbal task. ✓ *Normal/Large Sample?* Both sample sizes are small and both boxplots are skewed to the right with a high outlier. It is not appropriate to calculate a confidence interval for $\mu_1 - \mu_2$ using these data.

11. (a) *Tech:* 2.845 to 7.689 using df = 14.971; *Conservative approach:* df = $9 - 1 = 8$, $t^* = 1.860$; $(5.5 - 0.233) \pm 1.860 \sqrt{\dfrac{2.517^2}{9} + \dfrac{3.292^2}{9}}$

$\rightarrow$ 2.698 to 7.836 **(b)** Because 0 (no difference) is not a plausible value for the difference in means (2.845, 7.689), there is convincing evidence of a difference in the true mean percent change in polyphenols for subjects like the ones in this study under the two conditions.

13. (a) *Tech:* $-31{,}605$ to $10{,}205$ using df = 76.941; *Conservative approach:* df = $38 - 1 = 37$ (use df = 30), $t^* = 2.750$

$(47{,}400 - 58{,}100) \pm 2.750 \sqrt{\dfrac{29{,}400^2}{38} + \dfrac{41{,}900^2}{44}} \rightarrow -\$32{,}466$ to $\$11{,}066$ **(b)** Because 0 (no difference) is a plausible value for the difference in means ($-31{,}605$, $10{,}205$), there is not convincing evidence of a difference (Indiana − New Jersey) in the mean total family income for all residents in the two states.

15. S: 95% CI for $\mu_B - \mu_G$, where μ_B = the mean weight of all black squirrels in the large forest and $\mu_G = \ldots$ gray squirrels in the large forest. **P:** Two-sample t interval for $\mu_B - \mu_G$. *Random?* Independent random samples of black and gray squirrels. ✓ *Normal/Large Sample?* Both sample sizes are large. $n_B = 40 \geq 30$ and $n_G = 40 \geq 30$. ✓ **D:** *Tech:* 0.21 to 1.99 with df = 77.23; *Conservative approach:* df = $40 - 1 = 39$ (use df = 30), $t^* = 2.042$; $(20.3 - 19.2) \pm 2.042 \sqrt{\dfrac{2.1^2}{40} + \dfrac{1.9^2}{40}} \rightarrow$ 0.19 to 2.01 ounces. **C:** We are 95% confident that the interval from 0.21 to 1.99 ounces captures the difference in mean weight of all black squirrels and all gray squirrels in this forest. The interval suggests that the mean weight for all black squirrels is between 0.21 ounce and 1.99 ounces heavier than the mean weight for all gray squirrels in this forest.

17. S: 95% CI for $\mu_P - \mu_L$, where μ_P = the true mean number of words written by students like these who write notes with a pencil and paper and $\mu_L = \ldots$ who type notes on a laptop. **P:** Two-sample t interval for $\mu_P - \mu_L$. *Random?* Students were randomly assigned to write notes or type notes on a laptop. ✓ *Normal/Large Sample?* Both sample sizes are large. $n_P = 55 \geq 30$ and $n_L = 54 \geq 30$. ✓ **D:** *Tech:* -236.5 to -79.5 with df = 83.78; *Conservative approach:* df = $54 - 1 = 53$ (use df = 50), $t^* = 2.009$;

$(390.7-548.7)\pm 2.009\sqrt{\dfrac{143.9^2}{55}+\dfrac{252.7^2}{54}} \to -237.325$ to -78.675

words. **C:** We are 95% confident that the interval from -236.5 to -79.5 words captures the true difference in the mean number of words written using a pencil and paper or when typing on a laptop. The interval suggests that the true mean number of words written by students like these when taking notes with a pencil and paper is between 79.5 and 236.5 less than when typing notes on a laptop.

19. (a) S: 95% CI for $\mu_1-\mu_2$, where $\mu_1=$ the mean percent change in bone mineral content for all breastfeeding women and $\mu_2=\ldots$ women who are not pregnant or lactating. **P:** Two-sample t interval for $\mu_1-\mu_2$. *Random?* Independent random samples of breastfeeding women and women of similar age who were neither pregnant nor lactating. ✓ *Normal/Large Sample?* $n_1=47\geq 30$ and although $n_2=22<30$, the graph of the sample data shows no strong skewness or outliers. ✓ **D:** *Tech:* -4.817 to -2.983 with df $=66.204$; *Conservative approach:* df $=22-1=21$, $t^*=2.080$;

$(-3.59-0.31)\pm 2.080\sqrt{\dfrac{2.51^2}{47}+\dfrac{1.30^2}{22}} \to -4.855$ to -2.945 percent change. **C:** We are 95% confident that the interval from -4.817 to -2.983 captures the difference (Breastfeeding − Not pregnant or lactating) in the true mean percent change in BMC for these two populations of women. The interval suggests that breastfeeding women have a mean percent bone mineral loss that is between 2.983 and 4.817 percentage points greater than for women who are not pregnant or lactating. **(b)** No. This is an observational study. We cannot draw conclusions of cause and effect from an observational study.

21. (a) No, because our interval $(-0.4, 3.6)$ includes a difference of 0 (no difference) as a plausible value. **(b)** No. Zero is a plausible value for the difference in means $(-0.4, 3.6)$, but there are many plausible values other than 0 in the confidence interval.

23. (a) If we repeatedly took the 109 volunteers and randomly assigned them to the treatments of taking notes with a pencil and paper (55) or typing notes on a laptop (54), and each time constructed a 95% confidence interval in this same way, about 95% of the resulting intervals would capture the true difference in the mean number of words written for students who take notes using pencil and paper and those who take notes with a laptop. **(b)** A 99% confidence interval would be wider than a 95% confidence interval because increasing the confidence level increases the margin of error.

25. S: 95% CI for $\mu_{Off}-\mu_{On}$, where $\mu_{Off}=$ the mean diameter at breast height for all trees that are offset from the curb and $\mu_{On}=\ldots$ on the curb. **P:** Two-sample t interval for $\mu_{Off}-\mu_{On}$. *Random?* Independent random samples of trees that are offset from the curb and on the curb. ✓ *Normal/Large Sample?* Both sample sizes are large. $n_{Off}=337\geq 30$ and $n_{On}=7699\geq 30$. ✓ **D:** *Tech:* 1.072 to 3.334 with df $=355.493$. **C:** We are 95% confident that the interval from 1.072 to 3.334 inches captures the difference in mean diameter at breast height for all trees that are offset from the curb and all trees that are on the curb in New York City. The interval suggests that the mean diameter at breast height for all trees offset from the curb is between 1.072 and 3.334 inches larger than the mean diameter at breast height for all trees that are located on the curb.

27. (a) Yes. Because $n_D=40\geq 30$ and $n_C=30\geq 30$, the shape of the sampling distribution of $\bar{x}_D-\bar{x}_C$ is approximately normal. **(b)** $\mu_{\bar{x}_D-\bar{x}_C}=28-9.5=18.5$ pounds **(c)** $\sigma_{\bar{x}_D-\bar{x}_C}=\sqrt{\dfrac{14^2}{40}+\dfrac{2^2}{30}}=2.24$ pounds **(d)** We want to find $P(\bar{x}_D-\bar{x}_C\geq 20)$.

(i) $z=\dfrac{20-18.5}{2.24}=0.670$; *Tech:* Applet/normalcdf(lower: 0.670, upper: 1000, mean: 0, SD: 1) $=0.2514$; *Table A:* $1-0.7486=0.2514$. (ii) *Tech:* Applet/normalcdf(lower: 20, upper: 1000, mean: 18.5, SD: 2.24) $= 0.2515$.

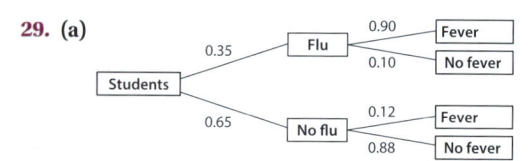

29. (a)

Students
— 0.35 → Flu — 0.90 → Fever
— 0.10 → No fever
— 0.65 → No flu — 0.12 → Fever
— 0.88 → No fever

$P(\text{student has a high fever})=(0.35)(0.90)+(0.65)(0.12)=0.393$

(b) $P(\text{flu}\mid\text{fever})=\dfrac{P(\text{flu and fever})}{P(\text{fever})}=\dfrac{(0.35)(0.90)}{0.393}=0.802$

(c) For any given student, $P(\text{flu and fever})=(0.35)(0.90)=0.315$. Let $X=$ the number of students in the sample of 5 who have both flu and fever. X is a binomial random variable with $n=5$ and $p=0.315$: $P(X=3)={}_5C_3(0.315)^3(1-0.315)^2=0.147$.

Answers to Section 9.4 Exercises

1. $H_0:\mu_1-\mu_2=0$

3. $t=\dfrac{(\bar{x}_1-\bar{x}_2)-0}{\sqrt{\dfrac{s_1^2}{n_1}+\dfrac{s_2^2}{n_2}}}$

5. (a) $H_0:\mu_1-\mu_2=0, H_a:\mu_1-\mu_2\neq 0$, where $\mu_1=$ the mean number of hours worked in a week by all employees in 2018 and $\mu_2=\ldots$ in 1978. **(b)** Yes. The evidence for H_a is: $\bar{x}_1-\bar{x}_2=41.28-40.81=0.47\neq 0$. **(c)** *Random?* Independent random samples of employees in 2018 and 1978. ✓ *Normal/Large Sample?* Both sample sizes are large. $n_1=1381\geq 30$ and $n_2=855\geq 30$. ✓

7. (a) $H_0:\mu_I-\mu_E=0, H_a:\mu_I-\mu_E\neq 0$, where $\mu_I=$ the true mean creativity score for students like these who are given internal reasons for writing and $\mu_E=\ldots$ external reasons for writing. **(b)** Yes. The evidence for H_a is: $\bar{x}_I-\bar{x}_E=19.883-15.739=4.144\neq 0$. **(c)** *Random?* College students were randomly assigned to receive lists of internal or external reasons for writing before writing poems. ✓ *Normal/Large Sample?* The sample sizes are small, but the dotplots don't show any outliers or strong skewness. ✓

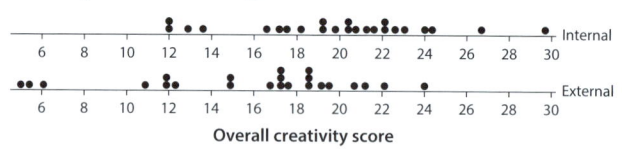

9. (a) $t=\dfrac{(41.28-40.81)-0}{\sqrt{\dfrac{6.59^2}{1381}+\dfrac{8.40^2}{855}}}=1.392$ **(b)** *Tech:* P-value $=0.1641$ with df $=1494.54$. *Conservative approach:* df $=$ smaller of $(1381-1, 855-1)=854$. Use df $=100$. Table B gives a P-value between $2(0.05)=0.10$ and $2(0.10)=0.20$.

11. (a) $t=\dfrac{(19.883-15.739)-0}{\sqrt{\dfrac{4.440^2}{24}+\dfrac{5.253^2}{23}}}=2.915$; *Tech:* P-value $=0.0056$ with df $=43.108$. *Conservative approach:* df $=$ smaller of $(24-1, 23-1)=22$. Table B gives a P-value between $2(0.0025)=0.005$ and $2(0.005)=0.01$. **(b)** Because the P-value of 0.0056 is less than $\alpha=0.05$, we reject H_0. We have convincing evidence that there is a difference in the true mean creativity scores for students like the ones in this study when given internal versus external reasons for writing.

13. S: $H_0:\mu_{ED}-\mu_{ND}=0, H_a:\mu_{ED}-\mu_{ND}>0$, where $\mu_{ED}=$ the mean annual income for the population of all college graduates and $\mu_{ND}=\ldots$ all college students who did not graduate. Use $\alpha=0.05$. The evidence for H_a is: $\bar{x}_{ED}-\bar{x}_{ND}=49,454.80-29,299.20=20,155.60>0$. **P:** Two-sample t test for $\mu_{ED}-\mu_{ND}$. *Random?* Independent random samples of college attendees who did and did not graduate. ✓ *Normal/Large Sample?* Both sample sizes are large. $n_{ED}=327\geq 30$ and $n_{ND}=173\geq 30$. ✓

D: $t = \dfrac{(49{,}454.80 - 29{,}299.20) - 0}{\sqrt{\dfrac{51{,}257.10^2}{327} + \dfrac{38{,}298.00^2}{173}}} = 4.960$; *Tech:* $t = 4.960$, *P*-value ≈ 0 with df = 442.7. *Conservative approach*: df = smaller of (327−1, 173−1) = 172. Use df = 100. Table B gives a *P*-value less than 0.0005. **C:** Because *P*-value ≈ 0 is less than $\alpha = 0.05$, we reject H_0. We have convincing evidence that the mean annual income is higher in the population of college graduates than in the population of college attendees who did not graduate.

15. S: $H_0: \mu_F - \mu_R = 0, H_a: \mu_F - \mu_R > 0$, where μ_F = true mean reduction in blood pressure (mmHg) for subjects like these who use fish oil and $\mu_R = \ldots$ use regular oil. Use $\alpha = 0.05$. The evidence for H_a is: $\bar{x}_F - \bar{x}_R = 6.571 - (-1.143) = 7.714 > 0$. **P:** Two-sample t test for $\mu_F - \mu_R$. *Random?* Subjects were randomly assigned to a diet with fish oil or a diet with regular oil. ✓ *Normal/Large Sample?* The sample sizes are small, but the dotplots don't show any outliers or strong skewness. ✓

Change in blood pressure (mmHg)

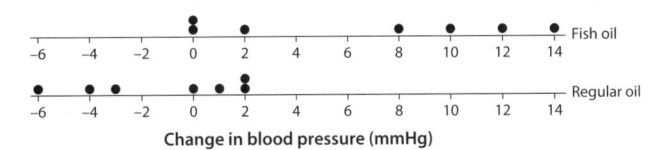

D: $t = \dfrac{[6.571 - (-1.143)] - 0}{\sqrt{\dfrac{5.855^2}{7} + \dfrac{3.185^2}{7}}} = 3.062$; *Tech:* $t = 3.062$, *P*-value = 0.0065 with df = 9.264. *Conservative approach:* df = smaller of (7−1, 7−1) = 6. Table B gives a *P*-value between 0.01 and 0.02. **C:** Because the *P*-value of 0.0065 is less than $\alpha = 0.05$, we reject H_0. We have convincing evidence that fish oil helps reduce blood pressure more than regular oil, on average, for men like the ones in this study.

17. (a) S: $H_0: \mu_1 - \mu_2 = 0, H_a: \mu_1 - \mu_2 > 0$, where μ_1 = true mean DRP score for students like these who do the reading activities and $\mu_2 = \ldots$ do not do the reading activities. Use $\alpha = 0.05$. The evidence for H_a is: $\bar{x}_1 - \bar{x}_2 = 51.48 - 41.52 = 9.96 > 0$. **P:** Two-sample t test for $\mu_1 - \mu_2$. *Random?* Students were randomly assigned to do new reading activities or just follow the standard curriculum. ✓ *Normal/Large Sample?* The sample sizes are small, but the boxplots don't show any outliers or strong skewness. ✓

D: $t = \dfrac{(51.48 - 41.52) - 0}{\sqrt{\dfrac{11.01^2}{21} + \dfrac{17.15^2}{23}}} = 2.312$; *Tech:* $t = 2.312$, *P*-value = 0.0132 with df = 37.859. *Conservative approach:* df = smaller of (21−1, 23−1) = 20. Table B gives a *P*-value between 0.01 and 0.02. **C:** Because the *P*-value of 0.0132 is less than $\alpha = 0.05$, we reject H_0. We have convincing evidence that the true mean DRP score is greater for students like these who do the new reading activities. **(b)** Yes. These data come from two groups (new reading activities or standard curriculum) in a randomized experiment and the results are statistically significant. **(c)** If the mean DRP score is actually the same for both treatments, we will have made a Type I error because we found convincing evidence that the true mean DRP score is higher for students like these who do the new reading activities.

19. (a) See Exercise 17 for the definition of parameters and check of conditions. **S:** 95% CI for $\mu_1 - \mu_2$. **P:** Two-sample t interval for $\mu_1 - \mu_2$. **D:** *Tech:* 1.238 to 18.683 with df = 37.859; *Conservative approach:* df = 21−1 = 20, $t^* = 2.086$; $(51.48 - 41.52) \pm 2.086\sqrt{\dfrac{11.01^2}{21} + \dfrac{17.15^2}{23}} \to 0.97$ to 18.95. **C:** We are 95% confident that the interval from 1.238 to 18.683 points captures the true difference in mean DRP score for students like these who do the reading activities and those who don't. The interval suggests that the true mean DRP score of students like these who do the reading activities is between 1.238 points and 18.683 points greater than the mean DRP score for students like these who do not do the reading activities. **(b)** The test only allows us to reject the null hypothesis. The confidence interval provides additional information that the significance test does not: a set of plausible values (1.238, 18.683) for the true difference in mean DRP score, $\mu_1 - \mu_2$, which does not include 0.

21. (a) Assuming that the difference in the mean number of hours worked by all U.S. employees in 1978 and 2018 is 0, there is a 0.1641 probability of getting a difference in sample means of 0.47 hour or more extreme (in either direction) by chance alone in random samples of 1381 employees in 2018 and 855 employees in 1978. **(b)** Because the *P*-value of 0.1641 is greater than $\alpha = 0.05$, we fail to reject H_0. We do not have convincing evidence of a difference in the mean number of hours worked per week for all U.S. employees in 1978 and 2018.

23. S: $H_0: \mu_1 - \mu_2 = 0, H_a: \mu_1 - \mu_2 \ne 0$, where μ_1 = the mean percentage of aluminum by weight for all vehicle windows that are float processed and $\mu_2 = \ldots$ all headlamps. Use $\alpha = 0.05$. The evidence for H_a is: $\bar{x}_1 - \bar{x}_2 = 1.2012 - 2.1228 = -0.9216 \ne 0$. **P:** Two-sample t test for $\mu_1 - \mu_2$. *Random?* Independent random samples of glass from vehicle windows that were float processed and headlamps. ✓ *Normal/Large Sample?* The sample sizes are small, but the dotplots don't show any outliers or strong skewness. ✓

Percent of aluminum by weight

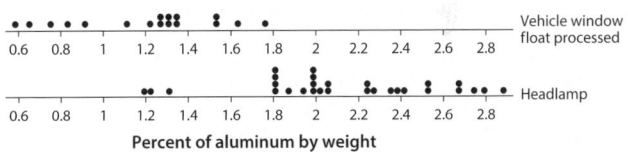

D: *Tech:* $t = 7.828$, *P*-value ≈ 0 with df = 40.159. **C:** Because the *P*-value of ≈ 0 is less than $\alpha = 0.05$, we reject H_0. We have convincing evidence of a difference in the mean amount of aluminum contained in all headlamps and all vehicle windows that are float processed.

25. (a) $H_0: \mu_1 - \mu_2 = 0, H_a: \mu_1 - \mu_2 > 0$, where μ_1 = the true mean number of putts required by subjects like the ones in this study after describing their putting and $\mu_2 = \ldots$ not describing their putting. **(b)** From the experiment, $\bar{x}_1 - \bar{x}_2 = 21.2 - 10.6 = 10.6$. From the 500 simulated differences, only 1 had a difference in means of 10.6 or greater. The estimated *P*-value is 1/500 = 0.002. **(c)** Because the estimated *P*-value = 0.002 is less than $\alpha = 0.05$, we reject H_0. The data give convincing evidence of overthinking it when putting by golfers like the ones in this study.

27. (a) Using the conservative df = 29, the critical value is $t^* = 1.699$. (Using df = 47.061, the critical value is $t^* = 1.678$.) The corresponding rejection region is $t > 1.699$ (or $t > 1.678$). **(b)** From Exercise 18, $t = 3.688$. Because $t = 3.688 > 1.699$ (or 1.678), we reject H_0. We have convincing evidence that the true mean number of mistakes found will be greater when people like these are told that the author is a non-native speaker.

29. (a) The treatments are: (1) 25% of all goods on sale, stated precisely (2) 50% of all goods on sale, stated precisely (3) 75% of all goods on sale, stated precisely (4) 100% of all goods on sale, stated precisely (5) 25% of all goods on sale, stated as a range (6) 50% of all goods on sale, stated as a range (7) 75% of all goods on sale, stated as a range (8) 100% of all goods on sale, stated as a range **(b)** (1) Label each volunteer with a different integer from 1 to 200. (2) Randomly generate 25 different integers from 1 to 200. The volunteers with these labels will be given treatment 1: 25% of all goods on sale, stated precisely. (3) Randomly generate an additional 25 different integers from 1 to 200, being careful to not choose volunteers who have already been selected. The volunteers with these labels will be given treatment 2: 50% of all goods on sale, stated precisely. (4) Continue in this manner until all of the volunteers are randomly assigned to the treatment groups. **(c)** The purpose of random assignment is to avoid confounding by creating eight groups of volunteers who are roughly equivalent at the beginning of the experiment. **(d)** The graph shows that as the percentage of goods on sale increases, the mean attractiveness of the sale decreases when a discount range is provided. The graph also shows that as the percentage of goods on sale increases, the mean attractiveness of the sale increases dramatically when the discount is stated precisely. When 25% of the goods are on sale, however, the mean attractiveness is greater when the discount is stated as a range than when it is stated precisely. When 50% of the goods are on sale, the mean attractiveness is about the same no matter whether the discount is stated precisely or as a range. When 75% or 100% of the goods are on sale, the mean attractiveness is greater when the discount is stated precisely.

Answers to Section 9.5 Exercises

1. Paired data result from recording two values of the same quantitative variable for each individual or for each pair of similar individuals.
3. mean difference ($\bar{x}_{\text{diff}}$); standard deviation of the differences (s_{diff})
5. True
7.

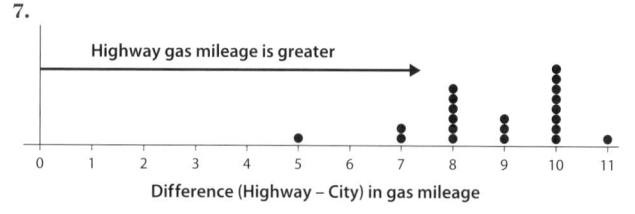

Highway gas mileage is greater

Difference (Highway − City) in gas mileage

All of the differences are positive, so the highway gas mileage exceeds the city gas mileage for all 21 vehicles. The differences vary from 5 mpg (Toyota Prius) to 11 mpg (Lincoln MKZ).
9. (a) Differences: 5, 246, −46, 121, 30, 55, 79, −94, −17, 95, 20, 14, 129, −39, 42 (b) Eleven of the differences are positive, so the transaction time for the regular lane exceeds the transaction time for the express lane in 11 out of the 15 visits to the supermarket. The differences vary from −94 to 246 seconds.
11. $\bar{x}_{\text{diff}} = 8.81$, $s_{\text{diff}} = 1.436$. The highway gas mileage for these 21 vehicles is 8.81 mpg higher than the city gas mileage, on average.
13. (a) $\bar{x}_{\text{diff}} = 42.667$, $s_{\text{diff}} = 84.019$ (b) The transaction time for the regular lane in these 15 visits to the supermarket is 42.667 seconds longer than the transaction time for the express lane, on average. (c) The difference (Regular lane − Express lane) in transaction time typically varies from the mean difference by about 84.019 seconds.
15. S: 95% CI for μ_{diff} = the true mean difference (Regular lane − Express lane) in transaction time (seconds). **P:** Paired t interval for μ_{diff}. *Random?* Paired data come from transaction times in the regular and express lanes for a random sample of 15 times during the week. ✓ *Normal/Large Sample?* The sample size is small ($n_{\text{diff}} = 15 < 30$), but the dotplot of the differences in Exercise 9 doesn't show any outliers or strong skewness. ✓ **D:** With 95% confidence and df = 15 − 1 = 14, $t^* = 2.145$. $42.667 \pm 2.145 \frac{84.019}{\sqrt{15}} \rightarrow -3.866$ to 89.2 seconds; *Tech:* (−3.862, 89.195) with df = 14. **C:** We are 95% confident that the interval from −3.862 seconds to 89.195 seconds captures the true mean difference (Regular lane − Express lane) in transaction time. This interval suggests that the transaction times are between 3.861 seconds shorter and 89.195 seconds longer, on average, when using the regular lane.
17. (a) **S:** 90% CI for μ_{diff} = the mean difference (After − Before) in reasoning scores for all preschool students who take 6 months of piano lessons. **P:** Paired t interval for μ_{diff}. *Random?* Paired data come from reasoning scores before and after 6 months of piano lessons for a random sample of 34 preschool children. ✓ *Normal/Large Sample?* The sample size is large: $n_{\text{diff}} = 34 \geq 30$. ✓ **D:** With 90% confidence and df = 34 − 1 = 33 (use df − 30), $t^* − 1.697$. $3.618 \pm 1.697 \frac{3.055}{\sqrt{34}}$; 2.729 to 4.507; *Tech:* (2.731, 4.505) with df = 33.
C: We are 90% confident that the interval from 2.731 to 4.505 captures the mean difference (After − Before) in reasoning scores for all preschool students who take 6 months of piano lessons. This interval suggests that the reasoning scores are between 2.731 and 4.505 greater, on average, after 6 months of piano lessons. (b) No. Although the confidence interval in part (a) does not include 0 as a plausible value for μ_{diff}, we cannot conclude that taking 6 months of piano lessons caused an increase in scores because this was an observational study, not an experiment.
19. (a)

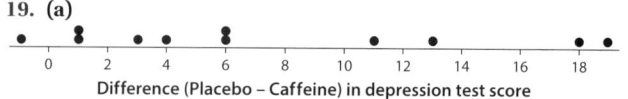

Difference (Placebo − Caffeine) in depression test score

The graph reveals that all but one of the volunteers were more depressed when they did not receive their normal caffeine intake. (b) **S:** 90% CI for μ_{diff} = the true mean change in depression test score (Placebo − Caffeine) for all caffeine-dependent subjects like these when they do not receive their normal caffeine intake. **P:** Paired t interval for μ_{diff}. *Random?* Paired data come from depression test scores with and without caffeine generated by the random assignment of 11 volunteers to a treatment order. ✓ *Normal/Large Sample?* The sample size is small ($n_{\text{diff}} = 11 < 30$), but the dotplot of the differences in in part (a) doesn't show any outliers or strong skewness. ✓ **D:** With 90% confidence and df = 11 − 1 = 10, $t^* = 1.812$. $7.364 \pm 1.812 \frac{6.918}{\sqrt{11}}$; 3.584 to 11.144; *Tech:* (3.583, 11.144) with df = 10. **C:** We are 90% confident that the interval from 3.583 to 11.144 captures the true mean change in depression scores for all caffeine-dependent subjects like these when they do not receive their normal caffeine intake. This interval suggests that when caffeine-dependent subjects like these do not receive their normal caffeine intake, their depression increases, on average, between 3.583 and 11.144 points. (c) Yes, because the volunteers were randomly assigned to a treatment order, and both endpoints of the interval are positive (3.583, 11.144), we have convincing evidence that caffeine deprivation causes an increase in the average depression test scores for subjects like these.
21. The Random condition would not be satisfied, as there was no random selection of car models. These 21 car models were the choice of the interested buyer.
23. (a) It was important for the researchers to randomly assign the order to the volunteers to avoid confounding. If the volunteers were all given the same treatment order, the order of the treatment might be confounded by the number of words the volunteers remembered. For example, they may improve slightly the second time around due to their experience. (b) *Random?* Paired data come from the number of words remembered with and without chewing gum generated by the random assignment of 30 volunteers to a treatment order. ✓ *Normal/Large Sample?* The sample size is large: $n_{\text{diff}} = 30 \geq 30$. ✓ (c) *Interval:* We are 95% confident that the interval from −0.67 to 1.54 words captures the true mean difference (Gum − No gum) in number of words remembered for subjects like these. This interval suggests that the true mean number of words remembered would be between 0.67 less and 1.54 more when chewing gum than when not chewing gum. *Level:* If we repeatedly took the 30 volunteers and randomly assigned them to the treatments of gum and no gum, and each time constructed a 95% confidence interval in this same way, about 95% of the resulting intervals would capture the true mean difference in the number of words recalled for volunteers like these. (d) Because zero is a plausible value for the true mean difference (−0.67, 1.54), we do not have convincing evidence that chewing gum helps subjects like the ones in this study with short-term memory.
25. S: 95% CI for μ_{diff} = the true mean difference (Iced − Hot) in coffee price for all establishments near this university. **P:** Paired t interval for μ_{diff}. *Random?* Paired data come from the cost of iced and hot coffee for a random sample of 41 establishments. ✓ *Normal/Large Sample?* The sample size is large: $n_{\text{diff}} = 41 \geq 30$. ✓ **D:** *Tech:* (0.4886, 0.8377) with df = 40. **C:** We are 95% confident that

the interval from $0.49 to $0.84 captures the true mean difference (Iced − Hot) in coffee price for all establishments near this university. This interval suggests that the price of an iced coffee is between $0.49 to $0.84 greater than the price of a hot coffee, on average, for all establishments near this university.

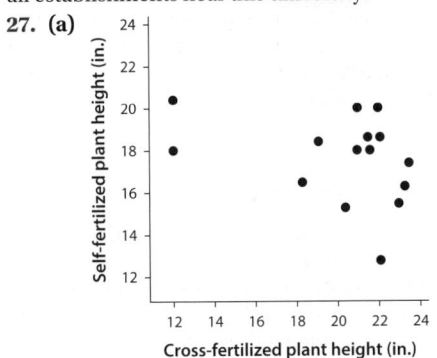

27. (a)

(b) No. There is a negative correlation ($r = -0.3379$) between cross-fertilized plant height and self-fertilized plant height for these 15 pairs of plants. **(c)** $\bar{x}_{cross} = 20.193$, $s_{cross} = 3.616$, $\bar{x}_{self} = 17.587$, $s_{self} = 2.038$ **(d)** The difference in the means $= 20.193 - 17.587 = 2.606$ inches. This is the same as the mean difference in height. **(e)** $s_{cross} = 3.616$, $s_{self} = 2.038$, $s_{diff} = 4.713$. In this case, the pairing did not help reduce the variability in height in this study.

29. (a) $H_0: \mu = 1.2$, $H_a: \mu > 1.2$, where $\mu =$ the mean weight of all tomatoes grown with the new fertilizer. **(b)** *Type II:* You do not find convincing evidence that the true mean weight for all tomatoes grown with the new fertilizer is greater than 1.2 pounds, when it really is greater than 1.2 pounds. *Consequence:* You don't use the fertilizer and tomatoes are smaller than they could have been. **(c)** Assuming that the mean weight for the population of all tomatoes grown using the new fertilizer is 1.2 pounds, there is a 0.062 probability of getting a sample mean of 1.33 pounds or greater by chance alone in a random sample of 35 tomatoes. **(d)** Because the *P*-value of $0.062 > \alpha = 0.05$, we fail to reject H_0. We do not have convincing evidence that the mean weight of all tomatoes grown using the new fertilizer is greater than 1.2 pounds.

Answers to Section 9.6 Exercises

1. $H_0: \mu_{diff} = 0$

3. $t = \dfrac{\bar{x}_{diff} - 0}{\dfrac{s_{diff}}{\sqrt{n_{diff}}}}$

5. S: $H_0: \mu_{diff} = 0$, $H_a: \mu_{diff} > 0$, where $\mu_{diff} =$ the true mean difference (Kiln − Regular) in barley yield (pounds per acre). Use $\alpha = 0.05$. The evidence for H_a is: $\bar{x}_{diff} = 33.727 > 0$. **P:** Paired t test for μ_{diff}. *Random?* Random assignment of one plot in each pair to get kiln-dried seeds and one to get regular seeds. ✓ *Normal/Large Sample?* The sample size is small ($n_{diff} = 11 < 30$), but the dotplot of the differences doesn't show any outliers or strong skewness. ✓

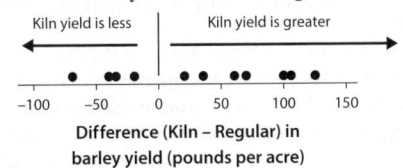

D: $t = \dfrac{33.727 - 0}{\dfrac{66.171}{\sqrt{11}}} = 1.690$; df $= 11 - 1 = 10$; *Table B:* The *P*-value is between 0.05 and 0.10; *Tech:* $t = 1.690$, *P*-value $= 0.0609$ with df $= 10$. **C:** Because the *P*-value of 0.0609 is greater than $\alpha = 0.05$, we fail to reject H_0. There is not convincing evidence that drying barley seeds in a kiln increases the yield of barley, on average.

7. (a) S: $H_0: \mu_{diff} = 0$, $H_a: \mu_{diff} < 0$, where $\mu_{diff} =$ the true mean difference (After − Before) in spin rate (rpm). Use $\alpha = 0.01$. The evidence for H_a is: $\bar{x}_{diff} = -74.114 < 0$. **P:** Paired t test for μ_{diff}. *Random?* Random sample of 44 pitchers. ✓ *Normal/Large Sample?* The number of differences in the sample is large: $n_{diff} = 44 \geq 30$. ✓

D: $t = \dfrac{-74.114 - 0}{\dfrac{79.948}{\sqrt{44}}} = -6.149$; df $= 44 - 1 = 43$; *Table B:* The *P*-value is less than 0.0005 (using df $= 40$); *Tech:* $t = -6.149$, *P*-value ≈ 0 with df $= 43$. **C:** Because the *P*-value of approximately 0 is less than $\alpha = 0.01$, we reject H_0. There is convincing evidence of a decrease in the average spin rate of all MLB pitchers from before to after the announcement of stricter enforcement of the sticky substances rules. **(b)** No, this was an observational study, not an experiment, so we don't know if the decrease in the average spin rate was caused by the announced changed in enforcement by MLB officials.

9. (a) Paired t procedures. The data come from a matched-pairs experiment with the two treatments (new fin or old fin) being randomly assigned to the left or right foot of each scuba diver. **(b)** Two-sample t procedures. The data come from two independent random samples: 60 white piranha in a recent year and 82 white piranha a decade ago. **(c)** Two-sample t procedures. The data come from two groups in a randomized experiment: 8 time periods assigned the new suit and 8 time periods assigned the black wet suit.

11. (a) Two-sample t procedures. The data come from two groups in a randomized experiment: 50 cars assigned Brand A tires and 50 cars assigned Brand B tires. **(b)** Paired t procedures. The data come from a matched-pairs experiment with the order of the two treatments (no music and music) being randomly assigned.

13. (a) There are two values of the same quantitative variable recorded for each patient with insomnia: sleep time before and after taking the drug. **(b) S:** $H_0: \mu_{diff} = 0$, $H_a: \mu_{diff} > 0$, where $\mu_{diff} =$ the true mean difference (Drug − No drug) in amount of sleep for patients with insomnia (hours). Use $\alpha = 0.05$. The evidence for H_a is: $\bar{x}_{diff} = 2.33 > 0$. **P:** Paired t test for μ_{diff}. *Random?* Random sample of 10 patients. ✓ *Normal/Large Sample?* The sample size is small, but the dotplot of the differences doesn't show any outliers or strong skewness. ✓

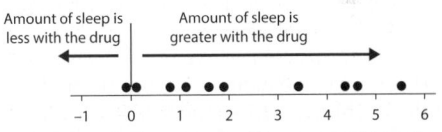

D: $t = \dfrac{2.33 - 0}{\dfrac{2.002}{\sqrt{10}}} = 3.680$; df $= 10 - 1 = 9$; *Table B:* The *P*-value is between 0.0025 and 0.005; *Tech:* $t = 3.68$, *P*-value $= 0.003$ with df $= 9$. **C:** Because the *P*-value of 0.003 is less than $\alpha = 0.05$, we reject H_0. There is convincing evidence that patients with insomnia would get more sleep, on average, when taking the drug. **(c)** No. There is no control group with which to compare results. It may be that the increase in sleep hours is due to the placebo effect. **(d)** If patients like these really do not get more sleep with the drug, we will have made a Type I error because we found convincing evidence that patients like these would get more sleep, on average, when taking the drug.

15. (a) See Exercise 13 for the definition of parameters and check of conditions. **S:** 90% CI for μ_{diff}. **P:** Paired t interval for μ_{diff}. **D:** With 90% confidence and df $= 10 - 1 = 9$, $t^* = 1.833$; $2.33 \pm 1.833 \dfrac{2.002}{\sqrt{10}} \rightarrow$ 1.170 to 3.490; *Tech:* 1.169 to 3.491 with df $= 9$. **C:** We are 95% confident that the interval from 1.169 hours to 3.491 hours captures the true mean difference (Drug − No drug) in amount of sleep. This interval suggests that patients with insomnia will sleep between 1.169 and

3.491 hours longer, on average, when taking the drug. **(b)** The test only allowed us to reject $H_0: \mu_{diff} = 0$ in favor of the one-sided alternative. The confidence interval gives us the entire set of plausible values for the true mean difference in amount of sleep for insomnia patients who use the drug: 1.169 to 3.491 hours (which does not include a difference of 0 as plausible).

17. (a) S: $H_0: \mu_{diff} = 0, H_a: \mu_{diff} \neq 0$, where μ_{diff} = the true mean difference (List – Sale) in price for all dwellings sold along the South Carolina coast during the third quarter of 2020. Use $\alpha = 0.01$. The evidence for H_a is: $\bar{x}_{diff} = 26,653.63 \neq 0$. **P:** Paired t test for μ_{diff}. *Random?* Random sample of 68 dwellings sold along the South Carolina coast during the third quarter of 2020. ✓ *Normal/Large Sample?* $n_{diff} = 68 \geq 30$. ✓ **D:** Tech: $t = 4.282$, P-value ≈ 0 with df = 67. **C:** Because the P-value of less than 0.001 is less than $\alpha = 0.01$, we reject H_0. There is convincing evidence of a nonzero mean difference in the list price and the sales price in the population of South Carolina coastal dwellings sold during the third quarter of 2020. **(b)** See part (a) for the definition of parameters and check of conditions. **S:** 99% CI for μ_{diff}. **P:** Paired t interval for μ_{diff}. **D:** Tech: $10,150.90 to $43,156.37 with df = 67. **C:** We are 99% confident that the interval from $10,150.90 to $43,156.37 captures the true mean difference (List – Sale) in price for the population of South Carolina coastal dwellings sold during the third quarter of 2020. This interval suggests that during the third quarter of 2020, coastal dwellings in South Carolina sold for $10,150.90 to $43,156.37 less than the listing price, on average.

19. (a) Paired t test using Gain. The data are paired by student (Try 1 and Try 2) and come from a random sample of 427 students who were coached. **(b) S:** $H_0: \mu_{diff} = 0, H_a: \mu_{diff} > 0$, where μ_{diff} = the true mean difference (Try 2 – Try 1) in SAT Verbal score for all students who are coached. Use $\alpha = 0.05$. The evidence for H_a is: $\bar{x}_{diff} = 29 > 0$. **P:** Paired t test for μ_{diff}. *Random?* Random sample of 427 students who were coached. ✓ *Normal/Large Sample?* $n_{diff} = 427 \geq 30$. ✓ **D:** $t = \dfrac{29 - 0}{\dfrac{59}{\sqrt{427}}} = 10.157$; df = 427 − 1 = 426; *Table B:* Use df = 100. The P-value is less than 0.0005. Tech: $t = 10.157$, P-value ≈ 0 with df = 426. **C:** Because the P-value of ≈ 0 is less than $\alpha = 0.05$, we reject H_0. There is convincing evidence that students who are coached increase their SAT scores, on average.

21. (a) $H_0: \mu_{diff} = 0, H_a: \mu_{diff} > 0$, where μ_{diff} = the true mean difference (New – Current) in the number of pain-free hours for all volunteers like these. **(b)** $\bar{x}_{diff} = 2.18$ hours, $s_{diff} = 3.762$ hours **(c)** It is not appropriate to carry out a paired t test in this case because the number of differences is small and a dotplot of the differences reveals that there is an outlier.

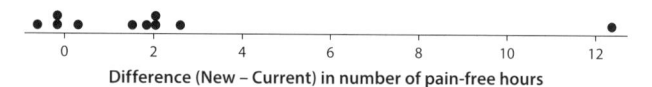

Difference (New – Current) in number of pain-free hours

(d) The probability of obtaining a mean difference of 2.18 or greater is about 7/500 = 0.014 based on the simulation. Because the estimated P-value of 0.014 is less than 0.05, we have convincing evidence that the new drug reduces arthritis pain more than the current drug, on average, for subjects like these.

23. (a) One-sample z interval for p **(b)** Paired t test for μ_{diff} **(c)** Two-sample z interval for $p_1 - p_2$ **(d)** Two-sample t test for $\mu_1 - \mu_2$

Answers to Section 9.7 Exercises

1. $H_0: \sigma_1 = \sigma_2$

3. $df_{Num} = n_L - 1$; $df_{Denom} = n_S - 1$

5. (a) $H_0: \sigma_A = \sigma_S, H_a: \sigma_A \neq \sigma_S$, where σ_A = the standard deviation of the amount of oxygen consumed for all crabs like the ones in the experiment that are exposed to ambient noise and $\sigma_S = \ldots$ exposed to ship noise. **(b)** $s_A = 16.549 \neq 14.319 = s_S$ **(c)** *Random?* The crabs were randomly assigned to the ambient noise and ship noise groups. ✓ *Normal?* Neither boxplot shows any strong skewness or outliers. ✓

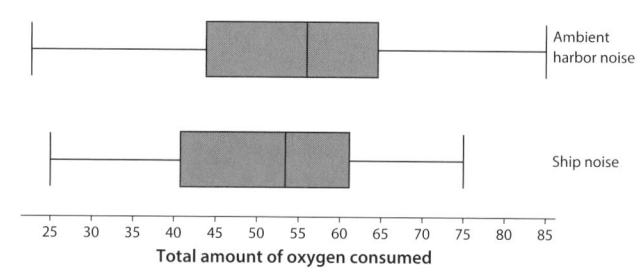

7. (a) $H_0: \sigma_R = \sigma_W, H_a: \sigma_R < \sigma_W$, where σ_R = the standard deviation of percent change in polyphenols for all healthy participants like these who drink half a bottle of red wine each day for two weeks and $\sigma_W = \ldots$ white wine each day for two weeks. **(b)** $s_R = 2.517 < 3.290 = s_W$ **(c)** *Random?* The subjects were randomly assigned to drink red or white wine. ✓ *Normal?* Neither boxplot shows any strong skewness or outliers. ✓

9. Despite the fact that the samples are selected randomly and the sample sizes are large, the Normal condition is not met because the distribution of total family income for Indiana shows an outlier and the distribution of total family income for New Jersey is clearly skewed to the right.

11. (a) $F = \dfrac{16.549^2}{14.319^2} = 1.336$ **(b)** $df_{Num} = 17 - 1 = 16$ and $df_{Denom} = 17 - 1 = 16$; Tech: P-value = $2(0.2845) = 0.5690$; *Table D:* P-value $> 2(0.10) = 0.20$.

13. (a) $F = \dfrac{3.29^2}{2.517^2} = 1.709$; $df_{Num} = 9 - 1 = 8$ and $df_{Denom} = 9 - 1 = 8$; Tech: P-value = 0.2326; *Table D:* P-value > 0.10. **(b)** Because the P-value of 0.2326 is greater than $\alpha = 0.05$, we fail to reject H_0. We do not have convincing evidence that the standard deviation of percent change in polyphenols is smaller for red wine than for white wine in people like those in the study.

15. S: $H_0: \sigma_W = \sigma_N, H_a: \sigma_W \neq \sigma_N$, where σ_W = the standard deviation of number of barbell curl repetitions for all subjects like the ones in the experiment who are given encouragement and $\sigma_N = \ldots$ not given encouragement. Use $\alpha = 0.01$. The evidence for H_a is: $s_W = 22.363 \neq 17.232 = s_N$. **P:** Two-sample F test comparing two standard deviations. *Random?* The subjects were randomly assigned to receive encouragement or no encouragement. ✓ *Normal?* Neither dotplot shows any strong skewness or outliers. ✓

Number of barbell curl repetitions

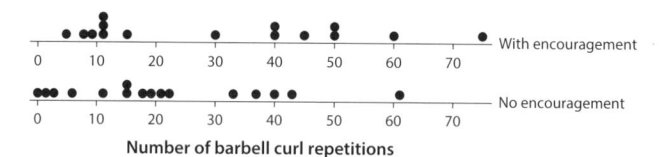

D: $F = \dfrac{22.363^2}{17.232^2} = 1.684$, $df_{Num} = 15 - 1 = 14$ and $df_{Denom} = 16 - 1 = 15$; Tech: $F = 1.684$, P-value = 0.3278; *Table D:* P-value $> 2(0.10) = 0.20$. **C:** Because the P-value of 0.3278 is greater than $\alpha = 0.01$, we fail to reject H_0. We do not have convincing evidence that the standard deviation of the number of barbell curl repetitions that people like the ones in this study can do differs depending on whether they receive encouragement or not.

17. S: $H_0: \sigma_A = \sigma_C, H_a: \sigma_A < \sigma_C$, where σ_A = the standard deviation of DRP score for all subjects like the ones in the experiment who do the activities and $\sigma_C = \ldots$ do not do the activities. Use $\alpha = 0.05$. The evidence for H_a is: $s_A = 11.01 < 17.15 = s_C$. **P:** Two-sample F test comparing two standard deviations. *Random?* The subjects were randomly assigned to do the activities or not do the activities. ✓ *Normal?* Neither boxplot shows any strong skewness or outliers. ✓

D: $F = \dfrac{17.15^2}{11.01^2} = 2.426$; $df_{Num} = 23 - 1 = 22$ and $df_{Denom} = 21 - 1 = 20$; *Tech:* $F = 2.426$, *P*-value = 0.0254; *Table D: P*-value is between 0.025 and 0.05. **C:** Because the *P*-value of 0.0254 is less than $\alpha = 0.05$, we reject H_0. We have convincing evidence that the standard deviation in DRP score is less for students like the ones in this study who do the activities than for students who do not do the activities.

19. (a) S: $H_0: \sigma_R^2 = \sigma_Y^2, H_a: \sigma_R^2 \neq \sigma_Y^2$, where $\sigma_R^2 = $ the variance of length of all *H. caribaea* red flowers on the island of Dominica and $\sigma_Y^2 = \ldots$ yellow flowers on the island of Dominica. Use $\alpha = 0.05$. The evidence for H_a is: $s_R^2 = 3.1915 \neq 0.9513 = s_Y^2$. **P:** Two-sample *F* test comparing two variances. *Random?* Independent random samples of *H. caribaea* red and *H. caribaea* yellow flowers. ✓ *Normal?* Neither dotplot shows any strong skewness or outliers. ✓

D: $F = \dfrac{3.1915}{0.9513} = 3.355$; $df_{Num} = 23 - 1 = 22$ and $df_{Denom} = 15 - 1 = 14$; *Tech:* $F = 3.355$, *P*-value = 0.0231; *Table D: P*-value is between $2(0.01) = 0.02$ and $2(0.025) = 0.05$. **C:** Because the *P*-value of 0.0231 is less than $\alpha = 0.05$, we reject H_0. We have convincing evidence of a difference in the variances of the lengths of these two species of flower on the island of Dominica. **(b)** If the variances of the lengths of these two species of flower are equal, we will have made a Type I error because we concluded that the variances are not equal.

21. (a) Assuming that the true standard deviation in the amounts of oxygen consumed by all crabs like these exposed to ambient harbor noise and ship noise is equal, there is a 0.5690 probability of getting a ratio of sample variances at least as surprising (in either direction) as 1.336 by chance alone in the random assignment of 34 crabs to ambient noise or ship noise. **(b)** Because the *P*-value of 0.5690 is greater than $\alpha = 0.05$, we fail to reject H_0. We do not have convincing evidence of a difference in standard deviations of oxygen consumption between crabs like the ones in the study exposed to ship noise and crabs like the ones in the study exposed to ambient harbor noise.

23. S: $H_0: \sigma_F = \sigma_M, H_a: \sigma_F \neq \sigma_M$, where $\sigma_F = $ the standard deviation of foot length for all female Australian possums and $\sigma_M = \ldots$ male Australian possums. Use $\alpha = 0.05$. The evidence for H_a is: $s_F = 4.911 \neq 3.982 = s_M$. **P:** Two-sample *F* test comparing two standard deviations. *Random?* Independent random samples of male and female Australian possums. ✓ *Normal?* Neither dotplot shows any strong skewness or outliers. ✓

D: *Tech:* $F = 1.521$, *P*-value = 0.1365. **C:** Because the *P*-value of 0.1365 is greater than $\alpha = 0.05$, we fail to reject H_0. We do not have convincing evidence of a difference in the standard deviations of the foot length of all male and female Australian possums.

25. (a) The 95% CI is

$$\dfrac{1}{F^*_{22,14}}\left(\dfrac{s_1^2}{s_2^2}\right) \text{ to } F^*_{14,22}\left(\dfrac{s_1^2}{s_2^2}\right) \to \dfrac{1}{2.814}\left(\dfrac{1.786^2}{0.975^2}\right) \text{ to } 2.528\left(\dfrac{1.786^2}{0.975^2}\right) \to$$

1.192 to 8.483. The 95% confidence interval for the ratio of the population standard deviations is $\sqrt{1.192}$ to $\sqrt{8.483}$; 1.092 to 2.913 **(b)** In the test, the *P*-value of 0.0231 is less than $\alpha = 0.05$, so we reject H_0. The 95% confidence interval does not contain $\dfrac{\sigma_R}{\sigma_Y} = 1$ as a plausible value, so we reject $H_0: \sigma_R = \sigma_Y$ at the $\alpha = 0.05$ level. Therefore, the decision about the two-sided test is consistent with the decision made based on the corresponding confidence interval. The confidence interval gives more information than the test, as it provides the entire set of plausible values for the ratio of the population standard deviations, $\dfrac{\sigma_R}{\sigma_Y}$.

27. (a) (i) $z = \dfrac{138 - 139.4}{0.58} = -2.414$; *Tech:* $P(Z < -2.414) = $ Applet/normalcdf(lower: -1000, upper: -2.414, mean: 0, SD 1) $= 0.0079$; *Table A:* $P(Z < -2.41) = 0.0080$ **(ii)** *Tech:* $P(X < 138) = $ Applet/normalcdf(lower: -1000, upper: 138, mean: 139.4, SD: 0.58) $= 0.0079$ **(b)** *Shape:* The sampling distribution is normal because the population distribution is normal. *Center:* $\mu_{\bar{x}} = 139.4$ ounces. *Variability:* $\sigma_{\bar{x}} = \dfrac{0.58}{\sqrt{5}} = 0.259$ ounce **(i)** $z = \dfrac{138 - 139.4}{0.259} = -5.405$; *Tech:* $P(Z < -5.405) = $ Applet/normalcdf(lower: -1000, upper: -5.405, mean: 0, SD: 1) ≈ 0; *Table A:* $P(Z < -5.41) < 0.0003$ **(ii)** *Tech:* $P(\bar{x} < 138) = $ Applet/normalcdf(lower: -1000, upper: 138, mean: 139.4, SD: 0.259) ≈ 0

Answers to Chapter 9 Review Exercises

1. S: $H_0: p_1 - p_2 = 0, H_a: p_1 - p_2 \neq 0$, where $p_1 = $ the true proportion of patients with RLS like the ones in this experiment who will experience much improved symptoms if they take pramipexole and $p_2 = \ldots$ take a placebo. Use $\alpha = 0.05$. The evidence for H_a is: $\hat{p}_1 - \hat{p}_2 = 0.819 - 0.543 = 0.276 \neq 0$. **P:** Two-sample *z* test for $p_1 - p_2$. *Random?* Patients were randomly assigned to take pramipexole or a placebo. ✓ *Large Counts?* $n_1\hat{p}_1 = 158$, $n_1(1 - \hat{p}_1) = 35$, $n_2\hat{p}_2 = 50$, and $n_2(1 - \hat{p}_2) = 42$ are all ≥ 10. ✓

D: $z = \dfrac{(0.819 - 0.543) - 0}{\sqrt{\dfrac{0.730(1 - 0.730)}{193} + \dfrac{0.730(1 - 0.730)}{92}}} = 4.907$; *Table A: P*-value $< 2(0.0003) = 0.0006$; *Tech:* $z = 4.89$, *P*-value ≈ 0. **C:** Because the *P*-value of approximately 0 is less than $\alpha = 0.05$, we reject H_0. We have convincing evidence that there is a difference in the true proportion of patients with RLS like the ones in this experiment who will experience much improved symptoms if they take pramipexole than if they take a placebo.

2. See Exercise 1 for the definition of parameters and check of conditions. **(a) S:** 95% CI for $p_1 - p_2$. **P:** Two-sample *z* interval for $p_1 - p_2$.

D: $(0.819 - 0.543) \pm 1.960\sqrt{\dfrac{0.819(1 - 0.819)}{193} + \dfrac{0.543(1 - 0.543)}{92}} \to 0.161$ to 0.391; *Tech:* 0.160 to 0.391. **C:** We are 95% confident that the interval from 0.160 to 0.391 captures the difference (Pramipexole − Placebo) in the true proportions of patients with RLS like these who will experience much improved symptoms. The interval suggests that the percentage of patients with RLS like these who would experience much improved symptoms when taking pramipexole is between 16.0 and 39.1 percentage points higher than the percentage who would experience much improved symptoms when taking a placebo. **(b)** The significance test allowed us to reject $H_0: p_1 - p_2 = 0$ in favor of the two-sided alternative. The confidence interval gives us the entire set of plausible values for the difference in the true proportion of patients with RLS like the ones in this experiment who will experience much improved symptoms when taking pramipexole than when taking a placebo: 0.160 to 0.391 (which does not include a difference of 0 as plausible).

3. S: 99% CI for $\mu_P - \mu_A$, where $\mu_P = $ the mean relative tail spine length for all *Daphnia pulicaria* with fish kairomone present and $\mu_A = \ldots$ fish kairomone absent. **P:** Two-sample *t* interval for $\mu_P - \mu_A$. *Random?* Independent random samples of *Daphnia pulicaria* with fish kairomone present and absent. ✓ *Normal/Large Sample?*

$n_P = 214 \geq 30$ and $n_A = 152 \geq 30$. ✓ **D:** *Tech:* 5.381 to 7.799 with df = 345.084; df = 152 − 1 = 151 (use df = 100), $t^* = 2.626$;

$(37.26 - 30.67) \pm 2.626 \sqrt{\dfrac{4.68^2}{214} + \dfrac{4.19^2}{152}} \rightarrow 5.364$ to 7.816. **C:** We are 95% confident that the interval from 5.381 to 7.799 captures the difference in mean relative tail spine length for all *Daphnia pulicaria* with fish kairomone present and absent. The interval suggests that the mean relative tail spine length for all *Daphnia pulicaria* with fish kairomone present is between 5.381 to 7.799 percentage points longer than the mean relative tail spine length for all *Daphnia pulicaria* with fish kairomone absent.

4. S: $H_0: \mu_1 - \mu_2 = 0, H_a: \mu_1 - \mu_2 > 0$, where μ_1 = true mean change in pulse rate for students like these who drink 12 ounces of cola with caffeine and $\mu_2 = \ldots$ caffeine-free cola. $\alpha = 0.01$. The evidence for H_a is: $\bar{x}_1 - \bar{x}_2 = 3.2 - 2.0 = 1.2 > 0$. **P:** Two-sample t test for $\mu_1 - \mu_2$. *Random?* Students were randomly assigned to drink cola with caffeine or caffeine-free cola. ✓ *Normal/Large Sample?* The sample sizes are small, but the dotplots don't show any outliers or strong skewness. ✓

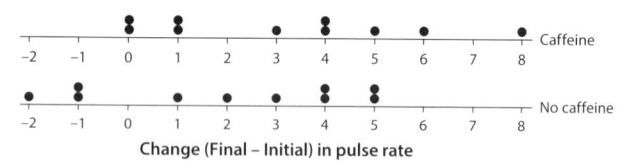

D: $t = \dfrac{(3.2 - 2.0) - 0}{\sqrt{\dfrac{2.70^2}{10} + \dfrac{2.625^2}{10}}} = 1.008$; *Tech:* $t = 1.008$, *P*-value = 0.1635 with df = 17.986. *Conservative approach:* df = smaller of $(10-1, 10-1) = 9$. *Table B:* *P*-value between 0.15 and 0.20. **C:** Because the *P*-value of 0.1635 is greater than $\alpha = 0.01$, we fail to reject H_0. We do not have convincing evidence that drinking caffeine increases pulse rates, on average, for people like the ones in this study.

5. (a) All but 4 of the differences are positive, so the reaction time before running exceeds the reaction time after running for 11 out of the 15 subjects. The differences vary from −29 milliseconds (student 2) to 50 milliseconds (student 4). **(b)** $\bar{x}_{\text{diff}} = 13.2$ milliseconds, $s_{\text{diff}} = 22.606$ milliseconds. The reaction time before running two laps for these 15 subjects is 13.2 milliseconds greater (slower) than their reaction time after running two laps, on average. **(c)** Yes, $\bar{x}_{\text{diff}} = 13.2 > 0$.

6. (a) S: 95% CI for μ_{diff} = the mean difference (Before − After) in reaction time (milliseconds) for all college students. **P:** Paired t interval for μ_{diff}. *Random?* Random sample of 15 college students. ✓ *Normal/Large Sample?* The sample size is small, but the dotplot of differences doesn't show any outliers or strong skewness. ✓ **D:** With 95% confidence and df = 15 − 1 = 14, $t^* = 2.145$. $13.2 \pm 2.145 \dfrac{22.606}{\sqrt{15}}$; 0.68 to 25.72; *Tech:* (0.681, 25.719) with df = 14. **C:** We are 95% confident that the interval from 0.681 to 25.719 captures the true mean difference (Before − After) in reaction time (milliseconds) for all college students. This interval suggests the reaction time before running two laps on the track is between 0.68 millisecond to 25.72 milliseconds slower (longer) than the reaction time after running two laps on the track, on average, for all college students. **(b)** Yes, because 0 is not a plausible value for the true mean difference (Before − After) in reaction time (milliseconds) for all college students and both endpoints of the interval are positive (0.681, 25.719), there is convincing evidence that exercise is associated with a decrease in the mean reaction time of all college students.

7. S: $H_0: \sigma_A = \sigma_B, H_a: \sigma_A \neq \sigma_B$, where σ_A = the standard deviation of annualized daily returns for all days over the past 5 years for stock A and $\sigma_B = \ldots$ stock B. Use $\alpha = 0.01$. The evidence for H_a is: $s_A = 12.9 \neq 9.6 = s_B$. **P:** Two-sample F test comparing two standard deviations. *Random?* Independent random samples of annualized daily returns for Stock A and Stock B. ✓ *Normal?* Graphs of the data suggest that it is reasonable to believe that the corresponding population distributions are normal. ✓

D: $F = \dfrac{12.9^2}{9.6^2} = 1.806$; $\text{df}_{\text{Num}} = 50 - 1 = 49$ and $\text{df}_{\text{Denom}} = 50 - 1 = 49$; *Tech:* $F = 1.806$, *P*-value = 0.0410; *Table D:* *P*-value is between $2(0.025) = 0.05$ and $2(0.05) = 0.10$. **C:** Because the *P*-value of 0.0410 is greater than $\alpha = 0.01$, we fail to reject H_0. We do not have convincing evidence of a difference in the standard deviations of the daily returns of these two stocks.

8. (a) S: $H_0: \mu_{\text{diff}} = 0, H_a: \mu_{\text{diff}} > 0$, where μ_{diff} = the true mean difference in test scores for students like these who get the neutral message. Use $\alpha = 0.05$. The evidence for H_a is: $\bar{x}_{\text{diff}} = 8.25 > 0$.

P: Paired t test for μ_{diff}. *Random?* Paired data come from pre- and post-test scores for a sample of 8 students who were randomly assigned to receive a neutral message. ✓ *Normal/Large Sample?* The sample size is small, but the dotplot shows no strong skewness and no outliers. ✓

D: $t = \dfrac{8.25 - 0}{\dfrac{3.694}{\sqrt{8}}} = 6.317$; *Tech:* $t = 6.318$, *P*-value ≈ 0 with df = 7.

Table B: *P*-value less than 0.0005. **C:** Because the *P*-value of approximately 0 is less than $\alpha = 0.05$, we reject H_0. We have convincing evidence that participating in the summer program improves the assessment test performance of students like the ones in this study who are only exposed to a neutral message.
(b) S: 95% CI for $\mu_1 - \mu_2$ where μ_1 = the true mean improvement in assessment test scores for students like these who are exposed to the subliminal message about getting better in math and μ_2 = the true mean improvement in assessment test scores for students like these who are only exposed to a neutral message. **P:** Two-sample t interval for $\mu_1 - \mu_2$. *Random?* Students were randomly assigned to a subliminal message about getting better in math or a neutral message. ✓ *Normal/Large Sample?* The sample sizes are small, but the dotplots show no strong skewness and no outliers. ✓

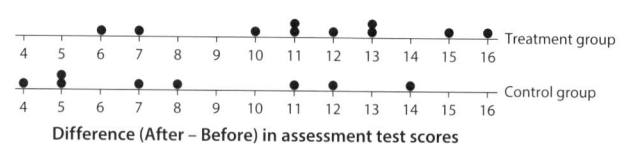

D: *Tech:* −0.383 to 6.683 with df = 13.919; df = smaller of (10 − 1, 8 − 1) = 7, $t^* = 2.365$; $(11.4 - 8.25) \pm 2.365 \sqrt{\dfrac{3.169^2}{10} + \dfrac{3.694^2}{8}} \rightarrow -0.743$ to 7.043. **C:** We are 95% confident that the interval from −0.383 to 6.683 captures the difference in mean improvement for students like these who receive the math message and students like these who receive the neutral message. The interval suggests that the mean improvement in score for students like these who receive the math message is between 0.383 less than and 6.683 more than the mean improvement in score for students like these who receive the neutral message. **(c)** We cannot generalize the results to all students who failed the test because our sample was not a random sample of all students who failed the test. It was a group of students who agreed to participate in the experiment.

Answers to Cumulative Review: Chapters 7–9

Section I: Multiple Choice

1. a
2. d
3. b
4. b
5. c
6. b
7. c
8. d
9. c
10. b
11. c
12. a
13. c
14. b
15. c
16. b
17. b
18. a
19. c
20. a

Section II: Free Response

1. (a) It is necessary to check this condition because when these counts are at least 10, we can be sure that the sampling distribution of $\hat{p}$ is approximately normal, which allows us to use the standard normal distribution to calculate z^*. **(b)** We are 95% confident that the interval from 0.105 to 0.255 captures p = the proportion of all members of the fitness club who would quit if the fee is raised to $50. **(c)** No. Some of the values in the confidence interval are greater than 20%, so the owner cannot be confident that fewer than 20% of the members will quit.

2. S: $H_0: \mu_{\text{diff}} = 0, H_a: \mu_{\text{diff}} < 0$, where μ_{diff} = the mean difference (After – Before) in weight (pounds) for all 5-week crash dieters. Use $\alpha = 0.05$. The evidence for H_a is: $\bar{x}_{\text{diff}} = -3.6 < 0$. **P:** Paired t test for μ_{diff}. *Random?* Random sample of 15 five-week crash dieters. ✓ *Normal/Large Sample?* The sample size is small ($n_{\text{diff}} = 15 < 30$), but the dotplot of the differences doesn't show any outliers or strong skewness. ✓

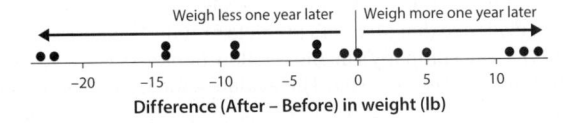

[Dotplot: Difference (After – Before) in weight (lb), ranging from –20 to 10, with "Weigh less one year later" on left and "Weigh more one year later" on right]

D: $t = \dfrac{-3.6 - 0}{\dfrac{11.525}{\sqrt{15}}} = -1.210$; df $= 15 - 1 = 14$; *Table B:* The P-value is between 0.10 and 0.15; *Tech:* $t = -1.210$, P-value = 0.1232 with df = 14. **C:** Because the P-value of 0.1232 is greater than $\alpha = 0.05$, we fail to reject H_0. There is not convincing evidence that 5-week crash dieters weigh less, on average, 1 year after finishing the diet.

3. (a) S: 90% CI for μ = the mean number of books read by all U.S. adults in the previous 12 months. **P:** One-sample t interval for μ. *Random?* Random sample of U.S. adults. ✓ *Normal/Large Sample?* $n = 1520 \geq 30$. ✓ **D:** With 90% confidence and df $= 1520 - 1 = 1519$, use df = 1000, so $t^* = 1.646$. $12 \pm 1.646 \dfrac{18}{\sqrt{1520}} \rightarrow 11.24$ to 12.76; *Tech:* 11.24 to 12.76 with df = 1519. **C:** We are 90% confident that the interval from 11.24 books to 12.76 books captures the mean number of books read by all U.S. adults in the previous 12 months. **(b)** If the Pew Research Center selects many random samples of U.S. adults and constructs a 90% confidence interval for each sample, about 90% of these intervals would capture the true mean number of books read by all U.S. adults in the previous 12 months. **(c)** It would not be appropriate to calculate a confidence interval for the population standard deviation in this setting because it is not reasonable to suggest that the distribution of number of books read is approximately normal. In a normal distribution, the minimum value should be at least 2 standard deviations below the mean. But in this distribution, the minimum value (0) is only 0.67 standard deviation below the mean.

4. (a) Because the subjects were randomly assigned to the treatments (AZT or placebo), we can draw conclusions about cause and effect. **(b)** $H_0: p_1 - p_2 = 0, H_a: p_1 - p_2 < 0$, where p_1 = proportion of all subjects like these who receive AZT and then develop AIDS and $p_2 = \ldots$ placebo and then develop AIDS. **(c)** *Type I:* We find convincing evidence that taking AZT decreases the proportion of infected people like the ones in this study who will develop AIDS in a given period of time, when in reality AZT does not decrease the proportion of infected people like the ones in this study who will develop AIDS in a given period of time. *Consequence:* We will use the drug broadly, when it is not effective. This wastes money and also is not an effective treatment for the patients. *Type II:* We fail to find convincing evidence that taking AZT decreases the proportion of infected people like the ones in this study who will develop AIDS in a given period of time, when in reality taking AZT *does* decrease the proportion of infected people like the ones in this study who will develop AIDS in a given period of time. *Consequence:* We will not use AZT broadly, when it actually is an effective treatment. This is a missed opportunity for effective care. **(d)** Answers will vary. Here is one possible answer: Researchers should use a significance level of $\alpha = 0.01$ to reduce the probability of making a Type I error. It is more serious to use the drug broadly when it is not an effective treatment for the patients. **(e) S:** $H_0: p_1 - p_2 = 0, H_a: p_1 - p_2 < 0$, where p_1 = proportion of all subjects like these who receive AZT and then develop AIDS and $p_2 = \ldots$ placebo and then develop AIDS. $\alpha = 0.01$. The evidence for H_a is: $\hat{p}_1 - \hat{p}_2 = 0.039 - 0.087 = -0.048 < 0$. **P:** Two-sample z test for $p_1 - p_2$. *Random?* Subjects were randomly assigned to take AZT or a placebo. ✓ *Large Counts?* 17, 418, 38, and 397 are all at least 10. ✓

D: $z = \dfrac{(0.039 - 0.087) - 0}{\sqrt{\dfrac{0.063(1 - 0.063)}{435} + \dfrac{0.063(1 - 0.063)}{435}}} = -2.914$; *Table A:* P-value = 0.0018; *Tech:* $z = -2.926$, P-value = 0.0017. **C:** Because the P-value of 0.0017 is less than $\alpha = 0.01$, we reject H_0. We have convincing evidence that taking AZT decreases the proportion of infected people like the ones in this study who will develop AIDS in a given period of time.

5. (a) S: 95% CI for $\mu_1 - \mu_2$, where μ_1 = the mean difference in electrical potential (millivolts) between the right hip and front feet for all mice with diabetes and $\mu_2 = \ldots$ normal mice. **P:** Two-sample t interval for $\mu_1 - \mu_2$. *Random?* Independent random samples of mice with diabetes and normal mice. ✓ *Normal/Large Sample?* The sample sizes are small, but graphs of the data for each group are consistent with normal population distributions. ✓ **D:** *Tech:* 0.633 to 5.503 with df = 38.460; df $= 18 - 1 = 17$, $t^* = 2.110$, $(13.090 - 10.022) \pm 2.110 \sqrt{\dfrac{4.839^2}{24} + \dfrac{2.915^2}{18}} \rightarrow 0.529$ to 5.607 millivolts.

C: We are 95% confident that the interval from 0.633 to 5.503 millivolts captures the true difference in the mean difference in electrical potential (millivolts) between the right hip and front feet for all mice with diabetes and all normal mice. The interval suggests that the mean difference in electrical potential (millivolts) for all mice with diabetes is between 0.633 and 5.503 millivolts more than the mean difference in electrical potential (millivolts) for all normal mice. **(b)** Assuming that the difference in the mean difference in electrical potential (millivolts) between the right hip and front feet for all mice with diabetes and normal mice is 0, there is a 0.015 probability of getting a difference in sample means of 3.068 millivolts or more extreme (in either direction) by chance alone in

random samples of 24 mice with diabetes and 18 normal mice. **(c)** In the test, the P-value of 0.015 is less than $\alpha = 0.05$, so we reject H_0. The 95% confidence interval does not contain 0 as a plausible value, so we reject $H_0: \mu_1 - \mu_2 = 0$ at the $\alpha = 0.05$ level. Therefore, the decision about the two-sided test is consistent with the decision made based on the corresponding confidence interval. The confidence interval gives more information than the test, as it provides the entire set of plausible values (0.633, 5.503) for the difference in the population means for normal mice and mice with diabetes.
6. S: $H_0: \sigma_1 = \sigma_2, H_a: \sigma_1 > \sigma_2$, where σ_1 = the standard deviation of the difference in electrical potential for all mice with diabetes and σ_2 = ... normal mice. Use $\alpha = 0.05$. The evidence for H_a is: $s_1 = 4.839 > 2.915 = s_2$. **P:** Two-sample F test comparing two standard deviations. *Random?* Independent random samples of mice with diabetes and normal mice. ✓ *Normal?* Graphs of the data for each group are consistent with normal population distributions. ✓
D: $F = \dfrac{4.839^2}{2.915^2} = 2.756$; $df_{Num} = 24 - 1 = 23$ and $df_{Denom} = 18 - 1 = 17$;
Tech: $F = 2.756$, *P*-value = 0.018; *Table D: P*-value is between 0.01 and 0.025. **C:** Because the *P*-value of 0.018 is less than $\alpha = 0.05$, we reject H_0. We have convincing evidence that the difference in electrical potential varies more (has a larger standard deviation) in mice with diabetes than in normal mice.

CHAPTER 10
Answers to Section 10.1 Exercises
1. categorical variable; population
3. $\chi^2 = \sum \dfrac{(\text{observed count} - \text{expected count})^2}{\text{expected count}}$
5. (a) H_0: The distribution of eye color/wing type is the same as what the biologists predict (9:3:3:1). H_a: ... different than ... **(b)** Red-eyed/straight-winged: 99/200 = 0.495; red-eyed/curly-winged: 42/200 = 0.210; white-eyed/straight-winged: 49/200 = 0.245; white-eyed/curly-winged: 10/200 = 0.050 **(c)** The sample proportions are different than 9/16 = 0.5625, 3/16 = 0.1875, 3/16 = 0.1875, and 1/16 = 0.0625.
7. (a) H_0: For Chicken McNuggets®, the proportions of bones, bells, boots, and balls are the same. H_a: ... not all the same. **(b)** Bone: 50/200 = 0.25; bell: 40/200 = 0.20; boot: 59/200 = 0.295; ball: 51/200 = 0.255 **(c)** The sample proportions are not all the same.
9. Red-eyed/straight-winged: 200(9/16) = 112.5; red-eyed/curly-winged: 200(3/16) = 37.5; white-eyed/straight-winged: 200(3/16) = 37.5; white-eyed/curly-winged: 200(1/16) = 12.5
11. Each shape has the same expected count: $200(0.25) = 50$
13. $\chi^2 = (99 - 112.5)^2/112.5 + (42 - 37.5)^2/37.5 + \cdots = 6.19$
15. $\chi^2 = (50 - 50)^2/50 + (40 - 50)^2/50 + \cdots = 3.64$
17. (a) H_0: The distribution of tree preference for nuthatches is the same as the distribution of tree type in the forest. H_a: ... different than ... **(b)** Douglas firs: 156(0.54) = 84.24; ponderosa pines: 156(0.40) = 62.4; other types: 156(0.06) = 9.36 **(c)** $\chi^2 = 7.418$
19. Because three simulated chi-square test statistics are greater than or equal to the observed chi-square test statistic of 7.418, the *P*-value is approximately 3/100 = 0.03. Because researchers are unlikely to get differences between the observed and expected counts this large by chance alone, these data provide convincing evidence that nuthatches do prefer particular types of trees when searching for seeds and insects.
21. (a) Using the empirical rule, < 50 g: 0.16; 50 to < 60 g: 0.34; 60 to < 70 g: 0.34; ≥ 70 g: 0.16 **(b)** < 50 g: 45(0.16) = 7.2; 50 to < 60 g: 45(0.34) = 15.3; 60 to < 70 g: 45(0.34) = 15.3; ≥ 70 g: 45(0.16) = 7.2 **(c)** $\chi^2 = 1.67$
23. Answers will vary. This poll is biased because it is based on voluntary response. Subscribers who are fearful of terrorism are probably more likely to respond. As a result, the sample percentage (49%) is likely greater than the percentage of all subscribers who are afraid there will be another terrorist attack.

Answers to Section 10.2 Exercises
1. False. To meet the Large Counts condition, the *expected* count in each category must be at least 5.
3. False. In a test for goodness of fit, df = number of categories – 1.
5. right
7. *Random?* Random sample of 387 U.S. residents with a landline phone. ✓ *Large Counts?* All expected counts (50.31, 135.45, 135.45, 65.79) are ≥ 5. ✓
9. *Random?* No, this was a census of one box. *Large Counts?* All expected counts are 3.85, which is not ≥ 5. Neither condition is met.
11. (a) *Table C:* 0.05 < *P*-value < 0.10; *Tech: P*-value = 0.0606 **(b)** *Using Table C: P*-value < 0.005; *Tech: P*-value = 0.0003
13. (a) $\chi^2 = 38.78$, df = 3; *Table C: P*-value < 0.005; *Tech: P*-value $\approx$ 0 **(b)** If the distribution of age among U.S. adult residents with a landline is the same as the distribution of age among the entire population of U.S. adult residents, there is about a 0 probability of getting an observed distribution of age this different or more different from the expected distribution by chance alone. **(c)** Because the *P*-value of approximately $0 < \alpha = 0.05$, we reject H_0. There is convincing evidence that the distribution of age among U.S. residents who have a landline is not the same as the distribution of age among the entire population of U.S. adult residents.
15. S: H_0: The distribution of car color in Oro Valley is the same as the distribution in North America. H_a: The distribution ... different than ... . Use $\alpha = 0.01$. The evidence for H_a is: The proportions in the sample (0.28, 0.127, 0.103, 0.153, 0.09, 0.097, 0.02, 0.13) are not all the same as the proportions in North America. **P:** Chi-square test for goodness of fit. *Random?* Random sample of 300 cars in Oro Valley. ✓ *Large Counts?* All expected counts (69, 54, 48, 45, 30, 27, 6, 21) are ≥ 5. ✓ **D:** $\chi^2 = 29.92$, df = 7; *Table C: P*-value < 0.005; *Tech: P*-value $\approx$ 0 **C:** Because the *P*-value of approximately $0 < \alpha = 0.01$, we reject H_0. There is convincing evidence that the distribution of car color in Oro Valley differs from the North American distribution.
17. S: H_0: Mendel's peas are produced in a ratio of 3:1 (smooth : wrinkled). H_a: ... are not produced .... Use $\alpha = 0.05$. The evidence for H_a is: The proportions in the sample (smooth: 423/556 = 0.761 and wrinkled: 133/556 = 0.239) are not the same as 3/4 = 0.75 and 1/4 = 0.25. **P:** Chi-square test for goodness of fit. *Random?* The problem states to assume the Random condition is met. ✓ *Large Counts?* All expected counts (417, 139) are ≥ 5. ✓ **D:** $\chi^2 = 0.35$, df = 1; *Table C: P*-value > 0.10; *Tech: P*-value $\approx$ 0.557. **C:** Because the *P*-value of $0.557 > \alpha = 0.05$, we fail to reject H_0. There is not convincing evidence that Mendel's predicted model is incorrect.
19. S: H_0: The company's invoices follow Benford's law. H_a: ... do not follow .... Use $\alpha = 0.05$. The evidence for H_a is: The proportions in the sample (61/250 = 0.244, 50/250 = 0.2, ..., 6/250 = 0.024) are not equal to those in Benford's law. **P:** Chi-square test for goodness of fit. *Random?* Random sample of 250 invoices from the company. ✓ *Large Counts?* All expected counts (75.25, 44, 31.25, 24.25, 19.75, 16.75, 14.5, 12.75, 11.5) are ≥ 5. ✓ **D:** $\chi^2 = 21.56$, df = 8; *Table C:* 0.005 < *P*-value < 0.01; *Tech: P*-value = 0.006. **C:** Because the *P*-value of $0.006 < \alpha = 0.05$, we reject H_0. There is convincing evidence that the company's invoices do not follow Benford's law.
21. (a) *Type I error:* The accountant concludes that the company is faking invoices, even though the company is really being honest. This decision would upset the company and may cost the accountant his job. *Type II error:* The accountant does not conclude that the company is faking invoices, even though the company is really faking invoices. This means the company gets away with being dishonest. If I were the accountant, a Type I error would be worse because I could lose my job. **(b)** The observed counts were much greater than expected for a first digit of 3 (43 > 31.25) and a first digit of 4 (34 > 24.25). The observed counts were much less than expected for a first digit of 1 (61 < 75.25) and a first digit of 7 (7 < 14.5).
23. Without combining several colors in a single "Other" category, the expected cell count for some of these colors might not be ≥ 5.

25. (a) Answers may vary. Here is one possible solution.

Digit	0	1	2	3	4	5	6	7	8	9
Frequency	23	18	17	21	20	15	22	23	19	22

(b) S: H_0: The random number generator gives each digit an equal chance of being generated. H_a: ... does not give.... Use $\alpha = 0.05$. The evidence for H_a is: The proportions in the sample ($23/200 = 0.115$, $18/200 = 0.09$, ..., and $22/200 = 0.11$) are not all equal. **P:** Chi-square test for goodness of fit. *Random?* We generated 200 random digits using a random number generator. ✓ *Large Counts?* All expected counts of 20 are ≥ 5. ✓ **D:** $\chi^2 = 3.3$, df $= 9$; *Table C*: P-value > 0.10; *Tech*: P-value $= 0.951$. **C:** Because the P-value of $0.951 > \alpha = 0.05$, we fail to reject H_0. There is not convincing evidence that the random number generator does not give each digit an equal chance of being generated. **(c)** $P(\text{Type I error}) = \alpha = 0.05$ **(d)** $P(\text{At least one student makes a Type I error}) = 1 - P(\text{no student makes a Type I error}) = 1 - 0.95^{25} = 0.7226$

27. (a) $P(\text{arthritis}) = \dfrac{43}{858} = 0.05$ **(b)** $P(\text{arthritis} \mid \text{elite}) = \dfrac{10}{71} = 0.141$

Answers to Section 10.3 Exercises

1. An association means that knowing the value of one variable helps us predict the value of the other variable.

3. Expected count $= \dfrac{(\text{row total})(\text{column total})}{\text{table total}}$

5. (a) H_0: There is no association between lighting condition and myopia status in the population of children. H_a: There is an association....
(b) [segmented bar chart: Relative frequency (%) vs Lighting condition (No light, Night light, Fully lit); Myopia status: No, Yes]
(c) The distribution of myopia status is not the same in all three lighting conditions for the members of the sample.

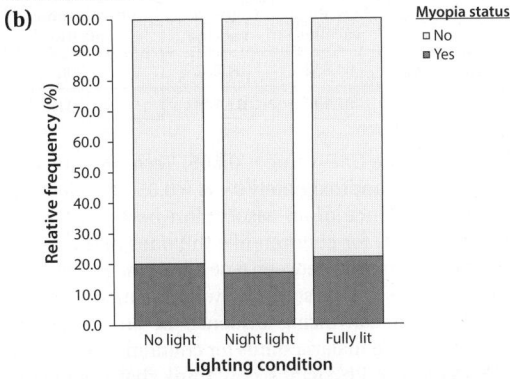

7. (a) H_0: There is no association between Twitter use and political views in the population of U.S. adults who self-identify as Democrats. H_a: There is an association....
(b) [segmented bar chart: Relative frequency (%) vs Use Twitter (No, Yes); Political views: Very liberal, Liberal, Moderate, Conservative, Very conservative]
(c) The distribution of political views is not the same for those who are and those who are not Twitter users for the members of the sample.

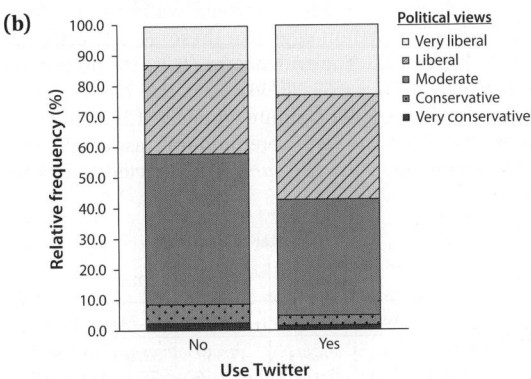

9.

		Lighting condition		
		No light	Night light	Fully lit
Myopia status	Yes	75.88	137.93	8.19
	No	341.12	620.07	36.81

11.

		Use Twitter	
		No	Yes
	Very conservative	117.87	48.13
	Conservative	222.25	90.75
Political views	Moderate	1990.99	813.01
	Liberal	1320.70	539.30
	Very liberal	663.19	270.81

13. $\chi^2 = (83 - 75.88)^2 / 75.88 + (129 - 137.93)^2 / 137.93 + \cdots = 2.012$

15. $\chi^2 = (127 - 117.87)^2 / 117.87 + (39 - 48.13)^2 / 48.13 + \cdots = 144.66$

17. (a) H_0: There is no association between weekly sauna frequency and SCD status for middle-aged men from eastern Finland. H_a: There is an association....
(b) [segmented bar chart: Relative frequency (%) vs Weekly sauna frequency (1 or fewer, 2–3, 4 or more); SCD: No, Yes]

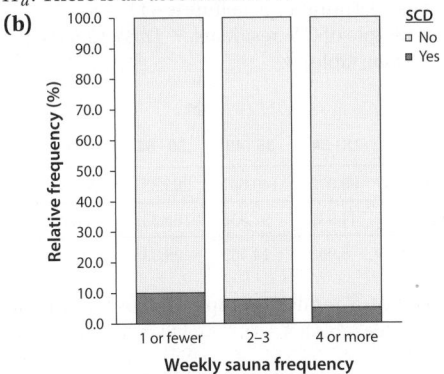

There is some evidence for H_a because the proportion of men who suffered SCD is not the same for each sauna frequency.
(c)

		Weekly sauna frequency		
		1 or fewer	2–3	4 or more
SCD	Yes	49.326	124.177	16.497
	No	551.674	1388.823	184.503

(d) $\chi^2 = 6.033$

19. Here is a segmented bar chart that shows no association between weekly sauna frequency and SCD status. The proportion of men who suffered SCD is the same for each sauna frequency.

[segmented bar chart: Relative frequency (%) vs Weekly sauna frequency (1 or fewer, 2–3, 4 or more); SCD: No, Yes]

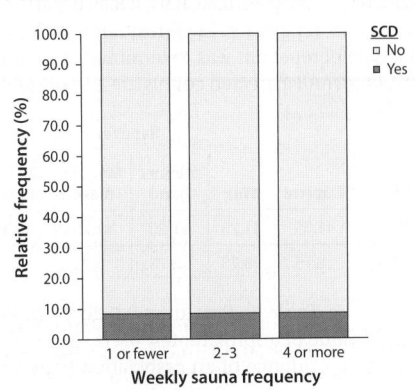

21. The distribution of wing length for male *Boloria chariclea* butterflies is roughly symmetric, whereas the distribution of wing length for female *Boloria chariclea* butterflies is skewed left with a high outlier. The center of the male distribution is much smaller than the center of the female distribution, while the amount of variability is roughly the same in each distribution. Because the female center is higher and there is very little overlap in the distributions, it appears that male and female butterflies of this species are different sizes.

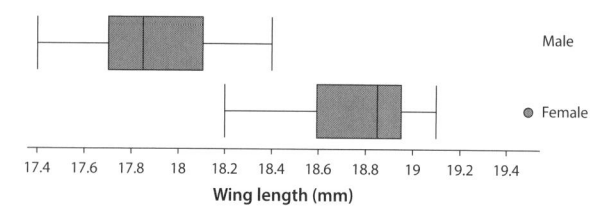

Answers to Section 10.4 Exercises

1. The Random condition is met if the data come from a random sample from the population of interest, from independent random samples from the populations of interest, or from groups in a randomized experiment.
3. df = (number of rows − 1)(number of columns − 1)
5. *Random?* Random sample of U.S. residents. ✓ *Large Counts?* All expected counts are ≥ 5 (see table). ✓

		Age			
		18–34	35–49	50–64	65+
Response	Support	68.072	140.681	285.333	361.914
	Oppose	44.930	92.856	188.333	238.880
	Don't know	6.998	14.463	29.333	37.206

7. *Random?* This was not a random sample. *Large Counts?* All expected counts are ≥ 5 (see table). ✓

		Class of travel		
		First	Second	Third
Survival status	Survived	116.817	95.557	229.606
	Died	202.183	165.423	397.394

9. (a) df = (5−1)(4−1) = 12; *Table C:* P-value > 0.10; *Tech:* P-value = 0.137 **(b)** df = (3−1)(2−1) = 2; *Table C:* P-value < 0.005; *Tech:* P-value ≈ 0.0002
11. df = (3−1)(4−1) = 6; *Table C:* P-value < 0.005; *Tech:* P-value ≈ 0
13. S: H_0: There is no association between type of repellent and shark response for surfboards like these. H_a: There is an association.... Use $\alpha = 0.05$. The evidence for H_a is: The proportion of trials that had a shark bite (0.96, 0.86, 0.88, 0.94, 0.898, 0.396) are not all the same for the various repellents. **P:** Chi-square test for association. *Random?* The type of repellent was determined at random for each trial. ✓ *Large Counts?* All expected counts are ≥ 5 (see table). ✓

		Repellent					
		Control	Wax	Magnet band	Magnet leash	Electric (Rpela)	Electric (Surf+)
Shark response	Bite	41.25	41.25	41.25	41.25	40.42	39.60
	No bite	8.75	8.75	8.75	8.75	8.58	8.40

D: $\chi^2 = 75.38$, df = 5; *Table C:* P-value < 0.005; *Tech:* P-value ≈ 0. **C:** Because the P-value of approximately $0 < \alpha = 0.05$, we reject H_0. There is convincing evidence of an association between repellent and shark response for surfboards like these.
15. S: H_0: There is no association between year and number of trips to the movie theater for U.S. adults. H_a: There is an association.... Use $\alpha = 0.01$. The evidence for H_a is: The segmented bar graph reveals that the distribution of number of trips to the movie theater is different in 2001 and 2019 for the members of these samples. **P:** Chi-square test for association. *Random?* The data come from two independent random samples of U.S. adults. ✓ *Large Counts?* All expected counts are ≥ 5 (see table). ✓

		Year	
		2001	2019
Number of trips to the movie theater	None	265.95	272.05
	1	85.02	86.98
	2	115.18	117.82
	3–5	225.41	230.59
	6–9	85.02	86.98
	10+	225.41	230.59

D: $\chi^2 = 19.66$, df = 5; *Table C:* P-value < 0.005; *Tech:* P-value = 0.0014. **C:** Because the P-value of $0.0014 < \alpha = 0.01$, we reject H_0. There is convincing evidence of an association between year and number of trips to the movie theater for U.S. adults.
17. (a) S: H_0: There is no association between type of vaccine and malaria status for children like the ones in the study. H_a: There is an association.... Use $\alpha = 0.05$. The evidence for H_a is: The proportion of children who had at least one episode of malaria (50/146 = 0.342, 39/146 = 0.267, 106/147 = 0.721) is not the same for the different vaccine groups. **P:** Chi-square test for association. *Random?* Children were randomly assigned into one of three treatment groups. ✓ *Large Counts?* All expected counts are ≥ 5 (see table). ✓

		Vaccine		
		Low-dose vaccine	High-dose vaccine	Control (rabies) vaccine
At least one episode of malaria?	Yes	64.852	64.852	65.296
	No	81.148	81.148	81.704

D: $\chi^2 = 70.31$, df = 2; *Table C:* P-value < 0.005; *Tech:* P-value ≈ 0. **C:** Because the P-value of approximately $0 < \alpha = 0.05$, we reject H_0. There is convincing evidence of an association between type of vaccine and malaria status for children like the ones in the study. **(b)** If there really is no association between type of vaccine and malaria status for children like these in the study, we will have made a Type I error because we found convincing evidence of an association between type of vaccine and malaria status for children like the ones in the study. *Consequence:* Physicians may think that the type of vaccine matters, when in reality the type of vaccine does not matter.
19. The high-dose malaria vaccine appears to be the most effective treatment for preventing malaria in children like the ones in the study, as its observed count of malaria cases (39) was the farthest below its expected count (64.852).
21. S: H_0: There is no association between water temperature and hatching status for python eggs like these. H_a: There is an association.... Use $\alpha = 0.05$. The evidence for H_a is: The proportion of eggs that hatched (16/27 = 0.59, 38/56 = 0.68, and 75/104 = 0.72) are different for the different temperature groups. **P:** Chi-square test for association. *Random?* The eggs were randomly assigned to the three temperature groups. ✓ *Large Counts?* All expected counts are ≥ 5 (see table). ✓

		Water temperature		
		Cold	Neutral	Hot
Hatched?	Yes	18.626	38.631	71.743
	No	8.374	17.369	32.257

D: $\chi^2 = 1.70$, df = 2; *Tech:* P-value = 0.427. **C:** Because the P-value of $0.427 > \alpha = 0.05$, we fail to reject H_0. There is not convincing evidence of an association between water temperature and hatching status for python eggs like these.

23. (a) $\chi^2 = 4.20$, df = 1; Table C: $0.025 < P\text{-value} < 0.05$; Tech: P-value = 0.0404 (b) $\hat{p}_W = \frac{27}{85} = 0.318$, $\hat{p}_L = \frac{15}{83} = 0.181$, $\hat{p}_C = \frac{27+15}{85+83} = 0.25$; $z = 2.049$; Table A: P-value = $2(0.0202) = 0.0404$; Tech: P-value = 0.0404 (c) The P-value for the two-sample z test (0.0404) is the same as the P-value from the chi-square test in part (a) (0.0404). The χ^2 test statistic is the square of the z test statistic $(2.049)^2 = 4.20 = \chi^2$.

25. We cannot use a chi-square test because these data are quantitative. A chi-square test for association requires categorical data. There are weights of nuts in each cell rather than counts of nuts in each cell.

27. (a) Paired t test. There are two values of the same quantitative variable recorded for each subject. (b) No. There was no control group, so we do not know if the scores increased due to the chess training or due to some other confounding variable, such as what the students were learning in school.

Answers to Section 10.5 Exercises

1. H_a: At least one mean is different.
3. Mean square = (sum of squares)/(degrees of freedom)
5. $H_0: \mu_N = \mu_{1yr} = \mu_{8yr}$, H_a: At least one mean is different, where μ_N, μ_{1yr}, and μ_{8yr} represent the mean numbers of trees per plot for all such plots in Borneo that had never been logged, all such plots that had been logged 1 year earlier, and all such plots that had been logged 8 years earlier, respectively.
7. $H_0: \mu_{IceCream} = \mu_{FrozenYogurt} = \mu_{Gelato} = \mu_{Sorbet} = \mu_{DairyFree}$, H_a: At least one mean is different, where $\mu_{Dessert}$ represents the mean amount of saturated fat per serving (g) in the specified dessert.
9. There is considerable overlap between the plots that were last logged 1 year earlier and 8 years earlier, suggesting that there may be little or no difference in the mean numbers of trees per plot for all such plots. However, the number of trees per plot for the plots in Borneo that had never been logged has less overlap and is centered at a much greater value than for the other two groups. The dotplots suggest that there may be at least one difference among the three population means.
11. The amount of saturated fat for the gelatos is centered at a much greater value than that for the frozen yogurts and ice creams. There is not much overlap between the ice creams and the gelatos and no overlap between the frozen yogurts and the gelatos. These boxplots suggest that there may be at least one difference among the five population means.
13. (a)

Source	df	SS	MS	F
Groups	2	455.6	227.80	12.62
Error	28	505.4	18.05	
Total	30	961.0		

(b) Because the F statistic of 12.62 is larger than 1, there is some evidence that at least one of the population means is different.

15. (a)

Source	df	SS	MS	F
Groups	4	84.68	21.17	1.31
Error	60	970.00	16.17	
Total	64	1054.68		

(b) Because the F statistic of 1.31 is larger than 1, there is some evidence that at least one of the population means is different.

17. (a) $H_0: \mu_A = \mu_B = \mu_C$, H_a: At least one mean is different, where μ_A, μ_B, and μ_C represent the mean amounts of lightness for all pieces of ramie fabric like those in the study when applying blue dye using Method A, B, and C, respectively. (b) The dotplots show almost no overlap between the Method B group and the Method C group, indicating that one of these methods may have a different mean amount of lightness.

(c)

Source	df	SS	MS	F
Groups	2	1.788	0.894	9.53
Error	21	1.97043	0.09383	
Total	23	3.75843		

(d) Because the F statistic of 9.53 is larger than 1, there is some evidence that at least one of the population means is different.

19. (a) Answers will vary. Here is one possibility.

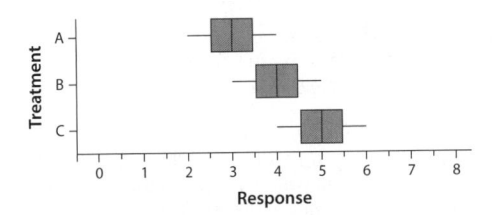

(b) Answers will vary. Here is one possibility.

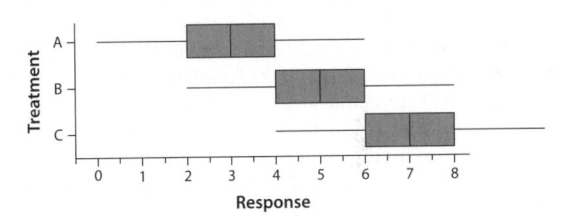

21. Because an F statistic of 12.62 or greater never occurred in this simulation, the P-value is approximately 0. There is convincing evidence that the mean number of trees per plot is not the same for all three levels of logging. It is extremely unlikely to get differences this big or bigger simply due to chance variation in the random assignment.

23. (a) $\bar{x} = \frac{5832}{48} = 121.5$ (b) SSG = 420 (c) SSE = 39,276

(d)

Source	df	SS	MS	F
Groups	2	420	210	0.24
Error	45	39276	872.8	
Total	47	39696		

25. (a) (i) $z = \frac{500-525}{110} = -0.227$; Tech: $P(Z > -0.227)$ = Applet/normalcdf(lower: -0.227, upper: 1000, mean: 0, SD: 1) = 0.5898, Table A: $P(Z > -0.23) = 1 - 0.4090 = 0.5910$ (ii) Tech: Applet/normalcdf(lower: 500, upper: 10,000, mean: 525, SD: 110) = 0.5899; Approximately 59% of these seeds weigh more than 500 mg. (b) (i) Tech: Applet/invNorm(area: 0.10, mean: 0, SD: 1) = -1.282; $-1.282 = \frac{x-525}{110}$; $x = 383.98$ mg; Table A: 0.10 area to the left gives $z = -1.28$; $-1.28 = \frac{x-525}{110}$; $x = 384.2$ mg; (ii) Tech: Applet/invNorm(area: 0.10, mean: 525, SD: 110) = 384.03 mg.

Answers to Section 10.6 Exercises

1. less than twice
3. True
5. *Random?* Assume a random sample of 214 specimens is equivalent to independent random samples of each type of glass. ✓ *Normal/Large Sample?* The dotplots for the tableware glass and the

vehicle window float-processed glass show skewness with possible outliers. Because the sample sizes for these groups are also less than 30, the Normal/Large Sample condition is not met. *Equal SD?* $0.00380 < 2(0.00192) = 0.00384$. ✓

7. *Random?* Three independent random samples of plots. ✓ *Normal/Large Sample?* The dotplots do not show strong skewness or outliers. ✓

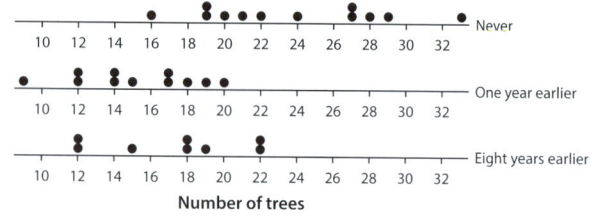

[Dotplots: Never, One year earlier, Eight years earlier; x-axis: Number of trees, 10–32]

Equal SD? $5.0655 < 2(3.3710) = 6.742$. ✓

9. S: $H_0: \mu_A = \mu_B = \mu_C = \mu_D = \mu_E = \mu_F = \mu_G = \mu_H = \mu_I = \mu_J$; H_a: At least one mean is different, where $\mu_{Fertilizer}$ is the mean growth (mm) for bean sprouts like the ones in the study using the specified fertilizer. Use $\alpha = 0.10$. The evidence for H_a is: The sample means (99.83, 95.25, ...) are not all the same. **P:** One-way ANOVA test for differences in means. *Random?* Each fertilizer was randomly assigned to 11 or 12 bean sprouts. ✓ *Normal/Large Sample?* The boxplots do not show strong skewness or outliers. ✓ *Equal SD?* $16.80 < 2(8.96) = 17.92$. ✓ **D:** The ANOVA table gives $F = 1.194$ and P-value = 0.3059. **C:** Because the P-value of $0.3059 > \alpha = 0.10$, we fail to reject H_0. There is not convincing evidence that the mean growth for bean sprouts like the ones in the study is not the same for all 10 fertilizers.

11. S: $H_0: \mu_N = \mu_{1yr} = \mu_{8yr}$, H_a: At least one mean is different, where μ_N, μ_{1yr}, and μ_{8yr} represent the mean number of species per plot for all plots in Borneo that had never been logged, all plots that had been logged 1 year earlier, and all plots that had been logged 8 years earlier, respectively. Use $\alpha = 0.05$. The evidence for H_a is: The sample means ($\bar{x}_{Never} = 17.50$, $\bar{x}_{1yr} = 12.64$, $\bar{x}_{8yr} = 14.88$) are not all the same. **P:** One-way ANOVA test for differences in means. *Random?* Three independent random samples of plots. ✓ *Normal/Large Sample?* The dotplots do not show strong skewness or outliers. ✓

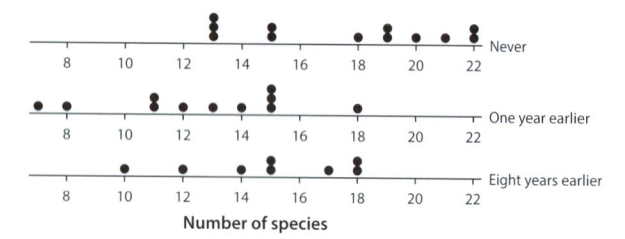

[Dotplots: Never, One year earlier, Eight years earlier; x-axis: Number of species, 8–22]

Equal SD? $3.529 < 2(2.850) = 5.700$. ✓
D: The ANOVA table gives $F = 6.35$ and P-value = 0.005.

Source	df	SS	MS	F	P-value
Groups	2	136.29	68.14	6.35	0.005
Error	28	300.42	10.73		
Total	30	436.71			

C: Because the P-value of $0.005 < \alpha = 0.05$, we reject H_0. There is convincing evidence that the mean number of species per plot is not the same for all three levels of logging in Borneo.

13. S: $H_0: \mu_{BlackFemale} = \mu_{BlackMale} = \mu_{WhiteFemale} = \mu_{WhiteMale}$, H_a: At least one mean is different, where μ_{Name} represents the mean rating given by counselors like the ones in the study to the specified type of name. Use $\alpha = 0.05$. The evidence for H_a is: The sample mean ratings (6.912, 8.879, 8.375, 8.433) are not the same for all four student names. **P:** One-way ANOVA test for differences in means. *Random?* Counselors were randomly assigned to one of the four groups. ✓ *Normal/Large Sample?* Three sample sizes are ≥ 30 and the remaining distribution (Black male) has no strong skewness or outliers. ✓ *Equal SD?* $1.712 < 2(1.090) = 2.180$. ✓ **D:** The ANOVA table gives $F = 10.58$ and P-value ≈ 0. **C:** Because the P-value of approximately $0 < \alpha = 0.05$, we reject H_0. There is convincing evidence that the mean rating given by counselors like the ones in the study is not the same for all four student names.

15. Fisher's LSD 95% interval for $\mu_{Never} - \mu_{1year}$: (2.06, 7.66); Fisher's LSD 95% interval for $\mu_{Never} - \mu_{8year}$: (−0.44, 5.69); Fisher's LSD 95% interval for $\mu_{1year} - \mu_{8year}$: (−5.36, 0.88). Because 0 is included in the second and third intervals, there is not convincing evidence that the mean number of species per plot is different for all plots logged 8 years earlier when compared to either all plots logged 1 year earlier or all plots that have never been logged. However, because 0 is not included in the first interval, there is convincing evidence that the mean number of species per plot is different for all plots that have never been logged when compared to all plots that have been logged 1 year earlier. In fact, the interval suggests that the mean for all plots that have never been logged is greater than the mean for all plots logged 1 year earlier.

17. Because 0 is included in the fourth, fifth, and sixth intervals, there is not convincing evidence that the mean rating given by counselors like the ones in the study is different for White females when compared to Black males and for White males when compared to either Black males or White females. However, because 0 is not included in the first, second, and third intervals, there is convincing evidence that the mean rating is different for Black females when compared to either Black males, White females, or White males. In fact, the intervals suggest that the mean rating for Black males, White females, and White males are all greater than the mean rating for Black females.

19. (a) S: $H_0: \mu_{ICF} = \mu_{Oral\&SS} = \mu_{Oral\&Theater} = \mu_{OralInfo}$, H_a: At least one mean is different, where $\mu_{Treatment}$ is the mean level of understanding for subjects like the ones in the study when given the information using the method in each treatment group. Use $\alpha = 0.05$. The evidence for H_a is: The sample mean level of understanding (4.413, 6.690, 7.034, 6.837) is not the same for all four treatment groups. **P:** One-way ANOVA test for differences in means. *Random?* Participants were randomly assigned to one of the treatment groups. ✓ *Normal/Large Sample?* All sample sizes are ≥ 30. ✓ *Equal SD?* $1.996 < 2(1.466) = 2.932$. ✓ **D:** The ANOVA table gives $F = 75.84$ and P-value ≈ 0. **C:** Because the P-value of approximately $0 < \alpha = 0.05$, we reject H_0. There is convincing evidence that the mean level of understanding for subjects like the ones in the study is not the same for all four treatments.
(b) Fisher's LSD 95% interval for $\mu_{ICF} - \mu_{Oral\&SS}$: (−2.654, −1.900)
Fisher's LSD 95% interval for $\mu_{ICF} - \mu_{Oral\&Theater}$: (−3.015, −2.227)
Fisher's LSD 95% interval for $\mu_{ICF} - \mu_{OralInformation}$: (−2.825, −2.023)
Fisher's LSD 95% interval for $\mu_{Oral\&SS} - \mu_{Oral\&Theater}$: (−0.724, 0.036)
Fisher's LSD 95% interval for $\mu_{Oral\&SS} - \mu_{OralInformation}$: (−0.535, 0.241)
Fisher's LSD 95% interval for $\mu_{Oral\&Theater} - \mu_{OralInformation}$: (−0.208, 0.602)

Because 0 is not included in the first three intervals, there is convincing evidence that the mean level of understanding is different between using the ICF only and any of the other three methods (oral and slideshow, oral and theater, oral information). In fact, the mean level of understanding is greater for oral information and slideshow, oral information and theater, and oral information as compared to receiving the information using an ICF alone. Because 0 is contained in the fourth, fifth, and sixth intervals, there is not convincing evidence that the mean level of understanding is different between using oral information, using oral information and slideshow, or using oral information and theater. **(c)** The mean level of understanding is better when using any of the three methods (oral information, oral information and slideshow, oral information and theater) compared to using the ICF alone.

21. *Normal/Large Sample?* Three of the distributions have sample sizes <30 and are strongly skewed to the right (see dotplots). This condition is not met.

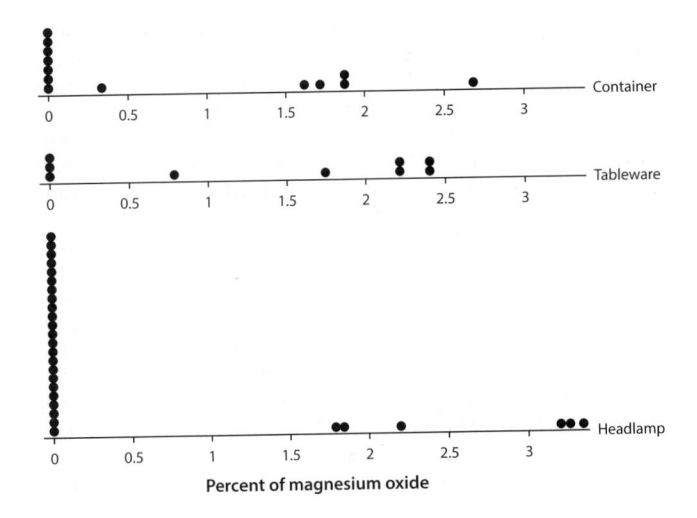

convincing evidence of a difference in the mean amount of caffeine per serving between Coke and Pepsi. **(c)** When using the pooled t test, the F statistic is equal to the square of the t statistic.

27. (a)

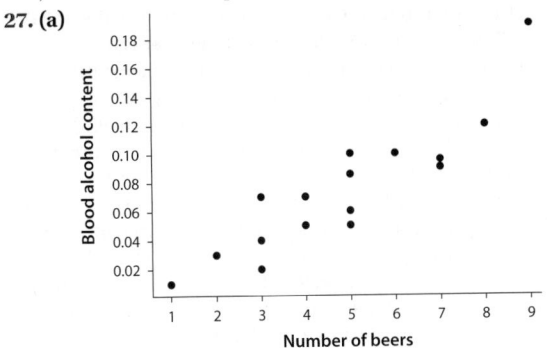

(b) The scatterplot suggests that there is a moderately strong, positive, linear relationship between the number of beers drunk and blood alcohol content for these students.

Equal SD? $1.21566 > 2(0.16279) = 0.32558$. This condition is not met.

23. (a) S: $H_0: \mu_{Applebee's} = \mu_{KFC} = \mu_{McDonald's}$, H_a: At least one mean is different, where $\mu_{Restaurant}$ is the mean amount of caffeine per serving of fountain cola at the restaurant chain. Use $\alpha = 0.05$. The evidence for H_a is: The sample mean level of caffeine per serving ($\bar{x}_A = 41.75$, $\bar{x}_M = 41.65$, $\bar{x}_K = 36.10$) is not the same for all three restaurant chains. **P:** One-way ANOVA test for differences in means. *Random?* Independent random samples of fountain cola. ✓ *Normal/Large Sample?* The boxplots do not indicate any strong skewness or outliers. ✓

Answers to Chapter 10 Review Exercises

1. S: H_0: The ratio of green (GG) to yellow-green (Gg) to albino (gg) tobacco plants is 1:2:1. H_a: The ratio is not 1:2:1. Use $\alpha = 0.01$. The evidence for H_a is: The proportions in the sample ($23/84 = 0.274$, $50/84 = 0.595$, $11/84 = 0.131$) are not the same as $1/4 = 0.25$, $2/4 = 0.50$, and $1/4 = 0.25$. **P:** Chi-square test for goodness of fit. *Random?* Pairs of yellow-green parent plants were randomly selected. ✓ *Large Counts?* All expected counts (21, 42, 21) are ≥ 5. ✓ **D:** $\chi^2 = 6.476$, df = 2; *Table C*: $0.025 <$ P-value is 0.05; *Tech*: P-value = 0.0392. **C:** Because the P-value of $0.0392 > \alpha = 0.01$, we fail to reject H_0. There is not convincing evidence that the true distribution of offspring is different from what the biologists predict.

2. (a) H_0: There is no association between wording and response for subjects like these. H_a: There is an association between wording and response for subjects like these.

(b)

		Wording		
		"Smashed into"	"Hit"	Control
Response	Yes	9.67	9.67	9.67
	No	40.33	40.33	40.33

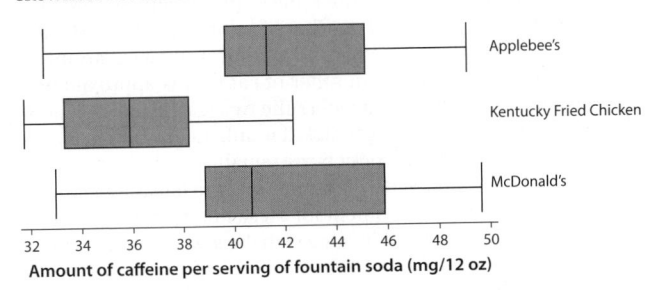

Equal SD? $4.719 < 2(3.155) = 6.310$. ✓
D: The ANOVA table gives $F = 12.344$ and P-value < 0.001.

Source	df	SS	MS	F	P-value
Groups	2	418.044	209.022	12.344	<0.001
Error	57	965.224	16.934		
Total	59	1383.267			

C: Because the P-value of less than $0.001 < \alpha = 0.05$, we reject H_0. There is convincing evidence that the mean amount of caffeine per serving is not the same at all three restaurant chains.
(b) Fisher's LSD 95% interval for $\mu_{Applebee's} - \mu_{KFC}$: $(3.042, 8.254)$
Fisher's LSD 95% interval for $\mu_{Applebee's} - \mu_{McDonald's}$: $(-2.507, 2.704)$
Fisher's LSD 95% interval for $\mu_{KFC} - \mu_{McDonald's}$: $(-8.155, -2.944)$
Because 0 is not included in the first and third intervals, there is convincing evidence that the mean amount of caffeine per serving is different between Kentucky Fried Chicken than for either Applebee's or McDonald's. In fact, the intervals suggest that the mean amount of caffeine is less for Kentucky Fried Chicken than for either Applebee's or McDonald's. Because 0 is included in the second interval, there is not convincing evidence that the mean amount of caffeine per serving is different between Applebee's and McDonald's. **(c)** From these data, Kentucky Fried Chicken has the least mean amount of caffeine per serving of fountain soda compared to either Applebee's or McDonald's. There appears to be little difference in the amount of caffeine per serving between Applebee's and McDonald's.
25. (a) $s_p = 4.079$; $t = 5.011$; df = 58, P-value ≈ 0 **(b)** Because the P-value of approximately $0 < \alpha = 0.05$, we reject H_0. There is

(c) *Random?* The subjects were randomly assigned to the three treatment groups. ✓ *Large Counts?* All expected counts are ≥ 5 (see table). ✓ **(d)** $\chi^2 = 7.78$, df = 2; *Table C*: $0.01 <$ P-value < 0.025; *Tech*: P-value = 0.020. Because the P-value of $0.020 < \alpha = 0.05$, we reject H_0. There is convincing evidence that there is an association between wording and response for subjects like these.

3. (a) S: $H_0: \mu_{Boiled} = \mu_{Filtered} = \mu_{None}$, H_a: At least one mean is different, where $\mu_{Treatment}$ is the mean change in cholesterol for subjects like the ones in the study after 81 days of the treatment. The evidence for H_a is: The sample mean change in cholesterol ($\bar{x}_B = 0.545$, $\bar{x}_F = 0.129$, $\bar{x}_N = -0.043$) is different for all three treatments. **P:** One-way ANOVA test for differences in means. *Random?* Subjects were randomly assigned to one of the three treatments. ✓ *Normal/Large Sample?* The boxplots do not indicate any strong skewness or outliers. ✓

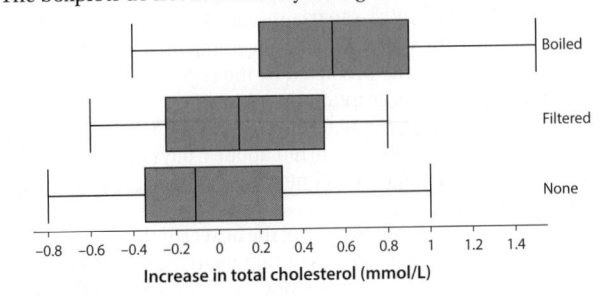

Equal SD? $0.523 < 2(0.434) = 0.868$.

D: The ANOVA table gives $F = 8.686$ and P-value $= 0.0005$.
C: Because the P-value of 0.0005 is $< \alpha = 0.05$, we reject H_0. There is convincing evidence that the mean change in cholesterol is not the same for all three treatments. **(b)** Because the results of the ANOVA test are significant, Fisher's LSD intervals are appropriate.
Fisher's LSD 95% interval for $\mu_{Boiled} - \mu_{Filtered}$: (0.126, 0.708)
Fisher's LSD 95% interval for $\mu_{Boiled} - \mu_{None}$: (0.297, 0.879)
Fisher's LSD 95% interval for $\mu_{Filtered} - \mu_{None}$: (−0.123, 0.466)
Because 0 is included in the third interval, there is not convincing evidence that the mean change in cholesterol is different for subjects like these who drink filtered coffee and those who do not drink coffee. However, because 0 is not included in the first and second intervals, there is convincing evidence that the mean change in cholesterol is different for subjects like these who drink boiled coffee as compared to drinking filtered coffee or not drinking coffee. In fact, the intervals suggest that the mean increase in total cholesterol for subjects like these who drink boiled coffee is greater than the mean increase for those who drink filtered coffee or those who do not drink coffee. **(c)** From these data, there is convincing evidence that drinking boiled coffee increases blood cholesterol levels for people like those in the study.
4. (a) S: H_0: There is no association between CVD event status and napping frequency for all adults in Lausanne; H_a: There is an association between CVD event status and napping frequency for all adults in Lausanne. Use $\alpha = 0.05$. The evidence for H_a is: The proportion of adults who had a CVD-related event (0.046, 0.018, 0.054, 0.076) is different for the four napping frequency categories. **P:** Chi-square test for association. *Random?* Random sample of adults from Lausanne. *Large Counts?* All expected counts are ≥ 5 (see table).

		Napping frequency			
		None	1–2 weekly	3–5 weekly	6–7 weekly
CVD event status	Yes	90.17	29.86	18.40	16.57
	No	1923.83	637.14	392.60	353.43

D: $\chi^2 = 20.28$, df $= 3$; *Table C:* P-value < 0.005; *Tech:* P-value $= 0.00015$.
C: Because the P-value of 0.00015 $< \alpha = 0.05$, we reject H_0. There is convincing evidence of an association between CVD event status and napping frequency for all adults in Lausanne. **(b)** No, this is an observational study, so we cannot draw a cause-and-effect conclusion. It is possible that people in Lausanne who nap 1 to 2 times weekly have a healthier lifestyle (better diet, more exercise, etc.) and this healthier lifestyle helps their CVD event status.

CHAPTER 11

Answers to Section 11.1 Exercises

1. $\hat{y}$ represents the predicted value of the response variable for a given value of x.
3. A residual is the difference $(y - \hat{y})$ between the actual value of y and the value of y predicted by the regression line for a given value of x.
5. The y intercept is the predicted value of y when x is 0.
7. The slope of the regression line is the amount by which the predicted value of y changes when x increases by one unit.
9. $\widehat{wait} = 33.35 + 13.29(4) = 86.51$ minutes
11. $\hat{y} = 33.1 - 4.69(6) = 4.96$ days $\approx$ April 5
13. (a) $\hat{y} = 33.35 + 13.29(2) = 59.93$ minutes; Residual $= 62 - 59.93 = 2.07$ minutes **(b)** The wait time between eruptions was 2.07 minutes longer than the wait time predicted by the regression line with $x =$ duration of the previous eruption (in minutes).
15. (a) $\hat{y} = 33.1 - 4.69(3.2) = 18.09$ days; Residual $= 11 - 18.09 = -7.09$ days **(b)** The first blossom occurred about 7 days earlier than the day predicted by the regression line with $x =$ the average temperature in March (°C).
17. (a) The predicted wait time until the next eruption increases by 13.29 minutes for each increase of 1 minute in duration of the previous eruption. **(b)** The y intercept does not have meaning because it does not make sense to have an eruption of duration 0 minutes.
19. (a) The predicted number of days in April until the first blossom decreases by 4.69 days for each increase of 1°C in the average temperature in March. **(b)** The y intercept does have meaning because it is possible to have an average March temperature of 0°C. If the average March temperature is 0°C, the predicted number of days in April until the first blossom is 33.1 days, which can be interpreted as May 3rd. However, because this prediction is an extrapolation, it may not be very accurate.
21. (a) $\hat{y} = 1425 - 19.87(49.4) = 443.42$ cubic feet per day; Residual $= 520 - 443.42 = 76.58$ cubic feet per day. This month had 76.58 cubic feet per day of gas consumption more than the consumption predicted by the regression line with $x =$ average temperature. **(b)** The predicted gas consumption decreases by 19.87 cubic feet per day for each increase of 1°F in the average monthly temperature. **(c)** The y intercept does have meaning because it is possible to have a monthly average temperature of 0°F. If a month has an average temperature of 0°F, the predicted gas consumption is 1425 cubic feet per day. However, because this prediction is an extrapolation, it may not be very accurate.
23. The slope of −19.87 tells us the predicted gas consumption decreases by 19.87 cubic feet per day for each increase of 1°F in the average monthly temperature. For an increase of 10°F in the average monthly temperature, the predicted gas consumption decreases by $(19.87)(10) = 198.7$ cubic feet per day.
25. (a) The predicted number of calories increases by 15 for each increase of 1 gram of fat. **(b)** The y intercept does have meaning because it is possible to have a product with 0 grams of fat. If a product has 0 grams of fat, the predicted number of calories is 118. **(c)** The slope of the least-squares regression line is greater for the drink products. However, the y intercept is greater for the food products. **(d)** The predicted number of calories per gram of fat is the same for food products and drink products when fat is approximately 6.50 grams and the number of calories is approximately 247 calories. This is the point where the two least-squares regression lines intersect and the predicted numbers of calories for the food products and drink products are equal.
27. (a) The correlation of 0.87 indicates that the linear relationship between calories and salt content for these hot dogs is moderately strong and positive. **(b)** The hot dog with the lowest calorie count is in the lower left and separated from the rest of the points in the horizontal direction, but is in the general pattern of the other points. Therefore, this point is influential and will make the correlation closer to 1. With this point, the correlation is more strongly positive.

Answers to Section 11.2 Exercises

1. The phrase "least-squares" reminds us that the least-squares regression line is the line that makes the sum of the *squared* residuals as small as possible (the *least*).
3. True. If an outlier has an x value that is equal to $\bar{x}$, the line will move toward the outlier (changing the y intercept), but the slope will remain the same.
5. Regression to the mean describes the tendency for extreme values of the explanatory variable to be paired with less extreme values of the response variable.
7. $\hat{y} = 300.04 + 2.83x$, where $x =$ amount of sugar (grams) and $y =$ calories
9. (a) Both data sets give $\hat{y} = 3 + 0.5x$. **(b)** While both data sets have the same least-squares regression line, the relationship in Data Set A is linear and the relationship in Data Set B is clearly curved.

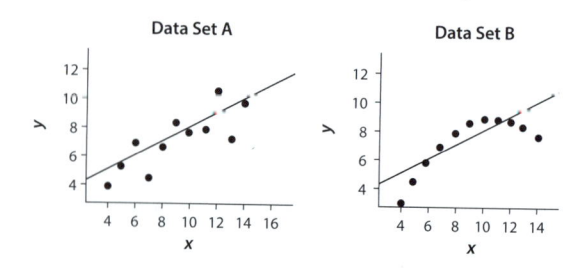

11. (a) Because the point for Maine is near $\bar{x}$ but above the rest of the points, it pulls the line up a little, which increases the y intercept but does not change the slope very much. **(b)** Because the point for Texas is slightly below the line on the right but still close to the line, it makes the slope of the regression line slightly less steep (less positive) and slightly increases the y intercept.

13. Because the point for Peanut M&M'S is near $\bar{x}$ but above the rest of the points, it pulls the line up a little, which increases the y intercept but does not change the slope very much.

15. (a) $6/6 = 100\%$ **(b)** $8/8 = 100\%$ **(c)** Among the teams with high winning percentages prior to the All-Star break, all of the teams had a lower winning percentage after the All-Star break. Likewise, among the teams with a low winning percentage prior to the All-Star break, all of the teams had a higher winning percentage after the All-Star break.

17. (a) $\hat{y} = x$, where x = midterm exam score and y = final exam score **(b)** $\hat{y} = 46.6 + 0.41(50) = 67.1$; $\hat{y} = 46.6 + 0.41(100) = 87.6$ **(c)** The student with a score of 50 on the midterm exam had a low score. The least-squares regression line predicts that this student's final exam score will also be low (67.1), but not as low as the score on the midterm exam. The student with a score of 100 on the midterm exam had a high score. The least-squares regression line predicts that this student's final exam score will also be high (87.6), but not as high as the score on the midterm exam.

19. (a) The scatterplot is shown here.

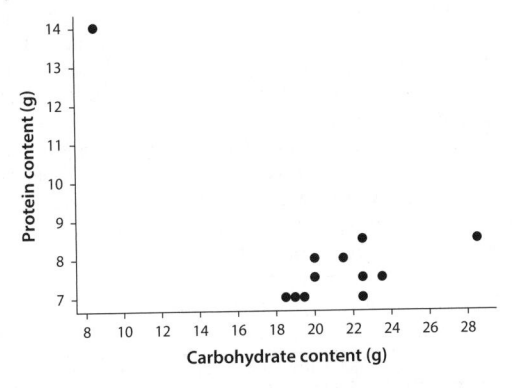

(b) $\hat{y} = 13.8981 - 0.28104x$, where x = carbohydrate content and y = protein content **(c)** Because the carbohydrate content in soybeans is much lower than that in the other types of beans and the protein content is much higher, the least-squares regression line has a negative slope. Without including soybeans, the least-squares regression line has a positive slope. The equation of the line is $\hat{y} = 4.772 + 0.130x$, where x = carbohydrate content (g) and y = protein content (g).

21. (a) The scatterplot is shown here.

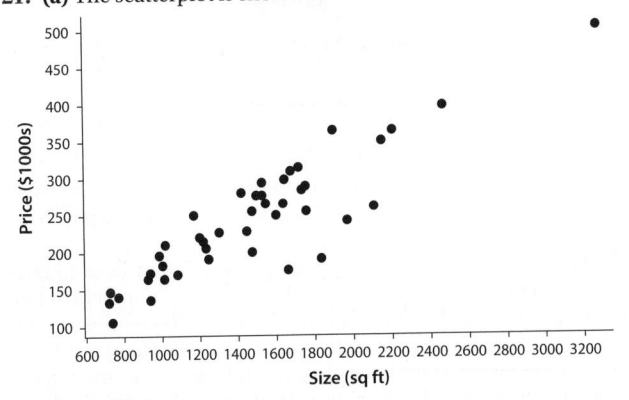

(b) $\hat{y} = 39.96205 + 0.1376x$, where x = size (square feet) and y = price (thousand \$) **(c)** $\hat{y} = 39.96205 + 0.1376(1633) = \$264,663$; Residual = $\$265,000 - \$264,663 = \$337$. The price of this house is \$337 more than the price predicted by the regression line with x = size. **(d)** The predicted price increases by \$137.60 for each increase of 1 square foot in size.

23. $\hat{y} = 31687.1 - 1754.03x$, where x = age (in years) and y = price (dollars)

25. (a) $\hat{y} = 4.774 + 0.135x$, where x = calories and y = amount of sugar (grams) **(b)** The slope and y intercept are not the same as in Exercise 7.

27. According to the scatterplot, the ideal value of k is 31. This is the value of k for which the sum of squared residuals is a minimum.

29. A histogram is shown here. The distribution of the weights of these diamonds is right-skewed, with several possible high outliers. Because of the skewness and outliers, we should use the median (5.4 mg) and IQR (5.5 mg) to describe the center and variability.

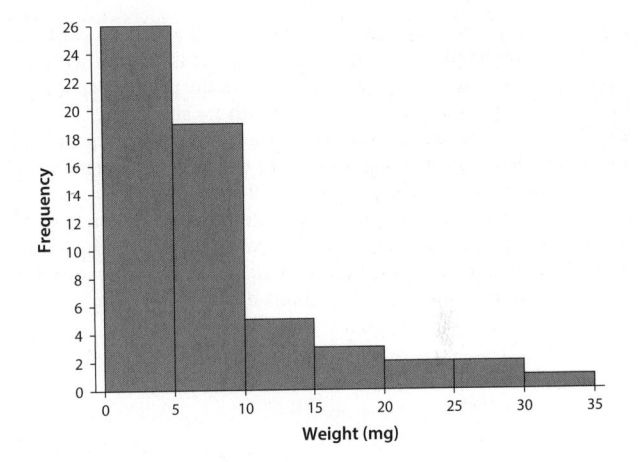

Answers to Section 11.3 Exercises

1. Is a line the right model to use or would a curve be better? If a line is the right model to use, how well does it make predictions?

3. True

5. The coefficient of determination measures the percent reduction in the sum of the squared residuals when using the least-squares regression line to make predictions, rather than the mean value of y. In other words, r^2 measures the percentage of the variability in the response variable that is accounted for by the least-squares regression line.

7. Because there is a leftover curved pattern (a U-shaped curve) in the residual plot, the least-squares regression line is not an appropriate model for relating the percentage of people who click on a link for a website to the website's position in the results of an internet search.

9. Because there is no leftover curved pattern in the residual plot, the least-squares regression line is an appropriate model for relating the battery life of a tablet to the price of the tablet.

11. The actual wait time until the next eruption of the Old Faithful geyser is typically about 6.49 minutes away from the wait time predicted by the least-squares regression line with x = duration of the previous eruption.

13. The actual number of days in April until the first blossom is typically about 3.02 days away from the number of days in April predicted by the least-squares regression line with x = the average temperature in March (°C).

15. 85.4% of the variability in wait time until the next eruption of the Old Faithful geyser is accounted for by the least-squares regression line with x = duration of the previous eruption.

17. 72.4% of the variability in number of days in April until the first blossom is accounted for by the least-squares regression line with x = the average temperature in March (°C).

19. (a) Because there is no leftover curved pattern in the residual plot, the least-squares regression line is an appropriate model for relating the percentage of the grass area burned to the number of wildebeest. **(b)** The actual percentage of the grass area burned is typically about 15.99 percentage points away from the percentage of the grass area burned predicted by the least-squares regression line with

x = wildebeest abundance. **(c)** 64.6% of the variability in the percentage of the grass area burned is accounted for by the least-squares regression line with x = wildebeest abundance.
21. (a)

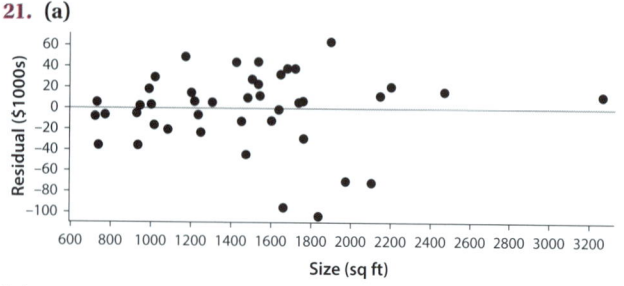

(b) $s = 36.395$; the actual price is typically about \$36,395 away from the price predicted by the least-squares regression line with x = size. **(c)** $r^2 = 0.794$; 79.4% of the variability in the price is accounted for by the least-squares regression line with x = size.
23. Because r^2 is expressed as a percentage, the change in the units of wait time will not affect the value of r^2. The change will affect the value of s. The value of s would be $6.49/60 = 0.108$ hour.
25. Answers will vary. Two possible scatterplots are shown here. In the first scatterplot, the association is clearly curved, but the points don't deviate much from the line ($s = 1.3$). In the second scatterplot, the association is linear, but the points deviate a lot from the line ($s = 4.5$).

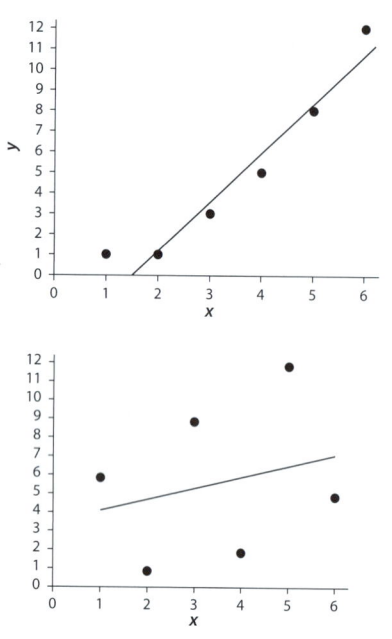

27. (a) $\hat{y} = -3.822 + 5.215x$ **(b)** $\hat{y} = -3.822 + 5.215(21) = 105.693$ floors **(c)** No, this would be an extrapolation, as a climb time of 35 minutes is far outside the interval of x values used to obtain the regression line. This prediction may not be very accurate.

Answers to Section 11.4 Exercises

1. False. The population least-squares regression line is $y = \beta_0 + \beta_1 x + \varepsilon$.
3. $b_1 \pm t^* SE_{b_1}$
5. *Random?* Each helicopter was randomly assigned a drop height. ✓ *Linear?* There is a linear association between drop height (cm) and drop time (sec) in the scatterplot. Also, there is no leftover curved pattern in the residual plot. ✓ *Equal SD?* In the residual plot, we do not see a clear < pattern or > pattern. ✓ *Normal?* A histogram of the residuals does not show strong skewness or outliers. ✓
7. *Random?* Random sample of 14 countries. ✓ *Linear?* The association between the income per person (\$1000s) and mortality rate (per 1000 live births) in the scatterplot does not appear to be linear. Also, the residual plot shows an obvious leftover curved pattern. This condition is not met. *Equal SD?* In the residual plot, the y values are more variable for smaller values of x. This condition is not met. *Normal?* A dotplot of the residuals shows a strong right skew. This condition is not met.
9. df $= 70 - 2 = 68$; *Tech:* $t^* = 1.995$, $0.0057244 \pm 1.995(0.0002018) \rightarrow$ 0.0053218 to 0.0061270; *Table B:* Using df $= 60$, $t^* = 2.000$, CI $= 0.0053208$ to 0.006128
11. S: 90% CI for $\beta_1 =$ the slope of the true least-squares regression line relating $y =$ number correct to $x =$ volume (decibels). **P:** t interval for the slope of a least-squares regression line. The conditions for inference are met. **D:** df $= 30 - 2 = 28$, $t^* = 1.701$, $-0.0483 \pm 1.701(0.0116) \rightarrow -0.0680$ to -0.0286 **C:** We are 90% confident that the interval from -0.0680 to -0.0286 captures the slope of the true least-squares regression line relating $y =$ number correct to $x =$ volume (decibels).
13. S: 95% CI for $\beta_1 =$ the slope of the population least-squares regression line relating $y =$ maximum speed (mph) to $x =$ height (ft). **P:** t interval for the slope of a least-squares regression line. *Random?* Random sample of 9 roller coasters. ✓ *Linear?* There is a linear association between height and speed in the scatterplot. Also, there is no leftover curved pattern in the residual plot. ✓ *Equal SD?* In the residual plot, we do not see a clear < pattern or > pattern. *Normal?* The dotplot of the residuals does not show strong skewness or outliers. ✓

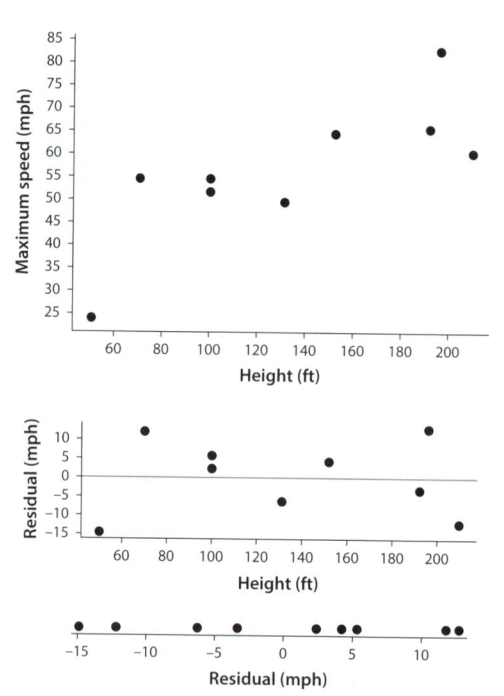

D: 0.062 to 0.367 with df $= 7$ **C:** We are 95% confident that the interval from 0.062 to 0.367 captures the slope of the population least-squares regression line relating $y =$ maximum speed (mph) to $x =$ height (ft).
15. S: 90% CI for $\beta_1 =$ the slope of the population least-squares regression line relating $y =$ calories burned to $x =$ number of steps for Josh. **P:** t interval for the slope of a least-squares regression line. *Random?* Random sample of 10 days. ✓ *Linear?* There is a linear association between number of steps and calories in the scatterplot. Also, there is no leftover curved pattern in the residual plot. ✓ *Equal SD?* In the residual plot, we do not see a clear < pattern or > pattern. ✓ *Normal?* The dotplot of the residuals does not show strong skewness or outliers. ✓

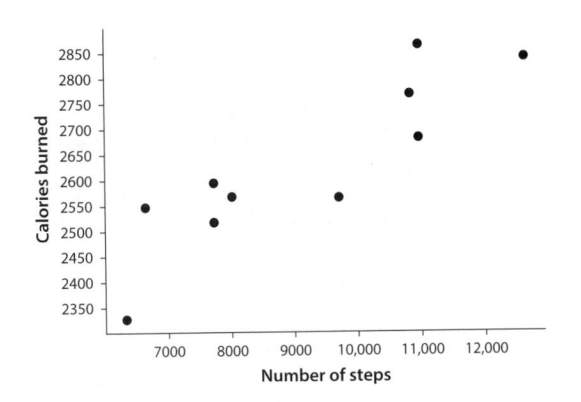

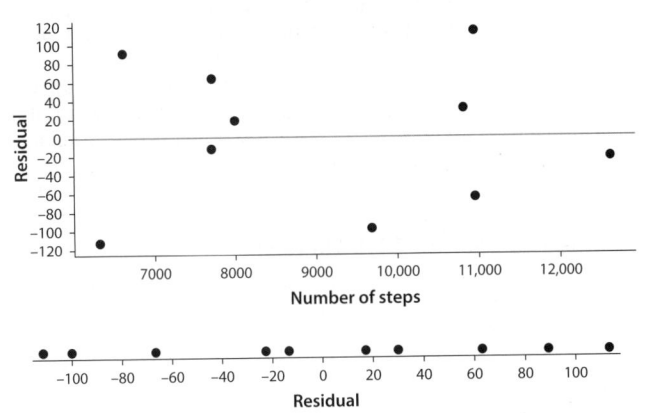

D: 0.044 to 0.091 with df = 8 **C:** We are 90% confident that the interval from 0.044 to 0.091 captures the slope of the population least-squares regression line relating y = calories burned to x = number of steps for Josh.

17. (a) S: 95% CI for β_1 = the slope of the true least-squares regression line relating y = average number of flowers to x = light intensity (μmol/m^2/sec). **P:** t interval for the slope of a least-squares regression line. *Random?* Two trays were randomly assigned to each light intensity. ✓ *Linear?* There is a linear association between light intensity and average number of flowers in the scatterplot. Also, there is no leftover curved pattern in the residual plot. ✓ *Equal SD?* In the residual plot, we do not see a clear < pattern or > pattern. ✓ *Normal?* The dotplot of the residuals does not show strong skewness or outliers. ✓

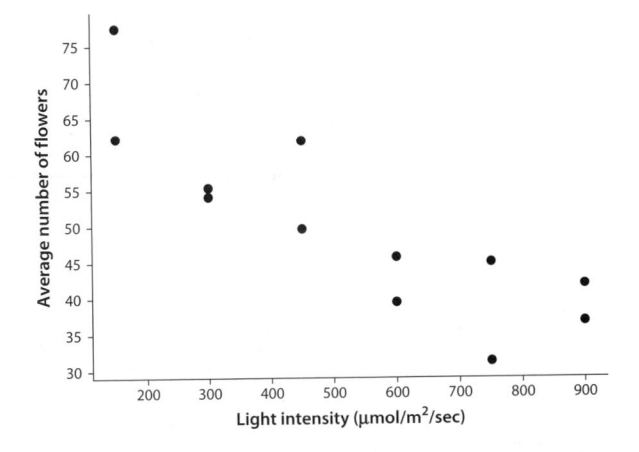

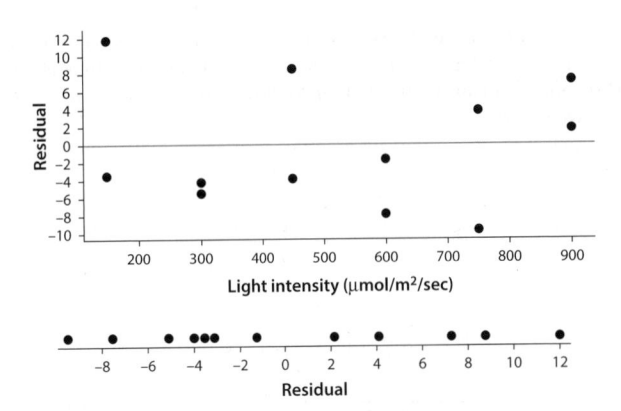

D: −0.059 to −0.023 with df = 10. **C:** We are 95% confident that the interval from −0.059 to −0.023 captures the slope of the true least-squares regression line relating y = average number of flowers to x = light intensity (μmol/m^2/sec). **(b)** The confidence interval from part (a) tells us that any value from −0.059 to −0.023 is a plausible value of the slope of the true least-squares regression line. Because all values in this interval are less than 0, there is convincing evidence that greater light intensity produces fewer flowers, on average.

19. S: 95% CI for β_1 = the slope of the population least-squares regression line relating y = price to x = age (years) **P:** t interval for the slope of a least-squares regression line. *Random?* Random sample of used Honda CRVs. ✓ *Linear?* From the scatterplot, there is a linear association between age (in years) and price ($) and there is no obvious leftover curved pattern in the residual plot. ✓ *Equal SD?* In the residual plot, we do not see a clear < pattern or > pattern. ✓ *Normal?* The dotplot of the residuals does not show strong skewness but does reveal some outliers. Due to the large sample size, it is not surprising to find outliers in the distribution of residuals when the Normal condition is met. Therefore, we can assume the Normal condition is met for these data. ✓

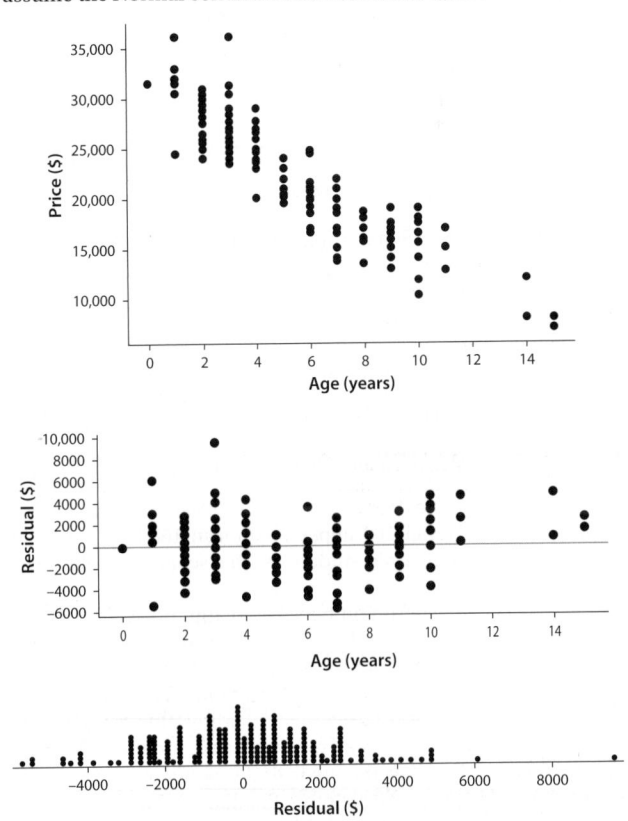

D: −1858.537 to −1649.524 with df =189 **C:** We are 95% confident that the interval from −1858.537 to −1649.524 captures the slope of the population least-squares regression line relating y = price ($) to x = age (years).

21. (a) Here is a table of income, mortality, log(income), and log(mortality).

Income	Mortality	log(income)	log(mortality)
38,003.90	4.4	4.5798	0.6435
2475.68	56.4	3.3937	1.7513
1202.53	127.5	3.0801	2.1055
1382.95	68.5	3.1408	1.8357
7858.97	21.3	3.8954	1.3284
1830.97	87.5	3.2627	1.9420
8199.03	26.3	3.9138	1.4200
4523.44	21.6	3.6555	1.3345
32,021.00	2.8	4.5054	0.4472
643.39	160.3	2.8085	2.2049
10,005.20	7.1	4.0002	0.8513
1493.53	84.0	3.1742	1.9243
4016.20	17.6	3.6038	1.2455
8826.90	14.5	3.9458	1.1614

(b) Here is the scatterplot of y = log(mortality) and x = log(income).

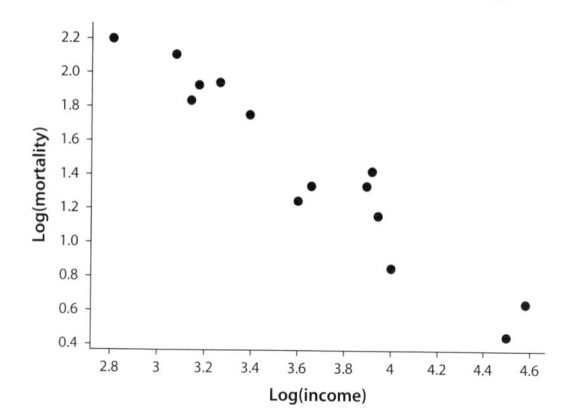

The relationship between x = log(income) and y = log(mortality) has a linear form.

23. (a) The average height of 15-year-old students at this school = 105 + 4.2(15) = 168 cm. **(b)** Because the conditions for the regression model were met, we know that the height of 15-year-olds varies approximately normally about the true regression line with a standard deviation of σ = 7 cm. (i) $z = \frac{180-168}{7} = 1.714$; *Tech:* P(z > 1.714) = Applet/normalcdf(lower: 1.714, upper: 1000, mean: 0, SD: 1) = 0.0433; *Table A:* P(z > 1.71) = 1 − 0.9564 = 0.0436 (ii) *Tech:* Applet/normalcdf(lower: 180, upper: 1000, mean: 168, SD: 7) = 0.0432. About 4.32% of 15-year-old students at this school are taller than 180 cm.

25. (a) This is an experiment because treatments were imposed on the students. Some students read the words first and then identified colors, while others identified colors first and read words second. **(b)** The purpose of randomizing the order is to eliminate the order of the treatment as a possible confounding variable. If all the subjects read the words first and then identified colors, it is possible that the color performance would improve simply because students were more familiar with the list the second time through. Then we would not know if any difference in performance was due to the order of the treatments or to the treatments themselves.

27. (a) Here is the scatterplot.

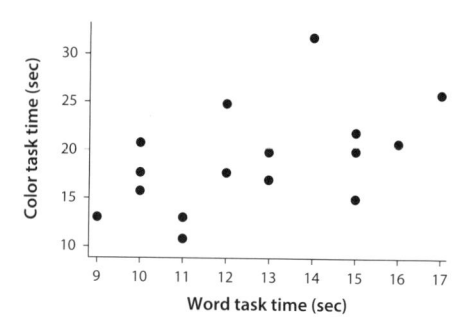

There is a moderate, positive, linear relationship between word task time (seconds) and color task time (seconds). **(b)** $\hat{y} = 4.887 + 1.132x$, where x = word task time (seconds) and y = color task time (seconds) **(c)** $\hat{y} = 4.887 + 1.132(9) = 15.075$ seconds; Residual = 13 − 15.075 = −2.075 seconds. This student was 2.075 seconds faster at identifying the colors than expected, based on the regression line using x = word task time (seconds).

Answers to Section 11.5 Exercises

1. False. Hypotheses for a test about a relationship between two quantitative variables may be expressed in words or in symbols.

3. $t = \dfrac{b_1 - 0}{SE_{b_1}}$

5. (a) $H_0: \beta_1 = 0$, $H_a: \beta_1 > 0$, where β_1 is the slope of the true least-squares regression line relating y = proportion of perch killed and x = number of perch in the pen. **(b)** The evidence for H_a is: $b_1 = 0.00857 > 0$.

7. (a) $H_0: \beta_1 = 0$, $H_a: \beta_1 < 0$, where β_1 is the slope of the population least-squares regression line relating y = heart disease death rate and x = per capital wine consumption in a country. **(b)** The evidence for H_a is: $b_1 = -22.97 < 0$.

9. $t = 3.489$, df = 16 − 2 = 14; *Table B*: The P-value is between 0.001 and 0.0025. *Tech*: P-value = 0.0018

11. $t = -6.452$, df = 19 − 2 = 17; *Table B*: The P-value is less than 0.0005. *Tech*: P-value ≈ 0

13. S: $H_0: \beta_1 = 0$, $H_a: \beta_1 < 0$, where β_1 is the slope of the population least-squares regression line relating y = birth rate (births per 1000 population) to x = female literacy rate (%) in a country. Use α = 0.05. The evidence for H_a is: $b_1 = -0.377 < 0$. **P:** t test for the slope of a least-squares regression line. *Random?* Random sample of 20 countries. ✓ *Linear?* There is a linear association between female literacy rate (%) and birth rate (births per 1000 population) in the scatterplot. Also, there is no obvious leftover curved pattern in the residual plot. ✓ *Equal SD?* In the residual plot, we do not see a clear < pattern or > pattern. *Normal?* A dotplot of the residuals does not show strong skewness or outliers. ✓

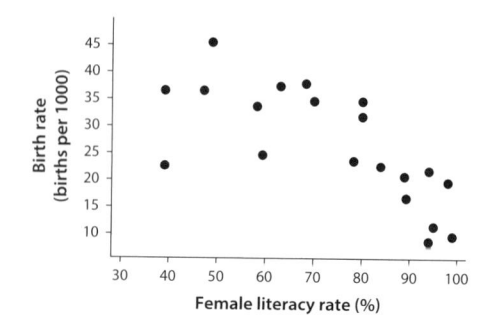

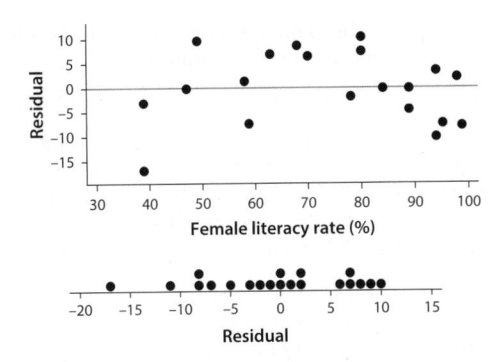

D: $b_1 = -0.377$, $t = -4.331$, P-value $= 0.0002$ with df $= 18$ **C:** Because the P-value of $0.0002 < \alpha = 0.05$, we reject H_0. There is convincing evidence that countries with higher female literacy rates have lower birth rates.

15. S: $H_0: \beta_1 = 0$, $H_a: \beta_1 \neq 0$, where β_1 is the slope of the population least-squares regression line relating $y =$ pulse rate (beats per minute) to $x =$ body temperature (°F). Use $\alpha = 0.01$. The evidence for H_a is: $b_1 = 4.944 \neq 0$. **P:** t test for the slope of a least-squares regression line. *Random?* Random sample of 20 students. ✓ *Linear?* There is a linear association between body temperature (°F) and pulse rate (beats per minute) in the scatterplot. Also, there is no leftover curved pattern in the residual plot. ✓ *Equal SD?* In the residual plot, we do not see a clear < pattern or > pattern. ✓ *Normal?* A dotplot of the residuals does not show strong skewness or outliers. ✓

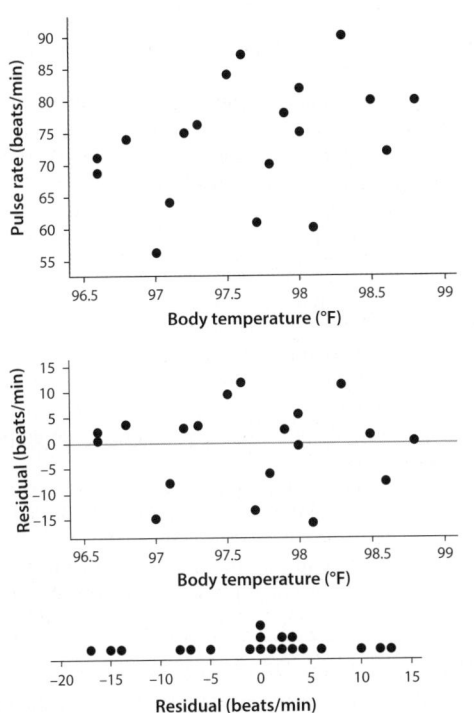

D: $b_1 = 4.944$, $t = 1.722$, P-value $= 0.102$ with df $= 18$ **C:** Because the P-value of $0.102 > \alpha = 0.01$, we fail to reject H_0. There is not convincing evidence of a linear association between body temperature and resting pulse rates for students at this university.

17. S: $H_0: \beta_1 = 0$, $H_a: \beta_1 > 0$, where β_1 is the slope of the population least-squares regression line relating $y =$ number of beetle larvae to $x =$ number of stumps in the plot. Use $\alpha = 0.05$. The evidence for H_a is: $b_1 = 11.8937 > 0$. **P:** t test for the slope of a least-squares regression line. *Random?* Random sample of 23 plots. ✓ *Linear?* There is a linear association between number of stumps and number of beetle larvae in the scatterplot. Also, there is no leftover curved pattern in the residual plot. ✓ *Equal SD?* In the residual plot, we do not see a clear < pattern or > pattern. ✓ *Normal?* A dotplot of the residuals does not show strong skewness or outliers. ✓

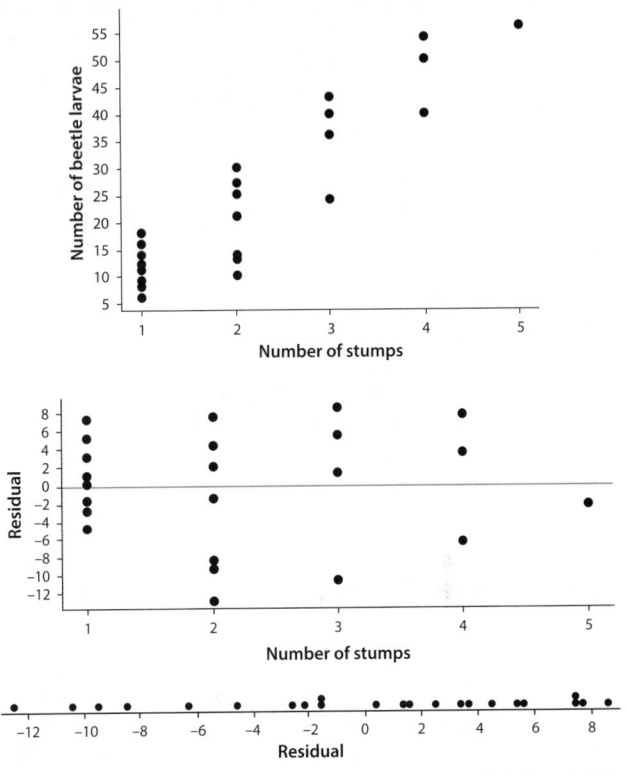

D: $b_1 = 11.8937$, $t = 10.467$, P-value ≈ 0 with df $= 21$ **C:** Because the P-value of approximately $0 < \alpha = 0.05$, we reject H_0. There is convincing evidence of a positive association between number of stumps and number of beetle larvae in this area.

19. Assuming there is no association between number of stumps and number of beetle larvae in the population of circular plots like these, there is ≈ 0 probability of getting a sample slope of 11.8937 or greater by chance alone.

21. (a) S: 95% CI for $\beta_1 =$ slope of the population least-squares regression line relating $y =$ pulse rate (beats per minute) to $x =$ body temperature (°F) **P:** t interval for the slope of a least-squares regression line **D:** -3.320 to 13.209 with df $= 18$ **C:** We are 99% confident that the interval from -3.320 to 13.209 captures the slope of the population least-squares regression line relating $y =$ pulse rate (beats per minute) to $x =$ body temperature (°F). **(b)** The confidence interval from part (a) tells us that any value from -3.320 to 13.209 is a plausible value of the slope of the population least-squares regression line. Because 0 is among these plausible values, there is not convincing evidence of a linear association between body temperature and resting pulse rates for students at this university.

23. S: $H_0: \beta_1 = 0$, $H_a: \beta_1 < 0$, where β_1 is the slope of the population least-squares regression line relating $y =$ price to $x =$ age (years). Use $\alpha = 0.05$. The evidence for H_a is: $b_1 = -1754.0303 < 0$. **P:** t test for the slope of a least-squares regression line. *Random?* Random sample of 191 used Honda CRVs. ✓ *Linear?* There is a linear association between age and price in the scatterplot. Also, there is no obvious leftover curved pattern in the residual plot. ✓ *Equal SD?* In the residual plot, we do not see a clear < pattern or > pattern. ✓ *Normal?* Due to the large sample size, it is not surprising to find outliers in the distribution of residuals when the Normal condition is met. Therefore, we can assume the Normal condition is met for these data. ✓

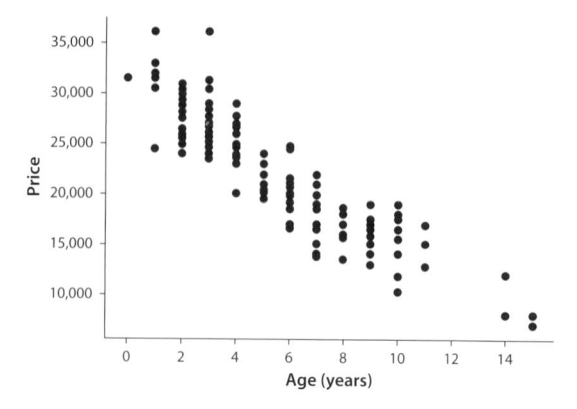

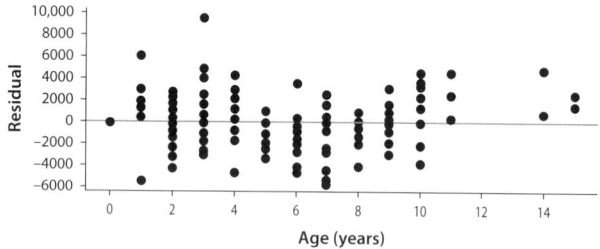

D: $b_1 = -1754.0303$, $t = -33.108$, P-value ≈ 0 with df $= 189$ **C:** Because the P-value of approximately $0 < \alpha = 0.05$, we reject H_0. There is convincing evidence of a negative linear association between $y =$ price and $x =$ age (years).

25. (a) $H_0: \rho = 0$, $H_a: \rho < 0$, where ρ is the population correlation coefficient between $x =$ female literacy rate (for women 15 years of age or older) and $y =$ birth rate (births per 1000 population).
(b) $r = -0.71434$

(c) $t = \dfrac{r - 0}{\sqrt{\dfrac{1 - r^2}{n - 2}}} = \dfrac{-0.71434 - 0}{\sqrt{\dfrac{1 - (-0.71434)^2}{20 - 2}}} = -4.331$. The value of the test statistic is equal to the value of the test statistic from Exercise 13.
(d) P-value $= 0.0002$ with df $= 18$. This is equal to the P-value from Exercise 13.

27. (a) S: H_0: There is no association between cry count and later aptitude score at age 3 for infants. H_a: There is an association between cry count and later aptitude score at age 3 for infants. Use $\alpha = 0.05$. The evidence for H_a is: The proportion of infants with a later aptitude score less than 115 ($10/15 = 0.667$ and $12/23 = 0.522$) is not the same for both cry count categories. **P:** Chi-square test for association. *Random?* Random sample of 38 infants. ✓ *Large Counts?* All expected counts are ≥ 5 (see the table). ✓

	Cry count	
	Less than 16	At least 16
Later aptitude score — Less than 115	8.684	13.316
Later aptitude score — At least 115	6.316	9.684

D: $\chi^2 = 0.7825$, df $= (2-1)(2-1) = 1$; *Table C: P*-value > 0.10; *Tech:* $\chi^2 = 0.782$, P-value $= 0.38$ with df $= 1$ **C:** Because the P-value of $0.38 > \alpha = 0.05$, we fail to reject H_0. There is not convincing evidence of an association between cry count and later aptitude score at age 3 for infants. **(b)** The magnitude of the values for later aptitude score and cry count are lost when turning the quantitative variables into the categorical variables needed for the chi-square test. As a result, the P-value for the chi-square test (0.38) is much larger than the P-value for the slope test (0.002), and the results of the chi-square test are not statistically significant.

29. (a) (i) $z = \dfrac{3 - 4.5}{0.9} = -1.667$ and $z = \dfrac{6 - 4.5}{0.9} = 1.667$. *Tech:* $P(-1.667 < z < 1.667) =$ Applet/normalcdf(lower: -1.667, upper: 1.667, mean: 0, SD: 1) $= 0.9045$; *Table A:* $P(-1.67 < z < 1.67) = 0.9525 - 0.0475 = 0.9050$ **(ii)** *Tech:* Applet/normalcdf(lower: 3, upper: 6, mean: 4.5, SD: 0.9) $= 0.9044$. **(b)** First quartile (i) *Tech:* Applet/invNorm(area: 0.25, mean: 0, SD: 1) gives $z = -0.674$. $-0.674 = \dfrac{x - 4.5}{0.9}$ gives $x = 3.8934$; *Table A:* A z-score of -0.67 gives the closest value (0.2514) to 0.25. $-0.67 = \dfrac{x - 4.5}{0.9}$; $x = (-0.67)(0.9) + 4.5 = 3.897$ (ii) *Tech:* invNorm(area: 0.25, mean: 4.5, SD: 0.9) $= 3.893$ minutes; Third quartile (i) *Tech:* Applet/invnorm(area: 0.75, mean: 0, SD: 1) gives $z = 0.674$. $0.674 = \dfrac{x - 4.5}{0.9}$ gives $x = 5.1066$; *Table A:* A z-score of 0.67 gives the closest value (0.7486) to 0.75. $0.67 = \dfrac{x - 4.5}{0.9}$; $x = (0.67)(0.9) + 4.5 = 5.103$ (ii) *Tech:* invNorm(area: 0.75, mean: 4.5, SD: 0.9) $= 5.107$. IQR $= 5.107 - 3.893 = 1.214$ minutes, high outlier $> Q_3 + 1.5 \times$ IQR $= 5.107 + 1.5(1.214) = 6.928$ minutes. Because 7 minutes > 6.928 minutes, a 7-minute shower would be classified as an outlier according to the $1.5 \times$ IQR rule. **(c)** *Shape:* The shape of the sampling distribution of $\bar{x}$ is approximately normal because the shape of the population distribution is normal. *Center:* $\mu_{\bar{x}} = 4.5$ minutes. *Variability:* $\sigma_{\bar{x}} = \dfrac{0.9}{\sqrt{10}} = 0.285$ minute (i) $z = \dfrac{5 - 4.5}{0.285} = 1.754$. *Tech:* $P(z > 1.754) =$ Applet/normalcdf(lower: 1.754, upper: 1000, mean: 0, SD: 1) $= 0.0397$; *Table A:* $P(z > 1.75) = 1 - 0.9599 = 0.0401$. (ii) *Tech:* Applet/normalcdf(lower: 5, upper: 1000, mean: 4.5, SD: 0.285) $= 0.0397$.

Answers to Section 11.6 Exercises

1. The population mean value of y at a given value of x and the value of y for a single observation at a given value of x.
3. The prediction interval will be wider for a given value of x.
5. We are 95% confident that the interval from 7.04 to 8.41 captures the average score on the math test for all volunteers like those in the study while listening to music at 45 decibels.
7. We are 90% confident that the interval from 198.23 to 231.02 captures the average death rate from heart disease (deaths per 100,000 people) for all countries for which the average per capita wine consumption is 2 L.
9. We are 95% confident that the interval from 4.46 to 10.99 captures the score on the math test for a single volunteer like those in the study while listening to music at 45 decibels.
11. We are 90% confident that the interval from 146.72 to 285.53 captures the heart disease death rate (deaths per 100,000 people) for a single country for which the average per capita wine consumption is 2 L.
13. (a) The width of both intervals will decrease because decreasing the confidence level from 95% to 90% decreases the margin of error. **(b)** Increasing the sample size from 30 to 40 will decrease the width of both intervals because increasing the sample size decreases the standard error, which decreases the margin of error. **(c)** Moving x^* from 45 decibels to 55 decibels will decrease the width of both intervals because 55 decibels is closer to the mean value of x than $x^* = 45$ decibels. This decreases the standard error, which decreases the margin of error.
15. (a) The width of both intervals will increase because increasing the confidence level from 90% to 95% increases the margin of error. **(b)** Decreasing the sample size from 19 countries to 15 countries will increase the width of both intervals because decreasing the sample size increases the standard error, which increases the margin of error. **(c)** Moving x^* from 2 L to 7 L will increase the width of both intervals because 7 L is farther from the mean value

of x than $x^* = 2$ L. This increases the standard error, which increases the margin of error.

17. (a) $\hat{y} = 201 + 24.03(45) = 1282.35$ cal/24 hr. **(b)** We are 98% confident that the interval from 1203 to 1362 captures the average metabolic rate for all individuals like those in the study with a lean body mass of 45 kg. **(c)** We are 98% confident that the interval from 1008 to 1557 captures the metabolic rate for a single individual like those in the study with a mean body mass of 45 kg. **(d)** Moving x^* from 45 kg to 38 kg will increase the width of both intervals because 38 kg is farther from the mean value of x than $x^* = 45$ kg. This increases the standard error, which increases the margin of error.

19. (a) $\hat{y} = 28.17 + 0.2143(175) = 65.6725$ mph **(b)** The wider interval (38.67, 92.70) is the 95% prediction interval, and the narrower interval (55.24, 76.13) is the 95% confidence interval. Because of the extra term of "1+" in the standard error of $\hat{y}$, the standard error of $\hat{y}$ is greater than the corresponding standard error of $\mu_{\hat{y}}$. This increases the margin of error for a prediction interval.

21. (a) We are 95% confident that the interval from $22,595.68 to $23,238.21 captures the mean price of all used Honda CRVs that are 5 years old. **(b)** We are 95% confident that the interval from $18,470.76 to $27,363.13 captures the price of a single used Honda CRV that is 5 years old.

23. (a) In both of the standard error formulas, s_x^2 appears in the denominator of the fractions. If we divide by a larger number, the overall result is a smaller standard error. And if the standard error is smaller, the width of the interval is also smaller. **(b)** A larger value of s_x^2 is preferred because a larger value of s_x^2 will result in a smaller standard error. And if the standard error is smaller, the width of the interval is also smaller. All other things being equal, an interval with a smaller width is always preferred. **(c)** When designing a study, be sure the variability in the x-values of the experimental units is as large as practically possible. In an observational study, there is much less control over the variablity of the x-values in the sample.

25. (a) Although there is considerable overlap in the distributions between the six times, the percent metamorphosed at time 0 is centered at a greater value than the other times, suggesting that at least one mean is different. **(b)** $H_0: \mu_0 = \mu_5 = \mu_{10} = \mu_{15} = \mu_{20} = \mu_{25}$, H_a: At least one mean is different, where μ_{time} represents the mean percentage of larvae that metamorphosed after time 0, 5, 10, 15, 20, and 25 minutes. **(c)** The final ANOVA table is

Source	df	SS	MS	F
Groups	5	0.6309	0.12618	5.97
Error	30	0.6346	0.02115	
Total	35	1.2655		

Because the F statistic of 5.97 is larger than 1, there is some evidence for at least one difference among the population means.

Answers to Chapter Review Exercises

1. (a) $\hat{y} = 130.49 - 0.0861(150) = 117.575$. I am not confident in this prediction because $x = 150$ is far outside the interval of x values used to obtain the regression line. **(b)** $\hat{y} = 130.49 - 0.0861(119) = 120.24$; Residual $= 104.01 - 120.24 = -16.23$. In 2019, the ice broke 16.23 days earlier than the date predicted by the regression line with x = number of years since 1900. **(c)** The predicted day that the ice will break decreases by 0.0861 day for each increase of 1 year since 1900. **(d)** Yes. In 1900, the predicted ice break day was 130.49 days into the year.

2. (a) The scatterplot for the average crawling age and average temperature is given. The equation of the least-squares regression line is $\widehat{age} = 35.7 - 0.078 temperature$.

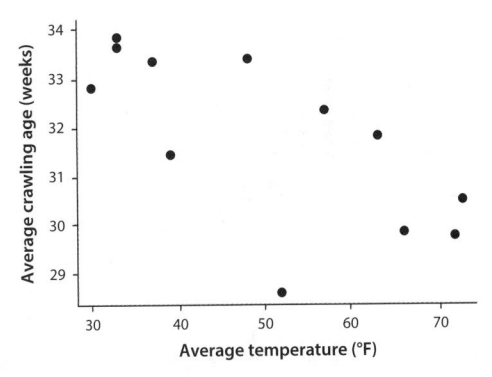

(b) The point for May has an x-value that is near $\bar{x}$ but a y-value that is far below the other points. This pulls the least-squares regression line down a little bit, which decreases the y intercept but does not change the slope very much.

3. (a) $5/8 = 62.5\%$ **(b)** $6/9 = 66.7\%$ **(c)** Among the quarterbacks with a high completion percentage during Season 1, more than half (62.5%) had a lower winning percentage during Season 2. Likewise, among the quarterbacks with a low completion percentage during Season 1, more than half (66.7%) had a higher winning percentage during Season 2.

4. (a) Because there is no leftover curved pattern in the residual plot, the least-squares regression line is an appropriate model for relating ice break day to the number of years since 1900. **(b)** The actual ice break day is typically about 6.1 days away from the ice break day predicted by the least-squares regression line with x = number of years since 1900. **(c)** 15% of the variability in the ice break day is accounted for by the least-squares regression line with x = number of years since 1900.

5. S: 95% CI for β_1 = the slope of the true least-squares regression line relating y = weight gain (ounces) to x = growth hormone dose (milligrams) **P:** t interval for the slope of a least-squares regression line. The conditions for inference are met. **D:** df $= 15 - 2 = 13$, $t^* = 2.160$, CI $= 2.6369$ to 7.0277 **C:** We are 95% confident that the interval from 2.6369 to 7.0277 captures the slope of the true least-squares regression line relating y = weight gain (ounces) to x = growth hormone dose (milligrams).

6. S: $H_0: \beta_1 = 0$, $H_a: \beta_1 < 0$, where β_1 is the slope of the population least-squares regression line relating y = saturated fat (grams) to x = amount of sugar (tsp). Use $\alpha = 0.05$. The evidence for H_a is: $b_1 = -0.3287 < 0$. **P:** t test for the slope of a least-squares regression line. *Random?* Random sample of 19 protein bars. ✓ *Linear?* There is a linear association between age and price in the scatterplot and there is no leftover curved pattern in the residual plot. ✓ *Equal SD?* In the residual plot, we do not see a clear < pattern or > pattern. ✓ *Normal?* A dotplot of the residuals does not show strong skewness or obvious outliers. ✓

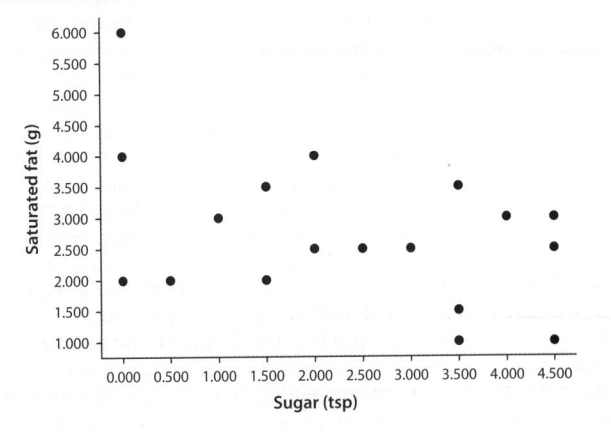

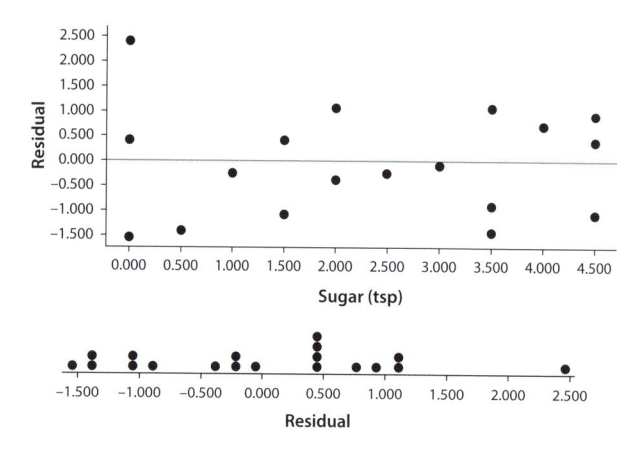

D: $b_1 = -0.3287$, $t = -2.022$, P-value = 0.030 with df = 17 **C:** Because the P-value of $0.030 < \alpha = 0.05$, we reject H_0. There is convincing evidence of a negative association between sugar and fat in protein bars.

7. (a) The wider interval, (0.808, 5.619), is the prediction interval. A prediction interval for an individual value of y will be wider than the corresponding confidence interval for the mean value of y. **(b)** Confidence interval for a mean response: We are 95% confident that the interval from 2.524 g to 3.902 g captures the mean fat content of all protein bars with one teaspoon of sugar. Prediction interval for an individual response: We are 95% confident that the interval from 0.808 g to 5.619 g captures the mean fat content of a single protein bar with one teaspoon of sugar.

CHAPTER 12

Answers to Section 12.1 Exercises

1. True
3. For the given values of the explanatory variables, use the multiple regression model to calculate the value of $\hat{y}$. Then calculate the residual using the actual value of y: residual = $y - \hat{y}$.
5. $\widehat{cGPA} = -0.1025 + 0.6974(3.05) + 0.0003(1140) = 2.3666$
7. $\hat{y} = 3.4022 + 0.0366(400) - 0.0251(300) = 10.5122$ wins
9. (a) $\widehat{cGPA} = -0.1025 + 0.6974(3.89) + 0.0003(1350) = 3.0154$; Residual = $3.73 - 3.0154 = 0.7146$ **(b)** The college GPA (after three semesters) for this student is 0.7146 greater than the value predicted by the multiple regression model with x_1 = high school GPA and x_2 = SAT score.
11. (a) $\hat{y} = 3.4022 + 0.0366(492) - 0.0251(355) = 12.4989$ wins; Residual = $11 - 12.4989 = -1.4989$ **(b)** The number of wins for the Tampa Bay Buccaneers during the 2020 season is 1.4989 wins less than the value predicted by the multiple regression model with x_1 = points scored and x_2 = points allowed.
13. $s = 0.71$; the actual college GPAs are typically about 0.71 point away from the college GPAs predicted by the multiple regression model using high school GPA and SAT score. $R^2 = 0.18$; 18% of the variability in the college GPAs is accounted for by the multiple regression model using high school GPA and SAT score.
15. $s = 1.375$; the actual number of wins is typically about 1.375 wins away from the number of wins predicted by the multiple regression model using points scored and points allowed. $R^2 = 0.855$; 85.5% of the variability in the number of wins is accounted for by the multiple regression model using points scored and points allowed.
17. (a) $\hat{y} = 33,957.20 - 1813.42(7) - 0.17875(7479) = \$19,926.39$ **(b)** Residual = $\$18,990 - \$19,926.39 = -\$936.39$; the price of this used Honda CRV is \$936.39 less than the value predicted by the multiple regression model with x_1 = age (years) and x_2 = annual miles (miles per year). **(c)** $s = 2129$; the actual prices are typically about \$2129 away from the prices predicted by the multiple regression model using age and annual miles. $R^2 = 0.868$; 86.8% of the variability in price is accounted for by the multiple regression model using age and annual miles.
19. (a) $\hat{y} = 82.954617 + 10.12442x_1 + 14.335391x_2 + 1.225832x_3$, where x_1 = saturated fat (grams), x_2 = added sugar (teaspoons), and x_3 = protein (grams) **(b)** $\hat{y} = 82.954617 + 10.12442(1.5) + 14.335391(0) + 1.225832(17) = 118.98$; Residual = $90 - 118.98 = -28.98$. The number of calories in this yogurt is 28.98 calories less than the number of calories predicted by the multiple regression model with x_1 = saturated fat (grams), x_2 = added sugar (teaspoons), and x_3 = protein (grams). **(c)** $s = 27.859$; the actual calories are typically about 27.859 calories away from the calories predicted by the multiple regression model using saturated fat, added sugar, and protein. $R^2 = 0.538$; 53.8% of the variability in the number of calories is accounted for by the multiple regression model using saturated fat, added sugar, and protein.
21. Because there is no leftover curved pattern in the residual plot, this multiple regression model is an appropriate model for predicting scoring average for LPGA golfers.
23. (a) $\hat{y} = 71.92 - 0.5667(1) - 0.7340(1) + 0.7980(-1) - 0.1942(1) = 69.6271$ **(b)** Because the coefficient for putting average has the largest magnitude, a change of 1 standard deviation in putting average leads to the biggest change in predicted scoring average. Because the coefficient for sand saves has the smallest magnitude, a change of 1 standard deviation in sand saves leads to the smallest change in predicted scoring average. From the scatterplots, the relationship between putting average and the residuals from the model using only average driving distance is the strongest; the relationship between sand saves and the residuals from the model using only average driving distance is the weakest.
25. (a) Because the shape of the probability distribution is skewed to the right, it is not possible to find the probability that a randomly selected orange has at least 10 seeds. **(b)** It is possible to find the probability using the sampling distribution of $\bar{x}$. $\mu_{\bar{x}} = 8.5$ seeds; $\sigma_{\bar{x}} = \dfrac{5.9}{\sqrt{30}} = 1.077$ seeds. Because $n = 30 \geq 30$, the sampling distribution of $\bar{x}$ is approximately normal. (i) $z = (10 - 8.5)/1.077 = 1.393$; Using technology: $P(z > 1.393)$ = Applet/normalcdf(lower: 1.393, upper: 1000, mean: 0, SD: 1) = 0.0818; Using Table A: $P(z > 1.39)$ = $1 - 0.9177 = 0.0823$. (ii) Using technology: normalcdf(lower: 10, upper: 1000, mean: 8.5, SD: 1.077) = 0.0818

Answers to Section 12.2 Exercises

1. 1, 0
3. False. The coefficient of an indicator describes the change in the predicted values of the response variable when the value of the indicator is equal to 1 after accounting for the other explanatory variables.
5. (a) The scatterplot shows that for any given light intensity, the average number of flowers tends to be greater for the trays that received the early light treatment. **(b)** $\widehat{flowers} = 71.31 - 0.0405(600) + 12.16(1) = 59.17$; Residual = $62.9 - 59.17 = 3.73$. The actual average number of flowers for this tray is 3.73 flowers greater than the value predicted by the multiple regression model with x_1 = light intensity and x_2 = early.
7. $\hat{y} = 109.8 + 372.82(1.023) - 118.25(1) = 372.94$ deer collisions; Residual = $143 - 372.94 = -229.94$ deer collisions. The actual number of collisions for this county is 229.94 collisions less than the number predicted by the multiple regression model with x_1 = annual miles driven (in billions) and x_2 = 5-year wolf presence.
9. (a) x_1: After accounting for whether the tray received the early light treatment or not, the predicted average number of flowers decreases by 0.0405 for each additional unit (micromoles per square meter per second) of light intensity. x_2: After accounting for light intensity, the predicted average number of flowers for the trays that received the early light treatment is 12.16 flowers greater than the predicted number of flowers for the trays that did not receive

the early light treatment. **(b)** Although it is possible for x_2 = early to equal 0, it does not make sense to set x_1 = light intensity equal to 0. The meadowfoam plants need light to grow.

11. (a) x_1: After accounting for 5-year wolf presence, the predicted number of deer collisions increases by 372.82 for each additional billion-mile increase in the annual number of miles driven. x_2: After accounting for annual miles driven, the predicted number of collisions in counties where wolves have been present in the previous 5 years is 118.25 less than the predicted number of collisions in counties where wolves have not been present in the previous 5 years. **(b)** The y intercept of 109.8 does not make sense. If $x_2 = 0$, then there would be no miles driven and no deer collisions, regardless of 5-year wolf presence.

13. (a) The model should include an interaction term because the slopes of the separate regression lines are different for beers and wines. **(b)** $\widehat{Calories} = 119.4 + 1.6314(10) - 126.63(0) + 28.75(10)(0) = 135.714$ calories **(c)** $\widehat{Calories} = 119.4 + 1.6314(Percent) - 126.63(1) + 28.75(Percent)(1) = -7.23 + 30.3814(Percent)$ **(d)** -126.63: The y intercept of the regression line for beer is 126.63 calories smaller than the y intercept of the regression line for wine. 28.75: The slope of the regression line for beer is 28.75 units greater than the slope of the regression line for wine.

15. (a) $\hat{y} = 64.16 + 465.44(1.023) - 52.01(1) - 139.12(1.023)(1) = 345.975$ collisions; Residual $= 143 - 345.975 = 143 - 345.975 = -202.975$. The actual number of deer collisions for this county is 202.975 collisions less than the number predicted by the multiple regression model with x_1 = annual miles driven (in billions) and x_2 = 5-year wolf presence. **(b)** $\hat{y} = 64.16 + 465.44x_1 - 52.01(0) - 139.12x_1(0) = 64.16 + 465.44x_1$ **(c)** By including the interaction term, the typical difference between the actual number of collisions and the predicted number of collisions decreased from 148.1 to 146.9. Although the decrease in s is relatively small, adding the interaction term did improve the ability to predict the number of deer collisions.

17. (a) $\hat{y} = 32,855.70 - 1757.65(6) - 0.2082(10171.5) + 2288.18(1) = \$22,480.27$; Residual $= \$20,990 - \$22,480.27 = -\$1490.27$. The actual price of this Honda CRV is $1490.27 less than the value predicted by the multiple regression model with x_1 = age, x_2 = annual miles, and x_3 = leather. **(b)** x_1: After accounting for annual miles and the presence of leather seats, the predicted price decreases by $1757.65 for each additional year in the age of the car. x_2: After accounting for age and the presence of leather seats, the predicted price decreases by $0.2082 for each additional mile per year increase in the annual miles driven. x_3: After accounting for the age and annual miles driven of the car, the predicted price of a Honda CRV with leather seats is $2288.18 greater than the predicted price of a Honda CRV without leather seats. **(c)** $s = 1810$; The actual prices are typically about $1810 away from the prices predicted by the multiple regression model using age, annual miles, and leather. $R^2 = 0.905$; 90.5% of the variability in price is accounted for by the multiple regression model using age, annual miles, and leather.

19. (a) By including the interaction term, the typical difference between the actual prices and the predicted prices decreased from $1810 to $1760. Although the decrease in s is relatively small, adding the interaction term did improve the ability to predict the price. **(b)** $\hat{y} = 32,131.80 - 1615.47(6) - 0.2126(10171.5) + 3703.99(1) - 289.59(6)(1) = \$22,242.97$; Residual $= \$20,990 - \$22,242.97 = -\$1252.97$. The actual price of this Honda CRV is $1252.97 less than the value predicted by the multiple regression model with x_1 = age, x_2 = annual miles, x_3 = leather, and the interaction term x_1x_3.

21. (a) $\hat{y} = 37.801813 + 10.970071x_1 + 15.083384x_2 + 4.003489x_3 + 41.01004x_4$, where x_1 = saturated fat (grams), x_2 = added sugar (teaspoons), x_3 = protein (grams), and x_4 = plant-based (1 = yes, 0 = no). **(b)** After accounting for the amount of saturated fat, added sugar, and protein, the predicted number of calories in plant-based yogurt is 41.01004 greater than the predicted number in non-plant-based yogurt. **(c)** By including the indicator variable, the typical difference between the actual number of calories and the predicted number of calories decreased from 27.859 calories to 24.953 calories. Based on this reduction, the indicator variable helped the model to improve the predictions.

23. $\hat{y} = 28.391577 + 15.728659x_1 + 14.021637x_2 + 4.091191x_3 + 68.215886x_4 - 12.786383x_1x_4$ where x_1 = saturated fat (grams), x_2 = added sugar (teaspoons), x_3 = protein (grams), x_4 = plant-based (1 = yes, 0 = no), and x_1x_4 is an interaction term. By including the interaction term, the typical difference between the actual number of calories and the predicted number of calories decreased from 24.953 calories to 21.509 calories. Based on this reduction, the interaction term helped to improve the predictions.

25. (a) $\hat{y} = 31,739.2 - 1706.85(5) - 0.2399(12500) + 1899.89(1) + 2943.07(0) + 5741.08(0) = \$22,106.09$ **(b)** x_5: After accounting for the age and annual miles of the car, the predicted price of a Honda CRV Touring is $5741.08 greater than the predicted price of a Honda CRV LX. **(c)** The predicted price increase for CRVs that have leather seats is $2943.07 - \$1899.89 = \$1043.18$. **(d)** In Exercise 17, the estimated value of leather is $2288.19, which is bigger than the estimated value of leather in part (c) ($1043.18). In part (c), the only difference between the EX and EX-L is leather seats. But in Exercise 17, leather = 0 includes both EX and LX, the lowest trim line. And leather = 1 includes both EX-L and Touring, the highest trim line. So, the difference between leather = 0 and leather = 1 includes more than just leather seats, which is why the estimate from Exercise 17 is greater than the estimate in part (c).

27. (a) Because it appears that separate models for each of the temperatures would have the same y-intercept, an indicator variable is not necessary. **(b)** $\hat{y} = 96.5357 - 14.837(2.5) + 8.926(2.5)(1) = 81.7582\%$

29. $\hat{y} = 10 + 0.15x_1 + 10x_2 - 0.03x_1x_2$

31. S: H_0: There is no difference in the distribution of color for name-brand (Haribo Gold) and store-brand (Great Value) gummy bears. H_a: There is a difference in the distribution of color for name-brand (Haribo Gold) and store-brand (Great Value) gummy bears. Use $\alpha = 0.05$. The evidence for H_a is: The proportion of red gummy bears is not the same for the two brands (Haribo: 0.367, Great Value: 0.341), etc. **P:** Chi-square test for association. *Random?* Independent random samples of each brand. ✓ *Large Counts?* 130.83, 218.17, 58.86, 98.14, 50.61, 84.39, 77.97, 130.03, 54.73, and 91.27 are ≥ 5. ✓ **D:** $\chi^2 = 1.81$, df = 4, P-value = 0.7698 **C:** Because the P-value of $0.7698 > \alpha = 0.05$, we fail to reject H_0. There is not convincing evidence of a difference in the distribution of color for name-brand and store-brand gummy bears.

Answers to Section 12.3 Exercises

1. H_a: At least one of the $\beta_i's \neq 0$
3. True
5. two
7. $H_0: \beta_1 = \beta_2 = \beta_3 = 0$; H_a: At least one of these $\beta_i's \neq 0$
9. $H_0: \beta_1 = \beta_2 = 0$; H_a: At least one of β_1 or $\beta_2 \neq 0$
11. Because the P-value of approximately $0 < \alpha = 0.05$, we reject H_0. There is convincing evidence that at least one of the explanatory variables [mouthpart damage, body length, exposure (yes/no) to Bd fungus] provides useful information about the length of the intestinal tract of tadpoles.
13. Because the P-value of $0.003 < \alpha = 0.05$, we reject H_0. There is convincing evidence that at least one of the explanatory variables (property value, monthly electricity cost) provides useful information about income.
15. Because the P-values for *Mouthpart Damage*, *Body*, and *Bd* are all $< \alpha = 0.05$, we have convincing evidence that each of these variables provides additional useful information about gut length.
17. Because the P-value for *Electric* $> \alpha = 0.05$, we do not have convincing evidence that this variable provides additional useful information about income. Because the P-value for *Property* $< \alpha = 0.05$, we have convincing evidence that this variable provides additional useful information about household income.

19. (a) $H_0: \beta_1 = \beta_2 = \beta_3 = 0$; H_a: At least one of these $\beta_i's \neq 0$. **(b)** Because the P-value of approximately $0 < \alpha = 0.05$, we reject H_0. There is convincing evidence that at least one of the explanatory variables (age, miles, leather) provides useful information about price. **(c)** Because the P-values for *age, miles,* and *leather* are all $< \alpha = 0.05$, we have convincing evidence that each of these variables provides additional useful information about price.

21. (a) $H_0: \beta_1 = \beta_2 = \beta_3 = 0$; H_a: At least one of these $\beta_i's \neq 0$. **(b)** *Random?* Random sample of Starbucks products. *Normal?* A dotplot of the residuals does not show strong skewness or outliers. ✓ *Linear?* There is no leftover curved pattern in the residual plot. ✓ *Equal SD?* In the residual plot, there is a slight > pattern, but the standard deviation on the left doesn't appear to be more than twice as big as the standard deviation on the right. So we can consider this condition met. ✓

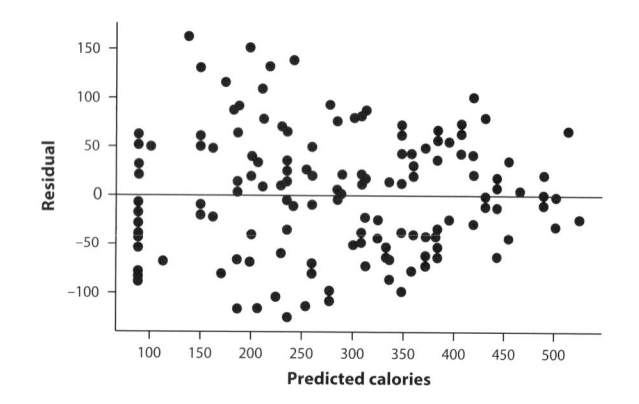

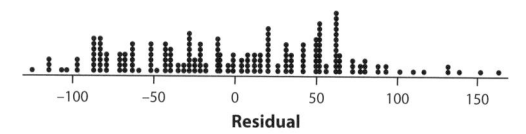

(c) From the ANOVA table, the P-value is approximately 0. Because the P-value of approximately $0 < \alpha = 0.05$, we reject H_0. There is convincing evidence that at least one of the explanatory variables (fat, food, fat*food) provides useful information about calories. **(d)** Because the F test is significant, we know that at least one of the coefficients is not zero, so we should look at the individual t tests to find out which one(s).

23. (a) For a 95% confidence interval with df = 187, $t^* = 1.973$. *Age:* $-1204.8 \pm 1.973(62.4) = (-1327.9152, -1081.6848)$. We are 95% confident that the interval from -1327.9152 to -1081.6848 captures the coefficient of age in the population regression model. *Leather:* $2216 \pm 1.973(238) = (1746.426, 2685.574)$. We are 95% confident that the interval from 1746.426 to 2685.574 captures the coefficient of leather in the population regression model. **(b)** *Age:* The 95% confidence interval does not include 0 as a plausible value, so we would reject $H_0: \beta_1 = 0$. We have convincing evidence that age provides additional useful information about price. *Leather:* The 95% confidence interval does not include 0 as a plausible value, so we would reject $H_0: \beta_3 = 0$. We have convincing evidence that leather provides additional useful information about price.

25. (a) 3 sizes × 4 quantities = 12 treatments **(b)** We will randomly assign 16 students to each of the first six treatment groups and 15 students to each of the next six treatment groups. Number the students from 1 to 186. Then use a random number generator to select 16 different numbers between 1 and 186 for group 1, an additional 16 different numbers for group 2, and so on for the first six groups. Continue using the random number generator to select 15 different numbers between 1 and 186 for groups 7, 8, 9, 10, and 11. The remaining 15 students are assigned to group 12. **(c)** Answers will vary. Variables to consider are the size of each student and whether or not the student makes healthy eating choices.

Answers to Chapter 12 Review Exercises

1. (a) $\widehat{calories} = 7.2629 + 12.9850(1.5) + 3.8457(7) = 53.6603$ **(b)** Residual $= 50 - 53.6603 = -3.6603$ calories. The actual number of calories is 3.6603 calories less than the value predicted by the multiple regression model, with x_1 = grams of saturated fat and x_2 = grams of protein. **(c)** $s = 4.46$; the actual number of calories is typically about 4.46 calories away from the number of calories predicted by the multiple regression model using saturated fat (g) and protein (g). $R^2 = 0.760$; 76% of the variability in the number of calories is accounted for by the multiple regression model using saturated fat and protein.

2. (a) $\hat{y} = 250.696 + 53.439(1.4) - 37.776(1) = 287.7346$; Residual $= 260 - 287.7346 = -27.7346$ g/mi. The actual CO_2 emissions are typically 27.7346 units less than the emissions predicted by the multiple regression model using x_1 = engine size and x_2 = front. **(b)** x_1: After accounting for whether the vehicle has front-wheel drive or not, the predicted CO_2 emissions increase 53.439 g/mi for each additional liter in engine size. x_2: After accounting for engine size, the predicted CO_2 emissions for front-wheel drive vehicles are 37.776 g/mi less than the predicted CO_2 emissions for vehicles without front-wheel drive. **(c)** The model does not need an interaction term because the slopes of the separate regression lines for vehicles with front-wheel drive and vehicles without front-wheel drive are approximately equal.

3. (a) $H_0: \beta_1 = \beta_2 = 0$; H_a: At least one of β_1 or $\beta_2 \neq 0$. Because the P-value of approximately $0 < \alpha = 0.05$, we reject H_0. There is convincing evidence that at least one of the explanatory variables (*EngSize, Front*) provides useful information about CO_2 emissions. **(b)** Because the P-value for *Front* and *EngSize* $< \alpha = 0.05$, we have convincing evidence that these variables provide additional useful information about CO_2 emissions. **(c)** Because the P-value for the interaction term *EngSize*Front* $> \alpha = 0.05$, we do not have convincing evidence that adding an interaction term is helpful.

4. (a) $\widehat{mass} = 18.0848 - 5.411(1.2) + 0.8484(1) = 12.44$ g **(b)** After accounting for humidity, the predicted nestling mass for nests that were exposed to fleas is 0.8484 g greater than the predicted nestling mass for nests that were not exposed to fleas. **(c)** 47.7% of the variability in the nestling mass is accounted for by the multiple regression model using humidity and exposed. **(d)** $H_0: \beta_1 = \beta_2 = 0$, H_a: At least one of β_1 or $\beta_2 \neq 0$. Because the P-value of approximately $0 < \alpha = 0.05$, we reject H_0. There is convincing evidence that at least one of the explanatory variables (*Humidity, Exposed*) provides useful information about nestling mass. **(e)** Because the P-value for *Humidity* $< \alpha = 0.05$, we have convincing evidence that this variable provides additional useful information about nestling mass. Because the P-value for *Exposed* $< \alpha = 0.05$, we have convincing evidence that this variable provides additional useful information about nestling mass.

Answers to Cumulative Review IV: Chapters 10–12

Section I: Multiple Choice

1. d
2. c
3. d
4. c
5. a
6. b
7. b
8. a
9. c
10. b
11. a
12. c
13. d
14. b
15. c
16. c

17. c
18. d
19. a
20. b

Section II: Free Response

1. S: H_0: There is no association between amount of exercise and whether or not a person has significant sleep disturbance for people like the ones in the study. H_a: There is an association between amount of exercise and whether or not a person has significant sleep disturbance for people like the ones in the study. Use $\alpha = 0.05$. The evidence for H_a is: The proportion of people who have a significant sleep disturbance ($41/92 = 0.446$, $46/151 = 0.305$, $33/99 = 0.333$, and $31/95 = 0.326$) is not the same for each exercise category. P: Chi-square test for association. *Random?* Participants were randomly assigned to one of four exercise treatments. ✓ *Large Counts?* All expected counts are ≥ 5 (see the table). ✓

		Amount of exercise			
		None	Light	Moderate	Heavy
Sleep disturbance	Yes	31.789	52.176	34.208	32.826
	No	60.211	98.824	64.792	62.174

D: $\chi^2 = \dfrac{(41-31.789)^2}{31.789} + \dfrac{(46-52.176)^2}{52.176} + \cdots + \dfrac{(64-62.174)^2}{62.174} = 5.415$; $df = (2-1)(4-1) = 3$; *Using Table C: P*-value is greater than 0.10. *Using technology:* $\chi^2 = 5.415$, *P*-value of 0.144 with df = 3. C: Because the *P*-value of $0.144 > \alpha = 0.05$, we fail to reject H_0. There is not convincing evidence of an association between amount of exercise and whether or not a person has significant sleep disturbance for people like the ones in the study.

2. (a) H_0: $\mu_{None} = \mu_{Light} = \mu_{Moderate} = \mu_{Heavy}$; H_a: At least one mean is different, where μ_{Amount} is the mean improvement in MOSSS score for the amount of exercise for people like the ones in the study. (b) *Random?* Participants were randomly assigned to one of four exercise treatments. ✓ *Normal/Large Sample?* Each treatment group has at least 30 participants. ✓ *Equal SD?* The sample standard deviations are $s_{None} = 9.9$, $s_{Light} = 10.2$, $s_{Moderate} = 10.0$, $s_{Heavy} = 9.6$ and $10.2 < 2(9.6) = 19.2$. ✓ (c) If the average MOSSS scores are the same for all four amounts of exercise, there is a 0.04 probability of getting sample means as different or more different than the sample means obtained in this study by chance alone. (d) Because the *P*-value of $0.04 < \alpha = 0.05$, we reject H_0. There is convincing evidence that the mean improvement in MOSSS scores is not the same for all four amounts of exercise for people like the ones in this study.

3. (a) H_0: $\beta_1 = 0$, H_a: $\beta_1 > 0$, where β_1 is the slope of the true least-squares regression line relating $y =$ improvement in MOSSS score and $x =$ amount of exercise (in KKW). (b) ANOVA tests for whether or not the mean improvement in MOSSS score is different for at least one level of exercise. The regression model specifies that the mean MOSSS score will increase as KWW increases, not just that at least one mean MOSSS score will be different. (c) Because the *P*-value of $0.02 < \alpha = 0.05$, we reject H_0. There is convincing evidence that improvements in MOSSS scores tend to get larger as the amount of exercise (KKW) increases for people like the ones in this study.

4. (a) $\widehat{ERA} = 6.12 - 0.22(14.3) = 2.974$; Residual $= 1.08 - 2.974 = -1.894$. The ERA for Jacob deGrom is 1.894 units less than the ERA predicted by the regression model with $x =$ strikeout rate. (b) The predicted ERA will decrease by 0.22 for each 1-unit increase in strikeout rate. The y-intercept does not have meaning because it does not make sense to have an MLB pitcher with a strikeout rate of 0. (c) *Random?* Random sample of pitchers. ✓ *Normal?* A histogram of the residuals does not show strong skewness or outliers. ✓ *Linear?* There is a linear association between strikeout rate and ERA. Also, there is no leftover curved pattern in the residual plot. ✓ *Equal SD?* In the residual plot, we do not see a clear < pattern or > pattern. ✓ (d) We are 95% confident that the interval from 0.355 to 5.684 captures the ERA of a pitcher with a strikeout rate of 14.3. (e) A 95% confidence interval is narrower than the corresponding 95% prediction interval. There is less variability in the estimate of the mean ERA among all pitchers with a strikeout rate of 14.3 than in the estimate of the ERA for a single pitcher with the same strikeout rate.

5. (a) $\widehat{ERA} = 2.73 - 0.19(14.3) + 1.81(0.6) + 0.25(1.1) = 1.374$; Residual $= 1.08 - 1.374 = -0.294$. The ERA for Jacob deGrom is 0.294 unit less than the ERA predicted by the regression model relating ERA to strikeout rate, homerun rate, and walk rate. (b) After accounting for homerun rate and walk rate, the predicted ERA will decrease by 0.19 for each 1-unit increase in strikeout rate. (c) H_0: $\beta_1 = \beta_2 = \beta_3 = 0$, H_a: At least one of these $\beta_i's \neq 0$. From the ANOVA table, the *P*-value is approximately 0. Because the *P*-value of approximately $0 < \alpha = 0.05$, we reject H_0. There is convincing evidence that at least one of the explanatory variables (strikeout rate, homerun rate, walk rate) provides useful information about ERA. (d) The *P*-values for strikeout rate, homerun rate, and walk rate are all $< \alpha = 0.05$. We have convincing evidence that each of these variables provides additional useful information about ERA in a model with the other two variables included.

CHAPTER 13

Answers to Section 13.1 Exercises

1. False; $X =$ number of negative differences when H_a: Population median difference < 0

3. ignored

5. hypothesized median

7. (a) H_0: Population median difference (Outbound − Return) $= 0$; H_a: Population median difference (Outbound − Return) > 0 (b) Calculating Outbound − Return: 33, 5, 25, 19, 12, 6, 11, 6, 1, 4, −5, 4. Because H_a tests for a population median difference > 0, the test statistic is $X = 11$, the number of positive differences.

9. (a) H_0: Population median difference (More reps/less weight − Fewer reps/more weight) $= 0$; H_a: Population median difference (More reps/less weight − Fewer reps/more weight) $\neq 0$ (b) Calculating More reps/less weight − Fewer reps/more weight: 2, −1, 1, −2, 1, 0, 2, −2, 3, 2. Because 6 of the differences are positive and 3 are negative and the difference of 0 is ignored, the test statistic is $X = 6$.

11. (a) *P*-value $= P(X \geq 11) = 0.0032$ (b) Because the *P*-value of $0.0032 < \alpha = 0.05$, we reject H_0. There is convincing evidence that the Dubai-to-Doha outbound flight typically takes longer.

13. (a) *P*-value $= 2P(X \geq 6) = 0.5078$ (b) Because the *P*-value of $0.5078 > \alpha = 0.05$, we fail to reject H_0. There is not convincing evidence that more reps with less weight tend to yield a different median number of curls for athletes like the ones in the study than would fewer reps with more weight.

15. S: H_0: Population median $= \$330{,}800$; H_a: Population median $< \$330{,}800$, where Population median is the median single-family home price in Huntingdon County, Pennsylvania. Use $\alpha = 0.10$. The evidence for H_a is: More than half the values (14 of 15) are less than $330,800. P: Sign test. *Random?* Random sample of Huntingdon County homes. ✓ D: $X = 14$; *P*-value $= P(X \geq 14) = 0.0005$. C: Because the *P*-value of $0.0005 < \alpha = 0.10$, we reject H_0. There is convincing evidence that the median single-family home price in Huntingdon County is less than the national median of $330,800.

17. S: H_0: Population median difference (L1 − L2) $= 0$, H_a: Population median difference (L1 − L2) $\neq 0$. Use $\alpha = 0.05$. The evidence for H_a is: More than half the differences (19 of 24) are positive. P: Sign test. *Random?* The order in which the subjects heard the two lists was randomized. ✓ D: $X = 19$; *P*-value $= 2P(X \geq 19) = 0.0066$. C: Because

the P-value of $0.0066 < \alpha = 0.05$, we reject H_0. There is convincing evidence that these two lists are not equally hard to hear.
19. (a) Because the sample size of 25 is not at least 30 and the distribution of differences (Human yawned – Human didn't yawn) is skewed to the right, the Normal/Large Sample condition is not met. **(b)** H_0: Population median difference (Human yawned – Human didn't yawn) = 0; H_a: Population median difference (Human yawned – Human didn't yawn) > 0. *Random?* The order in which the dog received the treatments was randomized. ✓ **(c)** The 15 differences of 0 will be ignored. $X = 8$; P-value $= P(X \geq 8) = 0.0547$ **(d)** Because the P-value of $0.0547 > \alpha = 0.05$, we do not reject H_0. There is not convincing evidence that dogs like the ones in this study tend to yawn more when in the presence of a yawning human than a non-yawning human.
21. S: H_0: $p = 0.5$, H_a: $p \neq 0.5$, where p is the true proportion of spins of a euro coin that would land heads. Use $\alpha = 0.05$. The evidence for H_a is: The sample proportion of spins that landed heads, $\hat{p} = \dfrac{140}{250} = 0.56$, is not equal to 0.5. **P:** Sign test. *Random?* We will consider the 250 spins of the euro coin to be a random sample of all such spins. ✓ **D:** $X = 140$; P-value $= 2P(X \geq 140) = 2(0.0332) = 0.0664$. **C:** Because the P-value of $0.0664 > \alpha = 0.05$, we do not reject H_0. There is not convincing evidence that the coin was minted asymmetrically.
23. (a) H_0: $\mu_d = 0$, H_a: $\mu_d > 0$, where μ_d is the mean difference (Traditional suit – Polyurethane suit) in 50-meter freestyle time for all swimmers like those in this study. **(b)** *Random?* The order in which each swimmer wore the two suits was randomly determined. ✓ *Normal/Large Sample?* The sample size is small, but there is no strong skewness or outliers in the dotplot.

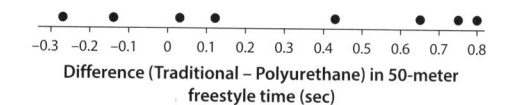

Difference (Traditional – Polyurethane) in 50-meter freestyle time (sec)

(c) Because the P-value of $0.042 < \alpha = 0.05$, we reject H_0. There is convincing evidence that polyurethane swimsuits help swimmers swim faster, on average. **(d)** H_0: Population median difference (Traditional suit – Polyurethane suit) = 0; H_a: Population median difference (Traditional suit – Polyurethane suit) > 0, where Population median difference is the median difference (Traditional suit – Polyurethane suit) in 50-meter freestyle time for all swimmers like those in this study. **(e)** *Random?* The order in which each swimmer wore the two suits was randomly determined. ✓ **(f)** $X = 6$; P-value $= P(X \geq 6) = 0.145$. Because the P-value of $0.145 > \alpha = 0.05$, we fail to reject H_0. There is not convincing evidence that polyurethane swimsuits help swimmers swim faster, on average. **(g)** Because the paired t test uses more information (that is, the actual values of the data) than the sign test, the paired t test is more likely to reject the null hypothesis and less likely to result in a Type II error than the sign test. This makes the paired t test a more powerful test than the sign test.
25. (a) A t-procedure is appropriate for this situation. All conditions are met. Independent random samples of healthy men and healthy women were selected, and the sample sizes are large ($n_1 = n_2 = 65 \geq 30$). **(b)** A t-procedure is not appropriate for this situation. Because the sample sizes are small ($n_1 = n_2 = 5 < 30$), we must check whether it's reasonable to believe that the population distributions are normal. The dotplot for the no group shows skewness with at least one potential outlier. This graph gives us reason to doubt the normality of the population.

Answers to Section 13.2 Exercises

1. absolute value
3. ignored
5. (a) The ranks of the differences are shown in the table. **(b)** The signed ranks are shown in the table.

Day	1	2	3	4	5	6	7	8	9	10	11	12
Outbound	75	42	62	63	54	46	52	50	42	46	43	52
Return	42	37	37	44	42	40	41	44	41	42	48	48
Difference	33	5	25	19	12	6	11	6	1	4	−5	4
|Difference|	33	5	25	19	12	6	11	6	1	4	5	4
Rank	12	4.5	11	10	9	6.5	8	6.5	1	2.5	4.5	2.5
Signed rank	12	4.5	11	10	9	6.5	8	6.5	1	2.5	−4.5	2.5

7. (a) The ranks of the differences are shown in the table. **(b)** The signed ranks are shown in the table.

Athlete	1	2	3	4	5	6	7	8	9	10
More reps/less weight	18	14	23	22	7	9	12	9	18	22
Fewer reps/more weight	16	15	22	24	6	9	10	11	15	20
Difference	2	−1	1	−2	1	0	2	−2	3	2
|Difference|	2	1	1	2	1	0	2	2	3	2
Rank	6	2	2	6	2	–	6	6	9	6
Signed rank	6	−2	2	−6	2	–	6	−6	9	6

9. (a) *Random?* The flights were randomly selected. ✓ *Symmetric?* The dotplot of the differences is reasonably symmetric. ✓

Difference (Outbound – Return) in time (min)

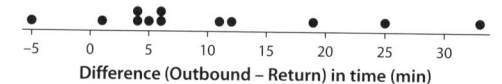

(b) $W^+ = 73.5$; $W^- = 4.5$; $W = 4.5$ **(c)** The P-value is less than 0.01. **(d)** Because the P-value of less than $0.01 < \alpha = 0.05$, we reject H_0. There is convincing evidence that the Dubai-to-Doha outbound flight typically takes longer.
11. (a) *Random?* The differences come from a randomized experiment. ✓ *Symmetric?* The dotplot of the differences is not strongly skewed. ✓

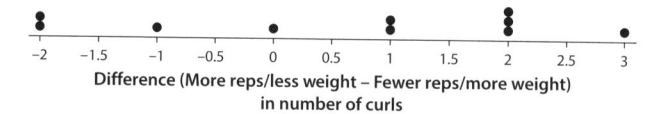

Difference (More reps/less weight – Fewer reps/more weight) in number of curls

(b) $W^+ = 31$; $W^- = 14$; $W = 14$ **(c)** The P-value is greater than 0.05. **(d)** Because the P-value of greater than $0.05 > \alpha = 0.05$, we fail to reject H_0. There is not convincing evidence that more reps with less weight tend to yield a different median number of curls for athletes like the ones in this study than would fewer reps with more weight.
13. S: H_0: Population median difference (Judge A – Judge B) in scoring dives = 0; H_a: Population median difference (Judge A – Judge B) in scoring dives $\neq 0$. Use $\alpha = 0.05$. The evidence for H_a is: The two judges differ in their diving scores more than half the time (8 of 14 times). **P:** Wilcoxon signed rank test *Random?* The differences come from a randomized experiment. ✓ *Symmetric?* The dotplot of the differences is reasonably symmetric ✓

Difference (Judge A – Judge B) in diving score

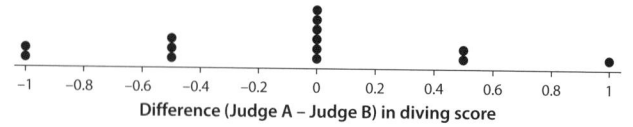

D: $W^+ = 13$; $W^- = 23$; $W = 13$. The P-value is greater than 0.05. **C:** Because the P-value $> \alpha = 0.05$, we fail to reject H_0. There is not convincing evidence that these two judges are judging the divers differently.
15. S: H_0: The median speed of drivers in this construction zone = 25 mph; H_a: The median speed of drivers in this construction zone > 25 mph. Use $\alpha = 0.05$. The evidence for H_a is: More than half (8 of 10)

of the drivers had speeds over 25 mph. **P:** Wilcoxon signed rank test. *Random?* Drivers were randomly selected. ✓ *Symmetric?* The dotplot of the differences is reasonably symmetric. ✓

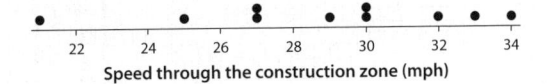

Speed through the construction zone (mph)

D: $W^+ = 41.5$; $W^- = 3.5$; $W = 3.5$. The *P*-value is between 0.01 and 0.05. **C:** Because the *P*-value $< \alpha = 0.05$, we reject H_0. There is convincing evidence that the median speed of drivers in this construction zone is greater than the posted speed limit of 25 mph.

17. (a) The difference, the ranks, and the signed ranks are shown in the table. **(b)** H_0: Population median = $1862; H_a: Population median < $1862, where Population median is the median apartment rent of all apartments in Chicago for which there was no application fee in May 2021. *Random?* A random sample of apartments was selected. ✓ *Symmetric?* The dotplot shows that the distribution of differences is reasonably symmetric. ✓

Rent for Chicago apartments with no application fee ($)

Apartment	1	2	3	4	5	6	7	8	9
Rental amount	975	1895	1500	1800	925	1600	1900	1600	870
Difference (Amount − $1862)	−887	33	−362	−62	−937	−262	38	−262	−992
|Difference|	887	33	362	62	937	262	38	262	992
Rank	7	1	6	3	8	4.5	2	4.5	9
Signed rank	−7	1	−6	−3	−8	−4.5	2	−4.5	−9

(c) $W^+ = 3$; $W^- = 42$; $W = 3$ **(d)** The *P*-value = 0.01. **(e)** Because the *P*-value of $0.01 < \alpha = 0.05$, we reject H_0. There is convincing evidence that the median rent of apartments in Chicago that do not have an application fee was less than $1862 in May 2021. The *P*-value for the Wilcoxon signed rank test is less than the *P*-value for the sign test. The conclusions from the two tests are different.

19. S: H_0: Population median weight gain (Terminal weight − Initial weight) = 0; H_a: Population median weight gain (Terminal weight − Initial weight) > 0, where Population median weight gain is the median weight gain (Terminal weight − Initial weight) of all first-year students at this university. Use $\alpha = 0.10$. The evidence for H_a is: More than half the weight gains (52 of 68) are positive. **P:** Wilcoxon signed rank test. Assume the conditions for inference are met. **D:** *Tech:* $W = 126.5$ and *P*-value < 0.001. **C:** Because the *P*-value of less than $0.001 < \alpha = 0.10$, we reject H_0. There is convincing evidence that the median weight gain of first-year students at this university is larger than 0.

21. (a) The least restrictive test is the sign test. The only condition required is the Random condition. The next most restrictive condition is for the Wilcoxon sign rank test. In addition to the Random condition, this test requires the population of differences to be symmetric. The most restrictive condition is for the *t* test. In addition to the Random condition, this test requires either the sample size to be large or the population distribution of differences to be normal. **(b)** $0.042 < 0.071 < 0.145$; *P*-value for *t* test $(0.042) <$ *P*-value for Wilcoxon signed rank test $(0.071) <$ *P*-value for sign test (0.145). This order goes from the most restrictive test to the least restrictive test.

23. S: H_0: There is no association between type of card and whether or not customers like the ones in this study leave a tip; H_a: There is an association between type of card and whether or not customers like the ones in this study leave a tip. Use $\alpha = 0.05$. The evidence for H_a is: The proportion of customers who leave a tip ($30/72 = 0.417$, $14/74 = 0.189$, $16/65 = 0.246$) is different for the three treatment groups. **P:** Chi-square test for association. *Random?* Customers were randomly assigned to one of three treatments. ✓ *Large Counts?* All expected counts are ≥ 5 (see the table). ✓

		Treatment		
		Joke card	Advertisement card	No card
Tip	Yes	20.474	21.043	18.483
	No	51.526	52.957	46.517

D: $\chi^2 = 9.953$; df $= (2-1)(3-1) = 2$; *P*-value $= 0.007$. **C:** Because the *P*-value of $0.007 < \alpha = 0.05$, we reject H_0. There is convincing evidence of an association between type of card and whether or not customers like the ones in this study leave a tip.

Answers to Section 13.3 Exercises

1. the two population (or treatment) medians are equal
3. False; both get the rank of 4.5.
5. (a) H_0: Median test score for people like these in the study using a computer for training = Median test score using a lecture video; H_a: Median test score for people like these in the study using a computer for training > Median test score using a lecture video. **(b)** The ranks for the observed values are shown in the table.

Method	Lecture	Lecture	Lecture	Lecture
Test score	0.33	1.99	2.41	3.24
Rank	1	2	3	4
Method	Lecture	Lecture	Computer	Lecture
Test score	4.56	4.86	5.07	5.22
Rank	5	6	7	8
Method	Computer	Lecture	Lecture	Lecture
Test score	5.45	5.64	5.7	5.98
Rank	9	10	11	12
Method	Lecture	Computer	Computer	Lecture
Test score	6.17	6.25	6.27	6.35
Rank	13	14	15	16
Method	Lecture	Lecture	Lecture	Lecture
Test score	6.44	6.6	6.66	6.75
Rank	17	18	19	20
Method	Computer	Computer	Computer	Computer
Test score	6.82	7.09	7.13	8.32
Rank	21	22	23	24
Method	Computer	Computer	Computer	Computer
Test score	8.35	8.39	8.52	8.64
Rank	25	26	27	28
Method	Computer	Computer	Computer	Computer
Test score	8.91	8.92	9.42	9.78
Rank	29	30	31	32

7. (a) H_0: Median height of all children with allergies = Median height of all children without allergies; H_a: Median height of all children with allergies $\neq$ Median height of all children without allergies. **(b)** The ranks for the observed values are shown in the table.

Group	No allergies	No allergies	No allergies	No allergies
Height (cm)	152	160	160	160
Rank	1	3	3	3
Group	Allergies	Allergies	Allergies	No allergies
Height (cm)	162.3	163	164	164
Rank	5	6	7.5	7.5
Group	Allergies	Allergies	No allergies	Allergies
Height (cm)	165	165	165.1	166
Rank	9.5	9.5	11	12
Group	Allergies	No allergies	Allergies	Allergies
Height (cm)	169	170	170.1	170.6
Rank	13	14	15	16
Group	Allergies	No allergies	No allergies	No allergies
Height (cm)	171	172.7	173	173
Rank	17	18	19.5	19.5
Group	Allergies	No allergies	No allergies	No allergies
Height (cm)	174	175	175.5	176
Rank	21	22	23	24
Group	No allergies	No allergies	No allergies	No allergies
Height (cm)	180	181	182	182
Rank	25	26	27.5	27.5
Group	Allergies	Allergies		
Height (cm)	183	183		
Rank	29.5	29.5		

9. (a) *Random?* The participants were randomly assigned to one of the treatments (computer, lecture). ✓ *Large Sample?* Let the computer group be Group 1; $n_1 = 16 \geq 5$, $n_2 = 16 \geq 5$. ✓ *Equal SD?* $s_L = 1.94$, $s_C = 1.44$, $1.94 < 2(1.44) = 2.88$ ✓ **(b)** $\mu_1 = \frac{16(16+16+1)}{2} = 264$, $\sigma_1 = \sqrt{\frac{16(16)(16+16+1)}{12}} = 26.533$, $R_1 = 363$, $z = \frac{363-264}{26.533} = 3.731$

(c) *Tech:* P-value = 0.0001 **(d)** Because the P-value of $0.0001 < \alpha = 0.05$, we reject H_0. There is convincing evidence that the median test score for people like those in this study who use a computer for training is larger than the median test score for people like those in this study who use a lecture video and readings for training.

11. (a) *Random?* It is reasonable to consider the two samples to be independent random samples of children with allergies and children without allergies. ✓ *Large Sample?* Let the allergy group be Group 1; $n_1 = 13 \geq 5$, $n_2 = 17 \geq 5$. ✓ *Equal SD?* $s_A = 6.86$, $s_{NA} = 9.02$, $9.02 < 2(6.86) = 13.72$ ✓ **(b)** $\mu_1 = \frac{13(13+17+1)}{2} = 201.5$, $\sigma_1 = \sqrt{\frac{13(17)(13+17+1)}{12}} = 23.894$, $R_1 = 190.5$, $z = \frac{190.5-201.5}{23.894} = -0.460$ **(c)** *Tech:* P-value = 2(0.3228) = 0.6456 **(d)** Because the P-value of $0.6456 > \alpha = 0.05$, we fail to reject H_0. There is not convincing evidence that all children with allergies have a different median height than all children without allergies.

13. S: H_0: Median number of steps for all days that the statistician walks the dog = Median number of steps for all days when the statistician does not walk the dog; H_a: Median number of steps for all days that the statistician walks the dog ≠ Median number of steps for all days when the statistician does not walk the dog. Use $\alpha = 0.05$. The evidence for H_a is: The median number of steps for the days that the statistician walks the dog = 7090 ≠ 5569.5 = median number of steps for the days that the statistician does not walk the dog. **P:** Wilcoxon rank sum test. *Random?* Random samples of days when they walked and when they didn't walk were selected. ✓ *Large Sample?* Let the walk group be Group 1; $n_1 = 11 \geq 5$, $n_2 = 12 \geq 5$. ✓ *Equal SD?* $s_W = 1934.93$, $s_{NW} = 3808.6$, $3808.6 < 2(1934.93) = 3869.86$ ✓ **D:** $\mu_1 = \frac{11(11+12+1)}{2} = 132$, $\sigma_1 = \sqrt{\frac{11(12)(11+12+1)}{12}} = 16.248$, $R_1 = 147$, $z = \frac{147-132}{16.248} = 0.923$ *Tech:* P-value = 0.3560. **C:** Because the P-value of $0.3560 > \alpha = 0.05$, we fail to reject H_0. There is not convincing evidence that the median number of steps for all days that the statistician walks the dog is different from the median number of steps for days when the statistician does not walk the dog.

15. S: H_0: Median post-test score for students like those in the study who are taught using the multimedia presentation = Median post-test score for students who are taught using the traditional lecture method; H_a: Median post-test score for students like those in the study who are taught using the multimedia presentation > Median post-test score for students who are taught using the traditional lecture method. Use $\alpha = 0.05$. The evidence for H_a is: The median post-test score for the multimedia group = 10.72 > 7.48 = median post-test score for the traditional lecture group. **P:** Wilcoxon rank sum test. *Random?* The researchers randomly assigned 32 students to one of two teaching methods. ✓ *Large Sample?* Let the multimedia group be Group 1; $n_1 = 13 \geq 5$, $n_2 = 19 \geq 5$. ✓ *Equal SD?* $s_M = 3.47$, $s_L = 2.45$, $3.47 < 2(2.45) = 4.90$ ✓ **D:** $\mu_1 = \frac{13(13+19+1)}{2} = 214.5$, $\sigma_1 = \sqrt{\frac{13(19)(13+19+1)}{12}} = 26.062$, $R_1 = 281$, $z = \frac{281-214.5}{26.062} = 2.552$; *Tech:* P-value = 0.0054. **C:** Because the P-value of $0.0054 < \alpha = 0.05$, we reject H_0. There is convincing evidence that the true median post-test score for students like those in the study who are taught using the multimedia presentation is greater than the true median post-test score for students like those in the study who are taught using the traditional lecture method.

17. (a) S: H_0: Median distance jumped when there is a target for children like the ones in the study = Median distance jumped when there is not a target; H_a: Median distance jumped when there is a target for children like the ones in the study > Median distance jumped when there is not a target. Use $\alpha = 0.05$. The evidence for H_a is: The median distance jumped for the target group = 168.5 cm > 157.0 cm = median distance jumped for the no target group. **P:** Wilcoxon rank sum test. *Random?* The teacher randomly assigned the children to one of two treatment groups (target, no target). ✓ *Large Sample?* Let the target group be Group 1; $n_1 = 14 \geq 5$, $n_2 = 15 \geq 5$. ✓ *Equal SD?* $s_T = 19.92$, $s_{NT} = 20.26$, $20.26 < 2(19.92) = 39.84$ ✓ **D:** $\mu_1 = \frac{14(14+15+1)}{2} = 210$, $\sigma_1 = \sqrt{\frac{14(15)(14+15+1)}{12}} = 22.913$, $R_1 = 234.5$, $z = \frac{234.5-210}{22.913} = 1.069$; *Tech:* P-value = 0.1425. **C:** Because the P-value of $0.1425 > \alpha = 0.05$, we fail to reject H_0. There is not convincing evidence that the median distance jumped will be longer when there is a target than when there is not a target for children like the ones in the study. **(b)** The random assignment was used to ensure that the effects of other variables (such as height, weight, and a child's natural ability to jump) are evenly spread among the two groups.

19. S: H_0: Median amount of calories consumed by children like those in the study who help their parents prepare the meal = Median amount of calories consumed by children who do not help their parents prepare the meal; H_a: Median amount of calories consumed by children like those in the study who help their parents prepare the meal ≠ Median amount of calories consumed by children who do not help their parents prepare the meal. Use $\alpha = 0.05$. The evidence for H_a is: The median amount of calories consumed by the children who helped their parents prepare the meal = 428.74 ≠ 361.02 = median amount of calories consumed by the children who did not help their parents prepare the meal. **P:** Wilcoxon rank sum test. *Random?* The child–parent pairs were randomly assigned to one of the treatment groups (child helps in the meal preparation, child does not help). ✓ *Large Sample?* Let the help group be Group 1; $n_1 = 25 \geq 5$, $n_2 = 22 \geq 5$. ✓ *Equal SD?* $s_H = 105.70$, $s_{NH} = 99.50$, $105.70 < 2(99.50) = 199.00$ ✓ **D:** *Tech:* $z = 2.281$, P-value = 0.023. **C:** Because the P-value of $0.023 < \alpha = 0.05$, we reject H_0. There is convincing evidence that there is a different true median amount of calories consumed by children like those in the study who help their parents prepare the meal compared to children like those in the study who do not help their parents prepare the meal.

21. (a) Let the calcium group be Group 1. Then $n_1 = 10$ and $n_2 = 11$.

$$\mu_1 = \frac{10(10+11+1)}{2} = 110, \sigma_1 = \sqrt{\frac{10(11)(10+11+1)}{12}} = 14.20$$

$R_1 = 122.5$, $z = \frac{122.5 - 110}{14.20} = 0.880$, P-value = $P(z > 0.880) = 0.1894$

The P-value increased from 0.1536 to 0.1894, but the conclusion doesn't change. **(b)** Using the value 12, $t = \frac{5-(-0.273)}{\sqrt{\frac{8.743^2}{10} + \frac{5.901^2}{11}}} = 1.604$;

df = 15.591; P-value = $P(t > 1.604) = 0.064$ **(c)** Using the value 22,

$t = \frac{5 - 0.636}{\sqrt{\frac{8.743^2}{10} + \frac{8.274^2}{11}}} = 1.172$; df = 18.552; P-value = $P(t > 1.172) = 0.128$. Using the larger value of 22, the P-value approximately doubled. **(d)** Because the Wilcoxon rank sum test uses the ranks of the data rather than the values of the data, the Wilcoxon rank sum test is more resistant to outliers. This is evidenced by the smaller change in the P-value for the Wilcoxon rank sum tests than for the two-sample t tests.

23. (a) *Randomized:* The four treatments (placebo, gingko biloba, acetazolamide, both medications) were randomly assigned to the 487 Western trekkers. *Double-blind:* Neither the 487 Western trekkers nor those interacting with them and measuring their responses knew who was receiving which treatment. *Placebo-controlled:* Some of the trekkers did not receive either of gingko biloba or acetazolamide. Rather, they received a pill that looks the same as the other treatment pills, providing a baseline for the occurrence of acute mountain sickness in people like those in the study who are not treated. **(b) S:** H_0: There is no association between medication given and the occurrence of acute mountain sickness in people like those in the study; H_a: There is an association between medication given and the occurrence of acute mountain sickness in people like those in the study. Use $\alpha = 0.05$. The evidence for H_a is: The proportion of hikers with an occurrence of acute mountain sickness ($40/119 = 0.336$, $14/118 = 0.119$, $43/124 = 0.347$, $18/126 = 0.143$) is not the same for each treatment group (placebo, acetazolamide, gingko biloba, both medications). **P:** Chi-square test for association. *Random?* Hikers were randomly assigned to one of four treatments. ✓ *Large Counts?* All expected counts are ≥ 5 (see the table). ✓ **D:**

		Placebo	Acetazo-lamide	Gingko biloba	Both	Total
Acute mountain sickness	Yes	28.101	27.864	29.281	29.754	115
	No	90.899	90.136	94.719	96.246	372
	Total	119	118	124	126	487

$\chi^2 = \frac{(40-28.101)^2}{28.101} + \frac{(14-27.864)^2}{27.864} + \cdots + \frac{(108-96.246)^2}{96.246} = 30.12$,

df = $(2-1)(4-1) = 3$, P-value ≈ 0.000. **C:** Because the P-value of approximately $0 < \alpha = 0.05$, we reject H_0. There is convincing evidence of an association between medication given and the occurrence of acute mountain sickness in people like those in the study.

Answers to Section 13.4 Exercises

1. all population (treatment) medians are equal; at least one median is different

3. True

5. (a) H_0: The true median amount of shrinkage (%) is the same for the three treatments; H_a: At least one true median is different. **(b)** Treatment 1: $R_1 = 207$, Treatment 2: $R_2 = 115$, Treatment 3: $R_3 = 56$

Treatment	Tr 3	Tr 3	Tr 3	Tr 2	Tr 3	Tr 3	Tr 2	Tr 3	Tr 3
Shrinkage (%)	32.3	32.3	33	33.6	34.8	34.8	35	35.2	35.6
Rank	1.5	1.5	3	4	5.5	5.5	7	8	9.5
Treatment	Tr 3	Tr 2	Tr 3	Tr 2	Tr 2	Tr 2	Tr 2	Tr 2	Tr 2
Shrinkage (%)	35.6	36.2	36.6	37.9	39.2	39.4	39.5	40.1	40.4
Rank	9.5	11	12	13	14	15	16	17	18
Treatment	Tr 1	Tr 1	Tr 1	Tr 1	Tr 1	Tr 1	Tr 1	Tr 1	Tr 1
Shrinkage (%)	42.8	43	43.5	43.8	45.2	45.6	45.7	46.2	47
Rank	19	20	21	22	23	24	25	26	27

7. (a) H_0: The true median amount of hexachlorobenzene (HCB) is the same at all three depths; H_a: At least one true median is different. **(b)** Surface: $R_1 = 106.5$, Mid-depth: $R_2 = 160$, Bottom: $R_3 = 198.5$

Depth	Mid-depth	Surface	Mid-depth	Bottom
HCB	3.55	3.74	3.77	3.89
Rank	1	2	3	4
Depth	Surface	Surface	Mid-depth	Surface
HCB	4	4.52	4.59	4.61
Rank	5	6	7	8
Depth	Surface	Mid-depth	Surface	Bottom
HCB	4.67	4.81	4.87	5.03
Rank	9	10	11	12
Depth	Surface	Surface	Bottom	Bottom
HCB	5.12	5.29	5.37	5.44
Rank	13	14	15	16.5
Depth	Bottom	Surface	Mid-depth	Surface
HCB	5.44	5.48	5.64	5.74
Rank	16.5	18	19	20.5
Depth	Mid-depth	Mid-depth	Bottom	Mid-depth
HCB	5.74	5.85	5.85	6.03
Rank	20.5	22.5	22.5	24
Depth	Bottom	Mid-depth	Mid-depth	Bottom
HCB	6.48	6.55	6.77	6.85
Rank	25	26	27	28
Depth	Bottom	Bottom		
HCB	6.88	7.16		
Rank	29	30		

9. (a) *Random?* The 27 pieces of wool were randomly assigned to one of the three treatments. ✓ *Large Sample?* $n_{Tr1} = 9 \geq 5$, $n_{Tr2} = 9 \geq 5$, $n_{Tr3} = 9 \geq 5$ ✓ *Equal SD?* The standard deviations are $s_{Tr1} = 1.513$, $s_{Tr2} = 2.434$, $s_{Tr3} = 1.558$; $2.434 < 2(1.513) = 3.026$. ✓

(b) $H = \frac{12}{27(28)}\left(\frac{(207)^2}{9} + \frac{(115)^2}{9} + \frac{(56)^2}{9}\right) - 3(28) = 20.427$

(c) df = 2; *Tech: P*-value ≈ 0 (d) Because the *P*-value of ≈ 0 < α = 0.05, we reject H_0. There is convincing evidence that the true median amount of shrinkage (%) is different for at least one treatment.

11. (a) *Random?* Water was randomly sampled at each of the three different depths. ✓ *Large Sample?* $n_{Surface} = 10 \geq 5$, $n_{Mid\text{-}depth} = 10 \geq 5$, $n_{Bottom} = 10 \geq 5$ ✓ *Equal SD?* The standard deviations are $s_{Surface} = 0.631$, $s_{Mid\text{-}depth} = 1.106$, $s_{Bottom} = 1.014$; $1.106 < 2(0.631) = 1.262$. ✓

(b) $H = \dfrac{12}{30(31)} \left(\dfrac{(106.5)^2}{10} + \dfrac{(160)^2}{10} + \dfrac{(198.5)^2}{10} \right) - 3(31) = 5.509$

(c) df = 2; *Tech: P*-value = 0.064 **(d)** Because the *P*-value of 0.064 > α = 0.01, we fail to reject H_0. There is not convincing evidence that the true median amount of hexachlorobenzene is different for at least one depth.

13. S: H_0: The population median salary for marketing managers is the same for the three regions (Northeast, Southeast, West); H_a: At least one of these regions has a different population median salary for marketing managers. Use α = 0.05. The evidence for H_a is: The median salaries for marketing managers in the three regions are not all the same (Median$_{NE}$ = \$91,760, Median$_{SE}$ = \$69,355, Median$_W$ = \$80,520). **P:** Kruskal-Wallis test. *Random?* Random samples of marketing managers were selected from the three regions. ✓ *Large Sample?* $n_{NE} = 9 \geq 5$, $n_{SE} = 8 \geq 5$, $n_W = 8 \geq 5$ ✓ *Equal SD?* The standard deviations are $s_{NE} = 21{,}380.92$, $s_{SE} = 13{,}070.51$, $s_W = 18{,}085.81$; $21{,}380.92 < 2(13{,}070.51) = 26{,}141.02$. ✓ **D:** NE: $R_1 = 151$, SE: $R_2 = 64$, W: $R_3 = 110$,

$H = \dfrac{12}{25(26)} \left(\dfrac{(151)^2}{9} + \dfrac{(64)^2}{8} + \dfrac{(110)^2}{8} \right) - 3(26) = 6.147$. H has a chi-square distribution with df = 3−1 = 2. *Tech*: *P*-value = 0.046. **C:** Because the *P*-value of 0.046 < α = 0.05, we reject H_0. There is convincing evidence that at least one of these regions has a different population median salary for marketing managers.

15. S: H_0: The population median salinity level in the Bimini Lagoon is the same in these three locations; H_a: The population median salinity level in the Bimini Lagoon is different in at least one of the locations. Use α = 0.05. The evidence for H_a is: The median salinity levels in the three locations are not all the same (Median$_{Site1}$ = 37.19, Median$_{Site2}$ = 39.98, Median$_{Site3}$ = 38.80). **P:** Kruskal-Wallis test. *Random?* Researchers took random samples of water at the three different locations. ✓ *Large Sample?* $n_{Site1} = 12 \geq 5$, $n_{Site2} = 8 \geq 5$, $n_{Site3} = 10 \geq 5$ ✓ *Equal SD?* The standard deviations are $s_{Site1} = 0.573$, $s_{Site2} = 0.531$, $s_{Site3} = 0.646$; $0.646 < 2(0.531) = 1.062$. ✓ **D:** Site 1: $R_1 = 83$, Site 2: $R_2 = 206$, Site 3: $R_3 = 176$,

$H = \dfrac{12}{30(31)} \left(\dfrac{(83)^2}{12} + \dfrac{(206)^2}{8} + \dfrac{(176)^2}{10} \right) - 3(31) = 22.822$. H has a chi-square distribution with df = 3−1 = 2. *Tech: P*-value ≈ 0. **C:** Because the *P*-value of ≈ 0 < α = 0.05, we reject H_0. There is convincing evidence that the median salinity level in the Bimini Lagoon is different in at least one of the locations.

17. (a) H_0: The true median amount of bacteria that grow is the same for all three temperatures (27°, 35°, 43°); H_a: The true median amount of bacteria that grow is different for at least one temperature. *Random?* Dishes of bacteria were randomly assigned to one of three temperatures. ✓ *Large Sample?* $n_{27°} = 10 \geq 5$, $n_{35°} = 10 \geq 5$, $n_{43°} = 10 \geq 5$ ✓ *Equal SD?* The standard deviations are $s_{27°} = 61.964$, $s_{35°} = 45.351$, $s_{43°} = 52.625$; $61.964 < 2(45.351) = 90.702$. ✓ **(b)** 27°: $R_1 = 102.5$, 35°: $R_2 = 197.5$, 43°: $R_3 = 165$,

$H = \dfrac{12}{30(31)} \left(\dfrac{(102.5)^2}{10} + \dfrac{(197.5)^2}{10} + \dfrac{(165)^2}{10} \right) - 3(31) = 6.016$

(c) H has a chi-square distribution with df = 3−1 = 2. *Tech: P*-value = 0.0494. **(d)** Use α = 0.05. Because the *P*-value of 0.0494 < α = 0.05, we reject H_0. There is convincing evidence that the median amount of bacteria that grow is different for at least one temperature.

19. S: H_0: The population median reading time is the same for all three types of electronic paper displays; H_a: The population median reading time is different for at least one of the three types of electronic paper displays. Use α = 0.05. The evidence for H_a is: The median reading times for the three types of paper displays are all different (Median$_1$ = 1233.115, Median$_2$ = 1085.92, Median$_3$ = 969.515). **P:** Kruskal-Wallis test. *Random?* Researchers randomly assigned 20 people to each of three different electronic paper displays. ✓ *Large Sample?* $n_1 = 20 \geq 5$, $n_2 = 20 \geq 5$, $n_3 = 20 \geq 5$ ✓ *Equal SD?* The standard deviations are $s_1 = 305.624$, $s_2 = 307.438$, $s_3 = 288.576$; $307.438 < 2(288.576) = 577.152$. ✓ **D:** *Tech: P*-value = 0.033, df = 2. **C:** Because the *P*-value of 0.033 < α = 0.05, we reject H_0. There is convincing evidence that the median reading time is different for at least one of the electronic paper displays.

21. (a) Blond: $R_1 = 137.5$, Brown: $R_2 = 52.5$.

$H = \dfrac{12}{19(20)} \left(\dfrac{(137.5)^2}{10} + \dfrac{(52.5)^2}{9} \right) - 3(20) = 9.375$.

H has a chi-square distribution with df = 2−1 = 1. *Tech: P*-value = 0.0022. **(b)** $z^2 = (-3.062)^2 = 9.375 = H$. The two *P*-values are equal.
23. $P(\text{all students will be on time}) = 0.9^{15} = 0.206$

Answers to Section 13.5 Exercises

1. The test statistic is the difference in two sample means $\bar{x}_1 - \bar{x}_2$.
3. False. Every time a randomization test is performed, a different approximate randomization distribution is created, leading to possibly slightly different *P*-values.
5. Take 1332 note cards. Write "VNV" (for "very negative view") on 73 of them and write "No VNV" (for "not very negative view") on the rest. Shuffle the cards and then deal 326 of them into one pile representing those U.S. adults who were surveyed in 2001, and deal the other 1006 into another pile representing those U.S. adults who were surveyed in 2021. Now count how many people with a very negative view are in each pile, compute the new sample proportion for each group, and compute the difference in sample proportions $\hat{p}_{2021} - \hat{p}_{2001}$.
7. Take 32 note cards. Write each of the 32 post-test scores on a card. Shuffle the cards and then deal 19 of them into one pile representing those who were taught using the traditional lecture method, and deal the other 13 into another pile representing those who were taught using the multimedia presentation. Now calculate the mean post-test score in each pile and compute the difference in sample means $\bar{x}_{Lecture} - \bar{x}_{Multimedia}$.
9. (a) $\hat{p}_{2021} - \hat{p}_{2001} = \dfrac{70}{1006} - \dfrac{3}{326} = 0.0604$ **(b)** There are 0 dots in the randomization distribution as large as or larger than 0.0604 or as small as or smaller than −0.0604. So, the *P*-value is approximately 0/100 = 0. **(c)** Because the *P*-value of approximately 0 < α = 0.05, we reject H_0. There is convincing evidence that the proportion of U.S. adults who had a very negative view of the computer industry in 2021 is different from the proportion of U.S. adults who had a very negative view of the computer industry in 2001.
11. (a) $\bar{x}_{Lecture} - \bar{x}_{Multimedia} = 8.3505 - 11.1 = -2.7495$ **(b)** There are 0 dots in the randomization distribution as low as or lower than −2.7495, so the *P*-value is approximately 0/100 = 0. **(c)** Assuming that the true mean post-test score for students like those in the study who were taught using the multimedia presentation is the same as the true mean post-test score for students like those in the study who were taught using the traditional lecture, there is approximately a 0 probability of getting a sample mean difference of −2.7495 or lower by chance alone.
13. S: H_0: $\mu_{Deprived} - \mu_{Unrestricted} = 0$, H_a: $\mu_{Deprived} - \mu_{Unrestricted} < 0$, where $\mu_{Deprived}$ = true mean improvement in visual discrimination

test time for subjects like the ones in the study who are deprived of sleep for an entire night and $\mu_{\text{Unrestricted}}$ = true mean improvement in visual discrimination test time for subjects like the ones in the study who are allowed unrestricted sleep. Use $\alpha = 0.05$. The evidence for H_a is: $\bar{x}_{\text{Deprived}} - \bar{x}_{\text{Unrestricted}} = 3.90 - 19.82 = -15.92 < 0$. **P:** Randomization test for the difference between two means. *Random?* Subjects were randomly assigned into the two groups. ✓ **D:** $\bar{x}_{\text{Deprived}} - \bar{x}_{\text{Unrestricted}} = 3.90 - 19.82 = -15.92$. There are 0 dots in the randomization distribution as low as or lower than -15.92, so the P-value is approximately $0/100 = 0$. **C:** Because the P-value of approximately $0 < \alpha = 0.05$, we reject H_0. There is convincing evidence that sleep deprivation decreases the true mean improvement time on the visual discrimination task compared to unrestricted sleep for subjects like the ones in the study.

15. S: H_0: $p_{\text{vibrations}} - p_{\text{control}} = 0$, H_a: $p_{\text{vibrations}} - p_{\text{control}} \neq 0$, where $p_{\text{vibrations}}$ = true proportion of all European seabass eggs placed in a vibration chamber that would hatch and p_{control} = true proportion of all European seabass eggs placed in similar environmental conditions with no vibrations that would hatch. Use $\alpha = 0.05$. The evidence for H_a is: $\hat{p}_{\text{vibrations}} - \hat{p}_{\text{control}} = 0.76 - 0.82 = -0.06 \neq 0$. **P:** Randomization test for the difference between two proportions. *Random?* Eggs were randomly assigned to the two treatment groups. ✓ **D:** $\hat{p}_{\text{vibrations}} - \hat{p}_{\text{control}} = -0.06$. There are 10 dots in the randomization distribution as low as or lower than -0.06 and 10 dots as large as or larger than 0.06, so the P-value is approximately $20/100 = 0.20$. **C:** Because the P-value of $\approx 0.20 > \alpha = 0.05$, we fail to reject H_0. There is not convincing evidence that vibration from a rocket launch will affect the proportion of European seabass eggs that hatch.

17. (a) S: H_0: $p_{\text{Ethiopia}} - p_{\text{Nigeria}} = 0$, H_a: $p_{\text{Ethiopia}} - p_{\text{Nigeria}} \neq 0$, where p_{Ethiopia} = true proportion of people from Ethiopia who would report that they trust scientists a little or a lot and p_{Nigeria} = true proportion of people from Nigeria who would report that they trust scientists a little or a lot. Use $\alpha = 0.05$. The evidence for H_a is: $\hat{p}_{\text{Ethiopia}} - \hat{p}_{\text{Nigeria}} = \frac{753}{1003} - \frac{439}{1002} = 0.313 \neq 0$. **P:** Randomization test for the difference between two proportions. *Random?* Independent random samples of people were selected from Nigeria and Ethiopia. ✓ **D:** $\hat{p}_{\text{Ethiopia}} - \hat{p}_{\text{Nigeria}} = \frac{753}{1003} - \frac{439}{1002} = 0.313$. There are 0 dots in the randomization distribution as low as or lower than -0.313 and 0 dots as large as or larger than 0.313, so the P-value is approximately $0/100 = 0$. **C:** Because the P-value of approximately $0 < \alpha = 0.05$, we reject H_0. There is convincing evidence that the proportion of people who trust scientists is different between Ethiopia and Nigeria. **(b)** *Random?* Independent random samples of people were selected from Nigeria and Ethiopia. ✓ *Large Counts?* 753; $1003 - 753 = 250$; 439; and $1002 - 439 = 563$ are all at least 10. ✓ **(c)** *Tech:* P-value ≈ 0. Because the P-value of approximately $0 < \alpha = 0.05$, we reject H_0. There is convincing evidence that the proportion of people who trust scientists is different between Ethiopia and Nigeria. **(d)** The randomization test and the two-sample z test do give similar P-values and conclusions.

19. (a) H_0: $\mu_{\text{Target}} - \mu_{\text{NoTarget}} = 0$, H_a: $\mu_{\text{Target}} - \mu_{\text{NoTarget}} > 0$, where μ_{Target} = true mean distance jumped for children like the ones in the study when there is a target and μ_{NoTarget} = true mean distance jumped for children like the ones in the study when there is not a target. **(b)** Take 29 note cards. Write each of the 29 distances on a card. Shuffle the cards and then deal 14 of them into one pile representing those children who had a target, and 15 into another pile representing those children who did not have a target. Now calculate the mean distance jumped in each pile and compute the difference in sample means $\bar{x}_{\text{Target}} - \bar{x}_{\text{NoTarget}}$. **(c)** $\bar{x}_{\text{Target}} - \bar{x}_{\text{NoTarget}} = 164.857 - 155.80 = 9.057$ **(d)** Answers will vary. Technology was used to create the following randomization distribution by simulating 1000 values of $\bar{x}_{\text{Target}} - \bar{x}_{\text{NoTarget}}$.

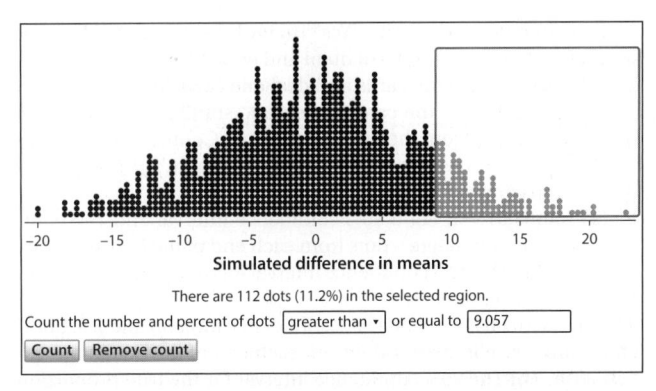

There are 112 dots in the randomization distribution as large as or larger than 9.057, so the P-value is approximately $112/1000 = 0.112$. **(e)** Because the P-value of $0.112 > \alpha = 0.05$, we fail to reject H_0. There is not convincing evidence that the mean distance jumped will be longer when there is a target than where there is not a target for children like the ones in the study.

21. (a) Take 40 note cards. Write the 20 temperature values on 20 of the note cards, and write the 20 pulse rate values on the remaining 20 note cards. Shuffle the two piles (temperature cards and pulse rate cards) separately. Select one card from the temperature pile and one card from the pulse rate pile. Record these values as (x, y) data. Repeat this process for a total of 20 (x, y) pairs. Calculate the sample slope b_1 of the least-squares regression line for these data. **(b)** There are 8 dots in the randomization distribution as low as or lower than -4.944 and 4 dots as high as or higher than 4.944, so the P-value is approximately $12/100 = 0.12$. **(c)** Because the P-value of $0.12 > \alpha = 0.05$, we fail to reject H_0. There is not convincing evidence of a relationship between pulse rate and temperature in this population.

23. The boxplots for the three groups are shown here.

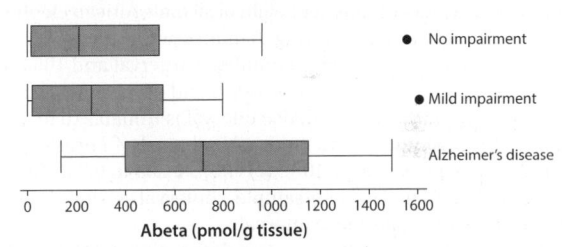

The distribution of Abeta measurements is skewed right for the group with no impairment and with mild impairment. Both groups have one high outlier. For these groups, the median, IQR, and range of the Abeta measurements are approximately equal. The distribution of Abeta measurements for the group with mild to moderate Alzheimer's disease is roughly symmetric with no outliers. The median Abeta measurement for this group is greater than the median measurement for both the group with no impairment and the group with mild impairment. The IQR is also greater for the group with mild to moderate Alzheimer's disease when compared to the IQR of the other two groups.

Answers to Section 13.6 Exercises

1. The data come from a random sample from the population of interest.

3. 5%; 5%; minimum; maximum

5. Take 28 note cards. Write each of the 28 copper percentages on a different card. Shuffle the cards and deal one card. Record the copper percentage on the card. Replace the card in the deck, shuffle again, and deal one card. Again, record the copper percentages on the card. Continue this process until 28 copper amounts have been written down. Compute $\bar{x}$ = the mean copper percentage for the 28 cards drawn.

7. Take 75 note cards. Write "Yes" (to indicate a squirrel that was indifferent to humans) on 3 of them and write "No" on the remaining 72 cards. Shuffle the cards and deal one card. Record the word on the card. Replace the card in the deck, shuffle again, and deal one card. Again, record the word on the card. Continue this process until 75 words have been written down. Compute $\hat{p}$ = the proportion of cards drawn that say "Yes."

9. (a) $100(0.05) = 5$. For a 90% confidence interval and 100 bootstrap samples, eliminate 5 dots from each end of the bootstrap distribution. (b) The 90% confidence interval for the true mean copper percentage is (0.31, 0.55).

11. (a) $200(0.005) = 1$. For a 99% confidence interval and 200 bootstrap samples, eliminate 1 dot from each end of the bootstrap distribution. (b) The 99% confidence interval for the true proportion of all squirrels in New York City's Central Park that are indifferent to humans is (0.000, 0.107).

13. S: 95% confidence interval for p = the true proportion of samples that these dogs would correctly identify. P: Bootstrap confidence interval for p. Random? Dogs were presented with 1012 randomized samples. ✓ D: $200(0.025) = 5$. For a 95% confidence interval and 200 bootstrap samples, eliminate 5 dots from each end of the bootstrap distribution. The bootstrap confidence interval is from 0.922 to 0.954. C: We are 95% confident that the interval from 0.922 to 0.954 captures the true proportion of samples that these dogs would correctly identify.

15. S: 90% confidence interval for μ = the true mean shoulder height of all male African elephants that lived through droughts during their first two years of life. P: Bootstrap confidence interval for μ. Random? The study used a random sample of 14 male African elephants that lived through droughts during their first two years of life. ✓ D: $100(0.05) = 5$. For a 90% confidence interval and 100 bootstrap samples, eliminate 5 dots from each end of the bootstrap distribution. The bootstrap confidence interval is from 234 cm to 256 cm. C: We are 90% confident that the interval from 234 cm to 256 cm captures the true mean shoulder height of all male African elephants that lived through droughts during their first two years of life.

17. (a) $100(0.05) = 5$. For a 90% confidence interval and 100 bootstrap samples, eliminate 5 dots from each end of the bootstrap distribution. The bootstrap confidence interval is from 0.210 to 0.254. (b) Random? Customers were randomly selected. ✓ Large Counts? $np = 170 \geq 10$, $n(1-p) = 568 \geq 10$ ✓ (c) Tech: 0.205 to 0.256 (d) The intervals are similar. The one-sample z-interval is slightly wider than the bootstrap confidence interval.

19. Answers will vary. For this bootstrap distribution, a 90% bootstrap confidence interval for μ is from 18.5 to 26.5. We are 90% confident that the interval from 18.5 to 26.5 captures the true mean amount of vitamin C in packages of CSB.

21. (a) Answers will vary. For this bootstrap distribution using 1000 bootstrap samples, a 90% bootstrap confidence interval for μ is from 232.7 cm to 258.9 cm. (b) From Exercise 15, the 90% bootstrap confidence interval using 100 bootstrap samples is from 234 cm to 256 cm. The two intervals are not the same. (c) The bootstrap interval using 1000 bootstrap samples is more trustworthy because we have a better estimate of the bootstrap distribution on which to base the confidence interval. With more bootstrap samples, we can better see the shape and the variability of the bootstrap distribution.

23. (a) First, take one pile of 10 note cards. For the people with naturally blond hair, write each of the 10 pain threshold scores on a different card. Shuffle the cards and deal one card. Record the score on the card. Replace the card in the deck, shuffle again, and deal one card. Again, record the score on the card. Continue this process until 10 scores have been written down. Compute $\bar{x}_{Blond}$ = the mean score for the 10 cards drawn. Next, take one pile of 9 note cards. For the people with naturally brown hair, write each of the 9 pain threshold scores on a different card. Shuffle the cards and deal one card. Record the score on the card. Replace the card in the deck, shuffle again, and deal one card. Again, record the score on the card. Continue this process until 9 scores have been written down. Compute $\bar{x}_{Brown}$ = the mean score for the 9 cards drawn. Finally, compute $\bar{x}_{Blond} - \bar{x}_{Brown}$. This simulates one possible value in a bootstrap distribution of $\bar{x}_{Blond} - \bar{x}_{Brown}$. (b) $200(0.025) = 5$. For a 95% confidence interval and 200 bootstrap samples, eliminate 5 dots from each end of the bootstrap distribution. The bootstrap confidence interval is from 8.0 to 22.5.

25. (a) (i) $z = \dfrac{(640 - 500)}{100} = 1.400$; Tech: $P(Z > 1.400)$ = Applet/normalcdf(lower: 1.4, upper: 1000, mean: 0, SD: 1) = 0.0808; Table A: $P(Z > 1.40) = 1 - 0.9192 = 0.0808$ (ii) $P(X > 640)$ = Applet/normalcdf (lower: 640, upper: 1000, mean: 500, SD: 100) = 0.0808
(b) (i) $z = \dfrac{(450 - 500)}{100} = -0.50$ and $z = \dfrac{(700 - 500)}{100} = 2.00$; Tech: $P(-0.50 < Z < 2.00)$ = Applet/normalcdf(lower: −0.50, upper: 2.00, mean: 0, SD: 1) = 0.6687; Table A: $P(-0.50 < Z < 2.00) = 0.9772 - 0.3085 = 0.6687$ (ii) $P(450 < X < 700)$ = Applet/normalcdf(lower: 450, upper: 700, mean: 500, SD: 100) = 0.6684

Answers to Chapter 13 Review Exercises

1. (a) Because the sample size is not large and the dotplot of the differences shows strong skewness, a paired t test for a mean difference is not appropriate. Because the dotplot of the differences shows strong skewness and is not reasonably symmetric, a Wilcoxon signed rank test is also not appropriate.

(b) S: H_0: Population median difference (Machete − Wooden tool) = 0; H_a: Population median difference (Machete − Wooden tool) > 0, where Population median difference refers to the true median difference (kg/hr) of manioc that all women like those in the study are able to dig. Use $\alpha = 0.05$. The evidence for H_a is: All the differences (5 of 5) are positive. P: Sign test. Random? The order in which the subjects used the two tools was randomized. ✓ D: Calculating Machete − Wooden tool: 80, 102, 90, 78, and 77. Because H_a tests for a Population median difference > 0, the test statistic is $X = 5$, the number of positive differences. P-value: $n = 5$, $p = 0.5$; $P(X \geq 5) = 0.03125$. C: Because the P-value of $0.03125 < \alpha = 0.05$, we reject H_0. There is convincing evidence that the women work better with the machete than with the traditional tool.

2. S: H_0: The median weight of corn grown using the new fertilizer = 8 ounces; H_a: The median weight of corn grown using the new fertilizer > 8 ounces. Use $\alpha = 0.05$. The evidence for H_a is: More than half (8 of 10) ears of corn weigh more than 8 ounces. P: Wilcoxon signed rank test. Random? Ears of corn were randomly selected. ✓ Symmetric? The dotplot of the differences is not strongly skewed. ✓

D:

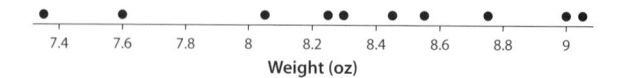

Weight	8.30	7.35	9.05	8.05	8.25	8.75	9.00	7.60	8.45	8.55
Difference (Weight − 8 oz)	0.30	−0.65	1.05	0.05	0.25	0.75	1.00	−0.40	0.45	0.55
|Difference|	0.30	0.65	1.05	0.05	0.25	0.75	1.00	0.40	0.45	0.55
Rank	3	7	10	1	2	8	9	4	5	6
Signed rank	3	−7	10	1	2	8	9	−4	5	6

$W^+ = 44$; $W^- = 11$; $W = 11$. Table E: $n = 10$, P-value > 0.05. Tech: P-value > 0.05. C: Because the P-value > 0.05, we fail to reject H_0. There is not convincing evidence that the fertilizer produces a median weight greater than 8 ounces.

3. S: H_0: Median coefficient of friction for all safety shoes that are not oil resistant = Median coefficient of friction for all safety shoes that are oil resistant; H_a: Median coefficient of friction for all safety shoes that are not oil resistant > Median coefficient of friction for all safety shoes that are oil resistant. Use $\alpha = 0.05$. The evidence for H_a is: The sample median coefficient of friction for safety shoes that are not oil resistant = 0.1525 > 0.131 = sample median coefficient of friction for safety shoes that are oil resistant. **P:** Wilcoxon rank sum test. *Random?* Independent random samples of safety shoes with each kind of outer soles were selected. ✓ *Large Sample?* Let the non-oil-resistant group be Group 1; $n_1 = 14 \geq 5$, $n_2 = 9 \geq 5$. ✓ *Equal SD?* The standard deviations are $s_{NotOR} = 0.49$, $s_{OR} = 0.28$; $0.49 < 2(0.28) = 0.56$. ✓

D:

Treatment	Not O.R.	Not O.R.	O.R	Not O.R.	Not O.R.	O.R	O.R	O.R
Coef of friction	0.068	0.073	0.101	0.103	0.105	0.112	0.119	0.12
Rank	1	2	3	4	5	6	7	8
Treatment	Not O.R.	O.R	O.R	Not O.R.	Not O.R.	Not O.R.	O.R	Not O.R.
Coef of friction	0.131	0.131	0.133	0.148	0.152	0.153	0.156	0.169
Rank	9.5	9.5	11	12	13	14	15	16
Treatment	Not O.R.	Not O.R.	O.R	O.R	Not O.R.	Not O.R.	Not O.R.	
Coef of friction	0.173	0.176	0.177	0.178	0.187	0.209	0.239	
Rank	17	18	19	20	21	22	23	

$$\mu_1 = \frac{14(14+9+1)}{2} = 168, \; \sigma_1 = \sqrt{\frac{14(9)(14+9+1)}{12}} = 15.875, \; R_1 = 177.5,$$

$$z = \frac{177.5 - 168}{15.875} = 0.598$$

Tech: P-value = 0.2749. **C:** Because the P-value of $0.2749 > \alpha = 0.05$, we fail to reject H_0. There is not convincing evidence that safety shoes with oil resistant outer soles are more slippery than safety shoes without oil resistant outer soles.

4. S: H_0: Median$_{Both}$ = Median$_{Dirt}$ = Median$_{Soaked}$; H_a: At least one of these treatments produces a different median failure pressure. Use $\alpha = 0.05$. The evidence for H_a is: The median failure pressure for the three treatments are not all the same (Median$_{Both}$ = 500.5 pounds, Median$_{Dirt}$ = 677.0 pounds, Median$_{Soaked}$ = 629.0 pounds). **P:** Kruskal-Wallis test. *Random?* Ropes were randomly assigned to one of the three treatments. ✓ *Large Sample?* $n_1 = 8 \geq 5$, $n_2 = 8 \geq 5$, $n_3 = 8 \geq 5$ ✓ *Equal SD?* The standard deviations are $s_{Both} = 45.35$, $s_{Dirt} = 44.77$, $s_{Soaked} = 23.44$; $45.35 < 2(23.44) = 46.88$. ✓

D:

Treatment	Both	Both	Both	Both	Both	Both	Both	Both
Failure pressure	415	417	486	486	515	520	520	524
Rank	1	2	3.5	3.5	5	6.5	6.5	8
Treatment	Soaked	Soaked	Soaked	Dirt	Soaked	Dirt	Dirt	Soaked
Failure pressure	600	609	615	620	621	624	625	637
Rank	9	10	11	12	13	14	15	16
Treatment	Soaked	Dirt	Soaked	Soaked	Dirt	Dirt	Dirt	Dirt
Failure pressure	640	655	660	664	699	702	721	722
Rank	17	18	19	20	21	22	23	24

Both: $R_1 = 36$, Dirt: $R_2 = 149$, Soaked: $R_3 = 115$,

$$H = \frac{12}{24(25)}\left(\frac{(36)^2}{8} + \frac{(149)^2}{8} + \frac{(115)^2}{8}\right) - 3(25) = 16.805.$$

H has a chi-square distribution with df $= 3 - 1 = 2$. *Tech:* P-value = 0.0002. **C:** Because the P-value of $0.0002 < \alpha = 0.05$, we reject H_0. There is convincing evidence that at least one of these treatments produces a different median failure pressure.

5. (a) Take 520 note cards. Write "Yes" on 149 of them and write "No" on the rest. Shuffle the cards and then deal 250 of them into one pile representing the random sample of Iowa residents, and 270 into another pile representing the random sample of Vermont residents. Now count how many residents who said "Yes" are in each pile, compute the sample proportion for each group, and compute the difference in sample proportions $\hat{p}_{Iowa} - \hat{p}_{Vermont}$. **(b) S:** H_0: $p_{Iowa} - p_{Vermont} = 0$, H_a: $p_{Iowa} - p_{Vermont} \neq 0$, where p_{Iowa} = population proportion of Iowa residents who stay in their vehicle while it warms up and $p_{Vermont}$ = population proportion of Vermont residents who stay in their vehicle while it warms up. Use $\alpha = 0.01$. The evidence for H_a is: $\hat{p}_{Iowa} - \hat{p}_{Vermont} = 0.260 - 0.311 = -0.051 \neq 0$. **P:** Randomization test for the difference between two proportions. *Random?* Random samples of Iowa residents and Vermont residents were selected. ✓ **D:** $\hat{p}_{Iowa} - \hat{p}_{Vermont} = 0.260 - 0.311 = -0.051$. There are at least 15 dots in the randomization distribution as low as or lower than -0.051 and at least 17 dots as large as or larger than 0.051, so the P-value is approximately $32/200 = 0.160$. **C:** Because the P-value of approximately $0.160 > \alpha = 0.01$, we fail to reject H_0. There is not convincing evidence of a difference between the population proportion of Iowa and Vermont residents who stay in their vehicle while it is warming up.

6. S: 95% confidence interval for μ = the true mean fuel efficiency (mpg) for this model of pickup truck. **P:** Bootstrap confidence interval for μ. *Random?* Random sample of 20 owners of this particular model of pickup truck. ✓ **D:** $200(0.025) = 5$. For a 95% confidence interval and 200 bootstrap samples, eliminate 5 dots from each end of the bootstrap distribution. The bootstrap confidence interval is from 17.3 mpg to 19.8 mpg. **C:** We are 95% confident that the interval from 17.3 mpg to 19.8 mpg captures the true mean fuel efficiency for this model of pickup truck.

7. (a) Because the 20 students were randomly assigned to one of the two treatments (caffeine, no caffeine) but were not paired in any way, the Wilcoxon rank sum test and a randomization test are appropriate tests for these data. **(b)** Wilcoxon rank sum test. **S:** H_0: Median$_{Caffeine}$ = Median$_{No\ caffeine}$; H_a: Median$_{Caffeine}$ > Median$_{No\ caffeine}$, where Median$_{Treatment}$ is the true median change in pulse rates for all students like those in the study who received the treatment (caffeine, no caffeine). Use $\alpha = 0.01$. The evidence for H_a is: The median change in pulse rates for the two treatments is not the same: (Median$_{Caffeine}$ = 3.5, Median$_{No\ caffeine}$ = 2.5). **P:** Wilcoxon rank sum test. *Random?* Students were randomly assigned to one of the two treatment groups (caffeine, no caffeine). ✓ *Large Sample?* Let the caffeine group be Group 1: $n_1 = 10 \geq 5$, $n_2 = 10 \geq 5$. ✓ *Equal SD?* $s_{Caffeine} = 2.700$; $s_{No\ caffeine} = 2.625$; $2.700 < 2(2.625) = 5.25$ ✓ **D:**

Treatment	No caffeine	No caffeine	No caffeine	Caffeine
Change in pulse rate	−2	−1	−1	0
Rank	1	2.5	2.5	4.5
Treatment	Caffeine	Caffeine	Caffeine	No caffeine
Change in pulse rate	0	1	1	1
Rank	4.5	7	7	7
Treatment	No caffeine	Caffeine	No caffeine	Caffeine
Change in pulse rate	2	3	3	4
Rank	9	10.5	10.5	13.5
Treatment	Caffeine	No caffeine	No caffeine	Caffeine
Change in pulse rate	4	4	4	5
Rank	13.5	13.5	13.5	17
Treatment	No caffeine	No caffeine	Caffeine	Caffeine
Change in pulse rate	5	5	6	8
Rank	17	17	19	20

$\mu_1 = \dfrac{10(10+10+1)}{2} = 105$, $\sigma_1 = \sqrt{\dfrac{10(10)(10+10+1)}{12}} = 13.229$,

$R_1 = 116.5$, $z = \dfrac{116.5-105}{13.229} = 0.869$, P-value $= 0.1924$. **C:** Because the P-value of $0.1924 > \alpha = 0.01$, we fail to reject H_0. There is not convincing evidence that drinking caffeine increases pulse rates, on average, for people like the ones in the study.

Glossary

addition rule for mutually exclusive events If A and B are mutually exclusive events, $P(A \text{ or } B) = P(A) + P(B)$.

alternative hypothesis (H_a) The claim that we are trying to find evidence *for* in a significance test.

ANOVA table Summary of the numerical information needed to conduct a significance test about differences among two or more population or treatment means. The typical format for the table is

Source	df	SS	MS	F
Groups	$df_G = I-1$	SSG	MSG	F
Error	$df_E = n-I$	SSE	MSE	
Total	$df_T = n-1$	SST		

where $SST = SSG + SSE$, each $MS = SS/df$, and $F = MSG/MSE$.

association A relationship between two variables in which knowing the value of one variable helps predict the value of the other. If knowing the value of one variable does not help predict the value of the other, there is no association between the variables.

bar chart A graph that shows each category as a bar. The heights of the bars show the category frequencies or relative frequencies.

Bayes' Theorem For any two events A and B with nonzero probabilities, the probability of event A given that event B has subsequently occurred is

$$P(A \mid B) = \frac{P(A) \cdot P(B \mid A)}{P(B)}$$

bias The design of a statistical study shows bias if it is very likely to underestimate or very likely to overestimate the value you want to know.

binomial distribution A distribution completely specified by two numbers: the number of trials n of the chance process and the probability p of success on each trial.

binomial random variable A variable measuring the count of successes X in a binomial setting. The possible values of X are $0, 1, 2, \ldots, n$.

binomial setting The situation that arises when we perform n independent trials of the same chance process and count the number of times that a particular outcome (called a "success") occurs. The four conditions for a binomial setting are: (1) **B**inary? The possible outcomes of each trial can be classified as "success" or "failure." (2) **I**ndependent? Trials are independent. That is, knowing the result of one trial must not tell us anything about the result of any other trial. (3) **N**umber? The number of trials n of the chance process is fixed in advance. (4) **S**ame probability? There is the same probability of success p on each trial.

block A group of experimental units that are known before the experiment to be similar in some way that is expected to affect the response to treatments.

bootstrap confidence interval A nonparametric confidence interval for a population parameter estimated with a bootstrap distribution.

bootstrap distribution A collection of sample statistics computed from many bootstrap samples.

bootstrap sample A random sample of the same size as the original sample that is selected with replacement from the original sample.

boxplot A visual representation of the five-number summary.

categorical variable A variable that takes values that are labels, which place each individual into a particular group, called a category.

census A study that collects data from every individual in the population.

central limit theorem (CLT) In an SRS of size n from any population with mean μ and standard deviation σ, when n is sufficiently large, the sampling distribution of the sample mean $\bar{x}$ is approximately normal.

chi-square distribution A distribution defined by a density curve that takes only non-negative values and is skewed to the right. A particular chi-square distribution is specified by its degrees of freedom.

chi-square test for association A significance test of the null hypothesis that there is no association between two categorical variables in the population of interest.

chi-square test for goodness of fit A significance test of the null hypothesis that a categorical variable has a specified distribution in the population of interest.

chi-square test statistic A measure of how different the observed counts are from the expected counts, relative to the expected counts. The formula for the statistic is

$$\chi^2 = \sum \frac{(\text{observed count} - \text{expected count})^2}{\text{expected count}}$$

where the sum is over all categories of the categorical variable.

cluster A group of individuals in the population that are located near each other.

cluster random sampling Selection of a sample by randomly choosing clusters and including each member of the selected clusters in the sample.

coefficient of determination r^2 A measure of the percent reduction in the sum of squared residuals when using the least-squares regression line to make predictions, rather than the mean value of y. In other words, r^2 measures the percentage of the variability in the response variable that is accounted for by the least-squares regression line.

combination A set of individuals chosen from some group in which the order of selection doesn't matter. The notation $_nC_r$ represents the number of different combinations of r individuals chosen from the entire group of n individuals.

complement The complement of event A, written as A^C, is the event that A does not occur.

complement rule The probability that an event does not occur is 1 minus the probability that the event does occur. In symbols, $P(A^C) = 1 - P(A)$, where event A^C is the complement of event A.

completely randomized design A design in which the experimental units are assigned to the treatments (or the treatments to the experimental units) completely at random.

conditional probability The probability that one event happens given that another event is known to have happened. The conditional probability that event B happens given that event A has happened is denoted by $P(B|A)$.

confidence interval A set of plausible values for a parameter based on sample data.

confidence level C Gives the long-run capture rate of confidence intervals calculated with $C\%$ confidence.

confounding When two variables are associated in such a way that their effects on a response variable cannot be distinguished from each other.

continuous random variable Variable that can take any value in a particular interval on the number line. The set of possible values for a continuous random variable is infinite.

control group A group used in an experiment to provide a baseline for comparing the effects of other treatments. Depending on the purpose of the experiment, a control group may be given an inactive treatment, an active treatment, or no treatment at all.

convenience sample A sample consisting of individuals from the population who are easy to reach.

correlation r A measure of the strength and direction of a linear relationship between two quantitative variables.

critical value A multiplier that makes the interval wide enough to have the stated capture rate.

density curve Models the probability distribution of a continuous random variable with a curve that is always on or above the horizontal axis and has an area of exactly 1 underneath it.

discrete random variable Variable that takes a set of individual values with gaps between them on the number line. The set of possible values for a discrete random variable can be finite or infinite.

distribution Tells what values a variable takes and how often it takes each value.

dotplot A graph showing each data value as a dot above its location on a number line.

double-blind An experiment in which neither the subjects nor those who interact with them and measure the response variable know which treatment a subject is receiving.

empirical rule In any normal distribution with mean μ and standard deviation σ, about 68% of the values fall within 1σ of the mean μ, about 95% of the values fall within 2σ of the mean μ, and about 99.7% of the values fall within 3σ of the mean μ.

event A set of outcomes from some chance process.

experiment A study in which researchers deliberately impose treatments (conditions) on individuals to measure their responses.

experimental unit The object to which a treatment is randomly assigned.

explanatory variable Variable that may help explain or predict changes in a response variable.

extrapolation Use of a regression line for prediction outside the interval of x values used to obtain the line. The further we extrapolate, the less reliable the predictions become.

F distribution A distribution defined by a density curve that takes only non-negative values and is skewed to the right. A particular F distribution is specified by its numerator degrees of freedom and its denominator degrees of freedom.

F test for a population multiple regression model A significance test of the null hypothesis that a population multiple regression model does not provide additional useful information about a response variable compared to a model that uses only the mean of the response variable.

factorial For any positive integer n, we define $n!$ (read "n factorial") as $n!=n(n-1)(n-2)\ldots 3\cdot 2\cdot 1$. That is, n factorial is the product of the integers starting with n and going down to 1.

first quartile Q_1 The median of the data values that are to the left of the median in the ordered list.

five-number summary The minimum, the first quartile Q_1, the median, the third quartile Q_3, and the maximum of a distribution of quantitative data.

frequency table A table showing the number of individuals having each data value.

general addition rule If A and B are any two events resulting from some chance process, $P(A \text{ or } B) = P(A) + P(B) - P(A \text{ and } B)$.

general multiplication rule For any chance process, the probability that events A and B both occur is $P(A \text{ and } B) = P(A) \cdot P(B|A)$.

histogram A graph that shows each interval as a bar. The heights of the bars show the frequencies or relative frequencies of values in each interval.

independent events Two events are independent if knowing whether or not one event has occurred does not change the probability that the other event will happen. In other words, events A and B are independent if $P(A|B) = P(A|B^C) = P(A)$. Alternatively, events A and B are independent if $P(B|A) = P(B|A^C) = P(B)$.

indicator variable A categorical variable with two possible outcomes. These outcomes are coded numerically, so they can be included in regression calculations. A success is reported as a "1" and a failure is recorded as a "0."

individual A person, animal, or thing described in a set of data.

interaction term In a multiple regression model, a term created by multiplying two explanatory variables and including the product as an additional explanatory variable in the model. A multiple regression model with two explanatory variables and an interaction term has the following form: $\hat{y} = b_0 + b_1 x_1 + b_2 x_2 + b_3 x_1 x_2$.

interquartile range (IQR) The distance between the first and third quartiles of a distribution. In symbols, $IQR = Q_3 - Q_1$.

intersection The event "A and B," consisting of all outcomes that are common to both events, denoted as $A \cap B$.

Kruskal-Wallis test A nonparametric significance test of the null hypothesis that two or more population medians are the same.

Large Counts condition Let X be the number of successes and $\hat{p}$ be the proportion of successes in a random sample of size n from a population with proportion of successes p. The Large Counts condition says that the shape of the probability distribution of X and the shape of the sampling distribution of $\hat{p}$ will be approximately normal when $np \geq 10$ and $n(1-p) \geq 10$. For a chi-square test, the Large Counts condition says that all expected counts are at least 5.

law of large numbers If we observe more and more repetitions of any chance process, the proportion of times that a specific outcome occurs approaches its probability.

least-squares regression line The line that makes the sum of the squared residuals as small as possible.

margin of error Describes how far, at most, we expect the point estimate to vary from the population parameter.

matched-pairs design An experimental design for comparing two treatments that uses blocks of size 2. In some matched-pairs designs, two very similar experimental units are paired and the two treatments are randomly assigned within each pair. In others, each experimental unit receives both treatments in a random order.

mean The average of all the individual data values in a distribution of quantitative data. To find the mean, add all the values and divide by the total number of data values.

mean (expected value) of a discrete random variable The average value of the variable over many, many repetitions of the same chance process.

median The midpoint of a distribution; the number such that about half the observations are smaller and about half are larger. To find the median, arrange the data values from smallest to largest. If the number n of data values is odd, the median is the middle value in the ordered list. If the number n of data values is even, use the average of the two middle values in the ordered list as the median.

multiple regression model A least-squares regression equation that relates a quantitative response variable to two or more explanatory variables. Multiple regression models are expressed in the form $\hat{y} = b_0 + b_1 x_1 + b_2 x_2 + \ldots$.

multiplication counting principle For a process involving multiple (r) steps, suppose that there are n_1 ways to do Step 1, n_2 ways to do Step 2, $\ldots$, and n_r ways to do Step r. The total number of different ways to complete the process is $n_1 \cdot n_2 \cdot \ldots \cdot n_r$.

multiplication rule for independent events If A and B are independent events, the probability that A and B both occur is $P(A \text{ and } B) = P(A) \cdot P(B)$.

mutually exclusive Two events A and B that have no outcomes in common and so can never occur together; that is, $P(A \text{ and } B) = 0$.

nonresponse When an individual chosen for a sample can't be contacted or refuses to participate.

normal curve A symmetric, single-peaked, bell-shaped density curve.

normal distribution Distribution described by a normal curve. It is completely specified by two numbers: its mean μ and its standard deviation σ.

normal probability plot A scatterplot of the ordered pair (data value, expected z-score) for each of the individuals in a quantitative data set. That is, the x-coordinate of each point is the actual data value and the y-coordinate is the expected z-score corresponding to the percentile of that data value in a standard normal distribution.

null hypothesis (H_0) The claim that we weigh evidence against in a significance test.

observational study A study that observes individuals and measures variables of interest, but does not attempt to influence the responses.

one-sample confidence interval for a standard deviation (variance) A confidence interval used to estimate a population standard deviation σ (or a population variance σ^2).

one-sample t interval for a mean A confidence interval used to estimate a population mean μ.

one-sample t test for a mean A significance test of the null hypothesis that a population mean μ is equal to a specified value.

one-sample test for a standard deviation (variance) A significance test of the null hypothesis that a population standard deviation σ (or a population variance σ^2) is equal to a specified value.

one-sample z interval for a proportion A confidence interval used to estimate a population proportion p.

one-sample z test for a proportion A significance test of the null hypothesis that a population proportion p is equal to a specified value.

one-sided An alternative hypothesis stating either that a parameter is greater than the null value or that the parameter is less than the null value.

one-sided test A significance test involving a one-sided alternative hypothesis.

one-way analysis of variance (ANOVA) test A significance test of the null hypothesis that two or more population (or treatment) means are the same.

outlier Individual value that falls outside the overall pattern in a distribution of quantitative data or a relationship between two quantitative variables.

paired data Data obtained by recording two values of the same quantitative variable for each individual or for each pair of similar individuals.

paired t interval for a mean difference A confidence interval used to estimate a population (true) mean difference for paired data.

paired t test for a mean difference A significance test of the null hypothesis that a population mean difference is equal to a specified value, usually 0.

parameter A number that describes some characteristic of the population.

percentile The percentage of values in a distribution that are less than the individual's data value.

permutation A distinct arrangement of some group of individuals where order matters.

pie chart A graph that shows each category as a sector of a circle. The areas of the sectors are proportional to the category frequencies or relative frequencies.

placebo A treatment that has no active ingredient but is otherwise like other treatments.

placebo effect Describes the fact that some subjects in an experiment will respond favorably to any treatment, even an inactive treatment (placebo).

point estimate A single-value estimate of a population parameter.

Poisson distribution The probability distribution of a Poisson random variable. Any Poisson distribution is completely specified by one number: the mean number of successes μ over the given interval of space or time.

Poisson random variable The count of successes X in a Poisson setting. The possible values of X are $0, 1, 2, \ldots$.

Poisson setting A situation in which we count the number of successes that occur over a given continuous interval of space or time. For the Poisson setting to apply, it must be possible to divide the given interval into small, equally sized subintervals for which the probability of two or more successes in any one subinterval is (essentially) 0; the probability of success in each subinterval is the same; and the occurrence of a success in any given subinterval is independent of the occurrence of a success in any other subinterval. These properties ensure that successes occur at random and at a constant rate over the interval of space or time.

population In a statistical study, the entire group of individuals we want information about.

population regression line A regression line calculated from every value in the population; also called a true regression line. The equation of a population regression line is $y = \beta_0 + \beta_1 x + \varepsilon$, where y is the value of the response variable for a given value of x, β_0 is the population y-intercept, β_1 is the population slope, and ε (the Greek letter epsilon) is the error term. We add the error term ε to the model as a reminder that individual values of the response variable will vary from values predicted by the model.

power The probability that a significance test will find convincing evidence for H_a when a specific alternative value of the parameter is true.

probability A number between 0 and 1 that describes the proportion of times a particular outcome of a chance process is expected to occur in a very large number of repetitions.

probability distribution The possible values of a random variable and their probabilities.

probability model A description of some chance process that consists of two parts: a list of all possible outcomes and the probability of each outcome.

P-value The probability, assuming the null hypothesis H_0 is true, of getting evidence for the alternative hypothesis H_a as strong as or stronger than the observed evidence by chance alone.

quantitative variable A variable that takes number values that are quantities — counts or measurements.

quartiles The quartiles of a distribution divide the ordered data set into four groups having roughly the same number of values. The first quartile is Q_1, the second quartile is the median, and the third quartile is Q_3.

random assignment In an experiment, assignment of the treatments to experimental units (or assignment of experimental units to treatments) using a chance process.

random sample A sample consisting of individuals from the population who are selected for the sample using a chance process.

random variable Variable that takes numerical values that describe the outcomes of a chance process.

randomization distribution The approximation of the sampling distribution of a statistic created by repeatedly simulating the possible values of the sample statistic while assuming the null hypothesis is true.

randomization test A nonparametric significance test that uses a randomization distribution to estimate the P-value.

range The distance between the minimum value and the maximum value in a distribution; range = maximum − minimum.

regression line A line that models how a response variable y changes as an explanatory variable x changes. Regression lines are expressed in the form $\hat{y} = b_0 + b_1 x$, where $\hat{y}$ is the predicted value of y for a given value of x.

regression to the mean The tendency for extreme values of the explanatory variable to be paired with less extreme values of the response variable.

relative frequency table Table that shows the proportion or percentage of individuals having each data value.

residual The difference between an actual value of y and the value of y predicted by the regression line; residual = actual y − predicted $y = y - \hat{y}$.

residual plot A scatterplot that plots the residuals on the vertical axis and the explanatory variable on the horizontal axis.

resistant A statistic that is not affected much by extreme observations.

response bias Bias that occurs when there is a consistent pattern of inaccurate responses to a survey question.

response variable Variable that measures an outcome of a study.

roughly symmetric A distribution is roughly symmetric if the right side of the graph (containing the half of the observations with larger values) is approximately a mirror image of the left side.

sample A subset of individuals in the population from which we collect data.

sample regression line A regression line calculated from a sample. The equation of a sample regression line is $\hat{y} = b_0 + b_1 x$, where $\hat{y}$ is the predicted y value (or estimated mean y value) for a given value of x, b_0 is the sample y-intercept, and b_1 is the sample slope.

sample space The list of all possible outcomes of a chance process.

sampling distribution The distribution of values taken by the statistic in all possible samples of the same size from the same population.

sampling distribution of the sample mean $\bar{x}$ The distribution of values taken by the sample mean $\bar{x}$ in all possible samples of the same size from the same population.

sampling distribution of the sample proportion $\hat{p}$ The distribution of values taken by the sample proportion $\hat{p}$ in all possible samples of the same size from the same population.

sampling variability The fact that different random samples of the same size from the same population produce different estimates.

scatterplot A graph that shows the relationship between two quantitative variables measured on the same individuals. The values of one variable appear on the horizontal axis, and the values of the other variable appear on the vertical axis. Each individual in the data set appears as a point in the graph.

segmented bar chart A graph that displays the distribution of a categorical variable as portions (segments) of a rectangle, with the area of each segment proportional to the percentage of individuals in the corresponding category.

side-by-side bar chart A graph that displays the distribution of a categorical variable for each value of another categorical variable. The bars are grouped together based on the values of one of the categorical variables and placed next to each other.

sign test A nonparametric significance test of the null hypothesis that a population median difference is equal to a specified value, usually 0.

significance level α The value that we use as a boundary to decide if an observed result is unlikely to happen by chance alone when the null hypothesis is true.

significance test A formal procedure for using observed data to decide between two competing claims (called hypotheses). The claims are usually statements about population parameters.

simple random sample (SRS) A sample of size n chosen in such a way that every group of n individuals in the population has an equal chance of being selected as the sample.

simulation Imitation of a chance process in a way that models real-world outcomes.

single-blind An experiment in which either the subjects or the people who interact with them and measure the response variable don't know which treatment a subject is receiving.

skewed to the left A distribution is skewed to the left if the left side of the graph (containing the half of the observations with smaller values) is much longer than the right side.

skewed to the right A distribution is skewed to the right if the right side of the graph (containing the half of the observations with larger values) is much longer than the left side.

slope In the regression equation $\hat{y} = b_0 + b_1 x$, the slope b_1 is the amount by which the predicted value of y changes when x increases by 1 unit.

standard deviation A measurement of the typical distance of the values in a distribution from the mean.

standard deviation of a discrete random variable A measurement of how much the values of the variable typically vary from the mean in many, many repetitions of the same chance process.

standard deviation of the residuals s A measurement of the size of a typical residual. That is, s measures the typical distance between the actual y values and the predicted y values.

standard error of $\hat{p}$ An estimate of the standard deviation of the sampling distribution of $\hat{p}$: $SE_{\hat{p}} = \sqrt{\dfrac{\hat{p}(1-\hat{p})}{n}}$. It estimates how much $\hat{p}$ typically varies from p.

standard error of $\bar{x}$ An estimate of the standard deviation of the sampling distribution of $\bar{x}$: $SE_{\bar{x}} = \dfrac{s_x}{\sqrt{n}}$. It estimates how much $\bar{x}$ typically varies from μ.

standard normal distribution The normal distribution with mean 0 and standard deviation 1.

standardized score (z-score) For an individual value in a distribution, the standardized score tells us how many standard deviations away from the mean the value falls, and in what direction. To find the standardized score (z-score), compute
$$z = \frac{\text{value} - \text{mean}}{\text{standard deviation}}.$$

standardized test statistic A measurement of how far a sample statistic is from what we would expect if the null hypothesis H_0 were true, in standard deviation units;
$$\text{standardized test statistic} = \frac{\text{statistic} - \text{null value}}{\text{standard deviation (error) of statistic}}.$$

statistic A number that describes some characteristic of a sample.

statistical problem-solving process The steps involved in solving statistics problems: (1) Ask questions, (2) Collect/consider data, (3) Analyze data, (4) Interpret results.

statistically significant When the observed results of a study are too unusual to be explained by chance alone we say that the result is statistically significant.

statistics The science and art of collecting, analyzing, and drawing conclusions from data.

stemplot A graph that shows each data value separated into two parts: a stem, which consists of the leftmost digit(s), and a leaf, the final digit. The stems are ordered from least to greatest and arranged in a vertical column. The leaves are arranged in increasing order out from the stems.

strata Groups of individuals in a population that share characteristics thought to be associated with the variables being measured in a study. The singular of strata is stratum.

stratified random sampling Selection of a sample by choosing a simple random sample from each stratum and combining those sub samples into one overall sample.

subjects Experimental units that are human beings.

systematic random sampling Selection of a sample from an ordered arrangement of the population by randomly selecting one of the first k individuals and choosing every kth individual thereafter.

t distribution A distribution described by a symmetric, single-peaked, bell-shaped density curve. Any t distribution is completely specified by its degrees of freedom (df).

t interval for the slope of a least-squares regression line A confidence interval used to estimate the slope of a population (true) regression line.

t test for an individual coefficient A significance test of the null hypothesis that an explanatory variable in a population multiple regression model does not provide additional useful information about the response variable, after accounting for other variables in the model.

t test for the slope of a least-squares regression line A significance test of the null hypothesis that there is no linear association between two quantitative variables.

test statistic for the sign test Compute the difference for each pair of observations. Classify each difference by its sign (ignore differences of 0). The test statistic for the sign test is $X =$ the number of negative differences when H_a: Population median difference < 0; X = the number of positive differences when H_a: Population median difference > 0; or X = the larger of the number of positive or negative differences when H_a: Population median difference ≠ 0.

test statistic W for the Wilcoxon signed rank test Let W^+ be the sum of the positive signed ranks, and let W^- be the sum of the absolute values of the negative signed ranks. The test statistic W for the Wilcoxon signed rank test is the smaller of W^+ nd W^-.

third quartile Q_3 The median of the data values that are to the right of the median in the ordered list.

timeplot A scatterplot with consecutive points connected by a line segment. The graph shows how a single quantitative variable changes over time.

treatment A specific condition applied to the individuals in an experiment.

tree diagram A figure that shows the sample space of a chance process involving multiple stages. The probability of each outcome is shown on the corresponding branch of the tree. All probabilities after the first stage are conditional probabilities.

two-sample F test comparing two standard deviations (variances) A significance test of the null hypothesis that the standard deviations (variances) for two populations or treatments are equal.

two-sample t interval for a difference between two means A confidence interval used to estimate a difference in the means of two populations or treatments.

two-sample t test for a difference between two means A significance test of the null hypothesis that the difference in the means of two populations or treatments is equal to a specified value (usually 0).

two-sample z interval for a difference between two proportions A confidence interval used to estimate a difference in the proportions of successes for two populations or treatments.

two-sample z test for a difference between two proportions A significance test of the null hypothesis that the difference in the proportions of successes for two populations or treatments is equal to 0.

two-sided An alternative hypothesis that states that the parameter is different from the null value (it could be either greater than or less than the null value).

two-sided test A significance test involving a two-sided alternative hypothesis.

two-way table A table of frequencies (or relative frequencies) that summarizes the relationship between two categorical variables for some group of individuals.

Type I error Error in which we reject H_0 when H_0 is true. That is, the data give convincing evidence that H_a is true when it really isn't.

Type II error Error in which we fail to reject H_0 when H_a is true. That is, the data do not give convincing evidence that H_a is true when it really is.

unbiased estimator A statistic used to estimate a parameter is an unbiased estimator if the mean of its sampling distribution is equal to the value of the parameter being estimated.

undercoverage The situation in which some members of the population are less likely to be chosen or cannot be chosen for the sample.

union The event "A or B", consisting of all outcomes that are in event A or event B, or both, denoted $A \cup B$.

variable Any attribute that can take different values for different individuals.

Venn diagram A figure that consists of one or more circles surrounded by a rectangle. Each circle represents an event. The region inside the rectangle represents the sample space of the chance process.

voluntary response sample A sample consisting of people who choose to be in the sample by responding to a general invitation.

Wilcoxon rank sum test A nonparametric significance test of the null hypothesis that two population (or treatment) medians are the same.

Wilcoxon signed rank test A nonparametric significance test of the null hypothesis that a population median difference is 0 or that a population median is equal to a specified value.

y-intercept In the regression equation $\hat{y} = b_0 + b_1 x$, the y-intercept b_0 is the predicted value of y when $x = 0$.

z-score *See* standardized score.

Notes and Data Sources

Note: The urls listed here were live when we researched this text.

CHAPTER 1

1. www.nytimes.com/2021/10/06/health/malaria-vaccine-who.html
2. www.plosmedicine.org/article/info%3Adoi%2F10.1371%2Fjournal.pmed.1001595
3. Adapted from C. Franklin et al. (2020), *Guidelines for Assessment and Instruction in Statistics Education* (American Statistical Association).
4. Roller coaster data from www.rcdb.com.
5. www.pewresearch.org/fact-tank/2021/01/25/though-not-especially-productive-in-passing-bills-the-116th-congress-set-new-marks-for-social-media-use/
6. John Mackenzie, "Family Dinner Linked to Better Grades for Teens: Survey Finds Regular Meal Time Yields Additional Benefits," ABC News, *World News Tonight,* September 13, 2005.
7. Signe Lund Mathiesen, Line Ahm Mielby, Derek Victor Byrne, and Qian Janice Wang, "Music to Eat By: A Systematic Investigation of the Relative Importance of Tempo and Articulation on Eating Time," *Appetite* 155 (2020): 104801.
8. Data obtained from The Numbers website: www.thenumbers.com/movie/records/All-Time-Worldwide-Box-Office on January 15, 2020.
9. B. Turnwald, D. Boles, & A. Crum (2017), "Association Between Indulgent Descriptions and Vegetable Consumption: Twisted Carrots and Dynamite Beets," *JAMA Internal Medicine.* doi:10.1001/jamainternmed.2017.1637
10. Katri Räikkönen, Anu-Katriina Pesonen, Anna-Liisa Järvenpää, Timo E. Strandberg, "Sweet Babies: Chocolate Consumption During Pregnancy and Infant Temperament at Six Months," *Early Human Development* 76, no. 2 (February 2004): 139–145.
11. Melissa G. Hunt, Rachel Marx, Courtney Lipson and Jordyn Young, "No More FOMO: Limiting Social Media Decreases Loneliness and Depression," *Journal of Social and Clinical Psychology* 37, no. 10 (2018): 751–768.
12. National Institute of Child Health and Human Development (NICHD), Study of Early Child Care and Youth Development. The article appears in the July 2003 issue of *Child Development.* The quotation is from the summary on the NICHD website: www.nichd.nih.gov.
13. Bill Hesselmar, Fei Sjöberg, Robert Saalman, Nils Åberg, Ingegerd Adlerberth & Agnes E. Wold (2013), "Pacifier Cleaning Practices and Risk of Allergy Development," *Pediatrics.* www.pediatrics.org/cgi/doi/10.1542/peds.2012-3345.
14. Martin Enserink, "Fraud and Ethics Charges Hit Stroke Drug Trial," *Science,* 274 (1996): 2004–2005.
15. bigstory.ap.org/content/coffee-buzz-study-finds-java-drinkers-live-longer
16. *Arizona Daily Star,* February 11, 2009.
17. Frederick Mosteller and David L. Wallace, *Inference and Disputed Authorship: The Federalist* (Reading, MA: Addison-Wesley, 1964). Other information obtained from en.wikipedia.org/wiki/ and www.constitution.org/fed/federa51.htm.
18. The advice columnist is Ann Landers.
19. www.huffpost.com/entry/women-sleep-better-with-dogs-than-with-human-partners-study_n_5c002dede4b0864f4f6b706f
20. Christy L. Hoffman, Kaylee Stutz & Terrie Vasilopoulos (2018), "An Examination of Adult Women's Sleep Quality and Sleep Routines in Relation to Pet Ownership and Bedsharing," *Anthrozoös* 31:6, 711–725. DOI: 10.1080/08927936.2018.1529354
21. www.econlib.org/archives/2009/02/parents_and_buy.html
22. www.pewresearch.org/methods/2017/05/15/what-low-response-rates-mean-for-telephone-surveys/
23. www.nytimes.com/1994/07/08/us/poll-on-doubt-of-holocaust-is-corrected.html
24. Data from Hailey Kiernan, Canyon del Oro High School.
25. www.cleaninginstitute.org/assets/1/AssetManager/2010%20Hand%20Washing%20Findings.pdf
26. *Arizona Daily Star,* April 18, 2016.
27. Bryan E. Porter and Thomas D. Berry, "A Nationwide Survey of Self-Reported Red Light Running: Measuring Prevalence, Predictors, and Perceived Consequences," *Accident Analysis and Prevention* 33 (2001): 735–741.
28. Mario A. Parada et al., "The validity of Self-Reported Seatbelt Use: Hispanic and Non-Hispanic Drivers in El Paso," *Accident Analysis and Prevention* 33 (2001): 139–143.
29. Data from Emma Merry, Canyon del Oro High School.
30. en.wikipedia.org/wiki/The_Literary_Digest
31. Bryan E. Porter and Thomas D. Berry, "A Nationwide Survey of Self-Reported Red Light Running: Measuring Prevalence, Predictors, and Perceived Consequences," *Accident Analysis and Prevention* 33 (2001): 735–741.
32. Ohio State University. "Night Lights Don't Lead to Nearsightedness, Study Suggests," *ScienceDaily,* March 9, 2000.
33. www.census.gov/content/dam/Census/library/publications/2020/acs/acsbr20-03.pdf and www.census.gov/acs/www/methodology/sample-size-and-data-quality/sample-size/index.php. The margin of error for this estimate is $842.
34. maristpoll.marist.edu/wp-content/uploads/2021/03/NPR_Marist-Poll_USA-NOS-and-Tables_202103291133.pdf
35. Ed O'Brien and Phoebe C. Ellsworth, "Saving the Last for Best: A Positivity Bias for End Experiences," *Psychological Science* 23, no. 2 (2012): 163–165.
36. Robert C. Parker and Patrick A. Glass, "Preliminary Results of Double-Sample Forest Inventory of Pine and Mixed Stands with High and Low-Density LiDAR," in Kristina F. Connoe (ed.), *Proceedings of the 12th Biennial Southern Silvicultural Research Conference* (Asheville, NC: U.S. Department of Agriculture, Forest Service, Southern Research Station, 2004), Table 1. The researchers actually sampled every tenth plot.
37. Gary S. Foster and Craig M. Eckert, "Up from the Grave: A Socio-Historical Reconstruction of an African American Community from Cemetery Data in the Rural Midwest," *Journal of Black Studies* 33 (2003): 468–489.
38. fivethirtyeight.com/features/how-much-football-is-even-in-a-football-broadcast/. Special thanks to Katherine Rowe for providing the raw data.
39. www.advfn.com/nyse/newyorkstockexchange.asp

40. sleepfoundation.org/sleep-polls-data/2015-sleep-and-pain
41. *San Gabriel Valley Tribune,* February 13, 2003.
42. www.smithsonianmag.com/ideas-innovations/How-Much-Do-Americans-Know-About-Science.html
43. Data courtesy of DeAnna McDonald, University of Arizona.
44. www.ppic.org/publication/california-voter-and-party-profiles/
45. tucson.com/news/local/saguaro-census-shows-more-giants-low-reproduction-in-namesake-park/article_c659cb2d-ddef-5371-b5a3-abe351f886ad.html. Thanks to Don Swann of the National Park Service for sharing additional information.
46. www.fivethirtyeight.com/features/donttakeyourvitamins
47. David O. Meltzer, Thomas J. Best, Hui Zhang, et al., "Association of Vitamin D Status and Other Clinical Characteristics with COVID-19 Test Results," *JAMA Network Open* 3, no. 9 (2020): e2019722. doi:10.1001/jamanetworkopen.2020.19722.
48. www.nbcwashington.com/news/health/ADHD_Linked_To_Lead_and_Mom_s_Smoking.html
49. *Journal of Clinical Endocrinology and Metabolism,* 2016. doi:10.1210/jc.2015-4013. Cited in *Nutrition Action,* April 2016.
50. Fernando P. Polack, et al., "Safety and Efficacy of the BNT162b2 mRNA Covid-19 Vaccine," *New England Journal of Medicine,* December 10, 2020. doi: 10.1056/NEJMoa2034577.
51. www.sciencedirect.com/science/article/pii/S0747563214001563
52. *Nutrition Action,* October 2013.
53. Katie Adolphus, Clare L. Lawton & Louise Dye (2019), "Associations Between Habitual School-Day Breakfast Consumption Frequency and Academic Performance in British Adolescents," *Frontiers in Public Health.* doi.org/10.3389/fpubh.2019.00283
54. Sheri Madigan, Dillon Browne, Nicole Racine, et al., "Association Between Screen Time and Children's Performance on a Developmental Screening Test," *JAMA Pediatrics* 173, no. 3 (2019): 244–250.
55. K. B. Suttle, Meredith A. Thomsen & Mary E. Power, "Species Interactions Reverse Grassland Responses to Changing Climate," *Science* 315 (2007): 640–642.
56. Information found online at ssw.unc.edu/about/news/careerstart_1-13-09.
57. Charlie Huveneers, Sasha Whitmarsh, Madeline Thiele, Lauren Meyer, Andrew Fox, Corey J.A. Bradshaw (2018), "Effectiveness of Five Personal Shark-Bite Deterrents for Surfers," *PeerJ.* doi:10.7717/peerj.5554.
58. *New York Times,* November 11, 2014. www.nytimes.com/2014/11/11/science/dead-jellyfish-are-more-nutrition-than-nuisance.html?_r=0.
59. foodpsychology.cornell.edu/OP/buffet_pricing
60. www.sciencedaily.com/releases/2014/11/141124081040.htm
61. Joel Brockner et al., "Layoffs, Equity Theory, and Work Performance: Further Evidence of the Impact of Survivor Guilt," *Academy of Management Journal* 29 (1986): 373–384.
62. Pam A. Mueller and Daniel M. Oppenheimer (2014), "The Pen Is Mightier Than the Keyboard: Advantages of Longhand over Laptop Note Taking," *Psychological Science.* doi: 10.1177/0956797614524581.
63. www.cnn.com/2018/12/19/health/reverse-cognitive-aging-exercise-diet-study/index.html
64. Faris M. Zuraikat, Liane S. Roe, Alissa D. Smethers & Barbara J. Rolls, "Doggy Bags and Downsizing: Packaging Uneaten Food to Go After a Meal Attenuates the Portion Size Effect in Women," *Appetite* 129 (2018): 162–170.
65. Naomi D. L. Fisher, Meghan Hughes, Marie Gerhard-Herman & Norman K. Hollenberg, "Flavonol-Rich Cocoa Induces Nitric Oxide–Dependent Vasodilation in Healthy Humans," *Journal of Hypertension* 21, no. 12 (2003): 2281–2286.
66. Christopher Anderson, "Measuring What Works in Health Care," *Science* 263 (1994): 1080–1082.
67. *Nutrition Action,* October 2013. Originally published in *Archives of Physical Medicine and Rehabilitation* 93 (2012): 1269.
68. www.nytimes.com/2008/03/05/health/research/05placebo.html?_r=0
69. *New York Times,* November 19, 2014. tinyurl.com/k8xm6rj.
70. www.sciencedirect.com/science/article/pii/S0360131512002254
71. David L. Strayer, Frank A. Drews & William A. Johnston, "Cell Phone–Induced Failures of Visual Attention During Simulated Driving," *Journal of Experimental Psychology: Applied* 9 (2003): 23–32.
72. Marianne Bertrand and Sendhil Mullainathan, *Are Emily and Greg More Employable Than Lakisha and Jamal? A Field Experiment on Labor Market Discrimination* (National Bureau of Economic Research, July 2003). www.nber.org/papers/w9873.
73. www.ocf.berkeley.edu/~broockma/kalla_broockman_donor_access_field_experiment.pdf
74. *Nutrition Action,* July/August 2014. Originally published in *Journal of the American Medical Association* 311 (2014): 2083.
75. *Nutrition Action,* March 2009.
76. H. R. Knapp and G. A. FitzGerald, "The antihypertensive effects of fish oil. A controlled study of polyunsaturated fatty acid supplements in essential hypertension," *New England Journal of Medicine,* 320 (1989): 1037–1043; cited in Fred Ramsey and Daniel Schafer, *The Statistical Sleuth* (Pacific Grove, CA: Duxbury Press, 2002), 23.
77. Bruce Barrett, Roger Brown, Dave Rakel, Marlon Mundt, Kerry Bone, Dip Phyto, Shari Barlow & Tola Ewers, "Echinacea for Treating the Common Cold," *Annals of Internal Medicine,* December 2010. doi.org/10.7326/0003-4819-153-12-201012210-00003.
78. www.nejm.org/doi/full/10.1056/NEJMoa1615869
79. www.plosone.org/article/info%3Adoi%2F10.1371%2Fjournal.pone.0016782
80. Steering Committee of the Physicians' Health Study Research Group, "Final Report on the Aspirin Component of the Ongoing Physicians' Health Study," *New England Journal of Medicine* 321 (1989): 129–135.
81. "Antibiotics No Better Than Placebo for Sinus Infections," in.news.yahoo.com/antibiotics-no-better-placebo-sinusinfections-074240018.html.
82. Carlos Vallbona et al., "Response of Pain to Static Magnetic Fields in Postpolio Patients, A Double Blind Pilot Study," *Archives of Physical Medicine and Rehabilitation* 78 (1997): 1200–1203.
83. www.nejm.org/doi/full/10.1056/NEJMoa0905471
84. www.nbc.com/the-voice/vote/rules
85. Mary O. Mundinger et al., "Primary Care Outcomes in Patients Treated by Nurse Practitioners or Physicians," *Journal of the American Medical Association* 238 (2000): 59–68.
86. E. C. Strain et al., "Caffeine Dependence Syndrome: Evidence from Case Histories and Experimental Evaluation," *Journal of the American Medical Association* 272 (1994): 1604–1607.

87. S. D. Brown, J. Duncan, D. Crabtree, D. Powell, M. Hudson & J. L.Allan, "We Are What We (Think We) Eat: The Effect of Expected Satiety on Subsequent Calorie Consumption," *Appetite*, September 2020. doi.org/10.1016/j.appet.2020.104717.

88. We obtained the tire pressure loss data from the *Consumer Reports* website: www.consumerreports.org/cro/news/2007/10/tires-nitrogen-air-loss-study/index.htm.

89. C. Proserpio, C. Invitti, S. Boesveldt, L. Pasqualinotto, M. Laureati, C. Cattaneo & E. Pagliarini, "Ambient Odor Exposure Affects Food Intake and Sensory Specific Appetite in Obese Women," *Frontiers in Psychology* 10 (2019): 7. doi: 10.3389/fpsyg.2019.00007.

90. Athanasios Koutsos, et al., "Two Apples a Day Lower Serum Cholesterol and Improve Cardiometabolic Biomarkers in Mildly Hypercholesterolemic Adults: A Randomized, Controlled, Crossover Trial," *American Journal of Clinical Nutrition* 111 (2020): 307–318.

91. www.bls.gov/cps

92. The sleep deprivation study is described in R. Stickgold, L. James, and J. Hobson, "Visual Discrimination Learning Requires Post-Training Sleep," *Nature Neuroscience* (2000): 1237–1238. We obtained the data from Allan Rossman, who got it courtesy of the authors.

93. academic.oup.com/ajcn/article/100/4/1182/4576550#110598138

94. ourworldindata.org/smoking-big-problem-in-brief

95. See the details on the website of the Office for Human Research Protections of the Department of Health and Human Services, hhs.gov/ohrp.

96. en.wikipedia.org/wiki/Tuskegee_Syphilis_Study

97. heart.bmj.com/content/105/23/1793

98. H. Scott et al., "Social media use and adolescent sleep patterns: cross-sectional findings from the UK millennium cohort study," *BMJ Open* 9 (2019): e031161.

99. *Nutrition Action*, December 2014. Originally published in *Acta Psychologica*, 153 (2014): 13.

100. Sarah J. Hardgrove and Stephen J. Livesley, "Applying Spent Coffee Grounds Directly to Urban Agriculture Soils Greatly Reduces Plant Growth," *Urban Forestry & Urban Greening*, 2016. dx.doi.org/10.1016/j.ufug.2016.02.015.

101. www.pnas.org/content/111/24/8788.full

102. en.wikipedia.org/wiki/Monster_Study

103. en.wikipedia.org/wiki/Stateville_Penitentiary_Malaria_Study

104. en.wikipedia.org/wiki/Willowbrook_State_School

105. Charles A. Nelson III et al., "Cognitive Recovery in Socially Deprived Young Children: The Bucharest Early Intervention Project," *Science* 318 (2007): 1937–1940.

106. Data from Lexie Gardner and Erica Chauvet, Waynesburg University.

107. R. Cialdini, et al., "Basking in Reflected Glory: Three (Football) Field Studies," *Journal of Personality and Social Psychology* 34, no. 3 (1976): 366–373.

108. *Nutrition Action*, March 2013. Originally published in *Circulation* 127 (2013): 188, www.ncbi.nlm.nih.gov/pubmed/23319811.

109. Chip Brown, "Cyrus Vance Jr.'s 'Moneyball' Approach to Crime," *New York Time Magazine*, December 7, 2014. www.nytimes.com/2014/12/07/magazine/cyrus-vance-jrs-moneyball-approach-to-crime.html?smid=fb-share&_r=0.

110. www.gallup.com/poll/180260/americans-rate-nurses-highest-honesty-ethical-standards.aspx

111. aem.asm.org/content/early/2016/08/15/AEM.01838-16.abstract?sid=61679ac7-4522-4b65-9284-04d7bfbbcdab

112. Julia Baudry, Karen E. Assmann, Mathilde Touvier, et al., "Association of Frequency of Organic Food Consumption With Cancer Risk: Findings From the NutriNet-Santé Prospective Cohort Study," *JAMA Internal Medicine* 178, no. 12 (2018): 1597–1606.

113. *Arizona Daily Star,* October 29, 2008.

114. www.ncbi.nlm.nih.gov/pubmed/29635503

115. hms.harvard.edu/news/e-readers-foil-good-nights-sleep

116. *Nutrition Action*, March 2014. Originally published in *JAMA Pediatrics*, 2013, doi: 10.1001/jamapediatrics.2013.4139.

CHAPTER 2

1. Data for 2020 from National Low Income Housing Coalition, "Out of Reach: The High Cost of Housing," reports.nlihc.org/oor.

2. This data set was obtained by taking a random sample of 3000 households from the 2019 American Community Survey data set: www.census.gov/programs-surveys/acs/microdata.html.

3. We actually selected a random sample of 100 students who completed the online survey during 2019. Then we cleaned the data to obtain the resulting sample of 75 students.

4. U.S. Census Bureau, *Current Population Survey, Annual Social and Economic Supplement,* April 21, 2021.

5. Regina A. Corso, "Few Hate Shopping For Clothes, But Love of It Varies by Country," *Harris Poll*, June 24, 2011.

6. Data from the CubeSat database, sites.google.com/a/slu.edu/swartwout/cubesat-database?authuser=0, accessed January 23, 2021.

7. The data on U.S. radio station formats in 2019 was obtained from statista.com.

8. Axalta Coating Systems, *Global Automotive 2020 Color Popularity Report.*

9. Found at www.toptenreviews.com, which claims to have compiled data "from a number of different reputable sources."

10. We got the idea for this exercise from David Lane's case study, "Who Is Buying iMacs?", which we found at onlinestatbook.com/case_studies_rvls/.

11. This exercise is based on information from www.minerandcostudio.com.

12. We obtained the squirrel data set from data.cityofnewyork.us/Environment/2018-Central-Park-Squirrel-Census-Squirrel-Data/vfnx-vebw/data.

13. Data for 2010 from the 2012 *Statistical Abstract of the United States,* U.S. Census Bureau, www.census.gov.

14. *The Asian Population: 2010,* www.census.gov, based on data collected in the 2010 U.S. Census.

15. news.gallup.com/poll/193874/half-americans-play-state-lotteries.aspx.

16. Data from the 2012 *Statistical Abstract of the United States.*

17. This data set was obtained by taking a random sample of 3000 households from the 2019 American Community Survey data set: www.census.gov/programs-surveys/acs/microdata.html.

18. COVID-19 case map accessed on January 24, 2022, from the *New York Times* coronavirus site: www.nytimes.com/interactive/2021/us/covid-cases.html.

19. Data obtained from skyscraperpage.com and Wikipedia.

20. Experiment based on an idea from the *Mythbusters* television show.

21. Survey data from the Visitor Use Study at www.nps.gov/yell.

22. *Nutrition Action,* "Weighing the Options: Do Extra Pounds Mean Extra Years?" (March 2013).
23. Elizabeth F. Loftus and John C. Palmer, "Reconstruction of Automobile Destruction: An Example of the Interaction Between Language and Memory," *Journal of Verbal Learning and Verbal Behavior* 13 (1974): 585–589, www.researchgate.net/publication/222307973_Reconstruction_of_Automobile_Destruction_An_Example_of_the_Interaction_Between_Language_and_Memory.
24. R. Shine et al., "The Influence of Nest Temperatures and Maternal Brooding on Hatchling Phenotypes in Water Pythons," *Ecology* 78 (1997): 1713–1721.
25. Squirrel data obtained from data.cityofnewyork.us/Environment/2018-Central-Park-Squirrel-Census-Squirrel-Data/vfnx-vebw/data.
26. N. Häusler, J. Haba-Rubio, R. Heinzer, et al., "Association of Napping with Incident Cardiovascular Events in a Prospective Cohort Study," *Heart* 105 (2019): 1793–1798.
27. Data from the CubeSat database, sites.google.com/a/slu.edu/swartwout/cubesat-database?authuser=0, accessed February 1, 2021.
28. K. Eagan, J. B. Lozano, S. Hurtado, and M. H. Case, *The American Freshman: National Norms, Fall 2013* (Los Angeles, CA: Higher Education Research Institute, UCLA, 2013).
29. poll.qu.edu/Poll-Release-Legacy?releaseid=2275.
30. Pew Research Center, "What It Takes to Truly Be 'One of Us'" (February 2017).
31. This data set was obtained by taking a random sample of 3000 households from the 2019 American Community Survey data set: www.census.gov/programs-surveys/acs/microdata.html.
32. This data set was obtained by taking a random sample of 3000 households from the 2019 American Community Survey data set: www.census.gov/programs-surveys/acs/microdata.html.
33. V. Rideout and M. B. Robb, *The Common Sense Census: Media Use by Tweens and Teens, 2019* (San Francisco, CA: Common Sense Media, 2019).
34. Siem Oppe and Frank De Charro, "The Effect of Medical Care by a Helicopter Trauma Team on the Probability of Survival and the Quality of Life of Hospitalized Victims," *Accident Analysis and Prevention* 33 (2001): 129–138. The authors give the data in this example as a "theoretical example" to illustrate the need for more elaborate analysis of actual data using severity scores for each victim.
35. Data from the 2015 report, "Images of Computer Science: Perceptions Among Students, Parents, and Educators in the U.S.," compiled by Google and Gallup: services.google.com/fh/files/misc/images-of-computer-science-report.pdf.
36. Data obtained from flintwaterstudy.org.
37. Data provided by the team manager for the youth football team of one of Mr. Starneses' grandsons.
38. R. R. Sokal and F. J. Rohlf, *Biometry* (New York, NY: Freeman Publishing, 1968), 109. Original data from R. R. Sokal and P. E. Hunter, "A Morphometric Analysis of DDT-Resistant and Non-resistant Housefly Strains," *Annals of the Entomological Society of America* 48 (1955): 499–507.
39. Data on household size were obtained using the Random Sampler tool from the Census at School websites for each of the two countries in 2013.
40. *Model Year 2020 Fuel Economy Guide,* www.fueleconomy.gov.
41. Data on the U.S. Women's National Soccer Team are from the U.S. Soccer Federation, www.ussoccer.com.
42. Frozen pizza data obtained from *Consumer Reports* magazine, January 2011.
43. Data collected by Doug Tyson in 2020.
44. The cereal data came from the Data and Story Library, dasl.datadescription.com/.
45. The original paper is T. M. Amabile, "Motivation and Creativity: Effects of Motivational Orientation on Creative Writers," *Journal of Personality and Social Psychology,* 48, no. 2 (February 1985): 393–399. The data for this exercise came from Fred L. Ramsey and Daniel W. Schafer, *The Statistical Sleuth,* 3rd ed. (Pacific Grove, CA: Brooks/Cole Cengage Learning, 2013).
46. Data from a student project by Rikki Schlott, Daniel Joseph, and Mandi Marom. Daniel was Interviewer 1 and Rikki was Interviewer 2.
47. *Nutrition Action Healthletter,* April 2016.
48. www.nutritionaction.com/daily/calories-in-food/how-many-calories-are-in-beer-wine-and-cocktails/.
49. D. B. Lindenmayer, K. L. Viggers, R. B. Cunningham, and C. F. Donnelly, "Morphological Variation Among Columns of the Mountain Brushtail Possum, *Trichosurus caninus Ogilby* (Phalangeridae: Marsupiala)," *Australian Journal of Zoology* 43 (1995): 449–458.
50. We obtained the guinea pig data from rdocumentation.org. The original source is listed as E. W. Crampton, "The Growth of the Odontoblast of the Incisor Teeth as a Criterion of Vitamin C Intake of the Guinea Pig," *Journal of Nutrition* 33, no. 5 (1947): 491–504, doi: 10.1093/jn/33.5.491.
51. Data for 2020 from U.S. Census Bureau, *Current Population Survey, 2020 Annual Social and Economic Supplement.* Data for 1980 were obtained from www.census.gov.
52. E. Anionwu et al., "Sickle-Cell Disease in a British Urban Community," *British Medical Journal* 282 (1981): 283–286.
53. From the Electronic Encyclopedia of Statistics Examples and Exercises (EESEE) story, "Acorn Size and Oak Tree Range."
54. USDA National Nutrient Database for Standard Reference 26 Software v.1.4
55. Data provided by Josh Tabor.
56. Data provided by Tim Brown.
57. Data from the U.S. Census Bureau, Population Estimates Program, 2020.
58. This activity is described in Pat Hopfensperger, Tim Jacobbe, Deborah Lurie, and Jerry Moreno, *Bridging the Gap Between Common Core State Standards and Teaching Statistics* (Alexandria, VA: American Statistical Association, 2012).
59. The finch data came from Fred L. Ramsey and Daniel W. Schafer, *The Statistical Sleuth,* 3rd ed. (Pacific Grove, CA: Brooks/Cole Cengage Learning, 2013).
60. National Center for Statistics and Analysis, *Seat Belt Use in 2019: Use Rates in the States and Territories,* Traffic Safety Facts Crash Stats. Report No. DOT HS 812 947 (Washington, DC: National Highway Traffic Safety Administration, April 2020).
61. Federal Highway Administration, *Highway Statistics, 2018* (Washington, DC: U.S. Department of Transportation, 2019).
62. www.nutritionaction.com/daily/calories-in-food/how-many-calories-are-in-beer-wine-and-cocktails/
63. D. B. Lindenmayer, K. L. Viggers, R. B. Cunningham, and C. F. Donnelly, "Morphological Variation Among Columns of the Mountain Brushtail Possum, *Trichosurus caninus Ogilby* (Phalangeridae: Marsupiala)," *Australian Journal of Zoology* 43 (1995): 449–458.
64. Data on earned run average from www.baseball-reference.com.
65. College tuition data from www.petersons.com.
66. Data on percentage of foreign-born residents in the states from www.statista.com, accessed January 28, 2021.

67. Data on annual income for college graduates and non-graduates obtained from the March 2021 Annual Social and Economic Supplement, downloaded from www.census.gov. We took a random sample of 500 people who had attended some college but earned no degree, or who had earned an associate or bachelor's degree.
68. CO_2 emissions data from Our World in Data website, ourworldindata.org/share-co2-emissions, accessed January 28, 2021.
69. Data on travel time to work from U.S. Census Bureau, 2015–2019 American Community Survey 5-Year Estimates.
70. Monthly stock returns from the website of Professor Kenneth French of Dartmouth University: mba.tuck.dartmouth.edu/pages/faculty/ken.french/data_library.html. A fine point: The data are the "excess returns" on stocks, which are actual returns less the small monthly returns on Treasury bills.
71. The cereal data came from the Data and Story Library, dasl.datadescription.com/.
72. Data from the Bureau of Labor Statistics, *Annual Demographic Supplement*, www.bls.gov.
73. Data on word lengths in Shakespeare's plays from C. B. Williams, *Style and Vocabulary: Numerological Studies* (Griffin, 1970). The data on word lengths in *Popular Science* magazine were collected by students as a class project.
74. S. M. Kerry, F. B. Micah, J. Plange-Rhule, J. B. Eastwood, and F. P. Cappuccio, "Blood Pressure and Body Mass Index in Lean Rural and Semi-urban subjects in West Africa," *Journal of Hypertension* 23, no. 9 (2005): 1645–1651.
75. Data from a student project in Josh Tabor's Introductory Statistics class.
76. Histogram of engagement ring costs from www.nytimes.com/2020/02/06/learning/whats-going-on-in-this-graph-engagement-ring-costs.html.
77. Fisher iris data from UCI Machine Learning Repository, archive.ics.uci.edu/ml/datasets/iris.
78. Data on the population distributions by age group in Australia and Ethiopia from the U.S. Census Bureau's International Database, www.census.gov, accessed October 4, 2021.
79. www.nutritionaction.com/daily/calories-in-food/how-many-calories-are-in-beer-wine-and-cocktails/
80. This data set was obtained by taking a random sample of 3000 households from the 2019 American Community Survey data set: www.census.gov/programs-surveys/acs/microdata.html.
81. Data on AP® exam scores from the College Board's AP® Central website: apcentral.collegeboard.com.
82. T. Laukkanen, H. Khan, F. Zaccardi, and J. A. Laukkanen, "Association Between Sauna Bathing and Fatal Cardiovascular and All-Cause Mortality Events," *JAMA Internal Medicine* 175, no. 4 (2015): 542–548, doi:10.1001/jamainternmed.2014.8187.
83. Data on income per person and life expectancy from the Gapminder Foundation, www.gapminder.org, accessed February 1, 2021.
84. Data on used car prices from autotrader.com. We searched for F-150 4 × 4's on sale within 50 miles of College Station, Texas.
85. Internet use data from internetworldstats.com, accessed October 11, 2021.
86. Time series graph inspired by one at ourworldindata.org/covid-cases, which cites Johns Hopkins University CSSE COVID-19 data as its source.
87. www.stat.columbia.edu/~gelman/research/published/golf.pdf.
88. Candy data from *Nutrition Action*, December 2009.
89. Data from the roller coaster database: www.rcdb.com.
90. Data from www.weatherbase.com.
91. Data on income and prayer from www.pewresearch.org/fact-tank/2019/05/01/with-high-levels-of-prayer-u-s-is-an-outlier-among-wealthy-nations/.
92. Data provided by Erica Chauvet from a class activity in her Introductory Statistics class at Waynesburg University.
93. Data on mortgage rates from fred.stlouisfed.org/.
94. Data on gold prices from goldprice.org.
95. Mauna Loa CO_2 data from www.esrl.noaa.gov/gmd/.
96. Data from www.starbucks.com/promo/nutrition.
97. Data from www.teamusa.org/road-to-rio-2016/team-usa/athletes. Thanks to Jeff Eicher for sharing on the AP® Statistics Teacher Community.
98. www.nutritionaction.com/daily/calories-in-food/how-many-calories-are-in-beer-wine-and-cocktails/
99. www.nutritionaction.com/daily/calories-in-food/how-many-calories-are-in-beer-wine-and-cocktails/
100. Data on internet use from the Internet/Broadband Fact Sheet compiled by the Pew Research Center, www.pewresearch.org.
101. Graph based on the bubble chart from the Gapminder Foundation, www.gapminder.org.
102. Table 1 of E. Thomassot et al., "Methane-Related Diamond Crystallization in the Earth's Mantle: Stable Isotopes Evidence from a Single Diamond-Bearing Xenolith," *Earth and Planetary Science Letters* 257 (2007): 362–371.
103. J. J. Valente, S. K. Nelson, J. W. Rivers, D. D. Roby, and M. G. Betts, "Experimental Evidence That Social Information Affects Habitat Selection in Marbled Murrelets," *Ornithology* 138, no. 2 (2021): ukaa086, doi.org/10.1093/ornithology/ukaa086.
104. Data for 2020 from National Low Income Housing Coalition, "The Gap: A Shortage of Affordable Housing," reports.nlihc.org/gap.
105. Data on winners of the World Canine Disc Championship for 1975–1999 from en.wikipedia.org/wiki/Frisbee_Dog_World_Championship and for 2000–2019 from skyhoundz.com/world-champions/.
106. Harris Interactive, "Cell Phones Key to Teens' Social Lives, 47% Can Text with Eyes Closed" (July 2008), www.marketingcharts.com.
107. From the EESEE story, "What Makes a Pre-teen Popular?"
108. S. M. Stigler, "Do Robust Estimators Work with Real Data?" *Annals of Statistics* 5 (1977): 1055–1078.
109. Data from a student project by Alex and Tempe in Mr. Starnes's Introductory Statistics class.
110. From the EESEE story, "Is It Tough to Crawl in March?"
111. Data compiled from a table of percentages in "Americans view higher education as key to the American dream," press release by the National Center for Public Policy and Higher Education, www.highereducation.org.
112. Data from the EESEE story, "Checkmating and Reading Skills."

CHAPTER 3

1. Water supply and snowfall data set provided by Patti Collings of Brigham Young University. Patti used this data set as the basis of an example she created for the Statistically Speaking video series, developed in partnership by Macmillan Publishing and Coast Learning Systems. Patti obtained the data from the Colorado Climate Center, climate.colostate.edu.

2. Data obtained from www.flintwaterstudy.org.

3. Population density data for 2020 from www.indexmundi.com, which cites its source as the CIA World Factbook, available online at www.cia.gov/the-world-factbook/.

4. Data from kitchencabinetkings.com.

5. *Model Year 2020 Fuel Economy Guide,* from the website www.fueleconomy.gov.

6. Copper data from a Hubday news release on March 29, 2021, which was highlighted in Tony Davis, "Hudbay Says It Found Copper That Could Result in Open Pit Mines on Santa Ritas' West Side," *Arizona Daily Star,* April 1, 2021.

7. Data on the U.S. Women's National Soccer Team are from the U.S. Soccer Federation website, www.ussoccer.com.

8. Data on birth rates obtained from www.indexmundi.com, which cites its source as the CIA World Factbook, available online at www.cia.gov/the-world-factbook/.

9. Graph from a story by Richard Florida at www.citylab.com, based on Carole Turley Voulgaris, Michael J. Smart, and Brian D. Taylor, "Tired of Commuting? Relationships Among Journeys to School, Sleep, and Exercise Among American Teenagers," *Journal of Planning Education and Research* 39, no. 2 (August 2017): 142–154.

10. The graph is based on data from an episode in the *Against All Odds: Inside Statistics* video series, at www.learner.org/series/against-all-odds-inside-statistics/.

11. From the American Community Survey, at the Census Bureau website, www.census.gov. The data are a subsample of the individuals in the ACS New York sample who had travel times greater than zero.

12. Frozen pizza data obtained from *Consumer Reports* magazine, January 2011.

13. Tom Lloyd et al., "Fruit Consumption, Fitness, and Cardiovascular Health in Female Adolescents: The Penn State Young Women's Health Study," *American Journal of Clinical Nutrition* 67 (1998): 624–630.

14. C. B. Williams, *Style and Vocabulary: Numerological Studies* (Griffin, 1970).

15. Zack Friedman, "Student Loan Debt Statistics in 2021: A Record $1.7 Trillion," February 20, 2021, www.forbes.com.

16. U.S. Census Bureau and U.S. Department of Housing and Urban Development, "New Residential Sales," March 23, 2021.

17. Data from Daren Starnes's Introductory Statistics class at The College of New Jersey.

18. Data from the article by Andrew Hurst, "Most Americans Are Concerned About Climate Change, But That Doesn't Mean They're Prepared for It," October 15, 2019, www.valuepenguin.com.

19. Population density data for 2020 from www.indexmundi.com, which cites its source as the CIA World Factbook, available online at www.cia.gov/the-world-factbook/.

20. Data from kitchencabinetkings.com.

21. Modified based on data from pirate.shu.edu/~wachsmut/Teaching/MATH1101/Descriptives/variability.html.

22. fivethirtyeight.com/features/how-much-football-is-even-in-a-football-broadcast/. Special thanks to Katherine Rowe for providing the raw data.

23. These data are a random sample from the values given in R. Sokal and J. Rohlf, *Introduction to Biostatistics* (Mineola, NY: Dover, 2009).

24. Data obtained from www.flintwaterstudy.org.

25. *Model Year 2020 Fuel Economy Guide,* from the website www.fueleconomy.gov.

26. Copper data from a Hudbay news release on March 29, 2021, which was highlighted in Tony Davis, "Hudbay Says It Found Copper That Could Result in Open Pit Mines on Santa Ritas' West Side," *Arizona Daily Star,* April 1, 2021.

27. Data on the U.S. Women's National Soccer Team are from the U.S. Soccer Federation website, www.ussoccer.com.

28. Data on 7-iron distances from Josh Tabor and Christine Franklin, *Statistical Reasoning in Sports*, 2nd ed. (New York, NY: Bedford, Freeman and Worth Publishers, 2019).

29. Data provided by David S. Moore, Purdue University (emeritus).

30. Instagram data collected by Luke Wilcox.

31. Data provided by Dr. Tim Brown.

32. From the *Electronic Encyclopedia of Statistics Examples and Exercises* (*EESEE*) story, "Acorn Size and Oak Tree Range."

33. Sidney Crosby data from www.hockey-reference.com.

34. wwwn.cdc.gov/nchs/nhanes/continuousnhanes/default.aspx?BeginYear=2017.

35. D. B. Lindenmayer, K. L. Viggers, R. B. Cunningham, and C. F. Donnelly, "Morphological Variation Among Columns of the Mountain Brushtail Possum, *Trichosurus caninus Ogilby* (Phalangeridae: Marsupiala)," *Australian Journal of Zoology* 43 (1995): 449–458.

36. Graph from a story by Richard Florida at www.citylab.com, based on Carole Turley Voulgaris, Michael J. Smart, and Brian D. Taylor, "Tired of Commuting? Relationships Among Journeys to School, Sleep, and Exercise Among American Teenagers," *Journal of Planning Education and Research* 39, no. 2 (August 2017): 142–154.

37. Apple stock price data obtained from the volatility calculator at www.buyupside.com on April 6, 2021.

38. Adapted from an idea suggested by Allan Rossman in his blog, "Ask Good Questions," at www.askgoodquestions.blog on August 12, 2019.

39. 2019 CIRP Freshman Survey; data from the Higher Education Research Institute, *The American Freshman: National Norms Fall 2019*, January 2020.

40. NBA scoring data from www.basketball-reference.com.

41. Burger King nutrition data accessed April 7, 2021, at www.bk.com/nutrition-explorer.

42. Data from a student project provided by Josh Tabor.

43. NBA scoring data from www.basketball-reference.com.

44. From the American Community Survey, at the Census Bureau website, www.census.gov. The data are a subsample of the individuals in the ACS New York sample who had travel times greater than zero.

45. Data provided by Dr. Tim Brown.

46. *Model Year 2020 Fuel Economy Guide,* from the website www.fueleconomy.gov.

47. Data provided by Josh Tabor.

48. Copper data from a Hudbay news release on March 29, 2021, which was highlighted in Tony Davis, "Hudbay Says It Found Copper That Could Result in Open Pit Mines on Santa Ritas' West Side," *Arizona Daily Star,* April 1, 2021.

49. Claim about number of texts per day from www.textrequest.com.

50. K. E. Flegal and M. C. Anderson, "Overthinking Skilled Motor Performance: Or Why Those Who Teach Can't Do," *Psychonomic Bulletin & Review* 15, no. 5 (2008): 927–932, doi: 10.3758/PBR.15.5.927. Thanks to Kristin Flegal for sharing the data from this study.

51. Nutrition data on McDonald's sandwiches obtained from www.mcdonalds.com/us/en-us/about-our-food/nutrition-calculator.html.

52. Joshua E. Stubbs and Toby C.B. Stubbs, "How to Lose Weight Well, According to *How to Lose Weight Well*," *Significance Magazine*, February 2021.

53. Freezer data from *Consumer Reports* magazine, May 2010.
54. Tablet ratings from www.consumerreports.org/cro/electronics-computers/computers-internet/tablets/tablet-ratings/ratings-overview/selector.htm.
55. The original paper is T. M. Amabile, "Motivation and Creativity: Effects of Motivational Orientation on Creative Writers," *Journal of Personality and Social Psychology* 48, no. 2 (1985): 393–399. The data for this exercise came from Fred L. Ramsey and Daniel W. Schafer, *The Statistical Sleuth*, 3rd ed. (Boston, MA: Brooks/Cole Cengage Learning, 2013).
56. We obtained the tire pressure loss data from the *Consumer Reports* website, www.consumerreports.org/tire-buying-maintenance/should-you-use-nitrogen-in-car-tires.
57. Data from a student project in fall 2014 provided by Doug Tyson.
58. Data on social media use by the 116th Congress from www.pewresearch.org/fact-tank/2021/01/25/though-not-especially-productive-in-passing-bills-the-116th-congress-set-new-marks-for-social-media-use/.
59. Air quality data for 2020 in the United States by county from aqs.epa.gov/aqsweb/airdata/download_files.html.
60. Data from the most recent Annual Demographic Supplement can be found at www.census.gov/cps.
61. These data are a random sample from the values given in R. Sokal and J. Rohlf, *Introduction to Biostatistics* (Mineola, NY: Dover, 2009).
62. Cherry blossom data from Yasuyuki Aono of Osaka Prefecture University, accessed at atmenv.envi.osakafu-u.ac.jp/aono/kyophenotemp4/.
63. www.prweek.com/article/1247089/paul-holmes-aarp-fooling-itself-misrepresenting-facts-gop-medicare-bill-members.
64. Data from the U.S. Census Bureau, Population Estimates Program, 2020.
65. Data obtained from www.flintwaterstudy.org.
66. National percentile rank tables from Assessment Technologies Institute, Inc. and National League for Nursing.
67. Values for the mean and standard deviation obtained from the Centers for Disease Control and Prevention's Growth Charts at www.cdc.gov.
68. Data on number of wins for each MLB team from www.baseball-reference.com.
69. The finch data came from Fred L. Ramsey and Daniel W. Schafer, *The Statistical Sleuth*, 3rd ed. (Boston, MA: Brooks/Cole Cengage Learning, 2013).
70. Median household incomes in 2019 from cps.ipums.org/cps/.
71. SAT summary statistics from reports.collegeboard.org/sat-suite-program-results/data-archive. ACT summary statistics from www.act.org.
72. Gymnastics score data from www.olympedia.org.
73. Stephen Jay Gould, "Entropic Homogeneity Isn't Why No One Hits .400 Anymore," *Discover* (August 1986): 60–66. Gould does not standardize but gives a speculative discussion instead.
74. Data provided by the team manager for the youth football team of one of Mr. Starneses' grandsons.
75. Data on birth rates obtained from www.indexmundi.com, which cites its source as the CIA World Factbook, available online at www.cia.gov/the-world-factbook/.
76. Beer and wine data from www.cspinet.org/article/which-alcoholic-beverages-have-most-or-least-calories.
77. Data on social media use by the 116th Congress from www.pewresearch.org/fact-tank/2021/01/25/though-not-especially-productive-in-passing-bills-the-116th-congress-set-new-marks-for-social-media-use/.
78. Information on bone density in the reference populations was found at www.courses.washington.edu/bonephys/opbmd.html.
79. Niels Juel-Nielsen, *Individual and Environment: Monozygotic Twins Reared Apart* (New York, NY: International Universities Press, 1980).
80. Data from a student project in Josh Tabor's statistics class.
81. Data on used car prices from autotrader.com. We searched for F-150 4×4's on sale within 50 miles of College Station, Texas.
82. You can also try the *Guess the Correlation* applet at www.rossmanchance.com/applets.
83. Data on boat registrations from www.flhsmv.gov/. Data on manatee deaths from myfwc.com/research/manatee/rescue-mortality-response/statistics/mortality/.
84. Data from Josh Tabor's experiment.
85. NBA data for the 2018 season from www.basketball-reference.com.
86. Based on a graph found at plotly.com.
87. Franz Messerli, "Chocolate Consumption, Cognitive Function, and Nobel Laureates," *New England Journal of Medicine* 367 (2012): 1562–1564.
88. Samuel Karelitz et al., "Relation of Crying Activity in Early Infancy to Speech and Intellectual Development at Age Three Years," *Child Development* 35 (1964): 769–777.
89. NBA data from www.basketball-reference.com.
90. From a graph in Bernt-Erik Saether, Steiner Engen, and Erik Mattysen, "Demographic Characteristics and Population Dynamical Patterns of Solitary Birds," *Science* 295 (2002): 2070–2073.
91. Chicago weather data from www.ncei.noaa.gov.
92. Example of a spurious correlation from www.tylervigen.com.
93. Example of a spurious correlation from www.tylervigen.com.
94. Data on wins and attendance for MLB teams from www.baseball-reference.com.
95. Data on GDP and math scores from www.gapminder.com.
96. M. Seddigh and G.D. Jolliff, "Light Intensity Effects on Meadowfoam Growth and Flowering," *Crop Science* 34 (1994): 497–503.
97. Data from a final project in Josh Tabor's introductory statistics class.
98. Scott DeCarlo, Michael Schubach, and Vladimir Naumovski, "A Decade of New Issues," *Forbes*, March 5, 2001, www.forbes.com.
99. Data on used car prices from autotrader.com. We searched for F-150 4×4's on sale within 50 miles of College Station, Texas.
100. Data from James R. Jordan.
101. Based on article by Matthew Russell, "Do Forests Have the Capacity for 1 Trillion Extra Trees?", *Significance Magazine*, December 2020.
102. Data from the roller coaster database: www.rcdb.com.
103. Data from www.weatherbase.com.
104. M. A. Houck et al., "Allometric Scaling in the Earliest Fossil Bird, *Archaeopteryx lithographica*," *Science* 247 (1990): 195–198. The authors conclude from a variety of evidence that all specimens represent the same species.
105. Cricket chirp data from mathbits.com/MathBits/TISection/Statistics2/linearREAL.html.
106. N. R. Draper and J. A. John, "Influential Observations and Outliers in Regression," *Technometrics* 23 (1981): 21–26.

107. G. A. Sacher and E. F. Staffelt, "Relation of Gestation Time to Brain Weight for Placental Mammals: Implications for the Theory of Vertebrate Growth," *American Naturalist* 108 (1974): 593–613. We found the data in Fred L. Ramsey and Daniel W. Schafer, *The Statistical Sleuth: A Course in Methods of Data Analysis* (Belmont, CA: Duxbury, 1997), 228.

108. Stand mixer data from *Consumer Reports,* November 2005.

109. Candy data from *Nutrition Action,* December 2009.

110. Based on article by Matthew Russell, "Do Forests Have the Capacity for 1 Trillion Extra Trees?", *Significance Magazine,* December 2020.

111. Beer and wine data from www.cspinet.org/article/which-alcoholic-beverages-have-most-or-least-calories.

112. Data from Defenders of Wildlife, *Is Our Tuna Family-Safe?*, 2006.

113. pediatrics.aappublications.org/content/early/2011/09/08/peds.2010-1919.

114. Water supply and snowfall data set provided by Patti Collings of Brigham Young University. Patti used this data set as the basis of an example she created for the Statistically Speaking video series, developed in partnership by Macmillan Publishing and Coast Learning Systems. Patti obtained the data from the Colorado Climate Center, climate.colostate.edu.

115. Data from a student project by Alex and Tempe in Mr. Starnes's introductory statistics class.

116. From the EESEE story, "Is It Tough to Crawl in March?"

117. Data set shared at statcrunch.com by Amanda Moulton, College of the Canyons.

118. M. S. Palmeirim, U. A. Mohammed, A. Ross, S. M. Ame, S. M. Ali, and J. Keiser, "Evaluation of Two Communication Tools, Slideshow and Theater, to Improve Participants' Understanding of a Clinical Trial in the Informed Consent Procedure on Pemba Island, Tanzania," *PLoS Neglected Tropical Diseases* 15, no. 5 (2021): e0009409, doi.org/10.1371/journal.pntd.0009409. We thank Jennifer Keiser and Marta Palmeirim for sharing the data with us.

119. Daniel I. Bolnick and Thomas J. Near, "Tempo of Hybrid Inviability in Centrarchid Fishes (teleostei: centrarchidae)," *Evolution* 59, no. 8 (2005): 1754–1767.

120. Danielle Ivory, Lauren Leatherby, and Robert Gebeloff, "Least Vaccinated U.S. Counties Have Something in Common: Trump Voters," *New York Times* April 17, 2021.

CHAPTER 4

1. Information obtained from the document "NCAA Drug-Testing Program 2020-21."

2. Hoang Nguyen, "One in Three Leave the Tap Running While Brushing Their Teeth," June 4, 2018, yougov.com.

3. Adeel Hassan and Audrey Carlsen, "How 'Crazy Rich' Asians Have Led to the Largest Income Gap in the U.S.," *New York Times,* August 17, 2018.

4. We obtained the color distribution for M&M'S® Milk Chocolate Candies from blogs.sas.com/content/iml/2017/02/20/proportion-of-colors-mandms.html for bags packaged at Mars, Incorporated's factory in Cleveland, Tennessee.

5. Distribution of blood types from the American Red Cross website, www.redcrossblood.org/donate-blood/blood-types.html.

6. Data for 2016 from the website of Statistics Canada, www.statcan.gc.ca.

7. U.S. Census Bureau, Current Population Survey, March and Annual Social and Economic Supplements, released December 2020.

8. Data for 2017 from the National Household Travel Survey website, nhts.ornl.gov.

9. Data on educational attainment in 2019 from nces.ed.gov/programs/coe/indicator_caa.asp.

10. The two-way table was based on Guicheng Zhang, Jeffery Spickett, Krassi Rumchev, Andy H. Lee, and Stephen Stick, "Snoring in Primary School Children and Domestic Environment: A Perth School Based Study," *Respiratory Research* 5, no. 1 (2004): 19, published online November 4, 2004.

11. Brooke Auxier and Monica Anderson, "Social Media Use in 2021," Pew Research Center, April 2021.

12. Exercise modified based on data provided in Gail Burrill, Two-*Way Tables: Introducing Probability Using Real Data*, paper presented at the Mathematics Education into the Twenty-First Century Project, Czech Republic, September 2003. Burrill cites as her source H. Kranendonk, P. Hopfensperger, and R. Scheaffer, *Exploring Probability* (New York, NY: Dale Seymour Publications, 1999).

13. R. Shine et al., "The Influence of Nest Temperatures and Maternal Brooding on Hatchling Phenotypes in Water Pythons," *Ecology* 78 (1997): 1713–1721.

14. Results based on a survey reported at statista.com.

15. Pew Research Center, January 2019 Core Trends Survey.

16. From the EESEE story, "What Makes a Pre-teen Popular?"

17. Pierre J. Meunier et al., "The Effects of Strontium Ranelate on the Risk of Vertebral Fracture in Women with Postmenopausal Osteoporosis," *New England Journal of Medicine* 350 (2004): 459–468.

18. Based on a graph in the article by Paul Hemez and Chanell Washington, "Percentage and Number of Children Living with Two Parents Has Dropped Since 1968," April 12, 2021, www.census.gov.

19. Data from the Yellowstone National Park Winter Use Survey, accessed at www.nps.gov/yell/learn/management/winter-use-management.htm.

20. Elizabeth F. Loftus and John C. Palmer, "Reconstruction of Automobile Destruction: An Example of the Interaction Between Language and Memory," *Journal of Verbal Learning and Verbal Behavior* 13 (1974): 585–589, www.researchgate.net/publication/222307973_Reconstruction_of_Automobile_Destruction_An_Example_of_the_Interaction_Between_Language_and_Memory.

21. Squirrel data from data.cityofnewyork.us/Environment/2018-Central-Park-Squirrel-Census-Squirrel-Data/vfnx-vebw/data.

22. R. Shine et al., "The Influence of Nest Temperatures and Maternal Brooding on Hatchling Phenotypes in Water Pythons," *Ecology* 78 (1997): 1713–1721.

23. Data on David Ortiz from www.baseball-reference.com.

24. Pew Research Center, January 2019 Core Trends Survey.

25. From the EESEE story, "What Makes a Pre-teen Popular?"

26. The data on superpower preference were obtained using the Random Sampler tool at www.amstat.org/censusatschool.

27. www.ncaa.org/about/resources/research/probability-competing-beyond-high-school

28. Data on smartphone use and broadband access from www.pewresearch.org/internet/fact-sheet/internet-broadband/ and www.pewresearch.org/internet/fact-sheet/mobile/, accessed May 16, 2021.

29. www.prnewswire.com/news-releases/powered-readers-americans-who-read-more-electronically-read-more-period-255609091.html
30. Thanks to Gerd Gigerenzer for suggesting this approach.
31. We got these data from the Energy Information Administration website, www.eia.gov.
32. From the National Institutes of Health's National Digestive Diseases Information Clearinghouse, www.niddk.nih.gov/health-information/digestive-diseases.
33. Data on Serena Williams's serve percentages from www.wtatennis.com.
34. Probabilities from trials with 2897 people known to be free of HIV antibodies and 673 people known to be infected are reported in J. Richard George, *Alternative Specimen Sources: Methods for Confirming Positives*, 1998 Conference on the Laboratory Science of HIV, found online at Centers for Disease Control and Prevention website, www.cdc.gov.
35. This exercise was inspired by the report at www.cbsnews.com/news/drug-tests-not-immune-from-false-positives.
36. Thomas F. Imperiale et al., "Multitarget Stool DNA Testing for Colorectal-Cancer Screening," *New England Journal of Medicine* 370 (2014): 1287–1297.
37. Todd Haugh and Suneal Bedi, "Just Because You Test Positive for Antibodies Doesn't Mean You Have Them," *New York Times*, May 13, 2020.
38. The probabilities given are realistic, according to the fundraising firm SCM Associates, scmassoc.com.
39. *Arizona Daily Star*, November 13, 2014, tucson.com/news/science/environment/tucson-country-club-area-uses-mostwater/article_fd2acb0d-6663-52ae-aa81-9e00bc4c7a76.html.
40. *Model Year 2020 Fuel Economy Guide*, www.fueleconomy.gov.
41. This is one of several tests discussed in Bernard M. Branson, *Rapid HIV Testing: 2005 Update*, presentation by Centers for Disease Control and Prevention, www.cdc.gov. The Malawi clinic result is reported by Bernard M. Branson, "Point-of-Care Rapid Tests for HIV Antibody," *Journal of Laboratory Medicine* 27 (2003): 288–295.
42. Data on school enrollment from U.S. Census Bureau, Current Population Survey, School Enrollment Supplement, October 2019. Data on population aged 55 and older from U.S. Census Bureau, Current Population Survey, Annual Social and Economic Supplement, 2019.
43. The probabilities in this exercise are taken from Tommy Bennett, "Expanded Horizons: Perfection," June 8, 2010, www.baseballprospectus.com.
44. Blood type information from www.redcrossblood.org/donate-blood/blood-types.html.
45. Information about Canada's Lotto 6/49 game from www.olg.ca/en/lottery/play-lotto-649-encore/odds-and-payouts.html.
46. Robert P. Dellavalle et al., "Going, Going, Gone: Lost Internet References," *Science* 302 (2003): 787–788.
47. Thanks to Corey Andreasen for suggesting the idea for this exercise.
48. Thanks to Michael Legacy for suggesting this exercise.
49. Research performed by Matthew O'Brien and Diego Bustos Arrieta, Canyon del Oro High School, under the supervision of Josh Tabor.
50. Nutritional information about bagels from www.einsteinbros.com.
51. Data on workplace injuries from Bureau of Labor Statistics, www.bls.gov.
52. This scenario is based on actual events reported by Mr. Starnes's youngest son, who at the time was one of the junior pilots. Details have been changed to protect confidential information, but the calculated probability is consistent with the original scenario.
53. Home run data from www.baseball-reference.com.
54. The table in this exercise was constructed using the search function at the GSS archive, sda.berkeley.edu/archive.htm.
55. The probability distribution was based on data found at www.statista.com/statistics/276506/change-in-us-car-demand-by-vehicle-type.
56. The table in this exercise was constructed using the search function at the GSS archive, sda.berkeley.edu/archive.htm.
57. www.fda.gov/news-events/press-announcements/coronavirus-covid-19-update-fda-authorizes-antigen-test-first-over-counter-fully-home-diagnostic
58. Results of the Harris Poll reported at www.statista.com/statistics/297156/united-states-common-superstitions-believe/.

CHAPTER 5

1. U.S. Supreme Court, *Strauder v. West Virginia*, 100 U.S. 303 (1879), supreme.justia.com/cases/federal/us/100/303/case.html.
2. U.S. Supreme Court, *Berghuis v. Smith*, Docket No. 08-1402, www.law.cornell.edu/supct/html/08-1402.ZO.html.
3. Thanks to Doug Tyson for suggesting this activity.
4. The Apgar score data came from National Center for Health Statistics, *Monthly Vital Statistics Reports* 30, no. 1 (May 6, 1981): suppl.
5. You can find a mathematical explanation of Benford's law in Ted Hill, "The First-Digit Phenomenon," *American Scientist* 86 (1996): 358–363; and Ted Hill, "The Difficulty of Faking Data," *Chance* 12, no. 3 (1999): 27–31. Applications in fraud detection are discussed in the second paper by Hill and in Mark A. Nigrini, "I've Got Your Number," *Journal of Accountancy* (May 1999), www.journalofaccountancy.com/issues/1999/may/nigrini.
6. The National Longitudinal Study of Adolescent Health interviewed a stratified random sample of 27,000 adolescents, then reinterviewed many of the subjects 6 years later, when most were aged 19 to 25. These data are from the Wave III reinterviews in 2000 and 2001, found at the website of the Carolina Population Center, www.cpc.unc.edu.
7. Probability distribution based on a sample of students from the U.S. Census at School database, www.amstat.org/censusatschool.
8. Aaron Smith, *The Best and Worst of Mobile Connectivity* (Pew Research Center, November 30, 2012), www.pewresearch.org.
9. M. Bravo, J. Montero, J. J. Bravo, P., Baca, J. C. Llodra, "Sealant and Fluoride Varnish in Caries: A Randomized Trial," *Journal of Dental Research* 84, no. 12 (2005): 1138–1143. doi: 10.1177/154405910508401209. PMID: 16304443.
10. You can find a mathematical explanation of Benford's law in Ted Hill, "The First-Digit Phenomenon," *American Scientist* 86 (1996): 358–363; and Ted Hill, "The Difficulty of Faking Data," *Chance* 12, no. 3 (1999): 27–31. Applications in fraud detection are discussed in the second paper by Hill and in Mark

A. Nigrini, "I've Got Your Number," *Journal of Accountancy* (May 1999), www.journalofaccountancy.com/issues/1999/may/nigrini.

11. The National Longitudinal Study of Adolescent Health interviewed a stratified random sample of 27,000 adolescents, then reinterviewed many of the subjects 6 years later, when most were aged 19 to 25. These data are from the Wave III reinterviews in 2000 and 2001, found at the website of the Carolina Population Center, www.cpc.unc.edu.

12. Probability distribution based on a sample of students from the U.S. Census at School database, www.amstat.org/censusatschool.

13. These exercises are based on an activity shared by Allan Rossman in his Ask Good Questions blog, askgoodquestions.blog/2020/11/02/70-batch-testing-part-2.

14. Emily Oster, "It's Hard to Know Where Gluten Sensitivity Stops and the Placebo Effect Begins," February 11, 2015, www.fivethirtyeight.com.

15. www.nielsen.com/us/en/press-releases/2020/super-bowl-liv-draws-nearly-100-million-tv-viewers-44-million-social-media-interactions

16. Ed O'Brien and Phoebe C. Ellsworth, "Saving the Last for Best: A Positivity Bias for End Experiences," *Psychological Science* 23, no. 2 (2011): 163–165.

17. Data from Emily Clymer.

18. We got the 9% figure from Pew Research Center, "Assessing the Representativeness of Public Opinion Surveys," May 15, 2012.

19. Office of Technology Assessment, *Scientific Validity of Polygraph Testing: A Research Review and Evaluation* (Washington, DC: Government Printing Office, 1983).

20. www.unitypoint.org/cedarrapids/maternity.aspx#:~:text=Last%20year%20nearly%202%2C400%20babies,their%20birthday%20at%20St.%20Luke's

21. www.math.hawaii.edu/~ramsey/Hurricane.html

22. www.worlddata.info/asia/japan/earthquakes.php

23. Mark R. Stromberg, "Subsurface Burrow Connections and Entrance Spatial Pattern of Prairie Dogs," *Southwestern Naturalist* 23, no. 2 (1978): 173–180.

24. Rate of collision claims from bankrate.com.

25. cosmosmagazine.com/space/earth-hit-by-17-meteors-a-day/#:~:text=Every%20year%2C%20the%20Earth%20is,places%20that%20catch%20more%20attention

26. www.futurity.org/african-forest-elephants-population-size-2352562/

27. www.worlddata.info/asia/japan/earthquakes.php

28. U.S. Supreme Court, *Berghuis v. Smith*, Docket No. 08-1402, www.law.cornell.edu/supct/html/08-1402.ZO.html.

29. We obtained the data on number of televisions per household from www.eia.gov/consumption/residential/index.php.

CHAPTER 6

1. www.nytimes.com/interactive/2015/02/17/upshot/what-do-people-actually-order-at-chipotle.html

2. D. B. Lindenmayer, K. L. Viggers, R. B. Cunningham, and C. F. Donnelly, "Morphological Variation Among Columns of the Mountain Brushtail Possum, *Trichosurus caninus Ogilby* (Phalangeridae: Marsupiala)," *Australian Journal of Zoology* 43 (1995): 449–458.

3. Ryan Wells and Allie Quiroz, Canyon del Oro High School.

4. www.pro-football-reference.com

5. www.lpga.com/statistics/driving/average-driving-distance?year=2021

6. www.ets.org

7. www.pro-football-reference.com/years/2020/receiving.htm

8. www.pewresearch.org/internet/wp-content/uploads/sites/9/2015/12/PI_2015-12-15_gaming-and-gamers_FINAL.pdf

9. stats.nba.com/players/bio

10. We found the information on birth weights of Norwegian children on the National Institute of Environmental Health Sciences website. The relevant article can be accessed here: www.ncbi.nlm.nih.gov/pubmed/1536353.

11. Braedon Day, Canyon del Oro High School.

12. www.fueleconomy.gov/feg/download.shtml

13. www.census.gov/data/tables/2019/demo/hispanic-origin/2019-cps.html

14. www.fueleconomy.gov/feg/download.shtml

15. www.wri.org

16. The data set was constructed based on information provided in P. D. Wood et al., "Plasma Lipoprotein Distributions in Male and Female Runners," in P. Milvey (Ed.), *The Marathon: Physiological, Medical, Epidemiological, and Psychological Studies* (New York, NY: New York Academy of Sciences, 1977).

17. S. M. Stigler, "Do Robust Estimators Work with Real Data?" *Annals of Statistics* 5 (1977): 1055–1078.

18. www.ofa.org/diseases/breed-statistics

19. www.coinnews.net/2021/01/22/u-s-mint-produces-14-77-billion-coins-for-circulation-in-2020/

20. www.nber.org/papers/w20689

21. www.census.gov/data/tables/time-series/demo/educational-attainment/cps-historical-time-series.html

22. www.pewresearch.org/internet/2021/04/07/social-media-use-in-2021/

23. www.insidehighered.com/views/2021/01/19/national-opinion-survey-shows-growing-public-support-helping-students-debt-opinion

24. www.mayoclinic.org/diseases-conditions/coronavirus/in-depth/herd-immunity-and-coronavirus/art-20486808

25. en.wikipedia.org/wiki/Giraffe

26. www.insidehighered.com/views/2021/01/19/national-opinion-survey-shows-growing-public-support-helping-students-debt-opinion

27. www.ultimatemotorcycling.com/2019/02/07/motorcycle-statistics-in-america-demographics-change-for-2018

28. www.pewresearch.org/fact-tank/2019/07/01/about-one-in-six-u-s-teachers-work-second-jobs-and-not-just-in-the-summer

29. www.kickstarter.com/help/stats

30. en.wikipedia.org/wiki/Language_demographics_of_Quebec

31. www.pewresearch.org/fact-tank/2017/08/24/jury-duty-is-rare-but-most-americans-see-it-as-part-of-good-citizenship

32. www.thelancet.com/journals/lancet/article/PIIS0140-6736(21)00943-0/fulltext

33. A. J. Wilcox and I. T. Russell, "Birthweight and Perinatal Mortality: I. On the Frequency Distribution of Birthweight," *International Journal of Epidemiology* 12 (1983): 314–318.

34. Based on a figure in Peter R. Grant, *Ecology and Evolution of Darwin's Finches* (Princeton, NJ: Princeton University Press, 1986).

35. David J. LeBlanc, Michael Sivak, and Scott Bogard, *Using Naturalistic Driving Data to Assess Variations in Fuel Efficiency Among Individual Drivers*, University of Michigan Transportation Research Institute, deepblue.lib.umich.edu/bitstream/handle/2027.42/78449/102705.pdf.
36. www.sleepfoundation.org/sites/default/files/inline-files/2014-NSF-Sleep-in-America-poll-summary-of-findings---FINAL-Updated-3-26-14-.pdf
37. www.nytimes.com/interactive/2015/02/17/upshot/what-do-people-actually-order-at-chipotle.html
38. www.dfw.state.or.us/MRP
39. J. D. Stewart et al., "Decreasing Body Lengths in North Atlantic Right Whales," *Current Biology* 31 (2021): 1–6, doi.org/10.1016/j.cub.2021.04.067.
40. pudding.cool/2019/10/laugh/
41. A. C. Gielen et al., "National Survey of Home Injuries During the Time of COVID-19: Who Is at Risk?" *Injury Epidemiology* 7 (2020): 63.
42. Based on an exercise from Ignacio Bello, Anton Kaul, and Jack R. Britton, *Topics in Contemporary Mathematics*, 10th ed. (Boston, MA: Cengage Learning, 2014).
43. www.highspeedinternet.com/resources/how-much-should-i-be-paying-for-high-speed-internet-resource

CHAPTER 7

1. Haipeng Shen, *Nonparametric Regression for Problems Involving Lognormal Distributions*, PhD thesis, University of Pennsylvania, 2003. Thanks to Haipeng Shen and Larry Brown for sharing the data.
2. Thanks to Floyd Bullard for sharing the idea for this activity.
3. www.gallup.com/poll/181217/americans-healthcare-low-wages-top-financial-problems.aspx
4. wwwn.cdc.gov/nchs/nhanes/continuousnhanes/default.aspx?BeginYear=2017. Data based on the second of three readings of systolic blood pressure.
5. www.cdc.gov/bloodpressure/about.htm
6. www.smithsonianmag.com/innovation/how-much-do-americans-know-about-science-27747364/
7. www.smithsonianmag.com/innovation/how-much-do-americans-know-about-science-27747364/
8. Robert Langkjaer-Bain, "The Murky Tale of Flint's Deceptive Water Data," *Significance Magazine*, April 2017. rss.onlinelibrary.wiley.com/doi/full/10.1111/j.1740-9713.2017.01016.x
9. www.census.gov/content/dam/Census/library/publications/2020/acs/acsbr20-03.pdf
10. www.atsdr.cdc.gov/csem/lead/docs/CSEM-Lead_toxicity_508.pdf
11. Morning Consult National Tracking Poll #200158, January 23–24, 2020.
12. Morning Consult National Tracking Poll #191255, December 13–15, 2019.
13. M. A. Wale, S. D. Simpson, and A. N. Radford, "Size-Dependent Physiological Responses of Shore Crabs to Single and Repeated Playback of Ship Noise," *Biology Letters* 9 (2013): 20121194. dx.doi.org/10.1098/rsbl.2012.1194.
14. www.basketball-reference.com
15. www.gallup.com/poll/181217/americans-healthcare-low-wages-top-financial-problems.aspx
16. NCHS Data Brief No. 322, October 2018.
17. Sungwoo Lim, Brett Wyker, Katherine Bartley, and Donna Eisenhower, "Measurement Error in Self-Reported Physical Activity Levels in New York City: Assessment and Correction," *American Journal of Epidemiology* 181, no. 9 (2015): 648–655, doi.org/10.1093/aje/kwu470. Confidence interval was inferred based on summary statistics in the study. Results from a follow-up study that had respondents use a fitness tracker showed that respondents overestimated their activity.
18. news.gallup.com/poll/284009/library-visits-outpaced-trips-movies-2019.aspx. Standard deviation estimated from summary statistics provided in supplement here: news.gallup.com/file/poll/284015/200123LeisureActivities.pdf.
19. Pew Research Center, "The Virtues and Downsides of Online Dating," February 6, 2020.
20. www.gallup.com
21. Data from 2013 Current Population Survey, found at www.eeps.com/zoo/acs/source/index.php.
22. Data from 2013 Current Population Survey, found at www.eeps.com/zoo/acs/source/index.php.
23. See also the *Simulating Confidence Intervals for a Population Parameter* applet at www.rossmanchance.com/applets.
24. fivethirtyeight.com/features/are-states-with-lower-income-tax-rates-better-at-winning-championships/
25. nces.ed.gov/programs/digest/d18/tables/dt18_104.10.asp
26. cdn.annenbergpublicpolicycenter.org/wp-content/uploads/Civics-survey-press-release-09-17-2014-for-PR-Newswire.pdf
27. This data set was obtained by taking a random sample of 3000 households from the March 2019 American Community Survey data set.
28. Una Osili, Jon Bergdoll, Andrea Pactor, Jacqueline Ackerman, and Peter Houston, "Charitable Crowdfunding: Who Gives, to What, and Why?" *ScholarWorks*, scholarworks.iupui.edu/handle/1805/25515.
29. aqs.epa.gov/aqsweb/airdata/download_files.html
30. www.nsf.org/blog/consumer/college-student-germ-study
31. P. Jendrny, C. Schulz, F. Twele, et al., "Scent Dog Identification of Samples from COVID-19 Patients: A Pilot Study," *BMC Infectious Diseases* 20, no. 536 (2020), doi.org/10.1186/s12879-020-05281-3.
32. morningconsult.com/2020/12/22/spoiler-culture-reality-tv-bachelorette
33. www.usatoday.com/picture-gallery/news/2015/04/07/usa-today-snapshots/6340793/
34. Michele L. Head, *Examining College Students' Ethical Values*, Consumer Science and Retailing honors project, Purdue University, 2003.
35. "Poll: Men, Women at Odds on Sexual Equality," Associated Press dispatch appearing in the *Lafayette (Ind.) Journal and Courier*, October 20, 1997.
36. www.indexmundi.com/facts/united-states/quick-facts/all-states/percent-of-population-under-18#table
37. sleepfoundation.org/sleep-polls-data/2015-sleep-and-pain
38. www.pewresearch.org/wp-content/uploads/2020/12/Phones-and-scams-methods-and-topline.pdf
39. www.forbes.com/sites/stevensavage/2016/02/08/inconvenient-truth-there-are-pesticide-residues-on-organics/
40. news.gallup.com/poll/163238/americans-reject-size-limit-soft-drinks-restaurants.aspx
41. poll.qu.edu/Poll-Release-Legacy?releaseid=2176
42. www.cnn.com

43. Data set randomly selected from full data set available at data.cityofnewyork.us/Environment/2015-Street-Tree-Census-Tree-Data/pi5s-9p35.
44. Data from *STAT2*, 2nd ed., 414.
45. Based on information in "NCAA 2003 National Study of Collegiate Sports Wagering and Associated Health Risks," www.ncaa.org.
46. V. Rideout and M. B. Robb, *The Common Sense Census: Media Use by Tweens and Teens, 2019* (San Francisco, CA: Common Sense Media, 2019).
47. www.nytimes.com/interactive/2019/02/13/upshot/engagement-rings-cost-two-weeks-pay.html
48. Mean and standard deviation estimated from graph at www.nytimes.com/interactive/2019/02/13/upshot/engagement-rings-cost-two-weeks-pay.html.
49. www.census.gov/library/stories/2019/09/us-median-household-income-up-in-2018-from-2017.html
50. news.gallup.com/poll/284009/library-visits-outpaced-trips-movies-2019.aspx
51. D. B. Lindenmayer, K. L. Viggers, R. B. Cunningham, and C. F. Donnelly, "Morphological Variation Among Columns of the Mountain Brushtail Possum, *Trichosurus caninus* Ogilby (Phalangeridae: Marsupiala)," *Australian Journal of Zoology* 43 (1995): 449–458.
52. Corey Heid, "Tootsie Pops: How Many Licks to the Chocolate?" *Significance Magazine* (October 2013): 47.
53. Alan S. Banks et al., "Juvenile hallux abducto valgus association with metatarsus adductus," *Journal of the American Podiatric Medical Association, 84* (1994): 219–224.
54. Nick Blanchard, Canyon del Oro High School.
55. M. Ann Laskey et al., "Bone Changes After 3 Mo of Lactation: Influence of Calcium Intake, Breast-Milk Output, and Vitamin D–Receptor Genotype," *American Journal of Clinical Nutrition, 67* (1998): 685–692.
56. news.gallup.com/poll/182261/four-americans-look-forward-checking-mail.aspx
57. fivethirtyeight.com/features/how-much-football-is-even-in-a-football-broadcast/. Special thanks to Katherine Rowe for providing the raw data.
58. Harry B. Meyers, *Investigations of the Life History of the Velvetleaf Seed Beetle,* Althaeus folkertsi Kingsolver, MS thesis, Purdue University, 1996.
59. data.cityofnewyork.us/City-Government/Citywide-Payroll-Data-Fiscal-Year/k397-673e
60. W. Alvarez and F. Asaro, "What Caused the Mass Extinction? An Extraterrestrial Impact," *Scientific American* 263 (4): 76–84; and E. Courtillot, "What Caused the Mass Extinction? A Volcanic Eruption," *Scientific American* 263 (4): 85–93.
61. These data are from *Results Report on the Vitamin C Pilot Program,* prepared by SUSTAIN (Sharing United States Technology to Aid in the Improvement of Nutrition) for the U.S. Agency for International Development. The report was used by the Committee on International Nutrition of the National Academy of Sciences/Institute of Medicine (NAS/IOM) to make recommendations on whether or not the vitamin C content of food commodities used in U.S. food aid programs should be increased. The program was directed by Peter Ranum and Françoise Chomé.
62. Data provided by Drina Iglesia, Purdue University. The data are part of a larger study reported in D. D. S. Iglesia, E. J. Cragoe, Jr., and J. W. Vanable, "Electric Field Strength and Epithelization in the Newt (*Notophthalmus viridescens*)," *Journal of Experimental Zoology* 274 (1996): 56–62.
63. M. A. Wale, S. D. Simpson, and A. N. Radford, "Size-Dependent Physiological Responses of Shore Crabs to Single and Repeated Playback of Ship Noise," *Biology Letters* 9 (2013): 20121194. dx.doi.org/10.1098/rsbl.2012.1194.
64. Data are from Phyllis Lee, Stirling University, and are related to P. Lee et al., "Enduring Consequences of Early Experiences: 40-Year Effects on Survival and Success Among African Elephants (*Loxodonta africana*)," *Biology Letters* 9 (2013): 20130011.
65. en.wikipedia.org/wiki/African_elephant#Size
66. Sample from the 2007 American Housing Survey at www.census.gov/programs-surveys/ahs/data/2007/ahs-2007-public-use-file--puf-.html.
67. Data set randomly selected from full data set available at data.cityofnewyork.us/Environment/2015-Street-Tree-Census-Tree-Data/pi5s-9p35.
68. www.ons.gov.uk/employmentandlabourmarket/peopleinwork/employmentandemployeetypes/bulletins/keystatisticsandquickstatisticsforlocalauthoritiesintheunitedkingdom/2013-12-04#national-statistics-socio-economic-classification-ns-sec; www.ons.gov.uk/peoplepopulationandcommunity/healthandsocialcare/healthandlifeexpectancies/bulletins/adultsmokinghabitsingreatbritain/2018
69. R. R. Sokol and F. J. Rohlf, *Biometry,* 2nd ed. (San Francisco: W. H. Freeman, 1981), 368.
70. These data are from *Results Report on the Vitamin C Pilot Program,* prepared by SUSTAIN (Sharing United States Technology to Aid in the Improvement of Nutrition) for the U.S. Agency for International Development. The report was used by the Committee on International Nutrition of the National Academy of Sciences/Institute of Medicine (NAS/IOM) to make recommendations on whether or not the vitamin C content of food commodities used in U.S. food aid programs should be increased. The program was directed by Peter Ranum and Françoise Chomé.
71. Nick Blanchard, Canyon del Oro High School.
72. Mean and standard deviation estimated from graph at www.nytimes.com/interactive/2019/02/13/upshot/engagement-rings-cost-two-weeks-pay.html.
73. Violetta N. Pivtoraiko, Eric E. Abrahamson, Sue E. Leurgans, Steven T. DeKosky, Elliott J. Mufson, and Milos D. Ikonomovic, "Cortical Pyroglutamate Amyloid-β Levels and Cognitive Decline in Alzheimer's Disease," *Neurobiology of Aging* 36 (2015): 12–19, Figure 1, panel d.
74. Raw data courtesy of Melanie L. Shoup and Gordon G. Gallup, Jr., "Men's Faces Convey Information About Their Bodies and Their Behavior: What You See Is What You Get," *Evolutionary Psychology* 6, no. 3 (2008): 469–479.
75. Subset of the data at dasl.datadescription.com/story/diet/, originally distributed with the Data Desk software package.
76. Random sample from a larger data set on the DASL website called CEO Compensation 2014, dasl.datadescription.com/datafile/ceo-compensation-2014.
77. wwwn.cdc.gov/nchs/nhanes/continuousnhanes/default.aspx?BeginYear=2017. Used the second of three readings of systolic blood pressure.
78. Robert Langkjaer-Bain, "The Murky Tale of Flint's Deceptive Water Data," *Significance Magazine,* April 2017. rss.onlinelibrary.wiley.com/doi/full/10.1111/j.1740-9713.2017.01016.x

79. Data originally from Emeritus (and late) Prof. Bob Black from Cornell College, Mount Vernon, Iowa.
80. fivethirtyeight.com/features/how-much-football-is-even-in-a-football-broadcast/. Special thanks to Katherine Rowe for providing the raw data.
81. M. A. Wale, S. D. Simpson, and A. N. Radford, "Size-Dependent Physiological Responses of Shore Crabs to Single and Repeated Playback of Ship Noise," *Biology Letters* 9 (2013): 20121194. dx.doi.org/10.1098/rsbl.2012.1194.
82. D. B. Lindenmayer, K. L. Viggers, R. B. Cunningham, and C. F. Donnelly, "Morphological Variation Among Columns of the Mountain Brushtail Possum, *Trichosurus caninus Ogilby* (Phalangeridae: Marsupiala)," *Australian Journal of Zoology* 43 (1995): 449–458.
83. W. Alvarez and F. Asaro, "What Caused the Mass Extinction? An Extraterrestrial Impact," *Scientific American* 263 (4): 76–84; and E. Courtillot, "What Caused the Mass Extinction? A Volcanic Eruption," *Scientific American* 263 (4): 85–93.
84. Data are from Phyllis Lee, Stirling University, and are related to P. Lee et al., "Enduring Consequences of Early Experiences: 40-Year Effects on Survival and Success Among African Elephants (*Loxodonta africana*)," *Biology Letters* 9 (2013): 20130011.
85. Data downloaded for SPO (South Pole) from the ESRL/GMD data page at www.esrl.noaa.gov/gmd/dv/data/.
86. Marla A. Sole and Sharon L. Weinberg, "What's Brewing? A Statistics Education Discovery Project," *Journal of Statistics Education* 25, no. 3 (2017): 137–144, www.tandfonline.com/doi/full/10.1080/10691898.2017.1395302.
87. Haipeng Shen, *Nonparametric Regression for Problems Involving Lognormal Distributions*, PhD thesis, University of Pennsylvania, 2003. Thanks to Haipeng Shen and Larry Brown for sharing the data.
88. www.washingtonpost.com/page/2010-2019/WashingtonPost/2014/03/23/National-Politics/Polling/question_13300.xml?uuid=07HdFrI_EeO4s0Sx0c1MHw
89. www.gapminder.org/ignorance/gms/ and www.ipsos.com/ipsos-mori/en-uk/mind-gap-ipsos-mori-survey-gapminder
90. realchristmastrees.org/dnn/News-Media/Industry-Statistics/Consumer-Survey
91. news.gallup.com/poll/260129/americans-reject-statehood.aspx
92. Data provided by Benni Delgado, Canyon del Oro High School.
93. www.caranddriver.com/news/a33457915/average-age-vehicles-on-road-12-years/

CHAPTER 8

1. P. A. Mackowiak, S. S. Wasserman, and M. M. Levine, "A Critical Appraisal of 98.6 Degrees F, the Upper Limit of the Normal Body Temperature, and Other Legacies of Carl Reinhold August Wunderlich," *Journal of the American Medical Association* 268 (1992): 1578–1580. We produced this data set with similar properties as the original temperature readings.
2. Center on Budget and Policy Priorities, "Tracking the COVID-19 Recession's Effects on Food, Housing, and Employment Hardships," August 9, 2021.
3. Results of this study were reported in the *San Gabriel Valley Tribune*, February 13, 2003.
4. Kevin M. Simmons, Daniel Sutter, and Roger Pielke, "Normalized Tornado Damage in the United States: 1950–2011," *Environmental Hazards* 12, no. 2 (2013): 132–147, sciencepolicy.colorado.edu/admin/publication_files/2012.31.pdf.
5. Colleen McClain, "Most Americans Don't Answer Cellphone Calls from Unknown Numbers," Pew Research Center, December 14, 2020.
6. Katie Hunt and Jacqueline Howard, "People with Blood Type O May Have Lower Risk of Covid-19 Infection and Severe Illness, Two New Studies Suggest," CNN, October 14, 2020.
7. The idea for this exercise was provided by Michael Legacy and Susan McGann.
8. Morning Consult National Tracking Poll #200473, April 22–24, 2020.
9. Robin Lake and Alvin Makori, "The Digital Divide Among Students During COVID-19: Who Has Access? Who Doesn't?" *The Lens*, June 16, 2020.
10. U.S. Department of Commerce, Census Bureau, Current Population Survey (CPS), October 2017.
11. Data from a project by Daniel Brown and Chad Porter.
12. U.S. Department of Agriculture, Agricultural Research Service, www.ars.usda.gov.
13. J. E. Schulenberg, M. E. Patrick, L. D. Johnston, P. M. O'Malley, J. G. Bachman, and R. A Miech, *Monitoring the Future National Survey Results on Drug Use, 1975–2020: Volume II, College Students and Adults Ages 19–60* (Ann Arbor, MI: Institute for Social Research, University of Michigan, 2021).
14. A. M. Gassó, J. R. Agustina, and E. Goméz-Durán, "Cross-Cultural Differences in Sexting Practices Between American and Spanish University Students," *International Journal of Environmental Research and Public Health* 18 (2021): 2058.
15. E. B. Stolzenberg, M. C. Aragon, E. Romo, V. Couch, D. McLennan, M. K. Eagan, and N. Kang, *The American Freshman: National Norms Fall 2019* (Los Angeles, CA: Higher Education Research Institute, University of California, Los Angeles, 2020).
16. www.newscientist.com/article/dn1748-euro-coin-accused-of-unfair-flipping/
17. This and similar results of Gallup polls are from the Gallup Organization website, www.gallup.com.
18. Andre Perrin and Monica Anderson, "Share of U.S. Adults Using Social Media, Including Facebook, Is Mostly Unchanged Since 2018," Pew Research Center, April 10, 2019.
19. American Community Survey data from www.census.gov/programs-surveys/acs/data.html.
20. wwwn.cdc.gov/nchs/nhanes/continuousnhanes/default.aspx?BeginYear=2017
21. *Model Year 2021 Fuel Economy Guide*, www.fueleconomy.gov.
22. Information about golden hamsters from animaldiversity.org.
23. Data provided by Tim Brown.
24. researchinuserexperience.wordpress.com/2003/08/12/a-comparison-of-two-computer-fonts-serif-versus-ornate-sans-serif/
25. This exercise is based on events that are real. The data and details have been altered to protect the privacy of the individuals involved.
26. Warren E. Leary, "Cell Phones: Questions But No Answers," *New York Times*, October 26, 1999.
27. American Community Survey data from www.census.gov/programs-surveys/acs/data.html.

28. wwwn.cdc.gov/nchs/nhanes/continuousnhanes/default.aspx?BeginYear=2017

29. blogs.edweek.org/edweek/curriculum/2015/03/homework_math_science_study.html, from this study: www.apa.org/pubs/journals/releases/edu-0000032.pdf

30. www.cbsnews.com/news/poll-who-can-get-ahead-in-the-u-s

31. The idea for this exercise was provided by Michael Legacy and Susan McGann.

32. news.gallup.com/poll/4831/americans-confident-safety-nations-food.aspx

33. See, for example, www.heart.org/en/health-topics/high-blood-pressure/the-facts-about-high-blood-pressure/all-about-heart-rate-pulse. Normal range is considered to be 60 to 100 beats/min. Heart rates are approximately normally distributed, so a range of 40 for the distribution leads to a standard deviation of approximately 6.67 beats/min (range is approximately 6 standard deviations).

34. Allen L. Shoemaker, "What's Normal? Temperature, Gender, and Heart Rate," *Journal of Statistics Education* 4, no. 2 (1996): 1–3. The data used here are a random sample of the data associated with the paper.

35. animaldiversity.org/accounts/Mesocricetus_auratus/. The range of the weights is listed as 0.89 ounce, which leads to a standard deviation of 0.15 ounce.

36. Allen L. Shoemaker, "What's Normal? Temperature, Gender, and Heart Rate," *Journal of Statistics Education* 4, no. 2 (1996): 1–3. The data used here are a random sample of the data associated with the paper.

37. Data provided by Tim Brown.

38. researchinuserexperience.wordpress.com/2003/08/12/a-comparison-of-two-computer-fonts-serif-versus-ornate-sans-serif/

39. www.ets.org/gre/score-users/scores/interpret-scores.html

40. Nick Blanchard, Canyon del Oro High School.

41. Data originally from Emeritus (and late) Prof. Bob Black from Cornell College, Mount Vernon, Iowa.

42. en.wikipedia.org/wiki/Red-tailed_hawk. The range of tail lengths is given as 188–258.7 mm. Dividing this range by 6 gives a standard deviation of 11.8 mm.

43. Fred L. Ramsey and Daniel W. Schafer, *The Statistical Sleuth*, 2nd ed. (Pacific Grove, CA, Duxbury, 2002), 405–407.

44. Data are from Phyllis Lee, Stirling University, and are related to P. Lee et al., "Enduring Consequences of Early Experiences: 40-Year Effects on Survival and Success Among African Elephants (*Loxodonta africana*)," *Biology Letters* 9 (2013): 20130011.

45. nces.ed.gov/programs/digest/d17/tables/dt17_226.40.asp

46. Marla A. Sole and Sharon L. Weinberg, "What's Brewing? A Statistics Education Discovery Project," *Journal of Statistics Education* 25, no. 3 (2017): 137–144, DOI: 10.1080/10691898.2017.1395302.

47. Data downloaded from dasl.datadescription.com/datafile/tuition-all-schools-2016/?_sfm_methods=Summarizing+Quantitative+Data&_sfm_cases=4+59943&sf_paged=3 on August 21, 2021. Original source is collegescorecard.ed.gov/data/.

48. P. A. Mackowiak, S. S. Wasserman, and M. M. Levine, "A Critical Appraisal of 98.6 Degrees F, the Upper Limit of the Normal Body Temperature, and Other Legacies of Carl Reinhold August Wunderlich," *Journal of the American Medical Association* 268 (1992): 1578–1580. Allen Shoemaker, from Calvin College, produced this data set with the same properties as the original temperature readings.

49. Myroslava Protsiv, Catherine Ley, Joanna Lankester, Trevor Hastie, and Julie Parsonnet, "Decreasing Human Body Temperature in the United States Since the Industrial Revolution," *eLife* 9 (January 7, 2020): e49555.

50. Jitter standard for Zoom calls from support.zoom.com.

CHAPTER 9

1. This case study is based on the story "Drive-Thru Competition" in the *Electronic Encyclopedia of Statistical Examples and Exercises* (EESEE). Updated data were obtained from the *QSR* magazine website, www.qsrmagazine.com/reports/2021-qsr-magazine-drive-thru-study.

2. www.prnewswire.com/news-releases/few-hate-shopping-for-clothes-but-love-of-it-varies-by-country-124480498.html

3. Benjamin W. Friedman et al., "Diazepam Is No Better Than Placebo When Added to Naproxen for Acute Low Back Pain," *Annals of Emergency Medicine*, February 7, 2017.

4. Instagram data from Pew Research Center, "Social Media Update 2016," November 2016, and Pew Research Center, "Social Media Use in 2021," April 2021.

5. Based on data on youth employment from the Bureau of Labor Statistics, www.bls.gov.

6. Clive G. Jones et al., "Chain Reactions Linking Acorns to Gypsy Moth Outbreaks and Lyme Disease Risk," *Science* 279 (1998): 1023–1026.

7. www.nejm.org/doi/full/10.1056/NEJMoa1615869

8. Joanne Ryan et al., "Randomized Placebo-Controlled Trial of the Effects of Aspirin on Dementia and Cognitive Decline," *Neurology* 95 (2020): e320–e331, doi:10.1212/WNL.0000000000009277.

9. Data on Christmas trees based on James Farmer et al., "Beauty Is in the Eye of the Tree Holder: Indiana Christmas Tree Consumer Survey," conducted by Indiana University.

10. iop.harvard.edu/youth-poll/spring-2021-harvard-youth-poll

11. www.ncbi.nlm.nih.gov/pubmed/29183701

12. Elray M. Roper and Charles G. Wright, "German Cockroach (Orthoptera: Blatellidae) Mortality on Various Surfaces Following Application of Diazinon," *Journal of Economic Entomology* 78, no. 4 (1985): 733–737, doi.org/10.1093/jee/78.4.733.

13. Michael J. Mack et al., "Transcatheter Aortic-Valve Replacement with a Balloon-Expandable Valve in Low-Risk Patients," *New England Journal of Medicine*, www.nejm.org/doi/full/10.1056/NEJMoa1814052.

14. Douglas E. Jorenby et al., "A Controlled Trial of Sustained-Release Bupropion, a Nicotine Patch, or Both for Smoking Cessation," *New England Journal of Medicine* 340 (1999): 685–691; doi: 10.1056/NEJM199903043400903.

15. Data set randomly selected from full data set available at data.cityofnewyork.us/Environment/2015-Street-Tree-Census-Tree-Data/pi5s-9p35.

16. data.cityofnewyork.us/Environment/2018-Central-Park-Squirrel-Census-Squirrel-Data/vfnx-vebw/data

17. W. E. Paulus et al., "Influence of Acupuncture on the Pregnancy Rate in Patients Who Undergo Assisted Reproductive Therapy," *Fertility and Sterility* 77, no. 4 (2002): 721–724.

18. www.gapminder.org/ignorance/gms/ and www.ipsos.com/ipsos-mori/en-uk/mind-gap-ipsos-mori-survey-gapminder

19. M. Frick et al., "Helsinki Heart Study: Primary-Prevention Trial with Gemfibrozil in Middle-Aged Men with Dyslipidemia," *New England Journal of Medicine* 317 (1987): 1237–1245, doi: 10.1056/NEJM198711123172001.

20. J. E. Schulenberg, M. E. Patrick, L. D. Johnston, P. M. O'Malley, J. G. Bachman, and R. A. Miech, *Monitoring the Future National Survey Results on Drug Use, 1975–2020: Volume II, College Students and Adults Ages 19-60* (Ann Arbor, MI: Institute for Social Research, University of Michigan, 2021).

21. Based on Deborah Roedder John and Ramnath Lakshmi-Ratan, "Age Differences in Children's Choice Behavior: The Impact of Available Alternatives," *Journal of Marketing Research* 29 (1992): 216–226.

22. The study is reported in William Celis III, "Study Suggests Head Start Helps Beyond School," *New York Times,* April 20, 1993. See www.highscope.org.

23. www.hakaimagazine.com/news/the-plan-to-rear-fish-on-the-moon/

24. Maureen Hack et al., "Outcomes in Young Adulthood for Very Low-Birth-Weight Infants," *New England Journal of Medicine* 346 (2002): 149–157. This exercise is simplified, in that the measures reported in this paper have been statistically adjusted for "sociodemographic status."

25. *Arizona Daily Star,* February 11, 2009.

26. Data from Emma Merry, Canyon del Oro High School.

27. George Du Toit et al., "Randomized Trial of Peanut Consumption in Infants at Risk for Peanut Allergy," *New England Journal of Medicine* 372 (February 2015): 803–813.

28. C. P. Cannon et al., "Intensive Versus Moderate Lipid Lowering with Statins After Acute Coronary Syndromes," *New England Journal of Medicine* 350 (2004): 1495–1504.

29. Data for this exercise were obtained from Wikipedia, en.wikipedia.org/wiki/Car_of_the_Year.

30. Data sourced from www.craigslist.com, February 15, 2020.

31. www.ncbi.nlm.nih.gov/pubmed/29635503

32. Shailija V. Nigdikar et al., "Consumption of Red Wine Polyphenols Reduces the Susceptibility of Low-Density Lipoproteins to Oxidation in Vivo," *American Journal of Clinical Nutrition* 68 (1998): 258–265.

33. Data from a student project by Daniel Flexas.

34. Kristen E. Flegal and Michael C. Anderson, "Overthinking Skilled Motor Performance: Or Why Those Who Teach Can't Do," *Psychonomic Bulletin & Review* 15, no. 5 (2008): 927–932, doi: 10.3758/PBR.15.5.927. Thanks to Kristin Flegal for sharing the data from this study.

35. cpb-us-w2.wpmucdn.com/sites.udel.edu/dist/6/132/files/2010/11/Psychological-Science-2014-Mueller-0956797614524581-1u0h0yu.pdf

36. Ethan J. Temeles and W. John Kress, "Adaptation in a Plant-Hummingbird Association," *Science* 300 (2003): 630–633. We thank Ethan J. Temeles for providing the data.

37. M. Ann Laskey et al., "Bone Changes After 3 Months of Lactation: Influence of Calcium Intake, Breast-Milk Output, and Vitamin D–Receptor Genotype," *American Journal of Clinical Nutrition* 67 (1998): 685–692.

38. *Electronic Encyclopedia of Statistical Examples and Exercises,* "Surgery in a Blanket."

39. Data set randomly selected from full data set available at data.cityofnewyork.us/Environment/2015-Street-Tree-Census-Tree-Data/pi5s-9p35.

40. American Community Survey data from www.census.gov/programs-surveys/acs/data.html.

41. This study is reported in Roseann M. Lyle et al., "Blood Pressure and Metabolic Effects of Calcium Supplementation in Normotensive White and Black Men," *Journal of the American Medical Association* 257 (1987): 1772–1776. The data were provided by Dr. Lyle.

42. www.cochrane.org/CD010037/HTN_extra-calcium-prevent-high-blood-pressure

43. Noel Cressie, *Statistics for Spatial Data* (Hoboken, NJ: Wiley, 1993).

44. General Social Survey results from gss.norc.org.

45. The data for this exercise came from Allan Rossman, George Cobb, Beth Chance, and John Holcomb's National Science Foundation project shared at the Joint Mathematics Meeting (JMM) 2008 in San Diego. Their original source was Robert Stickgold, LaTanya James, and J. Allan Hobson, "Visual Discrimination Learning Requires Sleep After Training," *Nature Neuroscience* 3 (2000): 1237–1238.

46. The original paper is T. M. Amabile, "Motivation and Creativity: Effects of Motivational Orientation on Creative Writers," *Journal of Personality and Social Psychology* 48, no. 2 (1985): 393–399. The data for this exercise came from Fred L. Ramsey and Daniel W. Schafer, *The Statistical Sleuth,* 3rd ed. (Boston, MA: Brooks/Cole Cengage Learning, 2013).

47. Gabriela S. Castellani, "The Effect of Cultural Values on Hispanics' Expectations About Service Quality," MS thesis, Purdue University, 2000.

48. Data on annual income for college graduates and non-graduates obtained from the March 2021 Annual Social and Economic Supplement, downloaded from www.census.gov. We took a random sample of 500 people who had attended some college but earned no degree, or who had earned an associate's or bachelor's degree.

49. Data from a final project in Josh Tabor's Introductory Statistics class.

50. H. R. Knapp and G. A. FitzGerald, "The Antihypertensive Effects of Fish Oil: A Controlled Study of Polyunsaturated Fatty Acid Supplements in Essential Hypertension," *New England Journal of Medicine* 320 (1989):1037–1043. Cited in Fred Ramsey and Daniel Schafer, *The Statistical Sleuth* (Belmont, CA: Duxbury Press, 2002), 23.

51. Data from a student project by Mary Ann McRae and Abigail O'Conner.

52. Adapted from Maribeth Cassidy Schmitt, "The Effects of an Elaborated Directed Reading Activity on the Metacomprehension Skills of Third Graders," Ph.D. dissertation, Purdue University.

53. Data obtained from archive.ics.uci.edu/ml/datasets.php. Original article is Ian W. Evett and Ernest J. Spiehler, "Rule Induction in Forensic Science" (Aldermaston, UK: Central Research Establishment, Home Office Forensic Science Service), 1987.

54. D. B. Lindenmayer, K. L. Viggers, R. B. Cunningham, and C. F. Donnelly, "Morphological Variation Among Columns of the Mountain Brushtail Possum, *Trichosurus caninus Ogilby* (Phalangeridae: Marsupiala)," *Australian Journal of Zoology* 43 (1995): 449–458.

55. Kristen E. Flegal and Michael C. Anderson, "Overthinking Skilled Motor Performance: Or Why Those Who Teach Can't Do," *Psychonomic Bulletin & Review* 15, no. 5 (2008): 927–932, doi: 10.3758/PBR.15.5.927. Thanks to Kristin Flegal for sharing the data from this study.

56. GDP data for European countries from the World Bank's website, data.worldbank.org, November 18, 2021.

57. Simplified from Sanjay K. Dhar, Claudia Gonzalez-Vallejo, and Dilip Soman, "Modeling the Effects of Advertised Price Claims: Tensile Versus Precise Pricing," *Marketing Science* 18 (1999): 154–177.

58. Data from Pennsylvania State University Stat 500 Applied Statistics online course, online.stat.psu.edu/statprogram/stat500.

59. C. Darwin, *The Effects of Cross and Self Fertilisation in the Vegetable Kingdom* (London, UK: John Murray, 1876).

60. *Model Year 2020 Fuel Economy Guide,* www.fueleconomy.gov.

61. United States Soccer Federation, www.ussoccer.com.

62. Data from a student project by Libby Foulk and Kathryn Hilton.

63. Data provided by Ramon Olivier.

64. F. H. Rauscher et al., "Music Training Causes Long-Term Enhancement of Preschool Children's Spatial–Temporal Reasoning," *Neurological Research* 19 (1997): 2–8.

65. E. C. Strain et al., "Caffeine Dependence Syndrome: Evidence from Case Histories and Experimental Evaluation," *Journal of the American Medical Association* 272 (1994): 1604–1607.

66. R. D. Stichler, G. G. Richey, and J. Mandel, "Measurement of Treadware of Commercial Tires," *Rubber Age* 73, no. 2 (May 1953).

67. Data on chewing gum and short-term memory from a study by Leila El-Ali and Valerie Pederson.

68. Data from a study by Sean Leader and Shelby Zismann.

69. Marla A. Sole and Sharon L. Weinberg, "What's Brewing? A Statistics Education Discovery Project," *Journal of Statistics Education* 25, no. 3 (2017): 137–144, doi: 10.1080/10691898.2017.1395302.

70. U.S. Census Bureau, Current Population Survey, 2020 Annual Social and Economic Supplement, Table 2 Educational Attainment of the Population 25 Years and Over, by Selected Characteristics: 2020.

71. Tire pressure loss data from *Consumer Reports* website, www.consumerreports.org/tire-buying-maintenance/should-you-use-nitrogen-in-car-tires.

72. W. S. Gosset, "The Probable Error of a Mean," *Biometrika* 6 (1908): 1–25.

73. Data provided by Ben Garcia and Maya Kraft.

74. www.nytimes.com/interactive/2021/07/19/upshot/major-league-baseball-spin-rate-shift.html

75. Data and Story Library, "Friday the 13th," dasl.datadescription.com.

76. The idea for this exercise was provided by Robert Hayden.

77. The idea for this exercise was provided by Robert Hayden.

78. We obtained the sleep data from the Data and Story Library (DASL) website, dasl.datadescription.com. They cite as a reference R. A. Fisher, *The Design of Experiments,* 3rd ed. (Edinburgh, UK: Oliver and Boyd, 1942), 27.

79. Data provided by Kirstyn Thwaits.

80. Data provided by Judy Starnes, who compiled the data from a local realtor.

81. Wayne J. Camera and Donald Powers, "Coaching and the SAT I," *TIP* (July 1999), www.siop.org/tip.

82. M. D. Venesky, S. M. Hanlon, K. Lynch, M. J. Parris, and J. R. Rohr, "Optimal Digestion Theory Does Not Predict the Effect of Pathogens on Intestinal Plasticity," *Biology Letters* 9 (2013): 20130038, dx.doi.org/10.1098/rsbl.2013.0038.

83. M. A. Wale, S. D. Simpson, and A. N. Radford, "Size-Dependent Physiological Responses of Shore Crabs to Single and Repeated Playback of Ship Noise," *Biology Letters* 9 (2013): 20121194, dx.doi.org/10.1098/rsbl.2012.1194.

84. The data for this exercise came from Allan Rossman, George Cobb, Beth Chance, and John Holcomb's National Science Foundation project shared at the Joint Mathematics Meeting (JMM) 2008 in San Diego. Their original source was Robert Stickgold, LaTanya James, and J. Allan Hobson, "Visual Discrimination Learning Requires Sleep After Training," *Nature Neuroscience* 3 (2000): 1237–1238.

85. Shailija V. Nigdikar et al., "Consumption of Red Wine Polyphenols Reduces the Susceptibility of Low-Density Lipoproteins to Oxidation in Vivo," *American Journal of Clinical Nutrition* 68 (1998): 258–265.

86. J. J. Halpern, "The Transaction Index: A Method for Standardizing Comparisons of Transaction Characteristics Across Different Contexts," *Group Decision and Negotiation* 6 (1997): 557–572, dasl.datadescription.com/datafile/buy-from-a-friend/?_sf_s=buy&_sfm_cases=4+59943.

87. Data from a student project by Daniel Flexas.

88. N. Teasdale, C. Bard, J. La Rue, and M. Fleury, "On the Cognitive Penetrability of Posture Control," *Experimental Aging Research* 19 (1993): 1–13. The data were obtained from the DASL Data and Story Library database, www.statsci.org/data/general/balaconc.html.

89. Adapted from Maribeth Cassidy Schmitt, "The Effects of an Elaborated Directed Reading Activity on the Metacomprehension Skills of Third Graders," Ph.D. dissertation, Purdue University.

90. Ethan J. Temeles and W. John Kress, "Adaptation in a Plant–Hummingbird Association," *Science* 300 (2003): 630–633. We thank Ethan J. Temeles for providing the data.

91. This exercise is based on M. Ann Laskey et al., "Bone Changes After 3 Months of Lactation: Influence of Calcium Intake, Breast-Milk Output, and Vitamin D–Receptor Genotype," *American Journal of Clinical Nutrition* 67 (1998): 685–692.

92. D. B. Lindenmayer, K. L. Viggers, R. B. Cunningham, and C. F. Donnelly, "Morphological Variation Among Columns of the Mountain Brushtail Possum, *Trichosurus caninus Ogilby* (Phalangeridae: Marsupiala)," *Australian Journal of Zoology* 43 (1995): 449–458.

93. Data obtained from archive.ics.uci.edu/ml/datasets.php. Original article is Ian W. Evett and Ernest J. Spiehler. "Rule Induction in Forensic Science" (Aldermaston, UK: Central Research Establishment, Home Office Forensic Science Service), 1987.

94. J. W. Winkelman et al., "Efficacy and Safety of Pramipexole in Restless Legs Syndrome," *Neurology* 67(6) (September 2006): 1034–039, doi: 10.1212/01.wnl.0000231513.23919.a1.

95. Piet Spaak and Maarten Boersma, "Tail Spine Length in the *Daphnia galeata* Complex: Costs and Benefits of Induction by Fish," *Aquatic Ecology* 31 (1997): 89–98.

96. Data from caffeine and pulse rate experiment provided by Josh Tabor.

97. Reaction time data from a student research project for Josh Tabor's statistics class.

98. Thanks to Michael Legacy for developing this exercise about stock volatility.

99. Data provided by Warren Page, New York City Technical College, from a study done by John Hudesman.

100. www.gallup.com/poll/159881/americans-call-term-limits-end-electoral-college.aspx

101. Data on crash dieting supplied by Michael Legacy.
102. Pew Research Center, "Book Reading 2016," September 2016. Standard deviation estimated from frequency table provided in the report.
103. The original randomized clinical trial testing the effectiveness of AZT in treating patients with AIDS was conducted by Burroughs Wellcome. Data for this exercise came from the original "Against All Odds" video series.
104. Data provided by Corinne Lim, Purdue University, from a student project supervised by Professor Joseph Vanable.

CHAPTER 10

1. Based on data from Pew Research Center, April 2021, "STEM Jobs See Uneven Progress in Increasing Gender, Racial, and Ethnic Diversity."
2. Biographical data from www.nhl.com.
3. McNugget information from www.thehits.co.nz/the-latest/mcdonalds-reveals-the-truth-about-chicken-mcnugget-shapes. Data from Carlos Poblano and Nathaniel Benavidez, Canyon del Oro High School.
4. R. W. Mannan and E. C. Meslow, "Bird Populations and Vegetation Characteristics in Managed and Old-Growth Forests, Northwestern Oregon," *Journal of Wildlife Management,* 48 (1984):1219–1238.
5. *New York Times Magazine,* August 2, 2015, p. 6.
6. ashpublications.org/bloodadvances/article/4/20/4990/463793/Reduced-prevalence-of-SARS-CoV-2-infection-in-ABO
7. www2.census.gov/library/publications/cen2010/briefs/c2010br-03.pdf
8. Cass Randall-Green, Canyon del Oro High School. North American percentages were obtained at www.ppg.com.
9. "Linkage Studies of the Tomato." *Transactions of the Canadian Institute,* 1931.
10. You can find a mathematical explanation of Benford's law in Ted Hill, "The First-Digit Phenomenon," *American Scientist* 86 (1996): 358–363; and Ted Hill, "The Difficulty of Faking Data," *Chance* 12, no. 3 (1999): 27–31. Applications in fraud detection are discussed in the second paper by Hill and in Mark A. Nigrini, "I've Got Your Number," *Journal of Accountancy,* May 1999, www.journalofaccountancy.com/issues/1999/may/nigrini.
11. The idea for this exercise came from a post to the AP® Statistics electronic discussion group by Joshua Zucker.
12. Constantine Daskalakis, "Tumor Growth Dataset," *TSHS Resources Portal* (2016), www.causeweb.org/tshs/tumor-growth/. One mouse observed for less than 2 weeks has been omitted.
13. H. Lindberg, H. Roos, and P. Gardsell, "Prevalence of Coxarthritis in Former Soccer Players," *Acta Orthopedica Scandinavica* 64 (1993): 165–167.
14. Janice E. Williams et al., "Anger Proneness Predicts Coronary Heart Disease Risk," *Circulation* 101 (2000): 63–95.
15. Ohio State University, "Night Lights Don't Lead to Nearsightedness, Study Suggests," *ScienceDaily,* March 9, 2000.
16. www.pewinternet.org/2015/12/15/gaming-and-gamers/
17. www.pewresearch.org/fact-tank/2020/02/03/democrats-on-twitter-more-liberal-less-focused-on-compromise-than-those-not-on-the-platform/
18. jamanetwork.com/journals/jamainternalmedicine/fullarticle/2130724
19. Andrea C. Gielen et al., "National Survey of Home Injuries During the Time of COVID-19: Who Is at Risk?" *Injury Epidemiology* 7 (2020): 63.
20. J. Bowden et al., "High-Arctic Butterflies Become Smaller with Rising Temperatures," *Biology Letters* 11, no. 10 (2015): 20150574.
21. Rachel K. Nelligan et al., "Effects of a Self-Directed Web-Based Strengthening Exercise and Physical Activity Program Supported by Automated Text Messages for People with Knee Osteoarthritis," *JAMA Internal Medicine,* pubmed.ncbi.nlm.nih.gov/33843948.
22. Pew Research Center, "What It Takes to Truly Be 'One of Us,'" February 2017.
23. The context of this example was inspired by C. M. Ryan et al., "The Effect of In-Store Music on Consumer Choice of Wine," *Proceedings of the Nutrition Society* 57 (1998): 1069A.
24. poll.qu.edu/Poll-Release-Legacy?releaseid=2275
25. www.pewresearch.org/fact-tank/2021/07/16/home-broadband-adoption-computer-ownership-vary-by-race-ethnicity-in-the-u-s/
26. Justin Yang et al., "Association Between Push-up Exercise Capacity and Future Cardiovascular Events Among Active Adult Men," *JAMA Network Open* 2, no. 2 (2019): e188341. jamanetwork.com/journals/jamanetworkopen/fullarticle/2724778.
27. Charlie Huveneers et al., "Effectiveness of Five Personal Shark-Bite Deterrents for Surfers," *Aquatic Biology,* August 31, 2018, peerj.com/articles/5554/.
28. Douglas E. Jorenby et al., "A Controlled Trial of Sustained Release Bupropion, a Nicotine Patch, or Both for Smoking Cessation," *New England Journal of Medicine* 340 (1990): 685–691.
29. news.gallup.com/poll/284009/library-visits-outpaced-trips-movies-2019.aspx
30. Jenna M. Dittmar et al., "Medieval Injuries: Skeletal Trauma as an Indicator of Past Living Conditions and Hazard Risk in Cambridge, England. *American Journal of Physical Anthropology* 175 (2021): 626–645, onlinelibrary.wiley.com/doi/10.1002/ajpa.24225.
31. Mehreen S. Datoo et al., "Efficacy of a Low-Dose Candidate Malaria Vaccine, R21 in Adjuvant Matrix-M, with Seasonal Administration to Children in Burkina Faso: A Randomised Controlled Trial," *Lancet* 397 (2021): 1809–1818, doi.org/10.1016/S0140-6736(21)00943-0.
32. Cameron Radford et al., "Artificial Eyespots on Cattle Reduce Predation by Large Carnivores," *Communications Biology* 3 (2020): 430. Study also described here: www.npr.org/2020/08/23/905181717/study-finds-painting-eyes-on-cows-butts-can-save-their-lives.
33. R. Shine et al., "The Influence of Nest Temperatures and Maternal Brooding on Hatchling Phenotypes in Water Pythons," *Ecology* 78 (1997): 1713–1721.
34. assets.morningconsult.com/wp-uploads/2021/05/27145135/2105101_crosstabs_ADWEEK_STREAMING_Adults_v1_NP.pdf
35. Robert B. Cialdini et al., "Basking in Reflected Glory: Three (Football) Field Studies," *Journal of Personality and Social Psychology* 34, no. 3 (1976): 366–373.
36. The idea for this exercise came from Bob Hayden.
37. Constantine Daskalakis, "Tumor Growth Dataset," *TSHS Resources Portal* (2016), www.causeweb.org/tshs/tumor-growth/. One mouse observed for 7 days was omitted. All other mice were observed for between 13 and 28 days.

38. C. H. Cannon, D. R. Peart, and M. Leighton, "Tree Species Diversity in Commercially Logged Bornean Rainforest," *Science* 281 (1998): 1366–1367. We thank Charles Cannon for providing the data.
39. The cereal data came from the Data and Story Library, lib.stat.cmu.edu/DASL/.
40. *Nutrition Action,* July/August 2020.
41. Data obtained from archive.ics.uci.edu/ml/datasets.php. The original article is Ian W. Evett and Ernest J. Spiehler, "Rule Induction in Forensic Science" (Aldermaston, UK: Central Research Establishment, Home Office Forensic Science Service).
42. Experiment conducted by Braedon Day, Canyon del Oro High School student.
43. Ilka Vosteen, Jonathan Gershenzon, and Grit Kunert, "Hoverfly Preference for High Honeydew Amounts Creates Enemy-Free Space for Aphids Colonizing Novel Host Plants," *Journal of Animal Ecology* 85 (2016): 1286–1297.
44. Data obtained from archive.ics.uci.edu/ml/datasets.php. The original article is Ian W. Evett and Ernest J. Spiehler, "Rule Induction in Forensic Science" (Aldermaston, UK: Central Research Establishment, Home Office Forensic Science Service).
45. The cereal data came from the Data and Story Library, lib.stat.cmu.edu/DASL/.
46. C. H. Cannon, D. R. Peart, and M. Leighton, "Tree Species Diversity in Commercially Logged Bornean Rainforest," *Science* 281 (1998): 1366–1367. We thank Charles Cannon for providing the data.
47. *Nutrition Action,* July/August 2020.
48. Data from Data and Story Library (DASL), dasl.datadescription.com/datafile/fertilizers/. Note that a low outlier of 58 in Treatment C was excluded from the analysis.
49. P. R. Jaffe, F. L. Parker, and D. J. Wilson, "Distribution of Toxic Substances in Rivers," *Journal of the Environmental Engineering Division* 108 (1982): 639–649.
50. Kathleen D. Vohs, Nicole L. Mead, and Miranda R. Goode, "The Psychological Consequences of Money," *Science* 314 (2006):1154–1156. We thank Kathleen Vohs for providing the data.
51. Dania V. Francis, Angela C. M. de Oliveira, and Carey Dimmitt, "Do School Counselors Exhibit Bias in Recommending Students for Advanced Coursework?" *B.E. Journal of Economic Analysis and Policy* 19, no. 4 (2019): 1–17.
52. Elisabeth Wells-Parker et al., "An Exploratory Study of the Relationship Between Road Rage and Crash Experience in a Representative Sample of U.S. Drivers," *Accident Analysis and Prevention* 34 (2002): 271–278.
53. M. S. Palmeirim, U. A. Mohammed, A. Ross, S. M. Ame, S. M. Ali, and J. Keiser, "Evaluation of Two Communication Tools, Slideshow and Theater, to Improve Participants' Understanding of a Clinical Trial in the Informed Consent Procedure on Pemba Island, Tanzania," *PLoS Neglected Tropical Diseases* 15, no. 5 (2021): e0009409, doi.org/10.1371/journal.pntd.0009409. We thank Jennifer Keiser and Marta Palmeirim for sharing the data with us.
54. Michael L. Dansinger, Joi Augustin Gleason, John L. Griffith, Harry P. Selker, and Ernst J. Schaefer, "Comparison of the Atkins, Ornish, Weight Watchers, and Zone Diets for Weight Loss and Heart Disease Risk Reduction: A Randomized Trial," *Journal of the American Medical Association* 293, no. 1 (2005): 43–53.
55. Data obtained from archive.ics.uci.edu/ml/datasets.php. The original article is Ian W. Evett and Ernest J. Spiehler, "Rule Induction in Forensic Science" (Aldermaston, UK: Central Research Establishment, Home Office Forensic Science Service).
56. Ethan J. Temeles and W. John Kress, "Adaptation in a Plant–Hummingbird Association," *Science* 300 (2003): 630–633. We thank Ethan J. Temeles for providing the data.
57. A. N. Garand and L. N. Bell, "Caffeine Content of Fountain and Private-Label Store Brand Carbonated Beverages," *Journal of the American Dietetic Association* 97, no. 2 (1997): 179–182.
58. James Hanley and Stanley Shapiro, "Sexual Activity and the Lifespan of Male Fruitflies: A Dataset That Gets Attention," *Journal of Statistics Education* 2, no. 1 (1994).
59. Data come from STAT2: Modeling with Regression and ANOVA. The original source is Violetta N. Pivtoraiko, Eric E. Abrahamson, Sue E. Leurgans, Steven T. DeKosky, Elliott J. Mufson, and Milos D. Ikonomovic, "Cortical Pyroglutamate Amyloid-β Levels and Cognitive Decline in Alzheimer's Disease," *Neurobiology of Aging* 36 (2015): 12–19. The data are read from Figure 1, panel d.
60. Based on data from Pew Research Center, "Women and Men in STEM Often at Odds of Workplace Equity," January 2018.
61. Elizabeth F. Loftus and John C. Palmer, "Reconstruction of Automobile Destruction: An Example of the Interaction Between Language and Memory," *Journal of Verbal Learning and Verbal Behavior* 13 (1974): 585–589, www.researchgate.net/publication/222307973_Reconstruction_of_Automobile_Destruction_An_Example_of_the_Interaction_Between_Language_and_Memory.
62. M. van Dusseldorp, et al., "Cholesterol-Raising Factor from Boiled Coffee Does Not Pass a Paper Filter," *Arteriosclerosis and Thrombosis* 11, no. 3 (1991). Data were generated to be consistent with summary statistics provided in the article.
63. heart.bmj.com/content/105/23/1793

CHAPTER 11

1. www.epa.gov/greenvehicles/fast-facts-transportation-greenhouse-gas-emissions
2. Data on used car prices from autotrader.com, September 8, 2012. We searched for F-150 4 × 4's on sale within 50 miles of College Station, Texas.
3. www.fs.usda.gov/pnw/pubs/pnw_rp283.pdf
4. Data from George W. Pierce, *The Songs of Insects* (Cambridge, MA, Harvard University Press, 1949), pp. 12–21.
5. Data provided by Dan Teague, North Carolina School of Science and Mathematics.
6. *Nutrition Action,* March 2019.
7. www.starbucks.com/promo/nutrition
8. *Consumer Reports,* June 1986, pp. 366–367.
9. www.pewresearch.org/internet/2013/08/07/51-of-u-s-adults-bank-online/
10. Tulin Yaklcl and Muhammet Arici, "Storage Stability of Aspartame in Orange Flavored Soft Drinks," *International*

Journal of Food Properties 16, no. 3 (2013): 698–705, doi: 10.1080/10942912.2011.565903.

11. N. R. Draper and J. A. John, "Influential Observations and Outliers in Regression," *Technometrics* 23 (1981): 21–26.

12. www.randomservices.org/random/data/Galton.html

13. www.lpga.com/tournaments/uswomensopenconductedbytheusga/results

14. *Nutrition Action,* December 2009.

15. jan.ucc.nau.edu/~dsk5/oldsite/S_AK/Arolik/Hu%20et%20al%20(2003)%20cyclic%20forcing%20200%20yr%20Alaska,%20Arolik.pdf

16. Based on an article by Matthew Russell, "Do Forests Have the Capacity for 1 Trillion Extra Trees?" *Significance Magazine* 17, no. 6 (2020): 8–9.

17. *The World Almanac and Book of Facts* (2009).

18. www.basketball-reference.com/leagues/NBA_2019_standings.html

19. Gary Smith, "Do Statistics Test Scores Regress Toward the Mean?" *Chance* 10, no. 4 (1997): 42–45.

20. www.baseball-reference.com

21. www.nutrition411.com/pdf/Bean%20Comparison%20Chart.pdf

22. Debora L. Arsenau, "Comparison of Diet Management Instruction for Patients with Non–Insulin Dependent Diabetes Mellitus: Learning Activity Package vs. Group Instruction," MS thesis, Purdue University, 1993.

23. Data set shared at statcrunch.com by Amanda Moulton, College of the Canyons.

24. *Nutrition Action,* July/August 2020.

25. David M. Fergusson and L. John Horwood, "Cannabis Use and Traffic Accidents in a Birth Cohort of Young Adults," *Accident Analysis and Prevention* 33 (2001): 703–711.

26. E. Thomassot et al., "Methane-Related Diamond Crystallization in the Earth's Mantle: Stable Isotopes Evidence from a Single Diamond-Bearing Xenolith," *Earth and Planetary Science Letters* 257 (2007): 362–371, Table 1.

27. www.advancedwebranking.com/cloud/ctrstudy/

28. *Consumer Reports,* January 2013.

29. Data from George W. Pierce, *The Songs of Insects* (Cambridge, MA, Harvard University Press, 1949), pp. 12–21.

30. *Nutrition Action,* March 2019.

31. From a graph in Craig Packer et al., "Ecological Change, Group Territoriality, and Population Dynamics in Serengeti Lions," *Science* 307 (2005): 390–393.

32. en.wikipedia.org/wiki/Figure_skating_at_the_2018_Winter_Olympics_%E2%80%93_Ladies%27_singles

33. ourworldindata.org/smoking-big-problem-in-brief

34. www.fs.usda.gov/pnw/pubs/pnw_rp283.pdf

35. The idea for this exercise came from Gloria Barrett, Floyd Bullard, and Dan Teague at the North Carolina School of Science and Math.

36. Data from Samuel Phillips, Purdue University.

37. www.gapminder.org

38. tucson.com/news/local/saguaro-census-shows-more-giants-low-reproduction-in-namesake-park/article_c659cb2d-ddef-5371-b5a3-abe351f886ad.html. Thanks to Don Swann of the National Park Service for sharing additional information.

39. Data from Nicole Enos and Elena Tesluk, Canyon del Oro High School.

40. Cheapest "wanna-get-away" fare on Southwest Airlines as of August 8, 2014.

41. www.rcdb.com

42. Data from Kerry Lane and Danielle Neal, Canyon del Oro High School.

43. www.wimbledon.com

44. M. Seddigh and G.D. Jolliff, "Light Intensity Effects on Meadowfoam Growth and Flowering," *Crop Science* 34 (1994): 497–503.

45. Samuel Karelitz et al., "Relation of Crying Activity in Early Infancy to Speech and Intellectual Development at Age Three Years," *Child Development* 35 (1964): 769–777.

46. Todd W. Anderson, "Predator Responses, Prey Refuges, and Density-Dependent Mortality of a Marine Fish," *Ecology* 81 (2001): 245–257.

47. M. H. Criqui, University of California, San Diego, reported in the *New York Times,* December 28, 1994.

48. www.gapminder.org

49. bcs.whfreeman.com/WebPub/Statistics/shared_resources/EESEE/BloodAlcoholContent/index.html

50. Data from James R. Jordan, Lawrenceville School.

51. Based on Marion E. Dunshee, "A Study of Factors Affecting the Amount and Kind of Food Eaten by Nursery School Children," *Child Development* 2 (1931): 163–183. This article gives the means, standard deviations, and correlation for 37 children but does not give the actual data.

52. Based on a plot in G. D. Martinsen, E. M. Driebe, and T. G. Whitham, "Indirect Interactions Mediated by Changing Plant Chemistry: Beaver Browsing Benefits Beetles," *Ecology,* 79 (1998): 192–200.

53. Thanks to Larry Green, Lake Tahoe Community College, for giving us permission to use several of the contexts from his website at www.ltcconline.net/greenl/java/Statistics/catStatProb/categorizingStatProblemsJavaScript.html.

54. www.wimbledon.com

55. Data from Nicole Enos and Elena Tesluk, Canyon del Oro High School.

56. Todd W. Anderson, "Predator Responses, Prey Refuges, and Density-Dependent Mortality of a Marine Fish," *Ecology* 81 (2001): 245–257.

57. M. H. Criqui, University of California, San Diego, reported in the *New York Times,* December 28, 1994.

58. bcs.whfreeman.com/WebPub/Statistics/shared_resources/EESEE/BloodAlcoholContent/index.html

59. Based on a plot in G. D. Martinsen, E. M. Driebe, and T. G. Whitham, "Indirect Interactions Mediated by Changing Plant Chemistry: Beaver Browsing Benefits Beetles," *Ecology* 79 (1998): 192–200.

60. www.rcdb.com

61. Based on Marion E. Dunshee, "A Study of Factors Affecting the Amount and Kind of Food Eaten by Nursery School Children," *Child Development* 2 (1931): 163–183. This article gives the means, standard deviations, and correlation for 37 children but does not give the actual data.

62. The data from J. Bowden et al., "High-Arctic Butterflies Become Smaller with Rising Temperatures," *Biology Letters* 11 (2015): 20150574.

63. A paper based on these data: P. J. Krug and R. K. Zimmer, "Larval Settlement: Chemical Markers for Tracing Production, Transport, and Distribution of a Waterborne Cue," *Marine Ecology Progress Series* 207 (2000): 283–296.

64. www.epa.gov/greenvehicles/fast-facts-transportation-greenhouse-gas-emissions

65. www.nenanaakiceclassic.com; dasl.datadescription.com/datafile/nenana-2017

66. From the EESEE story, "Is It Tough to Crawl in March?"
67. www.pro-football-reference.com
68. *Nutrition Action*, September 2019.
69. www.nature.com/articles/s41586-021-03550-y and portal.edirepository.org/nis/mapbrowse?packageid=edi.698.2

CHAPTER 12

1. Joshua D. Stewart et al., "Decreasing Body Lengths in North Atlantic Right Whales," *Current Biology* (2021), doi.org/10.1016/j.cub.2021.04.067. Whales younger than 2 years and older than 20 years were excluded to avoid curvature in the relationship between age and total length.
2. stats.washingtonpost.com/golf/averages.asp?tour=LPGA&rank=05. Data through August 16, 2021.
3. Data estimated from a graph in Kyle G. Ashton, Russell L. Burke, and James N. Layne, "Geographic Variation in Body and Clutch Size of Gopher Tortoises," *Copeia* 2 (2007): 355–363. We assumed that the sample was equivalent to a random sample for the purposes of inference in Section 12.3.
4. Results of the study are reported in P. F. Campbell and G. P. McCabe, "Predicting the Success of Freshmen in a Computer Science Major," *Communications of the ACM* 27 (1984): 1108–1113.
5. en.wikipedia.org/wiki/Refractive_index
6. archive.ics.uci.edu/ml/datasets/glass+identification
7. www.pro-football-reference.com
8. www.truecar.com/used-cars-for-sale/listings/honda/cr-v/location-tucson-az
9. Data set shared at statcrunch.com by Amanda Moulton, College of the Canyons.
10. Ben Dantzer et al., "Decoupling the Effects of Food and Density on Life-History Plasticity of Wild Animals Using Field Experiments: Insights from the Steward Who Sits in the Shadow of Its Tail, the North American Red Squirrel," *Journal of Animal Ecology,* doi: 10.1111/1365-2656.13341.
11. Anna D. Peterson and Laura Ziegler, "Building a Multiple Linear Regression Model with LEGO Brick Data," *Journal of Statistics and Data Science Education* (2021), doi: 10.1080/26939169.2021.1946450. Thanks to Laura Ziegler for sharing the data.
12. M. Seddigh and G. D. Jolliff, "Light Intensity Effects on Meadowfoam Growth and Flowering," *Crop Science* 34 (1994): 497–503.
13. Jennifer L. Raynor, Corbett A. Grainger, and Dominic P. Parker, "Wolves Make Roadways Safer, Generating Large Economic Returns to Predator Conservation," *PNAS* 118, no. 22 (2021): e2023251118, doi.org/10.1073/pnas.2023251118.
14. www.starbucks.com/promo/nutrition
15. www.truecar.com/used-cars-for-sale/listings/honda/cr-v/location-tucson-az
16. Data set shared at statcrunch.com by Amanda Moulton, College of the Canyons.
17. Tulin Yakici and Muhammet Arici, "Storage Stability of Aspartame in Orange Flavored Soft Drinks," *International Journal of Food Properties* 16, no. 3 (2013): 698–705, doi: 10.1080/10942912.2011.565903.
18. Details of the Carolina Abecedarian Project, including references to published work, can be found online at abc.fpg.unc.edu.
19. Data from Lexi Epperson and Courtney Johnson, Canyon del Oro High School.
20. M. D. Venesky, S. M. Hanlon, K. Lynch, M. J. Parris, and J. R. Rohr, "Optimal Digestion Theory Does Not Predict the Effect of Pathogens on Intestinal Plasticity," *Biology Letters* 9 (2013): 20130038, dx.doi.org/10.1098/rsbl.2013.0038.
21. Ben Dantzer et al., "Decoupling the Effects of Food and Density on Life-History Plasticity of Wild Animals Using Field Experiments: Insights from the Steward Who Sits in the Shadow of Its Tail, the North American Red Squirrel," *Journal of Animal Ecology,* doi: 10.1111/1365-2656.13341.
22. M. Seddigh and G. D. Jolliff, "Light Intensity Effects on Meadowfoam Growth and Flowering," *Crop Science* 34 (1994): 497–503.
23. www.truecar.com/used-cars-for-sale/listings/honda/cr-v/location-tucson-az
24. Data set shared at statcrunch.com by Amanda Moulton, College of the Canyons.
25. www.starbucks.com/promo/nutrition
26. We thank Ray and Pat Heithaus for providing data on the pine seedlings at the Brown Family Environmental Center.
27. www.researchgate.net/publication/332928434_Portion_size_effects_vary_The_size_of_food_units_is_a_bigger_problem_than_the_number
28. Joshua D. Stewart et al., "Decreasing Body Lengths in North Atlantic Right Whales," *Current Biology* (2021), doi.org/10.1016/j.cub.2021.04.067. Whales younger than 2 years and older than 20 years were excluded to avoid curvature in the relationship between age and total length.
29. *Nutrition Action Healthletter,* September 2021.
30. Data were estimated from a scatterplot in Philipp Heeb, Mathias Kolliker, and Heinz Richner, "Bird–Ectoparasite Interactions, Nest Humidity, and Ectoparasite Community Structure," *Ecology* 81 (2000): 958–968.
31. Scott W. Powers et al., "Trial of Amitriptyline, Topiramate, and Placebo for Pediatric Migraine," *New England Journal of Medicine* 376 (2017): L115–124, doi: 10.1056/NEJMoa1610384.
32. www.gapminder.com
33. Suma Thomas et al., "Effect of High-Dose Zinc and Ascorbic Acid Supplementation vs Usual Care on Symptom Length and Reduction Among Ambulatory Patients with SARS-CoV-2 Infection: The COVID A to Z Randomized Clinical Trial," *JAMA Network Open* 4, no. 2 (2021): e210369, doi:10.1001/jamanetworkopen.2021.0369.
34. www.baseball-reference.com/players/j/jeterde01.shtml
35. C. E. Kline et al., "Dose-Response Effects of Exercise Training on the Subjective Sleep Quality of Postmenopausal Women: Exploratory Analyses of a Randomised Controlled Trial," *BMJ Open* 2 (2012): e001044, doi:10.1136/bmjopen-2012-001044.
36. Means and standard deviations estimated to be consistent with values and graphs in the article.
37. www.baseball-reference.com

CHAPTER 13

1. www.statista.com/statistics/684723/adults-prone-to-headache-us
2. A. J. Vickers, R. W. Rees, C. E. Zollman, et al., "Acupuncture for Chronic Headache in Primary Care: Large, Pragmatic, Randomised Trial," *BMJ (Clinical Research Edition)* 328, no. 7442 (2004): 744. Data downloaded from www.causeweb.org/tshs/acupuncture.
3. data.cityofchicago.org/Parks-Recreation/Beach-E-coli-Predictions/xvsz-3xcj
4. R. J. Berger and J. M. Walker, "The Polygraphic Study of Sleep in the Tree Shrew," *Brain, Behavior, and Evolution* 5 (1972); 62.
5. www.rentcafe.com/average-rent-market-trends/us/il/chicago, accessed May 15, 2021.
6. Data taken from Chicago.craigslist.org on May 15, 2021.
7. Data obtained from flightaware.com on November 27, 2015.
8. fred.stlouisfed.org/series/MSPNHSUS
9. Prices collected from www.zillow.com.
10. Data obtained from finance.yahoo.com/quote/GOOG/history on May 16, 2021.
11. Data found in Ann R. Cannon, George W. Cobb, Bradley A. Hartlaub, et al., *STAT 2: Modeling with Regression and ANOVA*, 2nd ed. (New York, NY: W. H. Freeman and Company, 2019). Data originally came from F. Loven, *A Study of the Interlist Equivalency of the CID W-22 Word List Presented in Quiet and in Noise* (Unpublished master's thesis, University of Iowa, 1981).
12. Data found in Ann R. Cannon, George W. Cobb, Bradley A. Hartlaub, et al., *STAT 2: Modeling with Regression and ANOVA*, 2nd ed. (New York, NY: W. H. Freeman and Company, 2019). The original article is C. D. Scott and K. K. Chen, "Comparison of the Action of 1-Ethyl Theobromine and Caffeine in Animals and Man," *Journal of Pharmacological Experimental Therapy* 82 (1944): 89–97.
13. T. Romero, A. Konno, and T. Hasegawa, "Familiarity Bias and Physiological Responses in Contagious Yawning by Dogs Support Link to Empathy," *PLoS One* 8, no. 8 (2013): e71365, doi: 10.1371/journal.pone.0071365.
14. Data provided by Drina Iglesia, Department of Biological Sciences, Purdue University. The data are part of a larger study reported in D. D. S. Iglesia, E. J. Cragoe, Jr., and J. W. Vanable, "Electric Field Strength and Epithelization in the Newt (*Notophthalmus viridescens*)," *Journal of Experimental Zoology* 274 (1996): 56–62.
15. www.newscientist.com/article/dn1748-euro-coin-accused-of-unfair-flipping/#:~:text=Polish%20statisticians%20say%20the%20one,comes%20up%20heads%20more%20often
16. news.gallup.com/poll/349607/americans-expect-home-prices-rise-divided-buying.aspx
17. Data derived from an article in P. A. Mackowiak, S. S. Wasserman, and M. M. Levine, "A Critical Appraisal of 98.6 Degrees F, the Upper Limit of the Normal Body Temperature, and Other Legacies of Carl Reinhold August Wunderlich," *Journal of the American Medical Association* 268, no. 12 (1992):1578–1580. The given data appear in Allen L. Shoemaker, "What's Normal? Temperature, Gender, and Heart Rate," *Journal of Statistics Education* 4, no. 2 (1996), doi: 10.1080/10691898.1996.11910512.
18. N. Forbath, A. B. Kenshole, and G. Hetenyi, Jr., "Turnover Lactic Acid in Normal and Diabetic Dogs Calculated by Two Tracer Methods," *American Journal of Physiology* 212 (1967): 1179–1183.
19. W. S. Gosset, "The Probable Error of a Mean," *Biometrika* 6 (1908): 1–25.
20. www.cnn.com/2019/08/08/health/cat-name-study-trnd/index.html and www.nature.com/articles/s41598-019-40616-4#Abs1
21. Data obtained from flightaware.com on November 27, 2015.
22. www.deptofnumbers.com/income/new-york, accessed August 6, 2021.
23. Data collected from the NYC OpenData project, opendata.cityofnewyork.us. A random sample was taken from all of the salary data on the website.
24. Data found in Ann R. Cannon, George W. Cobb, Bradley A. Hartlaub, et al., *STAT 2: Modeling with Regression and ANOVA*, 2nd ed. (New York, NY: W. H. Freeman and Company, 2019). Data originally came from F. Loven, *A Study of the Interlist Equivalency of the CID W-22 Word List Presented in Quiet and in Noise* (Unpublished master's thesis, University of Iowa, 1981).
25. www.rentcafe.com/average-rent-market-trends/us/il/chicago, accessed May 15, 2021.
26. Data taken from Chicago.craigslist.org on May 15, 2021.
27. R. J. Berger and J. M. Walker, "The Polygraphic Study of Sleep in the Tree Shrew," *Brain, Behavior, and Evolution* 5 (1972): 62.
28. DASL website, dasl.datadescription.com/datafile/freshman-15/?_sfm_methods=Nonparametric+Methods&_sfm_cases=4+59943.
29. www.smithsonianmag.com/smart-news/human-body-temperature-getting-cooler-study-finds-180974006/#:~:text=In%201851%2C%20a%20German%20doctor,Celsius%2C%20or%2098.6%20degrees%20Fahrenheit
30. *Journal of Statistics Education* 4, no. 2 (July 1996), jse.amstat.org/jse_data_archive.htm.
31. James Hanley and Stanley Shapiro, "Sexual Activity and the Lifespan of Male Fruitflies: A Dataset That Gets Attention," *Journal of Statistics Education* 2, no. 1 (1994).
32. Nicholas Gueguen, "The Effects of a Joke on Tipping When It Is Delivered at the Same Time as the Bill," *Journal of Applied Social Psychology* 32 (2002): 1955–1963.
33. A. Kamimura, T. Takahishi, and Y. Watanabe, "Investigation of Topical Application of Procyanidin B-2 from Apple to Identify its Potential Use as a Hair Growing Agent," *Phytomedicine* 7, no. 6 (2000): 529–536. Found on the website users.stat.ufl.edu/~winner/datasets.html.
34. "Family Weekly," *Gainesville Sun*, February 5, 1978. J. T. McClave and F. H. Dietrich II, *Statistics* (San Francisco, CA: Dellen Publishing, 1991), Exercise 10.20. Data downloaded from www.statsci.org/data/oz/blonds.html.
35. This study is reported in Roseann M. Lyle et al., "Blood Pressure and Metabolic Effects of Calcium Supplementation in Normotensive White and Black Men," *Journal of the American Medical Association* 257 (1987): 1772–1776. The data were provided by Dr. Lyle.
36. J. Jung and Y. J. Ahn, "Effects of Interface on Procedural Skill Transfer in Virtual Training: Lifeboat Launching Operation Study," *Computer Animation & Virtual Worlds* 29 (2018): e1812. Data downloaded from users.stat.ufl.edu/~winner/datasets.html.

37. The data are a random subset of the Census at School data.

38. Data shared with the authors by Jeff Witmer, Oberlin College.

39. J. D. Klausner, S. Makonkawkeyoon, P. Akarasewi, et al., "The Effect of Thalidomide on the Pathogenesis of Human Immunodeficiency Virus Type 1 and *M. tuberculosis* Infection," *Journal of Acquired Immune Deficiency Syndrome and Human Retrovirology* 11 (1996): 247–257. Data downloaded from users.stat.ufl.edu/~winner/datasets.html.

40. J. Rusanganwa, "Multimedia as a Means to Enhance Teaching Technical Vocabulary to Physics Undergraduates in Rwanda," *English for Specific Purposes* 32 (2013): 36–44. Data downloaded from users.stat.ufl.edu/~winner/datasets.html.

41. This activity is described in Pat Hopfensperger, Tim Jaccobe, Deborah Lurie, and Jerry Moreno, *Bridging the Gap Between Common Core State Standards and Teaching Statistics* (American Statistical Association, 2012).

42. P. Gendreau et al., "Changes in EEG Alpha Frequency and Evoked Response Latency During Solitary Confinement," *Journal of Abnormal Psychology* 79 (1972): 54–59). Data accessed from dasl.datadescription.com/datafiles/?_sfm_methods=Comparing%20Two%20Groups&_sfm_cases = 4 + 59943.

43. K. nan der Horst, A. Ferrage, and A. Rytz, "Involving Children in Meal Preparation," *Appetite* 79 (2014): 18–24. Data downloaded from users.stat.ufl.edu/~winner/datasets.html.

44. dasl.datadescription.com/datafiles/?_sfm_methods=Comparing%20Two%20Groups&_sfm_cases=4+59943&sf_paged=2

45. J. H. Gertsch, B. Basnyat, E. W. Johnson, J. Onopa, and P. S. Holck, "Randomised, Double Blind, Placebo Controlled Comparison of Ginkgo Biloba and Acetazolamide for Prevention of Acute Mountain Sickness Among Himalayan Trekkers: The Prevention of High Altitude Illness Trial (PHAIT)," *BMJ* 328 (2004): 797. Data downloaded from users.stat.ufl.edu/~winner/datasets.html.

46. J. M. Smith and H. Misiak, "The Effect of Iris Color on Critical Flicker Frequency (Cff)," *Journal of General Psychology* (1973): 91–95. Downloaded from www.statsci.org/data/general/flicker.html.

47. B. Mehrgini, H. Memarian, M. B. Dusseault, A. Ghavidel, and M. Heydarizadeh, "Geomechanical Characteristics of Common Reservoir Caprocck in Iran (Gachsaran Formation), Experimental and Statistical Analysis," *Journal of Natural Gas Science and Engineering* 34 (2016): 898–907. Data downloaded from users.stat.ufl.edu/~winner/data.

48. Neal Binnie, "Using EDA, ANOVA and Regression to Optimise Some Microbiology Data," *Journal of Statistics Education* 12, no. 2 (2004), jse.amstat.org/v12n2/datasets.binnie.html.

49. J. Lindberg, "Relationship Between Various Surface Properties of Wool Fibers: Part II: Frictional Properties," *Textile Research Journal* 23 (1953): 225–237. Data downloaded from users.stat.ufl.edu/~winner/datasets.html.

50. D. J. Hand, F. Daly, A. D. Lunn, K. J. McConway, and E. Ostrowski, *A Handbook of Small Data Sets* (Chapman and Hall, 1994). Data downloaded from vincentarelbundock.github.io/Rdatasets/datasets.html.

51. P. R. Jaffe, F. L. Parker, and D. J. Wilson, "Distribution of Toxic Substances in Rivers," *Journal of the Environmental Engineering Division* 108 (1982): 639–649. Data downloaded from www.statsci.org/data/general/wolfrive.html.

52. S. Singh and S. Kumar, "Factorial Analysis of Lifting Task to Determine the Effect of Different Parameters and Interactions," *Journal of Manufacturing Technology Management* 23, no. 7 (2012): 947–953. Data downloaded from users.stat.ufl.edu/~winner/datasets.html.

53. dasl.datadescription.com/datafile/marketing-managers-salaries/?_sfm_methods=Analysis+of+Variance&_sfm_cases=4+59943&sf_paged=2

54. R. Till, *Statistical Methods for the Earth Scientist: An Introduction* (London. IL: Macmillan, 1974), 104. Data downloaded from vincentarelbundock.github.io/Rdatasets/doc/openintro/salinity.html.

55. A. Parenti, L. Guerrini, P. Masella, S. Spinelli, L. Calamai, and P. Spugnoli, "Comparison of Espresso Coffee Brewing Techniques," *Journal of Food Engineering* 121 (2014): 112–117. Data downloaded from users.stat.ufl.edu/~winner/datasets.html.

56. P. C. Chang, S. Y. Chou, and K. K. Shieh, "Reading Performance and Visual Fatigue When Using Electronic Displays in Long-Duration Reading Tasks Under Various Lighting Conditions," *Displays* 34 (2013): 208–214. Data downloaded from users.stat.ufl.edu/~winner/datasets.html.

57. D. Beis, K. Holzwarth, M. Flinders, M. Bader, M. Wöhr, and N. Alenina, "Brain Serotonin Deficiency Leads to Social Communication Deficits in Mice," *Biology Letters* 11 (2015): 20150057, dx.doi.org/10.1098/rsbl.2015.0057.

58. R. J. M. Engler, M. R. Nelson, M. M. Klote, et al., "Half- vs Full-Dose Trivalent Inactivated Influenza Vaccine (2004–2005): Age, Dose, and Sex Effects on Immune Responses," *Archives of Internal Medicine* 168, no. 22 (2008): 2405–2414, doi:10.1001/archinternmed.2008.513.

59. Data taken from two websites: news.gallup.com/poll/1591/Computers-Internet.aspx and news.gallup.com/poll/354749/internet-information-industries-new-image-lows.aspx.

60. J. E. Schulenberg, M. E. Patrick, L. D. Johnston, P. M. O'Malley, J. G. Bachman, and R. A. Miech, *Monitoring the Future National Survey Results on Drug Use, 1975–2020: Volume II, College Students and Adults Ages 19–60* (Ann Arbor, MI: Institute for Social Research, University of Michigan, 2021).

61. J. Rusanganwa, "Multimedia as a Means to Enhance Teaching Technical Vocabulary to Physics Undergraduates in Rwanda," *English for Specific Purposes* 32 (2013): 36–44. Data downloaded from users.stat.ufl.edu/~winner/datasets.html.

62. Data from a student project at Canyon del Oro High School, provided by Josh Tabor.

63. The data for this exercise came from Rossman, Cobb, Chance, & Holcomb's National Science Foundation project shared at the Joint Mathematics Meeting (JMM) 2008 in San Diego. Their original source was Robert Stickgold, LaTanya James, and J. Allan Hobson, "Visual Discrimination Learning Requires Sleep After Training," *Nature Neuroscience* 3 (2000): 1237–1238.

64. Based on Deborah Roedder John and Ramnath Lakshmi-Ratan, "Age Differences in Children's Choice Behavior: The Impact of Available Alternatives," *Journal of Marketing Research* 29 (1992): 216–226.

65. www.hakaimagazine.com/news/the-plan-to-rear-fish-on-the-moon/

66. A. J. Miller, D. E. Shaw, L. G. Veitch, and E. J. Smith, "Analyzing the Results of a Cloud-Seeding Experiment in Tasmania,"

Communications in Statistics: Theory and Methods 8, no. 10 (1979): 1017–1047, doi: 10.1080/03610927908827813.

67. wellcome.org/reports/wellcome-global-monitor-mental-health/2020?utm_source=twitter&utm_medium=o-spokesperson&utm_campaign=wgm&utm_content=gallup

68. "Family Weekly," *Gainesville Sun*, February 5, 1978. J. T. McClave and F. H. Dietrich II, *Statistics* (San Francisco, CA: Dellen Publishing, 1991), Exercise 10.20. Data downloaded from www.statsci.org/data/oz/blonds.html.

69. This activity is described in Pat Hopfensperger, Tim Jaccobe, Deborah Lurie, and Jerry Moreno, *Bridging the Gap Between Common Core State Standards and Teaching Statistics* (American Statistical Association, 2012).

70. Data description at leapstudy.co.uk/leap-0#.Y8B7F3bMJPZ

71. Data from James R. Jordan, Lawrenceville School.

72. news.gallup.com/poll/350048/retirees-experience-differs-nonretirees-outlook.aspx

73. Data are read from Figure 1, panel d, in Violetta N. Pivtoraiko, Eric E. Abrahamson, Sue E. Leurgans, Steven T. DeKosky, Elliott J. Mufson, and Milos D. Ikonomovic, "Cortical Pyroglutamate Amyloid-β Levels and Cognitive Decline in Alzheimer's Disease," *Neurobiology of Aging* 36 (2015): 12–19.

74. cdn.annenbergpublicpolicycenter.org/wp-content/uploads/Civics-survey-press-release-09-17-2014-for-PR-Newswire.pdf

75. fivethirtyeight.com/features/how-much-football-is-even-in-a-football-broadcast/. Special thanks to Katherine Rowe for providing the raw data.

76. Copper data from a Hudbay news release on March 29, 2021, which was highlighted in Tony Davis, "Hudbay Says It Found Copper That Could Result in Open Pit Mines on Santa Ritas' West Side," *Arizona Daily Star*, April 1, 2021.

77. "Charitable Crowdfunding: Who Gives, to What, and Why?," scholarworks.iupui.edu/handle/1805/25515.

78. M. A. Wale, S. D. Simpson, and A. N. Radford, "Size-Dependent Physiological Responses of Shore Crabs to Single and Repeated Playback of Ship Noise," *Biology Letters* 9 (2013): 20121194, dx.doi.org/10.1098/rsbl.2012.1194.

79. P. Jendrny, C. Schulz, F. Twele, et al., "Scent Dog Identification of Samples from COVID-19 Patients: A Pilot Study, *BMC Infectious Diseases* 20 (2020): 536, doi.org/10.1186/s12879-020-05281-3.

80. From the American Community Survey, at the Census Bureau website, www.census.gov. The data are a subsample of the individuals in the ACS New York sample who had travel times greater than zero.

81. Data are from Phyllis Lee, Stirling University, and are related to P. Lee et al., "Enduring Consequences of Early Experiences: 40-Year Effects on Survival and Success Among African Elephants (*Loxodonta africana*)," *Biology Letters* 9 (20130: 20130011.

82. www.usatoday.com/picture-gallery/news/2015/04/07/usa-today-snapshots/6340793/

83. These data are from "Results Report on the Vitamin C Pilot Program," prepared by SUSTAIN (Sharing United States Technology to Aid in the Improvement of Nutrition) for the U.S. Agency for International Development. The report was used by the Committee on International Nutrition of the National Academy of Sciences/Institute of Medicine to make recommendations on whether the vitamin C content of food commodities used in U.S. food aid programs should be increased. The program was directed by Peter Ranum and Françoise Chomé.

84. www.pewresearch.org/wp-content/uploads/2020/12/Phones-and-scams-methods-and-topline.pdf

85. "Family Weekly," *Gainesville Sun*, February 5, 1978. J. T. McClave and F. H. Dietrich II, *Statistics* (San Francisco, CA: Dellen Publishing, 1991), Exercise 10.20. Data downloaded from www.statsci.org/data/oz/blonds.html.

86. Data are read from Figure 1, panel d, in Violetta N. Pivtoraiko, Eric E. Abrahamson, Sue E. Leurgans, Steven T. DeKosky, Elliott J. Mufson, and Milos D. Ikonomovic, "Cortical Pyroglutamate Amyloid-β Levels and Cognitive Decline in Alzheimer's Disease," *Neurobiology of Aging* 36 (2015): 12–19

87. A. M. Hurtado and K. Hill, "Experimental Studies of Tool Efficiency Among Machinguenga Women and Implications for Root-Digging Foragers," *Journal of Anthropological Research* 45, no. 2 (1989): 207–217.

88. C. Aschan, M. Hirvonen, E. Rajamaki, and T. Mannelin, "Slip Resistance of Oil Resistant and Non-Oil Resistant Footwear Outsoles in Winter Conditions," *Safety Science* 43 (2005): 375–389.

89. Data description at users.stat.ufl.edu/~winner/data/ropestrength.txt, data subsetted from users.stat.ufl.edu/~winner/data/ropestrength.dat.

90. Data from caffeine and pulse rate experiment provided by Josh Tabor.

Index

1.5 × *IQR* rule, 148–149, 150
68-95-99.7 rule. *see* empirical rule

absolute values, 777–778, 779
accuracy, in sampling, 348
acorns, size of, 95–96
acupuncture, and chronic headaches, 767, 832–833
addition rule
   general, 213–215, 229, 239
   for mutually exclusive events, 209
age distribution, 158, 160
alternative hypothesis (H_a), **435**–436. *see also* hypotheses
   accepting/rejecting, 437–438
   one-sided, 435–436
   two-sided, 435–436
Alzheimer's disease, 423–424, 482–483, 485
American Community Survey (ACS), 62
amyloid-β, 423–424
analysis of variance, 625–630
   ANOVA table, 628–630, 675, 748–749
   hypotheses, 626
   means, using variation to compare, 626–628
   one-way analysis of variance (ANOVA) test, 626, 635–640, 674, 799
anonymity, 53
ANOVA. *see* one-way analysis of variance (ANOVA)
ANOVA table, 628–630, **629**, 675, 748–749
Apgar, Virginia, 264
Apgar score, 264, 265, 268
   mean of, 270–271
   standard deviation of, 272
aphid honeydew, 636–637, 638, 639–640
applets. *see* technology
aptitude tests, and crying infants, 702–703
areas, finding in normal distribution, 314–320, 315f
association, 76, 77, **606**–621
   between categorical variables, 606–612. *see also* significance test for relationship between categorical variables
   vs. causation, 78, 115, 612
   chi-square test for, 615–621
   conditional probability and, 607
   vs. correlation, 170, 176
   describing, 77
   independence and, 225

   linear, 170, 685, 686, 699
   in scatterplots, 114–115
athletes, drug testing for, 198, 257
attention-deficit/hyperactivity disorder (ADHD), maternal smoking and, 29
averaging, for sampling distribution of sample mean, 363–364
axes, in boxplots, 150

back pain treatment, 508–509
back-to-back stemplots, 95–96
bacteria, growing, 804–805
baldness, 788–789, 790–791, 810–811, 812–813
bar charts, 63–65, 64f
   definition of, **64**
   vs. histograms, 104
   interpreting, 65
   making, 64–65
   technology for, 66–67
barley seeds, drying, 777, 779
baseball, batting order in, 245
basketball players, free throws by, 434–435, 435f, 483–484
batteries, lifetime of, 347–348, 442, 443–444, 482
batting order, 245
Bayes' Theorem, **232**–234
beaches, *E. coli* measurements and, 768–770
Benford's law, 274, 605
bias, **9**
   margin of error and, 385
   mean and, 345
   nonresponse and, 11
   response, 11–12
   sample mean and, 359–360
   sample size and, 17
   sampling, 344–346
   in sampling distributions, 343–346
   sampling variability and, 16, 348, 348f
   sources of, 11–12
   in surveys, 11–12, 329–330
   unbiased estimators and, 344–346
   undercoverage and, 11
bimodal graphs, 86, 86f, 307f
binomial distributions, **280**–283, 281f
   analyzing, 281–282
   graphing, 282
   histogram for, 281f
   Large Counts condition and, 329, 330, 355
   normal approximations to, 327–329

   of sample count, 353–354
   of sample proportion, 353–354
   shape and histogram of, 328f
   shape of, 280–283, 281f, 327–329, 328f
binomial probability
   calculating, 279–280, 286–287
   complement rule for, 285
   formula for, 279–280, 285
   involving several values, 285–287
binomial random variables, 277–283
   analyzing, 285–289
   calculating, 285–287
   calculating probabilities for, 279–280
   definition of, **278**
   as discrete random variables, 278
   mean of, 287–289
   probability distribution for, 280–283
   standard deviation of, 288–289
binomial settings, **277**
   conditions for, 277–278
   identifying, 278
BINS (binomial settings mnemonic), 278
birth dates, of ice hockey players, 590–591, 592, 593–594, 597, 599–600
blackberry plant experiment, 43–46
blocks, **43**–47
   benefits of, 45–46
blood pressure
   calcium supplements for, 535, 536–537, 793–794
   confidence intervals for, 377
   measurement of, 378
   measuring, 379
blood types
   Covid-19 pandemic and, 600–601
   inheritance of, 278
body temperature, normal, 433, 498–499
bootstrap confidence interval, **825**–827
bootstrap distributions, **823**
bootstrap sample, **823**
bootstrapping, 821–828
   bootstrap confidence interval, 825–827
   bootstrap distributions, 822–823
   confidence intervals from bootstrap distributions, 823–825, 826

   four-step process for, 825–827
boundary values, 322–324, 322f, 438
boxplots (box-and-whisker plots), **149**, 411
   comparing distributions with, 152–153
   five-number summary for, 149
   limitations of, 151
   making and interpreting, 149–151
   parallel, 152–153
brand-name clothing, 506, 507
breast cancer, 232–234
butterfat in milk, 416–417, 419
buying behavior, and music, 618–619

cactuses, 24
caffeine experiment, pulse rate and, 30–31, 36, 36f, 38, 39f, 40
calcium, recommended intake of, 437, 438
calcium supplements, 535, 536–537, 793–794
calculator, graphing. *see* technology
cancer treatments, 626, 627–628, 629–630, 635–636
candy, 17–18
candy color distribution, 209, 589–590, 591–592, 597–598, 598f
carpet, age of, 747, 749, 750–751
cars. *see* vehicles
cases. *see* individuals
categorical data
   displaying, 62–67
   summarizing data, 62–63
categorical explanatory variables, 735–738
categorical variables, **3**, 606–612
   bar graphs and, 104
   describing relationship between, 76–78
   displaying relationship between, 74–76
   explanatory, 178
   relationship between, 72–78, 606–612. *see also* association
   response, 178
   significance test for distribution of, 589–594
   significance test for relationship between, 606–612
   summarizing data, 72–74
   technology for analysis of, 78

I-1

cats, attention of, 778–779, 780–781
causation
  vs. association, 78, 115, 612
  vs. correlation, 170–171
  establishment of, 51
census, **4**
center
  in dotplots, 87
  measuring, 130–135, 143. *see also* mean; median
  median as measure of, 768
  resistant measure of, 133
  of sampling distribution, 352–354
  of sampling distribution of sample mean, 359–360
  of sampling distribution of sample proportion, 352–354
  of sampling distribution of sample slope, 685
central limit theorem (CLT), **361**
Challenger Space Shuttle, 238
chance behavior, probability and, 200
charts. *see* graph(s)
Chicago
  cost of living in, 524–525, 526–527
  rent in, 772, 822–824
chi-square critical values, 417f
chi-square distribution, **415**, 417f, **598**–599
  density curves for, 415–416, 416f, 490, 490f, 598–599, 598f, 599f
  in Kruskal-Wallis test, 802
chi-square test for association, **615**–621
  conditions for, 615–616
  degrees of freedom for, 617
  four-step process for, 618–621
  P-values for, 611–612, 611f, 617–618
chi-square test for goodness of fit, **596**–603
  conditions for, 597
  degrees of freedom for, 598–599
  four-step process for, 600–603
  P-values for, 597–600, 599f
chi-square test statistic, **593**
  calculating, 593
  for distribution of categorical variable, 592–594, 594f, 598f
  observed and expected counts and, 592–594
  for relationship between categorical variables, 610–612
  sampling distribution of, 598
chronic headaches, and acupuncture, 767, 832–833
circle graphs. *see* pie charts

claims, testing, 17–18, 433–495
classroom seat location test scores and, 697–698
clothing
  brand names, 506, 507
  fit of, 425
CLT (central limit theorem), 361
cluster random sampling, **24**
clusters, **24**, 84
coefficient
  individual, *t* test for, 750–754
  of interaction term, 740
coefficient of determination (r^2), 675f, **675**
  calculating, 676–678
  in population multiple regression model, 752
  standard deviation of the residuals and, 676
coin tosses, 282
college students. *see also* teenagers
  age of, 239
  distance from home, 355–356
combinations, 249, **250**
  calculating, 250–251
  permutations and, 250
  probability and, 251–254
combined sample proportion, 515
complement, **209**, 216, 216f
complement rule, **209**
  for binomial probability, 285
completely randomized designs, 35–40, **37**
conclusions, in significance test, 438–439, 442–443
concussions, 84
conditional probability, 220, **221**
  association and, 607
  Bayes' Theorem and, 232–234
  calculating, 222–223
  independence and, 223–224, 278
  notation for, 221
  tree diagrams and, 230–232, 230f, 231f
  two-way tables and, 221–222
  vs. unconditional probability, 278
confidence interval for difference between means
  calculating, 525–527
  conditions for, 524–525
  constructing and interpreting, 527–530
  degrees of freedom for, 525–526, 536–537
  technology for, 529
confidence interval for difference between proportions, 505–510
  conditions for, 505–506
  constructing and interpreting, 508–510
  order of subtraction and, 508

confidence interval for mean difference, 549–551
confidence interval for population mean, 399–406, 409–412, 475–476
  calculating, 404–406
  conditions for, 400
  critical values for, 402–404
  four-step process for, 410–411
  Normal/Large Sample condition for, 400, 409–410
  Random condition for, 400
  t^* critical values and, 402–404
  two-sided tests and, 75–76
  with unknown standard deviation, 400–402, 402f
confidence interval for population proportion, 388–393
  calculating, 391–393
  conditions for, 388–389
  critical values for, 390
  four-step process for, 395–396
  margin of error for, 396–397
  sample size for, 396–397
confidence interval for slope, 688
  four-step process for, 689–692
confidence interval(s), 375–429
  from bootstrap distributions, 823–825, 826
  building, 378
  conditions for estimating, 416–417
  confidence levels and, 377, 383f
  controlling width of, 713–716, 714f
  critical values for, 390, 391, 403, 417–418, 417f. *see also* critical values
  decision making with, 378–379
  definition of, **376**
  for difference between means, 523–530
  four-step process for, 395–396, 410–411, 549–551
  interpreting, 377
  margin of error in, 378, 390, 396–397
  for mean response, 715–716
  for a mean response and prediction intervals in regression, 709–716
  for the mean *y* value at a given value of *x*, 709–711
  overview of, 376–379
  parameters vs. statistics and, 377
  plausible values and, 376
  point estimates and, 378, 390
  for population mean difference, 549–551
  population parameters and, 383

for population standard deviation, 416–417, 418–420, 423–426
for population variance, 418–420
sample size, effect of, 713–714
sample size for, 400, 409–410
vs. significance tests, 461, 475–476
size of, 384
for standard deviation, 424–426
standard error and, 391–393
for two population standard deviations or variances, 571
two-sided tests and, 461, 475–476
confidence levels, **377**, 383f
  interpreting, 382–383, 383f
  margin of error and, 384–385
confidentiality, 52, 53
confounding, 28–**29**
congressional members, 3–4
conservative degrees of freedom, 525, 536–537
contingency table. *see* two-way tables
continuity correction, 336
continuous random variables, **263**, 305
  density curves and, 305–307
  from measuring, 305
control group, **30**
convenience samples, **9**, 10
correlation (*r*), 167f, **167**
  vs. association, 176
  calculating, 176–177
  vs. causation, 170–171, 703
  estimating and interpreting, 167–170
  explanatory variables and, 178
  guessing, 168
  outliers and, 179–181
  properties of, 178–179
  response variables and, 178
  scatterplots and, 167–171, 167f, 168f, 180–181
  testing for, 703–704
  units of measure and, 178–179
correlation coefficient, calculating, 676–678
cost of living comparison, 524–527
counting, probability and, 251–254
counts
  expected. *see* expected counts
  observed. *see* observed counts
Covid-19 pandemic, 116, 600–601
  retiree income and, 813–814
credit card fraud, 375

critical values, **390**, 391
  in bootstrapping, 822
  for confidence interval for
    population mean, 402–404
  for confidence interval for
    population proportion,
    390
  for confidence interval for
    slope, 688
  for confidence interval for
    the mean y value at a given
    value of x, 710
  for confidence intervals
    for population standard
    deviation, 417–418, 417f
  for prediction intervals, 711
  t^*, 402–404, 688
cross-fertilization, 547, 548
Current Population Study
  (CPS), 25, 49, 63–64

Darwin, Charles, 547
data
  paired, 545–551. *see also*
    paired data
  quantitative. *see* quantitative
    data
data collection, 1–56
  blocking, 43–47
  cluster random sampling, 24
  completely randomized
    designs, 35–40
  ethics, 52–53
  experiments, 29–31
  inferences and, 49–50
  introduction to, 2–5
  observational studies, 4–5,
    28–29
  sampling, 8–12
  simple random sampling,
    15–19
  stratified random sampling,
    22–24
  systematic random sampling,
    25
data dredging, 478
data ethics, 52–53
decision making
  with confidence intervals,
    378–379
  significance test and,
    442–445
  degrees of freedom (df),
    403–404, 415
  in ANOVA table, 629
  for chi-square test for
    association, 617
  for chi-square test for
    goodness of fit, 598–599
  for confidence interval for
    difference between means,
    525–526, 536–537
  conservative, 525, 536–537
  for F distribution, 567
  in Kruskal-Wallis test, 802
  for significance test for mean,
    467–471, 467f, 468f

  for significance test for
    relationship between
    categorical variables, 617
  for t distribution, 467–468,
    536–537
denominator degrees of
  freedom, 567
density curves, **305**–307
  center of, 308
  for chi-square distributions,
    415–416, 416f, 490, 490f,
    598–599, 598f, 599f
  for normal distributions,
    307–309, 307f
  shape of, 307, 307f, 309
  for t distribution, 403f, 467,
    467f
  uniform distributions, 306
  variability, 308
df. *see* degrees of freedom (df)
diabetes, 28–30
dice, possible outcomes with,
  207–208, 207f
diet soda, sweetness of, 661–662
difference between means,
  523–530
  conditions for, 524–525, 555
  confidence intervals for,
    523–530
  estimating, 523
  four-step process for, 538–540
  vs. mean difference, 548–549
  mean of, 524
  sampling distribution of, 524
  shape of, 524
  significance tests for, 534–540
  standard deviation of, 524
  standard error of, 536, 549
  testing claim about, 534–540.
    see also significance test for
    difference between means
  two-sample t interval for, 527
difference between proportions
  confidence interval for,
    505–510
  critical values for, 506
  estimating, 505–506
  sampling distribution of, 505
  significance tests for, 514–520
  standard deviation for, 505, 507
  standard error of, 507
  testing claim about, 514–520.
    see also significance test
    for difference between
    proportions
  z statistic for, 515
discrete probability, displaying
  distributions, 268–269
discrete random variables,
  263–264
  analyzing, 268–274
  binomial random variables
    as, 278
  boundary value for, 265
  from counting, 305
  mean of, 269–271
  median of, 271

  probability distribution of,
    264–266, 268–269
  standard deviation for,
    271–274
discrimination, in hiring, 2
distractions
  driving and, 4, 37, 39
  multitasking, 36
distribution(s), **62**. *see also*
  specific types
  boxplots, comparing with,
    152–153
  center of, 130–135. *see also*
    mean; median
  chi-square, 598–599, 598f,
    599f
  dotplots, comparing with,
    86–88
  dotplots, describing with,
    86–88
  histograms, comparing with,
    105–106
  location, describing in,
    158–163. *see also*
    percentiles; z-scores
  location, measuring in,
    158–163
  locations, comparing in, 162
  mean of, 130, 132–133
  median of, 130–131, 159
  mode of, 130
  normal. *see* normal
    distributions
  percentiles and, 159
  probability. *see* probability
    distributions
  quartiles and, 142–144, 159
  sampling. *see* sampling
    distribution(s)
  shape, describing, 85
  skewness of, 157, 409–410
  standard normal. *see*
    standard normal
    distributions
  stemplots, comparing with,
    95–96
  summarizing with boxplots,
    152–153
  t, 467–471, 467f, 468f,
    536–537, 556
  z-scores and, 160–162
diversity, STEM workers and,
  588, 646
dogs, sleeping with, 10
dotplots, 36f, **82**, 133f
  for assessing normality of
    distributions, 411
  boxplots and, 150
  for comparing distributions,
    86–88
  vs. histograms, 101f
  inference for sampling
    and, 17
  interpreting, 83–84
  making, 83–84
  mean and median for, 133
  outliers and, 133–135

  parallel, 546f
  percentiles and, 159
  for quantitative data, 82–88,
    83f
  for sample mean
    distributions, 16
  technology for, 88
double-blind experiments, **31**
drinking water, zinc in, 545–548,
  546f, 550
drive-thrus, fast-food, 444–445,
  449, 451, 504–505, 577–578
driving, distractions in, 4, 37, 39
drug tests, false-negatives/
  false-positives in, 198, 257

E. coli measurements, 768–770
earnings
  and education, 105–106
  median income, 16–17
ebooks, 231–232
education, and earnings,
  105–106
educational levels, 64, 64f
effect size, 484
Electoral College, 93–94, 95
emergency room wait times,
  436
emissions, and engine size,
  651, 720
empirical rule, 309–311, **310**
employees, sampling, 353, 355
energy conservation, 30
engagement ring prices, 400,
  405
engine size
  and emissions, 651, 720
  and fuel efficiency, 739
English, opinions about
  speaking, 616, 618
equal-areas point, 307
equal standard deviation
  condition, 635–636
  for Kruskal-Wallis test, 802
  for population multiple
    regression model,
    747–748
  for significance test for slope,
    685, 700
  for Wilcoxon rank sum test,
    791
equally likely outcomes, 208
error. *see* margin of error;
  standard error
error term, 746
ethics, 52–53
event(s), **208**
  complement of, 209
  independent, 223–224,
    237–240, 278, 608
  mutually exclusive, 209–210,
    213, 239–240
  probability of, 208
exam scores
  classical music and, 46
  seating position and,
    697–698

expected counts
    calculating, 591–592
    for chi-square distribution, 598
    for chi-square test statistic, 592–594, 594f
    for distribution of categorical variables, 591–592
    for relationship between categorical variables, 609–610
expected value, of discrete random variable, 269–271, **270**
experimental unit, **29**
experiment(s), 4–**5**, 29–31
    blocking, 43–46
    matched pairs design, 46–47
    random assignment in, 31–32, 518–519. *see also* random assignment/sampling
    variability, sources of, 36
explanatory variables, **28**
    correlation and, 178
    indicator variables and, 737
    in multiple regression model, 727–729, 731
extrapolation, **653**
eye color, and flickering light, 799–800, 802

F distribution, 566–**567**, 567f, 567t–568t, 636
F statistic, 629, 630, 630f, 636
F test for a population multiple regression model, **749**–750
Facebook, rate of use, 215
factorials, **244**
false-negatives, 198, 232, 257
false-positives, 198, 232, 257
family dinners, 5
fast food
    drive-thrus, 444–445, 449, 451, 504–505, 577–578
    popularity of, 383, 385
Federalist Papers, 8–9
fertilization, and plants, 547, 548
financial markets, 25
first quartile (Q_1), **142**. *see also* interquartile range (IQR)
    equivalent percentile for, 160
    in five-number summary, 149
    resistance of, 143
first speech, age of, and mental aptitude, 663
Fisher's least significant difference (LSD) intervals, 637–640
fitted values, 747
five-number summary, **149**. *see also* boxplots
flickering light, and eye color, 799–800, 802
Flint water crisis, 82–83, 83f, 87, 130, 143, 159, 161

flu vaccine, and side effects, 809, 811
food consumption, 527–528
food insufficiency, 436
foot length, height and, 698–699, 700–701
football, actual game time in, 410–411, 825–826
four-step process
    for bootstrapping, 825–827
    for chi-square test for association, 618–621
    for chi-square test for goodness of fit, 600–603
    for confidence intervals, 395–396, 410–411, 549–551
    for difference between means, 538–540
    for Kruskal-Wallis test, 804–806
    for mean, 474–475
    for population standard deviation or variance, 492–495
    for proportion, 457–459
    for randomization tests, 813–816
    for relationship between categorical variables, 618–621
    for sign test, 771–772
    for significance test, 457–459, 474–475
    for slope, 701–705
    for Wilcoxon rank sum test, 793–795
    for Wilcoxon signed rank test, 781–784
free throws, 434–435, 435f
french fries, serving size of, 151
frequency histogram, 103f
frequency tables, **62**, 63
fuel mileage, 47, 332, 739

gaps, in graphs, 84
gas mileage, 134
general addition rule, 213–215, **214**, 229, 239
general multiplication rule, **229**–230, 231, 252
Gesell Adaptive Score, 663, 664f
Gladwell, Malcolm, 590
golden hamsters, weight of, 466, 468–469, 489–490, 490f
golf drive distance, 317, 319, 324, 726–729, 727f, 730
golfers, regression to the mean and, 665
Gosset, William S., 467, 777
government, branches of, 389, 391, 823, 824–825
Grand Canyon, visitors to, 116–117
graphing calculator. *see* technology

graph(s), 61–118
    advantages and disadvantages of, 411
    analyzing paired data with, 546–549, 546f
    assessing normality of distribution with, 411
    bar. *see* bar charts
    bimodal, 86, 86f
    of binomial distribution, 282
    boxplots, 149–151, 153, 411
    for categorial data, 62–67
    categorical variables, relationship between, 72–78
    clusters, 84
    dotplots, 82–88
    gaps in, 84
    histograms, 100–107
    misleading, 66–67
    multimodal, 86
    peaks in, 84
    for quantitative data, 82–88, 92–97, 100–107
    quantitative variables, relationship between, 112–118
    of sampling distributions, 337
    selecting, 411
    skewness in, 157, 409–410
    stemplots, 92–97
    unimodal, 86
    when to use, 411

H_0. *see* null hypothesis
H_a. *see* alternative hypothesis
hair
    color of and pain threshold, 789–790, 792–793
    growth of, 788–789, 790–791, 810–811, 812–813
hamsters, weight of, 466, 468–469, 489–490, 490f
hand washing, 12
handedness, 606–607, 607f, 609, 616
head circumference, 85
headaches, and acupuncture, 767, 832–833
healing time, 477
heart disease, and anger, 607–608, 610–612, 615, 617, 620–621
heart rate, 488–489, 491–492
    of shrews, 769–770, 771
height
    of fathers and sons, 664–665, 664f
    foot length and, 698–699, 700–701
    of ponderosa pine trees, 654–655, 672–676, 672f, 686–688, 709–712, 714, 714f
    of young men, 338
Helsinki Heart Study, 518–519
Hershey's Kisses, 17–18
high-income countries, 514–515, 516

histograms, **100**–107
    for assessing normality of distribution, 411
    vs. bar charts, 104
    of binomial distribution, 280–283, 281f, 328f
    density curves of. *see* density curves
    describing, 103–105
    distributions, comparing with, 105–106
    vs. dotplots, 101f
    making, 101–103
    normal density curve models, 307f
    normal probability plot, 331f
    of probability distribution, 268–269, 268f, 280–283, 281f
    for quantitative data, 100–107
    technology for, 106–107
HIV testing, 238
households, size of, 87
housing costs, in Chicago, 772, 822–824
hurricanes in Hawaii, 294–296
hybrid vehicles, 739
hypotheses
    accepting/rejecting, 437–438. *see also* significance test(s)
    alternative, 435–438
    for analysis of variance, 626
    for difference between means, 534–535
    for difference between proportions, 514–515
    for distribution of categorical variable, 555, 590–591
    independence and, 608
    for Kruskal-Wallis test, 799–801
    for mean difference, 555
    null, 435–436
    one-sided alternative, 435–436
    for population multiple regression model, 746–747
    for randomization tests, 809–810
    for relationship between categorical variables, 606–608
    for relationship between quantitative variables, 698–699
    for sign test, 769–770
    for slope, 698–699
    statement of, 435–436
    for two population standard deviations or variances, 565–566
    two-sided alternative, 435–436

ice breaking game, 476
ice hockey players, birth dates of, 590–591, 592, 593–594, 597, 599–600

immigration, 101–103, 103f, 104
income, globally, 514–515, 516
independent events, 223–224
    association between, 225
    conditional probability and, 223–224, 278
    hypotheses using, 608
    multiplication rule for, 237–240
    mutually exclusive, 239–240
indicator variables, **736**
    and interaction, 735–741
individual coefficient, t test for, 750–754
individuals, **3**–4
infants, crying and later aptitude, 702–703
inference, 376
    about slope, 684–687, 684f
    for multiple regression, 745–754
    for sampling, 17–18, 518–519, 537, 558–560
inference about a population, 49–**50**
inference about cause and effect, **50**–52
informed consent, 52, 53
Instagram, rate of use, 215
institutional review board, 52
interaction term, **739**–741
interquartile range (IQR), **142**, 151
    finding and interpreting, 142–143
    first quartile (Q_1), 142–144, 149, 159
    identifying outliers with, 148–149, 150
    resistant, 143
    third quartile (Q_3), 142–144, 149, 159
    when to use, 143
intersection, in Venn diagram, 216f, **216**
IQR. *see* interquartile range

James, LeBron, 148, 149, 150, 152
jury selection, 261, 300

Kruskal-Wallis test, 635, **799**
    conditions for, 802
    four-step process for, 804–806
    hypotheses, 799–801
    P-values, 801–803
    ranking data, 799–801
    technology, 805–806
    test statistic, 801–803

landfills, 15–16
Large Counts condition, **329**, 330, **355**–356, 451, 453
    in bootstrapping, 822, 824
    for chi-square goodness of fit test, 597, 598
    for chi-square test for association, 615–616

for confidence interval for difference between proportions, 505–506
for confidence interval for proportion, 388–389
for randomization distributions, 811
for randomization tests, 810
for significance test for difference between proportions, 514–515
for significance test for proportion, 448–449, 450
law of large numbers, 200
least-squares regression line, 658, 659f, **659**, 664f
    calculating equation of, 659–662
    confidence intervals for the slope of, 688
    inference for slope of, 684–687, 684f
    outliers and, 662–664
    population, 681–692, 682f
    regression to the mean and, 664–666
    sample, 584, 684
    scatterplots, 683f
    slope of. *see* slope of least-squares regression line
    sum of squared residuals and, 675f
left-handedness, 224–225
left-skewed distributions. *see* skewness
Legos, pricing, 737, 738, 740–741
license plate numbers, 244
life expectancy, 781–783
linear association, 170, 685, 686, 699, 747–748
linear regression, 651–720
    confidence intervals for mean response, 709–716
    least-squares regression line, 658–666
    population least-squares regression line, confidence intervals for, 681–692
    prediction intervals in, 709–716
    regression lines and, 652–655
    regression models, assessing, 671–678
    significance tests for slope of a population least-squares regression line, 697–705
location, measuring in a distribution, 158–163
lotteries, combinations and probability for, 250–251, 252–253
low-income countries, 514–515, 516

malaria prevention, 1, 56
mammograms, 232–234

Mann-Whitney U test, 788
margin of error, **378**
    bias and, 385
    in bootstrapping, 822
    calculating, 378
    confidence intervals and, 378, 390, 396–397
    confidence levels and, 384–385
    decreasing, 384
    factors affecting, 384–385
    factors not accounted for by, 385–386
    sample size and, 384–385, 396–397, 411
    vs. standard error, 391
matched pairs design, **46**–47, 558
mean difference
    conditions for, 549
    confidence interval for, 549–551
    vs. difference between means, 548–549
    estimating, 545–551
    mean of, 547–548
    paired data and, 545–551. *see also* paired data
    significance test for, 555–560
    standard deviation of, 547–548, 556
    standard error of, 549
mean square (MS), 629
mean(s), **132**–133. *see also* center
    as balance point, 133
    bias and, 345
    of binomial random variable, 287–289
    calculating, 143–144
    confidence intervals for, 399–406, 475–476
    in correlation calculation, 176, 178
    difference between, 523–530. *see also* difference between means
    of discrete random variable, 269–271
    estimating, 399–406
    finding from areas in normal distribution, 324–325
    interpreting, 133
    of mean difference, 547–548
    vs. median, 133–135
    as nonresistant measure, 133
    notation for, 132
    outliers and, 133, 180
    population. *see* population mean
    regression to, 664–666
    rounding of, 271
    sample. *see* sample mean of sampling distribution of
    of sampling distribution of difference between proportions, 505
    of sampling distribution of sample mean, 359–360

of sampling distribution of sample slope, 684, 685
significance test for, 474–478
standard deviation and, 140–141
testing claim for, 464–471. *see also* significance test for mean
trimmed, 137
using variation to compare, 626–628
when to use, 133–135
z-scores and, 160
measures of center, 130–135. *see also* mean; median
    calculating, 143–144
    choosing, 133–135
median, **130**–131
    calculating, 143–144
    of discrete random variable, 271
    of distributions, 130–131, 159
    equivalent percentile for, 159
    finding and interpreting, 131
    in five-number summary, 149
    vs. mean, 133–135
    as measure of center, 768
    with odd vs. even number of values, 130
    outliers and, 133
    as resistant measure of center, 133
    when to use, 133–135, 143
mental aptitude, and age of first speech, 663
middle-income countries, 514–515, 516
Milne, Joy, 262
Milne, Les, 262
mode, of distributions, 130
models
    probability, 207–208
    regression, assessing, 671–678
movie viewership, 360, 362
multicollinearity, 739
multimodal graphs, 86
multiple linear regression model, 728
multiple regression, 725–758
    idea of, 726–729
    indicator variables and interaction, 735–741
    inference for, 745–754
    introduction to, 726–732
multiple regression model, **728**
    assessing, 729–732, 752–753
    coefficients, interpreting, 738–739
    conditions for, 747–748
    explanatory variables and, 727–729, 731
    interaction term, 739–741
    vs. mean-only model, 746
    residuals and, 727–729
    technology for, 731–732
    testing, 746–750

multiplication counting principle, 243-**244**
multiplication rule(s)
general, 229-230, 231, 252
for independent events, **237**-240
multistage sampling, 25
multitasking, and learning, 36
multitasking experiment, 38
music, 5
amount of airplay, 464-465, 465f
and buying behavior, 618-619
mutually exclusive events, **209**-210, 213
addition rule for, 209-210
independence and, 239-240

National Hockey League players, birth dates of, 590-591, 592, 593-594, 597, 599-600
negative association, 114
New York, cost of living in, 524-525, 526-527
newspaper readership, 223
nitrogen-filled tires, 556-557
no association, 114
non-normal population, sampling from, 361
nonparametric methods, 767-833
bootstrapping, 821-828
Kruskal-Wallis test, 799-806
randomization tests, 809-817
sign test, 768-772
Wilcoxon rank sum test, 788-795
Wilcoxon signed rank test, 777-784
nonresponse, **11**
normal approximations to binomial distributions, 327-333
calculating probabilities, 330
continuity correction and, 336
finding probabilities involving sample proportion and, 355-356
Large Counts condition and, 329, 355-356
when to use, 330
Normal condition
for population multiple regression model, 747-748
for significance test for population standard deviation, 488-489
for significance test for slope, 685, 700
normal curve, **307**-309
drawing, 308
standard normal distribution and, 315
normal density curve models, 307f

normal distributions, **307**-309
density curves for, 307-309, 307f. *see also* density curves
drawing, 308
empirical rule and, 309-311
finding areas between two values in, 318-320
finding areas from values in, 314-320, 315f
finding areas in, 316
finding areas to the left in, 317
finding areas to the right in, 317-318
finding boundary values, 322-324, 322f
finding mean from areas in, 324-325
finding probabilities involving sample proportion and, 355-356
finding standard deviation from areas in, 324-325
finding values from areas in, 322-325
standard. *see* standard normal distributions
Normal/Large Sample condition
in bootstrapping, 822, 824, 826
for confidence interval for difference between means, 524-525
for confidence interval for mean, 400, 409-410
for confidence interval of population standard deviation, 423-424
for Kruskal-Wallis test, 802
for mean difference, 549
for one-way ANOVA test, 635-636
for significance test for difference between means, 535, 555
for significance test for mean, 464-465
for significance test for mean difference, 556-558
standardized test statistic and, 466
for two population standard deviations or variances, 565-566
for Wilcoxon rank sum test, 791, 794
normal population, sampling from, 360-362
normal probability plot, 330, 331f, **331**-333
normal quantile plot, 331
null hypothesis (H0), **435**-436. *see also* hypotheses
numerator degrees of freedom, 567, 567t-568t

numerical summaries
computing with technology, 143-144
for quantitative data, 129-189
numerical variables. *see* quantitative variables
nurse practitioners, 44, 46

observational studies, 4-**5**, 28-29
observational units. *see* individuals
observed counts, chi-square statistic and, 592-594
Old Faithful geyser, 682-684, 682f, 683f
On the Origin of Species (Darwin), 547
1.5 × *IQR* rule, 148-149, 150
one-sample confidence interval for a standard deviation, **424**-426
one-sample *t* interval for mean difference, 549
one-sample *t* interval for population mean, **410**-412. *see also* confidence interval for population mean
one-sample *t* test for a mean, **474**. *see also* significance test for mean
one-sample *t* test for mean difference, 557
one-sample test for a standard deviation (variance), **492**-495
one-sample *z* interval for population proportion, **395**. *see also* confidence interval for population proportion
one-sample *z* test for proportion, **457**. *see also* significance test for proportion
one-sided alternative hypothesis, 435-**436**
one-sided significance test, 457-458
one-sided tests, **457**-458
one-tailed test, 459
one-variable quantitative data
location in distribution, 158-163
normal distributions, finding areas from values, 314-320
one-way analysis of variance (ANOVA) test, **626**, 674, 799
conditions for, 635-636
Fisher's least significant difference intervals, 637-640
P-values, 636-637
technology, 639-640
online dating, 450-451, 452

outliers
1.5 × *IQR* rule, 148-149, 150
in boxplots, 149-151
correlation and, 179-181
describing effects of, 663
in dotplots, 87, 133-134
identifying with interquartile range, 148-149
least-squares regression line and, 662-664
mean and, 133, 180
median and, 133
in scatterplots, 180
in significance tests, 476
oxygen, level of in water, 728, 729, 730, 746, 750

pain threshold, and hair color, 789-790, 792-793
paired data, 545-551, **546**
analyzing graphically, 546-549
analyzing numerically, 547-548
confidence interval for mean difference and, 549-551
matched pairs design and, 558
significance test for mean difference and, 555-560
vs. two samples, 558-560
paired *t* interval, for mean difference, **549**
paired *t* test, for mean difference, **556**-558
parallel boxplots, 152-153
parallel dotplots, 546f
parameters, **336**-337
estimating, 375-429
plausible values for, 377
vs. statistics, 336-337
Parkinson's disease, odor from, 262
peaks, in graphs, 84
peanuts, jar weight of, 363
percentiles, **159**
describing, 159
finding and interpreting, 159-160
median and, 159
reporting as whole numbers, 159
z-scores and, 162
periodontal disease, 51-52
permutations, **244**-247
combinations and, 250
computing, 246-247
notation for, 244, 245
P-hacking, 478
Pick Six lotto game, 250-251
pictographs, 66
pie charts, 63-65, 64f
definition of, **64**
relative frequency, 65
technology for, 66-67
placebo, **29**
placebo effect, **31**

plants, fertilization and, 547, 548
plausible values, for parameter, 377
point estimate, **378**, 710, 711, 822
Poisson distribution, 292, **294**
Poisson probabilities
  calculating, 293–294
  formula for, 293
  involving several values, 294–295
  over different interval lengths, 295–298
  for random variables, 296–297
Poisson random variables, 292–298, **293**
Poisson setting, **292–293**
polls. *see* sampling
ponderosa pine trees, 654–655, 672–676, 672f, 686–688, 709–712, 714, 714f
pooled two-sample t procedures, 528
pop quizzes, grading, 277, 278, 280, 286, 289
population density, 138
population distribution, mean of, 345
population least-squares regression line, 681–692, 682f
population mean, 132. *see also* mean
  confidence intervals for, 399–406
  critical values for, 402–404
  difference between. *see* difference between means
  notation for, 132
  t distribution and, 467–471, 467f, 468f
  testing claims about, 464–471. *see also* significance test for mean
population multiple regression model, 746
  conditions for, 747–748
  F test, 749–750
  variables, selecting, 751–752
population parameters, 336–337
  confidence intervals and, 383
  plausible values for, 377
  vs. statistics, 336–337
population proportion
  confidence intervals for, 395–397. *see also* confidence
  estimating, 388–393
  interval for population proportion difference between. *see* difference between proportions
  one-sample z interval for, 395

one-sample z test for, 457
  testing claims about, 448–454. *see also* significance test for proportion
population regression line, 681–**682**, 683f
  slope of, 684, 684f
population standard deviation, 139, 404
  confidence intervals for, 416–417, 420, 423–426
  critical values for, 420
  estimating, 414–420
  significance tests for, 488–495, 490f
population(s), **4**
  comparing, 504–578
  normal, sampling from, 360–362
portion size, 527–528
positive association, 114
posterior probability, 232
potato chips, percentage of air in bags, 138, 143–144
power of a test, **481**–485
  increasing, 484
  Type I and II errors, 482, 484, 484f
precision, in sampling, 348
prediction intervals
  conditions for, 711
  for an individual y value at a given value of x, 711–713
  prediction intervals in regression, 709–716
predictions, using regression line, 652–653
presidential approval rating, 17
prior probability, 232
probability, 198–257, 200f
  basic rules of, 207–210
  binomial, 279–280, 285–287
  central limit theorem and, 361
  chance behavior and, 200
  combinations and, 251–254
  complement and, 209
  complement rule and, 209
  conditional, 220–225, 278, 607. *see also* conditional probability
  counting and, 251–254
  definition of, **200**
  estimating with simulations, 202–203
  of event, 208
  general multiplication rule and, 229–230
  idea of, 199–201
  interpreting, 201
  involving sample mean, 362–364
  involving sample proportion, 355–356
  multiplication counting principle and, 243–244

multiplication rule for independent events and, 237–240
myths about randomness and, 201
two-way tables for, 213–215
of Type I error, 445
unconditional, 278
Venn diagrams and, 215–217, 215f, 216f
probability distributions, **264**–266
  of binomial random variable, 280–283, 281f
  of discrete random variable, 264–266, 268–269
  displaying, 268–269
  shape and histogram of, 268–269, 268f, 280–283, 281f
  skewness of, 409–410
probability models, 207f, **207**–208
  Benford's law and, 274, 605
  equally likely outcomes and, 208
product pricing, 737, 738, 740–741
proportion(s). *see also* population proportion; sample proportion
  confidence intervals for. *see* confidence interval for population proportion
  difference between. *see* difference between proportions
  estimating, 388–393
  expected, 593
  testing claim about, 448–454. *see also* significance test for proportion
pulse rate, while standing vs. sitting, 559–560, 560f
P-values, **437**–438
  in AVOVA table, 629, 630
  calculating, 452, 594, 597–600
  from chi-square distribution, 598–599, 599f
  for chi-square test for association, 611–612, 611f, 617–618
  for chi-square test for goodness of fit, 597–600, 599f
  for chi-square test statistic, 594
  for chi-square tests for goodness of fit, 601–603
  for difference between means, 536–537
  for difference between proportions, 515–517
  interpreting, 437–438

for Kruskal-Wallis test, 801–803
in one-sided tests, 457–458
for one-way ANOVA test, 636–637
in population multiple regression model, 749, 750, 751–752
for randomization distributions, 811–813
for sign test, 770–771, 777
for significance test for mean, 467–471, 475–476
for significance test for mean difference, 556
for significance test for proportion, 452, 475
for significance test for slope, 699–701
in significance tests for population standard deviation, 489–492, 494
size of, 438
t distribution and, 467–471, 467f
for two population standard deviations or variances, 566–569, 571–572
in two-sided tests, 459–461
for Wilcoxon rank sum test, 790–793
for Wilcoxon signed rank test, 779–781

qualitative variables. *see* categorical variables
quantitative data
  displaying. *see* dotplots; graph(s); histograms; stemplots
  distribution of. *see* distribution(s)
  dotplots, 82–88
  histograms, 100–107
  numerical summaries, 129–189
  stemplots, 92–97
  summarizing, 149–151. *see also* boxplots
quantitative variables, **3**
  describing distributions, 86
  histograms and, 104
  relationship between, 112–118, 697–705. *see also* significance test for slope
quartiles, **142**–144, 159–160. *see also* interquartile range (IQR)
  boxplots and, 149
  resistance of, 143
quiz scores, 85

random assignment, **31**–32
  inference and, 518–519, 537, 558–560
  simulation and, 537

Random condition, 451
  in bootstrapping, 824
  for chi-square goodness of fit test, 597
  for chi-square test for association, 615–616
  for confidence interval for difference between means, 524–525
  for confidence interval for difference between proportions, 505–506
  for confidence interval for mean, 400
  for confidence interval for proportion, 388–389
  for Kruskal-Wallis test, 802
  for mean difference, 549
  for one-way ANOVA test, 635–636
  for population multiple regression model, 747–748
  for randomization tests, 810, 811–812
  for sign test, 770, 771, 777
  for significance test for difference between means, 535, 555
  for significance test for difference between proportions, 514–515
  for significance test for mean, 464–465
  for significance test for mean difference, 556–558
  for significance test for population standard deviation, 488–489
  for significance test for proportion, 448–449
  for significance test for slope, 685, 699
  for two population standard deviations or variances, 565–566
  for Wilcoxon rank sum test, 791
  for Wilcoxon signed rank test, 780
random number generators, 15
random samples, **10**
random sampling, 518–519.
  see also simple random sample (SRS)
  inference and, 537, 558–560
  in significance test for difference between proportions, 516, 518–519
  simulation and, 17, 537
random variables, **262**–266
  binomial, 277–283. *see also* binomial random variables
  continuous, 263, 305. *see also* continuous random variables
  discrete, 263–264, 268–274. *see also* discrete random variables
  Poisson, 292–298

randomization distributions, 809–811
  definition of, **810**
  P-values, 811–813
randomization tests, 809–817
  conditions for, 810, 811–812
  definition of, **810**
  four-step process for, 813–816
  hypotheses, 809–810
  randomization distributions, 809–811
  technology, 815–816
randomized designs, 35–40
randomness, myths about, 201
range, **138**–139
  interquartile. *see* interquartile range (*IQR*)
  nonresistance of, 139
ranks, 777–779
  for two samples or groups, 788–790
red squirrels, food availability and population size, 735–736, 737–738
regression line(s), **652**–655, 653f, 659f. *see also* least-squares regression line
  extrapolation with, 653
  interpreting, 654–655
  predictions, using regression line, 652–653
  residuals and, 653–654
regression models
  assessing, 671–678
  residual plots and, 671–673, 672f
regression to the mean, **664**–666
rejection region approach, 452–453
relative frequency tables, **62**, 63, 65, 73–74
rent, in Chicago, 772, 822–824
rental housing, 61, 123
replication, 32
residual plots, **671**
  interpreting, 672–675
  making, 676–678
  patterns in, 673, 685–686
  regression models and, 671–673, 672f, 747–748
residuals, 653f, **653**, 659f
  calculating and interpreting, 729
  in multiple regression model, 727–729
  regression lines and, 653–654, 659f
  squared, 674, 675, 675f
  standard deviation of, 673–675
  sum of squared, 674, 675, 675f
resistant measure of center, 133
response bias, **11**–12
response variables, **28**
  correlation and, 178

resting heart rate, 488–489, 491–492
retiree income, and Covid-19 pandemic, 813–814
right-skewed distributions. *see* skewness
rocks, density of, 801, 803
roller coasters, 3
roughly symmetric shape, **84**
roundoff error, 64

salt intake, 304, 367
sample count. *see also* counts
  binomial distribution of, 353–354
sample least-squares regression line, 682, 684
sample mean, 16
  center of, 359–360
  notation for, 132
  probabilities involving, 362–364
  sampling distribution of, 345, 347, 359–364. *see also* sampling distribution of sample mean
  shape of, 360–362
  standard error of, 404
  as unbiased estimator, 345
sample mean, 55, 254, 254f. *see* mean
sample proportion, 337
  binomial distribution of, 353–354
  combined, 515
  finding probabilities involving, 355–356
  P-values for, 450f, 451–454, 451f
  sampling distribution of, 352–356. *see also* sampling distribution of sample
  significance test for, 451–454, 451f. *see also* significance test for proportion
  standard error of, 391–393
  standardized test statistic for, 451–454, 451f
sample range, mean of sampling distribution of, 345
sample regression line, 681–**682**
  slope of, 684, 684f
sample size
  bias and, 17
  for confidence interval, 400, 409–410
  confidence intervals and, 713–714
  for margin of error, 384–385, 396–397, 411
sample space, **207**
sample standard deviation, 139, 391, 404
sample statistics, 336–337
sample surveys. *see* surveys
sample variance, 628

sampling distribution and, 415–416
sample(s), **4**
sampling, 8–12
  accuracy in, 348
  bias in, 11–12, 348. *see also* bias
  cluster random, 24
  inference for, 17–18, 537
  from non-normal population, 361
  nonresponse and, 11
  from normal population, 360–362
  precision in, 348
  random. *see* random sampling
  simple random, 15–19
  stratified random, 22–24
  systematic random, 25
  undercoverage and, 11
sampling distribution of difference between means, 524
sampling distribution of difference between proportions, 505
sampling distribution of sample count X. *see* sample count
sampling distribution of sample mean, **359**–364
  bias and, 359–360
  finding probabilities for, 362–364
  mean of, 359–360
  from normal population, 360–362
  shape of, 360–362, 362f
  standard deviation of, 359–360, 404
  using averages in, 363–364
sampling distribution of sample proportion, **352**–356
  center of, 352–354
  finding probabilities for, 355–356
  Large Counts condition and, 355–356
  mean of, 352–354
  shape of, 354–355
  standard deviation of, 352–354
  variability of, 352–354
sampling distribution of slope, 682–683
sampling distribution(s), 336–340, 338f, 344f, 346f, 347f
  bias, 343–346
  center of, 352–354
  definition of, **337**
  evaluating claims with, 339–340
  graphs of, 337
  population, 682
  of sample variance, 415–416
  shape of, 354–355

significance test for population proportion and, 449
terminology for, 337
unbiased estimators and, 344–346
variability, 346–348
variability of, 352–354
sampling frame, 11
sampling variability, **16–17**, 337, 346–348, 348f
bias and, 16–17, 348, 348f
sampling without replacement, 15
sandwich size, 474–475
scatterplots, **112–118**
correlation and, 167–171, 167f, 168f, 180–181
describing, 114–115
direction of, 114–115
form of, 114
linear association and, 170, 699
making, 113–114
outliers, 114
outliers and correlation in, 180–181
quantitative variables, relationship between, 112–118, 113f
regression line and, 653f, 659f, 664–665
strength of, 114
technology for, 117–118
timeplots, 115–118
SD. *see* standard deviation (SD)
seat location, test scores and, 697–698
segmented bar charts, **74**, 75–76, 75f
self-fertilization, 547, 548
self-selected samples. *see* voluntary response samples
shape
of binomial distribution, 280–283, 281f
of density curves, 307
describing, 84–86. *see also* graph(s)
of discrete probability distribution, 268–269
of distributions, 87
of sample mean, 360–362
of sampling distribution of difference between means, 524
of sampling distribution of difference between proportions, 505
of sampling distribution of sample mean, 360–362
of sampling distribution of sample proportion, 354–355
of sampling distribution of sample slope, 684, 685
skewness and. *see* skewness

shoe size, height and, 698–699, 700–701
shoes prices, 569–570
shopping, 66
shrews, heart rate and, 769–770, 771
side-by-side bar charts, **74**, 75f
side effects of flu vaccine, 809, 811
sign test, 768–772
conditions for, 770
definition of, **769**
four-step process for, 771–772
hypotheses, 769–770
median as measure of center, 768
for median of single population, 771
P-values, 770–771, 777
test statistic, 769–770
signed ranks, 777–779
significance level, **442–443**
Type I/Type II errors and, 445
significance test for difference between means, 534–540
conditions for, 534–535
four-step process for, 538–540
hypotheses statement for, 534–535
P-value for, 536–537
standard error for, 536–537
standardized test statistic for, 536–537
technology for, 539–540
significance test for difference between proportions, 514–520
conditions for, 514–515
hypotheses statement for, 514–515
P-value for, 515–517
randomized assignment in, 518–519
standardized test statistic for, 515–517
technology for, 519
significance test for distribution of categorical variable, 589–594
calculating expected counts in, 591–592
chi-square test statistic and, 592–594, 594f
four-step process for, 600–603
hypotheses statement for, 590–591
technology for, 601–603
significance test for mean, 474–478
conditions for, 464–465
confidence intervals and, 475–476
degrees of freedom for, 467–471
four-step process for, 474–475
P-value for, 467–471, 475–476

standardized test statistic for, 465–467
t distribution and, 467–471, 467f, 468f
technology for, 469–470
significance test for mean difference, 555–560
conditions for, 556–558
hypotheses statement for, 555
P-value for, 556
randomized vs. matched pairs design for, 558–560
standardized test statistic for, 556
significance test for proportion, 448–454
conditions for, 448–449
four-step process for, 457–459
one-sided, 457–458
P-value for, 452
sampling distribution and, 449
standardized test statistic and, 449–451
technology for, 453–454
two-sided, 459–461
significance test for relationship between categorical variables, 606–612
chi-square test for association and, 615–621
chi-square test statistic for, 610–612
conditions for, 615–616
degrees of freedom for, 617
expected counts in, 609–610
four-step process for, 618–621
hypotheses statement in, 606–608
P-values for, 617–618
technology for, 620–621
significance test for slope of least-squares regression line, 697–705
conditions for, 684–687, 684f, 699–701
four-step process for, 701–705
hypotheses statement for, 698–699
P-value for, 699–701
standardized test statistic for, 699–701
technology for, 704–705
significance test(s), **434–439**
applications of, 476–478
conclusions in, 438–439, 442–443
vs. confidence interval, 461, 475–476
data dredging and, 478
decision making and, 442–445
four-step process for, 457–459
hypothesis statement in, 435–436. *see also* hypotheses
multiple analyses and, 478
one-sided, 457–458

outliers and, 476
P-hacking and, 478
for population standard deviation or variance, 488–495, 490f
power of a test, 481–485
practical importance of, 477
P-values and, 437–438
reasoning in, 434–435
rejection region approach, 452–453
significance level in, 442–443
statistical significance in, 438, 477–478
for two population standard deviations or variances, 565–573
two-sided, 459–461
Type I/Type II errors in, 443–445
wise use of, 476–478
simple linear regression model, 728
simple random sample (SRS), **15–19**. *see also* random assignment/sampling
choosing, 15–16
random number generator for, 15
simulations, **202**
performing, 202–203
for random sampling, 17, 537
single-blind experiments, **31**
skewed to the left, **85**
skewed to the right, **84**
skewness
calculating, 157
of distributions, 157, 409–410
in normal probability plot, 332
sleep deprivation experiment, 49, 50
sleeping
with dogs, 10
survey of sleep habits, 395–396
slope, **654**–655
interaction term and, 739
of population regression line, 684, 684f
of sample regression line, 684, 684f
slope of least-squares regression line, 681–692
confidence interval for, 688
estimating, 681–692
sampling distribution of, 682–683
standard deviation of, 684, 688
standard error of, 688
t interval for, **689–692**
t test for, **701–705**
testing claim about, 697–705. *see also* significance test for slope of least-squares line

smartphones, rate of use, 229–230
smoking, 52f
　maternal, ADHD and, 29
social media, rate of use, 215
soda bottles, 4
sodium intake, 304
Space Shuttle *Challenger*, 238
splitting stems, 94, 94f
sports
　drug testing in, 198, 257
　regression to the mean in, 665–666
squared residuals, 674, 675, 675f
squirrels, food availability and population size, 735–736, 737–738
SRS. *see* simple random sample (SRS)
standard deviation (SD), **139**–141
　of binomial random variable, 288–289
　calculating and interpreting, 140, 346
　confidence interval for, 424–426
　in correlation calculation, 176, 178
　critical values and, 390
　of discrete random variable, 271–274
　of explanatory variable x, 710, 711
　finding from areas in normal distribution, 324–325
　greater than or equal to zero, 140–141
　of larger values, 140–141
　mean and, 140–141
　of mean difference, 547–548, 556
　nonresistance of, 141
　notation for, 139
　population, 139, 404
　of residuals, 673–675, 710, 711
　sample, 139, 391, 404
　of sampling distribution, 352–354
　of sampling distribution of difference between means, 524
　of sampling distribution of difference between proportions, 505, 507
　of sampling distribution of sample mean, 359–360
　of sampling distribution of sample proportion, 352–354
　of sampling distribution of sample slope, 684
　for significance test for mean difference, 556
　for significance test for slope, 685
　variance and, 139–141, 289
　variation from mean and, 141
standard deviation of the residuals, **673**–675
　calculating, 676–678
　coefficient of determination and, 675–678
standard error
　confidence intervals and, 391–393, 710
　of difference between means, 525, 536, 549
　of difference between proportions, 507
　vs. margin of error, 391
　of mean difference, 549
　prediction intervals and, 711
　of sample mean, 404
　of sample proportion, 391–393
　of slope, 688
standard normal curve, critical value for, 391
standard normal distributions, 314–317, 315f
　definition of, **315**
　finding areas between two values in, 318–320
　finding areas to the left in, 314–317
　finding areas to the right in, 317–318
　Wilcoxon rank sum test and, 790
standard normal distributions, 119, 128–131, 129f. *see* normal distributions
standardized scores. *see* z-scores (standardized scores)
standardized test statistic, 449–451
　calculating, 452, 465–467, 489–492, 515–517
　definition of, **450**
　for difference between means, 536–537
　for difference between proportions, 515–517
　for mean, 465–467
　for mean difference, 556
　for proportion, 451–454, 451f
　for *P*-value, 452
　for slope, 699, 700
　t distribution and, 467–471, 467f, 468f
　for two population standard deviations or variances, 566–569
statistical inference. *see* inference
statistical problem-solving process, **2**
statistical significance, **30**–40, 438
　determining, 442–443
　vs. practical importance, 477

statistics, **2**, **336**–337
　vs. parameters, 336–337
　summary. *see* five-number summary
stem-and-leaf plot. *see* stemplots
STEM workers, diversity of, 588, 646
stemplots (stem-and-leaf plots), 92–97, 94f
　back-to-back, 95–96
　comparing distributions with, 95–96
　describing, 95
　technology for, 96
stopping distance in cars, 308, 310, 315, 318, 322
strata, **22**
stratified random sampling, **22**–24
student loans, 339
Student's *t*, 467
subjects, **29**
sum of squared residuals, 674, 675, 675f
sum of squares error (SSE), 628, 675
sum of squares groups (SSG), 628
sum of squares (SS), 629
sum of squares total (SST), 628, 675
summarizing data
　five-number summary and, 149
　identifying outliers and, 148–149
　making and interpreting boxplots and, 151
sunscreen comparison, 558–559
surveys
　bias in, 11–12, 329–330
　nonresponse and, 11
　undercoverage and, 11
Survivor (television show), 457–458
sweet foods, 50, 51–52
symmetry, and Wilcoxon signed rank test, 780
systematic random sampling, **25**

t distribution, **402**–403, 403f, 467–471, 467f, 468f
　degrees of freedom for, 467–468, 536–537
　for mean difference, 556
　Student's, 467
t inference, 467
t interval
　for difference between means, 527
　for mean, 410
　for slope, 689–692
t statistic, 750
t test
　for mean, 474

　for sign test, 768, 770
　for slope, 701–705. *see also* significance test for slope of least-squares regression line
t test for an individual coefficient, **750**–754
*t** critical value, 402–404, 688
tables, two-way, 213–215, 214f, 221–222
tadpoles, gut length of, 566, 568–569
tattoos, proportion of students with, 397
technology
　analyzing binomial distributions, 281–282
　analyzing discrete random variables, 273
　for bar charts, 66–67
　bootstrap confidence intervals, 826–827
　calculating binomial probabilities, 286–287
　calculating combinations, 254
　calculating confidence interval for difference between means, 529
　calculating confidence interval for difference between proportions, 509–510
　calculating confidence interval for mean, 405–406
　calculating confidence interval for proportion, 392
　calculating correlation, 176–177, 181
　calculating equation of least-squares regression line, 659–662
　calculating factorials and permutation, 246–247
　calculating measures of center and variability, 143–144
　calculating probabilities for a Poisson random variable, 296–297
　calculating *t* interval for slope, 690–691
　categorical variables, analysis of, 78
　chi-square test for association, 620–621
　chi-square test for goodness of fit, 601–603
　choosing simple random sample, 15
　computing numerical summaries, 143–144
　confidence interval for difference between means, 529

confidence intervals for a mean response, 715–716
confidence intervals for population standard deviation, 420
critical values for population standard deviation, 420
for dotplots, 88
finding areas from values in normal distribution, 315–316
finding values from areas in normal distribution, 323
for histograms, 106–107
Kruskal-Wallis test, 805–806
making bar charts, 66–67
making boxplots, 153
making normal probability plots, 332–333
making pie charts, 66–67
making residual plots, 676–678
multiple regression model, assessing, 752–753
for multiple regression models, 731–732
one-way (ANOVA) test, 639–640
for prediction intervals, 712–713
for prediction intervals in regression, 715–716
P-values and chi-square tests for goodness of fit, 601–603
P-values and significance tests for a population mean, 469–470
P-values for a population standard deviation or deviance, 494
P-values for two population standard deviations or variance, 571–572
randomization tests, 815–816
sampling from non-normal population, 361
for scatterplots, 117–118
significance test for difference between means, 539–540
significance test for difference between proportions, 519
significance test for population mean, 469–470
significance test for population proportion, 453–454
significance tests for a population standard deviation or variance, 494
significance tests for two population standard deviations or variances, 571–572
simple random sampling with, 15, 18
for stemplots, 96
t test for slope, 704–705
Wilcoxon rank sum test, 795
Wilcoxon signed rank test, 783
teen romance, 558–559
teenagers
 calcium intake, 437, 438
 movie viewership, 360, 362
television viewing, ratings and, 457–458
temperature, normal body, 433
tennis, speed of serves, 711, 712, 714
test scores, seating position and, 697–698
test statistic. *see* standardized test statistic
test statistic for Kruskal-Wallis test, 801–803
test statistic for sign test, **769**
test statistic for Wilcoxon rank sum test, 790–793
test statistic W for Wilcoxon signed rank test, 779–781, **780**
testing claims. *see* significance test(s)
third quartile (Q_3), **142**. *see also* interquartile range (IQR)
 equivalent percentile for, 160
 in five-number summary, 149
timeplots, 115–118, **116**
tires, nitrogen-filled, 556–557
Titanic survivors, 73–74, 76, 77
transforming data. *see* data transformation
treatment, **29**
tree diagrams, 230f, **230**–232, 231f
trees
 comparative size of, 16, 538–539
 ponderosa pines, 654–655, 672–676, 672f, 686–688, 709–712, 714, 714f
trials, 277
trimmed mean, 137
trucks, value of, 113–114, 115, 652–653, 653f, 659, 659f, 661, 672f, 674, 675, 689–690
Tuskegee Syphilis Study, 53
Twitter, 3–4
two-sample test of standard deviations
 conditions for, 565–566
 confidence intervals for, 571
 hypotheses, 565–566
 P-values for, 566–569, 571–572
 significance tests for, 565–573
 standardized test statistic, 566–569
two-sample F test comparing two standard deviations (variances), **569**
two-sample t interval, for difference between means, **527**. *see also* confidence interval for difference between means
two-sample t test, for difference between means, **538**–540, 788, 794, 815. *see also* significance test for difference between means
two-sample z interval, for difference between proportions, **508**. *see also* confidence interval for difference between proportions
two-sample z test, for difference between proportions, **517**, 812, 815. *see also* significance test for difference between proportions
two-sided alternative hypothesis, 435–**436**
two-sided tests, **457**, 459–461
 confidence intervals and, 461, 475–476
two-way tables, 72–**73**, 213–215, 214f
 conditional probability and, 221–222
 relative frequency, 73–74
Type I error, **443**–445
 power and, 484, 484f
Type II error, **443**–445, 780
 power and, 482, 484, 484f

unbiased estimator, 344–346, **345**
undercoverage, **11**
uniform density curve, 305–306
unimodal graphs, 86, 307f
union, in Venn diagram, 216f, **216**
units of measure, correlation and, 178–179
upper-tail test, 777
U.S. Census Bureau, 49–50, 63–64
U.S. government, naming branches of, 389, 391
utility, 748

vaccines, 4, 31–32
values
 absolute, 777–778, 779
 fitted, 747
vaping, by college students, 460
variability
 calculating, 143
 in dotplots, 87
 interquartile range and, 142–144, 148
 measuring, 138–144
 range and, 138–139
 of sample mean, 359–360
 sampling, 16–17, 337, 346–348, 348f
 of sampling distribution, 352–354
 of sampling distribution of sample proportion, 352–354
 of sampling distribution of sample slope, 685
 in sampling distributions, 346–348
 variance and, 140
variables, **3**–4
 binomial random, 277–283
 categorical, 606–612. *see also* categorical variables
 explanatory, 178
 quantitative. *see* quantitative variables
 random, 262–266. *see also* random variables
 response, 178
 selecting for population multiple regression model, 751–752
variance, 140
 standard deviation and, 139–141, 289
 using to compare means, 626–628
vehicles. *see also* trucks, value of
 license plate numbers, 244
 nitrogen-filled tires, 556–558
 stopping distance, 308, 310, 315, 318, 322
Venn diagrams, 215f, **215**–217, 216f
video games, 558–559
vitamin D, 28–30
voluntary response samples, **9**, 10
 vs. nonresponse, 11
voting, 22

water, level of dissolved oxygen in, 728, 729, 730, 746, 750
water crisis in Flint, MI, 82–83, 83f, 87, 130, 143, 159, 161
water supply, annual, 129, 187–188
water temperature, 492–493
weight, of golden hamsters, 466, 468–469, 489–490, 490f
whales, growth of, 725, 758
whiskers, 150
Wilcoxon rank sum test, 788–795
 conditions for, 791–793
 definition of, **788**
 four-step process for, 793–795
 P-values, 790–793
 ranks for two samples or groups, 788–790
 technology, 795
 test statistic, 790–793

Wilcoxon signed rank test, 777–784
    conditions for, 780
    definition of, **779**
    four-step process for, 781–784
    P-values, 779–781
    ranks and signed ranks, 777–779
    technology, 783
    test statistic, 779–781
Wilcoxon-Mann-Whitney test, 788

y intercept, of regression line, **654**–655
Yellowstone National Park, snowmobile usage, 72–77, 75f

z interval, for a proportion, 395. *see also* confidence interval for population proportion.
z interval, for difference between proportions, 508. *see also* confidence interval for difference between proportions
z statistic, for a proportion, 452
z statistic, for difference between proportions, 515
z test, for a proportion, 457. *see also* significance test for proportion.
z test, for difference between proportions, 517. *see also* significance test for difference between proportions

z^* critical value, 391, 402, 402f
zinc, in drinking water, 545–548, 546f, 550
z-scores (standardized scores), 319, 322, 322f
    in correlation calculation, 178
    definition of, **160**
    finding and interpreting, 160–162, 324
    mean and, 160
    percentiles and, 162

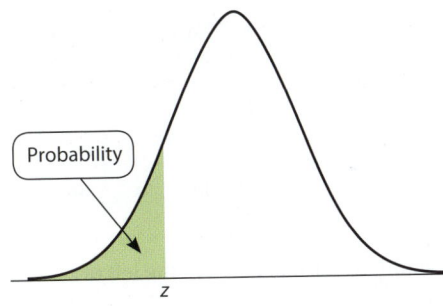

Table entry for z is the area under the standard normal curve to the left of z.

Table A Standard normal probabilities

z	.00	.01	.02	.03	.04	.05	.06	.07	.08	.09
−3.4	.0003	.0003	.0003	.0003	.0003	.0003	.0003	.0003	.0003	.0002
−3.3	.0005	.0005	.0005	.0004	.0004	.0004	.0004	.0004	.0004	.0003
−3.2	.0007	.0007	.0006	.0006	.0006	.0006	.0006	.0005	.0005	.0005
−3.1	.0010	.0009	.0009	.0009	.0008	.0008	.0008	.0008	.0007	.0007
−3.0	.0013	.0013	.0013	.0012	.0012	.0011	.0011	.0011	.0010	.0010
−2.9	.0019	.0018	.0018	.0017	.0016	.0016	.0015	.0015	.0014	.0014
−2.8	.0026	.0025	.0024	.0023	.0023	.0022	.0021	.0021	.0020	.0019
−2.7	.0035	.0034	.0033	.0032	.0031	.0030	.0029	.0028	.0027	.0026
−2.6	.0047	.0045	.0044	.0043	.0041	.0040	.0039	.0038	.0037	.0036
−2.5	.0062	.0060	.0059	.0057	.0055	.0054	.0052	.0051	.0049	.0048
−2.4	.0082	.0080	.0078	.0075	.0073	.0071	.0069	.0068	.0066	.0064
−2.3	.0107	.0104	.0102	.0099	.0096	.0094	.0091	.0089	.0087	.0084
−2.2	.0139	.0136	.0132	.0129	.0125	.0122	.0119	.0116	.0113	.0110
−2.1	.0179	.0174	.0170	.0166	.0162	.0158	.0154	.0150	.0146	.0143
−2.0	.0228	.0222	.0217	.0212	.0207	.0202	.0197	.0192	.0188	.0183
−1.9	.0287	.0281	.0274	.0268	.0262	.0256	.0250	.0244	.0239	.0233
−1.8	.0359	.0351	.0344	.0336	.0329	.0322	.0314	.0307	.0301	.0294
−1.7	.0446	.0436	.0427	.0418	.0409	.0401	.0392	.0384	.0375	.0367
−1.6	.0548	.0537	.0526	.0516	.0505	.0495	.0485	.0475	.0465	.0455
−1.5	.0668	.0655	.0643	.0630	.0618	.0606	.0594	.0582	.0571	.0559
−1.4	.0808	.0793	.0778	.0764	.0749	.0735	.0721	.0708	.0694	.0681
−1.3	.0968	.0951	.0934	.0918	.0901	.0885	.0869	.0853	.0838	.0823
−1.2	.1151	.1131	.1112	.1093	.1075	.1056	.1038	.1020	.1003	.0985
−1.1	.1357	.1335	.1314	.1292	.1271	.1251	.1230	.1210	.1190	.1170
−1.0	.1587	.1562	.1539	.1515	.1492	.1469	.1446	.1423	.1401	.1379
−0.9	.1841	.1814	.1788	.1762	.1736	.1711	.1685	.1660	.1635	.1611
−0.8	.2119	.2090	.2061	.2033	.2005	.1977	.1949	.1922	.1894	.1867
−0.7	.2420	.2389	.2358	.2327	.2296	.2266	.2236	.2206	.2177	.2148
−0.6	.2743	.2709	.2676	.2643	.2611	.2578	.2546	.2514	.2483	.2451
−0.5	.3085	.3050	.3015	.2981	.2946	.2912	.2877	.2843	.2810	.2776
−0.4	.3446	.3409	.3372	.3336	.3300	.3264	.3228	.3192	.3156	.3121
−0.3	.3821	.3783	.3745	.3707	.3669	.3632	.3594	.3557	.3520	.3483
−0.2	.4207	.4168	.4129	.4090	.4052	.4013	.3974	.3936	.3897	.3859
−0.1	.4602	.4562	.4522	.4483	.4443	.4404	.4364	.4325	.4286	.4247
−0.0	.5000	.4960	.4920	.4880	.4840	.4801	.4761	.4721	.4681	.4641

(*Continued*)

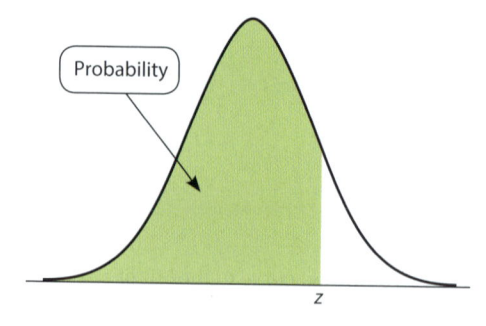

Table entry for z is the area under the standard normal curve to the left of z.

Table A Standard normal probabilities (continued)

z	.00	.01	.02	.03	.04	.05	.06	.07	.08	.09
0.0	.5000	.5040	.5080	.5120	.5160	.5199	.5239	.5279	.5319	.5359
0.1	.5398	.5438	.5478	.5517	.5557	.5596	.5636	.5675	.5714	.5753
0.2	.5793	.5832	.5871	.5910	.5948	.5987	.6026	.6064	.6103	.6141
0.3	.6179	.6217	.6255	.6293	.6331	.6368	.6406	.6443	.6480	.6517
0.4	.6554	.6591	.6628	.6664	.6700	.6736	.6772	.6808	.6844	.6879
0.5	.6915	.6950	.6985	.7019	.7054	.7088	.7123	.7157	.7190	.7224
0.6	.7257	.7291	.7324	.7357	.7389	.7422	.7454	.7486	.7517	.7549
0.7	.7580	.7611	.7642	.7673	.7704	.7734	.7764	.7794	.7823	.7852
0.8	.7881	.7910	.7939	.7967	.7995	.8023	.8051	.8078	.8106	.8133
0.9	.8159	.8186	.8212	.8238	.8264	.8289	.8315	.8340	.8365	.8389
1.0	.8413	.8438	.8461	.8485	.8508	.8531	.8554	.8577	.8599	.8621
1.1	.8643	.8665	.8686	.8708	.8729	.8749	.8770	.8790	.8810	.8830
1.2	.8849	.8869	.8888	.8907	.8925	.8944	.8962	.8980	.8997	.9015
1.3	.9032	.9049	.9066	.9082	.9099	.9115	.9131	.9147	.9162	.9177
1.4	.9192	.9207	.9222	.9236	.9251	.9265	.9279	.9292	.9306	.9319
1.5	.9332	.9345	.9357	.9370	.9382	.9394	.9406	.9418	.9429	.9441
1.6	.9452	.9463	.9474	.9484	.9495	.9505	.9515	.9525	.9535	.9545
1.7	.9554	.9564	.9573	.9582	.9591	.9599	.9608	.9616	.9625	.9633
1.8	.9641	.9649	.9656	.9664	.9671	.9678	.9686	.9693	.9699	.9706
1.9	.9713	.9719	.9726	.9732	.9738	.9744	.9750	.9756	.9761	.9767
2.0	.9772	.9778	.9783	.9788	.9793	.9798	.9803	.9808	.9812	.9817
2.1	.9821	.9826	.9830	.9834	.9838	.9842	.9846	.9850	.9854	.9857
2.2	.9861	.9864	.9868	.9871	.9875	.9878	.9881	.9884	.9887	.9890
2.3	.9893	.9896	.9898	.9901	.9904	.9906	.9909	.9911	.9913	.9916
2.4	.9918	.9920	.9922	.9925	.9927	.9929	.9931	.9932	.9934	.9936
2.5	.9938	.9940	.9941	.9943	.9945	.9946	.9948	.9949	.9951	.9952
2.6	.9953	.9955	.9956	.9957	.9959	.9960	.9961	.9962	.9963	.9964
2.7	.9965	.9966	.9967	.9968	.9969	.9970	.9971	.9972	.9973	.9974
2.8	.9974	.9975	.9976	.9977	.9977	.9978	.9979	.9979	.9980	.9981
2.9	.9981	.9982	.9982	.9983	.9984	.9984	.9985	.9985	.9986	.9986
3.0	.9987	.9987	.9987	.9988	.9988	.9989	.9989	.9989	.9990	.9990
3.1	.9990	.9991	.9991	.9991	.9992	.9992	.9992	.9992	.9993	.9993
3.2	.9993	.9993	.9994	.9994	.9994	.9994	.9994	.9995	.9995	.9995
3.3	.9995	.9995	.9995	.9996	.9996	.9996	.9996	.9996	.9996	.9997
3.4	.9997	.9997	.9997	.9997	.9997	.9997	.9997	.9997	.9997	.9998

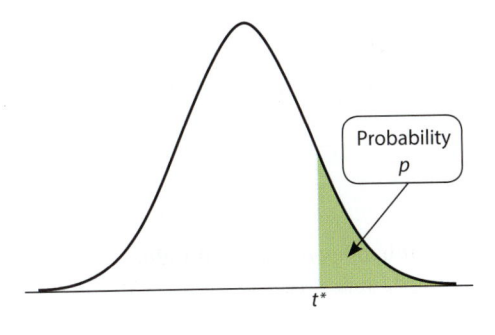

Table entry for p and C is the point t^* with probability p lying to its right and probability C lying between $-t^*$ and t^*.

Table B t distribution critical values

df	\.25	\.20	\.15	\.10	\.05	\.025	\.02	\.01	\.005	\.0025	\.001	\.0005
1	1.000	1.376	1.963	3.078	6.314	12.71	15.89	31.82	63.66	127.3	318.3	636.6
2	0.816	1.061	1.386	1.886	2.920	4.303	4.849	6.965	9.925	14.09	22.33	31.60
3	0.765	0.978	1.250	1.638	2.353	3.182	3.482	4.541	5.841	7.453	10.21	12.92
4	0.741	0.941	1.190	1.533	2.132	2.776	2.999	3.747	4.604	5.598	7.173	8.610
5	0.727	0.920	1.156	1.476	2.015	2.571	2.757	3.365	4.032	4.773	5.893	6.869
6	0.718	0.906	1.134	1.440	1.943	2.447	2.612	3.143	3.707	4.317	5.208	5.959
7	0.711	0.896	1.119	1.415	1.895	2.365	2.517	2.998	3.499	4.029	4.785	5.408
8	0.706	0.889	1.108	1.397	1.860	2.306	2.449	2.896	3.355	3.833	4.501	5.041
9	0.703	0.883	1.100	1.383	1.833	2.262	2.398	2.821	3.250	3.690	4.297	4.781
10	0.700	0.879	1.093	1.372	1.812	2.228	2.359	2.764	3.169	3.581	4.144	4.587
11	0.697	0.876	1.088	1.363	1.796	2.201	2.328	2.718	3.106	3.497	4.025	4.437
12	0.695	0.873	1.083	1.356	1.782	2.179	2.303	2.681	3.055	3.428	3.930	4.318
13	0.694	0.870	1.079	1.350	1.771	2.160	2.282	2.650	3.012	3.372	3.852	4.221
14	0.692	0.868	1.076	1.345	1.761	2.145	2.264	2.624	2.977	3.326	3.787	4.140
15	0.691	0.866	1.074	1.341	1.753	2.131	2.249	2.602	2.947	3.286	3.733	4.073
16	0.690	0.865	1.071	1.337	1.746	2.120	2.235	2.583	2.921	3.252	3.686	4.015
17	0.689	0.863	1.069	1.333	1.740	2.110	2.224	2.567	2.898	3.222	3.646	3.965
18	0.688	0.862	1.067	1.330	1.734	2.101	2.214	2.552	2.878	3.197	3.611	3.922
19	0.688	0.861	1.066	1.328	1.729	2.093	2.205	2.539	2.861	3.174	3.579	3.883
20	0.687	0.860	1.064	1.325	1.725	2.086	2.197	2.528	2.845	3.153	3.552	3.850
21	0.686	0.859	1.063	1.323	1.721	2.080	2.189	2.518	2.831	3.135	3.527	3.819
22	0.686	0.858	1.061	1.321	1.717	2.074	2.183	2.508	2.819	3.119	3.505	3.792
23	0.685	0.858	1.060	1.319	1.714	2.069	2.177	2.500	2.807	3.104	3.485	3.768
24	0.685	0.857	1.059	1.318	1.711	2.064	2.172	2.492	2.797	3.091	3.467	3.745
25	0.684	0.856	1.058	1.316	1.708	2.060	2.167	2.485	2.787	3.078	3.450	3.725
26	0.684	0.856	1.058	1.315	1.706	2.056	2.162	2.479	2.779	3.067	3.435	3.707
27	0.684	0.855	1.057	1.314	1.703	2.052	2.158	2.473	2.771	3.057	3.421	3.690
28	0.683	0.855	1.056	1.313	1.701	2.048	2.154	2.467	2.763	3.047	3.408	3.674
29	0.683	0.854	1.055	1.311	1.699	2.045	2.150	2.462	2.756	3.038	3.396	3.659
30	0.683	0.854	1.055	1.310	1.697	2.042	2.147	2.457	2.750	3.030	3.385	3.646
40	0.681	0.851	1.050	1.303	1.684	2.021	2.123	2.423	2.704	2.971	3.307	3.551
50	0.679	0.849	1.047	1.299	1.676	2.009	2.109	2.403	2.678	2.937	3.261	3.496
60	0.679	0.848	1.045	1.296	1.671	2.000	2.099	2.390	2.660	2.915	3.232	3.460
80	0.678	0.846	1.043	1.292	1.664	1.990	2.088	2.374	2.639	2.887	3.195	3.416
100	0.677	0.845	1.042	1.290	1.660	1.984	2.081	2.364	2.626	2.871	3.174	3.390
1000	0.675	0.842	1.037	1.282	1.646	1.962	2.056	2.330	2.581	2.813	3.098	3.300
z^*	0.674	0.841	1.036	1.282	1.645	1.960	2.054	2.326	2.576	2.807	3.091	3.291
	50%	60%	70%	80%	90%	95%	96%	98%	99%	99.5%	99.8%	99.9%

Confidence level C

Table entry for p is the point χ^2 with probability p lying to its right.

Table C Chi-square distribution critical values

df	\.995	\.99	\.975	\.95	\.90	\.10	\.05	\.025	\.01	\.005
1	—	—	0.001	0.004	0.016	2.706	3.841	5.024	6.635	7.879
2	0.010	0.020	0.051	0.103	0.211	4.605	5.991	7.378	9.210	10.597
3	0.072	0.115	0.216	0.352	0.584	6.251	7.815	9.348	11.345	12.838
4	0.207	0.297	0.484	0.711	1.064	7.779	9.488	11.143	13.277	14.860
5	0.412	0.554	0.831	1.145	1.610	9.236	11.071	12.833	15.086	16.750
6	0.676	0.872	1.237	1.635	2.204	10.645	12.592	14.449	16.812	18.548
7	0.989	1.239	1.690	2.167	2.833	12.017	14.067	16.013	18.475	20.278
8	1.344	1.646	2.180	2.733	3.490	13.362	15.507	17.535	20.090	21.955
9	1.735	2.088	2.700	3.325	4.168	14.684	16.919	19.023	21.666	23.589
10	2.156	2.558	3.247	3.940	4.865	15.987	18.307	20.483	23.209	25.188
11	2.603	3.053	3.816	4.575	5.578	17.275	19.675	21.920	24.725	26.757
12	3.074	3.571	4.404	5.226	6.304	18.549	21.026	23.337	26.217	28.299
13	3.565	4.107	5.009	5.892	7.042	19.812	22.362	24.736	27.688	29.819
14	4.075	4.660	5.629	6.571	7.790	21.064	23.685	26.119	29.141	31.319
15	4.601	5.229	6.262	7.261	8.547	22.307	24.996	27.488	30.578	32.801
16	5.142	5.812	6.908	7.962	9.312	23.542	26.296	28.845	32.000	34.267
17	5.697	6.408	7.564	8.672	10.085	24.769	27.587	30.191	33.409	35.718
18	6.265	7.015	8.231	9.390	10.865	25.989	28.869	31.526	34.805	37.156
19	6.844	7.633	8.907	10.117	11.651	27.204	30.144	32.852	36.191	38.582
20	7.434	8.260	9.591	10.851	12.443	28.412	31.410	34.170	37.566	39.997
21	8.034	8.897	10.283	11.591	13.240	29.615	32.671	35.479	38.932	41.401
22	8.643	9.542	10.982	12.338	14.042	30.813	33.924	36.781	40.289	42.796
23	9.260	10.196	11.689	13.091	14.848	32.007	35.172	38.076	41.638	44.181
24	9.886	10.856	12.401	13.848	15.659	33.196	36.415	39.364	42.980	45.559
25	10.520	11.524	13.120	14.611	16.473	34.382	37.652	40.646	44.314	46.928
26	11.160	12.198	13.844	15.379	17.292	35.563	38.885	41.923	45.642	48.290
27	11.808	12.879	14.573	16.151	18.114	36.741	40.113	43.194	46.963	49.645
28	12.461	13.565	15.308	16.928	18.939	37.916	41.337	44.461	48.278	50.993
29	13.121	14.257	16.047	17.708	19.768	39.087	42.557	45.722	49.588	52.336
30	13.787	14.954	16.791	18.493	20.599	40.256	43.773	46.979	50.892	53.672
40	20.707	22.164	24.433	26.509	29.051	51.805	55.758	59.342	63.691	66.766
50	27.991	29.707	32.357	34.764	37.689	63.167	67.505	71.420	76.154	79.490
60	35.534	37.485	40.482	43.188	46.459	74.397	79.082	83.298	88.379	91.952
70	43.275	45.442	48.758	51.739	55.329	85.527	90.531	95.023	100.425	104.215
80	51.172	53.540	57.153	60.391	64.270	96.578	101.879	106.629	112.329	116.321
90	59.196	61.754	65.647	69.126	73.291	107.565	113.145	118.136	124.116	128.299
100	67.328	70.065	74.222	77.929	82.358	118.498	124.342	129.561	135.807	140.169

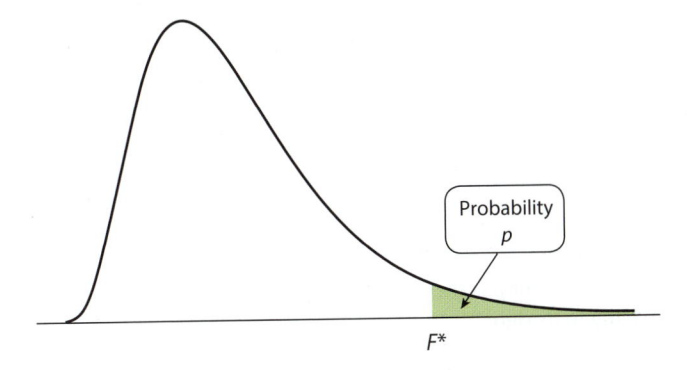

Table entry for p is the critical value F^* with probability p lying to its right.

Table D F distribution critical values

| | | | \multicolumn{9}{c}{Degrees of freedom in the numerator} |
Denom df	p	1	2	3	4	5	6	7	8	9
1	.100	39.86	49.50	53.59	55.83	57.24	58.20	58.91	59.44	59.86
	.050	161.45	199.50	215.71	224.58	230.16	233.99	236.77	238.88	240.54
	.025	647.79	799.50	864.16	899.58	921.85	937.11	948.22	956.66	963.28
	.010	4052.18	4999.50	5403.35	5624.58	5763.65	5858.99	5928.36	5981.07	6022.47
	.001	405284	500000	540379	562500	576405	585937	592873	598144	602284
2	.100	8.53	9.00	9.16	9.24	9.29	9.33	9.35	9.37	9.38
	.050	18.51	19.00	19.16	19.25	19.30	19.33	19.35	19.37	19.38
	.025	38.51	39.00	39.17	39.25	39.30	39.33	39.36	39.37	39.39
	.010	98.50	99.00	99.17	99.25	99.30	99.33	99.36	99.37	99.39
	.001	998.50	999.00	999.17	999.25	999.30	999.33	999.36	999.37	999.39
3	.100	5.54	5.46	5.39	5.34	5.31	5.28	5.27	5.25	5.24
	.050	10.13	9.55	9.28	9.12	9.01	8.94	8.89	8.85	8.81
	.025	17.44	16.04	15.44	15.10	14.88	14.73	14.62	14.54	14.47
	.010	34.12	30.82	29.46	28.71	28.24	27.91	27.67	27.49	27.35
	.001	167.03	148.50	141.11	137.10	134.58	132.85	131.58	130.62	129.86
4	.100	4.54	4.32	4.19	4.11	4.05	4.01	3.98	3.95	3.94
	.050	7.71	6.94	6.59	6.39	6.26	6.16	6.09	6.04	6.00
	.025	12.22	10.65	9.98	9.60	9.36	9.20	9.07	8.98	8.90
	.010	21.20	18.00	16.69	15.98	15.52	15.21	14.98	14.80	14.66
	.001	74.14	61.25	56.18	53.44	51.71	50.53	49.66	49.00	48.47
5	.100	4.06	3.78	3.62	3.52	3.45	3.40	3.37	3.34	3.32
	.050	6.61	5.79	5.41	5.19	5.05	4.95	4.88	4.82	4.77
	.025	10.01	8.43	7.76	7.39	7.15	6.98	6.85	6.76	6.68
	.010	16.26	13.27	12.06	11.39	10.97	10.67	10.46	10.29	10.16
	.001	47.18	37.12	33.20	31.09	29.75	28.83	28.16	27.65	27.24
6	.100	3.78	3.46	3.29	3.18	3.11	3.05	3.01	2.98	2.96
	.050	5.99	5.14	4.76	4.53	4.39	4.28	4.21	4.15	4.10
	.025	8.81	7.26	6.60	6.23	5.99	5.82	5.70	5.60	5.52
	.010	13.75	10.92	9.78	9.15	8.75	8.47	8.26	8.10	7.98
	.001	35.51	27.00	23.70	21.92	20.80	20.03	19.46	19.03	18.69
7	.100	3.59	3.26	3.07	2.96	2.88	2.83	2.78	2.75	2.72
	.050	5.59	4.74	4.35	4.12	3.97	3.87	3.79	3.73	3.68
	.025	8.07	6.54	5.89	5.52	5.29	5.12	4.99	4.90	4.82
	.010	12.25	9.55	8.45	7.85	7.46	7.19	6.99	6.84	6.72
	.001	29.25	21.69	18.77	17.20	16.21	15.52	15.02	14.63	14.33

(Continued)

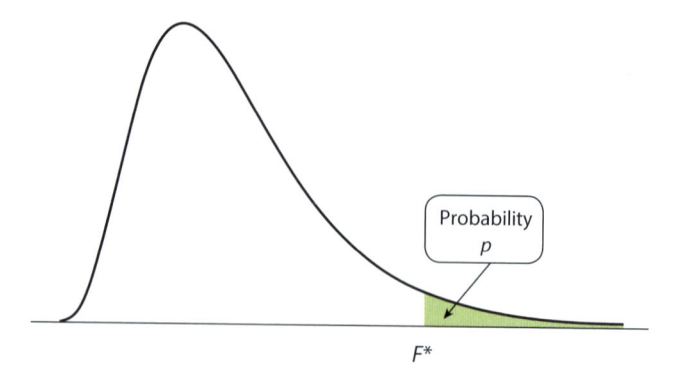

TABLE D F distribution critical values (continued)

Degrees of freedom in the numerator

		10	12	15	20	25	30	40	50	60	120	1000
		60.19	60.71	61.22	61.74	62.05	62.26	62.53	62.69	62.79	63.06	63.30
		241.88	243.91	245.95	248.01	249.26	250.10	251.14	251.77	252.20	253.25	254.19
	1	968.63	976.71	984.87	993.10	998.08	1001.41	1005.60	1008.12	1009.80	1014.02	1017.75
		6055.85	6106.32	6157.28	6208.73	6239.83	6260.65	6286.78	6302.52	6313.03	6339.39	6362.68
		605621	610668	615764	620908	624017	626099	628712	630285	631337	633972	636301
		9.39	9.41	9.42	9.44	9.45	9.46	9.47	9.47	9.47	9.48	9.49
		19.40	19.41	19.43	19.45	19.46	19.46	19.47	19.48	19.48	19.49	19.49
	2	39.40	39.41	39.43	39.45	39.46	39.46	39.47	39.48	39.48	39.49	39.50
		99.40	99.42	99.43	99.45	99.46	99.47	99.47	99.48	99.48	99.49	99.50
		999.40	999.42	999.43	999.45	999.46	999.47	999.47	999.48	999.48	999.49	999.50
		5.23	5.22	5.20	5.18	5.17	5.17	5.16	5.15	5.15	5.14	5.13
		8.79	8.74	8.70	8.66	8.63	8.62	8.59	8.58	8.57	8.55	8.53
Degrees of freedom in the denominator	3	14.42	14.34	14.25	14.17	14.12	14.08	14.04	14.01	13.99	13.95	13.91
		27.23	27.05	26.87	26.69	26.58	26.50	26.41	26.35	26.32	26.22	26.14
		129.25	128.32	127.37	126.42	125.84	125.45	124.96	124.66	124.47	123.97	123.53
		3.92	3.90	3.87	3.84	3.83	3.82	3.80	3.80	3.79	3.78	3.76
		5.96	5.91	5.86	5.80	5.77	5.75	5.72	5.70	5.69	5.66	5.63
	4	8.84	8.75	8.66	8.56	8.50	8.46	8.41	8.38	8.36	8.31	8.26
		14.55	14.37	14.20	14.02	13.91	13.84	13.75	13.69	13.65	13.56	13.47
		48.05	47.41	46.76	46.10	45.70	45.43	45.09	44.88	44.75	44.40	44.09
		3.30	3.27	3.24	3.21	3.19	3.17	3.16	3.15	3.14	3.12	3.11
		4.74	4.68	4.62	4.56	4.52	4.50	4.46	4.44	4.43	4.40	4.37
	5	6.62	6.52	6.43	6.33	6.27	6.23	6.18	6.14	6.12	6.07	6.02
		10.05	9.89	9.72	9.55	9.45	9.38	9.29	9.24	9.20	9.11	9.03
		26.92	26.42	25.91	25.39	25.08	24.87	24.60	24.44	24.33	24.06	23.82
		2.94	2.90	2.87	2.84	2.81	2.80	2.78	2.77	2.76	2.74	2.72
		4.06	4.00	3.94	3.87	3.83	3.81	3.77	3.75	3.74	3.70	3.67
	6	5.46	5.37	5.27	5.17	5.11	5.07	5.01	4.98	4.96	4.90	4.86
		7.87	7.72	7.56	7.40	7.30	7.23	7.14	7.09	7.06	6.97	6.89
		18.41	17.99	17.56	17.12	16.85	16.67	16.44	16.31	16.21	15.98	15.77
		2.70	2.67	2.63	2.59	2.57	2.56	2.54	2.52	2.51	2.49	2.47
		3.64	3.57	3.51	3.44	3.40	3.38	3.34	3.32	3.30	3.27	3.23
	7	4.76	4.67	4.57	4.47	4.40	4.36	4.31	4.28	4.25	4.20	4.15
		6.62	6.47	6.31	6.16	6.06	5.99	5.91	5.86	5.82	5.74	5.66
		14.08	13.71	13.32	12.93	12.69	12.53	12.33	12.20	12.12	11.91	11.72

TABLE D F distribution critical values (continued)

Degrees of freedom in the numerator

	p	1	2	3	4	5	6	7	8	9
8	.100	3.46	3.11	2.92	2.81	2.73	2.67	2.62	2.59	2.56
	.050	5.32	4.46	4.07	3.84	3.69	3.58	3.50	3.44	3.39
	.025	7.57	6.06	5.42	5.05	4.82	4.65	4.53	4.43	4.36
	.010	11.26	8.65	7.59	7.01	6.63	6.37	6.18	6.03	5.91
	.001	25.41	18.49	15.83	14.39	13.48	12.86	12.40	12.05	11.77
9	.100	3.36	3.01	2.81	2.69	2.61	2.55	2.51	2.47	2.44
	.050	5.12	4.26	3.86	3.63	3.48	3.37	3.29	3.23	3.18
	.025	7.21	5.71	5.08	4.72	4.48	4.32	4.20	4.10	4.03
	.010	10.56	8.02	6.99	6.42	6.06	5.80	5.61	5.47	5.35
	.001	22.86	16.39	13.90	12.56	11.71	11.13	10.70	10.37	10.11
10	.100	3.29	2.92	2.73	2.61	2.52	2.46	2.41	2.38	2.35
	.050	4.96	4.10	3.71	3.48	3.33	3.22	3.14	3.07	3.02
	.025	6.94	5.46	4.83	4.47	4.24	4.07	3.95	3.85	3.78
	.010	10.04	7.56	6.55	5.99	5.64	5.39	5.20	5.06	4.94
	.001	21.04	14.91	12.55	11.28	10.48	9.93	9.52	9.20	8.96
11	.100	3.23	2.86	2.66	2.54	2.45	2.39	2.34	2.30	2.27
	.050	4.84	3.98	3.59	3.36	3.20	3.09	3.01	2.95	2.90
	.025	6.72	5.26	4.63	4.28	4.04	3.88	3.76	3.66	3.59
	.010	9.65	7.21	6.22	5.67	5.32	5.07	4.89	4.74	4.63
	.001	19.69	13.81	11.56	10.35	9.58	9.05	8.66	8.35	8.12
12	.100	3.18	2.81	2.61	2.48	2.39	2.33	2.28	2.24	2.21
	.050	4.75	3.89	3.49	3.26	3.11	3.00	2.91	2.85	2.80
	.025	6.55	5.10	4.47	4.12	3.89	3.73	3.61	3.51	3.44
	.010	9.33	6.93	5.95	5.41	5.06	4.82	4.64	4.50	4.39
	.001	18.64	12.97	10.80	9.63	8.89	8.38	8.00	7.71	7.48
13	.100	3.14	2.76	2.56	2.43	2.35	2.28	2.23	2.20	2.16
	.050	4.67	3.81	3.41	3.18	3.03	2.92	2.83	2.77	2.71
	.025	6.41	4.97	4.35	4.00	3.77	3.60	3.48	3.39	3.31
	.010	9.07	6.70	5.74	5.21	4.86	4.62	4.44	4.30	4.19
	.001	17.82	12.31	10.21	9.07	8.35	7.86	7.49	7.21	6.98
14	.100	3.10	2.73	2.52	2.39	2.31	2.24	2.19	2.15	2.12
	.050	4.60	3.74	3.34	3.11	2.96	2.85	2.76	2.70	2.65
	.025	6.30	4.86	4.24	3.89	3.66	3.50	3.38	3.29	3.21
	.010	8.86	6.51	5.56	5.04	4.69	4.46	4.28	4.14	4.03
	.001	17.14	11.78	9.73	8.62	7.92	7.44	7.08	6.80	6.58
15	.100	3.07	2.70	2.49	2.36	2.27	2.21	2.16	2.12	2.09
	.050	4.54	3.68	3.29	3.06	2.90	2.79	2.71	2.64	2.59
	.025	6.20	4.77	4.15	3.80	3.58	3.41	3.29	3.20	3.12
	.010	8.68	6.36	5.42	4.89	4.56	4.32	4.14	4.00	3.89
	.001	16.59	11.34	9.34	8.25	7.57	7.09	6.74	6.47	6.26
16	.100	3.05	2.67	2.46	2.33	2.24	2.18	2.13	2.09	2.06
	.050	4.49	3.63	3.24	3.01	2.85	2.74	2.66	2.59	2.54
	.025	6.12	4.69	4.08	3.73	3.50	3.34	3.22	3.12	3.05
	.010	8.53	6.23	5.29	4.77	4.44	4.20	4.03	3.89	3.78
	.001	16.12	10.97	9.01	7.94	7.27	6.80	6.46	6.19	5.98
17	.100	3.03	2.64	2.44	2.31	2.22	2.15	2.10	2.06	2.03
	.050	4.45	3.59	3.20	2.96	2.81	2.70	2.61	2.55	2.49
	.025	6.04	4.62	4.01	3.66	3.44	3.28	3.16	3.06	2.98
	.010	8.40	6.11	5.19	4.67	4.34	4.10	3.93	3.79	3.68
	.001	15.72	10.66	8.73	7.68	7.02	6.56	6.22	5.96	5.75

(Continued)

TABLE D F distribution critical values (continued)

		\multicolumn{11}{c}{Degrees of freedom in the numerator}
		10	12	15	20	25	30	40	50	60	120	1000
Degrees of freedom in the denominator	8	2.54	2.50	2.46	2.42	2.40	2.38	2.36	2.35	2.34	2.32	2.30
		3.35	3.28	3.22	3.15	3.11	3.08	3.04	3.02	3.01	2.97	2.93
		4.30	4.20	4.10	4.00	3.94	3.89	3.84	3.81	3.78	3.73	3.68
		5.81	5.67	5.52	5.36	5.26	5.20	5.12	5.07	5.03	4.95	4.87
		11.54	11.19	10.84	10.48	10.26	10.11	9.92	9.80	9.73	9.53	9.36
	9	2.42	2.38	2.34	2.30	2.27	2.25	2.23	2.22	2.21	2.18	2.16
		3.14	3.07	3.01	2.94	2.89	2.86	2.83	2.80	2.79	2.75	2.71
		3.96	3.87	3.77	3.67	3.60	3.56	3.51	3.47	3.45	3.39	3.34
		5.26	5.11	4.96	4.81	4.71	4.65	4.57	4.52	4.48	4.40	4.32
		9.89	9.57	9.24	8.90	8.69	8.55	8.37	8.26	8.19	8.00	7.84
	10	2.32	2.28	2.24	2.20	2.17	2.16	2.13	2.12	2.11	2.08	2.06
		2.98	2.91	2.85	2.77	2.73	2.70	2.66	2.64	2.62	2.58	2.54
		3.72	3.62	3.52	3.42	3.35	3.31	3.26	3.22	3.20	3.14	3.09
		4.85	4.71	4.56	4.41	4.31	4.25	4.17	4.12	4.08	4.00	3.92
		8.75	8.45	8.13	7.80	7.60	7.47	7.30	7.19	7.12	6.94	6.78
	11	2.25	2.21	2.17	2.12	2.10	2.08	2.05	2.04	2.03	2.00	1.98
		2.85	2.79	2.72	2.65	2.60	2.57	2.53	2.51	2.49	2.45	2.41
		3.53	3.43	3.33	3.23	3.16	3.12	3.06	3.03	3.00	2.94	2.89
		4.54	4.40	4.25	4.10	4.01	3.94	3.86	3.81	3.78	3.69	3.61
		7.92	7.63	7.32	7.01	6.81	6.68	6.52	6.42	6.35	6.18	6.02
	12	2.19	2.15	2.10	2.06	2.03	2.01	1.99	1.97	1.96	1.93	1.91
		2.75	2.69	2.62	2.54	2.50	2.47	2.43	2.40	2.38	2.34	2.30
		3.37	3.28	3.18	3.07	3.01	2.96	2.91	2.87	2.85	2.79	2.73
		4.30	4.16	4.01	3.86	3.76	3.70	3.62	3.57	3.54	3.45	3.37
		7.29	7.00	6.71	6.40	6.22	6.09	5.93	5.83	5.76	5.59	5.44
	13	2.14	2.10	2.05	2.01	1.98	1.96	1.93	1.92	1.90	1.88	1.85
		2.67	2.60	2.53	2.46	2.41	2.38	2.34	2.31	2.30	2.25	2.21
		3.25	3.15	3.05	2.95	2.88	2.84	2.78	2.74	2.72	2.66	2.60
		4.10	3.96	3.82	3.66	3.57	3.51	3.43	3.38	3.34	3.25	3.18
		6.80	6.52	6.23	5.93	5.75	5.63	5.47	5.37	5.30	5.14	4.99
	14	2.10	2.05	2.01	1.96	1.93	1.91	1.89	1.87	1.86	1.83	1.80
		2.60	2.53	2.46	2.39	2.34	2.31	2.27	2.24	2.22	2.18	2.14
		3.15	3.05	2.95	2.84	2.78	2.73	2.67	2.64	2.61	2.55	2.50
		3.94	3.80	3.66	3.51	3.41	3.35	3.27	3.22	3.18	3.09	3.02
		6.40	6.13	5.85	5.56	5.38	5.25	5.10	5.00	4.94	4.77	4.62
	15	2.06	2.02	1.97	1.92	1.89	1.87	1.85	1.83	1.82	1.79	1.76
		2.54	2.48	2.40	2.33	2.28	2.25	2.20	2.18	2.16	2.11	2.07
		3.06	2.96	2.86	2.76	2.69	2.64	2.59	2.55	2.52	2.46	2.40
		3.80	3.67	3.52	3.37	3.28	3.21	3.13	3.08	3.05	2.96	2.88
		6.08	5.81	5.54	5.25	5.07	4.95	4.80	4.70	4.64	4.47	4.33
	16	2.03	1.99	1.94	1.89	1.86	1.84	1.81	1.79	1.78	1.75	1.72
		2.49	2.42	2.35	2.28	2.23	2.19	2.15	2.12	2.11	2.06	2.02
		2.99	2.89	2.79	2.68	2.61	2.57	2.51	2.47	2.45	2.38	2.32
		3.69	3.55	3.41	3.26	3.16	3.10	3.02	2.97	2.93	2.84	2.76
		5.81	5.55	5.27	4.99	4.82	4.70	4.54	4.45	4.39	4.23	4.08
	17	2.00	1.96	1.91	1.86	1.83	1.81	1.78	1.76	1.75	1.72	1.69
		2.45	2.38	2.31	2.23	2.18	2.15	2.10	2.08	2.06	2.01	1.97
		2.92	2.82	2.72	2.62	2.55	2.50	2.44	2.41	2.38	2.32	2.26
		3.59	3.46	3.31	3.16	3.07	3.00	2.92	2.87	2.83	2.75	2.66
		5.58	5.32	5.05	4.78	4.60	4.48	4.33	4.24	4.18	4.02	3.87

TABLE D F distribution critical values (continued)

Degrees of freedom in the numerator

	p	1	2	3	4	5	6	7	8	9
18	.100	3.01	2.62	2.42	2.29	2.20	2.13	2.08	2.04	2.00
	.050	4.41	3.55	3.16	2.93	2.77	2.66	2.58	2.51	2.46
	.025	5.98	4.56	3.95	3.61	3.38	3.22	3.10	3.01	2.93
	.010	8.29	6.01	5.09	4.58	4.25	4.01	3.84	3.71	3.60
	.001	15.38	10.39	8.49	7.46	6.81	6.35	6.02	5.76	5.56
19	.100	2.99	2.61	2.40	2.27	2.18	2.11	2.06	2.02	1.98
	.050	4.38	3.52	3.13	2.90	2.74	2.63	2.54	2.48	2.42
	.025	5.92	4.51	3.90	3.56	3.33	3.17	3.05	2.96	2.88
	.010	8.18	5.93	5.01	4.50	4.17	3.94	3.77	3.63	3.52
	.001	15.08	10.16	8.28	7.27	6.62	6.18	5.85	5.59	5.39
20	.100	2.97	2.59	2.38	2.25	2.16	2.09	2.04	2.00	1.96
	.050	4.35	3.49	3.10	2.87	2.71	2.60	2.51	2.45	2.39
	.025	5.87	4.46	3.86	3.51	3.29	3.13	3.01	2.91	2.84
	.010	8.10	5.85	4.94	4.43	4.10	3.87	3.70	3.56	3.46
	.001	14.82	9.95	8.10	7.10	6.46	6.02	5.69	5.44	5.24
21	.100	2.96	2.57	2.36	2.23	2.14	2.08	2.02	1.98	1.95
	.050	4.32	3.47	3.07	2.84	2.68	2.57	2.49	2.42	2.37
	.025	5.83	4.42	3.82	3.48	3.25	3.09	2.97	2.87	2.80
	.010	8.02	5.78	4.87	4.37	4.04	3.81	3.64	3.51	3.40
	.001	14.59	9.77	7.94	6.95	6.32	5.88	5.56	5.31	5.11
22	.100	2.95	2.56	2.35	2.22	2.13	2.06	2.01	1.97	1.93
	.050	4.30	3.44	3.05	2.82	2.66	2.55	2.46	2.40	2.34
	.025	5.79	4.38	3.78	3.44	3.22	3.05	2.93	2.84	2.76
	.010	7.95	5.72	4.82	4.31	3.99	3.76	3.59	3.45	3.35
	.001	14.38	9.61	7.80	6.81	6.19	5.76	5.44	5.19	4.99
23	.100	2.94	2.55	2.34	2.21	2.11	2.05	1.99	1.95	1.92
	.050	4.28	3.42	3.03	2.80	2.64	2.53	2.44	2.37	2.32
	.025	5.75	4.35	3.75	3.41	3.18	3.02	2.90	2.81	2.73
	.010	7.88	5.66	4.76	4.26	3.94	3.71	3.54	3.41	3.30
	.001	14.20	9.47	7.67	6.70	6.08	5.65	5.33	5.09	4.89
24	.100	2.93	2.54	2.33	2.19	2.10	2.04	1.98	1.94	1.91
	.050	4.26	3.40	3.01	2.78	2.62	2.51	2.42	2.36	2.30
	.025	5.72	4.32	3.72	3.38	3.15	2.99	2.87	2.78	2.70
	.010	7.82	5.61	4.72	4.22	3.90	3.67	3.50	3.36	3.26
	.001	14.03	9.34	7.55	6.59	5.98	5.55	5.23	4.99	4.80
25	.100	2.92	2.53	2.32	2.18	2.09	2.02	1.97	1.93	1.89
	.050	4.24	3.39	2.99	2.76	2.60	2.49	2.40	2.34	2.28
	.025	5.69	4.29	3.69	3.35	3.13	2.97	2.85	2.75	2.68
	.010	7.77	5.57	4.68	4.18	3.85	3.63	3.46	3.32	3.22
	.001	13.88	9.22	7.45	6.49	5.89	5.46	5.15	4.91	4.71
26	.100	2.91	2.52	2.31	2.17	2.08	2.01	1.96	1.92	1.88
	.050	4.23	3.37	2.98	2.74	2.59	2.47	2.39	2.32	2.27
	.025	5.66	4.27	3.67	3.33	3.10	2.94	2.82	2.73	2.65
	.010	7.72	5.53	4.64	4.14	3.82	3.59	3.42	3.29	3.18
	.001	13.74	9.12	7.36	6.41	5.80	5.38	5.07	4.83	4.64
27	.100	2.90	2.51	2.30	2.17	2.07	2.00	1.95	1.91	1.87
	.050	4.21	3.35	2.96	2.73	2.57	2.46	2.37	2.31	2.25
	.025	5.63	4.24	3.65	3.31	3.08	2.92	2.80	2.71	2.63
	.010	7.68	5.49	4.60	4.11	3.78	3.56	3.39	3.26	3.15
	.001	13.61	9.02	7.27	6.33	5.73	5.31	5.00	4.76	4.57

(Continued)

TABLE D F distribution critical values (continued)

Degrees of freedom in the numerator

Degrees of freedom in the denominator		10	12	15	20	25	30	40	50	60	120	1000
	18	1.98	1.93	1.89	1.84	1.80	1.78	1.75	1.74	1.72	1.69	1.66
		2.41	2.34	2.27	2.19	2.14	2.11	2.06	2.04	2.02	1.97	1.92
		2.87	2.77	2.67	2.56	2.49	2.44	2.38	2.35	2.32	2.26	2.20
		3.51	3.37	3.23	3.08	2.98	2.92	2.84	2.78	2.75	2.66	2.58
		5.39	5.13	4.87	4.59	4.42	4.30	4.15	4.06	4.00	3.84	3.69
	19	1.96	1.91	1.86	1.81	1.78	1.76	1.73	1.71	1.70	1.67	1.64
		2.38	2.31	2.23	2.16	2.11	2.07	2.03	2.00	1.98	1.93	1.88
		2.82	2.72	2.62	2.51	2.44	2.39	2.33	2.30	2.27	2.20	2.14
		3.43	3.30	3.15	3.00	2.91	2.84	2.76	2.71	2.67	2.58	2.50
		5.22	4.97	4.70	4.43	4.26	4.14	3.99	3.90	3.84	3.68	3.53
	20	1.94	1.89	1.84	1.79	1.76	1.74	1.71	1.69	1.68	1.64	1.61
		2.35	2.28	2.20	2.12	2.07	2.04	1.99	1.97	1.95	1.90	1.85
		2.77	2.68	2.57	2.46	2.40	2.35	2.29	2.25	2.22	2.16	2.09
		3.37	3.23	3.09	2.94	2.84	2.78	2.69	2.64	2.61	2.52	2.43
		5.08	4.82	4.56	4.29	4.12	4.00	3.86	3.77	3.70	3.54	3.40
	21	1.92	1.87	1.83	1.78	1.74	1.72	1.69	1.67	1.66	1.62	1.59
		2.32	2.25	2.18	2.10	2.05	2.01	1.96	1.94	1.92	1.87	1.82
		2.73	2.64	2.53	2.42	2.36	2.31	2.25	2.21	2.18	2.11	2.05
		3.31	3.17	3.03	2.88	2.79	2.72	2.64	2.58	2.55	2.46	2.37
		4.95	4.70	4.44	4.17	4.00	3.88	3.74	3.64	3.58	3.42	3.28
	22	1.90	1.86	1.81	1.76	1.73	1.70	1.67	1.65	1.64	1.60	1.57
		2.30	2.23	2.15	2.07	2.02	1.98	1.94	1.91	1.89	1.84	1.79
		2.70	2.60	2.50	2.39	2.32	2.27	2.21	2.17	2.14	2.08	2.01
		3.26	3.12	2.98	2.83	2.73	2.67	2.58	2.53	2.50	2.40	2.32
		4.83	4.58	4.33	4.06	3.89	3.78	3.63	3.54	3.48	3.32	3.17
	23	1.89	1.84	1.80	1.74	1.71	1.69	1.66	1.64	1.62	1.59	1.55
		2.27	2.20	2.13	2.05	2.00	1.96	1.91	1.88	1.86	1.81	1.76
		2.67	2.57	2.47	2.36	2.29	2.24	2.18	2.14	2.11	2.04	1.98
		3.21	3.07	2.93	2.78	2.69	2.62	2.54	2.48	2.45	2.35	2.27
		4.73	4.48	4.23	3.96	3.79	3.68	3.53	3.44	3.38	3.22	3.08
	24	1.88	1.83	1.78	1.73	1.70	1.67	1.64	1.62	1.61	1.57	1.54
		2.25	2.18	2.11	2.03	1.97	1.94	1.89	1.86	1.84	1.79	1.74
		2.64	2.54	2.44	2.33	2.26	2.21	2.15	2.11	2.08	2.01	1.94
		3.17	3.03	2.89	2.74	2.64	2.58	2.49	2.44	2.40	2.31	2.22
		4.64	4.39	4.14	3.87	3.71	3.59	3.45	3.36	3.29	3.14	2.99
	25	1.87	1.82	1.77	1.72	1.68	1.66	1.63	1.61	1.59	1.56	1.52
		2.24	2.16	2.09	2.01	1.96	1.92	1.87	1.84	1.82	1.77	1.72
		2.61	2.51	2.41	2.30	2.23	2.18	2.12	2.08	2.05	1.98	1.91
		3.13	2.99	2.85	2.70	2.60	2.54	2.45	2.40	2.36	2.27	2.18
		4.56	4.31	4.06	3.79	3.63	3.52	3.37	3.28	3.22	3.06	2.91
	26	1.86	1.81	1.76	1.71	1.67	1.65	1.61	1.59	1.58	1.54	1.51
		2.22	2.15	2.07	1.99	1.94	1.90	1.85	1.82	1.80	1.75	1.70
		2.59	2.49	2.39	2.28	2.21	2.16	2.09	2.05	2.03	1.95	1.89
		3.09	2.96	2.81	2.66	2.57	2.50	2.42	2.36	2.33	2.23	2.14
		4.48	4.24	3.99	3.72	3.56	3.44	3.30	3.21	3.15	2.99	2.84
	27	1.85	1.00	1.75	1.70	1.66	1.64	1.60	1.58	1.57	1.53	1.50
		2.20	2.13	2.06	1.97	1.92	1.88	1.84	1.81	1.79	1.73	1.68
		2.57	2.47	2.36	2.25	2.18	2.13	2.07	2.03	2.00	1.93	1.86
		3.06	2.93	2.78	2.63	2.54	2.47	2.38	2.33	2.29	2.20	2.11
		4.41	4.17	3.92	3.66	3.49	3.38	3.23	3.14	3.08	2.92	2.78

TABLE D F distribution critical values (continued)

Degrees of freedom in the numerator

	p	1	2	3	4	5	6	7	8	9
28	.100	2.89	2.50	2.29	2.16	2.06	2.00	1.94	1.90	1.87
	.050	4.20	3.34	2.95	2.71	2.56	2.45	2.36	2.29	2.24
	.025	5.61	4.22	3.63	3.29	3.06	2.90	2.78	2.69	2.61
	.010	7.64	5.45	4.57	4.07	3.75	3.53	3.36	3.23	3.12
	.001	13.50	8.93	7.19	6.25	5.66	5.24	4.93	4.69	4.50
29	.100	2.89	2.50	2.28	2.15	2.06	1.99	1.93	1.89	1.86
	.050	4.18	3.33	2.93	2.70	2.55	2.43	2.35	2.28	2.22
	.025	5.59	4.20	3.61	3.27	3.04	2.88	2.76	2.67	2.59
	.010	7.60	5.42	4.54	4.04	3.73	3.50	3.33	3.20	3.09
	.001	13.39	8.85	7.12	6.19	5.59	5.18	4.87	4.64	4.45
30	.100	2.88	2.49	2.28	2.14	2.05	1.98	1.93	1.88	1.85
	.050	4.17	3.32	2.92	2.69	2.53	2.42	2.33	2.27	2.21
	.025	5.57	4.18	3.59	3.25	3.03	2.87	2.75	2.65	2.57
	.010	7.56	5.39	4.51	4.02	3.70	3.47	3.30	3.17	3.07
	.001	13.29	8.77	7.05	6.12	5.53	5.12	4.82	4.58	4.39
40	.100	2.84	2.44	2.23	2.09	2.00	1.93	1.87	1.83	1.79
	.050	4.08	3.23	2.84	2.61	2.45	2.34	2.25	2.18	2.12
	.025	5.42	4.05	3.46	3.13	2.90	2.74	2.62	2.53	2.45
	.010	7.31	5.18	4.31	3.83	3.51	3.29	3.12	2.99	2.89
	.001	12.61	8.25	6.59	5.70	5.13	4.73	4.44	4.21	4.02
50	.100	2.81	2.41	2.20	2.06	1.97	1.90	1.84	1.80	1.76
	.050	4.03	3.18	2.79	2.56	2.40	2.29	2.20	2.13	2.07
	.025	5.34	3.97	3.39	3.05	2.83	2.67	2.55	2.46	2.38
	.010	7.17	5.06	4.20	3.72	3.41	3.19	3.02	2.89	2.78
	.001	12.22	7.96	6.34	5.46	4.90	4.51	4.22	4.00	3.82
60	.100	2.79	2.39	2.18	2.04	1.95	1.87	1.82	1.77	1.74
	.050	4.00	3.15	2.76	2.53	2.37	2.25	2.17	2.10	2.04
	.025	5.29	3.93	3.34	3.01	2.79	2.63	2.51	2.41	2.33
	.010	7.08	4.98	4.13	3.65	3.34	3.12	2.95	2.82	2.72
	.001	11.97	7.77	6.17	5.31	4.76	4.37	4.09	3.86	3.69
100	.100	2.76	2.36	2.14	2.00	1.91	1.83	1.78	1.73	1.69
	.050	3.94	3.09	2.70	2.46	2.31	2.19	2.10	2.03	1.97
	.025	5.18	3.83	3.25	2.92	2.70	2.54	2.42	2.32	2.24
	.010	6.90	4.82	3.98	3.51	3.21	2.99	2.82	2.69	2.59
	.001	11.50	7.41	5.86	5.02	4.48	4.11	3.83	3.61	3.44
200	.100	2.73	2.33	2.11	1.97	1.88	1.80	1.75	1.70	1.66
	.050	3.89	3.04	2.65	2.42	2.26	2.14	2.06	1.98	1.93
	.025	5.10	3.76	3.18	2.85	2.63	2.47	2.35	2.26	2.18
	.010	6.76	4.71	3.88	3.41	3.11	2.89	2.73	2.60	2.50
	.001	11.15	7.15	5.63	4.81	4.29	3.92	3.65	3.43	3.26
1000	.100	2.71	2.31	2.09	1.95	1.85	1.78	1.72	1.68	1.64
	.050	3.85	3.00	2.61	2.38	2.22	2.11	2.02	1.95	1.89
	.025	5.04	3.70	3.13	2.80	2.58	2.42	2.30	2.20	2.13
	.010	6.66	4.63	3.80	3.34	3.04	2.82	2.66	2.53	2.43
	.001	10.89	6.96	5.46	4.65	4.14	3.78	3.51	3.30	3.13

(Continued)

TABLE D F distribution critical values (continued)

		\multicolumn{11}{c	}{Degrees of freedom in the numerator}
		10	12	15	20	25	30	40	50	60	120	1000
Degrees of freedom in the denominator	28	1.84	1.79	1.74	1.69	1.65	1.63	1.59	1.57	1.56	1.52	1.48
		2.19	2.12	2.04	1.96	1.91	1.87	1.82	1.79	1.77	1.71	1.66
		2.55	2.45	2.34	2.23	2.16	2.11	2.05	2.01	1.98	1.91	1.84
		3.03	2.90	2.75	2.60	2.51	2.44	2.35	2.30	2.26	2.17	2.08
		4.35	4.11	3.86	3.60	3.43	3.32	3.18	3.09	3.02	2.86	2.72
	29	1.83	1.78	1.73	1.68	1.64	1.62	1.58	1.56	1.55	1.51	1.47
		2.18	2.10	2.03	1.94	1.89	1.85	1.81	1.77	1.75	1.70	1.65
		2.53	2.43	2.32	2.21	2.14	2.09	2.03	1.99	1.96	1.89	1.82
		3.00	2.87	2.73	2.57	2.48	2.41	2.33	2.27	2.23	2.14	2.05
		4.29	4.05	3.80	3.54	3.38	3.27	3.12	3.03	2.97	2.81	2.66
	30	1.82	1.77	1.72	1.67	1.63	1.61	1.57	1.55	1.54	1.50	1.46
		2.16	2.09	2.01	1.93	1.88	1.84	1.79	1.76	1.74	1.68	1.63
		2.51	2.41	2.31	2.20	2.12	2.07	2.01	1.97	1.94	1.87	1.80
		2.98	2.84	2.70	2.55	2.45	2.39	2.30	2.25	2.21	2.11	2.02
		4.24	4.00	3.75	3.49	3.33	3.22	3.07	2.98	2.92	2.76	2.61
	40	1.76	1.71	1.66	1.61	1.57	1.54	1.51	1.48	1.47	1.42	1.38
		2.08	2.00	1.92	1.84	1.78	1.74	1.69	1.66	1.64	1.58	1.52
		2.39	2.29	2.18	2.07	1.99	1.94	1.88	1.83	1.80	1.72	1.65
		2.80	2.66	2.52	2.37	2.27	2.20	2.11	2.06	2.02	1.92	1.82
		3.87	3.64	3.40	3.14	2.98	2.87	2.73	2.64	2.57	2.41	2.25
	50	1.73	1.68	1.63	1.57	1.53	1.50	1.46	1.44	1.42	1.38	1.33
		2.03	1.95	1.87	1.78	1.73	1.69	1.63	1.60	1.58	1.51	1.45
		2.32	2.22	2.11	1.99	1.92	1.87	1.80	1.75	1.72	1.64	1.56
		2.70	2.56	2.42	2.27	2.17	2.10	2.01	1.95	1.91	1.80	1.70
		3.67	3.44	3.20	2.95	2.79	2.68	2.53	2.44	2.38	2.21	2.05
	60	1.71	1.66	1.60	1.54	1.50	1.48	1.44	1.41	1.40	1.35	1.30
		1.99	1.92	1.84	1.75	1.69	1.65	1.59	1.56	1.53	1.47	1.40
		2.27	2.17	2.06	1.94	1.87	1.82	1.74	1.70	1.67	1.58	1.49
		2.63	2.50	2.35	2.20	2.10	2.03	1.94	1.88	1.84	1.73	1.62
		3.54	3.32	3.08	2.83	2.67	2.55	2.41	2.32	2.25	2.08	1.92
	100	1.66	1.61	1.56	1.49	1.45	1.42	1.38	1.35	1.34	1.28	1.22
		1.93	1.85	1.77	1.68	1.62	1.57	1.52	1.48	1.45	1.38	1.30
		2.18	2.08	1.97	1.85	1.77	1.71	1.64	1.59	1.56	1.46	1.36
		2.50	2.37	2.22	2.07	1.97	1.89	1.80	1.74	1.69	1.57	1.45
		3.30	3.07	2.84	2.59	2.43	2.32	2.17	2.08	2.01	1.83	1.64
	200	1.63	1.58	1.52	1.46	1.41	1.38	1.34	1.31	1.29	1.23	1.16
		1.88	1.80	1.72	1.62	1.56	1.52	1.46	1.41	1.39	1.30	1.21
		2.11	2.01	1.90	1.78	1.70	1.64	1.56	1.51	1.47	1.37	1.25
		2.41	2.27	2.13	1.97	1.87	1.79	1.69	1.63	1.58	1.45	1.30
		3.12	2.90	2.67	2.42	2.26	2.15	2.00	1.90	1.83	1.64	1.43
	1000	1.61	1.55	1.49	1.43	1.38	1.35	1.30	1.27	1.25	1.18	1.08
		1.84	1.76	1.68	1.58	1.52	1.47	1.41	1.36	1.33	1.24	1.11
		2.06	1.96	1.85	1.72	1.64	1.58	1.50	1.45	1.41	1.29	1.13
		2.34	2.20	2.06	1.90	1.79	1.72	1.61	1.54	1.50	1.35	1.16
		2.99	2.77	2.54	2.30	2.14	2.02	1.87	1.77	1.69	1.49	1.22

Table E Wilcoxon signed-rank test critical values

	Two-tailed test		One-tailed test	
n	$\alpha = .05$	$\alpha = .01$	$\alpha = .05$	$\alpha = .01$
5	—	—	0	—
6	0	—	2	—
7	2	—	3	0
8	3	0	5	1
9	5	1	8	3
10	8	3	10	5
11	10	5	13	7
12	13	7	17	10
13	17	10	21	12
14	21	13	25	16
15	25	16	30	19
16	30	19	35	23
17	35	23	41	28
18	40	27	47	33
19	46	32	53	37
20	52	37	60	43
21	58	42	67	49
22	66	48	75	55
23	73	54	83	62
24	81	61	91	69
25	89	68	100	76
26	98	75	110	84
27	107	83	119	93
28	116	91	130	101
29	126	100	140	110
30	137	109	151	120

Table values adapted from Table A.4 in *Nonparametric Statistical Methods* 2e by Myles Hollander and Douglas A. Wolfe, Wiley 1999.